THE AMERICAN PRACTICAL NAVIGATOR

BOWDITCH

NATIONAL GEOSPATIAL-INTELLIGENCE AGENCY

Skyhorse Publishing

All inquiries should be addressed to Skyhorse Publishing, 307 West 36th Street, 11th Floor, New York, NY 10018.

Skyhorse Publishing books may be purchased in bulk at special discounts for sales promotion, corporate gifts, fund-raising, or educational purposes. Special editions can also be created to specifications. For details, contact the Special Sales Department, Skyhorse Publishing, 307 West 36th Street, 11th Floor, New York, NY 10018 or info@skyhorsepublishing.com.

Visit our website at www.skyhorsepublishing.com.

10 9 8 7 6 5 4 3 2 1

Library of Congress Cataloging-in-Publication Data is available on file.
ISBN: 978-1-62087-796-8

Printed in the United States of America

TABLE OF CONTENTS

PART 7 — OCEANOGRAPHY

PART 8 — MARINE METEOROLOGY

NAVIGATION TABLES

MATHEMATICAL TABLES

CARTOGRAPHIC TABLES

PILOTING TABLES

CELESTIAL NAVIGATION TABLES

METEOROLOGICAL TABLES

GLOSSARIES

INDEX

Last painting by Gilbert Stuart (1828). Considered by the family of Bowditch to be the best of various paintings made, although it was unfinished when the artist died.

NATHANIEL BOWDITCH
(1773-1838)

Nathaniel Bowditch was born on March 26, 1773, in Salem, Mass., fourth of the seven children of shipmaster Habakkuk Bowditch and his wife, Mary.

Since the migration of William Bowditch from England to the Colonies in the 17th century, the family had resided at Salem. Most of its sons, like those of other families in this New England seaport, had gone to sea, and many of them became shipmasters. Nathaniel Bowditch himself sailed as master on his last voyage, and two of his brothers met untimely deaths while pursuing careers at sea.

Nathaniel Bowditch's father is said to have lost two ships at sea, and by late Revolutionary days he returned to the trade of cooper, which he had learned in his youth. This provided insufficient income to properly supply the needs of his growing family, who were often hungry and cold. For many years the nearly destitute family received an annual grant of 15 to 20 dollars from the Salem Marine Society. By the time Nathaniel had reached the age of 10, the family's poverty forced him to leave school and join his father in the cooper's trade to help support the family.

Nathaniel was unsuccessful as a cooper, and when he was about 12 years of age, he entered the first of two ship-chandlery firms by which he was employed. It was during the nearly 10 years he was so employed that his great mind first attracted public attention. From the time he began school Bowditch had an all-consuming interest in learning, particularly mathematics. By his middle teens he was recognized in Salem as an authority on that subject. Salem being primarily a shipping town, most of the inhabitants sooner or later found their way to the ship chandler, and news of the brilliant young clerk spread until eventually it came to the attention of the learned men of his day. Impressed by his desire to educate himself, they supplied him with books that he might learn of the discoveries of other men. Since many of the best books were written by Europeans, Bowditch first taught himself their languages. French, Spanish, Latin, Greek and German were among the two dozen or more languages and dialects he studied during his life. At the age of 16 he began the study of Newton's *Principia*, translating parts of it from the Latin. He even found an error in that classic text, and though lacking the confidence to announce it at the time, he later published his findings and had them accepted by the scientific community.

During the Revolutionary War a privateer out of Beverly, a neighboring town to Salem, had taken as one of its prizes an English vessel which was carrying the philosophical library of a famed Irish scholar, Dr. Richard Kirwan. The books were brought to the Colonies and there bought by a group of educated Salem men who used them to found the Philosophical Library Company, reputed to have been the best library north of Philadelphia at the time. In 1791, when Bowditch was 18, two Harvard-educated ministers, Rev. John Prince and Rev. William Bentley, persuaded the Company to allow Bowditch the use of its library. Encouraged by these two men and a third, Nathan Read, an apothecary and also a Harvard man, Bowditch studied the works of the great men who had preceded him, especially the mathematicians and the astronomers. By the time he became of age, this knowledge, acquired when not working long hours at the chandlery, had made young Nathaniel the outstanding mathematician in the Commonwealth, and perhaps in the country.

In the seafaring town of Salem, Bowditch was drawn to navigation early, learning the subject at the age of 13 from an old British sailor. A year later he began studying surveying, and in 1794 he assisted in a survey of the town. At 15 he devised an almanac reputed to have been of great accuracy. His other youthful accomplishments included the construction of a crude barometer and a sundial.

When Bowditch went to sea at the age of 21, it was as captain's writer and nominal second mate, the officer's berth being offered him because of his reputation as a scholar. Under Captain Henry Prince, the ship *Henry* sailed from Salem in the winter of 1795 on what was to be a year-long voyage to the Ile de Bourbon (now called Reunion) in the Indian Ocean.

Bowditch began his seagoing career when accurate time was not available to the average naval or merchant ship. A reliable marine chronometer had been invented some 60 years before, but the prohibitive cost, plus the long voyages without opportunity to check the error of the timepiece, made the large investment an impractical one. A system of determining longitude by "lunar distance," a method which did not require an accurate timepiece, was known, but this product of the minds of mathematicians and astronomers was so involved as to be beyond the capabilities of the uneducated seamen of that day. Consequently, ships were navigated by a combination of dead reckoning and parallel sailing (a system of sailing north or south to the latitude of the destination and then east or west to the destination). The navigational routine of the time was "lead, log, and lookout."

To Bowditch, the mathematical genius, computation of lunar distances was no mystery, of course, but he recognized the need for an easier method of working them in order to navigate ships more safely and efficiently. Through analysis and observation, he derived a new and simplified formula during his first trip.

John Hamilton Moore's *The Practical Navigator* was the leading navigational text when Bowditch first went to sea, and had been for many years. Early in his first voyage, however, the captain's writer-second mate began turning up errors in Moore's book, and before long he found it necessary to recompute some of the tables he most often used in working his sights. Bowditch recorded the errors he found, and by the end of his second voyage, made in the higher capacity of supercargo, the news of his findings in *The New Practical Navigator* had reached Edmund Blunt, a printer at Newburyport, Mass. At Blunt's request, Bowditch agreed to participate with other learned men in the preparation of an American edition of the thirteenth (1798) edition of Moore's work. The first American edition was published at Newburyport by Blunt in 1799. This edition corrected many of the errors that Moore had included.

Although most of the errors were of little significance to practical navigation because they were errors in the fifth and sixth places of logarithm tables, some errors were significant.The most significant mistake was listing the year 1800 as a leap year in the table of the sun's declination. The consequence was that Moore gave the declination for March 1, 1800, as 7°11'. Since the actual value was 7° 33', the calculation of a meridian altitude would be in error by 22 minutes of latitude, or 22 nautical miles.

Bowditch's principal contribution to the first American edition was his chapter "The Method of Finding the Longitude at Sea," which discussed his new method for computing lunar distances. Following publication of the first American edition, Blunt obtained Bowditch's services in checking the American and English editions for further errors. Blunt then published a second American edition of Moore's thirteenth edition in 1800. When preparing a third American edition for the press, Blunt decided that Bowditch had revised Moore's work to such an extent that Bowditch should be named as author. The title was changed to *The New American Practical Navigator* and the book was published in 1802 as a first edition. Bowditch vowed while writing this edition to "put down in the book nothing I can't teach the crew," and it is said that every member of his crew including the cook could take a lunar observation and plot the ship's position.

Bowditch made a total of five trips to sea, over a period of about nine years, his last as master and part owner of the three-masted *Putnam*. Homeward bound from a 13-month voyage to Sumatra and the Ile de France (now called Mauritius) the *Putnam* approached Salem harbor on December 25, 1803, during a thick fog without having had a celestial observation since noon on the 24th. Relying upon his dead reckoning, Bowditch conned his wooden-hulled ship to the entrance of the rocky harbor, where he had the good fortune to get a momentary glimpse of Eastern Point, Cape Ann, enough to confirm his position. The *Putnam* proceeded in, past such hazards as "Bowditch's Ledge" (named after a great-grandfather who had wrecked his ship on the rock more than a century before) and anchored safely at 1900 that evening. Word of the daring feat, performed when other masters were hove-to outside the harbor, spread along the coast and added greatly to Bowditch's reputation. He was, indeed, the "practical navigator."

His standing as a mathematician and successful shipmaster earned him a well-paid position ashore within a matter of weeks after his last voyage. He was installed as president of a Salem fire and marine insurance company at the age of 30, and during the 20 years he held that position the company prospered. In 1823 he left Salem to take a similar position with a Boston insurance firm, serving that company with equal success until his death.

From the time he finished the "*Navigator*" until 1814, Bowditch's mathematical and scientific pursuits consisted of studies and papers on the orbits of comets, applications of Napier's rules, magnetic variation, eclipses, calculations on tides, and the charting of Salem harbor. In that year, however, he turned to what he considered the greatest work of his life, the translation into English of *Mecanique Celeste*, by Pierre Laplace. *Mecanique Celeste* was a summary of all the then known facts about the workings of the heavens. Bowditch translated four of the five volumes before his death, and published them at his own expense. He gave many formula derivations which Laplace had not shown, and also included further discoveries following the time of publication. His work made this information available to American astronomers and enabled them to pursue their studies on the basis of that which was already known. Continuing his style of writing for the learner, Bowditch presented his English version of *Mecanique Celeste* in such a manner that the student of mathematics could easily trace the steps involved in reaching the most complicated conclusions.

Shortly after the publication of *The New American Practical Navigator*, Harvard College honored its author with the presentation of the honorary degree of Master of Arts, and in 1816 the college made him an honorary Doctor of Laws. From the time the Harvard graduates of Salem first assisted him in his studies, Bowditch had a great interest in that college, and in 1810 he was elected one of its Overseers, a position he held until 1826, when he was elected to the Corporation. During 1826-27 he was the leader of a small group of men who saved the school from financial disaster by forcing necessary economies on the college's reluctant president. At one time Bowditch was offered a Professorship in Mathematics at Harvard but this, as well as similar offers from West Point and the University of Virginia, he declined. In all his life he was never known to have made a public speech or to have addressed any large group of people.

Many other honors came to Bowditch in recognition of his astronomical, mathematical, and marine accomplishments. He became a member of the American Academy of Arts and Sciences, the East India Marine Society, the Royal Academy of Edinburgh, the Royal Society of London, the Royal Irish Academy, the American Philosophical Society, the Connecticut Academy of Arts and Sciences, the Boston Marine Society, the Royal Astronomical Society, the Palermo Academy of Science, and the Royal Academy of Berlin.

Nathaniel Bowditch outlived all of his brothers and sisters by nearly 30 years. He died on March 16, 1838, in his sixty-fifth year. The following eulogy by the Salem Marine Society indicates the regard in which this distinguished American was held by his contemporaries:

"In his death a public, a national, a human benefactor has departed. Not this community, nor our country only, but the whole world, has reason to do honor to his memory. When the voice of Eulogy shall be still, when the tear of Sorrow shall cease to flow, no monument will be needed to keep alive his memory among men; but as long as ships shall sail, the needle point to the north, and the stars go through their wonted courses in the heavens, the name of Dr. Bowditch will be revered as of one who helped his fellow-men in a time of need, who was and is a guide to them over the pathless ocean, and of one who forwarded the great interests of mankind."

THE NEW AMERICAN

PRACTICAL NAVIGATOR;

BEING AN

EPITOME OF NAVIGATION;

CONTAINING ALL THE TABLES NECESSARY TO BE USED WITH THE

NAUTICAL ALMANAC,

IN DETERMINING THE

LATITUDE;

AND THE

LONGITUDE BY LUNAR OBSERVATIONS;

AND

KEEPING A COMPLETE RECKONING AT SEA:

ILLUSTRATED BY

PROPER RULES AND EXAMPLES:

THE WHOLE EXEMPLIFIED IN A

JOURNAL,

KEPT FROM

BOSTON TO MADEIRA,

IN WHICH ALL THE RULES OF NAVIGATION ARE INTRODUCED:

ALSO

The Demonstration of the most useful Rules of Trigonometry: With many useful Problems in Mensuration, Surveying, and Gauging: And a Dictionary of Sea-Terms; with the Manner of performing the most common Evolutions at Sea.

TO WHICH ARE ADDED,

Some General Instructions and Information to Merchants, Masters of Vessels, and others concerned in Navigation, relative to Maritime Laws and Mercantile Customs.

FROM THE BEST AUTHORITIES.

ENRICHED WITH A NUMBER OF

NEW TABLES,

WITH ORIGINAL IMPROVEMENTS AND ADDITIONS, AND A LARGE VARIETY OF NEW AND IMPORTANT MATTER:

ALSO,

MANY THOUSAND ERRORS ARE CORRECTED,

WHICH HAVE APPEARED IN THE BEST SYSTEMS OF NAVIGATION YET PUBLISHED.

BY NATHANIEL BOWDITCH,

FELLOW OF THE AMERICAN ACADEMY OF ARTS AND SCIENCES.

ILLUSTRATED WITH COPPERPLATES.

First Edition.

PRINTED AT NEWBURYPORT, (MASS.) 1802,

BY

EDMUND M. BLUNT, (Proprietor)

FOR CUSHING & APPLETON, SALEM.

SOLD BY EVERY BOOKSELLER, SHIP-CHANDLER, AND MATHEMATICAL-INSTRUMENT-MAKER, IN THE UNITED STATES AND WEST-INDIES.

Original title page of *The New American Practical Navigator*, First Edition, published in 1802.

PREFACE

The Naval Observatory library in Washington, D.C., is unnaturally quiet. It is a large circular room, filled with thousands of books. Its acoustics are perfect; a mere whisper from the room's open circular balcony can be easily heard by those standing on the ground floor. A fountain in the center of the ground floor softly breaks the room's silence as its water stream gently splashes into a small pool. From this serene room, a library clerk will lead you into an antechamber, beyond which is a vault containing the Observatory's most rare books. In this vault, one can find an original 1802 first edition of the *New American Practical Navigator*.

One cannot hold this small, delicate, slipcovered book without being impressed by the nearly 200-year unbroken chain of publication that it has enjoyed. It sailed on U.S. merchantmen and Navy ships shortly after the quasi-war with France and during British impressment of merchant seamen that led to the War of 1812. It sailed on U.S. Naval vessels during operations against Mexico in the 1840's, on ships of both the Union and Confederate fleets during the Civil War, and with the U.S. Navy in Cuba in 1898. It went around the world with the Great White Fleet, across the North Atlantic to Europe during both World Wars, to Asia during the Korean and Vietnam Wars, and to the Middle East during Operation Desert Storm. It has circled the globe with countless thousands of merchant ships for 200 years.

As navigational requirements and procedures have changed throughout the years, *Bowditch* has changed with them. Originally devoted almost exclusively to celestial navigation, it now also covers a host of modern topics. It is as practical today as it was when Nathaniel Bowditch, master of the *Putnam*, gathered the crew on deck and taught them the mathematics involved in calculating lunar distances. It is that practicality that has been the publication's greatest strength, and that makes the publication as useful today as it was in the age of sail.

Seafarers have long memories. In no other profession is tradition more closely guarded. Even the oldest and most cynical acknowledge the special bond that connects those who have made their livelihood plying the sea. This bond is not comprised of a single strand; rather, it is a rich and varied tapestry that stretches from the present back to the birth of our nation and its seafaring culture. As this book is a part of that tapestry, it should not be lightly regarded; rather, it should be preserved, as much for its historical importance as for its practical utility.

Since antiquity, mariners have gathered available navigation information and put it into a text for others to follow. One of the first attempts at this involved volumes of Spanish and Portuguese navigational manuals translated into English between about 1550 to 1750. Writers and translators of the time "borrowed" freely in compiling navigational texts, a practice which continues today with works such as Sailing Directions and Pilots.

Colonial and early American navigators depended exclusively on English navigation texts because there were no American editions. The first American navigational text, *Orthodoxal Navigation*, was completed by Benjamin Hubbard in 1656. The first American navigation text published in America was Captain Thomas Truxton's *Remarks, Instructions, and Examples Relating to the Latitude and Longitude; also the Variation of the Compass, Etc., Etc.*, published in 1794.

The most popular navigational text of the late 18th century was John Hamilton Moore's *The New Practical Navigator*. Edmund M. Blunt, a Newburyport publisher, decided to issue a revised copy of this work for American navigators. Blunt convinced Nathaniel Bowditch, a locally famous mariner and mathematician, to revise and update *The New Practical Navigator*. Several other learned men assisted in this revision. Blunt's *The New Practical Navigator* was published in 1799. Blunt also published a second American edition of Moore's book in 1800.

By 1802, when Blunt was ready to publish a third edition, Nathaniel Bowditch and others had corrected so many errors in Moore's work that Blunt decided to issue the work as a first edition of the *New American Practical Navigator*. It is to that 1802 work that the current edition of the *American Practical Navigator* traces its pedigree.

The *New American Practical Navigator* stayed in the Bowditch and Blunt family until the government bought the copyright in 1867. Edmund M. Blunt published the book until 1833; upon his retirement, his sons, Edmund and George, took over publication.The elder Blunt died in 1862; his son Edmund followed in 1866. The next year, 1867, George Blunt sold the copyright to the government for $25,000. The government has published *Bowditch* ever since. George Blunt died in 1878.

Nathaniel Bowditch continued to correct and revise the book until his death in 1838. Upon his death, the editorial responsibility for the *American Practical Navigator* passed to his son, J. Ingersoll Bowditch. Ingersoll Bowditch continued editing the *Navigator* until George Blunt sold the copyright to the government. He outlived all of the principals involved in publishing and editing the *Navigator*, dying in 1889.

The U.S. government has published some 52 editions since acquiring the copyright to the book that has come to be known simply by its original author's name, "*Bowditch*." Since the government began production, the book has been known by its year of publishing, instead of by the edition number. During a revision in 1880 by Commander Phillip H. Cooper, USN, the name was changed to *American Practical Navigator*. Bowditch's original method of taking "lunars" was finally dropped from the book just after the turn of the 20th century. After several more revisions and printings through World Wars I and II, *Bowditch* was extensively revised for the 1958 edition and again in 1995.

Recognizing the limitations of the printed word, and that computers and electronic media permit us to think about the processes of both navigation and publishing in completely new ways, NIMA has, for the 2002 edition, produced the first official Compact Disk-Read Only Memory (CD-ROM) version of this work. This CD contains, in addition to the full text of the printed book, electronic enhancements and additions not possible in book form. Our goal is to put as much useful navigational information before the navigator as possible in the most understandable and readable format. We are only beginning to explore the possibilities of new technology in this area.

As much as it is a part of history, Bowditch is not a history book. As in past editions, dated material has been dropped and new methods, technologies and techniques added to keep pace with the rapidly changing world of navigation. The changes to this edition are intended to ensure that it remains the premier reference work for modern, practical marine navigation. This edition replaces but does not cancel former editions, which may be retained and consulted as to historical navigation methods not discussed herein.

PART 1, FUNDAMENTALS, includes an overview of the types and phases of marine navigation and the organizations which develop, support and regulate it. It includes chapters relating to the types, structure, use and limitations of nautical charts; a concise explanation of geodesy and chart datums; and a summary of various necessary navigational publications.

PART 2, PILOTING, emphasizes the practical aspects of navigating a vessel in restricted waters, using both traditional and electronic methods.

PART 3, ELECTRONIC NAVIGATION, explains the nature of radio waves and electronic navigation systems. Chapters deal with each of the several electronic methods of navigation--satellite, Loran C, and radar, with special emphasis on satellite navigation systems and electronic charts.

PART 4, CELESTIAL NAVIGATION, updates the former edition with more modern terminology, and discusses the use of calculators and computers for the solution of celestial navigation problems.

PART 5, NAVIGATIONAL MATHEMATICS, remains unchanged from the former edition.

PART 6, NAVIGATIONAL SAFETY, discusses recent developments in management of navigational resources, the changing role of the navigator, distress and safety communications, procedures for emergency navigation, and the increasingly complex web of navigation regulations.

PART 7, OCEANOGRAPHY, has been updated to reflect the latest science and terminology.

PART 8, MARINE WEATHER incorporates updated weather routing information and new cloud graphics.

The pronoun "he," used throughout this book as a reference to the navigator, refers to both genders.

The printed version of this volume may be corrected using the Notice to Mariners and Summary of Corrections. Suggestions and comments for changes and additions may be sent to:

NATIONAL IMAGERY AND MAPPING AGENCY
MARITIME SAFETY INFORMATION DIVISION
MAIL STOP D-44
4600 SANGAMORE RD.
BETHESDA, MARYLAND, 20816-5003
UNITED STATES OF AMERICA

This book could not have been produced without the expertise of dedicated personnel from many government organizations, among them: U.S. Coast Guard, U.S. Naval Academy, U.S. Naval Oceanographic Office, US Navy Fleet Training Center, the U.S. Naval Observatory, Office of the Navigator of the Navy, U.S. Merchant Marine Academy, U.S. Coast and Geodetic Survey, the National Ocean Service, and the National Weather Service. In addition to official government expertise, we must note the contributions of private organizations and individuals far too numerous to mention. Mariners worldwide can be grateful for the experience, dedication, and professionalism of the many people who generously gave their time in this effort. A complete list of contributors can be found in the "Contributor's Corner" of the CD-ROM version of this book.

THE EDITORS

CHAPTER 1

INTRODUCTION TO MARINE NAVIGATION

DEFINITIONS

100. The Art And Science Of Navigation

Marine navigation blends both science and art. A good navigator constantly thinks strategically, operationally, and tactically. He plans each voyage carefully. As it proceeds, he gathers navigational information from a variety of sources, evaluates this information, and determines his ship's position. He then compares that position with his voyage plan, his operational commitments, and his predetermined "dead reckoning" position. A good navigator anticipates dangerous situations well before they arise, and always stays "ahead of the vessel." He is ready for navigational emergencies at any time. He is increasingly a manager of a variety of resources--electronic, mechanical, and human. Navigation methods and techniques vary with the type of vessel, the conditions, and the navigator's experience. The navigator uses the methods and techniques best suited to the vessel, its equipment, and conditions at hand.

Some important elements of successful navigation cannot be acquired from any book or instructor. The science of navigation can be taught, but the art of navigation must be developed from experience.

101. Types of Navigation

Methods of navigation have changed throughout history. New methods often enhance the mariner's ability to complete his voyage safely and expeditiously, and make his job easier. One of the most important judgments the navigator must make involves choosing the best methods to use. Each method or type has advantages and disadvantages, while none is effective in all situations. Commonly recognized types of navigation are listed below.

- **Dead reckoning (DR)** determines position by advancing a known position for courses and distances. A position so determined is called a dead reckoning (DR) position. It is generally accepted that only course and speed determine the DR position. Correcting the DR position for leeway, current effects, and steering error result in an **estimated position (EP)**.

- **Piloting** involves navigating in restricted waters with frequent or constant determination of position relative to nearby geographic and hydrographic features.

- **Celestial navigation** involves reducing celestial measurements taken with a sextant to lines of position using calculators or computer programs, or by hand with almanacs and tables or using spherical trigonometry.

- **Radio navigation** uses radio waves to determine position through a variety of electronic devices.

- **Radar navigation** uses radar to determine the distance from or bearing of objects whose position is known. This process is separate from radar's use in collision avoidance.

- **Satellite navigation** uses radio signals from satellites for determining position.

Electronic systems and integrated bridge concepts are driving navigation system planning. Integrated systems take inputs from various ship sensors, electronically and automatically chart the position, and provide control signals required to maintain a vessel on a preset course. The navigator becomes a system manager, choosing system presets, interpreting system output, and monitoring vessel response.

In practice, a navigator synthesizes different methodologies into a single integrated system. He should never feel comfortable utilizing only one method when others are also available. Each method has advantages and disadvantages. The navigator must choose methods appropriate to each situation, and never rely completely on only one system.

With the advent of automated position fixing and electronic charts, modern navigation is almost completely an electronic process. The mariner is constantly tempted to rely solely on electronic systems. But electronic navigation systems are always subject to failure, and the professional mariner must never forget that the safety of his ship and crew may depend on skills that differ little from those practiced generations ago. Proficiency in conventional piloting and celestial navigation remains essential.

102. Phases of Navigation

Four distinct phases define the navigation process. The mariner should choose the system mix that meets the accuracy requirements of each phase.

- **Inland Waterway Phase**: Piloting in narrow canals, channels, rivers, and estuaries.
- **Harbor/Harbor Approach Phase**: Navigating to a harbor entrance through bays and sounds, and negotiating harbor approach channels.
- **Coastal Phase**: Navigating within 50 miles of the coast or inshore of the 200 meter depth contour.
- **Ocean Phase**: Navigating outside the coastal area in the open sea.

The navigator's position accuracy requirements, his fix interval, and his systems requirements differ in each phase. The following table can be used as a general guide for selecting the proper system(s).

	Inland	*Harbor/ Approach*	*Coastal*	*Ocean*
DR	X	X	X	X
Piloting	X	X	X	
Celestial			X	X
Radio		X	X	X
Radar	X	X	X	
Satellite	X*	X	X	X

*Table 102. The relationship of the types and phases of navigation. * With SA off and/or using DGPS*

NAVIGATION TERMS AND CONVENTIONS

103. Important Conventions and Concepts

Throughout the history of navigation, numerous terms and conventions have been established which enjoy worldwide recognition. The professional navigator, to gain a full understanding of his field, should understand the origin of certain terms, techniques, and conventions. The following section discusses some of the important ones.

Defining a **prime meridian** is a comparatively recent development. Until the beginning of the 19th century, there was little uniformity among cartographers as to the meridian from which to measure longitude. But it mattered little because there existed no method for determining longitude accurately.

Ptolemy, in the 2nd century AD, measured longitude eastward from a reference meridian 2 degrees west of the Canary Islands. In 1493, Pope Alexander VI established a line in the Atlantic west of the Azores to divide the territories of Spain and Portugal. For many years, cartographers of these two countries used this dividing line as the prime meridian. In 1570 the Dutch cartographer Ortelius used the easternmost of the Cape Verde Islands. John Davis, in his 1594 *The Seaman's Secrets*, used the Isle of Fez in the Canaries because there the variation was zero. Most mariners paid little attention to these conventions and often reckoned their longitude from several different capes and ports during a voyage.

The meridian of London was used as early as 1676, and over the years its popularity grew as England's maritime interests increased. The system of measuring longitude both east and west through 180° may have first appeared in the middle of the 18th century. Toward the end of that century, as the Greenwich Observatory increased in prominence, English cartographers began using the meridian of that observatory as a reference. The publication by the Observatory of the first British *Nautical Almanac* in 1767 further entrenched Greenwich as the prime meridian. An unsuccessful attempt was made in 1810 to establish Washington, D.C. as the prime meridian for American navigators and cartographers. In 1884, the meridian of Greenwich was officially established as the prime meridian. Today, all maritime nations have designated the Greenwich meridian the prime meridian, except in a few cases where local references are used for certain harbor charts.

Charts are graphic representations of areas of the Earth, in digital or graphic form, for use in marine or air navigation. Nautical charts, whether in digital or paper form, depict features of particular interest to the marine navigator. Charts have probably existed since at least 600 B.C. Stereographic and orthographic projections date from the 2nd century B.C. In 1569 Gerardus Mercator published a chart using the mathematical principle which now bears his name. Some 30 years later, Edward Wright published corrected mathematical tables for this projection, enabling other cartographers to produce charts on the Mercator projection. This projection is still the most widely used.

Sailing Directions or ***pilots*** have existed since at least the 6th century B.C. Continuous accumulation of navigational data, along with increased exploration and trade, led to increased production of volumes through the Middle Ages. "Routiers" were produced in France about 1500; the English referred to them as "rutters." In 1584 Lucas Waghenaer published the *Spieghel der Zeevaerdt (The Mariner's Mirror)*, which became the model for such publications for several generations of navigators. They were known as "Waggoners" by most sailors.

The **compass** was developed about 1000 years ago. The origin of the magnetic compass is uncertain, but

Norsemen used it in the 11th century, and Chinese navigators used the magnetic compass at least that early and probably much earlier. It was not until the 1870s that Lord Kelvin developed a reliable dry card marine compass. The fluid-filled compass became standard in 1906.

Variation was not understood until the 18th century, when Edmond Halley led an expedition to map lines of variation in the South Atlantic. **Deviation** was understood at least as early as the early 1600s, but adequate correction of compass error was not possible until Matthew Flinders discovered that a vertical iron bar could reduce certain types of errors. After 1840, British Astronomer Royal Sir George Airy and later Lord Kelvin developed combinations of iron masses and small magnets to eliminate most magnetic compass error.

The **gyrocompass** was made necessary by iron and steel ships. Leon Foucault developed the basic gyroscope in 1852. An American (Elmer Sperry) and a German (Anshutz Kampfe) both developed electrical gyrocompasses in the early years of the 20th century. Ring laser gyrocompasses and digital flux gate compasses are gradually replacing traditional gyrocompasses, while the magnetic compass remains an important backup device.

The **log** is the mariner's speedometer. Mariners originally measured speed by observing a chip of wood passing down the side of the vessel. Later developments included a wooden board attached to a reel of line. Mariners measured speed by noting how many knots in the line unreeled as the ship moved a measured amount of time; hence the term **knot.** Mechanical logs using either a small paddle wheel or a rotating spinner arrived about the middle of the 17th century. The taffrail log still in limited use today was developed in 1878. Modern logs use electronic sensors or spinning devices that induce small electric fields proportional to a vessel's speed. An engine revolution counter or shaft log often measures speed aboard large ships. Doppler speed logs are used on some vessels for very accurate speed readings. Inertial and satellite systems also provide highly accurate speed readings.

The Metric Conversion Act of 1975 and the Omnibus Trade and Competitiveness Act of 1988 established the **metric system** of weights and measures in the United States. As a result, the government is converting charts to the metric format. Notwithstanding the conversion to the metric system, the common measure of distance at sea is the **nautical mile**.

The current policy of the National Imagery and Mapping Agency (NIMA) and the National Ocean Service (NOS) is to convert new compilations of nautical, special purpose charts, and publications to the metric system. All digital charts use the metric system. This conversion began on January 2, 1970. Most modern maritime nations have also adopted the meter as the standard measure of depths and heights. However, older charts still on issue and the charts of some foreign countries may not conform to this standard.

The **fathom** as a unit of length or depth is of obscure origin. Posidonius reported a sounding of more than 1,000 fathoms in the 2nd century B.C. How old the unit was then is unknown. Many modern charts are still based on the fathom, as conversion to the metric system continues.

The sailings refer to various methods of mathematically determining course, distance, and position. They have a history almost as old as mathematics itself. Thales, Hipparchus, Napier, Wright, and others contributed the formulas that permit computation of course and distance by plane, traverse, parallel, middle latitude, Mercator, and great circle sailings.

104. The Earth

The Earth is an irregular oblate spheroid (a sphere flattened at the poles). Measurements of its dimensions and the amount of its flattening are subjects of geodesy. However, for most navigational purposes, assuming a spherical Earth introduces insignificant error. The Earth's axis of rotation is the line connecting the north and south geographic poles.

A great circle is the line of intersection of a sphere and a plane through its center. This is the largest circle that can be drawn on a sphere. The shortest line on the surface of a sphere between two points on the surface is part of a great circle. On the spheroidal Earth the shortest line is called a geodesic. A great circle is a near enough approximation to

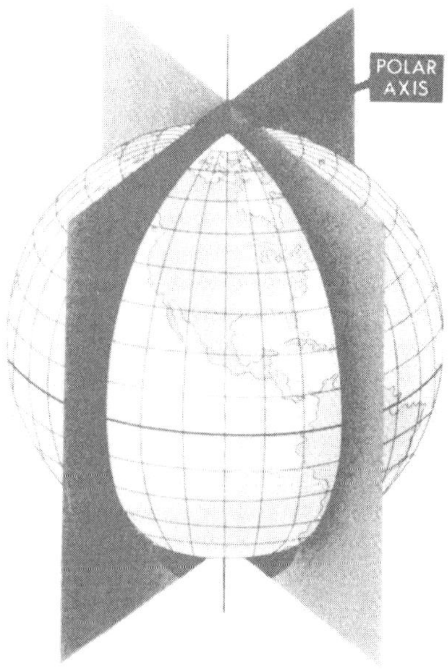

Figure 104a. The planes of the meridians at the polar axis.

a geodesic for most problems of navigation. A small circle is the line of intersection of a sphere and a plane which does not pass through the center. See Figure 104a.

The term **meridian** is usually applied to the **upper branch** of the half-circle from pole to pole which passes through a given point. The opposite half is called the **lower branch.**

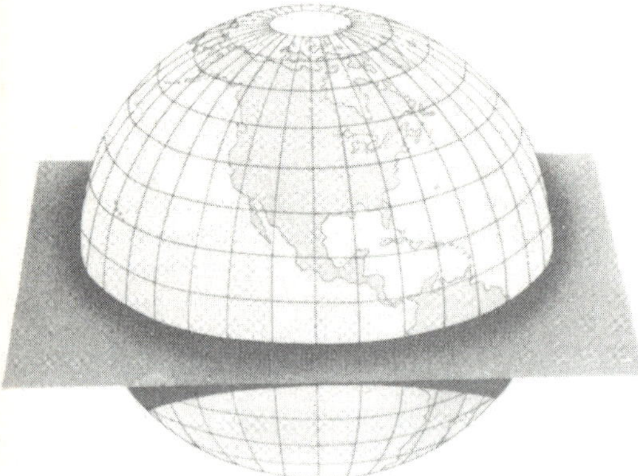

Figure 104b. The equator is a great circle midway between the poles.

A **parallel** or parallel of latitude is a circle on the surface of the Earth parallel to the plane of the equator. It connects all points of equal latitude. The equator is a great circle at latitude 0°. See Figure 104b. The poles are single points at latitude 90°. All other parallels are small circles.

105. Coordinates

Coordinates of latitude and longitude can define any position on Earth. **Latitude (L, lat.)** is the angular distance from the equator, measured northward or southward along a meridian from 0° at the equator to 90° at the poles. It is designated north (N) or south (S) to indicate the direction of measurement.

The **difference of latitude (*l*, DLat.)** between two places is the angular length of arc of any meridian between their parallels. It is the numerical difference of the latitudes if the places are on the same side of the equator; it is the sum of the latitudes if the places are on opposite sides of the equator. It may be designated north (N) or south (S) when appropriate. The middle or **mid-latitude (Lm)** between two places on the same side of the equator is half the sum of their latitudes. Mid-latitude is labeled N or S to indicate whether it is north or south of the equator.

The expression may refer to the mid-latitude of two places on opposite sides of the equator. In this case, it is equal to half the difference between the two latitudes and takes the name of the place farthest from the equator.

Longitude (l, long.) is the angular distance between the prime meridian and the meridian of a point on the Earth, measured eastward or westward from the prime meridian through 180°. It is designated east (E) or west (W) to indicate the direction of measurement.

The **difference of longitude (DLo)** between two places is the shorter arc of the parallel or the smaller angle at the pole between the meridians of the two places. If both places are on the same side (east or west) of Greenwich, DLo is the numerical difference of the longitudes of the two places; if on opposite sides, DLo is the numerical sum unless this exceeds 180°, when it is 360° minus the sum.

The distance between two meridians at any parallel of latitude, expressed in distance units, usually nautical miles, is called **departure (p, Dep.)**. It represents distance made good east or west as a craft proceeds from one point to another. Its numerical value between any two meridians decreases with increased latitude, while DLo is numerically the same at any latitude. Either DLo or p may be designated east (E) or west (W) when appropriate.

106. Distance on the Earth

Distance, as used by the navigator, is the length of the **rhumb line** connecting two places. This is a line making the same angle with all meridians. Meridians and parallels which also maintain constant true directions may be considered special cases of the rhumb line. Any other rhumb line spirals toward the pole, forming a **loxodromic curve** or **loxodrome**. See Figure 106. Distance along the great

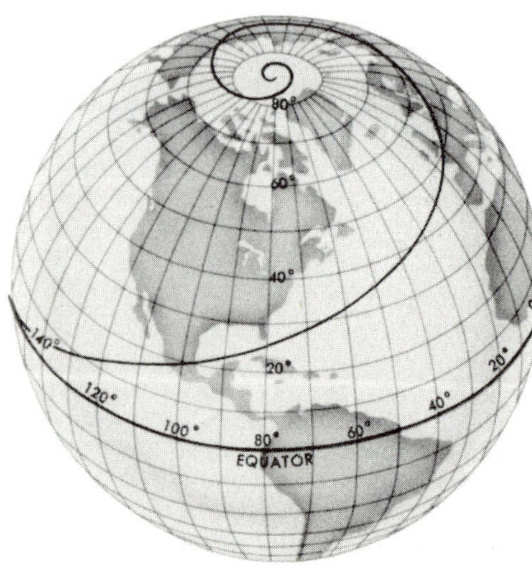

Figure 106. A loxodrome.

circle connecting two points is customarily designated **great-circle distance**. For most purposes, considering the nautical mile the length of one minute of latitude introduces no significant error

Speed (S) is rate of motion, or distance per unit of time. A **knot (kn.)**, the unit of speed commonly used in navigation, is a rate of 1 nautical mile per hour. The expression **speed of advance (SOA)** is used to indicate the speed to be made along the intended track. **Speed over the ground (SOG)** is the actual speed of the vessel over the surface of the Earth at any given time. To calculate **speed made good (SMG)** between two positions, divide the distance between the two positions by the time elapsed between the two positions.

107. Direction on the Earth

Direction is the position of one point relative to another. Navigators express direction as the angular difference in degrees from a reference direction, usually north or the ship's head. **Course (C, Cn)** is the horizontal direction in which a vessel is intended to be steered, expressed as angular distance from north clockwise through 360°. Strictly used, the term applies to direction through the water, not the direction intended to be made good over the ground.The course is often designated as true, magnetic, compass, or grid according to the reference direction.

Track made good (TMG) is the single resultant direction from the point of departure to point of arrival at any given time. **Course of advance (COA)** is the direction intended to be made good over the ground, and **course over ground (COG)** is the direction between a vessel's last fix and an EP. A **course line** is a line drawn on a chart extending in the direction of a course. It is sometimes convenient to express a course as an angle from either north or south, through 90° or 180°. In this case it is designated course angle (C) and should be properly labeled to indicate the origin (prefix) and direction of measurement (suffix). Thus, C N35°E = Cn 035° (000° + 35°), C N155°W = Cn 205° (360° - 155°), C S47°E = Cn 133° (180° - 47°). But Cn 260° may be either C N100°W or C S80°W, depending upon the conditions of the problem.

Track (TR) is the intended horizontal direction of travel with respect to the Earth. The terms intended track and trackline are used to indicate the path of intended travel. See Figure 107a. The track consists of one or a series of course lines, from the point of departure to the destination, along which one intends to proceed. A great circle which a vessel intends to follow is called a **great-circle track**, though it consists of a series of straight lines approximating a great circle

Heading (Hdg., SH) is the direction in which a vessel is pointed at any given moment, expressed as angular distance from 000° clockwise through 360°. It is easy to confuse heading and course. Heading constantly changes as a vessel yaws back and forth across the course due to sea, wind, and steering error.

Bearing (B, Brg.) is the direction of one terrestrial point from another, expressed as angular distance from 000° (North) clockwise through 360°. When measured through 90° or 180° from either north or south, it is called bearing angle (B). Bearing and azimuth are sometimes used interchangeably, but the latter more accurately refers to the horizontal direction of a point on the celestial sphere from a point on the Earth. A relative bearing is measured relative to the ship's heading from 000° (dead ahead) clockwise through 360°. However, it is sometimes conveniently measured right or left from 000° at the ship's head through 180°. This is particularly true when using the table for Distance of an Object by Two Bearings.

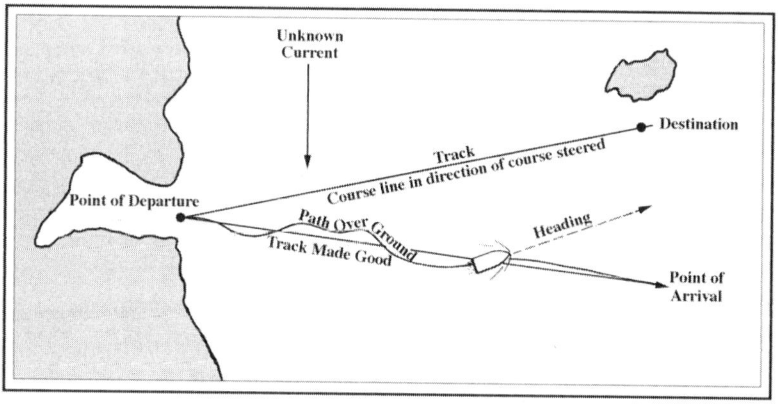

Figure 107a. Course line, track, track made good, and heading.

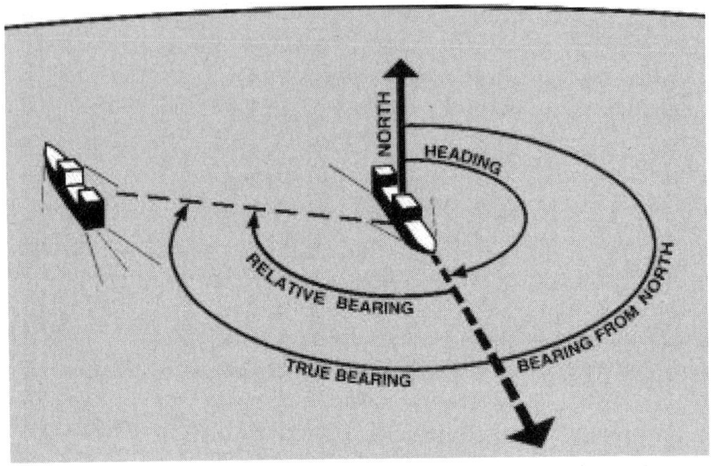

Figure 107b. Relative Bearing

To convert a relative bearing to a true bearing, add the true heading. See Figure 107b

True Bearing = Relative Bearing + True Heading.
Relative Bearing = True Bearing - True Heading.

108. Finding Latitude and Longitude

Navigators have made latitude observations for thousands of years. Accurate declination tables for the Sun have been published for centuries, enabling ancient seamen to compute latitude to within 1 or 2 degrees. Those who today determine their latitude by measuring the Sun at their meridian and the altitude of Polaris are using methods well known to 15th century navigators.

A method of finding longitude eluded mariners for centuries. Several solutions independent of time proved too cumbersome. Finding longitude by magnetic variation was tried, but found too inaccurate. The lunar distance method, which determines GMT by observing the Moon's position among the stars, became popular in the 1800s. However, the mathematics required by most of these processes were far above the abilities of the average seaman. It was apparent that the solution lay in keeping accurate time at sea.

In 1714, the British Board of Longitude was formed, offering a small fortune in reward to anyone who could provide a solution to the problem.

An Englishman, John Harrison, responded to the challenge, developing four chronometers between 1735 and 1760. The most accurate of these timepieces lost only 15 seconds on a 156 day round trip between London and Barbados. The Board, however, paid him only half the promised reward. The King finally intervened on Harrison's behalf, and at the age of 80 years Harrison received his full reward of £20,000.

Rapid chronometer development led to the problem of determining **chronometer error** aboard ship. **Time balls**, large black spheres mounted in port in prominent locations, were dropped at the stroke of noon, enabling any ship in harbor which could see the ball to determine chronometer error. By the end of the U.S. Civil War, telegraph signals were being used to key time balls. Use of radio signals to send time ticks to ships well offshore began in 1904, and soon worldwide signals were available.

109. The Navigational Triangle

Modern celestial navigators reduce their celestial observations by solving a **navigational triangle** whose points are the elevated pole, the celestial body, and the zenith of the observer. The sides of this triangle are the polar distance of the body (**codeclination**), its zenith distance (**coaltitude**), and the polar distance of the zenith (**colatitude** of the observer).

A spherical triangle was first used at sea in solving **lunar distance** problems. Simultaneous observations were made of the altitudes of the Moon and the Sun or a star near the ecliptic and the angular distance between the Moon and the other body. The zenith of the observer and the two celestial bodies formed the vertices of a triangle whose sides were the two coaltitudes and the angular distance between the bodies. Using a mathematical calculation the navigator "cleared" this distance of the effects of refraction and parallax applicable to each altitude. This corrected value was then used as an argument for entering the almanac. The almanac gave the true lunar distance from the Sun and several stars at 3 hour intervals. Previously, the navigator had set his watch or checked its error and rate with the local mean time determined by celestial observations. The local mean time of the watch, properly corrected, applied to the Greenwich mean time obtained from the lunar distance observation, gave the longitude.

The calculations involved were tedious. Few mariners could solve the triangle until Nathaniel Bowditch published his simplified method in 1802 in *The New American Practical Navigator*.

Reliable chronometers were available by1800, but their high cost precluded their general use aboard most ships. However, most navigators could determine their longitude using Bowditch's method. This eliminated the need for parallel sailing and the lost time associated with it. Tables for the lunar distance solution were carried in the American nautical almanac into the 20th century.

110. The Time Sight

The theory of the **time sight** had been known to mathematicians since the development of spherical trigonometry, but not until the chronometer was developed could it be used by mariners.

The time sight used the modern navigational triangle. The codeclination, or polar distance, of the body could be determined from the almanac. The zenith distance (coaltitude) was determined by observation. If the colatitude were known, three sides of the triangle were available. From these the meridian angle was computed. The comparison of this with the Greenwich hour angle from the almanac yielded the longitude.

The time sight was mathematically sound, but the navigator was not always aware that the longitude determined was only as accurate as the latitude, and together they merely formed a point on what is known today as a **line of position**. If the observed body was on the prime vertical, the line of position ran north and south and a small error in latitude generally had little effect on the longitude. But when the body was close to the meridian, a small error in latitude produced a large error in longitude.

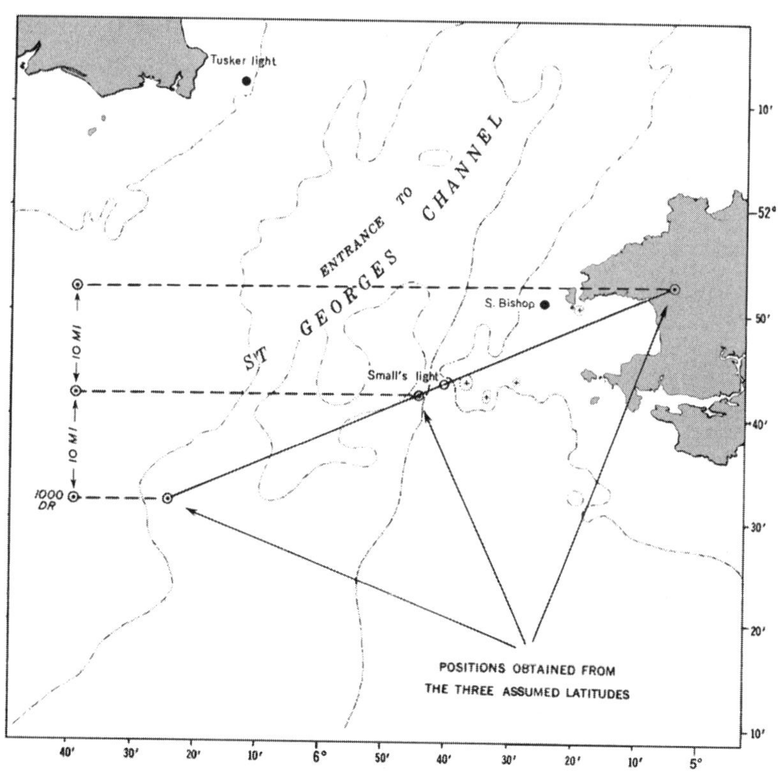

Figure 110. The first celestial line of position, obtained by Captain Thomas Sumner in 1837.

The line of position by celestial observation was unknown until discovered in 1837 by 30-year-old Captain Thomas H. Sumner, a Harvard graduate and son of a United States congressman from Massachusetts. The discovery of the "**Sumner line**," as it is sometimes called, was considered by Maury "the commencement of a new era in practical navigation." This was the turning point in the development of modern celestial navigation technique. In Sumner's own words, the discovery took place in this manner:

> *Having sailed from Charleston, S. C., 25th November, 1837, bound to Greenock, a series of heavy gales from the Westward promised a quick passage; after passing the Azores, the wind prevailed from the Southward, with thick weather; after passing Longitude 21° W, no observation was had until near the land; but soundings were had not far, as was supposed, from the edge of the Bank. The weather was now more boisterous, and very thick; and the wind still Southerly; arriving about midnight, 17th December, within 40 miles, by dead reckoning, of Tusker light; the wind hauled SE, true, making the Irish coast a lee shore; the ship was then kept close to the wind, and several tacks made to preserve her position as nearly as possible until daylight; when nothing being in sight, she was kept on ENE under short sail, with heavy gales; at about 10 AM an altitude of the Sun was observed, and the Chronometer time noted; but, having run so far without any observation, it was plain the Latitude by dead reckoning was liable to error, and could not be entirely relied on. Using, however, this Latitude, in finding the Longitude by Chronometer, it was found to put the ship 15' of Longitude E from her position by dead reckoning; which in Latitude 52° N is 9 nautical miles; this seemed to agree tolerably well with the dead reckoning; but feeling doubtful of the Latitude, the observation was tried with a Latitude 10' further N, finding this placed the ship ENE 27 nautical miles, of the former position, it was tried again with a Latitude 20' N of the dead reckoning; this also placed the ship still further ENE, and still 27 nautical miles further; these three positions were then seen to lie in the direction of Small's light. It then at once appeared that the observed altitude must have happened at all the three points, and at Small's light, and at the ship, at the same instant of time; and it followed, that Small's light must bear ENE, if the Chronometer was right. Having been convinced of this truth, the ship was kept on her course, ENE, the wind being still SE., and in less than an hour, Small's light was made bearing ENE 1/2 E, and close aboard.*

In 1843 Sumner published a book, A *New and Accurate Method of Finding a Ship's Position at Sea by Projection on Mercator's Chart*. He proposed solving a single time sight twice, using latitudes somewhat greater and somewhat less than that arrived at by dead reckoning, and joining the two positions obtained to form the line of position.

The Sumner method required the solution of two time sights to obtain each line of position. Many older navigators preferred not to draw the lines on their charts, but to fix their position mathematically by a method which Sumner had also devised and included in his book. This was a tedious but popular procedure.

111. Navigational Tables

Spherical trigonometry is the basis for solving every navigational triangle, and until about 80 years ago the navigator had no choice but to solve each triangle by tedious, manual computations.

Lord Kelvin, generally considered the father of modern navigational methods, expressed interest in a book of tables with which a navigator could avoid tedious trigonometric solutions. However, solving the many thousands of triangles involved would have made the project too costly. Computers finally provided a practical means of preparing tables. In 1936 the first volume of *Pub. No. 214* was made available; later, Pub. No. 249 was provided for air navigators. *Pub. No. 229, Sight Reduction Tables for Marine Navigation*, has replaced *Pub. No. 214*.

Electronic calculators are gradually replacing the tables. Scientific calculators with trigonometric functions can easily solve the navigational triangle. Navigational calculators readily solve celestial sights and perform a variety of voyage planning functions. Using a calculator generally gives more accurate lines of position because it eliminates the rounding errors inherent in tabular inspection and interpolation.

112. Development of Electronic Navigation

Perhaps the first application of electronics to navigation involved sending telegraphic time signals in 1865 to check chronometer error. Transmitting radio time signals for chronometer checks dates to 1904. Radio broadcasts providing navigational warnings, begun in 1907 by the U.S. Navy Hydrographic Office, helped increase the safety of navigation at sea.

By the latter part of World War I the directional properties of a loop antenna were successfully used in the radio direction finder. The first radiobeacon was installed in 1921. Early 20th century experiments by Behm and Langevin led to the U.S. Navy's development of the first practical echo sounder in 1922. Radar and hyperbolic systems grew out of WWII.

Today, electronics touches almost every aspect of navigation. Hyperbolic systems, satellite systems, and electronic charts all require an increasingly sophisticated electronics suite and the expertise to manage them. These systems' accuracy and ease of use make them invaluable assets to the navigator, but there is far more to using them than knowing which buttons to push.

113. Development of Radar

As early as 1904, German engineers were experimenting with reflected radio waves. In 1922 two American scientists, Dr. A. Hoyt Taylor and Leo C. Young, testing a communication system at the Naval Aircraft Radio Laboratory, noted fluctuations in the signals when ships passed between stations on opposite sides of the Potomac River. In 1935 the British began work on radar. In 1937 the USS Leary tested the first sea-going radar, and in 1940 United States and British scientists combined their efforts. When the British revealed the principle of the multicavity magnetron developed by J. T. Randall and H. A. H. Boot at the University of Birmingham in 1939, microwave radar became practical. In 1945, at the close of World War II, radar became available for commercial use.

114. Development of Hyperbolic Radio Aids

Various hyperbolic systems were developed beginning in World War II. These were outgrowths of the British GEE system, developed to help bombers navigate to and from their missions over Europe. Loran A was developed as a long-range marine navigation system. This was replaced by the more accurate Loran C system, deployed throughout much of the world. Various short range and regional hyperbolic systems have been developed by private industry for hydrographic surveying, offshore facilities positioning, and general navigation.

115. Other Electronic Systems

The underlying concept that led to development of satellite navigation dates to 1957 and the first launch of an artificial satellite into orbit. The first system, NAVSAT, has been replaced by the far more accurate and widely available **Global Positioning System (GPS)**, which has revolutionized all aspects of navigation

The first **inertial navigation system** was developed in 1942 for use in the V2 missile by the Peenemunde group under the leadership of Dr. Wernher von Braun. This system used two 2-degree-of-freedom gyroscopes and an integrating accelerometer to determine the missile velocity. By the end of World War II, the Peenemunde group had developed a stable platform with three single-degree-of-freedom gyroscopes and an integrating accelerometer. In 1958 an inertial navigation system was used to navigate the USS *Nautilus* under the ice to the North Pole.

NAVIGATION ORGANIZATIONS

116. Governmental Role

Navigation only a generation ago was an independent process, carried out by the mariner without outside assistance. With compass and charts, sextant and chronometer, he could independently travel anywhere in the world. The increasing use of electronic navigation systems has made the navigator dependent on many factors outside his control. Government organizations fund, operate, and regulate satellites, Loran, and other electronic systems. Governments are increasingly involved in regulation of vessel movements through traffic control systems and regulated areas. Understanding the governmental role in supporting and regulating navigation is vitally important to the mariner. In the United States, there are a number of official organizations which support the interests of navigators. Some have a policy-making role; others build and operate navigation systems. Many maritime nations have similar organizations performing similar functions. International organizations also play a significant role.

117. The Coast and Geodetic Survey

The **U.S. Coast and Geodetic Survey** was founded in 1807 when Congress passed a resolution authorizing a survey of the coast, harbors, outlying islands, and fishing banks of the United States. President Thomas Jefferson appointed Ferdinand Hassler, a Swiss immigrant and professor of mathematics at West Point, the first Director of the "Survey of the Coast." The survey became the "Coast Survey" in 1836.

The approaches to New York were the first sections of the coast charted, and from there the work spread northward and southward along the eastern seaboard. In 1844 the work was expanded and arrangements made to simultaneously chart the gulf and east coasts. Investigation of tidal conditions began, and in 1855 the first tables of tide predictions were published. The California gold rush necessitated a survey of the west coast, which began in 1850, the year California became a state. *Coast Pilots*, or *Sailing Directions*, for the Atlantic coast of the United States were privately published in the first half of the 19th century. In 1850 the Survey began accumulating data that led to federally produced *Coast Pilots*. The 1889 *Pacific Coast Pilot* was an outstanding contribution to the safety of west coast shipping.

In 1878 the survey was renamed "Coast and Geodetic Survey." In 1970 the survey became the "National Ocean Survey," and in 1983 it became the "National Ocean Service." The Office of Charting and Geodetic Services accomplished all charting and geodetic functions. In 1991 the name was changed back to the original "Coast and Geodetic Survey," organized under the National Ocean Service along with several other environmental offices. Today it provides the mariner with the charts and coast pilots of all waters of the United States and its possessions, and tide and tidal current tables for much of the world. Its administrative order requires the Coast and Geodetic Survey to plan and direct programs to produce charts and related information for safe navigation of U.S. waterways, territorial seas, and airspace. This work includes all activities related to the National Geodetic Reference System; surveying, charting, and data collection; production and distribution of charts; and research and development of new technologies to enhance these missions.

118. The National Imagery and Mapping Agency

In the first years of the newly formed United States of America, charts and instruments used by the Navy and merchant mariners were left over from colonial days or were obtained from European sources. In 1830 the U.S. Navy established a "Depot of Charts and Instruments" in Washington, D. C., as a storehouse from which available charts, pilots and sailing directions, and navigational instruments were issued to Naval ships. Lieutenant L. M. Goldsborough and one assistant, Passed Midshipman R. B. Hitchcock, constituted the entire staff.

The first chart published by the Depot was produced from data obtained in a survey made by Lieutenant Charles Wilkes, who had succeeded Goldsborough in 1834. Wilkes later earned fame as the leader of a United States expedition to Antarctica. From 1842 until 1861 Lieutenant Matthew Fontaine Maury served as Officer in Charge. Under his command the Depot rose to international prominence.

Maury decided upon an ambitious plan to increase the mariner's knowledge of existing winds, weather, and currents. He began by making a detailed record of pertinent matter included in old log books stored at the Depot. He then inaugurated a hydrographic reporting program among ship masters, and the thousands of reports received, along with the log book data, were compiled into the "*Wind and Current Chart of the North Atlantic*" in 1847. This is the ancestor of today's *Pilot Chart*.

The United States instigated an international conference in 1853 to interest other nations in a system of exchanging nautical information. The plan, which was Maury's, was enthusiastically adopted by other maritime nations. In 1854 the Depot was redesignated the "U.S. Naval Observatory and Hydrographical Office." At the outbreak of the American Civil War in 1861, Maury, a native of Virginia, resigned from the U.S. Navy and accepted a commission in the Confederate Navy. This effectively ended his career as a navigator, author, and oceanographer. At war's end, he fled the country, his reputation suffering from his embrace of the Confederate cause.

After Maury's return to the United States in 1868, he served as an instructor at the Virginia Military Institute. He continued at this position until his death in 1873. Since his death, his reputation as one of America's greatest hydrographers has been restored.

In 1866 Congress separated the Observatory and the Hydrographic Office, broadly increasing the functions of the latter. The Hydrographic Office was authorized to carry out surveys, collect information, and print every kind of nautical chart and publication "for the benefit and use of navigators generally."

The Hydrographic Office purchased the copyright of ***The New American Practical Navigator*** in 1867. The first *Notice to Mariners* appeared in 1869. Daily broadcast of navigational warnings was inaugurated in 1907. In 1912, following the sinking of the *Titanic*, the International Ice Patrol was established.

In 1962 the U.S. Navy Hydrographic Office was redesignated the U.S. Naval Oceanographic Office. In 1972 certain hydrographic functions of the latter office were transferred to the **Defense Mapping Agency Hydrographic Center**. In 1978 the **Defense Mapping Agency Hydrographic/Topographic Center (DMAHTC)** assumed hydrographic and topographic chart production functions. In 1996 the **National Imagery and Mapping Agency (NIMA)** was formed from DMA and certain other elements of the Department of Defense. NIMA continues to produce charts and publications and to disseminate maritime safety information in support of the U.S. military and navigators generally.

119. The United States Coast Guard

Alexander Hamilton established the **U.S. Coast Guard** as the Revenue Marine, later the Revenue Cutter Service, on August 4, 1790. It was charged with enforcing the customs laws of the new nation. A revenue cutter, the *Harriet Lane*, fired the first shot from a naval unit in the Civil War at Fort Sumter. The Revenue Cutter Service became the U.S. Coast Guard when combined with the Lifesaving Service in 1915. The Lighthouse Service was added in 1939, and the Bureau of Marine Inspection and Navigation was added in 1942. The Coast Guard was transferred from the Treasury Department to the Department of Transportation in 1967.

The primary functions of the Coast Guard include maritime search and rescue, law enforcement, and operation of the nation's aids to navigation system. In addition, the Coast Guard is responsible for port safety and security, merchant marine inspection, and marine pollution control. The Coast Guard operates a large and varied fleet of ships, boats, and aircraft in performing its widely ranging duties

Navigation systems operated by the Coast Guard include the system of some 40,000 lighted and unlighted beacons, buoys, and ranges in U.S. and territorial waters; the U.S. stations of the Loran C system; differential GPS (DGPS) services in the U.S.; and Vessel Traffic Services (VTS) in major ports and harbors of the U.S.

120. The United States Navy

The **U.S. Navy** was officially established in 1798. Its role in the development of navigational technology has been singular. From the founding of the Naval Observatory to the development of the most advanced electronics, the U.S. Navy has been a leader in developing devices and techniques designed to make the navigator's job safer and easier.

The development of almost every device known to navigation science has been deeply influenced by Naval policy. Some systems are direct outgrowths of specific Naval needs; some are the result of technological improvements shared with other services and with commercial maritime industry.

121. The United States Naval Observatory

One of the first observatories in the United States was built in 1831-1832 at Chapel Hill, N.C. The Depot of Charts and Instruments, established in 1830, was the agency from which the U.S. Navy Hydrographic Office and the **U.S. Naval Observatory** evolved 36 years later. In about 1835, under Lieutenant Charles Wilkes, the second Officer in Charge, the Depot installed a small transit instrument for rating chronometers.

The Mallory Act of 1842 provided for the establishment of a permanent observatory. The director was authorized to purchase everything necessary to continue astronomical study. The observatory was completed in 1844 and the results of its first observations were published two years later. Congress established the Naval Observatory as a separate agency in 1866. In 1873 a refracting telescope with a 26 inch aperture, then the world's largest, was installed. The observatory, located in Washington, D.C., has occupied its present site since 1893.

122. The Royal Greenwich Observatory

England had no early privately supported observatories such as those on the continent. The need for navigational advancement was ignored by Henry VIII and Elizabeth I, but in 1675 Charles II, at the urging of John Flamsteed, Jonas Moore, Le Sieur de Saint Pierre, and Christopher Wren, established the **Greenwich Royal Observatory**. Charles limited construction costs to £500, and appointed Flamsteed the first Astronomer Royal, at an annual salary of £100. The equipment available in the early years of the observatory consisted of two clocks, a "sextant" of 7 foot radius, a quadrant of 3 foot radius, two telescopes, and the star catalog published almost a century before by Tycho Brahe. Thirteen years passed before Flamsteed had an instrument with which he could determine his latitude accurately.

In 1690 a transit instrument equipped with a telescope and vernier was invented by Romer; he later added a vertical circle to the device. This enabled the astronomer to determine declination and right ascension at the same time. One of these instruments was added to the equipment at Greenwich in 1721, replacing the huge quadrant previously used. The development and perfection of the chronometer in the next hundred years added to the accuracy of observations.

Other national observatories were constructed in the years that followed: at Berlin in 1705, St. Petersburg in 1725, Palermo in 1790, Cape of Good Hope in 1820, Parramatta in New South Wales in 1822, and Sydney in 1855.

123. The International Hydrographic Organization

The **International Hydrographic Organization (IHO)** was originally established in 1921 as the International Hydrographic Bureau (IHB). The present name was adopted in 1970 as a result of a revised international agreement among member nations. However, the former name, International Hydrographic Bureau, was retained for the IHO's administrative body of three Directors and their staff at the organization's headquarters in Monaco.

The IHO sets forth hydrographic standards to be agreed upon by the member nations. All member states are urged and encouraged to follow these standards in their surveys, nautical charts, and publications. As these standards are uniformly adopted, the products of the world's hydrographic and oceanographic offices become more uniform. Much has been done in the field of standardization since the Bureau was founded.

The principal work undertaken by the IHO is:

- To bring about a close and permanent association between national hydrographic offices.
- To study matters relating to hydrography and allied sciences and techniques.
- To further the exchange of nautical charts and documents between hydrographic offices of member governments.
- To circulate the appropriate documents.
- To tender guidance and advice upon request, in particular to countries engaged in setting up or expanding their hydrographic service.
- To encourage coordination of hydrographic surveys with relevant oceanographic activities.
- To extend and facilitate the application of oceanographic knowledge for the benefit of navigators.
- To cooperate with international organizations and scientific institutions which have related objectives.

During the 19th century, many maritime nations established hydrographic offices to provide means for improving the navigation of naval and merchant vessels by providing nautical publications, nautical charts, and other navigational services. There were substantial differences in hydrographic procedures, charts, and publications. In 1889, an International Marine Conference was held at

Washington, D. C., and it was proposed to establish a "permanent international commission." Similar proposals were made at the sessions of the International Congress of Navigation held at St. Petersburg in 1908 and again in 1912.

In 1919 the hydrographers of Great Britain and France cooperated in taking the necessary steps to convene an international conference of hydrographers. London was selected as the most suitable place for this conference, and on July 24, 1919, the First International Conference opened, attended by the hydrographers of 24 nations. The object of the conference was "To consider the advisability of all maritime nations adopting similar methods in the preparation, construction, and production of their charts and all hydrographic publications; of rendering the results in the most convenient form to enable them to be readily used; of instituting a prompt system of mutual exchange of hydrographic information between all countries; and of providing an opportunity to consultations and discussions to be carried out on hydrographic subjects generally by the hydrographic experts of the world." This is still the major purpose of the International Hydrographic Organization.

As a result of the conference, a permanent organization was formed and statutes for its operations were prepared. The International Hydrographic Bureau, now the International Hydrographic Organization, began its activities in 1921 with 18 nations as members. The Principality of Monaco was selected because of its easy communication with the rest of the world and also because of the generous offer of Prince Albert I of Monaco to provide suitable accommodations for the Bureau in the Principality. There are currently 59 member governments. Technical assistance with hydrographic matters is available through the IHO to member states requiring it.

Many IHO publications are available to the general public, such as the International Hydrographic Review, International Hydrographic Bulletin, Chart Specifications of the IHO, Hydrographic Dictionary, and others. Inquiries should be made to the International Hydrographic Bureau, 7 Avenue President J. F. Kennedy, B.P. 445, MC98011, Monaco, CEDEX.

124. The International Maritime Organization

The **International Maritime Organization (IMO)** was established by United Nations Convention in 1948. The Convention actually entered into force in 1959, although an international convention on marine pollution was adopted in 1954. (Until 1982 the official name of the organization was the Inter-Governmental Maritime Consultative Organization.) It is the only permanent body of the U. N. devoted to maritime matters, and the only special U. N. agency to have its headquarters in the UK.

The governing body of the IMO is the **Assembly** of 137 member states, which meets every two years. Between Assembly sessions a Council, consisting of 32 member governments elected by the Assembly, governs the organization. Its work is carried out by the Maritime Safety Committee, with subcommittees for:

- Safety of Navigation
- Radiocommunications
- Life-saving
- Search and Rescue
- Training and Watchkeeping
- Carriage of Dangerous Goods
- Ship Design and Equipment
- Fire Protection
- Stability and Load Lines/Fishing Vessel Safety
- Containers and Cargoes
- Bulk Chemicals
- Marine Environment Protection Committee
- Legal Committee
- Technical Cooperation Committee
- Facilitation Committee

IMO is headed by the Secretary General, appointed by the council and approved by the Assembly. He is assisted by some 300 civil servants.

To achieve its objectives of coordinating international policy on marine matters, the IMO has adopted some 30 conventions and protocols, and adopted over 700 codes and recommendations. An issue to be adopted first is brought before a committee or subcommittee, which submits a draft to a conference. When the conference adopts the final text, it is submitted to member governments for ratification. Ratification by a specified number of countries is necessary for adoption; the more important the issue, the more countries must ratify. Adopted conventions are binding on member governments.

Codes and recommendations are not binding, but in most cases are supported by domestic legislation by the governments involved.

The first and most far-reaching convention adopted by the IMO was the Convention of **Safety of Life at Sea (SOLAS)** in 1960. This convention actually came into force in 1965, replacing a version first adopted in 1948. Because of the difficult process of bringing amendments into force internationally, none of subsequent amendments became binding. To remedy this situation, a new convention was adopted in 1974 and became binding in 1980. Among the regulations is V-20, requiring the carriage of up-to-date charts and publications sufficient for the intended voyage.

Other conventions and amendments were also adopted, such as the International Convention on Load Lines (adopted 1966, came into force 1968), a convention on the tonnage measurement of ships (adopted 1969, came into force 1982), The International Convention on Safe Containers (adopted 1972, came into force 1977), and the convention on **International Regulations for Preventing Collisions at Sea (COLREGS)** (adopted 1972, came into force 1977).

The 1972 COLREGS convention contained, among other provisions, a section devoted to Traffic Separation

Schemes, which became binding on member states after having been adopted as recommendations in prior years.

One of the most important conventions is the **International Convention for the Prevention of Pollution from Ships (MARPOL 73/78)**, which was first adopted in 1973, amended by Protocol in 1978, and became binding in 1983. This convention built on a series of prior conventions and agreements dating from 1954, highlighted by several severe pollution disasters involving oil tankers. The MARPOL convention reduces the amount of oil discharged into the sea by ships, and bans discharges completely in certain areas. A related convention known as the London Dumping Convention regulates dumping of hazardous chemicals and other debris into the sea.

The IMO also develops minimum performance standards for a wide range of equipment relevant to safety at sea. Among such standards is one for the **Electronic Chart Display and Information System (ECDIS)**, the digital display deemed the operational and legal equivalent of the conventional paper chart.

Texts of the various conventions and recommendations, as well as a catalog and publications on other subjects, are available from the Publications Section of the IMO at 4 Albert Embankment, London SE1 7SR, United Kingdom.

125. The International Association of Marine Aids to Navigation and Lighthouse Authorities

The **International Association of Marine Aids to Navigation and Lighthouse Authorities (formerly IALA)** brings together representatives of the aids to navigation services of more than 80 member countries for technical coordination, information sharing, and coordination of improvements to visual aids to navigation throughout the world. It was established in 1957 to provide a permanent organization to support the goals of the Technical Lighthouse Conferences, which had been convening since 1929. The General Assembly of IALA meets about every 4 years. The Council of 20 members meets twice a year to oversee the ongoing programs.

Five technical committees maintain the permanent programs:

- The Marine Marking Committee
- The Radionavigation Systems Committee
- The Vessel Traffic Services (VTS) Committee
- The Reliability Committee
- The Documentation Committee

IALA committees provide important documentation to the IHO and other international organizations, while the IALA Secretariat acts as a clearing house for the exchange of technical information, and organizes seminars and technical support for developing countries.

Its principle work since 1973 has been the implementation of the IALA Maritime Buoyage System, described in Chapter 5, Visual Aids to Navigation. This system replaced some 30 dissimilar buoyage systems in use throughout the world with 2 major systems.

IALA is based near Paris, France in Saint-Germaine-en-Laye.

126. The Radio Technical Commission for Maritime Services

The **Radio Technical Commission for Maritime Services** is a non-profit organization which serves as a focal point for the exchange of information and the development of recommendations and standards related to all aspects of maritime radiocommunications and radionavigation.

Specifically, RTCM:

- Promotes ideas and exchanges information on maritime radiocommunications and radionavigation.
- Facilitates the development and exchange of views among and between government and non-government interests both nationally and internationally.
- Conducts studies and prepares reports on maritime radiocommunications and radionavigation issues to improve efficiency and capabilities.

Both government and non-government organizations are members, coming from the U.S. and many other nations. The RTCM organization consists of a Board of Directors, and the Assembly consisting of all members, officers, staff, technical advisors, and working committees.

Working committees are formed as needed to develop official RTCM recommendations regarding technical standards and regulatory policies in the maritime field. Currently committees address such issues as maritime safety information, electronic charts, emergency position-indicating radiobeacons (EPIRB's), personal locator beacons, ship radars, differential GPS, GLONASS, and maritime survivor locator devices.

The RTCM headquarters office is in Alexandria, VA.

127. The National Marine Electronic Association

The **National Marine Electronic Association (NMEA)** is a professional trade association founded in 1957 whose purpose is to coordinate the efforts of marine electronics manufacturers, technicians, government agencies, ship and boat builders, and other interested groups. In addition to certifying marine electronics technicians and professionally recognizing outstanding achievements by corporate and individual members, the NMEA sets standards for the exchange of digital data by all manufacturers of marine electronic equipment. This allows the configuration of integrated navigation system using equipment from different manufacturers.

NMEA works closely with RTCM and other private organizations and with government agencies to monitor the status of laws and regulations affecting the marine electronics industry.

It also sponsors conferences and seminars, and publishes a number of guides and periodicals for members and the general public.

128. International Electrotechnical Commission

The **International Electrotechnical Commission (IEC)** was founded in 1906 as an outgrowth of the International Electrical Congress held at St. Louis, Missouri in 1904. Some 60 countries are active members. Its mission is to develop and promote standardization among all nations in the technical specifications of electrical and electronic equipment. These technologies include electronics, magnetics, electromagnetics, electroacoustics, multimedia, telecommunications, electrical energy production and distribution, and associated fields such as terminology and symbology, compatibility, performance standards, safety, and environmental factors.

By standardizing in these areas, the IEC seeks to promote more efficient markets, improve the quality of products and standards of performance, promote interoperability, increase production efficiency, and contribute to human health and safety and environmental protection.

Standards are published by the IEC in the form of official IEC documents after debate and input from the national committees. Standards thus represent a consensus of the views of many different interests. Adoption of a standard by any country is entirely voluntary. However, failure to adopt a standard may result in a technical barrier to trade, as goods manufactured to a proprietary standard in one country may be incompatible with the systems of others.

IEC standards are vital to the success of ECDIS and other integrated navigation systems because they help to ensure that systems from various manufacturers in different countries will be compatible and meet required specifications.

CHAPTER 2

GEODESY AND DATUMS IN NAVIGATION

GEODESY, THE BASIS OF CARTOGRAPHY

200. Definition

Geodesy is the science concerned with the exact positioning of points on the surface of the Earth. It also involves the study of variations of the Earth's gravity, the application of these variations to exact measurements on the Earth, and the study of the exact size and shape of the Earth. These factors were unimportant to early navigators because of the relative inaccuracy of their methods. The precision of today's navigation systems and the global nature of satellite and other long-range positioning methods demand a more complete understanding of geodesy by the navigator than has ever before been required.

201. The Shape of the Earth

The **topographic surface** is the actual surface of the earth, upon which geodetic measurements are made. These measurements are then reduced to the **geoid**. Marine navigation measurements are made on the ocean surface which approximates the geoid.

The **geoid** is a surface along which gravity is always equal and to which the direction of gravity is always perpendicular. The latter point is particularly significant because optical instruments containing leveling devices are commonly used to make geodetic measurements. When properly adjusted, the vertical axis of the instrument coincides exactly with the direction of gravity and is by definition perpendicular to the geoid. See Figure 201.

The geoid is that surface to which the oceans would conform over the entire Earth if free to adjust to the combined effect of the Earth's mass attraction and the centrifugal force of the Earth's rotation. Uneven distribution of the Earth's mass makes the geoidal surface irregular.

The geoid refers to the actual size and shape of the Earth, but such an irregular surface has serious limitations as a mathematical Earth model because:

- It has no complete mathematical expression.
- Small variations in surface shape over time introduce small errors in measurement.
- The irregularity of the surface would necessitate a prohibitive amount of computations.

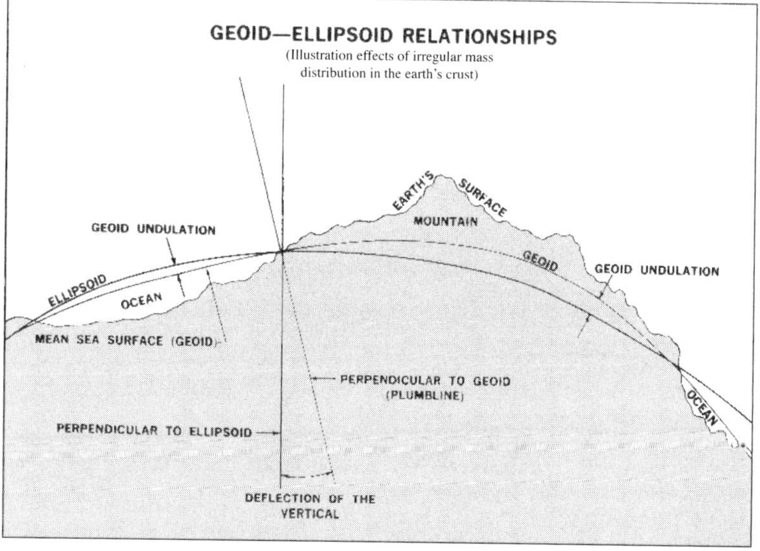

Figure 201. Geoid, ellipsoid, and topographic surface of the Earth, and deflection of the vertical due to differences in mass.

The surface of the geoid, with some exceptions, tends to rise under mountains and to dip above ocean basins.

For geodetic, mapping, and charting purposes, it is necessary to use a regular or geometric shape which closely approximates the shape of the geoid either on a local or global scale and which has a specific mathematical expression. This shape is called the **ellipsoid**.

The separations of the geoid and ellipsoid are called **geoidal heights**, **geoidal undulations**, or **geoidal separations**.

Natural irregularities in density and depths of the material making up the upper crust of the Earth also result in slight alterations of the direction of gravity. These alterations are reflected in the irregular shape of the geoid, the surface that is perpendicular to a plumb line.

Since the Earth is in fact flattened slightly at the poles and bulges somewhat at the equator, the geometric figure used in geodesy to most nearly approximate the shape of the Earth is the **oblate spheroid** or **ellipsoid of revolution**. This is the three dimensional shape obtained by rotating an ellipse about its minor axis.

202. Defining the Ellipsoid

An ellipsoid of revolution is uniquely defined by specifying two parameters. Geodesists, by convention, use the **semimajor axis** and **flattening**. The size is represented by the radius at the equator, the semimajor axis. The shape of the ellipsoid is given by the flattening, which indicates how closely an ellipsoid approaches a spherical shape. The flattening is the ratio of the difference between the semimajor and semiminor axes of the ellipsoid and the semimajor axis. See Figure 202. If a and b represent the semimajor and semiminor axes, respectively, of the ellipsoid, and f is the flattening,

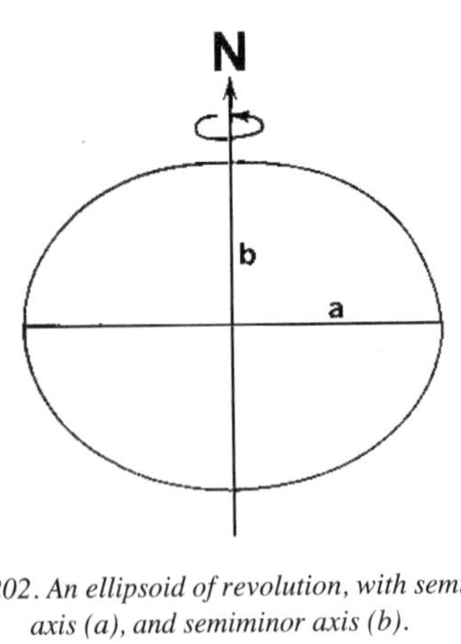

Figure 202. An ellipsoid of revolution, with semimajor axis (a), and semiminor axis (b).

$$f = \frac{a - b}{a}.$$

This ratio is about 1/300 for the Earth. The ellipsoidal Earth model has its minor axis parallel to the Earth's polar axis.

203. Ellipsoids and the Geoid as Reference Surfaces

Since the surface of the geoid is irregular and the surface of an ellipsoid is regular, no ellipsoid can provide more than an approximation of part of the geoidal surface. Figure 203 illustrates an example. A variety of ellipsoids are necessary to cover the entire earth.

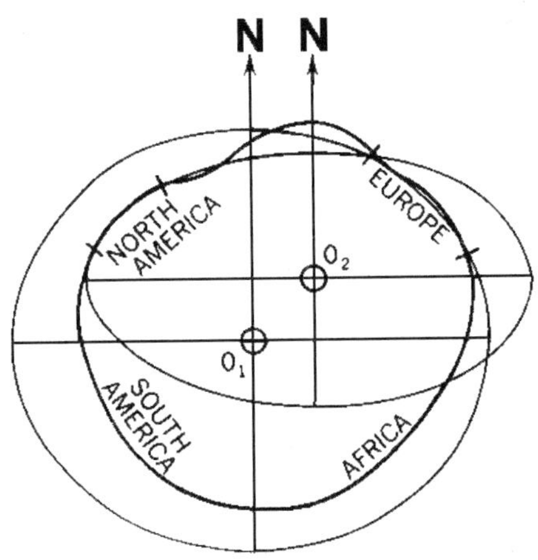

Figure 203. An ellipsoid which fits well in North America may not fit well in Europe, whose ellipsoid must have a different size, shape, and origin. Other ellipsoids are necessary for other areas

204. Coordinates

The **astronomic latitude** is the angle between a plumb line and the plane of the celestial equator. It is the latitude which results directly from observations of celestial bodies, uncorrected for deflection of the vertical component in the meridian (north-south) direction. Astronomic latitude applies only to positions on the Earth. It is reckoned from the astronomic equator (0°), north and south through 90°.

The **astronomic longitude** is the angle between the plane of the celestial meridian at a station and the plane of the celestial meridian at Greenwich. It is the longitude which results directly from observations of celestial bodies, uncorrected for deflection of the vertical component in the prime vertical (east-west) direction. These are the coordinates observed by the celestial navigator using a sextant and a very accurate clock based on the Earth's rotation.

Celestial observations by geodesists are made with optical instruments (theodolite, zenith camera, prismatic astrolabe) which all contain leveling devices. When properly adjusted, the vertical axis of the instrument coincides with the direction of gravity, which may not coincides with the plane of the meridian. Thus, geodetically derived astronomic positions are referenced to the geoid. The difference, from a navigational standpoint, is too small to be of concern.

The **geodetic latitude** is the angle which the normal to the ellipsoid at a station makes with the plane of the geodetic equator. In recording a geodetic position, it is essential that the geodetic datum on which it is based also be stated. A geodetic latitude differs from the corresponding astronomic latitude by the amount of the meridian component of the local deflection of the vertical.

The **geodetic longitude** is the angle between the plane of the geodetic meridian at a station and the plane of the geodetic meridian at Greenwich. A geodetic longitude differs from the corresponding astronomic longitude by the prime vertical component of the local deflection of the vertical divided by the cosine of the latitude. The geodetic coordinates are used for mapping.

The **geocentric latitude** is the angle at the center of the ellipsoid (used to represent the Earth) between the plane of the equator, and a straight line (or radius vector) to a point on the surface of the ellipsoid. This differs from geodetic latitude because the Earth is approximated more closely by a spheroid than a sphere and the meridians are ellipses, not perfect circles.

Both geocentric and geodetic latitudes refer to the reference ellipsoid and not the Earth. Since the parallels of latitude are considered to be circles, geodetic longitude is geocentric, and a separate expression is not used.

Because of the oblate shape of the ellipsoid, the length of a degree of geodetic latitude is not everywhere the same, increasing from about 59.7 nautical miles at the equator to about 60.3 nautical miles at the poles.

A **horizontal geodetic datum** usually consists of the astronomic and geodetic latitude, and astronomic and geodetic longitude of an initial point (origin); an azimuth of a line (direction); the parameters (radius and flattening) of the ellipsoid selected for the computations; and the geoidal separation at the origin. A change in any of these quantities affects every point on the datum.

For this reason, while positions within a given datum are directly and accurately relatable, those from different datums must be transformed to a common datum for consistency.

TYPES OF GEODETIC SURVEY

205. Triangulation

The most common type of geodetic survey is known as **triangulation**. Triangulation consists of the measurement of the angles of a series of triangles. The principle of triangulation is based on plane trigonometry. If the distance along one side of the triangle and the angles at each end are accurately measured, the other two sides and the remaining angle can be computed. In practice, all of the angles of every triangle are measured to provide precise measurements. Also, the latitude and longitude of one end of the measured side along with the length and direction (azimuth) of the side provide sufficient data to compute the latitude and longitude of the other end of the side.

The measured side of the base triangle is called a **baseline**. Measurements are made as carefully and accurately as possible with specially calibrated tapes or wires of Invar, an alloy with a very low coefficient of expansion. The tape or wires are checked periodically against standard measures of length.

To establish an arc of triangulation between two widely separated locations, the baseline may be measured and longitude and latitude determined for the initial points at each location. The lines are then connected by a series of adjoining triangles forming quadrilaterals extending from each end. All angles of the triangles are measured repeatedly to reduce errors. With the longitude, latitude, and azimuth of the initial points, similar data is computed for each vertex of the triangles, thereby establishing triangulation stations, or geodetic control stations. The coordinates of each of the stations are defined as geodetic coordinates.

Triangulation is extended over large areas by connecting and extending series of arcs to form a network or triangulation system. The network is adjusted so as to reduce observational errors to a minimum. A denser distribution of geodetic control is achieved by subdividing or filling in with other surveys.

There are four general classes or orders of triangulation. **First-order** (primary) triangulation is the most precise and exact type. The most accurate instruments and rigorous computation methods are used. It is costly and time-consuming, and is usually used to provide the basic framework of control data for an area, and the determination of the figure of the Earth. The most accurate first-order surveys furnish control points which can be interrelated with an accuracy ranging from 1 part in 25,000 over short distances to approximately 1 part in 100,000 for long distances.

Second-order triangulation furnishes points closer together than in the primary network. While second-order surveys may cover quite extensive areas, they are usually tied to a primary system where possible. The procedures are less exacting and the proportional error is 1 part in 10,000.

Third-order triangulation is run between points in a secondary survey. It is used to densify local control nets and position the topographic and hydrographic detail of the area. Error can amount to 1 part in 5,000.

The sole accuracy requirement for **fourth-order** triangulation is that the positions be located without any appreciable error on maps compiled on the basis of the control. Fourth-order control is done primarily as mapping control.

206. Trilateration, Traverse, And Vertical Surveying

Trilateration involves measuring the sides of a chain of triangles or other polygons. From them, the distance and direction from A to B can be computed. Figure 206 shows this process.

Traverse involves measuring distances and the angles between them without triangles for the purpose of computing the distance and direction from A to B. See Figure 206.

Vertical surveying is the process of determining elevations above mean sea-level. In geodetic surveys executed primarily for mapping, geodetic positions are referred to an ellipsoid, and the elevations of the positions are referred to the geoid. However, for satellite geodesy the geoidal heights must be considered to establish the correct height above the geoid.

Precise geodetic **leveling** is used to establish a basic network of vertical control points. From these, the height of other positions in the survey can be determined by supplementary methods. The mean sea-level surface used as a reference (vertical datum) is determined by averaging the hourly water heights for a specified period of time at specified tide gauges.

There are three leveling techniques: **differential**, **trigonometric**, and **barometric**. Differential leveling is the most accurate of the three methods. With the instrument locked in position, readings are made on two calibrated staffs held in an upright position ahead of and behind the instrument. The difference between readings is the difference in elevation between the points.

Trigonometric leveling involves measuring a vertical angle from a known distance with a theodolite and computing the elevation of the point. With this method, vertical measurement can be made at the same time horizontal angles are measured for triangulation. It is, therefore, a somewhat more economical method but less accurate than differential leveling. It is often the only mechanical method of establishing accurate elevation control in mountainous areas.

In barometric leveling, differences in height are determined by measuring the differences in atmospheric pressure at various elevations. Air pressure is measured by mercurial or aneroid barometer, or a boiling point thermometer. Although the accuracy of this method is not as great as either of the other two, it obtains relative heights very rapidly at points which are fairly far apart. It is used in reconnaissance and exploratory surveys where more accurate measurements will be made later or where a high degree of accuracy is not required.

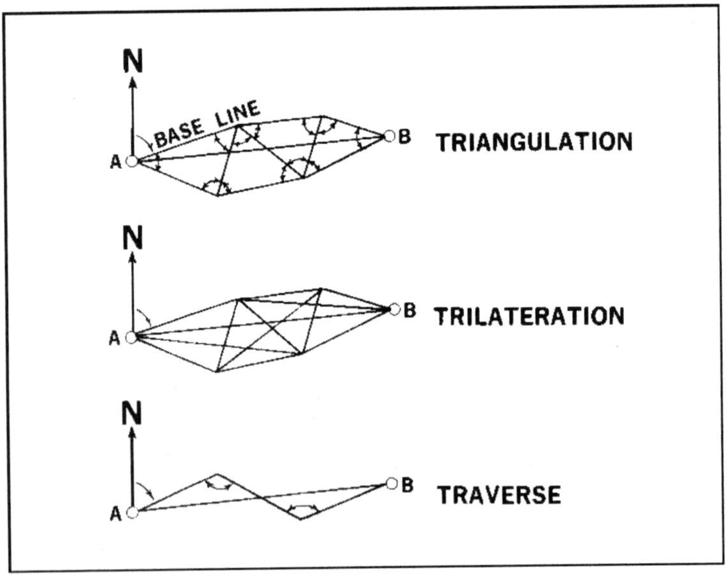

Figure 206. Triangulation, trilateration, and traverse.

DATUM CONNECTIONS

207. Definitions

A **datum** is defined as any numerical or geometrical quantity or set of such quantities which serves as a reference point from which to measure other quantities.

In geodesy, cartography, and navigation, two general types of datums must be considered: **horizontal datum** and **vertical datum**. The horizontal datum forms the basis for computations of horizontal position. The vertical datum provides the reference to measure heights or depths, and may be one of two types: **Vertical geodetic datum** is the reference used by surveyors to measure heights of topographic features, and by cartographers to portray them. This should not be confused with the various types of **tidal datums**, which are by definition vertical datums (and having no horizontal component), used to define the heights and depths of hydrographic features, such as water depths or bridge clearances. The vertical geodetic datum is derived from its mathematical expression, while the tidal datum is derived from actual tidal data. For a complete discussion of tidal datums, see Chapter 9.

This chapter will discuss only geodetic datums. For navigational purposes, vertical geodetic datums are quite unimportant, while horizontal geodetic datums and tidal datums are vital.

A horizontal datum may be defined at an origin point on the ellipsoid (local datum) such that the center of the ellipsoid coincides with the Earth's center of mass (geocentric datum). The coordinates for points in specific geodetic surveys and triangulation networks are computed from certain initial quantities, or datums.

208. Preferred Datums

In areas of overlapping geodetic triangulation networks, each computed on a different datum, the coordinates of the points given with respect to one datum will differ from those given with respect to the other. The differences can be used to derive transformation formulas. Datums are connected by developing transformation formulas at common points, either between overlapping control networks or by satellite connections.

Many countries have developed national datums which differ from those of their neighbors. Accordingly, national maps and charts often do not agree along national borders.

The **North American Datum, 1927** (NAD 27) has been used in the United States for about 60 years, but it is being replaced by datums based on the **World Geodetic System**. NAD 27 coordinates are based on the latitude and longitude of a triangulation station (the reference point) at Mead's Ranch in Kansas, the azimuth to a nearby triangulation station called Waldo, and the mathematical parameters of the Clarke Ellipsoid of 1866. Other datums throughout the world use different assumptions as to origin points and ellipsoids.

The origin of the **European Datum** is at Potsdam, Germany. Numerous national systems have been joined into a large datum based upon the International Ellipsoid of 1924 which was oriented by a modified astrogeodetic method. European, African, and Asian triangulation chains were connected, and African measurements from Cairo to Cape Town were completed. Thus, all of Europe, Africa, and Asia are molded into one great system. Through common survey stations, it was also possible to convert data from the Russian Pulkova, 1932 system to the European Datum, and as a result, the European Datum includes triangulation as far east as the 84th meridian. Additional ties across the Middle East have permitted connection of the Indian and European Datums.

The **Ordnance Survey of Great Britain 1936 Datum** has no point of origin. The data was derived as a best fit between retriangulation and original values of 11 points of the earlier Principal Triangulation of Great Britain (1783-1853).

Tokyo Datum has its origin in Tokyo. It is defined in terms of the Bessel Ellipsoid and oriented by a single astronomic station. Triangulation ties through Korea connect the Japanese datum with the Manchurian datum. Unfortunately, Tokyo is situated on a steep slope on the geoid, and the single-station orientation has resulted in large systematic geoidal separations as the system is extended from its initial point.

The **Indian Datum** is the preferred datum for India and several adjacent countries in Southeast Asia. It is computed on the Everest Ellipsoid with its origin at Kalianpur, in central India. It is largely the result of the untiring work of Sir George Everest (1790-1866), Surveyor General in India from 1830 to 1843. He is best known by the mountain named after him, but by far his most important legacy was the survey of the Indian subcontinent.

MODERN GEODETIC SYSTEMS

209. Development of the World Geodetic System

By the late 1950's the increasing range and sophistication of weapons systems had rendered local or national datums inadequate for military purposes; these new weapons required datums at least continental, if not global, in scope. In response to these requirements, the U.S. Department of Defense generated a geocentric (earth-centered) reference system to which different geodetic networks could be referred, and established compatibility

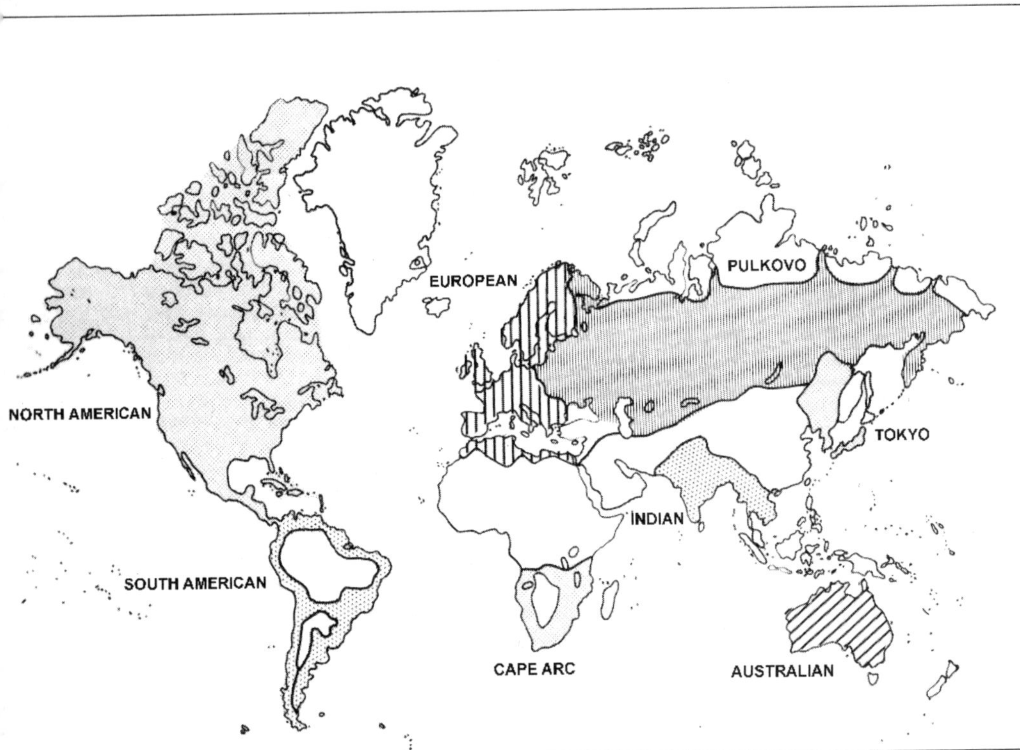

Figure 208. Major geodetic datum blocks.

between the coordinate systems. Efforts of the Army, Navy, and Air Force were combined, leading to the development of the DoD **World Geodetic System of 1960 (WGS 60)**.

In January 1966, a World Geodetic System Committee was charged with the responsibility for developing an improved WGS needed to satisfy mapping, charting, and geodetic requirements. Additional surface gravity observations, results from the extension of triangulation and trilateration networks, and large amounts of Doppler and optical satellite data had become available since the development of WGS 60. Using the additional data and improved techniques, the Committee produced **WGS 66** which served DoD needs following its implementation in 1967.

The same World Geodetic System Committee began work in 1970 to develop a replacement for WGS 66. Since the development of WGS 66, large quantities of additional data had become available from both Doppler and optical satellites, surface gravity surveys, triangulation and trilateration surveys, high precision traverses, and astronomic surveys.

In addition, improved capabilities had been developed in both computers and computer software. Continued research in computational procedures and error analyses had produced better methods and an improved facility for handling and combining data. After an extensive effort extending over a period of approximately three years, the Committee completed the development of the Department of Defense **World Geodetic System 1972 (WGS 72)**.

Further refinement of WGS 72 resulted in the new **World Geodetic System of 1984 (WGS 84),** now referred to as simply WGS. For surface navigation, WGS 60, 66, 72 and the new WGS 84 are essentially the same, so that positions computed on any WGS coordinates can be plotted directly on the others without correction.

The WGS system is not based on a single point, but many points, fixed with extreme precision by satellite fixes and statistical methods. The result is an ellipsoid which fits the real surface of the Earth, or geoid, far more accurately than any other. The WGS system is applicable worldwide. All regional datums can be referenced to WGS once a survey tie has been made.

210. The New North American Datum Of 1983

The Coast And Geodetic Survey of the National Ocean Service (NOS), NOAA, is responsible for charting United States waters. From 1927 to 1987, U.S. charts were based on NAD 27, using the Clarke 1866 ellipsoid. In 1989, the U.S. officially switched to **NAD 83** (navigationally equivalent to WGS) for all mapping and charting purposes, and all new NOS chart production is based on this new standard.

The grid of interconnected surveys which criss-crosses the United States consists of some 250,000 control points, each consisting of the latitude and longitude of the point, plus additional data such as elevation. Converting the NAD 27 coordinates to NAD 83 involved recomputing the position of each point based on the new NAD 83 datum. In addition to the 250,000 U.S. control points, several thousand more were added to tie in surveys from Canada, Mexico, and Central America.

Conversion of new edition charts to the new datums, either WGS 84 or NAD 83, involves converting reference points on each chart from the old datum to the new, and adjusting the latitude and longitude grid (known as the graticule) so that it reflects the newly plotted positions. This adjustment of the graticule is the only difference between charts which differ only in datum. All charted features remain in exactly the same relative positions.

The Global Positioning System (GPS) has transformed the science of surveying, enabling the establishment of precise ties to WGS in areas previously found to be too remote to survey to modern standards. As a result, new charts are increasingly precise as to position of features. The more recent a chart's date of publishing, the more likely it is that it will be accurate as to positions. Navigators should always refer to the title block of a chart to determine the date of the chart, the date of the surveys and sources used to compile it, and the datum on which it is based.

DATUMS AND NAVIGATION

211. Datum Shift

One of the most serious impacts of different datums on navigation occurs when a navigation system provides a fix based on a datum different from that used for the nautical chart. The resulting plotted position may be different from the actual location on that chart. This difference is known as a **datum shift**.

Modern electronic navigation systems have software installed that can output positions in a variety of datums, eliminating the necessity for applying corrections. All electronic charts produced by NIMA are compiled on WGS and are not subject to datum shift problems as long as the GPS receiver is outputting WGS position data to the display system. The same is true for NOAA charts of the U.S., which are compiled on NAD 83 datum, very closely related to WGS. GPS receivers, including the WRN-6, default to WGS, so that no action is necessary to use any U.S.-produced electronic charts.

To automate datum conversions, a number of datum transformation software programs have been written that will convert from any known datum to any other, in any location. MADTRAN and GEOTRANS-2 are two such programs. The amount of datum shift between two different datums is not linear. That is, the amount of shift is a function of the position of the observer, which must be specified for the shift to be computed. Varying differences of latitude and longitude between two different datums will be noted as one's location changes.

There are still a few NIMA-produced paper charts, and a number of charts from other countries, based on datums other than WGS. If the datum of these charts is noted in the title block of the chart, the WRN-6 and most other GPS receivers can be set to output position data in that datum, eliminating the datum shift problem. If the datum is not listed, extreme caution is necessary. An offset can sometimes be established if the ship's actual position can be determined with sufficient accuracy, and this offset applied to GPS positions in the local area. But remember that since a datum shift is not linear, this offset is only applicable locally.

Another effect on navigation occurs when shifting between charts that have been compiled using different datums. If a position is replotted on a chart of another datum using latitude and longitude, the newly plotted position will not match with respect to other charted features. The datum shift may be avoided by transferring positions using bearings and ranges to common points. If datum shift conversion notes for the applicable datums are given on the charts, positions defined by latitude and longitude may be replotted after applying the noted correction.

The positions given for chart corrections in the *Notice to Mariners* reflect the proper datum for each specific chart and edition number. Due to conversion of charts based on old datums to more modern ones, and the use of many different datums throughout the world, chart corrections intended for one edition of a chart may not be safely plotted on any other.

As noted, datum shifts are not constant throughout a given region, but vary according to how the differing datums fit together. For example, the NAD 27 to NAD 83 conversion resulted in changes in latitude of 40 meters in Miami, 11 meters in New York, and 20 meters in Seattle. Longitude changes for this conversion amounted to 22 meters in Miami, 35 meters in New York, and 93 meters in Seattle.

Most charts produced by NIMA and NOS show a

"datum note." This note is usually found in the title block or in the upper left margin of the chart. According to the year of the chart edition, the scale, and policy at the time of production, the note may say "World Geodetic System 1972 (WGS-72)", "World Geodetic System 1984 (WGS-84)", or "World Geodetic System (WGS)." A datum note for a chart for which satellite positions can be plotted without correction will read: "Positions obtained from satellite navigation systems referred to (Reference Datum) can be plotted directly on this chart."

NIMA reproductions of foreign charts will usually be in the datum or reference system of the producing country. In these cases a conversion factor is given in the following format: "Positions obtained from satellite navigation systems referred to the (Reference Datum) must be moved X.XX minutes (Northward/Southward) and X.XX minutes (Eastward/ Westward) to agree with this chart."

Some charts cannot be tied in to WGS because of lack of recent surveys. Currently issued charts of some areas are based on surveys or use data obtained in the age of sailing ships. The lack of surveyed control points means that they cannot be properly referenced to modern geodetic systems. In this case there may be a note that says: "Adjustments to WGS cannot be determined for this chart."

A few charts may have no datum note at all, but may carry a note which says: "From various sources to (year)." In these cases there is no way for the navigator to determine the mathematical difference between the local datum and WGS positions. However, if a radar or visual fix can be accurately determined, and an offset established as noted above. This offset can then be programmed into the GPS receiver.

To minimize problems caused by differing datums:

- Plot chart corrections only on the specific charts and editions for which they are intended. Each chart correction is specific to only one edition of a chart. When the same correction is made on two charts based on different datums, the positions for the same feature may differ slightly. This difference is equal to the datum shift between the two datums for that area.
- Try to determine the source and datum of positions of temporary features, such as drill rigs. In general they are given in the datum used in the area in question. Since these are precisely positioned using satellites, WGS is the normal datum. A datum correction, if needed, might be found on a chart of the area.
- Remember that if the datum of a plotted feature is not known, position inaccuracies may result. It is wise to allow a margin of error if there is any doubt about the datum.
- Know how the datum of the positioning system you are using (Loran, GPS, etc.) relates to your chart. GPS and other modern positioning systems use WGS datum. If your chart is on any other datum, you must program the system to use the chart's datum, or apply a datum correction when plotting GPS positions on the chart.

CHAPTER 3

NAUTICAL CHARTS

CHART FUNDAMENTALS

300. Definitions

A **nautical chart** represents part of the spherical earth on a plane surface. It shows water depth, the shoreline of adjacent land, prominent topographic features, aids to navigation, and other navigational information. It is a work area on which the navigator plots courses, ascertains positions, and views the relationship of the ship to the surrounding area. It assists the navigator in avoiding dangers and arriving safely at his destination.

Originally hand-drawn on sheepskin, traditional nautical charts have for generations been printed on paper. **Electronic charts** consisting of a digital data base and a display system are in use and are replacing paper charts aboard many vessels. An electronic chart is not simply a digital version of a paper chart; it introduces a new navigation methodology with capabilities and limitations very different from paper charts. The electronic chart is the legal equivalent of the paper chart if it meets certain International Maritime Organization specifications. See Chapter 14 for a complete discussion of electronic charts.

Should a marine accident occur, the nautical chart in use at the time takes on legal significance. In cases of grounding, collision, and other accidents, charts become critical records for reconstructing the event and assigning liability. Charts used in reconstructing the incident can also have tremendous training value.

301. Projections

Because a cartographer cannot transfer a sphere to a flat surface without distortion, he must project the surface of a sphere onto a **developable surface**. A developable surface is one that can be flattened to form a plane. This process is known as **chart projection**. If points on the surface of the sphere are projected from a single point, the projection is said to be **perspective** or **geometric**.

As the use of electronic charts becomes increasingly widespread, it is important to remember that the same cartographic principles that apply to paper charts apply to their depiction on video screens.

302. Selecting a Projection

Each projection has certain preferable features. However, as the area covered by the chart becomes smaller, the differences between various projections become less noticeable. On the largest scale chart, such as of a harbor, all projections are practically identical. Some desirable properties of a projection are:

1. True shape of physical features
2. Correct angular relationships
3. Equal area (Represents areas in proper proportions)
4. Constant scale values
5. Great circles represented as straight lines
6. Rhumb lines represented as straight lines

Some of these properties are mutually exclusive. For example, a single projection cannot be both conformal and equal area. Similarly, both great circles and rhumb lines cannot be represented on a single projection as straight lines.

303. Types of Projections

The type of developable surface to which the spherical surface is transferred determines the projection's classification. Further classification depends on whether the projection is centered on the equator (equatorial), a pole (polar), or some point or line between (oblique). The name of a projection indicates its type and its principal features.

Mariners most frequently use a **Mercator projection**, classified as a **cylindrical projection** upon a plane, the cylinder tangent along the equator. Similarly, a projection based upon a cylinder tangent along a meridian is called **transverse** (or inverse) **Mercator** or **transverse** (or inverse) **orthomorphic**. The Mercator is the most common projection used in maritime navigation, primarily because rhumb lines plot as straight lines.

In a **simple conic projection**, points on the surface of the earth are transferred to a tangent cone. In the **Lambert conformal projection**, the cone intersects the earth (a secant cone) at two small circles. In a **polyconic projection**, a series of tangent cones is used.

In an **azimuthal** or **zenithal projection**, points on the earth are transferred directly to a plane. If the origin of the projecting rays is the center of the earth, a **gnomonic projection** results; if it is the point opposite the plane's point of tangency, a **stereographic projection**; and if at infinity (the projecting lines being parallel to each other), an **orthographic projection**. The gnomonic, stereographic, and orthographic are **perspective projections**. In an **azimuthal equidistant projection**, which is not perspective, the scale of distances is constant along any radial line from the point of tangency. See Figure 303.

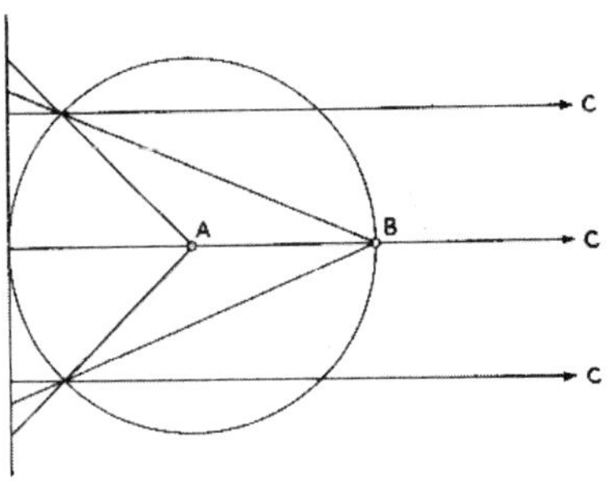

Figure 303. Azimuthal projections: A, gnomonic; B, stereographic; C, (at infinity) orthographic.

Cylindrical and **plane projections** are special conical projections, using heights infinity and zero, respectively.

A **graticule** is the network of latitude and longitude lines laid out in accordance with the principles of any projection.

304. Cylindrical Projections

If a cylinder is placed around the earth, tangent along the equator, and the planes of the meridians are extended, they intersect the cylinder in a number of vertical lines. See Figure 304. These parallel lines of projection are equidistant from each other, unlike the terrestrial meridians from which they are derived which converge as the latitude increases. On the earth, parallels of latitude are perpendicular to the meridians, forming circles of progressively smaller diameter as the latitude increases. On the cylinder they are shown perpendicular to the projected meridians, but because a cylinder is everywhere of the same diameter, the projected parallels are all the same size.

If the cylinder is cut along a vertical line (a meridian) and spread out flat, the meridians appear as equally spaced vertical lines; and the parallels appear as horizontal lines. The parallels' relative spacing differs in the various types of cylindrical projections.

If the cylinder is tangent along some great circle other than the equator, the projected pattern of latitude and longitude lines appears quite different from that described above, since the line of tangency and the equator no longer coincide. These projections are classified as **oblique** or **transverse projections**.

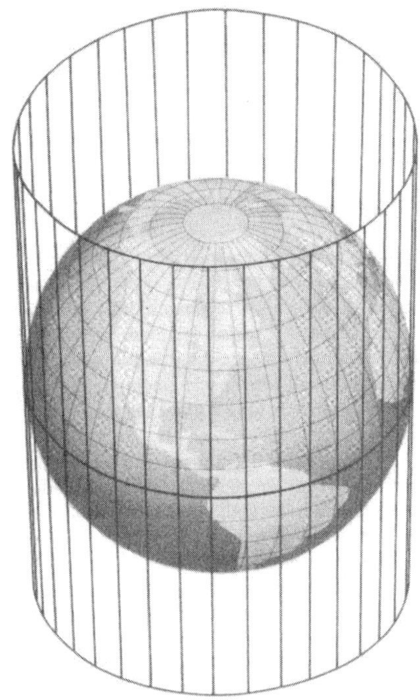

Figure 304. A cylindrical projection.

305. Mercator Projection

Navigators most often use the plane conformal projection known as the **Mercator projection**. The Mercator projection is not perspective, and its parallels can be derived mathematically as well as projected geometrically. Its distinguishing feature is that both the meridians and parallels are expanded at the same ratio with increased latitude. The expansion is equal to the secant of the latitude, with a small correction for the ellipticity of the earth. Since the secant of 90° is infinity, the projection cannot include the poles. Since the projection is conformal, expansion is the same in all directions and angles are correctly shown. Rhumb lines appear as straight lines, the directions of which can be measured directly on the chart. Distances can also be measured directly if the spread of latitude is small. Great circles, except meridians and the equator, appear as curved lines concave to the equator. Small areas appear in their correct shape but of increased size unless they are near the equator.

306. Meridional Parts

At the equator a degree of longitude is approximately equal in length to a degree of latitude. As the distance from the equator increases, degrees of latitude remain approximately the same, while degrees of longitude become

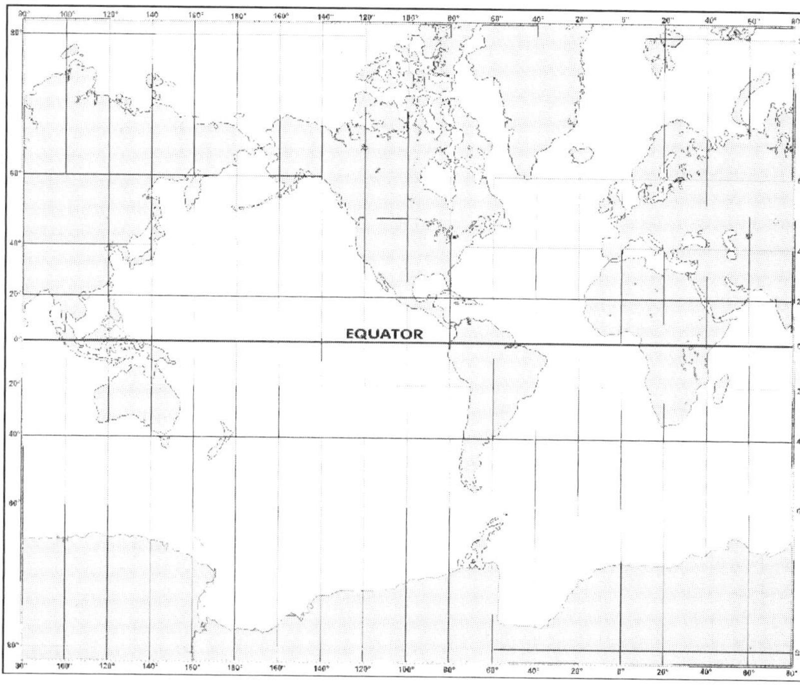

Figure 306. A Mercator map of the world.

progressively shorter. Since degrees of longitude appear everywhere the same length in the Mercator projection, it is necessary to increase the length of the meridians if the expansion is to be equal in all directions. Thus, to maintain the correct proportions between degrees of latitude and degrees of longitude, the degrees of latitude must be progressively longer as the distance from the equator increases. This is illustrated in Figure 306.

The length of a meridian, increased between the equator and any given latitude, expressed in minutes of arc at the equator as a unit, constitutes the number of meridional parts (M) corresponding to that latitude. Meridional parts, given in Table 6 for every minute of latitude from the equator to the pole, make it possible to construct a Mercator chart and to solve problems in Mercator sailing. These values are for the WGS ellipsoid of 1984.

307. Transverse Mercator Projections

Constructing a chart using Mercator principles, but with the cylinder tangent along a meridian, results in a **transverse Mercator** or **transverse orthomorphic projection**. The word "inverse" is used interchangeably with "transverse." These projections use a fictitious graticule similar to, but offset from, the familiar network of meridians and parallels. The tangent great circle is the fictitious equator. Ninety degrees from it are two fictitious poles. A group of great circles through these poles and perpendicular to the tangent great circle are the fictitious meridians, while a series of circles parallel to the plane of the tangent great circle form the fictitious parallels. The actual meridians and parallels appear as curved lines.

A straight line on the transverse or oblique Mercator projection makes the same angle with all fictitious meridians, but not with the terrestrial meridians. It is therefore a fictitious rhumb line. Near the tangent great circle, a straight line closely approximates a great circle. The projection is most useful in this area. Since the area of minimum distortion is near a meridian, this projection is useful for charts covering a large band of latitude and extending a relatively short distance on each side of the tangent meridian. It is sometimes used for star charts showing the evening sky at various seasons of the year. See Figure 307.

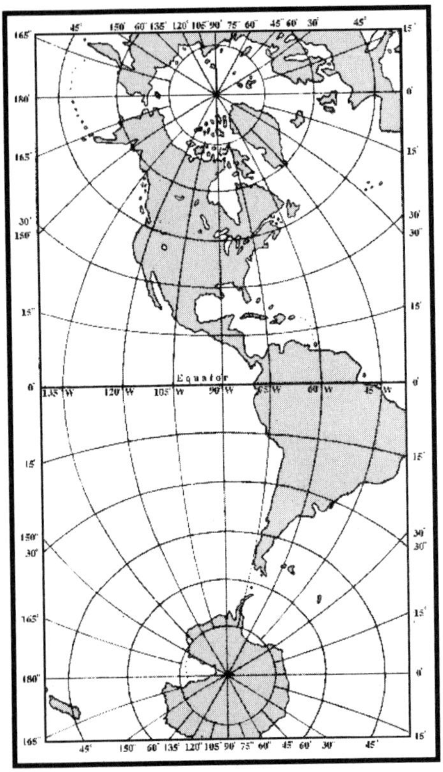

Figure 307. A transverse Mercator map of the Western Hemisphere.

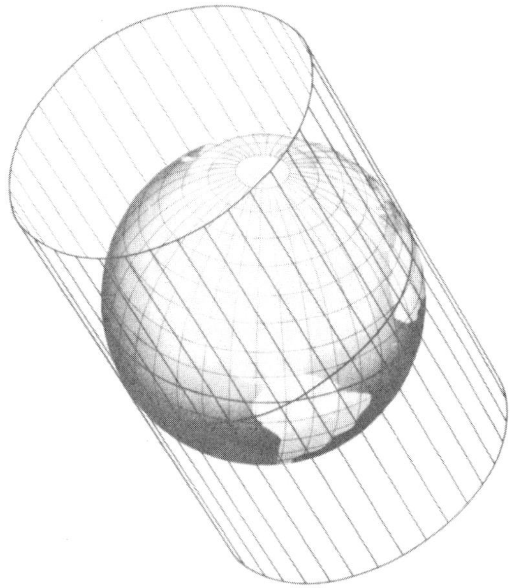

Figure 309a. An oblique Mercator projection.

308. Universal Transverse Mercator (UTM) Grid

The **Universal Transverse Mercator (UTM)** grid is a military grid superimposed upon a transverse Mercator graticule, or the representation of these grid lines upon any graticule. This grid system and these projections are often used for large-scale (harbor) nautical charts and military charts.

309. Oblique Mercator Projections

A Mercator projection in which the cylinder is tangent along a great circle other than the equator or a meridian is called an **oblique Mercator** or **oblique orthomorphic projection**. See Figure 309a and Figure 309b. This projection is used principally to depict an area in the near vicinity of an oblique great circle. Figure 309c, for example, shows the great circle joining Washington and Moscow. Figure 309d shows an oblique Mercator map with the great circle between these two centers as the tangent great circle or fictitious equator. The limits of the chart of Figure 309c are indicated in Figure 309d. Note the large variation in scale as the latitude changes.

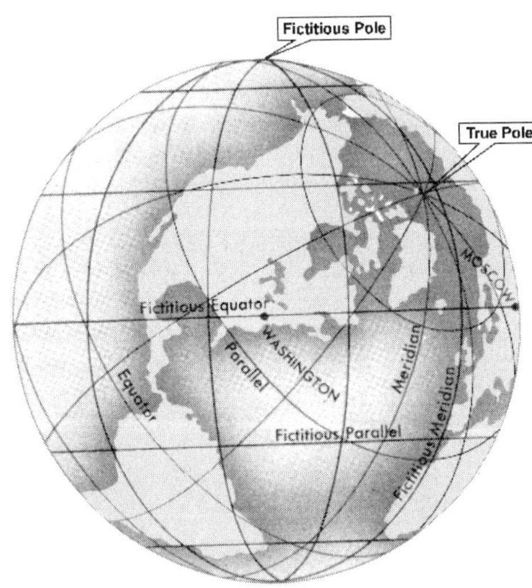

Figure 309b. The fictitious graticule of an oblique Mercator projection.

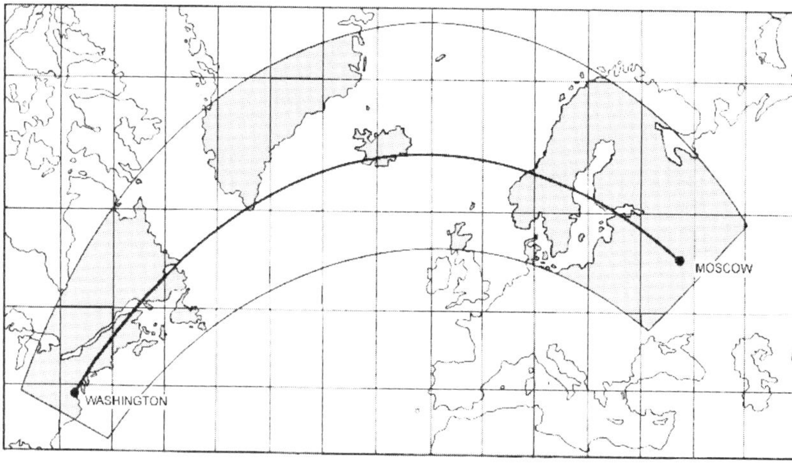

Figure 309c. The great circle between Washington and Moscow as it appears on a Mercator map.

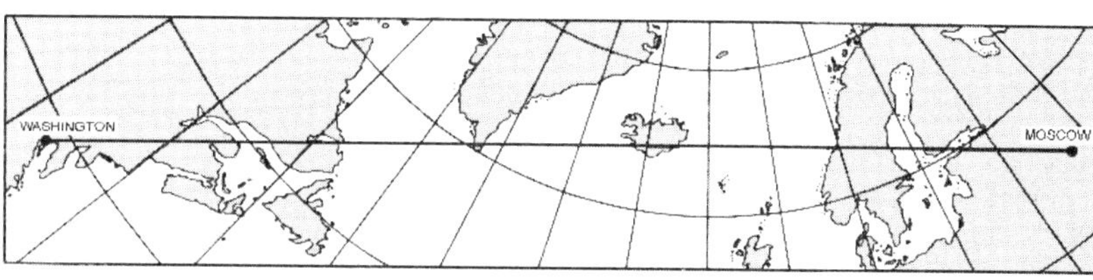

Figure 309d. An oblique Mercator map based upon a cylinder tangent along the great circle through Washington and Moscow. The map includes an area 500 miles on each side of the great circle. The limits of this map are indicated on the Mercator map of Figure 309c.

310. Rectangular Projection

A cylindrical projection similar to the Mercator, but with uniform spacing of the parallels, is called a **rectangular projection**. It is convenient for graphically depicting information where distortion is not important. The principal navigational use of this projection is for the star chart of the Air Almanac, where positions of stars are plotted by rectangular coordinates representing declination (ordinate) and sidereal hour angle (abscissa). Since the meridians are parallel, the parallels of latitude (including the equator and the poles) are all represented by lines of equal length.

311. Conic Projections

A **conic projection** is produced by transferring points from the surface of the earth to a cone or series of cones. This cone is then cut along an element and spread out flat to form the chart. When the axis of the cone coincides with the axis of the earth, then the parallels appear as arcs of circles, and the meridians appear as either straight or curved lines converging toward the nearer pole. Limiting the area covered to that part of the cone near the surface of the earth limits distortion. A parallel along which there is no distortion is called a **standard parallel**. Neither the transverse conic projection, in which the axis of the cone is in the equatorial plane, nor the oblique conic projection, in which the axis of the cone is oblique to the plane of the equator, is ordinarily used for navigation. They are typically used for illustrative maps.

Using cones tangent at various parallels, a secant (intersecting) cone, or a series of cones varies the appearance and features of a conic projection.

312. Simple Conic Projection

A conic projection using a single tangent cone is a **simple conic projection** (Figure 312a). The height of the cone increases as the latitude of the tangent parallel decreases. At the equator, the height reaches infinity and the cone becomes a cylinder. At the pole, its height is zero, and the cone becomes a plane. Similar to the Mercator projection, the simple conic projection is not perspective since only the meridians are projected geometrically, each becoming an element of the cone. When this projection is spread out flat to form a map, the meridians appear as straight lines converging at the apex of the cone. The standard parallel, where the cone is tangent to the earth, appears as the arc of a circle with its center at the apex of the cone. The other parallels are concentric circles. The distance along any meridian between consecutive parallels is in correct relation to the distance on the earth, and, therefore, can be derived mathematically. The pole is represented by a circle (Figure 312b). The scale is correct along any meridian and along the standard parallel. All other parallels are too great in length, with the error increasing with increased distance from the standard parallel. Since the scale is not the same in all directions about every point, the projection is neither a conformal nor equal-area projection. Its non-conformal nature is its principal disadvantage for navigation.

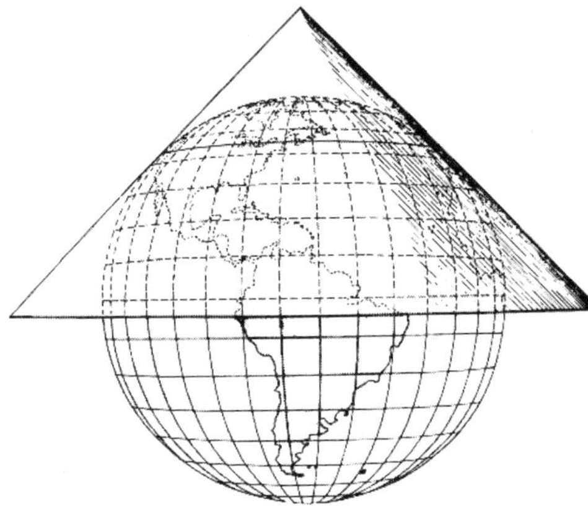

Figure 312a. A simple conic projection.

Since the scale is correct along the standard parallel and varies uniformly on each side, with comparatively little distortion near the standard parallel, this projection is useful for mapping an area covering a large spread of longitude and a comparatively narrow band of latitude. It was developed by Claudius Ptolemy in the second century A.D. to map just such an area: the Mediterranean Sea.

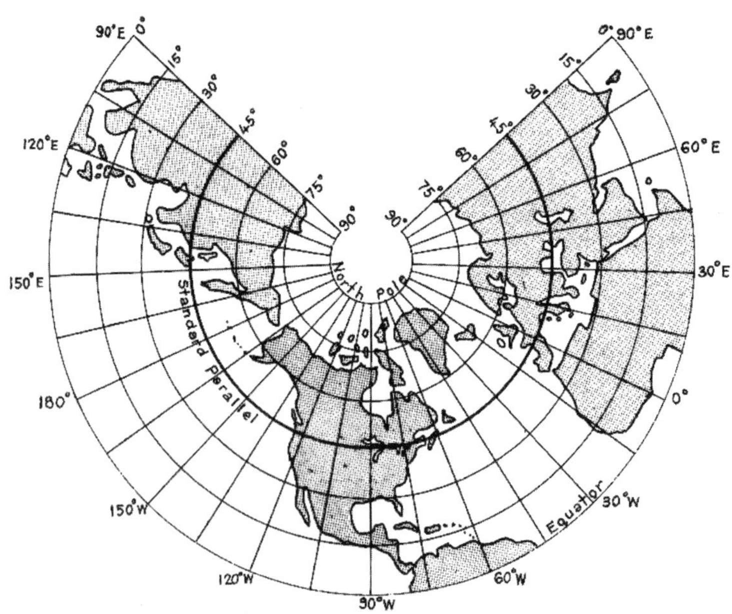

Figure 312b. A simple conic map of the Northern Hemisphere.

313. Lambert Conformal Projection

The useful latitude range of the simple conic projection can be increased by using a secant cone intersecting the earth at two standard parallels. See Figure 313. The area between the two standard parallels is compressed, and that beyond is expanded. Such a projection is called either a **secant conic** or **conic projection with two standard parallels**.

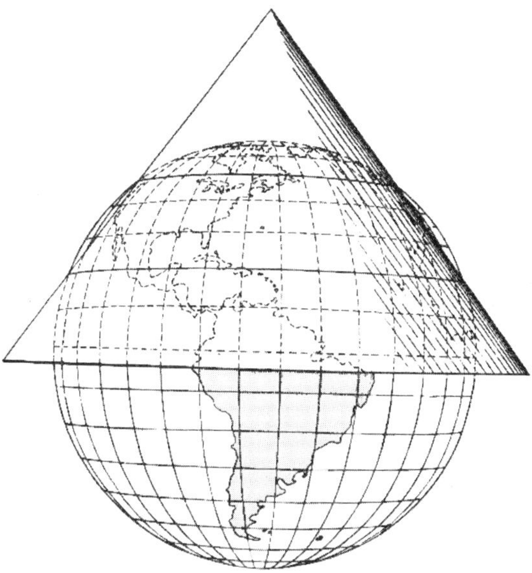

Figure 313. A secant cone for a conic projection with two standard parallels.

If in such a projection the spacing of the parallels is altered, such that the distortion is the same along them as along the meridians, the projection becomes conformal. This modification produces the **Lambert conformal projection**. If the chart is not carried far beyond the standard parallels, and if these are not a great distance apart, the distortion over the entire chart is small.

A straight line on this projection so nearly approximates a great circle that the two are nearly identical. Radio beacon signals travel great circles; thus, they can be plotted on this projection without correction. This feature, gained without sacrificing conformality, has made this projection popular for aeronautical charts because aircraft make wide use of radio aids to navigation. Except in high latitudes, where a slightly modified form of this projection has been used for polar charts, it has not replaced the Mercator projection for marine navigation.

314. Polyconic Projection

The latitude limitations of the secant conic projection can be minimized by using a series of cones. This results in a **polyconic projection**. In this projection, each parallel is the base of a tangent cone. At the edges of the chart, the area between parallels is expanded to eliminate gaps. The scale is correct along any parallel and along the central meridian of the projection. Along other meridians the scale increases with increased difference of longitude from the central meridian. Parallels appear as nonconcentric circles; meridians appear as curved lines converging toward the pole and concave to the central meridian.

The polyconic projection is widely used in atlases, particularly for areas of large range in latitude and reasonably large range in longitude, such as continents. However, since it is not conformal, this projection is not customarily used in navigation.

315. Azimuthal Projections

If points on the earth are projected directly to a plane surface, a map is formed at once, without cutting and flattening, or "developing." This can be considered a special case of a conic projection in which the cone has zero height.

The simplest case of the **azimuthal projection** is one in which the plane is tangent at one of the poles. The meridians are straight lines intersecting at the pole, and the parallels are concentric circles with their common center at the pole. Their spacing depends upon the method used to transfer points from the earth to the plane.

If the plane is tangent at some point other than a pole, straight lines through the point of tangency are great circles, and concentric circles with their common center at the point of tangency connect points of equal distance from that point. Distortion, which is zero at the point of tangency, increases along any great circle through this point. Along any circle whose center is the point of tangency, the distortion is constant. The bearing of any point from the point of tangency is correctly represented. It is for this reason that these projections are called **azimuthal**. They are also called **zenithal**. Several of the common azimuthal projections are perspective.

316. Gnomonic Projection

If a plane is tangent to the earth, and points are projected geometrically from the center of the earth, the result is a **gnomonic projection**. See Figure 316a. Since the projection is perspective, it can be demonstrated by placing a light at the center of a transparent terrestrial globe and holding a flat surface tangent to the sphere.

In an **oblique gnomonic projection** the meridians appear as straight lines converging toward the nearer pole. The parallels, except the equator, appear as curves (Figure 316b). As in all azimuthal projections, bearings from the point of tangency are correctly represented. The distance scale, however, changes rapidly. The projection is neither conformal nor equal area. Distortion is so great that shapes, as well as distances and areas, are very poorly represented, except near the point of tangency.

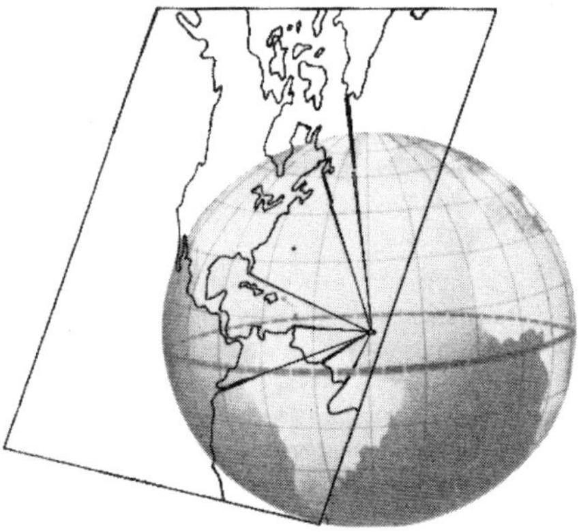

Figure 316a. An oblique gnomonic projection.

The usefulness of this projection rests upon the fact

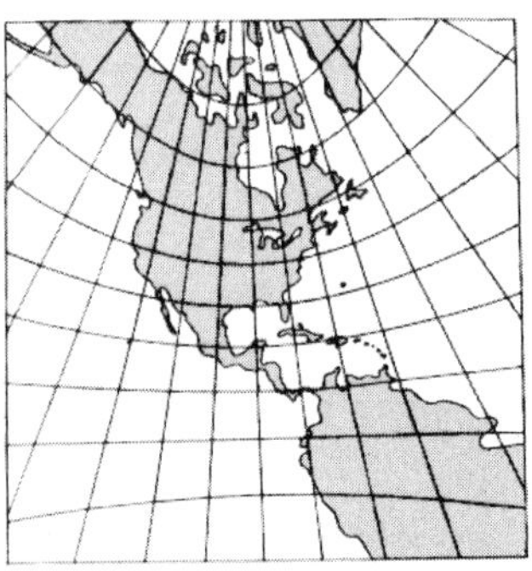

Figure 316b. An oblique gnomonic map with point of tangency at latitude 30°N, longitude 90°W.

that any great circle appears on the map as a straight line, giving charts made on this projection the common name **great-circle charts**.

Gnomonic charts are most often used for planning the great-circle track between points. Points along the determined track are then transferred to a Mercator projection. The great circle is then followed by following the rhumb lines from one point to the next. Computer programs which automatically calculate great circle routes between points and provide latitude and longitude of corresponding rhumb line endpoints are quickly making this use of the gnomonic chart obsolete.

317. Stereographic Projection

A **stereographic projection** results from projecting points on the surface of the earth onto a tangent plane, from a point on the surface of the earth opposite the point of tangency (Figure 317a). This projection is also called an **azimuthal orthomorphic projection**.

The scale of the stereographic projection increases with distance from the point of tangency, but it increases more slowly than in the gnomonic projection. The stereographic projection can show an entire hemisphere without excessive distortion (Figure 317b). As in other azimuthal

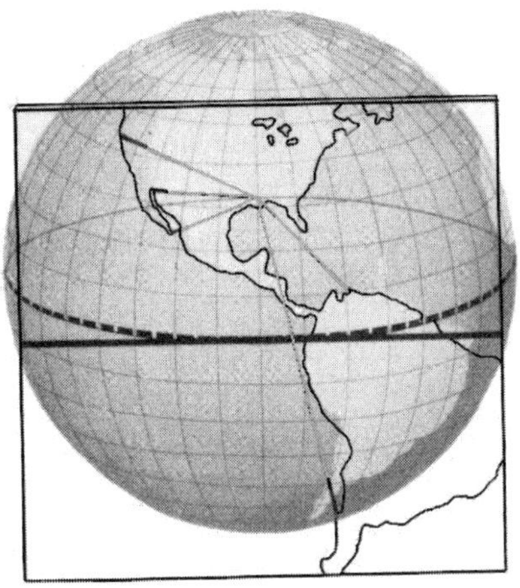

Figure 317a. An equatorial stereographic projection.

Figure 317b. A stereographic map of the Western Hemisphere.

projections, great circles through the point of tangency appear as straight lines. Other circles such as meridians and parallels appear as either circles or arcs of circles.

The principal navigational use of the stereographic projection is for charts of the polar regions and devices for mechanical or graphical solution of the navigational triangle. A **Universal Polar Stereographic (UPS)** grid, mathematically adjusted to the graticule, is used as a reference system.

318. Orthographic Projection

If terrestrial points are projected geometrically from infinity to a tangent plane, an **orthographic projection** results (Figure 318a). This projection is not conformal; nor does it result in an equal area representation. Its principal use is in navigational astronomy because it is useful for illustrating and solving the navigational triangle. It is also useful for illustrating celestial coordinates. If the plane is tangent at a point on the equator, the parallels (including the equator) appear as straight lines. The meridians would appear as ellipses, except that the meridian through the point of tangency would appear as a straight line and the one 90° away would appear as a circle (Figure 318b).

319. Azimuthal Equidistant Projection

An **azimuthal equidistant projection** is an azimuthal projection in which the distance scale along any great circle through the point of tangency is constant. If a pole is the point of tangency, the meridians appear as straight radial lines and the parallels as equally spaced concentric circles. If the plane is tangent at some point other than a pole, the concentric circles represent distances from the point of tangency. In this case, meridians and parallels appear as curves.

The projection can be used to portray the entire earth, the point 180° from the point of tangency appearing as the largest of the concentric circles. The projection is not conformal, equal area, or perspective. Near the point of tangency distortion is small, increasing with distance until shapes near the opposite side of the earth are unrecognizable (Figure 319).

The projection is useful because it combines the three features of being azimuthal, having a constant distance scale from the point of tangency, and permitting the entire earth to be shown on one map. Thus, if an important harbor or airport is selected as the point of tangency, the great-circle course, distance, and track from that point to any other point on the earth are quickly and accurately determined. For communication work with the station at the point of tangency, the path of an incoming signal is at once apparent if the direction of arrival has been determined and the direction to train a directional antenna can be determined easily. The projection is also used for polar charts and for the star finder, No. 2102D.

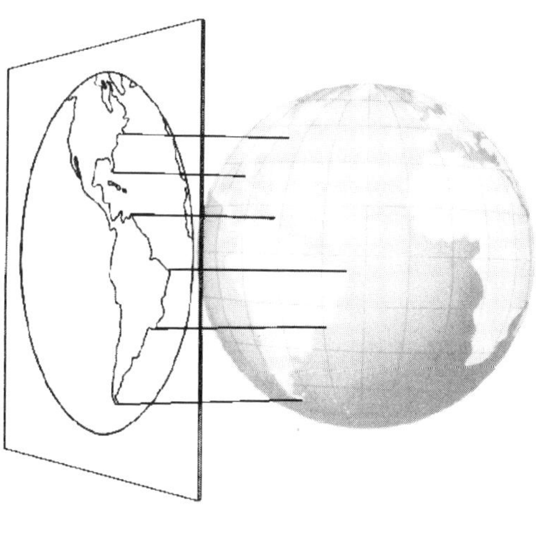

Figure 318a. An equatorial orthographic projection.

Figure 318b. An orthographic map of the Western Hemisphere.

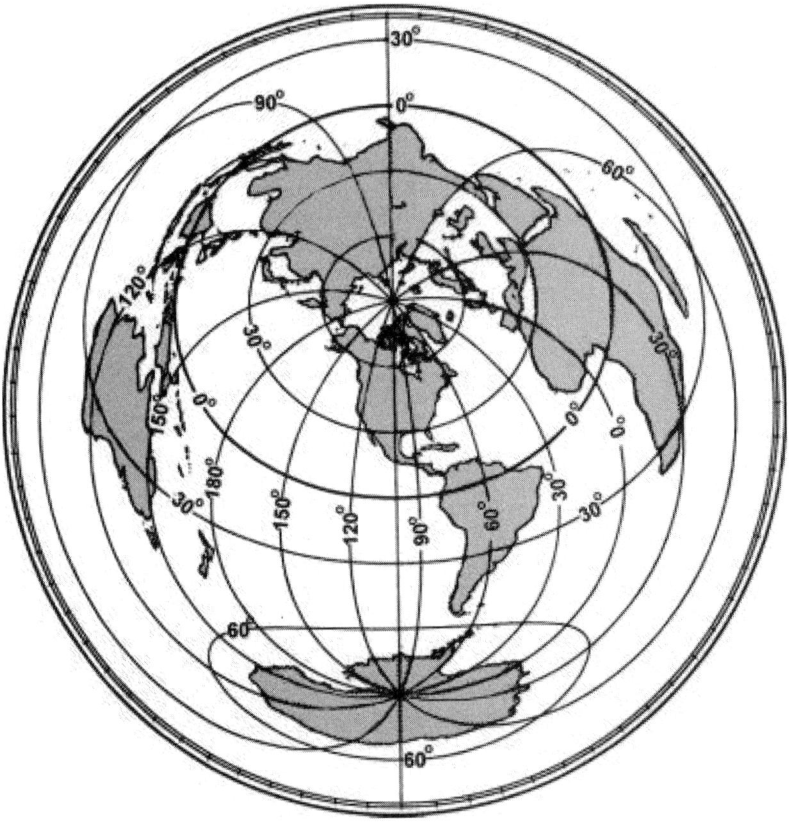

Figure 319. An azimuthal equidistant map of the world with the point of tangency latitude 40°N, longitude 100°W.

POLAR CHARTS

320. Polar Projections

Special consideration is given to the selection of projections for polar charts because the familiar projections become special cases with unique features.

In the case of cylindrical projections in which the axis of the cylinder is parallel to the polar axis of the earth, distortion becomes excessive and the scale changes rapidly. Such projections cannot be carried to the poles. However, both the transverse and oblique Mercator projections are used.

Conic projections with their axes parallel to the earth's polar axis are limited in their usefulness for polar charts because parallels of latitude extending through a full 360° of longitude appear as arcs of circles rather than full circles. This is because a cone, when cut along an element and flattened, does not extend through a full 360° without stretching or resuming its former conical shape. The usefulness of such projections is also limited by the fact that the pole appears as an arc of a circle instead of a point. However, by using a parallel very near the pole as the higher standard parallel, a conic projection with two standard parallels can be made. This requires little stretching to complete the circles of the parallels and eliminate that of the pole. Such a projection, called a **modified Lambert conformal** or **Ney's projection**, is useful for polar charts. It is particularly familiar to those accustomed to using the ordinary Lambert conformal charts in lower latitudes.

Azimuthal projections are in their simplest form when tangent at a pole. This is because the meridians are straight lines intersecting at the pole, and parallels are concentric circles with their common center at the pole. Within a few degrees of latitude of the pole they all look similar; however, as the distance becomes greater, the spacing of the parallels becomes distinctive in each projection. In the polar azimuthal equidistant it is uniform; in the polar stereographic it increases with distance from the pole until the equator is shown at a distance from the pole equal to twice the length of the radius of the earth; in the polar gnomonic the increase is considerably greater, becoming infinity at the equator; in the polar orthographic it decreases with distance from the pole (Figure 320). All of these but the last are used for polar charts.

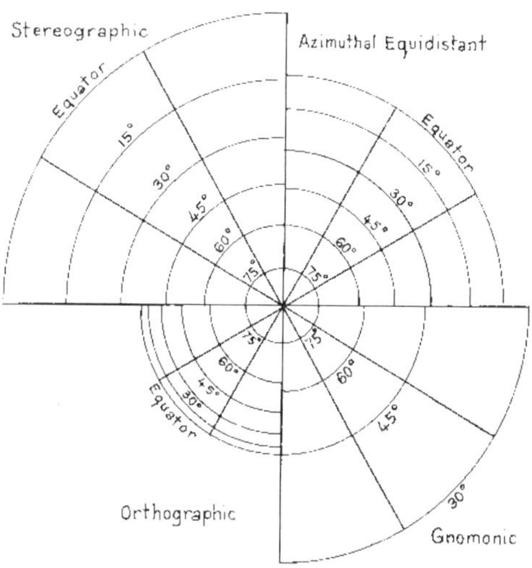

Figure 320. Expansion of polar azimuthal projections.

321. Selection of a Polar Projection

The principal considerations in the choice of a suitable projection for polar navigation are:

1. Conformality: When the projection represents angles correctly, the navigator can plot directly on the chart.
2. Great circle representation: Because great circles are more useful than rhumb lines at high altitudes, the projection should represent great circles as straight lines.
3. Scale variation: The projection should have a constant scale over the entire chart.
4. Meridian representation: The projection should show straight meridians to facilitate plotting and grid navigation
5. Limits: Wide limits reduce the number of projections needed to a minimum.

The projections commonly used for polar charts are the modified Lambert conformal, gnomonic, stereographic, and azimuthal equidistant. All of these projections are similar near the pole. All are essentially conformal, and a great circle on each is nearly a straight line.

As the distance from the pole increases, however, the distinctive features of each projection become important. The modified Lambert conformal projection is virtually conformal over its entire extent. The amount of its scale distortion is comparatively little if it is carried only to about 25° or 30° from the pole. Beyond this, the distortion increases rapidly. A great circle is very nearly a straight line anywhere on the chart. Distances and directions can be measured directly on the chart in the same manner as on a Lambert conformal chart. However, because this projection is not strictly conformal, and on it great circles are not exactly represented by straight lines, it is not suited for highly accurate work.

The polar gnomonic projection is the one polar projection on which great circles are exactly straight lines. However, a complete hemisphere cannot be represented upon a plane because the radius of 90° from the center would become infinity.

The polar stereographic projection is conformal over its entire extent, and a straight line closely approximates a great circle. See Figure 321. The scale distortion is not excessive for a considerable distance from the pole, but it is greater than that of the modified Lambert conformal projection.

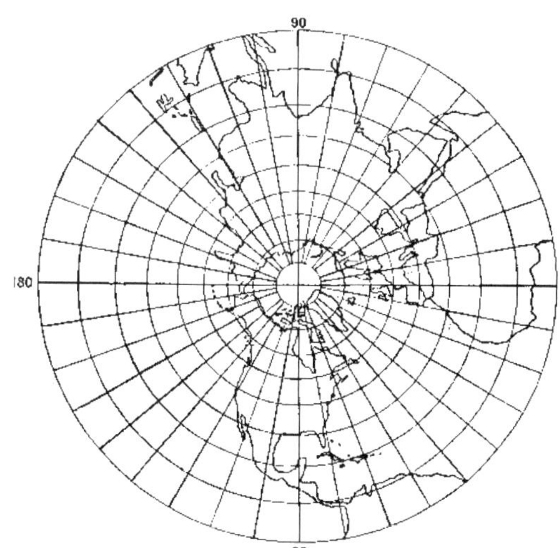

Figure 321. Polar stereographic projection.

The polar azimuthal equidistant projection is useful for showing a large area such as a hemisphere because there is no expansion along the meridians. However, the projection is not conformal and distances cannot be measured accurately in any but a north-south direction. Great circles other than the meridians differ somewhat from straight lines. The equator is a circle centered at the pole.

The two projections most commonly used for polar charts are the modified Lambert conformal and the polar stereographic. When a directional gyro is used as a directional reference, the track of the craft is approximately a great circle. A desirable chart is one on which a great circle is represented as a straight line with a constant scale and with angles correctly represented. These requirements are not met entirely by any single projection, but they are approximated by both the modified Lambert conformal and the polar stereographic. The scale is more nearly constant on the former, but the projection is not strictly conformal. The polar stereographic is conformal, and its maximum scale variation can be reduced by using a plane which intersects the earth at some parallel intermediate between the pole and the lowest parallel. The portion within this standard parallel is compressed, and that portion outside is expanded.

The selection of a suitable projection for use in polar regions depends upon mission requirements. These requirements establish the relative importance of various features. For a relatively small area, any of several projections is suitable. For a large area, however, the choice is more difficult. If grid directions are to be used, it is important that all units in related operations use charts on the same projection, with the same standard parallels, so that a single grid direction exists between any two points.

SPECIAL CHARTS

322. Plotting Sheets

Position plotting sheets are "charts" designed primarily for open ocean navigation, where land, visual aids to navigation, and depth of water are not factors in navigation. They have a latitude and longitude graticule, and they may have one or more compass roses. The meridians are usually unlabeled, so a plotting sheet can be used for any longitude. Plotting sheets on Mercator projection are specific to latitude, and the navigator should have enough aboard for all latitudes for his voyage. Plotting sheets are less expensive than charts.

A plotting sheet may be used in an emergency when charts have been lost or destroyed. Directions on how to construct plotting sheets suitable for emergency purposes are given in Chapter 26, Emergency Navigation.

323. Grids

No system exists for showing the surface of the earth on a plane without distortion. Moreover, the appearance of the surface varies with the projection and with the relation of that surface area to the point of tangency. One may want to identify a location or area simply by alpha-numeric rectangular coordinates. This is accomplished with a **grid**. In its usual form this consists of two series of lines drawn perpendicularly on the chart, marked by suitable alpha-numeric designations.

A grid may use the rectangular graticule of the Mercator projection or a set of arbitrary lines on a particular projection. **The World Geodetic Reference System (GEOREF)** is a method of designating latitude and longitude by a system of letters and numbers instead of by angular measure. It is not, therefore, strictly a grid. It is useful for operations extending over a wide area. Examples of the second type of grid are the **Universal Transverse Mercator (UTM)** grid, the **Universal Polar Stereographic (UPS)** grid, and the **Temporary Geographic Grid (TGG)**. Since these systems are used primarily by military forces, they are sometimes called military grids.

CHART SCALES

324. Types Of Scales

The **scale** of a chart is the ratio of a given distance on the chart to the actual distance which it represents on the earth. It may be expressed in various ways. The most common are:

1. A simple ratio or fraction, known as the **representative fraction**. For example, 1:80,000 or 1/80,000 means that one unit (such as a meter) on the chart represents 80,000 of the same unit on the surface of the earth. This scale is sometimes called the **natural** or **fractional** scale.
2. A **statement** that a given distance on the earth equals a given measure on the chart, or vice versa. For example, "30 miles to the inch" means that 1 inch on the chart represents 30 miles of the earth's surface. Similarly, "2 inches to a mile" indicates that 2 inches on the chart represent 1 mile on the earth. This is sometimes called the **numerical scale**.
3. A line or bar called a **graphic scale** may be drawn at a convenient place on the chart and subdivided into nautical miles, meters, etc. All charts vary somewhat in scale from point to point, and in some projections the scale is not the same in all directions about a single

point. A single subdivided line or bar for use over an entire chart is shown only when the chart is of such scale and projection that the scale varies a negligible amount over the chart, usually one of about 1:75,000 or larger. Since 1 minute of latitude is very nearly equal to 1 nautical mile, the latitude scale serves as an approximate graphic scale. On most nautical charts the east and west borders are subdivided to facilitate distance measurements.

On a Mercator chart the scale varies with the latitude. This is noticeable on a chart covering a relatively large distance in a north-south direction. On such a chart the border scale near the latitude in question should be used for measuring distances.

Of the various methods of indicating scale, the graphical method is normally available in some form on the chart. In addition, the scale is customarily stated on charts on which the scale does not change appreciably over the chart.

The ways of expressing the scale of a chart are readily interchangeable. For instance, in a nautical mile there are about 72,913.39 inches. If the natural scale of a chart is 1:80,000, one inch of the chart represents 80,000 inches of the earth, or a little more than a mile. To find the exact amount, divide the scale by the number of inches in a mile, or 80,000/72,913.39 = 1.097. Thus, a scale of 1:80,000 is the same as a scale of 1.097 (or approximately 1.1) miles to an inch. Stated another way, there are: 72,913.39/80,000 = 0.911 (approximately 0.9) inch to a mile. Similarly, if the scale is 60 nautical miles to an inch, the representative fraction is 1:(60 x 72,913.39) = 1:4,374,803.

A chart covering a relatively large area is called a **small-scale chart** and one covering a relatively small area is called a **large-scale chart**. Since the terms are relative, there is no sharp division between the two. Thus, a chart of scale 1:100,000 is large scale when compared with a chart of 1:1,000,000 but small scale when compared with one of 1:25,000.

As scale decreases, the amount of detail which can be shown decreases also. Cartographers selectively decrease the detail in a process called **generalization** when producing small scale charts using large scale charts as sources. The amount of detail shown depends on several factors, among them the coverage of the area at larger scales and the intended use of the chart.

325. Chart Classification by Scale

Charts are constructed on many different scales, ranging from about 1:2,500 to 1:14,000,000. Small-scale charts covering large areas are used for route planning and for offshore navigation. Charts of larger scale, covering smaller areas, are used as the vessel approaches land. Several methods of classifying charts according to scale are used in various nations. The following classifications of nautical charts are used by the National Ocean Service.

Sailing charts are the smallest scale charts used for planning, fixing position at sea, and for plotting the dead reckoning while proceeding on a long voyage. The scale is generally smaller than 1:600,000. The shoreline and topography are generalized and only offshore soundings, the principal navigational lights, outer buoys, and landmarks visible at considerable distances are shown.

General charts are intended for coastwise navigation outside of outlying reefs and shoals. The scales range from about 1:150,000 to 1:600,000.

Coastal charts are intended for inshore coastwise navigation, for entering or leaving bays and harbors of considerable width, and for navigating large inland waterways. The scales range from about 1:50,000 to 1:150,000.

Harbor charts are intended for navigation and anchorage in harbors and small waterways. The scale is generally larger than 1:50,000.

In the classification system used by NIMA, the sailing charts are incorporated in the general charts classification (smaller than about 1:150,000); those coast charts especially useful for approaching more confined waters (bays, harbors) are classified as approach charts. There is considerable overlap in these designations, and the classification of a chart is best determined by its use and by its relationship to other charts of the area. The use of insets complicates the placement of charts into rigid classifications.

CHART ACCURACY

326. Factors Relating to Accuracy

The accuracy of a chart depends upon the accuracy of the hydrographic surveys and other data sources used to compile it and the suitability of its scale for its intended use.

One can sometimes estimate the accuracy of a chart's surveys from the source notes given in the title of the chart. If the chart is based upon very old surveys, use it with caution. Many early surveys were inaccurate because of the technological limitations of the surveyor.

The number of soundings and their spacing indicates the completeness of the survey. Only a small fraction of the soundings taken in a thorough survey are shown on the chart, but sparse or unevenly distributed soundings indicate that the survey was probably not made in detail. See Figure 326a and Figure 326b. Large blank areas or absence of depth contours generally indicate lack of soundings in the area. Operate in an area with sparse sounding data only if required and then only with extreme caution. Run the echo sounder continuously and operate at a reduced speed.

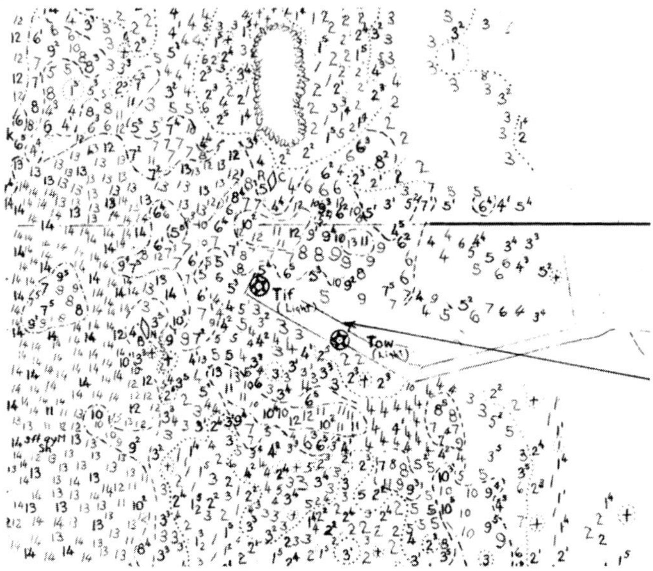

Figure 326a. Part of a "boat sheet," showing the soundings obtained in a survey.

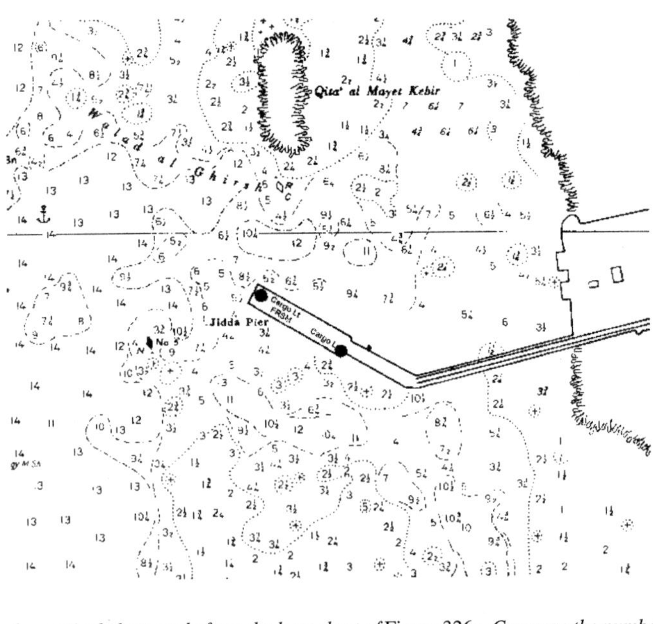

Figure 326b. Part of a nautical chart made from the boat sheet of Figure 326a. *Compare the number of soundings in the two figures.*

Sparse sounding information does not necessarily indicate an incomplete survey. Relatively few soundings are shown when there is a large number of depth contours, or where the bottom is flat, or gently and evenly sloping. Additional soundings are shown when they are helpful in indicating the uneven character of a rough bottom.

Even a detailed survey may fail to locate every rock or pinnacle. In waters where they might be located, the best method for finding them is a wire drag survey. Areas that have been dragged may be indicated on the chart by limiting lines and green or purple tint and a note added to show the effective depth at which the drag was operated.

Changes in bottom contours are relatively rapid in areas such as entrances to harbors where there are strong currents or heavy surf. Similarly, there is sometimes a tendency for dredged channels to shoal, especially if they are surrounded by sand or mud, and cross currents exist. Charts often contain notes indicating the bottom contours are known to change rapidly.

The same detail cannot be shown on a small-scale chart as on a large scale chart. On small-scale charts, detailed information is omitted or "generalized" in the areas covered by larger scale charts. The navigator should use the largest scale chart available for the area in which he is operating, especially when operating in the vicinity of hazards.

Charting agencies continually evaluate both the detail and the presentation of data appearing on a chart. Development of a new navigational aid may render previous charts inadequate. The development of radar, for example, required upgrading charts which lacked the detail required for reliable identification of radar targets.

After receiving a chart, the user is responsible for keeping it updated. Mariner's reports of errors, changes, and suggestions are useful to charting agencies. Even with modern automated data collection techniques, there is no substitute for on-sight observation of hydrographic conditions by experienced mariners. This holds true especially in less frequently traveled areas of the world.

CHART READING

327. Chart Dates

NOS charts have two dates. At the top center of the chart is the date of the first edition of the chart. In the lower left corner of the chart is the current edition number and date. This date shows the latest date through which *Notice to Mariners* were applied to the chart. Any subsequent change will be printed in the *Notice to Mariners*. Any notices which accumulate between the chart date and the announcement date in the *Notice to Mariners* will be given with the announcement. Comparing the dates of the first and current editions gives an indication of how often the chart is updated. Charts of busy areas are updated more frequently than those of less traveled areas. This interval may vary from 6 months to more than ten years for NOS charts. This update interval may be much longer for certain NIMA charts in remote areas.

New editions of charts are both demand and source driven. Receiving significant new information may or may not initiate a new edition of a chart, depending on the demand for that chart. If it is in a sparsely-traveled area, other priorities may delay a new edition for several years. Conversely, a new edition may be printed without the receipt of significant new data if demand for the chart is high and stock levels are low. *Notice to Mariners* corrections are always included on new editions.

NIMA charts have the same two dates as the NOS charts; the current chart edition number and date is given in the lower left corner. Certain NIMA charts are reproductions of foreign charts produced under joint agreements with a number of other countries. These charts, even though of recent date, may be based on foreign charts of considerably earlier date. Further, new editions of the foreign chart will not necessarily result in a new edition of the NIMA reproduction. In these cases, the foreign chart is the better chart to use.

328. Title Block

The chart title block should be the first thing a navigator looks at when receiving a new edition chart. Refer to Figure 328. The title itself tells what area the chart covers. The chart's scale and projection appear below the title. The chart will give both vertical and horizontal datums and, if necessary, a datum conversion note. Source notes or diagrams will list the date of surveys and other charts used in compilation.

329. Shoreline

The shoreline shown on nautical charts represents the line of contact between the land and water at a selected vertical datum. In areas affected by tidal fluctuations, this is usually the mean high-water line. In confined coastal waters of diminished tidal influence, a mean water level line may be used. The shoreline of interior waters (rivers, lakes) is usually a line representing a specified elevation above a

BALTIC SEA

GERMANY—NORTH COAST

DAHMESHÖVED TO WISMAR

From German Surveys

SOUNDINGS IN METERS

reduced to the approximate level of Mean Sea Level

HEIGHTS IN METERS ABOVE MEAN SEA LEVEL

MERCATOR PROJECTION

EUROPEAN DATUM

SCALE 1:50,000

Figure 328. A chart title block.

selected datum. A shoreline is symbolized by a heavy line. A broken line indicates that the charted position is approximate only. The nature of the shore may be indicated.

If the low water line differs considerably from the high water line, then a dotted line represents the low water line. If the bottom in this area is composed of mud, sand, gravel or stones, the type of material will be indicated. If the bottom is composed of coral or rock, then the appropriate symbol will be used. The area alternately covered and uncovered may be shown by a tint which is usually a combination of the land and water tint.

The apparent shoreline shows the outer edge of marine vegetation where that limit would appear as shoreline to the mariner. It is also used to indicate where marine vegetation prevents the mariner from defining the shoreline. A light line symbolizes this shoreline. A broken line marks the inner edge when no other symbol (such as a cliff or levee) furnishes such a limit. The combined land-water tint or the land tint marks the area between inner and outer limits.

330. Chart Symbols

Much of the information contained on charts is shown by symbols. These symbols are not shown to scale, but they indicate the correct position of the feature to which they refer. The standard symbols and abbreviations used on charts published by the United States of America are shown in ***Chart No. 1, Nautical Chart Symbols and Abbreviations***. See Figure 330.

Electronic chart symbols are, within programming and display limits, much the same as printed ones. The less expensive electronic charts have less extensive symbol libraries, and the screen's resolution may affect the presentation detail.

Most of the symbols and abbreviations shown in U.S. *Chart No. 1* agree with recommendations of the International Hydrographic Organization (IHO). The layout is explained in the general remarks section of *Chart No. 1*.

The symbols and abbreviations on any given chart may differ somewhat from those shown in *Chart No. 1*. In addition, foreign charts may use different symbology. When using a foreign chart, the navigator should have available the *Chart No. 1* from the country which produced the chart.

Chart No. 1 is organized according to subject matter, with each specific subject given a letter designator. The general subject areas are General, Topography, Hydrography, Aids and Services, and Indexes. Under each heading, letter designators further define subject areas, and individual numbers refer to specific symbols.

Information in *Chart No. 1* is arranged in columns. The first column contains the IHO number code for the symbol in question. The next two columns show the symbol itself, in NOS and NIMA formats. If the formats are the same, the two columns are combined into one. The next column is a text description of the symbol, term, or abbreviation. The next column contains the IHO standard symbol. The last column shows certain symbols used on foreign reproduction charts produced by NIMA.

331. Lettering

Except on some modified reproductions of foreign charts, cartographers have adopted certain lettering stan-

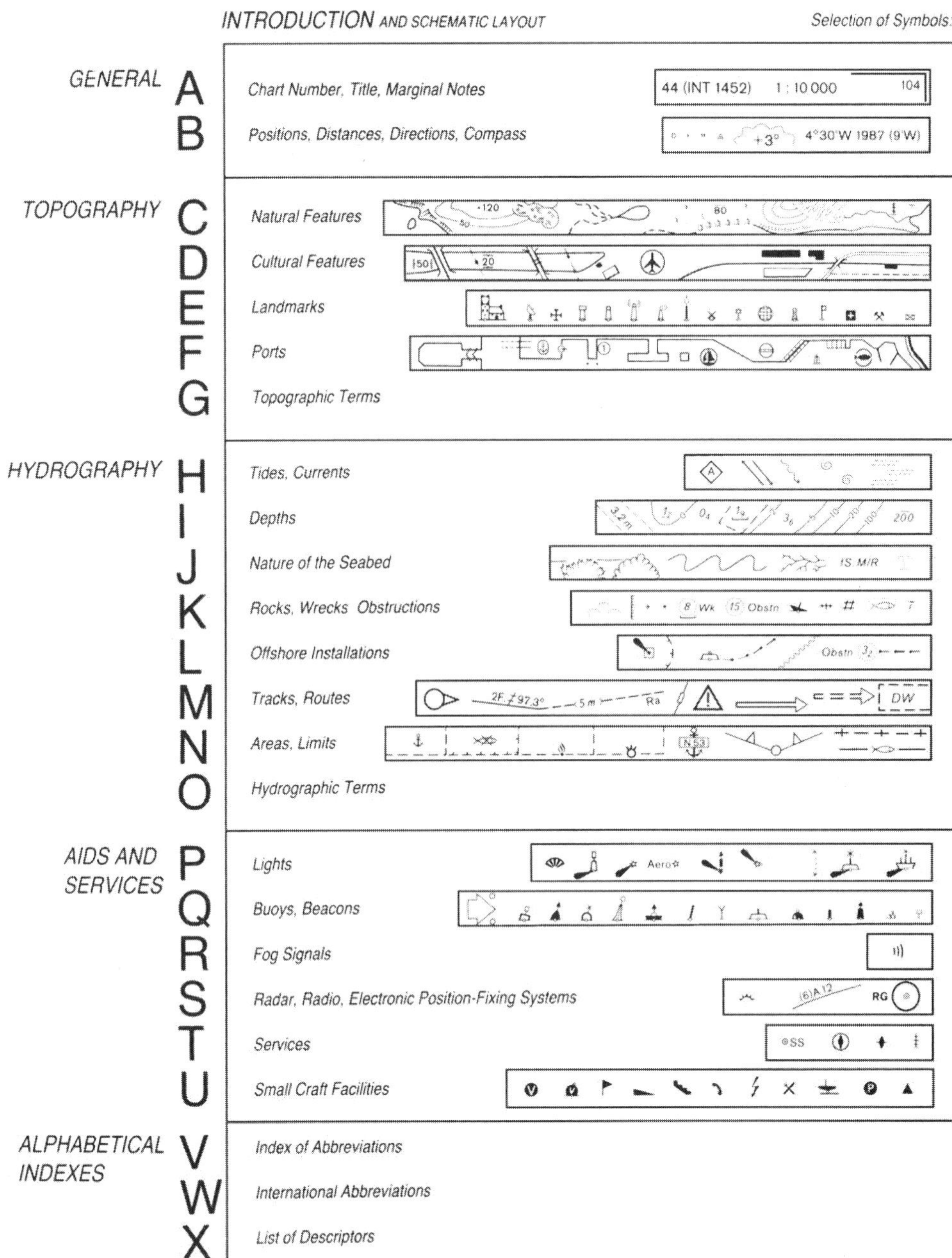

Figure 330. Contents of U.S. Chart No. 1.

dards. Vertical type is used for features which are dry at high water and not affected by movement of the water; slanting type is used for underwater and floating features.

There are two important exceptions to the two general rules listed above. Vertical type is not used to represent heights above the waterline, and slanting type is not used to indicate soundings, except on metric charts. Section 332 below discusses the conventions for indicating soundings.

Evaluating the type of lettering used to denote a feature, one can determine whether a feature is visible at high tide. For instance, a rock might bear the title "Rock" whether or not it extends above the surface. If the name is given in vertical letters, the rock constitutes a small islet; if in slanting type, the rock constitutes a reef, covered at high water.

332. Soundings

Charts show soundings in several ways. Numbers denote individual soundings. These numbers may be either vertical or slanting; both may be used on the same chart, distinguishing between data based upon different U.S. and foreign surveys, different datums, or smaller scale charts.

Large block letters at the top and bottom of the chart indicate the unit of measurement used for soundings. SOUNDINGS IN FATHOMS indicates soundings are in fathoms or fathoms and fractions. SOUNDINGS IN FATHOMS AND FEET indicates the soundings are in fathoms and feet. A similar convention is followed when the soundings are in meters or meters and tenths.

A **depth conversion scale** is placed outside the neatline on the chart for use in converting charted depths to feet, meters, or fathoms. "No bottom" soundings are indicated by a number with a line over the top and a dot over the line. This indicates that the spot was sounded to the depth indicated without reaching the bottom. Areas which have been wire dragged are shown by a broken limiting line, and the clear effective depth is indicated, with a characteristic symbol under the numbers. On NIMA charts a purple or green tint is shown within the swept area.

Soundings are supplemented by **depth contours**, lines connecting points of equal depth. These lines present a picture of the bottom. The types of lines used for various depths are shown in Section I of Chart No. 1. On some charts depth contours are shown in solid lines; the depth represented by each line is shown by numbers placed in breaks in the lines, as with land contours. Solid line depth contours are derived from intensively developed hydrographic surveys. A broken or indefinite contour is substituted for a solid depth contour whenever the reliability of the contour is questionable.

Depth contours are labeled with numerals in the unit of measurement of the soundings. A chart presenting a more detailed indication of the bottom configuration with fewer numerical soundings is useful when bottom contour navigating. Such a chart can be made only for areas which have undergone a detailed survey

Shoal areas often are given a blue tint. Charts designed to give maximum emphasis to the configuration of the bottom show depths beyond the 100-fathom curve over the entire chart by depth contours similar to the contours shown on land areas to indicate graduations in height. These are called **bottom contour** or **bathymetric charts**.

On electronic charts, a variety of other color schemes may be used, according to the manufacturer of the system. Color perception studies are being used to determine the best presentation.

The side limits of dredged channels are indicated by broken lines. The project depth and the date of dredging, if known, are shown by a statement in or along the channel. The possibility of silting is always present. Local authorities should be consulted for the controlling depth. NOS Charts frequently show controlling depths in a table, which is kept current by the *Notice to Mariners*.

The chart scale is generally too small to permit all soundings to be shown. In the selection of soundings, least depths are shown first. This conservative sounding pattern provides safety and ensures an uncluttered chart appearance. Steep changes in depth may be indicated by more dense soundings in the area. The limits of shoal water indicated on the chart may be in error, and nearby areas of undetected shallow water may not be included on the chart. Given this possibility, areas where shoal water is known to exist should be avoided. If the navigator must enter an area containing shoals, he must exercise extreme caution in avoiding shallow areas which may have escaped detection. By constructing a "safety range" around known shoals and ensuring his vessel does not approach the shoal any closer than the safety range, the navigator can increase his chances of successfully navigating through shoal water. Constant use of the echo sounder is also important.

Abbreviations listed in Section J of Chart No. 1 are used to indicate what substance forms the bottom. The meaning of these terms can be found in the Glossary of this volume. While in ages past navigators might actually navigate by knowing the bottom characteristics of certain local areas, today knowing the characteristic of the bottom is most important when anchoring.

333. Depths and Datums

Depths are indicated by soundings or explanatory notes. Only a small percentage of the soundings obtained in a hydrographic survey can be shown on a nautical chart. The least depths are generally selected first, and a pattern built around them to provide a representative indication of bottom relief. In shallow water, soundings may be spaced 0.2 to 0.4 inch apart. The spacing is gradually increased as water deepens, until a spacing of 0.8 to 1.0 inch is reached in deeper waters offshore. Where a sufficient number of soundings are available to permit adequate interpretation, depth curves are drawn in at selected intervals.

All depths indicated on charts are reckoned from a selected level of the water, called the **sounding datum**, (sometimes referred to as the **reference plane** to distinguish this term from the geodetic datum). The various

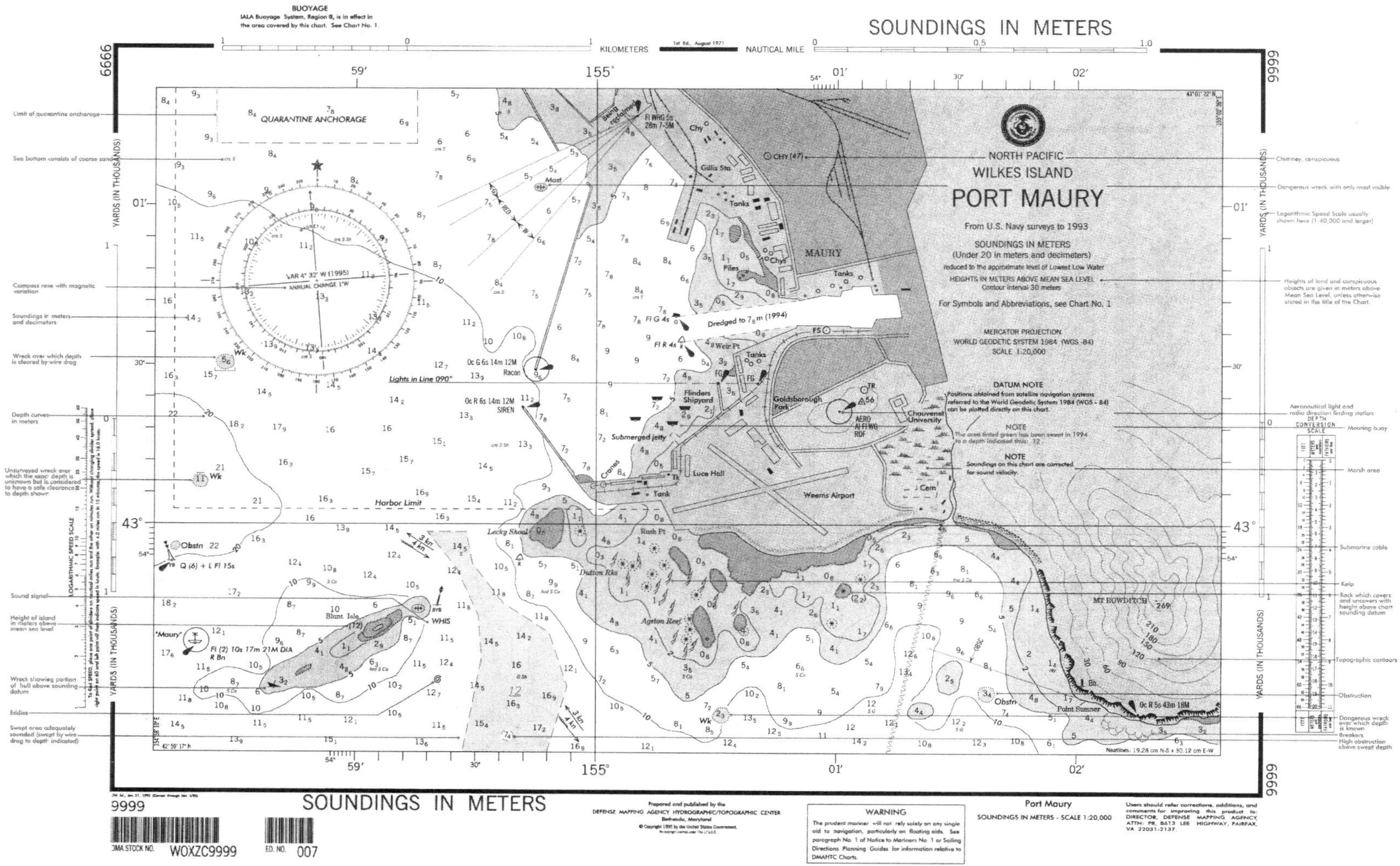

SOUNDINGS IN METERS
BUOYAGE
IALA Buoyage System, Region B, is in effect in the area covered by this chart. See Chart No. 1.
KILOMETERS
1st Ed., August 1971
NAUTICAL MILE
NORTH PACIFIC
WILKES ISLAND
PORT MAURY
From U.S. Navy surveys to 1993
SOUNDINGS IN METERS
(Under 20 in meters and decimeters)
reduced to the approximate level of Lowest Low Water
HEIGHTS IN METERS ABOVE MEAN SEA LEVEL
Contour interval 30 meters
For Symbols and Abbreviations, see Chart No. 1
MERCATOR PROJECTION
WORLD GEODETIC SYSTEM 1984 (WGS-84)
SCALE 1:20,000
DATUM NOTE
Positions obtained from satellite navigation systems referred to the World Geodetic System 1984 (WGS-84) can be plotted directly on this chart.
NOTE
The area tinted green has been swept in 1994 to a depth indicated thus: 12
NOTE
Soundings on this chart are corrected for sound velocity.
QUARANTINE ANCHORAGE
Lights in Line 090°
Harbor Limit
Dredged to 7.8m (1994)
Submerged jetty
Flinders Shipyard
Goldsborough Park
Chauvenet University
Luce Hall
Weems Airport
Cranes
Tank
Tanks
MAURY
Gillis Sta
Piles
Weir Pt
Rush Pt
Lecky Shoal
Dutton Rks
Agston Reef
Blunt Isle
MT BOWDITCH 269
Point Sumner
Obstn
Wk
'Maury'
Limit of quarantine anchorage
Sea bottom consists of coarse sand
Compass rose with magnetic variation
Soundings in meters and decimeters
Wreck over which depth is cleared by wire drag
Depth curves in meters
Unsurveyed wreck over which the exact depth is unknown but is considered to have a safe clearance to depth shown
Sound signal
Height of island in meters above mean sea level
Wreck showing portion of hull above sounding datum
Eddies
Swept area adequately sounded (swept by wire drag to depth indicated)
Chimney, conspicuous
Dangerous wreck with only mast visible
Logarithmic Speed Scale usually shown here (1:40,000 and larger)
Heights of land and conspicuous objects are given in meters above Mean Sea Level, unless otherwise stated in the title of the Chart.
Aeronautical light and radio direction finding station
Mooring buoy
Marsh area
Submarine cable
Kelp
Rock which covers and uncovers with height above chart sounding datum
Topographic contours
Obstruction
Dangerous wreck over which depth is known
Breakers
High obstruction above swept depth
DEPTH CONVERSION SCALE
LOGARITHMIC SPEED SCALE
YARDS (IN THOUSANDS)
Neatlines: 19.28 cm N-S x 30.12 cm E-W
9999
DMA STOCK NO. WOXZC9999
ED. NO. 007
Prepared and published by the DEFENSE MAPPING AGENCY HYDROGRAPHIC/TOPOGRAPHIC CENTER Bethesda, Maryland
© Copyright 1995 by the United States Government.
WARNING
The prudent mariner will not rely solely on any single aid to navigation, particularly on floating aids. See paragraph No. 1 of Notice to Mariners No. 1 or Sailing Directions Planning Guides for information relative to DMAHTC Charts.
Port Maury
SOUNDINGS IN METERS - SCALE 1:20,000
Users should refer corrections, additions, and comments for improving this product to: DIRECTOR, DEFENSE MAPPING AGENCY, ATTN: PR, 8613 LEE HIGHWAY, FAIRFAX, VA 22031-2137.

This nautical chart has been designed to promote safe navigation. The National Ocean Service encourages users to submit corrections, additions or comments for improving this chart to the Director, Coast and Geodetic Survey (N/CG22), National Ocean Service, NOAA, Silver Spring, Maryland 20910 - 3282.

SOUNDINGS IN FATHOMS AND FEET

(FATHOMS AND FEET TO 11 FATHOMS)

Formerly C&GS 4196

Nautical Chart Catalog No. 2, Panel B.

6666

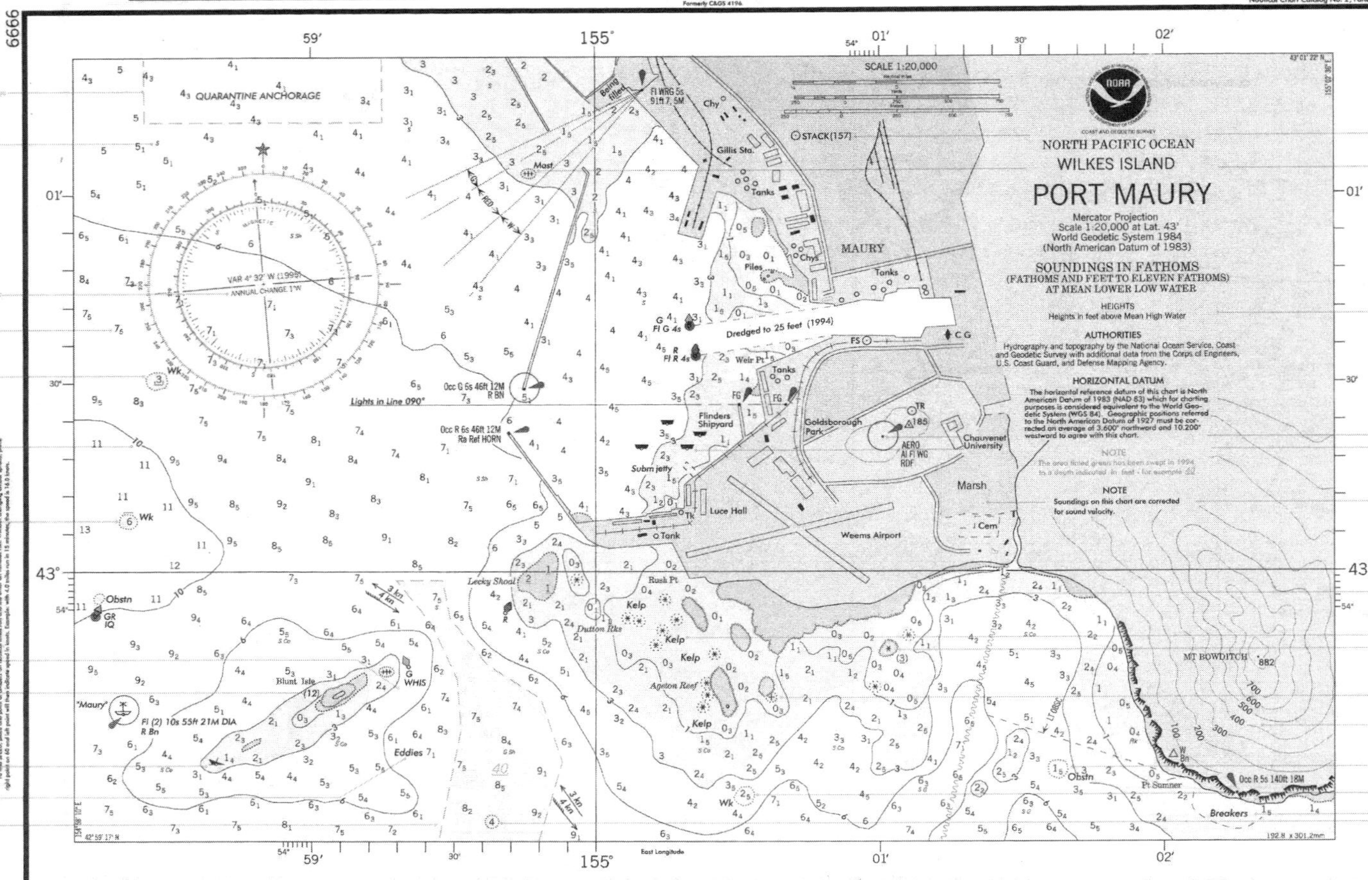

ED. NO. 007

DMA STOCK NO. WOXZC9999

7th Ed., Jan. 21, 1995 (Corrected through NM 3/95)

9999

CAUTION

This chart has been corrected from Notice to Mariners published weekly by the Defense Mapping Agency Hydrographic/Topographic Center and the Local Notices to Mariners issued periodically by each U.S. Coast Guard district to the date shown in the lower left hand corner.

Published at Washington, D.C.
U.S. DEPARTMENT OF COMMERCE
NATIONAL OCEANIC AND ATMOSPHERIC ADMINISTRATION
NATIONAL OCEAN SERVICE
COAST AND GEODETIC SURVEY

WARNING

The prudent mariner will not rely solely on any single aid to navigation, particularly on floating aids. See U.S. Coast Guard Light List and U.S. Coast Pilot for details.

Port Maury

SOUNDINGS IN FATHOMS - SCALE 1:20,000

9999

sounding datums are explained in Chapter 9, Tides and Tidal Currents. On charts produced from U.S. surveys, the sounding datum is selected with regard to the tides of the region. Depths shown are the least depths to be expected under average conditions. On charts compiled from foreign charts and surveys the sounding datum is that of the original authority. When it is known, the sounding datum used is stated on the chart. In some cases where the chart is based upon old surveys, particularly in areas where the range of tide is not great, the sounding datum may not be known.

For most National Ocean Service charts of the United States and Puerto Rico, the sounding datum is mean lower low water. Most NIMA charts are based upon mean low water, mean lower low water, or mean low water springs. The sounding datum for charts published by other countries varies greatly, but is usually lower than mean low water. On charts of the Baltic Sea, Black Sea, the Great Lakes, and other areas where tidal effects are small or without significance, the sounding datum adopted is an arbitrary height approximating the mean water level.

The sounding datum of the largest scale chart of an area is generally the same as the reference level from which height of tide is tabulated in the tide tables.

The chart datum is usually only an approximation of the actual mean value, because determination of the actual mean height usually requires a longer series of tidal observations than is usually available to the cartographer. In addition, the heights of the tide vary over time.

Since the chart datum is generally a computed mean or average height at some state of the tide, the depth of water at any particular moment may be less than shown on the chart. For example, if the chart datum is mean lower low water, the depth of water at lower low water will be less than the charted depth about as often as it is greater. A lower depth is indicated in the tide tables by a minus sign (–).

334. Heights

The shoreline shown on charts is generally mean high water. A light's height is usually reckoned from mean sea level. The heights of overhanging obstructions (bridges, power cables, etc.) are usually reckoned from mean high water. A high water reference gives the mariner the minimum clearance expected.

Since heights are usually reckoned from high water and depths from some form of low water, the reference levels are seldom the same. Except where the range of tide is very large, this is of little practical significance.

335. Dangers

Dangers are shown by appropriate symbols, as indicated in Section K of *Chart No. 1*.

A rock uncovered at mean high water may be shown as an islet. If an isolated, offlying rock is known to uncover at the sounding datum but to be covered at high water, the chart shows the appropriate symbol for a rock and gives the height above the sounding datum. The chart can give this height one of two ways. It can use a statement such as "Uncov 2 ft.," or it can indicate the number of feet the rock protrudes above the sounding datum, underline this value, and enclose it in parentheses (i.e. ($\underline{2}$)). A rock which does not uncover is shown by an enclosed figure approximating its dimensions and filled with land tint. It may be enclosed by a dotted depth curve for emphasis.

A tinted, irregular-line figure of approximately true dimensions is used to show a detached coral reef which uncovers at the chart datum. For a coral or rocky reef which is submerged at chart datum, the sunken rock symbol or an appropriate statement is used, enclosed by a dotted or broken line if the limits have been determined.

Several different symbols mark wrecks. The nature of the wreck or scale of the chart determines the correct symbol. A sunken wreck with less than 11 fathoms of water over it is considered dangerous and its symbol is surrounded by a dotted curve. The curve is omitted if the wreck is deeper than 11 fathoms. The safe clearance over a wreck, if known, is indicated by a standard sounding number placed at the wreck. If this depth was determined by a wire drag, the sounding is underscored by the wire drag symbol. An unsurveyed wreck over which the exact depth is unknown but a safe clearance depth is known is depicted with a solid line above the symbol.

Tide rips, eddies, and kelp are shown by symbol or legend. Piles, dolphins (clusters of piles), snags, and stumps are shown by small circles and a label identifying the type of obstruction. If such dangers are submerged, the letters "Subm" precede the label. Fish stakes and traps are shown when known to be permanent or hazardous to navigation.

336. Aids to Navigation

Aids to navigation are shown by symbols listed in Sections P through S of Chart No. 1. Abbreviations and additional descriptive text supplement these symbols. In order to make the symbols conspicuous, the chart shows them in size greatly exaggerated relative to the scale of the chart. "Position approximate" circles are used on floating aids to indicate that they have no exact position because they move around their moorings. For most floating aids, the position circle in the symbol marks the approximate location of the anchor or sinker. The actual aid may be displaced from this location by the scope of its mooring.

The type and number of aids to navigation shown on a chart and the amount of information given in their legends varies with the scale of the chart. Smaller scale charts may have fewer aids indicated and less information than larger scale charts of the same area.

Lighthouses and other navigation lights are shown as black dots with purple disks or as black dots with purple flare symbols. The center of the dot is the position of the light. Some modified facsimile foreign charts use a small star instead of a dot.

On large-scale charts the legend elements of lights are shown in the following order:

Legend	*Example*	*Meaning*
Characteristic	Fl(2)	group flashing; 2 flashes
Color	R	red
Period	10s	2 flashes in 10 seconds
Height	80m	80 meters
Range	19M	19 nautical miles
Designation	"6"	light number 6

The legend for this light would appear on the chart:

Fl(2) R 10s 80m 19M "6"

As chart scale decreases, information in the legend is selectively deleted to avoid clutter. The order of deletion is usually height first, followed by period, group repetition interval (e.g. (2)), designation, and range. Characteristic and color will almost always be shown.

Small triangles mark red daybeacons; small squares mark all others. On NIMA charts, pictorial beacons are used when the IALA buoyage system has been implemented. The center of the triangle marks the position of the aid. Except on Intracoastal Waterway charts and charts of state waterways, the abbreviation "Bn" is shown beside the symbol, along with the appropriate abbreviation for color if known. For black beacons the triangle is solid black and there is no color abbreviation. All beacon abbreviations are in vertical lettering.

Radiobeacons are indicated on the chart by a purple circle accompanied by the appropriate abbreviation indicating an ordinary radiobeacon (R Bn) or a radar beacon (Ramark or Racon, for example).

A variety of symbols, determined by both the charting agency and the types of buoys, indicate navigation buoys. IALA buoys (see Chapter 5, Short Range Aids to Navigation) in foreign areas are depicted by various styles of symbols with proper topmarks and colors; the position circle which shows the approximate location of the sinker is at the base of the symbol.

A mooring buoy is shown by one of several symbols as indicated in Chart No. 1. It may be labeled with a berth number or other information.

A buoy symbol with a horizontal line indicates the buoy has horizontal bands. A vertical line indicates vertical stripes; crossed lines indicate a checked pattern. There is no significance to the angle at which the buoy symbol appears on the chart. The symbol is placed so as to avoid interference with other features.

Lighted buoys are indicated by a purple flare from the buoy symbol or by a small purple disk centered on the position circle.

Abbreviations for light legends, type and color of buoy, designation, and any other pertinent information given near the symbol are in slanted type. The letter C, N, or S indicates a can, nun, or spar, respectively. Other buoys are assumed to be pillar buoys, except for special buoys such as spherical, barrel, etc. The number or letter designation of the buoy is given in quotation marks on NOS charts. On other charts they may be given without quotation marks or other punctuation.

Aeronautical lights included in the light lists are shown by the lighthouse symbol, accompanied by the abbreviation "AERO." The characteristics shown depend principally upon the effective range of other navigational lights in the vicinity and the usefulness of the light for marine navigation.

Directional ranges are indicated by a broken or solid line. The solid line, indicating that part of the range intended for navigation, may be broken at irregular intervals to avoid being drawn through soundings. That part of the range line drawn only to guide the eye to the objects to be kept in range is broken at regular intervals. The direction, if given, is expressed in degrees, clockwise from true north.

Sound signals are indicated by the appropriate word in capital letters (HORN, BELL, GONG, or WHIS) or an abbreviation indicating the type of sound. Sound signals of any type except submarine sound signals may be represented by three purple 45° arcs of concentric circles near the top of the aid. These are not shown if the type of signal is listed. The location of a sound signal which does not accompany a visual aid, either lighted or unlighted, is shown by a small circle and the appropriate word in vertical block letters.

Private aids, when shown, are marked "Priv" on NOS charts. Some privately maintained unlighted fixed aids are indicated by a small circle accompanied by the word "Marker," or a larger circle with a dot in the center and the word "MARKER." A privately maintained lighted aid has a light symbol and is accompanied by the characteristics and the usual indication of its private nature. Private aids should be used with caution.

A light sector is the sector or area bounded by two radii and the arc of a circle in which a light is visible or in which it has a distinctive color different from that of adjoining sectors. The limiting radii are indicated on the chart by dotted or dashed lines. Sector colors are indicated by words spelled out if space permits, or by abbreviations (W, R, etc.) if it does not. Limits of light sectors and arcs of visibility as observed from a vessel are given in the light lists, in clockwise order.

337. Land Areas

The amount of detail shown on the land areas of nautical charts depends upon the scale and the intended purpose of the chart. Contours, form lines, and shading indicate relief.

Contours are lines connecting points of equal elevation. Heights are usually expressed in feet (or in meters with means for conversion to feet). The interval between contours is uniform over any one chart, except that certain intermediate contours are sometimes shown by broken line. When contours are broken, their locations are approximate.

Form lines are approximations of contours used for the purpose of indicating relative elevations. They are used in areas where accurate information is not available in sufficient detail to permit exact location of contours. Elevations of individual form lines are not indicated on the chart.

Spot elevations are generally given only for summits or for tops of conspicuous landmarks. The heights of spot elevations and contours are given with reference to mean high water when this information is available.

When there is insufficient space to show the heights of islets or rocks, they are indicated by slanting figures enclosed in parentheses in the water area nearby.

338. Cities and Roads

Cities are shown in a generalized pattern that approximates their extent and shape. Street names are generally not charted except those along the waterfront on the largest scale charts. In general, only the main arteries and thoroughfares or major coastal highways are shown on smaller scale charts. Occasionally, highway numbers are given. When shown, trails are indicated by a light broken line. Buildings along the waterfront or individual ones back from the waterfront but of special interest to the mariner are shown on large-scale charts. Special symbols from Chart No. 1 are used for certain kinds of buildings. A single line with cross marks indicates both single and double track railroads. City electric railways are usually not charted. Airports are shown on small-scale charts by symbol and on large-scale charts by the shape of runways. The scale of the chart determines if single or double lines show breakwaters and jetties; broken lines show the submerged portion of these features.

339. Landmarks

Landmarks are shown by symbols in Chart No. 1.

A large circle with a dot at its center is used to indicate that the position is precise and may be used without reservation for plotting bearings. A small circle without a dot is used for landmarks not accurately located. Capital and lower case letters are used to identify an approximate landmark: "Mon," "Cup," or "Dome." The abbreviation "PA" (position approximate) may also appear. An accurate landmark is identified by all capital type ("MON," "CUP," "DOME").

When only one object of a group is charted, its name is followed by a descriptive legend in parenthesis, including the number of objects in the group, for example "(TALLEST OF FOUR)" or "(NORTHEAST OF THREE)."

340. Miscellaneous Chart Features

A measured nautical mile indicated on a chart is accurate to within 6 feet of the correct length. Most measured miles in the United States were made before 1959, when the United States adopted the International Nautical Mile. The new value is within 6 feet of the previous standard length of 6,080.20 feet. If the measured distance differs from the standard value by more than 6 feet, the actual measured distance is stated and the words "measured mile" are omitted.

Periods after abbreviations in water areas are omitted because these might be mistaken for rocks. However, a lower case i or j is dotted.

Commercial radio broadcasting stations are shown on charts when they are of value to the mariner either as landmarks or sources of direction-finding bearings.

Lines of demarcation between the areas in which international and inland navigation rules apply are shown only when they cannot be adequately described in notes on the chart.

Compass roses are placed at convenient locations on Mercator charts to facilitate the plotting of bearings and courses. The outer circle is graduated in degrees with zero at true north. The inner circle indicates magnetic north.

On many NIMA charts magnetic variation is given to the nearest 1' by notes in the centers of compass roses. the annual change is given to the nearest 1' to permit correction of the given value at a later date. On NOS charts, variation is to the nearest 15', updated at each new edition if over three years old. The current practice of NIMA is to give the magnetic variation to the nearest 1', but the magnetic information on new editions is only updated to conform with the latest five year epoch. Whenever a chart is reprinted, the magnetic information is updated to the latest epoch. On some smaller scale charts, the variation is given by isogonic lines connecting points of equal variation; usually a separate line represents each degree of variation. The line of zero variation is called the agonic line. Many plans and insets show neither compass roses nor isogonic lines, but indicate magnetic information by note. A local magnetic disturbance of sufficient force to cause noticeable deflection of the magnetic compass, called local attraction, is indicated by a note on the chart.

Currents are sometimes shown on charts with arrows giving the directions and figures showing speeds. The information refers to the usual or average conditions. According to tides and weather, conditions at any given time may differ considerably from those shown.

Review chart notes carefully because they provide important information. Several types of notes are used. Those in the margin give such information as chart number, publication notes, and identification of adjoining charts. Notes in connection with the chart title include information on scale, sources of data, tidal information, soundings, and cautions. Another class of notes covers such topics as local magnetic disturbance, controlling depths of channels, hazards to navigation, and anchorages.

A datum note will show the geodetic datum of the chart (Do not confuse with the sounding datum. See Chapter 2, Geodesy and Datums in Navigation.) It may also contain instructions on plotting positions from the WGS 84 or NAD 83 datums on the chart if such a conversion is needed.

Anchorage areas are labeled with a variety of magenta, black, or green lines depending on the status of the area. Anchorage berths are shown as purple circles, with the number or letter assigned to the berth inscribed within the circle. Caution notes are sometimes shown when there are specific anchoring regulations.

Spoil areas are shown within short broken black lines. Spoil areas are tinted blue on NOS charts and labeled. These areas contain no soundings and should be avoided.

Firing and bombing practice areas in the United States territorial and adjacent waters are shown on NOS and NIMA charts of the same area and comparable scale.

Danger areas established for short periods of time are not charted but are announced locally. Most military commands charged with supervision of gunnery and missile firing areas promulgate a weekly schedule listing activated danger areas. This schedule is subjected to frequent change; the mariner should always ensure he has the latest schedule prior to proceeding into a gunnery or missile firing area. Danger areas in effect for longer periods are published in the *Notice to Mariners*. Any aid to navigation established to mark a danger area or a fixed or floating target is shown on charts.

Traffic separation schemes are shown on standard nautical charts of scale 1:600,000 and larger and are printed in magenta.

A logarithmic time-speed-distance nomogram with an explanation of its application is shown on harbor charts.

Tidal information boxes are shown on charts of scales 1:200,000 and larger for NOS charts, and various scales on DMA charts, according to the source. See Figure 340a.

Tabulations of controlling depths are shown on some National Ocean Service harbor and coastal charts. See Figure 340b.

Study Chart No. 1 thoroughly to become familiar with all the symbols used to depict the wide variety of features on nautical charts.

TIDAL INFORMATION

Place	Position		Height above datum of soundings			
			Mean High Water		Mean Low Water	
	N. Lat.	E. Long.	Higher	Lower	Lower	Higher
			meters	meters	meters	meters
Olongapo	14°49'	120°17'	. . . 0.9 . . .	. . . 0.4 . . .	. . . 0.0 . . .	. . . 0.3 . . .

Figure 340a. Tidal box.

NANTUCKET HARBOR

Tabulated from surveys by the Corps of Engineers - report of June 1972 and surveys of Nov. 1971

Controlling depths in channels entering from seaward in feet at Mean Low Water					Project Dimensions		
Name of Channel	Left outside quarter	Middle half of channel	Right outside quarter	Date of Survey	Width (feet)	Length (naut. miles)	Depth M. L. W. (feet)
Entrance Channel	11.1	15.0	15.0	11 - 71	300	1.2	15

Note.-The Corps of Engineers should be consulted for changing conditions subsequent to the above.

Figure 340b. Tabulations of controlling depths.

REPRODUCTIONS OF FOREIGN CHARTS

341. Modified Facsimiles

Modified facsimile charts are modified reproductions of foreign charts produced in accordance with bilateral international agreements. These reproductions provide the mariner with up-to-date charts of foreign waters. Modified facsimile charts published by NIMA are, in general, reproduced with minimal changes, as listed below:

1. The original name of the chart may be removed and replaced by an anglicized version.
2. English language equivalents of names and terms on the original chart are printed in a suitable glossary on the reproduction, as appropriate.
3. All hydrographic information, except bottom characteristics, is shown as depicted on the original chart.
4. Bottom characteristics are as depicted in Chart No. 1, or as on the original with a glossary.
5. The unit of measurement used for soundings is shown in block letters outside the upper and lower neatlines.
6. A scale for converting charted depth to feet, meters, or fathoms is added.
7. Blue tint is shown from a significant depth curve to the shoreline.
8. Blue tint is added to all dangers enclosed by a dotted danger curve, dangerous wrecks, foul areas, obstructions, rocks awash, sunken rocks, and swept wrecks.
9. Caution notes are shown in purple and enclosed in a box.
10. Restricted, danger, and prohibited areas are usually outlined in purple and labeled appropriately.
11. Traffic separation schemes are shown in purple.
12. A note on traffic separation schemes, printed in black, is added to the chart.
13. Wire dragged (swept) areas are shown in purple or green.
14. Corrections are provided to shift the horizontal datum to the World Geodetic System (1984).

INTERNATIONAL CHARTS

342. International Chart Standards

The need for mariners and chart makers to understand and use nautical charts of different nations became increasingly apparent as the maritime nations of the world developed their own establishments for the compilation and publication of nautical charts from hydrographic surveys. Representatives of twenty-two nations formed a Hydrographic Conference in London in 1919. That conference resulted in the establishment of the **International Hydrographic Bureau (IHB)** in Monaco in 1921. Today, the IHB's successor, the **International Hydrographic Organization (IHO)** continues to provide international standards for the cartographers of its member nations. (See Chapter 1, Introduction to Marine Navigation, for a description of the IHO.)

Recognizing the considerable duplication of effort by member states, the IHO in 1967 moved to introduce the first **international chart**. It formed a committee of six member states to formulate specifications for two series of international charts. Eighty-three small-scale charts were approved; responsibility for compiling these charts has subsequently been accepted by the member states' Hydrographic Offices.

Once a Member State publishes an international chart, reproduction material is made available to any other Member State which may wish to print the chart for its own purposes.

International charts can be identified by the letters INT before the chart number and the International Hydrographic Organization seal in addition to other national seals which may appear.

CHART NUMBERING

343. The Chart Numbering System

NIMA and NOS use a system in which numbers are assigned in accordance with both the scale and geographical area of coverage of a chart. With the exception of certain charts produced for military use only, one- to five-digit numbers are used. With the exception of one-digit numbers, the first digit identifies the area; the number of digits establishes the scale range. The one-digit numbers are used for certain products in the chart system which are not actually charts.

Number of Digits	*Scale*
1	No Scale
2	1:9 million and smaller
3	1:2 million to 1:9 million
4	Special Purpose
5	1:2 million and larger

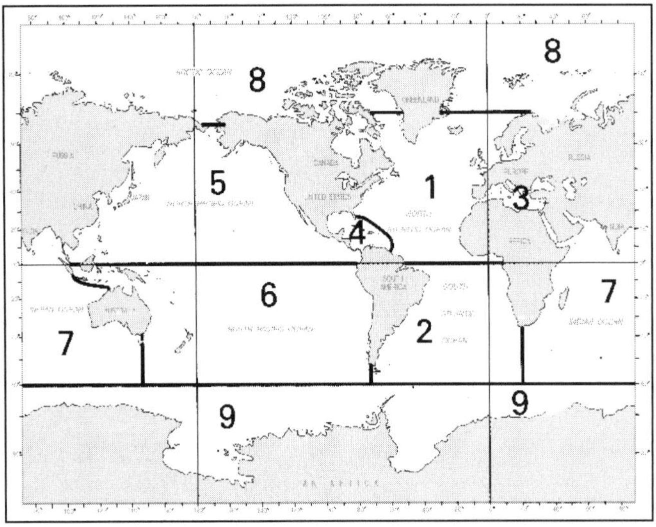

Figure 343a. Ocean basins with region numbers.

Two- and three-digit numbers are assigned to those small-scale charts which depict a major portion of an ocean basin or a large area. The first digit identifies the applicable ocean basin. See Figure 343a. Two-digit numbers are used for charts of scale 1:9,000,000 and smaller. Three-digit numbers are used for charts of scale 1:2,000,000 to 1:9,000,000.

Due to the limited sizes of certain ocean basins, no charts for navigational use at scales of 1:9,000,000 and smaller are published to cover these basins. The otherwise unused two-digit numbers (30 to 49 and 70 to 79) are assigned to special world charts.

One exception to the scale range criteria for three-digit numbers is the use of three-digit numbers for a series of position plotting sheets. They are of larger scale than 1:2,000,000 because they have application in ocean basins and can be used in all longitudes.

Four-digit numbers are used for non-navigational and special purpose charts, such as chart 5090, *Maneuvering Board*.

Five-digit numbers are assigned to those charts of scale 1:2,000,000 and larger that cover portions of the coastline rather than significant portions of ocean basins. These charts are based on the regions of the nautical chart index. See Figure 343b.

The first of the five digits indicates the region; the second digit indicates the subregion; the last three digits indicate the geographical sequence of the chart within the subregion. Many numbers have been left unused so that any future charts may be placed in their proper geographical sequence.

In order to establish a logical numbering system within the geographical subregions (for the 1:2,000,000 and larger-scale charts), a worldwide skeleton framework of coastal charts was laid out at a scale 1:250,000. This series was used as basic coverage except in areas where a coordinated series at about this scale already existed (such as the coast of Norway where a coordinated series of 1:200,000 charts was available).

Within each region, the geographical subregions are numbered counterclockwise around the continents, and within each subregion the basic series also is numbered counterclockwise around the continents. The basic coverage is assigned generally every 20th digit, except that the first 40 numbers in each subregion are reserved for smaller-scale coverage. Charts with scales larger than the basic coverage are assigned one of the 19 numbers following the number assigned to the sheet within which it falls. Figure 343c shows the numbering sequence in Iceland. Note the sequence of numbers around the coast, the direction of numbering, and the numbering of larger scale charts within the limits of smaller scales.

Five-digit numbers are also assigned to the charts produced by other hydrographic offices. This numbering system is applied to foreign charts so that they can be filed in logical sequence with the charts produced by the National Imagery and Mapping Agency and the National Ocean Service.

Certain exceptions to the standard numbering system have been made for charts intended for the military. Bottom contour charts depict parts of ocean basins. They are identified with a letter plus four digits according to a scheme best shown in the catalog, and are not available to civilian navigators.

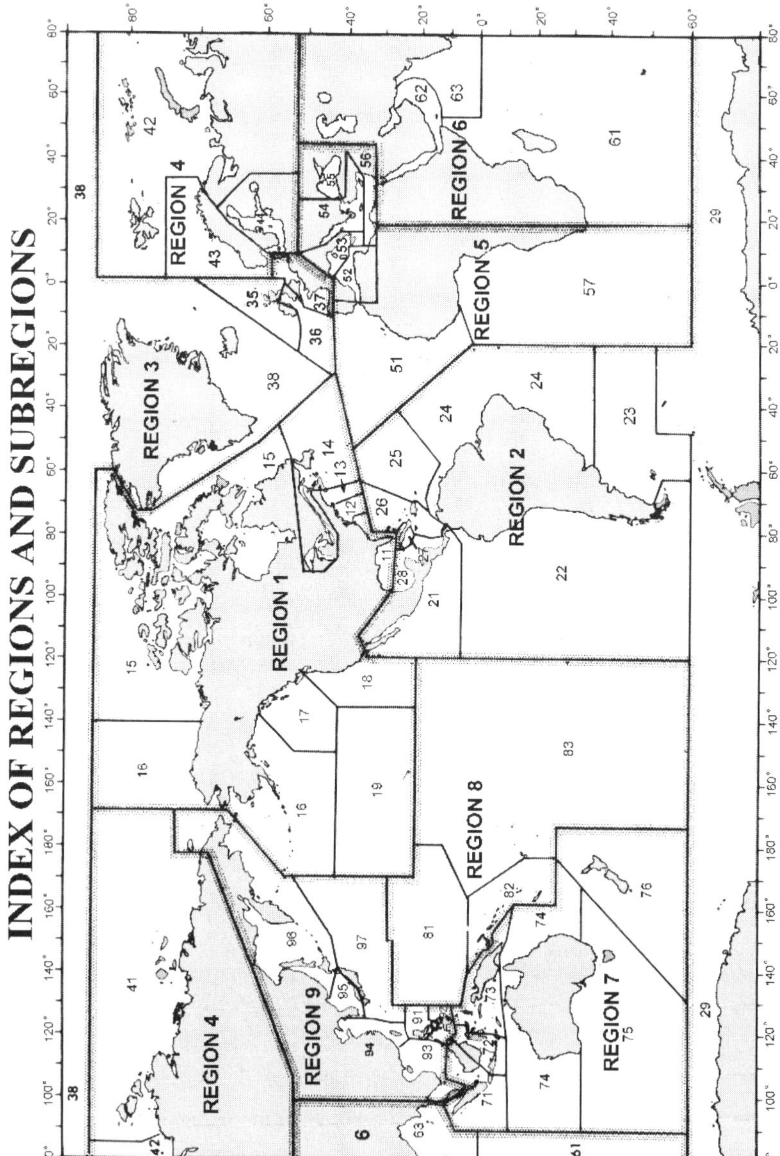

Figure 343b. Regions and subregions of the nautical chart index.

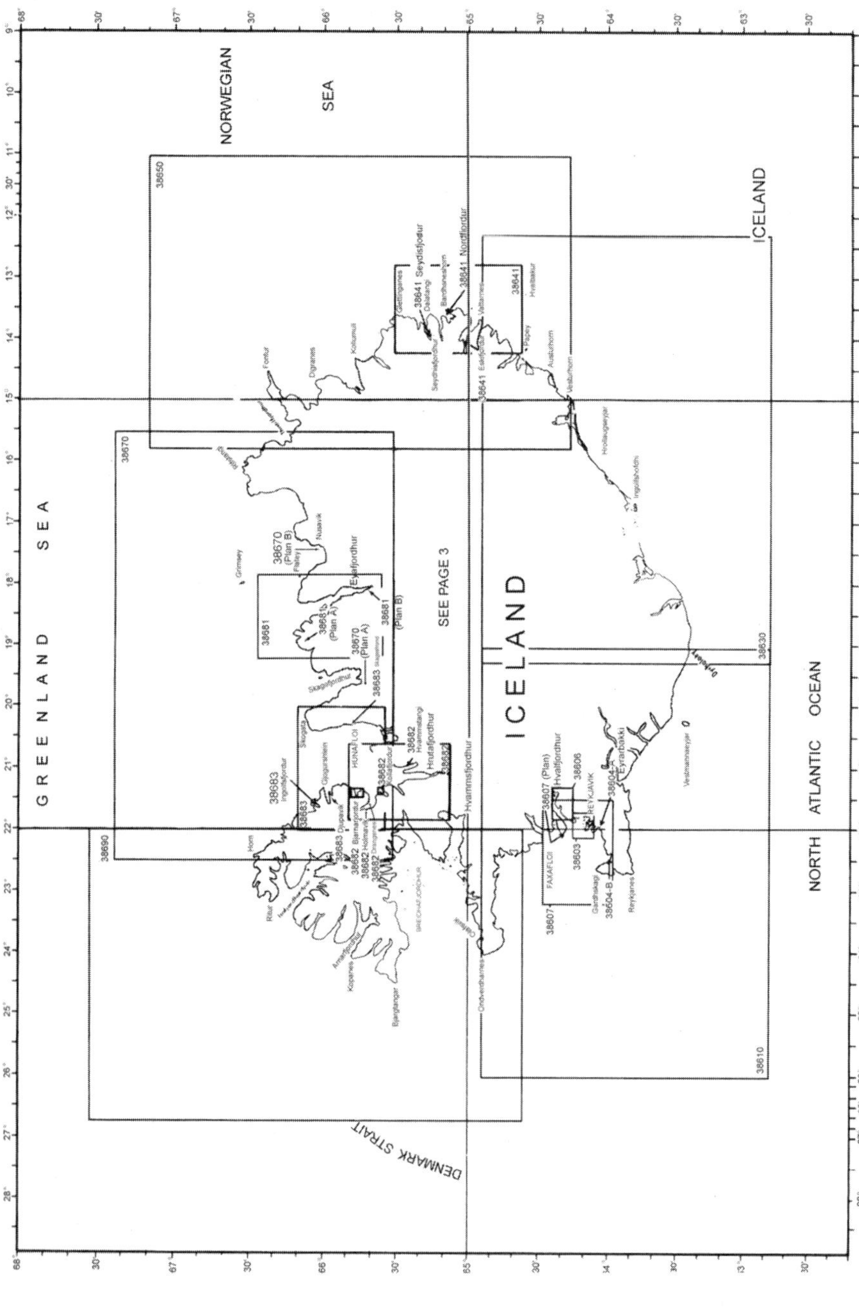

Figure 343c. Chart coverage of Iceland, illustrating the sequence and direction of the U.S. chart numbering system.

Combat charts have 6-digit numbers beginning with an "8." Neither is available to civilian navigators.

344. Catalogs and Stock Numbers

Chart catalogs provide information regarding not only chart coverage, but also a variety of special purpose charts and publications of interest. Keep a corrected chart catalog aboard ship for review by the navigator. The NIMA catalog contains operating area charts and other special products not available for civilian use, but does not contain any classified listings. The NOS catalogs contain all unclassified civilian-use NOS and NIMA charts. Military navigators receive their nautical charts and publications automatically; civilian navigators purchase them from chart sales agents.

The stock number and bar code are generally found in the lower left corner of a NIMA chart, and in the lower right corner of an NOS chart. The first two digits of the stock number refer to the region and subregion. These are followed by three letters, the first of which refers to the portfolio to which the chart belongs; the second two denote the type of chart: CO for coastal, HA for harbor and approach, and OA for military operating area charts. The last five digits are the actual chart number.

USING CHARTS

345. Preliminary Steps

Before using a new edition of a chart, verify its announcement in the *Notice to Mariners* and correct it with all applicable corrections. Read all the chart's notes; there should be no question about the meanings of symbols or the units in which depths are given. Since the latitude and longitude scales differ considerably on various charts, carefully note those on the chart to be used.

Place additional information on the chart as required. Arcs of circles might be drawn around navigational lights to indicate the limit of visibility at the height of eye of an observer on the bridge. Notes regarding other information from the light lists, tide tables, tidal current tables, and sailing directions might prove helpful.

346. Maintaining Charts

A mariner navigating on an uncorrected chart is courting disaster. The chart's print date reflects the latest *Notice to Mariners* used to update the chart; responsibility for maintaining it after this date lies with the user. The weekly *Notice to Mariners* contains information needed for maintaining charts. Radio broadcasts give advance notice of urgent corrections. Local *Notice to Mariners* should be consulted for inshore areas. The navigator must develop a system to keep track of chart corrections and to ensure that the chart he is using is updated with the latest correction. A convenient way of keeping this record is with a *Chart/Publication Correction Record Card* system. Using this system, the navigator does not immediately update every chart in his portfolio when he receives the *Notice to Mariners*. Instead, he constructs a card for every chart in his portfolio and notes the correction on this card. When the time comes to use the chart, he pulls the chart and chart's card, and he makes the indicated corrections on the chart. This system ensures that every chart is properly corrected prior to use.

A *Summary of Corrections*, containing a cumulative listing of previously published *Notice to Mariners* corrections, is published annually in 5 volumes by NIMA. Thus, to fully correct a chart whose edition date is several years old, the navigator needs only the Summary of Corrections for that region and the notices from that Summary forward; he does not need to obtain notices all the way back to the edition date. See Chapter 4, Nautical Publications, for a description of the *Summaries* and *Notice to Mariners*.

When a new edition of a chart is published, it is normally furnished automatically to U.S. Government vessels. It should not be used until it is announced as ready for use in the *Notice to Mariners*. Until that time, corrections in the Notice apply to the old edition and should not be applied to the new one. When it is announced, a new edition of a chart replaces an older one.

Commercial users and others who don't automatically receive new editions should obtain new editions from their sales agent. Occasionally, charts may be received or purchased several weeks in advance of their announcement in the *Notice to Mariners*. This is usually due to extensive rescheming of a chart region and the need to announce groups of charts together to avoid lapses in coverage. The mariner bears the responsibility for ensuring that his charts are the current edition. The fact that a new edition has been compiled and published often indicates that there have been extensive changes that cannot be made by hand corrections.

347. Using and Stowing Charts

Use and stow charts carefully. This is especially true with digital charts contained on electronic media. Keep optical and magnetic media containing chart data out of the sun, inside dust covers, and away from magnetic influences. Placing a disk in an inhospitable environment may destroy the data.

Make permanent corrections to paper charts in ink so that they will not be inadvertently erased. Pencil in all other markings so that they can be easily erased without damaging the chart. Lay out and label tracks on charts of frequently-traveled ports in ink. Draw lines and labels no larger than necessary. Do not obscure sounding data or other information when labeling a chart. When a voyage is completed, carefully erase the charts unless there has been a grounding or collision. In this case, preserve the charts without change because they will play a critical role in the investigation.

When not in use, stow charts flat in their proper portfolio. Minimize their folding and properly index them for easy retrieval.

348. Chart Lighting

Mariners often work in a red light environment because red light is least disturbing to night adapted vision. Such lighting seriously affects the appearance of a chart. Before using a chart in red light, test the effect red light has on its markings. Do not outline or otherwise indicate navigational hazards in red pencil because red markings disappear under red light.

349. Small-Craft Charts

NOS publishes a series of small craft charts sometimes called "strip charts." These charts depict segments of the Atlantic Intracoastal Waterway, the Gulf Intracoastal Waterway, and other inland routes used by yachtsmen, fishermen, and small commercial vessels for coastal travel. They are not "north-up" in presentation, but are aligned with the waterway they depict, whatever its orientation is. Most often they are used as a piloting aid for "eyeball" navigation and placed "course-up" in front of the helmsman, because the routes they show are too confined for taking and plotting fixes.

Although NOS small-craft charts are designed primarily for use aboard yachts, fishing vessels and other small craft, these charts, at scales of 1:80,000 and larger, are in some cases the only charts available depicting inland waters transited by large vessels. In other cases the small-craft charts may provide a better presentation of navigational hazards than the standard nautical chart because of better scale and more detail. Therefore, navigators should use these charts in areas where they provide the best coverage.

CHAPTER 4

NAUTICAL PUBLICATIONS

INTRODUCTION

400. Hardcopy vs. Softcopy Publications

The navigator uses many textual information sources when planning and conducting a voyage. These sources include notices to mariners, summary of corrections, sailing directions, light lists, tide tables, sight reduction tables, and almanacs. Historically, this information has been contained in paper or so-called "hardcopy" publications. But electronic methods of production and distribution of textual material are now commonplace, and will soon replace many of the navigator's familiar books. This volume's CD-ROM version is only one of many. Regardless of how technologically advanced we become, the printed word will always be an important method of communication. Only the means of access will change.

While it is still possible to obtain hard-copy printed publications, increasingly these texts are found on-line or in the form of Compact Disc-Read Only Memory (CD-ROM's). CD-ROM's are much less expensive than printed publications to reproduce and distribute, and on-line publications have no reproduction costs at all for the producer, and only minor costs to the user, if he chooses to print them at all. Also, a few CD-ROM's can hold entire libraries of information, making both distribution and on-board storage much easier.

The advantages of electronic publications go beyond their cost savings. They can be updated easier and more often, making it possible for mariners to have frequent or even continuous access to a maintained publications database instead of receiving new editions at infrequent intervals and entering hand corrections periodically. Generally, digital publications also provide links and search engines to quickly access related information.

Navigational publications are available from many sources. Military customers automatically receive or requisition most publications. The civilian navigator obtains his publications from a publisher's agent. Larger agents representing many publishers can completely supply a ship's chart and publication library. On-line publications produced by the U.S. government are available on the Web.

This chapter will refer generally to printed publications. If the navigator has access to this data electronically, his methods of access and use will differ somewhat, but the discussion herein applies equally to both electronic and hard-copy documents.

NAUTICAL TEXTS

401. *Sailing Directions*

National Imagery and Mapping Agency *Sailing Directions* consist of 37 ***Enroutes*** and 5 ***Planning Guides***. *Planning Guides* describe general features of ocean basins; *Enroutes* describe features of coastlines, ports, and harbors.

Sailing Directions are updated when new data requires extensive revision of an existing volume. These data are obtained from several sources, including pilots and foreign Sailing Directions.

One book comprises the *Planning Guide* and *Enroute* for Antarctica. This consolidation allows for a more effective presentation of material on this unique area.

The *Planning Guides* are relatively permanent; by contrast, *Sailing Directions (Enroute)* are frequently updated. Between updates, both are corrected by the *Notice to Mariners*.

402. *Sailing Directions (Planning Guide)*

Planning Guides assist the navigator in planning an extensive oceanic voyage. Each of the Guides provides useful information about all the countries adjacent to a particular ocean basin. The limits of the *Sailing Directions* in relation to the major ocean basins are shown in Figure 402.

Planning Guides are structured in the alphabetical order of countries contained within the region. Information pertaining to each country includes Buoyage Systems, Currency, Government, Industries, Holidays, Languages, Regulations, Firing Danger Areas, Mined Areas, Pilotage, Search and Rescue, Reporting Systems, Submarine Operating Areas, Time Zone, and the location of the U.S. Embassy.

403. *Sailing Directions (Enroute)*

Each volume of the *Sailing Directions (Enroute)*

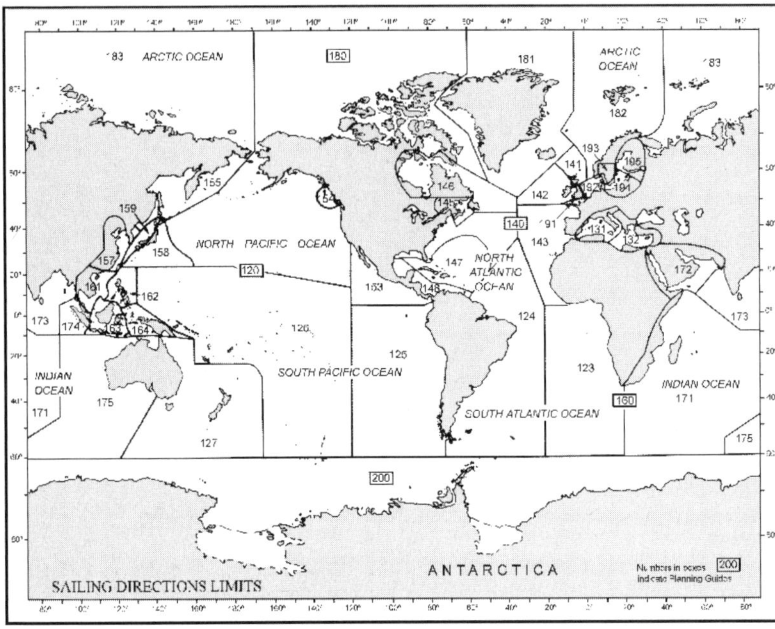

Figure 402. Sailing Directions limits in relation to the major ocean basins.

contains numbered sections along a coast or through a strait. Figure 403a illustrates this division. Each sector is sub-divided into paragraphs and discussed in turn. A preface with information about authorities, references, and conventions used in each book precedes the sector discussions. Each book also provides conversions between feet, fathoms, and meters, and an Information and Suggestion Sheet.

The Chart Information Graphic, the first item in each sector, is a graphic key for charts pertaining to that area. See Figure 403b. The graduation of the border scale of the chartlet enables navigators to identify the largest scale chart for a location and to find a feature listed in the Index-Gazetteer. These graphics are not maintained by *Notice to Mariners*; one should refer to the chart catalog for updated chart listings. Other graphics may contain special information on anchorages, significant coastal features, and navigation dangers.

A foreign terms glossary and a comprehensive Index-Gazetteer follow the sector discussions. The Index-Gazetteer is an alphabetical listing of described and charted features. The Index lists each feature by geographic coordinates and sector paragraph number.

U.S. military vessels have access to special files of data reported via official messages known as Port Visit After Action Reports. These reports, written in text form according to a standardized reporting format, give complete details of recent visits by U.S. military vessels to all foreign ports visited. Virtually every detail regarding navigation, services, supplies, official and unofficial contacts, and other matters is discussed in detail, making these reports an extremely useful adjunct to the *Sailing Directions*. These files are available to ".mil" users only, and may be accessed on the Web at: http://cnsl.spear.navy.mil, under the "Force Navigator" link. They are also available via DoD's classified Web.

404. *Coast Pilots*

The National Ocean Service publishes nine *United States Coast Pilots* to supplement nautical charts of U.S. waters. Information comes from field inspections, survey vessels, and various harbor authorities. Maritime officials and pilotage associations provide additional information. *Coast Pilots* provide more detailed information than *Sailing Directions* because *Sailing Directions* are intended exclusively for the oceangoing mariner. The *Notice to Mariners* updates *Coast Pilots*.

Each volume contains comprehensive sections on local operational considerations and navigation regulations. Following chapters contain detailed discussions of coastal navigation. An appendix provides information on obtaining additional weather information, communications services, and other data. An index and additional tables complete the volume.

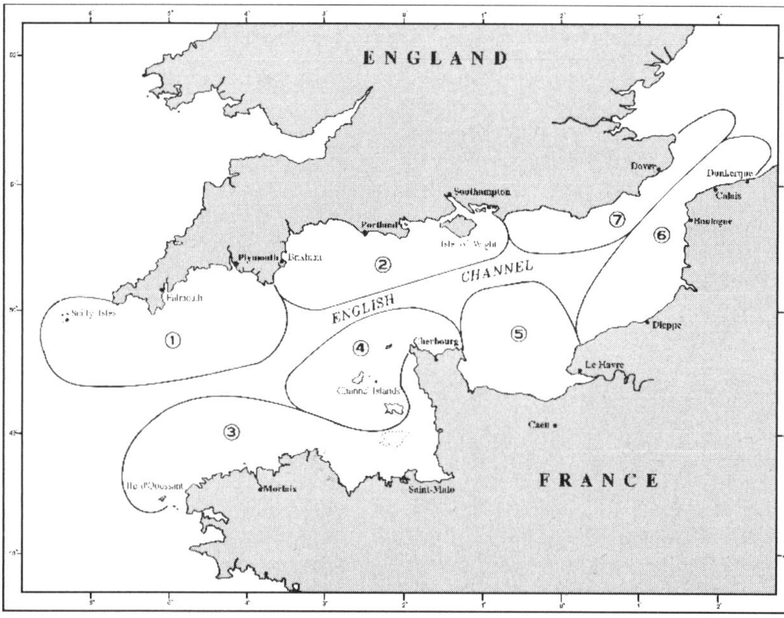

Figure 403a. Sector Limits graphic.

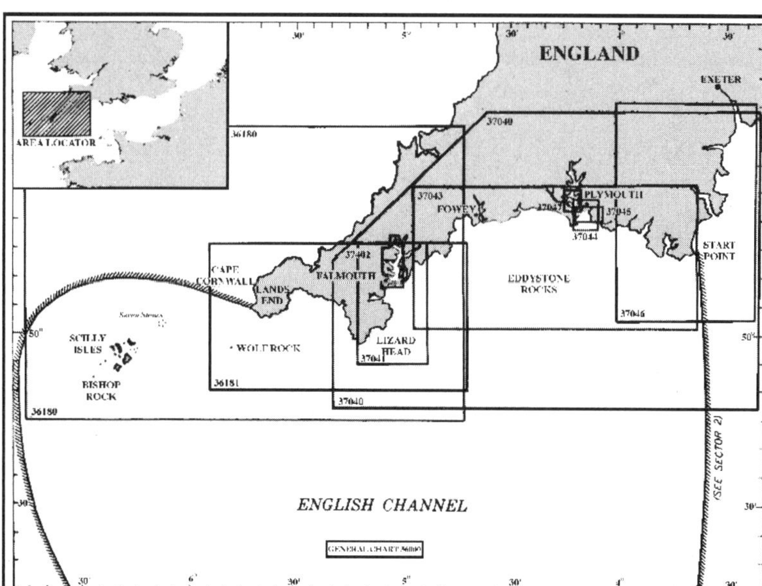

Additional chart coverage may be found in CATP2 Catalog of Nautical Charts.

Figure 403b. Chart Information graphic.

405. Other Nautical Texts

The government publishes several other nautical texts. NIMA, for example, publishes *Pub. 1310, Radar Navigation and Maneuvering Board Manual* and *Pub. 9, American Practical Navigator*.

The U.S. Coast Guard publishes *Navigation Rules* for international and inland waters. This publication, officially known as Commandant Instruction M16672.2d, contains the Inland Navigation Rules enacted in December 1980 and effective on all inland waters of the United States including the Great Lakes, as well as the *International Regulations for the Prevention of Collisions at Sea*, enacted in 1972 (1972 COLREGS). Mariners should ensure that they have the updated issue. The Coast Guard also publishes comprehensive user's manuals for the Loran and GPS navigation systems; *Navigation and Vessel Inspection Circulars*; and the *Chemical Data Guide for Bulk Shipment by Water*.

The Government Printing Office provides several publications on navigation, safety at sea, communications, weather, and related topics. Additionally, it publishes provisions of the Code of Federal Regulations (CFR) relating to maritime matters. A number of private publishers also provide maritime publications.

The International Maritime Organization, International Hydrographic Organization, and other governing international organizations provide information on international navigation regulations. Chapter 1 gives these organizations' addresses. Regulations for various Vessel Traffic Services (VTS), canals, lock systems, and other regulated waterways are published by the authorities which operate them. Nautical chart and publication sales agents are a good source of information about publications required for any voyage. Increasingly, many regulations, whether instituted by international or national governments, can be found online. This includes regulations for Vessel Traffic Services, Traffic Separation Schemes, special regulations for passage through major canal and lock systems, port and harbor regulations, and other information. A Web search can often find the textual information the navigator needs.

USING THE LIGHT LISTS

406. Light Lists

The United States publishes two different light lists. The U.S. Coast Guard publishes the *Light List* for lights in U.S. territorial waters; NIMA publishes the *List of Lights* for lights in foreign waters.

Light lists furnish detailed information about navigation lights and other navigation aids, supplementing the charts, *Coast Pilots*, and *Sailing Directions*. Consult the chart for the location and light characteristics of all navigation aids; consult the light lists to determine their detailed description.

The *Notice to Mariners* corrects both lists. Corrections which have accumulated since the print date are included in the *Notice to Mariners* as a *Summary of Corrections*. All of these summary corrections, and any corrections published subsequently, should be noted in the "Record of Corrections."

A navigator needs to know both the identity of a light and when he can expect to see it; he often plans the ship's track to pass within a light's range. If lights are not sighted when predicted, the vessel may be significantly off course and standing into danger.

A circle with a radius equal to the visible range of the light usually defines the area in which a light can be seen. On some bearings, however, obstructions may reduce the range. In this case, the obstructed arc might differ with height of eye and distance. Also, lights of different colors may be seen at different distances. Consider these facts both when identifying a light and predicting the range at which it can be seen.

Atmospheric conditions have a major effect on a light's range. Fog, haze, dust, smoke, or precipitation can obscure a light. Additionally, a light can be extinguished. Always report an extinguished light so maritime authorities can issue a warning and make repairs.

On a dark, clear night, the visual range is limited by either: (1) luminous intensity, or (2) curvature of the Earth. Regardless of the height of eye, one cannot see a weak light beyond a certain luminous range. Assuming light travels linearly, an observer located below the light's visible horizon cannot see it. The Distance to the Horizon table gives the distance to the horizon for various heights of eye. The light lists contain a condensed version of this table. Abnormal refraction patterns might change this range; therefore, one cannot exactly predict the range at which a light will be seen.

407. Finding Range and Bearing of a Light at Sighting

A light's **luminous range** is the maximum range at which an observer can see a light under existing visibility conditions. This luminous range ignores the elevation of the light, the observer's height of eye, the curvature of the Earth, and interference from background lighting. It is determined from the known **nominal range** and the existing visibility conditions. The nominal range is the maximum distance at which a light can be seen in weather conditions where visibility is 10 nautical miles.

The U.S. Coast Guard *Light List* usually lists a light's nominal range. Use the Luminous Range Diagram shown in the *Light List* and Figure 407a to convert this nominal range to luminous range. Remember that the luminous ranges obtained are approximate because of atmospheric or background lighting conditions. To use the Luminous Range

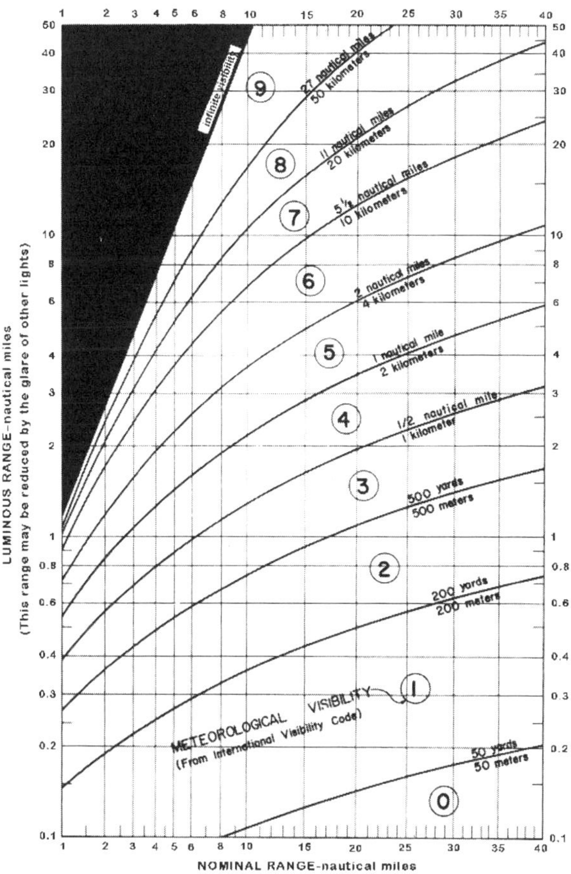

Figure 407a. Luminous Range Diagram.

Diagram, first estimate the meteorological visibility by the Meteorological Optical Range Table, Figure 407b. Next, enter the Luminous Range Diagram with the nominal range on the horizontal nominal range scale. Follow a vertical line until it intersects the curve or reaches the region on the diagram representing the meteorological visibility. Finally, follow a horizontal line from this point or region until it intersects the vertical luminous range scale.

Example 1: *The nominal range of a light as extracted from the Light List is 15 nautical miles.*

Required: *The luminous range when the meteorological visibility is (1) 11 nautical miles and (2) 1 nautical mile.*

Solution: *To find the luminous range when the meteorological visibility is 11 nautical miles, enter the Luminous Range Diagram with nominal range 15 nautical miles on the horizontal nominal range scale; follow a vertical line upward until it intersects the curve on the diagram representing a meteorological visibility of 11 nautical miles; from this point follow a horizontal line to the right until it intersects the vertical luminous range scale at 16 nautical miles. A similar procedure is followed to find the luminous range when the meteorological visibility is 1 nautical mile.*

Answers: *(1) 16 nautical miles; (2) 3 nautical miles.*

A light's **geographic range** depends upon the height of both the light and the observer. The sum of the observer's dis-

Code No.	Weather	Yards
0	Dense fog	Less than 50
1	Thick fog	50-200
2	Moderate fog	200-500
3	Light fog	500-1000
		Nautical Miles
4	Thin fog	1/2-1
5	Haze	1-2
6	Light Haze	2-5 1/2
7	Clear	5 1/2-11
8	Very Clear	11.0-27.0.
9	Exceptionally Clear	Over 27.0

From the International Visibility Code.

Figure 407b. Meteorological Optical Range Table.

tance to the visible horizon (based on his height of eye) plus the light's distance to the horizon (based on its height) is its geographic range. See Figure 407c. This illustration uses a light 150 feet above the water. Table 12, Distance of the Horizon, yields a value of 14.3 nautical miles for a height of 150 feet. Within this range, the light, if powerful enough and atmospheric conditions permit, is visible regardless of the height of eye of the observer. Beyond 14.3 nautical miles, the geographic range depends upon the observer's height of eye. Thus, by the Distance of the Horizon table mentioned above, an observer with height of eye of 5 feet can see the light on his horizon if he is 2.6 miles beyond the horizon of the light. The geographic range of the light is therefore 16.9 miles. For a height of 30 feet the distance is 14.3 + 6.4 = 20.7 miles. If the height of eye is 70 feet, the geographic range is 14.3 + 9.8 = 24.1 miles. A height of eye of 15 feet is often assumed when tabulating lights' geographic ranges.

To predict the bearing and range at which a vessel will initially sight a light first determine the light's geographic range. Compare the geographic range with the light's luminous range. The lesser of the two ranges is the range at which the light will first be sighted. Plot a visibility arc centered on the light and with a radius equal to the lesser of the geographic or luminous ranges. Extend the vessel's track until it intersects the visibility arc. The bearing from the intersection point to the light is the light's predicted bearing at first sighting.

If the extended track crosses the visibility arc at a small angle, a small lateral track error may result in large bearing and time prediction errors. This is particularly apparent if the vessel is farther from the light than predicted; the vessel may pass the light without sighting it. However, not sighting a light when predicted does not always indicate the vessel is farther from the light than expected. It could also mean that atmospheric conditions are affecting visibility.

Example 2: *The nominal range of a navigational light 120 feet above the chart datum is 20 nautical miles. The meteorological visibility is 27 nautical miles.*

Required: *The distance at which an observer at a height of eye of 50 feet can expect to see the light.*

Solution: *The maximum range at which the light may be seen is the lesser of the luminous or geographic ranges. At 120 feet the distance to the horizon, by table or formula, is 12.8 miles. Add 8.3 miles, the distance to the horizon for a height of eye of 50 feet to determine the geographic range. The geographic range, 21.1 miles, is less than the luminous range, 40 miles.*

Answer: *21 nautical miles. Because of various uncertainties, the range is rounded off to the nearest whole mile.*

When first sighting a light, an observer can determine if it is on the horizon by immediately reducing his height of eye. If the light disappears and then reappears when the observer returns to his original height, the light is on the horizon. This process is called **bobbing a light**.

If a vessel has considerable vertical motion due to rough seas, a light sighted on the horizon may alternately appear and disappear. Wave tops may also obstruct the light periodically. This may cause the characteristic to appear different than expected. The light's true characteristics can be ascertained either by closing the range to the light or by increasing the observer's height of eye.

If a light's range given in a foreign publication approximates the light's geographic range for a 15-foot observer's height of eye, one can assume that the printed range is the light's geographic range. Also assume that publication has listed the lesser of the geographic and nominal ranges. Therefore, if the light's listed range approximates the geographic range for an observer with a height of eye of 15 feet, then assume that the light's limiting range is the geographic range. Then, calculate the light's true geographic range using the actual observer's height of eye, not the assumed height of eye of 15 feet. This calculated true geographic range is the range at which the light will first be sighted.

Example 3: *The range of a light as printed on a foreign chart is 17 miles. The light is 120 feet above chart datum. The meteorological visibility is 10 nautical miles.*

Required: *The distance at which an observer at a height of eye of 50 feet can expect to see the light.*

Solution: *Calculate the geographic range of the light assuming a 15 foot observer's height of eye. At 120 feet the distance to the horizon is 12.8 miles. Add 4.5 miles (the distance to the horizon at a height of 15 feet) to 12.8 miles; this range is 17.3 miles. This approximates the range listed on the chart. Then assuming that the charted range is the*

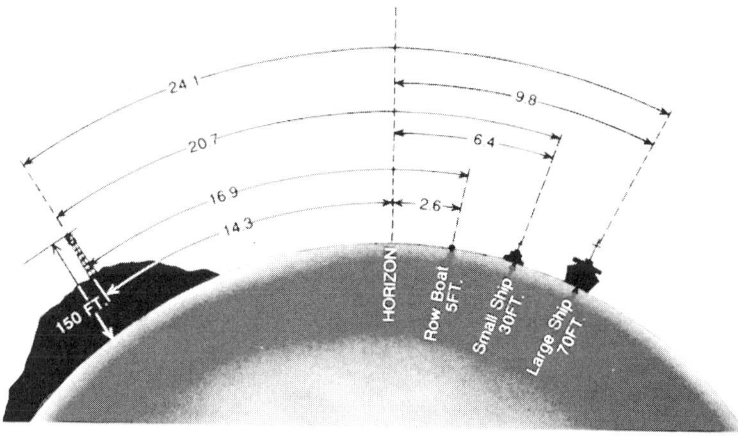

Figure 407c. Geographic Range of a light.

geographic range for a 15-foot observer height of eye and that the nominal range is the greater than this charted range, the predicted range is found by calculating the true geographic range with a 50 foot height of eye for the observer.

Answer: *The predicted range = 12.8 mi. + 8.3 mi. = 21.1 mi. The distance in excess of the charted range depends on the luminous intensity of the light and the meteorological visibility.*

408. USCG *Light Lists*

The U.S. Coast Guard *Light List* (7 volumes) gives information on lighted navigation aids, unlighted buoys, radiobeacons, radio direction finder calibration stations, daybeacons, racons, and Loran stations.

Each volume of the *Light List* contains aids to navigation in geographic order from north to south along the Atlantic coast, from east to west along the Gulf coast, and from south to north along the Pacific coast. It lists seacoast aids first, followed by entrance and harbor aids listed from seaward. Intracoastal Waterway aids are listed last in geographic order in the direction from New Jersey to Florida to the Texas/Mexico border.

The listings are preceded by a description of the aids to navigation system in the United States, luminous range diagram, geographic range tables, and other information.

409. NIMA *List of Lights, Radio Aids, and Fog Signals*

The National Imagery and Mapping Agency publishes the *List of Lights, Radio Aids, and Fog Signals* (usually referred to as the *List of Lights*, not to be confused with the Coast Guard's *Light List*). In addition to information on lighted aids to navigation and sound signals in foreign waters, the NIMA *List of Lights* provides information on storm signals, signal stations, racons, radiobeacons, radio direction finder calibration stations located at or near lights, and DGPS stations. For more details on radio navigational aids, consult *Pub. 117, Radio Navigational Aids*.

The NIMA *List of Lights* generally does not include information on buoys, although in certain instances, a large offshore buoy with a radio navigational aid may be listed. It does include certain aeronautical lights situated near the coast. However, these lights are not designed for marine navigation and are subject to unreported changes.

Foreign notices to mariners are the main correctional information source for the NIMA *Lists of Lights*; other sources, such as ship reports, are also used. Many aids to navigation in less developed countries may not be well maintained. They are subject to damage by storms and vandalism, and repairs may be delayed for long periods.

MISCELLANEOUS NAUTICAL PUBLICATIONS

410. NIMA *Radio Navigational Aids (Pub. 117)*

This publication is a selected list of worldwide radio stations which perform services to the mariner. Topics covered include radio direction finder and radar stations, radio time signals, radio navigation warnings, distress and safety communications, medical advice via radio, long-range navigation aids, the AMVER system, and interim procedures for U.S. vessels in the event of an outbreak of hostilities. *Pub. 117* is corrected via the

Notice to Mariners and is updated periodically with a new edition.

Though *Pub. 117* is essentially a list of radio stations providing vital maritime communication and navigation services, it also contains information which explains the capabilities and limitations of the various systems.

411. *Chart No. 1*

Chart No. 1 is not actually a chart but a book containing a key to chart symbols. Most countries which produce charts also produce such a list. The U.S. *Chart No. 1* contains a listing of chart symbols in four categories:

- Chart symbols used by the National Ocean Service
- Chart symbols used by NIMA
- Chart symbols recommended by the International Hydrographic Organization
- Chart symbols used on foreign charts reproduced by NIMA

Subjects covered include general features of charts, topography, hydrography, and aids to navigation. There is also a complete index of abbreviations and an explanation of the IALA buoyage system.

412. NIMA *World Port Index (Pub. 150)*

The *World Port Index* contains a tabular listing of thousands of ports throughout the world, describing their locations, characteristics, facilities, and services available. Information is arranged geographically; the index is arranged alphabetically.

Coded information is presented in columns and rows. This information supplements information in the *Sailing Directions*. The applicable volume of *Sailing Directions* and the number of the harbor chart are given in the *World Port Index*. The *Notice to Mariners* corrects this book.

413. NIMA *Distances Between Ports (Pub. 151)*

This publication lists the distances between major ports. Reciprocal distances between two ports may differ due to different routes chosen because of currents and climatic conditions. To reduce the number of listings needed, junction points along major routes are used to consolidate routes converging from different directions.

This book can be most effectively used for voyage planning in conjunction with the proper volume(s) of the *Sailing Directions (Planning Guide)*. It is corrected via the *Notice to Mariners*.

414. NIMA *International Code of Signals (Pub. 102)*

This book lists the signals to be employed by vessels at sea to communicate a variety of information relating to safety, distress, medical, and operational information. This publication became effective in 1969.

According to this code, each signal has a unique and complete meaning. The signals can be transmitted via Morse code light and sound, flag, radio telegraph and telephone, and semaphore. Since these methods of signaling are internationally recognized, differences in language between sender and receiver are immaterial; the message will be understood when decoded in the language of the receiver, regardless of the language of the sender. The *Notice to Mariners* corrects *Pub. 102*.

415. Almanacs

For celestial sight reduction, the navigator needs an **almanac** for ephemeris data. The *Nautical Almanac*, produced jointly by H.M. Nautical Almanac Office and the U.S. Naval Observatory, is the most common almanac used for celestial navigation. It also contains information on sunrise, sunset, moonrise, and moonset, as well as compact sight reduction tables. The *Nautical Almanac* is published annually.

The *Air Almanac* contains slightly less accurate ephemeris data for air navigation. It can be used for marine navigation if slightly reduced accuracy is acceptable.

Chapter 19 provides more detailed information on using the *Nautical Almanac*.

416. *Sight Reduction Tables*

Without a calculator or computer programmed for sight reduction, the navigator needs ***sight reduction tables*** to solve the celestial triangle. Two different sets of tables are commonly used at sea.

NIMA *Pub. 229, Sight Reduction Tables for Marine Navigation*, consists of six volumes of tables designed for use with the *Nautical Almanac* for solution of the celestial triangle by the **Marcq Saint Hilaire** or **intercept** method. The tabular data are the solutions of the navigational triangle of which two sides and the included angle are known and it is necessary to find the third side and adjacent angle.

Each volume of *Pub. 229* includes two 8 degree zones, comprising 15 degree bands from 0 to 90 degrees, with a 1° degree overlap between volumes. *Pub. 229* is a joint publication produced by the National Imagery and Mapping Agency, the U.S. Naval Observatory, and the Royal Greenwich Observatory.

Sight Reduction Tables for Air Navigation, Pub. 249, is also a joint production of the three organizations above. It is issued in three volumes. Volume 1 contains the values of the altitude and true azimuth of seven selected stars chosen to provide, for any given position and time, the best celestial observations. A new edition is issued every 5 years for the upcoming astronomical epoch. Volumes 2 (0° to 40°) and 3 (39° to 89°) provide for sights of the Sun, Moon, and planets.

417. Catalogs

A chart catalog is a valuable reference to the navigator for voyage planning, inventory control, and ordering. The catalog is used by military and civilian customers.

The navigator will see the NIMA nautical chart catalog as part of a larger suite of catalogs including aeronautical (Part 1), hydrographic (Part 2), and topographic (Part 3) products. Each Part consists of one or more volumes. Unclassified NIMA nautical charts are listed in Part 2, Volume 1.

This catalog contains comprehensive ordering instructions and information about the products listed. Also listed are addresses of all Map Support Offices, information on crisis support, and other special situations. The catalog is organized by geographic region corresponding to the chart regions 1 through 9. A special section of miscellaneous charts and publications is included. This section also lists products produced by NOS, the U.S. Army Corps of Engineers, U.S. Coast Guard, U.S. Naval Oceanographic Office, and some foreign publications from the United Kingdom and Canada.

The civilian navigator should also refer to catalogs produced by the National Ocean Service. For U.S. waters, NOS charts are listed in a series of large sheet "charts" showing a major region of the U.S. with individual chart graphics depicted. These catalogs also list charts showing titles and scales. They also list sales agents from whom the charts may be purchased.

NIMA products for the civilian navigator are listed by NOS in a series of regionalized catalogs similar to Part 2 Volume 1. These catalogs are also available through authorized NOS chart agents.

MARITIME SAFETY INFORMATION

418. *Notice to Mariners*

The ***Notice to Mariners*** is published weekly by the National Imagery and Mapping Agency (NIMA), prepared jointly with the National Ocean Service (NOS) and the U.S. Coast Guard. It advises mariners of important matters affecting navigational safety, including new hydrographic information, changes in channels and aids to navigation, and other important data. The information in the *Notice to Mariners* is formatted to simplify the correction of paper charts, sailing directions, light lists, and other publications produced by NIMA, NOS, and the U.S. Coast Guard.

It is the responsibility of users to decide which of their charts and publications require correction. Suitable records of *Notice to Mariners* should be maintained to facilitate the updating of charts and publications prior to use.

Information for the *Notice to Mariners* is contributed by: NIMA (Department of Defense) for waters outside the territorial limits of the United States; National Ocean Service (National Oceanic and Atmospheric Administration, Department of Commerce), which is charged with surveying and charting the coasts and harbors of the United States and its territories; the U.S. Coast Guard (Department of Transportation) which is responsible for, among other things, the safety of life at sea and the establishment and operation of aids to navigation; and the Army Corps of Engineers (Department of Defense), which is charged with the improvement of rivers and harbors of the United States. In addition, important contributions are made by foreign hydrographic offices and cooperating observers of all nationalities.

Over 60 countries which produce nautical charts also produce a notice to mariners. About one third of these are weekly, another third are bi-monthly or monthly, and the rest irregularly issued according to need. Much of the data in the U.S. *Notice to Mariners* is obtained from these foreign notices.

U.S. charts must be corrected only with a U.S. *Notice to Mariners*. Similarly, correct foreign charts using the foreign notice because chart datums often vary according to region and geographic positions are not the same for different datums.

The *Notice to Mariners* consists of a page of **Hydrograms** listing important items in the notice, a chart correction section organized by ascending chart number, a publications correction section, and a summary of broadcast navigation warnings and miscellaneous information.

Mariners are requested to cooperate in the correction of charts and publications by reporting all discrepancies between published information and conditions actually observed and by recommending appropriate improvements. A convenient reporting form is provided in the back of each *Notice to Mariners*.

Notice to Mariners No. 1 of each year contains important information on a variety of subjects which supplements information not usually found on charts and in navigational publications. This information is published as ***Special Notice to Mariners Paragraphs***. Additional items considered of interest to the mariner are also included in this *Notice*.

419. *Summary of Corrections*

A close companion to the *Notice to Mariners* is the

Summary of Corrections. The *Summary* is published in five volumes. Each volume covers a major portion of the Earth including several chart regions and their subregions. Volume 5 also includes special charts and publications corrected by the *Notice to Mariners*. Since the *Summaries* contain cumulative corrections, any chart, regardless of its print date, can be corrected with the proper volume of the *Summary* and all subsequent *Notice to Mariners*.

420. The Maritime Safety Information Website

The NIMA **Maritime Safety Information Website** provides worldwide remote query access to extensive menus of maritime safety information 24 hours a day. The Maritime Safety Information Website can be accessed via the NIMA Homepage (www.nima.mil) under the Safety of Navigation icon or directly at http://pollux.nss.nima.mil.

Databases made available for access, query and download include Chart Corrections, Publication Corrections, NIMA Hydrographic Catalog Corrections, Chart and Publication Reference Data (current edition number, dates, title, scale), NIMA *List of Lights*, U.S. Coast Guard *Light Lists*, World Wide Navigational Warning Service (WWNWS) Broadcast Warnings, Maritime Administration (MARAD) Advisories, Department of State Special Warnings, Mobile Offshore Drilling Units (MODUs), Anti-Shipping Activity Messages (ASAMs), *World Port Index*, and *Radio Navigational Aids*. Publications that are also made available as Portable Document Format (PDF) files include the U.S. *Notice to Mariners*, U.S. *Chart No. 1*, *The American Practical Navigator*, *International Code of Signals*, *Radio Navigational Aids*, *World Port Index*, *Distances Between Ports*, *Sight Reduction Tables for Marine Navigation*, *Sight Reduction Tables for Air Navigation*, and the *Radar Navigation and Maneuvering Board Manual*.

Navigators have online access to, and can download, all the information contained in the printed *Notice to Mariners* including chartlets. Information on this website is updated daily or weekly according to the *Notice to Mariners* production schedule. Broadcast Warnings, MARAD Advisories, ASAMs and MODUs are updated on a daily basis; the remaining data is updated on a weekly basis.

Certain files, for example U.S. Coast Guard *Light List* data, are entered directly into the database without editing and the accuracy of this information cannot be verified by NIMA staff. Also, drill rig locations are furnished by the companies which operate them. They are not required to provide these positions, and they cannot be verified. However, within these limitations, the Website can provide information 2 weeks sooner than the printed *Notice to Mariners*, because the paper *Notice* must be printed and mailed after the digital version is completed and posted on the Web.

Users can provide suggestions, changes, corrections or comments on any of the Maritime Safety Information Division products and services by submitting an online version of the Marine Information Report and Suggestion Sheet.

Access to the Maritime Safety Information Website is free, but the user must pay the applicable charges for internet service. Any questions concerning the Maritime Safety Information Website should be directed to the Maritime Safety Information Division, Attn.: NSS STAFF, Mail Stop D-44, NIMA, 4600 Sangamore Rd., Bethesda, MD, 20816-5003; telephone (1) 301-227-3296; fax (1) 301-227-4211; e-mail webmaster_nss@nima.mil.

421. *Local Notice to Mariners*

The ***Local Notice to Mariners*** is issued by each U.S. Coast Guard District to disseminate important information affecting navigational safety within that District. This Notice reports changes and deficiencies in aids to navigation maintained by the Coast Guard. Other marine information such as new charts, channel depths, naval operations, and regattas is included. Since temporary information of short duration is not included in the NIMA *Notice to Mariners*, the *Local Notice to Mariners* may be the only source for it. Since correcting information for U.S. charts in the NIMA *Notice* is obtained from the Coast Guard local notices, there is a lag of 1 or 2 weeks for NIMA *Notice* to publish a correction from this source.

The *Local Notice to Mariners* may be obtained free of charge by contacting the appropriate Coast Guard District Commander. Vessels operating in ports and waterways in several districts must obtain the *Local Notice to Mariners* from each district. See Figure 421 for a complete list of U.S. Coast Guard Districts.

422. Electronic Notice to Mariners

One major impediment to full implementation of electronic chart systems has been the issue of how to keep them up to date. The IMO, after reviewing the range standards which might be employed in the provision of updates to ECDIS charts, decided that the correction system must be "hands off" from the mariner's point of view. That is, the correction system could not rely on the ability of the mariner to enter individual correction data himself, as he would do on a paper chart. The process must be automated to maintain the integrity of the data and prevent errors in data entry by navigators.

National hydrographic offices which publish electronic charts must also publish corrections for them. The manner of doing so varies among the different types of systems. The corrections are applied to the data as the chart to be displayed is created, leaving the database unchanged.

Another possibility exists, and that is to simply reload the entire chart data file with updated information. This is not as crazy as it sounds when one considers the amount of data that can be stored on a single CD-ROM and the ease

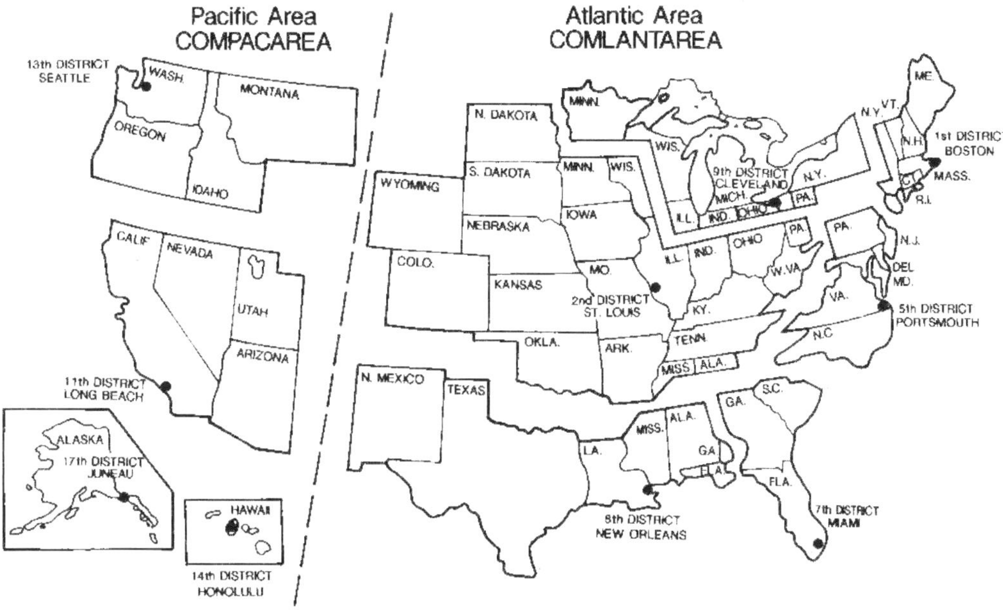

COMMANDER, FIRST COAST GUARD DISTRICT
408 ATLANTIC AVENUE
BOSTON, MA 02110-3350
PHONE: DAY 617-223-8338, NIGHT 617-223-8558

COMMANDER, SECOND COAST GUARD DISTRICT
1222 SPRUCE STREET
ST. LOUIS, MO 63103-2832
PHONE: DAY 314-539-3714, NIGHT 314-539-3709

COMMANDER, FIFTH COAST GUARD DISTRICT
FEDERAL BUILDING
431 CRAWFORD STREET
PORTSMOUTH, VA 23704-5004
PHONE: DAY 804-398-6486, NIGHT 804-398-6231

COMMANDER, SEVENTH COAST GUARD DISTRICT
BRICKELL PLAZA FEDERAL BUILDING
909 SE 1ST AVENUE, RM: 406
MIAMI, FL 33131-3050
PHONE: DAY 305-536-5621, NIGHT 305-536-5611

COMMANDER GREATER ANTILLES SECTION
U.S. COAST GUARD
P.O. BOX S-2029
SAN JUAN, PR 00903-2029
PHONE: 809-729-6870

COMMANDER, EIGHTH COAST GUARD DISTRICT
HALE BOGGS FEDERAL BUILDING
501 MAGAZINE STREET
NEW ORLEANS, LA 70130-3396
PHONE: DAY 504-589-6234, NIGHT 504-589-6225

COMMANDER, NINTH COAST GUARD DISTRICT
1240 EAST 9TH STREET
CLEVELAND, OH 44199-2060
PHONE: DAY 216-522-3991, NIGHT 216-522-3984

COMMANDER, ELEVENTH COAST GUARD DISTRICT
FEDERAL BUILDING
501 W. OCEAN BLVD.
LONG BEACH, CA 90822-5399
PHONE: DAY 310-980-4300, NIGHT 310-980-4400

COMMANDER, THIRTEENTH COAST GUARD DISTRICT
FEDERAL BUILDING
915 SECOND AVENUE
SEATTLE, WA 98174-1067
PHONE: DAY 206-220-7280, NIGHT 206-220-7004

COMMANDER, FOURTEENTH COAST GUARD DISTRICT
PRINCE KALANIANAOLE FEDERAL BLDG.
9TH FLOOR, ROOM 9139
300 ALA MOANA BLVD.
HONOLULU, HI 96850-4982
PHONE: DAY 808-541-2317, NIGHT 808-541-2500

COMMANDER, SEVENTEENTH COAST GUARD DISTRICT
P.O. BOX 25517
JUNEAU, AK 99802-5517
PHONE: DAY 907-463-2245, NIGHT 907-463-2000

Figure 421. U.S. Coast Guard Districts.

with which it can be reproduced. At present, these files are too large to be broadcast effectively, but with the proper bandwidth the concept of transferring entire chart portfolios worldwide via satellite or fiber-optic cable is entirely feasible.

Corrections to the DNC published by NIMA are being made by Vector Product Format Database Update (VDU). These are patch corrections and are available via the Web and by classified data links used by the Department of Defense.

Corrections to raster charts issued by NOAA are also available via the internet. To produce the patch, each chart is corrected and then compared, pixel by pixel, with the previous, uncorrected version. Any differences between the two must have been the result of a correction, so those files are saved and posted to a site for access by subscription users. The user accesses the site, downloads the compressed files, uncompresses them on his own terminal, and writes the patches onto his raster charts. He can then toggle between old and new versions to see exactly what has changed, and can view the patch by itself.

NOAA developed this process under an agreement with a commercial partner, which produces the CD-ROM containing chart data. The CD-ROM also contains *Coast Pilots*, *Light Lists*, *Tide Tables*, and *Tidal Current Tables*, thus comprising on one CD-ROM the entire suite of publications required by USCG regulations for certain classes of vessels. Additional information can be found at the NOAA Web site at: http://chartmaker.ncd.noaa.gov.

See Chapter 14 for a complete discussion on electronic charts and the means of correcting them.

CHAPTER 5

SHORT RANGE AIDS TO NAVIGATION

SHORT RANGE AIDS TO NAVIGATION

500. Terms and Definitions

Short range aids to navigation are those intended to be used visually or by radar while in inland, harbor and approach, and coastal navigation. The term encompasses lighted and unlighted beacons, ranges, leading lights, buoys, and their associated sound signals. Each short range aid to navigation, commonly referred to as a NAVAID, fits within a system designed to warn the mariner of dangers and direct him toward safe water. An aid's function determines its color, shape, light characteristic, and sound. This chapter explains the U.S. Aids to Navigation System as well as the IALA Maritime Buoyage System.

The placement and maintenance of marine aids to navigation in U.S. waters is the responsibility of the United States Coast Guard. The Coast Guard maintains lighthouses, radiobeacons, racons, sound signals, buoys, and daybeacons on the navigable waters of the United States, its territories, and possessions. Additionally, the Coast Guard exercises control over privately owned navigation aid systems.

A **beacon** is a stationary, visual navigation aid. Large lighthouses and small single-pile structures are both beacons. Lighted beacons are called **lights**; unlighted beacons are **daybeacons**. All beacons exhibit a **daymark** of some sort. In the case of a lighthouse, the color and type of structure are the daymarks. On small structures, these daymarks, consisting of colored geometric shapes called **dayboards**, often have lateral significance. The markings on lighthouses and towers convey no lateral significance.

FIXED LIGHTS

501. Major and Minor Lights

Lights vary from tall, high intensity coastal lights to battery-powered lanterns on single wooden piles. Immovable, highly visible, and accurately charted, fixed lights provide navigators with an excellent source for bearings. The structures are often distinctively colored to aid in identification. See Figure 501a.

A **major light** is a high-intensity light exhibited from a fixed structure or a marine site. Major lights include primary seacoast lights and secondary lights. **Primary seacoast lights** are major lights established for making landfall from sea and coastwise passages from headland to headland. **Secondary lights** are major lights established at harbor entrances and other locations where high intensity and reliability are required.

A **minor light** usually displays a light of low to moderate intensity. Minor lights are established in harbors, along channels, rivers, and in isolated locations. They usually have numbering, coloring, and light and sound characteristics that are part of the lateral system of buoyage.

Lighthouses are placed where they will be of most use: on prominent headlands, at harbor and port entrances, on isolated dangers, or at other points where mariners can best use them to fix their position. The lighthouse's principal purpose is to support a light at a considerable height above the water, thereby increasing its geographic range. Support equipment is often housed near the tower.

With few exceptions, all major lights operate automatically. There are also many automatic lights on smaller structures maintained by the Coast Guard or other attendants. Unmanned major lights may have emergency generators and automatic monitoring equipment to increase the light's reliability.

Light structures' appearances vary. Lights in low-lying areas usually are supported by tall towers; conversely, light structures on high cliffs may be relatively short. However its support tower is constructed, almost all lights are similarly generated, focused, colored, and characterized.

Some major lights use modern rotating or flashing lights, but many older lights use **Fresnel** lenses. These lenses consist of intricately patterned pieces of glass in a heavy brass framework. Modern Fresnel-type lenses are cast from high-grade plastic; they are much smaller and lighter than their glass counterparts.

A **buoyant beacon** provides nearly the positional accuracy of a light in a place where a buoy would normally be used. See Figure 501b. The buoyant beacon consists of a heavy sinker to which a pipe structure is tightly moored. A buoyancy chamber near the surface supports the pipe. The light, radar reflector, and other devices are located atop the pipe above the surface of the water. The pipe with its buoyancy chamber tends to remain upright even in severe weather and heavy currents, providing a smaller watch cir-

Figure 501a. Typical offshore light station.

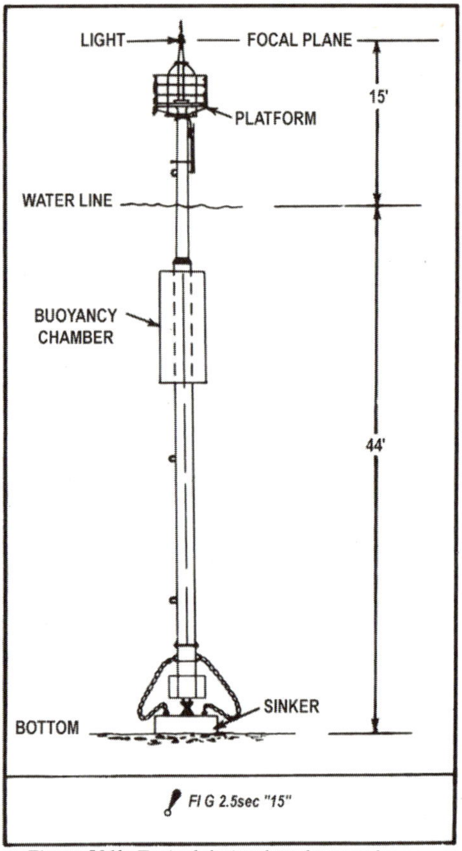

Figure 501b. Typical design for a buoyant beacon.

cle than a buoy. The buoyant beacon is most useful along narrow ship channels in relatively sheltered water.

502. Range Lights

Range lights are light pairs that indicate a specific line of position when they are in line. The higher rear light is placed behind the front light. When the mariner sees the lights vertically in line, he is on the range line. If the front light appears left of the rear light, the observer is to the right of the range line; if the front appears to the right of the rear, the observer is left of the range line. Range lights are sometimes equipped with high intensity lights for daylight use. These are effective for long channels in hazy conditions when dayboards might not be seen. The range light structures are usually also equipped with dayboards for ordinary daytime use. Some smaller ranges, primarily in the Intercoastal Waterway, rivers, and other inland waters, have just the dayboards with no lights. See Figure 502.

To enhance the visibility of range lights, the Coast Guard has developed 15-foot long lighted tubes called **light pipes**. They are mounted vertically, and the mariner sees them as vertical bars of light distinct from background lighting. Installation of light pipes is proceeding on several range markers throughout the country. The Coast Guard is also experimenting with long range sodium lights for areas requiring visibility greater than the light pipes can provide.

The output from a low pressure sodium light is almost entirely at one wavelength. This allows the use of an inexpensive band-pass filter to make the light visible even during the daytime. This arrangement eliminates the need for high intensity lights with their large power requirements.

Range lights are usually white, red, or green. They display various characteristics differentiating them from surrounding lights.

A **directional light** is a single light that projects a high intensity, special characteristic beam in a given direction. It is used in cases where a two-light range may not be practicable. A **directional sector light** is a directional light that emits two or more colored beams. The beams have a pre-

Figure 502. Range lights.

cisely oriented boundary between them. A normal application of a sector light would show three colored sections: red, white, and green. The white sector would indicate that the vessel is on the channel centerline; the green sector would indicate that the vessel is off the channel centerline in the direction of deep water; and the red sector would indicate that the vessel is off the centerline in the direction of shoal water.

503. Aeronautical Lights

Aeronautical lights may be the first lights observed at night when approaching the coast. Those situated near the coast and visible from sea are listed in the *List of Lights*. These lights are not listed in the Coast Guard *Light List*. They usually flash alternating white and green.

Aeronautical lights are sequenced geographically in the *List of Lights* along with marine navigation lights. However, since they are not maintained for marine navigation, they are subject to changes of which maritime authorities may not be informed. These changes will be published in *Notice to Airmen* but perhaps not in *Notice to Mariners*.

504. Bridge Lights

Navigational lights on bridges in the U.S. are prescribed by Coast Guard regulations. Red, green, and white lights mark bridges across navigable waters. Red lights mark piers and other parts of the bridge. Red lights are also used on drawbridges to show when they are in the closed position. Green lights mark open drawbridges and mark the centerline of navigable channels through fixed bridges. The position will vary according to the type of structure.

Infrequently-used bridges may be unlighted. In foreign waters, the type and method of lighting may be different from those normally found in the United States. Drawbridges which must be opened to allow passage operate upon sound and light signals given by the vessel and acknowledged by the bridge. These required signals are detailed in the Code of Federal Regulations and the applicable *Coast Pilot*. Certain bridges may also be equipped with sound signals and radar reflectors.

505. Shore Lights

Shore lights usually have a shore-based power supply. Lights on pilings, such as those found in the Intracoastal Waterway, are battery powered. Solar panels may be installed to enhance the light's power supply. The lights consist of a power source, a flasher to determine the characteristic, a lamp changer to replace burned-out lamps, and a focusing lens.

Various types of rotating lights are in use. They do not have flashers but remain continuously lit while a lens or reflector rotates around the horizon.

The aids to navigation system is carefully engineered

to provide the maximum amount of direction to the mariner for the least expense. Specially designed filaments and special grades of materials are used in the light to withstand the harsh marine environment.

The **flasher** electronically determines the characteristic by selectively interrupting the light's power supply according to the chosen cycle.

The **lamp changer** consists of several sockets arranged around a central hub. When the circuit is broken by a burned-out filament, a new lamp is rotated into position. Almost all lights have daylight switches which turn the light off at sunrise and on at dusk.

The **lens** for small lights may be one of several types. The common ones in use are omni-directional lenses of 155mm, 250mm, and 300mm diameter. In addition, lights using parabolic mirrors or focused-beam lenses are used in leading lights and ranges. The lamp filaments must be carefully aligned with the plane of the lens or mirror to provide the maximum output of light. The lens' size is chosen according to the type of platform, power source, and lamp characteristics. Additionally, environmental characteristics of the location are considered. Various types of light-condensing panels, reflex reflectors, or colored sector panels may be installed inside the lens to provide the proper characteristic. A specially reinforced 200mm lantern is used in locations where ice and breaking water are a hazard.

LIGHT CHARACTERISTICS

506. Characteristics

A light has distinctive **characteristics** which distinguish it from other lights or convey specific information by showing a distinctive sequence of light and dark intervals. Additionally, a light may display a distinctive color or color sequence. In the *Light Lists*, the dark intervals are referred to as **eclipses**.

An **occulting** light is a light totally eclipsed at regular intervals, the duration of light always being greater than the duration of darkness. A **flashing** light flashes on and off at regular intervals, the duration of light always being less than the duration of darkness. An **isophase** light flashes at regular intervals, the duration of light being equal to the duration of darkness.

Light phase characteristics (See Table 506) are the distinctive sequences of light and dark intervals or sequences in the variations of the luminous intensity of a light. The light phase characteristics of lights which change color do not differ from those of lights which do not change color. A light showing different colors alternately is described as an **alternating** light. The alternating characteristic may be used with other light phase characteristics.

TYPE	ABBREVIATION	GENERAL DESCRIPTION	ILLUSTRATION*
Fixed	F.	A continuous and steady light.	
Occulting	Oc.	The total duration of light in a period is longer than the total duration of darkness and the intervals of darkness (eclipses) are usually of equal duration. Eclipse regularly repeated.	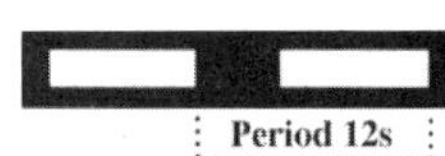
Group occulting	Oc.(2)	An occulting light for which a group of eclipses, specified in number, is regularly repeated.	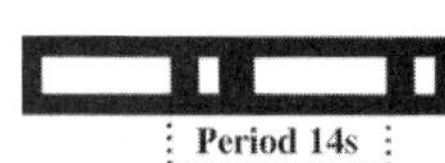
Composite group occulting	Oc.(2+1)	A light similar to a group occulting light except that successive groups in a period have different numbers of eclipses.	
Isophase	Iso	A light for which all durations of light and darkness are clearly equal.	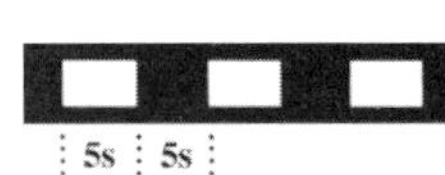

Table 506. Light phase characteristics.

TYPE	ABBREVIATION	GENERAL DESCRIPTION	ILLUSTRATION*
Flashing	Fl.	A light for which the total duration of light in a period is shorter than the total duration of darkness and the appearances of light (flashes) are usually of equal duration (at a rate of less than 50 flashes per minute).	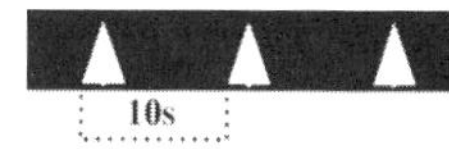
Long flashing	L.Fl.	A single flashing light for which an appearance of light of not less than 2 sec. duration (long flash) is regularly repeated.	
Group flashing	Fl.(3)	A flashing light for which a group of flashes, specified in number, is regularly repeated.	
Composite group flashing	Fl.(2+1)	A light similar to a group flashing light except that successive groups in a period have different numbers of flashes.	
Quick flashing	Q.	A light for which a flash is regularly repeated at a rate of not less than 50 flashes per minute but less than 80 flashes per minute.	
Group quick flashing	Q.(3)	A light for which a specified group of flashes is regularly repeated; flashes are repeated at a rate of not less than 50 flashes per minute but less than 80 flashes per minute.	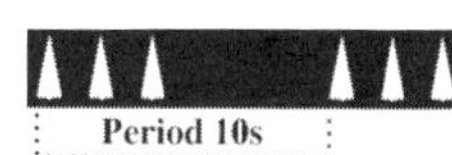
	Q.(9)		
	Q.(6)+L.Fl.		
Interrupted quick flashing	I.Q.	A light for which the sequence of quick flashes is interrupted by regularly repeated eclipses of constant and long duration.	
Very quick flashing	V.Q.	A light for which a flash is regularly repeated at a rate of not less than 80 flashes per minute but less than 160 flashes per minute.	

Table 506. Light phase characteristics.

TYPE	ABBREVIATION	GENERAL DESCRIPTION	ILLUSTRATION*
Group very quick flashing	V.Q.(3)	A light for which a specified group of very quick flashes is regularly repeated.	
	V.Q.(9)		
	V.Q.(6)+L.Fl.		
Interrupted very quick flashing	I.V.Q.	A light for which the sequence of very quick flashes is interrupted by regularly repeated eclipses of constant and long duration.	
Ultra quick flashing	U.Q.	A light for which a flash is regularly repeated at a rate of not less than 160 flashes per minute.	
Interrupted ultra quick flashing	I.U.Q.	A light for which the sequence of ultra quick flashes is interrupted by regularly repeated eclipses of constant and long duration.	
Morse code	Mo.(U)	A light for which appearances of light of two clearly different durations are grouped to represent a character or characters in Morse Code.	
Fixed and flashing	F.Fl.	A light for which a fixed light is combined with a flashing light of greater luminous intensity	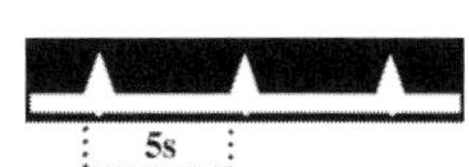
Alternate light	Al.	A light showing different colors alternately	* Periods shown are examples only.
NOTE: Alternating lights may be used in combined form with most of the previous types of lights			

Table 506. Light phase characteristics.

Light-sensitive switches extinguish most lighted navigation aids during daylight hours. However, owing to the various sensitivities of the light switches, all lights do not turn on or off at the same time. Mariners should account for this when identifying aids to navigation during twilight periods when some lighted aids are on while others are not.

507. Light Sectors

Sectors of colored glass or plastic are sometimes placed in the lanterns of certain lights to indicate dangerous waters. Lights so equipped show different colors when observed from different bearings. A sector changes the color of a light, but not its characteristic, when viewed from certain directions. For example, a four second flashing white light having a red sector will appear as a four second flashing red light when viewed from within the red sector.

Sectors may be only a few degrees in width or extend in a wide arc from deep water toward shore. Bearings referring to sectors are expressed in degrees true as observed from a vessel. In most cases, areas covered by red sectors should be avoided. The nature of the danger can be determined from the chart. In some cases a narrow sector may mark the best water across a shoal, or a turning point in a channel.

The transition from one color to another is not abrupt. The colors change through an arc of uncertainty of 2° or greater, depending on the optical design of the light. Therefore determining bearings by observing the color change is less accurate than obtaining a bearing with an azimuth circle.

508. Factors Affecting Range and Characteristics

The condition of the atmosphere has a considerable effect upon a light's range. Lights are sometimes obscured by fog, haze, dust, smoke, or precipitation. On the other hand, refraction may cause a light to be seen farther than under ordinary circumstances. A light of low intensity will be easily obscured by unfavorable conditions of the atmosphere. For this reason, the intensity of a light should always be considered when looking for it in thick weather. Haze and distance may reduce the apparent duration of a light's flash. In some conditions of the atmosphere, white lights may have a reddish hue. In clear weather green lights may have a more whitish hue.

Lights placed at higher elevations are more frequently obscured by clouds, mist, and fog than those near sea level. In regions where ice conditions prevail, an unattended light's lantern panes may become covered with ice or snow This may reduce the light's luminous range and change the light's observed color.

The distance from a light cannot be estimated by its apparent brightness. There are too many factors which can change the perceived intensity. Also, a powerful, distant light may sometimes be confused with a smaller, closer one with similar characteristics. Every light sighted should be carefully evaluated to determine if it is the one expected.

The presence of bright shore lights may make it difficult to distinguish navigational lights from background lighting. Lights may also be obscured by various shore obstructions, natural and man-made. The Coast Guard requests mariners to report these cases to the nearest Coast Guard station.

A light's **loom** is sometimes seen through haze or the reflection from low-lying clouds when the light is beyond its geographic range. Only the most powerful lights can generate a loom. The loom may be sufficiently defined to obtain a bearing. If not, an accurate bearing on a light beyond geographic range may sometimes be obtained by ascending to a higher level where the light can be seen, and noting a star directly over the light. The bearing of the star can then be obtained from the navigating bridge and the bearing to the light plotted indirectly.

At short distances, some of the brighter flashing lights may show a faint continuous light, or faint flashes, between regular flashes. This is due to reflections of a rotating lens on panes of glass in the lighthouse.

If a light is not sighted within a reasonable time after prediction, a dangerous situation may exist. Conversely, the light may simply be obscured or extinguished. The ship's position should immediately be fixed by other means to determine any possibility of danger.

The apparent characteristic of a complex light may change with the distance of the observer. For example, a light with a characteristic of fixed white and alternating flashing white and red may initially show as a simple flashing white light. As the vessel draws nearer, the red flash will become visible and the characteristic will apparently be alternating flashing white and red. Later, the fainter fixed white light will be seen between the flashes and the true characteristic of the light finally recognized as fixed white, alternating flashing white and red (F W Al W R). This is because for a given candlepower, white is the most visible color, green less so, and red least of the three. This fact also accounts for the different ranges given in the *Light Lists* for some multi-color sector lights. The same lamp has different ranges according to the color imparted by the sector glass.

A light may be **extinguished** due to weather, battery failure, vandalism, or other causes. In the case of unattended lights, this condition might not be immediately corrected. The mariner should report this condition to the nearest Coast Guard station. During periods of armed conflict, certain lights may be deliberately extinguished without notice. Offshore light stations should always be left well off the course whenever searoom permits.

BUOYS

509. Definitions and Types

Buoys are floating aids to navigation. They mark channels, indicate shoals and obstructions, and warn the mariner of dangers. Buoys are used where fixed aids would be uneconomical or impractical due to the depth of water. By their color, shape, topmark, number, and light characteristics, buoys indicate to the mariner how to avoid hazards and stay in safe water. The federal buoyage system in the U.S. is maintained by the Coast Guard.

There are many different sizes and types of buoys designed to meet a wide range of environmental conditions and user requirements. The size of a buoy is determined primarily by its location. In general, the smallest buoy which will stand up to local weather and current conditions is chosen.

There are five types of buoys maintained by the Coast Guard. They are:

1. Lateral marks
2. Isolated danger marks
3. Safe water marks
4. Special marks
5. Information/regulatory marks

These conform in general to the specifications of the **International Association of Lighthouse Authorities (IALA)** buoyage system.

A **lighted buoy** is a floating hull with a tower on which a light is mounted. Batteries for the light are in watertight pockets in the buoy hull or in watertight boxes mounted on the buoy hull. To keep the buoy in an upright position, a counterweight is attached to the hull below the water's surface. A radar reflector is built into the buoy tower.

The largest of the typical U.S. Coast Guard buoys can be moored in up to 190 feet of water, limited by the weight of chain the hull can support. The focal plane of the light is 15 to 20 feet high. The designed nominal visual range is 3.8 miles, and the radar range 4 miles. Actual conditions will cause these range figures to vary considerably.

The smallest buoys are designed for protected water. Some are made of plastic and weigh only 40 pounds. Specially designed buoys are used for fast current, ice, and other environmental conditions.

A variety of special purpose buoys are owned by other governmental organizations. Examples of these organizations include the St. Lawrence Seaway Development Corporation, NOAA, and the Department of Defense. These buoys are usually navigational marks or data collection buoys with traditional round, boat-shaped, or discus-shaped hulls.

A special class of buoy, the **Ocean Data Acquisition System (ODAS)** buoy, is moored or floats free in offshore waters. Positions are promulgated through radio warnings. These buoys are generally not large enough to cause damage to a large vessel in a collision, but should be given a wide berth regardless, as any loss would almost certainly result in the interruption of valuable scientific experiments. They are generally bright orange or yellow in color, with vertical stripes on moored buoys and horizontal bands on free-floating ones, and have a strobe light for night visibility.

Figure 509. Buoy showing counterweight.

Even in clear weather, the danger of collision with a buoy exists. If struck head-on, a large buoy can inflict severe damage to a large ship; it can sink a smaller one. Reduced visibility or heavy background lighting can contribute to the problem of visibility. The Coast Guard sometimes receives reports of buoys missing from station that were actually run down and sunk. Tugboats and towboats towing or pushing barges are particularly dangerous to buoys because of poor over-the-bow visibility when pushing or yawing during towing. The professional mariner must report *any* collision with a buoy to the nearest Coast Guard unit. Failure to do so may cause the next vessel to miss the channel or hit the obstruction marked by the buoy; it can also lead to fines and legal liability.

Routine on-station buoy maintenance consists of inspecting the mooring, cleaning the hull and superstructure, replacing the batteries, flasher, and lamps, checking wiring and venting systems, and verifying the buoy's exact position. Every few years, each buoy is replaced by a similar aid and returned to a Coast Guard maintenance facility for complete refurbishment.

The placement of a buoy depends on its purpose and its position on the chart. Most buoys are placed on their charted positions as accurately as conditions allow. However, if a buoy's purpose is to mark a shoal and the shoal is found to be in a different position than the chart shows, the buoy will be placed to properly mark the shoal, and not on its charted position.

510. Lights on Buoys

Buoy light systems consist of a **battery pack**, a **flasher** which determines the characteristic, a **lamp changer** which automatically replaces burned-out bulbs, a **lens** to focus the light, and a **housing** which supports the lens and protects the electrical equipment.

The **batteries** consist of 12-volt lead/acid type batteries electrically connected to provide sufficient power to run the proper flash characteristic and lamp size. These battery packs are contained in pockets in the buoy hull, accessible through water-tight bolted hatches or externally mounted boxes. Careful calculations based on light characteristics determine how much battery power to install.

The **flasher** determines the characteristic of the lamp. It is installed in the housing supporting the lens.

The **lamp changer** consists of several sockets arranged around a central hub. A new lamp rotates into position if the active one burns out.

Under normal conditions, the **lenses** used on buoys are 155mm in diameter at the base. 200 mm lenses are used where breaking waves or swells call for the larger lens. They are colored according to the charted characteristic of the buoy. As in shore lights, the lamp must be carefully focused so that the filament is directly in line with the focal plane of the lens. This ensures that the majority of the light produced is focused in a 360° horizontal fan beam. A buoy light has a relatively narrow vertical profile. Because the buoy rocks in the sea, the focal plane may only be visible for fractions of a second at great ranges. A realistic range for sighting buoy lights is 4-6 miles in good visibility and calm weather.

511. Sound Signals on Buoys

Lighted sound buoys have the same general configuration as lighted buoys but are equipped with either a bell, gong, whistle, or horn. **Bells** and **gongs** are sounded by tappers hanging from the tower that swing as the buoy rocks in the sea. Bell buoys produce only one tone; gong buoys produce several tones. The tone-producing device is mounted between the legs of the pillar or tower.

Whistle buoys make a loud moaning sound caused by the rising and falling motions of the buoy in the sea. A sound buoy equipped with an electronic **horn** will produce a pure tone at regular intervals regardless of the sea state. Unlighted sound buoys have the same general appearance as lighted buoys, but their underwater shape is designed to make them lively in all sea states.

512. Buoy Moorings

Buoys require **moorings** to hold them in position. Typically the mooring consists of **chain** and a large concrete or cast iron **sinker**. See Figure 512. Because buoys are subjected to waves, wind, and tides, the moorings must be deployed with chain lengths much greater than the water depth. The scope of chain will normally be about 3 times the water depth. The length of the mooring chain defines a **watch circle** within which the buoy can be expected to swing. It is for this reason that the charted buoy symbol has a "position approximate" circle to indicate its charted position, whereas a light position is shown by a dot at the exact location. Actual watch circles do not necessarily coincide with the "position approximate" circles which represent them.

Figure 512. A sinker used to anchor a buoy.

Over several years, the chain gradually wears out and must be replaced. The worn chain is often cast into the concrete of new sinkers.

513. Large Navigational Buoys

Large navigational buoys are moored in open water at approaches to certain major seacoast ports and monitored from shore stations by radio signals. These 40-foot diameter buoys (Figure 513) show lights from heights of about 36 feet above the water. Emergency lights automatically energize if the main light is extinguished. These buoys may also have a radiobeacon and sound signals.

514. Wreck Buoys

A **wreck buoy** usually cannot be placed directly over the wreck it is intended to mark because the buoy tender may not want to pass over a shallow wreck or risk fouling the buoy mooring. For this reason, a wreck buoy is usually

Figure 513. Large navigational buoy.

placed as closely as possible on the seaward or channelward side of a wreck. In some situations, two buoys may be used to mark the wreck, one lying off each end. The wreck may lie directly between them or inshore of a line between them, depending on the local situation. The *Local Notice to Mariners* should be consulted concerning details of the placement of wreck buoys on individual wrecks. Often it will also give particulars of the wreck and what activities may be in progress to clear it.

The charted position of a wreck buoy will usually be offset from the actual geographic position so that the wreck and buoy symbols do not coincide. Only on the largest scale chart will the actual and charted positions of both wreck and buoy be the same. Where they might overlap, it is the wreck symbol which occupies the exact charted position and the buoy symbol which is offset.

Wreck buoys are required to be placed by the owner of the wreck, but they may be placed by the Coast Guard if the owner is unable to comply with this requirement. In general, privately placed aids are not as reliable as Coast Guard aids.

Sunken wrecks are sometimes moved away from their buoys by storms, currents, freshets, or other causes. Just as shoals may shift away from the buoys placed to mark them, wrecks may shift away from wreck buoys.

515. Fallibility of Buoys

Buoys cannot be relied on to maintain their charted positions consistently. They are subject to a variety of hazards including severe weather, collision, mooring casualties, and electrical failure. Mariners should report discrepancies to the authority responsible for maintaining the aid.

The buoy symbol shown on charts indicates the approximate position of the sinker which secures the buoy to the seabed. The approximate position is used because of practical limitations in keeping buoys in precise geographical locations. These limitations include prevailing atmospheric and sea conditions, the slope and type of material making up the seabed, the scope of the mooring chain, and the fact that the positions of the buoys and the sinkers are not under continuous surveillance. The position of the buoy shifts around the area shown by the chart symbol due to the forces of wind and current.

A buoy may not be in its charted position because of changes in the feature it marks. For example, a buoy meant to mark a shoal whose boundaries are shifting might frequently be moved to mark the shoal accurately. A *Local Notice to Mariners* will report the change, and a *Notice to Mariners* chart correction may also be written. In some small channels which change often, buoys are not charted even when considered permanent; local knowledge is advised in such areas.

For these reasons, a mariner must not rely completely upon the position or operation of buoys, but should navigate using bearings of charted features, structures, and aids to navigation on shore. Further, a vessel attempting to pass too close aboard a buoy risks a collision with the buoy or the obstruction it marks.

BUOYAGE SYSTEMS

516. Lateral and Cardinal Systems

There are two major types of buoyage systems: the **lateral system** and the **cardinal system**. The lateral system is best suited for well-defined channels. The description of each buoy indicates the direction of danger relative to the course which is normally followed. In principle, the positions of marks in the lateral system are determined by the **general direction** taken by the mariner when approaching port from seaward. These positions may also be determined with reference to the main stream of flood current. The United States Aids to Navigation System is a lateral system.

The cardinal system is best suited for coasts with numerous isolated rocks, shoals, and islands, and for dangers in the open sea. The characteristic of each buoy indicates the approximate true bearing of the danger it marks. Thus, an eastern quadrant buoy marks a danger which lies to the west of the buoy. The following pages diagram the cardinal and lateral buoyage systems as found outside the United States.

517. The IALA Maritime Buoyage System

Although most of the major maritime nations have used either the lateral or the cardinal system for many years, details such as the buoy shapes and colors have varied from country to country. With the increase in maritime commerce between countries, the need for a uniform system of buoyage became apparent.

In 1889, an International Marine Conference held in Washington, D.C., recommended that in the lateral system, starboard hand buoys be painted red and port hand buoys black. Unfortunately, when lights for buoys were introduced some years later, some European countries placed red lights on the black port hand buoys to conform with the red lights marking the port side of harbor entrances, while in North America red lights were placed on red starboard hand buoys. In 1936, a League of Nations subcommittee recommended a coloring system opposite to the 1889 proposal.

The **International Association of Lighthouse Authorities (IALA)** is a non-governmental organization which consists of representatives of the worldwide community of aids to navigation services. It promotes information exchange and recommends improvements based on new technologies. In 1980, with the assistance of IMO and the IHO, the lighthouse authorities from 50 countries and representatives of 9 international organizations concerned with aids to navigation met and adopted the **IALA Maritime Buoyage System**. They established two regions, **Region A** and **Region B**, for the entire world. Region A roughly corresponds to the 1936 League of Nations system, and Region B to the older 1889 system.

Lateral marks differ between Regions A and B. Lateral marks in Region A use red and green colors by day and night to indicate port and starboard sides of channels, respectively. In Region B, these colors are reversed with red to starboard and green to port. In both systems, the conventional direction of buoyage is considered to be returning from sea, hence the phrase "red right returning" in IALA region B.

518. Types of Marks

The **IALA Maritime Buoyage System** applies to all fixed and floating marks, other than lighthouses, sector lights, range lights, daymarks, lightships and large navigational buoys, which indicate:

1. The side and center-lines of navigable channels
2. Natural dangers, wrecks, and other obstructions
3. Regulated navigation areas
4. Other important features

Most lighted and unlighted beacons other than range marks are included in the system. In general, beacon topmarks will have the same shape and colors as those used on buoys. The system provides five types of marks which may be used in any combination:

1. Lateral marks indicate port and starboard sides of channels.
2. Cardinal marks, named according to the four points of the compass, indicate that the navigable water lies to the named side of the mark.
3. Isolated danger marks erected on, or moored directly on or over, dangers of limited extent.
4. Safe water marks, such as midchannel buoys.

5. Special marks, the purpose of which is apparent from reference to the chart or other nautical documents.

Characteristics of Marks

The significance of a mark depends on one or more features:

1. By day—color, shape, and topmark
2. By night—light color and phase characteristics

Colors of Marks

The colors red and green are reserved for lateral marks, and yellow for special marks. The other types of marks have black and yellow or black and red horizontal bands, or red and white vertical stripes.

Shapes of Marks

There are five basic buoy shapes:

1. Can
2. Cone
3. Sphere
4. Pillar
5. Spar

In the case of can, conical, and spherical, the shapes have lateral significance because the shape indicates the correct side to pass. With pillar and spar buoys, the shape has no special significance.

The term "pillar" is used to describe any buoy which is smaller than a large navigation buoy (LNB) and which has a tall, central structure on a broad base; it includes beacon buoys, high focal plane buoys, and others (except spar buoys) whose body shape does not indicate the correct side to pass.

Topmarks

The IALA System makes use of **can, conical, spherical,** and **X-shaped** topmarks only. Topmarks on pillar and spar buoys are particularly important and will be used wherever practicable, but ice or other severe conditions may occasionally prevent their use.

Colors of Lights

Where marks are lighted, red and green lights are reserved for lateral marks, and yellow for special marks. The other types of marks have a white light, distinguished one from another by phase characteristic.

Phase Characteristics of Lights

Red and green lights may have any phase characteristic, as the color alone is sufficient to show on which side they should be passed. Special marks, when lighted, have a yellow light with any phase characteristic not reserved for white lights of the system. The other types of marks have clearly specified phase characteristics of white light: various quick-flashing phase characteristics for cardinal marks, group flashing (2) for isolated danger marks, and relatively long periods of light for safe water marks.

Some shore lights specifically excluded from the IALA System may coincidentally have characteristics corresponding to those approved for use with the new marks. Care is needed to ensure that such lights are not misinterpreted.

519. IALA Lateral Marks

Lateral marks are generally used for well-defined channels; they indicate the port and starboard hand sides of the route to be followed, and are used in conjunction with a **conventional direction of buoyage**.

This direction is defined in one of two ways:

1. **Local direction of buoyage** is the direction taken by the mariner when approaching a harbor, river estuary, or other waterway from seaward.
2. **General direction of buoyage** is determined by the buoyage authorities, following a clockwise direction around continental land-masses, given in sailing directions, and, if necessary, indicated on charts by a large open arrow symbol.

In some places, particularly straits open at both ends, the local direction of buoyage may be overridden by the general direction.

Along the coasts of the United States, the characteristics assume that proceeding "from seaward" constitutes a clockwise direction: a southerly direction along the Atlantic coast, a westerly direction along the Gulf of Mexico coast, and a northerly direction along the Pacific coast. On the Great Lakes, a westerly and northerly direction is taken as being "from seaward" (except on Lake Michigan, where a southerly direction is used). On the Mississippi and Ohio Rivers and their tributaries, the characteristics of aids to navigation are determined as proceeding from sea toward the head of navigation. On the Intracoastal Waterway, proceeding in a generally southerly direction along the Atlantic coast, and in a generally westerly direction along the gulf coast, is considered as proceeding "from seaward."

520. IALA Cardinal Marks

A **cardinal mark** is used in conjunction with the compass to indicate where the mariner may find the best navigable water. It is placed in one of the four quadrants (north, east, south, and west), bounded by the true bearings

NW-NE, NE-SE, SE-SW, and SW-NW, taken from the point of interest. A cardinal mark takes its name from the quadrant *in which it is placed*.

The mariner is safe if he passes north of a north mark, east of an east mark, south of a south mark, and west of a west mark.

A cardinal mark may be used to:

1. Indicate that the deepest water in an area is on the named side of the mark.
2. Indicate the safe side on which to pass a danger.
3. Emphasize a feature in a channel, such as a bend, junction, bifurcation, or end of a shoal.

Topmarks

Black double-cone topmarks are the most important feature, by day, of cardinal marks. The cones are vertically placed, one over the other. The arrangement of the cones is very logical: North is two cones with their points up (as in "north-up"). South is two cones, points down. East is two cones with bases together, and west is two cones with points together, which gives a wineglass shape. "West is a Wineglass" is a memory aid.

Cardinal marks carry topmarks whenever practicable, with the cones as large as possible and clearly separated.

Colors

Black and yellow horizontal bands are used to color a cardinal mark. The position of the black band, or bands, is related to the points of the black topmarks.

N	Points up	Black above yellow
S	Points down	Black below yellow
W	Points together	Black, yellow above and below
E	Points apart	Yellow, black above and below

Shape

The shape of a cardinal mark is not significant, but buoys must be pillars or spars.

Lights

When lighted, a cardinal mark exhibits a white light; its characteristics are based on a group of quick or very quick flashes which distinguish it as a cardinal mark and indicate its quadrant. The distinguishing quick or very quick flashes are:

North—Uninterrupted
East—three flashes in a group
South—six flashes in a group followed by a long flash
West—nine flashes in a group

As a memory aid, the number of flashes in each group can be associated with a clock face: 3 o'clock—E, 6 o'clock—S, and 9 o'clock—W.

The long flash (of not less than 2 seconds duration), immediately following the group of flashes of a south cardinal mark, is to ensure that its six flashes cannot be mistaken for three or nine.

The periods of the east, south, and west lights are, respectively, 10, 15, and 15 seconds if quick flashing; and 5, 10, and 10 seconds if very quick flashing.

Quick flashing lights flash at a rate between 50 and 79 flashes per minute, usually either 50 or 60. Very quick flashing lights flash at a rate between 80 and 159 flashes per minute, usually either 100 or 120.

It is necessary to have a choice of quick flashing or very quick flashing lights in order to avoid confusion if, for example, two north buoys are placed near enough to each other for one to be mistaken for the other.

521. IALA Isolated Danger Marks

An **isolated danger mark** is erected on, or moored on or above, an isolated danger of limited extent which has navigable water all around it. The extent of the surrounding navigable water is immaterial; such a mark can, for example, indicate either a shoal which is well offshore or an islet separated by a narrow channel from the coast.

Position

On a chart, the position of a danger is the center of the symbol or sounding indicating that danger; an isolated danger buoy may therefore be slightly displaced from its geographic position to avoid overprinting the two symbols. The smaller the scale, the greater this offset will be. At very large scales the symbol may be correctly charted.

Topmark

A black double-sphere topmark is, by day, the most important feature of an isolated danger mark. Whenever practicable, this topmark will be carried with the spheres as large as possible, disposed vertically, and clearly separated.

Color

Black with one or more red horizontal bands are the colors used for isolated danger marks.

Shape

The shape of an isolated danger mark is not significant, but a buoy will be a pillar or a spar.

Light

When lighted, a white flashing light showing a group of two flashes is used to denote an isolated danger mark. As a memory aid, associate two flashes with two balls in the topmark.

522. IALA Safe Water Marks

A **safe water mark** is used to indicate that there is navigable water all around the mark. Such a mark may be used as a center line, mid-channel, or landfall buoy.

Color

Red and white vertical stripes are used for safe water marks, and distinguish them from the black-banded, danger-marking marks.

Shape

Spherical, pillar, or spar buoys may be used as safe water marks.

Topmark

A single red spherical topmark will be carried, whenever practicable, by a pillar or spar buoy used as a safe water mark.

Lights

When lighted, safe water marks exhibit a white light. This light can be occulting, isophase, a single long flash, or Morse "A." If a long flash (i.e. a flash of not less than 2 seconds) is used, the period of the light will be 10 seconds. As a memory aid, remember a single flash and a single sphere topmark.

523. IALA Special Marks

A **special mark** may be used to indicate a special area or feature which is apparent by referring to a chart, sailing directions, or notices to mariners. Uses include:

1. Ocean Data Acquisition System (ODAS) buoys
2. Traffic separation marks
3. Spoil ground marks
4. Military exercise zone marks
5. Cable or pipeline marks, including outfall pipes
6. Recreation zone marks

Another function of a special mark is to define a channel within a channel. For example, a channel for deep draft vessels in a wide estuary, where the limits of the channel for normal navigation are marked by red and green lateral buoys, may have its boundaries or centerline marked by yellow buoys of the appropriate lateral shapes.

Color

Yellow is the color used for special marks.

Shape

The shape of a special mark is optional, but must not conflict with that used for a lateral or a safe water mark. For example, an outfall buoy on the port hand side of a channel could be can-shaped but not conical.

Topmark

When a topmark is carried it takes the form of a single yellow X.

Lights

When a light is exhibited it is yellow. It may show any phase characteristic except those used for the white lights of cardinal, isolated danger, and safe water marks. In the case of ODAS buoys, the phase characteristic used is group-flashing with a group of five flashes every 20 seconds.

524. IALA New Dangers

A newly discovered hazard to navigation not yet shown on charts, included in sailing directions, or announced by a *Notice to Mariners* is termed a **new danger**. The term covers naturally occurring and man-made obstructions.

Marking

A new danger is marked by one or more cardinal or lateral marks in accordance with the IALA system rules. If the danger is especially grave, at least one of the marks will be duplicated as soon as practicable by an identical mark until the danger has been sufficiently identified.

Lights

If a lighted mark is used for a new danger, it must exhibit a quick flashing or very quick flashing light. If a cardinal mark is used, it must exhibit a white light; if a lateral mark, a red or green light.

Racons

The duplicate mark may carry a Racon, Morse coded D, showing a signal length of 1 nautical mile on a radar display.

525. Chart Symbols and Abbreviations

Spar buoys and spindle buoys are represented by the same symbol; it is slanted to distinguish them from upright beacon symbols. The abbreviated description of the color of a buoy is given under the symbol. Where a buoy is colored in bands, the colors are indicated in sequence from the top. If the sequence of the bands is not known, or if the buoy is striped, the colors are indicated with the darker color first.

Topmarks

Topmark symbols are solid black except if the topmark is red.

Lights

The period of the light of a cardinal mark is determined by its quadrant and its flash characteristic (either quick-flashing or a very quick-flashing). The light's period is less important than its phase characteristic. Where space on charts is limited, the period may be omitted.

Light Flares

Magenta light-flares are normally slanted and inserted with their points adjacent to the position circles at the base of the symbols so the flare symbols do not obscure the topmark symbols.

Radar Reflectors

According to IALA rules, radar reflectors are not charted, for several reasons. First, all important buoys are fitted with radar reflectors. It is also necessary to reduce the size and complexity of buoy symbols and associated legends. Finally, it is understood that, in the case of cardinal buoys, buoyage authorities place the reflector so that it cannot be mistaken for a topmark.

The symbols and abbreviations of the IALA Maritime Buoyage System may be found in *U.S. Chart No. 1* and in foreign equivalents.

526. Description of the U.S. Aids to Navigation System

In the United States, the U.S. Coast Guard has incorporated the major features of the IALA system with the existing infrastructure of buoys and lights as explained below.

Colors

Under this system, green buoys mark a channel's port side and obstructions which must be passed by keeping the buoy on the port hand. Red buoys mark a channel's starboard side and obstructions which must be passed by keeping the buoy on the starboard hand.

Red and green horizontally banded **preferred channel buoys** mark junctions or bifurcations in a channel or obstructions which may be passed on either side. If the topmost band is green, the preferred channel will be followed by keeping the buoy on the port hand. If the topmost band is red, the preferred channel will be followed by keeping the buoy on the starboard hand.

Red and white vertically striped safe water buoys mark a fairway or mid-channel.

Reflective material is placed on buoys to assist in their detection at night with a searchlight. The color of the reflective material agrees with the buoy color. Red or green reflective material may be placed on preferred channel (junction) buoys; red if topmost band is red, or green if the topmost band is green. White reflective material is used on safe water buoys. Special purpose buoys display yellow reflective material. Warning or regulatory buoys display orange reflective horizontal bands and a warning symbol. Intracoastal Waterway buoys display a yellow reflective square, triangle, or horizontal strip along with the reflective material coincident with the buoy's function.

Shapes

Certain unlighted buoys are differentiated by shape. Red buoys and red and green horizontally banded buoys with the topmost band red are cone-shaped buoys called **nuns**. Green buoys and green and red horizontally banded buoys with the topmost band green are cylinder-shaped buoys called **cans**.

Unlighted red and white vertically striped buoys may be pillar shaped or spherical. Lighted buoys, sound buoys, and spar buoys are not differentiated by shape to indicate the side on which they should be passed. Their purpose is indicated not by shape but by the color, number, or light characteristics.

Numbers

All solid colored buoys are numbered, red buoys bearing even numbers and green buoys bearing odd numbers. (Note that this same rule applies in IALA System A also.) The numbers increase from seaward upstream or toward land. No other colored buoys are numbered; however, any buoy may have a letter for identification.

Light Colors

Red lights are used only on red buoys or red and green horizontally banded buoys with the topmost band red. Green lights are used only on the green buoys or green and red horizontally banded buoys with the topmost band green. White lights are used on both "safe water" aids showing a Morse Code "A" characteristic and on Information and Regulatory aids.

Light Characteristics

Lights on red buoys or green buoys, if not occulting

or isophase, will generally be regularly flashing (Fl). For ordinary purposes, the frequency of flashes will be not more than 50 flashes per minute. Lights with a distinct cautionary significance, such as at sharp turns or marking dangerous obstructions, will flash not less than 50 flashes but not more than 80 flashes per minute (quick flashing, Q). Lights on preferred channel buoys will show a series of group flashes with successive groups in a period having a different number of flashes - composite group flashing (or a quick light in which the sequence of flashes is interrupted by regularly repeated eclipses of constant and long duration). Lights on safe water buoys will always show a white Morse Code "A" (Short-Long) flash recurring at the rate of approximately eight times per minute.

Daylight Controls

Lighted buoys have a special device to energize the light when darkness falls and to de-energize the light when day breaks. These devices are not of equal sensitivity; therefore all lights do not come on or go off at the same time. Mariners should ensure correct identification of aids during twilight periods when some light aids to navigation are on while others are not.

Special Purpose Buoys

Buoys for special purposes are colored yellow. White buoys with orange bands are for informational or regulatory purposes. The shape of special purpose buoys has no significance. They are not numbered, but they may be lettered. If lighted, special purpose buoys display a yellow light usually with fixed or slow flash characteristics. Information and regulatory buoys, if lighted, display white lights.

BEACONS

527. Definition and Description

Beacons are fixed aids to navigation placed on shore or on pilings in relatively shallow water. If unlighted, the beacon is referred to as a **daybeacon**. A daybeacon is identified by the color, shape, and number of its **dayboard**. The simplest form of daybeacon consists of a single pile with a dayboard affixed at or near its top. See Figure 527. Daybeacons may be used to form an unlighted range.

Dayboards identify aids to navigation against daylight backgrounds. The size of the dayboard required to make the aid conspicuous depends upon the aid's intended range.

Most dayboards also display numbers or letters for identification. The numbers, letters, and borders of most dayboards have reflective tape to make them visible at night.

The detection, recognition, and identification distances vary widely for any particular dayboard. They depend upon the luminance of the dayboard, the Sun's position, and the local visibility conditions.

Figure 527. Daybeacon.

SOUND SIGNALS

528. Types of Sound Signals

Most lighthouses and offshore light platforms, as well as some minor light structures and buoys, are equipped with sound-producing devices to help the mariner in periods of low visibility. Charts and *Light Lists* contain the information required for positive identification. Buoys fitted with bells, gongs, or whistles actuated by wave motion may produce no sound when the sea is calm. Sound signals are not designed to identify the buoy or beacon for navigation purposes. Rather, they allow the mariner to pass clear of the buoy or beacon during low visibility.

Sound signals vary. The navigator must use the *Light List* to determine the exact length of each blast and silent interval. The various types of sound signals also differ in tone, facilitating recognition of the respective stations.

Diaphones produce sound with a slotted piston moved back and forth by compressed air. Blasts may consist of a high and low tone. These alternate-pitch signals are called "two-tone." Diaphones are not used by the Coast Guard, but the mariner may find them on some private navigation aids.

Horns produce sound by means of a disc diaphragm operated pneumatically or electrically. Duplex or triplex horn units of differing pitch produce a chime signal.

Sirens produce sound with either a disc or a cup-shaped rotor actuated electrically or pneumatically. Sirens are not used on U.S. navigation aids.

Whistles use compressed air emitted through a circumferential slot into a cylindrical bell chamber.

Bells and gongs are sounded with a mechanically operated hammer.

529. Limitations of Sound Signals

As aids to navigation, sound signals have serious limitations because sound travels through the air in an unpredictable manner.

It has been clearly established that:

1. Sound signals are heard at greatly varying distances and that the distance at which a sound signal can be heard may vary with the bearing and timing of the signal.
2. Under certain atmospheric conditions, when a sound signal has a combination high and low tone, it is not unusual for one of the tones to be inaudible. In the case of sirens, which produce a varying tone, portions of the signal may not be heard.
3. When the sound is screened by an obstruction, there are areas where it is inaudible.
4. Operators may not activate a remotely controlled sound aid for a condition unobserved from the controlling station.
5. Some sound signals cannot be immediately started.
6. The status of the vessel's engines and the location of the observer both affect the effective range of the aid.

These considerations justify the utmost caution when navigating near land in a fog. A navigator can never rely on sound signals alone; he should continuously man both the radar and fathometer. He should place lookouts in positions where the noises in the ship are least likely to interfere with hearing a sound signal. The aid upon which a sound signal rests is usually a good radar target, but collision with the aid or the danger it marks is always a possibility.

Emergency signals are sounded at some of the light and fog signal stations when the main and stand-by sound signals are inoperative. Some of these emergency sound signals are of a different type and characteristic than the main sound signal. The characteristics of the emergency sound signals are listed in the *Light List*.

The mariner should never assume:

1. That he is out of ordinary hearing distance because he fails to hear the sound signal.
2. That because he hears a sound signal faintly, he is far from it.
3. That because he hears it clearly, he is near it.
4. That the distance from and the intensity of a sound on any one occasion is a guide for any future occasion.
5. That the sound signal is not sounding because he does not hear it, even when in close proximity.
6. That the sound signal is in the direction the sound appears to come from.

MISCELLANEOUS U.S. SYSTEMS

530. Intracoastal Waterway Aids to Navigation

The Intracoastal Waterway (ICW) runs parallel to the Atlantic and Gulf of Mexico coasts from Manasquan Inlet on the New Jersey shore to the Texas/Mexican border. It follows rivers, sloughs, estuaries, tidal channels, and other natural waterways, connected with dredged channels where necessary. Some of the aids marking these waters are marked with yellow; otherwise, the marking of buoys and beacons follows the same system as that in other U.S. waterways.

Yellow symbols indicate that an aid marks the Intracoastal Waterway. Yellow triangles indicate starboard hand aids, and yellow squares indicate port hand aids when following the ICW's conventional direction of buoyage. Non-lateral aids such as safe water, isolated danger, and front range boards are marked with a horizontal yellow band. Rear range boards do not display the yellow band. At a junction with a federally-maintained waterway, the preferred channel mark will display a yellow triangle or square as appropriate. Junctions between the ICW and privately maintained waterways are not marked with preferred channel buoys.

531. Western Rivers System

Aids to navigation on the Mississippi River and its tributaries above Baton Rouge generally conform to the lateral system of buoyage in use in the rest of the U.S. The following differences are significant:

1. Buoys are not numbered.
2. The numbers on lights and daybeacons do not have lateral significance; they indicate the mileage from a designated point, normally the river mouth.
3. Flashing lights on the left side proceeding upstream show single green or white flashes while those on the right side show group flashing red or white flashes.
4. Diamond shaped crossing daymarks are used to indicate where the channel crosses from one side of the river to the other.

532. The Uniform State Waterway Marking System (USWMS)

This system was developed jointly by the U.S. Coast Guard and state boating administrators to assist the small craft operator in those state waters marked by participating states. The **USWMS** consists of two categories of aids to navigation. The first is a system of aids to navigation, generally compatible with the Federal lateral system of buoyage, supplementing the federal system in state waters. The other is a system of regulatory markers to warn small craft operators of dangers or to provide general information.

On a well-defined channel, red and black buoys are established in pairs called **gates**; the channel lies between the buoys. The buoy which marks the left side of the channel viewed looking upstream or toward the head of navigation is black; the buoy which marks the right side of the channel is red.

In an irregularly-defined channel, buoys may be staggered on alternate sides of the channel, but they are spaced at sufficiently close intervals to mark clearly the channel lying between them.

Where there is no well-defined channel or where a body of water is obstructed by objects whose nature or location is such that the obstruction can be approached by a vessel from more than one direction, aids to navigation having cardinal significance may be used. The aids conforming to the cardinal system consist of three distinctly colored buoys as follows:

1. A white buoy with a red top must be passed to the south or west of the buoy.
2. A white buoy with a black top must be passed to the north or east of the buoy.
3. A buoy showing alternate vertical red and white stripes indicates that an obstruction to navigation extends from the nearest shore to the buoy and that the vessel must not pass between the buoy and the nearest shore.

The shape of buoys has no significance under the USWMS.

Regulatory buoys are colored white with orange horizontal bands completely around them. One band is at the top of the buoy and a second band just above the waterline of the buoy so that both orange bands are clearly visible.

Geometric shapes colored orange are placed on the white portion of the buoy body. The authorized geometric shapes and meanings associated with them are as follows:

1. A vertical open faced diamond shape means danger.
2. A vertical open faced diamond shape with a cross centered in the diamond means that vessels are excluded from the marked area.
3. A circular shape means that vessels in the marked area are subject to certain operating restrictions.
4. A square or rectangular shape indicates that directions or information is written inside the shape.

Regulatory markers consist of square and rectangular shaped signs displayed from fixed structures. Each sign is white with an orange border. Geometric shapes with the same meanings as those displayed on buoys are centered on the sign boards. The geometric shape displayed on a regulatory marker tells the mariner if he should stay well clear of the marker or if he may approach the marker in order to read directions.

533. Private Aids to Navigation

A **private navigation aid** is any aid established and maintained by entities other than the Coast Guard.

The Coast Guard must approve the placement of private navigation aids. In addition, the District Engineer, U.S. Army Corps of Engineers, must approve the placement of any structure, including aids to navigation, in the navigable waters of the U.S.

Private aids to navigation are similar to the aids established and maintained by the U.S. Coast Guard; they are specially designated on the chart and in the *Light List*. In some cases, particularly on large commercial structures, the aids are the same type of equipment used by the Coast Guard. Although the Coast Guard periodically inspects some private navigation aids, the mariner should exercise special caution when using them.

In addition to private aids to navigation, numerous types of construction and anchor buoys are used in various oil drilling operations and marine construction. These buoys are not charted, as they are temporary, and may not be lighted well or at all. Mariners should give a wide berth to drilling and construction sites to avoid the possibility of fouling moorings. This is a particular danger in offshore oil fields, where large anchors are often used to stabilize the positions of drill rigs in deep water. Up to eight anchors may be placed at various positions as much as a mile from the drill ship. These positions may or may not be marked by buoys. Such operations in the U.S. are announced in the *Local Notice to Mariners*.

534. Protection by Law

It is unlawful to impair the usefulness of any navigation aid established and maintained by the United States. If any vessel collides with a navigation aid, it is the legal duty of the person in charge of the vessel to report the accident to the nearest U.S. Coast Guard station.

CHAPTER 6

COMPASSES

INTRODUCTION

600. Changes in Compass Technologies

This chapter discusses the major types of compasses available to the navigator, their operating principles, their capabilities, and limitations of their use. As with other aspects of navigation, technology is rapidly revolutionizing the field of compasses. Amazingly, after at least a millennia of constant use, it is now possible (however advisable it may or may not be aboard any given vessel) to dispense with the traditional magnetic compass.

For much of maritime history the only heading reference for navigators has been the magnetic compass. A great deal of effort and expense has gone into understanding the magnetic compass scientifically and making it as accurate as possible through elaborate compensation techniques.

The introduction of the electro-mechanical gyrocompass relegated the magnetic compass to backup status for many large vessels. Later came the development of inertial navigation systems based on gyroscopic principles. The interruption of electrical power to the gyrocompass or inertial navigator, mechanical failure, or its physical destruction would instantly elevate the magnetic compass to primary status for most vessels.

New technologies are both refining and replacing the magnetic compass as a heading reference and navigational tool. Although a magnetic compass for backup is certainly advisable, today's navigator can safely avoid nearly all of the effort and expense associated with the binnacle-mounted magnetic compass, its compensation, adjustment, and maintenance.

Similarly, electro-mechanical gyrocompasses are being supplanted by far lighter, cheaper, and more dependable ring laser gyrocompasses. These devices do not operate on the principle of the gyroscope (which is based on Newton's laws of motion), but instead rely on the principles of electromagnetic energy and wave theory.

Magnetic flux gate compasses, while relying on the earth's magnetic field for reference, have no moving parts and can compensate themselves, adjusting for both deviation and variation to provide true heading, thus completely eliminating the process of compass correction.

To the extent that one depends on the magnetic compass for navigation, it should be checked regularly and adjusted when observed errors exceed certain minimal limits, usually a few degrees for most vessels. Compensation of a magnetic compass aboard vessels expected to rely on it offshore during long voyages is best left to professionals. However, this chapter will present enough material for the competent navigator to do a passable job.

Whatever type of compass is used, it is advisable to check it periodically against an error free reference to determine its error. This may be done when steering along any range during harbor and approach navigation, or by aligning any two charted objects and finding the difference between their observed and charted bearings. When navigating offshore, the use of azimuths and amplitudes of celestial bodies will also suffice, a subject covered in Chapter 17.

MAGNETIC COMPASSES

601. The Magnetic Compass and Magnetism

The principle of the present day magnetic compass is no different from that of the compasses used by ancient mariners. The magnetic compass consists of a magnetized needle, or an array of needles, allowed to rotate in the horizontal plane. The superiority of present day magnetic compasses over ancient ones results from a better knowledge of the laws of magnetism which govern the behavior of the compass and from greater precision in design and construction.

Any magnetized piece of metal will have regions of concentrated magnetism called **poles**. Any such magnet will have at least two poles of opposite polarity. Magnetic force (flux) lines connect one pole of such a magnet with the other pole. The number of such lines per unit area represents the intensity of the magnetic field in that area.

If two magnets are placed close to each other, the like poles will repel each other and the unlike poles will attract each other.

Magnetism can be either **permanent** or **induced**. A bar having permanent magnetism will retain its magnetism when it is removed from a magnetizing field. A bar having induced magnetism will lose its magnetism when removed from the magnetizing field. Whether or not a bar will retain its magnetism on removal from the magnetizing field will depend on the strength of that field, the degree of hardness of the iron (retentivity), and upon the amount of physical stress applied to the bar while in the magnetizing field. The harder the iron, the more permanent will be the magnetism acquired.

602. Terrestrial Magnetism

Consider the Earth as a huge magnet surrounded by lines of magnetic flux connecting its two **magnetic poles**. These magnetic poles are near, but not coincidental with, the Earth's geographic poles. Since the north seeking end of a compass needle is conventionally called the **north pole**, or **positive pole**, it must therefore be attracted to a **south pole**, or **negative pole**.

Figure 602a illustrates the Earth and its surrounding magnetic field. The flux lines enter the surface of the Earth at different angles to the horizontal at different magnetic latitudes. This angle is called the **angle of magnetic dip**, θ, and increases from 0° at the magnetic equator to 90° at the magnetic poles. The total magnetic field is generally considered as having two components: H, the horizontal component; and Z, the vertical component. These components change as the angle θ changes, such that H is at its maximum at the magnetic equator and decreases in the direction of either pole, while Z is zero at the magnetic equator and increases in the direction of either pole.

Since the magnetic poles of the Earth do not coincide with the geographic poles, a compass needle in line with the Earth's magnetic field will not indicate true north, but magnetic north. The angular difference between the true meridian (great circle connecting the geographic poles) and the magnetic meridian (direction of the lines of magnetic flux) is called **variation**. This variation has different values at different locations on the Earth. These values of magnetic variation may be found on pilot charts and on the compass rose of navigational charts.

The poles are not geographically static. They are known to migrate slowly, so that variation for most areas undergoes a small annual change, the amount of which is also noted on charts. Figure 602b and Figure 602c show magnetic dip and variation for the world. Up-to-date information on geomagnetics is available at http://geomag.usgs.gov/dod.html.

603. Ship's Magnetism

A ship under construction or repair will acquire permanent magnetism due to hammering and vibration while sitting stationary in the Earth's magnetic field. After launching, the ship will lose some of this original magnetism as a result of vibration and pounding in varying magnetic fields, and will eventually reach a more or less stable magnetic condition. The magnetism which remains is the **permanent magnetism** of the ship.

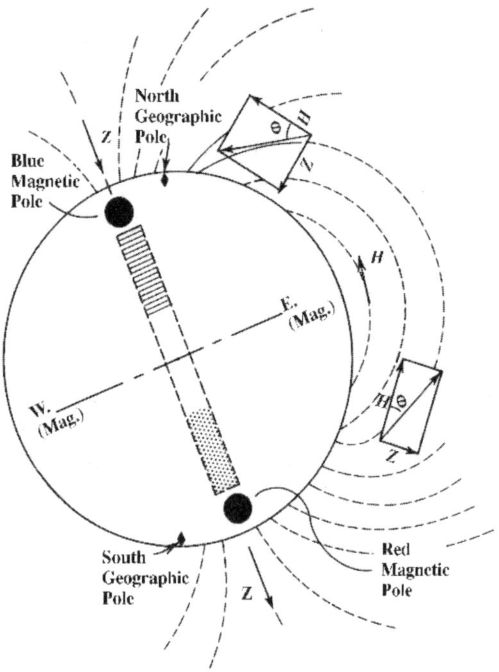

Figure 602a. Terrestrial magnetism.

In addition to its permanent magnetism, a ship acquires **induced magnetism** when placed in the Earth's magnetic field. The magnetism induced in any given piece of soft iron is a function of the field intensity, the alignment of the soft iron in that field, and the physical properties and dimensions of the iron. This induced magnetism may add to, or subtract from, the permanent magnetism already present in the ship, depending on how the ship is aligned in the magnetic field. The softer the iron, the more readily it will be magnetized by the Earth's magnetic field, and the more readily it will give up its magnetism when removed from that field.

The magnetism in the various structures of a ship, which tends to change as a result of cruising, vibration, or aging, but which does not alter immediately so as to be properly termed induced magnetism, is called **subpermanent magnetism**. This magnetism, at any instant, is part of the ship's permanent magnetism, and consequently must be corrected by permanent magnet correctors. It is the principal cause of deviation changes on a magnetic compass. Subsequent reference to permanent magnetism will refer to the apparent permanent magnetism which includes the existing permanent and subpermanent magnetism.

A ship, then, has a combination of permanent, subpermanent, and induced magnetism. Therefore, the ship's

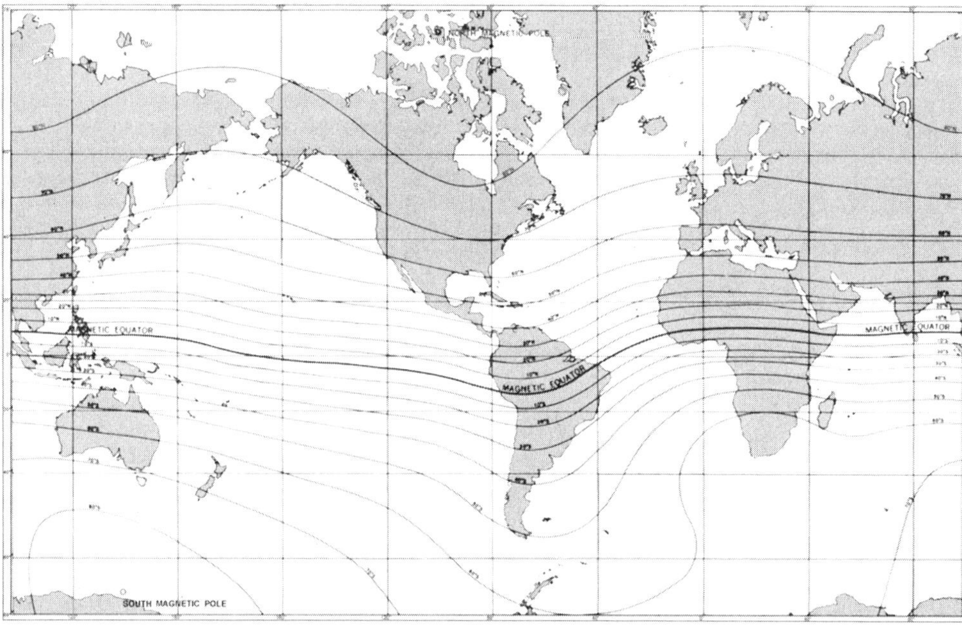

Figure 602b. Magnetic dip for the world.

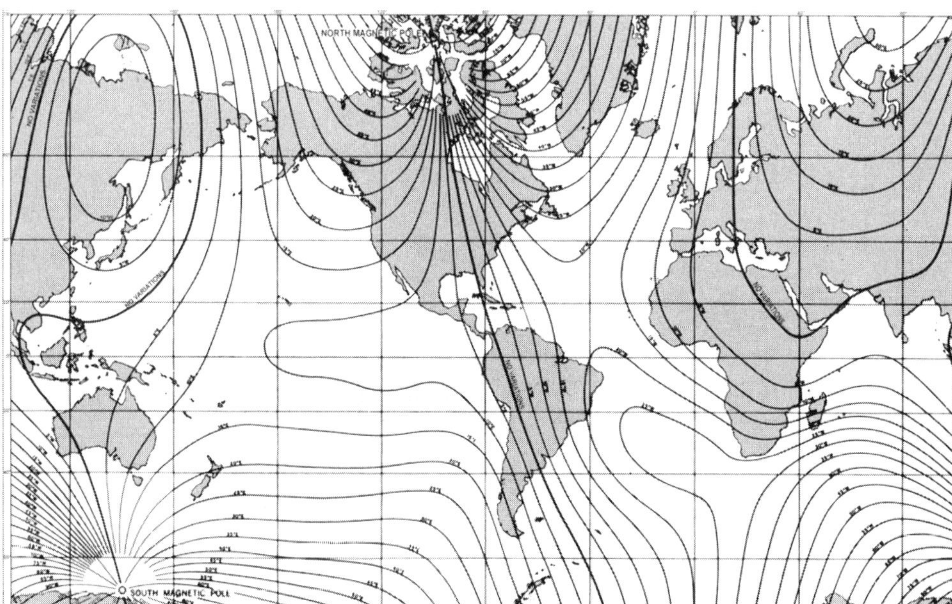

Figure 602c. Magnetic variation for the world.

apparent permanent magnetic condition is subject to change from deperming, shocks, welding, and vibration. The ship's induced magnetism will vary with the Earth's magnetic field strength and with the alignment of the ship in that field.

604. Magnetic Adjustment

A narrow rod of soft iron, placed parallel to the Earth's horizontal magnetic field, H, will have a north pole induced in the end toward the north geographic pole and a south pole induced in the end toward the south geographic pole. This same rod in a horizontal plane, but at right angles to the horizontal Earth's field, would have no magnetism induced in it, because its alignment in the magnetic field precludes linear magnetization, if the rod is of negligible cross section. Should the rod be aligned in some horizontal direction between those headings which create maximum and zero induction, it would be induced by an amount which is a function of the angle of alignment. However, if a similar rod is placed in a vertical position in northern latitudes so as to be aligned with the vertical Earth's field Z, it will have a south pole induced at the upper end and a north pole induced at the lower end. These polarities of vertical induced magnetization will be reversed in southern latitudes.

The amount of horizontal or vertical induction in such rods, or in ships whose construction is equivalent to combinations of such rods, will vary with the intensity of H and Z, heading, and heel of the ship.

The magnetic compass must be corrected for the vessel's permanent and induced magnetism so that its operation approximates that of a completely nonmagnetic vessel. Ship's magnetic conditions create magnetic compass deviations and sectors of sluggishness and unsteadiness. **Deviation** is defined as deflection right or left of the magnetic meridian caused by magnetic properties of the vessel. Adjusting the compass consists of arranging magnetic and soft iron **correctors** near the compass so that their effects are equal and opposite to the effects of the magnetic material in the ship.

The total permanent magnetic field effect at the compass may be broken into three components, mutually 90° to each other, as shown in Figure 604a.

The vertical permanent component tilts the compass card, and, when the ship rolls or pitches, causes oscillating deflections of the card. Oscillation effects which accompany roll are maximum on north and south compass headings, and those which accompany pitch are maximum on east and west compass headings.

The horizontal B and C components of permanent magnetism cause varying deviations of the compass as the ship swings in heading on an even keel. Plotting these deviations against compass heading yields the sine and cosine curves shown in Figure 604b. These deviation curves are called semicircular curves because they reverse direction by 180°.

A vector analysis is helpful in determining deviations

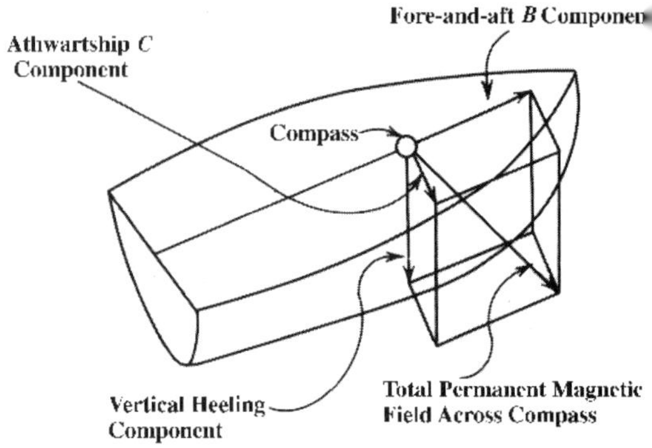

Figure 604a. Components of permanent magnetic field.

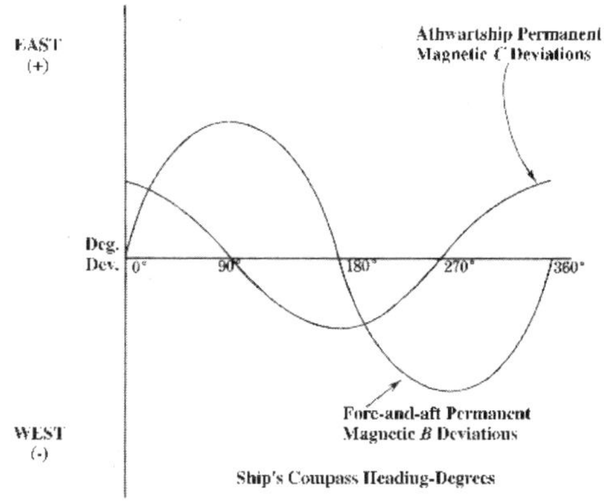

Figure 604b. Permanent magnetic deviation effects.

or the strength of deviating fields. For example, a ship as shown in Figure 604c on an east magnetic heading will subject its compass to a combination of magnetic effects; namely, the Earth's horizontal field H, and the deviating field B, at right angles to the field H. The compass needle will align itself in the resultant field which is represented by the vector sum of H and B, as shown. A similar analysis will reveal that the resulting directive force on the compass would be maximum on a north heading and minimum on a south heading because the deviations for both conditions are zero. The magnitude of the deviation caused by the permanent B magnetic field will vary with different values of H; hence, deviations resulting from permanent magnetic fields will vary with the magnetic latitude of the ship.

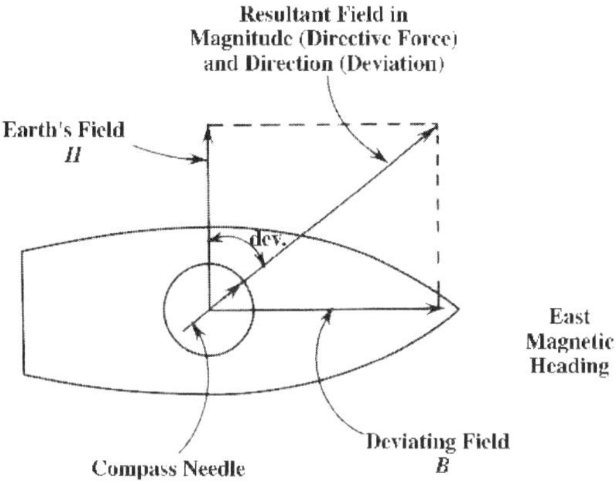

Figure 604c. General force diagram.

605. Effects of Induced Magnetism

Induced magnetism varies with the strength of the surrounding field, the mass of metal, and the alignment of the metal in the field. Since the intensity of the Earth's magnetic field varies over the Earth's surface, the induced magnetism in a ship will vary with latitude, heading, and heeling angle.

With the ship on an even keel, the resultant vertical induced magnetism, if not directed through the compass itself, will create deviations which plot as a semicircular deviation curve. This is true because the vertical induction changes magnitude and polarity only with magnetic latitude and heel, and not with heading of the ship. Therefore, as long as the ship is in the same magnetic latitude, its vertical induced pole swinging about the compass will produce the same effect on the compass as a permanent pole swinging about the compass.

The Earth's field induction in certain other unsymmetrical arrangements of horizontal soft iron create a constant A deviation curve. In addition to this magnetic A error, there are constant A deviations resulting from: (1) physical misalignments of the compass, pelorus, or gyro; (2) errors in calculating the Sun's azimuth, observing time, or taking bearings.

The nature, magnitude, and polarity of these induced effects are dependent upon the disposition of metal, the symmetry or asymmetry of the ship, the location of the binnacle, the strength of the Earth's magnetic field, and the angle of dip.

Certain heeling errors, in addition to those resulting from permanent magnetism, are created by the presence of both horizontal and vertical soft iron which experience changing induction as the ship rolls in the Earth's magnetic field. This part of the heeling error will change in magnitude proportional to changes of magnetic latitude of the ship. Oscillation effects associated with rolling are maximum on north and south headings, just as with the permanent magnetic heeling errors.

606. Adjustments and Correctors

Since some magnetic effects are functions of the vessel's magnetic latitude and others are not, each individual effect should be corrected independently. Furthermore, to make the corrections, we use (1) permanent magnet correctors to compensate for permanent magnetic fields at the compass, and (2) soft iron correctors to compensate for induced magnetism. The compass binnacle provides support for both the compass and its correctors. Typical large ship binnacles hold the following correctors:

1. Vertical permanent **heeling magnet** in the central vertical tube
2. Fore-and-aft **B permanent magnets** in their trays
3. Athwartship **C permanent magnets** in their trays
4. Vertical soft iron **Flinders bar** in its external tube
5. Soft iron **quadrantal spheres**

The heeling magnet is the only corrector which corrects for both permanent and induced effects. Therefore, it may need to be adjusted for changes in latitude if a vessel permanently changes its normal operating area. However, any movement of the heeling magnet will require readjustment of other correctors.

Fairly sophisticated magnetic compasses used on smaller commercial craft, larger yachts, and fishing vessels, may not have soft iron correctors or B and C permanent magnets. These compasses are adjusted by rotating magnets located inside the base of the unit, adjustable by small screws on the outside. A non-magnetic screwdriver is necessary to adjust these compasses. Occasionally one may find a permanent magnet corrector mounted near the compass, placed during the initial installation so as to remove a large, constant deviation before final adjustments are made. Normally, this remains in place for the life of the vessel.

Figure 606 summarizes all the various magnetic conditions in a ship, the types of deviation curves they create, the correctors for each effect, and headings on which each corrector is adjusted. When adjusting the compass, always apply the correctors symmetrically and as far away from the compass as possible. This preserves the uniformity of magnetic fields about the compass needle.

Occasionally, the permanent magnetic effects at the location of the compass are so large that they overcome the Earth's directive force, H. This condition will not only create sluggish and unsteady sectors, but may even freeze the compass to one reading or to one quadrant, regardless of the heading of the ship. Should the compass become so frozen, the polarity of the magnetism which must be attracting the compass needles is indicated; hence, correction may be effected simply by the application of permanent magnet

Coefficient	Type deviation curve	Compass headings of maximum deviation	Causes of such errors	Correctors for such errors	Magnetic or compass headings on which to apply correctors
A	Constant.	Same on all.	Human-error in calculations Physical-compass, gyro, pelorus alignment Magnetic-unsymmetrical arrangements of horiz. soft iron.	Check methods and calculations Check alignments Rare arrangement of soft iron rods.	Any.
B	Semicircular $\sin\phi$.	090° 270°	Fore-and-aft component of permanent magnetic field Induced magnetism in unsymmetrical vertical iron forward or aft of compass.	Fore-and-aft *B* magnets Flinders bar (forward or aft)	090° or 270°.
C	Semicircular $\cos\phi$.	000° 180°	Athwartship component of permanent magnetic field Induced magnetism in unsymmetrical vertical iron port or starboard of compass.	Athwartship *C* magnets Flinders bar (port or starboard)	000° or 180°.
D	Quadrantral $\sin 2\phi$.	045° 135° 225° 315°	Induced magnetism in all symmetrical arrangements of horizontal soft iron.	Spheres on appropriate axis. (athwartship for +D) (fore and aft for -D) *See sketch a*	045°, 135°, 225°, or 315°.
E	Quadrantral $\cos 2\phi$.	000° 090° 180° 270°	Induced magnetism in all unsymmetrical arrangements of horizontal soft iron.	Spheres on appropriate axis. (port fwd.-stb'd for +E) (stb'd fwd.-port aft for -E) *See sketch b*	000°, 090°, 180°, or 270°.
Heeling	Oscillations with roll or pitch. Deviations with constant list.	000°, 180° } roll 090°, 270° } pitch	Change in the horizontal component of the induced or permanent magnetic fields at the compass due to rolling or pitching of the ship.	Heeling magnet (must be readjusted for latitude changes).	090° or 270° with dip needle. 000° or 180° while rolling.

Deviation = $A + B\sin\phi + C\cos\phi + D\sin 2\phi + E\cos 2\phi$ (ϕ = compass heading)

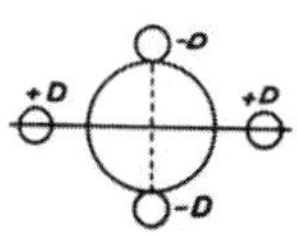

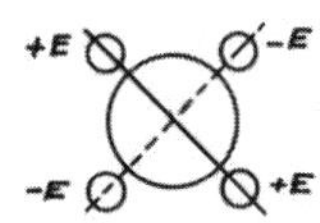

Figure 606. Summary of compass errors and adjustments.

correctors to neutralize this magnetism. Whenever such adjustments are made, the ship should be steered on a heading such that the unfreezing of the compass needles will be immediately evident. For example, a ship whose compass is frozen to a north reading would require fore-and-aft B corrector magnets with the positive ends forward in order to neutralize the existing negative pole which attracted the compass. If made on an east heading, such an adjustment would be evident when the compass card was freed to indicate an east heading.

607. Reasons for Correcting Compass

There are several reasons for correcting the errors of a magnetic compass, even if it is not the primary directional reference:

1. It is easier to use a magnetic compass if the deviations are small.
2. Even known and fully compensated deviation introduces error because the compass operates sluggishly and unsteadily when deviation is present.
3. Even though the deviations are compensated for, they will be subject to appreciable change as a function of heel and magnetic latitude.

Theoretically, it doesn't matter what the compass error is as long as it is known. But a properly adjusted magnetic compass is more accurate in all sea conditions, easier to steer by, and less subject to transient deviations which could result in deviations from the ship's chosen course.

Therefore, if a magnetic compass is installed and meant to be relied upon, it behooves the navigator to attend carefully to its adjustment. Doing so is known as "swinging ship."

608. Adjustment Check-off List

While a professional compass adjuster will be able to obtain the smallest possible error curve in the shortest time, many ship's navigators adjust the compass themselves with satisfactory results. Whether or not a "perfect" adjustment is necessary depends on the degree to which the magnetic compass will be relied upon in day-to-day navigation. If the magnetic compass is only used as a backup compass, removal of every last possible degree of error may not be worthwhile. If the magnetic compass is the only steering reference aboard, as is the case with many smaller commercial craft and fishing vessels, it should be adjusted as accurately as possible.

Prior to getting underway to swing ship, the navigator

must ensure that the process will proceed as expeditiously as possible by preparing the vessel and compass. The following tests and adjustment can be done at dockside, assuming that the compass has been installed and maintained properly. Initial installation and adjustment should be done by a professional compass technician during commissioning.

1. Check for bubbles in the compass bowl. Fluid may be added through the filling plug if necessary. Large bubbles indicate serious leakage, indicating that the compass should be taken to a professional compass repair facility for new gaskets.

2. Check for free movement of gimbals. Clean any dust or dirt from gimbal bearings and lubricate them as recommended by the maker.

3. Check for magnetization of the quadrantal spheres by moving them close to the compass and rotating them. If the compass needle moves more than 2 degrees, the spheres must be annealed to remove their magnetism. Annealing consists of heating the spheres to a dull red color in a non-magnetic area and allowing them to cool slowly to ambient temperature.

4. Check for magnetization of the Flinders bar by inverting it, preferably with the ship on an E/W heading. If the compass needle moves more than 2 degrees the Flinders bar must be annealed.

5. Synchronize the gyro repeaters with the master gyro so courses can be steered accurately.

6. Assemble past documentation relating to the compass and its adjustment. Have the ship's degaussing folder ready.

7. Ensure that every possible metallic object is stowed for sea. All guns, doors, booms, and other movable gear should be in its normal seagoing position. All gear normally turned on such as radios, radars, loudspeakers, etc. should be on while swinging ship.

8. Have the International Code flags Oscar-Quebec ready to fly.

Once underway to swing ship, the following procedures will expedite the process. Choose the best helmsman aboard and instruct him to steer each course as steadily and precisely as possible. Each course should be steered steadily for at least two minutes before any adjustments are made to remove Gaussin error. Be sure the gyro is set for the mean speed and latitude of the ship.

The navigator (or compass adjuster if one is employed) should have a pelorus and a table of azimuths prepared for checking the gyro, but the gyrocompass will be the primary steering reference. Normally the adjuster will request courses and move the magnets as he feels necessary, a process much more an art than a science. If a professional adjuster is not available, use the following sequence:

1. If there is a sea running, steer course 000° and adjust the heeling magnet to decrease oscillations to a minimum.

2. Come to course 090°. When steady on course 090°, for at least two minutes, insert, remove, or move fore-and-aft B magnets to remove ALL deviation.

3. Come to a heading of 180°. Insert, remove, or move athwartships C magnets to remove ALL deviation.

4. Come to 270° and move the B magnets to remove one half of the deviation.

5. Come to 000° and move the C magnets to remove one half of the deviation.

6. Come to 045° (or any intercardinal heading) and move the quadrantal spheres toward or away from the compass to minimize any error.

7. Come to 135° (or any intercardinal heading 90° from the previous course) and move the spheres in or out to remove one half of the observed error.

8. Steer the ship in turn on each cardinal and intercardinal heading around the compass, recording the error at each heading called for on the deviation card. If plotted, the errors should plot roughly as a sine curve about the 0° line.

If necessary, repeat steps 1-8. There is no average error, for each ship is different, but generally speaking, errors of more than a few degrees, or errors which seriously distort the sine curve, indicate a magnetic problem which should be addressed.

Once the compass has been swung, tighten all fittings and carefully record the placement of all magnets and correctors. Finally, swing for residual degaussed deviations with the degaussing circuits energized and record the deviations on the deviation card. Post this card near the chart table for ready reference by the navigation team.

Once properly adjusted, the magnetic compass deviations should remain constant until there is some change in the magnetic condition of the vessel resulting from magnetic treatment, shock, vibration, repair, or structural changes. Transient deviations are discussed below.

609. Sources of Transient Error

The ship must be in seagoing trim and condition to properly compensate a magnetic compass. Any movement of large metal objects or the energizing of any electrical equipment in the vicinity of the compass can cause errors. If in doubt about the effect of any such changes, temporarily move the gear or cycle power to the equipment while observing the compass card while on a steady heading. Preferably this should be done on two different headings 90° apart, since the compass might be affected on one heading and not on another.

Some magnetic items which cause deviations if placed too close to the compass are as follows:

1. Movable guns or weapon loads
2. Magnetic cargo
3. Hoisting booms
4. Cable reels
5. Metal doors in wheelhouse
6. Chart table drawers
7. Movable gyro repeater
8. Windows and ports
9. Signal pistols racked near compass
10. Sound powered telephones
11. Magnetic wheel or rudder mechanism
12. Knives or tools near binnacle
13. Watches, wrist bands, spectacle frames
14. Hat grommets, belt buckles, metal pencils
15. Heating of smoke stack or exhaust pipes
16. Landing craft

Some electrical items which cause variable deviations if placed too close to the compass are:

1. Electric motors
2. Magnetic controllers
3. Gyro repeaters
4. Nonmarried conductors
5. Loudspeakers
6. Electric indicators
7. Electric welding
8. Large power circuits
9. Searchlights or flashlights
10. Electrical control panels or switches
11. Telephone headsets
12. Windshield wipers
13. Rudder position indicators, solenoid type
14. Minesweeping power circuits
15. Engine order telegraphs
16. Radar equipment
17. Magnetically controlled switches
18. Radio transmitters
19. Radio receivers
20. Voltage regulators

Another source of transient deviation is the **retentive error**. This error results from the tendency of a ship's structure to retain induced magnetic effects for short periods of time. For example, a ship traveling north for several days, especially if pounding in heavy seas, will tend to retain some fore-and-aft magnetism induced under these conditions. Although this effect is transient, it may cause slightly incorrect observations or adjustments. This same type of error occurs when ships are docked on one heading for long periods of time. A short shakedown, with the ship on other headings, will tend to remove such errors. A similar sort of residual magnetism is left in many ships if the degaussing circuits are not secured by the correct reversal sequence.

A source of transient deviation somewhat shorter in duration than retentive error is known as **Gaussin error**. This error is caused by eddy currents set up by a changing number of magnetic lines of force through soft iron as the ship changes heading. Due to these eddy currents, the induced magnetism on a given heading does not arrive at its normal value until about 2 minutes after changing course.

Deperming and other magnetic treatment will change the magnetic condition of the vessel and therefore require compass readjustment. The decaying effects of deperming can vary. Therefore, it is best to delay readjustment for several days after such treatment. Since the magnetic fields used for such treatments are sometimes rather large at the compass locations, the Flinders bar, compass, and related equipment should be removed from the ship during these operations.

DEGAUSSING (MAGNETIC SILENCING) COMPENSATION

610. Degaussing

A steel vessel has a certain amount of **permanent magnetism** in its "hard" iron and **induced magnetism** in its "soft" iron. Whenever two or more magnetic fields occupy the same space, the total field is the vector sum of the individual fields. Thus, near the magnetic field of a vessel, the total field is the combined total of the Earth's field and the vessel's field. Not only does the Earth's field affect the vessel's, the vessel's field affects the Earth's field in its immediate vicinity.

Since certain types of explosive mines are triggered by the magnetic influence of a vessel passing near them, a vessel may use a degaussing system to minimize its magnetic field. One method of doing this is to neutralize each component of the field with an opposite field produced by electrical cables coiled around the vessel. These cables, when energized, counteract the permanent magnetism of the vessel, rendering it magnetically neutral. This has severe effects on magnetic compasses.

A unit sometimes used for measuring the strength of a magnetic field is the **gauss**. Reducing of the strength of a magnetic field decreases the number of gauss in that field. Hence, the process is called **degaussing**.

The magnetic field of the vessel is completely altered when the degaussing coils are energized, introducing large deviations in the magnetic compass. This deviation can be removed by introducing an equal and opposite force with energized coils near the compass. This is called **compass compensation**. When there is a possibility of confusion with compass adjustment to neutralize the effects of the natural magnetism of the vessel, the expression **degaussing compensation** is used. Since compensation may not be perfect, a small amount of deviation due to degaussing may remain on certain headings. This is the reason for swinging the ship with degaussing off and again with it on, and why there are two separate columns in the deviation table.

611. A Vessel's Magnetic Signature

A simplified diagram of the distortion of the Earth's magnetic field in the vicinity of a steel vessel is shown in Figure 611a. The field strength is directly proportional to the line spacing density. If a vessel passes over a device for detecting and recording the strength of the magnetic field, a certain pattern is traced. Figure 611b shows this pattern. Since the magnetic field of each vessel is different, each produces a distinctive trace. This distinctive trace is referred to as the vessel's **magnetic signature**.

Several **degaussing stations** have been established in major ports to determine magnetic signatures and recommend the currents needed in the various degaussing coils to render it magnetically neutral. Since a vessel's induced magnetism varies with heading and magnetic latitude, the current settings of the coils may sometimes need to be changed. A **degaussing folder** is provided to the vessel to indicate these changes and to document other pertinent information.

A vessel's permanent magnetism changes somewhat with time and the magnetic history of the vessel. Therefore, the data in the degaussing folder should be checked periodically at the magnetic station.

612. Degaussing Coils

For degaussing purposes, the total field of the vessel is divided into three components: (1) vertical, (2) horizontal fore-and-aft, and (3) horizontal athwartships. The positive (+) directions are considered downward, forward, and to port, respectively. These are the normal directions for a vessel headed north or east in north latitude.

Each component is opposed by a separate degaussing field just strong enough to neutralize it. Ideally, when this has been done, the Earth's field passes through the vessel smoothly and without distortion. The opposing degaussing fields are produced by direct current flowing in coils of wire. Each of the degaussing coils is placed so that the field it produces is directed to oppose one component of the ship's field.

The number of coils installed depends upon the magnetic characteristics of the vessel, and the degree of safety desired. The ship's permanent and induced magnetism may be neutralized separately so that control of induced magnetism can be varied as heading and latitude change, without disturbing the fields opposing the vessel's permanent field. The principal coils employed are the following:

Main (M) coil. The M coil is horizontal and completely encircles the vessel, usually at or near the waterline. Its function is to oppose the vertical component of the vessel's combined permanent and induced fields. Generally the induced field predominates. Current in the M-coil is varied or reversed according to the change of the induced component of the vertical field with latitude.

Forecastle (F) and **quarterdeck (Q) coils**. The F and Q coils are placed horizontally just below the forward and after thirds (or quarters), respectively, of the weather deck. These coils, in which current can be individually adjusted, remove much of the fore-and-aft component of the ship's permanent and induced fields. More commonly, the combined F and Q coils consist of two parts; one part the FP and QP coils, to take care of the permanent fore-and-aft field, and the other part, the FI and QI coils, to neutralize the induced fore-and-aft field. Generally, the forward and after coils of each type are connected in series, forming a split-coil installation and designated FP-QP coils and FI-QI coils. Current in the FP-QP coils is generally constant, but in the FI-QI coils is varied according to the heading and magnetic latitude of the vessel. In split-coil installations, the coil designations are often called simply the P-coil and I-coil.

Longitudinal (L) coil. Better control of the fore-and-aft components, but at greater installation expense, is provided by placing a series of vertical, athwartship coils along the length of the ship. It is the field, not the coils, which is longitudinal. Current in an L coil is varied as with the FI-QI coils. It is maximum on north and south headings, and zero on east and west headings.

Athwartship (A) coil. The A coil is in a vertical fore-and-aft plane, thus producing a horizontal athwartship field which neutralizes the athwartship component of the vessel's field. In most vessels, this component of the permanent field is small and can be ignored. Since the A-coil neutralizes the induced field, primarily, the current is changed with magnetic latitude and with heading, maximum on east or west headings, and zero on north or south headings.

The strength and direction of the current in each coil is indicated and adjusted at a control panel accessible to the navigator. Current may be controlled directly by rheostats at the control panel or remotely by push buttons which operate rheostats in the engine room.

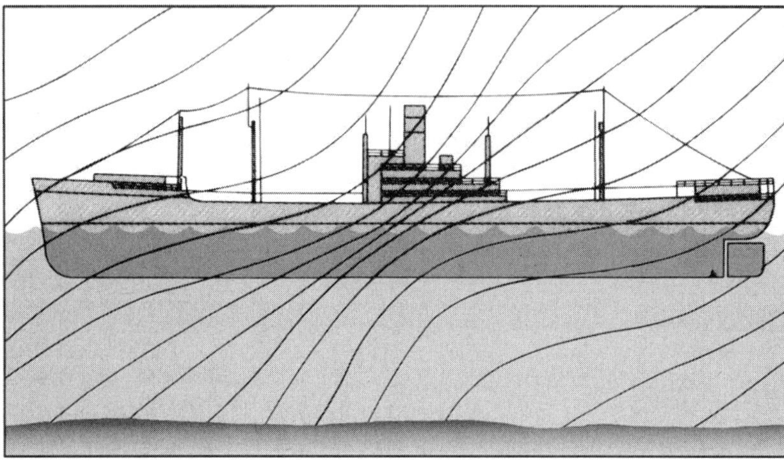

Figure 611a. Simplified diagram of distortion of Earth's magnetic field in the vicinity of a steel vessel.

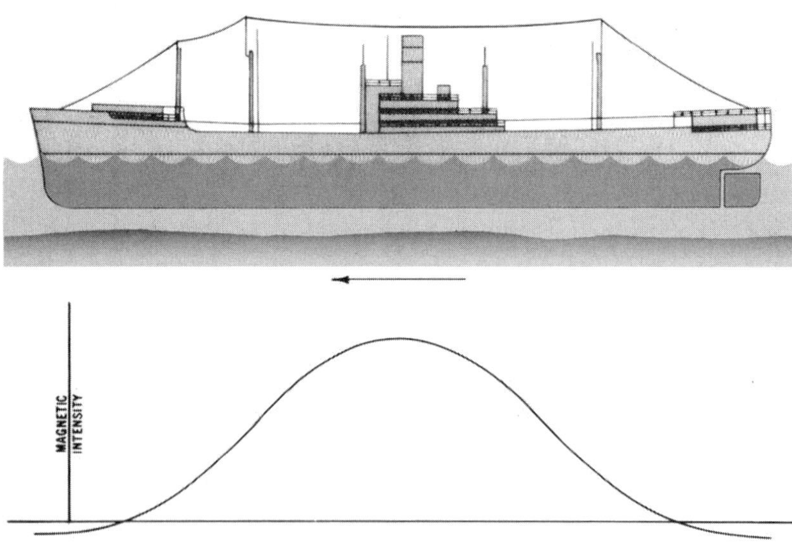

Figure 611b. A simplified signature of a vessel of Figure 611a.

Appropriate values of the current in each coil are determined at a degaussing station, where the various currents are adjusted until the vessel's magnetic signature is made as flat as possible. Recommended current values and directions for all headings and magnetic latitudes are set forth in the vessel's degaussing folder. This document is normally kept by the navigator, who must see that the recommended settings are maintained whenever the degaussing system is energized.

613. Securing The Degaussing System

Unless the degaussing system is properly secured, residual magnetism may remain in the vessel. During degaussing compensation and at other times, as recommended in the degaussing folder, the "reversal" method is used. The steps in the reversal process are as follows:

1. Start with maximum degaussing current used since the system was last energized.
2. Decrease current to zero and increase it in the opposite direction to the same value as in step 1.
3. Decrease the current to zero and increase it to three-fourths maximum value in the original direction.
4. Decrease the current to zero and increase it to one-half maximum value in the opposite direction.
5. Decrease the current to zero and increase it to one-fourth maximum value in the original direction.
6. Decrease the current to zero and increase it to one-eighth maximum value in the opposite direction.
7. Decrease the current to zero and open switch.

614. Magnetic Treatment Of Vessels

In some instances, degaussing can be made more effective by changing the magnetic characteristics of the vessel by a process known as **deperming**. Heavy cables are wound around the vessel in an athwartship direction, forming vertical loops around the longitudinal axis of the vessel. The loops are run beneath the keel, up the sides, and over the top of the weather deck at closely spaced equal intervals along the entire length of the vessel. Predetermined values of direct current are then passed through the coils. When the desired magnetic characteristics have been acquired, the cables are removed.

A vessel which does not have degaussing coils, or which has a degaussing system which is inoperative, can be given some temporary protection by a process known as **flashing**. A horizontal coil is placed around the outside of the vessel and energized with large predetermined values of direct current. When the vessel has acquired a vertical field of permanent magnetism of the correct magnitude and polarity to reduce to a minimum the resultant field below the vessel for the particular magnetic latitude involved, the cable is removed. This type protection is not as satisfactory as that provided by degaussing coils because it is not adjustable for various headings and magnetic latitudes, and also because the vessel's magnetism slowly readjusts following treatment.

During magnetic treatment all magnetic compasses and Flinders bars should be removed from the ship. Permanent adjusting magnets and quadrantal correctors are not materially affected, and need not be removed. If it is impractical to remove a compass, the cables used for magnetic treatment should be kept as far as practical from it.

615. Degaussing Effects

The degaussing of ships for protection against magnetic mines creates additional effects upon magnetic compasses, which are somewhat different from the permanent and induced magnetic effects. The degaussing effects are electromagnetic, and depend on:

1. Number and type of degaussing coils installed.
2. Magnetic strength and polarity of the degaussing coils.
3. Relative location of the different degaussing coils with respect to the binnacle.
4. Presence of masses of steel, which would tend to concentrate or distort magnetic fields in the vicinity of the binnacle.
5. The fact that degaussing coils are operated intermittently, with variable current values, and with different polarities, as dictated by necessary degaussing conditions.

616. Degaussing Compensation

The magnetic fields created by the degaussing coils would render the vessel's magnetic compasses useless unless compensated. This is accomplished by subjecting the compass to compensating fields along three mutually perpendicular axes. These fields are provided by small **compensating coils** adjacent to the compass. In nearly all installations, one of these coils, the **heeling coil**, is horizontal and on the same plane as the compass card, providing a vertical compensating field. Current in the heeling coil is adjusted until the vertical component of the total degaussing field is neutralized. The other compensating coils provide horizontal fields perpendicular to each other. Current is varied in these coils until their resultant field is equal and opposite to the horizontal component of the degaussing field. In early installations, these horizontal fields were directed fore-and-aft and athwartships by placing the coils around the Flinders bar and the quadrantal spheres. Compactness and other advantages are gained by placing the coils on perpendicular axes extending 045°-225° and 315°-135° relative to the heading. A frequently used compensating installation, called the **type K**, is shown in Figure 616. It consists of a heeling coil extending completely around the top of the binnacle, four **intercardinal coils**, and three control boxes. The intercardinal coils are named for their positions relative to the compass when the vessel is on a heading of north, and also for the compass headings on which the current in the coils is adjusted to the correct amount for compensation. The NE-SW coils operate together as one set, and the NW-SE coils operate as another. One control box is provided for each set, and one for the heeling coil.

The compass compensating coils are connected to the power supply of the degaussing coils, and the currents passing through the compensating coils are adjusted by series resistances so that the compensating field is equal to the degaussing field. Thus, a change in the degaussing currents is accompanied by a proportional change in the compensating currents. Each coil has a separate winding for each degaussing circuit it compensates.

Degaussing compensation is carried out while the vessel is moored at the shipyard where the degaussing coils are installed. This process is usually carried out by civilian professionals, using the following procedure:

Step 1. The compass is removed from its binnacle and a dip needle is installed in its place. The M coil and heeling coil are then energized, and the current in the heeling coil is adjusted until the dip needle indicates the correct value for the magnetic latitude of the vessel. The system is then secured by the reversing process.

Step 2. The compass is replaced in the binnacle. With auxiliary magnets, the compass card is deflected until the compass magnets are parallel to one of the compensating coils or set of coils used to produce a horizontal field. The compass magnets are then perpendicular to the field produced by that coil. One of the degaussing circuits producing a horizontal field, and its compensating winding, are then energized, and the current in the compensating winding is adjusted until the compass reading returns to the value it had before the degaussing circuit was energized. The system is then secured by the reversing process. The process is repeated with each additional circuit used to create a horizontal field. The auxiliary magnets are then removed.

Step 3. The auxiliary magnets are placed so that the compass magnets are parallel to the other compensating coils or set of coils used to produce a horizontal field. The procedure of step 2 is then repeated for each circuit producing a horizontal field.

When the vessel gets under way, it proceeds to a suitable maneuvering area. The vessel is then steered so that the compass magnets are parallel first to one compensating coil or set of coils, and then the other. Any needed adjustment is made in the compensating circuits to reduce the error to a

Figure 616. Type K degaussing compensation installation.

minimum. The vessel is then swung for residual deviation, first with degaussing off and then with degaussing on, and the correct current settings determined for each heading at the magnetic latitude of the vessel. From the values thus obtained, the "DG OFF" and "DG ON" columns of the deviation table are filled in. If the results indicate satisfactory compensation, a record is made of the degaussing coil settings and the resistance, voltages, and currents in the compensating coil circuits. The control boxes are then secured.

Under normal operating conditions, the settings do not need to be changed unless changes are made in the degaussing system, or unless an alteration is made in the length of the Flinders bar or the setting of the quadrantal spheres. However, it is possible for a ground to occur in the coils or control box if the circuits are not adequately protected from moisture. If this occurs, it should be reflected by a change in deviation with degaussing on, or by a decreased installation resistance. Under these conditions, compensation should be done again. If the compass will be used with degaussing on before the ship can be returned to a shipyard where the compensation can be made by experienced personnel, the compensation should be made at sea on the actual headings needed, rather than by deflection of the compass needles by magnets. More complete information related to this process is given in the degaussing folder.

If a vessel has been given magnetic treatment, its magnetic properties have changed, necessitating readjustment of each magnetic compass. This is best delayed for several days to permit the magnetic characteristics of the vessel to settle. If compensation cannot be delayed, the vessel should be swung again for residual deviation after a few days. Degaussing compensation should not be made until after compass adjustment has been completed.

GYROCOMPASSES

617. Principles of the Gyroscope

A gyroscope consists of a spinning wheel or rotor contained within gimbals which permit movement about three mutually perpendicular axes, known as the **horizontal axis**, the **vertical axis**, and the **spin axis**. When spun rapidly, assuming that friction is not considered, the gyroscope develops **gyroscopic inertia**, tending to remain spinning in the same plane indefinitely. The amount of gyroscopic inertia depends on the angular velocity, mass, and radius of the wheel or rotor.

When a force is applied to change alignment of the spin axis of a gyroscope, the resultant motion is perpendicular to the direction of the force. This tendency is known as precession. A force applied to the center of gravity of the gyroscope will move the entire system in the direction of the force. Only a force that tends to change the axis of rotation produces precession.

If a gyroscope is placed at the equator with its spin axis pointing east-west, as the earth turns on its axis, gyroscopic inertia will tend to keep the plane of rotation constant. To the observer, it is the gyroscope which is seen to rotate, not the earth. This effect is called the horizontal earth rate, and is maximum at the equator and zero at the poles. At points between, it is equal to the cosine of the latitude.

If the gyro is placed at a geographic pole with its spin axis horizontal, it will appear to rotate about its vertical axis. This is the vertical earth rate. At all points between the equator and the poles, the gyro appears to turn partly about its horizontal and partly about its vertical axis, being affected by both horizontal and vertical earth rates. In order to visualize these effects, remember that the gyro, at whatever latitude it is placed, is remaining aligned in space while the earth moves beneath it.

618. Gyrocompass Operation

The gyrocompass depends upon four natural phenomena: gyroscopic inertia, precession, earth's rotation, and gravity. To make a gyroscope into a gyrocompass, the wheel or rotor is mounted in a sphere, called the gyrosphere, and the sphere is then supported in a vertical ring. The whole is mounted on a base called the phantom. The gyroscope in a gyrocompass can be pendulous or non-pendulous, according to design. The rotor may weigh as little as half a kilogram to over 25 kg.

To make it seek and maintain true north, three things are necessary. First, the gyro must be made to stay on the plane of the meridian. Second, it must be made to remain horizontal. Third, it must stay in this position once it reaches it regardless of what the vessel on which it is mounted does or where it goes on the earth. To make it seek the meridian, a weight is added to the bottom of the vertical ring, causing it to swing on its vertical axis, and thus seek to align itself horizontally. It will tend to oscillate, so a second weight is added to the side of the sphere in which the rotor is contained, which dampens the oscillations until the gyro stays on the meridian. With these two weights, the only possible position of equilibrium is on the meridian with its spin axis horizontal.

To make the gyro seek north, a system of reservoirs filled with mercury, known as mercury ballistics, is used to apply a force against the spin axis. The ballistics, usually four in number, are placed so that their centers of gravity exactly coincide with the CG of the gyroscope. Precession then causes the spin axis to trace an ellipse, one ellipse taking about 84 minutes to complete. (This is the period of oscillation of a pendulum with an arm equal to the radius of the earth.) To dampen this oscillation, the force is applied, not in the vertical plane, but slightly to the east of the vertical plane. This causes the spin axis to trace a spiral instead of an ellipse and eventually settle on the meridian pointing north.

619. Gyrocompass Errors

The total of the all the combined errors of the gyrocompass is called **gyro error** and is expressed in degrees E or W, just like variation and deviation. But gyro error, unlike magnetic compass error, and being independent of Earth's magnetic field, will be constant in one direction; that is, an error of one degree east will apply to all bearings all around the compass.

The errors to which a gyrocompass is subject are speed error, latitude error, ballistic deflection error, ballistic damping error, quadrantal error, and gimballing error. Additional errors may be introduced by a malfunction or incorrect alignment with the centerline of the vessel.

Speed error is caused by the fact that a gyrocompass only moves directly east or west when it is stationary (on the rotating earth) or placed on a vessel moving exactly east or west. Any movement to the north or south will cause the compass to trace a path which is actually a function of the speed of advance and the amount of northerly or southerly heading. This causes the compass to tend to settle a bit off true north. This error is westerly if the vessel's course is northerly, and easterly if the course is southerly. Its magnitude depends on the vessel's speed, course, and latitude. This error can be corrected internally by means of a cosine cam mounted on the underside of the azimuth gear, which removes most of the error. Any remaining error is minor in amount and can be disregarded.

Tangent latitude error is a property only of gyros with mercury ballistics, and is easterly in north latitudes and westerly in south latitudes. This error is also corrected internally, by offsetting the lubber's line or with a small movable weight attached to the casing.

Ballistic deflection error occurs when there is a marked change in the north-south component of the speed. East-west accelerations have no effect. A change of course or speed also results in speed error in the opposite direction, and the two tend to cancel each other if the compass is properly designed. This aspect of design involves slightly offsetting the ballistics according to the operating latitude, upon which the correction is dependent. As latitude changes, the error becomes apparent, but can be minimized by adjusting the offset.

Ballistic damping error is a temporary oscillation introduced by changes in course or speed. During a change in course or speed, the mercury in the ballistic is subjected to centrifugal and acceleration/deceleration forces. This causes a torquing of the spin axis and subsequent error in the compass reading. Slow changes do not introduce enough error to be a problem, but rapid changes will. This error is counteracted by changing the position of the ballistics so that the true vertical axis is centered, thus not subject to error, but only when certain rates of turn or acceleration are exceeded.

Quadrantal error has two causes. The first occurs if the center of gravity of the gyro is not exactly centered in the phantom. This causes the gyro to tend to swing along its heavy axis as the vessel rolls in the sea. It is minimized by adding weight so that the mass is the same in all directions from the center. Without a long axis of weight, there is no tendency to swing in one particular direction. The second source of quadrantal error is more difficult to eliminate. As a vessel rolls in the sea, the apparent vertical axis is displaced, first to one side and then the other. The vertical axis of the gyro tends to align itself with the apparent vertical. On northerly or southerly courses, and on easterly or westerly courses, the compass precesses equally to both sides and the resulting error is zero. On intercardinal courses, the N-S and E-W precessions are additive, and a persistent error is introduced, which changes direction in different quadrants. This error is corrected by use of a second gyroscope called a floating ballistic, which stabilizes the mercury ballistic as the vessel rolls, eliminating the error. Another method is to use two gyros for the directive element, which tend to precess in opposite directions, neutralizing the error.

Gimballing error is caused by taking readings from the compass card when it is tilted from the horizontal plane. It applies to the compass itself and to all repeaters. To minimize this error, the outer ring of the gimbal of each repeater should be installed in alignment with the fore-and-aft line of the vessel. Of course, the lubber's line must be exactly centered as well.

620. Using the Gyrocompass

Since a gyrocompass is not influenced by magnetism, it is not subject to variation or deviation. Any error is constant and equal around the horizon, and can often be reduced to less than one degree, thus effectively eliminating it altogether. Unlike a magnetic compass, it can output a signal to repeaters spaced around the vessel at critical positions.

But it also requires a constant source of stable electrical power, and if power is lost, it requires several hours to settle on the meridian again before it can be used. This period can be reduced by aligning the compass with the meridian before turning on the power.

The directive force of a gyrocompass depends on the amount of precession to which it is subject, which in turn is dependent on latitude. Thus the directive force is maximum at the equator and decreases to zero at the poles. Vessels operating in high latitudes must construct error curves based on latitudes because the errors at high latitudes eventually overcome the ability of the compass to correct them.

The gyrocompass is typically located below decks as close as possible to the center of roll, pitch and yaw of the ship, thus minimizing errors caused by the ship's motion. Repeaters are located at convenient places throughout the ship, such as at the helm for steering, on the bridge wings for taking bearings, in after steering for emergency steering, and other places. The output can also be used to drive course recorders, autopilot systems, plotters, fire control systems, and stabilized radars. The repeaters should be checked regularly against the master to ensure they are all in alignment. The repeaters on the bridge wing used for taking bearings will likely be equipped with removable bearing circles, azimuth circles, and telescopic alidades, which allow one to sight a distant object and see its exact gyrocompass bearing.

ELECTRONIC COMPASSES

621. New Direction Sensing Technologies

The magnetic compass has serious limitations, chiefly that of being unable to isolate the earth's magnetic field from all others close enough to influence it. It also indicates magnetic north, whereas the mariner is most interested in true north. Most of the work involved with compensating a traditional magnetic compass involves neutralizing magnetic influences other than the earth's, a complicated and inexact process often involving more art than science. Residual error is almost always present even after compensation. Degaussing complicates the situation immensely.

The electro-mechanical gyrocompass has been the standard steering and navigational compass since the early 20th century, and has provided several generations of mariners a stable and reliable heading and bearing reference. However, it too has limitations: It is a large, expensive, heavy, sensitive device that must be mounted according to rather strict limitations. It requires a stable and uninterrupted supply of electrical power; it is sensitive to shock, vibration, and environmental changes; and it needs several hours to settle after being turned on.

Fortunately, several new technologies have been developed which promise to greatly reduce or eliminate the limitations of both the mechanical gyroscope and traditional magnetic compasses. Sometimes referred to as "electronic compasses," the digital flux gate magnetic compass and the ring laser gyrocompass are two such devices. They have the following advantages:

1. Solid state electronics, no moving parts
2. Operation at very low power
3. Easy backup power from independent sources
4. Standardized digital output
5. Zero friction, drift, or wear
6. Compact, lightweight, and inexpensive
7. Rapid start-up and self-alignment
8. Low sensitivity to vibration, shock, and temperature changes
9. Self-correcting

Both types are being installed on many vessels as the primary directional reference, enabling the decommissioning of the traditional magnetic compasses and the avoidance of periodic compensation and maintenance.

622. The Flux Gate Compass

The most widely used sensor for digital compasses is the flux-gate magnetometer, developed around 1928. Initially it was used for detecting submarines, for geophysical prospecting, and airborne mapping of earth's magnetic fields.

The most common type, called the second harmonic device, incorporates two coils, a primary and a secondary, both wrapped around a single highly permeable ferromagnetic core. In the presence of an external magnetic field, the core's magnetic induction changes. A signal applied to the primary winding causes the core to oscillate. The secondary winding emits a signal that is induced through the core from the primary winding. This induced signal is affected by changes in the permeability of the core and appears as an amplitude variation in the output of the sensing coil. The signal is then demodulated with a phase-sensitive detector and filtered to retrieve the magnetic field value. After being converted to a standardized digital format, the data can be output to numerous remote devices, including steering compasses, bearing compasses, emergency steering stations, and autopilots.

Since the influence of a ship's inherent magnetism is inversely proportional to the square of the distance to the compass, it is logical that if the compass could be located at some distance from the ship, the influence of the ship's magnetic field could be greatly reduced. One advantage of the flux gate compass is that the sensor can be located remotely from the readout device, allowing it to be placed at a position as far as possible from the hull and its contents, such as high up on a mast, the ideal place on most vessels.

A further advantage is that the digital signal can be processed mathematically, and algorithms written which can correct for observed deviation once the deviation table has been determined. Further, the "table," in digital format, can be found by merely steering the vessel in a full circle. Algorithms then determine and apply corrections that effectively flatten the usual sine wave pattern of deviation. The theoretical result is zero observed compass deviation.

Should there be an index error (which has the effect of skewing the entire sine wave below or above the zero degree axis of the deviation curve) this can be corrected with an index correction applied to all the readings. This problem is largely confined to asymmetric installations such as aircraft carriers. Similarly, a correction for variation can be applied, and with GPS input (so the system knows where it is with respect to the isogonic map) the variation correction can be applied automatically, thus rendering the output in true degrees, corrected for both deviation and variation.

It is important to remember that a flux gate compass is still a magnetic compass, and that it will be influenced by large changes to the ship's magnetic field. Compensation should be accomplished after every such change. Fortunately, as noted, compensation involves merely steering the vessel in a circle in accordance with the manufacturer's recommendations.

Flux-gate compasses from different manufacturers share some similar operational modes. Most of them will have the following:

SET COURSE MODE: A course can be set and "remembered" by the system, which then provides the helmsman a graphic steering aid, enabling him to see if the ship's head is right or left of the set course, as if on a digital "highway." Normal compass operation continues in the background.

DISPLAY RESPONSE DAMPING: In this mode, a switch is used to change the rate of damping and update of the display in response to changes in sea condition and vessel speed.

AUTO-COMPENSATION: This mode is used to determine the deviation curve for the vessel as it steams in a complete circle. The system will then automatically compute correction factors to apply around the entire compass, resulting in zero deviation at any given heading. This should be done after every significant change in the magnetic signature of the ship, and within 24 hours of entering restricted waters.

CONTINUOUS AUTO-COMPENSATION: This mode, which should normally be turned OFF in restricted waters and ON at sea, runs the compensation algorithm each time the ship completes a 360 degree turn in two minutes. A warning will flash on the display in the OFF mode.

PRE-SET VARIATION: In effect an index correction, pre-set variation allows the application of magnetic variation to the heading, resulting in a true output (assuming the unit has been properly compensated and aligned). Since variation changes according to one's location on the earth, it must be changed periodically to agree with the charted variation unless GPS input is provided. The GPS position input is used in an algorithm which computes the variation for the area and automatically corrects the readout.

U.S. Naval policy approves the use of flux gate compasses and the decommissioning, but not removal, of the traditional binnacle mounted compass, which should be clearly marked as "Out of Commission" once an approved flux gate compass is properly installed and tested.

623. The Ring Laser Gyrocompass

The ring laser had its beginnings in England, where in the 1890's two scientists, Joseph Larmor and Sir Oliver Lodge (also one of the pioneers of radio), debated the possibility of measuring rotation by a ring interferometer. Some 15 years later, a French physicist, Georges Sagnac, fully described the phenomenon which today bears his name, the Sagnac Effect. This principle states that if two beams of light are sent in opposite directions around a "ring" or polyhedron and steered so as to meet and combine, a standing wave will form around the ring. If the wave is observed from any point, and that point is then moved along the perimeter of the ring, the wave form will change in direct relationship to the direction and velocity of movement.

It wasn't until 1963 that W. Macek of Sperry-Rand Corporation tested and refined the concept into a useful research device. Initially, mirrors were used to direct light around a square or rectangular pattern. But such mirrors must be made and adjusted to exceptionally close tolerances to allow useful output, and must operate in a vacuum for best effect. Multilayer dielectric mirrors with a reflectivity of 99.9999 percent were developed. The invention of laser light sources and fiber-optics has enabled the production of small, light, and dependable ring laser gyros. Mirror-based devices continue to be used in physics research.

The ring laser gyrocompass (RLG) operates by measuring laser-generated light waves traveling around a fiber-optic ring. A beam splitter divides a beam of light into two counter-rotating waves, which then travel around the fiber-optic ring in opposite directions. The beams are then recombined and sent to an output detector. In the absence of rotation, the path lengths will be the same and the beams will recombine in phase. If the device has rotated, there will be a difference in the length of the paths of the two beams, resulting in a detectable phase difference in the combined signal. The signal will vary in amplitude depending on the amount of the phase shift. The amplitude is thus a measurement of the phase shift, and consequently, the rotation rate. This signal is processed into a digital readout in degrees. This readout, being digital, can then be sent to a variety of devices which need heading information, such as helm, autopilot, and electronic chart systems.

A single ring laser gyroscope can be used to provide a one-dimensional rotational reference, exactly what a compass needs. The usefulness of ring laser gyrocompasses stems from that fact that they share many of the same characteristics of flux gate compasses. They are compact, light, inexpensive, accurate, dependable, and robust. The ring laser device is also quite immune to magnetic influences which would send a traditional compass spinning hopelessly, and might adversely affect even the remotely mounted flux gate compass.

Ring laser gyroscopes can also serve as the stable elements in an inertial guidance system, using three gyros to represent the three degrees of freedom, thus providing both directional and position information. The principle of operation is the same as for mechanical inertial navigation devices, in that a single gyro can measure any rotation about its own axis. This implies that its orientation in space about its own axis will be known at all times. Three gyros arranged along three axes each at 90 degrees to the others can measure accelerations in three dimensional space, and thus track movement over time.

Inertial navigation systems based on ring lasers have been used in aircraft for a number of years, and are becoming increasingly common in maritime applications. Uses include navigation, radar and fire control systems, precise weapons stabilization, and stabilization of directional sensors such as satellite antennas.

CORRECTING AND UNCORRECTING THE COMPASS

624. Ship's Heading

Ship's heading is the angle, expressed in degrees clockwise from north, of the ship's fore-and-aft line with respect to the true meridian or the magnetic meridian. When this angle is referred to the true meridian, it is called a **true heading**. When this angle is referred to the magnetic meridian, it is called a **magnetic heading**. Heading, as indicated on a particular compass, is termed the ship's compass heading by that compass. It is essential to specify every heading as true (T), magnetic (M), or compass. Two abbreviations simplify recording of compass directions. The abbreviation PGC refers to "per gyro compass," and PSC refers to "per steering compass." The steering compass is the one being used by the helmsman or autopilot, regardless of type.

625. Variation And Deviation

Variation is the angle between the magnetic meridian and the true meridian at a given location. If the northerly part of the magnetic meridian lies to the right of the true meridian, the variation is easterly. Conversely, if this part is to the left of the true meridian, the variation is westerly. The local variation and its small **annual change** are noted on the compass rose of all navigational charts. Thus the true and magnetic headings of a ship differ by the local variation.

As previously explained, a ship's magnetic influence will generally cause the compass needle to deflect from the magnetic meridian. This angle of deflection is called **deviation**. If the north end of the needle points east of the magnetic meridian, the deviation is easterly; if it points west of the magnetic meridian, the deviation is westerly.

626. Heading Relationships

A summary of heading relationships follows:

1. **Deviation** is the difference between the compass heading and the magnetic heading.
2. **Variation** is the difference between the magnetic heading and the true heading.
3. The algebraic sum of deviation and variation is the **compass error**.

The following simple rules will assist in correcting and uncorrecting the compass:

1. Compass least, error east; compass best, error west.
2. When correcting, add easterly errors, subtract westerly errors (Remember: "Correcting Add East").
3. When uncorrecting, subtract easterly errors, add westerly errors.

Some typical correction operations follow:

Compass	Deviation	Magnetic	Variation	True
		-> +E, -W		
358°	5°E	003°	6°E	009°
120°	1°W	119°	3°E	122°
180°	6°E	186°	8°W	178°
240°	5°W	235°	7°W	228°
		+W, -E <-		

Figure 626. Examples of compass correcting.

Use the memory aid "Can Dead Men Vote Twice, At Elections" to remember the conversion process (Compass, Deviation, Magnetic, Variation, True; Add East). When converting compass heading to true heading, add easterly deviations and variations and subtract westerly deviations and variations.

The same rules apply to correcting gyrocompass errors, although gyro errors always apply in the same direction. That is, they are E or W all around the compass.

Complete familiarity with the correcting of compasses is essential for navigation by magnetic or gyro compass. The professional navigator who deals with them continually can do them in his head quickly and accurately.

CHAPTER 7

DEAD RECKONING

DEFINITION AND PURPOSE

700. Definition and Use

Dead reckoning is the process of determining one's present position by projecting course(s) and speed(s) from a known past position, and predicting a future position by projecting course(s) and speed(s) from a known present position. The DR position is only an approximate position because it does not allow for the effect of leeway, current, helmsman error, or compass error.

Dead reckoning helps in determining sunrise and sunset; in predicting landfall, sighting lights and predicting arrival times; and in evaluating the accuracy of electronic positioning information. It also helps in predicting which celestial bodies will be available for future observation. But its most important use is in projecting the position of the ship into the immediate future and avoiding hazards to navigation.

The navigator should carefully tend his DR plot, update it when required, use it to evaluate external forces acting on his ship, and consult it to avoid potential navigation hazards. A fix taken at each DR position will reveal the effects of current, wind, and steering error, and allow the navigator to stay on track by correcting for them.

The use of DR when an Electronic Charts Display and Information System (ECDIS) is the primary plotting method will vary with the type of system. An ECDIS allows the display of the ship's heading projected out to some future position as a function of time, the display of waypoint information, and progress toward each waypoint in turn.

Until ECDIS is proven to provide the level of safety and accuracy required, the use of a traditional DR plot on paper charts is a prudent backup, especially in restricted waters. The following procedures apply to DR plotting on the traditional paper chart.

CONSTRUCTING THE DEAD RECKONING PLOT

Maintain the DR plot directly on the chart in use. DR at least two fix intervals ahead while piloting. If transiting in the open ocean, maintain the DR at least four hours ahead of the last fix position. Maintaining the DR plot directly on the chart allows the navigator to evaluate a vessel's future position in relation to charted navigation hazards. It also allows the conning officer and captain to plan course and speed changes required to meet any operational commitments.

This section will discuss how to construct the DR plot.

701. Measuring Courses and Distances

To measure courses, use the chart's compass rose nearest to the chart area currently in use. Transfer course lines to and from the compass rose using parallel rulers, rolling rulers, or triangles. If using a parallel motion plotter (PMP), simply set the plotter at the desired course and plot that course directly on the chart. Transparent plastic navigation plotters that align with the latitude/longitude grid may also be used.

The navigator can measure direction at any convenient place on a Mercator chart because the meridians are parallel to each other and a line making an angle with any one makes the same angle with all others. One must measure direction on a conformal chart having nonparallel meridians at the meridian closest to the area of the chart in use. The only common nonconformal projection used is the gnomonic; a gnomonic chart usually contains instructions for measuring direction.

Compass roses may give both true and magnetic directions. True directions are on the outside of the rose; magnetic directions are on the inside. For most purposes, use true directions.

Measure distances using the chart's latitude scale. Although not technically true, assuming that one minute of latitude equals one nautical mile introduces no significant error. Since the Mercator chart's latitude scale expands as latitude increases, on small scale charts one must measure distances on the latitude scale closest to the area of interest, that is, at the same latitude, or directly to the side. On large scale charts, such as harbor charts, one can use either the latitude scale or the distance scale provided. To measure long distances on small-scale charts, break the distance into a number of segments and measure each segment at its mid-latitude.

702. Plotting and Labeling the Course Line and Positions

Draw a new **course line** whenever restarting the DR. Extend the course line from a fix in the direction of the ordered course. Above the course line place a capital C followed by the ordered course in degrees true. Below the course line, place a capital S followed by the speed in knots. Label all course lines and fixes immediately after plotting them because a conning officer or navigator can easily misinterpret an unlabeled line or position.

Enclose a fix from two or more Lines of Position (LOP's) by a small circle and label it with the time to the nearest minute, written horizontally. Mark a DR position with a semicircle and the time, written diagonally. Mark an **estimated position (EP)** by a small square and the time, written horizontally. Determining an EP is covered later in this chapter.

Express the time using four digits without punctuation, using either zone time or Greenwich Mean Time (GMT), according to procedure. Label the plot neatly, succinctly, and clearly.

Figure 702. A course line with labels.

Figure 702 illustrates this process. The navigator plots and labels the 0800 fix. The conning officer orders a course of 095°T and a speed of 15 knots. The navigator extends the course line from the 0800 fix in a direction of 095°T. He calculates that in one hour at 15 knots he will travel 15 nautical miles. He measures 15 nautical miles from the 0800 fix position along the course line and marks that point on the course line with a semicircle. He labels this DR with the time. Note that, by convention, he labels the fix time horizontally and the DR time diagonally.

THE RULES OF DEAD RECKONING

703. Plotting the DR

Plot the vessel's DR position:

1. At least every hour on the hour.
2. After every change of course or speed.
3. After every fix or running fix.
4. After plotting a single line of position.

Figure 703 illustrates applying these rules. Clearing the harbor at 0900, the navigator obtains a last visual fix. This is called **taking departure**, and the position determined is called the **departure**. At the 0900 departure, the conning officer orders a course of 090°T and a speed of 10 knots. The navigator lays out the 090°T course line from the departure.

At 1000, the navigator plots a DR position according to the rule requiring plotting a DR position at least every hour on the hour. At 1030, the conning officer orders a course change to 060°T. The navigator plots the 1030 DR position in accordance with the rule requiring plotting a DR position at every course and speed change. Note that the course line changes at 1030 to 060°T to conform to the new course. At 1100, the conning officer changes course back to 090°T. The navigator plots an 1100 DR due to the course change. Note that, regardless of the course change, an 1100 DR would have been required because of the "every hour on the hour" rule.

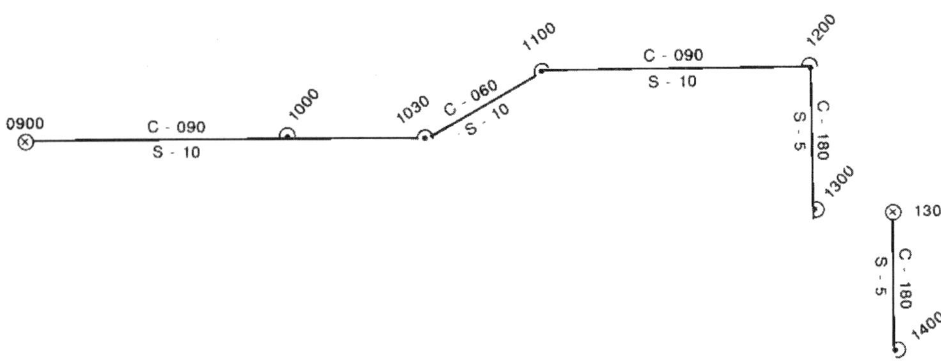

Figure 703. A typical dead reckoning plot.

At 1200, the conning officer changes course to 180°T and speed to 5 knots. The navigator plots the 1200 DR. At 1300, the navigator obtains a fix. Note that the fix position is offset to the east from the DR position. The navigator determines set and drift from this offset and applies this set and drift to any DR position from 1300 until the next fix to determine an estimated position. He also resets the DR to the fix; that is, he draws the 180°T course line from the 1300 fix, not the 1300 DR.

704. Resetting the DR

Reset the DR plot to each fix or running fix in turn. In addition, consider resetting the DR to an inertial estimated position, if an inertial system is installed.

If a navigator has not taken a fix for an extended period of time, the DR plot, not having been reset to a fix, will accumulate time-dependent errors. Over time that error may become so significant that the DR will no longer show the ship's position with acceptable accuracy. If the vessel is equipped with an inertial navigator, the navigator should consider resetting the DR to the inertial estimated position. Some factors to consider when making this determination are:

(1) Time since the last fix and availability of fix information. If it has been a short time since the last fix and fix information may soon become available, it may be advisable to wait for the next fix to reset the DR.

(2) Dynamics of the navigation situation. If, for example, a submerged submarine is operating in the Gulf Stream, fix information is available but operational considerations may preclude the submarine from going to periscope depth to obtain a fix. Similarly, a surface ship with an inertial navigator may be in a dynamic current and suffer a temporary loss of electronic fix equipment. In either case, the fix information will be available shortly but the dynamics of the situation call for a more accurate assessment of the vessel's position. Plotting an inertial EP and resetting the DR to that EP may provide the navigator with a more accurate assessment of the navigation situation.

(3) Reliability and accuracy of the fix source. If a submarine is operating under the ice, for example, only the inertial EP fixes may be available for weeks at a time. Given a high prior correlation between the inertial EP and highly accurate fix systems such as GPS, and the continued proper operation of the inertial navigator, the navigator may decide to reset the DR to the inertial EP.

DEAD RECKONING AND SHIP SAFETY

Properly maintaining a DR plot is important for ship safety. The DR allows the navigator to examine a future position in relation to a planned track. It allows him to anticipate charted hazards and plan appropriate action to avoid them. Recall that the DR position is only approximate. Using a concept called **fix expansion** compensates for the DR's inaccuracy and allows the navigator to use the DR more effectively to anticipate and avoid danger.

705. Fix Expansion

Often a ship steams in the open ocean for extended periods without a fix. This can result from any number of factors ranging from the inability to obtain celestial fixes to malfunctioning electronic navigation systems. Infrequent fixes are particularly common on submarines. Whatever the reason, in some instances a navigator may find himself in the position of having to steam many hours on DR alone.

The navigator must take precautions to ensure that all hazards to navigation along his path are accounted for by the approximate nature of a DR position. One method which can be used is **fix expansion**.

Fix expansion takes into account possible errors in the DR calculation caused by factors which tend to affect the vessel's actual course and speed over the ground. The navigator considers all such factors and develops an expanding "error circle" around the DR plot. One of the basic assumptions of fix expansion is that the various individual effects of current, leeway, and steering error combine to cause a cumulative error which increases over time, hence, the concept of expansion. While the errors may in fact cancel each other out, the worst case is that they will all be additive, and this is what the navigator must anticipate.

Errors considered in the calculation of fix expansion encompass all errors that can lead to DR inaccuracy. Some of the most important factors are current and wind, compass or gyro error, and steering error. Any method which attempts to determine an error circle must take these factors into account. The navigator can use the magnitude of set and drift calculated from his DR plot. See Article 707. He can obtain the current's estimated magnitude from pilot charts or weather reports. He can determine wind speed from weather instruments. He can determine compass error by comparison with an accurate standard or by obtaining an azimuth of the Sun. The navigator determines the effect each of these errors has on his course and speed over ground, and applies that error to the fix expansion calculation.

As noted previously, error is a function of time; it grows as the ship proceeds along the track without obtaining a fix. Therefore, the navigator must incorporate his calculated errors into an **error circle** whose radius grows with time. For example, assume the navigator calculates that all the various sources of error can create a cumulative position error of no more than 2 nm. Then his fix expansion error circle would grow at that rate; it would

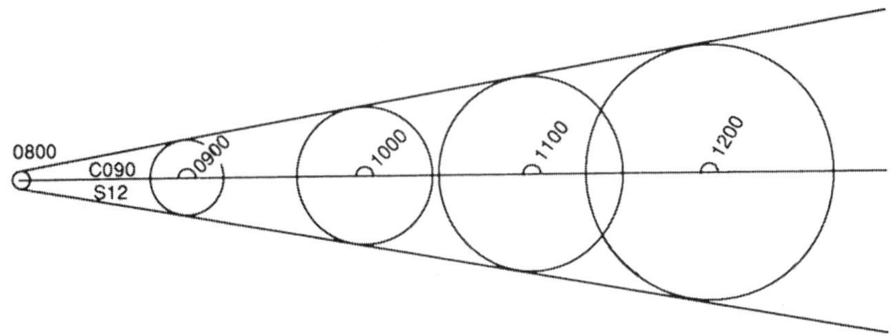

Figure 705. Fix expansion. All possible positions of the ship lie between the lines tangent to the expanding circles. Examine this area for dangers.

be 2 nm after the first hour, 4 nm after the second, and so on.

At what value should the navigator start this error circle? Recall that a DR is laid out from every fix. All fix sources have a finite absolute accuracy, and the initial error circle should reflect that accuracy. Assume, for example, that a satellite navigation system has an accuracy of 0.5 nm. Then the initial error circle around that fix should be set at 0.5 nm.

First, enclose the fix position in a circle, the radius of which is equal to the accuracy of the system used to obtain the fix. Next, lay out the ordered course and speed from the fix position. Then apply the fix expansion circle to the hourly DR's, increasing the radius of the circle by the error factor each time. In the example given above, the DR after one hour would be enclosed by a circle of radius 2.5 nm, after two hours 4.5 nm, and so on. Having encircled the four hour DR positions with the error circles, the navigator then draws two lines originating tangent to the original error circle and simultaneously tangent to the other error circles. The navigator then closely examines the area between the two tangent lines for hazards to navigation. This technique is illustrated in Figure 705.

The fix expansion encompasses the total area in which the vessel could be located (as long as all sources of error are considered). If any hazards are indicated within the cone, the navigator should be especially alert for those dangers. If, for example, the fix expansion indicates that the vessel may be standing into shoal water, continuously monitor the fathometer. Similarly, if the fix expansion indicates that the vessel might be approaching a charted obstruction, post extra lookouts.

The fix expansion may grow at such a rate that it becomes unwieldy. Obviously, if the fix expansion grows to cover too large an area, it has lost its usefulness as a tool for the navigator, and he should obtain a new fix by any available means.

DETERMINING AN ESTIMATED POSITION

An estimated position (EP) is a DR position corrected for the effects of leeway, steering error, and current. This section will briefly discuss the factors that cause the DR position to diverge from the vessel's actual position. It will then discuss calculating set and drift and applying these values to the DR to obtain an estimated position. It will also discuss determining the estimated course and speed made good.

706. Factors Affecting DR Position Accuracy

Tidal current is the periodic horizontal movement of the water's surface caused by the tide-affecting gravitational forces of the Moon and Sun. **Current** is the horizontal movement of the sea surface caused by meteorological, oceanographic, or topographical effects. From whatever its source, the horizontal motion of the sea's surface is an important dynamic force acting on a vessel.

Set refers to the current's direction, and **drift** refers to the current's speed. **Leeway** is the leeward motion of a vessel due to that component of the wind vector perpendicular to the vessel's track. Leeway and current combine to produce the most pronounced natural dynamic effects on a transiting vessel. Leeway especially affects sailing vessels and high-sided vessels.

In addition to these natural forces, relatively small helmsman and steering compass error may combine to cause additional error in the DR.

707. Calculating Set and Drift and Plotting an Estimated Position

It is difficult to quantify the errors discussed above individually. However, the navigator can easily quantify their cumulative effect by comparing simultaneous fix and DR positions. If there are no dynamic forces acting on the vessel and no steering error, the DR position and the fix position will coincide. However, they seldom do so. The fix is offset from the DR by the vector sum of all the errors.

Note again that this methodology provides no means to determine the magnitude of the individual errors. It simply provides the navigator with a measurable representation of their combined effect.

When the navigator measures this combined effect, he often refers to it as the "set and drift." Recall from above that these terms technically were restricted to describing current effects. However, even though the fix-to-DR offset is caused by effects in addition to the current, this text will follow the convention of referring to the offset as the set and drift.

The set is the direction from the DR to the fix. The drift is the distance in miles between the DR and the fix divided by the number of hours since the DR was last reset. This is true regardless of the number of changes of course or speed since the last fix. The prudent navigator calculates set and drift at every fix.

To calculate an EP, draw a vector from the DR position in the direction of the set, with the length equal to the product of the drift and the number of hours since the last reset. See Figure 707. From the 0900 DR position the navigator draws a set and drift vector. The end of that vector marks the 0900 EP. Note that the EP is enclosed in a square and labeled horizontally with the time. Plot and evaluate an EP with every DR position.

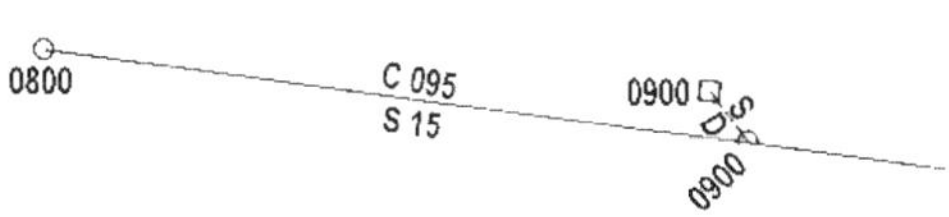

Figure 707. Determining an estimated position.

708. Estimated Course and Speed Made Good

The direction of a straight line from the last fix to the EP is the **estimated track made good**. The length of this line divided by the time between the fix and the EP is the **estimated speed made good**.

Solve for the estimated track and speed by using a vector diagram. See the example problems below and refer to Figure 708a.

Example 1: *A ship on course 080°, speed 10 knots, is steaming through a current having an estimated set of 140° and drift of 2 knots.*

Required: *Estimated track and speed made good.*

Solution: *See Figure 708a. From A, any convenient point, draw AB, the course and speed of the ship, in direction 080°, for a distance of 10 miles.*

From B draw BC, the set and drift of the current, in direction 140°, for a distance of 2 miles.

The direction and length of AC are the estimated track and speed made good.

Answers: *Estimated track made good 089°, estimated speed made good 11.2 knots.*

To find the course to steer at a given speed to make good a desired course, plot the current vector from the origin, A, instead of from B. See Figure 708b.

Example 2: *The captain desires to make good a course of 095° through a current having a set of 170° and a drift of 2.5 knots, using a speed of 12 knots.*

Required: *The course to steer and the speed made good.*

Solution: *See Figure 708b. From A, any convenient point, draw line AB extending in the direction of the course to be made good, 095°.*

From A draw AC, the set and drift of the current.

Using C as a center, swing an arc of radius CD, the speed through the water (12 knots), intersecting line AB at D.

Measure the direction of line CD, 083.5°. This is the course to steer.

Measure the length AD, 12.4 knots. This is the speed made good.

Answers: *Course to steer 083.5°, speed made good 12.4 knots.*

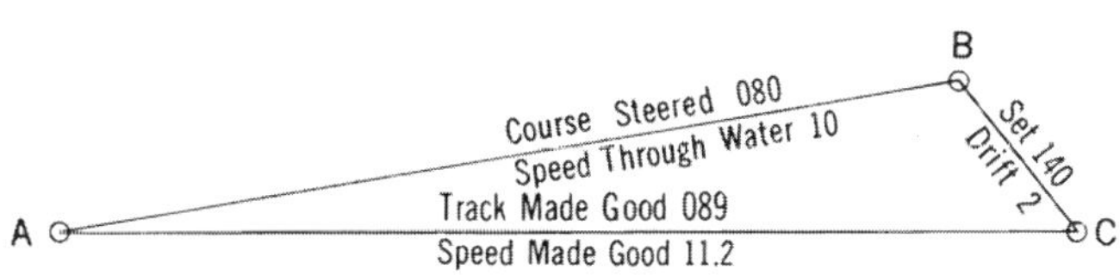

Figure 708a. Finding track and speed made good through a current.

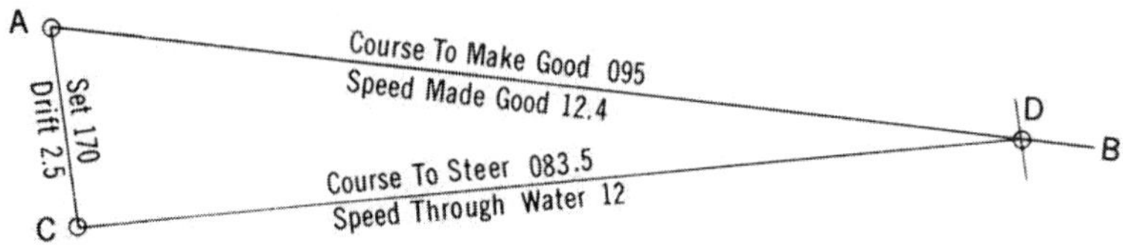

Figure 708b. Finding the course to steer at a given speed to make good a given course through a current.

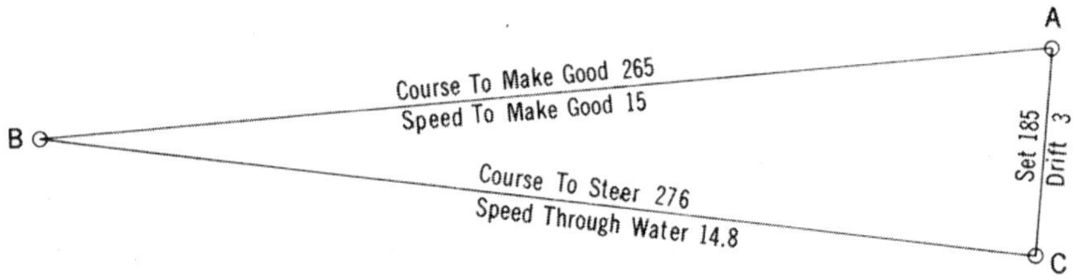

Figure 708c. Finding course to steer and speed to use to make good a given course and speed through the current.

To find the course to steer and the speed to use to make good a desired course and speed, proceed as follows: See Figure 708c.

Example 3: *The captain desires to make good a course of 265° and a speed of 15 knots through a current having a set of 185° and a drift of 3 knots.*

Required: *The course to steer and the speed to use.*

Solution: *See Figure 708c. From A, any convenient point, draw AB in the direction of the course to be made good, 265° and for length equal to the speed to be made good, 15 knots.*

From A draw AC, the set and drift of the current.

Draw a straight line from C to B. The direction of this line, 276°, is the required course to steer; and the length, 14.8 knots, is the required speed.

Answers: *Course to steer 276°, speed to use 14.8 kn.*

CHAPTER 8

PILOTING

DEFINITION AND PURPOSE

800. Introduction

Piloting involves navigating a vessel in restricted waters and fixing its position as precisely as possible at frequent intervals. More so than in other phases of navigation, proper preparation and attention to detail are important. This chapter will discuss a piloting methodology designed to ensure that procedures are carried out safely and efficiently. These procedures will vary from vessel to vessel according to the skills and composition of the piloting team. It is the responsibility of the navigator to choose the procedures applicable to his own situation, to train the piloting team in their execution, and to ensure that duties are carried out properly.

These procedures are written primarily from the perspective of the military navigator, with some notes included where civilian procedures might differ. This set of procedures is designed to minimize the chance of error and maximize safety of the ship.

The military navigation team will nearly always consist of several more people than are available to the civilian navigator. Therefore, the civilian navigator must streamline these procedures, eliminating certain steps, doing only what is essential to keep his ship in safe water.

The navigation of civilian vessels will therefore proceed differently than for military vessels. For example, while the military navigator might have bearing takers stationed at the gyro repeaters on the bridge wings for taking simultaneous bearings, the civilian navigator must often take and plot them himself. While the military navigator will have a bearing book and someone to record entries for each fix, the civilian navigator will simply plot the bearings on the chart as they are taken and not record them at all.

If the ship is equipped with an ECDIS, it is reasonable for the navigator to simply monitor the progress of the ship along the chosen track, visually ensuring that the ship is proceeding as desired, checking the compass, sounder and other indicators only occasionally. If a pilot is aboard, as is often the case in the most restricted of waters, his judgement can generally be relied upon explicitly, further easing the workload. But should the ECDIS fail, the navigator will have to rely on his skill in the manual and time-tested procedures discussed in this chapter.

While an ECDIS is the legal equivalent of a paper chart and can be used as the primary plot, an ECS, (non-ECDIS compliant electronic chart system) cannot be so used. An ECS may be considered as an additional resource used to ensure safe navigation, but cannot be relied upon for performing all the routine tasks associated with piloting. The individual navigator, with knowledge of his vessel, his crew, and the capabilities they possess, must make a professional judgement as to how the ECS can support his efforts to keep his ship in safe water. The navigator should always remember that reliance on any single navigation system courts disaster. An ECS does not relieve the navigator of maintaining a proper and legal plot on a paper chart.

PREPARATION

801. Plot Setup

The navigator's job begins well before getting underway. Much advance preparation is necessary to ensure a safe and efficient voyage. The following steps are representative:

Ensure the plotting station(s) have the following instruments:

- **Dividers:** Dividers are used to measure distances between points on the chart.

- **Compasses:** Compasses are used to plot range arcs for radar LOP's. **Beam compasses** are used when the range arc exceeds the spread of a conventional compass. Both should be available at both plots.

- **Plotters:** Several types of plotters are available. The preferred device for large vessels is the parallel motion plotter (PMP) used in conjunction with a drafting table. Otherwise, use a transparent protractor plotter, or triangles, parallel rulers or rolling rulers in conjunction with the chart's compass rose. Finally, the plotter can use a one arm protractor. The plotter should use the device with which he can work the most quickly and accurately.

- **Sharpened Pencils and Erasers:** Ensure an adequate supply of pencils is available.

- **Fischer Radar Plotting Templates:** Fischer plotting is covered in Chapter 13. The plotting templates for this technique should be stacked near the radar repeater.

- **Time-Speed-Distance Calculator:** Given two of the three unknowns (between time, speed, and distance), this calculator allows for rapid computation of the third.

- **Tide and Current Graphs:** Post the tide and current graphs near the primary plot for easy reference during the transit. Give a copy of the graphs to the conning officer and the captain.

Once the navigator verifies the above equipment is in place, he tapes down the charts on the chart table. If more than one chart is required for the transit, tape the charts in a stack such that the plotter works from the top to the bottom of the stack. This minimizes the time required to shift the chart during the transit. If the plotter is using a PMP, align the arm of the PMP with any meridian of longitude on the chart. While holding the PMP arm stationary, adjust the PMP to read 000.0°T. This procedure calibrates the PMP to the chart in use. Perform this alignment every time the piloting team shifts charts.

Be careful not to fold under any important information when folding the chart on the chart table. Ensure the chart's distance scale, the entire track, and all important warning information are visible.

Energize and test all electronic navigation equipment, if not already in operation. This includes the radar and the GPS receiver. Energize and test the fathometer. Ensure the entire electronic navigation suite is operating properly prior to entering restricted waters.

802. Preparing Charts and Publications

- **Assemble Required Publications.** These publications should include *Coast Pilots*, *Sailing Directions*, *USCG Light Lists*, *NIMA Lists of Lights*, *Tide Tables*, *Tidal Current Tables*, *Notice to Mariners*, and *Local Notice to Mariners*. Often, for military vessels, a port will be under the operational direction of a particular squadron; obtain that squadron's port Operation Order. Civilian vessels should obtain the port's harbor regulations. These publications will cover local regulations such as speed limits and bridge-to-bridge radio frequency monitoring requirements. Assemble and review the Broadcast Notice to Mariners file.

- **Select and Correct Charts.** Choose the largest scale chart available for the harbor approach or departure. Often, the harbor approach will be too long to be represented on only one chart. For example, three charts are required to cover the waters from the Naval Station in Norfolk to the entrance of the Chesapeake Bay. Therefore, obtain all the charts required to cover the entire passage. Using the *Notice to Mariners*, verify that these charts have been corrected through the latest *Notice to Mariners*. Check the *Local Notice to Mariners* and the Broadcast Notice to Mariners file to ensure the chart is fully corrected. Annotate on the chart or a chart correction card all the corrections that have been made; this will make it easier to verify the chart's correction status prior to its next use. Naval ships may need to prepare three sets of charts. One set is for the primary plot, the second set is for the secondary plot, and the third set is for the conning officer and captain. Civilian vessels will prepare one set.

- **Mark the Minimum Depth Contour:** Determine the minimum depth of water in which the vessel can safely operate and outline that depth contour on the chart. Do this step before doing any other harbor navigation planning. Highlight this outline in a bright color so that it clearly stands out. Carefully examine the area inside the contour and mark the isolated shoals less than the minimum depth which fall inside the marked contour. Determine the minimum depth in which the vessel can operate as follows:

  Minimum Depth = Ship's Draft – Height of Tide + Safety Margin + Squat. (See Article 804 and Article 818.)

  Remember that often the fathometer's transducer is not located at the section of the hull that extends the furthest below the waterline. Therefore, the indicated depth of water is that below the fathometer transducer, not the depth of water below the vessel's deepest draft.

- **Highlight Selected Visual Navigation Aids (NAVAIDS).** Circle, highlight and label the main navigational aids on the chart. Consult the applicable *Coast Pilot* or *Sailing Directions* to determine a port's best NAVAIDS if the piloting team has not visited the port previously. These aids can be lighthouses, piers, shore features, or tanks; any prominent feature that is displayed on the chart can be used as a NAVAID. Label critical buoys, such as those marking a harbor entrance or a traffic separation scheme. Verify charted lights against the *Light List* or the *List of Lights* to confirm the charted information is correct. This becomes most critical when attempting to identify a light at night. Label NAVAIDS succinctly and clearly. Ensure everyone in the navigation team refers to a NAVAID using the same terminology. This will reduce confusion between the bearing taker, the bearing recorder, and plotter.

- **Highlight Selected Radar NAVAIDS.** Highlight radar NAVAIDS with a triangle instead of a circle. If the NAVAID is suitable for either visual or radar piloting, it can be highlighted with either a circle or a triangle.

- **Plot the Departure/Approach Track.** This process is critical for ensuring safe pilotage. Consult the *Fleet Guide* and *Sailing Directions* for recommendations on the best track to use. Look for any information or regulations published by the local harbor authority. Lacking any of this information, locate a channel or safe route on the chart and plot the vessel's track. Most U.S. ports have well-defined channels marked with buoys. Carefully check the intended track to ensure a sufficient depth of water under the keel will exist for the entire passage. If the scale of the chart permits, lay the track out to the starboard side of the channel to allow for any vessel traffic proceeding in the opposite direction. Many channels are marked by natural or man-made ranges. The bearings of these ranges should be measured to the nearest 0.1° or noted from the *Light List*, and this value should be marked on the chart. Not only are ranges useful in keeping a vessel on track, they are invaluable for determining gyro error. See Article 807.

- **Label the Departure/Approach Track.** Label the track course to the nearest 0.5°. Similarly, label the distance of each track leg. Highlight the track courses for easy reference while piloting. Often a navigator might plan two separate tracks. One track would be for use during good visibility and the other for poor visibility. Considerations might include concern for the number of turns (fewer turns for poor visibility) or proximity to shoal water (smaller margin for error might be acceptable in good visibility). In this case, label both tracks as above and appropriately mark when to use each track.

- **Use Advance and Transfer to Find Turning Points.** The distance the vessel moves along its original course from the time the rudder is put over until the new course is reached is called **advance**. The distance the vessel moves perpendicular to the original course during the turn is called **transfer**.The track determined above does not account for these. See Figure 802a. Use the advance and transfer characteristics of the vessel to determine when the vessel must put its rudder over to gain the next course. From that point, fair in a curve between the original course and the new course. Mark the point on the original course where the vessel must put its rudder over as the **turning point**. See Figure 802b.

- **Plot Turn Bearings and Ranges.** A **turn bearing** is a predetermined bearing to a charted object from the track point at which the rudder must be put over in order to make a desired turn. In selecting a NAVAID for a turn bearing, find one as close to abeam as possible at the turning point, and if possible on the inside elbow of the turn. Account for advance and transfer and label the bearing to the nearest 0.1°. A **turn range** is similar, but taken as a radar range to a prominent object ahead or astern. Ideally, both can be used, one as a check against the other.

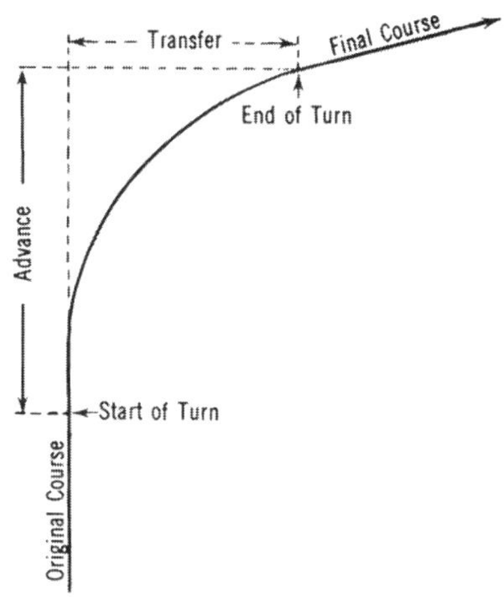

Figure 802a. Advance and transfer.

Example: *Figure 802b illustrates using advance and transfer to determine a turn bearing. A ship proceeding on course 100° is to turn 60° to the left to come on a range which will guide it up a channel. For a 60° turn and the amount of rudder used, the advance is 920 yards and the transfer is 350 yards.*

Required: *The bearing of flagpole "FP." when the rudder is put over.*

Solution:

1. *Extend the original course line, AB.*
2. *At a perpendicular distance of 350 yards, the transfer, draw a line A'B' parallel to the original course line AB. The point of intersection, C, of A'B' with the new course line is the place at which the turn is to be completed.*
3. *From C draw a perpendicular, CD, to the original course line, intersecting at D.*
4. *From D measure the advance, 920 yards, back along the original course line. This locates E, the point at which the turn should be started.*
5. *The direction of "FP." from E, 058°, is the bearing when the turn should be started.*

Answer: *Bearing 058°.*

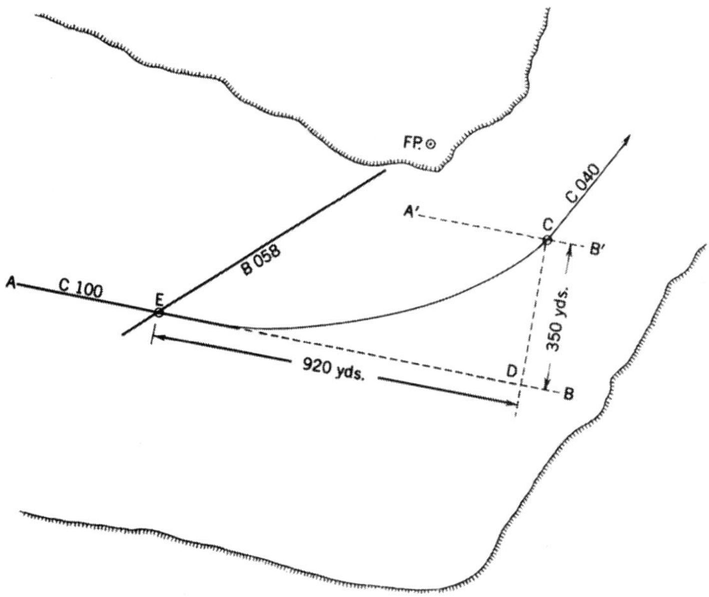

Figure 802b. Allowing for advance and transfer.

- **Plot a Slide Bar for Every Turn Bearing:** If the ship is off track immediately prior to a turn, a plotting technique known as the **slide bar** can quickly revise a turn bearing. See Figure 802c. A slide bar is a line drawn parallel to the new course through the turning point on the original course. The navigator can quickly determine a new turn bearing by dead reckoning ahead from the vessel's last fix position to where the DR intersects the slide bar. The revised turn bearing is simply the bearing from that intersection point to the turn bearing NAVAID. Draw the slide bar with a different color from that used for the track in order to see the slide bar clearly.

- **Label Distance to Go from Each Turn Point:** At each turning point, label the distance to go until either the ship moors (inbound) or the ship clears the harbor (outbound). For an inbound transit, a vessel's captain is usually more concerned about time of arrival, so assume a speed of advance and label each turn point with time to go until mooring.

- **Plot Danger Bearings:** Danger bearings warn a navigator he may be approaching a navigational hazard too closely. See Figure 802d. Vector AB indicates a vessel's intended track. This track passes close to the indicated shoal. Draw a line from the NAVAID H tangent to the shoal. The bearing of that tangent line measured from the ship's track is 074.0°T. In other words, as long as NAVAID H bears *less than* 074°T as the vessel proceeds down its track, the vessel will not ground on the shoal. Hatch the side of the bearing line on the side of the hazard and label the danger bearing NMT (no more than) 074.0°T. For an added margin of safety, the line does not have to be drawn exactly tangent to the shoal. Perhaps, in this case, the navigator might want to set an error margin and draw the danger bearing at 065°T from NAVAID H. Lay down a danger bearing from any appropriate NAVAID in the vicinity of any hazard to navigation. Ensure the track does not cross any danger bearing.

- **Plot Danger Ranges:** The danger range is analogous to the danger bearing. It is a standoff range from an object to prevent the vessel from approaching a hazard too closely.

- **Label Warning and Danger Soundings:** To determine the danger sounding, examine the vessel's proposed track and note the minimum expected sounding. The minimum expected sounding is the difference between the shallowest water expected on the transit and the vessel's maximum draft. Set 90% of this difference as the warning sounding and 80% of this difference as the danger sounding. There may be peculiarities about local conditions that will cause the navigator to choose another method of setting warning and danger soundings. Use the above method if no

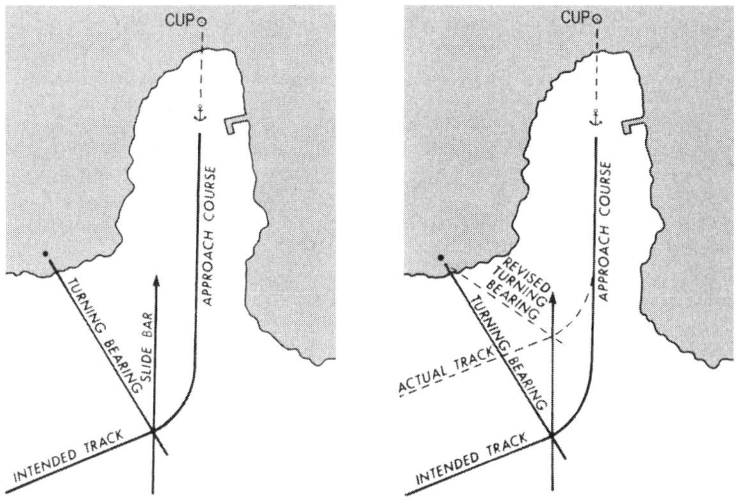

Figure 802c. The slide bar technique.

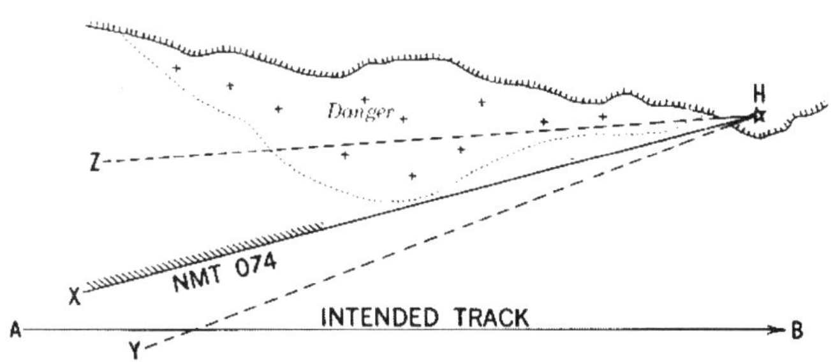

Figure 802d. A danger bearing, hatched on the dangerous side, labeled with the appropriate bearing.

other means is more suitable. For example: A vessel draws a maximum of 20 feet, and it is entering a channel dredged to a minimum depth of 50 feet. Set the warning and danger soundings at 0.9 (50ft. - 20ft) = 27ft and 0.8 (50ft. - 20ft.) = 24ft., respectively. Re-evaluate these soundings at different intervals along the track, when the minimum expected sounding may change. Carefully label the points along the track between which these warning and danger soundings apply.

- **Label Demarcation Line:** Clearly label the point on the ship's track where the Inland and International Rules of the Road apply. This is applicable only when piloting in U.S. ports.

- **Mark Speed Limits Where Applicable:** Often a harbor will have a local speed limit in the vicinity of piers, other vessels, or shore facilities. Mark these speed limits and the points between which they are applicable on the chart.

- **Mark the Point of Pilot Embarkation:** Some ports require vessels over a certain size to embark a pilot. If this is the case, mark the point on the chart where the pilot is to embark.

- **Mark the Tugboat Rendezvous Point:** If the vessel requires a tug to moor, mark the tug rendezvous point on the chart.

- **Mark the Chart Shift Point:** If more than one chart

will be required to complete the passage, mark the point where the navigator should shift to the next chart.

- **Harbor Communications:** Mark the point on the chart where the vessel must contact harbor control. Also mark the point where a vessel must contact its parent squadron to make an arrival report (military vessels only).
- **Tides and Currents:** Mark the points on the chart for which the tides and currents were calculated.

803. Records

Ensure the following records are assembled and personnel assigned to maintain them:

- **Bearing Record Book:** The bearing recorders for the primary and secondary plots should record all the bearings used on their plot during the entire transit. The books should clearly list what NAVAIDS are being used and what method of navigation was being used on their plot. In practice, the primary bearing book will contain mostly visual bearings and the secondary bearing book will contain mostly radar ranges and bearings.
- **Fathometer Log:** In restricted waters, monitor soundings continuously and record soundings every five minutes in the fathometer log. Record all fathometer settings that could affect the sounding display.
- **Deck Log:** This log is the legal record of the passage. Record all ordered course and speed changes. Record all the navigator's recommendations and whether the navigator concurs with the actions of the conning officer. Record all buoys passed, and the shift between different Rules of the Road. Record the name and embarkation of any pilot. Record who has the conn at all times. Record any casualty or important event. The deck log combined with the bearing log should constitute a complete record of the passage.

804. Tides and Currents

Determining the tidal and current conditions of the port is crucial. This process is covered in depth in Chapter 9. In order to anticipate early or late transit, plot a graph of the tidal range for the 24-hour period centered on the scheduled time of arrival or departure. Depending on a vessel's draft and the harbor's depth, some vessels may be able to transit only at high tide. If this is this case, it is critically important to determine the time and range of the tide correctly.

The magnitude and direction of the current will give the navigator some idea of the **set and drift** the vessel will experience during the transit. This will allow him to plan in advance for any potential current effects in the vicinity of navigational hazards.

While printed tide tables can be used for predicting and plotting tides, it is far more efficient to use a computer with appropriate software, or the internet, to compute tides and print out the graphs. These graphs can be posted on the bridge at the chart table for ready reference, and copies made for others involved in the piloting process. NOAA tide tables for the U.S. are available at the following site: http://co-ops.nos.noaa.gov/tp4days.html. Always remember that tide tables give predicted data, but that actual conditions may be quite different due to weather or other natural phenomena.

805. Weather

The navigator should obtain a weather report covering the route which he intends to transit. This will allow him to prepare for any adverse weather by stationing extra lookouts, adjusting speed for poor visibility, and preparing for radar navigation. If the weather is thick, consider standing off the harbor until it clears.

The navigator can receive weather information any number of ways. Military vessels may receive weather reports from their parent squadrons prior to coming into port. Marine band radio carries continuous weather reports. Many vessels are equipped with weather facsimile machines. Some navigators carry cellular phones to reach shoreside personnel and harbor control; these can also be used to get weather reports from NOAA weather stations. If the ship is using a weather routing service for the voyage, it should provide forecasts when asked. Finally, if the vessel has an internet connection, this is an ideal source of weather data. NOAA weather data can be obtained at: http://www.nws.noaa.gov. However he obtains the information, the navigator should have a good idea of the weather before entering piloting waters.

806. The Piloting Brief

Assemble the entire navigation team for a piloting brief prior to entering or leaving port. The vessel's captain and navigator should conduct the briefing. All navigation and bridge personnel should attend. The pilot, if he is already on board, should also attend. If the pilot is not onboard when the ship's company is briefed, the navigator should immediately brief him when he embarks. The pilot must know the ship's maneuvering characteristics before entering restricted waters. The briefing should cover, as a minimum, the following:

- **Detailed Coverage of the Track Plan:** Go over the planned route in detail. Use the prepared and approved chart as part of this brief. Concentrate especially on all the NAVAIDS and soundings which are being used to indicate danger. Cover the buoyage system in use and the port's major NAVAIDS. Point out the radar NAVAIDS for the radar operator. Often, a *Fleet Guide* or *Sailing Directions* will have pictures of a port's NAVAIDS. This is especially important for the piloting party that has never transited a particular port before. If no pictures are available, consider stationing a photographer to take some for submission to NIMA.

- **Harbor Communications:** Discuss the bridge-to bridge radio frequencies used to raise harbor control. Discuss what channel the vessel is supposed to monitor on its passage into port and the port's communication protocol.

- **Duties and Responsibilities:** Each member of the piloting team must have a thorough understanding of his duties and responsibilities. He must also understand how his part fits into the whole. The radar plotter, for example, must know if radar will be the primary or secondary source of fix information. The bearing recorder must know what fix interval the navigator is planning to use. Each person must be thoroughly briefed on his job; there is little time for questions once the vessel enters the channel.

807. Evolutions Prior to Piloting

The navigator should always accomplish the following evolutions prior to piloting:

- **Testing the Shaft on the Main Engines in the Astern Direction:** This ensures that the ship can answer a backing bell. If the ship is entering port, no special precautions are required prior to this test. If the ship is tied up at the pier preparing to get underway, exercise extreme caution to ensure no way is placed on the ship while testing the main engines.

- **Making the Anchor Ready for Letting Go:** Make the anchor ready for letting go and station a watchstander in direct communications with the bridge at the anchor windlass. Be prepared to drop anchor immediately when piloting if required to keep from drifting too close to a navigational hazard.

- **Calculate Gyro Error:** An error of greater than 1.0° T indicates a gyro problem which should be investigated prior to piloting. There are several ways to determine gyro error:

  1. Compare the gyro reading with a known accurate heading reference such as an inertial navigator. The difference in the readings is the gyro error.

  2. Mark the bearing of a charted range as the range NAVAID's come into line and compare the gyro bearing with the charted bearing. The difference is the gyro error.

  3. Prior to getting underway, plot a dockside fix using at least three lines of position. The three LOP's should intersect at a point. Their intersecting in a "cocked hat" indicates a gyro error. Incrementally adjust each visual bearing by the same amount and direction until the fix plots as a pinpoint. The total correction required to eliminate the cocked hat is the gyro error.

  4. Measure a celestial body's azimuth or amplitude, or Polaris' azimuth with the gyro, and then compare the measured value with a value computed from the *Sight Reduction Tables* or the *Nautical Almanac*. These methods are covered in detail in Chapter 17.

Report the magnitude and direction of the gyro error to the navigator and captain. The direction of the error is determined by the relative magnitude of the gyro reading and the value against which it is compared. When the compass is least, the error is east. Conversely, when the compass is best, the error is west. See Chapter 6.

808. Inbound Voyage Planning

The vessel's planned estimated time of arrival (ETA) at its mooring determines the vessel's course and speed to the harbor entrance. Arriving at the mooring site on time may be important in a busy port which operates its port services on a tight schedule. Therefore, it is important to plan the arrival accurately. Take the desired time of arrival at the mooring and subtract from that the time it will take to navigate to it from the entrance. The resulting time is when you must arrive at the harbor entrance. Next, measure the distance between the vessel's present location and the harbor entrance. Determine the speed of advance (SOA) the vessel will use to make the transit to the harbor. Use the distance to the harbor and the SOA to calculate what time to leave the present position to make the mooring ETA, or what speed must be made good to arrive on time.

Consider these factors which might affect this decision:

- **Weather:** This is the single most important factor in harbor approach planning because it directly affects the vessel's SOA. The thicker the weather, the more slowly the vessel must proceed. Therefore, if heavy fog or rain is in the forecast, the navigator must allow more time for the transit.

- **Mooring Procedures:** The navigator must take more than distance into account when calculating how long it will take him to pilot to his mooring. If the vessel needs a tug, that will increase the time needed. Similarly, picking up or dropping off a pilot adds time to the transit. It is better to allow a margin for error when trying to add up all the time delays caused by these procedures. It is always easier to avoid arriving early by slowing down than it is to make up lost time by speeding up.

- **Shipping Density:** Generally, the higher the shipping density entering and exiting the harbor, the longer it will take to proceed into the harbor entrance safely.

TRANSITION TO PILOTING

809. Stationing the Piloting Team

At the appropriate time, station the piloting team. Allow plenty of time to acclimate to the navigational situation and if at night, to the darkness. The number and type of personnel available for the piloting team depend on the vessel. A Navy warship, for example, has more people available for piloting than a merchant ship. Therefore, more than one of the jobs listed below may have to be filled by a single person. The piloting team should consist of:

- **The Captain:** The captain is ultimately responsible for the safe navigation of his vessel. His judgment regarding navigation is final. The piloting team acts to support the captain, advising him so he can make informed decisions on handling his vessel.

- **The Pilot:** The pilot is usually the only member of the piloting team not a member of the ship's company. The piloting team must understand the relationship between the pilot and the captain. The pilot is perhaps the captain's most important navigational advisor. Generally, the captain will follow his recommendations when navigating an unfamiliar harbor. The pilot, too, bears some responsibility for the safe passage of the vessel; he can be censured for errors of judgment which cause accidents. However, the presence of a pilot in no way relieves the captain of his ultimate responsibility for safe navigation. The piloting team works to support and advise the captain.

- **The Officer of the Deck (Conning Officer):** In Navy piloting teams, neither the pilot or the captain usually has the **conn**. The officer having the conn directs the ship's movements by rudder and engine orders. Another officer of the ship's company usually fulfills this function. The captain can take the conn immediately simply by issuing an order to the helm should an emergency arise. The conning officer of a merchant vessel can be either the pilot, the captain, or another watch officer. In any event, the officer having the conn must be clearly indicated in the ship's deck log at all times. Often a single officer will have the deck and the conn. However, sometimes a junior officer will take the conn for training. In this case, different officers will have the deck and the conn. The officer who retains the deck retains the responsibility for the vessel's safe navigation.

- **The Navigator:** The vessel's navigator is the officer directly responsible to the ship's captain for the safe navigation of the ship. He is the captain's principal navigational advisor. The piloting team works for him. He channels the required information developed by the piloting team to the ship's conning officer on recommended courses, speeds, and turns. He also carefully looks ahead for potential navigational hazards and makes appropriate recommendations. He is the most senior officer who devotes his effort exclusively to monitoring the navigation picture. The captain and the conning officer are concerned with all aspects of the passage, including contact avoidance and other necessary ship evolutions (making up tugs, maneuvering alongside a small boat for personnel transfers, engineering evolutions, and coordinating with harbor control via radio, for example). The navigator, on the other hand, focuses solely on safe navigation. It is his job to anticipate dangers, keep himself appraised of the navigation situation at all times, and manage the team.

- **Bearing Plotting Team:** This team consists, ideally, of three persons. The first person measures the bearings. The second person records the bearings in an official record book. The third person plots the bearings. The more quickly and accurately this process is completed, the sooner the navigator has an accurate picture of the ship's position. The bearing taker should be an experienced individual who has traversed the port before and who is familiar with the NAVAIDS. He should take his round of bearings as quickly as possible, beam bearings first, minimizing any time delay errors in the resulting fix. The plotter should also be an experienced individual who can quickly and accurately lay down the required bearings. The bearing recorder can be one of the junior members of the piloting team.

- **The Radar Operator:** The radar operator has one of the more difficult jobs of the team. The radar is as important for collision avoidance as it is for navigation. Therefore, this operator must often "time share" the radar between these two functions. Determining the amount of time spent on these functions falls within the judgment of the captain and the navigator. If the day is clear and the traffic heavy, the captain may want to use the radar mostly for collision avoidance. As the weather worsens, obscuring visual NAVAIDS, the importance of radar for safe navigation increases. The radar operator must be given clear guidance on how the captain and navigator want the radar to be operated.

- **Plot Supervisors:** On many military ships, the piloting team will consist of two plots: the primary plot and the secondary plot. The navigator should designate the type of navigation that will be employed on the primary plot. All other fix sources should be plotted on the secondary plot. The navigator can function as the primary plot supervisor. A senior, experienced individual should be employed as a secondary plot supervisor. The navigator should frequently compare the positions plotted on both plots as a check on the primary plot.

  There are three major reasons for maintaining a primary and secondary plot. First, as mentioned above, the secondary fix sources provide a good check on the accuracy of visual piloting. Large discrepancies between visual and radar positions may point out a problem with the visual fixes that the navigator might not otherwise suspect. Secondly, the navigator often must change the primary means of navigation during the transit. He may initially designate visual bearings as the primary fix method only to have a sudden storm or fog obscure the visual NAVAIDS. If he shifts the primary fix means to radar, he has a track history of the correlation between radar and visual fixes. Finally, the piloting team often must shift charts several times during the transit. When the old chart is taken off the plotting table and before the new chart is secured, there is a period of time when no chart is in use. Maintaining a secondary plot eliminates this complication. Ensure the secondary plot is not shifted prior to getting the new primary plot chart down on the chart table. In this case, there will always be a chart available on which to pilot. Do not consider the primary chart shifted until the new chart is properly secured and the plotter has transferred the last fix from the original chart onto the new chart.

- **Satellite Navigation Operator:** This operator normally works for the secondary plot supervisor. GPS accuracy with Selective Availability (SA) on is not sufficient for navigating restricted waters; but with SA off, GPS can support harbor navigation, in which case it should be considered as only one aid to navigation, not as a substitute for the entire process. If the team loses visual bearings in the channel and no radar NAVAIDS are available, GPS may be the most accurate fix source available. The navigator must have some data on the comparison between satellite positions and visual positions over the history of the passage to use satellite positions effectively. The only way to obtain this data is to plot satellite positions and compare these positions to visual positions throughout the harbor passage.

- **Fathometer Operator:** Run the fathometer continuously and station an operator to monitor it. Do not rely on audible alarms to key your attention to this critically important piloting tool. The fathometer operator must know the warning and danger soundings for the area the vessel is transiting. Most fathometers can display either total depth of water or depth under the keel. Set the fathometer to display depth under the keel. The navigator must check the sounding at each fix and compare that value to the charted sounding. A discrepancy between these values is cause for immediate action to take another fix and check the ship's position.

810. Harbor Approach (Inbound Vessels Only)

The piloting team must make the transition from coastal navigation to piloting smoothly as the vessel approaches restricted waters. There is no rigid demarcation between coastal navigation and piloting. Often visual NAVAIDS are visible miles from shore where Loran and GPS are easier to use. The navigator should take advantage of this overlap when approaching the harbor. Plotting Loran, GPS, and visual fixes concurrently ensures that the piloting team has correctly identified NAVAIDS and that the different types of systems are in agreement. Once the vessel is close enough to the shore such that sufficient NAVAIDS (at least three with sufficient bearing spread) become visible, the navigator should order visual bearings only for the primary plot and shift all other fixes to the secondary plot, unless the decision has been made to proceed with ECDIS as the primary system.

Take advantage of the coastal navigation and piloting overlap to shorten the fix interval gradually. The navigator must use his judgment in adjusting fix intervals. If the ship is steaming inbound directly towards the shore, set a fix interval such that two fix intervals lie between the vessel and the nearest danger. Upon entering restricted waters, the piloting team should be plotting visual fixes at three minute intervals.

Commercial vessels with GPS and/or Loran C, planning the harbor transit with a pilot, will approach a coast differently. The transition from ocean to coastal to harbor approach navigation will proceed as visual aids and radar targets appear and are plotted. With GPS or ECDIS operating and a waypoint set at the pilot station, only a few fixes are necessary to verify that the GPS position is correct. Once the pilot is aboard, the captain/pilot team may elect to navigate visually, depending on the situation.

TAKING FIXES WHILE PILOTING

Safe navigation while piloting requires frequent fixing of the ship's position. If ECDIS is the primary navigation system in use, this process is automatic, and the role of the navigator is to monitor the progress of the vessel, cross-check the position occasionally, and be alert for any indication that the system is not operating optimally.

If an ECS is in use, it should be considered only a supplement to the paper navigation plot, which legally must still be maintained. As long as the manual plot and the ECS plot are in agreement, the ECS is a valuable tool which shows the navigator where the ship is at any instant, not two or three minutes ago when the last fix was taken. It cannot legally take the place of the paper chart and the manual plot, but it can provide an additional measure of assurance that the ship is in safe water and alert the navigator to a developing dangerous situation before the next round of bearings or ranges.

The next several articles will discuss the three major manual methods used to fix a ship's position when piloting: crossing lines of position, copying satellite or Loran data, or advancing a single line of position. Using one method does not exclude using other methods. The navigator must obtain as much information as possible and employ as many of these methods as necessary.

811. Types of Fixes

While the intersection of two LOP's constitutes a **fix** under one definition, and only an estimated position by another, the prudent navigator will always use at least three LOP's if they are available, so that an error is apparent if they don't meet in a point. Some of the most commonly used methods of obtaining LOP's are discussed below:

- **Fix by Bearings:** The navigator can take and plot bearings from two or more charted objects. This is the most common and often the most accurate way to fix a vessel's position. Bearings may be taken directly to charted objects, or tangents of points of land. See Figure 811a. The intersection of these lines constitutes a fix. A position taken by bearings to buoys should not be considered a fix, but an estimated position (EP), because buoys swing about their watch circle and may be out of position.

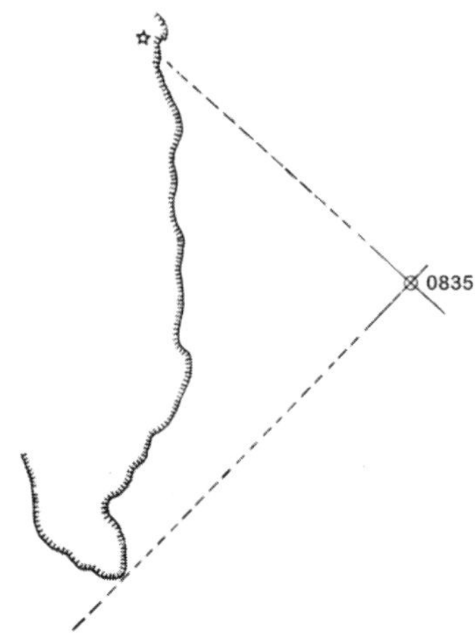

Figure 811a. A fix by two bearing lines.

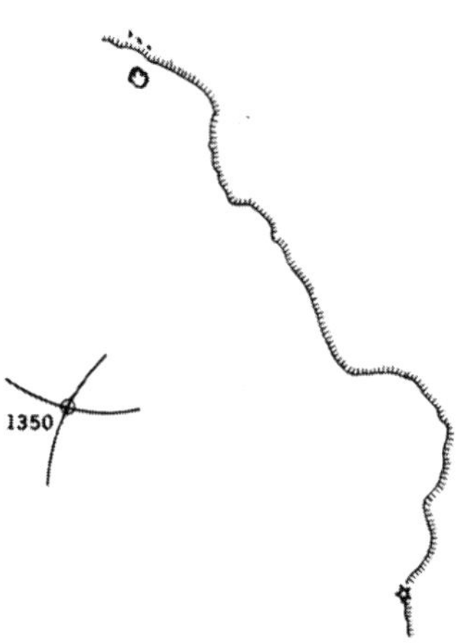

Figure 811b. A fix by two radar ranges.

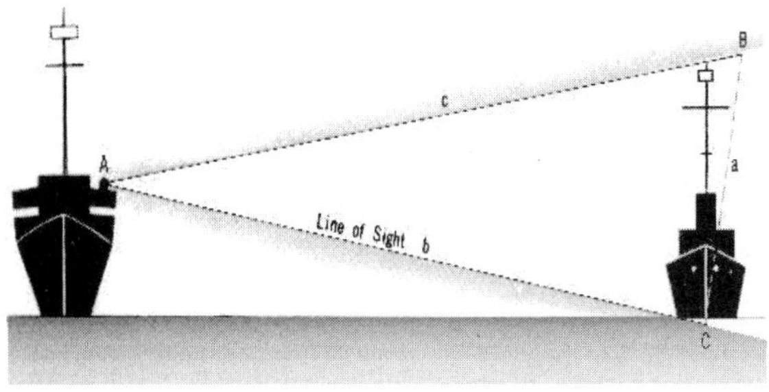

Figure 811c. Principle of stadimeter operation.

- **Fix by Ranges:** The navigator can plot a fix consisting of the intersection of two or more range arcs from charted objects. He can obtain an object's range in several ways:

 1. **Radar Ranges:** See Figure 811b. The navigator may take ranges to two fixed objects. The intersection of the range arcs constitutes a fix. He can plot ranges from any point on the radar scope which he can correlate on his chart. Remember that the shoreline of low-lying land may move many yards in an area of large tidal range, and swampy areas may be indistinct.

 2. **Stadimeter Ranges:** Given a known height of a NAVAID, one can use a stadimeter to determine its range. See Figure 811c for a representation of the geometry involved. Generally, stadimeters contain a height scale on which is set the height of the object. The observer then directs his line of sight through the stadimeter to the base of the object being observed. Finally, he adjusts the stadimeter's range index until the object's top reflection is "brought down" to the visible horizon. Read the object's range off the range index.

 3. **Sextant Vertical Angles:** Measure the vertical angle from the top of the NAVAID to the waterline below the NAVAID. Enter Table 16 to determine the distance of the NAVAID. The navigator must know the height of the NAVAID above sea level to use this table; it can be found in the *Light List*.

 4. **Sonar Ranges:** If the vessel is equipped with a sonar suite, the navigator can use sonar echoes to determine ranges to charted underwater objects. It may take some trial and error to set the active signal strength at a value that will give a strong return and still not cause excessive reverberation. Check local harbor restrictions on energizing active sonar. Avoid active sonar transmissions in the vicinity of divers.

- **Fix by Bearing and Range:** This is a hybrid fix of LOP's from a bearing and range to a single object. The radar is the only instrument that can give simultaneous range and bearing information to the same object. (A sonar system can also provide bearing and range information, but sonar bearings are far too inaccurate to use in piloting.) Therefore, with the radar, the navigator can obtain an instantaneous fix from only one NAVAID. This unique fix is shown in Figure 811d. This makes the radar an extremely useful tool for the piloting team. The radar's characteristics make it much more accurate determining range than determining

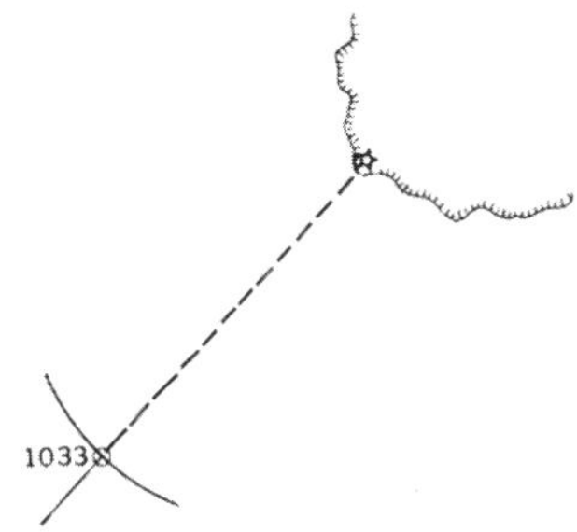

Figure 811d. A fix by range and bearing of a single object.

bearing; therefore, two radar ranges are preferable to a radar range and bearing.

- **Fix by Range Line and Distance:** When the vessel comes in line with a range, plot the bearing to the range (while checking compass error in the bargain) and cross this LOP with a distance from another NAVAID. Figure 811e shows this fix.

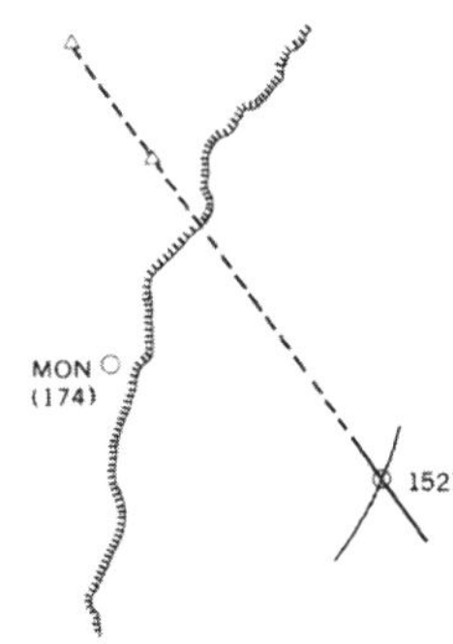

Figure 811e. A fix by a range and distance.

812. The Running Fix

When only one NAVAID is available from which to obtain bearings, use a technique known as the **running fix**. Use the following method:

1. Plot a bearing to a NAVAID (LOP 1).
2. Plot a second bearing to a NAVAID (either the same NAVAID or a different one) at a later time (LOP 2).
3. Advance LOP 1 to the time when LOP 2 was taken.
4. The intersection of LOP 2 and the advanced LOP 1 constitute the running fix.

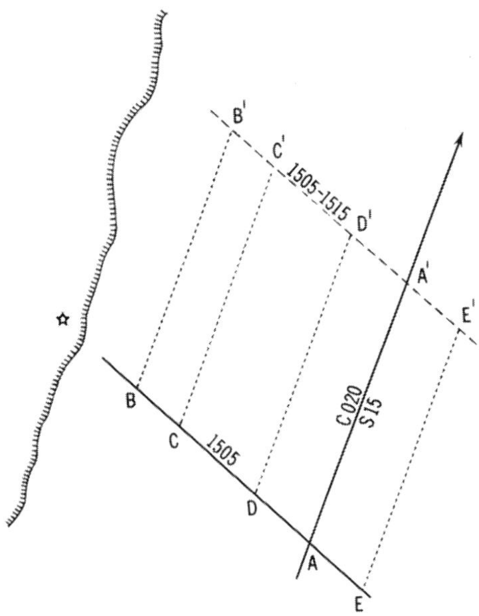

Figure 812a. Advancing a line of position.

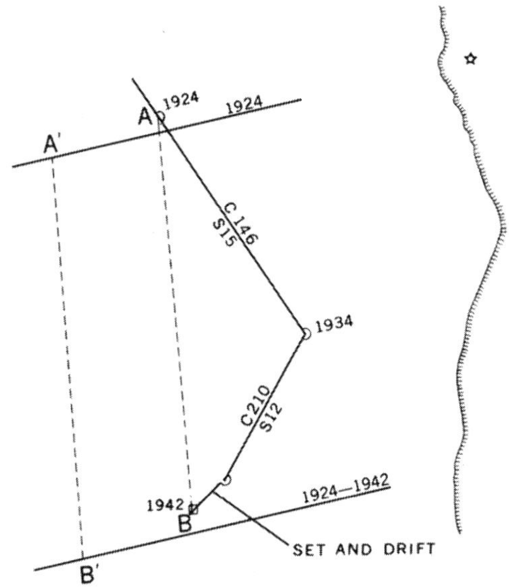

Figure 812b. Advancing a line of position with a change in course and speed, allowing for set and drift.

Figure 812a represents a ship proceeding on course 020°, speed 15 knots. At 1505, the plotter plots an LOP to a lighthouse bearing 310°. The ship can be at any point on this 1505 LOP. Some possible points are represented as points A, B, C, D, and E in Figure 812a. Ten minutes later the ship will have traveled 2.5 miles in direction 020°. If the ship was at A at 1505, it will be at A' at 1515. However, if the position at 1505 was B, the position at 1515 will be B'. A similar relationship exists between C and C', D and D', E and E'. Thus, if any point on the original LOP is moved a distance equal to the distance run in the direction of the motion, a line through this point parallel to the original line of position represents all possible positions of the ship at the later time. This process is called **advancing** a line of position. Moving a line back to an earlier time is called **retiring** a line of position.

When advancing a line of position, account for course changes, speed changes, and set and drift between the two bearing lines. Three methods of advancing an LOP are discussed below:

Method 1: See Figure 812b. To advance the 1924 LOP to 1942, first apply the best estimate of set and drift to the 1942 DR position and label the resulting position point B. Then, measure the distance between the dead reckoning position at 1924 (point A) and point B. Advance the LOP a distance equal to the distance between points A and B. Note that LOP A'B' is in the same direction as line AB.

Method 2: See Figure 812c. Advance the NAVAIDS position on the chart for the course and distance traveled by the vessel and draw the line of position from the NAVAIDS

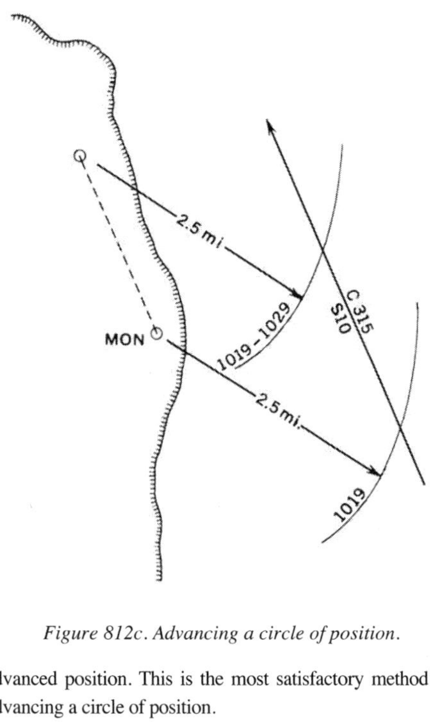

Figure 812c. Advancing a circle of position.

advanced position. This is the most satisfactory method for advancing a circle of position.

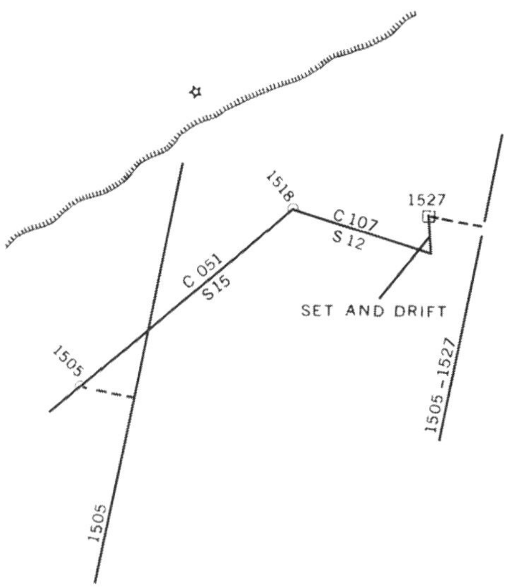

Figure 812d. Advancing a line of position by its relation to the dead reckoning.

Method 3: See Figure 812d. To advance the 1505 LOP to 1527, first draw a correction line from the 1505 DR position to the 1505 LOP. Next, apply a set and drift correction to the 1527 DR position. This results in a 1527 estimated position (EP). Then, draw from the 1527 EP a correction line of the same length and direction as the one drawn from the 1505 DR to the 1505 LOP. Finally, parallel the 1505 bearing to the end of the correction line as shown.

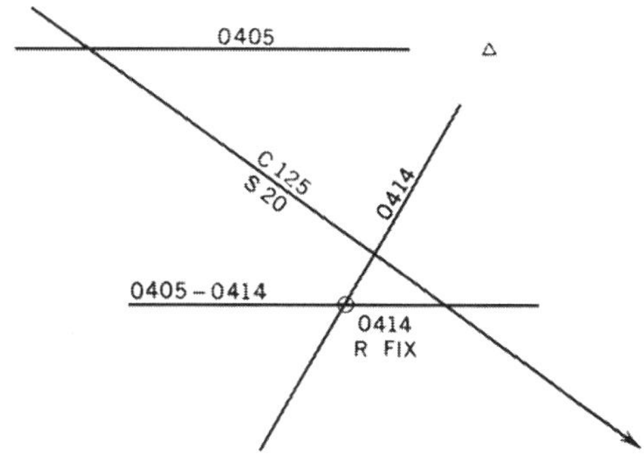

Figure 812e. A running fix by two bearings on the same object.

Label an advanced line of position with both the time of observation and the time to which the line is adjusted.

Figure 812e through Figure 812g demonstrate three running fixes. Figure 812e illustrates the case of obtaining a running fix with no change in course or speed between taking two bearings on the same NAVAID. Figure 812f illustrates a running fix with changes in a vessel's course and speed between taking two bearings on two different objects. Finally, Figure 812g illustrates a running fix obtained by advancing range circles of position using the second method discussed above.

PILOTING PROCEDURES

The previous section discussed the methods for fixing the ship's position. This section discusses integrating the manual fix methods discussed above, and the use of the fathometer, into a piloting procedure. The navigator must develop his piloting procedure to meet several requirements. He must obtain enough information to fix the position of the vessel without question. He must also plot and evaluate this information. Finally, he must relay his evaluations and recommendations to the vessel's conning officer. This section examines some considerations to ensure the navigator accomplishes all these requirements quickly and effectively. Of course, if ECDIS is the primary plot, manual methods as discussed here are for backup use.

813. Fix Type and Fix Interval

The preferred piloting fix is taken from visual bearings from charted fixed NAVAIDS. Plot visual bearings on the primary plot and plot all other fixes on the secondary plot. If poor visibility obscures visual NAVAIDS, shift to radar piloting on the primary plot. If neither visual or radar piloting is available, consider standing off until the visibility improves.

The interval between fixes in restricted waters should usually not exceed three minutes. Setting the fix interval at three minutes optimizes the navigator's ability to assimilate and evaluate all available information. He must relate it to charted navigational hazards and to his vessel's intended track. It should take a well trained plotting team no more than 30 seconds to measure, record, and plot three bearings to three separate NAVAIDS. The navigator should spend the majority of the fix interval time interpreting the information, evaluating the navigational situation, and making recommendations to the conning officer.

If three minutes goes by without a fix, inform the captain and try to plot a fix as soon as possible. If the delay was caused by a loss of visibility, shift to radar piloting. If the delay was caused by plotting error, take another fix. If the navigator cannot get a fix down on the plot for several more minutes, consider slowing or stopping the ship until its position can be fixed. Never continue a passage through

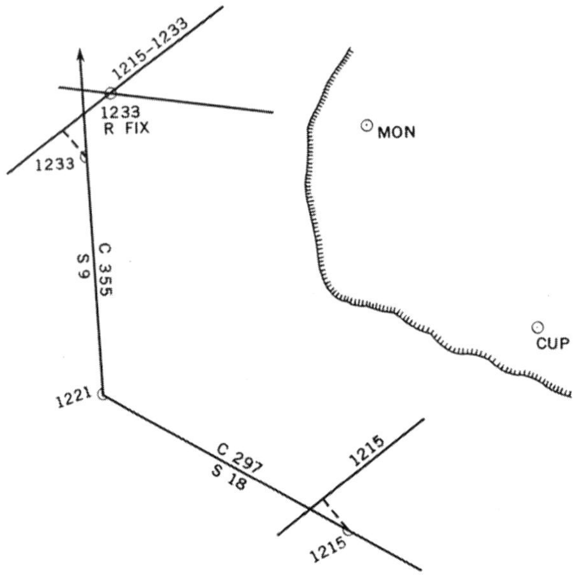

Figure 812f. A running fix with a change of course and speed between observations on separate landmarks.

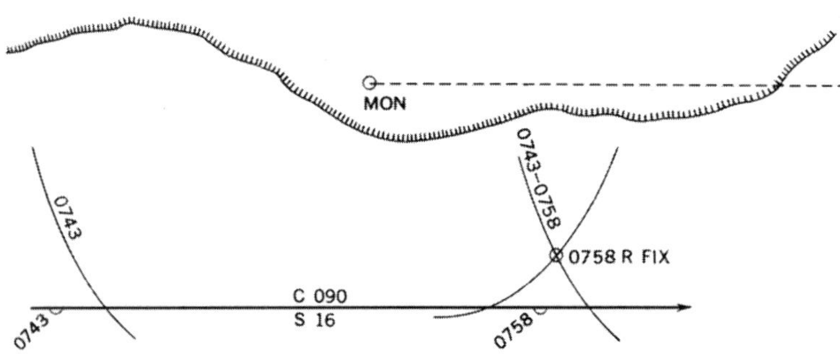

Figure 812g. A running fix by two circles of position.

restricted waters if the vessel's position is uncertain.

The secondary plot supervisor should maintain the same fix interval as the primary plot. Usually, this means he should plot a radar fix every three minutes. He should plot other fix sources (Loran and GPS fixes, for example) at an interval sufficient for making meaningful comparisons between fix sources. Every third fix interval, he should pass a radar fix to the primary plot for comparison with the visual fix. He should inform the navigator how well all the fix sources plotted on the secondary plot are tracking.

814. The Piloting Routine

Following a cyclic routine ensures the timely and efficient processing of data and forms a smoothly functioning piloting team. It quickly yields the information which the navigator needs to make informed recommendations to the conning officer and captain.

Repeat this routine at each fix interval beginning when the ship gets underway until it clears the harbor (outbound) or when the ship enters the harbor until it is moored (inbound).

The routine consists of the following steps:

1. Take, plot and label a fix.
2. Calculate set and drift from the DR position.
3. Reset the DR from the fix and DR two fixes ahead.

- **Plotting the Fix:** This involves coordination between the navigator, bearing taker(s), recorder, and plotter.

The navigator will call for each fix at the DR time. The bearing taker must measure his bearings as quickly as possible, beam bearings first, fore and aft last, on the navigator's mark. The recorder will write the bearings in the book, and the plotter will plot them immediately.

- **Labeling the Fix:** The plotter should clearly mark a visual fix with a circle or an electronic fix with a triangle. Clearly label the time of each fix. A visual running fix should be circled, marked "R Fix" and labeled with the time of the second LOP. Keep the chart neat and uncluttered when labeling fixes.

- **Dead Reckoning Two Fix Intervals Ahead:** After labeling the fix, the plotter should dead reckon the fix position ahead two fix intervals. The navigator should carefully check the area marked by this DR for any navigational hazards. If the ship is approaching a turn, update the turn bearing as discussed in Article 802.

- **Calculate Set and Drift at Every Fix:** Calculating set and drift is covered in Chapter 7. Calculate these values at every fix and inform the captain and conning officer. Compare the actual values of set and drift with the predicted values from the current graph discussed in Article 804. Evaluate how the current is affecting the vessel's position in relation to the track and recommend courses and speeds to regain the planned track. Because the navigator can determine set and drift only when comparing fixes and DR's plotted for the same time, take fixes exactly at the times for which a DR has been plotted. Repeat this routine at each fix interval beginning when the ship gets underway until it clears the harbor (outbound) or when the ship enters the harbor until she is moored (inbound).

- **Piloting Routine When Turning:** Modify the cyclic routine slightly when approaching a turn. Adjust the fix interval so that the plotting team has a fix plotted approximately one minute before a scheduled turn. This gives the navigator sufficient time to evaluate the position in relation to the planned track, DR ahead to the slide bar to determine a new turn bearing, relay the new turn bearing to the conning officer, and then monitor the turn bearing to mark the turn.

Approximately 30 seconds before the time to turn, train the alidade on the turn bearing NAVAID. Watch the bearing of the NAVAID approach the turn bearing. About 1° away from the turn bearing, announce to the conning officer: "Stand by to turn." Slightly before the turn bearing is indicated, report to the conning officer: "Mark the turn." Make this report slightly before the bearing is reached because it takes the conning officer a finite amount of time to acknowledge the report and order the helmsman to put over the rudder. Additionally, it takes a finite amount of time for the helmsman to turn the rudder and for the ship to start to turn. If the navigator waits until the turn bearing is indicated to report the turn, the ship will turn too late.

Once the ship is steady on the new course, immediately take another fix to evaluate the vessel's position in relation to the track. If the ship is not on the track after the turn, calculate and recommend a course to the conning officer to regain the track.

815. Using the Fathometer

Use the fathometer to determine whether the depth of water under the keel is sufficient to prevent the ship from grounding and to check the actual water depth with the charted water depth at the fix position. The navigator must compare the charted sounding at every fix position with the fathometer reading and report to the captain any discrepancies. Taking continuous soundings in restricted waters is mandatory.

See the discussion of calculating the warning and danger soundings in Article 802. If the warning sounding is received, then slow the ship, fix the ship's position more frequently, and proceed with extreme caution. Ascertain immediately where the ship is in the channel; if the minimum expected sounding was noted correctly, the warning sounding indicates the vessel may be leaving the channel and standing into shoal water. Notify the vessel's captain and conning officer immediately.

If the danger sounding is received, take immediate action to get the vessel back to deep water. Reverse the engines and stop the vessel's forward movement. Turn in the direction of the deepest water before the vessel loses steerageway. Consider dropping the anchor to prevent the ship from drifting aground. The danger sounding indicates that the ship has left the channel and is standing into immediate danger. It requires immediate corrective action by the ship's conning officer, navigator, and captain to avoid disaster.

Many underwater features are poorly surveyed. If a fathometer trace of a distinct underwater feature can be obtained along with accurate position information, send the fathometer trace and related navigational data to NIMA for entry into the Digital Bathymetric Data Base.

PILOTING TO AN ANCHORAGE

816. Choosing an Anchorage

Most U.S. Navy vessels receive instructions in their movement orders regarding the choice of anchorage. Merchant ships are often directed to specific anchorages by harbor authorities. However, lacking specific guidance, the mariner should choose his anchoring positions using the following criteria:

- **Depth of Water:** Choose an area that will provide sufficient depth of water through an entire range of tides. Water too shallow will cause the ship to go aground, and water too deep will allow the anchor to drag.

- **Type of Bottom:** Choose the bottom that will best hold the anchor. Avoid rocky bottoms and select sandy or muddy bottoms if they are available.

- **Proximity to navigational Hazards:** Choose an anchorage as far away as possible from known navigational hazards.

- **Proximity to Adjacent Ships:** Anchor well away from adjacent vessels; ensure that another vessel will not swing over your own anchor on a current or wind shift.

- **Proximity to Harbor Traffic Lanes:** Anchor clear of traffic lanes and ensure that the vessel will not swing into the channel on a current or wind shift.

- **Weather:** Choose an area with the weakest winds and currents.

- **Availability of NAVAIDS:** Choose an anchorage with several NAVAIDS available for monitoring the ship's position when anchored.

817. Navigational Preparations for Anchoring

It is usually best to follow an established procedure to ensure an accurate positioning of the anchor, even when anchoring in an open roadstead. The following procedure is representative. See Figure 817.

Locate the selected anchoring position on the chart. Consider limitations of land, current, shoals, and other vessels when determining the direction of approach. Where conditions permit, make the approach heading into the current. Close observation of any other anchored vessels will provide clues as to which way the ship will lie to her anchor. If wind and current are strong and from different directions, ships will lie to their anchors according to the balance between these two forces and the draft and trim of each ship. Different ships may lie at different headings in the same anchorage depending on the balance of forces affecting them.

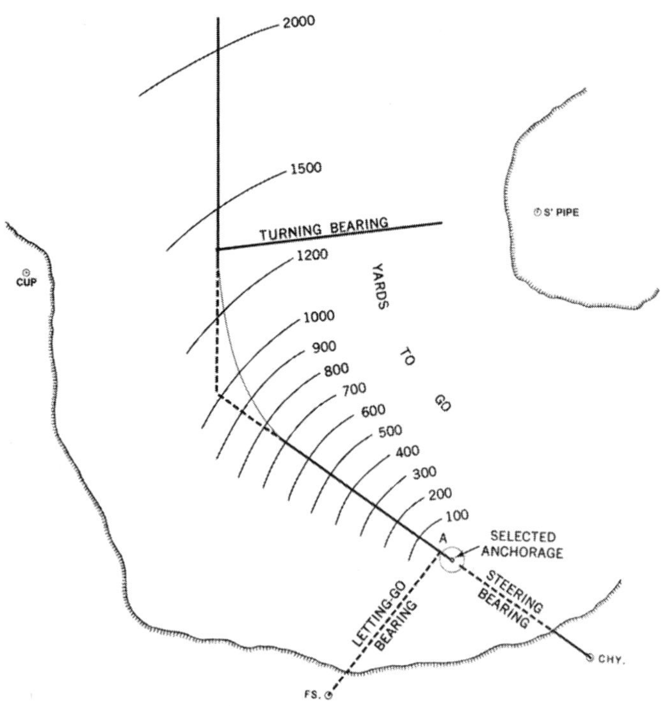

Figure 817. Anchoring.

Approach from a direction with a prominent NAVAID, preferably a range, available dead ahead to serve as a steering guide. If practicable, use a straight approach of at least 1200 yards to permit the vessel to steady on the required course. Draw in the approach track, allowing for advance and transfer during any turns. In Figure 817, the chimney was selected as this steering bearing. A turn range may also be used if a radar-prominent object can be found directly ahead or astern.

Next, draw a circle with the selected position of the anchor as the center, and with a radius equal to the distance between the hawsepipe and pelorus, alidade, or periscope used for measuring bearings. This circle is marked "A" in Figure 817. The intersection of this circle and the approach track is the position of the vessel's bearing-measuring instrument at the moment of letting the anchor go. Select a NAVAID which will be on the beam when the vessel is at the point of letting go the anchor. This NAVAID is marked "FS" in Figure 817. Determine what the bearing to that object will be when the ship is at the drop point and measure this bearing to the nearest 0.1°T. Label this bearing as the letting go bearing.

During the approach to the anchorage, plot fixes at frequent intervals. The navigator must advise the conning officer of any tendency of the vessel to drift from the desired track. The navigator must frequently report to the conning officer the distance to go, permitting adjustment of the speed so that the vessel will be dead in the water or have very slight sternway when the anchor is let go. To aid in determining the distance to the drop point, draw and label a number of range arcs as shown in Figure 817 representing distances to go to the drop point.

At the moment of letting the anchor go, take a fix and plot the vessel's exact position on the chart. This is important in the construction of the swing and drag circles discussed below. To draw these circles accurately, determine the position of the vessel at the time of letting go the anchor as accurately as possible.

Veer the anchor chain to a length equal to five to seven times the depth of water at the anchorage. The exact amount to veer is a function of both vessel type and severity of weather expected at the anchorage. When calculating the scope of anchor chain to veer, take into account the maximum height of tide.

Once the ship is anchored, construct two separate circles around the ship's position when the anchor was dropped. These circles are called the **swing circle** and the **drag circle**. Use the swing circle to check for navigational hazards and use the drag circle to ensure the anchor is holding.

The swing circle's radius is equal to the sum of the ship's length and the scope of the anchor chain released. This represents the maximum arc through which a ship can swing while riding at anchor if the anchor holds. Examine this swing circle carefully for navigational hazards, interfering contacts, and other anchored shipping. Use the lowest height of tide expected during the anchoring period when checking inside the swing circle for shoal water.

The drag circle's radius equals the sum of the hawsepipe to pelorus distance and the scope of the chain released. Any bearing taken to check on the position of the ship should, if the anchor is holding, fall within the drag circle. If a fix falls outside of that circle, then the anchor is dragging. If the vessel has a GPS or Loran system with an off-station alarm, set the alarm at the drag circle radius, or slightly more.

In some cases, the difference between the radii of the swing and drag circles will be so small that, for a given chart scale, there will be no difference between the circles when plotted. If that is the case, plot only the swing circle and treat that circle as both a swing and a drag circle. On the other hand, if there is an appreciable difference in radii between the circles when plotted, plot both on the chart. Which method to use falls within the sound judgment of the navigator.

When determining if the anchor is holding or dragging, the most crucial period is immediately after anchoring. Fixes should be taken frequently, at least every three minutes, for the first thirty minutes after anchoring. The navigator should carefully evaluate each fix to determine if the anchor is holding. If the anchor is holding, the navigator can then increase the fix interval. What interval to set falls within the judgment of the navigator, but the interval should not exceed 30 minutes. If an ECDIS, Loran, or GPS is available, use its off-station alarm feature for an additional safety factor.

NAVIGATIONAL ASPECTS OF SHIP HANDLING

818. Effects Of Banks, Channels, and Shallow Water

A ship moving through shallow water experiences pronounced effects from the proximity of the nearby bottom. Similarly, a ship in a channel will be affected by the proximity of the sides of the channel. These effects can easily cause errors in piloting which lead to grounding. The effects are known as **squat**, **bank cushion**, and **bank suction**. They are more fully explained in texts on shiphandling, but certain navigational aspects are discussed below.

Squat is caused by the interaction of the hull of the ship, the bottom, and the water between. As a ship moves through shallow water, some of the water it displaces rushes under the vessel to rise again at the stern. This causes a venturi effect, decreasing upward pressure on the hull. Squat makes the ship sink deeper in the water than normal and slows the vessel. The faster the ship moves through shallow water, the greater is this effect; groundings on both charted and uncharted shoals and rocks have occurred because of this phenomenon, when at reduced speed the ship could have safely cleared the dangers. When navigating in shallow water, the navigator must reduce speed to avoid squat. If bow and stern waves nearly perpendicular the direction of travel are noticed, and the vessel slows with no change in shaft speed, squat is occurring. Immediately slow the ship to counter it. Squatting occurs in deep water also, but is more pronounced and dangerous in shoal water. The large waves generated by a squatting ship also endanger shore facilities and other craft.

Bank cushion is the effect on a ship approaching a steep underwater bank at an oblique angle. As water is forced into the narrowing gap between the ship's bow and the shore, it tends to rise or pile up on the landward side, causing the ship to sheer away from the bank.

Bank suction occurs at the stern of a ship in a narrow channel. Water rushing past the ship on the landward side exerts less force than water on the opposite or open water side. This effect can actually be seen as a difference in draft readings from one side of the vessel to the other, and is similar to the venturi effect seen in squat. The stern of the ship is forced toward the bank. If the ship gets too close to the bank, it can be forced sideways into it. The same effect occurs between two vessels passing close to each other.

These effects increase as speed increases. Therefore, in shallow water and narrow channels, navigators should decrease speed to minimize these effects. Skilled pilots may use these effects to advantage in particular situations, but the average mariner's best choice is slow speed and careful attention to piloting.

ADVANCED PILOTING TECHNIQUES

819. Assuming Current Values to Set Safety Margins for Running Fixes

Current affects the accuracy of a running fix. Consider, for example, the situation of an unknown head current. In Figure 819a, a ship is proceeding along a coast, on course 250 ° speed 12 knots. At 0920 light A bears 190°, and at 0930 it bears 143°. If the earlier bearing line is advanced a distance of 2 miles (10 minutes at 12 knots) in the direction of the course, the running fix is as shown by the solid lines. However, if there is a head current of 2 knots, the ship is making good a speed of only 10 knots, and in 10 minutes will travel a distance of only 1 $^2/_3$ miles. If the first bearing line is advanced this distance, as shown by the broken line, the actual position of the ship is at B. This actual position is nearer the shore than the running fix actually plotted. A following current, conversely, would show a position too far from the shore from which the bearing was measured.

If the navigator assumes a following current when advancing his LOP, the resulting running fix will plot further from the NAVAID than the vessel's actual position. Conversely, if he assumes a head current, the running fix will plot closer to the NAVAID than the vessel's actual position. To ensure a margin of safety when plotting running fix bearings to a NAVAID on shore, always assume the current slows a vessel's speed over ground. This will cause the running fix to plot closer to the shore than the ship's actual position.

When taking the second running fix bearing from a different object, maximize the speed estimate if the second object is on the same side and farther forward, or on the opposite side and farther aft, than the first object was when observed.

All of these situations assume that danger is on the same side as the object observed first. If there is either a head or following current, a series of running fixes based upon a number of bearings of the same object will plot in a straight line parallel to the course line, as shown in Figure 819b. The plotted line will be too close to the object observed if there is a head current and too far out if there is a following current. The existence of the current will not be apparent unless the actual speed over the ground is known. The position of the plotted line relative to the dead reckoning course line is not a reliable guide.

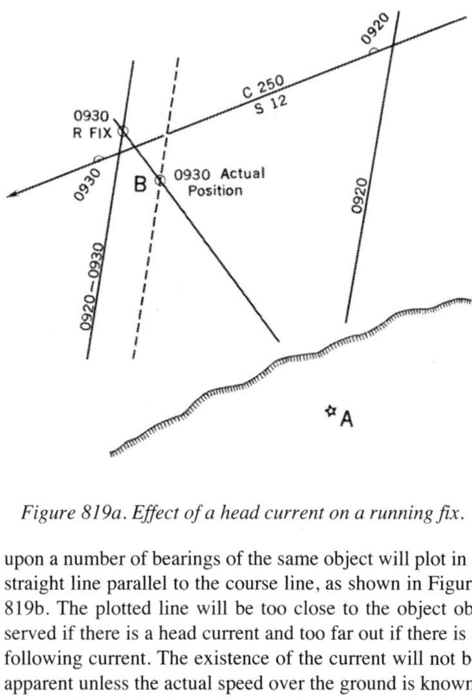

Figure 819a. Effect of a head current on a running fix.

820. Determining Track Made Good by Plotting Running Fixes

A current oblique to a vessel's course will also result in an incorrect running fix position. An oblique current can be detected by observing and plotting several bearings of the same object. The running fix obtained by advancing one

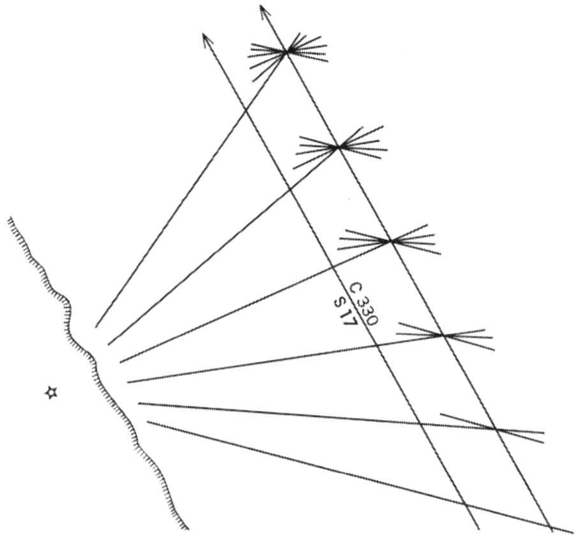

Figure 819b. A number of running fixes with a following current.

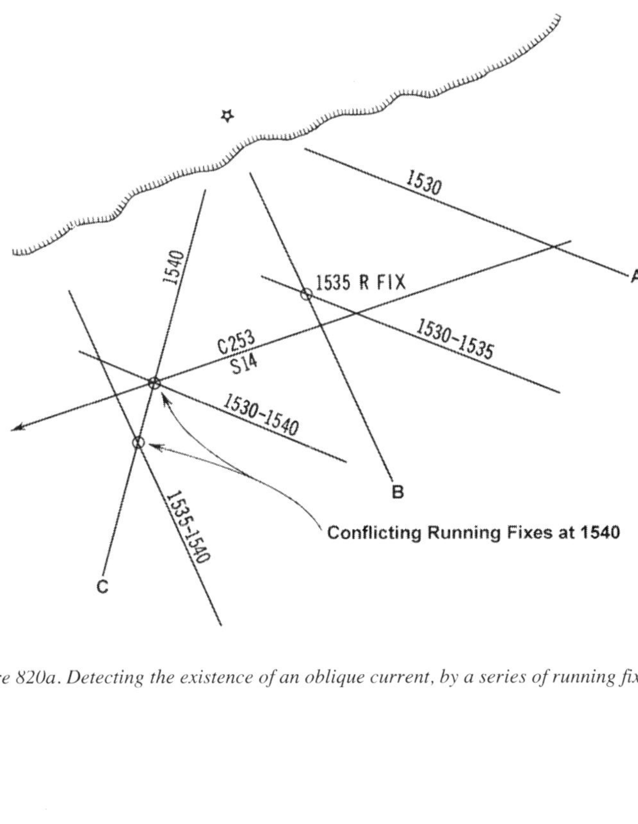

Figure 820a. Detecting the existence of an oblique current, by a series of running fixes.

bearing line to the time of the next one will not agree with the running fix obtained by advancing an earlier line. See Figure 820a. If bearings A, B, and C are observed at five-minute intervals, the running fix obtained by advancing B to the time of C will not be the same as that obtained by advancing A to the time of C, as shown in Figure 820a.

Whatever the current, the navigator can determine the direction of the track made good (assuming constant current and constant course and speed). Observe and plot three bearings of a charted object O. See Figure 820b. Through O draw XY in any direction. Using a convenient scale, determine points A and B so that OA and OB are proportional to the time intervals between the first and second bearings and the second and third bearings, respectively. From A and B draw lines parallel to the second bearing line, intersecting the first and third bearing lines at C and D, respectively. The direction of the line from C and D is the track made good.

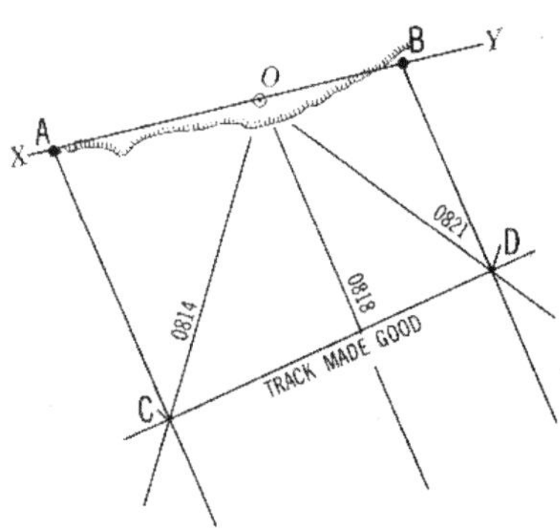

Figure 820b. Determining the track made good.

The distance of the line CD in Figure 820b from the track is in error by an amount proportional to the ratio of the speed made good to the speed assumed for the solution. If a good fix (not a running fix) is obtained at some time before the first bearing for the running fix, and the current has not changed, the track can be determined by drawing a line from the fix, in the direction of the track made good. The intersection of the track with any of the bearing lines is an actual position.

821. Fix by Distance of an Object by Two Bearings (Table 18)

Geometrical relationships can define a running fix. In Figure 821, the navigator takes a bearing on NAVAID D. The bearing is expressed as degrees right or left of course. Later, at B, he takes a second bearing to D; similarly, he takes a bearing at C, when the landmark is broad on the beam. The navigator knows the angles at A, B, and C and the distance run between points. The various triangles can be solved using Table 18. From this table, the navigator can calculate the lengths of segments AD, BD, and CD. He knows the range and bearing; he can then plot an LOP. He can then advance these LOP's to the time of taking the CD bearing to plot a running fix.

Enter the table with the difference between the course and first bearing (angle BAD in Figure 821) along the top of the table and the difference between the course and second bearing (angle CBD) at the left of the table. For each pair of angles listed, two numbers are given. To find the distance from the landmark at the time of the second bearing (BD), multiply the distance run between bearings (in nautical miles) by the first number from Table 18. To find the distance when the object is abeam (CD), multiply the distance run between A and B by the second number from the table. If the run between bearings is exactly 1 mile, the tabulated values are the distances sought.

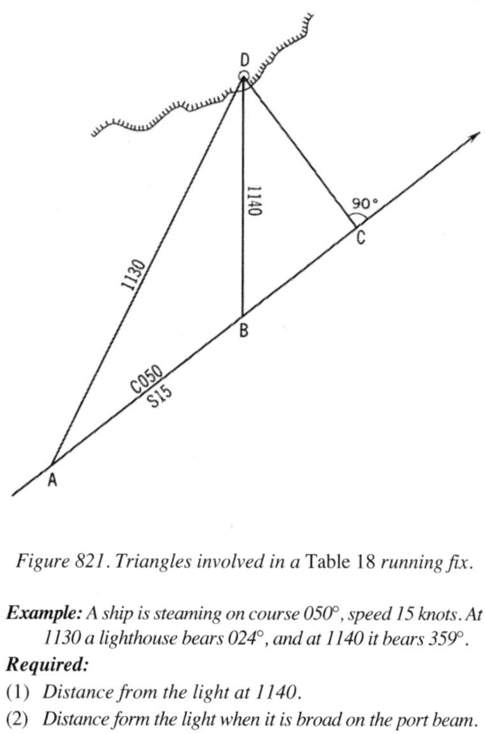

Figure 821. Triangles involved in a Table 18 *running fix.*

Example: *A ship is steaming on course 050°, speed 15 knots. At 1130 a lighthouse bears 024°, and at 1140 it bears 359°.*

Required:

(1) *Distance from the light at 1140.*

(2) *Distance form the light when it is broad on the port beam.*

Solution:

(1) *The difference between the course and the first bearing (050° – 24°) is 26°, and the difference between the course and the second bearing (050° + 360° - 359°) is 51°.*

(2) *From* Table 18, *the two numbers (factors are 1.04 and 0.81, found by interpolation.*

(3) *The distance run between bearings is 2.5 miles (10 minutes at 15 knots).*

(4) *The distance from the lighthouse at the time of the second bearing is 2.5 × 1.04 = 2.6 miles.*

(5) *The distance from the lighthouse when it is broad on the beam is 2.5 × 0.81 = 2.0 miles.*
Answer: *(1) D 2.6 mi., (2) D 2.0 mi.*

This method yields accurate results only if the helmsman has steered a steady course and the navigator uses the vessel's speed over ground.

MINIMIZING ERRORS IN PILOTING

822. Common Errors

Piloting requires a thorough familiarity with principles involved, constant alertness, and judgment. A study of groundings reveals that the cause of most is a failure to use or interpret available information. Among the more common errors are:

1. Failure to obtain or evaluate soundings
2. Mis-identification of aids to navigation
3. Failure to use available navigational aids effectively
4. Failure to correct charts
5. Failure to adjust a magnetic compass or keep a table of corrections
6. Failure to apply deviation
7. Failure to apply variation
8. Failure to check gyro and magnetic compass readings regularly
9. Failure to keep a dead reckoning plot
10. Failure to plot new information
11. Failure to properly evaluate information
12. Poor judgment
13. Failure to use information in charts and navigational publications
14. Poor navigation team organization
15. Failure to "keep ahead of the vessel"
16. Failure to have backup navigational methods in place
17. Failure to recognize degradation of electronically obtained LOP's or lat./long. positions

Some of the errors listed above are mechanical and some are matters of judgment. Conscientiously applying the principles and procedures of this chapter will go a long way towards eliminating many of the mechanical errors. However, the navigator must guard against the feeling that in following a checklist he has eliminated all sources of error. A navigator's judgment is just as important as his checklists.

823. Minimizing Errors with a Two Bearing Plot

When measuring bearings from two NAVAIDS, the fix error resulting from an error held constant for both observations is minimized if the angle of intersection of the bearings is 90°. If the observer in Figure 823a is located at point T and the bearings of a beacon and cupola are observed and plotted without error, the intersection of the bearing lines lies on the circumference of a circle passing through the beacon, cupola, and the observer. With constant error, the angular difference between the bearings of the beacon and the cupola is not affected. Thus, the angle formed at point F by the bearing lines plotted with constant error is equal to the angle formed at point T by the bearing lines plotted without error. From geometry it is known that angles having their apexes on the circumference of a circle and that are subtended by the same chord are equal. Since the angles at points T and F are equal and the angles are subtended by the same chord, the intersection at point F lies on the circumference of a circle passing through the beacon, cupola, and the observer.

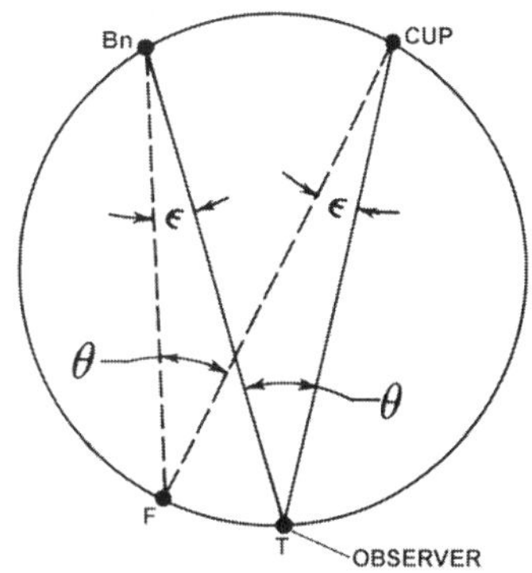

Figure 823a. Two-bearing plot.

Assuming only constant error in the plot, the direction of displacement of the two-bearing fix from the position of the observer is in accordance with the sign (or direction) of the constant error. However, a third bearing is required to determine the direction of the constant error.

Assuming only constant error in the plot, the two-bearing fix lies on the circumference of the circle passing through the two charted objects observed and the observer. The fix error, the length of the chord FT in Figure 823b, depends on the magnitude of the constant error $\in$, the distance between the charted objects, and the cosecant of the angle of cut, angle θ. In Figure 823b,

$$\text{The fix error} = \text{FT} = \frac{\text{BC} \csc\theta}{2}$$

where $\in$ is the magnitude of the constant error, BC is the length of the chord BC, and θ is the angle of the LOP's intersection.

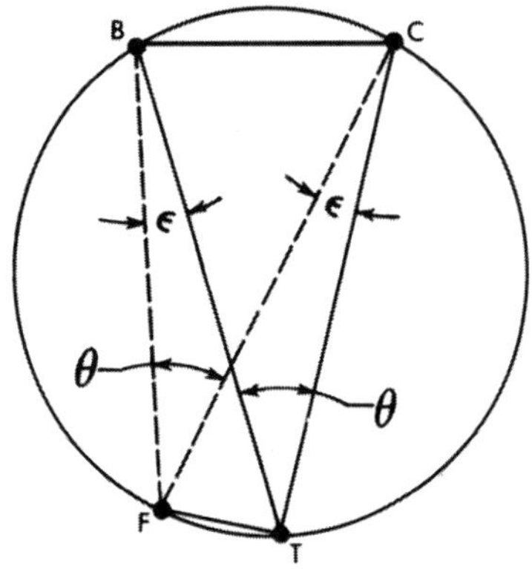

Figure 823b. Two-bearing plot with constant error.

Since the fix error is a function of the cosecant of the angle of intersection, it is least when the angle of intersection is 90°. As illustrated in Figure 823c, the error increases in accordance with the cosecant function as the angle of intersection decreases. The increase in the error becomes quite rapid after the angle of intersection has decreased to below about 30°. With an angle of intersection of 30°, the fix error is about twice that at 90°.

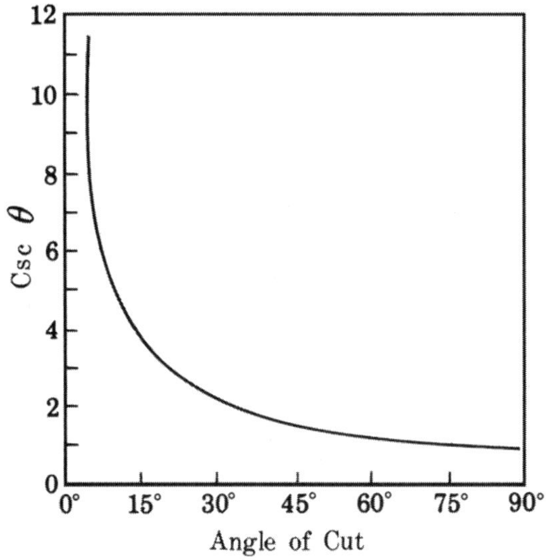

Figure 823c. Error of two-bearing plot.

824. Finding Compass Error by Trial and Error

If several fixes obtained by bearings on three objects produce triangles of error of about the same size, there might be a constant error in observing or plotting the bearings. If applying of a constant error to all bearings results in a pinpoint fix, apply such a correction to all subsequent fixes. Figure 824 illustrates this technique. The solid lines indicate the original plot, and the broken lines indicate each line of position moved 3° in a clockwise direction.

Employ this procedure carefully. Attempt to find and eliminate the error source. The error may be in the gyrocompass, the repeater, or the bearing transmission system. Compare the resulting fix positions with a satellite position, a radar position, or the charted sounding. A high degree of correlation between these three independent positioning systems and an "adjusted" visual fix is further confirmation of a constant bearing error.

TRAINING

825. Piloting Simulators

Civilian piloting training has traditionally been a function of both maritime academies and on-the-job experience. The latter is usually more valuable, because there is no substitute for experience in developing judgment. Military piloting training consists of advanced correspondence courses and formal classroom instruction combined with duties on the bridge. U.S. Navy Quartermasters frequently attend Ship's Piloting and Navigation (SPAN) trainers as a routine segment of shoreside training. Military vessels in general have a much clearer definition of responsibilities, as well as more people to carry them out, than civilian ships, so training is generally more thorough and targeted to specific skills.

Computer technology has made possible the development of computerized **ship simulators**, which allow piloting experience to be gained without risking accidents at sea and without incurring underway expenses. Simulators range from simple micro-computer-based software to a completely equipped ship's bridge with radar, engine controls, 360° horizon views, programmable sea motions, and the capability to simulate almost any navigational situation.

A different type of simulator consists of scale models of ships. The models, actually small craft of about 20-30 feet, have hull forms and power-to-weight ratios similar to various types of ships, primarily supertankers, and the operator pilots the vessel from a position such that his view is from the craft's "bridge." These are primarily used in training pilots and masters in docking maneuvers with exceptionally large vessels.

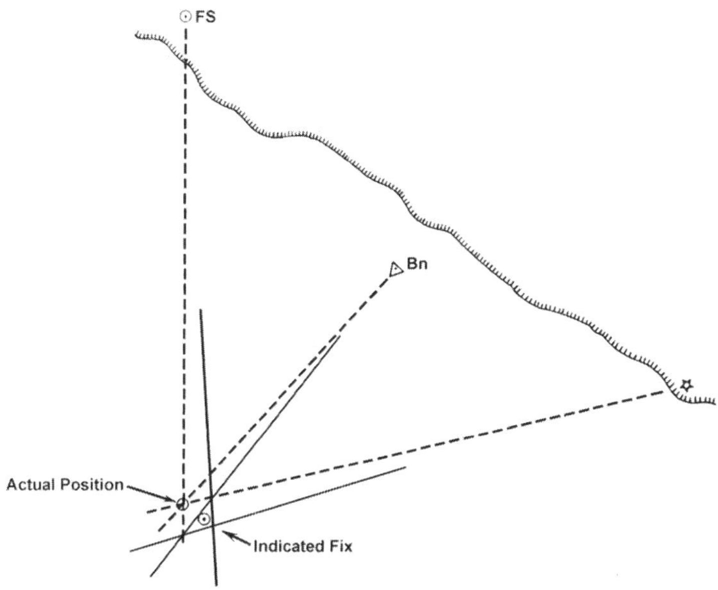

Figure 824. Adjusting a fix for constant error.

The first computer ship simulators came into use in the late 1970s. Several years later the U.S. Coast Guard began accepting a limited amount of simulator time as "sea time" for licensing purposes. They can simulate virtually any conditions encountered at sea or in piloting waters, including land, aids to navigation, ice, wind, fog, snow, rain, and lightning. The system can also be programmed to simulate hydrodynamic effects such as shallow water, passing vessels, current, and tugs.

Virtually any type of vessel can be simulated, including tankers, bulkers, container ships, tugs and barges, yachts, and military vessels. Similarly, any given navigational situation can be modeled, including passage through any chosen harbor, river, or passage, convoy operations, meeting and passing situations at sea and in harbors.

Simulators are used not only to train mariners, but also to test feasibility of port and harbor plans and visual aids to navigation system designs. This allows pilots to "navigate" simulated ships through simulated harbors before construction begins to test the adequacy of channels, turning basins, aids to navigation, and other factors.

A full-capability simulator consists of a ship's bridge which may have motion and noise/vibration inputs, a programmable visual display system which projects a simulated picture of the area surrounding the vessel in both daylight and night modes, image generators for the various inputs to the scenario such as video images and radar, a central data processor, a human factors monitoring system which may record and videotape bridge activities for later analysis, and a control station where instructors control the entire scenario.

Some simulators are part-task in nature, providing specific training in only one aspect of navigation such as radar navigation, collision avoidance, or night navigation.

While there is no substitute for on-the-job training, simulators are extremely cost effective systems which can be run for a fraction of the cost of an actual vessel. Further, they permit trainees to learn from mistakes with no possibility of an accident, they can model an infinite variety of scenarios, and they permit replay and reassessment of each maneuver.

CHAPTER 9

TIDES AND TIDAL CURRENTS

ORIGINS OF TIDES

900. Introduction

Tides are the periodic motion of the waters of the sea due to changes in the attractive forces of the Moon and Sun upon the rotating Earth. Tides can either help or hinder a mariner. A high tide may provide enough depth to clear a bar, while a low tide may prevent entering or leaving a harbor. Tidal current may help progress or hinder it, may set the ship toward dangers or away from them. By understanding tides and making intelligent use of predictions published in tide and tidal current tables and descriptions in sailing directions, the navigator can plan an expeditious and safe passage through tidal waters.

901. Tide and Current

The rise and fall of tide is accompanied by horizontal movement of the water called tidal current. It is necessary to distinguish clearly between tide and tidal current, for the relation between them is complex and variable. For the sake of clarity mariners have adopted the following definitions: Tide is the vertical rise and fall of the water, and tidal current is the horizontal flow. The tide rises and falls, the tidal current floods and ebbs. The navigator is concerned with the amount and time of the tide, as it affects access to shallow ports. The navigator is concerned with the time, speed, and direction of the tidal current, as it will affect his ship's position, speed, and course.

Tides are superimposed on nontidal rising and falling water levels, caused by weather, seismic events, or other natural forces. Similarly, tidal currents are superimposed upon non-tidal currents such as normal river flows, floods, and freshets.

902. Causes of Tides

The principal tidal forces are generated by the Moon and Sun. The Moon is the main tide-generating body. Due to its greater distance, the Sun's effect is only 46 percent of the Moon's. Observed tides will differ considerably from the tides predicted by equilibrium theory since size, depth, and configuration of the basin or waterway, friction, land masses, inertia of water masses, Coriolis acceleration, and other factors are neglected in this theory. Nevertheless, equilibrium theory is sufficient to describe the magnitude and distribution of the main tide-generating forces across the surface of the Earth.

Newton's universal law of gravitation governs both the orbits of celestial bodies and the tide-generating forces which occur on them. The force of gravitational attraction between any two masses, m_1 and m_2, is given by:

$$F = \frac{Gm_1m_2}{d^2}$$

where d is the distance between the two masses, and G is a constant which depends upon the units employed. This law assumes that m_1 and m_2 are point masses. Newton was able to show that homogeneous spheres could be treated as point masses when determining their orbits.

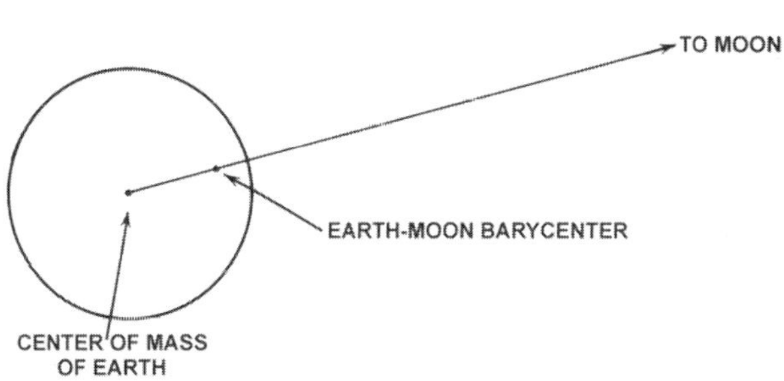

Figure 902a. Earth-Moon barycenter.

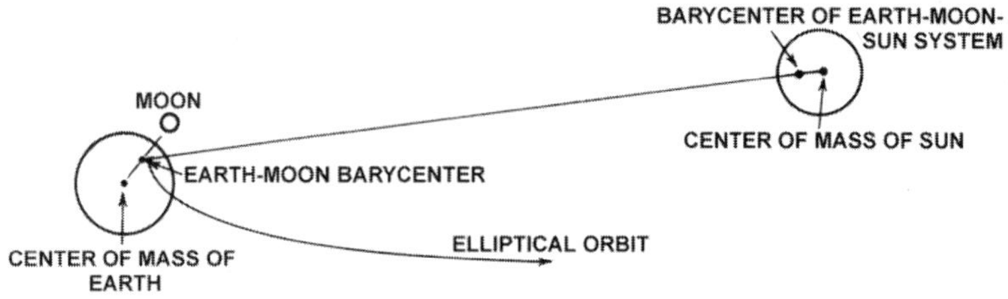

Figure 902b. Orbit of Earth-Moon barycenter (not to scale).

However, when computing differential gravitational forces, the actual dimensions of the masses must be taken into account.

Using the law of gravitation, it is found that the orbits of two point masses are conic sections about the **barycenter** of the two masses. If either one or both of the masses are homogeneous spheres instead of point masses, the orbits are the same as the orbits which would result if all of the mass of the sphere were concentrated at a point at the center of the sphere. In the case of the Earth-Moon system, both the Earth and the Moon describe elliptical orbits about their barycenter if both bodies are assumed to be homogeneous spheres and the gravitational forces of the Sun and other planets are neglected. The Earth-Moon barycenter is located 74/100 of the distance from the center of the Earth to its surface, along the line connecting the Earth's and Moon's centers. See Figure 902a.

Thus the center of mass of the Earth describes a very small ellipse about the Earth-Moon barycenter, while the center of mass of the Moon describes a much larger ellipse about the same barycenter. If the gravitational forces of the other bodies of the solar system are neglected, Newton's law of gravitation also predicts that the Earth-Moon barycenter will describe an orbit which is approximately elliptical about the barycenter of the Sun-Earth-Moon system. This barycentric point lies inside the Sun. See Figure 902b.

903. The Earth-Moon-Sun System

The fundamental tide-generating force on the Earth has two interactive but distinct components. The tide-generating forces are differential forces between the gravitational attraction of the bodies (Earth-Sun and Earth-Moon) and the centrifugal forces on the Earth produced by the Earth's orbit around the Sun and the Moon's orbit around the Earth. Newton's Law of Gravitation and his Second Law of Motion can be combined to develop formulations for the differential force at any point on the Earth, as the direction and magnitude are dependent on where you are on the Earth's surface. As a result of these differential forces, the tide generating forces F_{dm} (Moon) and F_{ds} (Sun) are inversely proportional to the cube of the distance between the bodies, where:

$$F_{dm} = \frac{GM_mR_e}{d_m^3}; \quad F_{ds} = \frac{GM_sR_e}{d_s^3}$$

where M_m is the mass of the Moon and M_s is the mass of the Sun, R_e is the radius of the Earth and d is the distance to the Moon or Sun. This explains why the tide-generating force of the Sun is only 46/100 of the tide-generating force of the Moon. Even though the Sun is much more massive, it is also much farther away.

Using Newton's second law of motion, we can calculate the differential forces generated by the Moon and the Sun affecting any point on the Earth. The easiest calculation is for the point directly below the Moon, known as the **sublunar point**, and the point on the Earth exactly opposite, known as the **antipode**. Similar calculations are done for the Sun.

If we assume that the entire surface of the Earth is covered with a uniform layer of water, the differential forces may be resolved into vectors perpendicular and parallel to the surface of the Earth to determine their effect. See Figure 903a.

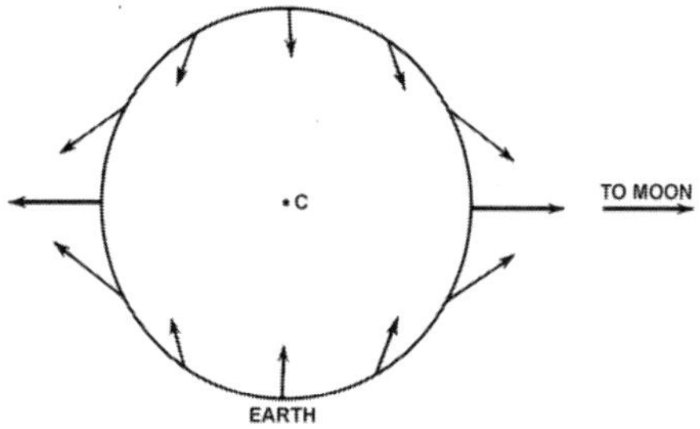

Figure 903a. Differential forces along a great circle connecting the sublunar point and antipode.

The perpendicular components change the mass on which they are acting, but do not contribute to the tidal effect. The horizontal components, parallel to the Earth's surface, have the effect of moving the water in a horizontal

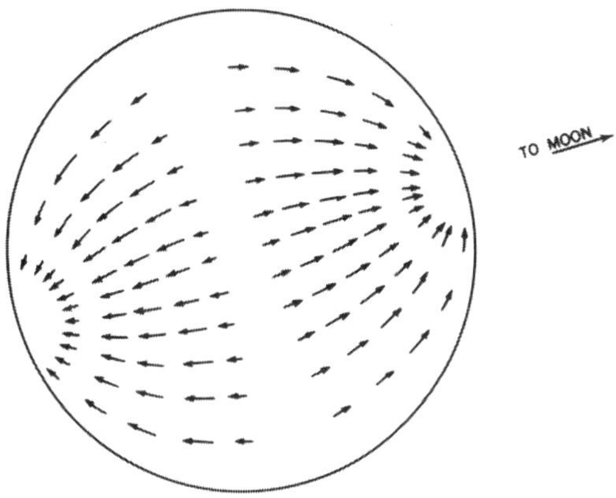

Figure 903b. Tractive forces across the surface of the Earth.

direction toward the sublunar and antipodal points until an equilibrium position is found. The *horizontal* components of the differential forces are the principal tide-generating forces. These are also called **tractive** forces. Tractive forces are zero at the sublunar and antipodal points and along the great circle halfway between these two points. Tractive forces are maximum along the small circles located 45° from the sublunar point and the antipode. Figure 903b shows the tractive forces across the surface of the Earth.

Equilibrium will be reached when a bulge of water has formed at the sublunar and antipodal points such that the tractive forces due to the Moon's differential gravitational forces on the mass of water covering the surface of the Earth are just balanced by the Earth's gravitational attraction (Figure 903c).

Now consider the effect of the rotation of the Earth. If the declination of the Moon is 0°, the bulges will lie on the equator. As the Earth rotates, an observer at the equator will note that the Moon transits approximately every 24 hours and 50 minutes. Since there are two bulges of water on the equator, one at the sublunar point and the other at the antipode, the observer will also see two high tides during this interval with one high tide occurring when the Moon is overhead and another high tide 12 hours 25 minutes later when the observer is at the antipode. He will also experience a low tide between each high tide. The theoretical range of these equilibrium tides at the equator will be less than 1 meter.

In theory, the heights of the two high tides should be equal at the equator. At points north or south of the equator, an observer would still experience two high and two low tides, but the heights of the high tides would not be as great as they are at the equator. The effects of the declination of the Moon are shown in Figure 903d, for three cases, A, B, and C.

A. When the Moon is on the plane of the equator, the forces are equal in magnitude at the two points on the same parallel of latitude and 180° apart in longitude.
B. When the Moon has north or south declination, the forces are unequal at such points and tend to cause an inequality in the two high waters and the two low waters each day.
C. Observers at points X, Y, and Z experience one high tide when the Moon is on their meridian, then another high tide 12 hours 25 minutes later when at X', Y', and Z'. The second high tide is the same at X' as at X. High tides at Y' and Z' are lower than high tides at Y and Z.

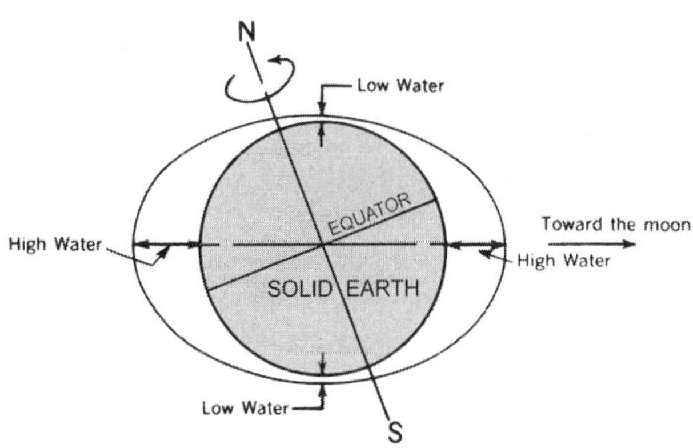

Figure 903c. Theoretical equilibrium configuration due to Moon's differential gravitational forces. One bulge of the water envelope is located at the sublunar point, the other bulge at the antipode.

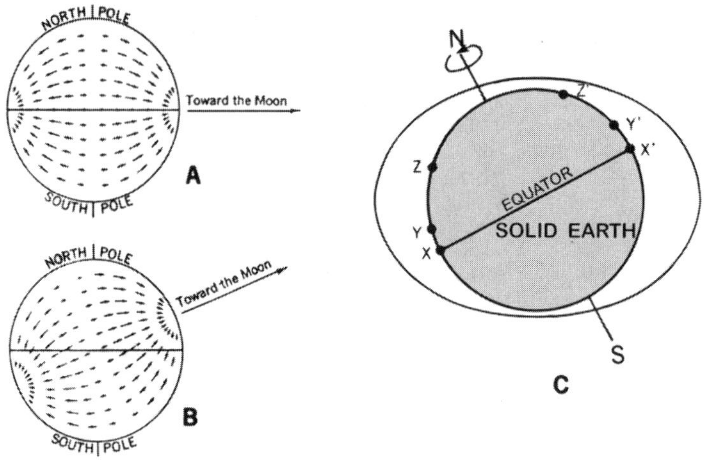

Figure 903d. Effects of the declination of the Moon.

The preceding discussion pertaining to the effects of the Moon is equally valid when discussing the effects of the Sun, taking into account that the magnitude of the solar effect is smaller. Hence, the tides will also vary according to the Sun's declination and its varying distance from the Earth. A second envelope of water representing the equilibrium tides due to the Sun would resemble the envelope shown in Figure 903c except that the heights of the high tides would be smaller, and the low tides correspondingly not as low. The theoretical tide at any place represents the combination of the effects of both the Moon and Sun.

FEATURES OF TIDES

904. General Features

At most places the tidal change occurs twice daily. The tide rises until it reaches a maximum height, called **high tide** or **high water**, and then falls to a minimum level called **low tide** or **low water**.

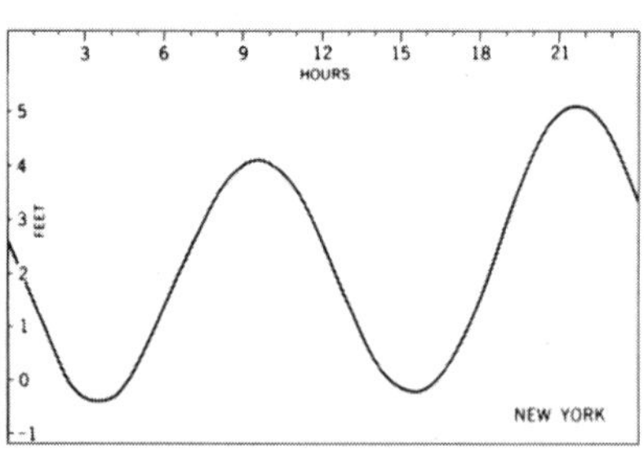

Figure 904. The rise and fall of the tide at New York, shown graphically.

The rate of rise and fall is not uniform. From low water, the tide begins to rise slowly at first, but at an increasing rate until it is about halfway to high water. The rate of rise then decreases until high water is reached, and the rise ceases.

The falling tide behaves in a similar manner. The period at high or low water during which there is no apparent change of level is called **stand**. The difference in height between consecutive high and low waters is the **range**.

Figure 904 is a graphical representation of the rise and fall of the tide at New York during a 24-hour period. The curve has the general form of a variable sine curve.

905. Types of Tide

A body of water has a natural period of oscillation, dependent upon its dimensions. None of the oceans is a single oscillating body; rather each one is made up of several separate oscillating basins. As such basins are acted upon by the tide-producing forces, some respond more readily to daily or diurnal forces, others to semidiurnal forces, and others almost equally to both. Hence, tides are classified as one of three types, semidiurnal, diurnal, or mixed, according to the characteristics of the tidal pattern.

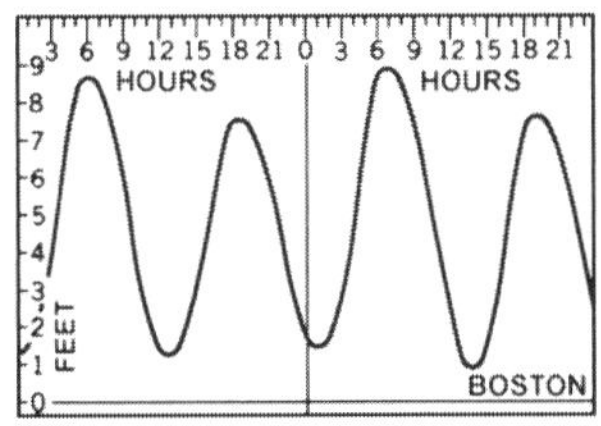

Figure 905a. Semidiurnal type of tide.

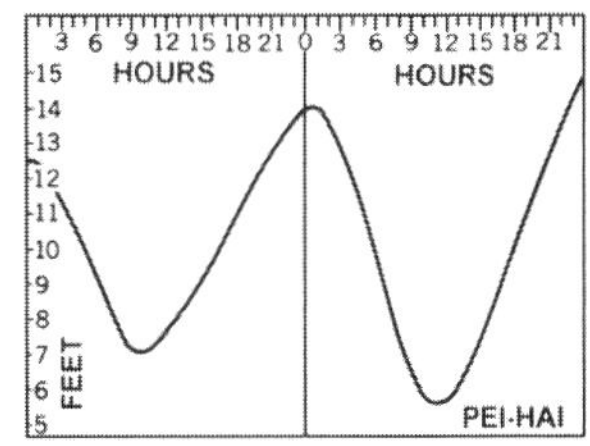

Figure 905b. Diurnal tide.

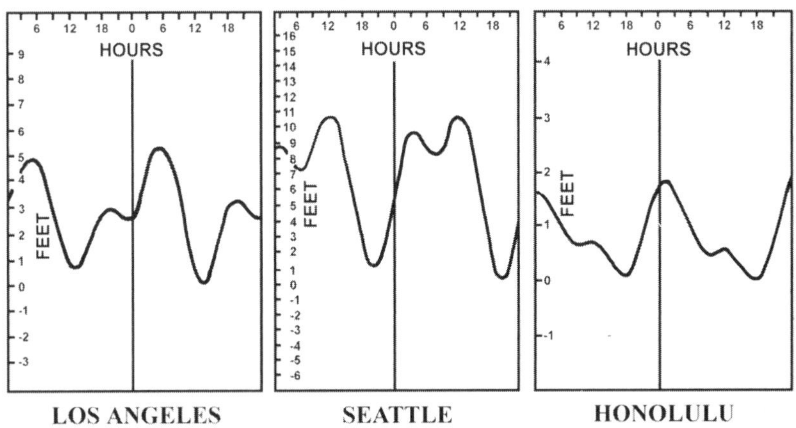

Figure 905c. Mixed tide.

In the **semidiurnal tide**, there are two high and two low waters each tidal day, with relatively small differences in the respective highs and lows. Tides on the Atlantic coast of the United States are of the semidiurnal type, which is illustrated in Figure 905a by the tide curve for Boston Harbor.

In the **diurnal tide**, only a single high and single low water occur each tidal day. Tides of the diurnal type occur along the northern shore of the Gulf of Mexico, in the Java Sea, the Gulf of Tonkin, and in a few other localities. The tide curve for Pei-Hai, China, illustrated in Figure 905b, is an example of the diurnal type.

In the **mixed tide**, the diurnal and semidiurnal oscillations are both important factors and the tide is characterized by a large inequality in the high water heights, low water heights, or in both. There are usually two high and two low waters each day, but occasionally the tide may become diurnal. Such tides are prevalent along the Pacific coast of the United States and in many other parts of the world. Examples of mixed types of tide are shown in Figure 905c. At Los Angeles, it is typical that the inequalities in the high and low waters are about the same. At Seattle the greater inequalities are typically in the low waters, while at Honolulu it is the high waters that have the greater inequalities.

906. Solar Tide

The natural period of oscillation of a body of water may accentuate either the solar or the lunar tidal oscillations. Though as a general rule the tides follow the Moon, the relative importance of the solar effect varies in different areas. There are a few places, primarily in the South Pacific and the Indonesian areas, where the solar oscillation is the more important, and at those places the high and low waters occur at about the same time each day. At Port Adelaide, Australia the solar and lunar semidiurnal oscillations are equal and nullify one another at neaps.

907. Special Tidal Effects

As a wave enters shallow water, its speed is decreased. Since the trough is shallower than the crest, it is retarded more, resulting in a steepening of the wave front. In a few estuaries, the advance of the low water trough is so much retarded that the crest of the rising tide overtakes the low, and advances upstream as a breaking wave called a **bore**. Bores that are large and dangerous at times of large tidal ranges may be mere ripples at those times of the month when the range is small. Examples occur in the Petitcodiac River in the Bay of Fundy, and at Haining, China, in the Tsientang Kaing. The tide tables indicate where bores occur.

Other special features are the **double low water** (as at Hoek Van Holland) and the **double high water** (as at Southampton, England). At such places there is often a slight fall or rise in the middle of the high or low water period. The practical effect is to create a longer period of stand at high or low tide. The tide tables list these and other peculiarities where they occur.

908. Variations in Range

Though the tide at a particular place can be classified as to type, it exhibits many variations during the month (Figure 908a). The range of the tide varies according to the intensity of the tide-producing forces, though there may be a lag of a day or two between a particular astronomic cause and the tidal effect.

The combined lunar-solar effect is obtained by adding the Moon's tractive forces vectorially to the Sun's tractive forces. The resultant tidal bulge will be predominantly lunar with modifying solar effects upon both the height of the tide and the direction of the tidal bulge. Special cases of interest occur during the times of new and full Moon (Figure 908b). With the Earth, Moon, and Sun lying approximately on the same line, the tractive forces of the Sun are acting in the same direction as the Moon's tractive forces (modified by declination effects). The resultant tides are called **spring tides**, whose ranges are greater than average.

Between the spring tides, the Moon is at first and third quarters. At those times, the tractive forces of the Sun are acting at approximately right angles to the Moon's tractive forces. The results are tides called **neap tides**, whose ranges are less than average.

With the Moon in positions between quadrature and new or full, the effect of the Sun is to cause the tidal bulge to either lag or precede the Moon (Figure 908c). These effects are called **priming** and **lagging** the tides.

Thus, when the Moon is at the point in its orbit nearest the Earth (at perigee), the lunar semidiurnal range is increased and **perigean tides** occur. When the Moon is farthest from the Earth (at apogee), the smaller **apogean tides** occur. When the Moon and Sun are in line and pulling together, as at new and full Moon, **spring tides** occur (the term spring has nothing to do with the season of year); when the Moon and Sun oppose each other, as at the quadratures, the smaller **neap tides** occur. When certain of these phenomena coincide, **perigean spring tides** and **apogean neap tides** occur.

These are variations in the semidiurnal portion of the tide. Variations in the diurnal portion occur as the Moon and Sun change declination. When the Moon is at its maximum semi-monthly declination (either north or south), **tropic tides** occur in which the diurnal effect is at a maximum. When it crosses the equator, the diurnal effect is a minimum and **equatorial tides** occur.

When the range of tide is increased, as at spring tides, there is more water available only at high tide; at low tide there is less, for the high waters rise higher and the low waters fall lower at these times. There is more water at neap low water than at spring low water. With tropic tides, there is usually more depth at one low water during the day than at the other. While it is desirable to know the meanings of these terms, the best way of determining the height of the tide at any place and time is to examine the tide predictions for the place as given in the tide tables, which take all these effects into account.

909. Tidal Cycles

Tidal oscillations go through a number of cycles. The shortest cycle, completed in about 12 hours and 25 minutes for a semidiurnal tide, extends from any phase of the tide to the next recurrence of the same phase. During a lunar day (averaging 24 hours and 50 minutes) there are two highs and two lows (two of the shorter cycles) for a semidiurnal tide. The Moon revolves around the Earth with respect to the Sun in a **synodical month** of about 29 1/2 days, commonly called the **lunar month**. The effect of the phase variation is completed in one-half of a synodical month or about 2 weeks as the Moon varies from new to full or full to new.

The effect of the Moon's declination is also repeated in one-half of a **tropical month** of 27 1/3 days, or about every 2 weeks. The cycle involving the Moon's distance requires an **anomalistic month** of about 27 1/2 days. The Sun's declination and distance cycles are respectively a half year and a year in length.

An important lunar cycle, called the **nodal period** or Metonic cycle (after Greek philosopher Meton, fifth century BC, who discovered the phenomenon) is 18.6 years (usually expressed in round figures as 19 years). For a tidal value, particularly a range, to be considered a true mean, it must be either based upon observations extended over this period of time, or adjusted to take account of variations known to occur during the nodal period.

The nodal period is the result of axis of the Moon's rotation being tilted 5 degrees with respect to the axis of the Earth's rotation. Since the Earth's axis is tilted 23.5 degrees with respect to the plane of its revolution around the sun, the combined effect is that the Moon's declination varies from 28.5 degrees to 18.5 degrees in a cycle lasting 18.6 years. For practical purposes, the nodal period can be con-

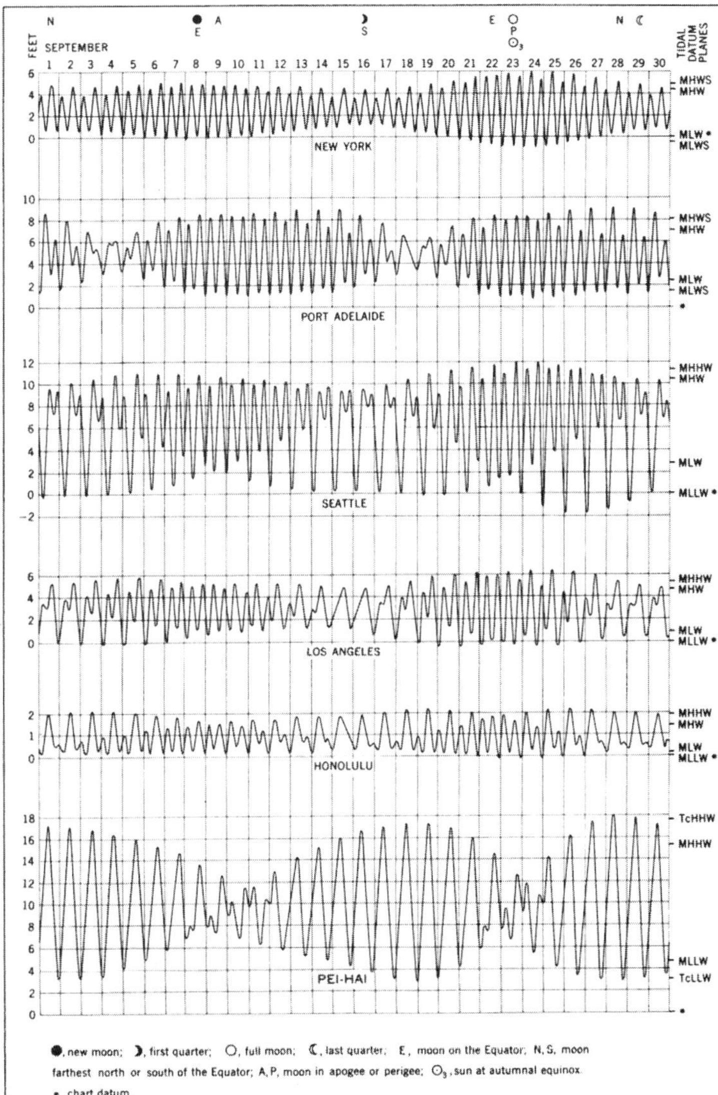

Figure 908a. Monthly tidal variations at various places.

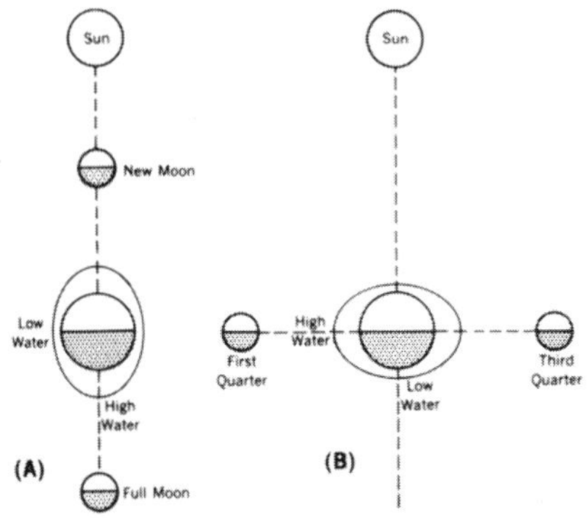

Figure 908b. (A) Spring tides occur at times of new and full Moon. Range of tide is greater than average since solar and lunar tractive forces act in same direction. (B) Neap ties occur at times of first and third quarters. Range of tide is less than average since solar and lunar tractive forces act at right angles.

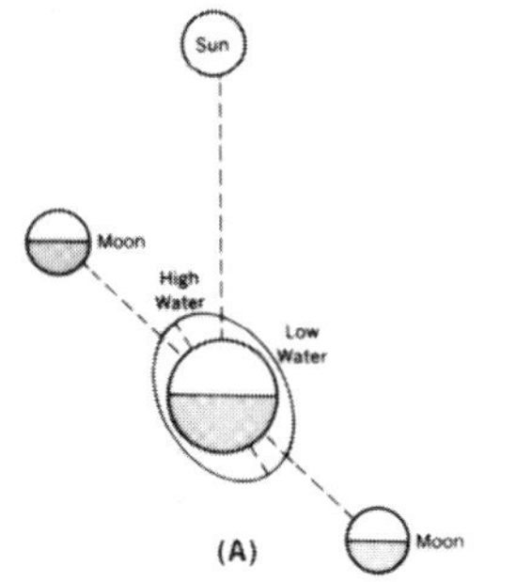

Priming occurs when moon is between new and first quarter and between full and third quarter. High tide occurs before transit moon.

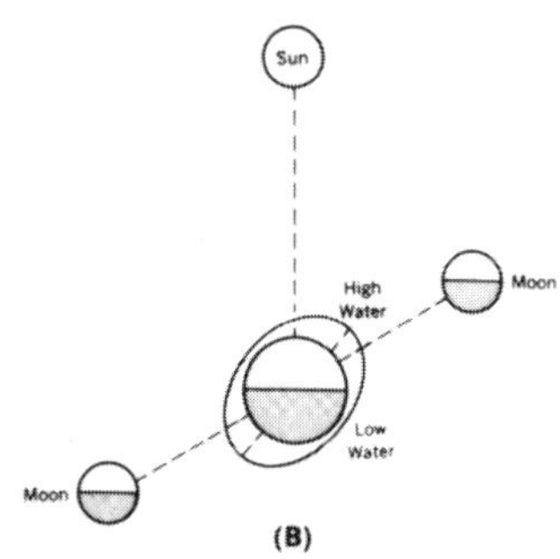

Lagging occurs when moon is between first quarter and full and between third quarter and new. High tide occurs after transit of moon.

Figure 908c. Priming and lagging the tides.

sidered as the time between the Sun and Moon appearing in precisely the same relative positions in the sky.

910. Time of Tide

Since the lunar tide-producing force has the greatest effect in producing tides at most places, the tides "follow the Moon." Because the Earth rotates, high water lags behind both upper and lower meridian passage of the Moon. The **tidal day**, which is also the lunar day, is the time between consecutive transits of the Moon, or 24 hours and 50 minutes on the average. Where the tide is largely semidiurnal in type, the **lunitidal interval** (the interval between the Moon's meridian transit and a particular phase of tide) is fairly constant throughout the month, varying somewhat with the tidal cycles. There are many places, however, where solar or diurnal oscillations are effective in upsetting this relationship. The interval generally given is the average elapsed time from the meridian transit (upper or lower) of the Moon until the next high tide. This may be called **mean high water lunitidal interval** or **corrected** (or **mean**) **establishment**. The **common establishment** is the average interval on days of full or new Moon, and approximates the mean high water lunitidal interval.

In the ocean, the tide may be in the nature of a progressive wave with the crest moving forward, a stationary or standing wave which oscillates in a seesaw fashion, or a combination of the two. Consequently, caution should be used in inferring the time of tide at a place from tidal data for nearby places. In a river or estuary, the tide enters from the sea and is usually sent upstream as a progressive wave so that the tide occurs progressively later at various places upstream.

TIDAL DATUMS

911. Low Water Datums

A tidal datum is a given average tide level from which heights of tides and overhead clearances are measured. It is a vertical datum, but is not the same as vertical geodetic datum, which is a mathematical quantity developed as part of a geodetic system used for horizontal positioning. There are a number of tidal levels of reference that are important to the mariner. See Figure 911.

The most important level of reference is the **sounding datum** shown on charts. The sounding datum is sometimes referred to as the reference plane to distinguish it from vertical geodetic datum. Since the tide rises and falls continually while soundings are being taken during a hy-

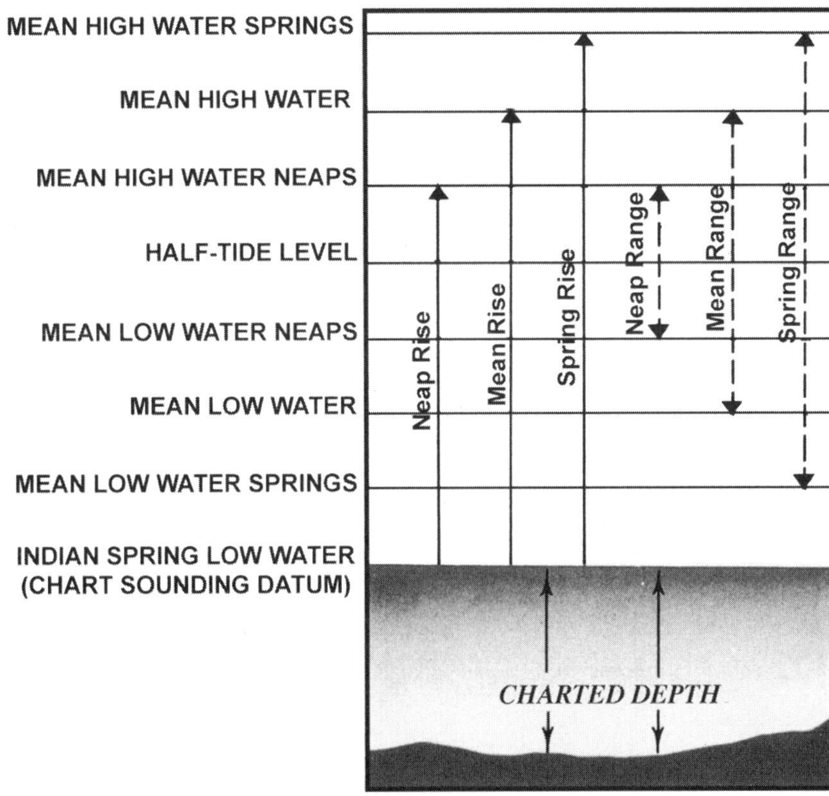

Figure 911. Variations in the ranges and heights of tide where the chart sounding datum is Indian Spring Low Water.

drographic survey, the tide is recorded during the survey so that soundings taken at all stages of the tide can be reduced to a common sounding datum. Soundings on charts show depths below a selected low water datum (occasionally mean sea level), and tide predictions in tide tables show heights above and below the same level. The depth of water available at any time is obtained by adding algebraically the height of the tide at the time in question to the charted depth.

By international agreement, the level used as chart datum should be low enough so that low waters do not fall very far below it. At most places, the level used is one determined from a mean of a number of low waters (usually over a 19 year period); therefore, some low waters can be expected to fall below it. The following are some of the datums in general use.

Mean low water (MLW) is the average height of all low waters at a given place. About half of the low waters fall below it, and half above.

Mean low water springs (MLWS), usually shortened to low water springs, is the average level of the low waters that occur at the times of spring tides.

Mean lower low water (MLLW) is the average height of the lower low waters of each tidal day.

Tropic lower low water (TcLLW) is the average height of the lower low waters (or of the single daily low waters if the tide becomes diurnal) that occur when the Moon is near maximum declination and the diurnal effect is most pronounced. This datum is not in common use as a tidal reference.

Indian spring low water (ISLW), sometimes called **Indian tide plane** or **harmonic tide plane**, is a low water datum that includes the spring effect of the semi-diurnal portion of the tide and the tropic effect of the diurnal portion. It is about the level of lower low water of mixed tides at the time that the Moon's maximum declination coincides with the time of new or full Moon.

Mean lower low water springs (MLLWS) is the average level of the lower of the two low waters on the days of spring tides.

Some lower datums used on charts are determined from tide observations and some are determined arbitrarily and later referred to the tide. Most of them fall close to one or the other of the following two datums.

Lowest normal low water is a datum that approximates the average height of monthly lowest low waters, discarding any tides disturbed by storms.

Lowest low water is an extremely low datum. It conforms generally to the lowest tide observed, or even somewhat lower. Once a tidal datum is established, it is sometimes retained for an indefinite period, even though it might differ slightly from a better determination from later observations. When this occurs, the established datum may be called **low water datum**, **lower low water datum**, etc. These datums are used in a limited area and primarily for river and harbor engineering purposes. Examples are Boston Harbor Low Water Datum and Columbia River Lower Low Water Datum.

Some sounding datums are based on the predicted tide rather than an average of observations. A British sounding datum that may be adopted internationally is the Lowest Astronomical Tide (LAT). LAT is the elevation of the lowest water level predicted in a 19-year period. Canadian coastal charts use a datum of Lower Low Water, Large Tide (LLWLT) which is the average of the lowest low waters, one from each of the 19 years of predictions.

Figure 911 illustrates variations in the ranges and heights of tides in a locality such as the Indian Ocean, where predicted and observed water levels are referenced to a chart sounding datum that will always cause them to be additive relative to the charted depth.

In areas where there is little or no tide, various other datums are used. For the Black Sea for instance, Mean Sea Level (MSL, sometimes referred to as Mean Water Level or MWL) is used, and is the average of the hourly heights observed over a period of time and adjusted to a 19-year period. In the United States, a Low Water Datum (LWD) is used in those coastal areas that have transitioned from tidal to non-tidal (e.g. Laguna Madre, Texas and Pamlico Sound, North Carolina) and is simply 0.5 foot below a computed MLW. For the Great Lakes, the United States and Canada use a separate LWD for each lake, which is designed to ensure that the actual water level is above the datum most of the time during the navigation season. Lake levels vary by several feet over a period of years.

Inconsistencies of terminology are found among charts of different countries and between charts issued at different times.

Large-scale charts usually specify the datum of soundings and may contain a tide note giving mean heights of the tide at one or more places on the chart. These heights are intended merely as a rough guide to the change in depth to be expected under the specified conditions. They should not be used for the prediction of heights on any particular day, which should be obtained from tide tables.

912. High Water Datums

Heights of terrestrial features are usually referred on nautical charts to a high water datum. This gives the mariner a margin of error when passing under bridges, overhead cables, and other obstructions. The one used on charts of the United States, its territories and possessions, and widely used elsewhere, is **mean high water (MHW)**, which is the average height of all high waters over a 19 year period. Any other high water datum in use on charts is likely to be higher than this. Other high water datums are **mean high water springs (MHWS)**, which is the average level of the high waters that occur at the time of spring tides; **mean higher high water (MHHW)**, which is the average height of the higher high waters of each tidal day; and **tropic higher high water (TcHHW)**, which is the average height of the higher high waters (or the single daily high waters if the tide becomes diurnal) that occur when the Moon is near maximum declination and the diurnal effect is most pronounced. A reference merely to "high water" leaves some doubt as to the specific level referred to, for the height of high water varies from day to day. Where the range is large, the variation during a 2 week period may be considerable.

Because there are periodic and apparent secular trends in sea level, a specific 19 year cycle (the **National Tidal Datum Epoch**) is issued for all United States datums. The National Tidal Datum Epoch officially adopted by the National Ocean Service is presently 1960 through 1978. The Epoch is reviewed for revision every 25 years.

TIDAL CURRENTS

913. Tidal and Nontidal Currents

Horizontal movement of water is called **current**. It may be either "tidal" and "nontidal." **Tidal current** is the periodic horizontal flow of water accompanying the rise and fall of the tide. **Nontidal current** includes all currents not due to the tidal movement. Nontidal currents include the permanent currents in the general circulatory system of the oceans as well as temporary currents arising from meteorological conditions. The current experienced at any time is usually a combination of tidal and nontidal currents.

914. General Features

Offshore, where the direction of flow is not restricted by any barriers, the tidal current is rotary; that is, it flows continuously, with the direction changing through all points of the compass during the tidal period. This rotation is caused by the Earth's rotation, and unless modified by local conditions, is clockwise in the Northern Hemisphere and counterclockwise in the Southern Hemisphere. The speed usually varies throughout the tidal cycle, passing through two maximums in approximately opposite directions, and two minimums about halfway between the maximums in time and direction. Rotary currents can be depicted as in Figure 914a, by a series of arrows representing the direction and speed of the current at each hour. This is sometimes called a **current rose**. Because of the elliptical pattern formed by the ends of the arrows, it is also referred to as a **current ellipse**.

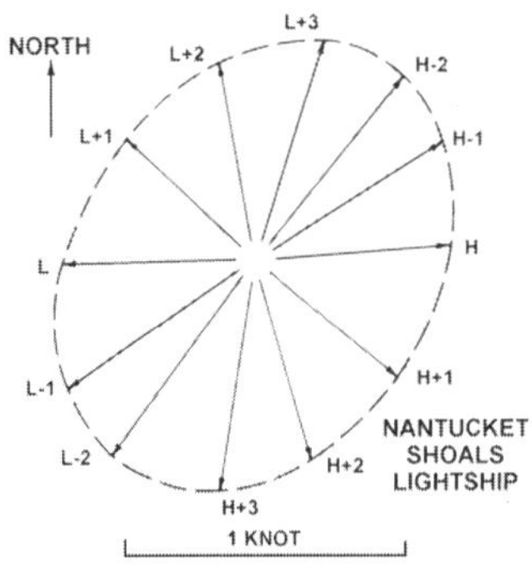

Figure 914a. Rotary tidal current. Times are hours before and after high and low tide at Nantucket Shoals. The bearing and length of each arrow represents the hourly direction and speed of the current.

In rivers or straits, or where the direction of flow is more or less restricted to certain channels, the tidal current is reversing; that is, it flows alternately in approximately opposite directions with an instant or short period of little or no current, called **slack water**, at each reversal of the current. During the flow in each direction, the speed varies from zero at the time of slack water to a maximum, called strength of flood or ebb, about midway between the slacks. Reversing currents can be indicated graphically, as in Figure 914b, by arrows that represent the speed of the current at each hour. The flood is usually depicted above the slack waterline and the ebb below it. The tidal current curve formed by the ends of the arrows has the same characteristic sine form as the tide curve. In illustrations and for certain other purposes it is convenient to omit the arrows and show only the curve.

A slight departure from the sine form is exhibited by the reversing current in a strait that connects two different tidal basins, such as the East River, New York. The tides at the two ends of a strait are seldom in phase or equal in range, and the current, called **hydraulic current**, is generated largely by the continuously changing difference in height of water at the two ends. The speed of a hydraulic current varies nearly as the square root of the difference in

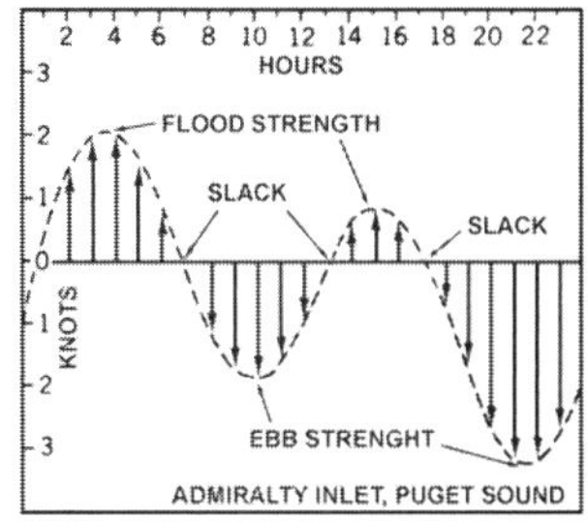

Figure 914b. Reversing tidal current.

height. The speed reaches a maximum more quickly and remains at strength for a longer period than shown in Figure 914b, and the period of weak current near the time of slack is considerably shortened.

The current direction, or **set**, is the direction toward which the current flows. The speed is sometimes called the **drift**. The term "velocity" is often used as the equivalent of "speed" when referring to current, although strictly speaking "velocity" implies direction as well as speed. The term "strength" is also used to refer to speed, but more often to greatest speed between consecutive slack waters. The movement toward shore or upstream is the **flood**, the movement away from shore or downstream is the **ebb**. In a purely semidiurnal current unaffected by nontidal flow, the flood and ebb each last about 6 hours and 13 minutes. But if there is either diurnal inequality or nontidal flow, the durations of flood and ebb may be quite unequal.

915. Types of Tidal Current

Tidal currents, like tides, may be of the **semidiurnal**, **diurnal**, or **mixed** type, corresponding to a considerable degree to the type of tide at the place, but often with a stronger semidiurnal tendency.

The tidal currents in tidal estuaries along the Atlantic coast of the United States are examples of the semidiurnal type of reversing current. Along the Gulf of Mexico coast, such as at Mobile Bay entrance, they are almost purely diurnal. At most places, however, the type is mixed to a greater or lesser degree. At Tampa and Galveston entrances there is only one flood and one ebb each day when the Moon is near its maximum declination, and two floods and two ebbs each day when the Moon is near the equator. Along the Pacific coast of the United States there are generally two floods and two ebbs every day, but one of the floods or ebbs has a greater speed and longer duration than the other, the inequality varying with the declination of the Moon.

The inequalities in the current often differ considerably from place to place even within limited areas, such as adjacent passages in Puget Sound and various passages between the Aleutian Islands. Figure 915a shows several types of re-

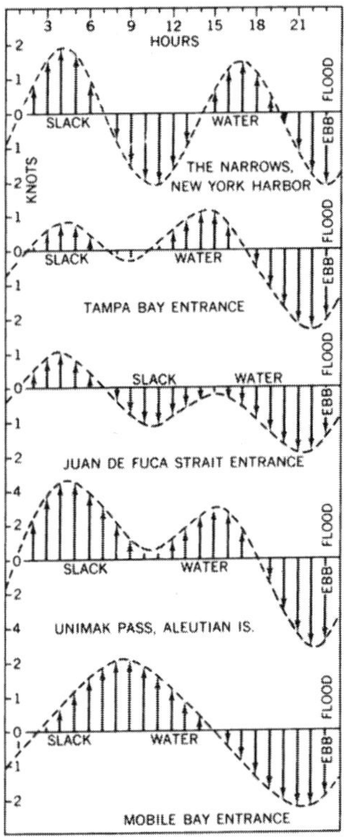

Figure 915a. Several types of reversing current. The pattern changes gradually from day to day, particularly for mixed types, passing through cycles.

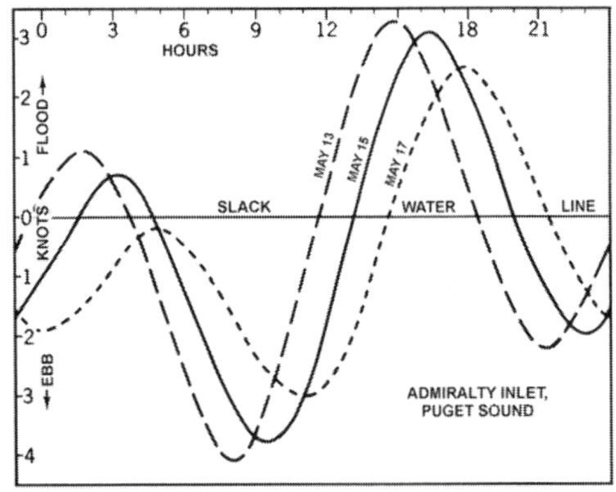

Figure 915b. Changes in a current of the mixed type. Note that each day as the inequality increases, the morning slacks draw together in time until on the 17th the morning flood disappears. On that day the current ebbs throughout the morning.

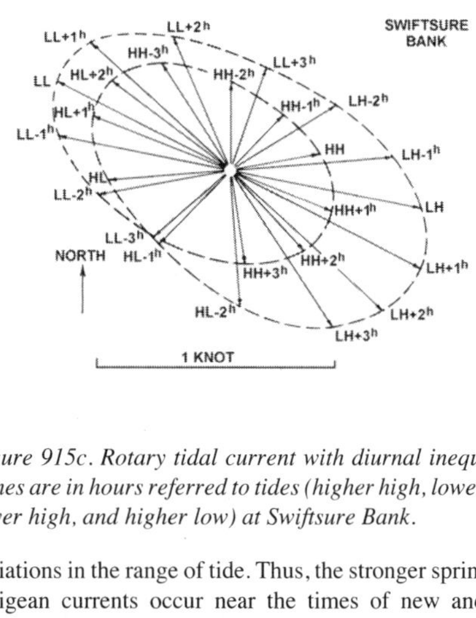

Figure 915c. Rotary tidal current with diurnal inequality. Times are in hours referred to tides (higher high, lower low, lower high, and higher low) at Swiftsure Bank.

versing current. Figure 915b shows how the flood disappears as the diurnal inequality increases at one station.

Offshore rotary currents that are purely semidiurnal repeat the elliptical pattern each tidal cycle of 12 hours and 25 minutes. If there is considerable diurnal inequality, the plotted hourly current arrows describe a set of two ellipses of different sizes during a period of 24 hours and 50 minutes, as shown in Figure 915c, and the greater the diurnal inequality, the greater the difference between the sizes of the two ellipses. In a completely diurnal rotary current, the smaller ellipse disappears and only one ellipse is produced in 24 hours and 50 minutes.

916. Tidal Current Periods and Cycles

Tidal currents have periods and cycles similar to those of the tides, and are subject to similar variations, but flood and ebb of the current do not necessarily occur at the same times as the rise and fall of the tide.

The speed at strength increases and decreases during the 2 week period, month, and year along with the variations in the range of tide. Thus, the stronger spring and perigean currents occur near the times of new and full Moon and near the times of the Moon's perigee, or at times of spring and perigean tides; the weaker neap and apogean currents occur at the times of neap and apogean tides; and tropic currents with increased diurnal speeds or with larger diurnal inequalities in speed occur at times of tropic tides; and equatorial currents with a minimum diurnal effect occur at times of equatorial tides.

As with the tide, a mean value represents an average obtained from a 19 year series. Since a series of current observations is usually limited to a few days, and seldom covers more than a month or two, it is necessary to adjust the observed values, usually by comparison with tides at a nearby place, to obtain such a mean.

917. Effect of Nontidal Flow

The current existing at any time is seldom purely tidal, but usually includes also a nontidal current that is due to drainage, oceanic circulation, wind, or other causes. The method in which tidal and nontidal currents combine is best explained graphically, as in Figure 917a and Figure 917b. The pattern of the tidal current remains unchanged, but the curve is shifted from the point or line from which the currents are measured, in the direction of the nontidal current, and by an amount equal to it. It is sometimes more convenient graphically merely to move the line or point of origin in the opposite direction. Thus, the speed of the current flowing in the direction of the nontidal current is increased by an amount equal to the magnitude of the nontidal current, and the speed of the current flowing in the opposite direction is decreased by an equal amount.

In Figure 917a, a nontidal current is represented both in direction and speed by the vector AO. Since this is greater than the speed of the tidal current in the opposite direction, the point A is outside the ellipse. The direction and speed of the combined tidal and nontidal currents at any time is represented by a vector from A to that point on the curve representing the given time, and can be scaled from the graph. The strongest and weakest currents may no longer be in the directions of the maximum and minimum of the tidal current. If the nontidal current is northwest at 0.3 knot, it may be represented by BO, and all hourly directions and speeds will then be measured from B. If it is 1.0 knot, it will be represented by AO and the actual resultant hourly directions and speeds will be measured from A, as shown by the arrows.

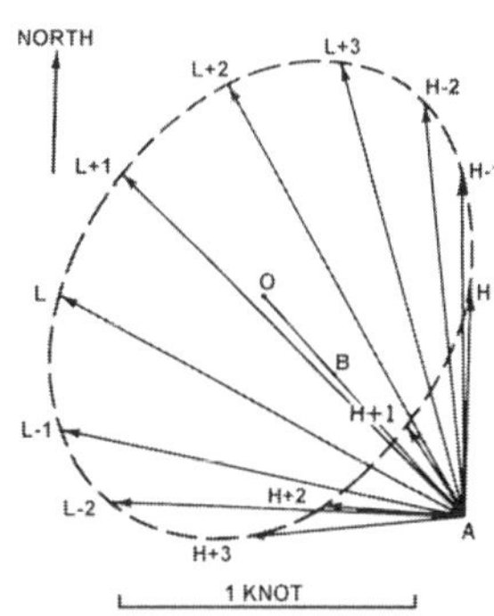

Figure 917a. Effect of nontidal current on the rotary tidal current of Figure 914a.

In a reversing current (Figure 917b), the effect is to advance the time of one slack, and to retard the following

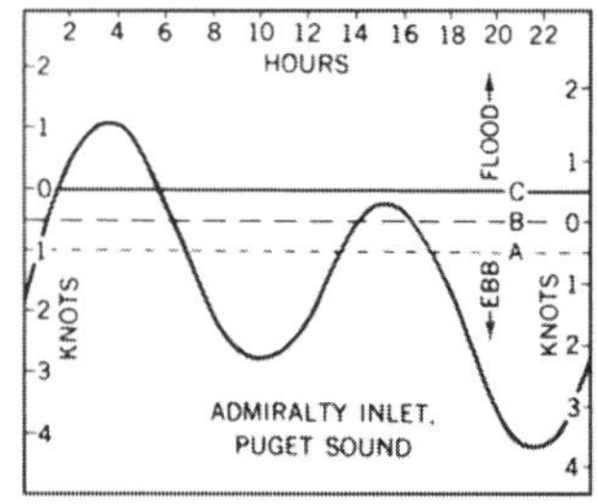

Figure 917b. Effect of nontidal current on the reversing tidal current of Figure 914b.

one. If the speed of the nontidal current exceeds that of the reversing tidal current, the resultant current flows continuously in one direction without coming to a slack. In this case, the speed varies from a maximum to a minimum and back to a maximum in each tidal cycle. In Figure 917b, the horizontal line A represents slack water if only tidal currents are present. Line B represents the effect of a 0.5 knot nontidal ebb, and line C the effect of a 1.0 knot nontidal ebb. With the condition shown at C there is only one flood each tidal day. If the nontidal ebb were to increase to approximately 2 knots, there would be no flood, two maximum ebbs and two minimum ebbs occurring during a tidal day.

918. Time of Tidal Current and Time of Tide

At many places where current and tide are both semidiurnal, there is a definite relationship between times of current and times of high and low water in the locality. Current atlases and notes on nautical charts often make use of this relationship by presenting for particular locations, the direction and speed of the current at each succeeding hour after high and low water, at a place for which tide predictions are available.

Where there is considerable diurnal inequality in tide or current, or where the type of current differs from the type of tide, the relationship is not constant, and it may be hazardous to try to predict the times of current from times of tide. Note the current curve for Unimak Pass in the Aleutians in Figure 915a. It shows the current as predicted in the tidal current tables. Predictions of high and low waters in the tide tables might have led one to expect the current to change from flood to ebb in the late morning, whereas actually the current continued to run flood with some strength at that time.

Since the relationship between times of tidal current and tide is not the same everywhere, and may be variable at the same place, one should exercise extreme caution in using general rules. The belief that slacks occur at local high and low tides and that the maximum flood and ebb occur when the tide is rising or falling most rapidly may be approximately true at the seaward entrance to, and in the upper reaches of, an inland tidal waterway. But generally this is not true in other parts of inland waterways. When an inland waterway is extensive or its entrance constricted, the slacks in some parts of the waterway often occur midway between the times of high and low tide. Usually in such waterways the relationship changes from place to place as one progresses upstream, slack water getting progressively closer in time to the local tide maximum until at the head of tidewater (the inland limit of water affected by a tide) the slacks occur at about the times of high and low tide.

919. Relationship Between Speed of Current and Range of Tide

The speed of the tidal current is not necessarily consistent with the range of tide. It may be the reverse. For example, currents are weak in the Gulf of Maine where the tides are large, and strong near Nantucket Island and in Nantucket Sound where the tides are small. However, at any one place the speed of the current at strength of flood and ebb varies during the month in about the same proportion as the range of tide, and this relationship can be used to determine the relative strength of currents on any given day.

920. Variation Across an Estuary

In inland tidal estuaries the time of tidal current varies across the channel from shore to shore. On the average, the current turns earlier near shore than in midstream, where the speed is greater. Differences of half an hour to an hour are not uncommon, but the difference varies and the relationship may be nullified by the effect of nontidal flow.

The speed of the current also varies across the channel, usually being greater in midstream or midchannel than near shore, but in a winding river or channel the strongest currents occur near the concave shore, or the outside corner of the curve. Near the opposite (convex) shore the currents are weak or eddying.

921. Variation with Depth

In tidal rivers the subsurface current acting on the lower portion of a ship's hull may differ considerably from the surface current. An appreciable subsurface current may be present when the surface movement appears to be practically slack, and the subsurface current may even be flowing with appreciable speed in the opposite direction to the surface current.

In a tidal estuary, particularly in the lower reaches where there is considerable difference in density from top to bottom, the flood usually begins earlier near the bottom than at the surface. The difference may be an hour or two, or as little as a few minutes, depending upon the estuary, the location in the estuary, and freshet conditions. Even when the freshwater runoff becomes so great as to prevent the surface current from flooding, it may still flood below the surface. The difference in time of ebb from surface to bottom is normally small but subject to variation with time and location.

The ebb speed at strength usually decreases gradually from top to bottom, but the speed of flood at strength often is stronger at subsurface depths than at the surface.

922. Tidal Current Observations

Observations of current are made with sophisticated electronic **current meters**. Current meters are suspended from a buoy or anchored to the bottom with no surface marker at all. Very sensitive current meters measure and record deep ocean currents; these are later recovered by triggering a release mechanism with a signal from the surface. Untended current meters either record data internally or send it by radio to a base station on ship or land. The period of observation varies from a few hours to as long as 6 months.

TIDE AND CURRENT PREDICTION

923. Tidal Height Predictions

To measure the height of tides, hydrographers select a reference level, sometimes referred to as the reference plane, or vertical datum. This vertical tidal datum is not the same as the vertical geodetic datum. Soundings shown on the largest scale charts are the vertical distances from this datum to the bottom. At any given time the actual depth is this charted depth plus the height of tide. In most places the reference level is some form of low water. But all low waters at a given place are not the same height, and the selected reference level is seldom the lowest tide occurring at the place. When lower tides occur, these are indicated in the tide tables by a negative sign. Thus, at a spot where the charted depth is 15 feet, the actual depth is 15 feet plus the tidal height. When the tide is three feet, the depth is 15 + 3 = 18 feet. When it is -1 foot, the depth is 15 - 1 = 14 feet. The actual depth can be less than the charted depth. In an area where there is a considerable range of tide (the difference between high water and low water), the height of tide might be an important consideration when using soundings to determine if the vessel is in safe water.

The heights given in the tide tables are predictions, and when assumed conditions vary considerably, the predictions shown may be considerably in error. Heights lower than predicted can be anticipated when the atmospheric pressure is higher than normal, or when there is a persistent strong offshore wind. The greater the range of tide, the less reliable are the predictions for both height and current.

924. Tidal Heights

The nature of the tide at any place can best be determined by observation. The predictions in tide tables and the tidal data on nautical charts are based upon detailed observations at specific locations, instead of theoretical predictions.

Tidal elevations are usually observed with a continuously recording gage. A year of observations is the minimum length desirable for determining the harmonic constants used in prediction. For establishing mean sea level and long-term changes in the relative elevations of land and sea, as well as for other special uses, observations have been made over periods of 20, 30, and even 120 years at important locations. Observations for a month or less will establish the type of tide and suffice for comparison with a longer series of observations to determine tidal differences and constants.

Mathematically, the variations in the lunar and solar tide-producing forces, such as those due to changing phase, distance, and declination, are considered as separate constituent forces, and the harmonic analysis of observations reveals the response of each constituent of the tide to its corresponding force. At any one place this response remains constant and is shown for each constituent by **harmonic constants** which are in the form of a phase angle for the time relation and an amplitude for the height. Harmonic constants are used in making technical studies of the tide and in tidal predictions on computers. The tidal predictions in most published tide tables are produced by computer.

925. Meteorological Effects

The foregoing discussion of tidal behavior assumes normal weather conditions. However, sea level is also affected by wind and atmospheric pressure. In general, onshore winds raise the level and offshore winds lower it, but the amount of change varies at different places. During periods of low atmospheric pressure, the water level tends to be higher than normal. For a stationary low, the increase in elevation can be found by the formula

$$R_0 = 0.01(1010 - P),$$

in which R_0 is the increase in elevation in meters and P is the atmospheric pressure in hectopascals. This is equal approximately to 1 centimeter per hectopascal depression, or about 13.6 inches per inch depression. For a moving low, the increase in elevation is given by the formula

$$R = \frac{R_0}{1 - \frac{C^2}{gh}}$$

in which R is the increase in elevation in feet, R_0 is the increase in meters for a stationary low, C is the rate of motion of the low in feet per second, g is the acceleration due to gravity (32.2 feet per second per second), and h is the depth of water in feet.

Where the range of tide is very small, the meteorological effect may sometimes be greater than the normal tide. Where a body of water is large in area but shallow, high winds can push the water from the windward to the lee shore, creating much greater local differences in water levels than occurs normally, and partially or completely masking the tides. The effect is dependent on the configuration and depth of the body of water relative to the wind direction, strength and duration.

926 Tidal Current Predictions

Tidal currents are due primarily to tidal action, but other causes are often present. The ***Tidal Current Tables*** give the best prediction of total current. Following heavy rains or a drought, a river's current prediction may be considerably in error. Set and drift may vary considerably over different parts of a harbor, because differences in bathymetry from place to place affect current. Since this is usually an area where small errors in a vessel's position are crucial, a knowledge of predicted currents, particularly in reduced visibility, is important. Strong currents occur mostly in narrow passages connecting larger bodies of water. Currents of more than 5 knots are sometimes encountered at the Golden Gate in San Francisco, and currents of more than 13 knots sometimes occur at Seymour Narrows, British Columbia.

In straight portions of rivers and channels, the strongest currents usually occur in the middle of the channel. In curved portions the swiftest currents (and deepest water) usually occur near the outer edge of the curve. Countercurrents and eddies may occur on either side of the main current of a river or narrow passage, especially near obstructions and in bights.

In general, the range of tide and the velocity of tidal current are at a minimum in the open ocean or along straight coasts. The greatest tidal effects are usually encountered in estuaries, bays, and other coastal indentations. A vessel proceeding along an indented coast may encounter a set toward or away from the shore; a similar set is seldom experienced along a straight coast.

PUBLICATIONS FOR PREDICTING TIDES AND CURRENTS

927. *Tide Tables*

Usually, tidal information is obtained from tide and tidal current tables, or from specialized computer software or calculators. However, if these are not available, or if they do not include information at a desired place, the mariner may be able to obtain locally the **mean high water lunitidal interval** or the **high water full and change**. The approximate time of high water can be found by adding either interval to the time of transit (either upper or lower) of the Moon. Low water occurs approximately 1/4 tidal day (about $6^h 12^m$) before and after the time of high water. The actual interval varies somewhat from day to day, but approximate results can be obtained in this manner. Similar information for tidal currents (**lunicurrent interval**) is seldom available.

The National Ocean Service (NOS) has traditionally published hard copy tide tables and tidal current tables. Tide and tidal current data continue to be updated by NOS, but hardcopy publication has been transferred to private companies working with NOS data, published on CD-ROM.

Tidal data for various parts of the world is published in 4 volumes by the National Ocean Service. These volumes are:

- Central and Western Pacific Ocean and Indian Ocean
- East Coast of North and South America (including Greenland)
- Europe and West Coast of Africa
- West Coast of North and South America (including the Hawaiian Islands)

A small separate volume, the Alaskan Supplement, is also published.

Each volume has 5 common tables:

- **Table 1** contains a complete list of the predicted times and heights of the tide for each day of the year at a number of places designated as **reference stations**.
- **Table 2** gives tidal differences and ratios which can be used to modify the tidal information for the reference stations to make it applicable to a relatively large number of **subordinate stations**.
- **Table 3** provides information for finding the approximate height of the tide at any time between high water and low water.
- **Table 4** is a sunrise-sunset table at five-day intervals for various latitudes from 76°N to 60°S (40°S in one volume).
- **Table 5** provides an adjustment to convert the local mean time of Table 4 to zone or standard time.

For the East Coast and West Coast volumes, each contains a Table 6, a moonrise and moonset table; Table 7 for conversion from feet to centimeters; Table 8, a table of estimated tide prediction accuracies; a glossary of terms; and an index to stations. Each table is preceded by a complete explanation. Sample problems are given where necessary. The inside back cover of each volume contains a calendar of critical astronomical data to help explain the variations of the tide during each month and throughout the year.

928. Tide Predictions for Reference Stations

For each day, the date and day of week are given, and the time and height of each high and low water are listed in chronological order. Although high and low waters are not labeled as such, they can be distinguished by the relative heights given immediately to the right of the times. If two high tides and two low tides occur each tidal day, the tide is semidiurnal. Since the tidal day is longer than the civil day (because of the revolution of the Moon eastward around the Earth), any given tide occurs later each day. Because of later times of corresponding tides from day to day, certain days have only one high water or only one low water.

929. Tide Predictions for Subordinate Stations

For each subordinate station listed, the following information is given:

1. **Number**. The stations are listed in geographical order and assigned consecutive numbers. Each volume contains an alphabetical station listing correlating the station with its consecutive number to assist in finding the entry in Table 2.
2. **Place**. The list of places includes both subordinate and reference stations; the latter are in bold type.
3. **Position**. The approximate latitude and longitude are given to assist in locating the station. The latitude is north or south, and the longitude east or west, depending upon the letters (N, S, E, W) next above the entry. These may not be the same as those at the top of the column.
4. **Differences**. The differences are to be applied to the predictions for the reference station, shown in capital letters above the entry. Time and height differences are given separately for high and low waters. Where differences are omitted, they are either unreliable or unknown.
5. **Ranges**. Various ranges are given, as indicated in the tables. In each case this is the difference in height between high water and low water for the tides indicated.
6. **Mean tide level**. This is the average between mean low and mean high water, measured from chart datum.

The **time difference** is the number of hours and minutes to be applied to the reference station time to find the time of the corresponding tide at the subordinate station. This interval is added if preceded by a plus sign (+) and subtracted if preceded by a minus sign (-). The results obtained by the application of the time differences will be in the zone time of the time meridian shown directly above the difference for the subordinate station. Special conditions occurring at a few stations are indicated by footnotes on the applicable pages. In some instances, the corresponding tide falls on a different date at reference and subordinate stations.

Height differences are shown in a variety of ways. For most entries, separate height differences in feet are given for high water and low water. These are applied to the height given for the reference station. In many cases a ratio is given for either high water or low water, or both. The height at the reference station is multiplied by this ratio to find the height at the subordinate station. For a few stations, both a ratio and difference are given. In this case the height at the reference station is first multiplied by the ratio, and the difference is then applied. An example is given in each volume of tide tables. Special conditions are indicated in the table or by footnote. For example, a footnote indicates that "Values for the Hudson River above George Washington Bridge are based upon averages for the six months May to October, when the fresh-water discharge is a minimum."

930. Finding Height of Tide at any Time

Table 3 provides means for determining the approximate height of tide at any time. It assumes that plotting height versus time yields a sine curve. Actual values may vary from this. The explanation of the table contains directions for both mathematical and graphical solutions. Though the mathematical solution is quicker, if the vessel's ETA changes significantly, it will have to be done for the new ETA. Therefore, if there is doubt about the ETA, the graphical solution will provide a plot of predictions for several hours and allow quick reference to the predicted height for any given time. This method will also quickly show at what time a given depth of water will occur. Figure 930a shows the OPNAV form used to calculate heights of tides. Figure 930b shows the importance of calculating tides in shallow water.

OPNAV 3530/40 (4-73)
HT OF TIDE

Date		
Location		
Time		
Ref Sta		
HW Time Diff		
LW Time Diff		
HW Ht Diff		
LW Ht Diff		
Ref Sta HW/LW Time		
HW/LW Time Diff		
Sub Sta HW/LW Time		
Ref Sta HW/LW Ht		
HW/LW Ht Diff		
Sub Sta HW/LW Ht		
Duration	Rise Fall	
Time Fm	Near Tide	
Range of Tide		
Ht of Neat Tide		
Corr Table 3		
Ht of Tide		
Charted Depth		
Depth of Water		
Draft		
Clearance		

Figure 930a. OPNAV 3530/40 Tide Form.

931. *Tidal Current Tables*

Tidal Current Tables are somewhat similar to *Tide Tables*, but the coverage is less extensive. NOS publishes 2 volumes on an annual basis: Atlantic Coast of North America, and Pacific Coast of North America and Asia. Each of the two volumes is arranged as follows:

Each volume also contains current diagrams and instructions for their use. Explanations and examples are given in each table.

- **Table 1** contains a complete list of predicted times of maximum currents and slack water, with the velocity of the maximum currents, for a number of reference stations.
- **Table 2** gives differences, ratios, and other information related to a relatively large number of subordinate

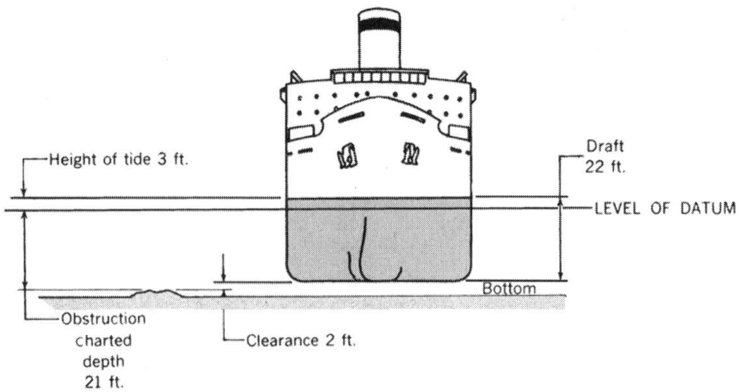

Figure 930b. Height of tide required to pass clear of charted obstruction.

stations.

- **Table 3** provides information to determine the current's velocity at any time between entries in tables 1 and 2.
- **Table 4** gives duration of slack, or the number of minutes the current does not exceed stated amounts, for various maximum velocities.
- **Table 5** (Atlantic Coast of North America only) gives information on rotary tidal currents.

The volumes also contain general descriptive information on wind-driven currents, combination currents, and information such as Gulf Stream currents for the east coast and coastal currents on the west coast.

932. Tidal Current Prediction for Reference Stations

For each day, the date and day of week are given; current information follows. If the cycle is repeated twice each tidal day, currents are semidiurnal. On most days there are four slack waters and four maximum currents, two floods (F) and two ebbs (E). However, since the tidal day is longer than the civil day, the corresponding condition occurs later each day, and on certain days there are only three slack waters or three maximum currents. At some places, the current on some days runs maximum flood twice, but ebbs only once, a minimum flood occurring in place of the second ebb. The tables show this information.

933. Tidal Current Predictions for Subordinate Stations

For each subordinate station listed in Table 2 of the tidal current tables, the following information is given:

1. **Number**: The stations are listed in geographical order and assigned consecutive numbers, as in the tide tables. Each volume contains an alphabetical station listing correlating the station with its consecutive number to assist in locating the entry in Table 2.
2. **Place**: The list of places includes both subordinate and reference stations, the latter given in bold type.
3. **Position**: The approximate latitude and longitude are given to assist in locating the station. The latitude is north or south and the longitude east or west as indicated by the letters (N, S, E, W) next above the entry. The current given is for the center of the channel unless another location is indicated by the station name.
4. **Time difference**: Two time differences are tabulated. One is the number of hours and minutes to be applied to the tabulated times of slack water at the reference station to find the times of slack waters at the subordinate station. The other time difference is applied to the times of maximum current at the reference station to find the times of the corresponding maximum current at the subordinate station. The intervals, which are added or subtracted in accordance with their signs, include any difference in time between the two stations, so that the answer is correct for the standard time of the subordinate station. Limited application and special conditions are indicated by footnotes.
5. **Velocity ratios**: Speed of the current at the subordinate station is the product of the velocity at the reference station and the tabulated ratio. Separate ratios may be given for flood and ebb currents. Special conditions are indicated by footnotes.
6. **Average Speeds and Directions**: Minimum and maximum velocities before flood and ebb are listed for each station, along with the true directions of the flow. Minimum velocity is not always 0.0 knots.

934. Finding Velocity of Tidal Current at any Time

Table 3 of the tidal current tables provides means for determining the approximate velocity at any time. Directions are given in an explanation preceding the table. Figure 934 shows the OPNAV form used for current prediction.

935. Duration of Slack Water

The predicted times of slack water listed in the tidal current tables indicate the instant of zero velocity. There is a period each side of slack water, however, during which the current is so weak that for practical purposes it may be considered negligible. Table 4 of the tidal current tables gives, for various maximum currents, the approximate period of time during which currents not exceeding 0.1 to 0.5 knots will be encountered. This period includes the last of the flood or ebb and the beginning of the following flood or ebb; that is, half of the duration will be before and half after the time of slack water.

When there is a difference between the velocities of the maximum flood and ebb preceding and following the slack for which the duration is desired, it will be sufficiently accurate to find a separate duration for each maximum velocity and average the two to determine the duration of the weak current.

Of the two sub-tables of Table 4, Table A is used for all places except those listed for Table B; Table B is used for just the places listed and the stations in Table 2 which are referred to them.

936. Additional Tide Prediction Publications

NOS also publishes a special *Regional Tide and Tidal Current Table for New York Harbor to Chesapeake Bay*, and a *Tidal Circulation and Water Level Forecast Atlas for Delaware River and Bay*.

937. *Tidal Current Charts*

Tidal Current charts present a comprehensive view of the hourly velocity of current in different bodies of water. They also provide a means for determining the current's velocity at various locations in these waters. The arrows show the direction of the current; the figures give the speed in knots at the time of spring tides. A weak current is defined as less than 0.1 knot. These charts depict the flow of the tidal current under normal weather conditions. Strong winds and freshets, however, may cause nontidal currents, considerably modifying the velocity indicated on the charts.

Tidal Current charts are provided (1994) for Boston Harbor, Charleston Harbor SC, Long Island Sound and Block Island Sound, Narragansett Bay, Narragansett Bay to Nantucket Sound, Puget Sound (Northern Part), Puget Sound (Southern Part), Upper Chesapeake Bay, and Tampa Bay.

OPNAV 3530/40 (4-73)
VEL OF CURRENT

Date	
Location	
Time	
Ref Sta	
Time Diff Stack Water	
Time Diff Max Current	
Vel Ratio Max Flood	
Vel Ratio Max Ebb	
Flood Dir	
Ebb Dir	
Ref Sta Stack Water Time	
Time Diff	
Local Sta Stack Water Time	
Ref Sta Max Current Time	
Time Diff	
Local Sta Max Current Time	
Ref Sta Max Current Vel	
Vel Ratio	
Local Sta Max Current Vel	
Int Between Slack and Desired Time	
Int Between Slack and Max Current	
Max Current	
Factor Table 3	
Velocity	
Direction	

Figure 934. OPNAV 3530/41 Current Form.

The tidal current's velocity varies from day to day as a function of the phase, distance, and declination of the Moon. Therefore, to obtain the velocity for any particular day and hour, the spring velocities shown on the charts must be modified by correction factors. A correction table given in the charts can be used for this purpose.

All of the charts except Narragansett Bay require the use of the annual *Tidal Current Tables*. Narragansett Bay requires use of the annual *Tide Tables*.

938. Current Diagrams

A current diagram is a graph showing the velocity of the current along a channel at different stages of the tidal current cycle. The current tables include diagrams for Martha's Vineyard and Nantucket Sounds (one diagram); East River, New York; New York Harbor; Delaware Bay and River (one diagram); and Chesapeake Bay. These diagrams are no longer published by NOS, but are available privately and remain useful as they are not ephemeral.

On Figure 938, each vertical line represents a given instant identified by the number of hours before or after slack water at The Narrows. Each horizontal line represents a distance from Ambrose Channel entrance, measured along the usually traveled route. The names along the left margin are placed at the correct distances from Ambrose Channel entrance. The current is for the center of the channel opposite these points. The intersection of any vertical line with any horizontal line represents a given moment in the current cycle at a given place in the channel. If this intersection is in a shaded area, the current is flooding; if in an unshaded area, it is ebbing. The velocity can be found by interpolation between the numbers given in the diagram. The given values are averages. To find the value at any time, multiply the velocity found from the diagram by the ratio of maximum velocity of the current involved to the maximum shown on the diagram. If the diurnal inequality is large, the accuracy can be improved by altering the width of the shaded area to fit conditions. The diagram covers 1 1/2 current cycles, so that the right 1/3 duplicates the left 1/3.

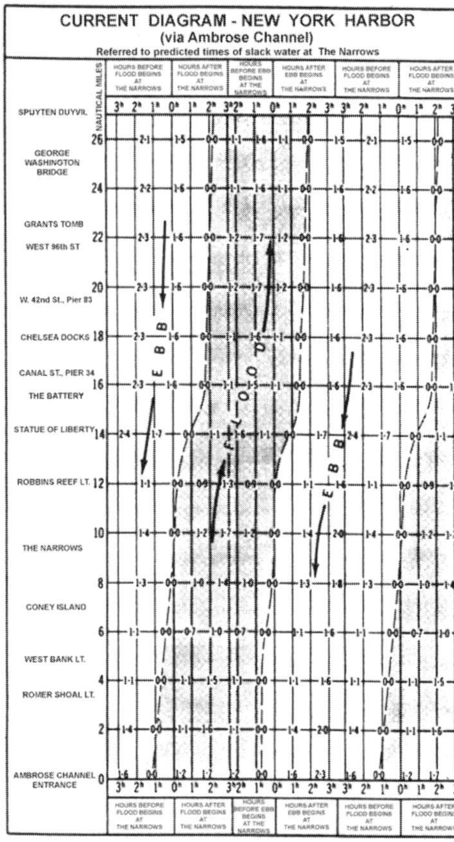

Figure 938. Current diagram for New York Harbor.

Use Table 1 or 2 to determine the current for a single station. The current diagrams are intended for use in either of two ways: to determine a favorable time for passage through the channel and to find the average current to be expected during a passage through the channel. For both of these uses, a number of "velocity lines" are provided. When the appropriate line is transferred to the correct part of the diagram, the current to be encountered during passage is indicated along the line.

If the transferred velocity line is partly in a flood current area, all ebb currents (those increasing the ship's velocity) are given a positive sign (+), and all flood currents a negative sign (-). A separate ratio should be determined for each current (flood or ebb), and applied to the entries for that current. In the Chesapeake Bay, it is common for an outbound vessel to encounter three or even four separate currents during passage. Under the latter condition, it is good practice to multiply each current taken from the diagram by the ratio for the current involved.

If the time of starting the passage is fixed, and the current during passage is desired, the starting time is identified in terms of the reference tidal cycle. The velocity line is then drawn through the intersection of this vertical time line and the horizontal line through the place. The average current is then determined in the same manner as when the velocity line is located as described above.

939. Computer Predictions

Until recently, tidal predictions were compiled only on mainframe or minicomputers and then put into hardcopy table form for the mariner. There are several types of commercial software available now for personal computers (PC's) that provide digital versions of the NOS tide tables and also graph the tidal heights. The tabular information and graphs can be printed for the desired locations for pre-voyage planning. There are also several types of specialized hand-held calculators and tide clocks that can be used to predict tides for local areas.

Newer versions of PC software use the actual harmonic constants available for locations, the prediction equation, and digital versions of Table 2 in the *Tide Tables* to produce even more products for the navigator's use. Since NOS has published the data, even inexpensive navigation electronics such as handheld GPS receivers and plotters for small craft navigation often include graphic tide tables.

Emerging applications include integration of tidal prediction with positioning systems and vessel traffic systems which are now moving towards full use of GPS. In addition, some electronic chart systems are already able to integrate tide prediction information. Many of these new systems will also use real-time water level and current information. Active research also includes providing predictions of total water level that will include not only the tidal prediction component, but also the weather-related component.

CHAPTER 10

RADIO WAVES

ELECTROMAGNETIC WAVE PROPAGATION

1000. Source of Radio Waves

Consider electric current as a flow of electrons along a conductor between points of differing potential. A **direct current** flows continuously in the same direction. This would occur if the polarity of the electromotive force causing the electron flow were constant, such as is the case with a battery. If, however, the current is induced by the relative motion between a conductor and a magnetic field, such as is the case in a rotating machine called a **generator**, then the resulting current changes direction in the conductor as the polarity of the electromotive force changes with the rotation of the generator's rotor. This is known as **alternating current**.

The energy of the current flowing through the conductor is either dissipated as heat (an energy loss proportional to both the current flowing through the conductor and the conductor's resistance) or stored in an electromagnetic field oriented symmetrically about the conductor. The orientation of this field is a function of the polarity of the source producing the current. When the current is removed from the wire, this electromagnetic field will, after a finite time, collapse back into the wire.

What would occur should the polarity of the current source supplying the wire be reversed at a rate which exceeds the finite amount of time required for the electromagnetic field to collapse back upon the wire? In this case, another magnetic field, proportional in strength but exactly opposite in magnetic orientation to the initial field, will be formed upon the wire. The initial magnetic field, its current source gone, cannot collapse back upon the wire because of the existence of this second electromagnetic field. Instead, it propagates out into space. This is the basic principle of a radio antenna, which transmits a wave at a frequency proportional to the rate of pole reversal and at a speed equal to the speed of light.

1001. Radio Wave Terminology

The magnetic field strength in the vicinity of a conductor is directly proportional to the magnitude of the current flowing through the conductor. Recall the discussion of alternating current above. A rotating generator produces current in the form of a sine wave. That is, the magnitude of the current varies as a function of the relative position of the rotating conductor and the stationary magnetic field used to induce the current. The current starts at zero, increases to a maximum as the rotor completes one quarter of its revolution, and falls to zero when the rotor completes one half of its revolution. The current then approaches a negative maximum; then it once again returns to zero. This cycle can be represented by a sine function.

The relationship between the current and the magnetic field strength induced in the conductor through which the current is flowing is shown in Figure 1001. Recall from the discussion above that this field strength is proportional to the magnitude of the current; that is, if the current is represented by a sine wave function, then so too will be the magnetic field strength resulting from that current. This characteristic shape of the field strength curve has led to the use of the term "wave" when referring to electromagnetic propagation. The maximum displacement of a peak from zero is called the **amplitude**. The forward side of any wave is called the **wave front**. For a non-directional antenna, each wave proceeds outward as an expanding sphere (or hemisphere).

One **cycle** is a complete sequence of values, as from crest to crest. The distance traveled by the energy during one cycle is the **wavelength**, usually expressed in metric units (meters, centimeters, etc.). The number of cycles repeated during unit time (usually 1 second) is the **frequency**. This is given in **hertz** (cycles per second). A kilohertz (kHz) is 1,000 cycles per second. A megahertz (MHz) is 1,000,000 cycles per second. Wavelength and frequency are inversely proportional.

The **phase** of a wave is the amount by which the cycle

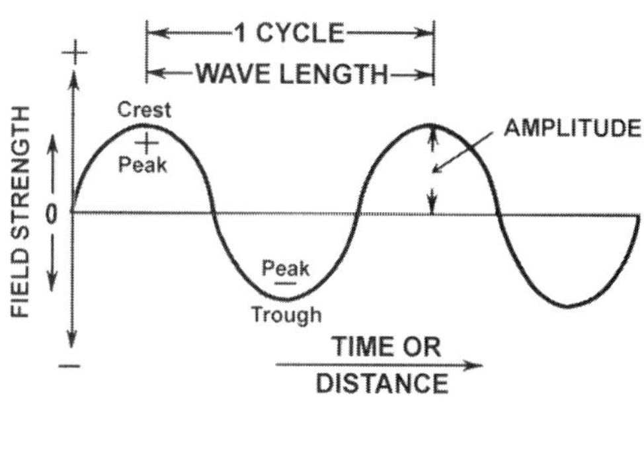

Figure 1001. Radio wave terminology.

has progressed from a specified origin. For most purposes it is stated in circular measure, a complete cycle being considered 360°. Generally, the origin is not important, principal interest being the phase relative to that of some other wave. Thus, two waves having crests 1/4 cycle apart are said to be 90° "out of phase." If the crest of one wave occurs at the trough of another, the two are 180° out of phase.

1002. The Electromagnetic Spectrum

The entire range of electromagnetic radiation frequencies is called the **electromagnetic spectrum**. The frequency range suitable for radio transmission, the **radio spectrum**, extends from 10 kilohertz to 300,000 megahertz. It is divided into a number of bands, as shown in Table 1002.

Below the radio spectrum, but overlapping it, is the audio frequency band, extending from 20 to 20,000 hertz. Above the radio spectrum are heat and infrared, the visible spectrum (light in its various colors), ultraviolet, X-rays, gamma rays, and cosmic rays. These are included in Table 1002. Waves shorter than 30 centimeters are usually called **microwaves**.

Within the frequencies from 1-40 gHz (1,000-40,000 MHz), additional bands are defined as follows:

L-band: 1-2 gHz (1,000-2,000 MHz)
S-band: 2-4 gHz (2,000-4,000 MHz
C-band: 4-8 gHz (4,000-8,000 MHz)
X-band: 8-12.5 gHz (8,000-12,500 MHz)
Lower K-band: 12.5-18 gHz (12,500-18,000 MHz)
Upper K-band: 26.5-40 gHz (26,500-40,000 MHz)

Marine radar systems commonly operate in the S and X bands, while satellite navigation system signals are found in the L-band.

The break of the K-band into lower and upper ranges is necessary because the resonant frequency of water vapor occurs in the middle region of this band, and severe absorption of radio waves occurs in this part of the spectrum.

Band	Abbreviation	Range of frequency	Range of wavelength
Audio frequency	AF	20 to 20,000 Hz	15,000,000 to 15,000 m
Radio frequency	RF	10 kHz to 300,000 MHz	30,000 m to 0.1 cm
Very low frequency	VLF	10 to 30 kHz	30,000 to 10,000 m
Low frequency	LF	30 to 300 kHz	10,000 to 1,000 m
Medium frequency	MF	300 to 3,000 kHz	1,000 to 100 m
High frequency	HF	3 to 30 MHz	100 to 10 m
Very high frequency	VHF	30 to 300 MHz	10 to 1 m
Ultra high frequency	UHF	300 to 3,000 MHz	100 to 10 cm
Super high frequency	SHF	3,000 to 30,000 MHz	10 to 1 cm
Extremely high frequency	EHF	30,000 to 300,000 MHz	1 to 0.1 cm
Heat and infrared*		10^6 to 3.9×10^8 MHz	0.03 to 7.6×10^{-5} cm
Visible spectrum*		3.9×10^8 to 7.9×10^8 MHz	7.6×10^{-5} to 3.8×10^{-5} cm
Ultraviolet*		7.9×10^8 to 2.3×10^{10} MHz	3.8×10^{-5} to 1.3×10^{-6} cm
X-rays*		2.0×10^9 to 3.0×10^{13} MHz	1.5×10^{-5} to 1.0×10^{-9} cm
Gamma rays*		2.3×10^{12} to 3.0×10^{14} MHz	1.3×10^{-8} to 1.0×10^{-10} cm
Cosmic rays*		$>4.8\times10^{15}$ MHz	$<6.2\times10^{-12}$ cm

* Values approximate.

Table 1002. Electromagnetic spectrum.

1003. Polarization

Radio waves produce both electric and magnetic fields. The direction of the electric component of the field is called the **polarization** of the electromagnetic field. Thus, if the electric component is vertical, the wave is said to be "vertically polarized," and if horizontal, "horizontally polarized."

A wave traveling through space may be polarized in any direction. One traveling along the surface of the Earth is always vertically polarized because the Earth, a conductor, short-circuits any horizontal component. The magnetic field and the electric field are always mutually perpendicular.

1004. Reflection

When radio waves strike a surface, the surface reflects them in the same manner as light waves. Radio waves of all frequencies are reflected by the surface of the Earth. The strength of the reflected wave depends upon angle of incidence (the angle between the incident ray and the horizontal), type of polarization, frequency, reflecting properties of the surface, and divergence of the reflected ray. Lower frequencies penetrate the earth's surface more than higher ones. At very low frequencies, usable radio signals can be received some distance below the surface of the sea.

A phase change occurs when a wave is reflected from the surface of the Earth. The amount of the change varies with the conductivity of the Earth and the polarization of the wave, reaching a maximum of 180° for a horizontally polarized wave reflected from sea water (considered to have infinite conductivity).

When direct waves (those traveling from transmitter to receiver in a relatively straight line, without reflection) and reflected waves arrive at a receiver, the total signal is the vector sum of the two. If the signals are in phase, they reinforce each other, producing a stronger signal. If there is a phase difference, the signals tend to cancel each other, the cancellation being complete if the phase difference is 180° and the two signals have the same amplitude. This interaction of waves is called **wave interference**.

A phase difference may occur because of the change of phase of a reflected wave, or because of the longer path it follows. The second effect decreases with greater distance between transmitter and receiver, for under these conditions the difference in path lengths is smaller.

At lower frequencies there is no practical solution to interference caused in this way. For VHF and higher frequencies, the condition can be improved by elevating the antenna, if the wave is vertically polarized. Additionally, interference at higher frequencies can be more nearly eliminated because of the greater ease of beaming the signal to avoid reflection.

Reflections may also occur from mountains, trees, and other obstacles. Such reflection is negligible for lower frequencies, but becomes more prevalent as frequency increases. In radio communication, it can be reduced by using directional antennas, but this solution is not always available for navigational systems.

Various reflecting surfaces occur in the atmosphere. At high frequencies, reflections take place from rain. At still higher frequencies, reflections are possible from clouds, particularly rain clouds. Reflections may even occur at a sharply defined boundary surface between air masses, as when warm, moist air flows over cold, dry air. When such a surface is roughly parallel to the surface of the Earth, radio waves may travel for greater distances than normal The principal source of reflection in the atmosphere is the ionosphere.

1005. Refraction

Refraction of radio waves is similar to that of light waves. Thus, as a signal passes from air of one density to that of a different density, the direction of travel is altered. The principal cause of refraction in the atmosphere is the difference in temperature and pressure occurring at various heights and in different air masses.

Refraction occurs at all frequencies, but below 30 MHz the effect is small as compared with ionospheric effects, diffraction, and absorption. At higher frequencies, refraction in the lower layer of the atmosphere extends the radio horizon to a distance about 15 percent greater than the visible horizon. The effect is the same as if the radius of the Earth were about one-third greater than it is and there were no refraction.

Sometimes the lower portion of the atmosphere becomes stratified. This stratification results in nonstandard temperature and moisture changes with height. If there is a marked temperature inversion or a sharp decrease in water vapor content with increased height, a horizontal radio duct may be formed. High frequency radio waves traveling horizontally within the duct are refracted to such an extent that they remain within the duct, following the curvature of the Earth for phenomenal distances. This is called **superrefraction**. Maximum results are obtained when both transmitting and receiving antennas are within the duct. There is a lower limit to the frequency affected by ducts. It varies from about 200 MHz to more than 1,000 MHz.

At night, surface ducts may occur over land due to cooling of the surface. At sea, surface ducts about 50 feet thick may occur at any time in the trade wind belt. Surface ducts 100 feet or more in thickness may extend from land out to sea when warm air from the land flows over the cooler ocean surface. Elevated ducts from a few feet to more than 1,000 feet in thickness may occur at elevations of 1,000 to 5,000 feet, due to the settling of a large air mass. This is a frequent occurrence in Southern California and certain areas of the Pacific Ocean.

A bending in the horizontal plane occurs when a groundwave crosses a coast at an oblique angle. This is due to a marked difference in the conducting and reflecting properties of the land and water over which the wave travels. The effect is known as **coastal refraction** or **land effect**.

1006. The Ionosphere

Since an atom normally has an equal number of negatively charged electrons and positively charged protons, it is electrically neutral. An **ion** is an atom or group of atoms which has become electrically charged, either positively or negatively, by the loss or gain of one or more electrons.

Loss of electrons may occur in a variety of ways. In the atmosphere, ions are usually formed by collision of atoms with rapidly moving particles, or by the action of cosmic rays or ultraviolet light. In the lower portion of the atmosphere, recombination soon occurs, leaving a small percentage of ions. In thin atmosphere far above the surface of the Earth, however, atoms are widely separated and a large number of ions may be present. The region of numerous positive and negative ions and unattached electrons is called the **ionosphere**. The extent of ionization depends upon the kinds of atoms present in the atmosphere, the density of the atmosphere, and the position relative to the Sun (time of day and season). After sunset, ions and electrons recombine faster than they are separated, decreasing the ionization of the atmosphere.

An electron can be separated from its atom only by the application of greater energy than that holding the electron. Since the energy of the electron depends primarily upon the kind of an atom of which it is a part, and its position relative to the nucleus of that atom, different kinds of radiation may cause ionization of different substances.

In the outermost regions of the atmosphere, the density is so low that oxygen exists largely as separate atoms, rather than combining as molecules as it does nearer the surface of the Earth. At great heights the energy level is low and ionization from solar radiation is intense. This is known as the **F layer**. Above this level the ionization decreases because of the lack of atoms to be ionized. Below this level it decreases because the ionizing agent of appropriate energy has already been absorbed. During daylight, two levels of maximum F ionization can be detected, the F_2 layer at about 125 statute miles above the surface of the Earth, and the F_1 layer at about 90 statute miles. At night, these combine to form a single F layer.

At a height of about 60 statute miles, the solar radiation not absorbed by the F layer encounters, for the first time, large numbers of oxygen molecules. A new maximum ionization occurs, known as the **E layer**. The height of this layer is quite constant, in contrast with the fluctuating F layer. At night the E layer becomes weaker by two orders of magnitude.

Below the E layer, a weak D layer forms at a height of about 45 statute miles, where the incoming radiation encounters ozone for the first time. The D layer is the principal source of absorption of HF waves, and of reflection of LF and VLF waves during daylight.

1007. The Ionosphere and Radio Waves

When a radio wave encounters a particle having an electric charge, it causes that particle to vibrate. The vibrating particle absorbs electromagnetic energy from the radio wave and radiates it. The net effect is a change of polarization and an alteration of the path of the wave. That portion of the wave in a more highly ionized region travels faster, causing the wave front to tilt and the wave to be directed toward a region of less intense ionization.

Refer to Figure 1007a, in which a single layer of the ionosphere is considered. Ray 1 enters the ionosphere at such an angle that its path is altered, but it passes through and proceeds outward into space. As the angle with the horizontal decreases, a critical value is reached where ray 2 is bent or reflected back toward the Earth. As the angle is still further decreased, such as at 3, the return to Earth occurs at a greater distance from the transmitter.

A wave reaching a receiver by way of the ionosphere is called a **skywave**. This expression is also appropriately applied to a wave reflected from an air mass boundary. In common usage, however, it is generally associated with the ionosphere. The wave which travels along the surface of the Earth is called a **groundwave**. At angles greater than the critical angle, no skywave signal is received. Therefore, there is a minimum distance from the transmitter at which skywaves can be received. This is called the **skip distance**, shown in Figure 1007a. If the groundwave extends out for less distance than the skip distance, a skip zone occurs, in which no signal is received.

The critical radiation angle depends upon the intensity of ionization, and the frequency of the radio wave. As the frequency increases, the angle becomes smaller. At frequencies greater than about 30 MHz virtually all of the energy penetrates through or is absorbed by the ionosphere. Therefore, at any given receiver there is a maximum usable frequency if skywaves are to be utilized. The strongest signals are received at or slightly below this frequency. There is also a lower practical frequency beyond which signals are too weak to be of value. Within this band the optimum frequency can be selected to give best results. It cannot be too near the maximum usable frequency because this frequency fluctuates with changes of intensity within the ionosphere. During magnetic storms the ionosphere density decreases. The maximum usable frequency decreases, and the lower usable frequency increases. The band of usable frequencies is thus narrowed. Under extreme conditions it may be completely eliminated, isolating the receiver and causing a radio blackout.

Skywave signals reaching a given receiver may arrive by any of several paths, as shown in Figure 1007b. A signal which undergoes a single reflection is called a "one-hop" signal, one which undergoes two reflections with a ground reflection between is called a "two-hop" signal, etc. A

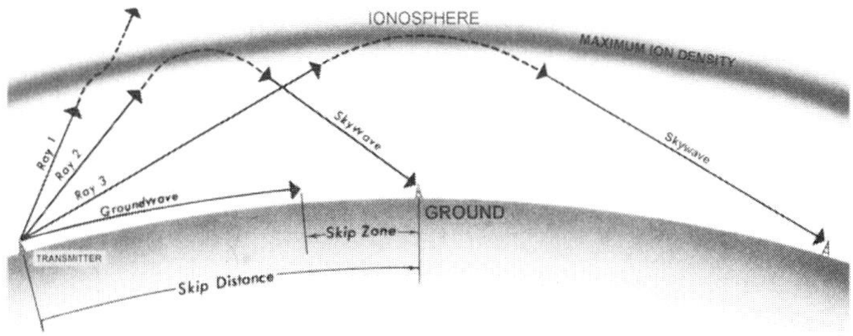

Figure 1007a. The effect of the ionosphere on radio waves.

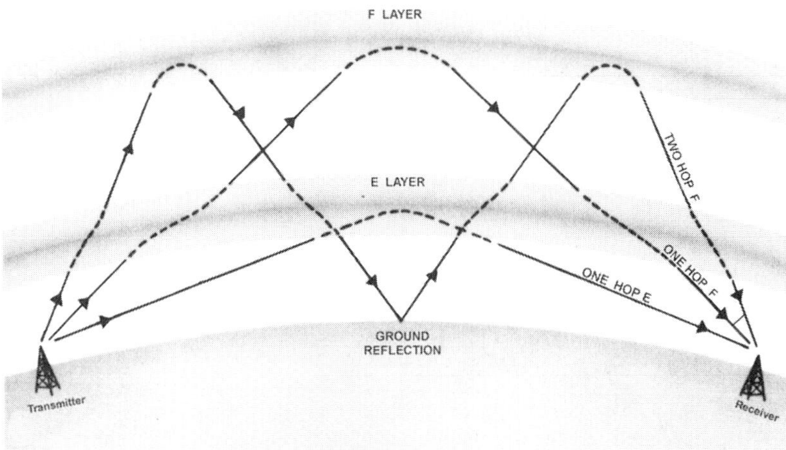

Figure 1007b. Various paths by which a skywave signal might be received.

"multihop" signal undergoes several reflections. The layer at which the reflection occurs is usually indicated, also, as "one-hop E," "two-hop F," etc.

Because of the different paths and phase changes occurring at each reflection, the various signals arriving at a receiver have different phase relationships. Since the density of the ionosphere is continually fluctuating, the strength and phase relationships of the various signals may undergo an almost continuous change. Thus, the various signals may reinforce each other at one moment and cancel each other at the next, resulting in fluctuations of the strength of the total signal received. This is called **fading**. This phenomenon may also be caused by interaction of components within a single reflected wave, or changes in its strength due to changes in the reflecting surface. Ionospheric changes are associated with fluctuations in the radiation received from the Sun, since this is the principal cause of ionization. Signals from the F layer are particularly erratic because of the rapidly fluctuating conditions within the layer itself.

The maximum distance at which a one-hop E signal can be received is about 1,400 miles. At this distance the signal leaves the transmitter in approximately a horizontal direction. A one-hop F signal can be received out to about 2,500 miles. At low frequencies groundwaves extend out for great distances.

A skywave may undergo a change of polarization during reflection from the ionosphere, accompanied by an alteration in the direction of travel of the wave. This is called **polarization error**. Near sunrise and sunset, when rapid changes are occurring in the ionosphere, reception may become erratic and polarization error a maximum. This is called **night effect**.

1008. Diffraction

When a radio wave encounters an obstacle, its energy is reflected or absorbed, causing a shadow beyond the obstacle. However, some energy does enter the shadow area because of diffraction. This is explained by Huygens' principle, which

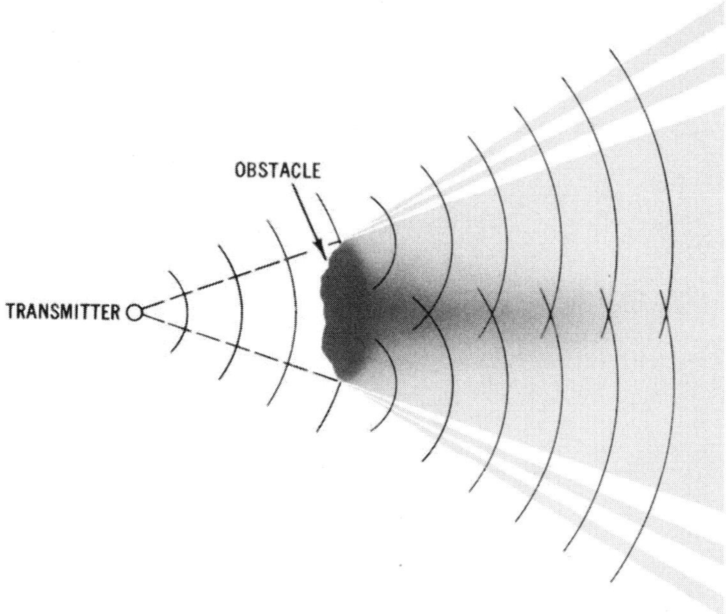

Figure 1008. Diffraction.

states that every point on the surface of a wave front is a source of radiation, transmitting energy in all directions ahead of the wave. No noticeable effect of this principle is observed until the wave front encounters an obstacle, which intercepts a portion of the wave. From the edge of the obstacle, energy is radiated into the shadow area, and also outside of the area. The latter interacts with energy from other parts of the wave front, producing alternate bands in which the secondary radiation reinforces or tends to cancel the energy of the primary radiation. Thus, the practical effect of an obstacle is a greatly reduced signal strength in the shadow area, and a disturbed pattern for a short distance outside the shadow area. This is illustrated in Figure 1008.

The amount of diffraction is inversely proportional to the frequency, being greatest at very low frequencies.

1009. Absorption and Scattering

The amplitude of a radio wave expanding outward through space varies inversely with distance, weakening with increased distance. The decrease of strength with distance is called **attenuation**. Under certain conditions the attenuation is greater than in free space.

A wave traveling along the surface of the Earth loses a certain amount of energy to the Earth. The wave is diffracted downward and absorbed by the Earth. As a result of this absorption, the remainder of the wave front tilts downward, resulting in further absorption by the Earth. Attenuation is greater over a surface which is a poor conductor. Relatively little absorption occurs over sea water, which is an excellent conductor at low frequencies, and low frequency groundwaves travel great distances over water.

A skywave suffers an attenuation loss in its encounter with the ionosphere. The amount depends upon the height and composition of the ionosphere as well as the frequency of the radio wave. Maximum ionospheric absorption occurs at about 1,400 kHz.

In general, atmospheric absorption increases with frequency. It is a problem only in the SHF and EHF frequency range. At these frequencies, attenuation is further increased by scattering due to reflection by oxygen, water vapor, water droplets, and rain in the atmosphere.

1010. Noise

Unwanted signals in a receiver are called **interference**. The intentional production of such interference to obstruct communication is called **jamming**. Unintentional interference is called **noise**.

Noise may originate within the receiver. Hum is usually the result of induction from neighboring circuits carrying alternating current. Irregular crackling or sizzling sounds may be caused by poor contacts or faulty components within the receiver. Stray currents in normal components cause some noise. This source sets the ultimate limit of sensitivity that can be achieved in a receiver. It is the same at any frequency.

Noise originating outside the receiver may be either man-made or natural. Man-made noises originate in electrical appliances, motor and generator brushes, ignition systems, and other sources of sparks which transmit electromagnetic signals that are picked up by the receiving antenna.

Natural noise is caused principally by discharge of static electricity in the atmosphere. This is called **atmospheric noise**, atmospherics, or static. An extreme example is a thunderstorm. An exposed surface may acquire a considerable charge of static electricity. This may be caused by friction of water or solid particles blown against or along such a surface. It may also be caused by splitting of a water droplet which strikes the surface, one part of the droplet requiring a positive charge and the other a negative charge. These charges may be transferred to the surface. The charge tends to gather at points and ridges of the conducting surface, and when it accumulates to a sufficient extent to overcome the insulating properties of the atmosphere, it discharges into the atmosphere. Under suitable conditions this becomes visible and is known as St. Elmo's fire, which is sometimes seen at mastheads, the ends of yardarms, etc.

Atmospheric noise occurs to some extent at all frequencies but decreases with higher frequencies. Above about 30 MHz it is not generally a problem.

1011. Antenna Characteristics

Antenna design and orientation have a marked effect upon radio wave propagation. For a single-wire antenna, strongest signals are transmitted along the perpendicular to the wire, and virtually no signal in the direction of the wire. For a vertical antenna, the signal strength is the same in all horizontal directions. Unless the polarization undergoes a change during transit, the strongest signal received from a vertical transmitting antenna occurs when the receiving antenna is also vertical.

For lower frequencies the radiation of a radio signal takes place by interaction between the antenna and the ground. For a vertical antenna, efficiency increases with greater length of the antenna. For a horizontal antenna, efficiency increases with greater distance between antenna and ground. Near-maximum efficiency is attained when this distance is one-half wavelength. This is the reason for elevating low frequency antennas to great heights. However, at the lowest frequencies, the required height becomes prohibitively great. At 10 kHz it would be about 8 nautical miles for a half-wavelength antenna. Therefore, lower frequency antennas are inherently inefficient. This is partly offset by the greater range of a low frequency signal of the same transmitted power as one of higher frequency.

At higher frequencies, the ground is not used, both conducting portions being included in a dipole antenna. Not only can such an antenna be made efficient, but it can also be made sharply directive, thus greatly increasing the strength of the signal transmitted in a desired direction.

The power received is inversely proportional to the square of the distance from the transmitter, assuming there is no attenuation due to absorption or scattering.

1012. Range

The range at which a usable signal is received depends upon the power transmitted, the sensitivity of the receiver, frequency, route of travel, noise level, and perhaps other factors. For the same transmitted power, both the groundwave and skywave ranges are greatest at the lowest frequencies, but this is somewhat offset by the lesser efficiency of antennas for these frequencies. At higher frequencies, only direct waves are useful, and the effective range is greatly reduced. Attenuation, skip distance, ground reflection, wave interference, condition of the ionosphere, atmospheric noise level, and antenna design all affect the distance at which useful signals can be received.

1013. Radio Wave Spectra

Frequency is an important consideration in radio wave propagation. The following summary indicates the principal effects associated with the various frequency bands, starting with the lowest and progressing to the highest usable radio frequency.

Very Low Frequency (VLF, 10 to 30 kHz): The VLF signals propagate between the bounds of the ionosphere and the Earth and are thus guided around the curvature of the Earth to great distances with low attenuation and excellent stability. Diffraction is maximum. Because of the long wavelength, large antennas are needed, and even these are inefficient, permitting radiation of relatively small amounts of power. Magnetic storms have little effect upon transmission because of the efficiency of the "Earth-ionosphere waveguide." During such storms, VLF signals may constitute the only source of radio communication over great distances. However, interference from atmospheric noise may be troublesome. Signals may be received from below the surface of the sea.

Low Frequency (LF, 30 to 300 kHz): As frequency is increased to the LF band and diffraction decreases, there is greater attenuation with distance, and range for a given power output falls off rapidly. However, this is partly offset by more efficient transmitting antennas. LF signals are most stable within groundwave distance of the transmitter. A wider bandwidth permits pulsed signals at 100 kHz. This allows separation of the stable groundwave pulse from the variable skywave pulse up to 1,500 km, and up to 2,000 km for overwater paths. The frequency for Loran C is in the LF band. This band is also useful for radio direction finding and time dissemination.

Medium Frequency (MF, 300 to 3,000 kHz): Groundwaves provide dependable service, but the range for a given power is reduced greatly. This range varies from about 400 miles at the lower portion of the band to about 15 miles at the upper end for a transmitted signal of 1 kilowatt. These values are influenced, however, by the power of the transmitter, the directivity and efficiency of the antenna, and the nature of the terrain over which signals travel. Elevating the antenna to obtain direct waves may improve the transmission. At the lower frequencies of the band, skywaves are available both day and night. As the frequency is increased, ionospheric absorption increases to a maximum at about 1,400 kHz. At higher frequencies the absorption decreases, permitting increased use of skywaves. Since the ionosphere changes with the hour, season, and sunspot cycle, the reliability of skywave signals is variable. By careful selection of frequency, ranges of as much as 8,000 miles with 1 kilowatt of transmitted power are possible, using multihop signals. However, the frequency selection is critical. If it is too high, the signals penetrate the ionosphere and are lost in space. If it is too low, signals are too weak. In general, skywave reception is equally good by day or night, but lower frequencies are needed at night. The standard broadcast band for commercial stations (535 to 1,605 kHz) is in the MF band.

High Frequency (HF, 3 to 30 MHz): As with higher medium frequencies, the groundwave range of HF signals is limited to a few miles, but the elevation of the antenna may increase the direct-wave distance of transmission. Also, the height of the antenna does have an important effect upon skywave transmission because the antenna has an "image" within the conducting Earth. The distance between antenna and image is related to the height of the antenna, and this distance is as critical as the distance between elements of an antenna system. Maximum usable frequencies fall generally within the HF band. By day this may be 10 to 30 MHz, but during the night it may drop to 8 to 10 MHz. The HF band is widely used for ship-to-ship and ship-to-shore communication.

Very High Frequency (VHF, 30 to 300 MHz): Communication is limited primarily to the direct wave, or the direct wave plus a ground-reflected wave. Elevating the antenna to increase the distance at which direct waves can be used results in increased distance of reception, even though some wave interference between direct and ground-reflected waves is present. Diffraction is much less than with lower frequencies, but it is most evident when signals cross sharp mountain peaks or ridges. Under suitable conditions, reflections from the ionosphere are sufficiently strong to be useful, but generally they are unavailable. There is relatively little interference from atmospheric noise in this band. Reasonably efficient directional antennas are possible with VHF. The VHF band is much used for communication.

Ultra High Frequency (UHF, 300 to 3,000 MHz): Skywaves are not used in the UHF band because the ionosphere is not sufficiently dense to reflect the waves, which pass through it into space. Groundwaves and ground-reflected waves are used, although there is some wave interference. Diffraction is negligible, but the radio horizon extends about 15 percent beyond the visible horizon, due principally to refraction. Reception of UHF signals is virtually free from fading and interference by atmospheric noise. Sharply directive antennas can be produced for transmission in this band, which is widely used for ship-to-ship and ship-to-shore communication.

Super High Frequency (SHF, 3,000 to 30,000 MHz): In the SHF band, also known as the microwave or as the centimeter wave band, there are no skywaves, transmission being entirely by direct and ground-reflected waves. Diffraction and interference by atmospheric noise are virtually nonexistent. Highly efficient, sharply directive antennas can be produced. Thus, transmission in this band is similar to that of UHF, but with the effects of shorter waves being greater. Reflection by clouds, water droplets, dust particles, etc., increases, causing greater scattering, increased wave interference, and fading. The SHF band is used for marine navigational radar.

Extremely High Frequency (EHF, 30,000 to 300,000 MHz): The effects of shorter waves are more pronounced in the EHF band, transmission being free from wave interference, diffraction, fading, and interference by atmospheric noise. Only direct and ground-reflected waves are available. Scattering and absorption in the atmosphere are pronounced and may produce an upper limit to the frequency useful in radio communication.

1014. Regulation of Frequency Use

While the characteristics of various frequencies are important to the selection of the most suitable one for any given purpose, these are not the only considerations. Confusion and extensive interference would result if every user had complete freedom of selection. Some form of regulation is needed. The allocation of various frequency bands to particular uses is a matter of international agreement. Within the United States, the Federal Communications Commission has responsibility for authorizing use of particular frequencies. In some cases a given frequency is allocated to several widely separated transmitters, but only under conditions which minimize interference, such as during daylight hours. Interference between stations is further reduced by the use of channels, each of a narrow band of frequencies. Assigned frequencies are separated by an arbitrary band of frequencies that are not authorized for use. In the case of radio aids to navigation and ship communications bands of several channels are allocated, permitting selection of band and channel by the user.

1015. Types of Radio Transmission

A series of waves transmitted at constant frequency and amplitude is called a continuous wave (CW). This cannot be heard except at the very lowest radio frequencies, when it may produce, in a receiver, an audible hum of high pitch.

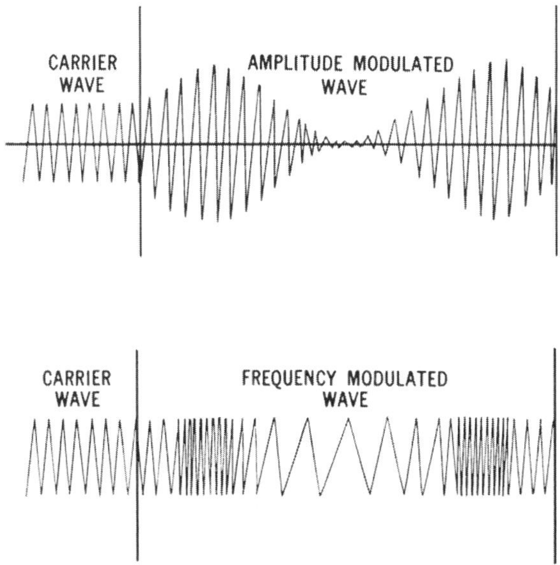

Figure 1015a. Amplitude modulation (upper figure) and frequency modulation (lower figure) by the same modulating wave.

Figure 1015b. Pulse modulation.

Although a continuous wave may be used directly, as in radiodirection finding or Decca, it is more commonly modified in some manner. This is called **modulation**. When this occurs, the continuous wave serves as a carrier wave for information. Any of several types of modulation may be used.

In **amplitude modulation (AM)** the amplitude of the carrier wave is altered in accordance with the amplitude of a modulating wave, usually of audio frequency, as shown in Figure 1015a. In the receiver the signal is demodulated by removing the modulating wave and converting it back to its original form. This form of modulation is widely used in voice radio, as in the standard broadcast band of commercial broadcasting.

If the frequency instead of the amplitude is altered in accordance with the amplitude of the impressed signal, as shown in Figure 1015a, **frequency modulation (FM)** occurs. This is used for commercial FM radio broadcasts and the sound portion of television broadcasts.

Pulse modulation (PM) is somewhat different, there being no impressed modulating wave. In this form of transmission, very short bursts of carrier wave are transmitted, separated by relatively long periods of "silence," during which there is no transmission. This type of transmission, illustrated in Figure 1015b, is used in some common radio navigational aids, including radar and Loran C.

1016. Transmitters

A radio transmitter consists essentially of (1) a power supply to furnish direct current, (2) an oscillator to convert direct current into radio-frequency oscillations (the carrier wave), (3) a device to control the generated signal, and (4) an amplifier to increase the output of the oscillator. For some transmitters a microphone is needed with a modulator and final amplifier to modulate the carrier wave. In addition, an antenna and ground (for lower frequencies) are needed to produce electromagnetic radiation. These components are illustrated in Figure 1016.

1017. Receivers

When a radio wave passes a conductor, a current is induced in that conductor. A radio receiver is a device which senses the power thus generated in an antenna, and transforms it into usable form. It is able to select signals of a single frequency (actually a narrow band of frequencies) from among the many which may reach the receiving antenna. The receiver is able to demodulate the signal and provide adequate amplification. The output of a receiver may be presented audibly by earphones or loudspeaker; or visually on a dial, cathode-ray tube, counter, or other

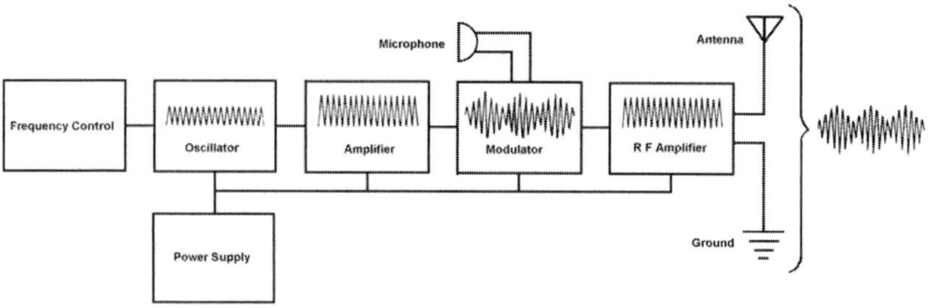

Figure 1016. Components of a radio transmitter.

display. Thus, the useful reception of radio signals requires three components: (1) an antenna, (2) a receiver, and (3) a display unit.

Radio receivers differ mainly in (1) frequency range, the range of frequencies to which they can be tuned; (2) selectivity, the ability to confine reception to signals of the desired frequency and avoid others of nearly the same frequency; (3) sensitivity, the ability to amplify a weak signal to usable strength against a background of noise; (4) stability, the ability to resist drift from conditions or values to which set; and (5) fidelity, the completeness with which the essential characteristics of the original signal are reproduced. Receivers may have additional features such as an automatic frequency control, automatic noise limiter, etc.

Some of these characteristics are interrelated. For instance, if a receiver lacks selectivity, signals of a frequency differing slightly from those to which the receiver is tuned may be received. This condition is called spillover, and the resulting interference is called crosstalk. If the selectivity is increased sufficiently to prevent spillover, it may not permit receipt of a great enough band of frequencies to obtain the full range of those of the desired signal. Thus, the fidelity may be reduced.

A transponder is a transmitter-receiver capable of accepting the challenge of an interrogator and automatically transmitting an appropriate reply.

U.S. RADIO NAVIGATION POLICY

1018. The Federal Radionavigation Plan

The ideal navigation system should provide three things to the user. First, it should be as accurate as necessary for the job it is expected to do. Second, it should be available 100% of the time, in all weather, at any time of day or night. Third, it should have 100% integrity, warning the user and shutting itself down when not operating properly. The mix of navigation systems in the U.S. is carefully chosen to provide maximum accuracy, availability, and integrity to all users, marine, aeronautical, and terrestrial, within the constraints of budget and practicality.

The Federal Radionavigation Plan (FRP) is produced by the U.S. Departments of Defense and Transportation. It establishes government policy on the mix of electronic navigation systems, ensuring consideration of national interests and efficient use of resources. It presents an integrated federal plan for all common-use civilian and military radionavigation systems, outlines approaches for consolidation of systems, provides information and schedules, defines and clarifies new or unresolved issues, and provides a focal point for user input. The FRP is a review of existing and planned radionavigation systems used in air, space, land, and marine navigation. It is available from the National Technical Information Service, Springfield, Virginia, 22161, http://www.ntis.gov.

The first edition of the FRP was released in 1980 as part of a presidential report to Congress. It marked the first time that a joint Department of Transportation/Department of Defense plan had been developed for systems used by both departments. The FRP has had international impact on navigation systems; it is distributed to the International Maritime Organization (IMO), the International Civil Aviation Organization (ICAO), the International Association of Lighthouse Authorities (IALA), and other international organizations.

During a national emergency, any or all of the systems may be temporarily discontinued by the federal government. The government's policy is to continue to operate radionavigation systems as long as the U.S. and its allies derive greater benefit than adversaries. Operating agencies may shut down systems or change signal formats and characteristics during such an emergency.

The plan is reviewed continually and updated

biennially. Industry, advisory groups, and other interested parties provide input. The plan considers governmental responsibilities for national security, public safety, and transportation system economy. It is the official source of radionavigation systems policy and planning for the United States. Systems covered by the FRP include GPS, DGPS, WAAS, LAAS, Loran C, TACAN, MLS, VOR/VOR-DME/VORTAC, and ILS.

1019. System Plans

In order to meet both civilian and military needs, the federal government has established a number of different navigation systems. Each system utilizes the latest technology available at the time of implementation and is upgraded as technology and resources permit. The FRP addresses the length of time each system should be part of the system mix. The 2001 FRP sets forth the following system policy guidelines:

RADIOBEACONS: All U.S. marine radiobeacons have been discontinued and most of the stations converted into DGPS sites.

LORAN C: Loran C provides navigation, location, and timing services for both civil and military air, land, and maritime users. It is slated for replacement by GPS, but due to the large number of users, is expected to remain in place indefinitely while its continuation is evaluated. Reasonable notice will be given if the decision is made to terminate it.

GPS: The Global Positioning System, or GPS, will be the nation's primary radionavigation system well into the next century. It is operated by the U.S. Air Force.

1020. Enhancements to GPS

Differential GPS (DGPS): The U.S. Coast Guard operates marine DGPS in U.S. coastal waters. DGPS is a system in which differences between observed and calculated GPS signals are broadcast to users using medium frequencies. DGPS service is available in all U.S. coastal waters including Hawaii, Alaska, and the Great Lakes. It will provide 4-20 meter continuous accuracy. A terrestrial DGPS system is being installed across the United States to bring differential GPS service to land areas.

Wide Area Augmentation System (WAAS): WAAS is a service of the FAA similar to DGPS, and is intended for cross-country and local air navigation, using a series of reference stations and broadcasting correction data through geostationary satellites. WAAS is not optimized for marine use, and while not certified for maritime navigation, may provide additional position accuracy if the signal is unobstructed. Accuracies of a few meters are possible, about the same as with DGPS.

Local Area Augmentation System (LAAS): LAAS is a precision positioning system provided by the FAA for local navigation in the immediate vicinity of airports so equipped. The correctional signals are broadcast on HF radio with a range of about 30 miles. LAAS is not intended or configured for marine use, but can provide extremely accurate position data in a local area.

1021. Factors Affecting Navigation System Mix

The navigator relies on simple, traditional gear, and on some of the most complex and expensive space-based electronic systems man has ever developed. The success of GPS as a robust, accurate, available, and flexible system is rapidly driving older systems off the scene. Several have met their demise already (Transit, Omega, and marine radiobeacons in the U.S.), and the days are numbered for others, as GPS assumes primacy in navigation technology.

In the U.S., the Departments of Defense and Transportation continually evaluate the components which make up the federally provided and maintained radionavigation system. Several factors influence the decision on the proper mix of systems; cost, military utility, accuracy requirements, and user requirements all drive the problem of allocating scarce resources to develop and maintain navigation systems. The decreasing cost of receivers and increasing accuracy of the Global Positioning System increase its attractiveness as the primary navigation method of the future for both military and civilian use, although there are issues of reliability to be addressed in the face of threats to jam or otherwise compromise the system.

Many factors influence the choice of navigation systems, which must satisfy an extremely diverse group of users. International agreements must be honored. The current investment in existing systems by both government and users must be considered. The full life-cycle cost of each system must be considered. No system will be phased out without consideration of these factors. The FRP recognizes that GPS may not meet the needs of all users; therefore, some systems are currently being evaluated independently of GPS. The goal is to meet all military and civilian requirements in the most efficient way possible.

RADIO DIRECTION FINDING

1022. Introduction

The simplest use of radio waves in navigation is radio direction finding, in which a medium frequency radio signal is broadcast from a station at a known location. This signal is omnidirectional, but a directional antenna on a vessel is used to determine the bearing of the station. This constitutes an LOP, which can be crossed with another LOP to determine a fix.

Once used extensively throughout the world, radiobeacons have been discontinued in the U.S. and many other areas. They are now chiefly used as homing devices by local fishermen, and very little of the ocean's surface is covered by any radiobeacon signal. Because of its limited range, limited availability, and inherent errors, radio direction finding is of limited usefulness to the professional navigator.

In the past, when radiobeacon stations were powerful and common enough for routine ocean navigation, correction of radio bearings was necessary to obtain the most accurate LOP's. The correction process accounted for the fact that, while radio bearings travel along great circles, they are most often plotted on Mercator charts. The relatively short range of those stations remaining has made this process obsolete. Once comprising a major part of NIMA *Pub. 117, Radio Navigational Aids*, radiobeacons are now listed in the back of each volume of the geographically appropriate *List of Lights*.

A Radio Direction Finding Station is one which the mariner can contact via radio and request a bearing. Most of these stations are for emergency use only, and a fee may be involved. These stations and procedures for use are listed in NIMA *Pub. 117, Radio Navigational Aids*.

1023. Using Radio Direction Finders

Depending upon the design of the RDF, the bearings of the radio transmissions are measured as relative bearings, or as both relative and true bearings. The most common type of marine radiobeacon transmits radio waves of approximately uniform strength in all directions. Except during calibration, radiobeacons operate continuously, regardless of weather conditions. Simple combinations of dots and dashes comprising Morse code letters are used for station identification. All radiobeacons superimpose the characteristic on a carrier wave, which is broadcast continuously during the period of transmission. A 10-second dash is incorporated in the characteristic signal to enable users of the aural null type of radio direction finder to refine the bearing.

Bearing measurement is accomplished with a directional antenna. Nearly all types of receiving antennas have some directional properties, but the RDF antenna is designed to be as directional as possible. Simple small craft RDF units usually have a ferrite rod antenna mounted directly on a receiver, with a 360 degree graduated scale. To get a bearing, align the unit to the vessel's course or to true north, and rotate the antenna back and forth to find the exact null point. The bearing to the station, relative or true according to the alignment, will be indicated on the dial. Some small craft RDF's have a portable hand-held combination ferrite rod and compass, with earphones to hear the null.

Two types of loop antenna are used in larger radio direction finders. In one of these, the crossed loop type, two loops are rigidly mounted in such manner that one is placed at 90 degrees to the other. The relative output of the two antennas is related to the orientation of each with respect to the direction of travel of the radio wave, and is measured by a device called a goniometer.

1024. Errors of Radio Direction Finders

RDF bearings are subject to certain errors. Quadrantal error occurs when radio waves arrive at a receiver and are influenced by the immediate shipboard environment.

A radio wave crossing a coastline at an oblique angle experiences a change of direction due to differences in conducting and reflecting properties of land and water known as coastal refraction, sometimes called land effect. It is avoided by not using, or regarding as of doubtful accuracy, bearings which cross a shoreline at an oblique angle.

In general, good radio bearings should not be in error by more than two or three degrees for distances under 150 nautical miles. However, conditions vary considerably, and skill is an important factor. By observing the technical instructions for the equipment and practicing frequently when results can be checked, one can develop skill and learn to what extent radio bearings can be relied upon under various conditions. Other factors affecting accuracy include range, the condition of the equipment, and the accuracy of calibration.

The strength of the signal determines the usable range of a radiobeacon. The actual useful range may vary considerably from the published range with different types of radio direction finders and during varying atmospheric conditions. The sensitivity of a radio direction finder determines the degree to which the full range of a radiobeacon can be utilized. Selectivity varies with the type of receiver and its condition.

CHAPTER 11

SATELLITE NAVIGATION

INTRODUCTION

1100. Development

The idea that led to development of the satellite navigation systems dates back to 1957 and the first launch of an artificial satellite into orbit, Russia's Sputnik I. Dr. William H. Guier and Dr. George C. Wieffenbach at the Applied Physics Laboratory of the Johns Hopkins University were monitoring the famous "beeps" transmitted by the passing satellite. They plotted the received signals at precise intervals, and noticed that a characteristic Doppler curve emerged. Since satellites generally follow fixed orbits, they reasoned that this curve could be used to describe the satellite's orbit. They then demonstrated that they could determine all of the orbital parameters for a passing satellite by Doppler observation of a single pass from a single fixed station. The Doppler shift apparent while receiving a transmission from a passing satellite proved to be an effective measuring device for establishing the satellite orbit.

Dr. Frank T. McClure, also of the Applied Physics Laboratory, reasoned in reverse: If the satellite orbit was known, Doppler shift measurements could be used to determine one's position on Earth. His studies in support of this hypothesis earned him the first National Aeronautics and Space Administration award for important contributions to space development.

In 1958, the Applied Physics Laboratory proposed exploring the possibility of an operational satellite Doppler navigation system. The Chief of Naval Operations then set forth requirements for such a system. The first successful launching of a prototype system satellite in April 1960 demonstrated the Doppler system's operational feasibility.

The **Navy Navigation Satellite System** (**NAVSAT**, also known as **TRANSIT**) was the first operational satellite navigation system. The system's accuracy was better than 0.1 nautical mile anywhere in the world, though its availability was somewhat limited. It was used primarily for the navigation of surface ships and submarines, but it also had some applications in air navigation. It was also used in hydrographic surveying and geodetic position determination.

The transit launch program ended in 1988 and the system was disestablished when the Global Positioning System became operational in 1996.

THE GLOBAL POSITIONING SYSTEM

1101. System Description

The Federal Radionavigation Plan has designated the NAVigation System using Timing And Ranging (NAVSTAR) Global Positioning System (GPS) as the primary navigation system of the U.S. government. GPS is a spaced-based radio positioning system which provides suitably equipped users with highly accurate position, velocity, and time data. It consists of three major segments: a **space segment**, a **control segment**, and a **user segment**.

The space segment comprises some 24 satellites. Spacing of the satellites in their orbits is arranged so that at least four satellites are in view to a user at any time, anywhere on the Earth. Each satellite transmits signals on two radio frequencies, superimposed on which are navigation and system data. Included in this data are predicted satellite ephemeris, atmospheric propagation correction data, satellite clock error information, and satellite health data. This segment normally consists of 21 operational satellites with three satellites orbiting as active spares. The satellites orbit at an altitude of 20,200 km, in six separate orbital planes, each plane inclined 55° relative to the equator. The satellites complete an orbit approximately once every 12 hours.

GPS satellites transmit **pseudorandom noise (PRN)** sequence-modulated radio frequencies, designated L1 (1575.42 MHz) and L2 (1227.60 MHz). The satellite transmits both a **Coarse Acquisition Code** (C/A code) and a **Precision Code** (P code). Both the P and C/A codes are transmitted on the L1 carrier; only the P code is transmitted on the L2 carrier. Superimposed on both the C/A and P codes is the navigation message. This message contains the satellite ephemeris data, atmospheric propagation correction data, and satellite clock bias.

GPS assigns a unique C/A code and a unique P code to each satellite. This practice, known as **code division multiple access (CDMA)**, allows all satellites the use of a common carrier frequency while still allowing the receiver to determine which satellite is transmitting. CDMA also allows for easy user identification of each GPS satellite. Since each satellite broadcasts using its own unique C/A and P code combination, it can be assigned a unique **PRN sequence number**. This number is how a satellite is identified when the GPS control system communicates with users about a particular GPS satellite.

The control segment includes a **master control station** (MCS), a number of monitor stations, and ground antennas located throughout the world. The master control station, located in Colorado Springs, Colorado, consists of equipment and facilities required for satellite monitoring, telemetry, tracking, commanding, control, uploading, and navigation message generation. The monitor stations, located in Hawaii, Colorado Springs, Kwajalein, Diego Garcia, and Ascension Island, passively track the satellites, accumulating ranging data from the satellites' signals and relaying them to the MCS. The MCS processes this information to determine satellite position and signal data accuracy, updates the navigation message of each satellite and relays this information to the ground antennas. The ground antennas then transmit this information to the satellites. The ground antennas, located at Ascension Island, Diego Garcia, and Kwajalein, are also used for transmitting and receiving satellite control information.

The user equipment is designed to receive and process signals from four or more orbiting satellites either simultaneously or sequentially. The processor in the receiver then converts these signals to navigation information. Since GPS is used in a wide variety of applications, from marine navigation to land surveying, these receivers can vary greatly in function and design.

1102. System Capabilities

GPS provides multiple users with accurate, continuous, worldwide, all-weather, common-grid, three-dimensional positioning and navigation information.

To obtain a navigation solution of position (latitude, longitude, and altitude) and time (four unknowns), four satellites must be used. The GPS user measures pseudorange and pseudorange rate by synchronizing and tracking the navigation signal from each of the four selected satellites. Pseudorange is the true distance between the satellite and the user plus an offset due to the user's clock bias. Pseudorange rate is the true slant range rate plus an offset due to the frequency error of the user's clock. By decoding the ephemeris data and system timing information on each satellite's signal, the user's receiver/processor can convert the pseudorange and pseudorange rate to three-dimensional position and velocity. Four measurements are necessary to solve for the three unknown components of position (or velocity) and the unknown user time (or frequency) bias.

The navigation accuracy that can be achieved by any user depends primarily on the variability of the errors in making pseudorange measurements, the instantaneous geometry of the satellites as seen from the user's location on Earth, and the presence of **Selective Availability** (SA). Selective Availability is discussed further below.

1103. Global Positioning System Concepts

GPS measures distances between satellites in orbit and a receiver on Earth, and computes spheres of position from those distances. The intersections of those spheres of position then determine the receiver's position.

The distance measurements described above are done by comparing timing signals generated simultaneously by the satellites' and receiver's internal clocks. These signals, characterized by a special wave form known as the pseudo-random code, are generated in phase with each other. The signal from the satellite arrives at the receiver following a time delay proportional to its distance traveled. This time delay is detected by the phase shift between the received pseudo-random code and the code generated by the receiver. Knowing the time required for the signal to reach the receiver from the satellite allows the receiver to calculate the distance from the satellite. The receiver, therefore, must be located on a sphere centered at the satellite with a radius equal to this distance measurement. The intersection of three spheres of position yields two possible points of receiver position. One of these points can be disregarded since it is hundreds of miles from the surface of the Earth. Theoretically, then, only three time measurements are required to obtain a fix from GPS.

In practice, however, a fourth measurement is required to obtain an accurate position from GPS. This is due to receiver clock error. Timing signals travel from the satellite to the receiver at the speed of light; even extremely slight timing errors between the clocks on the satellite and in the receiver will lead to tremendous range errors. The satellite's atomic clock is accurate to 10^{-9} seconds; installing a clock that accurate on a receiver would make the receiver prohibitively expensive. Therefore, receiver clock accuracy is sacrificed, and an additional satellite timing measurement is made. The fix error caused by the inaccuracies in the receiver clock is reduced by simultaneously subtracting a constant timing error from four satellite timing measurements until a pinpoint fix is reached.

Assuming that the satellite clocks are perfectly synchronized and the receiver clock's error is constant, the subtraction of that constant error from the resulting distance determinations will reduce the fix error until a "pinpoint" position is obtained. It is important to note here that the number of lines of position required to employ this technique is a function of the number of lines of position required to obtain a fix. GPS determines position in three dimensions; the presence of receiver clock error adds an additional unknown. Therefore, four timing measurements are required to solve for the resulting four unknowns.

1104. GPS Signal Coding

Two separate carrier frequencies carry the signal transmitted by a GPS satellite. The first carrier frequency (L1) transmits on 1575.42 MHz; the second (L2) transmits on 1227.60 MHz. The GPS signal consists of three separate messages: the P-code, transmitted on both L1 and L2; the C/A code, transmitted on L1 only; and a navigation data message. The P code and C/A code messages are divided into individual bits known as **chips**. The frequency at which bits are sent for each type of signal is known as the **chipping rate**. The chipping rate for the P-code is 10.23 MHz (10.23 $\times 10^6$ bits per second); for the C/A code, 1.023 MHz (1.023 $\times 10^6$ bits per second); and for the data message, 50 Hz (50 bits per second). The P and C/A codes **phase modulate** the carriers; the C/A code is transmitted at a phase angle of 90° from the P code. The periods of repetition for the C/A and P codes differ. The C/A code repeats once every millisecond; the P-code sequence repeats every seven days.

As stated above the GPS carrier frequencies are phase modulated. This is simply another way of saying that the digital "1's" and "0's" contained in the P and C/A codes are indicated along the carrier by a shift in the carrier phase. This is analogous to sending the same data along a carrier by varying its amplitude (amplitude modulation, or AM) or its frequency (frequency modulation, or FM). See Figure 1104a. In phase modulation, the frequency and the amplitude of the carrier are unchanged by the "information signal," and the digital information is transmitted by shifting the carrier's phase. The phase modulation employed by GPS is known as bi-phase shift keying (BPSK).

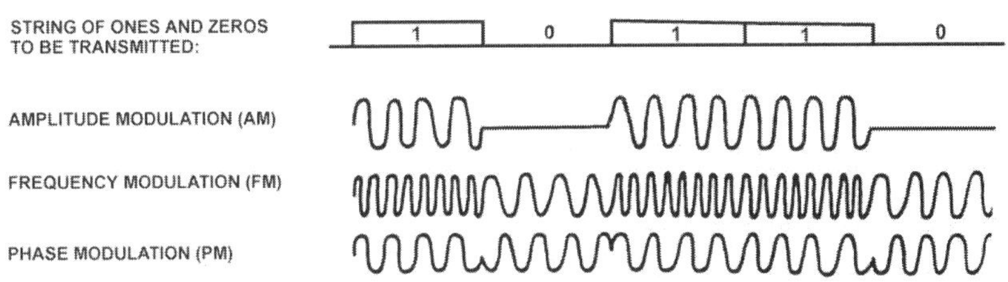

Figure 1104a. Digital data transmission with amplitude, frequency and phase modulation.

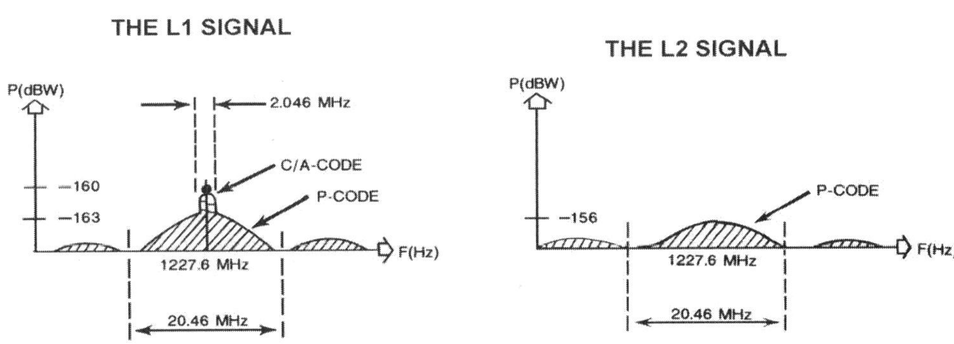

Figure 1104b. Modulation of the L1 and L2 carrier frequencies with the C/A and P code signals.

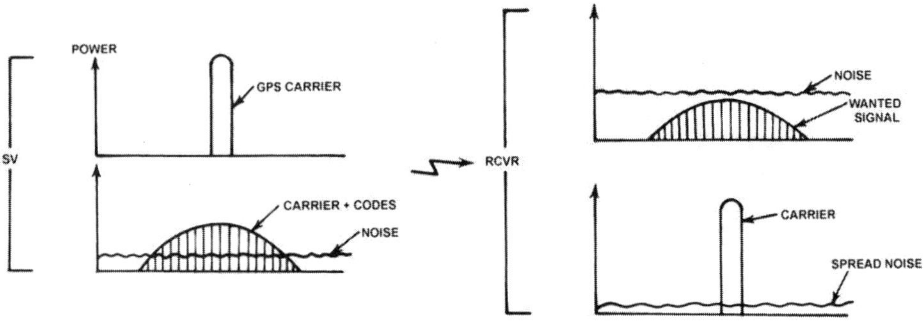

Figure 1104c. GPS signal spreading and recovery from satellite to receiver.

Due to this BPSK, the carrier frequency is "spread" about its center frequency by an amount equal to twice the "chipping rate" of the modulating signal. In the case of the P code, this spreading is equal to (2 × 10.23 MHz) = 20.46 MHz. For the C/A code, the spreading is equal to (2 × 1.023 MHz) = 2.046 MHz. See Figure 1104b. Note that the L1 carrier signal, modulated with both the P code and C/A code, is shaped differently from the L2 carrier, modulated with only the P code. This spreading of the carrier signal lowers the total signal strength below the thermal noise threshold present at the receiver. This effect is demonstrated in Figure 1104c. When the satellite signal is multiplied with the C/A and P codes generated by the receiver, the satellite signal will be collapsed into the original carrier frequency band. The signal power is then raised above the thermal noise level.

The navigation message is superimposed on both the P code and C/A code with a data rate of 50 bits per second (50 Hz.) The navigation message consists of 25 data frames, each frame consisting of 1500 bits. Each frame is divided into five subframes of 300 bits each. It will, therefore, take 30 seconds to receive one data frame and 12.5 minutes to receive all 25 frames. The navigation message contains GPS system time of transmission; a **handover word (HOW)**, allowing the transition between tracking the C/A code to the P code; ephemeris and clock data for the satellite being tracked; and almanac data for the satellites in orbit. It also contains coefficients for ionospheric delay models used by C/A receivers and coefficients used to calculate Universal Coordinated Time (UTC).

1105. The Correlation Process

The correlation process compares the signal received from the satellites with the signal generated by the receiver by comparing the square wave function of the received signal with the square wave function generated by the receiver. The computer logic of the receiver recognizes the square wave signals as either a +1 or a 0 depending on whether the signal is "on" or "off." The signals are processed and matched by using an **autocorrelation function**.

This process defines the necessity for a "pseudo-random code." The code must be repeatable (i.e., non-random) because it is in comparing the two signals that the receiver makes its distance calculations. At the same time, the code must be random for the correlation process to work; the randomness of the signals must be such that the matching process excludes all possible combinations except the combination that occurs when the generated signal is shifted a distance proportional to the received signal's time delay. These simultaneous requirements to be both repeatable (non-random) and random give rise to the description of "pseudo-random"; the signal has enough repeatability to enable the receiver to make the required measurement while simultaneously retaining enough randomness to ensure incorrect calculations are excluded.

1106. Precise Positioning Service and Standard Positioning Service

Two levels of navigational accuracy are provided by the GPS: the **Precise Positioning Service (PPS)** and the **Standard Positioning Service (SPS)**. GPS was designed, first and foremost, by the U.S. Department of Defense as a United States military asset; its extremely accurate positioning capability is an asset access to which the U.S. military may need to limit during time of war to prevent use by enemies. Therefore, the PPS is available only to authorized users, mainly the U.S. military and authorized allies. SPS, on the other hand, is available worldwide to anyone possessing a GPS receiver. Therefore PPS provides

SA/A-S Configuration	SIS Interface Conditions	PPS Users	SPS Users
SA Set to Zero A-S Off	P-Code, no errors C/A-Code, no errors	Full accuracy, spoofable	Full accuracy,* spoofable
SA at Non-Zero Value A-S Off	P-Code, errors C/A-Code, errors	Full accuracy, spoofable	Limited accuracy, spoofable
SA Set to Zero A-S On	Y-Code, no errors C/A-Code, no errors	Full accuracy, Not spoofable**	Full accuracy,*** spoofable
SA at Non-Zero Value A-S On	Y-Code, errors C/A-Code, errors	Full accuracy, Not spoofable**	Limited accuracy, spoofable

* "Full accuracy" defined as equivalent to a PPS-capable UE operated in a similar manner.
** Certain PPS-capable UE do not have P- or Y-code tracking abilities and remain spoofable despite A-S protection being applied
*** Assuming negligible accuracy degradation due to C/A-code operation (but more susceptible to jamming).

Figure 1106. Effect of SA and A-S on GPS accuracy.

a more accurate position than does SPS.

Two cryptographic methods are employed to deny PPS accuracy to civilian users: **selective availability (SA)** and **anti-spoofing (A-S)**. SA operates by introducing controlled errors into both the C/A and P code signals. SA can be programmed to degrade the signals' accuracy even further during time of war, denying a potential adversary the ability to use GPS to nominal SPS accuracy. SA introduces two errors into the satellite signal: (1) The **epsilon error**: an error in satellite ephemeris data in the navigation message; and (2) **clock dither**: error introduced in the satellite atomic clocks' timing. The presence of SA is the largest source of error present in an SPS GPS position measurement. The status of SA, whether off or on, can be checked at the USCG's NAVCEN Web site:

http://www.navcen.uscg.gov.

Anti-spoofing is designed to negate any hostile imitation of GPS signals. The technique alters the P code into another code, designated the Y code. The C/A code remains unaffected. The U.S. employs this technique to the satellite signals at random times and without warning; therefore, civilian users are unaware when this P code transformation takes place. Since anti-spoofing is applied only to the P code, the C/A code is not protected and can be spoofed.

Only users employing the proper cryptographic devices can defeat both SA and anti-spoofing. Without these devices, the user will be subject to the accuracy degradation of SA and will be unable to track the Y code.

GPS PPS receivers can use either the P code or the C/A code, or both, in determining position. Maximum accuracy is obtained by using the P code on both L1 and L2. The difference in propagation delay is then used to calculate ionospheric corrections. The C/A code is normally used to acquire the satellite signal and determine the approximate P code phase. Then, the receiver locks on the P code for precise positioning (subject to SA if not cryptographically equipped). Some PPS receivers possess a clock accurate enough to track and lock on the P code signal without initially tracking the C/A code. Some PPS receivers can track only the C/A code and disregard the P code entirely. Since the C/A code is transmitted on only one frequency, the dual frequency ionosphere correction methodology is unavailable and an ionospheric modeling procedure is required to calculate the required corrections.

SPS receivers, as mentioned above, provide positions with a degraded accuracy. The A-S feature denies SPS users access to the P code when transformed to the Y code. Therefore, the SPS user cannot rely on access to the P code to measure propagation delays between L1 and L2 and compute ionospheric delay corrections. Consequently, the typical SPS receiver uses only the C/A code because it is unaffected by A-S. Since C/A is transmitted only on L1, the dual frequency method of calculating ionospheric corrections is unavailable; an ionospheric modeling technique must be used. This is less accurate than the dual frequency method; this degradation in accuracy is accounted for in the 100-meter accuracy calculation. Figure 1106 presents the effect on SA and A-S on different types of GPS measurements.

1107. GPS Receiver Operations

In order for the GPS receiver to navigate, it has to track satellite signals, make pseudorange measurements, and collect navigation data.

A typical satellite tracking sequence begins with the receiver determining which satellites are available for it to track. Satellite visibility is determined by user-entered predictions of position, velocity, and time, and by almanac information stored internal to the receiver. If no stored almanac information exists, then the receiver must attempt to locate and lock onto the signal from any satellite in view. When the receiver is locked onto a satellite, it can demodulate the navigation message and read the almanac information about all the other satellites in the constellation. A carrier tracking loop tracks the carrier frequency while a code tracking loop tracks the C/A and P code signals. The two tracking loops operate together in an iterative process to acquire and track satellite signals.

The receiver's carrier tracking loop will locally generate an L1 carrier frequency which differs from the satellite produced L1 frequency due to a Doppler shift in the received frequency. This Doppler offset is proportional to the relative velocity along the line of sight between the satellite and the receiver, subject to a receiver frequency bias. The carrier tracking loop adjusts the frequency of the receiver-generated frequency until it matches the incoming frequency. This determines the relative velocity between the satellite and the receiver. The GPS receiver uses this relative velocity to calculate the velocity of the receiver. This velocity is then used to aid the code tracking loop.

The code tracking loop is used to make pseudorange measurements between the GPS receiver and the satellites. The receiver's tracking loop will generate a replica of the targeted satellite's C/A code with estimated ranging delay. In order to match the received signal with the internally generated replica, two things must be done: 1) The center frequency of the replica must be adjusted to be the same as the center frequency of the received signal; and 2) the phase of the replica code must be lined up with the phase of the received code. The center frequency of the replica is set by using the Doppler-estimated output of the carrier tracking loop. The receiver will then slew the code loop generated C/A code though a millisecond search window to correlate with the received C/A code and obtain C/A tracking.

Once the carrier tracking loop and the code tracking loop have locked onto the received signal and the C/A code has been stripped from the carrier, the navigation message is demodulated and read. This gives the receiver other information crucial to a pseudorange measurement. The navigation message also gives the receiver the handover word, the code that allows a GPS receiver to shift from C/A code tracking to P code tracking.

The handover word is required due to the long phase (seven days) of the P code signal. The C/A code repeats every millisecond, allowing for a relatively small search window. The seven day repeat period of the P code requires that the receiver be given the approximate P code phase to narrow its search window to a manageable time. The handover word provides this P code phase information. The handover word is repeated every subframe in a 30 bit long block of data in the navigation message. It is repeated in the second 30 second data block of each subframe. For some receivers, this handover word is unnecessary; they can acquire the P code directly. This normally requires the receiver to have a clock whose accuracy approaches that of an atomic clock. Since this greatly increases the cost of the receiver, most receivers for non-military marine use do not have this capability.

Once the receiver has acquired the satellite signals from four GPS satellites, achieved carrier and code tracking, and has read the navigation message, the receiver is ready to begin making pseudorange measurements. Recall that these measurements are termed *pseudorange* because a receiver clock offset makes them inaccurate; that is, they do not represent the true range from the satellite, only a range biased by a receiver clock error. This clock bias introduces a fourth unknown into the system of equations for which the GPS receiver must solve (the other three being the x coordinate, y coordinate, and z coordinate of the receiver position). Recall from the discussion in Article 1101 that the receiver solves this clock bias problem by making a fourth pseudorange measurement, resulting in a fourth equation to allow solving for the fourth unknown. Once the four equations are solved, the receiver has an estimate of the receiver's position in three dimensions and of GPS time. The receiver then converts this position into coordinates referenced to an Earth model based on the World Geodetic System (1984).

1108. User Range Errors and Geometric Dilution of Precision

There are two formal position accuracy requirements for GPS:

1) The PPS spherical position accuracy shall be 16 meters SEP (spherical error probable) or better.

2) The SPS user two dimensional position accuracy shall be 100 meters 2 drms or better.

Assume that a universal set of GPS pseudorange measurements results in a set of GPS position measurements. The accuracy of these measurements will conform to a normal (i.e. values symmetrically distributed around a mean of zero) probability function because the two most important factors affecting accuracy, the **geometric dilution of precision (GDOP)** and the **user equivalent range error (UERE)**, are continuously variable.

The UERE is the error in the measurement of the pseudoranges from each satellite to the user. The UERE is the product of several factors, including the clock stability, the predictability of the satellite's orbit, errors in the 50 Hz navigation message, the precision of the receiver's correlation process, errors due to atmospheric distortion and the calculations to compensate for it, and the quality of the satellite's signal. The UERE, therefore, is a random error which is the function of errors in both the satellites and the user's receiver.

The GDOP depends on the geometry of the satellites in relation to the user's receiver. It is independent of the quality of the broadcast signals and the user's receiver. Generally speaking, the GDOP measures the "spread" of the satellites around the receiver. The optimum case would be to have one satellite directly overhead and the other three spaced 120° around the receiver on the horizon. The worst GDOP would occur if the satellites were spaced closely together or in a line overhead.

There are special types of DOP's for each of the position and time solution dimensions; these particular DOP's combine to determine the GDOP. For the vertical dimension, the **vertical dilution of precision (VDOP)** describes the effect of satellite geometry on altitude calculations. The **horizontal dilution of precision (HDOP)** describes satellite geometry's effect on position (latitude and longitude) errors. These two DOP's combine

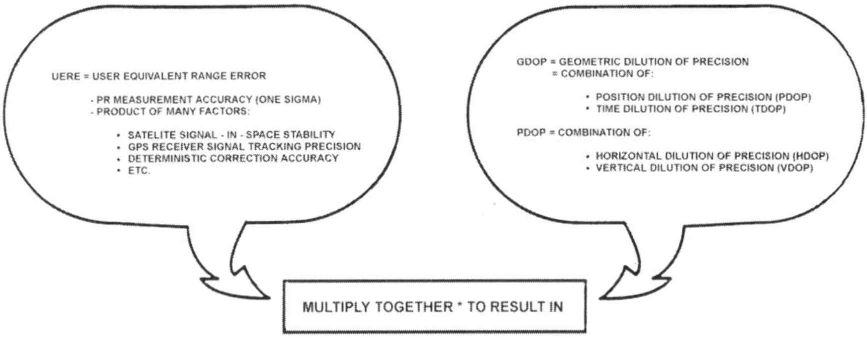

THREE-DIMENSIONAL USER NAVIGATION ERROR (UNE)**

UNE = UERE x PDOP

TWO-DIMENSIONAL USER HORIZONTAL NAVIGATION ERROR (UHNE)**

UHNE = UERE x HDOP

Figure 1108. Position and time error computations.

to determine the **position dilution of precision (PDOP)**. The PDOP combined with the **time dilution of precision (TDOP)** results in the GDOP. See Figure 1108.

1109. Ionospheric Delay Errors

Article 1108 covered errors in GPS positions due to errors inherent in the satellite signal (UERE) and the geometry of the satellite constellation (GDOP). Another major cause of accuracy degradation is the effect of the ionosphere on the radio frequency signals that comprise the GPS signal.

A discussion of a model of the Earth's atmosphere will be useful in understanding this concept. Consider the Earth as surrounded by three layers of atmosphere. The first layer, extending from the surface of the Earth to an altitude of approximately 10 km, is known as the troposphere. Above the troposphere and extending to an altitude of approximately 50 km is the stratosphere. Finally, above the stratosphere and extending to an altitude that varies as a function of the time of day is the **ionosphere**. Though radio signals are subjected to effects which degrade its accuracy in all three layers of this atmospheric model, the effects of the ionosphere are the most significant to GPS operation.

The ionosphere, as the name implies, is that region of the atmosphere which contains a large number of ionized molecules and a correspondingly high number of free electrons. These charged molecules have lost one or more electrons. No atom will loose an electron without an input of energy; the energy input that causes the ions to be formed in the ionosphere comes from the ultraviolet (U-V) radiation of the Sun. Therefore, the more intense the Sun's rays, the larger the number of free electrons which will exist in this region of the atmosphere.

The largest effect that this ionospheric effect has on GPS accuracy is a phenomenon known as **group time delay**. As the name implies, group time delay results in a delay in the time a signal takes to travel through a given distance. Obviously, since GPS relies on extremely accurate timing measurement of these signals between satellites and ground receivers, this group time delay can have a noticeable effect on the magnitude of GPS position error.

The group time delay is a function of several elements. It is inversely proportional to the square of the frequency at which the satellite transmits, and it is directly proportional to the atmosphere's **total electron content (TEC)**, a measure of the degree of the atmosphere's ionization. The general form of the equation describing the delay effect is:

$$\Delta t = \frac{(K \times TEC)}{f^2}$$

where

Δt = group time delay
f = operating frequency
K = constant

Since the Sun's U-V radiation ionizes the molecules in the upper atmosphere, it stands to reason that the time delay value will be highest when the Sun is shining and lowest at night. Experimental evidence has borne this out, showing that the value for TEC is highest around 1500 local time and lowest around 0500 local time. Therefore, the magnitude of the accuracy degradation caused by this effect will be highest during daylight operations. In addition to these daily variations, the magnitude of this time delay error also varies with the seasons; it is highest at the vernal equinox. Finally, this effect shows a solar cycle dependence. The greater the number of sunspots, the higher the TEC value and the greater the group time delay effect. The solar cycle typically follows an eleven year pattern. The next solar cycle will be at a minimum in 2006 and peak again in 2010.

Given that this ionospheric delay introduces a serious accuracy degradation into the system, how does GPS ac count for it? There are two methods used: (1) the dual frequency technique, and (2) the ionospheric delay method.

1110. Dual Frequency Correction Technique

As the term implies, the dual frequency technique requires the ability to acquire and track both the L1 and L2 frequency signals. Recall from the discussion in Article 1103 that the C/A and P codes are transmitted on carrier frequency L1, but only the P code is transmitted on L2. Recall also that only authorized operators with access to DOD cryptographic material are able to copy the P code. It follows, then, that only those authorized users are able to copy the L2 carrier frequency. Therefore, only those authorized users are able to use the dual frequency correction method. The dual frequency method measures the distance between the satellite and the user based on both the L1 and L2 carrier signal. These ranges will be different because the group time delay for each signal will be different. This is because of the frequency dependence of the time delay error. The range from the satellite to the user will be the true range combined with the range error caused by the time delay, as shown by the following equation:

$$R(f) = R_{actual} + \text{error term}$$

where R(f) is the range which differs from the actual range as a function of the carrier frequency. The dual frequency correction method takes two such range measurements, R(L1) and R(L2). Recall that the error term is a function of a constant divided by the square of the frequency. By combining the two range equations derived from the two frequency measurements, the constant term can be eliminated and one is left with an equation in which the true range is simply a function of the two carrier frequencies and the measured ranges R(L1) and R(L2). This method has two major advantages over the ionospheric model method. (1) It calculates corrections from real-time measured data; therefore, it is more accurate. (2) It alleviates the need to include ionospheric data on the navigation message. A significant portion of the data message is devoted to ionospheric correction data. If the receiver is dual frequency capable, then it does not need any of this data.

The vast majority of maritime users cannot copy dual frequency signals. For them, the ionospheric delay model provides the correction for the group time delay.

1111. The Ionospheric Delay Model

The ionospheric delay model mathematically models the diurnal ionospheric variation. The value for this time delay is determined from a cosinusoidal function into which coefficients representing the maximum value of the time delay (i.e., the amplitude of the cosine wave representing the delay function); the time of day; the period of the variation; and a minimum value of delay are introduced. This model is designed to be most accurate at the diurnal maximum. This is obviously a reasonable design consideration because it is at the time of day when the maximum diurnal time delay occurs that the largest magnitude of error appears. The coefficients for use in this delay model are transmitted to the receiver in the navigation data message. As stated in Article 1110, this method of correction is not as accurate as the dual frequency method; however, for the non-military user, it is the only method of correction available.

1112. Multipath Reflection Errors

Multipath reflection errors occur when the receiver detects parts of the same signal at two different times. The first reception is the direct path reception, the signal that is received directly from the satellite. The second reception is from a reflection of that same signal from the ground or any other reflective surface. The direct path signal arrives first, the reflected signal, having had to travel a longer distance to the receiver, arrives later. The GPS signal is designed to minimize this multipath error. The L1 and L2 frequencies used demonstrate a diffuse reflection pattern, lowering the signal strength of any reflection that arrives at the receiver. In addition, the receiver's antenna can be designed to reject a signal that it recognizes as a reflection. In addition to the properties of the carrier frequencies, the high data frequency of both the P and C/A codes and their resulting good correlation properties minimize the effect of multipath propagation.

The design features mentioned above combine to reduce the maximum error expected from multipath propagation to less than 20 feet.

DIFFERENTIAL GPS

1113. Differential GPS Concept

The discussions above make it clear that the Global Positioning System provides the most accurate positions available to navigators today. They should also make clear that the most accurate positioning information is available to only a small fraction of the using population: U.S. and allied military. For most open ocean navigation applications, the degraded accuracy inherent in selective availability and the inability to copy the precision code presents no serious hazard to navigation. A mariner seldom if ever needs greater than 100 meter accuracy in the middle of the ocean.

It is a different situation as the mariner approaches shore. Typically for harbor approaches and piloting, the mariner will shift to visual piloting. The increase in accuracy provided by this navigational method is required to ensure ship's safety. The 100 meter accuracy of GPS in this situation is not sufficient. Any mariner who has groped his way through a restricted channel in a thick fog will certainly appreciate the fact that even a degraded GPS position is available for them to plot. However, 100 meter accuracy is not sufficient to ensure ship's safety in most piloting situations. In this situation, the mariner needs P code accuracy. The problem then becomes how to obtain the accuracy of the Precise Positioning Service with due regard to the legitimate security concerns of the U.S. military. The answer to this seeming dilemma lies in the concept of **Differential GPS (DGPS)**.

Differential GPS is a system in which a receiver at an accurately surveyed position utilizes GPS signals to calculate timing errors and then broadcasts a correction signal to account for these errors. This is an extremely powerful concept. The errors which contribute to GPS accuracy degradation, ionospheric time delay and selective availability, are experienced simultaneously by both the DGPS receiver and a relatively close user's receiver. The extremely high altitude of the GPS satellites means that, as long as the DGPS receiver is within 100-200 km of the user's receiver, the user's receiver is close enough to take advantage of any DGPS correction signal.

The theory behind a DGPS system is straightforward. Located on an accurately surveyed site, the DGPS receiver already knows its location. It receives data which tell it where the satellite is. Knowing the two locations, it then calculates the theoretical time it should take for a satellite's signal to reach it. It then compares the time that it actually takes for the signal to arrive. This difference in time between the theoretical and the actual is the basis for the DGPS receiver's computation of a timing error signal; this difference in time is caused by all the errors to which the GPS signal is subjected; errors, except for receiver error and multipath error, to which both the DGPS and the user's receivers are simultaneously subject. The DGPS system then broadcasts a timing correction signal, the effect of which is to correct for selective availability, ionospheric delay, and all the other error sources the two receivers share in common.

For suitably equipped users, DGPS results in positions at least as accurate as those obtainable by the Precise Positioning Service. This capability is not limited to simply displaying the correct position for the navigator to plot. The DGPS position can be used as the primary input to an electronic chart system, providing an electronic readout of position accurate enough to pilot safely in the most restricted channel.

WAAS AND LAAS IN MARINE NAVIGATION

1114. WAAS/LAAS for Aeronautical Use

In 1994 the National Telecommunications and Information Administration (NTIA) produced a technical report for the Department of Transportation which concluded that the optimum mix of enhanced GPS systems for overall civilian use would consist of DGPS for marine and terrestrial use and a combined WAAS/LAAS system for air navigation.

The **Wide Area Augmentation System (WAAS)** concept is similar to the DGPS concept, except that correctional signals are sent from geostationary satellites via HF signals directly to the user's GPS receiver. This eliminates the need for a separate receiver and antenna, as is the case with DGPS. WAAS is intended for enroute air navigation, with 25 reference stations widely spaced across the United States, for coverage of the entire U.S. and parts of Mexico and Canada.

The **Local Area Augmentation System (LAAS)** is intended for precision airport approaches, with reference stations located at airports and broadcasting their correction message on VHF radio frequencies.

While many marine GPS receivers incorporate WAAS circuitry (but not the more accurate, shorter-range LAAS), WAAS is not optimized for surface navigation because the HF radio signals are line-of-sight and are transmitted from geostationary satellites. At low angles to the horizon, the WAAS signal may be blocked and the resulting GPS position accuracy significantly degraded with no warning. The DGPS signal, on the other hand, is a terrain-following signal that is unaffected by objects in its path. It simply flows around them and continues on unblocked.

The accuracy of WAAS and DGPS is comparable, on the order of a few meters. WAAS was designed to provide 7 meter accuracy 95% of the time. DGPS was designed to provide 10 meter accuracy 95% of the time, but in actual use one can expect about 1-3 meter accuracy when the user is within 100 miles of he DGPS transmitter. Over 100 miles, DGPS accuracy will commonly degrade by an additional 1 meter per 100 miles from the transmitter site. Both systems have been found in actual use to provide accuracies somewhat better than designed.

The WAAS signal, while not certified for use in the marine environment as is DGPS, can be a very useful navigational tool if its limitations are understood. In open waters of the continental U.S., the WAAS signal can be expected to be available and useful, provided the receiver has WAAS circuitry and is programmed to use the WAAS data. Outside the U.S., or in any area where tall buildings, trees, or other obstructions rise above the horizon, the WAAS signal may be blocked, and the resulting GPS fix could be in error by many meters. Since the highest accuracy is necessary in the most confined waters, WAAS should be used with extreme caution in these areas.

WAAS can enhance the navigator's situational awareness when available, but availability is not assured. Further, a marine receiver will provide no indication when WAAS data is not a part of the fix. [Aircraft GPS receivers may contain Receiver Autonomous Integrity Monitoring (RAIM) software, which does provide warning of WAAS satellite signal failure, and removes the affected signal from the fix solution.]

LAAS data, broadcast on VHF, is less subject to blocking, but is only available in selected areas near airports. Its range is about 30 miles. It is therefore not suitable for general marine navigational use.

NON-U.S. SATELLITE NAVIGATION SYSTEMS

1115. The Galileo System

Since the development of GPS, various European councils and commissions have expressed a need for a satellite navigation system independent of GPS. Economic studies have emphasized this need, and technological studies by the European Space Agency over several years have proven its feasibility. In early 2002 the European Union (EU) decided to fund the development of its new Galileo satellite navigation system. A great deal of preliminary scientific work has already been accomplished, which will enable the full deployment of Galileo over the next few years.

Several factors influenced the decision to develop Galileo, the primary one being that GPS is a U.S. military asset that can be degraded for civilian use on order of the U.S. Government (as is the Russian satellite navigation system GLONASS). Disruption of either system might leave European users without their primary navigation system at a critical time. In contrast, Galileo will be under civilian control and dedicated primarily to civilian use. It is important to note that since GPS has been operational, civilian uses are proliferating far more rapidly than anticipated, to the point that GPS planners are developing new frequencies and enhancements to GPS for civilian use (WAAS and LAAS), SA has been turned off (as of May 1, 2000), and the cost and size of receivers have plummeted.

Plans call for the Galileo constellation to consist of 30 satellites (27 usable and three spares) in three orbital planes, each inclined 56 degrees to the equator. The orbits are at an altitude of 23,616 km (about 12,750 nm). Galileo will be designed to serve higher latitudes than GPS, an additional factor in the EU decision, based on Scandinavian participation.

While U.S. GPS satellites are only launched one at a time, Galileo satellites are being designed with new miniaturization techniques that will allow several to be launched on the same rocket, a far more cost-efficient way to place them in orbit and maintain the constellation.

Galileo will also provide an important feature for civilian use that GPS does not: integrity monitoring. Currently, a civilian GPS user receives no indication that his unit is not receiving proper satellite signals, there being no provision for such notification in the code. However, Galileo will provide such a signal, alerting the user that the system is operating improperly.

The issue of compatibility with GPS is being addressed during ongoing development. Frequency sharing with GPS is under discussion, and it is reasonable to assume that a high degree of compatibility will exist when Galileo is operational. Manufacturers will undoubtedly offer a variety of systems which exploit the best technologies of both GPS and Galileo. Integration with existing shipboard electronic systems such as ECDIS and ECS will be ensured.

The benefit of Galileo for the navigator is that there will be two separate satellite navigation systems to rely on, providing not only redundancy, but also an increased degree of accuracy (for systems that can integrate both systems' signals). Galileo should be first available in 2005, and the full constellation is scheduled to be up by 2008.

1116. GLONASS

The Global Navigation Satellite System (GLONASS), under the control of the Russian military, has been in use since 1993, and is based on the same principles as GPS. The space segment consists of 24 satellites in three orbital planes, the planes separated by 120 degrees and the individual satellites by 45 degrees. The orbits are inclined to the equator at an angle of 64.8 degrees, and the orbital period is about 11hours, 15minutes at an altitude of 19,100 km (10,313 nm). The designed system fix accuracy for civilian use is 100 meters horizontal (95%), 150 meters vertical, and 15 cm/sec. in velocity. Military codes provide accuracies of some 10-20 meters horizontal.

The ground segment of GLONASS lies entirely within the former Soviet Union. Reliability has been an ongoing problem for the GLONASS system, but new satellite designs with longer life spans are addressing these concerns. The user segment consists of various types of receivers that provide position, time, and velocity information.

GLONASS signals are in the L-band, operating in 25 channels with 0.5625 MHz separation in 2 bands: from 1602.5625 MHz to 1615.5 MHz, and from 1240 to 1260 MHz.

CHAPTER 12

LORAN NAVIGATION

INTRODUCTION TO LORAN

1200. History and Role of Loran

The theory behind the operation of hyperbolic navigation systems was known in the late 1930's, but it took the urgency of World War II to speed development of the system into practical use. By early 1942, the British had an operating hyperbolic system in use designed to aid in long-range bomber navigation. This system, named Gee, operated on frequencies between 30 MHz and 80 MHz and employed "master" and "slave" transmitters spaced approximately 100 miles apart. The Americans were not far behind the British in development of their own system. By 1943, the U. S. Coast Guard was operating a chain of hyperbolic navigation transmitters that became Loran A (The term Loran was originally an acronym for LOng RAnge Navigation). By the end of the war, the network consisted of over 70 transmitters providing coverage over approximately 30% of the earth's surface.

In the late 1940's and early 1950's, experiments in low frequency Loran produced a longer range, more accurate system. Using the 90-110 kHz band, Loran developed into a 24-hour-a-day, all-weather radionavigation system named Loran C. From the late 1950's, Loran A and Loran C systems were operated in parallel until the mid 1970's when the U.S. Government began phasing out Loran A. The United States continued to operate Loran C in a number of areas around the world, including Europe, Asia, the Mediterranean Sea, and parts of the Pacific Ocean until the mid 1990's when it began closing its overseas Loran C stations or transferring them to the governments of the host countries. This was a result of the U.S. Department of Defense adopting the Global Positioning System (GPS) as its primary radionavigation service. In the United States, Loran serves the 48 contiguous states, their coastal areas and parts of Alaska. It provides navigation, location, and timing services for both civil and military air, land, and marine users. Loran systems are also operated in Canada, China, India, Japan, Northwest Europe, Russia, Saudi Arabia, and South Korea.

The future role of Loran depends on the radionavigation policies of the countries and international organizations that operate the individual chains. In the United States, the Federal Government plans to continue operating Loran in the short term while it evaluates the long-term need for the system. The U.S. Government will give users reasonable notice if it concludes that Loran is no longer needed or is not cost effective, so that users will have the opportunity to transition to alternative navigation aids and timing services.

Current information on the U.S. Loran system, including Notices to Mariners, may be obtained at the U.S. Coast Guard Navigation Center World Wide Web site at http://www.navcen.uscg.gov/.

LORAN C DESCRIPTION

1201. Summary of Operation

The Loran C (hereafter referred to simply as Loran) system consists of **transmitting stations**, which are placed several hundred miles apart and organized into **chains**. Within a Loran chain, one station is designated as the **master station** and the others as **secondary stations**. Every Loran chain contains at least one master station and two secondary stations in order to provide two lines of position.

The master and secondary stations transmit radio pulses at precise time intervals. A Loran receiver measures the **time difference** (TD) between when the vessel receives the master signal and when it receives each of the secondary signals. When this elapsed time is converted to distance, the locus of points having the same TD between the master and each secondary forms the hyperbolic LOP. The intersection of two or more of these LOP's produces a fix of the vessel's position.

There are two methods by which the navigator can convert this information into a geographic position. The first involves the use of a chart overprinted with a Loran **time delay lattice** consisting of hyperbolic TD lines spaced at convenient intervals. The navigator plots the displayed TD's by interpolating between the lattice lines printed on the chart, manually plots the fix where they intersect and then determines latitude and longitude. In the second method, computer algorithms in the receiver's software convert the TD's to latitude and longitude for display.

As with other computerized navigation receivers, a typical Loran receiver can accept and store **waypoints**.

Waypoints are sets of coordinates that describe either locations of navigational interest or points along a planned route. Waypoints may be entered by visiting the spot of interest and pressing the appropriate receiver control key, or by keying in the waypoint coordinates manually, either as a TD or latitude-longitude pair. If using waypoints to mark a planned route, the navigator can use the receiver to monitor the vessel's progress in relation to the track between each waypoint. By continuously providing parameters such as cross-track error, course over ground, speed over ground, and bearing and distance to next waypoint, the receiver continually serves as a check on the primary navigation plot.

1202. Components of the Loran System

For the marine navigator, the components of the Loran system consist of the land-based transmitting stations, the Loran **receiver** and **antenna**, the **Loran charts.** In addition to the master and secondary transmitting stations themselves, land-based Loran facilities also include the primary and secondary **system area monitor** sites, the **control station** and a precise time reference. The transmitters emit Loran signals at precisely timed intervals. The monitor sites and control stations continually measure and analyze the characteristics of the Loran signals received to detect any anomalies or out-of-specification conditions. Some transmitters serve only one function within a chain (i.e., either master or secondary). However, in many instances, one transmitter transmits signals for each of two adjacent chains. This practice is termed **dual rating**.

Loran receivers exhibit varying degrees of sophistication, but their signal processing is similar. The first processing stage consists of **search and acquisition**, during which the receiver searches for the signal from a particular Loran chain and establishes the approximate time reference of the master and secondaries with sufficient accuracy to permit subsequent settling and tracking.

After search and acquisition, the receiver enters the **settle** phase. In this phase, the receiver searches for and detects the front edge of the Loran pulse. After detecting the front edge of the pulse, it selects the correct cycle of the pulse to track.

Having selected the correct tracking cycle, the receiver begins the **tracking and lock** phase, in which the receiver maintains synchronization with the selected received signals. Once this phase is reached, the receiver displays either the time difference of the signals or the computed latitude and longitude.

1203. The Loran Signal

The Loran signal consists of a series of 100 kHz pulses sent first by the master station and then, in turn, by the secondary stations. Both the shape of the individual pulse and the pattern of the entire pulse sequence are shown in Figure 1203a. As compared to a carrier signal of constant amplitude, pulsed transmission allows the same signal range to be achieved with a lower average output power. Pulsed transmission also yields better signal identification properties and more precise timing of the signals.

The individual sinusoidal Loran pulse exhibits a steep rise to its maximum amplitude within 65 μsec of emission and an exponential decay to zero within 200 to 300 μsec. The signal frequency is nominally defined as 100 kHz; in actuality, the signal is designed such that 99% of the radiated power is contained in a 20 kHz band centered on 100 kHz.

The Loran receiver is programmed to track the signal on the cycle corresponding to the carrier frequency's third positive crossing of the x-axis. This occurrence, termed the **standard zero crossing**, is chosen for two reasons. First, it is late enough for the pulse to have built up sufficient signal strength for the receiver to detect it. Second, it is early enough in the pulse to ensure that the receiver is detecting the transmitting station's ground wave pulse and not its sky wave pulse. Sky wave pulses are affected by atmospheric refraction and if used unknowingly, would introduce large errors into positions determined by a Loran receiver. The pulse architecture described here reduces this major source of error.

Another important parameter of the pulse is the **envelope-to-cycle difference (ECD)**. This parameter indicates how propagation of the signal causes the pulse shape envelope (i.e., the imaginary line connecting the peak of each sinusoidal cycle) to shift in time relative to the zero crossings. The ECD is important because receivers use the precisely shaped pulse envelope to identify the correct zero crossing. Transmitting stations are required to keep the ECD within defined limits. Many receivers display the received ECD as well.

Next, individual pulses are combined into sequences. For the master signal, a series of nine pulses is transmitted, the first eight spaced 1000 μsec apart followed by a ninth transmitted 2000 μsec after the eighth. Secondary stations transmit a series of eight pulses, each spaced 1000 μsec apart. Secondary stations are given letter designations of U, W, X, Y, and Z; this letter designation indicates the order in which they transmit following the master. If a chain has two secondaries, they will be designated Y and Z. If a chain has three secondaries, they are X, Y and Z, and so on. Some exceptions to this general naming pattern exist (e.g., W, X and Y for some 3-secondary chains).

The spacing between the master signal and each of the secondary signals is governed by several parameters as illustrated in Figure 1203b. The general idea is that each of the signals must clear the entire chain coverage area before the next one is transmitted, so that no signal can be received out of order. The time required for the master signal to travel to the secondary station is defined as the average **baseline travel time (BTT)**, or **baseline length (BLL)**. To this time interval is added an additional delay defined as the **secondary coding delay (SCD)**, or simply **coding delay (CD).** The total of these two delays is termed the **emission delay**

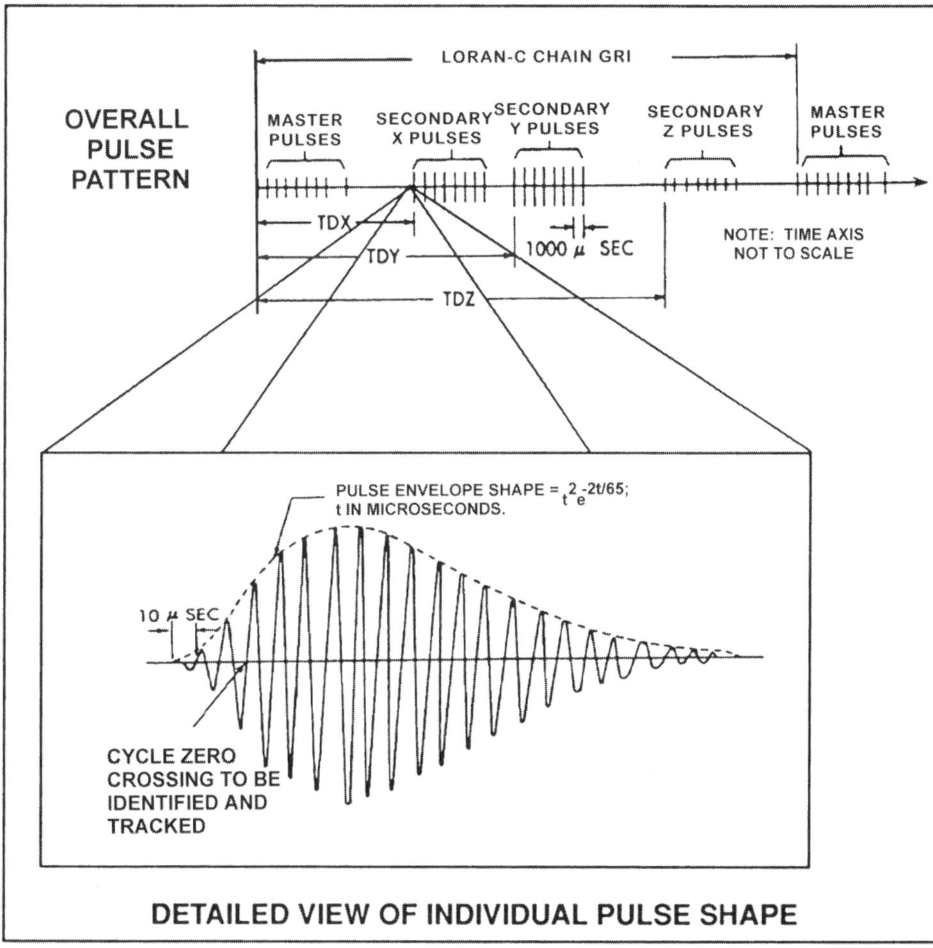

Figure 1203a. Pulse pattern and shape for Loran C transmission.

(ED), which is the exact time interval between the transmission of the master signal and the transmission of the secondary signal. Each secondary station has its own ED value. In order to ensure the proper sequence, the ED of secondary Y is longer than that of X, and the ED of Z is longer than that of Y.

Once the last secondary has transmitted, the master transmits again, and the cycle is repeated. The time to complete this cycle of transmission defines an important characteristic for the chain: the **group repetition interval (GRI)**. The group repetition interval divided by ten yields the chain's numeric designator. For example, the interval between successive transmissions of the master pulse group for the northeast U.S. chain is 99,600 μsec, just less than one tenth of a second. From the definition above, the GRI designator for this chain is defined as 9960. As mentioned previously, the GRI must be sufficiently large to allow the signals from the master and secondary stations in the chain to propagate fully throughout the region covered by the chain before the next cycle of pulses begins.

Two additional characteristics of the pulse group are **phase coding** and **blink coding**. In phase coding, the phase of the 100 kHz carrier signal is reversed from pulse to pulse in a preset pattern that repeats every two GRI's. Phase coding allows a receiver to remove skywave contamination from the groundwave signal. Loran C signals travel away

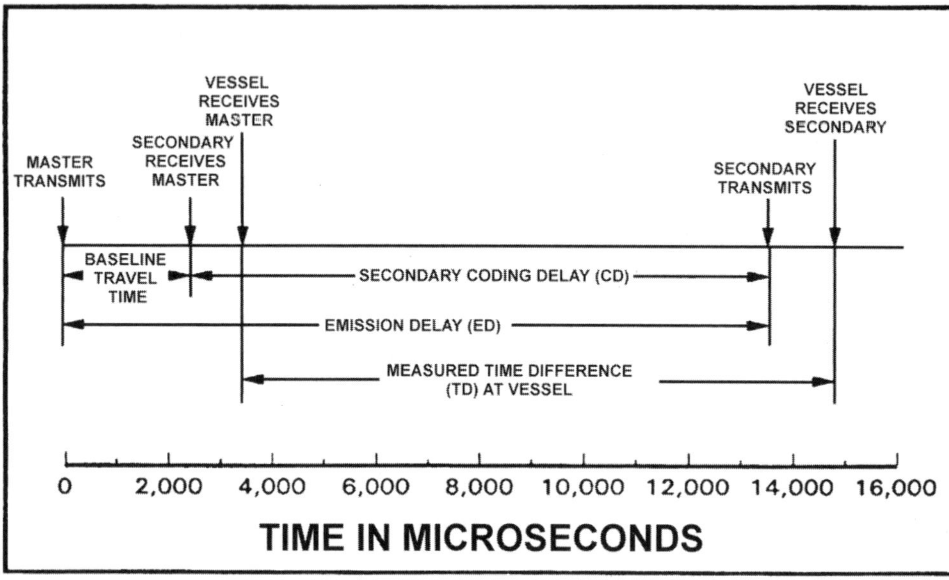

Figure 1203b. The time axis for Loran TD for point "A."

from a transmitting station in all possible directions. Groundwave is the Loran energy that travels along the surface of the earth. Skywave is Loran energy that travels up into the sky. The ionosphere reflects some of these skywaves back to the earth's surface. The skywave always arrives later than the groundwave because it travels a greater distance. The skywave of one pulse can thus contaminate the ground wave of the next pulse in the pulse group. Phase coding ensures that this skywave contamination will always "cancel out" when all the pulses of two consecutive GRI's are averaged together.

Blink coding provides integrity to the received Loran signal. When a signal from a secondary station is out of tolerance and therefore temporarily unsuitable for navigation, the affected secondary station will blink; that is, the first two pulses of the affected secondary station are turned off and on in a repeating cycle, 3.6 seconds off and 0.4 seconds on. The receiver detects this condition and displays it to the operator. When the blink indication is received, the operator should not use the affected secondary station. If a station's signal will be temporarily shut down for maintenance, the Coast Guard communicates this information in a *Notice to Mariners*. The U.S. Coast Guard Navigation Center posts these online at http://www.navcen.uscg.gov/. If a master station is out of tolerance, all secondaries in the affected chain will blink.

Two other concepts important to the understanding of Loran operation are the **baseline** and **baseline extension**. The geographic line connecting a master to a particular secondary station is defined as the station pair baseline. The baseline is, in other words, that part of a great circle on which lie all the points connecting the two stations. The extension of this line beyond the stations to encompass the points along this great circle not lying between the two stations defines the baseline extension. The optimal region for hyperbolic navigation occurs in the vicinity of the baseline, while the most care must be exercised in the regions near the baseline extension. These concepts are further developed in the next few articles.

1204. Loran Theory of Operation

In Loran navigation, the locus of points having a constant difference in distance between an observer and each of two transmitter stations defines a hyperbola, which is a line of position.

Assuming a constant speed of propagation of electromagnetic radiation in the atmosphere, the time difference in the arrival of electromagnetic radiation from the two transmitter sites is proportional to the distance between each of the transmitting sites, thus creating the hyperbola on the earth's surface. The following equations demonstrate this proportionality between distance and time:

Distance=Velocity x Time

or, using algebraic symbols

$d = c \times t$

Therefore, if the velocity (c) is constant, the distance between a vessel and each of two transmitting stations will be directly proportional to the time delay detected at the vessel between pulses of electromagnetic radiation transmitted from the two stations.

An example illustrates the concept. As shown in Figure 1204, let us assume that two Loran transmitting stations, a master and a secondary, are located along with an observer in a Cartesian coordinate system whose units are in nautical miles. We assume further that the master station, designated "M", is located at coordinates (x,y) = (-200,0) and the secondary, designated "X," is located at (x,y) = (+200,0). An observer with a receiver capable of detecting electromagnetic radiation is positioned at any point "A" whose coordinates are defined as (x_a,y_a).

Note that for mathematical convenience, these hyperbola labels have been normalized so that the hyperbola perpendicular to the baseline is labeled zero, with both negative and positive difference values. In actual practice, all Loran TD's are positive.

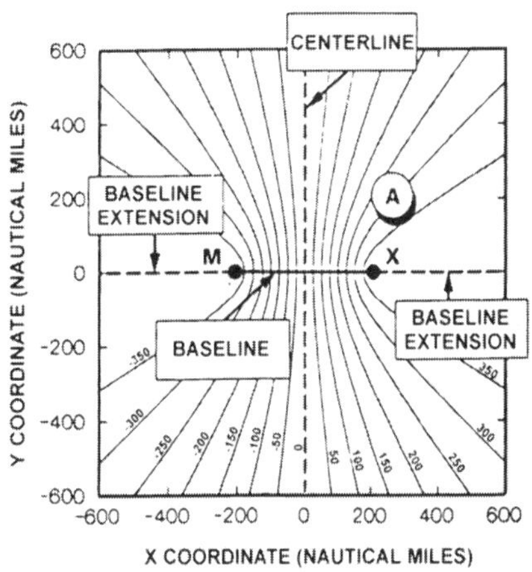

Figure 1204. Depiction of Loran LOP's.

The Pythagorean theorem can be used to determine the distance between the observer and the master station; similarly, one can obtain the distance between the observer and the secondary station:

$$\text{distance}_{am} = [(x_a + 200)^2 + y_a^2]^{0.5}$$

$$\text{distance}_{ax} = [(x_a - 200)^2 + y_a^2]^{0.5}$$

The difference between these distances (D) is:

$$D = \text{distance}_{am} - \text{distance}_{ax}$$

Substituting,

$$D = [(x_a + 200)^2 + y_a^2]^{0.5} - [(x_a - 200)^2 + y_a^2]^{0.5}$$

With the master and secondary stations in known geographic positions, the only unknowns are the two geographic coordinates of the observer.

Each hyperbolic line of position in Figure 1204 represents the locus of points for which (D) is held constant. For example, if the observer above were located at point A (271.9, 200) then the distance between that observer and the secondary station (the point designated "X" in Figure 1204a) would be 212.5 NM. In turn, the observer's distance from the master station would be 512.5 NM. The function D would simply be the difference of the two, or 300 NM. For every other point along the hyperbola passing through A, distance D has a value of 300 NM. Adjacent LOP's indicate where D is 250 NM or 350 NM.

To produce a fix, the observer must obtain a similar hyperbolic line of position generated by another master-secondary pair. Let us say another secondary station "Y" is placed at point (50,500). Mathematically, the observer will then have two equations corresponding to the M-X and M-Y TD pairs:

$$D_1 = [(x_a + 200)^2 + y_a^2]^{0.5} - [(x_a - 200)^2 + y_a^2]^{0.5}$$

$$D_2 = [(x_a + 200)^2 + y_a^2]^{0.5} - [(x_a - 50)^2 + (y_a - 500)^2]^{0.5}$$

Distances D_1 and D_2 are known because the time differences have been measured by the receiver and converted to these distances. The two remaining unknowns, x_a and y_a, may then be solved.

The above example is expressed in terms of distance in nautical miles. Because the navigator uses TD's to perform Loran hyperbolic navigation, let us rework the example for the M-X TD pair in terms of time rather than distance, adding timing details specific to Loran. Let us assume that electromagnetic radiation travels at the speed of light (one nautical mile traveled in 6.18 µsec). The distance from master station M to point A was 512.5 NM. From the relationship just defined between distance and time, it would take a signal (6.18 µsec/NM) × 512.5 NM = 3,167 µsec to travel from the master station to the observer at point A. At the arrival of this signal, the observer's Loran receiver would start the TD measurement. Recall from the general discussion above that a secondary station transmits after an emission delay equal to the sum of the baseline travel time and the secondary coding delay. In this example, the master and the secondary are 400 NM apart; therefore, the baseline travel time is (6.18 μsec/NM) × 400 NM = 2,472 μsec. Assuming a secondary coding delay of 11,000 μsec, the secondary station in this example would transmit (2,472 + 11,000) μsec or 13,472 μsec after the master station. The secondary signal then propagates over a distance 212.5 NM to reach point A, taking (6.18 μsec/NM) × 212.5 NM = 1,313 μsec to do so. Therefore, the total time from *transmission* of the master signal to the *reception* of the secondary signal by the observer at point A is (13,472 + 1,313) μsec = 14,785 μsec.

Recall, however, that the Loran receiver measures the time delay between *reception* of the master signal and the *reception* of the secondary signal. Therefore, the time quantity above must be corrected by subtracting the amount of time required for the signal to travel from the master transmitter to the observer at point A. This amount of time was 3,167 μsec. Therefore, the TD observed at point A in this hypothetical example would be (14,785 - 3,167) μsec or 11,618 μsec. Once again, this time delay is a function of the simultaneous differences in distance between the observer and the two transmitting stations, and it gives rise to a hyperbolic line of position which can be crossed with another LOP to fix the observer's position.

1205. Allowances for Non-Uniform Propagation Rates

The initial calculations above assumed the speed of light in free space; however, the actual speed at which electromagnetic radiation propagates on earth is reduced both by the atmosphere through which it travels and by the conductive surfaces—sea and land—over which it passes. The specified accuracy needed from Loran therefore requires three corrections to the propagation speed of the signal.

The reduction in propagation speed due to the atmosphere is represented by the first correction term: the **Primary Phase Factor (PF)**. Similarly, a **Secondary Phase Factor (SF)** accounts for the reduced propagation speed due to traveling over seawater. These two corrections are transparent to the operator since they are uniformly incorporated into all calculations represented on charts and in Loran receivers.

Because land surfaces have lower conductivity than seawater, the propagation speed of the Loran signal passing over land is further reduced as compared to the signal passing over seawater. A third and final correction, the **Additional Secondary Phase Factor (ASF)**, accounts for the delay due to the land conductivity when converting time delays to distances and then to geographic coordinates. Depending on the mariner's location, signals from some Loran transmitters may have traveled hundreds of miles over land and must be corrected to account for this non-seawater portion of the signal path. Of the three corrections mentioned in this article, this is the most complex and the most important one to understand, and is accordingly treated in detail in Article 1210.

LORAN ACCURACY

1206. Defining Accuracy

Specifications of Loran and other radionavigation systems typically refer to three types of accuracy: **absolute, repeatable** and **relative.**

Absolute accuracy, also termed predictable or geodetic accuracy, is the accuracy of a position with respect to the geographic coordinates of the earth. For example, if the navigator plots a position based on the Loran latitude and longitude (or based on Loran TD's) the difference between the Loran position and the actual position is a measure of the system's absolute accuracy.

Repeatable accuracy is the accuracy with which the navigator can return to a position whose coordinates have been measured previously with the same navigational system. For example, suppose a navigator were to travel to a buoy and note the TD's at that position. Later, suppose the navigator, wanting to return to the buoy, returns to the previously measured TD's. The resulting position difference between the vessel and the buoy is a measure of the system's repeatable accuracy.

Relative accuracy is the accuracy with which a user can measure position relative to that of another user of the same navigation system at the same time. If one vessel were to travel to the TD's determined by another vessel, the difference in position between the two vessels would be a measure of the system's relative accuracy.

The distinction between absolute and repeatable accuracy is the most important one to understand. With the correct application of ASF's and within the **coverage area** defined for each chain, the absolute accuracy of the Loran system varies from between 0.1 and 0.25 nautical miles. However, the repeatable accuracy of the system is much better, typically between 18 and 90 meters (approximately 60 to 300 feet) depending on one's location in the coverage area. If the navigator has been to an area previously and noted the TD's corresponding to different navigational aids (e.g., a buoy marking a harbor entrance), the high repeatable accuracy of the system enables location of the buoy in adverse weather. Similarly, selected TD data for various harbor navigational aids and other locations of interest have been collected and recorded and is generally commercially available. This information provides an excellent backup navigational source to conventional harbor approach navigation.

1207. Limitations to Loran Accuracy

There are limits on the accuracy of any navigational system, and Loran is no exception. Several factors that contribute to limiting the accuracy of Loran as a navigational aid are listed in Table 1207 and are briefly discussed in this article. Even though all these factors except operator error are included in the published accuracy of Loran, the mariner's aim should be to have a working knowledge of each one and minimize any that are under his control so as to obtain the best possible accuracy.

The geometry of LOP's used in a Loran fix is of prime importance to the mariner. Because understanding of this factor is so critical to proper Loran operation, the effects of crossing angles and gradients are discussed in detail in the Article 1208. The remaining factors are briefly explained as follows.

The age of the Coast Guard's Loran transmitting equipment varies from station to station. When some older types of equipment are switched from standby to active and vice versa, a slight timing shift as large as tens of nanoseconds may be seen. This is so small that it is undetectable by most marine receivers, but since all errors accumulate, it should be understood as part of the Loran "error budget."

The effects of actions to control chain timing are similar. The timing of each station in a chain is controlled based on data received at the primary system area monitor site. Signal timing errors are kept as near to zero as possible at the primary site, making the absolute accuracy of Loran generally the best in the vicinity of the primary site. Whenever, due to equipment casualty or to accomplish system maintenance, the control station shifts to the secondary system area monitor site, slight timing shifts may be introduced in parts of the coverage area.

Atmospheric noise, generally caused by lightning, reduces the **signal-to-noise ratio (SNR)** available at the receiver. This in turn degrades accuracy of the LOP. Man-made noise has a similar effect on accuracy. In rare cases, a man-made noise source whose carrier signal frequency or harmonics are near 100 kHz (such as the constant carrier control signals commonly used on high-tension power lines) may also interfere with lock-on and tracking of a Loran receiver. In general, Loran stations that are the closest to the user will have the highest SNR and will produce LOP's with the lowest errors. Geometry, however, remains a key factor in producing a good fix from combined LOP's. Therefore, the best LOP's for a fix may not all be from the very nearest stations.

The user should also be aware that the propagation speed of Loran changes with time as well. Temporal changes may be seasonal, due to snow cover or changing groundwater levels, or diurnal, due to atmospheric and surface changes from day to night. Seasonal changes may be as large as 1 μsec and diurnal changes as large as 0.2 μsec, but these vary with location and chain being used. Passing cold weather fronts may have temporary effects as well.

Disturbances on the sun's surface, most notably solar flares, disturb the earth's atmosphere as well. These Sudden Ionospheric Disturbances (SID's) increase attenuation of radio waves and thus disturb Loran signals and reduce SNR. Such a disturbance may interfere with Loran reception for periods of hours or even longer.

The factors above all relate to the propagated signal before it reaches the mariner. The remaining factors discussed below address the accuracy with which the mariner receives and interprets the signal.

Factor	Has effect on	
	Absolute Accuracy	Repeatable Accuracy
Crossing angles and gradients of the Loran LOP's	Yes	Yes
Stability of the transmitted signal (e.g., transmitter effect)	Yes	Yes
Loran chain control parameters (e.g., how closely actual ED is maintained to published ED, which system area monitor is being used, etc.)	Yes	Yes
Atmospheric and man-made ambient electronic noise	Yes	Yes
Factors with temporal variations in signal propagation speed (e.g., weather, seasonal effects, diurnal variations, etc.)	Yes	Yes
Sudden ionospheric disturbances	Yes	Yes
Receiver quality and sensitivity	Yes	Yes
Shipboard electric noise	Yes	Yes
Accuracy with which LOP's are printed on nautical charts	Yes	No
Accuracy of receiver's computer algorithms for coordinate conversion	Yes	No
Operator error	Yes	Yes

Table 1207. Selected Factors that Limit Loran Accuracy.

Receivers vary in precision, quality and sophistication. Some receivers display TD's to the nearest 0.1 µsec; others to 0.01 µsec. Internal processing also varies, whether in the analog "front end" or the digital computer algorithms that use the processed analog signal. By referencing the user manual, the mariner may gain an appreciation for the advantages and limitations of the particular model available, and may adjust operator settings to maximize performance.

The best receiver available may be hindered by a poor installation. Similarly, electronic noise produced by engine and drive machinery, various electric motors, other electronic equipment or even household appliances may hinder the performance of a Loran receiver. The mariner should consult documentation supplied with the receiver for proper installation. Generally, proper installation and placement of the receiver's components will mitigate these problems. In some cases, contacting the manufacturer or obtaining professional installation assistance may be appropriate.

The raw TD's obtained by the receiver must be corrected with ASF's and then translated to position. Whether the receiver performs this entire process or the mariner assists by translating TD's to position manually using a Loran overprinted chart, published accuracies take into account the small errors involved in this conversion process.

Finally, as in all endeavors, operator error when using Loran is always possible. This can be minimized with alertness, knowledge and practice.

1208. The Effects of Crossing Angles and Gradients

The hyperbolic nature of Loran requires the operator to pay special attention to the geometry of the fix, specifically to crossing angles and gradients, and to the possibility of fix ambiguity. We begin with crossing angles.

As discussed above, the TD's from any given master-secondary pair form a family of hyperbolas. Each hyperbola in this family can be considered a line of position; the vessel must be somewhere along that locus of points which forms the hyperbola. A typical family of hyperbolas is shown in Figure 1208a.

Now, suppose the hyperbolic family from the Master-Xray station pair shown in Figure 1204 were superimposed upon the family shown in Figure 1208a. The results would be the hyperbolic lattice shown in Figure 1208b.

As has been noted, Loran LOP's for various chains and secondaries are printed on nautical charts. Each of the sets of LOP's is given a separate color and is denoted by a characteristic set of symbols. For example, an LOP might be designated 9960-X-25750. The designation is read as follows: the chain GRI designator is 9960, the TD is for the Master-Xray pair (M-X), and the time difference along this LOP is 25750 µsec. The chart shows only a limited number of LOP's to reduce clutter on the chart. Therefore, if the observed time delay falls between two charted LOP's, interpolation between them is required to obtain the precise LOP. After having interpolated (if necessary) between two

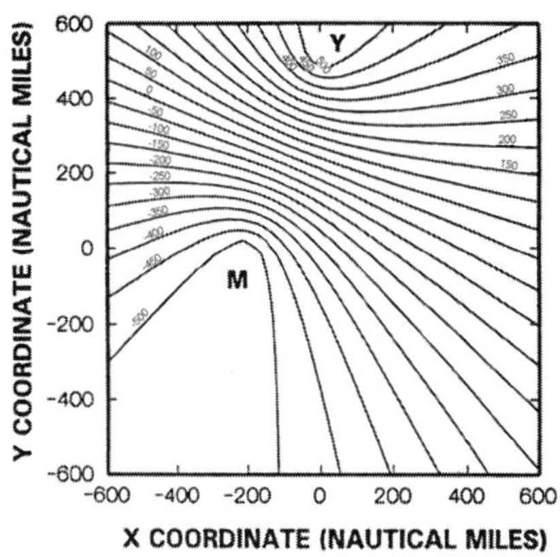

Figure 1208a. A family of hyperbolic lines generated by Loran signals.

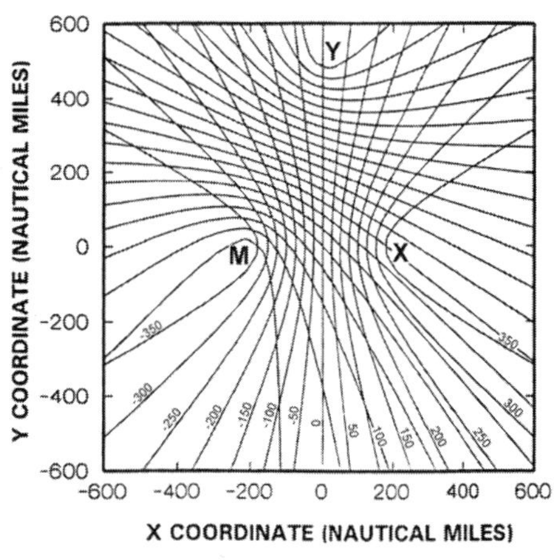

Figure 1208b. A hyperbolic lattice formed by station pairs M-X and M-Y.

TD measurements and plotted the resulting LOP's on the chart, the navigator marks the intersection of the LOP's and labels that intersection as the Loran fix. Note also in Figure 1208b the various angles at which the hyperbolas cross each other.

Figure 1208c shows graphically how error magnitude varies as a function of crossing angle. Assume that LOP 1

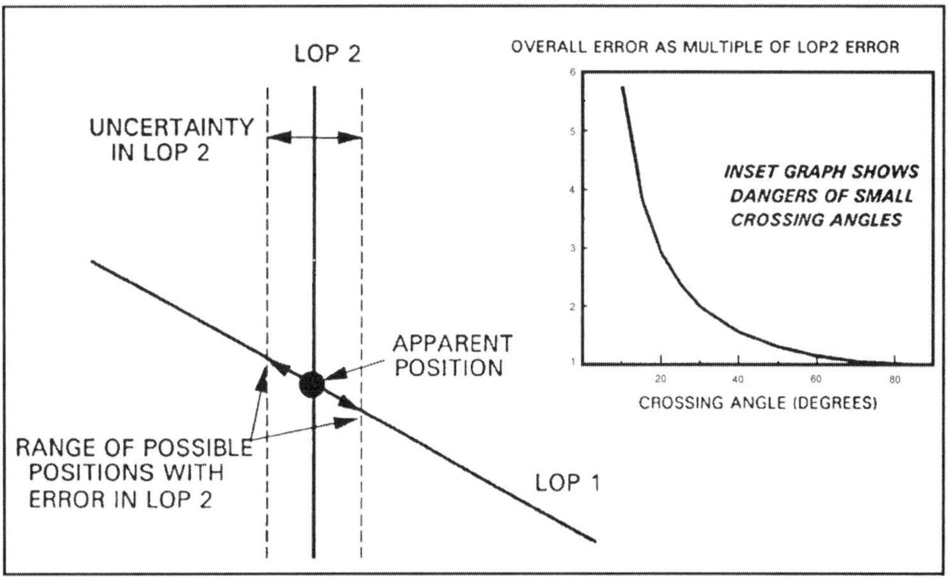

Figure 1208c. Error in Loran LOP's is magnified if the crossing angle is less than 90°.

is known to contain no error, while LOP 2 has an uncertainty as shown. As the crossing angle (i.e., the angle of intersection of the two LOP's) approaches 90°, range of possible positions along LOP 1 (i.e., the position uncertainty or fix error) approaches a minimum; conversely, as the crossing angle decreases, the position uncertainty increases; the line defining the range of uncertainty grows longer. This illustration demonstrates the desirability of choosing LOP's for which the crossing angle is as close to 90° as possible.

The relationship between crossing angle and fix uncertainty can be expressed mathematically:

$$\sin(x) = \frac{\text{LOP error}}{\text{fix uncertainty}}$$

where x is the crossing angle.

Rearranging algebraically,

$$\text{fix uncertainty} = \frac{\text{LOP error}}{\sin(x)}$$

Assuming that LOP error is constant, then position uncertainty is inversely proportional to the sine of the crossing angle. As the crossing angle increases from 0° to 90°, the sine of the crossing angle increases from 0 to 1. Therefore, the error is at a minimum when the crossing angle is 90°, and increases thereafter as the crossing angle decreases.

Understanding and proper use of TD gradients are also important to the navigator. The gradient is defined as the rate of change of distance with respect to TD. Put another way, this quantity is the ratio of the spacing between adjacent Loran TD's (usually expressed in feet or meters) and the difference in microseconds between these adjacent LOP's. For example, if at a particular location two printed TD lines differ by 20 μsec and are 6 NM apart, the gradient is.

$$\text{Gradient} = \frac{6\text{NM} \times 6076\text{ft/NM}}{20\mu\text{sec}} = 1822.8 \text{ ft/}\mu\text{sec}$$

The smaller the gradient, the smaller the distance error that results from any TD error. Thus, the best accuracy from Loran is obtained by using TD's whose gradient is the smallest possible (i.e. the hyperbolic lines are closest together). This occurs along the baseline. Gradients are much larger (i.e. hyperbolic lines are farther apart) in the vicinity of the baseline extension. Therefore, the user should select TD's having the smallest possible gradients.

Another Loran effect that can lead to navigational error in the vicinity of the baseline extension is fix ambiguity. Fix ambiguity results when one Loran LOP crosses another LOP in two separate places. Near the baseline extension, the "ends" of a hyperbola can wrap around so that they cross another LOP twice, once along the baseline, and again along the baseline extension. A third LOP would resolve the ambiguity.

Most Loran receivers have an ambiguity alarm to alert the navigator to this occurrence. However, both fix ambiguity and large gradients necessitate that the navigator avoid using a master-secondary pair when operating in the vicinity of that pair's baseline extension.

1209. Coverage Areas

The 0.25 NM absolute accuracy specified for Loran is valid within each chain's coverage area. This area, whose limits define the maximum range of Loran for a particular chain, is the region in which both accuracy and SNR criteria are met. The National Oceanographic and Atmospheric Administration (NOAA) has generally followed these coverage area limits when selecting where to print particular Loran TD lines on Loran overprinted charts. Coverage area diagrams of each chain are also available online from the U.S. Coast Guard's Navigation Center, currently at http://www.navcen.uscg.gov/ftp/loran/lgeninfo/h-book/loranappendixb.pdf. Other helpful information available at this FTP site includes the Loran C User's Handbook and the Loran C Signal Specification, two key sources of material in this chapter.

One caveat to remember when considering coverage areas is that the 0.25 NM accuracy criteria is modified inside the coverage area in the vicinity of the coastline due to ASF effects. The following article describes this more fully.

1210. Understanding Additional Secondary Factors (ASF's)

Mathematically, calculating the reduction in propagation speed of an electromagnetic signal passing over a land surface of known conductivity is relatively straightforward. In practice, however, determining this Loran ASF correction accurately for use in the real world can be complex.

There are at least four reasons for this complexity. First, the conductivity of ground varies from region to region, so the correction to be applied is different for every signal path. Moreover, ground conductivity data currently available do not take into account all the minor variations within each region. Second, methods used to compute ASF's vary. ASF's can be determined from either a mathematical model based on known approximate ground conductivities, or from empirical time delay measurements in various locations, or a combination of both. Methods incorporating empirical measurements tend to yield more accurate results. One receiver manufacturer may not use exactly the same correction method as another, and neither may use exactly the same method as those incorporated into time differences printed on a particular nautical chart. While such differences are minor, a user unaware of these differences may not obtain the best accuracy possible from Loran. Third, relatively large local variations in ASF variations that cannot fully be accounted for in current ASF models applied to the coverage area as a whole, may be observed in the region within 10 NM of the coast. Over the years, even empirically measured ASF's may change slightly in these areas with the addition of buildings, bridges and other structures to coastal areas. Fourth and finally, ASF's vary seasonally with changes in groundwater levels, snow pack depths and similar factors.

Designers of the Loran system, including Loran receiver manufacturers, have expended a great deal of effort to include ASF's in error calculations and to minimize these effects. Indeed, inaccuracies in ASF modeling are accounted for in published accuracy specifications for Loran. What then does the marine navigator need to know about ASF's beyond this? To obtain the 0.25 NM absolute accuracy advertised for Loran, the answer is clear. One must know *where* in the coverage area ASF's affect published accuracies, and one must know *when* ASF's are being incorporated, both in the receiver and on any chart in use.

With respect to *where* ASF's affect published accuracies, one must remember that local variations in the vicinity of the coastline are the most unpredictable of all ASF related effects because they are not adequately explained by current predictive ASF models. As a result, even though fixes determined by Loran may satisfy the 0.25 NM accuracy specification in these areas, such accuracy is not "guaranteed" for Loran within 10 NM of the coast. Users should also avoid relying solely on the lattice of Loran TD's in inshore areas.

With respect to *when* ASF's are being applied, one should realize that the default mode in most receivers combines ASF's with raw TD measurements. This is because the inclusion of ASF's is required in order to meet the 0.25 NM accuracy criteria. The navigator should verify which mode the receiver is in, and ensure the mode is not changed unknowingly. Similarly, current NOAA Loran overprinted charts of the U.S. incorporate ASF's, and in the chart's margin the following note appears:

> "Loran C correction tables published by the National Imagery and Mapping Agency or others should not be used with this chart. The lines of position shown have been adjusted based on survey data. Every effort has been made to meet the 0.25 nautical mile accuracy criteria established by the U.S. Coast Guard. Mariners are cautioned not to rely solely on the lattices in inshore waters."

The key point to remember there is that the "ASF included" and "ASF not included" modes must not be mixed. In other words, the receiver and any chart in use must handle ASF's in the same manner. If the receiver includes them, any chart in use must also include them. If operating on a chart that does not include ASF's—Loran coverage areas in another part of the world, for example—the receiver must be set to the same mode. If the navigator desires to correct

ASF's manually, tables for U.S. Loran chains are available at http://chartmaker.ncd.noaa.gov/mcd/loranc.htm. These documents also provide a fuller explanation of manual ASF corrections. When viewing ASF tables, remember that although the ASF correction for a single signal is always positive (indicating that the signal is always slowed and never speeded by its passage over land), the ASF correction for a time *difference* may be negative because two signal delays are included in the computation.

The U.S. Government does not guarantee the accuracy of ASF corrections incorporated into Loran receivers by their respective manufacturers. The prudent navigator will regularly check Loran TD's against charted LOP's when in a known position, and will compare Loran latitude and longitude readouts against other sources of position information. Ensuring the proper configuration and operation of the Loran receiver remains the navigator's responsibility.

Up to this point, our discussion has largely focused on correctly understanding and using Loran in order to obtain published accuracies. In some portions of the coverage areas, accuracy levels actually obtainable may be significantly better than these minimum published values. The following articles discuss practical techniques for maximizing the absolute, repeatable and relative accuracy of Loran.

1211. Maximizing Loran's Absolute Accuracy

Obtaining the best possible absolute accuracy from Loran rests primarily on the navigator's selection of TD's, particularly taking into account geometry, SNR and proximity to the baseline and baseline extension. As a vessel transits the coverage area, these factors gradually change and, except for SNR, are not visible on the display panel of the Loran receiver. Most receivers track an entire chain and some track multiple chains simultaneously, but the majority of installed marine receivers still use only two TD's to produce a latitude and longitude. Some receivers monitor these factors and may automatically select the best pair. The best way for the navigator, however, to monitor these factors is by referring to a Loran overprinted chart, even if not actually plotting fixes on it. The alert navigator will frequently reevaluate the selection of TD's during a transit and make adjustments as necessary.

Beyond this advice, two additional considerations may help the navigator maximize absolute accuracy. The first is the realization that Loran TD error is not evenly distributed over the coverage area. Besides the effects of transmitter station location on geometry and fix error, the locations of the primary and secondary monitor sites also have a discernible effect on TD error in the coverage area. As ASF's change daily and seasonally, the Loran control stations continually adjust the emission delay of each secondary station to keep it statistically at its nominal value as observed at the primary monitor site. What this means is that, on average, the Loran TD is more stable and more accurate in the absolute sense in the vicinity of the primary monitor site. The primary system area monitor for stations 9960-M, 9960-X and 9960-Y was placed at the entrance to New York harbor at Sandy Hook, New Jersey for just this reason. A switch by the control station to the secondary monitor site will shift the error distribution slightly within the coverage area, reducing it near the secondary site and slightly increasing it elsewhere. The locations of primary system area monitor sites can be found at the USCG NAVCEN web site.

The second consideration in maximizing absolute accuracy is that most Loran receivers may be manually calibrated using a feature variously called "bias," "offset," "homeport" or a similar term. When in homeport or another known location, the known latitude and longitude (or in some cases, the difference between the current Loran display and the known values) is entered into the receiver. This forces the receiver's position error to be zero at that particular point and time.

The limitation of this technique is that this correction becomes less accurate with the passage of time and with increasing distance away from the point used. Most published sources indicate the technique to be of value out to a distance of 10 to 100 miles of the point where the calibration was performed. This correction does not take into account local distortions of the Loran grid due to bridges, power lines or other such man-made structures. The navigator should evaluate experimentally the effectiveness of this technique in good weather conditions before relying on it for navigation at other times. The bias should also be adjusted regularly to account for seasonal Loran variations; using the same value throughout the year is not the most effective application of this technique. Also, entering an offset into a Loran receiver alters the apparent location of waypoints stored prior to establishing this correction.

Finally, receivers vary in how this feature is implemented. Some receivers save the offset when the receiver is turned off; others zero the correction when the receiver is turned on. Some receivers replace the internal ASF value with the offset, while others add it to the internal ASF values. Refer to the owner's manual for the receiver in use.

1212. Maximizing Loran's Repeatable Accuracy

Many users consider the high repeatable accuracy of Loran its most important characteristic. To obtain the best repeatable accuracy consistently, the navigator should use measured TD's rather than latitude and longitude values supplied by the receiver.

The reason for this lies in the ASF conversion process. Recall that Loran receivers use ASF's to correct TD's. Recall also that the ASF's are a function of the terrain over which the signal must pass to reach the receiver. Therefore, the ASF's for one station pair are different from the ASF's for another station pair because the signals from the different pairs must travel over different terrain to reach the receiver.

This consideration matters because a Loran receiver may not always use the same pairs of TD's to calculate a fix. Suppose a navigator marks the position of a channel buoy by recording its latitude and longitude using the TD pair selected automatically by the Loran receiver. If, on the return trip, the receiver is using a different TD pair, the latitude and longitude readings for the exact same buoy would be slightly different because the new TD pair would be using a different ASF value. By using previously-measured TD's and not previously-measured latitudes and longitudes, this ASF-introduced error is avoided. The navigator should also record the values of all secondary TD's at the waypoint and not just the ones used by the receiver at the time. When returning to the waypoint, other TD's will be available even if the previously used TD pair is not. Recording the time and date the waypoint is stored will also help evaluate the cyclical seasonal and diurnal variations that may have since occurred.

1213. Maximizing Loran's Relative Accuracy

The classical application of relative accuracy involves two users finding the same point on the earth's surface at the same time using the same navigation system. The maximum relative Loran accuracy would be theoretically be achieved by identical receivers, configured and installed identically on identical vessels, tracking the same TD's. In practice, the two most important factors are tracking the same TD's and ensuring that ASF's are being treated consistently between the two receivers. By attending to these, the navigator should obtain relative accuracy close to the theoretical maximum.

Another application of relative accuracy is the current practice of converting old Loran TD's into latitude and longitude for use with GPS and DGPS receivers. Several commercial firms sell software applications that perform this tedious task. One key question posed by these programs is whether or not the Loran TD's include ASF's. The difficulty in answering this question depends on how the Loran TD's were obtained, and of course an understanding of ASF's. If in doubt, the navigator can perform the conversion once by specifying "with" ASF's and once "without," and then carefully choosing which is the valid one, assisted by direct observation underway if needed.

To round out the discussion of Loran, the following article briefly describes present and possible future uses for this system beyond the well-known hyperbolic navigation mode.

NON-HYPERBOLIC USES OF LORAN C

1214. Precise Timing with Loran

Because Loran is fundamentally a precise timing system, a significant segment of the user community uses Loran for the propagation of Coordinated Universal Time (UTC). The accessibility of UTC at any desired location enables such applications as the synchronization of telephone and data networks. The U.S. Coast Guard makes every effort to ensure that each Loran master transmitter station emits its signal within 100 ns of UTC. Because the timing of each secondary station is relative to the master, its timing accuracy derives from that of the master.

The start of each Loran station's GRI periodically coincides with the start of the UTC second. This is termed the Time of Coincidence (TOC). The U.S. Naval Observatory publishes TOC's at http://tycho.usno.navy.mil/loran.html for the benefit of timing users. Because one Loran station is sufficient to provide an absolute timing reference, timing receivers do not typically rely on the hyperbolic mode or use TD's per se.

A noteworthy feature of Loran is that each transmitter station has an independent timing reference consisting of three modern cesium beam oscillators. Timing equipment at the transmitter stations constantly compares these signals and adjusts to minimize oscillator drift. The end result is a nationwide system with a large ensemble of independent timing sources. This strengthens the U.S. technology infrastructure. As another cross-check of Loran time, daily comparisons are made with UTC, as disseminated via GPS.

1215. Loran Time of Arrival (TOA) Mode

With the advent of the powerful digital processors and compact precise oscillators now embedded in user receivers, technical limitations that dictated Loran's hyperbolic architecture decades ago have been overcome. A receiver can now predict in real time the exact point in time a Loran station will transmit its signal, as well as the exact time the signal will be received at any assumed position.

An alternate receiver architecture that takes advantage of these capabilities uses Loran Time of Arrival (TOA) measurement, which are measured relative to UTC rather than to an arbitrary master station's transmission. A receiver operating in TOA mode can locate and track all Loran signals in view, prompting the descriptor "all in view" for this type of receiver. This architecture steps beyond the limitations of using only one Loran chain at a time. As a result, system availability can be improved across all the overlapping coverage areas. Coupled with advanced Receiver Autonomous Integrity Monitor (RAIM) algorithms, this architecture can also add an additional layer of integrity at the user level, independent of Loran blink.

One technical possibility arising out of this new capability is to control the time of transmission of each station independently with direct reference to UTC, rather than by using system area monitors. Such an arrangement could offer the advantage of more uniformly distributing Loran fix errors across the coverage areas. This could in turn more naturally configure Loran for use in an integrated navigation system.

1216. Loran in an Integrated Navigation System

An exponential worldwide increase in reliance on electronic navigation systems, most notably GPS, for positioning and timing has fueled a drive for more robust systems immune from accidental or intentional interference. Even a short outage of GPS, for example, would likely have severe safety and economic consequences for the United States and other nations.

In this environment, integrated navigation systems are attractive options as robust sources of position and time. The ideal integrated navigation system can tolerate the degradation or failure of any component system without degradation as a whole.

Loran offers several advantages to an integrated system based on GPS or DGPS. Although Loran relies on radio propagation and is thus similarly vulnerable to large-scale atmospheric events such as ionospheric disturbances, at 100 kHz it occupies a very different portion of the spectrum than the 1.2 GHz to 1.6 GHz band used by GPS. Loran is a high-power system whose low frequency requires a very large antenna for efficient propagation. Therefore, jamming Loran over a broad area is much more difficult than jamming GPS over the same area. Loran signals are present in urban and natural canyons and under foliage, where GPS signals may be partially or completely blocked. Loran's independent timing source also provides an additional degree of robustness to an integrated system. In short, the circumstances that cause failure or degradation of Loran are very different from those that cause failure or degradation of GPS or DGPS. When the absolute accuracy of Loran is continually calibrated by GPS, the repeatable accuracy of Loran could ensure near-GPS performance of an integrated system in several possible navigation and timing scenarios, for periods of several hours to a few days after a total loss of GPS.

1217. Loran as a Data Transfer Channel

The U.S. Coast Guard has practiced low data rate transmission using Loran signals during various periods since the 1970's. The two primary uses of this capability were Loran chain control and backup military communications. In all cases, the data superimposed on the Loran signal were transparent to the users, who were nearly universally unaware of this dual use.

In the late 1990's, the Northwest European Loran System (NELS) implemented a pulse-position modulation pattern termed Eurofix to provide differential GPS corrections via the Loran signal to certain areas in western and northern Europe. Eurofix successfully incorporated sophisticated data communications techniques to broadcast GPS corrections in real time while allowing traditional Loran users to operate without interruption.

Another possible use of a Loran data transfer channel is to broadcast GPS corrections provided by the U.S. Wide Area Augmentation System (WAAS), which was designed for the benefit of aircraft in the U.S. National Airspace System (NAS). Preliminary tests have shown modulated Loran signals could be successfully used to broadcast WAAS data.

CHAPTER 13

RADAR NAVIGATION

PRINCIPLES OF RADAR OPERATION

1300. Introduction

Radar determines distance to an object by measuring the time required for a radio signal to travel from a transmitter to the object and return. Such measurements can be converted into lines of position (LOP's) comprised of circles with radius equal to the distance to the object. Since marine radars use directional antennae, they can also determine an object's bearing. However, due to its design, a radar's bearing measurement is less accurate than its distance measurement. Understanding this concept is crucial to ensuring the optimal employment of the radar for safe navigation.

1301. Signal Characteristics

In most marine navigation applications, the radar signal is pulse modulated. Signals are generated by a timing circuit so that energy leaves the antenna in very short pulses. When transmitting, the antenna is connected to the transmitter but not the receiver. As soon as the pulse leaves, an electronic switch disconnects the antenna from the transmitter and connects it to the receiver. Another pulse is not transmitted until after the preceding one has had time to travel to the most distant target within range and return. Since the interval between pulses is long compared with the length of a pulse, strong signals can be provided with low average power. The duration or length of a single pulse is called **pulse length**, **pulse duration**, or **pulse width**. This pulse emission sequence repeats a great many times, perhaps 1,000 per second. This rate defines the **pulse repetition rate (PRR)**. The returned pulses are displayed on an indicator screen.

1302. The Display

The radar display is often referred to as the **plan position indicator (PPI)**. On a PPI, the sweep appears as a radial line, centered at the center of the scope and rotating in synchronization with the antenna. Any returned echo causes a brightening of the display screen at the bearing and range of the object. Because of a luminescent coating on the inside of the tube, the glow continues after the trace rotates past the target.

On a PPI, a target's actual range is proportional to its distance from the center of the scope. A moveable cursor helps to measure ranges and bearings. In the "heading-upward" presentation, which indicates relative bearings, the top of the scope represents the direction of the ship's head. In this unstabilized presentation, the orientation changes as the ship changes heading. In the stabilized "north-upward" presentation, gyro north is always at the top of the scope.

1303. The Radar Beam

The pulses of energy comprising the radar beam would form a single lobe-shaped pattern of radiation if emitted in free space. Figure 1303a shows this free space radiation pattern, including the undesirable minor lobes or side lobes associated with practical antenna design.

Although the radiated energy is concentrated into a relatively narrow main beam by the antenna, there is no clearly defined envelope of the energy radiated, although most of the energy is concentrated along the axis of the beam. With the rapid decrease in the amount of radiated energy in directions away from this axis, practical power limits may be used to define the dimensions of the radar beam.

A radar beam's horizontal and vertical beam widths are referenced to arbitrarily selected power limits. The most common convention defines beam width as the angular width between half power points. The half power point corresponds to a drop in 3 decibels from the maximum beam strength.

The definition of the decibel shows this halving of power at a decrease in 3 dB from maximum power. A decibel is simply the logarithm of the ratio of a final power level to a reference power level:

$$dB = 10 \log\left[\frac{P_1}{P_0}\right]$$

where P_1 is the final power level, and P_0 is a reference power level. When calculating the dB drop for a 50% reduction in power level, the equation becomes:

$$dB = 10 \log(.5)$$
$$dB = -3\ dB$$

The radiation diagram shown in Figure 1303b depicts relative values of power in the same plane existing at the same distances from the antenna or the origin of the radar beam. Maximum power is in the direction of the axis of the beam. Power values diminish rapidly in directions away from the axis. The beam width is taken as the angle between the half-power points.

The beam width depends upon the frequency or wavelength of the transmitted energy, antenna design, and the dimensions of the antenna. For a given antenna size (antenna aperture), narrower beam widths result from using shorter wavelengths. For a given wavelength, narrower beam widths result from using larger antennas.

With radar waves being propagated in the vicinity of the surface of the sea, the main lobe of the radar beam is composed of a number of separate lobes, as opposed to the single lobe-shaped pattern of radiation as emitted in free space. This phenomenon is the result of interference between radar waves directly transmitted, and those waves which are reflected from the surface of the sea. Radar waves strike the surface of the sea, and the indirect waves reflect off the surface of the sea. See Figure 1303c. These reflected waves either constructively or destructively interfere with the direct waves depending upon the waves' phase relationship.

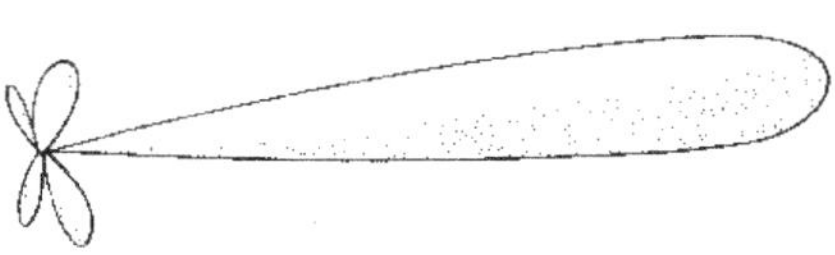

Figure 1303a. Freespace radiation pattern.

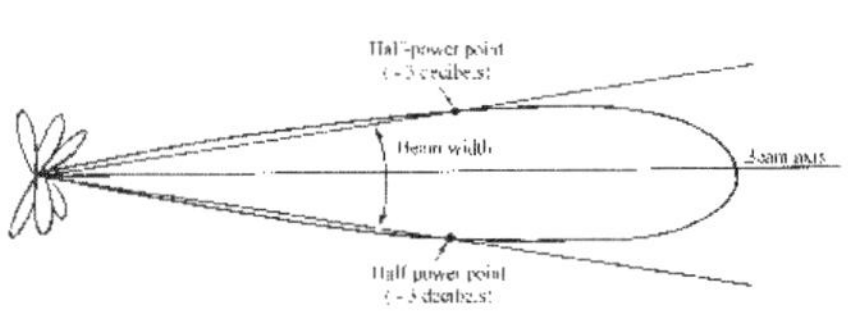

Figure 1303b. Radiation diagram.

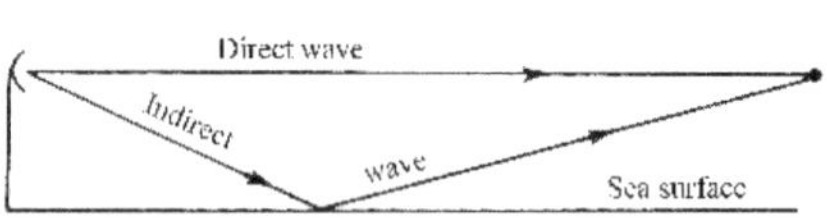

Figure 1303c. Direct and indirect waves.

1304. Diffraction and Attenuation

Diffraction is the bending of a wave as it passes an obstruction. Because of diffraction there is some illumination of the region behind an obstruction or target by the radar beam. Diffraction effects are greater at the lower frequencies. Thus, the radar beam of a lower frequency radar tends to illuminate more of the shadow region behind an obstruction than the beam of a radar of higher frequency or shorter wavelength.

Attenuation is the scattering and absorption of the energy in the radar beam as it passes through the atmosphere. It causes a decrease in echo strength. Attenuation is greater at the higher frequencies or shorter wavelengths.

While reflected echoes are much weaker than the transmitted pulses, the characteristics of their return to the source are similar to the characteristics of propagation. The strengths of these echoes are dependent upon the amount of transmitted energy striking the targets and the size and reflecting properties of the targets.

1305. Refraction

If the radar waves traveled in straight lines, the distance to the radar horizon would be dependent only on the power output of the transmitter and the height of the antenna. In other words, the distance to the radar horizon would be the same as that of the geometrical horizon for the antenna height. However, atmospheric density gradients bend radar rays as they travel to and from a target. This bending is called **refraction**.

The distance to the radar horizon does not limit the distance from which echoes may be received from targets. Assuming that adequate power is transmitted, echoes may be received from targets beyond the radar horizon if their reflecting surfaces extend above it. The distance to the radar horizon is the distance at which the radar rays pass tangent to the surface of the Earth.

The following formula, where h is the height of the antenna in feet, gives the theoretical distance to the radar horizon in nautical miles:

$$d = 1.22\sqrt{h} .$$

1306. Factors Affecting Radar Interpretation

Radar's value as a navigational aid depends on the navigator's understanding its characteristics and limitations. Whether measuring the range to a single reflective object or trying to discern a shoreline lost amid severe clutter, knowledge of the characteristics of the individual radar used are crucial. Some of the factors to be considered in interpretation are discussed below:

- **Resolution in Range.** In part A of Figure 1306a, a transmitted pulse has arrived at the second of two targets of insufficient size or density to absorb or reflect all of the energy of the pulse. While the pulse has traveled from the first to the second target, the echo from the first has traveled an equal distance in the opposite direction. At B, the transmitted pulse has continued on beyond the second target, and the two echoes are returning toward the transmitter. The distance between leading edges of the two echoes is twice the distance between targets. The correct distance will be shown on the scope, which is calibrated to show half the distance traveled out and back. At C the targets are closer together and the pulse length has been increased. The two echoes merge, and on the scope they will appear as a single, large target. At D the pulse length has been decreased, and the two echoes appear separated. The ability of a radar to separate targets close together on the same bearing is called **resolution in range**. It is related primarily to pulse length. The minimum distance between targets that can be distinguished as separate is half the pulse length. This (half the pulse length) is the apparent depth or thickness of a target presenting a flat perpendicular surface to the radar beam. Thus, several ships close together may appear as an island. Echoes from a number of small boats, piles, breakers, or even large ships close to the shore may blend with echoes from the shore, resulting in an incorrect indication of the position and shape of the shoreline.

- **Resolution in Bearing**. Echoes from two or more targets close together at the same range may merge to form a single, wider echo. The ability to separate targets close together at the same range is called **resolution in bearing**. Bearing resolution is a function of two variables: beam width and range to the targets. A narrower beam and a shorter distance to the objects both increase bearing resolution.

- **Height of Antenna and Target**. If the radar horizon is between the transmitting vessel and the target, the lower part of the target will not be visible. A large vessel may appear as a small craft, or a shoreline may appear at some distance inland.

- **Reflecting Quality and Aspect of Target**. Echoes from several targets of the same size may be quite different in appearance. A metal surface reflects radio waves more strongly than a wooden surface. A surface perpendicular to the beam returns a stronger echo than a non perpendicular one. A vessel seen broadside returns a stronger echo than one heading directly toward or away. Some surfaces absorb most radar energy rather that reflecting it.

- **Frequency**. As frequency increases, reflections occur from smaller targets.

Atmospheric noise, sea return, and precipitation complicate radar interpretation by producing **clutter**. Clutter is usually strongest near the vessel. Strong echoes can sometimes be detected by reducing receiver gain to eliminate weaker signals. By watching the repeater during several rotations of the antenna, the operator can often discriminate between clutter and a target even when the signal strengths from clutter and the target are equal. At each rotation, the signals from targets will remain relatively stationary on the display while those caused by clutter will appear at different locations on each sweep.

Another major problem lies in determining which features in the vicinity of the shoreline are actually represented by echoes shown on the repeater. Particularly in cases where a low lying shore is being scanned, there may be considerable uncertainty.

A related problem is that certain features on the shore will not return echoes because they are blocked from the radar beam by other physical features or obstructions. This factor in turn causes the chart-like image painted on the scope to differ from the chart of the area.

If the navigator is to be able to interpret the presentation on his radarscope, he must understand the characteristics of radar propagation, the capabilities of his radar set, the reflecting properties of different types of radar targets, and the ability to analyze his chart to determine which charted features are most likely to reflect the transmitted pulses or to be blocked. Experience gained during clear weather comparison between radar and visual images is invaluable.

Land masses are generally recognizable because of the steady brilliance of the relatively large areas painted on the PPI. Also, land should be at positions expected from the ship's navigational position. Although land masses are readily recognizable, the primary problem is the identification of specific land features. Identification of specific features can be quite difficult because of various factors, including distortion resulting from beam width and pulse length, and uncertainty as to just which charted features are reflecting the echoes.

Sand spits and smooth, clear beaches normally do not appear on the PPI at ranges beyond 1 or 2 miles because these targets have almost no area that can reflect energy back to the radar. Ranges determined from these targets are not reliable. If waves are breaking over a sandbar, echoes may be returned from the surf. Waves may, however, break well out from the actual shoreline, so that ranging on the surf may be misleading.

Mud flats and marshes normally reflect radar pulses only a little better than a sand spit. The weak echoes received at low tide disappear at high tide. Mangroves and other thick growth may produce a strong echo. Areas that are indicated as swamps on a chart, therefore, may return either strong or weak echoes, depending on the density type, and size of the vegetation growing in the area.

When sand dunes are covered with vegetation and are well back from a low, smooth beach, the apparent shoreline determined by radar appears as the line of the dunes rather than the true shoreline. Under some conditions, sand dunes may return strong echo signals because the combination of the vertical surface of the vegetation and the horizontal

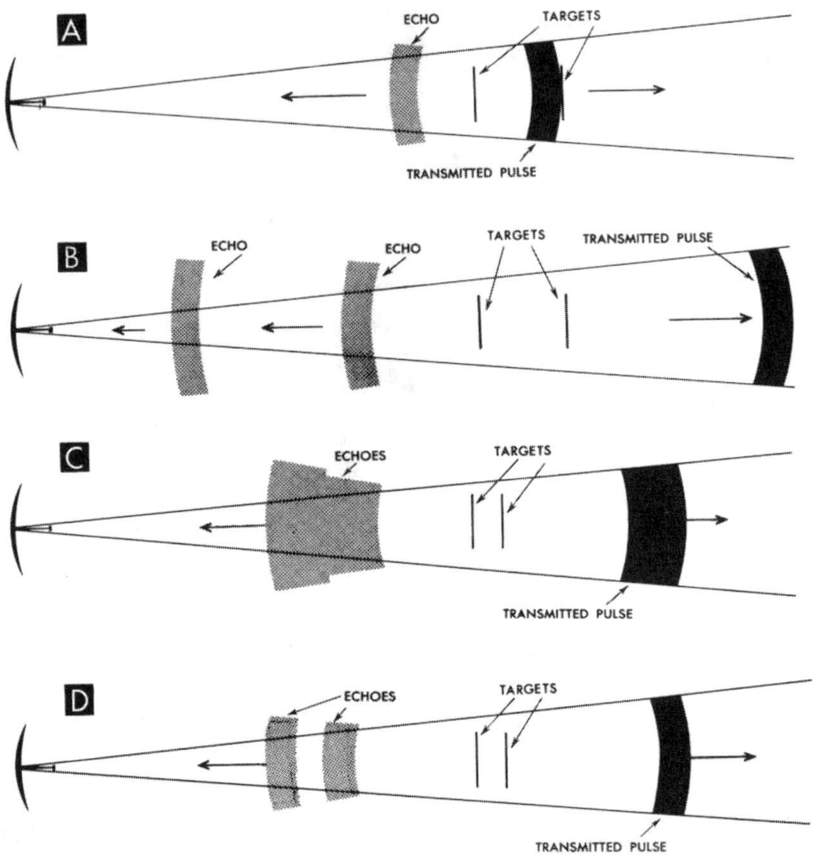

Figure 1306a. Resolution in range.

beach may form a sort of corner reflector.

Lagoons and inland lakes usually appear as blank areas on a PPI because the smooth water surface returns no energy to the radar antenna. In some instances, the sandbar or reef surrounding the lagoon may not appear on the PPI because it lies too low in the water.

Coral atolls and long chains of islands may produce long lines of echoes when the radar beam is directed perpendicular to the line of the islands. This indication is especially true when the islands are closely spaced. The reason is that the spreading resulting from the width of the radar beam causes the echoes to blend into continuous lines. When the chain of islands is viewed lengthwise, or obliquely, however, each island may produce a separate return. Surf breaking on a reef around an atoll produces a ragged, variable line of echoes.

One or two rocks projecting above the surface of the water, or waves breaking over a reef, may appear on the PPI.

If the land rises in a gradual, regular manner from the shoreline, no part of the terrain produces an echo that is stronger than the echo from any other part. As a result, a general haze of echoes appears on the PPI, and it is difficult to ascertain the range to any particular part of the land.

Blotchy signals are returned from hilly ground, because the crest of each hill returns a good echo although the valley beyond is in a shadow. If high receiver gain is used, the pattern may become solid except for the very deep shadows.

Low islands ordinarily produce small echoes. When thick palm trees or other foliage grow on the island, strong echoes often are produced because the horizontal surface of the water around the island forms a sort of corner reflector with the vertical surfaces of the trees. As a result, wooded islands give good echoes and can be detected at a much

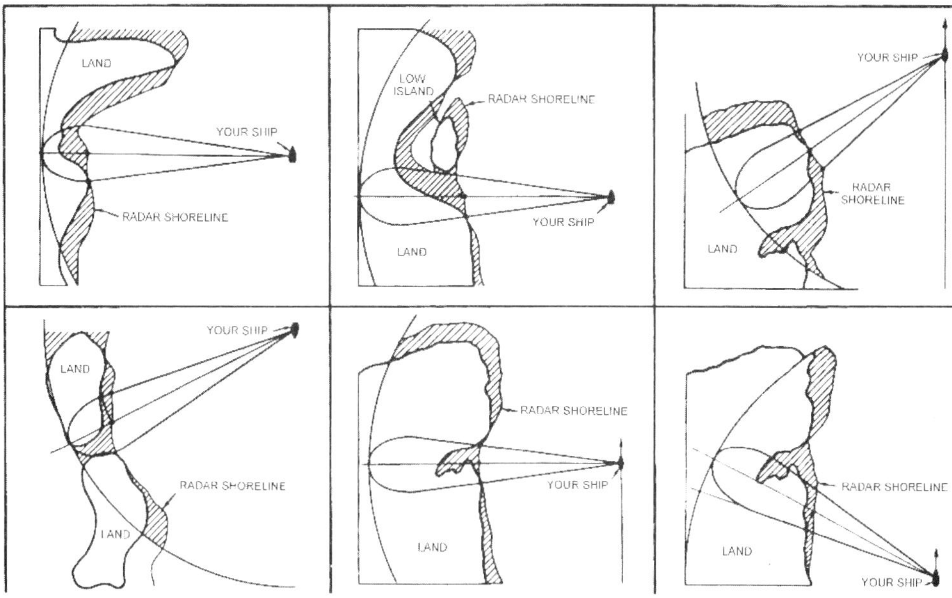

Figure 1306b. Effects of ship's position, beam width, and pulse length on radar shoreline.

greater range than barren islands.

Sizable land masses may be missing from the radar display because of certain features being blocked from the radar beam by other features. A shoreline which is continuous on the PPI display when the ship is at one position, may not be continuous when the ship is at another position and scanning the same shoreline. The radar beam may be blocked from a segment of this shoreline by an obstruction such as a promontory. An indentation in the shoreline, such as a cove or bay, appearing on the PPI when the ship is at one position, may not appear when the ship is at another position nearby. Thus, radar shadow alone can cause considerable differences between the PPI display and the chart presentation. This effect in conjunction with beam width and pulse length distortion of the PPI display can cause even greater differences.

The returns of objects close to shore may merge with the shoreline image on the PPI, because of distortion effects of horizontal beam width and pulse length. Target images on the PPI are distorted angularly by an amount equal to the effective horizontal beam width. Also, the target images always are distorted radially by an amount at least equal to one-half the pulse length (164 yards per microsecond of pulse length).

Figure 1306b illustrates the effects of ship's position, beam width, and pulse length on the radar shoreline. Because of beam width distortion, a straight, or nearly straight, shoreline often appears crescent-shaped on the PPI. This effect is greater with the wider beam widths. Note that this distortion increases as the angle between the beam axis and the shoreline decreases.

Figure 1306c illustrates the distortion effects of radar shadow, beam width, and pulse length. View A shows the actual shape of the shoreline and the land behind it. Note the steel tower on the low sand beach and the two ships at anchor close to shore. The heavy line in view B represents the shoreline on the PPI. The dotted lines represent the actual position and shape of all targets. Note in particular:

1. The low sand beach is not detected by the radar.
2. The tower on the low beach is detected, but it looks like a ship in a cove. At closer range the land would be detected and the cove-shaped area would begin to fill in; then the tower could not be seen without reducing the receiver gain.
3. The radar shadow behind both mountains. Distortion owing to radar shadows is responsible for more confusion than any other cause. The small island does not appear because it is in the radar shadow.
4. The spreading of the land in bearing caused by beam width distortion. Look at the upper shore of the peninsula. The shoreline distortion is greater to the west because the angle between the radar beam and the shore is smaller as the beam seeks out the more westerly shore.
5. Ship No. 1 appears as a small peninsula. Its return has merged with the land because of the beam width

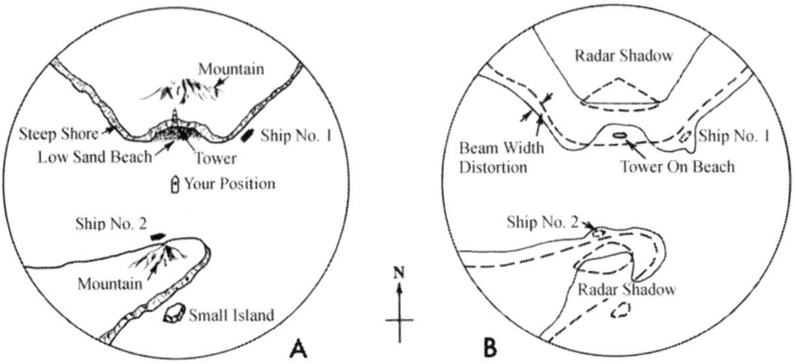

Figure 1306c. Distortion effects of radar shadow, beam width, and pulse length.

distortion.

6. Ship No. 2 also merges with the shoreline and forms a bump. This bump is caused by pulse length and beam width distortion. Reducing receiver gain might cause the ship to separate from land, provided the ship is not too close to the shore. The Fast Time Constant (FTC) control could also be used to attempt to separate the ship from land.

1307. Recognition of Unwanted Echoes

Indirect or false echoes are caused by reflection of the main lobe of the radar beam off ship's structures such as stacks and kingposts. When such reflection does occur, the echo will return from a legitimate radar contact to the antenna by the same indirect path. Consequently, the echo will appear on the PPI at the bearing of the reflecting surface. As shown in Figure 1307a, the indirect echo will appear on the PPI at the same range as the direct echo received, assuming that the additional distance by the indirect path is negligible.

Characteristics by which indirect echoes may be recognized are summarized as follows:

1. Indirect echoes will often occur in shadow sectors.
2. They are received on substantially constant bearings, although the true bearing of the radar contact may change appreciably.
3. They appear at the same ranges as the corresponding direct echoes.
4. When plotted, their movements are usually abnormal.
5. Their shapes may indicate that they are not direct echoes.

Side-lobe effects are readily recognized in that they produce a series of echoes (Figure 1307b) on each side of the main lobe echo at the same range as the latter. Semicircles, or even complete circles, may be produced. Because of the low energy of the side-lobes, these effects will normally occur only at the shorter ranges. The effects may be minimized or eliminated, through use of the gain and anti-clutter controls. Slotted wave guide antennas have largely eliminated the side-lobe problem.

Multiple echoes may occur when a strong echo is received from another ship at close range. A second or third or more echoes may be observed on the radarscope at double, triple, or other multiples of the actual range of the radar contact (Figure 1307c).

Second-trace echoes (multiple-trace echoes) are echoes received from a contact at an actual range greater than the radar range setting. If an echo from a distant target is received after the following pulse has been transmitted, the echo will appear on the radarscope at the correct bearing but not at the true range. Second-trace echoes are unusual, except under abnormal atmospheric conditions, or conditions under which super-refraction is present. Second-trace echoes may be recognized through changes in their positions on the radarscope in changing the pulse repetition rate (PRR); their hazy, streaky, or distorted shape; and the erratic movements on plotting.

As illustrated in Figure 1307d, a target return is detected on a true bearing of 090° at a distance of 7.5 miles. On changing the PRR from 2,000 to 1,800 pulses per second, the same target is detected on a bearing of 090° at a distance of 3 miles (Figure 1307e). The change in the position of the return indicates that the return is a second-trace echo. The actual distance of the target is the distance as indicated on the PPI plus half the distance the radar wave travels between pulses.

Electronic interference effects, such as may occur

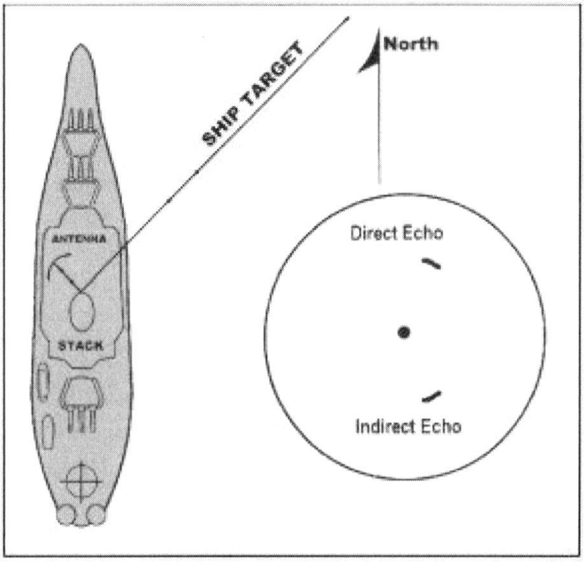

Figure 1307a. Indirect echo.

when near another radar operating in the same frequency band as that of the observer's ship, is usually seen on the PPI as a large number of bright dots either scattered at random or in the form of dotted lines extending from the center to the edge of the PPI.

Interference effects are greater at the longer radar range scale settings. The interference effects can be distinguished easily from normal echoes because they do not appear in the same places on successive rotations of the antenna.

Stacks, masts, samson posts, and other structures, may cause a reduction in the intensity of the radar beam beyond these obstructions, especially if they are close to the radar antenna. If the angle at the antenna subtended by the obstruction is more than a few degrees, the reduction of the intensity of the radar beam beyond the obstruction may produce a blind sector. Less reduction in the intensity of the beam beyond the obstructions may produce shadow sectors. Within a shadow sector, small targets at close range may not be detected, while larger targets at much greater ranges will appear.

Spoking appears on the PPI as a number of spokes or radial lines. Spoking is easily distinguished from interference effects because the lines are straight on all range-scale settings, and are lines rather than a series of dots.

The spokes may appear all around the PPI, or they may be confined to a sector. If spoking is confined to a narrow sector, the effect can be distinguished from a Ramark signal of similar appearance through observation of the steady relative bearing of the spoke in a situation where the bearing of the Ramark signal should change. Spoking indicates a need for maintenance or adjustment. The PPI display may appear as normal sectors alternating with dark sectors. This is usually due to the automatic frequency control being out of adjustment. The appearance of serrated range rings indicates a need for maintenance.

After the radar set has been turned on, the display may not spread immediately to the whole of the PPI because of static electricity inside the CRT. Usually, the static electricity effect, which produces a distorted PPI display, lasts no longer than a few minutes.

Hour-glass effect appears as either a constriction or expansion of the display near the center of the PPI. The expansion effect is similar in appearance to the expanded center display. This effect, which can be caused by a nonlinear time base or the sweep not starting on the indicator at the same instant as the transmission of the pulse, is most apparent when in narrow rivers or close to shore.

The echo from an overhead power cable can be wrongly identified as the echo from a ship on a steady bearing and decreasing range. Course changes to avoid the contact are ineffective; the contact remains on a steady bearing, decreasing range. This phenomenon is particularly apparent for the power cable spanning the Straits of Messina.

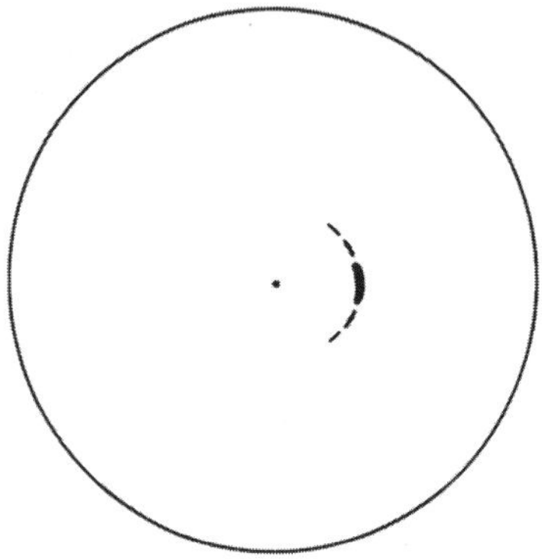

Figure 1307b. Side-lobe effects.

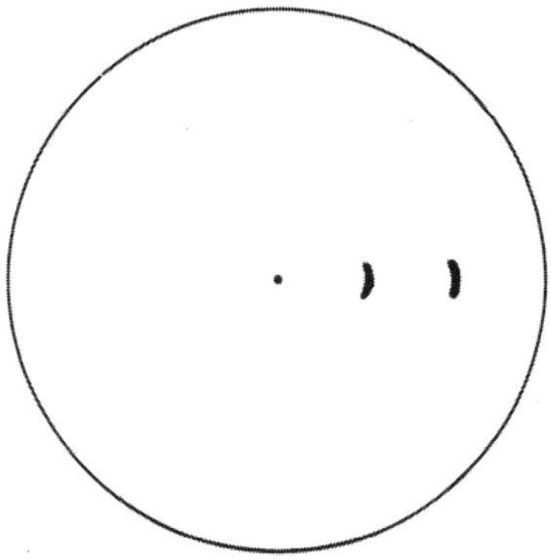

Figure 1307c. Multiple echoes.

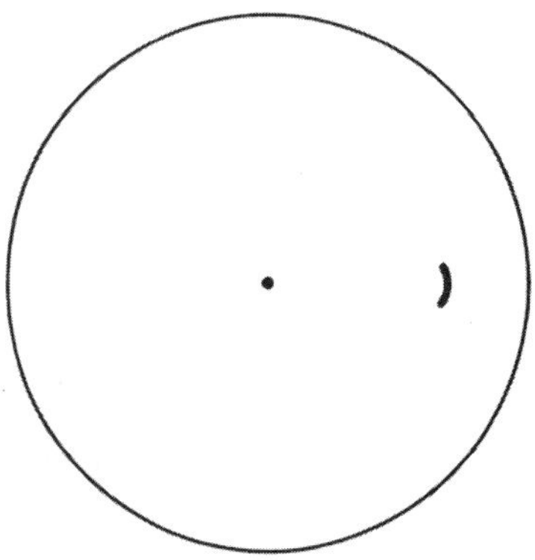

Figure 1307d. Second-trace echo on 12-mile range scale.

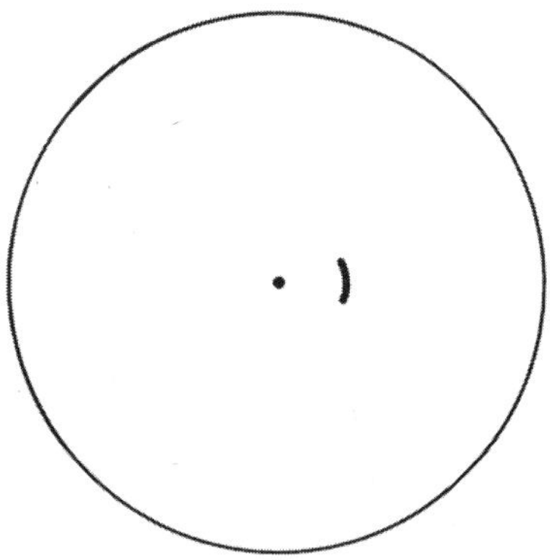

Figure 1307e. Position of second-trace echo on 12-mile range scale after changing PRR.

1308. Aids to Radar Navigation

Radar navigation aids help identify radar targets and increase echo signal strength from otherwise poor radar targets.

Buoys are particularly poor radar targets. Weak, fluctuating echoes received from these targets are easily lost in the sea clutter. To aid in the detection of these targets, **radar reflectors**, designated corner reflectors, may be used. These reflectors may be mounted on the tops of buoys or designed into the structure.

Each corner reflector, shown in Figure 1308a, consists of three mutually perpendicular flat metal surfaces. A radar wave striking any of the metal surfaces or plates will be reflected back in the direction of its source. Maximum energy will be reflected back to the antenna if the axis of the radar beam makes equal angles with all the metal surfaces. Frequently, corner reflectors are assembled in clusters to maximize the reflected signal.

Although radar reflectors are used to obtain stronger echoes from radar targets, other means are required for more positive identification of radar targets. **Radar beacons** are transmitters operating in the marine radar frequency band, which produce distinctive indications on the radarscopes of ships within range of these beacons. There are two general classes of these beacons: **racons**, which provide both bearing and range information to the target, and **ramarks** which provide bearing information only. However, if the ramark installation is detected as an echo on the radarscope, the range will be available also.

A racon is a radar transponder which emits a characteristic signal when triggered by a ship's radar. The signal may be emitted on the same frequency as that of the triggering radar, in which case it is superimposed on the ship's radar display automatically. The signal may be emitted on a separate frequency, in which case to receive the signal the ship's radar receiver must be tuned to the beacon frequency, or a special receiver must be used. In either case, the PPI will be blank except for the beacon signal. However, the only racons in service are "in band"

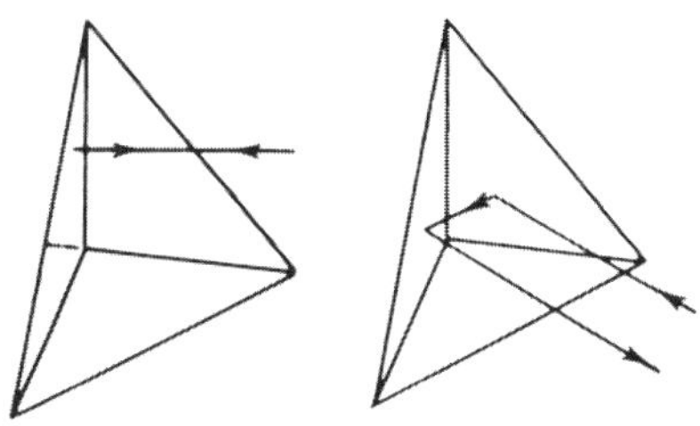

Figure 1308a. Corner reflectors.

beacons which transmit in one of the marine radar bands, usually only the 3-centimeter band.

The racon signal appears on the PPI as a radial line originating at a point just beyond the position of the radar beacon, or as a Morse code signal (Figure 1308b) displayed radially from just beyond the beacon.

A ramark is a radar beacon which transmits either continuously or at intervals. The latter method of transmission is used so that the PPI can be inspected without any clutter introduced by the ramark signal on the scope. The ramark signal as it appears on the PPI is a radial line from the center. The radial line may be a continuous narrow line, a broken line (Figure 1308c), a series of dots, or a series of dots and dashes.

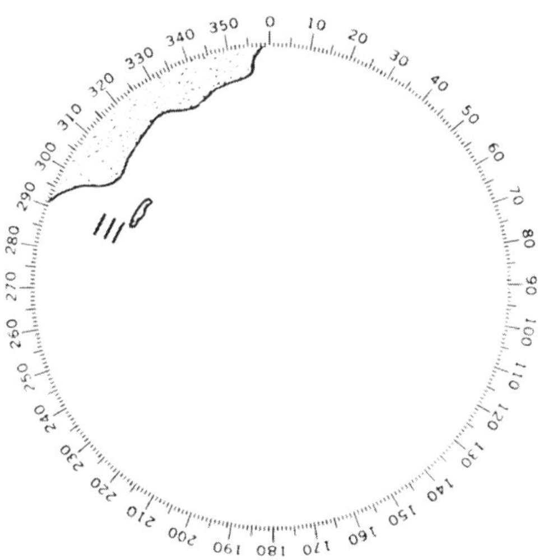

Figure 1308b. Coded racon signal.

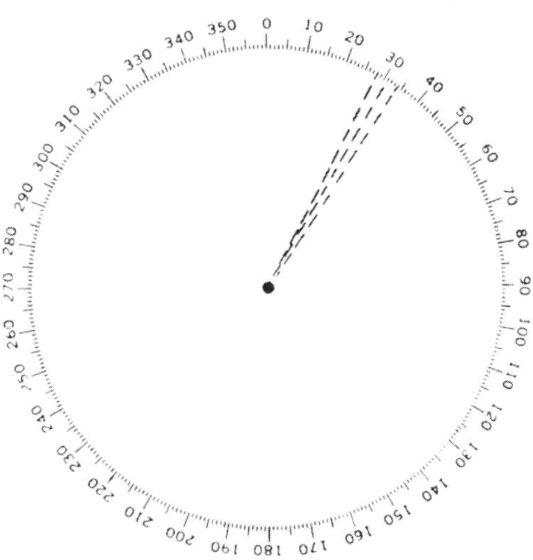

Figure 1308c. Ramark appears as broken radial line.

RADAR PILOTING

1309. Introduction

When navigating in restricted waters, a mariner most often relies on visual piloting to provide the accuracy required to ensure ship safety. Visual piloting, however, requires clear weather; often, mariners must navigate through fog. When weather conditions render visual piloting impossible on a vessel not equipped with ECDIS, radar navigation provides a method of fixing a vessel's position with sufficient accuracy to allow safe passage. See Chapter 8 for a detailed discussion of integrating radar into a piloting procedure.

1310. Fix by Radar Ranges

Since radar can more accurately determine ranges than bearings, the most accurate radar fixes result from measuring and plotting ranges to two or more objects. Measure objects directly ahead or astern first; measure objects closest to the beam last.

This procedure is the opposite to that recommended for taking visual bearings, where objects closest to the beam are measured first; however, both recommendations rest on the same principle. When measuring objects to determine a line of position, measure first those which have the greatest rate of change in the quantity being measured; measure last those which have the least rate of change. This minimizes measurement time delay errors. Since the range of those objects directly ahead or astern of the ship changes more rapidly than those objects located abeam, we measure ranges to objects ahead or astern first.

Record the ranges to the navigation aids used and lay the resulting range arcs down on the chart. Theoretically, these lines of position should intersect at a point coincident with the ship's position at the time of the fix.

Though verifying soundings is always a good practice in all navigation scenarios, its importance increases when piloting using only radar. Assuming proper operation of the fathometer, soundings give the navigator invaluable information on the reliability of his fixes.

1311. Fix by Range and Bearing to One Object

Visual piloting requires bearings from at least two objects; radar, with its ability to determine both bearing and range from one object, allows the navigator to obtain a fix where only a single navigation aid is available. An example of using radar in this fashion occurs in approaching a harbor whose entrance is marked with a single, prominent object such as Chesapeake Light at the entrance of the Chesapeake Bay. Well beyond the range of any land-based visual navigation aid, and beyond the visual range of the light itself, a shipboard radar can detect the light and provide bearings and ranges for the ship's piloting party. Care must be taken that fixes are not taken on any nearby stationary vessel.

This methodology is limited by the inherent inaccuracy associated with radar bearings; typically, a radar bearing is accurate to within about 5° of the true bearing. Therefore, the navigator must carefully evaluate the resulting position, possibly checking it with a sounding. If a visual bearing is available from the object, use that bearing instead of the radar bearing when laying down the fix. This illustrates the basic concept discussed above: radar ranges are inherently more accurate than radar bearings. One must also be aware that if the radar is gyro stabilized and there is a gyro error of more than a degree or so, radar bearings will be in error by that amount.

Prior to using this method, the navigator must ensure that he has correctly identified the object from which the bearing and range are to be taken. Using only one navigation aid for both lines of position can lead to disaster if the navigation aid is not properly identified.

1312. Fix Using Tangent Bearings and Range

This method combines bearings tangent to an object with a range measurement from some point on that object. The object must be large enough to provide sufficient bearing spread between the tangent bearings; often an island or peninsula works well. Identify some prominent feature of the object that is displayed on both the chart and the radar display. Take a range measurement from that feature and plot it on the chart. Then determine the tangent bearings to the feature and plot them on the chart.

Steep-sided features work the best. Tangents to low, sloping shorelines will seriously reduce accuracy, as will tangent bearings in areas of excessively high tides, which can change the location of the apparent shoreline by many meters.

1313. Fix by Radar Bearings

The inherent inaccuracy of radar bearings discussed above makes this method less accurate than fixing position by radar range. Use this method to plot a position quickly on the chart when approaching restricted waters to obtain an approximate ship's position for evaluating radar targets to use for range measurements. Unless no more accurate method is available, this method is not suitable while piloting in restricted waters.

1314. Fischer Plotting

In Fischer plotting, the navigator adjusts the scale of

the radar to match the scale of the chart in use. Then he places a clear plastic disk, sized to the radar, on the center of the radar screen and quickly traces the shape of land and location of any navigation aids onto the plastic overlay with a grease pencil. Taking the plastic with the tracings on it to the chart, he matches the features traced from the radar with the chart's features. A hole in the center of the plastic allows the navigator to mark the position of the ship at the time the tracing was done.

RASTER RADARS

1315. Basic Description

Conventional PPI-display radars use a circular **Cathode Ray Tube (CRT)** to direct an electron beam at a screen coated on the inside with phosphorus, which glows when illuminated by the beam. Internal circuitry forms the beam such that a "sweep" is indicated on the face of the PPI. This sweep is timed to coincide with the sweep of the radar's antenna. A return echo is added to the sweep signal so that the screen is more brightly illuminated at a point corresponding to the bearing and range of the target that returned the echo.

The raster radar also employs a cathode ray tube; however, the end of the tube upon which the picture is formed is rectangular, not circular as in the PPI display. The raster radar does not produce its picture from a circular sweep; it utilizes a liner scan in which the picture is "drawn," line by line, horizontally across the screen. As the sweep moves across the screen, the electron beam from the CRT illuminates the picture elements, or **pixels**, on the screen. A pixel is the smallest area of a display that can be excited individually.

In order to produce a sufficiently high resolution, larger raster radars require over 1 million pixels per screen combined with an update rate of 60 or more scans per second. Processing such a large number of pixel elements requires a rather sophisticated computer. One way to lower cost is to slow down the required processing speed. This speed can be lowered to approximately 30 frames per second before the picture develops a noticeable flicker, but the best radars have scan rates of at least 60 scans per second.

Further cost reduction can be gained by using an **interlaced display**. An interlaced display does not draw the entire picture in one pass. On the first pass, it draws every other line; it draws the remaining lines on the second pass. This type of display reduces the number of screens that have to be drawn per unit of time by a factor of two; however, if the two pictures are misaligned, the picture will appear to jitter.

CHAPTER 14

ELECTRONIC CHARTS

INTRODUCTION

1400. The Importance of Electronic Charts

Since the beginning of maritime navigation, the desire of the navigator has always been to answer a fundamental question: "Where, exactly, is my vessel?" To answer that question, the navigator was forced to continually take fixes on celestial bodies, on fixed objects ashore, or using radio signals, and plot the resulting lines of position as a fix on a paper chart. Only then could he begin to assess the safety of the ship and its progress toward its destination. He spent far more time taking fixes, working out solutions, and plotting the results than on making assessments, and the fix only told him where the ship was at the time that fix was taken, not where the vessel was some time later when the assessment was made. He was always "behind the vessel." On the high seas this is of little import. Near shore, it becomes vitally important.

Electronic charts automate the process of integrating real-time positions with the chart display and allow the navigator to continuously assess the position and safety of the vessel. Further, the GPS/DGPS fixes are far more accurate and taken far more often than any navigator ever could. A good piloting team is expected to take and plot a fix every three minutes. An electronic chart system can do it once per second to a standard of accuracy at least an order of magnitude better.

Electronic charts also allow the integration of other operational data, such as ship's course and speed, depth soundings, and radar data into the display. Further, they allow automation of alarm systems to alert the navigator to potentially dangerous situations well in advance of a disaster.

Finally, the navigator has a complete picture of the instantaneous situation of the vessel and all charted dangers in the area. With a radar overlay, the tactical situation with respect to other vessels is clear as well. This chapter will discuss the various types of electronic charts, the requirements for using them, their characteristics, capabilities and limitations.

1401. Terminology

Before understanding what an electronic chart is and what it does, one must learn a number of terms and definitions. We must first make a distinction between official and unofficial charts. Official charts are those, and only those, produced by a government hydrographic office (HO). Unofficial charts are produced by a variety of private companies and may or may not meet the same standards used by HO's for data accuracy, currency, and completeness.

An **electronic chart system (ECS)** is a commercial electronic chart system not designed to satisfy the regulatory requirements of the IMO Safety of Life at Sea (SOLAS) convention. ECS is an aid to navigation and when used on SOLAS regulated vessels is to be used in conjunctions with corrected paper charts.

An **electronic chart display and information system (ECDIS)** is an electronic chart system which satisfies the IMO SOLAS convention carriage requirements for corrected paper charts when used with an ENC or its functional equivalent (e.g. NIMA Digital Nautical Chart.)

An **electronic chart (EC)** is any digitized chart intended for display on a computerized navigation system.

An **electronic chart data base (ECDB)** is the digital database from which electronic charts are produced.

An **electronic navigational chart (ENC)** is an electronic chart issued by a national hydrographic authority designed to satisfy the regulatory requirements for chart carriage.

The **electronic navigation chart database (ENCDB)** is the hydrographic database from which the ENC is produced.

The **system electronic navigation chart (SENC)** is the database created by an ECDIS from the ENC data.

A **raster navigation chart (RNC)**is a raster-formatted chart produced by a national hydrographic office.

A **raster chart display system (RCDS)** is a system which displays official raster-formatted charts on an ECDIS system. Raster charts cannot take the place of paper charts because they lack key features required by the IMO, so that when an ECDIS uses raster charts it operates in the ECS mode.

Overscale and **underscale** refer to the display of electronic chart data at too large and too small a scale, respectively. In the case of overscale, the display is "zoomed in" too close, beyond the standard of accuracy to which the data was digitized. Underscale indicates that larger scale data is available for the area in question. ECDIS provides a warning in either case.

Raster chart data is a digitized "picture" of a chart comprised of millions of "picture elements" or "pixels." All data is in one layer and one format. The video display simply reproduces the picture from its digitized data file. With raster data, it is difficult to change individual elements of the chart since they are not separated in the data file. Raster data files tend to be large, since a data point with associated color and intensity values must be entered for every pixel on the chart.

Vector chart data is data that is organized into many separate files or layers. It contains graphics files and programs to produce certain symbols, points, lines, and areas with associated colors, text, and other chart elements. The programmer can change individual elements in the file and link elements to additional data. Vector files of a given area are a fraction the size of raster files, and at the same time much more versatile. The navigator can selectively display vector data, adjusting the display according to his needs. Vector data supports the computation of precise distances between features and can provide warnings when hazardous situations arise.

1402. Components of ECS's and ECDIS's

The terms ECS and ECDIS encompasses many possible combinations of equipment and software designed for a variety of navigational purposes. In general, the following components comprise an ECS or ECDIS.

- **Computer processor, software, and network:** These subsystems control the processing of information from the vessel's navigation sensors and the flow of information between various system components. Electronic positioning information from GPS or Loran C, contact information from radar, and digital compass data, for example, can be integrated with the electronic chart data.

- **Chart database:** At the heart of any ECS lies a database of digital charts, which may be in either raster or vector format. It is this dataset, or a portion of it, that produces the chart seen on the display screen.

- **System display:** This unit displays the electronic chart and indicates the vessel's position on it, and provides other information such as heading, speed, distance to the next waypoint or destination, soundings, etc. There are two modes of display, **relative** and **true**. In the relative mode the ship remains fixed in the center of the screen and the chart moves past it. This requires a lot of computer power, as all the screen data must be updated and re-drawn at each fix. In true mode, the chart remains fixed and the ship moves across it. The display may also be north-up or course-up, according to the availability of data from a heading sensor such as a digital compass.

- **User interface:** This is the user's link to the system. It allows the navigator to change system parameters, enter data, control the display, and operate the various functions of the system. Radar may be integrated with the ECDIS or ECS for navigation or collision avoidance, but is not required by SOLAS regulations.

1403. Legal Aspects of Using Electronic Charts

Requirements for carriage of charts are found in SOLAS Chapter V, which states in part: "All ships shall carry adequate and up-to-date charts... necessary for the intended voyage." As electronic charts have developed and the supporting technology has matured, regulations have been adopted internationally to set standards for what constitutes a "chart" in the electronic sense, and under what conditions such a chart will satisfy the chart carriage requirement.

An extensive body of rules and regulations controls the production of ECDIS equipment, which must meet certain high standards of reliability and performance. By definition, **only an ECDIS can replace a paper chart.** No system which is not an ECDIS relieves the navigator of the responsibility of maintaining a plot on a corrected paper chart. Neither can the presence of an electronic chart system substitute for good judgement, sea sense, and taking all reasonable precautions to ensure the safety of the vessel and crew.

An electronic chart system should be considered as an aid to navigation, one of many the navigator might have at his disposal to help ensure a safe passage. While possessing revolutionary capabilities, it must be considered as a tool, not an infallible answer to all navigational problems. The rule for the use of electronic charts is the same as for all other aids to navigation: The prudent navigator will never rely completely on any single one.

CAPABILITIES AND PERFORMANCE STANDARDS

1404. ECDIS Performance Standards

The specifications for ECDIS consist of a set of interrelated standards from three organizations, the International Maritime Organization (IMO), the International Hydrographic Organization (IHO), and the International Electrotechnical Commission (IEC). The IMO published a resolution in November 1995 to establish performance standards for the general functionality of ECDIS, and to define the conditions for its replacement of paper charts. It consisted of a 15-section annex and 5 original appendices. Appendix 6 was adopted in 1996 to define the backup requirements for ECDIS. Appendix 7 was adopted in 1998 to define the operation of ECDIS in a raster chart mode. Previous standards related only to vector data.

The IMO performance standards refer to IHO Special

Publication S-52 for specification of technical details pertaining to the ECDIS display. Produced in 1997, the 3rd edition of S-52 includes appendices specifying the issue, updating, display, color, and symbology of official electronic navigational charts (ENC), as well as a revised glossary of ECDIS-related terms. The IMO performance standards also refer to IEC International Standard 61174 for the requirements of type approval of an ECDIS. Published in 1998, the IEC standard defines the testing methods and required results for an ECDIS to be certified as compliant with IMO standards. Accordingly, the first ECDIS was given type approval by Germany's classification society (BSH) in 1999. Since then, several other makes of ECDIS have gained type approval by various classification societies.

The IMO performance standards specify the following general requirements: Display of government-authorized vector chart data including an updating capability; enable route planning, route monitoring, manual positioning, and continuous plotting of the ship's position; have a presentation as reliable and available as an official paper chart; provide appropriate alarms or indications regarding displayed information or malfunctions; and permit a mode of operation with raster charts similar to the above standards.

The performance standards also specify additional functions, summarized as follows:

- Display of system information in three selectable levels of detail
- Means to ensure correct loading of ENC data and updates
- Apply updates automatically to system display
- Protect chart data from any alteration
- Permit display of update content
- Store updates separately and keep records of application in system
- Indicate when user zooms too far in or out on a chart (over- or under-scale) or when a larger scale chart is available in memory
- Permit the overlay of radar image and ARPA information onto the display
- Require north-up orientation and true motion mode, but permit other combinations
- Use IHO-specified resolution, colors and symbols
- Use IEC-specified navigational elements and parameters (range & bearing marker, position fix, own ship's track and vector, waypoint, tidal information, etc.)
- Use specified size of symbols, letters and figures at scale specified in chart data
- Permit display of ship as symbol or in true scale
- Display route planning and other tasks
- Display route monitoring
- Permit display to be clearly viewed by more than one user in day or night conditions
- Permit route planning in straight and curved segments and adjustment of waypoints
- Display a route plan in addition to the route selected for monitoring
- Permit track limit selection and display an indication if track limit crosses a safety contour or a selected prohibited area
- Permit display of an area away from ship while continuing to monitor selected route
- Give an alarm at a selectable time prior to ship crossing a selected safety contour or prohibited area
- Plot ship's position using a continuous positioning system with an accuracy consistent with the requirements of safe navigation
- Identify selectable discrepancy between primary and secondary positioning system
- Provide an alarm when positioning system input is lost
- Provide an alarm when positioning system and chart are based on different geodetic datums
- Store and provide for replay the elements necessary to reconstruct navigation and verify chart data in use during previous 12 hours
- Record the track for entire voyage with at least four hour time marks
- Permit accurate drawing of ranges and bearings not limited by display resolution
- Require system connection to continuous position-fixing, heading and speed information
- Neither degrade nor be degraded by connection to other sensors
- Conduct on-board tests of major functions with alarm or indication of malfunction
- Permit normal functions on emergency power circuit
- Permit power interruptions of up to 45 seconds without system failure or need to reboot
- Enable takeover by backup unit to continue navigation if master unit fails,

Before an IMO-compliant ECDIS can replace paper charts on vessels governed by SOLAS regulations, the route of the intended voyage must be covered completely by ENC data, that ENC data must include the latest updates, the ECDIS installation must be IMO-compliant including the master-slave network with full sensor feed to both units, and the national authority of the transited waters must allow for paperless navigation through published regulations. The latter may also include requirements for certified training in the operational use of ECDIS.

The first type approval was earned in 1999 and since the finalization of the standards in 1998, many manufacturers of ECDIS equipment have gained such certification.

The certifying agency issues a certificate valid for two years. For renewal, a survey is conducted to ensure that systems, software versions, components and materials used comply with type-approved documents and to review possible changes in design of systems, software versions, components, materials performance, and make sure that such changes do not affect the type approval granted.

Manufacturers have been willing to provide type-approved ECDIS to vessel operators, but in a non-compliant installation. Without the geographical coverage of ENC data, the expensive dual-network installation required by ECDIS will not eliminate the requirement to carry a corrected portfolio of paper charts. These partial installations range from approved ECDIS software in a single PC, to ECDIS with its IEC-approved hardware. In these instances, plotting on paper charts continues to be the primary means of navigation. As more ENC data and updates become available, and as governments regulate paperless transits, vessel operators are upgrading their installations to meet full IMO compliance and to make ECDIS the primary means of navigation.

1405. ECS Standards

Although the IMO has declined to issue guidelines on ECS, the Radio Technical Commission for Maritime Services (RTCM) in the United States developed a voluntary, industry-wide standard for ECS. Published in December 1994, the RTCM Standard called for ECS to be capable of executing basic navigational functions, providing continuous plots of own ship position, and providing appropriate indicators with respect to information displayed. The RTCM ECS Standard allows the use of either raster or vector data, and includes the requirement for simple and reliable updating of information, or an indication that the electronic chart information has changed.

In November 2001, RTCM published Version 2.1 of the "RTCM Recommended Standards for Electronic Chart Systems." This updated version is intended to better define requirements applicable to various classes of vessels operating in a variety of areas. Three general classes of vessels are designated:

Large commercial vessels (oceangoing ships)
Small commercial vessels (tugs, research vessels. etc.)
Smaller craft (yachts, fishing boats, etc.)

The intent is that users, manufacturers, and regulatory authorities will have a means of differentiating between the needs of various vessels as relates to ECS. In concept, an ECS meeting the minimum requirements of the RTCM standard should reduce the risk of incidents and improve the efficiency of navigating for many types of vessels.

However, unlike IMO-compliant ECDIS, an ECS is not intended to comply with the up-to-date chart requirements of SOLAS. As such, an ECS must be considered as a single aid to navigation, and should always be used with a corrected chart from a government-authorized hydrographic office.

Initially, IMO regulations require the use of vector data in an ECDIS; raster data does not have the flexibility needed to do what the ECDIS must do. But it soon became clear that the hydrographic offices of the world would not be able to produce vector data for any significant part of the world for some years. Meanwhile, commercial interests were rasterizing charts as fast as they could for the emerging electronic chart market, and national hydrographic offices began rasterizing their own inventories to meet public demand. The result was a rather complete set of raster data for the most heavily travelled waters of the world, while production of man-power intensive vector data lagged far behind. IMO regulations were then amended to allow ECDIS to function in an RCDS mode using official raster data in conjunction with an appropriate portfolio of corrected paper charts. Nations may issue regulations authorizing the use of RCDS and define what constitutes an appropriate folio of paper charts for use in their waters.

In general, an ECS is not designed to read and display the S-57 format, and does not meet the performance standards of either ECDIS or RCDS. But an ECDIS can operate in ECS mode when using raster charts or when using non-S-57 vector charts. When a type-approved ECDIS is installed without being networked to a backup ECDIS, or when it is using non-official ENC data, or ENC data without updates, it can be said to be operating in an ECS mode, and as such cannot be used as a substitute for official, corrected paper charts.

1406. Display Characteristics

While manufacturers of electronic chart systems have designed their own proprietary colors and symbols, the IMO Performance Standard requires that all IMO approved ECDIS follow the International Hydrographic Organization (IHO) Color & Symbol Specifications. These specifications are embodied in the ECDIS Presentation Library. Their development was a joint effort between Canada and Germany during the 1990s. In order for ECDIS to enhance the safety of navigation, every detail of the display should be clearly visible, unambiguous in its meaning, and uncluttered by superfluous information. The unofficial ECS's continue to be free to develop independent of IHO control. In general they seek to emulate the look of the traditional paper chart.

To reduce clutter, the IMO Standard lays down a permanent display base of essentials such as depths, aids to navigation, shoreline, etc., making the remaining information selectable. The navigator may then select only what is essential for the navigational task at hand. A black-background display for night use provides good color contrast without compromising the mariner's night vision. Similarly, a "bright sun" color table is designed to output maximum luminance in order to be daylight visible, and the colors for details such as buoys are made as contrasting as possible.

The symbols for ECDIS are based on the familiar paper chart symbols, with some optional extras such as simplified buoy symbols that show up better at night. Since the ECDIS can be customized to each ship's requirements, new symbols were added such as a highlighted, mariner selectable, safety contour and a prominent isolated danger symbol.

The Presentation Library is a set of colors and symbols together with rules relating them to the digital data of the ENC, and procedures for handling special cases, such as priorities for the display of overlapping objects. Every feature in the ENC is first passed through the look-up table of the Presentation Library that either assigns a symbol or line style immediately, or, for complex cases, passes the object to a symbology procedure. Such procedures are used for objects like lights, which have so many variations that a look-up table for their symbolization would be too long. The Presentation Library includes a Chart 1, illustrating the symbology. Given the IHO S-57 data standards and S-52 display specifications, a waterway should look the same no matter which hydrographic office produced the ENC, and no matter which manufacturer built the ECDIS.

The overwhelming advantage of the vector-based ECDIS display is its ability to remove cluttering information not needed at a given time. By comparison, the paper chart and its raster equivalent is an unchangeable diagram. A second advantage is the ability to orient the display course-up when this is convenient, while the text remains screen-up.

Taking advantage of affordable yet high-powered computers, some ECDIS's now permit a split screen display, where mode of motion, orientation and scale are individually selectable on each panel. This permits, for example, a north-up small-scale overview in true motion alongside a course-up large-scale view in relative motion. Yet another display advantage occurs with zooming, in that symbols and text describing areas center themselves automatically in whatever part of the area appears on the screen. None of these functions are possible with raster charts.

The display operates by a set of rules, and data is arranged hierarchically, For example, where lines overlap, the less important line is not drawn. A more complex rule always places text at the same position relative to the object it applies to, no matter what else may be there. Since a long name or light description will often over-write another object, the only solution is to zoom in until the objects separate from each other. Note that because text causes so much clutter, and is seldom vital for safe navigation, it is written automatically when the object it refers to is on the display, but is an option under the "all other information" display level.

Flexibility in display scale requires some indication of distance to objects seen on the display. Some manufacturers use the rather restrictive but familiar radar range rings to provide this, while another uses a line symbol keyed to data's original scale. The ECDIS design also includes a one-mile scalebar at the side of the display, and an optionally displayed course and speed made good vector for own ship. There may be a heading line leading from the vessel's position indicating her future track for one minute, three minutes, or some other selectable time.

To provide the option of creating manual chart corrections, ECDIS includes a means of drawing lines, adding text and inserting stored objects on the display. These may be saved as user files, called up from a subdirectory, and edited on the display. Once loaded into the SENC, the objects may be selected or de-selected just as with other objects of the SENC.

Display options for ECDIS include transfer of ARPA-acquired targets and radar image overlay. IMO standards for ECDIS require that the operator be able to deselect the radar picture from the chart with a single operator action for fast "uncluttering" of the chart presentation.

1407. Units, Data Layers and Calculations

ECDIS uses the following units of measure:

- **Position:** Latitude and longitude will be shown in degrees, minutes, and decimal minutes, normally based on WGS-84 datum.
- **Depth:** Depths will be indicated in meters and decimeters.
- **Height:** Meters
- **Distance:** Nautical miles and tenths, or meters
- **Speed:** Knots and tenths

ECDIS requires data layers to establish a priority of data displayed. The minimum number of information categories required and their relative priority from highest to lowest are listed below:

- ECDIS warnings and messages
- Hydrographic office data
- *Notice to Mariners* information
- Hydrographic office cautions
- Hydrographic office color-fill area data
- Hydrographic office on demand data
- Radar information
- User's data
- Manufacturer's data
- User's color-fill area data
- Manufacturer's color-fill area data

As a minimum, an ECDIS system must be able to perform the following calculations and conversions:

- Geographical coordinates to display coordinates, and display coordinates to geographical coordinates.
- Transformation from local datum to WGS-84.
- True distance and azimuth between two geographical positions.
- Geographic position from a known position given distance and azimuth.
- Projection calculations such as great circle and rhumb line courses and distances.

1408. Warnings and Alarms

Appendix 5 of the IMO Performance Standard specifies that ECDIS must monitor the status of its systems continuously, and must provide alarms and indications for certain functions if a condition occurs that requires immediate attention. Indications may be either visual or audible. An alarm must be audible and may be visual as well.

An alarm is required for the following:

- Exceeding cross-track limits
- Crossing selected safety contour
- Deviation from route
- Position system failure
- Approaching a critical point
- Chart on different geodetic datum from positioning system

An alarm or indication is required for the following:

- Largest scale for alarm (indicates that presently loaded chart is too small a scale to activate anti-grounding feature)
- Area with special conditions (means a special type of chart is within a time or distance setting)
- Malfunction of ECDIS (means the master unit in a master-backup network has failed)

An indication is required for the following:

- Chart overscale (zoomed in too close)
- Larger scale ENC available
- Different reference units (charted depths not in meters)
- Route crosses safety contour
- Route crosses specified area activated for alarms
- System test failure

As these lists reveal, ECDIS has been programmed to constantly "know" what the navigation team should know, and to help the team to apply its experience and judgment through the adjustment of operational settings.

This automation in ECDIS has two important consequences: First, route or track monitoring does not replace situational awareness; it only enhances it. The alarm functions, while useful, are partial and have the potential to be in error, misinterpreted, ignored, or overlooked.

Secondly, situational awareness must now include, especially when ECDIS is used as the primary means of navigation, the processes and status of the electronic components of the system. This includes all attached sensors, the serial connections and communication ports and data interfaces, the computer processor and operating system, navigation and chart software, data storage devices, and power supply. Furthermore, these new responsibilities must still be balanced with the traditional matters of keeping a vigilant navigational watch.

ECDIS or not, the windows in the pilothouse are still the best tool for situational awareness. Paradoxically, ECDIS makes the navigator's job both simpler and more complex.

1409. ECDIS Outputs

During the past 12 hours of the voyage, ECDIS must be able to reconstruct the navigation and verify the official database used. Recorded at one minute intervals, the information includes:

- Own ship's past track including time, position, heading, and speed
- A record of official ENC used including source, edition, date, cell and update history

It is important to note that if ECDIS is turned off, such as for chart management or through malfunction, voyage recording ceases, unless a networked backup system takes over the functions of the master ECDIS. In that case, the voyage recording will continue, including an entry in the electronic log for all the alarms that were activated and reset during the switchover. Voyage files consist of logbook files, track files and target files. The file structure is based on the date and is automatically created at midnight for the time reference in use. If the computer system time is used for that purpose, the possibility exists for overwriting voyage files if the system time is manually set back. Allowing GPS time as the system reference avoids this pitfall.

In addition, ECDIS must be able to record the complete track for the entire voyage with time marks at least once every four hours. ECDIS should also have the capability to preserve the record of the previous 12 hours of the voyage. It is a requirement that the recorded information be inaccessible to alteration. Preserving voyage files should follow procedures for archiving data. Unless radar overlay data is being recorded, voyage files tend to be relatively small, permitting backup onto low-capacity media, and purging from system memory at regular intervals. (This form of backing up should not be confused with the network master-slave backup system.)

Adequate backup arrangements must be provided to ensure safe navigation in case of ECDIS failure. This includes provisions to take over ECDIS functions so that an ECDIS failure does not develop into a critical situation, and a means of safe navigation for the remaining part of the voyage in case of complete failure.

1410. Voyage Data Recorder (VDR)

The purpose of the voyage data recorder VDR is to provide accurate historical navigational data in the investigation of maritime incidents. It is additionally useful for system performance monitoring. A certified VDR configuration records all data points, as per IMO Resolution A.861(20) & EC Directive 1999/35/EC. Some of the voyage data can be relayed through ECDIS. A fully IEC compliant data capsule passes fire and immersion tests.

The implementation of a secure "black box" and comprehensive Voyage Data Recorder (VDR) is now a carriage requirement on passenger and Ro-Ro vessels over 3000 GT (1600 GRT) engaged in international passages. Existing vessels must be retrofitted by July 2004, and all vessels built after July 2002 must be fitted with a VDR. Retrofit regulations for other vessels built before July 2002 are still in development. Non-RO-RO passenger vessels built before July 2002 may be exempted from carriage where an operator can show that interfacing a VDR with the existing equipment on the ship is unreasonable and impracticable. The European Union requires that all RO-RO ferries or high speed craft engaged on a regular service in European waters (domestic or international) be fitted with a VDR if built before February 2003, and otherwise retrofitted by July 2004.

VDR features include:

- Radar video capture: Radar video is captured and compressed every 15 seconds to comply with IEC performance standards.
- I/O subsystem: To collect a wide variety of data types, a sensor interface unit provides signal conditioning for all analog, digital and serial inputs. All data is converted and transmitted to a data acquisition unit via an ethernet LAN.
- Audio compression: An audio module collects analog signals from microphone preamplifiers. The data is digitized and compressed to meet Lloyds of London 24-hour voice storage requirements.
- Integral uninterruptible power supply (UPS) IEC requires a UPS backup for all components of the data acquisition unit and for the data capsule to provide two hours continuous recording following a blackout.
- Hardened fixed data capsule: IEC 61996 compliant data capsules fitted with ethernet connections provide fast download as well as fast upload to satellite links.
- Remote data recovery and shoreside playback: Options available in several systems.
- Annual system certification: The IMO requires that the VDR system, including all sensors, be subjected to an annual performance test for certification.

DATA FORMATS

1411. Official Vector Data

How ECDIS operates depends on what type of chart data is used. ENC's (electronic navigational charts) and RNC's (raster nautical charts) are approved for use in ECDIS. By definition both ENC's and RNC's are issued under the authority of national hydrographic offices (HO's). ECDIS functions as a true ECDIS when used with corrected ENC data, but ECDIS operates in the less functional raster chart display system (RCDS) mode when using corrected RNC data. When ECDIS is used with non-official vector chart data (corrected or not), it operates in the ECS mode.

In vector charts, hydrographic data is comprised of a series of files in which different layers of information are stored or displayed. This form of "intelligent" spatial data is obtained by digitizing information from existing paper charts or by storing a list of instructions that define various position-referenced features or objects (e.g., buoys, lighthouses, etc.). In displaying vector chart data on ECDIS, the user has considerable flexibility and discretion regarding the amount of information that is displayed.

An ENC is vector data conforming to the IHO S-57 ENC product specification in terms of content, structure and format. An ENC contains all the chart information necessary for safe navigation and may contain supplementary information in addition to that contained in the paper chart. In general, an S-57 ENC is a structurally layered data set designed for a range of hydrographic applications. As defined in IHO S-57 Edition 3, the data is comprised of a series of points, lines, areas, features, and objects. The minimum size of a data set is a cell, which is a spherical rectangle (i.e., bordered by meridians and latitudes). Adjacent cells do not overlap. The scale of the data contained in the cell is dependent upon the navigational purpose (e.g., general, coastal, approach, harbor).

Under S-57, cells have a standard format but do not have a standard coverage size. Instead, cells are limited to 5mb of data. S-57 cells are normally copy protected and therefore require a permit before use is allowed. These permits are delivered as either a file containing the chart permits or as a code. In both cases the first step is to install the chart permit into the ECDIS. Some hydrographic offices deliver S-57 cells without copy protection and therefore permits are not required.

Any regional agency responsible for collecting and distributing S-57 data, such as PRIMAR for Northern Europe, will also maintain data consistency. National hydrographic offices are responsible for producing S-57 data for their own country area. Throughout the world HO's have been slow to produce sufficient quantities of ENC data. This is due to the fact that the standards evolved over several years, and that vector data is much harder to collect than raster data.

In 1996 the IHO S-57 data standard and IHO S-52 specifications for chart content and display were "frozen." It took three versions of S-57 before the issue was finally settled as to what actually comprises an ENC (i.e., ENC Product Specification) and what is required for updating (ENC Updating Profile). The ENC Test Dataset that the International Electrotechnical Commission (IEC) requires for use in conjunction with IEC Publication 61174 (IEC 1997) was finalized by IHO in 1998. It was not possible to conduct ECDIS type-approval procedures without a complete and validated IHO ENC Test Dataset.

Major areas of ENC coverage now include most of Canadian and Japanese waters, the Baltic and North Sea, and important waterways such as the Straits of Malacca, Singapore Strait, and the Straits of Magellan (Chile).

At the same time, many countries including the United States, are stepping up their production of ENC's where issues of port security require the collection of baseline data of submerged hazards. In the U.S., NOAA plans to complete its portfolio of large-scale charts of 42 ports in ENC format by mid-2003, with smaller scale chart completion by 2005. As the chart cells are completed, the data is being made available on the World Wide Web at no cost. Beginning in 2003, NOAA will post critical notice to mariner corrections without restrictions in monthly increments. At that point the status of NOAA's available ENC data will be changed from provisional to official.

ENC data is currently available from the HO's of most Northern European countries, Japan, Korea, Hong Kong, Singapore, Canada, Chile, and the United States, although the coverage and updating process is incomplete. Most ENC is available only through purchase, permits or licensing.

1412. Vector Data Formats Other Than IHO S-57

The largest of the non-S-57 format databases is the Digital Nautical Chart (DNC). The National Imagery and Mapping Agency (NIMA) produces the content and format for the DNC according to a military specification. This allows compatibility among all U.S. Defense Department assets. The DNC is a vector-based digital product that portrays significant maritime features in a format suitable for computerized marine navigation. The DNC is a general-purpose global database designed to support marine navigation and Geographic Information System (GIS) applications. DNC data is only available to the U.S. military and selected allies. It is designed to conform to the IMO Performance Standard and IHO specifications for ECDIS.

Several commercial manufacturers have developed vector databases beyond those that have been issued by official hydrographic offices. These companies are typically manufacturers of ECDIS or ECS equipment or have direct relationships with companies that do, and typically have developed data in proprietary format in order to provide options to raster charts in the absence of ENC data. HO-issued paper charts provide the source data for these formats, although in some cases non-official paper charts are used. In some cases, ECS manufacturers provide a regular updating and maintenance service for their vector data, resulting in added confidence and satisfaction among users. The manufacturer's source of the updates is through HO's. Hence, these two particular non-official formats allows for a very high degree of confidence and satisfaction among mariners using this data.

ECS systems sometimes apply rules of presentation similar to officially specified rules. Thus information is displayed or removed automatically according to scale level to manage clutter. The same indications pertinent to overscaling ENC apply to private vector data. Since the chart data is not ENC, the systems must display that non-official status when used in an ECDIS.

1413. Raster Data

Raster navigational chart (RNC) data is stored as picture elements (pixels). Each pixel is a minute component of the chart image with a defined color and brightness level. Raster-scanned images are derived by scanning paper charts to produce a digital photograph of the chart. Raster data are far easier to produce than vector data, but raster charts present many limitations to the user.

The official raster chart formats are:

ARCS (British Admiralty)
Seafarer (Australia)
BSB (U.S., NOAA/Maptech)

These charts are produced from the same raster process used to print paper charts. They are accurate representations of the original paper chart with every pixel geographically referenced. Where applicable, horizontal datum shifts are included with each chart to enable referencing to WGS84. This permits compatibility with information overlaid on the chart. *Note: Not all available charts have WGS84 shift information.* Extreme caution is necessary if the datum shift cannot be determined exactly.

Raster nautical charts require significantly larger

amounts of memory than vector charts. Whereas a world portfolio of more than 7500 vector charts may occupy about 500mb, a typical coastal region in raster format may consist of just 40 charts and occupy more than 1000mb of memory.

For practical purposes, most of a portfolio of raster charts should be left on the CD and not loaded into the ECDIS hard drive unless one is route planning or actually sailing in a given region. Of course, updates can only be performed on charts that are loaded onto the hard drive.

Certain non-official raster charts are produced that cover European and some South American waters. These are scanned from local paper charts. Additionally, some ECDIS and ECS manufacturers also produce raster charts in proprietary formats.

In 1998 the IMO's Maritime Safety Committee (MSC 70) adopted the Raster Chart Display System (RCDS) as Appendix 7 to the IMO Performance Standards. The IMO-IHO Harmonization Group on ECDIS (HGE) considered this issue for over three years. Where IHO S-57 Ed. 3 ENC data coverage is not available, raster data provided by official HO's can be used as an interim solution. But this RCDS mode does not have the full functionality of an otherwise IMO-compliant ECDIS using ENC data. Therefore, RCDS does not meet SOLAS requirements for carriage of paper charts, meaning that when ECDIS equipment is operated in the RCDS mode, it must be used together with an appropriate portfolio of corrected paper charts.

Some of the limitations of RCDS compared to ECDIS include:

- Chart features cannot be simplified or removed to suit a particular navigational circumstance or task.
- Orientation of the RCDS display to course-up may affect the readability of the chart text and symbols since these are fixed to the chart image in a north-up orientation.
- Depending on the source of the raster chart data, different colors may be used to show similar chart information, and there may be differences between colors used during day and night time.
- The accuracy of the raster chart data may be less than that of the position-fixing system being used.
- Unlike vector data, charted objects on raster charts do not support any underlying information.
- RNC data will not trigger automatic alarms. (However, some alarms can be generated by the RCDS from user-inserted information.).
- Soundings on raster charts may be in fathoms and feet, rather than meters.

The use of ECDIS in RCDS mode can only be considered as long as there is a backup folio of appropriate up-to-date paper charts.

INTEGRATED BRIDGE SYSTEMS

1414. Description

An Integrated Bridge System (IBS) is a combination of equipment and software which uses interconnected controls and displays to present a comprehensive suite of navigational information to the mariner. Rules from classification societies such as Det Norske Veritas (DNV) specify design criteria for bridge workstations. Their rules define tasks to be performed, and specify how and where equipment should be sited to enable those tasks to be performed. Equipment carriage requirements are specified for ships according to the requested class certification or notation. Publication IEC 61029 defines operational and performance requirements, methods of testing, and required test results for IBS.

Classification society rules address the total bridge system in four parts: technical system, human operator, man/machine interface, and operational procedures. The DNV classifies IBS with three certifications: NAUT-C covers bridge design; W1-OC covers bridge design, instrumentation and bridge procedures; W1 augments certain portions of W1-OC.

An IBS generally consists of at least:

- Dual ECDIS installation – one serving master and the other as backup and route planning station
- Dual radar/ARPA installation
- Conning display with a concentrated presentation of navigational information (the master ECDIS)
- DGPS positioning
- Ship's speed measuring system
- Auto-pilot and gyrocompass system
- Full GMDSS functionality

Some systems include full internal communications, and a means of monitoring fire control, shipboard status alarms, and machinery control. Additionally, functions for the loading and discharge of cargo may also be provided.

An IBS is designed to centralize the functions of monitoring collision and grounding risks, and to automate navigation and ship control. Control and display of component systems are not simply interconnected, but often share a proprietary language or code. Several instruments and indicators are considered essential for safe and efficient performance of tasks, and are easily readable at the navigation workstation, such as heading, rudder angle, depth, propeller speed or pitch, thruster azimuth and force, and speed and distance log.

Type approval by Det Norske Veritas for the DNV-W1-ANTS (Automatic Navigation and Track-Keeping

System) certification is given to ship bridge systems designed for one-man watch (W1) in an unbounded sea area. DNV also provides for the other two class notations, NAUT-C and W1-OC. The W1 specifications require the integration of:

- CDIS (providing the functions of safety-contour checks and alarms during voyage planning and execution)
- Manual and automatic steering system (including software for calculation, execution and adjustments to maintain a pre-planned route, and including rate of turn indicator)
- Automatic Navigation and Track-keeping System (ANTS)
- Conning information display
- Differential GPS (redundant)
- Gyrocompass (redundant)
- Radar (redundant) and ARPA
- Central alarm panel
- Wind measuring system
- Internal communications systems
- GMDSS
- Speed over ground (SOG) and speed through water (STW or Doppler log)
- Depth sounder (dual transducer >250m)
- Course alteration warnings and acknowledgment
- Provision to digitize paper charts for areas not covered by ENC data

The W1 classification requires that maneuvering information be made available on the bridge and presented as a pilot card, wheelhouse poster, and maneuvering booklet. The information should include characteristics of speed, stopping, turning, course change, low-speed steering, course stability, trials with the auxiliary maneuvering device, and man-overboard rescue maneuvers.

The W1-OC and W1 classifications specify responsibilities of ship owner and ship operator, qualifications, bridge procedures, and particular to W1, a requirement for operational safety standards. The W1 operational safety manual requires compliance with guidelines on bridge organization, navigational watch routines, operation and maintenance of navigational equipment, procedures for arrival and departure, navigational procedures for various conditions of confinement and visibility, and system fallback procedures. Both classifications also require compliance with a contingency and emergency manual, including organization, accident, security, evacuation, and other related issues.

MILITARY ECDIS

1415. ECDIS-N

In 1998, the U.S. Navy issued a policy letter for a naval version of ECDIS, ECDIS-N, and included a performance standard that not only conforms to the IMO Performance Standards, but extends it to meet unique requirements of the U.S. Department of Defense.

A major difference from an IMO-compliant ECDIS is the requirement that the ECDIS-N SENC must be the Digital Nautical Chart (DNC) issued by the National Imagery and Mapping Agency (NIMA). The DNC conforms to the U.S. DoD standard Vector Product Format (VPF), an implementation of the NATO DIGEST C Vector Relational Format. All of NIMA's nautical, aeronautical, and topographic vector databases are in VPF to ensure interoperability between DoD forces.

In the United States, NIMA produces the Digital Nautical Chart (DNC). It is a vector database of significant maritime features that can be used with shipboard integrated navigation systems such as ECDIS, ECDIS-N, or other types of geographic information systems. NIMA has been working closely with the U.S. Navy to help facilitate a transition from reliance on paper charts to electronic chart navigation using the DNC. The U.S. Navy plans to have all of its surface and sub-surface vessels using DNC's by 2004. NIMA has produced the DNC to support worldwide navigation requirements of the U.S. Navy and U.S. Coast Guard.

To ensure that the DNC data would not be manipulated or inadvertently altered when used by different military units, a decision was made to produce a specific data software product that must be used in a "direct read" capability. As such, a DNC is really a system electronic navigational chart (SENC) that contains specified data and display characteristics. Control of the SENC provides the military with interoperability across deployed systems, which is particularly important when integrated with military data layers.

1416. Navigation Sensor System Interface (NAVSSI)

The Navigation Sensor System Interface (NAVSSI) contains the U.S. Navy's version of ECDIS, and also has significant additional capabilities for the Navy's defense missions. NIMA's Vector Product Format (VPF) DNC's are used in conjunction with NAVSSI. NAVSSI performs three important functions:

- Navigation Safety: NAVSSI distributes real time navigation data to the navigation team members to ensure navigation safety.
- Weapons System Support: NAVSSI provides initial-

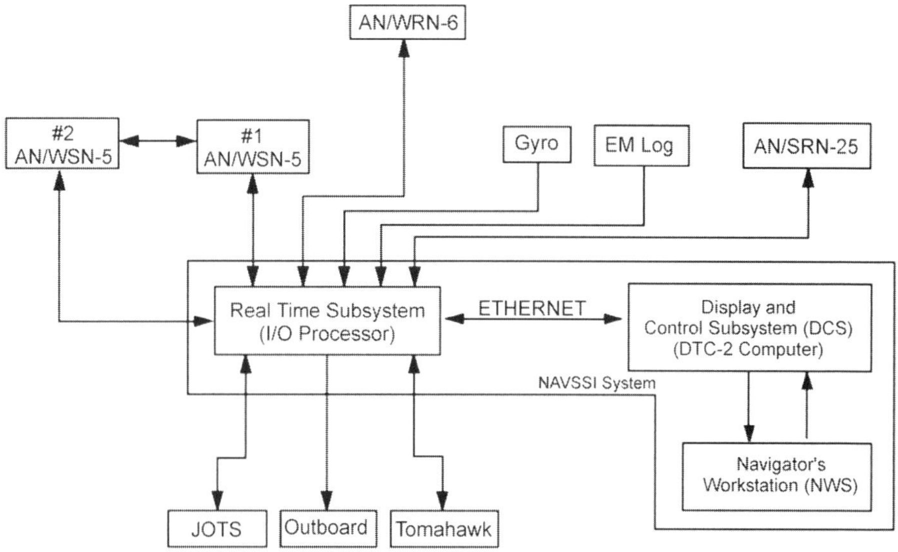

Figure 1416. Block diagram of NAVSSI.

ization data for weapons systems.

- Battlegroup Planning: NAVSSI provides a workstation for battlegroup planning.

The navigational function of NAVSSI, therefore, is only one of several tasks accomplished by the system. The navigational portion of NAVSSI complies with the IMO/IHO ECDIS standards for content and function.

The heart of NAVSSI is the Real Time Subsystem (RTS). The RTS receives, processes and distributes navigational data to the navigation display, weapons systems, and other networked vessels. This ensures that all elements of a battlegroup have the same navigational picture. Inputs come from GPS, Loran, inertial navigation systems, compass, and speed log. The bridge display consists of a monitor and control panel, while the RTS is mounted below decks. DNC's are contained in the **Display and Control Subsystem (DCS)** typically mounted in the chartroom with a monitor on the bridge. This is unlike many current commercial systems which house all hardware and software in a single unit on the bridge. A separate NAVSSI software package supports operator interface, waypoint capability, collision and grounding avoidance features, and other aspects of an ECDIS.

Figure 1416 illustrates a basic block diagram of the NAVSSI system. The RTS takes inputs from the inertial navigators, the GPS in PPS mode, the compass, the EM Log, and the SRN-25. The RTS distributes navigation information to the various tactical applications requiring navigation input, and it communicates via fiber optic network with the DCS. The DCS exchanges information with the Navigator's Workstation.

1417. The Digital Nautical Chart

NAVSSI uses the Digital Nautical Chart (DNC) as its chart database. The DNC is in Vector Product Format (VPF) and is based on the contents of the traditional paper harbor, approach, coastal and general charts produced by NIMA and NOS.

Horizontal datum is WGS 84 (NAD 83 in the U. S. is equivalent). There are three vertical datums. Topographic features are referenced to Mean Sea Level and the shore line is referenced to Mean High Water. Hydrography is referenced to a low water level suitable for the region. All measurements are metric.

The DNC portfolio consists of 29 CD-ROM's and provides global coverage between 84 degrees N and 81 degrees S. This comprises some 4,820 charts group into five libraries based on scale:

General: (>1:500K)
Coastal: (1:75K - 1: 500K)
Approach (1:25K - 1:75K)
Harbor (1 <1:50K)
Browse Index (1:3,100,000)

DNC data is layered together into 12 related feature classes:

- Cultural Landmarks
- Earth Cover
- Inland Waterways
- Relief
- Landcover
- Port Facilities
- Aids to Navigation
- Obstructions
- Hydrography
- Environment
- Limits
- Data Quality

Content is generally the same as on a paper chart. The data is stored in libraries; each library represents a different level of detail. The libraries are then stored on CD-ROM and organized as tiles according to the World Geodetic Reference System (GEOREF) tiling scheme.

A subset of the DNC is known as Tactical Ocean Data (TOD). TOD data is bathymetric in nature and intended for Naval operations.

There are 6 levels of TOD:

Level 0 - OPAREA charts
Level 1 - Bottom Contour
Level 2 - Bathymetric Navigation Planning Charts
Level 3 - Shallow Water
Level 4 - Hull Integrity Test Charts
Level 5 - Strategic Straits Charts

1418. Warship ECDIS (WECDIS)

A Warship ECDIS is an ECDIS approved by international authorities for warship use, which, while meeting the operating standards of ECDIS, may not conform exactly to ECDIS specifications.

Performance Standards for "Warship" ECDIS (WECDIS) were approved by the North Atlantic Treaty Organization (NATO) in 1999 and issued as STANAG 4564. The core functionality of WECDIS is an IMO-compliant ECDIS. Beyond the minimum performance requirements for ECDIS, WECDIS has the ability to use a variety of geospatial data from both civilian and military sources. For navigational data, WECDIS uses both IHO S-57 ENC data and data conforming to NATO Digital Geographic Information Exchange (DIGEST) Standards. This latter includes such products as Vector Product Format (VPF) and Digital Nautical Chart (DNC).

In addition to core navigation information (IHO S-57 ENC and VPF-DNC), WECDIS will also use Additional Navigation Information (ANI) provided by government hydrographic offices and military sources. Specific types of ANI data include Raster Navigational Charts (RNC's), such as Admiralty Raster Chart Service (ARCS) or NOAA's raster charts distributed and updated by Maptech, Inc. The ability to use different types of navigational data from a variety sources is often referred to as "multi-fuel."

CORRECTING ELECTRONIC CHARTS

1419. ECDIS Correction Systems

ECDIS software creates a database from the ENC data called the system electronic navigational chart (SENC) and from this selects information for display. The ECDIS software meanwhile receives and processes serial data from navigational sensors and displays that textual and graphical information simultaneously with the SENC information.

It is the SENC that is equivalent to up-to-date charts, as stated by the Performance Standards. As originally conceived, ECDIS was designed to use internationally standardized and officially produced vector data called the ENC (electronic navigational chart). Only when using ENC data can ECDIS create an SENC, and thereby function in the ECDIS mode.

Updates for ENC are installed into the ECDIS separate from the ENC data itself. For the mariner, this involves activating a special utility accompanying the ECDIS and following the on-screen prompts. Within this same utility, update content and update log files in textual form can be viewed. Once the ECDIS software itself is reactivated, the update information is accessed in conjunction with the ENC data and the SENC database is created.

Just as ENC and updates are transformed into the SENC, so too are other data types accessed and combined. The user has the option to add lines, objects, text and links to other files supported by application. Referred to in the Performance Standards as data added by the mariner, these notes function as layers on the displayed chart. The user can select all or parts of the layers for display to keep clutter to a minimum. The mariner's own layers, however, must be called into the SENC from stored memory. As a practical matter, not only must the mariner take care to associate file names with actual content, such as with manually created chart corrections, but also must realize that the files themselves do not have the tamper-proof status that ENC and official updates have.

Within the SENC resides all the information available for the display. The Presentation Library rules such as Standard Display and Display Base define what levels of information from the SENC can be shown.

An ENC updating profile is contained within the IHO S-57 Edition 3.0 specification. This enables the efficient addition, removal or replacement of any line, feature, object or area contained within the ENC dataset. Guidance on the means and process for ENC updating is provided in IHO S-52, Appendix 1. In terms of what is called for in the IMO Performance Standards, an ENC dataset being used in an ECDIS must also have an ENC updating service. This permits the ENC and the SENC to be corrected for the intended voyage, and thus achieves an important component of SOLAS compliance.

Accordingly, ECDIS must be capable of accepting official updates to the ENC data provided in conformity with IHO standard. Updated cells are stored in a file and transmitted by e-mail, floppy disk or CD-ROM, or satellite. For example, PRIMAR charts and updates are delivered on two CD's: the Base CD contains the PRIMAR database at the time indicated on the label and the second CD contains the updates for those charts. But the update CD also contains new charts issued since the base CD was printed. Since the operator must acquire the files and then initiate the update functions of the ECDIS software, this form of updating is referred to as semi-automatic.

Generally, ECDIS will reject updates if the update issuing authority is different from the cell issuing authority. It will also reject corrupted update files and files with an incorrect extension. ECDIS checks that updates are applied in the right sequence. If one update is missing the next update is rejected. An update CD-ROM should contain all available updates for all S57 cells. Generally, ECDIS will automatically run all updates in the right order for all cells.

For S-57 data, the content of updates in text form can be viewed from within the utility that permits the management of chart data. Generally it can only be run when ECDIS is terminated. ECDIS is also capable of showing or hiding S-57 updates on a given chart or cell. The update must first be installed via the chart utility. After restarting ECDIS, and after loading into the display the particular chart with the correction, the correction should be manually accepted. That enables the function in S-57 chart options to show or hide the symbol indicating the location of the correction.

NIMA DNC Corrections

NIMA has produced the DNC Vector Product Format Database Update (VDU) to support worldwide DNC navigation requirements of the U.S. Navy, the U.S. Coast Guard, and certain allies. NIMA does not distribute DNC to other than U.S. government agencies and foreign governments having data exchange agreements with NIMA. The DNC maintenance system will be able to apply new source materials such as bathymetry, imagery, *Notice to Mariners*, local notices, new foreign charts, etc. for inclusion in the DNC database.

The VDU system works by performing a binary comparison of the corrected chart with the previous version. The differences are then written to a binary "patch" file with instructions as to its exact location. The user then applies this patch by specifying the proper path and filename on his own ship. Every new change incorporates all previous changes, so the navigator is assured that, having received the latest change, he has all changes issued to date.

File sizes are small enough to support bandwidth limitations of ships at sea and requires only one-way communication. Patch files are posted every four weeks. Authorized commands may access DNC's and the associated VDU files through the NIMA Gateway:

OSIS http://osis.nima.mil/gidbe/index.htm
SIPRNET http://www.nima.smil.mil/products/dnc1
JWICS http://www.nima.is.gov/products/dnc1

The VDU patch files are posted to the World Wide Web monthly at:

http://www.nima.mil/dncpublic/

A separate layer within DNC provides the user with identification of where changes have been made during the updating process.

British RCS Corrections

For the British RCS system, updates for all 2700 charts affected by *Admiralty Notice to Mariners* are compiled and placed on a weekly ARCS Update CD-ROM. Applying the corrections is only semi-automatic (not fully automatic), but it is also error-free, and each CD-ROM provides cumulative updates. The CD-ROM's are available through chart agents.

NOAA Corrections

In the U.S., NOAA has contracted with Maptech, Inc. to provide updating of all NOS raster charts using information from the USCG, NIMA and the Canadian Hydrographic Service (CHS). Maptech uses a "patch technique" to update only those parts of a given chart identified as needing correction. The method compares the existing chart file and its corrected counterpart on a pixel-by-pixel basis. The software creates a "difference file" that is associated with the existing raster file to which it applies. This difference file is then compressed so that a typical patch contains only a few kilobytes of data. Ninety-nine percent are under 10kb. Typical downloads for a chart take 15 seconds to 5 minutes depending on modem speed.

The raster chart is updated as the patch file alters the pixels on the original chart. Update patches are available by download, and are cumulative for the all the charts packed on a given source folio CD. Further refinement will permit the separate storage of the RNC and update patches, so that as the patch is applied dynamically in real time, the user will be able to view the correction. The dynamic patching is similar to ENC updating in that the original chart data is not altered. Presently the service is a subscription service with weekly updates at a nominal cost. Information is available at http://chartmaker.ncd.noaa.gov.

Commercial Systems

There are a variety of ECS's available for small craft, often found aboard fishing vessels, tugs, research vessels, yachts, and other craft not large enough to need SOLAS equipment but wanting the best in navigation technology. Given that these systems comprise a single aid to navigation and do not represent a legal chart in any sense, it is probably not a critical point that correction systems for these products are not robust enough to support regular application of changes.

In fact, often the only way to make changes is to purchase new editions, although the more sophisticated ones allow the placement of electronic "notes" on the chart. The data is commonly stored on RAM chips of various types, and cannot be changed or without re-programming the chip from a CD-ROM or disk containing the data. If the data is on CD-ROM, a new CD-ROM is the update mechanism, and they are, for the most part, infrequently produced. Users of these systems are required to maintain a plot on a corrected paper chart.

USING ELECTRONIC CHARTS

1420. Digital Chart Accuracy

As is the case with any shipboard gear, the user must be aware of the capabilities and limitations of digital charts. The mariner should understand that nautical chart data displayed possess inherent accuracy limitations. Because digital charts are necessarily based primarily on paper charts, many of these limitations have migrated from the paper chart into the electronic chart. Electronic chart accuracy is, for the most part, dependent on the accuracy of the features being displayed and manipulated. While some ECDIS and ECS have the capability to use large-scale data produced from recent hydrographic survey operations (e.g., dredged channel limits or pier/terminal facilities) most raster and vector-based electronic chart data are derived from existing paper charts.

Twenty years ago, mariners were typically obtaining position fixes using radar ranges, visual bearings or Loran. Generally, these positioning methods were an order of magnitude less accurate than the horizontal accuracy of the survey information portrayed on the chart. For example, a three-line fix that results in an equilateral triangle with sides two millimeters in length at a chart scale of 1:20,000 represents a triangle with 40-meter sides in real-world coordinates.

A potential source of error is related to the system configuration, rather than the accuracy of electronic chart data being used. All ECDIS's and most ECS's enable the user to input the vessel's dimensions and GPS antenna location. On larger vessels, the relative position of the GPS antenna aboard the ship can be a source of error when viewing the "own-ship" icon next to a pier or wharf.

In U.S. waters, the Coast Guard's DGPS provides a horizontal accuracy of +/-10 meters (95 percent). However, with selective availability off, even the most basic GPS receiver in a non-differential mode may be capable of providing better than 10 meter horizontal accuracy. In actual operation, accuracies of 3-5 meters are being achieved. As a result, some mariners have reported that when using an electronic chart while moored alongside a pier, the vessel icon plots on top of the pier or out in the channel.

Similarly, some mariners transiting a range that marks the centerline of a channel report that the vessel icon plots along the edge or even outside of the channel. Mariners now expect, just as they did 20 years ago, that the horizontal accuracy of their charts will be as accurate as the positioning system available to them. Unfortunately, any electronic chart based on a paper chart, whether it is raster or vector, is not able to meet this expectation.

The overall horizontal accuracy of data portrayed on paper charts is a combination of the accuracy of the underlying source data and the accuracy of the chart compilation process. Most paper charts are generalized composite documents compiled from survey data that have been collected by various sources over a long period of time. A given chart might encompass one area that is based on a lead line and sextant hydrographic survey conducted in 1890, while another area of the same chart might have been surveyed in the year 2000 with a full-coverage shallow-water multibeam system. In the U.S., agencies have typically used the most accurate hydrographic survey instrumentation available at the time of the survey.

While survey positioning methods have changed over the years, standards have generally been such that surveys were conducted with a positioning accuracy of better than 0.75 millimeters at the scale of the chart. Therefore, on a 1:20,000-scale chart, the survey data was required to be accurate to 15 meters. Features whose positions originate in the local notice to mariners, reported by unknown source, are usually charted with qualifying notations like position approximate (PA) or position doubtful (PD). The charted positions of these features, if they do exist, may be in error by miles.

As of 2002, over 50 percent of the depth information found on U.S. charts is based on hydrographic surveys conducted before 1940. Surveys conducted many years ago with lead lines or single-beam echo sounders sampled only a tiny percentage of the ocean bottom. Hydrographers were unable to collect data between the sounding lines. Depending on the water depth, these lines may have been spaced at

50, 100, 200 or 400 meters. As areas are re-surveyed and full-bottom coverage is obtained, uncharted features, some dangerous to navigation, are discovered quite often. These features were either: 1) not detected on prior surveys, 2) objects such as wrecks that have appeared on the ocean bottom since the prior survey or 3) the result of natural changes that have occurred since the prior survey.

In a similar manner, the shoreline found on most U.S. charts is based on photogrammetric or plane table surveys that are more than 20 years old. In major commercial harbors, the waterfront is constantly changing. New piers, wharves, and docks are constructed and old facilities are demolished. Some of these man-made changes are added to the chart when the responsible authority provides as-built drawings. However, many changes are not reported and therefore do not appear on the chart. Natural erosion along the shoreline, shifting sand bars and spits, and geological subsidence and uplift also tend to render the charted shoreline inaccurate over time.

Another component of horizontal chart accuracy involves the chart compilation process. For example, in the U.S. before NOAA's suite of charts was scanned into raster format in 1994, all chart compilation was performed manually. Projection lines were constructed and drawn by hand and all plotting was done relative to these lines. Cartographers graphically reduced large scale surveys or engineering drawings to chart scale. Very often these drawings were referenced to state plane or other local coordinate systems. The data would then be converted to the horizontal datum of the chart (e.g., the North American 1927 (NAD 27) or the North American Datum 1983 (NAD 83). In the late 1980's and early 1990's, NOAA converted all of its charts to NAD 83. In accomplishing this task, averaging techniques were used and all of the projection lines were redrawn.

When NOAA scanned its charts and moved its cartographic production into a computer environment, variations were noted between manually constructed projection lines and those that were computer generated. All of the raster charts were adjusted or warped so that the manual projection lines conformed to the computer-generated projection. In doing so, all information displayed on the chart was moved or adjusted.

Similar processes take place during NIMA's digital chart production, but involving more complexity, since NIMA cartographers must work with a variety of different datums in use throughout the world, and with hydrographic data from hundreds of official and unofficial sources. While much of NIMA's incoming data was collected to IHO standards during hydrographic surveys, many sources are questionable at best, especially among the older data.

Today, when survey crews and contractors obtain DGPS positions on prominent shoreline features and compare those positions to the chart, biases may be found that are on the order of two millimeters at the scale of the chart (e.g., 20 meters on 1:10,000-scale chart). High accuracy aerial photography reveals similar discrepancies between the true shoreline and the charted shoreline. It stands to reason that other important features such as dredged channel limits and navigational aids also exhibit these types of biases. Unfortunately, on any given chart, the magnitude and the direction of these discrepancies will vary by unknown amounts in different areas of the chart. Therefore, no systematic adjustment can easily be performed that will improve the inherent accuracy of the paper or electronic chart.

Some mariners have the misconception that because charts can be viewed on a computer, the information has somehow become more accurate than it appears on paper. Some mariners believe that vector data is more accurate than paper or raster data. Clearly, if an electronic chart database is built by digitizing a paper chart, it can be no more accurate than the paper chart.

The most accurate way to create an ENC is to re-compile the chart from all of the original source material. Unfortunately, the process is far too labor intensive. In the U.S., NOAA has used original source material where possible to compile navigation critical information such as aids to navigation and channel limits. The remaining data are being digitized from the largest scale paper charts.

Once ENC's are compiled, they may be enhanced with higher-accuracy data over time. High-resolution shoreline data may be incorporated into the ENC's as new photogrammetric surveys are conducted. Likewise, depths from new hydrographic surveys will gradually supersede depths that originated from old surveys.

1421. Route Planning and Monitoring

Presumably, route planning takes place before the voyage begins, except in situations where major changes in the route are called for while the ship is underway. In either case, both ECDIS and ECS will allow the display of the smallest scale charts of the operating area and the selection of waypoints from those charts. ECDIS requires a warning that a chosen route crosses a safety contour or prohibited area; ECS will not necessarily do so. If the data is raster, this function is not possible. Once the waypoints are chosen, they can be saved as a route in a separate file for later reference and output to the autopilot.

It is a good idea to zoom in on each waypoint if the chart scale from which it is selected is very small, so that the navigational picture in the area can be seen at a reasonable scale. Also, if a great circle route is involved, the software may be able to enter the waypoints directly from the great circle route file. If not, they will have to be entered by hand.

During route monitoring, ECDIS must show own ship's position whenever the display covers that area. Although the navigator may chose to "look-ahead" while in route monitoring, it must be possible to return to own ship's position with a single operator action. Key information provided during route monitoring includes a continuous indication of vessel position, course, and speed. Additional information that ECDIS or ECS can provide includes distance right/left of intended track, time-to-turn, distance-to-turn, position and time of "wheel-over", and past track history.

As specified in Appendix 5 of the IMO Performance Standard, ECDIS must provide an indication of the condition of the system and its components. An alarm must be provided if there is a condition that requires immediate attention. An indication can be visual, while an alarm must either be audible, or both audible and visual.

The operator can control certain settings and functions, some of the most important of which are the parameters for certain alarms and indications, including:

- Cross-track error: Set the distance to either side of the track the vessel can stray before an alarm sounds. This will depend on the phase of navigation, weather, and traffic.
- Safety contour: Set the depth contour line which will alert the navigator that the vessel is approaching shallow water.
- Course deviation: Set the number of degrees off course the vessel's heading should be allowed to stray before an alarm sounds.
- Critical point approach: Set the distance before approaching each waypoint or other critical point that an alarm will sound.
- Datum: Set the datum of the positioning system to the datum of the chart, if different.

1422. Waypoints and Routes

In the route planning mode, the ECS or ECDIS will allow the entry of waypoints as coordinates of latitude and longitude, or the selection of waypoints by moving a cursor around on the charts. It will allow the creation and storage of numerous pre-defined routes, which can be combined in various ways to create complex voyages.

For example, one might define a route from the inner harbor to the outer harbor of a major port, a route for each of two or more channels to the sea, and several more for open sea routes to different destinations. These can then be combined in different ways to create comprehensive routes that will comprise entire dock-to-dock voyages. They may also be run in reverse for the return trip.

When selecting waypoints, take care to leave any aids to navigation marking the route well to one side of the course. Many navigational software programs contain databases listing the location of the aids to navigation in the United States and other countries. This list should NOT be used to create routes, because the accuracy of today's navigation systems is good enough that to do so invites a collision with any aid whose actual position is entered as a waypoint. Always leave a prudent amount of room between the waypoint and the aid.

Some published routes exist, also a feature of certain software programs. The wise navigator will not use these until he has verified the exact position of each waypoint using the best scale chart. Using pre-programmed routes from an unknown source is the same as letting someone else navigator your vessel. Such a route may pass over shoal water, under a bridge, or through an area that your own vessel might find hazardous. Always check each waypoint personally.

Many electronic chart systems will also allow the coupling of the navigation system to the autopilot. Technically, it is possible to turn the navigation of the vessel over to the autopilot almost as soon as the vessel is underway, allowing the autopilot to make the course changes according to each waypoint. While this may be possible for small craft in most inland, harbor and harbor approach situations, the larger the vessel, the less advisable this practice is, because autopilots do not take advance and transfer into account. The large ship under autopilot control will not anticipate the turn in a channel, and will not begin the turn until the antenna of the positioning system, presumably GPS and often located in the stern of the ship, is at the exact waypoint. By this time it is too late, for the turn should likely have been started at least two ship lengths previous. It is perfectly prudent to allow autopilot control of course changes for vessels in the open sea if the proper parameters for maximum rudder angle have been set.

1423. Training and Simulation

In 2001, the IMO issued guidelines for training with ECDIS simulation. The guidelines stipulate that ECDIS training should include simulation of live data streams, as well as ARPA and Automated Information System (AIS) target information, and a Voyage Data Recorder (VDR) interface. But the IMO has not specifically required ECDIS training other than as a general substitution in the Standards of Training, Certification, and Watchkeeping (STCW) 95 code for navigation with paper charts.

Also in 2001, the USCG approved the country's first STCW-compliant five day ECDIS training course in the U.S. Long-term STCW 95 training and education programs are presently in development. The two levels of competency defined by STCW are operational (OIC or 3rd mate / 2nd mate) and management (CCM or 1st officer / Master). It is likely that for mariners sailing since August 1998, training and education in navigation at both the OIC and CCM levels will include the five day competency-based ECDIS training course.

Accordingly, certified training in the operational use of ECDIS should consist of a five day course making use of simulation equipment for a real-time operating environment appropriate for tasks in navigation, watchkeeping and maneuvering. The primary goal is that the trainee should be able to smoothly operate the ECDIS equipment, use all of its navigational functions, select and assess all relevant information, respond correctly in the case of a malfunction, describe common errors of interpretation and describe potential errors of displayed data. The trainee should follow structured practice in the following: setting up and maintaining the display; operational use of electronic charts including updating, route monitoring, route planning, handling alarms; work with motion parameters and position correction; work with log records and voyage files; and operate interfaces with radar, ARPA, AIS transponders, and VDR's.

CHAPTER 15

NAVIGATIONAL ASTRONOMY

PRELIMINARY CONSIDERATIONS

1500. Definitions

The science of Astronomy studies the positions and motions of celestial bodies and seeks to understand and explain their physical properties. Navigational astronomy deals with their coordinates, time, and motions. The symbols commonly recognized in navigational astronomy are given in Table 1500.

Celestial Bodies

- ☉ Sun
- ☾ Moon
- ☿ Mercury
- ♀ Venus
- ⊕ Earth
- ♂ Mars
- ♃ Jupiter
- ♄ Saturn
- ♅ Uranus
- ♆ Neptune
- ♇ Pluto
- ☆ Star
- ☆-P Star-planet altitude correction (altitude)
- ☉̲ ☾̲ Lower limb
- ⊖ ☾ Center
- ☉̄ ☾̄ Upper limb
- ● New moon
- ◐ Crescent moon
- ◐ First quarter
- ◯ Gibbous moon
- ○ Full moon
- ◯ Gibbous moon
- ◑ Last quarter
- ◑ Crescent moon

Miscellaneous Symbols

- ʸ Years
- ᵐ Months
- ᵈ Days
- ʰ Hours
- ᵐ Minutes of time
- ˢ Seconds of time
- ▬ Remains below horizon
- ▭ Remains above horizon
- //// Twilight all night
- ✳ Interpolation impractical
- ° Degrees
- ′ Minutes of arc
- ″ Seconds of arc
- ☌ Conjunction
- ☍ Opposition
- □ Quadrature
- ☊ Ascending node
- ☋ Descending node

Signs of the Zodiac

- ♈ Aries (vernal equinox)
- ♉ Taurus
- ♊ Gemini
- ♋ Cancer (summer solstice)
- ♌ Leo
- ♍ Virgo
- ♎ Libra (autumnal equinox)
- ♏ Scorpius
- ♐ Sagittarius
- ♑ Capricornus (winter solstice)
- ♒ Aquarius
- ♓ Pisces

Table 1500. Astronomical symbols.

1501. The Celestial Sphere

Looking at the sky on a dark night, imagine that celestial bodies are located on the inner surface of a vast, Earth-centered sphere (Figure 1501). This model is useful since we are only interested in the relative positions and motions of celestial bodies on this imaginary surface. Understanding the concept of the celestial sphere is most important when discussing sight reduction in Chapter 20.

1502. Relative and Apparent Motion

Celestial bodies are in constant motion. There is no fixed position in space from which one can observe absolute motion. Since all motion is relative, the position of the observer must be noted when discussing planetary motion. From the Earth we see apparent motions of celestial bodies on the celestial sphere. In considering how planets follow their orbits around the Sun, we assume a hypothetical observer at some distant point in space. When discussing the rising or setting of a body on a local horizon, we must locate the observer at a particular point on the Earth because the setting Sun for one observer may be the rising Sun for another.

Motion on the celestial sphere results from the motions in space of both the celestial body and the Earth. Without special instruments, motions toward and away from the Earth cannot be discerned.

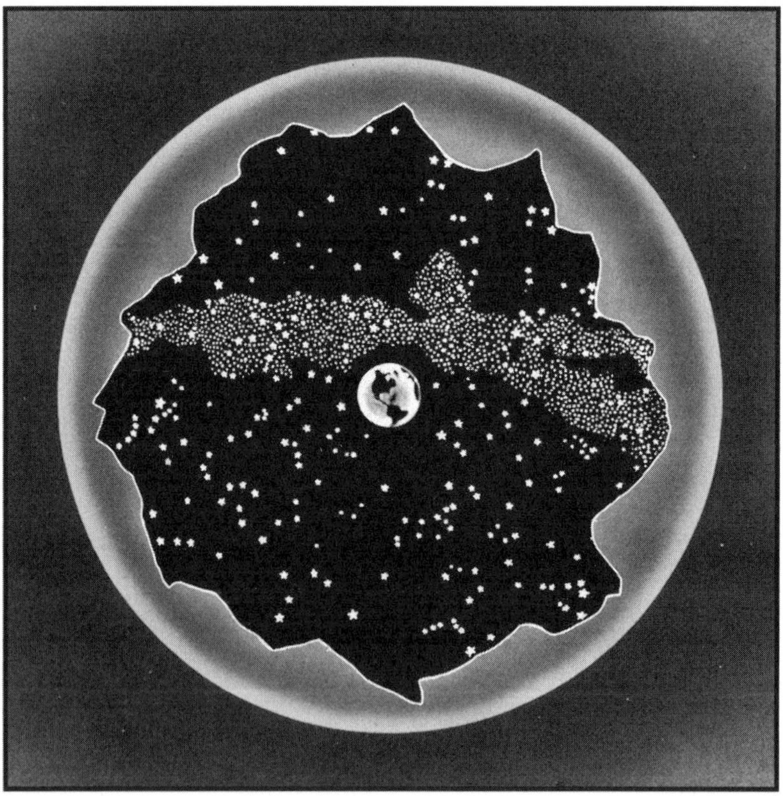

Figure 1501. The celestial sphere.

1503. Astronomical Distances

We can consider the celestial sphere as having an infinite radius because distances between celestial bodies are so vast. For an example in scale, if the Earth were represented by a ball one inch in diameter, the Moon would be a ball one-fourth inch in diameter at a distance of 30 inches, the Sun would be a ball nine feet in diameter at a distance of nearly a fifth of a mile, and Pluto would be a ball half an inch in diameter at a distance of about seven miles. The nearest star would be one-fifth of the actual distance to the Moon.

Because of the size of celestial distances, it is inconvenient to measure them in common units such as the mile or kilometer. The mean distance to our nearest neighbor, the Moon, is 238,855 miles. For convenience this distance is sometimes expressed in units of the equatorial radius of the Earth: 60.27 Earth radii.

Distances between the planets are usually expressed in terms of the **astronomical unit (AU)**, the mean distance between the Earth and the Sun. This is approximately 92,960,000 miles. Thus the mean distance of the Earth from the Sun is 1 AU. The mean distance of Pluto, the outermost known planet in our solar system, is 39.5 A.U. Expressed in astronomical units, the mean distance from the Earth to the Moon is 0.00257 A.U.

Distances to the stars require another leap in units. A commonly-used unit is the **light-year**, the distance light travels in one year. Since the speed of light is about 1.86×10^5 miles per second and there are about 3.16×10^7 seconds per year, the length of one light-year is about 5.88×10^{12} miles. The nearest stars, Alpha Centauri and its neighbor Proxima, are 4.3 light-years away. Relatively few stars are less than 100 light-years away. The nearest galaxies, the Clouds of Magellan, are 150,000 to 200,000 light years away. The most distant galaxies observed by astronomers are several billion light years away.

1504. Magnitude

The relative brightness of celestial bodies is indicated by a scale of stellar **magnitudes**. Initially, astronomers divided the stars into 6 groups according to brightness. The 20 brightest were classified as of the first magnitude, and the dimmest were of the sixth magnitude. In modern times, when it became desirable to define more precisely the limits of magnitude, a first magnitude star was considered 100 times brighter than one of the sixth magnitude. Since the fifth root of 100 is 2.512, this number is considered the **magnitude ratio**. A first magnitude star is 2.512 times as bright as a second magnitude star, which is 2.512 times as bright as a third magnitude star,. A second magnitude is $2.512 \times 2.512 = 6.310$ times as bright as a fourth magnitude star. A first magnitude star is 2.512^{20} times as bright as a star of the 21st magnitude, the dimmest that can be seen through a 200-inch telescope.

Brightness is normally tabulated to the nearest 0.1 magnitude, about the smallest change that can be detected by the unaided eye of a trained observer. All stars of magnitude 1.50 or brighter are popularly called "first magnitude" stars. Those between 1.51 and 2.50 are called "second magnitude" stars, those between 2.51 and 3.50 are called "third magnitude" stars, etc. Sirius, the brightest star, has a magnitude of –1.6. The only other star with a negative magnitude is Canopus, –0.9. At greatest brilliance Venus has a magnitude of about –4.4. Mars, Jupiter, and Saturn are sometimes of negative magnitude. The full Moon has a magnitude of about –12.6, but varies somewhat. The magnitude of the Sun is about –26.7.

THE UNIVERSE

1505. The Solar System

The **Sun**, the most conspicuous celestial object in the sky, is the central body of the solar system. Associated with it are at least nine principal **planets** and thousands of asteroids, comets, and meteors. Some planets have moons.

1506. Motions of Bodies of the Solar System

Astronomers distinguish between two principal motions of celestial bodies. Rotation is a spinning motion about an axis within the body, whereas revolution is the motion of a body in its orbit around another body. The body around which a celestial object revolves is known as that body's primary. For the satellites, the primary is a planet. For the planets and other bodies of the solar system, the primary is the Sun. The entire solar system is held together by the gravitational force of the Sun. The whole system revolves around the center of the Milky Way galaxy (Article 1515), and the Milky Way is in motion relative to its neighboring galaxies.

The hierarchies of motions in the universe are caused by the force of gravity. As a result of gravity, bodies attract each other in proportion to their masses and to the inverse square of the distances between them. This force causes the planets to go around the sun in nearly circular, elliptical orbits.

In each planet's orbit, the point nearest the Sun is called the **perihelion**. The point farthest from the Sun is called the **aphelion**. The line joining perihelion and aphelion is called the **line of apsides**. In the orbit of the Moon, the point nearest the Earth is called the **perigee**, and that point farthest from the Earth is called the **apogee**. Figure 1506 shows the orbit of the Earth (with exaggerated eccentricity), and the orbit of the Moon around the Earth.

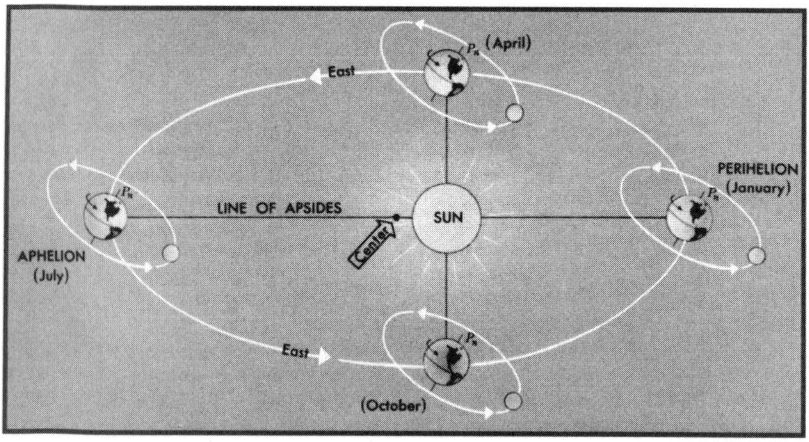

Figure 1506. Orbits of the Earth and Moon.

1507. The Sun

The Sun dominates our solar system. Its mass is nearly a thousand times that of all other bodies of the solar system combined. Its diameter is about 865,000 miles. Since it is a star, it generates its own energy through a thermonuclear reaction, thereby providing heat and light for the entire solar system.

The distance from the Earth to the Sun varies from 91,300,000 at perihelion to 94,500,000 miles at aphelion. When the Earth is at perihelion, which always occurs early in January, the Sun appears largest, 32.6' of arc in diameter. Six months later at aphelion, the Sun's apparent diameter is a minimum of 31.5'.

Observations of the Sun's surface (called the **photosphere**) reveal small dark areas called **sunspots**. These are areas of intense magnetic fields in which relatively cool gas (at 7000°F.) appears dark in contrast to the surrounding hotter gas (10,000°F.). Sunspots vary in size from perhaps 50,000 miles in diameter to the smallest spots that can be detected (a few hundred miles in diameter). They generally appear in groups. See Figure 1507. Large sunspots can be seen without a telescope if the eyes are protected.

Surrounding the photosphere is an outer **corona** of very hot but tenuous gas. This can only be seen during an eclipse of the Sun, when the Moon blocks the light of the photosphere.

The Sun is continuously emitting charged particles, which form the **solar wind**. As the solar wind sweeps past the Earth, these particles interact with the Earth's magnetic field. If the solar wind is particularly strong, the interaction can produce magnetic storms which adversely affect radio signals on the Earth. At such times the auroras are particularly brilliant and widespread.

The Sun is moving approximately in the direction of Vega at about 12 miles per second, or about two-thirds as fast as the Earth moves in its orbit around the Sun.

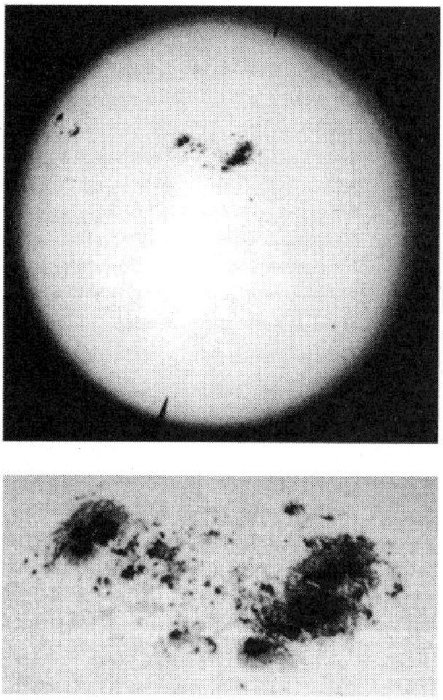

Figure 1507. Whole solar disk and an enlargement of the great spot group of April 7, 1947. Courtesy of Mt. Wilson and Palomar Observatories.

1508. The Planets

The principal bodies orbiting the Sun are called **planets**. Nine principal planets are known: Mercury, Venus, Earth, Mars, Jupiter, Saturn, Uranus, Neptune, and Pluto. Of these, only four are commonly used for celestial navigation: Venus, Mars, Jupiter, and Saturn.

Except for Pluto, the orbits of the planets lie in nearly the same plane as the Earth's orbit. Therefore, as seen from the Earth, the planets are confined to a strip of the celestial sphere near the **ecliptic**, which is the intersection of the mean plane of the Earth's orbit around the Sun with the celestial sphere.

The two planets with orbits smaller than that of the Earth are called **inferior planets**, and those with orbits larger than that of the Earth are called **superior planets**. The four planets nearest the Sun are sometimes called the inner planets, and the others the outer planets. Jupiter, Saturn, Uranus, and Neptune are so much larger than the others that they are sometimes classed as major planets. Uranus is barely visible to the unaided eye; Neptune and Pluto are not visible without a telescope.

Planets can be identified in the sky because, unlike the stars, they do not twinkle. The stars are so distant that they are point sources of light. Therefore the stream of light from a star is easily scattered in the atmosphere, causing the twinkling effect. The naked-eye planets, however, are close enough to present perceptible disks. The broader stream of light from a planet is not easily disrupted.

The orbits of many thousands of tiny minor planets or asteroids lie chiefly between the orbits of Mars and Jupiter. These are all too faint to be seen with the naked eye.

1509. The Earth

In common with other planets, the Earth **rotates** on its axis and **revolves** in its orbit around the Sun. These motions are the principal source of the daily apparent motions of other celestial bodies. The Earth's rotation also causes a deflection of water and air currents to the right in the Northern Hemisphere and to the left in the Southern Hemisphere. Because of the Earth's rotation, high tides on the open sea lag behind the meridian transit of the Moon.

For most navigational purposes, the Earth can be considered a sphere. However, like the other planets, the Earth is approximately an **oblate spheroid**, or **ellipsoid of revolution**, flattened at the poles and bulged at the equator. See Figure 1509. Therefore, the polar diameter is less than the equatorial diameter, and the meridians are slightly elliptical, rather than circular. The dimensions of the Earth are recomputed from time to time, as additional and more precise measurements become available. Since the Earth is not exactly an ellipsoid, results differ slightly when equally precise and extensive measurements are made on different parts of the surface.

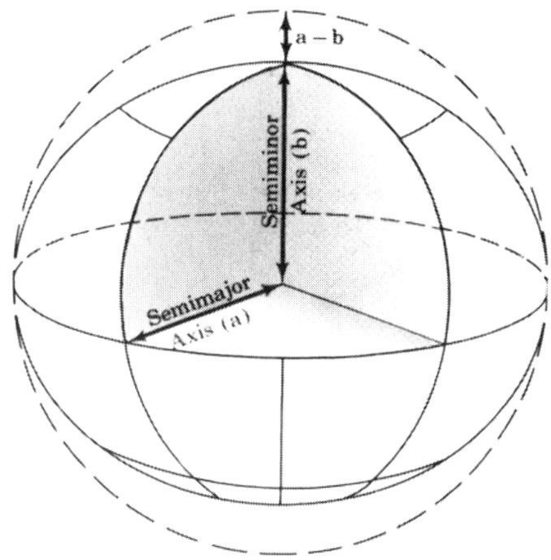

Figure 1509. Oblate spheroid or ellipsoid of revolution.

1510. Inferior Planets

Since Mercury and Venus are inside the Earth's orbit, they always appear in the neighborhood of the Sun. Over a period of weeks or months, they appear to oscillate back and forth from one side of the Sun to the other. They are seen either in the eastern sky before sunrise or in the western sky after sunset. For brief periods they disappear into the Sun's glare. At this time they are between the Earth and Sun (known as **inferior conjunction**) or on the opposite side of the Sun from the Earth (**superior conjunction**). On rare occasions at inferior conjunction, the planet will cross the face of the Sun as seen from the Earth. This is known as a **transit of the Sun**.

When Mercury or Venus appears most distant from the Sun in the evening sky, it is at greatest eastern elongation. (Although the planet is in the western sky, it is at its easternmost point from the Sun.) From night to night the planet will approach the Sun until it disappears into the glare of twilight. At this time it is moving between the Earth and Sun to inferior conjunction. A few days later, the planet will appear in the morning sky at dawn. It will gradually move away from the Sun to western elongation, then move back toward the Sun. After disappearing in the morning twilight, it will move behind the Sun to superior conjunction. After this it will reappear in the evening sky, heading toward eastern elongation.

Mercury is never seen more than about 28° from the Sun. For this reason it is not commonly used for navigation. Near greatest elongation it appears near the western horizon after sunset, or the eastern horizon before sunrise. At these times it resembles a first magnitude star and is sometimes reported as a new or strange object in the sky. The interval during which it appears as a morning or evening star can

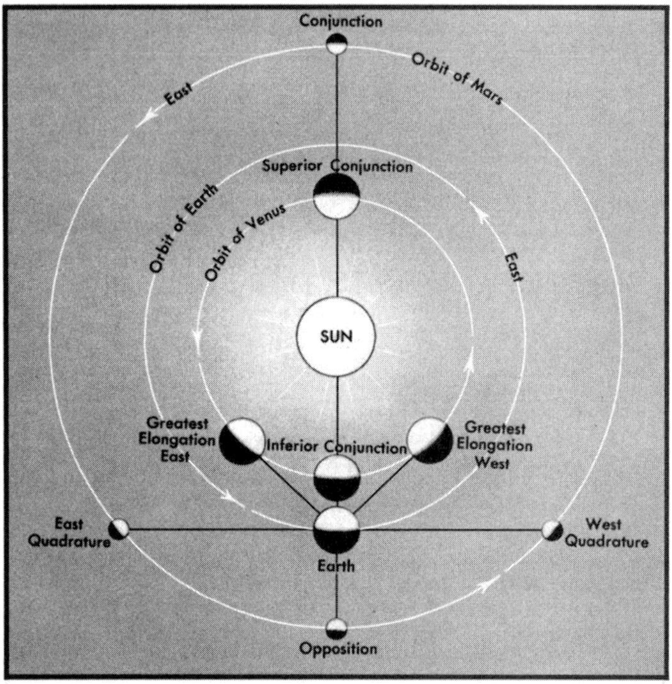

Figure 1510. Planetary configurations.

vary from about 30 to 50 days. Around inferior conjunction, Mercury disappears for about 5 days; near superior conjunction, it disappears for about 35 days. Observed with a telescope, Mercury is seen to go through phases similar to those of the Moon.

Venus can reach a distance of 47° from the Sun, allowing it to dominate the morning or evening sky. At maximum brilliance, about five weeks before and after inferior conjunction, it has a magnitude of about –4.4 and is brighter than any other object in the sky except the Sun and Moon. At these times it can be seen during the day and is sometimes observed for a celestial line of position. It appears as a morning or evening star for approximately 263 days in succession. Near inferior conjunction Venus disappears for 8 days; around superior conjunction it disappears for 50 days. When it transits the Sun, Venus can be seen by the naked eye as a small dot about the size of a group of Sunspots. Through strong binoculars or a telescope, Venus can be seen to go through a full set of phases.

1511. Superior Planets

As planets outside the Earth's orbit, the superior planets are not confined to the proximity of the Sun as seen from the Earth. They can pass behind the Sun (conjunction), but they cannot pass between the Sun and the Earth. Instead we see them move away from the Sun until they are opposite the Sun in the sky (**opposition**). When a superior planet is near conjunction, it rises and sets approximately with the Sun and is thus lost in the Sun's glare. Gradually it becomes visible in the early morning sky before sunrise. From day to day, it rises and sets earlier, becoming increasingly visible through the late night hours until dawn. Approaching opposition, the planet will rise in the late evening, until at opposition, it will rise when the Sun sets, be visible throughout the night, and set when the Sun rises.

Observed against the background stars, the planets normally move eastward in what is called **direct motion**. Approaching opposition, however, a planet will slow down, pause (at a stationary point), and begin moving westward (**retrograde motion**), until it reaches the next stationary point and resumes its direct motion. This is not because the planet is moving strangely in space. This relative, observed motion results because the faster moving Earth is catching up with and passing by the slower moving superior planet.

The superior planets are brightest and closest to the Earth at opposition. The interval between oppositions is known as the **synodic period**. This period is longest for the closest planet, Mars, and becomes increasingly shorter for the outer planets.

Unlike Mercury and Venus, the superior planets do not go through a full cycle of phases. They are always full or highly gibbous.

Mars can usually be identified by its orange color. It can become as bright as magnitude –2.8 but is more often between –1.0 and –2.0 at opposition. Oppositions occur at intervals of about 780 days. The planet is visible for about 330 days on either side of opposition. Near conjunction it is lost from view for about 120 days. Its two satellites can only be seen in a large telescope.

Jupiter, largest of the known planets, normally outshines Mars, regularly reaching magnitude –2.0 or brighter at opposition. Oppositions occur at intervals of about 400 days, with the planet being visible for about 180 days before and after opposition. The planet disappears for about 32 days at conjunction. Four satellites (of a total 16 currently known) are bright enough to be seen with binoculars. Their motions around Jupiter can be observed over the course of several hours.

Saturn, the outermost of the navigational planets, comes to opposition at intervals of about 380 days. It is visible for about 175 days before and after opposition, and disappears for about 25 days near conjunction. At opposition it becomes as bright as magnitude +0.8 to –0.2. Through good, high powered binoculars, Saturn appears as elongated because of its system of rings. A telescope is needed to examine the rings in any detail. Saturn is now known to have at least 18 satellites, none of which are visible to the unaided eye.

Uranus, **Neptune** and **Pluto** are too faint to be used for navigation; Uranus, at about magnitude 5.5, is faintly visible to the unaided eye.

1512. The Moon

The **Moon** is the only satellite of direct navigational interest. It revolves around the Earth once in about 27.3 days, as measured with respect to the stars. This is called the **sidereal month**. Because the Moon rotates on its axis with the same period with which it revolves around the Earth, the same side of the Moon is always turned toward the Earth. The cycle of phases depends on the Moon's revolution with respect to the Sun. This synodic month is approximately 29.53 days, but can vary from this average by up to a quarter of a day during any given month.

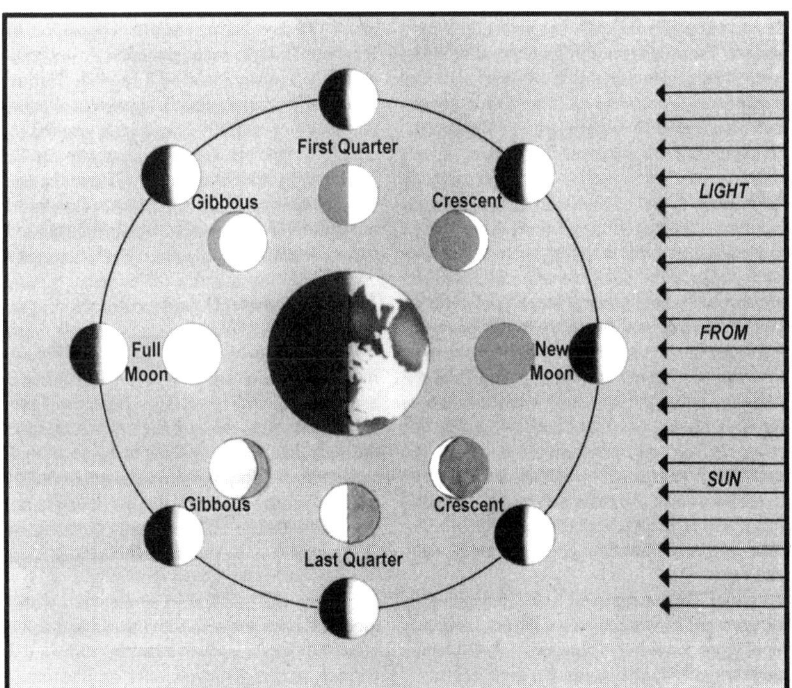

Figure 1512. Phases of the Moon. The inner figures of the Moon represent its appearance from the Earth.

When the Moon is in conjunction with the Sun (new Moon), it rises and sets with the Sun and is lost in the Sun's glare. The Moon is always moving eastward at about 12.2° per day, so that sometime after conjunction (as little as 16 hours, or as long as two days), the thin lunar crescent can be observed after sunset, low in the west. For the next couple of weeks, the Moon will **wax**, becoming more fully illuminated. From day to day, the Moon will rise (and set) later, becoming increasingly visible in the evening sky, until (about 7 days after new Moon) it reaches first quarter, when the Moon rises about noon and sets about midnight. Over the next week the Moon will rise later and later in the afternoon until full Moon, when it rises about sunset and dominates the sky throughout the night. During the next couple of weeks the Moon will **wane**, rising later and later at night. By last quarter (a week after full Moon), the Moon rises about midnight and sets at noon. As it approaches new Moon, the Moon becomes an increasingly thin crescent, and is seen only in the early morning sky. Sometime before conjunction (16 hours to 2 days before conjunction) the thin crescent will disappear in the glare of morning twilight.

At full Moon, the Sun and Moon are on opposite sides of the ecliptic. Therefore, in the winter the full Moon rises early, crosses the celestial meridian high in the sky, and sets late; as the Sun does in the summer. In the summer the full Moon rises in the southeastern part of the sky (Northern Hemisphere), remains relatively low in the sky, and sets along the southwestern horizon after a short time above the horizon.

At the time of the autumnal equinox, the part of the ecliptic opposite the Sun is most nearly parallel to the horizon. Since the eastward motion of the Moon is approximately along the ecliptic, the delay in the time of rising of the full Moon from night to night is less than at other times of the year. The full Moon nearest the autumnal equinox is called the **Harvest Moon**; the full Moon a month later is called the **Hunter's Moon**. See Figure 1512.

1513. Comets and Meteors

Although **comets** are noted as great spectacles of nature, very few are visible without a telescope. Those that become widely visible do so because they develop long, glowing tails. Comets are swarms of relatively small solid bodies held together by gravity. Around the nucleus, a gaseous head or coma and tail may form as the comet approaches the Sun. The tail is directed away from the Sun, so that it follows the head while the comet is approaching the Sun, and precedes the head while the comet is receding. The total mass of a comet is very small, and the tail is so thin that stars can easily be seen through it. In 1910, the Earth passed through the tail of Halley's comet without noticeable effect.

Compared to the well-ordered orbits of the planets, comets are erratic and inconsistent. Some travel east to west and some west to east, in highly eccentric orbits inclined at any angle to the ecliptic. Periods of revolution range from about 3 years to thousands of years. Some comets may speed away from the solar system after gaining velocity as they pass by Jupiter or Saturn.

The short-period comets long ago lost the gasses needed to form a tail. Long period comets, such as Halley's comet, are more likely to develop tails. The visibility of a comet depends very much on how close it approaches the Earth. In 1910, Halley's comet spread across the sky (Figure 1513). Yet when it returned in 1986, the Earth was not well situated to get a good view, and it was barely visible to the unaided eye.

Meteors, popularly called **shooting stars**, are tiny, solid bodies too small to be seen until heated to incandescence by air friction while passing through the Earth's atmosphere. A particularly bright meteor is called a **fireball**. One that explodes is called a **bolide**. A meteor that survives its trip through the atmosphere and lands as a solid particle is called a **meteorite**.

Vast numbers of meteors exist. An estimated average of some 1,000,000 meteors large enough to be seen enter the Earth's atmosphere each hour, and many times this number undoubtedly enter, but are too small to attract attention. The cosmic dust they create falls to earth in a constant shower.

Meteor showers occur at certain times of the year when the Earth passes through **meteor swarms**, the scattered remains of comets that have broken up. At these times the number of meteors observed is many times the usual number.

A faint glow sometimes observed extending upward approximately along the ecliptic before sunrise and after sunset has been attributed to the reflection of Sunlight from quantities of this material. This glow is called **zodiacal light**. A faint glow at that point of the ecliptic 180° from the Sun is called the **gegenschein** or **counterglow**.

1514. Stars

Stars are distant Suns, in many ways resembling our own. Like the Sun, stars are massive balls of gas that create their own energy through thermonuclear reactions.

Although stars differ in size and temperature, these differences are apparent only through analysis by astronomers. Some differences in color are noticeable to the unaided eye. While most stars appear white, some (those of lower temperature) have a reddish hue. In Orion, blue Rigel and red Betelgeuse, located on opposite sides of the belt, constitute a noticeable contrast.

The stars are not distributed uniformly around the sky. Striking configurations, known as **constellations**, were noted by ancient peoples, who supplied them with names and myths. Today astronomers use constellations—88 in all—to identify areas of the sky.

Under ideal viewing conditions, the dimmest star that can be seen with the unaided eye is of the sixth magnitude. In the entire sky there are about 6,000 stars of this

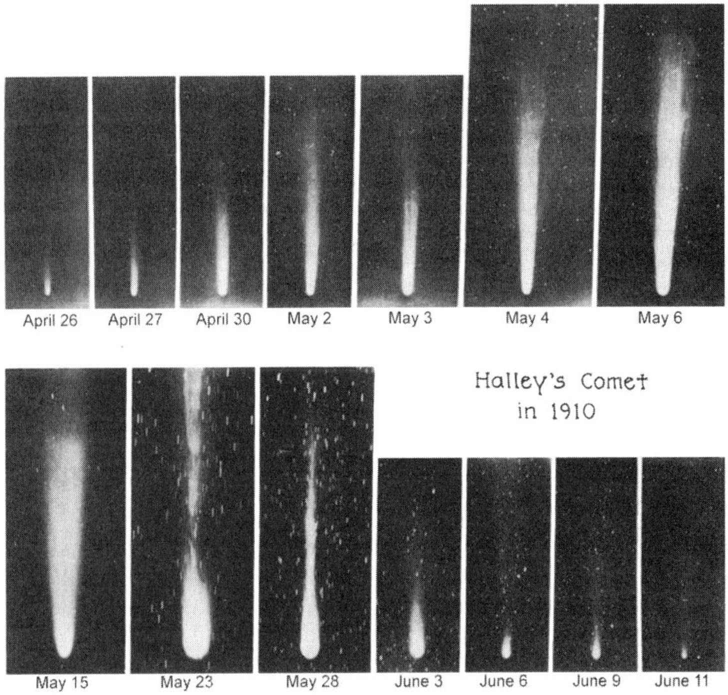

Figure 1513. Halley's Comet; fourteen views, made between April 26 and June 11, 1910. Courtesy of Mt. Wilson and Palomar Observatories.

magnitude or brighter. Half of these are below the horizon at any time. Because of the greater absorption of light near the horizon, where the path of a ray travels for a greater distance through the atmosphere, not more than perhaps 2,500 stars are visible to the unaided eye at any time. However, the average navigator seldom uses more than perhaps 20 or 30 of the brighter stars.

Stars which exhibit a noticeable change of magnitude are called **variable stars**. A star which suddenly becomes several magnitudes brighter and then gradually fades is called a **nova**. A particularly bright nova is called a **supernova**.

Two stars which appear to be very close together are called a **double star**. If more than two stars are included in the group, it is called a **multiple star**. A group of a few dozen to several hundred stars moving through space together is called an **open cluster**. The Pleiades is an example of an open cluster. There are also spherically symmetric clusters of hundreds of thousands of stars known as **globular clusters**. The globular clusters are all too distant to be seen with the naked eye.

A cloudy patch of matter in the heavens is called a **nebula**. If it is within the galaxy of which the Sun is a part, it is called a **galactic nebula**; if outside, it is called an **extragalactic nebula**.

Motion of a star through space can be classified by its vector components. That component in the line of sight is called **radial motion**, while that component across the line of sight, causing a star to change its apparent position relative to the background of more distant stars, is called **proper motion**.

1515. Galaxies

A **galaxy** is a vast collection of clusters of stars and clouds of gas. In a galaxy the stars tend to congregate in groups called **star clouds** arranged in long spiral arms. The spiral nature is believed due to revolution of the stars about the center of the galaxy, the inner stars revolving more rapidly than the outer ones (Figure 1515).

The Earth is located in the Milky Way galaxy, a slowly spinning disk more than 100,000 light years in diameter. All the bright stars in the sky are in the Milky Way. However, the most dense portions of the galaxy are seen as the great, broad band that glows in the summer nighttime sky. When we look toward the constellation Sagittarius, we are looking toward the

Figure 1515. Spiral nebula Messier 51, In Canes Venetici. Satellite nebula is NGC 5195.
Courtesy of Mt. Wilson and Palomar Observatories.

center of the Milky Way, 30,000 light years away.

Despite their size and luminance, almost all other galaxies are too far away to be seen with the unaided eye. An exception in the northern hemisphere is the Great Galaxy (sometimes called the Great Nebula) in Andromeda, which appears as a faint glow. In the southern hemisphere, the Large and Small Magellanic Clouds (named after Ferdinand Magellan) are the nearest known neighbors of the Milky Way. They are approximately 1,700,000 light years distant. The Magellanic Clouds can be seen as sizable glowing patches in the southern sky.

APPARENT MOTION

1516. Apparent Motion due to Rotation of the Earth

Apparent motion caused by the Earth's rotation is much greater than any other observed motion of celestial bodies. It is this motion that causes celestial bodies to appear to rise along the eastern half of the horizon, climb to maximum altitude as they cross the meridian, and set along the western horizon, at about the same point relative to due west as the rising point was to due east. This apparent motion along the daily path, or **diurnal circle**, of the body is approximately parallel to the plane of the equator. It would be exactly so if rotation of the Earth were the only motion and the axis of rotation of the Earth were stationary in space.

The apparent effect due to rotation of the Earth varies with the latitude of the observer. At the equator, where the equatorial plane is vertical (since the axis of rotation of the Earth is parallel to the plane of the horizon), bodies appear to rise and set vertically. Every celestial body is above the horizon approximately half the time. The celestial sphere as seen by an observer at the equator is called the right sphere, shown in Figure 1516a.

For an observer at one of the poles, bodies having constant declination neither rise nor set (neglecting precession of the equinoxes and changes in refraction), but circle the sky, always at the same altitude, making one complete trip around the horizon each day. At the North Pole the motion is clockwise, and at the South Pole it is counterclockwise. Approximately half the stars are always above the horizon and the other half never are. The parallel sphere at the poles is illustrated in Figure 1516b.

Between these two extremes, the apparent motion is a combination of the two. On this oblique sphere, illustrated in Figure 1516c, circumpolar celestial bodies remain above the horizon during the entire 24 hours, circling the elevated celestial pole each day. The stars of Ursa Major (the Big Dipper) and Cassiopeia are circumpolar for many observers in the United States.

An approximately equal part of the celestial sphere remains below the horizon during the entire day. For example, Crux is not visible to most observers in the United States. Other bodies rise obliquely along the eastern horizon, climb to maximum altitude at the celestial meridian, and set along the western horizon. The length of time above the horizon and the altitude at meridian transit vary with both the latitude of the observer and the declination of the body. At the polar circles of the Earth even the Sun becomes circumpolar. This is the land of the midnight Sun, where the Sun does not set during part of the summer and does not rise during part of the winter.

The increased obliquity at higher latitudes explains why days and nights are always about the same length in the tropics, and the change of length of the day becomes greater as latitude increases, and why twilight lasts longer in higher latitudes. Evening twilight starts at sunset, and morning twilight ends at sunrise. The darker limit of twilight occurs when the center of the Sun is a stated number of degrees below the celestial horizon. Three kinds of twilight are

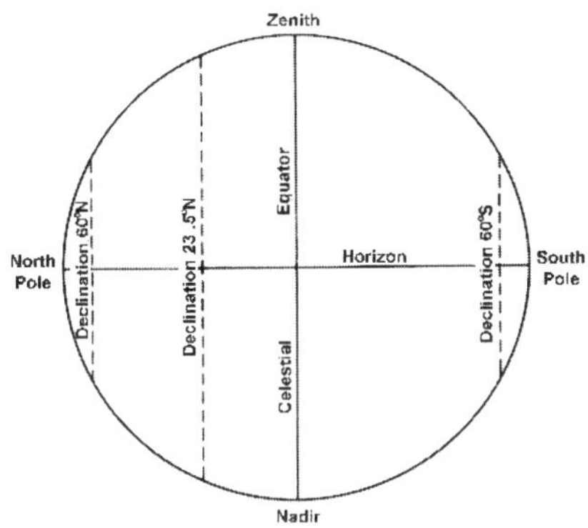

Figure 1516a. The right sphere.

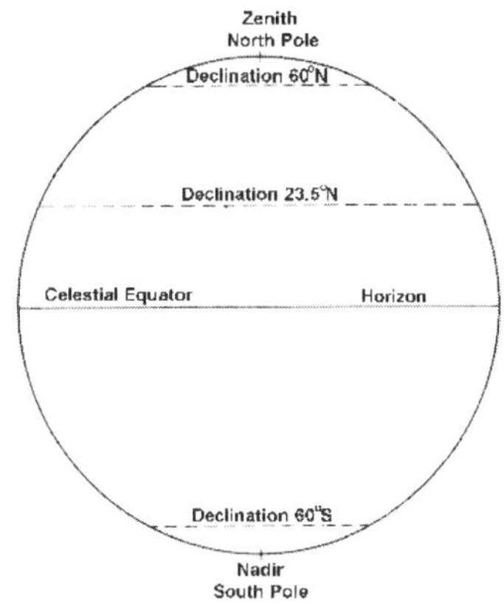

Figure 1516b. The parallel sphere.

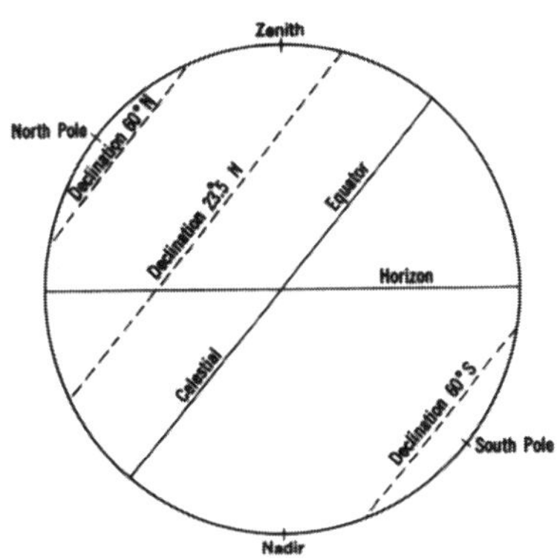

Figure 1516c. The oblique sphere at latitude 40°N.

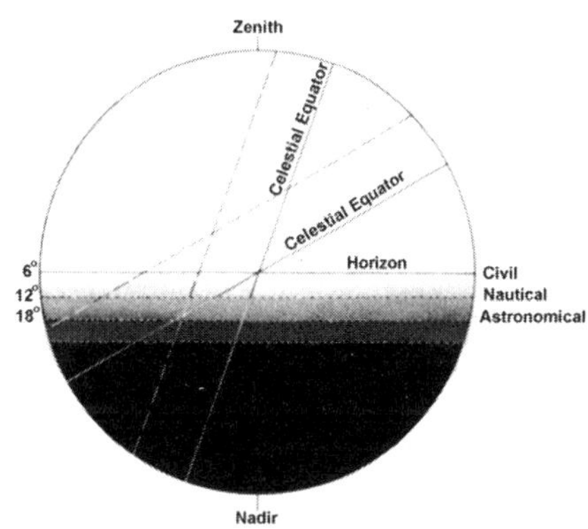

Figure 1516d. The various twilight at latitude 20°N and latitude 60°N.

Twilight	*Lighter limit*	*Darker limit*	*At darker limit*
civil	–0°50'	–6°	Horizon clear; bright stars visible
nautical	–0°50'	–12°	Horizon not visible
astronomical	–0°50'	–18°	Full night

Table 1516. Limits of the three twilights.

defined: civil, nautical and astronomical. See Table 1516.

The conditions at the darker limit are relative and vary considerably under different atmospheric conditions.

In Figure 1516d, the twilight band is shown, with the darker limits of the various kinds indicated. The nearly vertical celestial equator line is for an observer at latitude 20°N. The nearly horizontal celestial equator line is for an observer at latitude 60°N. The broken line in each case is the diurnal circle of the Sun when its declination is 15°N. The relative duration of any kind of twilight at the two latitudes is indicated by the portion of the diurnal circle between the horizon and the darker limit, although it is not directly proportional to the relative length of line shown since the projection is orthographic. The duration of twilight at the higher latitude is longer, proportionally, than shown. Note that complete darkness does not occur at latitude 60°N when the declination of the Sun is 15°N.

1517. Apparent Motion due to Revolution of the Earth

If it were possible to stop the rotation of the Earth so that the celestial sphere would appear stationary, the effects of the revolution of the Earth would become more noticeable. In one year the Sun would appear to make one complete trip around the Earth, from west to east. Hence, it would seem to move eastward a little less than 1° per day. This motion can be observed by watching the changing position of the Sun among the stars. But since both Sun and stars generally are not visible at the same time, a better way is to observe the constellations at the same time each night. On any night a star rises nearly four minutes earlier than on the previous night. Thus, the celestial sphere appears to shift westward nearly 1° each night, so that different constellations are associated with different seasons of the year.

Apparent motions of planets and the Moon are due to a combination of their motions and those of the Earth. If the rotation of the Earth were stopped, the combined apparent motion due to the revolutions of the Earth and other bodies would be similar to that occurring if both rotation and revolution of the Earth were stopped. Stars would appear nearly stationary in the sky but would undergo a small annual cycle of change due to aberration. The motion of the Earth in its orbit is sufficiently fast to cause the light from stars to appear to shift slightly in the direction of the Earth's motion. This is similar to the effect one experiences when walking in vertically-falling rain that appears to come from ahead due to the observer's own forward motion. The apparent direction of the light ray from the star is the vector difference of the motion of light and the motion of the Earth, similar to that of apparent wind on a moving vessel. This effect is most apparent for a body perpendicular to the line of travel of the Earth in its orbit, for which it reaches a maximum value of 20.5". The effect of aberration can be noted by comparing the coordinates (declination and sidereal hour angle) of various stars throughout the year. A change is observed in some bodies as the year progresses, but at the end of the year the values have returned almost to what they were at the beginning. The reason they do not return exactly is due to proper motion and precession of the equinoxes. It is also due to nutation, an irregularity in the motion of the Earth due to the disturbing effect of other celestial bodies, principally the Moon. Polar motion is a slight wobbling of the Earth about its axis of rotation and sometimes wandering of the poles. This motion, which does not exceed 40 feet from the mean position, produces slight variation of latitude and longitude of places on the Earth.

1518. Apparent Motion due to Movement of other Celestial Bodies

Even if it were possible to stop both the rotation and revolution of the Earth, celestial bodies would not appear stationary on the celestial sphere. The Moon would make one revolution about the Earth each sidereal month, rising in the west and setting in the east. The inferior planets would appear to move eastward and westward relative to the Sun, staying within the zodiac. Superior planets would appear to make one revolution around the Earth, from west to east, each sidereal period.

Since the Sun (and the Earth with it) and all other stars are in motion relative to each other, slow apparent motions would result in slight changes in the positions of the stars relative to each other. This space motion is, in fact, observed by telescope. The component of such motion across the line of sight, called proper motion, produces a change in the apparent position of the star. The maximum which has been observed is that of Barnard's Star, which is moving at the rate of 10.3 seconds per year. This is a tenth-magnitude star, not visible to the unaided eye. Of the 57 stars listed on the daily pages of the almanacs, Rigil Kentaurus has the greatest proper motion, about 3.7 seconds per year. Arcturus, with 2.3 seconds per year, has the greatest proper motion of the navigational stars in the Northern Hemisphere. In a few thousand years proper motion will be sufficient to materially alter some familiar configurations of stars, notably Ursa Major.

1519. The Ecliptic

The **ecliptic** is the path the Sun appears to take among the stars due to the annual revolution of the Earth in its orbit. It is considered a great circle of the celestial sphere, inclined at an angle of about 23°26' to the celestial equator, but undergoing a continuous slight change. This angle is called the **obliquity of the ecliptic**. This inclination is due to the fact that the axis of rotation of the Earth is not perpendicular to its orbit. It is this inclination which causes the Sun to appear to move north and south during the year, giving the Earth its seasons and changing lengths of periods of daylight.

Refer to Figure 1519a. The Earth is at perihelion early

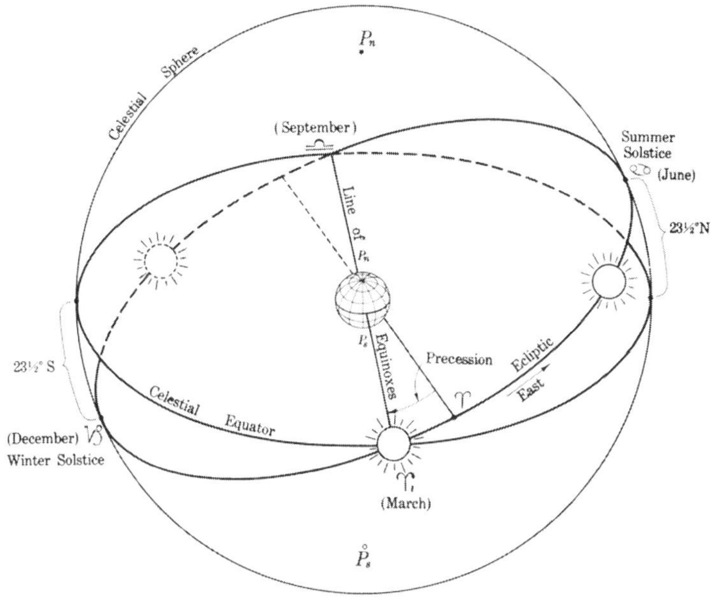

Figure 1519a. Apparent motion of the Sun in the ecliptic.

in January and at aphelion 6 months later. On or about June 21, about 10 or 11 days before reaching aphelion, the northern part of the Earth's axis is tilted toward the Sun. The north polar regions are having continuous Sunlight; the Northern Hemisphere is having its summer with long, warm days and short nights; the Southern Hemisphere is having winter with short days and long, cold nights; and the south polar region is in continuous darkness. This is the **summer solstice**. Three months later, about September 23, the Earth has moved a quarter of the way around the Sun, but its axis of rotation still points in about the same direction in space. The Sun shines equally on both hemispheres, and days and nights are the same length over the entire world. The Sun is setting at the North Pole and rising at the South Pole. The Northern Hemisphere is having its autumn, and the Southern Hemisphere its spring. This is the **autumnal equinox**. In another three months, on or about December 22, the Southern Hemisphere is tilted toward the Sun and conditions are the reverse of those six months earlier; the Northern Hemisphere is having its winter, and the Southern Hemisphere its summer. This is the **winter solstice**. Three months later, when both hemispheres again receive equal amounts of Sunshine, the Northern Hemisphere is having spring and the Southern Hemisphere autumn, the reverse of conditions six months before. This is the **vernal equinox**.

The word "equinox," meaning "equal nights," is applied because it occurs at the time when days and nights are of approximately equal length all over the Earth. The word "solstice," meaning "Sun stands still," is applied because the Sun stops its apparent northward or southward motion and momentarily "stands still" before it starts in the opposite direction. This action, somewhat analogous to the "stand" of the tide, refers to the motion in a north-south direction only, and not to the daily apparent revolution around the Earth. Note that it does not occur when the Earth is at perihelion or aphelion. Refer to Figure 1519a. At the time of the vernal equinox, the Sun is directly over the equator, crossing from the Southern Hemisphere to the Northern Hemisphere. It rises due east and sets due west, remaining above the horizon for approximately 12 hours. It is not exactly 12 hours because of refraction, semidiameter, and the height of the eye of the observer. These cause it to be above the horizon a little longer than below the horizon. Following the vernal equinox, the northerly declination increases, and the Sun climbs higher in the sky each day (at the latitudes of the United States), until the summer solstice, when a declination of about 23°26' north of the celestial equator is reached. The Sun then gradually retreats southward until it is again over the equator at the autumnal equinox, at about 23°26' south of the celestial equator at the winter solstice, and back over the celestial equator again at the next vernal equinox.

The Earth is nearest the Sun during the northern hemisphere winter. It is not the distance between the Earth and Sun that is responsible for the difference in temperature during the different seasons, but the altitude of the Sun in the sky and the length of time it remains above the horizon.

During the summer the rays are more nearly vertical, and hence more concentrated, as shown in Figure 1519b. Since the Sun is above the horizon more than half the time, heat is being added by absorption during a longer period than it is being lost by radiation. This explains the lag of the seasons. Following the longest day, the Earth continues to receive more heat than it dissipates, but at a decreasing proportion. Gradually the proportion decreases until a balance is reached, after which the Earth cools, losing more heat than it gains. This is analogous to the day, when the highest temperatures normally occur several hours after the Sun reaches maximum altitude at meridian transit. A similar lag occurs at other seasons of the year. Astronomically, the seasons begin at the equinoxes and solstices. Meteorologically, they differ from place to place.

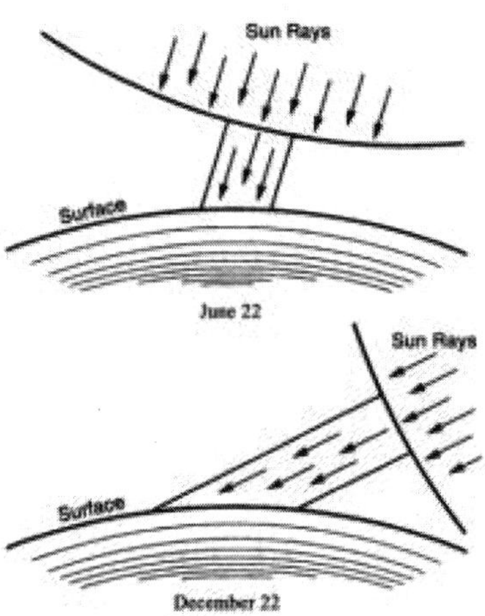

Figure 1519b. Sunlight in summer and winter. Winter sunlight is distributed over a larger area and shines fewer hours each day, causing less heat energy to reach the Earth.

Since the Earth travels faster when nearest the Sun, the northern hemisphere (astronomical) winter is shorter than its summer by about seven days.

Everywhere between the parallels of about 23°26'N and about 23°26'S the Sun is directly overhead at some time during the year. Except at the extremes, this occurs twice: once as the Sun appears to move northward, and the second time as it moves southward. This is the **torrid zone**. The northern limit is the **Tropic of Cancer**, and the southern limit is the **Tropic of Capricorn**. These names come from the constellations which the Sun entered at the solstices when the names were first applied more than 2,000 years ago. Today, the Sun is in the next constellation toward the west because of precession of the equinoxes. The parallels about 23°26' from the poles, marking the approximate limits of the circumpolar Sun, are called **polar circles**, the one in the Northern Hemisphere being the **Arctic Circle** and the one in the Southern Hemisphere the **Antarctic Circle**. The areas inside the polar circles are the north and south **frigid zones**. The regions between the frigid zones and the torrid zones are the north and south **temperate zones**.

The expression "vernal equinox" and associated expressions are applied both to the times and points of occurrence of the various phenomena. Navigationally, the vernal equinox is sometimes called the **first point of Aries** (symbol ♈) because, when the name was given, the Sun entered the constellation Aries, the ram, at this time. This point is of interest to navigators because it is the origin for measuring **sidereal hour angle**. The expressions March equinox, June solstice, September equinox, and December solstice are occasionally applied as appropriate, because the more common names are associated with the seasons in the Northern Hemisphere and are six months out of step for the Southern Hemisphere.

The axis of the Earth is undergoing a precessional motion similar to that of a top spinning with its axis tilted. In about 25,800 years the axis completes a cycle and returns to the position from which it started. Since the celestial equator is 90° from the celestial poles, it too is moving. The result is a slow westward movement of the equinoxes and solstices, which has already carried them about 30°, or one constellation, along the ecliptic from the positions they occupied when named more than 2,000 years ago. Since sidereal hour angle is measured from the vernal equinox, and declination from the celestial equator, the coordinates of celestial bodies would be changing even if the bodies themselves were stationary. This westward motion of the equinoxes along the ecliptic is called **precession of the equinoxes**. The total amount, called **general precession**, is about 50 seconds of arc per year. It may be considered divided into two components: precession in right ascension (about 46.10 seconds per year) measured along the celestial equator, and precession in declination (about 20.04" per year) measured perpendicular to the celestial equator. The annual change in the coordinates of any given star, due to precession alone, depends upon its position on the celestial sphere, since these coordinates are measured relative to the polar axis while the precessional motion is relative to the ecliptic axis.

Due to precession of the equinoxes, the celestial poles are slowly describing circles in the sky. The north celestial pole is moving closer to Polaris, which it will pass at a distance of approximately 28 minutes about the year 2102. Following this, the polar distance will increase, and eventually other stars, in their turn, will become the Pole Star.

The precession of the Earth's axis is the result of gravitational forces exerted principally by the Sun and Moon on the Earth's equatorial bulge. The spinning Earth responds to these forces in the manner of a gyroscope. Regression of the nodes introduces certain irregularities known as **nutation** in the precessional motion. See Figure 1519c.

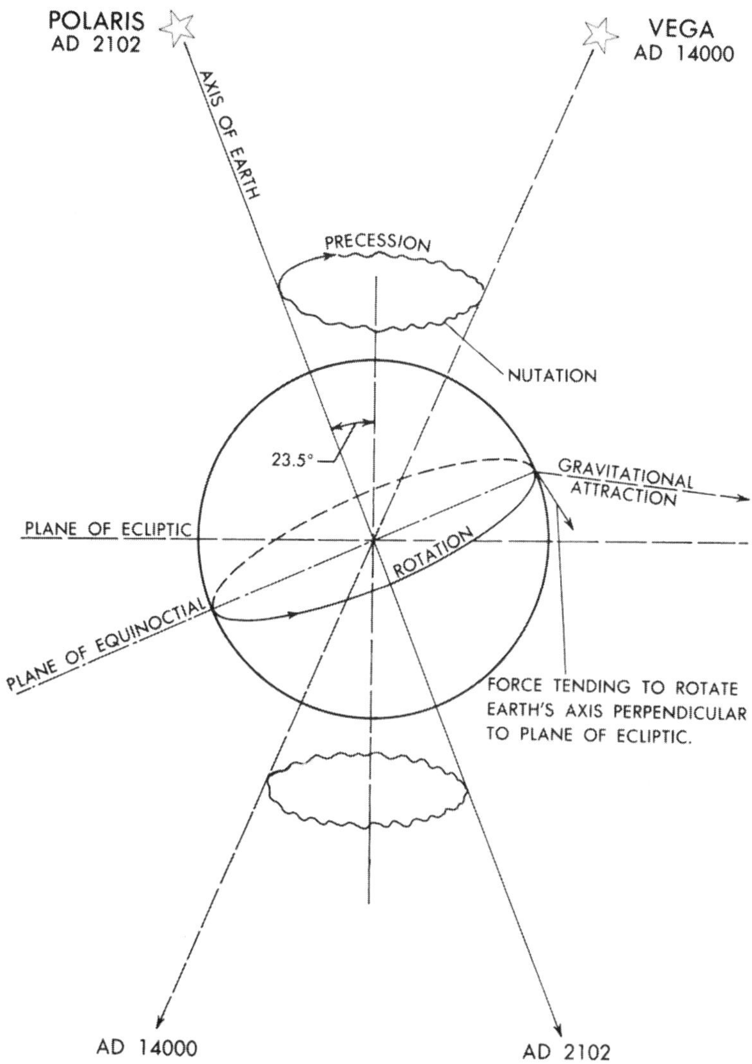

Figure 1519c. Precession and nutation.

1520. The Zodiac

The **zodiac** is a circular band of the sky extending 8° on each side of the ecliptic. The navigational planets and the Moon are within these limits. The zodiac is divided into 12 sections of 30° each, each section being given the name and symbol ("sign") of a constellation. These are shown in Figure 1520. The names were assigned more than 2,000 years ago, when the Sun entered Aries at the vernal equinox, Cancer at the summer solstice, Libra at the autumnal equinox, and Capricornus at the winter solstice. Because of precession, the zodiacal signs have shifted with respect to the constellations. Thus at the time of the vernal equinox, the Sun is said to be at the "first point of Aries," though it is in the constellation Pisces.

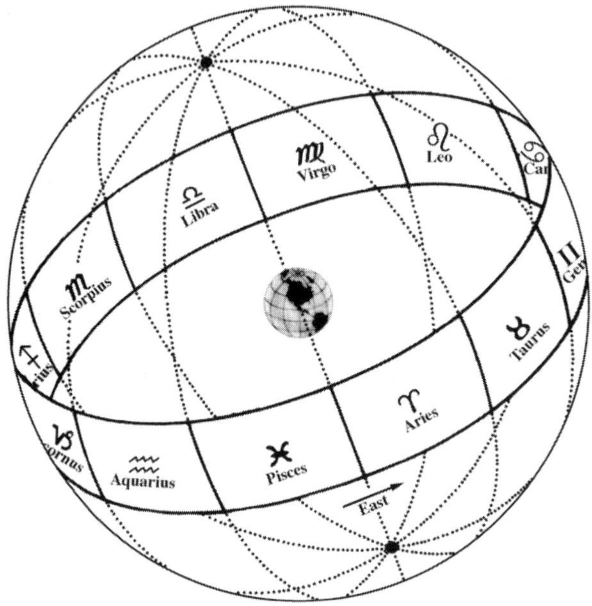

Figure 1520. The zodiac.

1521. Time and the Calendar

Traditionally, astronomy has furnished the basis for measurement of time, a subject of primary importance to the navigator. The **year** is associated with the revolution of the Earth in its orbit. The **day** is one rotation of the Earth about its axis.

The duration of one rotation of the Earth depends upon the external reference point used. One rotation relative to the Sun is called a **solar day**. However, rotation relative to the apparent Sun (the actual Sun that appears in the sky) does not provide time of uniform rate because of variations in the rate of revolution and rotation of the Earth. The error due to lack of uniform rate of revolution is removed by using a fictitious **mean Sun**. Thus, mean solar time is nearly equal to the average apparent solar time. Because the accumulated difference between these times, called the **equation of time**, is continually changing, the period of daylight is shifting slightly, in addition to its increase or decrease in length due to changing declination. Apparent and mean Suns seldom cross the celestial meridian at the same time. The earliest sunset (in latitudes of the United States) occurs about two weeks before the winter solstice, and the latest sunrise occurs about two weeks after winter solstice. A similar but smaller apparent discrepancy occurs at the summer solstice.

Universal Time is a particular case of the measure known in general as mean solar time. Universal Time is the mean solar time on the Greenwich meridian, reckoned in days of 24 mean solar hours beginning with 0 hours at midnight. Universal Time and sidereal time are rigorously related by a formula so that if one is known the other can be found. Universal Time is the standard in the application of astronomy to navigation.

If the vernal equinox is used as the reference, a **sidereal day** is obtained, and from it, **sidereal time**. This indicates the approximate positions of the stars, and for this reason it is the basis of star charts and star finders. Because of the revolution of the Earth around the Sun, a sidereal day is about 3 minutes 56 seconds shorter than a solar day, and there is one more sidereal than solar days in a year. One mean solar day equals 1.00273791 mean sidereal days. Because of precession of the equinoxes, one rotation of the Earth with respect to the stars is not quite the same as one rotation with respect to the vernal equinox. One mean solar day averages 1.0027378118868 rotations of the Earth with respect to the stars.

In tide analysis, the Moon is sometimes used as the reference, producing a **lunar day** averaging 24 hours 50 minutes (mean solar units) in length, and lunar time.

Since each kind of day is divided arbitrarily into 24 hours, each hour having 60 minutes of 60 seconds, the length of each of these units differs somewhat in the various kinds of time.

Time is also classified according to the terrestrial meridian used as a reference. **Local time** results if one's own meridian is used, **zone time** if a nearby reference meridian is used over a spread of longitudes, and **Greenwich** or **Universal Time** if the Greenwich meridian is used.

The period from one vernal equinox to the next (the cycle of the seasons) is known as the **tropical year**. It is approximately 365 days, 5 hours, 48 minutes, 45 seconds, though the length has been slowly changing for many centuries. Our calendar, the Gregorian calendar, approximates the tropical year with a combination of common years of 365 days and leap years of 366 days. A leap year is any year divisible by four, unless it is a century year, which must be divisible by 400 to be a leap year. Thus, 1700, 1800, and 1900 were not leap years, but 2000 was. A critical mistake was made by John Hamilton Moore in calling 1800 a leap year, causing an error in the tables in his book, *The Practical Navigator*. This error caused the loss of at least one ship and was later discovered by Nathaniel Bowditch while writing the first edition of *The New American Practical Navigator*.

See Chapter 18 for an in-depth discussion of time.

1522. Eclipses

If the orbit of the Moon coincided with the plane of the ecliptic, the Moon would pass in front of the Sun at every new Moon, causing a solar eclipse. At full Moon, the Moon would pass through the Earth's shadow, causing a lunar eclipse. Because of the Moon's orbit is inclined 5° with respect to the ecliptic, the Moon usually passes above or below the Sun at new Moon and above or below the Earth's shadow at full Moon. However, there are two points at which the plane of the Moon's orbit intersects the ecliptic. These are the **nodes** of the Moon's orbit. If the Moon passes one of these points at the same time as the Sun, a **solar eclipse** takes place. This is shown in Figure 1522.

The Sun and Moon are of nearly the same apparent size to an observer on the Earth. If the Moon is at perigee, the Moon's apparent diameter is larger than that of the Sun, and its shadow reaches the Earth as a nearly round dot only a few miles in diameter. The dot moves rapidly across the Earth, from west to east, as the Moon continues in its orbit. Within the dot, the Sun is completely hidden from view, and a total eclipse of the Sun occurs. For a considerable distance around the shadow, part of the surface of the Sun is obscured, and a **partial eclipse** occurs. In the line of travel of the shadow a partial eclipse occurs as the round disk of the Moon appears to move slowly across the surface of the Sun, hiding an ever-increasing part of it, until the total eclipse occurs. Because of the uneven edge of the mountainous Moon, the light is not cut off evenly. But several last illuminated portions appear through the valleys or passes between the mountain peaks. These are called **Baily's Beads**.

A total eclipse is a spectacular phenomenon. As the last light from the Sun is cut off, the solar **corona**, or envelope of thin, illuminated gas around the Sun becomes visible. Wisps of more dense gas may appear as **solar prominences**. The only light reaching the observer is that diffused by the atmosphere surrounding the shadow. As the Moon appears to continue on across the face of the Sun, the Sun finally emerges from the other side, first as Baily's Beads, and then as an ever widening crescent until no part of its surface is obscured by the Moon.

The duration of a total eclipse depends upon how nearly the Moon crosses the center of the Sun, the location of the shadow on the Earth, the relative orbital speeds of the Moon and Earth, and (principally) the relative apparent diameters of the Sun and Moon. The maximum length that can occur is a little more than seven minutes.

If the Moon is near apogee, its apparent diameter is less than that of the Sun, and its shadow does not quite reach the Earth. Over a small area of the Earth directly in line with the Moon and Sun, the Moon appears as a black disk almost covering the surface of the Sun, but with a thin ring of the Sun around its edge. This **annular eclipse** occurs a little more often than a total eclipse.

If the shadow of the Moon passes close to the Earth, but not directly in line with it, a partial eclipse may occur without a total or annular eclipse.

An eclipse of the Moon (or **lunar eclipse**) occurs when the Moon passes through the shadow of the Earth, as shown in Figure 1522. Since the diameter of the Earth is about $3^1/_2$ times that of the Moon, the Earth's shadow at the distance of the Moon is much larger than that of the Moon. A total eclipse of the Moon can last nearly $1^3/_4$ hours, and some part of the Moon may be in the Earth's shadow for almost 4 hours.

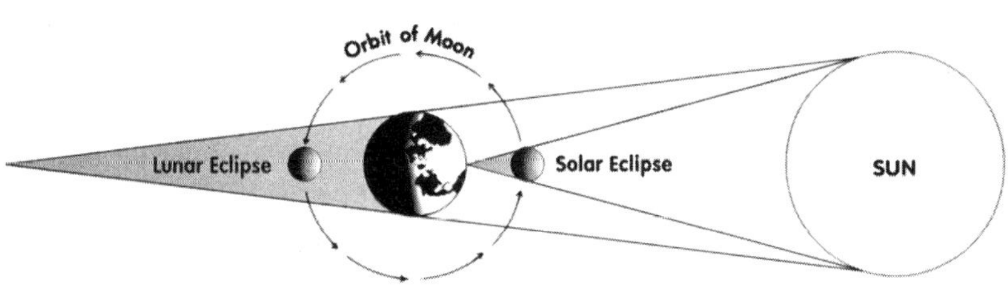

Figure 1522. Eclipses of the Sun and Moon.

During a total solar eclipse no part of the Sun is visible because the Moon is in the line of sight. But during a lunar eclipse some light does reach the Moon, diffracted by the atmosphere of the Earth, and hence the eclipsed full Moon is visible as a faint reddish disk. A lunar eclipse is visible over the entire hemisphere of the Earth facing the Moon. Anyone who can see the Moon can see the eclipse.

During any one year there may be as many as five eclipses of the Sun, and always there are at least two. There may be as many as three eclipses of the Moon, or none. The total number of eclipses during a single year does not exceed seven, and can be as few as two. There are more solar than lunar eclipses, but the latter can be seen more often because of the restricted areas over which solar eclipses are visible.

The Sun, Earth, and Moon are nearly aligned on the line of nodes twice each eclipse year of 346.6 days. This is less than a calendar year because of **regression of the nodes**. In a little more than 18 years the line of nodes returns to approximately the same position with respect to the Sun, Earth, and Moon. During an almost equal period, called the **saros**, a cycle of eclipses occurs. During the following saros the cycle is repeated with only minor differences.

COORDINATES

1523. Latitude And Longitude

Latitude and **longitude** are coordinates used to locate positions on the Earth. This article discusses three different definitions of these coordinates.

Astronomic latitude is the angle (ABQ, Figure 1523) between a line in the direction of gravity (AB) at a station and the plane of the equator (QQ'). **Astronomic longitude** is the angle between the plane of the celestial meridian at a station and the plane of the celestial meridian at Greenwich. These coordinates are customarily found by means of celestial observations. If the Earth were perfectly homogeneous and round, these positions would be consistent and satisfactory. However, because of deflection of the vertical due to uneven distribution of the mass of the Earth, lines of equal astronomic latitude and longitude are not circles, although the irregularities are small. In the United States the prime vertical component (affecting longitude) may be a little more than 18", and the meridional component (affecting latitude) as much as 25".

Geodetic latitude is the angle (ACQ, Figure 1523) between a normal to the spheroid (AC) at a station and the plane of the geodetic equator (QQ'). **Geodetic longitude** is the angle between the plane defined by the normal to the spheroid and the axis of the Earth and the plane of the geodetic meridian at Greenwich. These values are obtained when astronomical latitude and longitude are corrected for deflection of the vertical. These coordinates are used for charting and are frequently referred to as **geographic latitude** and **geographic longitude**, although these expressions are sometimes used to refer to astronomical latitude.

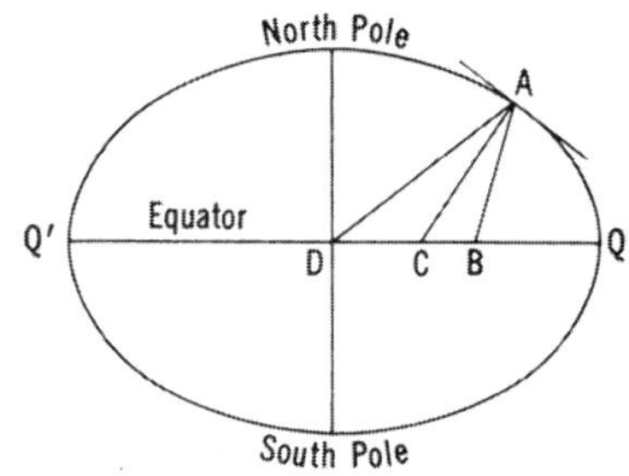

Figure 1523. Three kinds of latitude at point A.

Geocentric latitude is the angle (ADQ, Figure 1523) at the center of the ellipsoid between the plane of its equator (QQ') and a straight line (AD) to a point on the surface of the Earth. This differs from geodetic latitude because the Earth is a spheroid rather than a sphere, and the meridians are ellipses. Since the parallels of latitude are considered to be circles, geodetic longitude is geocentric, and a separate expression is not used. The difference between geocentric and geodetic latitudes is a maximum of about 11.6' at latitude 45°.

Because of the oblate shape of the ellipsoid, the length of a degree of geodetic latitude is not everywhere the same, increasing from about 59.7 nautical miles at the equator to about 60.3 nautical miles at the poles. The value of 60 nautical miles customarily used by the navigator is correct at about latitude 45°.

MEASUREMENTS ON THE CELESTIAL SPHERE

1524. Elements of the Celestial Sphere

The **celestial sphere** (Article 1501) is an imaginary sphere of infinite radius with the Earth at its center (Figure 1524a). The north and south celestial poles of this sphere are located by extension of the Earth's axis. The **celestial equator** (sometimes called **equinoctial**) is formed by projecting the plane of the Earth's equator to the celestial sphere. A **celestial meridian** is formed by the intersection of the plane of a terrestrial meridian and the celestial sphere. It is the arc of a great circle through the poles of the celestial sphere.

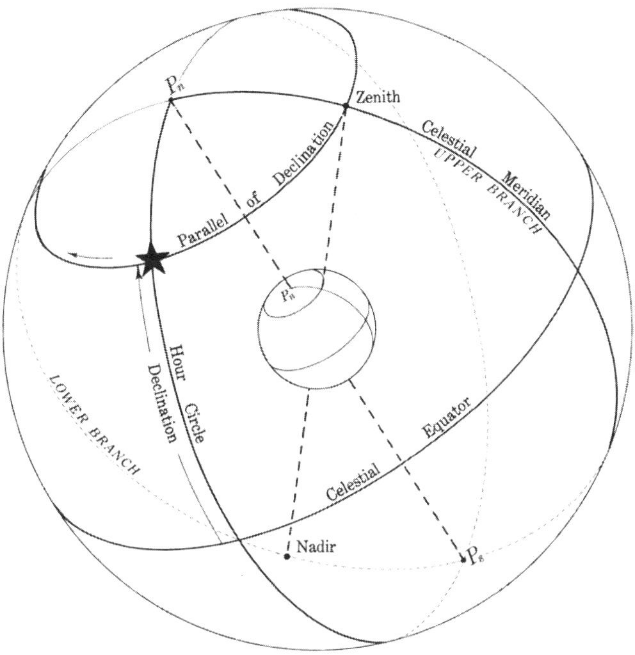

Figure 1524a. Elements of the celestial sphere. The celestial equator is the primary great circle.

The point on the celestial sphere vertically overhead of an observer is the **zenith**, and the point on the opposite side of the sphere vertically below him is the **nadir**. The zenith and nadir are the extremities of a diameter of the celestial sphere through the observer and the common center of the Earth and the celestial sphere. The arc of a celestial meridian between the poles is called the **upper branch** if it contains the zenith and the **lower branch** if it contains the nadir. The upper branch is frequently used in navigation, and references to a celestial meridian are understood to mean only its upper branch unless otherwise stated. Celestial meridians take the names of their terrestrial counterparts, such as 65° west.

An **hour circle** is a great circle through the celestial poles and a point or body on the celestial sphere. It is similar to a celestial meridian, but moves with the celestial sphere as it rotates about the Earth, while a celestial meridian remains fixed with respect to the Earth.

The location of a body on its hour circle is defined by the body's angular distance from the celestial equator. This distance, called **declination**, is measured north or south of the celestial equator in degrees, from 0° through 90°, similar to latitude on the Earth.

A circle parallel to the celestial equator is called a **parallel of declination**, since it connects all points of equal declination. It is similar to a parallel of latitude on the Earth. The path of a celestial body during its daily apparent revolution around the Earth is called its **diurnal circle**. It is not actually a circle if a body changes its declination. Since the declination of all navigational bodies is continually changing, the bodies are describing flat, spherical spirals as they circle the Earth. However, since the change is relatively slow, a diurnal circle and a parallel of declination are usually considered identical.

A point on the celestial sphere may be identified at the intersection of its parallel of declination and its hour circle. The parallel of declination is identified by the declination.

Two basic methods of locating the hour circle are in use. First, the angular distance west of a reference hour circle through a point on the celestial sphere, called the vernal equinox or first point of Aries, is called **sidereal hour angle (SHA)** (Figure 1524b). This angle, measured eastward from the vernal equinox, is called **right ascension** and is usually expressed in time units.

The second method of locating the hour circle is to indicate its angular distance west of a celestial meridian (Figure 1524c). If the Greenwich celestial meridian is used as the reference, the angular distance is called **Greenwich hour angle (GHA)**, and if the meridian of the observer, it is called **local hour angle (LHA)**. It is

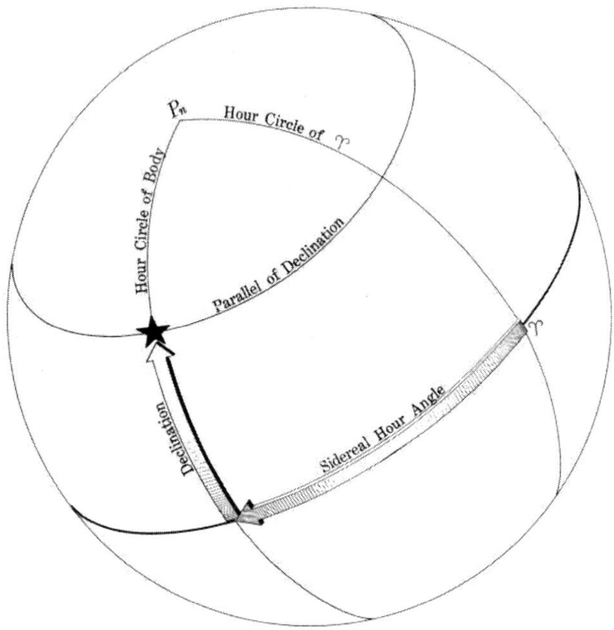

Figure 1524b. A point on the celestial sphere can be located by its declination and sidereal hour angle.

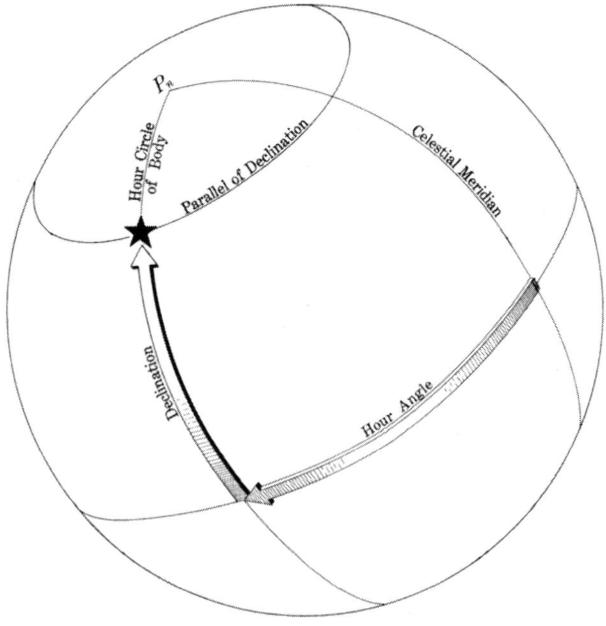

Figure 1524c. A point on the celestial sphere can be located by its declination and hour angle.

sometimes more convenient to measure hour angle either eastward or westward, as longitude is measured on the Earth, in which case it is called **meridian angle** (designated "t").

A point on the celestial sphere may also be located using altitude and azimuth coordinates based upon the horizon as the primary great circle instead of the celestial equator.

COORDINATE SYSTEMS

1525. The Celestial Equator System of Coordinates

The familiar graticule of latitude and longitude lines, expanded until it reaches the celestial sphere, forms the basis of the celestial equator system of coordinates. On the celestial sphere latitude becomes declination, while longitude becomes sidereal hour angle, measured from the vernal equinox.

Declination is angular distance north or south of the celestial equator (d in Figure 1525a). It is measured along an hour circle, from 0° at the celestial equator through 90° at the celestial poles. It is labeled N or S to indicate the direction of measurement. All points having the same declination lie along a parallel of declination.

Polar distance (p) is angular distance from a celestial pole, or the arc of an hour circle between the celestial pole and a point on the celestial sphere. It is measured along an hour circle and may vary from 0° to 180°, since either pole may be used as the origin of measurement. It is usually considered the complement of declination, though it may be either 90° – d or 90° + d, depending upon the pole used.

Local hour angle (LHA) is angular distance west of the local celestial meridian, or the arc of the celestial equator between the upper branch of the local celestial meridian and the hour circle through a point on the celestial sphere, measured westward from the local celestial meridian, through 360°. It is also the similar arc of the parallel of declination and the angle at the celestial pole, similarly measured. If the Greenwich (0°) meridian is used as the reference, instead of the local meridian, the expression **Greenwich hour angle (GHA)** is applied. It is sometimes convenient to measure the arc or angle in either an easterly or westerly direction from the local meridian, through 180°, when it is called **meridian angle (t)** and labeled E or W to indicate the direction of measurement. All bodies or other points having the same hour angle lie along the same hour circle.

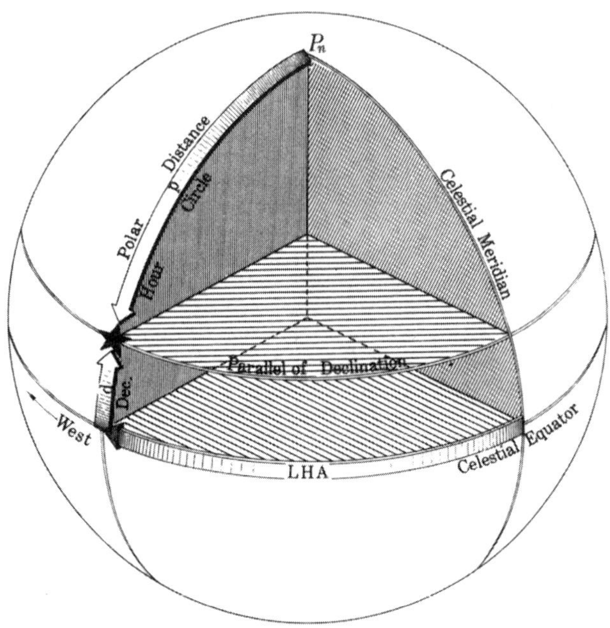

Figure 1525a. The celestial equator system of coordinates, showing measurements of declination, polar distance, and local hour angle.

Because of the apparent daily rotation of the celestial sphere, hour angle continually increases, but meridian angle increases from 0° at the celestial meridian to 180°W, which is also 180°E, and then decreases to 0° again. The rate of change for the mean Sun is 15° per hour. The rate of all other bodies except the Moon is within 3' of this value. The average rate of the Moon is about 15.5°.

As the celestial sphere rotates, each body crosses each branch of the celestial meridian approximately once a day. This crossing is called **meridian transit** (sometimes called culmination). It may be called **upper transit** to indicate crossing of the upper branch of the celestial meridian, and **lower transit** to indicate crossing of the lower branch.

The **time diagram** shown in Figure 1525b illustrates the relationship between the various hour angles and meridian angle. The circle is the celestial equator as seen from above the South Pole, with the upper branch of the observer's meridian (P_sM) at the top. The radius P_sG is the Greenwich meridian; P_s ♈ is the hour circle of the vernal equinox. The Sun's hour circle is to the east of the observer's meridian; the Moon's hour circle is to the west of the observer's meridian Note that when LHA is less than 180°, t is numerically the same and is labeled W, but that when LHA is greater than 180°, t = 360° – LHA and is labeled E. In Figure 1525b arc GM is the longitude, which in this case is west. The relationships shown apply equally to other arrangements of radii, except for relative magnitudes of the quantities involved.

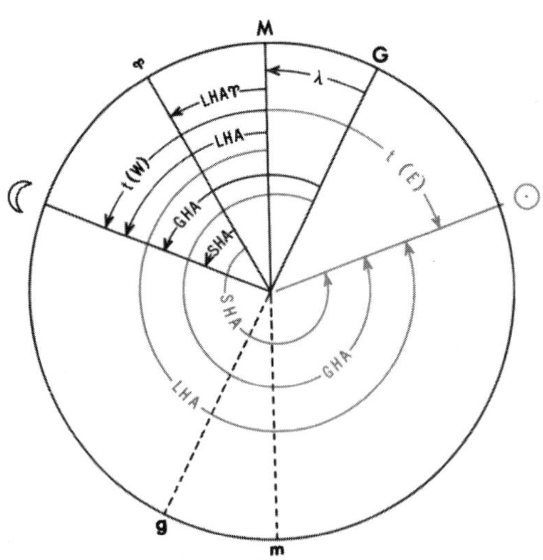

Figure 1525b. Time diagram.

1526. The Horizons

The second set of celestial coordinates with which the navigator is directly concerned is based upon the horizon as the primary great circle. However, since several different horizons are defined, these should be thoroughly understood before proceeding with a consideration of the horizon system of coordinates.

The line where Earth and sky appear to meet is called the **visible** or **apparent horizon**. On land this is usually an irregular line unless the terrain is level. At sea the visible horizon appears very regular and is often very sharp. However, its position relative to the celestial sphere depends primarily upon (1) the refractive index of the air and (2) the height of the observer's eye above the surface.

Figure 1526 shows a cross section of the Earth and celestial sphere through the position of an observer at A. A straight line through A and the center of the Earth O is the vertical of the observer and contains his zenith (Z) and nadir (Na). A plane perpendicular to the true vertical is a horizontal plane, and its intersection with the celestial sphere is a horizon. It is the **celestial horizon** if the plane passes through the center of the Earth, the **geoidal horizon** if it is tangent to the Earth, and the **sensible horizon** if it passes through the eye of the observer at A. Since the radius of the Earth is considered negligible with respect to that of the celestial sphere, these horizons become superimposed, and most measurements are referred only to the celestial horizon. This is sometimes called the **rational horizon**.

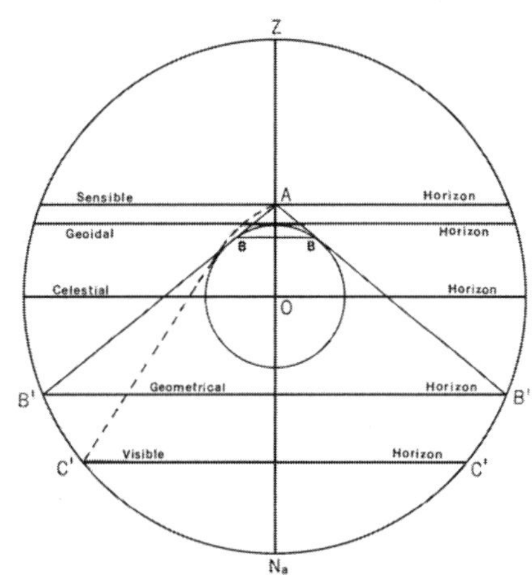

Figure 1526. The horizons used in navigation.

If the eye of the observer is at the surface of the Earth, his visible horizon coincides with the plane of the geoidal horizon; but when elevated above the surface, as at A, his eye becomes the vertex of a cone which is tangent to the

Earth at the small circle BB, and which intersects the celestial sphere in B'B', the **geometrical horizon**. This expression is sometimes applied to the celestial horizon.

Because of refraction, the visible horizon C'C' appears above but is actually slightly below the geometrical horizon as shown in Figure 1526. In Figure 1525b the Local hour angle, Greenwich hour angle, and sidereal hour angle are measured westward through 360°. Meridian angle (t) is measured eastward or westward through 180° and labeled E or W to indicate the direction of measurement.

For any elevation above the surface, the celestial horizon is usually above the geometrical and visible horizons, the difference increasing as elevation increases. It is thus possible to observe a body which is above the visible horizon but below the celestial horizon. That is, the body's altitude is negative and its zenith distance is greater than 90°.

1527. The Horizon System of Coordinates

This system is based upon the celestial horizon as the primary great circle and a series of secondary vertical circles which are great circles through the zenith and nadir of the observer and hence perpendicular to his horizon (Figure 1527a). Thus, the celestial horizon is similar to the equator, and the vertical circles are similar to meridians, but with one important difference. The celestial horizon and vertical circles are dependent upon the position of the observer and hence move with him as he changes position, while the primary and secondary great circles of both the geographical and celestial equator systems are independent of the observer. The horizon and celestial equator systems coincide for an observer at the geographical pole of the Earth and are mutually perpendicular for an observer on the equator. At all other places the two are oblique.

The vertical circle through the north and south points of the horizon passes through the poles of the celestial equator system of coordinates. One of these poles (having the same name as the latitude) is above the horizon and is called the **elevated pole**. The other, called the **depressed pole**, is below the horizon. Since this vertical circle is a great circle through the celestial poles, and includes the zenith of the observer, it is also a celestial meridian. In the horizon system it is called the **principal vertical circle**. The vertical circle through the east and west points of the horizon, and hence perpendicular to the principal vertical circle, is called the **prime vertical circle**, or simply the **prime vertical**.

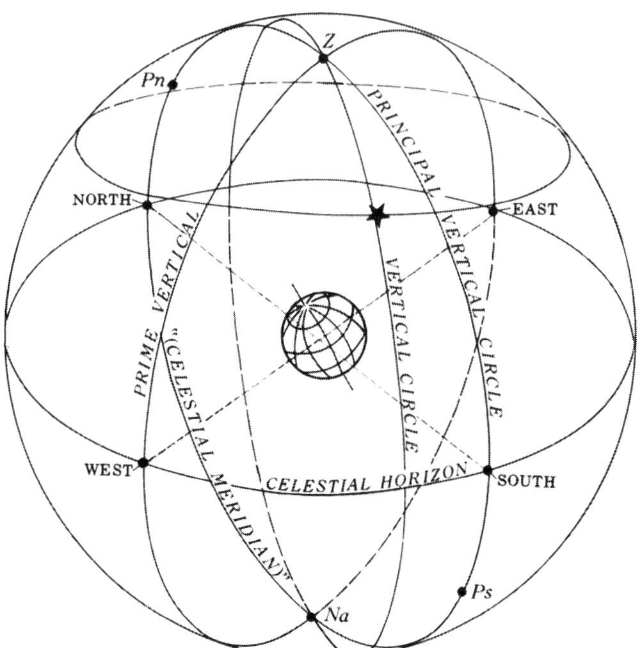

Figure 1527a. Elements of the celestial sphere. The celestial horizon is the primary great circle.

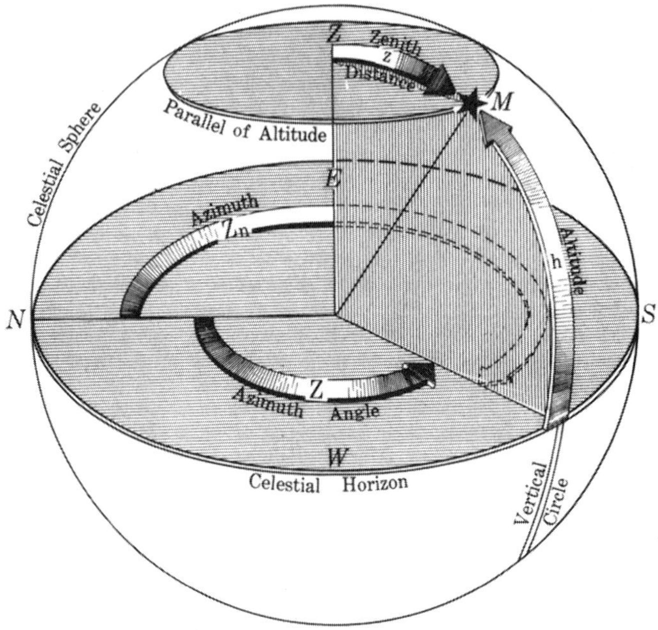

Figure 1527b. The horizon system of coordinates, showing measurement of altitude, zenith distance, azimuth, and azimuth angle.

Earth	*Celestial Equator*	*Horizon*	*Ecliptic*
equator	celestial equator	horizon	ecliptic
poles	celestial poles	zenith; nadir	ecliptic poles
meridians	hours circle; celestial meridians	vertical circles	circles of latitude
prime meridian	hour circle of Aries	principal or prime vertical circle	circle of latitude through Aries
parallels	parallels of declination	parallels of altitude	parallels of latitude
latitude	declination	altitude	celestial altitude
colatitude	polar distance	zenith distance	celestial colatitude
longitude	SHA; RA; GHA; LHA; t	azimuth; azimuth angle; amplitude	celestial longitude

Table 1527. The four systems of celestial coordinates and their analogous terms.

As shown in Figure 1527b, altitude is angular distance above the horizon. It is measured along a vertical circle, from 0° at the horizon through 90° at the zenith. Altitude measured from the visible horizon may exceed 90° because of the dip of the horizon, as shown in Figure 1526. Angular distance below the horizon, called negative altitude, is provided for by including certain negative altitudes in some tables for use in celestial navigation. All points having the same altitude lie along a parallel of altitude.

Zenith distance (z) is angular distance from the zenith, or the arc of a vertical circle between the zenith and a point on the celestial sphere. It is measured along a vertical circle from 0° through 180°. It is usually considered the complement of altitude. For a body above the celestial horizon it is equal to 90° – h and for a body below the celestial horizon it is equal to 90° – (– h) or 90° + h.

The horizontal direction of a point on the celestial sphere, or the bearing of the geographical position, is called **azimuth** or **azimuth angle** depending upon the method of measurement. In both methods it is an arc of the horizon (or parallel of altitude), or an angle at the zenith. It is **azimuth** (Zn) if measured clockwise through 360°, starting at the north point on the horizon, and **azimuth angle** (Z) if measured either clockwise or counterclockwise through 180°, starting at the north point of the horizon in north latitude and the south point of the horizon in south latitude.

The ecliptic system is based upon the ecliptic as the primary great circle, analogous to the equator. The points 90° from the ecliptic are the north and south ecliptic poles. The series of great circles through these poles, analogous to meridians, are circles of latitude. The circles parallel to the plane of the ecliptic, analogous to parallels on the Earth, are parallels of latitude or circles of longitude. Angular distance north or south of the ecliptic, analogous to latitude, is celestial latitude. Celestial longitude is measured eastward along the ecliptic through 360°, starting at the vernal equinox. This system of coordinates is of interest chiefly to astronomers.

The four systems of celestial coordinates are analogous to each other and to the terrestrial system, although each has distinctions such as differences in directions, units, and limits of measurement. Table 1527 indicates the analogous term or terms under each system.

1528. Diagram on the Plane of the Celestial Meridian

From an imaginary point outside the celestial sphere and over the celestial equator, at such a distance that the view would be orthographic, the great circle appearing as the outer limit would be a celestial meridian. Other celestial meridians would appear as ellipses. The celestial equator would appear as a diameter 90° from the poles, and parallels of declination as straight lines parallel to the equator. The view would be similar to an orthographic map of the Earth.

A number of useful relationships can be demonstrated by drawing a diagram on the plane of the celestial meridian showing this orthographic view. Arcs of circles can be substituted for the ellipses without destroying the basic relationships. Refer to Figure 1528a. In the lower diagram the circle represents the celestial meridian, QQ' the celestial equator, Pn and Ps the north and south celestial poles, respectively. If a star has a declination of 30° N, an angle of 30° can be measured from the celestial equator, as shown. It could be measured either to the right or left, and would have been toward the south pole if the declination had been south. The parallel of declination is a line through this point and parallel to the celestial equator. The star is somewhere on this line (actually a circle viewed on edge).

To locate the hour circle, draw the upper diagram so that Pn is directly above Pn of the lower figure (in line with the polar axis Pn-Ps), and the circle is of the same diameter as that of the lower figure. This is the plan view, looking down on the celestial sphere from the top. The circle is the celestial equator. Since the view is from above the north celestial pole, west is clockwise. The diameter QQ' is the celestial meridian shown as a circle in the lower diagram. If the right half is considered the upper branch, local hour angle is measured clockwise from this line to the hour circle, as shown. In this case the LHA is 80°. The intersection of the hour circle and celestial equator, point A, can be projected down to the lower diagram (point A') by a straight line parallel to the polar axis. The elliptical hour circle can be represented approximately by an arc of a circle through A', Pn, Ps. The center of this circle is somewhere along the celestial equator line QQ', extended if necessary. It is usually found by trial and error. The intersection of the hour circle and parallel of declination locates the star.

Since the upper diagram serves only to locate point A' in the lower diagram, the two can be combined. That is, the LHA arc can be drawn in the lower diagram, as shown, and point A projected upward to A'. In practice, the upper diagram is not drawn, being shown here for illustrative purposes.

In this example the star is on that half of the sphere toward the observer, or the western part. If LHA had been greater than 180°, the body would have been on the eastern or "back" side.

From the east or west point over the celestial horizon, the orthographic view of the horizon system of coordinates would be similar to that of the celestial equator system from a point over the celestial equator, since the celestial meridian is also the principal vertical circle. The horizon would appear as a diameter, parallels of altitude as straight lines parallel to the horizon, the zenith and nadir as poles 90° from the horizon, and vertical circles as ellipses through the zenith and nadir, except for the principal vertical circle, which would appear as a circle, and the prime vertical, which would appear as a diameter perpendicular to the horizon.

A celestial body can be located by altitude and azimuth in a manner similar to that used with the celestial equator system. If the altitude is 25°, this angle is measured from the horizon toward the zenith and the parallel of altitude is drawn as a straight line parallel to the horizon, as shown at hh' in the lower diagram of Figure 1528b. The plan view from above the zenith is shown in the upper diagram. If north is taken at the left, as shown, azimuths are measured clockwise from this point. In the figure the azimuth is 290° and the azimuth angle is N70°W. The vertical circle is located by measuring either arc. Point A thus located can be projected vertically downward to A' on the horizon of the lower diagram, and the vertical circle represented approximately by the arc of a circle through A' and the zenith and nadir. The center of this circle is on NS, extended if necessary. The body is at the intersection of the parallel of altitude and the vertical circle. Since the upper diagram serves only to locate A' on the lower diagram, the two can

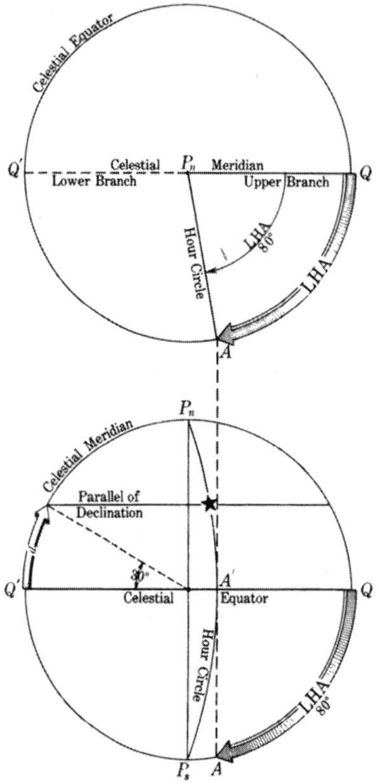

Figure 1528a. Measurement of celestial equator system of coordinates.

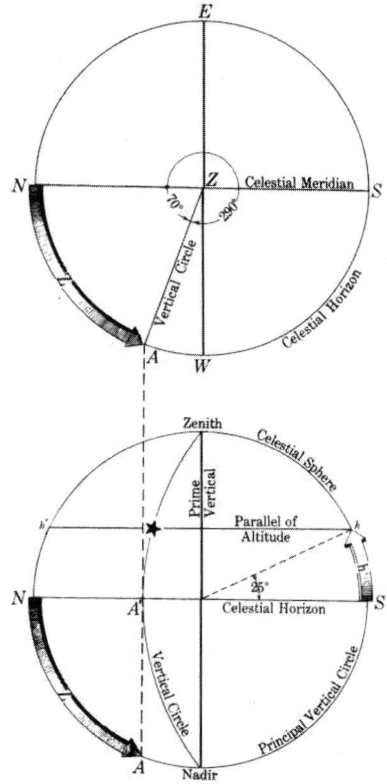

Figure 1528b. Measurement of horizon system of coordinates.

be combined, point A located on the lower diagram and projected upward to A', as shown. Since the body of the example has an azimuth greater than 180°, it is on the western or "front" side of the diagram.

Since the celestial meridian appears the same in both the celestial equator and horizon systems, the two diagrams can be combined and, if properly oriented, a body can be located by one set of coordinates, and the coordinates of the other system can be determined by measurement.

Refer to Figure 1528c, in which the black lines represent the celestial equator system, and the gray lines the horizon system. By convention, the zenith is shown at the top and the north point of the horizon at the left. The west point on the horizon is at the center, and the east point directly behind it. In the figure the latitude is 37°N. Therefore, the zenith is 37° north of the celestial equator. Since the zenith is established at the top of the diagram, the equator can be found by measuring an arc of 37° toward the south, along the celestial meridian. If the declination is 30°N and the LHA is 80°, the body can be located as shown by the black lines, and described above.

The altitude and azimuth can be determined by the reverse process to that described above. Draw a line hh' through the body and parallel to the horizon, NS. The altitude, 25°, is found by measurement, as shown. Draw the arc of a circle through the body and the zenith and nadir. From A', the intersection of this arc with the horizon, draw a vertical line intersecting the circle at A. The azimuth, N70°W, is found by measurement, as shown. The prefix N is applied to agree with the latitude. The body is left (north) of ZNa, the prime vertical circle. The suffix W applies because the LHA, 80°, shows that the body is west of the meridian.

If altitude and azimuth are given, the body is located by means of the gray lines. The parallel of declination is then drawn parallel to QQ', the celestial equator, and the declination determined by measurement. Point L' is located by drawing the arc of a circle through Pn, the star, and Ps. From L' a line is drawn perpendicular to QQ', locating L. The meridian angle is then found by measurement. The declination is known to be north because the body is between

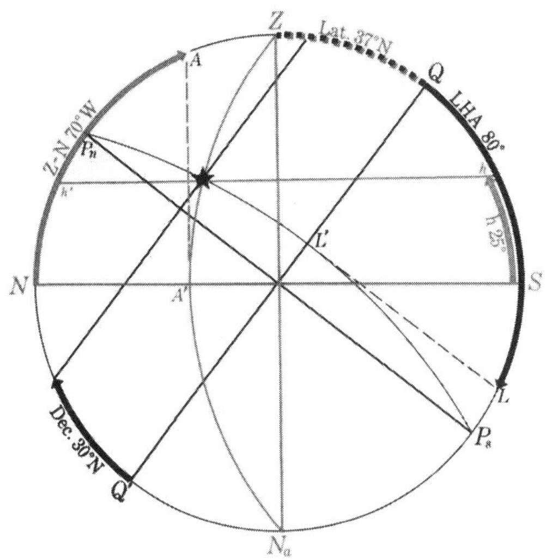

Figure 1528c. Diagram on the plane of the celestial meridian.

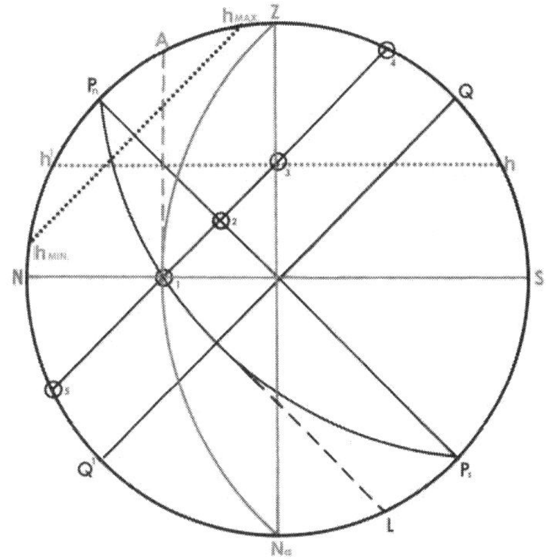

Figure 1528d. A diagram on the plane of the celestial meridian for lat. 45°N.

the celestial equator and the north celestial pole. The meridian angle is west, to agree with the azimuth, and hence LHA is numerically the same.

Since QQ' and PnPs are perpendicular, and ZNa and NS are also perpendicular, arc NPn is equal to arc ZQ. That is, the altitude of the elevated pole is equal to the declination of the zenith, which is equal to the latitude. This relationship is the basis of the method of determining latitude by an observation of Polaris.

The diagram on the plane of the celestial meridian is useful in approximating a number of relationships. Consider Figure 1528d. The latitude of the observer (NPn or ZQ) is 45°N. The declination of the Sun (Q4) is 20°N. Neglecting the change in declination for one day, note the following: At sunrise, position 1, the Sun is on the horizon (NS), at the "back" of the diagram. Its altitude, h, is 0°. Its azimuth angle, Z, is the arc NA, N63°E. This is prefixed N to agree with the latitude and suffixed E to agree with the meridian angle of the Sun at sunrise. Hence, Zn = 063°. The amplitude, A, is the arc ZA, E27°N. The meridian angle, t, is the arc QL, 110°E. The suffix E is applied because the Sun is east of the meridian at rising. The LHA is 360° – 110° = 250°.

As the Sun moves upward along its parallel of declination, its altitude increases. It reaches position 2 at about 0600, when t = 90°E. At position 3 it is on the prime vertical, ZNa. Its azimuth angle, Z, is N90°E, and Zn = 090°. The altitude is Nh' or Sh, 27°.

Moving on up its parallel of declination, it arrives at position 4 on the celestial meridian about noon-when t and LHA are both 0°, by definition. On the celestial meridian a body's azimuth is 000° or 180°. In this case it is 180° because the body is south of the zenith. The maximum altitude occurs at meridian transit. In this case the arc S4 represents the maximum altitude, 65°. The zenith distance, z, is the arc Z4, 25°. A body is not in the zenith at meridian transit unless its declination's magnitude and name are the same as the latitude.

Continuing on, the Sun moves downward along the "front" or western side of the diagram. At position 3 it is again on the prime vertical. The altitude is the same as when previously on the prime vertical, and the azimuth angle is numerically the same, but now measured toward the west. The azimuth is 270°. The Sun reaches position 2 six hours after meridian transit and sets at position 1. At this point, the azimuth angle is numerically the same as at sunrise, but westerly, and Zn = 360° – 63° = 297°. The amplitude is W27°N.

After sunset the Sun continues on downward, along its parallel of declination, until it reaches position 5, on the lower branch of the celestial meridian, about midnight. Its negative altitude, arc N5, is now greatest, 25°, and its azimuth is 000°. At this point it starts back up along the "back" of the diagram, arriving at position 1 at the next sunrise, to start another cycle.

Half the cycle is from the crossing of the 90° hour circle (the PnPs line, position 2) to the upper branch of the celestial meridian (position 4) and back to the PnPs line (position 2). When the declination and latitude have the same name (both north or both south), more than half the parallel of declination (position 1 to 4 to 1) is above the horizon, and the body is above the horizon more than half the time, crossing the 90° hour circle above the horizon. It rises and sets on the same side of the prime vertical as the elevated pole. If the declination is of the same name but numerically smaller than the latitude, the body crosses the prime vertical above the horizon. If the declination and latitude have the same name and are numerically equal, the body is in the zenith at upper transit. If the declination is of the same name but numerically greater than the latitude, the body crosses the upper branch of the celestial meridian between the zenith and elevated pole and does not cross the prime vertical. If the declination is of the same name as the latitude and complementary to it (d + L = 90°), the body is on the horizon at lower transit and does not set. If the declination is of the same name as the latitude and numerically greater than the colatitude, the body is above the horizon during its entire daily cycle and has maximum and minimum altitudes. This is shown by the black dotted line in Figure 1528d.

If the declination is 0° at any latitude, the body is above the horizon half the time, following the celestial equator QQ', and rises and sets on the prime vertical. If the declination is of contrary name (one north and the other south), the body is above the horizon less than half the time and crosses the 90° hour circle below the horizon. It rises and sets on the opposite side of the prime vertical from the elevated pole. If the declination is of contrary name and numerically smaller than the latitude, the body crosses the prime vertical below the horizon. If the declination is of contrary name and numerically equal to the latitude, the body is in the nadir at lower transit. If the declination is of contrary name and complementary to the latitude, the body is on the horizon at upper transit. If the declination is of contrary name and numerically greater than the colatitude, the body does not rise.

All of these relationships, and those that follow, can be derived by means of a diagram on the plane of the celestial meridian. They are modified slightly by atmospheric refraction, height of eye, semidiameter, parallax, changes in declination, and apparent speed of the body along its diurnal circle.

It is customary to keep the same orientation in south latitude, as shown in Figure 1528e. In this illustration the latitude is 45°S, and the declination of the body is 15°N. Since Ps is the elevated pole, it is shown above the southern horizon, with both SPs and ZQ equal to the latitude, 45°. The body rises at position 1, on the opposite side of the prime vertical from the elevated pole. It moves upward along its parallel of declination to position 2, on the upper branch of the celestial meridian, bearing north; and then it moves downward along the "front" of the diagram to position 1, where it sets. It remains above the horizon for less than half the time because declination and latitude are of contrary name. The azimuth at rising is arc NA, the amplitude ZA, and the azimuth angle SA. The altitude circle at meridian transit is shown at hh'.

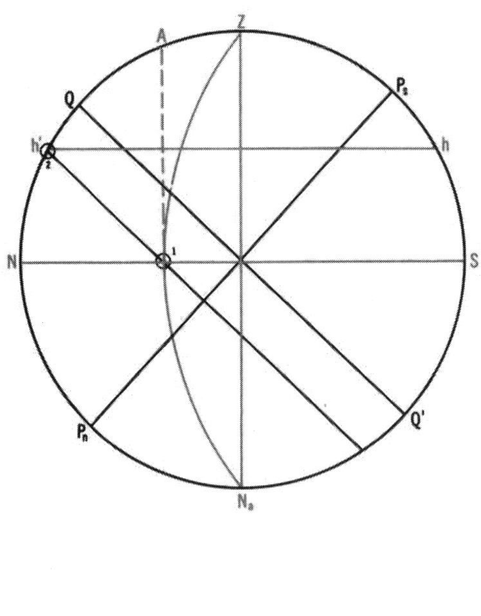

Figure 1528e. A diagram on the plane of the celestial meridian for lat. 45°S.

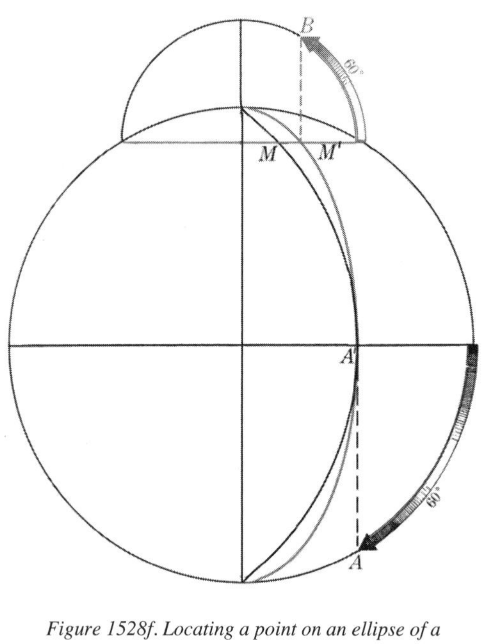

Figure 1528f. Locating a point on an ellipse of a diagram on the plane of the celestial meridian.

A diagram on the plane of the celestial meridian can be used to demonstrate the effect of a change in latitude. As the latitude increases, the celestial equator becomes more nearly parallel to the horizon. The colatitude becomes smaller increasing the number of circumpolar bodies and those which neither rise nor set. It also increases the difference in the length of the days between summer and winter. At the poles celestial bodies circle the sky, parallel to the horizon. At the equator the 90° hour circle coincides with the horizon. Bodies rise and set vertically; and are above the horizon half the time. At rising and setting the amplitude is equal to the declination. At meridian transit the altitude is equal to the codeclination. As the latitude changes name, the same-contrary name relationship with declination reverses. This accounts for the fact that one hemisphere has winter while the other is having summer.

NAVIGATIONAL COORDINATES

Coordinate	Symbol	Measured from	Measured along	Direction	Measured to	Units	Precision	Maximum value	Labels
latitude	L, lat.	equator	meridian	N, S	parallel	°, ′	0′.1	90°	N, S
colatitude	colat.	poles	meridian	S, N	parallel	°, ′	0′.1	90°	—
longitude	λ, long.	prime meridian	parallel	E, W	local meridian	°, ′	0′.1	180°	E, W
declination	d, dec.	celestial equator	hour circle	N, S	parallel of declination	°, ′	0′.1	90°	N, S
polar distance	p	elevated pole	hour circle	S, N	parallel of declination	°, ′	0′.1	180°	—
altitude	h	horizon	vertical circle	up	parallel of altitude	°, ′	0′.1	90°*	—
zenith distance	z	zenith	vertical circle	down	parallel of altitude	°, ′	0′.1	180°	—
azimuth	Zn	north	horizon	E	vertical circle	°	0°.1	360°	—
azimuth angle	Z	north, south	horizon	E, W	vertical circle	°	0°.1	180° or 90°	N, S...E, W
amplitude	A	east, west	horizon	N, S	body	°	0°.1	90°	E, W...N, S
Greenwich hour angle	GHA	Greenwich celestial meridian	parallel of declination	W	hour circle	°, ′	0′.1	360°	—
local hour angle	LHA	local celestial meridian	parallel of declination	W	hour circle	°, ′	0′.1	360°	—
meridian angle	t	local celestial meridian	parallel of declination	E, W	hour circle	°, ′	0′.1	180°	E, W
sidereal hour angle	SHA	hour circle of vernal equinox	parallel of declination	W	hour circle	°, ′	0′.1	360°	—
right ascension	RA	hour circle of vernal equinox	parallel of declination	E	hour circle	h, m, s	1^s	24^h	—
Greenwich mean time	GMT	lower branch Greenwich celestial meridian	parallel of declination	W	hour circle mean Sun	h, m, s	1^s	24^h	—
local mean time	LMT	lower branch local celestial meridian	parallel of declination	W	hour circle mean Sun	h, m, s	1^s	24^h	—
zone time	ZT	lower branch zone celestial meridian	parallel of declination	W	hour circle mean Sun	h, m, s	1^s	24^h	—
Greenwich apparent time	GAT	lower branch Greenwich celestial meridian	parallel of declination	W	hour circle apparent Sun	h, m, s	1^s	24^h	—
local apparent time	LAT	lower branch local celestial meridian	parallel of declination	W	hour circle apparent Sun	h, m, s	1^s	24^h	—
Greenwich sidereal time	GST	Greenwich celestial meridian	parallel of declination	W	hour circle vernal equinox	h, m, s	1^s	24^h	—
local sidereal time	LST	local celestial meridian	parallel of declination	W	hour circle vernal equinox	h, m, s	1^s	24^h	—

*When measured from celestial horizon.

Figure 1528g. Navigational Coordinates.

The error arising from showing the hour circles and vertical circles as arcs of circles instead of ellipses increases with increased declination or altitude. More accurate results can be obtained by measurement of azimuth on the parallel of altitude instead of the horizon, and of hour angle on the parallel of declination instead of the celestial equator. Refer to Figure 1528f. The vertical circle shown is for a body having an azimuth angle of S60°W. The arc of a circle is shown in black, and the ellipse in gray. The black arc is obtained by measurement around the horizon, locating A' by means of A, as previously described. The intersection of this arc with the altitude circle at 60° places the body at M. If a semicircle is drawn with the altitude circle as a diameter, and the azimuth angle measured around this, to B, a perpendicular to the hour circle locates the body at M', on the ellipse. By this method the altitude circle, rather than the horizon, is, in effect, rotated through 90° for the measurement. This refinement is seldom used because actual values are usually found mathematically, the diagram on the plane of the meridian being used primarily to indicate relationships.

With experience, one can visualize the diagram on the plane of the celestial meridian without making an actual drawing. Devices with two sets of spherical coordinates, on either the orthographic or stereographic projection, pivoted at the center, have been produced commercially to provide a mechanical diagram on the plane of the celestial meridian. However, since the diagram's principal use is to illustrate certain relationships, such a device is not a necessary part of the navigator's equipment.

Figure 1528g summarizes navigation coordinate systems.

1529. The Navigational Triangle

A triangle formed by arcs of great circles of a sphere is called a **spherical triangle**. A spherical triangle on the celestial sphere is called a **celestial triangle**. The spherical triangle of particular significance to navigators is called the **navigational triangle**, formed by arcs of a *celestial meridian*, an *hour circle*, and a *vertical circle*. Its vertices are the *elevated pole*, the *zenith*, and a *point on the celestial sphere* (usually a celestial body). The terrestrial counterpart is also called a navigational triangle, being formed by arcs of two meridians and the great circle connecting two places on the Earth, one on each meridian. The vertices are the two places and a pole. In great-circle sailing these places are the point of departure and the destination. In celestial navigation they are the **assumed position (AP)** of the observer and the **geographical position (GP)** of the body (the point having the body in its zenith). The GP of the Sun is sometimes called the **subsolar point**, that of the Moon the **sublunar point**, that of a satellite (either natural or artificial) the **subsatellite point**, and that of a star its **substellar** or **subastral point**. When used to solve a celestial observation, either the celestial or terrestrial triangle may be called the **astronomical triangle**.

The navigational triangle is shown in Figure 1529a on a diagram on the plane of the celestial meridian. The Earth is at the center, O. The star is at M, dd' is its parallel of declination, and hh' is its altitude circle.

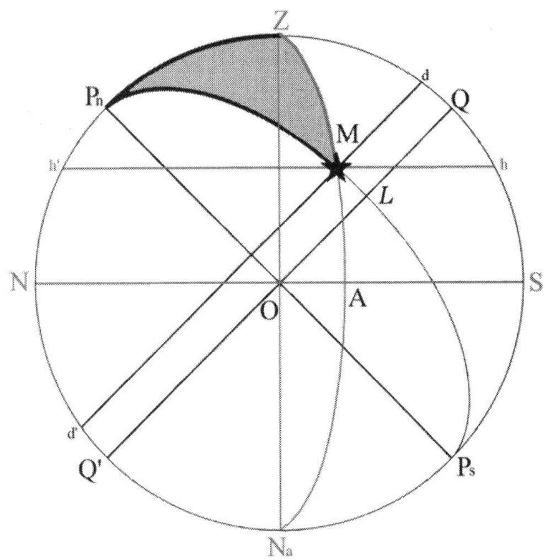

Figure 1529a. The navigational triangle.

In the figure, arc QZ of the celestial meridian is the latitude of the observer, and PnZ, one side of the triangle, is the colatitude. Arc AM of the vertical circle is the altitude of the body, and side ZM of the triangle is the zenith distance, or coaltitude. Arc LM of the hour circle is the declination of the body, and side PnM of the triangle is the polar distance, or codeclination.

The angle at the elevated pole, ZPnM, having the hour circle and the celestial meridian as sides, is the meridian angle, t. The angle at the zenith, PnZM, having the vertical circle and that arc of the celestial meridian, which includes the elevated pole, as sides, is the azimuth angle. The angle at the celestial body, ZMPn, having the hour circle and the vertical circle as sides, is the parallactic angle (X) (sometimes called the position angle), which is not generally used by the navigator.

A number of problems involving the navigational triangle are encountered by the navigator, either directly or indirectly. Of these, the most common are:

1. Given latitude, declination, and meridian angle, to find altitude and azimuth angle. This is used in the reduction of a celestial observation to establish a line of position.
2. Given latitude, altitude, and azimuth angle, to find declination and meridian angle. This is used to identify an unknown celestial body.

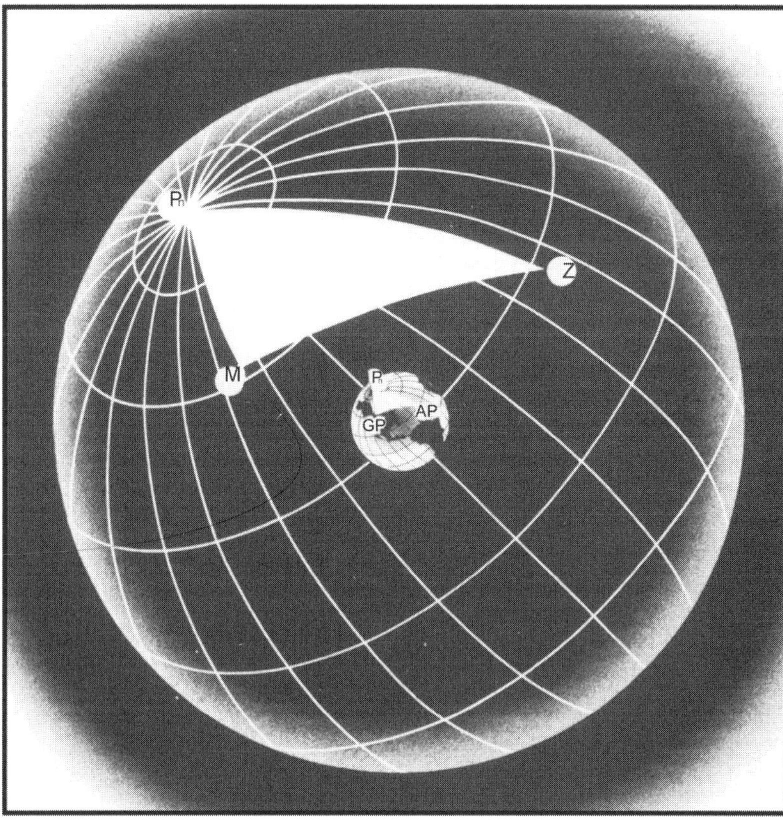

Figure 1529b. The navigational triangle in perspective.

3. Given meridian angle, declination, and altitude, to find azimuth angle. This may be used to find azimuth when the altitude is known.
4. Given the latitude of two places on the Earth and the difference of longitude between them, to find the initial great-circle course and the great-circle distance. This involves the same parts of the triangle as in 1, above, but in the terrestrial triangle, and hence is defined differently.

Both celestial and terrestrial navigational triangles are shown in perspective in Figure 1529b.

IDENTIFICATION OF STARS AND PLANETS

1530. Introduction

A basic requirement of celestial navigation is the ability to identify the bodies observed. This is not difficult because relatively few stars and planets are commonly used for navigation, and various aids are available to assist in their identification. See Figure 1530a and Figure 1532a.

Navigational calculators or computer programs can identify virtually any celestial body observed, given inputs of DR position, azimuth, and altitude. In fact, a complete round of sights can be taken and solved without knowing the names of a single observed body. Once the data is entered, the computer identifies the bodies, solves the sights,

NAVIGATIONAL STARS AND THE PLANETS

Name	Pronunciation	Bayer name	Origin of name	Meaning of name	Distance*
Acamar	ă'kȧ·mă̤r	θ Eridani	Arabic	another form of Achernar	120
Achernar	ă'kēr·nă̤r	α Eridani	Arabic	end of the river (Eridanus)	72
Acrux	ā'krŭks	α Crucis	Modern	coined from Bayer name	220
Adhara	ȧ·dā'rȧ	ε Canis Majoris	Arabic	the virgin(s)	350
Aldebaran	ăl dĕb'ȧ·răn	α Tauri	Arabic	follower (of the Pleiades)	64
Alioth	ăl'ĭ·ŏth	ε Ursa Majoris	Arabic	another form of Capella	49
Alkaid	ăl·kād'	η Ursa Majoris	Arabic	leader of the daughters of the bier	190
Al Na'ir	ăl·nâr'	α Gruis	Arabic	bright one (of the fish's tail)	90
Alnilam	ăl'nĭ·lăm	ε Orionis	Arabic	string of pearls	410
Alphard	ăl'fă̤rd	α Hydrae	Arabic	solitary star of the serpent	200
Alphecca	ăl·fĕk'ȧ	α Corona Borealis	Arabic	feeble one (in the crown)	76
Alpheratz	ăl·fē'răts	α Andromeda	Arabic	the horse's navel	120
Altair	ăl·târ'	α Aquilae	Arabic	flying eagle or vulture	16
Ankaa	ă̤n'kă̤	α Phoenicis	Arabic	coined name	93
Antares	ăn·tā'rēz	α Scorpii	Greek	rival of Mars (in color)	250
Arcturus	ă̤rk·tū'rŭs	α Bootis	Greek	the bear's guard	37
Atria	ăt'rĭ·ȧ	α Trianguli Australis	Modern	coined from Bayer name	130
Avior	ā'vĭ·ôr	ε Carinae	Modern	coined name	350
Bellatrix	bĕ·lā'trĭks.	γ Orionis	Latin	female warrior	250
Betelgeuse	bĕt'ĕl·jūz	α Orionis	Arabic	the arm pit (of Orion)	300
Canopus	kȧ·nō'pŭs	α Carinae	Greek	city of ancient Egypt	230
Capella	kȧ·pĕl'ȧ	α Aurigae	Latin	little she-goat	46
Deneb	dĕn'ĕb	α Cygni	Arabic	tail of the hen	600
Denebola	dĕ·nĕb'ō·lȧ	β Leonis	Arabic	tail of the lion	42
Diphda	dĭf'dȧ	β Ceti	Arabic	the second frog (Fomalhaut was once the first)	57
Dubhe	dŭb'ĕ	α Ursa Majoris	Arabic	the bear's back	100
Elnath	ĕl'năth	β Tauri	Arabic	one butting with horns	130
Eltanin	ĕl·tā'nĭn	γ Draconis	Arabic	head of the dragon	150
Enif	ĕn'ĭf	ε Pegasi	Arabic	nose of the horse	250
Fomalhaut	fō'măl·ôt	α Piscis Austrini	Arabic	mouth of the southern fish	23
Gacrux	gā'krŭks	γ Crucis	Modern	coined from Bayer name	72
Gienah	jē'nȧ	γ Corvi	Arabic	right wing of the raven	136
Hadar	hā'dă̤r	β Centauri	Modern	leg of the centaur	200
Hamal	hăm'ăl	α Arietis	Arabic	full-grown lamb	76
Kaus Australis	kôs ôs·trā'lĭs	ε Sagittarii	Ar., L.	southern part of the bow	163
Kochab	kō'kăb	β Ursa Minoris	Arabic	shortened form of "north star" (named when it was that, c. 1500 BC–AD 300)	100
Markab	mă̤r'kăb	α Pegasi	Arabic	saddle (of Pegasus)	100
Menkar	mĕn'kă̤r	α Ceti	Arabic	nose (of the whale)	1,100
Menkent	mĕn'kĕnt	θ Centauri	Modern	shoulder of the centaur	55
Miaplacidus	mī'ȧ·plăs'ĭ·dŭs	β Carinae	Ar., L.	quiet or still waters	86
Mirfak	mĭr'făk	α Persei	Arabic	elbow of the Pleiades	130
Nunki	nŭn'kē	σ Sagittarii	Bab.	constellation of the holy city (Eridu)	150
Peacock	pē'kŏk	α Pavonis	Modern	coined from English name of constellation	250
Polaris	pō·lā'rĭs	α Ursa Minoris	Latin	the pole (star)	450
Pollux	pŏl'ŭks	β Geminorum	Latin	Zeus' other twin son (Castor, α Geminorum, is first twin)	33
Procyon	prō'sĭ·ŏn	α Canis Minoris	Greek	before the dog (rising before the dog star, Sirius)	11
Rasalhague	răs'ă̤l·hā'gwē	α Ophiuchi	Arabic	head of the serpent charmer	67
Regulus	rĕg'ū·lŭs	α Leonis	Latin	the prince	67
Rigel	rī'jĕl	β Orionis	Arabic	foot (left foot of Orion)	500
Rigil Kentaurus	rī'jĭl kĕn·tô'rŭs	α Centauri	Arabic	foot of the centaur	4.3
Sabik	sā'bĭk	η Ophiuchi	Arabic	second winner or conqueror	69
Schedar	shĕd'ȧr	α Cassiopeiae	Arabic	the breast (of Cassiopeia)	360
Shaula	shō'lȧ	λ Scorpii	Arabic	cocked-up part of the scorpion's tail	200
Sirius	sĭr'ĭ·ŭs	α Canis Majoris	Greek	the scorching one (popularly, the dog star)	8.6
Spica	spī'kȧ	α Virginis	Latin	the ear of corn	155
Suhail	sōō·hāl'	λ Velorum	Arabic	shortened form of Al Suhail, one Arabic name for Canopus	200
Vega	vē'gȧ	α Lyrae	Arabic	the falling eagle or vulture	27
Zubenelgenubi	zōō·bĕn'ĕl·jē·nū'bē	α Librae	Arabic	southern claw (of the scorpion)	66

PLANETS

Name	Pronunciation	Origin of name	Meaning of name
Mercury	mûr'kū·rĭ	Latin	god of commerce and gain
Venus	vē'nŭs	Latin	goddess of love
Earth	ûrth	Mid. Eng.	—
Mars	mărz	Latin	god of war
Jupiter	jōō'pĭ·tēr	Latin	god of the heavens, identified with the Greek Zeus, chief of the Olympian gods
Saturn	săt'ērn	Latin	god of seed-sowing
Uranus	ū'rȧ·nŭs	Greek	the personification of heaven
Neptune	nĕp'tūn	Latin	god of the sea
Pluto	plōō'tō	Greek	god of the lower world (Hades)

Guide to pronunciations:
fāte, ădd, finȧl, lȧst, ȧbound, ärm; bē, ĕnd, camēl, readēr; īce, bĭt, anĭmal; ōver, pôetic, hŏt, lôrd, mōōn; tūbe, ûnite, tŭb, circŭs, ûrn

*Distances in light-years. One light-year equals approximately 63,300 AU, or 5,880,000,000,000 miles. Authorities differ on distances of the stars; the values given are representative.

Figure 1530a. Navigational stars and the planets.

and plots the results. In this way, the navigator can learn the stars by observation instead of by rote memorization.

No problem is encountered in the identification of the Sun and Moon. However, the planets can be mistaken for stars. A person working continually with the night sky recognizes a planet by its changing position among the relatively fixed stars. The planets are identified by noting their positions relative to each other, the Sun, the Moon, and the stars. They remain within the narrow limits of the zodiac, but are in almost constant motion relative to the stars. The magnitude and color may be helpful. The information needed is found in the *Nautical Almanac*. The "Planet Notes" near the front of that volume are particularly useful. Planets can also be identified by planet diagram, star finder, sky diagram, or by computation.

1531. Stars

The *Nautical Almanac* lists full navigational information on 19 first magnitude stars and 38 second magnitude stars, plus Polaris. Abbreviated information is listed for 115 more. Additional stars are listed in the *Astronomical Almanac* and in various star catalogs. About 6,000 stars of the sixth magnitude or brighter (on the entire celestial sphere) are visible to the unaided eye on a clear, dark night.

Stars are designated by one or more of the following naming systems:

- **Common Name:** Most names of stars, as now used, were given by the ancient Arabs and some by the Greeks or Romans. One of the stars of the *Nautical Almanac*, Nunki, was named by the Babylonians. Only a relatively few stars have names. Several of the stars on the daily pages of the almanacs had no name prior to 1953.

- **Bayer's Name:** Most bright stars, including those with names, have been given a designation consisting of a Greek letter followed by the possessive form of the name of the constellation, such as α Cygni (Deneb, the brightest star in the constellation Cygnus, the swan). Roman letters are used when there are not enough Greek letters. Usually, the letters are assigned in order of brightness within the constellation; however, this is not always the case. For example, the letter designations of the stars in Ursa Major or the Big Dipper are assigned in order from the outer rim of the bowl to the end of the handle. This system of star designation was suggested by John Bayer of Augsburg, Germany, in 1603. All of the 173 stars included in the list near the back of the *Nautical Almanac* are listed by Bayer's name, and, when applicable, their common name.

- **Flamsteed's Number:** This system assigns numbers to stars in each constellation, from west to east in the order in which they cross the celestial meridian. An example is 95 Leonis, the 95th star in the constellation Leo. This system was suggested by John Flamsteed (1646-1719).

- **Catalog Number:** Stars are sometimes designated by the name of a star catalog and the number of the star as given in the catalog, such as A. G. Washington 632. In these catalogs, stars are listed in order from west to east, without regard to constellation, starting with the hour circle of the vernal equinox. This system is used primarily for fainter stars having no other designation. Navigators seldom have occasion to use this system.

1532. Star Charts

It is useful to be able to identify stars by relative position. A **star chart** (Figure 1532a and Figure 1532b) is helpful in locating these relationships and others which may be useful. This method is limited to periods of relatively clear, dark skies with little or no overcast. Stars can also be identified by the *Air Almanac* **sky diagrams**, a **star finder**, *Pub. No. 249*, or by computation by hand or calculator.

Star charts are based upon the celestial equator system of coordinates, using declination and sidereal hour angle (or right ascension). The zenith of the observer is at the intersection of the parallel of declination equal to his latitude, and the hour circle coinciding with his celestial meridian. This hour circle has an SHA equal to 360° – LHA ♈ (or RA = LHA ♈). The horizon is everywhere 90° from the zenith. A **star globe** is similar to a terrestrial sphere, but with stars (and often constellations) shown instead of geographical positions. The *Nautical Almanac* includes instructions for using this device. On a star globe the celestial sphere is shown as it would appear to an observer outside the sphere. Constellations appear reversed. Star charts may show a similar view, but more often they are based upon the view from inside the sphere, as seen from the Earth. On these charts, north is at the top, as with maps, but east is to the left and west to the right. The directions seem correct when the chart is held overhead, with the top toward the north, so the relationship is similar to the sky.

The *Nautical Almanac* has four star charts. The two principal ones are on the polar azimuthal equidistant projection, one centered on each celestial pole. Each chart extends from its pole to declination 10° (same name as pole). Below each polar chart is an auxiliary chart on the Mercator projection, from 30°N to 30°S. On any of these charts, the zenith can be located as indicated, to determine which stars are overhead. The horizon is 90° from the zenith. The charts can also be used to determine the location of a star relative to surrounding stars.

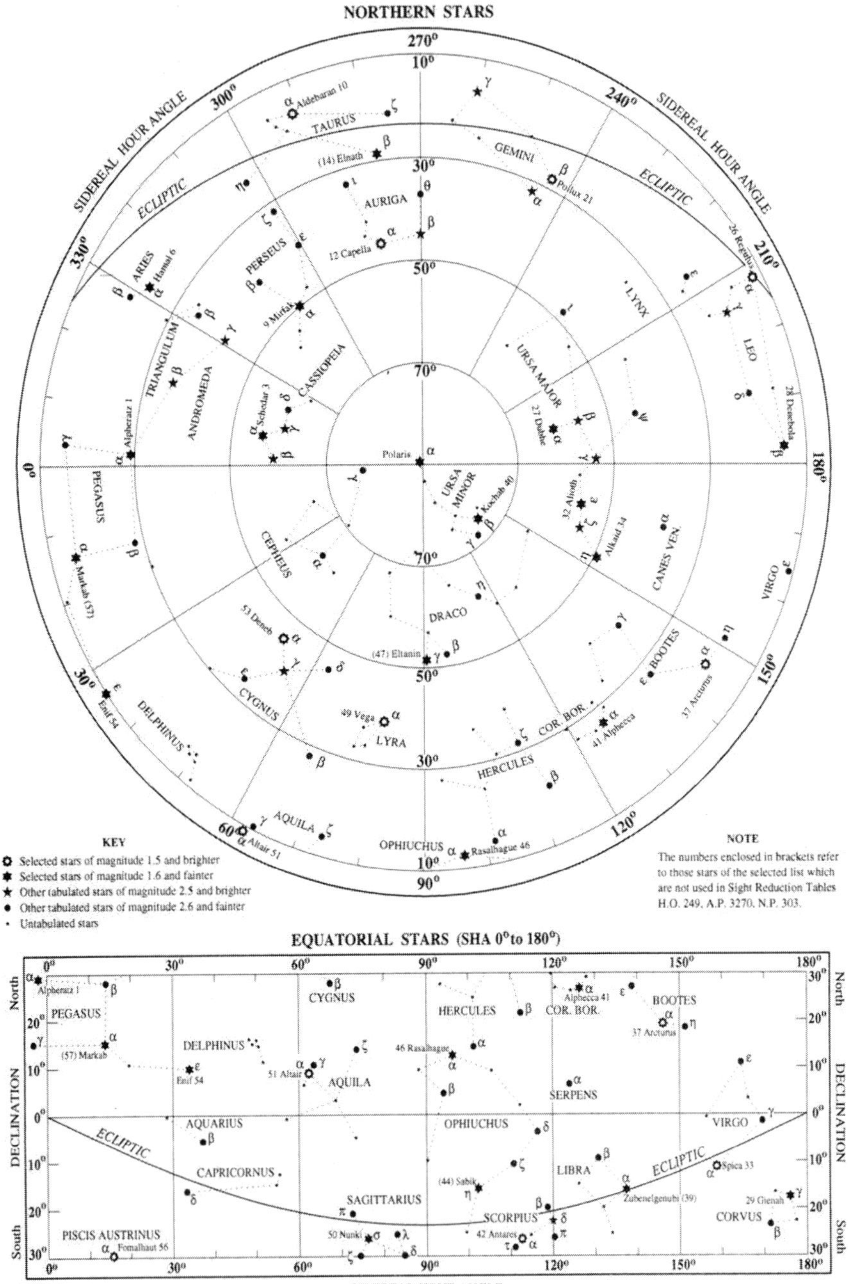

Figure 1532a. Star chart from Nautical Almanac.

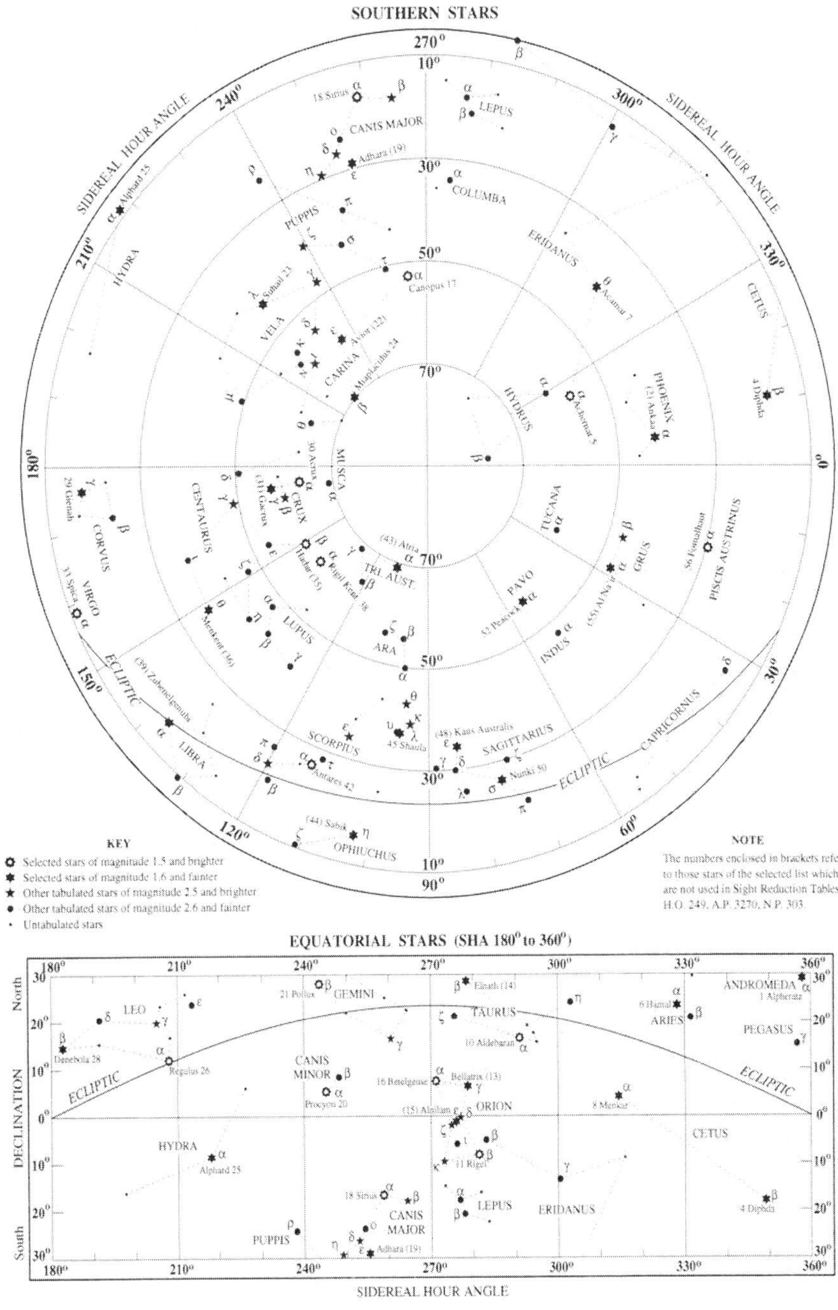

Figure 1532b. Star chart from Nautical Almanac.

	Fig. 1534	Fig.1535	Fig. 1536	Fig. 1537
Local sidereal time	0000	0600	1200	1800
LMT 1800	Dec. 21	Mar. 22	June 22	Sept. 21
LMT 2000	Nov. 21	Feb. 20	May 22	Aug. 21
LMT 2200	Oct. 21	Jan. 20	Apr. 22	July 22
LMT 0000	Sept. 22	Dec. 22	Mar. 23	June 22
LMT 0200	Aug. 22	Nov. 22	Feb. 21	May 23
LMT 0400	July 23	Oct. 22	Jan 21	Apr. 22
LMT 0600	June 22	Sept. 21	Dec. 22	Mar. 23

Table 1532. Locating the zenith on the star diagrams.

The star charts shown in Figure 1533 through Figure 1536, on the transverse Mercator projection, are designed to assist in learning Polaris and the stars listed on the daily pages of the *Nautical Almanac*. Each chart extends about 20° beyond each celestial pole, and about 60° (four hours) each side of the central hour circle (at the celestial equator). Therefore, they do not coincide exactly with that half of the celestial sphere above the horizon at any one time or place. The zenith, and hence the horizon, varies with the position of the observer on the Earth. It also varies with the rotation of the Earth (apparent rotation of the celestial sphere). The charts show all stars of fifth magnitude and brighter as they appear in the sky, but with some distortion toward the right and left edges.

The overprinted lines add certain information of use in locating the stars. Only Polaris and the 57 stars listed on the daily pages of the *Nautical Almanac* are named on the charts. The almanac star charts can be used to locate the additional stars given near the back of the *Nautical Almanac* and the *Air Almanac*. Dashed lines connect stars of some of the more prominent constellations. Solid lines indicate the celestial equator and useful relationships among stars in different constellations. The celestial poles are marked by crosses, and labeled. By means of the celestial equator and the poles, one can locate his zenith approximately along the mid hour circle, when this coincides with his celestial meridian, as shown in Table 1532. At any time earlier than those shown in Table 1532 the zenith is to the right of center, and at a later time it is to the left, approximately one-quarter of the distance from the center to the outer edge (at the celestial equator) for each hour that the time differs from that shown. The stars in the vicinity of the North Pole can be seen in proper perspective by inverting the chart, so that the zenith of an observer in the Northern Hemisphere is up from the pole.

1533. Stars in the Vicinity of Pegasus

In autumn the evening sky has few first magnitude stars. Most are near the southern horizon of an observer in the latitudes of the United States. A relatively large number of second and third magnitude stars seem conspicuous, perhaps because of the small number of brighter stars. High in the southern sky three third magnitude stars and one second magnitude star form a square with sides nearly 15° of arc in length. This is Pegasus, the winged horse.

Only Markab at the southwestern corner and Alpheratz at the northeastern corner are listed on the daily pages of the *Nautical Almanac*. Alpheratz is part of the constellation Andromeda, the princess, extending in an arc toward the northeast and terminating at Mirfak in Perseus, legendary rescuer of Andromeda.

A line extending northward through the eastern side of the square of Pegasus passes through the leading (western) star of M-shaped (or W-shaped) Cassiopeia, the legendary mother of the princess Andromeda. The only star of this constellation listed on the daily pages of the *Nautical Almanac* is Schedar, the second star from the leading one as the configuration circles the pole in a counterclockwise direction. If the line through the eastern side of the square of Pegasus is continued on toward the north, it leads to second magnitude Polaris, the North Star (less than 1° from the north celestial pole) and brightest star of Ursa Minor, the Little Dipper. Kochab, a second magnitude star at the other end of Ursa Minor, is also listed in the almanacs. At this season Ursa Major is low in the northern sky, below the celestial pole. A line extending from Kochab through Polaris leads to Mirfak, assisting in its identification when Pegasus and Andromeda are near or below the horizon.

Deneb, in Cygnus, the swan, and Vega are bright, first magnitude stars in the northwestern sky. The line through the eastern side of the square of Pegasus approximates the hour circle of the vernal equinox, shown at Aries on the celestial equator to the south. The Sun is at Aries on or about March 21, when it crosses the celestial equator from south to north. If the line through the eastern side of Pegasus is extended southward and curved slightly toward the east, it leads to second magnitude Diphda. A longer and straighter line southward through the western side of Pegasus leads to first magnitude Fomalhaut. A line extending northeasterly from Fomalhaut through Diphda leads to Menkar, a third magnitude star, but the brightest in its vicinity. Ankaa, Diphda, and Fomalhaut form an isosceles triangle, with the apex at Diphda. Ankaa is near or below the southern horizon of observers in latitudes of the United States. Four stars farther south than Ankaa may be visible when on the celes-

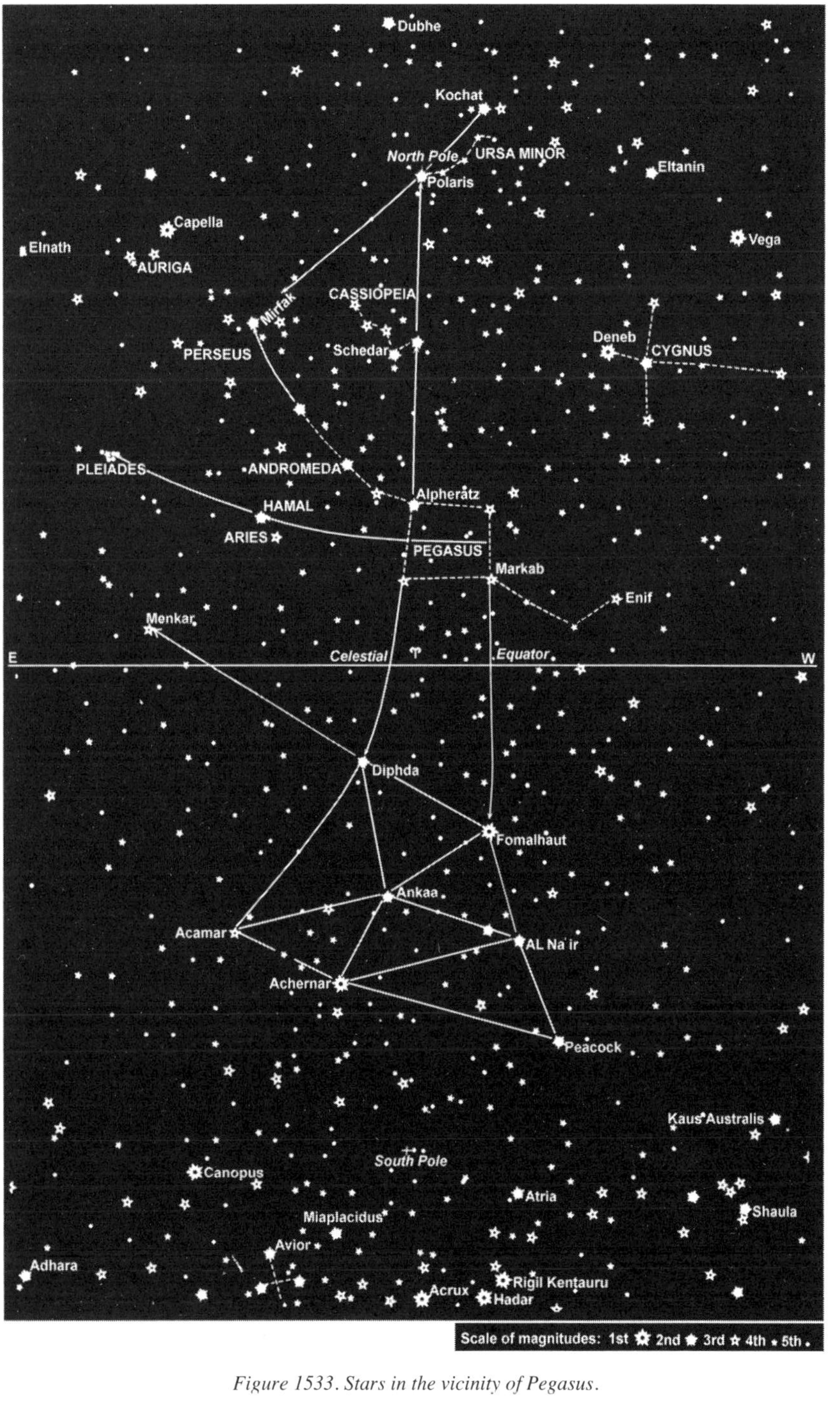

Figure 1533. Stars in the vicinity of Pegasus.

tial meridian, just above the horizon of observers in latitudes of the extreme southern part of the United States. These are Acamar, Achernar, Al Na'ir, and Peacock. These stars, with each other and with Ankaa, Fomalhaut, and Diphda, form a series of triangles as shown in Figure 1533. Almanac stars near the bottom of Figure 1533 are discussed in succeeding articles.

Two other almanac stars can be located by their positions relative to Pegasus. These are Hamal in the constellation Aries, the ram, east of Pegasus, and Enif, west of the southern part of the square, identified in Figure 1533. The line leading to Hamal, if continued, leads to the Pleiades (the Seven Sisters), not used by navigators for celestial observations, but a prominent figure in the sky, heralding the approach of the many conspicuous stars of the winter evening sky.

1534. Stars in the Vicinity of Orion

As Pegasus leaves the meridian and moves into the western sky, Orion, the hunter, rises in the east. With the possible exception of Ursa Major, no other configuration of stars in the entire sky is as well known as Orion and its immediate surroundings. In no other region are there so many first magnitude stars.

The belt of Orion, nearly on the celestial equator, is visible in virtually any latitude, rising and setting almost on the prime vertical, and dividing its time equally above and below the horizon. Of the three second magnitude stars forming the belt, only Alnilam, the middle one, is listed on the daily pages of the *Nautical Almanac*.

Four conspicuous stars form a box around the belt. Rigel, a hot, blue star, is to the south. Betelgeuse, a cool, red star lies to the north. Bellatrix, bright for a second magnitude star but overshadowed by its first magnitude neighbors, is a few degrees west of Betelgeuse. Neither the second magnitude star forming the southeastern corner of the box, nor any star of the dagger, is listed on the daily pages of the *Nautical Almanac*.

A line extending eastward from the belt of Orion, and curving toward the south, leads to Sirius, the brightest star in the entire heavens, having a magnitude of –1.6. Only Mars and Jupiter at or near their greatest brilliance, the Sun, Moon, and Venus are brighter than Sirius. Sirius is part of the constellation Canis Major, the large hunting dog of Orion. Starting at Sirius a curved line extends northward through first magnitude Procyon, in Canis Minor, the small hunting dog; first magnitude Pollux and second magnitude Castor (not listed on the daily pages of the *Nautical Almanac*), the twins of Gemini; brilliant Capella in Auriga, the charioteer; and back down to first magnitude Aldebaran, the follower, which trails the Pleiades, the seven sisters. Aldebaran, brightest star in the head of Taurus, the bull, may also be found by a curved line extending northwestward from the belt of Orion. The V-shaped figure forming the outline of the head and horns of Taurus points toward third magnitude Menkar. At the summer solstice the Sun is between Pollux and Aldebaran.

If the curved line from Orion's belt southeastward to Sirius is continued, it leads to a conspicuous, small, nearly equilateral triangle of three bright second magnitude stars of nearly equal brilliancy. This is part of Canis Major. Only Adhara, the westernmost of the three stars, is listed on the daily pages of the *Nautical Almanac*. Continuing on with somewhat less curvature, the line leads to Canopus, second brightest star in the heavens and one of the two stars having a negative magnitude (–0.9). With Suhail and Miaplacidus, Canopus forms a large, equilateral triangle which partly encloses the group of stars often mistaken for Crux. The brightest star within this triangle is Avior, near its center. Canopus is also at one apex of a triangle formed with Adhara to the north and Suhail to the east, another triangle with Acamar to the west and Achernar to the southwest, and another with Achernar and Miaplacidus. Acamar, Achernar, and Ankaa form still another triangle toward the west. Because of chart distortion, these triangles do not appear in the sky in exactly the relationship shown on the star chart. Other daily-page almanac stars near the bottom of Figure 1534 are discussed in succeeding articles.

In the winter evening sky, Ursa Major is east of Polaris, Ursa Minor is nearly below it, and Cassiopeia is west of it. Mirfak is northwest of Capella, nearly midway between it and Cassiopeia. Hamal is in the western sky. Regulus and Alphard are low in the eastern sky, heralding the approach of the configurations associated with the evening skies of spring.

1535. Stars in the Vicinity of Ursa Major

As if to enhance the splendor of the sky in the vicinity of Orion, the region toward the east, like that toward the west, has few bright stars, except in the vicinity of the south celestial pole. However, as Orion sets in the west, leaving Capella and Pollux in the northwestern sky, a number of good navigational stars move into favorable positions for observation.

Ursa Major, the great bear, appears prominently above the north celestial pole, directly opposite Cassiopeia, which appears as a "W" just above the northern horizon of most observers in latitudes of the United States. Of the seven stars forming Ursa Major, only Dubhe, Alioth, and Alkaid are listed on the daily pages of the *Nautical Almanac*. See Figure 1535.

The two second magnitude stars forming the outer part of the bowl of Ursa Major are often called the pointers because a line extending northward (down in spring evenings) through them points to Polaris. Ursa Minor, the Little Bear, contains Polaris at one end and Kochab at the other. Relative to its bowl, the handle of Ursa Minor curves in the opposite direction to that of Ursa Major.

A line extending southward through the pointers, and curving somewhat toward the west, leads to first magnitude Regulus, brightest star in Leo, the lion. The head,

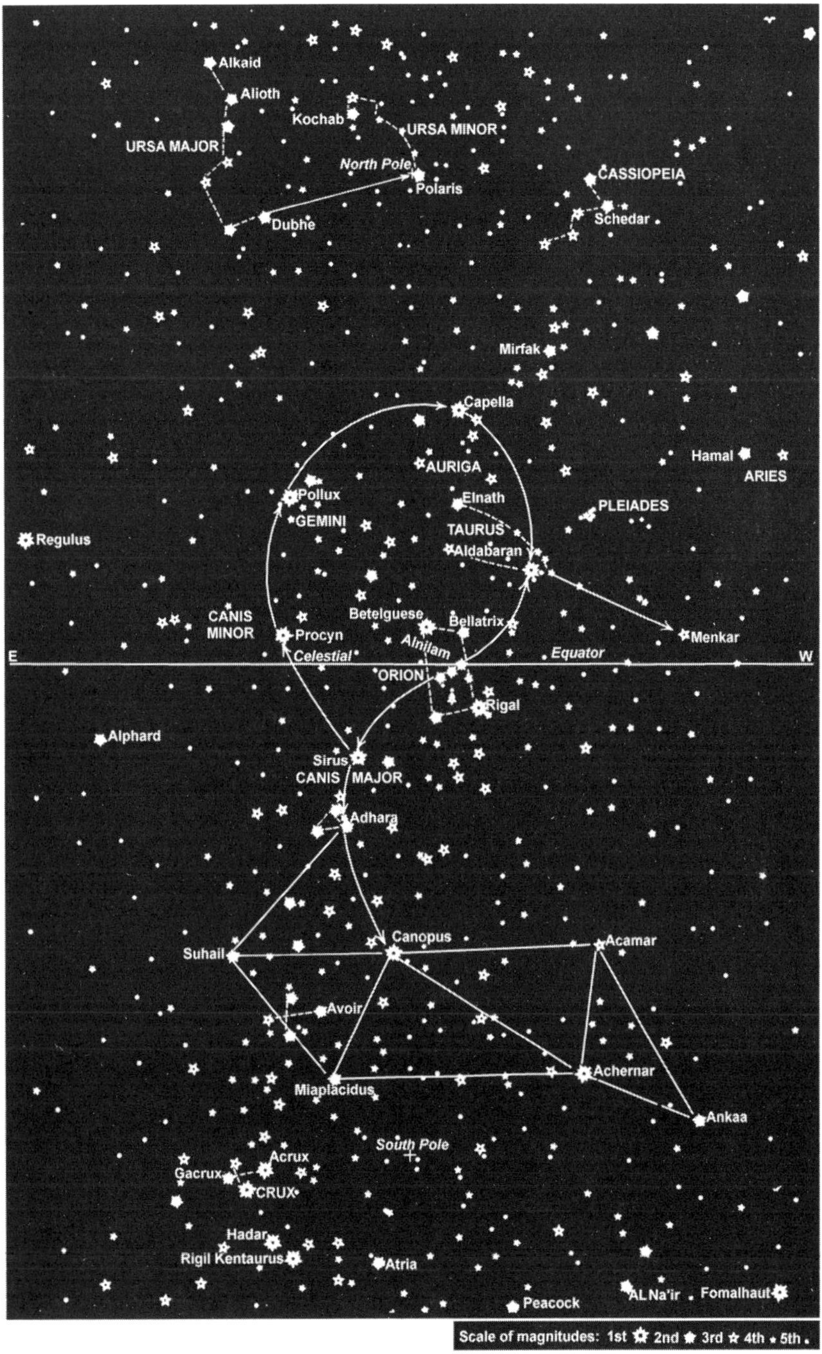

Figure 1534. Stars in the vicinity of Orion.

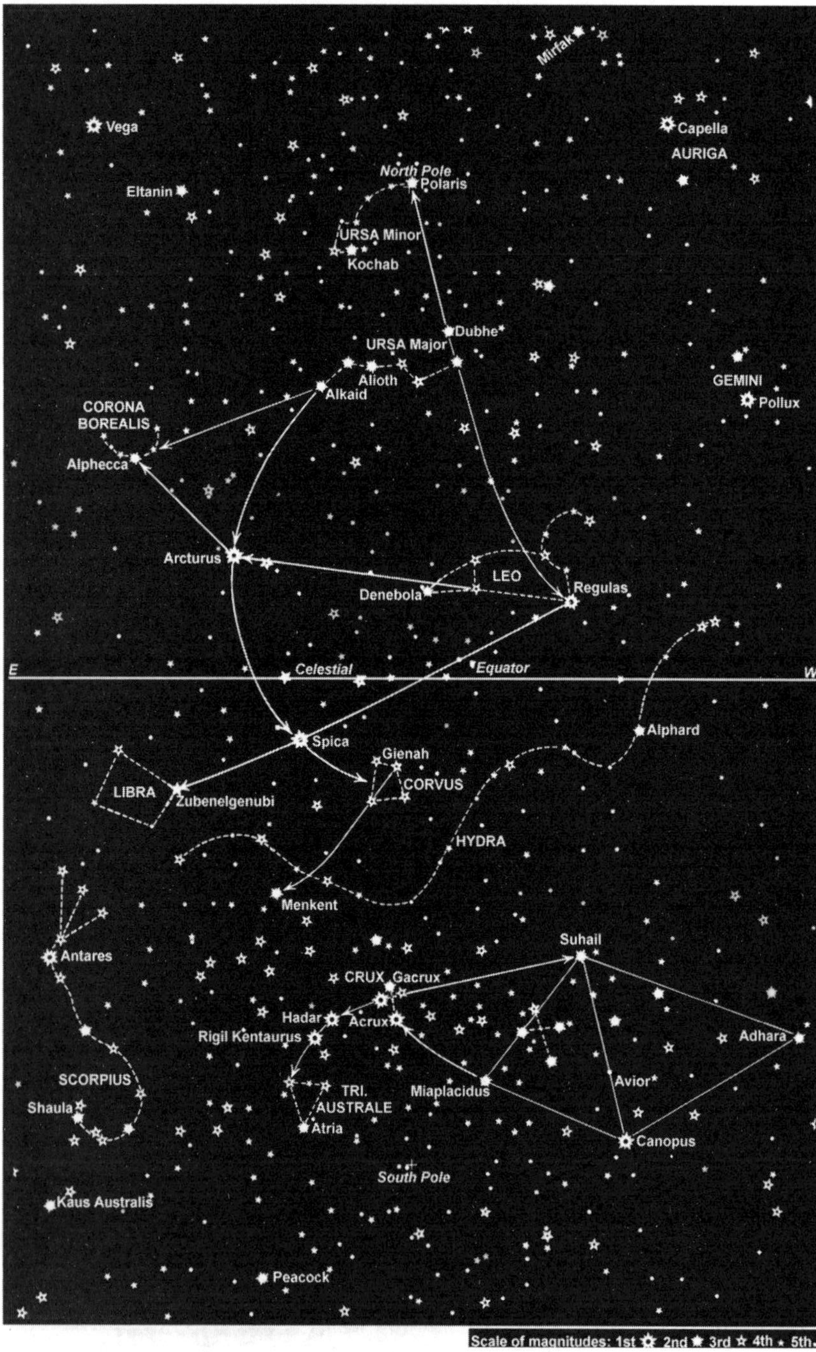

Figure 1535. Stars in the vicinity of Ursa Major.

shoulders, and front legs of this constellation form a sickle, with Regulus at the end of the handle. Toward the east is second magnitude Denebola, the tail of the lion. On toward the southwest from Regulus is second magnitude Alphard, brightest star in Hydra, the sea serpent. A dark sky and considerable imagination are needed to trace the long, winding body of this figure.

A curved line extending the arc of the handle of Ursa Major leads to first magnitude Arcturus. With Alkaid and Alphecca, brightest star in Corona Borealis, the Northern Crown, Arcturus forms a large, inconspicuous triangle. If the arc through Arcturus is continued, it leads next to first magnitude Spica and then to Corvus, the crow. The brightest star in this constellation is Gienah, but three others are nearly as bright. At autumnal equinox, the Sun is on the celestial equator, about midway between Regulus and Spica.

A long, slightly curved line from Regulus, east-southeasterly through Spica, leads to Zubenelgenubi at the southwestern corner of an inconspicuous box-like figure called Libra, the scales.

Returning to Corvus, a line from Gienah, extending diagonally across the figure and then curving somewhat toward the east, leads to Menkent, just beyond Hydra.

Far to the south, below the horizon of most northern hemisphere observers, a group of bright stars is a prominent feature of the spring sky of the Southern Hemisphere. This is Crux, the Southern Cross. Crux is about 40° south of Corvus. The "false cross" to the west is often mistaken for Crux. Acrux at the southern end of Crux and Gacrux at the northern end are listed on the daily pages of the *Nautical Almanac*.

The triangles formed by Suhail, Miaplacidus, and Canopus, and by Suhail, Adhara, and Canopus, are west of Crux. Suhail is in line with the horizontal arm of Crux. A line from Canopus, through Miaplacidus, curved slightly toward the north, leads to Acrux. A line through the east-west arm of Crux, eastward and then curving toward the south, leads first to Hadar and then to Rigil Kentaurus, both very bright stars. Continuing on, the curved line leads to small Triangulum Australe, the Southern Triangle, the easternmost star of which is Atria.

1536. Stars in the Vicinity of Cygnus

As the celestial sphere continues in its apparent westward rotation, the stars familiar to a spring evening observer sink low in the western sky. By midsummer, Ursa Major has moved to a position to the left of the north celestial pole, and the line from the pointers to Polaris is nearly horizontal. Ursa Minor, is standing on its handle, with Kochab above and to the left of the celestial pole. Cassiopeia is at the right of Polaris, opposite the handle of Ursa Major. See Figure 1536.

The only first magnitude star in the western sky is Arcturus, which forms a large, inconspicuous triangle with Alkaid, the end of the handle of Ursa Major, and Alphecca, the brightest star in Corona Borealis, the Northern Crown.

The eastern sky is dominated by three very bright stars. The westernmost of these is Vega, the brightest star north of the celestial equator, and third brightest star in the heavens, with a magnitude of 0.1. With a declination of a little less than 39°N, Vega passes through the zenith along a path across the central part of the United States, from Washington in the east to San Francisco on the Pacific coast. Vega forms a large but conspicuous triangle with its two bright neighbors, Deneb to the northeast and Altair to the southeast. The angle at Vega is nearly a right angle. Deneb is at the end of the tail of Cygnus, the swan. This configuration is sometimes called the Northern Cross, with Deneb at the head. To modern youth it more nearly resembles a dive bomber, while it is still well toward the east, with Deneb at the nose of the fuselage. Altair has two fainter stars close by, on opposite sides. The line formed by Altair and its two fainter companions, if extended in a northwesterly direction, passes through Vega, and on to second magnitude Eltanin. The angular distance from Vega to Eltanin is about half that from Altair to Vega. Vega and Altair, with second magnitude Rasalhague to the west, form a large equilateral triangle. This is less conspicuous than the Vega-Deneb-Altair triangle because the brilliance of Rasalhague is much less than that of the three first magnitude stars, and the triangle is overshadowed by the brighter one.

Far to the south of Rasalhague, and a little toward the west, is a striking configuration called Scorpius, the scorpion. The brightest star, forming the head, is red Antares. At the tail is Shaula.

Antares is at the southwestern corner of an approximate parallelogram formed by Antares, Sabik, Nunki, and Kaus Australis. With the exception of Antares, these stars are only slightly brighter than a number of others nearby, and so this parallelogram is not a striking figure. At winter solstice the Sun is a short distance northwest of Nunki.

Northwest of Scorpius is the box-like Libra, the scales, of which Zubenelgenubi marks the southwest corner.

With Menkent and Rigil Kentaurus to the southwest, Antares forms a large but unimpressive triangle. For most observers in the latitudes of the United States, Antares is low in the southern sky, and the other two stars of the triangle are below the horizon. To an observer in the Southern Hemisphere Crux is to the right of the south celestial pole, which is not marked by a conspicuous star. A long, curved line, starting with the now-vertical arm of Crux and extending northward and then eastward, passes successively through Hadar, Rigil Kentaurus, Peacock, and Al Na'ir.

Fomalhaut is low in the southeastern sky of the southern hemisphere observer, and Enif is low in the eastern sky at nearly any latitude. With the appearance of these stars it is not long before Pegasus will appear over the eastern horizon during the evening, and as the winged horse climbs evening by

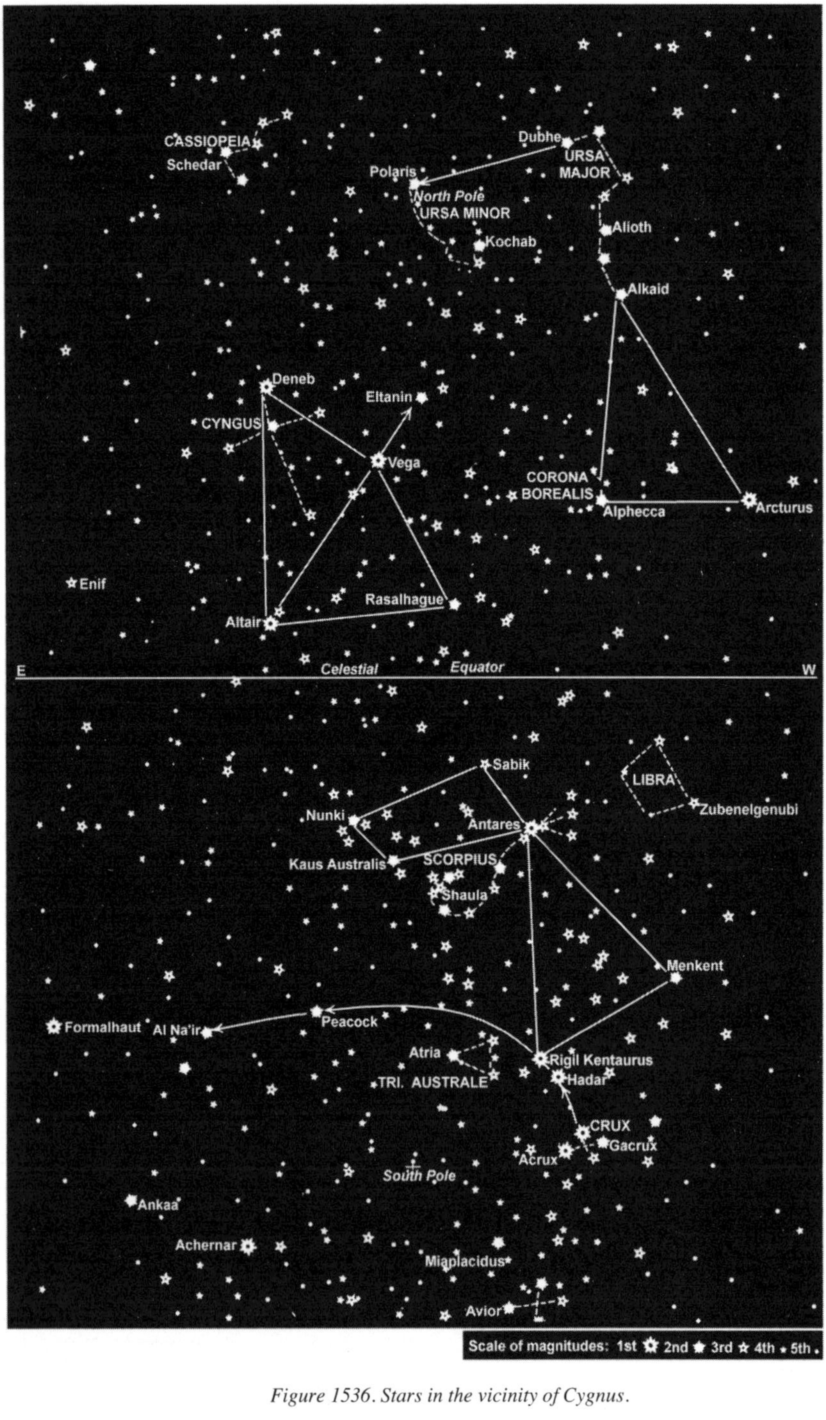

Figure 1536. Stars in the vicinity of Cygnus.

evening to a position higher in the sky, a new annual cycle approaches.

1537. Planet Diagram

The planet diagram in the *Nautical Almanac* shows, for any date, the LMT of meridian passage of the Sun, for the five planets Mercury, Venus, Mars, Jupiter, and Saturn, and of each 30° of SHA. The diagram provides a general picture of the availability of planets and stars for observation, and thus shows:

1. Whether a planet or star is too close to the Sun for observation.
2. Whether a planet is a morning or evening star.
3. Some indication of the planet's position during twilight.
4. The proximity of other planets.
5. Whether a planet is visible from evening to morning twilight.

A band 45 minutes wide is shaded on each side of the curve marking the LMT of meridian passage of the Sun. Any planet and most stars lying within the shaded area are too close to the Sun for observation.

When the meridian passage occurs at midnight, the body is in opposition to the Sun and is visible all night; planets may be observable in both morning and evening twilights. As the time of meridian passage decreases, the body ceases to be observable in the morning, but its altitude above the eastern horizon during evening twilight gradually increases; this continues until the body is on the meridian at twilight. From then onwards the body is observable above the western horizon and its altitude at evening twilight gradually decreases; eventually the body comes too close to the Sun for observation. When the body again becomes visible, it is seen as a morning star low in the east. Its altitude at twilight increases until meridian passage occurs at the time of morning twilight. Then, as the time of meridian passage decreases to 0^h, the body is observable in the west in the morning twilight with a gradually decreasing altitude, until it once again reaches opposition.

Only about one-half the region of the sky along the ecliptic, as shown on the diagram, is above the horizon at one time. At sunrise (LMT about 6^h) the Sun and, hence, the region near the middle of the diagram, are rising in the east; the region at the bottom of the diagram is setting in the west. The region half way between is on the meridian. At sunset (LMT about 18^h) the Sun is setting in the west; the region at the top of the diagram is rising in the east. Marking the planet diagram of the *Nautical Almanac* so that east is at the top of the diagram and west is at the bottom can be useful to interpretation.

If the curve for a planet intersects the vertical line connecting the date graduations below the shaded area, the planet is a morning star; if the intersection is above the shaded area, the planet is an evening star.

A similar planet location diagram in the *Air Almanac* represents the region of the sky along the ecliptic within which the Sun, Moon, and planets always move; it shows, for each date, the Sun in the center and the relative positions of the Moon, the five planets Mercury, Venus, Mars, Jupiter, Saturn and the four first magnitude stars Aldebaran, Antares, Spica, and Regulus, and also the position on the ecliptic which is north of Sirius (i.e. Sirius is 40° south of this point). The first point of Aries is also shown for reference. The magnitudes of the planets are given at suitable intervals along the curves. The Moon symbol shows the correct phase. A straight line joining the date on the left-hand side with the same date of the right-hand side represents a complete circle around the sky, the two ends of the line representing the point 180° from the Sun; the intersections with the curves show the spacing of the bodies along the ecliptic on the date. The time scale indicates roughly the local mean time at which an object will be on the observer's meridian.

At any time only about half the region on the diagram is above the horizon. At sunrise the Sun (and hence the region near the middle of the diagram), is rising in the east and the region at the end marked "West" is setting in the west; the region half-way between these extremes is on the meridian, as will be indicated by the local time (about 6^h). At the time of sunset (local time about 18^h) the Sun is setting in the west, and the region at the end marked "East" is rising in the east. The diagram should be used in conjunction with the Sky Diagrams.

1538. Finding Stars for a Fix

Various devices have been invented to help an observer find individual stars. The most widely used is the **Star Finder and Identifier**, formerly published by the U.S. Navy Hydrographic Office as *No. 2102D*. It is no longer issued, having been replaced officially by the STELLA computer program, but it is still available commercially. A navigational calculator or computer program is much quicker, more accurate, and less tedious.

In fact, the process of identifying stars is no longer necessary because the computer or calculator does it automatically. The navigator need only take sights and enter the required data. The program identifies the bodies, solves for the LOP's for each, combines them into the best fix, and displays the lat./long. position. Most computer programs also print out a plotted fix, just as the navigator might have drawn by hand.

The data required by the calculator or program consists of the DR position, the sextant altitude of the body, the time, and the azimuth of the body. The name of the body is not necessary because there will be only one possible body meeting those conditions, which the computer will identify.

Computer sight reduction programs can also automatically predict twilight on a moving vessel and create a plot of the sky at the vessel's twilight location (or any location, at any time). This plot will be free of the distortion inherent in the mechanical star finders and will show all bodies, even planets, Sun, and Moon, in their correct relative orientation centered on the observer's zenith. It will also indicate which stars provide the best geometry for a fix.

Computer sight reduction programs or celestial navigation calculators are especially useful when the sky is only briefly visible thorough broken cloud cover. The navigator can quickly shoot any visible body without having to identify it by name, and let the computer do the rest.

1539. Identification by Computation

If the altitude and azimuth of the celestial body, and the approximate latitude of the observer, are known, the navigational triangle can be solved for meridian angle and declination. The meridian angle can be converted to LHA, and this to GHA. With this and GHA ♈ at the time of observation, the SHA of the body can be determined. With SHA and declination, one can identify the body by reference to an almanac. Any method of solving a spherical triangle, with two sides and the included angle being given, is suitable for this purpose. A large-scale, carefully-drawn diagram on the plane of the celestial meridian, using the refinement shown in Figure 1528f, should yield satisfactory results.

Although no formal star identification tables are included in *Pub. No. 229*, a simple approach to star identification is to scan the pages of the appropriate latitudes, and observe the combination of arguments which give the altitude and azimuth angle of the observation. Thus the declination and LHA ★ are determined directly. The star's SHA is found from SHA ★ = LHA ★ – LHA ♈. From these quantities the star can be identified from the *Nautical Almanac*.

Another solution is available through an interchange of arguments using the nearest integral values. The procedure consists of entering *Pub. No. 229* with the observer's latitude (same name as declination), with the observed azimuth angle (converted from observed true azimuth as required) as LHA and the observed altitude as declination, and extracting from the tables the altitude and azimuth angle respondents. The extracted altitude becomes the body's declination; the extracted azimuth angle (or its supplement) is the meridian angle of the body. Note that the tables are always entered with latitude of same name as declination. In north latitudes the tables can be entered with true azimuth as LHA.

If the respondents are extracted from above the C-S Line on a right-hand page, the name of the latitude is actually contrary to the declination. Otherwise, the declination of the body has the same name as the latitude. If the azimuth angle respondent is extracted from above the C-S Line, the supplement of the tabular value is the meridian angle, t, of the body. If the body is east of the observer's meridian, LHA = 360° – t; if the body is west of the meridian, LHA = t.

CHAPTER 16

INSTRUMENTS FOR CELESTIAL NAVIGATION

THE MARINE SEXTANT

1600. Description and Use

The marine sextant measures the angle between two points by bringing the direct image from one point and a double-reflected image from the other into coincidence. Its principal use is to measure the altitudes of celestial bodies above the visible sea horizon. It may also be used to measure vertical angles to find the range from an object of known height. Sometimes it is turned on its side and used for measuring the angular distance between two terrestrial objects.

A marine sextant can measure angles up to approximately 120°. Originally, the term "sextant" was applied to the navigator's double-reflecting, altitude-measuring instrument only if its arc was 60° in length, or 1/6 of a circle, permitting measurement of angles from 0° to 120°. In modern usage the term is applied to all modern navigational altitude-measuring instruments regardless of angular range or principles of operation.

1601. Optical Principles of a Sextant

When a plane surface reflects a light ray, the angle of reflection equals the angle of incidence. The angle between the first and final directions of a ray of light that has undergone double reflection in the same plane is twice the angle the two reflecting surfaces make with each other (Figure 1601).

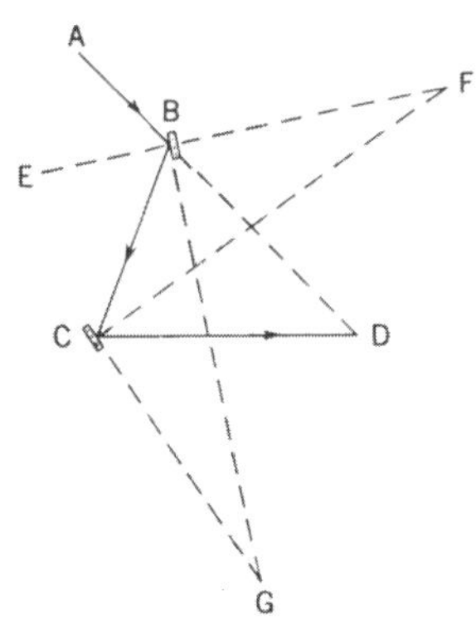

Figure 1601. Optical principle of the marine sextant.

In Figure 1601, AB is a ray of light from a celestial body. The index mirror of the sextant is at B, the horizon glass at C, and the eye of the observer at D. Construction lines EF and CF are perpendicular to the index mirror and horizon glass, respectively. Lines BG and CG are parallel to these mirrors. Therefore, angles BFC and BGC are equal because their sides are mutually perpendicular. Angle BGC is the inclination of the two reflecting surfaces. The ray of light AB is reflected at mirror B, proceeds to mirror C, where it is again reflected, and then continues on to the eye of the observer at D. Since the angle of reflection is equal to the angle of incidence,

ABE = EBC, and ABC = 2EBC.
BCF = FCD, and BCD = 2BCF.

Since an exterior angle of a triangle equals the sum of the two non adjacent interior angles,

ABC = BDC+BCD, and EBC = BFC+BCF.
Transposing,
BDC = ABC-BCD, and BFC = EBC-BCF.

Substituting 2EBC for ABC, and 2BCF for BCD in the first of these equations,

BDC = 2EBC-2BCF, or BDC=2 (EBC-BCF).

Since BFC=EBC - BCF, and BFC = BGC, therefore

BDC = 2BFC = 2BGC.

That is, BDC, the angle between the first and last directions of the ray of light, is equal to 2BGC, twice the angle of inclination of the reflecting surfaces. Angle BDC is the altitude of the celestial body.

If the two mirrors are parallel, the incident ray from any observed body must be parallel to the observer's line of sight through the horizon glass. In that case, the body's altitude would be zero. The angle that these two reflecting surfaces make with each other is one-half the observed angle. The graduations on the arc reflect this half angle relationship between the angle observed and the mirrors' angle.

1602. Micrometer Drum Sextant

Figure 1602 shows a modern marine sextant, called a **micrometer drum sextant**. In most marine sextants, brass or aluminum comprise the **frame**, A. Frames come in various designs; most are similar to this. Teeth mark the outer edge of the **limb**, B; each tooth marks one degree of altitude. The altitude graduations, C, along the limb, mark the **arc**. Some sextants have an arc marked in a strip of brass, silver, or platinum inlaid in the limb.

The **index arm**, D, is a movable bar of the same material as the frame. It pivots about the center of curvature of the limb. The **tangent screw**, E, is mounted perpendicularly on the end of the index arm, where it engages the teeth of the limb. Because the observer can move the index arm through the length of the arc by rotating the tangent screw, this is sometimes called an "endless tangent screw." The **release**, F, is a spring-actuated clamp that keeps the tangent screw engaged with the limb's teeth. The observer can disengage the tangent screw and move the index arm along the limb for rough adjustment. The end of the tangent screw mounts a **micrometer drum**, G, graduated in minutes of altitude. One complete turn of the drum moves the index arm one degree along the arc. Next to the micrometer drum and fixed on the index arm is a **vernier**, H, that reads in fractions of a minute. The vernier shown is graduated into ten parts, permitting readings to $^1/_{10}$ of a minute of arc (0.1'). Some sextants have verniers graduated into only five parts, permitting readings to 0.2'.

The **index mirror**, I, is a piece of silvered plate glass mounted on the index arm, perpendicular to the plane of the instrument, with the center of the reflecting surface directly over the pivot of the index arm. The **horizon glass**, J, is a piece of optical glass silvered on its half nearer the frame. It is mounted on the frame, perpendicular to the plane of the sextant. The index mirror and horizon glass are mounted so that their surfaces are parallel when the micrometer drum is set at 0°, if the instrument is in perfect adjustment. **Shade glasses**, K, of varying darkness are mounted on the sextant's frame in front of the index mirror and horizon glass. They can be moved into the line of sight as needed to reduce the intensity of light reaching the eye.

The **telescope**, L, screws into an adjustable collar in line with the horizon glass and parallel to the plane of the instrument. Most modern sextants are provided with only one telescope. When only one telescope is provided, it is of the "erect image type," either as shown or with a wider "object glass" (far end of telescope), which generally is shorter in length and gives a greater field of view. The second telescope, if provided, may be the "inverting type." The inverting telescope, having one lens less than the erect type, absorbs less light, but at the expense of producing an inverted image. A small colored glass cap is sometimes provided, to be placed over the "eyepiece" (near end of telescope) to reduce glare. With this in place, shade glasses are generally not needed. A "peep sight," or clear tube which serves to direct the line of sight of the observer when no telescope is used, may be fitted.

Sextants are designed to be held in the right hand. Some have a small light on the index arm to assist in reading altitudes. The batteries for this light are fitted inside a recess in the **handle**, M. Not clearly shown in Figure 1602 are the **tangent screw**, E, and the three legs.

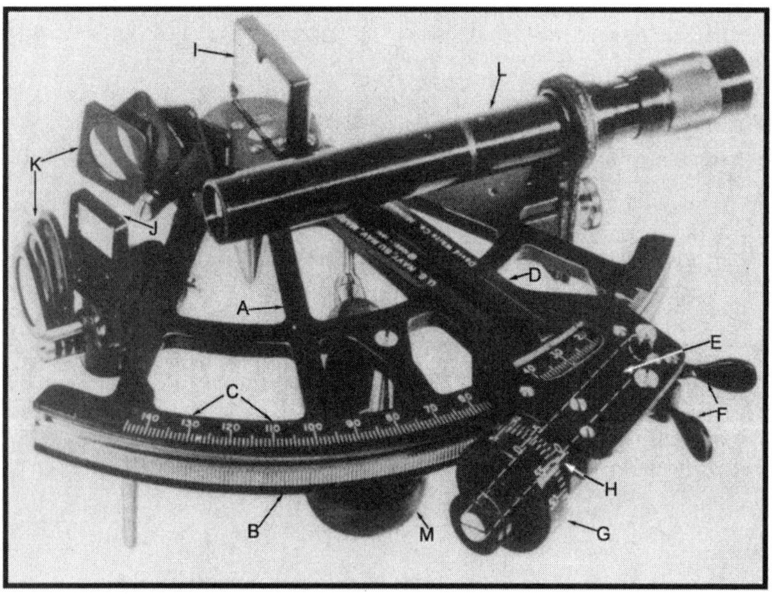

Figure 1602. U.S. Navy Mark 2 micrometer drum sextant.

There are two basic designs commonly used for mounting and adjusting mirrors on marine sextants. On the U.S. Navy Mark 3 and certain other sextants, the mirror is mounted so that it can be moved against retaining or mounting springs within its frame. Only one perpendicular adjustment screw is required. On the U.S. Navy Mark 2 and other sextants the mirror is fixed within its frame. Two perpendicular adjustment screws are required. One screw must be loosened before the other screw bearing on the same surface is tightened.

1603. Vernier Sextant

Most recent marine sextants are of the micrometer drum type, but at least two older-type sextants are still in use. These differ from the micrometer drum sextant principally in the manner in which the final reading is made. They are called **vernier sextants**.

The **clamp screw vernier sextant** is the older of the two. In place of the modern release clamp, a clamp screw is fitted on the underside of the index arm. To move the index arm, the clamp screw is loosened, releasing the arm. When the arm is placed at the approximate altitude of the body being observed, the clamp screw is tightened. Fixed to the clamp screw and engaged with the index arm is a long tangent screw. When this screw is turned, the index arm moves slowly, permitting accurate setting. Movement of the index arm by the tangent screw is limited to the length of the screw (several degrees of arc). Before an altitude is measured, this screw should be set to the approximate mid-point of its range. The final reading is made on a vernier set in the index arm below the arc. A small microscope or magnifying glass fitted to the index arm is used in making the final reading.

The **endless tangent screw vernier sextant** is identical to the micrometer drum sextant, except that it has no drum, and the fine reading is made by a vernier along the arc, as with the clamp screw vernier sextant. The release is the same as on the micrometer drum sextant, and teeth are cut into the underside of the limb which engage with the endless tangent screw.

1604. Sextant Sun Sights

For a Sun sight, hold the sextant vertically and direct the sight line at the horizon directly below the Sun. After moving suitable shade glasses into the line of sight, move the index arm outward along the arc until the reflected image appears in the horizon glass near the direct view of the horizon. Rock the sextant slightly to the right and left to ensure it is perpendicular. As you rock the sextant, the image of the Sun appears to move in an arc, and you may have to turn slightly to prevent the image from moving off the horizon glass.

The sextant is vertical when the Sun appears at the bottom of the arc. This is the correct position for making the observation. The Sun's reflected image appears at the center of the horizon glass; one half appears on the silvered part, and the other half appears on the clear part. Move the index arm with the drum or vernier slowly until the Sun appears to be resting exactly on the horizon, tangent to the lower limb. The novice observer needs practice to determine the exact point of tangency. Beginners often err by bringing the image down too far.

Some navigators get their most accurate observations by letting the body contact the horizon by its own motion, bringing it slightly below the horizon if rising, and above if setting. At the instant the horizon is tangent to the disk, the navigator notes the time. The sextant altitude is the uncorrected reading of the sextant.

1605. Sextant Moon Sights

When observing the Moon, follow the same procedure as for the Sun. Because of the phases of the Moon, the upper limb of the Moon is observed more often than that of the Sun. When the terminator (the line between light and dark areas) is nearly vertical, be careful in selecting the limb to shoot. Sights of the Moon are best made during either daylight hours or that part of twilight in which the Moon is least luminous. At night, false horizons may appear below the Moon because the Moon illuminates the water below it.

1606. Sextant Star and Planet Sights

While the relatively large Sun and Moon are easy to find in the sextant, stars and planets can be more difficult to locate because the field of view is so narrow. One of three methods may help locate a star or planet:

Method 1. Set the index arm and micrometer drum on 0° and direct the line of sight at the body to be observed. Then, while keeping the reflected image of the body in the mirrored half of the horizon glass, swing the index arm out and rotate the frame of the sextant down. Keep the reflected image of the body in the mirror until the horizon appears in the clear part of the horizon glass. Then, make the observation. When there is little contrast between brightness of the sky and the body, this procedure is difficult. If the body is "lost" while it is being brought down, it may not be recovered without starting over again.

Method 2. Direct the line of sight at the body while holding the sextant upside down. Slowly move the index arm out until the horizon appears in the horizon glass. Then invert the sextant and take the sight in the usual manner.

Method 3. Determine in advance the approximate altitude and azimuth of the body by a star finder such as No. 2102D. Set the sextant at the indicated altitude and face in the direction of the azimuth. The image of the body should appear in the horizon glass with a little searching.

When measuring the altitude of a star or planet, bring its center down to the horizon. Stars and planets have no discernible upper or lower limb; you must observe the center of the point of light. Because stars and planets have no discernible limb and because their visibility may be limited, the method of letting a star or planet intersect the horizon by its own motion is not recommended. As with the Sun and Moon, however, "rock the sextant" to establish perpendicularity.

1607. Taking a Sight

Unless you have a navigation calculator or computer that will identify bodies automatically, predict expected altitudes and azimuths for up to eight bodies when preparing to take celestial sights. Choose the stars and planets that give the best bearing spread. Try to select bodies with a predicted altitude between 30° and 70°. Take sights of the brightest stars first in the evening; take sights of the brightest stars last in the morning.

Occasionally, fog, haze, or other ships in a formation may obscure the horizon directly below a body which the navigator wishes to observe. If the arc of the sextant is sufficiently long, a **back sight** might be obtained, using the opposite point of the horizon as the reference. For this the observer faces away from the body and observes the supplement of the altitude. If the Sun or Moon is observed in this manner, what appears in the horizon glass to be the lower limb is in fact the upper limb, and vice versa. In the case of the Sun, it is usually preferable to observe what appears to be the upper limb. The arc that appears when rocking the sextant for a back sight is inverted; that is, the highest point indicates the position of perpendicularity.

If more than one telescope is furnished with the sextant, the erecting telescope is used to observe the Sun. A wider field of view is present if the telescope is not used. The collar into which the sextant telescope fits may be adjusted in or out, in relation to the frame. When moved in, more of the mirrored half of the horizon glass is visible to the navigator, and a star or planet is more easily observed when the sky is relatively bright. Near the darker limit of twilight, the telescope can be moved out, giving a broader view of the clear half of the glass, and making the less distinct horizon more easily discernible. If both eyes are kept open until the last moments of an observation, eye strain will be lessened. Practice will permit observations to be made quickly, reducing inaccuracy due to eye fatigue.

When measuring an altitude, have an assistant note and record the time if possible, with a "stand-by" warning when the measurement is almost ready, and a "mark" at the moment a sight is made. If a flashlight is needed to see the comparing watch, the assistant should be careful not to interfere with the navigator's night vision.

If an assistant is not available to time the observations, the observer holds the watch in the palm of his left hand, leaving his fingers free to manipulate the tangent screw of the sextant. After making the observation, he notes the time as quickly as possible. The delay between completing the altitude observation and noting the time should not be more than one or two seconds.

1608. Reading the Sextant

Reading a micrometer drum sextant is done in three steps. The degrees are read by noting the position of the arrow on the index arm in relation to the arc. The minutes are read by noting the position of the zero on the vernier with relation to the graduations on the micrometer drum. The fraction of a minute is read by noting which mark on the vernier most nearly coincides with one of the graduations on the micrometer drum. This is similar to reading the time with the hour, minute, and second hands of a watch. In both, the relationship of one part of the reading to the others should be kept in mind. Thus, if the hour hand of a watch were about on "4," one would know that the time was about four o'clock. But if the minute hand were on "58," one would know that the time was 0358 (or 1558), not 0458 (or 1658). Similarly, if the arc indicated a reading of about 40°, and 58' on the micrometer drum were opposite zero on the vernier, one would know that the reading was 39° 58', not 40°58'. Similarly, any doubt as to the correct minute can be removed by noting the fraction of a minute from the position of the vernier. In Figure 1608a the reading is 29° 42.5'. The arrow on the index mark is between 29° and 30°, the zero on the vernier is between 42' and 43', and the 0.5' graduation on the vernier coincides with one of the graduations on the micrometer drum.

The principle of reading a vernier sextant is the same, but the reading is made in two steps. Figure 1608b shows a typical altitude setting. Each degree on the arc of this sextant is graduated into three parts, permitting an initial reading by the reference mark on the index arm to the nearest 20' of arc. In this illustration the reference mark lies between 29°40' and 30°00', indicating a reading between these values. The reading for the fraction of 20' is made using the vernier, which is engraved on the index arm and has the small reference mark as its zero graduation. On this vernier, 40 graduations coincide with 39 graduations on the arc. Each graduation on the vernier is equivalent to 1/40 of one graduation of 20' on the arc, or 0.5', or 30". In the illustration, the vernier graduation representing 2 1/2' (2'30") most nearly coincides with one of the graduations on the arc. Therefore, the reading is 29°42'30", or 29°42.5', as before. When a vernier of this type is used, any doubt as to which mark on the vernier coincides with a graduation on the arc can usually be resolved by noting the position of the vernier mark on each side of the one that seems to be in coincidence.

Negative readings, such as a negative index correction, are made in the same manner as positive readings; the various figures are added algebraically. Thus, if the three parts of a micrometer drum reading are (-)1°, 56' and 0.3', the total reading is (-)1° + 56' + 0.3' = (-)3.7'.

1609. Developing Observational Skill

A well-constructed marine sextant is capable of measuring angles with an instrument error not exceeding 0.1'. Lines of position from altitudes of this accuracy would not be

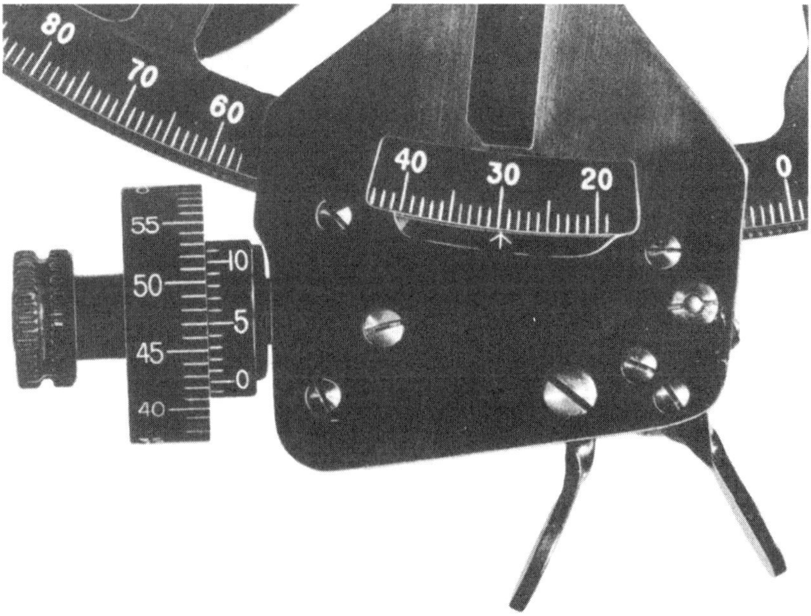

Figure 1608a. Micrometer drum sextant set at 29° 42.5'.

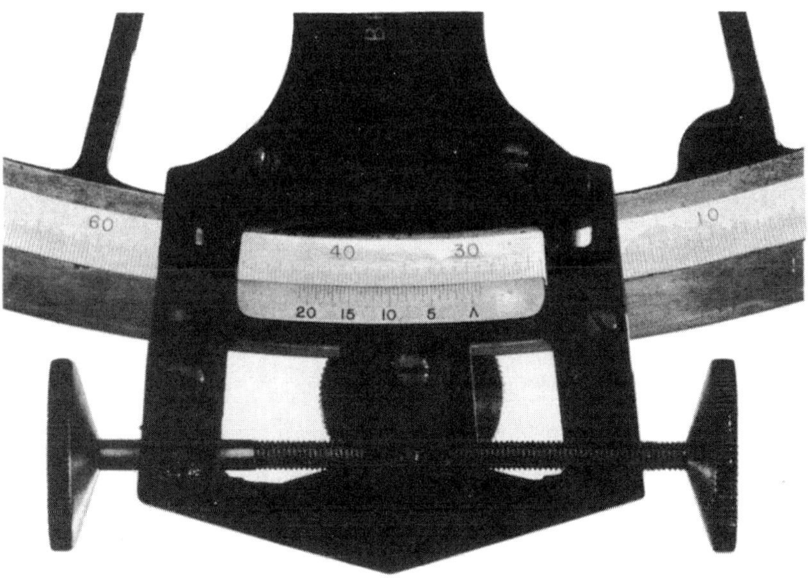

Figure 1608b. Vernier sextant set at 29°42'30".

in error by more than about 200 yards. However, there are various sources of error, other than instrumental, in altitudes measured by sextant. One of the principal sources is the observer.

The first fix a student celestial navigator plots is likely to be disappointing. Most navigators require a great amount of practice to develop the skill necessary for consistently good observations. But practice alone is not sufficient. Good technique should be developed early and refined throughout the navigator's career. Many good pointers can be obtained from experienced navigators, but each develops his own technique, and a practice that proves successful for one observer may not help another. Also, an experienced navigator is not necessarily a good observer. Navigators have a natural tendency to judge the accuracy of their observations by the size of the figure formed when the lines of position are plotted. Although this is some indication, it is an imperfect one, because it does not indicate errors of individual observations, and may not reflect constant errors. Also, it is a compound of a number of errors, some of which are not subject to the navigator's control.

Lines of position from celestial observations should be compared often with good positions obtained by electronics or piloting. Common sources of error are:

1. The sextant may not be rocked properly.
2. Tangency may not be judged accurately.
3. A false horizon may have been used.
4. Subnormal refraction (dip) might be present.
5. The height of eye may be wrong.
6. Time might be in error.
7. The index correction may have been determined incorrectly.
8. The sextant might be out of adjustment.
9. An error may have been made in the computation.

Generally, it is possible to correct observation technique errors, but occasionally a personal error will persist. This error might vary as a function of the body observed, degree of fatigue of the observer, and other factors. For this reason, a personal error should be applied with caution.

To obtain greater accuracy, take a number of closely-spaced observations. Plot the resulting altitudes versus time and fair a curve through the points. Unless the body is near the celestial meridian, this curve should be a straight line. Use this graph to determine the altitude of the body at any time covered by the graph. It is best to use a point near the middle of the line. Using a navigational calculator or computer program to reduce sights will yield greater accuracy because of the rounding errors inherent in the use of sight reduction tables, and because many more sights can be reduced in a given time, thus averaging out errors.

A simpler method involves making observations at equal intervals. This procedure is based upon the assumption that, unless the body is on the celestial meridian, the change in altitude should be equal for equal intervals of time. Observations can be made at equal intervals of altitude or time. If time intervals are constant, the mid time and the average altitude are used as the observation. If altitude increments are constant, the average time and mid altitude are used.

If only a small number of observations is available, reduce and plot the resulting lines of position; then adjust them to a common time. The average position of the line might be used, but it is generally better practice to use the middle line. Reject any observation considered unreliable when determining the average.

1610. Care of the Sextant

A sextant is a rugged instrument. However, careless handling or neglect can cause it irreparable harm. If you drop it, take it to an instrument repair shop for testing and inspection. When not using the sextant, stow it in a sturdy and sufficiently padded case. Keep the sextant away from excessive heat and dampness. Do not expose it to excessive vibration. Do not leave it unattended when it is out of its case. Do not hold it by its limb, index arm, or telescope. Lift it only by its frame or handle. Do not lift it by its arc or index bar.

Next to careless handling, moisture is the sextant's greatest enemy. Wipe the mirrors and the arc after each use. If the mirrors get dirty, clean them with lens paper and a small amount of alcohol. Clean the arc with ammonia; never use a polishing compound. When cleaning, do not apply excessive pressure to any part of the instrument.

Silica gel kept in the sextant case will help keep the instrument free from moisture and preserve the mirrors. Occasionally heat the silica gel to remove the absorbed moisture.

Rinse the sextant with fresh water if sea water gets on it. Wipe the sextant gently with a soft cotton cloth and dry the optics with lens paper.

Glass optics do not transmit all the light received because glass surfaces reflect a small portion of light incident on their face. This loss of light reduces the brightness of the object viewed. Viewing an object through several glass optics affects the perceived brightness and makes the image indistinct. The reflection also causes glare which obscures the object being viewed. To reduce this effect to a minimum, the glass optics are treated with a thin, fragile, anti-reflection coating. Therefore, apply only light pressure when polishing the coated optics. Blow loose dust off the lens before wiping them so grit does not scratch the lens.

Occasionally, oil and clean the tangent screw and the teeth on the side of the limb. Use the oil provided with the sextant or an all-purpose light machine oil. Occasionally set the index arm of an endless tangent screw at one extremity of the limb, oil it lightly, and then rotate the tangent screw over the length of the arc. This will clean the teeth and spread oil over them. When stowing a sextant for a long period, clean it thoroughly, polish and oil it, and protect its arc with a thin coat of petroleum jelly. If the mirrors need re-silvering, take the sextant to an instrument shop.

1611. Non Adjustable Sextant Errors

The non-adjustable sextant errors are prismatic error, graduation error, and centering error. The higher the quality of the instrument, the less these error will be.

Prismatic error occurs when the faces of the shade glasses and mirrors are not parallel. Error due to lack of parallelism in the shade glasses may be called **shade error**. The navigator can determine shade error in the shade glasses near the index mirror by comparing an angle measured when a shade glass is in the line of sight with the same angle measured when the glass is not in the line of sight. In this manner, determine and record the error for each shade glass. Before using a combination of shade glasses, determine their combined error. If certain observations require additional shading, use the colored telescope eyepiece cover. This does not introduce an error because direct and reflected rays are traveling together when they reach the cover and are, therefore, affected equally by any lack of parallelism of its two sides.

Graduation errors occur in the arc, micrometer drum, and vernier of a sextant which is improperly cut or incorrectly calibrated. Normally, the navigator cannot determine whether the arc of a sextant is improperly cut, but the principle of the vernier makes it possible to determine the existence of graduation errors in the micrometer drum or vernier. This is a useful guide in detecting a poorly made instrument. The first and last markings on any vernier should align perfectly with one less graduation on the adjacent micrometer drum.

Centering error results if the index arm does not pivot at the exact center of the arc's curvature. Calculate centering error by measuring known angles after removing all adjustable errors. Use horizontal angles accurately measured with a theodolite as references for this procedure. Several readings by both theodolite and sextant should minimize errors. If a theodolite is not available, use calculated angles between the lines of sight to stars as the reference, comparing these calculated values with the values determined by the sextant. To minimize refraction errors, select stars at about the same altitude and avoid stars near the horizon. The same shade glasses, if any, used for determining index error should be used for measuring centering error.

The manufacturer normally determines the magnitude of all three non-adjustable errors and reports them to the user as **instrument error**. The navigator should apply the correction for this error to each sextant reading.

1612. Adjustable Sextant Error

The navigator should measure and remove the following adjustable sextant errors in the order listed:

1. **Perpendicularity Error:** Adjust first for perpendicularity of the index mirror to the frame of the sextant. To test for perpendicularity, place the index arm at about 35° on the arc and hold the sextant on its side with the index mirror up and toward the eye. Observe the direct and reflected views of the sextant arc, as illustrated in Figure 1612a. If the two views are not joined in a straight line, the index mirror is not perpendicular. If the reflected image is above the direct view, the mirror is inclined forward. If the reflected image is below the direct view, the mirror is inclined backward. Make the adjustment using two screws behind the index mirror.

2. **Side Error:** An error resulting from the horizon glass not being perpendicular is called **side error**. To test for side error, set the index arm at zero and direct the line of sight at a star. Then rotate the tangent screw back and forth so that the reflected image passes alternately above and below the direct view. If, in changing from one position to the other, the reflected image passes directly over the unreflected image, no side error exists. If it passes to one side, side error exists. Figure 1612b illustrates observations without side error (left) and with side error (right). Whether the sextant reads zero when the true and reflected images are in coincidence is immaterial for this test. An alternative method is to observe a vertical line, such as one edge of the mast of another vessel (or the sextant can be held on its side and the horizon used). If the direct and reflected portions do not form a continuous line, the horizon glass is not perpendicular to the frame of the sextant. A third method involves holding the sextant vertically, as in observing the altitude of a celestial body. Bring the reflected image of the horizon into coincidence with the direct view until it appears as a continuous line across the horizon glass. Then tilt the sextant right or left. If the horizon still appears continuous, the horizon glass is perpendicular to the frame, but if the reflected portion appears above or below the part seen directly, the glass is not perpendicular. Make the appropriate adjustment using two screws behind the horizon glass.

3. **Collimation Error:** If the line of sight through the telescope is not parallel to the plane of the instrument, a **collimation error** will result. Altitudes measured will be greater than their actual values. To check for parallelism of the telescope, insert it in its collar and observe two stars 90° or more apart. Bring the reflected image of one into coincidence with the direct view of the other near either the right or left edge of the field of view (the upper or lower edge if the sextant is horizontal). Then tilt the sextant so that the stars appear near the opposite edge. If they remain in coincidence, the telescope is parallel to the frame; if they separate, it is not. An alternative method involves placing the telescope in its collar and then laying the sextant on a flat table. Sight along the frame of the sextant and have an assistant place a mark on the opposite bulkhead, in line with the frame. Place another mark above the first, at a distance equal to the distance from the center of the telescope to the frame. This second line should be in the center of the field

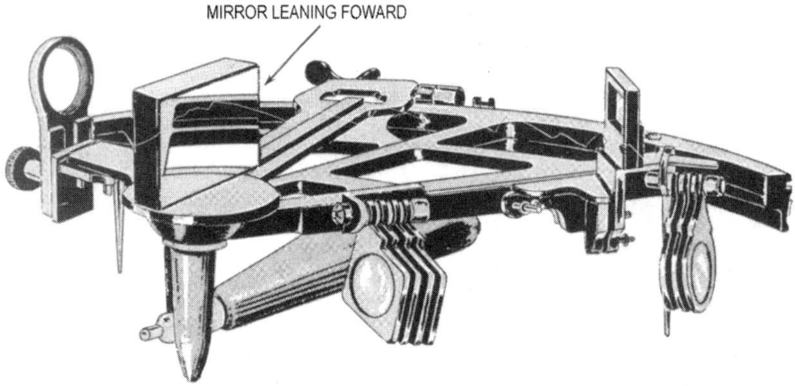

Figure 1612a. Testing the perpendicularity of the index mirror. Here the mirror is not perpendicular.

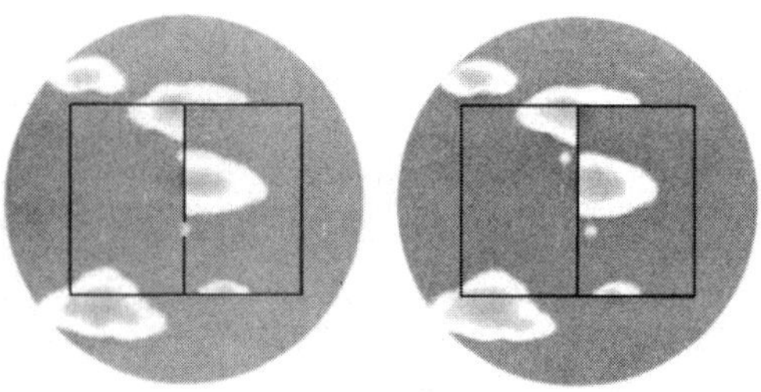

Figure 1612b. Testing the perpendicularity of the horizon glass. On the left, side error does not exist. At the right, side error does exist.

of view of the telescope if the telescope is parallel to the frame. Adjust the collar to correct for non-parallelism.

4. **Index Error:** Index error is the error remaining after the navigator has removed perpendicularity error, side error, and collimation error. The index mirror and horizon glass not being parallel when the index arm is set exactly at zero is the major cause of index error. To test for parallelism of the mirrors, set the instrument at zero and direct the line of sight at the horizon. Adjust the sextant reading as necessary to cause both images of the horizon to come into line. The sextant's reading when the horizon comes into line is the index error. If the index error is positive, subtract it from each sextant reading. If the index error is negative, add it to each sextant reading.

1613. Selecting a Sextant

Carefully match the selected sextant to its required uses. For occasional small craft or student use, a plastic sextant may be adequate. A plastic sextant may also be appropriate for an emergency navigation kit. Accurate offshore navigation requires a quality metal instrument. For ordinary use in measuring altitudes of celestial bodies, an arc of 90° or slightly more is sufficient. If back sights or determining horizontal angles are often required, purchase one with a longer arc. An experienced mariner or nautical instrument technician can provide valuable advice on the purchase of a sextant.

1614. The Artificial Horizon

Measurement of altitude requires an exact horizontal reference, normally provided at sea by the visible horizon. If the horizon is not clearly visible, however, a different horizontal reference is required. Such a reference is commonly termed an **artificial horizon**. If it is attached to, or part of, the sextant, altitudes can be measured at sea, on land, or in the air, whenever celestial bodies are available for observations.

An external artificial horizon can be improvised by a carefully levelled mirror or a pan of dark liquid. To use an external artificial horizon, stand or sit so that the celestial body is reflected in the mirror or liquid, and is also visible in direct view. With the sextant, bring the double-reflected image into coincidence with the image appearing in the liquid. For a lower limb observation of the Sun or the Moon, bring the bottom of the double-reflected image into coincidence with the top of the image in the liquid. For an upper-limb observation, bring the opposite sides into coincidence. If one image covers the other, the observation is of the center of the body.

After the observation, apply the index correction and any other instrumental correction. Then take *half* the remaining angle and apply all other corrections except dip (height of eye) correction, since this is not applicable. If the center of the Sun or Moon is observed, omit the correction for semidiameter.

1615. Artificial Horizon Sextants

Various types of artificial horizons have been used, including a bubble, gyroscope, and pendulum. Of these, the bubble has been most widely used. This type of instrument is fitted as a backup system to inertial and other positioning systems in a few aircraft, fulfilling the requirement for a self-contained, non-emitting system. On land, a skilled observer using a 2-minute averaging bubble or pendulum sextant can measure altitudes to an accuracy of perhaps 2', (2 miles). This, of course, refers to the accuracy of measurement only, and does not include additional errors such as abnormal refraction, deflection of the vertical, computing and plotting errors, etc. In steady flight through smooth air the error of a 2-minute observation is increased to perhaps 5 to 10 miles.

At sea, with virtually no roll or pitch, results should approach those on land. However, even a gentle roll causes large errors. Under these conditions observational errors of 10-16 miles are not unreasonable. With a moderate sea, errors of 30 miles or more are common. In a heavy sea, any useful observations are virtually impossible to obtain. Single altitude observations in a moderate sea can be in error by a matter of degrees.

When the horizon is obscured by ice or haze, polar navigators can sometimes obtain better results with an artificial-horizon sextant than with a marine sextant. Some artificial-horizon sextants have provision for making observations with the natural horizon as a reference, but results are not generally as satisfactory as by marine sextant. Because of their more complicated optical systems, and the need for providing a horizontal reference, artificial-horizon sextants are generally much more costly to manufacture than marine sextants.

Altitudes observed by artificial-horizon sextants are subject to the same errors as those observed by marine sextant, except that the dip (height of eye) correction does not apply. Also, when the center of the Sun or Moon is observed, no correction for semidiameter is required.

CHRONOMETERS

1616. The Marine Chronometer

The spring-driven **marine chronometer** is a precision timepiece used aboard ship to provide accurate time for celestial observations. A chronometer differs from a spring-driven watch principally in that it contains a variable lever device to maintain even pressure on the mainspring, and a special balance designed to compensate for temperature variations.

A spring-driven chronometer is set approximately to Greenwich mean time (GMT) and is not reset until the instrument is overhauled and cleaned, usually at three-year intervals. The difference between GMT and chronometer time (C) is carefully determined and applied as a correction to all chronometer readings. This difference, called chronometer error (CE), is **fast** (F) if chronometer time is later than GMT, and **slow** (S) if earlier. The amount by which chronometer error changes in 1 day is called **chronometer rate**. An erratic rate indicates a defective instrument requiring repair.

The principal maintenance requirement is regular winding at about the same time each day. At maximum intervals of about three years, a spring-driven chronometer should be sent to a chronometer repair shop for cleaning and overhaul.

1617. Quartz Crystal Marine Chronometers

Quartz crystal marine chronometers have replaced spring-driven chronometers aboard many ships because of their greater accuracy. They are maintained on GMT directly from radio time signals. This eliminates chronometer error (CE) and watch error (WE) corrections. Should the second hand be in error by a readable amount, it can be reset electrically.

The basic element for time generation is a quartz crystal oscillator. The quartz crystal is temperature compensated and is hermetically sealed in an evacuated envelope. A calibrated adjustment capability is provided to adjust for the aging of the crystal.

The chronometer is designed to operate for a minimum of 1 year on a single set of batteries. A good marine chronometer has a built-in push button battery test meter. The meter face is marked to indicate when the battery should be replaced. The chronometer continues to operate and keep the correct time for at least 5 minutes while the batteries are changed. The chronometer is designed to accommodate the gradual voltage drop during the life of the batteries while maintaining accuracy requirements.

1618. Watches

A chronometer should not be removed from its case to time sights. Observations may be timed and ship's clocks set with a **comparing watch**, which is set to chronometer time (GMT, also known as UT) and taken to the bridge wing for recording sight times. In practice, a wrist watch coordinated to the nearest second with the chronometer will be adequate.

A stop watch, either spring wound or digital, may also be used for celestial observations. In this case, the watch is started at a known GMT by chronometer, and the elapsed time of each sight added to this to obtain GMT of the sight.

All chronometers and watches should be checked regularly with a radio time signal. Times and frequencies of radio time signals are listed in *NIMA Pub. 117, Radio Navigational Aids*.

1619. Navigational Calculators

While not considered "instruments" in the strict sense of the word, certainly one of the professional navigator's most useful tools is the navigational calculator or computer program. Calculators eliminate several potential sources of error in celestial navigation, and permit the solution of many more sights in much less time, making it possible to refine a celestial position much more accurately than is practical using mathematical or tabular methods.

Calculators also save space and weight, a valuable consideration on many craft. One small calculator can replace several heavy and expensive volumes of tables, and is inexpensive enough that there is little reason not to carry a spare for backup use should the primary one fail. The pre-programmed calculators are at least as robust in construction, probably more so, than the sextant itself, and properly cared for, will last a lifetime with no maintenance except new batteries from time to time.

If the vessel carries a computer for other ship's chores such as inventory control or personnel administration, there is little reason not to use it for celestial navigation. Freeware or inexpensive programs are available which take up little hard disk space and allow rapid solution of all types of celestial navigation problems. Typically they will also take care of route planning, sailings, tides, weather routing, electronic charts, and numerous other tasks.

Using a calculator or sight reduction program, it is possible to take and solve half a dozen or more sights in a fraction of the time it would normally take to shoot two or three and solve them by hand. This will increase the accuracy of the fix by averaging out errors in taking the sights. The computerized solution is always more accurate than tabular methods because it is free of rounding errors.

CHAPTER 17

AZIMUTHS AND AMPLITUDES

INTRODUCTION

1700. Checking Compass Error

The navigator must constantly be concerned about the accuracy of the ship's primary and backup compasses, and should check them regularly. A regularly annotated compass log book will allow the navigator to notice a developing error before it becomes a serious problem.

As long as at least two different types of compass (e.g. mechanical gyro and flux gate, or magnetic and ring laser gyro) are consistent with each other, one can be reasonably sure that there is no appreciable error in either system. Since different types of compasses depend on different scientific principles and are not subject to the same error sources, their agreement indicates almost certainly that no error is present.

A navigational compass can be checked against the heading reference of an inertial navigation system if one is installed. One can also refer to the ship's indicated GPS track as long as current and leeway are not factors, so that the ship's COG and heading are in close agreement.

The navigator's only completely independent directional reference (because it is extra-terrestrial and not man-made) is the sky. The primary compass should be checked occasionally by comparing the observed and calculated azimuths and amplitudes of a celestial body. The difference between the observed and calculated values is the compass error. This chapter discusses these procedures.

Theoretically, these procedures work with any celestial body. However, the Sun and Polaris are used most often when measuring azimuths, and the rising or setting Sun when measuring amplitudes.

While errors can be computed to the nearest tenth of a degree or so, it is seldom possible to steer a ship that accurately, especially when a sea is running, and it is reasonable to round calculations to the nearest half or perhaps whole degree for most purposes.

Various hand-held calculators and computer programs are available to relieve the tedium and errors of tabular and mathematical methods of calculating azimuths and amplitudes. Naval navigators will find the STELLA program useful in this regard. Chapter 20 discusses this program in greater detail.

AZIMUTHS

1701. Compass Error by Azimuth of the Sun

Mariners may use *Pub 229, Sight Reduction Tables for Marine Navigation* to compute the Sun's azimuth. They compare the computed azimuth to the azimuth measured with the compass to determine compass error. In computing an azimuth, interpolate the tabular azimuth angle for the difference between the table arguments and the actual values of declination, latitude, and local hour angle. Do this triple interpolation of the azimuth angle as follows:

1. Enter the *Sight Reduction Tables* with the nearest integral values of declination, latitude, and local hour angle. For each of these arguments, extract a base azimuth angle.

2. Reenter the tables with the same latitude and LHA arguments but with the declination argument 1° greater or less than the base declination argument, depending upon whether the actual declination is greater or less than the base argument. Record the difference between the respondent azimuth angle and the base azimuth angle and label it as the azimuth angle difference (Z Diff.).

3. Reenter the tables with the base declination and LHA arguments, but with the latitude argument 1° greater or less than the base latitude argument, depending upon whether the actual (usually DR) latitude is greater or less than the base argument. Record the Z Diff. for the increment of latitude.

4. Reenter the tables with the base declination and latitude arguments, but with the LHA argument 1° greater or less than the base LHA argument, depending upon whether the actual LHA is greater or less than the base argument. Record the Z Diff. for the increment of LHA.

5. Correct the base azimuth angle for each increment.

	Actual	Base Arguments	Base Z	Tab* Z	Z Diff.	Increments	Correction (Z Diff x Inc.÷ 60)
Dec.	20°13.8' N	20°	97.8°	96.4°	−1.4°	13.8'	−0.3°
DR Lat.	33°24.0' N	33°(Same)	97.8°	98.9°	+1.1°	24.0'	+0.4°
LHA	316°41.2'	317°	97.8°	97.1°	− 0.7°	18.8'	−0.2°
						Total Corr.	−0.1°

Base Z	97.8°
Corr.	(−) 0.1°
Z	N 97.7° E
Zn	097.7°
Zn pgc	096.5°
Gyro Error	1.2° E

*Respondent for the two base arguments and 1° change from third base argument, in vertical order of Dec., DR Lat., and LHA.

Figure 1701. Azimuth by Pub. No. 229.

Example:

In DR latitude 33° 24.0'N, the azimuth of the Sun is 096.5° pgc. At the time of the observation, the declination of the Sun is 20° 13.8'N; the local hour angle of the Sun is 316° 41.2'. Determine compass error.

Solution:

See Figure 1701 Enter the actual value of declination, DR latitude, and LHA. Round each argument to the nearest whole degree. In this case, round the declination and the latitude down to the nearest whole degree. Round the LHA up to the nearest whole degree. Enter the Sight Reduction Tables with these whole degree arguments and extract the base azimuth value for these rounded off arguments. Record the base azimuth value in the table.

As the first step in the triple interpolation process, increase the value of declination by 1° (to 21°) because the actual declination value was greater than the base declination. Enter the Sight Reduction Tables with the following arguments: (1) Declination = 21°; (2) DR Latitude = 33°; (3) LHA = 317°. Record the tabulated azimuth for these arguments.

As the second step in the triple interpolation process, increase the value of latitude by 1° to 34° because the actual DR latitude was greater than the base latitude. Enter the Sight Reduction Tables with the following arguments: (1) Declination = 20°; (2) DR Latitude = 34°; (3) LHA = 317°. Record the tabulated azimuth for these arguments.

As the third and final step in the triple interpolation process, decrease the value of LHA to 316° because the actual LHA value was smaller than the base LHA. Enter the Sight Reduction Tables with the following arguments: (1) Declination = 20°; (2) DR Latitude = 33°; (3) LHA = 316°. Record the tabulated azimuth for these arguments.

Calculate the Z Difference by subtracting the base azimuth from the tabulated azimuth. Be careful to carry the correct sign.

Z Difference = Tab Z - Base Z

Next, determine the increment for each argument by taking the difference between the actual values of each argument and the base argument. Calculate the correction for each of the three argument interpolations by multiplying the increment by the Z difference and dividing the resulting product by 60.

The sign of each correction is the same as the sign of the corresponding Z difference used to calculate it. In the above example, the total correction sums to -0.1'. Apply this value to the base azimuth of 97.8° to obtain the true azimuth 97.7°. Compare this to the compass reading of 096.5° pgc. The compass error is 1.2°E, which can be rounded to 1° for steering and logging purposes.

AZIMUTH OF POLARIS

1702. Compass Error By Azimuth Of Polaris

The Polaris tables in the *Nautical Almanac* list the azimuth of Polaris for latitudes between the equator and 65° N. Figure 2012 in Chapter 20 shows this table. Compare a compass bearing of Polaris to the tabular value of Polaris to determine compass error. The entering arguments for the table are LHA of Aries and observer latitude.

Example:

On March 17, 2001, at L 33°15.0' N and λ 045°00.0'W, at 02-00-00 GMT, Polaris bears 358.6° pgc. Calculate the compass error.

Date	*17 March 2001*
Time (GMT)	*02-00-00*
GHA Aries	*204° 43.0'*
Longitude	*045° 00.0'W*
LHA Aries	*159° 43.0'*

Solution:

Enter the azimuth section of the Polaris table with the

calculated LHA of Aries. In this case, go to the column for LHA Aries between 160° and 169°. Follow that column down and extract the value for the given latitude. Since the increment between tabulated values is so small, visual interpolation is sufficient. In this case, the azimuth for Polaris for the given LHA of Aries and the given latitude is 359.3°.

Tabulated Azimuth	*359.2°T*
Compass Bearing	*358.6°C*
Error	*0.6°E*

AMPLITUDES

1703. Amplitudes

A celestial body's amplitude angle is the complement of its azimuth angle. At the moment that a body rises or sets, the amplitude angle is the arc of the horizon between the body and the East/West point of the horizon where the observer's prime vertical intersects the horizon (at 90°), which is also the point where the plane of the equator intersects the horizon (at an angle numerically equal to the observer's co-latitude). See Figure 1703.

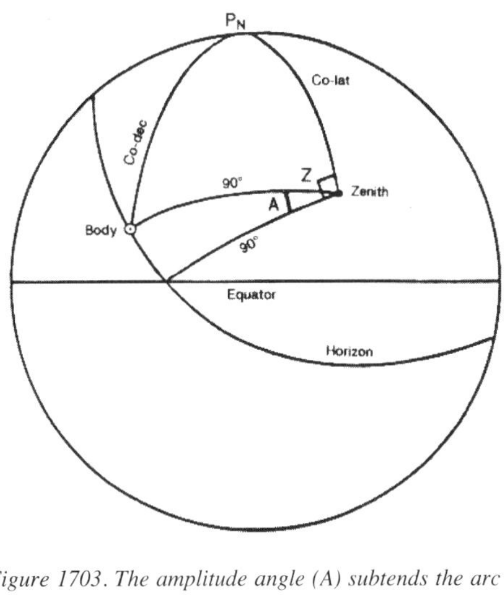

Figure 1703. The amplitude angle (A) subtends the arc of the horizon between the body and the point where the prime vertical and the equator intersect the horizon. Note that it is the compliment of the azimuth angle (Z).

In practical navigation, a bearing (psc or pgc) of a body can be observed when it is on either the celestial or the visible horizon. To determine compass error, simply convert the computed amplitude angle to true degrees and compare it with the observed compass bearing.

The angle is computed by the formula:

$$\sin A = \sin Dec / \cos Lat.$$

This formula gives the angle at the instant the body is on the celestial horizon. It does not contain an altitude term because the body's computed altitude is zero at this instant.

The angle is prefixed E if the body is rising and W if it is setting. This is the only angle in celestial navigation referenced FROM East or West, i.e. from the prime vertical. A body with northerly declination will rise and set North of the prime vertical. Likewise, a body with southerly declination will rise and set South of the prime vertical. Therefore, the angle is suffixed N or S to agree with the name of the body's declination. A body whose declination is zero rises and sets exactly on the prime vertical.

The Sun is on the celestial horizon when its lower limb is approximately two thirds of a diameter above the visible horizon. The Moon is on the celestial horizon when its upper limb is on the visible horizon. Stars and planets are on the celestial horizon when they are approximately one Sun diameter above the visible horizon.

When observing a body on the visible horizon, a correction from Table 23 must be applied. This correction accounts for the slight change in bearing as the body moves between the visible and celestial horizons. It reduces the bearing on the visible horizon to the celestial horizon, from which the table is computed.

For the Sun, stars, and planets, apply this correction to the observed bearing in the direction away from the elevated pole. For the moon, apply one half of the correction toward the elevated pole. Note that the algebraic sign of the correction does not depend upon the body's declination, but only on the observer's latitude. Assuming the body is the Sun the rule for applying the correction can be outlined as follows:

Observer's Lat.	Rising/Setting	Observed bearing
North	Rising	Add to
North	Setting	Subtract from
South	Rising	Subtract from
South	Setting	Add to

The following two articles demonstrate the procedure for obtaining the amplitude of the Sun on both the celestial and visible horizons.

1704. Amplitude of the Sun on the Celestial Horizon

Example:

The DR latitude of a ship is 51° 24.6' N. The navigator observes the setting Sun on the celestial horizon. Its decli-

Actual	Base	Base Amp.	Tab. Amp.	Diff.	Inc.	Correction
L=51.4°N	51°	32.0°	32.8°	+0.8°	0.4	+0.3°
dec=19.67°N	19.5°	32.0°	32.9°	+0.9°	0.3	+0.3°
					Total	+0.6°

Figure 1704. Interpolation in Table 22 *for Amplitude.*

nation is N 19° 40.4'. Its observed bearing is 303° pgc.

Required:

Gyro error.

Solution:

Interpolate in Table 22 for the Sun's calculated amplitude as follows. See Figure 1704. The actual values for latitude and declination are L = 51.4° N and dec. = N 19.67°. Find the tabulated values of latitude and declination closest to these actual values. In this case, these tabulated values are L = 51° and dec. = 19.5°. Record the amplitude corresponding to these base values, 32.0°, as the base amplitude.

Next, holding the base declination value constant at 19.5°, increase the value of latitude to the next tabulated value: N 52°. Note that this value of latitude was increased because the actual latitude value was greater than the base value of latitude. Record the tabulated amplitude for L = 52° and dec. = 19.5°: 32.8°. Then, holding the base latitude value constant at 51°, increase the declination value to the next tabulated value: 20°. Record the tabulated amplitude for L = 51° and dec. = 20°: 32.9°.

The latitude's actual value (51.4°) is 0.4 of the way between the base value (51°) and the value used to determine the tabulated amplitude (52°). The declination's actual value (19.67°) is 0.3 of the way between the base value (19.5°) and the value used to determine the tabulated amplitude (20.0°). To determine the total correction to base amplitude, multiply these increments (0.4 and 0.3) by the respective difference between the base and tabulated values (+0.8 and +0.9, respectively) and sum the products. The total correction is +0.6°. Add the total correction (+0.6°) to the base amplitude (32.0°) to determine the final amplitude (32.6°) which will be converted to a true bearing.

Because of its northerly declination (in this case), the Sun was 32.6° north of west when it was on the celestial horizon. Therefore its true bearing was 302.6° (270° + 32.6°) at this moment. Comparing this with the gyro bearing of 303° gives an error of 0.4°W, which can be rounded to 1/2°W.

1705. Amplitude of the Sun on the Visible Horizon

In higher latitudes, amplitude observations should be made when the body is on the visible horizon because the value of the correction is large enough to cause significant error if the observer misjudges the exact position of the celestial horizon. The observation will yield precise results whenever the visible horizon is clearly defined.

Example:

Observer's DR latitude is 59°47'N, Sun's declination is 5°11.3'S. At sunrise the Sun is observed on the visible horizon bearing 098.5° pgc.

Required:

Compass error.

Solution:

Given this particular latitude and declination, the amplitude angle is E100.4°S, so that the Sun's true bearing is 100.4° at the moment it is on the celestial horizon, that is, when its Hc is precisely 0°. Applying the Table 23 correction to the observed bearing using the rules given in Article 1703, the Sun would have been bearing 099.7° pgc had the observation been made when the Sun was on the celestial horizon. Therefore, the gyro error is 0.7°E.

1706. Amplitude by Calculation

As an alternative to using Table 22 and Table 23, a visible horizon amplitude observation can be solved by the "altitude azimuth" formula, because azimuth and amplitude angles are complimentary, and the co-functions of complimentary angles are equal; i.e., cosine Z = sine A.

Sine A = [SinD - (sin L sin H)] / (cos L cos H)

For shipboard observations, the Sun's (computed) altitude is negative 0.7° when it is on the visible horizon. Using the same entities as in Article 1705, the amplitude angle is computed as follows:

Sin A = [sin 5.2°- (sin 59.8° X sin -0.7°)] / (cos 59.8° X cos 0.7°)

CHAPTER 18

TIME

TIME IN NAVIGATION

1800. Solar Time

The Earth's rotation on its axis causes the Sun and other celestial bodies to appear to move across the sky from east to west each day. If a person located on the Earth's equator measured the time interval between two successive transits overhead of a very distant star, he would be measuring the period of the Earth's rotation. If he then made a similar measurement of the Sun, the resulting time would be about 4 minutes longer. This is due to the Earth's motion around the Sun, which continuously changes the apparent place of the Sun among the stars. Thus, during the course of a day the Sun appears to move a little to the east among the stars, so that the Earth must rotate on its axis through more than 360° in order to bring the Sun overhead again.

See Figure 1800. If the Sun is on the observer's meridian when the Earth is at point A in its orbit around the Sun, it will not be on the observer's meridian after the Earth has rotated through 360° because the Earth will have moved along its orbit to point B. Before the Sun is again on the observer's meridian, the Earth must turn a little more on its axis. The Sun will be on the observer's meridian again when the Earth has moved to point C in its orbit. Thus, during the course of a day the Sun appears to move eastward with respect to the stars.

The apparent positions of the stars are commonly reckoned with reference to an imaginary point called the **vernal equinox**, the intersection of the celestial equator and the ecliptic. The period of the Earth's rotation measured with respect to the vernal equinox is called a **sidereal day**. The period with respect to the Sun is called an **apparent solar day**.

When measuring time by the Earth's rotation, using the actual position of the Sun, or the apparent Sun, results in **apparent solar time**. Use of the apparent Sun as a time reference results in time of non-constant rate for at least three reasons. First, revolution of the Earth in its orbit is not constant. Second, time is measured along the celestial equator and the path of the real Sun is not along the celestial equator. Rather, its path is along the ecliptic, which is tilted at an angle of 23° 27' with respect to the celestial equator. Third, rotation of the Earth on its axis is not constant.

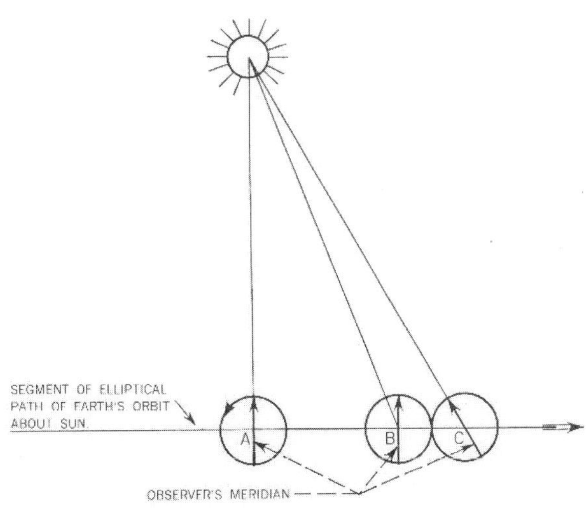

Figure 1800. Apparent eastward movement of the Sun with respect to the stars.

To obtain a constant rate of time, we replace the apparent Sun with a fictitious **mean Sun**. This mean Sun moves eastward along the celestial equator at a uniform speed equal to the average speed of the apparent Sun along the ecliptic. This mean Sun, therefore, provides a uniform measure of time which approximates the average apparent time. The speed of the mean Sun along the celestial equator is 15° per hour of mean solar time.

1801. Equation of Time

Mean solar time, or **mean time** as it is commonly called, is sometimes ahead of and sometimes behind apparent solar time. This difference, which never exceeds about 16.4 minutes, is called the **equation of time**.

The navigator most often deals with the equation of time when determining the time of **upper meridian passage** of the Sun. The Sun transits the observer's upper meridian at **local apparent noon**. Were it not for the difference in rate between the mean and apparent Sun, the Sun would be on the observer's meridian when the mean Sun indicated 1200 local time. The apparent solar time of upper meridian passage, however, is offset from exactly 1200 mean solar time. This time difference, the equation of time at meridian transit, is listed on the right hand daily pages of the *Nautical Almanac*.

The sign of the equation of time is negative if the time of Sun's meridian passage is earlier than 1200 and positive if later than 1200. Therefore: Apparent Time = Mean Time + (equation of time).

Example 1: *Determine the time of the Sun's meridian passage (Local Apparent Noon) on June 16, 1994.*

Solution: *See Figure 2008 in Chapter 20, the Nautical Almanac's right hand daily page for June 16, 1994. The equation of time is listed in the bottom right hand corner of the page. There are two ways to solve the problem, depending on the accuracy required for the value of meridian passage. The time of the Sun at meridian passage is given to the nearest minute in the "Mer. Pass."column. For June 16, 1994, this value is 1201.*

To determine the exact time of meridian passage, use the value given for the equation of time. This value is listed immediately to the left of the "Mer. Pass." column on the daily pages. For June 16, 1994, the value is given as 00^m37^s. *Use the "*12^h*" column because the problem asked for meridian passage at LAN. The value of meridian passage from the "Mer. Pass." column indicates that meridian passage occurs* <u>*after*</u> *1200; therefore, add the 37 second correction to 1200 to obtain the exact time of meridian passage. The exact time of meridian passage for June 16, 1994, is* $12^h00^m37^s$.

The equation of time's maximum value approaches 16^m22^s in November.

If the *Almanac* lists the time of meridian passage as 1200, proceed as follows. Examine the equations of time listed in the *Almanac* to find the dividing line marking where the equation of time changes between positive and negative values. Examine the trend of the values near this dividing line to determine the correct sign for the equation of time.

Example 2: *See Figure 1801. Determine the time of the upper meridian passage of the Sun on April 16, 1995.*

Solution: *From Figure 1801, upper meridian passage of the Sun on April 16, 1995, is given as 1200. The dividing line between the values for upper and lower meridian passage on April 16th indicates that the sign of the equation of time changes between lower meridian passage and upper meridian passage on this date; the question, therefore, becomes: does it become positive or negative? Note that on April 18, 1995, upper meridian passage is given as 1159, indicating that on April 18, 1995, the equation of time is positive. All values for the equation of time on the same side of the dividing line as April 18th are positive. Therefore, the equation of time for upper meridian passage of the Sun on April 16, 1995 is (+)* 00^m05^s. *Upper meridian passage, therefore, takes place at* $11^h59^m55^s$.

Day	SUN Eqn. of Time 00^h	SUN Eqn. of Time 12^h	SUN Mer. Pass.	MOON Mer. Pass. Upper	MOON Mer. Pass. Lower	MOON Age	MOON Phase
	m s	m s	h m	h m	h m	d	
16	00 02	00 05	12 00	00 26	12 55	16	○
17	00 13	00 20	12 00	01 25	13 54	17	
18	00 27	00 33	11 59	02 25	14 55	18	

Figure 1801. The equation of time for April 16, 17, 18, 1995.

To calculate latitude and longitude at LAN, the navigator seldom requires the time of meridian passage to accuracies greater than one minute. Therefore, use the time listed under the "Mer. Pass." column to estimate LAN unless extraordinary accuracy is required.

1802. Fundamental Systems of Time

Atomic time is defined by the Systeme International (SI) second, with a duration of 9,192,631,770 cycles of radiation corresponding to the transition between two hyperfine levels of the ground state of cesium 133. **International Atomic Time** (**TAI**) is an international time scale based on the average of a number of atomic clocks.

Universal time (**UT**) is counted from 0 hours at midnight, with a duration of one mean solar day, averaging out minor variations in the rotation of the Earth.

UT0 is the rotational time of a particular place of observation, observed as the diurnal motion of stars or extraterrestrial radio sources.

UT1 is computed by correcting UT0 for the effect of polar motion on the longitude of the observer, and varies because of irregularities in the Earth's rotation.

Coordinated Universal Time, or **UTC**, used as a standard reference worldwide for certain purposes, is kept within one second of TAI by the introduction of leap seconds. It differs from TAI by an integral number of seconds, but is always kept within 0.9 seconds of TAI.

Dynamical time has replaced ephemeris time in theoretical usage, and is based on the orbital motions of the Earth, Moon, and planets.

Terrestrial time (**TT**), also known as **Terrestrial Dynamical Time** (**TDT**), is defined as 86,400 seconds on the geoid.

Sidereal time is the hour angle of the vernal equinox, and has a unit of duration related to the period of the Earth's rotation with respect to the stars.

Delta T is the difference between UT1 and TDT.

Dissemination of time is an inherent part of various electronic navigation systems. The U.S. Naval Observatory Master Clock is used to coordinate Loran signals, and GPS signals have a time reference encoded in the data message. GPS time is normally within 15 nanoseconds with SA off, about 70 nanoseconds with SA on. One nanosecond (one one-billionth of a second) of time is roughly equivalent to one foot on the Earth for the GPS system.

1803. Time and Arc

One day represents one complete rotation of the Earth. Each day is divided into 24 hours of 60 minutes; each minute has 60 seconds.

Time of day is an indication of the phase of rotation of the Earth. That is, it indicates how much of a day has elapsed, or what part of a rotation has been completed. Thus, at zero hours the day begins. One hour later, the Earth has turned through 1/24 of a day, or 1/24 of 360°, or 360° ÷ 24 = 15°

Smaller intervals can also be stated in angular units; since 1 hour or 60 minutes is equivalent to 15° of arc, 1 minute of time is equivalent to 15° ÷ 60 = 0.25° = 15' of arc, and 1 second of time is equivalent to 15' ÷ 60 = 0.25' = 15" of arc.

Summarizing in table form:

1^d	$=24^h$	=360°
60^m	$=1^h$	=15°
4^m	= 1°	=60'
60^s	$= 1^m$	= 15'
4^s	= 1'	= 60"
1^s	= 15"	= 0.25'

Therefore any time interval can be expressed as an equivalent amount of rotation, and vice versa. Interconversion of these units can be made by the relationships indicated above.

To convert time to arc:

1. Multiply the hours by 15 to obtain degrees of arc.
2. Divide the minutes of time by four to obtain degrees.
3. Multiply the remainder of step 2 by 15 to obtain minutes of arc.
4. Divide the seconds of time by four to obtain minutes of arc
5. Multiply the remainder by 15 to obtain seconds of arc.
6. Add the resulting degrees, minutes, and seconds.

Example 1: *Convert $14^h21^m39^s$ to arc.*

Solution:

(1)	$14^h \times 15$	= 210° 00' 00"
(2)	$21^m \div 4$	= 005° 00' 00" (remainder 1)
(3)	1×15	= 000° 15' 00"
(4)	$39^s \div 4$	= 000° 09' 00" (remainder 3)
(5)	3×15	= 000° 00' 45"
(6)	$14^h21^m39^s$	= 215° 24' 45"

To convert arc to time:

1. Divide the degrees by 15 to obtain hours.
2. Multiply the remainder from step 1 by four to obtain minutes of time.
3. Divide the minutes of arc by 15 to obtain minutes of time.
4. Multiply the remainder from step 3 by four to obtain seconds of time.
5. Divide the seconds of arc by 15 to obtain seconds of time.
6. Add the resulting hours, minutes, and seconds.

Example 2: *Convert 215° 24' 45" to time units.*

Solution:

(1)	215° ÷ 15	$= 14^h00^m00^s$	remainder 5
(2)	5×4	$= 00^h20^m00^s$	
(3)	24' ÷ 15	$= 00^h01^m00^s$	remainder 9
(4)	9×4	$= 00^h00^m36^s$	
(5)	45" ÷ 15	$= 00^h00^m03^s$	
(6)	215° 24' 45"	$= 14^h21^m39^s$	

Solutions can also be made using arc to time conversion tables in the almanacs. In the *Nautical Almanac*, the table given near the back of the volume is in two parts, permitting separate entries with degrees, minutes, and quarter minutes of arc. This table is arranged in this manner because the navigator converts arc to time more often than the reverse.

Example 3: *Convert 334°18'22" to time units, using the Nautical Almanac arc to time conversion table.*

Solution:

Convert the 22" to the nearest quarter minute of arc for solution to the nearest second of time. Interpolate if more precise results are required.

334° 00.00ᵐ	=	$22^h16^m00^s$
000° 18.25ᵐ	=	$00^h01^m13^s$
334° 18' 22"	=	$22^h17^m13^s$

1804. Time and Longitude

Suppose the Sun were directly over a certain point on the Earth at a given time. An hour later the Earth would have turned through 15°, and the Sun would then be directly over a meridian 15° farther west. Thus, any difference of longitude between two points is a measure of the angle through which the Earth must rotate to separate them. Therefore, places east of an observer have later time, and those west have earlier time, and the difference is exactly equal to the difference in longitude, expressed in *time* units. The difference in time between two places is equal to the difference of longitude between their meridians, expressed in units of time instead of arc.

1805. The Date Line

Since time grows later toward the east and earlier toward the west of an observer, time at the lower branch of one's meridian is 12 hours earlier or later, depending upon the direction of reckoning. A traveler circling the Earth gains or loses an entire day depending on the direction of travel, and only for a single instant of time, at precisely Greenwich noon, is it the same date around the earth. To prevent the date from being in error and to provide a starting place for each new day, a **date line** is fixed by informal agreement. This line coincides with the 180th meridian over most of its length. In crossing this line, the date is altered by one day. If a person is traveling eastward from east longitude to west longitude, time is becoming later, and when the date line is crossed the date becomes 1 day earlier. At any instant the date immediately to the west of the date line (east longitude) is 1 day later than the date immediately to the east of the line. When solving celestial problems, we convert local time to Greenwich time and then convert this to local time on the opposite side of the date line.

1806. Zone Time

At sea, as well as ashore, watches and clocks are normally set to some form of **zone time (ZT)**. At sea the nearest meridian exactly divisible by 15° is usually used as the **time meridian** or **zone meridian**. Thus, within a time zone extending 7.5° on each side of the time meridian the time is the same, and time in consecutive zones differs by exactly one hour. The time is changed as convenient, usually at a whole hour, when crossing the boundary between zones. Each time zone is identified by the number of times the longitude of its zone meridian is divisible by 15°, *positive in west longitude* and *negative in east longitude*. This number and its sign, called the **zone description (ZD)**, is the number of whole hours that are added to or subtracted from the zone time to obtain Greenwich Mean Time (GMT). The mean Sun is the celestial reference point for zone time. See Figure 1806.

Converting ZT to GMT, a positive ZT is added and a negative one subtracted; converting GMT to ZT, a positive ZD is subtracted, and a negative one added.

Example: *The GMT is* $15^h27^m09^s$.

Required: *(1) ZT at long. 156°24.4' W.*
(2) ZT at long. 039°04.8' E.

Solutions:

(1)	*GMT*	$15^h27^m09^s$
	ZD	$+10^h$ *(rev.)*
	ZT	$05^h27^m09^s$
(2)	*GMT*	$15^h27^m09^s$
	ZD	-03^h *(rev.)*
	ZT	$18^h27^m09^s$

1807. Chronometer Time

Chronometer time (C) is time indicated by a chronometer. Since a chronometer is set approximately to GMT and not reset until it is overhauled and cleaned about every 3 years, there is nearly always a **chronometer error (CE)**, either fast (F) or slow (S). The change in chronometer error in 24 hours is called **chronometer rate**, or **daily rate**, and designated gaining or losing. With a consistent rate of 1^s per day for three years, the chronometer error would total approximately 18^m. Since chronometer error is subject to change, it should be determined from time to time, preferably daily at sea. Chronometer error is found by radio time signal, by comparison with another timepiece of known error, or by applying chronometer rate to previous readings of the same instrument. It is recorded to the nearest whole or half second. Chronometer rate is recorded to the nearest 0.1 second.

Example: *At GMT 1200 on May 12 the chronometer reads* $12^h04^m21^s$. *At GMT 1600 on May 18 it reads* $4^h04^m25^s$.

Required: . *1. Chronometer error at 1200 GMT May 12.*
2. Chronometer error at 1600 GMT May 18.
3. Chronometer rate.
4. Chronometer error at GMT 0530, May 27.

TIME ZONE CHART

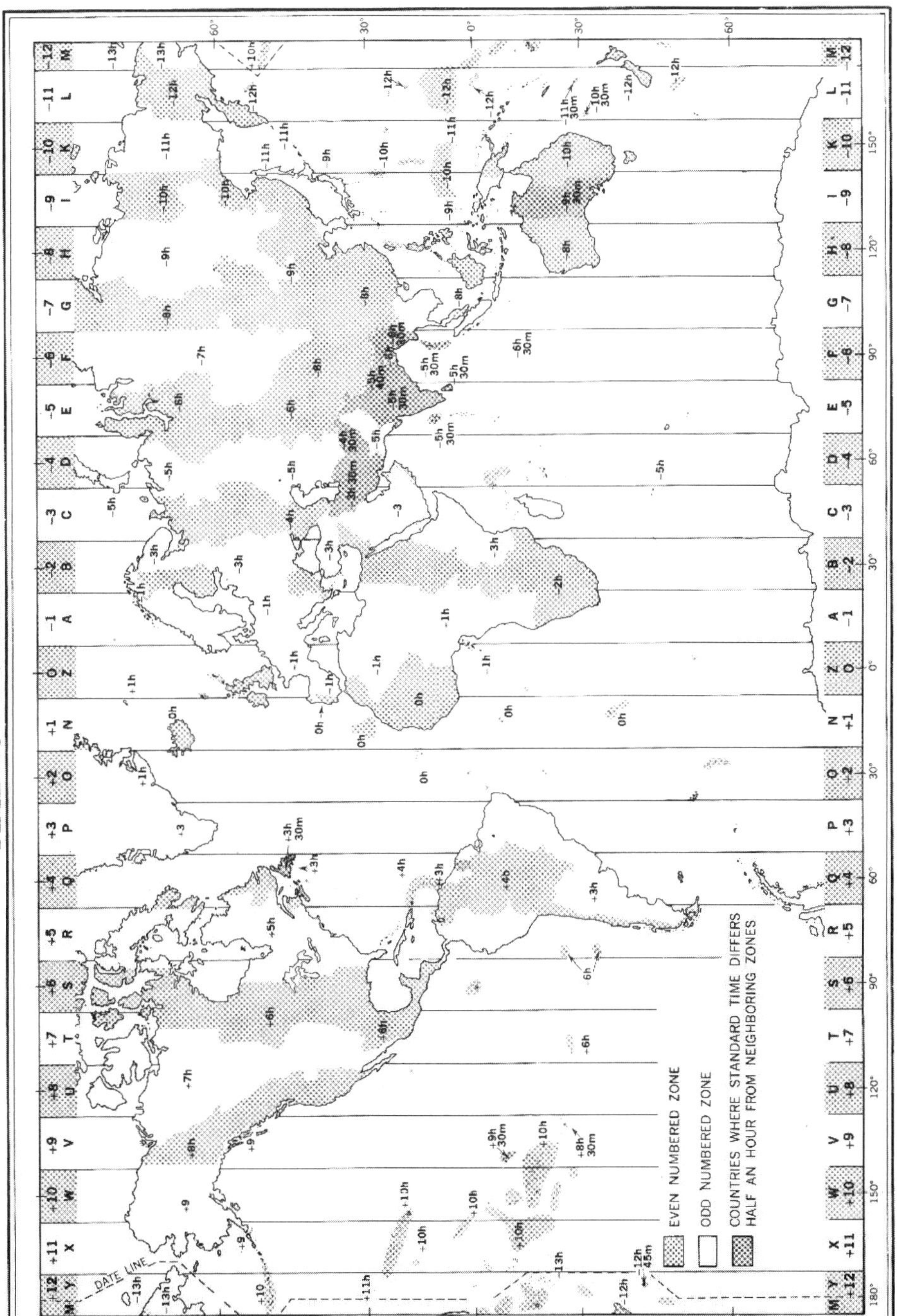

Figure 1806. Time Zone Chart.

Solutions:

1.	*GMT*	*$12^h00^m00^s$*	*May 12*
	C	*$12^h04^m21^s$*	
	CE	*(F)4^m21^s*	
2.	*GMT*	*$16^h00^m00^s$*	*May 18*
	C	*04 04 25*	
	CE	*(F)4^m25^s*	
3.	*GMT*	*18^d16^h*	
	GMT	*12^d12h*	
	diff.	*$06^d04^h = 6.2^d$*	
	CE	*(F)4^m21^s*	*1200 May 12*
	CE	*(F)4^m25^s*	*1600 May 18*
	diff.	*4^s (gained)*	
	daily rate	*0.6^s (gain)*	
4.	*GMT*	*$27^d05^h30^m$*	
	GMT	*$18^d16^h00^m$*	
	diff.	*$08^d13^h30^m$ (8.5^d)*	
	CE	*(F)4^m25^s*	*1600 May 18*
	corr.	*(+)0^m05^s*	*diff. × rate*
	CE	*(F)4^m30^s*	*0530 May 27*

Because GMT is on a 24-hour basis and chronometer time on a 12-hour basis, a 12-hour ambiguity exists. This is ignored in finding chronometer error. However, if chronometer error is applied to chronometer time to find GMT, a 12-hour error can result. This can be resolved by mentally applying the zone description to local time to obtain approximate GMT. A time diagram can be used for resolving doubt as to approximate GMT and Greenwich date. If the Sun for the kind of time used (mean or apparent) is between the lower branches of two time meridians (as the standard meridian for local time, and the Greenwich meridian for GMT), the date at the place farther east is one day later than at the place farther west.

1808. Watch Time

Watch time (WT) is usually an approximation of zone time, except that for timing celestial observations it is easiest to set a comparing watch to GMT. If the watch has a second-setting hand, the watch can be set exactly to ZT or GMT, and the time is so designated. If the watch is not set exactly to one of these times, the difference is known as **watch error (WE)**, labeled fast (F) or slow (S) to indicate whether the watch is ahead of or behind the correct time.

If a watch is to be set exactly to ZT or GMT, set it to some whole minute slightly ahead of the correct time and stopped. When the set time arrives, start the watch and check it for accuracy.

The GMT may be in error by 12^h, but if the watch is graduated to 12 hours, this will not be reflected. If a watch with a 24-hour dial is used, the actual GMT should be determined.

To determine watch error compare the reading of the watch with that of the chronometer at a selected moment. This may also be at some selected GMT. Unless a watch is graduated to 24 hours, its time is designated am before noon and pm after noon.

Even though a watch is set to zone time approximately, its error on GMT can be determined and used for timing observations. In this case the 12-hour ambiguity in GMT should be resolved, and a time diagram used to avoid error. This method requires additional work, and presents a greater probability of error, without compensating advantages.

If a stopwatch is used for timing observations, it should be started at some convenient GMT, such as a whole 5^m or 10^m. The time of each observation is then the GMT plus the watch time. Digital stopwatches and wristwatches are ideal for this purpose, as they can be set from a convenient GMT and read immediately after the altitude is taken.

1809. Local Mean Time

Local mean time (LMT), like zone time, uses the mean Sun as the celestial reference point. It differs from zone time in that the local meridian is used as the terrestrial reference, rather than a zone meridian. Thus, the local mean time at each meridian differs from every other meridian, the difference being equal to the difference of longitude expressed in time units. At each zone meridian, including 0°, LMT and ZT are identical.

In navigation the principal use of LMT is in rising, setting, and twilight tables. The problem is usually one of converting the LMT taken from the table to ZT. At sea, the difference between the times is normally not more than 30^m, and the conversion is made directly, without finding GMT as an intermediate step. This is done by applying a correction equal to the difference of longitude. If the observer is west of the time meridian, the correction is added, and if east of it, the correction is subtracted. If Greenwich time is desired, it is found from ZT.

Where there is an irregular zone boundary, the longitude may differ by more than 7.5° (30^m) from the time meridian.

If LMT is to be corrected to daylight saving time, the difference in longitude between the local and time meridian can be used, or the ZT can first be found and then increased by one hour.

Conversion of ZT (including GMT) to LMT is the same as conversion in the opposite direction, except that the sign of difference of longitude is reversed. This problem is not normally encountered in navigation.

1810. Sidereal Time

Sidereal time uses the first point of Aries (vernal equinox) as the celestial reference point. Since the Earth revolves around the Sun, and since the direction of the Earth's rotation and revolution are the same, it completes a rotation with respect to the stars in less time (about $3^m56.6^s$ of mean solar units) than with respect to the Sun, and during one revolution about the Sun (1 year) it makes one complete rotation more with respect to the stars than with the Sun. This accounts for the daily shift of the stars nearly 1° westward each night. Hence, sidereal days are shorter than solar days, and its hours, minutes, and seconds are correspondingly shorter. Because of nutation, sidereal time is not quite constant in rate. Time based upon the average rate is called **mean sidereal time**, when it is to be distinguished from the slightly irregular sidereal time. The ratio of mean solar time units to mean sidereal time units is 1:1.00273791.

A navigator very seldom uses sidereal time. Astronomers use it to regulate mean time because its celestial reference point remains almost fixed in relation to the stars.

1811. Time And Hour Angle

Both time and hour angle are a measure of the phase of rotation of the Earth, since both indicate the angular distance of a celestial reference point west of a terrestrial reference meridian. Hour angle, however, applies to any point on the celestial sphere. Time might be used in this respect, but only the apparent Sun, mean Sun, the first point of Aries, and occasionally the Moon, are commonly used.

Hour angles are usually expressed in arc units, and are measured from the upper branch of the celestial meridian. Time is customarily expressed in time units. Sidereal time is measured from the upper branch of the celestial meridian, like hour angle, but solar time is measured from the lower branch. Thus, LMT = LHA mean Sun plus or minus 180°, LAT = LHA apparent Sun plus or minus 180°, and LST = LHA Aries.

As with time, local hour angle (LHA) at two places differs by their difference in longitude, and LHA at longitude 0° is called Greenwich hour angle (GHA). In addition, it is often convenient to express hour angle in terms of the shorter arc between the local meridian and the body. This is similar to measurement of longitude from the Greenwich meridian. Local hour angle measured in this way is called meridian angle (t), which is labeled east or west, like longitude, to indicate the direction of measurement. A westerly meridian angle is numerically equal to LHA, while an easterly meridian angle is equal to 360° – LHA. LHA = t (W), and LHA = 360° – t (E). Meridian angle is used in the solution of the navigational triangle.

Example: *Find LHA and t of the Sun at GMT $3^h24^m16^s$ on June 1, 1975, for long. 118°48.2' W.*

Solution:

GMT	*$3^h24^m16^s$*	*June 1*
3^h	*225°35.7'*	
24^m16^s	*6°04.0'*	
GHA	*231°39.7'*	
λ	*118°48.2'* W	
LHA	*112°51.5'*	
t	*112°51.5' W*	

RADIO DISSEMINATION OF TIME SIGNALS

1812. Dissemination Systems

Of the many systems for time and frequency dissemination, the majority employ some type of radio transmission, either in dedicated time and frequency emissions or established systems such as radionavigation systems. The most accurate means of time and frequency dissemination today is by the mutual exchange of time signals through communication (commonly called Two-Way) and by the mutual observation of navigation satellites (commonly called Common View).

Radio time signals can be used either to perform a clock's function or to set clocks. When using a radio wave instead of a clock, however, new considerations evolve. One is the delay time of approximately 3 microseconds per kilometer it takes the radio wave to propagate and arrive at the reception point. Thus, a user 1,000 kilometers from a transmitter receives the time signal about 3 milliseconds later than the on-time transmitter signal. If time is needed to better than 3 milliseconds, a correction must be made for the time it takes the signal to pass through the receiver.

In most cases standard time and frequency emissions as received are more than adequate for ordinary needs. However, many systems exist for the more exacting scientific requirements.

1813. Characteristic Elements of Dissemination Systems

A number of common elements characterize most time and frequency dissemination systems. Among the more important elements are accuracy, ambiguity, repeatability, coverage, availability of time signal, reliability, ease of use, cost to the user, and the number of users served. No single system incorporates all desired characteristics. The relative importance of these characteristics will vary from one user to the next, and the solution for one user may not be satisfactory to another. These common elements are discussed in the following examination of a hypothetical radio signal.

Consider a very simple system consisting of an unmodulated 10-kHz signal as shown in Figure 1813. This signal, leaving the transmitter at 0000 UTC, will reach the receiver at a later time equivalent to the propagation delay. The user must know this delay because the accuracy of his knowledge of time can be no better than the degree to which the delay is known. Since all cycles of the signal are identical, the signal is ambiguous and the user must somehow decide which cycle is the "on time" cycle. This means, in the case of the hypothetical 10-kHz signal, that the user must know the time to ± 50 microseconds (half the period of the signal). Further, the user may desire to use this system, say once a day, for an extended period of time to check his clock or frequency standard. However, if the delay varies from one day to the next without the user knowing, accuracy will be limited by the lack of repeatability.

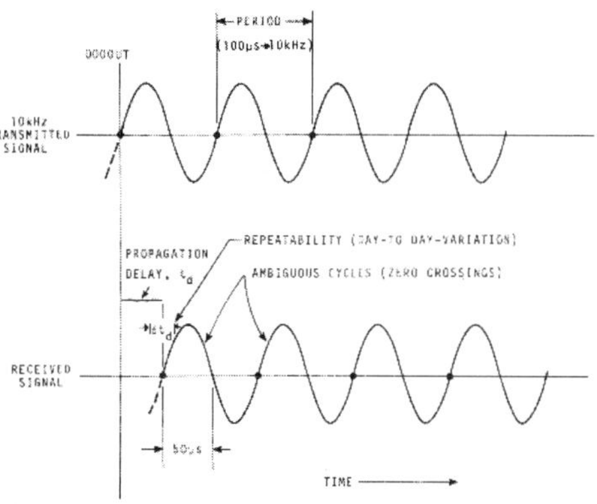

Figure 1813. Single tone time dissemination.

Many users are interested in making time coordinated measurements over large geographic areas. They would like all measurements to be referenced to one time system to eliminate corrections for different time systems used at scattered or remote locations. This is a very important practical consideration when measurements are undertaken in the field. In addition, a one-reference system, such as a single time broadcast, increases confidence that all measurements can be related to each other in some known way. Thus, the coverage of a system is an important concept. Another important characteristic of a timing system is the percent of time available. The man on the street who has to keep an appointment needs to know the time perhaps to a minute or so. Although requiring only coarse time information, he wants it on demand, so he carries a wristwatch that gives the time 24 hours a day. On the other hand, a user who needs time to a few microseconds employs a very good clock which only needs an occasional update, perhaps only once or twice a day. An additional characteristic of time and frequency dissemination is reliability, i.e., the likelihood that a time signal will be available when scheduled. Propagation fade-out can sometimes prevent reception of HF signals.

1814. Radio Wave Propagation Factors

Radio has been used to transmit standard time and frequency signals since the early 1900's. As opposed to the physical transfer of time via portable clocks, the transfer of information by radio entails propagation of electromagnetic energy from a transmitter to a distant receiver.

In a typical standard frequency and time broadcast, the signals are directly related to some master clock and are transmitted with little or no degradation in accuracy. In a vacuum and with a noise-free background, the signals should be received at a distant point essentially as transmitted, except for a constant path delay with the radio wave propagating near the speed of light (299,773 kilometers per second). The propagation media, including the Earth, atmosphere, and ionosphere, as well as physical and electrical characteristics of transmitters and receivers, influence the stability and accuracy of received radio signals, dependent upon the frequency of the transmission and length of signal path. Propagation delays are affected in varying degrees by extraneous radiations in the propagation media, solar disturbances, diurnal effects, and weather conditions, among others.

Radio dissemination systems can be classified in a number of different ways. One way is to divide those carrier frequencies low enough to be reflected by the ionosphere (below 30 MHz) from those sufficiently high to penetrate the ionosphere (above 30 MHz). The former can be observed at great distances from the transmitter but suffer from ionospheric propagation anomalies that limit accuracy; the latter are restricted to line-of-sight applications but show little or no signal deterioration caused by propagation anomalies. The most accurate systems tend to be those which use the higher, line-of-sight frequencies, while broadcasts of the lower carrier frequencies show the greatest number of users.

1815. Standard Time Broadcasts

The World Administrative Radio Council (WARC) has allocated certain frequencies in five bands for standard frequency and time signal emission. For such dedicated standard frequency transmissions, the International Radio Consultative Committee (CCIR) recommends that carrier frequencies be maintained so that the average daily fractional frequency deviations from the internationally designated standard for measurement of time interval should not exceed 1 X 10^{-10}. The U.S. Naval Observatory Time Service Announcement Series 1, No. 2, gives characteristics of standard time signals assigned to allocated bands, as reported by the CCIR.

1816. Time Signals

The usual method of determining chronometer error and daily rate is by radio time signals, popularly called **time ticks**. Most maritime nations broadcast time signals several times daily from one or more stations, and a vessel equipped with radio receiving equipment normally has no difficulty in obtaining a time tick anywhere in the world. Normally, the time transmitted is maintained virtually uniform with respect to atomic clocks. The Coordinated Universal Time (UTC) as received by a vessel may differ from (GMT) by as much as 0.9 second.

The majority of radio time signals are transmitted automatically, being controlled by the standard clock of an astronomical observatory or a national measurement standards laboratory. Absolute reliance may be placed on these signals because they are required to be accurate to at least 0.001^s as transmitted.

Other radio stations, however, have no automatic transmission system installed, and the signals are given by hand. In this instance the operator is guided by the standard clock at the station. The clock is checked by astronomical observations or radio time signals and is normally correct to 0.25 second.

At sea, a spring-driven chronometer should be checked daily by radio time signal, and in port daily checks should be maintained, or begun at least three days prior to departure, if conditions permit. Error and rate are entered in the chronometer record book (or record sheet) each time they are determined.

The various time signal systems used throughout the world are discussed in NIMA *Pub. 117, Radio Navigational Aids*, and volume 5 of *Admiralty List of Radio Signals*. Only the United States signals are discussed here.

The National Institute of Standards and Technology (NIST) broadcasts continuous time and frequency reference signals from WWV, WWVH, WWVB, and the GOES satellite system. Because of their wide coverage and relative simplicity, the HF services from WWV and WWVH are used extensively for navigation.

Station WWV broadcasts from Fort Collins, Colorado at the internationally allocated frequencies of 2.5, 5.0, 10.0, 15.0, and 20.0 MHz; station WWVH transmits from Kauai, Hawaii on the same frequencies with the exception of 20.0 MHz. The broadcast signals include standard time and frequencies, and various voice announcements. Details of these broadcasts are given in NIST *Special Publication 432, NIST Frequency and Time Dissemination Services*. Both HF emissions are directly controlled by cesium beam frequency standards with periodic reference to the NIST atomic frequency and time standards.

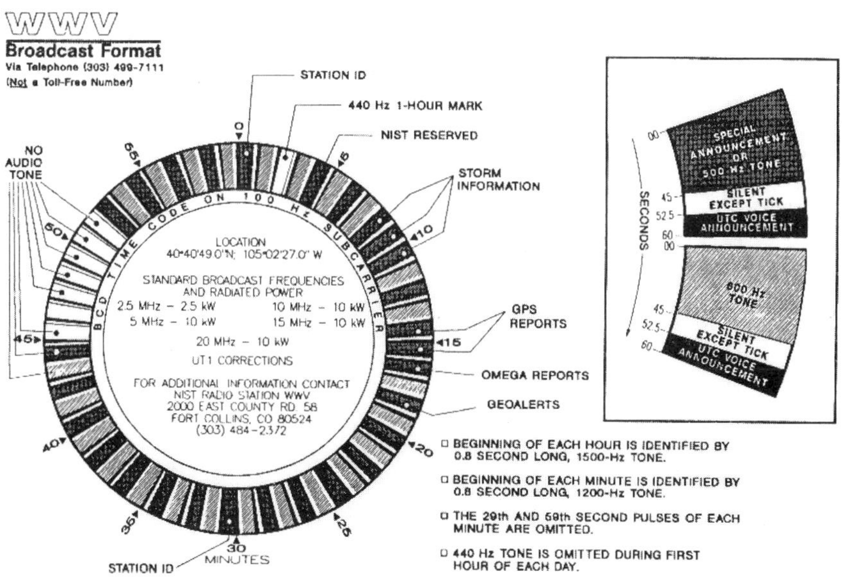

Figure 1816a. Broadcast format of station WWV.

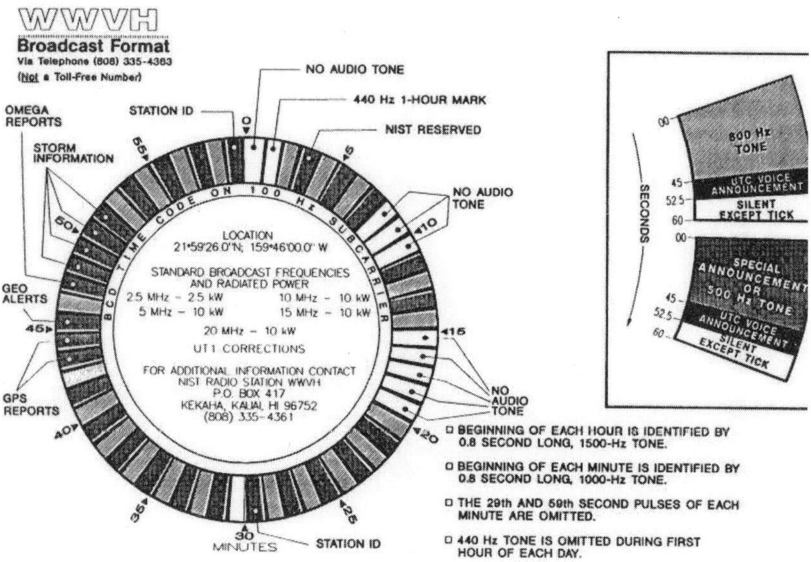

Figure 1816b. Broadcast format of station WWVH.

The time ticks in the WWV and WWVH emissions are shown in Figure 1816a and Figure 1816b. The 1-second UTC markers are transmitted continuously by WWV and WWVH, except for omission of the 29th and 59th marker each minute. With the exception of the beginning tone at each minute (800 milliseconds) all 1-second markers are of 5 milliseconds duration. Each pulse is preceded by 10 milliseconds of silence and followed by 25 milliseconds of silence. Time voice announcements are given also at 1-minute intervals. All time announcements are UTC.

Pub. No. 117, Radio Navigational Aids, should be referred to for further information on time signals.

1817. Leap-Second Adjustments

By international agreement, UTC is maintained within about 0.9 seconds of the celestial navigator's time scale, UT1. The introduction of **leap seconds** allows a clock to keep approximately in step with the Sun. Because of the variations in the rate of rotation of the Earth, however, the occurrences of the leap seconds are not predictable in detail.

The Central Bureau of the International Earth Rotation Service (IERS) decides upon and announces the introduction of a leap second. The IERS announces the new leap second at least several weeks in advance. A positive or negative leap

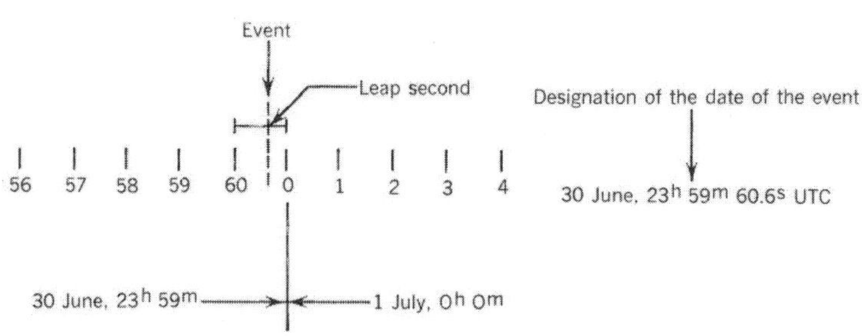

Figure 1817a. Dating of event in the vicinity of a positive leap second.

second is introduced the last second of a UTC month, but first preference is given to the end of December and June, and second preference is given to the end of March and September. A positive leap second begins at $23^h59^m60^s$ and ends at $00^h00^m00^s$ of the first day of the following month. In the case of a negative leap second, $23^h59^m58^s$ is followed one second later by $00^h00^m00^s$ of the first day of the following month.

The dating of events in the vicinity of a leap second is effected in the manner indicated in Figure 1817a and Figure 1817b.

Whenever leap second adjustments are to be made to UTC, mariners are advised by messages from NIMA.

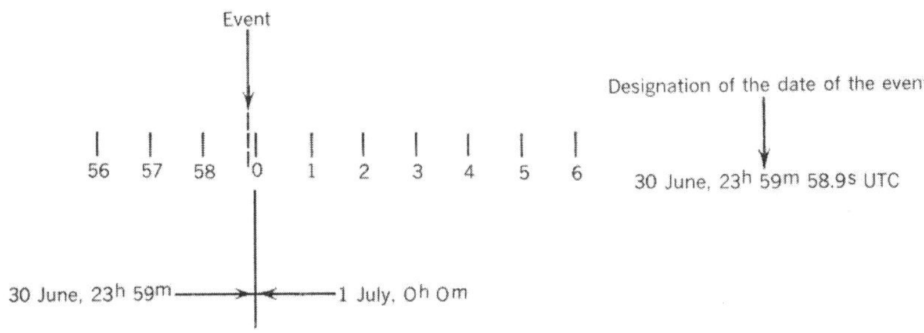

Figure 1817b. Dating of event in the vicinity of a negative leap second.

CHAPTER 19

THE ALMANACS

PURPOSE OF ALMANACS

1900. Introduction

Celestial navigation requires accurate predictions of the geographic positions of the celestial bodies observed. These predictions are available from three almanacs published annually by the United States Naval Observatory and H. M. Nautical Almanac Office, Royal Greenwich Observatory.

The ***Astronomical Almanac*** precisely tabulates celestial data for the exacting requirements found in several scientific fields. Its precision is far greater than that required by celestial navigation. Even if the *Astronomical Almanac* is used for celestial navigation, it will not necessarily result in more accurate fixes due to the limitations of other aspects of the celestial navigation process.

The ***Nautical Almanac*** contains the astronomical information specifically needed by marine navigators. Information is tabulated to the nearest 0.1' of arc and 1 second of time. GHA and declination are available for the Sun, Moon, planets, and 173 stars, as well as corrections necessary to reduce the observed values to true.

The ***Air Almanac*** was originally intended for air navigators, but is used today mostly by a segment of the maritime community. In general, the information is similar to the *Nautical Almanac*, but is given to a precision of 1' of arc and 1 second of time, at intervals of 10 minutes (values for the Sun and Aries are given to a precision of 0.1'). This publication is suitable for ordinary navigation at sea, but lacks the precision of the *Nautical Almanac*, and provides GHA and declination for only the 57 commonly used navigation stars.

The ***Multi-Year Interactive Computer Almanac*** (MICA) is a computerized almanac produced by the U.S. Naval Observatory. This and other web-based calculators are available from: http://aa.usno.navy.mil. The Navy's **STELLA** program, found aboard all seagoing naval vessels, contains an interactive almanac as well. A variety of privately produced electronic almanacs are available as computer programs or installed in pocket calculators. These invariably are associated with sight reduction software which replaces tabular and mathematical sight reduction methods.

FORMAT OF THE NAUTICAL AND AIR ALMANACS

1901. *Nautical Almanac*

The major portion of the *Nautical Almanac* is devoted to hourly tabulations of Greenwich Hour Angle (GHA) and declination, to the nearest 0.1' of arc. On each set of facing pages, information is listed for three consecutive days. On the left-hand page, successive columns list GHA of Aries (♈), and both GHA and declination of Venus, Mars, Jupiter, and Saturn, followed by the Sidereal Hour Angle (SHA) and declination of 57 stars. The GHA and declination of the Sun and Moon, and the horizontal parallax of the Moon, are listed on the right-hand page. Where applicable, the quantities v and d are given to assist in interpolation. The quantity v is the difference between the actual change of GHA in 1 hour and a constant value used in the interpolation tables, while d is the change in declination in 1 hour. Both v and d are listed to the nearest 0.1'.

To the right of the Moon data is listed the Local Mean Time (LMT) of sunrise, sunset, and beginning and ending of nautical and civil twilight for latitudes from 72°N to 60°S. The LMT of moonrise and moonset at the same latitudes is listed for each of the three days for which other information is given, and for the following day. Magnitude of each planet at UT 1200 of the middle day is listed at the top of the column. The UT of transit across the celestial meridian of Greenwich is listed as "Mer. Pass.". The value for the first point of Aries for the middle of the three days is listed to the nearest 0.1' at the bottom of the Aries column. The time of transit of the planets for the middle day is given to the nearest whole minute, with SHA (at UT 0000 of the middle day) to the nearest 0.1', below the list of stars. For the Sun and Moon, the time of transit to the nearest whole minute is given for each day. For the Moon, both upper and lower transits are given. This information is tabulated below the rising, setting, and twilight information. Also listed, are the equation of time for 0^h and 12^h, and the age and phase of the Moon. Equation of time is listed, without sign, to the nearest whole second. Age is given to the nearest whole day. Phase is given by symbol.

The main tabulation is preceded by a list of religious and civil holidays, phases of the Moon, a calendar, information on eclipses occurring during the year, and notes and a diagram giving information on the planets.

The main tabulation is followed by explanations and examples. Next are four pages of standard times (zone descriptions). Star charts are next, followed by a list of 173 stars in order of increasing SHA. This list includes the stars given on the daily pages. It gives the SHA and declination each month, and the magnitude. Stars are listed by Bayer's name and also by popular name where applicable. Following the star list are the Polaris tables. These tables give the azimuth and the corrections to be applied to the observed altitude to find the latitude.

Following the Polaris table is a section that gives formulas and examples for the entry of almanac data, the calculations that reduce a sight, and a method of solution for position, all for use with a calculator or microcomputer. This is followed by concise sight reduction tables, with instructions and examples, for use when a calculator or traditional sight reduction tables are not available. Tabular precision of the concise tables is one minute of arc.

Next is a table for converting arc to time units. This is followed by a 30-page table called "Increments and Corrections," used for interpolation of GHA and declination. This table is printed on tinted paper for quick location. Then come tables for interpolating for times of rise, set, and twilight; followed by two indices of the 57 stars listed on the daily pages, one index in alphabetical order, and the other in order of decreasing SHA.

Sextant altitude corrections are given at the front and back of the almanac. Tables for the Sun, stars, and planets, and a dip table, are given on the inside front cover and facing page, with an additional correction for nonstandard temperature and atmospheric pressure on the following page. Tables for the Moon, and an abbreviated dip table, are given on the inside back cover and facing page. Corrections for the Sun, stars, and planets for altitudes greater than 10°, and the dip table, are repeated on one side of a loose bookmark. The star indices are repeated on the other side.

1902. *Air Almanac*

As in the *Nautical Almanac*, the major portion of the *Air Almanac* is devoted to a tabulation of GHA and declination. However, in the *Air Almanac* values are listed at intervals of 10 minutes, to a precision of 0.1' for the Sun and Aries, and to a precision of 1' for the Moon and the planets. Values are given for the Sun, first point of Aries (GHA only), the three navigational planets most favorably located for observation, and the Moon. The magnitude of each planet listed is given at the top of its column, and the percentage of the Moon's disc illuminated, waxing (+) or waning(-), is given at the bottom of each page. Values for the first 12 hours of the day are given on the right-hand page, and those for the second half of the day on the back. In addition, each page has a table of the Moon's parallax in altitude, and below this the semidiameter of the Sun, and both the semidiameter and age of the Moon. Each daily page includes the LMT of moonrise and moonset; and a difference column to find the time of moonrise and moonset at any longitude.

Critical tables for interpolation for GHA are given on the inside front cover, which also has an alphabetical listing of the stars, with the number, magnitude, SHA, and declination of each. The same interpolation table and star list are printed on a flap which follows the daily pages. This flap also contains a star chart, a star index in order of decreasing SHA, and a table for interpolation of the LMT of moonrise and moonset for longitude.

Following the flap are instructions for the use of the almanac; a list of symbols and abbreviations in English, French, and Spanish; a list of time differences between Greenwich and other places; sky diagrams; a planet location diagram; star recognition diagrams for periscopic sextants; sunrise, sunset, and civil twilight tables; rising, setting, and depression graphs; semiduration graphs of Sunlight, twilight, and Moonlight in high latitudes; percentage of the Moon illuminated at 6 and 18 hours UT daily; a list of 173 stars by number and Bayer's name (also popular name where there is one), giving the SHA and declination each month (to a precision of 0.1'), and the magnitude; tables for interpolation of GHA Sun and GHA ♈ ; a table for converting arc to time; a single Polaris correction table; an aircraft standard dome refraction table; a refraction correction table; a Coriolis correction table; and on the inside back cover, a correction table for dip of the horizon.

USING THE ALMANACS

1903. Entering Arguments

The time used as an entering argument in the almanacs is 12^h + GHA of the mean Sun and is denoted by UT, formerly referred to as GMT and so referred to in this book to avoid confusion. This scale may differ from the broadcast time signals by an amount which, if ignored, will introduce an error of up to 0.2' in longitude determined from astronomical observations. The difference arises because the time argument depends on the variable rate of rotation of the Earth while the broadcast time signals are now based on atomic time. Step adjustments of exactly one second are made to the time signals as required (primarily at 24h on December 31 and June 30) so that the difference between the time signals and UT, as used in the almanacs, may not exceed 0.9^s. If observations to a precision of better than 1^s are required, corrections must be obtained from coding in the signal, or from other sources. The correction may be applied to each of the times of observation. Alternatively, the longitude, when determined from observations, may be corrected by the corresponding amount shown in Table 1903.

The main contents of the almanacs consist of data from

Correction to time signals	*Correction to longitude*
-0.7^s to -0.9^s	0.2' to east
-0.6^s to -0.3^s	0.1' to east
-0.2^s to $+0.2^s$	no correction
$+0.3^s$ to $+0.6^s$	0.1' to west
$+0.7^s$ to $+0.9^s$	0.2' to west

Table 1903. Corrections to time.

which the GHA and the declination of all the bodies used for navigation can be obtained for any instant of UT. The LHA can then be obtained with the formula:

LHA = GHA + east longitude.
LHA = GHA - west longitude.

For the Sun, Moon, and the four navigational planets, the GHA and declination are tabulated directly in the *Nautical Almanac* for each hour of GMT throughout the year; in the *Air Almanac*, the values are tabulated for each whole 10 m of GMT. For the stars, the SHA is given, and the GHA is obtained from:

GHA Star = GHA ♈ + SHA Star.

The SHA and declination of the stars change slowly and may be regarded as constant over periods of several days or even months if lesser accuracy is required. The SHA and declination of stars tabulated in the *Air Almanac* may be considered constant to a precision of 1.5' to 2' for the period covered by each of the volumes providing the data for a whole year, with most data being closer to the smaller value. GHA ♈ , or the GHA of the first point of Aries (the vernal equinox), is tabulated for each hour in the *Nautical Almanac* and for each whole 10^m in the *Air Almanac*. Permanent tables list the appropriate increments to the tabulated values of GHA and declination for the minutes and seconds of time.

In the *Nautical Almanac*, the permanent table for increments also includes corrections for *v*, the difference between the actual change of GHA in one hour and a constant value used in the interpolation tables; and *d*, the change in declination in one hour.

In the *Nautical Almanac*, *v* is always positive unless a negative sign (-) is shown. This occurs only in the case of Venus. For the Sun, the tabulated values of GHA have been adjusted to reduce to a minimum the error caused by treating *v* as negligible; there is no *v* tabulated for the Sun.

No sign is given for tabulated values of *d*, which is positive if declination is increasing, and negative if decreasing. The sign of a *v* or *d* value is also given to the related correction.

In the *Air Almanac*, the tabular values of the GHA of the Moon are adjusted so that use of an interpolation table based on a fixed rate of change gives rise to negligible error; no such adjustment is necessary for the Sun and planets. The tabulated declination values, except for the Sun, are those for the middle of the interval between the time indicated and the next following time for which a value is given, making interpolation unnecessary. Thus, it is always important to take out the GHA and declination for the time immediately *before* the time of observation.

In the *Air Almanac*, GHA ♈ and the GHA and declination of the Sun are tabulated to a precision of 0.1'. If these values are extracted with the tabular precision, the "Interpolation of GHA" table on the inside front cover (and flap) should not be used; use the "Interpolation of GHA Sun" and "Interpolation of GHA Aries' tables, as appropriate. These tables are found immediately preceding the Polaris Table.

1904. Finding GHA and Declination of the Sun

Nautical Almanac: Enter the daily page table with the whole hour before the given GMT, unless the exact time is a whole hour, and take out the tabulated GHA and declination. Also record the d value given at the bottom of the declination column. Next, enter the increments and corrections table for the number of minutes of GMT. If there are seconds, use the next earlier whole minute. On the line corresponding to the seconds of GMT, extract the value from the Sun-Planets column. Add this to the value of GHA from the daily page. This is GHA of the Sun. Next, enter the correction table for the same minute with the d value and take out the correction. Give this the sign of the d value and apply it to the declination from the daily page. This is the declination.

The correction table for GHA of the Sun is based upon a rate of change of 15° per hour, the average rate during a year. At most times the rate differs slightly. The slight error is minimized by adjustment of the tabular values. The d value is the amount that the declination changes between 1200 and 1300 on the middle day of the three shown.

Air Almanac: Enter the daily page with the whole 10^m preceding the given GMT, unless the time is itself a whole 10^m, and extract the GHA. The declination is extracted without interpolation from the same line as the tabulated GHA or, in the case of planets, the top line of the block of six. If the values extracted are rounded to the nearest minute, next enter the "Interpolation of GHA" table on the inside front cover (and flap), using the "Sun, etc." entry column, and take out the value for the remaining minutes and seconds of GMT. If the entry time is an exact tabulated value, use the correction listed half a line above the entry time. Add this correction to the GHA taken from the daily page. This is GHA. No adjustment of declination is needed. If the values are extracted with a precision of 0.1', the table for interpolating the GHA of the Sun to a precision of 0.1' must be used. Again no adjustment of declination is needed.

1905. Finding GHA and Declination of the Moon

Nautical Almanac: Enter the daily page table with the whole hour before the given GMT, unless this time is itself a whole hour, and extract the tabulated GHA and declination. Record the corresponding v and d values tabulated on the same line, and determine the sign of the d value. The v value of the Moon is always positive (+) and is not marked in the almanac. Next, enter the increments and corrections table for the minutes of GMT, and on the line for the seconds of GMT, take the GHA correction from the Moon column. Then, enter the correction table for the same minute with the v value, and extract the correction. Add both of these corrections to the GHA from the daily page. This is GHA of the Moon. Then, enter the same correction table with the d value and extract the correction. Give this correction the sign of the d value and apply it to the declination from the daily page. This is declination.

The correction table for GHA of the Moon is based upon the minimum rate at which the Moon's GHA increases, 14°19.0' per hour. The v correction adjusts for the actual rate. The v value is the difference between the minimum rate and the actual rate during the hour following the tabulated time. The d value is the amount that the declination changes during the hour following the tabulated time.

Air Almanac: Enter the daily page with the whole 10^m next preceding the given GMT, unless this time is a whole 10^m, and extract the tabulated GHA and the declination without interpolation. Next, enter the "Interpolation of GHA" table on the inside front cover, using the "Moon" entry column, and extract the value for the remaining minutes and seconds of GMT. If the entry time is an exact tabulated value, use the correction given half a line above the entry time. Add this correction to the GHA taken from the daily page to find the GHA at the given time. No adjustment of declination is needed.

The declination given in the table is correct for the time 5 minutes later than tabulated, so that it can be used for the 10-minute interval without interpolation, to an accuracy to meet most requirements. Declination changes much more slowly than GHA. If greater accuracy is needed, it can be obtained by interpolation, remembering to allow for the 5 minutes.

1906. Finding GHA and Declination of a Planet

Nautical Almanac: Enter the daily page table with the whole hour before the given GMT, unless the time is a whole hour, and extract the tabulated GHA and declination. Record the v value given at the bottom of each of these columns. Next, enter the increments and corrections table for the minutes of GMT, and on the line for the seconds of GMT, take the GHA correction from the Sun-planets column. Next, enter the correction table with the v value and extract the correction, giving it the sign of the v value. Add the first correction to the GHA from the daily page, and apply the second correction in accordance with its sign. This is GHA. Then enter the correction table for the same minute with the d value, and extract the correction. Give this correction the sign of the d value, and apply it to the declination from the daily page to find the declination at the given time.

The correction table for GHA of planets is based upon the mean rate of the Sun, 15° per hour. The v value is the difference between 15° and the change of GHA of the planet between 1200 and 1300 on the middle day of the three shown. The d value is the amount the declination changes between 1200 and 1300 on the middle day. Venus is the only body listed which ever has a negative v value.

Air Almanac: Enter the daily page with the whole 10^m before the given GMT, unless this time is a whole 10^m, and extract the tabulated GHA and declination, without interpolation. The tabulated declination is correct for the time 30^m later than tabulated, so interpolation during the hour following tabulation is not needed for most purposes. Next, enter the "Interpolation of GHA" table on the inside front cover, using the "Sun, etc." column, and take out the value for the remaining minutes and seconds of GMT. If the entry time is an exact tabulated value, use the correction half a line above the entry time. Add this correction to the GHA from the daily page to find the GHA at the given time. No adjustment of declination is needed.

1907. Finding GHA and Declination of a Star

If the GHA and declination of each navigational star were tabulated separately, the almanacs would be several times their present size. But since the sidereal hour angle and the declination are nearly constant over several days (to the nearest 0.1') or months (to the nearest 1'), separate tabulations are not needed. Instead, the GHA of the first point of Aries, from which SHA is measured, is tabulated on the daily pages, and a single listing of SHA and declination is given for each double page of the *Nautical Almanac*, and for an entire volume of the *Air Almanac*. Finding the GHA ♈ is similar to finding the GHA of the Sun, Moon, and planets.

Nautical Almanac: Enter the daily page table with the whole hour before the given GMT, unless this time is a whole hour, and extract the tabulated GHA of Aries. Also record the tabulated SHA and declination of the star from the listing on the left-hand daily page. Next, enter the increments and corrections table for the minutes of GMT, and, on the line for the seconds of GMT, extract the GHA correction from the Aries column. Add this correction and the SHA of the star to the GHA ♈ on the daily page to find the GHA of the star at the given time. No adjustment of declination is needed.

The SHA and declination of 173 stars, including Polaris and the 57 listed on the daily pages, are given for the middle of each month. For a star not listed on the daily pages, this is the only almanac source of this information. Interpolation in this table is not necessary for ordinary purposes of navigation, but is sometimes needed for precise results.

Air Almanac: Enter the daily page with the whole 10^m before the given GMT, unless this is a whole 10^m, and extract the tabulated GHA ♈. Next, enter the "Interpolation of GHA" table on the inside front cover, using the "Sun, etc." entry column, and extract the value for the remaining minutes and seconds of GMT. If the entry time is an exact tabulated value, use the correction given half a line above the entry time. From the tabulation at the left side of the same page, extract the SHA and declination of the star. Add the GHA from the daily page and the two values taken from the inside front cover to find the GHA at the given time. No adjustment of declination is needed.

RISING, SETTING, AND TWILIGHT

1908. Rising, Setting, and Twilight

In both *Air* and *Nautical Almanacs*, the times of sunrise, sunset, moonrise, moonset, and twilight information, at various latitudes between 72°N and 60°S, is listed to the nearest whole minute. By definition, rising or setting occurs when the upper limb of the body is on the visible horizon, assuming standard refraction for zero height of eye. Because of variations in refraction and height of eye, computation to a greater precision than 1 minute of time is not justified.

In high latitudes, some of the phenomena do not occur during certain periods. Symbols are used in the almanacs to indicate:

1. Sun or Moon does not set, but remains continuously above the horizon, indicated by an open rectangle.
2. Sun or Moon does not rise, but remains continuously below the horizon, indicated by a solid rectangle.
3. Twilight lasts all night, indicated by 4 slashes (////).

The *Nautical Almanac* makes no provision for finding the times of rising, setting, or twilight in polar regions. The *Air Almanac* has graphs for this purpose.

In the *Nautical Almanac*, sunrise, sunset, and twilight tables are given only once for the middle of the three days on each page opening. For navigational purposes this information can be used for all three days. Both almanacs have moonrise and moonset tables for each day.

The tabulations are in LMT. On the zone meridian, this is the zone time (ZT). For every 15' of longitude the observer's position differs from the zone meridian, the zone time of the phenomena differs by 1^m, being later if the observer is west of the zone meridian, and earlier if east of the zone meridian. The LMT of the phenomena varies with latitude of the observer, declination of the body, and hour angle of the body relative to the mean Sun.

The UT of the phenomenon is found from LMT by the formula:

UT = LMT + W Longitude
UT = LMT - E Longitude.

To use this formula, convert the longitude to time using the table on page i or by computation, and add or subtract as indicated. Apply the zone description (ZD) to find the zone time of the phenomena.

Sunrise and sunset are also tabulated in the tide tables (from 76°N to 60°S).

1909. Finding Times of Sunrise and Sunset

To find the time of sunrise or sunset in the *Nautical Almanac*, enter the table on the daily page, and extract the LMT for the latitude next smaller than your own (unless it is exactly the same). Apply a correction from Table I on almanac page xxxii to interpolate for altitude, determining the sign by inspection. Then convert LMT to ZT using the difference of longitude between the local and zone meridians.

For the *Air Almanac*, the procedure is the same as for the *Nautical Almanac*, except that the LMT is taken from the tables of sunrise and sunset instead of from the daily page, and the latitude correction is by linear interpolation.

The tabulated times are for the Greenwich meridian. Except in high latitudes near the time of the equinoxes, the time of sunrise and sunset varies so little from day to day that no interpolation is needed for longitude. In high latitudes interpolation is not always possible. Between two tabulated entries, the Sun may in fact cease to set. In this case, the time of rising and setting is greatly influenced by small variations in refraction and changes in height of eye.

1910. Twilight

Morning twilight ends at sunrise, and evening twilight begins at sunset. The time of the darker limit can be found from the almanacs. The time of the darker limits of both civil and nautical twilights (center of the Sun 6° and 12°, respectively, below the celestial horizon) is given in the *Nautical Almanac*. The *Air Almanac* provides tabulations of civil twilight from 60°S to 72°N. The brightness of the sky at any given depression of the Sun below the horizon may vary considerably from day to day, depending upon the amount of cloudiness, haze, and other atmospheric conditions. In general, the most effective period for observing stars and planets occurs when the center of the Sun is between about 3° and 9° below the celestial horizon. Hence, the darker limit of civil twilight occurs at about the mid-point of this period. At the darker limit of nautical twilight, the horizon is generally too dark for good observations.

At the darker limit of astronomical twilight (center of the Sun 18° below the celestial horizon), full night has set in. The time of this twilight is given in the *Astronomical Almanac*. Its approximate value can be determined by extrapolation in the *Nautical Almanac*, noting that the duration of the different kinds of twilight is proportional to the number of degrees of depression for the center of the Sun. More precise determination of the time at which the center of the Sun is any given number of degrees below the celestial horizon can be determined by a large-scale diagram on the plane of the celestial meridian, or by computation. Duration of twilight in latitudes higher than 65°N is given in a graph in the *Air Almanac*.

In both *Nautical* and *Air Almanacs*, the method of finding the darker limit of twilight is the same as that for sunrise and sunset.

Sometimes in high latitudes the Sun does not rise but twilight occurs. This is indicated in the *Air Almanac* by a solid black rectangle symbol in the sunrise and sunset column. To find the time of beginning of morning twilight, subtract half the duration of twilight as obtained from the duration of twilight graph from the time of meridian transit of the Sun; and for the time of ending of evening twilight, add it to the time of meridian transit. The LMT of meridian transit never differs by more than 16.4^m (approximately) from 1200. The actual time on any date can be determined from the almanac.

1911. Moonrise and Moonset

Finding the time of moonrise and moonset is similar to finding the time of sunrise and sunset, with one important difference. Because of the Moon's rapid change of declination, and its fast eastward motion relative to the Sun, the time of moonrise and moonset varies considerably from day to day. These changes of position on the celestial sphere are continuous, as moonrise and moonset occur successively at various longitudes around the Earth. Therefore, the change in time is distributed over all longitudes. For precise results, it would be necessary to compute the time of the phenomena at any given place by lengthy complex calculation. For ordinary purposes of navigation, however, it is sufficiently accurate to interpolate between consecutive moonrises or moonsets at the Greenwich meridian. Since apparent motion of the Moon is westward, relative to an observer on the Earth, interpolation in west longitude is between the phenomenon on the given date and the following one. In east longitude it is between the phenomenon on the given date and the preceding one.

To find the time of moonrise or moonset in the *Nautical Almanac*, enter the daily-page table with latitude, and extract the LMT for the tabulated latitude next smaller than the observer's latitude (unless this is an exact tabulated value). Apply a correction from table I of almanac page xxxii to interpolate for latitude, determining the sign of the correction by inspection. Repeat this procedure for the day following the given date, if in west longitude; or for the day preceding, if in east longitude. Using the difference between these two times, and the longitude, enter table II of the almanac on the same page and take out the correction. Apply this correction to the LMT of moonrise or moonset at the Greenwich meridian on the given date to find the LMT at the position of the observer. The sign to be given the correction is such as to make the corrected time fall between the times for the two dates between which interpolation is being made. This is nearly always positive (+) in west longitude and negative (-) in east longitude. Convert the corrected LMT to ZT.

To find the time of moonrise or moonset by the *Air Almanac* for the given date, determine LMT for the observer's latitude at the Greenwich meridian in the same manner as with the *Nautical Almanac*, except that linear interpolation is made directly from the main tables, since no interpolation table is provided. Extract, also, the value from the "Diff." column to the right of the moonrise and moonset column, interpolating if necessary. This "Diff." is the half-daily difference. The error introduced by this approximation is generally not more than a few minutes, although it increases with latitude. Using this difference, and the longitude, enter the "Interpolation of moonrise, moonset" table on flap F4 of the *Air Almanac* and extract the correction. The *Air Almanac* recommends taking the correction from this table without interpolation. The results thus obtained are sufficiently accurate for ordinary purposes of navigation. If greater accuracy is desired, the correction can be taken by interpolation. However, since the "Diff." itself is an approximation, the *Nautical Almanac* or computation should be used if accuracy is a consideration. Apply the correction to the LMT of moonrise or moonset at the Greenwich meridian on the given date to find the LMT at the position of the observer. The correction is positive (+) for west longitude, and negative (-) for east longitude, unless the "Diff." on the daily page is preceded by the negative sign (-), when the correction is negative (-) for west longitude, and positive (+) for east longitude. If the time is near midnight, record the date at each step, as in the *Nautical Almanac* solution.

As with the Sun, there are times in high latitudes when interpolation is inaccurate or impossible. At such periods, the times of the phenomena themselves are uncertain, but an approximate answer can be obtained by the Moonlight graph in the *Air Almanac*, or by computation. With the Moon, this condition occurs when the Moon rises or sets at one latitude, but not at the next higher tabulated latitude, as with the Sun. It also occurs when the Moon rises or sets on one day, but not on the preceding or following day. This latter condition is indicated in the *Air Almanac* by the symbol * in the "Diff." column.

Because of the eastward revolution of the Moon around the Earth, there is one day each synodical month ($29^1/_2$ days) when the Moon does not rise, and one day when it does not set. These occur near last quarter and first quarter, respectively. Since this day is not the same at all latitudes or at all longitudes, the time of moonrise or moonset found from the almanac may occasionally be the preceding or succeeding one to that desired. When interpolating near midnight, caution will prevent an error.

The effect of the revolution of the Moon around the Earth is to cause the Moon to rise or set later from day to day. The daily retardation due to this effect does not differ greatly from 50^m. However, the change in declination of the Moon may increase or decrease this effect. This effect increases with latitude, and in extreme conditions it may be greater than the effect due to revolution of the Moon. Hence, the interval between successive moonrises or moonsets is more erratic in high latitudes than in low latitudes. When the two effects act in the same direction, daily differences can be quite large. When they act in opposite directions, they are small, and when the effect due to change in declination is larger than that due to revolution, the Moon sets *earlier* on succeeding days.

This condition is reflected in the *Air Almanac* by a negative "Diff." If this happens near the last quarter or first quarter, two moonrises or moonsets might occur on the same day, one a few minutes after the day begins, and the other a few minutes before it ends, as on June 8, 2002, where two moonrises occur at latitude 72°. Interpolation for longitude is always made between consecutive moonrises or moonsets, regardless of the days on which they fall.

Beyond the northern limits of the almanacs the values can be obtained from a series of graphs given near the back of the *Air Almanac*. For high latitudes, graphs are used instead of tables because graphs give a clearer picture of conditions, which may change radically with relatively little change in position or date. Under these conditions interpolation to practical precision is simpler by graph than by table. In those parts of the graph which are difficult to read, the times of the phenomena's occurrence are uncertain, being altered considerably by a relatively small change in refraction or height of eye.

On all of these graphs, any given latitude is represented by a horizontal line and any given date by a vertical line. At the intersection of these two lines the duration is read from the curves, interpolating by eye between curves.

The "Semiduration of Sunlight" graph gives the number of hours between sunrise and meridian transit or between meridian transit and sunset. The dot scale near the top of the graph indicates the LMT of meridian transit, the time represented by the minute dot nearest the vertical dateline being used. If the intersection occurs in the area marked "Sun above horizon," the Sun does not set; and if in the area marked "Sun below horizon," the Sun does not rise.

The "Duration of Twilight" graph gives the number of hours between the beginning of morning civil twilight (center of Sun 6° below the horizon) and sunrise, or between sunset and the end of evening civil twilight. If the Sun does not rise, but twilight occurs, the time taken from the graph is half the total length of the single twilight period, or the number of hours from beginning of morning twilight to LAN, or from LAN to end of evening twilight. If the intersection occurs in the area marked "continuous twilight or Sunlight," the center of the Sun does not move more than 6° below the horizon, and if in the area marked "no twilight nor Sunlight," the Sun remains more than 6° below the horizon throughout the entire day.

The "Semiduration of Moonlight" graph gives the number of hours between moonrise and meridian transit or between meridian transit and moonset. The dot scale near the top of the graph indicates the LMT of meridian transit, each dot representing one hour. The phase symbols indicate the date on which the principal Moon phases occur, the open circle indicating full Moon and the dark circle indicating new Moon. If the intersection of the vertical dateline and the horizontal latitude line falls in the "Moon above horizon" or "Moon below horizon" area, the Moon remains above or below the horizon, respectively, for the entire 24 hours of the day.

If approximations of the times of moonrise and moonset are sufficient, the semiduration of Moonlight is taken for the time of meridian passage and can be used without adjustment. When an estimated time of rise falls on the preceding day, that phenomenon may be recalculated using the meridian passage and semiduration for the day following. When an estimated time of set falls on the following day, that phenomenon may be recalculated using meridian passage and semiduration for the preceding day. For more accurate results (seldom justified), the times on the required date and the adjacent date (the following date in W longitude and the preceding date in E longitude) should be determined, and an interpolation made for longitude, as in any latitude, since the intervals given are for the Greenwich meridian.

Sunlight, twilight, and Moonlight graphs are not given for south latitudes. Beyond latitude 65°S, the northern hemisphere graphs can be used for determining the semiduration or duration, by using the vertical dateline for a day when the declination has the same numerical value but opposite sign. The time of meridian transit and the phase of the Moon are determined as explained above, using the correct date. Between latitudes 60°S and 65°S, the solution is made by interpolation between the tables and the graphs.

Other methods of solution of these phenomena are available. The *Tide Tables* tabulate sunrise and sunset from latitude 76°N to 60°S. Semiduration or duration can be determined graphically using a diagram on the plane of the celestial meridian, or by computation. When computation is used, solution is made for the meridian angle at which the required negative altitude occurs. The meridian angle expressed in time units is the semiduration in the case of sunrise, sunset, moonrise, and moonset; and the semiduration of the combined Sunlight and twilight, or the time from meridian transit at which morning twilight begins or evening twilight ends. For sunrise and sunset the altitude used is (-)50'. Allowance for height of eye can be made by algebraically subtracting (numerically adding) the dip correction from this altitude. The altitude used for twilight is (-)6°, (-)12°, or (-)18° for civil, nautical, or astronomical twilight, respectively. The altitude used for moonrise and moonset is -34' - SD + HP, where SD is semidiameter and HP is horizontal parallax, from the daily pages of the *Nautical Almanac*.

1912. Rising, Setting, and Twilight on a Moving Craft

Instructions to this point relate to a fixed position on the Earth. Aboard a moving craft the problem is complicated somewhat by the fact that time of occurrence depends upon the position of the craft, which itself depends on the time. At ship speeds, it is generally sufficiently accurate to make an approximate mental solution and use the position of the vessel at this time to make a more accurate solution. If greater accuracy is required, the position at the time indicated in the second solution can be used for a third solution. If desired, this process can be repeated until the same answer is obtained from two consecutive solutions. However, it is generally sufficient to alter the first solution by 1^m for each 15' of longitude that the position of the craft differs from that used in the solution, adding if west of the estimated position, and subtracting if east of it. In applying this rule, use both longitudes to the nearest 15'. The first solution is the **first estimate**; the second solution is the **second estimate**.

CHAPTER 20

SIGHT REDUCTION

BASIC PROCEDURES

2000. Computer Sight Reduction

The purely mathematical process of sight reduction is an ideal candidate for computerization, and a number of different hand-held calculators and computer programs have been developed to relieve the tedium of working out sights by tabular or mathematical methods. The civilian navigator can choose from a wide variety of hand-held calculators and computer programs which require only the entry of the DR position, altitude and azimuth of the body, and GMT. It is not even necessary to know the name of the body because the computer can figure out what it must be based on the entered data. Calculators and computers provide more accurate solutions than tabular and mathematical methods because they can be based on actual values rather than theoretical assumptions and do not have inherent rounding errors.

U.S. Naval navigators have access to a program called **STELLA** (System To Estimate Latitude and Longitude Astronomically; do not confuse with a commercial astronomy program with the same name). STELLA was developed by the Astronomical Applications Department of the U.S. Naval Observatory based on a Navy requirement. The algorithms used in STELLA provide an accuracy of one arc-second on the Earth's surface, a distance of about 30 meters. While this accuracy is far better than can be obtained using a sextant, it does support possible naval needs for automated navigation systems based on celestial objects. These algorithms take into account the oblateness of the Earth, movement of the vessel during sight-taking, and other factors not fully addressed by traditional methods.

STELLA can perform almanac functions, position updating/DR estimations, celestial body rise/set/transit calculations, compass error calculations, sight planning, and sight reduction. On-line help and user's guide are included, and it is a component of the Block III NAVSSI. Because STELLA logs all entered data for future reference, it is authorized to replace the Navy Navigation Workbook. STELLA is now an allowance list requirement for Naval ships, and is available from:

Superintendent
U.S. Naval Observatory
Code: AA/STELLA
3450 Massachusetts Ave. NW
Washington, DC, 20392-5420

or on the Navigator of the Navy Web site at

http://www.navigator.navy.mil/navigator/surface.html.

2001. Tabular Sight Reduction

The remainder of this chapter concentrates on sight reduction using the *Nautical Almanac* and *Pub. No. 229, Sight Reduction Tables for Marine Navigation*. The method explained here is only one of many methods of reducing a sight. The *Nautical Almanac* contains directions for solving sights using its own concise sight reduction tables or calculators, along with examples for the current year

Reducing a celestial sight to obtain a line of position using the tables consists of six steps:

1. Correct the sextant altitude (hs) to obtain observed altitude (ho).
2. Determine the body's GHA and declination (dec.).
3. Select an assumed position (AP) and find its local hour angle (LHA).
4. Compute altitude and azimuth for the AP.
5. Compare the computed and observed altitudes.
6. Plot the line of position.

The introduction to each volume of *Pub. 229* contains information: (1) discussing use of the publication for a variety of special celestial navigation techniques; (2) discussing interpolation, explaining the double second difference interpolation required in some sight reductions, and providing tables to facilitate the interpolation process; and (3) discussing the publication's use in solving problems of great circle sailings. Prior to using *Pub. 229*, carefully read this introductory material.

Celestial navigation involves determining a circular line of position based on an observer's distance from a celestial body's geographic position (GP). Should the observer determine both a body's GP and his distance from the GP, he would have enough information to plot a line of position; he would be somewhere on a circle whose center was the GP and whose radius equaled his distance from that GP. That circle, from all points on which a body's measured altitude would be equal, is a **circle of equal altitude**. There is a direct proportionality between a body's altitude as measured by an observer and the distance of its GP from that observer; the lower the altitude, the farther away the GP.

Therefore, when an observer measures a body's altitude he obtains an indirect measure of the distance between himself and the body's GP. Sight reduction is the process of converting that indirect measurement into a line of position.

Sight reduction reduces the problem of scale to manageable size. Depending on a body's altitude, its GP could be thousands of miles from the observer's position. The size of a chart required to plot this large distance would be impractical. To eliminate this problem, the navigator does not plot this line of position directly. Indeed, he does not plot the GP at all. Rather, he chooses an **assumed position (AP)** near, but usually not coincident with, his DR position. The navigator chooses the AP's latitude and longitude to correspond to the entering arguments of LHA and latitude used in *Pub. 229*. From *Pub. 229*, the navigator computes what the body's altitude would have been had it been measured from the AP. This yields the **computed altitude (h_c)**. He then compares this computed value with the **observed altitude (h_o)** obtained at his actual position. The difference between the computed and observed altitudes is directly proportional to the distance between the circles of equal altitude for the assumed position and the actual position. *Pub. 229* also gives the direction from the GP to the AP. Having selected the assumed position, calculated the distance between the circles of equal altitude for that AP and his actual position, and determined the direction from the assumed position to the body's GP, the navigator has enough information to plot a line of position (LOP).

To plot an LOP, plot the assumed position on either a chart or a plotting sheet. From the *Sight Reduction Tables*, determine: 1) the altitude of the body for a sight taken at the AP and 2) the direction from the AP to the GP. Then, determine the difference between the body's calculated altitude at this AP and the body's measured altitude. This difference represents the difference in radii between the equal altitude circle passing through the AP and the equal altitude circle passing through the actual position. Plot this difference from the AP either towards or away from the GP along the axis between the AP and the GP. Finally, draw the circle of equal altitude representing the circle with the body's GP at the center and with a radius equal to the distance between the GP and the navigator's actual position.

One final consideration simplifies the plotting of the equal altitude circle. Recall that the GP is usually thousands of miles away from the navigator's position. The equal altitude circle's radius, therefore, can be extremely large. Since this radius is so large, the navigator can approximate the section close to his position with a straight line drawn perpendicular to the line connecting the AP and the GP. This straight line approximation is good only for sights at relatively low altitudes. The higher the altitude, the shorter the distance between the GP and the actual position, and the smaller the circle of equal altitude. The shorter this distance, the greater the inaccuracy introduced by this approximation.

2002. Selection of the Assumed Position (AP)

Use the following arguments when entering *Pub. 229* to compute altitude (h_c) and azimuth:

1. Latitude (L)
2. Declination (d or Dec.)
3. Local hour angle (LHA)

Latitude and LHA are functions of the assumed position. Select an AP longitude resulting in a whole degree of LHA and an AP latitude equal to that whole degree of latitude closest to the DR position. Selecting the AP in this manner eliminates interpolation for LHA and latitude in *Pub. 229*.

2003. Comparison of Computed and Observed Altitudes

The difference between the computed altitude (h_c) and the observed altitude (h_o) is the **altitude intercept** (a).

The altitude intercept is the difference in the length of the radii of the circles of equal altitude passing through the AP and the observer's actual position. The position having the greater altitude is on the circle of smaller radius and is closer to the observed body's GP. In Figure 2004, the AP is shown on the inner circle. Therefore, h_c is greater than h_o.

Express the altitude intercept in nautical miles and label it T or A to indicate whether the line of position is toward or away from the GP, as measured from the AP.

A useful aid in remembering the relation between h_o, h_c, and the altitude intercept is: $\underline{H}_o$ $\underline{M}_o$ $\underline{T}_o$ for H_o More Toward. Another is C-G-A: Computed Greater Away, remembered as Coast Guard Academy. In other words, if h_o is greater than h_c, the line of position intersects a point measured from the AP towards the GP a distance equal to the altitude intercept. Draw the LOP through this intersection point perpendicular to the axis between the AP and GP.

2004. Plotting the Line of Position

Plot the line of position as shown in Figure 2004. Plot the AP first; then plot the azimuth line from the AP toward or away from the GP. Then, measure the altitude intercept along this line. At the point on the azimuth line equal to the intercept distance, draw a line perpendicular to the azimuth line. This perpendicular represents that section of the circle of equal altitude passing through the navigator's actual position. This is the line of position.

A navigator often takes sights of more than one celestial body when determining a celestial fix. After plotting the lines of position from these several sights, advance the resulting LOP's along the track to the time of the last sight and label the resulting fix with the time of this last sight.

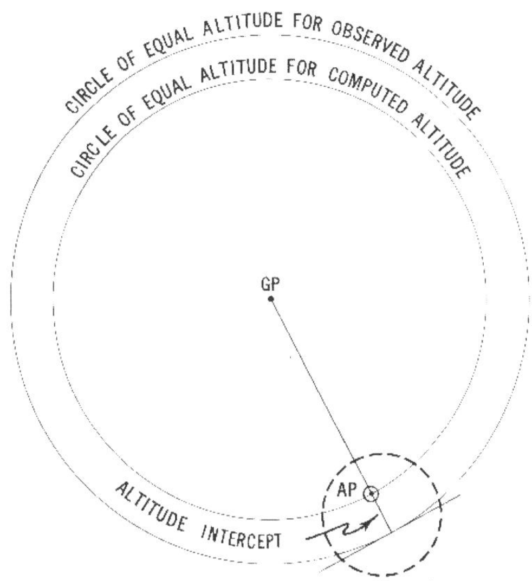

Figure 2004. The basis for the line of position from a celestial observation.

2005. Sight Reduction Procedures

Just as it is important to understand the theory of sight reduction, it is also important to develop a practical procedure to reduce celestial sights consistently and accurately. Sight reduction involves several consecutive steps, the accuracy of each completely dependent on the accuracy of the steps that went before. *Sight reduction tables* have, for the most part, reduced the mathematics involved to simple addition and subtraction. However, careless errors will render even the most skillfully measured sights inaccurate. The navigator using tabular or mathematical techniques must work methodically to reduce careless errors.

Naval navigators will most likely use OPNAV 3530, U.S. Navy Navigation Workbook, which contains pre-formatted pages with "strip forms" to guide the navigator through sight reduction. A variety of commercially-produced forms are also available. Pick a form and learn its method thoroughly. With familiarity will come increasing understanding, speed and accuracy.

Figure 2005 represents a functional and complete worksheet designed to ensure a methodical approach to any sight reduction problem. The recommended procedure discussed below is not the only one available; however, the navigator who uses it can be assured that he has considered every correction required to obtain an accurate fix.

SECTION ONE consists of two parts: (1) Correcting sextant altitude to obtain apparent altitude; and (2) Correcting the apparent altitude to obtain the observed altitude.

Body: Enter the name of the body whose altitude you have measured. If using the Sun or the Moon, indicate which limb was measured.

Index Correction: This is determined by the characteristics of the individual sextant used. Chapter 16 discusses determining its magnitude and algebraic sign.

Dip: The dip correction is a function of the height of eye of the observer. It is always negative; its magnitude is determined from the Dip Table on the inside front cover of the *Nautical Almanac*.

Sum: Enter the algebraic sum of the dip correction and the index correction.

Sextant Altitude: Enter the altitude of the body measured by the sextant.

Apparent Altitude: Apply the correction determined above to the measured altitude and enter the result as the apparent altitude.

Altitude Correction: Every observation requires an altitude correction. This correction is a function of the apparent altitude of the body. The *Almanac* contains tables for determin-

SECTION ONE: OBSERVED ALTITUDE

Body		
Index Correction		
Dip (height of eye)		
Sum		
Sextant Altitude (h_s)		
Apparent Altitude (h_a)		
Altitude Correction		
Mars or Venus Additional Correction		
Additional Correction		
Horizontal Parallax Correction		
Moon Upper Limb Correction		
Correction to Apparent Altitude (h_a)		
Observed Altitude (h_o)		

SECTION TWO: GMT TIME AND DATE

Date		
DR Latitude		
DR Longitude		
Observation Time		
Watch Error		
Zone Time		
Zone Description		
Greenwich Mean Time		
Date GMT		

SECTION THREE: LOCAL HOUR ANGLE AND DECLINATION

Tabulated GHA and *v* Correction Factor		
GHA Increment		
Sidereal Hour Angle (SHA) or *v* Correction		
GHA		
+ or - 360° if needed		
Assumed Longitude (-W, +E)		
Local Hour Angle (LHA)		
Tabulated Declination and *d* Correction Factor		
d Correction		
True Declination		
Assumed Latitude		

SECTION FOUR: ALTITUDE INTERCEPT AND AZIMUTH

Declination Increment and *d* Interpolation Factor		
Computed Altitude (Tabulated)		
Double Second Difference Correction		
Total Correction		
Computed Altitude (h_c)		
Observed Altitude (h_o)		
Altitude Intercept		
Azimuth Angle		
True Azimuth		

Figure 2005. Complete sight reduction form.

ing these corrections. For the Sun, planets, and stars, these tables are located on the inside front cover and facing page. For the Moon, these tables are located on the back inside cover and preceding page.

Mars or Venus Additional Correction: As the name implies, this correction is applied to sights of Mars and Venus. The correction is a function of the planet measured, the time of year, and the apparent altitude. The inside front cover of the *Almanac* lists these corrections.

Additional Correction: Enter this additional correction from Table A-4 located at the front of the *Nautical Almanac* when obtaining a sight under non-standard atmospheric temperature and pressure conditions. This correction is a function of atmospheric pressure, temperature, and apparent altitude.

Horizontal Parallax Correction: This correction is unique to reducing Moon sights. Obtain the H.P. correction value from the daily pages of the *Almanac*. Enter the H.P correction table at the back of the *Almanac* with this value. The H.P correction is a function of the limb of the Moon used (upper or lower), the apparent altitude, and the H.P. correction factor. The H.P. correction is always added to the apparent altitude.

Moon Upper Limb Correction: Enter -30' for this correction if the sight was of the upper limb of the Moon.

Correction to Apparent Altitude: Sum the altitude correction, the Mars or Venus additional correction, the additional correction, the horizontal parallax correction, and the Moon's upper limb correction. Be careful to determine and carry the algebraic sign of the corrections and their sum correctly. Enter this sum as the correction to the apparent altitude.

Observed Altitude: Apply the Correction to Apparent Altitude algebraically to the apparent altitude. The result is the observed altitude.

SECTION TWO determines the Greenwich Mean Time (GMT; referred to in the *Almanac*s as Universal time or UT) and GMT date of the sight.

Date: Enter the local time zone date of the sight.

DR Latitude: Enter the dead reckoning latitude of the vessel.

DR Longitude: Enter the dead reckoning longitude of the vessel.

Observation Time: Enter the local time of the sight as recorded on the ship's chronometer or other timepiece.

Watch Error: Enter a correction for any known watch error.

Zone Time: Correct the observation time with watch error to determine zone time.

Zone Description: Enter the zone description of the time zone indicated by the DR longitude. If the longitude is west of the Greenwich Meridian, the zone description is positive. Conversely, if the longitude is east of the Greenwich Meridian, the zone description is negative. The zone description represents the correction necessary to convert local time to Greenwich Mean Time.

Greenwich Mean Time: Add to the zone description the zone time to determine Greenwich Mean Time.

Date: Carefully evaluate the time correction applied above and determine if the correction has changed the date. Enter the GMT date.

SECTION THREE determines two of the three arguments required to enter *Pub. 229*: Local Hour Angle (LHA) and Declination. This section employs the principle that a celestial body's LHA is the algebraic sum of its Greenwich Hour Angle (GHA) and the observer's longitude. Therefore, the basic method employed in this section is: (1) Determine the body's GHA; (2) Determine an assumed longitude; (3) Algebraically combine the two quantities, remembering to subtract a western assumed longitude from GHA and to add an eastern longitude to GHA; and (4) Extract the declination of the body from the appropriate *Almanac* table, correcting the tabular value if required.

Tabulated GHA and (2) *v* Correction Factor:

For the Sun, the Moon, or a planet, extract the value for the whole hour of GHA corresponding to the sight. For example, if the sight was obtained at 13-50-45 GMT, extract the GHA value for 1300. For a star sight reduction, extract the value of the GHA of Aries (GHA ♈), again using the value corresponding to the whole hour of the time of the sight.

For a planet or Moon sight reduction, enter the *v* correction value. This quantity is not applicable to a Sun or star sight. The *v* correction for a planet sight is found at the bottom of the column for each particular planet. The *v* correction factor for the Moon is located directly beside the tabulated hourly GHA values. The *v* correction factor for the Moon is always positive. If a planet's *v* correction factor is listed without sign, it is positive. If listed with a negative sign, the planet's *v* correction factor is negative. This *v* correction factor is not the magnitude of the *v* correction; it is used later to enter the Increments and Correction table to determine the magnitude of the correction.

GHA Increment: The GHA increment serves as an interpolation factor, correcting for the time that the sight differed from the whole hour. For example, in the sight at 13-50-45 discussed above, this increment correction accounts for the 50 minutes and 45 seconds after the whole hour at which the sight was taken. Obtain this correction value from the Increments and Corrections tables in the *Almanac*. The entering arguments for these tables are the minutes and seconds after the hour at which the sight was taken and the body sighted. Extract the proper correction from the applicable table and enter the correction.

Sidereal Hour Angle or *v* Correction: If reducing a star sight, enter the star's Sidereal Hour Angle (SHA). The SHA is found in the star column of the daily pages of the *Almanac*. The SHA combined with the GHA of Aries results in the star's GHA. The SHA entry is applicable only to a star. If reducing a planet or Moon sight, obtain the *v* correction from the Increments and Corrections Table. The correction is a function of only the *v* correction factor; its magnitude is the same for both the Moon and the planets.

GHA: A star's GHA equals the sum of the Tabulated GHA of Aries, the GHA Increment, and the star's SHA. The Sun's GHA equals the sum of the Tabulated GHA and the GHA Increment. The GHA of the Moon or a planet equals the sum of the Tabulated GHA, the GHA Increment, and the *v* correction.

+ or – 360° (if needed): Since the LHA will be determined from subtracting or adding the assumed longitude to the GHA, adjust the GHA by 360° if needed to facilitate the addition or subtraction.

Assumed Longitude: If the vessel is west of the prime meridian, the assumed longitude will be subtracted from the GHA to determine LHA. If the vessel is east of the prime meridian, the assumed longitude will be added to the GHA to determine the LHA. Select the assumed longitude to meet the following two criteria: (1) When added or subtracted (as applicable) to the GHA determined above, a whole degree of LHA will result; and (2) It is the longitude closest to that DR longitude that meets criterion (1).

Local Hour Angle (LHA): Combine the body's GHA with the assumed longitude as discussed above to determine the body's LHA.

Tabulated Declination and *d* Correction factor: (1) Obtain the tabulated declination for the Sun, the Moon, the stars, or the planets from the daily pages of the *Almanac*. The declination values for the stars are given for the entire three day period covered by the daily page of the *Almanac*. The values for the Sun, Moon, and planets are listed in hourly increments. For these bodies, enter the declination value for the whole hour of the sight. For example, if the sight is at 12-58-40, enter the tabulated declination for 1200. (2) There is no *d* correction factor for a star sight. There are *d* correction factors for Sun, Moon, and planet sights. Similar to the v correction factor discussed above, the *d* correction factor does not equal the magnitude of the *d* correction; it provides the argument to enter the Increments and Corrections tables in the *Almanac*. The sign of the *d* correction factor, which determines the sign of the *d* correction, is determined by the trend of declination values, not the trend of *d* values. The *d* correction factor is simply an interpolation factor; therefore, to determine its sign, look at the declination values for the hours that frame the time of the sight. For example, suppose the sight was taken on a certain date at 12-30-00. Compare the declination value for 1200 and 1300 and determine if the declination has increased or decreased. If it has increased, the *d* correction factor is positive. If it has decreased, the *d* correction factor is negative.

***d* correction:** Enter the Increments and Corrections table with the *d* correction factor discussed above. Extract the proper correction, being careful to retain the proper sign.

True Declination: Combine the tabulated declination and the *d* correction to obtain the true declination.

Assumed Latitude: Choose as the assumed latitude that whole value of latitude closest to the vessel's DR latitude. If the assumed latitude and declination are both north or both south, label the assumed latitude "Same." If one is north and the other is south, label the assumed latitude "Contrary."

SECTION FOUR uses the arguments of assumed latitude, LHA, and declination determined in Section Three to enter *Pub. 229* to determine azimuth and computed altitude. Then, Section Four compares computed and observed altitudes to calculate the altitude intercept. From this the LOP is plotted.

Declination Increment and *d* Interpolation Factor: Note that two of the three arguments used to enter *Pub. 229*, LHA and latitude, are whole degree values. Section Three does not determine the third argument, declination, as a whole degree. Therefore, the navigator must interpolate in *Pub. 229* for declination, given whole degrees of LHA and latitude. The first steps of Section Four involve this interpolation for declination. Since declination values are tabulated every whole degree in *Pub. 229*, the declination increment is the minutes and tenths of the true declination. For example, if the true declination is 13° 15.6', then the declination increment is 15.6'.

Pub. 229 also lists a *d* Interpolation Factor. This is the magnitude of the difference between the two successive tabulated values for declination that frame the true declination. Therefore, for the hypothetical declination listed above, the tabulated *d* interpolation factor listed in the table would be the difference between declination values given for 13° and 14°. If the declination increases between these two values, *d* is positive. If the declination decreases between these two values, *d* is negative.

Computed Altitude (Tabulated): Enter *Pub. 229* with the following arguments: (1) LHA from Section Three; (2) assumed latitude from Section Three; (3) the whole degree value of the true declination. For example, if the true declination were 13° 15.6', then enter *Pub. 229* with 13° as the value for declination. Record the tabulated computed altitude.

Double Second Difference Correction: Use this correction when linear interpolation of declination for computed altitude is not sufficiently accurate due to the non-linear change in the computed altitude as a function of declination. The need for double second difference interpolation is indicated by the *d* interpolation factor appearing in italic type followed by a small dot. When this procedure must be employed, refer to detailed instructions in the introduction to *Pub. 229*.

Total Correction: The total correction is the sum of the double second difference (if required) and the interpolation corrections. Calculate the interpolation correction by dividing the declination increment by 60' and multiply the resulting quotient by the *d* interpolation factor.

Computed Altitude (h_c): Apply the total correction, being careful to carry the correct sign, to the tabulated computed altitude. This yields the computed altitude.

Observed Altitude (h_o): Enter the observed altitude from Section One.

Altitude Intercept: Compare h_c and h_o. Subtract the smaller from the larger. The resulting difference is the magnitude of the altitude intercept. If h_o is greater than h_c, then label the altitude intercept "Toward." If h_c is greater than h_o, then label the altitude intercept "Away."

Azimuth Angle: Obtain the azimuth angle (Z) from *Pub. 229*, using the same arguments which determined tabulated computed altitude. Visual interpolation is sufficiently accurate.

True Azimuth: Calculate the true azimuth (Z_n) from the azimuth angle (Z) as follows:

a) If in northern latitudes:

LHA > 180°, then $Z_n = Z$
LHA < 180°, then $Z_n = 360° - Z$

b) If in southern latitudes:

LHA > 180°, then $Z_n = 180° - Z$
LHA < 180°, then $Z_n = 180° + Z$

SIGHT REDUCTION

The section above discussed the basic theory of sight reduction and presented a method to be followed when reducing sights. This section puts that method into practice in reducing sights of a star, the Sun, the Moon, and planets.

2006. Reducing Star Sights to a Fix

On May 16, 1995, at the times indicated, the navigator takes and records the following sights:

Star	Sextant Altitude	Zone Time
Kochab	47° 19.1'	20-07-43
Spica	32° 34.8'	20-11-26

Height of eye is 48 feet and index correction (IC) is +2.1'. The DR latitude for both sights is 39° N. The DR longitude for the Spica sight is 157° 10'W. The DR longitude for the Kochab sight is 157° 08.0'W. Determine the intercept and azimuth for both sights. See Figure 2006.

First, convert the sextant altitudes to observed altitudes. Reduce the Spica sight first:

Body	Spica
Index Correction	+2.1'
Dip (height 48 ft)	-6.7'
Sum	-4.6'
Sextant Altitude (h_s)	32° 34.8'
Apparent Altitude (h_a)	32° 30.2'
Altitude Correction	-1.5'
Additional Correction	0
Horizontal Parallax	0
Correction to h_a	-1.5'
Observed Altitude (h_o)	32° 28.7'

Determine the sum of the index correction and the dip correction. Go to the inside front cover of the *Nautical Almanac* to the table entitled "DIP." This table lists dip corrections as a function of height of eye measured in either feet or meters. In the above problem, the observer's height of eye is 48 feet. The heights of eye are tabulated in intervals, with the correction corresponding to each interval listed between the interval's endpoints. In this case, 48 feet lies between the tabulated 46.9 to 48.4 feet interval; the corresponding correction for this interval is -6.7'. Add the IC and the dip correction, being careful to carry the correct sign. The sum of the corrections here is -4.6'. Apply this correction to the sextant altitude to obtain the apparent altitude (h_a).

Next, apply the altitude correction. Find the altitude correction table on the inside front cover of the *Nautical Almanac* next to the dip table. The altitude correction varies as a function of both the type of body sighted (Sun, star, or planet) and the body's apparent altitude. For the problem above, enter the star altitude correction table. Again, the correction is given within an altitude interval; h_a in this case was 32° 30.2'. This value lies between the tabulated endpoints 32° 00.0' and 33° 45.0'. The correction corresponding to this interval is -1.5'. Applying this correction to h_a yields an observed altitude of 32° 28.7'.

Having calculated the observed altitude, determine the time and date of the sight in Greenwich Mean Time:

Date	16 May 1995
DR Latitude	39° N
DR Longitude	157° 10' W
Observation Time	20-11-26
Watch Error	0
Zone Time	20-11-26
Zone Description	+10
GMT	06-11-26
GMT Date	17 May 1995

Record the observation time and then apply any watch error to determine zone time. Then, use the DR longitude at the time of the sight to determine time zone description. In this case, the DR longitude indicates a zone description of +10 hours. Add the zone description to the zone time to obtain GMT. It is important to carry the correct date when applying this correction. In this case, the +10 correction made it 06-11-26 GMT on May 17, when the date in the local time zone was May 16.

After calculating both the observed altitude and the GMT

time, enter the daily pages of the *Nautical Almanac* to calculate the star's Greenwich Hour Angle (GHA) and declination.

Tab GHA ♈	324° 28.4'
GHA Increment	2° 52.0'
SHA	158° 45.3'
GHA	486° 05.7'
+/- 360°	not required
Assumed Longitude	157° 05.7'
LHA	329°
Tabulated Dec/*d*	S 11° 08.4'/n.a.
d Correction	—
True Declination	S 11° 08.4'
Assumed Latitude	N 39° contrary

First, record the GHA of Aries from the May 17, 1995 daily page: 324° 28.4'.

Next, determine the incremental addition for the minutes and seconds after 0600 from the Increments and Corrections table in the back of the *Nautical Almanac*. The increment for 11 minutes and 26 seconds is 2° 52'.

Then, calculate the GHA of the star. Remember:

GHA (star) = GHA ♈ + SHA (star)

The *Nautical Almanac* lists the SHA of selected stars on each daily page. The SHA of Spica on May 17, 1995: 158° 45.3'.

Pub. 229's entering arguments are whole degrees of LHA and assumed latitude. Remember that LHA = GHA - west longitude or GHA + east longitude. Since in this example the vessel is in west longitude, subtract its assumed longitude from the GHA of the body to obtain the LHA. Assume a longitude meeting the criteria listed in Article 2005.

From those criteria, the assumed longitude must end in 05.7 minutes so that, when subtracted from the calculated GHA, a whole degree of LHA will result. Since the DR longitude was 157° 10.0', then the assumed longitude ending in 05.7' closest to the DR longitude is 157° 05.7'. Subtracting this assumed longitude from the calculated GHA of the star yields an LHA of 329°.

The next value of concern is the star's true declination. This value is found on the May 17th daily page next to the star's SHA. Spica's declination is S 11° 08.4'. There is no d correction for a star sight, so the star's true declination equals its tabulated declination. The assumed latitude is determined from the whole degree of latitude closest to the DR latitude at the time of the sight. In this case, the assumed latitude is N 39°. It is marked "contrary" because the DR latitude is north while the star's declination is south.

The following information is known: (1) the assumed position's LHA (329°) and assumed latitude (39°N contrary name); and (2) the body's declination (S11° 08.4').

Find the page in the *Sight Reduction Table* corresponding to an LHA of 329° and an assumed latitude of N 39°, with latitude contrary to declination. Enter this table with the body's whole degree of declination. In this case, the body's whole degree of declination is 11°. This declination corresponds to a tabulated altitude of 32° 15.9'. This value is for a declination of 11°; the true declination is 11° 08.4'. Therefore, interpolate to determine the correction to add to the tabulated altitude to obtain the computed altitude.

The difference between the tabulated altitudes for 11° and 12° is given in *Pub. 229* as the value d; in this case, d = -53.0. Express as a ratio the declination increment (in this case, 8.4') and the total interval between the tabulated declination values (in this case, 60') to obtain the percentage of the distance between the tabulated declination values represented by the declination increment. Next, multiply that percentage by the increment between the two values for computed altitude. In this case:

$$\frac{8.4}{60} \times (-53.0) = -7.4$$

Subtract 7.4' from the tabulated altitude to obtain the final computed altitude: H_c = 32° 08.5'.

Dec Inc / + or - d	8.4' / -53.0
h_c (tabulated)	32° 15.9'
Correction (+ or -)	-7.4'
h_c (computed)	32° 08.5'

It will be valuable here to review exactly what h_o and h_c represent. Recall the methodology of the altitude-intercept method. The navigator first measures and corrects an altitude for a celestial body. This corrected altitude, h_o, corresponds to a circle of equal altitude passing through the navigator's actual position whose center is the geographic position (GP) of the body. The navigator then determines an assumed position (AP) near, but not coincident with, his actual position; he then calculates an altitude for an observer at that assumed position (AP).The circle of equal altitude passing through this assumed position is concentric with the circle of equal altitude passing through the navigator's actual position. The difference between the body's altitude at the assumed position (h_c) and the body's observed altitude (h_o) is equal to the differences in radii length of the two corresponding circles of equal altitude. In the above problem, therefore, the navigator knows that the equal altitude circle passing through his actual position is:
away from the equal altitude circle passing through his assumed position. Since h_o is greater than h_c, the navigator knows that the radius of the equal altitude circle passing through his actual position is less than

$$\begin{aligned} h_o &= 32°28.7' \\ -h_c &= \underline{32°08.5'} \\ &\quad 20.2\text{ NM} \end{aligned}$$

the radius of the equal altitude circle passing through the assumed position. The only remaining question is: in what direction from the assumed position is the body's actual GP. *Pub. 229* also provides this final piece of information. This is the value for Z tabulated with the h_c and d values discussed above. In this case, enter *Pub. 229* as before, with LHA, assumed latitude, and declination. Visual interpolation is sufficient. Extract the value Z = 143.3°. The relation between Z and Z_n, the true azimuth, is as follows:

In northern latitudes:

LHA > 180°, then $Z_n = Z$
LHA < 180°, then $Z_n = 360° - Z$

In southern latitudes:

LHA > 180°, then $Z_n = 180° - Z$
LHA < 180°, then $Z_n = 180° + Z$

In this case, LHA > 180° and the vessel is in northern latitude. Therefore, Z_n = Z = 143.3°T. The navigator now has enough information to plot a line of position.

The values for the reduction of the Kochab sight follow:

Body	Kochab
Index Correction	+2.1'
Dip Correction	-6.7'
Sum	-4.6'
h_s	47° 19.1'
h_a	47° 14.5'
Altitude Correction	-.9'
Additional Correction	not applicable
Horizontal Parallax	not applicable
Correction to h_a	-9'
h_o	47° 13.6'
Date	16 May 1995
DR latitude	39°N
DR longitude	157° 08.0' W
Observation Time	20-07-43
Watch Error	0
Zone Time	20-07-43
Zone Description	+10
GMT	06-07-43
GMT Date	17 May 1995
Tab GHA ♈	324° 28.4'
GHA Increment	1° 56.1'
SHA	137° 18.5'
GHA	463° 43.0'
+/- 360°	not applicable
Assumed Longitude	156° 43.0'
LHA	307°
Tab Dec / *d*	N74° 10.6' / n.a.
d Correction	not applicable
True Declination	N74° 10.6'
Assumed Latitude	39°N (same)
Dec Inc / + or - d	10.6' / -24.8
h_c	47° 12.6'
Total Correction	-4.2'
h_c (computed)	47° 08.4'
h_o	47° 13.6'
a (intercept)	5.2 towards
Z	018.9°
Z_n	018.9°

2007. Reducing a Sun Sight

The example below points out the similarities between reducing a Sun sight and reducing a star sight. It also demonstrates the additional corrections required for low altitude (<10°) sights and sights taken during non-standard temperature and pressure conditions.

On June 16, 1994, at 05-15-23 local time, at DR position L 30°N λ 45°W, a navigator takes a sight of the Sun's upper limb. The navigator has a height of eye of 18 feet, the temperature is 88° F, and the atmospheric pressure is 982 mb. The sextant altitude is 3° 20.2'. There is no index error. Determine the observed altitude. See Figure 2007.

Apply the index and dip corrections to h_s to obtain h_a. Because h_a is less than 10°, use the special altitude correction table for sights between 0° and 10° located on the right inside front page of the *Nautical Almanac*.

Enter the table with the apparent altitude, the limb of the Sun used for the sight, and the period of the year. Interpolation for the apparent altitude is not required. In this case, the table yields a correction of -29.4'. The correction's algebraic sign is found at the head of each group of entries and at every change of sign.

The additional correction is required because of the non-standard temperature and atmospheric pressure under which the sight was taken. The correction for these non-standard conditions is found in the *Additional Corrections* table located on page A4 in the front of the *Nautical Almanac*.

First, enter the *Additional Corrections* table with the temperature and pressure to determine the correct zone letter: in this case, zone L. Then, locate the correction in the L column corresponding to the apparent altitude of 3° 16.1'. Interpolate between the table arguments of 3° 00.0' and 3° 30.0' to determine the additional correction: +1.4'. The total correction to the apparent altitude is the sum of the altitude and additional corrections: -28.0'. This results in an h_o of 2° 48.1'.

Next, determine the Sun's GHA and declination.

1995 MAY 16, 17, 18 (TUES., WED., THURS.)

UT (GMT)	ARIES	VENUS −3.9		MARS +0.7		JUPITER −2.5		SATURN +1.3		STARS		
d h	G.H.A. ° ′	G.H.A. ° ′	Dec. ° ′	G.H.A. ° ′	Dec. ° ′	G.H.A. ° ′	Dec. ° ′	G.H.A. ° ′	Dec. ° ′	Name	S.H.A. ° ′	Dec. ° ′
16 00	233 14.4	205 51.6	N 9 30.5	84 34.3	N14 31.2	342 02.6	S21 28.7	239 13.9	S 4 40.8	Acamar	315 29.1	S40 19.4
01	248 16.9	220 51.2	31.6	99 35.8	30.8	357 05.3	28.7	254 16.2	40.7	Achernar	335 37.4	S57 15.5
02	263 19.4	235 50.8	32.7	114 37.3	30.3	12 08.1	28.7	269 18.5	40.6	Acrux	173 24.0	S63 04.7
03	278 21.8	250 50.4	· · 33.8	129 38.8	· · 29.9	27 10.9	· · 28.6	284 20.7	· · 40.6	Adhara	255 23.5	S28 58.3
04	293 24.3	265 50.0	34.9	144 40.3	29.5	42 13.7	28.6	299 23.0	40.5	Aldebaran	291 05.3	N16 29.9
05	308 26.8	280 49.6	36.0	159 41.7	29.1	57 16.4	28.5	314 25.3	40.4			
06	323 29.2	295 49.2	N 9 37.1	174 43.2	N14 28.7	72 19.2	S21 28.5	329 27.6	S 4 40.4	Alioth	166 32.3	N55 59.2
07	338 31.7	310 48.8	38.2	189 44.7	28.3	87 22.0	28.5	344 29.9	40.3	Alkaid	153 09.2	N49 20.3
T 08	353 34.2	325 48.4	39.2	204 46.2	27.9	102 24.7	28.4	359 32.1	40.2	Al Na'ir	28 00.8	S46 58.7
U 09	8 36.6	340 48.0	· · 40.3	219 47.7	· · 27.5	117 27.5	· · 28.4	14 34.4	· · 40.2	Alnilam	276 00.5	S 1 12.5
E 10	23 39.1	355 47.6	41.4	234 49.2	27.1	132 30.3	28.4	29 36.7	40.1	Alphard	218 09.5	S 8 38.6
S 11	38 41.5	10 47.1	42.5	249 50.6	26.6	147 33.1	28.3	44 39.0	40.0			
D 12	53 44.0	25 46.7	N 9 43.6	264 52.1	N14 26.2	162 35.8	S21 28.3	59 41.3	S 4 40.0	Alphecca	126 22.2	N26 43.8
A 13	68 46.5	40 46.3	44.7	279 53.6	25.8	177 38.6	28.3	74 43.6	39.9	Alpheratz	357 57.8	N29 03.8
Y 14	83 48.9	55 45.9	45.8	294 55.1	25.4	192 41.4	28.2	89 45.8	39.8	Altair	62 21.3	N 8 51.4
15	98 51.4	70 45.5	· · 46.9	309 56.6	· · 25.0	207 44.2	· · 28.2	104 48.1	· · 39.8	Ankaa	353 29.4	S42 19.7
16	113 53.9	85 45.1	47.9	324 58.0	24.6	222 46.9	28.2	119 50.4	39.7	Antares	112 42.6	S26 25.3
17	128 56.3	100 44.7	49.0	339 59.5	24.2	237 49.7	28.1	134 52.7	39.6			
18	143 58.8	115 44.3	N 9 50.1	355 01.0	N14 23.8	252 52.5	S21 28.1	149 55.0	S 4 39.6	Arcturus	146 07.8	N19 12.4
19	159 01.3	130 43.9	51.2	10 02.5	23.3	267 55.3	28.1	164 57.3	39.5	Atria	107 56.1	S69 01.0
20	174 03.7	145 43.5	52.3	25 03.9	22.9	282 58.0	28.0	179 59.5	39.4	Avior	234 23.8	S59 30.1
21	189 06.2	160 43.1	· · 53.4	40 05.4	· · 22.5	298 00.8	· · 28.0	195 01.8	· · 39.4	Bellatrix	278 46.9	N 6 20.6
22	204 08.7	175 42.7	54.5	55 06.9	22.1	313 03.6	28.0	210 04.1	39.3	Betelgeuse	271 16.3	N 7 24.2
23	219 11.1	190 42.3	55.5	70 08.4	21.7	328 06.3	27.9	225 06.4	39.2			
17 00	234 13.6	205 41.9	N 9 56.6	85 09.8	N14 21.3	343 09.1	S21 27.9	240 08.7	S 4 39.1	Canopus	264 02.6	S52 41.9
01	249 16.0	220 41.5	57.7	100 11.3	20.9	358 11.9	27.8	255 11.0	39.1	Capella	280 55.0	N45 59.5
02	264 18.5	235 41.1	58.8	115 12.8	20.5	13 14.7	27.8	270 13.2	39.0	Deneb	49 40.6	N45 15.7
03	279 21.0	250 40.7	9 59.9	130 14.3	· · 20.0	28 17.4	· · 27.8	285 15.5	· · 38.9	Denebola	182 47.4	N14 35.8
04	294 23.4	265 40.3	10 01.0	145 15.7	19.6	43 20.2	27.7	300 17.8	38.9	Diphda	349 09.8	S18 00.7
05	309 25.9	280 39.8	02.0	160 17.2	19.2	58 23.0	27.7	315 20.1	38.8			
06	324 28.4	295 39.4	N10 03.1	175 18.7	N14 18.8	73 25.8	S21 27.7	330 22.4	S 4 38.7	Dubhe	194 08.2	N61 46.7
W 07	339 30.8	310 39.0	04.2	190 20.2	18.4	88 28.6	27.6	345 24.7	38.7	Elnath	278 30.2	N28 36.1
E 08	354 33.3	325 38.6	05.3	205 21.6	18.0	103 31.3	27.6	0 26.9	38.6	Eltanin	90 52.0	N51 29.3
D 09	9 35.8	340 38.2	· · 06.4	220 23.1	· · 17.5	118 34.1	· · 27.6	15 29.2	· · 38.5	Enif	34 00.5	N 9 51.2
N 10	24 38.2	355 37.8	07.4	235 24.6	17.1	133 36.9	27.5	30 31.5	38.5	Fomalhaut	15 39.1	S29 38.6
E 11	39 40.7	10 37.4	08.5	250 26.0	16.7	148 39.7	27.5	45 33.8	38.4			
S 12	54 43.1	25 37.0	N10 09.6	265 27.5	N14 16.3	163 42.4	S21 27.5	60 36.1	S 4 38.3	Gacrux	172 15.6	S57 05.5
D 13	69 45.6	40 36.6	10.7	280 29.0	15.9	178 45.2	27.4	75 38.4	38.3	Gienah	176 06.1	S17 31.2
A 14	84 48.1	55 36.2	11.8	295 30.5	15.5	193 48.0	27.4	90 40.7	38.2	Hadar	149 06.6	S60 21.2
Y 15	99 50.5	70 35.7	· · 12.8	310 31.9	· · 15.0	208 50.8	· · 27.4	105 42.9	· · 38.1	Hamal	328 16.4	N23 26.3
16	114 53.0	85 35.3	13.9	325 33.4	14.6	223 53.5	27.3	120 45.2	38.1	Kaus Aust.	84 01.5	S34 23.0
17	129 55.5	100 34.9	15.0	340 34.9	14.2	238 56.3	27.3	135 47.5	38.0			
18	144 57.9	115 34.5	N10 16.1	355 36.3	N14 13.8	253 59.1	S21 27.2	150 49.8	S 4 37.9	Kochab	137 18.5	N74 10.6
19	160 00.4	130 34.1	17.2	10 37.8	13.4	269 01.9	27.2	165 52.1	37.9	Markab	13 52.0	N15 10.8
20	175 02.9	145 33.7	18.2	25 39.3	13.0	284 04.6	27.2	180 54.4	37.8	Menkar	314 29.6	N 4 04.2
21	190 05.3	160 33.3	· · 19.3	40 40.7	· · 12.5	299 07.4	· · 27.1	195 56.7	· · 37.7	Menkent	148 23.3	S36 21.0
22	205 07.8	175 32.8	20.4	55 42.2	12.1	314 10.2	27.1	210 58.9	37.7	Miaplacidus	221 42.6	S69 42.4
23	220 10.3	190 32.4	21.5	70 43.7	11.7	329 13.0	27.1	226 01.2	37.6			
18 00	235 12.7	205 32.0	N10 22.5	85 45.1	N14 11.3	344 15.8	S21 27.0	241 03.5	S 4 37.5	Mirfak	309 00.4	N49 50.6
01	250 15.2	220 31.6	23.6	100 46.6	10.9	359 18.5	27.0	256 05.8	37.5	Nunki	76 14.9	S26 18.0
02	265 17.6	235 31.2	24.7	115 48.1	10.5	14 21.3	27.0	271 08.1	37.4	Peacock	53 40.4	S56 44.7
03	280 20.1	250 30.8	· · 25.8	130 49.5	· · 10.0	29 24.1	· · 26.9	286 10.4	· · 37.3	Pollux	243 44.6	N28 02.2
04	295 22.6	265 30.4	26.8	145 51.0	09.6	44 26.9	26.9	301 12.7	37.3	Procyon	245 14.1	N 5 14.0
05	310 25.0	280 29.9	27.9	160 52.5	09.2	59 29.6	26.9	316 15.0	37.2			
06	325 27.5	295 29.5	N10 29.0	175 53.9	N14 08.8	74 32.4	S21 26.8	331 17.2	S 4 37.1	Rasalhague	96 18.8	N12 33.8
07	340 30.0	310 29.1	30.0	190 55.4	08.4	89 35.2	26.8	346 19.5	37.1	Regulus	207 58.0	N11 59.3
T 08	355 32.4	325 28.7	31.1	205 56.9	07.9	104 38.0	26.8	1 21.8	37.0	Rigel	281 25.5	S 8 12.6
H 09	10 34.9	340 28.3	· · 32.2	220 58.3	· · 07.5	119 40.8	· · 26.7	16 24.1	· · 36.9	Rigil Kent.	140 09.6	S60 49.0
U 10	25 37.4	355 27.9	33.3	235 59.8	07.1	134 43.5	26.7	31 26.4	36.9	Sabik	102 27.8	S15 43.1
R 11	40 39.8	10 27.4	34.3	251 01.2	06.7	149 46.3	26.6	46 28.7	36.8			
S 12	55 42.3	25 27.0	N10 35.4	266 02.7	N14 06.3	164 49.1	S21 26.6	61 31.0	S 4 36.8	Schedar	349 56.4	N56 30.5
D 13	70 44.8	40 26.6	36.5	281 04.2	05.8	179 51.9	26.6	76 33.3	36.7	Shaula	96 40.0	S37 05.9
A 14	85 47.2	55 26.2	37.5	296 05.6	05.4	194 54.7	26.5	91 35.6	36.6	Sirius	258 45.9	S16 42.8
Y 15	100 49.7	70 25.8	· · 38.6	311 07.1	· · 05.0	209 57.4	· · 26.5	106 37.8	· · 36.6	Spica	158 45.3	S11 08.4
16	115 52.1	85 25.3	39.7	326 08.6	04.6	225 00.2	26.5	121 40.1	36.5	Suhail	223 02.5	S43 25.2
17	130 54.6	100 24.9	40.7	341 10.0	04.2	240 03.0	26.4	136 42.4	36.4			
18	145 57.1	115 24.5	N10 41.8	356 11.5	N14 03.7	255 05.8	S21 26.4	151 44.7	S 4 36.4	Vega	80 47.8	N38 46.7
19	160 59.5	130 24.1	42.9	11 12.9	03.3	270 08.6	26.4	166 47.0	36.3	Zuben'ubi	137 20.2	S16 01.4
20	176 02.0	145 23.6	44.0	26 14.4	02.9	285 11.3	26.3	181 49.3	36.2		S.H.A. ° ′	Mer. Pass. h m
21	191 04.5	160 23.2	· · 45.0	41 15.9	· · 02.5	300 14.1	· · 26.3	196 51.6	· · 36.2			
22	206 06.9	175 22.8	46.1	56 17.3	02.0	315 16.9	26.2	211 53.9	36.1	Venus	331 28.3	10 17
23	221 09.4	190 22.4	47.2	71 18.8	01.6	330 19.7	26.2	226 56.2	36.0	Mars	210 56.3	18 18
Mer. Pass.	h m 8 21.7	v −0.4	d 1.1	v 1.5	d 0.4	v 2.8	d 0.0	v 2.3	d 0.1	Jupiter	108 55.5	1 07
										Saturn	5 55.1	7 58

Figure 2006. Left hand daily page of the Nautical Almanac for May 17, 1995.

Body	Sun UL
Index Correction	0
Dip Correction (18 ft)	-4.1'
Sum	-4.1'
h_s	3° 20.2'
h_a	3° 16.1'
Altitude Correction	-29.4'
Additional Correction	+1.4'
Horizontal Parallax	0
Correction to h_a	-28.0'
h_o	2° 48.1'

Date	June 16, 1994
DR Latitude	N30° 00.0'
DR Longitude	W045° 00.0'
Observation Time	05-15-23
Watch Error	0
Zone Time	05-15-23
Zone Description	+03
GMT	08-15-23
Date GMT	June 16, 1994
Tab GHA / *v*	299° 51.3' / n.a.
GHA Increment	3° 50.8'
SHA or *v* correction	not applicable
GHA	303°42.1'
Assumed Longitude	44° 42.1' W
LHA	259°
Tab Declination / *d*	N23° 20.5' / +0.1'
d Correction	0.0
True Declination	N23° 20.5'
Assumed Latitude	N30° (same)

Again, this process is similar to the star sights reduced above. Notice, however, that SHA, a quantity unique to star sight reduction, is not used in Sun sight reduction.

Determining the Sun's GHA is less complicated than determining a star's GHA. The Nautical Almanac's daily pages list the Sun's GHA in hourly increments. In this case, the Sun's GHA at 0800 GMT on June 16, 1994 is 299° 51.3'. The *v* correction is not applicable for a Sun sight; therefore, applying the increment correction yields the Sun's GHA. In this case, the GHA is 303° 42.1'.

Determining the Sun's LHA is similar to determining a star's LHA. In determining the Sun's declination, however, an additional correction not encountered in the star sight, the *d* correction, must be considered. The bottom of the Sun column on the daily pages of the *Nautical Almanac* lists the *d* value. This is an interpolation factor for the Sun's declination. The sign of the *d* factor is not given; it must be determined by noting from the *Almanac* if the Sun's declination is increasing or decreasing throughout the day. If it is increasing, the factor is positive; if it is decreasing, the factor is negative. In the above problem, the Sun's declination is increasing throughout the day. Therefore, the *d* factor is +0.1.

Having obtained the *d* factor, enter the 15 minute increment and correction table. Under the column labeled "*v* or *d* corr^n," find the value for *d* in the left hand column. The corresponding number in the right hand column is the correction; apply it to the tabulated declination. In this case, the correction corresponding to a *d* value of +0.1 is 0.0'.

Correction (+ or -)	+10.8'
Computed Altitude (h_c)	2° 39.6'
Observed Altitude (h_o)	2° 48.1'
Intercept	8.5 NM (towards)
Z	064.7°
Z_n	064.7°

The final step will be to determine h_c and Z_n. Enter *Pub. 229* with an LHA of 259°, a declination of N23° 20.5', and an assumed latitude of 30°N.

Declination Increment / + or - *d*	20.5' / +31.5
Tabulated Altitude	2° 28.8'

2008. Reducing a Moon Sight

The Moon is easy to identify and is often visible during the day. However, the Moon's proximity to the Earth requires applying additional corrections to h_a to obtain h_o. This article will cover Moon sight reduction.

At 10-00-00 GMT, June 16, 1994, the navigator obtains a sight of the Moon's upper limb. H_s is 26° 06.7'. Height of eye is 18 feet; there is no index error. Determine h_o, the Moon's GHA, and the Moon's declination. See Figure 2008.

This example demonstrates the extra corrections required for obtaining h_o for a Moon sight. Apply the index and dip corrections in the same manner as for star and Sun sights. The altitude correction comes from tables located on the inside back covers of the *Nautical Almanac*.

In this case, the apparent altitude was 26° 02.6'. Enter the altitude correction table for the Moon with the above apparent altitude. Interpolation is not required. The correction is +60.5'. The additional correction in this case is not applicable because the sight was taken under standard temperature and pressure conditions.

The horizontal parallax correction is unique to Moon sights. The table for determining this HP correction is on the back inside cover of the *Nautical Almanac*. First, go to the daily page for June 16 at 10-00-00 GMT. In the column for the Moon, find the HP correction factor corresponding to 10-00-00. Its value is 58.4. Take this value to the HP correction table on the inside back cover of the *Almanac*. Notice that the HP correction columns line up vertically with the Moon altitude correction table columns. Find the HP correction column directly under the altitude correction table heading corresponding to the apparent altitude. Enter that column with the HP correction factor from the daily pages. The column has two sets of figures listed under "U" and "L" for upper and lower limb, respectively. In this case, trace down the "U" column until it intersects with the HP

1994 JUNE 15, 16, 17 (WED., THURS., FRI.)

UT (GMT)	ARIES	VENUS −4.0		MARS +1.2		JUPITER −2.3		SATURN +1.0		STARS		
d h	G.H.A.	G.H.A.	Dec.	G.H.A.	Dec.	G.H.A.	Dec.	G.H.A.	Dec.	Name	S.H.A.	Dec.
	° ′	° ′	° ′	° ′	° ′	° ′	° ′	° ′	° ′		° ′	° ′
15 00	263 03.0	140 48.0	N22 09.1	218 59.0	N16 08.6	49 42.0	S12 03.8	278 39.3	S 8 31.9	Acamar	315 29.5	S40 19.5
01	278 05.4	155 47.4	08.5	233 59.7	09.1	64 44.6	03.8	293 41.7	31.9	Achernar	335 37.6	S57 15.6
02	293 07.9	170 46.7	07.9	249 00.3	09.7	79 47.2	03.7	308 44.1	31.9	Acrux	173 25.1	S63 04.5
03	308 10.4	185 46.1	· · 07.3	264 01.0	· · 10.2	94 49.7	· · 03.7	323 46.6	· · 31.9	Adhara	255 24.1	S28 58.0
04	323 12.8	200 45.5	06.7	279 01.6	10.8	109 52.3	03.7	338 49.0	31.9	Aldebaran	291 06.1	N16 29.8
05	338 15.3	215 44.8	06.0	294 02.2	11.3	124 54.9	03.6	353 51.4	31.9			
06	353 17.8	230 44.2	N22 05.4	309 02.9	N16 11.9	139 57.5	S12 03.6	8 53.9	S 8 31.9	Alioth	166 33.0	N55 59.6
W 07	8 20.2	245 43.6	04.8	324 03.5	12.4	155 00.1	03.6	23 56.3	31.9	Alkaid	153 09.8	N49 20.6
E 08	23 22.7	260 42.9	04.2	339 04.2	13.0	170 02.7	03.5	38 58.7	31.9	Al Na'ir	28 01.4	S46 58.9
D 09	38 25.2	275 42.3	· · 03.5	354 04.8	· · 13.5	185 05.3	· · 03.5	54 01.1	· · 31.9	Alnilam	276 01.2	S 1 12.4
N 10	53 27.6	290 41.7	02.9	9 05.5	14.0	200 07.9	03.5	69 03.6	31.9	Alphard	218 10.3	S 8 38.3
E 11	68 30.1	305 41.0	02.3	24 06.1	14.6	215 10.4	03.4	84 06.0	31.9			
S 12	83 32.6	320 40.4	N22 01.6	39 06.8	N16 15.1	230 13.0	S12 03.4	99 08.4	S 8 31.9	Alphecca	126 22.7	N26 44.1
D 13	98 35.0	335 39.8	01.0	54 07.4	15.7	245 15.6	03.4	114 10.9	31.9	Alpheratz	357 58.3	N29 03.5
A 14	113 37.5	350 39.2	22 00.4	69 08.1	16.2	260 18.2	03.3	129 13.3	31.9	Altair	62 21.8	N 8 51.3
Y 15	128 39.9	5 38.5	21 59.7	84 08.7	· · 16.8	275 20.8	· · 03.3	144 15.7	· · 31.9	Ankaa	353 29.8	S42 19.9
16	143 42.4	20 37.9	59.1	99 09.4	17.3	290 23.4	03.3	159 18.2	31.9	Antares	112 43.4	S26 25.2
17	158 44.9	35 37.3	58.5	114 10.0	17.8	305 26.0	03.3	174 20.6	31.9			
18	173 47.3	50 36.7	N21 57.8	129 10.7	N16 18.4	320 28.5	S12 03.2	189 23.0	S 8 31.9	Arcturus	146 08.5	N19 12.7
19	188 49.8	65 36.0	57.2	144 11.3	18.9	335 31.1	03.2	204 25.5	31.9	Atria	107 57.4	S69 01.1
20	203 52.3	80 35.4	56.6	159 12.0	19.5	350 33.7	03.2	219 27.9	31.9	Avior	234 24.3	S59 29.8
21	218 54.7	95 34.8	· · 55.9	174 12.6	· · 20.0	5 36.3	· · 03.1	234 30.3	· · 31.9	Bellatrix	278 47.6	N 6 20.6
22	233 57.2	110 34.2	55.3	189 13.3	20.6	20 38.9	03.1	249 32.8	31.9	Betelgeuse	271 17.0	N 7 24.3
23	248 59.7	125 33.5	54.6	204 13.9	21.1	35 41.5	03.1	264 35.2	31.9			
16 00	264 02.1	140 32.9	N21 54.0	219 14.6	N16 21.6	50 44.1	S12 03.0	279 37.6	S 8 31.9	Canopus	264 03.0	S52 41.7
01	279 04.6	155 32.3	53.4	234 15.2	22.2	65 46.6	03.0	294 40.1	31.9	Capella	280 56.0	N45 59.5
02	294 07.1	170 31.7	52.7	249 15.9	22.7	80 49.2	03.0	309 42.5	31.9	Deneb	49 40.8	N45 15.6
03	309 09.5	185 31.1	· · 52.1	264 16.5	· · 23.2	95 51.8	· · 03.0	324 44.9	· · 31.9	Denebola	182 48.1	N14 36.2
04	324 12.0	200 30.4	51.4	279 17.2	23.8	110 54.4	02.9	339 47.4	31.9	Diphda	349 10.3	S18 00.9
05	339 14.4	215 29.8	50.8	294 17.8	24.3	125 57.0	02.9	354 49.8	31.8			
06	354 16.9	230 29.2	N21 50.1	309 18.5	N16 24.9	140 59.5	S12 02.9	9 52.2	S 8 31.8	Dubhe	194 09.2	N61 47.0
07	9 19.4	245 28.6	49.5	324 19.1	25.4	156 02.1	02.8	24 54.7	31.8	Elnath	278 31.0	N28 36.1
T 08	24 21.8	260 28.0	48.8	339 19.8	25.9	171 04.7	02.8	39 57.1	31.8	Eltanin	90 52.2	N51 29.4
H 09	39 24.3	275 27.4	· · 48.2	354 20.4	· · 26.5	186 07.3	· · 02.8	54 59.5	· · 31.8	Enif	34 00.9	N 9 51.0
U 10	54 26.8	290 26.8	47.5	9 21.1	27.0	201 09.9	02.7	70 02.0	31.8	Fomalhaut	15 39.6	S29 38.8
R 11	69 29.2	305 26.1	46.9	24 21.7	27.6	216 12.5	02.7	85 04.4	31.8			
S 12	84 31.7	320 25.5	N21 46.2	39 22.3	N16 28.1	231 15.0	S12 02.7	100 06.8	S 8 31.8	Gacrux	172 16.6	S57 05.3
D 13	99 34.2	335 24.9	45.6	54 23.0	28.6	246 17.6	02.7	115 09.3	31.8	Gienah	176 06.9	S17 30.9
A 14	114 36.6	350 24.3	44.9	69 23.6	29.2	261 20.2	02.6	130 11.7	31.8	Hadar	149 07.7	S60 21.1
Y 15	129 39.1	5 23.7	· · 44.3	84 24.3	· · 29.7	276 22.8	· · 02.6	145 14.1	· · 31.8	Hamal	328 17.1	N23 26.1
16	144 41.5	20 23.1	43.6	99 24.9	30.2	291 25.3	02.6	160 16.6	31.8	Kaus Aust.	84 02.3	S34 23.1
17	159 44.0	35 22.5	42.9	114 25.6	30.8	306 27.9	02.5	175 19.0	31.8			
18	174 46.5	50 21.9	N21 42.3	129 26.2	N16 31.3	321 30.5	S12 02.5	190 21.4	S 8 31.8	Kochab	137 18.6	N74 10.9
19	189 48.9	65 21.3	41.6	144 26.9	31.8	336 33.1	02.5	205 23.9	31.8	Markab	13 52.4	N15 10.5
20	204 51.4	80 20.6	41.0	159 27.5	32.4	351 35.7	02.5	220 26.3	31.8	Menkar	314 30.2	N 4 04.1
21	219 53.9	95 20.0	· · 40.3	174 28.2	· · 32.9	6 38.2	· · 02.4	235 28.7	· · 31.8	Menkent	148 24.1	S36 20.8
22	234 56.3	110 19.4	39.6	189 28.8	33.4	21 40.8	02.4	250 31.2	31.8	Miaplacidus	221 43.2	S69 42.1
23	249 58.8	125 18.8	39.0	204 29.5	34.0	36 43.4	02.4	265 33.6	31.8			
17 00	265 01.3	140 18.2	N21 38.3	219 30.1	N16 34.5	51 46.0	S12 02.4	280 36.0	S 8 31.8	Mirfak	309 01.2	N49 50.3
01	280 03.7	155 17.6	37.6	234 30.8	35.0	66 48.5	02.3	295 38.5	31.8	Nunki	76 15.6	S26 18.1
02	295 06.2	170 17.0	37.0	249 31.4	35.6	81 51.1	02.3	310 40.9	31.8	Peacock	53 41.1	S56 44.9
03	310 08.7	185 16.4	· · 36.3	264 32.1	· · 36.1	96 53.7	· · 02.3	325 43.4	· · 31.8	Pollux	243 45.4	N28 02.3
04	325 11.1	200 15.8	35.6	279 32.7	36.6	111 56.3	02.2	340 45.8	31.8	Procyon	245 14.9	N 5 14.2
05	340 13.6	215 15.2	35.0	294 33.3	37.2	126 58.8	02.2	355 48.2	31.8			
06	355 16.0	230 14.6	N21 34.3	309 34.0	N16 37.7	142 01.4	S12 02.2	10 50.7	S 8 31.8	Rasalhague	96 19.3	N12 33.9
07	10 18.5	245 14.0	33.6	324 34.6	38.2	157 04.0	02.2	25 53.1	31.8	Regulus	207 58.8	N11 59.6
08	25 21.0	260 13.4	32.9	339 35.3	38.8	172 06.6	02.1	40 55.5	31.8	Rigel	281 26.1	S 8 12.6
F 09	40 23.4	275 12.8	· · 32.3	354 35.9	· · 39.3	187 09.1	· · 02.1	55 58.0	· · 31.8	Rigil Kent.	140 10.7	S60 48.9
R 10	55 25.9	290 12.2	31.6	9 36.6	39.8	202 11.7	02.1	71 00.4	31.8	Sabik	102 28.5	S15 43.0
I 11	70 28.4	305 11.6	30.9	24 37.2	40.4	217 14.3	02.1	86 02.9	31.8			
D 12	85 30.8	320 11.0	N21 30.2	39 37.9	N16 40.9	232 16.9	S12 02.0	101 05.3	S 8 31.8	Schedar	349 56.9	N56 30.2
A 13	100 33.3	335 10.4	29.6	54 38.5	41.4	247 19.4	02.0	116 07.7	31.8	Shaula	96 40.8	S37 05.9
Y 14	115 35.8	350 09.8	28.9	69 39.2	41.9	262 22.0	02.0	131 10.2	31.8	Sirius	258 46.6	S16 42.6
15	130 38.2	5 09.2	· · 28.2	84 39.8	· · 42.5	277 24.6	· · 02.0	146 12.6	· · 31.8	Spica	158 46.1	S11 08.1
16	145 40.7	20 08.6	27.5	99 40.5	43.0	292 27.1	01.9	161 15.1	31.8	Suhail	223 03.2	S43 24.9
17	160 43.2	35 08.0	26.8	114 41.1	43.5	307 29.7	01.9	176 17.5	31.8			
18	175 45.6	50 07.4	N21 26.2	129 41.7	N16 44.1	322 32.3	S12 01.9	191 19.9	S 8 31.8	Vega	80 48.2	N38 46.8
19	190 48.1	65 06.8	25.5	144 42.4	44.6	337 34.9	01.9	206 22.4	31.8	Zuben'ubi	137 20.9	S16 01.2
20	205 50.5	80 06.2	24.8	159 43.0	45.1	352 37.4	01.8	221 24.8	31.8		S.H.A.	Mer. Pass.
21	220 53.0	95 05.6	· · 24.1	174 43.7	· · 45.6	7 40.0	· · 01.8	236 27.3	· · 31.8		° ′	h m
22	235 55.5	110 05.1	23.4	189 44.3	46.2	22 42.6	01.8	251 29.7	31.8	Venus	236 30.8	14 38
23	250 57.9	125 04.5	22.7	204 45.0	46.7	37 45.1	01.8	266 32.1	31.8	Mars	315 12.5	9 23
Mer. Pass.	h m 6 22.8	v −0.6	d 0.7	v 0.6	d 0.5	v 2.6	d 0.0	v 2.4	d 0.0	Jupiter	146 41.9	20 34
										Saturn	15 35.5	5 21

Figure 2007. Left hand daily page of the Nautical Almanac for June 16, 1994.

Body	Moon (UL)
Index Correction	0.0'
Dip (18 feet)	-4.1'
Sum	-4.1'
Sextant Altitude (h_s)	26° 06.7'
Apparent Altitude (h_a)	26° 02.6'
Altitude Correction	+60.5'
Additional Correction	0.0'
Horizontal Parallax (58.4)	+4.0'
Moon Upper Limb Correction	-30.0'
Correction to h_a	+34.5'
Observed Altitude (h_o)	26° 37.1'

correction factor of 58.4. Interpolating between 58.2 and 58.5 yields a value of +4.0' for the horizontal parallax correction.

The final correction is a constant -30.0' correction to h_a applied only to sights of the Moon's upper limb. This correction is always negative; apply it only to sights of the Moon's upper limb, not its lower limb. The total correction to h_a is the sum of all the corrections; in this case, this total correction is +34.5 minutes.

To obtain the Moon's GHA, enter the daily pages in the Moon column and extract the applicable data just as for a star or Sun sight. Determining the Moon's GHA requires an additional correction, the *v* correction.

GHA Moon and *v*	245° 45.1' and +11.3
GHA Increment	0° 00.0'
v Correction	+0.1'
GHA	245° 45.2'

First, record the GHA of the Moon for 10-00-00 on June 16, 1994, from the daily pages of the *Nautical Almanac*. Record also the *v* correction factor; in this case, it is +11.3. The *v* correction factor for the Moon is always positive. The increment correction is, in this case, zero because the sight was recorded on the even hour. To obtain the *v* correction, go to the tables of increments and corrections. In the 0 minute table in the *v* or *d* correction columns, find the correction that corresponds to a *v* = 11.3. The table yields a correction of +0.1'. Adding this correction to the tabulated GHA gives the final GHA as 245° 45.2'.

Finding the Moon's declination is similar to finding the declination for the Sun or stars. Go to the daily pages for June 16, 1994; extract the Moon's declination and *d* factor.

Tabulated Declination / *d*	S 00° 13.7' / +12.1
d Correction	+0.1'
True Declination	S 00° 13.8'

The tabulated declination and the *d* factor come from the *Nautical Almanac's* daily pages. Record the declination and *d* correction and go to the increment and correction pages to extract the proper correction for the given *d* factor. In this case, go to the correction page for 0 minutes. The correction corresponding to a *d* factor of +12.1 is +0.1. It is important to extract the correction with the correct algebraic sign. The *d* correction may be positive or negative depending on whether the Moon's declination is increasing or decreasing in the interval covered by the *d* factor. In this case, the Moon's declination at 10-00-00 GMT on 16 June was S 00° 13.7'; at 11-00-00 on the same date the Moon's declination was S 00° 25.8'. Therefore, since the declination was increasing over this period, the *d* correction is positive. Do not determine the sign of this correction by noting the trend in the *d* factor. In other words, had the *d* factor for 11-00-00 been a value less than 12.1, that would not indicate that the *d* correction should be negative. Remember that the *d* factor is analogous to an interpolation factor; it provides a correction to declination. Therefore, the trend in declination values, not the trend in *d* values, controls the sign of the *d* correction. Combine the tabulated declination and the *d* correction factor to determine the true declination. In this case, the Moon's true declination is S 00° 13.8'.

Having obtained the Moon's GHA and declination, calculate LHA and determine the assumed latitude. Enter the *Sight Reduction Table* with the LHA, assumed latitude, and calculated declination. Calculate the intercept and azimuth in the same manner used for star and Sun sights.

2009. Reducing a Planet Sight

There are four navigational planets: Venus, Mars, Jupiter, and Saturn. Reducing a planet sight is similar to reducing a Sun or star sight, but there are a few important differences. This Article will cover the procedure for determining h_o, the GHA and the declination for a planet sight.

On July 27, 1995, at 09-45-20 GMT, you take a sight of Mars. H_s is 33° 20.5'. The height of eye is 25 feet, and the index correction is +0.2'. Determine h_o, GHA, and declination. See Figure 2009.

The table above demonstrates the similarity between reducing planet sights and reducing sights of the Sun and stars. Calculate and apply the index and dip corrections exactly as for any other sight. Take the resulting apparent altitude and enter the altitude correction table for the stars and planets on the inside front cover of the *Nautical Almanac*.

In this case, the altitude correction for 33° 15.8' results in a correction of -1.5'. The additional correction is not applicable because the sight was taken at standard temperature and pressure; the horizontal parallax correction is not applicable to a planet sight. All that remains is the correction specific to Mars or Venus. The altitude correction table in the *Nautical Almanac* also contains this correction. Its magnitude is a function of the body sighted (Mars or Venus), the time of year, and the body's apparent altitude. Entering this table with the data for this problem yields a correction of +0.1'. Applying these cor-

1994 JUNE 15, 16, 17 (WED., THURS., FRI.)

UT (GMT)	SUN G.H.A.	SUN Dec.	MOON G.H.A.	v	MOON Dec.	d	H.P.
d h	° ′	° ′	° ′	′	° ′	′	′
15 00	179 55.6	N23 17.2	112 24.4	11.8	N 6 28.5	11.4	57.6
01	194 55.5	17.4	126 55.2	11.8	6 17.1	11.3	57.6
02	209 55.4	17.5	141 26.0	11.8	6 05.8	11.4	57.6
03	224 55.2	. . 17.6	155 56.8	11.7	5 54.4	11.4	57.7
04	239 55.1	17.7	170 27.5	11.8	5 43.0	11.5	57.7
05	254 55.0	17.8	184 58.3	11.8	5 31.5	11.5	57.7
06	269 54.8	N23 17.9	199 29.1	11.7	N 5 20.0	11.6	57.7
W 07	284 54.7	18.0	213 59.8	11.8	5 08.4	11.5	57.8
E 08	299 54.6	18.1	228 30.6	11.7	4 56.9	11.7	57.8
D 09	314 54.4	. . 18.2	243 01.3	11.8	4 45.2	11.6	57.8
N 10	329 54.3	18.3	257 32.1	11.7	4 33.6	11.7	57.8
E 11	344 54.2	18.5	272 02.8	11.7	4 21.9	11.7	57.9
S 12	359 54.0	N23 18.6	286 33.5	11.7	N 4 10.2	11.8	57.9
D 13	14 53.9	18.7	301 04.2	11.7	3 58.4	11.7	57.9
A 14	29 53.8	18.8	315 34.9	11.7	3 46.7	11.8	57.9
Y 15	44 53.6	. . 18.9	330 05.6	11.7	3 34.9	11.9	58.0
16	59 53.5	19.0	344 36.3	11.6	3 23.0	11.8	58.0
17	74 53.4	19.1	359 06.9	11.6	3 11.2	11.9	58.0
18	89 53.2	N23 19.2	13 37.5	11.7	N 2 59.3	11.9	58.0
19	104 53.1	19.3	28 08.2	11.6	2 47.4	12.0	58.1
20	119 53.0	19.4	42 38.8	11.6	2 35.4	11.9	58.1
21	134 52.8	. . 19.5	57 09.4	11.5	2 23.5	12.0	58.1
22	149 52.7	19.6	71 39.9	11.6	2 11.5	12.0	58.1
23	164 52.6	19.7	86 10.5	11.5	1 59.5	12.0	58.2
16 00	179 52.4	N23 19.8	100 41.0	11.5	N 1 47.5	12.1	58.2
01	194 52.3	19.9	115 11.5	11.5	1 35.4	12.0	58.2
02	209 52.2	20.0	129 42.0	11.5	1 23.4	12.1	58.2
03	224 52.0	. . 20.1	144 12.5	11.5	1 11.3	12.1	58.3
04	239 51.9	20.1	158 43.0	11.4	0 59.2	12.1	58.3
05	254 51.7	20.2	173 13.4	11.4	0 47.1	12.1	58.3
06	269 51.6	N23 20.3	187 43.8	11.4	N 0 35.0	12.2	58.3
07	284 51.5	20.4	202 14.2	11.3	0 22.8	12.1	58.4
T 08	299 51.3	20.5	216 44.5	11.3	N 0 10.7	12.2	58.4
H 09	314 51.2	. . 20.6	231 14.8	11.3	S 0 01.5	12.2	58.4
U 10	329 51.1	20.7	245 45.1	11.3	0 13.7	12.1	58.4
R 11	344 50.9	20.8	260 15.4	11.2	0 25.8	12.2	58.5
S 12	359 50.8	N23 20.9	274 45.6	11.3	S 0 38.0	12.2	58.5
D 13	14 50.7	21.0	289 15.9	11.1	0 50.2	12.2	58.5
A 14	29 50.5	21.1	303 46.0	11.2	1 02.4	12.2	58.5
Y 15	44 50.4	. . 21.1	318 16.2	11.1	1 14.6	12.2	58.6
16	59 50.3	21.2	332 46.3	11.1	1 26.8	12.3	58.6
17	74 50.1	21.3	347 16.4	11.0	1 39.1	12.2	58.6
18	89 50.0	N23 21.4	1 46.4	11.0	S 1 51.3	12.2	58.6
19	104 49.9	21.5	16 16.4	11.0	2 03.5	12.2	58.7
20	119 49.7	21.6	30 46.4	10.9	2 15.7	12.2	58.7
21	134 49.6	. . 21.6	45 16.3	10.9	2 27.9	12.2	58.7
22	149 49.5	21.7	59 46.2	10.9	2 40.1	12.2	58.7
23	164 49.3	21.8	74 16.1	10.8	2 52.3	12.2	58.8
17 00	179 49.2	N23 21.9	88 45.9	10.8	S 3 04.5	12.2	58.8
01	194 49.1	22.0	103 15.7	10.7	3 16.7	12.2	58.8
02	209 48.9	22.0	117 45.4	10.7	3 28.9	12.2	58.8
03	224 48.8	. . 22.1	132 15.1	10.7	3 41.1	12.2	58.9
04	239 48.6	22.2	146 44.8	10.6	3 53.3	12.1	58.9
05	254 48.5	22.3	161 14.4	10.6	4 05.4	12.2	58.9
06	269 48.4	N23 22.3	175 44.0	10.5	S 4 17.6	12.1	58.9
07	284 48.2	22.4	190 13.5	10.5	4 29.7	12.1	59.0
08	299 48.1	22.5	204 43.0	10.4	4 41.8	12.1	59.0
F 09	314 48.0	. . 22.6	219 12.4	10.4	4 53.9	12.1	59.0
R 10	329 47.8	22.6	233 41.8	10.3	5 06.0	12.1	59.0
I 11	344 47.7	22.7	248 11.1	10.3	5 18.1	12.1	59.1
D 12	359 47.6	N23 22.8	262 40.4	10.2	S 5 30.2	12.0	59.1
A 13	14 47.4	22.9	277 09.6	10.2	5 42.2	12.0	59.1
Y 14	29 47.3	22.9	291 38.8	10.1	5 54.2	12.0	59.1
15	44 47.2	. . 23.0	306 07.9	10.1	6 06.2	11.9	59.2
16	59 47.0	23.1	320 37.0	10.0	6 18.1	12.0	59.2
17	74 46.9	23.1	335 06.0	10.0	6 30.1	11.9	59.2
18	89 46.8	N23 23.2	349 35.0	9.9	S 6 42.0	11.9	59.2
19	104 46.6	23.3	4 03.9	9.9	6 53.9	11.8	59.3
20	119 46.5	23.3	18 32.8	9.8	7 05.7	11.8	59.3
21	134 46.3	. . 23.4	33 01.6	9.7	7 17.5	11.8	59.3
22	149 46.2	23.5	47 30.3	9.7	7 29.3	11.8	59.3
23	164 46.1	23.5	61 59.0	9.6	7 41.1	11.7	59.4
	S.D. 15.8	d 0.1	S.D. 15.8		15.9		16.1

Lat.	Twilight Naut.	Twilight Civil	Sunrise	Moonrise 15	Moonrise 16	Moonrise 17	Moonrise 18
°	h m	h m	h m	h m	h m	h m	h m
N 72	▭	▭	▭	09 54	11 50	13 49	15 56
N 70	▭	▭	▭	10 01	11 49	13 40	15 36
68	▭	▭	▭	10 06	11 48	13 33	15 21
66	▭	▭	▭	10 11	11 47	13 26	15 09
64	////	////	01 33	10 14	11 47	13 21	14 59
62	////	////	02 10	10 18	11 46	13 17	14 50
60	////	00 52	02 36	10 20	11 46	13 13	14 43
N 58	////	01 41	02 56	10 23	11 45	13 10	14 37
56	////	02 11	03 13	10 25	11 45	13 07	14 31
54	00 48	02 33	03 27	10 27	11 45	13 04	14 26
52	01 33	02 51	03 39	10 29	11 44	13 02	14 21
50	02 00	03 06	03 50	10 30	11 44	13 00	14 17
45	02 46	03 35	04 13	10 34	11 44	12 55	14 08
N 40	03 16	03 58	04 31	10 37	11 43	12 51	14 01
35	03 39	04 16	04 46	10 39	11 43	12 48	13 54
30	03 58	04 31	04 59	10 42	11 42	12 45	13 49
20	04 27	04 56	05 20	10 45	11 42	12 40	13 39
N 10	04 49	05 16	05 39	10 49	11 41	12 35	13 31
0	05 08	05 34	05 57	10 52	11 41	12 31	13 23
S 10	05 25	05 51	06 14	10 55	11 40	12 27	13 16
20	05 41	06 09	06 33	10 59	11 40	12 23	13 07
30	05 58	06 28	06 54	11 03	11 40	12 18	12 58
35	06 06	06 38	07 06	11 05	11 39	12 15	12 53
40	06 16	06 50	07 20	11 07	11 39	12 12	12 47
45	06 26	07 03	07 37	11 10	11 39	12 08	12 40
S 50	06 38	07 19	07 58	11 14	11 38	12 04	12 32
52	06 43	07 27	08 08	11 15	11 38	12 02	12 28
54	06 49	07 35	08 19	11 17	11 38	12 00	12 24
56	06 55	07 44	08 31	11 19	11 38	11 58	12 19
58	07 02	07 54	08 46	11 21	11 38	11 55	12 14
S 60	07 09	08 06	09 03	11 24	11 37	11 52	12 09

Lat.	Sunset	Twilight Civil	Twilight Naut.	Moonset 15	Moonset 16	Moonset 17	Moonset 18
°	h m	h m	h m	h m	h m	h m	h m
N 72	▭	▭	▭	23 38	23 25	23 12	22 54
N 70	▭	▭	▭	23 35	23 29	23 23	23 16
68	▭	▭	▭	23 34	23 33	23 32	23 32
66	▭	▭	▭	23 32	23 36	23 40	23 46
64	22 29	////	////	23 31	23 38	23 47	23 57
62	21 52	////	////	23 29	23 40	23 52	24 07
60	21 26	23 10	////	23 28	23 42	23 57	24 15
N 58	21 05	22 21	////	23 27	23 44	24 02	00 02
56	20 49	21 51	////	23 26	23 45	24 06	00 06
54	20 34	21 29	23 15	23 26	23 47	24 09	00 09
52	20 22	21 11	22 29	23 25	23 48	24 13	00 13
50	20 11	20 56	22 01	23 24	23 49	24 15	00 15
45	19 49	20 26	21 16	23 23	23 52	24 22	00 22
N 40	19 31	20 04	20 45	23 22	23 54	24 27	00 27
35	19 16	19 45	20 22	23 21	23 55	24 32	00 32
30	19 03	19 30	20 04	23 20	23 57	24 36	00 36
20	18 41	19 05	19 35	23 18	24 00	00 00	00 43
N 10	18 22	18 45	19 12	23 17	24 02	00 02	00 50
0	18 04	18 27	18 53	23 15	24 05	00 05	00 55
S 10	17 47	18 10	18 36	23 14	24 07	00 07	01 01
20	17 28	17 52	18 20	23 12	24 09	00 09	01 08
30	17 07	17 34	18 03	23 10	24 12	00 12	01 15
35	16 55	17 23	17 55	23 09	24 13	00 13	01 19
40	16 41	17 11	17 45	23 08	24 15	00 15	01 24
45	16 24	16 58	17 35	23 07	24 17	00 17	01 29
S 50	16 03	16 42	17 23	23 05	24 20	00 20	01 36
52	15 53	16 34	17 18	23 04	24 21	00 21	01 39
54	15 42	16 26	17 12	23 04	24 22	00 22	01 42
56	15 30	16 17	17 06	23 03	24 23	00 23	01 46
58	15 15	16 07	16 59	23 02	24 25	00 25	01 50
S 60	14 58	15 55	16 52	23 00	24 27	00 27	01 55

Day	SUN Eqn. of Time 00ʰ	SUN Eqn. of Time 12ʰ	SUN Mer. Pass.	MOON Mer. Pass. Upper	MOON Mer. Pass. Lower	Age	Phase
	m s	m s	h m	h m	h m	d	
15	00 17	00 24	12 00	17 04	04 39	06	◐
16	00 30	00 37	12 01	17 53	05 28	07	
17	00 43	00 49	12 01	18 43	06 18	08	

Figure 2008. Right hand daily page of the Nautical Almanac for June 16, 1994.

Body	Mars
Index Correction	+0.2'
Dip Correction (25 feet)	-4.9'
Sum	-4.7'
h_s	33° 20.5'
h_a	33° 15.8'
Altitude Correction	-1.5'
Additional Correction	Not applicable
Horizontal Parallax	Not applicable
Additional Correction for Mars	+0.1'
Correction to h_a	-1.4'
h_o	33° 14.4'

rections to h_a results in an h_o of 33° 14.4'.

Tabulated GHA / *v*	256°10.6' / 1.1
GHA Increment	11° 20.0'
v correction	+0.8'
GHA	267°31.4'

The only difference between determining the Sun's GHA and a planet's GHA lies in applying the *v* correction. Calculate this correction from the *v* or *d* correction section of the Increments and Correction table in the *Nautical Almanac*.
Find the *v* factor at the bottom of the planets' GHA columns on the daily pages of the *Nautical Almanac*. For Mars on July 27, 1995, the *v* factor is 1.1. If no algebraic sign precedes the *v* factor, add the resulting correction to the tabulated GHA. Subtract the resulting correction only when a negative sign precedes the *v* factor. Entering the *v* or *d* correction table corresponding to 45 minutes yields a correction of 0.8'. Remember, because no sign preceded the *v* factor on the daily pages, add this correction to the tabulated GHA. The final GHA is 267°31.4'.

Tabulated Declination / *d*	S 01° 06.1' / 0.6
d Correction	+0.5'
True Declination	S 01° 06.6'

Read the tabulated declination directly from the daily pages of the *Nautical Almanac*. The *d* correction factor is listed at the bottom of the planet column; in this case, the factor is 0.6. Note the trend in the declination values for the planet; if they are increasing during the day, the correction factor is positive. If the planet's declination is decreasing during the day, the correction factor is negative. Next, enter the *v* or *d* correction table corresponding to 45 minutes and extract the correction for a *d* factor of 0.6. The correction in this case is +0.5'.

From this point, reducing a planet sight is exactly the same as reducing a Sun sight.

MERIDIAN PASSAGE

This section covers determining both latitude and longitude at the meridian passage of the Sun, or Local Apparent Noon (LAN). Determining a vessel's latitude at LAN requires calculating the Sun's zenith distance and declination and combining them according to the rules discussed below.

Latitude at LAN is a special case of the navigational triangle where the Sun is on the observer's meridian and the triangle becomes a straight north/south line. No "solution" is necessary, except to combine the Sun's zenith distance and its declination according to the rules discussed below.

Longitude at LAN is a function of the time elapsed since the Sun passed the Greenwich meridian. The navigator must determine the time of LAN and calculate the GHA of the Sun at that time. The following examples demonstrates these processes.

2010. Latitude at Meridian Passage

At 1056 ZT, May 16, 1995, a vessel's DR position is L 40° 04.3'N and λ 157° 18.5' W. The ship is on course 200°T at a speed of ten knots. (1) Calculate the first and second estimates of Local Apparent Noon. (2) The navigator actually observes LAN at 12-23-30 zone time. The sextant altitude at LAN is 69° 16.0'. The index correction is +2.1' and the height of eye is 45 feet. Determine the vessel's latitude.

First, determine the time of meridian passage from the daily pages of the *Nautical Almanac*. In this case, the meridian passage for May 16, 1995, is 1156. That is, the Sun crosses the central meridian of the time zone at 1156 ZT and the observer's local meridian at 1156 local time. Next, determine the vessel's DR longitude for the time of meridian passage. In this case, the vessel's 1156 DR longitude is 157° 23.0' W. Determine the time zone in which this DR longitude falls and record the longitude of that time zone's central meridian. In this case, the central meridian is 150° W. Enter the Conversion of Arc to Time table in the *Nautical Almanac* with the difference between the DR longitude and the central meridian longitude. The conversion for 7° of arc is 28^m of time, and the conversion for 23' of arc is 1^m32^s of time. Sum these two times. If the DR position is west of the central meridian (as it is in this case), add this time to the time of tabulated meridian passage. If the longitude difference is to the east of the central meridian, subtract this time from the tabulated meridian passage. In this case, the DR position is west of the central meridian. Therefore, add 29 minutes and 32 seconds to 1156, the tabulated time of meridian passage. The estimated time of LAN is 12-25-32 ZT.

This first estimate for LAN does not take into account the vessel's movement. To calculate the second estimate of LAN, first determine the DR longitude for the time of first estimate of LAN (12 25 32 ZT). In this case, that longitude would be 157° 25.2' W. Then, calculate the difference between the longitude of the 12-25-32 DR position and the central meridian longitude. This would be 7° 25.2'. Again, enter the arc to time conversion table and calculate the time difference corresponding to this

1995 JULY 27, 28, 29 (THURS., FRI., SAT.)

UT (GMT) d h	ARIES G.H.A.	VENUS −3.9 G.H.A.	VENUS Dec.	MARS +1.3 G.H.A.	MARS Dec.	JUPITER −2.3 G.H.A.	JUPITER Dec.	SATURN +1.0 G.H.A.	SATURN Dec.
	° ′	° ′	° ′	° ′	° ′	° ′	° ′	° ′	° ′
27 00	304 12.4	185 23.5	N21 31.7	121 00.7	S 1 00.4	60 23.8	S20 36.7	308 27.9	S 4 15.6
01	319 14.9	200 22.7	31.2	136 01.8	01.0	75 26.3	36.7	323 30.4	15.7
02	334 17.3	215 21.9	30.7	151 02.9	01.7	90 28.8	36.7	338 33.0	15.7
03	349 19.8	230 21.1	· · 30.2	166 04.0	· · 02.3	105 31.3	· · 36.7	353 35.5	· · 15.8
04	4 22.3	245 20.4	29.7	181 05.1	02.9	120 33.8	36.7	8 38.1	15.8
05	19 24.7	260 19.6	29.2	196 06.2	03.6	135 36.4	36.7	23 40.6	15.8
06	34 27.2	275 18.8	N21 28.7	211 07.3	S 1 04.2	150 38.9	S20 36.7	38 43.1	S 4 15.9
07	49 29.7	290 18.0	28.2	226 08.4	04.8	165 41.4	36.7	53 45.7	15.9
T 08	64 32.1	305 17.2	27.7	241 09.5	05.4	180 43.9	36.7	68 48.2	16.0
H 09	79 34.6	320 16.4	· · 27.1	256 10.6	· · 06.1	195 46.4	· · 36.7	83 50.7	· · 16.0
U 10	94 37.1	335 15.6	26.6	271 11.7	06.7	210 48.9	36.7	98 53.3	16.1
R 11	109 39.5	350 14.8	26.1	286 12.8	07.3	225 51.5	36.7	113 55.8	16.1
S 12	124 42.0	5 14.0	N21 25.6	301 13.9	S 1 08.0	240 54.0	S20 36.7	128 58.4	S 4 16.1
D 13	139 44.4	20 13.2	25.1	316 15.0	08.6	255 56.5	36.7	144 00.9	16.2
A 14	154 46.9	35 12.4	24.5	331 16.1	09.2	270 59.0	36.7	159 03.4	16.2
Y 15	169 49.4	50 11.6	· · 24.0	346 17.2	· · 09.8	286 01.5	· · 36.7	174 06.0	· · 16.3
16	184 51.8	65 10.8	23.5	1 18.3	10.5	301 04.0	36.7	189 08.5	16.3
17	199 54.3	80 10.0	23.0	16 19.4	11.1	316 06.6	36.7	204 11.1	16.4
18	214 56.8	95 09.2	N21 22.5	31 20.6	S 1 11.7	331 09.1	S20 36.7	219 13.6	S 4 16.4
19	229 59.2	110 08.4	21.9	46 21.7	12.4	346 11.6	36.7	234 16.1	16.4
20	245 01.7	125 07.7	21.4	61 22.8	13.0	1 14.1	36.7	249 18.7	16.5
21	260 04.2	140 06.9	· · 20.9	76 23.9	· · 13.6	16 16.6	· · 36.7	264 21.2	· · 16.5
22	275 06.6	155 06.1	20.3	91 25.0	14.2	31 19.1	36.7	279 23.8	16.6
23	290 09.1	170 05.3	19.8	106 26.1	14.9	46 21.6	36.7	294 26.3	16.6
28 00	305 11.6	185 04.5	N21 19.3	121 27.2	S 1 15.5	61 24.1	S20 36.7	309 28.8	S 4 16.7
01	320 14.0	200 03.7	18.8	136 28.3	16.1	76 26.7	36.7	324 31.4	16.7
02	335 16.5	215 02.9	18.2	151 29.4	16.8	91 29.2	36.7	339 33.9	16.8
03	350 18.9	230 02.1	· · 17.7	166 30.5	· · 17.4	106 31.7	· · 36.7	354 36.5	· · 16.8
04	5 21.4	245 01.3	17.1	181 31.6	18.0	121 34.2	36.7	9 39.0	16.8
05	20 23.9	260 00.6	16.6	196 32.7	18.6	136 36.7	36.7	24 41.5	16.9
06	35 26.3	274 59.8	N21 16.1	211 33.8	S 1 19.3	151 39.2	S20 36.7	39 44.1	S 4 16.9
07	50 28.8	289 59.0	15.5	226 34.9	19.9	166 41.7	36.7	54 46.6	17.0
08	65 31.3	304 58.2	15.0	241 36.0	20.5	181 44.2	36.7	69 49.2	17.0
F 09	80 33.7	319 57.4	· · 14.5	256 37.1	· · 21.2	196 46.7	· · 36.7	84 51.7	· · 17.1
R 10	95 36.2	334 56.6	13.9	271 38.2	21.8	211 49.2	36.7	99 54.3	17.1
I 11	110 38.7	349 55.8	13.4	286 39.3	22.4	226 51.8	36.7	114 56.8	17.2
D 12	125 41.1	4 55.1	N21 12.8	301 40.4	S 1 23.0	241 54.3	S20 36.7	129 59.3	S 4 17.2
A 13	140 43.6	19 54.3	12.3	316 41.5	23.7	256 56.8	36.7	145 01.9	17.2
Y 14	155 46.1	34 53.5	11.7	331 42.6	24.3	271 59.3	36.7	160 04.4	17.3
15	170 48.5	49 52.7	· · 11.2	346 43.6	· · 24.9	287 01.8	· · 36.7	175 07.0	· · 17.3
16	185 51.0	64 51.9	10.6	1 44.7	25.6	302 04.3	36.7	190 09.5	17.4
17	200 53.4	79 51.1	10.1	16 45.8	26.2	317 06.8	36.7	205 12.1	17.4
18	215 55.9	94 50.4	N21 09.5	31 46.9	S 1 26.8	332 09.3	S20 36.7	220 14.6	S 4 17.5
19	230 58.4	109 49.6	09.0	46 48.0	27.5	347 11.8	36.7	235 17.1	17.5
20	246 00.8	124 48.8	08.4	61 49.1	28.1	2 14.3	36.7	250 19.7	17.6
21	261 03.3	139 48.0	· · 07.9	76 50.2	· · 28.7	17 16.8	· · 36.7	265 22.2	· · 17.6
22	276 05.8	154 47.2	07.3	91 51.3	29.3	32 19.3	36.7	280 24.8	17.6
23	291 08.2	169 46.4	06.8	106 52.4	30.0	47 21.8	36.7	295 27.3	17.7
29 00	306 10.7	184 45.7	N21 06.2	121 53.5	S 1 30.6	62 24.3	S20 36.8	310 29.9	S 4 17.7
01	321 13.2	199 44.9	05.7	136 54.6	31.2	77 26.8	36.8	325 32.4	17.8
02	336 15.6	214 44.1	05.1	151 55.7	31.9	92 29.3	36.8	340 34.9	17.8
03	351 18.1	229 43.3	· · 04.5	166 56.8	· · 32.5	107 31.8	· · 36.8	355 37.5	· · 17.9
04	6 20.5	244 42.5	04.0	181 57.9	33.1	122 34.3	36.8	10 40.0	17.9
05	21 23.0	259 41.8	03.4	196 59.0	33.8	137 36.8	36.8	25 42.6	18.0
06	36 25.5	274 41.0	N21 02.9	212 00.1	S 1 34.4	152 39.3	S20 36.8	40 45.1	S 4 18.0
07	51 27.9	289 40.2	02.3	227 01.2	35.0	167 41.8	36.8	55 47.7	18.1
S 08	66 30.4	304 39.4	01.7	242 02.3	35.6	182 44.4	36.8	70 50.2	18.1
A 09	81 32.9	319 38.7	· · 01.2	257 03.4	· · 36.3	197 46.9	· · 36.8	85 52.8	· · 18.1
T 10	96 35.3	334 37.9	00.6	272 04.5	36.9	212 49.4	36.8	100 55.3	18.2
U 11	111 37.8	349 37.1	21 00.0	287 05.6	37.5	227 51.9	36.8	115 57.9	18.2
R 12	126 40.3	4 36.3	N20 59.4	302 06.7	S 1 38.2	242 54.4	S20 36.8	131 00.4	S 4 18.3
D 13	141 42.7	19 35.6	58.9	317 07.8	38.8	257 56.9	36.8	146 02.9	18.3
A 14	156 45.2	34 34.8	58.3	332 08.9	39.4	272 59.4	36.8	161 05.5	18.4
Y 15	171 47.7	49 34.0	· · 57.7	347 10.0	· · 40.1	288 01.8	· · 36.8	176 08.0	· · 18.4
16	186 50.1	64 33.2	57.2	2 11.0	40.7	303 04.3	36.8	191 10.6	18.5
17	201 52.6	79 32.5	56.6	17 12.1	41.3	318 06.8	36.8	206 13.1	18.5
18	216 55.0	94 31.7	N20 56.0	32 13.2	S 1 42.0	333 09.3	S20 36.8	221 15.7	S 4 18.6
19	231 57.5	109 30.9	55.4	47 14.3	42.6	348 11.8	36.8	236 18.2	18.6
20	247 00.0	124 30.1	54.9	62 15.4	43.2	3 14.3	36.8	251 20.8	18.6
21	262 02.4	139 29.4	· · 54.3	77 16.5	· · 43.8	18 16.8	· · 36.8	266 23.3	· · 18.7
22	277 04.9	154 28.6	53.7	92 17.6	44.5	33 19.3	36.8	281 25.9	18.7
23	292 07.4	169 27.8	53.1	107 18.7	45.1	48 21.8	36.8	296 28.4	18.8
Mer. Pass.	h m 3 38.6	v −0.8	d 0.5	v 1.1	d 0.6	v 2.5	d 0.0	v 2.5	d 0.0

STARS Name	S.H.A.	Dec.
	° ′	° ′
Acamar	315 28.7	S40 19.1
Achernar	335 36.7	S57 15.2
Acrux	173 24.6	S63 04.8
Adhara	255 23.4	S28 58.0
Aldebaran	291 05.0	N16 29.9
Alioth	166 32.7	N55 59.3
Alkaid	153 09.6	N49 20.4
Al Na'ir	28 00.2	S46 58.7
Alnilam	276 00.3	S 1 12.3
Alphard	218 09.6	S 8 38.4
Alphecca	126 22.3	N26 44.0
Alpheratz	357 57.2	N29 04.0
Altair	62 21.0	N 8 51.6
Ankaa	353 28.8	S42 19.5
Antares	112 42.5	S26 25.3
Arcturus	146 08.0	N19 12.5
Atria	107 56.1	S69 01.3
Avior	234 24.1	S59 29.8
Bellatrix	278 46.7	N 6 20.7
Betelgeuse	271 16.1	N 7 24.3
Canopus	264 02.6	S52 41.6
Capella	280 54.7	N45 59.4
Deneb	49 40.1	N45 16.0
Denebola	182 47.6	N14 35.9
Diphda	349 09.2	S18 00.4
Dubhe	194 08.7	N61 46.6
Elnath	278 29.9	N28 36.1
Eltanin	90 51.9	N51 29.7
Enif	34 00.0	N 9 51.5
Fomalhaut	15 38.5	S29 38.5
Gacrux	172 16.1	S57 05.6
Gienah	176 06.3	S17 31.1
Hadar	149 07.0	S60 21.3
Hamal	328 15.9	N23 26.4
Kaus Aust.	84 01.3	S34 23.1
Kochab	137 19.4	N74 10.8
Markab	13 51.5	N15 11.0
Menkar	314 29.2	N 4 04.4
Menkent	148 23.4	S36 21.0
Miaplacidus	221 43.3	S69 42.1
Mirfak	308 59.8	N49 50.5
Nunki	76 14.6	S26 18.0
Peacock	53 39.8	S56 44.8
Pollux	243 44.5	N28 02.1
Procyon	245 14.1	N 5 14.1
Rasalhague	96 18.7	N12 34.0
Regulus	207 58.1	N11 59.3
Rigel	281 25.2	S 8 12.4
Rigil Kent.	140 10.0	S60 49.2
Sabik	102 27.7	S15 43.1
Schedar	349 55.6	N56 30.6
Shaula	96 39.8	S37 06.0
Sirius	258 45.9	S16 42.6
Spica	158 45.5	S11 08.3
Suhail	223 02.7	S43 25.0
Vega	80 47.7	N38 47.1
Zuben'ubi	137 20.3	S16 01.4

	S.H.A.	Mer. Pass.
	° ′	h m
Venus	239 52.9	11 40
Mars	176 15.6	15 53
Jupiter	116 12.6	19 51
Saturn	4 17.3	3 22

Figure 2009. Left hand daily page of the Nautical Almanac for July 27, 1995.

Date	16 May 1995
DR Latitude (1156 ZT)	39° 55.0' N
DR Longitude (1156 ZT)	157° 23.0' W
Central Meridian	150° W
d Longitude (arc)	7° 23' W
d Longitude (time)	+29 min. 32 sec
Meridian Passage (LMT)	1156
ZT (first estimate)	12-25-32
DR Longitude (12-25-32)	157° 25.2'
d Longitude (arc)	7° 25.2'
d Longitude (time)	+29 min. 41 sec
Meridian Passage	1156
ZT (second estimate)	12-25-41
ZT (actual transit)	12-23-30 local
Zone Description	+10
GMT	22-23-30
Date (GMT)	16 May 1995
Tabulated Declination / *d*	N 19° 09.0' / +0.6
d correction	+0.2'
True Declination	N 19° 09.2'
Index Correction	+2.1'
Dip (48 ft)	-6.7'
Sum	-4.6'
h_s (at LAN)	69° 16.0'
h_a	69° 11.4'
Altitude Correction	+15.6'
89° 60'	89° 60.0'
h_o	69° 27.0'
Zenith Distance	N 20° 33.0'
True Declination	N 19° 09.2'
Latitude	39° 42.2'

longitude difference. The correction for 7° of arc is 28' of time, and the correction for 25.2' of arc is 1'41" of time. Finally, apply this time correction to the original tabulated time of meridian passage (1156 ZT). The resulting time, 12-25-41 ZT, is the second estimate of LAN.

Solving for latitude requires that the navigator calculate two quantities: the Sun's declination and the Sun's zenith distance. First, calculate the Sun's true declination at LAN. The problem states that LAN is 12-28-30. (Determining the exact time of LAN is covered in Article 2011.) Enter the time of observed LAN and add the correct zone description to determine GMT. Determine the Sun's declination in the same manner as in the sight reduction problem in Article 2006. In this case, the tabulated declination was N 19° 19.1', and the d correction +0.2'. The true declination, therefore, is N 19° 19.3'.

Next, calculate zenith distance. Recall from Navigational Astronomy that zenith distance is simply 90° - observed altitude. Therefore, correct h_s to obtain h_a; then correct h_a to obtain h_o. Then, subtract h_o from 90° to determine the zenith distance. Name the zenith distance North or South depending on the relative position of the observer and the Sun's declination. If the observer is to the north of the Sun's declination, name the zenith distance north. Conversely, if the observer is to the south of the Sun's declination, name the zenith distance south. In this case, the DR latitude is N 39° 55.0' and the Sun's declination is N 19° 19.3'. The observer is to the north of the Sun's declination; therefore, name the zenith distance north. Next, compare the names of the zenith distance and the declination. If their names are the same (i.e., both are north or both are south), add the two values together to obtain the latitude. This was the case in this problem. Both the Sun's declination and zenith distance were north; therefore, the observer's latitude is the sum of the two.

If the name of the body's zenith distance is contrary to the name of the Sun's declination, then subtract the smaller of the two quantities from the larger, carrying for the name of the difference the name of the larger of the two quantities. The result is the observer's latitude. The following examples illustrate this process.

Zenith Distance	N 25°	Zenith Distance	S 50°
True Declination	S 15°	True Declination	N10°
Latitude	N 10°	Latitude	S 40°

2011. Longitude at Meridian Passage

Determining a vessel's longitude at LAN is straightforward. In the western hemisphere, the Sun's GHA at LAN equals the vessel's longitude. In the eastern hemisphere, subtract the Sun's GHA from 360° to determine longitude. The difficult part lies in determining the precise moment of meridian passage.

Determining the time of meridian passage presents a problem because the Sun appears to hang for a finite time at its local maximum altitude. Therefore, noting the time of maximum sextant altitude is not sufficient for determining the precise time of LAN. Two methods are available to obtain LAN with a precision sufficient for determining longitude: (1) the graphical method and (2) the calculation method. The graphical method is discussed first below.

See Figure 2011. For about 30 minutes before the estimated time of LAN, measure and record several sextant altitudes and their corresponding times. Continue taking sights for about 30 minutes after the Sun has descended from the maximum recorded altitude. Increase the sighting frequency near the meridian passage. One sight every 20-30 seconds should yield good results near meridian passage; less frequent sights are required before and after.

Plot the resulting data on a graph of sextant altitude versus time and draw a fair curve through the plotted data. Next, draw a series of horizontal lines across the curve formed by the data points. These lines will intersect the faired curve at two different points. The x coordinates of the points where these lines intersect the faired curve represent the two different times when the Sun's altitude was equal (one time when the Sun was ascending; the other time when the Sun was descending). Draw three such lines, and ensure the lines have sufficient vertical separation. For each line, average the two times where it intersects the faired curve. Finally, average the three resulting times to obtain a final value

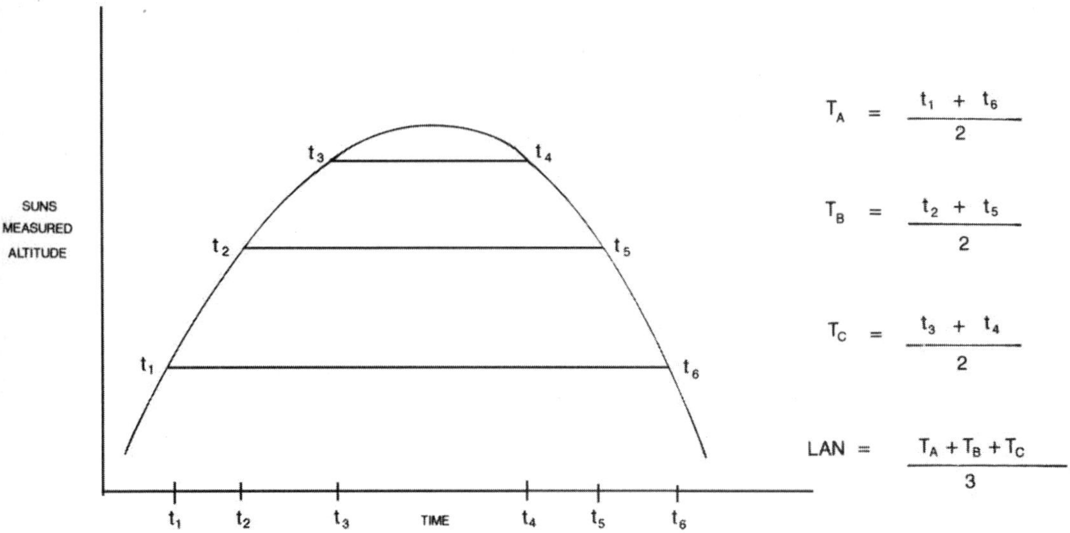

Figure 2011. Time of LAN.

for the time of LAN. From the *Nautical Almanac*, determine the Sun's GHA at that time; this is your longitude in the western hemisphere. In the eastern hemisphere, subtract the Sun's GHA from 360° to determine longitude. For a quicker but less exact time, simply drop a perpendicular from the apex of the curve and read the time along the time scale.

The second method of determining LAN is similar to the first. Estimate the time of LAN as discussed above, Measure and record the Sun's altitude as the Sun approaches its maximum altitude. As the Sun begins to descend, set the sextant to correspond to the altitude recorded just before the Sun's reaching its maximum altitude. Note the time when the Sun is again at that altitude. Average the two times. Repeat this procedure with two other altitudes recorded before LAN, each time presetting the sextant to those altitudes and recording the corresponding times that the Sun, now on its descent, passes through those altitudes. Average these corresponding times. Take a final average among the three averaged times; the result will be the time of meridian passage. Determine the vessel's longitude by determining the Sun's GHA at the exact time of LAN.

LATITUDE BY POLARIS

2012. Latitude by Polaris

Since Polaris is always within about 1° of the North Pole, the altitude of Polaris, with a few minor corrections, equals the latitude of the observer. This relationship makes Polaris an extremely important navigational star in the northern hemisphere.

The corrections are necessary because Polaris orbits in a small circle around the pole. When Polaris is at the exact same altitude as the pole, the correction is zero. At two points in its orbit it is in a direct line with the observer and the pole, either nearer than or beyond the pole. At these points the corrections are maximum. The following example illustrates converting a Polaris sight to latitude.

At 23-18-56 GMT, on April 21, 1994, at DR Lat. 50° 23.8' N, λ=37° 14.0' W, the observed altitude of Polaris (h_o) is 49° 31.6'. Find the vessel's latitude.

To solve this problem, use the equation:

$$\text{Latitude} = h_o - 1° + A_0 + A_1 + A_2$$

where h_o is the sextant altitude (h_s) corrected as in any other star sight; 1° is a constant; and A_0, A_1, and A_2 are correction factors from the Polaris tables found in the *Nautical Almanac*. These three correction factors are always positive. One needs the following information to enter the tables: LHA of Aries, DR latitude, and the month of the year. Therefore:

Enter the Polaris table with the calculated LHA of Aries

POLARIS (POLE STAR) TABLES, 1994

FOR DETERMINING LATITUDE FROM SEXTANT ALTITUDE AND FOR AZIMUTH

LHA ARIES	120°–129°	130°–139°	140°–149°	150°–159°	160°–169°	170°–179°	180°–189°	190°–199°	200°–209°	210°–219°	220°–229°	230°–239°
	a_0	a_0	a_0	a_0	a_0	a_0	a_0	a_0	a_0	a_0	a_0	a_0
°	° ′	° ′	° ′	° ′	° ′	° ′	° ′	° ′	° ′	° ′	° ′	° ′
0	0 53·9	1 01·8	1 09·7	1 17·2	1 24·1	1 30·3	1 35·5	1 39·6	1 42·5	1 44·1	1 44·3	1 43·2
1	54·7	02·6	10·4	17·9	24·8	30·9	36·0	40·0	42·7	44·2	44·3	43·0
2	55·5	03·4	11·2	18·6	25·4	31·4	36·4	40·3	42·9	44·3	44·2	42·8
3	56·3	04·2	12·0	19·3	26·1	32·0	36·9	40·6	43·1	44·3	44·1	42·6
4	57·1	05·0	12·7	20·0	26·7	32·5	37·3	40·9	43·3	44·4	44·0	42·4
5	0 57·8	1 05·8	1 13·5	1 20·7	1 27·3	1 33·0	1 37·7	1 41·2	1 43·5	1 44·4	1 43·9	1 42·1
6	58·6	06·6	14·2	21·4	27·9	33·5	38·1	41·5	43·6	44·4	43·8	41·9
7	0 59·4	07·3	15·0	22·1	28·5	34·1	38·5	41·8	43·8	44·4	43·7	41·6
8	1 00·2	08·1	15·7	22·8	29·1	34·6	38·9	42·0	43·9	44·4	43·5	41·3
9	01·0	08·9	16·4	23·5	29·7	35·0	39·3	42·3	44·0	44·4	43·4	41·0
10	1 01·8	1 09·7	1 17·2	1 24·1	1 30·3	1 35·5	1 39·6	1 42·5	1 44·1	1 44·3	1 43·2	1 40·7
Lat.	a_1	a_1	a_1	a_1	a_1	a_1	a_1	a_1	a_1	a_1	a_1	a_1
°	′	′	′	′	′	′	′	′	′	′	′	′
0	0·2	0·2	0·3	0·3	0·4	0·4	0·5	0·6	0·6	0·6	0·6	0·6
10	·3	·3	·3	·4	·4	·5	·5	·6	·6	·6	·6	·6
20	·3	·4	·4	·4	·4	·5	·5	·6	·6	·6	·6	·6
30	·4	·4	·4	·5	·5	·5	·5	·6	·6	·6	·6	·6
40	0·5	0·5	0·5	0·5	0·5	0·6	0·6	0·6	0·6	0·6	0·6	0·6
45	·5	·5	·5	·6	·6	·6	·6	·6	·6	·6	·6	·6
50	·6	·6	·6	·6	·6	·6	·6	·6	·6	·6	·6	·6
55	·7	·7	·7	·7	·6	·6	·6	·6	·6	·6	·6	·6
60	·8	·8	·7	·7	·7	·7	·6	·6	·6	·6	·6	·6
62	0·8	0·8	0·8	0·8	0·7	0·7	0·7	0·6	0·6	0·6	0·6	0·6
64	·9	·9	·8	·8	·8	·7	·7	·6	·6	·6	·6	·6
66	0·9	0·9	·9	·8	·8	·7	·7	·6	·6	·6	·6	·6
68	1·0	1·0	0·9	0·9	0·8	0·8	0·7	0·7	0·6	0·6	0·6	0·6
Month	a_2	a_2	a_2	a_2	a_2	a_2	a_2	a_2	a_2	a_2	a_2	a_2
	′	′	′	′	′	′	′	′	′	′	′	′
Jan.	0·6	0·6	0·6	0·5	0·5	0·5	0·4	0·4	0·4	0·4	0·4	0·4
Feb.	·8	·8	·7	·7	·6	·6	·5	·5	·4	·4	·4	·3
Mar.	0·9	0·9	0·9	·8	·8	·7	·6	·6	·5	·5	·4	·4
Apr.	1·0	1·0	1·0	0·9	0·9	0·8	0·8	0·7	0·7	0·6	0·5	0·5
May	0·9	1·0	1·0	1·0	1·0	0·9	·9	·9	·8	·8	·7	·6
June	·8	0·9	0·9	0·9	0·9	1·0	·9	·9	·9	·9	·8	·8
July	0·7	0·7	0·8	0·8	0·8	0·9	0·9	0·9	0·9	0·9	0·9	0·9
Aug.	·5	·5	·6	·6	·7	·7	·8	·8	·8	·9	·9	·9
Sept.	·3	·4	·4	·5	·5	·6	·6	·7	·7	·7	·8	·8
Oct.	0·3	0·3	0·3	0·3	0·3	0·4	0·4	0·5	0·5	0·6	0·6	0·7
Nov.	·2	·2	·2	·2	·2	·2	·2	·3	·3	·4	·5	·5
Dec.	0·3	0·2	0·2	0·2	0·1	0·1	0·1	0·2	0·2	0·2	0·3	0·4
Lat.	AZIMUTH											
°	°	°	°	°	°	°	°	°	°	°	°	°
0	359·2	359·2	359·3	359·3	359·4	359·5	359·6	359·7	359·8	0·0	0·1	0·2
20	359·2	359·2	359·2	359·3	359·4	359·5	359·6	359·7	359·8	0·0	0·1	0·3
40	359·0	359·0	359·1	359·1	359·2	359·3	359·5	359·6	359·8	0·0	0·1	0·3
50	358·8	358·8	358·9	359·0	359·1	359·2	359·4	359·6	359·8	0·0	0·2	0·4
55	358·7	358·7	358·7	358·8	359·0	359·1	359·3	359·5	359·7	0·0	0·2	0·4
60	358·5	358·5	358·6	358·7	358·8	359·0	359·2	359·5	359·7	0·0	0·2	0·5
65	358·2	358·2	358·3	358·4	358·6	358·8	359·1	359·4	359·6	359·9	0·3	0·6

Figure 2012. Excerpt from the Polaris Tables.

(162° 03.5'). See Figure 2012. The first correction, A_0, is a function solely of the LHA of Aries. Enter the table column indicating the proper range of LHA of Aries; in this case, enter the 160°-169° column. The numbers on the left hand side of the A_0 correction table represent the whole degrees of LHA ♈ ; interpolate to determine the proper A_0 correction. In this case, LHA ♈ was 162° 03.5'. The A_0 correction for LHA = 162° is 1° 25.4' and the A_0 correction for LHA = 163° is 1° 26.1'. The A_0 correction for 162° 03.5' is 1° 25.4'.

LHA ♈	162° 03.5'
A_0 (162° 03.5')	+1° 25.4'
A_1 (L = 50°N)	+0.6'
A_2 (April)	+0.9'
Sum	1° 26.9'
Constant	-1° 00.0'
Observed Altitude	49° 31.6'
Total Correction	+26.9'
Latitude	N 49° 58.5'

Tabulated GHA ♈ (2300 hrs.)	194° 32.7'
Increment (18-56)	4° 44.8'
GHA ♈	199° 17.5'
DR Longitude (-W +E)	37° 14.0'

To calculate the A_1 correction, enter the A_1 correction table with the DR latitude, being careful to stay in the 160°-169° LHA column. There is no need to interpolate here; simply choose the latitude that is closest to the vessel's DR latitude. In this case, L is 50°N. The A_1 correction corresponding to an LHA range of 160°-169° and a latitude of 50°N is + 0.6'.

Finally, to calculate the A_2 correction factor, stay in the 160°-169° LHA ♈ column and enter the A_2 correction table. Follow the column down to the month of the year; in this case, it is April. The correction for April is + 0.9'.

Sum the corrections, remembering that all three are always positive. Subtract 1° from the sum to determine the total correction; then apply the resulting value to the observed altitude of Polaris. This is the vessel's latitude.

THE DAY'S WORK IN CELESTIAL NAVIGATION

2013. Celestial Navigation Daily Routine

The navigator need not follow the entire celestial routine if celestial navigation is not the primary navigation method. It is appropriate to use only the steps of the celestial day's work that are necessary to provide a meaningful check on the primary fix source and maintain competency in celestial techniques.

The list of procedures below provides a complete daily celestial routine to follow. This sequence works equally well for all sight reduction methods, whether tabular, mathematical, computer program, or celestial navigation calculator. See Figure 2013 for an example of a typical day's celestial plot.

1. Before dawn, compute the time of morning twilight and plot the dead reckoning position for that time.
2. At morning twilight, take and reduce celestial observations for a fix. At sunrise take an amplitude of the Sun for a compass check.
3. Mid-morning, wind the chronometer and determine chronometer error with a radio time tick.
4. Mid-morning, reduce a Sun sight for a morning Sun line.
5. Calculate an azimuth of the Sun for a compass check, if no amplitude was taken at sunrise.
6. At LAN, obtain a Sun line and advance the morning Sun line for the noon fix. Compute a longitude determined at LAN for an additional LOP.
7. Mid afternoon, again take and reduce a Sun sight. This is primarily for use with an advanced noon Sun line, or with a Moon or Venus line if the skies are overcast during evening twilight.
8. Calculate an azimuth of the Sun for a compass check at about the same time as the afternoon Sun observation. The navigator may replace this azimuth with an amplitude observation at sunset.
9. During evening twilight, reduce celestial observations for a fix.
10. Be alert at all times for the moon or brighter planets which may be visible during daylight hours for additional LOP's, and Polaris at twilight for a latitude line.

Chapter 7, Chapter 17, and Chapter 20 contain detailed explanations of the procedures required to carry out the various functions of this routine.

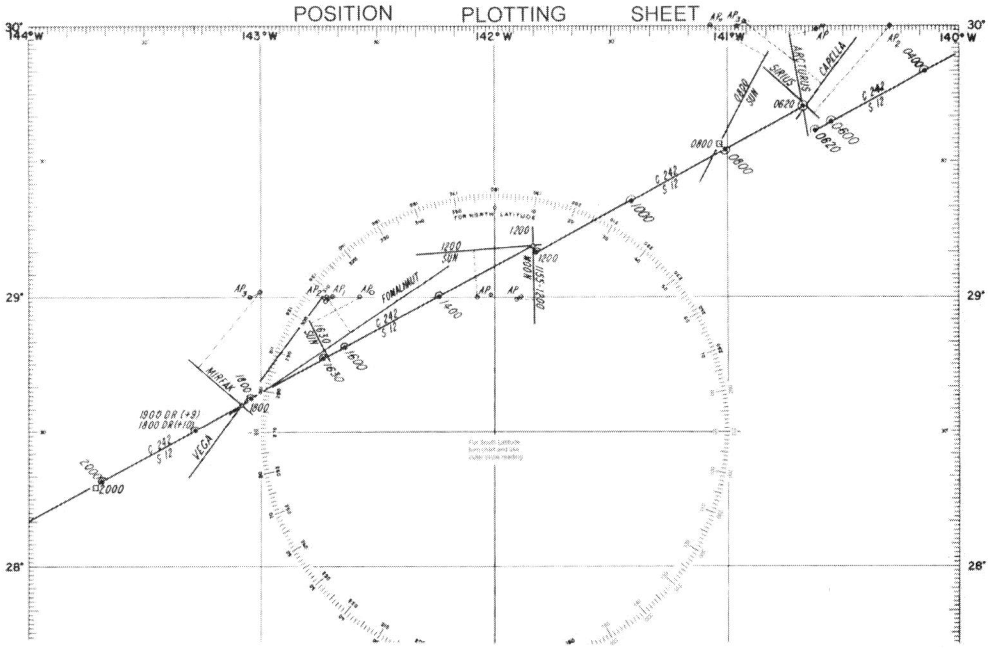

Figure 2013. Typical celestial plot at sea.

CHAPTER 21

NAVIGATIONAL MATHEMATICS

GEOMETRY

2100. Definition

Geometry deals with the properties, relations, and measurement of lines, surfaces, solids, and angles. **Plane geometry** deals with plane figures, and **solid geometry** deals with three–dimensional figures.

A **point**, considered mathematically, is a place having position but no extent. It has no length, breadth, or thickness. A point in motion produces a **line**, which has length, but neither breadth nor thickness. A **straight or right line** is the shortest distance between two points in space. A line in motion in any direction except along itself produces a **surface**, which has length and breadth, but not thickness. A **plane surface** or **plane** is a surface without curvature. A straight line connecting any two of its points lies wholly within the plane. A plane surface in motion in any direction except within its plane produces a **solid**, which has length, breadth, and thickness. **Parallel** lines or surfaces are those which are everywhere equidistant. **Perpendicular** lines or surfaces are those which meet at right or 90° angles. A perpendicular may be called a **normal**, particularly when it is perpendicular to the tangent to a curved line or surface at the point of tangency. All points equidistant from the ends of a straight line are on the perpendicular bisector of that line. The shortest distance from a point to a line is the length of the perpendicular between them.

2101. Angles

An **angle** is formed by two straight lines which meet at a point. It is measured by the arc of a circle intercepted between the two lines forming the angle, the center of the circle being at the point of intersection. In Figure 2101, the angle formed by lines *AB* and *BC*, may be designated "angle *B*," "angle *ABC*," or "angle *CBA*"; or by Greek letter as "angle α." The three letter designation is preferred if there is more than one angle at the point. When three letters are used, the middle one should always be that at the **vertex** of the angle.

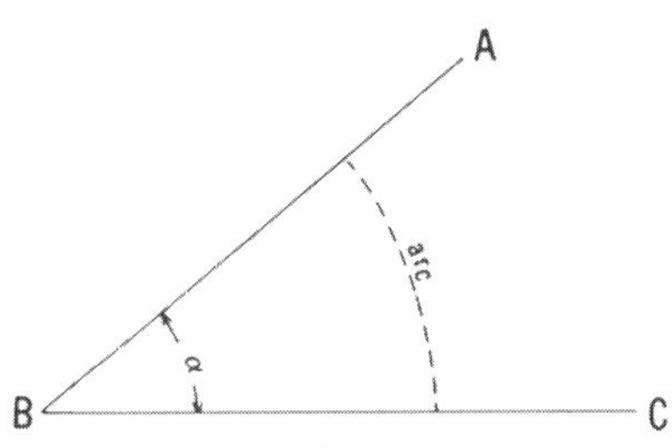

Figure 2101. An angle.

An **acute angle** is one less than a right angle (90°).

A **right angle** is one whose sides are perpendicular (90°).

An **obtuse angle** is one greater than a right angle (90°) but less than 180°.

A **straight angle** is one whose sides form a continuous straight line (180°).

A **reflex angle** is one greater than a straight angle (180°) but less than a circle (360°). Any two lines meeting at a point form two angles, one less than a straight angle of 180° (unless exactly a straight angle) and the other greater than a straight angle.

An **oblique angle** is any angle not a multiple of 90°.

Two angles whose sum is a right angle (90°) are **complementary angles**, and either is the **complement** of the other.

Two angles whose sum is a straight angle (180°) are **supplementary angles**, and either is the **supplement** of the other.

Two angles whose sum is a circle (360°) are **explementary angles**, and either is the **explement** of the other. The two angles formed when any two lines terminate at a common point are explementary.

If the sides of one angle are perpendicular to those of another, the two angles are either equal or supplementary. Also, if the sides of one angle are parallel to those of another, the two angles are either equal or supplementary.

When two straight lines intersect, forming four angles, the two opposite angles, called **vertical angles**, are equal. Angles which have the same vertex and lie on opposite sides of a common side are **adjacent angles**. Adjacent angles formed by intersecting lines are supplementary, since each pair of adjacent angles forms a straight angle.

A **transversal** is a line that intersects two or more other lines. If two or more parallel lines are cut by a transversal, groups of adjacent and vertical angles are formed,

A **dihedral angle** is the angle between two intersecting planes.

2102. Triangles

A **plane triangle** is a closed figure formed by three straight lines, called **sides**, which meet at three points called **vertices**. The vertices are labeled with capital letters and the sides with lowercase letters, as shown in Figure 2102a.

An **equilateral triangle** is one with its three sides equal in length. It must also be **equiangular**, with its three angles equal.

An **isosceles triangle** is one with two equal sides, called **legs**. The angles opposite the legs are equal. A line which **bisects** (divides into two equal parts) the unequal angle of an isosceles triangle is the perpendicular bisector of the opposite side, and divides the triangle into two equal right triangles.

A **scalene triangle** is one with no two sides equal. In such a triangle, no two angles are equal.

An **acute triangle** is one with three acute angles.

A **right triangle** is one having a right angle. The side opposite the right angle is called the **hypotenuse**. The other two sides may be called **legs**. A plane triangle can have only one right angle.

An **obtuse triangle** is one with an obtuse angle. A plane triangle can have only one obtuse angle.

An **oblique triangle** is one which does not contain a right angle.

The **altitude** of a triangle is a line or the distance from any vertex perpendicular to the opposite side.

A **median** of a triangle is a line from any vertex to the center of the opposite side. The three medians of a triangle meet at a point called the **centroid** of the triangle. This point divides each median into two parts, that part between the centroid and the vertex being twice as long as the other part.

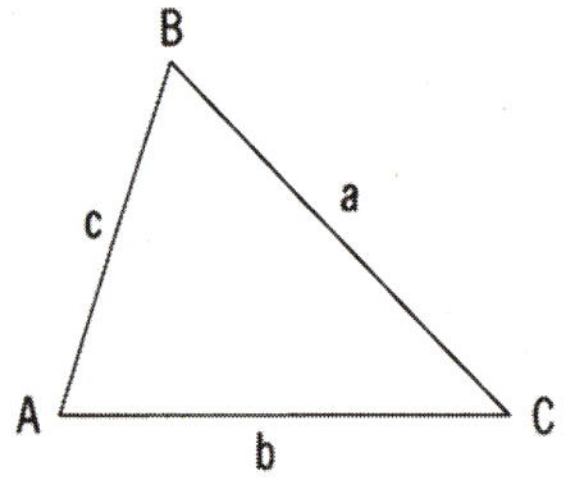

Figure 2102a. A triangle.

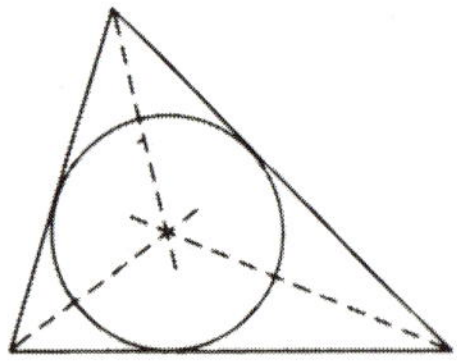

Figure 2102b. A circle inscribed in a triangle.

Lines bisecting the three *angles* of a triangle meet at a point which is equidistant from the three sides, which is the center of the **inscribed circle**, as shown in Figure 2102b. This point is of particular interest to navigators because it is the point theoretically taken as the fix when three lines of position of equal weight and having only random errors do not meet at a common point. In practical navigation, the point is found visually, not by construction, and other factors often influence the chosen fix position.

The perpendicular bisectors of the three *sides* of a triangle meet at a point which is equidistant from the three vertices, which is the center of the **circumscribed circle**, the circle through the three vertices and the smallest circle which can be drawn enclosing the triangle. The center of a circumscribed circle is *within* an acute triangle, *on the hypotenuse* of a right triangle, and *outside* an obtuse triangle.

A line connecting the mid–points of two sides of a triangle is always parallel to the third side and half as long. Also, a line parallel to one side of a triangle and intersecting the other two sides divides these sides proportionally. This principle can be used to divide a line into any number of equal or proportional parts.

The sum of the angles of a plane triangle is always 180°. Therefore, the sum of the acute angles of a right triangle is 90°, and the angles are complementary. If one side of a triangle is extended, the **exterior angle** thus formed is supplementary to the adjacent **interior angle** and is therefore equal to the sum of the two non adjacent angles. If two angles of one triangle are equal to two angles of another triangle, the third angles are also equal, and the triangles are **similar**. If the area of one triangle is equal to the area of another, the triangles are **equal**. Triangles having equal bases and altitudes also have equal areas. Two figures are **congruent** if one can be placed over the other to make an exact fit. Congruent figures are both similar and equal. If any side of one triangle is equal to any side of a similar triangle, the triangles are congruent. For example, if two right triangles have equal sides, they are congruent; if two right triangles have two corresponding sides equal, they are congruent. Triangles are congruent only if the sides and angles are equal.

The sum of two sides of a plane triangle is always greater than the third side; their difference is always less than the third side.

The area of a triangle is equal to 1/2 of the area of the polygon formed from its base and height. This can be stated algebraically as:

$$\text{Area of plane triangle A} = \frac{bh}{2}$$

The square of the hypotenuse of a right triangle is equal

to the sum of the squares of the other two sides, or $a^2 + b^2 = c^2$. Therefore the length of the hypotenuse of plane right triangle can be found by the formula:

$$c = \sqrt{a^2 + b^2}$$

2103. Circles

A **circle** is a plane, closed curve, all points of which are equidistant from a point within, called the **center**.

The distance around a circle is called the **circumference**. Technically the length of this line is the **perimeter**, although the term "circumference" is often used. An arc is part of a circumference. A major arc is more than a semicircle (180°), a **minor** are is less than a semicircle (180°). A **semi–circle** is half a circle (180°), a **quadrant** is a quarter of a circle (90°), a **quintant** is a fifth of a circle (72°), a **sextant** is a sixth of a circle (60°), an **octant** is an eighth of a circle (45°). Some of these names have been applied to instruments used by navigators for measuring altitudes of celestial bodies because of the part of a circle used for the length of the arc of the instrument.

Concentric circles have a common center. A **radius** (plural **radii**) or **semidiameter** is a straight line connecting the center of a circle with any point on its circumference.

A **diameter** of a circle is a straight line passing through its center and terminating at opposite sides of the circumference. It divides a circle into two equal parts. The ratio of the length of the circumference of any circle to the length of its diameter is 3.14159+, or π (the Greek letter pi), a relationship that has many useful applications.

A **sector** is that part of a circle bounded by two radii and an arc. The angle formed by two radii is called a **central angle**. Any pair of radii divides a circle into sectors, one less than a semicircle (180°) and the other greater than a semicircle (unless the two radii form a diameter).

A **chord** is a straight line connecting any two points on the circumference of a circle. Chords equidistant from the center of a circle are equal in length.

A **segment** is the part of a circle bounded by a chord and the intercepted arc. A chord divides a circle into two segments, one less than a semicircle (180°), and the other greater than a semicircle (unless the chord is a diameter). A diameter perpendicular to a chord bisects it, its arc, and its segments. Either pair of vertical angles formed by intersecting chords has a combined number of degrees equal to the sum of the number of degrees in the two arcs intercepted by the two angles.

An **inscribed angle** is one whose vertex is on the circumference of a circle and whose sides are chords. It has half as many degrees as the arc it intercepts. Hence, an angle inscribed in a semicircle is a right angle if its sides terminate at the ends of the diameter forming the semicircle.

A **secant** of a circle is a line intersecting the circle, or a chord extended beyond the circumference.

A **tangent** to a circle is a straight line, in the plane of the circle, which has only one point in common with the circumference. A tangent is perpendicular to the radius at the **point of tangency**. Two tangents from a common point to opposite sides of a circle are equal in length, and a line from the point to the center of the circle bisects the angle formed by the two tangents. An angle formed outside a circle by the intersection of two tangents, a tangent and a secant, or two secants has *half* as many degrees as the *difference* between the two intercepted arcs. An angle formed by a tangent and a chord, with the apex at the point of tangency, has half as many degrees as the arc it intercepts. A **common tangent** is one tangent to more than one circle. Two circles are tangent to each other if they touch at one point only. If of different sizes, the smaller circle may be either inside or outside the larger one.

Parallel lines intersecting a circle intercept equal arcs.

If A = area; r = radius; d = diameter; C = circumference; s = linear length of an arc; a = angular length of an arc, or the angle it subtends at the center of a circle, in degrees; b = angular length of an arc, or the angle it subtends at the center of a circle, in radians:

$$\text{Area of circle } A = \pi r^2 = \frac{\pi d^2}{4}$$

$$\text{Circumference of a circle } C = 2\pi r = \pi d = 2\pi \text{ rad}$$

$$\text{Area of sector} = \frac{\pi r^2 a}{360} = \frac{r^2 b}{2} = \frac{rs}{2}$$

$$\text{Area of segment} = \frac{r^2(b - \sin a)}{2}$$

2104. Spheres

A **sphere** is a solid bounded by a surface every point of which is equidistant from a point within called the **center**. It may also be formed by rotating a circle about any diameter.

A **radius** or **semidiameter** of a sphere is a straight line connecting its center with any point on its surface. A **diameter** of a sphere is a straight line through its center and terminated at both ends by the surface of the sphere.

The intersection of a plane and the surface of a sphere is a circle, a **great circle** if the plane passes through the center of the sphere, and a **small circle** if it does not. The shorter arc of the great circle between two points on the surface of a sphere is the shortest distance, on the surface of the sphere, between the points. Every great circle of a sphere bisects every other great circle of that sphere. The **poles** of a circle on a sphere are the extremities of the sphere's diameter which is perpendicular to the plane of the circle. All points on the circumference of the circle are equidistant from either of its poles. In the ease of a great circle, *both* poles are 90° from any point on the circumference of the circle. Any great circle may be considered a **primary**, particularly when it serves as the origin of measurement of a coordinate. The great circles through its poles are called **secondary**. Secondaries are perpendicular to their primary.

A **spherical triangle** is the figure formed on the surface of a sphere by the intersection of three great circles. The lengths of the sides of a spherical triangle are measured in degrees, minutes, and seconds, as the angular lengths of the arcs forming them. The sum of the three sides is always less than 360°. The sum of the three angles is always *more* than 180° and *less* than 540°.

A **lune** is the part of the surface of a sphere bounded by halves of two great circles.

2105. Coordinates

Coordinates are magnitudes used to define a position. Many different types of coordinates are used. Important navigational ones are described below.

If a position is known to be on a given *line*, only one magnitude (coordinate) is needed to identify the position if an origin is stated or understood.

If a position is known to be on a given *surface*, two magnitudes (coordinates) are needed to define the position.

If nothing is known regarding a position other than that it exists in space, three magnitudes (coordinates) are needed to define its position.

Each coordinate requires an origin, either stated or implied. If a position is known to be on a given plane, it might be defined by means of its distance from each of two intersecting lines, called **axes**. These are called **rectangular coordinates**. In Figure 2105, OY is called the **ordinate**, and OX is called the **abscissa**. Point O is the **origin**, and lines OX and OY the axes (called the X and Y axes, respectively). Point A is at position x,y. If the axes are not perpendicular but the lines x and y are drawn parallel to the axes, **oblique coordinates** result. Either type are called **Cartesian coordinates**. A three–dimensional system of Cartesian coordinates, with X Y, and Z axes, is called **space coordinates**.

Another system of plane coordinates in common usage consists of the *direction* and *distance* from the origin (called the **pole**). A line extending in the direction indicated is called a **radius vector**. Direction and distance from a fixed point constitute **polar coordinates**, sometimes called the rho–theta (the Greek ρ, to indicate distance, and the Greek θ, to indicate direction) system. An example of its use is the radar scope.

Spherical coordinates are used to define a position on the surface of a sphere or spheroid by indicating angular distance from a primary great circle and a reference secondary great circle. Examples used in navigation are latitude and longitude, altitude and azimuth, and declination and hour angle.

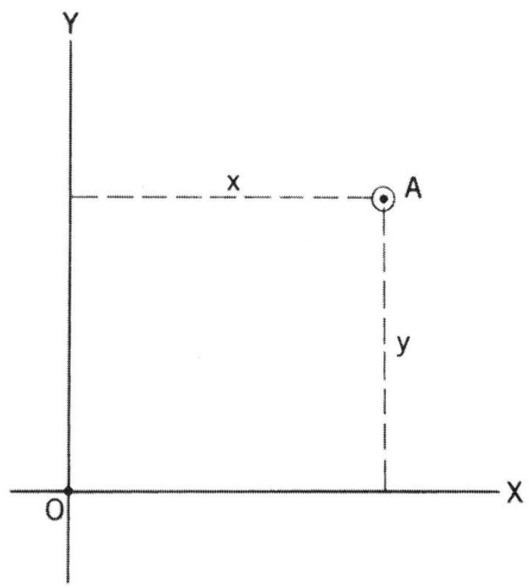

Figure 2105. Rectangular coordinates.

TRIGONOMETRY

2106. Definitions

Trigonometry deals with the relations among the angles and sides of triangles. **Plane trigonometry** deals with plane triangles, those on a plane surface. Spherical trigonometry deals with spherical triangles, which are drawn on the surface of a sphere. In navigation, the common methods of celestial sight reduction use spherical triangles on the surface of the Earth. For most navigational purposes, the Earth is assumed to be a sphere, though it is somewhat flattened.

2107. Angular Measure

A circle may be divided into 360 **degrees** (°), which is the **angular length** of its circumference. Each degree may be divided into 60 **minutes** ('), and each minute into 60 **seconds** ("). The angular measure of an arc is usually expressed in these units. By this system a right angle or quadrant has 90° and a straight angle or semicircle 180°. In marine navigation, altitudes, latitudes, and longitudes are usually expressed in degrees, minutes, and tenths (27°14.4'). Azimuths are usually expressed in degrees and tenths (164.7°). The system of degrees, minutes, and seconds indicated above is the **sexagesimal system**. In the **centesimal system,** used chiefly in France, the circle is divided into 400 **centesimal degrees** (sometimes called **grades**) each of which is divided into 100 centesimal minutes of 100 centesimal seconds each.

A **radian** is the angle subtended at the center of a circle by an arc having a linear length equal to the radius of the circle. A circle (360°) = 2π radians, a semicircle (180°) = π radians, a right angle (90°) = π/2 radians. The length of the arc of a circle is equal to the radius multiplied by the angle subtended in radians.

2108. Trigonometric Functions

Trigonometric functions are the various proportions or ratios of the sides of a plane right triangle, defined in relation to one of the acute angles. In Figure 2108a, let θ be any acute angle. From any point R on line OA, draw a line perpendicular to OB at F. From any other point R' on OA, draw a line perpendicular to OB at F'. Then triangles OFR and OF'R' are similar right triangles because all their corresponding *angles* are equal. Since in any pair of similar triangles the ratio of any two sides of one triangle is equal to the ratio of the corresponding two sides of the other triangle,

$$\frac{RF}{OF} = \frac{R'F'}{OF'} = \frac{RF}{OR} = \frac{R'F'}{OR'} \quad \text{and} \quad \frac{OF}{OR} = \frac{OF'}{OR'}.$$

No matter where the point R is located on OA, the ratio between the lengths of any two sides in the triangle OFR has a constant value. Hence, for any value of the acute angle θ, there is a fixed set of values for the ratios of the various sides of the triangle. These ratios are defined as follows:

$$\text{sine } \theta = \sin\theta = \frac{\text{side opposite}}{\text{hypotenuse}}$$

$$\text{cosine } \theta = \cos\theta = \frac{\text{side adjacent}}{\text{hypotenuse}}$$

$$\text{tangent } \theta = \tan\theta = \frac{\text{side opposite}}{\text{side adjacent}}$$

$$\text{cosecant } \theta = \csc\theta = \frac{\text{hypotenuse}}{\text{side opposite}}$$

$$\text{secant } \theta = \sec\theta = \frac{\text{hypotenuse}}{\text{side adjacent}}$$

$$\text{cotangent } \theta = \cot\theta = \frac{\text{side adjacent}}{\text{side opposite}}$$

Of these six principal functions, the second three are the reciprocals of the first three; therefore

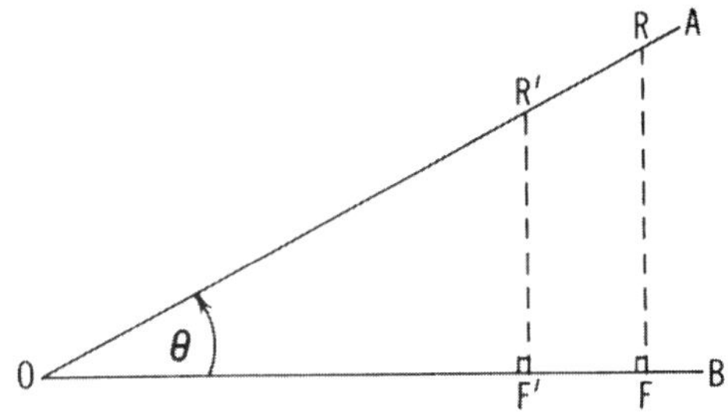

Figure 2108a. Similar right triangles.

$$\sin\theta = \frac{1}{\csc\theta} \qquad \csc\theta = \frac{1}{\sin\theta}$$

$$\cos\theta = \frac{1}{\sec\theta} \qquad \sec\theta = \frac{1}{\cos\theta}$$

$$\tan\theta = \frac{1}{\cot\theta} \qquad \cot\theta = \frac{1}{\tan\theta}$$

In Figure 2108b, A, B, and C are the angles of a plane right triangle, with the right angle at C. The sides are labeled a, b, c, with opposite angles labeled A, B, and C respectively.

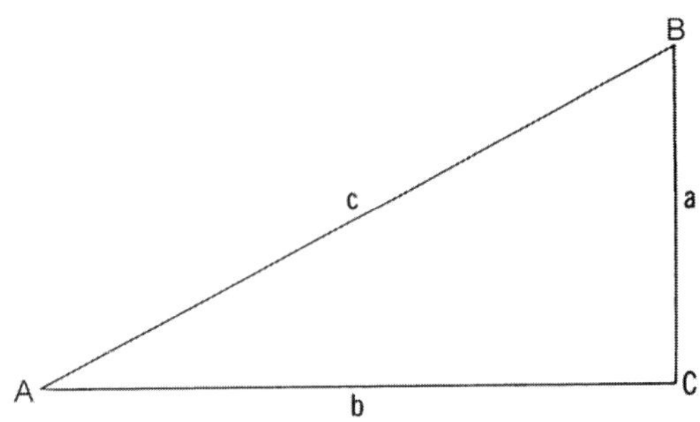

Figure 2108b. A right triangle.

The six principal trigonometric functions of angle B are:

$$\sin B = \frac{b}{c} = \cos A = \cos(90° - B)$$

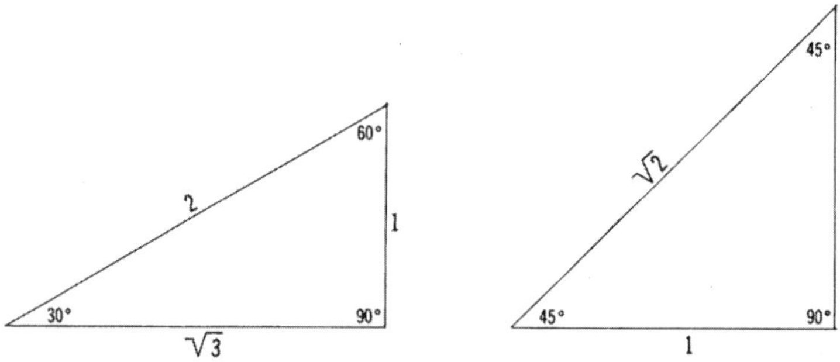

Figure 2108c. Numerical relationship of sides of 30°, 60°, and 45° triangles.

$$\cos B = \frac{a}{c} = \sin A = \sin(90^\circ - B)$$

$$\tan B = \frac{b}{a} = \cot A = \cot(90^\circ - B)$$

$$\cot B = \frac{a}{b} = \tan A = \tan(90^\circ - B)$$

$$\sec B = \frac{c}{a} = \csc A = \csc(90^\circ - B)$$

$$\csc B = \frac{c}{b} = \sec A = \sec(90^\circ - B)$$

Function	*30°*	*45°*	*60°*
sine	$\frac{1}{2}$	$\frac{1}{\sqrt{2}} = \frac{1}{2}\sqrt{2}$	$\frac{\sqrt{3}}{2} = \frac{1}{2}\sqrt{3}$
cosine	$\frac{\sqrt{3}}{2} = \frac{1}{2}\sqrt{3}$	$\frac{1}{\sqrt{2}} = \frac{1}{2}\sqrt{2}$	$\frac{1}{2}$
tangent	$\frac{1}{\sqrt{3}} = \frac{1}{3}\sqrt{3}$	$\frac{1}{1} = 1$	$\frac{\sqrt{3}}{1} = \sqrt{3}$
cotangen t	$\frac{\sqrt{3}}{1} = \sqrt{3}$	$\frac{1}{1} = 1$	$\frac{1}{\sqrt{3}} = \frac{1}{3}\sqrt{3}$
secant	$\frac{2}{\sqrt{3}} = \frac{2}{3}\sqrt{3}$	$\frac{\sqrt{2}}{1} = \sqrt{2}$	$\frac{2}{1} = 2$
cosecant	$\frac{2}{1} = 2$	$\frac{\sqrt{2}}{1} = \sqrt{2}$	$\frac{2}{\sqrt{3}} = \frac{2}{3}\sqrt{3}$

Table 2108d. Values of various trigonometric functions for angles 30°, 45°, and 60°.

Since A and B are *complementary*, these relations show that the sine of an angle is the cosine of its complement, the tangent of an angle is the cotangent of its complement, and the secant of an angle is the cosecant of its complement. Thus, the co–function of an angle is the function of its complement.

$$\sin(90^\circ - A) = \cos A$$
$$\cos(90^\circ - A) = \sin A$$
$$\tan(90^\circ - A) = \cot A$$
$$\csc(90^\circ - A) = \sec A$$
$$\sec(90^\circ - A) = \csc A$$
$$\cot(90^\circ - A) = \tan A$$

The numerical value of a trigonometric function is sometimes called the **natural function** to distinguish it from the logarithm of the function, called the **logarithmic function**.

Since the relationships of 30°, 60°, and 45° right triangles are as shown in Figure 2108c, certain values of the basic functions can be stated exactly, as shown in Table 2108d.

2109. Functions in Various Quadrants

To make the definitions of the trigonometric functions more general to include those angles greater than 90°, the functions are defined in terms of the rectangular Cartesian coordinates of point R of Figure 2108a, due regard being given to the sign of the function. In Figure 2109a, OR is assumed to be a *unit* radius. By convention the sign of OR is always positive. This radius is imagined to rotate in a counterclockwise direction through 360° from the horizontal position at 0°, the positive direction along the X axis. Ninety degrees (90°) is the positive direction along the Y axis. The angle between the original position of the radius and its position at any time increases from 0° to 90° in the first quadrant (I), 90° to 180° in the second quadrant (II), 180° to 270° in the third quadrant (III), and 270° to 360° in the fourth quadrant (IV).

The numerical value of the sine of an angle is equal to the projection of the unit radius on the Y–axis. According to the definition given in Article 2108, the sine of angle in the first quadrant of Figure 2109a is $\frac{+y}{+OR}$. If the radius OR is equal to one, sin θ =+y. Since +y is equal to the projection of the unit radius OR on the Y axis, the sine function of an angle in the first quadrant defined in terms of rectangular Cartesian coordinates does not contradict the definition in Article 2108. In Figure 2109a,

$$\sin\theta = +y$$

$$\sin(180°-\theta) = +y = \sin\theta$$

$$\sin(180°+\theta) = -y = -\sin\theta$$

$$\sin(360°-\theta) = -y = \sin(-\theta) = -\sin\theta$$

The numerical value of the cosine of an angle is equal to the projection of the unit radius on the X axis. In Figure 2109a,

$$\cos\theta = +x$$
$$\cos(180°-\theta) = -x = -\cos\theta$$
$$\cos(180°+\theta) = -x = -\cos\theta$$
$$\cos(360°-\theta) = +x = \cos(-\theta) = \cos\theta$$

The numerical value of the tangent of an angle is equal to the ratio of the projections of the unit radius on the Y and X axes. In Figure 2109a,

$$\tan\theta = \frac{+y}{+x}$$

$$(180°-\theta) = \frac{+y}{-x} = -\tan\theta$$

$$\tan(180°+\theta) = \frac{-y}{-x} = \tan\theta$$

$$\tan(360°-\theta) = \frac{-y}{+x} = \tan(-\theta) = -\tan\theta$$

The cosecant, secant, and cotangent functions of angles in the various quadrants are similarly determined.

$$\csc\theta = \frac{1}{+y}$$

$$\csc(180°-\theta) = \frac{1}{+y} = \csc\theta$$

$$\csc(180°+\theta) = \frac{1}{-y} = -\csc\theta$$

$$\csc(360°-\theta) = \frac{1}{-y} = \csc(-\theta) = -\csc\theta$$

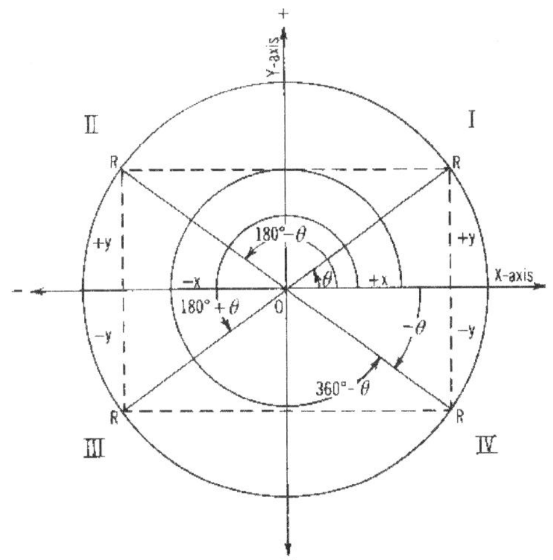

Figure 2109a. The functions in various quadrants, mathematical convention.

$$\sec\theta = \frac{1}{+x}$$

$$\sec(180°-\theta) = \frac{1}{-x} = -\sec\theta$$

$$\sec(180°+\theta) = \frac{1}{-x} = -\sec\theta$$

$$\sec(360°-\theta) = \frac{1}{+x} = \sec(-\theta) = \sec\theta$$

$$\cot\theta = \frac{+x}{+y}$$

$$\cot(180°-\theta) = \frac{-x}{+y} = -\cot\theta$$

$$\cot(180°+\theta) = \frac{-x}{-y} = \cot\theta$$

$$\cot(360°-\theta) = \frac{+x}{-y} = \cot(-\theta) = -\cot\theta.$$

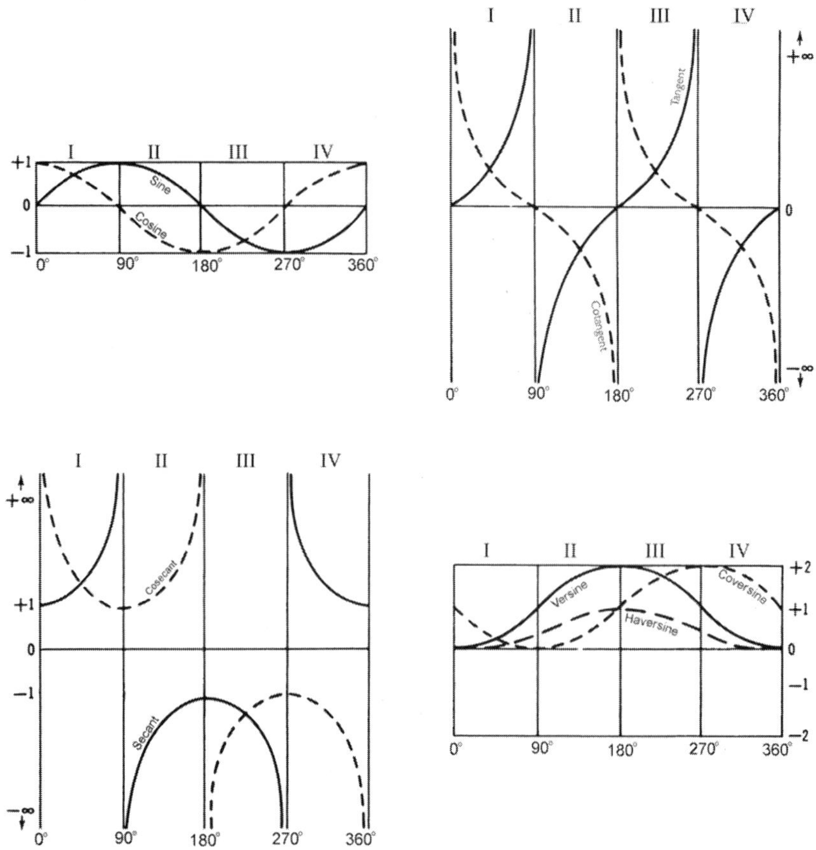

Figure 2109b. Graphic representation of values of trigonometric functions in various quadrants.

The signs of the functions in the four different quadrants are shown below:

	I	II	III	IV
sine and cosecant	+	+	-	-
cosine and secant	+	-	-	+
tangent and cotangent	+	-	+	-

The numerical values vary by quadrant as shown above:

	I	II	III	IV
sin	0 to +1	+1 to 0	0 to −1	−1 to 0
csc	+∞ to +1	+1 to 0	−∞ to −1	−1 to −∞
cos	+1 to 0	0 to −1	−1 to 0	0 to +1
sec	+1 to +∞	−∞ to −1	−1 to −∞	+∞ to +1

	I	II	III	IV
tan	0 to +∞	−∞ to 0	0 to +∞	−∞ to 0
cot	+∞ to 0	0 to −∞	+∞ to 0	0 to −∞

These relationships are shown graphically in Figure 2109b.

2110. Trigonometric Identities

A **trigonometric identity** is an equality involving trigonometric functions of θ which is true for all values of θ, except those values for which one of the functions is not defined or for which a denominator in the equality is equal to zero. The **fundamental identities** are those identities from which other identities can be derived.

$$\sin\theta = \frac{1}{\csc\theta} \qquad \csc\theta = \frac{1}{\sin\theta}$$

$$\cos\theta = \frac{1}{\sec\theta} \qquad \sec\theta = \frac{1}{\cos\theta}$$

$$\tan\theta = \frac{1}{\cot\theta} \qquad \cot\theta = \frac{1}{\tan\theta}$$

$$\tan\theta = \frac{\sin\theta}{\cos\theta} \qquad \cot\theta = \frac{\cos\theta}{\sin\theta}$$

$$\sin^2\theta + \cos^2\theta = 1 \qquad \tan^2\theta + 1 = \sec^2\theta$$

$$\sin(90° - \theta) = \cos\theta \qquad \csc(90° - \theta) = \sec\theta$$
$$\cos(90° - \theta) = \sin\theta \qquad \sec(90° - \theta) = \csc\theta$$
$$\tan(90° - \theta) = \cot\theta \qquad \cot(90° - \theta) = \tan\theta$$

$$\sin(-\theta) = -\sin\theta \qquad \csc(-\theta) = -\csc\theta$$
$$\cos(-\theta) = \cos\theta \qquad \sec(-\theta) = \sec\theta$$
$$\tan(-\theta) = -\tan\theta \qquad \cot(-\theta) = -\cot\theta$$

$$\sin(90 + \theta) = \cos\theta \qquad \csc(90 + \theta) = \sec\theta$$
$$\cos(90 + \theta) = -\sin\theta \qquad \sec(90 + \theta) = -\csc\theta$$
$$\tan(90 + \theta) = -\cot\theta \qquad \cot(90 + \theta) = -\tan\theta$$

$$\sin(180° - \theta) = \sin\theta \qquad \csc(180° - \theta) = \csc\theta$$
$$\cos(180° - \theta) = -\cos\theta \qquad \sec(180° - \theta) = -\sec\theta$$
$$\tan(180° - \theta) = -\tan\theta \qquad \cot(180° - \theta) = -\cot\theta$$

$$\sin(180° + \theta) = -\sin\theta \qquad \csc(180° + \theta) = -\csc\theta$$
$$\cos(180° + \theta) = -\cos\theta \qquad \sec(180° + \theta) = -\sec\theta$$
$$\tan(180° + \theta) = \tan\theta \qquad \cot(180° + \theta) = \cot\theta$$

$$\sin(360° - \theta) = -\sin\theta \qquad \csc(360° - \theta) = -\csc\theta$$
$$\cos(360° - \theta) = \cos\theta \qquad \sec(360° - \theta) = \sec\theta$$
$$\tan(360° - \theta) = -\tan\theta \qquad \cot(360° - \theta) = -\cot\theta$$

2111. Inverse Trigonometric Functions

An angle having a given trigonometric function may be indicated in any of several ways. Thus, sin y = x, y = arc sin x, and $y = \sin^{-1} x$ have the same meaning. The superior "–1" is not an exponent in this case. In each case, y is "the angle whose sine is x." In this case, y is the **inverse sine** of x. Similar relationships hold for all trigonometric functions.

SOLVING TRIANGLES

A triangle is composed of six parts: three angles and three sides. The angles may be designated A, B, and C; and the sides opposite these angles as *a*, *b*, and *c*, respectively. In general, when any three parts are known, the other three parts can be found, unless the known parts are the three angles.

2112. Right Plane Triangles

In a right plane triangle it is only necessary to substitute numerical values in the appropriate formulas representing the basic trigonometric functions and solve. Thus, if *a* and *b* are known,

$$\tan A = \frac{a}{b}$$
$$B = 90° - A$$
$$c = a \csc A$$

Similarly, if *c* and *B* are given,

$$A = 90° - B$$
$$a = c \sin A$$
$$b = c \cos A$$

2113. Oblique Plane Triangles

When solving an oblique plane triangle, it is often desirable to draw a rough sketch of the triangle approximately to scale, as shown in Figure 2113. The following laws are helpful in solving such triangles:

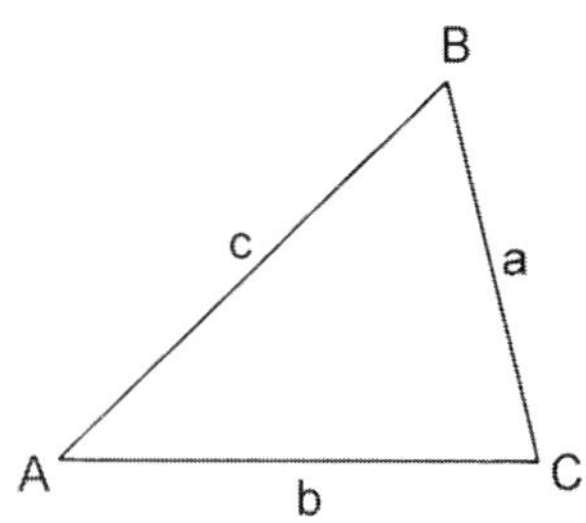

Figure 2113. An oblique plane triangle.

Known	*To find*	*Formula*	*Comments*
a, b, c	A	$\cos A = \frac{c^2 + b^2 - a^2}{2bc}$	Cosine law
a, b, A	B	$\sin B = \frac{b \sin A}{a}$	Sine law. Two solutions if $b>a$
	C	$C = 180° - (A + B)$	$A + B + C = 180°$
	c	$c = \frac{a \sin C}{\sin A}$	Sine law
a, b, C	A	$\tan A = \frac{a \sin C}{b - a \cos C}$	
	B	$B = 180° - (A + C)$	$A + B + C = 180°$
	c	$c = \frac{a \sin C}{\sin A}$	Sine law
a, A, B	b	$b = \frac{a \sin B}{\sin A}$	Sine law
	C	$C = 180° - (A + B)$	$A + B + C = 180°$
	c	$c = \frac{a \sin C}{\sin A}$	Sine law

Table 2113. Formulas for solving oblique plane triangles.

Law of sines: $\frac{a}{\sin A} = \frac{b}{\sin B} = \frac{c}{\sin C}$

Law of cosines: $a^2 = b^2 + c^2 - 2bc \cos A$.

The unknown parts of oblique plane triangles can be computed by the formulas in Table 2113, among others. By reassignment of letters to sides and angles, these formulas can be used to solve for all unknown parts of oblique plane triangles.

SPHERICAL TRIGONOMETRY

2114. Napier's Rules

Right spherical triangles can be solved with the aid of **Napier's Rules of Circular Parts**. If the right angle is omitted, the triangle has five parts: two angles and three sides, as shown in Figure 2114a. Since the right angle is already known, the triangle can be solved if any two other parts are known. If the two sides forming the right angle, and the *complements* of the other three parts are used, these elements (called "parts" in the rules) can be arranged in five sectors of a circle in the same order in which they occur in the triangle, as shown in Figure 2114b. Considering any part as the middle part, the two parts nearest it in the diagram are considered the adjacent parts, and the two farthest from it the opposite parts.

Napier's Rules state: The sine of a middle part equals the product of (1) the tangents of the adjacent parts or (2) the cosines of the opposite parts.

In the use of these rules, the co–function of a complement can be given as the function of the element. Thus, the cosine of co–A is the same as the sine of A. From these rules the following formulas can be derived:

$$\sin a = \tan b \cot B = \sin c \sin A$$

$$\sin b = \tan a \cot A = \sin c \sin B$$

$$\cos c = \cot A \cot B = \cos a \cos b$$

$$\cos A = \tan b \cot c = \cos a \sin B$$

$$\cos B = \tan a \cot c = \cos b \sin A$$

The following rules apply:

1. An oblique angle and the side opposite are in the same quadrant.
2. Side c (the hypotenuse) is less then 90° when a and

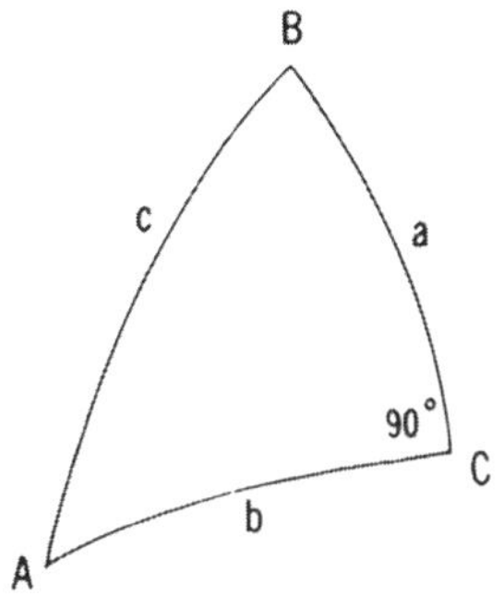

Figure 2114a. Parts of a right spherical triangle as used in Napier's rules.

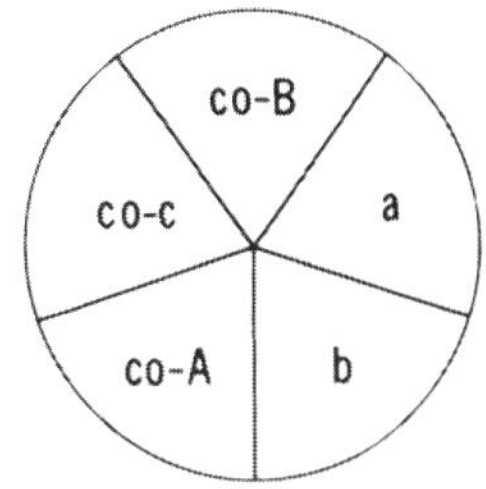

Figure 2114b. Diagram for Napier's Rules of Circular Parts.

b are in the same quadrant, and more than 90° when *a* and *b* are in different quadrants.

If the known parts are an angle and its opposite side, two solutions are possible.

A **quadrantal spherical triangle** is one having one side of 90°. A **biquadrantal spherical triangle** has two sides of 90°. A **triquadrantal spherical triangle** has three sides of 90°. A biquadrantal spherical triangle is isosceles and has two right angles opposite the 90° sides. A triquadrantal spherical triangle is equilateral, has three right angles, and bounds an octant (one–eighth) of the surface of the sphere. A quadrantal spherical triangle can be solved by Napier's rules provided any two elements in addition to the 90° side are known. The 90° side is omitted and the other parts are arranged in order in a five–sectored circle, using the complements of the three parts farthest from the 90° side. In the case of a quadrantal triangle, rule 1 above is used, and rule 2 restated: angle C (the angle opposite the side of 90°) is *more* than 90° when A and B are in the same quadrant, and *less* than 90° when A and B are in different quadrants. If the rule requires an angle of more than 90° and the solution produces an angle of less than 90°, subtract the solved angle from 180°.

2115. Oblique Spherical Triangles

An oblique spherical triangle can be solved by dropping a perpendicular from one of the apexes to the opposite side, subtended if necessary, to form two right spherical triangles. It can also be solved by the following formulas in Table 2115, reassigning the letters as necessary.

Known	*To find*	*Formula*	*Comments*
a, b, C	A	$\tan A = \dfrac{\sin D \tan C}{\sin (b - D)}$	$\tan D = \tan a \cos C$
	B	$\sin B = \dfrac{\sin C \sin b}{\sin c}$	
c, A, B	C	$\cos C = \sin A \sin B \cos c \quad \cos A \cos B$	
	a	$\tan a = \dfrac{\tan c \sin E}{\sin (B + E)}$	$\tan E = \tan A \cos c$
	b	$\tan b = \dfrac{\tan c \sin F}{\sin (A + F)}$	$\tan F = \tan B \cos c$

Table 2115. Formulas for solving oblique spherical triangles.

Known	*To find*	*Formula*	*Comments*
a, b, A	c	$\sin(c + G) = \frac{\cos a \sin G}{\cos b}$	$\cot G = \cos A \tan b$ Two solutions
	B	$\sin B = \frac{\sin A \sin b}{\sin a}$	Two solutions
	C	$\sin(C + H) = \sin H \tan b \cot a$	$\tan H = \tan A \cos b$ Two solutions
a, A, B	C	$\sin(C - K) = \frac{\cos A \sin K}{\cos B}$	$\cot K = \tan B \cos a$ Two solutions
	b	$\sin b = \frac{\sin a \sin B}{\sin A}$	Two solutions
	c	$\sin(c - M) = \cot A \tan B \sin M$	$\tan M = \cos B \tan a$ Two solutions

Table 2115. Formulas for solving oblique spherical triangles.

CHAPTER 22

CALCULATIONS AND CONVERSIONS

INTRODUCTION

2200. Purpose and Scope

This chapter discusses the use of calculators and computers in navigation and summarizes the formulas the navigator depends on during voyage planning, piloting, celestial navigation, and various related tasks. To fully utilize this chapter, the navigator should be competent in basic mathematics including algebra and trigonometry (See Chapter 21, Navigational Mathematics), and be familiar with the use of a basic scientific calculator. The navigator should choose a calculator based on personal needs, which may vary greatly from person to person according to individual abilities and responsibilities.

2201. Use of Calculators in Navigation

Any common calculator can be used in navigation, even one providing only the four basic arithmetic functions of addition, subtraction, multiplication, and division. Any good scientific calculator can be used for sight reduction, sailings, and other tasks. However, the use of a computer program or handheld calculator specifically designed for navigation will greatly reduce the workload of the navigator, reduce the possibility of errors, and increase the accuracy of results over those obtained by hand calculation.

Calculations of position based on celestial observations are becoming increasingly obsolete as GPS takes its place as a dependable position reference for all modes of navigation. This is especially true since handheld, battery-powered GPS units have become less expensive, and can provide a worldwide backup position reference to more sophisticated systems with far better accuracy and reliability than celestial.

However, for those who still use celestial techniques, a celestial navigation calculator or computer program can improve celestial positions by easily solving numerous sights, and by reducing mathematical and tabular errors inherent in the manual sight reduction process. They can also provide weighted plots of the LOP's from any number of celestial bodies, based on the navigator's subjective analysis of each sight, and calculate the best fix with lat./long. readout.

On a vessel with a laptop or desktop computer convenient to the bridge, a good choice would be a comprehensive computer program to handle all navigational functions such as sight reduction, sailings, tides, and other tasks, backed up by a handheld navigational calculator for basic calculations should the computer fail. Handheld calculators are dependable enough that the navigator can expect to never have to solve celestial sights, sailings, and other problems by tables or calculations.

In using a calculator for any navigational task, it important to remember that the accuracy of the result, even if carried to many decimal places, is only as good as the least accurate entry. If a sextant observation is taken to an accuracy of only a minute, that is the best accuracy of the final solution, regardless of a calculator's ability to solve to 12 decimal places. See Chapter 23, Navigational Errors, for a discussion of the sources of error in navigation.

Some basic calculators require the conversion of degrees, minutes and seconds (or tenths) to decimal degrees before solution. A good navigational calculator, however, should permit entry of degrees, minutes and tenths of minutes directly, and should do conversions at will. Though many non-navigational computer programs have an on-screen calculator, these are generally very simple versions with only the four basic arithmetical functions. They are thus too simple for many navigational problems. Conversely, a good navigational computer program requires no calculator per se, since the desired answer is calculated automatically from the entered data.

The following articles discuss calculations involved in various aspects of navigation.

2202. Calculations of Piloting

- **Hull speed in knots** is found by:

  $$S = 1.34\sqrt{\text{waterline length}}\ \text{(in feet)}.$$

  This is an approximate value which varies with hull shape.

- **Nautical and U.S. survey miles** can be interconverted by the relationships:

  1 nautical mile = 1.15077945 U.S. survey miles.

  1 U.S. survey mile = 0.86897624 nautical miles.

- **The speed of a vessel over a measured mile** can be calculated by the formula:

$$S = \frac{3600}{T}$$

where S is the speed in knots and T is the time in seconds.

- **The distance traveled at a given speed** is computed by the formula:

$$D = \frac{ST}{60}$$

where D is the distance in nautical miles, S is the speed in knots, and T is the time in minutes.

- **Distance to the visible horizon in nautical miles** can be calculated using the formula:

$$D = 1.17\sqrt{h_f}\text{ , or}$$

$$D = 2.07\sqrt{h_m}$$

depending upon whether the height of eye of the observer above sea level is in feet (h_f) or in meters (h_m).

- **Dip of the visible horizon in minutes of arc** can be calculated using the formula:

$$D = 0.97'\sqrt{h_f}\text{ , or}$$

$$D = 1.76'\sqrt{h_m}$$

depending upon whether the height of eye of the observer above sea level is in feet (h_f) or in meters (h_m)

- **Distance to the radar horizon** in nautical miles can be calculated using the formula:

$$D = 1.22\sqrt{h_f}\text{ , or}$$

$$D = 2.21\sqrt{h_m}$$

depending upon whether the height of the antenna above sea level is in feet (h_f) or in meters (h_m).

- **Dip of the sea short of the horizon** can be calculated using the formula:

$$Ds = 60\tan^{-1}\left(\frac{h_f}{6076.1\,d_s} + \frac{d_s}{8268}\right)$$

where Ds is the dip short of the horizon in minutes of arc; h_f is the height of eye of the observer above sea level, in feet and d_s is the distance to the waterline of the object in nautical miles.

- **Distance by vertical angle between the waterline and the top of an object** is computed by solving the right triangle formed between the observer, the top of the object, and the waterline of the object by simple trigonometry. This assumes that the observer is at sea level, the Earth is flat between observer and object, there is no refraction, and the object and its waterline form a right angle. For most cases of practical significance, these assumptions produce no large errors.

$$D = \sqrt{\frac{\tan^2 a}{0.0002419^2} + \frac{H-h}{0.7349}} - \frac{\tan a}{0.0002419}$$

where D is the distance in nautical miles, a is the corrected vertical angle, H is the height of the top of the object above sea level, and h is the observer's height of eye in feet. The constants (0.0002419 and 0.7349) account for refraction.

2203. Tide Calculations

- **The rise and fall of a diurnal tide** can be roughly calculated from the following table, which shows the fraction of the total range the tide rises or falls during flood or ebb.

Hour	Amount of flood/ebb
1	1/12
2	2/12
3	3/12
4	3/12
5	2/12
6	1/12

2204. Calculations of Celestial Navigation

Unlike sight reduction by tables, sight reduction by calculator permits the use of nonintegral values of latitude of the observer, and LHA and declination of the celestial body. Interpolation is not needed, and the sights can be readily reduced from any assumed position. Simultaneous, or nearly simultaneous, observations can be reduced using a single assumed position. Using the observer's DR or MPP for the assumed longitude usually provides a better representation of the circle of equal altitude, particularly at high observed altitudes.

- **The dip correction** is computed in the *Nautical Almanac* using the formula:

$$D = 0.97\sqrt{h}$$

where dip is in minutes of arc and h is height of eye in feet. This correction includes a factor for refraction. The *Air Almanac* uses a different formula intended for air navigation. The differences are of no significance in practical navigation.

- **The computed altitude** (Hc) is calculated using the basic formula for solution of the undivided navigational triangle:

$$\sin h = \sin L \sin d + \cos L \cos d \cos LHA,$$

in which h is the altitude to be computed (Hc), L is the latitude of the assumed position, d is the declination of the celestial body, and LHA is the local hour angle of the body. Meridian angle (t) can be substituted for LHA in the basic formula.
Restated in terms of the inverse trigonometric function:

$$Hc = \sin^{-1}[(\sin L \ \sin d) + (\cos L \cos d \ \cos LHA)].$$

When latitude and declination are of contrary name, declination is treated as a negative quantity. No special sign convention is required for the local hour angle, as in the following azimuth angle calculations.

- **The azimuth angle** (Z) can be calculated using the altitude azimuth formula if the altitude is known. The formula stated in terms of the inverse trigonometric function is:

$$Z = \cos^{-1}\left(\frac{\sin d - (\sin L \sin Hc)}{(\cos L \cos Hc)}\right)$$

If the altitude is unknown or a solution independent of altitude is required, the azimuth angle can be calculated using the time azimuth formula:

$$Z = \tan^{-1}\left(\frac{\sin LHA}{(\cos L \tan d) - (\sin L \cos LHA)}\right)$$

The sign conventions used in the calculations of both azimuth formulas are as follows: (1) if latitude and declination are of contrary name, declination is treated as a negative quantity; (2) if the local hour angle is greater than 180°, it is treated as a negative quantity.
If the azimuth angle as calculated is negative, add 180° to obtain the desired value.

- **Amplitudes** can be computed using the formula:

$$A = \sin^{-1}(\sin d \sec L)$$

this can be stated as

$$A = \sin^{-1}\left(\frac{\sin d}{\cos L}\right)$$

where A is the arc of the horizon between the prime vertical and the body, L is the latitude at the point of observation, and d is the declination of the celestial body.

2205. Calculations of the Sailings

- **Plane sailing** is based on the assumption that the meridian through the point of departure, the parallel through the destination, and the course line form a plane right triangle, as shown in Figure 2205.

From this: $\cos C = \frac{l}{D}$, $\sin C = \frac{p}{D}$, and $\tan C = \frac{p}{l}$.

From this: $l = D \cos C$, $D = l \sec C$, and $p = D \sin C$.

From this, given course and distance (C and D), the difference of latitude (l) and departure (p) can be found, and given the latter, the former can be found, using simple trigonometry. See Chapter 24.

- **Traverse sailing** combines plane sailings with two or more courses, computing course and distance along a series of rhumb lines. See Chapter 24.

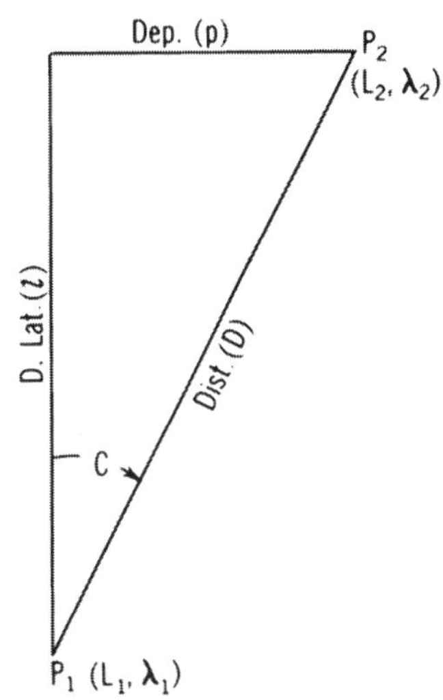

Figure 2205. The plane sailing triangle.

- **Parallel sailing** consists of interconverting departure and difference of longitude. Refer to Figure 2205.

$$DLo = p \sec L, \text{ and } p = DLo \cos L$$

- **Mid-latitude sailing** combines plane and parallel sailing, with certain assumptions. The mean latitude (Lm) is half of the arithmetical sum of the latitudes of two places on the same side of the equator. For places on opposite sides of the equator, the N and S portions are solved separately.

In mid-latitude sailing:

DLo = p sec Lm, and p= DLo cos Lm

- **Mercator Sailing** problems are solved graphically on a Mercator chart. For mathematical Mercator solutions the formulas are:

$$\tan C = \frac{DLo}{m} \text{ or } DLo = m \tan C$$

where m is the meridional part from Table 6 in the Tables Part of this volume. Following solution of the course angle by Mercator sailing, the distance is by the plane sailing formula:

$$D = l \sec C.$$

- **Great-circle solutions for distance and initial course angle** can be calculated from the formulas:

$$D = \cos^{-1}[(\sin L_1 \sin L_2 + \cos L_1 \cos L_2 \cos DLo)],$$

and

$$C = \tan^{-1}\left(\frac{\sin DLo}{(\cos L_1 \tan L_2) - (\sin L_1 \cos DLo)}\right)$$

where D is the great-circle distance, C is the initial great-circle course angle, L_1 is the latitude of the point of departure, L_2 is the latitude of the destination, and DLo is the difference of longitude of the points of departure and destination. If the name of the latitude of the destination is contrary to that of the point of departure, it is treated as a negative quantity.

- **The latitude of the vertex**, L_v, is always numerically equal to or greater than L_1 or L_2. If the initial course angle C is less than 90°, the vertex is toward L_2, but if C is greater than 90°, the nearer vertex is in the opposite direction. The vertex nearer L_1 has the same name as L_1.

The latitude of the vertex can be calculated from the formula:

$$L_v = \cos^{-1}(\cos L_1 \sin C)$$

The difference of longitude of the vertex and the point of departure (DLo_v) can be calculated from the formula:

$$DLo_v = \sin^{-1}\left(\frac{\cos C}{\sin L_v}\right).$$

The distance from the point of departure to the vertex (D_v) can be calculated from the formula:

$$D_v = \sin^{-1}(\cos L_1 \sin DLo_v).$$

- **The latitudes of points on the great-circle track** can be determined for equal DLo intervals each side of the vertex (DLo_{vx}) using the formula:

$$L_x = \tan^{-1}(\cos DLo_{vx} \tan L_v)$$

The DLo_v and D_v of the nearer vertex are never greater than 90°. However, when L_1 and L_2 are of contrary name, the other vertex, 180° away, may be the better one to use in the solution for points on the great-circle track if it is nearer the mid point of the track.

The method of selecting the longitude (or DLo_{vx}), and determining the latitude at which the great-circle crosses the selected meridian, provides shorter legs in higher latitudes and longer legs in lower latitudes. Points at desired distances or desired equal intervals of distance on the great-circle from the vertex (D_{vx}) can be calculated using the formulas:

$$L_x = \sin^{-1}[\sin L_v \cos D_{vx}],$$

and

$$DLo_{vx} = \sin^{-1}\left(\frac{\sin D_{vx}}{\cos L_x}\right).$$

A calculator which converts rectangular to polar coordinates provides easy solutions to plane sailings. However, the user must know whether the difference of latitude corresponds to the calculator's X-coordinate or to the Y-coordinate.

2206. Calculations Of Meteorology And Oceanography

- **Converting thermometer scales** between centigrade, Fahrenheit, and Kelvin scales can be done using the following formulas:

$$C° = \frac{5(F° - 32°)}{9},$$

$$F° = \frac{9}{5}C° + 32°, \text{ and}$$

$$K° = C° + 273.15°.$$

- **Maximum length of sea waves** can be found by the formula:

$$W = 1.5\sqrt{\text{fetch in nautical miles}}.$$

- **Wave height** = 0.026 S^2 where S is the wind speed in knots.
- **Wave speed** in knots

$$= 1.34\sqrt{\text{wavelength in feet}}, \text{ or}$$

$$= 3.03 \times \text{wave period in seconds}.$$

UNIT CONVERSION

Use the conversion tables that appear on the following pages to convert between different systems of units. Conversions followed by an asterisk are exact relationships.

MISCELLANEOUS DATA

Area

1 square inch	= 6.4516 square centimeters*
1 square foot	= 144 square inches*
	= 0.09290304 square meter*
	= 0.000022957 acre
1 square yard	= 9 square feet*
	= 0.83612736 square meter
1 square (statute) mile	= 27,878,400 square feet*
	= 640 acres*
	= 2.589988110336 square kilometers*
1 square centimeter	= 0.1550003 square inch
	= 0.00107639 square foot
1 square meter	= 10.76391 square feet
	= 1.19599005 square yards
1 square kilometer	= 247.1053815 acres
	= 0.38610216 square statute mile
	= 0.29155335 square nautical mile

Astronomy

1 mean solar unit	= 1.00273791 sidereal units
1 sidereal unit	= 0.99726957 mean solar units
1 microsecond	= 0.000001 second*
1 second	= 1,000,000 microseconds*
	= 0.01666667 minute
	= 0.00027778 hour
	= 0.00001157 day
1 minute	= 60 seconds*
	= 0.01666667 hour
	= 0.00069444 day
1 hour	= 3,600 seconds*
	= 60 minutes*
	= 0.04166667 day
1 mean solar day	= $24^h03^m56^s.55536$ of mean sidereal time
	= 1 rotation of Earth with respect to Sun (mean)*
	= 1.00273791 rotations of Earth with respect to vernal equinox (mean)
	= 1.0027378118868 rotations of Earth with respect to stars (mean)

1 mean sidereal day	= $23^{h}56^{m}04^{s}09054$ of mean solar time
1 sidereal month	= 27.321661 days
	= $27^{d}07^{h}43^{m}11^{s}.5$
1 synodical month	= 29.530588 days
	= $29^{d}12^{h}44^{m}02^{s}.8$
1 tropical (ordinary) year	= 31,556,925.975 seconds
	= 525,948.766 minutes
	= 8,765.8128 hours
	= $365^{d}.24219879 - 0^{d}.0000000614(t-1900)$,
	where t = the year (date)
	= $365^{d}05^{h}48^{m}46^{s}$ (–) $0^{s}.0053(t-1900)$
1 sidereal year	= $365^{d}.25636042 + 0.0000000011(t-1900)$,
	where t = the year (date)
	= $365^{d}06^{h}09^{m}09^{s}.5$ (+) $0^{s}.0001(t-1900)$
1 calendar year (common)	= 31,536,000 seconds*
	= 525,600 minutes*
	= 8,760 hours*
	= 365 days*
1 calendar year (leap)	= 31,622,400 seconds*
	= 527,040 minutes*
	= 8,784 hours*
	= 366 days
1 light-year	= 9,460,000,000,000 kilometers
	= 5,880,000,000,000 statute miles
	= 5,110,000,000,000 nautical miles
	= 63,240 astronomical units
	= 0.3066 parsecs
1 parsec	= 30,860,000,000,000 kilometers
	= 19,170,000,000,000 statute miles
	= 16,660,000,000,000 nautical miles
	= 206,300 astronomical units
	= 3.262 light years
1 astronomical unit	= 149,600,000 kilometers
	= 92,960,000 statute miles
	= 80,780,000 nautical miles
	= $499^{s}.012$ light-time
	= mean distance, Earth to Sun
Mean distance, Earth to Moon	= 384,400 kilometers
	= 238,855 statute miles
	= 207,559 nautical miles
Mean distance, Earth to Sun	= 149,600,000 kilometers
	= 92,957,000 statute miles
	= 80,780,000 nautical miles
	= 1 astronomical unit
Sun's diameter	= 1,392,000 kilometers
	= 865,000 statute miles
	= 752,000 nautical miles
Sun's mass	= 1,987,000,000,000,000,000,000,000,000,000,000 grams
	= 2,200,000,000,000,000,000,000,000,000 short tons
	= 2,000,000,000,000,000,000,000,000,000 long tons
Speed of Sun relative to neighboring stars	= 19.4 kilometers per second
	= 12.1 statute miles per second
	= 10.5 nautical miles per second
Orbital speed of Earth	= 29.8 kilometers per second
	= 18.5 statute miles per second
	= 16.1 nautical miles per second
Obliquity of the ecliptic	= $23°27'08''.26 - 0''.4684\ (t-1900)$,
	where t = the year (date)
General precession of the equinoxes	= $50''.2564 + 0''.000222\ (t-1900)$, per year,
	where t = the year (date)
Precession of the equinoxes in right ascension	= $46''.0850 + 0''.000279\ (t-1900)$, per year,
	where t = the year (date)

Precession of the equinoxes in declination	= 20″.0468 – 0″.000085 (t–1900), per year, where t = the year (date)
Magnitude ratio	= 2.512
	= $\sqrt[5]{100}$*

Charts

Nautical miles per inch	= reciprocal of natural scale ÷ 72,913.39
Statute miles per inch	= reciprocal of natural scale ÷ 63,360*
Inches per nautical mile	= 72,913.39 × natural scale
Inches per statute mile	= 63,360 × natural scale*
Natural scale	= 1:72,913.39 × nautical miles per inch
	= 1:63,360 × statute miles per inch*

Earth

Acceleration due to gravity (standard)	= 980.665 centimeters per second per second
	= 32.1740 feet per second per second
Mass-ratio—Sun/Earth	= 332,958
Mass-ratio—Sun/(Earth & Moon)	= 328,912
Mass-ratio—Earth/Moon	= 81.30
Mean density	= 5.517 grams per cubic centimeter
Velocity of escape	= 6.94 statute miles per second
Curvature of surface	= 0.8 foot per nautical mile

World Geodetic System (WGS) Ellipsoid of 1984

Equatorial radius (a)	= 6,378,137 meters
	= 3,443.918 nautical miles
Polar radius (b)	= 6,356,752.314 meters
	= 3432.372 nautical miles
Mean radius (2a + b)/3	= 6,371,008.770 meters
	= 3440.069 nautical miles
Flattening or ellipticity (f = 1 – b/a)	= 1/298.257223563
	= 0.003352811
Eccentricity ($e = (2f - f^2)^{1/2}$)	= 0.081819191
Eccentricity squared (e^2)	= 0.006694380

Length

1 inch	= 25.4 millimeters*
	= 2.54 centimeters*
1 foot (U.S.)	= 12 inches*
	= 1 British foot
	= $^1/_3$ yard*
	= 0.3048 meter*
	= $^1/_6$ fathom*
1 foot (U.S. Survey)	= 0.30480061 meter
1 yard	= 36 inches*
	= 3 feet*
	= 0.9144 meter*
1 fathom	= 6 feet*
	= 2 yards*
	= 1.8288 meters*
1 cable	= 720 feet*
	= 240 yards*
	= 219.4560 meters*
1 cable (British)	= 0.1 nautical mile
1 statute mile	= 5,280 feet*
	= 1,760 yards*
	= 1,609.344 meters*
	= 1.609344 kilometers*
	= 0.86897624 nautical mile
1 nautical mile	= 6,076.11548556 feet
	= 2,025.37182852 yards
	= 1,852 meters*

	= 1.852 kilometers*
	= 1.150779448 statute miles
1 meter	= 100 centimeters*
	= 39.370079 inches
	= 3.28083990 feet
	= 1.09361330 yards
	= 0.54680665 fathom
	= 0.00062137 statute mile
	= 0.00053996 nautical mile
1 kilometer	= 3,280.83990 feet
	= 1,093.61330 yards
	= 1,000 meters*
	= 0.62137119 statute mile
	= 0.53995680 nautical mile

Mass

1 ounce	= 437.5 grains*
	= 28.349523125 grams*
	= 0.0625 pound*
	= 0.028349523125 kilogram*
1 pound	= 7,000 grains*
	= 16 ounces*
	= 0.45359237 kilogram*
1 short ton	= 2,000 pounds*
	= 907.18474 kilograms*
	= 0.90718474 metric ton*
	= 0.8928571 long ton
1 long ton	= 2,240 pounds*
	= 1,016.0469088 kilograms*
	= 1.12 short tons*
	= 1.0160469088 metric tons*
1 kilogram	= 2.204623 pounds
	= 0.00110231 short ton
	= 0.0009842065 long ton
1 metric ton	= 2,204.623 pounds
	= 1,000 kilograms*
	= 1.102311 short tons
	= 0.9842065 long ton

Mathematics

π	= 3.14159265358979323846264338327950288841971
π^2	= 9.8696044011
$\sqrt{\pi}$	= 1.7724538509
Base of Naperian logarithms (e)	= 2.718281828459
Modulus of common logarithms ($\log_{10}e$)	= 0.4342944819032518
1 radian	= 206,264.″80625
	= 3,437′.7467707849
	= 57°.2957795131
	= 57°17′44″.80625
1 circle	= 1,296,000″*
	= 21,600′*
	= 360°*
	= 2π radians*
180°	= π radians*
1°	= 3600″*
	= 60′*
	= 0.0174532925199432957666 radian
1′	= 60″*
	= 0.000290888208665721596 radian
1″	= 0.0000048481368110953599333 radian
Sine of 1′	= 0.00029088820456342460
Sine of 1″	= 0.00000484813681107637

Meteorology

Atmosphere (dry air)

Nitrogen	= 78.08%	99.99% (Nitrogen, Oxygen, Argon, Carbon dioxide)
Oxygen	= 20.95%	
Argon	= 0.93%	
Carbon dioxide	= 0.03%	
Neon	= 0.0018%	
Helium	= 0.000524%	
Krypton	= 0.0001%	
Hydrogen	= 0.00005%	
Xenon	= 0.0000087%	
Ozone	= 0 to 0.000007% (increasing with altitude)	
Radon	= 0.000000000000000006% (decreasing with altitude)	

Standard atmospheric pressure at sea level	= 1,013.250 dynes per square centimeter
	= 1,033.227 grams per square centimeter
	= 1,033.227 centimeters of water
	= 1,013.250 hectopascals (millibars)*
	= 760 millimeters of mercury
	= 76 centimeters of mercury
	= 33.8985 feet of water
	= 29.92126 inches of mercury
	= 14.6960 pounds per square inch
	= 1.033227 kilograms per square centimeter
	= 1.013250 bars*
Absolute zero	= (–)273.16°C
	= (–)459.69°F

Pressure

1 dyne per square centimeter	= 0.001 hectopascal (millibar)*
	= 0.000001 bar*
1 gram per square centimeter	= 1 centimeter of water
	= 0.980665 hectopascal (millibar)*
	= 0.07355592 centimeter of mercury
	= 0.0289590 inch of mercury
	= 0.0142233 pound per square inch
	= 0.001 kilogram per square centimeter*
	= 0.000967841 atmosphere
1 hectopascal (millibar)	= 1,000 dynes per square centimeter*
	= 1.01971621 grams per square centimeter
	= 0.7500617 millimeter of mercury
	= 0.03345526 foot of water
	= 0.02952998 inch of mercury
	= 0.01450377 pound per square inch
	= 0.001 bar*
	= 0.00098692 atmosphere
1 millimeter of mercury	= 1.35951 grams per square centimeter
	= 1.3332237 hectopascals (millibars)
	= 0.1 centimeter of mercury*
	= 0.04460334 foot of water
	= 0.039370079 inch of mercury
	= 0.01933677 pound per square inch
	= 0.001315790 atmosphere
1 centimeter of mercury	= 10 millimeters of mercury*
1 inch of mercury	= 34.53155 grams per square centimeter
	= 33.86389 hectopascals (millibars)
	= 25.4 millimeters of mercury*
	= 1.132925 feet of water
	= 0.4911541 pound per square inch
	= 0.03342106 atmosphere
1 centimeter of water	= 1 gram per square centimeter
	= 0.001 kilogram per square centimeter
1 foot of water	= 30.48000 grams per square centimeter
	= 29.89067 hectopascals (millibars)
	= 2.241985 centimeters of mercury
	= 0.882671 inch of mercury
	= 0.4335275 pound per square inch
	= 0.02949980 atmosphere
1 pound per square inch	= 68,947.57 dynes per square centimeter
	= 70.30696 grams per square centimeter

	= 70.30696 centimeters of water
	= 68.94757 hectopascals (millibars)
	= 51.71493 millimeters of mercury
	= 5.171493 centimeters of mercury
	= 2.306659 feet of water
	= 2.036021 inches of mercury
	= 0.07030696 kilogram per square centimeter
	= 0.06894757 bar
	= 0.06804596 atmosphere
1 kilogram per square centimeter	= 1,000 grams per square centimeter*
	= 1,000 centimeters of water
1 bar	= 1,000,000 dynes per square centimeter*
	= 1,000 hectopascals (millibars)*

Speed

1 foot per minute	= 0.01666667 foot per second
	= 0.00508 meter per second*
1 yard per minute	= 3 feet per minute*
	= 0.05 foot per second*
	= 0.03409091 statute mile per hour
	= 0.02962419 knot
	= 0.01524 meter per second*
1 foot per second	= 60 feet per minute*
	= 20 yards per minute*
	= 1.09728 kilometers per hour*
	= 0.68181818 statute mile per hour
	= 0.59248380 knot
	= 0.3048 meter per second*
1 statute mile per hour	= 88 feet per minute*
	= 29.33333333 yards per minute
	= 1.609344 kilometers per hour*
	= 1.46666667 feet per second
	= 0.86897624 knot
	= 0.44704 meter per second*
1 knot	= 101.26859143 feet per minute
	= 33.75619714 yards per minute
	= 1.852 kilometers per hour*
	= 1.68780986 feet per second
	= 1.15077945 statute miles per hour
	= 0.51444444 meter per second
1 kilometer per hour	= 0.62137119 statute mile per hour
	= 0.53995680 knot
1 meter per second	= 196.85039340 feet per minute
	= 65.6167978 yards per minute
	= 3.6 kilometers per hour*
	= 3.28083990 feet per second
	= 2.23693632 statute miles per hour
	= 1.94384449 knots
Light in vacuum	= 299,792.5 kilometers per second
	= 186,282 statute miles per second
	= 161,875 nautical miles per second
	= 983.570 feet per microsecond
Light in air	= 299,708 kilometers per second
	= 186,230 statute miles per second
	= 161,829 nautical miles per second
	= 983.294 feet per microsecond
Sound in dry air at 59°F or 15°C and standard sea level pressure	= 1,116.45 feet per second
	= 761.22 statute miles per hour
	= 661.48 knots
	= 340.29 meters per second
Sound in 3.485 percent saltwater at 60°F	= 4,945.37 feet per second
	= 3,371.85 statute miles per hour

	= 2,930.05 knots
	= 1,507.35 meters per second

Volume

1 cubic inch	= 16.387064 cubic centimeters*
	= 0.016387064 liter*
	= 0.004329004 gallon
1 cubic foot	= 1,728 cubic inches*
	= 28.316846592 liters*
	= 7.480519 U.S. gallons
	= 6.228822 imperial (British) gallons
	= 0.028316846592 cubic meter*
1 cubic yard	= 46,656 cubic inches*
	= 764.554857984 liters*
	= 201.974026 U.S. gallons
	= 168.1782 imperial (British) gallons
	= 27 cubic feet*
	= 0.764554857984 cubic meter*
1 milliliter	= 0.06102374 cubic inch
	= 0.0002641721 U.S. gallon
	= 0.00021997 imperial (British) gallon
1 cubic meter	= 264.172035 U.S. gallons
	= 219.96878 imperial (British) gallons
	= 35.31467 cubic feet
	= 1.307951 cubic yards
1 quart (U.S.)	= 57.75 cubic inches*
	= 32 fluid ounces*
	= 2 pints*
	= 0.9463529 liter
	= 0.25 gallon*
1 gallon (U.S.)	= 3,785.412 milliliters
	= 231 cubic inches*
	= 0.1336806 cubic foot
	= 4 quarts*
	= 3.785412 liters
	= 0.8326725 imperial (British) gallon
1 liter	= 1,000 milliliters
	= 61.02374 cubic inches
	= 1.056688 quarts
	= 0.2641721 gallon
1 register ton	= 100 cubic feet*
	= 2.8316846592 cubic meters*
1 measurement ton	= 40 cubic feet*
	= 1 freight ton*
1 freight ton	= 40 cubic feet*
	= 1 measurement ton*

Volume-Mass

1 cubic foot of seawater	= 64 pounds
1 cubic foot of freshwater	= 62.428 pounds at temperature of maximum density (4°C = 39°.2F)
1 cubic foot of ice	= 56 pounds
1 displacement ton	= 35 cubic feet of seawater*
	= 1 long ton

Prefixes to Form Decimal Multiples and Sub-Multiples of International System of Units (SI)

Multiplying factor		Prefix	Symbol
1 000 000 000 000	$= 10^{12}$	tera	T
1 000 000 000	$= 10^{9}$	giga	G
1 000 000	$= 10^{6}$	mega	M
1 000	$= 10^{3}$	kilo	k
100	$= 10^{2}$	hecto	h
10	$= 10^{1}$	deka	da
0. 1	$= 10^{-1}$	deci	d
0. 01	$= 10^{-2}$	centi	c
0. 001	$= 10^{-3}$	milli	m
0. 000 001	$= 10^{-6}$	micro	μ
0. 000 000 001	$= 10^{-9}$	nano	n
0. 000 000 000 001	$= 10^{-12}$	pico	p
0. 000 000 000 000 001	$= 10^{-15}$	femto	f
0. 000 000 000 000 000 001	$= 10^{-18}$	atto	a

CHAPTER 23

NAVIGATIONAL ERRORS

DEFINING NAVIGATIONAL ERRORS

2300. Introduction

Navigation is an increasingly exact science. Electronic positioning systems give the navigator a greater certainty than ever that his position is correct within a few meters. However, the navigator makes certain assumptions which would be unacceptable in purely scientific work.

For example, when the navigator uses his latitude graduations as a mile scale to compute a great-circle course and distance, he neglects the flattening of the Earth at the poles. When the navigator plots a visual bearing on a Mercator chart, he uses a rhumb line to represent a great circle. When he plots a celestial line of position, he substitutes a rhumb line for a small circle. When he interpolates in sight reduction tables, he assumes a linear (constant-rate) change between tabulated values. All of these assumptions introduce errors.

There are so many approximations in navigation that there is a natural tendency for some of them to cancel others. However, if the various small errors in a particular fix all have the same sign, the error might be significant. The navigator must recognize the limitations of his positioning systems and understand the sources of position error.

The errors inherent in the use of various types of navigation systems are included in the chapters relating to those systems. This chapter discusses errors in general terms.

2301. Definitions

Error is the difference between a specific value and the correct or standard value. As used here, it does not include mistakes, but is related to lack of perfection. for example, an altitude determined by a marine sextant is corrected for a standard amount of refraction. But if the actual refraction at the time of observation varies from the standard, the value taken from the table is in error by the difference between standard and actual values. This error will be compounded with others in the observed altitude. Similarly, a depth determined by an echo sounder is in error, among other things, by the difference between the actual speed of sound waves in the water and the speed used for calibration of the instrument.

The navigator studying sources of error is concerned primarily with the deviation from standard values. Generalized corrections can be applied for standard values of error. It is the deviation from standard, as well as mistakes, that produce inaccurate results in navigation.

A **mistake** is a blunder, such as an incorrect reading of an instrument, the taking of a wrong value from a table, a data entry error, or plotting a reciprocal bearing.

A **standard** is a value or quantity established by custom, agreement, or authority as a basis for comparison. Frequently, a standard is chosen as a model which approximates a mean or average condition. However, the distinction between the standard value and the actual value at any time should not be forgotten. Thus, a standard atmosphere has been established in which the temperature, pressure, and density are precisely specified for each altitude. Actual conditions, however, are generally different from those defined by the standard atmosphere. Similarly, the values for dip given in the almanacs are considered standard by those who use them, but actual dip may be appreciably different due to non-standard atmospheric conditions.

Accuracy is the degree of conformance with the correct value, while **precision** is a measure of refinement of a value. Thus, an altitude determined by a marine sextant might be stated to the nearest 0.1', and yet be accurate only to the nearest 1.0' if the horizon is indistinct.

There are three types of accuracy with respect to navigation systems. The first is absolute accuracy, sometimes referred to as predictable or geodetic accuracy. This is the accuracy of a position with respect to the true geographic coordinates according to the particular datum being used. Repeatable accuracy is the accuracy with which a navigation system can return to a previously identified position. Relative accuracy is a measure of the ability of two different receivers of the same type to define a position at the same time.

2302. Systematic and Random Errors

Systematic errors are those which follow some rule by which they can be predicted. **Random errors**, on the other hand, are unpredictable. The laws of probability govern random errors.

If a navigator takes several measurements that are subject to random error and graphs the results, the error values would be normally distributed around a mean, or average, value. Suppose, for example, that a navigator takes 500 celestial observations. Table 2302 shows the frequency of each error in the measurement, and Figure 2302 shows a plot of these errors. The curve's height at any point represents the percentage of observations that can be expected to have the error indicated at that point. The probability of any similar observation having any given error is the proportion of the number of observations having this error to the total number of observations. Thus, the probability of an observation having an error of -3' is:

$$\frac{40}{500} = \frac{1}{12.5} = 0.08(8\%)$$

An important characteristic of a probability distribution is the **standard deviation**. For a normal error curve, square each error, sum the squares, and divide the sum by one less than the total number of measurements. Finally, take the square root of the quotient. In the illustration, the standard deviation is:

$$\sqrt{\frac{4474}{499}} = \sqrt{8.966} = 2.99$$

Error	No. of obs.	Percent of obs.
- 10′	0	0.0
- 9′	1	0.2
- 8′	2	0.4
- 7′	4	0.8
- 6′	9	1.8
- 5′	17	3.4
- 4′	28	5.6
- 3′	40	8.0
- 2′	53	10.6
- 1′	63	12.6
0	66	13.2
+ 1′	63	12.6
+ 2′	53	10.6
+ 3′	40	8.0
+ 4′	28	5.6
+ 5′	17	3.4
+ 6′	9	1.8
+ 7′	4	0.8
+ 8′	2	0.4
+ 9′	1	0.2
+10′	0	0.0
0	500	100.0

Table 2302. Normal distribution of random errors.

One standard deviation on either side of the mean defines the area under the probability curve in which lie 67 percent of all errors. Two standard deviations encompass 95 percent of all errors, and three standard deviations encompass 99 percent of all errors.

The normalized curve of any type of random error is symmetrical about the line representing zero error. This means that in the normalized plot every positive error is matched by a negative error of the same magnitude. The average of all readings is zero. Increasing the number of readings increases the probability that the errors will fit the normalized curve.

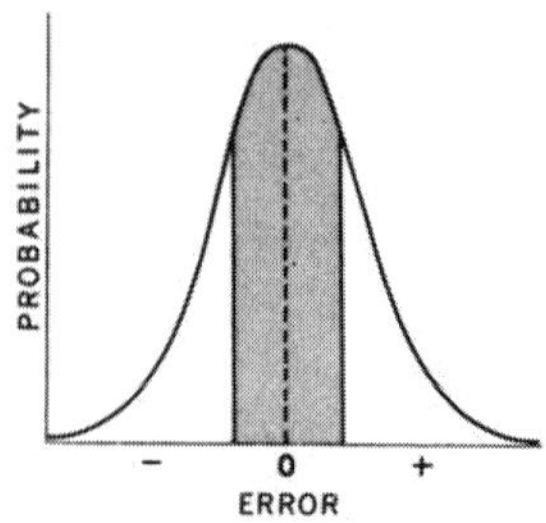

Figure 2302. Normal curve of random error with 50 percent of area shaded. Limits of shaded area indicate probable error.

When both systematic and random errors are present in a process, increasing the number of readings decreases the residual random error but does not decrease the systematic error. For example, if a number of phase-difference readings are made at a fixed point, the average of all the readings should be a good approximation of the true value if there is no systematic error. But increasing the number of readings will not correct a systematic error. If a constant error is combined with a normal random error, the error curve will have the correct shape but will be offset from the zero value.

2303. Navigation System Accuracy

In a navigation system, predictability is the measure of the accuracy with which the system can define the position in terms of geographical coordinates; repeatability is the measure of the accuracy with which the system permits the user to return to a position as defined only in terms of the coordinates peculiar to that system. For example, the distance specified for the repeatable accuracy of a system, such as Loran C, is the distance between two Loran C positions established using the same stations and time-difference readings at different times. The correlation between the geographical coordinates and the system coordinates may or may not be known.

Relative accuracy is the accuracy with which a user can determine his position relative to another user of the same navigation system, at the same time. Hence, a system with high relative accuracy provides good rendezvous capability for the users of the system. The correlation between the geographical coordinates and the system coordinates is not relevant.

2304. Most Probable Position

Some navigators have been led by simplified definitions and explanations to conclude that the line of position is almost infallible and that a good fix has very little error.

A more realistic concept is that of the most probable position (MPP), which recognizes the probability of error in all navigational information and determines position by an evaluation of all available information.

Suppose a vessel were to start from an accurate position and proceed on dead reckoning. If course and speed over the bottom were of equal accuracy, the uncertainty of dead reckoning positions would increase equally in all directions, with either distance or elapsed time (for any one speed these would be directly proportional, and therefore either could be used). A circle of uncertainty would grow around the dead reckoning position as the vessel proceeded. If the navigator had full knowledge of the distribution and nature of the errors of course and speed, and the necessary knowledge of statistical analysis, he could compute the radius of a circle of uncertainty, using the 50 percent, 95 percent, or other probabilities. This technique is known as fix expansion when done graphically. See Chapter 7 for a more detailed discussion of fix expansion.

In ordinary navigation, statistical computation is not practicable. However, the navigator might estimate at any time the likely error of his dead reckoning or estimated position. With practice, considerable skill in making this estimate is possible. He would take into account, too, the fact that the area of uncertainty might better be represented by an ellipse than a circle, with the major axis along the course line if the estimated error of the speed were greater than that of the course and the minor axis along the course line if the estimated error of the course were greater. He would recognize, too, that the size of the area of uncertainty would not grow in direct proportion to the distance or elapsed time, because disturbing factors, such as wind and current, could not be expected to remain of constant magnitude and direction. Also, he would know that the starting point of the dead reckoning might not be completely free from error.

The navigator can combine an LOP with either a dead reckoning or estimated position to determine an MPP. Determining the accuracy of the dead reckoning and estimated positions from which an MPP is determined is primarily a judgment call by the navigator. See Figure 2304a.

If a fix is obtained from two lines of position, the area of uncertainty is a circle if the lines are perpendicular and have equal error. If one is considered more accurate than the other, the area is an ellipse. As shown in Figure 2304b, it is also an ellipse if the likely error of each is equal and the lines cross at an oblique angle. If the errors are unequal, the major axis of the ellipse is more nearly in line with the line of position having the smaller likely error.

If a fix is obtained from three or more lines of position

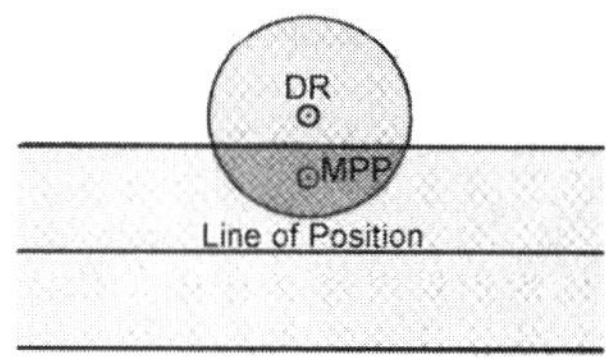

Figure 2304a. A most probable position based upon a dead reckoning position and line of position having equal probable errors.

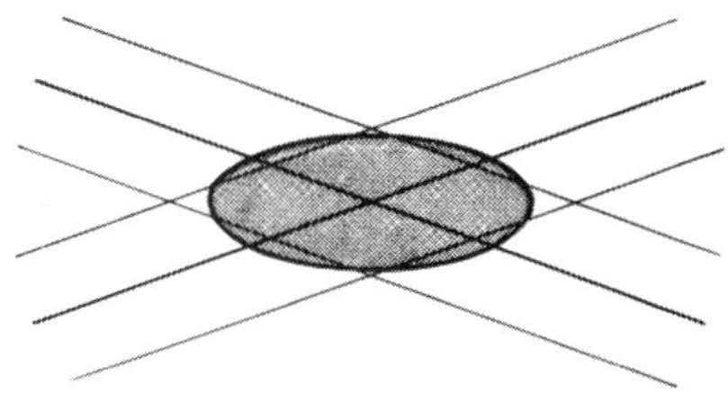

Figure 2304b. Ellipse of uncertainty with lines of positions of equal probable errors crossing at an oblique angle.

with a total bearing spread greater 180°, and the error of each line is normally distributed and equal to that of the others, the most probable position is the point within the figure equidistant from the sides. If the lines are of unequal error, the distance of the most probable position from each line of position varies as a function of the accuracy of each LOP.

Systematic errors are treated differently. Generally, the navigator tries to discover the errors and eliminate them or compensate for them. In the case of a position determined by three or more lines of position resulting from readings with constant error, the error might be eliminated by finding and applying that correction which will bring all lines through a common point.

Lines of position which are known to be of uncertain accuracy might better be considered as "bands of position", with a width of twice the possible amount of error. Intersecting bands of position define areas of position. It is most probable that the vessel is near the center of the area, but the navigator must realize that he could be anywhere within the area, and must navigate accordingly.

2305. Mistakes

The recognition of a mistake, as contrasted with an error, is not always easy, since a mistake is random, may have any magnitude, and may be either positive or negative. A large mistake should be readily apparent if the navigator is alert and has an understanding of the size of error to be reasonably expected. A small mistake is usually not detected unless the work is checked.

If results by two methods are compared, such as a dead reckoning position and a line of position, exact agreement is unlikely. But, if the discrepancy is unreasonably large, a mistake is a logical conclusion. If the 99.9 percent areas of the two results just touch, it is possible that no mistake has been made. However, the probability of either one having so great an error is remote if the errors are normal. The probability of both having 99.9 percent error of opposite sign at the same instant is extremely small. Perhaps a reasonable standard is that unless the most accurate result lies within the 95 percent area of the least accurate result, the possibility of a mistake should be investigated.

2306. Conclusion

A navigator need not understand the mathematical theory of error probability to navigate his ship safely. However, he must understand that his systems and processes are subject to numerous errors, some random, some systematic, and all potentially dangerous.

From a practical standpoint, the best policy is to never trust any single aid to navigation system explicitly in all situations. Some backup method must be used regularly to check for errors in the primary system. As the navigator becomes more and more confident in using a certain system or method, it is easy to also become dependent on that one system. He must understand his systems' limitations and use this understanding to bring his ship safely into harbor.

CHAPTER 24

THE SAILINGS

INTRODUCTION

2400. Introduction

Dead reckoning involves the determination of one's present or future position by projecting the ship's course and distance run from a known position. A closely related problem is that of finding the course and distance from one known point to another. For short distances, these problems are easily solved directly on charts, but for trans-oceanic distances, a purely mathematical solution is often a better method. Collectively, these methods are called **The Sailings**.

Navigational computer programs and calculators commonly contain algorithms for computing all of the problems of the sailings. For those situations when a calculator is not available, this chapter discusses hand calculation methods and tabular solutions. Navigators can also refer to NIMA *Pub. 151, Distances Between Ports*, for distances along normal ocean routes.

Because most commonly used formulas for the sailings are based on rules of spherical trigonometry and assume a perfectly spherical Earth, there may be inherent errors in the calculated answers. Errors of a few miles over distances of a few thousand can be expected. These will generally be much less than errors due to currents, steering error, and leeway.

To increase the accuracy of these calculations, one would have to take into account the oblateness of the Earth. Formulas exist which account for oblateness, reducing these errors to less than the length of the typical vessel using them, but far larger errors can be expected on any voyage of more than a few day's duration.

2401. Rhumb Lines and Great Circles

The principal advantage of a rhumb line is that it maintains constant true direction. A ship following the rhumb line between two places does not change its true course. A rhumb line makes the same angle with all meridians it crosses and appears as a straight line on a Mercator chart. For any other case, the difference between the rhumb line and the great circle connecting two points increases (1) as the latitude increases, (2) as the difference of latitude between the two points decreases, and (3) as the difference of longitude increases.

A great circle is the intersection of the surface of a sphere and a plane passing through the center of the sphere. It is the largest circle that can be drawn on the surface of the sphere, and is the shortest distance along the surface between any two points. Any two points are connected by only one great circle unless the points are **antipodal** (180° apart on the Earth), and then an infinite number of great circles passes through them. Every great circle bisects every other great circle. Thus, except for the equator, every great circle lies exactly half in the Northern Hemisphere and half in the Southern Hemisphere. Any two points 180° apart on a great circle have the same latitude numerically, but contrary names, and are 180° apart in longitude. The point of greatest latitude is called the **vertex**. For each great circle, there is a vertex in each hemisphere, 180° apart in longitude. At these points the great circle is tangent to a parallel of latitude, and its direction is due east-west. On each side of these vertices, the direction changes progressively until the intersection with the equator is reached, 90° in longitude away, where the great circle crosses the equator at an angle equal to the latitude of the vertex.

On a Mercator chart, a great circle appears as a sine curve extending equal distances each side of the equator. The rhumb line connecting any two points of the great circle on the same side of the equator is a chord of the curve. Along any intersecting meridian the great circle crosses at a higher latitude than the rhumb line. If the two points are on opposite sides of the equator, the direction of curvature of the great circle relative to the rhumb line changes at the equator. The rhumb line and great circle may intersect each other, and if the points are equal distances on each side of the equator, the intersection takes place at the equator.

Great circle sailing takes advantage of the shorter distance along the great circle between two points, rather than the longer rhumb line. The arc of the great circle between the points is called the **great circle track**. If it could be followed exactly, the destination would be dead ahead throughout the voyage (assuming course and heading were the same). The rhumb line appears the more direct route on a Mercator chart because of chart distortion. The great circle crosses meridians at higher latitudes, where the distance between them is less. This is why the great circle route is shorter than the rhumb line.

The decision as to whether or not to use great circle sailing depends upon the conditions. The savings in distance should be worth the additional effort, and of course the great circle route cannot cross land, nor should it carry the vessel into dangerous waters. **Composite sailing** (see Article 2402 and Article 2410) may save time and distance over the rhumb line track without leading the vessel into danger.

Since a great circle other than a meridian or the equator is a curved line whose true direction changes continually, the navigator does not attempt to follow it exactly. Instead, he selects a number of waypoints along the great circle, constructs rhumb lines between the waypoints, and steers along these rhumb lines.

2402. Kinds of Sailings

There are seven types of sailings:

1. **Plane sailing** solves problems involving a single course and distance, difference of latitude, and departure, in which the Earth is regarded as a plane surface. This method, therefore, provides solution for latitude of the point of arrival, but not for longitude. To calculate the longitude, the spherical sailings are necessary. Plane sailing is not intended for distances of more than a few hundred miles.
2. **Traverse sailing** combines the plane sailing solutions when there are two or more courses and determines the equivalent course and distance made good by a vessel steaming along a series of rhumb lines.
3. **Parallel sailing** is the interconversion of departure and difference of longitude when a vessel is proceeding due east or due west.
4. **Middle- (or mid-) latitude sailing** uses the mean latitude for converting departure to difference of longitude when the course is not due east or due west.
5. **Mercator sailing** provides a mathematical solution of the plot as made on a Mercator chart. It is similar to plane sailing, but uses meridional difference and difference of longitude in place of difference of latitude and departure.
6. **Great circle sailing** involves the solution of courses, distances, and points along a great circle between two points.
7. **Composite sailing** is a modification of great circle sailing to limit the maximum latitude, generally to avoid ice or severe weather near the poles.

2403. Terms and Definitions

In solutions of the sailings, the following quantities are used:

1. **Latitude (L)**. The latitude of the point of departure is designated L_1; that of the destination, L_2; middle (mid) or mean latitude, L_m; latitude of the vertex of a great circle, L_v; and latitude of any point on a great circle, L_x.
2. **Mean latitude (L_m)**. Half the arithmetical sum of the latitudes of two places on the same side of the equator.
3. **Middle or mid latitude (L_m)**. The latitude at which the arc length of the parallel separating the meridians passing through two specific points is exactly equal to the departure in proceeding from one point to the other by mid-latitude sailing. The mean latitude is used when there is no practicable means of determining the middle latitude.
4. **Difference of latitude (l or DLat.)**.
5. **Meridional parts (M)**. The meridional parts of the point of departure are designated M_1, and of the point of arrival or the destination, M_2.
6. **Meridional difference (m)**.
7. **Longitude (λ)**. The longitude of the point of departure is designated λ_1; that of the point of arrival or the destination, λ_2; of the vertex of a great circle, λ_v; and of any point on a great circle, λ_x.
8. **Difference of longitude (DLo)**.
9. **Departure (p or Dep.)**.
10. **Course** or **course angle (Cn or C)**.
11. **Distance (D or Dist.)**.

GREAT CIRCLE SAILING

2404. Great Circle Sailing by Chart

The graphic solution of great circle problems involves the use of two charts. NIMA publishes several gnomonic projections covering the principal navigable waters of the world. On these **great circle charts**, any straight line is a great circle. The chart, however, is not conformal; therefore, the navigator cannot directly measure directions and distances as on a Mercator chart.

The usual method of using a gnomonic chart is to plot the route and pick points along the track every 5° of longitude using the latitude and longitude scales in the immediate vicinity of each point. These points are then transferred to a Mercator chart and connected by rhumb lines. The course and distance for each leg can then be measured, and the points entered as waypoints in an electronic chart system, GPS, or Loran C. See Figure 2404..

2405. Great Circle Sailing by *Sight Reduction Tables*

Any method of solving a spherical triangle can be used for solving great circle sailing problems. The point of departure replaces the assumed position of the observer, the destination replaces the geographical position of the body, the difference of longitude replaces the meridian angle or local hour angle, the initial course angle replaces the azimuth angle, and the great

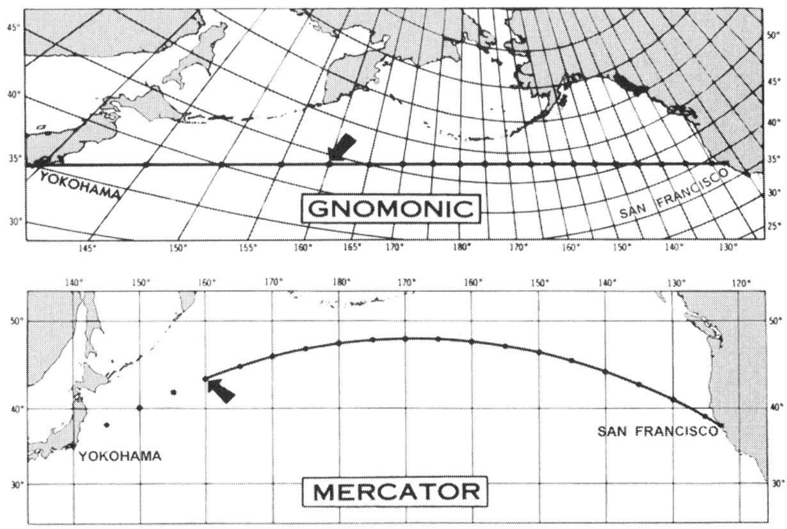

Figure 2404. Constructing a great circle track on a Mercator projection.

circle distance replaces the zenith distance (90° - altitude). See Figure 2405. Therefore, any table of azimuths (if the entering values are meridian angle, declination, and latitude) can be used for determining initial great circle course. Tables which solve for altitude, such as *Pub. No. 229*, can be used for determining great circle distance. The required distance is 90° - altitude.

In inspection tables such as *Pub. No. 229*, the given combination of L_1, L_2, and DLo may not be tabulated. In this case reverse the name of L_2 and use 180° - DLo for entering the table. The required course angle is then 180° minus the tabulated azimuth, and distance is 90° plus the altitude. If neither combination can be found, solution cannot be made by that method. By interchanging L_1 and L_2, one can find the supplement of the final course angle.

Solution by table often provides a rapid approximate check, but accurate results usually require triple interpolation. Except for *Pub. No. 229*, inspection tables do not provide a solution for points along the great circle. *Pub. No. 229* provides solutions for these points only if interpolation is not required.

By entering *Pub. No. 229* with the latitude of the point of departure as latitude, latitude of destination as declination, and difference of longitude as LHA, the tabular altitude and azimuth angle may be extracted and converted to great circle distance and course. As in sight reduction, the tables are entered according to whether the name of the latitude of the point of departure is the same as or contrary to the name of the latitude of the destination (declination). If the values correspond to those of a celestial body above the celestial horizon, 90° minus the arc of the tabular altitude becomes the distance; the tabular azimuth angle becomes the initial great circle course angle. If the respondents correspond to those of a celestial body below the celestial horizon, the arc of the tabular altitude plus 90° becomes the distance; the supplement of the tabular azimuth angle becomes the initial great circle course angle.

When the Contrary/Same (CS) Line is crossed in either direction, the altitude becomes negative; the body lies below the celestial horizon. For example: If the tables are entered with the LHA (DLo) at the bottom of a right-hand page and declination (L_2) such that the respondents lie above the CS Line, the CS Line has been crossed. Then the distance is 90° plus the tabular altitude; the initial course angle is the supplement of the tabular azimuth angle. Similarly, if the tables are entered with the LHA (DLo) at the top of a right-hand page and the respondents are found below the CS Line, the distance is 90° plus the tabular altitude; the initial course angle is the supplement of the tabular azimuth angle. If the tables are entered with the LHA (DLo) at the bottom of a right-hand page and the name of L_2 is contrary to L_1, the respondents are found in the column for L_1 on the facing page. In this case, the CS Line has been crossed; the distance is 90° plus the tabular altitude; the initial course angle is the supplement of the

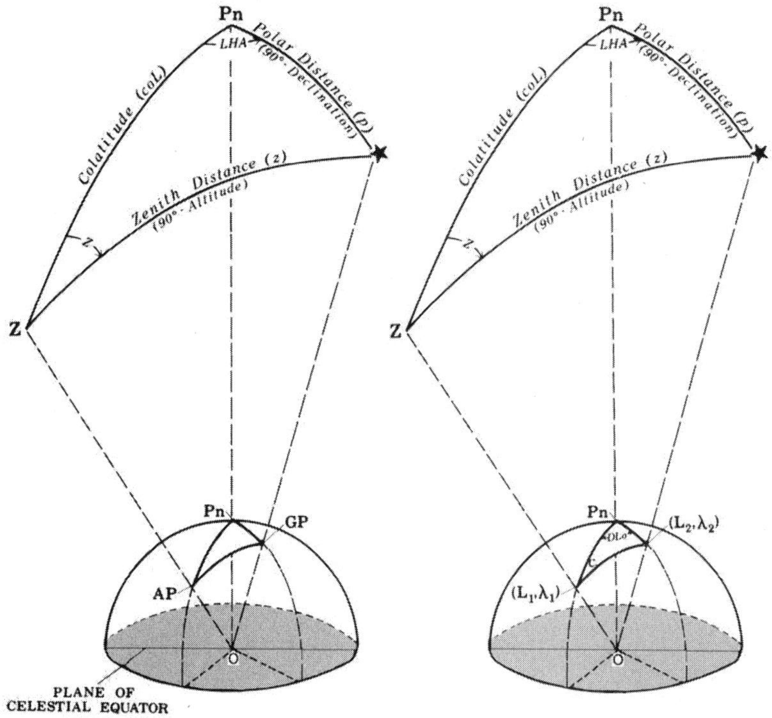

Figure 2405. Adapting the astronomical triangle to the navigational triangle of great circle sailing.

tabular azimuth angle.

The tabular azimuth angle, or its supplement, is prefixed N or S for the latitude of the point of departure and suffixed E or W depending upon the destination being east or west of the point of departure.

If all entering arguments are integral degrees, the distance and course angle are obtained directly from the tables without interpolation. If the latitude of the destination is nonintegral, interpolation for the additional minutes of latitude is done as in correcting altitude for any declination increment; if the latitude of departure or difference of longitude is nonintegral, the additional interpolation is done graphically.

Since the latitude of destination becomes the declination entry, and all declinations appear on every page, the great circle solution can always be extracted from the volume which covers the latitude of the point of departure.

Example 1: *Using Pub. No. 229, find the distance and initial great circle course from lat. 32°S, long. 116°E to lat. 30°S, long. 31°E.*

Solution: *Refer to Figure 2405. The point of departure (lat. 32°S, long. 116°E) replaces the AP of the observer; the destination (lat. 30°S, long. 31°E) replaces the GP of the celestial body; the difference of longitude (DLo 85°) replaces local hour angle (LHA) of the body.*

Enter Pub. No. 229, Volume 3 with lat. 32° (Same Name), LHA 85°, and declination 30°. The respondents correspond to a celestial body above the celestial horizon. Therefore, 90° minus the tabular altitude (90° - 19°12.4' = 70°47.6') becomes the distance; the tabular azimuth angle (S66.0°W) becomes the initial great circle course angle, prefixed S for the latitude of the point of departure and suffixed W due to the destination being west of the point of departure.

Answer:

D = 4248 nautical miles

C = S66.0°W = 246.0°.

Example 2: *Using Pub. No. 229, find the distance and initial great circle course from lat. 38°N, long. 122°W to lat. 24°S, long. 151°E.*

Solution: *Refer to Figure 2405. The point of departure (lat. 38°N, long. 122°W) replaces the AP of the observer; the destination (lat. 24°S, long. 151°E) replaces the GP of the celestial body; the difference of longitude (DLo 87°) replaces local hour angle (LHA) of the body*

Enter Pub. No. 229 Volume 3 with lat. 38° (Contrary Name), LHA 87°, and declination 24°. The respondents correspond to those of a celestial body below the celestial horizon. Therefore, the tabular altitude plus 90° (12°17.0' + 90° = 102°17.0') becomes the distance; the supplement of tabular azimuth angle (180° - 69.0° = 111.0°) becomes the initial great circle course angle, prefixed N for the latitude of the point of departure and suffixed W since the destination is west of the point of departure.

Note that the data is extracted from across the CS Line from the entering argument (LHA 87°), indicating that the corresponding celestial body would be below the celestial horizon.

Answer:
D = 6137 nautical miles
C = N111.0°W = 249°.

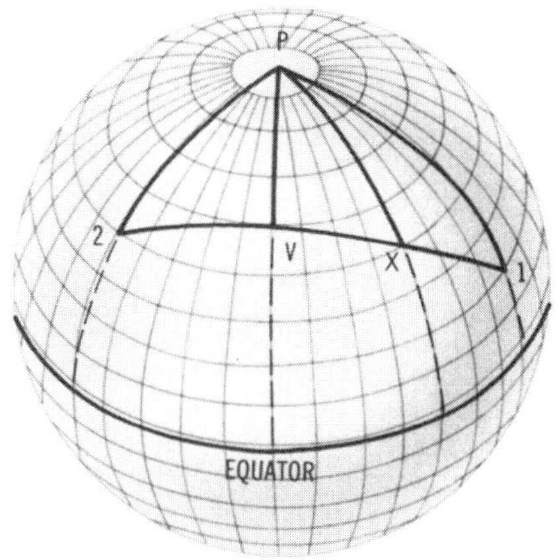

Figure 2406. The navigational triangle and great circle sailing.

2406. Great Circle Sailing by Computation

In Figure 2406, 1 is the point of departure, 2 the destination, P the pole nearer 1, 1-X-V-2 the great circle through 1 and 2, V the vertex, and X any point on the great circle. The arcs P1, PX, PV, and P2 are the colatitudes of points 1, X, V, and 2, respectively. If 1 and 2 are on opposite sides of the equator, P2 is 90°+ L_2. The length of arc 1-2 is the great circle distance between 1 and 2. Arcs 1-2, P1, and P2 form a spherical triangle. The angle at 1 is the initial great circle course from 1 to 2, that at 2 the supplement of the final great circle course (or the initial course from 2 to 1), and that at P the DLo between 1 and 2.

Great circle sailing by computation usually involves solving for the initial great circle course, the distance, latitude/longitude (and sometimes the distance) of the vertex, and the latitude and longitude of various points (X) on the great circle. The computation for initial course and the distance involves solution of an oblique spherical triangle, and any method of solving such a triangle can be used. If 2 is the geographical position (GP) of a celestial body (the point at which the body is in the zenith), this triangle is solved in celestial navigation, except that 90° - D (the altitude) is desired instead of D. The solution for the vertex and any point X usually involves the solution of right spherical triangles.

2407. Points Along the Great Circle

If the latitude of the point of departure and the initial great circle course angle are integral degrees, points along the great circle are found by entering the tables with the latitude of departure as the latitude argument (always Same Name), the initial great circle course angle as the LHA argument, and 90° minus distance to a point on the great circle as the declination argument. The latitude of the point on the great circle and the difference of longitude between that point and the point of departure are the tabular altitude and azimuth angle, respectively. If, however, the respondents are extracted from across the CS Line, the tabular altitude corresponds to a latitude on the side of the equator opposite from that of the point of departure; the tabular azimuth angle is the supplement of the difference of longitude.

Example 1: *Find a number of points along the great circle from latitude 38°N, longitude 125°W when the initial great circle course angle is N111°W.*

Solution: *Entering the tables with latitude 38° (Same Name), LHA 111°, and with successive* *declinations of 85°, 80°, 75°, etc., the latitudes and differences in longitude from 125°W are found as tabular altitudes and azimuth angles respectively:*

Answer:

D (NM)	*300*	*600*	*900*	*3600*
D (arc)	*5°*	*10°*	*15°*	*60°*
dec	*85°*	*80°*	*75°*	*30°*
Lat.	*36.1° N*	*33.9° N*	*31.4° N*	*3.6° N*
Dep.	*125° W*	*125° W*	*125° W*	*125° W*
DLo	*5.8°*	*11.3°*	*16.5°*	*54.1°*
Long	*130.8°W*	*136.3°W*	*141.5°W*	*179.1°*

Example 2: *Find a number of points along the great circle track from latitude 38°N, long. 125°W when the initial great circle course angle is N 69° W.*

Solution: *Enter the tables with latitude 38° (Same Name), LHA 69°, and with successive declinations as shown. Find the latitudes and differences of longitude from 125°W as tabular altitudes and azimuth angles, respectively:*

Answer:

D (NM.)	*300*	*600*	*900*	*6600*
D (arc)	*5°*	*10°*	*15°*	*110°*
dec	*85°*	*80°*	*75°*	*20°*
Lat.	*39.6° N*	*40.9° N*	*41.9° N*	*3.1° N*
Dep.	*125° W*	*125° W*	*125° W*	*125° W*
DLo	*6.1°*	*12.4°*	*18.9°*	*118.5°*
Long	*131.1°W*	*137.4°W*	*143.9°W*	*116.5°E*

2408. Finding the Vertex

Using *Pub. No. 229* to find the approximate position of the vertex of a great circle track provides a rapid check on the solution by computation. This approximate solution is also useful for voyage planning purposes.

Using the procedures for finding points along the great circle, inspect the column of data for the latitude of the point of departure and find the maximum value of tabular altitude. This maximum tabular altitude and the tabular azimuth angle correspond to the latitude of the vertex and the difference of longitude of the vertex and the point of departure.

Example 1: *Find the vertex of the great circle track from lat. 38°N, long. 125°W when the initial great circle course angle is N69°W.*

Solution: *Enter Pub. No. 229 with lat. 38° (Same Name), LHA 69°, and inspect the column for lat. 38° to find the maximum tabular altitude. The maximum altitude is 42°38.1' at a distance of 1500 nautical miles (90° - 65° = 25°) from the point of departure. The corresponding tabular azimuth angle is 32.4°. Therefore, the difference of longitude of vertex and point of departure is 32.4°.*

Answer:

Latitude of vertex = 42°38.1'N.
Longitude of vertex = 125° + 32.4° = 157.4°W.

2409. Altering a Great Circle Track to Avoid Obstructions

Land, ice, or severe weather may prevent the use of great circle sailing for some or all of one's route. One of the principal advantages of the solution by great circle chart is that any hazards become immediately apparent. The pilot charts are particularly useful in this regard. Often a relatively short run by rhumb line is sufficient to reach a point from which the great circle track can be followed. Where a choice is possible, the rhumb line selected should conform as nearly as practicable to the direct great circle.

If the great circle route passes too near a navigation hazard, it may be necessary to follow a great circle to the vicinity of the hazard, one or more rhumb lines along the edge of the hazard, and another great circle to the destination. Another possible solution is the use of composite sailing; still another is the use of two great circles, one from the point of departure to a point near the maximum latitude of unobstructed water and the second from this point to the destination.

2410. Composite Sailing

When the great circle would carry a vessel to a higher latitude than desired, a modification of great circle sailing called **composite sailing** may be used to good advantage. The composite track consists of a great circle from the point of departure and tangent to the limiting parallel, a course line along the parallel, and a great circle tangent to the limiting parallel and through the destination.

Solution of composite sailing problems is most easily made with a great circle chart. For this solution, draw lines from the point of departure and the destination, tangent to the limiting parallel. Then measure the coordinates of various selected points along the composite track and transfer them to a Mercator chart, as in great circle sailing. Composite sailing problems can also be solved by computation, using the equation:

$$\cos DLo_{VX} = \tan L_x \cot L_v$$

The point of departure and the destination are used successively as point X. Solve the two great circles at each end of the limiting parallel, and use parallel sailing along the limiting parallel. Since both great circles have vertices at the same parallel, computation for C, D, and DLo_{vx} can be made by considering them parts of the same great circle with L_1, L_2, and L_v as given and DLo = DLo_{v1} + DLo_{v2}. The total distance is the sum of the great circle and parallel distances.

TRAVERSE TABLES

2411. Using Traverse Tables

Traverse tables can be used in the solution of any of the sailings except great circle and composite. They consist of the tabulation of the solutions of plane right triangles. Because the solutions are for integral values of the course angle and the distance, interpolation for intermediate values may be required. Through appropriate interchanges of the headings of the columns, solutions for other than plane sailing can be made. For the solution of the plane right triangle, any value N in the distance (Dist.) column is the hypotenuse; the value opposite in the difference of latitude (D. Lat.) column is the product of N and the cosine of the acute angle; and the other number opposite in the departure (Dep.) column is the product of N and the sine of the acute angle. Or, the number in the D. Lat. column is the value of the side adjacent, and the number in the Dep. column is the value of the side opposite the acute angle. Hence, if the acute angle is the course angle, the side adjacent in the D. Lat. column is meridional difference m; the side opposite in the Dep. column is DLo. If the acute angle is the midlatitude of the formula p = DLo cos Lm, then DLo is any value N in the Dist. column, and the departure is the value N × cos L_m in the D. Lat. column.

The examples below clarify the use of the traverse tables for plane, traverse, parallel, mid latitude, and Mercator sailings.

2412. Plane Sailing

In plane sailing the figure formed by the meridian through the point of departure, the parallel through the point of arrival, and the course line is considered a plane right triangle. This is illustrated in Figure 2412a. P_1 and P_2 are the points of departure and arrival, respectively. The course angle and the three sides are as labeled. From this triangle:

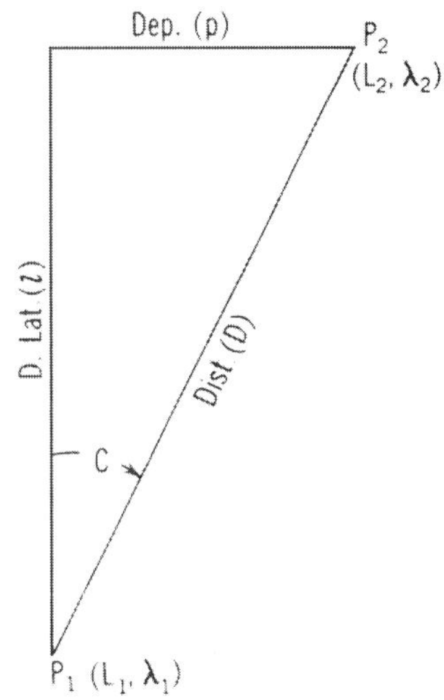

Figure 2412a. The plane sailing triangle.

$$\cos C = \frac{l}{D} \qquad \sin C = \frac{p}{D} \qquad \tan C = \frac{p}{l}.$$

From the first two of these formulas the following relationships can be derived:

$$l = D \cos C \qquad D = l \sec C \qquad p = D \sin C.$$

Label l as N or S, and p as E or W, to aid in identification of the quadrant of the course. Solutions by calculations and traverse tables are illustrated in the following examples:

Example 1: *A vessel steams 188.0 miles on course 005°.*

Required: *(1) (a) Difference of latitude and (b) departure by computation. (2) (a) difference of latitude and (b) departure by traverse table.*

Solution:

(1) (a) Difference of latitude by computation:

diff latitude $= D \times \cos C$
$= 188.0 \text{ miles} \times \cos(005°)$
$= 187.3 \text{ arc min}$
$= 3° \, 07.3' \text{ N}$

(1) (b) Departure by computation:

departure $= D \times \sin C$

$= 188.0 \text{ miles} \times \sin(005°)$

$= 16.4 \text{ miles}$

Answer:

Diff. Lat. = 3° 07.3' N

departure = 16.4 miles

(2) Difference of latitude and departure by traverse table:

Refer to Figure 2412b. Enter the traverse table and find course 005° at the top of the page. Using the column headings at the top of the table, opposite 188 in the Dist. column extract D. Lat. 187.3 and Dep. 16.4.

(a) D. Lat. = 187.3' N.

(b) Dep. = 16.4 mi. E.

Example 2: *A ship has steamed 136.0 miles north and 203.0 miles west.*

Required: *(1) (a) Course and (b) distance by computation. (2) (a) course and (b) distance by traverse table.*

Solution:

$$C = \arctan\frac{203.0}{136.0}$$

(1) (a) Course by computation:

$$C = \arctan\frac{departure}{diff.\ \text{lat.}}$$

$$C = N\ 56°\ 10.8'\ W$$

$$C = 304° (\text{to nearest degree})$$

Draw the course vectors to determine the correct course. In this case the vessel has gone north 136 miles and west 203 miles. The course, therefore, must have been between 270° and 360°. No solution other than 304° is reasonable.

(1) (b) Distance by computation:

$D = diff.\ latitude \times \sec C$

$= 136 \text{ miles} \times \sec(304°)$

$= 136 \text{ miles} \times 1.8$

$= 244.8 \text{ miles}$

Answer:

C = 304°

D = 244.8 miles

355° | 005° / 185° | 175° TABLE 4 Traverse 5° Table 355° | 005° / 185° | 175°

Dist.	D. Lat.	Dep.	Dist.	D. Lat.	Dep.	Dist.	D. Lat.	Dep.	Dist.	D. Lat.	Dep.	Dist.	D. Lat.	Dep.
1	1.0	0.1	61	60.8	5.3	121	120.5	10.5	181	180.3	15.8	241	240.1	21.0
2	2.0	0.2	62	61.8	5.4	22	121.5	10.6	82	181.3	15.9	42	241.1	21.1
3	3.0	0.3	63	62.8	5.5	23	122.5	10.7	83	182.3	15.9	43	242.1	21.2
4	4.0	0.3	64	63.8	5.6	24	123.5	10.8	84	183.3	16.0	44	243.1	21.3
5	5.0	0.4	65	64.8	5.7	25	124.5	10.9	85	184.3	16.1	45	244.1	21.4
6	6.0	0.5	66	65.7	5.8	26	125.5	11.0	86	185.3	16.2	46	245.1	21.4
7	7.0	0.6	67	66.7	5.8	27	126.5	11.1	87	186.3	16.3	47	246.1	21.5
8	8.0	0.7	68	67.7	5.9	28	127.5	11.2	88	187.3	16.4	48	247.1	21.6
9	9.0	0.8	69	68.7	6.0	29	128.5	11.2	89	188.3	16.5	49	248.1	21.7
10	10.0	0.9	70	69.7	6.1	30	129.5	11.3	90	189.3	16.6	50	249.0	21.8
60	59.8	5.2	20	119.5	10.5	80	179.3	15.7	40	239.1	20.9	300	298.9	26.1
Dist.	Dep.	D. Lat.	Dist.	Dep.	D. Lat.	Dist.	Dep.	D. Lat.	Dist.	Dep.	D. Lat.	Dist.	Dep.	D. Lat.

275° | 085° / 265° | 095° 85°

Dist.	D. Lat.	Dep.
N.	N x Cos.	N x Sin.
Hypotenuse	Side Adj.	Side Opp.

Figure 2412b. Extract from Table 4.

326° | 034° / 214° | 146°				TABLE 4 Traverse 34° Table									326° | 034° / 214° | 146°	
Dist.	D. Lat.	Dep.	Dist.	D. Lat.	Dep.	Dist.	D. Lat.	Dep.	Dist.	D. Lat.	Dep.	Dist.	D. Lat.	Dep.
1	0.8	0.6	61	50.6	34.1	121	100.3	67.7	181	150.1	101.2	241	199.8	134.8
2	1.7	1.1	62	51.4	34.7	22	101.1	68.2	82	150.9	101.8	42	200.6	135.3
3	2.5	1.7	63	52.2	35.2	23	102.0	68.8	83	151.7	102.3	43	201.5	135.9
4	3.3	2.2	64	53.1	35.8	24	102.8	69.3	84	152.5	102.9	44	202.3	136.4
5	4.1	2.8	65	53.9	36.3	25	103.6	69.9	85	153.4	103.5	45	203.1	137.0
60	49.7	33.6	20	99.5	67.1	80	149.2	100.7	40	199.0	134.2	300	248.7	167.8
Dist.	Dep.	D. Lat.	Dist.	Dep.	D. Lat.	Dist.	Dep.	D. Lat.	Dist.	Dep.	D. Lat.	Dist.	Dep.	D. Lat.
304° | 056° / 236° | 124°				56°										

Dist.	D. Lat.	Dep.
N.	N x Cos.	N x Sin.
Hypotenuse	Side Adj.	Side Opp.

Figure 2412c. Extract from Table 4.

(2) Solution by traverse table:

Refer to Figure 2412c. Enter the table and find 136 and 203 beside each other in the columns labeled D. Lat. and Dep., respectively. This occurs most nearly on the page for course angle 56°. Therefore, the course is 304°. Interpolating for intermediate values, the corresponding number in the Dist. column is 244.3 miles.

Answer:

(a) C = 304°
(b) D = 244.3 mi.

2413. Traverse Sailing

A **traverse** is a series of courses or a track consisting of a number of course lines, such as might result from a sailing vessel beating into the wind. **Traverse sailing** is the finding of a single equivalent course and distance.

Though the problem can be solved graphically on the chart, traverse tables provide a mathematical solution. The distance to the north or south and to the east or west on each course is tabulated, the algebraic sum of difference of latitude and departure is found, and converted to course and distance.

Example: *A ship steams as follows: course 158°, distance 15.5 miles; course 135°, distance 33.7 miles; course 259°, distance 16.1 miles; course 293°, distance 39.0 miles; course 169°, distance 40.4 miles.*

Required: *Equivalent single (1) course (2) distance.*

Solution: *Solve each leg as a plane sailing and tabulate each solution as follows: For course 158°, extract the values for D. Lat. and Dep. opposite 155 in the Dist. column. Then, divide the values by 10 and round them off to the nearest tenth. Repeat the procedure for each leg.*

Course degrees	*Dist. mi.*	*N mi.*	*S mi.*	*E mi.*	*W mi.*
158	*15.5*		*14.4*	*5.8*	
135	*33.7*		*23.8*	*23.8*	
259	*16.1*		*3.1*		*15.8*
293	*39.0*	*15.2*			*35.9*
169	*40.4*		*39.7*	*7.7*	
Subtotals		*15.2*	*81.0*	*37.3*	*51.7*
			-15.2		*-37.3*
N/S Total			*65.8 S*		*14.4 W*

Thus, the latitude difference is S 65.8 miles and the departure is W 14.4 miles. Convert this to a course and distance using the formulas discussed in Article 2413.

Answer:

(1) C = 192.3°
(2) D = 67.3 miles.

2414. Parallel Sailing

Parallel sailing consists of the interconversion of departure and difference of longitude. It is the simplest form of spherical sailing. The formulas for these transformations are:

$$DLo = p \sec L \qquad p = DLo \cos L$$

Example 1: *The DR latitude of a ship on course 090° is 49°30' N. The ship steams on this course until the longitude changes 3°30'.*

Required: *The departure by (1) computation and (2) traverse table.*

Solution:

(1) Solution by computation:

$DLo = 3° 30'$
$DLo = 210$ arc min
$p = DLo \times \cos L$
$p = 210$ arc minutes $\times \cos(49.5°)$
$p = 136.4$ miles

Answer:

p = 136.4 miles

(2) Solution by traverse table:

Refer to Figure 2414a. Enter the traverse table with latitude as course angle and substitute DLo as the heading of the Dist. column and Dep. as the heading of the D. Lat. column. Since the table is computed for integral degrees of course angle (or latitude), the tabulations in the pages for 49° and 50° must be interpolated for the intermediate value (49°30'). The departure for latitude 49° and DLo 210' is 137.8 miles. The departure for latitude 50° and DLo 210' is 135.0 miles. Interpolating for the intermediate latitude, the departure is 136.4 miles.

Answer:

p = 136.4 miles

Example 2: *The DR latitude of a ship on course 270° is 38°15'S. The ship steams on this course for a distance of 215.5 miles.*

Required: *The change in longitude by (1) computation and (2) traverse table.*

Solution:

(1) Solution by computation

$DLo = 215.5$ arc min $\times \sec(38.25°)$
$DLo = 215.5$ arc min $\times 1.27$
$DLo = 274.4$ minutes of arc (west)
$DLo = 4° 34.4'$ W

Answer:

DLo = 4° 34.4' W

(2) Solution by traverse table

Refer to Figure 2414b. Enter the traverse tables with latitude as course angle and substitute DLo as the heading of the Dist. column and Dep. as the heading of the D. Lat. column. As the table is computed for integral degrees of course angle (or latitude), the tabulations in the pages for 38° and 39° must be interpolated for the minutes of latitude. Corresponding to Dep. 215.5 miles in the former is DLo 273.5', and in the latter DLo 277.3'. Interpolating for minutes of latitude, the DLo is 274.4'W.

Answer:

DLo = 4° 34.4'

2415. Middle-Latitude Sailing

Middle-latitude sailing combines plane sailing and parallel sailing. Plane sailing is used to find difference of latitude and departure when course and distance are known, or vice versa. Parallel sailing is used to interconvert departure and difference of longitude. The mean latitude (L_m) is normally used for want of a practical means of determining the middle latitude, or the latitude at which the arc length of the parallel separating the meridians passing through two specific points is exactly equal to the departure in proceeding from one point to the other. The formulas for these transformations are:

$$DLo = p \sec L_m \qquad p = DLo \cos L_m.$$

319° | 041° / 221° | 139°					TABLE 4 Traverse 41° Table								319° | 041° / 221° | 139°	
Dist.	D. Lat.	Dep.	Dist.	D. Lat.	Dep.	Dist.	D. Lat.	Dep.	Dist.	D. Lat.	Dep.	Dist.	D. Lat.	Dep.
21	15.8	13.8	81	61.1	53.1	141	106.4	92.5	201	151.7	131.9	261	197.0	171.2
22	16.6	14.4	82	61.9	53.8	42	107.2	93.2	02	152.5	132.5	62	197.7	171.9
23	17.4	15.1	83	62.6	54.5	43	107.9	93.8	03	153.2	133.2	63	198.5	172.5
24	18.1	15.7	84	63.4	55.1	44	108.7	94.5	04	154.0	133.8	64	199.2	173.2
25	18.9	16.4	85	64.2	55.8	45	109.4	95.1	05	154.7	134.5	65	200.0	173.9
26	19.6	17.1	86	64.9	56.4	46	110.2	95.8	06	155.5	135.1	66	200.8	174.5
27	20.4	17.7	87	65.7	57.1	47	110.9	96.4	07	156.2	135.8	67	201.5	175.2
28	21.1	18.4	88	66.4	57.7	48	111.7	97.1	08	157.0	136.5	68	202.3	175.8
29	21.9	19.0	89	67.2	58.4	49	112.5	97.8	09	157.7	137.1	69	203.0	176.5
30	22.6	19.7	90	67.9	59.0	50	113.2	98.4	10	158.5	137.8	70	203.8	177.1
Dist.	Dep.	D. Lat.	Dist.	Dep.	D. Lat.	Dist.	Dep.	D. Lat.	Dist.	Dep.	D. Lat.	Dist.	Dep.	D. Lat.
311° | 049° / 229° | 131°					49°									

Dist.	D. Lat.	Dep.
D Lo	Dep.	
	m	D Lo

*Figure 2414a. Extract from*Table 4.

322° | 038° / 218° | 142°					TABLE 4 Traverse 38° Table								322° | 038° / 218° | 142°	
Dist.	D. Lat.	Dep.	Dist.	D. Lat.	Dep.	Dist.	D. Lat.	Dep.	Dist.	D. Lat.	Dep.	Dist.	D. Lat.	Dep.
31	24.4	19.1	91	71.7	56.0	151	119.0	93.0	211	166.3	129.9	271	213.6	166.8
32	25.2	19.7	92	72.5	56.6	52	119.8	93.6	12	167.1	130.5	72	214.3	167.5
33	26.0	20.3	93	73.3	57.3	53	120.6	94.2	13	167.8	131.1	73	215.1	168.1
34	26.8	20.9	94	74.1	57.9	54	121.4	94.8	14	168.6	131.8	74	215.9	168.7
35	27.6	21.5	95	74.9	58.5	55	122.1	95.4	15	169.4	132.4	75	216.7	169.3

Dist.	D. Lat.	Dep.
D Lo	Dep.	
	m	D Lo

Figure 2414b. Extract from Table 4.

The mean latitude (L_m) is half the arithmetic sum of the latitudes of two places on the same side of the equator. It is labeled N or S to indicate its position north or south of the equator. If a course line crosses the equator, solve each course line segment separately.

Example 1: *A vessel steams 1,253 miles on course 070° from lat. 15°17.0' N, long. 151°37.0' E.*

Required: *Latitude and longitude of the point of arrival by (1) computation and (2) traverse table.*

Solution:

(1) Solution by computation:

$l = D \cos C$; $p = D \sin C$; and $DLo = p \sec L_m$.

D = 1253.0 miles.

C = 070°

l = 428.6' N

p = 1177.4 miles E

L_1 = 15°17.0' N

l = 7°08.6' N

L_2 = 22°25.6' N

L_m = 18°51.3' N

$DLo = 1244.2' E$
$\lambda_1 = 151°37.0' E$
$DLo = 20°44.2' E$
$\lambda_2 = 172° 21.2' E$

Answer:

$L_2 = 22° 25.6' N$
$\lambda_2 = 172° 21.2' E$

(2) Solution by traverse tables:

Refer to Figure 2415a. Enter the traverse table with course 070° and distance 1,253 miles. Because a number as high as 1,253 is not tabulated in the Dist. column, obtain the values for D. Lat. and Dep. for a distance of 125.3 miles and multiply them by 10. Interpolating between the tabular distance arguments yields D. Lat. = 429' and Dep. = 1,178 miles. Converting the D. Lat. value to degrees of latitude yields 7° 09.0'. The point of arrival's latitude, therefore, is 22° 26' N. This results in a mean latitude of 18° 51.5' N.

Reenter the table with the mean latitude as course angle and substitute DLo as the heading of the Dist. column and Dep. as the heading of the D. Lat. column. Since the table is computed for integral degrees of course angle (or latitude), the tabulations in the pages for 18° and 19° must be interpolated for the minutes of L_m. In the 18° table, interpolate for DLo between the departure values of 117.0 miles and 117.9 miles. This results in a DLo value of 123.9. In the 19° table, interpolate for DLo between the departure values of 117.2 and 118.2. This yields a DLo value of 124.6.

Having obtained the DLo values corresponding to mean latitudes of 18° and 19°, interpolate for the actual value of the mean latitude: 18° 51.5' N. This yields the value of DLo: 124.5. Multiply this final value by ten to obtain DLo = 1245 minutes = 20° 45' E.

Add the changes in latitude and longitude to the original position's latitude and longitude to obtain the final position.

Answer:

$L_2 = 22° 26' N$
$\lambda_2 = 172° 22.0' E$

Example 2: *A vessel at lat. 8°48.9'S, long. 89°53.3'W is to proceed to lat. 17°06.9'S, long. 104°51.6'W.*

Required: *Course and distance by (1) computation and (2) traverse table.*

Solution:

(1) Solution by computation:

$$p = DLo \cos L_m; \quad \tan C = \frac{p}{l}; \quad \text{and} \quad D = l \sec C$$

$DLo = 14° 58.3'$
$DLo = 898.3'$
$L_m = 12° 57.9' S$
$p = 893.8 \text{ arc min} \times \cos(12° 57.9')$
$p = 875.4 \text{ arc min}$

$l = 17.1° - 8.8°$
$l = 8.3°$
$l = 498 \text{ arc min}$

$$C = \arctan\frac{875.4 \text{arc min}}{498 \text{arc min}}$$

$C = S\ 60.4°\ W$
$C = 240.4°$

$D = 498 \text{ arc min} \times \sec(60.4°)$
$D = 1008.2 \text{ miles}$

Answer:

$C = 240.4°$
$D = 1008.2$ miles

The labels (N, S, E, W) of l, p, and C are determined by noting the direction of motion or the relative positions of the two places.

(2) Solution by traverse tables:

Refer to Figure 2415b. Enter the traverse table with the mean latitude as course angle and substitute DLo as the heading of the Dist. column and Dep. as the heading of the D. Lat. column. Since the table is computed for integral values of course angle (or latitude), it is usually necessary to extract the value of departure for values just less and just greater than the L_m and then interpolate for the minutes of Lm. In this case where L_m is almost 13°, enter the table with L_m 13° and DLo 898.3' to find Dep. 875 miles. The departure is found for DLo 89.9', and then multiplied by 10.

Reenter the table to find the numbers 875 and 498

TABLE 4

Traverse **20°** Table

340° | 020° / 200° | 160°

Dist.	D. Lat.	Dep.	Dist.	D. Lat.	Dep.	Dist.	D. Lat.	Dep.	Dist.	D. Lat.	Dep.	Dist.	D. Lat.	Dep.
1	0.9	0.3	61	57.3	20.9	121	113.7	41.4	181	170.1	61.9	241	226.5	82.4
2	1.9	0.7	62	58.3	21.2	22	114.6	41.7	82	171.0	62.2	42	227.4	82.8
3	2.8	1.0	63	59.2	21.5	23	115.6	42.1	83	172.0	62.6	43	228.3	83.1
4	3.8	1.4	64	60.1	21.9	24	116.5	42.4	84	172.9	62.9	44	229.3	83.5
5	4.7	1.7	65	61.1	22.2	25	117.5	42.8	85	173.8	63.3	45	230.2	83.8
6	5.6	2.1	66	62.0	22.6	26	118.4	43.1	86	174.8	63.6	46	231.2	84.1
7	6.6	2.4	67	63.0	22.9	27	119.3	43.4	87	175.7	64.0	47	232.1	84.5
Dist.	Dep.	D. Lat.	Dist.	Dep.	D. Lat.	Dist.	Dep.	D. Lat.	Dist.	Dep.	D. Lat.	Dist.	Dep.	D. Lat.

290° | 070° / 250° | 110°

70°

Dist.	D. Lat.	Dep.
N.	N x Cos.	N x Sin.
Hypotenuse	Side Adj.	Side Opp.

TABLE 4

Traverse **18°** Table

342° | 018° / 198° | 162°

Dist.	D. Lat.	Dep.	Dist.	D. Lat.	Dep.	Dist.	D. Lat.	Dep.	Dist.	D. Lat.	Dep.	Dist.	D. Lat.	Dep.
1	1.0	0.3	61	58.0	18.9	121	115.1	37.4	181	172.1	55.9	241	229.2	74.5
2	1.9	0.6	62	59.0	19.2	22	116.0	37.7	82	173.1	56.2	42	230.2	74.8
3	2.9	0.9	63	59.9	19.5	23	117.0	38.0	83	174.0	56.6	43	231.1	75.1
4	3.8	1.2	64	60.9	19.8	24	117.9	38.3	84	175.0	56.9	44	232.1	75.4
5	4.8	1.5	65	61.8	20.1	25	118.9	38.6	85	175.9	57.2	45	233.0	75.7
Dist.	Dep.	D. Lat.	Dist.	Dep.	D. Lat.	Dist.	Dep.	D. Lat.	Dist.	Dep.	D. Lat.	Dist.	Dep.	D. Lat.

288° | 072° / 252° | 108°

72°

Dist.	D. Lat.	Dep.
D Lo	Dep.	
	m	D Lo

TABLE 4

Traverse **19°** Table

341° | 019° / 199° | 161°

Dist.	D. Lat.	Dep.	Dist.	D. Lat.	Dep.	Dist.	D. Lat.	Dep.	Dist.	D. Lat.	Dep.	Dist.	D. Lat.	Dep.
1	0.9	0.3	61	57.7	19.9	121	114.4	39.4	181	171.1	58.9	241	227.9	78.5
2	1.9	0.7	62	58.6	20.2	22	115.4	39.7	82	172.1	59.3	42	228.8	78.8
3	2.8	1.0	63	59.6	20.5	23	116.3	40.0	83	173.0	59.6	43	229.8	79.1
4	3.8	1.3	64	60.5	20.8	24	117.2	40.4	84	174.0	59.9	44	230.7	79.4
5	4.7	1.6	65	61.5	21.2	25	118.2	40.7	85	174.9	60.2	45	231.7	79.8
Dist.	Dep.	D. Lat.	Dist.	Dep.	D. Lat.	Dist.	Dep.	D. Lat.	Dist.	Dep.	D. Lat.	Dist.	Dep.	D. Lat.

289° | 071° / 251° | 109°

71°

Dist.	D. Lat.	Dep.
D Lo	Dep.	
	m	D Lo

Figure 2415a. Extracts from the Table 4.

347° | 013° / 193° | 167°				TABLE 4 / Traverse 13° Table									347° | 013° / 193° | 167°	
Dist.	D. Lat.	Dep.	Dist.	D. Lat	Dep.	Dist.	D. Lat.	Dep.	Dist.	D. Lat.	Dep.	Dist.	D. Lat.	Dep.
27	26.3	6.1	87	84.8	19.6	47	143.2	33.1	07	201.7	46.6	67	260.2	60.1
28	27.3	6.3	88	85.7	19.8	48	144.2	33.3	08	202.7	46.8	68	261.1	60.3
29	28.3	6.5	89	86.7	20.0	49	145.2	33.5	09	203.6	47.0	69	262.1	60.5
30	29.2	6.7	90	87.7	20.2	50	146.2	33.7	10	204.6	47.2	70	263.1	60.7
Dist.	Dep.	D. Lat.	Dist.	Dep.	D. Lat.	Dist.	Dep.	D. Lat.	Dist.	Dep.	D. Lat.	Dist.	Dep.	D. Lat.
283° | 077° / 257° | 103°				77°										

Dist.	D. Lat.	Dep.
D Lo	Dep.	
	m	D Lo

Figure 2415b. Extract from Table 4.

beside each other in the columns labeled Dep. and D. Lat., respectively. Because these high numbers are not tabulated, divide them by 10, and find 87.5 and 49.8. This occurs most nearly on the page for course angle 60°. Interpolating for intermediate values, the corresponding number in the Dist. column is about 100.5. Multiplying this by 10, the distance is about 1005 miles.

Answer:

C = 240°
D = 1005 miles.

The labels (N, S, E, W) of l, p, DLo, and C are determined by noting the direction of motion or the relative positions of the two places.

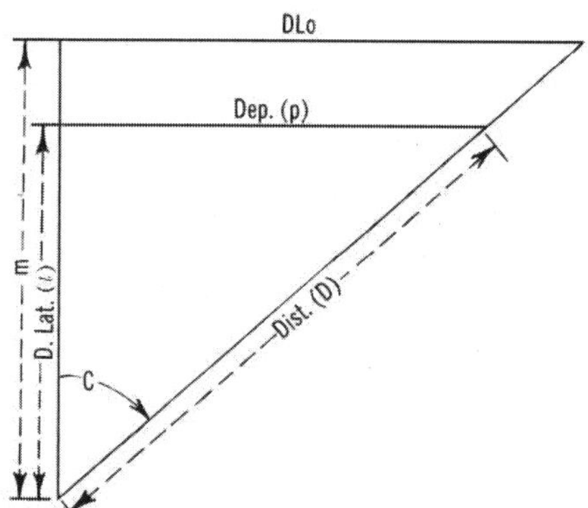

Figure 2416a

2416. Mercator Sailing

Mercator sailing problems can be solved graphically on a Mercator chart. For mathematical solution, the formulas of Mercator sailing are:

$$\tan C = \frac{DLo}{m} \qquad DLo = m \tan C.$$

After solving for course angle by Mercator sailing, solve for distance using the plane sailing formula:

$$D = l \sec C$$

Example 1: *A ship at lat. 32°14.7'N, long. 66°28.9'W is to head for a point near Chesapeake Light, lat. 36°58.7'N, long. 75°42.2'W.*

Required: *Course and distance by (1) computation and (2) traverse table.*

Solution:
(1) Solution by computation:

$$\tan C = \frac{DLo}{m}, \text{ and } D = l \sec C.$$

First calculate the meridional difference by entering Table 6 and interpolating for the meridional parts for the original and final latitudes. The meridional difference is the difference between these two val-

ues. Having calculated the meridional difference, simply solve for course and distance from the equations above.

M_2 *(36° 58.7' N) = 2377.1*
M_1 *(32° 14.7' N) = 2033.4*
m = 343.7

λ_2 *= 075° 42.2' W*
λ_1 *= 066° 28.9' W*
DLo = 9° 13.3' W
DLo = 553.3' W

C = arctan (553.3÷343.7')
C = N 58.2° W
C = 301.8°

L_2 *= 36° 58.7' N*
L_1 *= 32° 14.7' N*
l = 4° 44.0' N
l = 284.0'

D = 284.0 arc min × sec (58.2°)
D = 537.4 miles

Answer:
C = 301.8°
D = 538.2 miles
(2) Solution by traverse table:

Refer to Figure 2416b. Substitute m as the heading of the D. Lat. column and DLo as the heading of the Dep. column. Inspect the table for the numbers 343.7 and 553.3 in the columns relabeled m and DLo, respectively.

Because a number as high as 343.7 is not tabulated in the m column, it is necessary to divide m and DLo by 10. Then inspect to find 34.4 and 55.3 abreast in the m and DLo columns, respectively. This occurs most nearly on the page for course angle 58° or course 302°.

Reenter the table with course 302° to find Dist. for D. Lat. 284.0'. This distance is 536 miles.

Answer:
C = 302°
D = 536 miles

Example 2: *A ship at lat. 75°31.7' N, long. 79°08.7'W, in Baffin Bay, steams 263.5 miles on course 155°.*

Required: *Latitude and longitude of point of arrival by (1) computation and (2) traverse table.*

Solution:
(1) Solution by computation:
l = D cos C; and DLo = m tan C

328° | 032° / 212° | 148° — TABLE 4 — Traverse **32°** Table — 328° | 032° / 212° | 148°

Dist.	D. Lat.	Dep.	Dist.	D. Lat.	Dep.	Dist.	D. Lat.	Dep.	Dist.	D. Lat.	Dep.	Dist.	D. Lat.	Dep.
1	0.8	0.5	61	51.7	32.3	121	102.6	64.1	181	153.5	95.9	241	204.4	127.7
2	1.7	1.1	62	52.6	32.9	22	103.5	64.7	82	154.3	96.4	42	205.2	128.2
3	2.5	1.6	63	53.4	33.4	23	104.3	65.2	83	155.2	97.0	43	206.1	128.8
4	3.4	2.1	64	54.3	33.9	24	105.2	65.7	84	156.0	97.5	44	206.9	129.3
5	4.2	2.6	65	55.1	34.4	25	106.0	66.2	85	156.9	98.0	45	207.8	129.8
6	5.1	3.2	66	56.0	35.0	26	106.9	66.8	86	157.7	98.6	46	208.6	130.4
351	297.7	186.0	411	348.5	217.8	471	399.4	249.6	531	450.3	281.4	591	501.2	313.2
52	298.5	186.5	12	349.4	218.3	72	400.3	250.1	32	451.2	281.9	92	502.0	313.7
53	299.4	187.1	13	350.2	218.9	73	401.1	250.7	33	452.0	282.4	93	502.9	314.2
54	300.2	187.6	14	351.1	219.4	74	402.0	251.2	34	452.9	283.0	94	503.7	314.8
55	301.1	188.1	15	351.9	219.9	75	402.8	251.7	35	453.7	283.5	95	504.6	315.3
56	301.9	188.7	16	352.8	220.4	76	403.7	252.2	36	454.6	284.0	96	505.4	315.8
57	302.8	189.2	17	353.6	221.0	77	404.5	252.8	37	455.4	284.6	97	506.3	316.4
58	303.6	189.7	18	354.5	221.5	78	405.4	253.3	38	456.2	285.1	98	507.1	316.9
59	304.4	190.2	19	355.3	222.0	79	406.2	253.8	39	457.1	285.6	99	508.0	317.4
60	305.3	190.8	20	356.2	222.6	80	407.1	254.4	40	457.9	286.2	600	508.8	318.0
Dist.	Dep.	D. Lat.	Dist.	Dep.	D. Lat.	Dist.	Dep.	D. Lat.	Dist.	Dep.	D. Lat.	Dist.	Dep.	D. Lat.

302° | 058° / 238° | 122° — **58°**

Dist.	D. Lat.	Dep.
D Lo	Dep.	
	m	D Lo

Figure 2416b. Extract from Table 4 *composed of parts of left and right hand pages for course angle 58°.*

D	=	*263.5 mi.*
C	=	*155°*
l	=	*238.8 ' S*
l	=	*3° 58.8 ' S*
L_1	=	*75°31.7' N*
l	=	*3° 58.8' S*
L_2	=	*71°32.9' N*
M_1	=	*7072.4*
M_2	=	*6226.1*
m	=	*846.3*
DLo	=	*394.6' E*
DLo	=	*6°34.6' E*
λ_1	=	*79°08.7' W*
DLo	=	*6°34.6' E*
λ_2	=	*072°34.1' W*

The labels (N, S, E, W) of l, DLo, and C are determined by noting the direction of motion or the relative positions of the two places.

Answer:

L_2 *= 71° 32.9'*

λ_2 *= 072° 34.1'*

(2) Solution by traverse table:

Refer to Figure 2416c. Enter the traverse table with course 155° and Dist. 263.5 miles to find D. Lat. 238.8'. The latitude of the point of arrival is found by subtracting the D. Lat. from the latitude of the point of departure. Determine the meridional difference by Table Table 4 (m = 846.3).

Reenter the table with course 155° to find the DLo corresponding to m = 846.3. Substitute meridional difference m as the heading of the D. Lat. column and DLo as the heading of the Dep. column. Because a number as high as 846.3 is not tabulated in the m column, divide m by 10 and then inspect the m column for a value of 84.6. Interpolating as necessary, the latter value is opposite DLo 39.4'. The DLo is 394' (39.4' × 10). The longitude of the point of arrival is found by applying the DLo to the longitude of the point of departure.

Answer:

L_2 *= 71°32.9' N.*

λ_2 *= 72°34.7' W.*

TABLE 4

Traverse **25°** Table

335° | 025° / 205° | 155°

Dist.	D. Lat.	Dep.	Dist.	D. Lat.	Dep.	Dist.	D. Lat.	Dep.	Dist.	D. Lat.	Dep.	Dist.	D. Lat.	Dep.
21	19.0	8.9	81	73.4	34.2	141	127.8	59.6	201	182.2	84.9	261	236.5	110.3
22	19.9	9.3	82	74.3	34.7	42	128.7	60.0	02	183.1	85.4	62	237.5	110.7
23	20.8	9.7	83	75.2	35.1	43	129.6	60.4	03	184.0	85.8	63	238.4	111.1
24	21.8	10.1	84	76.1	35.5	44	130.5	60.9	04	184.9	86.2	64	239.3	111.6
25	22.7	10.6	85	77.0	35.9	45	131.4	61.3	05	185.8	86.6	65	240.2	112.0
33	29.9	13.9	93	84.3	39.3	53	138.7	64.7	13	193.0	90.0	73	247.4	115.4
34	30.8	14.4	94	85.2	39.7	54	139.6	65.1	14	193.9	90.4	74	248.3	115.8
Dist.	Dep.	D. Lat.	Dist.	Dep.	D. Lat.	Dist.	Dep.	D. Lat.	Dist.	Dep.	D. Lat.	Dist.	Dep.	D. Lat.

295° | 065° / 245° | 115°

65°

Dist.	D. Lat.	Dep.
D Lo	Dep.	
	m	D Lo

Figure 2416c. Extract from Table 4.

2417. Additional Problems

Example: *A vessel steams 117.3 miles on course 214°.*
Required: *(1) Difference of latitude, (2) departure, by plane sailing.*
Answers: *(1) l 97.2'S, (2) p 65.6 mi. W.*

Example: *A steamer is bound for a port 173.3 miles south and 98.6 miles east of the vessel's position*
Required: *(1) Course, (2) distance, by plane sailing.*
Answers: *(1) C 150.4°; (2) D 199.4 mi. by computation, 199.3 mi. by traverse table.*

Example: *A ship steams as follows: course 359°, distance 28.8 miles; course 006°, distance 16.4 miles; course 266°, distance 4.9 miles; course 144°, distance 3.1 miles; course 333°, distance 35.8 miles; course 280°, distance 19.3 miles.*
Required: *(1) Course, (2) distance, by traverse sailing.*
Answers: *(1) C 334.4°, (2) D 86.1 mi.*

Example: *The 1530 DR position of a ship is lat. 44°36.3'N, long. 31°18.3'W. The ship is on course 270°, speed 17 knots.*
Required: *The 2000 DR position, by parallel sailing.*
Answer: *2000 DR: L 44°36.3'N, λ 33°05.7'W.*

Example: *A ship at lat. 33°53.3'S, long. 18°23.1'E, leaving Cape Town, heads for a destination near Ambrose Light, lat. 40°27.1'N, long. 73°49.4'W.*
Required: *(1) Course and (2) distance, by Mercator sailing.*
Answers: *(1) C 310.9°; (2) D 6,811.5 mi. by computation, 6,812.8 mi. by traverse table.*

Example: *A ship at lat. 15°03.7'N, long. 151°26.8'E steams 57.4 miles on course 035°.*
Required: *(1) Latitude and (2) longitude of the point of arrival, by Mercator sailing.*
Answers: *(1) L 15°50.7'N; (2) λ 152°00.7'E.*

CHAPTER 25

NAVIGATION PROCESSES

INTRODUCTION

2500. Understanding the Process of Navigation

Navigation is comprised of a number of different processes. Some are done in a set order, some randomly, some almost constantly, others only infrequently. It is in choosing using these processes that an individual navigator's experience and judgment are most crucial. Compounding this subject's difficulty is the fact that there are no set rules regarding the optimum employment of navigational systems and techniques. Optimum use of navigational systems varies as a function of the type of vessel, the quality of the navigational equipment on board, and the experience and skill of the navigator and all the members of his team.

For the watch officer, ensuring the ship's safety always takes priority over completing operational commitments and carrying out the ship's routine. Navigation is his primary responsibility. Any ambiguity about the position of the vessel which constitutes a danger must be resolved immediately. The best policy is to prevent ambiguity by using all the tools available and continually checking different sources of position information to see that they agree. This includes the routine use of several different navigational techniques, both as operational checks and to maintain skills which might be needed in an emergency. Any single navigational system constitutes a single point of failure, which must be backed up with another source to ensure the safety of the vessel.

It is also the navigator's responsibility to ensure that he and all members of his team are properly trained and ready in all respects for their duties, and that he is familiar with the operation of all gear and systems for which he is responsible. He must also ensure that all digital and/or hardcopy charts and publications are updated with information from the *Notice to Mariners*, and that all essential navigational gear is in operating condition.

Navigating a vessel is a dynamic process. Schedules, missions, and weather often change. Planning a voyage is a process that begins well before the ship gets underway. Executing that plan does not end until the ship ties up at the pier at its final destination. While it is possible to over plan a voyage, it is a more serious error to under plan it. Carefully planning a route, preparing required charts and publications, and using various methods to monitor the ship's position as the trip proceeds are fundamental to safe navigation and are the marks of a professional navigator.

This chapter will examine navigational processes, the means by which a navigator manages all of the resources at his command to ensure a safe and efficient voyage.

BRIDGE RESOURCE MANAGEMENT

2501. The Navigator as Manager

The development of computers and navigational technologies driven by them has led to an evolution—some might say revolution—in the role of the navigator. Increasingly, the navigator is the manager of a combination of systems of varying complexity, which are used to direct the course of the ship and ensure its safety. The navigator is thus becoming less concerned with the direct control of the ship and more concerned with managing the systems and people which do so under his direction. The navigator must become competent and comfortable with the management of advanced technology and human resources, especially in stressful situations.

A modern ship's navigational suite might include an integrated bridge system with a comprehensive ship and voyage management software package, an ECDIS replacing paper charts and including radar overlay, dual interswitched X- and S-band ARPA radars, autopilot linked to digital flux gate and ring laser gyrocompasses linked to the ECDIS, integrated GPS/DGPS and Loran C positioning system, numerous environmental sensors, digital depth sounder, and Doppler speed log. The communications suite might include a GMDSS workstation with NAVTEX receiver, a weatherfax and computer weather routing system, SATCOM terminal, several installed and portable VHF radios, an internal telephone exchange, a public address and alarm system, and sound powered telephones. As all this technology is coming aboard, crew size is decreasing, placing increased responsibility on each member of the team.

Thus, the modern navigator is becoming a manager of resources, both electronic and human. Of course, he has always been so, but today's systems are far more complex, and the consequences of a navigational error far more serious, than ever before. The prudent navigator will therefore be familiar with the techniques of Bridge Resource Management (BRM), by which he can supervise the numerous complex tasks involved with maintaining navigational control of his vessel.

Bridge Team Management refers to the management of the human resources available to the navigator—helmsman, lookout, engine room watch, etc.—and how to ensure that all members contribute to the goal of a safe and efficient voyage.

Bridge Resource Management (BRM) is the study of the resources available to the navigator and the exploitation of them in order to conduct safe and efficient voyages. The terms "bridge resource management" and "bridge team management" are not precisely defined. For most, bridge resources consist of the complete suite of assets available to the navigator including electronic and human, while bridge team management refers only to human assets, except for the pilot, who is normally not considered a member of the team.

The resources available will vary according to the size of the ship, its mission, its crew, its shoreside management, funding, and numerous other variables. No two vessels are alike in resources, for even if two ships of a single class are alike in every physical respect, the people who man them will be different, and people are the most important resource the navigator has.

Effective Bridge Resource Management requires:

- Clearly defined navigational goals
- Defined procedures—a system—for achieving goals
- Means to achieve the goals
- Measures of progress toward goals
- Constant awareness of the situation tactically, operationally, and strategically
- Clearly defined accountability and responsibility
- Open communication throughout the system
- External support

2502. Watch Conditions

Whenever the navigational situation demands more resources than are immediately available to the navigator, a dangerous condition exists. This can be dealt with in two ways. First, the navigator can call up additional resources, such as by adding a bow lookout or an additional watch officer. Second, he can lower the navigational demands to the point where his available resources are able to cope, perhaps by reducing speed, changing course, heaving to, or anchoring.

Some conditions that increase the demands on the navigator include:

- Fog
- Heavy traffic
- Entering a channel, harbor or restricted area
- Heavy weather
- Fire, flooding, or other emergency

These and many other situations can increase the demands on the time and energy of the navigator, and cause him to need additional resources—another watch officer, a bow lookout, a more experienced helmsman—to take some of the workload and rebalance the amount of work to be done with the people available to do it.

There is no strict legal direction as to the assignment of personnel on watch. Various rules and regulations establish certain factors which must be addressed, but the responsibility for using the available people to meet them rests with the watch officer. Laws and admiralty cases have established certain requirements relating to the position and duties of the lookout, safe speed under certain conditions, mode of steering, and the use of radar. The maritime industry has established certain standards known as **Watch Conditions** to help define the personnel and procedures to be used under various situations.

Watch Condition I indicates unrestricted maneuverability, weather clear, little or no traffic, and all systems operating normally. In this condition, depending on the size and type of vessel and its mission, often a single licensed person can handle the bridge watch.

Watch Condition II applies to situations where visibility is somewhat restricted, and maneuverability is constrained by hydrography and other traffic. This condition may require additional navigational resources, such as a lookout, helmsman, or another licensed watch officer.

Watch Condition III reflects a condition where navigation is seriously constrained by poor visibility, close quarters (as in bays, sounds, or approach channels), and heavy traffic.

Watch Condition IV is the most serious, occurring when visibility is poor, maneuvering is tightly constrained (as in channels and inner harbors), and traffic is heavy.

Any watch condition can change almost momentarily due to planned or unforeseen events. Emergency drills or actual emergencies on one's own or other nearby vessels can quickly overwhelm the unprepared bridge team.

Under each of these conditions, the navigator must manage his resources effectively and efficiently, calling in extra help when necessary, assigning personnel as needed to jobs for which they are qualified and ready to perform. He must consider the peculiarities of his ship and its people, including considerations of vessel design and handling characteristics, personalities and qualifications of individuals, and the needs of the situation.

2503. Laws Relating to Bridge Resources

Numerous laws and regulations relate to the navigation of ships, particularly in less than ideal conditions. Title 33 of the Code of Federal Regulations (CFR) specifies bridge visibility parameters. Title 46 CFR and IMO standards relate to medical fitness. Public Law 101-380 specifies the maximum hours of work permitted, while 46 CFR specifies the minimum hours of rest required. Competency and certification are addressed by 46 CFR and STCW 95. Charts, publications, and navigational equipment are the subject of 33 CFR, which also specifies tests required before getting underway. This code also requires reporting of certain dangerous conditions aboard the vessel.

Various U.S. state and local regulations also apply to the duties and responsibilities of the bridge team, and numerous regulations and admiralty case law relate indirectly to bridge resource management.

2504. Pilots

One of the navigator's key resources in the harbor and harbor approaches is the pilot, a professional shiphandler with encyclopedic knowledge of a local port and harbor area. His presence is very often required by local regulation or law. He is not considered, by the common definition, to be a member of the bridge team, but he is an extremely important bridge resource. He remains, except in certain defined areas, an advisor to the captain, who retains full responsibility for the safety of the ship. Only in the Suez Canal and Panama Canal are pilots given full navigational responsibility.

As an important navigational resource, the pilot requires management, and as a professional navigator, he deserves respect. The balance of these two elements is the responsibility of the captain, who manages the Master-Pilot Exchange (MPX).

The explicit purpose of the MPX is to tell the pilot the particulars of the ship: its draft, condition of engines and navigational equipment, and special conditions or characteristics which might affect the pilot's ability to understand how the ship will handle in close quarters. However, simply relating the ship's characteristics and condition does not constitute a proper MPX, which must be more comprehensive.

The implicit purpose of the MPX is to establish a rapport with the pilot so that a mental model of the transit can be agreed on and shared with the bridge team. Thus, the MPX is not an event but a process, which will ensure that everyone responsible for navigating the vessel shares the same plan for the transit.

Some ships prepare a pilot card that lists the essential vessel parameters for the pilot's ready reference. The pilot himself may use a checklist to ensure that all required areas of concern are covered. The pilot may or may not require a signature on his own forms, and may or may not be requested or allowed to sign ship's forms. These are matters of local law and custom that must be respected.

Often, among the pilot's first words upon boarding will be a perfunctory recommendation to the captain to take up a certain course and speed. The captain then gives the appropriate orders to the bridge team. As the vessel gathers way, the rest of the MPX can proceed. As time permits, the pilot can be engaged in conversation about the events and hazards to be expected during the transit, such as turning points, shoal areas, weather and tides, other ship traffic, tugs and berthing arrangements, status of ground tackle, and other matters of concern. This information should be shared with the bridge team. At any time during the transit, the captain should bring up matters of concern to the pilot for discussion. Communication is the vital link between pilot and master that ensures a safe transit.

2505. Managing the Bridge Team

Shipboard personnel organization is among the most hierarchical to be found. Orders are given and expected to be obeyed down the chain of command without hesitation or question, especially in military vessels. While this operational style defines responsibilities clearly, it does not take advantage of the entire knowledge base held by the bridge team, which increasingly consists of a number of highly trained people with a variety of skills, abilities, and perceptions.

While the captain may have the explicit right to issue orders without discussion or consultation (and in most routine situations it is appropriate to do so), in unusual, dangerous, and stressful situations it is often better to consult other members of the team. Communication, up and down, is the glue that holds the bridge team together and ensures that all resources are effectively used. Many serious groundings could have been prevented by the simple exchange of information from crew to captain, information which, for reasons of tradition and mindless obedience to protocol, was not shared or was ignored.

A classic case of failure to observe principles of bridge team management occurred in 1950 when the USS Missouri, fully loaded and making over 12 knots at high tide, grounded hard on Thimble Shoals in Chesapeake Bay. The Captain ignored the advice of his Executive Officer, berated the helmsman for speaking out of turn, and failed to order a right turn into Thimble Shoals Channel. It took more than two weeks to free the ship.

Most transportation accidents are caused by human error, usually resulting from a combination of circumstances, and almost always involving a communications failure. Analysis of numerous accidents across a broad range of transportation fields reveals certain facts about human behavior in a dynamic team environment:

- Better decisions result from input by many individuals

- Success or failure of a team depends on their ability to communicate and cooperate
- More ideas present more opportunities for success and simultaneously limit failure
- Effective teams can share workloads and reduce stress, thus reducing stress-caused errors
- All members make mistakes; no one has all the right answers
- Effective teams usually catch mistakes before they happen, or soon after, and correct them

These facts argue for a more inclusive and less hierarchical approach to bridge team management than has been traditionally followed. The captain/navigator should include input from bridge team members when constructing the passage plan and during the pre-voyage conference, and should share his views openly when making decisions, especially during stressful situations. He should look for opportunities to instruct less experienced team members by involving them in debate and decisions regarding the voyage. This ensures that all team members know what is expected and share the same mental model of the transit.

Effective bridge teams do not just happen. They are the result of planning, education, training, practice, drills, open communication, honest responses, and management support. All of these attributes can and should be taught, and a number of professional schools and courses are dedicated to this subject. The U.S. Coast Guard web site at http://www.uscg.mil/STCW/m-achome.htm lists courses in Bridge Resource Management and other subjects that will help the navigator manage resources effectively.

2506. Standards of Training, Certification, and Watchkeeping (STCW)

From a personnel standpoint, the management of a military vessel is a very different proposition than a commercial vessel. Procedures are more formalized, lines of communication and responsibility are more structured, and the process of navigation is more highly organized. Military personnel are trained for navigational duties through a variety of required on-the-job and school-based training programs. The watch officer on a military vessel can generally be assured that the ratings under his command have passed the proper tests for their rank and are trained for their jobs. (Experience is another matter.)

The commercial captain has had, until recently, little such assurance. Training programs and certification for deck personnel only minimally addressed routine duties of bridge watchstanders, concentrating on emergency procedures and deck-related skills. The IMO's International Convention on Standards of Training, Certification and Watchkeeping for Seafarers (STCW) of 1978 set certain qualifications for masters, mates, and watch personnel. It entered into force in 1984, and the United States became a party to this convention in 1991.

Between 1984 and 1992, significant limitations to the 1978 conventions became apparent. Vague requirements, lack of clear standards, limited oversight and control, and failure to address modern issues of watchkeeping were all seen as problems meriting a review of the 1978 agreement. Prior to this, the IMO had concentrated mostly on construction and equipment of ships. This new review, spearheaded by the U.S., was to concentrate on the human element, which in fact is the cause of most marine casualties. Three serious maritime casualties in which human factors played a part spurred the leadership of the IMO to immediate action, and in 1995, a year sooner than initially planned, the new convention was signed, and entered into force on February 1, 1997.

The U.S. Coast Guard immediately began the process of changing the regulations related to issuing licenses to U.S. maritime personnel to comply with the new guidelines. Mariners licensed under the 1978 convention had until February 1, 2002, to renew their documents under the old rule. All others would have to comply with the new standards. This date was subsequently amended to allow more time for compliance.

The provisions of STCW 95 strongly address the human element of bridge team management. They mandate maximum duty hours, minimum rest periods, and training requirements for specific navigational and communications systems such as ARPA and GMDSS. They require that officers understand and comply with the principles of bridge resource management. They require not merely that people be trained in certain procedures and operations, but that they demonstrate competence therein.

The competencies relating to navigation required of unlicensed personnel relate to general watchstanding duties. Such personnel must not only pass training, but must demonstrate competency in the use of magnetic and gyrocompasses for steering and course changes, response to standard helm commands, change from automatic to hand steering and back, responsibilities of the lookout, and proper watch relief procedures.

Competence may be demonstrated at sea or in approved simulators, and must be documented by Designated Examiners (DE's) who provide documentation which will allow the examinee to be certified under the provisions of STCW 95.

VOYAGE PLANNING

2507. The Passage Plan

Before each voyage begins, the navigator should develop a detailed mental model of how the entire voyage is to proceed sequentially, from getting underway to mooring. This mental model will include charting courses, forecasting the weather and tides, checking *Sailing Directions* and *Coast Pilots*, and projecting the various future events—landfalls, narrow passages, and course changes—that will transpire during the voyage. This mental model becomes the standard by which he will measure progress toward the goal of a safe and efficient voyage, and it is manifested in a passage plan.

The passage plan is a comprehensive, step by step description of how the voyage is to proceed from berth to berth, including undocking, departure, enroute, approach and mooring at the destination. The passage plan should be communicated to the navigation team in a pre-voyage conference in order to ensure that all members of the team share the same mental model of the entire trip. This differs from the more detailed piloting brief discussed in Chapter 8, though it may be held in conjunction with it, and may be a formal or informal process.

Differences of opinion must be addressed. For example, one watch officer might consider a one mile minimum passing distance appropriate, while the captain prefers to pass no closer than two miles. These kinds of differences must be reconciled before the voyage begins, and the passage plan is the appropriate forum in which to do so.

Thus, each member of the navigation team will be able to assess the vessel's situation at any time and make a judgement as to whether or not additional bridge resources are necessary. Passage planning procedures are specified in Title 33 of the U.S. Code, IMO Resolutions, and a number of professional books and publications. There are some fifty elements of a comprehensive passage plan depending on the size and type of vessel, each applicable according to the individual situation.

Passage planning software can greatly simplify the process and ensure that nothing important is overlooked. A good passage planning software program will include great circle waypoint/distance calculators, tide and tidal current predictors, celestial navigational calculators, consumables estimators for fuel, oil, water, and stores, and other useful applications.

As the voyage proceeds, the navigator must maintain situational awareness to continually assess the progress of the ship as measured against the passage plan and the mental model of the voyage. Situational awareness consists of perceiving, comprehending, and comparing what is known at any given time with the mental model and passage plan. Both individual and team situational awareness are necessary for a safe voyage, and the former must be established by all members of the bridge team before the latter is possible.

The enemies of situational awareness are complacency, ignorance, personal bias, fatigue, stress, illness, and any other condition which prevents the navigator and his team members from clearly seeing and assessing the situation.

2508. Constructing a Voyage Track

Coastwise passages of a few hundred miles or less can be laid out directly on charts, either electronic or paper. Over these distances, it is reasonable to ignore great circle routes and plot voyages directly on Mercator charts.

For trans-oceanic voyages, construct the track using a navigational computer, a great circle (gnomonic) chart, or the sailings. It is best to use a navigational computer or calculator if one is available to save time and to eliminate the plotting errors inherent in transferring the track from a gnomonic to a Mercator projection. Because they solve problems mathematically, computers and calculators also eliminate rounding errors inherent in the tables, providing more accurate solutions.

To use a navigational computer for voyage planning, the navigator simply enters the two endpoints of his planned voyage or major legs thereof in the appropriate spaces. The program may ask for track segment intervals every X number of degrees. It then computes waypoints along the great circle track between the two endpoints, determines each track leg's distance and, given a speed of advance, calculates the time the vessel can expect to pass each waypoint. The waypoints may be saved as a route, viewed on screen, and sent to the autopilot. On paper charts, construct the track on an appropriate Mercator chart by plotting the computer-generated waypoints and the tracks between them.

After adjusting the track as necessary to pass well clear of any hazard, choose a speed of advance (SOA) that ensures the ship will arrive on time at its destination or at any required point. If the time of arrival is open-ended, that is, not specifically required, choose a reasonable average SOA. Given an SOA, mark the track with the vessel's first few planned hourly positions. In the Navy, these planned positions are points of intended movement (PIM's). The SOA chosen for each track leg is the PIM speed. Merchant vessels usually refer to them as waypoints.

An operation order often assigns a naval vessel to an operating area. In that case, plan a track from the departure to the edge of the operating area to ensure that the vessel arrives at the operating area on time. Following a planned track inside the assigned area may be impossible because of the dynamic nature of an exercise. In that case, carefully examine the entire operating area for navigational hazards. If simply transiting through the area, the ship should still follow a planned and approved track.

2509. Following a Voyage Plan

Complete the planning discussed in Article 2508 prior to leaving port. Once the ship is transiting, frequently compare the ship's actual position to the planned position and adjust the ship's course and speed to compensate for any deviations. Order courses and speeds to keep the vessel on track without significant deviation.

Often a vessel will have its operational commitments changed after it gets underway. If this happens, it will be necessary to begin the voyage planning process anew.

VOYAGE PREPARATION

2510. Equipment Inventory

Prior to getting the ship underway, the navigator should inventory all navigational equipment, charts, and publications. He should develop a checklist of navigational equipment specific to his vessel and check that all required equipment is onboard and in operating order. The navigator should have all applicable *Sailing Directions*, pilot charts, and navigation charts covering his planned route. He should also have all charts and *Sailing Directions* covering ports at which his vessel may call. He should have all the equipment and publications required to support all appropriate navigational methods. Finally, he must have all technical documentation required to support the operation of his electronic navigation suite.

It is important to complete this inventory well before the departure date and obtain all missing items before sailing.

2511. Chart Preparation

Just as the navigator must prepare charts for piloting, he must also prepare his small scale charts for an open ocean transit. The following is the minimum chart preparation required for an open ocean or offshore coastal transit.

Correcting the Chart: Correct all applicable charts through the latest *Notice to Mariners*, *Local Notice to Mariners*, and Broadcast Notice to Mariners. Ensure the chart to be used is the latest announced edition.

Plotting the Track: Mark the track course above the track line with a "C" followed by the course. Similarly, mark each track leg's distance under the course line with a "D" followed by the distance in nautical miles.

Calculating Minimum Expected, Danger, and Warning Soundings: Chapter 8 discusses calculating minimum expected, danger and warning soundings. Determining these soundings is particularly important for ships passing a shoal close aboard. Set these soundings to warn the conning officer that he is passing close to the shoal. Mark the minimum expected sounding, the warning sounding, and the danger sounding clearly on the chart and indicate the section of the track for which they are applicable.

Marking Allowed Operating Areas: (Military vessels) Often an operation order assigns a naval vessel to an operating area for a specific period of time. There may be operational restrictions placed on the ship while within this area. For example, a surface ship assigned to an operating area may be ordered not to exceed a certain speed for the duration of an exercise. When assigned an operating area, clearly mark that area on the chart. Label it with the time the vessel must remain in the area and what, if any, operational restrictions it must follow. The conning officer and the captain should be able to glean the entire navigational situation from the chart alone without reference to the directive from which the chart was constructed. Therefore, put all operationally important information directly on the chart.

Marking Chart Shift Points: Mark the chart points where the navigator must shift to the next chart, and note the next chart number.

Examining Either Side of Track: Highlight any shoal water or other navigational hazard near the planned track. This will alert the conning officer as he approaches a possible danger.

NAVIGATION ROUTINE AT SEA

2512. Fix Frequency

If ECDIS is in use, fix frequency is not an issue. The ship's position will be displayed on the chart once per second, and the navigator need only monitor the process. If only an ECS is available, more careful attention is necessary since ECS cannot substitute for a paper chart. Nevertheless, it is reasonable to plot fixes at less frequent intervals when using an ECS, checking the system with a hand-plotted fix at prudent intervals.

Assuming that an electronic chart system is not available and hand-plotted fixes are the order of the day, adjust the fix interval to ensure that the vessel remains at least two fixes from the nearest danger. Choose a fix interval that provides a sufficient safety margin from all charted hazards.

Table 2512 below lists recommended fix intervals as a function of the phase of navigation:

	Harbor/Appr.	Coastal	Ocean
Frequency	3 min. or less	3-15 min.	30 min.

Table 2512. Recommended fix intervals.

Use all available fix information. With the advent of accurate satellite navigational systems, it is especially tempting to disregard this maxim. However, the experienced navigator never feels comfortable relying solely on one particular system. Supplement the satellite position with positions from Loran, celestial fixes, radar lines of position, soundings, or visual observations. Evaluate the accuracy of the various fix methods against the satellite position.

Use an inertial navigator if one is available. The inertial navigator may actually produce estimated positions more accurate than non-GPS based fix positions. Inertial navigators are completely independent of any external input. Therefore, they are invaluable for maintaining an accurate ship's position during periods when external fix sources are unreliable or unavailable.

Always check a position determined by a fix, inertial navigator, or DR by comparing the charted sounding at the position with the fathometer reading. If the soundings do not correlate, investigate the discrepancy.

Chapter 7 covers the importance of maintaining a proper DR. It bears repeating here. Determine the difference between the fix and the DR positions at every fix and use this information to calculate an EP from every DR. Constant application of set and drift to the DR is crucial if the vessel must pass a known navigational hazard close aboard.

2513. Fathometer Operations

While the science of hydrography has made tremendous advances in the last few years, these developments have yet to translate into significantly more accurate soundings on charts. Further, mariners often misunderstand the concept of an electronic chart, erroneously thinking that the conversion of a chart to electronic format indicates that updated hydrographic information has been used to compile it. This is rarely the case. In fact, most electronic charts are simply digitized versions of the paper charts, newly compiled but based on the same sounding databases, which in some cases are more than a century old.

While busy ports and harbors tend to be surveyed and dredged at regular intervals, in less travelled areas it is common for the navigator to find significant differences between the observed and charted soundings. If in doubt about the date of the soundings, refer to the title block of the chart, where information regarding the data used to compile it may be found.

Standardized rules and procedures for the use of the depth sounder are advisable and prudent. Table 2513 suggests a set of guidelines for depth sounder use on a typical ship.

Water Depth	Sounding Interval
< 10 m	Monitor continuously.
10 m - < 100 m	Every 15 minutes.
100 m - < 300 m	Every 30 minutes.
> 300 m	Every hour.

Table 2513. Fathometer operating guidelines.

2514. Compass Checks

Determine gyro compass error at least once daily and before each transit of restricted waters. Check the gyro compass reading against the inertial navigator if one is installed. If the vessel does not have an inertial navigator, check gyro error using a flux gate magnetic or ring laser gyro compass, or by using the celestial techniques discussed in Chapter 17.

The magnetic compass, if operational, should be adjusted regularly and a deviation table prepared and posted as required (See Chapter 6). If the magnetic compass has been deactivated in favor of a digital flux gate magnetic, ring laser gyro, or other type of electronic compass, the electronic compass should be checked to ensure that it is operating within manufacturer's specifications, and that all remote repeaters are in agreement. Note that the electronic compass must not be in the ADJUST mode when in restricted waters.

2515. Night Orders and Standing Orders

The Night Order Book is the vehicle by which the captain informs the officer of the deck of his orders for operating the ship. It may be in hardcopy or softcopy format. The Night Order Book, despite its name, can contain orders for the entire 24 hour period for which the Captain or Commanding Officer issues it.

The navigator may write the Night Orders pertaining to navigation. Such orders include assigned operating areas, maximum speeds allowed, required positions with respect to PIM or DR, and, regarding submarines, the maximum depth at which the ship can operate. Each department head should include in the Night Order book the evolutions he wants to accomplish during the night that would normally require the captain's permission. The captain can add further orders and directions as required.

The Officer of the Deck or mate on watch must not follow the Night Orders blindly. Circumstances under which the captain signed the Orders may have changed, rendering some evolutions impractical or impossible. The Officer of the Deck, when exercising his judgment on completing ordered evolutions, must always inform the captain of any deviation from the Night Orders as soon as such a deviation occurs.

While Night Orders are in effect only for the 24 hours after they are written, Standing Orders are continuously in force. The captain sets the ship's navigation policy in these orders. He sets required fix intervals, intervals for fathometer operations, minimum CPA's, and other general navigation and collision avoidance requirements.

2516. Watch Relief Procedures

When a watch officer relieves as Officer of the Deck or mate on watch, he assumes the responsibility for the safe navigation of the ship. He becomes the Captain's direct representative, and is directly responsible for the safety of the ship and the lives of its crew. He must prepare himself carefully prior to assuming these responsibilities. A checklist developed specifically for each vessel can serve as a reminder that all watch relief procedures have been followed. The following list contains those items that, as a minimum, the relieving watch officer must check prior to assuming the navigation watch.

- **Conduct a Pre-Watch Tour:** The relieving watch officer should tour the ship prior to his watch. He should familiarize himself with any maintenance in progress, and check for general cleanliness and stowage. He should see that any loose gear that could pose a safety hazard in rough seas is secured.

- **Check the Position Log and Chart:** Check the type and accuracy of the ship's last fix. Verify that the navigation watch has plotted the last fix properly. Ensure there is a properly constructed DR plot on the chart. Examine the DR track for any potential navigational hazards. Check ship's position with respect to the PIM or DR. Ensure that the ship is in the correct operating area, if applicable. Check to ensure that the navigation watch has properly applied fix expansion if necessary.

- **Check the Fathometer Log:** Ensure that previous watches have taken soundings at required intervals and that the navigation watch took a sounding at the last fix. Verify that the present sounding matches the charted sounding at the vessel's position.

- **Check the Compass Record Log:** Verify that the navigation watch has conducted compass checks at the proper intervals. Verify that gyro error is less than 1° and that all repeaters agree within 1° with the master gyro.

- **Read the Night Orders:** Check the Night Order Book for the captain's directions for the duration of the watch.

- **Check Planned Operations and Evolutions:** For any planned operations or evolutions, verify that the ship meets all prerequisites and that all watchstanders have reviewed the operation order or plan. If the operation is a complicated one, consider holding an operations brief with applicable watchstanders prior to assuming the watch.

- **Check the Broadcast Schedule:** Read any message traffic that could have a bearing on the upcoming watch. Find out when the last safety and operational messages were received. Determine if there are any required messages to be sent during the watch (e.g. position reports, weather reports, Amver messages).

- **Check the Contact Situation:** Check the radar picture (and sonar contacts if so equipped). Determine which contact has the nearest CPA and what maneuvers, if any, might be required to open the CPA. Find out from the off-going watch officer if there have been any bridge-to-bridge communications with any vessels in the area. Check that no CPA will be less than the minimum set by the Standing Orders.

- **Review Watchstander Logs:** Review the log entries for all watchstanders. Note any out-of-specification readings or any trends in log readings indicating that a system will soon fail.

After conducting these checks, the relieving watch officer should report that he is ready to relieve the watch. The watch officer should brief the relieving watch officer on the following:

- Present course and speed
- Present depth (submarines only)
- Evolutions planned or in progress
- Status of the engineering plant
- Status of any out-of-commission equipment
- Orders not noted in the Night Order Book
- Status of cargo
- Hazardous operations planned or in progress
- Routine maintenance planned or in progress
- Planned ship's drills
- Any individuals working aloft, or in a tank or hold
- Any tank cleaning operations in progress

If the relieving watch officer has no questions following this brief, he should relieve the watch and announce to the rest of the bridge team that he has the deck and the conn. The change of watch should be noted in the ship's deck log.

Watch officers should not relieve the watch in the middle of an evolution or when casualty procedures are being carried out. This ensures that there is watchstander continuity when carrying out a specific evolution or combating a casualty. Alternatively, the on-coming watch officer might relieve only the conn, leaving the deck watch with the off-going officer until the situation is resolved.

CHAPTER 26

EMERGENCY NAVIGATION

BASIC TECHNIQUES OF EMERGENCY NAVIGATION

2600. Planning for Emergencies

Increasing reliance on electronic navigation and communication systems has dramatically changed the perspective of emergency navigation. While emergency navigation once concentrated on long-distance lifeboat navigation, today it is far more likely that a navigator will suffer failure of his ship's primary electronic navigation systems than that he will be forced to navigate a lifeboat. In the unlikely event that he must abandon ship, his best course of action is to remain as close to the scene as possible, for this is where rescuers will concentrate their search efforts. Leaving the scene of a disaster radically decreases the chance of rescue, and there is little excuse for failure to notify rescue authorities with worldwide communications and maritime safety systems available at little cost. See Chapter 28 for further discussion of these systems.

In the event of failure or destruction of electronic systems when the vessel itself is not in danger, navigational equipment and methods may need to be improvised. This is especially true with ECDIS and electronic charts. The navigator of a paperless ship, whose primary method of navigation is ECDIS, must assemble enough backup paper charts, equipment, and knowledge to complete his voyage in the event of a major computer system failure. A navigator who keeps a couple of dozen paper charts and a spare handheld GPS receiver under his bunk will be a hero in such an event. If he has a sextant and celestial calculator or tables and the knowledge to use them, so much the better.

No navigator should ever become completely dependent on electronic methods. The navigator who regularly navigates by blindly pushing buttons and reading the coordinates from "black boxes" will not be prepared to use basic principles to improvise solutions in an emergency.

For offshore voyaging, the professional navigator should become thoroughly familiar with the theory of celestial navigation. He should be able to identify the most useful stars and know how to solve various types of sights. He should be able to construct a plotting sheet with a protractor and improvise a sextant. He should know how to solve sights using tables or a navigational calculator. For the navigator prepared with such knowledge the situation is never hopeless. Some method of navigation is always available to one who understands certain basic principles.

The modern ship's regular suite of navigation gear consists of many complex electronic systems. Though they may possess a limited backup power supply, most depend on an uninterrupted supply of ship's electrical power. The failure of that power due to breakdown, fire, or hostile action can instantly render the unprepared navigator helpless. This discussion is intended to provide the navigator with the information needed to navigate a vessel in the absence of the regular suite of navigational gear. Training and preparation for a navigational emergency are essential. This should consist of regular practice in the techniques discussed herein while the regular navigation routine is in effect in order to establish confidence in emergency procedures.

2601. Emergency Navigation Kit

The navigator should assemble a kit containing equipment for emergency navigation. This kit should contain:

1. At least one proven and personally tested handheld GPS receiver with waypoints and routes entered, and with plenty of spare batteries.
2. A small, magnetic hand-bearing compass such as is used in small craft navigation, to be used if all other compasses fail.
3. A minimal set of paper charts for the voyage at hand, ranging from small-scale to coastal to approach and perhaps harbor, for the most likely scenarios. A *pilot chart* for the ocean basin in question makes a good small scale chart for offshore use.
4. A notebook or journal suitable for use as a deck log and for computations, plus maneuvering boards, graph paper, and position plotting sheets.
5. Pencils, erasers, a straightedge, protractor or plotter, dividers and compasses, and a knife or pencil sharpener.
6. A timepiece. The optimum timepiece is a quartz crystal chronometer, but any high-quality digital wristwatch will suffice if it is synchronized with the ship's chronometer. A portable radio capable of receiving time signals, together with a good wristwatch, will also suffice.
7. A marine sextant. (An inexpensive plastic sextant will

suffice.) Several types are available commercially. The emergency sextant should be used periodically so its limitations and capabilities are fully understood.

8. A celestial navigation calculator and spare batteries, or a current *Nautical Almanac* and this book or a similar text. Another year's almanac can be used for stars and the Sun without serious error by emergency standards. Some form of long-term almanac might be copied or pasted in the notebook.
9. Tables. Some form of table might be needed for reducing celestial observations if the celestial calculator fails. The *Nautical Almanac* produced by the U.S. Naval Observatory contains detailed procedures for calculator sight reduction and a compact *sight reduction table*.
10. Flashlight. Check the batteries periodically and include extra batteries and bulbs in the kit.
11. Portable radio. A handheld VHF transceiver approved by the Federal Communications Commission for emergency use can establish communications with rescue authorities. A small portable radio may be used as a radio direction finder or for receiving time signals.
12. An Emergency Position Indicating Radiobeacon (EPIRB) and a Search and Rescue Transponder (SART) are absolutely essential. (See Chapter 28).

2602. Most Probable Position

In the event of failure of primary electronic navigation systems, the navigator may need to establish the **most probable position** (MPP) of the vessel. Usually there is little doubt as to the position. The most recent fix updated with a DR position will be adequate. But when conflicting information or information of questionable reliability is received, the navigator must determine the MPP.

When complete positional information is lacking, or when the available information is questionable, the most probable position might be determined from the intersection of a single line of position and a DR, from a line of soundings, from lines of position which are somewhat inconsistent, or from a dead reckoning position with a correction for set and drift. Continue a dead reckoning plot from one fix to another because the DR plot often provides the best estimate of the MPP.

A series of estimated positions may not be consistent because of the continual revision of the estimate as additional information is received. However, it is good practice to plot all MPP's, and sometimes to maintain a separate EP plot based upon the best estimate of track and speed made good. This could indicate whether the present course is a safe one (See Chapter 23).

2603. Plotting Sheets

If plotting sheets are not available, a Mercator plotting sheet can be constructed through either of two alternative methods based upon a graphical solution of the secant of the latitude, which approximates the expansion of latitude.

First method (Figure 2603a):

Step one: Draw a series of equally spaced vertical lines at any spacing desired. These are the meridians; label them at any desired interval, such as 1', 2', 5', 10', 30', 1°, etc.

Step two: Draw and label a horizontal line through the center of the sheet to represent the parallel of the mid-latitude of the area.

Step three: Through any convenient point, such as the intersection of the central meridian and the parallel of the mid-latitude, draw a line making an angle with the horizontal equal to the mid-latitude. In Figure 2603a this angle is 35°.

Step four: Draw in and label additional parallels. The length of the oblique line between meridians is the perpendicular distance between parallels, as shown by the broken arc. The number of minutes of arc between parallels is the same as that between the meridians.

Step five: Graduate the oblique line into convenient units. If 1' is selected, this scale serves as both a latitude and mile scale. It can also be used as a longitude scale by measuring horizontally from a meridian instead of obliquely along the line.

The meridians may be shown at the desired interval and the mid-parallel may be printed and graduated in units of longitude. In using the sheet it is necessary only to label the meridians and draw the oblique line. From it determine the interval used to draw in and label additional parallels. If the central meridian is graduated, the oblique line need not be.

Second method (Figure 2603b):

Step one: At the center of the sheet draw a circle with a radius equal to 1° (or any other convenient unit) of latitude at the desired scale. If a sheet with a compass rose is available, as in Figure 2603b, the compass rose can be used as the circle and will prove useful for measuring directions. It need not limit the scale of the chart, as an additional concentric circle can be drawn, and desired graduations extended to it.

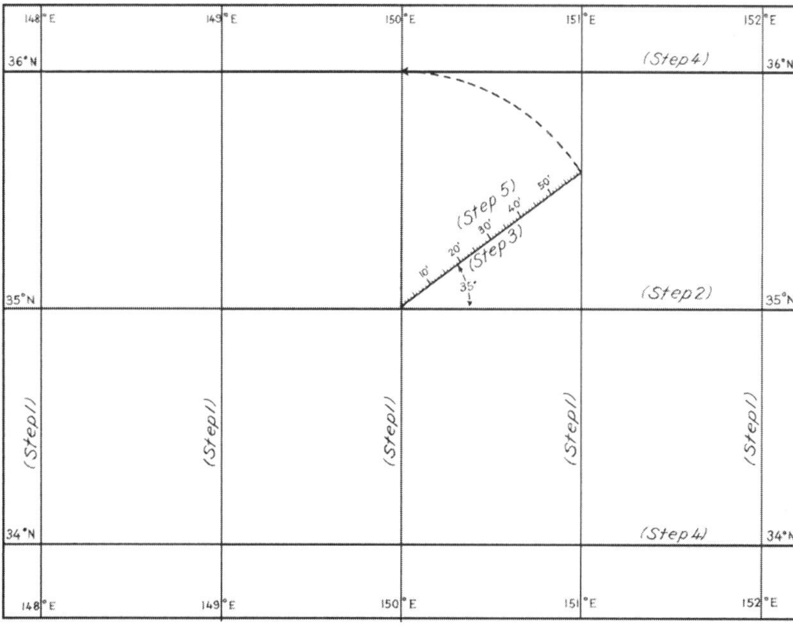

Figure 2603a. Small area plotting sheet with selected longitude scale.

Step two: Draw horizontal lines through the center of the circle and tangent at the top and bottom. These are parallels of latitude; label them accordingly, at the selected interval (as every 1°, 30', etc.).

Step three: From the center of the circle draw a line making an angle with the horizontal equal to the mid-latitude. In Figure 2603b this angle is 40°.

Step four: Draw in and label the meridians. The first is a vertical line through the center of the circle. The second is a vertical line through the intersection of the oblique line and the circle. Additional meridians are drawn the same distance apart as the first two.

Step five: Graduate the oblique line into convenient units. If 1' is selected, this scale serves as a latitude and mile scale. It can also be used as a longitude scale by measuring horizontally from a meridian, instead of obliquely along the line.

In the second method, the parallels may be shown at the desired interval, and the central meridian may be printed and graduated in units of latitude. In using the sheet it is necessary only to label the parallels, draw the oblique line, and from it determine the interval and draw in and label additional meridians. If the central meridian is graduated, as shown in Figure 2603b, the oblique line need not be.

The same result is produced by either method. The first method, starting with the selection of the longitude scale, is particularly useful when the longitude limits of the plotting sheet determine the scale. When the latitude coverage is more important, the second method may be preferable. In either method a simple compass rose might be printed.

Both methods use a constant relationship of latitude to longitude over the entire sheet and both fail to allow for the ellipticity of the Earth. For practical navigation these are not important considerations.

2604. Dead Reckoning

Of the various types of navigation, dead reckoning alone is always available in some form. In an emergency it is of more than average importance. With electronic systems out of service, keep a close check on speed, direction, and distance made good. Carefully evaluate the effects of wind and current. Long voyages with accurate landfalls have been successfully completed by this method alone. This is not meant to minimize the importance of other methods of determining position. However, a good dead reckoning position may actually be more accurate than one determined from several inexact LOP's. If the means of determining direction and distance (the elements of dead reckoning)

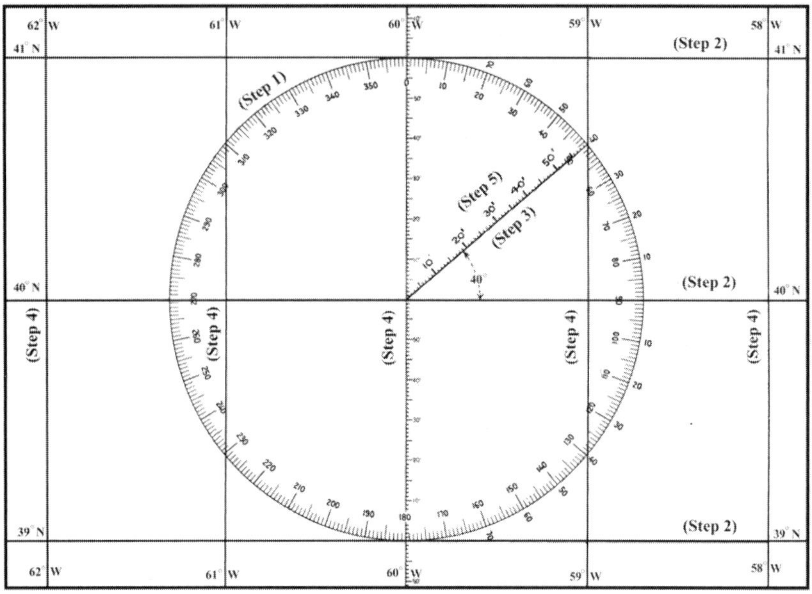

Figure 2603b. Small area plotting sheet with selected latitude scale.

Angle	0		18		31		41		49		56		63		69		75		81		87		90
Factor		1.0		0.9		0.8		0.7		0.6		0.5		0.4		0.3		0.2		0.1		0.0	

Table 2604. Simplified traverse table.

are accurate, it may be best to adjust the dead reckoning only after a confident fix.

Plotting can be done directly on a *pilot chart* or plotting sheet. If this proves too difficult, or if an independent check is desired, some form of mathematical reckoning may be useful. Table 2604, a simplified traverse table, can be used for this purpose. To find the difference or change of latitude in minutes, enter the table with course angle, reckoned from north or south toward the east or west. Multiply the distance run in miles by the factor. To find the departure in miles, enter the table with the complement of the course angle. Multiply the distance run in miles by the factor. To convert departure to difference of longitude in minutes, enter the table with mid-latitude and divide the departure by the factor.

Example: *A vessel travels 26 miles on course 205°, from Lat. 41°44'N, Long. 56°21'W.*

Required: *Latitude and longitude of the point of arrival.*

Solution: *The course angle is 205° - 180° = S25°W, and the complement is 90° - 25° = 65°. The factors corresponding to these angles are 0.9 and 0.4, respectively. The difference of latitude is 26 × 0.9 = 23' (to the nearest minute) and the departure is 26 × 0.4 = 10 NM. Since the course is in the southwestern quadrant in the Northern Hemisphere, the latitude of the point of arrival is 41°44'N -23' = 41°21'N. The factor corresponding to the mid-latitude 41°32'N is 0.7. The difference of longitude is 10 ÷ 0.7 = 14'. The longitude of the point of arrival is 56°21'W + 14 = 56°35'W.*

Answer: *Lat. 41°21'N, Long. 56°35'W.*

2605. Deck Log

At the onset of a navigational emergency, a navigation log should be started if a deck log is not already being maintained. The date and time of the casualty should be the first entry, followed by navigational information such as ship's position, status of all navigation systems, the decisions made, and the reasons for them.

The best determination of the position of the casualty should be recorded, followed by a full account of courses, distances, positions, winds, currents, and leeway. No important navigational information should be left to memory.

2606. Direction

Direction is one of the elements of dead reckoning. A deviation table for each compass, including any lifeboat compasses, should already have been determined. In the event of destruction or failure of the gyrocompass and bridge magnetic compass, lifeboat compasses can be used.

If an almanac, accurate Greenwich time, and the necessary tables are available, the azimuth of any celestial body can be computed and this value compared with an azimuth measured by the compass. If it is difficult to observe the compass azimuth, select a body dead ahead and note the compass heading. The difference between the computed and observed azimuths is compass error on that heading. This is of more immediate value than deviation, but if the latter is desired, it can be determined by applying variation to the compass error.

Several unique astronomical situations occur, permitting determination of azimuth without computation:

Polaris: Polaris is always within 2° of true north for observers between the equator and about 60° North. When Polaris is directly above or below the celestial pole, its azimuth is true north at any latitude. This occurs when the trailing star of either Cassiopeia or the Big Dipper is directly above or below Polaris. When these two stars form a horizontal line with Polaris, the maximum correction applies. Below about 50° latitude, this correction is 1°, and between 50° and 65°, it is 2°. If Cassiopeia is to the right of Polaris, the azimuth is 001° (002° above 50°N), and if Cassiopeia is to the left of Polaris, the azimuth is 359° (358° above 50°N).

The south celestial pole is located approximately at the intersection of a line through the longer axis of the Southern Cross with a line from the northernmost star of Triangulum Australe, perpendicular to the line joining the other two stars of the triangle. No conspicuous star marks this spot.

Meridian Transit: Any celestial body bears due north or south at meridian transit, either upper or lower. This is the moment of maximum (or minimum) altitude of the body. However, since the altitude at this time is nearly constant during a considerable change of azimuth, the instant of meridian transit may be difficult to determine. If time and an almanac are available, and the longitude is known, the time of transit can be computed. It can also be graphed as a curve on graph paper and the time of meridian transit determined with sufficient accuracy for emergency purposes.

Body on Prime Vertical: If any method is available for determining when a body is on the prime vertical (due east or west), the compass azimuth at this time can be observed. Table 20, Meridian Angle and Altitude of a Body on the Prime Vertical Circle provides this information. Any body on the celestial equator (declination 0°) is on the prime vertical at the time of rising or setting. For the Sun this occurs at the time of the equinoxes. The star Mintaka (δ Orionis), the leading star of Orion's belt, has a declination of approximately 0.3°S and can be considered on the celestial equator. For an observer near the equator, such a body is always nearly east or west. Because of refraction and dip, the azimuth should be noted when the center of the Sun or a star is a little more than one Sun diameter (half a degree) above the horizon. The Moon should be observed when its upper limb is on the horizon.

Body at Rising or Setting: Except for the Moon, the azimuth angle of a body is almost the same at rising as at setting, except that the former is toward the east and the latter toward the west. If the azimuth is measured both at rising and setting, true south (or north) is midway between the two observed values, and the difference between this value and 180° (or 000°) is the compass error. Thus, if the compass azimuth of a body is 073° at rising, and 277° at setting, true south (180°) is $\frac{073° + 277°}{2} = 175°$ by compass, and the compass error is 5°E. This method may be in error if the vessel is moving rapidly in a northerly or southerly direction. If the declination and latitude are known, the true azimuth of any body at rising or setting can be determined by means of a diagram on the plane of the celestial meridian or by computation. For this purpose, the body (except the Moon) should be considered as rising or setting when its center is a little more than one Sun diameter (half a degree) above the horizon, because of refraction and dip.

Finding direction by the relationship of the Sun to the hands of a watch is sometimes advocated, but the limitations of this method prevent its practical use at sea.

A simple technique can be used for determining deviation. Find an object that is easily visible and that floats, but will not drift too fast in the wind. A life preserver, or several tied together, will suffice. Throw this marker overboard, and steer the vessel steadily in the exact opposite direction to the chosen course. At a distance of perhaps half a mile, or more if the marker is still clearly in view, execute a Williamson turn, or turn the vessel 180° in the smallest practical radius, and head back toward the marker. The magnetic course will be midway between the course toward the object and the reciprocal of the course away from the object. Thus, if the boat is on compass course 151° while heading away from the object, and 337° while returning, the magnetic course is midway between 337° and 151° + 180° = 331°, or

$$\frac{337° + 331°}{2} = 334° .$$

Since 334° magnetic is the same as 337° by compass, the deviation on this heading is 3°W.

If a compass is not available, any celestial body can be used to steer by, if its diurnal apparent motion is considered. A reasonably straight course can be steered by noting the direction of the wind, the movement of the clouds, the direction of the waves, or by watching the wake of the vessel. The angle between the centerline and the wake is an indication of the amount of leeway.

A body having a declination the same as the latitude of the destination is directly over the destination once each day, when its hour angle equals the longitude, measured westward through 360°. At this time it should be dead ahead if the vessel is following the great circle leading directly to the destination. Inspect the almanac to find a body with a suitable declination.

EMERGENCY CELESTIAL NAVIGATION

2607. Almanacs

Almanac information, particularly declination and Greenwich Hour Angle of bodies, is important to celestial navigation. If the only copy available is for a previous year, it can be used for the Sun, Aries (♈), and stars without serious error by emergency standards. However, for greater accuracy, proceed as follows:

For declination of the Sun, enter the almanac with a time that is earlier than the correct time by $5^h 49^m$ multiplied by the number of years between the date of the almanac and the correct date, adding 24 hours for each February 29th that occurs between the dates. If the date is February 29th, use March 1 and reduce by one the number of 24 hour periods added. For GHA of the Sun or Aries, determine the value for the correct time, adjusting the minutes and tenths of arc to agree with that at the time for which the declination is determined. Since the adjustment never exceeds half a degree, care should be used when the value is near a whole degree, to prevent the value from being in error by 1°.

If no almanac is available, a rough approximation of the declination of the Sun can be obtained as follows: Count the days from the given date to the nearer solstice (June 21st or December 22nd). Divide this by the number of days from that solstice to the equinox (March 21st or September 23rd), using the equinox that will result in the given date being between it and the solstice. Multiply the result by 90°. Enter Table 2604 with the angle so found and extract the factor. Multiply this by 23.45° to find the declination.

Example 1: *The date is August 24th.*

Required: *The approximate declination of the Sun.*

Solution: *The number of days from the given date to the nearer solstice (June 21) is 64. There are 94 days between June 21 and September 23. Dividing and multiplying by 90°,*

$$\frac{64}{94} \times 90° = 61.3'$$

The factor from Table 2604 is 0.5. The declination is 23.45° × 0.5 = 11.7°. We know it is north because of the date.

Answer: *Dec. 11.7°N.*

The accuracy of this solution can be improved by considering the factor of Table 2604 as the value for the mid-angle between the two limiting ones (except that 1.00 is correct for 0° and 0.00 is correct for 90°), and interpolating to one additional decimal. In this instance the interpolation would be between 0.50 at 59.5° and 0.40 at 66°. The interpolated value is 0.47, giving a declination of 11.0°N. Still greater accuracy can be obtained by using a table of natural cosines instead of Table 2604. By natural cosine, the value is 11.3°N.

If the latitude is known, the declination of any body can be determined by observing a meridian altitude. It is usually best to make a number of observations shortly before and after transit, plot the values on graph paper, letting the ordinate (vertical scale) represent altitude, and the abscissa (horizontal scale) the time. The altitude is found by fairing a curve or drawing an arc of a circle through the points, and taking the highest value. A meridian altitude problem is then solved in reverse.

Example 2: *The latitude of a vessel is 40°16'S. The Sun is observed on the meridian, bearing north. The observed altitude is 36°29'.*

Required: *Declination of the Sun.*

Solution: *The zenith distance is 90° - 36°29' = 53°31'. The Sun is 53°31' north of the observer, or 13°15' north of the equator. Hence, the declination is 13°15' N.*

Answer: *Dec. 13°15' N.*

The GHA of Aries can be determined approximately by considering it equal to GMT (in angular units) on September 23rd. To find GHA Aries on any other date, add 1° for each day following September 23rd. The value is approximately 90° on December 22nd, 180° on March 21st and 270° on June 21st. The values found can be in error by as much as several degrees, and so should not be used if better information is available. An approximate check is provided by the great circle through Polaris, Caph (the leading star of Cassiopeia), and the eastern side of the square of Pegasus. When this great circle coincides with the meridian, LHA ♈ is approximately 0°. The hour angle of a body is equal to its SHA plus the hour angle of Aries. If an error of up to 4°, or a little more, is acceptable, the GHA of the Sun can be considered equal to GMT ± 180° (12^h).

For more accurate results, one can make a table of the equation of time from the *Nautical Almanac* perhaps at five- or ten-day intervals, and include this in the emergency navigation kit. The equation of time is applied according to its sign to GMT ± 180° to find GHA.

2608. Altitude Measurement

With a sextant, altitudes are measured in the usual manner. If in a small boat or raft, it is a good idea to make a number of observations and average both the altitudes and times, or plot on graph paper the altitudes versus time. The rougher the sea, the more important this process becomes, which tends to average out errors caused by rough weather observations.

The improvisations which may be made in the absence of a sextant are so varied that in virtually any circumstances a little ingenuity will produce a device to measure altitude. The results obtained with any improvised method will be approximate at best, but if a number of observations are averaged, the accuracy can be improved. A measurement, however approximate, is better than an estimate. Two general types of improvisation are available:

1. Circle. Any circular degree scale, such as a maneuvering board, compass rose, protractor, or plotter can be used to measure altitude or zenith distance directly. This is the principle of the ancient astrolabe. A maneuvering board or compass rose can be mounted on a flat board. A protractor or plotter may be used directly. There are a number of variations of the technique of using such a device. Some of them are:

A peg or nail is placed at the center of the circle as seen in Figure 2608a. A weight is hung from the 90° graduation, and a string for holding the device is attached at the 270° graduation. When it is held with the weight acting as a plumb bob, the 0° - 180° line is horizontal. In this position the board is turned in azimuth until it is in line with the Sun. The intersection of the shadow of the center peg with the arc of the circle indicates the altitude of the center of the Sun.

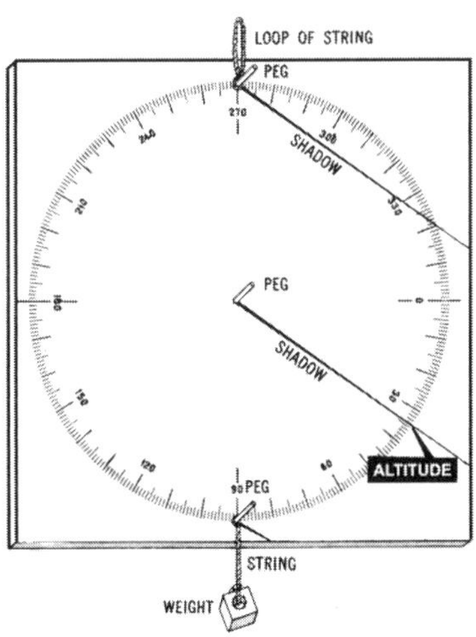

Figure 2608a. Improvised astrolabe; shadow method.

The weight and loop can be omitted and pegs placed at the 0° and 180° points of the circle. While one observer sights along the line of pegs to the horizon, an assistant notes the altitude.

The weight can be attached to the center pin, and the three pins (0°, center, 180°) aligned with the celestial body. The reading is made at the point where the string holding the weight crosses the scale. The reading thus obtained is the zenith distance unless the graduations are labeled to indicate altitude. This method, illustrated in Figure 2608b, is used for bodies other than the Sun.

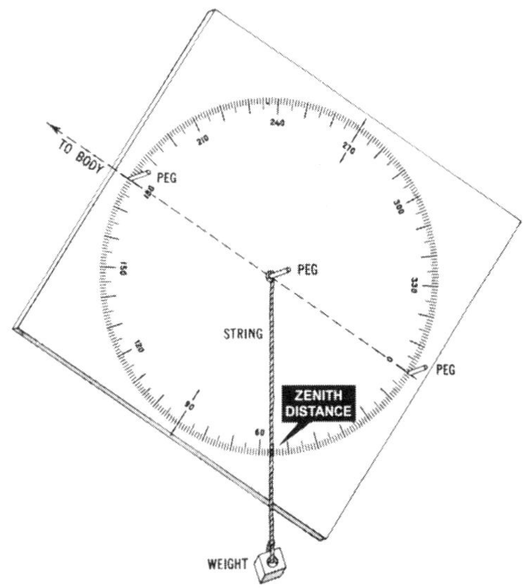

Figure 2608b. Improvised astrolabe; direct sighting method.

Whatever the technique, reverse the device for half the readings of a series to minimize errors of construction. Generally, the circle method produces more accurate results than the right triangle method, described below.

2. Right triangle. A cross-staff can be used to establish one or more right triangles, which can be solved by measuring the angle representing the altitude, either directly or by reconstructing the triangle. Another way of determining the altitude is to measure two sides of the triangle and divide one by the other to determine one of the trigonometric functions. This procedure, of course, requires a source of information on the values of trigonometric functions corresponding to various angles. If the cosine is found, Table 2604 can be used. The tabulated factors can be considered correct to one additional decimal for the value midway between the limited values (except that 1.00 is the correct value for 0° and 0.00 is the correct value for 90°) without serious error by emergency standards. Interpolation can then be made between such values.

By either protractor or table, most devices can be graduated in advance so that angles can be read directly. There are many variations of the right triangle method. Some of these are described below.

Two straight pieces of wood can be attached to each other in such a way that the shorter one can be moved along the longer, the two always being perpendicular to each other. The shorter piece is attached at its center. One end of the longer arm is held to the eye. The shorter arm is moved until its top edge is in line with the celestial body, and its bottom edge is in line with the horizon. Thus, two right triangles are formed, each representing half the altitude. See Figure 2608c.For low altitudes, only one of the triangles is used, the long arm being held in line with the horizon. The length of half the short arm, divided by the length of that part of the long arm between the eye and the intersection with the short arm, is the tangent of half the altitude (the whole altitude if only one right triangle is used). The cosine can be found by dividing that part of the long arm between the eye and the intersection with the short arm by the slant distance from the eye to one end of the short arm. Graduations consist of a series of marks along the long arm indicating settings for various angles. The device should be inverted for alternate readings of a series.

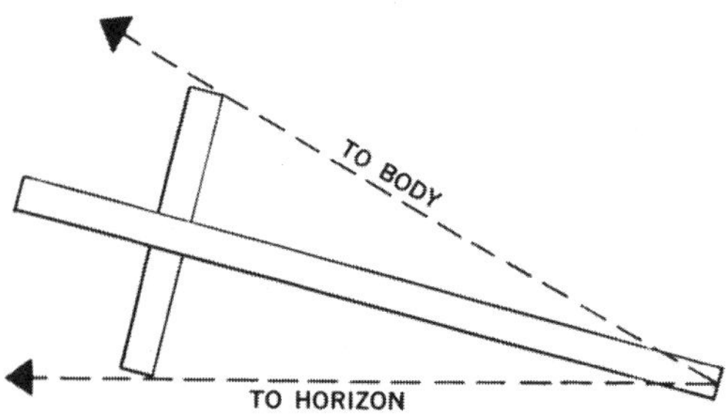

Figure 2608c. Improvised cross-staff.

A rule or any stick can be held at arm's length. The top of the rule is placed in line with the celestial body being observed, and the top of the thumb is placed in line with the horizon. The rule is held vertically. The length of rule above the thumb, divided by the distance from the eye to the top of the thumb, is the tangent of the angle observed. The cosine can be found by dividing the distance from the eye to the top of the thumb by the distance from the eye to the top of the rule. If the rule is tilted toward the eye until the minimum of rule is used, the distance from the eye to the middle of the rule is substituted for the distance from the eye to the top of the thumb, half the length of the rule above the thumb is used, and the angle found is multiplied by 2. Graduations consist of marks on the rule or stick indicating various altitudes. For the average observer each inch of rule will subtend an angle of about 2.3°, assuming an eye-to-ruler distance of 25 inches. This relationship is good to a maximum altitude of about 20°.

The accuracy of this relationship can be checked by comparing the measurement against known angles in the sky. Angular distances between stars can be computed by sight reduction methods, including *Pub. No. 229*, by using the declination of one star as the latitude of the assumed position, and the difference between the hour angles (or SHA's) of the two bodies as the local hour angle. The angular distance is the complement of the computed altitude. The angular distances between some well-known star pairs are: end stars of Orion's belt, 2.7°; pointers of the Big Dipper, 5.4°, Rigel to Orion's belt, 9.0°; eastern side of the great square of Pegasus, 14.0°; Dubhe (the pointer nearer Polaris) and Mizar (the second star in the Big Dipper, counting from the end of the handle), 19.3°.

The angle between the lines of sight from each eye is, at arm's length, about 6°. By holding a pencil or finger horizontally, and placing the head on its side, one can estimate an angle of about 6° by closing first one eye and then the other, and noting how much the pencil or finger appears to move in the sky.

The length of the shadow of a peg or nail mounted perpendicular to a horizontal board can be used as one side of an altitude triangle. The other sides are the height of the peg and the slant distance from the top of the peg to the end of the shadow. The height of the peg, divided by the length of the shadow, is the tangent of the altitude of the center of the Sun. The length of the shadow, divided by the slant distance, is the cosine. Graduations consist of a series of concentric circles indicating various altitudes, the peg being at the common center. The device is kept horizontal by floating it in a bucket of water. Half the readings of a series are taken with the board turned 180° in azimuth.

Two pegs or nails can be mounted perpendicular to a board, with a weight hung from the one farther from the eye. The board is held vertically and the two pegs aligned with the body being observed. A finger is then placed over the string holding the weight, to keep it in position as the board is turned on its side. A perpendicular line is dropped from the peg nearer the eye, to the string. The body's altitude is the acute angle nearer the eye. For alternate readings of a series, the board should be inverted. Graduations consist of a series of marks indicating the position of the string at various altitudes.

As the altitude decreases, the triangle becomes smaller. At the celestial horizon it becomes a straight line. No instrument is needed to measure the altitude when either the upper or lower limb is tangent to the horizon, as the sextant altitude is then 0°.

2609. Sextant Altitude Corrections

If altitudes are measured by a marine sextant, the usual sextant altitude corrections apply. If the center of the Sun or Moon is observed, either by sighting at the center or by shadow, the lower-limb corrections should be applied, as usual, and an additional correction of minus 16' applied. If the upper limb is observed, use minus 32'. If a weight is used as a plumb bob, or if the length of a shadow is measured, omit the dip (height of eye) correction.

If an almanac is not available for corrections, each source of error can be corrected separately, as follows:

If a sextant is used, the **index correction** should be determined and applied to all observations, or the sextant adjusted to

Altitude	5°		6°		7°		8°		10°		12°		15°		21°		33°		63°		90°
Refraction		9'		8'		7'		6'		5'		4'		3'		2'		1'		0	

Table 2609. Simplified refraction table.

eliminate index error.

Refraction is given to the nearest minute of arc in Table 2609. The value for a horizon observation is 34'. If the nearest 0.1° is sufficiently accurate, as with an improvised method of observing altitude, a correction of 0.1° should be applied for altitudes between 5° and 18°, and no correction applied for greater altitudes. Refraction applies to all observations, and is always minus.

Dip, in minutes of arc, is approximately equal to the square root of the height of eye, in feet. The dip correction applies to all observations in which the horizon is used as the horizontal reference. It is always a minus. If 0.1° accuracy is acceptable, no dip correction is needed for height of eye in a small boat.

The **semidiameter** of the Sun and Moon is approximately 16' of arc. The correction does not apply to other bodies or to observations of the center of the Sun and Moon, by whatever method, including shadow. The correction is positive if the lower limb is observed, and negative if the upper limb is observed.

For emergency accuracy, **parallax** is applied to observations of the Moon only. An approximate value, in minutes of arc, can be found by multiplying 57' by the factor from Table 2604, entering that table with altitude. For more accurate results, the factors can be considered correct to one additional decimal for the altitude midway between the limiting values (except that 1.00 is correct for 0° and 0.00 is correct for 90°), and the values for other altitudes can be found by interpolation. This correction is always positive.

For observations of celestial bodies on the horizon, the total correction for zero height of eye is:

Sun: Lower limb: (–)18', upper limb: (–)50'.
Moon: Lower limb: (+)39', upper limb: (+)7'.

Planet/Star: (-)34°.

Dip should be added algebraically to these values. Since the sextant altitude is zero, the observed altitude is equal to the total correction.

2610. Sight Reduction

Sight reduction tables should be used, if available. If not, use the compact *sight reduction tables* found in the *Nautical Almanac*. If trigonometric tables and the necessary formulas are available, they will serve the purpose. Speed in solution is seldom a factor in a liferaft, but might be important aboard ship, particularly in hostile areas. If tables but no formulas are available, determine the mathematical knowledge possessed by the crew. Someone may be able to provide the missing information. If the formulas are available, but no tables, approximate natural values of the various trigonometric functions can be obtained graphically. Graphical solution of the navigational triangle can be made by the orthographic method explained Chapter 15, Navigational Astronomy. A maneuvering board might prove helpful in the graphical solution for either trigonometric functions or altitude and azimuth. Very careful work will be needed for useful results by either method. Unless proper navigational equipment is available, better results might be obtained by making separate determinations of latitude and longitude.

2611. Finding Latitude

Several methods are available for determining latitude; none requires accurate time.

Latitude can be determined using a **meridian altitude** of any body, if its declination is known. If accurate time, knowledge of the longitude, and an almanac are available, the observation can be made at the correct moment, as determined in advance. However, if any of these are lacking, or if an accurate altitude measuring instrument is unavailable, it is better to make a number of altitude observations before and after meridian transit. Then plot altitude versus time on graph paper, and the highest (or lowest, for lower transit) altitude is scaled from a curve faired through the plotted points. At small boat speeds, this procedure is not likely to introduce a significant error. The time used for plotting the observations need not be accurate, as elapsed time between observations is all that is needed, and this is not of critical accuracy. Any altitudes that are not consistent with others of the series should be discarded.

Latitude by Polaris is explained in Chapter 20, Sight Reduction. In an emergency, only the first correction is of practical significance. If suitable tables are not available, this correction can be estimated. The trailing star of Cassiopeia (ε Cassiopeiae) and Polaris have almost exactly the same SHA. The trailing star of the Big Dipper (Alkaid) is nearly opposite Polaris and ε Cassiopeiae. These three stars, ε Cassiopeiae, Polaris, and Alkaid, form a line through the N. Celestial Pole (approximately). When this line is horizontal, there is no correction. When it is vertical, the maximum correction of 56' applies. It should be added to the observed altitude if Alkaid is at the top, and subtracted if ε Cassiopeiae is at the top. For any other position, estimate the angle this line makes with the vertical, and multiply the maximum correction (56') by the factor from Table 2604, adding if Alkaid is higher than ε Cassiopeiae, and subtracting if it is lower. See Figure 2611. For more accurate results, the factor from Table 2604 can be considered accurate to one additional decimal for the mid-value between those tabulated (except that 1.00 is correct for 0° and 0.00 for 90°). Other values can be found by interpolation.

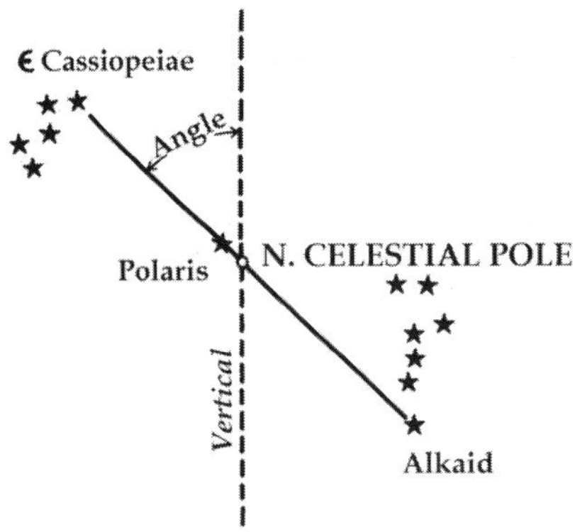

Figure 2611. Relative positions of ε Cassiopeiae, Polaris, and Alkaid with respect to the north celestial pole.

The **length of the day** varies with latitude. Hence, latitude can be determined if the elapsed time between sunrise and sunset can be accurately observed. Correct the observed length of day by adding 1 minute for each 15' of longitude traveled toward the east and subtracting 1 minute for each 15' of longitude traveled toward the west. The latitude determined by length of day is the value for the time of meridian transit. Since meridian transit occurs approximately midway between sunrise and sunset, half the interval may be observed and doubled. If a sunrise and sunset table is not available, the length of daylight can be determined graphically using a diagram on the plane of the celestial meridian, as explained in Chapter 15. A maneuvering board is useful for this purpose. This method cannot be used near the time of the equinoxes and is of little value near the equator. The Moon can be used if moonrise and moonset tables are available. However, with the Moon, the half-interval method is of insufficient accuracy, and allowance should be made for the longitude correction.

The declination of a **body in zenith** is equal to the latitude of the observer. If no means are available to measure altitude, the position of the zenith can be determined by holding a weighted string overhead.

2612. Finding Longitude

Unlike latitude, determining longitude requires accurate Greenwich time. All such methods consist of noting the Greenwich time at which a phenomenon occurs locally. In addition, a table indicating the time of occurrence of the same phenomenon at Greenwich, or equivalent information, is needed. Three methods may be used to determine longitude.

When a body is on the local celestial meridian, its GHA is the same as the longitude of the observer if in west longitude, or 360 - λ in east longitude. Thus, if the GMT of local **time of transit** is determined and a table of Greenwich Hour Angles (or time of transit of the Greenwich meridian) is available, longitude can be computed. If only the equation of time is available, the method can be used with the Sun. This is the reverse of the problem of finding the time of transit of a body. The time of transit is not always apparent. If a curve is made of altitude versus time, as suggested previously, the time corresponding to the highest altitude is used in finding longitude. Under some conditions, it may be preferable to observe an altitude before meridian transit, and then again after meridian transit when the body has returned to the same altitude as at the first observation. Meridian transit occurs midway between these two times. A body in the zenith is on the celestial meridian. If accurate azimuth measurement is available, note the time when the azimuth is 000° or 180°.

The difference between the observed GMT of sunrise or sunset and the LMT tabulated in the almanac is the longitude in time units, which can then be converted to angular measure. If the *Nautical Almanac* is used, this information is tabulated for each third day only. Greater accuracy can be obtained if interpolation is used for determining intermediate values. Moonrise or moonset can be used if the tabulated LMT is corrected for longitude. Planets and stars can be used if the time of rising or setting can be determined. This can be computed, or approximated using a diagram on the plane of the celestial meridian (See Chapter 15, Navigational Astronomy).

Either of these methods can be used in reverse to set a watch that has run down or to check the accuracy of a watch if the longitude is known. In the case of a meridian transit, the time at the instant of transit is not necessary.

Simply start the watch and measure the altitude several times before and after transit, or at equal altitudes before and after transit. Note the times of these observations and find the exact watch time of meridian transit. The difference between this time and the correct time of transit is the correction factor by which to reset the watch.

CHAPTER 27

NAVIGATION REGULATIONS

SHIP ROUTING

2700. Purpose and Types of Routing Systems

Navigation, once independent throughout the world, is an increasingly regulated activity. The consequences of collision or grounding for a large, modern ship carrying tremendous quantities of high-value, perhaps dangerous cargo are so severe that authorities have instituted many types of regulations and control systems to minimize the chances of loss. These range from informal and voluntary systems to closely controlled systems requiring strict compliance with numerous regulations. The regulations may concern navigation, communications, equipment, procedures, personnel, and many other aspects of ship management. This chapter will be concerned primarily with navigation regulations and procedures.

There are many types of vessel traffic rules. However, the cornerstone of all these are the *Navigation Rules: International-Inland*. The International Rules (Title 33 U.S.C. Chap. 30) were formalized in the Convention of the International Regulations for the Preventing of Collisions at Sea of 1972 (COLREGS '72) and became effective on July 15, 1977. Following the signing of the Convention, an effort was made to unify and update the various domestic navigation rules. This effort culminated in the enactment of the Inland Navigation Rules Act of 1980.

The Inland Navigation Rules (Title 33 U.S.C. Chap. 34) recodified parts of the Motorboat Act of 1940 and a large body of existing navigational practices, pilot rules, interpretive rules previously referred to as the Great Lakes Rules, Inland Rules and Western River Rules. The effective date for the Inland Navigation Rules was December 24, 1981, except for the Great Lakes where the effective date was March 1, 1983.

The International Rules apply to vessels on waters outside of the established lines of demarcation (COLREGS Demarcation Lines, 33 C.F.R. §80). These lines are depicted on U.S. charts with dashed lines, and generally run between major headlands and prominent points of land at the entrance to coastal rivers and harbors. The Inland Navigation Rules apply to waters inside the lines of demarcation. It is important to note that with the exception of Annex V to the Inland Rules, the International and Inland Navigation Rules are very similar in both content and format.

Much information relating to maritime regulations may be found on the World Wide Web, and any common search engine can turn up increasing amounts of documents posted for mariners to access. As more and more regulatory information is posted to new Web sites and bandwidth increases, mariners will have easier access to the numerous rules with which they must comply.

2701. Terminology

There are several specific types of regulatory systems. For commonly used open ocean routes where risk of collision is present, the use of **Recommended Routes** separates ships going in opposite directions. In areas where ships converge at headlands, straits, and major harbors, **Traffic Separation Schemes** (TSS's) have been instituted to separate vessels and control crossing and meeting situations. **Vessel Traffic Services** (VTS's), sometimes used in conjunction with a TSS, are found in many of the major ports of the world. While TSS's are often found offshore in international waters, VTS's are invariably found closer to shore, in national waters. Environmentally sensitive areas may be protected by **Areas to be Avoided** which prevent vessels of a certain size or carrying certain cargoes from navigating within specified boundaries. In confined waterways such as canals, lock systems, and rivers leading to major ports, local navigation regulations often control ship movement.

The following terms relate to ship's routing:

Routing System: Any system of routes or routing measures designed to minimize the possibility of collisions between ships, including TSS's, two-way routes, recommended tracks, areas to be avoided, inshore traffic zones, precautionary areas, and deep-water routes.

Traffic Separation Scheme: A routing measure which separates opposing traffic flow with traffic lanes.

Separation Zone or Line: An area or line which separates opposing traffic, separates traffic from adjacent areas, or separates different classes of ships from one another.

Traffic Lane: An area within which one-way traffic is established.

Roundabout: A circular traffic lane used at junctions of several routes, within which traffic moves counterclockwise around a separation point or zone.

Inshore Traffic Zone: The area between a traffic separation scheme and the adjacent coast, usually designated for coastal traffic.

Two-Way Route: A two-way track for guidance of ships through hazardous areas.

Recommended Route: A route established for convenience of ship navigation, often marked with centerline buoys.

Recommended Track: A route, generally found to be free of dangers, which ships are advised to follow to avoid possible hazards nearby.

Deep-Water Route: A route surveyed and chosen for the passage of deep-draft vessels through shoal areas.

Precautionary Area: A defined area within which ships must use particular caution and should follow the recommended direction of traffic flow.

Area to be Avoided: An area within which navigation by certain classes of ships is prohibited because of particular navigational dangers or environmentally sensitive natural features. They are depicted on charts by dashed or composite lines. The smallest may cover less than a mile in extent; the largest may cover hundreds of square miles. Notes on the appropriate charts and in pilots and *Sailing Directions* tell which classes of ships are excluded from the area.

Established Direction of Traffic Flow: The direction in which traffic within a lane must travel.

Recommended Direction of Traffic Flow: The direction in which traffic is recommended to travel.

There are various methods by which ships may be separated using Traffic Separation Schemes. The simplest scheme might consist of just one method. More complex schemes will use several different methods together in a coordinated pattern to route ships to and from several areas at once. Schemes may be just a few miles in extent, or cover relatively large sea areas.

2702. Recommended Routes and Tracks

Recommended Routes across the North Atlantic have been followed since 1898, when the risk of collision between increasing numbers of ships became too great, particularly at junction points. The International Convention for the Safety of Life at Sea (SOLAS) codifies the use of certain routes. These routes vary with the seasons, with winter and summer tracks chosen so as to avoid iceberg-prone areas. These routes are often shown on charts, particularly small scale ones, and are generally used to calculate distances between ports in tables.

Recommended Routes consist of single tracks, either one-way or two-way. Two-way routes show the best water through confined areas such as among islands and reefs. Ships following these routes can expect to meet other vessels head-on and engage in normal passings. One-way routes are generally found in areas where many ships are on similar or opposing courses. They are intended to separate opposing traffic so that most maneuvers are overtaking situations instead of the more dangerous meeting situation.

2703. Charting Routing Systems

Routing Systems and TSS's are depicted on nautical charts in magenta (purple) or black as the primary color. Zones are shown by purple tint, limits are shown by composite lines such as are used in other maritime limits, and lines are dashed. Arrows are outlined or dashed-lined depending on use. Deep-water routes are marked with the designation "DW" in bold purple letters, and the least depth may be indicated.

Recommended Routes and recommended tracks are generally indicated on charts by black lines, with arrowheads indicating the desired direction of traffic. Areas to be Avoided are depicted on charts by dashed lines or composite lines, either point to point straight lines or as a circle centered on a feature in question such as a rock or island.

Note that not all ship's routing measures are charted. U.S. charts generally depict recommended routes only on charts made directly from foreign charts. Special provisions applying to a scheme may be mentioned in notes on the chart and are usually discussed in detail in the *Sailing Directions*. In the U.S., the boundaries and routing scheme's general location and purpose are set forth in the Code of Federal Regulations and appear in the *Coast Pilot*.

TRAFFIC SEPARATION SCHEMES

2704. Traffic Separation Schemes (TSS)

In 1961, representatives from England, France, and Germany met to discuss ways to separate traffic in the congested Straits of Dover and subsequently in other congested areas. Their proposals were submitted to the

International Maritime Organization (IMO) and were adopted in general form. IMO expanded on the proposals and has since instituted a system of **Traffic Separation Schemes (TSS)** throughout the world.

The IMO is the only international body responsible for establishing and recommending measures for ship's routing in international waters. It does not attempt to regulate traffic within the territorial waters of any nation.

In deciding whether or not to adopt a TSS, IMO considers the aids to navigation system in the area, the state of hydrographic surveys, the scheme's adherence to accepted standards of routing, and the International Rules of the Road. The selection and development of TSS's are the responsibility of individual governments, who may seek IMO adoption of their plans, especially if the system extends into international waters.

Governments may develop and implement TSS's not adopted by the IMO, but in general only IMO-adopted schemes are charted. Rule 10 of the International Regulations for Preventing Collisions at Sea (Rules of the Road) addresses the subject of TSS's. This rule specifies the actions to be taken by various classes of vessels in and near traffic schemes.

Traffic separation schemes adopted by the IMO are listed in *Ship's Routing*, a publication of the IMO, 4 Albert Embankment, London SE1 7SR, United Kingdom http://www.imo.org. Because of differences in datums, chartlets in this publication which depict the various schemes must not be used either for navigation or to chart the schemes on navigational charts. The *Notice to Mariners* should be consulted for charting details.

2705. Methods and Depiction

A number of different methods of separating traffic have been developed, using various zones, lines, and defined areas. One or more methods may be employed in a given traffic scheme to direct and control converging or passing traffic. These are discussed below. Refer to definitions in Article 2701 and Figure 2705.

Method 1. Separation of opposing streams of traffic by separation zones or lines. In this method, typically a central separation zone is established within which ships are not to navigate. The central zone is bordered by traffic lanes with established directions of traffic flow. The lanes are bounded on the outside by limiting lines.

Method 2. Separation of opposing streams of traffic by natural features or defined objects. In this method islands, rocks, or other features may be used to separate traffic. The feature itself becomes the separation zone.

Method 3. Separation of through traffic from local traffic by provision of Inshore Traffic Zones. Inshore traffic zones provide an area within which local traffic may travel at will without interference from through traffic in the lanes. Inshore zones are separated from traffic lanes by separation zones or lines.

Method 4. Division of traffic from several different directions into sectors. This approach is used at points of convergence such as pilot stations and major entrances.

Method 5. Routing traffic through junctions of two or more major shipping routes. The exact design of the scheme in this method varies with conditions. It may be a circular or rectangular precautionary area, a roundabout, or a junction of two routes with crossing routes and directions of flow well defined.

2706. Use of Traffic Separation Schemes

A TSS is not officially approved for use until adopted by the IMO. Once adopted, it is implemented at a certain time and date and announced in the *Notice to Mariners* and by other means. The *Notice to Mariners* will also describe the scheme's general location and purpose, and give specific directions in the chart correction section on plotting the various zones and lines which define it. These corrections usually apply to several charts. Because the charts may range in scale from quite small to very large, the corrections for each should be followed closely. The positions for the various features may be slightly different from chart to chart due to differences in rounding off positions or chart datum.

Use of TSS's by all ships is recommended but not always required. In the event of a collision, vessel compliance with the TSS is a factor in assigning liability in admiralty courts. TSS's are intended for use in all weather, both day and night. Adequate aids to navigation are a part of all TSS's. There is no special right of one ship over another in TSS's because the Rules of the Road apply in all cases. Deep-water routes should be avoided by ships that do not need them to keep them clear for deep-draft vessels. Ships need not keep strictly to the courses indicated by the arrows, but are free to navigate as necessary within their lanes to avoid other traffic. The signal "YG" is provided in the International Code of Signals to indicate to another ship: "You appear not to be complying with the traffic separation scheme." TSS's are discussed in detail in the *Sailing Directions* for the areas where they are found.

Certain special rules adopted by IMO apply in constricted areas such as the Straits of Malacca and Singapore, the English Channel and Dover Strait, and in the Gulf of Suez. These regulations are summarized in the appropriate *Sailing Directions (Planning Guides)*. For a complete summary of worldwide ships' routing measures, the IMO publication Ship's Routing should be obtained. See Article 2704.

Routing term	Symbol	Description	Applications
1 Established direction of traffic flow		Outlined arrow	Traffic separation schemes and deep-water routes (when part of a traffic lane)
2 Recommended direction of traffic flow		Dashed outlined arrow	Precautionary areas, two-way routes, recommended routes and deep-water routes
3 Separation lines		Tint, 3 mm wide	Traffic separation schemes and between traffic separation schemes and inshore traffic zone
4 Separation zones		Tint, may be any shape	Traffic separation schemes and between traffic separation schemes and inshore traffic zones
5 Limits of restricted areas (charting term)		T-Shaped dashes	Areas to be avoided and defined ends of inshore traffic zones
6 General maritime limits (charting term)		Dashed line	Traffic separation schemes, precautionary areas, two-way routes and deep-water routes
7 Recommended tracks: one-way two-way		Dashed lines with arrowheads (colour black)	Generally reserved for use by charting authorities
8 Recommended routes		Dashed line and dashed outlined arrows	Recommended routes
9 Precautionary areas	⚠	Precautionary symbol	Precautionary areas

Figure 2705. Traffic separation scheme symbology. On charts the symbols are usually in magenta.

VESSEL TRAFFIC SERVICES (VTS)

2707. Description and Purpose

The purpose of Vessel Traffic Services (VTS) is to provide interactive monitoring and navigational advice for vessels in particularly confined and busy waterways. There are two main types of VTS, surveilled and non-surveilled. Surveilled systems consist of one or more land-based radar sites which transmit their signals to a central location where operators monitor and, to a certain extent, control traffic flows. Ships contact the VTS authority at predetermined, charted calling -in points. Non-surveilled systems consist of one or more calling-in points at which ships are required to report their identity, course, speed, and other data to the monitoring authority. At least 18 countries in the world now operate vessel traffic services of some sort, including most major maritime nations.

Vessel Traffic Services in the U.S. are implemented under the authority of the Ports and Waterways Safety Act of 1972 (Public Law 92-340 as amended) and the St. Lawrence Seaway Act (Public Law 358). They encompass a wide range of techniques and capabilities aimed at preventing vessel allisions, collisions, and groundings in the approach, harbor and inland waterway phases of navigation. They are also designed to expedite ship movements, increase transportation system capacity, and improve all-weather operating capability. Automatic Identification Systems (AIS) may be integrated into VTS operations.

A VHF-FM communications network forms the basis of most VTS's. Transiting vessels make position reports to an operations center by radiotelephone and are in turn provided with accurate, complete, and timely navigational safety information. The addition of a network of radars for surveillance and computer-assisted tracking and tagging, similar to that used in air traffic control, allows the VTS to play a more significant role in marine traffic management. This decreases vessel congestion and critical encounter situations, and lessens the probability of a marine casualty resulting in environmental damage. Surveilled VTS's are found in many large ports and harbors where congestion is a safety and operational hazard. Less sophisticated services have been established in other areas in response to hazardous navigational conditions according to the needs and resources of the authorities.

An important and rapidly developing technology is the ship Automated Information System (AIS). AIS is similar to the transponder in an aircraft, which sends out a radio signal containing information such as the name of the vessel, course, speed, etc. This data appears as a text tag, attached to the radar blip, on systems designed to receive and process the signals. It enhances the ability of VTS operators to monitor and control shipping in busy ports.

2708. Development of U.S. VTS's

Since the early 1960's the U.S. Coast Guard has been investigating various concepts by which navigational safety can be improved in the harbor and harbor approach phases. Equipment installations in various ports for this investigation have included shore-based radar, closed-circuit television (LLL-CCTV), VHF-FM communications, broadcast television, and computer driven electronic situation displays.

In 1962 an experimental installation called **Ratan** (Radar and Television Aid to Navigation) was completed in New York Harbor. In this system a radar at Sandy Hook, New Jersey, scanned the approaches to the harbor. The radar video, formatted by a scan conversion storage tube, was broadcast by a television band UHF transmitter. This enabled mariners to observe on commercial television sets the presentation on the radarscope at Sandy Hook. The mariner could identify his vessel on the television screen by executing a turn and by observing the motions of the targets. The high persistency created by the scan converter provided target "tails" which aided in observing target movement. This Ratan experiment was discontinued primarily because of allocation of the commercial television frequency spectrum for other purposes.

In January 1970 the Coast Guard established a harbor radar facility in San Francisco to gather data on vessel traffic patterns. The information was used to determine parameters for new equipment procurements. The initial installation consisted of standard marine X-band (3-centimeter) search radars located on Point Bonita and Yerba Buena Island in San Francisco Bay. Radar video was relayed from these two radar sites to a manned center co-located with the San Francisco Marine Exchange. When the parameter definition work was completed, VHF-FM communications equipment was added to enable communications throughout the harbor area. This experimental system, previously called Harbor Advisory Radar (HAR) was designated in August 1972 as an operational Vessel Traffic System (VTS); a continuous radar watch with advisory radio broadcasts to traffic in the harbor was provided. This change from HAR to VTS coincided with the effective date of the Ports and Waterways Safety Act of 1972, authorizing the U.S. Coast Guard to install and operate such systems in United States waters to increase vessel safety and protect the environment.

In late 1972 improved developmental radar systems were installed side by side with the operational system, operated by a new research evaluation center at Yerba Buena Island. Redundant operator-switchable transceivers provided 50 kW peak power and incorporated receivers with large dynamic ranges of automatic gain control giving considerable protection against receiver saturation by interfering signals and interference by rain and sea clutter. Parabolic antennas with apertures of 27 feet (8.2 meters) and beam widths of 0.3 degrees improved the radar system accuracy. Variable pulse lengths (50 and 200 nanoseconds), three pulse repetition rates (1000, 2500, and 4000 pps), two receiver bandwidths (22 MHz and 2 MHz), and three antenna polarizations (horizontal, vertical, and circular) were provided to evaluate the optimum parameters for future procurements.

After a period of extensive engineering evaluation, the radar system was accepted in May 1973 as an operational replacement for the equipment installed earlier at the HAR.

In 1980 an analysis indicated that a modified version of the Coast Guard standard shipboard radar would meet all the VTS standard operating requirements. Additionally, it was more cost effective to procure and maintain than the specially designed, non-standard radar. After a period of evaluation at VTS San Francisco and with certain technical modifications, the standard radar was accepted for VTS use. The radar includes a tracking system which enhances the radar capability by allowing the VTS to track up to 20 targets automatically. The PPI can operate in an environment that is half as bright as a normal room with an option for a TV type display that can operate under any lighting conditions.

The new radar was installed in VTS Prince William Sound in August, 1984. VTS Houston-Galveston's radar was replaced in January, 1985. VTS San Francisco's radars were replaced in May, 1985. VTS New York reopened in late 1990.

2709. U.S. Operational Systems

VTS New York became operational in December 1990. It had been open previously but was closed in 1988 due to a change in funding priorities.

This VTS has the responsibility of coordinating vessel traffic movements in the busy ports of New York and New Jersey. The VTS New York area includes the entrance to the harbor via Ambrose and Sandy Hook Channels, through the Verrazano Narrows Bridge to the Brooklyn Bridge in the East River, to the Holland Tunnel in the Hudson River, and the Kill Van Kull, including Newark Bay.

The current operation uses surveillance data provided by several radar sites and three closed circuit TV sites. VTS communications are on VHF-FM channels 12 and 14.

VTS San Francisco was commissioned in August of 1972. When the original radar system became operational in May 1973, the control center for VTS San Francisco was shifted to the Yerba Buena Island. This center was designated a Vessel Traffic Center (VTC).

As of early 1985, the major components of the system include a Vessel Traffic Center at Yerba Buena Island, two high resolution radars, a VHF-FM communications network, a traffic separation scheme, and a vessel movement reporting system (VMRS). Channels 12 and 14 are the working frequencies. In 1985, all existing radar equipment was replaced with the standard Coast Guard radar.

VTS San Francisco also operates an Offshore Vessel Movement Reporting System (OVMRS). The OVMRS is completely voluntary and operates using a broadcast system with information provided by participants.

VTS Puget Sound became operational in September 1972 as the second Vessel Traffic Service. It collected vessel movement report data and provided traffic advisories by means of a VHF-FM communications network. In this early service a VMRS was operated in conjunction with a Traffic Separation Scheme (TSS), without radar surveillance. Operational experience gained from this service and VTS San Francisco soon proved the expected need for radar surveillance in those services with complex traffic flow.

In 1973 radar coverage in critical areas of Puget Sound was provided. Efforts to develop a production generation of radar equipment for future port development were initiated. To satisfy the need for immediate radar coverage, redundant military grade Coast Guard shipboard radar transceivers were installed at four Coast Guard light stations along the Admiralty Inlet part of Puget Sound. Combination microwave radio link and radar antenna towers were installed at each site. Radar video and azimuth data, in a format similar to that used with VTS San Francisco, were relayed by broad band video links to the VTC in Seattle. At that center, standard Navy shipboard repeaters were used for operator display. Although the resolution parameters and display accuracy of the equipment were less than those of the VTS San Francisco equipment, the use of a shorter range scale (8 nautical miles) and overlapping coverage resulted in very satisfactory operation. In December 1980 additional radar surveillance was added in the Strait of Juan De Fuca and Rosario Strait, as well as increased surveillance of the Seattle area, making a total of 10 remote radar sites.

The communications equipment was upgraded in July 1991 to be capable of a two frequency, four sector system. Channels 5A and 14 are the frequencies for VTS Puget Sound. A total of 13 communication sites are in operation (3 extended area sites, 10 low level sites). The three extended area sites allow the VTS the ability to communicate in a large area when needed. The low level sites can be used in conjunction with one another without interference, and have greatly reduced congestion on the frequency. VTS Puget Sound now covers the Strait of Juan de Fuca, Rosario Strait, Admiralty Inlet, and Puget Sound south as far as Olympia.

The major components of the system include the Vessel Traffic Center at Pier 36 in Seattle, a VHF-FM communications network, a traffic separation scheme, radar surveillance of about 80% of the VTS area, and a Vessel Movement Reporting System. Regulations are in effect which require certain classes of vessels to participate in the system and make movement reports at specified points. The traffic separation scheme in the Strait of Juan de Fuca was extended as far west as Cape Flattery in March 1975 in cooperation with Canada and was formally adopted by the International Maritime Organization in 1982.

Under an agreement between the United States and Canada, regulations for the Strait of Juan de Fuca took effect in 1984. The Cooperative Vessel Traffic Management System (CVTMS) divides responsibility among the two Canadian VTS's and VTS Puget Sound.

VTS Houston-Galveston became operational in February 1975 as the third U.S. Vessel Traffic Service. The operating area is the Houston Ship Channel from the sea buoy to the Turning Basin (a distance of 53 miles) and the side channels to Galveston, Texas City, Bayport, and the Intracoastal Waterway. The area contains approximately 70 miles of restricted waterways. The greater part of the Houston Ship Channel is 400 feet wide with depths of 36-40 feet. Several bends in the channel are in excess of 90 degrees.

The major components of the system include the VTC at Galena Park, Houston; a VHF-FM communications network; low light level, closed circuit television (LLL-

CCTV) surveillance covering approximately three miles south of Morgan's Point west through the ship channel to City Dock #27 in Houston; a Vessel Movement Reporting System; and a radar surveillance system covering lower Galveston Bay approaches, Bolivar Roads, and Lower Galveston Bay.

A second radar was installed in 1994. This radar provides surveillance coverage between the Texas City channel and Morgan's Point.

VTS Prince William Sound is required by The Trans-Alaska Pipeline Authorization Act (Public Law 93-153), pursuant to authority contained in Title 1 of the Ports and Waterways Safety Act of 1972 (86 Stat. 424, Public Law 92-340).

The southern terminus of the pipeline is on the south shoreline of Port Valdez, at the Alyeska Pipeline Service Company tanker terminal. Port Valdez is at the north end of Prince William Sound, and Cape Hinchinbrook is at the south entrance.

Geographically, the area is comprised of deep open waterways surrounded by mountainous terrain. The only constrictions to navigation are at Cape Hinchinbrook, the primary entrance to Prince William Sound, and at Valdez Narrows, the entrance to Port Valdez.

The vessel traffic center is located in Valdez. The system is composed of two radars, two major microwave data relay systems, and a VMRS which covers Port Valdez, Prince William Sound, and Gulf of Alaska. There is also a vessel traffic separation scheme from Cape Hinchinbrook to Valdez Arm.

The Coast Guard is installing a dependent surveillance system to improve its ability to track tankers transiting Prince William Sound. To extend radar coverage the length of the traffic lanes in Prince William Sound would require several radars at remote, difficult-to-access sites and an extensive data relay network. As an alternative to radar, the Coast Guard is installing a dependent surveillance system that will require vessels to carry position and identification reporting equipment. The ability to supplement radar with dependent surveillance will bridge the gap in areas where conditions dictate some form of surveillance and where radar coverage is impractical. Once the dependent surveillance information is returned to the vessel traffic center, it will be integrated with radar data and presented to the watchstander on an electronic chart display.

2710. Vessel Traffic Management and Information Systems

An emerging concept is that of Vessel Traffic Management and Information Services (VTMIS) wherein a VTS is only part of a larger and much more comprehensive information exchange. Under this concept, not only can vessel traffic be managed from the standpoint of navigation safety and efficiency, but also tugs, pilots, line handlers, intermodal shipping operators, port authorities, customs and immigration, law enforcement, and disaster response agencies and others can use vessel transit information to enhance the delivery of their services.

A VTS need not be part of a VTMIS, but it is logical that no port needing the latter would be without the former. It is important to note that VTMIS is a service, not a system, and requires no particular set of equipment or software. VTMIS development and installations are proceeding in several busy ports and waterways worldwide, and mariners can expect this concept to be implemented in many more areas in the future.

AUTOMATIC IDENTIFICATION SYSTEMS

2711. Development and Purpose

The Automatic Identification System (AIS) is a shipboard transponder that operates in the maritime VHF band, transmitting detailed information about a particular vessel and its operation. Similarly equipped vessels and shore stations can receive and display this information on an ECDIS, making it possible for each to know the identity, course, speed, condition, and other vital information about the others.

Each AIS consists of a VHF transmitter, two receivers, a VHF DSC receiver, and a link to shipboard display and information systems. Positional and timing information is generally derived from GPS or other electronic navigation systems, and includes differential GPS in coastal and inland waters. Heading and speed information come from shipboard sensors.

Once a signal broadcast from a vessel is received aboard another vessel or a shore station, it is processed and symbolized with basic information on the navigation display. It is then possible to query the symbol for additional information, such as name, tonnage, dimensions, draft, cargo, etc. This allows navigators to know the exact identity of nearby vessels with which a risk of collision exists, and to call them by name to agree on procedures for meeting, passing, crossing, or overtaking. Other information might consist of destination, ETA, rate of turn, and other data. It also allows VTS and other authorities to know the identity of each ship in their system.

AIS is capable of handling over 2,000 reports per minute, with updates every two seconds. It is intended to replace DSC based transponder systems currently in operation. Operation is autonomous and continuous, and each station automatically synchronizes itself with all others in range.

As with all VHF transmissions, AIS range depends on

the antenna height, and since the wavelength is slightly longer than radar, AIS signals tend to cross land and other obstructions moderately well. The use of shore-based repeater stations can greatly enhance the range and strength of the signals. At sea, ranges of about 20 miles can be expected.

In May of 1998, the Marine Safety Committee of the IMO formally adopted the Performance Standards for a Universal Shipborne Automatic Identification System. These standards dictate that AIS systems must meet three requirements:

1. Operate in a ship-to-ship mode for collision avoidance.
2. Operate in a ship-to-shore mode for traffic management.
3. Carry specified data about the ship and its cargo.

The goal of the IMO in publishing these standards is to have one AIS as the worldwide standard, so that all vessels and countries may benefit. Originally envisioned as operating in a ship-to-shore mode for vessel tracking by VTS and harbor authorities, the concept has evolved into a "4-s" system: ship-to-shore/ship-to-ship, available for collision avoidance as well as traffic control. AIS transponders use a frequency available worldwide and the system has sufficient capacity to operate in the busiest ports. Eventually, it is likely that all SOLAS and many other types of vessels will be required to carry an AIS transponder.

The integration of AIS technology into the world's port and harbor control systems is ongoing, while the integration into ECDIS and other ship-to-ship systems is in the developmental stage. It is reasonable to expect that in the future, all commercial ocean-going vessels, most commercial coastal craft, and many other vessels will be able, if not required, to use AIS.

2712. Classes and Capabilities of AIS's

There are two classes of AIS transponders. The Class A unit meets all IMO requirements, while the Class B, intended for smaller vessels or those not requiring the more capable Class A device, lacks some of the IMO-required features, but still provides vital data.

The Class A AIS broadcasts the following data every 2-10 seconds while underway, and every three minutes at anchor, at a power of 12.5 watts:

- MMSI number, a unique identification number
- Navigation status: underway, anchored, not under command, etc.
- Rate of turn, right or left, to 720 degrees per minute
- Speed over ground
- Course over ground
- Position accuracy; GPS, DGPS and whether RAIM is in operation
- Lat. and long. to 1/10,000 minute
- True heading, derived from gyro if installed
- Time of report

In addition, the Class A AIS will transmit every six minutes:

- MMSI number as above, links data above to vessel
- IMO number, a unique identifier related to ship's construction
- International call sign
- Name of ship, to 20 characters
- Type of ship and cargo, from list of types
- Dimensions of ship, to nearest meter
- Location on ship of reference point for position reports
- Source of fix information: GPS, Loran, DR, undefined, etc.
- Draft of ship, to 0.1 meter; air draft is not defined
- Destination, to 20 characters
- ETA: month, day, hour, and minute in UTC

Class B AIS capabilities are not yet specifically defined, but in general the Class B units will report less often, leave out certain information such as IMO number, destination, rate of turn, draft, and status, and are not required to transmit textual safety messages.

AIS has the potential of eventually replacing racons, since shore stations can transmit data on aids to navigation for display through the AIS system. This would enable aids to navigation to appear with appropriate text data on the display, instead of as simple unidentified blips.

IMO requirements specify various classes of ships that must commence use of AIS by certain dates under a phased schedule. In general, by 2007 all vessels operating under SOLAS V must have AIS equipment. Additionally, in the U.S., all vessels subject to the Bridge-to-bridge Radiotelephone Act may be required to carry AIS equipment. The U.S. Coast Guard will define the requirements for certification of U.S. vessels.

REGULATED WATERWAYS

2713. Purpose and Authorities

In confined waterways not considered international waters, local authorities may establish certain regulations for the safe passage of ships and operate waterway systems consisting of locks, canals, channels, and ports. This generally occurs in especially busy or highly developed waterways which form the major constrictions on international shipping routes. The Panama Canal, St. Lawrence Seaway, and the Suez Canal represent systems of this type.

Nearly all ports and harbors have a body of regulations concerning the operation of vessels within the port limits, particularly if locks and other structures are part of the system. The regulations covering navigation through these areas are typically part of a much larger body of regulations relating to assessment and payment of tariffs and tolls, vessel condition and equipment, personnel, communications equipment, and many other factors. In general, the larger the investment in the system, the larger the body of regulations which control it will be.

Where a waterway separates two countries, a joint authority may be established to administer the regulations, collect tolls, and operate the system, as in the St. Lawrence Seaway.

Copies of the regulations are usually required to be aboard each vessel in transit. These regulations are available from the authority in charge or an authorized agent. Summaries of the regulations are contained in the appropriate volumes of the *Sailing Directions (Enroute)*.

CHAPTER 28

MARITIME SAFETY SYSTEMS

MARITIME SAFETY AND THE NAVIGATOR

2800. Introduction

The navigator's chief responsibility is the safety of the vessel and its crew. Fulfilling this duty consists mostly of ascertaining the ship's position and directing its course so as to avoid dangers. But accidents can happen to the most cautious, and the most prudent of navigators may experience an emergency which requires outside assistance. Distress incidents at sea are more likely to be resolved without loss of vessel and life if they are reported immediately. The more information that rescue authorities have, and the sooner they have it, the more likely it is that the outcome of a distress at sea will be favorable.

Global distress communication systems, ship reporting systems, emergency radiobeacons, and other technologies have greatly enhanced mariners' safety. Therefore, it is critical that mariners understand the purpose, functions, and limitations of maritime safety systems.

The mariner's direct high-seas link to shoreside rescue authorities is the Global Maritime Distress and Safety System (GMDSS), which was developed to both simplify and improve the dependability of communications for all ships at sea. GMDSS nicely compliments the operation of the U.S. Coast Guard's Amver system, which tracks participating ships worldwide and directs them as needed to distress incidents. GMDSS and Amver rely on radiotelephone or satellite communications for passing information. But even with normal communications disabled, a properly equipped vessel has every prospect of rapid rescue or aid if it carries a SOLAS-required Emergency Position Indicating Radiobeacon (EPIRB) and a Search and Rescue radar Transponder (SART). These systems are the subject of this chapter.

GLOBAL MARITIME DISTRESS AND SAFETY SYSTEM

2801. Introduction and Background

The **Global Maritime Distress and Safety System (GMDSS)** represents a significant improvement in maritime safety over the previous system of short range and high seas radio transmissions. Its many parts include satellite as well as advanced terrestrial communications systems. Operational service of the GMDSS began on February 1, 1992, with full implementation accomplished by February 1, 1999.

GMDSS was adopted in 1988 by amendments to the Conference of Contracting Governments to the International Convention for the Safety of Life at Sea (SOLAS), 1974. This was the culmination of more than a decade of work by the International Maritime Organization (IMO) in conjunction with the International Telecommunications Union (ITU), International Hydrographic Organization (IHO), World Meteorological Organization (WMO), Inmarsat (International Maritime Satellite Organization), and others.

GMDSS offers the greatest advancement in maritime safety since the enactment of regulations following the Titanic disaster in 1912. It is an automated ship-to-ship, shore-to-ship and ship-to-shore communications system covering distress alerting and relay, the provision of **maritime safety information (MSI)**, and routine communications. Satellite and advanced terrestrial systems are incorporated into a communications network to promote and improve safety of life and property at sea throughout the world. The equipment required on board ships depends not on their tonnage, but rather on the area in which the vessel operates. This is fundamentally different from the previous system, which based requirements on vessel size alone. The greatest benefit of the GMDSS is that it vastly reduces the chances of ships sinking without a trace, and enables search and rescue (SAR) operations to be launched without delay and directed to the exact site of a maritime disaster.

2802. Ship Carriage Requirements

By the terms of the SOLAS Convention, the GMDSS provisions apply to cargo ships of 300 gross tons and over and ships carrying more than 12 passengers on international voyages. Unlike previous shipboard carriage regulations that specified equipment according to size of vessel, the GMDSS carriage requirements stipulate equipment according to the area in which the vessel operates. These sea areas are designated as follows:

Sea Area A1 An area within the radiotelephone coverage of at least one VHF coast station in which continuous Digital Selective Calling is available, as may be defined by a Contracting Government to the 1974 SOLAS Convention. This area extends from the coast to about 20 miles offshore.

Sea Area A2 An area, excluding sea area A1, within the radiotelephone coverage of at least one MF coast station in which continuous DSC alerting is available, as may be defined by a Contracting Government. The general area is from the A1 limit out to about 100 miles offshore.

Sea Area A3 An area, excluding sea areas A1 and A2, within the coverage of an Inmarsat geostationary satellite in which continuous alerting is available. This area is from about 70°N to 70°S.

Sea Area A4 All areas outside of sea areas A1, A2 and A3. This area includes the polar regions, where geostationary satellite coverage is not available.

Ships at sea must be capable of the following functional GMDSS requirements:

1. Ship-to-shore distress alerting
2. Shore-to-ship distress alerting
3. Ship-to-ship distress alerting
4. SAR coordination
5. On-scene communications
6. Transmission and receipt of emergency locating signals
7. Transmission and receipt of MSI
8. General radio communications
9. Bridge-to-bridge communications

To meet the requirements of the functional areas above the following is a list of the minimum communications equipment needed for all ships:

1. VHF radio capable of transmitting and receiving DSC on channel 70, and radio telephony on channels 6, 13 and 16
2. Radio receiver capable of maintaining a continuous Digital Selective Calling (DSC) watch on channel 70 VHF
3. Search and rescue transponders (SART), a minimum of two, operating in the 9 GHz band
4. Receiver capable of receiving NAVTEX broadcasts anywhere within NAVTEX range
5. Receiver capable of receiving SafetyNET anywhere NAVTEX is not available
6. Satellite emergency position indicating radiobeacon (EPIRB), manually activated and float-free self-activated
7. Two-way handheld VHF radios (two sets minimum on 300-500 gross tons cargo vessels and three sets minimum on cargo vessels of 500 gross tons and upward and on all passenger ships)

Additionally, each sea area has its own requirements under GMDSS which are as follows:

Sea Area A1

1. General VHF radio telephone capability
2. Free-floating satellite EPIRB
3. Capability of initiating a distress alert from a navigational position using DSC on either VHF, HF or MF; manually activated EPIRB; or Ship Earth Station (SES)

Sea Areas A1 and A2

1. Radio telephone MF radiotelephony or direct printing 2182 kHz, and DSC on 2187.5 kHz
2. Equipment capable of maintaining a continuous DSC watch on 2187.5 kHz
3. General working radio communications in the MF band (1605-4000 kHz), or Inmarsat SES
4. Capability of initiating a distress alert by HF (using DSC), manual activation of an EPIRB, or Inmarsat SES

Sea Areas A1, A2 and A3

1. Radio telephone MF 2182 kHz and DSC 2187.5 kHz.
2. Equipment capable of maintaining a continuous DSC watch on 2187.5 kHz
3. Inmarsat-A, -B or -C (class 2) or Fleet 77 SES Enhanced Group Call (EGC), or HF as required for sea area A4
4. Capability of initiating a distress alert by two of the following:
   a. Inmarsat-A, -B or -C (class 2)or Fleet 77 SES
   b. Manually activated EPIRB
   c. HF/DSC radio communication

Sea Area A4

1. HF/MF receiving and transmitting equipment for band 1605-27500 kHz using DSC, radiotelephone and direct printing
2. Equipment capable of selecting any safety and

distress DSC frequency for band 4000-27500 kHz, maintaining DSC watch on 2187.5, 8414.5 kHz and at least one additional safety and distress DSC frequency in the band

3. Capability of initiating a distress alert from a navigational position via the Polar Orbiting System on 406 MHz (manual activation of 406 MHz satellite EPIRB)

2803. The Inmarsat System

Inmarsat (International Maritime Satellite Organization), a key player within GMDSS, is an international corporation comprising over 75 international partners providing maritime safety communications for ships at sea. Inmarsat provides the space segment necessary for improving distress communications, efficiency and management of ships, as well as public correspondence services.

The basic components of the Inmarsat system include the Inmarsat **space segment, Land Earth Stations (LES)**, also referred to as **Coast Earth Stations (CES)**, and mobile **Ship Earth Stations (SES)**.

The Inmarsat space segment consists of 11 geostationary satellites. Four operational Inmarsat satellites provide primary coverage, four additional satellites (including satellites leased from the European Space Agency (ESA) and the International Telecommunications Satellite Organization (INTELSAT)) serve as spares and three remaining leased satellites serve as back-ups.

The polar regions are not visible to the operational satellites but coverage is available from about 75°N to 75°S. Satellite coverage (Figure 2803) is divided into four overlapping regions:

1. Atlantic Ocean - East (AOR-E)
2. Atlantic Ocean - West (AOR-W)
3. Pacific Ocean (POR)
4. Indian Ocean (IOR)

The LES's provide the link between the Space Segment and the land-based national/international fixed communications networks. These communications networks are funded and operated by the authorized communications authorities of a participating nation. This network links registered information providers to the LES. The data then travels from the LES to the Inmarsat **Network Coordination Station (NCS)** and then down to the SES's on ships at sea. The SES's provide two-way communications between ship and shore. **Inmarsat-A**, the original Inmarsat system, operates at a transfer rate of up to 64k bits per second and is telephone, telex and facsimile (fax) capable. The similarly sized **Inmarsat-B** system uses digital technology, also at rates to 64kbps. Fleet 77 service is also digital and operates at up to 64kbps.

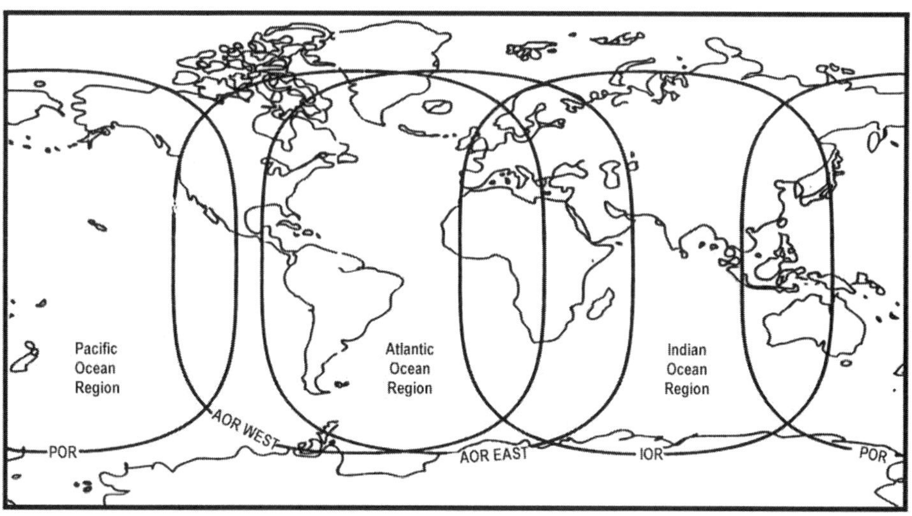

Figure 2803. The four regions of Inmarsat coverage.

Inmarsat-C provides a **store and forward** data messaging capability (but no voice) at 600 bits per second and was designed specifically to meet the GMDSS requirements for receiving MSI data on board ship. These units are small, lightweight and use an omni-directional antenna.

2804. Maritime Safety Information (MSI)

Major categories of MSI for both NAVTEX and SafetyNET are:

1. Navigational warnings
2. Meteorological warnings
3. Ice reports
4. Search and rescue information
5. Meteorological forecasts
6. Pilot service messages (not in the U.S.)
7. Electronic navigation system messages (i.e., LORAN, GPS, DGPS, etc.)

Broadcasts of MSI in NAVTEX international service are in English, but may be in languages other than English to meet requirements of the host government.

2805. SafetyNET

SafetyNET is a broadcast service of Inmarsat-C's **Enhanced Group Call (EGC)** system. The EGC system (Figure 2805) is a method used to specifically address particular regions or groups of ships. Its unique addressing capabilities allow messages to be sent to all vessels in both fixed geographical areas or to predetermined groups of ships. SafetyNET is a service designated by the IMO through which ships receive maritime safety information. The other service under the EGC system, called **FleetNET**, is used by commercial companies to communicate directly and privately with their individual fleets.

SafetyNET is an international shore to ship satellite-based service for the promulgation of distress alerts, navigational warnings, meteorological warnings and forecasts, and other safety messages. It fulfills an integral role in GMDSS as developed by the IMO. The ability to receive SafetyNET messages is required for all SOLAS ships that sail beyond coverage of NAVTEX (approximately 200 miles from shore).

SafetyNET can direct a message to a given geographic area based on EGC addressing. The area may be fixed, as in the case of a NAVAREA or weather forecast area, or it may be uniquely defined by the originator. This is particularly useful for messages such as local storm warnings or focussed shore to ship distress alerts.

SafetyNET messages can be originated by a **Registered Information Provider** anywhere in the world and broadcast to the appropriate ocean area through an Inmarsat-C LES. Messages are broadcast according to their priority (i.e. Distress, Urgent, Safety, and Routine).

Virtually all navigable waters of the world are covered by the operational satellites in the Inmarsat system. Each satellite broadcasts EGC traffic on a designated channel. Any ship sailing within the coverage area of an Inmarsat satellite will be able to receive all the SafetyNET messages broadcast over this channel. The EGC channel is optimized to enable the signal to be monitored by SES's dedicated to the reception of EGC messages. This capability can be built into other standard SES's. It is a feature of satellite communications that reception is not generally affected by the position of the ship within the ocean region, atmospheric conditions, or time of day.

Messages can be transmitted either to geographic areas (area calls) or to groups of ships (group calls):

1. **Area calls** can be to a fixed area such as one of the 16 NAVAREA's or to a temporary geographic area selected by the originator. Area calls will be received automatically by any ship whose receiver has been set to one or more fixed areas.

2. **Group calls** will be received automatically by any ship whose receiver acknowledges the unique group identity associated with a particular message.

Reliable delivery of messages is ensured by forward error correction techniques. Experience has demonstrated that the transmission link is generally error-free and low error reception is achieved under normal circumstances.

Given the vast ocean coverage by satellite, some form of discrimination and selectivity in printing the various messages is required. Area calls are received by all ships within the ocean region coverage of the satellite; however, they will be printed only by those receivers that recognize the fixed area or the geographic position in the message. The message format includes a **preamble** that enables the microprocessor in a ship's receiver to decide to print those MSI messages that relate to the present position, intended route or a fixed area programmed by the operator. This preamble also allows suppression of certain types of MSI that are not relevant to a particular ship. As each message will also have a unique identity, the reprinting of messages already received correctly is automatically suppressed.

MSI is promulgated by various information providers around the world. Messages for transmission through the SafetyNET service will, in many cases, be the result of coordination between authorities. Information providers will be authorized by IMO to broadcast via SafetyNET. Authorized information providers are:

1. National hydrographic offices for navigational warnings
2. National weather services for meteorological warnings and forecasts

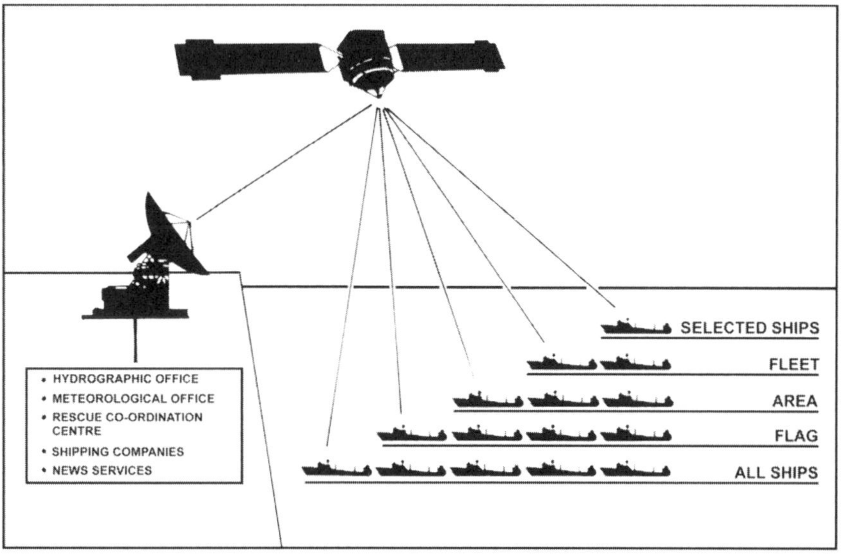

Figure 2805. SafetyNET EGC concept.

3. Rescue Coordination Centers (RCC's) for ship-to-shore distress alerts and other urgent information
4. In the U.S., the International Ice Patrol (IIP) for North Atlantic ice hazards

Each information provider prepares their SafetyNET messages with certain characteristics recognized by the EGC service. These characteristics, known as "C" codes are combined into a generalized message header format as follows: C1:C2:C3:C4:C5. Each "C" code controls a different broadcast criterion and is assigned a numerical value according to available options. A sixth "C" code, "C0" may be used to indicate the ocean region (i.e., AOR-E, AOR-W, POR, IOR) when sending a message to an LES which operates in more than one ocean region. Because errors in the header format of a message may prevent its being broadcast, MSI providers must install an Inmarsat SafetyNET receiver to monitor the broadcasts it originates. This also ensures quality control.

The "C" codes are transparent to the mariner, but are used by information providers to identify various transmitting parameters. C1 designates the message priority, either distress to urgent, safety, or routine. MSI messages will always be at least at the safety level. C2 is the service code or type of message (for example, long range NAVAREA warning or coastal NAVTEX warning). It also tells the receiver the length of the address (the C3 code) it will need to decode. C3 is the address code. It can be the two digit code for the NAVAREA number for instance, or a 10 digit number to indicate a circular area for a meteorological warning. C4 is the repetition code which instructs the LES when to send the message to the NCS for actual broadcast. A six minute echo (repeat) may also be used to ensure that an urgent (unscheduled) message has been received by all ships affected. C5 is a constant and represents a presentation code, International Alphabet number 5, "00".

Broadcasts of MSI in the international SafetyNET service must be in English, but may be supplemented by other languages to meet requirements of the host government.

2806. NAVTEX

NAVTEX is a maritime radio warning system consisting of a series of coast stations transmitting radio teletype (standard narrow-band direct printing, called **Sitor** for Simplex Telex Over Radio) safety messages on the internationally standard medium frequency of 518 kHz. It is a GMDSS requirement for the reception of MSI in coastal and local waters. Coast stations transmit during previously arranged time slots to minimize mutual interference. Routine messages are normally broadcast four times daily. Urgent messages are broadcast upon receipt, provided that an adjacent station is not transmitting. Since the broadcast uses the medium frequency band, a typical station service radius ranges from 100 to 500 NM day and night (although a 200 mile rule of thumb is applied in the U.S.). Interference from or receipt of stations further away occasionally occurs at night.

Each NAVTEX message broadcast contains a four-character header describing: identification of station (first character), message content or type (second character), and message serial number (third and fourth characters). This header allows the microprocessor in the shipboard receiver to screen messages from only those stations relevant to the user, messages of subject categories needed by the user and messages not previously received by the user. Messages so screened are printed as they are received, to be read by the mariner when convenient. All other messages are suppressed. Suppression of unwanted messages is becoming more and more a necessity to the mariner as the number of messages, including rebroadcast messages, increases yearly. With NAVTEX, a mariner will not find it necessary to listen to, or sift through, a large number of non-relevant data to obtain the information necessary for safe navigation.

The NAVTEX receiver is a small unit with an internal printer, which takes a minimum of room on the bridge. Its antenna is also of modest size, needing only a receive capability.

2807. Digital Selective Calling (DSC)

Digital Selective Calling (DSC) is a system of digitized radio communications which allows messages to be targeted to all stations or to specific stations, allows for unattended and automated receipt and storage of messages for later retrieval, and permits the printing of messages in hardcopy form. All DCS calls automatically include error-checking signals and the identity of the calling unit. Digital codes allow DSC stations to transmit and receive distress messages, transmit and receive acknowledgments of distress messages, relay distress messages, make urgent and safety calls, and initiate routine message traffic.

Each unit has a MAYDAY button which allows the instant transmittal of a distress message to all nearby ships and shore stations. The location of the distress will be automatically indicated if the unit is connected to a GPS or Loran C receiver. Each unit must be registered with the Coast Guard and have unique identifier programmed into it. Distress alerts can be sent on only one or as many as six channels consecutively on some units.

Listening watch on 2182 kHz ended with implementation of GMDSS in 1999. When DSC has been implemented worldwide, the traditional listening watch on Channel 16 VHF will no longer be necessary. The introduction of DSC throughout the world is expected to take to take a number of years.

There are four basic types of DSC calls:

- Distress
- Urgent
- Safety
- Routine

Distress calls are immediately received by rescue authorities for action, and all vessels receiving a distress call are alerted by an audible signal.

Each DSC unit has a unique Maritime Mobile Service Identity (MMSI) code number, which is attached to all outgoing messages. The MMSI number is a nine-digit number to identify individual vessels, groups of vessels, and coast stations. Ship stations will have a leading number consisting of 3 digits which identify the country in which the ship is registered, followed by a unique identifying number for the vessel. A group of vessels will have a leading zero, followed by a unique number for that group. A coast station will have 2 leading zeros followed by a code number. Other codes may identify all stations, or all stations in a particular geographic area.

DSC frequencies are found in the VHF, MF and HF bands. Within each band except VHF, one frequency is allocated for distress, urgent, and safety messages. Other frequencies are reserved for routine calls. In the VHF band, only one channel is available, Channel 70 (156.525 MHz), which is used for all calls. In the MF band, 2187.5 kHz and 2189.5 kHz are reserved for distress/safety, and 2177 kHz for ship to ship and ship to shore calls.

2808. Using DSC

A distress call consists of a Format Specifier--Distress; the MMSI code; the nature of the distress (selected from a list: fire/explosion, flooding, collision, grounding, listing, sinking, disabled/adrift, or abandoning ship; defaults to Undesignated); the time of the call, and the format for subsequent communications (radiotelephone or NDBP). Once activated, a distress signal is repeated automatically every few minutes until an acknowledgment is received or the function is switched off. As soon as an acknowledgment is received by the vessel in distress, it must commence communications with appropriate an message by radiotelephone or NDBP according to the format:

"MAYDAY"
MMSI CODE NUMBER AND CALL SIGN
NAME OF VESSEL
POSITION
NATURE OF DISTRESS
TYPE OF ASSISTANCE NEEDED
OTHER INFORMATION

Routine calls should be made on a channel reserved for non-distress traffic. Once made, a call should not be repeated, since the receiving station either received the call and stored it, or did not receive it because it was not in service. At least 5 minutes should elapse between calls by vessels on the first attempt, then at 15 minute minimum intervals.

To initiate a routine ship to shore or ship to ship call to a specific station, the following procedures are typical (consult the operator's manual for the equipment for specific directions):

- Select the appropriate frequency
- Select or enter the MMSI number of the station to be called
- Select the category of the call
- Select subsequent communications method (R/T, NDBP)
- Select proposed working channel (coast stations will indicate vacant channel in acknowledgment)
- Select end-of-message signal (RQ for acknowledgment required)
- Press <CALL>

The digital code is broadcast. The receiving station may acknowledge receipt either manually or automatically, at which point the working channel can be agreed on and communications begin.

Watchkeeping using DSC consists of keeping the unit ON while in the appropriate Sea Area. DSC watch frequencies are VHF Channel 70, 2187.5 kHz, 8414.5 kHz, and one HF frequency selected according to the time of day and season. Coast stations maintaining a watch on DCS channels are listed in NIMA *Pub. 117* and other lists of radio stations.

AMVER

2809. The Automated Mutual-Assistance Vessel Rescue System (Amver)

The purpose of ship reporting systems is to monitor vessels' positions at sea so that a response to any high-seas emergency can be coordinated among those nearest and best able to help. It is important that complete information be made available to search and rescue (SAR) coordinators immediately so that the right type of assistance can be sent to the scene with the least possible delay.

For example, a medical emergency at sea might require a doctor; a ship reporting system can find the nearest vessel with a doctor aboard. A sinking craft might require a vessel to rescue the crew, and perhaps another to provide a lee. A ship reporting system allows SAR coordinators to quickly assemble the required assets to complete the rescue.

The International Convention for the Safety of Life at Sea (SOLAS) obligates the master of any vessel who becomes aware of a distress incident to proceed to the emergency and assist until other aid is at hand or until released by the distressed vessel. Other international treaties and conventions impose the same requirement.

By maintaining a database of information as to the particulars of each participating vessel, and monitoring their positions as their voyages proceed, the Amver coordinator can quickly ascertain which vessels are closest and best able to respond to any maritime distress incident. They can also release vessels that might feel obligated to respond from their legal obligation to do so, allowing them to proceed on their way without incurring liability for not responding. International agreements ensure that no costs are incurred by a participating vessel

Several ship reporting systems are in operation throughout the world. The particulars of each system are given in publications of the International Maritime Organization (IMO). Masters of vessels making offshore passages are requested to always participate these systems when in the areas covered by them. The only worldwide system in operation is the U.S. Coast Guard's Amver system.

Amver is an international maritime mutual assistance program that coordinates search and rescue efforts around the world. It is voluntary, free of charge, and endorsed by the IMO. Merchant ships of all nations are encouraged to file a sailing plan, periodic position reports, and a final report at the end of each voyage, to the Amver Center located in the U.S. Coast Guard Operations Systems Center in Martinsburg, WV. Reports can be sent via e-mail, Inmarsat-C, Amver/SEAS "compressed message" format, Sat-C format, HF radiotelex, HF radio or telefax message. Most reports can be sent at little or no cost to the ship.

Data from these reports is protected as "commercial proprietary" business information, and is released by U.S. Coast Guard only to recognized national SAR authorities and only for the purposes of SAR in an actual distress. Information concerning the predicted location and SAR characteristics of each vessel is available upon request to recognized SAR agencies of any nation or to vessels needing assistance. Predicted locations are disclosed only for reasons related to marine safety.

The Amver computer uses a dead reckoning system to predict the positions of participating ships at any time during their voyage. Benefits to participating vessels and companies include:

- Improved chances of timely assistance in an emergency.
- Reduced number of calls for ships not favorably located.
- Reduced lost time for vessels responding.
- Added safety for crews in the event of an overdue vessel.

Amver participants can also act as the eyes and ears of SAR authorities to verify the authenticity of reports, reducing the strain on SAR personnel and facilities. Amver is designed to compliment computer and communications technologies, including GMDSS systems that provide distress alerting, and GPS positioning systems. These technologies can reduce or entirely eliminate the search aspect of search and rescue (since the precise location of the distress can be known), allowing SAR authorities to concentrate immediately on the response.

The Amver Sailing Plan provides information on the port of departure, destination, course, speed, navigational method, waypoints, communications capabilities, and the presence of onboard medical personnel. The database contains information on the ship's official name and registry, call sign, type of ship, tonnage, propulsion, maximum speed, and ownership. Changes in any of this data should be reported to Amver at the earliest opportunity.

Amver participants bound for U.S. ports enjoy an additional benefit: Amver messages which include the necessary information are considered to meet the requirements of 33 CFR 161 (Notice of Arrival).

2810. The Amver Communications Network

The following methods are recommended for ships to transmit information to Amver:

1. **Electronic mail** (e-mail) via the Internet: The Amver internet e-mail address is amvermsg@amver.com. If a ship already has an inexpensive means of sending e-mail to an internet address, this is the preferred method. The land-based portion of an e-mail message is free, but there may be a charge for any ship-to-shore portion. Reports should be sent in the body of the message, not as attachments.

2. **Amver/SEAS Compressed Message** via Inmarsat-C through certain Land Earth Stations (LES's): Ships equipped with an Inmarsat-C transceiver with floppy drive and capability to transmit a binary file (The ship's GMDSS Inmarsat-C transceiver can be used); and ships equipped with an IBM-compatible computer with hard drive, 286 or better processor, VGA graphics interface, and Amver/SEAS software; may send combined Amver/Weather Observation messages free of charge via TELENOR-USA Land Earth Stations at:

001 Atlantic Ocean Region-West (AOR-W)-Southbury
101 Atlantic Ocean Region East (AOR-E)-Southbury
201 Pacific Ocean Region (POR)-Santa Paula
321 Indian Ocean Region (IOR)-Assaguel

Amver/SEAS software can be downloaded free of charge from http://dbcp.nos.noaa.gov/seas.html.

3. **HF Radiotelex Service** of the U.S. Coast Guard Communication Stations; see full instructions at:

http://www.navcen.uscg.mil/marcomms/cgcomms/call.htm

4. **HF Radio** at no cost via Coast Guard contractual agreements with Globe Wireless Super Station Network, or Mobile Marine Radio (WLO) (under Telaurus Communications Inc.).

5. **Telex**: Amver Address (0) 230 127594 AMVERNYK

6. **Telefax**: To the USCG Operations Systems Center at: +1 304 264 2505. Telefax should be used only if other means are unavailable.

The **Amver Bulletin** provides information on the operation of the Amver System of general interest to the mariner and up-to-date information on the Amver communications network.

2811. Amver Participation

Instructions guiding participation in the Amver System are available from the Amver User's Manual published in the following languages: Chinese, Danish, Dutch, English, French, German, Greek, Italian, Japanese, Korean, Norwegian, Polish, Portuguese, Russian, Spanish and Swedish. This manual is available free from:

Amver Maritime Relations Office
USCG Battery Park Building
1 South Street
New York, NY, USA, 10004-1499
Telephone: (212) 668-7764
Fax: (212) 668-7684

or from:

Commander, Pacific Area
United States Coast Guard
Government Island
Alameda, CA 94501

The manual may also be obtained from Coast Guard District Offices, Marine Safety Offices, and Captain of the Port Offices in all major U.S. ports. Requests should indicate the language desired if other than English.

SAR operational procedures are contained in the International Aeronautical and Maritime Search and Rescue (IAMSAR) Manual published jointly by the IMO and the ICAO. Volume III of this manual is required aboard SOLAS vessels.

To enroll in Amver, a ship must first complete a SAR Questionnaire (SAR-Q). Participation involves filing four types of reports:

1. Sailing Plan
2. Position Report
3. Deviation Report
4. Final Report

The **Sailing Plan** is sent before leaving port, and indicates the departure time and date, destination, route and waypoints, speed, and navigational method.

The **Position Report** is sent after the first 24 hours to confirm departure as planned and conformance with the reported Sailing Plan. An additional report is requested every 48 hours to verify the DR plot being kept in the Amver computer.

A **Deviation Report** should be sent whenever a change of route is made, or a change to course or speed due to weather, heavy seas, casualty, or any other action that would render the computerized DR inaccurate.

A **Final Report** should be sent at the destination port. The system then removes the vessel from the DR plot and logs the total time the ship was participating.

Vessels that travel certain routes on a recurring basis may be automatically tracked for successive voyages as long as delays in regular departures are reported. The system may also be used to track vessels sailing under special circumstances such as tall ships, large ocean tows, research vessel operations, factory fishing vessels, etc. At any given time nearly 3,000 vessels worldwide are being plotted by Amver, and the number of persons rescued as a direct result of Amver operations is in the hundreds each year.

2812. Amver Reporting Requirements

The U.S. Maritime Administration (MARAD) regulations state that certain U.S. flag vessels and foreign flag "War Risk" vessels must report and regularly update their voyages to the Amver Center. This reporting is required of the following: (a) U.S. flag vessels of 1,000 tons or greater, operating in foreign commerce; (b) foreign flag vessels of 1,000 gross tons or greater, for which an Interim War Risk Insurance Binder has been issued under the provisions of Title XII, Merchant Marine Act, 1936.

2813. The Surface Picture (SURPIC)

When a maritime distress is reported to SAR authorities, the Amver computer is queried to produce a Surface Picture (SURPIC) in the vicinity of the distress. Several different types of SURPIC are available, and they can be generated for any specified time. The SURPIC output is a text file containing the names of all vessels meeting the criteria requested, plus a subset of the information recorded in the database about each vessel. See Figure 2813. A graphic display can be brought up for RCC use, and the data can be sent immediately to other SAR authorities worldwide. The information provided by the SURPIC includes the position of all vessels in the requested area, their courses, speeds, estimated time to reach the scene of the distress, and the amount of deviation from its course required for each vessel if it were to divert. RCC staff can then direct the best-placed, best-equipped vessel to respond.

Four types of SURPIC can be generated:

A **Radius SURPIC** may be requested for any radius from 50 to 500 miles. A sample request might read:

"REQUEST 062100Z RADIUS SURPIC OF DOCTORSHIPS WITHIN 800 MILES OF 43.6N 030.2W FOR MEDICAL EVALUATION M/V SEVEN SEAS."

The **Rectangular SURPIC** is obtained by specifying the date, time, and two latitudes and two longitudes. As with the Radius SURPIC, the controller can limit the types of ships to be listed. There is no maximum or minimum size limitation on a Rectangular SURPIC.

A sample Area SURPIC request is as follows:

"REQUEST 151300Z AREA SURPIC OF WESTBOUND SHIPS FROM 43N TO 31N LATITUDE AND FROM 130W TO 150W LONGITUDE FOR SHIP DISTRESS M/V EVENING SUN LOCATION 37N, 140W."

The **Snapshot** or **Trackline SURPIC** is obtained by specifying the date and time, two points (P1 and P2), whether the trackline should be rhumb line or great circle, what the half-width (D) coverage should be (in nautical miles), and whether all ships are desired or only those meeting certain parameters (e.g. doctor on board).

Name	*Call sign*	*Position*	*Course*	*Speed*	*SAR data*	*Destination and ETA*	
CHILE MARU	JAYU	26.2 N 179.9E	C294	12.5K	H 1 6 R T X Z	KOBE	11
CPA 258 DEG. 012 MI. 032000Z							
WILYAMA	LKBD	24.8N 179.1W	C106	14.0K	H X R T V X Z	BALBOA	21
CPA 152 DEG. 092 MI. 032000Z							
PRES CLEVELAND	WITM	25.5N 177.0W	C284	19.3K	H 2 4 R D T X Z S	YKHAMA	08
CPA 265 WILL PASS WITHIN 10 MI. 040430Z							
AENEAS	GMRT	25.9N 176.9E	C285	16.0K	H 8 R N V X Z	YKHAMA	10
CPA 265 DEG. 175 MI. 03200Z							

Figure 2813. Radius SURPIC as received by a rescue center.

A Snapshot Trackline SURPIC request might look like:

"REQUEST 310100Z GREAT CIRCLE TRACKLINE SURPIC OF ALL SHIPS WITHIN 50 MILES OF A LINE FROM 20.1N 150.2W TO 21.5N 158.0W FOR AIRCRAFT PRECAUTION."

A **Moving Point SURPIC** is defined by the starting and ending points of a vessel's trackline, the estimated departure time of the vessel, and the varying time of the SURPIC. This SURPIC is useful when a vessel is overdue at her destination. If the vessel's trackline can be accurately estimated, a SURPIC can generated for increments of time along the trackline, and a list can be generated of ships that might have sighted the missing ship.

2814. Uses of Amver Information

After evaluating the circumstances of a reported distress, The RCC can select the best available vessel to divert to the scene. In many cases a participating ship will be asked only to change course for a few hours or take a slightly different route to their destination, in order to provide a lookout in a certain area. RCC coordinators strive to use participating ships efficiently, and release them as soon as possible.

An example of the use of a Radius SURPIC is depicted in Figure 2814. In this situation rescue authorities believe that a ship in distress, or her survivors, might be found in the rectangular area. The RCC requests a SURPIC of all eastbound ships within 100 miles of a position well west of the rectangular area. With this list, the RCC staff prepares a modified route for each of four ships which will comprise a "search team" to cover the entire area, while adding only a few miles to each ship's route. Messages to each ship specify the exact route to follow and what to look for enroute.

Each ship contacted may be asked to sail a rhumb line between two specified points, one at the beginning of the search area and one at the end. By carefully assigning ships to areas of needed coverage, very little time need be lost from the sailing schedule of each cooperating ship. Those ships joining the search would report their positions every few hours to the Rescue Coordination Center, together with weather data and any significant sightings. In order to achieve saturation coverage, a westbound SURPIC at the eastern end of the search area would also be used.

The Trackline SURPIC is most commonly used as a precautionary measure for aircraft. Occasionally a plane loses of one or more of its engines. A Trackline SURPIC, provided from the point of difficulty to the destination, provides the pilot with the added assurance of knowing the positions of vessels beneath him and that they have been alerted. While the chance of an airliner experiencing such an emergency is extremely remote, SURPIC's have been used successfully to save the lives of pilots of general aviation aircraft on oceanic flights.

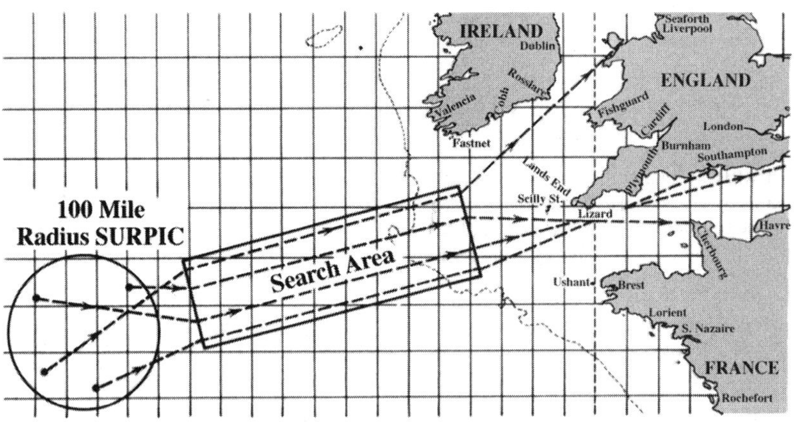

Figure 2814. Example of the use of a radius SURPIC to locate ships to search a rectangular area.

EMERGENCY POSITION INDICATING RADIOBEACONS (EPIRB'S)

2815. Description And Capabilities

Emergency Position Indicating Radiobeacons (EPIRB's) are designed to save lives by automatically alerting rescue authorities and indicating the distress location. EPIRB types are described below (Figure 2815a):

121.5/243 MHz EPIRB's (Class A, B, S): These are the most common and least expensive type of EPIRB, designed to be detected by overflying commercial or military aircraft.

The IMO and the International Civil Aviation Organization (ICAO) have announced plans to eventually terminate the processing distress signals for 121.5/243 MHz EPIRBS. Support for Class A, B, and S EPIRB's will be discontinued at some unannounced time in the future due to the high number of false alarms and the superiority of other systems.

Satellites were designed to detect these EPIRB's, but are limited for the following reasons:

1. Satellite detection range is limited for these EPIRB's (satellites must be within line of sight of both the EPIRB and a ground terminal for detection to occur).

2. EPIRB design and frequency congestion cause a high false alarm rate (over 99%); consequently, confirmation is required before SAR forces deploy.

3. EPIRB's manufactured before October 1988 may have design or construction problems (e.g. some models will leak and cease operating when immersed in water) or may not be detectable by satellite.

406 MHz EPIRB's (Category I, II): The 406 MHz EPIRB was designed to operate with satellites. Its signal allows authorities to locate the EPIRB much more accurately than 121.5/243 MHz devices and identify the individual vessel anywhere in the world. There is no range limitation. These devices also include a 121.5 MHz homing signal, allowing aircraft and rescue vessels to quickly locate the vessel in distress once underway. These are the only type of EPIRB which must be tested by Coast Guard-approved independent laboratories before they can be sold for use in the United States.

An automatically activated, float-free version of this EPIRB has been required on SOLAS vessels (cargo ships over 300 tons and passenger ships on international voyages) since August 1, 1993. The Coast Guard requires U.S. commercial fishing vessels to carry this device, and requires the same for other U.S. commercial uninspected vessels which travel more than 3 miles offshore.

Inmarsat-E EPIRB's: Inmarsat-E EPIRB's operate on 1.6 GHz (L-band) and transmit a distress signal to Inmarsat geostationary satellites, which includes a registered identity similar to that of the 406 MHz EPIRB, and a location derived from a GPS navigational satellite receiver inside the EPIRB. Inmarsat-E EPIRB's may be detected anywhere in the world between 70°N and 70°S. Since geostationary satellites are used, alerts are transmitted almost instantly to a RCC associated with the Inmarsat CES receiving the alert. The distress alert transmitted by an Inmarsat-E EPIRB is received by two CES's in each ocean region, giving 100 percent duplication for each ocean region in case of failures or outages associated with any of the CES's. Alerts received over the Inmarsat Atlantic Ocean Regions are routed to the U.S. Coast Guard Atlantic Area command center in Portsmouth, and alerts received over the Inmarsat Pacific Ocean Region are routed to the U.S. Coast Guard Pacific Area command center in Alameda. This type of EPIRB is designated for use in the GMDSS, but it is not sold in the United States or approved for use by U.S. flag vessels.

Type	Frequency	Description
Class A	121.5/243 MHz	Float-free, automatic activating, detectable by aircraft and satellite. Coverage limited (see Figure 2815b).
Class B	121.5/243 MHz	Manually activated version of Class A.
Class S	121.5/243 MHz	Similar to Class B, except that it floats, or is an integral part of a survival craft.
Category I	406 MHz	Float-free, automatically activated. Detectable by satellite anywhere in the world.
Category II	406 MHz	Similar to Category I, except manually activated.
Inmarsat-E	1646 MHz	Float-free, automatically activated EPIRB. Detectable by Inmarsat geostationary satellite.

Figure 2815a. EPIRB classifications.

Feature	*406 MHz EPIRB*	*121.5/243 MHz EPIRB*
Frequencies	406.025 MHz (locating) 121.500 MHz (homing)	 243.000 MHz (military)
Primary Function	Satellite alerting, locating, identification of distressed vessels.	Transmission of distress signal to passing aircraft and ships.
Distress Confirmation	Positive identification of coded beacon; each beacon signal is a coded, unique signal with registration data (vessel name, description, and telephone number ashore, assisting in confirmation).	Virtually impossible; no coded information, beacons often incompatible with satellites; impossible to know if signals are from EPIRB, ELT, or non-beacon source.
Signal	Pulse digital, providing accurate beacon location and vital information on distressed vessel.	Continuous signal allows satellite locating at reduced accuracy; close range homing.
Signal Quality	Excellent; exclusive use of 406 MHz for distress beacons; no problems with false alerts from non-beacon sources.	Relatively poor; high number of false alarms caused by other transmitters in the 121.5 MHz band.
Satellite Coverage	Global coverage, worldwide detection; satellite retains beacon data until next Earth station comes into view.	Both beacon and LUT must be within coverage of satellite; detection limited to line of sight.
Operational Time	48 hrs. at -20°C.	48 hrs. at -20°C.
Output Power	5 watts at 406 MHz, 0.025 watts at 121.5 MHz.	0.1 watts average.
Strobe Light	High intensity strobe helps in visually locating search target.	None.
Location Accuracy (Search Area) and Time Required	1 to 3 miles (10.8 sq. miles); accurate position on first satellite overflight enables rapid SAR response, often within 30 min.	10 to 20 miles (486 sq. miles); SAR forces must wait for second system alert to determine final position before responding (1 to 3 hr. delay).

Figure 2815b. Comparison of 121.5/406 MHz and 121.5/243 MHz EPIRB's.

Mariners should be aware of the differences between capabilities of 121.5/243 MHz and 121.5/406 MHz EPIRB's, as they have implications for alerting and locating of distress sites, as well as response by SAR forces. See Figure 2815b. The advantages of 121.5/406 MHz devices are substantial, and are further enhanced by EPIRB-transmitted registration data on the carrying vessel. Owners of 121.5/406 MHz EPIRB's furnish registration information about their vessel, type of survival gear, and emergency points of contact ashore, all of which greatly enhance the quality of the response. The database for U.S. vessels is maintained by the National Oceanographic and Atmospheric Administration, and is accessed worldwide by SAR authorities to facilitate SAR response.

2816. Testing EPIRB's

EPIRB owners should periodically check for water tightness, battery expiration date, and signal presence. FCC rules allow Class A, B, and S EPIRB's to be turned on briefly (for three audio sweeps, or 1 second only) during the first 5 minutes of any hour. Signal presence can be detected by an FM radio tuned to 99.5 MHz, or an AM radio tuned to any vacant frequency and located close to an EPIRB. All 121.5/406 MHz EPIRB's have a self-test function that should be used in accordance with manufacturers' instructions at least monthly.

2817. The COSPAS/SARSAT System

COSPAS is a Russian acronym for "Space System for Search of Distressed Vessels"; SARSAT signifies "Search And Rescue Satellite-Aided Tracking." COSPAS-SARSAT is an international satellite-based search and rescue system established by the U.S., Russia, Canada, and France to locate emergency radiobeacons transmitting on the frequencies 121.5, 243, and 406 MHz. Since its inception, the COSPAS-SARSAT system (SARSAT satellite only) has contributed to saving over 13,000 lives.

The USCG receives data from MRCC stations and SAR Points of Contact (SPOC). See Figure 2817.

Country	*Location*	*Designator*
Australia	Canberra	AUMCC
Brazil	San Paulo	BBMCC
Canada	Trenton	CMCC
Chile	Santiago	CHMCC
France	Toulouse	FMCC
Hong Kong	Hong Kong	HKMCC
India	Bangalore	INMCC
Indonesia	Jakarta	IONCC
ITDC	Taipei	TAMCC
Japan	Tokyo	JAMCC
Norway	Bodo	NMCC
Pakistan	Lahore	PAMCC*
Singapore	Singapore	SIMCC
Spain	Maspalomas	SPMCC
South Africa		SAMCC
Russian Federation	Moscow	CMC
United Kingdom	Plymouth	UKMCC
United States	Suitland	USMCC

* Status Unknown

Figure 2817. Participants in COSPASS/SARSAT system.

2818. Operation of The COSPAS/SARSAT System

If an EPIRB is activated, COSPAS/SARSAT picks up the signal, locates the source and passes the information to a land station. From there, the information is relayed to Rescue Coordination Centers, rescue vessels and nearby ships. This constitutes a one-way only communications system, from the EPIRB via the satellite to the rescuers. It employs low altitude, near polar orbiting satellites and by exploiting the Doppler principle, locates the 406 MHz EPIRB within about two miles. Due to the low polar orbit, there may by a delay in receiving the distress message unless the footprint of the satellite is simultaneously in view with a monitoring station. However, unlike SafetyNET, worldwide coverage is provided.

As a satellite approaches a transmitting EPIRB, the frequency of the signals it receives is higher than that being transmitted; when the satellite has passed the EPIRB, the received frequency is lower. This creates a notable Doppler shift. Calculations which take into account the Earth's rotation and other factors then determine the location of the EPIRB.

Each 406 MHz EPIRB incorporates a unique identification code. Once the satellite receives the beacon's signals, the Doppler shift is measured and the beacon's digital data is recovered from the signal. The information is time-lagged, formatted as digital data and transferred to the repeater downlink for real time transmission to a local user terminal. The digital data coded into each 406 MHz EPIRB's memory indicates the identity of the vessel to SAR authorities. They can then refer to the EPIRB registration database for information about the type of vessel, survival gear carried aboard, whom to contact in an emergency, etc. The data includes a maritime identification digit (MID, a three digit number identifying the administrative country) and either a ship station identifier (SSI, a 6 digit number assigned to specific ships), a ship radio call sign or a serial number to identify the ship in distress.

With the Inmarsat-E satellite EPIRB's, coverage does not extend to very high latitudes, but within the coverage area the satellite connection is instantaneous. However, to establish the EPIRB's geographic position, an interface with a GPS receiver or other sensor is needed.

2819. Alarm, Warning, and Alerting Signals

For MF (i.e. 2182 kHz), the signal consists of either (1) a keyed emission modulated by a tone of 1280 Hz to 1320 Hz with alternating periods of emission and silence of 1 to 1.2 seconds each; or (2) the radiotelephone alarm signal followed by Morse code B (— • • •) and/or the call sign of the transmitting ship, sent by keying a carrier modulated by a tone of 1300 Hz or 2200 Hz. For VHF (i.e. 121.5 MHz and 243 MHz), the signal characteristics are in accordance with the specifications of Appendix 37A of the ITU Radio Regulations. For 156.525 MHz and UHF (i.e. 406 MHz to 406.1 MHz and 1645.5 MHz to 1646.5 MHz), the signal characteristics are in accordance with CCIR recommendations.

The purpose of these signals is to help determine the position of survivors for SAR operations. They indicate that one or more persons are in distress, may no longer be aboard a ship or aircraft, and may not have a receiver available.

SEARCH AND RESCUE RADAR TRANSPONDERS

2820. Operational Characteristics

Operating much like a RACON, the Search and Rescue Radar Transponder (SART) is a passive rescue device which, when it senses the pulse from a radar operating in the 9 gHz frequency band, emits a series of pulses in response, which alerts the radar operator that some sort of maritime distress is in progress. Further, the SART signal allows the radar operator to home in on the exact location of the SART. The SART can be activated manually, or will activate automatically when placed in water.

The SART signal appears on the radar screen as a series of 12 blips, each 0.64 nautical miles apart. As the vessel or aircraft operating the radar approaches the SART location, the blips change to concentric arcs, and within about a mile of the SART become concentric circles, centered on the SART.

Because the SART actively responds to radar pulses, it also informs its user, with an audible or visual signal, that it is being triggered. This alerts the user in distress that there is an operating radar in the vicinity, whereupon they may send up flares or initiate other actions to indicate their position.

Approved SART's operate in standby mode for at least 96 hours and actively for at least 8 hours. Because the SART signal is stronger than any surrounding radar returns, it will be easily sensed by any nearby radar. But because it is much weaker than the radar, its own range is the limiting factor in detection.

2821. Factors Affecting SART Range

SART range is affected by three main factors. First, The type of radar and how it is operated is most important. Larger vessels with powerful, high-gain antennae, set higher above sea level, will trigger and detect the SART signal sooner than low-powered radars set closer to sea level. The radar should be set to a range of 12 or 6 miles for best indication of a SART's signal, and should not have too narrow a receive bandwidth, which might reduce the strength of the received signal.

Second, weather is a factor in SART range. A flat calm might cause multipath propagation and distort the SART's signal. Heavy seas may cause the SART signal to be received intermittently as the transponder falls into the troughs of the seas. Careful adjustment of the sea and rain clutter controls will maximize the SART's received signal strength.

Third, the height of the SART will greatly affect the range, because the signal obeys the normal rules for radio waves in its spectrum and does not follow the curvature of the earth, except for a small amount of refraction. Tests indicate that a SART floating in the sea will have a range of about 2 nautical miles when triggered by a radar mounted 15 meters above sea level. At a height of 1 meter, range increases to about 5 miles. To an aircraft actively searching for a SART at an altitude of 3.000 feet, the range increases to about 40 miles.

2822. Operating the Radar for SART Detection

Only an X-band (3 cm) radar can trigger and sense a SART. An S-Band (10 cm) radar will neither trigger nor detect a SART. Normally, an X-band radar will sense a SART at about 8 nm. When triggered by an incoming radar signal, the SART will transmit a return signal across the entire 3 cm radar frequency band. The first signal is a rapid 0.4 microsecond sweep, followed by a 7.5 microsecond sweep, repeated 12 times. This will cause a series of 12 blips on the radar, spaced 0.64 nm apart. See Figure 2822a.

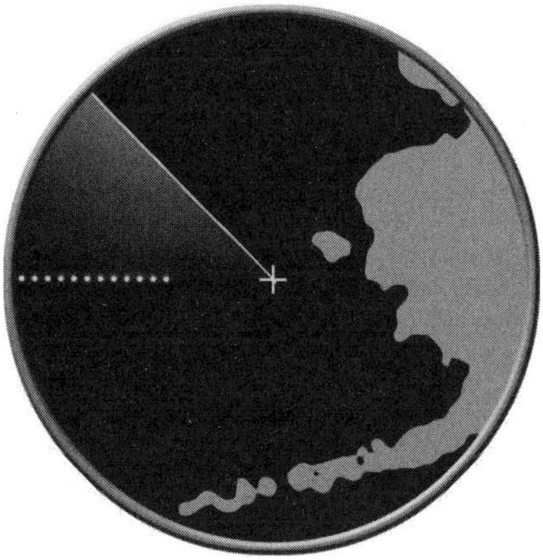

Figure 2822a. SART 12-dot blip code

For best reception, the radar should be set to medium bandwidth and to the 12 or 6 mile range. Too narrow a bandwidth will cause the SART signal to be weakened, as the radar is not sensing the entire SART pulse. The radar operator's manual should be consulted for these settings. Less expensive radars may not be able to change settings.

As the range to the SART decreases to about 1 nm, the initial 0.4 microsecond sweeps may become visible as weaker and smaller dots on the radar screen. When first sensed, the first blip will appear about 0.6 miles beyond the actual location of the SART. As range decreases, the blips will become centered on the SART.

As the SART is approached more closely, the blips ap-

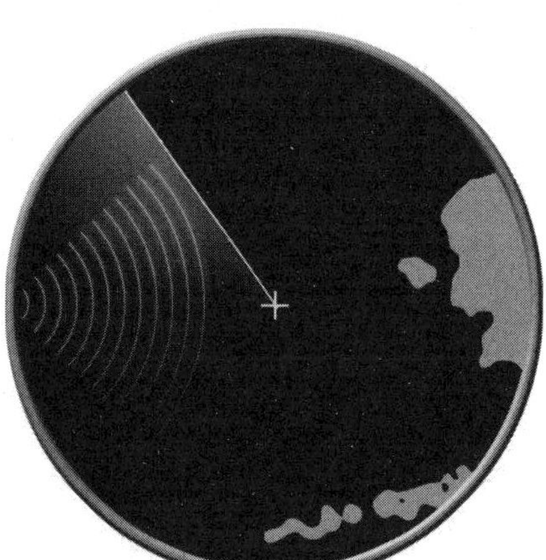

Figure 2822b. SART arcs

pearing on the radar become concentric arcs centered on the SART itself. The arcs are actually caused by the radar return of side lobes associated with the radar signal. While use of the sea return or clutter control may decrease or eliminate these arcs, it is often best to retain them, as they indicate the proximity of the SART. See Figure 2822b. Eventually the arcs become rings centered on the SART, as in Figure 2822c.

On some radars it may be possible to detune the radar signal in situations where heavy clutter or sea return obscures the SART signal. With the Automatic Frequency Control (AFC) on, the SART signal may become more visible, but the radar should be returned to normal operation as soon as possible. The gain control should usually be set to normal level for best detection, with the sea clutter control at its minimum and rain clutter control in normal position for the ambient conditions.

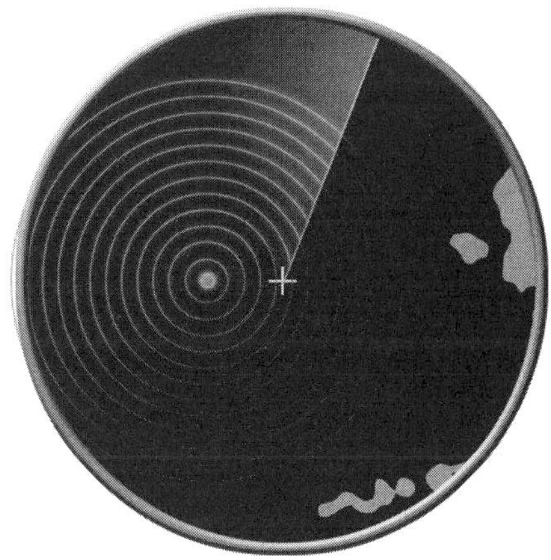

Figure 2822c. SART rings

CHAPTER 29

HYDROGRAPHY

2900. Introduction

Hydrography is the science of measurement and description of the features which affect marine navigation, including water depths, shorelines, tides, currents, bottom types, and undersea obstructions. Cartography transforms the scientific data collected by hydrographers into data usable by the mariner, and is the final step in a long process which leads from raw data to a usable chart.

The mariner, in addition to being the primary user of hydrographic data, is also an important source of data used in the production and correction of nautical charts. This chapter discusses the processes involved in producing a nautical chart, whether in digital or paper form, from the initial planning of a hydrographic survey to the final printing. It is important to note that digital charts are no more accurate than the paper charts and other sources from which they are produced. "Digital" does not mean "more accurate," for in most cases the digitized data comes from the same sources that the paper charts use.

With this information, the mariner can better understand the information presented on charts, evaluate hydrographic information which comes to his attention, and report discrepancies in a form that will be most useful to charting agencies.

BASICS OF HYDROGRAPHIC SURVEYING

2901. Planning the Survey

The basic sources of data used to produce nautical charts are hydrographic surveys. Much additional information is included, but the survey is central to the compilation of a chart. A survey begins long before actual data collection starts. Some elements which must be decided are:

- Exact area of the survey.
- Type of survey (reconnaissance or standard), scaled to meet standards of charts to be produced.
- Scope of the survey (short or long term).
- Platforms available (ships, launches, aircraft, leased vessels, cooperative agreements).
- Support work required (aerial or satellite photography, geodetics, tides).
- Limiting factors (budget, politics, geographic or operational constraints, positioning system limitations, logistics).

Once these issues are decided, all information available in the survey area is reviewed. This includes aerial photography, satellite data, topographic maps, existing nautical charts, geodetic information, tidal information, and anything else affecting the survey. The survey planners then compile sound velocity information, climatology, water clarity data, any past survey data, and information from light lists, *Sailing Directions*, and *Notices to Mariners*. Tidal information is thoroughly reviewed and tide gauge locations chosen. Local vertical control data is reviewed to see if it meets the expected accuracy standards so the tide gauges can be linked to the vertical datum used for the survey. Horizontal control is reviewed to check for accuracy and discrepancies and to determine sites for local positioning systems to be used in the survey.

Line spacing refers to the distance between tracks to be run by the survey vessel. It is chosen to provide the best coverage of the area using the equipment available. Line spacing is a function of the depth of water, the sound footprint of the collection equipment to be used, and the complexity of the bottom. Once line spacing is chosen, the hydrographer can compute the total miles of survey track to be run and have an idea of the time required for the survey, factoring in the expected weather and other possible delays. The scale of the survey, orientation to the shorelines in the area, and the method of positioning determine line spacing. Planned tracks are laid out so that there will be no gaps between sound lines and sufficient overlaps between individual survey areas.

Wider lines are run at right angles to the primary survey development to verify data repeatability. These are called **cross check lines**.

Other tasks to be completed with the survey include bottom sampling, seabed coring, production of sonar pictures of the seabed, gravity and magnetic measurements (on deep ocean surveys), and sound velocity measurements in the water column.

2902. Echo Sounders in Hydrographic Surveying

Echo sounders were developed in the early 1920s, and compute the depth of water by measuring the time it takes for a pulse of sound to travel from the source to the sea bottom and return. A device called a **transducer** converts electrical energy into sound energy and vice versa. For basic hydrographic surveying, the transducer is mounted permanently in the bottom of the survey vessel, which then follows the planned trackline, generating soundings along the track.

The major difference between different types of echo sounders is in the frequencies they use. Transducers can be classified according to their beam width, frequency, and power rating. The sound radiates from the transducer in a cone, with about 50% actually reaching to sea bottom. **Beam width** is determined by the frequency of the pulse and the size of the transducer. In general, lower frequencies produce a wider beam, and at a given frequency, a smaller transducer will produce a wider beam. Lower frequencies also penetrate deeper into the water, but have less resolution in depth. Higher frequencies have greater resolution in depth, but less range, so the choice is a trade-off. Higher frequencies also require a smaller transducer. A typical low frequency transducer operates at 12 kHz and a high frequency one at 200 kHz.

The formula for depth determined by an echo sounder is:

$$D = \frac{V \times T}{2} + K + D_r$$

where D is depth from the water surface, V is the average velocity of sound in the water column, T is round-trip time for the pulse, K is the system index constant, and D_r is the depth of the transducer below the surface (which may not be the same as vessel draft). V, D_r, and T can be only generally determined, and K must be determined from periodic calibration. In addition, T depends on the distinctiveness of the echo, which may vary according to whether the sea bottom is hard or soft. V will vary according to the density of the water, which is determined by salinity, temperature, and pressure, and may vary both in terms of area and time. In practice, average sound velocity is usually measured on site and the same value used for an entire survey unless variations in water mass are expected. Such variations could occur in areas of major currents or river outflows. While V is a vital factor in deep water surveys, it is normal practice to reflect the echo sounder signal off a plate suspended under the ship at typical depths for the survey areas in shallow waters. The K parameter, or index constant, refers to electrical or mechanical delays in the circuitry, and also contains any constant correction due to the change in sound velocity between the upper layers of water and the average used for the whole project. Further, vessel speed is factored in and corrections are computed for settlement and squat, which affect transducer depth. Vessel roll, pitch, and heave are also accounted for. Finally, the observed tidal data is recorded in order to correct the soundings during processing.

Tides are accurately measured during the entire survey so that all soundings can be corrected for tide height and thus reduced to the chosen vertical datum. Tide corrections eliminate the effect of the tides on the charted waters and ensure that the soundings portrayed on the chart are the minimum available to the mariner at the sounding datum. Observed, not predicted, tides are used to account for both astronomically and meteorologically induced water level changes during the survey.

2903. Collecting Survey Data

While sounding data is being collected along the planned tracklines by the survey vessel(s), a variety of other related activities are taking place. A large-scale **boat sheet** is produced with many thousands of individual soundings plotted. A complete navigation journal is kept of the survey vessel's position, course and speed. Side-scan sonar may be deployed to investigate individual features and identify rocks, wrecks, and other dangers. Divers may also be sent down to investigate unusual objects. Time is the single parameter which links the ship's position with the various echograms, sonograms, journals, and boat sheets that make up the hydrographic data package.

2904. Processing Hydrographic Data

During processing, echogram data and navigational data are combined with tidal data and vessel/equipment corrections to produce **reduced soundings**. This reduced data is combined on a plot of the vessel's actual track with the boat sheet data to produce a **smooth sheet**. A contour overlay is usually made to test the logic of all the data shown. All anomolous depths are rechecked in either the survey records or in the field. If necessary, sonar data are then overlayed to analyze individual features as related to depths. It may take dozens of smooth sheets to cover the area of a complete survey. The smooth sheets are then ready for cartographers, who will choose representative soundings manually or using automated systems from thousands shown, to produce a nautical chart. Documentation of the process is such that any individual sounding on any chart can be traced back to its original uncorrected value. See Figure 2904.

The process is increasingly computerized, such that all the data from an entire survey can be collected and reduced to a selected set of soundings ready for incorporation into an electronic chart, without manual processes of any kind. Only the more advanced maritime nations have this capability, but less developed nations often borrow advanced technology from them under cooperative hydrographic agreements.

2905. Automated Hydrographic Surveying

The evolution of echo sounders has followed the same

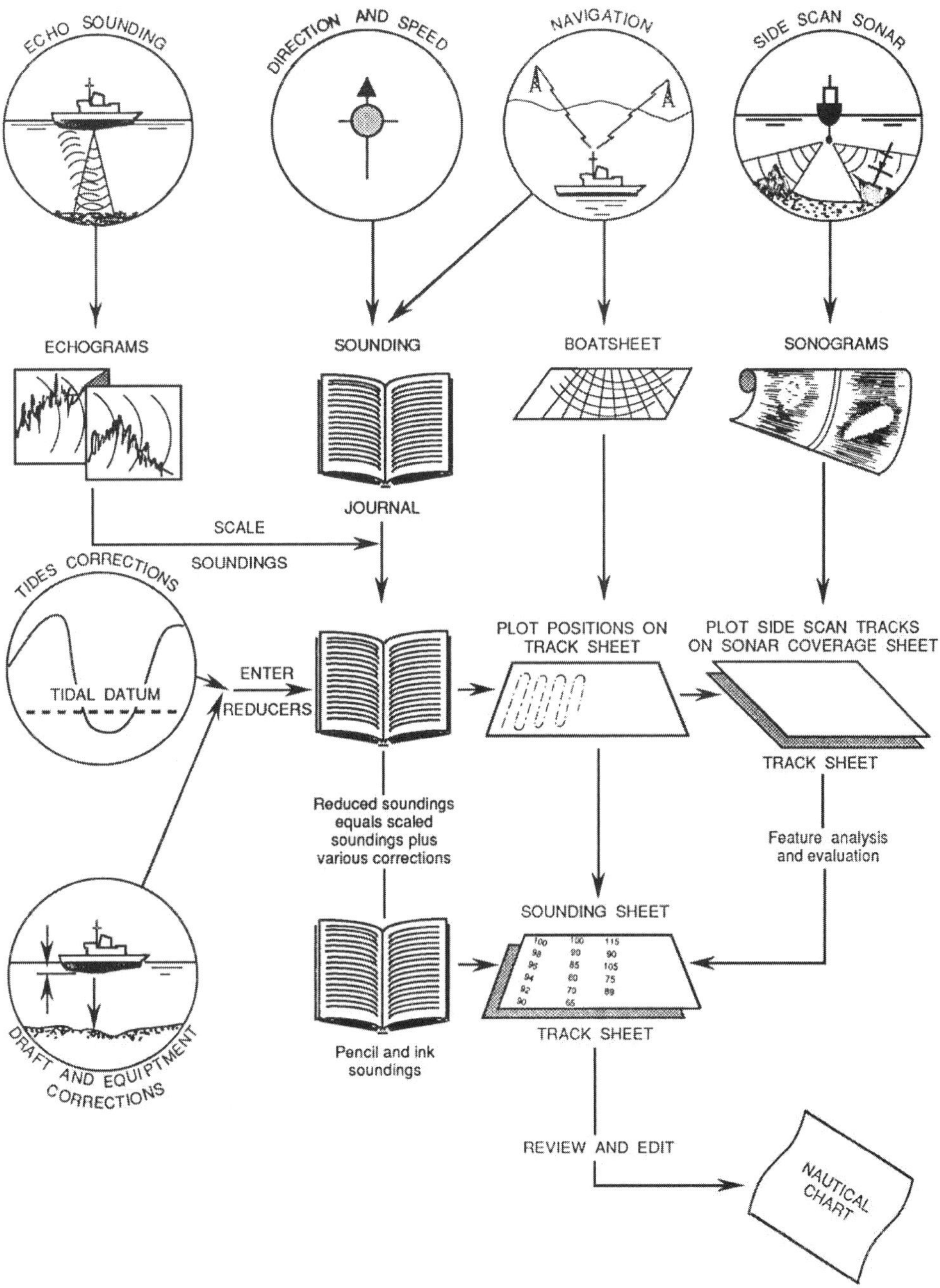

Figure 2904. The process of hydrographic surveying.

pattern of technological innovation seen in other areas. In the 1940s low frequency/wide beam sounders were developed for ships to cover larger ocean areas in less time with some loss of resolution. Boats used smaller sounders which usually required visual monitoring of the depth. Later, narrow beam sounders gave ship systems better resolution using higher frequencies, but with a corresponding loss of area. These were then combined into dual-frequency systems. All echo sounders, however, used a single transducer, which limited surveys to single lines of soundings. For boat equipment, automatic recording became standard.

The last three decades have seen the development of multiple-transducer, multiple-frequency sounding systems which are able to scan a wide area of seabed. Two general types are in use. Open waters are best surveyed using an array of transducers spread out athwartships across the hull of the survey vessel. They may also be deployed from an array towed behind the vessel at some depth to eliminate corrections for vessel heave, roll, and pitch. Typically, as many as 16 separate transducers are arrayed, sweeping an arc of 90°. The area covered by these **swath survey systems** is thus a function of water depth. See Figure 2905. In shallow water, track lines must be much closer together than in deep water. This is fine with hydrographers, because shallow waters need more closely spaced data to provide an accurate portrayal of the bottom on charts. The second type of multiple beam system uses an array of vertical beam transducers rigged out on poles abeam the survey vessel with transducers spaced to give overlapping coverage for the general water depth. This is an excellent configuration for very shallow water, providing very densely spaced soundings from which an accurate picture of the bottom can be made for harbor and small craft charts. The width of the swath of this system is fixed by the distance between the two outermost transducers and is not dependent on water depth.

Airborne Laser Hydrography (ALH) uses laser light to conduct hydrographic surveys from aircraft. It is particularly suitable in areas of complex hydrography containing numerous rocks, shoals, and obstructions dangerous to survey vessels. The technology has developed and matured since the 1970's, and in some areas of the world up to 50% of the hydrographic surveying is done with lasers. Survey rates of some 65 square km per hour are possible, at about a quarter of the cost of comparable vessel surveys. Data density is variable, ranging down to some 1-2 meters square, and depths from one half to over 70 meters have been successfully surveyed.

The technology uses laser light generators mounted in the bottom of a fixed or rotary wing aircraft. Two colors are used, one which reflects off the surface of the sea and back to the aircraft, and a different color which penetrates to the seabed before reflecting back to the aircraft. The difference in the time of reception of the two beams is a function of the water depth. This data is correlated with position data obtained from GPS, adjusted for tides, and added to a bathymetric database from which subsets of data are drawn for compilation of nautical charts.

Obviously water clarity has a great deal to do with the success of ALH, but even in most areas of murky water, seasonal or meteorological variations often allow sufficient penetration of the laser to conduct surveys. Some 80% of the earth's shallow waters are suitable for ALH.

In addition to hydrographic uses, ALH data finds application in coastal resource management, maritime boundaries, environmental studies, submarine pipeline construction, and oil and gas exploration.

HYDROGRAPHIC REPORTS

2906. Chart Accuracies

The chart resulting from a hydrographic survey can be no more accurate than that survey; the survey's accuracy, in turn, is limited by the positioning system used. For many older charts, the positioning system controlling data collection involved using two sextants to measure horizontal angles between surveyed points established ashore. The accuracy of this method, and to a lesser extent the accuracy of modern, shore based electronic positioning methods, deteriorates rapidly with distance. In the past this often determined the maximum scale which could be considered for the final chart. With the advent of the Global Positioning System (GPS) and enhancements such as DGPS and WAAS, the mariner can often now navigate with greater accuracy than could the hydrographic surveyor who collected the chart's source data. Therefore, one must exercise care not to take shoal areas or other hazards closer aboard than necessary because they may not be exactly where they are charted. This is especially true in less-travelled waters.

This is in addition to the caution the mariner must exercise to be sure that his navigation system and chart are on the same datum. The potential danger to the mariner increases with digital charts because by zooming in, he can increase the chart scale beyond what can be supported by the source data. The constant and automatic update of the vessel's position on the chart display can give the navigator a false sense of security, causing him to rely on the accuracy of a chart when the source data from which the chart was compiled cannot support the scale of the chart displayed.

2907. Navigational and Oceanographic Information

Mariners at sea, because of their professional skills and location, represent a unique data collection capability unobtainable by any government agency. Provision of high quality navigational and oceanographic information by government agencies requires active participation by mariners in data collection and reporting. Examples of the type of information required are reports of obstructions, shoals or

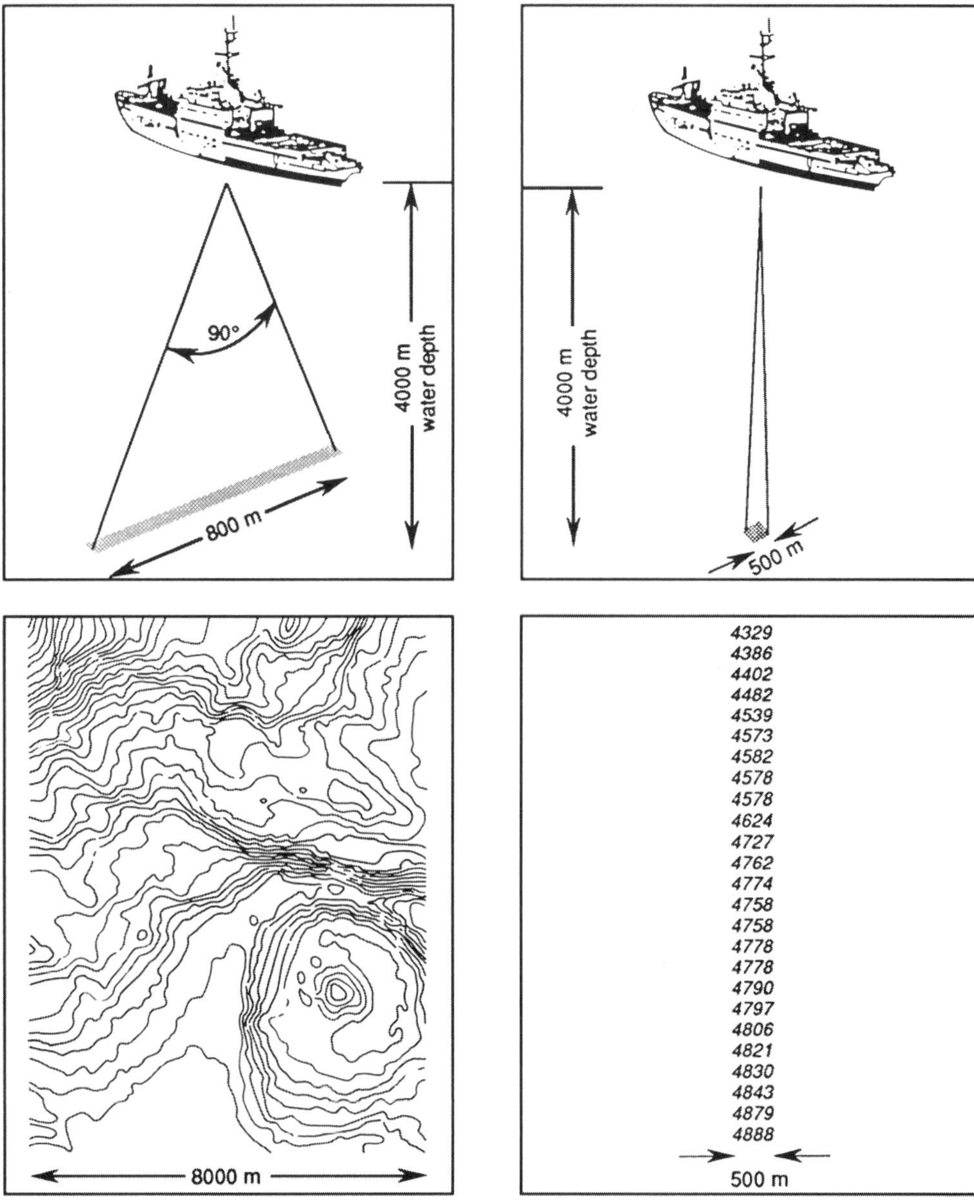

Figure 2905. Swath versus single-transducer surveys.

hazards to navigation, unusual sea ice or icebergs, unusual soundings, currents, geophysical phenomena such as magnetic disturbances and subsurface volcanic eruptions, and marine pollution. In addition, detailed reports of harbor conditions and facilities in both busy and out-of-the-way ports and harbors helps charting agencies keep their products current. The responsibility for collecting hydrographic data by U.S. Naval vessels is detailed in various directives and instructions. Civilian mariners, because they often travel to a wider range of ports, also have an opportunity to contribute substantial amounts of valuable information.

2908. Responsibility for Information

The National Imagery and Mapping Agency (NIMA), the U.S. Naval Oceanographic Office (NAVOCEANO), the U.S. Coast Guard and NOAA's Coast and Geodetic Survey (C&GS) are the primary agencies which receive, process, and disseminate marine information in the U.S.

NIMA produces charts, *Notice to Mariners*, and other nautical materials for the U.S. military services and for navigators in general for waters outside the U.S.

NAVOCEANO conducts hydrographic and oceanographic surveys of primarily foreign or international waters, and disseminates information to naval forces, government agencies, and civilians.

The Coast and Geodetic Survey (C&GS) conducts hydrographic and oceanographic surveys and provides charts for marine and air navigation in the coastal waters of the United States and its territories.

The U.S. Coast Guard is charged with protecting safety of life and property at sea, maintaining aids to navigation, law enforcement, and improving the quality of the marine environment. In the execution of these duties, the Coast Guard collects, analyzes, and disseminates navigational and oceanographic data.

Modern technology allows navigators to easily contribute to the body of hydrographic and oceanographic information.

Navigational reports are divided into four categories:

1. Safety Reports
2. Sounding Reports
3. Marine Data Reports
4. Port Information Reports

The seas and coastlines continually change through the actions of man and nature. Improvements realized over the years in the nautical products published by NIMA, the National Ocean Service (NOS), and U.S. Coast Guard have been made possible in part by the reports and constructive criticism of seagoing observers, both naval and merchant marine. NIMA and NOS continue to rely to a great extent on the personal observations of those who have seen the changes and can compare charts and publications with actual conditions. In addition, many ocean areas and a significant portion of the world's coastal waters have never been adequately surveyed for the purpose of producing modern nautical charts.

Information from all sources is evaluated and used in the production and maintenance of NIMA, NOS and Coast Guard charts and publications. Information from surveys, while originally accurate, is subject to continual change. As it is impossible for any hydrographic office to conduct continuous worldwide surveys, U.S. charting authorities depend on reports from mariners to provide a steady flow of valuable information from all parts of the globe.

After careful analysis of a report and comparison with all other data concerning the same area or subject, the organization receiving the information takes appropriate action. If the report is of sufficient urgency to affect the immediate safety of navigation, the information will be broadcast as a SafetyNET or NAVTEX message. Each report is compared with others and contributes in the compilation, construction, or correction of charts and publications. It is only through the constant flow of new information that charts and publications can be kept accurate and up-to-date.

2909. Safety Reports

Safety reports are those involving navigational safety which must be reported and disseminated by message. The types of dangers to navigation which will be discussed in this article include ice, floating derelicts, wrecks, shoals, volcanic activity, mines, and other hazards to shipping.

1. Ice—Mariners encountering ice, icebergs, bergy bits, or growlers in the North Atlantic should report to Commander, International Ice Patrol, Groton, CT through a U.S. Coast Guard Communications Station. Direct printing radio teletype (SITOR) is available through USCG Communications Stations Boston or Portsmouth.

Satellite telephone calls may be made to the Ice Patrol office in Groton, Connecticut throughout the season at (203) 441-2626 (Ice Patrol Duty Officer). Messages can also be sent through the Coast Guard Operations Center, Boston at (617) 223-8555.

When sea ice is observed, the concentration, thickness, and position of the leading edge should be reported. The size, position, and, if observed, rate and direction of drift, along with the local weather and sea surface temperature, should be reported when icebergs, bergy bits, or growlers are encountered.

Ice sightings should also be included in the regular synoptic ship weather report, using the five-figure group following the indicator for ice. This will assure the widest distribution to all interested ships and persons. In addition, sea surface temperature and weather reports should be made to COMINTICEPAT every 6 hours by vessels within latitude 40°N and 52°N and longitude 38°W and 58°W, if a routine weather report is not made to METEO Washington.

2. Floating Derelicts—All observed floating and drifting dangers to navigation that could damage the hull or propellers of a vessel at sea should be immediately reported by radio. The report should include a brief description of the danger, the date, time (GMT) and the location as exactly as can be determined (latitude and longitude).

3.Wrecks/Man-Made Obstructions—Information is needed to assure accurate charting of wrecks, man-made obstructions, other objects dangerous to surface and submerged navigation, and repeatable sonar contacts that may be of interest to the U.S. Navy. Man-made obstructions not in use or abandoned are particularly hazardous if unmarked and should be reported immediately. Examples include abandoned wellheads and pipelines, submerged platforms and pilings, and disused oil structures. Ship sinkings, strandings, disposals, or salvage data are also reportable, along with any large amounts of debris, particularly metallic.

Accuracy, especially in position, is vital. Therefore, the date and time of the observation, as well as the method used in establishing the position, and an estimate of the fix accuracy should be included. Reports should also include the depth of water, preferably measured by soundings (in fathoms or meters). If known, the name, tonnage, cargo, and cause of casualty should be provided.

Data concerning wrecks, man-made obstructions, other sunken objects, and any salvage work should be as complete as possible. Additional substantiating information is encouraged.

4. Shoals—When a vessel discovers an uncharted or erroneously charted shoal or an area that is dangerous to navigation, all essential details should be immediately reported to NIMA NAVSAFETY BETHESDA MD via radio. An uncharted depth of 300 fathoms or less is considered an urgent danger to submarine navigation. Immediately upon receipt of any message reporting dangers to navigation, NIMA may issue an appropriate NAVAREA warning. The information must appear on published charts as "reported" until sufficient substantiating evidence (i.e. clear and properly annotated echograms and navigation logs, and any other supporting information) is received.

Therefore, originators of shoal reports are requested to verify and forward all substantiating evidence to NIMA at the earliest opportunity. Clear and properly annotated echograms and navigation logs are especially important in verifying or disproving shoal reports.

5. Volcanic Activity—On occasion, volcanic eruptions may occur beneath the surface of the water. These submarine eruptions may occur more frequently and be more widespread than has been suspected in the past. Sometimes the only evidence of a submarine eruption is a noticeable discoloration of the water, a marked rise in sea surface temperature, or floating pumice. Mariners witnessing submarine activity have reported steams with a foul sulfurous odor rising from the sea surface, and strange sounds heard through the hull, including shocks resembling a sudden grounding. A subsea volcanic eruption may be accompanied by rumbling and hissing as hot lava meets the cold sea.

In some cases, reports of discolored water at the sea surface have been investigated and found to be the result of newly formed volcanic cones on the sea floor. These cones can grow rapidly and within a few years constitute a hazardous shoal.

It is imperative that mariners report evidence of volcanic activity immediately to NIMA by message. Additional substantiating information is encouraged.

6. Mines—All mines or objects resembling mines should be considered armed and dangerous. An immediate radio report to NIMA should include (if possible):

1. Greenwich Mean Time (UT) and date
2. Position of mine, and how near it was approached
3. Size, shape, color, condition of paint, and presence of marine growth
4. Presence or absence of horns or rings
5. Certainty of identification

2910. Instructions for Safety Report Messages

The International Convention for the Safety of Life at Sea (1974), which is applicable to all U.S. flag ships, states "The master of every ship which meets with dangerous ice, dangerous derelict, or any other direct danger to navigation, or a tropical storm, or encounters subfreezing air temperatures associated with gale force winds causing severe ice accretion on superstructures, or winds of force 10 or above on the Beaufort scale for which no storm warning has been received, is bound to communicate the information by all means at his disposal to ships in the vicinity, and also to the competent authorities at the first point on the coast with which he can communicate."

The transmission of information regarding ice, derelicts, tropical storms, or any other direct danger to navigation is obligatory. The form in which the information is sent is not obligatory. It may be transmitted either in plain language (preferably English) or by any means of International Code of Signals (wireless telegraphy section). It should be sent to all vessels in the area and to the first station with which communication can be made, with the request that it be transmitted to the appropriate authority. A vessel will not be charged for radio messages to government authorities reporting dangers to navigation.

Each radio report of a danger to navigation should answer briefly three questions:

1. What? A description of the object or phenomenon
2. Where? Latitude and longitude
3. When? Greenwich Mean Time (GMT) and date

Examples:

Ice

SECURITE. ICE: LARGE BERG SIGHTED DRIFTING SW AT 0.5 KT 4605N, 4410W, AT 0800 GMT, MAY 15.

Derelicts

SECURITE. DERELICT: OBSERVED WOODEN 25 METER DERELICT ALMOST SUBMERGED AT 4406N, 1243W AT 1530 GMT, APRIL 21.

The report should be addressed to one of the following shore authorities as appropriate:

1. U.S. Inland Waters—Commander of the Local Coast Guard District
2. Outside U.S. Waters—NIMA NAVSAFETY BETHESDA MD

Whenever possible, messages should be transmitted via the nearest government radio station. If it is impractical to use a government station, a commercial station may be used. U.S. government navigational warning messages should invariably be sent through U.S. radio stations, government or commercial, and never through foreign stations. Detailed instructions for reporting via radio are contained in NIMA *Pub. 117, Radio Navigational Aids*.

OCEANIC SOUNDING REPORTS

2911. Sounding Reports

Acquisition of reliable sounding data from all ocean areas of the world is a continuing effort of NIMA, NAVOCEANO, and NOS. There are vast ocean areas where few soundings have ever been acquired. Much of the bathymetric data shown on charts has been compiled from information submitted by mariners. Continued cooperation in observing and submitting sounding data is absolutely necessary to enable the compilation of accurate charts. Compliance with sounding data collection procedures by merchant ships is voluntary, but for U.S. Naval vessels compliance is required under various fleet directives.

2912. Areas Where Soundings are Needed

Prior to a voyage, navigators can determine the importance of recording sounding data by checking the charts for the route. Indications that soundings may be particularly useful are:

1. Old sources listed on source diagram or note
2. Absence of soundings in large areas
3. Presence of soundings, but only along well-defined lines with few or no soundings between tracks
4. Legends such as "Unexplored area"

2913. Fix Accuracy

A realistic goal of open ocean positioning for sounding reports is a few meters using GPS or Loran C. Depths of 300 fathoms or less should always be reported regardless of the fix accuracy. When such depths are uncharted or erroneously charted, they should be reported by message to NIMA NAVSAFETY BETHESDA MD, giving the best available positioning accuracy. Echograms and other supporting information should then be forwarded by mail to NIMA.

The accuracy goal noted above has been established to enable NIMA to create a high quality data base which will support the compilation of accurate nautical charts. It is particularly important that reports contain the navigator's best estimate of his fix accuracy and that the positioning system being used (GPS, Loran C, etc.) be identified.

2914. False Shoals

Many poorly identified shoals and banks shown on charts are probably based on encounters with the **Deep Scattering Layer (DSL)**, ambient noise, or, on rare occasions, submarine earthquakes. While each appears real enough at the time of its occurrence, a knowledge of the events that normally accompany these incidents may prevent erroneous data from becoming a charted feature.

The DSL is found in most parts of the world. It consists of a concentration of marine life which descends from near the surface at sunrise to an approximate depth of 200 fathoms during the day. It returns near the surface at sunset. Although at times the DSL may be so concentrated that it will completely mask the bottom, usually the bottom return can be identified at its normal depth at the same time the DSL is being recorded.

Ambient noise or interference from other sources can cause erroneous data. This interference may come from equipment on board the ship, from another transducer being operated close by, or from waterborne noise. Most of these returns can be readily identified on the echo sounder records and should cause no major problems. However, on occasion they may be so strong and consistent as to appear as the true bottom.

Finally, a volcanic disturbance beneath the ship or in the immediate vicinity may give erroneous indications of a shoal. The experience has at times been described as similar to running aground or striking a submerged object. Regardless of whether the feature is an actual shoal or a submarine eruption, the positions, date/time, and other information should be promptly reported to NIMA.

2915. Doubtful Hydrographic Data

Navigators are requested to assist in confirming and charting actual shoals and the removal from the charts of doubtful data which was erroneously reported.

The classification or confidence level assigned to doubtful hydrographic data is indicated by the following standard abbreviations:

Abbreviation	*Meaning*
Rep (date)	Reported (year)
E.D.	Existence Doubtful
P.A.	Position Approximate
P.D.	Position Doubtful

Many of these reported features are sufficiently deep that a ship can safely navigate across the area. Confirmation of the existence of the feature will result in proper charting. On the other hand, properly collected and annotated sounding reports of the area may enable cartographers to accumulate sufficient evidence to justify the removal of the erroneous sounding from the database.

2916. Preparation of Sounding Reports

The procedures for preparing sounding reports have been designed to minimize the efforts of the shipboard observers, yet provide essential information. Submission of plotted sounding tracks is not required. Annotated echograms and navigation logs are preferred. The procedure for collecting sounding reports is for the ship to operate a recording echo sounder while transiting an area where soundings are desired. Fixes and course changes are recorded in the log, and the event marker is used to note these events on the echogram. Both the log and echogram can then be sent to NIMA whenever convenient. From this data, the track will be reconstructed and the soundings keyed to logged times.

The following annotations or information should be clearly written on the echogram to ensure maximum use of the recorded depths:

1. **Ship's name**—At the beginning and end of each roll or portion of the echogram.
2. **Date**—Date, noted as local or GMT, on each roll or portion of a roll.
3. **Time**—The echogram should be annotated at the beginning of the sounding run, regularly thereafter (hourly is best), at every scale change, and at all breaks in the echogram record. Accuracy of these time marks is critical for correlation with ship's position.
4. **Time Zone**—Greenwich Mean Time (GMT) should be used if possible. In the event local zone times are used, annotate echogram whenever clocks are reset and identify zone time in use. It is most important that the echogram and navigation log use the same time basis.
5. **Phase or scale changes**—If echosounder does not indicate scale setting on echogram automatically, clearly label all depth phase (or depth scale) changes and the exact time they occur. Annotate the upper and lower limits of the echogram if necessary.

Figure 2916a and Figure 2916b illustrate the data necessary to reconstruct a sounding track. If ship operations dictate that only periodic single ping soundings can be obtained, the depths may be recorded in the Remarks column. Cartographers always prefer an annotated echogram over single soundings. The navigation log is vital to the reconstruction of a sounding track. Without the position information from the log, the echogram is virtually useless.

The data received from these reports is digitized and becomes part of the digital bathymetric data library of NIMA, from which new charts are compiled. Even in areas where numerous soundings already exist, sounding reports allow valuable cross-checking to verify existing data and more accurately portray the sea floor. Keep in mind that many soundings seen on currently issued charts, and in the sounding database used to make digital charts, were taken when navigation was still largely an art. Soundings accurate to modern GPS standards are helpful to our Naval forces and particularly to the submarine fleet, and are also useful to geologists, geophysicists, and other scientific disciplines.

A report of oceanic soundings should contain:

1. All pertinent information about the ship, sounding system, transducer, etc.
2. A detailed Navigation Log
3. The echo sounding trace, properly annotated

Each page of the report should be clearly marked with the ship's name and date, so that it can be identified if it becomes separated. Mail the report to:

NIMA/PTNM
MS D-44
4600 Sangamore Rd.
Bethesda, MD 20816-5003

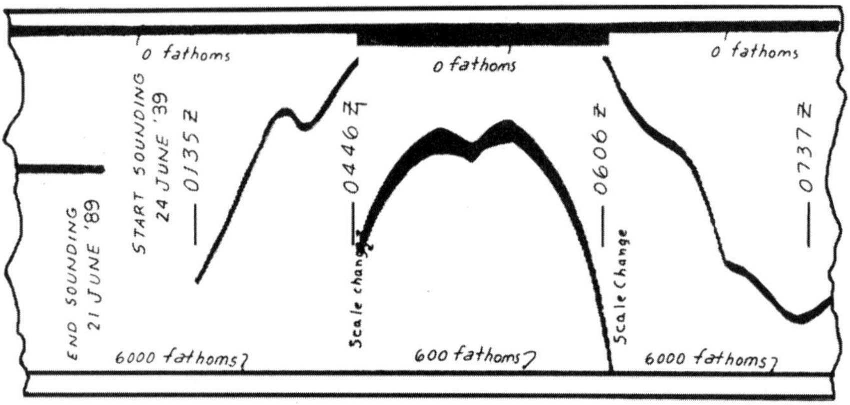

Figure 2916a. Annotated echo sounding record.

NAVIGATION LOG							REMARKS
DATE	TIME (GMT)	LAT.	LONG.	NAV. FIX	COURSE	SPEED	
11/2/83	0221	29°41'N	124°10'E	LORAN	093°	12.3	
	0340				097°	12.3	CHANGE COURSE
	0400	29°40'N	124°35'E	NOON FIX	097°	12.3	
	0728	29°35'N	125°22'E	LORAN	097°	12.3	
	0810				VARIOUS	8.2	REDUCE SPEED - MANUVERING TO AVOID FISHING BOATS
	0826	29°34'N	125°35.5'E	LORAN	097°	12.3	RESUME COURSE AND SPEED
	1011	29°32'N	125°56'E	EVENING STARS	097°	12.3	
	1620	29°23'N	127°22'E	LORAN	102°	12.4	CHANGE COURSE
	2230	29°06.2'N	128°48.5'E	RADAR/STAR	102°	12.5	
	2305				102°	10.1	REDUCE SPEED

Figure 2916b. Typical navigation log for hydrographic reporting.

OTHER HYDROGRAPHIC REPORTS

2917. Marine Information Reports

Marine Information Reports are reports of items of navigational interest such as the following:

1. Discrepancies in published information
2. Changes in aids to navigation
3. Electronic navigation reports
4. Satellite navigation reports
5. Radar navigation reports
6. Magnetic disturbances

Any information believed to be useful to charting authorities or other mariners should be reported. Depending on the type of report, certain information is absolutely critical for a correct evaluation. The following general suggestions are offered to assist in reporting information that will be of maximum value:

1. The geographical position included in the report may be used to correct charts. Accordingly, it should be fixed by the most exact method available, and more than one if possible.
2. If geographical coordinates are used to report position, they should be as exact as circumstances permit. Reference should be made to paper charts by number, edition number, and edition date.
3. The report should state the method used to fix the position and an estimate of fix accuracy.
4. When reporting a position within sight of charted objects, the position may be expressed as bearings and ranges from them. Bearings should preferably

be reported as true and expressed in degrees.

5. Always report the limiting bearings from the ship toward the light when describing the sectors in which a light is either visible or obscured. Although this is just the reverse of the form used to locate objects, it is the standard method used on NIMA nautical charts and in light lists.
6. A report prepared by one person should, if possible, be checked by another.

In most cases marine information can be adequately reported on one of the various forms printed by NIMA or NOS. It may be more convenient to annotate information directly on the affected chart and mail it to NIMA. As an example, it may be useful to sketch uncharted or erroneously charted shoals, buildings, or geological features directly on the chart. Appropriate supporting information should also be provided. NIMA forwards reports as necessary to NOS, NAVOCEANO, or U.S. Coast Guard.

Reports by letter or e-mail are just as acceptable as those prepared on regular forms. A letter report will often allow more flexibility in reporting details, conclusions, or recommendations concerning the observation. When reporting on the regular forms, use additional sheets if necessary to complete the details of an observation.

Reports are required concerning any errors in information published on nautical charts or in nautical publications. The reports should be as accurate and complete as possible. This will result in corrections to the information, including the issuance of a *Notice to Mariners* when appropriate.

Report all changes, defects, establishment or discontinuance of navigational aids and the source of the information. Check your report against the *List of Lights*, *Pub. 117, Radio Navigational Aids*, and the largest scale chart of the area. If a new, uncharted light has been established, report the light and its characteristics in a format similar to that carried in light lists. For changes and defects, report only elements that differ with light lists. If it is a lighted aid, identify by number. Defective aids to navigation in U.S. waters should be reported immediately to the Commander of the local Coast Guard District.

2918. Electronic Navigation System Reports

Electronic navigation systems have become an integral part of modern navigation. Reports on propagation anomalies or any unusual reception while using the electronic navigation system are desired.

Information should include:

1. Type of system
2. Type of antenna
3. Nature and description of the reception
4. Date and time
5. Position of ship
6. Manufacturer and model of receiver

2919. Radar Navigation Reports

Reports of any unusual reception or anomalous propagation by radar systems caused by atmospheric conditions are especially desirable. Comments concerning the use of radar in piloting, with the locations and description of good radar targets, are particularly needed. Reports should include:

1. Type of radar, frequency, antenna height and type.
2. Manufacturer and model of the radar
3. Date, time and duration of observed anomaly
4. Position
5. Weather and sea conditions

Radar reception problems caused by atmospheric parameters are contained in four groups. In addition to the previously listed data, reports should include the following specific data for each group:

1. Unexplained echoes—Description of echo, apparent velocity and direction relative to the observer, and range
2. Unusual clutter—Extent and Sector
3. Extended detection ranges—Surface or airborne target, and whether point or distributed target, such as a coastline or landmass
4. Reduced detection ranges—Surface or airborne target, and whether point or distributed target, such as a coastline or landmass

2920. Magnetic Disturbances

Magnetic anomalies, the result of a variety of causes, exist in many parts of the world. NIMA maintains a record of such magnetic disturbances and whenever possible attempts to find an explanation. A better understanding of this phenomenon can result in more detailed charts which will be of greater value to the mariner.

The report of a magnetic disturbance should be as specific as possible. For instance: "Compass quickly swung 190° to 170°, remained offset for approximately 3 minutes and slowly returned." Include position, ship's course, speed, date, and time.

Whenever the readings of the standard magnetic compass are unusual, an azimuth check should be made as soon as possible and this information included in a report to NIMA.

PORT INFORMATION REPORTS

2921. Importance of Port Information Reports

Port Information Reports provide essential information obtained during port visits which can be used to update and improve coastal, approach, and harbor charts as well as nautical publications including *Sailing Directions*, *Coast Pilots*, and *Fleet Guides*. Engineering drawings, hydrographic surveys and port plans showing new construction affecting charts and publications are especially valuable.

Items involving navigation safety should be reported by message or e-mail. Items which are not of immediate urgency, as well as additional supporting information may be submitted by the *Sailing Directions* Information and Suggestion Sheet found in the front of each volume of *Sailing Directions*, or the *Notice to Mariners* Marine Information Report and Suggestion Sheet found in the back of each *Notice to Mariners*. Reports by letter are completely acceptable and may permit more reporting flexibility.

In some cases it may be more convenient and more effective to annotate information directly on a chart and mail it to NIMA. As an example, new construction, such as new port facilities, pier or breakwater modifications, etc., may be drawn on a chart in cases where a written report would be inadequate.

Specific reporting requirements exist for U.S. Navy ships visiting foreign ports. These reports are primarily intended to provide information for use in updating the Navy Port Directories. A copy of the navigation information resulting from port visits should be provided directly to NIMA by including NIMA NAVSAFETY BETHESDA MD as an INFO addressee on messages containing hydrographic information.

2922. What to Report

Coastal features and landmarks are almost constantly changing. What may at one time have been a major landmark may now be obscured by new construction, destroyed, or changed by the elements. *Sailing Directions (Enroute)* and *Coast Pilots* utilize a large number of photographs and line sketches. Photographs, particularly a series of overlapping views showing the coastline, landmarks, and harbor entrances are very useful.

Especially convenient are e-mailed pictures taken with a digital camera. Use the highest resolution possible and e-mail the picture(s) with description of the feature and the exact Lat./Long. where the picture was taken to: navsafety@nima.mil. There is also an increasing need for video clips on VHS or other media of actual entrances to ports and harbors.

The following questions are suggested as a guide in preparing reports on coastal areas that are not included or that differ from the *Sailing Directions* and *Coast Pilots*.

Approach

1. What is the first landfall sighted?
2. Describe the value of soundings, GPS, LORAN, radar and other positioning systems in making a landfall and approaching the coast. Are depths, curves, and coastal dangers accurately charted?
3. Are prominent points, headlands, landmarks, and aids to navigation adequately described in *Sailing Directions* and *Coast Pilots*? Are they accurately charted?
4. Do land hazes, fog or local showers often obscure the prominent features of the coast?
5. Do discolored water and debris extend offshore? How far? Were tidal currents or rips experienced along the coasts or in approaches to rivers or bays?
6. Are any features of special value as radar targets?

Tides and Currents

1. Are the published tide and current tables accurate?
2. Does the tide have any special effect such as river bore? Is there a local phenomenon, such as double high or low water or interrupted rise and fall?
3. Was any special information on tides obtained from local sources?
4. What is the set and drift of tidal currents along coasts, around headlands, among islands, in coastal indentations?
5. Are tidal currents reversing or rotary? If rotary, do they rotate in a clockwise or counterclockwise direction?
6. Do subsurface currents affect the maneuvering of surface craft? If so, describe.
7. Are there any countercurrents, eddies, overfalls, or tide rips in the area? If so, where?

River and Harbor Entrances

1. What is the depth of water over the bar, and is it subject to change? Was a particular stage of tide necessary to permit crossing the bar?
2. What is the least depth in the channel leading from sea to berth?
3. If the channel is dredged, when and to what depth and width? Is the channel subject to silting?
4. What is the maximum draft, length and width of a vessel that can enter port?
5. If soundings were taken, what was the stage of tide? If the depth information was received from other sources, what were they?
6. What was the date and time of water depth observations?

Hills, Mountains, and Peaks

1. Are hills and mountains conical, flat-topped, or of any particular shape?
2. At what range are they visible in clear weather?
3. Are they snowcapped throughout the year?
4. Are they cloud covered at any particular time?
5. Are the summits and peaks adequately charted? Can accurate distances and/or bearings be obtained by sextant, pelorus, or radar?
6. What is the quality of the radar return?

Pilotage

1. Where is the signal station located?
2. Where does the pilot board the vessel? Are special arrangements necessary before a pilot boards?
3. Is pilotage compulsory? Is it advisable?
4. Will a pilot direct a ship in at night, during foul weather, or during periods of low visibility?
5. Where does the pilot boat usually lie?
6. Does the pilot boat change station during foul weather?
7. Describe the radiotelephone communication facilities available at the pilot station or pilot boat. What is the call sign, frequency, and the language spoken?

General

1. What cautionary advice, additional data, and information on outstanding features should be given to a mariner entering the area for the first time?
2. At any time did a question or need for clarification arise while using NIMA, NOS, or Coast Guard products?
3. Were charted land contours useful while navigating using radar? Indicate the charts and their edition numbers.
4. Would it be useful to have radar targets or topographic features that aid in identification or position plotting described or portrayed in the *Sailing Directions* and *Coast Pilots*?

Photographs

Use overlapping photographs to create panoramic views of wide features or areas. On the back of the photograph (negatives should accompany the required information), indicate the camera position by bearing and distance from a charted object if possible, name of the vessel, the date, time of exposure, height of eye (camera) and stage of tide. All features of navigational value should be clearly and accurately identified on an overlay, if time permits. Bearings and distances (from the vessel) of uncharted features identified on the print should be included. If photographs are digital and sent electronically, include this information in the e-mail message and add the photographs as attachments. Digital photographs can be sent via e-mail, on floppy disks or CR-ROM's.

Radarscope Photography

Because of the value of radar as an aid to navigation, NIMA desires radarscope photographs. Guidelines for radar settings for radarscope photography are given in *Pub. 1310, Radar Navigation and Maneuvering Board Manual.* Such photographs, reproduced in the *Sailing Directions* and *Fleet Guides*, supplement textual information concerning critical navigational areas and assist the navigator in correlating the radarscope presentation with the chart. To be of the greatest value, radarscope photographs should be taken at landfalls, sea buoys, harbor approaches, major turns in channels, constructed areas and other places where they will most aid the navigator. Two prints of each photograph are needed; one should be unmarked, the other annotated.

Examples of desired photographs are images of fixed and floating navigational aids of various sizes and shapes as observed under different sea and weather conditions, and images of sea return and precipitation of various intensities. There should be two photographs of this type of image, one without the use of special anti-clutter circuits and another showing the remedial effects of these. Photographs of actual icebergs, growlers, and bergy bits under different sea conditions, correlated with photographs of their radarscope images are also desired.

Radarscope photographs should include the following annotations:

1. Wavelength
2. Antenna height and rotation rate
3. Range-scale setting and true bearing
4. Antenna type (parabolic, slotted waveguide)
5. Weather and sea conditions, including tide
6. Manufacturer's model identification
7. Position at time of observation
8. Identification of target by *Light List*, *List of Lights*, or chart
9. Camera and exposure data

Other desired annotations include:

1. Beam width between half-power points
2. Pulse repetition rate
3. Pulse duration (width).
4. Antenna aperture (width)
5. Peak power
6. Polarization
7. Settings of radar operating controls, particularly use of special circuits
8. Characteristics of display (stabilized or unstabilized), diameter, etc.

Port Regulations and Restrictions

Sailing Directions (Planning Guides) are concerned with pratique, pilotage, signals, pertinent regulations, warning areas, and navigational aids. The following questions are suggested as a guide to the requested data.

1. Is this a port of entry for overseas vessels?
2. If not a port of entry, where must a vessel go for customs entry and pratique?
3. Where do customs, immigration, and health officials board?
4. What are the normal working hours of officials?
5. Will the officials board vessels after working hours? Are there overtime charges for after-hour services?
6. If the officials board a vessel underway, do they remain on board until the vessel is berthed?
7. Were there delays? If so, give details.
8. Were there any restrictions placed on the vessel?
9. Was a copy of the Port Regulations received from the local officials?
10. What verbal instructions were received from the local officials?
11. What preparations prior to arrival would expedite formalities?
12. Are there any unwritten requirements peculiar to the port?
13. What are the speed regulations?
14. What are the dangerous cargo regulations?
15. What are the flammable cargo and fueling regulations?
16. Are there special restrictions on blowing tubes, pumping bilges, oil pollution, fire warps, etc.?
17. Are the restricted and anchorage areas correctly shown on charts, and described in the *Sailing Directions* and *Coast Pilots*?
18. What is the reason for the restricted areas: gunnery, aircraft operating, waste disposal, etc.?
19. Are there specific hours of restrictions, or are local blanket notices issued?
20. Is it permissible to pass through, but not anchor in, restricted areas?
21. Do fishing boats, stakes, nets, etc., restrict navigation?
22. What are the heights of overhead cables, bridges, and pipelines?
23. What are the locations of submarine cables, their landing points, and markers?
24. Are there ferry crossings or other areas of heavy local traffic?
25. What is the maximum draft, length, and breadth of a vessel that can enter?

Port Installations

Much of the port information which appears in the *Sailing Directions* and *Coast Pilots* is derived from visit reports and port brochures submitted by mariners. Comments and recommendations on entering ports are needed so that corrections to these publications can be made.

If extra copies of local port plans, diagrams, regulations, brochures, photographs, etc. can be obtained, send them to NIMA. It is not essential that they be printed in English. Local pilots, customs officials, company agents, etc., are usually good information sources.

The following list may be used as a check-off list when submitting a letter report:

General

1. Name of the port
2. Date of observation and report
3. Name and type of vessel
4. Gross tonnage
5. Length (overall)
6. Breadth (extreme)
7. Draft (fore and aft)
8. Name of captain and observer
9. U.S. mailing address for acknowledgment

Tugs and Locks

1. Are tugs available or obligatory? What is their power?
2. If there are locks, what is the maximum size and draft of a vessel that can be locked through?

Cargo Handling Facilities

1. What are the capacities of the largest stationary, mobile, and floating cranes available? How was this information obtained?
2. What are the capacities, types, and number of lighters and barges available?
3. Is special cargo handling equipment available (e.g. grain elevators, coal and ore loaders, fruit or sugar conveyors, etc.)?
4. If cargo is handled from anchorage, what methods are used? Where is the cargo loaded? Are storage facilities available there?

Supplies

1. Are fuel oils, diesel oils, and lubricating oils available? If so, in what quantity?

Berths

1. What are the dimensions of the pier, wharf, or basin used?
2. What are the depths alongside? How were they

obtained?
3. Describe berth or berths for working containers or roll-on/roll-off cargo.
4. Does the port have berth for working deep draft tankers? If so, describe.
5. Are both dry and refrigerated storage available?
6. Are any unusual methods used when docking? Are special precautions necessary at berth?

Medical, Consular, and Other Services

1. Is there a hospital or the services of a doctor and dentist available?
2. Is there a United States consulate? Where is it located? If none, where is the nearest?

Anchorages

1. What are the limits of the anchorage areas?
2. In what areas is anchoring prohibited?
3. What is the depth, character of the bottom, types of holding ground, and swinging room available?
4. What are the effects of weather, sea, swell, tides, and currents on the anchorages?
5. Where is the special quarantine anchorage?
6. Are there any unusual anchoring restrictions?

Repairs and Salvage

1. What are the capacities of drydocks and marine railways, if available?
2. What repair facilities are available? Are there repair facilities for electrical and electronic equipment?
3. Are divers and diving gear available?
4. Are there salvage tugs available? What is the size and operating radius?
5. Are any special services (e.g. compass compensation or degaussing) available?

MISCELLANEOUS HYDROGRAPHIC REPORTS

2923. Ocean Current Reports

The set and drift of ocean currents are of great concern to the navigator. Only with the correct current information can the shortest and most efficient voyages be planned. As with all forces of nature, most currents vary considerably with time at a given location. Therefore, it is imperative that NIMA receive ocean current reports on a continuous basis.

The general surface currents along the principal trade routes of the world are well known. However, in other less traveled areas the current has not been well defined because of a lack of information. Detailed current reports from these areas are especially valuable.

An urgent need exists for more inshore current reports along all coasts of the world because data is scarce. Furthermore, information from deep draft ships is needed as this type of vessel is significantly influenced by the deeper layer of surface currents.

The CURRENT REPORT form, NAVOCEANO 3141/6, is designed to facilitate passing information to NAVOCEANO so that all mariners may benefit. The form is self-explanatory and can be used for ocean or coastal current information. Reports by the navigator will contribute significantly to accurate current information for nautical charts, current atlases, *Pilot Charts*, *Sailing Directions* and other special charts and publications.

2924. Route Reports

Route Reports enable NIMA, through its *Sailing Directions (Planning Guides)*, to make recommendations for ocean passages based upon the actual experience of mariners. Of particular importance are reports of routes used by very large ships and from any ship in regions where, from experience and familiarity with local conditions, mariners have devised routes that differ from the "preferred track." In addition, because of the many and varied local conditions which must be taken into account, coastal route information is urgently needed for updating both *Sailing Directions* and *Coast Pilots*.

A Route Report should include a comprehensive summary of the voyage with reference to currents, dangers, weather, and the draft of the vessel. If possible, each report should answer the following questions and should include any other data that may be considered pertinent to the particular route. All information should be given in sufficient detail to assure accurate conclusions and appropriate recommendations. Some questions to be answered are:

1. Why was the route selected?
2. Were anticipated conditions met during the voyage?

CHAPTER 30

THE OCEANS

INTRODUCTION

3000. The Importance of Oceanography

Oceanography is the scientific study of the oceans. It includes a study of their physical, chemical, and geological forms, and biological features. Thus, it embraces the widely separated fields of geography, geology, chemistry, physics, and biology, along with their many subdivisions, such as sedimentation, ecology, bacteriology, biochemistry, hydrodynamics, acoustics, and optics.

The oceans cover 70.8 percent of the surface of the Earth. The Atlantic covers 16.2 percent, the Pacific 32.4 percent (3.2 percent more than the land area of the entire Earth), the Indian Ocean 14.4 percent, and marginal and adjacent areas (of which the largest is the Arctic Ocean) 7.8 percent. Their extent alone makes them an important subject for study. However, greater incentive lies in their use for transportation, their influence upon weather and climate, and their potential as a source of power, food, fresh water, minerals, and organic substances.

3001. Origin of the Oceans

The structure of the continents is fundamentally different from that of the oceans. The rocks underlying the ocean floors are more dense than those underlying the continents. According to one theory, all the Earth's crust floats on a central liquid core, and the portions that make up the continents, being lighter, float with a higher freeboard. Thus, the thinner areas, composed of heavier rock, form natural basins where water has collected.

The shape of the oceans is constantly changing due to continental drift. The surface of the Earth consists of many different "**plates**." These plates are joined along **fracture** or **fault lines**. There is constant and measurable movement of these plates at rates of 0.02 meters per year or more.

The origin of the water in the oceans is unclear. Although some geologists have postulated that all the water existed as vapor in the atmosphere of the primeval Earth, and that it fell in great torrents of rain as soon as the Earth cooled sufficiently, another school holds that the atmosphere of the original hot Earth was lost, and that the water gradually accumulated as it was given off in steam by volcanoes, or worked to the surface in hot springs.

Most of the water on the Earth's crust is now in the oceans–about 1,370,000,000 cubic kilometers, or about 85 percent of the total. The mean depth of the ocean is 3,795 meters, and the total area is 360,000,000 square kilometers.

CHEMISTRY OF THE OCEANS

3002. Chemical Description

Oceanographic chemistry may be divided into three main parts: the chemistry of (1) seawater, (2) marine sediments, and (3) organisms living in the sea. The first is of particular interest to the navigator.

Chemical properties of seawater are usually determined by analyzing samples of water obtained at various locations and depths. Samples of water from below the surface are obtained with special bottles designed for this purpose. The open bottles are mounted in a rosette which is attached to the end of a wire cable which contains insulated electrical wires. The rosette is lowered to the depth of the deepest sample, and a bottle is closed electronically. As the rosette is raised to the surface, other bottles are closed at the desired depths. Sensors have also been developed to measure a few chemical properties of sea water continuously.

Physical properties of seawater are dependent primarily upon salinity, temperature, and pressure. However, factors like motion of the water and the amount of suspended matter affect such properties as color and transparency, conduction of heat, absorption of radiation, etc.

3003. Salinity

Salinity is a measure of the amount of dissolved solid material in the water. It has been defined as the total amount of solid material in grams contained in one kilogram of seawater when carbonate has been converted to oxide, bromine and iodine replaced by chlorine, and all organic material completely oxidized. It is usually expressed as parts per thousand (by weight), for example the average salinity of sea water is 35 grams per kilogram which would be written "35 ppt" or "35 ‰". Historically the determination of salinity was a slow and difficult process, while the amount of chlorine ions (plus the chlorine equivalent of the bromine and iodine), called **chlorinity**, could be determined easily and accurately by titration with silver nitrate. From chlorinity, the salinity was determined by a relation based upon the measured ratio of chlorinity to total dissolved substances:

$$\text{Salinity} = 1.80655 \times \text{Chlorinity}$$

This is now called the absolute salinity, (S_A). With titration techniques, salinity could be determined to about 0.02 parts per thousand.

This definition of salinity has now been replaced by the **Practical Salinity Scale**, (S). Using this scale, the salinity of a seawater sample is defined as the ratio between the conductivity of the sample and the conductivity of a standard potassium chloride (KCl) sample.

As salinity on the practical scale is defined to be conservative with respect to addition and removal of water, the entire salinity range is accessible through precise weight dilution or evaporation without additional definitions. Since practical salinity is a ratio, it has no physical units but is designated **practical salinity units**, or **psu**. The Practical Salinity Scale, combined with modern conductivity cells and bench salinometers, provides salinity measurements which are almost an order of magnitude more accurate and precise, about 0.003 psu, than titration. Numerically, absolute salinity and salinity are nearly equal.

It has also been found that electrical conductivity is better related to density than chlorinity. Since one of the main reasons to measure salinity is to deduce the density, this favors the Practical Salinity Scale as well.

Salinity generally varies between about 33 and 37 psu. However, when the water has been diluted, as near the mouth of a river or after a heavy rainfall, the salinity is somewhat less; and in areas of excessive evaporation, the salinity may be as high as 40 psu. In certain confined bodies of water, notably the Great Salt Lake in Utah, and the Dead Sea in Asia Minor, the salinity is several times this maximum.

3004. Temperature

Temperature in the ocean varies widely, both horizontally and with depth. Maximum values of about 32°C are encountered at the surface in the Persian Gulf in summer, and the lowest possible values of about –2°C (the usual minimum freezing point of seawater) occur in polar regions.

Except in the polar regions, the vertical distribution of temperature in the sea nearly everywhere shows a decrease of temperature with depth. Since colder water is denser (assuming the same salinity), it sinks below warmer water. This results in a temperature distribution just opposite to that of the Earth's crust, where temperature increases with depth below the surface of the ground.

In the sea there is usually a mixed layer of isothermal water below the surface, where the temperature is the same as that of the surface. This layer is caused by two physical processes: wind mixing, and convective overturning as surface water cools and becomes more dense. The layer is best developed in the Arctic and Antarctic regions, and in seas like the Baltic and Sea of Japan during the winter, where it may extend to the bottom of the ocean. In the Tropics, the wind-mixed layer may exist to a depth of 125 meters, and may exist throughout the year. Below this layer is a zone of rapid temperature decrease, called the **thermocline**. At a depth greater than 400 meters, the temperature everywhere is below 15°C. In the deeper layers, fed by cooled waters that have sunk from the surface in the Arctic and Antarctic, temperatures as low as –2°C exist.

In the colder regions the cooling creates the convective overturning and isothermal water in the winter; but in the summer a seasonal thermocline is created as the upper water becomes warmer. A typical curve of temperature at various depths is shown in Figure 3010a. Temperature is commonly measured with either a platinum or copper resistance thermometer or a thermistor (devices that measure the change in conductivity of a semiconductor with change in temperature).

The **CTD (conductivity-temperature-depth)** is an instrument that generates continuous signals as it is lowered into the ocean; temperature is determined by means of a platinum resistance thermometer, salinity by conductivity, and depth by pressure. These signals are transmitted to the surface through a cable and recorded. Accuracy of temperature measurement is 0.005°C and resolution an order of magnitude better.

A method commonly used to measure upper ocean temperature profiles from a vessel which is underway is the **expendable bathythermograph (XBT)**. The XBT uses a thermistor and is connected to the vessel by a fine wire. The wire is coiled inside the probe, and as the probe freefalls in the ocean, the wire pays out. Depth is determined by elapsed time and a known sink rate. Depth range is determined by the amount of wire stored in the probe; the most common model has a depth range of 450 meters. At the end of the drop, the wire breaks and the probe falls to the ocean bottom. One instrument of this type is dropped from an aircraft; the data is relayed to the aircraft from a buoy to which the wire of the XBT is attached. The accuracy and precision of an XBT is about 0.1°C.

3005. Pressure

The appropriate international standard (SI) unit for pressure in oceanography is 1 $kPa = 10^3\ Pa$ where Pa is a Pascal and is equal to one Newton per square meter. A more commonly used unit is a bar, which is nearly equal to 1 atmosphere (atmospheric pressure is measured with a barometer and may be read as hectopascals). Water pressure is expressed in terms of decibars, 10 of these being equal to 1 bar. One decibar is equal to nearly 1 $^1/_2$ pounds per square inch. This unit is convenient because it is very nearly the pressure exerted by 1 meter of water. Thus, the pressure in decibars is approximately the same as the depth in meters, the unit of depth.

Although virtually all of the physical properties of seawater are affected to a measurable extent by pressure, the effect is not as great as those of salinity and temperature. Pressure is of particular importance to submarines, directly because of the stress it induces on the hull and structures, and indirectly because of its effect upon buoyancy.

3006. Density

Density is mass per unit of volume. The appropriate SI unit is kilograms per cubic meter. The density of seawater depends upon salinity, temperature, and pressure. At constant temperature and pressure, density varies with salinity. A temperature of 0°C and atmospheric pressure are considered standard for density determination. The effects of thermal expansion and compressibility are used to determine the density at other temperatures and pressures. Slight density changes at the surface generally do not affect the draft or trim of a ship, though a noticeable change may occur as a ship travels from salt to fresh water. But density changes at a particular subsurface pressure affect the buoyancy of submarines because they are ballasted to be neutrally buoyant. For oceanographers, density is important because of its relationship to ocean currents.

Open ocean values of density range from about 1,021 kilograms per cubic meter at the surface to about 1,070 kilograms per cubic meter at 10,000 meters depth. As a matter of convenience, it is usual in oceanography to define a density anomaly which is equal to the density minus 1,000 kilograms per cubic meter. Thus, when an oceanographer speaks of seawater with a density of 25 kilograms per cubic meter, the actual density is 1,025 kilograms per cubic meter.

The greatest changes in density of seawater occur at the surface, where the water is subject to influences not present at depths. At the surface, density is decreased by precipitation, run-off from land, melting ice, or heating. When the surface water becomes less dense, it tends to float on top of the more dense water below. There is little tendency for the water to mix, and so the condition is one of stability. The density of surface water is increased by evaporation, formation of sea ice, and by cooling. If the surface water becomes more dense than that below, convection currents cause vertical mixing. The more dense surface water sinks and mixes with less dense water below. The resultant layer of water is of intermediate density. This process continues until the density of the mixed layer becomes less than that of the water below. The convective circulation established as part of this process can create very deep uniform mixed layers.

If the surface water becomes sufficiently dense, it sinks all the way to the bottom. If this occurs in an area where horizontal flow is unobstructed, the water which has descended spreads to other regions, creating a dense bottom layer. Since the greatest increase in density occurs in polar regions, where the air is cold and great quantities of ice form, the cold, dense polar water sinks to the bottom and then spreads to lower latitudes. In the Arctic Ocean region, the cold, dense water is confined by the Bering Strait and the underwater ridge from Greenland to Iceland to Europe. In the Antarctic, however, there are no similar geographic restrictions and large quantities of very cold, dense water formed there flow to the north along the ocean bottom. This process has continued for a sufficiently long period of time that the entire ocean floor is covered with this dense water, thus explaining the layer of cold water at great depths in all the oceans.

In some respects, oceanographic processes are similar to those occurring in the atmosphere. Masses of water of uniform characteristics are analogous to air masses.

3007. Compressibility

Seawater is nearly incompressible, its coefficient of compressibility being only 0.000046 per bar under standard conditions. This value changes slightly with changes in temperature or salinity. The effect of compression is to force the molecules of the substance closer together, causing it to become more dense. Even though the compressibility is low, its total effect is considerable because of the amount of water involved. If the compressibility of seawater were zero, sea level would be about 90 feet higher than it is now.

Compressibility is inversely proportional to temperature, i.e., cold water is more compressible than warm water. Waters which flow into the North Atlantic from the Mediterranean and Greenland Seas are equal in density, but because the water from the Greenland Sea is colder, it is more compressible and therefore becomes denser at depth. These waters from the Greenland Sea are therefore found beneath those waters which derive their properties from the Mediterranean.

3008. Viscosity

Viscosity is resistance to flow. Seawater is slightly more viscous than freshwater. Its viscosity increases with greater salinity, but the effect is not nearly as marked as that occurring with decreasing temperature. The rate is not uniform, becoming greater as the temperature decreases. Because of the effect of temperature upon viscosity, an incompressible object might sink at a faster rate in warm surface water than in colder water below. However, for most objects, this effect may be more than offset by the compressibility of the object.

The actual relationships existing in the ocean are considerably more complex than indicated by the simple explanation here, because of turbulent motion within the sea. The disturbing effect is called **eddy viscosity**.

3009. Specific Heat

Specific Heat is the amount of heat required to raise the temperature of a unit mass of a substance a stated amount. In oceanography, specific heat is stated, in SI units, as the number of Joules needed to raise 1 kilogram of a given substance 1°C. Specific heat at constant pressure is usually the quantity desired when liquids are involved, but occasionally the specific heat at constant volume is required. The ratio of these two quantities is directly related to the speed of sound in seawater.

The specific heat of seawater decreases slightly as salinity increases. However, it is much greater than that of land. The ocean is a giant storage area for heat. It can absorb large quantities of heat with very little change in temperature. This is partly due to the high specific heat of water and partly due to mixing in the ocean that distributes the heat throughout a layer. Land has a lower specific heat and, in addition, all heat is lost or gained from a thin layer at the surface; there is no mixing. This accounts for the greater temperature range of land and the atmosphere above it, resulting in monsoons, and the familiar land and sea breezes of tropical and temperate regions.

3010. Sound Speed

The speed of sound in sea water is a function of its density, compressibility and, to a minor extent, the ratio of specific heat at constant pressure to that at constant volume. As these properties depend on the temperature, salinity and pressure (depth) of sea water, it is customary to relate the speed of sound directly to the water temperature, salinity and pressure. An increase in any of these three properties causes an increase in the sound speed; the converse is true also. Figure 3010a portrays typical mid-ocean profiles of temperature and salinity; the resultant sound speed profile is shown in Figure 3010b.

The speed of sound changes by 3 to 5 meters per second per °C temperature change, by about 1.3 meters per second per psu salinity change and by about 1.7 meters per second per 100 m depth change. A simplified formula adapted from Wilson's (1960) equation for the computation of the sound speed in sea water is:

$$U = 1449 + 4.6T - 0.055T^2 + 0.0003T^3 + 1.39(S - 35) + 0.017D$$

where U is the speed (m/s), T is the temperature (°C), S is the salinity (psu), and D is depth (m).

3011. Thermal Expansion

One of the more interesting differences between salt and fresh water relates to thermal expansion. Saltwater continues to become more dense as it cools to the freezing point; freshwater reaches maximum density at 4°C and then expands (becomes less dense) as the water cools to 0°C and freezes. This means that the convective

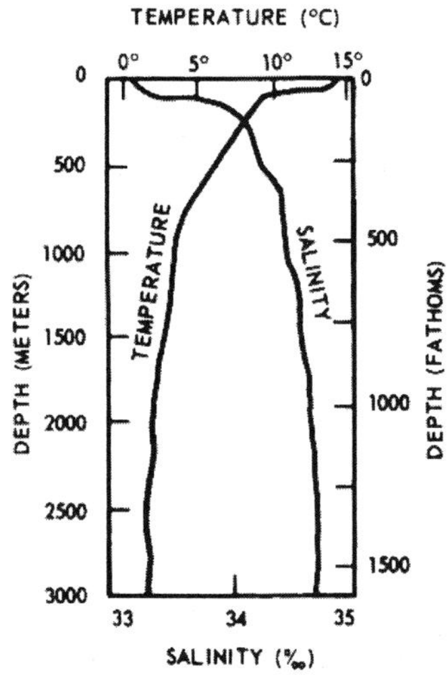

Figure 3010a. Typical variation of temperature and salinity with depth for a mid-latitude location.

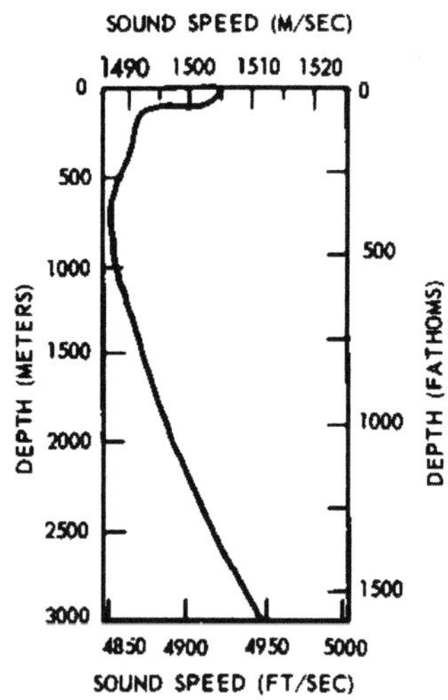

Figure 3010b. Resultant sound speed profile based on the temperature and salinity profile in Figure 3010a.

mixing of freshwater stops at 4°C; freezing proceeds very rapidly beyond that point. The rate of expansion with increased temperature is greater in seawater than in fresh water. Thus, at temperature 15°C, and atmospheric pressure, the coefficient of thermal expansion is 0.000151 per degree Celsius for freshwater, and 0.000214 per degree Celsius for average seawater. The coefficient of thermal expansion increases not only with greater salinity, but also with increased temperature and pressure. At a salinity of 35 psu, the coefficient of surface water increases from 0.000051 per degree Celsius at 0°C to 0.000334 per degree Celsius at 31°C. At a constant temperature of 0°C and a salinity of 34.85 psu, the coefficient increases to 0.000276 per degree Celsius at a pressure of 10,000 decibars (a depth of approximately 10,000 meters).

3012. Thermal Conductivity

In water, as in other substances, one method of heat transfer is by conduction. Freshwater is a poor conductor of heat, having a coefficient of thermal conductivity of 582 Joules per second per meter per degree Celsius. For seawater it is slightly less, but increases with greater temperature or pressure.

However, if turbulence is present, which it nearly always is to some extent, the processes of heat transfer are altered. The effect of turbulence is to increase greatly the rate of heat transfer. The "eddy" coefficient used in place of the still-water coefficient is so many times larger, and so dependent upon the degree of turbulence, that the effects of temperature and pressure are not important.

3013. Electrical Conductivity

Water without impurities is a very poor conductor of electricity. However, when salt is in solution in water, the salt molecules are ionized and become carriers of electricity. (What is commonly called freshwater has many impurities and is a good conductor of electricity; only pure distilled water is a poor conductor.) Hence, the electrical conductivity of seawater is directly proportional to the number of salt molecules in the water. For any given salinity, the conductivity increases with an increase in temperature.

3014. Radioactivity

Although the amount of radioactive material in seawater is very small, this material is present in marine sediments to a greater extent than in the rocks of the Earth's crust. This is probably due to precipitation of radium or other radioactive material from the water. The radioactivity of the top layers of sediment is less than that of deeper layers. This may be due to absorption of radioactive material in the soft tissues of marine organisms.

3015. Transparency

The two basic processes that alter the underwater distribution of light are absorption and scattering. Absorption is a change of light energy into other forms of energy; scattering entails a change in direction of the light, but without loss of energy. If seawater were purely absorbing, the loss of light with distance would be given by $I_x = I_0 e^{-ax}$ where I_x is the intensity of light at distance x, I_0 is the intensity of light at the source, and "a" is the absorption coefficient in the same units with which distance is measured. In a pure scattering medium, the transmission of light is governed by the same power law only in this case the exponential term is $I_0 e^{-bx}$, where "b" is the volume scattering coefficient. The attenuation of light in the ocean is defined as the sum of absorption and scattering so that the attenuation coefficient, c, is given by c = a + b. In the ocean, the attenuation of light with depth depends not only on the wavelength of the light but also the clarity of the water. The clarity is mostly controlled by biological activity although at the coast, sediments transported by rivers or resuspended by wave action can strongly attenuate light.

Attenuation in the sea is measured with a **transmissometer**. Transmissometers measure the attenuation of light over a fixed distance using a monochromatic light source which is close to red in color. Transmissometers are designed for in situ use and are usually attached to a CTD.

Since sunlight is critical for almost all forms of plant life in the ocean, oceanographers developed a simple method to measure the penetration of sunlight in the sea using a white disk 31 centimeters (a little less than 1 foot) in diameter which is called a **Secchi disk**. This is lowered into the sea, and the depth at which it disappears is recorded. In coastal waters the depth varies from about 5 to 25 meters. Offshore, the depth is usually about 45 to 60 meters. The greatest recorded depth at which the disk has disappeared is 79 meters in the eastern Weddell Sea. These depths, D, are sometimes reported as a diffuse attenuation (or "extinction") coefficient, k, where k = 1.7/D and the penetration of sunlight is given by $I_z = I_0 e^{-kz}$ where z is depth and I_0 is the energy of the sunlight at the ocean's surface.

3016. Color

The color of seawater varies considerably. Water of the Gulf Stream is a deep indigo blue, while a similar current off Japan was named Kuroshio (Black Stream) because of the dark color of its water. Along many coasts the water is green. In certain localities a brown or brownish-red water has been observed. Colors other than blue are caused by biological sources, such as plankton, or by suspended sediments from river runoff.

Offshore, some shade of blue is common, particularly in tropical or subtropical regions. It is due to scattering of sunlight by minute particles suspended in the water, or by molecules of the water itself. Because of its short wavelength, blue light is more effectively scattered than light of longer waves. Thus, the ocean appears blue for the same reason that the sky does. The green color often seen near the coast is a mixture of the blue due to scattering of light and a stable soluble yellow pigment associated with phytoplankton. Brown or brownish-red water receives its color from large quantities of certain types of **algae**, microscopic plants in the sea, or from river runoff.

3017. Bottom Relief

Compared to land, relatively little is known of relief below the surface of the sea. The development of an effective echo sounder in 1922 greatly simplified the determination of bottom depth. Later, a recording echo sounder was developed to permit the continuous tracing of a bottom profile. The latest sounding systems employ an array of echosounders aboard a single vessel, which continuously sound a wide swath of ocean floor. This has contributed immensely to our knowledge of bottom relief. By this means, many undersea mountain ranges, volcanoes, rift valleys, and other features have been discovered.

Along most of the coasts of the continents, the bottom slopes gradually downward to a depth of about 130 meters or somewhat less, where it falls away more rapidly to greater depths. This **continental shelf** averages about 65 kilometers in width, but varies from nothing to about 1400 kilometers, the widest part being off the Siberian Arctic coast. A similar shelf extending outward from an island or group of islands is called an **island shelf**. At the outer edge of the shelf, the steeper slope of 2° to 4° is called the **continental slope**, or the **island slope**, according to whether it surrounds a continent or a group of islands. The shelf itself is not uniform, but has numerous hills, ridges, terraces, and canyons, the largest being comparable in size to the Grand Canyon.

The relief of the ocean floor is comparable to that of land. Both have steep, rugged mountains, deep canyons, rolling hills, plains, etc. Most of the ocean floor is considered to be made up of a number of more-or-less circular or oval depressions called **basins**, surrounded by walls (**sills**) of lesser depth.

A wide variety of submarine features has been identified and defined. Some of these are shown in Figure 3017. Detailed definitions and descriptions of such features can be found in Kennett (1982) or Fairbridge (1966). The term **deep** may be used for a very deep part of the ocean, generally that part deeper than 6,000 meters.

The average depth of water in the oceans is 3795 meters (2,075 fathoms), as compared to an average height of land above the sea of about 840 meters. The greatest known depth is 11,524 meters, in the Marianas Trench in the Pacific. The highest known land is Mount Everest, 8,840 meters. About 23 percent of the ocean is shallower than 3,000 meters, about 76 percent is between 3,000 and 6,000 meters, and a little more than 1 percent is deeper than 6,000 meters.

3018. Marine Sediments

The ocean floor is composed of material deposited through the ages. This material consists principally of (1) earth and rocks washed into the sea by streams and waves, (2) volcanic ashes and lava, and (3) the remains of marine organisms. Lesser amounts of land material are carried into the sea by glaciers, blown out to sea by wind, or deposited by chemical means. This latter process is responsible for the **manganese nodules** that cover some parts of the ocean floor. In the ocean, the material is transported by ocean currents, waves, and ice. Near shore the material is deposited at the rate of about 8 centimeters in 1,000 years, while in the deep water offshore the rate is only about 1 centimeter in 1,000 years. Marine deposits in water deep enough to be relatively free from wave action are subject to little erosion. Recent studies have shown that some bottom currents are strong enough to move sediments. There are **turbidity currents**, similar to land slides, that move large masses of sediments. Turbidity currents have been known to rip apart large transoceanic cables on the ocean bottom. Because of this and the slow rate of deposit, marine sediments provide a better geological record than does the land.

Marine sediments are composed of individual particles of all sizes from the finest clay to large boulders. In general, the inorganic deposits near shore are relatively coarse (sand, gravel, shingle, etc.), while those in deep water are much finer (clay). In some areas the siliceous remains of marine organisms or calcareous deposits of either organic or inorganic origin predominate on the ocean floor.

A wide range of colors is found in marine sediments. The lighter colors (white or a pale tint) are usually associated with coarse-grained quartz or limestone deposits. Darker colors (red, blue, green, etc.) are usually found in mud having a predominance of some mineral substance, such as an oxide of iron or manganese. Black mud is often found in an area that is little disturbed, such as at the bottom of an inlet or in a depression without free access to other areas.

Marine sediments are studied primarily through bottom samples. Samples of surface deposits are obtained by means of a "snapper" (for mud, sand, etc.) or "dredge" (usually for rocky material). If a sample of material below the bottom surface is desired, a "coring" device is used. This device consists essentially of a tube driven into the bottom by weights or explosives. A sample obtained in this way preserves the natural order of the various layers. Samples of more than 100 feet in depth have been obtained using coring devices.

3019. Satellite Oceanography

Weather satellites are able to observe ocean surface temperatures in cloud free regions by using infrared sensors. Although these sensors are only able to penetrate a few

Figure 3017. Ocean basin features.

millimeters into the ocean, the temperatures that they yield are representative of upper ocean conditions except when the air is absolutely calm during daylight hours. For cloud covered regions, it is usually possible to wait a few days for the passage of a cold front and then use a sequence of infrared images to map the ocean temperature over a region. The patterns of warm and cold water yield information on ocean currents, the existence of fronts and eddies, and the temporal and spatial scales of ocean processes.

Other satellite sensors are capable of measuring ocean color, ice coverage, ice age, ice edge, surface winds and seas, ocean currents, and the shape of the surface of the ocean. (The latter is controlled by gravity and ocean circulation patterns. See Chapter 2.) The perspective provided by these satellites is a global one and in some cases they yield sufficient quantities of data that synoptic charts of the ocean surface, similar to weather maps and pilot charts, can be provided to the mariner for use in navigation.

The accuracy of satellite observations of the ocean surface depends, in many cases, on calibration procedures which use observations of sea surface conditions provided by mariners. These observations include marine weather observations, expendable bathythermograph soundings, and currents measured by electromagnetic logs or acoustic Doppler current profilers. Care and diligence in these observations will improve the accuracy and the quality of satellite data.

3020. Synoptic Oceanography

Oceanographic data provided by ships, buoys, and satellites are analyzed by the Naval Oceanographic Office and the National Meteorological Center. These data are utilized in computer models both to provide a synoptic view of ocean conditions and to predict how these conditions will change in the future. These products are available to the mariner via radio or satellite.

CHAPTER 31

OCEAN CURRENTS

TYPES AND CAUSES OF CURRENTS

3100. Definitions

The movement of ocean water is one of the two principal sources of discrepancy between dead reckoned and actual positions of vessels. Water in motion is called a current; the direction toward which it moves is called **set**, and its speed is called **drift**. Modern shipping speeds have lessened the impact of currents on a typical voyage, and since electronic navigation allows continuous adjustment of course, there is less need to estimate current set and drift before setting the course to be steered. Nevertheless, a knowledge of ocean currents can be used in cruise planning to reduce transit times, and current models are an integral part of ship routing systems.

Oceanographers have developed a number of methods of classifying currents in order to facilitate descriptions of their physics and geography. Currents may be referred to according to their forcing mechanism as either **wind driven** or **thermohaline**. Alternatively, they may be classified according to their depth (surface, intermediate, deep or bottom). The surface circulation of the world's oceans is mostly wind driven. Thermohaline currents are driven by differences in heat and salt and are associated with the sinking of dense water at high latitudes; the currents driven by thermohaline forces are typically subsurface. Note that this classification scheme is not unambiguous; the circumpolar current, which is wind driven, extends from the surface to the bottom.

A **periodic current** is one for which the speed or direction changes cyclically at somewhat regular intervals, such as a tidal current. A **seasonal current** is one which changes in speed or direction due to seasonal winds. The mean circulation of the ocean consists of semi-permanent currents which experience relatively little periodic or seasonal change.

A **coastal current** flows roughly parallel to a coast, outside the surf zone, while a **longshore current** is one parallel to a shore, inside the surf zone, generated by waves striking the beach at an angle. Any current some distance from the shore may be called an **offshore current**, and one close to the shore an **inshore current**.

General information on ocean currents is available from NOAA's National Ocean Data Center at: http://www.nodc.noaa.gov. Satellite graphics and other data can be found at: http://wwwo2c.nesdis.noaa.gov.

3101. Causes of Ocean Currents

The primary generating forces are wind and differences in water density caused by variations in heat and salinity Currents generated by these forces are modified by such factors as depth of water, underwater topography including shape of the basin in which the current is running, extent and location of land, and deflection by the rotation of the Earth.

3102. Wind Driven Currents

The stress of wind blowing across the sea causes a surface layer of water to move. Due to the low viscosity of water, this stress is not directly communicated to the ocean interior, but is balanced by the Coriolis force within a relatively thin surface layer, 10-200m thick. This layer is called the **Ekman layer** and the motion of this layer is called the **Ekman transport**. Because of the deflection by the Coriolis force, the Ekman transport is not in the direction of the wind, but is 90° to the right in the Northern Hemisphere and 90° toward the left in the Southern Hemisphere. The amount of water flowing in this layer depends only upon the wind and the Coriolis force and is independent of the depth of the Ekman layer and the viscosity of the water.

The large scale convergence or divergence of Ekman transport serves to drive the general ocean circulation. Consider the case of the Northern Hemisphere subtropics. To the south lie easterly winds with associated northward Ekman transport. To the north lie westerly winds with southward Ekman transport. The convergence of these Ekman transports is called **Ekman pumping** and results in a thickening of the upper ocean and a increase in the depth of the thermocline. The resulting subsurface pressure gradients, balanced by the Coriolis force, give rise to the anticyclonic subtropical gyres found at mid latitudes in each ocean basin. In subpolar regions, Ekman suction produces cyclonic gyres.

These wind driven gyres are not symmetrical. Along the western boundary of the oceans, currents are narrower, stronger, and deeper, often following a meandering course. These currents are sometimes called a **stream**. In contrast, currents in mid-ocean and at the eastern boundary, are often broad, shallow and slow-moving. Sometimes these are called **drift currents**.

Within the Ekman layer, the currents actually form a

spiral. At the surface, the difference between wind direction and surface wind-current direction varies from about 15° along shallow coastal areas to a maximum of 45° in the deep oceans. As the motion is transmitted to successively deep layers, the Coriolis force continues to deflect the current. At the bottom of the Ekman layer, the current flows in the opposite direction to the surface current. This shift of current directions with depth, combined with the decrease in velocity with depth, is called the **Ekman spiral**.

The velocity of the surface current is the sum of the velocities of the Ekman, geostrophic, tidal, and other currents. The Ekman surface current or wind drift current depends upon the speed of the wind, its constancy, the length of time it has blown, and other factors. In general, however, wind drift current is about 2 percent of the wind speed, or a little less, in deep water where the wind has been blowing steadily for at least 12 hours.

3103. Currents Related to Density Differences

The density of water varies with salinity, temperature, and pressure. At any given depth, the differences in density are due only to differences in temperature and salinity. With sufficient data, maps showing geographical density distribution at a certain depth can be drawn, with lines connecting points of equal density. These lines would be similar to isobars on a weather map and serve an analogous purpose, showing areas of high density and those of low density. In an area of high density, the water surface is lower than in an area of low density, the maximum difference in height being about 1 meter in 100 km. Because of this difference, water tends to flow from an area of higher water (low density) to one of lower water (high density). But due to rotation of the Earth, it is deflected by the Coriolis force or toward the right in the Northern Hemisphere, and toward the left in the Southern Hemisphere. This balance, between subsurface pressure fields and the Coriolis force, is called **geostrophic equilibrium**. At a given latitude, the greater the density gradient (rate of change with distance), the faster the geostrophic current.

OCEANIC CIRCULATION

3104. Introduction

A number of ocean currents flow with great persistence, setting up a circulation that continues with relatively little change throughout the year. Because of the influence of wind in creating current, there is a relationship between this oceanic circulation and the general circulation of the atmosphere. The oceanic circulation is shown on the chart following this page (winter N. hemisphere), with the names of the major ocean currents. Some differences in opinion exist regarding the names and limits of some of the currents, but those shown are representative. Speed may vary somewhat with the season. This is particularly noticeable in the Indian Ocean and along the South China coast, where currents are influenced to a marked degree by the monsoons.

3105. Southern Ocean Currents

The Southern Ocean has no meridional boundaries and its waters are free to circulate around the world. It serves as a conveyor belt for the other oceans, exchanging waters between them. The northern boundary of the Southern Ocean is marked by the Subtropical Convergence zone. This zone marks the transition from the temperate region of the ocean to the polar region and is associated with the surfacing of the main thermocline. This zone is typically found at 40°S but varies with longitude and season.

In the Antarctic, the circulation is generally from west to east in a broad, slow-moving current extending completely around Antarctica. This is called the **Antarctic Circumpolar Current** or the **West Wind Drift**, and it is formed partly by the strong westerly wind in this area, and partly by density differences. This current is augmented by the Brazil and Falkland Currents in the Atlantic, the East Australia Current in the Pacific, and the Agulhas Current in the Indian Ocean. In return, part of it curves northward to form the Cape Horn, Falkland, and most of the Benguela Currents in the Atlantic, and the Peru Current in the Pacific.

In a narrow zone next to the Antarctic continent, a westward flowing coastal current is usually found. This current is called the **East Wind Drift** because it is attributed to the prevailing easterly winds which occur there.

3106. Atlantic Ocean Currents

The trade winds set up a system of equatorial currents which at times extends over as much as 50° of latitude or more. There are two westerly flowing currents conforming generally with the areas of trade winds, separated by a weaker, easterly flowing countercurrent.

The **North Equatorial Current** originates to the northward of the Cape Verde Islands and flows almost due west at an average speed of about 0.7 knot.

The **South Equatorial Current** is more extensive. It starts off the west coast of Africa, south of the Gulf of Guinea, and flows in a generally westerly direction at an average speed of about 0.6 knot. However, the speed gradually increases until it may reach a value of 2.5 knots, or more, off the east coast of South America. As the current approaches Cabo de Sao Roque, the eastern extremity of South America, it divides, the southern part curving toward the south along the coast of Brazil, and the

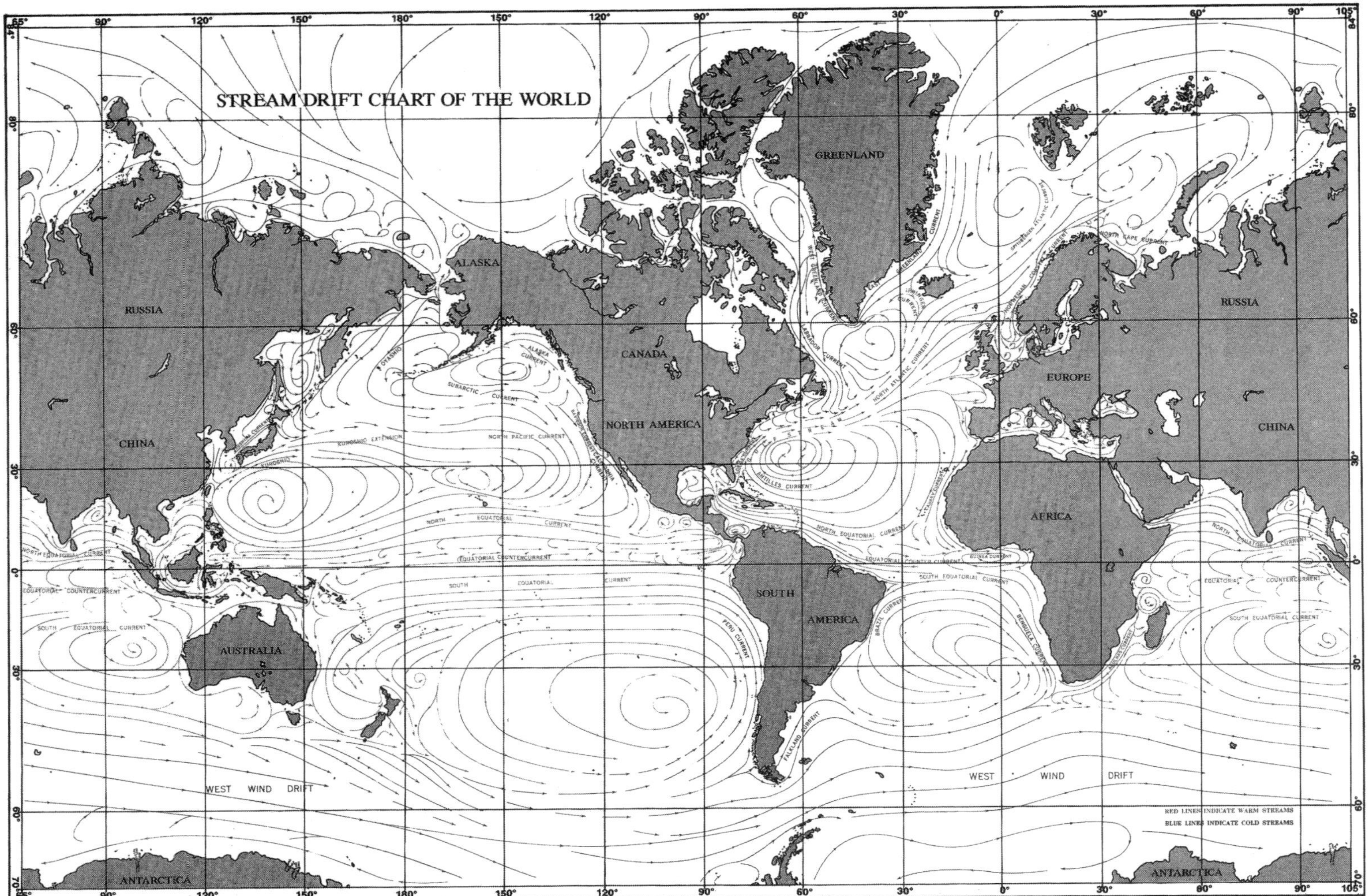
STREAM DRIFT CHART OF THE WORLD
GREENLAND
ALASKA
RUSSIA
CANADA
NORTH AMERICA
CHINA
EUROPE
AFRICA
SOUTH AMERICA
AUSTRALIA
ANTARCTICA
OYASHIO
KUROSHIO
KUROSHIO EXTENSION
ALASKA CURRENT
SUBARCTIC CURRENT
NORTH PACIFIC CURRENT
NORTH EQUATORIAL CURRENT
EQUATORIAL COUNTERCURRENT
SOUTH EQUATORIAL CURRENT
PERU CURRENT
LABRADOR CURRENT
WEST GREENLAND CURRENT
EAST GREENLAND CURRENT
IRMINGER CURRENT
NORTH ATLANTIC CURRENT
NORWEGIAN CURRENT
NORTH CAPE CURRENT
ANTILLES CURRENT
BRAZIL CURRENT
FALKLAND CURRENT
BENGUELA CURRENT
GUINEA CURRENT
WEST WIND DRIFT
RED LINES INDICATE WARM STREAMS
BLUE LINES INDICATE COLD STREAMS

northern part being deflected northward by the continent of South America.

Between the North and South Equatorial Currents, the weaker **North Equatorial Countercurrent** sets toward the east in the general vicinity of the doldrums. This is fed by water from the two westerly flowing equatorial currents, particularly the South Equatorial Current. The extent and strength of the Equatorial Countercurrent changes with the seasonal variations of the wind. It reaches a maximum during July and August, when it extends from about 50° west longitude to the Gulf of Guinea. During its minimum, in December and January, it is of very limited extent, the western portion disappearing altogether.

That part of the South Equatorial Current flowing along the northern coast of South America which does not feed the Equatorial Countercurrent unites with the North Equatorial Current at a point west of the Equatorial Countercurrent. A large part of the combined current flows through various passages between the Windward Islands and into the Caribbean Sea. It sets toward the west, and then somewhat north of west, finally arriving off the Yucatan peninsula. From there, the water enters the Gulf of Mexico and forms the **Loop Current**; the path of the Loop Current is variable with a 13-month period. It begins by flowing directly from Yucatan to the Florida Straits, but gradually grows to flow anticyclonically around the entire Eastern Gulf; it then collapses, again following the direct path from Yucatan to the Florida Straits, with the loop in the Eastern Gulf becoming a separate eddy which slowly flows into the Western Gulf.

Within the Straits of Florida, the Loop Current feeds the beginnings of the most remarkable of American ocean currents, the **Gulf Stream**. Off the southeast coast of Florida this current is augmented by the **Antilles Current** which flows along the northern coasts of Puerto Rico, Hispaniola, and Cuba. Another current flowing eastward of the Bahamas joins the stream north of these islands.

The Gulf Stream follows generally along the east coast of North America, flowing around Florida, northward and then northeastward toward Cape Hatteras, and then curving toward the east and becoming broader and slower. After passing the Grand Banks, it turns more toward the north and becomes a broad drift current flowing across the North Atlantic. The part in the Straits of Florida is sometimes called the **Florida Current**.

A tremendous volume of water flows northward in the Gulf Stream. It can be distinguished by its deep indigo-blue color, which contrasts sharply with the dull green of the surrounding water. It is accompanied by frequent squalls. When the Gulf Stream encounters the cold water of the Labrador Current, principally in the vicinity of the Grand Banks, there is little mixing of the waters. Instead, the junction is marked by a sharp change in temperature. The line or surface along which this occurs is called the **cold wall**. When the warm Gulf Stream water encounters cold air, evaporation is so rapid that the rising vapor may be visible as frost smoke.

Investigations have shown that the current itself is much narrower and faster than previously supposed, and considerably more variable in its position and speed. The maximum current off Florida ranges from about 2 to 4 knots. Northward, the speed is generally less, and it decreases further after the current passes Cape Hatteras. As the stream meanders and shifts position, eddies sometimes break off and continue as separate, circular flows until they dissipate. Boats in the Newport-Bermuda sailing yacht race have been known to be within sight of each other and be carried in opposite directions by different parts of the same current. This race is generally won by the boat which catches an eddy just right. As the current shifts position, its extent does not always coincide with the area of warm, blue water. When the sea is relatively smooth, the edges of the current are marked by ripples.

A recirculation region exists adjacent to and southeast of the Gulf Stream. The flow of water in the recirculation region is opposite to that in the Gulf Stream and surface currents are much weaker, generally less than half a knot.

As the Gulf Stream continues eastward and northeastward beyond the Grand Banks, it gradually widens and decreases speed until it becomes a vast, slow-moving current known as the **North Atlantic Current**, in the general vicinity of the prevailing westerlies. In the eastern part of the Atlantic it divides into the **Northeast Drift Current** and the **Southeast Drift Current**.

The Northeast Drift Current continues in a generally northeasterly direction toward the Norwegian Sea. As it does so, it continues to widen and decrease speed. South of Iceland it branches to form the **Irminger Current** and the **Norway Current**. The Irminger Current curves toward the north and northwest to join the East Greenland Current southwest of Iceland. The Norway Current continues in a northeasterly direction along the coast of Norway. Part of it, the **North Cape Current**, rounds North Cape into the Barents Sea. The other part curves toward the north and becomes known as the **Spitsbergen Current**. Before reaching Svalbard (Spitsbergen), it curves toward the west and joins the cold **East Greenland Current** flowing southward in the Greenland Sea. As this current flows past Iceland, it is further augmented by the Irminger Current.

Off Kap Farvel, at the southern tip of Greenland, the East Greenland Current curves sharply to the northwest following the coastline. As it does so, it becomes known as the **West Greenland Current**, and its character changes from that of an intense western boundary current to a weaker eastern boundary current. This current continues along the west coast of Greenland, through Davis Strait, and into Baffin Bay.

In Baffin Bay the West Greenland Current generally follows the coast, curving westward off Kap York to form the southerly flowing **Labrador Current**. This cold current flows southward off the coast of Baffin Island, through Davis Strait, along the coast of Labrador and Newfoundland, to the Grand Banks, carrying with it large quantities of ice. Here it encounters the warm water of the Gulf Stream, creating the cold wall. Some of the cold water flows southward along the east coast of North America, inshore of the Gulf Stream, as far as Cape Hatteras. The remainder curves toward the east and flows along the northern edge of the North Atlantic and Northeast Drift Currents, gradually merging with them.

The **Southeast Drift Current** curves toward the east, southeast, and then south as it is deflected by the coast of Europe. It flows past the Bay of Biscay, toward southeastern Europe and the Canary Islands, where it continues as the **Canary Current**. In the vicinity of the Cape Verde Islands, this current divides, part of it curving toward the west to help form the **North Equatorial Current**, and part of it curving toward the east to follow the coast of Africa into the Gulf of Guinea, where it is known as the **Guinea Current**. This current is augmented by the **North Equatorial Countercurrent** and, in summer, it is strengthened by monsoon winds. It flows in close proximity to the South Equatorial Current, but in the opposite direction. As it curves toward the south, still following the African coast, it merges with the South Equatorial Current.

The clockwise circulation of the North Atlantic leaves a large central area between the recirculation region and the Canary Current which has no well-defined currents. This area is known as the **Sargasso Sea**, from the large quantities of sargasso or gulfweed encountered there.

That branch of the South Equatorial Current which curves toward the south off the east coast of South America, follows the coast as the warm, highly-saline **Brazil Current**, which in some respects resembles a weak Gulf Stream. Off Uruguay it encounters the colder, less-salty **Falkland** or **Malvinas Current** forming a sharp meandering front in which eddies may form. The two currents curve toward the east to form the broad, slow-moving, **South Atlantic Current** in the general vicinity of the prevailing westerlies and the front dissipates somewhat. This current flows eastward to a point west of the Cape of Good Hope, where it curves northward to follow the west coast of Africa as the strong **Benguela Current**, augmented somewhat by part of the **Agulhas Current** flowing around the southern part of Africa from the Indian Ocean. As it continues northward, the current gradually widens and slows. At a point east of St. Helena Island it curves westward to continue as part of the South Equatorial Current, thus completing the counterclockwise circulation of the South Atlantic. The Benguela Current is also augmented somewhat by the West Wind Drift, a current which flows easterly around Antarctica. As the West Wind Drift flows past Cape Horn, that part in the immediate vicinity of the cape is called the **Cape Horn Current**. This current rounds the cape and flows in a northerly and northeasterly direction along the coast of South America as the Falkland or Malvinas Current.

3107. Pacific Ocean Currents

Pacific Ocean currents follow the general pattern of those in the Atlantic. The **North Equatorial Current** flows westward in the general area of the northeast trades, and the **South Equatorial Current** follows a similar path in the region of the southeast trades. Between these two, the weaker **North Equatorial Countercurrent** sets toward the east, just north of the equator.

After passing the Mariana Islands, the major part of the North Equatorial Current curves somewhat toward the northwest, past the Philippines and Taiwan. Here it is deflected further toward the north, where it becomes known as the **Kuroshio**, and then toward the northeast past the Nansei Shoto and Japan, and on in a more easterly direction. Part of the Kuroshio, called the **Tsushima Current**, flows through Tsushima Strait, between Japan and Korea, and the Sea of Japan, following generally the northwest coast of Japan. North of Japan it curves eastward and then southeastward to rejoin the main part of the Kuroshio. The limits and volume of the Kuroshio are influenced by the monsoons, being augmented during the season of southwesterly winds, and diminished when the northeasterly winds are prevalent.

The Kuroshio (Japanese for "Black Stream") is so named because of the dark color of its water. It is sometimes called the **Japan Current**. In many respects it is similar to the Gulf Stream of the Atlantic. Like that current, it carries large quantities of warm tropical water to higher latitudes, and then curves toward the east as a major part of the general clockwise circulation in the Northern Hemisphere. As it does so, it widens and slows, continuing on between the Aleutians and the Hawaiian Islands, where it becomes known as the **North Pacific Current**.

As this current approaches the North American continent, most of it is deflected toward the right to form a clockwise circulation between the west coast of North America and the Hawaiian Islands called the **California Current**. This part of the current has become so broad that the circulation is generally weak. Near the coast, the southeastward flow intensifies and average speeds are about 0.8 knot. But the flow pattern is complex, with offshore directed jets often found near more prominent capes, and poleward flow often found over the upper slope and outer continental shelf. It is strongest near land. Near the southern end of Baja California, this current curves sharply to the west and broadens to form the major portion of the North Equatorial Current.

During the winter, a weak countercurrent flows northwestward, inshore of the southeastward flowing California Current, along the west coast of North America from Baja California to Vancouver Island. This is called the **Davidson Current**.

Off the west coast of Mexico, south of Baja California the current flows southeastward during the winter as a continuation of part of the California Current. During the summer, the current in this area is northwestward as a continuation of the North Equatorial Countercurrent.

As in the Atlantic, there is in the Pacific a counterclockwise circulation to the north of the clockwise circulation. Cold water flowing southward through the western part of Bering Strait between Alaska and Siberia, is joined by water circulating counterclockwise in the Bering Sea to form the **Oyashio**. As the current leaves the strait, it curves toward the right and flows southwesterly along the coast of Siberia and the Kuril Islands. This current brings quantities of sea ice, but no icebergs. When it encounters the Kuroshio, the Oyashio curves southward and then eastward, the greater portion joining the Kuroshio and North Pacific Current.

The northern branch of the North Pacific Current curves in a counterclockwise direction to form the **Alaska Current**, which generally follows the coast of Canada and Alaska. When the Alaska Current turns to the southwest and flows along the Kodiak Island and the Alaska Peninsula, its character changes to that of a western boundary current and it is called the Alaska Stream. When this westward flow arrives off the Aleutian Islands, it is less intense and becomes known as the **Aleutian Current**. Part of it flows along the southern side of these islands to about the 180th meridian, where it curves in a counterclockwise direction and becomes an easterly flowing current, being augmented by the northern part of the Oyashio. The other part of the Aleutian Current flows through various openings between the Aleutian Islands, into the Bering Sea. Here it flows in a general counterclockwise direction. The southward flow along the Kamchatka peninsula is called the **Kamchatka Current** which feeds the southerly flowing Oyashio. Some water flows northward from the Bering Sea through the eastern side of the Bering Strait, into the Arctic Ocean.

The **South Equatorial Current**, extending in width between about 4°N latitude and 10°S, flows westward from South America to the western Pacific. After this current crosses the 180th meridian, the major part curves in a counterclockwise direction, entering the Coral Sea, and then curving more sharply toward the south along the east coast of Australia, where it is known as the **East Australian Current**. The East Australian Current is the weakest of the subtropical western boundary currents and separates from the Australian coast near 34°S. The path of the current from Australia to New Zealand is known as the **Tasman Front**, which marks the boundary between the warm water of the Coral Sea and the colder water of the Tasman Sea. The continuation of the East Australian Current east of New Zealand is the **East Auckland Current**. The East Auckland Current varies seasonally: in winter, it separates from the shelf and flows eastward, merging with the West Wind Drift, while in winter it follows the New Zealand shelf southward as the **East Cape Current** until it reaches Chatham Rise where it turns eastward, thence merging with the West Wind Drift.

Near the southern extremity of South America, most of this current flows eastward into the Atlantic, but part of it curves toward the left and flows generally northward along the west coast of South America as the **Peru Current** or **Humboldt Current**. Occasionally a set directly toward land is encountered. At about Cabo Blanco, where the coast falls away to the right, the current curves toward the left, past the Galapagos Islands, where it takes a westerly set and constitutes the major portion of the South Equatorial Current, thus completing the counterclockwise circulation of the South Pacific.

During the northern hemisphere summer, a weak northern branch of the South Equatorial Current, known as the **New Guinea Coastal Current**, continues on toward the west and northwest along both the southern and northeastern coasts of New Guinea. The southern part flows through Torres Strait, between New Guinea and Australia, into the Arafura Sea. Here, it gradually loses its identity, part of it flowing on toward the west as part of the South Equatorial Current of the Indian Ocean, and part of it following the coast of Australia and finally joining the easterly flowing West Wind Drift. The northern part of New Guinea Coastal Current both curves in a clockwise direction to help form the Pacific Equatorial Countercurrent and off Mindanao turns southward to form a southward flowing boundary current called the **Mindanao Current**. During the northern hemisphere winter, the New Guinea Coastal Current may reverse direction for a few months.

3108. Indian Ocean Currents

Indian Ocean currents follow generally the pattern of the Atlantic and Pacific but with differences caused principally by the monsoons, the more limited extent of water in the Northern Hemisphere, and by limited communication with the Pacific Ocean along the eastern boundary. During the northern hemisphere winter, the **North Equatorial Current** and **South Equatorial Current** flow toward the west, with the weaker, eastward **Equatorial Countercurrent** flowing between them, as in the Atlantic and Pacific (but somewhat south of the equator). But during the northern hemisphere summer, both the North Equatorial Current and the Equatorial Countercurrent are replaced by the **Southwest Monsoon Current**, which flows eastward and southeastward across the Arabian Sea and the Bay of Bengal. Near Sumatra, this current curves in a clockwise direction and flows westward, augmenting the South Equatorial Current, and setting up a clockwise circulation in the northern part of the Indian Ocean. Off the coast of Somalia, the **Somali Current** reverses direction during the northern hemisphere summer with northward currents reaching speeds of 5 knots or more. Twice a year, around May and November, westerly winds along the equator result in an eastward **Equatorial Jet** which feeds warm water towards Sumatra.

As the South Equatorial Current approaches the coast of Africa, it curves toward the southwest, part of it flowing through the Mozambique Channel between Madagascar and the mainland, and part flowing along the east coast of Madagascar. At the southern end of this island the two join to form the strong **Agulhas Current**, which is analogous to the Gulf Stream. This current, when opposed by strong winds from Southern Ocean storms, creates dangerously large seas.

South of South Africa, the Agulhas Current retroflects, and most of the flow curves sharply southward and then eastward to join the West Wind Drift; this junction is often marked by a broken and confused sea, made much worse by westerly storms. A small part of the Agulhas Current rounds the southern end of Africa and helps form the **Benguela Current**; occasionally, strong eddies are formed in the retroflection region and these too move into the Southeastern Atlantic.

The eastern boundary currents in the Indian Ocean are quite different from those found in the Atlantic and Pacific. The seasonally reversing **South Java Current** has strongest westward flow during August when monsoon winds are easterly and the Equatorial jet is inactive. Along the coast of Australia, a vigorous poleward flow, the **Leeuwin Current**, runs against the prevailing winds.

3109. Arctic Currents

The waters of the North Atlantic enter the Arctic Ocean between Norway and Svalbard. The currents flow easterly, north of Siberia, to the region of the Novosibirskiye Ostrova, where they turn northerly across the North Pole, and continue down the Greenland coast to form the **East Greenland Current**. On the American side of the Arctic basin, there is a weak, continuous clockwise flow centered in the vicinity of 80°N, 150°W. A current north through Bering Strait along the American coast is balanced by an outward southerly flow along the Siberian coast, which eventually becomes part of the **Kamchatka Current**. Each of the main islands or island groups in the Arctic, as far as is known, seems to have a clockwise nearshore circulation around it. The Barents Sea, Kara Sea, and Laptev Sea each have a weak counterclockwise circulation. A similar but weaker counterclockwise current system appears to exist in the East Siberian Sea.

OCEANIC CURRENT PHENOMENA

3110. Ocean Eddies and Rings

Eddies with horizontal diameters varying from 50-150 km have their own pattern of surface currents. These features may have either a warm or a cold core and currents flow around this core, either cyclonically for cold cores or anticyclonically for warm cores. The most intense of these features are called **rings** and are formed by the pinching off of meanders of western boundary currents such as the Gulf Stream. Maximum speed associated with these features is about 2 knots. Rings have also been observed to pinch off from the Agulhas retroflexion and to then drift to the northwest into the South Atlantic. Similarly, strong anticyclonic eddies are occasionally spawned by the loop current into the Western Gulf Mexico.

In general, mesoscale variability is strongest in the region of western boundary currents and in the Circumpolar Current. The strength of mesoscale eddies is greatly reduced at distances of 200-400 km from these strong boundary currents, because mean currents are generally weaker in these regions. The eddies may be sufficiently strong to reverse the direction of the surface currents.

3111. Undercurrents

At the equator and along some ocean boundaries, shallow undercurrents exist, flowing in a direction counter to that at the surface. These currents may affect the operation of submarines or trawlers. The most intense of these flows, called the Pacific **Equatorial Undercurrent**, is found at the equator in the Pacific. It is centered at a depth of 150m to the west of the Galapagos, is about 4 km wide, and eastward speeds of up to 1.5 m/s have been observed. Equatorial Undercurrents are also observed in the Atlantic and Indian Ocean, but they are somewhat weaker. In the Atlantic, the Equatorial Undercurrent is found to the east of 24°W and in the Indian Ocean, it appears to be seasonal.

Undercurrents also exist along ocean boundaries. They seem to be most ubiquitous at the eastern boundary of oceans. Here they are found at depths of 100-200m, may be 100 km wide, and have maximum speeds of 0.5 m/s.

3112. Ocean Currents and Climate

Many of the ocean currents exert a marked influence upon the climate of the coastal regions along which they flow. Thus, warm water from the Gulf Stream, continuing as the North Atlantic, Northeast Drift, and Irminger Currents, arrives off the southwest coast of Iceland, warming it to the extent that Reykjavik has a higher average winter temperature than New York City, far to the south. Great Britain and Labrador are about the same latitude, but the climate of Great Britain is much milder because of the relatively warm currents. The west coast of the United States is cooled in the summer by the California Current, and warmed in the winter by the Davidson Current. Partly as a result of this circulation, the range of monthly average temperature is comparatively small.

Currents exercise other influences besides those on temperature. The pressure pattern is affected materially, as air over a cold current contracts as it is cooled, and that over a warm current expands. As air cools above a cold ocean current, fog is likely to form. Frost smoke occurs over a warm current which flows into a colder region. Evaporation is greater from warm water than from cold water, adding to atmospheric moisture.

3113. Ocean Current Observations

Historically, our views of the surface circulation of the ocean have been shaped by reports of ocean currents provided by mariners. As mentioned at the start of this chapter, these observations consist of reports of the difference between the dead reckoning and the observed position of the vessel. These observations were routinely collected until the start of World War II.

Today, two observation systems are generally used for surface current studies. The first utilizes autonomous free-drifting buoys which are tracked by satellite or relay their position via satellite. These buoys consist of either a spherical or cylindrical surface float which is about 0.5m in diameter with a drogue at a depth of about 35m. The second system utilizes **acoustic Doppler current profilers**. These profilers utilize hull mounted transducers, operate at a frequency of 150 kHz, and have pulse repetition rates of about 1 second. They can penetrate to about 300m, and, where water is shallower than this depth, track the bottom. Merchant and naval vessels are increasingly being outfitted with acoustic Doppler current profilers which, when operated with the Global Positioning System, provide accurate observations of currents.

CHAPTER 32

WAVES, BREAKERS AND SURF

OCEAN WAVES

3200. Introduction

Ocean waves, the most easily observed phenomenon at sea, are probably the least understood by the average seaman. More than any other single factor, ocean waves are likely to cause a navigator to change course or speed to avoid damage to ship and cargo. Wind-generated ocean waves have been measured at more than 100 feet high, and tsunamis, caused by earthquakes, far higher. A mariner with knowledge of basic facts concerning waves is able to use them to his advantage, avoid hazardous conditions, and operate with a minimum of danger if such conditions cannot be avoided. See Chapter 37, Weather Routing, for details on how to avoid areas of severe waves.

3201. Causes of Waves

Waves on the surface of the sea are caused principally by wind, but other factors, such as submarine earthquakes, volcanic eruptions, and the tide, also cause waves. If a breeze of less than 2 knots starts to blow across smooth water, small wavelets called **ripples** form almost instantaneously. When the breeze dies, the ripples disappear as suddenly as they formed, the level surface being restored by surface tension of the water. If the wind speed exceeds 2 knots, more stable **gravity waves** gradually form, and progress with the wind.

While the generating wind blows, the resulting waves may be referred to as **sea**. When the wind stops or changes direction, waves that continue on without relation to local winds are called **swell**.

Unlike wind and current, waves are not deflected appreciably by the rotation of the Earth, but move in the direction in which the generating wind blows. When this wind ceases, friction and spreading cause the waves to be reduced in height, or attenuated, as they move. However, the reduction takes place so slowly that swell often continues until it reaches some obstruction, such as a shore.

The Fleet Numerical Meteorology and Oceanography Center produces synoptic analyses and predictions of ocean wave heights using a spectral numerical model. The wave information consists of heights and directions for different periods and wavelengths. Verification of projected data has proven the model to be very good. Information from the model is provided to the U.S. Navy on a routine basis and is a vital input to the Optimum Track Ship Routing program.

3202. Wave Characteristics

Ocean waves are very nearly in the shape of an inverted cycloid, the figure formed by a point inside the rim of a wheel rolling along a level surface. This shape is shown in Figure 3202a. The highest parts of waves are called **crests**, and the intervening lowest parts, troughs. Since the crests are steeper and narrower than the troughs, the mean or still water level is a little lower than halfway between the crests and troughs. The vertical distance between trough and crest is called **wave height**, labeled H in Figure 3202a. The horizontal distance between successive crests, measured in the direction of travel, is called **wavelength**, labeled L. The time interval between passage of successive crests at a stationary point is called **wave period** (P). Wave height, length, and period depend upon a number of factors, such as the wind speed, the length of time it has blown, and its **fetch** (the straight distance it has traveled over the surface). Table 3202 indicates the relationship between wind speed, fetch, length of time the wind blows, wave height, and wave period in deep water.

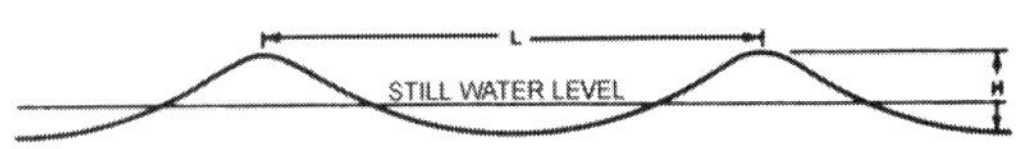

Figure 3202a. A typical sea wave.

If the water is deeper than one-half the wavelength (L), this length in feet is theoretically related to period (P) in seconds by the formula:

$$L = 5.12\,P^2.$$

The actual value has been found to be a little less than this for swell, and about two-thirds the length determined by this formula for sea. When the waves leave the generating area and continue as free waves, the wavelength and period continue to increase, while the height decreases. The rate of change gradually decreases.

The speed (S) of a free wave in deep water is nearly independent of its height or steepness. For swell, its

Fetch	BEAUFORT NUMBER 3			4			5			6			7			8			9			10			11			Fetch
	T	H	P	T	H	P	T	H	P	T	H	P	T	H	P	T	H	P	T	H	P	T	H	P	T	H	P	
10	4.4	1.8	2.1	3.7	2.6	2.4	3.2	3.5	2.8	2.7	5.0	3.1	2.5	6.0	3.4	2.3	7.3	3.9	2.0	8.0	4.1	1.9	10.0	4.2	1.8	10.0	5.0	10
20	7.1	2.0	2.5	6.2	3.2	2.9	5.4	4.9	3.3	4.7	7.0	3.8	4.2	8.6	4.3	3.9	10.0	4.4	3.5	12.0	5.0	3.2	14.0	5.2	3.0	16.0	5.9	20
30	9.8	2.0	2.8	8.3	3.8	3.3	7.2	5.8	3.7	6.2	8.0	4.2	5.8	10.0	4.6	5.2	12.1	5.0	4.7	15.8	5.5	4.4	18.0	6.0	4.1	19.8	6.3	30
40	12.0	2.0	3.0	10.3	3.9	3.6	8.9	6.2	4.1	7.8	9.0	4.6	7.1	11.2	4.9	6.5	14.0	5.4	5.8	17.7	5.9	5.4	21.0	6.3	5.1	22.5	6.7	40
50	14.0	2.0	3.2	12.4	4.0	3.8	11.0	6.5	4.4	9.1	9.8	4.8	8.4	12.2	5.2	7.7	15.7	5.6	6.9	19.8	6.3	6.4	23.0	6.7	6.1	25.0	7.1	50
60	16.0	2.0	3.5	14.0	4.0	4.0	12.0	6.8	4.6	10.2	10.3	5.1	9.6	13.2	5.5	8.7	17.0	6.0	8.0	21.0	6.5	7.4	25.0	7.0	7.0	27.5	7.5	60
70	18.0	2.0	3.7	15.8	4.0	4.1	13.5	7.0	4.8	11.9	10.8	5.4	10.5	13.9	5.7	9.9	18.0	6.4	9.0	22.5	6.8	8.3	26.5	7.3	7.8	29.5	7.7	70
80	20.0	2.0	3.8	17.0	4.0	4.2	15.0	7.2	4.9	13.0	11.0	5.6	12.0	14.5	6.0	11.0	18.9	6.6	10.0	24.0	7.1	9.3	28.0	7.7	8.6	31.5	7.9	80
90	23.6	2.0	3.9	18.8	4.0	4.3	16.5	7.3	5.1	14.1	11.2	5.8	13.0	15.0	6.3	12.0	20.0	6.7	11.0	25.0	7.2	10.2	30.0	7.9	9.5	34.0	8.2	90
100	27.1	2.0	4.0	20.0	4.0	4.4	17.5	7.3	5.3	15.1	11.4	6.0	14.0	15.5	6.5	12.8	20.5	6.9	11.9	26.5	7.6	11.0	32.0	8.1	10.3	35.0	8.5	100
120	31.1	2.0	4.2	22.4	4.1	4.7	20.0	7.8	5.4	17.0	11.7	6.2	15.9	16.0	6.7	14.5	21.5	7.3	13.1	27.5	7.9	12.3	33.5	8.4	11.5	37.5	8.8	120
140	36.6	2.0	4.5	25.8	4.2	4.9	22.5	7.9	5.8	19.1	11.9	6.4	17.6	16.2	7.0	16.0	22.0	7.6	14.8	29.0	8.3	13.9	35.5	8.8	13.0	40.0	9.2	140
160	43.2	2.0	4.9	28.4	4.2	5.2	24.3	7.9	6.0	21.1	12.0	6.6	19.5	16.5	7.3	18.0	23.0	8.0	16.4	30.5	8.7	15.1	37.0	9.1	14.5	42.5	9.6	160
180	50.0	2.0	4.9	30.9	4.3	5.4	27.0	8.0	6.2	23.1	12.1	6.8	21.3	17.0	7.5	19.9	23.5	8.3	18.0	31.5	9.0	16.5	38.5	9.5	16.0	44.5	10.0	180
200				33.5	4.3	5.6	29.0	8.0	6.4	25.4	12.2	7.1	23.1	17.5	7.7	21.5	23.5	8.5	19.3	32.5	9.2	18.1	40.0	9.8	17.1	46.0	10.3	200
220				36.5	4.4	5.8	31.1	8.0	6.6	27.2	12.3	7.2	25.0	17.9	8.0	22.9	24.0	8.8	20.9	34.0	9.6	19.1	41.5	10.1	18.2	47.5	10.6	220
240				39.2	4.4	5.9	33.1	8.0	6.8	29.0	12.4	7.3	26.8	17.9	8.2	24.4	24.5	9.0	22.0	34.5	9.8	20.5	43.0	10.3	19.5	49.0	10.8	240
260				41.9	4.4	6.0	34.9	8.0	6.9	30.5	12.6	7.5	28.0	18.0	8.4	26.0	25.0	9.2	23.5	34.5	10.0	21.8	44.0	10.6	20.9	50.5	11.1	260
280				44.5	4.4	6.2	36.8	8.0	7.0	32.4	12.9	7.8	29.5	18.0	8.5	27.7	25.0	9.4	25.0	35.0	10.2	23.0	45.0	10.9	22.0	51.5	11.3	280
300				47.0	4.4	6.3	38.5	8.0	7.1	34.1	13.1	8.0	31.5	18.0	8.7	29.0	25.0	9.5	26.3	35.0	10.4	24.3	45.0	11.1	23.2	53.0	11.6	300
320							40.5	8.0	7.2	36.0	13.3	8.2	33.0	18.0	8.9	30.2	25.0	9.6	27.6	35.5	10.6	25.5	45.5	11.2	24.5	54.0	11.8	320
340							42.4	8.0	7.3	37.6	13.4	8.3	34.2	18.0	9.0	31.6	25.0	9.8	29.0	36.0	10.8	26.7	46.0	11.4	25.5	55.0	12.0	340
360							44.2	8.0	7.4	38.8	13.4	8.4	35.7	18.1	9.1	33.0	25.0	9.9	30.0	36.5	10.9	27.7	46.5	11.6	26.6	55.0	12.2	360
380							46.1	8.0	7.5	40.2	13.5	8.5	37.1	18.2	9.3	34.2	25.5	10.0	31.3	37.0	11.1	29.1	47.0	11.8	27.7	55.5	12.4	380
400							48.0	8.0	7.7	42.2	13.5	8.6	38.8	18.4	9.5	35.6	26.0	10.2	32.5	37.0	11.2	30.2	47.5	12.0	28.9	56.0	12.6	400
420							50.0	8.0	7.8	43.5	13.6	8.7	40.0	18.7	9.6	36.9	26.5	10.3	33.7	37.5	11.4	31.5	47.5	12.2	29.6	56.5	12.7	420
440							52.0	8.0	7.9	44.7	13.7	8.8	41.3	18.8	9.7	38.1	27.0	10.4	34.8	37.5	11.5	32.5	48.0	12.3	30.9	57.0	12.9	440
460							54.0	8.0	8.0	46.2	13.7	8.9	42.8	19.0	9.8	39.5	27.5	10.6	36.0	37.5	11.7	33.5	48.5	12.5	31.8	57.5	13.1	460
480							56.0	8.0	8.1	47.8	13.7	9.0	44.0	19.0	9.9	41.0	27.5	10.8	37.0	37.5	11.8	34.5	49.0	12.6	32.7	57.5	13.2	480
500							58.0	8.0	8.2	49.2	13.8	9.1	45.5	19.1	10.1	42.1	27.5	10.9	38.3	38.0	11.9	35.5	49.0	12.7	33.9	58.0	13.4	500
550										53.0	13.8	9.3	48.5	19.5	10.3	44.9	27.5	11.1	41.0	38.5	12.2	38.2	50.0	13.0	36.5	59.0	13.7	550
600										56.3	13.8	9.5	51.8	19.7	10.5	47.7	27.5	11.3	43.6	39.0	12.5	40.3	50.0	13.3	38.7	60.0	14.0	600
650													55.0	19.8	10.7	50.3	27.5	11.6	46.4	39.5	12.8	43.0	50.0	13.7	41.0	60.0	14.2	650
700													58.5	19.8	11.0	53.2	27.5	11.8	49.0	40.0	13.1	45.4	50.5	14.0	43.5	60.5	14.5	700
750																56.2	27.5	12.1	51.0	40.0	13.3	48.0	51.0	14.2	45.8	61.0	14.8	750
800																59.2	27.5	12.3	53.8	40.0	13.5	50.6	51.5	14.5	47.8	61.5	15.0	800
850																			56.2	40.0	13.8	52.5	52.0	14.6	50.0	62.0	15.2	850
900																			58.2	40.0	14.0	54.6	52.0	14.9	52.0	62.5	15.5	900
950																						57.2	52.0	15.1	54.0	63.0	15.7	950
1000																						59.3	52.0	15.3	56.3	63.0	16.0	1000

Table 3202. Minimum Time (T) in hours that wind must blow to form waves of H significant height (in feet) and P period (in seconds). Fetch in nautical miles.

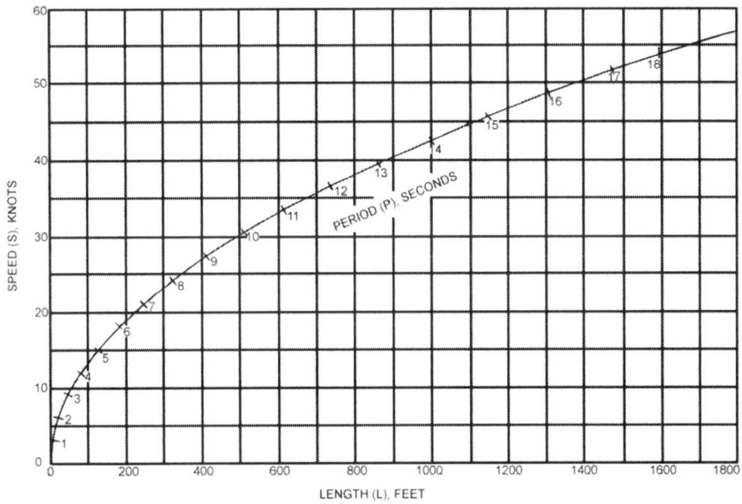

Figure 3202b. Relationship between speed, length, and period of waves in deep water, based upon the theoretical relationship between period and length.

relationship in knots to the period (P) in seconds is given by the formula

$$S = 3.03P.$$

The relationship for sea is not known.

The theoretical relationship between speed, wavelength, and period is shown in Figure 3202b. As waves continue on beyond the generating area, the period, wavelength, and speed remain the same. Because the waves of each period have different speeds they tend to sort themselves by periods as they move away from the generating area. The longer period waves move at a greater speed and move ahead. At great enough distances from a storm area the waves will have sorted themselves into sets based on period.

All waves are attenuated as they propagate but the short period waves attenuate faster, so that far from a storm only the longer waves remain.

The time needed for a wave system to travel a given distance is double that which would be indicated by the speed of individual waves. This is because each leading wave in succession gradually disappears and transfers its energy to following wave. The process occurs such that the whole wave *system* advances at a speed which is just half that of each individual wave. This process can easily be seen in the bow wave of a vessel. The speed at which the wave system advances is called **group velocity**.

Because of the existence of many independent wave systems at the same time, the sea surface acquires a complex and irregular pattern. Since the longer waves overrun the shorter ones, the resulting interference adds to the complexity of the pattern. The process of interference, illustrated in Figure 3202c, is duplicated many times in the sea; it is the principal reason that successive waves are not of the same height. The irregularity of the surface may be further accentuated by the presence of wave systems crossing at an angle to each other, producing peak-like rises.

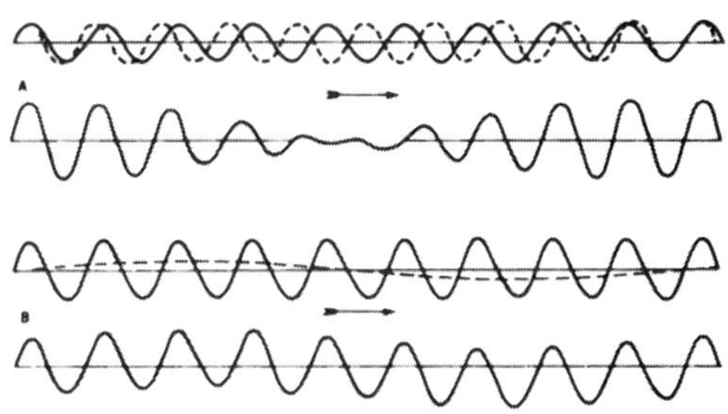

Figure 3202c. Interference. The upper part of A shows two waves of equal height and nearly equal length traveling in the same direction. The lower part of A shows the resulting wave pattern. In B similar information is shown for short waves and long swell.

In reporting average wave heights, the mariner has a tendency to neglect the lower ones. It has been found that the reported value is about the average for the highest one-third. This is sometimes called the "significant" wave height. The approximate relationship between this height and others, is as follows:

Wave	*Relative height*
Average	0.64
Significant	1.00
Highest 10 percent	1.29
Highest	1.87

3203. Path of Water Particles in a Wave

As shown in Figure 3203, a particle of water on the surface of the ocean follows a somewhat circular orbit as a wave passes, but moves very little in the direction of motion of the wave. The common wave producing this action is called an **oscillatory wave**. As the crest passes, the particle moves forward, giving the water the appearance of moving with the wave. As the trough passes, the motion is in the opposite direction. The radius of the circular orbit decreases with depth, approaching zero at a depth equal to about half the wavelength. In shallower water the orbits become more elliptical, and in very shallow water the vertical motion disappears almost completely.

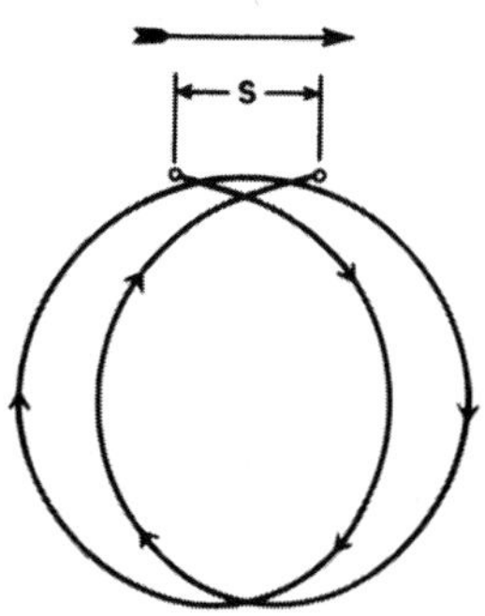

Figure 3203. Orbital motion and displacement, s, of a particle on the surface of deep water during two wave periods.

Since the speed is greater at the top of the orbit than at the bottom, the particle is not at exactly its original point following passage of a wave, but has moved slightly in the wave's direction of motion. However, since this advance is small in relation to the vertical displacement, a floating object is raised and lowered by passage of a wave, but moved little from its original position. If this were not so, a slow moving vessel might experience considerable difficulty in making way against a wave train. In Figure 3203 the forward displacement is greatly exaggerated.

3204. Effects of Current and Ice on Waves

A following current increases wavelengths and decreases wave heights. An opposing current has the opposite effect, decreasing the length and increasing the height. This effect can be dangerous in certain areas of the world where a stream current opposes waves generated by severe weather. An example of this effect is off the coast of South Africa, where the Agulhas current is often opposed by westerly storms, creating steep, dangerous seas. A strong opposing current may cause the waves to break, as in the case of **overfalls** in tidal currents. The extent of wave alteration is dependent upon the ratio of the still-water wave speed to the speed of the current.

Moderate ocean currents running at oblique angles to wave directions appear to have little effect, but strong tidal currents perpendicular to a system of waves have been observed to completely destroy them in a short period of time.

When ice crystals form in seawater, internal friction is greatly increased. This results in smoothing of the sea surface. The effect of pack ice is even more pronounced. A vessel following a lead through such ice may be in smooth water even when a gale is blowing and heavy seas are beating against the outer edge of the pack. Hail or torrential rain is also effective in flattening the sea, even in a high wind.

3205. Waves and Shallow Water

When a wave encounters shallow water, the movement of the water is restricted by the bottom, resulting in reduced wave speed. In deep water wave speed is a function of period. In shallow water, the wave speed becomes a function of depth. The shallower the water, the slower the wave speed. As the wave speed slows, the period remains the same, so the wavelength becomes shorter. Since the energy in the waves remains the same, the shortening of wavelengths results in increased heights. This process is called **shoaling**. If the wave approaches a shallow area at an angle, each part is slowed successively as the depth decreases. This causes a change in direction of motion, or **refraction**, the wave tending to change direction parallel to the depth curves. The effect is similar to the refraction of light and other forms of radiant energy.

As each wave slows, the next wave behind it, in deeper water, tends to catch up. As the wavelength decreases, the height generally becomes greater. The lower part of a wave, being nearest the bottom, is slowed more than the top. This may cause the wave to become unstable, the faster-moving top falling forward or breaking. Such a wave is called a **breaker**, and a series of breakers is **surf**.

Swell passing over a shoal but not breaking undergoes a decrease in wavelength and speed, and an increase in height, which may be sudden and dramatic, depending on the steepness of the seafloor's slope. This **ground swell**

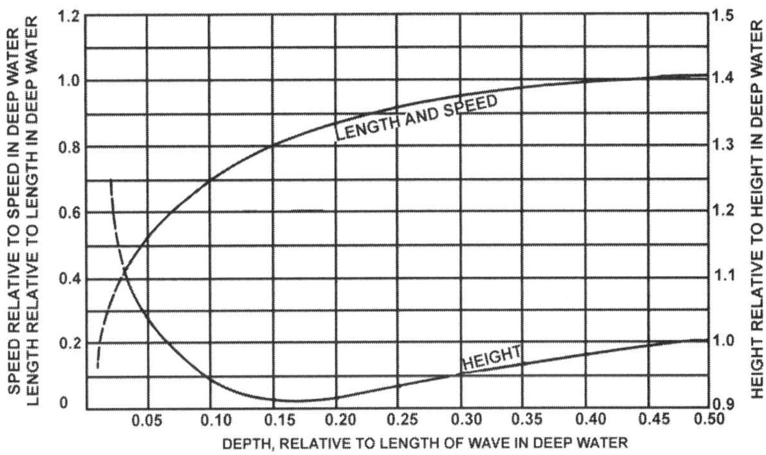

Figure 3205. Alteration of the characteristics of waves crossing a shoal.

may cause heavy rolling if it is on the beam and its period is the same as the period of roll of a vessel, even though the sea may appear relatively calm. It may also cause a **rage sea**, when the swell waves encounter water shoal enough to make them break. Rage seas are dangerous to small craft, particularly approaching from seaward, as the vessel can be overwhelmed by enormous breakers in perfectly calm weather. The swell waves, of course, may have been generated hundreds of miles away. In the open ocean they are almost unnoticed due to their very long period and wavelength. Figure 3205 illustrates the approximate alteration of the characteristics of waves as they cross a shoal.

3206. Energy Of Waves

The potential energy of a wave is related to the vertical distance of each particle from its still-water position. Therefore potential energy moves with the wave. In contrast, the kinetic energy of a wave is related to the speed of the particles, distributed evenly along the entire wave.

The amount of kinetic energy in a wave is tremendous. A 4-foot, 10-second wave striking a coast expends more than 35,000 horsepower per mile of beach. For each 56 miles of coast, the energy expended equals the power generated at Hoover Dam. An increase in temperature of the water in the relatively narrow surf zone in which this energy is expended would seem to be indicated, but no pronounced increase has been measured. Apparently, any heat that may be generated is dissipated to the deeper water beyond the surf zone.

3207. Wave Measurement Aboard Ship

With suitable equipment and adequate training, reliable measurements of the height, length, period, and speed of waves can be made. However, the mariner's estimates of height and length often contain relatively large errors. There is a tendency to underestimate the heights of low waves, and overestimate the heights of high ones. There are numerous accounts of waves 75 to 80 feet high, or even higher, although waves more than 55 feet high are very rare. Wavelength is usually underestimated. The motions of the vessel from which measurements are made contribute to such errors.

Height. Measurement of wave height is particularly difficult. A microbarograph can be used if the wave is long enough or the vessel small enough to permit the vessel to ride from crest to trough. If the waves are approaching from dead ahead or dead astern, this requires a wavelength at least twice the length of the vessel. For most accurate results the instrument should be placed at the center of roll and pitch, to minimize the effects of these motions. Wave height can often be estimated with reasonable accuracy by comparing it with freeboard of the vessel. This is less accurate as wave height and vessel motion increase. If a point of observation can be found at which the top of a wave is in line with the horizon when the observer is in the trough, the wave height is equal to height of eye. However, if the vessel is rolling or pitching, this height at the moment of observation may be difficult to determine. The highest wave ever reliably reported was 112 feet observed from the USS Ramapo in 1933.

Length. The dimensions of the vessel can be used to determine wavelength. Errors are introduced by perspective and disturbance of the wave pattern by the vessel. These errors are minimized if observations are made from maximum height. Best results are obtained if the sea is from dead ahead or dead astern.

Period. If allowance is made for the motion of the vessel, wave period can be determined by measuring the interval between passages of wave crests past the observer. The relative motion of the vessel can be eliminated by timing the passage of successive wave crests past a patch of foam or a floating object at some distance from the vessel. Accuracy of results can be improved by averaging several observations.

Speed. Speed can be determined by timing the passage of the wave between measured points along the side of the ship, if corrections are applied for the direction of travel for the wave and the speed of the ship.

The length, period, and speed of waves are interrelated by the relationships indicated previously. There is no definite mathematical relationship between wave height and length, period, or speed.

3208. Tsunamis

A **Tsunami** is an ocean wave produced by sudden, large-scale motion of a portion of the ocean floor or the shore, such as a volcanic eruption, earthquake (sometimes called seaquake if it occurs at sea), or landslide. If they are caused by a submarine earthquake, they are usually called **seismic sea waves**. The point directly above the disturbance, at which the waves originate, is called the **epicenter**. Either a tsunami or a storm tide that overflows the land is popularly called a **tidal wave**, although it bears no relation to the tide.

If a volcanic eruption occurs below the surface of the sea, the escaping gases cause a quantity of water to be pushed upward in the shape of a dome. The same effect is caused by the sudden rising of a portion of the bottom. As this water settles back, it creates a wave which travels at high speed across the surface of the ocean.

Tsunamis are a series of waves. Near the epicenter, the first wave may be the highest. At greater distances, the highest wave usually occurs later in the series, commonly between the third and the eighth wave. Following the maximum, they again become smaller, but the tsunami may be detectable for several days.

In deep water the wave height of a tsunami is probably never greater than 2 or 3 feet. Since the wavelength is usually considerably more than 100 miles, the wave is not conspicuous at sea. In the Pacific, where most tsunamis occur, the wave period varies between about 15 and 60 minutes, and the speed in deep water is more than 400 knots. The approximate speed can be computed by the formula:

$$S = 0.6\sqrt{gd} = 3.4\sqrt{d}$$

where S is the speed in knots, g is the acceleration due to gravity (32.2 feet per second per second), and d is the depth of water in feet. This formula is applicable to any wave in water having a depth of less than half the wavelength. For most ocean waves it applies only in shallow water, because of the relatively short wavelength.

When a tsunami enters shoal water, it undergoes the same changes as other waves. The formula indicates that speed is proportional to depth of water. Because of the great speed of a tsunami when it is in relatively deep water, the slowing is relatively much greater than that of an ordinary wave crested by wind. Therefore, the increase in height is also much greater. The size of the wave depends upon the nature and intensity of the disturbance. The height and destructiveness of the wave arriving at any place depends upon its distance from the epicenter, topography of the ocean floor, and the coastline. The angle at which the wave arrives, the shape of the coastline, and the topography along the coast and offshore, all have an effect. The position of the shore is also a factor, as it may be sheltered by intervening land, or be in a position where waves have a tendency to converge, either because of refraction or reflection, or both.

Tsunamis 50 feet in height or higher have reached the shore, inflicting widespread damage. On April 1, 1946, seismic sea waves originating at an epicenter near the Aleutians spread over the entire Pacific. Scotch Cap Light on Unimak Island, 57 feet above sea level, was completely destroyed and its keepers killed. Traveling at an average speed of 490 miles per hour, the waves reached the Hawaiian Islands in 4 hours and 34 minutes, where they arrived as waves 50 feet above the high water level, and flooded a strip of coast more than 1,000 feet wide at some places. They left a death toll of 173 and property damage of $25 million. Less destructive waves reached the shores of North and South America, as well as Australia, 6,700 miles from the epicenter.

After this disaster, a tsunami warning system was set up in the Pacific, even though destructive waves are relatively rare (averaging about one in 20 years in the Hawaiian Islands). This system monitors seismic disturbances throughout the Pacific basin and predicts times and heights of tsunamis. Warnings are immediately sent out if a disturbance is detected.

In addition to seismic sea waves, earthquakes below the surface of the sea may produce a longitudinal pressure wave that travels upward at the speed of sound. When a ship encounters such a wave, it is felt as a sudden shock which may be so severe that the crew thinks the vessel has struck bottom.

3209. Storm Tides

In relatively tideless seas like the Baltic and Mediterranean, winds cause the chief fluctuations in sea level. Elsewhere, the astronomical tide usually masks these variations. However, under exceptional conditions, either severe extra-tropical storms or tropical cyclones can produce changes in sea level that exceed the normal range of tide. Low sea level is of little concern except to coastal shipping, but a rise above ordinary high-water mark, particularly when it is accompanied by high waves, can result in a catastrophe.

Although, like tsunamis, these storm tides or storm surges are popularly called tidal waves, they are not associated with the tide. They consist of a single wave crest and hence have no period or wavelength.

Three effects in a storm induce a rise in sea level. The first is wind stress on the sea surface, which results in a piling-up of water (sometimes called "wind set-up"). The second effect is the convergence of wind-driven currents, which elevates the sea surface along the convergence line. In shallow water, bottom friction and the effects of local topography cause this elevation to persist and may even intensify it. The low atmospheric pressure that accompanies severe storms causes the third effect, which is sometimes referred to as the "inverted barometer" as the sea surface rises into the low pressure area. An inch of mercury is equivalent to about 13.6 inches of water, and the adjustment of the sea surface to the reduced pressure can amount to several feet at equilibrium.

All three of these causes act independently, and if they happen to occur simultaneously, their effects are additive. In addition, the wave can be intensified or amplified by the effects of local topography. Storm tides may reach heights of 20 feet or more, and it is estimated that they cause three-fourths of the deaths attributed to hurricanes.

3210. Standing Waves and Seiches

Previous articles in this chapter have dealt with progressive waves which appear to move regularly with time. When two systems of progressive waves having the same period travel in opposite directions across the same area, a series of **standing waves** may form. These appear to remain stationary.

Another type of standing wave, called a **seiche**, sometimes occurs in a confined body of water. It is a long wave, usually having its crest at one end of the confined space, and its trough at the other. Its period may be anything from a few minutes to an hour or more, but somewhat less than the tidal period. Seiches are usually attributed to strong winds or sudden changes in atmospheric pressure.

3211. Tide-Generated Waves

There are, in general, two regions of high tide separated by two regions of low tide, and these regions move progressively westward around the Earth as the moon revolves in its orbit. The high tides are the crests of these tide waves, and the low tides are the troughs. The wave is not noticeable at sea, but becomes apparent along the coasts, particularly in funnel-shaped estuaries. In certain river mouths, or estuaries of particular configuration, the incoming wave of high water overtakes the preceding low tide, resulting in a steep, breaking wave which progresses upstream in a surge called a **bore**.

3212. Internal Waves

Thus far, the discussion has been confined to waves on the surface of the sea, the boundary between air and water. Internal waves, or boundary waves, are created below the surface, at the boundaries between water strata of different densities. The density differences between adjacent water strata in the sea are considerably less than that between sea and air. Consequently, internal waves are much more easily formed than surface waves, and they are often much larger. The maximum height of wind waves on the surface is about 60 feet, but internal wave heights as great as 300 feet have been encountered.

Internal waves are detected by a number of observations of the vertical temperature distribution, using recording devices such as the bathythermograph. They have periods as short as a few minutes, and as long as 12 or 24 hours, these greater periods being associated with the tides.

A slow-moving ship, operating in a freshwater layer having a depth approximating the draft of the vessel, may produce short-period internal waves. This may occur off rivers emptying into the sea, or in polar regions in the vicinity of melting ice. Under suitable conditions, the normal propulsion energy of the ship is expended in generating and maintaining these internal waves and the ship appears to "stick" in the water, becoming sluggish and making little headway. The phenomenon, known as **dead water**, disappears when speed is increased by a few knots.

The full significance of internal waves has not yet been determined, but it is known that they may cause submarines to rise and fall like a ship at the surface, and they may also affect sound transmission in the sea.

3213. Waves and Ships

The effects of waves on a ship vary considerably with the type of ship, its course and speed, and the condition of the sea. A short vessel has a tendency to ride up one side of a wave and down the other side, while a larger vessel may tend to ride through the waves on an even keel. If the waves are of such length that the bow and stern of a vessel are alternately riding in successive crests and troughs, the vessel is subject to heavy sagging and hogging stresses, and under extreme conditions may break in two. A change of heading may reduce the danger. Because of the danger from sagging and hogging, a small vessel is sometimes better able to ride out a storm than a large one.

If successive waves strike the side of a vessel at the same phase of successive rolls, relatively small waves can cause heavy rolling. The same effect, if applied to the bow or stern in time with the natural period of pitch, can cause heavy pitching. A change of either heading or speed can quickly reduce the effect.

A wave having a length twice that of a ship places that ship in danger of falling off into the trough of the sea, particularly if it is a slow-moving vessel. The effect is especially pronounced if the sea is broad on the bow or broad on the quarter. An increase in speed reduces the hazard.

3214. Using Oil to Calm Breaking Waves

Historically oil was used to calm breaking waves, and was useful to vessels when lowering or hoisting boats in rough weather. Its effect was greatest in deep water, where a small quantity sufficed if the oil were made to spread to windward of the vessel. Oil increases the surface tension of the water, lessening the tendency for waves to break.

BREAKERS AND SURF

3215. Refraction

As explained previously, waves are slowed in shallow water, causing refraction if the waves approach the beach at an angle. Along a perfectly straight beach, with uniform shoaling, the wave fronts tend to become parallel to the shore. Any irregularities in the coastline or bottom contours, however, affect the refraction, causing irregularities. In the case of a ridge perpendicular to the beach, for instance, the shoaling is more rapid, causing greater refraction towards the ridge. The waves tend to align themselves with the bottom contours. Waves on both sides of the ridge have a component of motion toward the ridge. This convergence of wave energy toward the ridge causes an increase in wave or breaker height. A submarine canyon or valley perpendicular to the beach, on the other hand, produces divergence, with a decrease in wave or breaker height. These effects are illustrated in Figure 3215. Bends in the coast line have a similar effect, convergence occurring at a point, and divergence if the coast is concave to the sea. Points act as focal areas for wave energy and experience large breakers. Concave bays have small breakers because the energy is spread out as the waves approach the beach.

Under suitable conditions, currents also cause refraction. This is of particular importance at entrances of tidal estuaries. When waves encounter a current running in the opposite direction, they become higher and shorter. This results in a choppy sea, often with breakers. When waves move in the same direction as current, they decrease in height, and become longer. Refraction occurs when waves encounter a current at an angle.

Refraction diagrams, useful in planning amphibious operations, can be prepared with the aid of nautical charts or aerial photographs. When computer facilities are available, computer programs can be used to produce refraction diagrams quickly and accurately.

3216. Classes Of Breakers

In deep water, swell generally moves across the surface as somewhat regular, smooth undulations. When shoal water is reached, the wave period remains the same, but the speed decreases. The amount of decrease is negligible until the depth of water becomes about one-half the wavelength, when the waves begin to "feel" bottom. There is a slight decrease in wave height, followed by a rapid increase, if the waves are traveling perpendicular to a straight coast with a uniformly sloping bottom. As the waves become higher and shorter, they also become steeper, and the crest narrows. When the speed of the crest becomes greater than that of the wave, the front face of the wave becomes steeper than the rear face. This process continues at an accelerating rate as the depth of water decreases. If the wave becomes too unstable, it topples forward to form a breaker.

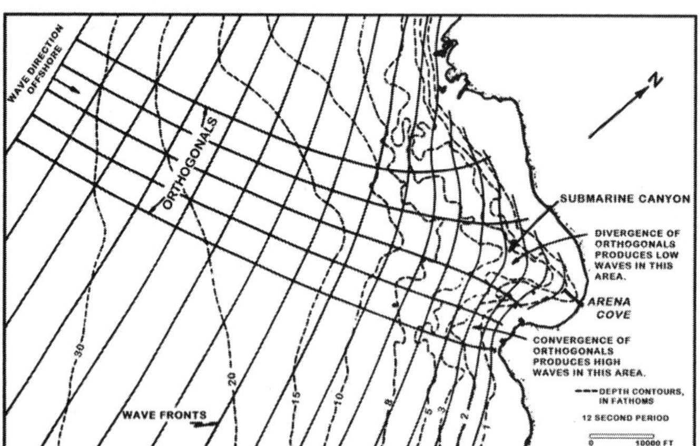

Figure 3215. The effect of bottom topography in causing wave convergence and wave divergence. Courtesy of Robert L. Wiegel, Council on Wave Research, University of California.

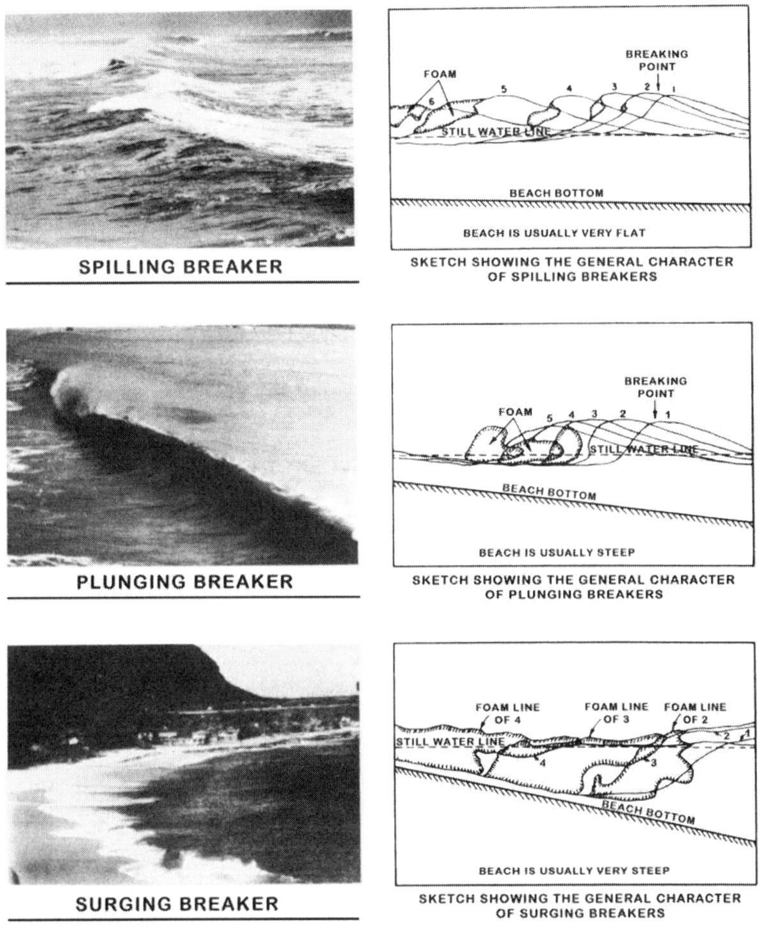

Figure 3216. The three types of breakers.
Courtesy of Robert L. Wiegel, Council on Wave Research, University of California.

There are three general classes of breakers. A **spilling breaker** breaks gradually over a considerable distance. A **plunging breaker** tends to curl over and break with a single crash. A **surging breaker** peaks up, but surges up the beach without spilling or plunging. It is classed as a breaker even though it does not actually break. The type of breaker which forms is determined by the steepness of the beach and the steepness of the wave before it reaches shallow water, as illustrated in Figure 3216.

Long waves break in deeper water, and have a greater breaker height. A steep beach also increases breaker height. The height of breakers is less if the waves approach the beach at an acute angle. With a steeper beach slope there is greater tendency of the breakers to plunge or surge. Following the uprush of water onto a beach after the breaking of a wave, the seaward backrush occurs. The returning water is called **backwash**. It tends to further slow the bottom of a wave, thus increasing its tendency to break. This effect is greater as either the speed or depth of the backwash increases. The still water depth at the point of breaking is approximately 1.3 times the average breaker height.

Surf varies with both position along the beach and time. A change in position often means a change in bottom contour, with the refraction effects discussed before. At the same point, the height and period of waves vary considerably from wave to wave. A group of high waves is usually followed by several lower ones. Therefore, passage through surf can usually be made most easily immediately following a series of higher waves.

Since surf conditions are directly related to height of the waves approaching a beach, and to the configuration of the bottom, the state of the surf at any time can be predicted if one has the necessary information and knowledge of the principles involved. Height of the sea and swell can be predicted from wind data, and information on bottom configuration can sometimes be obtained from the largest scale nautical chart. In addition, the area of lightest surf along a beach can be predicted if details of the bottom configuration are available. Surf predictions may, however, be significantly in error due to the presence of swell from unknown storms hundreds of miles away.

3217. Currents in the Surf Zone

In and adjacent to the surf zone, currents are generated by waves approaching the bottom contours at an angle, and by irregularities in the bottom.

Waves approaching at an angle produce a **longshore current** parallel to the beach, inside of the surf zone. Longshore currents are most common along straight beaches. Their speeds increase with increasing breaker height, decreasing wave period, increasing angle of breaker line with the beach, and increasing beach slope. Speed seldom exceeds 1 knot, but sustained speeds as high as 3 knots have been recorded. Longshore currents are usually constant in direction. They increase the danger of landing craft broaching to.

Where the bottom is sandy a good distance offshore, one or more **sand bars** typically form. The innermost bar will break in even small waves, and will isolate the longshore current. The second bar, if one forms, will break only in heavier weather, and the third, if present, only in storms. It is possible to move parallel to the coast in small craft in relatively deep water in the area between these bars, between the lines of breakers.

3218. Rip Currents

As explained previously, wave fronts advancing over nonparallel bottom contours are refracted to cause convergence or divergence of the energy of the waves. Energy concentrations in areas of convergence form barriers to the returning backwash, which is deflected along the beach to areas of less resistance. Backwash accumulates at weak points, and returns seaward in concentrations, forming **rip currents** through the surf. At these points the large volume of returning water has an easily seen retarding effect upon the incoming waves, thus adding to the condition causing the rip current. The waves on one or both sides of the rip, having greater energy and not being retarded by the concentration of backwash, advance faster and farther up the beach. From here, they move along the beach as feeder currents. At some point of low resistance, the water flows seaward through the surf, forming the neck of the rip current. Outside the breaker line the current widens and slackens, forming the head. The various parts of a rip current are shown in Figure 3218.

Rip currents may also be caused by irregularities in the beach face. If a beach indentation causes an uprush to advance farther than the average, the backrush is delayed and this in turn retards the next incoming foam line (the front of a wave as it advances shoreward after breaking) at that point. The foam line on each side of the retarded point continues in its advance, however, and tends to fill in the retarded area, producing a rip current.

Rip currents are dangerous for swimmers, but may provide a clear path to the beach for small craft, as they tend to scour out the bottom and break through any sand bars that have formed. Rip currents also change location over time as conditions change.

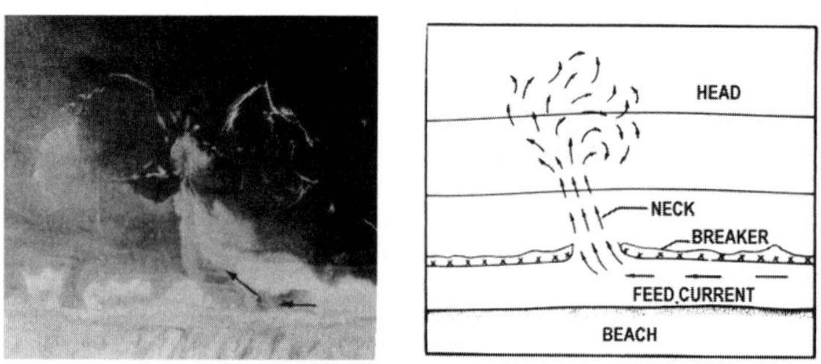

Figure 3218. A rip current (left) and a diagram of its parts (right).
Courtesy of Robert L. Wiegel, Council on Wave Research, University of California.

3219. Beach Sediments

In the surf zone, large amounts of sediment are suspended in the water. When the water's motion decreases, the sediments settle to the bottom. The water motion can be either waves or currents. Promontories or points are rocky because the large breakers scour the points and small sediments are suspended in the water and carried away. Bays tend to have sandy beaches because of the smaller waves.

In the winter when storms create large breakers and surf, the waves erode beaches and carry the particles offshore where offshore sand bars form; sandy beaches tend to be narrower in stormy seasons. In the summer the waves gradually move the sand back to the beaches and the offshore sand bars decrease; then sandy beaches tend to be wider.

Longshore currents move large amounts of sand along the coast. These currents deposit sand on the upcurrent side of a jetty or pier, and erode the beach on the downcurrent side. Groins are sometimes built to impede the longshore flow of sediments and preserve beaches for recreational use. As with jetties, the downcurrent side of each groin will have the best water for approaching the beach.

CHAPTER 33

ICE NAVIGATION

INTRODUCTION

3300. Ice and the Navigator

Sea ice has posed a problem to the navigator since antiquity. During a voyage from the Mediterranean to England and Norway sometime between 350 B.C. and 300 B.C., Pytheas of Massalia sighted a strange substance which he described as "neither land nor air nor water" floating upon and covering the northern sea over which the summer Sun barely set. Pytheas named this lonely region Thule, hence Ultima Thule (farthest north or land's end). Thus began over 20 centuries of polar exploration.

Ice is of direct concern to the navigator because it restricts and sometimes controls his movements; it affects his dead reckoning by forcing frequent changes of course and speed; it affects piloting by altering the appearance or obliterating the features of landmarks; it hinders the establishment and maintenance of aids to navigation; it affects the use of electronic equipment by affecting propagation of radio waves; it produces changes in surface features and in radar returns from these features; it affects celestial navigation by altering the refraction and obscuring the horizon and celestial bodies either directly or by the weather it influences, and it affects charts by introducing several plotting problems.

Because of his direct concern with ice, the prospective polar navigator must acquaint himself with its nature and extent in the area he expects to navigate. In addition to this volume, books, articles, and reports of previous polar operations and expeditions will help acquaint the polar navigator with the unique conditions at the ends of the Earth.

3301. Formation of Ice

As it cools, water contracts until the temperature of maximum density is reached. Further cooling results in expansion. The maximum density of fresh water occurs at a temperature of 4.0°C, and freezing takes place at 0°C. The addition of salt lowers both the temperature of maximum density and, to a lesser extent, that of freezing. These relationships are shown in Figure 3301. The two lines meet at a salinity of 24.7 parts per thousand, at which maximum density occurs at the freezing temperature of –1.3°C. At this and greater salinities, the temperature of maximum density of sea water is coincident with the freezing point temperature, i. e., the density increases as the temperature gets colder. At a salinity of 35 parts per thousand, the approximate average for the oceans, the freezing point is –1.88°C.

As the density of surface seawater increases with decreasing temperature, convective density-driven currents are induced bringing warmer, less dense water to the surface. If the polar seas consisted of water with constant salinity, the entire water column would have to be cooled to the freezing point in this manner before ice would begin to form. This is not the case, however, in the polar regions where the vertical salinity distribution is such that the surface waters are underlain at shallow depth by waters of higher salinity. In this instance density currents form a shallow mixed layer which subsequently cannot mix with the deep layer of warmer but saltier water. Ice will then begin forming at the water surface when density currents cease and the surface water reaches its freezing point. In shoal water, however, the mixing process can be sufficient to extend the freezing temperature from the surface to the bottom. Ice crystals can, therefore, form at any depth in this case. Because of their decreased density, they tend to rise to the surface, unless they form at the bottom and attach themselves there. This ice, called anchor ice, may continue to grow as additional ice freezes to that already formed.

3302. Land Ice

Ice of land origin is formed on land by the freezing of freshwater or the compacting of snow as layer upon layer adds to the pressure on that beneath.

Under great pressure, ice becomes slightly plastic, and is forced downward along an inclined surface. If a large area is relatively flat, as on the Antarctic plateau, or if the outward flow is obstructed, as on Greenland, an **ice cap** forms and remains essentially permanent. The thickness of these ice caps ranges from nearly 1 kilometer on Greenland to as much as 4.5 kilometers on the Antarctic Continent. Where ravines or mountain passes permit flow of the ice, a **glacier** is formed. This is a mass of snow and ice which continuously flows to lower levels, exhibiting many of the characteristics of rivers of water. The flow may be more than 30 meters per day, but is generally much less. When a glacier reaches a comparatively level area, it spreads out. When a glacier flows into the sea, the buoyant force of the water breaks off pieces from time to time, and these float away as **icebergs**. Icebergs may be described as dome shaped, sloping or pinnacled (Figure 3302a), tabular (Figure 3302b), glacier, or weathered.

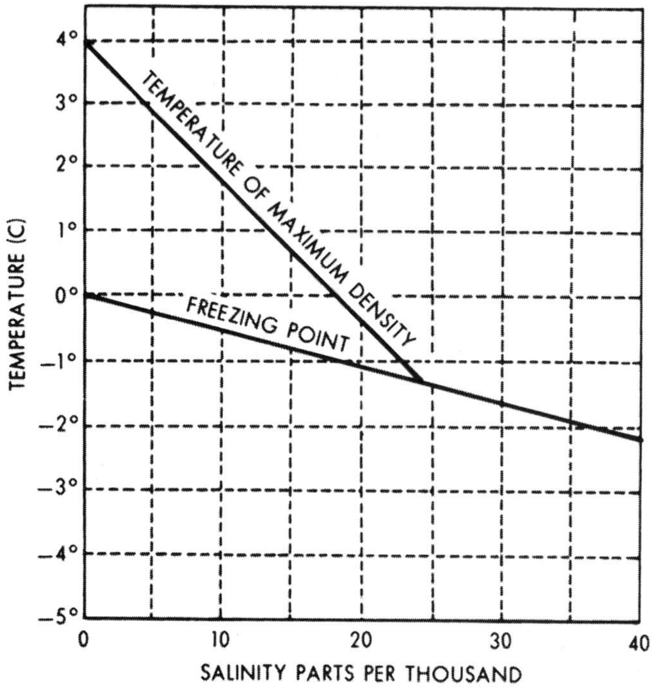

Figure 3301. Relationship between temperature of maximum density and freezing point for water of varying salinity.

A floating iceberg seldom melts uniformly because of lack of uniformity in the ice itself, differences in the temperature above and below the waterline, exposure of one side to the Sun, strains, cracks, mechanical erosion, etc. The inclusion of rocks, silt, and other foreign matter further accentuates the differences. As a result, changes in equilibrium take place, which may cause the berg to periodically tilt or capsize. Parts of it may break off or **calve**, forming separate smaller bergs. A relatively large piece of floating ice, generally extending 1 to 5 meters above the sea surface and normally about 100 to 300 square meters in area, is called a **bergy bit**. A smaller piece of ice large enough to inflict serious damage to a vessel is called a **growler** because of the noise it sometimes makes as it bobs up and down in the sea. Growlers extend less than 1 meter above the sea surface and normally occupy an area of about 20 square meters. Bergy bits and growlers are usually pieces calved from icebergs, but they may be the remains of a mostly melted iceberg.

One danger from icebergs is their tendency to break or capsize. Soon after a berg is calved, while remaining in far northern waters, 60–80% of its bulk is submerged. But as the berg drifts into warmer waters, the underside can sometimes melt faster than the exposed portion, especially in very cold weather. As the mass of the submerged portion deteriorates, the berg becomes increasingly unstable, and it may eventually roll over. Icebergs that have not yet capsized have a jagged and possibly dirty appearance. A recently capsized berg will be smooth, clean, and curved in appearance. Previous waterlines at odd angles can sometimes be seen after one or more capsizings.

The stability of a berg can sometimes be noted by its reaction to ocean swells. The livelier the berg, the more unstable it is. It is extremely dangerous for a vessel to approach an iceberg closely, even one which appears stable, because in addition to the danger from capsizing, unseen cracks can cause icebergs to split in two or calve off large chunks.

Another danger is from underwater extensions, called **rams**, which are usually formed due to melting or erosion above the waterline at a faster rate than below. Rams may also extend from a vertical ice cliff, also known as an **ice front**, which forms the seaward face of a massive ice sheet or floating glacier; or from an **ice wall**, which is the ice cliff forming the seaward margin of a glacier which is aground. In addition to rams, large portions of an iceberg may extend well beyond the waterline at greater depths.

Strangely, icebergs may be helpful to the mariner in some ways. The melt water found on the surface of icebergs is a source of freshwater, and in the past some daring sea-

Figure 3302a. Pinnacled iceberg.

Figure 3302b. A tabular iceberg.

men have made their vessels fast to icebergs which, because they are affected more by currents than the wind, have proceeded to tow them out of the ice pack.

Icebergs can be used as a navigational aid in extreme latitudes where charted depths may be in doubt or non-existent. Since an iceberg (except a large tabular berg) must be at least as deep in the water as it is high to remain upright, a grounded berg can provide an estimate of the minimum water depth at its location. Water depth will be at least equal to the exposed height of the grounded iceberg. Grounded bergs remain stationary while current and wind move sea ice past them. Drifting ice may pile up against the upcurrent side of a grounded berg.

3303. Sea Ice

Sea ice forms by the freezing of seawater and accounts for 95 percent of all ice encountered. The first indication of the formation of new sea ice (up to 10 centimeters in thickness) is the development of small individual, needle-like crystals of ice, called **spicules**, which become suspended in the top few centimeters of seawater. These spicules, also known as **frazil ice**, give the sea surface an oily appearance. **Grease ice** is formed when the spicules coagulate to form a soupy layer on the surface, giving the sea a matte appearance. The next stage in sea ice formation occurs when **shuga**, an accumulation of spongy white ice lumps a few centimeters across, develops from grease ice. Upon further freezing, and depending upon wind exposure, seas, and salinity, shuga and grease ice develop into **nilas**, an elastic crust of high salinity, up to 10 centimeters in thickness, with a matte surface, or into **ice rind**, a brittle, shiny crust of low salinity with a thickness up to approximately 5 centimeters. A layer of 5 centimeters of freshwater ice is brittle but strong enough to support the weight of a heavy man. In contrast, the same thickness of newly formed sea ice will support not more than about 10 percent of this weight, although its strength varies with the temperatures at which it is formed; very cold ice supports a greater weight than warmer ice. As it ages, sea ice becomes harder and more brittle.

New ice may also develop from slush which is formed when snow falls into seawater which is near its freezing point, but colder than the melting point of snow. The snow does not melt, but floats on the surface, drifting with the wind into beds. If the temperature then drops below the freezing point of the seawater, the slush freezes quickly into a soft ice similar to shuga.

Sea ice is exposed to several forces, including currents, waves, tides, wind, and temperature variations. In its early stages, its plasticity permits it to conform readily to virtually any shape required by the forces acting upon it. As it becomes older, thicker, more brittle, and exposed to the influence of wind and wave action, new ice usually separates into circular pieces from 30 centimeters to 3 meters in diameter and up to approximately 10 centimeters in thickness with raised edges due to individual pieces striking against each other. These circular pieces of ice are called **pancake ice** (Figure 3303) and may break into smaller pieces with strong wave motion. Any single piece of relatively flat sea ice less than 20 meters across is called an **ice cake**. With continued low temperatures, individual ice cakes and pancake ice will, depending on wind or wave motion, either freeze together to form a continuous sheet or unite into pieces of ice 20 meters or more across. These larger pieces are then called **ice floes**, which may further freeze together to form an ice covered area greater than 10 kilometers across known as an **ice field**. In wind sheltered areas thickening ice usually forms a continuous sheet before it can develop into the characteristic ice cake form. When sea ice reaches a thickness of between 10 to 30 centimeters it is referred to as **gray** and **gray-white ice**, or collectively as **young ice**, and is the transition stage between nilas and first-year ice. First-year ice usually attains a thickness of between 30 centimeters and 2 meters in its first winter's growth.

Figure 3303. Pancake ice, with an iceberg in the background.

Sea ice may grow to a thickness of 10 to 13 centimeters within 48 hours, after which it acts as an insulator between the ocean and the atmosphere progressively slowing its further growth. However, sea ice may grow to a thickness of between 2 to 3 meters in its first winter. Ice which has survived at least one summer's melt is classified as **old ice**. If it has survived only one summer's melt it may be referred to as **second-year ice**, but this term is seldom used today. Old ice which has attained a thickness of 3 meters or more and has survived at least two summers' melt is known as **multiyear ice** and is almost salt free. This term is increasingly used to refer to any ice more than one season old. Old ice can be recognized by a bluish tone to its surface color in contrast to the greenish tint of first-year ice, but it is often covered with snow. Another sign of old ice is a smoother, more rounded appearance due to melting/refreezing and weathering.

Greater thicknesses in both first and multiyear ice are attained through the deformation of the ice resulting from the movement and interaction of individual floes. Deformation processes occur after the development of new and young ice and are the direct consequence of the effects of winds, tides, and currents. These processes transform a relatively flat sheet of ice into pressure ice which has a rough surface. **Bending**, which is the first stage in the formation of pressure ice, is the upward or downward motion of thin and very plastic ice. Rarely, **tenting** occurs when bending produces an upward displacement of ice forming a flat sided arch with a cavity beneath. More frequently, however, **rafting** takes place as one piece of ice overrides another. When pieces of first-year ice are piled haphazardly over one another forming a wall or line of broken ice, referred to as a **ridge**, the process is known as **ridging**. Pressure ice with topography consisting of numerous mounds or hillocks is called **hummocked ice**, each mound being called a **hummock**.

The motion of adjacent floes is seldom equal. The rougher the surface, the greater is the effect of wind, since each piece extending above the surface acts as a sail. Some ice floes are in rotary motion as they tend to trim themselves into the wind. Since ridges extend below as well as above the surface, the deeper ones are influenced more by deep water currents. When a strong wind blows in the same direction for a considerable period, each floe exerts pressure on the next one, and as the distance increases, the pressure becomes tremendous. Ridges on sea ice are generally about 1 meter high and 5 meters deep, but under considerable pressure may attain heights of 20 meters and depths of 50 meters in extreme cases.

The alternate melting and growth of sea ice, combined with the continual motion of various floes that results in separation as well as consolidation, causes widely varying conditions within the ice cover itself. The mean areal density, or concentration, of pack ice in any given area is expressed in tenths. Concentrations range from:

Open water (total concentration of all ice is < one tenth)
Very open pack (1-3 tenths concentration)
Open pack (4-6 tenths concentration)
Close pack (7-8 tenths concentration)
Very close pack (9-10 to <10-10 concentration)
Compact or consolidated pack (100% coverage)

The extent to which an ice cover of varying concentrations can be penetrated by a vessel varies from place to place and with changing weather conditions. With a concentration of 1 to 3 tenths in a given area, an unreinforced vessel can generally navigate safely, but the danger of receiving heavy damage is always present. When the concentration increases to between 3 and 5 tenths, the area becomes only occasionally accessible to an unreinforced vessel, depending upon the wind and current. With concentrations of 5 to 7 tenths, the area becomes accessible only to ice strengthened vessels, which on occasion will require icebreaker assistance. Navigation in areas with concentrations of 7 tenths or more should only be attempted by icebreakers.

Within the ice cover, openings may develop resulting from a number of deformation processes. Long, jagged **cracks** may appear first in the ice cover or through a single floe. When these cracks part and reach lengths of a few meters to many kilometers, they are referred to as **fractures**. If they widen further to permit passage of a ship, they are called **leads**. In winter, a thin coating of new ice may cover the water within a lead, but in summer the water usually remains ice-free until a shift in the movement forces the two sides together again. A lead ending in a pressure ridge or other impenetrable barrier is a **blind lead**.

A lead between pack ice and shore is a **shore lead**, and one between pack and fast ice is a **flaw lead**. Navigation in these two types of leads is dangerous, because if the pack ice closes with the fast ice, the ship can be caught between the two, and driven aground or caught in the shear zone between.

Before a lead refreezes, lateral motion generally occurs between the floes, so that they no longer fit and unless the pressure is extreme, numerous large patches of open water remain. These nonlinear shaped openings enclosed in ice are called **polynyas**. Polynyas may contain small fragments of floating ice and may be covered with miles of new and young ice. **Recurring polynyas** occur in areas where upwelling of relatively warmer water occurs periodically. These areas are often the site of historical native settlements, where the polynyas permit fishing and hunting at times before regular seasonal ice breakup. Thule, Greenland, is an example.

Sea ice which is formed *in situ* from seawater or by the freezing of pack ice of any age to the shore and which remains attached to the coast, to an ice wall, to an ice front, or between shoals is called **fast ice**. The width of this fast ice varies considerably and may extend for a few meters or several hundred kilometers. In bays and other sheltered areas, fast ice, often augmented by annual snow accumulations and the seaward extension of land ice, may attain a thickness of over 2 meters above the sea surface. When a floating sheet of ice grows to this or a greater thickness and extends over a great horizontal distance, it is called an **ice shelf**. Massive ice shelves, where the ice thickness reaches several hundred meters, are found in both the Arctic and Antarctic.

The majority of the icebergs found in the Antarctic do not originate from glaciers, as do those found in the Arctic, but are calved from the outer edges of broad expanses of shelf ice. Icebergs formed in this manner are called **tabular icebergs**, having a box like shape with horizontal dimensions measured in kilometers, and heights above the sea surface approaching 60 meters. See Figure 3302b. The largest Antarctic ice shelves are found in the Ross and Weddell Seas. The expression "tabular iceberg" is not applied to bergs which break off from Arctic ice shelves; similar formations there are called **ice islands**. These originate when shelf ice, such as that found on the northern coast of Greenland and in the bays of Ellesmere Island, breaks up. As a rule, Arctic ice islands are not as large as the tabular icebergs found in the Antarctic. They attain a thickness of up to 55 meters and on the average extend 5 to 7 meters above the sea surface. Both tabular icebergs and ice islands possess a gently rolling surface. Because of their deep draft, they are influenced much more by current than wind. Arctic ice islands have been used as floating scientific platforms from which polar research has been conducted.

3304. Thickness of Sea Ice

Sea ice has been observed to grow to a thickness of almost

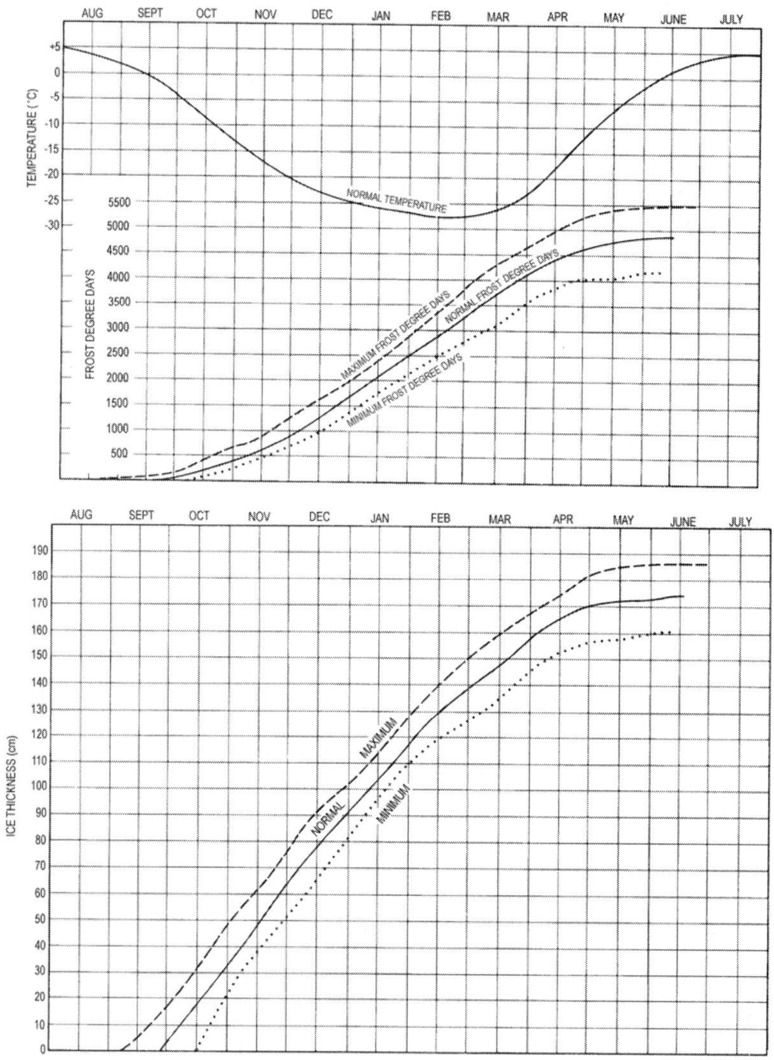

Figure 3304a. Relationship between accumulated frost degree days and theoretical ice thickness at Point Barrow, Alaska.

3 meters during its first year. However, the thickness of first-year ice that has not undergone deformation does not generally exceed 2 meters. In coastal areas where the melting rate is less than the freezing rate, the thickness may increase during succeeding winters, being augmented by compacted and frozen snow, until a maximum thickness of about 3.5 to 4.5 meters may eventually be reached. Old sea ice may also attain a thickness of over 4 meters in this manner, or when summer melt water from its surface or from snow cover runs off into the sea and refreezes under the ice where the seawater temperature is below the freezing point of the fresher melt water.

The growth of sea ice is dependent upon a number of meteorological and oceanographic parameters. Such parameters include air temperature, initial ice thickness, snow depth, wind speed, seawater salinity and density, and the specific heats of sea ice and seawater. Investigations, how-

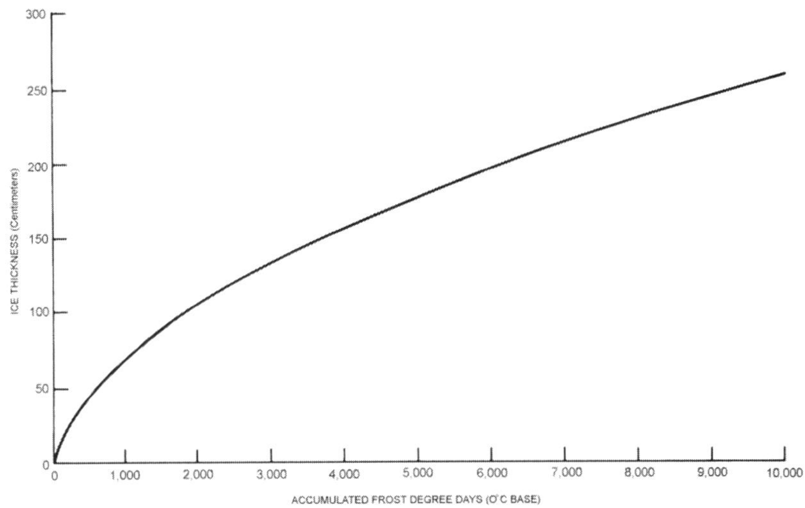

Figure 3304b. Relationship between accumulated frost degree days (°C) and ice thickness (cm).

ever, have shown that the most influential parameters affecting sea ice growth are air temperature, wind speed, snow depth and initial ice thickness. Many complex equations have been formulated to predict ice growth using these four parameters. However, except for the first two, these parameters are not routinely observed for remote polar locations.

Field measurements suggest that reasonable growth estimates can be obtained from air temperature data alone.Various empirical formulae have been developed based on this premise. All appear to perform better under thin ice conditions when the temperature gradient through the ice is linear, generally true for ice less than 100 centimeters thick. Differences in predicted thicknesses between models generally reflect differences in environmental parameters (snowfall, heat content of the underlying water column, etc.) at the measurement site. As a result, such equations must be considered partially site specific and their general use approached with caution. For example, applying an equation derived from central Arctic data to coastal conditions or to Antarctic conditions could lead to substantial errors. For this reason Zubov's formula is widely cited as it represents an average of many years of observations from the Russian Arctic:

$$h^2 + 50h = 8\phi$$

where h is the ice thickness in centimeters for a given day and ϕ is the cumulative number of frost degree days in degrees Celsius since the beginning of the freezing season.

A frost degree day is defined as a day with a mean temperature of 1° below an arbitrary base. The base most commonly used is the freezing point of freshwater (0°C). If, for example, the mean temperature on a given day is 5° below freezing, then five frost degree days are noted for that day. These frost degree days are then added to those noted the next day to obtain an accumulated value, which is then added to those noted the following day. This process is repeated daily throughout the ice growing season. Temperatures usually fluctuate above and below freezing for several days before remaining below freezing. Therefore, frost degree day accumulations are initiated on the first day of the period when temperatures remain below freezing. The relationship between frost degree day accumulations and theoretical ice growth curves at Point Barrow, Alaska is shown in Figure 3304a. Similar curves for other Arctic stations are contained in publications available from the U.S. Naval Oceanographic Office and the National Ice Center. Figure 3304b graphically depicts the relationship between accumulated frost degree days (°C) and ice thickness in centimeters.

During winter, the ice usually becomes covered with snow, which insulates the ice beneath and tends to slow down its rate of growth. This thickness of snow cover varies considerably from region to region as a result of differing climatic conditions. Its depth may also vary widely within very short distances in response to variable winds and ice topography. While this snow cover persists, about 80 to 85 percent of the incoming radiation is reflected back to space. Eventually, however, the snow begins to melt, as the air temperature rises above 0°C in early summer and the resulting freshwater forms puddles on the surface. These puddles absorb about 90 percent of the incoming radiation and rapidly enlarge as they melt the surrounding snow or ice. Eventually the puddles penetrate to the bottom surface of the floes forming **thawholes**. This slow process is characteristic of ice in the Arctic Ocean and seas where movement is restricted by the coastline or islands. Where ice is free to drift into warmer waters (e.g., the Antarctic, East Greenland, and the Labrador Sea), decay is accelerated in response to wave erosion as well as warmer air and sea temperatures.

3305. Salinity of Sea Ice

Sea ice forms first as salt-free crystals near the surface of the sea. As the process continues, these crystals are joined together and, as they do so, small quantities of brine are trapped within the ice. On the average, new ice 15 centimeters thick contains 5 to 10 parts of salt per thousand. With lower temperatures, freezing takes place faster. With faster freezing, a greater amount of salt is trapped in the ice.

Depending upon the temperature, the trapped brine may either freeze or remain liquid, but because its density is greater than that of the pure ice, it tends to settle down through the pure ice. As it does so, the ice gradually freshens, becoming clearer, stronger, and more brittle. At an age of 1 year, sea ice is sufficiently fresh that its melt water, if found in puddles of sufficient size, and not contaminated by spray from the sea, can be used to replenish the freshwater supply of a ship. However, ponds of sufficient size to water ships are seldom found except in ice of great age, and then much of the meltwater is from snow which has accumulated on the surface of the ice. When sea ice reaches an age of about 2 years, virtually all of the salt has been eliminated. Icebergs, having formed from precipitation, contain no salt, and uncontaminated melt water obtained from them is fresh.

The settling out of the brine gives sea ice a honeycomb structure which greatly hastens its disintegration when the temperature rises above freezing. In this state, when it is called **rotten ice**, much more surface is exposed to warm air and water, and the rate of melting is increased. In a day's time, a floe of apparently solid ice several inches thick may disappear completely.

3306. Density of Ice

The density of freshwater ice at its freezing point is 0.917gm/cm^3. Newly formed sea ice, due to its salt content, is more dense, 0.925 gm/cm^3 being a representative value. The density decreases as the ice freshens. By the time it has shed most of its salt, sea ice is less dense than freshwater ice, because ice formed in the sea contains more air bubbles. Ice having no salt but containing air to the extent of 8 percent by volume (an approximately maximum value for sea ice) has a density of 0.845 gm/cm^3.

The density of land ice varies over even wider limits. That formed by freezing of freshwater has a density of 0.917gm/cm^3, as stated above. Much of the land ice, however, is formed by compacting of snow. This results in the entrapping of relatively large quantities of air. **Névé**, a snow which has become coarse grained and compact through temperature change, forming the transition stage to glacier ice, may have an air content of as much as 50 percent by volume. By the time the ice of a glacier reaches the sea, its density approaches that of freshwater ice. A sample taken from an iceberg on the Grand Banks had a density of 0.899gm/cm^3.

When ice floats, part of it is above water and part is below the surface. The percentage of the mass below the surface can be found by dividing the average density of the ice by the density of the water in which it floats. Thus, if an iceberg of density 0.920 floats in water of density 1.028 (corresponding to a salinity of 35 parts per thousand and a temperature of –1°C), 89.5 percent of its mass will be below the surface.

The height to draft ratio for a blocky or tabular iceberg probably varies fairly closely about 1:5. This average ratio was computed for icebergs south of Newfoundland by considering density values and a few actual measurements, and by seismic means at a number of locations along the edge of the Ross Ice Shelf near Little America Station. It was also substantiated by density measurements taken in a nearby hole drilled through the 256-meter thick ice shelf. The height to draft ratios of icebergs become significant when determining their drift.

3307. Drift of Sea Ice

Although surface currents have some affect upon the drift of pack ice, the principal factor is wind. Due to Coriolis force, ice does not drift in the direction of the wind, but varies from approximately 18° to as much as 90° from this direction, depending upon the force of the surface wind and the ice thickness. In the Northern Hemisphere, this drift is to the right of the direction toward which the wind blows, and in the Southern Hemisphere it is toward the left. Although early investigators computed average angles of approximately 28° or 29° for the drift of close multiyear pack ice, large drift angles were usually observed with low, rather than high, wind speeds. The relationship between surface wind speed, ice thickness, and drift angle was derived theoretically for the drift of consolidated pack under equilibrium (a balance of forces acting on the ice) conditions, and shows that the drift angle increases with increasing ice thickness and decreasing surface wind speed. See Figure 3307. A slight increase also occurs with higher latitude.

Since the cross-isobar deflection of the surface wind over the oceans is approximately 20°, the deflection of the ice varies, from approximately along the isobars to as much as 70° to the right of the isobars, with low pressure on the left and high pressure on the right in the Northern Hemisphere. The positions of the low and high pressure areas are, of course, reversed in the Southern Hemisphere.

The rate of drift depends upon the roughness of the sur-

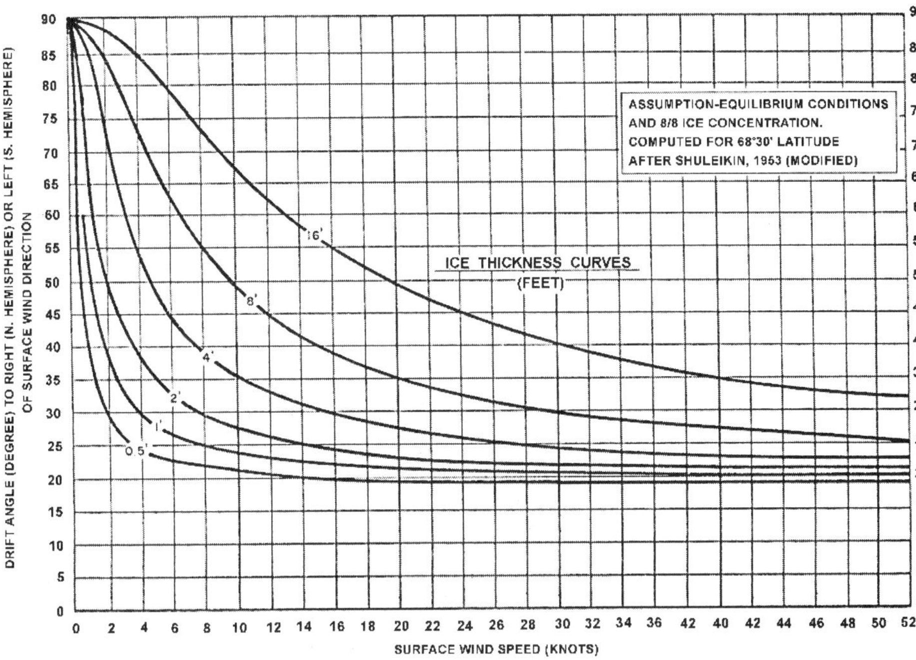

Figure 3307. Ice drift direction for varying wind speed and ice thickness.

face and the concentration of the ice. Percentages vary from approximately 0.25 percent to almost 8 percent of the surface wind speed as measured approximately 6 meters above the ice surface. Low concentrations of heavily ridged or hummocked floes drift faster than high concentrations of lightly ridged or hummocked floes with the same wind speed. Sea ice of 8 to 9 tenths concentrations and six tenths hummocking or close multiyear ice will drift at approximately 2 percent of the surface wind speed. Additionally, the response factors of 1 and 5 tenths ice concentrations, respectively, are approximately three times and twice the magnitude of the response factor for 9 tenths ice concentrations with the same extent of surface roughness. Isolated ice floes have been observed to drift as fast as 10 percent to 12 percent of strong surface winds.

The rates at which sea ice drifts have been quantified through empirical observation. The drift angle, however, has been determined theoretically for 10 tenths ice concentration. This relationship presently is extended to the drift of all ice concentrations, due to the lack of basic knowledge of the dynamic forces that act upon, and result in redistribution of sea ice, in the polar regions.

3308. Iceberg Drift

Icebergs extend a considerable distance below the surface and have relatively small "sail areas" compared to their subsurface mass. Therefore, the near-surface current is thought to be primarily responsible for drift; however, observations have shown that wind can be the dominant force that governs iceberg drift at a particular location or time. Also, the current and wind may contribute nearly equally to the resultant drift.

Two other major forces which act on a drifting iceberg are the Coriolis force and, to a lesser extent, the pressure gradient force which is caused by gravity owing to a tilt of the sea surface, and is important only for iceberg drift in a major current. Near-surface currents are generated by a variety of factors such as horizontal pressure gradients owing to density variations in the water, rotation of the Earth, gravitational attraction of the Moon, and slope of the sea surface. Not only does wind act directly on an iceberg, it also acts indirectly by generating waves and a surface current in about the same direction as the wind. Because of inertia, an iceberg may continue to move from the influence of wind for some time after the wind stops or changes direction.

The relative influence of currents and winds on the

Iceberg type	*Height to draft ratio*
Blocky or tabular	1:5
Rounded or domed	1:4
Picturesque or Greenland (sloping)	1:3
Pinnacled or ridged	1:2
Horned, winged, dry dock, or spired (weathered)	1:1

Table 3308a. Height to draft ratios for various types of icebergs.

Wind Speed (knots)	*Ice Speed/Wind Speed (percent)*		*Drift Angle (degrees)*	
	Small Berg	*Med. Berg*	*Small Berg*	*Med. Berg*
10	3.6	2.2	12	69
20	3.8	3.1	14	55
30	4.1	3.4	17	36
40	4.4	3.5	19	33
50	4.5	3.6	23	32
60	4.9	3.7	24	31

Table 3308b. Drift of iceberg as percentage of wind speed.

drift of an iceberg varies according to the direction and magnitude of the forces acting on its sail area and subsurface cross-sectional area. The resultant force therefore involves the proportions of the iceberg above and below the sea surface in relation to the velocity and depth of the current, and the velocity and duration of the wind. Studies tend to show that, generally, where strong currents prevail, the current is dominant. In regions of weak currents, however, winds that blow for a number of hours in a steady direction materially affect the drift of icebergs. Generally, it can be stated that currents tend to have a greater effect on deep-draft icebergs, while winds tend to have a greater effect on shallow-draft icebergs.

As icebergs waste through melting, erosion, and calving, observations indicate the height to draft ratio may approach 1:1 during their last stage of decay, when they are referred to as a dry dock, winged, horned, or pinnacle icebergs. The height to draft ratios found for icebergs in their various stages are presented in Table 3308a. Since wind tends to have a greater effect on shallow than on deep-draft icebergs, the wind can be expected to exert increasing influence on iceberg drift as wastage increases.

Simple equations which precisely define iceberg drift cannot be formulated at present because of the uncertainty in the water and air drag coefficients associated with iceberg motion. Values for these parameters not only vary from iceberg to iceberg, but they probably change for the same iceberg over its period of wastage.

Present investigations utilize an analytical approach, facilitated by computer calculations, in which the air and water drag coefficients are varied within reasonable limits. Combinations of these drag values are then used in several increasingly complex water models that try to duplicate observed iceberg trajectories. The results indicate that with a wind-generated current, Coriolis force, and a uniform wind, but without a gradient current, small and medium icebergs will drift with the percentages of the wind as given in Table 3308b. The drift will be to the right in the Northern Hemisphere and to the left in the Southern Hemisphere.

When gradient currents are introduced, trajectories vary considerably depending on the magnitude of the wind and current, and whether they are in the same or opposite direction. When a 1-knot current and wind are in the same direction, drift is to the right of both wind and current with drift angles increasing linearly from approximately 5° at 10 knots to 22° at 60 knots. When the wind and a 1-knot current are in opposite directions, drift is to the left of the current, with the angle increasing from approximately 3° at 10 knots, to 20° at 30 knots, and to 73° at 60 knots. As a limiting case for increasing wind speeds, drift may be approximately normal (to the right) to the wind direction. This indicates that the wind driven current is clearly dominating the drift. In general, the various models used demonstrated that a combination of the wind and current was responsible for the drift of icebergs.

3309. Extent of Ice in the Sea

When an area of sea ice, no matter what form it takes or how it is disposed, is described, it is referred to as **pack ice**. In both polar regions the pack ice is a very dynamic feature, with wide deviations in its extent dependent upon changing oceanographic and meteorological phenomena. In winter the Arctic pack extends over the entire Arctic Ocean, and for a varying distance outward from it; the limits recede considerably during the warmer summer months. The average positions of the seasonal absolute and mean maximum and minimum extents of sea ice in the Arctic region are plotted in Figure 3309a. Each year a large portion of the ice from the Arctic Ocean moves outward between Greenland and Spitsbergen (Fram Strait) into the North Atlantic Ocean and is replaced by new ice. Because of this constant annual removal and replacement of sea ice, relatively little of the Arctic pack ice is more than 10 years old.

Ice covers a large portion of the Antarctic waters and is probably the greatest single factor contributing to the isolation of the Antarctic Continent. During the austral

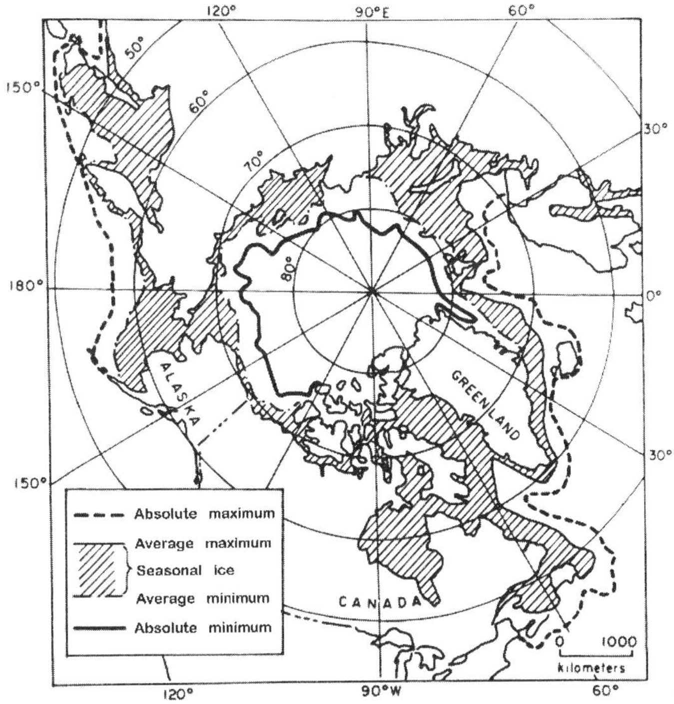

Figure 3309a. Average maximum and minimum extent of Arctic sea ice.

winter (June through September), ice completely surrounds the continent, forming an almost impassable barrier that extends northward on the average to about 54°S in the Atlantic and to about 62°S in the Pacific. Disintegration of the pack ice during the austral summer months of December through March allows the limits of the ice edge to recede considerably, opening some coastal areas of the Antarctic to navigation. The seasonal absolute and mean maximum and minimum positions of the Antarctic ice limit are shown in Figure 3309b.

Historical information on sea ice conditions for specific localities and time periods can be found in publications of the Naval Ice Center/National Ice Center and the National Imagery and Mapping Agency (NIMA). National Ice Center (NIC) publications include sea ice annual atlases (1972 to present for Eastern Arctic, Western Arctic and Antarctica), sea ice climatologies, and forecasting guides. NIC sea ice annual atlases include years 1972 to the present for all Arctic and Antarctic seas. NIC ice climatologies describe multiyear statistics for ice extent and coverage. NIC forecasting guides cover procedures for the production of short-term (daily, weekly), monthly, and seasonal predictions. NIMA publications include sailing directions, which describe localized ice conditions and the effect of ice on polar navigation.

3310. Icebergs in the North Atlantic

Sea level glaciers exist on a number of landmasses bordering the northern seas, including Alaska, Greenland, Svalbard (Spitsbergen), Zemlya Frantsa-Iosifa (Franz Josef Land), Novaya Zemlya, and Severnaya Zemlya (Nicholas II Land). Except in Greenland and Franz Josef Land, the rate of calving is relatively slow, and the few icebergs produced melt near their points of formation. Many of those produced along the western coast of Greenland, however, are eventually carried into the shipping lanes of the North Atlantic, where they constitute a major menace to ships. Those calved from Franz Josef Land glaciers drift southwest in the Barents Sea to the vicinity of Bear Island

Generally the majority of icebergs produced along the east coast of Greenland remain near their source. However, a small number of bergy bits, growlers, and small icebergs are transported south from this region by the East Greenland Current around Kap Farvel at the southern tip of Greenland and then northward by the West Greenland Current into Davis Strait to the vicinity of 67°N. Relatively few of these icebergs menace shipping, but some are carried

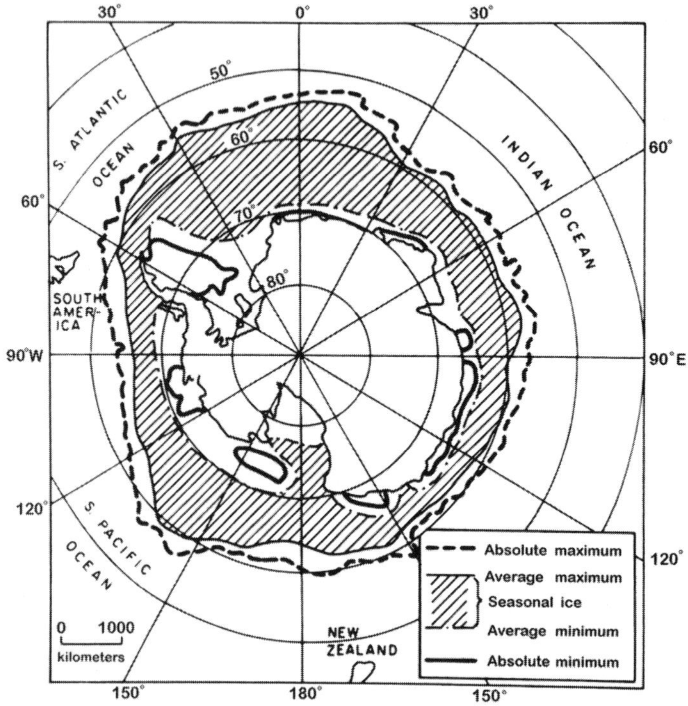

Figure 3309b. Average maximum and minimum extent of Antarctic sea ice.

to the south and southeast of Kap Farvel by a counter-clockwise current gyre centered near 57°N and 43°W.

The main source of the icebergs encountered in the North Atlantic is the west coast of Greenland between 67°N and 76°N, where approximately 10,000–15,000 icebergs are calved each year. In this area there are about 100 low-lying coastal glaciers, 20 of them being the principal producers of icebergs. Of these 20 major glaciers, 2 located in Disko Bugt between 69°N and 70°N are estimated to contribute 28 percent of all icebergs appearing in Baffin Bay and the Labrador Sea. The West Greenland Current carries icebergs from this area northward and then westward until they encounter the south flowing Labrador Current. West Greenland icebergs generally spend their first winter locked in the Baffin Bay pack ice; however, a large number can also be found within the sea ice extending along the entire Labrador coast by late winter.

During the next spring and summer they are transported farther southward by the Labrador Current. The general drift patterns of icebergs that are prevalent in the eastern portion of the North American Arctic are shown in Figure 3310a. Observations over a 101-year period show that an average of 479 icebergs per year reach latitudes south of 48°N, with approximately 10 percent of this total carried south of the Grand Banks (43°N) before they melt. Icebergs may be encountered during any part of the year, but in the Grand Banks area they are most numerous during spring. The maximum monthly average of iceberg sightings below 48°N occurs during April, May and June, with May having the highest average of 147.

It has been suggested that the distribution of the Davis Strait-Labrador Sea pack ice influences the melt rate of the icebergs as they drift south. Sea ice will decrease iceberg erosion by damping waves and holding surface water temperatures below 0°C, so as the areal extent of the sea ice increases the icebergs will tend to survive longer. Stronger than average northerly or northeasterly winds during late winter and spring will enhance sea ice drift to the south, which also may lengthen iceberg lifetimes. There are also large inter-annual variations in the number of icebergs calved from Greenland's glaciers, so the problem of forecasting the length and severity of an iceberg season is exceedingly complex.

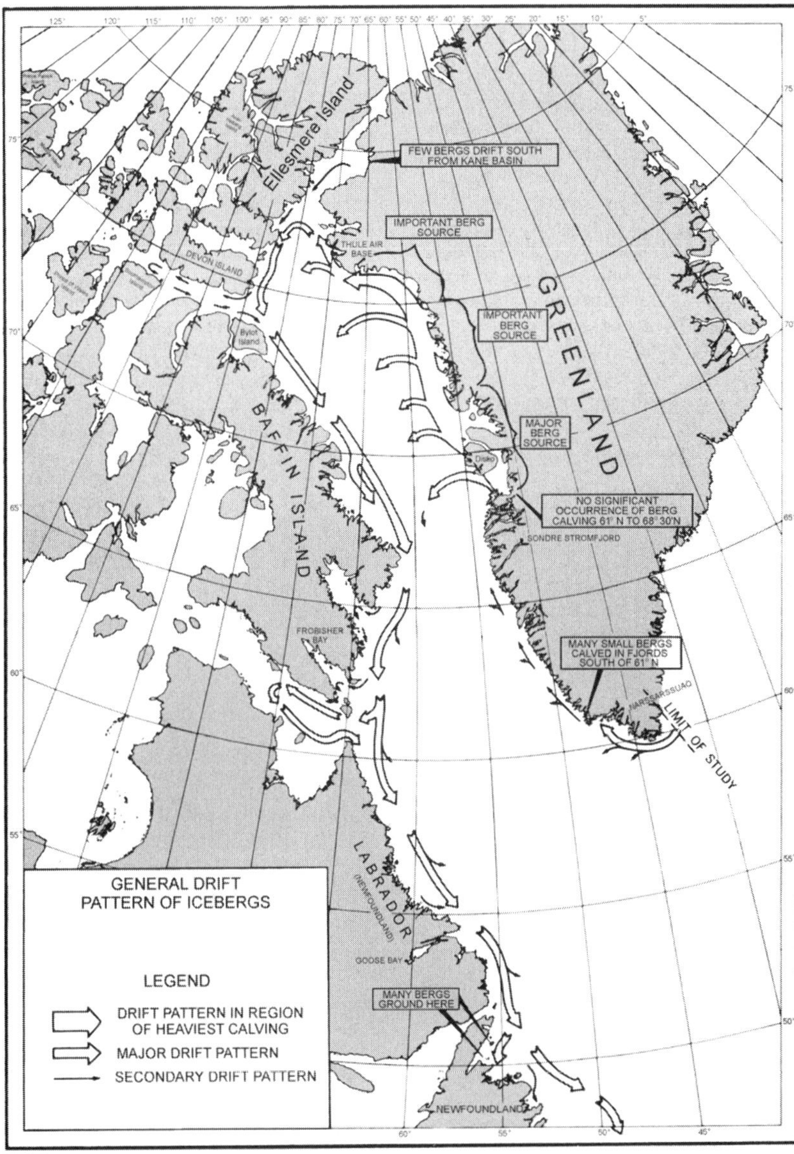

Figure 3310a. General drift patterns of icebergs in Baffin Bay, Davis Strait, and Labrador Sea.

The variation from average conditions is considerable. More than 2,202 icebergs have been sighted south of latitude 48°N in a single year (1984), while in 1966 not a single iceberg was encountered in this area. In the years of 1940 and 1958, only one iceberg was observed south of 48°N. The length of the iceberg "season" as defined by the International Ice Patrol also varies considerably, from a maximum of 203 days in 1992 to the minimum in 1999, when there was no formal ice season. The average length of the ice season is about 130 days. Although this variation has not

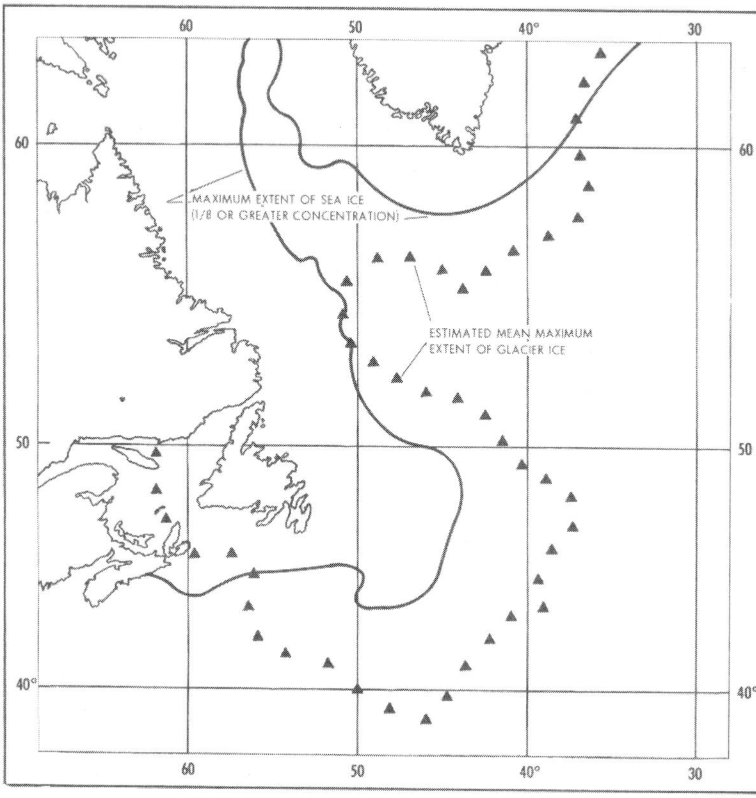

Figure 3310b. Average iceberg and pack ice limits during the month of May.

been fully explained, it is apparently related to wind and ocean current conditions, to the distribution of pack ice in Davis Strait, and to the amount of pack ice off Labrador.

Average iceberg and pack ice limits in this area during May are shown in Figure 3310b. Icebergs have been observed in the vicinity of Bermuda, the Azores, and within 400 to 500 kilometers of Great Britain.

Pack ice may also be found in the North Atlantic, some having been brought south by the Labrador Current and some coming through Cabot Strait after having formed in the Gulf of St. Lawrence.

3311. The International Ice Patrol

The International Ice Patrol was established in 1914 by the *International Convention for the Safety of Life at Sea* (SOLAS), held in 1913 as a result of the sinking of the RMS *Titanic* in 1912. The *Titanic* struck an iceberg on its maiden voyage and sank with the loss of 1,513 lives. In accordance with the agreement reached at the SOLAS conventions of 1960 and 1974, the International Ice Patrol is conducted by the U.S. Coast Guard, which is responsible for the observation and dissemination of information concerning ice conditions in the North Atlantic. Information on ice conditions for the Gulf of St. Lawrence and the coastal waters of Newfoundland and Labrador, including the Strait of Belle Isle, is provided by ECAREG Canada (Eastern Canada Traffic System), through any Coast Guard Radio Station, from the month of December through late June. Sea ice data for these areas can also be obtained from the Ice Operations Officer, located at Dartmouth, Nova Scotia, via Sydney, Halifax, or St. John's marine radio.

During the war years of 1916-18 and 1941-45, the Ice Patrol was suspended. Aircraft were added to the patrol force following World War II, and today perform the majority of the reconnaissance work. During each ice season, aerial reconnaissance surveys are made in the vicinity of the Grand Banks off Newfoundland to determine the southeastern, southern, and southwestern limit of the seaward extent of icebergs. The U.S. Coast Guard aircraft use Side-Looking Airborne Radar (SLAR) as well as Forward-Looking Airborne Radar (FLAR) to help detect and identify icebergs in this notoriously fog-ridden area. Reports of ice sightings are also requested and collected from ships transiting the Grand Banks area. When reporting ice, vessels are requested to detail the concentration and stage of development of sea ice, number of icebergs, the bearing of the principal sea ice edge, and the present ice situation and trend over the preceding three hours. These five parameters are part of the ICE group of the ship synoptic code which is addressed in more detail in Article 3416 on ice observation. In addition to ice reports, masters who do not issue routine weather reports are urged to make sea surface temperature and weather reports to the Ice Patrol every six hours when within latitudes 40° to 52°N and longitudes 38° to 58°W (the Ice Patrol Operations Area). Ice reports may be sent at no charge using INMARSAT Code 42.

International Ice Patrol activities are directed from an Operations Center at Avery Point, Groton, Connecticut. The Ice Patrol gathers all sightings and puts them into a computer model which analyzes and predicts iceberg drift and deterioration. Due to the large size of the Ice Patrol's operating area, icebergs are usually seen only once. The model predictions are crucial to setting the limits of all known ice. The fundamental model force balance is between iceberg acceleration and accelerations due to air and water drag, the Coriolis acceleration, and a sea surface slope term. The model is driven primarily by a water current that combines a depth- and time-independent geostrophic (mean) current with a depth- and time-dependent current driven by the wind (Ekman flow).

Environmental parameters for the model, including sea surface temperature, wave height and period, and wind, are obtained from the U.S. Navy's Fleet Numerical Meteorology and Oceanography Center (FNMOC) in Monterey, California every 12 hours. The International Ice Patrol also deploys from 12–15 World Ocean Circulation Experiment (WOCE) drifting buoys per year, and uses the buoy drifts to alter the climatological mean (geostrophic) currents used by the model in the immediate area of the buoys. The buoy drift data have been archived at the National Oceanographic Data Center (NODC) and are available for use by researchers. Sea surface temperature, wave height and wave period are the main factors that determine the rate of iceberg deterioration. Ship observations of these variables are extremely important because the accuracy of the deterioration model depends on accurate input data.

The results from the iceberg drift and deterioration model are used to compile bulletins that are issued twice daily during the ice season by radio communications from Boston, Massachusetts; St. John's, Newfoundland; and other radio stations. Bulletins are also available over INMARSAT. When icebergs are sighted outside the known limits of ice, special safety broadcasts are issued in between the regularly scheduled bulletins. Iceberg positions in the ice bulletins are updated for drift and deterioration at 12-hour intervals. A radio-facsimile chart is also broadcast twice a day throughout the ice season. A summary of broadcast times and frequencies is found in *Pub. 117, Radio Navigational Aids*, and on the International Ice Patrol Web site, http://www.uscg.mil/lantarea/iip/home.html.

The Ice Patrol, in addition to patrolling possible iceberg areas, conducts oceanographic surveys, maintains up-to-date records of the currents in its area of operation to aid in predicting the drift of icebergs, and studies iceberg conditions in general.

3312. Ice Detection

Safe navigation in the polar seas depends on a number of factors, not the least of which is accurate knowledge of the location and amount of sea ice that lies between the mariner and his destination. Sophisticated electronic equipment, such as radar, sonar, and the visible, infrared, and microwave radiation sensors on board satellites, have added to our ability to detect and thus avoid ice.

As a ship proceeds into higher latitudes, the first ice encountered is likely to be in the form of icebergs, because such large pieces require a longer time to disintegrate. Icebergs can easily be avoided if detected soon enough. The distance at which an iceberg can be seen visually depends upon meteorological visibility, height of the iceberg, source and condition of lighting, and the observer. On a clear day with excellent visibility, a large iceberg might be sighted at a distance of 20 miles. With a low-lying haze around the horizon, this distance will be reduced. In light fog or drizzle this distance is further reduced, down to near zero in heavy fog.

In a dense fog an iceberg may not be perceptible until it is close aboard where it will appear in the form of a luminous, white object if the Sun is shining; or as a dark, somber mass with a narrow streak of blackness at the waterline if the Sun is not shining. If the layer of fog is not too thick, an iceberg may be sighted from aloft sooner than from a point lower on the vessel, but this does not justify omitting a bow lookout. The diffusion of light in a fog will produce a **blink**, or area of whiteness, above and at the sides of an iceberg which will appear to increase the apparent size of its mass.

On dark, clear nights icebergs may be seen at a distance of from 1 to 3 miles, appearing either as white or black objects with occasional light spots where waves break against it. Under such conditions of visibility growlers are a greater menace to vessels; the vessel's speed should be reduced and a sharp lookout maintained.

The Moon may either help or hinder, depending upon its phase and position relative to ship and iceberg. A full Moon in the direction of the iceberg interferes with its detection, while Moonlight from behind the observer may produce a blink which renders the iceberg visible for a greater distance, as much as 3 or more miles. A clouded sky at night, through which the Moonlight is intermittent, also renders ice detection difficult. A night sky with heavy passing clouds may also dim or obscure any object which has been sighted, and fleecy cumulus and cumulonimbus clouds often may give the appearance of blink from icebergs.

If an iceberg is in the process of disintegration, its presence may be detected by a cracking sound as a piece breaks off, or by a thunderous roar as a large piece falls into the water. These sounds are unlikely to be heard due to shipboard noise. The appearance of small pieces of ice in the water often indicates the presence of an iceberg nearby. In calm weather these pieces may form a curved line with the parent iceberg on the concave side. Some of the pieces broken from an iceberg are themselves large enough to be a menace to ships.

As the ship moves closer towards areas known to contain sea ice, one of the most reliable signs that pack ice is being approached is the absence of swell or wave motion in a fresh breeze or a sudden flattening of the sea, especially from leeward. The observation of icebergs is not a good indication that pack ice will be encountered soon, since icebergs may be found at great distances from pack ice. If the sea ice is approached from windward, it is usually compacted and the edge will be sharply defined. However, if it is approached from leeward, the ice is likely to be loose and somewhat scattered, often in long narrow arms.

Another reliable sign of the approach of pack ice not yet in sight is the appearance of a pattern, or **sky map**, on the horizon or on the underside of distant, extensive cloud areas, created by the varying amounts of light reflected from different materials on the sea or Earth's surface. A bright white glare, or **snow blink**, will be observed above a snow covered surface. When the reflection on the underside of clouds is caused by an accumulation of distant ice, the glare is a little less bright and is referred to as an **ice blink**. A relatively dark pattern is reflected on the underside of clouds when it is over land that is not snow covered. This is known as a **land sky**. The darkest pattern will occur when the clouds are above an open water area, and is called a **water sky**. A mariner experienced in recognizing these sky maps will find them useful in avoiding ice or searching out openings which may permit his vessel to make progress through an ice field.

Another indication of the presence of sea ice is the formation of thick bands of fog over the ice edge, as moisture condenses from warm air when passing over the colder ice. An abrupt change in air or sea temperature or seawater salinity is *not* a reliable sign of the approach of icebergs or pack ice.

The presence of certain species of animals and birds can also indicate that pack ice is in close proximity. The sighting of walruses, seals, or polar bears in the Arctic should warn the mariner that pack ice is close at hand. In the Antarctic, the usual precursors of sea ice are penguins, terns, fulmars, petrels, and skuas.

Ice presents only about 1/60th of the radar return of a vessel of the same cross sectional area, and has a reflection coefficient of 0.33. But when visibility becomes limited, radar can prove to be a valuable tool. Although many icebergs will be observed visually on clear days before there is a return on the radarscope, radar under bad weather conditions will detect the average iceberg at a range of about 8 to 10 miles.

The intensity of the return is a function of the nature of the iceberg's exposed surface (slope, surface roughness); however, it is unusual to find an iceberg which will not produce a detectable echo. Ice is not frequency-sensitive; both S- and X-band radars provide the same detectability. The detectability of ice and seawater is almost identical.

In spring in the North Atlantic, especially on the Grand Banks and just when the danger from ice is greatest, atmospheric conditions often produce subnormal radar propagation, shortening the range at which ice can be detected. Large, vertical-sided tabular icebergs of the Antarctic and Arctic ice islands are usually detected by radar at ranges of 15 to 30 miles; a range of 37 miles has been reported.

Whereas a large iceberg is almost always detected by radar in time to be avoided, a growler large enough to be a serious menace to a vessel may be lost in the sea return and escape detection. Growlers cannot usually be detected at ranges greater than four miles, and are lost in a sea greater than four feet. If an iceberg or growler is detected by radar, tracking is sometimes necessary to distinguish it from a rock, islet, or another ship.

Radar can be of great assistance to experienced radar observers. Smooth sea ice, like smooth water, returns little or no echo, but small floes of rough, hummocky sea ice capable of inflicting damage to a ship can be detected in a smooth sea at a range of about 2 to 4 miles. The return may be similar to sea return, but the same echoes appear at each sweep. A lead in smooth ice is clearly visible on a radarscope, even though a thin coating of new ice may have formed in the opening. A light covering of snow obliterating many of the features to the eye has little effect upon a radar return. The ranges at which ice can be detected by radar are somewhat dependent upon refraction, which is sometimes quite abnormal in polar regions.

Experience in interpretation is gained through comparing various radar returns with actual observations. The most effective use of radar in ice detection and navigation is constant surveillance by trained and experienced operators.

Echoes from the ship's whistle or horn may sometimes indicate the presence of icebergs and can give an indication of direction. If the time interval between the sound and its echo is measured, the distance in meters can be determined by multiplying the number of seconds by 168. However, echoes are unreliable because only ice with a large vertical area facing the ship returns enough echo to be heard. Once

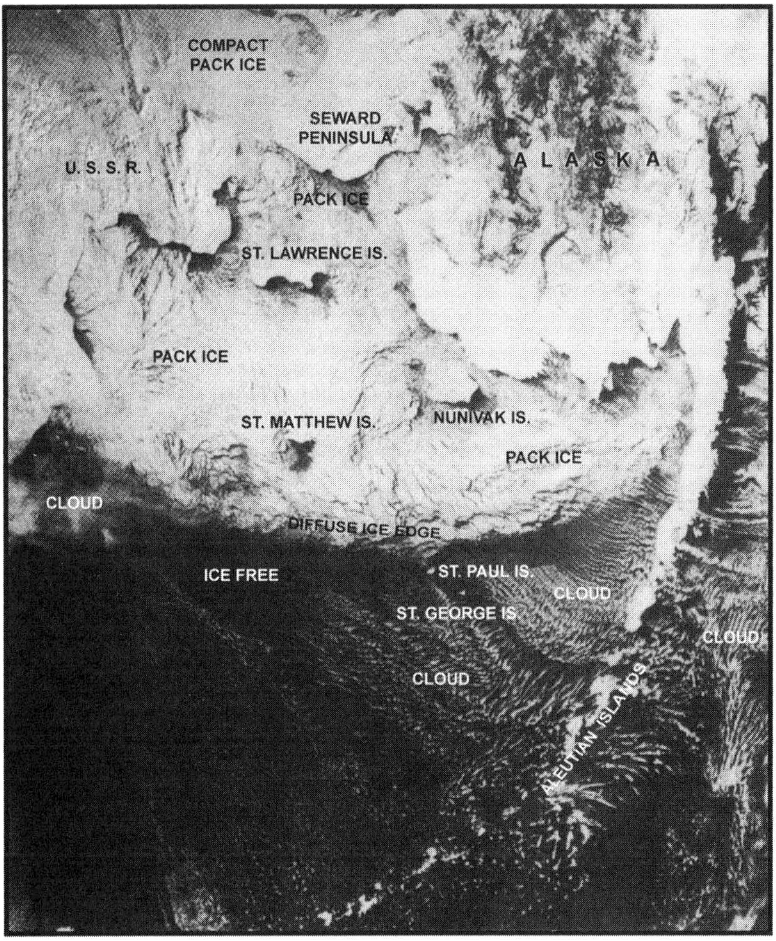

Figure 3312a. Example of satellite imagery with a resolution of 0.9 kilometer.

an echo is heard, a distinct pattern of horn blasts (not a Navigational Rules signal) should be made to confirm that the echo is not another vessel.

At relatively short ranges, sonar is sometimes helpful in locating ice. The initial detection of icebergs may be made at a distance of about 3 miles or more, but usually considerably less. Growlers may be detected at a distance of $^1/_2$ to 2 miles, and even smaller pieces may be detected in time to avoid them.

Ice in the polar regions is best detected and observed from the air, either from aircraft or by satellite. Fixed-winged aircraft have been utilized extensively for obtaining detailed aerial ice reconnaissance information since the early 1930's. Some ships, particularly icebreakers, proceeding into high latitudes carry helicopters, which are invaluable in locating leads and determining the relative navigability of different portions of the ice pack. Ice reports from personnel at Arctic and Antarctic coastal shore stations can also prove valuable to the polar mariner.

The enormous ice reconnaissance capabilities of meteorological satellites were confirmed within hours of the launch by the National Aeronautics and Space Administration (NASA) of the first experimental meteorological satellite, TIROS I, on April 1, 1960. With the advent of the polar-orbiting meteorological satellites during the mid and late 1960's, the U.S. Navy initiated an operational satellite ice reconnaissance program which could observe ice and its

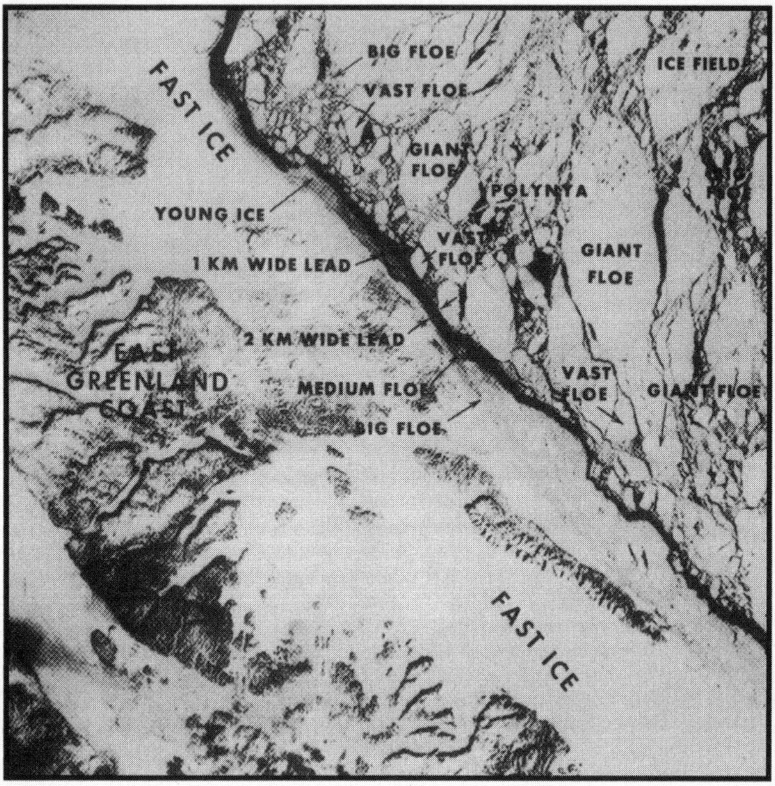

Figure 3312b. Example of satellite imagery with a resolution of 80 meters.

movement in any region of the globe on a daily basis, depending upon solar illumination. Since then, improvements in satellite sensor technology have provided a capability to make detailed global observations of ice properties under all weather and lighting conditions. The current suite of airborne and satellite sensors employed by the National Ice Center include: aerial reconnaissance including visual and Side-Looking Airborne Radar (SLAR), TIROS AVHRR visual and infrared, Defense Meteorological Satellite Program (DMSP) Operational Linescan System (OLS) visual and infrared, all-weather passive microwave from the DMSP Special Sensor Microwave Imager (SSM/I) and the ERS-1 Synthetic Aperture Radar (SAR). Examples of satellite imagery of ice covered waters are shown in Figure 3312a and Figure 3312b.

3313. Operations in Ice

Operations in ice-prone regions necessarily require considerable advanced planning and many more precautionary measures than those taken prior to a typical open ocean voyage. The crew, large or small, of a polar-bound vessel should be thoroughly indoctrinated in the fundamentals of polar operations, utilizing the best information sources available. The subjects covered should include training in ship handling in ice, polar navigation, effects of low temperatures on materials and equipment, damage control procedures, communications problems inherent in polar regions, polar meteorology, sea ice terminology, ice observing and reporting procedures (including classification and codes) and polar survival. Training materials should consist of reports on previous Arctic and Antarctic voyages, sailing directions, ice atlases, training films on polar operations, and U.S. Navy service manuals detailing the recommended procedures to follow during high latitude missions. Various sources of information can be obtained from the Director, National Ice Center, 4251 Suitland Road, Washington, D.C., 20395 and from the Office of Polar Programs, National Science Foundation, 4201 Wilson Blvd., Arlington, VA 22230.

The preparation of a vessel for polar operations is of extreme importance and the considerable experience gained from previous operations should be drawn upon to bring the ship to optimum operating condition. At the very least, operations conducted in ice-infested waters require that the vessel's hull and propulsion system undergo certain modifications.

The bow and waterline of the forward part of the vessel should be heavily reinforced. Similar reinforcement should also be considered for the propulsion spaces of the vessel. Cast iron propellers and those made of a bronze alloy do not possess the strength necessary to operate safely in ice. Therefore, it is strongly recommended that propellers made of these materials be replaced by steel. Other desirable features are the absence of vertical sides, deep placement of the propellers, a blunt bow, metal guards to protect propellers from ice damage, and lifeboats for 150 percent of personnel aboard. The complete list of desirable features depends upon the area of operations, types of ice to be encountered, length of stay in the vicinity of ice, anticipated assistance by icebreakers, and possibly other factors. Strength requirements and the minimum thicknesses deemed necessary for the vessel's frames and additional plating to be used as reinforcement, as well as other procedures needed to outfit a vessel for ice operations, can be obtained from the American Bureau of Shipping. For a more definitive and complete guide to the ice strengthening of ships, the mariner may desire to consult the procedures outlined in Rules for Ice Strengthening of Ships, from the Board of Navigation, Helsinki, Finland.

Equipment necessary to meet the basic needs of the crew and to insure the successful and safe completion of the polar voyage should not be overlooked. A minimum list of essential items should consist of polar clothing and footwear, 100% u/v protective sunglasses, food, vitamins, medical supplies, fuel, storage batteries, antifreeze, explosives, detonators, fuses, meteorological supplies, and survival kits containing sleeping bags, trail rations, firearms, ammunition, fishing gear, emergency medical supplies, and a repair kit.

The vessel's safety depends largely upon the thoroughness of advance preparations, the alertness and skill of its crew, and their ability to make repairs if damage is incurred. Spare propellers, rudder assemblies, and patch materials, together with the equipment necessary to effect emergency repairs of structural damage should be carried. Examples of repair materials needed include quick setting cement, oakum, canvas, timbers, planks, pieces of steel of varying shapes, welding equipment, clamps, and an assortment of nuts, bolts, washers, screws, and nails.

Ice and snow accumulation on the vessel poses a definite capsize hazard. Mallets, baseball bats, ax handles, and scrapers to aid in the removal of heavy accumulations of ice, together with snow shovels and stiff brooms for snow removal should be provided. A live steam line may be useful in removing ice from superstructures.

Navigation in polar waters is at best difficult and, during poor conditions, impossible, except using satellite or inertial systems. Environmental conditions encountered in high latitudes such as fog, storms, compass anomalies, atmospheric effects, and, of course, ice, hinder polar operations. Also, deficiencies in the reliability and detail of hydrographic and geographical information presented on polar navigation charts, coupled with a distinct lack of reliable bathymetry, current, and tidal data, add to the problems of polar navigation. Much work is being carried out in polar regions to improve the geodetic control, triangulation, and quality of hydrographic and topographic information necessary for accurate polar charts. However, until this massive task is completed, the only resource open to the polar navigator, especially during periods of poor environmental conditions, is to rely upon the basic principles of navigation and adapt them to unconventional methods when abnormal situations arise.

Upon the approach to pack ice, a careful decision is needed to determine the best action. Often it is possible to go around the ice, rather than through it. Unless the pack is quite loose, this action usually gains rather than loses time. When skirting an ice field or an iceberg, do so to windward, if a choice is available, to avoid projecting tongues of ice or individual pieces that have been blown away from the main body of ice.

When it becomes necessary to enter pack ice, a thorough examination of the distribution and extent of the ice conditions should be made beforehand from the highest possible location. Aircraft (particularly helicopters) and direct satellite readouts are of great value in determining the nature of the ice to be encountered. The most important features to be noted include the location of open water, such as leads and polynyas, which may be manifested by water sky; icebergs; and the presence or absence of both ice under pressure and rotten ice. Some protection may be offered the propeller and rudder assemblies by trimming the vessel down by the stern slightly (not more than 2–3 feet) prior to entering the ice; however, this precaution usually impairs the maneuvering characteristics of most vessels not specifically built for ice breaking.

Selecting the point of entry into the pack should be done with great care; and if the ice boundary consists of closely packed ice or ice under pressure, it is advisable to skirt the edge until a more desirable point of entry is located. Seek areas with low ice concentrations, areas of rotten ice or those containing navigable leads, and if possible enter from leeward on a course perpendicular to the ice edge. It is also advisable to take into consideration the direction and force of the wind, and the set and drift of the prevailing currents when determining the point of entry and the course followed thereafter. Due to wind induced wave action, ice floes close to the periphery of the ice pack will take on a bouncing motion which can be quite hazardous to the hull of thin-skinned vessels. In addition, note that pack ice will drift slightly to the right of the true wind in the Northern Hemisphere and to the left in the Southern Hemisphere, and that leads opened by the force of the wind will appear perpendicular to the wind direction. If a suitable entry point cannot be located due to less than favorable conditions, patience may be called for. Unfavorable conditions generally improve over a short period of time by a change in the wind, tide, or sea state.

Once in the pack, always try to work with the ice, not against it, and keep moving, but do not rush. Respect the ice but do not fear it. Proceed at slow speed at first, staying in open water or in areas of weak ice if possible. The vessel's speed may be safely increased after it has been ascertained how well it handles under the varying ice conditions encountered. It is better to make good progress in the general direction desired than to fight large thick floes in the exact direction to be made good. However, avoid the temptation to proceed far to one side of the intended track; it is almost always better to back out and seek a more penetrable area. During those situations when it becomes necessary to back, always do so with extreme caution and *with the rudder amidships*. If the ship is stopped by ice, the first command should be "rudder amidships," given while the screw is still turning. This will help protect the propeller when backing and prevent ice jamming between rudder and hull. If the rudder becomes ice-jammed, man after steering, establish communications, and *do not* give any helm commands until the rudder is clear. A quick full-ahead burst may clear it. If it does not, try going to "hard rudder" *in the same direction slowly* while turning full or flank speed ahead.

Ice conditions may change rapidly while a vessel is working in pack ice, necessitating quick maneuvering. Conventional vessels, even if ice strengthened, are not built for ice breaking. The vessel should be conned to first attempt to place it in leads or polynyas, giving due consideration to wind conditions. The age, thickness, and size of ice which can be navigated depends upon the type, size, hull strength, and horsepower of the vessel employed. If contact with an ice floe is unavoidable, never strike it a glancing blow. This maneuver may cause the ship to veer off in a direction which will swing the stern into the ice. If possible, seek weak spots in the floe and hit it head-on at slow speed. Unless the ice is rotten or very young, do not attempt to break through the floe, but rather make an attempt to swing it aside as speed is slowly increased. Keep clear of corners and projecting points of ice, but do so without making sharp turns which may throw the stern against the ice, resulting in a damaged propeller, propeller shaft, or rudder. The use of full rudder in non-emergency situations is not recommended because it may swing either the stern or mid-section of the vessel into the ice. This does not preclude use of alternating full rudder (swinging the rudder) aboard ice-breakers as a technique for penetrating heavy ice.

Offshore winds may open relatively ice free navigable coastal leads, but such leads should not be entered without benefit of icebreaker escort. If it becomes necessary to enter coastal leads, narrow straits, or bays, an alert watch should be maintained since a shift in the wind may force drifting ice down upon the vessel. An increase in wind on the windward side of a prominent point, grounded iceberg, or land ice tongue extending into the sea will also endanger a vessel. It is wiser to seek out leads toward the windward side of the main body of the ice pack. In the event that the vessel is under imminent danger of being trapped close to shore by pack ice, immediately attempt to orient the vessel's bow seaward. This will help to take advantage of the little maneuvering room available in the open water areas found between ice floes. Work carefully through these areas, easing the ice floes aside while maintaining a close watch on the general movement of the ice pack.

If the vessel is completely halted by pack ice, it is best to keep the rudder amidships, and the propellers turning at slow speed. The wash of the propellers will help to clear ice away from the stern, making it possible to back down safely. When the vessel is stuck fast, an attempt first should be made to free the vessel by going full speed astern. If this maneuver proves ineffective, it may be possible to get the vessel's stern to move slightly, thereby causing the bow to shift, by quickly shifting the rudder from one side to the other while going full speed ahead. Another attempt at going astern might then free the vessel. The vessel may also be freed by either transferring water from ballast tanks, causing the vessel to list, or by alternately flooding and emptying the fore and aft tanks. A heavy weight swung out on the cargo boom might give the vessel enough list to break free. If all these methods fail, the utilization of deadmen (2– to 4–meter lengths of timber buried in holes out in the ice and to which a vessel is moored) and ice anchors (a stockless, single fluked hook embedded in the ice) may be helpful. With a deadman or ice anchors attached to the ice astern, the vessel may be warped off the ice by winching while the engines are going full astern. If all the foregoing methods fail, explosives placed in holes cut nearly to the bottom of the ice approximately 10 to 12 meters off the beam of the vessel and detonated while the engines are working full astern might succeed in freeing the vessel. A vessel may also be sawed out of the ice if the air temperature is above the freezing point of seawater.

When a vessel becomes so closely surrounded by ice that all steering control is lost and it is unable to move, it is **beset**. It may then be carried by the drifting pack into shallow water or areas containing thicker ice or icebergs with their accompanying dangerous underwater projections. If ice forcibly presses itself against the hull, the vessel is said to be **nipped**, whether or not damage is sustained. When this occurs, the gradually increasing pressure may be capable of holing the vessel's bottom or crushing the sides. When a vessel is beset or nipped, freedom may be achieved through the careful maneuvering procedures, the physical efforts of the crew, or by the use of explosives similar to those previously detailed. Under severe conditions the mariner's best ally may be patience since there will be many times when nothing can be done to improve the vessel's plight until there is a change in meteorological conditions. It may be well to preserve fuel and perform any needed repairs to the vessel and its engines. Damage to the vessel while it is beset is usually attributable to collisions or pressure exerted between the vessel's hull, propellers, or rudder assembly, and the sharp corners of ice floes. These collisions can be minimized greatly by attempting to align the vessel in such a manner as to insure that the pressure from the surrounding pack ice is distributed as evenly as possible over the hull. This is best accomplished when medium or large ice floes encircle the vessel.

In the vicinity of icebergs, either in or outside of the pack ice, a sharp lookout should be kept and all icebergs given a wide berth. The commanding officers and masters of all vessels, irrespective of their size, should treat all icebergs with great respect. The best locations for lookouts are generally in a crow's nest, rigged in the foremast or housed in a shelter built specifically for a bow lookout in the eyes of a vessel. Telephone communications between these sites and the navigation bridge on larger vessels will prove invaluable. It is dangerous to approach close to an iceberg of any size because of the possibility of encountering underwater extensions, and because icebergs that are disintegrating may suddenly capsize or readjust their masses to new positions of equilibrium. In periods of low visibility the utmost caution is needed at all times. Vessel speed should be reduced and the watch prepared for quick maneuvering. Radar becomes an effective but not infallible tool, and does not negate the need for trained lookouts.

Since icebergs may have from eight to nine-tenths of their masses below the water surface, their drift is generally influenced more by currents than winds, particularly under light wind conditions. The drift of pack ice, on the other hand, is usually dependent upon the wind. Under these conditions, icebergs within the pack may be found moving at a different rate and in a different direction from that of the pack ice. In regions of strong currents, icebergs should always be given a wide berth because they may travel upwind under the influence of contrary currents, breaking heavy pack in their paths and endangering vessels unable to work clear. In these situations, open water will generally be found to leeward of the iceberg, with piled up pack ice to windward. Where currents are weak and a strong wind predominates, similar conditions will be observed as the wind driven ice pack overtakes an iceberg and piles up to windward with an open water area lying to leeward.

Under ice, submarine operations require knowledge of prevailing and expected sea ice conditions to ensure maximum operational efficiency and safety. The most important ice features are the frequency and extent of downward projections (bummocks and ice keels) from the underside of the ice canopy (pack ice and enclosed water areas from the point of view of the submariner), the distribution of thin ice areas through which submarines can attempt to surface, and the probable location of the outer pack edge where submarines can remain surfaced during emergencies to rendezvous with surface ship or helicopter units.

Bummocks are the subsurface counterpart of hummocks, and **ice keels** are similarly related to ridges. When the physical nature of these ice features is considered, it is apparent that ice keels may have considerable horizontal extent, whereas individual bummocks can be expected to have little horizontal extent. In shallow water lanes to the Arctic Basin, such as the Bering Strait and the adjoining portions of the Bering Sea and Chukchi Sea, deep bummocks and ice keels may leave little vertical room for submarine passage. Widely separated bummocks may be circumnavigated but make for a hazardous passage. Extensive ice areas, with numerous bummocks or ice keels which cross the lane may effectively block both surface and submarine passage into the Arctic Basin.

Bummocks and ice keels may extend downward approximately five times their vertical extent above the ice surface. Therefore, observed ridges of approximately 10 meters may extend as much as 50 meters below sea level. Because of the direct relation of the frequency and vertical extent between these surface features and their subsurface counterparts, aircraft ice reconnaissance should be conducted over a planned submarine cruise track before under ice operations commence.

Skylights are thin places (usually less than 1 meter thick) in the ice canopy, and appear from below as relatively light translucent patches in dark surroundings. The undersurface of a skylight is usually flat; not having been subjected to great pressure. Skylights are called large if big enough for a submarine to attempt to surface through them; that is, have a linear extent of at least 120 meters. Skylights smaller than 120 meters are referred to as small. An ice canopy along a submarine's track that contains a number of large skylights or other features such as leads and polynyas, which permit a submarine to surface more frequently than 10 times in 30 miles, is called **friendly ice**. An ice canopy containing no large skylights or other features which permit a submarine to surface is called **hostile ice**.

3314. Great Lakes Ice

Large vessels have been navigating the Great Lakes since the early 1760's. This large expanse of navigable water has since become one of the world's busiest waterways. Due to the northern geographical location of the Great Lakes Basin and its susceptibility to Arctic outbreaks of polar air during winter, the formation of ice plays a major disruptive role in the region's economically vital marine industry. Because of the relatively large size of the five Great Lakes, the ice cover which forms on them is affected by the wind and currents to a greater degree than on smaller lakes.

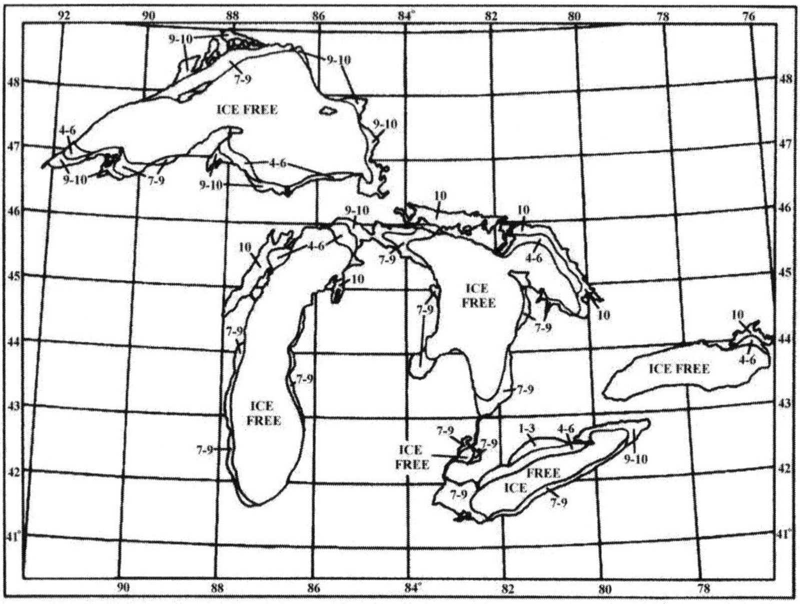

Figure 3314a. Great Lakes ice cover during a mild winter.

The Great Lakes' northern location results in a long ice growth season, which in combination with the effect of wind and current, imparts to their ice covers some of the characteristics and behavior of an Arctic ice pack.

Since the five Great Lakes extend over a distance of approximately 800 kilometers in a north-south direction, each lake is influenced differently by various meteorological phenomena. These, in combination with the fact that each lake also possesses different geographical characteristics, affect the extent and distribution of their ice covers.

The largest, deepest, and most northern of the Great Lakes is Lake Superior. Initial ice formation normally begins at the end of November or early December in harbors and bays along the north shore, in the western portion of the lake and over the shallow waters of Whitefish Bay. As the season progresses, ice forms and thickens in all coastal areas of the lake perimeter prior to extending offshore. This formation pattern can be attributed to a maximum depth in excess of 400 meters and an associated large heat storage capacity that hinders early ice formation in the center of the lake. During a normal winter, ice not under pressure ranges in thickness from 45–85 centimeters. During severe winters, maximum thicknesses are reported to approach 100 centimeters. Winds and currents acting upon the ice have been known to cause ridging with heights approaching 10 meters. During normal years, maximum ice cover extends over approximately 75% of the lake surface with heaviest ice conditions occurring by early March. This value increases to 95% coverage during severe winters and decreases to less than 20% coverage during a mild winter. Winter navigation is most difficult in the southeastern portion of the lake due to heavy ridging and compression of the ice under the influence of prevailing westerly winds. Breakup normally starts near the end of March with ice in a state of advanced deterioration by the middle of April. Under normal conditions, most of the lake is ice-free by the first week of May.

Lake Michigan extends in a north-south direction over 490 kilometers and possesses the third largest surface area of the five Great Lakes. Depths range from 280 meters in the center of the lake to 40 meters in the shipping lanes through the Straits of Mackinac, and less in passages between island groups. During average years, ice formation first occurs in the shallows of Green Bay and extends eastward along the northern coastal areas into the Straits of Mackinac during the second half of December and early January. Ice formation and accumulation proceeds southward with coastal ice found throughout the southern perimeter of the lake by late January. Normal ice thicknesses range from 10–20 centimeters in the south to 40–60

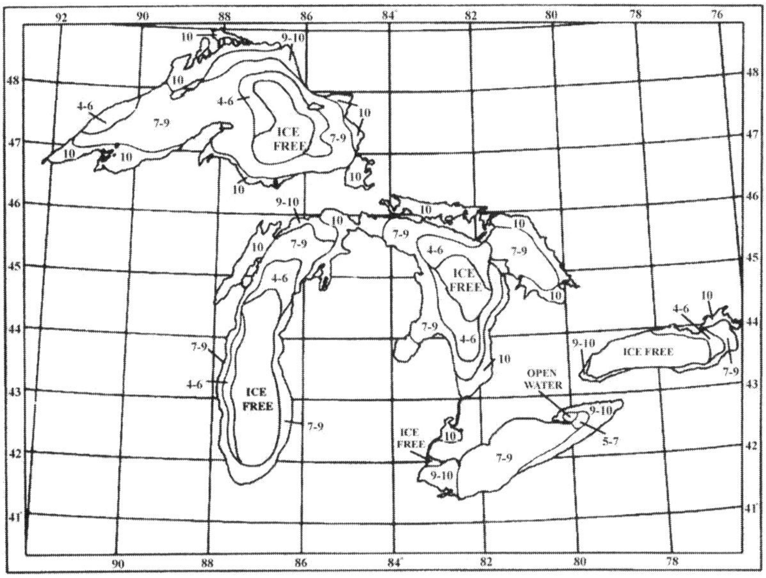

Figure 3314b. Great Lakes ice cover during a normal winter.

centimeters in the north. During normal years, maximum ice cover extends over approximately 40% of the lake surface with heaviest conditions occurring in late February and early March. Ice coverage increases to 85–90% during a severe winter and decreases to only 10–15% during a mild year. Coverage of 100% occurs, but rarely. Throughout the winter, ice formed in mid-lake areas tends to drift eastward because of prevailing westerly winds. This movement of ice causes an area in the southern central portion of the lake to remain ice-free throughout a normal winter. Extensive ridging of ice around the island areas adjacent to the Straits of Mackinac presents the greatest hazard to year-round navigation on this lake. Due to an extensive length and north-south orientation, ice formation and deterioration often occur simultaneously in separate regions of this lake. Ice break-up normally begins by early March in southern areas and progresses to the north by early April. Under normal conditions, only 5–10% of the lake surface is ice covered by mid-April with lingering ice in Green Bay and the Straits of Mackinac completely melting by the end of April.

Lake Huron, the second largest of the Great Lakes, has maximum depths of 230 meters in the central basin west of the Bruce peninsula and 170 meters in Georgian Bay. The pattern of ice formation in Lake Huron is similar to the north-south progression described in Lake Michigan. Initial ice formation normally begins in the North Channel and along the eastern coast of Saginaw and Georgian Bays by mid-December. Ice rapidly expands into the western and southern coastal areas before extending out into the deeper portions of the lake by late January. Normal ice thicknesses are 45–75 centimeters. During severe winters, maximum ice thicknesses often exceed 100 centimeters with windrows of ridged ice achieving thicknesses of up to 10 meters. During normal years, maximum ice cover occurs in late February with 60% coverage in Lake Huron and nearly 95% coverage in Georgian Bay. These values increase to 85–90% in Lake Huron and nearly 100% in Georgian Bay during severe winters. The percent of lake surface area covered by ice decreases to 20–25% for both bodies of water during mild years. During the winter, ice as a hazard to navigation is of greatest concern in the St. Mary's River/North Channel area and the Straits of Mackinac. Ice break-up normally begins in mid-March in southern coastal areas with melting conditions rapidly spreading northward by early April. A recurring threat to navigation is the southward drift and accumulation of melting ice at the entrance of the St. Clair river. Under normal conditions, the lake becomes ice free by the first week of May.

The shallowest and most southern of the Great Lakes is Lake Erie. Although the maximum depth nears 65 meters in the eastern portion of the lake, an overall mean depth of only 20 meters results in the rapid accumulation of ice over

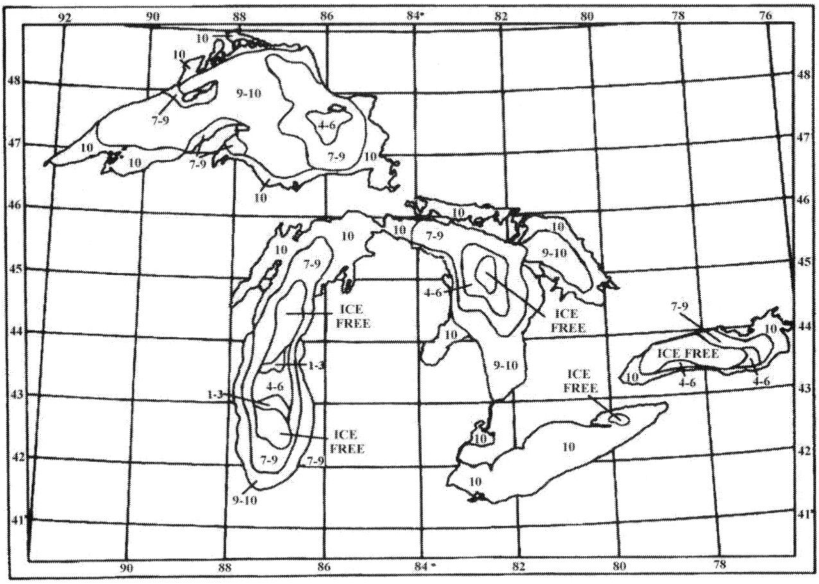

Figure 3314c. Great Lakes ice cover during a severe winter.

a short period of time with the onset of winter. Initial ice formation begins in the very shallow western portion of the lake in mid-December with ice rapidly extending eastward by early January. The eastern portion of the lake does not normally become ice covered until late January. During a normal winter, ice thicknesses range from 25–45 centimeters in Lake Erie. During the period of rapid ice growth, prevailing winds and currents routinely move existing ice to the northeastern end of the lake. This accumulation of ice under pressure is often characterized by ridging with maximum heights of 8–10 meters. During a severe winter, initial ice formation may begin in late November with maximum seasonal ice thicknesses exceeding 70 centimeters. Since this lake reacts rapidly to changes in air temperature, the variability of percent ice cover is the greatest of the five Great Lakes. During normal years, ice cover extends over approximately 90–95% of the lake surface by mid to late February. This value increases to nearly 100% during a severe winter and decreases to 30% ice coverage during a mild year. Lake St. Clair, on the connecting waterway to Lake Huron, is normally consolidated from the middle of January until early March. Ice break-up normally begins in the western portion of Lake Erie in early March with the lake becoming mostly ice-free by the middle of the month. The exception to this rapid deterioration is the extreme eastern end of the lake where ice often lingers until early May.

Lake Ontario has the smallest surface area and second greatest mean depth of the Great Lakes. Depths range from 245 meters in the southeastern portion of the lake to 55 meters in the approaches to the St. Lawrence River. Like Lake Superior, a large mean depth gives Lake Ontario a large heat storage capacity which, in combination with a small surface area, causes Lake Ontario to respond slowly to changing meteorological conditions. As a result, this lake produces the smallest amount of ice cover found on any of the Great Lakes. Initial ice formation normally begins from the middle to late December in the Bay of Quinte and extends to the western coastal shallows near the mouth of the St. Lawrence River by early January. By the first half of February, Lake Ontario is almost 20% ice covered with shore ice lining the perimeter of the lake. During normal years, ice cover extends over approximately 25% of the lake's surface by the second half of February. During this period of maximum ice coverage, ice is typically concentrated in the northeastern portion of the lake by prevailing westerly winds and currents. Ice coverage can extend over 50–60% of the lake surface during a severe winter and less than 10% during a mild year. Level lake ice thicknesses normally fall within the 20–60 centimeter range with occasional reports exceeding 70 centimeters during severe years. Ice break-up normally begins in early March with the lake generally becoming ice-free by mid-April.

The maximum ice cover distribution attained by each of the Great lakes for mild, normal and severe winters is shown in Figure 3314a, Figure 3314b and Figure 3314c. It should be noted that although the average maximum ice cover for each lake appears on the same chart, the actual occurrence of each distribution takes place during the time periods described within the preceding narratives.

Information concerning ice analyses and forecasts for the Great Lakes can be obtained from the Director, National Ice Center, 4251 Suitland Road, Washington D.C. 20395 and the National Weather Service Forecast Office located at Cleveland Hopkins International Airport, Cleveland, Ohio, 44135. Ice climatological information can be obtained from the Great Lakes Environmental Research Laboratory, 2205 Commonwealth Blvd., Ann Arbor, Michigan, 48105 (http://www.glerl.noaa.gov).

ICE INFORMATION SERVICES

3315. Importance of Ice Information

Advance knowledge of ice conditions to be encountered and how these conditions will change over specified time periods are invaluable for both the planning and operational phases of a voyage to the polar regions. Branches of the United States Federal Government responsible for providing operational ice products and services for safety of navigation include the Departments of Defense (U.S. Navy), Commerce (NOAA), and Transportation (U.S. Coast Guard). Manpower and resources from these agencies comprise the National Ice Center (NIC), which replaced the Navy/NOAA Joint Ice Center. The NIC provides ice products and services to U.S. Government military and civilian interests. Routine and tailored ice products of the NIC shown in Table 3317 can be separated into two categories: a) analyses which describe current ice conditions and b) forecasts which define the expected changes in the existing ice cover over a specified time period.

The content of sea ice analyses is directly dependent upon the planned use of the product, the required level of detail, and the availability of on-site ice observations and/or remotely-sensed data. Ice analyses are produced by blending relatively small numbers of visual ice observations from ships, shore stations and fixed wing aircraft with increasing amounts of remotely sensed data. These data include aircraft and satellite imagery in the visual, infrared, passive microwave and radar bands. The efficient receipt and accurate interpretation of these data are critical to producing a near real-time (24–48 hour old) analysis or "picture" of the ice cover. In general, global and regional scale ice analyses depict ice edge location, ice concentrations within the pack and the ice stages of development or thickness. Local scale ice analyses emphasize the location of thin ice covered or open water leads/polynyas, areas of heavy compression, frequency of ridging, and the presence or absence of dangerous multiyear ice and/or icebergs. The parameters defined in this tactical scale analysis are considered critical to both safety of navigation and the efficient routing of ships through the sea ice cover.

3316. Ice Forecasts and Observations

Sea ice forecasts are routinely separated into four temporal classes: short-term (24–72 hour), weekly (5–7 days), monthly (15–30 days) and seasonal (60–90 days) forecasts. Short-term forecasts are generally paired with local-scale ice analyses and focus on changes in the ice cover based on ice drift, ice formation and ablation, and divergent/convergent processes. Of particular importance are the predicted location of the ice edge and the presence or absence of open water polynyas and coastal/flaw leads. The accurate prediction of the location of these ice features are important for both ice avoidance and ice exploitation purposes.

Similar but with less detail, weekly ice forecasts also emphasize the change in ice edge location and concentration areas within the pack. The National Ice Center presently employs several prediction models to produce both short-term and weekly forecasts. These include empirical models which relate ice drift with geostrophic winds and a coupled dynamic/thermodynamic model called the Polar Ice Prediction System (PIPS). Unlike earlier models, the latter accounts for the effects of ice thickness, concentration, and growth on ice drift.

Monthly ice forecasts predict changes in overall ice extent and are based upon the predicted trends in air temperatures, projected paths of transiting low pressure systems, and continuity of ice conditions.

Seasonal or 90 day ice forecasts predict seasonal ice severity and the projected impact on annual shipping operations. Of particular interest to the National Ice Center are seasonal forecasts for the Alaskan North Slope, Baffin Bay for the annual resupply of Thule, Greenland, and Ross Sea/McMurdo Sound in Antarctica. Seasonal forecasts are also important to Great Lakes and St. Lawrence Seaway shipping interests.

Ice services provided to U.S. Government agencies upon request include aerial reconnaissance for polar shipping operations, ship visits for operational briefing and training, and optimum track ship routing (OTSR) recommendations through ice-infested seas. Commercial operations interested in ice products may obtain routinely produced ice products from the National Ice Center as well as ice analyses and forecasts for Alaskan waters from the National Weather Service Forecast Office in Anchorage, Alaska. Specific information on request procedures, types of ice products, ice services, methods of product dissemination and ship weather support is contained in the publication "Environmental Services for Polar Operations" prepared and distributed by the Director, National Ice Center, 4251 Suitland Road, Washington, D.C., 20395.

The U.S. Coast Guard has an additional responsibility, separate from the National Ice Center, for providing icebreaker support for polar operations and the administration and operations of the International Ice Patrol (IIP). Inquiries for further information on these subjects should be sent to Commandant (G–OPN–1), 2100 Second Street S.W., Washington D.C. 20593.

Other countries which provide sea ice information services are as follows: Arctic – Canada, Denmark (Greenland), Japan (Seas of Okhotsk, Japan and Bo Hai), Iceland, Norway, Russia, and the United Kingdom; Antarctic – Argentina, Australia, Chile, Germany, Japan, and Russia; and Baltic – Finland, Germany, Sweden and Russia. Except for the United States, the ice information services of all countries place specific focus upon ice conditions in territorial seas or waters adjacent to claims on the Antarctic continent. The National Ice Center of the United States is the only organization which provides global ice products and services. Names and locations of foreign sea ice service organizations can be found in "Sea Ice Information Services in the World," WMO Publication No. 574.

Mariners operating in and around sea ice can contribute substantially to increasing the knowledge of synoptic ice conditions, and therefore the accuracy of subsequent ice products by routinely taking and distributing ice observations. The code normally used by personnel trained only to take meteorological observations consists of a five character group appended to the World Meteorological Organization (WMO) weather reporting code: FM 13–X SHIP –Report of Surface Observation from a Sea Station. The five digit ICE group has the following format: ICE + $c_is_ib_iD_iz_i$. In general, the symbols represent:

c	=	total concentration of sea ice.
s	=	stage of development of sea ice.
b	=	ice of land origin (number of icebergs, growlers and bergy bits).
D	=	bearing of principal ice edge.
z	=	present situation and trend of conditions over three preceding hours.

The complete format and tables for the code are described in the WMO publication "Manual on Codes", Volume 1, WMO No. 306. This publication is available from the Secretariat of the World Meteorological Organization, Geneva, Switzerland, or on the Web at: http://www.wmo.ch/.

A more complete and detailed reporting code (ICEOB) has been in use since 1972 by vessels reporting to the U.S. National Ice Center. 1993 revisions to this code and the procedures for use are described in the "Ice Observation Handbook" prepared and distributed by the Director, National Ice Center, 4251 Suitland Road, Washington D.C., 20395.

All ice observation codes make use of special nomenclature which is precisely defined in several languages by the WMO publication "Sea Ice Nomenclature", WMO No. 259, TP 145. This publication, available from the Secretariat of the WMO, contains descriptive definitions along with photography of most ice features. This publication is very useful for vessels planning to submit ice observations.

3317. Distribution of Ice Products and Services

The following is intended as a brief overview of the distribution methods for NIC products and services. For detailed information the user should consult the publications discussed in Article 3316 or refer specific inquiries to Director, National Ice Center, 4251 Suitland Road, FOB #4, Room 2301, Washington, D.C. 20395 or call (301) 763–1111 or –2000. The Web address is http://www.natice.noaa.gov/home.htm. Facsimile inquiries can be phoned to (301) 763–1366 and will generally be answered by mail, therefore addresses must be included. NIC ice product distribution methods are as follows:

1. Autopolling: Customer originated menu-driven facsimile product distribution system. Call (301) 763–3190/3191 for menu directions or (301) 763–5972 for assignment of Personal Identification Number (PIN).
2. Autodin: Alphanumeric message transmission to U.S. Government organizations or vessels. Address is NAVICECEN Suitland MD.
3. OMNET/SCIENCENET: electronic mail and bulletin board run by OMNET, Inc. (617) 265–9230. Product request messages may be sent to mailbox NATIONAL.ICE.CTR. Ice products are routinely posted on bulletin board SEA.ICE.
4. INTERNET: Product requests may be forwarded to electronic mail address which is available by request from the NIC at (301) 763–5972.
5. Mail Subscription: For weekly Arctic and Antarctic sea ice analysis charts from the National Climatic Data Center, NESDIS, NOAA, 37 Battery Park Ave., Asheville, NC, 28801–2733. Call (704) 271–4800 with requests for ice products. The NESDIS Web address is http://www.nesdis.noaa.gov.
6. Mail: Annual ice atlases and multiyear ice climatologies are available either from the National Ice Center (if in stock) or from the National Technical Information Service, 5285 Port Royal Road, Springfield, VA, 22161, on the Web at http://www.ntis.gov. Call (703)

487–4600 for sales service desk. Digital files (in SIGRID format) of weekly NIC ice analyses may be obtained from the National Snow and Ice Data Center, CIRES, Box 449, University of Colorado, Boulder, Colorado 80309, on the Web at http://nsidc.org. Call (303) 492–5171 for information.

NAVAL ICE CENTER PRODUCTS

PRODUCT	*FREQUENCY*	*FORMAT*
GLOBAL SCALE		
Eastern Arctic Analysis/Fcst	Wed	Fax Chart
Western Arctic Analysis/Fcst	Tue	Fax Chart
Antarctic Analysis	Thu	Fax Chart
South Ice Limit-East Arctic	Wed	Posted to OMNET
South Ice Limit-West Arctic	Tue	Posted to OMNET
North Ice Limit-Antarctic	Mon	Fax Chart
30 Day Forecast-East Arctic	1st & 15th of month	Fax Chart
30 Day Forecast-West Arctic	1st & 15th of month	Fax Chart
East Arctic Seasonal Outlook	Annually (15 May)	Booklet
West Arctic Seasonal Outlook	Annually (15 May)	Booklet
REGIONAL SCALE		
Alaska Regional Analysis	Tue & Fri	Fax Chart
Great Lakes Analysis	15 Dec–01 May (Mon, Wed, Fri)	Fax Chart
30 Day Forecast-Gt Lakes	15 Nov–15 Apr (1st & 15th of Mo.)	Fax Chart
St. Mary's River Analysis	01 Jan– 01 May (Mon, Wed, Fri)	Fax Chart
Ross Sea/McMurdo Sound	Annually	Booklet
Seasonal Outlook	(30 Oct)	Booklet
Gt. Lakes Seasonal Outlook	Annually (1 Dec)	Fax Chart
LOCAL SCALE		
Large-Scale Analysis-User-Defined Area	Thrice Weekly	Fax Chart

Table 3317. Products produced by National Ice Center.

CHAPTER 34

WEATHER ELEMENTS

GENERAL DESCRIPTION OF THE ATMOSPHERE

3400. Introduction

Weather is the state of the Earth's atmosphere with respect to temperature, humidity, precipitation, visibility, cloudiness, and other factors. **Climate** refers to the average long-term meteorological conditions of a place or region.

All weather may be traced to the effect of the Sun on the Earth. Most changes in weather involve large-scale horizontal motion of air. Air in motion is called **wind**. This motion is produced by differences of atmospheric pressure, which are attributable both to differences of temperature and the nature of the motion itself.

Weather is of vital importance to the mariner. The wind and state of the sea affect dead reckoning. Reduced visibility limits piloting. The state of the atmosphere affects electronic navigation and radio communication. If the skies are overcast, celestial observations are not available; and under certain conditions refraction and dip are disturbed. When wind was the primary motive power, knowledge of the areas of favorable winds was of great importance. Modern vessels are still affected considerably by wind and sea.

3401. The Atmosphere

The **atmosphere** is a relatively thin shell of air, water vapor, and suspended particulates surrounding the Earth. Air is a mixture of gases and, like any gas, is elastic and highly compressible. Although extremely light, it has a definite weight which can be measured. A cubic foot of air at standard sea-level temperature and pressure weighs 1.22 ounces, or about $^1/_{817}$th the weight of an equal volume of water. Because of this weight, the atmosphere exerts a pressure upon the surface of the Earth of about 15 pounds per square inch.

As altitude increases, air pressure decreases due to the decreased weight of air above. With less pressure, the density decreases. More than three-fourths of the air is concentrated within a layer averaging about 7 statute miles thick, called the **troposphere**. This is the region of most "weather," as the term is commonly understood.

The top of the troposphere is marked by a thin transition zone called the **tropopause**, immediately above which is the **stratosphere**. Beyond this lie several other layers having distinctive characteristics. The average height of the tropopause ranges from about 5 miles or less at high latitudes to about 10 miles at low latitudes.

The **standard atmosphere** is a conventional vertical structure of the atmosphere characterized by a standard sea-level pressure of 1013.25 hectopascals of mercury (29.92 inches) and a sea-level air temperature of 15° C (59° F). The temperature decreases with height at the **standard lapse rate**, a uniform 2° C (3.6° F) per thousand feet to 11 kilometers (36,089 feet), and above that remains constant at –56.5° C (-69.7° F).

The **jet stream** refers to relatively strong (greater than 60 knots) quasi-horizontal winds, usually concentrated within a restricted layer of the atmosphere. Research has indicated that the jet stream is important in relation to the sequence of weather. There are two commonly known jet streams. The **sub-tropical jet stream (STJ)** occurs in the region of 30°N during the northern hemisphere winter, decreasing in summer. The core of highest winds in the STJ is found at about 12km altitude (40,000 feet) in the region of 70°W, 40°E, and 150°E, although considerable variability is common. The **polar frontal jet stream (PFJ)** is found in middle to upper-middle latitudes and is discontinuous and variable. Maximum jet stream winds have been measured by weather balloons at 291 knots.

3402. General Circulation Of The Atmosphere

The heat required to warm the air is supplied originally by the Sun. As radiant energy from the Sun arrives at the Earth, about 29 percent is reflected back into space by the Earth and its atmosphere, 19 percent is absorbed by the atmosphere, and the remaining 52 percent is absorbed by the surface of the Earth. Much of the Earth's absorbed heat is radiated back into space. Earth's radiation is in comparatively long waves relative to the short-wave radiation from the Sun because it emanates from a cooler body. Long-wave radiation, readily absorbed by the water vapor in the air, is primarily responsible for the warmth of the atmosphere near the Earth's surface. Thus, the atmosphere acts much like the glass on the roof of a greenhouse. It allows part of the incoming solar radiation to reach the surface of the Earth but is heated by the terrestrial radiation passing outward. Over the entire Earth and for long periods of time, the total outgoing energy must be equivalent to the incoming energy (minus any converted to another form and retained), or the temperature of the Earth and its atmosphere would steadily increase or decrease. In local areas, or over relatively short periods of time, such a balance is not required, and in fact does not exist, resulting in changes such as those occurring from one year to another, in different seasons and in different parts of the day.

The more nearly perpendicular the rays of the Sun strike the surface of the Earth, the more heat energy per unit area is received at that place. Physical measurements show that in the tropics, more heat per unit area is received than is radiated away, and that in polar regions, the opposite is true. Unless there were some process to transfer heat from the tropics to polar regions, the tropics would be much warmer than they are, and the polar regions would be much colder. Atmospheric motions bring about the required transfer of heat. The oceans also participate in the process, but to a lesser degree.

If the Earth had a uniform surface and did not rotate on its axis, with the Sun following its normal path across the sky (solar heating increasing with decreasing latitude), a simple circulation would result, as shown in Figure 3402a. However, the surface of the Earth is far from uniform, being covered with an irregular distribution of land and water. Additionally, the Earth rotates about its axis so that the portion heated by the Sun continually changes. In addition, the axis of rotation is tilted so that as the Earth moves along its orbit about the Sun, seasonal changes occur in the exposure of specific areas to the Sun's rays, resulting in variations in the heat balance of these areas. These factors, coupled with others, result in constantly changing large-scale movements of air. For example, the rotation of the Earth exerts an apparent force, known as **Coriolis force**, which diverts the air from a direct path between high and low pressure areas. The diversion of the air is toward the right in the Northern Hemisphere and toward the left in the Southern Hemisphere. At some distance above the surface of the Earth, the wind tends to blow along lines connecting points of equal pressure called **isobars**. The wind is called a **geostrophic wind** if it blows parallel to the isobars. This normally occurs when the isobars are straight (great circles). However, isobars curve around highs and lows, and the air is not generally able to maintain itself parallel to these. The resulting cross-isobar flow is called a **gradient wind**. Near the surface of the Earth, friction tends to divert the wind from the isobars toward the center of low pressure. At sea, where friction is less than on land, the wind follows the isobars more closely.

A simplified diagram of the general circulation pattern is shown in Figure 3402b. Figure 3402c and Figure 3402d give a generalized picture of the world's pressure distribution and wind systems as actually observed.

A change in pressure with horizontal distance is called a **pressure gradient**. It is maximum along a normal (perpendicular) to the isobars. A force results which is called **pressure gradient force** and is always directed from high to low pressure. Speed of the wind is approximately proportional to this pressure gradient.

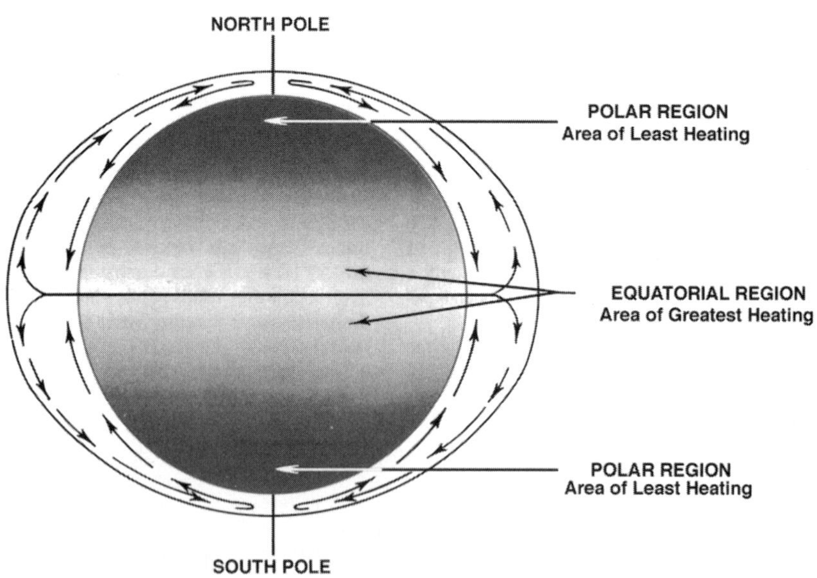

Figure 3402a. Ideal atmospheric circulation for a uniform and non-rotating Earth.

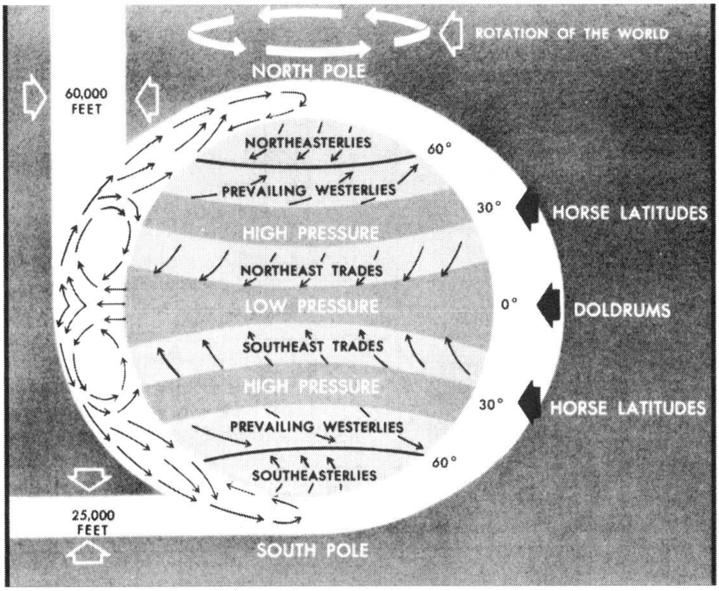

Figure 3402b. Simplified diagram of the general circulation of the atmosphere.

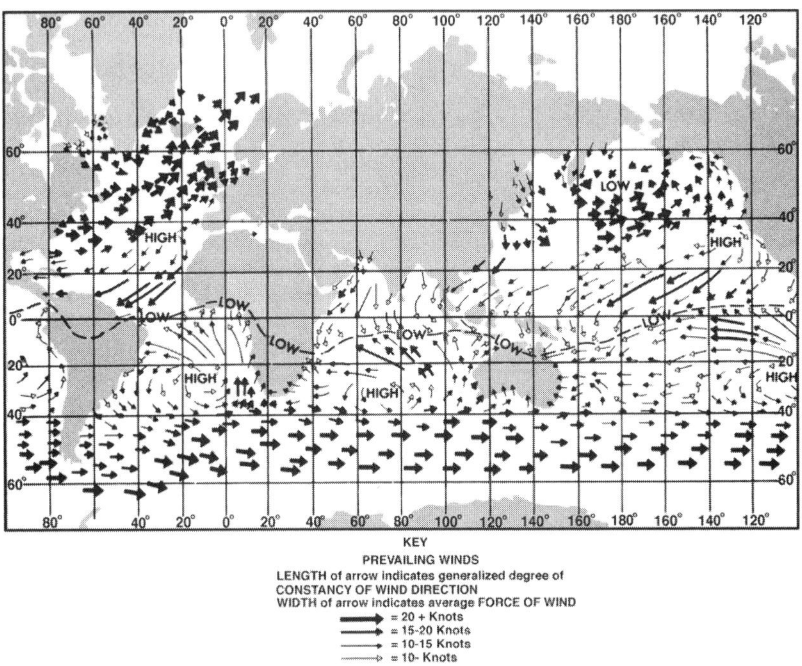

Figure 3402c. Generalized pattern of actual surface winds in January and February.

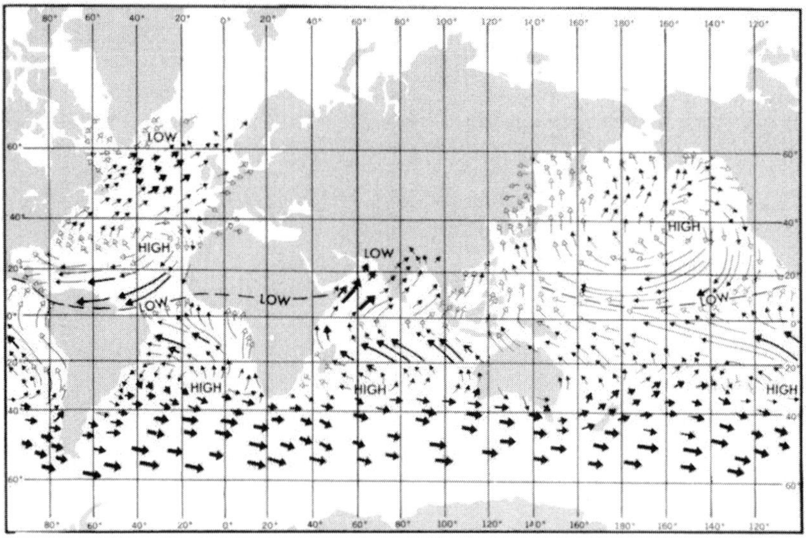

Figure 3402d. Generalized pattern of actual surface winds in July and August. (See key with Figure 3402c.*)*

MAJOR WIND PATTERNS

3403. The Doldrums

A belt of low pressure at the Earth's surface near the equator known as the **doldrums** occupies a position approximately midway between high pressure belts at about latitude 30° to 35° on each side. Except for significant intradiurnal changes, the atmospheric pressure along the equatorial low is almost uniform. With minimal pressure gradient, wind speeds are light and directions are variable. Hot, sultry days are common. The sky is often overcast, and showers and thundershowers are relatively frequent. In these atmospherically unstable areas, brief periods of strong wind occur.

The doldrums occupy a thin belt near the equator, the eastern part in both the Atlantic and Pacific being wider than the western part. However, both the position and extent of the belt vary with longitude and season. During all seasons in the Northern Hemisphere, the belt is centered in the eastern Atlantic and Pacific; however, there are wide excursions of the doldrum regions at longitudes with considerable landmass. On the average, the position is at 5°N, frequently called the **meteorological equator**.

3404. The Trade Winds

The trade winds at the surface blow from the belts of high pressure toward the equatorial belts of low pressure. Because of the rotation of the Earth, the moving air is deflected toward the west. Therefore, the trade winds in the Northern Hemisphere are from the northeast and are called the **northeast trades**, while those in the Southern Hemisphere are from the southeast and are called the **southeast trades**. The trade-wind directions are best defined over eastern ocean areas.

The trade winds are generally considered among the most constant of winds, blowing for days or even weeks with little change of direction or speed. However, at times they weaken or shift direction, and there are regions where the general pattern is disrupted. A notable example is found in the island groups of the South Pacific, where the trades are practically nonexistent during January and February. Their best development is attained in the South Atlantic and in the South Indian Ocean. In general, they are stronger during the winter than during the summer season.

In July and August, when the belt of equatorial low pressure moves to a position some distance north of the equator, the southeast trades blow across the equator, into the Northern Hemisphere, where the Earth's rotation diverts them toward the right, causing them to be southerly and southwesterly winds. The "southwest monsoons" of the African and Central American coasts originate partly in these diverted southeast trades.

Cyclones from the middle latitudes rarely enter the regions of the trade winds, although tropical cyclones originate within these areas.

3405. The Horse Latitudes

Along the poleward side of each trade-wind belt, and corresponding approximately with the belt of high pressure in each hemisphere, is another region with weak pressure gradients and correspondingly light, variable winds. These are called the **horse latitudes**, apparently so named because becalmed sailing ships threw horses overboard in this region when water supplies ran short. The weather is generally good although low clouds are common. Compared to the doldrums, periods of stagnation in the horse latitudes are less persistent. The difference is due primarily to the rising currents of warm air in the equatorial low, which carry large amounts of moisture. This moisture condenses as the air cools at higher levels, while in the horse latitudes the air is apparently descending and becoming less humid as it is warmed at lower heights.

3406. The Prevailing Westerlies

On the poleward side of the high pressure belt in each hemisphere, the atmospheric pressure again diminishes. The currents of air set in motion along these gradients toward the poles are diverted by the Earth's rotation toward the east, becoming southwesterly winds in the Northern Hemisphere and northwesterly in the Southern Hemisphere. These two wind systems are known as the **prevailing westerlies** of the temperate zones.

In the Northern Hemisphere this relatively simple pattern is distorted considerably by secondary wind circulations, due primarily to the presence of large landmasses. In the North Atlantic, between latitudes 40° and 50°, winds blow from some direction between south and northwest during 74 percent of the time, being somewhat more persistent in winter than in summer. They are stronger in winter, too, averaging about 25 knots (Beaufort 6) as compared with 14 knots (Beaufort 4) in the summer.

In the Southern Hemisphere the westerlies blow throughout the year with a steadiness approaching that of the trade winds. The speed, though variable, is generally between 17 and 27 knots (Beaufort 5 and 6). Latitudes 40°S to 50°S, where these boisterous winds occur, are called the **roaring forties**. These winds are strongest at about latitude 50°S.

The greater speed and persistence of the westerlies in the Southern Hemisphere are due to the difference in the atmospheric pressure pattern, and its variations, from the Northern Hemisphere. In the comparatively landless Southern Hemisphere, the average yearly atmospheric pressure diminishes much more rapidly on the poleward side of the high pressure belt, and has fewer irregularities due to continental interference, than in the Northern Hemisphere.

3407. Polar Winds

Partly because of the low temperatures near the geographical poles of the Earth, the surface pressure tends to remain higher than in surrounding regions, since cold air is more dense than warm air. Consequently, the winds blow outward from the poles, and are deflected westward by the rotation of the Earth, to become **northeasterlies** in the Arctic, and **southeasterlies** in the Antarctic. Where the polar easterlies meet the prevailing westerlies, near 50°N and 50°S on the average, a discontinuity in temperature and wind exists. This discontinuity is called the **polar front**. Here the warmer low-latitude air ascends over the colder polar air creating a zone of cloudiness and precipitation.

In the Arctic, the general circulation is greatly modified by surrounding landmasses. Winds over the Arctic Ocean are somewhat variable, and strong surface winds are rarely encountered.

In the Antarctic, on the other hand, a high central landmass is surrounded by water, a condition which augments, rather than diminishes, the general circulation. The high pressure, although weaker than in the horse latitudes, is stronger than in the Arctic, and of great persistence especially in eastern Antarctica. The cold air from the plateau areas moves outward and downward toward the sea and is deflected toward the west by the Earth's rotation. The winds remain strong throughout the year, frequently attaining hurricane force near the base of the mountains. These are some of the strongest surface winds encountered anywhere in the world, with the possible exception of those in well-developed tropical cyclones.

3408. Modifications of the General Circulation

The general circulation of the atmosphere is greatly modified by various conditions. The high pressure in the horse latitudes is not uniformly distributed around the belts, but tends to be accentuated at several points, as shown in Figure 3402c and Figure 3402d. These semi-permanent highs remain at about the same places with great persistence.

Semi-permanent lows also occur in various places, the most prominent ones being west of Iceland, and over the Aleutians (winter only) in the Northern Hemisphere, and in the Ross Sea and Weddell Sea in the Antarctic areas. The regions occupied by these semi-permanent lows are sometimes called the graveyards of the lows, since many lows move directly into these areas and lose their identity as they merge with and reinforce the semi-permanent lows. The low pressure in these areas is maintained largely by the migratory lows which stall there, with topography also important, especially in Antarctica.

Another modifying influence is land, which undergoes greater temperature changes than does the sea. During the

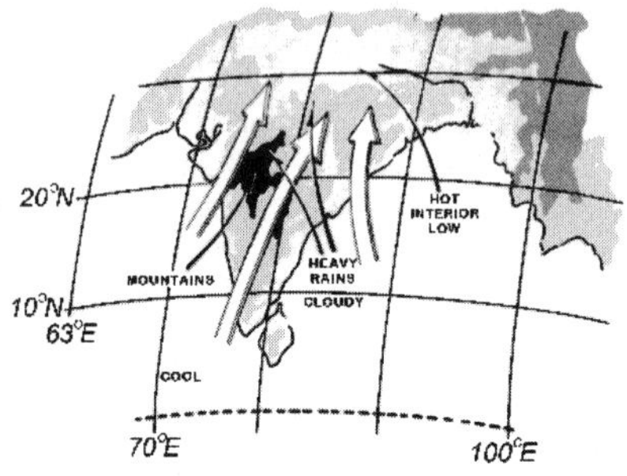

Figure 3408a. The summer monsoon.

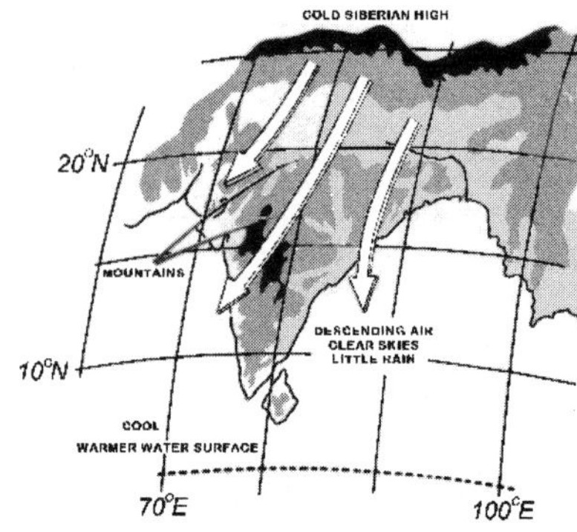

Figure 3408b. The winter monsoon.

summer, a continent is warmer than its adjacent oceans. Therefore, low pressures tend to prevail over the land. If a climatological belt of high pressure encounters a continent, its pattern is distorted or interrupted, whereas a belt of low pressure is intensified over the same area. In winter, the opposite effect takes place, belts of high pressure being intensified over land and those of low pressure being weakened.

The most striking example of a wind system produced by the alternate heating and cooling of a landmass is the **monsoon** (seasonal wind) of the China Sea and Indian Ocean. A portion of this effect is shown in Figure 3408a and Figure 3408b. In the summer, low pressure prevails over the warm continent of Asia, and relatively higher pressure prevails over the adjacent, cooler sea. Between these two systems the wind blows in a nearly steady direction. The lower portion of the pattern is in the Southern Hemisphere, extending to about 10° south latitude. Here the rotation of the Earth causes a deflection to the left, resulting in southeasterly winds. As they cross the equator, the deflection is in the opposite direction, causing them to curve toward the right, becoming southwesterly winds. In the winter, the positions of high and low pressure areas are interchanged, and the direction of flow is reversed.

In the China Sea, the summer monsoon blows from the southwest, usually from May to September. The strong winds are accompanied by heavy squalls and thunderstorms, the rainfall being much heavier than during the winter monsoon. As the season advances, squalls and rain become less frequent. In some places the wind becomes a light breeze which is unsteady in direction, or stops altogether, while in other places it continues almost undiminished, with changes in direction or calms being infrequent. The winter monsoon blows from the northeast, usually from October to April. It blows with a steadiness similar to that of the trade winds, often attaining the speed of a moderate gale (28–33 knots). Skies are generally clear during this season, and there is relatively little rain.

The general circulation is further modified by winds of cyclonic origin and various local winds. Some common local winds are listed by local name below.

Abroholos	A squall frequent from May through August between Cabo de Sao Tome and Cabo Frio on the coast of Brazil.
Bali wind	A strong east wind at the eastern end of Java.
Barat	A heavy northwest squall in Manado Bay on the north coast of the island of Celebes, prevalent from December to February.
Barber	A strong wind carrying damp snow or sleet and spray that freezes upon contact with objects, especially the beard and hair.
Bayamo	A violent wind blowing from the land on the south coast of Cuba, especially near the Bight of Bayamo.
Bentu de Soli	An east wind on the coast of Sardinia.
Bora	A cold, northerly wind blowing from the Hungarian basin into the Adriatic Sea. See also FALL WIND.
Borasco	A thunderstorm or violent squall, especially in the Mediterranean.
Brisa, Briza	1. A northeast wind which blows on the coast of South America or an east wind which blows on Puerto Rico during the trade wind season. 2. The northeast monsoon in the Philippines.

Brisote The northeast trade wind when it is blowing stronger than usual on Cuba.

Brubu A name for a squall in the East Indies.

Bull's Eye Squall A fair weather squall characteristic of the ocean off the coast of South Africa. It is named for the peculiar appearance of the small isolated cloud marking the top of the invisible vortex of the storm.

Cape Doctor The strong southeast wind which blows on the South African coast. Also called the DOCTOR.

Caver, Kaver A gentle breeze in the Hebrides.

Chubasco A violent squall with thunder and lightning, encountered during the rainy season along the west coast of Central America.

Churada A severe rain squall in the Mariana Islands during the northeast monsoon. They occur from November to April or May, especially from January through March.

Cierzo See MISTRAL.

Contrastes Winds a short distance apart blowing from opposite quadrants, frequent in the spring and fall in the western Mediterranean.

Cordonazo The "Lash of St. Francis." Name applied locally to southerly hurricane winds along the west coast of Mexico. It is associated with tropical cyclones in the southeastern North Pacific Ocean. These storms may occur from May to November, but ordinarily affect the coastal areas most severely near or after the Feast of St. Francis, October 4.

Coromell A night land breeze prevailing from November to May at La Paz, near the southern extremity of the Gulf of California.

Doctor 1. A cooling sea breeze in the Tropics. 2. See HARMATTAN. 3. The strong SE wind which blows on the south African coast. Usually called CAPE DOCTOR.

Elephanta A strong southerly or southeasterly wind which blows on the Malabar coast of India during the months of September and October and marks the end of the southwest monsoon.

Etesian A refreshing northerly summer wind of the Mediterranean, especially over the Aegean Sea.

Gregale A strong northeast wind of the central Mediterranean.

Harmattan The dry, dusty trade wind blowing off the Sahara Desert across the Gulf of Guinea and the Cape Verde Islands. Sometimes called the DOCTOR, because of its supposed healthful properties.

Knik Wind A strong southeast wind in the vicinity of Palmer, Alaska, most frequent in the winter.

Kona Storm A storm over the Hawaiian Islands, characterized by strong southerly or southwesterly winds and heavy rains.

Leste A hot, dry, easterly wind of the Madeira and Canary Islands.

Levanter A strong easterly wind of the Mediterranean, especially in the Strait of Gibraltar, attended by cloudy, foggy, and sometimes rainy weather especially in winter.

Levantera A persistent east wind of the Adriatic, usually accompanied by cloudy weather.

Levanto A hot southeasterly wind which blows over the Canary Islands.

Leveche A warm wind in Spain, either a foehn or a hot southerly wind in advance of a low pressure area moving from the Sahara Desert. Called a SIROCCO in other parts of the Mediterranean area.

Maestro A northwesterly wind with fine weather which blows, especially in summer, in the Adriatic. It is most frequent on the western shore. This wind is also found on the coasts of Corsica and Sardinia.

Matanuska Wind A strong, gusty, northeast wind which occasionally occurs during the winter in the vicinity of Palmer, Alaska.

Mistral A cold, dry wind blowing from the north over the northwest coast of the Mediterranean Sea, particularly over the Gulf of Lions. Also called CIERZO. See also FALL WIND.

Nashi, N'aschi A northeast wind which occurs in winter on the Iranian coast of the Persian Gulf, especially near the entrance to the gulf, and also on the Makran coast. It is probably associated with an outflow from the central Asiatic anticyclone which extends over the high land of Iran. It is similar in character but less severe than the BORA.

Norte A strong cold northeasterly wind which blows in Mexico and on the shores of the Gulf of Mexico. It results from an outbreak of cold air from the north. It is the Mexican extension of a norther.

Papagayo A violent northeasterly fall wind on the Pacific coast of Nicaragua and Guatemala. It consists of the cold air mass of a *norte* which has overridden the mountains of Central America. See also TEHUANTEPECER.

Pampero A fall wind of the Argentine coast.

Santa Ana A strong, hot, dry wind blowing out into San Pedro Channel from the southern California desert through Santa Ana Pass.

Shamal A summer northwesterly wind blowing over Iraq and the Persian Gulf, often strong during the day, but decreasing at night.

Sharki A southeasterly wind which sometimes blows in the Persian Gulf.

Sirocco A warm wind of the Mediterranean area, either a foehn or a hot southerly wind in advance of a low pressure area moving from the Sahara or Arabian deserts. Called LEVECHE in Spain.

Squamish A strong and often violent wind occurring in many of the fjords of British Columbia. Squamishes occur in those fjords oriented in a northeast-southwest or east-west direction where cold polar air can be funneled westward. They are notable in Jervis, Toba, and Bute inlets and in Dean Channel and Portland Canal. Squamishes lose their strength when free of the confining fjords and are not noticeable 15 to 20 miles offshore.

Suestado A storm with southeast gales, caused by intense cyclonic activity off the coasts of Argentina and Uruguay, which affects the southern part of the coast of Brazil in the winter.

Sumatra A squall with violent thunder, lightning, and rain, which blows at night in the Malacca Straits, especially during the southwest monsoon. It is intensified by strong mountain breezes.

Taku Wind A strong, gusty, east-northeast wind, occurring in the vicinity of Juneau, Alaska, between October and March. At the mouth of the Taku River, after which it is named, it sometimes attains hurricane force.

Tehuantepecer A violent squally wind from north or north-northeast in the Gulf of Tehuantepec (south of southern Mexico) in winter. It originates in the Gulf of Mexico as a norther which crosses the isthmus and blows through the gap between the Mexican and Guatamalan mountains. It may be felt up to 100 miles out to sea. See also PAPAGAYO.

Tramontana A northeasterly or northerly winter wind off the west coast of Italy. It is a fresh wind of the fine weather mistral type.

Vardar A cold fall wind blowing from the northwest down the Vardar valley in Greece to the Gulf of Salonica. It occurs when atmospheric pressure over eastern Europe is higher than over the Aegean Sea, as is often the case in winter. Also called VARDARAC.

Warm Braw A foehn wind in the Schouten Islands north of New Guinea.

White Squall A sudden, strong gust of wind coming up without warning, noted by whitecaps or white, broken water; usually seen in whirlwind form in clear weather in the tropics.

Williwaw A sudden blast of wind descending from a mountainous coast to the sea, in the Strait of Magellan or the Aleutian Islands.

AIR MASSES

3409. Types of Air Masses

Because of large differences in physical characteristics of the Earth's surface, particularly the oceanic and continental contrasts, the air overlying these surfaces acquires differing values of temperature and moisture. The processes of radiation and convection in the lower portions of the troposphere act in differing characteristic manners for a number of well-defined regions of the Earth. The air overlying these regions acquires characteristics common to the particular area, but contrasting to those of other areas. Each distinctive part of the atmosphere, within which common characteristics prevail over a reasonably large area, is called an **air mass**.

Air masses are named according to their source regions. Four regions are generally recognized: (1) equatorial (E), the doldrums area between the north and south trades; (2) tropical (T), the trade wind and lower temperate regions; (3) polar (P), the higher temperate latitudes; and (4) Arctic or Antarctic (A), the north or south polar regions of ice and snow. This classification is a general indication of relative temperature, as well as latitude of origin.

Air masses are further classified as maritime (m) or continental (c), depending upon whether they form over water or land. This classification is an indication of the relative moisture content of the air mass. Tropical air might be designated maritime tropical (mT) or continental tropical (cT). Similarly, polar air may be either maritime polar (mP) or continental polar (cP). Arctic/Antarctic air, due to the predominance of landmasses and ice fields in the high latitudes, is rarely maritime Arctic (mA). Equatorial air is found exclusively over the ocean surface and is designated neither (cE) nor (mE), but simply (E).

A third classification sometimes applied to tropical and polar air masses indicates whether the air mass is warm (w) or cold (k) relative to the underlying surface. Thus, the symbol mTw indicates maritime tropical air which is warmer than the underlying surface, and cPk indicates continental polar air which is colder than the underlying surface. The w and k classifications are primarily indications of stability (i.e., change of temperature with increasing height). If the air is cold relative to the surface, the lower portion of the air mass will be heated, resulting in instability (temperature markedly decreases with increasing height) as the warmer air tends to rise by convection. Conversely, if the air is warm relative to the surface, the lower portion of the air mass is cooled, tending to remain close to the surface. This is a stable condition (temperature increases with increasing height).

Two other types of air masses are sometimes recognized. These are monsoon (M), a transitional form between cP and E; and superior (S), a special type formed in the free atmosphere by the sinking and consequent warming of air aloft.

3410. Fronts

As air masses move within the general circulation, they travel from their source regions to other areas dominated by air having different characteristics. This leads to a zone of separation between the two air masses, called a **frontal zone** or **front**, across which temperature, humidity, and wind speed and direction change rapidly. Fronts are represented on weather maps by lines; a cold front is shown with pointed barbs, a warm front with rounded barbs, and an occluded front with both, alternating. A stationary front is shown with pointed and rounded barbs alternating and on opposite sides of the line with the pointed barbs away from the colder air. The front may take on a wave-like character, becoming a "frontal wave."

Before the formation of frontal waves, the isobars (lines of equal atmospheric pressure) tend to run parallel to the fronts. As a wave is formed, the pattern is distorted somewhat, as shown in Figure 3410a. In this illustration, colder air is north of warmer air. In Figures 3510a–3510d isobars are drawn at 4-hectopascal intervals.

The wave tends to travel in the direction of the general circulation, which in the temperate latitudes is usually in an easterly and slightly poleward direction.

Along the leading edge of the wave, warmer air is replacing colder air. This is called the **warm front**. The trailing edge is the **cold front**, where colder air is underrunning and displacing warmer air.

The warm air, being less dense, tends to ride up greatly over the colder air it is replacing. Partly because of the replacement of cold, dense air with warm, light air, the pressure decreases. Since the slope is gentle, the upper part of a warm frontal surface may be many hundreds of miles ahead of the surface portion. The decreasing pressure, indicated by a "falling barometer," is often an indication of the approach of such a wave. In a slow-moving, well-developed wave, the barometer may begin to fall several days before the wave arrives. Thus, the amount and nature of the change of atmospheric pressure between observations, called pressure tendency, is of assistance in predicting the approach of such a system.

The advancing cold air, being more dense, tends to ride under the warmer air at the cold front, lifting it to greater heights. The slope here is such that the upper-air portion of the cold front is behind the surface position relative to its motion. After a cold front has passed, the pressure increases, giving a rising barometer.

In the first stages, these effects are not marked, but as the wave continues to grow, they become more pronounced, as shown in Figure 3410b. As the amplitude of the wave increases, pressure near the center usually decreases, and the low is said to "deepen." As it deepens, its forward speed generally decreases.

The approach of a well-developed warm front (i.e.,

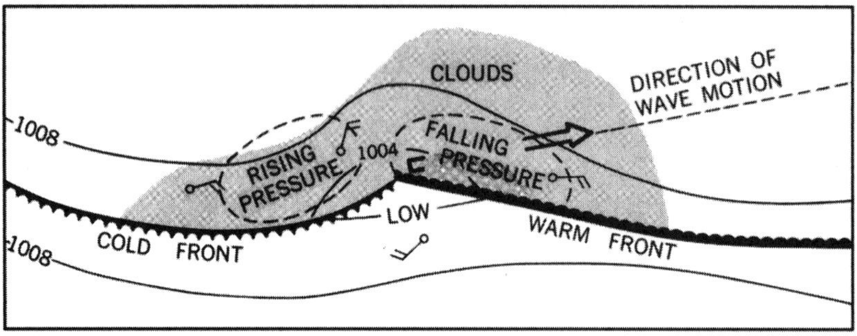

Figure 3410a. First stage in the development of a frontal wave (top view).

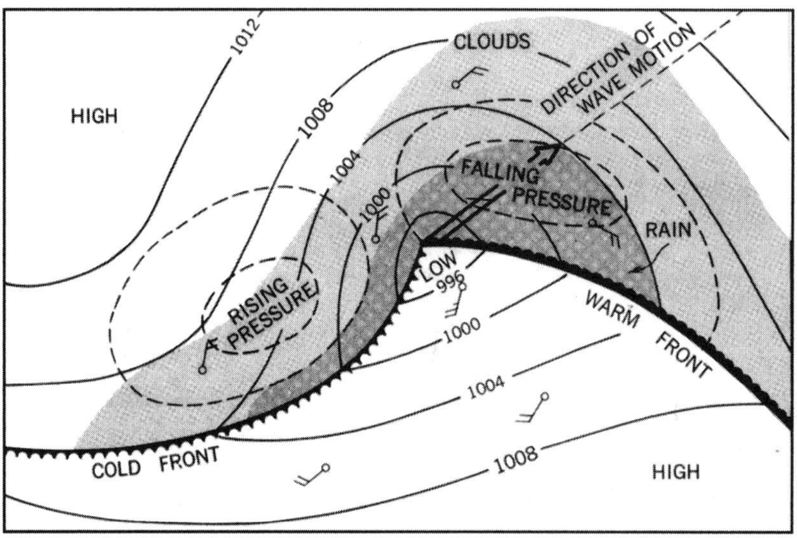

Figure 3410b. A fully developed frontal wave (top view).

when the warm air is mT) is usually heralded not only by falling pressure, but also by a more-or-less regular sequence of clouds. First, cirrus appear. These give way successively to cirrostratus, altostratus, altocumulus, and nimbostratus. Brief showers may precede the steady rain accompanying the nimbostratus.

As the warm front passes, the temperature rises, the wind shifts clockwise (in the Northern Hemisphere), and the steady rain stops. Drizzle may fall from low-lying stratus clouds, or there may be fog for some time after the wind shift. During passage of the warm sector between the warm front and the cold front, there is little change in temperature or pressure. However, if the wave is still growing and the low deepening, the pressure might slowly decrease. In the warm sector the skies are generally clear or partly cloudy, with cumulus or stratocumulus clouds most frequent. The warm air is usually moist, and haze or fog may often be present.

As the faster moving, steeper cold front passes, the wind veers (shifts clockwise in the Northern Hemisphere counter-clockwise in the Southern Hemisphere), the temperature falls rapidly, and there are often brief and sometimes violent squalls with showers, frequently accompanied by thunder

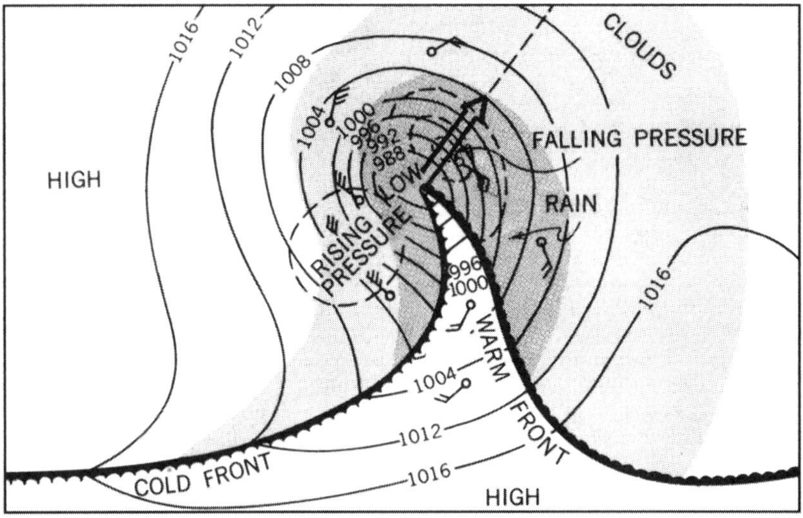

Figure 3410c. A frontal wave nearing occlusion (top view).

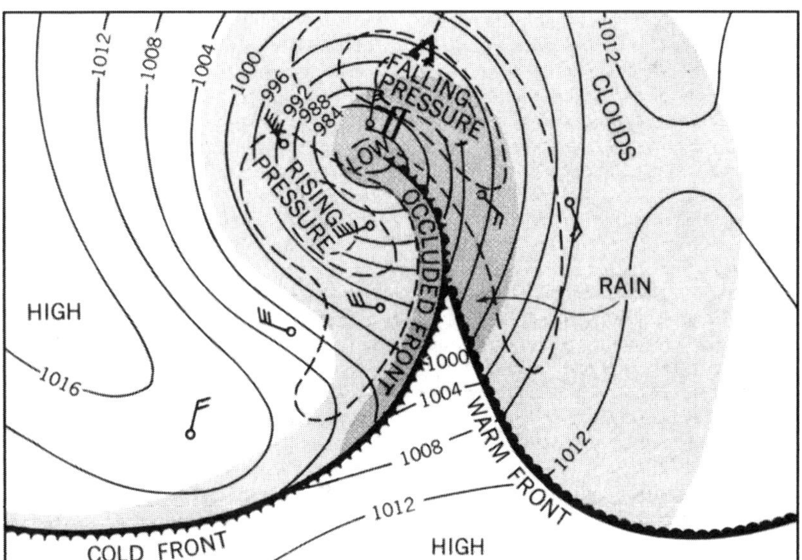

Figure 3410d. An occluded front (top view).

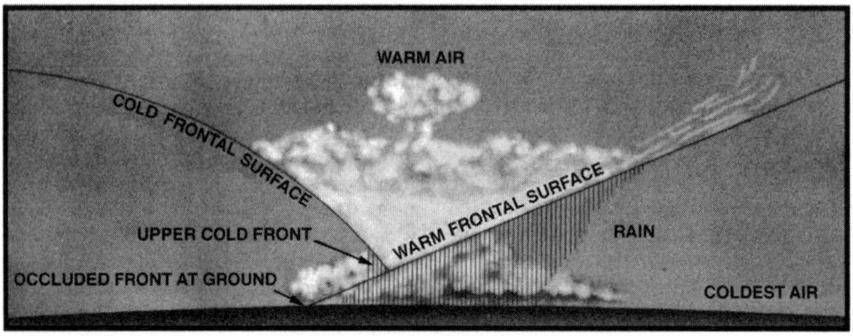

Figure 3410e. An occluded front (cross section).

and lightning. Clouds are usually of the convective type. A cold front usually coincides with a well-defined wind-shift line (a line along which the wind shifts abruptly from southerly or southwesterly to northerly or northwesterly in the Northern Hemisphere, and from northerly or northwesterly to southerly or southwesterly in the Southern Hemisphere). At sea a series of brief showers accompanied by strong, shifting winds may occur along or some distance (up to 200 miles) ahead of a cold front. These are called **squalls** (in common nautical use, the term squall may be additionally applied to any severe local storm accompanied by gusty winds, precipitation, thunder, and lightning), and the line along which they occur is called a **squall line**.

Because of its greater speed and steeper slope, which may approach or even exceed the vertical near the Earth's surface (due to friction), a cold front and its associated weather pass more quickly than a warm front. After a cold front passes, the pressure rises, often quite rapidly, the visibility usually improves, and the clouds tend to diminish. Clear, cool or cold air replaces the warm hazy air.

As the wave progresses and the cold front approaches the slower moving warm front, the low becomes deeper and the warm sector becomes smaller, as shown in Figure 3410c.

Finally, the faster moving cold front overtakes the warm front (Figure 3410d), resulting in an **occluded front** at the surface, and an upper front aloft (Figure 3410e). When the two parts of the cold air mass meet, the warmer portion tends to rise above the colder part. The warm air continues to rise until the entire frontal system dissipates.

As the warmer air is replaced by colder air, the pressure gradually rises, a process called **filling**. This usually occurs within a few days after an occluded front forms. Finally, there results a cold low, or simply a low pressure system across which little or no gradient in temperature and moisture can be found.

The sequence of weather associated with a low depends greatly upon the observer's location with respect to the path of the center. That described above assumes that the low center passes poleward of the observer. If the low center passes south of the observer, between the observer and the equator, the abrupt weather changes associated with the passage of fronts are not experienced. Instead, the change from the weather characteristically found ahead of a warm front, to that behind a cold front, takes place gradually, the exact sequence dictated by distance from the center, and the severity and age of the low.

Although each low generally follows this pattern, no two are ever exactly alike. Other centers of low pressure and high pressure, and the air masses associated with them, even though they may be 1,000 miles or more away, influence the formation and motion of individual low centers and their accompanying weather. Particularly, a high stalls or diverts a low. This is true of temporary highs as well as semi-permanent highs, but not to as great a degree.

3411. Cyclones and Anticyclones

An area of relatively low pressure, generally circular, is called a **cyclone**. Its counterpart for high pressure is called an **anticyclone**. These terms are used particularly in connection with the winds associated with such centers. Wind tends to blow from an area of high pressure to one of low pressure, but due to rotation of the Earth, wind is deflected toward the right in the Northern Hemisphere and toward the left in the Southern Hemisphere.

Because of the rotation of the Earth, therefore, the circulation tends to be counterclockwise around areas of low pressure and clockwise around areas of high pressure in the Northern Hemisphere, and the speed is proportional to the spacing of isobars. In the Southern Hemisphere, the direction of circulation is reversed. Based upon this condition, a general rule, known as Buys Ballot's Law, or the Baric Wind Law, can be stated:

If an observer in the Northern Hemisphere faces away from the surface wind, the low pressure is toward his left; the high pressure is toward his right.

If an observer in the Southern Hemisphere faces away from the surface wind, the low pressure is toward his right; the high pressure is toward his left.

In a general way, these relationships apply in the case

of the general distribution of pressure, as well as to temporary local pressure systems.

The reason for the wind shift along a front is that the isobars have an abrupt change of direction along these lines. Since the direction of the wind is directly related to the direction of isobars, any change in the latter results in a shift in the wind direction.

In the Northern Hemisphere, the wind shifts toward the right (clockwise) when either a warm or cold front passes. In the Southern Hemisphere, the shift is toward the left (counterclockwise). When an observer is on the poleward side of the path of a frontal wave, wind shifts are reversed (i.e., to the left in the Northern Hemisphere and to the right in the Southern Hemisphere).

In an anticyclone, successive isobars are relatively far apart, resulting in light winds. In a cyclone, the isobars are more closely spaced. With a steeper pressure gradient, the winds are stronger.

Since an anticyclonic area is a region of outflowing winds, air is drawn into it from aloft. Descending air is warmed, and as air becomes warmer, its capacity for holding uncondensed moisture increases. Therefore, clouds tend to dissipate. Clear skies are characteristic of an anticyclone, although scattered clouds and showers are sometimes encountered.

In contrast, a cyclonic area is one of converging winds. The resulting upward movement of air results in cooling, a condition favorable to the formation of clouds and precipitation. More or less continuous rain and generally stormy weather are usually associated with a cyclone.

Between the two hemispheric belts of high pressure associated with the horse latitudes, called subtropical anticyclones, cyclones form only occasionally over certain areas at sea, generally in summer and fall. Tropical cyclones (hurricanes and typhoons) are usually quite violent.

In the areas of the prevailing westerlies in temperate latitudes, migratory cyclones (lows) and anticyclones (highs) are a common occurrence. These are sometimes called extratropical cyclones and extratropical anticyclones to distinguish them from the more violent tropical cyclones. Formation occurs over sea and land. The lows intensify as they move poleward; the highs weaken as they move equatorward. In their early stages, cyclones are elongated, as shown in Figure 3410a, but as their life cycle proceeds, they become more nearly circular (Figure 3410b, Figure 3410c, and Figure 3410d).

LOCAL WEATHER PHENOMENA

3412. Local Winds

In addition to the winds of the general circulation and those associated with migratory cyclones and anticyclones, there are numerous local winds which influence the weather in various places.

The most common are the land and sea breezes, caused by alternate heating and cooling of land adjacent to water. The effect is similar to that which causes the monsoons, but on a much smaller scale, and over shorter periods. By day the land is warmer than the water, and by night it is cooler. This effect occurs along many coasts during the summer. Between about 0900 and 1100 local time the temperature of the land becomes greater than that of the adjacent water. The lower levels of air over the land are warmed, and the air rises, drawing in cooler air from the sea. This is the **sea breeze**. Late in the afternoon, when the Sun is low in the sky, the temperature of the two surfaces equalizes and the breeze stops. After sunset, as the land cools below the sea temperature, the air above it is also cooled. The contracting cool air becomes more dense, increasing the pressure near the surface. This results in an outflow of winds to the sea. This is the **land breeze**, which blows during the night and dies away near sunrise. Since the atmospheric pressure changes associated with this cycle are not great, the accompanying winds generally do not exceed gentle to moderate breezes. The circulation is usually of limited extent, reaching a distance of perhaps 20 miles inland, and not more than 5 or 6 miles offshore, and to a height of a few hundred feet. In the doldrums and subtropics, this process is repeated with great regularity throughout most of the year. As the latitude increases, it becomes less prominent, being masked by winds of migratory cyclones and anticyclones. However, the effect often may be present to reinforce, retard, or deflect stronger prevailing winds.

Varying conditions of topography produce a large variety of local winds throughout the world. Winds tend to follow valleys, and to be deflected from high banks and shores. In mountain areas wind flows in response to temperature distribution and gravity. An **anabolic wind** is one that blows up an incline, usually as a result of surface heating. A **katabatic wind** is one which blows down an incline. There are two types, foehn and fall wind.

The foehn (fān) is a warm dry wind which initiates from horizontally moving air encountering a mountain barrier. As it blows upward to clear the mountains, it is cooled below the dew point, resulting in clouds and rain on the windward side. As the air continues to rise, its rate of cooling is reduced because the condensing water vapor gives off heat to the surrounding atmosphere. After crossing the mountain barrier, the air flows downward along the leeward slope, being warmed by compression as it descends to lower levels. Since it loses less heat on the ascent than it gains during descent, and since it has lost its moisture during ascent, it arrives at the bottom of the mountains as very warm, dry air. This accounts for the warm, arid regions along the eastern side of the Rocky Mountains and in similar areas. In the Rocky Mountain region this wind is known by the name **chinook**. It may occur at any season of the year, at any hour of the day or night, and have any speed from a gentle breeze to a gale. It may last for several days, or for a very short period. Its effect is most marked in winter, when it may cause the temperature to rise as much as 20°F to 30°F within 15 minutes, and cause snow and ice to melt within a few hours. On the west coast of the United States, a foehn wind, given the name **Santa Ana**, blows through a pass and down a valley of that name in Southern California. This wind is frequently very strong and may endanger small craft immediately off the coast.

A cold wind blowing down an incline is called a **fall wind**. Although it is warmed somewhat during descent, as is the foehn, it remains cold relative to the surrounding air. It occurs when cold air is dammed up in great quantity on the windward side of a mountain and then spills over suddenly, usually as an overwhelming surge down the other side. It is usually quite violent, sometimes reaching hurricane force. A different name for this type wind is given at each place where it is common. The **tehuantepecer** of the Mexican and Central American coast, the **pampero** of the Argentine coast, the **mistral** of the western Mediterranean, and the **bora** of the eastern Mediterranean are examples of this wind.

Many other local winds common to certain areas have been given distinctive names. A **blizzard** is a violent, intensely cold wind laden with snow mostly or entirely picked up from the ground, although the term is often used popularly to refer to any heavy snowfall accompanied by strong wind. A **dust whirl** is a rotating column of air about 100 to 300 feet in height, carrying dust, leaves, and other light material. This wind, which is similar to a waterspout at sea, is given various local names such as dust devil in southwestern United States and desert devil in South Africa. A **gust** is a sudden, brief increase in wind speed, followed by a slackening, or the violent wind or squall that accompanies a thunderstorm. A puff of wind or a light breeze affecting a small area, such as would cause patches of ripples on the surface of water, is called a **cat's paw**.

3413. Waterspouts

A **waterspout** is a small, whirling storm over ocean or inland waters. Its chief characteristic is a tall, funnel-shaped cloud; when fully developed it is usually attached to the base of a cumulus cloud. See Figure 3413. The water in a waterspout is mostly confined to its lower portion, and may be either salt spray drawn up by the sea surface, or freshwater resulting from condensation due to the lowered pressure in the center of the vortex creating the spout. The air in waterspouts may rotate clockwise or counterclockwise, depending on the manner of formation. They are found most frequently in tropical regions, but are not uncommon in higher latitudes.

There are two types of waterspouts: those derived from violent convective storms over land moving seaward, called tornadoes, and those formed over the sea and which are associated with fair or foul weather. The latter type is most common, lasts a maximum of 1 hour, and has variable strength. Many waterspouts are no stronger than dust whirlwinds, which they resemble; at other times they are strong enough to destroy small craft or to cause damage to larger vessels, although modern ocean-going vessels have little to fear.

Waterspouts vary in diameter from a few feet to several hundred feet, and in height from a few hundred feet to several thousand feet. Sometimes they assume fantastic shapes; in early stages of development an elongated hour glass shape between cloud and sea is common. Since a waterspout is often inclined to the vertical, its actual length may be much greater than indicated by its height.

Figure 3413. Waterspouts.

3414. Deck Ice

Ships traveling through regions where the air temperature is below freezing may acquire thick deposits of ice as a result of salt spray freezing on the rigging, deckhouses, and deck areas. This accumulation of ice is called **ice accretion**. Also, precipitation may freeze to the superstructure and exposed areas of the vessel, increasing the load of ice. See Figure 3414.

On small vessels in heavy seas and freezing weather, deck ice may accumulate very rapidly and increase the topside weight enough to capsize the vessel. Accumulations of more than 2 cm per hour are classified as heavy freezing spray. Fishing vessels with outriggers, A-frames, and other top hamper are particularly susceptible.

Figure 3414. Deck ice.

RESTRICTED VISIBILITY

3415. Fog

Fog is a cloud whose base is at the surface of the Earth, and is composed of droplets of water or ice crystals (ice fog) formed by condensation or crystallization of water vapor in the air.

Radiation fog forms over low-lying land on clear, calm nights. As the land radiates heat and becomes cooler, it cools the air immediately above the surface. This causes a temperature inversion to form, the temperature increasing with height. If the air is cooled to its dew point, fog forms. Often, cooler and more dense air drains down surrounding slopes to heighten the effect. Radiation fog is often quite shallow, and is usually densest at the surface. After sunrise the fog may "lift" and gradually dissipate, usually being entirely gone by noon. At sea the temperature of the water undergoes little change between day and night, and so radiation fog is seldom encountered more than 10 miles from shore.

Advection fog forms when warm, moist air blows over a colder surface and is cooled below its dew point. It is most commonly encountered at sea, may be quite dense, and often persists over relatively long periods. Advection fog is common over cold ocean currents. If the wind is strong enough to thoroughly mix the air, condensation may take place at some distance above the surface of the Earth, forming low stratus clouds rather than fog.

Off the coast of California, seasonal winds create an offshore current which displaces the warm surface water, causing an upwelling of colder water. Moist Pacific air is transported along the coast in the same wind system, and is cooled by the relatively cold water. Advection fog results. In the coastal valleys, fog is sometimes formed when moist air blown inland during the afternoon is cooled by radiation during the night.

When very cold air moves over warmer water, wisps of visible water vapor may rise from the surface as the water "steams," In extreme cases this **frost smoke**, or **Arctic sea smoke**, may rise to a height of several hundred feet, the portion near the surface constituting a dense fog which obscures the horizon and surface objects, but usually leaves the sky relatively clear.

Haze consists of fine dust or salt particles in the air, too small to be individually apparent, but in sufficient number to reduce horizontal visibility and cast a bluish or yellowish veil over the landscape, subduing its colors and making objects appear indistinct. This is sometimes called **dry haze** to distinguish it from **damp haze**, which consists of small water droplets or moist particles in the air, smaller and more scattered than light fog. In international meteorological practice, the term "haze" is used to refer to a condition of atmospheric obscurity caused by dust and smoke.

Mist is synonymous with drizzle in the United States but is often considered as intermediate between haze and fog in its properties. Heavy mist can reduce visibility to a mile or less.

A mixture of smoke and fog is called **smog**. Normally it is not a problem in navigation except in severe cases accompanied by an offshore wind from the source, when it may reduce visibility to 2–4 miles.

ATMOSPHERIC EFFECTS ON LIGHT RAYS

3416. Mirage

Light is refracted as it passes through the atmosphere. When refraction is normal, objects appear slightly elevated, and the visible horizon is farther from the observer than it otherwise would be. Since the effects are uniformly progressive, they are not apparent to the observer. When refraction is not normal, some form of mirage may occur. A **mirage** is an optical phenomenon in which objects appear distorted, displaced (raised or lowered), magnified, multiplied, or inverted due to varying atmospheric refraction which occurs when a layer of air near the Earth's surface differs greatly in density from surrounding air. This may occur when there is a rapid and sometimes irregular change of temperature or humidity with height.

If there is a temperature inversion (increase of temperature with height), particularly if accompanied by a rapid decrease in humidity, the refraction is greater than normal. Objects appear elevated, and the visible horizon is farther away. Objects which are normally below the horizon become visible. This is called **looming**. If the upper portion of an object is raised much more than the bottom part, the object appears taller than usual, an effect called **towering**. If the lower part of an object is raised more than the upper part, the object appears shorter, an effect called **stooping**. When the refraction is greater than normal, a **superior mirage** may occur. An inverted image is seen above the object, and sometimes an erect image appears over the inverted one, with the bases of the two images touching. Greater than normal refraction usually occurs when the water is much colder than the air above it.

If the temperature decrease with height is much greater than normal, refraction is less than normal, or may even cause bending in the opposite direction. Objects appear lower than normal, and the visible horizon is closer to the observer. This is called **sinking**. Towering or stooping may occur if conditions are suitable. When the refraction is reversed, an **inferior mirage** may occur. A ship or an island appears to be floating in the air above a shimmering horizon, possibly with an inverted image beneath it. Conditions suitable to the formation of an inferior mirage occur when the surface is much warmer than the air above it. This usually requires a heated landmass, and therefore is more common near the coast than at sea.

When refraction is not uniformly progressive, objects may appear distorted, taking an almost endless variety of shapes. The Sun when near the horizon is one of the objects most noticeably affected. A **fata morgana** is a complex mirage characterized by marked distortion, generally in the vertical. It may cause objects to appear towering, magnified, and at times even multiplied.

3417. Sky Coloring

White light is composed of light of all colors. Color is related to wavelength, the visible spectrum varying from about 0.000038 to 0.000076 centimeters. The characteristics of each color are related to its wavelength (or frequency). The shorter the wavelength, the greater the amount of bending when light is refracted. It is this principle that permits the separation of light from celestial bodies into a **spectrum** ranging from red, through orange, yellow, green, and blue, to violet, with long-wave **infrared** being slightly outside the visible range at one end and short-wave **ultraviolet** being slightly outside the visible range at the other end. Light of shorter wavelength is scattered and diffracted more than that of longer wavelength.

Light from the Sun and Moon is white, containing all colors. As it enters the Earth's atmosphere, a certain amount of it is scattered. The blue and violet, being of shorter wavelength than other colors, are scattered most. Most of the violet light is absorbed in the atmosphere. Thus, the scattered blue light is most apparent, and the sky appears blue. At great heights, above most of the atmosphere, it appears black.

When the Sun is near the horizon, its light passes through more of the atmosphere than when higher in the sky, resulting in greater scattering and absorption of blue and green light, so that a larger percentage of the red and orange light penetrates to the observer. For this reason the Sun and Moon appear redder at this time, and when this light falls upon clouds, they appear colored. This accounts for the colors at sunset and sunrise. As the setting Sun approaches the horizon, the sunset colors first appear as faint tints of yellow and orange. As the Sun continues to set, the colors deepen. Contrasts occur, due principally to difference in height of clouds. As the Sun sets, the clouds become a deeper red, first the lower clouds and then the higher ones, and finally they fade to a gray.

When there is a large quantity of smoke, dust, or other material in the sky, unusual effects may be observed. If the material in the atmosphere is of suitable substance and quantity to absorb the longer wave red, orange, and yellow radiation, the sky may have a greenish tint, and even the Sun or Moon may appear green. If the green light, too, is absorbed, the Sun or Moon may appear blue. A green Moon or blue Moon is most likely to occur when the Sun is slightly below the horizon and the longer wavelength light from the Sun is absorbed, resulting in green or blue light being cast upon the atmosphere in front of the Moon. The effect is most apparent if the Moon is on the same side of the sky as the Sun.

3418. Rainbows

The **rainbow**, that familiar arc of concentric colored bands seen when the Sun shines on rain, mist, spray, etc., is caused by refraction, internal reflection, and diffraction of sunlight by the drops of water. The center of the arc is a point 180° from the Sun, in the direction of a line from the Sun passing through the observer. The radius of the brightest rainbow is 42°. The colors are visible because of the difference in the amount of refraction of the different colors making up white light, the light being spread out to form a spectrum. Red is on the outer side and blue and violet on the inner side, with orange, yellow, and green between, in that order from red.

Sometimes a secondary rainbow is seen outside the primary one, at a radius of about 50°. Very rarely, a third can be seen. The order of colors of this rainbow is reversed. On rare occasions a faint rainbow is seen on the same side as the Sun. The radius of this rainbow and the order of colors are the same as those of the primary rainbow.

A similar arc formed by light from the Moon (a lunar rainbow) is called a **Moonbow**. The colors are usually very faint. A faint, white arc of about 39° radius is occasionally seen in fog opposite the Sun. This is called a **fogbow**, although its origin is controversial, some considering it a halo.

3419. Halos

Refraction, or a combination of refraction and reflection, of light by ice crystals in the atmosphere may cause a **halo** to appear. The most common form is a ring of light of radius 22° or 46° with the Sun or Moon at the center. Cirrostratus clouds are a common source of atmospheric ice crystals. Occasionally a faint, white circle with a radius of 90° appears around the Sun. This is called a **Hevelian halo**. It is probably caused by refraction and internal reflection of the Sun's light by bipyramidal ice crystals. A halo formed by refraction is usually faintly colored like a rainbow, with red nearest the celestial body, and blue farthest from it.

A brilliant rainbow-colored arc of about a quarter of a circle with its center at the zenith, and the bottom of the arc about 46° above the Sun, is called a **circumzenithal arc**. Red is on the outside of the arc, nearest the Sun. It is produced by the refraction and dispersion of the Sun's light striking the top of prismatic ice crystals in the atmosphere. It usually lasts for only about 5 minutes, but may be so brilliant as to be mistaken for an unusually bright rainbow. A similar arc formed 46° below the Sun, with red on the upper side, is called a **circumhorizontal arc**. Any arc tangent to a heliocentric halo (one surrounding the Sun) is called a **tangent arc**. As the Sun increases in elevation, such arcs tangent to the halo of 22° gradually bend their ends toward each other. If they meet, the elongated curve enclosing the circular halo is called a **circumscribed halo**. The inner edge is red.

A halo consisting of a faint, white circle through the Sun and parallel to the horizon is called a **parhelic circle**. A similar one through the Moon is called a **paraselenic circle**. They are produced by reflection of Sunlight or Moonlight from vertical faces of ice crystals.

A **parhelion (plural: parhelia)** is a form of halo consisting of an image of the Sun at the same altitude and some distance from it, usually 22°, but occasionally 46°. A similar phenomenon occurring at an angular distance of 120° (sometimes 90° or 140°) from the Sun is called a **paranthelion**. One at an angular distance of 180°, a rare occurrence, is called an **anthelion**, although this term is also used to refer to a luminous, colored ring or glory sometimes seen around the shadow of one's head on a cloud or fog bank. A parhelion is popularly called a **mock Sun** or **Sun dog**. Similar phenomena in relation to the Moon are called **paraselene** (popularly a **mock Moon** or **Moon dog**), **parantiselene**, and **antiselene**. The term parhelion should not be confused with perihelion, the orbital point nearest the Sun when the Sun is the center of attraction.

A **Sun pillar** is a glittering shaft of white or reddish light occasionally seen extending above and below the Sun, usually when the Sun is near the horizon. A phenomenon similar to a Sun pillar, but observed in connection with the Moon, is called a **Moon pillar**. A rare form of halo in which horizontal and vertical shafts of light intersect at the Sun is called a **Sun cross**. It is probably due to the simultaneous occurrence of a Sun pillar and a parhelic circle.

3420. Corona

When the Sun or Moon is seen through altostratus clouds, its outline is indistinct, and it appears surrounded by a glow of light called a **corona**. This is somewhat similar in appearance to the corona seen around the Sun during a solar eclipse. When the effect is due to clouds, however, the glow may be accompanied by one or more rainbow-colored rings of small radii, with the celestial body at the center. These can be distinguished from a halo by their much smaller radii and also by the fact that the order of the colors is reversed, red being on the inside, nearest the body, in the case of the halo, and on the outside, away from the body, in the case of the corona.

A corona is caused by diffraction of light by tiny droplets of water. The radius of a corona is inversely proportional to the size of the water droplets. A large corona indicates small droplets. If a corona decreases in size, the water droplets are becoming larger and the air more humid. This may be an indication of an approaching rainstorm. The glow portion of a corona is called an **aureole**.

3421. The Green Flash

As light from the Sun passes through the atmosphere, it is refracted. Since the amount of bending is slightly different for each color, separate images of the Sun are formed in each color of the spectrum. The effect is similar to that of imperfect color printing, in which the various colors are slightly out of register. However, the difference is so slight that the effect is not usually noticeable. At the horizon, where refraction is maximum, the greatest difference, which occurs between violet at one end of the spectrum and red at the other, is about 10 seconds of arc. At latitudes of the United States, about 0.7 second of time is needed for the Sun to change altitude by this amount when it is near the horizon. The red image, being bent least by refraction, is first to set and last to rise. The shorter wave blue and violet colors are scattered most by the atmosphere, giving it its characteristic blue color. Thus, as the Sun sets, the green image may be the last of the colored images to drop out of sight. If the red, orange, and yellow images are below the horizon, and the blue and violet light is scattered and absorbed, the upper rim of the green image is the only part seen, and the Sun appears green. This is the **green flash**. The shade of green varies, and occasionally the blue image is seen, either separately or following the green flash (at sunset). On rare occasions the violet image is also seen. These colors may also be seen at sunrise, but in reverse order. They are occasionally seen when the Sun disappears behind a cloud or other obstruction.

The phenomenon is not observed at each sunrise or sunset, but under suitable conditions is far more common than generally supposed. Conditions favorable to observation of the green flash are a sharp horizon, clear atmosphere, a temperature inversion, and a very attentive observer. Since these conditions are more frequently met when the horizon is formed by the sea than by land, the phenomenon is more common at sea. With a sharp sea horizon and clear atmosphere, an attentive observer may see the green flash at as many as 50 percent of sunsets and sunrises, although a telescope may be needed for some of the observations.

Duration of the green flash (including the time of blue and violet flashes) of as long as 10 seconds has been reported, but such length is rare. Usually it lasts for a period of about $^1/_2$ to $2^1/_2$ seconds, with about $1^1/_4$ seconds being average. This variability is probably due primarily to changes in the index of refraction of the air near the horizon.

Under favorable conditions, a momentary green flash has been observed at the setting of Venus and Jupiter. A telescope improves the chances of seeing such a flash from a planet, but is not a necessity.

3422. Crepuscular Rays

Crepuscular rays are beams of light from the Sun passing through openings in the clouds, and made visible by illumination of dust in the atmosphere along their paths. Actually, the rays are virtually parallel, but because of perspective, appear to diverge. Those appearing to extend downward are popularly called **backstays of the Sun**, or the Sun drawing water. Those extending upward and across the sky, appearing to converge toward a point 180° from the Sun, are called **anticrepuscular rays**.

THE ATMOSPHERE AND RADIO WAVES

3423. Atmospheric Electricity

Radio waves traveling through the atmosphere exhibit many of the properties of light, being refracted, reflected, diffracted, and scattered. These effects are discussed in greater detail in Chapter 10, Radio Waves.

Various conditions induce the formation of electrical charges in the atmosphere. When this occurs, there is often a difference of electron charge between various parts of the atmosphere, and between the atmosphere and Earth or terrestrial objects. When this difference exceeds a certain minimum value, depending upon the conditions, the static electricity is discharged, resulting in phenomena such as lightning or St. Elmo's fire.

Lightning is the discharge of electricity from one part of a thundercloud to another, between different clouds, or between a cloud and the Earth or a terrestrial object.

Enormous electrical stresses build up within thunderclouds, and between such clouds and the Earth. At some point the resistance of the intervening air is overcome. At first the process is a progressive one, probably starting as a brush discharge (St. Elmo's fire), and growing by ionization. The breakdown follows an irregular path along the line of least resistance. A hundred or more individual discharges may be necessary to complete the path between points of opposite polarity. When this "leader stroke" reaches its destination, a heavy "main stroke" immediately follows in the opposite direction. This main stroke is the visible lightning, which may be tinted any color, depending upon the nature of the gases through which it passes. The illumination is due to the high degree of ionization of the air, which causes many of the atoms to become excited and emit radiation.

Thunder, the noise that accompanies lightning, is caused by the heating and ionizing of the air by lightning, which results in rapid expansion of the air along its path and the sending out of a compression wave. Thunder may be heard at a distance of as much as 15 miles, but generally does not carry that far. The elapsed time between the flash of lightning and reception of the accompanying sound of thunder is an indication of the distance, because of the difference in travel time of light and sound. Since the former is comparatively instantaneous, and the speed of sound is about 1,117 feet per second, the approximate distance in nautical miles is equal to the elapsed time in seconds, divided by 5.5. If the thunder accompanying lightning cannot be heard due to its distance, the lightning is called **heat lightning**.

St. Elmo's fire is a luminous discharge of electricity from pointed objects such as the masts and antennas of ships, lightning rods, steeples, mountain tops, blades of grass, human hair, arms, etc., when there is a considerable difference in the electrical charge between the object and the air. It appears most frequently during a storm. An object from which St. Elmo's fire emanates is in danger of being struck by lightning, since this discharge may be the initial phase of the leader stroke. Throughout history those who have not understood St. Elmo's fire have regarded it with superstitious awe, considering it a supernatural manifestation. This view is reflected in the name **corposant** (from "corpo santo," meaning "body of a saint") sometimes given this phenomenon.

The **aurora** is a luminous glow appearing in varied forms in the thin atmosphere high above the Earth in high latitudes. It closely follows solar flare activity, and is believed caused by the excitation of atoms of oxygen and hydrogen, and molecules of nitrogen (N_2). Auroras extend across hundreds of miles of sky, in colored sheets, folds, and rays, constantly changing in form and color. On occasion they are seen in temperate or even more southern latitudes. The maximum occurrence is at about 64–70° of geomagnetic latitude. These are called the **auroral zones** in both northern and southern regions.

The aurora of the northern regions is the **Aurora Borealis,** or **northern lights**, and that of the southern region the **Aurora Australis**, or **southern lights**. The term **polar lights** is occasionally used to refer to either.

In the northern zone, there is an apparent horizontal motion to the westward in the evening and eastward in the morning; a general southward motion occurs during the course of the night.

Variation in auroral activity occurs in sequence with the 11-year Sunspot cycle, and also with the 27-day period of the Sun's synodical rotation. Daily occurrence is greatest near midnight.

WEATHER ANALYSIS AND FORECASTING

3424. Forecasting Weather

The prediction of weather at some future time is based upon an understanding of weather processes and observations of present conditions. Thus, when there is a certain sequence of cloud types, rain usually can be expected to follow. If the sky is cloudless, more heat will be received from the Sun by day, and more heat will be radiated outward from the warm Earth by night than if the sky is overcast. If the wind is from a direction that transports warm, moist air over a colder surface, fog can be expected. A falling barometer indicates the approach of a "low," probably accompanied by stormy weather. Thus, before meteorology passed from an "art" to "science," many individuals learned to interpret certain atmospheric phenomena in terms of future weather, and to make reasonably accurate forecasts for short periods into the future.

With the establishment of weather observation stations, continuous and accurate weather information became available. As observations expanded and communication techniques improved, knowledge of simultaneous conditions over wider areas became available. This made possible the collection of "synoptic" reports at civilian and military forecast centers.

Individual observations are made at stations on shore and aboard vessels at sea. Observations aboard merchant ships at sea are made and transmitted on a voluntary and cooperative basis. The various national meteorological services supply shipmasters with blank forms, printed instructions, and other materials essential to the making, recording, and interpreting of observations. Any shipmaster can render a particularly valuable service by reporting all unusual or non-normal weather occurrences.

Symbols and numbers are used to indicate on a synoptic chart, popularly called a weather map, the conditions at each observation station. Isobars are drawn through lines of equal atmospheric pressure, fronts are located and symbolically marked (See Figure 3425), areas of precipitation and fog are indicated, etc.

Ordinarily, weather maps for surface observations are prepared every 6 (sometimes 3) hours. In addition, synoptic charts for selected heights are prepared every 12 (sometimes 6) hours. Knowledge of conditions aloft is of value in establishing the three-dimensional structure and motion of the atmosphere as input to the forecast.

With the advent of the computer, highly sophisticated numerical models have been developed to analyze and forecast weather patterns. The civil and military weather centers prepare and disseminate vast numbers of weather charts (analyses and prognoses) daily to assist local forecasters in their efforts to provide users with accurate weather forecasts. The accuracy of the forecast decreases with the length of the forecast period. A 12-hour forecast is likely to be more reliable than a 24-hour forecast. Long term forecasts for 2 weeks or a month in advance are limited to general statements. For example, a prediction may be made about which areas will have temperatures above or below normal, and how precipitation will compare with normal, but no attempt is made to state that rainfall will occur at a certain time and place.

Forecasts are issued for various areas. The national meteorological services of most maritime nations, including the United States, issue forecasts for ocean areas and warnings of approaching storms. The efforts all nations are coordinated by the World Meteorological Organization.

3425. Weather Forecast Dissemination

Dissemination of weather information is carried out in a number of ways. Forecasts and warnings are made available by various means including television and radio broadcast, satellite broadcast, telephone, and the internet. Visual storm warnings are displayed in various ports, and storm warnings are broadcast by radio.

The Global Maritime Distress and Safety System (GMDSS) was established to provide more effective and efficient emergency and safety communications, and to disseminate Maritime Safety Information (MSI) to all ships on the world's oceans regardless of location or atmospheric conditions. MSI includes navigational warnings, meteorological warnings and forecasts, and other urgent safety related information. GMDSS goals are defined in the International Convention for the The Safety Of Life At Sea (SOLAS), and affects vessels over 300 gross tons and passenger vessels of any size. The U.S. National Weather Service participates directly in the GMDSS by preparing meteorological forecasts and warnings for broadcast via NAVTEX and SafetyNET.

NAVTEX is an international automated medium frequency (518 kHz) direct-printing service for delivery of navigational and meteorological warnings and forecasts, as well as urgent marine safety information to ships. It was developed to provide a low-cost, simple, and automated means of receiving this information aboard ships at sea within approximately 200 nautical miles of shore. NAVTEX stations in the U.S. are operated by the U.S. Coast Guard.

Inmarsat-C SafetyNET is an internationally adopted, automated satellite system for promulgating weather forecasts and warnings, marine navigational warnings and other safety related information to all types of vessels.

Radiofax, also known as HF FAX, radiofacsimile or weatherfax, is a means of broadcasting graphic weather maps and other graphic images via HF radio. HF radiofax is also known as WEFAX, although this term is generally used to refer to the reception of weather charts and imagery via satellite. Maps are received using a dedicated radiofax receiver or a single sideband shortwave receiver connected to an external facsimile recorder or PC equipped with a radiofax interface and application software. Inexpensive internet access to weather charts at sea awaits the establishment of the bandwidth to quickly transmit large graphic files.

Information on dissemination of marine weather information may be found in NIMA *Pub. 117, Radio Navigational Aids*, the *Admiralty List of Signals Volumes III* and *V*, and the IMO publication, *Master Plan of Shore Based Facilities for the GMDSS*. Information on day and night visual storm warnings is given in the various volumes of *Sailing Directions*, both *Enroutes* and *Planning Guides*.

Through the use of codes, a simplified version of synoptic weather charts is transmitted to various stations ashore and afloat. Rapid transmission of completed maps is accomplished by facsimile. This system is based upon detailed scanning, by a photoelectric detector, of illuminated black and white copy. The varying degrees of light intensity are converted to electric energy, which is transmitted to the receiver and converted back to a black and white presentation. The proliferation of both commercial and restricted computer bulletin board systems having weather information has also greatly increased the accessibility of environmental data.

Complete information on dissemination of weather information by radio is provided in *Selected Worldwide Marine Weather Broadcasts*, published jointly by the National Weather Service and the Naval Meteorology and Oceanography Command. This publication lists broadcast schedules and weather codes. Information on day and night visual storm warnings is given in the various volumes of *Sailing Directions*, both *Enroutes* and *Planning Guides*.

3426. Interpreting Weather

The factors which determine weather are numerous and varied. Ever-increasing knowledge regarding them makes possible a continually improving weather service. However, the ability to forecast is acquired through study and long practice, and therefore the services of a trained meteorologist should be utilized whenever available.

The value of a forecast is increased if one has access to the information upon which it is based, and understands the principles and processes involved. It is sometimes as important to know the various types of weather which may be experienced as it is to know which of several possibilities is most likely to occur.

At sea, reporting stations are unevenly distributed, sometimes leaving relatively large areas with incomplete reports, or none at all. Under these conditions, the locations of highs, lows, fronts, etc., are imperfectly known, and their very existence may even be in doubt. At such times the mariner who can interpret the observations made from his own vessel may be able to predict weather for the next several hours more reliably than a trained meteorologist ashore.

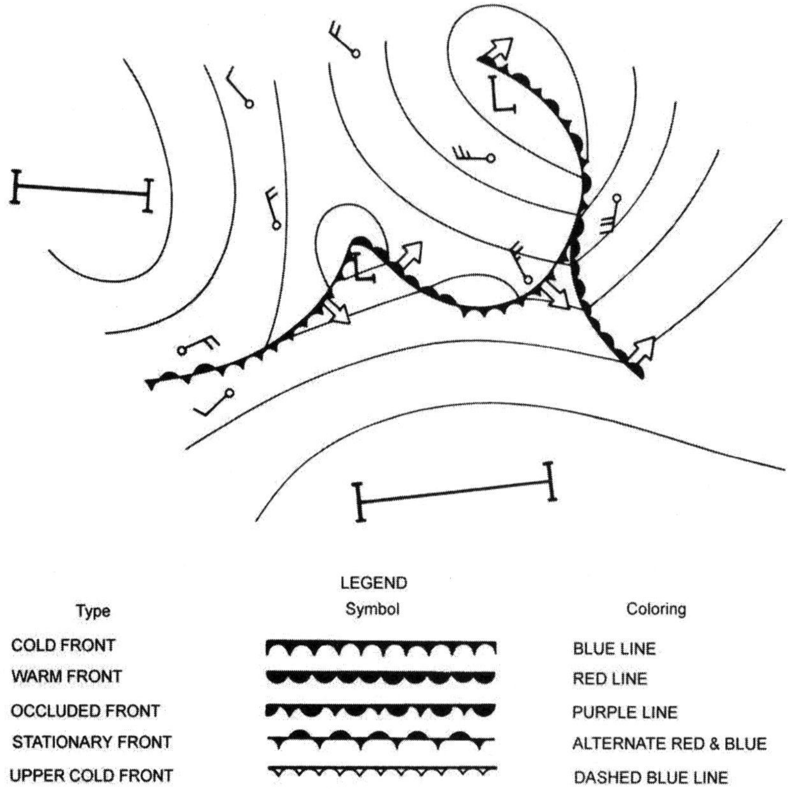

Figure 3425. Designation of fronts on weather maps.

CHAPTER 35

TROPICAL CYCLONES

DESCRIPTION AND CAUSES

3500. Introduction

A **tropical cyclone** is a cyclone originating in the tropics or subtropics. Although it generally resembles the extratropical cyclone of higher latitudes, there are important differences, the principal one being the concentration of a large amount of energy into a relatively small area. Tropical cyclones are infrequent in comparison with middle and high latitude storms, but they have a record of destruction far exceeding that of any other type of storm. Because of their fury, and because they are predominantly oceanic, they merit special attention by mariners.

A tropical storm may have a deceptively small size, and beautiful weather may be experienced only a few hundred miles from the center. The rapidity with which the weather can deteriorate with approach of the storm, and the violence of the fully developed tropical cyclone, are difficult to imagine if they have not been experienced.

On his second voyage to the New World, Columbus encountered a tropical storm. Although his vessels suffered no damage, this experience proved valuable during his fourth voyage when his ships were threatened by a fully developed hurricane. Columbus read the signs of an approaching storm from the appearance of a southeasterly swell, the direction of the high cirrus clouds, and the hazy appearance of the atmosphere. He directed his vessels to shelter. The commander of another group, who did not heed the signs, lost most of his ships and more than 500 men perished.

3501. Definitions

Tropical cyclones are classified by form and intensity as they increase in size.

A **tropical disturbance** is a discrete system of apparently organized convection, generally 100 to 300 miles in diameter, having a nonfrontal migratory character, and having maintained its identity for 24 hours or more. It may or may not be associated with a detectable disturbance of the wind field. It has no strong winds and no closed isobars i.e., isobars that completely enclose the low.

At its next stage of development it becomes a **tropical depression**. A tropical depression has one or more closed isobars and some rotary circulation at the surface. The highest sustained (1-minute mean) surface wind speed is 33 knots.

The next stage is **tropical storm**. A tropical storm has closed isobars and a distinct rotary circulation. The highest sustained (1-minute mean) surface wind speed is 34 to 63 knots.

When fully developed, a **hurricane** or **typhoon** has closed isobars, a strong and very pronounced rotary circulation, and a sustained (1-minute mean) surface wind speed of 64 knots or higher.

3502. Areas of Occurrence

Tropical cyclones occur almost entirely in six distinct areas, four in the Northern Hemisphere and two in the Southern Hemisphere, as shown in Figure 3502. The name by which the tropical cyclone is commonly known varies somewhat with the locality.

1. North Atlantic. A tropical cyclone with winds of 64 knots or greater is called a **hurricane**.
2. Eastern North Pacific. The name **hurricane** is used as in the North Atlantic.
3. Western North Pacific. A fully developed storm with winds of 64 knots or greater is called a **typhoon** or, locally in the Philippines, a **baguio**.
4. North Indian Ocean. A tropical cyclone with winds of 34 knots or greater is called a **cyclonic storm**.
5. South Indian Ocean. A tropical cyclone with winds of 34 knots or greater is called a **cyclone**.
6. Southwest Pacific and Australian Area. The name **cyclone** is used as in the South Indian Ocean. A severe tropical cyclone originating in the Timor Sea and moving southwest and then southeast across the interior of northwestern Australia is called a **willy-willy**.

Tropical cyclones have not been observed in the South Atlantic or in the South Pacific east of 140°W.

3503. Origin, Season and Frequency

See Figure 3503a and Figure 3503b. Origin, season, and frequency of occurrence of the tropical cyclones in the six areas are as follows:

North Atlantic: Tropical cyclones can affect the entire North Atlantic Ocean in any month. However, they are mostly a threat south of about 35°N from June through November; August, September, and October are the

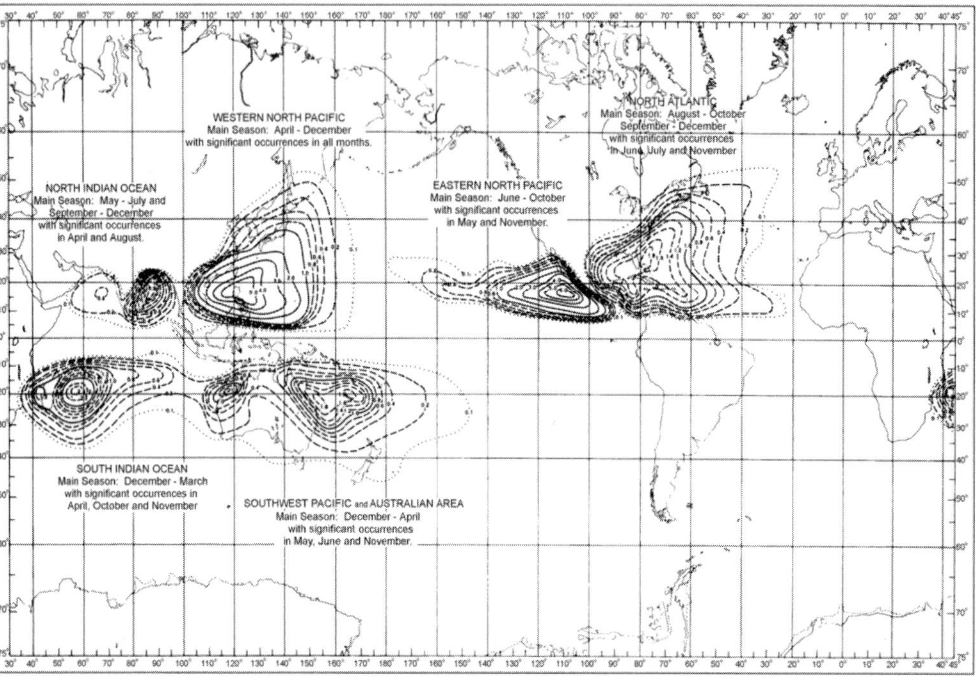

Figure 3502. Areas in which tropical cyclones occur. The average number of tropical cyclones per 5° square has been analyzed for this figure. The main season for intense tropical storm activity is also shown for each major basin.

months of highest incidence. See Figure 3503b. About 9 or 10 tropical cyclones (tropical storms and hurricanes) form each season; 5 or 6 reach hurricane intensity (winds of 64 knots and higher). A few hurricanes have generated winds estimated as high as 200 knots. Early and late season storms usually develop west of 50°W; during August and September, this spawning ground extends to the Cape Verde Islands. These storms usually move westward or west northwestward at speeds of less than 15 knots in the lower latitudes. After moving into the northern Caribbean or Greater Antilles regions, they usually either move toward the Gulf of Mexico or recurve and accelerate in the North Atlantic. Some will recurve after reaching the Gulf of Mexico, while others will continue westward to a landfall in Texas or Mexico.

Eastern North Pacific: The season is from June through October, although a storm can form in any month. An average of 15 tropical cyclones form each year with about 6 reaching hurricane strength. The most intense storms are often the early- and late-season ones; these form close to the coast and far south. Mid season storms form anywhere in a wide band from the Mexican-Central American coast to the Hawaiian Islands. August and September are the months of highest incidence. These storms differ from their North Atlantic counterparts in that they are usually smaller in size. However, they can be just as intense.

Western North Pacific: More tropical cyclones form in the tropical western North Pacific than anywhere else in the world. More than 25 tropical storms develop each year, and about 18 become typhoons. These typhoons are the largest and most intense tropical cyclones in the world. Each year an average of five generate maximum winds over 130 knots; circulations covering more than 600 miles in diameter are not uncommon. Most of these storms form east of the Philippines, and move across the Pacific toward the Philippines, Japan, and China; a few storms form in the South China Sea. The season extends from April through December. However, tropical cyclones are more common in the off-season months in this area than anywhere else. The peak of the season is July through October, when nearly 70 percent of all typhoons develop. There is a noticeable seasonal shift in storm tracks in this region. From July through September, storms move north of the Philippines and recurve, while early- and late-season typhoons move on a more westerly track through the Philippines before recurving.

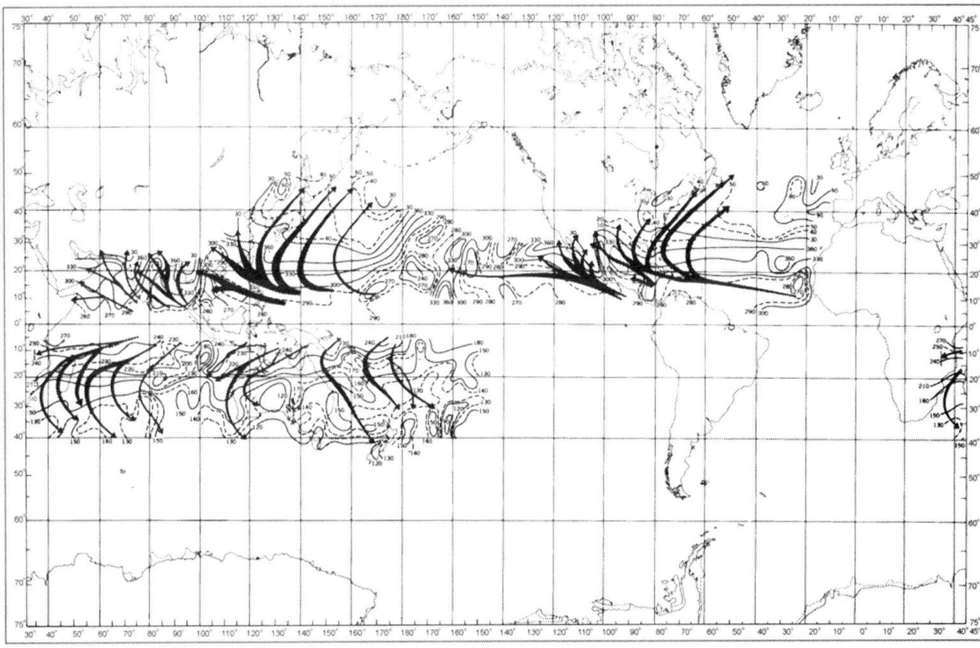

Figure 3503a. Storm tracks.The width of the arrow indicates the approximate frequency of storms; the wider the arrow the higher the frequency. Isolines on the base map show the resultant direction toward which storms moved. Data for the entire year has been summarized for this figure.

AREA AND STAGE	JAN	FEB	MAR	APR	MAY	JUN	JUL	AUG	SEP	OCT	NOV	DEC	ANNUAL
NORTH ATLANTIC													
TROPICAL STORMS	*	*	*	*	0.1	0.4	0.3	1.0	1.5	1.2	0.4	*	4.2
HURRICANES	*	*	*	*	*	0.3	0.4	1.5	2.7	1.3	0.3	*	5.2
TROPICAL STORMS AND HURRICANES	*	*	*	*	*0.2*	*0.7*	*0.8*	*2.5*	*4.3*	*2.5*	*0.7*	*0.1*	*9.4*

	JAN	FEB	MAR	APR	MAY	JUN	JUL	AUG	SEP	OCT	NOV	DEC	ANNUAL
EASTERN NORTH PACIFIC													
TROPICAL STORMS	*	*	*	*	*	1.5	2.8	2.3	2.3	1.2	0.3	*	9.3
HURRICANES	*	*	*	*	0.3	0.6	0.9	2.0	1.8	1.0	*	*	5.8
TROPICAL STORMS AND HURRICANES	*	*	*	*	*0.3*	*2.0*	*3.6*	*4.5*	*4.1*	*2.2*	*0.3*	*	*15.2*

	JAN	FEB	MAR	APR	MAY	JUN	JUL	AUG	SEP	OCT	NOV	DEC	ANNUAL
WESTERN NORTH PACIFIC													
TROPICAL STORMS	0.2	0.3	0.3	0.2	0.4	0.5	1.2	1.8	1.5	1.0	0.8	0.6	7.5
TYPHOONS	0.3	0.2	0.2	0.7	0.9	1.2	2.7	4.0	4.1	3.3	2.1	0.7	17.8
TROPICAL STORMS AND TYPHOONS	*0.4*	*0.4*	*0.5*	*0.9*	*1.3*	*1.8*	*3.9*	*5.8*	*5.6*	*4.3*	*2.9*	*1.3*	*25.3*

	JAN	FEB	MAR	APR	MAY	JUN	JUL	AUG	SEP	OCT	NOV	DEC	ANNUAL
SOUTHWEST PACIFIC AND AUSTRALIAN AREA													
TROPICAL STORMS	2.7	2.8	2.4	1.3	0.3	0.2	*	*	*	0.1	0.4	1.5	10.9
HURRICANES	0.7	1.1	1.3	0.3	*	*	0.1	0.1	*	*	0.3	0.5	3.8
TROPICAL STORMS AND HURRICANES	*3.4*	*4.1*	*3.7*	*1.7*	*0.3*	*0.2*	*0.1*	*0.1*	*	*0.1*	*0.7*	*2.0*	*14.8*

	JAN	FEB	MAR	APR	MAY	JUN	JUL	AUG	SEP	OCT	NOV	DEC	ANNUAL
SOUTHWEST INDIAN OCEAN													
TROPICAL STORMS	2.0	2.2	1.7	0.6	0.2	*	*	*	*	0.3	0.3	0.8	7.4
HURRICANES	1.3	1.1	0.8	0.4	*	*	*	*	*	*	*	0.5	3.8
TROPICAL STORMS AND HURRICANES	*3.2*	*3.3*	*2.5*	*1.1*	*0.2*	*	*	*	*	*0.3*	*0.4*	*1.4*	*11.2*

* Less than 0.05 [1]Winds ≥ 48 Kts.

Monthly values cannot be combined because single storms overlapping two months were counted once in each month and once in the annual.

Figure 3503b. Monthly and annual average number of storms per year for each area.

AREA AND STAGE	JAN	FEB	MAR	APR	MAY	JUN	JUL	AUG	SEP	OCT	NOV	DEC	ANNUAL
NORTH INDIAN OCEAN													
TROPICAL STORMS	0.1	*	*	0.1	0.3	0.5	0.5	0.4	0.4	0.6	0.5	0.3	3.5
CYCLONES[1]	*	*	*	0.1	0.5	0.2	0.1	*	0.1	0.4	0.6	0.2	2.2
TROPICAL STORMS AND CYCLONES[1]	*0.1*	*	*0.1*	*0.3*	*0.7*	*0.7*	0.6	*0.4*	*0.5*	*1.0*	*1.1*	*0.5*	*5.7*

* Less than 0.05 [1]Winds ≥ 48 Kts.

Monthly values cannot be combined because single storms overlapping two months were counted once in each month and once in the annual.

Figure 3503b. Monthly and annual average number of storms per year for each area.

North Indian Ocean—Tropical cyclones develop in the Bay of Bengal and Arabian Sea during the spring and fall. Tropical cyclones in this area form between latitudes 8°N and 15°N, except from June through September, when the little activity that does occur is confined north of about 15°N. These storms are usually short-lived and weak; however, winds of 130 knots have been encountered. They often develop as disturbances along the Intertropical Convergence Zone (ITCZ); this inhibits summertime development, since the ITCZ is usually over land during this monsoon season. However, it is sometimes displaced southward, and when this occurs, storms will form over the monsoon-flooded plains of Bengal. On the average, six cyclonic storms form each year. These include two storms that generate winds of 48 knots or greater. Another 10 tropical cyclones never develop beyond tropical depressions. The Bay of Bengal is the area of highest incidence. However, it is not unusual for a storm to move across southern India and reintensify in the Arabian Sea. This is particularly true during October, the month of highest incidence during the tropical cyclone season. It is also during this period that torrential rains from these storms, dumped over already rain-soaked areas, cause disastrous floods.

South Indian Ocean—Over the waters west of 100°E, to the east African coast, an average of 11 tropical cyclones (tropical storms and hurricanes) form each season, and about 4 reach hurricane intensity. The season is from December through March, although it is possible for a storm to form in any month. Tropical cyclones in this region usually form south of 10°S. The latitude of recurvature usually migrates from about 20°S in January to around 15°S in April. After crossing 30°S, these storms sometimes become intense extratropical lows.

Southwest Pacific and Australian Area—These tropical waters spawn an annual average of 15 tropical cyclones 4, of which reach hurricane intensity. The season extends from about December through April, although storms can form in any month. Activity is widespread in January and February, and it is in these months that tropical cyclones are most likely to affect Fiji, Samoa, and the other eastern islands. Tropical cyclones usually form in the waters from 105°E to 160°W, between 5° and 20°S. Storms affecting northern and western Australia often develop in the Timor or Arafura Sea, while those that affect the east coast form in the Coral Sea. These storms are often small, but can develop winds in excess of 130 knots. New Zealand is sometimes reached by decaying Coral Sea storms, and occasionally by an intense hurricane. In general, tropical cyclones in this region move southwestward and then recurve southeastward.

ANATOMY OF TROPICAL CYCLONES

3504. Formation

Hurricane formation was once believed to result from an intensification of convective forces which produce the towering cumulonimbus clouds of the doldrums. This view of hurricane generation held that surface heating caused warm moist air to ascend convectively to levels where condensation produced cumulonimbus clouds, which, after an inexplicable drop in atmospheric pressure, coalesced and were spun into a cyclonic motion by Coriolis force.

This hypothesis left much unexplained. Although some hurricanes develop from disturbances beginning in the doldrums, very few reach maturity in that region. Also, the high incidence of seemingly ideal convective situations does not match the low incidence of Atlantic hurricanes. Finally, the hypothesis did not explain the drop in atmospheric pressure, so essential to development of hurricane-force winds.

There is still no exact understanding of the triggering mechanism involved in hurricane generation, the balance of conditions needed to generate hurricane circulation, and the relationships between large- and small-scale atmospheric processes. But scientists today, treating the hurricane system as an atmospheric heat engine, present a more comprehensive and convincing view.

They begin with a starter mechanism in which either internal or external forces intensify the initial disturbance. The initial disturbance becomes a region into which low-level air from the surrounding area begins to flow, accelerating the convection already occurring inside the disturbance. The vertical circulation becomes increasingly well organized as water vapor in the ascending moist layer is condensed (releasing large amounts of heat energy to drive the wind system), and as the system is swept into a counterclockwise cyclonic spiral. But this incipient hurricane would soon fill up because of inflow at lower levels, unless the chimney in which converging air surges upward is provided the exhaust mechanism of high-altitude winds.

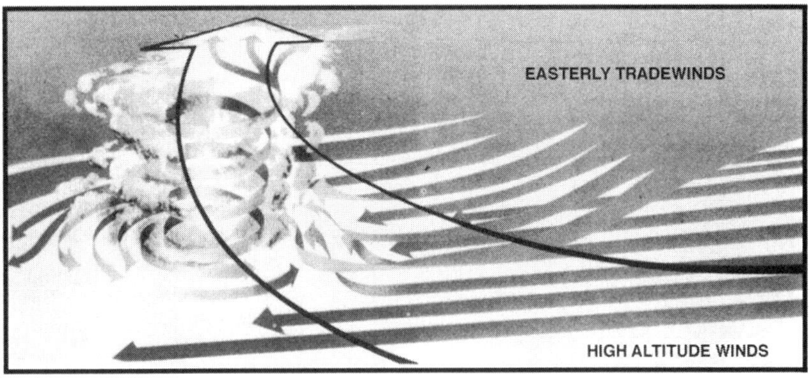

Figure 3504. Pumping action of high-altitude winds.

These high-altitude winds pump ascending air out of the cyclonic system, into a high-altitude anticyclone, which transports the air well away from the disturbance, before sinking occurs. See Figure 3504. Thus, a large scale vertical circulation is set up, in which low-level air is spiraled up the cyclonic twisting of the disturbance, and, after a trajectory over the sea, returned to lower altitudes some distance from the storm. This pumping action-and the heat released by the ascending air may account for the sudden drop of atmospheric pressure at the surface, which produces the steep pressure gradient along which winds reach hurricane proportions.

It is believed that the interaction of low-level and high-altitude wind systems determines the intensity the hurricane will attain. If less air is pumped out than converges at low levels, the system will fill and die out. If more is pumped out than flows in, the circulation will be sustained and will intensify.

Scientists have found that any process which increases the rate of low-level inflow is favorable for hurricane development, provided the inflowing air carries sufficient heat and moisture to fuel the hurricane's power system. It has also been shown that air above the developing disturbance, at altitudes between 20,000 and 40,000 feet, increases 1° to 3°F in temperature about 24 hours before the disturbance develops into a hurricane. But it is not known whether low-level inflow and high-level warming cause hurricanes. They could very well be measurable symptoms of another effect which actually triggers the storm's increase to hurricane intensity.

The view of hurricanes as atmospheric engines is necessarily a general one. The exact role of each contributor is not completely understood. The engine seems to be both inefficient and unreliable; a myriad of delicate conditions must be satisfied for the atmosphere to produce a hurricane. Their relative infrequency indicates that many potential hurricanes dissipate before developing into storms.

3505. Portrait of a Hurricane

In the early life of the hurricane, the spiral covers an area averaging 100 miles in diameter with winds of 64 knots and greater, and spreads gale-force winds over a 400-mile diameter. The cyclonic spiral is marked by heavy cloud bands from which torrential rains fall, separated by areas of light rain or no rain at all. These spiral bands ascend in decks of cumulus and cumulonimbus clouds to the convective limit of cloud formation, where condensing water vapor is swept off as ice-crystal wisps of cirrus clouds. See Figure 3505. Thunderstorm electrical activity is observed in these bands, both as lightning and as tiny electrostatic discharges.

In the lower few thousand feet, air flows in through the cyclone, and is drawn upward through ascending columns of air near the center. The size and intensity decrease with altitude, the cyclonic circulation being gradually replaced above 40,000 feet by an anticyclonic circulation centered hundreds of miles away, which is the exhaust system of the hurricane heat engine.

At lower levels, where the hurricane is more intense, winds on the rim of the storm follow a wide pattern, like the slower currents around the edge of a whirlpool; and, like those currents, these winds accelerate as they approach the center of the vortex. The outer band has light winds at the rim of the storm, perhaps no more than 25 knots; within 30 miles of the center, winds may have velocities exceeding 130 knots. The inner band is the region of maximum wind velocity, where the storm's worst winds are felt, and where ascending air is chimneyed upward, releasing heat to drive the storm. In most hurricanes, these winds reach 85 knots, and more than 170 knots in severe storms.

In the hurricane, winds flow toward the low pressure in the warm, comparatively calm core. There, converging air is whirled upward by convection, the mechanical thrusting of other converging air, and the pumping action of high-altitude circulations. This spiral is marked by the thick

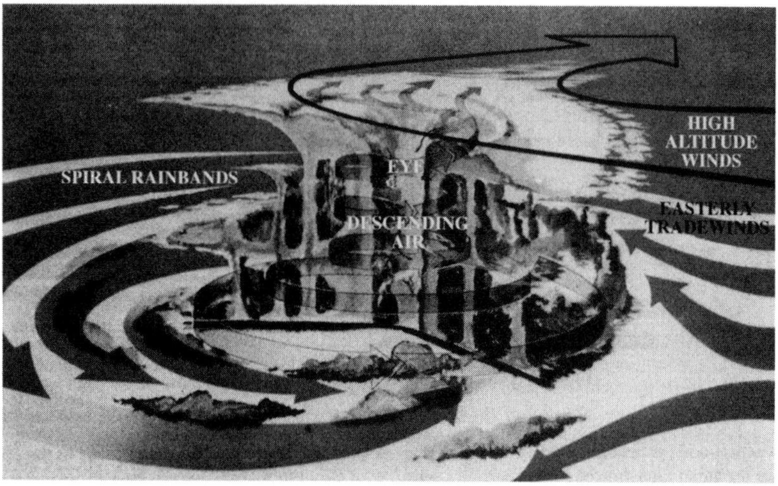

Figure 3505. Cutaway view of a hurricane greatly exaggerated in vertical dimension. Actual hurricanes are less than 50,000 feet high and may have a diameter of several hundred miles.

cloud walls curling inward toward the storm center, releasing heavy precipitation and enormous quantities of heat energy. At the center, surrounded by a band in which this strong vertical circulation is greatest, is the **eye** of the hurricane.

On the average, eye diameter is about 14 miles, although diameters of 25 miles are not unusual. From the heated tower of maximum winds and cumulonimbus clouds, winds diminish rapidly to something less than 15 miles per hour in the eye; at the opposite wall, winds increase again, but come from the opposite direction because of the cyclonic circulation of the storm. This sudden transformation of storm into comparative calm, and from calm into violence from another quarter is spectacular. The eye's abrupt existence in the midst of opaque rain squalls and hurricane winds, the intermittent bursts of blue sky and sunlight through light clouds in the core of the cyclone, and the galleried walls of cumulus and cumulonimbus clouds are unforgettable.

Every hurricane is individual, and the more or less orderly circulation described here omits the extreme variability and instability within the storm system. Pressure and temperature gradients fluctuate wildly across the storm as the hurricane maintains its erratic life. If it is an August storm, its average life expectancy is 12 days; if a July or November storm, it lives an average of 8 days.

3506. Life of a Tropical Cyclone

Reports from ships in the vicinity of an **easterly wave** (a westward-moving trough of low pressure embedded in deep easterlies) may indicate that the atmospheric pressure in the region has fallen more than 5 hectopascals (hPa) in the past 24 hours. This is cause for alarm, because in the Tropics pressure varies little; the normal diurnal pressure change is only about 3 hPa. Satellite pictures may indicate thickening middle and high clouds. Squalls are reported ahead of the easterly wave, and wind reports indicate a cyclonic circulation is forming. The former easterly wave, now classified a tropical disturbance, is moving westward at 10 knots under the canopy of a large high-pressure system aloft. Sea surface temperatures in the vicinity are in the 28°-30°C range.

Within 48 hours winds increase to 25 knots near the center of definite circulation, and central pressure has dropped below 1000 hPa. The disturbance is now classified as a tropical depression. Soon the circulation extends out to 100 miles and upward to 20,000 feet. Winds near the center increase to gale force, central pressure falls below 990 hPa, and towering cumulonimbus clouds shield a developing eye; a tropical storm has developed.

Satellite photographs now reveal a tightly organized tropical cyclone, and reconnaissance reports indicate maximum winds of 80 knots around a central pressure of 980 hPa; a hurricane has developed. A ship to the right (left in the Southern Hemisphere) of the hurricane's center (looking toward the direction of storm movement) reports 30-foot seas. The hurricane is rapidly maturing as it continues westward.

A few days later the hurricane reaches its peak. The satellite photographs show a textbook picture (Figure 3506), as 120-knot winds roar around a 940-hPa pressure center; hurricane-force winds extend 50 miles in all directions, and seas are reported up to 40 feet. There is no

Figure 3506. Satellite photograph of a hurricane.

further deepening now, but the hurricane begins to expand. In 2 days, gales extend out to 200 miles, and hurricane winds out to 75 miles. Then the hurricane slows and begins to recurve; this turning marks the beginning of its final phase.

The hurricane accelerates, and, upon reaching temperate latitudes, it begins to lose its tropical characteristics. The circulation continues to expand, but now cold air is intruding. (Cold air, cold water, dry air aloft, and land aid in the decay of a tropical cyclone.) The winds gradually abate as the concentrated storm disintegrates. The warm core survives for a few more days before the transformation to a large extratropical low-pressure system is complete.

Not all tropical cyclones follow this average pattern. Most falter in the early stages, some dissipate over land, and others remain potent for several weeks.

FORECASTING AND PREDICTING TROPICAL CYCLONES

3507. Weather Broadcasts and Radiofacsimile

The marine weather broadcast and radiofacsimile weather maps are the most important tools for avoiding tropical cyclones. These broadcasts, covering all tropical areas, provide information about the tropical cyclone's location, maximum winds and seas, and future conditions expected.

The U S. Navy, the National Oceanic and Atmospheric Administration, and the U.S. Air Force have developed a highly effective surveillance system for the tropical cyclone-prone areas of the world. Routine and special weather reports enable accurate detection, location, and tracking of tropical cyclones. International cooperation is effective. These reports originate from land stations, ships at sea, aircraft, weather satellite imagery, and specially instrumented weather reconnaissance aircraft of National Oceanic and Atmospheric Administration and the U.S. Air Force. Data buoys, both moored and drifting, provide another source of information.

The tropical warning services have three principal functions:

1. Collection and analysis of data
2. Preparation of timely and accurate warnings
3. The distribution of advisories

To provide timely and accurate information and warnings regarding tropical cyclones, the oceans have been divided into overlapping geographical areas of responsibility.

For detailed information on the areas of responsibility of the countries participating in the international forecasting and warning program, and radio aids, refer to Selected Worldwide Marine Weather Broadcasts, published jointly by the Naval Meteorology and Oceanography Command and the National Weather Service.

Although the areas of forecasting responsibility are fairly well defined for the Department of Defense, the international and domestic civilian system provides many overlaps and is dependent upon qualitative factors. For example, when a tropical storm or hurricane is traveling westward and crosses 35°W longitude, the continued issuance of forecasts and warnings to the general public, shipping interests, etc., becomes the responsibility of the National Hurricane Center of the National Weather Service at Miami, Florida. When a tropical storm or hurricane crosses 35°W longitude traveling from west to east, the National Hurricane Center ceases to issue formal public advisories, but will issue marine bulletins on any dangerous tropical cyclone in the North Atlantic, if it is of importance or constitutes a threat to shipping and other interests. These advisories are included in National Weather Service Marine Bulletins broadcast to ships over radio station NAM Norfolk, Virginia. Special advisories may be issued at any time. In the Atlantic Ocean, Department of Defense responsibility rests with the Naval Atlantic Meteorology and Oceanography Center in Norfolk, Virginia.

In the eastern Pacific east of longitude 140°W, responsibility for the issuance of tropical storm and hurricane advisories and warnings for the general public, merchant shipping, and other interests rests with the National Weather Service Eastern Pacific Hurricane Center, San Francisco, California. The Department of Defense responsibility rests with the Naval Pacific Meteorology and Oceanography Center, Pearl Harbor, Hawaii. Formal advisories and warnings are issued daily and are included in the marine bulletins broadcast by radio stations KFS, NMC, and NMQ.

In the central Pacific (between the meridian and longitude 140°W), the civilian responsibility rests with the National Weather Service Central Pacific Hurricane Center, Honolulu, Hawaii. Department of Defense responsibility rests with the Naval Pacific Meteorology and Oceanography Center in Pearl Harbor. Formal tropical storm and hurricane advisories and warnings are issued daily and are included in the marine bulletins broadcast by radio station NMO and NRV.

Tropical cyclone messages contain position of the storm, intensity, direction and speed of movement, and a description of the area of strong winds. Included is a forecast of future movement and intensity. When the storm is likely to affect any land area, details on when and where it will be felt, and data on tides, rain, floods, and maximum winds are also included. Figure 3507 provides an example of a marine advisory issued by the National Hurricane Center.

The Naval Pacific Meteorology and Oceanography Center Center-West/Joint Typhoon Warning Center (NPMOC-W/JTWC) in Guam is responsible for all U.S. tropical storm and typhoon advisories and warnings from the 180th meridian westward to the mainland of Asia. A secondary area of responsibility extends westward to longitude 90°E. Whenever a tropical cyclone is observed in the western North Pacific area, serially numbered warnings, bearing an "immediate" precedence are broadcast from the NPMOC-W/JTWC at 0000, 0600, 1200, and 1800 GMT.

The responsibility for issuing gale and storm warnings for the Indian Ocean, Arabian Sea, Bay of Bengal, Western Pacific, and South Pacific rests with many countries. In general, warnings of approaching tropical cyclones will include the following information: storm type, central pressure given in hPa, wind speed observed within the storm, storm location, speed and direction of movement, the extent of the affected area, visibility, and the state of the sea, as well as any other pertinent information received. All storm warning messages commence with the international call sign "TTT."

NOAA/NATIONAL HURRICANE CENTER MARINE ADVISORY NUMBER 13 HURRICANE LADY 0400Z SEPTEMBER 21 20--.

HURRICANE WARNINGS ARE DISPLAYED FROM KEY LARGO TO CAPE KENNEDY. GALE WARNINGS ARE DISPLAYED FROM KEY WEST TO JACKSONVILLE AND FROM FLORIDA BAY TO CEDAR KEY.

HURRICANE CENTER LOCATED NEAR LATITUDE 25.5 NORTH LONGITUDE 78.5 WEST AT 21/0400Z. POSITION EXCELLENT ACCURATE WITHIN 10 MILES BASED ON AIR FORCE RECONNAISSANCE AND SYNOPTIC REPORTS.

PRESENT MOVEMENT TOWARD THE WEST NORTHWEST OR 285 DEGREES AT 10 KT. MAX SUSTAINED WINDS OF 100 KT NEAR CENTER WITH GUSTS TO 160 KT.
MAX WINDS OVER INLAND AREAS 35 KT.
RAD OF 65 KT WINDS 90 NE 60 SE 80 SW 90 NW QUAD.
RAD OF 50 KT WINDS 120 NE 70 SE 90 SW 120 NW QUAD.
RAD OF 30 KT WINDS 210 NE 210 SE 210 SW 210 NW QUAD.
REPEAT CENTER LOCATED 25.5N 78.3W AT 21/0400Z.

12 HOUR FORECAST VALID 21/1600Z LATITUDE 26.0N LONGITUDE 80.5W.
MAX WINDS OF 100 KT NEAR CENTER WITH GUSTS TO 160 KT.
MAX WINDS OVER INLAND AREAS 65 KT.
RADIUS OF 50 KT WINDS 120 NE 70 SE 90 SW 120 NW QUAD.
24 HOUR FORECAST VALID 22/0400Z LATITUDE 26.0N LONGITUDE 83.0W.
MAX WINDS OF 75 KT NEAR CENTER WITH GUSTS TO 120 KT.
MAX WINDS OVER INLAND AREAS 45 KT.
RADIUS OF 50 KT WINDS 120 NE 120 SE 120 SW 120 NW QUAD.

STORM TIDE OF 9 TO 12 FT SOUTHEAST FLA COAST GREATER MIAMI AREA TO THE PALM BEACHES.

NEXT ADVISORY AT 21/1000Z.

Figure 3507. Example of marine advisory issued by the National Hurricane Center.

These warnings are broadcast on specified radio frequency bands immediately upon receipt of the information and at specific intervals thereafter. Generally, the broadcast interval is every 6 to 8 hours, depending upon receipt of new information.

Bulletins and forecasts are excellent guides to the present and future behavior of the tropical cyclone, and a plot should be kept of all positions.

AVOIDING TROPICAL CYCLONES

3508. Approach and Passage of a Tropical Cyclone

An early indication of the approach of a tropical cyclone is the presence of a long swell. In the absence of a tropical cyclone, the crests of swell in the deep waters of the Atlantic pass at the rate of perhaps eight per minute. Swell generated by a hurricane is about twice as long, the crests passing at the rate of perhaps four per minute. Swell may be observed several days before arrival of the storm.

When the storm center is 500 to 1,000 miles away, the barometer usually rises a little, and the skies are relatively clear. Cumulus clouds, if present at all, are few in number and their vertical development appears suppressed. The barometer usually appears restless, pumping up and down a few hundredths of an inch.

As the tropical cyclone comes nearer, a cloud sequence begins which resembles that associated with the approach of a warm front in middle latitudes. Snow-white, fibrous "mare's tails" (cirrus) appear when the storm is about 300 to 600 miles away. Usually these seem to converge, more or less, in the direction from which the storm is approaching. This convergence is particularly apparent at about the time of sunrise and sunset.

Shortly after the cirrus appears, but sometimes before, the barometer starts a long, slow fall. At first the fall is so gradual that it only appears to alter somewhat the normal

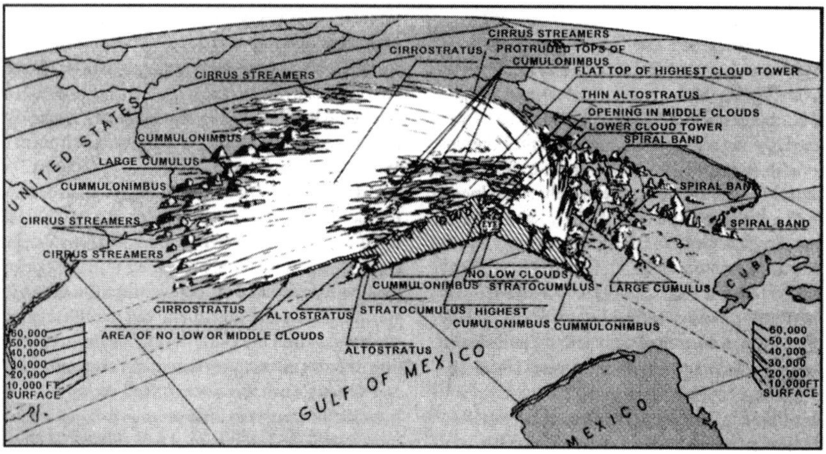

Figure 3508. Typical hurricane cloud formations.

daily cycle (two maxima and two minima in the Tropics). As the rate of fall increases, the daily pattern is completely lost in the more or less steady fall.

The cirrus becomes more confused and tangled, and then gradually gives way to a continuous veil of cirrostratus. Below this veil, altostratus forms, and then stratocumulus. These clouds gradually become more dense, and as they do so, the weather becomes unsettled. A fine, mist-like rain begins to fall, interrupted from time to time by rain showers. The barometer has fallen perhaps a tenth of an inch.

As the fall becomes more rapid, the wind increases in gustiness, and its speed becomes greater, reaching perhaps 22 to 40 knots (Beaufort 6-8). On the horizon appears a dark wall of heavy cumulonimbus, called the **bar** of the storm. This is the heavy bank of clouds comprising the main mass of the cyclone. Portions of this heavy cloud become detached from time to time, and drift across the sky, accompanied by rain squalls and wind of increasing speed. Between squalls, the cirrostratus can be seen through breaks in the stratocumulus.

As the bar approaches, the barometer falls more rapidly and wind speed increases. The seas, which have been gradually mounting, become tempestuous. Squall lines, one after the other, sweep past in ever increasing number and intensity.

With the arrival of the bar, the day becomes very dark, squalls become virtually continuous, and the barometer falls precipitously, with a rapid increase in wind speed. The center may still be 100 to 200 miles away in a fully developed tropical cyclone. As the center of the storm comes closer, the ever-stronger wind shrieks through the rigging, and about the superstructure of the vessel. As the center approaches, rain falls in torrents. The wind fury increases. The seas become mountainous. The tops of huge waves are blown off to mingle with the rain and fill the air with water. Visibility is virtually zero in blinding rain and spray. Even the largest and most seaworthy vessels become virtually unmanageable, and may sustain heavy damage. Less sturdy vessels may not survive. Navigation virtually stops as safety of the vessel becomes the only consideration. The awesome fury of this condition can only be experienced. Words are inadequate to describe it.

If the eye of the storm passes over the vessel, the winds suddenly drop to a breeze as the wall of the eye passes. The rain stops, and the skies clear sufficiently to permit the Sun or stars to shine through holes in the comparatively thin cloud cover. Visibility improves. Mountainous seas approach from all sides in complete confusion. The barometer reaches its lowest point, which may be $1^1/_2$ or 2 inches below normal in fully developed tropical cyclones. As the wall on the opposite side of the eye arrives, the full fury of the wind strikes as suddenly as it ceased, but from the opposite direction. The sequence of conditions that occurred during approach of the storm is reversed, and passes more quickly, as the various parts of the storm are not as wide in the rear of a storm as on its forward side.

Typical cloud formations associated with a hurricane are shown in Figure 3508.

3509. Locating the Center of a Tropical Cyclone

If intelligent action is to be taken to avoid the full fury of a tropical cyclone, early determination of its location and direction of travel relative to the vessel is essential. The bulletins and forecasts are an excellent general guide, but they are not infallible, and may be sufficiently in error to induce

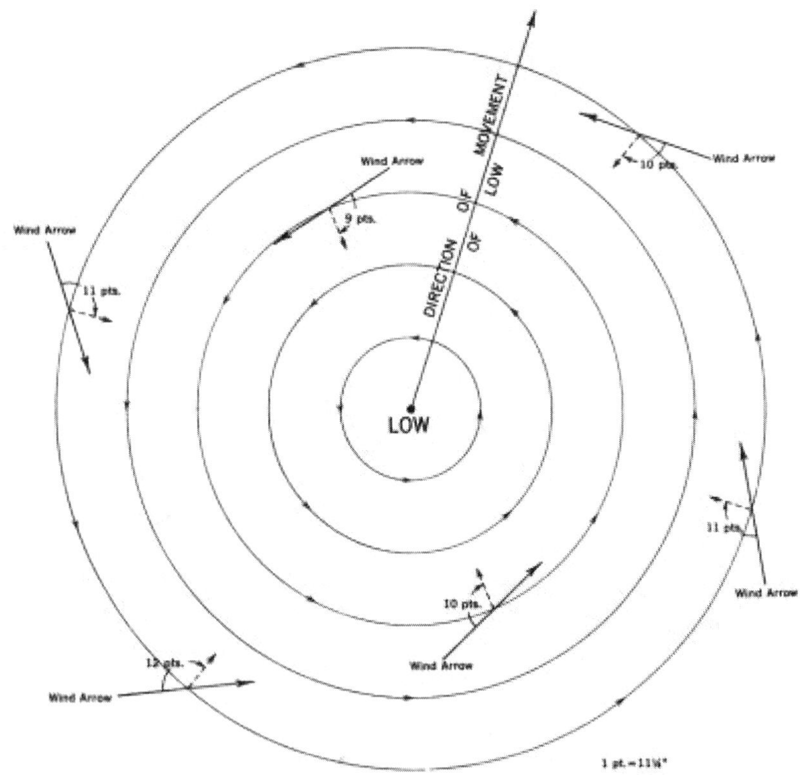

Figure 3509a. Approximate relationship of wind to isobars and storm center in the Northern Hemisphere.

a mariner in a critical position to alter course so as to unwittingly increase the danger to his vessel. Often it is possible, using only those observations made aboard ship, to obtain a sufficiently close approximation to enable the vessel to maneuver to the best advantage.

The presence of an exceptionally long swell is usually the first visible indication of the existence of a tropical cyclone. In deep water it approaches from the general direction of origin (the position of the storm center when the swell was generated). However, in shoaling water this is a less reliable indication because the direction is changed by refraction, the crests being more nearly parallel to the bottom contours.

When the cirrus clouds appear, their point of convergence provides an indication of the direction of the storm center. If the storm is to pass well to one side of the observer, the point of convergence shifts slowly in the direction of storm movement. If the storm center will pass near the observer, this point remains steady. When the bar becomes visible, it appears to rest upon the horizon for several hours. The darkest part of this cloud is in the direction of the storm center. If the storm is to pass to one side, the bar appears to drift slowly along the horizon. If the storm is heading directly toward the observer, the position of the bar remains fixed. Once within the area of the dense, low clouds, one should observe their direction of movement, which is almost exactly along the isobars, with the center of the storm being 90° from the direction of cloud movement (left of direction of movement in the Northern Hemisphere, and right in the Southern Hemisphere).

The winds are probably the best guide to the direction of the center of a tropical cyclone. The circulation is cyclonic, but because of the steep pressure gradient near the center, the winds there blow with greater violence and are more nearly circular than in extratropical cyclones.

According to **Buys Ballot's law**, an observer whose back is to the wind has the low pressure on his left in the Northern Hemisphere, and on his right in the Southern Hemisphere. If the wind followed circular isobars exactly, the center would be exactly 90° from behind when facing away from the wind. However, the track of the wind is usually inclined somewhat toward the center, so that the angle from

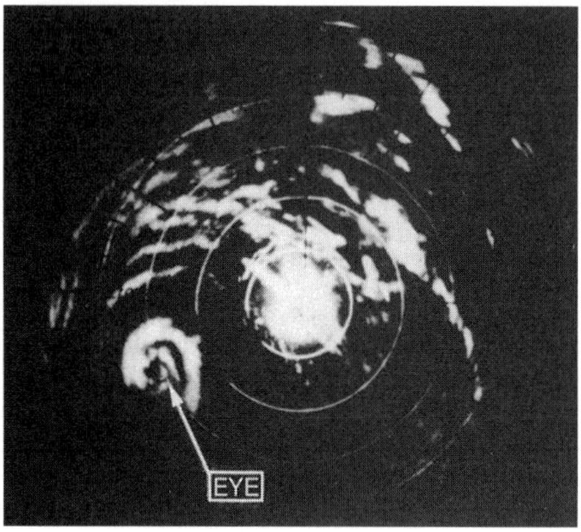

Figure 3509b. Radar PPI presentation of a tropical cyclone.

dead astern varies between perhaps 90° to 135°. The inclination varies in different parts of the same storm. It is least in front of the storm, and greatest in the rear, since the actual wind is the vector sum of the pressure gradient and the motion of the storm along the track. A good average is perhaps 110° in front, and 120-135° in the rear. These values apply when the storm center is still several hundred miles away. Closer to the center, the wind blows more nearly along the isobars, the inclination being reduced by one or two points at the wall of the eye. Since wind direction usually shifts temporarily during a squall, its direction at this time should not be used for determining the position of the center. The approximate relationship of wind to isobars and storm center in the Northern Hemisphere is shown in Figure 3509a.

When the center is within radar range, it will probably be visible on the scope. However, since the radar return is predominantly from the rain, results can be deceptive, and other indications should not be neglected. Figure 3509b shows a radar PPI presentation of a tropical cyclone. If the eye is out of range, the spiral bands (Figure 3509b) may indicate its direction from the vessel. Tracking the eye or upwind portion of the spiral bands enables determining the direction and speed of movement; this should be done for at least 1 hour because the eye tends to oscillate. The tracking of individual cells, which tend to move tangentially around the eye, for 15 minutes or more, either at the end of the band or between bands, will provide an indication of the wind speed in that area of the storm.

Distance from the storm center is more difficult to determine than direction. Radar is perhaps the best guide. However, the rate of fall of the barometer is some indication.

3510. Statistical Analysis of Barometric Pressure

The lowest-sea-level pressure ever recorded was 877 hPa in typhoon Ida, on September 24, 1958. The observation was taken by a reconnaissance aircraft dropsonde, some 750 miles east of Luzon, Philippines. This observation was obtained again in typhoon Nora on October 6, 1973. The lowest barometric reading of record for the United States is 892.3 hPa, obtained during a hurricane at Lower Matecumbe Key, Florida, in September 1935. In hurricane Camille in 1969, a 905 hPa pressure was measured by reconnaissance aircraft. During a 1927 typhoon, the S.S. Sapoeroea recorded a pressure of 886.6 hPa, the lowest sea-level pressure reported from a ship. Pressure has been observed to drop more than 33 hPa per hour, with a pressure gradient amounting to a change of 3.7 hPa per mile.

A method for alerting the mariner to possible tropical cyclone formation involves a statistical comparison of observed weather parameters with the climatology (30 year averaged conditions) for those parameters. Significant fluctuations away from these average conditions could mean the onset of severe weather. One such statistical method involves a comparison of mean surface pressure in the tropics with the standard deviation (s.d.) of surface pressure. Any significant deviation from the norm could indicate proximity to a tropical cyclone. Analysis shows that surface pressure can be expected to be lower than the mean minus 1 s.d. less than 16% of the time, lower than the mean minus 1.5 s.d. less than 7% of the time, and lower than the mean minus 2 s.d. less than 3% of the time. Comparison of the observed pressure with the mean will indicate how unusual the present conditions are.

As an example, assume the mean surface pressure in the South China Sea to be about 1005 mb during August with a s.d. of about 2 mb. Therefore, surface pressure can be expected to fall below 1003 mb about 16% of the time and below 1000 mb about 7% of the time. Ambient pressure any lower than that would alert the mariner to the possible onset of heavy weather. Charts showing the mean surface pressure and the s.d. of surface pressure for various global regions can be found in the U.S. Navy Marine Climatic Atlas of the World.

3511. Maneuvering to Avoid the Storm Center

The safest procedure with respect to tropical cyclones is to avoid them. If action is taken sufficiently early, this is simply a matter of setting a course that will take the vessel well to one side of the probable track of the storm, and then continuing to plot the positions of the storm center as given in the weather bulletins, revising the course as needed.

However, this is not always possible. If the ship is found to be within the storm area, the proper action to take depends in part upon its position relative to the storm center and its direction of travel. It is customary to divide the circular area of the storm into two parts.

In the Northern Hemisphere, that part to the right of the storm track (facing in the direction toward which the storm is moving) is called the **dangerous semicircle**. It is considered dangerous because (1) the actual wind speed is greater than that due to the pressure gradient alone, since it is augmented by the forward motion of the storm, and (2) the direction of the wind and sea is such as to carry a vessel into the path of the storm (in the forward part of the semicircle).

The part to the left of the storm track is called the **less dangerous semicircle**, or **navigable semicircle**. In this part, the wind is decreased by the forward motion of the storm, and the wind blows vessels away from the storm track (in the forward part). Because of the greater wind speed in the dangerous semicircle, the seas are higher than in the less dangerous semicircle. In the Southern Hemisphere, the dangerous semicircle is to the left of the storm track, and the less dangerous semicircle is to the right of the storm track.

A plot of successive positions of the storm center should indicate the semicircle in which a vessel is located. However, if this is based upon weather bulletins, it may not be a reliable guide because of the lag between the observations upon which the bulletin is based and the time of reception of the bulletin, with the ever-present possibility of a change in the direction of the storm. The use of radar eliminates this lag at short range, but the return may not be a true indication of the center. Perhaps the most reliable guide is the wind. Within the cyclonic circulation, a wind shifting to the right in the northern hemisphere and to the left in the southern hemisphere indicates the vessel is probably in the dangerous semicircle. A steady wind shift opposite to this indicates the vessel is probably in the less dangerous semicircle.

However, if a vessel is underway, its own motion should be considered. If it is outrunning the storm or pulling rapidly toward one side (which is not difficult during the early stages of a storm, when its speed is low), the opposite effect occurs. This should usually be accompanied by a rise in atmospheric pressure, but if motion of the vessel is nearly along an isobar, this may not be a reliable indication. If in doubt, the safest action is usually to stop long enough to define the proper semicircle. The loss in time may be more than offset by the minimizing of the possibility of taking the wrong action, increasing the danger to the vessel. If the wind direction remains steady (for a vessel which is stopped), with increasing speed and falling barometer, the vessel is in or near the path of the storm. If it remains steady with decreasing speed and rising barometer, the vessel is near the storm track, behind the center.

The first action to take if the ship is within the cyclonic circulation is to determine the position of his vessel with respect to the storm center. While the vessel can still make considerable way through the water, a course should be selected to take it as far as possible from the center. If the vessel can move faster than the storm, it is a relatively simple matter to outrun the storm if sea room permits. But when the storm is faster, the solution is not as simple. In this case, the vessel, if ahead of the storm, will approach nearer to the center. The problem is to select a course that will produce the greatest possible minimum distance. This is best determined by means of a relative movement plot, as shown in the following example solved on a maneuvering board.

Example: *A tropical cyclone is estimated to be moving in direction 320° at 19 knots. Its center bears 170°, at an estimated distance of 200 miles from a vessel which has a maximum speed of 12 knots.*

Required:

(1) *The course to steer at 12 knots to produce the greatest possible minimum distance between the vessel and the storm center.*

(2) *The distance to the center at nearest approach.*

(3) *Elapsed time until nearest approach.*

Solution: *(Figure 3511) Consider the vessel remaining at the center of the plot throughout the solution, as on a radar PPI.*

(1) To locate the position of the storm center relative to the vessel, plot point C at a distance of 200 miles (scale 20:1) in direction 170° from the center of the diagram. From the center of the diagram, draw RA, the speed vector of the storm center, in direction 320°, speed 19 knots (scale 2:1). From A draw a line tangent to the 12-knot speed circle (labeled 6 at scale 2:1) on the side opposite the storm center. From the center of the diagram, draw a perpendicular to this tangent line, locating point B. The line RB is the required speed vector for the vessel. Its direction, 011°, is the required course.

(2) The path of the storm center relative to the vessel will be along a line from C in the direction BA, if both storm and vessel maintain course and speed. The point of nearest approach will be at D, the foot of a perpendicular from the center of the diagram. This distance, at scale 20:1, is 187 miles.

(3) The length of the vector BA (14.8 knots) is the speed of the storm with respect to the vessel. Mark this on the lowest scale of the nomogram at the bottom of the diagram. The relative distance CD is 72 miles, by measurement. Mark this (scale 10:1) on the middle scale at the bottom of the diagram. Draw a line between the two points and extend it to intersect the top scale at 29.2 (292 at 10:1 scale). The elapsed time is therefore 292 minutes, or 4 hours 52 minutes.

Answers: *(1) C 011°, (2) D 187 mi., (3)* $4^h\ 52^m$.
The storm center will be dead astern at its nearest approach.

As a general rule, for a vessel in the Northern Hemisphere, safety lies in placing the wind on the starboard bow in the dangerous semicircle and on the starboard quarter in the less dangerous semicircle. If on the storm track ahead of the storm, the wind should be put about 160° on the starboard quarter until the vessel is well within the less dangerous semicircle, and the rule for that semicircle then followed. In the Southern Hemisphere the same rules

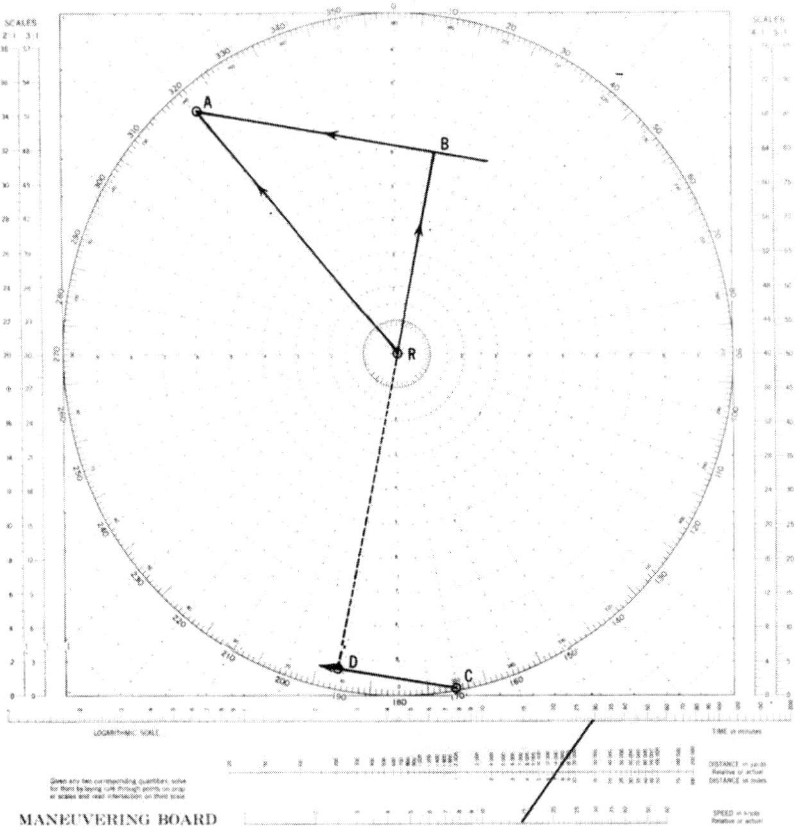

Figure 3511. Determining the course to avoid the storm center.

hold, but with respect to the port side. With a faster than average vessel, the wind can be brought a little farther aft in each case. However, as the speed of the storm increases along its track, the wind should be brought farther forward. If land interferes with what would otherwise be the best maneuver, the solution should be altered to fit the circumstances.

If the vessel is faster than the storm, it is possible to overtake it. In this case, the only action usually needed is to slow enough to let the storm pull ahead.

In all cases, one should be alert to changes in the direction of movement of the storm center, particularly in the area where the track normally curves toward the pole. If the storm maintains its direction and speed, the ship's course should be maintained as the wind shifts.

If it becomes necessary for a vessel to heave to, the characteristics of the vessel should be considered. A power vessel is concerned primarily with damage by direct action of the sea. A good general rule is to heave to with head to the sea in the dangerous semicircle, or stern to the sea in the less dangerous semicircle. This will result in greatest amount of headway away from the storm center, and least amount of leeway toward it. If a vessel handles better with the sea astern or on the quarter, it may be placed in this position in the less dangerous semicircle or in the rear half of the dangerous semicircle, but never in the forward half of the dangerous semicircle. It has been reported that when the wind reaches hurricane speed and the seas become confused, some ships ride out the storm best if the engines are stopped, and the vessel is left to seek its own position, or lie ahull. In this way, it is said, the ship rides with the storm instead of fighting against it.

In a sailing vessel attempting to avoid a storm center, one should steer courses as near as possible to those prescribed above for power vessels. However, if it becomes necessary for such a vessel to heave to, the wind is of greater concern than the sea. A good general rule always is to heave to on whichever tack permits the shifting wind to draw aft. In the Northern Hemisphere, this is the starboard tack in the dangerous semicircle, and the port tack in the less dangerous semicircle. In the Southern Hemisphere these are reversed.

While each storm requires its own analysis, and frequent or continual resurvey of the situation, the general rules for a steamer may be summarized as follows:

Northern Hemisphere

Right or dangerous semicircle: Bring the wind on the starboard bow (045° relative), hold course and make as much way as possible. If necessary, heave to with head to the sea.

Left or less dangerous semicircle: Bring the wind on the starboard quarter (135° relative), hold course and make as much way as possible. If necessary, heave to with stern to the sea.

On storm track, ahead of center: Bring the wind 2 points on the starboard quarter (about 160° relative), hold course and make as much way as possible. When well within the less dangerous semicircle, maneuver as indicated above.

On storm track, behind center: Avoid the center by the best practicable course, keeping in mind the tendency of tropical cyclones to curve northward and eastward.

Southern Hemisphere

Left or dangerous semicircle: Bring the wind on the port bow (315° relative), hold course and make as much way as possible. If necessary, heave to with head to the sea.

Right or less dangerous semicircle: Bring the wind on the port quarter (225° relative), hold course and make as much way as possible. If necessary, heave to with stern to the sea.

On storm track, ahead of center: Bring the wind about 200° relative, hold course and make as much way as possible. When well within the less dangerous semicircle, maneuver as indicated above.

On storm track, behind center: Avoid the center by the best practicable course, keeping in mind the tendency of tropical cyclones to curve southward and eastward.

It is possible, particularly in temperate latitudes after the storm has recurved, that the dangerous semicircle is the left one in the Northern Hemisphere (right one in the Southern Hemisphere). This can occur if a large high lies north of the storm and causes a tightening of the pressure gradient in the region.

The *Typhoon Havens Handbook* for the Western Pacific and Indian Oceans is published by the Naval Oceanographic and Atmospheric Research Lab (NOARL) Monterey, California, as an aid to captains and commanding officers of ships in evaluating a typhoon situation, and to assist them in deciding whether to sortie, to evade, to remain in port, or to head for the shelter of a specific harbor.

CONSEQUENCES OF TROPICAL CYCLONES

3512. High Winds and Flooding

The high winds of a tropical cyclone inflict widespread damage when such a storm leaves the ocean and crosses land. Aids to navigation may be blown out of position or destroyed. Craft in harbors, often lifted by the storm surge, break moorings or drag anchor and are blown ashore and against obstructions. Ashore, trees are blown over, houses are damaged, power lines are blown down, etc. The greatest damage usually occurs in the dangerous semicircle a short distance from the center, where the strongest winds occur. As the storm continues on across land, its fury subsides faster than it would if it had remained over water.

Wind instruments are usually incapable of measuring the 175 to 200 knot winds of the more intense hurricanes. Even if the instrument holds up, often the supporting structure is destroyed. Doppler radar may be effective in determining wind speeds, but may also be blown away.

Wind gusts, which are usually 30 to 50 percent higher than sustained winds, add significantly to the destructiveness of the tropical cyclone. Many tropical cyclones that reach hurricane intensity develop winds of more than 90 knots sometime during their lives, but few develop winds of more than 130 knots.

Tropical cyclones have produced some of the world's heaviest rainfalls. While average amounts range from 6 to 10 inches, totals near 100 inches over a 4-day period have been observed. A 24-hour world's record of 73.62 inches fell at Reunion Island during a tropical cyclone in 1952. Forward movement of the storm and land topography have a considerable influence on rainfall totals. Torrential rains can occur when a storm moves against a mountain range; this is common in the Philippines and Japan, where even weak tropical depressions produce considerable rainfall. A 24-hour total of 46 inches was recorded in the Philippines during a typhoon in 1911. As hurricane Camille crossed southern Virginia's Blue Ridge Mountains in August of 1969, there was nearly 30 inches of rain in about 8 hours. This caused some of the most disastrous floods in the state's history.

Flooding is an extremely destructive by-product of the tropical cyclone's torrential rains. Whether an area will be flooded depends on the physical characteristics of the drainage basin, rate and accumulation of precipitation, and river stages at the time the rains begin. When heavy rains fall over flat terrain, the countryside may lie under water for a month or so, and while buildings, furnishings, and underground power lines may be damaged, there are usually few fatalities. In mountainous or hill country, disastrous floods develop rapidly and can cause a great loss of life.

There have been occasional reports in tropical cyclones of waves greater than 40 feet in height, and numerous reports in the 30- to 40-foot category. However, in tropical cyclones, strong winds rarely persist for a sufficiently long time or over a large enough area to permit enormous wave heights to develop. The direction and speed of the wind changes more rapidly in tropical cyclones than in extratropical storms. Thus, the maximum duration and fetch for any wind condition is often less in tropical cyclones than in extratropical storms, and the waves accompanying any given local wind conditions are generally not so high as those expected, with similar local wind conditions, in the high-latitude storms. In hurricane Camille, significant waves of 43 feet were recorded; an extreme wave height reached 72 feet.

Exceptional conditions may arise when waves of certain dimensions travel within the storm at a speed equal to the storm's speed, thus, in effect, extending the duration and fetch of the wave and significantly increasing its height. This occurs most often to the right of the track in the Northern Hemisphere (left of the track in the Southern Hemisphere). Another condition that may give rise to exceptional wave heights is the intersection of waves from two or more distinct directions. This may lead to a zone of confused seas in which the heights of some waves will equal the sums of each individual wave train. This process can occur in any quadrant of the storm, so it should not be assumed that the highest waves will always be encountered to the right of the storm track in the Northern Hemisphere (left of the track in the Southern Hemisphere).

When these waves move beyond the influence of the generating winds, they become known as **swell**. They are recognized by their smooth, undulating form, in contrast to the steep, ragged crests of wind waves. This swell, particularly that generated by the right side of the storm, can travel a thousand miles or more and may produce tides 3 or 4 feet above normal along several hundred miles of coastline. It may also produce tremendous surf over offshore reefs which normally are calm.

When a tropical cyclone moves close to a coast, wind often causes a rapid rise in water level, and along with the falling pressure may produce a **storm surge**. This surge is usually confined to the right of the track in the Northern Hemisphere (left of the track in the Southern Hemisphere) and to a relatively small section of the coastline. It most often occurs with the approach of the storm, but in some cases, where a surge moves into a long channel, the effect may be delayed. Occasionally, the greatest rise in water is observed on the opposite side of the track, when northerly winds funnel into a partially landlocked harbor. The surge could be 3 feet or less, or it could be 20 feet or more, depending on the combination of factors involved.

There have been reports of a "hurricane wave," described as a "wall of water," which moves rapidly toward the coastline. Authenticated cases are rare, but some of the world's greatest natural disasters have occurred as a result of this wave, which may be a rapidly rising and abnormally high storm surge. In India, such a disaster occurred in 1876, between Calcutta and Chittagong, and drowned more than 100,000 persons.

Along the coast, greater damage may be inflicted by water than by the wind. There are at least four sources of water damage. First, the unusually high seas generated by the storm winds pound against shore installations and craft in their way. Second, the continued blowing of the wind toward land causes the water level to increase perhaps 3 to 10 feet above its normal level. This **storm tide**, which may begin when the storm center is 500 miles or even farther from the shore, gradually increases until the storm passes. The highest storm tides are caused by a slow-moving tropical cyclone of large diameter, because both of these effects result in greater duration of wind in the same direction. The effect is greatest in a partly enclosed body of water, such as the Gulf of Mexico, where the concave coastline does not readily permit the escape of water. It is least on small islands, which present little obstruction to the flow of water. Third, the furious winds which blow around the wall of the eye create a ridge of water called a **storm wave**, which strikes the coast and often inflicts heavy damage. The effect is similar to that of a seismic sea wave, caused by an earthquake in the ocean floor. Both of these waves are popularly called **tidal waves**. Storm waves of 20 feet or more have occurred. About 3 or 4 feet of this wave might be due to the decrease in atmospheric pressure as the sea surface is drawn up into the low pressure area, and the rest to winds. Like the damage caused by wind, damage due to high seas, the storm surge and tide, and the storm wave is greatest in the dangerous semicircle, near the center. The fourth source of water damage is the heavy rain that accompanies a tropical cyclone. This causes floods that add to the damage caused in other ways.

There have been many instances of tornadoes occurring within the circulation of tropical cyclones. Most of these have been associated with tropical cyclones of the North Atlantic Ocean and have occurred in the West Indies and along the gulf and Atlantic coasts of the United States. They are usually observed in the forward semicircle or along the advancing periphery of the storm. These tornadoes are usually short-lived and less intense than those that occur in the midwestern United States.

When proceeding along a shore recently visited by a tropical cyclone, a navigator should remember that time is required to restore aids to navigation which have been blown out of position or destroyed. In some instances the aid may remain but its light, sound apparatus, or radiobeacon may be inoperative. Landmarks may have been damaged or destroyed, and in some instances the coastline and hydrography may be changed.

CHAPTER 36

WEATHER OBSERVATIONS

BASIC WEATHER OBSERVATIONS

3600. Introduction

Weather forecasts are based upon information acquired by observations made at a large number of stations. Ashore, these stations are located so as to provide adequate coverage of the area of interest. The observations at sea are made by mariners, buoys, and satellites. Since the number of observations at sea is small compared to the number ashore, marine observations are of great importance. Data recorded by designated vessels are sent by radio or satellite to national meteorological centers ashore, where they are calculated into the computer forecast models for the development of synoptic charts, which are then used to prepare local and global forecasts. The complete set of weather data gathered at sea is then sent to the appropriate meteorological services for use in the preparation of weather atlases and in marine climatological studies.

Weather observations are normally taken on the major synoptic hours (0000, 0600, 1200, and 1800 UTC), but three-hourly intermediate observations are necessary on the Great Lakes, within 200 nautical miles from the United States or Canadian coastline, or within 300 nautical miles of a named tropical cyclone. Even with satellite imagery, actual reports are needed to confirm developing patterns and provide accurate temperature, pressure, and other measurements. Forecasts can be no better than the data received.

3601. Atmospheric Pressure

The sea of air surrounding the Earth exerts a pressure of about 14.7 pounds per square inch on the surface of the Earth. This **atmospheric pressure**, sometimes called **barometric pressure**, varies from place to place, and at the same place it varies over time.

Atmospheric pressure is one of the most basic elements of a meteorological observation. When the pressure at each station is plotted on a synoptic chart, lines of equal atmospheric pressure, called **isobars**, indicate the areas of high and low pressure. These are useful in making weather predictions, because certain types of weather are characteristic of each type of area, and the wind patterns over large areas can be deduced from the isobars.

Atmospheric pressure is measured with a **barometer**. The earliest known barometer was the **mercurial barometer,** invented by Evangelista Torricelli in 1643. In its simplest form, it consists of a glass tube a little more than 30 inches in length and of uniform internal diameter. With one end closed, the tube is filled with mercury, and inverted into a cup of mercury. The mercury in the tube falls until the column is just supported by the pressure of the atmosphere on the open cup, leaving a vacuum at the upper end of the tube. The height of the column indicates atmospheric pressure, with greater pressures supporting higher columns of mercury.

The **aneroid barometer** has a partly evacuated, thin metal cell which is compressed by atmospheric pressure. Slight changes in air pressure cause the cell to expand or contract, while a system of levers magnifies and converts this motion to a reading on a gauge or recorder.

The early mercurial barometers were calibrated to indicate the height, usually in inches or millimeters, of the column of mercury needed to balance the column of air above the point of measurement. While units of inches and millimeters are still widely used, many modern barometers are calibrated to indicate the centimeter-gram-second unit of pressure, the **hectopascal (hPa)**, formerly known as the millibar. The hectopascal is equal to 1,000 dynes per square centimeter. A dyne is the force required to accelerate a mass of one gram at the rate of one centimeter per second per second. 1,000 hPa = 100,000 Pascal = 14.50 pounds per square inch = 750.0 mm Hg = 0.9869 atmosphere. A reading in any of the three units of measurement can be converted to the equivalent reading in any of the other units by using Table 34 or the conversion factors. However, the pressure reading should always be reported in hPa.

3602. The Aneroid Barometer

The **aneroid barometer** (Figure 3602) measures the force exerted by atmospheric pressure on a partly evacuated, thin metal element called a sylphon cell (aneroid capsule). A small spring is used, either internally or externally, to partly counteract the tendency of the atmospheric pressure to crush the cell. Atmospheric pressure is indicated directly by a scale and a pointer connected to the cell by a combination of levers. The linkage provides considerable magnification of the slight motion of the cell, to permit readings to higher precision than could be obtained without it. An aneroid barometer should be mounted permanently. Prior to installation, the barometer should be carefully set.

Figure 3602. An aneroid barometer.

U.S. ships of the **Voluntary Observation Ship (VOS)** program are set to sea level pressure. Other vessels may be set to station pressure and corrected for height as necessary. An adjustment screw is provided for this purpose. The error of the instrument is determined by comparison with a mercurial barometer, Digiquartz barometer, or a standard precision aneroid barometer. If a qualified meteorologist is not available to make this adjustment, adjust by first removing only one half the apparent error. Then tap the case gently to assist the linkage to adjust itself, and repeat the adjustment. If the remaining error is not more than half a hPa (0.015 inch), no attempt should be made to remove it by further adjustment. Instead, a correction should be applied to the readings. The accuracy of this correction should be checked from time to time.

3603. The Barograph

The **barograph** (Figure 3603) is a recording barometer. In principle it is the same as a non-recording aneroid barometer except that the pointer carries a pen at its outer end, and a slowly rotating cylinder around which a chart is wrapped replaces the scale. A clock mechanism inside the cylinder rotates it so that a continuous line is traced on the chart to indicate the pressure at any time. The barograph is usually mounted on a shelf or desk in a room open to the atmosphere, in a location which minimizes the effect of the ship's vibration. Shock absorbing material such as sponge rubber may be placed under the instrument to minimize vibration. The pen should be checked and the inkwell filled each time the chart is changed.

A **marine microbarograph** is a precision barograph using greater magnification and an expanded chart. It is designed to maintain its precision through the conditions encountered aboard ship. Two sylphon cells are used, one mounted over the other in tandem. Minor fluctuations due to shocks or vibrations are eliminated by damping. Since oil filled dashpots are used for this purpose, the instrument should never be inverted. The dashpots of the marine microbarograph should be kept filled with dashpot oil to within three-eighths inch of the top. The marine microbarograph is fitted with a valve so it can be vented to the outside for more accurate pressure readings.

Ship motions are compensated by damping and spring loading which make it possible for the microbarograph to be tilted up to 22° without varying more than 0.3 hPa from the true reading. Microbarographs have been almost entirely replaced by standard barographs.

Both instruments require checking from time to time to insure correct indication of pressure. The position of the pen is adjusted by a small knob provided for this purpose. The adjustment should be made in stages, eliminating half the apparent error, tapping the case to insure linkage adjustment to the new setting, and then repeating the process.

Figure 3603. A marine barograph

3604. Adjusting Barometer Readings

Atmospheric pressure as indicated by a barometer or barograph may be subject to several errors.

Instrument error: Inaccuracy due to imperfection or incorrect adjustment can be determined by comparison with a standard precision instrument. The National Weather Service provides a comparison service. In major U.S. ports, a Port Meteorological Officer (PMO) carries a portable precision aneroid barometer or a digital barometer for barometer comparisons on board ships which participate in the VOS program. The portable barometer is compared with station barometers before and after a ship visit. If a barometer is taken to a National Weather Service shore station, the comparison can be made there. The correct sea level pressure can also be obtained by telephone. The shipboard barometer should be corrected for height, as explained below, before comparison with this value. If there is reason to believe that the barometer is in error, it should be compared with a standard, and if an error is found, the barometer should be adjusted to the correct reading, or a correction applied to all readings.

Height error: The atmospheric pressure reading at the height of the barometer is called the **station pressure** and is subject to a height correction in order to correct it to sea level. Isobars adequately reflect wind conditions and geographic distribution of pressure only when they are drawn for pressure at constant height (or the varying height at which a constant pressure exists). On synoptic charts it is customary to show the equivalent pressure at sea level, called **sea level pressure**. This is found by applying a correction to station pressure. The correction depends upon the height of the barometer and the average temperature of the air between this height and the surface. The outside air temperature taken aboard ship is sufficiently accurate for this purpose. This is an important correction that should be applied to all readings of any type of barometer. See Table 31 for this correction. Of special note on the Great Lakes, each Lake is at a different height above sea level, so an additional correction is needed.

Temperature error: Barometers are calibrated at a standard temperature of 32° F.Modern aneroid barometers compensate for temperature changes by using different metals having unequal coefficients of linear expansion.

3605. Temperature

Temperature is a measure of heat energy, measured in degrees. Several different temperature scales are in use.

On the **Fahrenheit (F)** scale, pure water freezes at 32° and boils at 212°.

On the **Celsius (C)** scale, commonly used with the metric system, the freezing point of pure water is 0° and the boiling point is 100° .This scale has been known by various names in different countries. In the United States it was formerly called the centigrade scale. The Ninth General Conference of Weights and Measures, held in France in 1948, adopted the name Celsius to be consistent with the naming of other temperature scales after their inventors, and to avoid the use of different names in different countries. On the original Celsius scale, invented in 1742 by a Swedish astronomer named Anders Celsius, numbering was the reverse of the modern scale, 0° representing the boiling point of water, and 100° its freezing point.

Temperature of one scale can be easily converted to another because of the linear mathematical relationship between them. Note that the sequence of calculation is slightly different; algebraic rules must be followed.

$$C = \frac{5}{9}(F - 32), \text{ or} \qquad C = \frac{F - 32}{1.8}$$

$$F = \frac{9}{5}C + 32, \text{ or} \qquad F = 1.8C + 32$$

$$K = C + 273.15$$

$$R = F + 459.69$$

A temperature of –40° is the same by either the Celsius or Fahrenheit scale. Similar formulas can be made for conversion of other temperature scale readings. The Conversion Table for Thermometer Scales (Table 29) gives the equivalent values of Fahrenheit, Celsius, and Kelvin temperatures.

The intensity or degree of heat (temperature) should not be confused with the amount of heat. If the temperature of air or some other substance is to be increased by a given number of degrees, the amount of heat that must be added depends on the mass of the substance. Also, because of differences in their specific heat, equal amounts of different substances require the addition of unequal amounts of heat to raise their temperatures by equal amounts. The units used for measurement of heat are the **British thermal unit (BTU,** the amount of heat needed to raise the temperature of 1 pound of water 1° Fahrenheit), and the **calorie** (the amount of heat needed to raise the temperature of 1 gram of water 1° Celsius).

3606. Temperature Measurement

Temperature is measured with a **thermometer**. Most thermometers are based upon the principle that materials expand with an increase of temperature, and contract as temperature decreases. In its most common form, a thermometer consists of a bulb filled with mercury or a glycol based fluid, which is connected to a tube of very small cross sectional area. The fluid only partly fills the tube. In the remainder is a vacuum. Air is driven out by boiling the fluid, and the top of the tube is then sealed. As the fluid expands or contracts with changing temperature, the length of the fluid column in the tube changes.

Sea surface temperature observations are used in the forecasting of fog, and furnish important information about the development and movement of tropical cyclones. Commercial fishermen are interested in the sea surface temperature as an aid in locating certain species of fish. There are several methods of determining seawater temperature. These include engine room intake readings, condenser intake readings, thermistor probes attached to the hull, and readings from buckets recovered from over the side. Although the condenser intake method is not a true measure of surface water temperature, the error is generally small.

If the surface temperature is desired, a sample should be obtained by bucket, preferably made of canvas, from a forward position well clear of any discharge lines. The sample should be taken immediately to a place where it is sheltered from wind and Sun. The water should then be stirred with the thermometer, keeping the bulb submerged, until a constant reading is obtained.

A considerable variation in sea surface temperature can be experienced in a relatively short distance of travel. This is especially true when crossing major ocean currents such as the Gulf Stream and the Kuroshio Current. Significant variations also occur where large quantities of fresh water are discharged from rivers. A clever navigator will note these changes as an indication of when to allow for set and drift in dead reckoning.

3607. Humidity

Humidity is a measure of the atmosphere's water vapor content. **Relative humidity** is the ratio, stated as a percentage, of the pressure of water vapor present in the atmosphere to the saturation vapor pressure at the same temperature.

As air temperature decreases, the relative humidity increases. At some point, saturation takes place, and any further cooling results in condensation of some of the moisture. The temperature at which this occurs is called the dew point, and the moisture deposited upon objects is called dew if it forms in the liquid state, or frost if it forms as ice crystals.

The same process causes moisture to form on the outside of a container of cold liquid, the liquid cooling the air in the immediate vicinity of the container until it reaches the dew point. When moisture is deposited on man-made objects, it is sometimes called **sweat**. It occurs whenever the temperature of a surface is lower than the dew point of air in contact with it. It is of particular concern to the mariner because of its effect upon instruments, and possible damage to ship or cargo. Lenses of optical instruments may sweat, usually with such small droplets that the surface has a "frosted" appearance. When this occurs, the instrument is said to "fog" or "fog up," and is useless until the moisture is removed. Damage is often caused by corrosion or direct water damage when pipes sweat and drip, or when the inside of the shell plates of a vessel sweat. Cargo may sweat if it is cooler than the dew point of the air.

Clouds and fog form from condensation of water on minute particles of dust, salt, and other material in the air.

Each particle forms a nucleus around which a droplet of water forms. If air is completely free from solid particles on which water vapor may condense, the extra moisture remains vaporized, and the air is said to be **supersaturated**.

Relative humidity and dew point are measured with a **hygrometer**. The most common type, called a **psychrometer**, consists of two thermometers mounted together on a single strip of material. One of the thermometers is mounted a little lower than the other, and has its bulb covered with muslin. When the muslin covering is thoroughly moistened and the thermometer well ventilated, evaporation cools the bulb of the thermometer, causing it to indicate a lower reading than the other. A **sling psychrometer** is ventilated by whirling the thermometers. The difference between the dry-bulb and wet-bulb temperatures is used to enter **psychrometric tables** (Table 35 and Table 36) to find the relative humidity and dew point. If the wet-bulb temperature is above freezing, reasonably accurate results can be obtained by a psychrometer consisting of dry- and wet-bulb thermometers mounted so that air can circulate freely around them without special ventilation. This type of installation is common aboard ship.

Example: *The dry-bulb temperature is 65°F, and the wet-bulb temperature is 61°F.*

Required: *(1) Relative humidity, (2) dew point.*

Solution: *The difference between readings is 4°. Entering Table 35 with this value, and a dry-bulb temperature of 65°, the relative humidity is found to be 80 percent. From Table 36 the dew point is 58°.*

Answers: *(1) Relative humidity 80 percent, (2) dew point 58°.*

Also in use aboard many ships is the **electric psychrometer**. This is a hand held, battery operated instrument with two mercury thermometers for obtaining dry- and wet-bulb temperature readings. It consists of a plastic housing that holds the thermometers, batteries, motor, and fan.

3608. Wind Measurement

Wind measurement consists of determination of the direction and speed of the wind. Direction is measured by a **wind vane**, and speed by an **anemometer**. Several types of wind speed and direction sensors are available, using vanes to indicate wind direction and rotating cups or propellers for speed sensing. Many ships have reliable wind instruments installed, and inexpensive wind instruments are available for even the smallest yacht. If no anemometer is available, wind speed can be estimated by its effect upon the sea and nearby objects. The direction can be computed accurately, even on a fast moving vessel, by maneuvering board or Table 30.

3609. True and Apparent Wind

An observer aboard a vessel proceeding through still air experiences an apparent wind which is from dead ahead and has an apparent speed equal to the speed of the vessel. Thus, if the actual or true wind is zero and the speed of the vessel is 10 knots, the apparent wind is from dead ahead at 10 knots. If the true wind is from dead ahead at 15 knots, and the speed of the vessel is 10 knots, the apparent wind is 15 + 10 = 25 knots from dead ahead. If the vessel reverses course, the apparent wind is 15 – 10 = 5 knots, from dead astern.

The **apparent wind** is the vector sum of the true wind and the *reciprocal* of the vessel's course and speed vector. Since wind vanes and anemometers measure apparent wind, the usual problem aboard a vessel equipped with an anemometer is to convert apparent wind to true wind. There are several ways of doing this. Perhaps the simplest is by the graphical solution illustrated in the following example:

Example 1: *A ship is proceeding on course 240° at a speed of 18 knots. The apparent wind is from 040° relative at 30 knots.*

Required: *The direction and speed of the true wind.*

Solution: *(Figure 3609) First starting from the center of a maneuvering board, plot the ship's vector "er," at 240°, length 18 knots (using the 3–1 scale). Next plot the relative wind's vector from r, in a direction of 100° (the reciprocal of 280°) length 30 knots. The true wind is from the center to the end of this vector or line "ew."*

Alternatively, you can plot the ship's vector from the center, then plot the relative wind's vector <u>toward</u> the center, and see the true wind's vector from the end of this line to the end of the ship's vector. Use parallel rulers to transfer the wind vector to the center for an accurate reading.

Answer: *True wind is from 315° at 20 knots.*

On a moving ship, the direction of the true wind is always on the same side and aft of the direction of the apparent wind. The faster the ship moves, the more the apparent wind draws ahead of the true wind.

A solution can also be made in the following manner without plotting: On a maneuvering board, label the circles 5, 10, 15, 20, etc., from the center, and draw vertical lines tangent to these circles. Cut out the 5:1 scale and discard that part having graduations greater than the maximum speed of the vessel. Keep this sheet for all solutions. (For durability, the two parts can be mounted on cardboard or other suitable material.) To find true wind, spot in point 1 by eye. Place the zero of the 5:1 scale on this point and align the scale (inverted) using the vertical lines. Locate point 2 at the speed of the vessel as indicated on the 5:1 scale. It is always vertically below point 1. Read the relative direction and the speed of the true wind, using eye interpolation if needed.

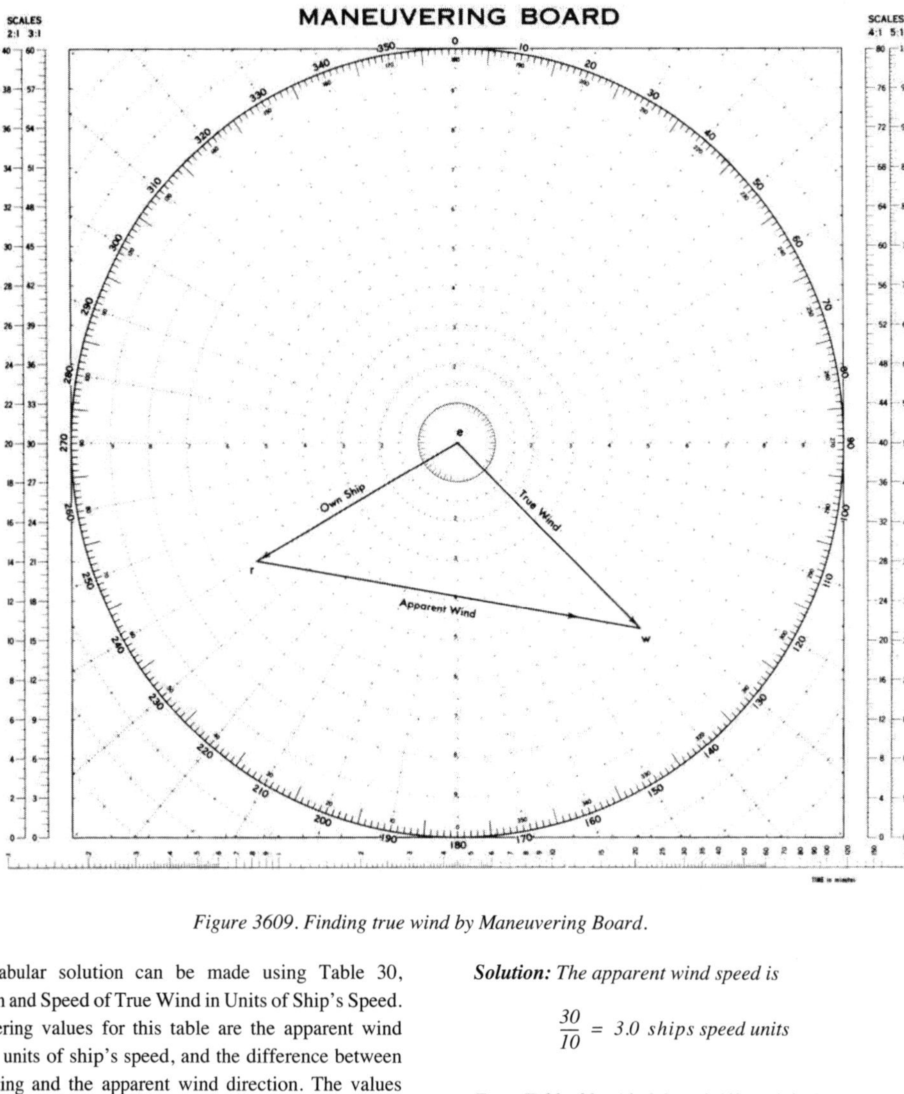

Figure 3609. Finding true wind by Maneuvering Board.

A tabular solution can be made using Table 30, Direction and Speed of True Wind in Units of Ship's Speed. The entering values for this table are the apparent wind speed in units of ship's speed, and the difference between the heading and the apparent wind direction. The values taken from the table are the relative direction (right or left) of the true wind, and the speed of the true wind in units of ship's speed. If a vessel is proceeding at 12 knots, 6 knots constitutes one-half (0.5) unit, 12 knots one unit, 18 knots 1.5 units, 24 knots two units, etc.

Example 2: *A ship is proceeding on course 270° at a speed of 10 knots. The apparent wind is from 10° off the port bow, speed 30 knots.*

Required: *The relative direction, true direction, and speed of the true wind by table.*

Solution: *The apparent wind speed is*

$$\frac{30}{10} = 3.0 \text{ ships speed units}$$

Enter Table 30 with 3.0 and 10° and find the relative direction of the true wind to be 15° off the port bow (345° relative), and the speed to be 2.02 times the ship's speed, or 20 knots, approximately. The true direction is 345° + 270° (-360) = 255°.

Answers: *True wind from 345° relative = 255° true, at 20 knots.*

One can also find apparent wind from the true wind, course or speed required to produce an apparent wind from a given direction or speed, or course and speed to produce an apparent wind of a given speed from a given direction.

Such problems often arise in aircraft carrier operations and in some rescue situations. See *Pub. 1310, The Radar Navigation and Maneuvering Board Manual*, for more detailed information.

When wind speed and direction are determined by the appearance of the sea, the result is true speed and direction. Waves move in the same direction as the generating wind, and are not deflected by Earth's rotation. If a wind vane is used, the direction of the apparent wind thus determined can be used with the speed of the true wind to determine the direction of the true wind by vector diagram.

WIND AND WAVES

3610. Effects of Wind on the Sea

There is a direct relationship between the speed of the wind and the state of the sea. This is useful in predicting the sea conditions to be anticipated when future wind speed forecasts are available. It can also be used to estimate the speed of the wind, which may be necessary when an anemometer is not available.

Wind speeds are usually grouped in accordance with the *Beaufort Scale of Wind Force*, devised in 1806 by English Admiral Sir Francis Beaufort (1774-1857). As adopted in 1838, Beaufort numbers ranged from 0 (calm) to 12 (hurricane). The Beaufort wind scale and sea state photographs at the end of this chapter can be used to estimate wind speed. These pictures (courtesy of the Meteorological Service of Canada) represent the results of a project carried out on board the Canadian Ocean Weather Ships VANCOUVER and QUADRA at Ocean Weather Station PAPA (50°N., 145°W), between April 1976 and May 1981. The aim of the project was to collect color photographs of the sea surface as it appears under the influence of the various ranges of wind speed, as defined by The Beaufort Scale. The photographs represent as closely as possible steady state sea conditions over many hours for each Beaufort wind force. They were taken from heights ranging from 12-17 meters above the sea surface; anemometer height was 28 meters.

3611. Estimating the Wind at Sea

When there is not a functioning anemometer, observers on board ships will usually determine the speed of the wind by estimating Beaufort force. Through experience, ships' officers have developed various methods of estimating this force. The effect of the wind on the observer himself, the ship's rigging, flags, etc., is used as a guide, but estimates based on these indications give the relative wind which must be corrected for the motion of the ship before an estimate of the true wind speed can be obtained.

The most common method involves the appearance of the sea surface. The state of the sea disturbance, i.e. the dimensions of the waves, the presence of white caps, foam, or spray, depends principally on three factors:

1. **The wind speed**. The higher the speed of the wind, the greater is the sea disturbance.
2. **The wind's duration**. At any point on the sea, the disturbance will increase the longer the wind blows at a given speed, until a maximum state of disturbance is reached.
3. **The fetch**. This is the length of the stretch of water over which the wind acts on the sea surface from the same direction.

For a given wind speed and duration, the longer the fetch, the greater is the sea disturbance. If the fetch is short, such as a few miles, the disturbance will be relatively small no matter how great the wind speed is or how long it has been blowing.

Swell waves are not considered when estimating wind speed and direction. Only those waves raised by the wind blowing at the time are of any significance.

A wind of a given Beaufort force will, therefore, produce a characteristic appearance of the sea surface provided that it has been blowing for a sufficient length of time, and over a sufficiently long fetch.

In practice, the mariner observes the sea surface, noting the size of the waves, the white caps, spindrift, etc., and then finds the criterion which best describes the sea surface as observed. This criterion is associated with a Beaufort number, for which a corresponding mean wind speed and range in knots are given. Since meteorological reports require that wind speeds be reported in knots, the mean speed for the Beaufort number may be reported, or an experienced observer may judge that the sea disturbance is such that a higher or lower speed within the range for the force is more accurate.

This method should be used with caution. The sea conditions described for each Beaufort force are "steady-state" conditions; i.e. the conditions which result when the wind has been blowing for a relatively long time, and over a great stretch of water. However, at any particular time at sea the duration of the wind or the fetch, or both, may not have been great enough to produce these "steady-state" conditions. When a high wind springs up suddenly after previously calm or near calm conditions, it will require some hours, depending on the strength of the wind, to generate waves of maximum height. The height of the waves increases rapidly in the first few hours after the commencement of the blow, but increases at a much slower rate later on.

At the beginning of the fetch (such as at a coastline when the wind is offshore) after the wind has been blowing for a long time, the waves are quite small near shore, and in-

Beaufort force of wind.	Theoretical maximum wave height (ft) unlimited duration and fetch.	Duration of winds (hours), with unlimited fetch, to produce percent of maximum wave height indicated.			Fetch (nautical miles), with unlimited duration of blow, to produce percent of maximum wave height indicated.		
		50%	75%	90%	50%	75%	90%
3	2	1.5	5	8	3	13	25
5	8	3.5	8	12	10	30	60
7	20	5.5	12	21	22	75	150
9	40	7	16	25	55	150	280
11	70	9	19	32	85	200	450

Table 3611. Duration of winds and length of fetches required for various wind forces.

crease in height rapidly over the first 50 miles or so of the fetch. Farther offshore, the rate of increase in height with distance slows down, and after 500 miles or so from the beginning of the fetch, there is little or no increase in height.

Table 3611 illustrates the duration of winds and the length of fetches required for various wind forces to build seas to 50 percent, 75 percent, and 90 percent of their theoretical maximum heights.

The theoretical maximum wave heights represent the average heights of the highest third of the waves, as these waves are most significant.

It is clear that winds of force 5 or less can build seas to 90 percent of their maximum height in less than 12 hours, provided the fetch is long enough. Higher winds require a much greater time, force 11 winds requiring 32 hours to build waves to 90 percent of their maximum height. The times given in Table 3611 represent those required to build waves starting from initially calm sea conditions. If waves are already present at the onset of the blow, the times would be somewhat less, depending on the initial wave heights and their direction relative to the direction of the wind which has sprung up.

The first consideration when using the sea criterion to estimate wind speed, therefore, is to decide whether the wind has been blowing long enough from the same direction to produce a steady state sea condition. If not, then it is possible that the wind speed may be underestimated.

Experience has shown that the appearance of white-caps, foam, spindrift, etc. reaches a steady state condition before the height of the waves attain their maximum value. It is a safe assumption that the appearance of the sea (such as white-caps, etc.) will reach a steady state in the time required to build the waves to 50-75 percent of their maximum height. Thus, from Table 3611 it is seen that a force 5 wind could require 8 hours at most to produce a characteristic appearance of the sea surface.

A second consideration when using the sea criteria is the amount of the fetch over which the wind has been blowing to produce the present state of the sea. On the open sea, unless the mariner has the latest synoptic weather map available, the length of the fetch will not be known. It will be seen from Table 3611 though, that only relatively short fetches are required for the lower wind forces to generate their characteristic seas. On the open sea, the fetches associated with most storms and other weather systems are usually long enough so that even winds up to force 9 can build seas up to 90 percent or more of their maximum height, providing the wind blows from the same direction long enough.

When navigating close to a coast or in restricted waters, however, it may be necessary to make allowances for the shorter stretches of water over which the wind blows. For example, referring to Table 3611, if the ship is 22 miles from a coast, and an offshore wind with an actual speed of force 7 is blowing, the waves at the ship will never attain more than 50 percent of their maximum height for this speed no matter how long the wind blows. Hence, if the sea criteria were used under these conditions without consideration of the short fetch, the wind speed would be underestimated. With an offshore wind, the sea criteria may be used with confidence if the distance to the coast is greater than the values given in the extreme right-hand column of Table 3611, provided that the wind has been blowing offshore for a sufficient length of time.

3612. Wind Speed Calculating Factors

Tidal and Other Currents: A wind blowing against the tide or a strong non-tidal current causes higher, steeper waves having a shorter period than normal, which may result in an overestimate of the wind speed if the estimation is made by wave height alone. On the other hand, a wind blowing in the same direction as a tide or strong current causes less sea disturbance than normal, with longer period waves, which may result in underestimating the wind speed.

Shallow Water: Waves running from deep water into

shallow water increase in steepness, hence their tendency to break. Therefore, with an onshore wind there will naturally be more whitecaps over shallow waters than over the deeper water farther offshore. It is only over relatively deep water that the sea criteria can be used with confidence.

Swell: Swell is the name given to waves, generally of considerable length, which were raised in some distant area and which have moved into the vicinity of the ship, or to waves raised nearby that continue after the wind has abated or changed direction. The direction of swell waves is usually different from the direction of the wind and the sea waves. Swell waves are not considered when estimating wind speed and direction. Only those waves raised by the wind blowing at the time are used for estimation. The wind-driven waves show a greater tendency to break when superimposed on the crests of swell, and hence, more whitecaps may be formed than if the swell were absent. Under these conditions, the use of the sea criteria may result in a slight overestimate of the wind speed.

Precipitation: Heavy rain has a damping or smoothing effect on the sea surface that is mechanical in character. Since the sea surface will therefore appear less disturbed than would be the case without the rain, the wind speed may be underestimated unless the smoothing effect is taken into account.

Ice: Even small concentrations of ice floating on the sea surface will dampen waves considerably, and concentrations averaging greater than about seven-tenths will eliminate waves altogether. Young sea ice, which in the early stages of formation has a thick soupy consistency and later takes on a rubbery appearance, is very effective in dampening waves. Consequently, the sea criteria cannot be used with any degree of confidence when sea ice is present. In higher latitudes, the presence of an ice field some distance to windward of the ship may be suspected if, when the ship is not close to any coast, the wind is relatively strong but the seas abnormally underdeveloped. The edge of the ice field acts like a coastline, and the short fetch between the ice and the ship is not sufficient for the wind to fully develop the seas.

Wind Shifts: Following a rapid change in the direction of the wind, as occurs at the passage of a cold front, the new wind will flatten out to a great extent the waves which were present before the wind shift. This happens because the direction of the wind after the shift may differ by 90° or more from the direction of the waves, which does not change. Hence, the wind may oppose the progress of the waves and quickly dampen them out. At the same time, the new wind begins to generate its own waves on top of this dissipating swell, and it is not long before the cross pattern of waves gives the sea a "choppy" or confused appearance. It is during the first few hours following the wind shift that the appearance of the sea surface may not provide a reliable indication of wind speed. The wind is normally stronger than the sea would indicate, as old waves are being flattened out, and the new wave pattern develops.

Night Observations: On a dark night, when it is impossible to see the sea clearly, the observer may estimate the apparent wind from its effect on the ship's rigging, flags, etc., or simply the "feel" of the wind.

CLOUDS

3613. Cloud Formation

Clouds consist of innumerable tiny droplets of water, or ice crystals, formed by condensation of water vapor around microscopic particles in the air. **Fog** is a cloud in contact with the surface of the Earth.

The shape, size, height, thickness, and nature of a cloud all depend upon the conditions under which it is formed. Therefore, clouds are indicators of various processes occurring in the atmosphere. The ability to recognize different types, and a knowledge of the conditions associated with them, are useful in predicting future weather.

Although the variety of clouds is virtually endless, they may be classified by type. Clouds are grouped into three families according to common characteristics and the altitude of their bases. The families are High, Middle, and Low clouds. As shown in Table 3613, the altitudes of the cloud bases vary depending on the latitude in which they are located. Large temperature changes cause most of this latitudinal variation.

Cloud Group	Tropical Regions	Temperate Regions	Polar Regions
High	6,000 to 18,000m (20,000 to 60,000ft)	5,000 to 13,000m (16,000 to 43,000ft)	3,000 to 8,000m (10,000 to 26,000ft)
Middle	2,000 to 8,000m (6,500 to 26,000ft)	2,000 to 7,000m (6,500 to 23,000ft)	2,000 to 4,000m (6,500 to 13,000ft)
Low	surface to 2,000m (0 to 6,500ft)	surface to 2,000m (0 to 6,500ft)	surface to 2,000m (0 to 6,500ft)

Table 3613. Approximate height of cloud bases above the surface for various locations

High clouds are composed principally of ice crystals. As shown in Table 3613, the air temperatures in the tropic regions that are low enough to freeze all liquid water usually occur above 6000 meters, but in the polar regions these temperatures are found at altitudes as low as 3000 meters. **Middle clouds** are composed largely of water droplets, although the higher ones have a tendency toward ice particles. **Low clouds** are composed entirely of water droplets. Clouds types cannot be sufficiently distinguished just by their base altitudes, so within these 3 families are 10 principal cloud types. The names of these are composed of various combinations and forms of the following basic words, all from Latin:

Cirrus, meaning “curl, lock, or tuft of hair.”

Cumulus, meaning “heap, a pile, an accumulation.”

Stratus, meaning “spread out, flatten, cover with a layer.”

Alto, meaning “high, upper air.”

Nimbus, meaning “rainy cloud.”

Individual cloud types recognize certain characteristics, variations, or combinations of these. The 10 principal cloud types and their commonly used symbols are:

3614. High Clouds

Cirrus (Ci) (Figure 3614a through Figure 3614f) are detached high clouds of delicate and fibrous appearance, without shading, generally white in color, often of a silky appearance. Their fibrous and feathery appearance is caused by their composition of ice crystals. Cirrus appear in varied forms, such as isolated tufts; long, thin lines across the sky; branching, feather-like plumes; curved wisps which may end in tufts, and other shapes. These clouds may be arranged in parallel bands which cross the sky in great circles, and appear to converge toward a point on the horizon. This may indicate the general direction of a low pressure area. Cirrus may be brilliantly colored at sunrise and sunset. Because of their height, they become illuminated before other clouds in the morning, and remain lighted after others at sunset. Cirrus are generally associated with fair weather, but if they are followed by lower and thicker clouds, they are often the forerunner of rain or snow.

Figure 3614a. Dense Cirrus in patches or sheaves, not increasing, or Cirrus like cumuliform tufts.

Figure 3614b. Cirrus filaments, strands, hooks, not expanding.

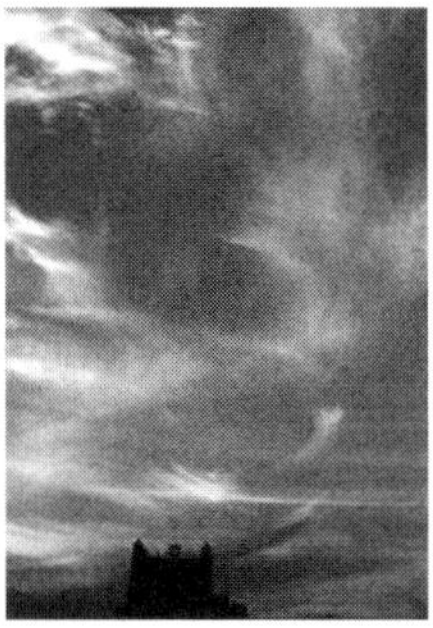

Figure 3614c. Cirrus filaments, strands, hooks, not expanding.

Figure 3614d. Dense Cirrus in patches or sheaves, not increasing, or Cirrus like cumuliform tufts.

Figure 3614e. Dense Cirrus, often the anvil remaining from Cumulonimbus.

Figure 3614f. Dense Cirrus, often the anvil remaining from Cumulonimbus.

Cirrostratus (Cs) (Figure 3614g through Figure 3614p) are thin, whitish, high clouds sometimes covering the sky completely and giving it a milky appearance and at other times presenting, more or less distinctly, a formation like a tangled web. The thin veil is not sufficiently dense to blur the outline of the Sun or Moon. However, the ice crystals of which the cloud is composed refract the light passing through to form halos with the Sun or Moon at the center. As cirrus begins to thicken, it will change into cirrostratus. In this form it is popularly known as "mares' tails." If it continues to thicken and lower, with the ice crystals melting to form water droplets, the cloud formation is known as altostratus. When this occurs, rain may normally be expected within 24 hours. The more brush-like the cirrus when the sky appears, the stronger the wind at the level of the cloud.

Figure 3614g. Cirrus hooks or filaments, increasing and becoming denser.

Figure 3614i. Cirrus bands and/or Cirrostratus, increasing, growing denser, veil below 45.

Figure 3614h. Cirrus hooks or filaments, increasing and becoming denser.

Figure 3614j. Cirrus bands and/or Cirrostratus, increasing, growing denser, veil below 45.

Figure 3614k. Cirrus bands and/or Cirrostratus, increasing, growing denser, veil below 45.

Figure 3614l. Cirrus bands and/or Cirrostratus, increasing, growing denser, veil below 45.

Figure 3614m. Cirrostratus covering the whole sky.

Figure 3614n. Cirrostratus covering the whole sky.

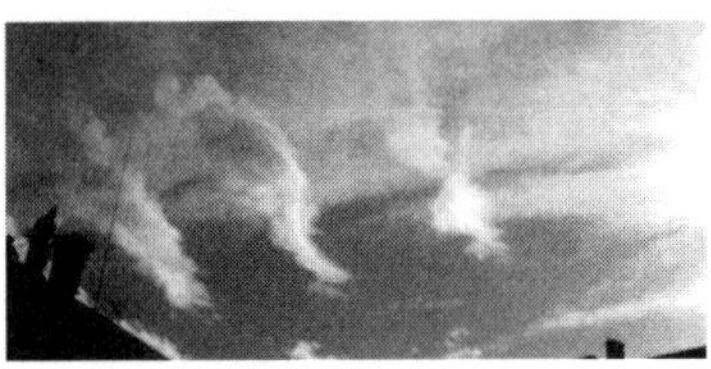

Figure 3614o. Cirrostratus, not increasing, not covering the whole sky.

Figure 3614p. Cirrostratus, not increasing, not covering the whole sky.

Cirrocumulus (Cc) (Figure 3614q and Figure 3614r) are high clouds composed of small white flakes or scales, or of very small globular masses, usually without shadows and arranged in groups of lines, or more often in ripples resembling sand on the seashore. One form of cirrocumulus is popularly known as "mackerel sky" because the pattern resembles the scales on the back of a mackerel. Like cirrus, cirrocumulus are composed of ice crystals and are generally associated with fair weather, but may precede a storm if they thicken and lower. They may turn gray and appear hard before thickening.

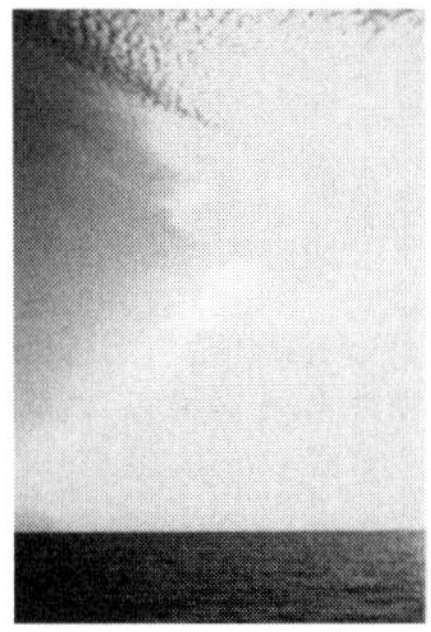

Figure 3614q. Cirrocumulus alone, and/or Cirrus and Cirrostratus.

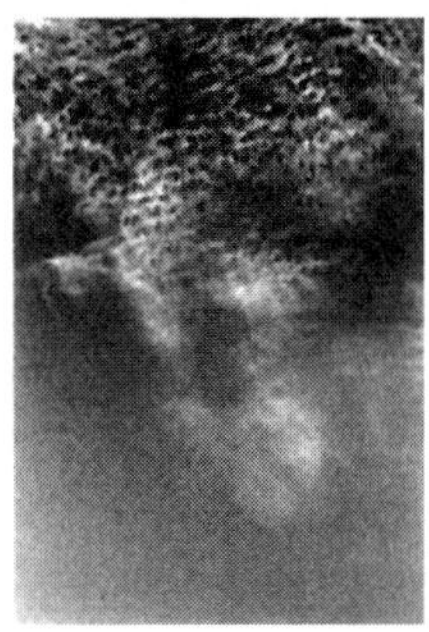

Figure 3614r. Cirrocumulus alone, and/or Cirrus and Cirrostratus.

3615. Middle Level Clouds

Altostratus (As) (Figure 3615a through Figure 3615d) are middle level clouds having the appearance of a grayish or bluish, fibrous veil or sheet. The Sun or Moon, when seen through these clouds, appears as if it were shining through ground glass with a corona around it. Halos are not formed. If these clouds thicken and lower, or if low, ragged "scud" or rain clouds (nimbostratus) form below them, continuous rain or snow may be expected within a few hours.

Figure 3615a. Altostratus, semitransparent, Sun or Moon dimly visible.

Figure 3615b. Altostratus, semitransparent, Sun or Moon dimly visible.

Figure 3615c. Altostratus, dense enough to hide Sun or Moon, or nimbostratus.

Figure 3615d. Altostratus, dense enough to hide Sun or Moon, or nimbostratus

Altocumulus (Ac) (Figure 3615e through Figure 3615r) are middle level clouds consisting of a layer of large, ball-like masses that tend to merge together. The balls or patches may vary in thickness and color from dazzling white to dark gray, but they are more or less regularly arranged. They may appear as distinct patches similar to cirrocumulus, but can be distinguished by having individual patches which are generally larger, showing distinct shadows in some places. They are often mistaken for stratocumulus. If altocumulus thickens and lowers, it may produce thundery weather and showers, but it does not bring prolonged bad weather. Sometimes the patches merge to form a series of big rolls resembling ocean waves, with streaks of blue sky between. Because of perspective, the rolls appear to run together near the horizon. These regular parallel bands differ from cirrocumulus because they occur in larger masses with shadows. Altocumulus move in the direction of the short dimension of the rolls, like ocean waves. Sometimes altocumulus appear briefly in the form shown in Figure 3615o and Figure 3615p, sometimes before a thunderstorm. They are generally arranged in a line with a flat horizontal base, giving the impression of turrets on a castle. The turreted tops may look like miniature cumulus and possess considerable depth and great length. These clouds usually indicate a change to chaotic, thundery skies.

Figure 3615e. Altocumulus, semitransparent, cloud elements change slowly, one level.

Figure 3615f. Altocumulus, semitransparent, cloud elements change slowly, one level.

Figure 3615g. Altocumulus patches, semitransparent, multilevel, cloud elements changing, also Altocumulus Lenticular

Figure 3615h. Altocumulus patches, semitransparent, multilevel, cloud elements changing, also Altocumulus Lenticular

Figure 3615i. Altocumulus, one or more bands or layers, expanding, thickening.

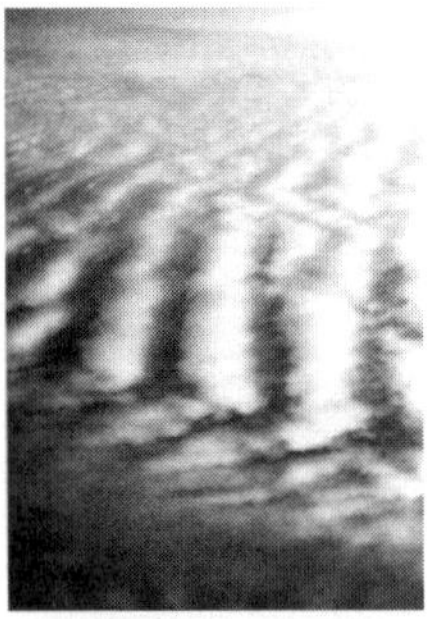

Figure 3615j. Altocumulus, one or more bands or layers, expanding, thickening.

Figure 3615k. Altocumulus from the spreading of Cumulus or Cumulonimbus.

Figure 3615l. Altocumulus from the spreading of Cumulus or Cumulonimbus.

Figure 3615m. Altocumulus, one or more layers, mainly opaque, not expanding, or Altocumulus with Altostratus or Nimbostratus.

Figure 3615n. Altocumulus, one or more layers, mainly opaque, not expanding, or Altocumulus with Altostratus or Nimbostratus.

Figure 3615o. Altocumulus with tower or turret like sproutings.

Figure 3615p. Altocumulus with tower or turret-like sproutings.

Figure 3615q. Altocumulus of a chaotic sky, usually with heavy broken cloud sheets at different levels.

Figure 3615r. Altocumulus of a chaotic sky, usually with heavy broken cloud sheets at different levels.

3616. Low Clouds

Cumulus (Cu) (Figure 3616a through Figure 3616d) are dense clouds with vertical development formed by rising air which is cooled as it reaches greater heights. They have a horizontal base and dome-shaped upper surfaces, with protuberances extending above the dome. Cumulus appear in patches, never covering the entire sky. When vertical development is not great, the clouds resemble tufts of cotton or wool, being popularly called "woolpack" clouds. The horizontal bases of such clouds may not be noticeable. These are called "fair weather" cumulus because they commonly accompany stable air and good weather. However, they may merge with altocumulus, or may grow to cumulonimbus before a thunderstorm. Since cumulus are formed by updrafts, they are accompanied by turbulence, causing "bumpiness" in the air. The extent of turbulence is proportional to the vertical extent of the clouds. Cumulus are marked by strong contrasts of light and dark.

Figure 3616a. Cumulus with very little vertical extent.

Figure 3616b. Cumulus with very little vertical extent.

Figure 3616c. Cumulus with moderate or greater vertical extent.

Figure 3616d. Cumulus with moderate or greater vertical extent.

Stratocumulus (Sc) (Figure 3616e through Figure 3616h) are low level clouds appearing as soft, gray, roll-shaped masses. They may be shaped in long, parallel rolls similar to altocumulus, moving forward with the wind. The motion is in the direction of their short dimension, like ocean waves. These clouds, which vary greatly in altitude, are the final product of the characteristic daily change taking place in cumulus clouds. They are usually followed by clear skies during the night.

Figure 3616e. Stratocumulus from the spreading out of Cumulus.

Figure 3616f. Stratocumulus from the spreading out of Cumulus.

Figure 3616g. Stratocumulus not formed from the spreading out of Cumulus.

Figure 3616h. Stratocumulus not formed from the spreading out of Cumulus.

Stratus (St) (Figure 3616i through Figure 3616l) is a low cloud in a uniform layer resembling fog. Often the base is not more than 1,000 feet high. A veil of thin stratus gives the sky a hazy appearance. Stratus is often quite thick, permitting so little sunlight to penetrate that it appears dark to an observer below. From above it is white. Light mist may descend from stratus. Strong wind sometimes breaks stratus into shreds called "fractostratus."

Figure 3616i. Stratus in a sheet or layer.

Figure 3616j. Stratus in a sheet or layer.

Figure 3616k. Stratus Fractus and/or Cumulus fractus of bad weather.

Figure 3616l. Stratus Fractus and/or Cumulus fractus of bad weather.

Nimbostratus (Ns) (Figure 3616m and Figure 3616n) is a low, dark, shapeless cloud layer, usually nearly uniform, but sometimes with ragged, wet-looking bases. Nimbostratus is the typical rain cloud. The precipitation which falls from this cloud is steady or intermittent, but not showery.

Figure 3616m. Nimbostratus formed from lowering Altostratus.

Figure 3616n. Nimbostratus formed from lowering Altostratus.

Cumulonimbus (Cb) (Figure 3616o through Figure 3616r) is a massive cloud with great vertical development, rising in mountainous towers to great heights. The upper part consists of ice crystals, and often spreads out in the shape of an anvil which may be seen at such distances that the base may be below the horizon. Cumulonimbus often produces showers of rain, snow, or hail, frequently accompanied by lightning and thunder. Because of this, the cloud is often popularly called a "thundercloud" or "thunderhead." The base is horizontal, but as showers occur it lowers and becomes ragged.

Figure 3616o. Cumulonimbus, tops not fibrous, outline not completely sharp, no anvil.

Figure 3616p. Cumulonimbus, tops not fibrous, outline not completely sharp, no anvil.

Figure 3616q. Cumulonimbus with fibrous top, often with an anvil.

Figure 3616r. Cumulonimbus with fibrous top, often with an anvil.

3617. Cloud Height Measurement

At sea, cloud heights are often determined by estimation. This is a difficult task, particularly at night.

The height of the base of clouds formed by vertical development (any form of cumulus), if formed in air that has risen from the surface of the Earth, can be determined by psychrometer. This is because the height to which the air must rise before condensation takes place is proportional to the difference between surface air temperature and the dew point. At sea, this difference multiplied by 126.3 gives the height in meters. That is, for every degree difference between surface air temperature and the dew point, the air must rise 126.3 meters before condensation will take place. Thus, if the dry-bulb temperature is 26.8°C, and the wet-bulb temperature is 25.0°C, the dew point is 24°C, or 2.8°C lower than the surface air temperature. The height of the cloud base is 2.8 × 126.3 = 354 meters.

OTHER OBSERVATIONS

3618. Visibility Measurement

Visibility is the horizontal distance at which prominent objects can be seen and identified by the unaided eye. It is usually measured directly by the human eye. Ashore, the distances of various buildings, trees, lights, and other objects can be used as a guide in estimating the visibility. At sea, however, such an estimate is difficult to make with accuracy. Other ships and the horizon may be of some assistance. See Table 12, Distance of the Horizon.

Ashore, visibility is sometimes measured by a **transmissometer**, a device which measures the transparency of the atmosphere by passing a beam of light over a known short distance, and comparing it with a reference light.

3619. Upper Air Observations

Upper air information provides the third dimension to the weather map. Unfortunately, the equipment necessary to obtain such information is quite expensive, and the observations are time consuming. Consequently, the network of observing stations is quite sparse compared to that for surface observations, particularly over the oceans and in isolated land areas. Where facilities exist, upper air observations are made by means of unmanned balloons, in conjunction with theodolites, radiosondes, radar, and radio direction finders.

3620. New Technologies in Weather Observing

Shipboard, upper air, buoy, radar, and satellite observations are the foundation for the development of accurate forecast computer models, both in the short and long term. New techniques such as Doppler radar, satellite analysis, and the integration of data from many different sites into complex computer algorithms provide a method of predicting storm tracks with a high degree of accuracy. Tornadoes, line squalls, individual thunderstorms, and entire storm systems can be continuously tracked and their paths predicted with unprecedented accuracy. At sea, the mariner has immediate access to this data through facsimile transmission of synoptic charts and actual satellite photographs, and through radio or communications satellite contact with weather routing services, or through internet providers.

Automated weather stations and buoy systems provide regular transmissions of meteorological and oceanographic information by radio. Some of these buoys or stations can be accessed via the telephone. For further information, visit the National Data Buoy Center's web site at http:// www.ndbc.noaa.gov. These buoys and stations are generally located at isolated and relatively inaccessible locations from which weather and ocean data are of great importance. Depending on the type of system used, the elements usually measured include wind direction and speed, atmospheric pressure, air and sea surface temperature, spectral wave data, and a temperature profile from the sea surface to a predetermined depth.

Regardless of advances in the technology of observing and forecasting, the shipboard weather report remains the cornerstone upon which the accuracy of many forecasts is based.

3621. Recording Observations

Instructions for recording weather observations aboard vessels of the United States Navy are given in NAVMETOCCOMINST 3144.1 (series). For information on obtaining a copy of this instruction, visit http://cnmoc.navy.mil. Instructions for recording observations aboard merchant vessels are given in the National Weather Service Observing Handbook No. 1, Marine Surface Observations. Contact the local Port Meteorological Officer (PMO) or the VOS program lead at http://www.vos.noaa.gov.

Force 0: Wind Speed less than 1 knot.
Sea: Sea like a mirror.

Force 1: Wind Speed 1-3 knots.
Sea: Wave height 0.1m (.25ft); Ripples with appearance of scales, no foam crests.

Force 2: Wind Speed 4-6 knots.
Sea: Wave height 0.2-0.3 m (0.5-1 ft); Small wavelets, crests of glassy appearance, not breaking.

Force 3: Wind Speed 7-10 knots.
Sea: Wave height 0.6-1m (2-3 ft); Large wavelets, crests begin to break, scattered whitecaps.

Force 4: Wind Speed 11-16 knots.
Sea: Wave height 1-1.5 m (3.5-5 ft); Small waves becoming longer, numerous whitecaps.

Force 5: Wind Speed 17-21 knots.
Sea: Wave height 2-2.5 m (6-8 ft); Moderate waves, taking longer form, many whitecaps, some spray.

Force 6: Wind Speed 22-27 knots.
Sea: Wave height 3-4 m (9.5-13 ft); Larger waves forming, whitecaps everywhere, more spray.

Force 7: Wind Speed 28-33 knots.
Sea: Wave height 4-5.5 m (13.5-19 ft); Sea heaps up, white foam from breaking waves begins to be blown in streaks along direction of wind.

Force 8: Wind Speed 34-40 knots.
Sea: Wave height 5.5-7.5 m (18-25 ft); Moderately high waves of greater length, edges of crests begin to break into spindrift, foam is blown in well marked streaks.

Force 9: Wind Speed 41-47 knots.
Sea: Wave height 7-10 m (23-32 ft); High waves, sea begins to roll, dense streaks of foam along wind direction, spray may reduce visibility.

Force 10: Wind Speed 48-55 knots (storm).
Sea: Wave height 9-12.5 m (29-41 ft); Very high waves with overhanging crests, sea takes white appearance as foam is blown in very dense streaks, rolling is heavy and shocklike, visibility is reduced.

Force 11: Wind Speed 56-63 knots.
Sea: Wave height 11.5-16 m (37-52 ft); Exceptionally high waves, sea covered with white foam patches, visibility still more reduced.

Force 12: Wind Speed 64-71 knots.
Sea: Wave height more than 16 m (52 ft); Air filled with foam, sea completely white with driving spray, visibility greatly reduced.

CHAPTER 37

WEATHER ROUTING

PRINCIPLES OF WEATHER ROUTING

3700. Introduction

Ship weather routing develops an optimum track for ocean voyages based on forecasts of weather, sea conditions, and a ship's individual characteristics for a particular transit. Within specified limits of weather and sea conditions, the term optimum is used to mean maximum safety and crew comfort, minimum fuel consumption, minimum time underway, or any desired combination of these factors. The purpose of this chapter is to acquaint the mariner with the basic philosophy and procedures of ship weather routing as an aid to understanding the routing agency's recommendations.

The mariner's first resources for route planning in relation to weather are the *Pilot Chart Atlases*, the *Sailing Directions (Planning Guides)*, and other climatological sources such as historical weather data tables. These publications give climatic data, such as wind speed and direction, wave height frequencies and ice limits, for the major ocean basins of the world. They may recommend specific routes based on probabilities, but not on specific conditions.

The ship routing agency, acting as an advisory service, attempts to avoid or reduce the effects of specific adverse weather and sea conditions by issuing initial route recommendations prior to sailing. It recommends track changes while underway (diversions), and weather advisories to alert the commanding officer or master about approaching unfavorable weather and sea conditions which cannot be effectively avoided by a diversion. Adverse weather and sea conditions are defined as those conditions which will cause damage, significant speed reduction, or time loss.

The initial route recommendation is based on a survey of weather and sea forecasts between the point of departure and the destination. It takes into account the type of vessel, hull type, speed capability, safety considerations, cargo, and loading conditions. The vessel's progress is continually monitored, and if adverse weather and sea conditions are forecast along the vessel's current track, a recommendation for a diversion or a weather advisory is transmitted. By this process of initial route selection and continued monitoring of progress for possible changes in the forecast weather and sea conditions along a route, it is possible to maximize both speed and safety.

In providing for optimum sailing conditions, the advisory service also attempts to reduce transit time by avoiding the adverse conditions which may be encountered on a shorter route, or if the forecasts permit, diverting to a shorter track to take advantage of favorable weather and sea conditions. A significant advantage of weather routing accrues when: (1) the passage is relatively long, about 1,500 miles or more; (2) the waters are navigationally unrestricted, so that there is a choice of routes; and (3) weather is a factor in determining the route to be followed.

Use of this advisory service in no way relieves the commanding officer or master of responsibility for prudent seamanship and safe navigation. There is no intent by the routing agency to inhibit the exercise of professional judgment and prerogatives of commanding officers and masters.

The techniques of ship routing and access to the advice are increasingly less expensive, and are thus being made available to coastal vessels, smaller commercial craft, and even yachts.

3701. Historical Perspective

The advent of extended range forecasting and the development of selective climatology, along with powerful computer modeling techniques, have made ship routing systems possible. The ability to effectively advise ships to take advantage of favorable weather was hampered previously by forecast limitations and the lack of an effective communications system.

Development work in the area of data accumulation and climatology has a long history. Benjamin Franklin, as deputy postmaster general of the British Colonies in North America, produced a chart of the Gulf Stream from information supplied by masters of New England whaling ships. This first mapping of the Gulf Stream helped improve the mail packet service between the British Colonies and England. In some passages the sailing time was reduced by as much as 14 days over routes previously sailed.

In the mid-19th century, Matthew Fontaine Maury compiled large amounts of atmospheric and oceanographic data from ships' log books. For the first time, a climatology of ocean weather and currents of the world was available to the mariner. This information was used by Maury to develop seasonally recommended routes for sailing ships and early steam powered vessels in the latter half of the 19th century. In many cases, Maury's charts were proved correct by the savings in transit time. Average transit time on the New York to California via Cape Horn route was reduced from 183 days to 139 days with the use of his recommended seasonal routes.

In the 1950's the concept of ship weather routing was put into operation by several private meteorological groups and by the U.S. Navy. By applying the available surface and upper air forecasts to transoceanic shipping, it was possible to effectively avoid much heavy weather while generally sailing shorter routes than previously. The development of computers, the internet and communications technology has made weather routing available to nearly everyone afloat.

3702. System Types

Optimum Track Ship Routing (OTSR), the ship routing service of the U.S. Navy, utilizes short range and extended range forecasting techniques in route selection and surveillance procedures. The short range dynamic forecasts of 3 to 5 days are derived from meteorological equations. These forecasts are computed at least twice daily from a data base of northern hemisphere surface and upper air observations, and include surface pressure, upper air constant pressure heights, and the spectral wave values. A significant increase in data input, particularly from satellite information over ocean areas, can extend the time period for which these forecasts are useful.

Selective climatology has been effective in predicting average conditions months in advance during such events as El Nino and La Nina. Such predictions do not represent forecasting, but can indicate the likelihood of certain conditions prevailing.

For extended range forecasting, generally 3 to 14 days, a computer searches a library of historical northern hemisphere surface pressure and 500 hPa analyses for an analogous weather pattern. This is an attempt at selective climatology by matching the current weather pattern with past weather patterns and providing a logical sequence-of-events forecast for the 10 to 14 day period following the dynamic forecast. It is performed for both the Atlantic and Pacific Oceans using climatological data for the entire period of data stored in the computer. For longer ocean transits, monthly values of wind, seas, fog, and ocean currents are used to further extend the time range.

Aviation was first in applying the principle of minimum time tracks (MTT) to a changing wind field. But the problem of finding an MTT for a specific flight is much simpler than for a transoceanic ship passage because an aircraft's transit time is much shorter than a ship's. Thus, marine minimum time tracks require significantly longer range forecasts to develop an optimum route.

Automation has enabled ship routing agencies to develop realistic minimum time tracks. Computation of minimum time tracks makes use of:

1. A navigation system to compute route distance, time enroute, estimated times of arrival (ETA's), and to provide 6 hourly DR synoptic positions for the range of the dynamic forecasts for the ship's current track.
2. A surveillance system to survey wind, seas, fog, and ocean currents obtained from the dynamic and climatological fields.
3. An environmental constraint system imposed as part of the route selection and surveillance process. Constraints are the upper limits of wind and seas desired for the transit. They are determined by the ship's loading, speed capability, and vulnerability. The constraint system is an important part of the route selection process and acts as a warning system when the weather and sea forecast along the present track exceeds predetermined limits.
4. Ship speed characteristics used to approximate ship's speed of advance (SOA) while transiting the forecast sea states.

Criteria for route selection reflect a balance between the captain's desired levels of speed, safety, comfort, and consideration of operations such as fleet maneuvers, fishing, towing, etc.

Ship weather routing services are being offered by many nations. These include Japan, United Kingdom, Russia, Netherlands, Germany, and the United States. Also, several private firms provide routing services to shipping industry clients. Several PC-based software applications have become available, making weather routing available to virtually everyone at sea.

There are two general types of routing services available. The first uses techniques similar to the Navy's OTSR system to forecast conditions and compute routing recommendations, which are then broadcast to the vessel. The second assembles and processes weather and sea condition data and transmits this to ships at sea for on-board processing and generation of route recommendations. The former system allows for greater computer power to be applied to the routing task because powerful computers are available ashore. The latter system allows greater flexibility to the ship's master in changing parameters, evaluating various scenarios, selecting routes, and displaying data.

3703. Ship and Cargo Considerations

Ship and cargo characteristics have a significant influence on the application of ship weather routing. Ship size, speed capability, and type of cargo are important considerations in the route selection process prior to sailing and the surveillance procedure while underway. A ship's characteristics identify its vulnerability to adverse conditions and its ability to avoid them.

Generally, ships with higher speed capability and lighter loads will have shorter routes and be better able to maintain near normal SOA's than ships with lower speed capability or heavy cargoes. Some routes are unique be-

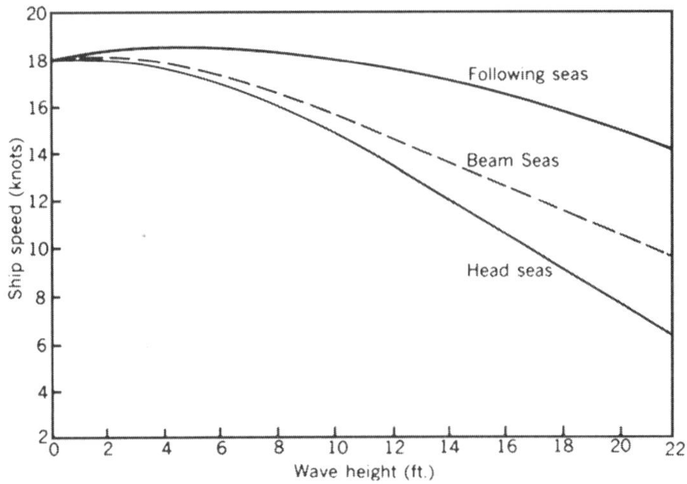

Figure 3703. Performance curves for head, beam, and following seas.

cause of the type of ship or cargo. Avoiding one element of weather to reduce pounding or rolling may be of prime importance. For example, a 20 knot ship with a heavy deck cargo may be severely hampered in its ability to maintain a 20 knot SOA in any seas exceeding moderate head or beam seas because of the possibility of damage resulting from the deck load's characteristics. A similar ship with a stable cargo under the deck is not as vulnerable and may be able to maintain the 20 knot SOA in conditions which would drastically slow the deck-loaded vessel. In towing operations, a tug is more vulnerable to adverse weather and sea conditions, not only in consideration of the tow, but also because of its already limited speed capability. Its slow speed adds to the difficulty of avoiding adverse weather and sea conditions.

Ship performance curves (speed curves) are used to estimate the ship's SOA while transiting the forecast sea states. The curves indicate the effect of head, beam, and following seas of various significant wave heights on the ship's speed. Figure 3703 is a performance curve prepared for a commercial 18-knot vessel. Each vessel will have its own performance curves, which vary widely according to hull type, length, beam, shape, power, and tonnage. Recommendations for sailing vessels must account for wind speed, wind angle, and vessel speed.

With the speed curves it is possible to determine just how costly a diversion will be in terms of the required distance and time. A diversion may not be necessary where the duration of the adverse conditions is limited. In this case, it may be better to ride out the weather and seas knowing that a diversion, even if able to maintain the normal SOA, will not overcome the increased distance and time required.

At other times, the diversion track is less costly because it avoids an area of adverse weather and sea conditions, while being able to maintain normal SOA even though the distance to destination is increased. Based on input data for environmental conditions and ship's behavior, route selection and surveillance techniques seek to achieve the optimum balance between time, distance, and acceptable environmental and seakeeping conditions. Although speed performance curves are an aid to the ship routing agency, the response by mariners to deteriorating weather and sea conditions is not uniform. Some reduce speed voluntarily or change heading sooner than others when unfavorable conditions are encountered. Certain waves with characteristics such that the ship's bow and stern are in successive crests and troughs present special problems for the mariner. Being nearly equal to the ship's length, such wavelengths may induce very dangerous stresses. The degree of hogging and sagging and the associated danger may be more apparent to the mariner than to the ship routing agency. Therefore, adjustment in course and speed for a more favorable ride must be initiated by the commanding officer or master when this situation is encountered.

3704. Environmental Factors

Environmental factors of importance to ship weather routing are those elements of the atmosphere and ocean that may produce a change in the status of a ship transit. In ship routing, consideration is given to wind, seas, fog, ice, and ocean currents. While all of the environmental factors are important for route selection and surveillance, optimum routing is normally considered attained if the effects of wind and seas can be optimized.

Wind: The effect of wind speed on ship performance is difficult to determine. In light winds (less than 20-knots), ships lose speed in headwinds and gain speed slightly in following winds. For higher wind speeds, ship speed is reduced in both head and following winds. This is due to the increased wave action, which even in following seas results in increased drag from steering corrections, and indicates the importance of sea conditions in determining ship performance. In dealing with wind, it is also necessary to know the ship's sail area. High winds will have a greater adverse effect on a large, fully loaded container ship or car carrier than a fully loaded tanker of similar length. This effect is most noticeable when docking, but the effect of beam winds over several days at sea can also be considerable. For sailing vessels, the wind is critical and accurate forecasts are vital to a successful voyage.

Wave Height: Wave height is the major factor affecting ship performance. Wave action is responsible for ship motions which reduce propeller thrust and cause increased drag from steering corrections. The relationship of ship speed to wave direction and height is similar to that of wind. Head seas reduce ship speed, while following seas increase ship speed slightly to a certain point, beyond which they retard it. In heavy seas, exact performance may be difficult to predict because of the adjustments to course and speed for shiphandling and comfort. Although the effect of sea and swell is much greater for large commercial vessels than is wind speed and direction, it is difficult to separate the two in ship routing.

In an effort to provide a more detailed description of the actual and forecast sea state, the U.S. Navy Fleet Numerical Meteorology and Oceanography Center, Monterey, California, produces the Global Spectral Ocean Wave Model (GSOWM) for use by the U.S. Navy's Optimum Track Ship Routing (OTSR) service. This model provides energy values from 12 different directions (30° sectors) and 15 frequency bands for wave periods from 6 to 26 seconds with the total wave energy propagated throughout the grid system as a function of direction and frequency. It is based on the analyzed and forecast planetary boundary layer model wind fields, and is produced for both the Northern and Southern Hemispheres out to 72 hours. For OTSR purposes, primary and secondary waves are derived from the spectral wave program, where the primary wave train has the principal energy (direction and frequency), and the secondary has to be 20 percent of the primary.

Fog: Fog, while not directly affecting ship performance, should be avoided as much as feasible, in order to maintain normal speed in safe conditions. Extensive areas of fog during summertime can be avoided by selecting a lower latitude route than one based solely upon wind and seas. Although the route may be longer, transit time may be less due to not having to reduce speed in reduced visibility. In addition, crew fatigue due to increased watchkeeping vigilance can be reduced.

North Wall Effect: During the Northern Hemisphere fall and winter, the waters to the north of the Gulf Stream in the North Atlantic are at their coldest, while the Gulf Stream itself remains at a constant relatively warm temperature. After passage of a strong cold front or behind a developing coastal low pressure system, Arctic air is sometimes drawn off the Mid-Atlantic coast of the United States and out over the warm waters of the Gulf Stream by northerly winds. This cold air is warmed as it passes over the Gulf Stream, resulting in rapid and intense deepening of the low pressure system and higher than normal surface winds. Higher waves and confused seas result from these winds. When these winds oppose the northeast set of the current, the result is increased wave heights and a shortening of the wave period. If the opposing current is sufficiently strong, the waves will break. These phenomena are collectively called the "North Wall Effect," referring to the region of most dramatic temperature change between the cold water to the north and the warm Gulf Stream water to the south. The most dangerous aspect of this phenomenon is that the strong winds and extremely high, steep waves occur in a limited area and may develop without warning. Thus, a ship that is laboring in near-gale force northerly winds and rough seas, proceeding on a northerly course, can suddenly encounter storm force winds and dangerously high breaking seas. Numerous ships have foundered off the North American coast in the approximate position of the Gulf Stream's North Wall. A similar phenomenon occurs in the North Pacific near the Kuroshio Current and off the Southeast African coast near the Agulhas Current.

Ocean Currents: Ocean currents do not present a significant routing problem, but they can be a determining factor in route selection and diversion. This is especially true when the points of departure and destination are at relatively low latitudes. The important considerations to be evaluated are the difference in distance between a great-circle route and a route selected for optimum current, with the expected increase in SOA from the following current, and the decreased probability of a diversion for weather and seas at the lower latitude. For example, it has proven beneficial to remain equatorward of approximately 22°N for westbound passages between the Canal Zone and southwest Pacific ports. For eastbound passages, if the maximum latitude on a great-circle track from the southwest Pacific to the Canal Zone is below 24°N, a route passing near the axis of the Equatorial Countercurrent is practical because the increased distance is offset by favorable current. Direction and speed of ocean currents are more predictable than wind and seas, but some variability can be expected. Major ocean currents can be disrupted for several days by very intense weather systems such as hurricanes and by global phenomena such as El Nino.

Ice: The problem of ice is twofold: floating ice (icebergs) and deck ice. If possible, areas of icebergs or pack ice should be avoided because of the difficulty of detection and the potential for collision. Deck ice may be more difficult to contend with from a ship routing point of view because it is caused by freezing weather associated with a large weather system. While mostly a nuisance factor on large ships, it causes significant problems with the stability of small ships.

Latitude: Generally, the higher the latitude of a route, even in the summer, the greater are the problems with the environment. Certain operations should benefit from seasonal planning as well as optimum routing. For example, towing operations north of about 40° latitude should be avoided in non-summer months if possible.

3705. Synoptic Weather Considerations

A ship routing agency should direct its forecasting skills to avoiding or limiting the effect of weather and seas associated with extratropical low pressure systems in the mid and higher latitudes and the tropical systems in low latitude. Seasonal or monsoon weather is also a factor in route selection and diversion in certain areas.

Despite the amount of attention and publicity given to tropical cyclones, mid-latitude low pressure systems generally present more difficult problems to a ship routing agency. This is primarily due to the fact that major ship traffic is sailing in the latitudes of the migrating low pressure systems, and the amount of potential exposure to intense weather systems, especially in winter, is much greater.

Low pressure systems weaker than gale intensity (winds less than 34 knots) are not a severe problem for most ships. However, a relatively weak system may generate prolonged periods of rough seas which may hamper normal work aboard ship. Ship weather routing can frequently limit rough conditions to short periods of time and provide more favorable conditions for most of the transit. Relatively small vessels, tugs with tows, low powered ships, yachts, and ships with sensitive cargoes can be significantly affected by weather systems weaker than gale intensity. Using a routing agency can enhance both safety and efficiency.

Gales (winds 34 to 47 knots) and storms (winds greater than 48 knots) in the open sea can generate very rough or high seas, particularly when an adverse current such as the Gulf Stream is involved. This can force a reduction in speed in order to gain a more comfortable and safe ride. Because of the extensive geographic area covered by a well developed low pressure system, once ship's speed is reduced the ability to improve the ship's situation is severely hampered. Thus, exposure to potential damage and danger is greatly increased. The vessel in such conditions may be forced to slow down just when it is necessary to speed up to avoid even worse conditions.

A recommendation for a diversion by a routing agency well in advance of the intense weather and associated seas will limit the duration of exposure of the vessel. If effective, ship speed will not be reduced and satisfactory progress will be maintained even though the remaining distance to destination is increased. Overall transit time is usually shorter than if no track change had been made and the ship had remained in heavy weather. In some cases diversions are made to avoid adverse weather conditions and shorten the track at the same time. Significant savings in time and costs can result.

In very intense low pressure systems, with high winds and long duration over a long fetch, seas will be generated and propagated as swell over considerable distances. Even on a diversion, it is difficult to effectively avoid all unfavorable conditions. Generally, original routes for transoceanic passages, issued by the U.S. Navy's ship routing service, are equatorward of the 10% frequency isoline for gale force winds for the month of transit, as interpreted from the U.S. Navy's Marine Climatic Atlas of the World. These are shown in Figure 3705a and Figure 3705b for the Pacific. To avoid the area of significant gale activity in the Atlantic from October to April, the latitude of transit is generally in the lower thirties.

The areas, seasons, and the probability of development of tropical cyclones are fairly well defined in climatological publications. In long range planning, considerable benefit can be gained by limiting the exposure to the potential hazards of tropical systems.

In the North Pacific, avoid areas with the greatest probability of tropical cyclone formation. Avoiding existing tropical cyclones with a history of 24 hours or more of 6-hourly warnings is in most cases relatively straightforward. However, when transiting the tropical cyclone generating area, the ship under routing may provide the first report of environmental conditions indicating that a new disturbance is developing. In the eastern North Pacific the generating area for a high percentage of tropical cyclones is relatively compact (Figure 3705c). Remain south of a line from lat. 9°N, long. 90°W to lat. 14°N, long. 115°W. In the western North Pacific it is advisable to hold north of 22°N when no tropical systems are known to exist. See Figure 3705d.

In the Atlantic, sail near the axis of the Bermuda high or northward to avoid the area of formation of tropical cyclones. Of course, avoiding an existing tropical cyclone takes precedence over avoiding a general area of potential development.

It has proven equally beneficial to employ similar considerations for routing in the monsoon areas of the Indian Ocean and the South China Sea. This is accomplished by providing routes and diversions that generally avoid the areas of high frequency of gale force winds and associated heavy seas, as much as feasible. Ships can then remain in satisfactory conditions with limited increases in route distance.

Depending upon the points of departure and destination, there are many combinations of routes that can be used when

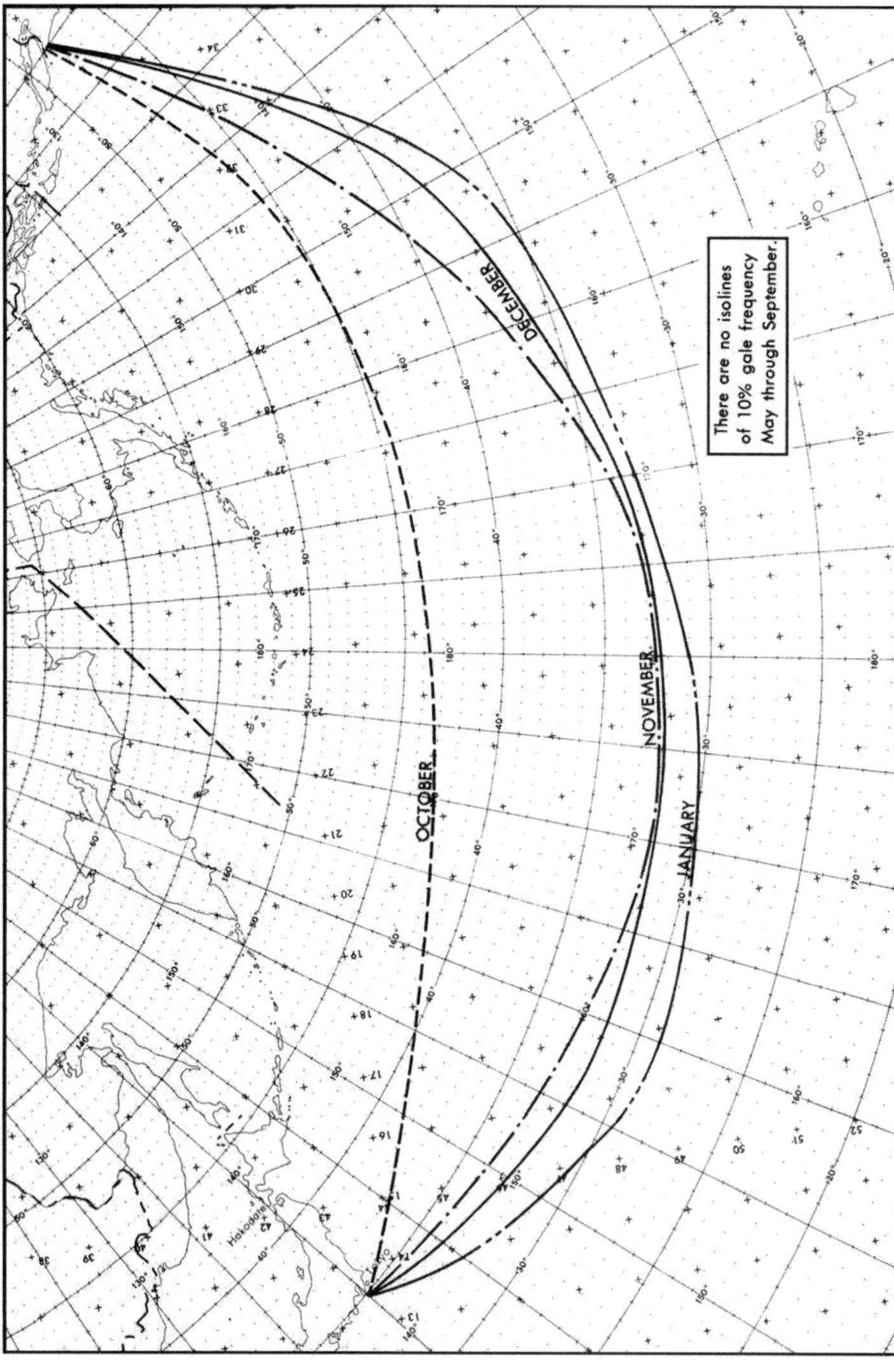

Figure 3705a. Generalized 10% frequency isolines of gale force winds for October through January.

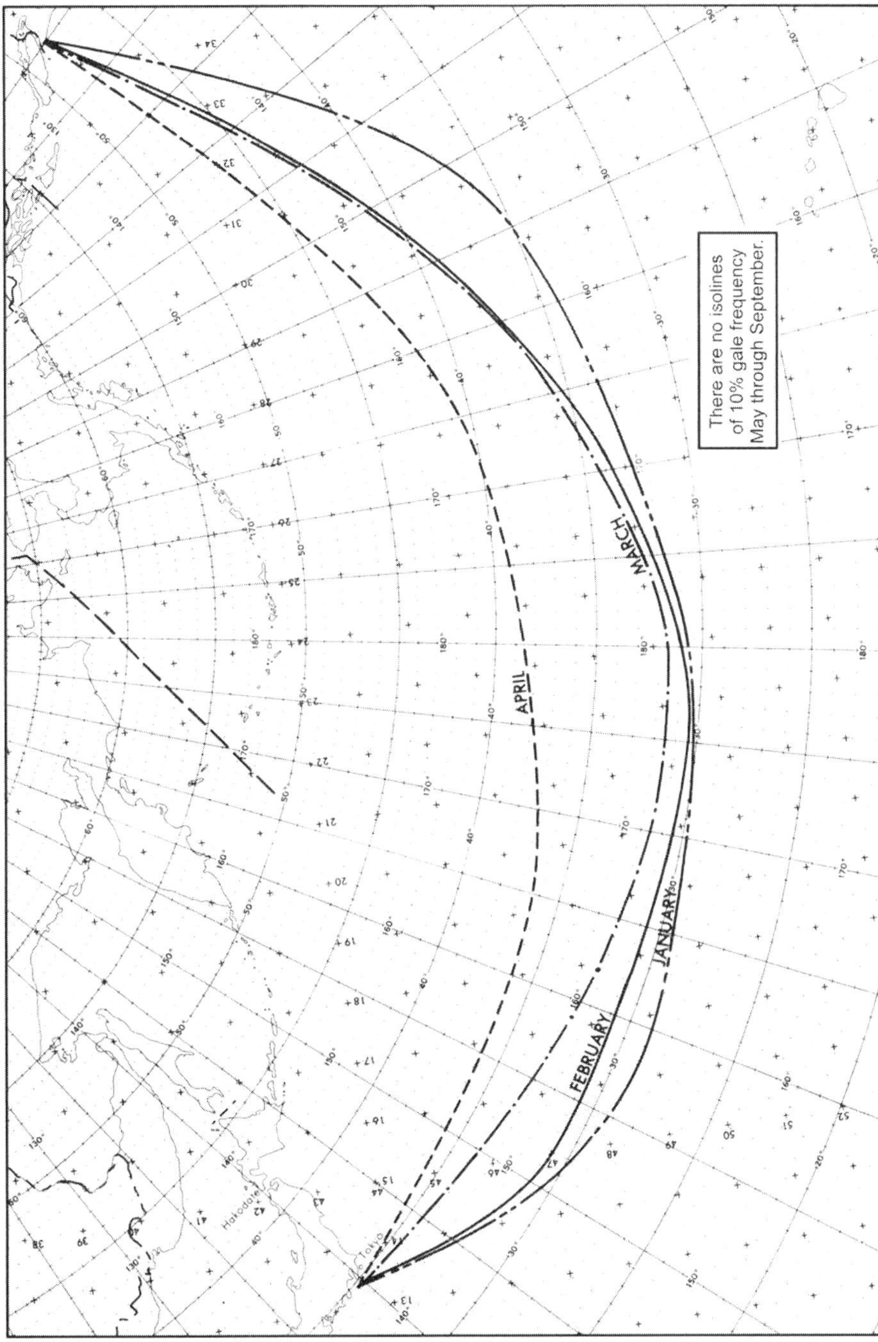

Figure 3705b. Generalized 10% frequency isolines of gale force winds for January through April.

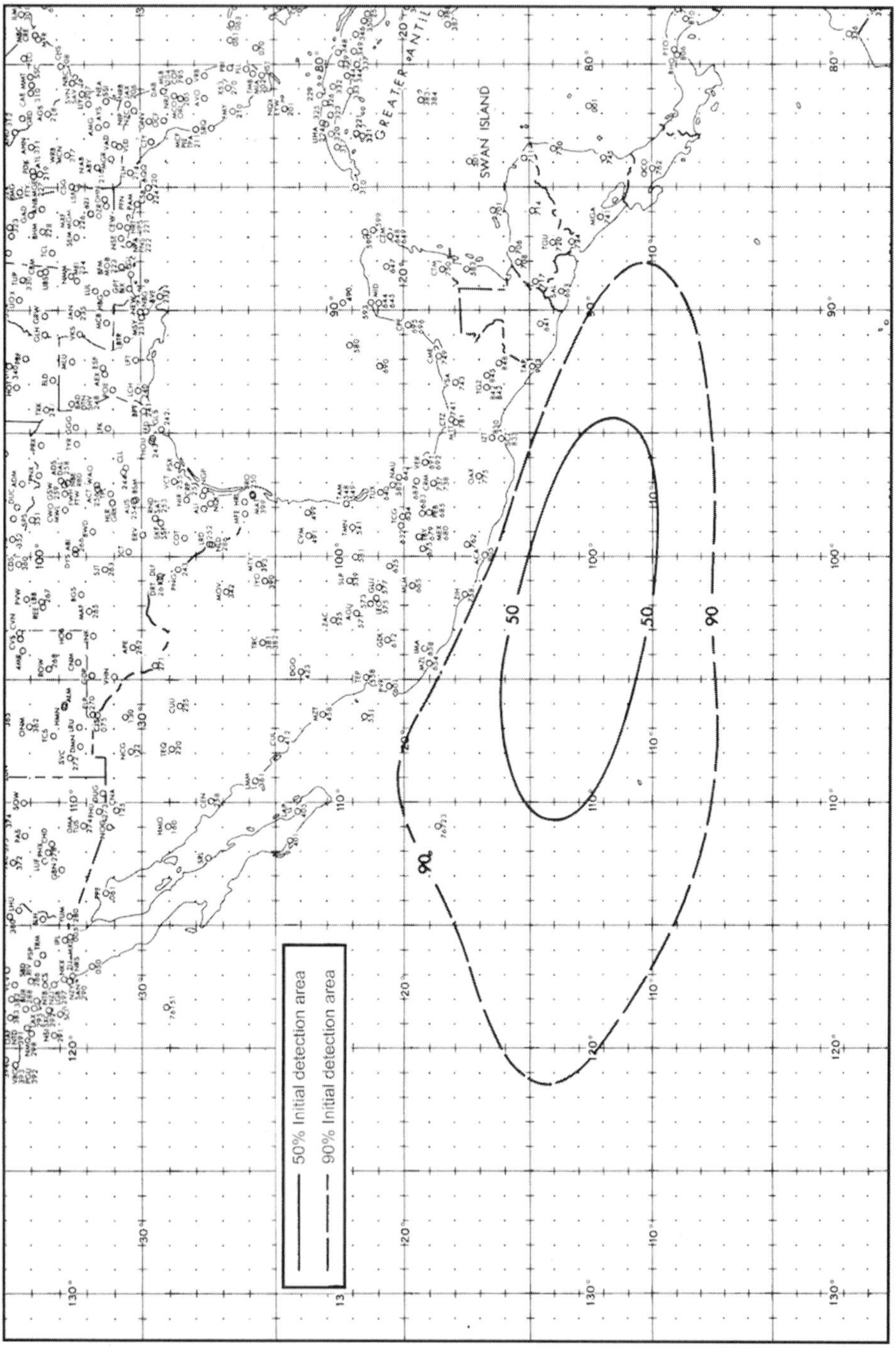

Figure 3705c. Area of initial detection of high percentage of tropical cyclones which later developed to tropical storm or typhoon intensity, 1957-1974.

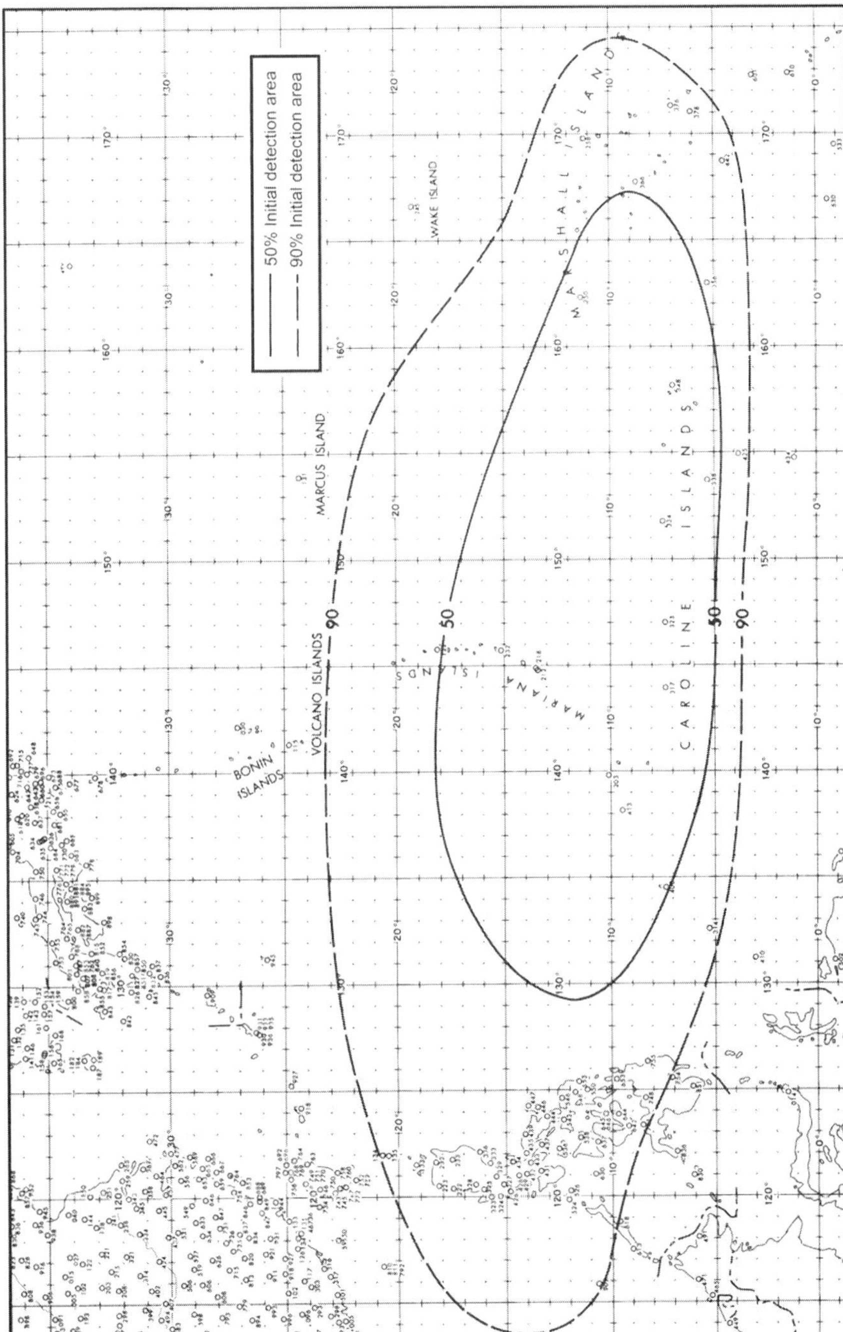

Figure 3705d. Area of initial detection of high percentage of tropical cyclones which later developed to tropical storm or typhoon intensity, 1946-1973.

transiting the northern Indian Ocean (Arabian Sea, Bay of Bengal) and the South China Sea. For example, in the Arabian Sea during the summer monsoon, routes to and from the Red Sea, the western Pacific, and the eastern Indian Ocean should hold equatorward. Ships proceeding to the Persian Gulf during this period are held farther south and west to put the heaviest seas on the quarter or stern when transiting the Arabian Sea. Eastbound ships departing the Persian Gulf may proceed generally east southeast toward the Indian sub-continent, then south, to pass north and east of the highest southwesterly seas in the Arabian Sea. Westbound ships out of the Persian Gulf for the Cape of Good Hope appear to have little choice in routes unless considerable distance is added to the transit by passing east of the highest seas. In the winter monsoon, routes to or from the Red Sea for the western Pacific and the Indian Ocean are held farther north in the Arabian Sea to avoid the highest seas. Ships proceeding to the Persian Gulf from the western Pacific and eastern Indian Ocean may hold more eastward when proceeding north in the Arabian Sea. Ships departing the Persian Gulf area will have considerably less difficulty than during the summer monsoon. Similar considerations can be given when routing ships proceeding to and from the Bay of Bengal.

In the South China Sea, transits via the Palawan Passage are recommended when strong, opposing wind and seas are forecast. This is especially true during the winter monsoon. During periods when the major monsoon flow is slack, ships can use the shortest track as conditions permit.

3706. Special Weather and Environmental Considerations

In addition to the synoptic weather considerations in ship weather routing, there are special environmental problems that can be avoided by following recommendations and advisories of ship routing agencies. These problems generally cover a smaller geographic area and are seasonal in nature, but are still important to ship routing.

In the North Atlantic, because of heavy shipping traffic, frequent poor visibility in rain or fog, and restricted navigation, particularly east of Dover Strait, some mariners prefer to transit to or from the North Sea via Pentland Firth, passing north of the British Isles rather than via the English Channel.

Weather routed ships generally avoid the area of dense fog with low visibility in the vicinity of the Grand Banks off Newfoundland and the area east of Japan north of 35°N. Fishing vessels in these two areas provide an added hazard to safe navigation. This condition exists primarily from June through September. Arctic supply ships en route from the U.S. east coast to the Davis Strait-Baffin Bay area in the summer frequently transit via Cabot Strait and the Strait of Belle Isle, where navigation aids are available and icebergs are generally grounded.

Icebergs are a definite hazard in the North Atlantic from late February through June, and occasionally later. The hazard of floating ice is frequently combined with restricted visibility in fog. International Ice Patrol reports and warnings are incorporated into the planning of routes to safely avoid dangerous iceberg areas. It is usually necessary to hold south of at least 45°N until well southeast of Newfoundland. The U.S. Navy ship routing office at the Naval Atlantic Meteorology and Oceanography Center in Norfolk maintains a safety margin of at least 100 miles from icebergs reported by the International Ice Patrol. Also, in a severe winter, the Denmark Strait may be closed by ice.

In the northern hemisphere winter, a strong high pressure system moving southeast out of the Rocky Mountains brings cold air down across Central America and the western Gulf of Mexico producing gale force winds in the Gulf of Tehuantepec. This fall wind is similar to the pampero, mistral, and bora of other areas of the world. An adjustment to ship's track can successfully avoid the highest seas associated with the "Tehuantepecer." For transits between the Canal Zone and northwest Pacific ports, little additional distance is required to avoid this area (in winter) by remaining south of at least 12°N when crossing 97°W. While avoiding the highest seas, some unfavorable swell conditions may be encountered south of this line. Ships transiting between the Panama Canal and North American west coast ports can stay close along the coast of the Gulf of Tehuantepec to avoid heavy seas during gale conditions, but may still encounter high offshore winds.

In the summer, the semi-permanent high pressure systems over the world's oceans produce strong equatorward flow along the west coasts of continents. This feature is most pronounced off the coast of California and Portugal in the Northern Hemisphere and along Chile, western Australia, and southwest Africa in the Southern Hemisphere. Very rough seas are generated and are considered a definite factor in route selection or diversion when transiting these areas.

3707. Types of Recommendations and Advisories

An **initial route recommendation** is issued to a ship or routing authority normally 48 to 72 hours prior to sailing, and the process of surveillance begins. Surveillance is a continuous process, maintained until the ship arrives at its destination. Initial route recommendations are a composite representation of experience, climatology, weather and sea state forecasts, the vessel's mission and operational concerns, and the vessel's seagoing characteristics. A planning route provides a best estimate of a realistic route for a specific transit period. Such routes are provided when estimated dates of departure (EDD's) are given to the routing agency well in advance of departure, usually a week to several months. Long range planning routes are based more on seasonal and climatological expectations than the current weather situation. While planning routes are an attempt to make extended range (more than a week) or long range (more than a month) forecasts, these recommendations are likely to be revised near the time of departure to reflect the current weather pattern. An initial route recommendation is more closely related to the current weather patterns by using the latest dynamic forecasts than are the planning route recommendations. These, too, are subject to revision prior to sailing, if weather and sea conditions warrant.

Adjustment of departure time is a recommendation for delay in departure, or early departure if feasible, and is intended to avoid or significantly reduce the adverse weather and seas forecast on the first portion of the route, if sailing on the original EDD. The initial route is not revised, only the timing of the ship's transit through an area with currently unfavorable weather conditions. Adjusting the departure time is an effective method of avoiding a potentially hazardous situation where there is no optimum route for sailing at the originally scheduled time. A go/no go recommendation may be made to vessels engaged in special missions such as speed record attempts or heavy-lift voyages.

A **diversion** is an underway adjustment in track and is intended to avoid or limit the effect of adverse weather conditions forecast to be encountered along the ship's current track, or to take advantage of favorable conditions along another route. Ship's speed is expected to be reduced by the encounter with the heavy weather. In most cases the distance to destination is increased in attempting to avoid the adverse weather, but this is partially overcome by being able to maintain a nearly normal SOA.

Adjustment of SOA is a recommendation for slowing or increasing the ship's speed as much as practicable, in an attempt to avoid an adverse weather situation by adjusting the timing of the encounter. This is also an effective means of maintaining maximum ship operating efficiency, while not diverting from the present ship's track. By adjusting the SOA, a major weather system can sometimes be avoided with no increase in distance. The development of fast ships (SOA greater than 30 knots) gives the ship routing agency the potential to "make the ship's weather" by adjusting the ship's speed and track for encounter with favorable weather conditions.

Evasion is a recommendation to the vessel to take independent action to avoid, as much as possible, a potentially dangerous weather system. The ship routing meteorologist may recommend a general direction for safe evasion but does not specify an exact track. The recommendation for evasion is an indication that the weather and sea conditions have deteriorated to a point where shiphandling and safety are the primary considerations and progress toward destination has been temporarily suspended, or is at least of secondary consideration.

A **weather advisory** is a transmission sent to the ship advising the commanding officer or master of expected adverse conditions, their duration, and geographic extent. It is initiated by the ship routing agency as a service and an aid to the ship. The best example of a situation for which a forecast is helpful is when the ship is currently in good weather but adverse weather is expected within 24 hours for which a diversion has not been recommended, or a diversion where adverse weather conditions are still expected. This type of advisory may include a synoptic weather discussion, and a wind, seas, or fog forecast.

The ability of the routing agency to achieve optimum conditions for the ship is aided by the commanding officer or master adjusting course and speed where necessary for an efficient and safe ride. At times, the local sea conditions may dictate that the commanding officer or master take independent action.

3708. Southern Hemisphere Routing

Available data on which to base analyses and forecasts is generally very limited in the Southern Hemisphere, although this situation is improving with the increasing availability of remotely sensed data. Weather and other environmental information obtained from satellites is contributing greatly to an improvement in southern hemisphere forecast products.

Passages south of the Cape of Good Hope and Cape Horn should be timed to avoid heavy weather as much as possible, since intense and frequent low pressure systems are common in these areas. In particular, near the southeast coasts of Africa and South America, intense low pressure systems form in the lee of relatively high terrain near the coasts of both continents. Winter transits south of Cape Horn are difficult, since the time required for transit is longer than the typical interval between storms. Remaining equatorward of about 35°S as much as practicable will limit exposure to adverse conditions. If the frequency of lows passing these areas is once every three or four days, the probability of encountering heavy weather is high.

Tropical cyclones in the Southern Hemisphere present a significant problem because of the sparse surface and upper air observations from which forecasts can be made. Satellites provide the most reliable means by which to obtain accurate positions of tropical systems, and also give the first indication of tropical cyclone formation.

In the Southern Hemisphere, OTSR and other ship weather routing services are available, but are hampered by sparse data reports from which reliable short and extended range forecasts can be produced. Strong climatological consideration is usually given to any proposed southern hemisphere transit, but satellite data is increasingly available to enhance short and extended range forecasts. OTSR procedures for the Northern Hemisphere can be instituted in the Southern Hemisphere whenever justified by basic data input and available forecast models.

3709. Communications

A vital part of a ship routing service is communication between the ship and the routing agency. Reports from the ship show the progress and ability to proceed in existing conditions. Weather reports from the ship enrich the basic data on which analyses are based and forecasts derived, assisting both the reporting ship and others in the vicinity.

Despite all efforts to achieve the best forecasts possible, the quality of forecasts does not always warrant maintaining the route selected. In the U.S. Navy's ship routing program, experience shows that one-third of the ships using OTSR receive some operational or weather-dependent change while underway.

The routing agency needs reports of the ship's position and the ability to transmit recommendations for track change or weather advisories to the ship. The ship needs both send and receive capability for the required information. Information on seakeeping changes initiated by the ship is desirable in a coordinated effort to provide optimum transit conditions. New satellite communications services are making possible the transmission of larger amounts of data than possible through traditional radio messages, a development which supports systems using on-board analysis to generate routes.

3710. Benefits

The benefits of ship weather routing services are primarily in time and cost reductions and increased safety. The savings in operating costs are derived from reductions in transit time, heavy weather encounters, fuel consumption, cargo and hull damage, and more efficient scheduling of dockside activities. The savings are further increased by fewer emergency repairs, more efficient use of personnel, improved topside working conditions, lower insurance rates as preferred risks under weather routing, and ultimately, extended ship operating life.

An effective routing service maximizes safety by greatly reducing the probability of severe or catastrophic damage to the ship, and injury of crew members. The efficiency and health of the crew is also enhanced by avoiding heavy weather. This is especially important on modern, automated ships with reduced crews and smaller craft such as fishing vessels and yachts.

3711. Conclusion

The success of ship weather routing depends upon the validity of forecasts and the routing agency's ability to make appropriate route recommendations and diversions. Anticipated improvements in a routing agency's recommendations will come from advancements in meteorology, technology, and the application of ocean wave forecast models. Advancements in mathematical meteorology, coupled with the continued application of forecast computer models, will extend the time range and accuracy of the dynamic and statistical forecasts. Additionally, a better understanding of the problems encountered by the mariner and their implications while offshore will assist the routing agency in making appropriate recommendations.

Technological advancements in the areas of satellite and automated communications and onboard ship response systems will increase the amount and type of information to and from the ship with fewer delays. Mariners will have a better quality of meteorological information, and the meteorologists will have a better understanding of the problems, constraints, and priorities of the vessel's masters. Ship response and performance data included with the ship's weather report will provide the routing agency with real-time information with which to ascertain the actual state of the ship. Being able to predict a ship's response in most weather and sea conditions will result in improved routing procedures.

Shipboard and anchored wave measuring devices contribute to the development of ocean wave analysis and forecast models. Shipboard seakeeping instrumentation, with input of measured wave conditions and predetermined ship response data for the particular hull, enables a master or commanding officer to adjust course and speed for actual conditions.

Modern ship designs, exotic cargoes, and sophisticated transport methods require individual attention to each ship's areas of vulnerability. Any improvement in the description of sea conditions by ocean wave models will improve the output from ship routing and seakeeping systems.

Advanced planning of a proposed transit, combined with the study of expected weather conditions, both before and during the voyage, as is done by ship routing agencies, and careful on board attention to seakeeping (with instrumentation if available) provide the greatest opportunity to achieve the goal of optimum environmental conditions for ocean transit.

EXPLANATION OF NAVIGATION TABLES

Mathematical Tables

Table 1. Logarithms of Numbers – The first page of this table gives the complete common logarithm (characteristic and mantissa) of numbers 1 through 250. Succeeding pages give the mantissa only of the common logarithm of any number. Values are given for four significant digits of entering values, the first three being in the left-hand column, and the fourth at the heading of one of the other columns. Thus, the mantissa of a three-digit number is given in the column headed 0, on the line with the given number; while the mantissa of a four-digit number is given in the column headed by the fourth digit, on the line with the first three digits. As an example, the mantissa of 328 is 51587, while that of 3.284 is 51640. For additional digits, interpolation should be used. The difference between each tabulated mantissa and the next larger tabulated mantissa is given in the "d" column to the right of the smaller mantissa. This difference can be used to enter the appropriate proportional parts ("Prop. parts") auxiliary table to interpolate for the fifth digit of the given number. If an accuracy of more than five significant digits is to be preserved in a computation, a table of logarithms to additional decimal places should be used. For a number of one or two digits, use the first page of the table or add zeros to make three digits. That is, the mantissa of 3, 30, and 300 is the same, 47712. Interpolation on the first page of the table is not recommended. The second part should be used for values not listed on the first page.

Table 2. Natural Trigonometric Functions – This table gives the values of natural sines, cosecants, tangents, cotangents, secants, and cosines of angles from 0° to 180°, at intervals of 1'. For angles between 0° and 45° use the column labels at the top and the minutes at the left; for angles between 45° and 90° use the column labels at the bottom and the minutes at the right; for angles between 90° and 135° use the column labels at the bottom and the minutes at the left; and for angles between 135° and 180° use the column labels at the top and the minutes at the right. These combinations are indicated by the arrows accompanying the figures representing the number of degrees. For angles between 180° and 360°, subtract 180° and proceed as indicated above to obtain the numerical values of the various functions.

Differences between consecutive entries are shown in the "Diff. 1'" column to the right of each column of values of a trigonometric function, as an aid to interpolation. These differences are one-half line out of step with the numbers to which they apply, as in a critical table. Each difference applies to the values half a line above and half a line below. To determine the correction to apply to the value for the smaller entering angle, multiply the difference by the number of tenths of a minute (or seconds ÷ 60) of the entering angle. Note whether the function is increasing or decreasing, and add or subtract the correction as appropriate, so that the interpolated value lies between the two values between which interpolation is made.

Table 3. Logarithms of Trigonometric Functions – This table gives the common logarithms (+10) of sines, cosecants, tangents, cotangents, secants, and cosines of angles from 0° to 180°, at intervals of 1'. For angles between 0° and 45° use the column labels at the top and the minutes at the left; for angles between 45° and 90° use the column labels at the bottom and the minutes at the right; for angles between 90° and 135° use the column labels at the bottom and the minutes at the left; and for angles between 135° and 180° use the column labels at the top and the minutes at the right. These combinations are indicated by the arrows accompanying the figures representing the number of degrees. For angles between 180° and 360°, subtract 180° and proceed as indicated above to obtain the numerical values of the various functions.

Differences between consecutive entries are shown in the "Diff. 1'" columns, except that one difference column is used for both sines and cosecants, another for both tangents and cotangents, and a third for both secants and cosines. These differences, given as an aid to interpolation, are one-half line out of step with the numbers to which they apply, as in a critical table. Each difference applies to the values half a line above and half a line below. To determine the correction to apply to the value for the smaller entering angle, multiply the difference by the number of tenths of a minute (or seconds ÷ 60) of the entering angle. Note whether the function is increasing or decreasing, and add or subtract the correction as appropriate, so that the interpolated value lies between the two values between which interpolation is made.

Table 4. Traverse Table – This table can be used in the solution of any of the sailings except great-circle and composite. In providing the values of the difference of latitude and departure corresponding to distances up to 600 miles and for courses for every degree of the compass, Table 4 is essentially a tabulation of the solutions of plane right triangles. Since the solutions are for integral values of the acute angle and the distance, interpolation for intermediate values may be required. Through appropriate interchanges of the headings of the columns, solutions for other than plane sailings can be made. The interchanges of the headings of the different columns are summarized at the foot of each table opening.

The distance, difference of latitude, and departure columns are labeled Dist., D. Lat., and Dep., respectively.

For solution of a plane right triangle, any number N in the distance column is the hypotenuse; the number opposite in the difference of latitude column is N times the cosine of the acute angle; and the other number opposite in the departure column is N times the sine of the acute angle. Or, the number in the column labeled D. Lat. is the value of the side adjacent and the number in the column labeled Dep. is the value of the side opposite the acute angle.

Cartographic Tables

Table 5. Natural and Numerical Chart Scales – This table gives the numerical scale equivalents for various natural or fractional chart scales. The scale of a chart is the ratio of a given distance on the chart to the actual distance which it represents on the earth. The scale may be expressed as a simple ratio or fraction, known as the **natural scale**. For example, 1:80,000 or $\frac{1}{80000}$ means that one unit (such as an inch) on the chart represents 80,000 of the same unit on the surface of the earth. The scale may also be expressed as a statement of that distance on the earth shown as one unit (usually an inch) on the chart, or vice versa. This is the **numerical scale**.

The table was computed using 72,913.39 inches per nautical mile and 63,360 inches per statute mile.

Table 6. Meridional Parts – In this table the meridional parts used in the construction of Mercator charts and in Mercator sailing are tabulated to one decimal place for each minute of latitude from the equator to the poles.

The table was computed using the formula:

$$M = a \log_e 10 \log \tan\left(45 + \frac{L}{2}\right) - a\left(e^2 \sin L + \frac{e^4}{3}\sin^3 L + \frac{e^6}{5}\sin^5 L + \ldots\right),$$

in which M is the number of meridional parts between the equator and the given latitude, a is the equatorial radius of the earth, expressed in minutes of arc of the equator, or

$$a = \frac{21600}{2\pi} = 3437.74677078(\log = 3.5362739),$$

$\log_e$ is the natural (Naperian) logarithm, using the base e = 2.71828182846,

$$\log_e 10 = 2.3025851 \quad (\log = 0.36221569)$$

L is the latitude,

e is eccentricity of the earth, or $\sqrt{2f - f^2} = 0.0818188$ (log=8.087146894 –10)

f is earth's flattening, or $f = \frac{1}{298.26} = 0.00335278$ (log = 7.4745949 – 10)

Using these values,

$$a \log_e 10 = 7915.704468 \ (\log = 3.8984896)$$

$$ae^2 = 23.01336332 \ (\log = 1.3619801)$$

$$\frac{ae^4}{3} = 0.05135291 (\log = 8.28943495 - 10)$$

$$\frac{ae^6}{5} = 0.00020626 (\log = 6.6855774 - 10)$$

Hence, the formula becomes

$$M = 7915.704468 \log \tan\left(45° + \frac{L}{2}\right) - 23.0133633 \sin L - 0.051353 \sin^3 L - 0.000206 \sin^5 L \ldots$$

The constants used in this derivation and in the table are based upon the World Geodetic System (WGS) ellipsoid of 1972.

Table 7. Length of a Degree of Latitude and Longitude – This table gives the length of one degree of latitude and longitude at intervals of 1° from the equator to the poles. In the case of latitude, the values given are the lengths of the arcs extending half a degree on each side of the tabulated latitudes. Lengths are given in nautical miles, statute miles, feet, and meters.

The values were computed in meters, using the World Geodetic System ellipsoid of 1972, and converted to other units. The following formulas were used:

$$M = 111{,}132.92 - 559.82 \cos 2L + 1.175 \cos 4L - 0.0023 \cos 6L + \ldots$$

$$P = 111{,}412.84 \cos L - 93.5 \cos 3L + 0.118 \cos 5L - \ldots$$

in which M is the length of 1° of the meridian (latitude), L is the latitude, and P is the length of 1° of the parallel (longitude).

Piloting Tables

Table 8. Conversion Table for Meters, Feet, and

Fathoms – The number of feet and fathoms corresponding to a given number of meters, and vice versa, can be taken directly from this table for any value of the entering argument from 1 to 120. The entering value can be multiplied by any power of 10, including negative powers, if the corresponding values of the other units are multiplied by the same power. Thus, 420 meters are equivalent to 1378.0 feet, and 11.2 fathoms are equivalent to 20.483 meters.

The table was computed by means of the relationships:

1 meter = 39.370079 inches,

1 foot = 12 inches,

1 fathom = 6 feet.

Table 9. Conversion Table for Nautical and Statute Miles – This table gives the number of statute miles corresponding to any whole number of nautical miles from 1 to 100, and the number of nautical miles corresponding to any whole number of statute miles within the same range. The entering value can be multiplied by any power of 10, including negative powers, if the corresponding value of the other unit is multiplied by the same power. Thus, 2,700 nautical miles are equivalent to 3,107.1 statute miles, and 0.3 statute mile is equivalent to 0.2607 nautical mile.

The table was computed using the conversion factors:

1 nautical mile = 1.15077945 statute miles,

1 statute mile = 0.86897624 nautical mile.

Table 10. Speed Table for Measured Mile – To find the speed of a vessel on a measured nautical mile in a given number of minutes and seconds of time, enter this table at the top or bottom with the number of minutes, and at either side with the number of seconds. The number taken from the table is speed in knots. Accurate results can be obtained by interpolating to the nearest 0.1 second.

This table was computed by means of the formula:

$S = \frac{3600}{T}$, in which S is speed in knots, and T is elapsed time in seconds.

Table 11. Speed, Time, and Distance Table – To find the distance steamed at any given speed between 0.5 and 40 knots in any given number of minutes from 1 to 60, enter this table at the top with the speed, and at the left with the number of minutes. The number taken from the table is the distance in nautical miles. If hours are substituted for minutes, the tabulated distance should be multiplied by 60; if seconds are substituted for minutes, the tabulated distance should be divided by 60.

The table was computed by means of the formula:

$D = \frac{ST}{60}$, in which D is distance in nautical miles, S is speed in knots, and T is elapsed time in minutes.

Table 12. Distance of the Horizon – This table gives the distance in nautical and statute miles of the visible sea horizon for various heights of eye in feet and meters. The actual distance varies somewhat as refraction changes. However, the error is generally less than that introduced by nonstandard atmospheric conditions. Also the formula used contains an approximation which introduces a small error at the greatest heights tabulated.

The table was computed using the formula:

$$D = \sqrt{\frac{2r_o h_f}{6076.1\ \beta_o}}$$

in which D is the distance to the horizon in nautical miles; ro is the mean radius of the earth, 3440.1 nautical miles; h_f is the height of eye in feet; and β_o (0.8279) accounts for terrestrial refraction.

This formula simplifies to: $D\ (nm) = 1.169\sqrt{h_f}$

$$(\text{statute miles}) = 1.345\sqrt{h_f}$$

Table 13. Geographic Range – This table gives the geographic range or the maximum distance at which the curvature of the earth permits a light to be seen from a particular height of eye without regard to the luminous intensity of the light. The geographic range depends upon the height of both the light and the eye of the observer.

The table was computed using the formula:

$$D = 1.17\sqrt{H} + 1.17\sqrt{h}$$

in which D is the geographic range in nautical miles, H is the height in feet of the light above sea level, and h is the height in feet of the eye of the observer above sea level.

Table 14. Dip of the Sea Short of the Horizon – If land, another vessel, or other obstruction is between the observer and the sea horizon, use the waterline of the obstruction as the horizontal reference for altitude measurements, and substitute dip from this table for the dip of the horizon (height of eye correction) given in the *Nautical Almanac*. The values below the bold rules are for normal dip, the visible horizon being between the observer and the obstruction.

The table was computed with the formula:

$$D_s = 60 \tan^{-1}\left(\frac{h_f}{6076.1 d_s} + \frac{\beta_o d_s}{2r_o}\right)$$

in which D_s is the dip short of the sea horizon, in minutes of arc; h_f is the height of eye of the observer above sea level in feet; β_o (0.8321) accounts for terrestrial refraction; r_o is the mean radius of the earth, 3440.1 nautical miles; and d_s is the distance to the waterline of the obstruction in nautical miles.

Table 15. Distance by Vertical Angle Measured Between Sea Horizon and Top of Object Beyond Sea Horizon – This table tabulates the distance to an object of known height above sea level when the object lies beyond the horizon. The vertical angle between the top of the object and the visible horizon is measured with a sextant and corrected for index error and dip only. The table is entered with the difference in the height of the object and the height of eye of the observer and the corrected vertical angle; and the distance in nautical miles is taken directly from the table. An error may be introduced if refraction differs from the standard value used in the computation of the table.

The table was computed using the formula:

$$D = \sqrt{\left(\frac{\tan\alpha}{0.0002419}\right)^2 + \frac{H-h}{0.7349}} - \frac{\tan\alpha}{0.0002419}$$

in which D is the distance in nautical miles, α is the corrected vertical angle, H is the height of the top of the object above sea level in feet, and h is the height of eye of the observer above sea level in feet. The constants 0.0002419 and 0.7349 account for terrestrial refraction.

Table 16. Distance by Vertical Angle Measured Between Waterline at Object and Top of Object – This table tabulates the angle subtended by an object of known height lying at a particular distance within the observer's visible horizon or vice versa.

The table provides the solution of a plane right triangle having its right angle at the base of the observed object and its altitude coincident with the vertical dimension of the observed object. The solutions are based upon the following simplifying assumptions: (1) the eye of the observer is at sea level, (2) the sea surface between the observer and the object is flat, (3) atmospheric refraction is negligible, and (4) the waterline at the object is vertically below the peak of the object. The error due to the height of eye of the observer does not exceed 3 percent of the distance-off for sextant angles less than 20° and heights of eye less than one-third of the object height. The error due to the waterline not being below the peak of the object does not exceed 3 percent of the distance-off when the height of eye is less than one-third of the object height and the offset of the waterline from the base of the object is less than one-tenth of the distance-off. Errors due to earth's curvature and atmospheric refraction are negligible for cases of practical interest.

Table 17. Distance by Vertical Angle Measured Between Waterline at Object and Sea Horizon Beyond Object – This table tabulates the distance to an object lying within or short of the horizon when the height of eye of the observer is known. The vertical angle between the waterline at the object and the visible (sea) horizon beyond is measured and corrected for index error. The table is entered with the corrected vertical angle and the height of eye of the observer *in nautical miles*; the distance in yards is taken directly from the table

The table was computed from the formula:

$$\tan h_s = (A - B) \div (1 + AB) \text{ where}$$

$$A = \frac{h}{d_s} + \frac{\beta_o d_s}{2r_o} \text{ and}$$

$$B = \sqrt{2\beta_o h / r_o}$$

in which β_o (0.8279) accounts for terrestrial refraction, r_o is the mean radius of the earth, 3440.1 nautical miles; h is the height of eye of the observer in feet; h_s is the observed vertical angle corrected for index error; and d_s is the distance to the waterline of the object in nautical miles.

Table 18. Distance of an Object by Two Bearings – To determine the distance of an object as a vessel on a steady course passes it, observe the difference between the course and two bearings of the object, and note the time interval between bearings. Enter this table with the two differences. Multiply the distance run between bearings by the number in the first column to find the distance of the object at the time of the second bearing, and by the number in the second column to find the distance when abeam.

The table was computed by solving plane oblique and right triangles.

Celestial Navigation Tables

Table 19. Table of Offsets – This table gives the corrections to the straight line of position (LOP) as drawn on a chart or plotting sheet to provide a closer approximation to the arc of the circle of equal altitude, a small circle of radius equal to the zenith distance.

In adjusting the straight LOP to obtain a closer approximation of the arc of the circle of equal altitude, points on the LOP are offset at right angles to the LOP in the direction of the celestial body. The arguments for entering the table are the distance from the DR to the foot of the perpendicular and the altitude of the body.

The table was computed using the formulas:

$$R = 3438' \cot h$$

$$\sin\theta = D/R$$

$$X = R(1 - \cos\theta),$$

in which X is the offset, R is the radius of a circle of equal altitude for altitude h, and D is the distance from the intercept to the point on the LOP to be offset.

Table 20. Meridian Angle and Altitude of a Body on the Prime Vertical Circle – A celestial body having a declination of contrary name to the latitude does not cross the prime vertical above the celestial horizon, its nearest approach being at rising or setting.

If the declination and latitude are of the same name, and the declination is numerically greater, the body does not cross the prime vertical, but makes its nearest approach (in azimuth) when its meridian angle, east or west, and altitude are as shown in this table, these values being given in italics above the heavy line. At this time the body is stationary in azimuth.

If the declination and latitude are of the same name and numerically equal, the body passes through the zenith as it crosses both the celestial meridian and the prime vertical, as shown in the table.

If the declination and latitude are of the same name, and the declination is numerically less, the body crosses the prime vertical when its meridian angle, east or west, and altitude are as tabulated in vertical type below the heavy line.

The table is entered with declination of the celestial body and the latitude of the observer. Computed altitudes are given, with no allowance made for refraction, dip, parallax, etc. The tabulated values apply to any celestial body, but values are not given for declination greater than 23° because the tabulated information is generally desired for the sun only.

The table was computed using the following formulas, derived by Napier's rules:

Nearest approach (in azimuth) to the prime vertical:

$$\csc h = \sin d \ \csc L$$

$$\sec t = \tan d \ \cot L$$

On the prime vertical:

$$\sin h = \sin d \ \csc L$$

$$\cos t = \tan d \ \cot L$$

In these formulas, h is the altitude, d is the declination, L is the latitude, t is the meridian angle.

Table 21. Latitude and Longitude Factors – The latitude obtained by an ex-meridian sight is inaccurate if the longitude used in determining the meridian angle is incorrect. Similarly, the longitude obtained by solution of a time sight is inaccurate if the latitude used in the solution is incorrect, unless the celestial body is on the prime vertical. This table gives the errors resulting from unit errors in the assumed values used in the computations. There are two columns for each tabulated value of latitude. The first gives the latitude factor, f, which is the error in minutes of latitude for a one-minute error of longitude. The second gives the longitude factor, F, which is the error in minutes of longitude for a one-minute error of latitude. In each case, the total error is the factor multiplied by the number of minutes error in the assumed value. Although the factors were originally intended for use in correcting ex-meridian altitudes and time-sight longitudes, they have other uses as well.

The azimuth angle used for entering the table can be measured from either the north or south, through 90°; or it may be measured from the elevated pole, through 180°. If the celestial body is in the southeast (090°– 180°) or northwest (270°– 360°) quadrant, the f correction is applied to the northward if the correct longitude is east of that used in the solution, and to the southward if the correct longitude is west of that used; while the F correction is applied to the eastward if the correct latitude is north of that used in the solution, and to the westward if the correct latitude is south of that used. If the body is in the northeast (000°– 090°) or southwest (180°– 270°) quadrant, the correction is applied in the opposite direction. These rules apply in both north and south latitude.

The table was computed using the formulas:

$$f = \cos L \tan Z = \frac{1}{\sec L \cot Z} = \frac{1}{F}$$

$$F = \sec L \cot Z = \frac{1}{\cos L \tan Z} = \frac{1}{f}$$

in which f is the tabulated latitude factor, L is the latitude, Z is the azimuth angle, and F is the tabulated longitude factor.

Table 22. Amplitudes – This table lists amplitudes of celestial bodies at rising and setting. Enter with the declination of the body and the latitude of the observer. The value taken from the table is the amplitude when the *center* of the body is on the *celestial* horizon. For the sun, this occurs when the lower limb is a little more than half a diameter above the visible horizon. For the moon it occurs when the upper limb is about on the horizon. Use the prefix E if the body is rising, and W if it is setting; use the suffix N or S to agree with the declination of the body. Table 23 can be used with reversed sign to correct the tabulations to the values for the visible horizon.

The table was computed using the following formula, derived by Napier's rules:

$$\sin A = \sec L \ \sin d$$

in which A is the amplitude, L is the latitude of the observer, and d is the declination of the celestial body.

Table 23. Correction of Amplitude Observed on the Visible Horizon – This table contains a correction to be applied to the amplitude observed when the center of a celestial body is on the visible horizon, to obtain the corresponding amplitude when the center of the body is on the celestial horizon. For the sun, a planet, or a star, apply the correction in the direction away from the elevated pole, thus *increasing* the *azimuth angle*. For the moon apply *half* the correction *toward* the elevated pole. This correction can be applied in the opposite direction to a value taken from Table 22 to find the corresponding amplitude when the center of a celestial body is on the visible horizon. The table was computed for a height of eye of 41 feet. For other heights normally encountered, the error is too small to be of practical significance in ordinary navigation.

The values in the table were determined by computing the azimuth angle when the center of the celestial body is on the visible horizon, converting this to amplitude, and determining the difference between this value and the corresponding value from Table 22. Computation of azimuth angle was made for an altitude of (–)0°42.0' determined as follows:

Azimuth angle was computed by means of the formula:

$$\cos Z = \frac{\sin d - \sin h \sin L}{\cos h \cos L}$$

in which Z is the azimuth angle, d is the declination of the celestial body, h is the altitude (–0°42.0'), and L is the latitude of the observer.

Table 24. Altitude Factors – In one minute of time from meridian transit the altitude of a celestial body changes by the amount shown in this table if the altitude is between 6° and 86°, the latitude is not more than 60°, and the declination is not more than 63°. The values taken from this table are used to enter Table 25 for solving reduction to the meridian (ex-meridian) problems.

For upper transit, use the left-hand pages if the declination and latitude are of the same name (both north or both south) and the right-hand pages if of contrary name. For lower transit, use the values below the heavy lines on the last three contrary-name pages. When a factor is taken from this part of the table, the correction from table 25 is *subtracted* from the observed altitude to obtain the corresponding meridian altitude. All other corrections are added.

The table was computed using the formula:

$$a = 1.9635'' \cos L \cos d \csc (L \sim d)$$

in which a is the change of altitude in one minute from meridian transit (the tabulated value), L is the latitude of the observer, and d is the declination of the celestial body.

This formula can be used to compute values outside the limits of the table, but is not accurate if the altitude is greater than 86°.

Table 25. Change of Altitude in Given Time from Meridian Transit – Enter this table with the altitude factor from table 24 and the meridian angle, in either arc or time units, and take out the difference between the altitude at the given time and the altitude at meridian transit. Enter the table separately with whole numbers and tenths of a, interpolating for t if necessary, and add the two values to obtain the total difference. This total can be applied as a correction to observed altitude to obtain the corresponding meridian altitude, adding for upper transit and subtracting for lower transit.

The table was computed using the formulas:

$$C = \frac{at^2}{60}$$

in which C is the tabulated difference to be used as a correction to observed altitude in minutes of arc; a is the altitude factor from table 24 in seconds of arc; and t is the meridian angle in minutes of time.

This formula should not be used for determining values beyond the limits of the table unless reduced accuracy is acceptable.

Table 26. Time Zones, Descriptions, and Suffixes – The zone description and the single letter of the alphabet designating a time zone and sometimes used as a suffix to zone time for all time zones are given in this table.

Table 27. Altitude Correction for Air Temperature – This table provides a correction to be applied to the altitude of a celestial body when the air temperature varies from the 50° F used for determining mean refraction with the *Nautical Almanac*. For maximum accuracy, apply index correction and dip to sextant altitude first, obtaining apparent (rectified) altitude for use in entering this table. Enter the table with altitude and air temperature in degrees Fahrenheit. Apply the correction in accordance with its tabulated sign to altitude.

The table was computed using formula:

$$\text{Correction} = R_m\left(1 - \frac{510}{460 + T}\right)$$

in which R_m is mean refraction and T is temperature in degrees Fahrenheit.

Table 28. Altitude Correction for Atmospheric Pressure – This table provides a correction to be applied to the altitude of a celestial body when the atmospheric pressure varies from the 29.83 inches (1010 millibars) used for determining mean refraction using the *Nautical Almanac*. For most accurate results, apply index correction and dip to sextant altitude first, obtaining apparent (rectified) altitude for use in entering this table. Enter the table with altitude and atmospheric pressure. Apply the correction to altitude, *adding* if the pressure is less than 29.83 inches and *subtracting* if it is more than 29.83 inches. The table was computed by means of the formula:

$$\text{Correction} = R_m\left(1 - \frac{P}{29.83}\right)$$

in which R_m is mean refraction and P is atmospheric pressure in inches of mercury.

Meteorological Tables

Table 29. Conversion Table for Thermometer Scales – Enter this table with temperature Fahrenheit, F; Celsius (centigrade), C; or Kelvin, K; and take out the corresponding readings on the other two temperature scales.

On the Fahrenheit scale, the freezing temperature of pure water at standard sea level pressure is 32°, and the boiling point under the same conditions is 212°. The corresponding temperatures are 0° and 100° on the Celsius scale and 273.15° and 373.15°, respectively, on the Kelvin scale. The value of (–) 273.15° C for absolute zero, the starting point of the Kelvin scale, is the value recognized officially by the National Institute of Standards and Technology (NIST).

The formulas are:

$$C = 5/9(F \times 32°) = K - 273.15°$$

$$F = 9/5C + 32° = 9/5\ K - 459.67°$$

$$K = 5/9(F - 459.67°) = C + 273.15°$$

Table 30. Direction and Speed of True Wind – This table converts apparent wind to true wind. To use the table, divide the apparent wind in knots by the vessel's speed in knots. This gives the apparent wind speed in units of ship's speed. Enter the table with this value and the difference between the heading and the apparent wind direction. The values taken from the table are (1) the difference between the heading and the true wind direction, and (2) the speed of the true wind in units of ship's speed. The true wind is on the same side as the apparent wind, and from a point farther aft.

To convert wind speed in units of ship's speed to speed in knots, multiply by the vessel's speed in knots. The steadiness of the wind and the accuracy of its measurement are seldom sufficient to warrant interpolation in this table. If speed of the true wind and relative direction of the apparent wind are known, enter the column for direction of the apparent wind, and find the speed of the true wind in units of ship's speed. The number to the left is the relative direction of the true wind. The number on the same line in the side columns is the speed of the apparent wind in units of ship's speed. Two solutions are possible if speed of the true wind is less than ship's speed.

The table was computed by solving the triangle involved in a graphical solution, using the formulas:

$$\tan \alpha = \frac{\sin B_A}{S_A - \cos B_A}$$

$$B_T = B_A + \alpha$$

$$S_T = \frac{\sin B_A}{\sin \alpha}$$

in which α is an auxiliary angle, B_A is the difference between the heading and the apparent wind direction, S_A is the speed of the apparent wind in units of ship's speed, B_T is the difference between the heading and the true wind direction, and S_T is the speed of the true wind in units of ship's speed.

Table 31. Correction of Barometer Reading for Height Above Sea Level – If simultaneous barometer readings at different heights are to be of maximum value in weather analysis, they should be converted to the corresponding readings at a standard height, usually sea level. To convert the observed barometer reading to this level, enter this table with the outside temperature and the height of the barometer above sea level. The height of a barometer is the height of its sensitive element; in the case of a mercurial barometer, this is the height of the free surface of mercury in the cistern. The correction taken from this table applies to the readings of any type barometer, and is always *added* to the observed readings, unless the barometer is below sea level.

The table was computed using the formula:

$$C = 29.92126\left(1 - \frac{1}{\text{antilog}\left(\frac{0.0081350H}{T + 0.00178308H}\right)}\right)$$

in which C is the correction in inches of mercury, H is the height of the barometer above sea level in feet, and T is the mean temperature, in degrees Rankine (degrees Fahrenheit plus 459.67°), of the air between the barometer and sea level. At sea the outside air temperature is sufficiently accurate for this purpose.

Table 32. Correction of Barometer Reading for Gravity – The height of the column of a mercury barometer is affected by the force of gravity, which changes with latitude and is approximately equal along any parallel of latitude. The average gravitational force at latitude 45°32'40" is used as the standard for calibration. This table provides a correction to convert the observed reading at any other latitude to the corresponding value at latitude 45°32'40". Enter the table with the latitude, take out the correction, and apply in accordance with the sign given. This correction does not apply to aneroid barometers.

The correction was computed using the formula:

$$C = B(-0.002637\cos 2L + 0.000006\cos^2 2L - 0.000050).$$

in which C is the correction in inches, B is the observed reading of the barometer (corrected for temperature and instrumental errors) in inches of mercury, and L is the latitude. This table was computed for a standard height of 30 inches.

Table 33. Correction of Barometer Reading for Temperature – Because of the difference in expansion of the mercury column of a mercurial barometer and that of the brass scale by which the height is measured, a correction should be applied to the reading when the temperature differs from the standard used for calibration of the instrument. To find the correction, enter this table with the temperature in degrees Fahrenheit and the barometer reading. Apply the correction in accordance with the sign given. This correction does not apply to aneroid barometers.

The standard temperature used for calibration is 32° F for the mercury, and 62° F for the brass. The correction was computed using the formula:

$$C = -B\frac{m(T-32°) - l(T-62°)}{1 + m(T-32°)}$$

in which C is the correction in inches, B is the observed reading of the barometer in inches of mercury, m is the coefficient of cubical expansion of mercury = 0.0001010 cubic inches per degree F, l is the coefficient of linear expansion of brass = 0.0000102 inches per degree F, and T is the temperature of the attached thermometer in degrees F. Substituting the values for m and l and simplifying:

$$C = -B\frac{T - 28.630°}{1.1123T + 10978°}$$

The minus sign before B indicates that the correction is negative if the temperature is more than 28.630°.

Table 34. Conversion Table for hecto-Pascals (Millibars), Inches of Mercury, and Millimeters of Mercury – The reading of a barometer in inches or millimeters of mercury corresponding to a given reading in hecto-Pascals can be found directly from this table.

The formula for the pressure in hecto-Pascals is:

$$P = \frac{B_m D_g}{1000}$$

in which P is the atmospheric pressure in hecto-Pascals, B_m is the height of the column of mercury in millimeters, D is the density of mercury = 13.5951 grams per cubic centimeter, and g is the standard value of gravity = 980.665 dynes. Substituting numerical values:

$$P = 1.33322B_m, \text{ and}$$

$$B_m = \frac{P}{1.33322} = 0.750064P$$

Since one millimeter = 0.750064 inches

$$B_i = \frac{0.03937P}{1.33322} = 0.0295300P,$$

in which B_i is the height of the column of mercury in inches.

Table 35. Relative Humidity – To determine the relative humidity of the atmosphere, enter this table with the dry-bulb (air) temperature (F), and the *difference* between the dry-bulb and wet-bulb temperatures (F). The value taken from the table is the approximate percentage of relative humidity. If the dry-bulb and wet-bulb temperatures are the same, relative humidity is 100 percent.

The table was computed using the formula:

$$R = \frac{100_e}{e_w}$$

in which R is the approximate relative humidity in percent, e is the ambient vapor pressure, and e_w is the saturation vapor pressure over water at dry-bulb temperature. Professor Ferrel's psychrometric formula was used for computation of e:

$$e' - \left(0.000367P(t - t')\left(1 + \frac{t - 32°}{1571}\right)\right)$$

in which e is the ambient vapor pressure in millibars, e' is the saturation vapor pressure in millibars at wet-bulb temperature with respect to water, P is the atmospheric pressure (the millibar equivalent of 30 inches of mercury is used for this table), t is the dry-bulb temperature in degrees Fahrenheit, and t' is the wet-bulb temperature in degrees Fahrenheit.

The values of e_w were taken from the International Meteorological Organization Publication Number 79, 1951, table 2, pages 82–83.

Table 36. Dew Point – To determine the dew point, enter this table with the dry-bulb (air) temperature (F), and the *difference* between the dry-bulb and wet-bulb temperatures (F). The value taken from the table is the dew point in degrees Fahrenheit. If the dry-bulb and wet-bulb temperatures are the same, the air is at or below the dew point.

TABLE 1
Logarithms of Numbers

1–250

No.	Log	No.	Log	No.	Log	No.	Log	No.	Log
1	0. 00000	51	1. 70757	101	2. 00432	151	2. 17898	201	2. 30320
2	0. 30103	52	1. 71600	102	2. 00860	152	2. 18184	202	2. 30535
3	0. 47712	53	1. 72428	103	2. 01284	153	2. 18469	203	2. 30750
4	0. 60206	54	1. 73239	104	2. 01703	154	2. 18752	204	2. 30963
5	0. 69897	55	1. 74036	105	2. 02119	155	2. 19033	205	2. 31175
6	0. 77815	56	1. 74819	106	2. 02531	156	2. 19312	206	2. 31387
7	0. 84510	57	1. 75587	107	2. 02938	157	2. 19590	207	2. 31597
8	0. 90309	58	1. 76343	108	2. 03342	158	2. 19866	208	2. 31806
9	0. 95424	59	1. 77085	109	2. 03743	159	2. 20140	209	2. 32015
10	1. 00000	60	1. 77815	110	2. 04139	160	2. 20412	210	2. 32222
11	1. 04139	61	1. 78533	111	2. 04532	161	2. 20683	211	2. 32428
12	1. 07918	62	1. 79239	112	2. 04922	162	2. 20952	212	2. 32634
13	1. 11394	63	1. 79934	113	2. 05308	163	2. 21219	213	2. 32838
14	1. 14613	64	1. 80618	114	2. 05690	164	2. 21484	214	2. 33041
15	1. 17609	65	1. 81291	115	2. 06070	165	2. 21748	215	2. 33244
16	1. 20412	66	1. 81954	116	2. 06446	166	2. 22011	216	2. 33445
17	1. 23045	67	1. 82607	117	2. 06819	167	2. 22272	217	2. 33646
18	1. 25527	68	1. 83251	118	2. 07188	168	2. 22531	218	2. 33846
19	1. 27875	69	1. 83885	119	2. 07555	169	2. 22789	219	2. 34044
20	1. 30103	70	1. 84510	120	2. 07918	170	2. 23045	220	2. 34242
21	1. 32222	71	1. 85126	121	2. 08279	171	2. 23300	221	2. 34439
22	1. 34242	72	1. 85733	122	2. 08636	172	2. 23553	222	2. 34635
23	1. 36173	73	1. 86332	123	2. 08991	173	2. 23805	223	2. 34830
24	1. 38021	74	1. 86923	124	2. 09342	174	2. 24055	224	2. 35025
25	1. 39794	75	1. 87506	125	2. 09691	175	2. 24304	225	2. 35218
26	1. 41497	76	1. 88081	126	2. 10037	176	2. 24551	226	2. 35411
27	1. 43136	77	1. 88649	127	2. 10380	177	2. 24797	227	2. 35603
28	1. 44716	78	1. 89209	128	2. 10721	178	2. 25042	228	2. 35793
29	1. 46240	79	1. 89763	129	2. 11059	179	2. 25285	229	2. 35984
30	1. 47712	80	1. 90309	130	2. 11394	180	2. 25527	230	2. 36173
31	1. 49136	81	1. 90849	131	2. 11727	181	2. 25768	231	2. 36361
32	1. 50515	82	1. 91381	132	2. 12057	182	2. 26007	232	2. 36549
33	1. 51851	83	1. 91908	133	2. 12385	183	2. 26245	233	2. 36736
34	1. 53148	84	1. 92428	134	2. 12710	184	2. 26482	234	2. 36922
35	1. 54407	85	1. 92942	135	2. 13033	185	2. 26717	235	2. 37107
36	1. 55630	86	1. 93450	136	2. 13354	186	2. 26951	236	2. 37291
37	1. 56820	87	1. 93952	137	2. 13672	187	2. 27184	237	2. 37475
38	1. 57978	88	1. 94448	138	2. 13988	188	2. 27416	238	2. 37658
39	1. 59106	89	1. 94939	139	2. 14301	189	2. 27646	239	2. 37840
40	1. 60206	90	1. 95424	140	2. 14613	190	2. 27875	240	2. 38021
41	1. 61278	91	1. 95904	141	2. 14922	191	2. 28103	241	2. 38202
42	1. 62325	92	1. 96379	142	2. 15229	192	2. 28330	242	2. 38382
43	1. 63347	93	1. 96848	143	2. 15534	193	2. 28556	243	2. 38561
44	1. 64345	94	1. 97313	144	2. 15836	194	2. 28780	244	2. 38739
45	1. 65321	95	1. 97772	145	2. 16137	195	2. 29003	245	2. 38917
46	1. 66276	96	1. 98227	146	2. 16435	196	2. 29226	246	2. 39094
47	1. 67210	97	1. 98677	147	2. 16732	197	2. 29447	247	2. 39270
48	1. 68124	98	1. 99123	148	2. 17026	198	2. 29667	248	2. 39445
49	1. 69020	99	1. 99564	149	2. 17319	199	2. 29885	249	2. 39620
50	1. 69897	100	2. 00000	150	2. 17609	200	2. 30103	250	2. 39794

TABLE 1
Logarithms of Numbers

1000–1500

No.	0	d	1	d	2	d	3	d	4	d	5	d	6	d	7	d	8	d	9	d
100	00000	43	00043	44	00087	43	00130	43	00173	44	00217	43	00260	43	00303	43	00346	43	00389	43
101	00432	43	00475	43	00518	43	00561	43	00604	43	00647	42	00689	43	00732	43	00775	42	00817	43
102	00860	43	00903	42	00945	43	00988	42	01030	42	01072	43	01115	42	01157	42	01199	43	01242	42
103	01284	42	01326	42	01368	42	01410	42	01452	42	01494	42	01536	42	01578	42	01620	42	01662	41
104	01703	42	01745	42	01787	41	01828	42	01870	42	01912	41	01953	42	01995	41	02036	42	02078	41
105	02119	41	02160	42	02202	41	02243	41	02284	41	02325	41	02366	41	02407	42	02449	41	02490	41
106	02531	41	02572	40	02612	41	02653	41	02694	41	02735	41	02776	40	02816	41	02857	41	02898	40
107	02938	41	02979	40	03019	41	03060	40	03100	41	03141	40	03181	41	03222	40	03262	40	03302	40
108	03342	41	03383	40	03423	40	03463	40	03503	40	03543	40	03583	40	03623	40	03663	40	03703	40
109	03743	39	03782	40	03822	40	03862	40	03902	39	03941	40	03981	40	04021	39	04060	40	04100	39
110	04139	40	04179	39	04218	40	04258	39	04297	39	04336	40	04376	39	04415	39	04454	39	04493	39
111	04532	39	04571	39	04610	40	04650	39	04689	38	04727	39	04766	39	04805	39	04844	39	04883	39
112	04922	39	04961	38	04999	39	05038	39	05077	38	05115	39	05154	38	05192	39	05231	38	05269	39
113	05308	38	05346	39	05385	38	05423	38	05461	39	05500	38	05538	38	05576	38	05614	38	05652	38
114	05690	39	05729	38	05767	38	05805	38	05843	38	05881	37	05918	38	05956	38	05994	38	06032	38
115	06070	38	06108	37	06145	38	06183	38	06221	37	06258	38	06296	37	06333	38	06371	37	06408	38
116	06446	37	06483	38	06521	37	06558	37	06595	38	06633	37	06670	37	06707	37	06744	37	06781	38
117	06819	37	06856	37	06893	37	06930	37	06967	37	07004	37	07041	37	07078	37	07115	36	07151	37
118	07188	37	07225	37	07262	36	07298	37	07335	37	07372	36	07408	37	07445	37	07482	36	07518	37
119	07555	36	07591	37	07628	36	07664	36	07700	37	07737	36	07773	36	07809	37	07846	36	07882	36
120	07918	36	07954	36	07990	37	08027	36	08063	36	08099	36	08135	36	08171	36	08207	36	08243	36
121	08279	35	08314	36	08350	36	08386	36	08422	36	08458	35	08493	36	08529	36	08565	35	08600	36
122	08636	36	08672	35	08707	36	08743	35	08778	36	08814	35	08849	35	08884	36	08920	35	08955	36
123	08991	35	09026	35	09061	35	09096	36	09132	35	09167	35	09202	35	09237	35	09272	35	09307	35
124	09342	35	09377	35	09412	35	09447	35	09482	35	09517	35	09552	35	09587	34	09621	35	09656	35
125	09691	35	09726	34	09760	35	09795	35	09830	34	09864	35	09899	35	09934	34	09968	35	10003	34
126	10037	35	10072	34	10106	34	10140	35	10175	34	10209	34	10243	35	10278	34	10312	34	10346	34
127	10380	35	10415	34	10449	34	10483	34	10517	34	10551	34	10585	34	10619	34	10653	34	10687	34
128	10721	34	10755	34	10789	34	10823	34	10857	33	10890	34	10924	34	10958	34	10992	33	11025	34
129	11059	34	11093	33	11126	34	11160	33	11193	34	11227	34	11261	33	11294	33	11327	34	11361	33
130	11394	34	11428	33	11461	33	11494	34	11528	33	11561	33	11594	34	11628	33	11661	33	11694	33
131	11727	33	11760	33	11793	33	11826	34	11860	33	11893	33	11926	33	11959	33	11992	32	12024	33
132	12057	33	12090	33	12123	33	12156	33	12189	33	12222	32	12254	33	12287	33	12320	32	12352	33
133	12385	33	12418	32	12450	33	12483	33	12516	32	12548	33	12581	32	12613	33	12646	32	12678	32
134	12710	33	12743	32	12775	33	12808	32	12840	32	12872	33	12905	32	12937	32	12969	32	13001	32
135	13033	33	13066	32	13098	32	13130	32	13162	32	13194	32	13226	32	13258	32	13290	32	13322	32
136	13354	32	13386	32	13418	32	13450	31	13481	32	13513	32	13545	32	13577	32	13609	31	13640	32
137	13672	32	13704	31	13735	32	13767	32	13799	31	13830	32	13862	31	13893	32	13925	31	13956	32
138	13988	31	14019	32	14051	31	14082	32	14114	31	14145	31	14176	32	14208	31	14239	31	14270	31
139	14301	32	14333	31	14364	31	14395	31	14426	31	14457	32	14489	31	14520	31	14551	31	14582	31
140	14613	31	14644	31	14675	31	14706	31	14737	31	14768	31	14799	30	14829	31	14860	31	14891	31
141	14922	31	14953	30	14983	31	15014	31	15045	31	15076	30	15106	31	15137	31	15168	30	15198	31
142	15229	30	15259	31	15290	30	15320	31	15351	30	15381	31	15412	30	15442	31	15473	30	15503	31
143	15534	30	15564	30	15594	31	15625	30	15655	30	15685	30	15715	31	15746	30	15776	30	15806	30
144	15836	30	15866	31	15897	30	15927	30	15957	30	15987	30	16017	30	16047	30	16077	30	16107	30
145	16137	30	16167	30	16197	30	16227	29	16256	30	16286	30	16316	30	16346	30	16376	30	16406	29
146	16435	30	16465	30	16495	29	16524	30	16554	30	16584	29	16613	30	16643	30	16673	29	16702	30
147	16732	29	16761	30	16791	29	16820	30	16850	29	16879	30	16909	29	16938	29	16967	30	16997	29
148	17026	30	17056	29	17085	29	17114	29	17143	30	17173	29	17202	29	17231	29	17260	29	17289	30
149	17319	29	17348	29	17377	29	17406	29	17435	29	17464	29	17493	29	17522	29	17551	29	17580	29
150	17609	29	17638	29	17667	29	17696	29	17725	29	17754	28	17782	29	17811	29	17840	29	17869	29
No.	0	d	1	d	2	d	3	d	4	d	5	d	6	d	7	d	8	d	9	d

Prop. parts

	44	43	42	41	40	39	38	37	36	35	34	33
1	4	4	4	4	4	4	4	4	4	4	3	3
2	9	9	8	8	8	8	8	7	7	7	7	7
3	13	13	13	12	12	12	11	11	11	10	10	10
4	18	17	17	16	16	16	15	15	14	14	14	13
5	22	22	21	20	20	20	19	18	18	18	17	16
6	26	26	25	25	24	23	23	22	22	21	20	20
7	31	30	29	29	28	27	27	26	25	24	24	23
8	35	34	34	33	32	31	30	30	29	28	27	26
9	40	39	38	37	36	35	34	33	32	32	31	30

TABLE 1
Logarithms of Numbers

1500–2000

No.	0	d	1	d	2	d	3	d	4	d	5	d	6	d	7	d	8	d	9	d
150	17609	29	17638	29	17667	29	17696	29	17725	29	17754	28	17782	29	17811	29	17840	29	17869	29
151	17898	28	17926	29	17955	29	17984	29	18013	28	18041	29	18070	29	18099	28	18127	29	18156	28
152	18184	29	18213	28	18241	29	18270	28	18298	29	18327	28	18355	29	18384	28	18412	29	18441	28
153	18469	29	18498	28	18526	28	18554	29	18583	28	18611	28	18639	28	18667	29	18696	28	18724	28
154	18752	28	18780	28	18808	29	18837	28	18865	28	18893	28	18921	28	18949	28	18977	28	19005	28
155	19033	28	19061	28	19089	28	19117	28	19145	28	19173	28	19201	28	19229	28	19257	28	19285	27
156	19312	28	19340	28	19368	28	19396	28	19424	27	19451	28	19479	28	19507	28	19535	27	19562	28
157	19590	28	19618	27	19645	28	19673	27	19700	28	19728	28	19756	27	19783	28	19811	27	19838	28
158	19866	27	19893	28	19921	27	19948	28	19976	27	20003	27	20030	28	20058	27	20085	27	20112	28
159	20140	27	20167	27	20194	28	20222	27	20249	27	20276	27	20303	27	20330	28	20358	27	20385	27
160	20412	27	20439	27	20466	27	20493	27	20520	28	20548	27	20575	27	20602	27	20629	27	20656	27
161	20683	27	20710	27	20737	26	20763	27	20790	27	20817	27	20844	27	20871	27	20898	27	20925	27
162	20952	26	20978	27	21005	27	21032	27	21059	26	21085	27	21112	27	21139	26	21165	27	21192	27
163	21219	26	21245	27	21272	27	21299	26	21325	27	21352	26	21378	27	21405	26	21431	27	21458	26
164	21484	27	21511	26	21537	27	21564	26	21590	27	21617	26	21643	26	21669	27	21696	26	21722	26
165	21748	27	21775	26	21801	26	21827	27	21854	26	21880	26	21906	26	21932	26	21958	27	21985	26
166	22011	26	22037	26	22063	26	22089	26	22115	26	22141	26	22167	27	22194	26	22220	26	22246	26
167	22272	26	22298	26	22324	26	22350	26	22376	25	22401	26	22427	26	22453	26	22479	26	22505	26
168	22531	26	22557	26	22583	25	22608	26	22634	26	22660	26	22686	26	22712	25	22737	26	22763	26
169	22789	25	22814	26	22840	26	22866	25	22891	26	22917	26	22943	25	22968	26	22994	25	23019	26
170	23045	25	23070	26	23096	25	23121	26	23147	25	23172	26	23198	25	23223	26	23249	25	23274	26
171	23300	25	23325	25	23350	26	23376	25	23401	25	23426	26	23452	25	23477	25	23502	26	23528	25
172	23553	25	23578	25	23603	26	23629	25	23654	25	23679	25	23704	25	23729	25	23754	25	23779	26
173	23805	25	23830	25	23855	25	23880	25	23905	25	23930	25	23955	25	23980	25	24005	25	24030	25
174	24055	25	24080	25	24105	25	24130	25	24155	25	24180	24	24204	25	24229	25	24254	25	24279	25
175	24304	25	24329	24	24353	25	24378	25	24403	25	24428	24	24452	25	24477	25	24502	25	24527	24
176	24551	25	24576	25	24601	24	24625	25	24650	24	24674	25	24699	25	24724	24	24748	25	24773	24
177	24797	25	24822	24	24846	25	24871	24	24895	25	24920	24	24944	25	24969	24	24993	25	25018	24
178	25042	24	25066	25	25091	24	25115	24	25139	25	25164	24	25188	24	25212	25	25237	24	25261	24
179	25285	25	25310	24	25334	24	25358	24	25382	24	25406	25	25431	24	25455	24	25479	24	25503	24
180	25527	24	25551	24	25575	25	25600	24	25624	24	25648	24	25672	24	25696	24	25720	24	25744	24
181	25768	24	25792	24	25816	24	25840	24	25864	24	25888	24	25912	23	25935	24	25959	24	25983	24
182	26007	24	26031	24	26055	24	26079	23	26102	24	26126	24	26150	24	26174	24	26198	23	26221	24
183	26245	24	26269	24	26293	23	26316	24	26340	24	26364	23	26387	24	26411	24	26435	23	26458	24
184	26482	23	26505	24	26529	24	26553	23	26576	24	26600	23	26623	24	26647	23	26670	24	26694	23
185	26717	24	26741	23	26764	24	26788	23	26811	23	26834	24	26858	23	26881	24	26905	23	26928	23
186	26951	24	26975	23	26998	23	27021	24	27045	23	27068	23	27091	23	27114	24	27138	23	27161	23
187	27184	23	27207	24	27231	23	27254	23	27277	23	27300	23	27323	23	27346	24	27370	23	27393	23
188	27416	23	27439	23	27462	23	27485	23	27508	23	27531	23	27554	23	27577	23	27600	23	27623	23
189	27646	23	27669	23	27692	23	27715	23	27738	23	27761	23	27784	23	27807	23	27830	22	27852	23
190	27875	23	27898	23	27921	23	27944	23	27967	22	27989	23	28012	23	28035	23	28058	23	28081	22
191	28103	23	28126	23	28149	22	28171	23	28194	23	28217	23	28240	22	28262	23	28285	22	28307	23
192	28330	23	28353	22	28375	23	28398	23	28421	22	28443	23	28466	22	28488	23	28511	22	28533	23
193	28556	22	28578	23	28601	22	28623	23	28646	22	28668	23	28691	22	28713	22	28735	23	28758	22
194	28780	23	28803	22	28825	22	28847	23	28870	22	28892	22	28914	23	28937	22	28959	22	28981	22
195	29003	23	29026	22	29048	22	29070	22	29092	23	29115	22	29137	22	29159	22	29181	22	29203	23
196	29226	22	29248	22	29270	22	29292	22	29314	22	29336	22	29358	22	29380	23	29403	22	29425	22
197	29447	22	29469	22	29491	22	29513	22	29535	22	29557	22	29579	22	29601	22	29623	22	29645	22
198	29667	21	29688	22	29710	22	29732	22	29754	22	29776	22	29798	22	29820	22	29842	21	29863	22
199	29885	22	29907	22	29929	22	29951	22	29973	21	29994	22	30016	22	30038	22	30060	21	30081	22
200	30103	22	30125	21	30146	22	30168	22	30190	21	30211	22	30233	22	30255	21	30276	22	30298	22
No.	0	d	1	d	2	d	3	d	4	d	5	d	6	d	7	d	8	d	9	d

Prop. parts

	32	31	30	29	28	27	26	25	24	23	22	21
1	3	3	3	3	3	3	3	2	2	2	2	2
2	6	6	6	6	6	5	5	5	5	5	4	4
3	10	9	9	9	8	8	8	8	7	7	7	6
4	13	12	12	12	11	11	10	10	10	9	9	8
5	16	16	15	14	14	14	13	12	12	12	11	10
6	19	19	18	17	17	16	16	15	14	14	13	13
7	22	22	21	20	20	19	18	18	17	16	15	15
8	26	25	24	23	22	22	21	20	19	18	18	17
9	29	28	27	26	25	24	23	22	22	21	20	19

TABLE 1
Logarithms of Numbers

2000–2500

No.	0	d	1	d	2	d	3	d	4	d	5	d	6	d	7	d	8	d	9	d
200	30103	22	30125	21	30146	22	30168	22	30190	21	30211	22	30233	22	30255	21	30276	22	30298	22
201	30320	21	30341	22	30363	21	30384	22	30406	22	30428	21	30449	22	30471	21	30492	22	30514	21
202	30535	22	30557	21	30578	22	30600	21	30621	22	30643	21	30664	21	30685	22	30707	21	30728	22
203	30750	21	30771	21	30792	22	30814	21	30835	21	30856	22	30878	21	30899	21	30920	22	30942	21
204	30963	21	30984	22	31006	21	31027	21	31048	21	31069	22	31091	21	31112	21	31133	21	31154	21
205	31175	22	31197	21	31218	21	31239	21	31260	21	31281	21	31302	21	31323	22	31345	21	31366	21
206	31387	21	31408	21	31429	21	31450	21	31471	21	31492	21	31513	21	31534	21	31555	21	31576	21
207	31597	21	31618	21	31639	21	31660	21	31681	21	31702	21	31723	21	31744	21	31765	20	31785	21
208	31806	21	31827	21	31848	21	31869	21	31890	21	31911	20	31931	21	31952	21	31973	21	31994	21
209	32015	20	32035	21	32056	21	32077	21	32098	20	32118	21	32139	21	32160	21	32181	20	32201	21
210	32222	21	32243	20	32263	21	32284	21	32305	20	32325	21	32346	20	32366	21	32387	21	32408	20
211	32428	21	32449	20	32469	21	32490	20	32510	21	32531	21	32552	20	32572	21	32593	20	32613	21
212	32634	20	32654	21	32675	20	32695	20	32715	21	32736	20	32756	21	32777	20	32797	21	32818	20
213	32838	20	32858	21	32879	20	32899	20	32919	21	32940	20	32960	20	32980	21	33001	20	33021	20
214	33041	21	33062	20	33082	20	33102	20	33122	21	33143	20	33163	20	33183	20	33203	21	33224	20
215	33244	20	33264	20	33284	20	33304	21	33325	20	33345	20	33365	20	33385	20	33405	20	33425	20
216	33445	20	33465	21	33486	20	33506	20	33526	20	33546	20	33566	20	33586	20	33606	20	33626	20
217	33646	20	33666	20	33686	20	33706	20	33726	20	33746	20	33766	20	33786	20	33806	20	33826	20
218	33846	20	33866	19	33885	20	33905	20	33925	20	33945	20	33965	20	33985	20	34005	20	34025	19
219	34044	20	34064	20	34084	20	34104	20	34124	19	34143	20	34163	20	34183	20	34203	20	34223	19
220	34242	20	34262	20	34282	19	34301	20	34321	20	34341	20	34361	19	34380	20	34400	20	34420	19
221	34439	20	34459	20	34479	19	34498	20	34518	19	34537	20	34557	20	34577	19	34596	20	34616	19
222	34635	20	34655	19	34674	20	34694	19	34713	20	34733	20	34753	19	34772	20	34792	19	34811	19
223	34830	20	34850	19	34869	20	34889	19	34908	20	34928	19	34947	20	34967	19	34986	19	35005	20
224	35025	19	35044	20	35064	19	35083	19	35102	20	35122	19	35141	19	35160	20	35180	19	35199	19
225	35218	20	35238	19	35257	19	35276	19	35295	20	35315	19	35334	19	35353	19	35372	20	35392	19
226	35411	19	35430	19	35449	19	35468	20	35488	19	35507	19	35526	19	35545	19	35564	19	35583	20
227	35603	19	35622	19	35641	19	35660	19	35679	19	35698	19	35717	19	35736	19	35755	19	35774	19
228	35793	20	35813	19	35832	19	35851	19	35870	19	35889	19	35908	19	35927	19	35946	19	35965	19
229	35984	19	36003	18	36021	19	36040	19	36059	19	36078	19	36097	19	36116	19	36135	19	36154	19
230	36173	19	36192	19	36211	18	36229	19	36248	19	36267	19	36286	19	36305	19	36324	18	36342	19
231	36361	19	36380	19	36399	19	36418	18	36436	19	36455	19	36474	19	36493	18	36511	19	36530	19
232	36549	19	36568	18	36586	19	36605	19	36624	18	36642	19	36661	19	36680	18	36698	19	36717	19
233	36736	18	36754	19	36773	18	36791	19	36810	19	36829	18	36847	19	36866	18	36884	19	36903	19
234	36922	18	36940	19	36959	18	36977	19	36996	18	37014	19	37033	18	37051	19	37070	18	37088	19
235	37107	18	37125	19	37144	18	37162	19	37181	18	37199	19	37218	18	37236	18	37254	19	37273	18
236	37291	19	37310	18	37328	18	37346	19	37365	18	37383	18	37401	19	37420	18	37438	19	37457	18
237	37475	18	37493	18	37511	19	37530	18	37548	18	37566	19	37585	18	37603	18	37621	18	37639	19
238	37658	18	37676	18	37694	18	37712	19	37731	18	37749	18	37767	18	37785	18	37803	19	37822	18
239	37840	18	37858	18	37876	18	37894	18	37912	19	37931	18	37949	18	37967	18	37985	18	38003	18
240	38021	18	38039	18	38057	18	38075	18	38093	19	38112	18	38130	18	38148	18	38166	18	38184	18
241	38202	18	38220	18	38238	18	38256	18	38274	18	38292	18	38310	18	38328	18	38346	18	38364	18
242	38382	17	38399	18	38417	18	38435	18	38453	18	38471	18	38489	18	38507	18	38525	18	38543	18
243	38561	17	38578	18	38596	18	38614	18	38632	18	38650	18	38668	18	38686	17	38703	18	38721	18
244	38739	18	38757	18	38775	17	38792	18	38810	18	38828	18	38846	17	38863	18	38881	18	38899	18
245	38917	17	38934	18	38952	18	38970	17	38987	18	39005	18	39023	18	39041	17	39058	18	39076	18
246	39094	17	39111	18	39129	17	39146	18	39164	18	39182	17	39199	18	39217	18	39235	17	39252	18
247	39270	17	39287	18	39305	17	39322	18	39340	18	39358	17	39375	18	39393	17	39410	18	39428	17
248	39445	18	39463	17	39480	18	39498	17	39515	18	39533	17	39550	18	39568	17	39585	17	39602	18
249	39620	17	39637	18	39655	17	39672	18	39690	17	39707	17	39724	18	39742	17	39759	18	39777	17
250	39794	17	39811	18	39829	17	39846	17	39863	18	39881	17	39898	17	39915	18	39933	17	39950	17
No.	0	d	1	d	2	d	3	d	4	d	5	d	6	d	7	d	8	d	9	d

Prop. parts

	22	21	20	19	18	17
1	2	2	2	2	2	2
2	4	4	4	4	4	3
3	7	6	6	6	5	5
4	9	8	8	8	7	7
5	11	10	10	10	9	8
6	13	13	12	11	11	10
7	15	15	14	13	13	12
8	18	17	16	15	14	14
9	20	19	18	17	16	15

TABLE 1
Logarithms of Numbers

2500–3000

No.	0	d	1	d	2	d	3	d	4	d	5	d	6	d	7	d	8	d	9	d
250	39794	17	39811	18	39829	17	39846	17	39863	18	39881	17	39898	17	39915	18	39933	17	39950	17
251	39967	18	39985	17	40002	17	40019	18	40037	17	40054	17	40071	17	40088	18	40106	17	40123	17
252	40140	17	40157	18	40175	17	40192	17	40209	17	40226	17	40243	18	40261	17	40278	17	40295	17
253	40312	17	40329	17	40346	18	40364	17	40381	17	40398	17	40415	17	40432	17	40449	17	40466	17
254	40483	17	40500	18	40518	17	40535	17	40552	17	40569	17	40586	17	40603	17	40620	17	40637	17
255	40654	17	40671	17	40688	17	40705	17	40722	17	40739	17	40756	17	40773	17	40790	17	40807	17
256	40824	17	40841	17	40858	17	40875	17	40892	17	40909	17	40926	17	40943	17	40960	16	40976	17
257	40993	17	41010	17	41027	17	41044	17	41061	17	41078	17	41095	16	41111	17	41128	17	41145	17
258	41162	17	41179	17	41196	16	41212	17	41229	17	41246	17	41263	17	41280	16	41296	17	41313	17
259	41330	17	41347	16	41363	17	41380	17	41397	17	41414	16	41430	17	41447	17	41464	17	41481	16
260	41497	17	41514	17	41531	16	41547	17	41564	17	41581	16	41597	17	41614	17	41631	16	41647	17
261	41664	17	41681	16	41697	17	41714	17	41731	16	41747	17	41764	16	41780	17	41797	17	41814	16
262	41830	17	41847	16	41863	17	41880	16	41896	17	41913	16	41929	17	41946	17	41963	16	41979	17
263	41996	16	42012	17	42029	16	42045	17	42062	16	42078	17	42095	16	42111	16	42127	17	42144	16
264	42160	17	42177	16	42193	17	42210	16	42226	17	42243	16	42259	16	42275	17	42292	16	42308	17
265	42325	16	42341	16	42357	17	42374	16	42390	16	42406	17	42423	16	42439	16	42455	17	42472	16
266	42488	16	42504	17	42521	16	42537	16	42553	17	42570	16	42586	16	42602	17	42619	16	42635	16
267	42651	16	42667	17	42684	16	42700	16	42716	16	42732	17	42749	16	42765	16	42781	16	42797	16
268	42813	17	42830	16	42846	16	42862	16	42878	16	42894	17	42911	16	42927	16	42943	16	42959	16
269	42975	16	42991	17	43008	16	43024	16	43040	16	43056	16	43072	16	43088	16	43104	16	43120	16
270	43136	16	43152	17	43169	16	43185	16	43201	16	43217	16	43233	16	43249	16	43265	16	43281	16
271	43297	16	43313	16	43329	16	43345	16	43361	16	43377	16	43393	16	43409	16	43425	16	43441	16
272	43457	16	43473	16	43489	16	43505	16	43521	16	43537	16	43553	16	43569	15	43584	16	43600	16
273	43616	16	43632	16	43648	16	43664	16	43680	16	43696	16	43712	15	43727	16	43743	16	43759	16
274	43775	16	43791	16	43807	16	43823	15	43838	16	43854	16	43870	16	43886	16	43902	15	43917	16
275	43933	16	43949	16	43965	16	43981	15	43996	16	44012	16	44028	16	44044	15	44059	16	44075	16
276	44091	16	44107	15	44122	16	44138	16	44154	16	44170	15	44185	16	44201	16	44217	15	44232	16
277	44248	16	44264	15	44279	16	44295	16	44311	15	44326	16	44342	16	44358	15	44373	16	44389	15
278	44404	16	44420	16	44436	15	44451	16	44467	16	44483	15	44498	16	44514	15	44529	16	44545	15
279	44560	16	44576	16	44592	15	44607	16	44623	15	44638	16	44654	15	44669	16	44685	15	44700	16
280	44716	15	44731	16	44747	15	44762	16	44778	15	44793	16	44809	15	44824	16	44840	15	44855	16
281	44871	15	44886	16	44902	15	44917	15	44932	16	44948	15	44963	16	44979	15	44994	16	45010	15
282	45025	15	45040	16	45056	15	45071	15	45086	16	45102	15	45117	16	45133	15	45148	15	45163	16
283	45179	15	45194	15	45209	16	45225	15	45240	15	45255	16	45271	15	45286	15	45301	16	45317	15
284	45332	15	45347	15	45362	16	45378	15	45393	15	45408	15	45423	16	45439	15	45454	15	45469	15
285	45484	16	45500	15	45515	15	45530	15	45545	16	45561	15	45576	15	45591	15	45606	15	45621	16
286	45637	15	45652	15	45667	15	45682	15	45697	15	45712	16	45728	15	45743	15	45758	15	45773	15
287	45788	15	45803	15	45818	16	45834	15	45849	15	45864	15	45879	15	45894	15	45909	15	45924	15
288	45939	15	45954	15	45969	15	45984	16	46000	15	46015	15	46030	15	46045	15	46060	15	46075	15
289	46090	15	46105	15	46120	15	46135	15	46150	15	46165	15	46180	15	46195	15	46210	15	46225	15
290	46240	15	46255	15	46270	15	46285	15	46300	15	46315	15	46330	15	46345	14	46359	15	46374	15
291	46389	15	46404	15	46419	15	46434	15	46449	15	46464	15	46479	15	46494	15	46509	14	46523	15
292	46538	15	46553	15	46568	15	46583	15	46598	15	46613	14	46627	15	46642	15	46657	15	46672	15
293	46687	15	46702	14	46716	15	46731	15	46746	15	46761	15	46776	14	46790	15	46805	15	46820	15
294	46835	15	46850	14	46864	15	46879	15	46894	15	46909	14	46923	15	46938	15	46953	14	46967	15
295	46982	15	46997	15	47012	14	47026	15	47041	15	47056	14	47070	15	47085	15	47100	14	47114	15
296	47129	15	47144	15	47159	14	47173	15	47188	14	47202	15	47217	15	47232	14	47246	15	47261	15
297	47276	14	47290	15	47305	14	47319	15	47334	15	47349	14	47363	15	47378	14	47392	15	47407	15
298	47422	14	47436	15	47451	14	47465	15	47480	14	47494	15	47509	15	47524	14	47538	15	47553	14
299	47567	15	47582	14	47596	15	47611	14	47625	15	47640	14	47654	15	47669	14	47683	15	47698	14
300	47712	15	47727	14	47741	15	47756	14	47770	14	47784	15	47799	14	47813	15	47828	14	47842	15
No.	0	d	1	d	2	d	3	d	4	d	5	d	6	d	7	d	8	d	9	d

Prop. parts

	18	17	16	15	14
1	2	2	2	2	1
2	4	3	3	3	3
3	5	5	5	4	4
4	7	7	6	6	6
5	9	8	8	8	7
6	11	10	10	9	8
7	13	12	11	10	10
8	14	14	13	12	11
9	16	15	14	14	13

TABLE 1
Logarithms of Numbers

3000–3500

No.	0	d	1	d	2	d	3	d	4	d	5	d	6	d	7	d	8	d	9	d
300	47712	15	47727	14	47741	15	47756	14	47770	14	47784	15	47799	14	47813	15	47828	14	47842	15
301	47857	14	47871	14	47885	15	47900	14	47914	15	47929	14	47943	15	47958	14	47972	14	47986	15
302	48001	14	48015	14	48029	15	48044	14	48058	15	48073	14	48087	14	48101	15	48116	14	48130	14
303	48144	15	48159	14	48173	14	48187	15	48202	14	48216	14	48230	14	48244	15	48259	14	48273	14
304	48287	15	48302	14	48316	14	48330	14	48344	15	48359	14	48373	14	48387	14	48401	15	48416	14
305	48430	14	48444	14	48458	15	48473	14	48487	14	48501	14	48515	15	48530	14	48544	14	48558	14
306	48572	14	48586	15	48601	14	48615	14	48629	14	48643	14	48657	14	48671	15	48686	14	48700	14
307	48714	14	48728	14	48742	14	48756	14	48770	15	48785	14	48799	14	48813	14	48827	14	48841	14
308	48855	14	48869	14	48883	14	48897	14	48911	15	48926	14	48940	14	48954	14	48968	14	48982	14
309	48996	14	49010	14	49024	14	49038	14	49052	14	49066	14	49080	14	49094	14	49108	14	49122	14
310	49136	14	49150	14	49164	14	49178	14	49192	14	49206	14	49220	14	49234	14	49248	14	49262	14
311	49276	14	49290	14	49304	14	49318	14	49332	14	49346	14	49360	14	49374	14	49388	14	49402	13
312	49415	14	49429	14	49443	14	49457	14	49471	14	49485	14	49499	14	49513	14	49527	14	49541	13
313	49554	14	49568	14	49582	14	49596	14	49610	14	49624	14	49638	13	49651	14	49665	14	49679	14
314	49693	14	49707	14	49721	13	49734	14	49748	14	49762	14	49776	14	49790	13	49803	14	49817	14
315	49831	14	49845	14	49859	13	49872	14	49886	14	49900	14	49914	13	49927	14	49941	14	49955	14
316	49969	13	49982	14	49996	14	50010	14	50024	13	50037	14	50051	14	50065	14	50079	13	50092	14
317	50106	14	50120	13	50133	14	50147	14	50161	13	50174	14	50188	14	50202	13	50215	14	50229	14
318	50243	13	50256	14	50270	14	50284	13	50297	14	50311	14	50325	13	50338	14	50352	13	50365	14
319	50379	14	50393	13	50406	14	50420	13	50433	14	50447	14	50461	13	50474	14	50488	13	50501	14
320	50515	14	50529	13	50542	14	50556	13	50569	14	50583	13	50596	14	50610	13	50623	14	50637	14
321	50651	13	50664	14	50678	13	50691	14	50705	13	50718	14	50732	13	50745	14	50759	13	50772	14
322	50786	13	50799	14	50813	13	50826	14	50840	13	50853	13	50866	14	50880	13	50893	14	50907	13
323	50920	14	50934	13	50947	14	50961	13	50974	13	50987	14	51001	13	51014	14	51028	13	51041	14
324	51055	13	51068	13	51081	14	51095	13	51108	13	51121	14	51135	13	51148	14	51162	13	51175	13
325	51188	14	51202	13	51215	13	51228	14	51242	13	51255	13	51268	14	51282	13	51295	13	51308	14
326	51322	13	51335	13	51348	14	51362	13	51375	13	51388	14	51402	13	51415	13	51428	13	51441	14
327	51455	13	51468	13	51481	14	51495	13	51508	13	51521	13	51534	14	51548	13	51561	13	51574	13
328	51587	14	51601	13	51614	13	51627	13	51640	14	51654	13	51667	13	51680	13	51693	13	51706	14
329	51720	13	51733	13	51746	13	51759	13	51772	14	51786	13	51799	13	51812	13	51825	13	51838	13
330	51851	14	51865	13	51878	13	51891	13	51904	13	51917	13	51930	13	51943	14	51957	13	51970	13
331	51983	13	51996	13	52009	13	52022	13	52035	13	52048	13	52061	14	52075	13	52088	13	52101	13
332	52114	13	52127	13	52140	13	52153	13	52166	13	52179	13	52192	13	52205	13	52218	13	52231	13
333	52244	13	52257	13	52270	14	52284	13	52297	13	52310	13	52323	13	52336	13	52349	13	52362	13
334	52375	13	52388	13	52401	13	52414	13	52427	13	52440	13	52453	13	52466	13	52479	13	52492	12
335	52504	13	52517	13	52530	13	52543	13	52556	13	52569	13	52582	13	52595	13	52608	13	52621	13
336	52634	13	52647	13	52660	13	52673	13	52686	13	52699	12	52711	13	52724	13	52737	13	52750	13
337	52763	13	52776	13	52789	13	52802	13	52815	12	52827	13	52840	13	52853	13	52866	13	52879	13
338	52892	13	52905	12	52917	13	52930	13	52943	13	52956	13	52969	13	52982	12	52994	13	53007	13
339	53020	13	53033	13	53046	12	53058	13	53071	13	53084	13	53097	13	53110	12	53122	13	53135	13
340	53148	13	53161	12	53173	13	53186	13	53199	13	53212	12	53224	13	53237	13	53250	13	53263	12
341	53275	13	53288	13	53301	13	53314	12	53326	13	53339	13	53352	12	53364	13	53377	13	53390	13
342	53403	12	53415	13	53428	13	53441	12	53453	13	53466	13	53479	12	53491	13	53504	13	53517	12
343	53529	13	53542	13	53555	12	53567	13	53580	13	53593	12	53605	13	53618	13	53631	12	53643	13
344	53656	12	53668	13	53681	13	53694	12	53706	13	53719	13	53732	12	53744	13	53757	12	53769	13
345	53782	12	53794	13	53807	13	53820	12	53832	13	53845	12	53857	13	53870	12	53882	13	53895	13
346	53908	12	53920	13	53933	12	53945	13	53958	12	53970	13	53983	12	53995	13	54008	12	54020	13
347	54033	12	54045	13	54058	12	54070	13	54083	12	54095	13	54108	12	54120	13	54133	12	54145	13
348	54158	12	54170	13	54183	12	54195	13	54208	12	54220	13	54233	12	54245	13	54258	12	54270	13
349	54283	12	54295	12	54307	13	54320	12	54332	13	54345	12	54357	13	54370	12	54382	12	54394	13
350	54407	12	54419	13	54432	12	54444	12	54456	13	54469	12	54481	13	54494	12	54506	12	54518	13
No.	0	d	1	d	2	d	3	d	4	d	5	d	6	d	7	d	8	d	9	d

Prop. parts

	15	14	13	12
1	2	1	1	1
2	3	3	3	2
3	4	4	4	4
4	6	6	5	5
5	8	7	6	6
6	9	8	8	7
7	10	10	9	8
8	12	11	10	10
9	14	13	12	11

TABLE 1
Logarithms of Numbers

3500–4000

No.	0	d	1	d	2	d	3	d	4	d	5	d	6	d	7	d	8	d	9	d
350	54407	12	54419	13	54432	12	54444	12	54456	13	54469	12	54481	13	54494	12	54506	12	54518	13
351	54531	12	54543	12	54555	13	54568	12	54580	13	54593	12	54605	12	54617	13	54630	12	54642	12
352	54654	13	54667	12	54679	12	54691	13	54704	12	54716	12	54728	13	54741	12	54753	12	54765	12
353	54777	13	54790	12	54802	12	54814	13	54827	12	54839	12	54851	13	54864	12	54876	12	54888	12
354	54900	13	54913	12	54925	12	54937	12	54949	13	54962	12	54974	12	54986	12	54998	13	55011	12
355	55023	12	55035	12	55047	13	55060	12	55072	12	55084	12	55096	12	55108	13	55121	12	55133	12
356	55145	12	55157	12	55169	13	55182	12	55194	12	55206	12	55218	12	55230	12	55242	13	55255	12
357	55267	12	55279	12	55291	12	55303	12	55315	13	55328	12	55340	12	55352	12	55364	12	55376	12
358	55388	12	55400	13	55413	12	55425	12	55437	12	55449	12	55461	12	55473	12	55485	12	55497	12
359	55509	13	55522	12	55534	12	55546	12	55558	12	55570	12	55582	12	55594	12	55606	12	55618	12
360	55630	12	55642	12	55654	12	55666	12	55678	13	55691	12	55703	12	55715	12	55727	12	55739	12
361	55751	12	55763	12	55775	12	55787	12	55799	12	55811	12	55823	12	55835	12	55847	12	55859	12
362	55871	12	55883	12	55895	12	55907	12	55919	12	55931	12	55943	12	55955	12	55967	12	55979	12
363	55991	12	56003	12	56015	12	56027	11	56038	12	56050	12	56062	12	56074	12	56086	12	56098	12
364	56110	12	56122	12	56134	12	56146	12	56158	12	56170	12	56182	12	56194	11	56205	12	56217	12
365	56229	12	56241	12	56253	12	56265	12	56277	12	56289	12	56301	11	56312	12	56324	12	56336	12
366	56348	12	56360	12	56372	12	56384	12	56396	11	56407	12	56419	12	56431	12	56443	12	56455	12
367	56467	11	56478	12	56490	12	56502	12	56514	12	56526	12	56538	11	56549	12	56561	12	56573	12
368	56585	12	56597	11	56608	12	56620	12	56632	12	56644	12	56656	11	56667	12	56679	12	56691	12
369	56703	11	56714	12	56726	12	56738	12	56750	11	56761	12	56773	12	56785	12	56797	11	56808	12
370	56820	12	56832	12	56844	11	56855	12	56867	12	56879	12	56891	11	56902	12	56914	12	56926	11
371	56937	12	56949	12	56961	11	56972	12	56984	12	56996	12	57008	11	57019	12	57031	12	57043	11
372	57054	12	57066	12	57078	11	57089	12	57101	12	57113	11	57124	12	57136	12	57148	11	57159	12
373	57171	12	57183	11	57194	12	57206	11	57217	12	57229	12	57241	11	57252	12	57264	12	57276	11
374	57287	12	57299	11	57310	12	57322	12	57334	11	57345	12	57357	11	57368	12	57380	12	57392	11
375	57403	12	57415	11	57426	12	57438	11	57449	12	57461	12	57473	11	57484	12	57496	11	57507	12
376	57519	11	57530	12	57542	11	57553	12	57565	11	57576	12	57588	12	57600	11	57611	12	57623	11
377	57634	12	57646	11	57657	12	57669	11	57680	12	57692	11	57703	12	57715	11	57726	12	57738	11
378	57749	12	57761	11	57772	12	57784	11	57795	12	57807	11	57818	12	57830	11	57841	11	57852	12
379	57864	11	57875	12	57887	11	57898	12	57910	11	57921	12	57933	11	57944	11	57955	12	57967	11
380	57978	12	57990	11	58001	12	58013	11	58024	11	58035	12	58047	11	58058	12	58070	11	58081	11
381	58092	12	58104	11	58115	12	58127	11	58138	11	58149	12	58161	11	58172	12	58184	11	58195	11
382	58206	12	58218	11	58229	11	58240	12	58252	11	58263	11	58274	12	58286	11	58297	12	58309	11
383	58320	11	58331	12	58343	11	58354	11	58365	12	58377	11	58388	11	58399	11	58410	12	58422	11
384	58433	11	58444	12	58456	11	58467	11	58478	12	58490	11	58501	11	58512	12	58524	11	58535	11
385	58546	11	58557	12	58569	11	58580	11	58591	11	58602	12	58614	11	58625	11	58636	11	58647	12
386	58659	11	58670	11	58681	11	58692	12	58704	11	58715	11	58726	11	58737	12	58749	11	58760	11
387	58771	11	58782	12	58794	11	58805	11	58816	11	58827	11	58838	12	58850	11	58861	11	58872	11
388	58883	11	58894	12	58906	11	58917	11	58928	11	58939	11	58950	11	58961	12	58973	11	58984	11
389	58995	11	59006	11	59017	11	59028	12	59040	11	59051	11	59062	11	59073	11	59084	11	59095	11
390	59106	12	59118	11	59129	11	59140	11	59151	11	59162	11	59173	11	59184	11	59195	12	59207	11
391	59218	11	59229	11	59240	11	59251	11	59262	11	59273	11	59284	11	59295	11	59306	12	59318	11
392	59329	11	59340	11	59351	11	59362	11	59373	11	59384	11	59395	11	59406	11	59417	11	59428	11
393	59439	11	59450	11	59461	11	59472	11	59483	11	59494	12	59506	11	59517	11	59528	11	59539	11
394	59550	11	59561	11	59572	11	59583	11	59594	11	59605	11	59616	11	59627	11	59638	11	59649	11
395	59660	11	59671	11	59682	11	59693	11	59704	11	59715	11	59726	11	59737	11	59748	11	59759	11
396	59770	10	59780	11	59791	11	59802	11	59813	11	59824	11	59835	11	59846	11	59857	11	59868	11
397	59879	11	59890	11	59901	11	59912	11	59923	11	59934	11	59945	11	59956	10	59966	11	59977	11
398	59988	11	59999	11	60010	11	60021	11	60032	11	60043	11	60054	11	60065	11	60076	10	60086	11
399	60097	11	60108	11	60119	11	60130	11	60141	11	60152	11	60163	10	60173	11	60184	11	60195	11
400	60206	11	60217	11	60228	11	60239	10	60249	11	60260	11	60271	11	60282	11	60293	11	60304	10
No.	0	d	1	d	2	d	3	d	4	d	5	d	6	d	7	d	8	d	9	d

Prop. parts

	13	12	11	10
1	1	1	1	1
2	3	2	2	2
3	4	4	3	3
4	5	5	4	4
5	6	6	6	5
6	8	7	7	6
7	9	8	8	7
8	10	10	9	8
9	12	11	10	9

TABLE 1
Logarithms of Numbers

4000–4500

No.	0	d	1	d	2	d	3	d	4	d	5	d	6	d	7	d	8	d	9	d
400	60206	11	60217	11	60228	11	60239	10	60249	11	60260	11	60271	11	60282	11	60293	11	60304	10
401	60314	11	60325	11	60336	11	60347	11	60358	11	60369	10	60379	11	60390	11	60401	11	60412	11
402	60423	10	60433	11	60444	11	60455	11	60466	11	60477	10	60487	11	60498	11	60509	11	60520	11
403	60531	10	60541	11	60552	11	60563	11	60574	10	60584	11	60595	11	60606	11	60617	10	60627	11
404	60638	11	60649	11	60660	10	60670	11	60681	11	60692	11	60703	10	60713	11	60724	11	60735	11
405	60746	10	60756	11	60767	11	60778	10	60788	11	60799	11	60810	11	60821	10	60831	11	60842	11
406	60853	10	60863	11	60874	11	60885	10	60895	11	60906	11	60917	10	60927	11	60938	11	60949	10
407	60959	11	60970	11	60981	10	60991	11	61002	11	61013	10	61023	11	61034	11	61045	10	61055	11
408	61066	11	61077	10	61087	11	61098	11	61109	10	61119	11	61130	10	61140	11	61151	11	61162	10
409	61172	11	61183	11	61194	10	61204	11	61215	10	61225	11	61236	11	61247	10	61257	11	61268	10
410	61278	11	61289	11	61300	10	61310	11	61321	10	61331	11	61342	10	61352	11	61363	11	61374	10
411	61384	11	61395	10	61405	11	61416	10	61426	11	61437	11	61448	10	61458	11	61469	10	61479	11
412	61490	10	61500	11	61511	10	61521	11	61532	10	61542	11	61553	10	61563	11	61574	10	61584	11
413	61595	11	61606	10	61616	11	61627	10	61637	11	61648	10	61658	11	61669	10	61679	11	61690	10
414	61700	11	61711	10	61721	10	61731	11	61742	10	61752	11	61763	10	61773	11	61784	10	61794	11
415	61805	10	61815	11	61826	10	61836	11	61847	10	61857	11	61868	10	61878	10	61888	11	61899	10
416	61909	11	61920	10	61930	11	61941	10	61951	11	61962	10	61972	10	61982	11	61993	10	62003	11
417	62014	10	62024	10	62034	11	62045	10	62055	11	62066	10	62076	10	62086	11	62097	10	62107	11
418	62118	10	62128	10	62138	11	62149	10	62159	11	62170	10	62180	10	62190	11	62201	10	62211	10
419	62221	11	62232	10	62242	10	62252	11	62263	10	62273	11	62284	10	62294	10	62304	11	62315	10
420	62325	10	62335	11	62346	10	62356	10	62366	11	62377	10	62387	10	62397	11	62408	10	62418	10
421	62428	11	62439	10	62449	10	62459	10	62469	11	62480	10	62490	10	62500	11	62511	10	62521	10
422	62531	11	62542	10	62552	10	62562	10	62572	11	62583	10	62593	10	62603	10	62613	11	62624	10
423	62634	10	62644	11	62655	10	62665	10	62675	10	62685	11	62696	10	62706	10	62716	10	62726	11
424	62737	10	62747	10	62757	10	62767	11	62778	10	62788	10	62798	10	62808	10	62818	11	62829	10
425	62839	10	62849	10	62859	11	62870	10	62880	10	62890	10	62900	10	62910	11	62921	10	62931	10
426	62941	10	62951	10	62961	11	62972	10	62982	10	62992	10	63002	10	63012	10	63022	11	63033	10
427	63043	10	63053	10	63063	10	63073	10	63083	11	63094	10	63104	10	63114	10	63124	10	63134	10
428	63144	11	63155	10	63165	10	63175	10	63185	10	63195	10	63205	10	63215	10	63225	11	63236	10
429	63246	10	63256	10	63266	10	63276	10	63286	10	63296	10	63306	11	63317	10	63327	10	63337	10
430	63347	10	63357	10	63367	10	63377	10	63387	10	63397	10	63407	10	63417	11	63428	10	63438	10
431	63448	10	63458	10	63468	10	63478	10	63488	10	63498	10	63508	10	63518	10	63528	10	63538	10
432	63548	10	63558	10	63568	11	63579	10	63589	10	63599	10	63609	10	63619	10	63629	10	63639	10
433	63649	10	63659	10	63669	10	63679	10	63689	10	63699	10	63709	10	63719	10	63729	10	63739	10
434	63749	10	63759	10	63769	10	63779	10	63789	10	63799	10	63809	10	63819	10	63829	10	63839	10
435	63849	10	63859	10	63869	10	63879	10	63889	10	63899	10	63909	10	63919	10	63929	10	63939	10
436	63949	10	63959	10	63969	10	63979	9	63988	10	63998	10	64008	10	64018	10	64028	10	64038	10
437	64048	10	64058	10	64068	10	64078	10	64088	10	64098	10	64108	10	64118	10	64128	9	64137	10
438	64147	10	64157	10	64167	10	64177	10	64187	10	64197	10	64207	10	64217	10	64227	10	64237	9
439	64246	10	64256	10	64266	10	64276	10	64286	10	64296	10	64306	10	64316	10	64326	9	64335	10
440	64345	10	64355	10	64365	10	64375	10	64385	10	64395	9	64404	10	64414	10	64424	10	64434	10
441	64444	10	64454	10	64464	9	64473	10	64483	10	64493	10	64503	10	64513	10	64523	9	64532	10
442	64542	10	64552	10	64562	10	64572	10	64582	9	64591	10	64601	10	64611	10	64621	10	64631	9
443	64640	10	64650	10	64660	10	64670	10	64680	9	64689	10	64699	10	64709	10	64719	10	64729	9
444	64738	10	64748	10	64758	10	64768	9	64777	10	64787	10	64797	10	64807	9	64816	10	64826	10
445	64836	10	64846	10	64856	9	64865	10	64875	10	64885	10	64895	9	64904	10	64914	10	64924	9
446	64933	10	64943	10	64953	10	64963	9	64972	10	64982	10	64992	10	65002	9	65011	10	65021	10
447	65031	9	65040	10	65050	10	65060	10	65070	9	65079	10	65089	10	65099	9	65108	10	65118	10
448	65128	9	65137	10	65147	10	65157	10	65167	9	65176	10	65186	10	65196	9	65205	10	65215	10
449	65225	9	65234	10	65244	10	65254	9	65263	10	65273	10	65283	9	65292	10	65302	10	65312	9
450	65321	10	65331	10	65341	9	65350	10	65360	9	65369	10	65379	10	65389	9	65398	10	65408	10
No.	0	d	1	d	2	d	3	d	4	d	5	d	6	d	7	d	8	d	9	d

Prop. parts

	11	10	9
1	1	1	1
2	2	2	2
3	3	3	3
4	4	4	4
5	6	5	4
6	7	6	5
7	8	7	6
8	9	8	7
9	10	9	8

TABLE 1

Logarithms of Numbers

4500–5000

No.	0	d	1	d	2	d	3	d	4	d	5	d	6	d	7	d	8	d	9	d
450	65321	10	65331	10	65341	9	65350	10	65360	9	65369	10	65379	10	65389	9	65398	10	65408	10
451	65418	9	65427	10	65437	10	65447	9	65456	10	65466	9	65475	10	65485	10	65495	9	65504	10
452	65514	9	65523	10	65533	10	65543	9	65552	10	65562	9	65571	10	65581	10	65591	9	65600	10
453	65610	9	65619	10	65629	10	65639	9	65648	10	65658	9	65667	10	65677	9	65686	10	65696	10
454	65706	9	65715	10	65725	9	65734	10	65744	9	65753	10	65763	9	65772	10	65782	10	65792	9
455	65801	10	65811	9	65820	10	65830	9	65839	10	65849	9	65858	10	65868	9	65877	10	65887	9
456	65896	10	65906	10	65916	9	65925	10	65935	9	65944	10	65954	9	65963	10	65973	9	65982	10
457	65992	9	66001	10	66011	9	66020	10	66030	9	66039	10	66049	9	66058	10	66068	9	66077	10
458	66087	9	66096	10	66106	9	66115	9	66124	10	66134	9	66143	10	66153	9	66162	10	66172	9
459	66181	10	66191	9	66200	10	66210	9	66219	10	66229	9	66238	9	66247	10	66257	9	66266	10
460	66276	9	66285	10	66295	9	66304	10	66314	9	66323	9	66332	10	66342	9	66351	10	66361	9
461	66370	10	66380	9	66389	9	66398	10	66408	9	66417	10	66427	9	66436	9	66445	10	66455	9
462	66464	10	66474	9	66483	9	66492	10	66502	9	66511	10	66521	9	66530	9	66539	10	66549	9
463	66558	9	66567	10	66577	9	66586	10	66596	9	66605	9	66614	10	66624	9	66633	9	66642	10
464	66652	9	66661	10	66671	9	66680	9	66689	10	66699	9	66708	9	66717	10	66727	9	66736	9
465	66745	10	66755	9	66764	9	66773	10	66783	9	66792	9	66801	10	66811	9	66820	9	66829	10
466	66839	9	66848	9	66857	10	66867	9	66876	9	66885	9	66894	10	66904	9	66913	9	66922	10
467	66932	9	66941	9	66950	10	66960	9	66969	9	66978	9	66987	10	66997	9	67006	9	67015	10
468	67025	9	67034	9	67043	9	67052	10	67062	9	67071	9	67080	9	67089	10	67099	9	67108	9
469	67117	10	67127	9	67136	9	67145	9	67154	10	67164	9	67173	9	67182	9	67191	10	67201	9
470	67210	9	67219	9	67228	9	67237	10	67247	9	67256	9	67265	9	67274	10	67284	9	67293	9
471	67302	9	67311	10	67321	9	67330	9	67339	9	67348	9	67357	10	67367	9	67376	9	67385	9
472	67394	9	67403	10	67413	9	67422	9	67431	9	67440	9	67449	10	67459	9	67468	9	67477	9
473	67486	9	67495	9	67504	10	67514	9	67523	9	67532	9	67541	9	67550	10	67560	9	67569	9
474	67578	9	67587	9	67596	9	67605	9	67614	10	67624	9	67633	9	67642	9	67651	9	67660	9
475	67669	10	67679	9	67688	9	67697	9	67706	9	67715	9	67724	9	67733	9	67742	10	67752	9
476	67761	9	67770	9	67779	9	67788	9	67797	9	67806	9	67815	10	67825	9	67834	9	67843	9
477	67852	9	67861	9	67870	9	67879	9	67888	9	67897	9	67906	10	67916	9	67925	9	67934	9
478	67943	9	67952	9	67961	9	67970	9	67979	9	67988	9	67997	9	68006	9	68015	9	68024	10
479	68034	9	68043	9	68052	9	68061	9	68070	9	68079	9	68088	9	68097	9	68106	9	68115	9
480	68124	9	68133	9	68142	9	68151	9	68160	9	68169	9	68178	9	68187	9	68196	9	68205	10
481	68215	9	68224	9	68233	9	68242	9	68251	9	68260	9	68269	9	68278	9	68287	9	68296	9
482	68305	9	68314	9	68323	9	68332	9	68341	9	68350	9	68359	9	68368	9	68377	9	68386	9
483	68395	9	68404	9	68413	9	68422	9	68431	9	68440	9	68449	9	68458	9	68467	9	68476	9
484	68485	9	68494	8	68502	9	68511	9	68520	9	68529	9	68538	9	68547	9	68556	9	68565	9
485	68574	9	68583	9	68592	9	68601	9	68610	9	68619	9	68628	9	68637	9	68646	9	68655	9
486	68664	9	68673	8	68681	9	68690	9	68699	9	68708	9	68717	9	68726	9	68735	9	68744	9
487	68753	9	68762	9	68771	9	68780	9	68789	8	68797	9	68806	9	68815	9	68824	9	68833	9
488	68842	9	68851	9	68860	9	68869	9	68878	8	68886	9	68895	9	68904	9	68913	9	68922	9
489	68931	9	68940	9	68949	9	68958	8	68966	9	68975	9	68984	9	68993	9	69002	9	69011	9
490	69020	8	69028	9	69037	9	69046	9	69055	9	69064	9	69073	9	69082	8	69090	9	69099	9
491	69108	9	69117	9	69126	9	69135	9	69144	8	69152	9	69161	9	69170	9	69179	9	69188	9
492	69197	8	69205	9	69214	9	69223	9	69232	9	69241	8	69249	9	69258	9	69267	9	69276	9
493	69285	9	69294	8	69302	9	69311	9	69320	9	69329	9	69338	8	69346	9	69355	9	69364	9
494	69373	8	69381	9	69390	9	69399	9	69408	9	69417	8	69425	9	69434	9	69443	9	69452	9
495	69461	8	69469	9	69478	9	69487	9	69496	8	69504	9	69513	9	69522	9	69531	8	69539	9
496	69548	9	69557	9	69566	8	69574	9	69583	9	69592	9	69601	8	69609	9	69618	9	69627	9
497	69636	8	69644	9	69653	9	69662	9	69671	8	69679	9	69688	9	69697	8	69705	9	69714	9
498	69723	9	69732	8	69740	9	69749	9	69758	9	69767	8	69775	9	69784	9	69793	8	69801	9
499	69810	9	69819	8	69827	9	69836	9	69845	9	69854	8	69862	9	69871	9	69880	8	69888	9
500	69897	9	69906	8	69914	9	69923	9	69932	8	69940	9	69949	9	69958	8	69966	9	69975	9
No.	0	d	1	d	2	d	3	d	4	d	5	d	6	d	7	d	8	d	9	d

Prop. parts

	10
1	1
2	2
3	3
4	4
5	5
6	6
7	7
8	8
9	9

	9
1	1
2	2
3	3
4	4
5	4
6	5
7	6
8	7
9	8

	8
1	1
2	2
3	2
4	3
5	4
6	5
7	6
8	6
9	7

TABLE 1

Logarithms of Numbers

5000–5500

No.	0	d	1	d	2	d	3	d	4	d	5	d	6	d	7	d	8	d	9	d
500	69897	9	69906	8	69914	9	69923	9	69932	8	69940	9	69949	9	69958	8	69966	9	69975	9
501	69984	8	69992	9	70001	9	70010	8	70018	9	70027	9	70036	8	70044	9	70053	9	70062	8
502	70070	9	70079	9	70088	8	70096	9	70105	9	70114	8	70122	9	70131	9	70140	8	70148	9
503	70157	8	70165	9	70174	9	70183	8	70191	9	70200	9	70209	8	70217	9	70226	8	70234	9
504	70243	9	70252	8	70260	9	70269	9	70278	8	70286	9	70295	8	70303	9	70312	9	70321	8
505	70329	9	70338	8	70346	9	70355	9	70364	8	70372	9	70381	8	70389	9	70398	8	70406	9
506	70415	9	70424	8	70432	9	70441	8	70449	9	70458	9	70467	8	70475	9	70484	8	70492	9
507	70501	8	70509	9	70518	8	70526	9	70535	9	70544	8	70552	9	70561	8	70569	9	70578	8
508	70586	9	70595	8	70603	9	70612	9	70621	8	70629	9	70638	8	70646	9	70655	8	70663	9
509	70672	8	70680	9	70689	8	70697	9	70706	8	70714	9	70723	8	70731	9	70740	9	70749	8
510	70757	9	70766	8	70774	9	70783	8	70791	9	70800	8	70808	9	70817	8	70825	9	70834	8
511	70842	9	70851	8	70859	9	70868	8	70876	9	70885	8	70893	9	70902	8	70910	9	70919	8
512	70927	8	70935	9	70944	8	70952	9	70961	8	70969	9	70978	8	70986	9	70995	8	71003	9
513	71012	8	71020	9	71029	8	71037	9	71046	8	71054	9	71063	8	71071	8	71079	9	71088	8
514	71096	9	71105	8	71113	9	71122	8	71130	9	71139	8	71147	8	71155	9	71164	8	71172	9
515	71181	8	71189	9	71198	8	71206	8	71214	9	71223	8	71231	9	71240	8	71248	9	71257	8
516	71265	8	71273	9	71282	8	71290	9	71299	8	71307	8	71315	9	71324	8	71332	9	71341	8
517	71349	8	71357	9	71366	8	71374	9	71383	8	71391	8	71399	9	71408	8	71416	9	71425	8
518	71433	8	71441	9	71450	8	71458	8	71466	9	71475	8	71483	9	71492	8	71500	8	71508	9
519	71517	8	71525	8	71533	9	71542	8	71550	9	71559	8	71567	8	71575	9	71584	8	71592	8
520	71600	9	71609	8	71617	8	71625	9	71634	8	71642	8	71650	9	71659	8	71667	8	71675	9
521	71684	8	71692	8	71700	9	71709	8	71717	8	71725	9	71734	8	71742	8	71750	9	71759	8
522	71767	8	71775	9	71784	8	71792	8	71800	9	71809	8	71817	8	71825	9	71834	8	71842	8
523	71850	8	71858	9	71867	8	71875	8	71883	9	71892	8	71900	8	71908	9	71917	8	71925	8
524	71933	8	71941	9	71950	8	71958	8	71966	9	71975	8	71983	8	71991	8	71999	9	72008	8
525	72016	8	72024	8	72032	9	72041	8	72049	8	72057	9	72066	8	72074	8	72082	8	72090	9
526	72099	8	72107	8	72115	8	72123	9	72132	8	72140	8	72148	8	72156	9	72165	8	72173	8
527	72181	8	72189	9	72198	8	72206	8	72214	8	72222	8	72230	9	72239	8	72247	8	72255	8
528	72263	9	72272	8	72280	8	72288	8	72296	8	72304	9	72313	8	72321	8	72329	8	72337	9
529	72346	8	72354	8	72362	8	72370	8	72378	9	72387	8	72395	8	72403	8	72411	8	72419	9
530	72428	8	72436	8	72444	8	72452	8	72460	9	72469	8	72477	8	72485	8	72493	8	72501	8
531	72509	9	72518	8	72526	8	72534	8	72542	8	72550	8	72558	9	72567	8	72575	8	72583	8
532	72591	8	72599	8	72607	9	72616	8	72624	8	72632	8	72640	8	72648	8	72656	9	72665	8
533	72673	8	72681	8	72689	8	72697	8	72705	8	72713	9	72722	8	72730	8	72738	8	72746	8
534	72754	8	72762	8	72770	9	72779	8	72787	8	72795	8	72803	8	72811	8	72819	8	72827	8
535	72835	8	72843	9	72852	8	72860	8	72868	8	72876	8	72884	8	72892	8	72900	8	72908	8
536	72916	9	72925	8	72933	8	72941	8	72949	8	72957	8	72965	8	72973	8	72981	8	72989	8
537	72997	9	73006	8	73014	8	73022	8	73030	8	73038	8	73046	8	73054	8	73062	8	73070	8
538	73078	8	73086	8	73094	8	73102	9	73111	8	73119	8	73127	8	73135	8	73143	8	73151	8
539	73159	8	73167	8	73175	8	73183	8	73191	8	73199	8	73207	8	73215	8	73223	8	73231	8
540	73239	8	73247	8	73255	8	73263	9	73272	8	73280	8	73288	8	73296	8	73304	8	73312	8
541	73320	8	73328	8	73336	8	73344	8	73352	8	73360	8	73368	8	73376	8	73384	8	73392	8
542	73400	8	73408	8	73416	8	73424	8	73432	8	73440	8	73448	8	73456	8	73464	8	73472	8
543	73480	8	73488	8	73496	8	73504	8	73512	8	73520	8	73528	8	73536	8	73544	8	73552	8
544	73560	8	73568	8	73576	8	73584	8	73592	8	73600	8	73608	8	73616	8	73624	8	73632	8
545	73640	8	73648	8	73656	8	73664	8	73672	7	73679	8	73687	8	73695	8	73703	8	73711	8
546	73719	8	73727	8	73735	8	73743	8	73751	8	73759	8	73767	8	73775	8	73783	8	73791	8
547	73799	8	73807	8	73815	8	73823	7	73830	8	73838	8	73846	8	73854	8	73862	8	73870	8
548	73878	8	73886	8	73894	8	73902	8	73910	8	73918	8	73926	7	73933	8	73941	8	73949	8
549	73957	8	73965	8	73973	8	73981	8	73989	8	73997	8	74005	8	74013	7	74020	8	74028	8
550	74036	8	74044	8	74052	8	74060	8	74068	8	74076	8	74084	8	74092	7	74099	8	74107	8
No.	0	d	1	d	2	d	3	d	4	d	5	d	6	d	7	d	8	d	9	d

Prop. parts

	9
1	1
2	2
3	3
4	4
5	4
6	5
7	6
8	7
9	8

	8
1	1
2	2
3	2
4	3
5	4
6	5
7	6
8	6
9	7

	7
1	1
2	1
3	2
4	3
5	4
6	4
7	5
8	6
9	6

TABLE 1
Logarithms of Numbers

5500–6000

No.	0	d	1	d	2	d	3	d	4	d	5	d	6	d	7	d	8	d	9	d
550	74036	8	74044	8	74052	8	74060	8	74068	8	74076	8	74084	8	74092	7	74099	8	74107	8
551	74115	8	74123	8	74131	8	74139	8	74147	8	74155	7	74162	8	74170	8	74178	8	74186	8
552	74194	8	74202	8	74210	8	74218	7	74225	8	74233	8	74241	8	74249	8	74257	8	74265	8
553	74273	7	74280	8	74288	8	74296	8	74304	8	74312	8	74320	7	74327	8	74335	8	74343	8
554	74351	8	74359	8	74367	7	74374	8	74382	8	74390	8	74398	8	74406	8	74414	7	74421	8
555	74429	8	74437	8	74445	8	74453	8	74461	7	74468	8	74476	8	74484	8	74492	8	74500	7
556	74507	8	74515	8	74523	8	74531	8	74539	8	74547	7	74554	8	74562	8	74570	8	74578	8
557	74586	7	74593	8	74601	8	74609	8	74617	7	74624	8	74632	8	74640	8	74648	8	74656	7
558	74663	8	74671	8	74679	8	74687	8	74695	7	74702	8	74710	8	74718	8	74726	7	74733	8
559	74741	8	74749	8	74757	7	74764	8	74772	8	74780	8	74788	8	74796	7	74803	8	74811	8
560	74819	8	74827	7	74834	8	74842	8	74850	8	74858	7	74865	8	74873	8	74881	8	74889	7
561	74896	8	74904	8	74912	8	74920	7	74927	8	74935	8	74943	7	74950	8	74958	8	74966	8
562	74974	7	74981	8	74989	8	74997	8	75005	7	75012	8	75020	8	75028	7	75035	8	75043	8
563	75051	8	75059	7	75066	8	75074	8	75082	7	75089	8	75097	8	75105	8	75113	7	75120	8
564	75128	8	75136	7	75143	8	75151	8	75159	7	75166	8	75174	8	75182	7	75189	8	75197	8
565	75205	8	75213	7	75220	8	75228	8	75236	7	75243	8	75251	8	75259	7	75266	8	75274	8
566	75282	7	75289	8	75297	8	75305	7	75312	8	75320	8	75328	7	75335	8	75343	8	75351	7
567	75358	8	75366	8	75374	7	75381	8	75389	8	75397	7	75404	8	75412	8	75420	7	75427	8
568	75435	7	75442	8	75450	8	75458	7	75465	8	75473	8	75481	7	75488	8	75496	8	75504	7
569	75511	8	75519	7	75526	8	75534	8	75542	7	75549	8	75557	8	75565	7	75572	8	75580	7
570	75587	8	75595	8	75603	7	75610	8	75618	8	75626	7	75633	8	75641	7	75648	8	75656	8
571	75664	7	75671	8	75679	7	75686	8	75694	8	75702	7	75709	8	75717	7	75724	8	75732	8
572	75740	7	75747	8	75755	7	75762	8	75770	8	75778	7	75785	8	75793	7	75800	8	75808	7
573	75815	8	75823	8	75831	7	75838	8	75846	7	75853	8	75861	7	75868	8	75876	8	75884	7
574	75891	8	75899	7	75906	8	75914	7	75921	8	75929	8	75937	7	75944	8	75952	7	75959	8
575	75967	7	75974	8	75982	7	75989	8	75997	8	76005	7	76012	8	76020	7	76027	8	76035	7
576	76042	8	76050	7	76057	8	76065	7	76072	8	76080	7	76087	8	76095	8	76103	7	76110	8
577	76118	7	76125	8	76133	7	76140	8	76148	7	76155	8	76163	7	76170	8	76178	7	76185	8
578	76193	7	76200	8	76208	7	76215	8	76223	7	76230	8	76238	7	76245	8	76253	7	76260	8
579	76268	7	76275	8	76283	7	76290	8	76298	7	76305	8	76313	7	76320	8	76328	7	76335	8
580	76343	7	76350	8	76358	7	76365	8	76373	7	76380	8	76388	7	76395	8	76403	7	76410	8
581	76418	7	76425	8	76433	7	76440	8	76448	7	76455	7	76462	8	76470	7	76477	8	76485	7
582	76492	8	76500	7	76507	8	76515	7	76522	8	76530	7	76537	8	76545	7	76552	7	76559	8
583	76567	7	76574	8	76582	7	76589	8	76597	7	76604	8	76612	7	76619	7	76626	8	76634	7
584	76641	8	76649	7	76656	8	76664	7	76671	7	76678	8	76686	7	76693	8	76701	7	76708	8
585	76716	7	76723	7	76730	8	76738	7	76745	8	76753	7	76760	8	76768	7	76775	7	76782	8
586	76790	7	76797	8	76805	7	76812	7	76819	8	76827	7	76834	8	76842	7	76849	7	76856	8
587	76864	7	76871	8	76879	7	76886	7	76893	8	76901	7	76908	8	76916	7	76923	7	76930	8
588	76938	7	76945	8	76953	7	76960	7	76967	8	76975	7	76982	7	76989	8	76997	7	77004	8
589	77012	7	77019	7	77026	8	77034	7	77041	7	77048	8	77056	7	77063	7	77070	8	77078	7
590	77085	8	77093	7	77100	7	77107	8	77115	7	77122	7	77129	8	77137	7	77144	7	77151	8
591	77159	7	77166	7	77173	8	77181	7	77188	7	77195	8	77203	7	77210	7	77217	8	77225	7
592	77232	8	77240	7	77247	7	77254	8	77262	7	77269	7	77276	7	77283	8	77291	7	77298	7
593	77305	8	77313	7	77320	7	77327	8	77335	7	77342	7	77349	8	77357	7	77364	7	77371	8
594	77379	7	77386	7	77393	8	77401	7	77408	7	77415	7	77422	8	77430	7	77437	7	77444	8
595	77452	7	77459	7	77466	8	77474	7	77481	7	77488	7	77495	8	77503	7	77510	7	77517	8
596	77525	7	77532	7	77539	7	77546	8	77554	7	77561	7	77568	8	77576	7	77583	7	77590	7
597	77597	8	77605	7	77612	7	77619	8	77627	7	77634	7	77641	7	77648	8	77656	7	77663	7
598	77670	7	77677	8	77685	7	77692	7	77699	7	77706	8	77714	7	77721	7	77728	7	77735	8
599	77743	7	77750	7	77757	7	77764	8	77772	7	77779	7	77786	7	77793	8	77801	7	77808	7
600	77815	7	77822	8	77830	7	77837	7	77844	7	77851	8	77859	7	77866	7	77873	7	77880	7
No.	0	d	1	d	2	d	3	d	4	d	5	d	6	d	7	d	8	d	9	d

Prop. parts

	8
1	1
2	2
3	2
4	3
5	4
6	5
7	6
8	6
9	7

	7
1	1
2	1
3	2
4	3
5	4
6	4
7	5
8	6
9	6

TABLE 1
Logarithms of Numbers

6000–6500

No.	0	d	1	d	2	d	3	d	4	d	5	d	6	d	7	d	8	d	9	d
600	77815	7	77822	8	77830	7	77837	7	77844	7	77851	8	77859	7	77866	7	77873	7	77880	7
601	77887	8	77895	7	77902	7	77909	7	77916	8	77924	7	77931	7	77938	7	77945	7	77952	8
602	77960	7	77967	7	77974	7	77981	7	77988	8	77996	7	78003	7	78010	7	78017	8	78025	7
603	78032	7	78039	7	78046	7	78053	8	78061	7	78068	7	78075	7	78082	7	78089	8	78097	7
604	78104	7	78111	7	78118	7	78125	7	78132	8	78140	7	78147	7	78154	7	78161	7	78168	8
605	78176	7	78183	7	78190	7	78197	7	78204	7	78211	8	78219	7	78226	7	78233	7	78240	7
606	78247	7	78254	8	78262	7	78269	7	78276	7	78283	7	78290	7	78297	8	78305	7	78312	7
607	78319	7	78326	7	78333	7	78340	7	78347	8	78355	7	78362	7	78369	7	78376	7	78383	7
608	78390	8	78398	7	78405	7	78412	7	78419	7	78426	7	78433	7	78440	7	78447	8	78455	7
609	78462	7	78469	7	78476	7	78483	7	78490	7	78497	7	78504	8	78512	7	78519	7	78526	7
610	78533	7	78540	7	78547	7	78554	7	78561	8	78569	7	78576	7	78583	7	78590	7	78597	7
611	78604	7	78611	7	78618	7	78625	8	78633	7	78640	7	78647	7	78654	7	78661	7	78668	7
612	78675	7	78682	7	78689	7	78696	8	78704	7	78711	7	78718	7	78725	7	78732	7	78739	7
613	78746	7	78753	7	78760	7	78767	7	78774	7	78781	8	78789	7	78796	7	78803	7	78810	7
614	78817	7	78824	7	78831	7	78838	7	78845	7	78852	7	78859	7	78866	7	78873	7	78880	8
615	78888	7	78895	7	78902	7	78909	7	78916	7	78923	7	78930	7	78937	7	78944	7	78951	7
616	78958	7	78965	7	78972	7	78979	7	78986	7	78993	7	79000	7	79007	7	79014	7	79021	8
617	79029	7	79036	7	79043	7	79050	7	79057	7	79064	7	79071	7	79078	7	79085	7	79092	7
618	79099	7	79106	7	79113	7	79120	7	79127	7	79134	7	79141	7	79148	7	79155	7	79162	7
619	79169	7	79176	7	79183	7	79190	7	79197	7	79204	7	79211	7	79218	7	79225	7	79232	7
620	79239	7	79246	7	79253	7	79260	7	79267	7	79274	7	79281	7	79288	7	79295	7	79302	7
621	79309	7	79316	7	79323	7	79330	7	79337	7	79344	7	79351	7	79358	7	79365	7	79372	7
622	79379	7	79386	7	79393	7	79400	7	79407	7	79414	7	79421	7	79428	7	79435	7	79442	7
623	79449	7	79456	7	79463	7	79470	7	79477	7	79484	7	79491	7	79498	7	79505	6	79511	7
624	79518	7	79525	7	79532	7	79539	7	79546	7	79553	7	79560	7	79567	7	79574	7	79581	7
625	79588	7	79595	7	79602	7	79609	7	79616	7	79623	7	79630	7	79637	7	79644	6	79650	7
626	79657	7	79664	7	79671	7	79678	7	79685	7	79692	7	79699	7	79706	7	79713	7	79720	7
627	79727	7	79734	7	79741	7	79748	6	79754	7	79761	7	79768	7	79775	7	79782	7	79789	7
628	79796	7	79803	7	79810	7	79817	7	79824	7	79831	6	79837	7	79844	7	79851	7	79858	7
629	79865	7	79872	7	79879	7	79886	7	79893	7	79900	6	79906	7	79913	7	79920	7	79927	7
630	79934	7	79941	7	79948	7	79955	7	79962	7	79969	6	79975	7	79982	7	79989	7	79996	7
631	80003	7	80010	7	80017	7	80024	6	80030	7	80037	7	80044	7	80051	7	80058	7	80065	7
632	80072	7	80079	6	80085	7	80092	7	80099	7	80106	7	80113	7	80120	7	80127	7	80134	6
633	80140	7	80147	7	80154	7	80161	7	80168	7	80175	7	80182	6	80188	7	80195	7	80202	7
634	80209	7	80216	7	80223	6	80229	7	80236	7	80243	7	80250	7	80257	7	80264	7	80271	6
635	80277	7	80284	7	80291	7	80298	7	80305	7	80312	6	80318	7	80325	7	80332	7	80339	7
636	80346	7	80353	6	80359	7	80366	7	80373	7	80380	7	80387	6	80393	7	80400	7	80407	7
637	80414	7	80421	7	80428	6	80434	7	80441	7	80448	7	80455	7	80462	6	80468	7	80475	7
638	80482	7	80489	7	80496	6	80502	7	80509	7	80516	7	80523	7	80530	6	80536	7	80543	7
639	80550	7	80557	7	80564	6	80570	7	80577	7	80584	7	80591	7	80598	6	80604	7	80611	7
640	80618	7	80625	7	80632	6	80638	7	80645	7	80652	7	80659	6	80665	7	80672	7	80679	7
641	80686	7	80693	6	80699	7	80706	7	80713	7	80720	6	80726	7	80733	7	80740	7	80747	7
642	80754	6	80760	7	80767	7	80774	7	80781	6	80787	7	80794	7	80801	7	80808	6	80814	7
643	80821	7	80828	7	80835	6	80841	7	80848	7	80855	7	80862	6	80868	7	80875	7	80882	7
644	80889	6	80895	7	80902	7	80909	7	80916	6	80922	7	80929	7	80936	7	80943	6	80949	7
645	80956	7	80963	6	80969	7	80976	7	80983	7	80990	6	80996	7	81003	7	81010	7	81017	6
646	81023	7	81030	7	81037	6	81043	7	81050	7	81057	7	81064	6	81070	7	81077	7	81084	6
647	81090	7	81097	7	81104	7	81111	6	81117	7	81124	7	81131	6	81137	7	81144	7	81151	7
648	81158	6	81164	7	81171	7	81178	6	81184	7	81191	7	81198	6	81204	7	81211	7	81218	6
649	81224	7	81231	7	81238	7	81245	6	81251	7	81258	7	81265	6	81271	7	81278	7	81285	6
650	81291	7	81298	7	81305	6	81311	7	81318	7	81325	6	81331	7	81338	7	81345	6	81351	7
No.	0	d	1	d	2	d	3	d	4	d	5	d	6	d	7	d	8	d	9	d

Prop. parts

	8
1	1
2	2
3	2
4	3
5	4
6	5
7	6
8	6
9	7

	7
1	1
2	1
3	2
4	3
5	4
6	4
7	5
8	6
9	6

	6
1	1
2	1
3	2
4	2
5	3
6	4
7	4
8	5
9	5

TABLE 1
Logarithms of Numbers

6500–7000

No.	0	d	1	d	2	d	3	d	4	d	5	d	6	d	7	d	8	d	9	d
650	81291	7	81298	7	81305	6	81311	7	81318	7	81325	6	81331	7	81338	7	81345	6	81351	7
651	81358	7	81365	6	81371	7	81378	7	81385	6	81391	7	81398	7	81405	6	81411	7	81418	7
652	81425	6	81431	7	81438	7	81445	6	81451	7	81458	7	81465	6	81471	7	81478	7	81485	6
653	81491	7	81498	7	81505	6	81511	7	81518	7	81525	6	81531	7	81538	6	81544	7	81551	7
654	81558	6	81564	7	81571	7	81578	6	81584	7	81591	7	81598	6	81604	7	81611	6	81617	7
655	81624	7	81631	6	81637	7	81644	7	81651	6	81657	7	81664	7	81671	6	81677	7	81684	6
656	81690	7	81697	7	81704	6	81710	7	81717	6	81723	7	81730	7	81737	6	81743	7	81750	7
657	81757	6	81763	7	81770	6	81776	7	81783	7	81790	6	81796	7	81803	6	81809	7	81816	7
658	81823	6	81829	7	81836	6	81842	7	81849	7	81856	6	81862	7	81869	6	81875	7	81882	7
659	81889	6	81895	7	81902	6	81908	7	81915	6	81921	7	81928	7	81935	6	81941	7	81948	6
660	81954	7	81961	7	81968	6	81974	7	81981	6	81987	7	81994	6	82000	7	82007	7	82014	6
661	82020	7	82027	6	82033	7	82040	6	82046	7	82053	7	82060	6	82066	7	82073	6	82079	7
662	82086	6	82092	7	82099	6	82105	7	82112	7	82119	6	82125	7	82132	6	82138	7	82145	6
663	82151	7	82158	6	82164	7	82171	7	82178	6	82184	7	82191	6	82197	7	82204	6	82210	7
664	82217	6	82223	7	82230	6	82236	7	82243	6	82249	7	82256	7	82263	6	82269	7	82276	6
665	82282	7	82289	6	82295	7	82302	6	82308	7	82315	6	82321	7	82328	6	82334	7	82341	6
666	82347	7	82354	6	82360	7	82367	6	82373	7	82380	7	82387	6	82393	7	82400	6	82406	7
667	82413	6	82419	7	82426	6	82432	7	82439	6	82445	7	82452	6	82458	7	82465	6	82471	7
668	82478	6	82484	7	82491	6	82497	7	82504	6	82510	7	82517	6	82523	7	82530	6	82536	7
669	82543	6	82549	7	82556	6	82562	7	82569	6	82575	7	82582	6	82588	7	82595	6	82601	6
670	82607	7	82614	6	82620	7	82627	6	82633	7	82640	6	82646	7	82653	6	82659	7	82666	6
671	82672	7	82679	6	82685	7	82692	6	82698	7	82705	6	82711	7	82718	6	82724	6	82730	7
672	82737	6	82743	7	82750	6	82756	7	82763	6	82769	7	82776	6	82782	7	82789	6	82795	7
673	82802	6	82808	6	82814	7	82821	6	82827	7	82834	6	82840	7	82847	6	82853	7	82860	6
674	82866	6	82872	7	82879	6	82885	7	82892	6	82898	7	82905	6	82911	7	82918	6	82924	6
675	82930	7	82937	6	82943	7	82950	6	82956	7	82963	6	82969	6	82975	7	82982	6	82988	7
676	82995	6	83001	7	83008	6	83014	6	83020	7	83027	6	83033	7	83040	6	83046	6	83052	7
677	83059	6	83065	7	83072	6	83078	7	83085	6	83091	6	83097	7	83104	6	83110	7	83117	6
678	83123	6	83129	7	83136	6	83142	7	83149	6	83155	6	83161	7	83168	6	83174	7	83181	6
679	83187	6	83193	7	83200	6	83206	7	83213	6	83219	6	83225	7	83232	6	83238	7	83245	6
680	83251	6	83257	7	83264	6	83270	6	83276	7	83283	6	83289	7	83296	6	83302	6	83308	7
681	83315	6	83321	6	83327	7	83334	6	83340	7	83347	6	83353	6	83359	7	83366	6	83372	6
682	83378	7	83385	6	83391	7	83398	6	83404	6	83410	7	83417	6	83423	6	83429	7	83436	6
683	83442	6	83448	7	83455	6	83461	6	83467	7	83474	6	83480	7	83487	6	83493	6	83499	7
684	83506	6	83512	6	83518	7	83525	6	83531	6	83537	7	83544	6	83550	6	83556	7	83563	6
685	83569	6	83575	7	83582	6	83588	6	83594	7	83601	6	83607	6	83613	7	83620	6	83626	6
686	83632	7	83639	6	83645	6	83651	7	83658	6	83664	6	83670	7	83677	6	83683	6	83689	7
687	83696	6	83702	6	83708	7	83715	6	83721	6	83727	7	83734	6	83740	6	83746	7	83753	6
688	83759	6	83765	6	83771	7	83778	6	83784	6	83790	7	83797	6	83803	6	83809	7	83816	6
689	83822	6	83828	7	83835	6	83841	6	83847	6	83853	7	83860	6	83866	6	83872	7	83879	6
690	83885	6	83891	6	83897	7	83904	6	83910	6	83916	7	83923	6	83929	6	83935	7	83942	6
691	83948	6	83954	6	83960	7	83967	6	83973	6	83979	6	83985	7	83992	6	83998	6	84004	7
692	84011	6	84017	6	84023	6	84029	7	84036	6	84042	6	84048	7	84055	6	84061	6	84067	6
693	84073	7	84080	6	84086	6	84092	6	84098	7	84105	6	84111	6	84117	6	84123	7	84130	6
694	84136	6	84142	6	84148	7	84155	6	84161	6	84167	6	84173	7	84180	6	84186	6	84192	6
695	84198	7	84205	6	84211	6	84217	6	84223	7	84230	6	84236	6	84242	6	84248	7	84255	6
696	84261	6	84267	6	84273	7	84280	6	84286	6	84292	6	84298	7	84305	6	84311	6	84317	6
697	84323	7	84330	6	84336	6	84342	6	84348	6	84354	7	84361	6	84367	6	84373	6	84379	7
698	84386	6	84392	6	84398	6	84404	6	84410	7	84417	6	84423	6	84429	6	84435	7	84442	6
699	84448	6	84454	6	84460	6	84466	7	84473	6	84479	6	84485	6	84491	6	84497	7	84504	6
700	84510	6	84516	6	84522	6	84528	7	84535	6	84541	6	84547	6	84553	6	84559	7	84566	6
No.	0	d	1	d	2	d	3	d	4	d	5	d	6	d	7	d	8	d	9	d

Prop. parts

	7
1	1
2	1
3	2
4	3
5	4
6	4
7	5
8	6
9	6

	6
1	1
2	1
3	2
4	2
5	3
6	4
7	4
8	5
9	5

TABLE 1
Logarithms of Numbers

7000–7500

No.	0	d	1	d	2	d	3	d	4	d	5	d	6	d	7	d	8	d	9	d
700	84510	6	84516	6	84522	6	84528	7	84535	6	84541	6	84547	6	84553	6	84559	7	84566	6
701	84572	6	84578	6	84584	6	84590	7	84597	6	84603	6	84609	6	84615	6	84621	7	84628	6
702	84634	6	84640	6	84646	6	84652	6	84658	7	84665	6	84671	6	84677	6	84683	6	84689	7
703	84696	6	84702	6	84708	6	84714	6	84720	6	84726	7	84733	6	84739	6	84745	6	84751	6
704	84757	6	84763	7	84770	6	84776	6	84782	6	84788	6	84794	6	84800	7	84807	6	84813	6
705	84819	6	84825	6	84831	6	84837	7	84844	6	84850	6	84856	6	84862	6	84868	6	84874	6
706	84880	7	84887	6	84893	6	84899	6	84905	6	84911	6	84917	7	84924	6	84930	6	84936	6
707	84942	6	84948	6	84954	6	84960	7	84967	6	84973	6	84979	6	84985	6	84991	6	84997	6
708	85003	6	85009	7	85016	6	85022	6	85028	6	85034	6	85040	6	85046	6	85052	6	85058	7
709	85065	6	85071	6	85077	6	85083	6	85089	6	85095	6	85101	6	85107	7	85114	6	85120	6
710	85126	6	85132	6	85138	6	85144	6	85150	6	85156	7	85163	6	85169	6	85175	6	85181	6
711	85187	6	85193	6	85199	6	85205	6	85211	6	85217	7	85224	6	85230	6	85236	6	85242	6
712	85248	6	85254	6	85260	6	85266	6	85272	6	85278	7	85285	6	85291	6	85297	6	85303	6
713	85309	6	85315	6	85321	6	85327	6	85333	6	85339	6	85345	7	85352	6	85358	6	85364	6
714	85370	6	85376	6	85382	6	85388	6	85394	6	85400	6	85406	6	85412	6	85418	7	85425	6
715	85431	6	85437	6	85443	6	85449	6	85455	6	85461	6	85467	6	85473	6	85479	6	85485	6
716	85491	6	85497	6	85503	6	85509	7	85516	6	85522	6	85528	6	85534	6	85540	6	85546	6
717	85552	6	85558	6	85564	6	85570	6	85576	6	85582	6	85588	6	85594	6	85600	6	85606	6
718	85612	6	85618	7	85625	6	85631	6	85637	6	85643	6	85649	6	85655	6	85661	6	85667	6
719	85673	6	85679	6	85685	6	85691	6	85697	6	85703	6	85709	6	85715	6	85721	6	85727	6
720	85733	6	85739	6	85745	6	85751	6	85757	6	85763	6	85769	6	85775	6	85781	7	85788	6
721	85794	6	85800	6	85806	6	85812	6	85818	6	85824	6	85830	6	85836	6	85842	6	85848	6
722	85854	6	85860	6	85866	6	85872	6	85878	6	85884	6	85890	6	85896	6	85902	6	85908	6
723	85914	6	85920	6	85926	6	85932	6	85938	6	85944	6	85950	6	85956	6	85962	6	85968	6
724	85974	6	85980	6	85986	6	85992	6	85998	6	86004	6	86010	6	86016	6	86022	6	86028	6
725	86034	6	86040	6	86046	6	86052	6	86058	6	86064	6	86070	6	86076	6	86082	6	86088	6
726	86094	6	86100	6	86106	6	86112	6	86118	6	86124	6	86130	6	86136	5	86141	6	86147	6
727	86153	6	86159	6	86165	6	86171	6	86177	6	86183	6	86189	6	86195	6	86201	6	86207	6
728	86213	6	86219	6	86225	6	86231	6	86237	6	86243	6	86249	6	86255	6	86261	6	86267	6
729	86273	6	86279	6	86285	6	86291	6	86297	6	86303	5	86308	6	86314	6	86320	6	86326	6
730	86332	6	86338	6	86344	6	86350	6	86356	6	86362	6	86368	6	86374	6	86380	6	86386	6
731	86392	6	86398	6	86404	6	86410	5	86415	6	86421	6	86427	6	86433	6	86439	6	86445	6
732	86451	6	86457	6	86463	6	86469	6	86475	6	86481	6	86487	6	86493	6	86499	5	86504	6
733	86510	6	86516	6	86522	6	86528	6	86534	6	86540	6	86546	6	86552	6	86558	6	86564	6
734	86570	6	86576	5	86581	6	86587	6	86593	6	86599	6	86605	6	86611	6	86617	6	86623	6
735	86629	6	86635	6	86641	5	86646	6	86652	6	86658	6	86664	6	86670	6	86676	6	86682	6
736	86688	6	86694	6	86700	5	86705	6	86711	6	86717	6	86723	6	86729	6	86735	6	86741	6
737	86747	6	86753	6	86759	5	86764	6	86770	6	86776	6	86782	6	86788	6	86794	6	86800	6
738	86806	6	86812	5	86817	6	86823	6	86829	6	86835	6	86841	6	86847	6	86853	6	86859	5
739	86864	6	86870	6	86876	6	86882	6	86888	6	86894	6	86900	6	86906	5	86911	6	86917	6
740	86923	6	86929	6	86935	6	86941	6	86947	6	86953	5	86958	6	86964	6	86970	6	86976	6
741	86982	6	86988	6	86994	5	86999	6	87005	6	87011	6	87017	6	87023	6	87029	6	87035	5
742	87040	6	87046	6	87052	6	87058	6	87064	6	87070	5	87075	6	87081	6	87087	6	87093	6
743	87099	6	87105	6	87111	5	87116	6	87122	6	87128	6	87134	6	87140	6	87146	5	87151	6
744	87157	6	87163	6	87169	6	87175	6	87181	5	87186	6	87192	6	87198	6	87204	6	87210	6
745	87216	5	87221	6	87227	6	87233	6	87239	6	87245	6	87251	5	87256	6	87262	6	87268	6
746	87274	6	87280	6	87286	5	87291	6	87297	6	87303	6	87309	6	87315	5	87320	6	87326	6
747	87332	6	87338	6	87344	5	87349	6	87355	6	87361	6	87367	6	87373	6	87379	5	87384	6
748	87390	6	87396	6	87402	6	87408	5	87413	6	87419	6	87425	6	87431	6	87437	5	87442	6
749	87448	6	87454	6	87460	6	87466	5	87471	6	87477	6	87483	6	87489	6	87495	5	87500	6
750	87506	6	87512	6	87518	5	87523	6	87529	6	87535	6	87541	6	87547	5	87552	6	87558	6
No.	0	d	1	d	2	d	3	d	4	d	5	d	6	d	7	d	8	d	9	d

Prop. parts

	7
1	1
2	1
3	2
4	3
5	4
6	4
7	5
8	6
9	6

	6
1	1
2	1
3	2
4	2
5	3
6	4
7	4
8	5
9	5

	5
1	0
2	1
3	2
4	2
5	2
6	3
7	4
8	4
9	4

TABLE 1
Logarithms of Numbers

7500–8000

No.	0	d	1	d	2	d	3	d	4	d	5	d	6	d	7	d	8	d	9	d
750	87506	6	87512	6	87518	5	87523	6	87529	6	87535	6	87541	6	87547	5	87552	6	87558	6
751	87564	6	87570	6	87576	5	87581	6	87587	6	87593	6	87599	5	87604	6	87610	6	87616	6
752	87622	6	87628	5	87633	6	87639	6	87645	6	87651	5	87656	6	87662	6	87668	6	87674	5
753	87679	6	87685	6	87691	6	87697	6	87703	5	87708	6	87714	6	87720	6	87726	5	87731	6
754	87737	6	87743	6	87749	5	87754	6	87760	6	87766	6	87772	5	87777	6	87783	6	87789	6
755	87795	5	87800	6	87806	6	87812	6	87818	5	87823	6	87829	6	87835	6	87841	5	87846	6
756	87852	6	87858	6	87864	5	87869	6	87875	6	87881	6	87887	5	87892	6	87898	6	87904	6
757	87910	5	87915	6	87921	6	87927	6	87933	5	87938	6	87944	6	87950	5	87955	6	87961	6
758	87967	6	87973	5	87978	6	87984	6	87990	6	87996	5	88001	6	88007	6	88013	5	88018	6
759	88024	6	88030	6	88036	5	88041	6	88047	6	88053	5	88058	6	88064	6	88070	6	88076	5
760	88081	6	88087	6	88093	5	88098	6	88104	6	88110	6	88116	5	88121	6	88127	6	88133	5
761	88138	6	88144	6	88150	6	88156	5	88161	6	88167	6	88173	5	88178	6	88184	6	88190	5
762	88195	6	88201	6	88207	6	88213	5	88218	6	88224	6	88230	5	88235	6	88241	6	88247	5
763	88252	6	88258	6	88264	6	88270	5	88275	6	88281	6	88287	5	88292	6	88298	6	88304	5
764	88309	6	88315	6	88321	5	88326	6	88332	6	88338	5	88343	6	88349	6	88355	5	88360	6
765	88366	6	88372	5	88377	6	88383	6	88389	6	88395	5	88400	6	88406	6	88412	5	88417	6
766	88423	6	88429	5	88434	6	88440	6	88446	5	88451	6	88457	6	88463	5	88468	6	88474	6
767	88480	5	88485	6	88491	6	88497	5	88502	6	88508	5	88513	6	88519	6	88525	5	88530	6
768	88536	6	88542	5	88547	6	88553	6	88559	5	88564	6	88570	6	88576	5	88581	6	88587	6
769	88593	5	88598	6	88604	6	88610	5	88615	6	88621	6	88627	5	88632	6	88638	5	88643	6
770	88649	6	88655	5	88660	6	88666	6	88672	5	88677	6	88683	6	88689	5	88694	6	88700	5
771	88705	6	88711	6	88717	5	88722	6	88728	6	88734	5	88739	6	88745	5	88750	6	88756	6
772	88762	5	88767	6	88773	6	88779	5	88784	6	88790	5	88795	6	88801	6	88807	5	88812	6
773	88818	6	88824	5	88829	6	88835	5	88840	6	88846	6	88852	5	88857	6	88863	5	88868	6
774	88874	6	88880	5	88885	6	88891	6	88897	5	88902	6	88908	5	88913	6	88919	6	88925	5
775	88930	6	88936	5	88941	6	88947	6	88953	5	88958	6	88964	5	88969	6	88975	6	88981	5
776	88986	6	88992	5	88997	6	89003	6	89009	5	89014	6	89020	5	89025	6	89031	6	89037	5
777	89042	6	89048	5	89053	6	89059	5	89064	6	89070	6	89076	5	89081	6	89087	5	89092	6
778	89098	6	89104	5	89109	6	89115	5	89120	6	89126	5	89131	6	89137	6	89143	5	89148	6
779	89154	5	89159	6	89165	5	89170	6	89176	6	89182	5	89187	6	89193	5	89198	6	89204	5
780	89209	6	89215	6	89221	5	89226	6	89232	5	89237	6	89243	5	89248	6	89254	6	89260	5
781	89265	6	89271	5	89276	6	89282	5	89287	6	89293	5	89298	6	89304	6	89310	5	89315	6
782	89321	5	89326	6	89332	5	89337	6	89343	5	89348	6	89354	6	89360	5	89365	6	89371	5
783	89376	6	89382	5	89387	6	89393	5	89398	6	89404	5	89409	6	89415	6	89421	5	89426	6
784	89432	5	89437	6	89443	5	89448	6	89454	5	89459	6	89465	5	89470	6	89476	5	89481	6
785	89487	5	89492	6	89498	6	89504	5	89509	6	89515	5	89520	6	89526	5	89531	6	89537	5
786	89542	6	89548	5	89553	6	89559	5	89564	6	89570	5	89575	6	89581	5	89586	6	89592	5
787	89597	6	89603	6	89609	5	89614	6	89620	5	89625	6	89631	5	89636	6	89642	5	89647	6
788	89653	5	89658	6	89664	5	89669	6	89675	5	89680	6	89686	5	89691	6	89697	5	89702	6
789	89708	5	89713	6	89719	5	89724	6	89730	5	89735	6	89741	5	89746	6	89752	5	89757	6
790	89763	5	89768	6	89774	5	89779	6	89785	5	89790	6	89796	5	89801	6	89807	5	89812	6
791	89818	5	89823	6	89829	5	89834	6	89840	5	89845	6	89851	5	89856	6	89862	5	89867	6
792	89873	5	89878	5	89883	6	89889	5	89894	6	89900	5	89905	6	89911	5	89916	6	89922	5
793	89927	6	89933	5	89938	6	89944	5	89949	6	89955	5	89960	6	89966	5	89971	6	89977	5
794	89982	6	89988	5	89993	5	89998	6	90004	5	90009	6	90015	5	90020	6	90026	5	90031	6
795	90037	5	90042	6	90048	5	90053	6	90059	5	90064	5	90069	6	90075	5	90080	6	90086	5
796	90091	6	90097	5	90102	6	90108	5	90113	6	90119	5	90124	5	90129	6	90135	5	90140	6
797	90146	5	90151	6	90157	5	90162	6	90168	5	90173	6	90179	5	90184	5	90189	6	90195	5
798	90200	6	90206	5	90211	6	90217	5	90222	5	90227	6	90233	5	90238	6	90244	5	90249	6
799	90255	5	90260	6	90266	5	90271	5	90276	6	90282	5	90287	6	90293	5	90298	6	90304	5
800	90309	5	90314	6	90320	5	90325	6	90331	5	90336	6	90342	5	90347	5	90352	6	90358	5
No.	0	d	1	d	2	d	3	d	4	d	5	d	6	d	7	d	8	d	9	d

Prop. parts

	6
1	1
2	1
3	2
4	2
5	3
6	4
7	4
8	5
9	5

	5
1	0
2	1
3	2
4	2
5	2
6	3
7	4
8	4
9	4

TABLE 1
Logarithms of Numbers

8000–8500

No.	0	d	1	d	2	d	3	d	4	d	5	d	6	d	7	d	8	d	9	d
800	90309	5	90314	6	90320	5	90325	6	90331	5	90336	6	90342	5	90347	5	90352	6	90358	5
801	90363	6	90369	5	90374	6	90380	5	90385	5	90390	6	90396	5	90401	6	90407	5	90412	5
802	90417	6	90423	5	90428	6	90434	5	90439	6	90445	5	90450	5	90455	6	90461	5	90466	6
803	90472	5	90477	5	90482	6	90488	5	90493	6	90499	5	90504	5	90509	6	90515	5	90520	6
804	90526	5	90531	5	90536	6	90542	5	90547	6	90553	5	90558	5	90563	6	90569	5	90574	6
805	90580	5	90585	5	90590	6	90596	5	90601	6	90607	5	90612	5	90617	6	90623	5	90628	6
806	90634	5	90639	5	90644	6	90650	5	90655	5	90660	6	90666	5	90671	6	90677	5	90682	5
807	90687	6	90693	5	90698	5	90703	6	90709	5	90714	6	90720	5	90725	5	90730	6	90736	5
808	90741	6	90747	5	90752	5	90757	6	90763	5	90768	5	90773	6	90779	5	90784	5	90789	6
809	90795	5	90800	6	90806	5	90811	5	90816	6	90822	5	90827	5	90832	6	90838	5	90843	6
810	90849	5	90854	5	90859	6	90865	5	90870	5	90875	6	90881	5	90886	5	90891	6	90897	5
811	90902	5	90907	6	90913	5	90918	6	90924	5	90929	5	90934	6	90940	5	90945	5	90950	6
812	90956	5	90961	5	90966	6	90972	5	90977	5	90982	6	90988	5	90993	5	90998	6	91004	5
813	91009	5	91014	6	91020	5	91025	5	91030	6	91036	5	91041	5	91046	6	91052	5	91057	5
814	91062	6	91068	5	91073	5	91078	6	91084	5	91089	5	91094	6	91100	5	91105	5	91110	6
815	91116	5	91121	5	91126	6	91132	5	91137	5	91142	6	91148	5	91153	5	91158	6	91164	5
816	91169	5	91174	6	91180	5	91185	5	91190	6	91196	5	91201	5	91206	6	91212	5	91217	5
817	91222	6	91228	5	91233	5	91238	5	91243	6	91249	5	91254	5	91259	6	91265	5	91270	5
818	91275	6	91281	5	91286	5	91291	6	91297	5	91302	5	91307	5	91312	6	91318	5	91323	5
819	91328	6	91334	5	91339	5	91344	6	91350	5	91355	5	91360	5	91365	6	91371	5	91376	5
820	91381	6	91387	5	91392	5	91397	6	91403	5	91408	5	91413	5	91418	6	91424	5	91429	5
821	91434	6	91440	5	91445	5	91450	5	91455	6	91461	5	91466	5	91471	6	91477	5	91482	5
822	91487	5	91492	6	91498	5	91503	5	91508	6	91514	5	91519	5	91524	5	91529	6	91535	5
823	91540	5	91545	6	91551	5	91556	5	91561	5	91566	6	91572	5	91577	5	91582	5	91587	6
824	91593	5	91598	5	91603	6	91609	5	91614	5	91619	5	91624	6	91630	5	91635	5	91640	5
825	91645	6	91651	5	91656	5	91661	5	91666	6	91672	5	91677	5	91682	5	91687	6	91693	5
826	91698	5	91703	6	91709	5	91714	5	91719	5	91724	6	91730	5	91735	5	91740	5	91745	6
827	91751	5	91756	5	91761	5	91766	6	91772	5	91777	5	91782	5	91787	6	91793	5	91798	5
828	91803	5	91808	6	91814	5	91819	5	91824	5	91829	5	91834	6	91840	5	91845	5	91850	5
829	91855	6	91861	5	91866	5	91871	5	91876	6	91882	5	91887	5	91892	5	91897	6	91903	5
830	91908	5	91913	5	91918	6	91924	5	91929	5	91934	5	91939	5	91944	6	91950	5	91955	5
831	91960	5	91965	6	91971	5	91976	5	91981	5	91986	5	91991	6	91997	5	92002	5	92007	5
832	92012	6	92018	5	92023	5	92028	5	92033	5	92038	6	92044	5	92049	5	92054	5	92059	6
833	92065	5	92070	5	92075	5	92080	5	92085	6	92091	5	92096	5	92101	5	92106	5	92111	6
834	92117	5	92122	5	92127	5	92132	5	92137	6	92143	5	92148	5	92153	5	92158	5	92163	6
835	92169	5	92174	5	92179	5	92184	5	92189	6	92195	5	92200	5	92205	5	92210	5	92215	6
836	92221	5	92226	5	92231	5	92236	5	92241	6	92247	5	92252	5	92257	5	92262	5	92267	6
837	92273	5	92278	5	92283	5	92288	5	92293	5	92298	6	92304	5	92309	5	92314	5	92319	5
838	92324	6	92330	5	92335	5	92340	5	92345	5	92350	5	92355	6	92361	5	92366	5	92371	5
839	92376	5	92381	6	92387	5	92392	5	92397	5	92402	5	92407	5	92412	6	92418	5	92423	5
840	92428	5	92433	5	92438	5	92443	6	92449	5	92454	5	92459	5	92464	5	92469	5	92474	6
841	92480	5	92485	5	92490	5	92495	5	92500	5	92505	6	92511	5	92516	5	92521	5	92526	5
842	92531	5	92536	6	92542	5	92547	5	92552	5	92557	5	92562	5	92567	5	92572	6	92578	5
843	92583	5	92588	5	92593	5	92598	5	92603	6	92609	5	92614	5	92619	5	92624	5	92629	5
844	92634	5	92639	6	92645	5	92650	5	92655	5	92660	5	92665	5	92670	5	92675	6	92681	5
845	92686	5	92691	5	92696	5	92701	5	92706	5	92711	5	92716	6	92722	5	92727	5	92732	5
846	92737	5	92742	5	92747	5	92752	6	92758	5	92763	5	92768	5	92773	5	92778	5	92783	5
847	92788	5	92793	6	92799	5	92804	5	92809	5	92814	5	92819	5	92824	5	92829	5	92834	6
848	92840	5	92845	5	92850	5	92855	5	92860	5	92865	5	92870	5	92875	6	92881	5	92886	5
849	92891	5	92896	5	92901	5	92906	5	92911	5	92916	5	92921	6	92927	5	92932	5	92937	5
850	92942	5	92947	5	92952	5	92957	5	92962	5	92967	6	92973	5	92978	5	92983	5	92988	5
No.	0	d	1	d	2	d	3	d	4	d	5	d	6	d	7	d	8	d	9	d

Prop. parts

	6
1	1
2	1
3	2
4	2
5	3
6	4
7	4
8	5
9	5

	5
1	0
2	1
3	2
4	2
5	2
6	3
7	4
8	4
9	4

TABLE 1
Logarithms of Numbers

8500–9000

No.	0	d	1	d	2	d	3	d	4	d	5	d	6	d	7	d	8	d	9	d
850	92942	5	92947	5	92952	5	92957	5	92962	5	92967	6	92973	5	92978	5	92983	5	92988	5
851	92993	5	92998	5	93003	5	93008	5	93013	5	93018	6	93024	5	93029	5	93034	5	93039	5
852	93044	5	93049	5	93054	5	93059	5	93064	5	93069	6	93075	5	93080	5	93085	5	93090	5
853	93095	5	93100	5	93105	5	93110	5	93115	5	93120	5	93125	6	93131	5	93136	5	93141	5
854	93146	5	93151	5	93156	5	93161	5	93166	5	93171	5	93176	5	93181	5	93186	6	93192	5
855	93197	5	93202	5	93207	5	93212	5	93217	5	93222	5	93227	5	93232	5	93237	5	93242	5
856	93247	5	93252	6	93258	5	93263	5	93268	5	93273	5	93278	5	93283	5	93288	5	93293	5
857	93298	5	93303	5	93308	5	93313	5	93318	5	93323	5	93328	6	93334	5	93339	5	93344	5
858	93349	5	93354	5	93359	5	93364	5	93369	5	93374	5	93379	5	93384	5	93389	5	93394	5
859	93399	5	93404	5	93409	5	93414	6	93420	5	93425	5	93430	5	93435	5	93440	5	93445	5
860	93450	5	93455	5	93460	5	93465	5	93470	5	93475	5	93480	5	93485	5	93490	5	93495	5
861	93500	5	93505	5	93510	5	93515	5	93520	6	93526	5	93531	5	93536	5	93541	5	93546	5
862	93551	5	93556	5	93561	5	93566	5	93571	5	93576	5	93581	5	93586	5	93591	5	93596	5
863	93601	5	93606	5	93611	5	93616	5	93621	5	93626	5	93631	5	93636	5	93641	5	93646	5
864	93651	5	93656	5	93661	5	93666	5	93671	5	93676	6	93682	5	93687	5	93692	5	93697	5
865	93702	5	93707	5	93712	5	93717	5	93722	5	93727	5	93732	5	93737	5	93742	5	93747	5
866	93752	5	93757	5	93762	5	93767	5	93772	5	93777	5	93782	5	93787	5	93792	5	93797	5
867	93802	5	93807	5	93812	5	93817	5	93822	5	93827	5	93832	5	93837	5	93842	5	93847	5
868	93852	5	93857	5	93862	5	93867	5	93872	5	93877	5	93882	5	93887	5	93892	5	93897	5
869	93902	5	93907	5	93912	5	93917	5	93922	5	93927	5	93932	5	93937	5	93942	5	93947	5
870	93952	5	93957	5	93962	5	93967	5	93972	5	93977	5	93982	5	93987	5	93992	5	93997	5
871	94002	5	94007	5	94012	5	94017	5	94022	5	94027	5	94032	5	94037	5	94042	5	94047	5
872	94052	5	94057	5	94062	5	94067	5	94072	5	94077	5	94082	4	94086	5	94091	5	94096	5
873	94101	5	94106	5	94111	5	94116	5	94121	5	94126	5	94131	5	94136	5	94141	5	94146	5
874	94151	5	94156	5	94161	5	94166	5	94171	5	94176	5	94181	5	94186	5	94191	5	94196	5
875	94201	5	94206	5	94211	5	94216	5	94221	5	94226	5	94231	5	94236	4	94240	5	94245	5
876	94250	5	94255	5	94260	5	94265	5	94270	5	94275	5	94280	5	94285	5	94290	5	94295	5
877	94300	5	94305	5	94310	5	94315	5	94320	5	94325	5	94330	5	94335	5	94340	5	94345	4
878	94349	5	94354	5	94359	5	94364	5	94369	5	94374	5	94379	5	94384	5	94389	5	94394	5
879	94399	5	94404	5	94409	5	94414	5	94419	5	94424	5	94429	4	94433	5	94438	5	94443	5
880	94448	5	94453	5	94458	5	94463	5	94468	5	94473	5	94478	5	94483	5	94488	5	94493	5
881	94498	5	94503	4	94507	5	94512	5	94517	5	94522	5	94527	5	94532	5	94537	5	94542	5
882	94547	5	94552	5	94557	5	94562	5	94567	4	94571	5	94576	5	94581	5	94586	5	94591	5
883	94596	5	94601	5	94606	5	94611	5	94616	5	94621	5	94626	4	94630	5	94635	5	94640	5
884	94645	5	94650	5	94655	5	94660	5	94665	5	94670	5	94675	5	94680	5	94685	4	94689	5
885	94694	5	94699	5	94704	5	94709	5	94714	5	94719	5	94724	5	94729	5	94734	4	94738	5
886	94743	5	94748	5	94753	5	94758	5	94763	5	94768	5	94773	5	94778	5	94783	4	94787	5
887	94792	5	94797	5	94802	5	94807	5	94812	5	94817	5	94822	5	94827	5	94832	4	94836	5
888	94841	5	94846	5	94851	5	94856	5	94861	5	94866	5	94871	5	94876	4	94880	5	94885	5
889	94890	5	94895	5	94900	5	94905	5	94910	5	94915	4	94919	5	94924	5	94929	5	94934	5
890	94939	5	94944	5	94949	5	94954	5	94959	4	94963	5	94968	5	94973	5	94978	5	94983	5
891	94988	5	94993	5	94998	4	95002	5	95007	5	95012	5	95017	5	95022	5	95027	5	95032	4
892	95036	5	95041	5	95046	5	95051	5	95056	5	95061	5	95066	5	95071	4	95075	5	95080	5
893	95085	5	95090	5	95095	5	95100	5	95105	4	95109	5	95114	5	95119	5	95124	5	95129	5
894	95134	5	95139	4	95143	5	95148	5	95153	5	95158	5	95163	5	95168	5	95173	4	95177	5
895	95182	5	95187	5	95192	5	95197	5	95202	5	95207	4	95211	5	95216	5	95221	5	95226	5
896	95231	5	95236	4	95240	5	95245	5	95250	5	95255	5	95260	5	95265	5	95270	4	95274	5
897	95279	5	95284	5	95289	5	95294	5	95299	4	95303	5	95308	5	95313	5	95318	5	95323	5
898	95328	4	95332	5	95337	5	95342	5	95347	5	95352	5	95357	4	95361	5	95366	5	95371	5
899	95376	5	95381	5	95386	4	95390	5	95395	5	95400	5	95405	5	95410	5	95415	4	95419	5
900	95424	5	95429	5	95434	5	95439	5	95444	4	95448	5	95453	5	95458	5	95463	5	95468	4
No.	0	d	1	d	2	d	3	d	4	d	5	d	6	d	7	d	8	d	9	d

Prop. parts

	6
1	1
2	1
3	2
4	2
5	3
6	4
7	4
8	5
9	5

	5
1	0
2	1
3	2
4	2
5	2
6	3
7	4
8	4
9	4

	4
1	0
2	1
3	1
4	2
5	2
6	2
7	3
8	3
9	4

TABLE 1
Logarithms of Numbers

9000–9500

No.	0	d	1	d	2	d	3	d	4	d	5	d	6	d	7	d	8	d	9	d
900	95424	5	95429	5	95434	5	95439	5	95444	4	95448	5	95453	5	95458	5	95463	5	95468	4
901	95472	5	95477	5	95482	5	95487	5	95492	5	95497	4	95501	5	95506	5	95511	5	95516	5
902	95521	4	95525	5	95530	5	95535	5	95540	5	95545	5	95550	4	95554	5	95559	5	95564	5
903	95569	5	95574	4	95578	5	95583	5	95588	5	95593	5	95598	4	95602	5	95607	5	95612	5
904	95617	5	95622	4	95626	5	95631	5	95636	5	95641	5	95646	4	95650	5	95655	5	95660	5
905	95665	5	95670	4	95674	5	95679	5	95684	5	95689	5	95694	4	95698	5	95703	5	95708	5
906	95713	5	95718	4	95722	5	95727	5	95732	5	95737	5	95742	4	95746	5	95751	5	95756	5
907	95761	5	95766	4	95770	5	95775	5	95780	5	95785	4	95789	5	95794	5	95799	5	95804	5
908	95809	4	95813	5	95818	5	95823	5	95828	4	95832	5	95837	5	95842	5	95847	5	95852	4
909	95856	5	95861	5	95866	5	95871	4	95875	5	95880	5	95885	5	95890	5	95895	4	95899	5
910	95904	5	95909	5	95914	4	95918	5	95923	5	95928	5	95933	5	95938	4	95942	5	95947	5
911	95952	5	95957	4	95961	5	95966	5	95971	5	95976	4	95980	5	95985	5	95990	5	95995	4
912	95999	5	96004	5	96009	5	96014	5	96019	4	96023	5	96028	5	96033	5	96038	4	96042	5
913	96047	5	96052	5	96057	4	96061	5	96066	5	96071	5	96076	4	96080	5	96085	5	96090	5
914	96095	4	96099	5	96104	5	96109	5	96114	4	96118	5	96123	5	96128	5	96133	4	96137	5
915	96142	5	96147	5	96152	4	96156	5	96161	5	96166	5	96171	4	96175	5	96180	5	96185	5
916	96190	4	96194	5	96199	5	96204	5	96209	4	96213	5	96218	5	96223	4	96227	5	96232	5
917	96237	5	96242	4	96246	5	96251	5	96256	5	96261	4	96265	5	96270	5	96275	5	96280	4
918	96284	5	96289	5	96294	4	96298	5	96303	5	96308	5	96313	4	96317	5	96322	5	96327	5
919	96332	4	96336	5	96341	5	96346	4	96350	5	96355	5	96360	5	96365	4	96369	5	96374	5
920	96379	5	96384	4	96388	5	96393	5	96398	4	96402	5	96407	5	96412	5	96417	4	96421	5
921	96426	5	96431	4	96435	5	96440	5	96445	5	96450	4	96454	5	96459	5	96464	4	96468	5
922	96473	5	96478	5	96483	4	96487	5	96492	5	96497	4	96501	5	96506	5	96511	4	96515	5
923	96520	5	96525	5	96530	4	96534	5	96539	5	96544	4	96548	5	96553	5	96558	4	96562	5
924	96567	5	96572	5	96577	4	96581	5	96586	5	96591	4	96595	5	96600	5	96605	4	96609	5
925	96614	5	96619	5	96624	4	96628	5	96633	5	96638	4	96642	5	96647	5	96652	4	96656	5
926	96661	5	96666	4	96670	5	96675	5	96680	5	96685	4	96689	5	96694	5	96699	4	96703	5
927	96708	5	96713	4	96717	5	96722	5	96727	4	96731	5	96736	5	96741	4	96745	5	96750	5
928	96755	4	96759	5	96764	5	96769	5	96774	4	96778	5	96783	5	96788	4	96792	5	96797	5
929	96802	4	96806	5	96811	5	96816	4	96820	5	96825	5	96830	4	96834	5	96839	5	96844	4
930	96848	5	96853	5	96858	4	96862	5	96867	5	96872	4	96876	5	96881	5	96886	4	96890	5
931	96895	5	96900	4	96904	5	96909	5	96914	4	96918	5	96923	5	96928	4	96932	5	96937	5
932	96942	4	96946	5	96951	5	96956	4	96960	5	96965	5	96970	4	96974	5	96979	5	96984	4
933	96988	5	96993	4	96997	5	97002	5	97007	4	97011	5	97016	5	97021	4	97025	5	97030	5
934	97035	4	97039	5	97044	5	97049	4	97053	5	97058	5	97063	4	97067	5	97072	5	97077	4
935	97081	5	97086	4	97090	5	97095	5	97100	4	97104	5	97109	5	97114	4	97118	5	97123	5
936	97128	4	97132	5	97137	5	97142	4	97146	5	97151	4	97155	5	97160	5	97165	4	97169	5
937	97174	5	97179	4	97183	5	97188	4	97192	5	97197	5	97202	4	97206	5	97211	5	97216	4
938	97220	5	97225	5	97230	4	97234	5	97239	4	97243	5	97248	5	97253	4	97257	5	97262	5
939	97267	4	97271	5	97276	4	97280	5	97285	5	97290	4	97294	5	97299	5	97304	4	97308	5
940	97313	4	97317	5	97322	5	97327	4	97331	5	97336	4	97340	5	97345	5	97350	4	97354	5
941	97359	5	97364	4	97368	5	97373	4	97377	5	97382	5	97387	4	97391	5	97396	4	97400	5
942	97405	5	97410	4	97414	5	97419	5	97424	4	97428	5	97433	4	97437	5	97442	5	97447	4
943	97451	5	97456	4	97460	5	97465	5	97470	4	97474	5	97479	4	97483	5	97488	5	97493	4
944	97497	5	97502	4	97506	5	97511	5	97516	4	97520	5	97525	4	97529	5	97534	5	97539	4
945	97543	5	97548	4	97552	5	97557	5	97562	4	97566	5	97571	4	97575	5	97580	5	97585	4
946	97589	5	97594	4	97598	5	97603	4	97607	5	97612	5	97617	4	97621	5	97626	4	97630	5
947	97635	5	97640	4	97644	5	97649	4	97653	5	97658	5	97663	4	97667	5	97672	4	97676	5
948	97681	4	97685	5	97690	5	97695	4	97699	5	97704	4	97708	5	97713	4	97717	5	97722	5
949	97727	4	97731	5	97736	4	97740	5	97745	4	97749	5	97754	5	97759	4	97763	5	97768	4
950	97772	5	97777	5	97782	4	97786	5	97791	4	97795	5	97800	4	97804	5	97809	4	97813	5
No.	0	d	1	d	2	d	3	d	4	d	5	d	6	d	7	d	8	d	9	d

Prop. parts

	5
1	0
2	1
3	2
4	2
5	2
6	3
7	4
8	4
9	4

	4
1	0
2	1
3	1
4	2
5	2
6	2
7	3
8	3
9	4

TABLE 1
Logarithms of Numbers

9500–10000

No.	0	d	1	d	2	d	3	d	4	d	5	d	6	d	7	d	8	d	9	d
950	97772	5	97777	5	97782	4	97786	5	97791	4	97795	5	97800	4	97804	5	97809	4	97813	5
951	97818	5	97823	4	97827	5	97832	4	97836	5	97841	4	97845	5	97850	5	97855	4	97859	5
952	97864	4	97868	5	97873	4	97877	5	97882	4	97886	5	97891	5	97896	4	97900	5	97905	4
953	97909	5	97914	4	97918	5	97923	5	97928	4	97932	5	97937	4	97941	5	97946	4	97950	5
954	97955	4	97959	5	97964	4	97968	5	97973	5	97978	4	97982	5	97987	4	97991	5	97996	4
955	98000	5	98005	4	98009	5	98014	5	98019	4	98023	5	98028	4	98032	5	98037	4	98041	5
956	98046	4	98050	5	98055	4	98059	5	98064	4	98068	5	98073	5	98078	4	98082	5	98087	4
957	98091	5	98096	4	98100	5	98105	4	98109	5	98114	4	98118	5	98123	4	98127	5	98132	5
958	98137	4	98141	5	98146	4	98150	5	98155	4	98159	5	98164	4	98168	5	98173	4	98177	5
959	98182	4	98186	5	98191	4	98195	5	98200	4	98204	5	98209	5	98214	4	98218	5	98223	4
960	98227	5	98232	4	98236	5	98241	4	98245	5	98250	4	98254	5	98259	4	98263	5	98268	4
961	98272	5	98277	4	98281	5	98286	4	98290	5	98295	4	98299	5	98304	4	98308	5	98313	5
962	98318	4	98322	5	98327	4	98331	5	98336	4	98340	5	98345	4	98349	5	98354	4	98358	5
963	98363	4	98367	5	98372	4	98376	5	98381	4	98385	5	98390	4	98394	5	98399	4	98403	5
964	98408	4	98412	5	98417	4	98421	5	98426	4	98430	5	98435	4	98439	5	98444	4	98448	5
965	98453	4	98457	5	98462	4	98466	5	98471	4	98475	5	98480	4	98484	5	98489	4	98493	5
966	98498	4	98502	5	98507	4	98511	5	98516	4	98520	5	98525	4	98529	5	98534	4	98538	5
967	98543	4	98547	5	98552	4	98556	5	98561	4	98565	5	98570	4	98574	5	98579	4	98583	5
968	98588	4	98592	5	98597	4	98601	4	98605	5	98610	4	98614	5	98619	4	98623	5	98628	4
969	98632	5	98637	4	98641	5	98646	4	98650	5	98655	4	98659	5	98664	4	98668	5	98673	4
970	98677	5	98682	4	98686	5	98691	4	98695	5	98700	4	98704	5	98709	4	98713	4	98717	5
971	98722	4	98726	5	98731	4	98735	5	98740	4	98744	5	98749	4	98753	5	98758	4	98762	5
972	98767	4	98771	5	98776	4	98780	4	98784	5	98789	4	98793	5	98798	4	98802	5	98807	4
973	98811	5	98816	4	98820	5	98825	4	98829	5	98834	4	98838	5	98843	4	98847	4	98851	5
974	98856	4	98860	5	98865	4	98869	5	98874	4	98878	5	98883	4	98887	5	98892	4	98896	4
975	98900	5	98905	4	98909	5	98914	4	98918	5	98923	4	98927	5	98932	4	98936	5	98941	4
976	98945	4	98949	5	98954	4	98958	5	98963	4	98967	5	98972	4	98976	5	98981	4	98985	4
977	98989	5	98994	4	98998	5	99003	4	99007	5	99012	4	99016	5	99021	4	99025	4	99029	5
978	99034	4	99038	5	99043	4	99047	5	99052	4	99056	5	99061	4	99065	4	99069	5	99074	4
979	99078	5	99083	4	99087	5	99092	4	99096	4	99100	5	99105	4	99109	5	99114	4	99118	5
980	99123	4	99127	4	99131	5	99136	4	99140	5	99145	4	99149	5	99154	4	99158	4	99162	5
981	99167	4	99171	5	99176	4	99180	5	99185	4	99189	4	99193	5	99198	4	99202	5	99207	4
982	99211	5	99216	4	99220	4	99224	5	99229	4	99233	5	99238	4	99242	5	99247	4	99251	4
983	99255	5	99260	4	99264	5	99269	4	99273	4	99277	5	99282	4	99286	5	99291	4	99295	5
984	99300	4	99304	4	99308	5	99313	4	99317	5	99322	4	99326	4	99330	5	99335	4	99339	5
985	99344	4	99348	4	99352	5	99357	4	99361	5	99366	4	99370	4	99374	5	99379	4	99383	5
986	99388	4	99392	4	99396	5	99401	4	99405	5	99410	4	99414	5	99419	4	99423	4	99427	5
987	99432	4	99436	5	99441	4	99445	4	99449	5	99454	4	99458	5	99463	4	99467	4	99471	5
988	99476	4	99480	4	99484	5	99489	4	99493	5	99498	4	99502	4	99506	5	99511	4	99515	5
989	99520	4	99524	4	99528	5	99533	4	99537	5	99542	4	99546	4	99550	5	99555	4	99559	5
990	99564	4	99568	4	99572	5	99577	4	99581	4	99585	5	99590	4	99594	5	99599	4	99603	4
991	99607	5	99612	4	99616	5	99621	4	99625	4	99629	5	99634	4	99638	4	99642	5	99647	4
992	99651	5	99656	4	99660	4	99664	5	99669	4	99673	4	99677	5	99682	4	99686	5	99691	4
993	99695	4	99699	5	99704	4	99708	4	99712	5	99717	4	99721	5	99726	4	99730	4	99734	5
994	99739	4	99743	4	99747	5	99752	4	99756	4	99760	5	99765	4	99769	5	99774	4	99778	4
995	99782	5	99787	4	99791	4	99795	5	99800	4	99804	4	99808	5	99813	4	99817	5	99822	4
996	99826	4	99830	5	99835	4	99839	4	99843	5	99848	4	99852	4	99856	5	99861	4	99865	5
997	99870	4	99874	4	99878	5	99883	4	99887	4	99891	5	99896	4	99900	4	99904	5	99909	4
998	99913	4	99917	5	99922	4	99926	4	99930	5	99935	4	99939	5	99944	4	99948	4	99952	5
999	99957	4	99961	4	99965	5	99970	4	99974	4	99978	5	99983	4	99987	4	99991	5	99996	4
1000	00000	4	00004	5	00009	4	00013	4	00017	5	00022	4	00026	4	00030	5	00035	4	00039	4
No.	0	d	1	d	2	d	3	d	4	d	5	d	6	d	7	d	8	d	9	d

Prop. parts

	5
1	0
2	1
3	2
4	2
5	2
6	3
7	4
8	4
9	4

	4
1	0
2	1
3	1
4	2
5	2
6	2
7	3
8	3
9	4

TABLE 2

Natural Trigonometric Functions

0°→ ↓ ′	sin	Diff. 1′	csc	Diff. 1′	tan	Diff. 1′	cot	Diff. 1′	sec	Diff. 1′	cos	Diff. 1′	← 179° ↓ ′
0	0.00000		∞		0.00000		∞		1.00000		1.00000		60
1	.00029	29	3437.75	— —	.00029	29	3437.75	— —	.00000	0	.00000	0	59
2	.00058	29	1718.87	1718.88	.00058	29	1718.87	1718.88	.00000	0	.00000	0	58
3	.00087	29	1145.92	572.958	.00087	29	1145.92	572.958	.00000	0	.00000	0	57
4	.00116	29	859.437	286.479	.00116	29	859.436	286.479	.00000	0	.00000	0	56
5	0.00145	29	687.550	171.887	0.00145	29	687.549	171.887	1.00000	0	1.00000	0	55
6	.00175	30	572.958	114.592	.00175	30	572.957	114.592	.00000	0	.00000	0	54
7	.00204	29	491.107	81.851	.00204	29	491.106	81.851	.00000	0	.00000	0	53
8	.00233	29	429.719	61.388	.00233	29	429.718	61.388	.00000	0	.00000	0	52
9	.00262	29	381.972	47.746	.00262	29	381.971	47.747	.00000	0	.00000	0	51
10	0.00291	29	343.775	38.197	0.00291	29	343.774	38.197	1.00000	0	1.00000	0	50
11	.00320	29	312.523	31.252	.00320	29	312.521	31.252	.00001	1	.99999	1	49
12	.00349	29	286.479	26.043	.00349	29	286.478	26.044	.00001	0	.99999	0	48
13	.00378	29	264.443	22.037	.00378	29	264.441	22.037	.00001	0	.99999	0	47
14	.00407	29	245.554	18.889	.00407	29	245.552	18.889	.00001	0	.99999	0	46
15	0.00436	29	229.184	16.370	0.00436	29	229.182	16.370	1.00001	0	0.99999	0	45
16	.00465	29	214.860	14.324	.00465	29	214.858	14.324	.00001	0	.99999	0	44
17	.00495	30	202.221	12.639	.00495	30	202.219	12.639	.00001	0	.99999	0	43
18	.00524	29	190.987	11.234	.00524	29	190.984	11.235	.00001	0	.99999	0	42
19	.00553	29	180.935	10.052	.00553	29	180.932	10.052	.00002	1	.99998	1	41
20	0.00582	29	171.888	9.047	0.00582	29	171.885	9.047	1.00002	0	0.99998	0	40
21	.00611	29	163.703	8.185	.00611	29	163.700	8.185	.00002	0	.99998	0	39
22	.00640	29	156.262	7.441	.00640	29	156.259	7.441	.00002	0	.99998	0	38
23	.00669	29	149.468	6.794	.00669	29	149.465	6.794	.00002	0	.99998	0	37
24	.00698	29	143.241	6.228	.00698	29	143.237	6.228	.00002	0	.99998	0	36
25	0.00727	29	137.511	5.730	0.00727	29	137.507	5.730	1.00003	1	0.99997	1	35
26	.00756	29	132.222	5.289	.00756	29	132.219	5.289	.00003	0	.99997	0	34
27	.00785	29	127.325	4.897	.00785	29	127.321	4.897	.00003	0	.99997	0	33
28	.00814	29	122.778	4.547	.00815	30	122.774	4.547	.00003	0	.99997	0	32
29	.00844	30	118.544	4.234	.00844	29	118.540	4.234	.00004	1	.99996	1	31
30	0.00873	29	114.593	3.951	0.00873	29	114.589	3.952	1.00004	0	0.99996	0	30
31	.00902	29	110.897	3.696	.00902	29	110.892	3.697	.00004	0	.99996	0	29
32	.00931	29	107.431	3.465	.00931	29	107.426	3.466	.00004	0	.99996	0	28
33	.00960	29	104.176	3.255	.00960	29	104.171	3.256	.00005	1	.99995	1	27
34	.00989	29	101.112	3.064	.00989	29	101.107	3.064	.00005	0	.99995	0	26
35	0.01018	29	98.2230	2.8888	0.01018	29	98.2179	2.8890	1.00005	0	0.99995	0	25
36	.01047	29	95.4947	2.7283	.01047	29	95.4895	2.7285	.00005	0	.99995	0	24
37	.01076	29	92.9139	2.5808	.01076	29	92.9085	2.5810	.00006	1	.99994	1	23
38	.01105	29	90.4689	2.4450	.01105	29	90.4633	2.4452	.00006	0	.99994	0	22
39	.01134	29	88.1492	2.3196	.01135	30	88.1436	2.3198	.00006	0	.99994	0	21
40	0.01164	30	85.9456	2.2036	0.01164	29	85.9398	2.2038	1.00007	1	0.99993	1	20
41	.01193	29	83.8495	2.0961	.01193	29	83.8435	2.0963	.00007	0	.99993	0	19
42	.01222	29	81.8531	1.9963	.01222	29	81.8470	1.9965	.00007	0	.99993	0	18
43	.01251	29	79.9497	1.9035	.01251	29	79.9434	1.9036	.00008	1	.99992	1	17
44	.01280	29	78.1327	1.8169	.01280	29	78.1263	1.8171	.00008	0	.99992	0	16
45	0.01309	29	76.3966	1.7362	0.01309	29	76.3900	1.7363	1.00009	1	0.99991	1	15
46	.01338	29	74.7359	1.6607	.01338	29	74.7292	1.6608	.00009	0	.99991	0	14
47	.01367	29	73.1458	1.5900	.01367	29	73.1390	1.5902	.00009	0	.99991	0	13
48	.01396	29	71.6221	1.5238	.01396	29	71.6151	1.5239	.00010	1	.99990	1	12
49	.01425	29	70.1605	1.4616	.01425	29	70.1533	1.4617	.00010	0	.99990	0	11
50	0.01454	29	68.7574	1.4031	0.01455	30	68.7501	1.4033	1.00011	1	0.99989	1	10
51	.01483	29	67.4093	1.3481	.01484	29	67.4019	1.3482	.00011	0	.99989	0	9
52	.01513	30	66.1130	1.2962	.01513	29	66.1055	1.2964	.00011	0	.99989	0	8
53	.01542	29	64.8657	1.2473	.01542	29	64.8580	1.2475	.00012	1	.99988	1	7
54	.01571	29	63.6646	1.2011	.01571	29	63.6567	1.2013	.00012	0	.99988	0	6
55	0.01600	29	62.5072	1.1574	0.01600	29	62.4992	1.1576	1.00013	1	0.99987	1	5
56	.01629	29	61.3911	1.1161	.01629	29	61.3829	1.1162	.00013	0	.99987	0	4
57	.01658	29	60.3141	1.0769	.01658	29	60.3058	1.0771	.00014	1	.99986	1	3
58	.01687	29	59.2743	1.0398	.01687	29	59.2659	1.0399	.00014	0	.99986	0	2
59	.01716	29	58.2698	1.0046	.01716	29	58.2612	1.0047	.00015	1	.99985	1	1
60	0.01745	29	57.2987	0.9711	0.01746	30	57.2900	0.9712	1.00015	0	0.99985	0	0
↑ 90°→	cos	Diff. 1′	sec	Diff. 1′	cot	Diff. 1′	tan	Diff. 1′	csc	Diff. 1′	sin	Diff. 1′ ← 89°	↑

TABLE 2

Natural Trigonometric Functions

1°→ ↓ ′	sin	Diff. 1′	csc	Diff. 1′	tan	Diff. 1′	cot	Diff. 1′	sec	Diff. 1′	cos	Diff. 1′	← 178° ↓ ′
0	0.01745		57.2987		0.01746		57.2900		1.00015		0.99985		60
1	.01774	29	56.3595	9392	.01775	29	56.3506	9394	.00016	1	.99984	1	59
2	.01803	29	55.4505	9089	.01804	29	55.4415	9091	.00016	0	.99984	0	58
3	.01832	29	54.5705	8801	.01833	29	54.5613	8802	.00017	1	.99983	1	57
4	.01862	30	53.7179	8526	.01862	29	53.7086	8527	.00017	0	.99983	0	56
5	0.01891	29	52.8916	8263	0.01891	29	52.8821	8265	1.00018	1	0.99982	1	55
6	.01920	29	52.0903	8013	.01920	29	52.0807	8014	.00018	0	.99982	0	54
7	.01949	29	51.3129	7774	.01949	29	51.3032	7775	.00019	1	.99981	1	53
8	.01978	29	50.5584	7545	.01978	29	50.5485	7547	.00020	1	.99980	1	52
9	.02007	29	49.8258	7326	.02007	29	49.8157	7328	.00020	0	.99980	0	51
10	0.02036	29	49.1141	7117	0.02036	29	49.1039	7118	1.00021	1	0.99979	1	50
11	.02065	29	48.4224	6917	.02066	30	48.4121	6918	.00021	0	.99979	0	49
12	.02094	29	47.7500	6724	.02095	29	47.7395	6726	.00022	1	.99978	1	48
13	.02123	29	47.0960	6540	.02124	29	47.0853	6542	.00023	1	.99977	1	47
14	.02152	29	46.4596	6363	.02153	29	46.4489	6365	.00023	0	.99977	0	46
15	0.02181	29	45.8403	6194	0.02182	29	45.8294	6195	1.00024	1	0.99976	1	45
16	.02211	30	45.2372	6031	.02211	29	45.2261	6032	.00024	0	.99976	0	44
17	.02240	29	44.6498	5874	.02240	29	44.6386	5875	.00025	1	.99975	1	43
18	.02269	29	44.0775	5723	.02269	29	44.0661	5725	.00026	1	.99974	1	42
19	.02298	29	43.5196	5578	.02298	29	43.5081	5580	.00026	0	.99974	0	41
20	0.02327	29	42.9757	5439	0.02328	30	42.9641	5440	1.00027	1	0.99973	1	40
21	.02356	29	42.4452	5305	.02357	29	42.4335	5306	.00028	1	.99972	1	39
22	.02385	29	41.9277	5175	.02386	29	41.9158	5177	.00028	0	.99972	0	38
23	.02414	29	41.4227	5051	.02415	29	41.4106	5052	.00029	1	.99971	1	37
24	.02443	29	40.9296	4930	.02444	29	40.9174	4932	.00030	1	.99970	1	36
25	0.02472	29	40.4482	4814	0.02473	29	40.4358	4816	1.00031	1	0.99969	1	35
26	.02501	29	39.9780	4702	.02502	29	39.9655	4704	.00031	0	.99969	0	34
27	.02530	29	.5185	4594	.02531	29	.5059	4596	.00032	1	.99968	1	33
28	.02560	30	39.0696	4490	.02560	29	39.0568	4491	.00033	1	.99967	1	32
29	.02589	29	38.6307	4389	.02589	29	38.6177	4390	.00034	1	.99966	1	31
30	0.02618	29	38.2016	4291	0.02619	30	38.1885	4293	1.00034	0	0.99966	0	30
31	.02647	29	37.7818	4197	.02648	29	37.7686	4198	.00035	1	.99965	1	29
32	.02676	29	37.3713	4106	.02677	29	37.3579	4107	.00036	1	.99964	1	28
33	.02705	29	36.9695	4017	.02706	29	36.9560	4019	.00037	1	.99963	1	27
34	.02734	29	.5763	3932	.02735	29	.5627	3933	.00037	0	.99963	0	26
35	0.02763	29	36.1914	3849	0.02764	29	36.1776	3851	1.00038	1	0.99962	1	25
36	.02792	29	35.8145	3769	.02793	29	35.8006	3770	.00039	1	.99961	1	24
37	.02821	29	.4454	3691	.02822	29	.4313	3693	.00040	1	.99960	1	23
38	.02850	29	35.0838	3616	.02851	29	35.0695	3617	.00041	1	.99959	1	22
39	.02879	29	34.7295	3543	.02881	30	34.7151	3544	.00041	0	.99959	0	21
40	0.02908	29	34.3823	3472	0.02910	29	34.3678	3473	1.00042	1	0.99958	1	20
41	.02938	30	34.0420	3403	.02939	29	34.0273	3405	.00043	1	.99957	1	19
42	.02967	29	33.7083	3336	.02968	29	33.6935	3338	.00044	1	.99956	1	18
43	.02996	29	.3812	3272	.02997	29	.3662	3273	.00045	1	.99955	1	17
44	.03025	29	33.0603	3209	.03026	29	33.0452	3210	.00046	1	.99954	1	16
45	0.03054	29	32.7455	3148	0.03055	29	32.7303	3149	1.00047	1	0.99953	1	15
46	.03083	29	.4367	3088	.03084	29	.4213	3090	.00048	1	.99952	1	14
47	.03112	29	32.1337	3031	.03114	30	32.1181	3032	.00048	0	.99952	0	13
48	.03141	29	31.8362	2974	.03143	29	31.8205	2976	.00049	1	.99951	1	12
49	.03170	29	.5442	2920	.03172	29	.5284	2921	.00050	1	.99950	1	11
50	0.03199	29	31.2576	2867	0.03201	29	31.2416	2868	1.00051	1	0.99949	1	10
51	.03228	29	30.9761	2815	.03230	29	30.9599	2816	.00052	1	.99948	1	9
52	.03257	29	.6996	2765	.03259	29	.6833	2766	.00053	1	.99947	1	8
53	.03286	29	.4280	2716	.03288	29	.4116	2717	.00054	1	.99946	1	7
54	.03316	30	30.1612	2668	.03317	29	30.1446	2670	.00055	1	.99945	1	6
55	0.03345	29	29.8990	2622	0.03346	29	29.8823	2623	1.00056	1	0.99944	1	5
56	.03374	29	.6414	2577	.03376	30	.6245	2578	.00057	1	.99943	1	4
57	.03403	29	.3881	2532	.03405	29	.3711	2534	.00058	1	.99942	1	3
58	.03432	29	29.1392	2490	.03434	29	29.1220	2491	.00059	1	.99941	1	2
59	.03461	29	28.8944	2448	.03463	29	28.8771	2449	.00060	1	.99940	1	1
60	0.03490	29	28.6537	2407	0.03492	29	28.6363	2408	1.00061	1	0.99939	1	0
↑ 91°→	cos	Diff. 1′	sec	Diff. 1′	cot	Diff. 1′	tan	Diff. 1′	csc	Diff. 1′	sin	Diff. 1′ ← 88°	↑

TABLE 2
Natural Trigonometric Functions

2°→ ↓ ′	sin	Diff. 1′	csc	Diff. 1′	tan	Diff. 1′	cot	Diff. 1′	sec	Diff. 1′	cos	Diff. 1′	← 177° ↓ ′
0	0.03490		28.6537		0.03492		28.6363		1.00061		0.99939		60
1	.03519	29	.4170	2367	.03521	29	.3994	2369	.00062	1	.99938	1	59
2	.03548	29	28.1842	2328	.03550	29	28.1664	2330	.00063	1	.99937	1	58
3	.03577	29	27.9551	2290	.03579	29	27.9372	2292	.00064	1	.99936	1	57
4	.03606	29	.7298	2253	.03609	30	.7117	2255	.00065	1	.99935	1	56
5	0.03635	29	27.5080	2217	0.03638	29	27.4899	2219	1.00066	1	0.99934	1	55
6	.03664	29	.2898	2182	.03667	29	.2715	2184	.00067	1	.99933	1	54
7	.03693	29	27.0750	2148	.03696	29	27.0566	2149	.00068	1	.99932	1	53
8	.03723	30	26.8636	2114	.03725	29	26.8450	2116	.00069	1	.99931	1	52
9	.03752	29	.6555	2081	.03754	29	.6367	2083	.00070	1	.99930	1	51
10	0.03781	29	26.4505	2049	0.03783	29	26.4316	2051	1.00072	2	0.99929	1	50
11	.03810	29	.2487	2018	.03812	29	.2296	2020	.00073	1	.99927	2	49
12	.03839	29	26.0499	1988	.03842	30	26.0307	1989	.00074	1	.99926	1	48
13	.03868	29	25.8542	1958	.03871	29	25.8348	1959	.00075	1	.99925	1	47
14	.03897	29	.6613	1928	.03900	29	.6418	1930	.00076	1	.99924	1	46
15	0.03926	29	25.4713	1900	0.03929	29	25.4517	1901	1.00077	1	0.99923	1	45
16	.03955	29	.2841	1872	.03958	29	.2644	1873	.00078	1	.99922	1	44
17	.03984	29	25.0997	1845	.03987	29	25.0798	1846	.00079	1	.99921	1	43
18	.04013	29	24.9179	1818	.04016	29	24.8978	1819	.00081	2	.99919	2	42
19	.04042	29	.7387	1792	.04046	30	.7185	1793	.00082	1	.99918	1	41
20	0.04071	29	24.5621	1766	0.04075	29	24.5418	1768	1.00083	1	0.99917	1	40
21	.04100	29	.3880	1741	.04104	29	.3675	1742	.00084	1	.99916	1	39
22	.04129	29	.2164	1716	.04133	29	.1957	1718	.00085	1	.99915	1	38
23	.04159	30	24.0471	1692	.04162	29	24.0263	1694	.00087	2	.99913	2	37
24	.04188	29	23.8802	1669	.04191	29	23.8593	1670	.00088	1	.99912	1	36
25	0.04217	29	23.7156	1646	0.04220	29	23.6945	1647	1.00089	1	0.99911	1	35
26	.04246	29	.5533	1623	.04250	30	.5321	1625	.00090	1	.99910	1	34
27	.04275	29	.3932	1601	.04279	29	.3718	1603	.00091	1	.99909	1	33
28	.04304	29	.2352	1580	.04308	29	.2137	1581	.00093	2	.99907	2	32
29	.04333	29	23.0794	1558	.04337	29	23.0577	1560	.00094	1	.99906	1	31
30	0.04362	29	22.9256	1538	0.04366	29	22.9038	1539	1.00095	1	0.99905	1	30
31	.04391	29	.7739	1517	.04395	29	.7519	1519	.00097	2	.99904	1	29
32	.04420	29	.6241	1497	.04424	29	.6020	1499	.00098	1	.99902	2	28
33	.04449	29	.4764	1478	.04454	30	.4541	1479	.00099	1	.99901	1	27
34	.04478	29	.3305	1459	.04483	29	.3081	1460	.00100	1	.99900	1	26
35	0.04507	29	22.1865	1440	0.04512	29	22.1640	1441	1.00102	2	0.99898	2	25
36	.04536	29	22.0444	1421	.04541	29	22.0217	1423	.00103	1	.99897	1	24
37	.04565	29	21.9041	1403	.04570	29	21.8813	1405	.00104	1	.99896	1	23
38	.04594	29	.7656	1385	.04599	29	.7426	1387	.00106	2	.99894	2	22
39	.04623	29	.6288	1368	.04628	29	.6056	1369	.00107	1	.99893	1	21
40	0.04653	30	21.4937	1351	0.04658	30	21.4704	1352	1.00108	1	0.99892	1	20
41	.04682	29	.3603	1334	.04687	29	.3369	1335	.00110	2	.99890	2	19
42	.04711	29	.2285	1318	.04716	29	.2049	1319	.00111	1	.99889	1	18
43	.04740	29	21.0984	1301	.04745	29	21.0747	1303	.00113	2	.99888	1	17
44	.04769	29	20.9698	1286	.04774	29	20.9460	1287	.00114	1	.99886	2	16
45	0.04798	29	20.8428	1270	0.04803	29	20.8188	1271	1.00115	1	0.99885	1	15
46	.04827	29	.7174	1255	.04833	30	.6932	1256	.00117	2	.99883	2	14
47	.04856	29	.5934	1240	.04862	29	.5691	1241	.00118	1	.99882	1	13
48	.04885	29	.4709	1225	.04891	29	.4465	1226	.00120	2	.99881	1	12
49	.04914	29	.3499	1210	.04920	29	.3253	1212	.00121	1	.99879	2	11
50	0.04943	29	20.2303	1196	0.04949	29	20.2056	1198	1.00122	1	0.99878	1	10
51	.04972	29	20.1121	1182	.04978	29	20.0872	1184	.00124	2	.99876	2	9
52	.05001	29	19.9952	1168	.05007	29	19.9702	1170	.00125	1	.99875	1	8
53	.05030	29	.8798	1155	.05037	30	.8546	1156	.00127	2	.99873	2	7
54	.05059	29	.7656	1142	.05066	29	.7403	1143	.00128	1	.99872	1	6
55	0.05088	29	19.6528	1128	0.05095	29	19.6273	1130	1.00130	2	0.99870	2	5
56	.05117	29	.5412	1116	.05124	29	.5156	1117	.00131	1	.99869	1	4
57	.05146	29	.4309	1103	.05153	29	.4051	1105	.00133	2	.99867	2	3
58	.05175	29	.3218	1091	.05182	29	.2959	1092	.00134	1	.99866	1	2
59	.05205	30	.2140	1078	.05212	30	.1879	1080	.00136	2	.99864	2	1
60	0.05234	29	19.1073	1066	0.05241	29	19.0811	1068	1.00137	1	0.99863	1	0
↑ 92°→	cos	Diff. 1′	sec	Diff. 1′	cot	Diff. 1′	tan	Diff. 1′	csc	Diff. 1′	sin	Diff. 1′ ←	↑ 87°

TABLE 2
Natural Trigonometric Functions

3°→ ↓ ′	sin	Diff. 1′	csc	Diff. 1′	tan	Diff. 1′	cot	Diff. 1′	sec	Diff. 1′	cos	Diff. 1′	← 176° ↓ ′
0	0.05234		19.1073		0.05241		19.0811		1.00137		0.99863		60
1	.05263	29	19.0019	1055	.05270	29	18.9755	1056	.00139	2	.99861	2	59
2	.05292	29	18.8975	1043	.05299	29	.8711	1045	.00140	1	.99860	1	58
3	.05321	29	.7944	1032	.05328	29	.7678	1033	.00142	2	.99858	2	57
4	.05350	29	.6923	1020	.05357	29	.6656	1022	.00143	1	.99857	1	56
5	0.05379	29	18.5914	1009	0.05387	30	18.5645	1011	1.00145	2	0.99855	2	55
6	.05408	29	.4915	999	.05416	29	.4645	1000	.00147	2	.99854	1	54
7	.05437	29	.3927	988	.05445	29	.3655	989	.00148	1	.99852	2	53
8	.05466	29	.2950	977	.05474	29	.2677	979	.00150	2	.99851	1	52
9	.05495	29	.1983	967	.05503	29	.1708	968	.00151	1	.99849	2	51
10	0.05524	29	18.1026	957	0.05533	30	18.0750	958	1.00153	2	0.99847	2	50
11	.05553	29	18.0079	947	.05562	29	17.9802	948	.00155	2	.99846	1	49
12	.05582	29	17.9142	937	.05591	29	.8863	938	.00156	1	.99844	2	48
13	.05611	29	.8215	927	.05620	29	.7934	929	.00158	2	.99842	2	47
14	.05640	29	.7298	918	.05649	29	.7015	919	.00159	1	.99841	1	46
15	0.05669	29	17.6389	908	0.05678	29	17.6106	910	1.00161	2	0.99839	2	45
16	.05698	29	.5490	899	.05708	30	.5205	900	.00163	2	.99838	1	44
17	.05727	29	.4600	890	.05737	29	.4314	891	.00164	1	.99836	2	43
18	.05756	29	.3720	881	.05766	29	.3432	882	.00166	2	.99834	2	42
19	.05785	29	.2848	872	.05795	29	.2558	873	.00168	2	.99833	1	41
20	0.05814	29	17.1984	863	0.05824	29	17.1693	865	1.00169	1	0.99831	2	40
21	.05844	30	.1130	855	.05854	30	17.0837	856	.00171	2	.99829	2	39
22	.05873	29	17.0283	846	.05883	29	16.9990	848	.00173	2	.99827	2	38
23	.05902	29	16.9446	838	.05912	29	.9150	839	.00175	2	.99826	1	37
24	.05931	29	.8616	830	.05941	29	.8319	831	.00176	1	.99824	2	36
25	0.05960	29	16.7794	822	0.05970	29	16.7496	823	1.00178	2	0.99822	2	35
26	.05989	29	.6981	814	.05999	29	.6681	815	.00180	2	.99821	1	34
27	.06018	29	.6175	806	.06029	30	.5874	807	.00182	2	.99819	2	33
28	.06047	29	.5377	798	.06058	29	.5075	799	.00183	1	.99817	2	32
29	.06076	29	.4587	790	.06087	29	.4283	792	.00185	2	.99815	2	31
30	0.06105	29	16.3804	783	0.06116	29	16.3499	784	1.00187	2	0.99813	2	30
31	.06134	29	.3029	775	.06145	29	.2722	777	.00189	2	.99812	1	29
32	.06163	29	.2261	768	.06175	30	.1952	769	.00190	1	.99810	2	28
33	.06192	29	.1500	761	.06204	29	.1190	762	.00192	2	.99808	2	27
34	.06221	29	16.0746	754	.06233	29	16.0435	755	.00194	2	.99806	2	26
35	0.06250	29	15.9999	747	0.06262	29	15.9687	748	1.00196	2	0.99804	2	25
36	.06279	29	.9260	740	.06291	29	.8945	741	.00198	2	.99803	1	24
37	.06308	29	.8527	733	.06321	30	.8211	734	.00200	2	.99801	2	23
38	.06337	29	.7801	726	.06350	29	.7483	728	.00201	1	.99799	2	22
39	.06366	29	.7081	720	.06379	29	.6762	721	.00203	2	.99797	2	21
40	0.06395	29	15.6368	713	0.06408	29	15.6048	714	1.00205	2	0.99795	2	20
41	.06424	29	.5661	707	.06437	29	.5340	708	.00207	2	.99793	2	19
42	.06453	29	.4961	700	.06467	30	.4638	702	.00209	2	.99792	1	18
43	.06482	29	.4267	694	.06496	29	.3943	695	.00211	2	.99790	2	17
44	.06511	29	.3579	688	.06525	29	.3254	689	.00213	2	.99788	2	16
45	0.06540	29	15.2898	682	0.06554	29	15.2571	683	1.00215	2	0.99786	2	15
46	.06569	29	.2222	676	.06584	30	.1893	677	.00216	1	.99784	2	14
47	.06598	29	.1553	670	.06613	29	.1222	671	.00218	2	.99782	2	13
48	.06627	29	.0889	664	.06642	29	15.0557	665	.00220	2	.99780	2	12
49	.06656	29	15.0231	658	.06671	29	14.9898	659	.00222	2	.99778	2	11
50	0.06685	29	14.9579	652	0.06700	29	14.9244	654	1.00224	2	0.99776	2	10
51	.06714	29	.8932	647	.06730	30	.8596	648	.00226	2	.99774	2	9
52	.06743	29	.8291	641	.06759	29	.7954	642	.00228	2	.99772	2	8
53	.06773	30	.7656	635	.06788	29	.7317	637	.00230	2	.99770	2	7
54	.06802	29	.7026	630	.06817	29	.6685	631	.00232	2	.99768	2	6
55	0.06831	29	14.6401	625	0.06847	30	14.6059	626	1.00234	2	0.99766	2	5
56	.06860	29	.5782	619	.06876	29	.5438	621	.00236	2	.99764	2	4
57	.06889	29	.5168	614	.06905	29	.4823	616	.00238	2	.99762	2	3
58	.06918	29	.4559	609	.06934	29	.4212	610	.00240	2	.99760	2	2
59	.06947	29	.3955	604	.06963	29	.3607	605	.00242	2	.99758	2	1
60	0.06976	29	14.3356	599	0.06993	30	14.3007	600	1.00244	2	0.99756	2	0
↑ 93°→	cos	Diff. 1′	sec	Diff. 1′	cot	Diff. 1′	tan	Diff. 1′	csc	Diff. 1′	sin	Diff. 1′ ←	↑ 86°

TABLE 2
Natural Trigonometric Functions

4°→ ↓ ′	sin	Diff. 1′	csc	Diff. 1′	tan	Diff. 1′	cot	Diff. 1′	sec	Diff. 1′	cos	Diff. 1′	← 175° ↓ ′
0	0.06976	29	14.3356	594	0.06993	29	14.3007	595	1.00244	2	0.99756	2	60
1	.07005	29	.2762	589	.07022	29	.2411	590	.00246	2	.99754	2	59
2	.07034	29	.2173	584	.07051	29	.1821	586	.00248	2	.99752	2	58
3	.07063	29	.1589	579	.07080	30	.1235	581	.00250	2	.99750	2	57
4	.07092	29	.1010	575	.07110	29	.0655	576	.00252	2	.99748	2	56
5	0.07121	29	14.0435	570	0.07139	29	14.0079	571	1.00254	3	0.99746	2	55
6	.07150	29	13.9865	565	.07168	29	13.9507	567	.00257	2	.99744	2	54
7	.07179	29	.9300	561	.07197	30	.8940	562	.00259	2	.99742	2	53
8	.07208	29	.8739	556	.07227	29	.8378	558	.00261	2	.99740	2	52
9	.07237	29	.8183	552	.07256	29	.7821	553	.00263	2	.99738	2	51
10	0.07266	29	13.7631	547	0.07285	29	13.7267	549	1.00265	2	0.99736	2	50
11	.07295	29	.7084	543	.07314	30	.6719	544	.00267	2	.99734	3	49
12	.07324	29	.6541	539	.07344	29	.6174	540	.00269	2	.99731	2	48
13	.07353	29	.6002	534	.07373	29	.5634	536	.00271	3	.99729	2	47
14	.07382	29	.5468	530	.07402	29	.5098	532	.00274	2	.99727	2	46
15	0.07411	29	13.4937	526	0.07431	30	13.4566	528	1.00276	2	0.99725	2	45
16	.07440	29	.4411	522	.07461	29	.4039	523	.00278	2	.99723	2	44
17	.07469	29	.3889	518	.07490	29	.3515	519	.00280	2	.99721	2	43
18	.07498	29	.3371	514	.07519	29	.2996	515	.00282	2	.99719	3	42
19	.07527	29	.2857	510	.07548	30	.2480	511	.00284	3	.99716	2	41
20	0.07556	29	13.2347	506	0.07578	29	13.1969	508	1.00287	2	0.99714	2	40
21	.07585	29	.1841	502	.07607	29	.1461	504	.00289	2	.99712	2	39
22	.07614	29	.1339	498	.07636	29	.0958	500	.00291	2	.99710	2	38
23	.07643	29	.0840	495	.07665	30	13.0458	496	.00293	3	.99708	3	37
24	.07672	29	13.0346	491	.07695	29	12.9962	492	.00296	2	.99705	2	36
25	0.07701	29	12.9855	487	0.07724	29	12.9469	489	1.00298	2	0.99703	2	35
26	.07730	29	.9368	484	.07753	29	.8981	485	.00300	2	.99701	2	34
27	.07759	29	.8884	480	.07782	30	.8496	481	.00302	3	.99699	3	33
28	.07788	29	.8404	476	.07812	29	.8014	478	.00305	2	.99696	2	32
29	.07817	29	.7928	473	.07841	29	.7536	474	.00307	2	.99694	2	31
30	0.07846	29	12.7455	469	0.07870	29	12.7062	471	1.00309	3	0.99692	3	30
31	.07875	29	.6986	466	.07899	30	.6591	467	.00312	2	.99689	2	29
32	.07904	29	.6520	462	.07929	29	.6124	464	.00314	2	.99687	2	28
33	.07933	29	.6057	459	.07958	29	.5660	461	.00316	2	.99685	2	27
34	.07962	29	.5598	456	.07987	30	.5199	457	.00318	3	.99683	3	26
35	0.07991	29	12.5142	452	0.08017	29	12.4742	454	1.00321	2	0.99680	2	25
36	.08020	29	.4690	449	.08046	29	.4288	451	.00323	3	.99678	2	24
37	.08049	29	.4241	446	.08075	29	.3838	447	.00326	2	.99676	3	23
38	.08078	29	.3795	443	.08104	30	.3390	444	.00328	2	.99673	2	22
39	.08107	29	.3352	440	.08134	29	.2946	441	.00330	3	.99671	3	21
40	0.08136	29	12.2913	436	0.08163	29	12.2505	438	1.00333	2	0.99668	2	20
41	.08165	29	.2476	433	.08192	29	.2067	435	.00335	2	.99666	2	19
42	.08194	29	.2043	430	.08221	30	.1632	432	.00337	3	.99664	3	18
43	.08223	29	.1612	427	.08251	29	.1201	429	.00340	2	.99661	2	17
44	.08252	29	.1185	424	.08280	29	.0772	426	.00342	3	.99659	2	16
45	0.08281	29	12.0761	421	0.08309	30	12.0346	423	1.00345	2	0.99657	3	15
46	.08310	29	12.0340	418	.08339	29	11.9923	420	.00347	3	.99654	2	14
47	.08339	29	11.9921	415	.08368	29	.9504	417	.00350	2	.99652	3	13
48	.08368	29	.9506	413	.08397	30	.9087	414	.00352	2	.99649	2	12
49	.08397	29	.9093	410	.08427	29	.8673	411	.00354	3	.99647	3	11
50	0.08426	29	11.8684	407	0.08456	29	11.8262	408	1.00357	2	0.99644	2	10
51	.08455	29	.8277	404	.08485	29	.7853	406	.00359	3	.99642	3	9
52	.08484	29	.7873	401	.08514	30	.7448	403	.00362	2	.99639	2	8
53	.08513	29	.7471	399	.08544	29	.7045	400	.00364	3	.99637	2	7
54	.08542	29	.7073	396	.08573	29	.6645	397	.00367	2	.99635	3	6
55	0.08571	29	11.6677	393	0.08602	30	11.6248	395	1.00369	3	0.99632	2	5
56	.08600	29	.6284	391	.08632	29	.5853	392	.00372	2	.99630	3	4
57	.08629	29	.5893	388	.08661	29	.5461	389	.00374	3	.99627	2	3
58	.08658	29	.5505	385	.08690	30	.5072	387	.00377	2	.99625	3	2
59	.08687	29	.5120	383	.08720	29	.4685	384	.00379	3	.99622	3	1
60	0.08716		11.4737		0.08749		11.4301		1.00382		0.99619		0
↑ 94°→	cos	Diff. 1′	sec	Diff. 1′	cot	Diff. 1′	tan	Diff. 1′	csc	Diff. 1′	sin	Diff. 1′	← 85° ↑

TABLE 2
Natural Trigonometric Functions

5°→ ↓ ′	sin	Diff. 1′	csc	Diff. 1′	tan	Diff. 1′	cot	Diff. 1′	sec	Diff. 1′	cos	Diff. 1′	← 174° ↓ ′
0	0.08716	29	11.4737	380	0.08749	29	11.4301	382	1.00382	3	0.99619	2	60
1	.08745	29	.4357	378	.08778	29	.3919	379	.00385	2	.99617	3	59
2	.08774	29	.3979	375	.08807	30	.3540	377	.00387	3	.99614	2	58
3	.08803	28	.3604	373	.08837	29	.3163	374	.00390	2	.99612	3	57
4	.08831	29	.3231	370	.08866	29	.2789	372	.00392	3	.99609	2	56
5	0.08860	29	11.2861	368	0.08895	30	11.2417	369	1.00395	2	0.99607	3	55
6	.08889	29	.2493	365	.08925	29	.2048	367	.00397	3	.99604	2	54
7	.08918	29	.2128	363	.08954	29	.1681	365	.00400	3	.99602	3	53
8	.08947	29	.1765	361	.08983	30	.1316	362	.00403	2	.99599	3	52
9	.08976	29	.1404	358	.09013	29	.0954	360	.00405	3	.99596	2	51
10	0.09005	29	11.1045	356	0.09042	29	11.0594	358	1.00408	3	0.99594	3	50
11	.09034	29	.0689	354	.09071	30	11.0237	355	.00411	2	.99591	3	49
12	.09063	29	11.0336	352	.09101	29	10.9882	353	.00413	3	.99588	2	48
13	.09092	29	10.9984	349	.09130	29	.9529	351	.00416	3	.99586	3	47
14	.09121	29	.9635	347	.09159	30	.9178	349	.00419	2	.99583	3	46
15	0.09150	29	10.9288	345	0.09189	29	10.8829	346	1.00421	3	0.99580	2	45
16	.09179	29	.8943	343	.09218	29	.8483	344	.00424	3	.99578	3	44
17	.09208	29	.8600	341	.09247	30	.8139	342	.00427	2	.99575	3	43
18	.09237	29	.8260	338	.09277	29	.7797	340	.00429	3	.99572	2	42
19	.09266	29	.7921	336	.09306	29	.7457	338	.00432	3	.99570	3	41
20	0.09295	29	10.7585	334	0.09335	30	10.7119	336	1.00435	3	0.99567	3	40
21	.09324	29	.7251	332	.09365	29	.6783	334	.00438	2	.99564	2	39
22	.09353	29	.6919	330	.09394	29	.6450	332	.00440	3	.99562	3	38
23	.09382	29	.6589	328	.09423	30	.6118	329	.00443	3	.99559	3	37
24	.09411	29	.6261	326	.09453	29	.5789	327	.00446	3	.99556	3	36
25	0.09440	29	10.5935	324	0.09482	29	10.5462	325	1.00449	2	0.99553	2	35
26	.09469	29	.5611	322	.09511	30	.5136	323	.00451	3	.99551	3	34
27	.09498	29	.5289	320	.09541	29	.4813	321	.00454	3	.99548	3	33
28	.09527	29	.4969	318	.09570	30	.4491	320	.00457	3	.99545	3	32
29	.09556	29	.4650	316	.09600	29	.4172	318	.00460	3	.99542	2	31
30	0.09585	29	10.4334	314	0.09629	29	10.3854	316	1.00463	2	0.99540	3	30
31	.09614	28	.4020	312	.09658	30	.3538	314	.00465	3	.99537	3	29
32	.09642	29	.3708	310	.09688	29	.3224	312	.00468	3	.99534	3	28
33	.09671	29	.3397	309	.09717	29	.2913	310	.00471	3	.99531	3	27
34	.09700	29	.3089	307	.09746	30	.2602	308	.00474	3	.99528	2	26
35	0.09729	29	10.2782	305	0.09776	29	10.2294	306	1.00477	3	0.99526	3	25
36	.09758	29	.2477	303	.09805	29	.1988	305	.00480	2	.99523	3	24
37	.09787	29	.2174	301	.09834	30	.1683	303	.00482	3	.99520	3	23
38	.09816	29	.1873	300	.09864	29	.1381	301	.00485	3	.99517	3	22
39	.09845	29	.1573	298	.09893	30	.1080	299	.00488	3	.99514	3	21
40	0.09874	29	10.1275	296	0.09923	29	10.0780	297	1.00491	3	0.99511	3	20
41	.09903	29	.0979	294	.09952	29	.0483	296	.00494	3	.99508	2	19
42	.09932	29	.0685	293	.09981	30	10.0187	2940	.00497	3	.99506	3	18
43	.09961	29	.0392	291	.10011	29	9.98931	2923	.00500	3	.99503	3	17
44	.09990	29	10.0101	2891	.10040	29	.96007	2907	.00503	3	.99500	3	16
45	0.10019	29	9.98123	2876	0.10069	30	9.93101	2890	1.00506	3	0.99497	3	15
46	.10048	29	.95248	2859	.10099	29	.90211	2873	.00509	3	.99494	3	14
47	.10077	29	.92389	2841	.10128	30	.87338	2857	.00512	3	.99491	3	13
48	.10106	29	.89547	2826	.10158	29	.84482	2840	.00515	3	.99488	3	12
49	.10135	29	.86722	2810	.10187	29	.81641	2824	.00518	3	.99485	3	11
50	0.10164	28	9.83912	2793	0.10216	30	9.78817	2809	1.00521	3	0.99482	3	10
51	.10192	29	.81119	2778	.10246	29	.76009	2792	.00524	3	.99479	3	9
52	.10221	29	.78341	2761	.10275	30	.73217	2777	.00527	3	.99476	3	8
53	.10250	29	.75579	2747	.10305	29	.70441	2760	.00530	3	.99473	3	7
54	.10279	29	.72833	2730	.10334	29	.67680	2746	.00533	3	.99470	3	6
55	0.10308	29	9.70103	2716	0.10363	30	9.64935	2730	1.00536	3	0.99467	3	5
56	.10337	29	.67387	2700	.10393	29	.62205	2714	.00539	3	.99464	3	4
57	.10366	29	.64687	2684	.10422	30	.59490	2700	.00542	3	.99461	3	3
58	.10395	29	.62002	2670	.10452	29	.56791	2684	.00545	3	.99458	3	2
59	.10424	29	.59332	2656	.10481	29	.54106	2670	.00548	3	.99455	3	1
60	0.10453		9.56677		0.10510		9.51436		1.00551		0.99452		0
↑ 95°→	cos	Diff. 1′	sec	Diff. 1′	cot	Diff. 1′	tan	Diff. 1′	csc	Diff. 1′	sin	Diff. 1′	← 84° ↑

TABLE 2
Natural Trigonometric Functions

6° → ↓ ′	sin	Diff. 1′	csc	Diff. 1′	tan	Diff. 1′	cot	Diff. 1′	sec	Diff. 1′	cos	Diff. 1′	← 173° ↓ ′
0	0.10453	29	9.56677	2640	0.10510	30	9.51436	2654	1.00551	3	0.99452	3	60
1	.10482	29	.54037	2626	.10540	29	.48781	2640	.00554	3	.99449	3	59
2	.10511	29	.51411	2611	.10569	30	.46141	2626	.00557	3	.99446	3	58
3	.10540	29	.48800	2597	.10599	29	.43515	2611	.00560	3	.99443	3	57
4	.10569	28	.46203	2582	.10628	29	.40904	2598	.00563	3	.99440	3	56
5	0.10597	29	9.43620	2569	0.10657	30	9.38307	2583	1.00566	3	0.99437	3	55
6	.10626	29	.41052	2554	.10687	29	.35724	2570	.00569	4	.99434	3	54
7	.10655	29	.38497	2540	.10716	30	.33155	2556	.00573	3	.99431	3	53
8	.10684	29	.35957	2527	.10746	29	.30599	2541	.00576	3	.99428	4	52
9	.10713	29	.33430	2513	.10775	30	.28058	2528	.00579	3	.99424	3	51
10	0.10742	29	9.30917	2500	0.10805	29	9.25530	2514	1.00582	3	0.99421	3	50
11	.10771	29	.28417	2487	.10834	29	.23016	2500	.00585	3	.99418	3	49
12	.10800	29	.25931	2472	.10863	30	.20516	2488	.00588	4	.99415	3	48
13	.10829	29	.23459	2460	.10893	29	.18028	2474	.00592	3	.99412	3	47
14	.10858	29	.20999	2447	.10922	30	.15554	2460	.00595	3	.99409	3	46
15	0.10887	29	9.18553	2433	0.10952	29	9.13093	2448	1.00598	3	0.99406	4	45
16	.10916	29	.16120	2420	.10981	30	.10646	2434	.00601	3	.99402	3	44
17	.10945	28	.13699	2408	.11011	29	.08211	2422	.00604	4	.99399	3	43
18	.10973	29	.11292	2394	.11040	30	.05789	2410	.00608	3	.99396	3	42
19	.11002	29	.08897	2382	.11070	29	.03379	2397	.00611	3	.99393	3	41
20	0.11031	29	9.06515	2370	0.11099	29	9.00983	2384	1.00614	3	0.99390	4	40
21	.11060	29	.04146	2358	.11128	30	8.98598	2371	.00617	4	.99386	3	39
22	.11089	29	9.01788	2344	.11158	29	.96227	2360	.00621	3	.99383	3	38
23	.11118	29	8.99444	2332	.11187	30	.93867	2348	.00624	3	.99380	3	37
24	.11147	29	.97111	2320	.11217	29	.91520	2336	.00627	3	.99377	3	36
25	0.11176	29	8.94791	2309	0.11246	30	8.89185	2322	1.00630	4	0.99374	4	35
26	.11205	29	.92482	2297	.11276	29	.86862	2311	.00634	3	.99370	3	34
27	.11234	29	.90186	2284	.11305	30	.84551	2300	.00637	3	.99367	3	33
28	.11263	28	.87901	2272	.11335	29	.82252	2288	.00640	4	.99364	4	32
29	.11291	29	.85628	2261	.11364	30	.79964	2276	.00644	3	.99360	3	31
30	0.11320	29	8.83367	2250	0.11394	29	8.77689	2264	1.00647	3	0.99357	3	30
31	.11349	29	.81118	2239	.11423	29	.75425	2252	.00650	4	.99354	3	29
32	.11378	29	.78880	2227	.11452	30	.73172	2241	.00654	3	.99351	4	28
33	.11407	29	.76653	2216	.11482	29	.70931	2230	.00657	3	.99347	3	27
34	.11436	29	.74438	2204	.11511	30	.68701	2219	.00660	4	.99344	3	26
35	0.11465	29	8.72234	2192	0.11541	29	8.66482	2208	1.00664	3	0.99341	4	25
36	.11494	29	.70041	2181	.11570	30	.64275	2197	.00667	4	.99337	3	24
37	.11523	29	.67859	2170	.11600	29	.62078	2186	.00671	3	.99334	3	23
38	.11552	28	.65688	2160	.11629	30	.59893	2174	.00674	3	.99331	4	22
39	.11580	29	.63528	2150	.11659	29	.57718	2163	.00677	4	.99327	3	21
40	0.11609	29	8.61379	2139	0.11688	30	8.55555	2152	1.00681	3	0.99324	4	20
41	.11638	29	.59241	2128	.11718	29	.53402	2142	.00684	4	.99320	3	19
42	.11667	29	.57113	2118	.11747	30	.51259	2131	.00688	3	.99317	3	18
43	.11696	29	.54996	2107	.11777	29	.49128	2121	.00691	4	.99314	4	17
44	.11725	29	.52889	2097	.11806	30	.47007	2110	.00695	3	.99310	3	16
45	0.11754	29	8.50793	2086	0.11836	29	8.44896	2100	1.00698	3	0.99307	4	15
46	.11783	29	.48707	2076	.11865	30	.42795	2090	.00701	4	.99303	3	14
47	.11812	28	.46632	2066	.11895	29	.40705	2080	.00705	3	.99300	3	13
48	.11840	29	.44566	2056	.11924	30	.38625	2070	.00708	4	.99297	4	12
49	.11869	29	.42511	2046	.11954	29	.36555	2060	.00712	3	.99293	3	11
50	0.11898	29	8.40466	2036	0.11983	30	8.34496	2050	1.00715	4	0.99290	4	10
51	.11927	29	.38431	2026	.12013	29	.32446	2040	.00719	3	.99286	3	9
52	.11956	29	.36405	2016	.12042	30	.30406	2030	.00722	4	.99283	4	8
53	.11985	29	.34390	2006	.12072	29	.28376	2020	.00726	4	.99279	3	7
54	.12014	29	.32384	1997	.12101	30	.26355	2010	.00730	3	.99276	4	6
55	0.12043	28	8.30388	1987	0.12131	29	8.24345	2001	1.00733	4	0.99272	3	5
56	.12071	29	.28402	1977	.12160	30	.22344	1991	.00737	3	.99269	4	4
57	.12100	29	.26425	1968	.12190	29	.20352	1981	.00740	4	.99265	3	3
58	.12129	29	.24457	1958	.12219	30	.18370	1972	.00744	3	.99262	4	2
59	.12158	29	.22500	1949	.12249	29	.16398	1963	.00747	4	.99258	3	1
60	0.12187		8.20551		0.12278		8.14435		1.00751		0.99255		0
↑ 96° →	cos	Diff. 1′	sec	Diff. 1′	cot	Diff. 1′	tan	Diff. 1′	csc	Diff. 1′	sin	Diff. 1′	↑ ← 83°

TABLE 2
Natural Trigonometric Functions

7° → ↓ ′	sin	Diff. 1′	csc	Diff. 1′	tan	Diff. 1′	cot	Diff. 1′	sec	Diff. 1′	cos	Diff. 1′	← 172° ↓ ′
0	0.12187	29	8.20551	1940	0.12278	30	8.14435	1953	1.00751	4	0.99255	4	60
1	.12216	29	.18612	1930	.12308	30	.12481	1944	.00755	3	.99251	3	59
2	.12245	29	.16681	1920	.12338	29	.10536	1936	.00758	4	.99248	4	58
3	.12274	28	.14760	1911	.12367	30	.08600	1927	.00762	3	.99244	4	57
4	.12302	29	.12849	1902	.12397	29	.06674	1918	.00765	4	.99240	3	56
5	0.12331	29	8.10946	1893	0.12426	30	8.04756	1909	1.00769	4	0.99237	4	55
6	.12360	29	.09052	1886	.12456	29	.02848	1900	.00773	3	.99233	3	54
7	.12389	29	.07167	1877	.12485	30	8.00948	1890	.00776	4	.99230	4	53
8	.12418	29	.05291	1868	.12515	29	7.99058	1882	.00780	4	.99226	4	52
9	.12447	29	.03423	1859	.12544	30	.97176	1873	.00784	3	.99222	3	51
10	0.12476	28	8.01565	1850	0.12574	29	7.95302	1864	1.00787	4	0.99219	4	50
11	.12504	29	7.99714	1841	.12603	30	.93438	1857	.00791	4	.99215	4	49
12	.12533	29	.97873	1832	.12633	29	.91582	1848	.00795	4	.99211	3	48
13	.12562	29	.96040	1824	.12662	30	.89734	1840	.00799	3	.99208	4	47
14	.12591	29	.94216	1817	.12692	30	.87895	1830	.00802	4	.99204	4	46
15	0.12620	29	7.92399	1808	0.12722	29	7.86064	1822	1.00806	4	0.99200	3	45
16	.12649	29	.90592	1800	.12751	30	.84242	1814	.00810	3	.99197	4	44
17	.12678	28	.88792	1791	.12781	29	.82428	1806	.00813	4	.99193	4	43
18	.12706	29	.87001	1782	.12810	30	.80622	1798	.00817	4	.99189	3	42
19	.12735	29	.85218	1774	.12840	29	.78825	1790	.00821	4	.99186	4	41
20	0.12764	29	7.83443	1767	0.12869	30	7.77035	1781	1.00825	3	0.99182	4	40
21	.12793	29	.81677	1759	.12899	30	.75254	1773	.00828	4	.99178	3	39
22	.12822	29	.79918	1750	.12929	29	.73480	1766	.00832	4	.99175	4	38
23	.12851	29	.78167	1742	.12958	30	.71715	1758	.00836	4	.99171	4	37
24	.12880	28	.76424	1736	.12988	29	.69957	1750	.00840	4	.99167	4	36
25	0.12908	29	7.74689	1728	0.13017	30	7.68208	1741	1.00844	4	0.99163	3	35
26	.12937	29	.72962	1720	.13047	29	.66466	1734	.00848	3	.99160	4	34
27	.12966	29	.71242	1711	.13076	30	.64732	1727	.00851	4	.99156	4	33
28	.12995	29	.69530	1704	.13106	30	.63005	1719	.00855	4	.99152	4	32
29	.13024	29	.67826	1697	.13136	29	.61287	1711	.00859	4	.99148	4	31
30	0.13053	28	7.66130	1690	0.13165	30	7.59575	1703	1.00863	4	0.99144	3	30
31	.13081	29	.64441	1681	.13195	29	.57872	1697	.00867	4	.99141	4	29
32	.13110	29	.62759	1674	.13224	30	.56176	1689	.00871	4	.99137	4	28
33	.13139	29	.61085	1667	.13254	30	.54487	1681	.00875	3	.99133	4	27
34	.13168	29	.59418	1660	.13284	29	.52806	1673	.00878	4	.99129	4	26
35	0.13197	29	7.57759	1652	0.13313	30	7.51132	1667	1.00882	4	0.99125	3	25
36	.13226	28	.56107	1644	.13343	29	.49465	1660	.00886	4	.99122	4	24
37	.13254	29	.54462	1638	.13372	30	.47806	1652	.00890	4	.99118	4	23
38	.13283	29	.52825	1630	.13402	30	.46154	1646	.00894	4	.99114	4	22
39	.13312	29	.51194	1623	.13432	29	.44509	1638	.00898	4	.99110	4	21
40	0.13341	29	7.49571	1617	0.13461	30	7.42871	1630	1.00902	4	0.99106	4	20
41	.13370	29	.47955	1610	.13491	30	.41240	1623	.00906	4	.99102	4	19
42	.13399	28	.46346	1602	.13521	29	.39616	1617	.00910	4	.99098	4	18
43	.13427	29	.44743	1596	.13550	30	.37999	1610	.00914	4	.99094	3	17
44	.13456	29	.43148	1589	.13580	29	.36389	1603	.00918	4	.99091	4	16
45	0.13485	29	7.41560	1581	0.13609	30	7.34786	1597	1.00922	4	0.99087	4	15
46	.13514	29	.39978	1574	.13639	30	.33190	1590	.00926	4	.99083	4	14
47	.13543	29	.38403	1569	.13669	29	.31600	1582	.00930	4	.99079	4	13
48	.13572	28	.36835	1561	.13698	30	.30018	1576	.00934	4	.99075	4	12
49	.13600	29	.35274	1554	.13728	30	.28442	1570	.00938	4	.99071	4	11
50	0.13629	29	7.33719	1549	0.13758	29	7.26873	1562	1.00942	4	0.99067	4	10
51	.13658	29	.32171	1541	.13787	30	.25310	1557	.00946	4	.99063	4	9
52	.13687	29	.30630	1534	.13817	29	.23754	1550	.00950	4	.99059	4	8
53	.13716	28	.29095	1529	.13846	30	.22204	1543	.00954	4	.99055	4	7
54	.13744	29	.27566	1521	.13876	30	.20661	1537	.00958	4	.99051	4	6
55	0.13773	29	7.26044	1516	0.13906	29	7.19125	1530	1.00962	4	0.99047	4	5
56	.13802	29	.24529	1510	.13935	30	.17594	1523	.00966	4	.99043	4	4
57	.13831	29	.23019	1502	.13965	30	.16071	1518	.00970	5	.99039	4	3
58	.13860	29	.21517	1497	.13995	29	.14553	1511	.00975	4	.99035	4	2
59	.13889	28	.20020	1490	.14024	30	.13042	1504	.00979	4	.99031	4	1
60	0.13917		7.18530		0.14054		7.11537		1.00983		0.99027		0
↑ 97° →	cos	Diff. 1′	sec	Diff. 1′	cot	Diff. 1′	tan	Diff. 1′	csc	Diff. 1′	sin	Diff. 1′	↑ ← 82°

TABLE 2
Natural Trigonometric Functions

8° → ↓ ′	sin	Diff. 1′	csc	Diff. 1′	tan	Diff. 1′	cot	Diff. 1′	sec	Diff. 1′	cos	Diff. 1′	← 171° ↓ ′
0	0.13917		7.18530		0.14054		7.11537		1.00983		0.99027		60
1	.13946	29	.17046	1484	.14084	30	.10038	1499	.00987	4	.99023	4	59
2	.13975	29	.15568	1478	.14113	29	.08546	1492	.00991	4	.99019	4	58
3	.14004	29	.14096	1471	.14143	30	.07059	1487	.00995	4	.99015	4	57
4	.14033	29	.12630	1466	.14173	30	.05579	1480	.00999	4	.99011	4	56
5	0.14061	28	7.11171	1460	0.14202	29	7.04105	1474	1.01004	5	0.99006	5	55
6	.14090	29	.09717	1453	.14232	30	.02637	1469	.01008	4	.99002	4	54
7	.14119	29	.08269	1448	.14262	30	7.01174	1462	.01012	4	.98998	4	53
8	.14148	29	.06828	1441	.14291	29	6.99718	1457	.01016	4	.98994	4	52
9	.14177	29	.05392	1436	.14321	30	.98268	1450	.01020	4	.98990	4	51
10	0.14205	28	7.03962	1430	0.14351	30	6.96823	1444	1.01024	4	0.98986	4	50
11	.14234	29	.02538	1423	.14381	30	.95385	1439	.01029	5	.98982	4	49
12	.14263	29	7.01120	1419	.14410	29	.93952	1432	.01033	4	.98978	4	48
13	.14292	29	6.99708	1412	.14440	30	.92525	1428	.01037	4	.98973	5	47
14	.14320	28	.98301	1407	.14470	30	.91104	1421	.01041	4	.98969	4	46
15	0.14349	29	6.96900	1400	0.14499	29	6.89688	1416	1.01046	5	0.98965	4	45
16	.14378	29	.95505	1396	.14529	30	.88278	1410	.01050	4	.98961	4	44
17	.14407	29	.94115	1390	.14559	30	.86874	1404	.01054	4	.98957	4	43
18	.14436	29	.92731	1384	.14588	29	.85475	1399	.01059	5	.98953	4	42
19	.14464	28	.91352	1379	.14618	30	.84082	1393	.01063	4	.98948	5	41
20	0.14493	29	6.89979	1372	0.14648	30	6.82694	1388	1.01067	4	0.98944	4	40
21	.14522	29	.88612	1368	.14678	30	.81312	1382	.01071	4	.98940	4	39
22	.14551	29	.87250	1362	.14707	29	.79936	1377	.01076	5	.98936	4	38
23	.14580	29	.85893	1357	.14737	30	.78564	1371	.01080	4	.98931	5	37
24	.14608	28	.84542	1351	.14767	30	.77199	1366	.01084	4	.98927	4	36
25	0.14637	29	6.83196	1346	0.14796	29	6.75838	1360	1.01089	5	0.98923	4	35
26	.14666	29	.81856	1340	.14826	30	.74483	1356	.01093	4	.98919	4	34
27	.14695	29	.80521	1336	.14856	30	.73133	1350	.01097	4	.98914	5	33
28	.14723	28	.79191	1330	.14886	30	.71789	1344	.01102	5	.98910	4	32
29	.14752	29	.77866	1324	.14915	29	.70450	1340	.01106	4	.98906	4	31
30	0.14781	29	6.76547	1320	0.14945	30	6.69116	1334	1.01111	5	0.98902	4	30
31	.14810	29	.75233	1314	.14975	30	.67787	1329	.01115	4	.98897	5	29
32	.14838	28	.73924	1310	.15005	30	.66463	1323	.01119	4	.98893	4	28
33	.14867	29	.72620	1303	.15034	29	.65144	1319	.01124	5	.98889	4	27
34	.14896	29	.71321	1299	.15064	30	.63831	1313	.01128	4	.98884	5	26
35	0.14925	29	6.70027	1293	0.15094	30	6.62523	1309	1.01133	5	0.98880	4	25
36	.14954	29	.68738	1289	.15124	30	.61219	1303	.01137	4	.98876	4	24
37	.14982	28	.67454	1283	.15153	29	.59921	1299	.01142	5	.98871	5	23
38	.15011	29	.66176	1279	.15183	30	.58627	1293	.01146	4	.98867	4	22
39	.15040	29	.64902	1273	.15213	30	.57339	1289	.01151	5	.98863	4	21
40	0.15069	29	6.63633	1269	0.15243	30	6.56055	1283	1.01155	4	0.98858	5	20
41	.15097	28	.62369	1264	.15272	29	.54777	1279	.01160	5	.98854	4	19
42	.15126	29	.61110	1260	.15302	30	.53503	1273	.01164	4	.98849	5	18
43	.15155	29	.59855	1254	.15332	30	.52234	1269	.01169	5	.98845	4	17
44	.15184	29	.58606	1250	.15362	30	.50970	1264	.01173	4	.98841	4	16
45	0.15212	28	6.57361	1244	0.15391	29	6.49710	1260	1.01178	5	0.98836	5	15
46	.15241	29	.56121	1240	.15421	30	.48456	1254	.01182	4	.98832	4	14
47	.15270	29	.54886	1236	.15451	30	.47206	1250	.01187	5	.98827	5	13
48	.15299	29	.53655	1230	.15481	30	.45961	1246	.01191	4	.98823	4	12
49	.15327	28	.52429	1226	.15511	30	.44720	1240	.01196	5	.98818	5	11
50	0.15356	29	6.51208	1221	0.15540	29	6.43484	1236	1.01200	4	0.98814	4	10
51	.15385	29	.49991	1217	.15570	30	.42253	1231	.01205	5	.98809	5	9
52	.15414	29	.48779	1212	.15600	30	.41026	1227	.01209	4	.98805	4	8
53	.15442	28	.47572	1208	.15630	30	.39804	1222	.01214	5	.98800	5	7
54	.15471	29	.46369	1202	.15660	30	.38587	1218	.01219	5	.98796	4	6
55	0.15500	29	6.45171	1199	0.15689	29	6.37374	1213	1.01223	4	0.98791	5	5
56	.15529	29	.43977	1193	.15719	30	.36165	1209	.01228	5	.98787	4	4
57	.15557	28	.42787	1190	.15749	30	.34961	1204	.01233	5	.98782	5	3
58	.15586	29	.41602	1186	.15779	30	.33761	1200	.01237	4	.98778	4	2
59	.15615	29	.40422	1180	.15809	30	.32566	1196	.01242	5	.98773	5	1
60	0.15643	28	6.39245	1177	0.15838	29	6.31375	1190	1.01247	5	0.98769	4	0
↑ 98° →	cos	Diff. 1′	sec	Diff. 1′	cot	Diff. 1′	tan	Diff. 1′	csc	Diff. 1′	sin	Diff. 1′ ←	↑ 81°

TABLE 2
Natural Trigonometric Functions

9° → ↓ ′	sin	Diff. 1′	csc	Diff. 1′	tan	Diff. 1′	cot	Diff. 1′	sec	Diff. 1′	cos	Diff. 1′	← 170° ↓ ′
0	0.15643		6.39245		0.15838		6.31375		1.01247		0.98769		60
1	.15672	29	.38073	1171	.15868	30	.30189	1187	.01251	4	.98764	5	59
2	.15701	29	.36906	1168	.15898	30	.29007	1182	.01256	5	.98760	4	58
3	.15730	29	.35743	1163	.15928	30	.27829	1178	.01261	5	.98755	5	57
4	.15758	28	.34584	1159	.15958	30	.26655	1173	.01265	4	.98751	4	56
5	0.15787	29	6.33429	1154	0.15988	30	6.25486	1170	1.01270	5	0.98746	5	55
6	.15816	29	.32279	1150	.16017	29	.24321	1166	.01275	5	.98741	5	54
7	.15845	29	.31133	1147	.16047	30	.23160	1160	.01279	4	.98737	4	53
8	.15873	28	.29991	1141	.16077	30	.22003	1157	.01284	5	.98732	5	52
9	.15902	29	.28853	1138	.16107	30	.20851	1152	.01289	5	.98728	4	51
10	0.15931	29	6.27719	1133	0.16137	30	6.19703	1149	1.01294	5	0.98723	5	50
11	.15959	28	.26590	1130	.16167	30	.18559	1144	.01298	4	.98718	5	49
12	.15988	29	.25464	1126	.16196	29	.17419	1140	.01303	5	.98714	4	48
13	.16017	29	.24343	1121	.16226	30	.16283	1136	.01308	5	.98709	5	47
14	.16046	29	.23226	1118	.16256	30	.15151	1131	.01313	5	.98704	5	46
15	0.16074	28	6.22113	1113	0.16286	30	6.14023	1128	1.01317	4	0.98700	4	45
16	.16103	29	.21004	1110	.16316	30	.12899	1123	.01322	5	.98695	5	44
17	.16132	29	.19898	1106	.16346	30	.11779	1120	.01327	5	.98690	5	43
18	.16160	28	.18797	1101	.16376	30	.10664	1116	.01332	5	.98686	4	42
19	.16189	29	.17700	1098	.16405	29	.09552	1111	.01337	5	.98681	5	41
20	0.16218	29	6.16607	1093	0.16435	30	6.08444	1108	1.01342	5	0.98676	5	40
21	.16246	28	.15517	1090	.16465	30	.07340	1104	.01346	4	.98671	5	39
22	.16275	29	.14432	1086	.16495	30	.06240	1100	.01351	5	.98667	4	38
23	.16304	29	.13350	1081	.16525	30	.05143	1097	.01356	5	.98662	5	37
24	.16333	29	.12273	1078	.16555	30	.04051	1092	.01361	5	.98657	5	36
25	0.16361	28	6.11199	1073	0.16585	30	6.02962	1089	1.01366	5	0.98652	5	35
26	.16390	29	.10129	1070	.16615	30	.01878	1084	.01371	5	.98648	4	34
27	.16419	29	.09062	1067	.16645	30	6.00797	1080	.01376	5	.98643	5	33
28	.16447	28	.08000	1062	.16674	29	5.99720	1078	.01381	5	.98638	5	32
29	.16476	29	.06941	1059	.16704	30	.98646	1073	.01386	5	.98633	5	31
30	0.16505	29	6.05886	1056	0.16734	30	5.97576	1070	1.01391	5	0.98629	4	30
31	.16533	28	.04834	1051	.16764	30	.96510	1066	.01395	4	.98624	5	29
32	.16562	29	.03787	1048	.16794	30	.95448	1062	.01400	5	.98619	5	28
33	.16591	29	.02743	1043	.16824	30	.94390	1059	.01405	5	.98614	5	27
34	.16620	29	.01702	1040	.16854	30	.93335	1054	.01410	5	.98609	5	26
35	0.16648	28	6.00666	1037	0.16884	30	5.92283	1051	1.01415	5	0.98604	5	25
36	.16677	29	5.99633	1033	.16914	30	.91236	1048	.01420	5	.98600	4	24
37	.16706	29	.98603	1030	.16944	30	.90191	1044	.01425	5	.98595	5	23
38	.16734	28	.97577	1026	.16974	30	.89151	1040	.01430	5	.98590	5	22
39	.16763	29	.96555	1022	.17004	30	.88114	1037	.01435	5	.98585	5	21
40	0.16792	29	5.95536	1019	0.17033	29	5.87080	1033	1.01440	5	0.98580	5	20
41	.16820	28	.94521	1016	.17063	30	.86051	1030	.01445	5	.98575	5	19
42	.16849	29	.93509	1011	.17093	30	.85024	1027	.01450	5	.98570	5	18
43	.16878	29	.92501	1009	.17123	30	.84001	1022	.01455	5	.98565	5	17
44	.16906	28	.91496	1004	.17153	30	.82982	1020	.01460	5	.98561	4	16
45	0.16935	29	5.90495	1001	0.17183	30	5.81966	1017	1.01466	6	0.98556	5	15
46	.16964	29	.89497	998	.17213	30	.80953	1012	.01471	5	.98551	5	14
47	.16992	28	.88502	994	.17243	30	.79944	1010	.01476	5	.98546	5	13
48	.17021	29	.87511	991	.17273	30	.78938	1006	.01481	5	.98541	5	12
49	.17050	29	.86524	988	.17303	30	.77936	1002	.01486	5	.98536	5	11
50	0.17078	28	5.85539	984	0.17333	30	5.76937	1000	1.01491	5	0.98531	5	10
51	.17107	29	.84558	981	.17363	30	.75941	996	.01496	5	.98526	5	9
52	.17136	29	.83581	978	.17393	30	.74949	992	.01501	5	.98521	5	8
53	.17164	28	.82606	974	.17423	30	.73960	990	.01506	5	.98516	5	7
54	.17193	29	.81635	971	.17453	30	.72974	986	.01512	6	.98511	5	6
55	0.17222	29	5.80667	968	0.17483	30	5.71992	982	1.01517	5	0.98506	5	5
56	.17250	28	.79703	964	.17513	30	.71013	980	.01522	5	.98501	5	4
57	.17279	29	.78742	961	.17543	30	.70037	976	.01527	5	.98496	5	3
58	.17308	29	.77783	959	.17573	30	.69064	972	.01532	5	.98491	5	2
59	.17336	28	.76829	954	.17603	30	.68094	970	.01537	5	.98486	5	1
60	0.17365	29	5.75877	951	0.17633	30	5.67128	967	1.01543	6	0.98481	5	0
↑ 99° →	cos	Diff. 1′	sec	Diff. 1′	cot	Diff. 1′	tan	Diff. 1′	csc	Diff. 1′	sin	Diff. 1′ ←	↑ 80°

TABLE 2
Natural Trigonometric Functions

10° → ↓ ′	sin	Diff. 1′	csc	Diff. 1′	tan	Diff. 1′	cot	Diff. 1′	sec	Diff. 1′	cos	Diff. 1′	← 169° ↓ ′
0	0.17365		5.75877		0.17633		5.67128		1.01543		0.98481		60
1	.17393	28	.74929	949	.17663	30	.66165	963	.01548	5	.98476	5	59
2	.17422	29	.73983	946	.17693	30	.65205	960	.01553	5	.98471	5	58
3	.17451	29	.73041	942	.17723	30	.64248	957	.01558	5	.98466	5	57
4	.17479	28	.72102	939	.17753	30	.63295	953	.01564	6	.98461	5	56
5	0.17508	29	5.71166	936	0.17783	30	5.62344	950	1.01569	5	0.98455	6	55
6	.17537	29	.70234	932	.17813	30	.61397	948	.01574	5	.98450	5	54
7	.17565	28	.69304	930	.17843	30	.60452	944	.01579	5	.98445	5	53
8	.17594	29	.68377	927	.17873	30	.59511	941	.01585	6	.98440	5	52
9	.17623	29	.67454	923	.17903	30	.58573	939	.01590	5	.98435	5	51
10	0.17651	28	5.66533	920	0.17933	30	5.57638	936	1.01595	5	0.98430	5	50
11	.17680	29	.65616	918	.17963	30	.56706	932	.01601	6	.98425	5	49
12	.17708	28	.64701	914	.17993	30	.55777	930	.01606	5	.98420	5	48
13	.17737	29	.63790	911	.18023	30	.54851	927	.01611	5	.98414	6	47
14	.17766	29	.62881	909	.18053	30	.53927	923	.01616	5	.98409	5	46
15	0.17794	28	5.61976	906	0.18083	30	5.53007	920	1.01622	6	0.98404	5	45
16	.17823	29	.61073	902	.18113	30	.52090	918	.01627	5	.98399	5	44
17	.17852	29	.60174	900	.18143	30	.51176	914	.01633	6	.98394	5	43
18	.17880	28	.59277	897	.18173	30	.50264	911	.01638	5	.98389	5	42
19	.17909	29	.58383	893	.18203	30	.49356	909	.01643	5	.98383	6	41
20	0.17937	28	5.57493	890	0.18233	30	5.48451	906	1.01649	6	0.98378	5	40
21	.17966	29	.56605	888	.18263	30	.47548	902	.01654	5	.98373	5	39
22	.17995	29	.55720	886	.18293	30	.46648	900	.01659	5	.98368	5	38
23	.18023	28	.54837	882	.18323	30	.45751	897	.01665	6	.98362	6	37
24	.18052	29	.53958	880	.18353	30	.44857	894	.01670	5	.98357	5	36
25	0.18081	29	5.53081	877	0.18384	31	5.43966	891	1.01676	6	0.98352	5	35
26	.18109	28	.52208	873	.18414	30	.43077	889	.01681	5	.98347	5	34
27	.18138	29	.51337	870	.18444	30	.42192	886	.01687	6	.98341	6	33
28	.18166	28	.50468	869	.18474	30	.41309	882	.01692	5	.98336	5	32
29	.18195	29	.49603	866	.18504	30	.40429	880	.01698	6	.98331	5	31
30	0.18224	29	5.48740	862	0.18534	30	5.39552	878	1.01703	5	0.98325	6	30
31	.18252	28	.47881	860	.18564	30	.38677	874	.01709	6	.98320	5	29
32	.18281	29	.47023	858	.18594	30	.37805	871	.01714	5	.98315	5	28
33	.18309	28	.46169	854	.18624	30	.36936	870	.01720	6	.98310	5	27
34	.18338	29	.45317	851	.18654	30	.36070	867	.01725	5	.98304	6	26
35	0.18367	29	5.44468	850	0.18684	30	5.35206	863	1.01731	6	0.98299	5	25
36	.18395	28	.43622	847	.18714	30	.34345	860	.01736	5	.98294	5	24
37	.18424	29	.42778	843	.18745	31	.33487	859	.01742	6	.98288	6	23
38	.18452	28	.41937	840	.18775	30	.32631	856	.01747	5	.98283	5	22
39	.18481	29	.41099	839	.18805	30	.31778	853	.01753	6	.98277	6	21
40	0.18509	28	5.40263	836	0.18835	30	5.30928	850	1.01758	5	0.98272	5	20
41	.18538	29	.39430	833	.18865	30	.30080	848	.01764	6	.98267	5	19
42	.18567	29	.38600	830	.18895	30	.29235	846	.01769	5	.98261	6	18
43	.18595	28	.37772	828	.18925	30	.28393	842	.01775	6	.98256	5	17
44	.18624	29	.36947	826	.18955	30	.27553	840	.01781	6	.98250	6	16
45	0.18652	28	5.36124	822	0.18986	31	5.26715	838	1.01786	5	0.98245	5	15
46	.18681	29	.35304	820	.19016	30	.25880	834	.01792	6	.98240	5	14
47	.18710	29	.34486	818	.19046	30	.25048	832	.01798	6	.98234	6	13
48	.18738	28	.33671	816	.19076	30	.24218	830	.01803	5	.98229	5	12
49	.18767	29	.32859	812	.19106	30	.23391	828	.01809	6	.98223	6	11
50	0.18795	28	5.32049	810	0.19136	30	5.22566	824	1.01815	6	0.98218	5	10
51	.18824	29	.31241	808	.19166	30	.21744	822	.01820	5	.98212	6	9
52	.18852	28	.30436	806	.19197	31	.20925	820	.01826	6	.98207	5	8
53	.18881	29	.29634	802	.19227	30	.20107	818	.01832	6	.98201	6	7
54	.18910	29	.28833	800	.19257	30	.19293	814	.01837	5	.98196	5	6
55	0.18938	28	5.28036	798	0.19287	30	5.18480	812	1.01843	6	0.98190	6	5
56	.18967	29	.27241	796	.19317	30	.17671	810	.01849	6	.98185	5	4
57	.18995	28	.26448	792	.19347	30	.16863	808	.01854	5	.98179	6	3
58	.19024	29	.25658	790	.19378	31	.16058	804	.01860	6	.98174	5	2
59	.19052	28	.24870	788	.19408	30	.15256	802	.01866	6	.98168	6	1
60	0.19081	29	5.24084	786	0.19438	30	5.14455	800	1.01872	6	0.98163	5	0
↑ 100° →	cos	Diff. 1′	sec	Diff. 1′	cot	Diff. 1′	tan	Diff. 1′	csc	Diff. 1′	sin	Diff. 1′ ←	↑ 79°

TABLE 2
Natural Trigonometric Functions

11° → ↓ ′	sin	Diff. 1′	csc	Diff. 1′	tan	Diff. 1′	cot	Diff. 1′	sec	Diff. 1′	cos	Diff. 1′	← 168° ↓ ′
0	0.19081		5.24084		0.19438		5.14455		1.01872		0.98163		60
1	.19109	28	.23301	783	.19468	30	.13658	798	.01877	5	.98157	6	59
2	.19138	29	.22521	780	.19498	30	.12862	796	.01883	6	.98152	5	58
3	.19167	29	.21742	779	.19529	31	.12069	793	.01889	6	.98146	6	57
4	.19195	28	.20966	776	.19559	30	.11279	790	.01895	6	.98140	6	56
5	0.19224	29	5.20193	773	0.19589	30	5.10490	789	1.01901	6	0.98135	5	55
6	.19252	28	.19421	771	.19619	30	.09704	786	.01906	5	.98129	6	54
7	.19281	29	.18652	769	.19649	30	.08921	783	.01912	6	.98124	5	53
8	.19309	28	.17886	767	.19680	31	.08139	781	.01918	6	.98118	6	52
9	.19338	29	.17121	764	.19710	30	.07360	780	.01924	6	.98112	6	51
10	0.19366	28	5.16359	762	0.19740	30	5.06584	777	1.01930	6	0.98107	5	50
11	.19395	29	.15599	760	.19770	30	.05809	774	.01936	6	.98101	6	49
12	.19423	28	.14842	758	.19801	31	.05037	772	.01941	5	.98096	5	48
13	.19452	29	.14087	756	.19831	30	.04267	770	.01947	6	.98090	6	47
14	.19481	29	.13334	752	.19861	30	.03499	768	.01953	6	.98084	6	46
15	0.19509	28	5.12583	750	0.19891	30	5.02734	766	1.01959	6	0.98079	5	45
16	.19538	29	.11835	749	.19921	30	.01971	763	.01965	6	.98073	6	44
17	.19566	28	.11088	747	.19952	31	.01210	760	.01971	6	.98067	6	43
18	.19595	29	.10344	744	.19982	30	5.00451	759	.01977	6	.98061	6	42
19	.19623	28	.09602	741	.20012	30	4.99695	757	.01983	6	.98056	5	41
20	0.19652	29	5.08863	740	0.20042	30	4.98940	754	1.01989	6	0.98050	6	40
21	.19680	28	.08125	738	.20073	31	.98188	752	.01995	6	.98044	6	39
22	.19709	29	.07390	736	.20103	30	.97438	750	.02001	6	.98039	5	38
23	.19737	28	.06657	733	.20133	30	.96690	748	.02007	6	.98033	6	37
24	.19766	29	.05926	730	.20164	31	.95945	746	.02013	6	.98027	6	36
25	0.19794	28	5.05197	729	0.20194	30	4.95201	743	1.02019	6	0.98021	6	35
26	.19823	29	.04471	727	.20224	30	.94460	741	.02025	6	.98016	5	34
27	.19851	28	.03746	724	.20254	30	.93721	740	.02031	6	.98010	6	33
28	.19880	29	.03024	722	.20285	31	.92984	738	.02037	6	.98004	6	32
29	.19908	28	.02303	720	.20315	30	.92249	734	.02043	6	.97998	6	31
30	0.19937	29	5.01585	719	0.20345	30	4.91516	732	1.02049	6	0.97992	6	30
31	.19965	28	.00869	717	.20376	31	.90785	730	.02055	6	.97987	5	29
32	.19994	29	5.00155	714	.20406	30	.90056	729	.02061	6	.97981	6	28
33	.20022	28	4.99443	711	.20436	30	.89330	727	.02067	6	.97975	6	27
34	.20051	29	.98733	710	.20466	30	.88605	724	.02073	6	.97969	6	26
35	0.20079	28	4.98025	708	0.20497	31	4.87882	722	1.02079	6	0.97963	6	25
36	.20108	29	.97320	706	.20527	30	.87162	720	.02085	6	.97958	5	24
37	.20136	28	.96616	703	.20557	30	.86444	719	.02091	6	.97952	6	23
38	.20165	29	.95914	701	.20588	31	.85727	717	.02097	6	.97946	6	22
39	.20193	28	.95215	700	.20618	30	.85013	714	.02103	6	.97940	6	21
40	0.20222	29	4.94517	698	0.20648	30	4.84300	712	1.02110	7	0.97934	6	20
41	.20250	28	.93821	696	.20679	31	.83590	710	.02116	6	.97928	6	19
42	.20279	29	.93128	693	.20709	30	.82882	709	.02122	6	.97922	6	18
43	.20307	28	.92436	691	.20739	30	.82175	707	.02128	6	.97916	6	17
44	.20336	29	.91746	690	.20770	31	.81471	704	.02134	6	.97910	6	16
45	0.20364	28	4.91058	688	0.20800	30	4.80769	702	1.02140	6	0.97905	5	15
46	.20393	29	.90373	686	.20830	30	.80068	700	.02146	6	.97899	6	14
47	.20421	28	.89689	683	.20861	31	.79370	699	.02153	7	.97893	6	13
48	.20450	29	.89007	681	.20891	30	.78673	697	.02159	6	.97887	6	12
49	.20478	28	.88327	680	.20921	30	.77978	694	.02165	6	.97881	6	11
50	0.20507	29	4.87649	678	0.20952	31	4.77286	692	1.02171	6	0.97875	6	10
51	.20535	28	.86973	677	.20982	30	.76595	690	.02178	7	.97869	6	9
52	.20563	28	.86299	674	.21013	31	.75906	689	.02184	6	.97863	6	8
53	.20592	29	.85627	672	.21043	30	.75219	687	.02190	6	.97857	6	7
54	.20620	28	.84956	670	.21073	30	.74534	686	.02196	6	.97851	6	6
55	0.20649	29	4.84288	669	0.21104	31	4.73851	683	1.02203	7	0.97845	6	5
56	.20677	28	.83621	667	.21134	30	.73170	681	.02209	6	.97839	6	4
57	.20706	29	.82956	664	.21164	30	.72490	680	.02215	6	.97833	6	3
58	.20734	28	.82294	662	.21195	31	.71813	678	.02221	6	.97827	6	2
59	.20763	29	.81633	660	.21225	30	.71137	676	.02228	7	.97821	6	1
60	0.20791	28	4.80973	660	0.21256	31	4.70463	673	1.02234	6	0.97815	6	0
↑ 101° →	cos	Diff. 1′	sec	Diff. 1′	cot	Diff. 1′	tan	Diff. 1′	csc	Diff. 1′	sin	Diff. 1′ ←	↑ 78°

TABLE 2
Natural Trigonometric Functions

12°→ ↓ ′	sin	Diff. 1′	csc	Diff. 1′	tan	Diff. 1′	cot	Diff. 1′	sec	Diff. 1′	cos	Diff. 1′	← 167° ↓ ′
0	0.20791	29	4.80973	658	0.21256	30	4.70463	672	1.02234	6	0.97815	6	60
1	.20820	28	.80316	656	.21286	30	.69791	670	.02240	7	.97809	6	59
2	.20848	29	.79661	653	.21316	31	.69121	669	.02247	6	.97803	6	58
3	.20877	28	.79007	651	.21347	30	.68452	667	.02253	6	.97797	6	57
4	.20905	28	.78355	650	.21377	31	.67786	664	.02259	7	.97791	7	56
5	0.20933	29	4.77705	649	0.21408	30	4.67121	662	1.02266	6	0.97784	6	55
6	.20962	28	.77057	647	.21438	31	.66458	661	.02272	7	.97778	6	54
7	.20990	29	.76411	644	.21469	30	.65797	660	.02279	6	.97772	6	53
8	.21019	28	.75766	642	.21499	30	.65138	658	.02285	6	.97766	6	52
9	.21047	29	.75123	641	.21529	31	.64480	656	.02291	7	.97760	6	51
10	0.21076	28	4.74482	640	0.21560	30	4.63825	654	1.02298	6	0.97754	6	50
11	.21104	28	.73843	638	.21590	31	.63171	652	.02304	7	.97748	6	49
12	.21132	29	.73205	636	.21621	30	.62518	650	.02311	6	.97742	7	48
13	.21161	28	.72569	634	.21651	31	.61868	649	.02317	6	.97735	6	47
14	.21189	29	.71935	632	.21682	30	.61219	648	.02323	7	.97729	6	46
15	0.21218	28	4.71303	630	0.21712	31	4.60572	646	1.02330	6	0.97723	6	45
16	.21246	29	.70673	629	.21743	30	.59927	643	.02336	7	.97717	6	44
17	.21275	28	.70044	628	.21773	31	.59283	641	.02343	6	.97711	6	43
18	.21303	28	.69417	626	.21804	30	.58641	640	.02349	7	.97705	7	42
19	.21331	29	.68791	623	.21834	30	.58001	639	.02356	6	.97698	6	41
20	0.21360	28	4.68167	622	0.21864	31	4.57363	637	1.02362	7	0.97692	6	40
21	.21388	29	.67545	620	.21895	30	.56726	636	.02369	6	.97686	6	39
22	.21417	28	.66925	619	.21925	31	.56091	633	.02375	7	.97680	7	38
23	.21445	29	.66307	617	.21956	30	.55458	631	.02382	6	.97673	6	37
24	.21474	28	.65690	616	.21986	31	.54826	630	.02388	7	.97667	6	36
25	0.21502	28	4.65074	613	0.22017	30	4.54196	629	1.02395	7	0.97661	6	35
26	.21530	29	.64461	611	.22047	31	.53568	627	.02402	6	.97655	7	34
27	.21559	28	.63849	610	.22078	30	.52941	626	.02408	7	.97648	6	33
28	.21587	29	.63238	609	.22108	31	.52316	623	.02415	6	.97642	6	32
29	.21616	28	.62630	608	.22139	30	.51693	621	.02421	7	.97636	6	31
30	0.21644	28	4.62023	606	0.22169	31	4.51071	620	1.02428	7	0.97630	7	30
31	.21672	29	.61417	603	.22200	31	.50451	619	.02435	6	.97623	6	29
32	.21701	28	.60813	602	.22231	30	.49832	617	.02441	7	.97617	6	28
33	.21729	29	.60211	600	.22261	31	.49215	616	.02448	6	.97611	7	27
34	.21758	28	.59611	599	.22292	30	.48600	613	.02454	7	.97604	6	26
35	0.21786	28	4.59012	598	0.22322	31	4.47986	612	1.02461	7	0.97598	6	25
36	.21814	29	.58414	596	.22353	30	.47374	610	.02468	6	.97592	7	24
37	.21843	28	.57819	594	.22383	31	.46764	609	.02474	7	.97585	6	23
38	.21871	28	.57224	592	.22414	30	.46155	608	.02481	7	.97579	6	22
39	.21899	29	.56632	591	.22444	31	.45548	606	.02488	6	.97573	7	21
40	0.21928	28	4.56041	590	0.22475	30	4.44942	604	1.02494	7	0.97566	6	20
41	.21956	29	.55451	588	.22505	31	.44338	602	.02501	7	.97560	7	19
42	.21985	28	.54863	587	.22536	31	.43735	601	.02508	7	.97553	6	18
43	.22013	28	.54277	584	.22567	30	.43134	600	.02515	6	.97547	6	17
44	.22041	29	.53692	583	.22597	31	.42534	598	.02521	7	.97541	7	16
45	0.22070	28	4.53109	581	0.22628	30	4.41936	597	1.02528	7	0.97534	6	15
46	.22098	28	.52527	580	.22658	31	.41340	594	.02535	7	.97528	7	14
47	.22126	29	.51947	579	.22689	30	.40745	593	.02542	6	.97521	6	13
48	.22155	28	.51368	578	.22719	31	.40152	591	.02548	7	.97515	7	12
49	.22183	29	.50791	576	.22750	31	.39560	590	.02555	7	.97508	6	11
50	0.22212	28	4.50216	574	0.22781	30	4.38969	589	1.02562	7	0.97502	6	10
51	.22240	28	.49642	572	.22811	31	.38381	588	.02569	7	.97496	7	9
52	.22268	29	.49069	571	.22842	30	.37793	586	.02576	6	.97489	6	8
53	.22297	28	.48498	570	.22872	31	.37207	584	.02582	7	.97483	7	7
54	.22325	28	.47928	569	.22903	31	.36623	582	.02589	7	.97476	6	6
55	0.22353	29	4.47360	567	0.22934	30	4.36040	581	1.02596	7	0.97470	7	5
56	.22382	28	.46793	566	.22964	31	.35459	580	.02603	7	.97463	6	4
57	.22410	28	.46228	563	.22995	31	.34879	579	.02610	7	.97457	7	3
58	.22438	29	.45664	562	.23026	30	.34300	578	.02617	7	.97450	6	2
59	.22467	28	.45102	560	.23056	31	.33723	576	.02624	6	.97444	7	1
60	0.22495		4.44541		0.23087		4.33148		1.02630		0.97437		0
↑ 102°→	cos	Diff. 1′	sec	Diff. 1′	cot	Diff. 1′	tan	Diff. 1′	csc	Diff. 1′	sin	Diff. 1′←	↑ 77°

TABLE 2
Natural Trigonometric Functions

13°→ ↓ ′	sin	Diff. 1′	csc	Diff. 1′	tan	Diff. 1′	cot	Diff. 1′	sec	Diff. 1′	cos	Diff. 1′	← 166° ↓ ′
0	0.22495	28	4.44541	560	0.23087	30	4.33148	574	1.02630	7	0.97437	7	60
1	.22523	29	.43982	558	.23117	31	.32573	572	.02637	7	.97430	6	59
2	.22552	28	.43424	557	.23148	31	.32001	571	.02644	7	.97424	7	58
3	.22580	28	.42867	556	.23179	30	.31430	570	.02651	7	.97417	6	57
4	.22608	29	.42312	553	.23209	31	.30860	569	.02658	7	.97411	7	56
5	0.22637	28	4.41759	552	0.23240	31	4.30291	567	1.02665	7	0.97404	6	55
6	.22665	28	.41206	550	.23271	30	.29724	566	.02672	7	.97398	7	54
7	.22693	29	.40656	550	.23301	31	.29159	564	.02679	7	.97391	7	53
8	.22722	28	.40106	548	.23332	31	.28595	562	.02686	7	.97384	6	52
9	.22750	28	.39558	547	.23363	30	.28032	561	.02693	7	.97378	7	51
10	0.22778	29	4.39012	546	0.23393	31	4.27471	560	1.02700	7	0.97371	6	50
11	.22807	28	.38466	543	.23424	31	.26911	559	.02707	7	.97365	7	49
12	.22835	28	.37923	542	.23455	30	.26352	558	.02714	7	.97358	7	48
13	.22863	29	.37380	541	.23485	31	.25795	556	.02721	7	.97351	6	47
14	.22892	28	.36839	540	.23516	31	.25239	554	.02728	7	.97345	7	46
15	0.22920	28	4.36299	539	0.23547	31	4.24685	553	1.02735	7	0.97338	7	45
16	.22948	29	.35761	537	.23578	30	.24132	551	.02742	7	.97331	6	44
17	.22977	28	.35224	536	.23608	31	.23580	550	.02749	7	.97325	7	43
18	.23005	28	.34689	534	.23639	31	.23030	549	.02756	7	.97318	7	42
19	.23033	29	.34154	532	.23670	30	.22481	548	.02763	7	.97311	7	41
20	0.23062	28	4.33622	531	0.23700	31	4.21933	547	1.02770	7	0.97304	6	40
21	.23090	28	.33090	530	.23731	31	.21387	544	.02777	7	.97298	7	39
22	.23118	28	.32560	529	.23762	31	.20842	543	.02784	7	.97291	7	38
23	.23146	29	.32031	528	.23793	30	.20298	542	.02791	8	.97284	6	37
24	.23175	28	.31503	527	.23823	31	.19756	540	.02799	7	.97278	7	36
25	0.23203	28	4.30977	524	0.23854	31	4.19215	540	1.02806	7	0.97271	7	35
26	.23231	29	.30452	523	.23885	31	.18675	539	.02813	7	.97264	7	34
27	.23260	28	.29929	522	.23916	30	.18137	538	.02820	7	.97257	6	33
28	.23288	28	.29406	520	.23946	31	.17600	536	.02827	7	.97251	7	32
29	.23316	29	.28885	520	.23977	31	.17064	534	.02834	8	.97244	7	31
30	0.23345	28	4.28366	519	0.24008	31	4.16530	533	1.02842	7	0.97237	7	30
31	.23373	28	.27847	518	.24039	30	.15997	531	.02849	7	.97230	7	29
32	.23401	28	.27330	516	.24069	31	.15465	530	.02856	7	.97223	6	28
33	.23429	29	.26814	514	.24100	31	.14934	530	.02863	7	.97217	7	27
34	.23458	28	.26300	513	.24131	31	.14405	528	.02870	8	.97210	7	26
35	0.23486	28	4.25787	511	0.24162	31	4.13877	527	1.02878	7	0.97203	7	25
36	.23514	28	.25275	510	.24193	30	.13350	526	.02885	7	.97196	7	24
37	.23542	29	.24764	510	.24223	31	.12825	524	.02892	7	.97189	7	23
38	.23571	28	.24255	509	.24254	31	.12301	522	.02899	8	.97182	6	22
39	.23599	28	.23746	507	.24285	31	.11778	521	.02907	7	.97176	7	21
40	0.23627	29	4.23239	506	0.24316	31	4.11256	520	1.02914	7	0.97169	7	20
41	.23656	28	.22734	504	.24347	30	.10736	520	.02921	7	.97162	7	19
42	.23684	28	.22229	503	.24377	31	.10216	518	.02928	8	.97155	7	18
43	.23712	28	.21726	501	.24408	31	.09699	517	.02936	7	.97148	7	17
44	.23740	29	.21224	500	.24439	31	.09182	516	.02943	7	.97141	7	16
45	0.23769	28	4.20723	500	0.24470	31	4.08666	514	1.02950	8	0.97134	7	15
46	.23797	28	.20224	499	.24501	31	.08152	513	.02958	7	.97127	7	14
47	.23825	28	.19725	498	.24532	30	.07639	511	.02965	7	.97120	7	13
48	.23853	29	.19228	496	.24562	31	.07127	510	.02972	8	.97113	7	12
49	.23882	28	.18733	494	.24593	31	.06616	510	.02980	7	.97106	6	11
50	0.23910	28	4.18238	493	0.24624	31	4.06107	509	1.02987	7	0.97100	7	10
51	.23938	28	.17744	492	.24655	31	.05599	508	.02994	8	.97093	7	9
52	.23966	29	.17252	491	.24686	31	.05092	506	.03002	7	.97086	7	8
53	.23995	28	.16761	490	.24717	30	.04586	504	.03009	8	.97079	7	7
54	.24023	28	.16271	489	.24747	31	.04081	503	.03017	7	.97072	7	6
55	0.24051	28	4.15782	488	0.24778	31	4.03578	502	1.03024	8	0.97065	7	5
56	.24079	29	.15295	487	.24809	31	.03076	501	.03032	7	.97058	7	4
57	.24108	28	.14809	486	.24840	31	.02574	500	.03039	7	.97051	7	3
58	.24136	28	.14323	484	.24871	31	.02074	499	.03046	8	.97044	7	2
59	.24164	28	.13839	482	.24902	31	.01576	498	.03054	7	.97037	7	1
60	0.24192		4.13357		0.24933		4.01078		1.03061		0.97030		0
↑ 103°→	cos	Diff. 1′	sec	Diff. 1′	cot	Diff. 1′	tan	Diff. 1′	csc	Diff. 1′	sin	Diff. 1′←	↑ 76°

TABLE 2
Natural Trigonometric Functions

14°→ ↓ ′	sin	Diff. 1′	csc	Diff. 1′	tan	Diff. 1′	cot	Diff. 1′	sec	Diff. 1′	cos	Diff. 1′	← 165° ↓ ′
0	0.24192		4.13357		0.24933		4.01078		1.03061		0.97030		60
1	.24220	28	.12875	481	.24964	31	.00582	497	.03069	8	.97023	7	59
2	.24249	29	.12394	480	.24995	31	4.00086	496	.03076	7	.97015	8	58
3	.24277	28	.11915	480	.25026	31	3.99592	494	.03084	8	.97008	7	57
4	.24305	28	.11437	479	.25056	30	.99099	492	.03091	7	.97001	7	56
5	0.24333	28	4.10960	478	0.25087	31	3.98607	491	1.03099	8	0.96994	7	55
6	.24362	29	.10484	476	.25118	31	.98117	490	.03106	7	.96987	7	54
7	.24390	28	.10009	474	.25149	31	.97627	490	.03114	8	.96980	7	53
8	.24418	28	.09535	473	.25180	31	.97139	489	.03121	7	.96973	7	52
9	.24446	28	.09063	472	.25211	31	.96651	488	.03129	8	.96966	7	51
10	0.24474	28	4.08591	471	0.25242	31	3.96165	487	1.03137	8	0.96959	7	50
11	.24503	29	.08121	470	.25273	31	.95680	486	.03144	7	.96952	7	49
12	.24531	28	.07652	470	.25304	31	.95196	483	.03152	8	.96945	7	48
13	.24559	28	.07184	469	.25335	31	.94713	482	.03159	7	.96937	8	47
14	.24587	28	.06717	467	.25366	31	.94232	481	.03167	8	.96930	7	46
15	0.24615	28	4.06251	466	0.25397	31	3.93751	480	1.03175	8	0.96923	7	45
16	.24644	29	.05786	464	.25428	31	.93271	480	.03182	7	.96916	7	44
17	.24672	28	.05322	463	.25459	31	.92793	479	.03190	8	.96909	7	43
18	.24700	28	.04860	462	.25490	31	.92316	478	.03197	7	.96902	7	42
19	.24728	28	.04398	461	.25521	31	.91839	477	.03205	8	.96894	8	41
20	0.24756	28	4.03938	460	0.25552	31	3.91364	476	1.03213	8	0.96887	7	40
21	.24784	28	.03479	460	.25583	31	.90890	474	.03220	7	.96880	7	39
22	.24813	29	.03020	459	.25614	31	.90417	473	.03228	8	.96873	7	38
23	.24841	28	.02563	458	.25645	31	.89945	471	.03236	8	.96866	7	37
24	.24869	28	.02107	457	.25676	31	.89474	470	.03244	8	.96858	8	36
25	0.24897	28	4.01652	456	0.25707	31	3.89004	470	1.03251	7	0.96851	7	35
26	.24925	28	.01198	453	.25738	31	.88536	469	.03259	8	.96844	7	34
27	.24954	29	.00745	452	.25769	31	.88068	468	.03267	8	.96837	7	33
28	.24982	28	4.00293	451	.25800	31	.87601	467	.03275	8	.96829	8	32
29	.25010	28	3.99843	450	.25831	31	.87136	466	.03282	7	.96822	7	31
30	0.25038	28	3.99393	450	0.25862	31	3.86671	464	1.03290	8	0.96815	7	30
31	.25066	28	.98944	449	.25893	31	.86208	463	.03298	8	.96807	8	29
32	.25094	28	.98497	448	.25924	31	.85745	462	.03306	8	.96800	7	28
33	.25122	28	.98050	447	.25955	31	.85284	461	.03313	7	.96793	7	27
34	.25151	29	.97604	446	.25986	31	.84824	460	.03321	8	.96786	7	26
35	0.25179	28	3.97160	444	0.26017	31	3.84364	460	1.03329	8	0.96778	8	25
36	.25207	28	.96716	443	.26048	31	.83906	459	.03337	8	.96771	7	24
37	.25235	28	.96274	442	.26079	31	.83449	458	.03345	8	.96764	7	23
38	.25263	28	.95832	441	.26110	31	.82992	457	.03353	8	.96756	8	22
39	.25291	28	.95392	440	.26141	31	.82537	456	.03360	7	.96749	7	21
40	0.25320	29	3.94952	440	0.26172	31	3.82083	454	1.03368	8	0.96742	7	20
41	.25348	28	.94514	439	.26203	31	.81630	453	.03376	8	.96734	8	19
42	.25376	28	.94076	438	.26235	32	.81177	452	.03384	8	.96727	7	18
43	.25404	28	.93640	437	.26266	31	.80726	451	.03392	8	.96719	8	17
44	.25432	28	.93204	436	.26297	31	.80276	450	.03400	8	.96712	7	16
45	0.25460	28	3.92770	434	0.26328	31	3.79827	450	1.03408	8	0.96705	7	15
46	.25488	28	.92337	433	.26359	31	.79378	449	.03416	8	.96697	8	14
47	.25516	28	.91904	432	.26390	31	.78931	448	.03424	8	.96690	7	13
48	.25545	29	.91473	431	.26421	31	.78485	447	.03432	8	.96682	8	12
49	.25573	28	.91042	430	.26452	31	.78040	446	.03439	7	.96675	7	11
50	0.25601	28	3.90613	430	0.26483	31	3.77595	444	1.03447	8	0.96667	8	10
51	.25629	28	.90184	429	.26515	32	.77152	443	.03455	8	.96660	7	9
52	.25657	28	.89756	428	.26546	31	.76709	442	.03463	8	.96653	7	8
53	.25685	28	.89330	427	.26577	31	.76268	441	.03471	8	.96645	8	7
54	.25713	28	.88904	426	.26608	31	.75828	440	.03479	8	.96638	7	6
55	0.25741	28	3.88479	424	0.26639	31	3.75388	440	1.03487	8	0.96630	8	5
56	.25769	28	.88056	423	.26670	31	.74950	439	.03495	8	.96623	7	4
57	.25798	29	.87633	422	.26701	31	.74512	438	.03503	8	.96615	8	3
58	.25826	28	.87211	421	.26733	32	.74075	437	.03511	8	.96608	7	2
59	.25854	28	.86790	420	.26764	31	.73640	436	.03520	9	.96600	8	1
60	0.25882	28	3.86370	420	0.26795	31	3.73205	434	1.03528	8	0.96593	7	0
↑ 104°→	cos	Diff. 1′	sec	Diff. 1′	cot	Diff. 1′	tan	Diff. 1′	csc	Diff. 1′	sin	Diff. 1′←	↑ 75°

TABLE 2
Natural Trigonometric Functions

15°→ ↓ ′	sin	Diff. 1′	csc	Diff. 1′	tan	Diff. 1′	cot	Diff. 1′	sec	Diff. 1′	cos	Diff. 1′	← 164° ↓ ′
0	0.25882		3.86370		0.26795		3.73205		1.03528		0.96593		60
1	.25910	28	.85951	419	.26826	31	.72771	433	.03536	8	.96585	8	59
2	.25938	28	.85533	419	.26857	31	.72338	432	.03544	8	.96578	7	58
3	.25966	28	.85116	418	.26888	31	.71907	431	.03552	8	.96570	8	57
4	.25994	28	.84700	417	.26920	32	.71476	430	.03560	8	.96562	8	56
5	0.26022	28	3.84285	416	0.26951	31	3.71046	430	1.03568	8	0.96555	7	55
6	.26050	28	.83871	414	.26982	31	.70616	430	.03576	8	.96547	8	54
7	.26079	29	.83457	413	.27013	31	.70188	429	.03584	8	.96540	7	53
8	.26107	28	.83045	412	.27044	31	.69761	428	.03592	8	.96532	8	52
9	.26135	28	.82633	411	.27076	32	.69335	427	.03601	9	.96524	8	51
10	0.26163	28	3.82223	410	0.27107	31	3.68909	426	1.03609	8	0.96517	7	50
11	.26191	28	.81813	410	.27138	31	.68485	424	.03617	8	.96509	8	49
12	.26219	28	.81404	409	.27169	31	.68061	423	.03625	8	.96502	7	48
13	.26247	28	.80996	408	.27201	32	.67638	422	.03633	8	.96494	8	47
14	.26275	28	.80589	407	.27232	31	.67217	421	.03642	9	.96486	8	46
15	0.26303	28	3.80183	407	0.27263	31	3.66796	420	1.03650	8	0.96479	7	45
16	.26331	28	.79778	406	.27294	31	.66376	420	.03658	8	.96471	8	44
17	.26359	28	.79374	404	.27326	32	.65957	420	.03666	8	.96463	8	43
18	.26387	28	.78970	403	.27357	31	.65538	419	.03674	8	.96456	7	42
19	.26415	28	.78568	402	.27388	31	.65121	418	.03683	9	.96448	8	41
20	0.26443	28	3.78166	401	0.27419	31	3.64705	417	1.03691	8	0.96440	8	40
21	.26471	28	.77765	400	.27451	32	.64289	416	.03699	8	.96433	7	39
22	.26500	29	.77365	400	.27482	31	.63874	414	.03708	9	.96425	8	38
23	.26528	28	.76966	399	.27513	31	.63461	413	.03716	8	.96417	8	37
24	.26556	28	.76568	399	.27545	32	.63048	412	.03724	8	.96410	7	36
25	0.26584	28	3.76171	398	0.27576	31	3.62636	412	1.03732	8	0.96402	8	35
26	.26612	28	.75775	397	.27607	31	.62224	411	.03741	9	.96394	8	34
27	.26640	28	.75379	396	.27638	31	.61814	410	.03749	8	.96386	8	33
28	.26668	28	.74984	394	.27670	32	.61405	410	.03757	8	.96379	7	32
29	.26696	28	.74591	393	.27701	31	.60996	409	.03766	9	.96371	8	31
30	0.26724	28	3.74198	392	0.27732	31	3.60588	408	1.03774	8	0.96363	8	30
31	.26752	28	.73806	392	.27764	32	.60181	407	.03783	9	.96355	8	29
32	.26780	28	.73414	391	.27795	31	.59775	407	.03791	8	.96347	8	28
33	.26808	28	.73024	390	.27826	31	.59370	406	.03799	8	.96340	7	27
34	.26836	28	.72635	390	.27858	32	.58966	404	.03808	9	.96332	8	26
35	0.26864	28	3.72246	389	0.27889	31	3.58562	403	1.03816	8	0.96324	8	25
36	.26892	28	.71858	388	.27921	32	.58160	402	.03825	9	.96316	8	24
37	.26920	28	.71471	387	.27952	31	.57758	401	.03833	8	.96308	8	23
38	.26948	28	.71085	387	.27983	31	.57357	400	.03842	9	.96301	7	22
39	.26976	28	.70700	386	.28015	32	.56957	400	.03850	8	.96293	8	21
40	0.27004	28	3.70315	384	0.28046	31	3.56557	400	1.03858	8	0.96285	8	20
41	.27032	28	.69931	383	.28077	31	.56159	399	.03867	9	.96277	8	19
42	.27060	28	.69549	382	.28109	32	.55761	398	.03875	8	.96269	8	18
43	.27088	28	.69167	382	.28140	31	.55364	397	.03884	9	.96261	8	17
44	.27116	28	.68785	381	.28172	32	.54968	397	.03892	8	.96253	8	16
45	0.27144	28	3.68405	380	0.28203	31	3.54573	396	1.03901	9	0.96246	7	15
46	.27172	28	.68025	380	.28234	31	.54179	394	.03909	8	.96238	8	14
47	.27200	28	.67647	379	.28266	32	.53785	393	.03918	9	.96230	8	13
48	.27228	28	.67269	378	.28297	31	.53393	392	.03927	9	.96222	8	12
49	.27256	28	.66892	378	.28329	32	.53001	391	.03935	8	.96214	8	11
50	0.27284	28	3.66515	377	0.28360	31	3.52609	391	1.03944	9	0.96206	8	10
51	.27312	28	.66140	376	.28391	31	.52219	390	.03952	8	.96198	8	9
52	.27340	28	.65765	374	.28423	32	.51829	390	.03961	9	.96190	8	8
53	.27368	28	.65391	373	.28454	31	.51441	389	.03969	8	.96182	8	7
54	.27396	28	.65018	373	.28486	32	.51053	388	.03978	9	.96174	8	6
55	0.27424	28	3.64645	372	0.28517	31	3.50666	388	1.03987	9	0.96166	8	5
56	.27452	28	.64274	371	.28549	32	.50279	387	.03995	8	.96158	8	4
57	.27480	28	.63903	370	.28580	31	.49894	386	.04004	9	.96150	8	3
58	.27508	28	.63533	370	.28612	32	.49509	384	.04013	9	.96142	8	2
59	.27536	28	.63164	370	.28643	31	.49125	384	.04021	8	.96134	8	1
60	0.27564	28	3.62796	369	0.28675	32	3.48741	383	1.04030	9	0.96126	8	0
↑ 105°→	cos	Diff. 1′	sec	Diff. 1′	cot	Diff. 1′	tan	Diff. 1′	csc	Diff. 1′	sin	Diff. 1′←	↑ 74°

TABLE 2

Natural Trigonometric Functions

16° → ↓ ′	sin	Diff. 1′	csc	Diff. 1′	tan	Diff. 1′	cot	Diff. 1′	sec	Diff. 1′	cos	Diff. 1′	← 163° ↓ ′
0	0.27564		3.62796		0.28675		3.48741		1.04030		0.96126		60
1	.27592	28	.62428	368	.28706	31	.48359	382	.04039	9	.96118	8	59
2	.27620	28	.62061	367	.28738	32	.47977	381	.04047	8	.96110	8	58
3	.27648	28	.61695	367	.28769	31	.47596	380	.04056	9	.96102	8	57
4	.27676	28	.61330	366	.28801	32	.47216	380	.04065	9	.96094	8	56
5	0.27704	28	3.60965	364	0.28832	31	3.46837	380	1.04073	8	0.96086	8	55
6	.27731	27	.60601	363	.28864	32	.46458	379	.04082	9	.96078	8	54
7	.27759	28	.60238	363	.28895	31	.46080	378	.04091	9	.96070	8	53
8	.27787	28	.59876	362	.28927	32	.45703	378	.04100	9	.96062	8	52
9	.27815	28	.59514	361	.28958	31	.45327	377	.04108	8	.96054	8	51
10	0.27843	28	3.59154	360	0.28990	32	3.44951	376	1.04117	9	0.96046	8	50
11	.27871	28	.58794	360	.29021	31	.44576	374	.04126	9	.96037	9	49
12	.27899	28	.58434	360	.29053	32	.44202	374	.04135	9	.96029	8	48
13	.27927	28	.58076	359	.29084	31	.43829	373	.04144	9	.96021	8	47
14	.27955	28	.57718	358	.29116	32	.43456	372	.04152	8	.96013	8	46
15	0.27983	28	3.57361	358	0.29147	31	3.43084	371	1.04161	9	0.96005	8	45
16	.28011	28	.57005	357	.29179	32	.42713	371	.04170	9	.95997	8	44
17	.28039	28	.56649	356	.29210	31	.42343	370	.04179	9	.95989	8	43
18	.28067	28	.56294	354	.29242	32	.41973	370	.04188	9	.95981	8	42
19	.28095	28	.55940	354	.29274	32	.41604	369	.04197	9	.95972	9	41
20	0.28123	28	3.55587	353	0.29305	31	3.41236	369	1.04206	9	0.95964	8	40
21	.28150	27	.55234	352	.29337	32	.40869	368	.04214	8	.95956	8	39
22	.28178	28	.54883	351	.29368	31	.40502	367	.04223	9	.95948	8	38
23	.28206	28	.54531	351	.29400	32	.40136	366	.04232	9	.95940	8	37
24	.28234	28	.54181	350	.29432	32	.39771	366	.04241	9	.95931	9	36
25	0.28262	28	3.53831	350	0.29463	31	3.39406	364	1.04250	9	0.95923	8	35
26	.28290	28	.53482	349	.29495	32	.39042	363	.04259	9	.95915	8	34
27	.28318	28	.53134	349	.29526	31	.38679	363	.04268	9	.95907	8	33
28	.28346	28	.52787	348	.29558	32	.38317	362	.04277	9	.95898	9	32
29	.28374	28	.52440	347	.29590	32	.37955	361	.04286	9	.95890	8	31
30	0.28402	28	3.52094	347	0.29621	31	3.37594	360	1.04295	9	0.95882	8	30
31	.28429	27	.51748	346	.29653	32	.37234	360	.04304	9	.95874	8	29
32	.28457	28	.51404	344	.29685	32	.36875	360	.04313	9	.95865	9	28
33	.28485	28	.51060	343	.29716	31	.36516	359	.04322	9	.95857	8	27
34	.28513	28	.50716	343	.29748	32	.36158	359	.04331	9	.95849	8	26
35	0.28541	28	3.50374	342	0.29780	32	3.35800	358	1.04340	9	0.95841	8	25
36	.28569	28	.50032	341	.29811	31	.35443	357	.04349	9	.95832	9	24
37	.28597	28	.49691	341	.29843	32	.35087	357	.04358	9	.95824	8	23
38	.28625	28	.49350	340	.29875	32	.34732	356	.04367	9	.95816	8	22
39	.28652	27	.49010	340	.29906	31	.34377	354	.04376	9	.95807	9	21
40	0.28680	28	3.48671	340	0.29938	32	3.34023	353	1.04385	9	0.95799	8	20
41	.28708	28	.48333	339	.29970	32	.33670	353	.04394	9	.95791	8	19
42	.28736	28	.47995	338	.30001	31	.33317	352	.04403	9	.95782	9	18
43	.28764	28	.47658	338	.30033	32	.32965	351	.04413	10	.95774	8	17
44	.28792	28	.47321	337	.30065	32	.32614	351	.04422	9	.95766	8	16
45	0.28820	28	3.46986	336	0.30097	32	3.32264	350	1.04431	9	0.95757	9	15
46	.28847	27	.46651	336	.30128	31	.31914	350	.04440	9	.95749	8	14
47	.28875	28	.46316	334	.30160	32	.31565	350	.04449	9	.95740	9	13
48	.28903	28	.45983	333	.30192	32	.31216	349	.04458	9	.95732	8	12
49	.28931	28	.45650	333	.30224	32	.30868	348	.04468	10	.95724	8	11
50	0.28959	28	3.45317	332	0.30255	31	3.30521	348	1.04477	9	0.95715	9	10
51	.28987	28	.44986	331	.30287	32	.30174	347	.04486	9	.95707	8	9
52	.29015	28	.44655	331	.30319	32	.29829	346	.04495	9	.95698	9	8
53	.29042	27	.44324	330	.30351	32	.29483	346	.04504	9	.95690	8	7
54	.29070	28	.43995	330	.30382	31	.29139	344	.04514	10	.95681	9	6
55	0.29098	28	3.43666	330	0.30414	32	3.28795	343	1.04523	9	0.95673	8	5
56	.29126	28	.43337	329	.30446	32	.28452	343	.04532	9	.95664	9	4
57	.29154	28	.43010	328	.30478	32	.28109	342	.04541	9	.95656	8	3
58	.29182	28	.42683	328	.30509	31	.27767	341	.04551	10	.95647	9	2
59	.29209	27	.42356	327	.30541	32	.27426	341	.04560	9	.95639	8	1
60	0.29237	28	3.42030	326	0.30573	32	3.27085	340	1.04569	9	0.95630	9	0
↑ 106° →	cos	Diff. 1′	sec	Diff. 1′	cot	Diff. 1′	tan	Diff. 1′	csc	Diff. 1′	sin	Diff. 1′	↑ ← 73°

TABLE 2

Natural Trigonometric Functions

17° → ↓ ′	sin	Diff. 1′	csc	Diff. 1′	tan	Diff. 1′	cot	Diff. 1′	sec	Diff. 1′	cos	Diff. 1′	← 162° ↓ ′
0	0.29237		3.42030		0.30573		3.27085		1.04569		0.95630		60
1	.29265	28	.41705	326	.30605	32	.26745	340	.04578	9	.95622	8	59
2	.29293	28	.41381	324	.30637	32	.26406	340	.04588	10	.95613	9	58
3	.29321	28	.41057	323	.30669	32	.26067	339	.04597	9	.95605	8	57
4	.29348	27	.40734	323	.30700	31	.25729	339	.04606	9	.95596	9	56
5	0.29376	28	3.40411	322	0.30732	32	3.25392	338	1.04616	10	0.95588	8	55
6	.29404	28	.40089	321	.30764	32	.25055	337	.04625	9	.95579	9	54
7	.29432	28	.39768	321	.30796	32	.24719	337	.04635	10	.95571	8	53
8	.29460	28	.39448	320	.30828	32	.24383	336	.04644	9	.95562	9	52
9	.29487	27	.39128	320	.30860	32	.24049	334	.04653	9	.95554	8	51
10	0.29515	28	3.38808	320	0.30891	31	3.23714	334	1.04663	10	0.95545	9	50
11	.29543	28	.38489	319	.30923	32	.23381	333	.04672	9	.95536	9	49
12	.29571	28	.38171	319	.30955	32	.23048	332	.04682	10	.95528	8	48
13	.29599	28	.37854	318	.30987	32	.22715	332	.04691	9	.95519	9	47
14	.29626	27	.37537	317	.31019	32	.22384	331	.04700	9	.95511	8	46
15	0.29654	28	3.37221	317	0.31051	32	3.22053	331	1.04710	10	0.95502	9	45
16	.29682	28	.36905	316	.31083	32	.21722	330	.04719	9	.95493	9	44
17	.29710	28	.36590	314	.31115	32	.21392	330	.04729	10	.95485	8	43
18	.29737	27	.36276	314	.31147	32	.21063	330	.04738	9	.95476	9	42
19	.29765	28	.35962	313	.31178	31	.20734	329	.04748	10	.95467	9	41
20	0.29793	28	3.35649	313	0.31210	32	3.20406	329	1.04757	9	0.95459	8	40
21	.29821	28	.35336	312	.31242	32	.20079	328	.04767	10	.95450	9	39
22	.29849	28	.35025	311	.31274	32	.19752	327	.04776	9	.95441	9	38
23	.29876	27	.34713	311	.31306	32	.19426	327	.04786	10	.95433	8	37
24	.29904	28	.34403	310	.31338	32	.19100	326	.04795	9	.95424	9	36
25	0.29932	28	3.34092	310	0.31370	32	3.18775	324	1.04805	10	0.95415	9	35
26	.29960	28	.33783	310	.31402	32	.18451	324	.04815	10	.95407	8	34
27	.29987	27	.33474	309	.31434	32	.18127	323	.04824	9	.95398	9	33
28	.30015	28	.33166	309	.31466	32	.17804	323	.04834	10	.95389	9	32
29	.30043	28	.32858	308	.31498	32	.17481	322	.04843	9	.95380	9	31
30	0.30071	28	3.32551	308	0.31530	32	3.17159	321	1.04853	10	0.95372	8	30
31	.30098	27	.32244	307	.31562	32	.16838	321	.04863	10	.95363	9	29
32	.30126	28	.31939	306	.31594	32	.16517	320	.04872	9	.95354	9	28
33	.30154	28	.31633	306	.31626	32	.16197	320	.04882	10	.95345	9	27
34	.30182	28	.31328	304	.31658	32	.15877	320	.04891	9	.95337	8	26
35	0.30209	27	3.31024	304	0.31690	32	3.15558	320	1.04901	10	0.95328	9	25
36	.30237	28	.30721	303	.31722	32	.15240	319	.04911	10	.95319	9	24
37	.30265	28	.30418	302	.31754	32	.14922	318	.04920	9	.95310	9	23
38	.30292	27	.30115	302	.31786	32	.14605	318	.04930	10	.95301	9	22
39	.30320	28	.29814	301	.31818	32	.14288	317	.04940	10	.95293	8	21
40	0.30348	28	3.29512	301	0.31850	32	3.13972	317	1.04950	10	0.95284	9	20
41	.30376	28	.29212	300	.31882	32	.13656	316	.04959	9	.95275	9	19
42	.30403	27	.28912	300	.31914	32	.13341	314	.04969	10	.95266	9	18
43	.30431	28	.28612	300	.31946	32	.13027	314	.04979	10	.95257	9	17
44	.30459	28	.28313	299	.31978	32	.12713	313	.04989	10	.95248	9	16
45	0.30486	27	3.28015	299	0.32010	32	3.12400	313	1.04998	9	0.95240	8	15
46	.30514	28	.27717	298	.32042	32	.12087	312	.05008	10	.95231	9	14
47	.30542	28	.27420	298	.32074	32	.11775	312	.05018	10	.95222	9	13
48	.30570	28	.27123	297	.32106	32	.11464	311	.05028	10	.95213	9	12
49	.30597	27	.26827	297	.32139	33	.11153	310	.05038	10	.95204	9	11
50	0.30625	28	3.26531	296	0.32171	32	3.10842	310	1.05047	9	0.95195	9	10
51	.30653	28	.26237	294	.32203	32	.10532	310	.05057	10	.95186	9	9
52	.30680	27	.25942	294	.32235	32	.10223	310	.05067	10	.95177	9	8
53	.30708	28	.25648	293	.32267	32	.09914	309	.05077	10	.95168	9	7
54	.30736	28	.25355	293	.32299	32	.09606	309	.05087	10	.95159	9	6
55	0.30763	27	3.25062	292	0.32331	32	3.09298	308	1.05097	10	0.95150	9	5
56	.30791	28	.24770	292	.32363	32	.08991	308	.05107	10	.95142	8	4
57	.30819	28	.24478	291	.32396	33	.08685	307	.05116	9	.95133	9	3
58	.30846	27	.24187	291	.32428	32	.08379	306	.05126	10	.95124	9	2
59	.30874	28	.23897	290	.32460	32	.08073	306	.05136	10	.95115	9	1
60	0.30902	28	3.23607	290	0.32492	32	3.07768	304	1.05146	10	0.95106	9	0
↑ 107° →	cos	Diff. 1′	sec	Diff. 1′	cot	Diff. 1′	tan	Diff. 1′	csc	Diff. 1′	sin	Diff. 1′	↑ ← 72°

TABLE 2
Natural Trigonometric Functions

18° → ↓	sin	Diff. 1′	csc	Diff. 1′	tan	Diff. 1′	cot	Diff. 1′	sec	Diff. 1′	cos	Diff. 1′	← 161° ↓
′													′
0	0.30902		3.23607		0.32492		3.07768		1.05146		0.95106		60
1	.30929	27	.23317	290	.32524	32	.07464	304	.05156	10	.95097	9	59
2	.30957	28	.23028	289	.32556	32	.07160	303	.05166	10	.95088	9	58
3	.30985	28	.22740	289	.32588	32	.06857	303	.05176	10	.95079	9	57
4	.31012	27	.22452	288	.32621	33	.06554	302	.05186	10	.95070	9	56
5	0.31040	28	3.22165	288	0.32653	32	3.06252	302	1.05196	10	0.95061	9	55
6	.31068	28	.21878	287	.32685	32	.05950	301	.05206	10	.95052	9	54
7	.31095	27	.21592	287	.32717	32	.05649	301	.05216	10	.95043	9	53
8	.31123	28	.21306	286	.32749	32	.05349	300	.05226	10	.95033	10	52
9	.31151	28	.21021	286	.32782	33	.05049	300	.05236	10	.95024	9	51
10	0.31178	27	3.20737	284	0.32814	32	3.04749	300	1.05246	10	0.95015	9	50
11	.31206	28	.20453	284	.32846	32	.04450	299	.05256	10	.95006	9	49
12	.31233	27	.20169	283	.32878	32	.04152	299	.05266	10	.94997	9	48
13	.31261	28	.19886	283	.32911	33	.03854	298	.05276	10	.94988	9	47
14	.31289	28	.19604	282	.32943	32	.03556	298	.05286	10	.94979	9	46
15	0.31316	27	3.19322	281	0.32975	32	3.03260	297	1.05297	11	0.94970	9	45
16	.31344	28	.19040	281	.33007	32	.02963	297	.05307	10	.94961	9	44
17	.31372	28	.18759	280	.33040	33	.02667	296	.05317	10	.94952	9	43
18	.31399	27	.18479	280	.33072	32	.02372	296	.05327	10	.94943	9	42
19	.31427	28	.18199	280	.33104	32	.02077	294	.05337	10	.94933	10	41
20	0.31454	27	3.17920	280	0.33136	32	3.01783	294	1.05347	10	0.94924	9	40
21	.31482	28	.17641	279	.33169	33	.01489	293	.05357	10	.94915	9	39
22	.31510	28	.17363	279	.33201	32	.01196	293	.05367	10	.94906	9	38
23	.31537	27	.17085	278	.33233	32	.00903	292	.05378	11	.94897	9	37
24	.31565	28	.16808	278	.33266	33	.00611	292	.05388	10	.94888	9	36
25	0.31593	28	3.16531	277	0.33298	32	3.00319	291	1.05398	10	0.94878	10	35
26	.31620	27	.16255	277	.33330	32	3.00028	291	.05408	10	.94869	9	34
27	.31648	28	.15979	276	.33363	33	2.99738	290	.05418	10	.94860	9	33
28	.31675	27	.15704	276	.33395	32	.99447	290	.05429	11	.94851	9	32
29	.31703	28	.15429	274	.33427	32	.99158	290	.05439	10	.94842	9	31
30	0.31730	27	3.15155	274	0.33460	33	2.98868	290	1.05449	10	0.94832	10	30
31	.31758	28	.14881	273	.33492	32	.98580	289	.05459	10	.94823	9	29
32	.31786	28	.14608	273	.33524	32	.98292	289	.05470	11	.94814	9	28
33	.31813	27	.14335	272	.33557	33	.98004	288	.05480	10	.94805	9	27
34	.31841	28	.14063	272	.33589	32	.97717	288	.05490	10	.94795	10	26
35	0.31868	27	3.13791	271	0.33621	32	2.97430	287	1.05501	11	0.94786	9	25
36	.31896	28	.13520	271	.33654	33	.97144	287	.05511	10	.94777	9	24
37	.31923	27	.13249	270	.33686	32	.96858	286	.05521	10	.94768	9	23
38	.31951	28	.12979	270	.33718	32	.96573	286	.05532	11	.94758	10	22
39	.31979	28	.12709	270	.33751	33	.96288	284	.05542	10	.94749	9	21
40	0.32006	27	3.12440	270	0.33783	32	2.96004	284	1.05552	10	0.94740	9	20
41	.32034	28	.12171	269	.33816	33	.95721	283	.05563	11	.94730	10	19
42	.32061	27	.11903	269	.33848	32	.95437	283	.05573	10	.94721	9	18
43	.32089	28	.11635	268	.33881	33	.95155	282	.05584	11	.94712	9	17
44	.32116	27	.11367	268	.33913	32	.94872	282	.05594	10	.94702	10	16
45	0.32144	28	3.11101	267	0.33945	32	2.94591	281	1.05604	10	0.94693	9	15
46	.32171	27	.10834	267	.33978	33	.94309	281	.05615	11	.94684	9	14
47	.32199	28	.10568	266	.34010	32	.94028	280	.05625	10	.94674	10	13
48	.32227	28	.10303	266	.34043	33	.93748	280	.05636	11	.94665	9	12
49	.32254	27	.10038	264	.34075	32	.93468	280	.05646	10	.94656	9	11
50	0.32282	28	3.09774	264	0.34108	33	2.93189	280	1.05657	11	0.94646	10	10
51	.32309	27	.09510	263	.34140	32	.92910	279	.05667	10	.94637	9	9
52	.32337	28	.09246	263	.34173	33	.92632	279	.05678	11	.94627	10	8
53	.32364	27	.08983	263	.34205	32	.92354	278	.05688	10	.94618	9	7
54	.32392	28	.08721	262	.34238	33	.92076	278	.05699	11	.94609	9	6
55	0.32419	27	3.08459	262	0.34270	32	2.91799	278	1.05709	10	0.94599	10	5
56	.32447	28	.08197	261	.34303	33	.91523	277	.05720	11	.94590	9	4
57	.32474	27	.07936	261	.34335	32	.91246	277	.05730	10	.94580	10	3
58	.32502	28	.07675	260	.34368	33	.90971	276	.05741	11	.94571	9	2
59	.32529	27	.07415	260	.34400	32	.90696	276	.05751	10	.94561	10	1
60	0.32557	28	3.07155	260	0.34433	33	2.90421	274	1.05762	11	0.94552	9	0
↑ 108° →	cos	Diff. 1′	sec	Diff. 1′	cot	Diff. 1′	tan	Diff. 1′	csc	Diff. 1′	sin	Diff. 1′	↑ ← 71°

TABLE 2
Natural Trigonometric Functions

19° → ↓	sin	Diff. 1′	csc	Diff. 1′	tan	Diff. 1′	cot	Diff. 1′	sec	Diff. 1′	cos	Diff. 1′	← 160° ↓
′													′
0	0.32557		3.07155		0.34433		2.90421		1.05762		0.94552		60
1	.32584	27	.06896	260	.34465	32	.90147	274	.05773	11	.94542	10	59
2	.32612	28	.06637	259	.34498	33	.89873	273	.05783	10	.94533	9	58
3	.32639	27	.06379	259	.34530	32	.89600	273	.05794	11	.94523	10	57
4	.32667	28	.06121	258	.34563	33	.89327	272	.05805	11	.94514	9	56
5	0.32694	27	3.05864	258	0.34596	33	2.89055	272	1.05815	10	0.94504	10	55
6	.32722	28	.05607	257	.34628	32	.88783	271	.05826	11	.94495	9	54
7	.32749	27	.05350	257	.34661	33	.88511	271	.05836	10	.94485	10	53
8	.32777	28	.05094	257	.34693	32	.88240	270	.05847	11	.94476	9	52
9	.32804	27	.04839	256	.34726	33	.87970	270	.05858	11	.94466	10	51
10	0.32832	28	3.04584	256	0.34758	32	2.87700	270	1.05869	11	0.94457	9	50
11	.32859	27	.04329	254	.34791	33	.87430	270	.05879	10	.94447	10	49
12	.32887	28	.04075	254	.34824	33	.87161	270	.05890	11	.94438	9	48
13	.32914	27	.03821	253	.34856	32	.86892	269	.05901	11	.94428	10	47
14	.32942	28	.03568	253	.34889	33	.86624	269	.05911	10	.94418	10	46
15	0.32969	27	3.03315	252	0.34922	33	2.86356	268	1.05922	11	0.94409	9	45
16	.32997	28	.03062	252	.34954	32	.86089	268	.05933	11	.94399	10	44
17	.33024	27	.02810	251	.34987	33	.85822	267	.05944	11	.94390	9	43
18	.33051	27	.02559	251	.35020	33	.85555	267	.05955	11	.94380	10	42
19	.33079	28	.02308	251	.35052	32	.85289	267	.05965	10	.94370	10	41
20	0.33106	27	3.02057	250	0.35085	33	2.85023	266	1.05976	11	0.94361	9	40
21	.33134	28	.01807	250	.35118	33	.84758	266	.05987	11	.94351	10	39
22	.33161	27	.01557	250	.35150	32	.84494	264	.05998	11	.94342	9	38
23	.33189	28	.01308	250	.35183	33	.84229	264	.06009	11	.94332	10	37
24	.33216	27	.01059	249	.35216	33	.83965	263	.06020	11	.94322	10	36
25	0.33244	28	3.00810	249	0.35248	32	2.83702	263	1.06030	10	0.94313	9	35
26	.33271	27	.00562	249	.35281	33	.83439	262	.06041	11	.94303	10	34
27	.33298	27	.00315	248	.35314	33	.83176	262	.06052	11	.94293	10	33
28	.33326	28	3.00067	248	.35346	32	.82914	262	.06063	11	.94284	9	32
29	.33353	27	2.99821	247	.35379	33	.82653	261	.06074	11	.94274	10	31
30	0.33381	28	2.99574	247	0.35412	33	2.82391	261	1.06085	11	0.94264	10	30
31	.33408	27	.99329	246	.35445	33	.82130	260	.06096	11	.94254	10	29
32	.33436	28	.99083	246	.35477	32	.81870	260	.06107	11	.94245	9	28
33	.33463	27	.98838	246	.35510	33	.81610	260	.06118	11	.94235	10	27
34	.33490	27	.98594	244	.35543	33	.81350	260	.06129	11	.94225	10	26
35	0.33518	28	2.98349	244	0.35576	33	2.81091	260	1.06140	11	0.94215	10	25
36	.33545	27	.98106	243	.35608	32	.80833	259	.06151	11	.94206	9	24
37	.33573	28	.97862	243	.35641	33	.80574	259	.06162	11	.94196	10	23
38	.33600	27	.97619	242	.35674	33	.80316	258	.06173	11	.94186	10	22
39	.33627	27	.97377	242	.35707	33	.80059	258	.06184	11	.94176	10	21
40	0.33655	28	2.97135	242	0.35740	33	2.79802	258	1.06195	11	0.94167	9	20
41	.33682	27	.96893	241	.35772	32	.79545	257	.06206	11	.94157	10	19
42	.33710	28	.96652	241	.35805	33	.79289	257	.06217	11	.94147	10	18
43	.33737	27	.96411	240	.35838	33	.79033	256	.06228	11	.94137	10	17
44	.33764	27	.96171	240	.35871	33	.78778	256	.06239	11	.94127	10	16
45	0.33792	28	2.95931	240	0.35904	33	2.78523	254	1.06250	11	0.94118	9	15
46	.33819	27	.95691	240	.35937	33	.78269	254	.06261	11	.94108	10	14
47	.33846	27	.95452	240	.35969	32	.78014	254	.06272	11	.94098	10	13
48	.33874	28	.95213	239	.36002	33	.77761	253	.06283	11	.94088	10	12
49	.33901	27	.94975	239	.36035	33	.77507	253	.06295	12	.94078	10	11
50	0.33929	28	2.94737	238	0.36068	33	2.77254	252	1.06306	11	0.94068	10	10
51	.33956	27	.94500	238	.36101	33	.77002	252	.06317	11	.94058	10	9
52	.33983	27	.94263	238	.36134	33	.76750	252	.06328	11	.94049	9	8
53	.34011	28	.94026	237	.36167	33	.76498	251	.06339	11	.94039	10	7
54	.34038	27	.93790	237	.36199	32	.76247	251	.06350	11	.94029	10	6
55	0.34065	27	2.93554	236	0.36232	33	2.75996	250	1.06362	12	0.94019	10	5
56	.34093	28	.93318	236	.36265	33	.75746	250	.06373	11	.94009	10	4
57	.34120	27	.93083	236	.36298	33	.75496	250	.06384	11	.93999	10	3
58	.34147	27	.92849	234	.36331	33	.75246	250	.06395	11	.93989	10	2
59	.34175	28	.92614	234	.36364	33	.74997	250	.06407	12	.93979	10	1
60	0.34202	27	2.92380	233	0.36397	33	2.74748	249	1.06418	11	0.93969	10	0
↑ 109° →	cos	Diff. 1′	sec	Diff. 1′	cot	Diff. 1′	tan	Diff. 1′	csc	Diff. 1′	sin	Diff. 1′	↑ ← 70°

TABLE 2
Natural Trigonometric Functions

20°→ ↓ ′	sin	Diff. 1′	csc	Diff. 1′	tan	Diff. 1′	cot	Diff. 1′	sec	Diff. 1′	cos	Diff. 1′	← 159° ↓ ′
0	0.34202		2.92380		0.36397		2.74748		1.06418		0.93969		60
1	.34229	27	.92147	233	.36430	33	.74499	249	.06429	11	.93959	10	59
2	.34257	28	.91914	233	.36463	33	.74251	249	.06440	11	.93949	10	58
3	.34284	27	.91681	232	.36496	33	.74004	248	.06452	12	.93939	10	57
4	.34311	27	.91449	232	.36529	33	.73756	248	.06463	11	.93929	10	56
5	0.34339	28	2.91217	231	0.36562	33	2.73509	247	1.06474	11	0.93919	10	55
6	.34366	27	.90986	231	.36595	33	.73263	247	.06486	12	.93909	10	54
7	.34393	27	.90754	231	.36628	33	.73017	247	.06497	11	.93899	10	53
8	.34421	28	.90524	230	.36661	33	.72771	246	.06508	11	.93889	10	52
9	.34448	27	.90293	230	.36694	33	.72526	246	.06520	12	.93879	10	51
10	0.34475	27	2.90063	230	0.36727	33	2.72281	244	1.06531	11	0.93869	10	50
11	.34503	28	.89834	230	.36760	33	.72036	244	.06542	11	.93859	10	49
12	.34530	27	.89605	230	.36793	33	.71792	244	.06554	12	.93849	10	48
13	.34557	27	.89376	229	.36826	33	.71548	243	.06565	11	.93839	10	47
14	.34584	27	.89148	229	.36859	33	.71305	243	.06577	12	.93829	10	46
15	0.34612	28	2.88920	229	0.36892	33	2.71062	243	1.06588	11	0.93819	10	45
16	.34639	27	.88692	228	.36925	33	.70819	242	.06600	12	.93809	10	44
17	.34666	27	.88465	228	.36958	33	.70577	242	.06611	11	.93799	10	43
18	.34694	28	.88238	227	.36991	33	.70335	241	.06622	11	.93789	10	42
19	.34721	27	.88011	227	.37024	33	.70094	241	.06634	12	.93779	10	41
20	0.34748	27	2.87785	227	0.37057	33	2.69853	241	1.06645	11	0.93769	10	40
21	.34775	27	.87560	226	.37090	33	.69612	240	.06657	12	.93759	10	39
22	.34803	28	.87334	226	.37123	33	.69371	240	.06668	11	.93748	11	38
23	.34830	27	.87109	224	.37157	34	.69131	240	.06680	12	.93738	10	37
24	.34857	27	.86885	224	.37190	33	.68892	240	.06691	11	.93728	10	36
25	0.34884	27	2.86661	224	0.37223	33	2.68653	240	1.06703	12	0.93718	10	35
26	.34912	28	.86437	223	.37256	33	.68414	239	.06715	12	.93708	10	34
27	.34939	27	.86213	223	.37289	33	.68175	239	.06726	11	.93698	10	33
28	.34966	27	.85990	223	.37322	33	.67937	239	.06738	12	.93688	10	32
29	.34993	27	.85767	222	.37355	33	.67700	238	.06749	11	.93677	11	31
30	0.35021	28	2.85545	222	0.37388	33	2.67462	238	1.06761	12	0.93667	10	30
31	.35048	27	.85323	221	.37422	34	.67225	237	.06773	12	.93657	10	29
32	.35075	27	.85102	221	.37455	33	.66989	237	.06784	11	.93647	10	28
33	.35102	27	.84880	221	.37488	33	.66752	237	.06796	12	.93637	10	27
34	.35130	28	.84659	220	.37521	33	.66516	236	.06807	11	.93626	11	26
35	0.35157	27	2.84439	220	0.37554	33	2.66281	236	1.06819	12	0.93616	10	25
36	.35184	27	.84219	220	.37588	34	.66046	236	.06831	12	.93606	10	24
37	.35211	27	.83999	220	.37621	33	.65811	234	.06842	11	.93596	10	23
38	.35239	28	.83780	220	.37654	33	.65576	234	.06854	12	.93585	11	22
39	.35266	27	.83561	220	.37687	33	.65342	234	.06866	12	.93575	10	21
40	0.35293	27	2.83342	219	0.37720	33	2.65109	233	1.06878	12	0.93565	10	20
41	.35320	27	.83124	219	.37754	34	.64875	233	.06889	11	.93555	10	19
42	.35347	27	.82906	218	.37787	33	.64642	232	.06901	12	.93544	11	18
43	.35375	28	.82688	218	.37820	33	.64410	232	.06913	12	.93534	10	17
44	.35402	27	.82471	218	.37853	33	.64177	232	.06925	12	.93524	10	16
45	0.35429	27	2.82254	217	0.37887	34	2.63945	231	1.06936	11	0.93514	10	15
46	.35456	27	.82037	217	.37920	33	.63714	231	.06948	12	.93503	11	14
47	.35484	28	.81821	217	.37953	33	.63483	231	.06960	12	.93493	10	13
48	.35511	27	.81605	216	.37986	33	.63252	230	.06972	12	.93483	10	12
49	.35538	27	.81390	216	.38020	34	.63021	230	.06984	12	.93472	11	11
50	0.35565	27	2.81175	216	0.38053	33	2.62791	230	1.06995	11	0.93462	10	10
51	.35592	27	.80960	214	.38086	33	.62561	230	.07007	12	.93452	10	9
52	.35619	27	.80746	214	.38120	34	.62332	230	.07019	12	.93441	11	8
53	.35647	28	.80531	214	.38153	33	.62103	230	.07031	12	.93431	10	7
54	.35674	27	.80318	213	.38186	33	.61874	229	.07043	12	.93420	11	6
55	0.35701	27	2.80104	213	0.38220	34	2.61646	229	1.07055	12	0.93410	10	5
56	.35728	27	.79891	213	.38253	33	.61418	229	.07067	12	.93400	10	4
57	.35755	27	.79679	212	.38286	33	.61190	228	.07079	12	.93389	11	3
58	.35782	27	.79466	212	.38320	34	.60963	228	.07091	12	.93379	10	2
59	.35810	28	.79254	211	.38353	33	.60736	228	.07103	12	.93368	11	1
60	0.35837	27	2.79043	211	0.38386	33	2.60509	227	1.07114	11	0.93358	10	0
↑ 110°→	cos	Diff. 1′	sec	Diff. 1′	cot	Diff. 1′	tan	Diff. 1′	csc	Diff. 1′	sin	Diff. 1′←	69° ↑

TABLE 2
Natural Trigonometric Functions

21°→ ↓ ′	sin	Diff. 1′	csc	Diff. 1′	tan	Diff. 1′	cot	Diff. 1′	sec	Diff. 1′	cos	Diff. 1′	← 158° ↓ ′
0	0.35837		2.79043		0.38386		2.60509		1.07114		0.93358		60
1	.35864	27	.78832	211	.38420	34	.60283	227	.07126	12	.93348	10	59
2	.35891	27	.78621	210	.38453	33	.60057	226	.07138	12	.93337	11	58
3	.35918	27	.78410	210	.38487	34	.59831	226	.07150	12	.93327	10	57
4	.35945	27	.78200	210	.38520	33	.59606	226	.07162	12	.93316	11	56
5	0.35973	28	2.77990	210	0.38553	33	2.59381	224	1.07174	12	0.93306	10	55
6	.36000	27	.77780	210	.38587	34	.59156	224	.07186	12	.93295	11	54
7	.36027	27	.77571	210	.38620	33	.58932	224	.07199	13	.93285	10	53
8	.36054	27	.77362	209	.38654	34	.58708	223	.07211	12	.93274	11	52
9	.36081	27	.77154	209	.38687	33	.58484	223	.07223	12	.93264	10	51
10	0.36108	27	2.76945	209	0.38721	34	2.58261	223	1.07235	12	0.93253	11	50
11	.36135	27	.76737	208	.38754	33	.58038	222	.07247	12	.93243	10	49
12	.36162	27	.76530	208	.38787	33	.57815	222	.07259	12	.93232	11	48
13	.36190	28	.76323	208	.38821	34	.57593	222	.07271	12	.93222	10	47
14	.36217	27	.76116	207	.38854	33	.57371	221	.07283	12	.93211	11	46
15	0.36244	27	2.75909	207	0.38888	34	2.57150	221	1.07295	12	0.93201	10	45
16	.36271	27	.75703	207	.38921	33	.56928	221	.07307	12	.93190	11	44
17	.36298	27	.75497	206	.38955	34	.56707	220	.07320	13	.93180	10	43
18	.36325	27	.75292	206	.38988	33	.56487	220	.07332	12	.93169	11	42
19	.36352	27	.75086	206	.39022	34	.56266	220	.07344	12	.93159	10	41
20	0.36379	27	2.74881	204	0.39055	33	2.56046	220	1.07356	12	0.93148	11	40
21	.36406	27	.74677	204	.39089	34	.55827	220	.07368	12	.93137	11	39
22	.36434	28	.74473	204	.39122	33	.55608	219	.07380	12	.93127	10	38
23	.36461	27	.74269	203	.39156	34	.55389	219	.07393	13	.93116	11	37
24	.36488	27	.74065	203	.39190	34	.55170	219	.07405	12	.93106	10	36
25	0.36515	27	2.73862	203	0.39223	33	2.54952	219	1.07417	12	0.93095	11	35
26	.36542	27	.73659	202	.39257	34	.54734	219	.07429	12	.93084	11	34
27	.36569	27	.73456	202	.39290	33	.54516	218	.07442	13	.93074	10	33
28	.36596	27	.73254	202	.39324	34	.54299	218	.07454	12	.93063	11	32
29	.36623	27	.73052	201	.39357	33	.54082	218	.07466	12	.93052	11	31
30	0.36650	27	2.72850	201	0.39391	34	2.53865	217	1.07479	13	0.93042	10	30
31	.36677	27	.72649	201	.39425	34	.53648	217	.07491	12	.93031	11	29
32	.36704	27	.72448	201	.39458	33	.53432	217	.07503	12	.93020	11	28
33	.36731	27	.72247	200	.39492	34	.53217	216	.07516	13	.93010	10	27
34	.36758	27	.72047	200	.39526	34	.53001	216	.07528	12	.92999	11	26
35	0.36785	27	2.71847	200	0.39559	33	2.52786	216	1.07540	12	0.92988	11	25
36	.36812	27	.71647	200	.39593	34	.52571	214	.07553	13	.92978	10	24
37	.36839	27	.71448	200	.39626	33	.52357	214	.07565	12	.92967	11	23
38	.36867	28	.71249	200	.39660	34	.52142	214	.07578	13	.92956	11	22
39	.36894	27	.71050	199	.39694	34	.51929	213	.07590	12	.92945	11	21
40	0.36921	27	2.70851	199	0.39727	33	2.51715	213	1.07602	12	0.92935	10	20
41	.36948	27	.70653	199	.39761	34	.51502	213	.07615	13	.92924	11	19
42	.36975	27	.70455	198	.39795	34	.51289	212	.07627	12	.92913	11	18
43	.37002	27	.70258	198	.39829	34	.51076	212	.07640	13	.92902	11	17
44	.37029	27	.70061	198	.39862	33	.50864	212	.07652	12	.92892	10	16
45	0.37056	27	2.69864	197	0.39896	34	2.50652	211	1.07665	13	0.92881	11	15
46	.37083	27	.69667	197	.39930	34	.50440	211	.07677	12	.92870	11	14
47	.37110	27	.69471	197	.39963	33	.50229	211	.07690	13	.92859	11	13
48	.37137	27	.69275	196	.39997	34	.50018	211	.07702	12	.92849	10	12
49	.37164	27	.69079	196	.40031	34	.49807	210	.07715	13	.92838	11	11
50	0.37191	27	2.68884	196	0.40065	34	2.49597	210	1.07727	12	0.92827	11	10
51	.37218	27	.68689	196	.40098	33	.49386	210	.07740	13	.92816	11	9
52	.37245	27	.68494	194	.40132	34	.49177	210	.07752	12	.92805	11	8
53	.37272	27	.68299	194	.40166	34	.48967	210	.07765	13	.92794	11	7
54	.37299	27	.68105	194	.40200	34	.48758	210	.07778	13	.92784	10	6
55	0.37326	27	2.67911	193	0.40234	34	2.48549	209	1.07790	12	0.92773	11	5
56	.37353	27	.67718	193	.40267	33	.48340	209	.07803	13	.92762	11	4
57	.37380	27	.67525	193	.40301	34	.48132	209	.07816	13	.92751	11	3
58	.37407	27	.67332	192	.40335	34	.47924	209	.07828	12	.92740	11	2
59	.37434	27	.67139	192	.40369	34	.47716	208	.07841	13	.92729	11	1
60	0.37461	27	2.66947	192	0.40403	34	2.47509	208	1.07853	12	0.92718	11	0
↑ 111°→	cos	Diff. 1′	sec	Diff. 1′	cot	Diff. 1′	tan	Diff. 1′	csc	Diff. 1′	sin	Diff. 1′←	68° ↑

TABLE 2
Natural Trigonometric Functions

22° → ↓ ′	sin	Diff. 1′	csc	Diff. 1′	tan	Diff. 1′	cot	Diff. 1′	sec	Diff. 1′	cos	Diff. 1′	← 157° ↓ ′
0	0.37461		2.66947		0.40403		2.47509		1.07853		0.92718		60
1	.37488	27	.66755	192	.40436	33	.47302	208	.07866	13	.92707	11	59
2	.37515	27	.66563	191	.40470	34	.47095	207	.07879	13	.92697	10	58
3	.37542	27	.66371	191	.40504	34	.46888	207	.07892	13	.92686	11	57
4	.37569	27	.66180	191	.40538	34	.46682	207	.07904	12	.92675	11	56
5	0.37595	26	2.65989	190	0.40572	34	2.46476	206	1.07917	13	0.92664	11	55
6	.37622	27	.65799	190	.40606	34	.46270	206	.07930	13	.92653	11	54
7	.37649	27	.65609	190	.40640	34	.46065	206	.07943	13	.92642	11	53
8	.37676	27	.65419	190	.40674	34	.45860	206	.07955	12	.92631	11	52
9	.37703	27	.65229	190	.40707	33	.45655	204	.07968	13	.92620	11	51
10	0.37730	27	2.65040	190	0.40741	34	2.45451	204	1.07981	13	0.92609	11	50
11	.37757	27	.64851	190	.40775	34	.45246	204	.07994	13	.92598	11	49
12	.37784	27	.64662	189	.40809	34	.45043	203	.08006	12	.92587	11	48
13	.37811	27	.64473	189	.40843	34	.44839	203	.08019	13	.92576	11	47
14	.37838	27	.64285	189	.40877	34	.44636	203	.08032	13	.92565	11	46
15	0.37865	27	2.64097	188	0.40911	34	2.44433	203	1.08045	13	0.92554	11	45
16	.37892	27	.63909	188	.40945	34	.44230	202	.08058	13	.92543	11	44
17	.37919	27	.63722	188	.40979	34	.44027	202	.08071	13	.92532	11	43
18	.37946	27	.63535	188	.41013	34	.43825	202	.08084	13	.92521	11	42
19	.37973	27	.63348	187	.41047	34	.43623	201	.08097	13	.92510	11	41
20	0.37999	26	2.63162	187	0.41081	34	2.43422	201	1.08109	12	0.92499	11	40
21	.38026	27	.62976	187	.41115	34	.43220	201	.08122	13	.92488	11	39
22	.38053	27	.62790	186	.41149	34	.43019	201	.08135	13	.92477	11	38
23	.38080	27	.62604	186	.41183	34	.42819	200	.08148	13	.92466	11	37
24	.38107	27	.62419	186	.41217	34	.42618	200	.08161	13	.92455	11	36
25	0.38134	27	2.62234	186	0.41251	34	2.42418	200	1.08174	13	0.92444	11	35
26	.38161	27	.62049	184	.41285	34	.42218	200	.08187	13	.92432	12	34
27	.38188	27	.61864	184	.41319	34	.42019	200	.08200	13	.92421	11	33
28	.38215	27	.61680	184	.41353	34	.41819	200	.08213	13	.92410	11	32
29	.38241	26	.61496	183	.41387	34	.41620	200	.08226	13	.92399	11	31
30	0.38268	27	2.61313	183	0.41421	34	2.41421	199	1.08239	13	0.92388	11	30
31	.38295	27	.61129	183	.41455	34	.41223	199	.08252	13	.92377	11	29
32	.38322	27	.60946	183	.41490	35	.41025	199	.08265	13	.92366	11	28
33	.38349	27	.60763	182	.41524	34	.40827	198	.08278	13	.92355	11	27
34	.38376	27	.60581	182	.41558	34	.40629	198	.08291	13	.92343	12	26
35	0.38403	27	2.60399	182	0.41592	34	2.40432	198	1.08305	14	0.92332	11	25
36	.38430	27	.60217	181	.41626	34	.40235	198	.08318	13	.92321	11	24
37	.38456	26	.60035	181	.41660	34	.40038	197	.08331	13	.92310	11	23
38	.38483	27	.59853	181	.41694	34	.39841	197	.08344	13	.92299	11	22
39	.38510	27	.59672	181	.41728	34	.39645	197	.08357	13	.92287	12	21
40	0.38537	27	2.59491	180	0.41763	35	2.39449	197	1.08370	13	0.92276	11	20
41	.38564	27	.59311	180	.41797	34	.39253	196	.08383	13	.92265	11	19
42	.38591	27	.59130	180	.41831	34	.39058	196	.08397	14	.92254	11	18
43	.38617	26	.58950	180	.41865	34	.38863	196	.08410	13	.92243	11	17
44	.38644	27	.58771	180	.41899	34	.38668	194	.08423	13	.92231	12	16
45	0.38671	27	2.58591	180	0.41933	34	2.38473	194	1.08436	13	0.92220	11	15
46	.38698	27	.58412	180	.41968	35	.38279	194	.08449	13	.92209	11	14
47	.38725	27	.58233	179	.42002	34	.38084	194	.08463	14	.92198	11	13
48	.38752	27	.58054	179	.42036	34	.37891	193	.08476	13	.92186	12	12
49	.38778	26	.57876	179	.42070	34	.37697	193	.08489	13	.92175	11	11
50	0.38805	27	2.57698	179	0.42105	35	2.37504	193	1.08503	14	0.92164	11	10
51	.38832	27	.57520	178	.42139	34	.37311	193	.08516	13	.92152	12	9
52	.38859	27	.57342	178	.42173	34	.37118	192	.08529	13	.92141	11	8
53	.38886	27	.57165	178	.42207	34	.36925	192	.08542	13	.92130	11	7
54	.38912	26	.56988	178	.42242	35	.36733	192	.08556	14	.92119	11	6
55	0.38939	27	2.56811	177	0.42276	34	2.36541	191	1.08569	13	0.92107	12	5
56	.38966	27	.56634	177	.42310	34	.36349	191	.08582	13	.92096	11	4
57	.38993	27	.56458	177	.42345	35	.36158	191	.08596	14	.92085	11	3
58	.39020	27	.56282	177	.42379	34	.35967	191	.08609	13	.92073	12	2
59	.39046	26	.56106	176	.42413	34	.35776	190	.08623	14	.92062	11	1
60	0.39073	27	2.55930	176	0.42447	34	2.35585	190	1.08636	13	0.92050	12	0
↑ 112° →	cos	Diff. 1′	sec	Diff. 1′	cot	Diff. 1′	tan	Diff. 1′	csc	Diff. 1′	sin	Diff. 1′ ←	↑ 67°

TABLE 2
Natural Trigonometric Functions

23° → ↓ ′	sin	Diff. 1′	csc	Diff. 1′	tan	Diff. 1′	cot	Diff. 1′	sec	Diff. 1′	cos	Diff. 1′	← 156° ↓ ′
0	0.39073		2.55930		0.42447		2.35585		1.08636		0.92050		60
1	.39100	27	.55755	176	.42482	35	.35395	190	.08649	13	.92039	11	59
2	.39127	27	.55580	174	.42516	34	.35205	190	.08663	14	.92028	11	58
3	.39153	26	.55405	174	.42551	35	.35015	190	.08676	13	.92016	12	57
4	.39180	27	.55231	174	.42585	34	.34825	190	.08690	14	.92005	11	56
5	0.39207	27	2.55057	174	0.42619	34	2.34636	190	1.08703	13	0.91994	11	55
6	.39234	27	.54883	173	.42654	35	.34447	190	.08717	14	.91982	12	54
7	.39260	26	.54709	173	.42688	34	.34258	189	.08730	13	.91971	11	53
8	.39287	27	.54536	173	.42722	34	.34069	189	.08744	14	.91959	12	52
9	.39314	27	.54363	173	.42757	35	.33881	189	.08757	13	.91948	11	51
10	0.39341	27	2.54190	172	0.42791	34	2.33693	189	1.08771	14	0.91936	12	50
11	.39367	26	.54017	172	.42826	35	.33505	188	.08784	13	.91925	11	49
12	.39394	27	.53845	172	.42860	34	.33317	188	.08798	14	.91914	11	48
13	.39421	27	.53672	172	.42894	34	.33130	188	.08811	13	.91902	12	47
14	.39448	27	.53500	171	.42929	35	.32943	188	.08825	14	.91891	11	46
15	0.39474	26	2.53329	171	0.42963	34	2.32756	187	1.08839	14	0.91879	12	45
16	.39501	27	.53157	171	.42998	35	.32570	187	.08852	13	.91868	11	44
17	.39528	27	.52986	171	.43032	34	.32383	187	.08866	14	.91856	12	43
18	.39555	27	.52815	170	.43067	35	.32197	187	.08880	14	.91845	11	42
19	.39581	26	.52645	170	.43101	34	.32012	186	.08893	13	.91833	12	41
20	0.39608	27	2.52474	170	0.43136	35	2.31826	186	1.08907	14	0.91822	11	40
21	.39635	27	.52304	170	.43170	34	.31641	186	.08920	13	.91810	12	39
22	.39661	26	.52134	170	.43205	35	.31456	186	.08934	14	.91799	11	38
23	.39688	27	.51965	170	.43239	34	.31271	184	.08948	14	.91787	12	37
24	.39715	27	.51795	170	.43274	35	.31086	184	.08962	14	.91775	12	36
25	0.39741	26	2.51626	170	0.43308	34	2.30902	184	1.08975	13	0.91764	11	35
26	.39768	27	.51457	169	.43343	35	.30718	184	.08989	14	.91752	12	34
27	.39795	27	.51289	169	.43378	35	.30534	183	.09003	14	.91741	11	33
28	.39822	27	.51120	169	.43412	34	.30351	183	.09017	14	.91729	12	32
29	.39848	26	.50952	169	.43447	35	.30167	183	.09030	13	.91718	11	31
30	0.39875	27	2.50784	168	0.43481	34	2.29984	183	1.09044	14	0.91706	12	30
31	.39902	27	.50617	168	.43516	35	.29801	182	.09058	14	.91694	12	29
32	.39928	26	.50449	168	.43550	34	.29619	182	.09072	14	.91683	11	28
33	.39955	27	.50282	168	.43585	35	.29437	182	.09086	14	.91671	12	27
34	.39982	27	.50115	167	.43620	35	.29254	182	.09099	13	.91660	11	26
35	0.40008	26	2.49948	167	0.43654	34	2.29073	181	1.09113	14	0.91648	12	25
36	.40035	27	.49782	167	.43689	35	.28891	181	.09127	14	.91636	12	24
37	.40062	27	.49616	167	.43724	35	.28710	181	.09141	14	.91625	11	23
38	.40088	26	.49450	166	.43758	34	.28528	181	.09155	14	.91613	12	22
39	.40115	27	.49284	166	.43793	35	.28348	180	.09169	14	.91601	12	21
40	0.40141	26	2.49119	166	0.43828	35	2.28167	180	1.09183	14	0.91590	11	20
41	.40168	27	.48954	166	.43862	34	.27987	180	.09197	14	.91578	12	19
42	.40195	27	.48789	164	.43897	35	.27806	180	.09211	14	.91566	12	18
43	.40221	26	.48624	164	.43932	35	.27626	180	.09224	13	.91555	11	17
44	.40248	27	.48459	164	.43966	34	.27447	180	.09238	14	.91543	12	16
45	0.40275	27	2.48295	164	0.44001	35	2.27267	180	1.09252	14	0.91531	12	15
46	.40301	26	.48131	164	.44036	35	.27088	180	.09266	14	.91519	12	14
47	.40328	27	.47967	163	.44071	35	.26909	179	.09280	14	.91508	11	13
48	.40355	27	.47804	163	.44105	34	.26730	179	.09294	14	.91496	12	12
49	.40381	26	.47640	163	.44140	35	.26552	179	.09308	14	.91484	12	11
50	0.40408	27	2.47477	163	0.44175	35	2.26374	179	1.09323	15	0.91472	12	10
51	.40434	26	.47314	162	.44210	35	.26196	179	.09337	14	.91461	11	9
52	.40461	27	.47152	162	.44244	34	.26018	178	.09351	14	.91449	12	8
53	.40488	27	.46989	162	.44279	35	.25840	178	.09365	14	.91437	12	7
54	.40514	26	.46827	162	.44314	35	.25663	178	.09379	14	.91425	12	6
55	0.40541	27	2.46665	161	0.44349	35	2.25486	178	1.09393	14	0.91414	11	5
56	.40567	26	.46504	161	.44384	35	.25309	177	.09407	14	.91402	12	4
57	.40594	27	.46342	161	.44418	34	.25132	177	.09421	14	.91390	12	3
58	.40621	27	.46181	161	.44453	35	.24956	177	.09435	14	.91378	12	2
59	.40647	26	.46020	160	.44488	35	.24780	177	.09449	14	.91366	12	1
60	0.40674	27	2.45859	160	0.44523	35	2.24604	176	1.09464	15	0.91355	11	0
↑ 113° →	cos	Diff. 1′	sec	Diff. 1′	cot	Diff. 1′	tan	Diff. 1′	csc	Diff. 1′	sin	Diff. 1′ ←	↑ 66°

TABLE 2

Natural Trigonometric Functions

24° → ↓ ′	sin	Diff. 1′	csc	Diff. 1′	tan	Diff. 1′	cot	Diff. 1′	sec	Diff. 1′	cos	Diff. 1′	← 155° ↓ ′
0	0.40674		2.45859		0.44523		2.24604		1.09464		0.91355		60
1	.40700	26	.45699	160	.44558	35	.24428	176	.09478	14	.91343	12	59
2	.40727	27	.45539	160	.44593	35	.24252	176	.09492	14	.91331	12	58
3	.40753	26	.45378	160	.44627	34	.24077	176	.09506	14	.91319	12	57
4	.40780	27	.45219	160	.44662	35	.23902	176	.09520	14	.91307	12	56
5	0.40806	26	2.45059	160	0.44697	35	2.23727	174	1.09535	15	0.91295	12	55
6	.40833	27	.44900	160	.44732	35	.23553	174	.09549	14	.91283	12	54
7	.40860	27	.44741	160	.44767	35	.23378	174	.09563	14	.91272	11	53
8	.40886	26	.44582	159	.44802	35	.23204	174	.09577	14	.91260	12	52
9	.40913	27	.44423	159	.44837	35	.23030	173	.09592	15	.91248	12	51
10	0.40939	26	2.44264	159	0.44872	35	2.22857	173	1.09606	14	0.91236	12	50
11	.40966	27	.44106	159	.44907	35	.22683	173	.09620	14	.91224	12	49
12	.40992	26	.43948	159	.44942	35	.22510	173	.09635	15	.91212	12	48
13	.41019	27	.43790	158	.44977	35	.22337	172	.09649	14	.91200	12	47
14	.41045	26	.43633	158	.45012	35	.22164	172	.09663	14	.91188	12	46
15	0.41072	27	2.43476	158	0.45047	35	2.21992	172	1.09678	15	0.91176	12	45
16	.41098	26	.43318	158	.45082	35	.21819	172	.09692	14	.91164	12	44
17	.41125	27	.43162	157	.45117	35	.21647	172	.09707	15	.91152	12	43
18	.41151	26	.43005	157	.45152	35	.21475	171	.09721	14	.91140	12	42
19	.41178	27	.42848	157	.45187	35	.21304	171	.09735	14	.91128	12	41
20	0.41204	26	2.42692	157	0.45222	35	2.21132	171	1.09750	15	0.91116	12	40
21	.41231	27	.42536	157	.45257	35	.20961	171	.09764	14	.91104	12	39
22	.41257	26	.42380	156	.45292	35	.20790	171	.09779	15	.91092	12	38
23	.41284	27	.42225	156	.45327	35	.20619	170	.09793	14	.91080	12	37
24	.41310	26	.42070	156	.45362	35	.20449	170	.09808	15	.91068	12	36
25	0.41337	27	2.41914	156	0.45397	35	2.20278	170	1.09822	14	0.91056	12	35
26	.41363	26	.41760	154	.45432	35	.20108	170	.09837	15	.91044	12	34
27	.41390	27	.41605	154	.45467	35	.19938	170	.09851	14	.91032	12	33
28	.41416	26	.41450	154	.45502	35	.19769	170	.09866	15	.91020	12	32
29	.41443	27	.41296	154	.45538	36	.19599	170	.09880	14	.91008	12	31
30	0.41469	26	2.41142	154	0.45573	35	2.19430	170	1.09895	15	0.90996	12	30
31	.41496	27	.40988	153	.45608	35	.19261	170	.09909	14	.90984	12	29
32	.41522	26	.40835	153	.45643	35	.19092	169	.09924	15	.90972	12	28
33	.41549	27	.40681	153	.45678	35	.18923	169	.09939	15	.90960	12	27
34	.41575	26	.40528	153	.45713	35	.18755	169	.09953	14	.90948	12	26
35	0.41602	27	2.40375	152	0.45748	35	2.18587	169	1.09968	15	0.90936	12	25
36	.41628	26	.40222	152	.45784	36	.18419	168	.09982	14	.90924	12	24
37	.41655	27	.40070	152	.45819	35	.18251	168	.09997	15	.90911	13	23
38	.41681	26	.39918	152	.45854	35	.18084	168	.10012	15	.90899	12	22
39	.41707	26	.39766	152	.45889	35	.17916	168	.10026	14	.90887	12	21
40	0.41734	27	2.39614	151	0.45924	35	2.17749	168	1.10041	15	0.90875	12	20
41	.41760	26	.39462	151	.45960	36	.17582	167	.10056	15	.90863	12	19
42	.41787	27	.39311	151	.45995	35	.17416	167	.10071	15	.90851	12	18
43	.41813	26	.39159	151	.46030	35	.17249	167	.10085	14	.90839	12	17
44	.41840	27	.39008	151	.46065	35	.17083	167	.10100	15	.90826	13	16
45	0.41866	26	2.38857	150	0.46101	36	2.16917	167	1.10115	15	0.90814	12	15
46	.41892	26	.38707	150	.46136	35	.16751	166	.10130	15	.90802	12	14
47	.41919	27	.38556	150	.46171	35	.16585	166	.10144	14	.90790	12	13
48	.41945	26	.38406	150	.46206	35	.16420	166	.10159	15	.90778	12	12
49	.41972	27	.38256	150	.46242	36	.16255	166	.10174	15	.90766	12	11
50	0.41998	26	2.38107	150	0.46277	35	2.16090	166	1.10189	15	0.90753	13	10
51	.42024	26	.37957	150	.46312	35	.15925	164	.10204	15	.90741	12	9
52	.42051	27	.37808	150	.46348	36	.15760	164	.10218	14	.90729	12	8
53	.42077	26	.37658	150	.46383	35	.15596	164	.10233	15	.90717	12	7
54	.42104	27	.37509	149	.46418	35	.15432	164	.10248	15	.90704	13	6
55	0.42130	26	2.37361	149	0.46454	36	2.15268	163	1.10263	15	0.90692	12	5
56	.42156	26	.37212	149	.46489	35	.15104	163	.10278	15	.90680	12	4
57	.42183	27	.37064	149	.46525	36	.14940	163	.10293	15	.90668	12	3
58	.42209	26	.36916	149	.46560	35	.14777	163	.10308	15	.90655	13	2
59	.42235	26	.36768	148	.46595	35	.14614	163	.10323	15	.90643	12	1
60	0.42262	27	2.36620	148	0.46631	36	2.14451	162	1.10338	15	0.90631	12	0
↑ 114° →	cos	Diff. 1′	sec	Diff. 1′	cot	Diff. 1′	tan	Diff. 1′	csc	Diff. 1′	sin	Diff. 1′ ←	65° ↑

TABLE 2

Natural Trigonometric Functions

25° → ↓ ′	sin	Diff. 1′	csc	Diff. 1′	tan	Diff. 1′	cot	Diff. 1′	sec	Diff. 1′	cos	Diff. 1′	← 154° ↓ ′
0	0.42262		2.36620		0.46631		2.14451		1.10338		0.90631		60
1	.42288	26	.36473	148	.46666	35	.14288	162	.10353	15	.90618	13	59
2	.42315	27	.36325	148	.46702	36	.14125	162	.10368	15	.90606	12	58
3	.42341	26	.36178	148	.46737	35	.13963	162	.10383	15	.90594	12	57
4	.42367	26	.36031	147	.46772	35	.13801	162	.10398	15	.90582	12	56
5	0.42394	27	2.35885	147	0.46808	36	2.13639	161	1.10413	15	0.90569	13	55
6	.42420	26	.35738	147	.46843	35	.13477	161	.10428	15	.90557	12	54
7	.42446	26	.35592	147	.46879	36	.13316	161	.10443	15	.90545	12	53
8	.42473	27	.35446	147	.46914	35	.13154	161	.10458	15	.90532	13	52
9	.42499	26	.35300	146	.46950	36	.12993	161	.10473	15	.90520	12	51
10	0.42525	26	2.35154	146	0.46985	35	2.12832	160	1.10488	15	0.90507	13	50
11	.42552	27	.35009	146	.47021	36	.12671	160	.10503	15	.90495	12	49
12	.42578	26	.34863	146	.47056	35	.12511	160	.10518	15	.90483	12	48
13	.42604	26	.34718	146	.47092	36	.12350	160	.10533	15	.90470	13	47
14	.42631	27	.34573	144	.47128	36	.12190	160	.10549	16	.90458	12	46
15	0.42657	26	2.34429	144	0.47163	35	2.12030	160	1.10564	15	0.90446	12	45
16	.42683	26	.34284	144	.47199	36	.11871	160	.10579	15	.90433	13	44
17	.42709	26	.34140	144	.47234	35	.11711	160	.10594	15	.90421	12	43
18	.42736	27	.33996	144	.47270	36	.11552	160	.10609	15	.90408	13	42
19	.42762	26	.33852	143	.47305	35	.11392	160	.10625	16	.90396	12	41
20	0.42788	26	2.33708	143	0.47341	36	2.11233	159	1.10640	15	0.90383	13	40
21	.42815	27	.33565	143	.47377	36	.11075	159	.10655	15	.90371	12	39
22	.42841	26	.33422	143	.47412	35	.10916	159	.10670	15	.90358	13	38
23	.42867	26	.33278	143	.47448	36	.10758	159	.10686	16	.90346	12	37
24	.42894	27	.33135	142	.47483	35	.10600	159	.10701	15	.90334	12	36
25	0.42920	26	2.32993	142	0.47519	36	2.10442	159	1.10716	15	0.90321	13	35
26	.42946	26	.32850	142	.47555	36	.10284	158	.10731	15	.90309	12	34
27	.42972	26	.32708	142	.47590	35	.10126	158	.10747	16	.90296	13	33
28	.42999	27	.32566	142	.47626	36	.09969	158	.10762	15	.90284	12	32
29	.43025	26	.32424	141	.47662	36	.09811	158	.10777	15	.90271	13	31
30	0.43051	26	2.32282	141	0.47698	36	2.09654	158	1.10793	16	0.90259	12	30
31	.43077	26	.32140	141	.47733	35	.09498	157	.10808	15	.90246	13	29
32	.43104	27	.31999	141	.47769	36	.09341	157	.10824	16	.90233	13	28
33	.43130	26	.31858	141	.47805	36	.09184	157	.10839	15	.90221	12	27
34	.43156	26	.31717	140	.47840	35	.09028	157	.10854	15	.90208	13	26
35	0.43182	26	2.31576	140	0.47876	36	2.08872	157	1.10870	16	0.90196	12	25
36	.43209	27	.31436	140	.47912	36	.08716	156	.10885	15	.90183	13	24
37	.43235	26	.31295	140	.47948	36	.08560	156	.10901	16	.90171	12	23
38	.43261	26	.31155	140	.47984	36	.08405	156	.10916	15	.90158	13	22
39	.43287	26	.31015	140	.48019	35	.08250	156	.10932	16	.90146	12	21
40	0.43313	26	2.30875	140	0.48055	36	2.08094	156	1.10947	15	0.90133	13	20
41	.43340	27	.30735	140	.48091	36	.07939	154	.10963	16	.90120	13	19
42	.43366	26	.30596	140	.48127	36	.07785	154	.10978	15	.90108	12	18
43	.43392	26	.30457	140	.48163	36	.07630	154	.10994	16	.90095	13	17
44	.43418	26	.30318	140	.48198	35	.07476	154	.11009	15	.90082	13	16
45	0.43445	27	2.30179	139	0.48234	36	2.07321	154	1.11025	16	0.90070	12	15
46	.43471	26	.30040	139	.48270	36	.07167	154	.11041	16	.90057	13	14
47	.43497	26	.29901	139	.48306	36	.07014	153	.11056	15	.90045	12	13
48	.43523	26	.29763	139	.48342	36	.06860	153	.11072	16	.90032	13	12
49	.43549	26	.29625	139	.48378	36	.06706	153	.11087	15	.90019	13	11
50	0.43575	26	2.29487	138	0.48414	36	2.06553	153	1.11103	16	0.90007	12	10
51	.43602	27	.29349	138	.48450	36	.06400	153	.11119	16	.89994	13	9
52	.43628	26	.29211	138	.48486	36	.06247	152	.11134	15	.89981	13	8
53	.43654	26	.29074	138	.48521	35	.06094	152	.11150	16	.89968	13	7
54	.43680	26	.28937	138	.48557	36	.05942	152	.11166	16	.89956	12	6
55	0.43706	26	2.28800	138	0.48593	36	2.05790	152	1.11181	15	0.89943	13	5
56	.43733	27	.28663	137	.48629	36	.05637	152	.11197	16	.89930	13	4
57	.43759	26	.28526	137	.48665	36	.05485	152	.11213	16	.89918	12	3
58	.43785	26	.28390	137	.48701	36	.05333	151	.11229	16	.89905	13	2
59	.43811	26	.28253	137	.48737	36	.05182	151	.11244	15	.89892	13	1
60	0.43837	26	2.28117	137	0.48773	36	2.05030	151	1.11260	16	0.89879	13	0
↑ 115° →	cos	Diff. 1′	sec	Diff. 1′	cot	Diff. 1′	tan	Diff. 1′	csc	Diff. 1′	sin	Diff. 1′ ←	64° ↑

TABLE 2
Natural Trigonometric Functions

26°→ ↓	sin	Diff. 1′	csc	Diff. 1′	tan	Diff. 1′	cot	Diff. 1′	sec	Diff. 1′	cos	Diff. 1′	← 153° ↓
′													′
0	0.43837		2.28117		0.48773		2.05030		1.11260		0.89879		60
1	.43863	26	.27981	136	.48809	36	.04879	151	.11276	16	.89867	12	59
2	.43889	26	.27845	136	.48845	36	.04728	151	.11292	16	.89854	13	58
3	.43916	27	.27710	136	.48881	36	.04577	150	.11308	16	.89841	13	57
4	.43942	26	.27574	136	.48917	36	.04426	150	.11323	15	.89828	13	56
5	0.43968	26	2.27439	136	0.48953	36	2.04276	150	1.11339	16	0.89816	12	55
6	.43994	26	.27304	136	.48989	36	.04125	150	.11355	16	.89803	13	54
7	.44020	26	.27169	134	.49026	37	.03975	150	.11371	16	.89790	13	53
8	.44046	26	.27035	134	.49062	36	.03825	150	.11387	16	.89777	13	52
9	.44072	26	.26900	134	.49098	36	.03675	150	.11403	16	.89764	13	51
10	0.44098	26	2.26766	134	0.49134	36	2.03526	150	1.11419	16	0.89752	12	50
11	.44124	26	.26632	134	.49170	36	.03376	150	.11435	16	.89739	13	49
12	.44151	27	.26498	133	.49206	36	.03227	150	.11451	16	.89726	13	48
13	.44177	26	.26364	133	.49242	36	.03078	150	.11467	16	.89713	13	47
14	.44203	26	.26230	133	.49278	36	.02929	149	.11483	16	.89700	13	46
15	0.44229	26	2.26097	133	0.49315	37	2.02780	149	1.11499	16	0.89687	13	45
16	.44255	26	.25963	133	.49351	36	.02631	149	.11515	16	.89674	13	44
17	.44281	26	.25830	133	.49387	36	.02483	149	.11531	16	.89662	12	43
18	.44307	26	.25697	132	.49423	36	.02335	149	.11547	16	.89649	13	42
19	.44333	26	.25565	132	.49459	36	.02187	149	.11563	16	.89636	13	41
20	0.44359	26	2.25432	132	0.49495	36	2.02039	148	1.11579	16	0.89623	13	40
21	.44385	26	.25300	132	.49532	37	.01891	148	.11595	16	.89610	13	39
22	.44411	26	.25167	132	.49568	36	.01743	148	.11611	16	.89597	13	38
23	.44437	26	.25035	132	.49604	36	.01596	148	.11627	16	.89584	13	37
24	.44464	27	.24903	131	.49640	36	.01449	148	.11643	16	.89571	13	36
25	0.44490	26	2.24772	131	0.49677	37	2.01302	148	1.11659	16	0.89558	13	35
26	.44516	26	.24640	131	.49713	36	.01155	147	.11675	16	.89545	13	34
27	.44542	26	.24509	131	.49749	36	.01008	147	.11691	16	.89532	13	33
28	.44568	26	.24378	131	.49786	37	.00862	147	.11708	17	.89519	13	32
29	.44594	26	.24247	131	.49822	36	.00715	147	.11724	16	.89506	13	31
30	0.44620	26	2.24116	130	0.49858	36	2.00569	147	1.11740	16	0.89493	13	30
31	.44646	26	.23985	130	.49894	36	.00423	147	.11756	16	.89480	13	29
32	.44672	26	.23855	130	.49931	37	.00277	146	.11772	16	.89467	13	28
33	.44698	26	.23724	130	.49967	36	2.00131	146	.11789	17	.89454	13	27
34	.44724	26	.23594	130	.50004	37	1.99986	146	.11805	16	.89441	13	26
35	0.44750	26	2.23464	130	0.50040	36	1.99841	146	1.11821	16	0.89428	13	25
36	.44776	26	.23334	130	.50076	36	.99695	146	.11838	17	.89415	13	24
37	.44802	26	.23205	130	.50113	37	.99550	146	.11854	16	.89402	13	23
38	.44828	26	.23075	130	.50149	36	.99406	144	.11870	16	.89389	13	22
39	.44854	26	.22946	130	.50185	36	.99261	144	.11886	16	.89376	13	21
40	0.44880	26	2.22817	130	0.50222	37	1.99116	144	1.11903	17	0.89363	13	20
41	.44906	26	.22688	129	.50258	36	.98972	144	.11919	16	.89350	13	19
42	.44932	26	.22559	129	.50295	37	.98828	144	.11936	17	.89337	13	18
43	.44958	26	.22430	129	.50331	36	.98684	144	.11952	16	.89324	13	17
44	.44984	26	.22302	129	.50368	37	.98540	143	.11968	16	.89311	13	16
45	0.45010	26	2.22174	129	0.50404	36	1.98396	143	1.11985	17	0.89298	13	15
46	.45036	26	.22045	129	.50441	37	.98253	143	.12001	16	.89285	13	14
47	.45062	26	.21918	128	.50477	36	.98110	143	.12018	17	.89272	13	13
48	.45088	26	.21790	128	.50514	37	.97966	143	.12034	16	.89259	13	12
49	.45114	26	.21662	128	.50550	36	.97823	143	.12051	17	.89245	14	11
50	0.45140	26	2.21535	128	0.50587	37	1.97681	142	1.12067	16	0.89232	13	10
51	.45166	26	.21407	128	.50623	36	.97538	142	.12083	16	.89219	13	9
52	.45192	26	.21280	128	.50660	37	.97395	142	.12100	17	.89206	13	8
53	.45218	26	.21153	127	.50696	36	.97253	142	.12117	17	.89193	13	7
54	.45243	25	.21026	127	.50733	37	.97111	142	.12133	16	.89180	13	6
55	0.45269	26	2.20900	127	0.50769	36	1.96969	142	1.12150	17	0.89167	13	5
56	.45295	26	.20773	127	.50806	37	.96827	141	.12166	16	.89153	14	4
57	.45321	26	.20647	127	.50843	37	.96685	141	.12183	17	.89140	13	3
58	.45347	26	.20521	127	.50879	36	.96544	141	.12199	16	.89127	13	2
59	.45373	26	.20395	126	.50916	37	.96402	141	.12216	17	.89114	13	1
60	0.45399	26	2.20269	126	0.50953	37	1.96261	141	1.12233	17	0.89101	13	0
↑ 116°→	cos	Diff. 1′	sec	Diff. 1′	cot	Diff. 1′	tan	Diff. 1′	csc	Diff. 1′	sin	Diff. 1′	← 63° ↑

TABLE 2
Natural Trigonometric Functions

27°→ ↓	sin	Diff. 1′	csc	Diff. 1′	tan	Diff. 1′	cot	Diff. 1′	sec	Diff. 1′	cos	Diff. 1′	← 152° ↓
′													′
0	0.45399		2.20269		0.50953		1.96261		1.12233		0.89101		60
1	.45425	26	.20143	126	.50989	36	.96120	141	.12249	16	.89087	14	59
2	.45451	26	.20018	126	.51026	37	.95979	140	.12266	17	.89074	13	58
3	.45477	26	.19892	126	.51063	37	.95838	140	.12283	17	.89061	13	57
4	.45503	26	.19767	126	.51099	36	.95698	140	.12299	16	.89048	13	56
5	0.45529	26	2.19642	126	0.51136	37	1.95557	140	1.12316	17	0.89035	13	55
6	.45554	25	.19517	124	.51173	37	.95417	140	.12333	17	.89021	14	54
7	.45580	26	.19393	124	.51209	36	.95277	140	.12349	16	.89008	13	53
8	.45606	26	.19268	124	.51246	37	.95137	140	.12366	17	.88995	13	52
9	.45632	26	.19144	124	.51283	37	.94997	140	.12383	17	.88981	14	51
10	0.45658	26	2.19019	124	0.51319	36	1.94858	140	1.12400	17	0.88968	13	50
11	.45684	26	.18895	124	.51356	37	.94718	140	.12416	16	.88955	13	49
12	.45710	26	.18772	123	.51393	37	.94579	140	.12433	17	.88942	13	48
13	.45736	26	.18648	123	.51430	37	.94440	140	.12450	17	.88928	14	47
14	.45762	26	.18524	123	.51467	37	.94301	139	.12467	17	.88915	13	46
15	0.45787	25	2.18401	123	0.51503	36	1.94162	139	1.12484	17	0.88902	13	45
16	.45813	26	.18277	123	.51540	37	.94023	139	.12501	17	.88888	14	44
17	.45839	26	.18154	123	.51577	37	.93885	139	.12518	17	.88875	13	43
18	.45865	26	.18031	122	.51614	37	.93746	139	.12534	16	.88862	13	42
19	.45891	26	.17909	122	.51651	37	.93608	139	.12551	17	.88848	14	41
20	0.45917	26	2.17786	122	0.51688	37	1.93470	139	1.12568	17	0.88835	13	40
21	.45942	25	.17663	122	.51724	36	.93332	138	.12585	17	.88822	13	39
22	.45968	26	.17541	122	.51761	37	.93195	138	.12602	17	.88808	14	38
23	.45994	26	.17419	122	.51798	37	.93057	138	.12619	17	.88795	13	37
24	.46020	26	.17297	122	.51835	37	.92920	138	.12636	17	.88782	13	36
25	0.46046	26	2.17175	121	0.51872	37	1.92782	138	1.12653	17	0.88768	14	35
26	.46072	26	.17053	121	.51909	37	.92645	138	.12670	17	.88755	13	34
27	.46097	25	.16932	121	.51946	37	.92508	137	.12687	17	.88741	14	33
28	.46123	26	.16810	121	.51983	37	.92371	137	.12704	17	.88728	13	32
29	.46149	26	.16689	121	.52020	37	.92235	137	.12721	17	.88715	13	31
30	0.46175	26	2.16568	121	0.52057	37	1.92098	137	1.12738	17	0.88701	14	30
31	.46201	26	.16447	120	.52094	37	.91962	137	.12755	17	.88688	13	29
32	.46226	25	.16326	120	.52131	37	.91826	137	.12772	17	.88674	14	28
33	.46252	26	.16206	120	.52168	37	.91690	137	.12789	17	.88661	13	27
34	.46278	26	.16085	120	.52205	37	.91554	136	.12807	18	.88647	14	26
35	0.46304	26	2.15965	120	0.52242	37	1.91418	136	1.12824	17	0.88634	13	25
36	.46330	26	.15845	120	.52279	37	.91282	136	.12841	17	.88620	14	24
37	.46355	25	.15725	120	.52316	37	.91147	136	.12858	17	.88607	13	23
38	.46381	26	.15605	120	.52353	37	.91012	136	.12875	17	.88593	14	22
39	.46407	26	.15485	120	.52390	37	.90876	136	.12892	17	.88580	13	21
40	0.46433	26	2.15366	120	0.52427	37	1.90741	134	1.12910	18	0.88566	14	20
41	.46458	25	.15246	120	.52464	37	.90607	134	.12927	17	.88553	13	19
42	.46484	26	.15127	120	.52501	37	.90472	134	.12944	17	.88539	14	18
43	.46510	26	.15008	120	.52538	37	.90337	134	.12961	17	.88526	13	17
44	.46536	26	.14889	119	.52575	37	.90203	134	.12979	18	.88512	14	16
45	0.46561	25	2.14770	119	0.52613	38	1.90069	134	1.12996	17	0.88499	13	15
46	.46587	26	.14651	119	.52650	37	.89935	134	.13013	17	.88485	14	14
47	.46613	26	.14533	119	.52687	37	.89801	133	.13031	18	.88472	13	13
48	.46639	26	.14414	119	.52724	37	.89667	133	.13048	17	.88458	14	12
49	.46664	25	.14296	119	.52761	37	.89533	133	.13065	17	.88445	13	11
50	0.46690	26	2.14178	119	0.52798	37	1.89400	133	1.13083	18	0.88431	14	10
51	.46716	26	.14060	118	.52836	38	.89266	133	.13100	17	.88417	14	9
52	.46742	26	.13942	118	.52873	37	.89133	133	.13117	17	.88404	13	8
53	.46767	25	.13825	118	.52910	37	.89000	133	.13135	18	.88390	14	7
54	.46793	26	.13707	118	.52947	37	.88867	132	.13152	17	.88377	13	6
55	0.46819	26	2.13590	118	0.52985	38	1.88734	132	1.13170	18	0.88363	14	5
56	.46844	25	.13473	118	.53022	37	.88602	132	.13187	17	.88349	14	4
57	.46870	26	.13356	118	.53059	37	.88469	132	.13205	18	.88336	13	3
58	.46896	26	.13239	117	.53096	37	.88337	132	.13222	17	.88322	14	2
59	.46921	25	.13122	117	.53134	38	.88205	132	.13239	17	.88308	14	1
60	0.46947	26	2.13005	117	0.53171	37	1.88073	132	1.13257	18	0.88295	13	0
↑ 117°→	cos	Diff. 1′	sec	Diff. 1′	cot	Diff. 1′	tan	Diff. 1′	csc	Diff. 1′	sin	Diff. 1′	← 62° ↑

TABLE 2
Natural Trigonometric Functions

28° → ↓ ′	sin	Diff. 1′	csc	Diff. 1′	tan	Diff. 1′	cot	Diff. 1′	sec	Diff. 1′	cos	Diff. 1′	← 151° ↓ ′
0	0.46947		2.13005		0.53171		1.88073		1.13257		0.88295		60
1	.46973	26	.12889	117	.53208	37	.87941	131	.13275	18	.88281	14	59
2	.46999	26	.12773	117	.53246	38	.87809	131	.13292	17	.88267	14	58
3	.47024	25	.12657	117	.53283	37	.87677	131	.13310	18	.88254	13	57
4	.47050	26	.12540	117	.53320	37	.87546	131	.13327	17	.88240	14	56
5	0.47076	26	2.12425	116	0.53358	38	1.87415	131	1.13345	18	0.88226	14	55
6	.47101	25	.12309	116	.53395	37	.87283	131	.13362	17	.88213	13	54
7	.47127	26	.12193	116	.53432	37	.87152	131	.13380	18	.88199	14	53
8	.47153	26	.12078	116	.53470	38	.87021	130	.13398	18	.88185	14	52
9	.47178	25	.11963	116	.53507	37	.86891	130	.13415	17	.88172	13	51
10	0.47204	26	2.11847	116	0.53545	38	1.86760	130	1.13433	18	0.88158	14	50
11	.47229	25	.11732	116	.53582	37	.86630	130	.13451	18	.88144	14	49
12	.47255	26	.11617	114	.53620	38	.86499	130	.13468	17	.88130	14	48
13	.47281	26	.11503	114	.53657	37	.86369	130	.13486	18	.88117	13	47
14	.47306	25	.11388	114	.53694	37	.86239	130	.13504	18	.88103	14	46
15	0.47332	26	2.11274	114	0.53732	38	1.86109	130	1.13521	17	0.88089	14	45
16	.47358	26	.11159	114	.53769	37	.85979	130	.13539	18	.88075	14	44
17	.47383	25	.11045	114	.53807	38	.85850	130	.13557	18	.88062	13	43
18	.47409	26	.10931	114	.53844	37	.85720	130	.13575	18	.88048	14	42
19	.47434	25	.10817	113	.53882	38	.85591	130	.13593	18	.88034	14	41
20	0.47460	26	2.10704	113	0.53920	38	1.85462	130	1.13610	17	0.88020	14	40
21	.47486	26	.10590	113	.53957	37	.85333	130	.13628	18	.88006	14	39
22	.47511	25	.10477	113	.53995	38	.85204	129	.13646	18	.87993	13	38
23	.47537	26	.10363	113	.54032	37	.85075	129	.13664	18	.87979	14	37
24	.47562	25	.10250	113	.54070	38	.84946	129	.13682	18	.87965	14	36
25	0.47588	26	2.10137	113	0.54107	37	1.84818	129	1.13700	18	0.87951	14	35
26	.47614	26	.10024	112	.54145	38	.84689	129	.13718	18	.87937	14	34
27	.47639	25	.09911	112	.54183	38	.84561	129	.13735	17	.87923	14	33
28	.47665	26	.09799	112	.54220	37	.84433	129	.13753	18	.87909	14	32
29	.47690	25	.09686	112	.54258	38	.84305	128	.13771	18	.87896	13	31
30	0.47716	26	2.09574	112	0.54296	38	1.84177	128	1.13789	18	0.87882	14	30
31	.47741	25	.09462	112	.54333	37	.84049	128	.13807	18	.87868	14	29
32	.47767	26	.09350	112	.54371	38	.83922	128	.13825	18	.87854	14	28
33	.47793	26	.09238	111	.54409	38	.83794	128	.13843	18	.87840	14	27
34	.47818	25	.09126	111	.54446	37	.83667	128	.13861	18	.87826	14	26
35	0.47844	26	2.09014	111	0.54484	38	1.83540	128	1.13879	18	0.87812	14	25
36	.47869	25	.08903	111	.54522	38	.83413	128	.13897	18	.87798	14	24
37	.47895	26	.08791	111	.54560	38	.83286	127	.13915	18	.87784	14	23
38	.47920	25	.08680	111	.54597	37	.83159	127	.13934	19	.87770	14	22
39	.47946	26	.08569	111	.54635	38	.83033	127	.13952	18	.87756	14	21
40	0.47971	25	2.08458	110	0.54673	38	1.82906	127	1.13970	18	0.87743	13	20
41	.47997	26	.08347	110	.54711	38	.82780	127	.13988	18	.87729	14	19
42	.48022	25	.08236	110	.54748	37	.82654	127	.14006	18	.87715	14	18
43	.48048	26	.08126	110	.54786	38	.82528	127	.14024	18	.87701	14	17
44	.48073	25	.08015	110	.54824	38	.82402	126	.14042	18	.87687	14	16
45	0.48099	26	2.07905	110	0.54862	38	1.82276	126	1.14061	19	0.87673	14	15
46	.48124	25	.07795	110	.54900	38	.82150	126	.14079	18	.87659	14	14
47	.48150	26	.07685	110	.54938	38	.82025	126	.14097	18	.87645	14	13
48	.48175	25	.07575	110	.54975	37	.81899	126	.14115	18	.87631	14	12
49	.48201	26	.07465	110	.55013	38	.81774	126	.14134	19	.87617	14	11
50	0.48226	25	2.07356	110	0.55051	38	1.81649	126	1.14152	18	0.87603	14	10
51	.48252	26	.07246	110	.55089	38	.81524	126	.14170	18	.87589	14	9
52	.48277	25	.07137	110	.55127	38	.81399	124	.14188	18	.87575	14	8
53	.48303	26	.07027	110	.55165	38	.81274	124	.14207	19	.87561	14	7
54	.48328	25	.06918	110	.55203	38	.81150	124	.14225	18	.87546	15	6
55	0.48354	26	2.06809	109	0.55241	38	1.81025	124	1.14243	18	0.87532	14	5
56	.48379	25	.06701	109	.55279	38	.80901	124	.14262	19	.87518	14	4
57	.48405	26	.06592	109	.55317	38	.80777	124	.14280	18	.87504	14	3
58	.48430	25	.06483	109	.55355	38	.80653	124	.14299	19	.87490	14	2
59	.48456	26	.06375	109	.55393	38	.80529	123	.14317	18	.87476	14	1
60	0.48481	25	2.06267	109	0.55431	38	1.80405	123	1.14335	18	0.87462	14	0
↑ 118° →	cos	Diff. 1′	sec	Diff. 1′	cot	Diff. 1′	tan	Diff. 1′	csc	Diff. 1′	sin	Diff. 1′ ←	↑ 61°

TABLE 2
Natural Trigonometric Functions

29° → ↓ ′	sin	Diff. 1′	csc	Diff. 1′	tan	Diff. 1′	cot	Diff. 1′	sec	Diff. 1′	cos	Diff. 1′	← 150° ↓ ′
0	0.48481		2.06267		0.55431		1.80405		1.14335		0.87462		60
1	.48506	25	.06158	109	.55469	38	.80281	123	.14354	19	.87448	14	59
2	.48532	26	.06050	109	.55507	38	.80158	123	.14372	18	.87434	14	58
3	.48557	25	.05942	108	.55545	38	.80034	123	.14391	19	.87420	14	57
4	.48583	26	.05835	108	.55583	38	.79911	123	.14409	18	.87406	14	56
5	0.48608	25	2.05727	108	0.55621	38	1.79788	123	1.14428	19	0.87391	15	55
6	.48634	26	.05619	108	.55659	38	.79665	123	.14446	18	.87377	14	54
7	.48659	25	.05512	108	.55697	38	.79542	122	.14465	19	.87363	14	53
8	.48684	25	.05405	108	.55736	39	.79419	122	.14483	18	.87349	14	52
9	.48710	26	.05298	108	.55774	38	.79296	122	.14502	19	.87335	14	51
10	0.48735	25	2.05191	108	0.55812	38	1.79174	122	1.14521	19	0.87321	14	50
11	.48761	26	.05084	107	.55850	38	.79051	122	.14539	18	.87306	15	49
12	.48786	25	.04977	107	.55888	38	.78929	122	.14558	19	.87292	14	48
13	.48811	25	.04870	107	.55926	38	.78807	122	.14576	18	.87278	14	47
14	.48837	26	.04764	107	.55964	38	.78685	122	.14595	19	.87264	14	46
15	0.48862	25	2.04657	107	0.56003	39	1.78563	121	1.14614	19	0.87250	14	45
16	.48888	26	.04551	107	.56041	38	.78441	121	.14632	18	.87235	15	44
17	.48913	25	.04445	107	.56079	38	.78319	121	.14651	19	.87221	14	43
18	.48938	25	.04339	106	.56117	38	.78198	121	.14670	19	.87207	14	42
19	.48964	26	.04233	106	.56156	39	.78077	121	.14689	19	.87193	14	41
20	0.48989	25	2.04128	106	0.56194	38	1.77955	121	1.14707	18	0.87178	15	40
21	.49014	25	.04022	106	.56232	38	.77834	121	.14726	19	.87164	14	39
22	.49040	26	.03916	106	.56270	38	.77713	121	.14745	19	.87150	14	38
23	.49065	25	.03811	106	.56309	39	.77592	120	.14764	19	.87136	14	37
24	.49090	25	.03706	106	.56347	38	.77471	120	.14782	18	.87121	15	36
25	0.49116	26	2.03601	106	0.56385	38	1.77351	120	1.14801	19	0.87107	14	35
26	.49141	25	.03496	104	.56424	39	.77230	120	.14820	19	.87093	14	34
27	.49166	25	.03391	104	.56462	38	.77110	120	.14839	19	.87079	14	33
28	.49192	26	.03286	104	.56501	39	.76990	120	.14858	19	.87064	15	32
29	.49217	25	.03182	104	.56539	38	.76869	120	.14877	19	.87050	14	31
30	0.49242	25	2.03077	104	0.56577	38	1.76749	120	1.14896	19	0.87036	14	30
31	.49268	26	.02973	104	.56616	39	.76629	120	.14914	18	.87021	15	29
32	.49293	25	.02869	104	.56654	38	.76510	120	.14933	19	.87007	14	28
33	.49318	25	.02765	104	.56693	39	.76390	120	.14952	19	.86993	14	27
34	.49344	26	.02661	103	.56731	38	.76271	120	.14971	19	.86978	15	26
35	0.49369	25	2.02557	103	0.56769	38	1.76151	120	1.14990	19	0.86964	14	25
36	.49394	25	.02453	103	.56808	39	.76032	120	.15009	19	.86949	15	24
37	.49419	25	.02349	103	.56846	38	.75913	120	.15028	19	.86935	14	23
38	.49445	26	.02246	103	.56885	39	.75794	120	.15047	19	.86921	14	22
39	.49470	25	.02143	103	.56923	38	.75675	119	.15066	19	.86906	15	21
40	0.49495	25	2.02039	103	0.56962	39	1.75556	119	1.15085	19	0.86892	14	20
41	.49521	26	.01936	103	.57000	38	.75437	119	.15105	20	.86878	14	19
42	.49546	25	.01833	102	.57039	39	.75319	119	.15124	19	.86863	15	18
43	.49571	25	.01730	102	.57078	39	.75200	119	.15143	19	.86849	14	17
44	.49596	25	.01628	102	.57116	38	.75082	119	.15162	19	.86834	15	16
45	0.49622	26	2.01525	102	0.57155	39	1.74964	119	1.15181	19	0.86820	14	15
46	.49647	25	.01422	102	.57193	38	.74846	119	.15200	19	.86805	15	14
47	.49672	25	.01320	102	.57232	39	.74728	118	.15219	19	.86791	14	13
48	.49697	25	.01218	102	.57271	39	.74610	118	.15239	20	.86777	14	12
49	.49723	26	.01116	102	.57309	38	.74492	118	.15258	19	.86762	15	11
50	0.49748	25	2.01014	102	0.57348	39	1.74375	118	1.15277	19	0.86748	14	10
51	.49773	25	.00912	101	.57386	38	.74257	118	.15296	19	.86733	15	9
52	.49798	25	.00810	101	.57425	39	.74140	118	.15315	19	.86719	14	8
53	.49824	26	.00708	101	.57464	39	.74022	118	.15335	20	.86704	15	7
54	.49849	25	.00607	101	.57503	39	.73905	118	.15354	19	.86690	14	6
55	0.49874	25	2.00505	101	0.57541	38	1.73788	118	1.15373	19	0.86675	15	5
56	.49899	25	.00404	101	.57580	39	.73671	117	.15393	20	.86661	14	4
57	.49924	25	.00303	101	.57619	39	.73555	117	.15412	19	.86646	15	3
58	.49950	26	.00202	101	.57657	38	.73438	117	.15431	19	.86632	14	2
59	.49975	25	.00101	100	.57696	39	.73321	117	.15451	20	.86617	15	1
60	0.50000	25	2.00000	100	0.57735	39	1.73205	117	1.15470	19	0.86603	14	0
↑ 119° →	cos	Diff. 1′	sec	Diff. 1′	cot	Diff. 1′	tan	Diff. 1′	csc	Diff. 1′	sin	Diff. 1′ ←	↑ 60°

TABLE 2
Natural Trigonometric Functions

30°→ ↓ ′	sin	Diff. 1′	csc	Diff. 1′	tan	Diff. 1′	cot	Diff. 1′	sec	Diff. 1′	cos	Diff. 1′	← 149° ↓ ′
0	0.50000		2.00000		0.57735		1.73205		1.15470		0.86603		60
1	.50025	25	1.99899	100	.57774	39	.73089	117	.15489	19	.86588	15	59
2	.50050	25	.99799	100	.57813	39	.72973	117	.15509	20	.86573	15	58
3	.50076	26	.99698	100	.57851	38	.72857	117	.15528	19	.86559	14	57
4	.50101	25	.99598	100	.57890	39	.72741	116	.15548	20	.86544	15	56
5	0.50126	25	1.99498	100	0.57929	39	1.72625	116	1.15567	19	0.86530	14	55
6	.50151	25	.99398	100	.57968	39	.72509	116	.15587	20	.86515	15	54
7	.50176	25	.99298	100	.58007	39	.72393	116	.15606	19	.86501	14	53
8	.50201	25	.99198	100	.58046	39	.72278	116	.15626	20	.86486	15	52
9	.50227	26	.99098	100	.58085	39	.72163	116	.15645	19	.86471	15	51
10	0.50252	25	1.98998	100	0.58124	39	1.72047	116	1.15665	20	0.86457	14	50
11	.50277	25	.98899	100	.58162	38	.71932	116	.15684	19	.86442	15	49
12	.50302	25	.98799	100	.58201	39	.71817	116	.15704	20	.86427	15	48
13	.50327	25	.98700	100	.58240	39	.71702	114	.15724	20	.86413	14	47
14	.50352	25	.98601	100	.58279	39	.71588	114	.15743	19	.86398	15	46
15	0.50377	25	1.98502	100	0.58318	39	1.71473	114	1.15763	20	0.86384	14	45
16	.50403	26	.98403	99	.58357	39	.71358	114	.15782	19	.86369	15	44
17	.50428	25	.98304	99	.58396	39	.71244	114	.15802	20	.86354	15	43
18	.50453	25	.98205	99	.58435	39	.71129	114	.15822	20	.86340	14	42
19	.50478	25	.98107	99	.58474	39	.71015	114	.15841	19	.86325	15	41
20	0.50503	25	1.98008	99	0.58513	39	1.70901	114	1.15861	20	0.86310	15	40
21	.50528	25	.97910	99	.58552	39	.70787	113	.15881	20	.86295	15	39
22	.50553	25	.97811	99	.58591	39	.70673	113	.15901	20	.86281	14	38
23	.50578	25	.97713	99	.58631	40	.70560	113	.15920	19	.86266	15	37
24	.50603	25	.97615	99	.58670	39	.70446	113	.15940	20	.86251	15	36
25	0.50628	25	1.97517	98	0.58709	39	1.70332	113	1.15960	20	0.86237	14	35
26	.50654	26	.97420	98	.58748	39	.70219	113	.15980	20	.86222	15	34
27	.50679	25	.97322	98	.58787	39	.70106	113	.16000	20	.86207	15	33
28	.50704	25	.97224	98	.58826	39	.69992	113	.16019	19	.86192	15	32
29	.50729	25	.97127	98	.58865	39	.69879	113	.16039	20	.86178	14	31
30	0.50754	25	1.97029	98	0.58905	40	1.69766	112	1.16059	20	0.86163	15	30
31	.50779	25	.96932	98	.58944	39	.69653	112	.16079	20	.86148	15	29
32	.50804	25	.96835	98	.58983	39	.69541	112	.16099	20	.86133	15	28
33	.50829	25	.96738	98	.59022	39	.69428	112	.16119	20	.86119	14	27
34	.50854	25	.96641	97	.59061	39	.69316	112	.16139	20	.86104	15	26
35	0.50879	25	1.96544	97	0.59101	40	1.69203	112	1.16159	20	0.86089	15	25
36	.50904	25	.96448	97	.59140	39	.69091	112	.16179	20	.86074	15	24
37	.50929	25	.96351	97	.59179	39	.68979	112	.16199	20	.86059	15	23
38	.50954	25	.96255	97	.59218	39	.68866	112	.16219	20	.86045	14	22
39	.50979	25	.96158	97	.59258	40	.68754	111	.16239	20	.86030	15	21
40	0.51004	25	1.96062	97	0.59297	39	1.68643	111	1.16259	20	0.86015	15	20
41	.51029	25	.95966	97	.59336	39	.68531	111	.16279	20	.86000	15	19
42	.51054	25	.95870	97	.59376	40	.68419	111	.16299	20	.85985	15	18
43	.51079	25	.95774	96	.59415	39	.68308	111	.16319	20	.85970	15	17
44	.51104	25	.95678	96	.59454	39	.68196	111	.16339	20	.85956	14	16
45	0.51129	25	1.95583	96	0.59494	40	1.68085	111	1.16359	20	0.85941	15	15
46	.51154	25	.95487	96	.59533	39	.67974	111	.16380	21	.85926	15	14
47	.51179	25	.95392	96	.59573	40	.67863	111	.16400	20	.85911	15	13
48	.51204	25	.95296	96	.59612	39	.67752	111	.16420	20	.85896	15	12
49	.51229	25	.95201	96	.59651	39	.67641	110	.16440	20	.85881	15	11
50	0.51254	25	1.95106	96	0.59691	40	1.67530	110	1.16460	20	0.85866	15	10
51	.51279	25	.95011	96	.59730	39	.67419	110	.16481	21	.85851	15	9
52	.51304	25	.94916	94	.59770	40	.67309	110	.16501	20	.85836	15	8
53	.51329	25	.94821	94	.59809	39	.67198	110	.16521	20	.85821	15	7
54	.51354	25	.94726	94	.59849	40	.67088	110	.16541	20	.85806	15	6
55	0.51379	25	1.94632	94	0.59888	39	1.66978	110	1.16562	21	0.85792	14	5
56	.51404	25	.94537	94	.59928	40	.66867	110	.16582	20	.85777	15	4
57	.51429	25	.94443	94	.59967	39	.66757	110	.16602	20	.85762	15	3
58	.51454	25	.94349	94	.60007	40	.66647	110	.16623	21	.85747	15	2
59	.51479	25	.94254	94	.60046	39	.66538	110	.16643	20	.85732	15	1
60	0.51504	25	1.94160	94	0.60086	40	1.66428	110	1.16663	20	0.85717	15	0
↑ 120°→	cos	Diff. 1′	sec	Diff. 1′	cot	Diff. 1′	tan	Diff. 1′	csc	Diff. 1′	sin	Diff. 1′←	↑ 59°

TABLE 2
Natural Trigonometric Functions

31°→ ↓ ′	sin	Diff. 1′	csc	Diff. 1′	tan	Diff. 1′	cot	Diff. 1′	sec	Diff. 1′	cos	Diff. 1′	← 148° ↓ ′
0	0.51504		1.94160		0.60086		1.66428		1.16663		0.85717		60
1	.51529	25	.94066	93	.60126	40	.66318	110	.16684	21	.85702	15	59
2	.51554	25	.93973	93	.60165	39	.66209	110	.16704	20	.85687	15	58
3	.51579	25	.93879	93	.60205	40	.66099	110	.16725	21	.85672	15	57
4	.51604	25	.93785	93	.60245	40	.65990	110	.16745	20	.85657	15	56
5	0.51628	24	1.93692	93	0.60284	39	1.65881	110	1.16766	21	0.85642	15	55
6	.51653	25	.93598	93	.60324	40	.65772	110	.16786	20	.85627	15	54
7	.51678	25	.93505	93	.60364	40	.65663	109	.16806	20	.85612	15	53
8	.51703	25	.93412	93	.60403	39	.65554	109	.16827	21	.85597	15	52
9	.51728	25	.93319	93	.60443	40	.65445	109	.16848	21	.85582	15	51
10	0.51753	25	1.93226	92	0.60483	40	1.65337	109	1.16868	20	0.85567	15	50
11	.51778	25	.93133	92	.60522	39	.65228	109	.16889	21	.85551	16	49
12	.51803	25	.93040	92	.60562	40	.65120	109	.16909	20	.85536	15	48
13	.51828	25	.92947	92	.60602	40	.65011	109	.16930	21	.85521	15	47
14	.51852	24	.92855	92	.60642	40	.64903	109	.16950	20	.85506	15	46
15	0.51877	25	1.92762	92	0.60681	39	1.64795	109	1.16971	21	0.85491	15	45
16	.51902	25	.92670	92	.60721	40	.64687	109	.16992	21	.85476	15	44
17	.51927	25	.92578	92	.60761	40	.64579	108	.17012	20	.85461	15	43
18	.51952	25	.92486	92	.60801	40	.64471	108	.17033	21	.85446	15	42
19	.51977	25	.92394	92	.60841	40	.64363	108	.17054	21	.85431	15	41
20	0.52002	25	1.92302	91	0.60881	40	1.64256	108	1.17075	21	0.85416	15	40
21	.52026	24	.92210	91	.60921	40	.64148	108	.17095	20	.85401	15	39
22	.52051	25	.92118	91	.60960	39	.64041	108	.17116	21	.85385	16	38
23	.52076	25	.92027	91	.61000	40	.63934	108	.17137	21	.85370	15	37
24	.52101	25	.91935	91	.61040	40	.63826	108	.17158	21	.85355	15	36
25	0.52126	25	1.91844	91	0.61080	40	1.63719	108	1.17178	20	0.85340	15	35
26	.52151	25	.91752	91	.61120	40	.63612	108	.17199	21	.85325	15	34
27	.52175	24	.91661	91	.61160	40	.63505	107	.17220	21	.85310	15	33
28	.52200	25	.91570	91	.61200	40	.63398	107	.17241	21	.85294	16	32
29	.52225	25	.91479	91	.61240	40	.63292	107	.17262	21	.85279	15	31
30	0.52250	25	1.91388	90	0.61280	40	1.63185	107	1.17283	21	0.85264	15	30
31	.52275	25	.91297	90	.61320	40	.63079	107	.17304	21	.85249	15	29
32	.52299	24	.91207	90	.61360	40	.62972	107	.17325	21	.85234	15	28
33	.52324	25	.91116	90	.61400	40	.62866	107	.17346	21	.85218	16	27
34	.52349	25	.91026	90	.61440	40	.62760	107	.17367	21	.85203	15	26
35	0.52374	25	1.90935	90	0.61480	40	1.62654	107	1.17388	21	0.85188	15	25
36	.52399	25	.90845	90	.61520	40	.62548	106	.17409	21	.85173	15	24
37	.52423	24	.90755	90	.61561	41	.62442	106	.17430	21	.85157	16	23
38	.52448	25	.90665	90	.61601	40	.62336	106	.17451	21	.85142	15	22
39	.52473	25	.90575	90	.61641	40	.62230	106	.17472	21	.85127	15	21
40	0.52498	25	1.90485	90	0.61681	40	1.62125	106	1.17493	21	0.85112	15	20
41	.52522	24	.90395	90	.61721	40	.62019	106	.17514	21	.85096	16	19
42	.52547	25	.90305	90	.61761	40	.61914	106	.17535	21	.85081	15	18
43	.52572	25	.90216	90	.61801	40	.61808	106	.17556	21	.85066	15	17
44	.52597	25	.90126	90	.61842	41	.61703	106	.17577	21	.85051	15	16
45	0.52621	24	1.90037	90	0.61882	40	1.61598	106	1.17598	21	0.85035	16	15
46	.52646	25	.89948	90	.61922	40	.61493	106	.17620	22	.85020	15	14
47	.52671	25	.89858	90	.61962	40	.61388	104	.17641	21	.85005	15	13
48	.52696	25	.89769	90	.62003	41	.61283	104	.17662	21	.84989	16	12
49	.52720	24	.89680	89	.62043	40	.61179	104	.17683	21	.84974	15	11
50	0.52745	25	1.89591	89	0.62083	40	1.61074	104	1.17704	21	0.84959	15	10
51	.52770	25	.89503	89	.62124	41	.60970	104	.17726	22	.84943	16	9
52	.52794	24	.89414	89	.62164	40	.60865	104	.17747	21	.84928	15	8
53	.52819	25	.89325	89	.62204	40	.60761	104	.17768	21	.84913	15	7
54	.52844	25	.89237	89	.62245	41	.60657	104	.17790	22	.84897	16	6
55	0.52869	25	1.89148	89	0.62285	40	1.60553	104	1.17811	21	0.84882	15	5
56	.52893	24	.89060	89	.62325	40	.60449	104	.17832	21	.84866	16	4
57	.52918	25	.88972	89	.62366	41	.60345	103	.17854	22	.84851	15	3
58	.52943	25	.88884	89	.62406	40	.60241	103	.17875	21	.84836	15	2
59	.52967	24	.88796	88	.62446	40	.60137	103	.17896	21	.84820	16	1
60	0.52992	25	1.88708	88	0.62487	41	1.60033	103	1.17918	22	0.84805	15	0
↑ 121°→	cos	Diff. 1′	sec	Diff. 1′	cot	Diff. 1′	tan	Diff. 1′	csc	Diff. 1′	sin	Diff. 1′←	↑ 58°

TABLE 2
Natural Trigonometric Functions

32°→ ↓ ′	sin	Diff. 1′	csc	Diff. 1′	tan	Diff. 1′	cot	Diff. 1′	sec	Diff. 1′	cos	Diff. 1′	← 147° ↓ ′
0	0.52992		1.88708		0.62487		1.60033		1.17918		0.84805		60
1	.53017	25	.88620	88	.62527	40	.59930	103	.17939	21	.84789	16	59
2	.53041	24	.88532	88	.62568	41	.59826	103	.17961	22	.84774	15	58
3	.53066	25	.88445	88	.62608	40	.59723	103	.17982	21	.84759	15	57
4	.53091	25	.88357	88	.62649	41	.59620	103	.18004	22	.84743	16	56
5	0.53115	24	1.88270	88	0.62689	40	1.59517	103	1.18025	21	0.84728	15	55
6	.53140	25	.88183	88	.62730	41	.59414	103	.18047	22	.84712	16	54
7	.53164	24	.88095	88	.62770	40	.59311	102	.18068	21	.84697	15	53
8	.53189	25	.88008	88	.62811	41	.59208	102	.18090	22	.84681	16	52
9	.53214	25	.87921	88	.62852	41	.59105	102	.18111	21	.84666	15	51
10	0.53238	24	1.87834	87	0.62892	40	1.59002	102	1.18133	22	0.84650	16	50
11	.53263	25	.87748	87	.62933	41	.58900	102	.18155	22	.84635	15	49
12	.53288	25	.87661	87	.62973	40	.58797	102	.18176	21	.84619	16	48
13	.53312	24	.87574	87	.63014	41	.58695	102	.18198	22	.84604	15	47
14	.53337	25	.87488	87	.63055	41	.58593	102	.18220	22	.84588	16	46
15	0.53361	24	1.87401	87	0.63095	40	1.58490	102	1.18241	21	0.84573	15	45
16	.53386	25	.87315	87	.63136	41	.58388	102	.18263	22	.84557	16	44
17	.53411	25	.87229	87	.63177	41	.58286	102	.18285	22	.84542	15	43
18	.53435	24	.87142	87	.63217	40	.58184	101	.18307	22	.84526	16	42
19	.53460	25	.87056	87	.63258	41	.58083	101	.18328	21	.84511	15	41
20	0.53484	24	1.86970	86	0.63299	41	1.57981	101	1.18350	22	0.84495	16	40
21	.53509	25	.86885	86	.63340	41	.57879	101	.18372	22	.84480	15	39
22	.53534	25	.86799	86	.63380	40	.57778	101	.18394	22	.84464	16	38
23	.53558	24	.86713	86	.63421	41	.57676	101	.18416	22	.84448	16	37
24	.53583	25	.86627	86	.63462	41	.57575	101	.18437	21	.84433	15	36
25	0.53607	24	1.86542	86	0.63503	41	1.57474	101	1.18459	22	0.84417	16	35
26	.53632	25	.86457	86	.63544	41	.57372	101	.18481	22	.84402	15	34
27	.53656	24	.86371	86	.63584	40	.57271	101	.18503	22	.84386	16	33
28	.53681	25	.86286	86	.63625	41	.57170	100	.18525	22	.84370	16	32
29	.53705	24	.86201	86	.63666	41	.57069	100	.18547	22	.84355	15	31
30	0.53730	25	1.86116	86	0.63707	41	1.56969	100	1.18569	22	0.84339	16	30
31	.53754	24	.86031	84	.63748	41	.56868	100	.18591	22	.84324	15	29
32	.53779	25	.85946	84	.63789	41	.56767	100	.18613	22	.84308	16	28
33	.53804	25	.85861	84	.63830	41	.56667	100	.18635	22	.84292	16	27
34	.53828	24	.85777	84	.63871	41	.56566	100	.18657	22	.84277	15	26
35	0.53853	25	1.85692	84	0.63912	41	1.56466	100	1.18679	22	0.84261	16	25
36	.53877	24	.85608	84	.63953	41	.56366	100	.18701	22	.84245	16	24
37	.53902	25	.85523	84	.63994	41	.56265	100	.18723	22	.84230	15	23
38	.53926	24	.85439	84	.64035	41	.56165	100	.18745	22	.84214	16	22
39	.53951	25	.85355	84	.64076	41	.56065	100	.18767	22	.84198	16	21
40	0.53975	24	1.85271	84	0.64117	41	1.55966	100	1.18790	23	0.84182	16	20
41	.54000	25	.85187	84	.64158	41	.55866	100	.18812	22	.84167	15	19
42	.54024	24	.85103	83	.64199	41	.55766	100	.18834	22	.84151	16	18
43	.54049	25	.85019	83	.64240	41	.55666	100	.18856	22	.84135	16	17
44	.54073	24	.84935	83	.64281	41	.55567	100	.18878	22	.84120	15	16
45	0.54097	24	1.84852	83	0.64322	41	1.55467	100	1.18901	23	0.84104	16	15
46	.54122	25	.84768	83	.64363	41	.55368	100	.18923	22	.84088	16	14
47	.54146	24	.84685	83	.64404	41	.55269	100	.18945	22	.84072	16	13
48	.54171	25	.84601	83	.64446	42	.55170	100	.18967	22	.84057	15	12
49	.54195	24	.84518	83	.64487	41	.55071	100	.18990	23	.84041	16	11
50	0.54220	25	1.84435	83	0.64528	41	1.54972	99	1.19012	22	0.84025	16	10
51	.54244	24	.84352	83	.64569	41	.54873	99	.19034	22	.84009	16	9
52	.54269	25	.84269	83	.64610	41	.54774	99	.19057	23	.83994	15	8
53	.54293	24	.84186	82	.64652	42	.54675	99	.19079	22	.83978	16	7
54	.54317	24	.84103	82	.64693	41	.54576	99	.19102	23	.83962	16	6
55	0.54342	25	1.84020	82	0.64734	41	1.54478	99	1.19124	22	0.83946	16	5
56	.54366	24	.83938	82	.64775	41	.54379	99	.19146	22	.83930	16	4
57	.54391	25	.83855	82	.64817	42	.54281	99	.19169	23	.83915	15	3
58	.54415	24	.83773	82	.64858	41	.54183	99	.19191	22	.83899	16	2
59	.54440	25	.83690	82	.64899	41	.54085	99	.19214	23	.83883	16	1
60	0.54464	24	1.83608	82	0.64941	42	1.53986	99	1.19236	22	0.83867	16	0
↑ 122°→	cos	Diff. 1′	sec	Diff. 1′	cot	Diff. 1′	tan	Diff. 1′	csc	Diff. 1′	sin	Diff. 1′←	↑ 57°

TABLE 2
Natural Trigonometric Functions

33°→ ↓ ′	sin	Diff. 1′	csc	Diff. 1′	tan	Diff. 1′	cot	Diff. 1′	sec	Diff. 1′	cos	Diff. 1′	← 146° ↓ ′
0	0.54464		1.83608		0.64941		1.53986		1.19236		0.83867		60
1	.54488	24	.83526	82	.64982	41	.53888	99	.19259	23	.83851	16	59
2	.54513	25	.83444	82	.65024	42	.53791	98	.19281	22	.83835	16	58
3	.54537	24	.83362	82	.65065	41	.53693	98	.19304	23	.83819	16	57
4	.54561	24	.83280	81	.65106	41	.53595	98	.19327	23	.83804	15	56
5	0.54586	25	1.83198	81	0.65148	42	1.53497	98	1.19349	22	0.83788	16	55
6	.54610	24	.83116	81	.65189	41	.53400	98	.19372	23	.83772	16	54
7	.54635	25	.83034	81	.65231	42	.53302	98	.19394	22	.83756	16	53
8	.54659	24	.82953	81	.65272	41	.53205	98	.19417	23	.83740	16	52
9	.54683	24	.82871	81	.65314	42	.53107	98	.19440	23	.83724	16	51
10	0.54708	25	1.82790	81	0.65355	41	1.53010	98	1.19463	23	0.83708	16	50
11	.54732	24	.82709	81	.65397	42	.52913	98	.19485	22	.83692	16	49
12	.54756	24	.82627	81	.65438	41	.52816	98	.19508	23	.83676	16	48
13	.54781	25	.82546	81	.65480	42	.52719	97	.19531	23	.83660	16	47
14	.54805	24	.82465	81	.65521	41	.52622	97	.19553	22	.83645	15	46
15	0.54829	24	1.82384	80	0.65563	42	1.52525	97	1.19576	23	0.83629	16	45
16	.54854	25	.82303	80	.65604	41	.52429	97	.19599	23	.83613	16	44
17	.54878	24	.82222	80	.65646	42	.52332	97	.19622	23	.83597	16	43
18	.54902	24	.82142	80	.65688	42	.52235	97	.19645	23	.83581	16	42
19	.54927	25	.82061	80	.65729	41	.52139	97	.19668	23	.83565	16	41
20	0.54951	24	1.81981	80	0.65771	42	1.52043	97	1.19691	23	0.83549	16	40
21	.54975	24	.81900	80	.65813	42	.51946	97	.19713	22	.83533	16	39
22	.54999	24	.81820	80	.65854	41	.51850	97	.19736	23	.83517	16	38
23	.55024	25	.81740	80	.65896	42	.51754	97	.19759	23	.83501	16	37
24	.55048	24	.81659	80	.65938	42	.51658	97	.19782	23	.83485	16	36
25	0.55072	24	1.81579	80	0.65980	42	1.51562	96	1.19805	23	0.83469	16	35
26	.55097	25	.81499	80	.66021	41	.51466	96	.19828	23	.83453	16	34
27	.55121	24	.81419	80	.66063	42	.51370	96	.19851	23	.83437	16	33
28	.55145	24	.81340	80	.66105	42	.51275	96	.19874	23	.83421	16	32
29	.55169	24	.81260	80	.66147	42	.51179	96	.19897	23	.83405	16	31
30	0.55194	25	1.81180	80	0.66189	42	1.51084	96	1.19920	23	0.83389	16	30
31	.55218	24	.81101	80	.66230	41	.50988	96	.19944	24	.83373	16	29
32	.55242	24	.81021	80	.66272	42	.50893	96	.19967	23	.83356	17	28
33	.55266	24	.80942	80	.66314	42	.50797	96	.19990	23	.83340	16	27
34	.55291	25	.80862	80	.66356	42	.50702	96	.20013	23	.83324	16	26
35	0.55315	24	1.80783	80	0.66398	42	1.50607	96	1.20036	23	0.83308	16	25
36	.55339	24	.80704	80	.66440	42	.50512	96	.20059	23	.83292	16	24
37	.55363	24	.80625	80	.66482	42	.50417	94	.20083	24	.83276	16	23
38	.55388	25	.80546	79	.66524	42	.50322	94	.20106	23	.83260	16	22
39	.55412	24	.80467	79	.66566	42	.50228	94	.20129	23	.83244	16	21
40	0.55436	24	1.80388	79	0.66608	42	1.50133	94	1.20152	23	0.83228	16	20
41	.55460	24	.80309	79	.66650	42	.50038	94	.20176	24	.83212	16	19
42	.55484	24	.80231	79	.66692	42	.49944	94	.20199	23	.83195	17	18
43	.55509	25	.80152	79	.66734	42	.49849	94	.20222	23	.83179	16	17
44	.55533	24	.80074	79	.66776	42	.49755	94	.20246	24	.83163	16	16
45	0.55557	24	1.79995	79	0.66818	42	1.49661	94	1.20269	23	0.83147	16	15
46	.55581	24	.79917	79	.66860	42	.49566	94	.20292	23	.83131	16	14
47	.55605	24	.79839	79	.66902	42	.49472	94	.20316	24	.83115	16	13
48	.55630	25	.79761	79	.66944	42	.49378	94	.20339	23	.83098	17	12
49	.55654	24	.79682	79	.66986	42	.49284	93	.20363	24	.83082	16	11
50	0.55678	24	1.79604	78	0.67028	42	1.49190	93	1.20386	23	0.83066	16	10
51	.55702	24	.79527	78	.67071	43	.49097	93	.20410	24	.83050	16	9
52	.55726	24	.79449	78	.67113	42	.49003	93	.20433	23	.83034	16	8
53	.55750	24	.79371	78	.67155	42	.48909	93	.20457	24	.83017	17	7
54	.55775	25	.79293	78	.67197	42	.48816	93	.20480	23	.83001	16	6
55	0.55799	24	1.79216	78	0.67239	42	1.48722	93	1.20504	24	0.82985	16	5
56	.55823	24	.79138	78	.67282	43	.48629	93	.20527	23	.82969	16	4
57	.55847	24	.79061	78	.67324	42	.48536	93	.20551	24	.82953	16	3
58	.55871	24	.78984	78	.67366	42	.48442	93	.20575	24	.82936	17	2
59	.55895	24	.78906	78	.67409	43	.48349	93	.20598	23	.82920	16	1
60	0.55919	24	1.78829	78	0.67451	42	1.48256	93	1.20622	24	0.82904	16	0
↑ 123°→	cos	Diff. 1′	sec	Diff. 1′	cot	Diff. 1′	tan	Diff. 1′	csc	Diff. 1′	sin	Diff. 1′←	↑ 56°

TABLE 2
Natural Trigonometric Functions

34°→ ↓ ′	sin	Diff. 1′	csc	Diff. 1′	tan	Diff. 1′	cot	Diff. 1′	sec	Diff. 1′	cos	Diff. 1′	← 145° ↓ ′
0	0.55919		1.78829		0.67451		1.48256		1.20622		0.82904		60
1	.55943	24	.78752	78	.67493	42	.48163	92	.20645	23	.82887	17	59
2	.55968	25	.78675	77	.67536	43	.48070	92	.20669	24	.82871	16	58
3	.55992	24	.78598	77	.67578	42	.47977	92	.20693	24	.82855	16	57
4	.56016	24	.78521	77	.67620	42	.47885	92	.20717	24	.82839	16	56
5	0.56040	24	1.78445	77	0.67663	43	1.47792	92	1.20740	23	0.82822	17	55
6	.56064	24	.78368	77	.67705	42	.47699	92	.20764	24	.82806	16	54
7	.56088	24	.78291	77	.67748	43	.47607	92	.20788	24	.82790	16	53
8	.56112	24	.78215	77	.67790	42	.47514	92	.20812	24	.82773	17	52
9	.56136	24	.78138	77	.67832	42	.47422	92	.20836	24	.82757	16	51
10	0.56160	24	1.78062	77	0.67875	43	1.47330	92	1.20859	23	0.82741	16	50
11	.56184	24	.77986	77	.67917	42	.47238	92	.20883	24	.82724	17	49
12	.56208	24	.77910	77	.67960	43	.47146	92	.20907	24	.82708	16	48
13	.56232	24	.77833	77	.68002	42	.47053	92	.20931	24	.82692	16	47
14	.56256	24	.77757	77	.68045	43	.46962	91	.20955	24	.82675	17	46
15	0.56280	24	1.77681	76	0.68088	43	1.46870	91	1.20979	24	0.82659	16	45
16	.56305	25	.77606	76	.68130	42	.46778	91	.21003	24	.82643	16	44
17	.56329	24	.77530	76	.68173	43	.46686	91	.21027	24	.82626	17	43
18	.56353	24	.77454	76	.68215	42	.46595	91	.21051	24	.82610	16	42
19	.56377	24	.77378	76	.68258	43	.46503	91	.21075	24	.82593	17	41
20	0.56401	24	1.77303	76	0.68301	43	1.46411	91	1.21099	24	0.82577	16	40
21	.56425	24	.77227	76	.68343	42	.46320	91	.21123	24	.82561	16	39
22	.56449	24	.77152	76	.68386	43	.46229	91	.21147	24	.82544	17	38
23	.56473	24	.77077	76	.68429	43	.46137	91	.21171	24	.82528	16	37
24	.56497	24	.77001	76	.68471	42	.46046	91	.21195	24	.82511	17	36
25	0.56521	24	1.76926	76	0.68514	43	1.45955	91	1.21220	25	0.82495	16	35
26	.56545	24	.76851	76	.68557	43	.45864	91	.21244	24	.82478	17	34
27	.56569	24	.76776	74	.68600	43	.45773	90	.21268	24	.82462	16	33
28	.56593	24	.76701	74	.68642	42	.45682	90	.21292	24	.82446	16	32
29	.56617	24	.76626	74	.68685	43	.45592	90	.21316	24	.82429	17	31
30	0.56641	24	1.76552	74	0.68728	43	1.45501	90	1.21341	25	0.82413	16	30
31	.56665	24	.76477	74	.68771	43	.45410	90	.21365	24	.82396	17	29
32	.56689	24	.76402	74	.68814	43	.45320	90	.21389	24	.82380	16	28
33	.56713	24	.76328	74	.68857	43	.45229	90	.21414	25	.82363	17	27
34	.56736	23	.76253	74	.68900	43	.45139	90	.21438	24	.82347	16	26
35	0.56760	24	1.76179	74	0.68942	42	1.45049	90	1.21462	24	0.82330	17	25
36	.56784	24	.76105	74	.68985	43	.44958	90	.21487	25	.82314	16	24
37	.56808	24	.76031	74	.69028	43	.44868	90	.21511	24	.82297	17	23
38	.56832	24	.75956	74	.69071	43	.44778	90	.21535	24	.82281	16	22
39	.56856	24	.75882	74	.69114	43	.44688	90	.21560	25	.82264	17	21
40	0.56880	24	1.75808	73	0.69157	43	1.44598	90	1.21584	24	0.82248	16	20
41	.56904	24	.75734	73	.69200	43	.44508	90	.21609	25	.82231	17	19
42	.56928	24	.75661	73	.69243	43	.44418	90	.21633	24	.82214	17	18
43	.56952	24	.75587	73	.69286	43	.44329	90	.21658	25	.82198	16	17
44	.56976	24	.75513	73	.69329	43	.44239	90	.21682	24	.82181	17	16
45	0.57000	24	1.75440	73	0.69372	43	1.44149	90	1.21707	25	0.82165	16	15
46	.57024	24	.75366	73	.69416	44	.44060	90	.21731	24	.82148	17	14
47	.57047	23	.75293	73	.69459	43	.43970	90	.21756	25	.82132	16	13
48	.57071	24	.75219	73	.69502	43	.43881	90	.21781	25	.82115	17	12
49	.57095	24	.75146	73	.69545	43	.43792	90	.21805	24	.82098	17	11
50	0.57119	24	1.75073	73	0.69588	43	1.43703	90	1.21830	25	0.82082	16	10
51	.57143	24	.75000	73	.69631	43	.43614	90	.21855	25	.82065	17	9
52	.57167	24	.74927	73	.69675	44	.43525	90	.21879	24	.82048	17	8
53	.57191	24	.74854	72	.69718	43	.43436	89	.21904	25	.82032	16	7
54	.57215	24	.74781	72	.69761	43	.43347	89	.21929	25	.82015	17	6
55	0.57238	23	1.74708	72	0.69804	43	1.43258	89	1.21953	24	0.81999	16	5
56	.57262	24	.74635	72	.69847	43	.43169	89	.21978	25	.81982	17	4
57	.57286	24	.74562	72	.69891	44	.43080	89	.22003	25	.81965	17	3
58	.57310	24	.74490	72	.69934	43	.42992	89	.22028	25	.81949	16	2
59	.57334	24	.74417	72	.69977	43	.42903	89	.22053	25	.81932	17	1
60	0.57358	24	1.74345	72	0.70021	44	1.42815	89	1.22077	24	0.81915	17	0
↑ 124°→	cos	Diff. 1′	sec	Diff. 1′	cot	Diff. 1′	tan	Diff. 1′	csc	Diff. 1′	sin	Diff. 1′←	↑ 55°

TABLE 2
Natural Trigonometric Functions

35°→ ↓ ′	sin	Diff. 1′	csc	Diff. 1′	tan	Diff. 1′	cot	Diff. 1′	sec	Diff. 1′	cos	Diff. 1′	← 144° ↓ ′
0	0.57358		1.74345		0.70021		1.42815		1.22077		0.81915		60
1	.57381	23	.74272	72	.70064	43	.42726	89	.22102	25	.81899	16	59
2	.57405	24	.74200	72	.70107	43	.42638	89	.22127	25	.81882	17	58
3	.57429	24	.74128	72	.70151	44	.42550	89	.22152	25	.81865	17	57
4	.57453	24	.74056	72	.70194	43	.42462	89	.22177	25	.81848	17	56
5	0.57477	24	1.73983	72	0.70238	44	1.42374	89	1.22202	25	0.81832	16	55
6	.57501	24	.73911	72	.70281	43	.42286	89	.22227	25	.81815	17	54
7	.57524	23	.73840	71	.70325	44	.42198	88	.22252	25	.81798	17	53
8	.57548	24	.73768	71	.70368	43	.42110	88	.22277	25	.81782	16	52
9	.57572	24	.73696	71	.70412	44	.42022	88	.22302	25	.81765	17	51
10	0.57596	24	1.73624	71	0.70455	43	1.41934	88	1.22327	25	0.81748	17	50
11	.57619	23	.73552	71	.70499	44	.41847	88	.22352	25	.81731	17	49
12	.57643	24	.73481	71	.70542	43	.41759	88	.22377	25	.81714	17	48
13	.57667	24	.73409	71	.70586	44	.41672	88	.22402	25	.81698	16	47
14	.57691	24	.73338	71	.70629	43	.41584	88	.22428	26	.81681	17	46
15	0.57715	24	1.73267	71	0.70673	44	1.41497	88	1.22453	25	0.81664	17	45
16	.57738	23	.73195	71	.70717	44	.41409	88	.22478	25	.81647	17	44
17	.57762	24	.73124	71	.70760	43	.41322	88	.22503	25	.81631	16	43
18	.57786	24	.73053	71	.70804	44	.41235	88	.22528	25	.81614	17	42
19	.57810	24	.72982	71	.70848	44	.41148	88	.22554	26	.81597	17	41
20	0.57833	23	1.72911	70	0.70891	43	1.41061	88	1.22579	25	0.81580	17	40
21	.57857	24	.72840	70	.70935	44	.40974	87	.22604	25	.81563	17	39
22	.57881	24	.72769	70	.70979	44	.40887	87	.22629	25	.81546	17	38
23	.57904	23	.72698	70	.71023	44	.40800	87	.22655	26	.81530	16	37
24	.57928	24	.72628	70	.71066	43	.40714	87	.22680	25	.81513	17	36
25	0.57952	24	1.72557	70	0.71110	44	1.40627	87	1.22706	26	0.81496	17	35
26	.57976	24	.72487	70	.71154	44	.40540	87	.22731	25	.81479	17	34
27	.57999	23	.72416	70	.71198	44	.40454	87	.22756	25	.81462	17	33
28	.58023	24	.72346	70	.71242	44	.40367	87	.22782	26	.81445	17	32
29	.58047	24	.72275	70	.71285	43	.40281	87	.22807	25	.81428	17	31
30	0.58070	23	1.72205	70	0.71329	44	1.40195	87	1.22833	26	0.81412	16	30
31	.58094	24	.72135	70	.71373	44	.40109	87	.22858	25	.81395	17	29
32	.58118	24	.72065	70	.71417	44	.40022	87	.22884	26	.81378	17	28
33	.58141	23	.71995	70	.71461	44	.39936	87	.22909	25	.81361	17	27
34	.58165	24	.71925	70	.71505	44	.39850	87	.22935	26	.81344	17	26
35	0.58189	24	1.71855	70	0.71549	44	1.39764	86	1.22960	25	0.81327	17	25
36	.58212	23	.71785	70	.71593	44	.39679	86	.22986	26	.81310	17	24
37	.58236	24	.71715	70	.71637	44	.39593	86	.23012	26	.81293	17	23
38	.58260	24	.71646	70	.71681	44	.39507	86	.23037	25	.81276	17	22
39	.58283	23	.71576	70	.71725	44	.39421	86	.23063	26	.81259	17	21
40	0.58307	24	1.71506	70	0.71769	44	1.39336	86	1.23089	26	0.81242	17	20
41	.58330	23	.71437	70	.71813	44	.39250	86	.23114	25	.81225	17	19
42	.58354	24	.71368	70	.71857	44	.39165	86	.23140	26	.81208	17	18
43	.58378	24	.71298	70	.71901	44	.39079	86	.23166	26	.81191	17	17
44	.58401	23	.71229	70	.71946	45	.38994	86	.23192	26	.81174	17	16
45	0.58425	24	1.71160	70	0.71990	44	1.38909	86	1.23217	25	0.81157	17	15
46	.58449	24	.71091	70	.72034	44	.38824	86	.23243	26	.81140	17	14
47	.58472	23	.71022	70	.72078	44	.38738	86	.23269	26	.81123	17	13
48	.58496	24	.70953	69	.72122	44	.38653	86	.23295	26	.81106	17	12
49	.58519	23	.70884	69	.72167	45	.38568	84	.23321	26	.81089	17	11
50	0.58543	24	1.70815	69	0.72211	44	1.38484	84	1.23347	26	0.81072	17	10
51	.58567	24	.70746	69	.72255	44	.38399	84	.23373	26	.81055	17	9
52	.58590	23	.70677	69	.72299	44	.38314	84	.23398	25	.81038	17	8
53	.58614	24	.70609	69	.72344	45	.38229	84	.23424	26	.81021	17	7
54	.58637	23	.70540	69	.72388	44	.38145	84	.23450	26	.81004	17	6
55	0.58661	24	1.70472	69	0.72432	44	1.38060	84	1.23476	26	0.80987	17	5
56	.58684	23	.70403	69	.72477	45	.37976	84	.23502	26	.80970	17	4
57	.58708	24	.70335	69	.72521	44	.37891	84	.23529	27	.80953	17	3
58	.58731	23	.70267	69	.72565	44	.37807	84	.23555	26	.80936	17	2
59	.58755	24	.70198	69	.72610	45	.37722	84	.23581	26	.80919	17	1
60	0.58779	24	1.70130	69	0.72654	44	1.37638	84	1.23607	26	0.80902	17	0
↑ 125°→	cos	Diff. 1′	sec	Diff. 1′	cot	Diff. 1′	tan	Diff. 1′	csc	Diff. 1′	sin	Diff. 1′←	↑ 54°

TABLE 2
Natural Trigonometric Functions

36° → ↓ ′	sin	Diff. 1′	csc	Diff. 1′	tan	Diff. 1′	cot	Diff. 1′	sec	Diff. 1′	cos	Diff. 1′	← 143° ↓ ′
0	0.58779		1.70130		0.72654		1.37638		1.23607		0.80902		60
1	.58802	23	.70062	69	.72699	45	.37554	84	.23633	26	.80885	17	59
2	.58826	24	.69994	69	.72743	44	.37470	84	.23659	26	.80867	18	58
3	.58849	23	.69926	68	.72788	45	.37386	84	.23685	26	.80850	17	57
4	.58873	24	.69858	68	.72832	44	.37302	83	.23711	26	.80833	17	56
5	0.58896	23	1.69790	68	0.72877	45	1.37218	83	1.23738	27	0.80816	17	55
6	.58920	24	.69723	68	.72921	44	.37134	83	.23764	26	.80799	17	54
7	.58943	23	.69655	68	.72966	45	.37050	83	.23790	26	.80782	17	53
8	.58967	24	.69587	68	.73010	44	.36967	83	.23816	26	.80765	17	52
9	.58990	23	.69520	68	.73055	45	.36883	83	.23843	27	.80748	17	51
10	0.59014	24	1.69452	68	0.73100	45	1.36800	83	1.23869	26	0.80730	18	50
11	.59037	23	.69385	68	.73144	44	.36716	83	.23895	26	.80713	17	49
12	.59061	24	.69318	68	.73189	45	.36633	83	.23922	27	.80696	17	48
13	.59084	23	.69250	68	.73234	45	.36549	83	.23948	26	.80679	17	47
14	.59108	24	.69183	68	.73278	44	.36466	83	.23975	27	.80662	17	46
15	0.59131	23	1.69116	68	0.73323	45	1.36383	83	1.24001	26	0.80644	18	45
16	.59154	23	.69049	68	.73368	45	.36300	83	.24028	27	.80627	17	44
17	.59178	24	.68982	67	.73413	45	.36217	83	.24054	26	.80610	17	43
18	.59201	23	.68915	67	.73457	44	.36134	83	.24081	27	.80593	17	42
19	.59225	24	.68848	67	.73502	45	.36051	82	.24107	26	.80576	17	41
20	0.59248	23	1.68782	67	0.73547	45	1.35968	82	1.24134	27	0.80558	18	40
21	.59272	24	.68715	67	.73592	45	.35885	82	.24160	26	.80541	17	39
22	.59295	23	.68648	67	.73637	45	.35802	82	.24187	27	.80524	17	38
23	.59318	23	.68582	67	.73681	44	.35719	82	.24213	26	.80507	17	37
24	.59342	24	.68515	67	.73726	45	.35637	82	.24240	27	.80489	18	36
25	0.59365	23	1.68449	67	0.73771	45	1.35554	82	1.24267	27	0.80472	17	35
26	.59389	24	.68382	67	.73816	45	.35472	82	.24293	26	.80455	17	34
27	.59412	23	.68316	67	.73861	45	.35389	82	.24320	27	.80438	17	33
28	.59436	24	.68250	67	.73906	45	.35307	82	.24347	27	.80420	18	32
29	.59459	23	.68183	67	.73951	45	.35224	82	.24373	26	.80403	17	31
30	0.59482	23	1.68117	67	0.73996	45	1.35142	82	1.24400	27	0.80386	17	30
31	.59506	24	.68051	67	.74041	45	.35060	82	.24427	27	.80368	18	29
32	.59529	23	.67985	66	.74086	45	.34978	82	.24454	27	.80351	17	28
33	.59552	23	.67919	66	.74131	45	.34896	82	.24481	27	.80334	17	27
34	.59576	24	.67853	66	.74176	45	.34814	81	.24508	27	.80316	18	26
35	0.59599	23	1.67788	66	0.74221	45	1.34732	81	1.24534	26	0.80299	17	25
36	.59622	23	.67722	66	.74267	46	.34650	81	.24561	27	.80282	17	24
37	.59646	24	.67656	66	.74312	45	.34568	81	.24588	27	.80264	18	23
38	.59669	23	.67591	66	.74357	45	.34487	81	.24615	27	.80247	17	22
39	.59693	24	.67525	66	.74402	45	.34405	81	.24642	27	.80230	17	21
40	0.59716	23	1.67460	66	0.74447	45	1.34323	81	1.24669	27	0.80212	18	20
41	.59739	23	.67394	66	.74492	45	.34242	81	.24696	27	.80195	17	19
42	.59763	24	.67329	66	.74538	46	.34160	81	.24723	27	.80178	17	18
43	.59786	23	.67264	66	.74583	45	.34079	81	.24750	27	.80160	18	17
44	.59809	23	.67198	66	.74628	45	.33998	81	.24777	27	.80143	17	16
45	0.59832	23	1.67133	66	0.74674	46	1.33916	81	1.24804	27	0.80125	18	15
46	.59856	24	.67068	66	.74719	45	.33835	81	.24832	28	.80108	17	14
47	.59879	23	.67003	66	.74764	45	.33754	81	.24859	27	.80091	17	13
48	.59902	23	.66938	64	.74810	46	.33673	81	.24886	27	.80073	18	12
49	.59926	24	.66873	64	.74855	45	.33592	81	.24913	27	.80056	17	11
50	0.59949	23	1.66809	64	0.74900	45	1.33511	80	1.24940	27	0.80038	18	10
51	.59972	23	.66744	64	.74946	46	.33430	80	.24967	27	.80021	17	9
52	.59995	23	.66679	64	.74991	45	.33349	80	.24995	28	.80003	18	8
53	.60019	24	.66615	64	.75037	46	.33268	80	.25022	27	.79986	17	7
54	.60042	23	.66550	64	.75082	45	.33187	80	.25049	27	.79968	18	6
55	0.60065	23	1.66486	64	0.75128	46	1.33107	80	1.25077	28	0.79951	17	5
56	.60089	24	.66421	64	.75173	45	.33026	80	.25104	27	.79934	17	4
57	.60112	23	.66357	64	.75219	46	.32946	80	.25131	27	.79916	18	3
58	.60135	23	.66292	64	.75264	45	.32865	80	.25159	28	.79899	17	2
59	.60158	23	.66228	64	.75310	46	.32785	80	.25186	27	.79881	18	1
60	0.60182	24	1.66164	64	0.75355	45	1.32704	80	1.25214	28	0.79864	17	0
↑ 126° →	cos	Diff. 1′	sec	Diff. 1′	cot	Diff. 1′	tan	Diff. 1′	csc	Diff. 1′	sin	Diff. 1′ ←	53° ↑

TABLE 2
Natural Trigonometric Functions

37° → ↓ ′	sin	Diff. 1′	csc	Diff. 1′	tan	Diff. 1′	cot	Diff. 1′	sec	Diff. 1′	cos	Diff. 1′	← 142° ↓ ′
0	0.60182		1.66164		0.75355		1.32704		1.25214		0.79864		60
1	.60205	23	.66100	64	.75401	46	.32624	80	.25241	27	.79846	18	59
2	.60228	23	.66036	64	.75447	46	.32544	80	.25269	28	.79829	17	58
3	.60251	23	.65972	63	.75492	45	.32464	80	.25296	27	.79811	18	57
4	.60274	23	.65908	63	.75538	46	.32384	80	.25324	28	.79793	18	56
5	0.60298	24	1.65844	63	0.75584	46	1.32304	80	1.25351	27	0.79776	17	55
6	.60321	23	.65780	63	.75629	45	.32224	80	.25379	28	.79758	18	54
7	.60344	23	.65717	63	.75675	46	.32144	80	.25406	27	.79741	17	53
8	.60367	23	.65653	63	.75721	46	.32064	80	.25434	28	.79723	18	52
9	.60390	23	.65589	63	.75767	46	.31984	80	.25462	28	.79706	17	51
10	0.60414	24	1.65526	63	0.75812	45	1.31904	80	1.25489	27	0.79688	18	50
11	.60437	23	.65462	63	.75858	46	.31825	80	.25517	28	.79671	17	49
12	.60460	23	.65399	63	.75904	46	.31745	80	.25545	28	.79653	18	48
13	.60483	23	.65335	63	.75950	46	.31666	80	.25572	27	.79635	18	47
14	.60506	23	.65272	63	.75996	46	.31586	80	.25600	28	.79618	17	46
15	0.60529	23	1.65209	63	0.76042	46	1.31507	80	1.25628	28	0.79600	18	45
16	.60553	24	.65146	63	.76088	46	.31427	80	.25656	28	.79583	17	44
17	.60576	23	.65083	63	.76134	46	.31348	80	.25683	27	.79565	18	43
18	.60599	23	.65020	63	.76180	46	.31269	80	.25711	28	.79547	18	42
19	.60622	23	.64957	62	.76226	46	.31190	80	.25739	28	.79530	17	41
20	0.60645	23	1.64894	62	0.76272	46	1.31110	80	1.25767	28	0.79512	18	40
21	.60668	23	.64831	62	.76318	46	.31031	80	.25795	28	.79494	18	39
22	.60691	23	.64768	62	.76364	46	.30952	80	.25823	28	.79477	17	38
23	.60714	23	.64705	62	.76410	46	.30873	79	.25851	28	.79459	18	37
24	.60738	24	.64643	62	.76456	46	.30795	79	.25879	28	.79441	18	36
25	0.60761	23	1.64580	62	0.76502	46	1.30716	79	1.25907	28	0.79424	17	35
26	.60784	23	.64518	62	.76548	46	.30637	79	.25935	28	.79406	18	34
27	.60807	23	.64455	62	.76594	46	.30558	79	.25963	28	.79388	18	33
28	.60830	23	.64393	62	.76640	46	.30480	79	.25991	28	.79371	17	32
29	.60853	23	.64330	62	.76686	46	.30401	79	.26019	28	.79353	18	31
30	0.60876	23	1.64268	62	0.76733	47	1.30323	79	1.26047	28	0.79335	18	30
31	.60899	23	.64206	62	.76779	46	.30244	79	.26075	28	.79318	17	29
32	.60922	23	.64144	62	.76825	46	.30166	79	.26104	29	.79300	18	28
33	.60945	23	.64081	62	.76871	46	.30087	79	.26132	28	.79282	18	27
34	.60968	23	.64019	62	.76918	47	.30009	79	.26160	28	.79264	18	26
35	0.60991	23	1.63957	61	0.76964	46	1.29931	79	1.26188	28	0.79247	17	25
36	.61015	24	.63895	61	.77010	46	.29853	79	.26216	28	.79229	18	24
37	.61038	23	.63834	61	.77057	47	.29775	79	.26245	29	.79211	18	23
38	.61061	23	.63772	61	.77103	46	.29696	79	.26273	28	.79193	18	22
39	.61084	23	.63710	61	.77149	46	.29618	78	.26301	28	.79176	17	21
40	0.61107	23	1.63648	61	0.77196	47	1.29541	78	1.26330	29	0.79158	18	20
41	.61130	23	.63587	61	.77242	46	.29463	78	.26358	28	.79140	18	19
42	.61153	23	.63525	61	.77289	47	.29385	78	.26387	29	.79122	18	18
43	.61176	23	.63464	61	.77335	46	.29307	78	.26415	28	.79105	17	17
44	.61199	23	.63402	61	.77382	47	.29229	78	.26443	28	.79087	18	16
45	0.61222	23	1.63341	61	0.77428	46	1.29152	78	1.26472	29	0.79069	18	15
46	.61245	23	.63279	61	.77475	47	.29074	78	.26500	28	.79051	18	14
47	.61268	23	.63218	61	.77521	46	.28997	78	.26529	29	.79033	18	13
48	.61291	23	.63157	61	.77568	47	.28919	78	.26557	28	.79016	17	12
49	.61314	23	.63096	61	.77615	47	.28842	78	.26586	29	.78998	18	11
50	0.61337	23	1.63035	61	0.77661	46	1.28764	78	1.26615	29	0.78980	18	10
51	.61360	23	.62974	61	.77708	47	.28687	78	.26643	28	.78962	18	9
52	.61383	23	.62913	60	.77754	46	.28610	78	.26672	29	.78944	18	8
53	.61406	23	.62852	60	.77801	47	.28533	78	.26701	29	.78926	18	7
54	.61429	23	.62791	60	.77848	47	.28456	78	.26729	28	.78908	18	6
55	0.61451	22	1.62730	60	0.77895	47	1.28379	78	1.26758	29	0.78891	17	5
56	.61474	23	.62669	60	.77941	46	.28302	78	.26787	29	.78873	18	4
57	.61497	23	.62609	60	.77988	47	.28225	77	.26815	28	.78855	18	3
58	.61520	23	.62548	60	.78035	47	.28148	77	.26844	29	.78837	18	2
59	.61543	23	.62487	60	.78082	47	.28071	77	.26873	29	.78819	18	1
60	0.61566	23	1.62427	60	0.78129	47	1.27994	77	1.26902	29	0.78801	18	0
↑ 127° →	cos	Diff. 1′	sec	Diff. 1′	cot	Diff. 1′	tan	Diff. 1′	csc	Diff. 1′	sin	Diff. 1′ ←	52° ↑

TABLE 2

Natural Trigonometric Functions

38° ↓ ′	sin	Diff. 1′	csc	Diff. 1′	tan	Diff. 1′	cot	Diff. 1′	sec	Diff. 1′	cos	Diff. 1′	← 141° ↓ ′
0	0.61566		1.62427		0.78129		1.27994		1.26902		0.78801		60
1	.61589	23	.62366	60	.78175	46	.27917	77	.26931	29	.78783	18	59
2	.61612	23	.62306	60	.78222	47	.27841	77	.26960	29	.78765	18	58
3	.61635	23	.62246	60	.78269	47	.27764	77	.26988	28	.78747	18	57
4	.61658	23	.62185	60	.78316	47	.27688	77	.27017	29	.78729	18	56
5	0.61681	23	1.62125	60	0.78363	47	1.27611	77	1.27046	29	0.78711	18	55
6	.61704	23	.62065	60	.78410	47	.27535	77	.27075	29	.78694	17	54
7	.61726	22	.62005	60	.78457	47	.27458	77	.27104	29	.78676	18	53
8	.61749	23	.61945	60	.78504	47	.27382	77	.27133	29	.78658	18	52
9	.61772	23	.61885	60	.78551	47	.27306	77	.27162	29	.78640	18	51
10	0.61795	23	1.61825	60	0.78598	47	1.27230	77	1.27191	29	0.78622	18	50
11	.61818	23	.61765	60	.78645	47	.27153	77	.27221	30	.78604	18	49
12	.61841	23	.61705	60	.78692	47	.27077	77	.27250	29	.78586	18	48
13	.61864	23	.61646	60	.78739	47	.27001	77	.27279	29	.78568	18	47
14	.61887	23	.61586	60	.78786	47	.26925	76	.27308	29	.78550	18	46
15	0.61909	22	1.61526	60	0.78834	48	1.26849	76	1.27337	29	0.78532	18	45
16	.61932	23	.61467	60	.78881	47	.26774	76	.27366	29	.78514	18	44
17	.61955	23	.61407	60	.78928	47	.26698	76	.27396	30	.78496	18	43
18	.61978	23	.61348	60	.78975	47	.26622	76	.27425	29	.78478	18	42
19	.62001	23	.61288	60	.79022	47	.26546	76	.27454	29	.78460	18	41
20	0.62024	23	1.61229	60	0.79070	48	1.26471	76	1.27483	29	0.78442	18	40
21	.62046	22	.61170	60	.79117	47	.26395	76	.27513	30	.78424	18	39
22	.62069	23	.61111	60	.79164	47	.26319	76	.27542	29	.78405	19	38
23	.62092	23	.61051	60	.79212	48	.26244	76	.27572	30	.78387	18	37
24	.62115	23	.60992	60	.79259	47	.26169	76	.27601	29	.78369	18	36
25	0.62138	23	1.60933	60	0.79306	47	1.26093	76	1.27630	29	0.78351	18	35
26	.62160	22	.60874	60	.79354	48	.26018	76	.27660	30	.78333	18	34
27	.62183	23	.60815	59	.79401	47	.25943	76	.27689	29	.78315	18	33
28	.62206	23	.60756	59	.79449	48	.25867	76	.27719	30	.78297	18	32
29	.62229	23	.60698	59	.79496	47	.25792	76	.27748	29	.78279	18	31
30	0.62251	22	1.60639	59	0.79544	48	1.25717	76	1.27778	30	0.78261	18	30
31	.62274	23	.60580	59	.79591	47	.25642	76	.27807	29	.78243	18	29
32	.62297	23	.60521	59	.79639	48	.25567	74	.27837	30	.78225	18	28
33	.62320	23	.60463	59	.79686	47	.25492	74	.27867	30	.78206	19	27
34	.62342	22	.60404	59	.79734	48	.25417	74	.27896	29	.78188	18	26
35	0.62365	23	1.60346	59	0.79781	47	1.25343	74	1.27926	30	0.78170	18	25
36	.62388	23	.60287	59	.79829	48	.25268	74	.27956	30	.78152	18	24
37	.62411	23	.60229	59	.79877	48	.25193	74	.27985	29	.78134	18	23
38	.62433	22	.60171	59	.79924	47	.25118	74	.28015	30	.78116	18	22
39	.62456	23	.60112	59	.79972	48	.25044	74	.28045	30	.78098	18	21
40	0.62479	23	1.60054	59	0.80020	48	1.24969	74	1.28075	30	0.78079	19	20
41	.62502	23	.59996	59	.80067	47	.24895	74	.28105	30	.78061	18	19
42	.62524	22	.59938	59	.80115	48	.24820	74	.28134	29	.78043	18	18
43	.62547	23	.59880	59	.80163	48	.24746	74	.28164	30	.78025	18	17
44	.62570	23	.59822	58	.80211	48	.24672	74	.28194	30	.78007	18	16
45	0.62592	22	1.59764	58	0.80258	47	1.24597	74	1.28224	30	0.77988	19	15
46	.62615	23	.59706	58	.80306	48	.24523	74	.28254	30	.77970	18	14
47	.62638	23	.59648	58	.80354	48	.24449	74	.28284	30	.77952	18	13
48	.62660	22	.59590	58	.80402	48	.24375	74	.28314	30	.77934	18	12
49	.62683	23	.59533	58	.80450	48	.24301	74	.28344	30	.77916	18	11
50	0.62706	23	1.59475	58	0.80498	48	1.24227	74	1.28374	30	0.77897	19	10
51	.62728	22	.59418	58	.80546	48	.24153	73	.28404	30	.77879	18	9
52	.62751	23	.59360	58	.80594	48	.24079	73	.28434	30	.77861	18	8
53	.62774	23	.59302	58	.80642	48	.24005	73	.28464	30	.77843	18	7
54	.62796	22	.59245	58	.80690	48	.23931	73	.28495	31	.77824	19	6
55	0.62819	23	1.59188	58	0.80738	48	1.23858	73	1.28525	30	0.77806	18	5
56	.62842	23	.59130	58	.80786	48	.23784	73	.28555	30	.77788	18	4
57	.62864	22	.59073	58	.80834	48	.23710	73	.28585	30	.77769	19	3
58	.62887	23	.59016	58	.80882	48	.23637	73	.28615	30	.77751	18	2
59	.62909	22	.58959	58	.80930	48	.23563	73	.28646	31	.77733	18	1
60	0.62932	23	1.58902	58	0.80978	48	1.23490	73	1.28676	30	0.77715	18	0
↑ 128° →	cos	Diff. 1′	sec	Diff. 1′	cot	Diff. 1′	tan	Diff. 1′	csc	Diff. 1′	sin	Diff. 1′	← 51° ↑

TABLE 2

Natural Trigonometric Functions

39° ↓ ′	sin	Diff. 1′	csc	Diff. 1′	tan	Diff. 1′	cot	Diff. 1′	sec	Diff. 1′	cos	Diff. 1′	← 140° ↓ ′
0	0.62932		1.58902		0.80978		1.23490		1.28676		0.77715		60
1	.62955	23	.58845	58	.81027	49	.23416	73	.28706	30	.77696	19	59
2	.62977	22	.58788	57	.81075	48	.23343	73	.28737	31	.77678	18	58
3	.63000	23	.58731	57	.81123	48	.23270	73	.28767	30	.77660	18	57
4	.63022	22	.58674	57	.81171	48	.23196	73	.28797	30	.77641	19	56
5	0.63045	23	1.58617	57	0.81220	49	1.23123	73	1.28828	31	0.77623	18	55
6	.63068	23	.58560	57	.81268	48	.23050	73	.28858	30	.77605	18	54
7	.63090	22	.58503	57	.81316	48	.22977	73	.28889	31	.77586	19	53
8	.63113	23	.58447	57	.81364	48	.22904	73	.28919	30	.77568	18	52
9	.63135	22	.58390	57	.81413	49	.22831	73	.28950	31	.77550	18	51
10	0.63158	23	1.58333	57	0.81461	48	1.22758	72	1.28980	30	0.77531	19	50
11	.63180	22	.58277	57	.81510	49	.22685	72	.29011	31	.77513	18	49
12	.63203	23	.58221	57	.81558	48	.22612	72	.29042	31	.77494	19	48
13	.63225	22	.58164	57	.81606	48	.22539	72	.29072	30	.77476	18	47
14	.63248	23	.58108	57	.81655	49	.22467	72	.29103	31	.77458	18	46
15	0.63271	23	1.58051	57	0.81703	48	1.22394	72	1.29133	30	0.77439	19	45
16	.63293	22	.57995	57	.81752	49	.22321	72	.29164	31	.77421	18	44
17	.63316	23	.57939	57	.81800	48	.22249	72	.29195	31	.77402	19	43
18	.63338	22	.57883	57	.81849	49	.22176	72	.29226	31	.77384	18	42
19	.63361	23	.57827	57	.81898	49	.22104	72	.29256	30	.77366	18	41
20	0.63383	22	1.57771	57	0.81946	48	1.22031	72	1.29287	31	0.77347	19	40
21	.63406	23	.57715	56	.81995	49	.21959	72	.29318	31	.77329	18	39
22	.63428	22	.57659	56	.82044	49	.21886	72	.29349	31	.77310	19	38
23	.63451	23	.57603	56	.82092	48	.21814	72	.29380	31	.77292	18	37
24	.63473	22	.57547	56	.82141	49	.21742	72	.29411	31	.77273	19	36
25	0.63496	23	1.57491	56	0.82190	49	1.21670	72	1.29442	31	0.77255	18	35
26	.63518	22	.57436	56	.82238	48	.21598	72	.29473	31	.77236	19	34
27	.63540	22	.57380	56	.82287	49	.21526	72	.29504	31	.77218	18	33
28	.63563	23	.57324	56	.82336	49	.21454	72	.29535	31	.77199	19	32
29	.63585	22	.57269	56	.82385	49	.21382	71	.29566	31	.77181	18	31
30	0.63608	23	1.57213	56	0.82434	49	1.21310	71	1.29597	31	0.77162	19	30
31	.63630	22	.57158	56	.82483	49	.21238	71	.29628	31	.77144	18	29
32	.63653	23	.57103	56	.82531	48	.21166	71	.29659	31	.77125	19	28
33	.63675	22	.57047	56	.82580	49	.21094	71	.29690	31	.77107	18	27
34	.63698	23	.56992	56	.82629	49	.21023	71	.29721	31	.77088	19	26
35	0.63720	22	1.56937	56	0.82678	49	1.20951	71	1.29752	31	0.77070	18	25
36	.63742	22	.56881	56	.82727	49	.20879	71	.29784	32	.77051	19	24
37	.63765	23	.56826	56	.82776	49	.20808	71	.29815	31	.77033	18	23
38	.63787	22	.56771	56	.82825	49	.20736	71	.29846	31	.77014	19	22
39	.63810	23	.56716	56	.82874	49	.20665	71	.29877	31	.76996	18	21
40	0.63832	22	1.56661	54	0.82923	49	1.20593	71	1.29909	32	0.76977	19	20
41	.63854	22	.56606	54	.82972	49	.20522	71	.29940	31	.76959	18	19
42	.63877	23	.56551	54	.83022	50	.20451	71	.29971	31	.76940	19	18
43	.63899	22	.56497	54	.83071	49	.20379	71	.30003	32	.76921	19	17
44	.63922	23	.56442	54	.83120	49	.20308	71	.30034	31	.76903	18	16
45	0.63944	22	1.56387	54	0.83169	49	1.20237	71	1.30066	32	0.76884	19	15
46	.63966	22	.56332	54	.83218	49	.20166	71	.30097	31	.76866	18	14
47	.63989	23	.56278	54	.83268	50	.20095	71	.30129	32	.76847	19	13
48	.64011	22	.56223	54	.83317	49	.20024	71	.30160	31	.76828	19	12
49	.64033	22	.56169	54	.83366	49	.19953	70	.30192	32	.76810	18	11
50	0.64056	23	1.56114	54	0.83415	49	1.19882	70	1.30223	31	0.76791	19	10
51	.64078	22	.56060	54	.83465	50	.19811	70	.30255	32	.76772	19	9
52	.64100	22	.56005	54	.83514	49	.19740	70	.30287	32	.76754	18	8
53	.64123	23	.55951	54	.83564	50	.19669	70	.30318	31	.76735	19	7
54	.64145	22	.55897	54	.83613	49	.19599	70	.30350	32	.76717	18	6
55	0.64167	22	1.55843	54	0.83662	49	1.19528	70	1.30382	32	0.76698	19	5
56	.64190	23	.55789	54	.83712	50	.19457	70	.30413	31	.76679	19	4
57	.64212	22	.55734	54	.83761	49	.19387	70	.30445	32	.76661	18	3
58	.64234	22	.55680	54	.83811	50	.19316	70	.30477	32	.76642	19	2
59	.64256	22	.55626	54	.83860	49	.19246	70	.30509	32	.76623	19	1
60	0.64279	23	1.55572	53	0.83910	50	1.19175	70	1.30541	32	0.76604	19	0
↑ 129° →	cos	Diff. 1′	sec	Diff. 1′	cot	Diff. 1′	tan	Diff. 1′	csc	Diff. 1′	sin	Diff. 1′	← 50° ↑

TABLE 2
Natural Trigonometric Functions

40°→ ↓	sin	Diff. 1′	csc	Diff. 1′	tan	Diff. 1′	cot	Diff. 1′	sec	Diff. 1′	cos	Diff. 1′	← 139° ↓
′													′
0	0.64279		1.55572		0.83910		1.19175		1.30541		0.76604		60
1	.64301	22	.55518	53	.83960	50	.19105	70	.30573	32	.76586	18	59
2	.64323	22	.55465	53	.84009	49	.19035	70	.30605	32	.76567	19	58
3	.64346	23	.55411	53	.84059	50	.18964	70	.30636	31	.76548	19	57
4	.64368	22	.55357	53	.84108	49	.18894	70	.30668	32	.76530	18	56
5	0.64390	22	1.55303	53	0.84158	50	1.18824	70	1.30700	32	0.76511	19	55
6	.64412	22	.55250	53	.84208	50	.18754	70	.30732	32	.76492	19	54
7	.64435	23	.55196	53	.84258	50	.18684	70	.30764	32	.76473	19	53
8	.64457	22	.55143	53	.84307	49	.18614	70	.30796	32	.76455	18	52
9	.64479	22	.55089	53	.84357	50	.18544	70	.30829	33	.76436	19	51
10	0.64501	22	1.55036	53	0.84407	50	1.18474	70	1.30861	32	0.76417	19	50
11	.64524	23	.54982	53	.84457	50	.18404	70	.30893	32	.76398	19	49
12	.64546	22	.54929	53	.84507	50	.18334	70	.30925	32	.76380	18	48
13	.64568	22	.54876	53	.84556	49	.18264	70	.30957	32	.76361	19	47
14	.64590	22	.54822	53	.84606	50	.18194	70	.30989	32	.76342	19	46
15	0.64612	22	1.54769	53	0.84656	50	1.18125	70	1.31022	33	0.76323	19	45
16	.64635	23	.54716	53	.84706	50	.18055	70	.31054	32	.76304	19	44
17	.64657	22	.54663	53	.84756	50	.17986	70	.31086	32	.76286	18	43
18	.64679	22	.54610	53	.84806	50	.17916	70	.31119	33	.76267	19	42
19	.64701	22	.54557	53	.84856	50	.17846	70	.31151	32	.76248	19	41
20	0.64723	22	1.54504	52	0.84906	50	1.17777	70	1.31183	32	0.76229	19	40
21	.64746	23	.54451	52	.84956	50	.17708	70	.31216	33	.76210	19	39
22	.64768	22	.54398	52	.85006	50	.17638	70	.31248	32	.76192	18	38
23	.64790	22	.54345	52	.85057	51	.17569	70	.31281	33	.76173	19	37
24	.64812	22	.54292	52	.85107	50	.17500	70	.31313	32	.76154	19	36
25	0.64834	22	1.54240	52	0.85157	50	1.17430	70	1.31346	33	0.76135	19	35
26	.64856	22	.54187	52	.85207	50	.17361	70	.31378	32	.76116	19	34
27	.64878	22	.54134	52	.85257	50	.17292	70	.31411	33	.76097	19	33
28	.64901	23	.54082	52	.85308	51	.17223	70	.31443	32	.76078	19	32
29	.64923	22	.54029	52	.85358	50	.17154	70	.31476	33	.76059	19	31
30	0.64945	22	1.53977	52	0.85408	50	1.17085	69	1.31509	33	0.76041	18	30
31	.64967	22	.53924	52	.85458	50	.17016	69	.31541	32	.76022	19	29
32	.64989	22	.53872	52	.85509	51	.16947	69	.31574	33	.76003	19	28
33	.65011	22	.53820	52	.85559	50	.16878	69	.31607	33	.75984	19	27
34	.65033	22	.53768	52	.85609	50	.16809	69	.31640	33	.75965	19	26
35	0.65055	22	1.53715	52	0.85660	51	1.16741	69	1.31672	32	0.75946	19	25
36	.65077	22	.53663	52	.85710	50	.16672	69	.31705	33	.75927	19	24
37	.65100	23	.53611	52	.85761	51	.16603	69	.31738	33	.75908	19	23
38	.65122	22	.53559	52	.85811	50	.16535	69	.31771	33	.75889	19	22
39	.65144	22	.53507	52	.85862	51	.16466	69	.31804	33	.75870	19	21
40	0.65166	22	1.53455	51	0.85912	50	1.16398	69	1.31837	33	0.75851	19	20
41	.65188	22	.53403	51	.85963	51	.16329	69	.31870	33	.75832	19	19
42	.65210	22	.53351	51	.86014	51	.16261	69	.31903	33	.75813	19	18
43	.65232	22	.53299	51	.86064	50	.16192	69	.31936	33	.75794	19	17
44	.65254	22	.53247	51	.86115	51	.16124	69	.31969	33	.75775	19	16
45	0.65276	22	1.53196	51	0.86166	51	1.16056	69	1.32002	33	0.75756	19	15
46	.65298	22	.53144	51	.86216	50	.15987	69	.32035	33	.75738	18	14
47	.65320	22	.53092	51	.86267	51	.15919	69	.32068	33	.75719	19	13
48	.65342	22	.53041	51	.86318	51	.15851	69	.32101	33	.75700	19	12
49	.65364	22	.52989	51	.86368	50	.15783	69	.32134	33	.75680	20	11
50	0.65386	22	1.52938	51	0.86419	51	1.15715	69	1.32168	34	0.75661	19	10
51	.65408	22	.52886	51	.86470	51	.15647	69	.32201	33	.75642	19	9
52	.65430	22	.52835	51	.86521	51	.15579	68	.32234	33	.75623	19	8
53	.65452	22	.52784	51	.86572	51	.15511	68	.32267	33	.75604	19	7
54	.65474	22	.52732	51	.86623	51	.15443	68	.32301	34	.75585	19	6
55	0.65496	22	1.52681	51	0.86674	51	1.15375	68	1.32334	33	0.75566	19	5
56	.65518	22	.52630	51	.86725	51	.15308	68	.32368	34	.75547	19	4
57	.65540	22	.52579	51	.86776	51	.15240	68	.32401	33	.75528	19	3
58	.65562	22	.52527	51	.86827	51	.15172	68	.32434	33	.75509	19	2
59	.65584	22	.52476	51	.86878	51	.15104	68	.32468	34	.75490	19	1
60	0.65606	22	1.52425	51	0.86929	51	1.15037	68	1.32501	33	0.75471	19	0
↑ 130°→	cos	Diff. 1′	sec	Diff. 1′	cot	Diff. 1′	tan	Diff. 1′	csc	Diff. 1′	sin	Diff. 1′	↑ ←49°

TABLE 2
Natural Trigonometric Functions

41°→ ↓	sin	Diff. 1′	csc	Diff. 1′	tan	Diff. 1′	cot	Diff. 1′	sec	Diff. 1′	cos	Diff. 1′	← 138° ↓
′													′
0	0.65606		1.52425		0.86929		1.15037		1.32501		0.75471		60
1	.65628	22	.52374	50	.86980	51	.14969	68	.32535	34	.75452	19	59
2	.65650	22	.52323	50	.87031	51	.14902	68	.32568	33	.75433	19	58
3	.65672	22	.52273	50	.87082	51	.14834	68	.32602	34	.75414	19	57
4	.65694	22	.52222	50	.87133	51	.14767	68	.32636	34	.75395	19	56
5	0.65716	22	1.52171	50	0.87184	51	1.14699	68	1.32669	33	0.75375	20	55
6	.65738	22	.52120	50	.87236	52	.14632	68	.32703	34	.75356	19	54
7	.65759	21	.52069	50	.87287	51	.14565	68	.32737	34	.75337	19	53
8	.65781	22	.52019	50	.87338	51	.14498	68	.32770	33	.75318	19	52
9	.65803	22	.51968	50	.87389	51	.14430	68	.32804	34	.75299	19	51
10	0.65825	22	1.51918	50	0.87441	52	1.14363	68	1.32838	34	0.75280	19	50
11	.65847	22	.51867	50	.87492	51	.14296	68	.32872	34	.75261	19	49
12	.65869	22	.51817	50	.87543	51	.14229	68	.32905	33	.75241	20	48
13	.65891	22	.51766	50	.87595	52	.14162	68	.32939	34	.75222	19	47
14	.65913	22	.51716	50	.87646	51	.14095	67	.32973	34	.75203	19	46
15	0.65935	22	1.51665	50	0.87698	52	1.14028	67	1.33007	34	0.75184	19	45
16	.65956	21	.51615	50	.87749	51	.13961	67	.33041	34	.75165	19	44
17	.65978	22	.51565	50	.87801	52	.13894	67	.33075	34	.75146	19	43
18	.66000	22	.51515	50	.87852	51	.13828	67	.33109	34	.75126	20	42
19	.66022	22	.51465	50	.87904	52	.13761	67	.33143	34	.75107	19	41
20	0.66044	22	1.51415	50	0.87955	51	1.13694	67	1.33177	34	0.75088	19	40
21	.66066	22	.51364	50	.88007	52	.13627	67	.33211	34	.75069	19	39
22	.66088	22	.51314	50	.88059	52	.13561	67	.33245	34	.75050	19	38
23	.66109	21	.51265	50	.88110	51	.13494	67	.33279	34	.75030	20	37
24	.66131	22	.51215	50	.88162	52	.13428	67	.33314	35	.75011	19	36
25	0.66153	22	1.51165	50	0.88214	52	1.13361	67	1.33348	34	0.74992	19	35
26	.66175	22	.51115	50	.88265	51	.13295	67	.33382	34	.74973	19	34
27	.66197	22	.51065	50	.88317	52	.13228	67	.33416	34	.74953	20	33
28	.66218	21	.51015	50	.88369	52	.13162	67	.33451	35	.74934	19	32
29	.66240	22	.50966	50	.88421	52	.13096	67	.33485	34	.74915	19	31
30	0.66262	22	1.50916	50	0.88473	52	1.13029	67	1.33519	34	0.74896	19	30
31	.66284	22	.50866	50	.88524	51	.12963	67	.33554	35	.74876	20	29
32	.66306	22	.50817	50	.88576	52	.12897	67	.33588	34	.74857	19	28
33	.66327	21	.50767	50	.88628	52	.12831	67	.33622	34	.74838	19	27
34	.66349	22	.50718	50	.88680	52	.12765	67	.33657	35	.74818	20	26
35	0.66371	22	1.50669	50	0.88732	52	1.12699	67	1.33691	34	0.74799	19	25
36	.66393	22	.50619	50	.88784	52	.12633	67	.33726	35	.74780	19	24
37	.66414	21	.50570	50	.88836	52	.12567	66	.33760	34	.74760	20	23
38	.66436	22	.50521	50	.88888	52	.12501	66	.33795	35	.74741	19	22
39	.66458	22	.50471	50	.88940	52	.12435	66	.33830	35	.74722	19	21
40	0.66480	22	1.50422	50	0.88992	52	1.12369	66	1.33864	34	0.74703	19	20
41	.66501	21	.50373	50	.89045	53	.12303	66	.33899	35	.74683	20	19
42	.66523	22	.50324	50	.89097	52	.12238	66	.33934	35	.74664	19	18
43	.66545	22	.50275	50	.89149	52	.12172	66	.33968	34	.74644	20	17
44	.66566	21	.50226	50	.89201	52	.12106	66	.34003	35	.74625	19	16
45	0.66588	22	1.50177	49	0.89253	52	1.12041	66	1.34038	35	0.74606	19	15
46	.66610	22	.50128	49	.89306	53	.11975	66	.34073	35	.74586	20	14
47	.66632	22	.50079	49	.89358	52	.11909	66	.34108	35	.74567	19	13
48	.66653	21	.50030	49	.89410	52	.11844	66	.34142	34	.74548	19	12
49	.66675	22	.49981	49	.89463	53	.11778	66	.34177	35	.74528	20	11
50	0.66697	22	1.49933	49	0.89515	52	1.11713	66	1.34212	35	0.74509	19	10
51	.66718	21	.49884	49	.89567	52	.11648	66	.34247	35	.74489	20	9
52	.66740	22	.49835	49	.89620	53	.11582	66	.34282	35	.74470	19	8
53	.66762	22	.49787	49	.89672	52	.11517	66	.34317	35	.74451	19	7
54	.66783	21	.49738	49	.89725	53	.11452	66	.34352	35	.74431	20	6
55	0.66805	22	1.49690	49	0.89777	52	1.11387	66	1.34387	35	0.74412	19	5
56	.66827	22	.49641	49	.89830	53	.11321	66	.34423	36	.74392	20	4
57	.66848	21	.49593	49	.89883	53	.11256	66	.34458	35	.74373	19	3
58	.66870	22	.49544	49	.89935	52	.11191	66	.34493	35	.74353	20	2
59	.66891	21	.49496	49	.89988	53	.11126	66	.34528	35	.74334	19	1
60	0.66913	22	1.49448	49	0.90040	52	1.11061	64	1.34563	35	0.74314	20	0
↑ 131°→	cos	Diff. 1′	sec	Diff. 1′	cot	Diff. 1′	tan	Diff. 1′	csc	Diff. 1′	sin	Diff. 1′	↑ ←48°

TABLE 2
Natural Trigonometric Functions

42° → ↓ ′	sin	Diff. 1′	csc	Diff. 1′	tan	Diff. 1′	cot	Diff. 1′	sec	Diff. 1′	cos	Diff. 1′	← 137° ↓ ′
0	0.66913		1.49448		0.90040		1.11061		1.34563		0.74314		60
1	.66935	22	.49399	49	.90093	53	.10996	64	.34599	36	.74295	19	59
2	.66956	21	.49351	49	.90146	53	.10931	64	.34634	35	.74276	19	58
3	.66978	22	.49303	49	.90199	53	.10867	64	.34669	35	.74256	20	57
4	.66999	21	.49255	49	.90251	52	.10802	64	.34704	35	.74237	19	56
5	0.67021	22	1.49207	49	0.90304	53	1.10737	64	1.34740	36	0.74217	20	55
6	.67043	22	.49159	49	.90357	53	.10672	64	.34775	35	.74198	19	54
7	.67064	21	.49111	48	.90410	53	.10607	64	.34811	36	.74178	20	53
8	.67086	22	.49063	48	.90463	53	.10543	64	.34846	35	.74159	19	52
9	.67107	21	.49015	48	.90516	53	.10478	64	.34882	36	.74139	20	51
10	0.67129	22	1.48967	48	0.90569	53	1.10414	64	1.34917	35	0.74120	19	50
11	.67151	22	.48919	48	.90621	52	.10349	64	.34953	36	.74100	20	49
12	.67172	21	.48871	48	.90674	53	.10285	64	.34988	35	.74080	20	48
13	.67194	22	.48824	48	.90727	53	.10220	64	.35024	36	.74061	19	47
14	.67215	21	.48776	48	.90781	54	.10156	64	.35060	36	.74041	20	46
15	0.67237	22	1.48728	48	0.90834	53	1.10091	64	1.35095	35	0.74022	19	45
16	.67258	21	.48681	48	.90887	53	.10027	64	.35131	36	.74002	20	44
17	.67280	22	.48633	48	.90940	53	.09963	64	.35167	36	.73983	19	43
18	.67301	21	.48586	48	.90993	53	.09899	64	.35203	36	.73963	20	42
19	.67323	22	.48538	48	.91046	53	.09834	64	.35238	35	.73944	19	41
20	0.67344	21	1.48491	48	0.91099	53	1.09770	64	1.35274	36	0.73924	20	40
21	.67366	22	.48443	48	.91153	54	.09706	64	.35310	36	.73904	20	39
22	.67387	21	.48396	48	.91206	53	.09642	64	.35346	36	.73885	19	38
23	.67409	22	.48349	48	.91259	53	.09578	64	.35382	36	.73865	20	37
24	.67430	21	.48301	48	.91313	54	.09514	63	.35418	36	.73846	19	36
25	0.67452	22	1.48254	48	0.91366	53	1.09450	63	1.35454	36	0.73826	20	35
26	.67473	21	.48207	48	.91419	53	.09386	63	.35490	36	.73806	20	34
27	.67495	22	.48160	48	.91473	54	.09322	63	.35526	36	.73787	19	33
28	.67516	21	.48113	48	.91526	53	.09258	63	.35562	36	.73767	20	32
29	.67538	22	.48066	48	.91580	54	.09195	63	.35598	36	.73747	20	31
30	0.67559	21	1.48019	48	0.91633	53	1.09131	63	1.35634	36	0.73728	19	30
31	.67580	21	.47972	47	.91687	54	.09067	63	.35670	36	.73708	20	29
32	.67602	22	.47925	47	.91740	53	.09003	63	.35707	37	.73688	20	28
33	.67623	21	.47878	47	.91794	54	.08940	63	.35743	36	.73669	19	27
34	.67645	22	.47831	47	.91847	53	.08876	63	.35779	36	.73649	20	26
35	0.67666	21	1.47784	47	0.91901	54	1.08813	63	1.35815	36	0.73629	20	25
36	.67688	22	.47738	47	.91955	54	.08749	63	.35852	37	.73610	19	24
37	.67709	21	.47691	47	.92008	53	.08686	63	.35888	36	.73590	20	23
38	.67730	21	.47644	47	.92062	54	.08622	63	.35924	36	.73570	20	22
39	.67752	22	.47598	47	.92116	54	.08559	63	.35961	37	.73551	19	21
40	0.67773	21	1.47551	47	0.92170	54	1.08496	63	1.35997	36	0.73531	20	20
41	.67795	22	.47504	47	.92224	54	.08432	63	.36034	37	.73511	20	19
42	.67816	21	.47458	47	.92277	53	.08369	63	.36070	36	.73491	20	18
43	.67837	21	.47411	47	.92331	54	.08306	63	.36107	37	.73472	19	17
44	.67859	22	.47365	47	.92385	54	.08243	63	.36143	36	.73452	20	16
45	0.67880	21	1.47319	47	0.92439	54	1.08179	63	1.36180	37	0.73432	20	15
46	.67901	21	.47272	47	.92493	54	.08116	63	.36217	37	.73413	19	14
47	.67923	22	.47226	47	.92547	54	.08053	63	.36253	36	.73393	20	13
48	.67944	21	.47180	47	.92601	54	.07990	63	.36290	37	.73373	20	12
49	.67965	21	.47134	47	.92655	54	.07927	62	.36327	37	.73353	20	11
50	0.67987	22	1.47087	47	0.92709	54	1.07864	62	1.36363	36	0.73333	20	10
51	.68008	21	.47041	47	.92763	54	.07801	62	.36400	37	.73314	19	9
52	.68029	21	.46995	47	.92817	54	.07738	62	.36437	37	.73294	20	8
53	.68051	22	.46949	47	.92872	55	.07676	62	.36474	37	.73274	20	7
54	.68072	21	.46903	47	.92926	54	.07613	62	.36511	37	.73254	20	6
55	0.68093	21	1.46857	46	0.92980	54	1.07550	62	1.36548	37	0.73234	20	5
56	.68115	22	.46811	46	.93034	54	.07487	62	.36585	37	.73215	19	4
57	.68136	21	.46765	46	.93088	54	.07425	62	.36622	37	.73195	20	3
58	.68157	21	.46719	46	.93143	55	.07362	62	.36659	37	.73175	20	2
59	.68179	22	.46674	46	.93197	54	.07299	62	.36696	37	.73155	20	1
60	0.68200	21	1.46628	46	0.93252	55	1.07237	62	1.36733	37	0.73135	20	0
↑ 132° →	cos	Diff. 1′	sec	Diff. 1′	cot	Diff. 1′	tan	Diff. 1′	csc	Diff. 1′	sin	Diff. 1′	↑ ← 47°

TABLE 2
Natural Trigonometric Functions

43° → ↓ ′	sin	Diff. 1′	csc	Diff. 1′	tan	Diff. 1′	cot	Diff. 1′	sec	Diff. 1′	cos	Diff. 1′	← 136° ↓ ′
0	0.68200		1.46628		0.93252		1.07237		1.36733		0.73135		60
1	.68221	21	.46582	46	.93306	54	.07174	62	.36770	37	.73116	19	59
2	.68242	21	.46537	46	.93360	54	.07112	62	.36807	37	.73096	20	58
3	.68264	22	.46491	46	.93415	55	.07049	62	.36844	37	.73076	20	57
4	.68285	21	.46445	46	.93469	54	.06987	62	.36881	37	.73056	20	56
5	0.68306	21	1.46400	46	0.93524	55	1.06925	62	1.36919	38	0.73036	20	55
6	.68327	21	.46354	46	.93578	54	.06862	62	.36956	37	.73016	20	54
7	.68349	22	.46309	46	.93633	55	.06800	62	.36993	37	.72996	20	53
8	.68370	21	.46263	46	.93688	55	.06738	62	.37030	37	.72976	20	52
9	.68391	21	.46218	46	.93742	54	.06676	62	.37068	38	.72957	19	51
10	0.68412	21	1.46173	46	0.93797	55	1.06613	62	1.37105	37	0.72937	20	50
11	.68434	22	.46127	46	.93852	55	.06551	62	.37143	38	.72917	20	49
12	.68455	21	.46082	46	.93906	54	.06489	62	.37180	37	.72897	20	48
13	.68476	21	.46037	46	.93961	55	.06427	62	.37218	38	.72877	20	47
14	.68497	21	.45992	46	.94016	55	.06365	62	.37255	37	.72857	20	46
15	0.68518	21	1.45946	46	0.94071	55	1.06303	61	1.37293	38	0.72837	20	45
16	.68539	21	.45901	46	.94125	54	.06241	61	.37330	37	.72817	20	44
17	.68561	22	.45856	46	.94180	55	.06179	61	.37368	38	.72797	20	43
18	.68582	21	.45811	46	.94235	55	.06117	61	.37406	38	.72777	20	42
19	.68603	21	.45766	44	.94290	55	.06056	61	.37443	37	.72757	20	41
20	0.68624	21	1.45721	44	0.94345	55	1.05994	61	1.37481	38	0.72737	20	40
21	.68645	21	.45676	44	.94400	55	.05932	61	.37519	38	.72717	20	39
22	.68666	21	.45631	44	.94455	55	.05870	61	.37556	37	.72697	20	38
23	.68688	22	.45587	44	.94510	55	.05809	61	.37594	38	.72677	20	37
24	.68709	21	.45542	44	.94565	55	.05747	61	.37632	38	.72657	20	36
25	0.68730	21	1.45497	44	0.94620	55	1.05685	61	1.37670	38	0.72637	20	35
26	.68751	21	.45452	44	.94676	56	.05624	61	.37708	38	.72617	20	34
27	.68772	21	.45408	44	.94731	55	.05562	61	.37746	38	.72597	20	33
28	.68793	21	.45363	44	.94786	55	.05501	61	.37784	38	.72577	20	32
29	.68814	21	.45319	44	.94841	55	.05439	61	.37822	38	.72557	20	31
30	0.68835	21	1.45274	44	0.94896	55	1.05378	61	1.37860	38	0.72537	20	30
31	.68857	22	.45229	44	.94952	56	.05317	61	.37898	38	.72517	20	29
32	.68878	21	.45185	44	.95007	55	.05255	61	.37936	38	.72497	20	28
33	.68899	21	.45141	44	.95062	55	.05194	61	.37974	38	.72477	20	27
34	.68920	21	.45096	44	.95118	56	.05133	61	.38012	38	.72457	20	26
35	0.68941	21	1.45052	44	0.95173	55	1.05072	61	1.38051	39	0.72437	20	25
36	.68962	21	.45007	44	.95229	56	.05010	61	.38089	38	.72417	20	24
37	.68983	21	.44963	44	.95284	55	.04949	61	.38127	38	.72397	20	23
38	.69004	21	.44919	44	.95340	56	.04888	61	.38165	38	.72377	20	22
39	.69025	21	.44875	44	.95395	55	.04827	61	.38204	39	.72357	20	21
40	0.69046	21	1.44831	44	0.95451	56	1.04766	61	1.38242	38	0.72337	20	20
41	.69067	21	.44787	44	.95506	55	.04705	60	.38280	38	.72317	20	19
42	.69088	21	.44742	44	.95562	56	.04644	60	.38319	39	.72297	20	18
43	.69109	21	.44698	44	.95618	56	.04583	60	.38357	38	.72277	20	17
44	.69130	21	.44654	44	.95673	55	.04522	60	.38396	39	.72257	20	16
45	0.69151	21	1.44610	43	0.95729	56	1.04461	60	1.38434	38	0.72236	21	15
46	.69172	21	.44567	43	.95785	56	.04401	60	.38473	39	.72216	20	14
47	.69193	21	.44523	43	.95841	56	.04340	60	.38512	39	.72196	20	13
48	.69214	21	.44479	43	.95897	56	.04279	60	.38550	38	.72176	20	12
49	.69235	21	.44435	43	.95952	55	.04218	60	.38589	39	.72156	20	11
50	0.69256	21	1.44391	43	0.96008	56	1.04158	60	1.38628	39	0.72136	20	10
51	.69277	21	.44347	43	.96064	56	.04097	60	.38666	38	.72116	20	9
52	.69298	21	.44304	43	.96120	56	.04036	60	.38705	39	.72095	21	8
53	.69319	21	.44260	43	.96176	56	.03976	60	.38744	39	.72075	20	7
54	.69340	21	.44217	43	.96232	56	.03915	60	.38783	39	.72055	20	6
55	0.69361	21	1.44173	43	0.96288	56	1.03855	60	1.38822	39	0.72035	20	5
56	.69382	21	.44129	43	.96344	56	.03794	60	.38860	38	.72015	20	4
57	.69403	21	.44086	43	.96400	56	.03734	60	.38899	39	.71995	20	3
58	.69424	21	.44042	43	.96457	57	.03674	60	.38938	39	.71974	21	2
59	.69445	21	.43999	43	.96513	56	.03613	60	.38977	39	.71954	20	1
60	0.69466	21	1.43956	43	0.96569	56	1.03553	60	1.39016	39	0.71934	20	0
↑ 133° →	cos	Diff. 1′	sec	Diff. 1′	cot	Diff. 1′	tan	Diff. 1′	csc	Diff. 1′	sin	Diff. 1′	↑ ← 46°

TABLE 2
Natural Trigonometric Functions

44°→ ↓ ′	sin	Diff. 1′	csc	Diff. 1′	tan	Diff. 1′	cot	Diff. 1′	sec	Diff. 1′	cos	Diff. 1′	← 135° ↓ ′
0	0.69466		1.43956		0.96569		1.03553		1.39016		0.71934		60
1	.69487	21	.43912	43	.96625	56	.03493	60	.39055	39	.71914	20	59
2	.69508	21	.43869	43	.96681	56	.03433	60	.39095	40	.71894	20	58
3	.69529	21	.43826	43	.96738	57	.03372	60	.39134	39	.71873	21	57
4	.69549	20	.43783	43	.96794	56	.03312	60	.39173	39	.71853	20	56
5	0.69570	21	1.43739	43	0.96850	56	1.03252	60	1.39212	39	0.71833	20	55
6	.69591	21	.43696	43	.96907	57	.03192	60	.39251	39	.71813	20	54
7	.69612	21	.43653	43	.96963	56	.03132	60	.39291	40	.71792	21	53
8	.69633	21	.43610	43	.97020	57	.03072	60	.39330	39	.71772	20	52
9	.69654	21	.43567	43	.97076	56	.03012	60	.39369	39	.71752	20	51
10	0.69675	21	1.43524	43	0.97133	57	1.02952	60	1.39409	40	0.71732	20	50
11	.69696	21	.43481	42	.97189	56	.02892	60	.39448	39	.71711	21	49
12	.69717	21	.43438	42	.97246	57	.02832	60	.39487	39	.71691	20	48
13	.69737	20	.43395	42	.97302	56	.02772	60	.39527	40	.71671	20	47
14	.69758	21	.43352	42	.97359	57	.02713	60	.39566	39	.71650	21	46
15	0.69779	21	1.43309	42	0.97416	57	1.02653	60	1.39606	40	0.71630	20	45
16	.69800	21	.43267	42	.97472	56	.02593	60	.39646	40	.71610	20	44
17	.69821	21	.43224	42	.97529	57	.02533	60	.39685	39	.71590	20	43
18	.69842	21	.43181	42	.97586	57	.02474	60	.39725	40	.71569	21	42
19	.69862	20	.43139	42	.97643	57	.02414	60	.39764	39	.71549	20	41
20	0.69883	21	1.43096	42	0.97700	57	1.02355	60	1.39804	40	0.71529	20	40
21	.69904	21	.43053	42	.97756	56	.02295	60	.39844	40	.71508	21	39
22	.69925	21	.43011	42	.97813	57	.02236	60	.39884	40	.71488	20	38
23	.69946	21	.42968	42	.97870	57	.02176	60	.39924	40	.71468	20	37
24	.69966	20	.42926	42	.97927	57	.02117	60	.39963	39	.71447	21	36
25	0.69987	21	1.42883	42	0.97984	57	1.02057	60	1.40003	40	0.71427	20	35
26	.70008	21	.42841	42	.98041	57	.01998	60	.40043	40	.71407	20	34
27	.70029	21	.42799	42	.98098	57	.01939	60	.40083	40	.71386	21	33
28	.70049	20	.42756	42	.98155	57	.01879	60	.40123	40	.71366	20	32
29	.70070	21	.42714	42	.98213	58	.01820	60	.40163	40	.71345	21	31
30	0.70091	21	1.42672	42	0.98270	57	1.01761	60	1.40203	40	0.71325	20	30
31	.70112	21	.42630	42	.98327	57	.01702	60	.40243	40	.71305	20	29
32	.70132	20	.42587	42	.98384	57	.01642	60	.40283	40	.71284	21	28
33	.70153	21	.42545	42	.98441	57	.01583	60	.40324	41	.71264	20	27
34	.70174	21	.42503	42	.98499	58	.01524	60	.40364	40	.71243	21	26
35	0.70195	21	1.42461	42	0.98556	57	1.01465	60	1.40404	40	0.71223	20	25
36	.70215	20	.42419	42	.98613	57	.01406	60	.40444	40	.71203	20	24
37	.70236	21	.42377	41	.98671	58	.01347	59	.40485	41	.71182	21	23
38	.70257	21	.42335	41	.98728	57	.01288	59	.40525	40	.71162	20	22
39	.70277	20	.42293	41	.98786	58	.01229	59	.40565	40	.71141	21	21
40	0.70298	21	1.42251	41	0.98843	57	1.01170	59	1.40606	41	0.71121	20	20
41	.70319	21	.42209	41	.98901	58	.01112	59	.40646	40	.71100	21	19
42	.70339	20	.42168	41	.98958	57	.01053	59	.40687	41	.71080	20	18
43	.70360	21	.42126	41	.99016	58	.00994	59	.40727	40	.71059	21	17
44	.70381	21	.42084	41	.99073	57	.00935	59	.40768	41	.71039	20	16
45	0.70401	20	1.42042	41	0.99131	58	1.00876	59	1.40808	40	0.71019	20	15
46	.70422	21	.42001	41	.99189	58	.00818	59	.40849	41	.70998	21	14
47	.70443	21	.41959	41	.99247	58	.00759	59	.40890	41	.70978	20	13
48	.70463	20	.41918	41	.99304	57	.00701	59	.40930	40	.70957	21	12
49	.70484	21	.41876	41	.99362	58	.00642	59	.40971	41	.70937	20	11
50	0.70505	21	1.41835	41	0.99420	58	1.00583	59	1.41012	41	0.70916	21	10
51	.70525	20	.41793	41	.99478	58	.00525	59	.41053	41	.70896	20	9
52	.70546	21	.41752	41	.99536	58	.00467	59	.41093	40	.70875	21	8
53	.70567	21	.41710	41	.99594	58	.00408	59	.41134	41	.70855	20	7
54	.70587	20	.41669	41	.99652	58	.00350	59	.41175	41	.70834	21	6
55	0.70608	21	1.41627	41	0.99710	58	1.00291	59	1.41216	41	0.70813	21	5
56	.70628	20	.41586	41	.99768	58	.00233	59	.41257	41	.70793	20	4
57	.70649	21	.41545	41	.99826	58	.00175	59	.41298	41	.70772	21	3
58	.70670	21	.41504	41	.99884	58	.00116	59	.41339	41	.70752	20	2
59	.70690	20	.41463	41	.99942	58	.00058	59	.41380	41	.70731	21	1
60	0.70711	21	1.41421	41	1.00000	58	1.00000	59	1.41421	41	0.70711	20	0
↑ 134°→	cos	Diff. 1′	sec	Diff. 1′	cot	Diff. 1′	tan	Diff. 1′	csc	Diff. 1′	sin	Diff. 1′	↑ ← 45°

TABLE 3
Common Logarithms of Trigonometric Functions (offset +10)

0°→ ↓	sin	Diff. 1′	csc	tan	Diff. 1′	cot	sec	Diff. 1′	cos	←179° ↓
0	∞		∞	∞		∞	10.00000		10.00000	60
1	6.46373	— —	13.53627	6.46373	— —	13.53627	.00000	0	.00000	59
2	.76476	30103	.23524	.76476	30103	.23524	.00000	0	.00000	58
3	6.94085	17609	.05915	6.94085	17609	.05915	.00000	0	.00000	57
4	7.06579	12494	.93421	7.06579	12494	.93421	.00000	0	.00000	56
5	7.16270	9691	12.83730	7.16270	9691	12.83730	10.00000	0	10.00000	55
6	.24188	7918	.75812	.24188	7918	.75812	.00000	0	.00000	54
7	.30882	6694	.69118	.30882	6694	.69118	.00000	0	.00000	53
8	.36682	5800	.63318	.36682	5800	.63318	.00000	0	.00000	52
9	.41797	5115	.58203	.41797	5115	.58203	.00000	0	.00000	51
10	7.46373	4576	12.53627	7.46373	4576	12.53627	10.00000	0	10.00000	50
11	.50512	4139	.49488	.50512	4139	.49488	.00000	0	.00000	49
12	.54291	3779	.45709	.54291	3779	.45709	.00000	0	.00000	48
13	.57767	3476	.42233	.57767	3476	.42233	.00000	0	.00000	47
14	.60985	3218	.39015	.60986	3219	.39014	.00000	0	.00000	46
15	7.63982	2997	12.36018	7.63982	2996	12.36018	10.00000	0	10.00000	45
16	.66784	2802	.33216	.66785	2803	.33215	.00000	0	.00000	44
17	.69417	2633	.30583	.69418	2633	.30582	.00001	1	.99999	43
18	.71900	2483	.28100	.71900	2482	.28100	.00001	0	.99999	42
19	.74248	2348	.25752	.74248	2348	.25752	.00001	0	.99999	41
20	7.76475	2227	12.23525	7.76476	2228	12.23524	10.00001	0	9.99999	40
21	.78594	2119	.21406	.78595	2119	.21405	.00001	0	.99999	39
22	.80615	2021	.19385	.80615	2020	.19385	.00001	0	.99999	38
23	.82545	1930	.17455	.82546	1931	.17454	.00001	0	.99999	37
24	.84393	1848	.15607	.84394	1848	.15606	.00001	0	.99999	36
25	7.86166	1773	12.13834	7.86167	1773	12.13833	10.00001	0	9.99999	35
26	.87870	1704	.12130	.87871	1704	.12129	.00001	0	.99999	34
27	.89509	1639	.10491	.89510	1639	.10490	.00001	0	.99999	33
28	.91088	1579	.08912	.91089	1579	.08911	.00001	0	.99999	32
29	.92612	1524	.07388	.92613	1524	.07387	.00002	1	.99998	31
30	7.94084	1472	12.05916	7.94086	1473	12.05914	10.00002	0	9.99998	30
31	.95508	1424	.04492	.95510	1424	.04490	.00002	0	.99998	29
32	.96887	1379	.03113	.96889	1379	.03111	.00002	0	.99998	28
33	.98223	1336	.01777	.98225	1336	.01775	.00002	0	.99998	27
34	7.99520	1297	.00480	7.99522	1297	.00478	.00002	0	.99998	26
35	8.00779	1259	11.99221	8.00781	1259	11.99219	10.00002	0	9.99998	25
36	.02002	1223	.97998	.02004	1223	.97996	.00002	0	.99998	24
37	.03192	1190	.96808	.03194	1190	.96806	.00003	1	.99997	23
38	.04350	1158	.95650	.04353	1159	.95647	.00003	0	.99997	22
39	.05478	1128	.94522	.05481	1128	.94519	.00003	0	.99997	21
40	8.06578	1100	11.93422	8.06581	1100	11.93419	10.00003	0	9.99997	20
41	.07650	1072	.92350	.07653	1072	.92347	.00003	0	.99997	19
42	.08696	1046	.91304	.08700	1047	.91300	.00003	0	.99997	18
43	.09718	1022	.90282	.09722	1022	.90278	.00003	0	.99997	17
44	.10717	999	.89283	.10720	998	.89280	.00004	1	.99996	16
45	8.11693	976	11.88307	8.11696	976	11.88304	10.00004	0	9.99996	15
46	.12647	954	.87353	.12651	955	.87349	.00004	0	.99996	14
47	.13581	934	.86419	.13585	934	.86415	.00004	0	.99996	13
48	.14495	914	.85505	.14500	915	.85500	.00004	0	.99996	12
49	.15391	896	.84609	.15395	895	.84605	.00004	0	.99996	11
50	8.16268	877	11.83732	8.16273	878	11.83727	10.00005	1	9.99995	10
51	.17128	860	.82872	.17133	860	.82867	.00005	0	.99995	9
52	.17971	843	.82029	.17976	843	.82024	.00005	0	.99995	8
53	.18798	827	.81202	.18804	828	.81196	.00005	0	.99995	7
54	.19610	812	.80390	.19616	812	.80384	.00005	0	.99995	6
55	8.20407	797	11.79593	8.20413	797	11.79587	10.00006	1	9.99994	5
56	.21189	782	.78811	.21195	782	.78805	.00006	0	.99994	4
57	.21958	769	.78042	.21964	769	.78036	.00006	0	.99994	3
58	.22713	755	.77287	.22720	756	.77280	.00006	0	.99994	2
59	.23456	743	.76544	.23462	742	.76538	.00006	0	.99994	1
60	8.24186	730	11.75814	8.24192	730	11.75808	10.00007	1	9.99993	0
↑ 90°→	cos	Diff. 1′	sec	cot	Diff. 1′	tan	csc	Diff. 1′	sin	← 89° ↑

TABLE 3
Common Logarithms of Trigonometric Functions (offset +10)

1°→ ↓	sin	Diff. 1′	csc	tan	Diff. 1′	cot	sec	Diff. 1′	cos	←178° ↓
0	8.24186		11.75814	8.24192		11.75808	10.00007		9.99993	60
1	.24903	717	.75097	.24910	718	.75090	.00007	0	.99993	59
2	.25609	706	.74391	.25616	706	.74384	.00007	0	.99993	58
3	.26304	695	.73696	.26312	696	.73688	.00007	0	.99993	57
4	.26988	684	.73012	.26996	684	.73004	.00008	1	.99992	56
5	8.27661	673	11.72339	8.27669	673	11.72331	10.00008	0	9.99992	55
6	.28324	663	.71676	.28332	663	.71668	.00008	0	.99992	54
7	.28977	653	.71023	.28986	654	.71014	.00008	0	.99992	53
8	.29621	644	.70379	.29629	643	.70371	.00008	0	.99992	52
9	.30255	634	.69745	.30263	634	.69737	.00009	1	.99991	51
10	8.30879	624	11.69121	8.30888	625	11.69112	10.00009	0	9.99991	50
11	.31495	616	.68505	.31505	617	.68495	.00009	0	.99991	49
12	.32103	608	.67897	.32112	607	.67888	.00010	1	.99990	48
13	.32702	599	.67298	.32711	599	.67289	.00010	0	.99990	47
14	.33292	590	.66708	.33302	591	.66698	.00010	0	.99990	46
15	8.33875	583	11.66125	8.33886	584	11.66114	10.00010	0	9.99990	45
16	.34450	575	.65550	.34461	575	.65539	.00011	1	.99989	44
17	.35018	568	.64982	.35029	568	.64971	.00011	0	.99989	43
18	.35578	560	.64422	.35590	561	.64410	.00011	0	.99989	42
19	.36131	553	.63869	.36143	553	.63857	.00011	0	.99989	41
20	8.36678	547	11.63322	8.36689	546	11.63311	10.00012	1	9.99988	40
21	.37217	539	.62783	.37229	540	.62771	.00012	0	.99988	39
22	.37750	533	.62250	.37762	533	.62238	.00012	0	.99988	38
23	.38276	526	.61724	.38289	527	.61711	.00013	1	.99987	37
24	.38796	520	.61204	.38809	520	.61191	.00013	0	.99987	36
25	8.39310	514	11.60690	8.39323	514	11.60677	10.00013	0	9.99987	35
26	.39818	508	.60182	.39832	509	.60168	.00014	1	.99986	34
27	.40320	502	.59680	.40334	502	.59666	.00014	0	.99986	33
28	.40816	496	.59184	.40830	496	.59170	.00014	0	.99986	32
29	.41307	491	.58693	.41321	491	.58679	.00015	1	.99985	31
30	8.41792	485	11.58208	8.41807	486	11.58193	10.00015	0	9.99985	30
31	.42272	480	.57728	.42287	480	.57713	.00015	0	.99985	29
32	.42746	474	.57254	.42762	475	.57238	.00016	1	.99984	28
33	.43216	470	.56784	.43232	470	.56768	.00016	0	.99984	27
34	.43680	464	.56320	.43696	464	.56304	.00016	0	.99984	26
35	8.44139	459	11.55861	8.44156	460	11.55844	10.00017	1	9.99983	25
36	.44594	455	.55406	.44611	455	.55389	.00017	0	.99983	24
37	.45044	450	.54956	.45061	450	.54939	.00017	0	.99983	23
38	.45489	445	.54511	.45507	446	.54493	.00018	1	.99982	22
39	.45930	441	.54070	.45948	441	.54052	.00018	0	.99982	21
40	8.46366	436	11.53634	8.46385	437	11.53615	10.00018	0	9.99982	20
41	.46799	433	.53201	.46817	432	.53183	.00019	1	.99981	19
42	.47226	427	.52774	.47245	428	.52755	.00019	0	.99981	18
43	.47650	424	.52350	.47669	424	.52331	.00019	0	.99981	17
44	.48069	419	.51931	.48089	420	.51911	.00020	1	.99980	16
45	8.48485	416	11.51515	8.48505	416	11.51495	10.00020	0	9.99980	15
46	.48896	411	.51104	.48917	412	.51083	.00021	1	.99979	14
47	.49304	408	.50696	.49325	408	.50675	.00021	0	.99979	13
48	.49708	404	.50292	.49729	404	.50271	.00021	0	.99979	12
49	.50108	400	.49892	.50130	401	.49870	.00022	1	.99978	11
50	8.50504	396	11.49496	8.50527	397	11.49473	10.00022	0	9.99978	10
51	.50897	393	.49103	.50920	393	.49080	.00023	1	.99977	9
52	.51287	390	.48713	.51310	390	.48690	.00023	0	.99977	8
53	.51673	386	.48327	.51696	386	.48304	.00023	0	.99977	7
54	.52055	382	.47945	.52079	383	.47921	.00024	1	.99976	6
55	8.52434	379	11.47566	8.52459	380	11.47541	10.00024	0	9.99976	5
56	.52810	376	.47190	.52835	376	.47165	.00025	1	.99975	4
57	.53183	373	.46817	.53208	373	.46792	.00025	0	.99975	3
58	.53552	369	.46448	.53578	370	.46422	.00026	1	.99974	2
59	.53919	367	.46081	.53945	367	.46055	.00026	0	.99974	1
60	8.54282	363	11.45718	8.54308	363	11.45692	10.00026	0	9.99974	0
↑ 91°→	cos	Diff. 1′	sec	cot	Diff. 1′	tan	csc	Diff. 1′	sin	← 88° ↑

TABLE 3

Common Logarithms of Trigonometric Functions (offset +10)

2° → ′	sin	Diff. 1′	csc	tan	Diff. 1′	cot	sec	Diff. 1′	cos	← 177° ′
0	8.54282	360	11.45718	8.54308	361	11.45692	10.00026	1	9.99974	60
1	.54642	357	.45358	.54669	358	.45331	.00027	0	.99973	59
2	.54999	355	.45001	.55027	355	.44973	.00027	1	.99973	58
3	.55354	351	.44646	.55382	352	.44618	.00028	0	.99972	57
4	.55705	349	.44295	.55734	349	.44266	.00028	1	.99972	56
5	8.56054	346	11.43946	8.56083	346	11.43917	10.00029	0	9.99971	55
6	.56400	343	.43600	.56429	344	.43571	.00029	1	.99971	54
7	.56743	341	.43257	.56773	341	.43227	.00030	0	.99970	53
8	.57084	337	.42916	.57114	338	.42886	.00030	1	.99970	52
9	.57421	336	.42579	.57452	336	.42548	.00031	0	.99969	51
10	8.57757	332	11.42243	8.57788	333	11.42212	10.00031	1	9.99969	50
11	.58089	330	.41911	.58121	330	.41879	.00032	0	.99968	49
12	.58419	328	.41581	.58451	328	.41549	.00032	1	.99968	48
13	.58747	325	.41253	.58779	326	.41221	.00033	0	.99967	47
14	.59072	323	.40928	.59105	323	.40895	.00033	0	.99967	46
15	8.59395	320	11.40605	8.59428	321	11.40572	10.00033	1	9.99967	45
16	.59715	318	.40285	.59749	319	.40251	.00034	0	.99966	44
17	.60033	316	.39967	.60068	316	.39932	.00034	1	.99966	43
18	.60349	313	.39651	.60384	314	.39616	.00035	1	.99965	42
19	.60662	311	.39338	.60698	311	.39302	.00036	0	.99964	41
20	8.60973	309	11.39027	8.61009	310	11.38991	10.00036	1	9.99964	40
21	.61282	307	.38718	.61319	307	.38681	.00037	0	.99963	39
22	.61589	305	.38411	.61626	305	.38374	.00037	1	.99963	38
23	.61894	302	.38106	.61931	303	.38069	.00038	0	.99962	37
24	.62196	301	.37804	.62234	301	.37766	.00038	1	.99962	36
25	8.62497	298	11.37503	8.62535	299	11.37465	10.00039	0	9.99961	35
26	.62795	296	.37205	.62834	297	.37166	.00039	1	.99961	34
27	.63091	294	.36909	.63131	295	.36869	.00040	0	.99960	33
28	.63385	293	.36615	.63426	292	.36574	.00040	1	.99960	32
29	.63678	290	.36322	.63718	291	.36282	.00041	0	.99959	31
30	8.63968	288	11.36032	8.64009	289	11.35991	10.00041	1	9.99959	30
31	.64256	287	.35744	.64298	287	.35702	.00042	0	.99958	29
32	.64543	284	.35457	.64585	285	.35415	.00042	1	.99958	28
33	.64827	283	.35173	.64870	284	.35130	.00043	1	.99957	27
34	.65110	281	.34890	.65154	281	.34846	.00044	0	.99956	26
35	8.65391	279	11.34609	8.65435	280	11.34565	10.00044	1	9.99956	25
36	.65670	277	.34330	.65715	278	.34285	.00045	0	.99955	24
37	.65947	276	.34053	.65993	276	.34007	.00045	1	.99955	23
38	.66223	274	.33777	.66269	274	.33731	.00046	0	.99954	22
39	.66497	272	.33503	.66543	273	.33457	.00046	1	.99954	21
40	8.66769	270	11.33231	8.66816	271	11.33184	10.00047	1	9.99953	20
41	.67039	269	.32961	.67087	269	.32913	.00048	0	.99952	19
42	.67308	267	.32692	.67356	268	.32644	.00048	1	.99952	18
43	.67575	266	.32425	.67624	266	.32376	.00049	0	.99951	17
44	.67841	263	.32159	.67890	264	.32110	.00049	1	.99951	16
45	8.68104	263	11.31896	8.68154	263	11.31846	10.00050	1	9.99950	15
46	.68367	260	.31633	.68417	261	.31583	.00051	0	.99949	14
47	.68627	259	.31373	.68678	260	.31322	.00051	1	.99949	13
48	.68886	258	.31114	.68938	258	.31062	.00052	0	.99948	12
49	.69144	256	.30856	.69196	257	.30804	.00052	1	.99948	11
50	8.69400	254	11.30600	8.69453	255	11.30547	10.00053	1	9.99947	10
51	.69654	253	.30346	.69708	254	.30292	.00054	0	.99946	9
52	.69907	252	.30093	.69962	252	.30038	.00054	1	.99946	8
53	.70159	250	.29841	.70214	251	.29786	.00055	1	.99945	7
54	.70409	249	.29591	.70465	249	.29535	.00056	0	.99944	6
55	8.70658	247	11.29342	8.70714	248	11.29286	10.00056	1	9.99944	5
56	.70905	246	.29095	.70962	246	.29038	.00057	1	.99943	4
57	.71151	244	.28849	.71208	245	.28792	.00058	0	.99942	3
58	.71395	243	.28605	.71453	244	.28547	.00058	1	.99942	2
59	.71638	242	.28362	.71697	243	.28303	.00059	1	.99941	1
60	8.71880		11.28120	8.71940		11.28060	10.00060		9.99940	0
92° →	cos	Diff. 1′	sec	cot	Diff. 1′	tan	csc	Diff. 1′	sin	← 87°

TABLE 3

Common Logarithms of Trigonometric Functions (offset +10)

3° → ′	sin	Diff. 1′	csc	tan	Diff. 1′	cot	sec	Diff. 1′	cos	← 176° ′
0	8.71880	240	11.28120	8.71940	241	11.28060	10.00060	0	9.99940	60
1	.72120	239	.27880	.72181	239	.27819	.00060	1	.99940	59
2	.72359	238	.27641	.72420	239	.27580	.00061	1	.99939	58
3	.72597	237	.27403	.72659	237	.27341	.00062	0	.99938	57
4	.72834	235	.27166	.72896	236	.27104	.00062	1	.99938	56
5	8.73069	234	11.26931	8.73132	234	11.26868	10.00063	1	9.99937	55
6	.73303	232	.26697	.73366	234	.26634	.00064	0	.99936	54
7	.73535	232	.26465	.73600	232	.26400	.00064	1	.99936	53
8	.73767	230	.26233	.73832	231	.26168	.00065	1	.99935	52
9	.73997	229	.26003	.74063	229	.25937	.00066	0	.99934	51
10	8.74226	228	11.25774	8.74292	229	11.25708	10.00066	1	9.99934	50
11	.74454	226	.25546	.74521	227	.25479	.00067	1	.99933	49
12	.74680	226	.25320	.74748	226	.25252	.00068	0	.99932	48
13	.74906	224	.25094	.74974	225	.25026	.00068	1	.99932	47
14	.75130	223	.24870	.75199	224	.24801	.00069	1	.99931	46
15	8.75353	222	11.24647	8.75423	222	11.24577	10.00070	1	9.99930	45
16	.75575	220	.24425	.75645	222	.24355	.00071	0	.99929	44
17	.75795	220	.24205	.75867	220	.24133	.00071	1	.99929	43
18	.76015	219	.23985	.76087	219	.23913	.00072	1	.99928	42
19	.76234	217	.23766	.76306	219	.23694	.00073	1	.99927	41
20	8.76451	216	11.23549	8.76525	217	11.23475	10.00074	0	9.99926	40
21	.76667	216	.23333	.76742	216	.23258	.00074	1	.99926	39
22	.76883	214	.23117	.76958	215	.23042	.00075	1	.99925	38
23	.77097	213	.22903	.77173	214	.22827	.00076	1	.99924	37
24	.77310	212	.22690	.77387	213	.22613	.00077	0	.99923	36
25	8.77522	211	11.22478	8.77600	211	11.22400	10.00077	1	9.99923	35
26	.77733	210	.22267	.77811	211	.22189	.00078	1	.99922	34
27	.77943	209	.22057	.78022	210	.21978	.00079	1	.99921	33
28	.78152	208	.21848	.78232	209	.21768	.00080	0	.99920	32
29	.78360	208	.21640	.78441	208	.21559	.00080	1	.99920	31
30	8.78568	206	11.21432	8.78649	206	11.21351	10.00081	1	9.99919	30
31	.78774	205	.21226	.78855	206	.21145	.00082	1	.99918	29
32	.78979	204	.21021	.79061	205	.20939	.00083	0	.99917	28
33	.79183	203	.20817	.79266	204	.20734	.00083	1	.99917	27
34	.79386	202	.20614	.79470	203	.20530	.00084	1	.99916	26
35	8.79588	201	11.20412	8.79673	202	11.20327	10.00085	1	9.99915	25
36	.79789	201	.20211	.79875	201	.20125	.00086	1	.99914	24
37	.79990	199	.20010	.80076	201	.19924	.00087	0	.99913	23
38	.80189	199	.19811	.80277	199	.19723	.00087	1	.99913	22
39	.80388	197	.19612	.80476	198	.19524	.00088	1	.99912	21
40	8.80585	197	11.19415	8.80674	198	11.19326	10.00089	1	9.99911	20
41	.80782	196	.19218	.80872	196	.19128	.00090	1	.99910	19
42	.80978	195	.19022	.81068	196	.18932	.00091	0	.99909	18
43	.81173	194	.18827	.81264	195	.18736	.00091	1	.99909	17
44	.81367	193	.18633	.81459	194	.18541	.00092	1	.99908	16
45	8.81560	192	11.18440	8.81653	193	11.18347	10.00093	1	9.99907	15
46	.81752	192	.18248	.81846	192	.18154	.00094	1	.99906	14
47	.81944	190	.18056	.82038	192	.17962	.00095	1	.99905	13
48	.82134	190	.17866	.82230	190	.17770	.00096	0	.99904	12
49	.82324	189	.17676	.82420	190	.17580	.00096	1	.99904	11
50	8.82513	188	11.17487	8.82610	189	11.17390	10.00097	1	9.99903	10
51	.82701	187	.17299	.82799	188	.17201	.00098	1	.99902	9
52	.82888	187	.17112	.82987	188	.17013	.00099	1	.99901	8
53	.83075	186	.16925	.83175	186	.16825	.00100	1	.99900	7
54	.83261	185	.16739	.83361	186	.16639	.00101	1	.99899	6
55	8.83446	184	11.16554	8.83547	185	11.16453	10.00102	0	9.99898	5
56	.83630	183	.16370	.83732	184	.16268	.00102	1	.99898	4
57	.83813	183	.16187	.83916	184	.16084	.00103	1	.99897	3
58	.83996	181	.16004	.84100	182	.15900	.00104	1	.99896	2
59	.84177	181	.15823	.84282	182	.15718	.00105	1	.99895	1
60	8.84358		11.15642	8.84464		11.15536	10.00106		9.99894	0
93° →	cos	Diff. 1′	sec	cot	Diff. 1′	tan	csc	Diff. 1′	sin	← 86°

TABLE 3

Common Logarithms of Trigonometric Functions (offset +10)

4° → ↓	sin	Diff. 1′	csc	tan	Diff. 1′	cot	sec	Diff. 1′	cos	← 175° ↓
′										′
0	8.84358	181	11.15642	8.84464	182	11.15536	10.00106	1	9.99894	60
1	.84539	179	.15461	.84646	180	.15354	.00107	1	.99893	59
2	.84718	179	.15282	.84826	180	.15174	.00108	1	.99892	58
3	.84897	178	.15103	.85006	179	.14994	.00109	0	.99891	57
4	.85075	177	.14925	.85185	178	.14815	.00109	1	.99891	56
5	8.85252	177	11.14748	8.85363	177	11.14637	10.00110	1	9.99890	55
6	.85429	176	.14571	.85540	177	.14460	.00111	1	.99889	54
7	.85605	175	.14395	.85717	176	.14283	.00112	1	.99888	53
8	.85780	175	.14220	.85893	176	.14107	.00113	1	.99887	52
9	.85955	173	.14045	.86069	174	.13931	.00114	1	.99886	51
10	8.86128	173	11.13872	8.86243	174	11.13757	10.00115	1	9.99885	50
11	.86301	173	.13699	.86417	174	.13583	.00116	1	.99884	49
12	.86474	171	.13526	.86591	172	.13409	.00117	1	.99883	48
13	.86645	171	.13355	.86763	172	.13237	.00118	1	.99882	47
14	.86816	171	.13184	.86935	171	.13065	.00119	1	.99881	46
15	8.86987	169	11.13013	8.87106	171	11.12894	10.00120	1	9.99880	45
16	.87156	169	.12844	.87277	170	.12723	.00121	0	.99879	44
17	.87325	169	.12675	.87447	169	.12553	.00121	1	.99879	43
18	.87494	167	.12506	.87616	169	.12384	.00122	1	.99878	42
19	.87661	168	.12339	.87785	168	.12215	.00123	1	.99877	41
20	8.87829	166	11.12171	8.87953	167	11.12047	10.00124	1	9.99876	40
21	.87995	166	.12005	.88120	167	.11880	.00125	1	.99875	39
22	.88161	165	.11839	.88287	166	.11713	.00126	1	.99874	38
23	.88326	164	.11674	.88453	165	.11547	.00127	1	.99873	37
24	.88490	164	.11510	.88618	165	.11382	.00128	1	.99872	36
25	8.88654	163	11.11346	8.88783	165	11.11217	10.00129	1	9.99871	35
26	.88817	163	.11183	.88948	163	.11052	.00130	1	.99870	34
27	.88980	162	.11020	.89111	163	.10889	.00131	1	.99869	33
28	.89142	162	.10858	.89274	163	.10726	.00132	1	.99868	32
29	.89304	160	.10696	.89437	161	.10563	.00133	1	.99867	31
30	8.89464	161	11.10536	8.89598	162	11.10402	10.00134	1	9.99866	30
31	.89625	159	.10375	.89760	160	.10240	.00135	1	.99865	29
32	.89784	159	.10216	.89920	160	.10080	.00136	1	.99864	28
33	.89943	159	.10057	.90080	160	.09920	.00137	1	.99863	27
34	.90102	158	.09898	.90240	159	.09760	.00138	1	.99862	26
35	8.90260	157	11.09740	8.90399	158	11.09601	10.00139	1	9.99861	25
36	.90417	157	.09583	.90557	158	.09443	.00140	1	.99860	24
37	.90574	156	.09426	.90715	157	.09285	.00141	1	.99859	23
38	.90730	155	.09270	.90872	157	.09128	.00142	1	.99858	22
39	.90885	155	.09115	.91029	156	.08971	.00143	1	.99857	21
40	8.91040	155	11.08960	8.91185	155	11.08815	10.00144	1	9.99856	20
41	.91195	154	.08805	.91340	155	.08660	.00145	1	.99855	19
42	.91349	153	.08651	.91495	155	.08505	.00146	1	.99854	18
43	.91502	153	.08498	.91650	153	.08350	.00147	1	.99853	17
44	.91655	152	.08345	.91803	154	.08197	.00148	1	.99852	16
45	8.91807	152	11.08193	8.91957	153	11.08043	10.00149	1	9.99851	15
46	.91959	151	.08041	.92110	152	.07890	.00150	2	.99850	14
47	.92110	151	.07890	.92262	152	.07738	.00152	1	.99848	13
48	.92261	150	.07739	.92414	151	.07586	.00153	1	.99847	12
49	.92411	150	.07589	.92565	151	.07435	.00154	1	.99846	11
50	8.92561	149	11.07439	8.92716	150	11.07284	10.00155	1	9.99845	10
51	.92710	149	.07290	.92866	150	.07134	.00156	1	.99844	9
52	.92859	148	.07141	.93016	149	.06984	.00157	1	.99843	8
53	.93007	147	.06993	.93165	148	.06835	.00158	1	.99842	7
54	.93154	147	.06846	.93313	149	.06687	.00159	1	.99841	6
55	8.93301	147	11.06699	8.93462	147	11.06538	10.00160	1	9.99840	5
56	.93448	146	.06552	.93609	147	.06391	.00161	1	.99839	4
57	.93594	146	.06406	.93756	147	.06244	.00162	1	.99838	3
58	.93740	145	.06260	.93903	146	.06097	.00163	1	.99837	2
59	.93885	145	.06115	.94049	146	.05951	.00164	2	.99836	1
60	8.94030		11.05970	8.94195		11.05805	10.00166		9.99834	0
94° → ↑	cos	Diff. 1′	sec	cot	Diff. 1′	tan	csc	Diff. 1′	sin	← 85° ↑

TABLE 3

Common Logarithms of Trigonometric Functions (offset +10)

5° → ↓	sin	Diff. 1′	csc	tan	Diff. 1′	cot	sec	Diff. 1′	cos	← 174° ↓
′										′
0	8.94030	144	11.05970	8.94195	145	11.05805	10.00166	1	9.99834	60
1	.94174	143	.05826	.94340	145	.05660	.00167	1	.99833	59
2	.94317	144	.05683	.94485	145	.05515	.00168	1	.99832	58
3	.94461	142	.05539	.94630	143	.05370	.00169	1	.99831	57
4	.94603	143	.05397	.94773	144	.05227	.00170	1	.99830	56
5	8.94746	141	11.05254	8.94917	143	11.05083	10.00171	1	9.99829	55
6	.94887	142	.05113	.95060	142	.04940	.00172	1	.99828	54
7	.95029	141	.04971	.95202	142	.04798	.00173	2	.99827	53
8	.95170	140	.04830	.95344	142	.04656	.00175	1	.99825	52
9	.95310	140	.04690	.95486	141	.04514	.00176	1	.99824	51
10	8.95450	139	11.04550	8.95627	140	11.04373	10.00177	1	9.99823	50
11	.95589	139	.04411	.95767	141	.04233	.00178	1	.99822	49
12	.95728	139	.04272	.95908	139	.04092	.00179	1	.99821	48
13	.95867	138	.04133	.96047	140	.03953	.00180	1	.99820	47
14	.96005	138	.03995	.96187	138	.03813	.00181	2	.99819	46
15	8.96143	137	11.03857	8.96325	139	11.03675	10.00183	1	9.99817	45
16	.96280	137	.03720	.96464	138	.03536	.00184	1	.99816	44
17	.96417	136	.03583	.96602	137	.03398	.00185	1	.99815	43
18	.96553	136	.03447	.96739	138	.03261	.00186	1	.99814	42
19	.96689	136	.03311	.96877	136	.03123	.00187	1	.99813	41
20	8.96825	135	11.03175	8.97013	137	11.02987	10.00188	2	9.99812	40
21	.96960	135	.03040	.97150	135	.02850	.00190	1	.99810	39
22	.97095	134	.02905	.97285	136	.02715	.00191	1	.99809	38
23	.97229	134	.02771	.97421	135	.02579	.00192	1	.99808	37
24	.97363	133	.02637	.97556	135	.02444	.00193	1	.99807	36
25	8.97496	133	11.02504	8.97691	134	11.02309	10.00194	2	9.99806	35
26	.97629	133	.02371	.97825	134	.02175	.00196	1	.99804	34
27	.97762	132	.02238	.97959	133	.02041	.00197	1	.99803	33
28	.97894	132	.02106	.98092	133	.01908	.00198	1	.99802	32
29	.98026	131	.01974	.98225	133	.01775	.00199	1	.99801	31
30	8.98157	131	11.01843	8.98358	132	11.01642	10.00200	2	9.99800	30
31	.98288	131	.01712	.98490	132	.01510	.00202	1	.99798	29
32	.98419	130	.01581	.98622	131	.01378	.00203	1	.99797	28
33	.98549	130	.01451	.98753	131	.01247	.00204	1	.99796	27
34	.98679	129	.01321	.98884	131	.01116	.00205	2	.99795	26
35	8.98808	129	11.01192	8.99015	130	11.00985	10.00207	1	9.99793	25
36	.98937	129	.01063	.99145	130	.00855	.00208	1	.99792	24
37	.99066	128	.00934	.99275	130	.00725	.00209	1	.99791	23
38	.99194	128	.00806	.99405	129	.00595	.00210	2	.99790	22
39	.99322	128	.00678	.99534	128	.00466	.00212	1	.99788	21
40	8.99450	127	11.00550	8.99662	129	11.00338	10.00213	1	9.99787	20
41	.99577	127	.00423	.99791	128	.00209	.00214	1	.99786	19
42	.99704	126	.00296	8.99919	127	.00081	.00215	2	.99785	18
43	.99830	126	.00170	9.00046	128	.99954	.00217	1	.99783	17
44	8.99956	126	.00044	.00174	127	.99826	.00218	1	.99782	16
45	9.00082	125	10.99918	9.00301	126	10.99699	10.00219	1	9.99781	15
46	.00207	125	.99793	.00427	126	.99573	.00220	2	.99780	14
47	.00332	124	.99668	.00553	126	.99447	.00222	1	.99778	13
48	.00456	125	.99544	.00679	126	.99321	.00223	1	.99777	12
49	.00581	123	.99419	.00805	125	.99195	.00224	1	.99776	11
50	9.00704	124	10.99296	9.00930	125	10.99070	10.00225	2	9.99775	10
51	.00828	123	.99172	.01055	124	.98945	.00227	1	.99773	9
52	.00951	123	.99049	.01179	124	.98821	.00228	1	.99772	8
53	.01074	122	.98926	.01303	124	.98697	.00229	2	.99771	7
54	.01196	122	.98804	.01427	123	.98573	.00231	1	.99769	6
55	9.01318	122	10.98682	9.01550	123	10.98450	10.00232	1	9.99768	5
56	.01440	121	.98560	.01673	123	.98327	.00233	2	.99767	4
57	.01561	121	.98439	.01796	122	.98204	.00235	1	.99765	3
58	.01682	121	.98318	.01918	122	.98082	.00236	1	.99764	2
59	.01803	120	.98197	.02040	122	.97960	.00237	2	.99763	1
60	9.01923		10.98077	9.02162		10.97838	10.00239		9.99761	0
95° → ↑	cos	Diff. 1′	sec	cot	Diff. 1′	tan	csc	Diff. 1′	sin	← 84° ↑

TABLE 3

Common Logarithms of Trigonometric Functions (offset +10)

6° → ′	sin	Diff. 1′	csc	tan	Diff. 1′	cot	sec	Diff. 1′	cos	← 173° ′
0	9.01923	120	10.98077	9.02162	121	10.97838	10.00239	1	9.99761	60
1	.02043	120	.97957	.02283	121	.97717	.00240	1	.99760	59
2	.02163	120	.97837	.02404	121	.97596	.00241	2	.99759	58
3	.02283	119	.97717	.02525	120	.97475	.00243	1	.99757	57
4	.02402	118	.97598	.02645	121	.97355	.00244	1	.99756	56
5	9.02520	119	10.97480	9.02766	119	10.97234	10.00245	2	9.99755	55
6	.02639	118	.97361	.02885	120	.97115	.00247	1	.99753	54
7	.02757	117	.97243	.03005	119	.96995	.00248	1	.99752	53
8	.02874	118	.97126	.03124	118	.96876	.00249	2	.99751	52
9	.02992	117	.97008	.03242	119	.96758	.00251	1	.99749	51
10	9.03109	117	10.96891	9.03361	118	10.96639	10.00252	1	9.99748	50
11	.03226	116	.96774	.03479	118	.96521	.00253	2	.99747	49
12	.03342	116	.96658	.03597	117	.96403	.00255	1	.99745	48
13	.03458	116	.96542	.03714	118	.96286	.00256	2	.99744	47
14	.03574	116	.96426	.03832	116	.96168	.00258	1	.99742	46
15	9.03690	115	10.96310	9.03948	117	10.96052	10.00259	1	9.99741	45
16	.03805	115	.96195	.04065	116	.95935	.00260	2	.99740	44
17	.03920	114	.96080	.04181	116	.95819	.00262	1	.99738	43
18	.04034	115	.95966	.04297	116	.95703	.00263	1	.99737	42
19	.04149	113	.95851	.04413	115	.95587	.00264	2	.99736	41
20	9.04262	114	10.95738	9.04528	115	10.95472	10.00266	1	9.99734	40
21	.04376	114	.95624	.04643	115	.95357	.00267	2	.99733	39
22	.04490	113	.95510	.04758	115	.95242	.00269	1	.99731	38
23	.04603	112	.95397	.04873	114	.95127	.00270	2	.99730	37
24	.04715	113	.95285	.04987	114	.95013	.00272	1	.99728	36
25	9.04828	112	10.95172	9.05101	113	10.94899	10.00273	1	9.99727	35
26	.04940	112	.95060	.05214	114	.94786	.00274	2	.99726	34
27	.05052	112	.94948	.05328	113	.94672	.00276	1	.99724	33
28	.05164	111	.94836	.05441	112	.94559	.00277	2	.99723	32
29	.05275	111	.94725	.05553	113	.94447	.00279	1	.99721	31
30	9.05386	111	10.94614	9.05666	112	10.94334	10.00280	2	9.99720	30
31	.05497	110	.94503	.05778	112	.94222	.00282	1	.99718	29
32	.05607	110	.94393	.05890	112	.94110	.00283	1	.99717	28
33	.05717	110	.94283	.06002	111	.93998	.00284	2	.99716	27
34	.05827	110	.94173	.06113	111	.93887	.00286	1	.99714	26
35	9.05937	109	10.94063	9.06224	111	10.93776	10.00287	2	9.99713	25
36	.06046	109	.93954	.06335	110	.93665	.00289	1	.99711	24
37	.06155	109	.93845	.06445	111	.93555	.00290	2	.99710	23
38	.06264	108	.93736	.06556	110	.93444	.00292	1	.99708	22
39	.06372	109	.93628	.06666	109	.93334	.00293	2	.99707	21
40	9.06481	108	10.93519	9.06775	110	10.93225	10.00295	1	9.99705	20
41	.06589	107	.93411	.06885	109	.93115	.00296	2	.99704	19
42	.06696	108	.93304	.06994	109	.93006	.00298	1	.99702	18
43	.06804	107	.93196	.07103	108	.92897	.00299	2	.99701	17
44	.06911	107	.93089	.07211	109	.92789	.00301	1	.99699	16
45	9.07018	106	10.92982	9.07320	108	10.92680	10.00302	2	9.99698	15
46	.07124	107	.92876	.07428	108	.92572	.00304	1	.99696	14
47	.07231	106	.92769	.07536	107	.92464	.00305	2	.99695	13
48	.07337	105	.92663	.07643	108	.92357	.00307	1	.99693	12
49	.07442	106	.92558	.07751	107	.92249	.00308	2	.99692	11
50	9.07548	105	10.92452	9.07858	106	10.92142	10.00310	1	9.99690	10
51	.07653	105	.92347	.07964	107	.92036	.00311	2	.99689	9
52	.07758	105	.92242	.08071	106	.91929	.00313	1	.99687	8
53	.07863	105	.92137	.08177	106	.91823	.00314	2	.99686	7
54	.07968	104	.92032	.08283	106	.91717	.00316	1	.99684	6
55	9.08072	104	10.91928	9.08389	106	10.91611	10.00317	2	9.99683	5
56	.08176	104	.91824	.08495	105	.91505	.00319	1	.99681	4
57	.08280	103	.91720	.08600	105	.91400	.00320	2	.99680	3
58	.08383	103	.91617	.08705	105	.91295	.00322	1	.99678	2
59	.08486	103	.91514	.08810	104	.91190	.00323	2	.99677	1
60	9.08589		10.91411	9.08914		10.91086	10.00325		9.99675	0
96° →	cos	Diff. 1′	sec	cot	Diff. 1′	tan	csc	Diff. 1′	sin	← 83°

TABLE 3

Common Logarithms of Trigonometric Functions (offset +10)

7° → ′	sin	Diff. 1′	csc	tan	Diff. 1′	cot	sec	Diff. 1′	cos	← 172° ′
0	9.08589	103	10.91411	9.08914	105	10.91086	10.00325	1	9.99675	60
1	.08692	103	.91308	.09019	104	.90981	.00326	2	.99674	59
2	.08795	102	.91205	.09123	104	.90877	.00328	2	.99672	58
3	.08897	102	.91103	.09227	103	.90773	.00330	1	.99670	57
4	.08999	102	.91001	.09330	104	.90670	.00331	2	.99669	56
5	9.09101	101	10.90899	9.09434	103	10.90566	10.00333	1	9.99667	55
6	.09202	102	.90798	.09537	103	.90463	.00334	2	.99666	54
7	.09304	101	.90696	.09640	102	.90360	.00336	1	.99664	53
8	.09405	101	.90595	.09742	103	.90258	.00337	2	.99663	52
9	.09506	100	.90494	.09845	102	.90155	.00339	2	.99661	51
10	9.09606	101	10.90394	9.09947	102	10.90053	10.00341	1	9.99659	50
11	.09707	100	.90293	.10049	101	.89951	.00342	2	.99658	49
12	.09807	100	.90193	.10150	102	.89850	.00344	1	.99656	48
13	.09907	99	.90093	.10252	101	.89748	.00345	2	.99655	47
14	.10006	100	.89994	.10353	101	.89647	.00347	2	.99653	46
15	9.10106	99	10.89894	9.10454	101	10.89546	10.00349	1	9.99651	45
16	.10205	99	.89795	.10555	101	.89445	.00350	2	.99650	44
17	.10304	98	.89696	.10656	100	.89344	.00352	1	.99648	43
18	.10402	99	.89598	.10756	100	.89244	.00353	2	.99647	42
19	.10501	98	.89499	.10856	100	.89144	.00355	2	.99645	41
20	9.10599	98	10.89401	9.10956	100	10.89044	10.00357	1	9.99643	40
21	.10697	98	.89303	.11056	99	.88944	.00358	2	.99642	39
22	.10795	98	.89205	.11155	99	.88845	.00360	2	.99640	38
23	.10893	97	.89107	.11254	99	.88746	.00362	1	.99638	37
24	.10990	97	.89010	.11353	99	.88647	.00363	2	.99637	36
25	9.11087	97	10.88913	9.11452	99	10.88548	10.00365	2	9.99635	35
26	.11184	97	.88816	.11551	98	.88449	.00367	1	.99633	34
27	.11281	96	.88719	.11649	98	.88351	.00368	2	.99632	33
28	.11377	97	.88623	.11747	98	.88253	.00370	1	.99630	32
29	.11474	96	.88526	.11845	98	.88155	.00371	2	.99629	31
30	9.11570	96	10.88430	9.11943	97	10.88057	10.00373	2	9.99627	30
31	.11666	95	.88334	.12040	98	.87960	.00375	1	.99625	29
32	.11761	96	.88239	.12138	97	.87862	.00376	2	.99624	28
33	.11857	95	.88143	.12235	97	.87765	.00378	2	.99622	27
34	.11952	95	.88048	.12332	96	.87668	.00380	2	.99620	26
35	9.12047	95	10.87953	9.12428	97	10.87572	10.00382	1	9.99618	25
36	.12142	94	.87858	.12525	96	.87475	.00383	2	.99617	24
37	.12236	95	.87764	.12621	96	.87379	.00385	2	.99615	23
38	.12331	94	.87669	.12717	96	.87283	.00387	1	.99613	22
39	.12425	94	.87575	.12813	96	.87187	.00388	2	.99612	21
40	9.12519	93	10.87481	9.12909	95	10.87091	10.00390	2	9.99610	20
41	.12612	94	.87388	.13004	95	.86996	.00392	1	.99608	19
42	.12706	93	.87294	.13099	95	.86901	.00393	2	.99607	18
43	.12799	93	.87201	.13194	95	.86806	.00395	2	.99605	17
44	.12892	93	.87108	.13289	95	.86711	.00397	2	.99603	16
45	9.12985	93	10.87015	9.13384	94	10.86616	10.00399	1	9.99601	15
46	.13078	93	.86922	.13478	95	.86522	.00400	2	.99600	14
47	.13171	92	.86829	.13573	94	.86427	.00402	2	.99598	13
48	.13263	92	.86737	.13667	94	.86333	.00404	1	.99596	12
49	.13355	92	.86645	.13761	93	.86239	.00405	2	.99595	11
50	9.13447	92	10.86553	9.13854	94	10.86146	10.00407	2	9.99593	10
51	.13539	91	.86461	.13948	93	.86052	.00409	2	.99591	9
52	.13630	92	.86370	.14041	93	.85959	.00411	1	.99589	8
53	.13722	91	.86278	.14134	93	.85866	.00412	2	.99588	7
54	.13813	91	.86187	.14227	93	.85773	.00414	2	.99586	6
55	9.13904	90	10.86096	9.14320	92	10.85680	10.00416	2	9.99584	5
56	.13994	91	.86006	.14412	92	.85588	.00418	1	.99582	4
57	.14085	90	.85915	.14504	93	.85496	.00419	2	.99581	3
58	.14175	91	.85825	.14597	91	.85403	.00421	2	.99579	2
59	.14266	90	.85734	.14688	92	.85312	.00423	2	.99577	1
60	9.14356		10.85644	9.14780		10.85220	10.00425		9.99575	0
97° →	cos	Diff. 1′	sec	cot	Diff. 1′	tan	csc	Diff. 1′	sin	← 82°

TABLE 3

Common Logarithms of Trigonometric Functions (offset +10)

8° → ′ ↓	sin	Diff. 1′	csc	tan	Diff. 1′	cot	sec	Diff. 1′	cos	← 171° ′ ↓
0	9.14356	89	10.85644	9.14780	92	10.85220	10.00425	1	9.99575	60
1	.14445	90	.85555	.14872	91	.85128	.00426	2	.99574	59
2	.14535	89	.85465	.14963	91	.85037	.00428	2	.99572	58
3	.14624	90	.85376	.15054	91	.84946	.00430	2	.99570	57
4	.14714	89	.85286	.15145	91	.84855	.00432	2	.99568	56
5	9.14803	88	10.85197	9.15236	91	10.84764	10.00434	1	9.99566	55
6	.14891	89	.85109	.15327	90	.84673	.00435	2	.99565	54
7	.14980	89	.85020	.15417	91	.84583	.00437	2	.99563	53
8	.15069	88	.84931	.15508	90	.84492	.00439	2	.99561	52
9	.15157	88	.84843	.15598	90	.84402	.00441	2	.99559	51
10	9.15245	88	10.84755	9.15688	89	10.84312	10.00443	1	9.99557	50
11	.15333	88	.84667	.15777	90	.84223	.00444	2	.99556	49
12	.15421	87	.84579	.15867	89	.84133	.00446	2	.99554	48
13	.15508	88	.84492	.15956	90	.84044	.00448	2	.99552	47
14	.15596	87	.84404	.16046	89	.83954	.00450	2	.99550	46
15	9.15683	87	10.84317	9.16135	89	10.83865	10.00452	2	9.99548	45
16	.15770	87	.84230	.16224	88	.83776	.00454	1	.99546	44
17	.15857	87	.84143	.16312	89	.83688	.00455	2	.99545	43
18	.15944	86	.84056	.16401	88	.83599	.00457	2	.99543	42
19	.16030	86	.83970	.16489	88	.83511	.00459	2	.99541	41
20	9.16116	87	10.83884	9.16577	88	10.83423	10.00461	2	9.99539	40
21	.16203	86	.83797	.16665	88	.83335	.00463	2	.99537	39
22	.16289	85	.83711	.16753	88	.83247	.00465	2	.99535	38
23	.16374	86	.83626	.16841	87	.83159	.00467	1	.99533	37
24	.16460	85	.83540	.16928	88	.83072	.00468	2	.99532	36
25	9.16545	86	10.83455	9.17016	87	10.82984	10.00470	2	9.99530	35
26	.16631	85	.83369	.17103	87	.82897	.00472	2	.99528	34
27	.16716	85	.83284	.17190	87	.82810	.00474	2	.99526	33
28	.16801	85	.83199	.17277	86	.82723	.00476	2	.99524	32
29	.16886	84	.83114	.17363	87	.82637	.00478	2	.99522	31
30	9.16970	85	10.83030	9.17450	86	10.82550	10.00480	2	9.99520	30
31	.17055	84	.82945	.17536	86	.82464	.00482	1	.99518	29
32	.17139	84	.82861	.17622	86	.82378	.00483	2	.99517	28
33	.17223	84	.82777	.17708	86	.82292	.00485	2	.99515	27
34	.17307	84	.82693	.17794	86	.82206	.00487	2	.99513	26
35	9.17391	83	10.82609	9.17880	85	10.82120	10.00489	2	9.99511	25
36	.17474	84	.82526	.17965	86	.82035	.00491	2	.99509	24
37	.17558	83	.82442	.18051	85	.81949	.00493	2	.99507	23
38	.17641	83	.82359	.18136	85	.81864	.00495	2	.99505	22
39	.17724	83	.82276	.18221	85	.81779	.00497	2	.99503	21
40	9.17807	83	10.82193	9.18306	85	10.81694	10.00499	2	9.99501	20
41	.17890	83	.82110	.18391	84	.81609	.00501	2	.99499	19
42	.17973	82	.82027	.18475	85	.81525	.00503	2	.99497	18
43	.18055	82	.81945	.18560	84	.81440	.00505	1	.99495	17
44	.18137	83	.81863	.18644	84	.81356	.00506	2	.99494	16
45	9.18220	82	10.81780	9.18728	84	10.81272	10.00508	2	9.99492	15
46	.18302	81	.81698	.18812	84	.81188	.00510	2	.99490	14
47	.18383	82	.81617	.18896	83	.81104	.00512	2	.99488	13
48	.18465	82	.81535	.18979	84	.81021	.00514	2	.99486	12
49	.18547	81	.81453	.19063	83	.80937	.00516	2	.99484	11
50	9.18628	81	10.81372	9.19146	83	10.80854	10.00518	2	9.99482	10
51	.18709	81	.81291	.19229	83	.80771	.00520	2	.99480	9
52	.18790	81	.81210	.19312	83	.80688	.00522	2	.99478	8
53	.18871	81	.81129	.19395	83	.80605	.00524	2	.99476	7
54	.18952	81	.81048	.19478	83	.80522	.00526	2	.99474	6
55	9.19033	80	10.80967	9.19561	82	10.80439	10.00528	2	9.99472	5
56	.19113	80	.80887	.19643	82	.80357	.00530	2	.99470	4
57	.19193	80	.80807	.19725	82	.80275	.00532	2	.99468	3
58	.19273	80	.80727	.19807	82	.80193	.00534	2	.99466	2
59	.19353	80	.80647	.19889	82	.80111	.00536	2	.99464	1
60	9.19433		10.80567	9.19971		10.80029	10.00538		9.99462	0
↑ 98° →	cos	Diff. 1′	sec	cot	Diff. 1′	tan	csc	Diff. 1′	sin	← 81° ↑

TABLE 3

Common Logarithms of Trigonometric Functions (offset +10)

9° → ′ ↓	sin	Diff. 1′	csc	tan	Diff. 1′	cot	sec	Diff. 1′	cos	← 170° ′ ↓
0	9.19433	80	10.80567	9.19971	82	10.80029	10.00538	2	9.99462	60
1	.19513	79	.80487	.20053	81	.79947	.00540	2	.99460	59
2	.19592	80	.80408	.20134	82	.79866	.00542	2	.99458	58
3	.19672	79	.80328	.20216	81	.79784	.00544	2	.99456	57
4	.19751	79	.80249	.20297	81	.79703	.00546	2	.99454	56
5	9.19830	79	10.80170	9.20378	81	10.79622	10.00548	2	9.99452	55
6	.19909	79	.80091	.20459	81	.79541	.00550	2	.99450	54
7	.19988	79	.80012	.20540	81	.79460	.00552	2	.99448	53
8	.20067	78	.79933	.20621	80	.79379	.00554	2	.99446	52
9	.20145	78	.79855	.20701	81	.79299	.00556	2	.99444	51
10	9.20223	79	10.79777	9.20782	80	10.79218	10.00558	2	9.99442	50
11	.20302	78	.79698	.20862	80	.79138	.00560	2	.99440	49
12	.20380	78	.79620	.20942	80	.79058	.00562	2	.99438	48
13	.20458	77	.79542	.21022	80	.78978	.00564	2	.99436	47
14	.20535	78	.79465	.21102	80	.78898	.00566	2	.99434	46
15	9.20613	78	10.79387	9.21182	79	10.78818	10.00568	3	9.99432	45
16	.20691	77	.79309	.21261	80	.78739	.00571	2	.99429	44
17	.20768	77	.79232	.21341	79	.78659	.00573	2	.99427	43
18	.20845	77	.79155	.21420	79	.78580	.00575	2	.99425	42
19	.20922	77	.79078	.21499	79	.78501	.00577	2	.99423	41
20	9.20999	77	10.79001	9.21578	79	10.78422	10.00579	2	9.99421	40
21	.21076	77	.78924	.21657	79	.78343	.00581	2	.99419	39
22	.21153	76	.78847	.21736	78	.78264	.00583	2	.99417	38
23	.21229	77	.78771	.21814	79	.78186	.00585	2	.99415	37
24	.21306	76	.78694	.21893	78	.78107	.00587	2	.99413	36
25	9.21382	76	10.78618	9.21971	78	10.78029	10.00589	2	9.99411	35
26	.21458	76	.78542	.22049	78	.77951	.00591	2	.99409	34
27	.21534	76	.78466	.22127	78	.77873	.00593	3	.99407	33
28	.21610	75	.78390	.22205	78	.77795	.00596	2	.99404	32
29	.21685	76	.78315	.22283	78	.77717	.00598	2	.99402	31
30	9.21761	75	10.78239	9.22361	77	10.77639	10.00600	2	9.99400	30
31	.21836	76	.78164	.22438	78	.77562	.00602	2	.99398	29
32	.21912	75	.78088	.22516	77	.77484	.00604	2	.99396	28
33	.21987	75	.78013	.22593	77	.77407	.00606	2	.99394	27
34	.22062	75	.77938	.22670	77	.77330	.00608	2	.99392	26
35	9.22137	74	10.77863	9.22747	77	10.77253	10.00610	2	9.99390	25
36	.22211	75	.77789	.22824	77	.77176	.00612	3	.99388	24
37	.22286	75	.77714	.22901	76	.77099	.00615	2	.99385	23
38	.22361	74	.77639	.22977	77	.77023	.00617	2	.99383	22
39	.22435	74	.77565	.23054	76	.76946	.00619	2	.99381	21
40	9.22509	74	10.77491	9.23130	76	10.76870	10.00621	2	9.99379	20
41	.22583	74	.77417	.23206	77	.76794	.00623	2	.99377	19
42	.22657	74	.77343	.23283	76	.76717	.00625	3	.99375	18
43	.22731	74	.77269	.23359	76	.76641	.00628	2	.99372	17
44	.22805	73	.77195	.23435	75	.76565	.00630	2	.99370	16
45	9.22878	74	10.77122	9.23510	76	10.76490	10.00632	2	9.99368	15
46	.22952	73	.77048	.23586	75	.76414	.00634	2	.99366	14
47	.23025	73	.76975	.23661	76	.76339	.00636	2	.99364	13
48	.23098	73	.76902	.23737	75	.76263	.00638	3	.99362	12
49	.23171	73	.76829	.23812	75	.76188	.00641	2	.99359	11
50	9.23244	73	10.76756	9.23887	75	10.76113	10.00643	2	9.99357	10
51	.23317	73	.76683	.23962	75	.76038	.00645	2	.99355	9
52	.23390	72	.76610	.24037	75	.75963	.00647	2	.99353	8
53	.23462	73	.76538	.24112	74	.75888	.00649	3	.99351	7
54	.23535	72	.76465	.24186	75	.75814	.00652	2	.99348	6
55	9.23607	72	10.76393	9.24261	74	10.75739	10.00654	2	9.99346	5
56	.23679	73	.76321	.24335	75	.75665	.00656	2	.99344	4
57	.23752	71	.76248	.24410	74	.75590	.00658	2	.99342	3
58	.23823	72	.76177	.24484	74	.75516	.00660	3	.99340	2
59	.23895	72	.76105	.24558	74	.75442	.00663	2	.99337	1
60	9.23967		10.76033	9.24632		10.75368	10.00665		9.99335	0
↑ 99° →	cos	Diff. 1′	sec	cot	Diff. 1′	tan	csc	Diff. 1′	sin	← 80° ↑

TABLE 3

Common Logarithms of Trigonometric Functions (offset +10)

10°→ ′ ↓	sin	Diff. 1′	csc	tan	Diff. 1′	cot	sec	Diff. 1′	cos	←169° ′ ↓
0	9.23967	72	10.76033	9.24632	74	10.75368	10.00665	2	9.99335	60
1	.24039	71	.75961	.24706	73	.75294	.00667	2	.99333	59
2	.24110	71	.75890	.24779	74	.75221	.00669	3	.99331	58
3	.24181	72	.75819	.24853	73	.75147	.00672	2	.99328	57
4	.24253	71	.75747	.24926	74	.75074	.00674	2	.99326	56
5	9.24324	71	10.75676	9.25000	73	10.75000	10.00676	2	9.99324	55
6	.24395	71	.75605	.25073	73	.74927	.00678	3	.99322	54
7	.24466	70	.75534	.25146	73	.74854	.00681	2	.99319	53
8	.24536	71	.75464	.25219	73	.74781	.00683	2	.99317	52
9	.24607	70	.75393	.25292	73	.74708	.00685	2	.99315	51
10	9.24677	71	10.75323	9.25365	72	10.74635	10.00687	3	9.99313	50
11	.24748	70	.75252	.25437	73	.74563	.00690	2	.99310	49
12	.24818	70	.75182	.25510	72	.74490	.00692	2	.99308	48
13	.24888	70	.75112	.25582	73	.74418	.00694	2	.99306	47
14	.24958	70	.75042	.25655	72	.74345	.00696	3	.99304	46
15	9.25028	70	10.74972	9.25727	72	10.74273	10.00699	2	9.99301	45
16	.25098	70	.74902	.25799	72	.74201	.00701	2	.99299	44
17	.25168	69	.74832	.25871	72	.74129	.00703	3	.99297	43
18	.25237	70	.74763	.25943	72	.74057	.00706	2	.99294	42
19	.25307	69	.74693	.26015	71	.73985	.00708	2	.99292	41
20	9.25376	69	10.74624	9.26086	72	10.73914	10.00710	2	9.99290	40
21	.25445	69	.74555	.26158	71	.73842	.00712	3	.99288	39
22	.25514	69	.74486	.26229	72	.73771	.00715	2	.99285	38
23	.25583	69	.74417	.26301	71	.73699	.00717	2	.99283	37
24	.25652	69	.74348	.26372	71	.73628	.00719	3	.99281	36
25	9.25721	69	10.74279	9.26443	71	10.73557	10.00722	2	9.99278	35
26	.25790	68	.74210	.26514	71	.73486	.00724	2	.99276	34
27	.25858	69	.74142	.26585	70	.73415	.00726	3	.99274	33
28	.25927	68	.74073	.26655	71	.73345	.00729	2	.99271	32
29	.25995	68	.74005	.26726	71	.73274	.00731	2	.99269	31
30	9.26063	68	10.73937	9.26797	70	10.73203	10.00733	3	9.99267	30
31	.26131	68	.73869	.26867	70	.73133	.00736	2	.99264	29
32	.26199	68	.73801	.26937	71	.73063	.00738	2	.99262	28
33	.26267	68	.73733	.27008	70	.72992	.00740	3	.99260	27
34	.26335	68	.73665	.27078	70	.72922	.00743	2	.99257	26
35	9.26403	67	10.73597	9.27148	70	10.72852	10.00745	3	9.99255	25
36	.26470	68	.73530	.27218	70	.72782	.00748	2	.99252	24
37	.26538	67	.73462	.27288	69	.72712	.00750	2	.99250	23
38	.26605	67	.73395	.27357	70	.72643	.00752	3	.99248	22
39	.26672	67	.73328	.27427	69	.72573	.00755	2	.99245	21
40	9.26739	67	10.73261	9.27496	70	10.72504	10.00757	2	9.99243	20
41	.26806	67	.73194	.27566	69	.72434	.00759	3	.99241	19
42	.26873	67	.73127	.27635	69	.72365	.00762	2	.99238	18
43	.26940	67	.73060	.27704	69	.72296	.00764	3	.99236	17
44	.27007	66	.72993	.27773	69	.72227	.00767	2	.99233	16
45	9.27073	67	10.72927	9.27842	69	10.72158	10.00769	2	9.99231	15
46	.27140	66	.72860	.27911	69	.72089	.00771	3	.99229	14
47	.27206	67	.72794	.27980	69	.72020	.00774	2	.99226	13
48	.27273	66	.72727	.28049	68	.71951	.00776	3	.99224	12
49	.27339	66	.72661	.28117	69	.71883	.00779	2	.99221	11
50	9.27405	66	10.72595	9.28186	68	10.71814	10.00781	2	9.99219	10
51	.27471	66	.72529	.28254	69	.71746	.00783	3	.99217	9
52	.27537	65	.72463	.28323	68	.71677	.00786	2	.99214	8
53	.27602	66	.72398	.28391	68	.71609	.00788	3	.99212	7
54	.27668	66	.72332	.28459	68	.71541	.00791	2	.99209	6
55	9.27734	65	10.72266	9.28527	68	10.71473	10.00793	3	9.99207	5
56	.27799	65	.72201	.28595	67	.71405	.00796	2	.99204	4
57	.27864	66	.72136	.28662	68	.71338	.00798	2	.99202	3
58	.27930	65	.72070	.28730	68	.71270	.00800	3	.99200	2
59	.27995	65	.72005	.28798	67	.71202	.00803	2	.99197	1
60	9.28060		10.71940	9.28865		10.71135	10.00805		9.99195	0
↑ 100°→	cos	Diff. 1′	sec	cot	Diff. 1′	tan	csc	Diff. 1′	sin	← 79° ↑

TABLE 3

Common Logarithms of Trigonometric Functions (offset +10)

11°→ ′ ↓	sin	Diff. 1′	csc	tan	Diff. 1′	cot	sec	Diff. 1′	cos	←168° ′ ↓
0	9.28060	65	10.71940	9.28865	68	10.71135	10.00805	3	9.99195	60
1	.28125	65	.71875	.28933	67	.71067	.00808	2	.99192	59
2	.28190	64	.71810	.29000	67	.71000	.00810	3	.99190	58
3	.28254	65	.71746	.29067	67	.70933	.00813	2	.99187	57
4	.28319	65	.71681	.29134	67	.70866	.00815	3	.99185	56
5	9.28384	64	10.71616	9.29201	67	10.70799	10.00818	2	9.99182	55
6	.28448	64	.71552	.29268	67	.70732	.00820	3	.99180	54
7	.28512	65	.71488	.29335	67	.70665	.00823	2	.99177	53
8	.28577	64	.71423	.29402	66	.70598	.00825	3	.99175	52
9	.28641	64	.71359	.29468	67	.70532	.00828	2	.99172	51
10	9.28705	64	10.71295	9.29535	66	10.70465	10.00830	3	9.99170	50
11	.28769	64	.71231	.29601	67	.70399	.00833	2	.99167	49
12	.28833	63	.71167	.29668	66	.70332	.00835	3	.99165	48
13	.28896	64	.71104	.29734	66	.70266	.00838	2	.99162	47
14	.28960	64	.71040	.29800	66	.70200	.00840	3	.99160	46
15	9.29024	63	10.70976	9.29866	66	10.70134	10.00843	2	9.99157	45
16	.29087	63	.70913	.29932	66	.70068	.00845	3	.99155	44
17	.29150	64	.70850	.29998	66	.70002	.00848	2	.99152	43
18	.29214	63	.70786	.30064	66	.69936	.00850	3	.99150	42
19	.29277	63	.70723	.30130	65	.69870	.00853	2	.99147	41
20	9.29340	63	10.70660	9.30195	66	10.69805	10.00855	3	9.99145	40
21	.29403	63	.70597	.30261	65	.69739	.00858	2	.99142	39
22	.29466	63	.70534	.30326	65	.69674	.00860	3	.99140	38
23	.29529	62	.70471	.30391	66	.69609	.00863	2	.99137	37
24	.29591	63	.70409	.30457	65	.69543	.00865	3	.99135	36
25	9.29654	62	10.70346	9.30522	65	10.69478	10.00868	2	9.99132	35
26	.29716	63	.70284	.30587	65	.69413	.00870	3	.99130	34
27	.29779	62	.70221	.30652	65	.69348	.00873	3	.99127	33
28	.29841	62	.70159	.30717	65	.69283	.00876	2	.99124	32
29	.29903	63	.70097	.30782	64	.69218	.00878	3	.99122	31
30	9.29966	62	10.70034	9.30846	65	10.69154	10.00881	2	9.99119	30
31	.30028	62	.69972	.30911	64	.69089	.00883	3	.99117	29
32	.30090	61	.69910	.30975	65	.69025	.00886	2	.99114	28
33	.30151	62	.69849	.31040	64	.68960	.00888	3	.99112	27
34	.30213	62	.69787	.31104	64	.68896	.00891	3	.99109	26
35	9.30275	61	10.69725	9.31168	65	10.68832	10.00894	2	9.99106	25
36	.30336	62	.69664	.31233	64	.68767	.00896	3	.99104	24
37	.30398	61	.69602	.31297	64	.68703	.00899	2	.99101	23
38	.30459	62	.69541	.31361	64	.68639	.00901	3	.99099	22
39	.30521	61	.69479	.31425	64	.68575	.00904	3	.99096	21
40	9.30582	61	10.69418	9.31489	63	10.68511	10.00907	2	9.99093	20
41	.30643	61	.69357	.31552	64	.68448	.00909	3	.99091	19
42	.30704	61	.69296	.31616	63	.68384	.00912	2	.99088	18
43	.30765	61	.69235	.31679	64	.68321	.00914	3	.99086	17
44	.30826	61	.69174	.31743	63	.68257	.00917	3	.99083	16
45	9.30887	60	10.69113	9.31806	64	10.68194	10.00920	2	9.99080	15
46	.30947	61	.69053	.31870	63	.68130	.00922	3	.99078	14
47	.31008	60	.68992	.31933	63	.68067	.00925	3	.99075	13
48	.31068	61	.68932	.31996	63	.68004	.00928	2	.99072	12
49	.31129	60	.68871	.32059	63	.67941	.00930	3	.99070	11
50	9.31189	61	10.68811	9.32122	63	10.67878	10.00933	3	9.99067	10
51	.31250	60	.68750	.32185	63	.67815	.00936	2	.99064	9
52	.31310	60	.68690	.32248	63	.67752	.00938	3	.99062	8
53	.31370	60	.68630	.32311	62	.67689	.00941	3	.99059	7
54	.31430	60	.68570	.32373	63	.67627	.00944	2	.99056	6
55	9.31490	59	10.68510	9.32436	62	10.67564	10.00946	3	9.99054	5
56	.31549	60	.68451	.32498	63	.67502	.00949	3	.99051	4
57	.31609	60	.68391	.32561	62	.67439	.00952	2	.99048	3
58	.31669	59	.68331	.32623	62	.67377	.00954	3	.99046	2
59	.31728	60	.68272	.32685	62	.67315	.00957	3	.99043	1
60	9.31788		10.68212	9.32747		10.67253	10.00960		9.99040	0
↑ 101°→	cos	Diff. 1′	sec	cot	Diff. 1′	tan	csc	Diff. 1′	sin	← 78° ↑

TABLE 3
Common Logarithms of Trigonometric Functions (offset +10)

12°→ ′ ↓	sin	Diff. 1′	csc	tan	Diff. 1′	cot	sec	Diff. 1′	cos	←167° ′ ↓
0	9.31788	59	10.68212	9.32747	63	10.67253	10.00960	2	9.99040	60
1	.31847	60	.68153	.32810	62	.67190	.00962	3	.99038	59
2	.31907	59	.68093	.32872	61	.67128	.00965	3	.99035	58
3	.31966	59	.68034	.32933	62	.67067	.00968	2	.99032	57
4	.32025	59	.67975	.32995	62	.67005	.00970	3	.99030	56
5	9.32084	59	10.67916	9.33057	62	10.66943	10.00973	3	9.99027	55
6	.32143	59	.67857	.33119	61	.66881	.00976	2	.99024	54
7	.32202	59	.67798	.33180	62	.66820	.00978	3	.99022	53
8	.32261	58	.67739	.33242	61	.66758	.00981	3	.99019	52
9	.32319	59	.67681	.33303	62	.66697	.00984	3	.99016	51
10	9.32378	59	10.67622	9.33365	61	10.66635	10.00987	2	9.99013	50
11	.32437	58	.67563	.33426	61	.66574	.00989	3	.99011	49
12	.32495	58	.67505	.33487	61	.66513	.00992	3	.99008	48
13	.32553	59	.67447	.33548	61	.66452	.00995	3	.99005	47
14	.32612	58	.67388	.33609	61	.66391	.00998	2	.99002	46
15	9.32670	58	10.67330	9.33670	61	10.66330	10.01000	3	9.99000	45
16	.32728	58	.67272	.33731	61	.66269	.01003	3	.98997	44
17	.32786	58	.67214	.33792	61	.66208	.01006	3	.98994	43
18	.32844	58	.67156	.33853	60	.66147	.01009	2	.98991	42
19	.32902	58	.67098	.33913	61	.66087	.01011	3	.98989	41
20	9.32960	58	10.67040	9.33974	60	10.66026	10.01014	3	9.98986	40
21	.33018	57	.66982	.34034	61	.65966	.01017	3	.98983	39
22	.33075	58	.66925	.34095	60	.65905	.01020	2	.98980	38
23	.33133	57	.66867	.34155	60	.65845	.01022	3	.98978	37
24	.33190	58	.66810	.34215	61	.65785	.01025	3	.98975	36
25	9.33248	57	10.66752	9.34276	60	10.65724	10.01028	3	9.98972	35
26	.33305	57	.66695	.34336	60	.65664	.01031	2	.98969	34
27	.33362	58	.66638	.34396	60	.65604	.01033	3	.98967	33
28	.33420	57	.66580	.34456	60	.65544	.01036	3	.98964	32
29	.33477	57	.66523	.34516	60	.65484	.01039	3	.98961	31
30	9.33534	57	10.66466	9.34576	59	10.65424	10.01042	3	9.98958	30
31	.33591	56	.66409	.34635	60	.65365	.01045	2	.98955	29
32	.33647	57	.66353	.34695	60	.65305	.01047	3	.98953	28
33	.33704	57	.66296	.34755	59	.65245	.01050	3	.98950	27
34	.33761	57	.66239	.34814	60	.65186	.01053	3	.98947	26
35	9.33818	56	10.66182	9.34874	59	10.65126	10.01056	3	9.98944	25
36	.33874	57	.66126	.34933	59	.65067	.01059	3	.98941	24
37	.33931	56	.66069	.34992	59	.65008	.01062	2	.98938	23
38	.33987	56	.66013	.35051	60	.64949	.01064	3	.98936	22
39	.34043	57	.65957	.35111	59	.64889	.01067	3	.98933	21
40	9.34100	56	10.65900	9.35170	59	10.64830	10.01070	3	9.98930	20
41	.34156	56	.65844	.35229	59	.64771	.01073	3	.98927	19
42	.34212	56	.65788	.35288	59	.64712	.01076	3	.98924	18
43	.34268	56	.65732	.35347	58	.64653	.01079	2	.98921	17
44	.34324	56	.65676	.35405	59	.64595	.01081	3	.98919	16
45	9.34380	56	10.65620	9.35464	59	10.64536	10.01084	3	9.98916	15
46	.34436	55	.65564	.35523	58	.64477	.01087	3	.98913	14
47	.34491	56	.65509	.35581	59	.64419	.01090	3	.98910	13
48	.34547	55	.65453	.35640	58	.64360	.01093	3	.98907	12
49	.34602	56	.65398	.35698	59	.64302	.01096	3	.98904	11
50	9.34658	55	10.65342	9.35757	58	10.64243	10.01099	3	9.98901	10
51	.34713	56	.65287	.35815	58	.64185	.01102	2	.98898	9
52	.34769	55	.65231	.35873	58	.64127	.01104	3	.98896	8
53	.34824	55	.65176	.35931	58	.64069	.01107	3	.98893	7
54	.34879	55	.65121	.35989	58	.64011	.01110	3	.98890	6
55	9.34934	55	10.65066	9.36047	58	10.63953	10.01113	3	9.98887	5
56	.34989	55	.65011	.36105	58	.63895	.01116	3	.98884	4
57	.35044	55	.64956	.36163	58	.63837	.01119	3	.98881	3
58	.35099	55	.64901	.36221	58	.63779	.01122	3	.98878	2
59	.35154	55	.64846	.36279	57	.63721	.01125	3	.98875	1
60	9.35209		10.64791	9.36336		10.63664	10.01128		9.98872	0
↑ 102°→	cos	Diff. 1′	sec	cot	Diff. 1′	tan	csc	Diff. 1′	sin	← 77° ↑

TABLE 3
Common Logarithms of Trigonometric Functions (offset +10)

13°→ ′ ↓	sin	Diff. 1′	csc	tan	Diff. 1′	cot	sec	Diff. 1′	cos	←166° ′ ↓
0	9.35209	54	10.64791	9.36336	58	10.63664	10.01128	3	9.98872	60
1	.35263	55	.64737	.36394	58	.63606	.01131	2	.98869	59
2	.35318	55	.64682	.36452	57	.63548	.01133	3	.98867	58
3	.35373	54	.64627	.36509	57	.63491	.01136	3	.98864	57
4	.35427	54	.64573	.36566	58	.63434	.01139	3	.98861	56
5	9.35481	55	10.64519	9.36624	57	10.63376	10.01142	3	9.98858	55
6	.35536	54	.64464	.36681	57	.63319	.01145	3	.98855	54
7	.35590	54	.64410	.36738	57	.63262	.01148	3	.98852	53
8	.35644	54	.64356	.36795	57	.63205	.01151	3	.98849	52
9	.35698	54	.64302	.36852	57	.63148	.01154	3	.98846	51
10	9.35752	54	10.64248	9.36909	57	10.63091	10.01157	3	9.98843	50
11	.35806	54	.64194	.36966	57	.63034	.01160	3	.98840	49
12	.35860	54	.64140	.37023	57	.62977	.01163	3	.98837	48
13	.35914	54	.64086	.37080	57	.62920	.01166	3	.98834	47
14	.35968	54	.64032	.37137	56	.62863	.01169	3	.98831	46
15	9.36022	53	10.63978	9.37193	57	10.62807	10.01172	3	9.98828	45
16	.36075	54	.63925	.37250	56	.62750	.01175	3	.98825	44
17	.36129	53	.63871	.37306	57	.62694	.01178	3	.98822	43
18	.36182	54	.63818	.37363	56	.62637	.01181	3	.98819	42
19	.36236	53	.63764	.37419	57	.62581	.01184	3	.98816	41
20	9.36289	53	10.63711	9.37476	56	10.62524	10.01187	3	9.98813	40
21	.36342	53	.63658	.37532	56	.62468	.01190	3	.98810	39
22	.36395	54	.63605	.37588	56	.62412	.01193	3	.98807	38
23	.36449	53	.63551	.37644	56	.62356	.01196	3	.98804	37
24	.36502	53	.63498	.37700	56	.62300	.01199	3	.98801	36
25	9.36555	53	10.63445	9.37756	56	10.62244	10.01202	3	9.98798	35
26	.36608	52	.63392	.37812	56	.62188	.01205	3	.98795	34
27	.36660	53	.63340	.37868	56	.62132	.01208	3	.98792	33
28	.36713	53	.63287	.37924	56	.62076	.01211	3	.98789	32
29	.36766	53	.63234	.37980	55	.62020	.01214	3	.98786	31
30	9.36819	52	10.63181	9.38035	56	10.61965	10.01217	3	9.98783	30
31	.36871	53	.63129	.38091	56	.61909	.01220	3	.98780	29
32	.36924	52	.63076	.38147	55	.61853	.01223	3	.98777	28
33	.36976	52	.63024	.38202	55	.61798	.01226	3	.98774	27
34	.37028	53	.62972	.38257	56	.61743	.01229	3	.98771	26
35	9.37081	52	10.62919	9.38313	55	10.61687	10.01232	3	9.98768	25
36	.37133	52	.62867	.38368	55	.61632	.01235	3	.98765	24
37	.37185	52	.62815	.38423	56	.61577	.01238	3	.98762	23
38	.37237	52	.62763	.38479	55	.61521	.01241	3	.98759	22
39	.37289	52	.62711	.38534	55	.61466	.01244	3	.98756	21
40	9.37341	52	10.62659	9.38589	55	10.61411	10.01247	3	9.98753	20
41	.37393	52	.62607	.38644	55	.61356	.01250	4	.98750	19
42	.37445	52	.62555	.38699	55	.61301	.01254	3	.98746	18
43	.37497	52	.62503	.38754	54	.61246	.01257	3	.98743	17
44	.37549	51	.62451	.38808	55	.61192	.01260	3	.98740	16
45	9.37600	52	10.62400	9.38863	55	10.61137	10.01263	3	9.98737	15
46	.37652	51	.62348	.38918	54	.61082	.01266	3	.98734	14
47	.37703	52	.62297	.38972	55	.61028	.01269	3	.98731	13
48	.37755	51	.62245	.39027	55	.60973	.01272	3	.98728	12
49	.37806	52	.62194	.39082	54	.60918	.01275	3	.98725	11
50	9.37858	51	10.62142	9.39136	54	10.60864	10.01278	3	9.98722	10
51	.37909	51	.62091	.39190	55	.60810	.01281	4	.98719	9
52	.37960	51	.62040	.39245	54	.60755	.01285	3	.98715	8
53	.38011	51	.61989	.39299	54	.60701	.01288	3	.98712	7
54	.38062	51	.61938	.39353	54	.60647	.01291	3	.98709	6
55	9.38113	51	10.61887	9.39407	54	10.60593	10.01294	3	9.98706	5
56	.38164	51	.61836	.39461	54	.60539	.01297	3	.98703	4
57	.38215	51	.61785	.39515	54	.60485	.01300	3	.98700	3
58	.38266	51	.61734	.39569	54	.60431	.01303	3	.98697	2
59	.38317	51	.61683	.39623	54	.60377	.01306	4	.98694	1
60	9.38368		10.61632	9.39677		10.60323	10.01310		9.98690	0
↑ 103°→	cos	Diff. 1′	sec	cot	Diff. 1′	tan	csc	Diff. 1′	sin	← 76° ↑

TABLE 3

Common Logarithms of Trigonometric Functions (offset +10)

14° → ↓	sin	Diff. 1′	csc	tan	Diff. 1′	cot	sec	Diff. 1′	cos	← 165° ↓
′										′
0	9.38368	50	10.61632	9.39677	54	10.60323	10.01310	3	9.98690	60
1	.38418	51	.61582	.39731	54	.60269	.01313	3	.98687	59
2	.38469	50	.61531	.39785	53	.60215	.01316	3	.98684	58
3	.38519	51	.61481	.39838	54	.60162	.01319	3	.98681	57
4	.38570	50	.61430	.39892	53	.60108	.01322	3	.98678	56
5	9.38620	50	10.61380	9.39945	54	10.60055	10.01325	4	9.98675	55
6	.38670	51	.61330	.39999	53	.60001	.01329	3	.98671	54
7	.38721	50	.61279	.40052	54	.59948	.01332	3	.98668	53
8	.38771	50	.61229	.40106	53	.59894	.01335	3	.98665	52
9	.38821	50	.61179	.40159	53	.59841	.01338	3	.98662	51
10	9.38871	50	10.61129	9.40212	54	10.59788	10.01341	3	9.98659	50
11	.38921	50	.61079	.40266	53	.59734	.01344	4	.98656	49
12	.38971	50	.61029	.40319	53	.59681	.01348	3	.98652	48
13	.39021	50	.60979	.40372	53	.59628	.01351	3	.98649	47
14	.39071	50	.60929	.40425	53	.59575	.01354	3	.98646	46
15	9.39121	49	10.60879	9.40478	53	10.59522	10.01357	3	9.98643	45
16	.39170	50	.60830	.40531	53	.59469	.01360	4	.98640	44
17	.39220	50	.60780	.40584	52	.59416	.01364	3	.98636	43
18	.39270	49	.60730	.40636	53	.59364	.01367	3	.98633	42
19	.39319	50	.60681	.40689	53	.59311	.01370	3	.98630	41
20	9.39369	49	10.60631	9.40742	53	10.59258	10.01373	4	9.98627	40
21	.39418	49	.60582	.40795	52	.59205	.01377	3	.98623	39
22	.39467	50	.60533	.40847	53	.59153	.01380	3	.98620	38
23	.39517	49	.60483	.40900	52	.59100	.01383	3	.98617	37
24	.39566	49	.60434	.40952	53	.59048	.01386	4	.98614	36
25	9.39615	49	10.60385	9.41005	52	10.58995	10.01390	3	9.98610	35
26	.39664	49	.60336	.41057	52	.58943	.01393	3	.98607	34
27	.39713	49	.60287	.41109	52	.58891	.01396	3	.98604	33
28	.39762	49	.60238	.41161	53	.58839	.01399	4	.98601	32
29	.39811	49	.60189	.41214	52	.58786	.01403	3	.98597	31
30	9.39860	49	10.60140	9.41266	52	10.58734	10.01406	3	9.98594	30
31	.39909	49	.60091	.41318	52	.58682	.01409	3	.98591	29
32	.39958	48	.60042	.41370	52	.58630	.01412	4	.98588	28
33	.40006	49	.59994	.41422	52	.58578	.01416	3	.98584	27
34	.40055	48	.59945	.41474	52	.58526	.01419	3	.98581	26
35	9.40103	49	10.59897	9.41526	52	10.58474	10.01422	4	9.98578	25
36	.40152	48	.59848	.41578	51	.58422	.01426	3	.98574	24
37	.40200	49	.59800	.41629	52	.58371	.01429	3	.98571	23
38	.40249	48	.59751	.41681	52	.58319	.01432	3	.98568	22
39	.40297	49	.59703	.41733	51	.58267	.01435	4	.98565	21
40	9.40346	48	10.59654	9.41784	52	10.58216	10.01439	3	9.98561	20
41	.40394	48	.59606	.41836	51	.58164	.01442	3	.98558	19
42	.40442	48	.59558	.41887	52	.58113	.01445	4	.98555	18
43	.40490	48	.59510	.41939	51	.58061	.01449	3	.98551	17
44	.40538	48	.59462	.41990	51	.58010	.01452	3	.98548	16
45	9.40586	48	10.59414	9.42041	52	10.57959	10.01455	4	9.98545	15
46	.40634	48	.59366	.42093	51	.57907	.01459	3	.98541	14
47	.40682	48	.59318	.42144	51	.57856	.01462	3	.98538	13
48	.40730	48	.59270	.42195	51	.57805	.01465	4	.98535	12
49	.40778	47	.59222	.42246	51	.57754	.01469	3	.98531	11
50	9.40825	48	10.59175	9.42297	51	10.57703	10.01472	3	9.98528	10
51	.40873	48	.59127	.42348	51	.57652	.01475	4	.98525	9
52	.40921	47	.59079	.42399	51	.57601	.01479	3	.98521	8
53	.40968	48	.59032	.42450	51	.57550	.01482	3	.98518	7
54	.41016	47	.58984	.42501	51	.57499	.01485	4	.98515	6
55	9.41063	48	10.58937	9.42552	51	10.57448	10.01489	3	9.98511	5
56	.41111	47	.58889	.42603	50	.57397	.01492	3	.98508	4
57	.41158	47	.58842	.42653	51	.57347	.01495	4	.98505	3
58	.41205	47	.58795	.42704	51	.57296	.01499	3	.98501	2
59	.41252	48	.58748	.42755	50	.57245	.01502	4	.98498	1
60	9.41300		10.58700	9.42805		10.57195	10.01506		9.98494	0
↑ 104° →	cos	Diff. 1′	sec	cot	Diff. 1′	tan	csc	Diff. 1′	sin	← 75° ↑

TABLE 3

Common Logarithms of Trigonometric Functions (offset +10)

15° → ↓	sin	Diff. 1′	csc	tan	Diff. 1′	cot	sec	Diff. 1′	cos	← 164° ↓
′										′
0	9.41300	47	10.58700	9.42805	51	10.57195	10.01506	3	9.98494	60
1	.41347	47	.58653	.42856	50	.57144	.01509	3	.98491	59
2	.41394	47	.58606	.42906	51	.57094	.01512	4	.98488	58
3	.41441	47	.58559	.42957	50	.57043	.01516	3	.98484	57
4	.41488	47	.58512	.43007	50	.56993	.01519	4	.98481	56
5	9.41535	47	10.58465	9.43057	51	10.56943	10.01523	3	9.98477	55
6	.41582	46	.58418	.43108	50	.56892	.01526	3	.98474	54
7	.41628	47	.58372	.43158	50	.56842	.01529	4	.98471	53
8	.41675	47	.58325	.43208	50	.56792	.01533	3	.98467	52
9	.41722	46	.58278	.43258	50	.56742	.01536	4	.98464	51
10	9.41768	47	10.58232	9.43308	50	10.56692	10.01540	3	9.98460	50
11	.41815	46	.58185	.43358	50	.56642	.01543	4	.98457	49
12	.41861	47	.58139	.43408	50	.56592	.01547	3	.98453	48
13	.41908	46	.58092	.43458	50	.56542	.01550	3	.98450	47
14	.41954	47	.58046	.43508	50	.56492	.01553	4	.98447	46
15	9.42001	46	10.57999	9.43558	49	10.56442	10.01557	3	9.98443	45
16	.42047	46	.57953	.43607	50	.56393	.01560	4	.98440	44
17	.42093	47	.57907	.43657	50	.56343	.01564	3	.98436	43
18	.42140	46	.57860	.43707	49	.56293	.01567	4	.98433	42
19	.42186	46	.57814	.43756	50	.56244	.01571	3	.98429	41
20	9.42232	46	10.57768	9.43806	49	10.56194	10.01574	4	9.98426	40
21	.42278	46	.57722	.43855	50	.56145	.01578	3	.98422	39
22	.42324	46	.57676	.43905	49	.56095	.01581	4	.98419	38
23	.42370	46	.57630	.43954	50	.56046	.01585	3	.98415	37
24	.42416	45	.57584	.44004	49	.55996	.01588	3	.98412	36
25	9.42461	46	10.57539	9.44053	49	10.55947	10.01591	4	9.98409	35
26	.42507	46	.57493	.44102	49	.55898	.01595	3	.98405	34
27	.42553	46	.57447	.44151	50	.55849	.01598	4	.98402	33
28	.42599	45	.57401	.44201	49	.55799	.01602	3	.98398	32
29	.42644	46	.57356	.44250	49	.55750	.01605	4	.98395	31
30	9.42690	45	10.57310	9.44299	49	10.55701	10.01609	3	9.98391	30
31	.42735	46	.57265	.44348	49	.55652	.01612	4	.98388	29
32	.42781	45	.57219	.44397	49	.55603	.01616	3	.98384	28
33	.42826	46	.57174	.44446	49	.55554	.01619	4	.98381	27
34	.42872	45	.57128	.44495	49	.55505	.01623	4	.98377	26
35	9.42917	45	10.57083	9.44544	48	10.55456	10.01627	3	9.98373	25
36	.42962	46	.57038	.44592	49	.55408	.01630	4	.98370	24
37	.43008	45	.56992	.44641	49	.55359	.01634	3	.98366	23
38	.43053	45	.56947	.44690	48	.55310	.01637	4	.98363	22
39	.43098	45	.56902	.44738	49	.55262	.01641	3	.98359	21
40	9.43143	45	10.56857	9.44787	49	10.55213	10.01644	4	9.98356	20
41	.43188	45	.56812	.44836	48	.55164	.01648	3	.98352	19
42	.43233	45	.56767	.44884	49	.55116	.01651	4	.98349	18
43	.43278	45	.56722	.44933	48	.55067	.01655	3	.98345	17
44	.43323	44	.56677	.44981	48	.55019	.01658	4	.98342	16
45	9.43367	45	10.56633	9.45029	49	10.54971	10.01662	4	9.98338	15
46	.43412	45	.56588	.45078	48	.54922	.01666	3	.98334	14
47	.43457	45	.56543	.45126	48	.54874	.01669	4	.98331	13
48	.43502	44	.56498	.45174	48	.54826	.01673	3	.98327	12
49	.43546	45	.56454	.45222	49	.54778	.01676	4	.98324	11
50	9.43591	44	10.56409	9.45271	48	10.54729	10.01680	3	9.98320	10
51	.43635	45	.56365	.45319	48	.54681	.01683	4	.98317	9
52	.43680	44	.56320	.45367	48	.54633	.01687	4	.98313	8
53	.43724	45	.56276	.45415	48	.54585	.01691	3	.98309	7
54	.43769	44	.56231	.45463	48	.54537	.01694	4	.98306	6
55	9.43813	44	10.56187	9.45511	48	10.54489	10.01698	3	9.98302	5
56	.43857	44	.56143	.45559	47	.54441	.01701	4	.98299	4
57	.43901	45	.56099	.45606	48	.54394	.01705	4	.98295	3
58	.43946	44	.56054	.45654	48	.54346	.01709	3	.98291	2
59	.43990	44	.56010	.45702	48	.54298	.01712	4	.98288	1
60	9.44034		10.55966	9.45750		10.54250	10.01716		9.98284	0
↑ 105° →	cos	Diff. 1′	sec	cot	Diff. 1′	tan	csc	Diff. 1′	sin	← 74° ↑

TABLE 3
Common Logarithms of Trigonometric Functions (offset +10)

16°→ ↓ ′	sin	Diff. 1′	csc	tan	Diff. 1′	cot	sec	Diff. 1′	cos	←163° ↓ ′
0	9.44034	44	10.55966	9.45750	47	10.54250	10.01716	3	9.98284	60
1	.44078	44	.55922	.45797	48	.54203	.01719	4	.98281	59
2	.44122	44	.55878	.45845	47	.54155	.01723	4	.98277	58
3	.44166	44	.55834	.45892	48	.54108	.01727	3	.98273	57
4	.44210	43	.55790	.45940	47	.54060	.01730	4	.98270	56
5	9.44253	44	10.55747	9.45987	48	10.54013	10.01734	4	9.98266	55
6	.44297	44	.55703	.46035	47	.53965	.01738	3	.98262	54
7	.44341	44	.55659	.46082	48	.53918	.01741	4	.98259	53
8	.44385	43	.55615	.46130	47	.53870	.01745	4	.98255	52
9	.44428	44	.55572	.46177	47	.53823	.01749	3	.98251	51
10	9.44472	44	10.55528	9.46224	47	10.53776	10.01752	4	9.98248	50
11	.44516	43	.55484	.46271	48	.53729	.01756	4	.98244	49
12	.44559	43	.55441	.46319	47	.53681	.01760	3	.98240	48
13	.44602	44	.55398	.46366	47	.53634	.01763	4	.98237	47
14	.44646	43	.55354	.46413	47	.53587	.01767	4	.98233	46
15	9.44689	44	10.55311	9.46460	47	10.53540	10.01771	3	9.98229	45
16	.44733	43	.55267	.46507	47	.53493	.01774	4	.98226	44
17	.44776	43	.55224	.46554	47	.53446	.01778	4	.98222	43
18	.44819	43	.55181	.46601	47	.53399	.01782	3	.98218	42
19	.44862	43	.55138	.46648	46	.53352	.01785	4	.98215	41
20	9.44905	43	10.55095	9.46694	47	10.53306	10.01789	4	9.98211	40
21	.44948	44	.55052	.46741	47	.53259	.01793	3	.98207	39
22	.44992	43	.55008	.46788	47	.53212	.01796	4	.98204	38
23	.45035	42	.54965	.46835	46	.53165	.01800	4	.98200	37
24	.45077	43	.54923	.46881	47	.53119	.01804	4	.98196	36
25	9.45120	43	10.54880	9.46928	47	10.53072	10.01808	3	9.98192	35
26	.45163	43	.54837	.46975	46	.53025	.01811	4	.98189	34
27	.45206	43	.54794	.47021	47	.52979	.01815	4	.98185	33
28	.45249	43	.54751	.47068	46	.52932	.01819	4	.98181	32
29	.45292	42	.54708	.47114	46	.52886	.01823	3	.98177	31
30	9.45334	43	10.54666	9.47160	47	10.52840	10.01826	4	9.98174	30
31	.45377	42	.54623	.47207	46	.52793	.01830	4	.98170	29
32	.45419	43	.54581	.47253	46	.52747	.01834	4	.98166	28
33	.45462	42	.54538	.47299	47	.52701	.01838	3	.98162	27
34	.45504	43	.54496	.47346	46	.52654	.01841	4	.98159	26
35	9.45547	42	10.54453	9.47392	46	10.52608	10.01845	4	9.98155	25
36	.45589	43	.54411	.47438	46	.52562	.01849	4	.98151	24
37	.45632	42	.54368	.47484	46	.52516	.01853	3	.98147	23
38	.45674	42	.54326	.47530	46	.52470	.01856	4	.98144	22
39	.45716	42	.54284	.47576	46	.52424	.01860	4	.98140	21
40	9.45758	43	10.54242	9.47622	46	10.52378	10.01864	4	9.98136	20
41	.45801	42	.54199	.47668	46	.52332	.01868	3	.98132	19
42	.45843	42	.54157	.47714	46	.52286	.01871	4	.98129	18
43	.45885	42	.54115	.47760	46	.52240	.01875	4	.98125	17
44	.45927	42	.54073	.47806	46	.52194	.01879	4	.98121	16
45	9.45969	42	10.54031	9.47852	45	10.52148	10.01883	4	9.98117	15
46	.46011	42	.53989	.47897	46	.52103	.01887	3	.98113	14
47	.46053	42	.53947	.47943	46	.52057	.01890	4	.98110	13
48	.46095	41	.53905	.47989	46	.52011	.01894	4	.98106	12
49	.46136	42	.53864	.48035	45	.51965	.01898	4	.98102	11
50	9.46178	42	10.53822	9.48080	46	10.51920	10.01902	4	9.98098	10
51	.46220	42	.53780	.48126	45	.51874	.01906	4	.98094	9
52	.46262	41	.53738	.48171	46	.51829	.01910	3	.98090	8
53	.46303	42	.53697	.48217	45	.51783	.01913	4	.98087	7
54	.46345	41	.53655	.48262	45	.51738	.01917	4	.98083	6
55	9.46386	42	10.53614	9.48307	46	10.51693	10.01921	4	9.98079	5
56	.46428	41	.53572	.48353	45	.51647	.01925	4	.98075	4
57	.46469	42	.53531	.48398	45	.51602	.01929	4	.98071	3
58	.46511	41	.53489	.48443	46	.51557	.01933	4	.98067	2
59	.46552	42	.53448	.48489	45	.51511	.01937	3	.98063	1
60	9.46594		10.53406	9.48534		10.51466	10.01940		9.98060	0
↑ 106°→	cos	Diff. 1′	sec	cot	Diff. 1′	tan	csc	Diff. 1′	sin	← 73° ↑

TABLE 3
Common Logarithms of Trigonometric Functions (offset +10)

17°→ ↓ ′	sin	Diff. 1′	csc	tan	Diff. 1′	cot	sec	Diff. 1′	cos	←162° ↓ ′
0	9.46594	41	10.53406	9.48534	45	10.51466	10.01940	4	9.98060	60
1	.46635	41	.53365	.48579	45	.51421	.01944	4	.98056	59
2	.46676	41	.53324	.48624	45	.51376	.01948	4	.98052	58
3	.46717	41	.53283	.48669	45	.51331	.01952	4	.98048	57
4	.46758	42	.53242	.48714	45	.51286	.01956	4	.98044	56
5	9.46800	41	10.53200	9.48759	45	10.51241	10.01960	4	9.98040	55
6	.46841	41	.53159	.48804	45	.51196	.01964	4	.98036	54
7	.46882	41	.53118	.48849	45	.51151	.01968	3	.98032	53
8	.46923	41	.53077	.48894	45	.51106	.01971	4	.98029	52
9	.46964	41	.53036	.48939	45	.51061	.01975	4	.98025	51
10	9.47005	40	10.52995	9.48984	45	10.51016	10.01979	4	9.98021	50
11	.47045	41	.52955	.49029	44	.50971	.01983	4	.98017	49
12	.47086	41	.52914	.49073	45	.50927	.01987	4	.98013	48
13	.47127	41	.52873	.49118	45	.50882	.01991	4	.98009	47
14	.47168	41	.52832	.49163	44	.50837	.01995	4	.98005	46
15	9.47209	40	10.52791	9.49207	45	10.50793	10.01999	4	9.98001	45
16	.47249	41	.52751	.49252	44	.50748	.02003	4	.97997	44
17	.47290	40	.52710	.49296	45	.50704	.02007	4	.97993	43
18	.47330	41	.52670	.49341	44	.50659	.02011	3	.97989	42
19	.47371	40	.52629	.49385	45	.50615	.02014	4	.97986	41
20	9.47411	41	10.52589	9.49430	44	10.50570	10.02018	4	9.97982	40
21	.47452	40	.52548	.49474	45	.50526	.02022	4	.97978	39
22	.47492	41	.52508	.49519	44	.50481	.02026	4	.97974	38
23	.47533	40	.52467	.49563	44	.50437	.02030	4	.97970	37
24	.47573	40	.52427	.49607	45	.50393	.02034	4	.97966	36
25	9.47613	41	10.52387	9.49652	44	10.50348	10.02038	4	9.97962	35
26	.47654	40	.52346	.49696	44	.50304	.02042	4	.97958	34
27	.47694	40	.52306	.49740	44	.50260	.02046	4	.97954	33
28	.47734	40	.52266	.49784	44	.50216	.02050	4	.97950	32
29	.47774	40	.52226	.49828	44	.50172	.02054	4	.97946	31
30	9.47814	40	10.52186	9.49872	44	10.50128	10.02058	4	9.97942	30
31	.47854	40	.52146	.49916	44	.50084	.02062	4	.97938	29
32	.47894	40	.52106	.49960	44	.50040	.02066	4	.97934	28
33	.47934	40	.52066	.50004	44	.49996	.02070	4	.97930	27
34	.47974	40	.52026	.50048	44	.49952	.02074	4	.97926	26
35	9.48014	40	10.51986	9.50092	44	10.49908	10.02078	4	9.97922	25
36	.48054	40	.51946	.50136	44	.49864	.02082	4	.97918	24
37	.48094	39	.51906	.50180	43	.49820	.02086	4	.97914	23
38	.48133	40	.51867	.50223	44	.49777	.02090	4	.97910	22
39	.48173	40	.51827	.50267	44	.49733	.02094	4	.97906	21
40	9.48213	39	10.51787	9.50311	44	10.49689	10.02098	4	9.97902	20
41	.48252	40	.51748	.50355	43	.49645	.02102	4	.97898	19
42	.48292	40	.51708	.50398	44	.49602	.02106	4	.97894	18
43	.48332	39	.51668	.50442	43	.49558	.02110	4	.97890	17
44	.48371	40	.51629	.50485	44	.49515	.02114	4	.97886	16
45	9.48411	39	10.51589	9.50529	43	10.49471	10.02118	4	9.97882	15
46	.48450	40	.51550	.50572	44	.49428	.02122	4	.97878	14
47	.48490	39	.51510	.50616	43	.49384	.02126	4	.97874	13
48	.48529	39	.51471	.50659	44	.49341	.02130	4	.97870	12
49	.48568	39	.51432	.50703	43	.49297	.02134	5	.97866	11
50	9.48607	40	10.51393	9.50746	43	10.49254	10.02139	4	9.97861	10
51	.48647	39	.51353	.50789	44	.49211	.02143	4	.97857	9
52	.48686	39	.51314	.50833	43	.49167	.02147	4	.97853	8
53	.48725	39	.51275	.50876	43	.49124	.02151	4	.97849	7
54	.48764	39	.51236	.50919	43	.49081	.02155	4	.97845	6
55	9.48803	39	10.51197	9.50962	43	10.49038	10.02159	4	9.97841	5
56	.48842	39	.51158	.51005	43	.48995	.02163	4	.97837	4
57	.48881	39	.51119	.51048	44	.48952	.02167	4	.97833	3
58	.48920	39	.51080	.51092	43	.48908	.02171	4	.97829	2
59	.48959	39	.51041	.51135	43	.48865	.02175	4	.97825	1
60	9.48998		10.51002	9.51178		10.48822	10.02179		9.97821	0
↑ 107°→	cos	Diff. 1′	sec	cot	Diff. 1′	tan	csc	Diff. 1′	sin	← 72° ↑

TABLE 3

Common Logarithms of Trigonometric Functions (offset +10)

18° → ′ ↓	sin	Diff. 1′	csc	tan	Diff. 1′	cot	sec	Diff. 1′	cos	← 161° ↓ ′
0	9.48998	39	10.51002	9.51178	43	10.48822	10.02179	4	9.97821	60
1	.49037	39	.50963	.51221	43	.48779	.02183	5	.97817	59
2	.49076	39	.50924	.51264	42	.48736	.02188	4	.97812	58
3	.49115	38	.50885	.51306	43	.48694	.02192	4	.97808	57
4	.49153	39	.50847	.51349	43	.48651	.02196	4	.97804	56
5	9.49192	39	10.50808	9.51392	43	10.48608	10.02200	4	9.97800	55
6	.49231	38	.50769	.51435	43	.48565	.02204	4	.97796	54
7	.49269	39	.50731	.51478	42	.48522	.02208	4	.97792	53
8	.49308	39	.50692	.51520	43	.48480	.02212	4	.97788	52
9	.49347	38	.50653	.51563	43	.48437	.02216	5	.97784	51
10	9.49385	39	10.50615	9.51606	42	10.48394	10.02221	4	9.97779	50
11	.49424	38	.50576	.51648	43	.48352	.02225	4	.97775	49
12	.49462	38	.50538	.51691	43	.48309	.02229	4	.97771	48
13	.49500	39	.50500	.51734	42	.48266	.02233	4	.97767	47
14	.49539	38	.50461	.51776	43	.48224	.02237	4	.97763	46
15	9.49577	38	10.50423	9.51819	42	10.48181	10.02241	5	9.97759	45
16	.49615	39	50385	.51861	42	.48139	.02246	4	.97754	44
17	.49654	38	50346	.51903	43	.48097	.02250	4	.97750	43
18	.49692	38	50308	.51946	42	.48054	.02254	4	.97746	42
19	.49730	38	50270	.51988	43	.48012	.02258	4	.97742	41
20	9.49768	38	10.50232	9.52031	42	10.47969	10.02262	4	9.97738	40
21	.49806	38	.50194	.52073	42	.47927	.02266	5	.97734	39
22	.49844	38	.50156	.52115	42	.47885	.02271	4	.97729	38
23	.49882	38	.50118	.52157	43	.47843	.02275	4	.97725	37
24	.49920	38	.50080	.52200	42	.47800	.02279	4	.97721	36
25	9.49958	38	10.50042	9.52242	42	10.47758	10.02283	4	9.97717	35
26	.49996	38	.50004	.52284	42	.47716	.02287	5	.97713	34
27	.50034	38	.49966	.52326	42	.47674	.02292	4	.97708	33
28	.50072	38	.49928	.52368	42	.47632	.02296	4	.97704	32
29	.50110	38	.49890	.52410	42	.47590	.02300	4	.97700	31
30	9.50148	37	10.49852	9.52452	42	10.47548	10.02304	5	9.97696	30
31	.50185	38	.49815	.52494	42	.47506	.02309	4	.97691	29
32	.50223	38	.49777	.52536	42	.47464	.02313	4	.97687	28
33	.50261	37	.49739	.52578	42	.47422	.02317	4	.97683	27
34	.50298	38	.49702	.52620	41	.47380	.02321	5	.97679	26
35	9.50336	38	10.49664	9.52661	42	10.47339	10.02326	4	9.97674	25
36	.50374	37	.49626	.52703	42	.47297	.02330	4	.97670	24
37	.50411	38	.49589	.52745	42	.47255	.02334	4	.97666	23
38	.50449	37	.49551	.52787	42	.47213	.02338	5	.97662	22
39	.50486	37	.49514	.52829	41	.47171	.02343	4	.97657	21
40	9.50523	38	10.49477	9.52870	42	10.47130	10.02347	4	9.97653	20
41	.50561	37	.49439	.52912	41	.47088	.02351	4	.97649	19
42	.50598	37	.49402	.52953	42	.47047	.02355	5	.97645	18
43	.50635	38	.49365	.52995	42	.47005	.02360	4	.97640	17
44	.50673	37	.49327	.53037	41	.46963	.02364	4	.97636	16
45	9.50710	37	10.49290	9.53078	42	10.46922	10.02368	4	9.97632	15
46	.50747	37	.49253	.53120	41	.46880	.02372	5	.97628	14
47	.50784	37	.49216	.53161	41	.46839	.02377	4	.97623	13
48	.50821	37	.49179	.53202	42	.46798	.02381	4	.97619	12
49	.50858	38	.49142	.53244	41	.46756	.02385	5	.97615	11
50	9.50896	37	10.49104	9.53285	42	10.46715	10.02390	4	9.97610	10
51	.50933	37	.49067	.53327	41	.46673	.02394	4	.97606	9
52	.50970	37	.49030	.53368	41	.46632	.02398	5	.97602	8
53	.51007	36	.48993	.53409	41	.46591	.02403	4	.97597	7
54	.51043	37	.48957	.53450	42	.46550	.02407	4	.97593	6
55	9.51080	37	10.48920	9.53492	41	10.46508	10.02411	5	9.97589	5
56	.51117	37	.48883	.53533	41	.46467	.02416	4	.97584	4
57	.51154	37	.48846	.53574	41	.46426	.02420	4	.97580	3
58	.51191	36	.48809	.53615	41	.46385	.02424	5	.97576	2
59	.51227	37	.48773	.53656	41	.46344	.02429	4	.97571	1
60	9.51264		10.48736	9.53697		10.46303	10.02433		9.97567	0
↑ 108° →	cos	Diff. 1′	sec	cot	Diff. 1′	tan	csc	Diff. 1′	sin	← 71° ↑

TABLE 3

Common Logarithms of Trigonometric Functions (offset +10)

19° → ′ ↓	sin	Diff. 1′	csc	tan	Diff. 1′	cot	sec	Diff. 1′	cos	← 160° ↓ ′
0	9.51264	37	10.48736	9.53697	41	10.46303	10.02433	4	9.97567	60
1	.51301	37	.48699	.53738	41	.46262	.02437	5	.97563	59
2	.51338	36	.48662	.53779	41	.46221	.02442	4	.97558	58
3	.51374	37	.48626	.53820	41	.46180	.02446	4	.97554	57
4	.51411	36	.48589	.53861	41	.46139	.02450	5	.97550	56
5	9.51447	37	10.48553	9.53902	41	10.46098	10.02455	4	9.97545	55
6	.51484	36	.48516	.53943	41	.46057	.02459	5	.97541	54
7	.51520	37	.48480	.53984	41	.46016	.02464	4	.97536	53
8	.51557	36	.48443	.54025	40	.45975	.02468	4	.97532	52
9	.51593	36	.48407	.54065	41	.45935	.02472	5	.97528	51
10	9.51629	37	10.48371	9.54106	41	10.45894	10.02477	4	9.97523	50
11	.51666	36	.48334	.54147	40	.45853	.02481	4	.97519	49
12	.51702	36	.48298	.54187	41	.45813	.02485	5	.97515	48
13	.51738	36	.48262	.54228	41	.45772	.02490	4	.97510	47
14	.51774	37	.48226	.54269	40	.45731	.02494	5	.97506	46
15	9.51811	36	10.48189	9.54309	41	10.45691	10.02499	4	9.97501	45
16	.51847	36	.48153	.54350	40	.45650	.02503	5	.97497	44
17	.51883	36	.48117	.54390	41	.45610	.02508	4	.97492	43
18	.51919	36	.48081	.54431	40	.45569	.02512	4	.97488	42
19	.51955	36	.48045	.54471	41	.45529	.02516	5	.97484	41
20	9.51991	36	10.48009	9.54512	40	10.45488	10.02521	4	9.97479	40
21	.52027	36	.47973	.54552	41	.45448	.02525	5	.97475	39
22	.52063	36	.47937	.54593	40	.45407	.02530	4	.97470	38
23	.52099	36	.47901	.54633	40	.45367	.02534	5	.97466	37
24	.52135	36	.47865	.54673	41	.45327	.02539	4	.97461	36
25	9.52171	36	10.47829	9.54714	40	10.45286	10.02543	4	9.97457	35
26	.52207	35	.47793	.54754	40	.45246	.02547	5	.97453	34
27	.52242	36	.47758	.54794	41	.45206	.02552	4	.97448	33
28	.52278	36	.47722	.54835	40	.45165	.02556	5	.97444	32
29	.52314	36	.47686	.54875	40	.45125	.02561	4	.97439	31
30	9.52350	35	10.47650	9.54915	40	10.45085	10.02565	5	9.97435	30
31	.52385	36	.47615	.54955	40	.45045	.02570	4	.97430	29
32	.52421	35	.47579	.54995	40	.45005	.02574	5	.97426	28
33	.52456	36	.47544	.55035	40	.44965	.02579	4	.97421	27
34	.52492	35	.47508	.55075	40	.44925	.02583	5	.97417	26
35	9.52527	36	10.47473	9.55115	40	10.44885	10.02588	4	9.97412	25
36	.52563	35	.47437	.55155	40	.44845	.02592	5	.97408	24
37	.52598	36	.47402	.55195	40	.44805	.02597	4	.97403	23
38	.52634	35	.47366	.55235	40	.44765	.02601	5	.97399	22
39	.52669	36	.47331	.55275	40	.44725	.02606	4	.97394	21
40	9.52705	35	10.47295	9.55315	40	10.44685	10.02610	5	9.97390	20
41	.52740	35	.47260	.55355	40	.44645	.02615	4	.97385	19
42	.52775	36	.47225	.55395	39	.44605	.02619	5	.97381	18
43	.52811	35	.47189	.55434	40	.44566	.02624	4	.97376	17
44	.52846	35	.47154	.55474	40	.44526	.02628	5	.97372	16
45	9.52881	35	10.47119	9.55514	40	10.44486	10.02633	4	9.97367	15
46	.52916	35	.47084	.55554	39	.44446	.02637	5	.97363	14
47	.52951	35	.47049	.55593	40	.44407	.02642	5	.97358	13
48	.52986	35	.47014	.55633	40	.44367	.02647	4	.97353	12
49	.53021	35	.46979	.55673	39	.44327	.02651	5	.97349	11
50	9.53056	36	10.46944	9.55712	40	10.44288	10.02656	4	9.97344	10
51	.53092	34	.46908	.55752	39	.44248	.02660	5	.97340	9
52	.53126	35	.46874	.55791	40	.44209	.02665	4	.97335	8
53	.53161	35	.46839	.55831	39	.44169	.02669	5	.97331	7
54	.53196	35	.46804	.55870	40	.44130	.02674	4	.97326	6
55	9.53231	35	10.46769	9.55910	39	10.44090	10.02678	5	9.97322	5
56	.53266	35	.46734	.55949	40	.44051	.02683	5	.97317	4
57	.53301	35	.46699	.55989	39	.44011	.02688	4	.97312	3
58	.53336	34	.46664	.56028	39	.43972	.02692	5	.97308	2
59	.53370	35	.46630	.56067	40	.43933	.02697	4	.97303	1
60	9.53405		10.46595	9.56107		10.43893	10.02701		9.97299	0
↑ 109° →	cos	Diff. 1′	sec	cot	Diff. 1′	tan	csc	Diff. 1′	sin	← 70° ↑

TABLE 3

Common Logarithms of Trigonometric Functions (offset +10)

20° → ↓ ′	sin	Diff. 1′	csc	tan	Diff. 1′	cot	sec	Diff. 1′	cos	← 159° ↓ ′
0	9.53405	35	10.46595	9.56107	39	10.43893	10.02701	5	9.97299	60
1	.53440	35	.46560	.56146	39	.43854	.02706	5	.97294	59
2	.53475	34	.46525	.56185	39	.43815	.02711	4	.97289	58
3	.53509	35	.46491	.56224	40	.43776	.02715	5	.97285	57
4	.53544	34	.46456	.56264	39	.43736	.02720	4	.97280	56
5	9.53578	35	10.46422	9.56303	39	10.43697	10.02724	5	9.97276	55
6	.53613	34	.46387	.56342	39	.43658	.02729	5	.97271	54
7	.53647	35	.46353	.56381	39	.43619	.02734	4	.97266	53
8	.53682	34	.46318	.56420	39	.43580	.02738	5	.97262	52
9	.53716	35	.46284	.56459	39	.43541	.02743	5	.97257	51
10	9.53751	34	10.46249	9.56498	39	10.43502	10.02748	4	9.97252	50
11	.53785	34	.46215	.56537	39	.43463	.02752	5	.97248	49
12	.53819	35	.46181	.56576	39	.43424	.02757	5	.97243	48
13	.53854	34	.46146	.56615	39	.43385	.02762	4	.97238	47
14	.53888	34	.46112	.56654	39	.43346	.02766	5	.97234	46
15	9.53922	35	10.46078	9.56693	39	10.43307	10.02771	5	9.97229	45
16	.53957	34	.46043	.56732	39	.43268	.02776	4	.97224	44
17	.53991	34	.46009	.56771	39	.43229	.02780	5	.97220	43
18	.54025	34	.45975	.56810	39	.43190	.02785	5	.97215	42
19	.54059	34	.45941	.56849	38	.43151	.02790	4	.97210	41
20	9.54093	34	10.45907	9.56887	39	10.43113	10.02794	5	9.97206	40
21	.54127	34	.45873	.56926	39	.43074	.02799	5	.97201	39
22	.54161	34	.45839	.56965	39	.43035	.02804	4	.97196	38
23	.54195	34	.45805	.57004	38	.42996	.02808	5	.97192	37
24	.54229	34	.45771	.57042	39	.42958	.02813	5	.97187	36
25	9.54263	34	10.45737	9.57081	39	10.42919	10.02818	4	9.97182	35
26	.54297	34	.45703	.57120	38	.42880	.02822	5	.97178	34
27	.54331	34	.45669	.57158	39	.42842	.02827	5	.97173	33
28	.54365	34	.45635	.57197	38	.42803	.02832	5	.97168	32
29	.54399	34	.45601	.57235	39	.42765	.02837	4	.97163	31
30	9.54433	33	10.45567	9.57274	38	10.42726	10.02841	5	9.97159	30
31	.54466	34	.45534	.57312	39	.42688	.02846	5	.97154	29
32	.54500	34	.45500	.57351	38	.42649	.02851	4	.97149	28
33	.54534	33	.45466	.57389	39	.42611	.02855	5	.97145	27
34	.54567	34	.45433	.57428	38	.42572	.02860	5	.97140	26
35	9.54601	34	10.45399	9.57466	38	10.42534	10.02865	5	9.97135	25
36	.54635	33	.45365	.57504	39	.42496	.02870	4	.97130	24
37	.54668	34	.45332	.57543	38	.42457	.02874	5	.97126	23
38	.54702	33	.45298	.57581	38	.42419	.02879	5	.97121	22
39	.54735	34	.45265	.57619	39	.42381	.02884	5	.97116	21
40	9.54769	33	10.45231	9.57658	38	10.42342	10.02889	4	9.97111	20
41	.54802	34	.45198	.57696	38	.42304	.02893	5	.97107	19
42	.54836	33	.45164	.57734	38	.42266	.02898	5	.97102	18
43	.54869	34	.45131	.57772	38	.42228	.02903	5	.97097	17
44	.54903	33	.45097	.57810	39	.42190	.02908	5	.97092	16
45	9.54936	33	10.45064	9.57849	38	10.42151	10.02913	4	9.97087	15
46	.54969	34	.45031	.57887	38	.42113	.02917	5	.97083	14
47	.55003	33	.44997	.57925	38	.42075	.02922	5	.97078	13
48	.55036	33	.44964	.57963	38	.42037	.02927	5	.97073	12
49	.55069	33	.44931	.58001	38	.41999	.02932	5	.97068	11
50	9.55102	34	10.44898	9.58039	38	10.41961	10.02937	4	9.97063	10
51	.55136	33	.44864	.58077	38	.41923	.02941	5	.97059	9
52	.55169	33	.44831	.58115	38	.41885	.02946	5	.97054	8
53	.55202	33	.44798	.58153	38	.41847	.02951	5	.97049	7
54	.55235	33	.44765	.58191	38	.41809	.02956	5	.97044	6
55	9.55268	33	10.44732	9.58229	38	10.41771	10.02961	4	9.97039	5
56	.55301	33	.44699	.58267	37	.41733	.02965	5	.97035	4
57	.55334	33	.44666	.58304	38	.41696	.02970	5	.97030	3
58	.55367	33	.44633	.58342	38	.41658	.02975	5	.97025	2
59	.55400	33	.44600	.58380	38	.41620	.02980	5	.97020	1
60	9.55433		10.44567	9.58418		10.41582	10.02985		9.97015	0
↑ 110° →	cos	Diff. 1′	sec	cot	Diff. 1′	tan	csc	Diff. 1′	sin	← 69° ↑

TABLE 3

Common Logarithms of Trigonometric Functions (offset +10)

21° → ↓ ′	sin	Diff. 1′	csc	tan	Diff. 1′	cot	sec	Diff. 1′	cos	← 158° ↓ ′
0	9.55433	33	10.44567	9.58418	37	10.41582	10.02985	5	9.97015	60
1	.55466	33	.44534	.58455	38	.41545	.02990	5	.97010	59
2	.55499	33	.44501	.58493	38	.41507	.02995	4	.97005	58
3	.55532	32	.44468	.58531	38	.41469	.02999	5	.97001	57
4	.55564	33	.44436	.58569	37	.41431	.03004	5	.96996	56
5	9.55597	33	10.44403	9.58606	38	10.41394	10.03009	5	9.96991	55
6	.55630	33	.44370	.58644	37	.41356	.03014	5	.96986	54
7	.55663	32	.44337	.58681	38	.41319	.03019	5	.96981	53
8	.55695	33	.44305	.58719	38	.41281	.03024	5	.96976	52
9	.55728	33	.44272	.58757	37	.41243	.03029	5	.96971	51
10	9.55761	32	10.44239	9.58794	38	10.41206	10.03034	4	9.96966	50
11	.55793	33	.44207	.58832	37	.41168	.03038	5	.96962	49
12	.55826	32	.44174	.58869	38	.41131	.03043	5	.96957	48
13	.55858	33	.44142	.58907	37	.41093	.03048	5	.96952	47
14	.55891	32	.44109	.58944	37	.41056	.03053	5	.96947	46
15	9.55923	33	10.44077	9.58981	38	10.41019	10.03058	5	9.96942	45
16	.55956	32	.44044	.59019	37	.40981	.03063	5	.96937	44
17	.55988	33	.44012	.59056	38	.40944	.03068	5	.96932	43
18	.56021	32	.43979	.59094	37	.40906	.03073	5	.96927	42
19	.56053	32	.43947	.59131	37	.40869	.03078	5	.96922	41
20	9.56085	33	10.43915	9.59168	37	10.40832	10.03083	5	9.96917	40
21	.56118	32	.43882	.59205	38	.40795	.03088	5	.96912	39
22	.56150	32	.43850	.59243	37	.40757	.03093	4	.96907	38
23	.56182	33	.43818	.59280	37	.40720	.03097	5	.96903	37
24	.56215	32	.43785	.59317	37	.40683	.03102	5	.96898	36
25	9.56247	32	10.43753	9.59354	37	10.40646	10.03107	5	9.96893	35
26	.56279	32	.43721	.59391	38	.40609	.03112	5	.96888	34
27	.56311	32	.43689	.59429	37	.40571	.03117	5	.96883	33
28	.56343	32	.43657	.59466	37	.40534	.03122	5	.96878	32
29	.56375	33	.43625	.59503	37	.40497	.03127	5	.96873	31
30	9.56408	32	10.43592	9.59540	37	10.40460	10.03132	5	9.96868	30
31	.56440	32	.43560	.59577	37	.40423	.03137	5	.96863	29
32	.56472	32	.43528	.59614	37	.40386	.03142	5	.96858	28
33	.56504	32	.43496	.59651	37	.40349	.03147	5	.96853	27
34	.56536	32	.43464	.59688	37	.40312	.03152	5	.96848	26
35	9.56568	31	10.43432	9.59725	37	10.40275	10.03157	5	9.96843	25
36	.56599	32	.43401	.59762	37	.40238	.03162	5	.96838	24
37	.56631	32	.43369	.59799	36	.40201	.03167	5	.96833	23
38	.56663	32	.43337	.59835	37	.40165	.03172	5	.96828	22
39	.56695	32	.43305	.59872	37	.40128	.03177	5	.96823	21
40	9.56727	32	10.43273	9.59909	37	10.40091	10.03182	5	9.96818	20
41	.56759	31	.43241	.59946	37	.40054	.03187	5	.96813	19
42	.56790	32	.43210	.59983	36	.40017	.03192	5	.96808	18
43	.56822	32	.43178	.60019	37	.39981	.03197	5	.96803	17
44	.56854	32	.43146	.60056	37	.39944	.03202	5	.96798	16
45	9.56886	31	10.43114	9.60093	37	10.39907	10.03207	5	9.96793	15
46	.56917	32	.43083	.60130	36	.39870	.03212	5	.96788	14
47	.56949	31	.43051	.60166	37	.39834	.03217	5	.96783	13
48	.56980	32	.43020	.60203	37	.39797	.03222	6	.96778	12
49	.57012	32	.42988	.60240	36	.39760	.03228	5	.96772	11
50	9.57044	31	10.42956	9.60276	37	10.39724	10.03233	5	9.96767	10
51	.57075	32	.42925	.60313	36	.39687	.03238	5	.96762	9
52	.57107	31	.42893	.60349	37	.39651	.03243	5	.96757	8
53	.57138	31	.42862	.60386	36	.39614	.03248	5	.96752	7
54	.57169	32	.42831	.60422	37	.39578	.03253	5	.96747	6
55	9.57201	31	10.42799	9.60459	36	10.39541	10.03258	5	9.96742	5
56	.57232	32	.42768	.60495	37	.39505	.03263	5	.96737	4
57	.57264	31	.42736	.60532	36	.39468	.03268	5	.96732	3
58	.57295	31	.42705	.60568	37	.39432	.03273	5	.96727	2
59	.57326	32	.42674	.60605	36	.39395	.03278	5	.96722	1
60	9.57358		10.42642	9.60641		10.39359	10.03283		9.96717	0
↑ 111° →	cos	Diff. 1′	sec	cot	Diff. 1′	tan	csc	Diff. 1′	sin	← 68° ↑

TABLE 5
Common Logarithms of Trigonometric Functions (offset +10)

22° → ′	sin	Diff. 1′	csc	tan	Diff. 1′	cot	sec	Diff. 1′	cos	← 157° ′
0	9.57358	31	10.42642	9.60641	36	10.39359	10.03283	6	9.96717	60
1	.57389	31	.42611	.60677	37	.39323	.03289	5	.96711	59
2	.57420	31	.42580	.60714	36	.39286	.03294	5	.96706	58
3	.57451	31	.42549	.60750	36	.39250	.03299	5	.96701	57
4	.57482	32	.42518	.60786	37	.39214	.03304	5	.96696	56
5	9.57514	31	10.42486	9.60823	36	10.39177	10.03309	5	9.96691	55
6	.57545	31	.42455	.60859	36	.39141	.03314	5	.96686	54
7	.57576	31	.42424	.60895	36	.39105	.03319	5	.96681	53
8	.57607	31	.42393	.60931	36	.39069	.03324	6	.96676	52
9	.57638	31	.42362	.60967	37	.39033	.03330	5	.96670	51
10	9.57669	31	10.42331	9.61004	36	10.38996	10.03335	5	9.96665	50
11	.57700	31	.42300	.61040	36	.38960	.03340	5	.96660	49
12	.57731	31	.42269	.61076	36	.38924	.03345	5	.96655	48
13	.57762	31	.42238	.61112	36	.38888	.03350	5	.96650	47
14	.57793	31	.42207	.61148	36	.38852	.03355	5	.96645	46
15	9.57824	31	10.42176	9.61184	36	10.38816	10.03360	6	9.96640	45
16	.57855	30	.42145	.61220	36	.38780	.03366	5	.96634	44
17	.57885	31	.42115	.61256	36	.38744	.03371	5	.96629	43
18	.57916	31	.42084	.61292	36	.38708	.03376	5	.96624	42
19	.57947	31	.42053	.61328	36	.38672	.03381	5	.96619	41
20	9.57978	30	10.42022	9.61364	36	10.38636	10.03386	6	9.96614	40
21	.58008	31	.41992	.61400	36	.38600	.03392	5	.96608	39
22	.58039	31	.41961	.61436	36	.38564	.03397	5	.96603	38
23	.58070	31	.41930	.61472	36	.38528	.03402	5	.96598	37
24	.58101	30	.41899	.61508	36	.38492	.03407	5	.96593	36
25	9.58131	31	10.41869	9.61544	35	10.38456	10.03412	6	9.96588	35
26	.58162	30	.41838	.61579	36	.38421	.03418	5	.96582	34
27	.58192	31	.41808	.61615	36	.38385	.03423	5	.96577	33
28	.58223	30	.41777	.61651	36	.38349	.03428	5	.96572	32
29	.58253	31	.41747	.61687	35	.38313	.03433	5	.96567	31
30	9.58284	30	10.41716	9.61722	36	10.38278	10.03438	6	9.96562	30
31	.58314	31	.41686	.61758	36	.38242	.03444	5	.96556	29
32	.58345	30	.41655	.61794	36	.38206	.03449	5	.96551	28
33	.58375	31	.41625	.61830	35	.38170	.03454	5	.96546	27
34	.58406	30	.41594	.61865	36	.38135	.03459	6	.96541	26
35	9.58436	31	10.41564	9.61901	35	10.38099	10.03465	5	9.96535	25
36	.58467	30	.41533	.61936	36	.38064	.03470	5	.96530	24
37	.58497	30	.41503	.61972	36	.38028	.03475	5	.96525	23
38	.58527	30	.41473	.62008	35	.37992	.03480	6	.96520	22
39	.58557	31	.41443	.62043	36	.37957	.03486	5	.96514	21
40	9.58588	30	10.41412	9.62079	35	10.37921	10.03491	5	9.96509	20
41	.58618	30	.41382	.62114	36	.37886	.03496	6	.96504	19
42	.58648	30	.41352	.62150	35	.37850	.03502	5	.96498	18
43	.58678	31	.41322	.62185	36	.37815	.03507	5	.96493	17
44	.58709	30	.41291	.62221	35	.37779	.03512	5	.96488	16
45	9.58739	30	10.41261	9.62256	36	10.37744	10.03517	6	9.96483	15
46	.58769	30	.41231	.62292	35	.37708	.03523	5	.96477	14
47	.58799	30	.41201	.62327	35	.37673	.03528	5	.96472	13
48	.58829	30	.41171	.62362	36	.37638	.03533	6	.96467	12
49	.58859	30	.41141	.62398	35	.37602	.03539	5	.96461	11
50	9.58889	30	10.41111	9.62433	35	10.37567	10.03544	5	9.96456	10
51	.58919	30	.41081	.62468	36	.37532	.03549	6	.96451	9
52	.58949	30	.41051	.62504	35	.37496	.03555	5	.96445	8
53	.58979	30	.41021	.62539	35	.37461	.03560	5	.96440	7
54	.59009	30	.40991	.62574	35	.37426	.03565	6	.96435	6
55	9.59039	30	10.40961	9.62609	36	10.37391	10.03571	5	9.96429	5
56	.59069	29	.40931	.62645	35	.37355	.03576	5	.96424	4
57	.59098	30	.40902	.62680	35	.37320	.03581	6	.96419	3
58	.59128	30	.40872	.62715	35	.37285	.03587	5	.96413	2
59	.59158	30	.40842	.62750	35	.37250	.03592	5	.96408	1
60	9.59188		10.40812	9.62785		10.37215	10.03597		9.96403	0
↑ 112° →	cos	Diff. 1′	sec	cot	Diff. 1′	tan	csc	Diff. 1′	sin	← 67° ↑

TABLE 5
Common Logarithms of Trigonometric Functions (offset +10)

23° → ′	sin	Diff. 1′	csc	tan	Diff. 1′	cot	sec	Diff. 1′	cos	← 156° ′
0	9.59188	30	10.40812	9.62785	35	10.37215	10.03597	6	9.96403	60
1	.59218	29	.40782	.62820	35	.37180	.03603	5	.96397	59
2	.59247	30	.40753	.62855	35	.37145	.03608	5	.96392	58
3	.59277	30	.40723	.62890	36	.37110	.03613	6	.96387	57
4	.59307	29	.40693	.62926	35	.37074	.03619	5	.96381	56
5	9.59336	30	10.40664	9.62961	35	10.37039	10.03624	6	9.96376	55
6	.59366	30	.40634	.62996	35	.37004	.03630	5	.96370	54
7	.59396	29	.40604	.63031	35	.36969	.03635	5	.96365	53
8	.59425	30	.40575	.63066	35	.36934	.03640	6	.96360	52
9	.59455	29	.40545	.63101	34	.36899	.03646	5	.96354	51
10	9.59484	30	10.40516	9.63135	35	10.36865	10.03651	6	9.96349	50
11	.59514	29	.40486	.63170	35	.36830	.03657	5	.96343	49
12	.59543	30	.40457	.63205	35	.36795	.03662	5	.96338	48
13	.59573	29	.40427	.63240	35	.36760	.03667	6	.96333	47
14	.59602	30	.40398	.63275	35	.36725	.03673	5	.96327	46
15	9.59632	29	10.40368	9.63310	35	10.36690	10.03678	6	9.96322	45
16	.59661	29	.40339	.63345	34	.36655	.03684	5	.96316	44
17	.59690	30	.40310	.63379	35	.36621	.03689	6	.96311	43
18	.59720	29	.40280	.63414	35	.36586	.03695	5	.96305	42
19	.59749	29	.40251	.63449	35	.36551	.03700	6	.96300	41
20	9.59778	30	10.40222	9.63484	35	10.36516	10.03706	5	9.96294	40
21	.59808	29	.40192	.63519	34	.36481	.03711	5	.96289	39
22	.59837	29	.40163	.63553	35	.36447	.03716	6	.96284	38
23	.59866	29	.40134	.63588	35	.36412	.03722	5	.96278	37
24	.59895	29	.40105	.63623	34	.36377	.03727	6	.96273	36
25	9.59924	30	10.40076	9.63657	35	10.36343	10.03733	5	9.96267	35
26	.59954	29	.40046	.63692	34	.36308	.03738	6	.96262	34
27	.59983	29	.40017	.63726	35	.36274	.03744	5	.96256	33
28	.60012	29	.39988	.63761	35	.36239	.03749	6	.96251	32
29	.60041	29	.39959	.63796	34	.36204	.03755	5	.96245	31
30	9.60070	29	10.39930	9.63830	35	10.36170	10.03760	6	9.96240	30
31	.60099	29	.39901	.63865	34	.36135	.03766	5	.96234	29
32	.60128	29	.39872	.63899	35	.36101	.03771	6	.96229	28
33	.60157	29	.39843	.63934	34	.36066	.03777	5	.96223	27
34	.60186	29	.39814	.63968	35	.36032	.03782	6	.96218	26
35	9.60215	29	10.39785	9.64003	34	10.35997	10.03788	5	9.96212	25
36	.60244	29	.39756	.64037	35	.35963	.03793	6	.96207	24
37	.60273	29	.39727	.64072	34	.35928	.03799	5	.96201	23
38	.60302	29	.39698	.64106	34	.35894	.03804	6	.96196	22
39	.60331	28	.39669	.64140	35	.35860	.03810	5	.96190	21
40	9.60359	29	10.39641	9.64175	34	10.35825	10.03815	6	9.96185	20
41	.60388	29	.39612	.64209	34	.35791	.03821	5	.96179	19
42	.60417	29	.39583	.64243	35	.35757	.03826	6	.96174	18
43	.60446	28	.39554	.64278	34	.35722	.03832	6	.96168	17
44	.60474	29	.39526	.64312	34	.35688	.03838	5	.96162	16
45	9.60503	29	10.39497	9.64346	35	10.35654	10.03843	6	9.96157	15
46	.60532	29	.39468	.64381	34	.35619	.03849	5	.96151	14
47	.60561	28	.39439	.64415	34	.35585	.03854	6	.96146	13
48	.60589	29	.39411	.64449	34	.35551	.03860	5	.96140	12
49	.60618	28	.39382	.64483	34	.35517	.03865	6	.96135	11
50	9.60646	29	10.39354	9.64517	35	10.35483	10.03871	6	9.96129	10
51	.60675	29	.39325	.64552	34	.35448	.03877	5	.96123	9
52	.60704	28	.39296	.64586	34	.35414	.03882	6	.96118	8
53	.60732	29	.39268	.64620	34	.35380	.03888	5	.96112	7
54	.60761	28	.39239	.64654	34	.35346	.03893	6	.96107	6
55	9.60789	29	10.39211	9.64688	34	10.35312	10.03899	6	9.96101	5
56	.60818	28	.39182	.64722	34	.35278	.03905	5	.96095	4
57	.60846	29	.39154	.64756	34	.35244	.03910	6	.96090	3
58	.60875	28	.39125	.64790	34	.35210	.03916	5	.96084	2
59	.60903	28	.39097	.64824	34	.35176	.03921	6	.96079	1
60	9.60931		10.39069	9.64858		10.35142	10.03927		9.96073	0
↑ 113° →	cos	Diff. 1′	sec	cot	Diff. 1′	tan	csc	Diff. 1′	sin	← 66° ↑

TABLE 3

Common Logarithms of Trigonometric Functions (offset +10)

24° → ′	sin	Diff. 1′	csc	tan	Diff. 1′	cot	sec	Diff. 1′	cos	← 155° ′
0	9.60931	29	10.39069	9.64858	34	10.35142	10.03927	6	9.96073	60
1	.60960	28	.39040	.64892	34	.35108	.03933	5	.96067	59
2	.60988	28	.39012	.64926	34	.35074	.03938	6	.96062	58
3	.61016	29	.38984	.64960	34	.35040	.03944	6	.96056	57
4	.61045	28	.38955	.64994	34	.35006	.03950	5	.96050	56
5	9.61073	28	10.38927	9.65028	34	10.34972	10.03955	6	9.96045	55
6	.61101	28	.38899	.65062	34	.34938	.03961	5	.96039	54
7	.61129	29	.38871	.65096	34	.34904	.03966	6	.96034	53
8	.61158	28	.38842	.65130	34	.34870	.03972	6	.96028	52
9	.61186	28	.38814	.65164	33	.34836	.03978	5	.96022	51
10	9.61214	28	10.38786	9.65197	34	10.34803	10.03983	6	9.96017	50
11	.61242	28	.38758	.65231	34	.34769	.03989	6	.96011	49
12	.61270	28	.38730	.65265	34	.34735	.03995	5	.96005	48
13	.61298	28	.38702	.65299	34	.34701	.04000	6	.96000	47
14	.61326	28	.38674	.65333	33	.34667	.04006	6	.95994	46
15	9.61354	28	10.38646	9.65366	34	10.34634	10.04012	6	9.95988	45
16	.61382	29	.38618	.65400	34	.34600	.04018	5	.95982	44
17	.61411	27	.38589	.65434	33	.34566	.04023	6	.95977	43
18	.61438	28	.38562	.65467	34	.34533	.04029	6	.95971	42
19	.61466	28	.38534	.65501	34	.34499	.04035	5	.95965	41
20	9.61494	28	10.38506	9.65535	33	10.34465	10.04040	6	9.95960	40
21	.61522	28	.38478	.65568	34	.34432	.04046	6	.95954	39
22	.61550	28	.38450	.65602	34	.34398	.04052	6	.95948	38
23	.61578	28	.38422	.65636	33	.34364	.04058	5	.95942	37
24	.61606	28	.38394	.65669	34	.34331	.04063	6	.95937	36
25	9.61634	28	10.38366	9.65703	33	10.34297	10.04069	6	9.95931	35
26	.61662	27	.38338	.65736	34	.34264	.04075	5	.95925	34
27	.61689	28	.38311	.65770	33	.34230	.04080	6	.95920	33
28	.61717	28	.38283	.65803	34	.34197	.04086	6	.95914	32
29	.61745	28	.38255	.65837	33	.34163	.04092	6	.95908	31
30	9.61773	27	10.38227	9.65870	34	10.34130	10.04098	5	9.95902	30
31	.61800	28	.38200	.65904	33	.34096	.04103	6	.95897	29
32	.61828	28	.38172	.65937	34	.34063	.04109	6	.95891	28
33	.61856	27	.38144	.65971	33	.34029	.04115	6	.95885	27
34	.61883	28	.38117	.66004	34	.33996	.04121	6	.95879	26
35	9.61911	28	10.38089	9.66038	33	10.33962	10.04127	5	9.95873	25
36	.61939	27	.38061	.66071	33	.33929	.04132	6	.95868	24
37	.61966	28	.38034	.66104	34	.33896	.04138	6	.95862	23
38	.61994	27	.38006	.66138	33	.33862	.04144	6	.95856	22
39	.62021	28	.37979	.66171	33	.33829	.04150	6	.95850	21
40	9.62049	27	10.37951	9.66204	34	10.33796	10.04156	5	9.95844	20
41	.62076	28	.37924	.66238	33	.33762	.04161	6	.95839	19
42	.62104	27	.37896	.66271	33	.33729	.04167	6	.95833	18
43	.62131	28	.37869	.66304	33	.33696	.04173	6	.95827	17
44	.62159	27	.37841	.66337	34	.33663	.04179	6	.95821	16
45	9.62186	28	10.37814	9.66371	33	10.33629	10.04185	5	9.95815	15
46	.62214	27	.37786	.66404	33	.33596	.04190	6	.95810	14
47	.62241	27	.37759	.66437	33	.33563	.04196	6	.95804	13
48	.62268	28	.37732	.66470	33	.33530	.04202	6	.95798	12
49	.62296	27	.37704	.66503	34	.33497	.04208	6	.95792	11
50	9.62323	27	10.37677	9.66537	33	10.33463	10.04214	6	9.95786	10
51	.62350	27	.37650	.66570	33	.33430	.04220	5	.95780	9
52	.62377	28	.37623	.66603	33	.33397	.04225	6	.95775	8
53	.62405	27	.37595	.66636	33	.33364	.04231	6	.95769	7
54	.62432	27	.37568	.66669	33	.33331	.04237	6	.95763	6
55	9.62459	27	10.37541	9.66702	33	10.33298	10.04243	6	9.95757	5
56	.62486	27	.37514	.66735	33	.33265	.04249	6	.95751	4
57	.62513	28	.37487	.66768	33	.33232	.04255	6	.95745	3
58	.62541	27	.37459	.66801	33	.33199	.04261	6	.95739	2
59	.62568	27	.37432	.66834	33	.33166	.04267	5	.95733	1
60	9.62595		10.37405	9.66867		10.33133	10.04272		9.95728	0
↑ 114° →	cos	Diff. 1′	sec	cot	Diff. 1′	tan	csc	Diff. 1′	sin	← 65° ↑

TABLE 3

Common Logarithms of Trigonometric Functions (offset +10)

25° → ′	sin	Diff. 1′	csc	tan	Diff. 1′	cot	sec	Diff. 1′	cos	← 154° ′
0	9.62595	27	10.37405	9.66867	33	10.33133	10.04272	6	9.95728	60
1	.62622	27	.37378	.66900	33	.33100	.04278	6	.95722	59
2	.62649	27	.37351	.66933	33	.33067	.04284	6	.95716	58
3	.62676	27	.37324	.66966	33	.33034	.04290	6	.95710	57
4	.62703	27	.37297	.66999	33	.33001	.04296	6	.95704	56
5	9.62730	27	10.37270	9.67032	33	10.32968	10.04302	6	9.95698	55
6	.62757	27	.37243	.67065	33	.32935	.04308	6	.95692	54
7	.62784	27	.37216	.67098	33	.32902	.04314	6	.95686	53
8	.62811	27	.37189	.67131	32	.32869	.04320	6	.95680	52
9	.62838	27	.37162	.67163	33	.32837	.04326	6	.95674	51
10	9.62865	27	10.37135	9.67196	33	10.32804	10.04332	5	9.95668	50
11	.62892	26	.37108	.67229	33	.32771	.04337	6	.95663	49
12	.62918	27	.37082	.67262	33	.32738	.04343	6	.95657	48
13	.62945	27	.37055	.67295	32	.32705	.04349	6	.95651	47
14	.62972	27	.37028	.67327	33	.32673	.04355	6	.95645	46
15	9.62999	27	10.37001	9.67360	33	10.32640	10.04361	6	9.95639	45
16	.63026	26	.36974	.67393	33	.32607	.04367	6	.95633	44
17	.63052	27	.36948	.67426	32	.32574	.04373	6	.95627	43
18	.63079	27	.36921	.67458	33	.32542	.04379	6	.95621	42
19	.63106	27	.36894	.67491	33	.32509	.04385	6	.95615	41
20	9.63133	26	10.36867	9.67524	32	10.32476	10.04391	6	9.95609	40
21	.63159	27	.36841	.67556	33	.32444	.04397	6	.95603	39
22	.63186	27	.36814	.67589	33	.32411	.04403	6	.95597	38
23	.63213	26	.36787	.67622	32	.32378	.04409	6	.95591	37
24	.63239	27	.36761	.67654	33	.32346	.04415	6	.95585	36
25	9.63266	26	10.36734	9.67687	32	10.32313	10.04421	6	9.95579	35
26	.63292	27	.36708	.67719	33	.32281	.04427	6	.95573	34
27	.63319	26	.36681	.67752	33	.32248	.04433	6	.95567	33
28	.63345	27	.36655	.67785	32	.32215	.04439	6	.95561	32
29	.63372	26	.36628	.67817	33	.32183	.04445	6	.95555	31
30	9.63398	27	10.36602	9.67850	32	10.32150	10.04451	6	9.95549	30
31	.63425	26	.36575	.67882	33	.32118	.04457	6	.95543	29
32	.63451	27	.36549	.67915	32	.32085	.04463	6	.95537	28
33	.63478	26	.36522	.67947	33	.32053	.04469	6	.95531	27
34	.63504	27	.36496	.67980	32	.32020	.04475	6	.95525	26
35	9.63531	26	10.36469	9.68012	32	10.31988	10.04481	6	9.95519	25
36	.63557	26	.36443	.68044	33	.31956	.04487	6	.95513	24
37	.63583	27	.36417	.68077	32	.31923	.04493	7	.95507	23
38	.63610	26	.36390	.68109	33	.31891	.04500	6	.95500	22
39	.63636	26	.36364	.68142	32	.31858	.04506	6	.95494	21
40	9.63662	27	10.36338	9.68174	32	10.31826	10.04512	6	9.95488	20
41	.63689	26	.36311	.68206	33	.31794	.04518	6	.95482	19
42	.63715	26	.36285	.68239	32	.31761	.04524	6	.95476	18
43	.63741	26	.36259	.68271	32	.31729	.04530	6	.95470	17
44	.63767	27	.36233	.68303	33	.31697	.04536	6	.95464	16
45	9.63794	26	10.36206	9.68336	32	10.31664	10.04542	6	9.95458	15
46	.63820	26	.36180	.68368	32	.31632	.04548	6	.95452	14
47	.63846	26	.36154	.68400	32	.31600	.04554	6	.95446	13
48	.63872	26	.36128	.68432	33	.31568	.04560	6	.95440	12
49	.63898	26	.36102	.68465	32	.31535	.04566	7	.95434	11
50	9.63924	26	10.36076	9.68497	32	10.31503	10.04573	6	9.95427	10
51	.63950	26	.36050	.68529	32	.31471	.04579	6	.95421	9
52	.63976	26	.36024	.68561	32	.31439	.04585	6	.95415	8
53	.64002	26	.35998	.68593	33	.31407	.04591	6	.95409	7
54	.64028	26	.35972	.68626	32	.31374	.04597	6	.95403	6
55	9.64054	26	10.35946	9.68658	32	10.31342	10.04603	6	9.95397	5
56	.64080	26	.35920	.68690	32	.31310	.04609	7	.95391	4
57	.64106	26	.35894	.68722	32	.31278	.04616	6	.95384	3
58	.64132	26	.35868	.68754	32	.31246	.04622	6	.95378	2
59	.64158	26	.35842	.68786	32	.31214	.04628	6	.95372	1
60	9.64184		10.35816	9.68818		10.31182	10.04634		9.95366	0
↑ 115° →	cos	Diff. 1′	sec	cot	Diff. 1′	tan	csc	Diff. 1′	sin	← 64° ↑

TABLE 3

Common Logarithms of Trigonometric Functions (offset +10)

26° → ′ ↓	sin	Diff. 1′	csc	tan	Diff. 1′	cot	sec	Diff. 1′	cos	← 153° ′ ↑
0	9.64184	26	10.35816	9.68818	32	10.31182	10.04634	6	9.95366	60
1	.64210	26	.35790	.68850	32	.31150	.04640	6	.95360	59
2	.64236	26	.35764	.68882	32	.31118	.04646	6	.95354	58
3	.64262	26	.35738	.68914	32	.31086	.04652	7	.95348	57
4	.64288	25	.35712	.68946	32	.31054	.04659	6	.95341	56
5	9.64313	26	10.35687	9.68978	32	10.31022	10.04665	6	9.95335	55
6	.64339	26	.35661	.69010	32	.30990	.04671	6	.95329	54
7	.64365	26	.35635	.69042	32	.30958	.04677	6	.95323	53
8	.64391	26	.35609	.69074	32	.30926	.04683	7	.95317	52
9	.64417	25	.35583	.69106	32	.30894	.04690	6	.95310	51
10	9.64442	26	10.35558	9.69138	32	10.30862	10.04696	6	9.95304	50
11	.64468	26	.35532	.69170	32	.30830	.04702	6	.95298	49
12	.64494	25	.35506	.69202	32	.30798	.04708	6	.95292	48
13	.64519	26	.35481	.69234	32	.30766	.04714	7	.95286	47
14	.64545	26	.35455	.69266	32	.30734	.04721	6	.95279	46
15	9.64571	25	10.35429	9.69298	31	10.30702	10.04727	6	9.95273	45
16	.64596	26	.35404	.69329	32	.30671	.04733	6	.95267	44
17	.64622	25	.35378	.69361	32	.30639	.04739	7	.95261	43
18	.64647	26	.35353	.69393	32	.30607	.04746	6	.95254	42
19	.64673	25	.35327	.69425	32	.30575	.04752	6	.95248	41
20	9.64698	26	10.35302	9.69457	31	10.30543	10.04758	6	9.95242	40
21	.64724	25	.35276	.69488	32	.30512	.04764	7	.95236	39
22	.64749	26	.35251	.69520	32	.30480	.04771	6	.95229	38
23	.64775	25	.35225	.69552	32	.30448	.04777	6	.95223	37
24	.64800	26	.35200	.69584	31	.30416	.04783	6	.95217	36
25	9.64826	25	10.35174	9.69615	32	10.30385	10.04789	7	9.95211	35
26	.64851	26	.35149	.69647	32	.30353	.04796	6	.95204	34
27	.64877	25	.35123	.69679	31	.30321	.04802	6	.95198	33
28	.64902	25	.35098	.69710	32	.30290	.04808	7	.95192	32
29	.64927	26	.35073	.69742	32	.30258	.04815	6	.95185	31
30	9.64953	25	10.35047	9.69774	31	10.30226	10.04821	6	9.95179	30
31	.64978	25	.35022	.69805	32	.30195	.04827	6	.95173	29
32	.65003	26	.34997	.69837	31	.30163	.04833	7	.95167	28
33	.65029	25	.34971	.69868	32	.30132	.04840	6	.95160	27
34	.65054	25	.34946	.69900	32	.30100	.04846	6	.95154	26
35	9.65079	25	10.34921	9.69932	31	10.30068	10.04852	7	9.95148	25
36	.65104	26	.34896	.69963	32	.30037	.04859	6	.95141	24
37	.65130	25	.34870	.69995	31	.30005	.04865	6	.95135	23
38	.65155	25	.34845	.70026	32	.29974	.04871	7	.95129	22
39	.65180	25	.34820	.70058	31	.29942	.04878	6	.95122	21
40	9.65205	25	10.34795	9.70089	32	10.29911	10.04884	6	9.95116	20
41	.65230	25	.34770	.70121	31	.29879	.04890	7	.95110	19
42	.65255	26	.34745	.70152	32	.29848	.04897	6	.95103	18
43	.65281	25	.34719	.70184	31	.29816	.04903	7	.95097	17
44	.65306	25	.34694	.70215	32	.29785	.04910	6	.95090	16
45	9.65331	25	10.34669	9.70247	31	10.29753	10.04916	6	9.95084	15
46	.65356	25	.34644	.70278	31	.29722	.04922	7	.95078	14
47	.65381	25	.34619	.70309	32	.29691	.04929	6	.95071	13
48	.65406	25	.34594	.70341	31	.29659	.04935	6	.95065	12
49	.65431	25	.34569	.70372	32	.29628	.04941	7	.95059	11
50	9.65456	25	10.34544	9.70404	31	10.29596	10.04948	6	9.95052	10
51	.65481	25	.34519	.70435	31	.29565	.04954	7	.95046	9
52	.65506	25	.34494	.70466	32	.29534	.04961	6	.95039	8
53	.65531	25	.34469	.70498	31	.29502	.04967	6	.95033	7
54	.65556	24	.34444	.70529	31	.29471	.04973	7	.95027	6
55	9.65580	25	10.34420	9.70560	32	10.29440	10.04980	6	9.95020	5
56	.65605	25	.34395	.70592	31	.29408	.04986	7	.95014	4
57	.65630	25	.34370	.70623	31	.29377	.04993	6	.95007	3
58	.65655	25	.34345	.70654	31	.29346	.04999	6	.95001	2
59	.65680	25	.34320	.70685	32	.29315	.05005	7	.94995	1
60	9.65705		10.34295	9.70717		10.29283	10.05012		9.94988	0
↑ 116° →	cos	Diff 1′	sec	cot	Diff. 1′	tan	csc	Diff. 1′	sin	← 63° ↑

TABLE 3

Common Logarithms of Trigonometric Functions (offset +10)

27° → ′ ↓	sin	Diff. 1′	csc	tan	Diff. 1′	cot	sec	Diff. 1′	cos	← 152° ′ ↑
0	9.65705	24	10.34295	9.70717	31	10.29283	10.05012	6	9.94988	60
1	.65729	25	.34271	.70748	31	.29252	.05018	7	.94982	59
2	.65754	25	.34246	.70779	31	.29221	.05025	6	.94975	58
3	.65779	25	.34221	.70810	31	.29190	.05031	7	.94969	57
4	.65804	24	.34196	.70841	32	.29159	.05038	6	.94962	56
5	9.65828	25	10.34172	9.70873	31	10.29127	10.05044	7	9.94956	55
6	.65853	25	.34147	.70904	31	.29096	.05051	6	.94949	54
7	.65878	24	.34122	.70935	31	.29065	.05057	7	.94943	53
8	.65902	25	.34098	.70966	31	.29034	.05064	6	.94936	52
9	.65927	25	.34073	.70997	31	.29003	.05070	7	.94930	51
10	9.65952	24	10.34048	9.71028	31	10.28972	10.05077	6	9.94923	50
11	.65976	25	.34024	.71059	31	.28941	.05083	6	.94917	49
12	.66001	24	.33999	.71090	31	.28910	.05089	7	.94911	48
13	.66025	25	.33975	.71121	32	.28879	.05096	6	.94904	47
14	.66050	25	.33950	.71153	31	.28847	.05102	7	.94898	46
15	9.66075	24	10.33925	9.71184	31	10.28816	10.05109	6	9.94891	45
16	.66099	25	.33901	.71215	31	.28785	.05115	7	.94885	44
17	.66124	24	.33876	.71246	31	.28754	.05122	7	.94878	43
18	.66148	25	.33852	.71277	31	.28723	.05129	6	.94871	42
19	.66173	24	.33827	.71308	31	.28692	.05135	7	.94865	41
20	9.66197	24	10.33803	9.71339	31	10.28661	10.05142	6	9.94858	40
21	.66221	25	.33779	.71370	31	.28630	.05148	7	.94852	39
22	.66246	24	.33754	.71401	30	.28599	.05155	6	.94845	38
23	.66270	25	.33730	.71431	31	.28569	.05161	7	.94839	37
24	.66295	24	.33705	.71462	31	.28538	.05168	6	.94832	36
25	9.66319	24	10.33681	9.71493	31	10.28507	10.05174	7	9.94826	35
26	.66343	25	.33657	.71524	31	.28476	.05181	6	.94819	34
27	.66368	24	.33632	.71555	31	.28445	.05187	7	.94813	33
28	.66392	24	.33608	.71586	31	.28414	.05194	7	.94806	32
29	.66416	25	.33584	.71617	31	.28383	.05201	6	.94799	31
30	9.66441	24	10.33559	9.71648	31	10.28352	10.05207	7	9.94793	30
31	.66465	24	.33535	.71679	30	.28321	.05214	6	.94786	29
32	.66489	24	.33511	.71709	31	.28291	.05220	7	.94780	28
33	.66513	24	.33487	.71740	31	.28260	.05227	6	.94773	27
34	.66537	25	.33463	.71771	31	.28229	.05233	7	.94767	26
35	9.66562	24	10.33438	9.71802	31	10.28198	10.05240	7	9.94760	25
36	.66586	24	.33414	.71833	30	.28167	.05247	6	.94753	24
37	.66610	24	.33390	.71863	31	.28137	.05253	7	.94747	23
38	.66634	24	.33366	.71894	31	.28106	.05260	6	.94740	22
39	.66658	24	.33342	.71925	30	.28075	.05266	7	.94734	21
40	9.66682	24	10.33318	9.71955	31	10.28045	10.05273	7	9.94727	20
41	.66706	25	.33294	.71986	31	.28014	.05280	6	.94720	19
42	.66731	24	.33269	.72017	31	.27983	.05286	7	.94714	18
43	.66755	24	.33245	.72048	30	.27952	.05293	7	.94707	17
44	.66779	24	.33221	.72078	31	.27922	.05300	6	.94700	16
45	9.66803	24	10.33197	9.72109	31	10.27891	10.05306	7	9.94694	15
46	.66827	24	.33173	.72140	30	.27860	.05313	7	.94687	14
47	.66851	24	.33149	.72170	31	.27830	.05320	6	.94680	13
48	.66875	24	.33125	.72201	30	.27799	.05326	7	.94674	12
49	.66899	23	.33101	.72231	31	.27769	.05333	7	.94667	11
50	9.66922	24	10.33078	9.72262	31	10.27738	10.05340	6	9.94660	10
51	.66946	24	.33054	.72293	30	.27707	.05346	7	.94654	9
52	.66970	24	.33030	.72323	31	.27677	.05353	7	.94647	8
53	.66994	24	.33006	.72354	30	.27646	.05360	6	.94640	7
54	.67018	24	.32982	.72384	31	.27616	.05366	7	.94634	6
55	9.67042	24	10.32958	9.72415	30	10.27585	10.05373	7	9.94627	5
56	.67066	24	.32934	.72445	31	.27555	.05380	6	.94620	4
57	.67090	23	.32910	.72476	30	.27524	.05386	7	.94614	3
58	.67113	24	.32887	.72506	31	.27494	.05393	7	.94607	2
59	.67137	24	.32863	.72537	30	.27463	.05400	7	.94600	1
60	9.67161		10.32839	9.72567		10.27433	10.05407		9.94593	0
↑ 117° →	cos	Diff. 1′	sec	cot	Diff. 1′	tan	csc	Diff. 1′	sin	← 62° ↑

TABLE 3

Common Logarithms of Trigonometric Functions (offset +10)

28°→ ′	sin	Diff. 1′	csc	tan	Diff. 1′	cot	sec	Diff. 1′	cos	←151° ′
0	9.67161	24	10.32839	9.72567	31	10.27433	10.05407	6	9.94593	60
1	.67185	23	.32815	.72598	30	.27402	.05413	7	.94587	59
2	.67208	24	.32792	.72628	31	.27372	.05420	7	.94580	58
3	.67232	24	.32768	.72659	30	.27341	.05427	6	.94573	57
4	.67256	24	.32744	.72689	31	.27311	.05433	7	.94567	56
5	9.67280	23	10.32720	9.72720	30	10.27280	10.05440	7	9.94560	55
6	.67303	24	.32697	.72750	30	.27250	.05447	7	.94553	54
7	.67327	23	.32673	.72780	31	.27220	.05454	6	.94546	53
8	.67350	24	.32650	.72811	30	.27189	.05460	7	.94540	52
9	.67374	24	.32626	.72841	31	.27159	.05467	7	.94533	51
10	9.67398	23	10.32602	9.72872	30	10.27128	10.05474	7	9.94526	50
11	.67421	24	.32579	.72902	30	.27098	.05481	6	.94519	49
12	.67445	23	.32555	.72932	31	.27068	.05487	7	.94513	48
13	.67468	24	.32532	.72963	30	.27037	.05494	7	.94506	47
14	.67492	23	.32508	.72993	30	.27007	.05501	7	.94499	46
15	9.67515	24	10.32485	9.73023	31	10.26977	10.05508	7	9.94492	45
16	.67539	23	.32461	.73054	30	.26946	.05515	6	.94485	44
17	.67562	24	.32438	.73084	30	.26916	.05521	7	.94479	43
18	.67586	23	.32414	.73114	30	.26886	.05528	7	.94472	42
19	.67609	24	.32391	.73144	31	.26856	.05535	7	.94465	41
20	9.67633	23	10.32367	9.73175	30	10.26825	10.05542	7	9.94458	40
21	.67656	24	.32344	.73205	30	.26795	.05549	6	.94451	39
22	.67680	23	.32320	.73235	30	.26765	.05555	7	.94445	38
23	.67703	23	.32297	.73265	30	.26735	.05562	7	.94438	37
24	.67726	24	.32274	.73295	31	.26705	.05569	7	.94431	36
25	9.67750	23	10.32250	9.73326	30	10.26674	10.05576	7	9.94424	35
26	.67773	23	.32227	.73356	30	.26644	.05583	7	.94417	34
27	.67796	24	.32204	.73386	30	.26614	.05590	6	.94410	33
28	.67820	23	.32180	.73416	30	.26584	.05596	7	.94404	32
29	.67843	23	.32157	.73446	30	.26554	.05603	7	.94397	31
30	9.67866	24	10.32134	9.73476	31	10.26524	10.05610	7	9.94390	30
31	.67890	23	.32110	.73507	30	.26493	.05617	7	.94383	29
32	.67913	23	.32087	.73537	30	.26463	.05624	7	.94376	28
33	.67936	23	.32064	.73567	30	.26433	.05631	7	.94369	27
34	.67959	23	.32041	.73597	30	.26403	.05638	7	.94362	26
35	9.67982	24	10.32018	9.73627	30	10.26373	10.05645	6	9.94355	25
36	.68006	23	.31994	.73657	30	.26343	.05651	7	.94349	24
37	.68029	23	.31971	.73687	30	.26313	.05658	7	.94342	23
38	.68052	23	.31948	.73717	30	.26283	.05665	7	.94335	22
39	.68075	23	.31925	.73747	30	.26253	.05672	7	.94328	21
40	9.68098	23	10.31902	9.73777	30	10.26223	10.05679	7	9.94321	20
41	.68121	23	.31879	.73807	30	.26193	.05686	7	.94314	19
42	.68144	23	.31856	.73837	30	.26163	.05693	7	.94307	18
43	.68167	23	.31833	.73867	30	.26133	.05700	7	.94300	17
44	.68190	23	.31810	.73897	30	.26103	.05707	7	.94293	16
45	9.68213	24	10.31787	9.73927	30	10.26073	10.05714	7	9.94286	15
46	.68237	23	.31763	.73957	30	.26043	.05721	6	.94279	14
47	.68260	23	.31740	.73987	30	.26013	.05727	7	.94273	13
48	.68283	22	.31717	.74017	30	.25983	.05734	7	.94266	12
49	.68305	23	.31695	.74047	30	.25953	.05741	7	.94259	11
50	9.68328	23	10.31672	9.74077	30	10.25923	10.05748	7	9.94252	10
51	.68351	23	.31649	.74107	30	.25893	.05755	7	.94245	9
52	.68374	23	.31626	.74137	29	.25863	.05762	7	.94238	8
53	.68397	23	.31603	.74166	30	.25834	.05769	7	.94231	7
54	.68420	23	.31580	.74196	30	.25804	.05776	7	.94224	6
55	9.68443	23	10.31557	9.74226	30	10.25774	10.05783	7	9.94217	5
56	.68466	23	.31534	.74256	30	.25744	.05790	7	.94210	4
57	.68489	23	.31511	.74286	30	.25714	.05797	7	.94203	3
58	.68512	22	.31488	.74316	29	.25684	.05804	7	.94196	2
59	.68534	23	.31466	.74345	30	.25655	.05811	7	.94189	1
60	9.68557		10.31443	9.74375		10.25625	10.05818		9.94182	0
↑ 118°→	cos	Diff. 1′	sec	cot	Diff. 1′	tan	csc	Diff. 1′	sin	← 61° ↑

TABLE 3

Common Logarithms of Trigonometric Functions (offset +10)

29°→ ′	sin	Diff. 1′	csc	tan	Diff. 1′	cot	sec	Diff. 1′	cos	←150° ′
0	9.68557	23	10.31443	9.74375	30	10.25625	10.05818	7	9.94182	60
1	.68580	23	.31420	.74405	30	.25595	.05825	7	.94175	59
2	.68603	22	.31397	.74435	30	.25565	.05832	7	.94168	58
3	.68625	23	.31375	.74465	29	.25535	.05839	7	.94161	57
4	.68648	23	.31352	.74494	30	.25506	.05846	7	.94154	56
5	9.68671	23	10.31329	9.74524	30	10.25476	10.05853	7	9.94147	55
6	.68694	22	.31306	.74554	29	.25446	.05860	7	.94140	54
7	.68716	23	.31284	.74583	30	.25417	.05867	7	.94133	53
8	.68739	23	.31261	.74613	30	.25387	.05874	7	.94126	52
9	.68762	22	.31238	.74643	30	.25357	.05881	7	.94119	51
10	9.68784	23	10.31216	9.74673	29	10.25327	10.05888	7	9.94112	50
11	.68807	22	.31193	.74702	30	.25298	.05895	7	.94105	49
12	.68829	23	.31171	.74732	30	.25268	.05902	8	.94098	48
13	.68852	23	.31148	.74762	29	.25238	.05910	7	.94090	47
14	.68875	22	.31125	.74791	30	.25209	.05917	7	.94083	46
15	9.68897	23	10.31103	9.74821	30	10.25179	10.05924	7	9.94076	45
16	.68920	22	.31080	.74851	29	.25149	.05931	7	.94069	44
17	.68942	23	.31058	.74880	30	.25120	.05938	7	.94062	43
18	.68965	22	.31035	.74910	29	.25090	.05945	7	.94055	42
19	.68987	23	.31013	.74939	30	.25061	.05952	7	.94048	41
20	9.69010	22	10.30990	9.74969	29	10.25031	10.05959	7	9.94041	40
21	.69032	23	.30968	.74998	30	.25002	.05966	7	.94034	39
22	.69055	22	.30945	.75028	30	.24972	.05973	7	.94027	38
23	.69077	23	.30923	.75058	29	.24942	.05980	8	.94020	37
24	.69100	22	.30900	.75087	30	.24913	.05988	7	.94012	36
25	9.69122	22	10.30878	9.75117	29	10.24883	10.05995	7	9.94005	35
26	.69144	23	.30856	.75146	30	.24854	.06002	7	.93998	34
27	.69167	22	.30833	.75176	29	.24824	.06009	7	.93991	33
28	.69189	23	.30811	.75205	30	.24795	.06016	7	.93984	32
29	.69212	22	.30788	.75235	29	.24765	.06023	7	.93977	31
30	9.69234	22	10.30766	9.75264	30	10.24736	10.06030	7	9.93970	30
31	.69256	23	.30744	.75294	29	.24706	.06037	8	.93963	29
32	.69279	22	.30721	.75323	30	.24677	.06045	7	.93955	28
33	.69301	22	.30699	.75353	29	.24647	.06052	7	.93948	27
34	.69323	22	.30677	.75382	29	.24618	.06059	7	.93941	26
35	9.69345	23	10.30655	9.75411	30	10.24589	10.06066	7	9.93934	25
36	.69368	22	.30632	.75441	29	.24559	.06073	7	.93927	24
37	.69390	22	.30610	.75470	30	.24530	.06080	8	.93920	23
38	.69412	22	.30588	.75500	29	.24500	.06088	7	.93912	22
39	.69434	22	.30566	.75529	29	.24471	.06095	7	.93905	21
40	9.69456	23	10.30544	9.75558	30	10.24442	10.06102	7	9.93898	20
41	.69479	22	.30521	.75588	29	.24412	.06109	7	.93891	19
42	.69501	22	.30499	.75617	30	.24383	.06116	8	.93884	18
43	.69523	22	.30477	.75647	29	.24353	.06124	7	.93876	17
44	.69545	22	.30455	.75676	29	.24324	.06131	7	.93869	16
45	9.69567	22	10.30433	9.75705	30	10.24295	10.06138	7	9.93862	15
46	.69589	22	.30411	.75735	29	.24265	.06145	8	.93855	14
47	.69611	22	.30389	.75764	29	.24236	.06153	7	.93847	13
48	.69633	22	.30367	.75793	29	.24207	.06160	7	.93840	12
49	.69655	22	.30345	.75822	30	.24178	.06167	7	.93833	11
50	9.69677	22	10.30323	9.75852	29	10.24148	10.06174	7	9.93826	10
51	.69699	22	.30301	.75881	29	.24119	.06181	8	.93819	9
52	.69721	22	.30279	.75910	29	.24090	.06189	7	.93811	8
53	.69743	22	.30257	.75939	30	.24061	.06196	7	.93804	7
54	.69765	22	.30235	.75969	29	.24031	.06203	8	.93797	6
55	9.69787	22	10.30213	9.75998	29	10.24002	10.06211	7	9.93789	5
56	.69809	22	.30191	.76027	29	.23973	.06218	7	.93782	4
57	.69831	22	.30169	.76056	30	.23944	.06225	7	.93775	3
58	.69853	22	.30147	.76086	29	.23914	.06232	8	.93768	2
59	.69875	22	.30125	.76115	29	.23885	.06240	7	.93760	1
60	9.69897		10.30103	9.76144		10.23856	10.06247		9.93753	0
↑ 119°→	cos	Diff. 1′	sec	cot	Diff. 1′	tan	csc	Diff. 1′	sin	← 60° ↑

TABLE 3

Common Logarithms of Trigonometric Functions (offset +10)

30° → ↓	sin	Diff. 1′	csc	tan	Diff. 1′	cot	sec	Diff. 1′	cos	← 149° ↓
′										′
0	9.69897	22	10.30103	9.76144	29	10.23856	10.06247	7	9.93753	60
1	.69919	22	.30081	.76173	29	.23827	.06254	8	.93746	59
2	.69941	22	.30059	.76202	29	.23798	.06262	7	.93738	58
3	.69963	21	.30037	.76231	30	.23769	.06269	7	.93731	57
4	.69984	22	.30016	.76261	29	.23739	.06276	7	.93724	56
5	9.70006	22	10.29994	9.76290	29	10.23710	10.06283	8	9.93717	55
6	.70028	22	.29972	.76319	29	.23681	.06291	7	.93709	54
7	.70050	22	.29950	.76348	29	.23652	.06298	7	.93702	53
8	.70072	21	.29928	.76377	29	.23623	.06305	8	.93695	52
9	.70093	22	.29907	.76406	29	.23594	.06313	7	.93687	51
10	9.70115	22	10.29885	9.76435	29	10.23565	10.06320	7	9.93680	50
11	.70137	22	.29863	.76464	29	.23536	.06327	8	.93673	49
12	.70159	21	.29841	.76493	29	.23507	.06335	7	.93665	48
13	.70180	22	.29820	.76522	29	.23478	.06342	8	.93658	47
14	.70202	22	.29798	.76551	29	.23449	.06350	7	.93650	46
15	9.70224	21	10.29776	9.76580	29	10.23420	10.06357	7	9.93643	45
16	.70245	22	.29755	.76609	30	.23391	.06364	8	.93636	44
17	.70267	21	.29733	.76639	29	.23361	.06372	7	.93628	43
18	.70288	22	.29712	.76668	29	.23332	.06379	7	.93621	42
19	.70310	22	.29690	.76697	28	.23303	.06386	8	.93614	41
20	9.70332	21	10.29668	9.76725	29	10.23275	10.06394	7	9.93606	40
21	.70353	22	.29647	.76754	29	.23246	.06401	8	.93599	39
22	.70375	21	.29625	.76783	29	.23217	.06409	7	.93591	38
23	.70396	22	.29604	.76812	29	.23188	.06416	7	.93584	37
24	.70418	21	.29582	.76841	29	.23159	.06423	8	.93577	36
25	9.70439	22	10.29561	9.76870	29	10.23130	10.06431	7	9.93569	35
26	.70461	21	.29539	.76899	29	.23101	.06438	8	.93562	34
27	.70482	22	.29518	.76928	29	.23072	.06446	7	.93554	33
28	.70504	21	.29496	.76957	29	.23043	.06453	8	.93547	32
29	.70525	22	.29475	.76986	29	.23014	.06461	7	.93539	31
30	9.70547	21	10.29453	9.77015	29	10.22985	10.06468	7	9.93532	30
31	.70568	22	.29432	.77044	29	.22956	.06475	8	.93525	29
32	.70590	21	.29410	.77073	28	.22927	.06483	7	.93517	28
33	.70611	22	.29389	.77101	29	.22899	.06490	8	.93510	27
34	.70633	21	.29367	.77130	29	.22870	.06498	7	.93502	26
35	9.70654	21	10.29346	9.77159	29	10.22841	10.06505	8	9.93495	25
36	.70675	22	.29325	.77188	29	.22812	.06513	7	.93487	24
37	.70697	21	.29303	.77217	29	.22783	.06520	8	.93480	23
38	.70718	21	.29282	.77246	28	.22754	.06528	7	.93472	22
39	.70739	22	.29261	.77274	29	.22726	.06535	8	.93465	21
40	9.70761	21	10.29239	9.77303	29	10.22697	10.06543	7	9.93457	20
41	.70782	21	.29218	.77332	29	.22668	.06550	8	.93450	19
42	.70803	21	.29197	.77361	29	.22639	.06558	7	.93442	18
43	.70824	22	.29176	.77390	28	.22610	.06565	8	.93435	17
44	.70846	21	.29154	.77418	29	.22582	.06573	7	.93427	16
45	9.70867	21	10.29133	9.77447	29	10.22553	10.06580	8	9.93420	15
46	.70888	21	.29112	.77476	29	.22524	.06588	7	.93412	14
47	.70909	22	.29091	.77505	28	.22495	.06595	8	.93405	13
48	.70931	21	.29069	.77533	29	.22467	.06603	7	.93397	12
49	.70952	21	.29048	.77562	29	.22438	.06610	8	.93390	11
50	9.70973	21	10.29027	9.77591	28	10.22409	10.06618	7	9.93382	10
51	.70994	21	.29006	.77619	29	.22381	.06625	8	.93375	9
52	.71015	21	.28985	.77648	29	.22352	.06633	7	.93367	8
53	.71036	22	.28964	.77677	29	.22323	.06640	8	.93360	7
54	.71058	21	.28942	.77706	28	.22294	.06648	8	.93352	6
55	9.71079	21	10.28921	9.77734	29	10.22266	10.06656	7	9.93344	5
56	.71100	21	.28900	.77763	28	.22237	.06663	8	.93337	4
57	.71121	21	.28879	.77791	29	.22209	.06671	7	.93329	3
58	.71142	21	.28858	.77820	29	.22180	.06678	8	.93322	2
59	.71163	21	.28837	.77849	28	.22151	.06686	7	.93314	1
60	9.71184		10.28816	9.77877		10.22123	10.06693		9.93307	0
↑ 120° →	cos	Diff. 1′	sec	cot	Diff. 1′	tan	csc	Diff. 1′	sin	← 59° ↑

TABLE 3

Common Logarithms of Trigonometric Functions (offset +10)

31° → ↓	sin	Diff. 1′	csc	tan	Diff. 1′	cot	sec	Diff. 1′	cos	← 148° ↓
′										′
0	9.71184	21	10.28816	9.77877	29	10.22123	10.06693	8	9.93307	60
1	.71205	21	.28795	.77906	29	.22094	.06701	8	.93299	59
2	.71226	21	.28774	.77935	28	.22065	.06709	7	.93291	58
3	.71247	21	.28753	.77963	29	.22037	.06716	8	.93284	57
4	.71268	21	.28732	.77992	28	.22008	.06724	7	.93276	56
5	9.71289	21	10.28711	9.78020	29	10.21980	10.06731	8	9.93269	55
6	.71310	21	.28690	.78049	28	.21951	.06739	8	.93261	54
7	.71331	21	.28669	.78077	29	.21923	.06747	7	.93253	53
8	.71352	21	.28648	.78106	29	.21894	.06754	8	.93246	52
9	.71373	20	.28627	.78135	28	.21865	.06762	8	.93238	51
10	9.71393	21	10.28607	9.78163	29	10.21837	10.06770	7	9.93230	50
11	.71414	21	.28586	.78192	28	.21808	.06777	8	.93223	49
12	.71435	21	.28565	.78220	29	.21780	.06785	8	.93215	48
13	.71456	21	.28544	.78249	28	.21751	.06793	7	.93207	47
14	.71477	21	.28523	.78277	29	.21723	.06800	8	.93200	46
15	9.71498	21	10.28502	9.78306	28	10.21694	10.06808	8	9.93192	45
16	.71519	20	.28481	.78334	29	.21666	.06816	7	.93184	44
17	.71539	21	.28461	.78363	28	.21637	.06823	8	.93177	43
18	.71560	21	.28440	.78391	28	.21609	.06831	8	.93169	42
19	.71581	21	.28419	.78419	29	.21581	.06839	7	.93161	41
20	9.71602	20	10.28398	9.78448	28	10.21552	10.06846	8	9.93154	40
21	.71622	21	.28378	.78476	29	.21524	.06854	8	.93146	39
22	.71643	21	.28357	.78505	28	.21495	.06862	7	.93138	38
23	.71664	21	.28336	.78533	29	.21467	.06869	8	.93131	37
24	.71685	20	.28315	.78562	28	.21438	.06877	8	.93123	36
25	9.71705	21	10.28295	9.78590	28	10.21410	10.06885	7	9.93115	35
26	.71726	21	.28274	.78618	29	.21382	.06892	8	.93108	34
27	.71747	20	.28253	.78647	28	.21353	.06900	8	.93100	33
28	.71767	21	.28233	.78675	29	.21325	.06908	8	.93092	32
29	.71788	21	.28212	.78704	28	.21296	.06916	7	.93084	31
30	9.71809	20	10.28191	9.78732	28	10.21268	10.06923	8	9.93077	30
31	.71829	21	.28171	.78760	29	.21240	.06931	8	.93069	29
32	.71850	20	.28150	.78789	28	.21211	.06939	8	.93061	28
33	.71870	21	.28130	.78817	28	.21183	.06947	7	.93053	27
34	.71891	20	.28109	.78845	29	.21155	.06954	8	.93046	26
35	9.71911	21	10.28089	9.78874	28	10.21126	10.06962	8	9.93038	25
36	.71932	20	.28068	.78902	28	.21098	.06970	8	.93030	24
37	.71952	21	.28048	.78930	29	.21070	.06978	8	.93022	23
38	.71973	21	.28027	.78959	28	.21041	.06986	7	.93014	22
39	.71994	20	.28006	.78987	28	.21013	.06993	8	.93007	21
40	9.72014	20	10.27986	9.79015	28	10.20985	10.07001	8	9.92999	20
41	.72034	21	.27966	.79043	29	.20957	.07009	8	.92991	19
42	.72055	20	.27945	.79072	28	.20928	.07017	7	.92983	18
43	.72075	21	.27925	.79100	28	.20900	.07024	8	.92976	17
44	.72096	20	.27904	.79128	28	.20872	.07032	8	.92968	16
45	9.72116	21	10.27884	9.79156	29	10.20844	10.07040	8	9.92960	15
46	.72137	20	.27863	.79185	28	.20815	.07048	8	.92952	14
47	.72157	20	.27843	.79213	28	.20787	.07056	8	.92944	13
48	.72177	21	.27823	.79241	28	.20759	.07064	7	.92936	12
49	.72198	20	.27802	.79269	28	.20731	.07071	8	.92929	11
50	9.72218	20	10.27782	9.79297	29	10.20703	10.07079	8	9.92921	10
51	.72238	21	.27762	.79326	28	.20674	.07087	8	.92913	9
52	.72259	20	.27741	.79354	28	.20646	.07095	8	.92905	8
53	.72279	20	.27721	.79382	28	.20618	.07103	8	.92897	7
54	.72299	21	.27701	.79410	28	.20590	.07111	8	.92889	6
55	9.72320	20	10.27680	9.79438	28	10.20562	10.07119	7	9.92881	5
56	.72340	20	.27660	.79466	29	.20534	.07126	8	.92874	4
57	.72360	21	.27640	.79495	28	.20505	.07134	8	.92866	3
58	.72381	20	.27619	.79523	28	.20477	.07142	8	.92858	2
59	.72401	20	.27599	.79551	28	.20449	.07150	8	.92850	1
60	9.72421		10.27579	9.79579		10.20421	10.07158		9.92842	0
↑ 121° →	cos	Diff. 1′	sec	cot	Diff. 1′	tan	csc	Diff. 1′	sin	← 58° ↑

TABLE 3

Common Logarithms of Trigonometric Functions (offset +10)

32°→ ↓	sin	Diff. 1′	csc	tan	Diff. 1′	cot	sec	Diff. 1′	cos	←147° ↓
′										′
0	9.72421	20	10.27579	9.79579	28	10.20421	10.07158	8	9.92842	60
1	.72441	20	.27559	.79607	28	.20393	.07166	8	.92834	59
2	.72461	21	.27539	.79635	28	.20365	.07174	8	.92826	58
3	.72482	20	.27518	.79663	28	.20337	.07182	8	.92818	57
4	.72502	20	.27498	.79691	28	.20309	.07190	7	.92810	56
5	9.72522	20	10.27478	9.79719	28	10.20281	10.07197	8	9.92803	55
6	.72542	20	.27458	.79747	29	.20253	.07205	8	.92795	54
7	.72562	20	.27438	.79776	28	.20224	.07213	8	.92787	53
8	.72582	20	.27418	.79804	28	.20196	.07221	8	.92779	52
9	.72602	20	.27398	.79832	28	.20168	.07229	8	.92771	51
10	9.72622	21	10.27378	9.79860	28	10.20140	10.07237	8	9.92763	50
11	.72643	20	.27357	.79888	28	.20112	.07245	8	.92755	49
12	.72663	20	.27337	.79916	28	.20084	.07253	8	.92747	48
13	.72683	20	.27317	.79944	28	.20056	.07261	8	.92739	47
14	.72703	20	.27297	.79972	28	.20028	.07269	8	.92731	46
15	9.72723	20	10.27277	9.80000	28	10.20000	10.07277	8	9.92723	45
16	.72743	20	.27257	.80028	28	.19972	.07285	8	.92715	44
17	.72763	20	.27237	.80056	28	.19944	.07293	8	.92707	43
18	.72783	20	.27217	.80084	28	.19916	.07301	8	.92699	42
19	.72803	20	.27197	.80112	28	.19888	.07309	8	.92691	41
20	9.72823	20	10.27177	9.80140	28	10.19860	10.07317	8	9.92683	40
21	.72843	20	.27157	.80168	27	.19832	.07325	8	.92675	39
22	.72863	20	.27137	.80195	28	.19805	.07333	8	.92667	38
23	.72883	19	.27117	.80223	28	.19777	.07341	8	.92659	37
24	.72902	20	.27098	.80251	28	.19749	.07349	8	.92651	36
25	9.72922	20	10.27078	9.80279	28	10.19721	10.07357	8	9.92643	35
26	.72942	20	.27058	.80307	28	.19693	.07365	8	.92635	34
27	.72962	20	.27038	.80335	28	.19665	.07373	8	.92627	33
28	.72982	20	.27018	.80363	28	.19637	.07381	8	.92619	32
29	.73002	20	.26998	.80391	28	.19609	.07389	8	.92611	31
30	9.73022	19	10.26978	9.80419	28	10.19581	10.07397	8	9.92603	30
31	.73041	20	.26959	.80447	27	.19553	.07405	8	.92595	29
32	.73061	20	.26939	.80474	28	.19526	.07413	8	.92587	28
33	.73081	20	.26919	.80502	28	.19498	.07421	8	.92579	27
34	.73101	20	.26899	.80530	28	.19470	.07429	8	.92571	26
35	9.73121	19	10.26879	9.80558	28	10.19442	10.07437	8	9.92563	25
36	.73140	20	.26860	.80586	28	.19414	.07445	9	.92555	24
37	.73160	20	.26840	.80614	28	.19386	.07454	8	.92546	23
38	.73180	20	.26820	.80642	27	.19358	.07462	8	.92538	22
39	.73200	19	.26800	.80669	28	.19331	.07470	8	.92530	21
40	9.73219	20	10.26781	9.80697	28	10.19303	10.07478	8	9.92522	20
41	.73239	20	.26761	.80725	28	.19275	.07486	8	.92514	19
42	.73259	19	.26741	.80753	28	.19247	.07494	8	.92506	18
43	.73278	20	.26722	.80781	27	.19219	.07502	8	.92498	17
44	.73298	20	.26702	.80808	28	.19192	.07510	8	.92490	16
45	9.73318	19	10.26682	9.80836	28	10.19164	10.07518	9	9.92482	15
46	.73337	20	.26663	.80864	28	.19136	.07527	8	.92473	14
47	.73357	20	.26643	.80892	27	.19108	.07535	8	.92465	13
48	.73377	19	.26623	.80919	28	.19081	.07543	8	.92457	12
49	.73396	20	.26604	.80947	28	.19053	.07551	8	.92449	11
50	9.73416	19	10.26584	9.80975	28	10.19025	10.07559	8	9.92441	10
51	.73435	20	.26565	.81003	27	.18997	.07567	8	.92433	9
52	.73455	19	.26545	.81030	28	.18970	.07575	9	.92425	8
53	.73474	20	.26526	.81058	28	.18942	.07584	8	.92416	7
54	.73494	19	.26506	.81086	27	.18914	.07592	8	.92408	6
55	9.73513	20	10.26487	9.81113	28	10.18887	10.07600	8	9.92400	5
56	.73533	19	.26467	.81141	28	.18859	.07608	8	.92392	4
57	.73552	20	.26448	.81169	27	.18831	.07616	8	.92384	3
58	.73572	19	.26428	.81196	28	.18804	.07624	9	.92376	2
59	.73591	20	.26409	.81224	28	.18776	.07633	8	.92367	1
60	9.73611		10.26389	9.81252		10.18748	10.07641		9.92359	0
↑ 122°→	cos	Diff. 1′	sec	cot	Diff. 1′	tan	csc	Diff. 1′	sin ←	57° ↑

TABLE 3

Common Logarithms of Trigonometric Functions (offset +10)

33°→ ↓	sin	Diff. 1′	csc	tan	Diff. 1′	cot	sec	Diff. 1′	cos	←146° ↓
′										′
0	9.73611	19	10.26389	9.81252	27	10.18748	10.07641	8	9.92359	60
1	.73630	20	.26370	.81279	28	.18721	.07649	8	.92351	59
2	.73650	19	.26350	.81307	28	.18693	.07657	8	.92343	58
3	.73669	20	.26331	.81335	27	.18665	.07665	9	.92335	57
4	.73689	19	.26311	.81362	28	.18638	.07674	8	.92326	56
5	9.73708	19	10.26292	9.81390	28	10.18610	10.07682	8	9.92318	55
6	.73727	20	.26273	.81418	27	.18582	.07690	8	.92310	54
7	.73747	19	.26253	.81445	28	.18555	.07698	9	.92302	53
8	.73766	19	.26234	.81473	27	.18527	.07707	8	.92293	52
9	.73785	20	.26215	.81500	28	.18500	.07715	8	.92285	51
10	9.73805	19	10.26195	9.81528	28	10.18472	10.07723	8	9.92277	50
11	.73824	19	.26176	.81556	27	.18444	.07731	9	.92269	49
12	.73843	20	.26157	.81583	28	.18417	.07740	8	.92260	48
13	.73863	19	.26137	.81611	27	.18389	.07748	8	.92252	47
14	.73882	19	.26118	.81638	28	.18362	.07756	9	.92244	46
15	9.73901	20	10.26099	9.81666	27	10.18334	10.07765	8	9.92235	45
16	.73921	19	.26079	.81693	28	.18307	.07773	8	.92227	44
17	.73940	19	.26060	.81721	27	.18279	.07781	8	.92219	43
18	.73959	19	.26041	.81748	28	.18252	.07789	9	.92211	42
19	.73978	19	.26022	.81776	27	.18224	.07798	8	.92202	41
20	9.73997	20	10.26003	9.81803	28	10.18197	10.07806	8	9.92194	40
21	.74017	19	.25983	.81831	27	.18169	.07814	9	.92186	39
22	.74036	19	.25964	.81858	28	.18142	.07823	8	.92177	38
23	.74055	19	.25945	.81886	27	.18114	.07831	8	.92169	37
24	.74074	19	.25926	.81913	28	.18087	.07839	9	.92161	36
25	9.74093	20	10.25907	9.81941	27	10.18059	10.07848	8	9.92152	35
26	.74113	19	.25887	.81968	28	.18032	.07856	8	.92144	34
27	.74132	19	.25868	.81996	27	.18004	.07864	9	.92136	33
28	.74151	19	.25849	.82023	28	.17977	.07873	8	.92127	32
29	.74170	19	.25830	.82051	27	.17949	.07881	8	.92119	31
30	9.74189	19	10.25811	9.82078	28	10.17922	10.07889	9	9.92111	30
31	.74208	19	.25792	.82106	27	.17894	.07898	8	.92102	29
32	.74227	19	.25773	.82133	28	.17867	.07906	8	.92094	28
33	.74246	19	.25754	.82161	27	.17839	.07914	9	.92086	27
34	.74265	19	.25735	.82188	27	.17812	.07923	8	.92077	26
35	9.74284	19	10.25716	9.82215	28	10.17785	10.07931	9	9.92069	25
36	.74303	19	.25697	.82243	27	.17757	.07940	8	.92060	24
37	.74322	19	.25678	.82270	28	.17730	.07948	8	.92052	23
38	.74341	19	.25659	.82298	27	.17702	.07956	9	.92044	22
39	.74360	19	.25640	.82325	27	.17675	.07965	8	.92035	21
40	9.74379	19	10.25621	9.82352	28	10.17648	10.07973	9	9.92027	20
41	.74398	19	.25602	.82380	27	.17620	.07982	8	.92018	19
42	.74417	19	.25583	.82407	28	.17593	.07990	8	.92010	18
43	.74436	19	.25564	.82435	27	.17565	.07998	9	.92002	17
44	.74455	19	.25545	.82462	27	.17538	.08007	8	.91993	16
45	9.74474	19	10.25526	9.82489	28	10.17511	10.08015	9	9.91985	15
46	.74493	19	.25507	.82517	27	.17483	.08024	8	.91976	14
47	.74512	19	.25488	.82544	27	.17456	.08032	9	.91968	13
48	.74531	18	.25469	.82571	28	.17429	.08041	8	.91959	12
49	.74549	19	.25451	.82599	27	.17401	.08049	9	.91951	11
50	9.74568	19	10.25432	9.82626	27	10.17374	10.08058	8	9.91942	10
51	.74587	19	.25413	.82653	28	.17347	.08066	9	.91934	9
52	.74606	19	.25394	.82681	27	.17319	.08075	8	.91925	8
53	.74625	19	.25375	.82708	27	.17292	.08083	9	.91917	7
54	.74644	18	.25356	.82735	27	.17265	.08092	8	.91908	6
55	9.74662	19	10.25338	9.82762	28	10.17238	10.08100	9	9.91900	5
56	.74681	19	.25319	.82790	27	.17210	.08109	8	.91891	4
57	.74700	19	.25300	.82817	27	.17183	.08117	9	.91883	3
58	.74719	18	.25281	.82844	27	.17156	.08126	8	.91874	2
59	.74737	19	.25263	.82871	28	.17129	.08134	9	.91866	1
60	9.74756		10.25244	9.82899		10.17101	10.08143		9.91857	0
↑ 123°→	cos	Diff. 1′	sec	cot	Diff. 1′	tan	csc	Diff. 1′	sin ←	56° ↑

TABLE 3

Common Logarithms of Trigonometric Functions (offset +10)

34° → ↓ ′	sin	Diff. 1′	csc	tan	Diff. 1′	cot	sec	Diff. 1′	cos	← 145° ↓ ′
0	9.74756	19	10.25244	9.82899	27	10.17101	10.08143	8	9.91857	60
1	.74775	19	.25225	.82926	27	.17074	.08151	9	.91849	59
2	.74794	18	.25206	.82953	27	.17047	.08160	8	.91840	58
3	.74812	19	.25188	.82980	28	.17020	.08168	9	.91832	57
4	.74831	19	.25169	.83008	27	.16992	.08177	8	.91823	56
5	9.74850	18	10.25150	9.83035	27	10.16965	10.08185	9	9.91815	55
6	.74868	19	.25132	.83062	27	.16938	.08194	8	.91806	54
7	.74887	19	.25113	.83089	28	.16911	.08202	9	.91798	53
8	.74906	18	.25094	.83117	27	.16883	.08211	8	.91789	52
9	.74924	19	.25076	.83144	27	.16856	.08219	9	.91781	51
10	9.74943	18	10.25057	9.83171	27	10.16829	10.08228	9	9.91772	50
11	.74961	19	.25039	.83198	27	.16802	.08237	8	.91763	49
12	.74980	19	.25020	.83225	27	.16775	.08245	9	.91755	48
13	.74999	18	.25001	.83252	28	.16748	.08254	8	.91746	47
14	.75017	19	.24983	.83280	27	.16720	.08262	9	.91738	46
15	9.75036	18	10.24964	9.83307	27	10.16693	10.08271	9	9.91729	45
16	.75054	19	.24946	.83334	27	.16666	.08280	8	.91720	44
17	.75073	18	.24927	.83361	27	.16639	.08288	9	.91712	43
18	.75091	19	.24909	.83388	27	.16612	.08297	8	.91703	42
19	.75110	18	.24890	.83415	27	.16585	.08305	9	.91695	41
20	9.75128	19	10.24872	9.83442	28	10.16558	10.08314	9	9.91686	40
21	.75147	18	.24853	.83470	27	.16530	.08323	8	.91677	39
22	.75165	19	.24835	.83497	27	.16503	.08331	9	.91669	38
23	.75184	18	.24816	.83524	27	.16476	.08340	9	.91660	37
24	.75202	19	.24798	.83551	27	.16449	.08349	8	.91651	36
25	9.75221	18	10.24779	9.83578	27	10.16422	10.08357	9	9.91643	35
26	.75239	19	.24761	.83605	27	.16395	.08366	9	.91634	34
27	.75258	18	.24742	.83632	27	.16368	.08375	8	.91625	33
28	.75276	18	.24724	.83659	27	.16341	.08383	9	.91617	32
29	.75294	19	.24706	.83686	27	.16314	.08392	9	.91608	31
30	9.75313	18	10.24687	9.83713	27	10.16287	10.08401	8	9.91599	30
31	.75331	19	.24669	.83740	28	.16260	.08409	9	.91591	29
32	.75350	18	.24650	.83768	27	.16232	.08418	9	.91582	28
33	.75368	18	.24632	.83795	27	.16205	.08427	8	.91573	27
34	.75386	19	.24614	.83822	27	.16178	.08435	9	.91565	26
35	9.75405	18	10.24595	9.83849	27	10.16151	10.08444	9	9.91556	25
36	.75423	18	.24577	.83876	27	.16124	.08453	9	.91547	24
37	.75441	18	.24559	.83903	27	.16097	.08462	8	.91538	23
38	.75459	19	.24541	.83930	27	.16070	.08470	9	.91530	22
39	.75478	18	.24522	.83957	27	.16043	.08479	9	.91521	21
40	9.75496	18	10.24504	9.83984	27	10.16016	10.08488	8	9.91512	20
41	.75514	19	.24486	.84011	27	.15989	.08496	9	.91504	19
42	.75533	18	.24467	.84038	27	.15962	.08505	9	.91495	18
43	.75551	18	.24449	.84065	27	.15935	.08514	9	.91486	17
44	.75569	18	.24431	.84092	27	.15908	.08523	8	.91477	16
45	9.75587	18	10.24413	9.84119	27	10.15881	10.08531	9	9.91469	15
46	.75605	19	.24395	.84146	27	.15854	.08540	9	.91460	14
47	.75624	18	.24376	.84173	27	.15827	.08549	9	.91451	13
48	.75642	18	.24358	.84200	27	.15800	.08558	9	.91442	12
49	.75660	18	.24340	.84227	27	.15773	.08567	8	.91433	11
50	9.75678	18	10.24322	9.84254	26	10.15746	10.08575	9	9.91425	10
51	.75696	18	.24304	.84280	27	.15720	.08584	9	.91416	9
52	.75714	19	.24286	.84307	27	.15693	.08593	9	.91407	8
53	.75733	18	.24267	.84334	27	.15666	.08602	9	.91398	7
54	.75751	18	.24249	.84361	27	.15639	.08611	8	.91389	6
55	9.75769	18	10.24231	9.84388	27	10.15612	10.08619	9	9.91381	5
56	.75787	18	.24213	.84415	27	.15585	.08628	9	.91372	4
57	.75805	18	.24195	.84442	27	.15558	.08637	9	.91363	3
58	.75823	18	.24177	.84469	27	.15531	.08646	9	.91354	2
59	.75841	18	.24159	.84496	27	.15504	.08655	9	.91345	1
60	9.75859		10.24141	9.84523		10.15477	10.08664		9.91336	0
↑ 124° →	cos	Diff. 1′	sec	cot	Diff. 1′	tan	csc	Diff. 1′	sin	← 55° ↑

TABLE 3

Common Logarithms of Trigonometric Functions (offset +10)

35° → ↓ ′	sin	Diff. 1′	csc	tan	Diff. 1′	cot	sec	Diff. 1′	cos	← 144° ↓ ′
0	9.75859	18	10.24141	9.84523	27	10.15477	10.08664	8	9.91336	60
1	.75877	18	.24123	.84550	26	.15450	.08672	9	.91328	59
2	.75895	18	.24105	.84576	27	.15424	.08681	9	.91319	58
3	.75913	18	.24087	.84603	27	.15397	.08690	9	.91310	57
4	.75931	18	.24069	.84630	27	.15370	.08699	9	.91301	56
5	9.75949	18	10.24051	9.84657	27	10.15343	10.08708	9	9.91292	55
6	.75967	18	.24033	.84684	27	.15316	.08717	9	.91283	54
7	.75985	18	.24015	.84711	27	.15289	.08726	8	.91274	53
8	.76003	18	.23997	.84738	26	.15262	.08734	9	.91266	52
9	.76021	18	.23979	.84764	27	.15236	.08743	9	.91257	51
10	9.76039	18	10.23961	9.84791	27	10.15209	10.08752	9	9.91248	50
11	.76057	18	.23943	.84818	27	.15182	.08761	9	.91239	49
12	.76075	18	.23925	.84845	27	.15155	.08770	9	.91230	48
13	.76093	18	.23907	.84872	27	.15128	.08779	9	.91221	47
14	.76111	18	.23889	.84899	26	.15101	.08788	9	.91212	46
15	9.76129	17	10.23871	9.84925	27	10.15075	10.08797	9	9.91203	45
16	.76146	18	.23854	.84952	27	.15048	.08806	9	.91194	44
17	.76164	18	.23836	.84979	27	.15021	.08815	9	.91185	43
18	.76182	18	.23818	.85006	27	.14994	.08824	9	.91176	42
19	.76200	18	.23800	.85033	26	.14967	.08833	9	.91167	41
20	9.76218	18	10.23782	9.85059	27	10.14941	10.08842	9	9.91158	40
21	.76236	17	.23764	.85086	27	.14914	.08851	8	.91149	39
22	.76253	18	.23747	.85113	27	.14887	.08859	9	.91141	38
23	.76271	18	.23729	.85140	26	.14860	.08868	9	.91132	37
24	.76289	18	.23711	.85166	27	.14834	.08877	9	.91123	36
25	9.76307	17	10.23693	9.85193	27	10.14807	10.08886	9	9.91114	35
26	.76324	18	.23676	.85220	27	.14780	.08895	9	.91105	34
27	.76342	18	.23658	.85247	26	.14753	.08904	9	.91096	33
28	.76360	18	.23640	.85273	27	.14727	.08913	9	.91087	32
29	.76378	17	.23622	.85300	27	.14700	.08922	9	.91078	31
30	9.76395	18	10.23605	9.85327	27	10.14673	10.08931	9	9.91069	30
31	.76413	18	.23587	.85354	26	.14646	.08940	9	.91060	29
32	.76431	17	.23569	.85380	27	.14620	.08949	9	.91051	28
33	.76448	18	.23552	.85407	27	.14593	.08958	9	.91042	27
34	.76466	18	.23534	.85434	26	.14566	.08967	10	.91033	26
35	9.76484	17	10.23516	9.85460	27	10.14540	10.08977	9	9.91023	25
36	.76501	18	.23499	.85487	27	.14513	.08986	9	.91014	24
37	.76519	18	.23481	.85514	26	.14486	.08995	9	.91005	23
38	.76537	17	.23463	.85540	27	.14460	.09004	9	.90996	22
39	.76554	18	.23446	.85567	27	.14433	.09013	9	.90987	21
40	9.76572	18	10.23428	9.85594	26	10.14406	10.09022	9	9.90978	20
41	.76590	17	.23410	.85620	27	.14380	.09031	9	.90969	19
42	.76607	18	.23393	.85647	27	.14353	.09040	9	.90960	18
43	.76625	17	.23375	.85674	26	.14326	.09049	9	.90951	17
44	.76642	18	.23358	.85700	27	.14300	.09058	9	.90942	16
45	9.76660	17	10.23340	9.85727	27	10.14273	10.09067	9	9.90933	15
46	.76677	18	.23323	.85754	26	.14246	.09076	9	.90924	14
47	.76695	17	.23305	.85780	27	.14220	.09085	9	.90915	13
48	.76712	18	.23288	.85807	27	.14193	.09094	10	.90906	12
49	.76730	17	.23270	.85834	26	.14166	.09104	9	.90896	11
50	9.76747	18	10.23253	9.85860	27	10.14140	10.09113	9	9.90887	10
51	.76765	17	.23235	.85887	26	.14113	.09122	9	.90878	9
52	.76782	18	.23218	.85913	27	.14087	.09131	9	.90869	8
53	.76800	17	.23200	.85940	27	.14060	.09140	9	.90860	7
54	.76817	18	.23183	.85967	26	.14033	.09149	9	.90851	6
55	9.76835	17	10.23165	9.85993	27	10.14007	10.09158	10	9.90842	5
56	.76852	18	.23148	.86020	26	.13980	.09168	9	.90832	4
57	.76870	17	.23130	.86046	27	.13954	.09177	9	.90823	3
58	.76887	17	.23113	.86073	27	.13927	.09186	9	.90814	2
59	.76904	18	.23096	.86100	26	.13900	.09195	9	.90805	1
60	9.76922		10.23078	9.86126		10.13874	10.09204		9.90796	0
↑ 125° →	cos	Diff. 1′	sec	cot	Diff. 1′	tan	csc	Diff. 1′	sin	← 54° ↑

TABLE 3

Common Logarithms of Trigonometric Functions (offset +10)

36°→ ↓	sin	Diff. 1′	csc	tan	Diff. 1′	cot	sec	Diff. 1′	cos	←143° ↓
′										′
0	9.76922	17	10.23078	9.86126	27	10.13874	10.09204	9	9.90796	60
1	.76939	18	.23061	.86153	26	.13847	.09213	10	.90787	59
2	.76957	17	.23043	.86179	27	.13821	.09223	9	.90777	58
3	.76974	17	.23026	.86206	26	.13794	.09232	9	.90768	57
4	.76991	18	.23009	.86232	27	.13768	.09241	9	.90759	56
5	9.77009	17	10.22991	9.86259	26	10.13741	10.09250	9	9.90750	55
6	.77026	17	.22974	.86285	27	.13715	.09259	10	.90741	54
7	.77043	18	.22957	.86312	26	.13688	.09269	9	.90731	53
8	.77061	17	.22939	.86338	27	.13662	.09278	9	.90722	52
9	.77078	17	.22922	.86365	27	.13635	.09287	9	.90713	51
10	9.77095	17	10.22905	9.86392	26	10.13608	10.09296	10	9.90704	50
11	.77112	18	.22888	.86418	27	.13582	.09306	9	.90694	49
12	.77130	17	.22870	.86445	26	.13555	.09315	9	.90685	48
13	.77147	17	.22853	.86471	27	.13529	.09324	9	.90676	47
14	.77164	17	.22836	.86498	26	.13502	.09333	10	.90667	46
15	9.77181	18	10.22819	9.86524	27	10.13476	10.09343	9	9.90657	45
16	.77199	17	.22801	.86551	26	.13449	.09352	9	.90648	44
17	.77216	17	.22784	.86577	26	.13423	.09361	9	.90639	43
18	.77233	17	.22767	.86603	27	.13397	.09370	10	.90630	42
19	.77250	18	.22750	.86630	26	.13370	.09380	9	.90620	41
20	9.77268	17	10.22732	9.86656	27	10.13344	10.09389	9	9.90611	40
21	.77285	17	.22715	.86683	26	.13317	.09398	10	.90602	39
22	.77302	17	.22698	.86709	27	.13291	.09408	9	.90592	38
23	.77319	17	.22681	.86736	26	.13264	.09417	9	.90583	37
24	.77336	17	.22664	.86762	27	.13238	.09426	9	.90574	36
25	9.77353	17	10.22647	9.86789	26	10.13211	10.09435	10	9.90565	35
26	.77370	17	.22630	.86815	27	.13185	.09445	9	.90555	34
27	.77387	18	.22613	.86842	26	.13158	.09454	9	.90546	33
28	.77405	17	.22595	.86868	26	.13132	.09463	10	.90537	32
29	.77422	17	.22578	.86894	27	.13106	.09473	9	.90527	31
30	9.77439	17	10.22561	9.86921	26	10.13079	10.09482	9	9.90518	30
31	.77456	17	.22544	.86947	27	.13053	.09491	10	.90509	29
32	.77473	17	.22527	.86974	26	.13026	.09501	9	.90499	28
33	.77490	17	.22510	.87000	27	.13000	.09510	10	.90490	27
34	.77507	17	.22493	.87027	26	.12973	.09520	9	.90480	26
35	9.77524	17	10.22476	9.87053	26	10.12947	10.09529	9	9.90471	25
36	.77541	17	.22459	.87079	27	.12921	.09538	10	.90462	24
37	.77558	17	.22442	.87106	26	.12894	.09548	9	.90452	23
38	.77575	17	.22425	.87132	26	.12868	.09557	9	.90443	22
39	.77592	17	.22408	.87158	27	.12842	.09566	10	.90434	21
40	9.77609	17	10.22391	9.87185	26	10.12815	10.09576	9	9.90424	20
41	.77626	17	.22374	.87211	27	.12789	.09585	10	.90415	19
42	.77643	17	.22357	.87238	26	.12762	.09595	9	.90405	18
43	.77660	17	.22340	.87264	26	.12736	.09604	10	.90396	17
44	.77677	17	.22323	.87290	27	.12710	.09614	9	.90386	16
45	9.77694	17	10.22306	9.87317	26	10.12683	10.09623	9	9.90377	15
46	.77711	17	.22289	.87343	26	.12657	.09632	10	.90368	14
47	.77728	16	.22272	.87369	27	.12631	.09642	9	.90358	13
48	.77744	17	.22256	.87396	26	.12604	.09651	10	.90349	12
49	.77761	17	.22239	.87422	26	.12578	.09661	9	.90339	11
50	9.77778	17	10.22222	9.87448	27	10.12552	10.09670	10	9.90330	10
51	.77795	17	.22205	.87475	26	.12525	.09680	9	.90320	9
52	.77812	17	.22188	.87501	26	.12499	.09689	10	.90311	8
53	.77829	17	.22171	.87527	27	.12473	.09699	9	.90301	7
54	.77846	16	.22154	.87554	26	.12446	.09708	10	.90292	6
55	9.77862	17	10.22138	9.87580	26	10.12420	10.09718	9	9.90282	5
56	.77879	17	.22121	.87606	27	.12394	.09727	10	.90273	4
57	.77896	17	.22104	.87633	26	.12367	.09737	9	.90263	3
58	.77913	17	.22087	.87659	26	.12341	.09746	10	.90254	2
59	.77930	16	.22070	.87685	26	.12315	.09756	9	.90244	1
60	9.77946		10.22054	9.87711		10.12289	10.09765		9.90235	0
↑ 126°→	cos	Diff. 1′	sec	cot	Diff. 1′	tan	csc	Diff. 1′	sin	←53° ↑

TABLE 3

Common Logarithms of Trigonometric Functions (offset +10)

37°→ ↓	sin	Diff. 1′	csc	tan	Diff. 1′	cot	sec	Diff. 1′	cos	←142° ↓
′										′
0	9.77946	17	10.22054	9.87711	27	10.12289	10.09765	10	9.90235	60
1	.77963	17	.22037	.87738	26	.12262	.09775	9	.90225	59
2	.77980	17	.22020	.87764	26	.12236	.09784	10	.90216	58
3	.77997	16	.22003	.87790	27	.12210	.09794	9	.90206	57
4	.78013	17	.21987	.87817	26	.12183	.09803	10	.90197	56
5	9.78030	17	10.21970	9.87843	26	10.12157	10.09813	9	9.90187	55
6	.78047	16	.21953	.87869	26	.12131	.09822	10	.90178	54
7	.78063	17	.21937	.87895	27	.12105	.09832	9	.90168	53
8	.78080	17	.21920	.87922	26	.12078	.09841	10	.90159	52
9	.78097	16	.21903	.87948	26	.12052	.09851	10	.90149	51
10	9.78113	17	10.21887	9.87974	26	10.12026	10.09861	9	9.90139	50
11	.78130	17	.21870	.88000	27	.12000	.09870	10	.90130	49
12	.78147	16	.21853	.88027	26	.11973	.09880	9	.90120	48
13	.78163	17	.21837	.88053	26	.11947	.09889	10	.90111	47
14	.78180	17	.21820	.88079	26	.11921	.09899	10	.90101	46
15	9.78197	16	10.21803	9.88105	26	10.11895	10.09909	9	9.90091	45
16	.78213	17	.21787	.88131	27	.11869	.09918	10	.90082	44
17	.78230	16	.21770	.88158	26	.11842	.09928	9	.90072	43
18	.78246	17	.21754	.88184	26	.11816	.09937	10	.90063	42
19	.78263	17	.21737	.88210	26	.11790	.09947	10	.90053	41
20	9.78280	16	10.21720	9.88236	26	10.11764	10.09957	9	9.90043	40
21	.78296	17	.21704	.88262	27	.11738	.09966	10	.90034	39
22	.78313	16	.21687	.88289	26	.11711	.09976	10	.90024	38
23	.78329	17	.21671	.88315	26	.11685	.09986	9	.90014	37
24	.78346	16	.21654	.88341	26	.11659	.09995	10	.90005	36
25	9.78362	17	10.21638	9.88367	26	10.11633	10.10005	10	9.89995	35
26	.78379	16	.21621	.88393	27	.11607	.10015	9	.89985	34
27	.78395	17	.21605	.88420	26	.11580	.10024	10	.89976	33
28	.78412	16	.21588	.88446	26	.11554	.10034	10	.89966	32
29	.78428	17	.21572	.88472	26	.11528	.10044	9	.89956	31
30	9.78445	16	10.21555	9.88498	26	10.11502	10.10053	10	9.89947	30
31	.78461	17	.21539	.88524	26	.11476	.10063	10	.89937	29
32	.78478	16	.21522	.88550	27	.11450	.10073	9	.89927	28
33	.78494	16	.21506	.88577	26	.11423	.10082	10	.89918	27
34	.78510	17	.21490	.88603	26	.11397	.10092	10	.89908	26
35	9.78527	16	10.21473	9.88629	26	10.11371	10.10102	10	9.89898	25
36	.78543	17	.21457	.88655	26	.11345	.10112	9	.89888	24
37	.78560	16	.21440	.88681	26	.11319	.10121	10	.89879	23
38	.78576	16	.21424	.88707	26	.11293	.10131	10	.89869	22
39	.78592	17	.21408	.88733	26	.11267	.10141	10	.89859	21
40	9.78609	16	10.21391	9.88759	27	10.11241	10.10151	9	9.89849	20
41	.78625	17	.21375	.88786	26	.11214	.10160	10	.89840	19
42	.78642	16	.21358	.88812	26	.11188	.10170	10	.89830	18
43	.78658	16	.21342	.88838	26	.11162	.10180	10	.89820	17
44	.78674	17	.21326	.88864	26	.11136	.10190	9	.89810	16
45	9.78691	16	10.21309	9.88890	26	10.11110	10.10199	10	9.89801	15
46	.78707	16	.21293	.88916	26	.11084	.10209	10	.89791	14
47	.78723	16	.21277	.88942	26	.11058	.10219	10	.89781	13
48	.78739	17	.21261	.88968	26	.11032	.10229	10	.89771	12
49	.78756	16	.21244	.88994	26	.11006	.10239	9	.89761	11
50	9.78772	16	10.21228	9.89020	26	10.10980	10.10248	10	9.89752	10
51	.78788	17	.21212	.89046	27	.10954	.10258	10	.89742	9
52	.78805	16	.21195	.89073	26	.10927	.10268	10	.89732	8
53	.78821	16	.21179	.89099	26	.10901	.10278	10	.89722	7
54	.78837	16	.21163	.89125	26	.10875	.10288	10	.89712	6
55	9.78853	16	10.21147	9.89151	26	10.10849	10.10298	9	9.89702	5
56	.78869	17	.21131	.89177	26	.10823	.10307	10	.89693	4
57	.78886	16	.21114	.89203	26	.10797	.10317	10	.89683	3
58	.78902	16	.21098	.89229	26	.10771	.10327	10	.89673	2
59	.78918	16	.21082	.89255	26	.10745	.10337	10	.89663	1
60	9.78934		10.21066	9.89281		10.10719	10.10347		9.89653	0
↑ 127°→	cos	Diff. 1′	sec	cot	Diff. 1′	tan	csc	Diff. 1′	sin	←52° ↑

TABLE 5

Common Logarithms of Trigonometric Functions (offset +10)

38° → ′	sin	Diff. 1′	csc	tan	Diff. 1′	cot	sec	Diff. 1′	cos	← 141° ′
0	9.78934	16	10.21066	9.89281	26	10.10719	10.10347	10	9.89653	60
1	.78950	17	.21050	.89307	26	.10693	.10357	10	.89643	59
2	.78967	16	.21033	.89333	26	.10667	.10367	9	.89633	58
3	.78983	16	.21017	.89359	26	.10641	.10376	10	.89624	57
4	.78999	16	.21001	.89385	26	.10615	.10386	10	.89614	56
5	9.79015	16	10.20985	9.89411	26	10.10589	10.10396	10	9.89604	55
6	.79031	16	.20969	.89437	26	.10563	.10406	10	.89594	54
7	.79047	16	.20953	.89463	26	.10537	.10416	10	.89584	53
8	.79063	16	.20937	.89489	26	.10511	.10426	10	.89574	52
9	.79079	16	.20921	.89515	26	.10485	.10436	10	.89564	51
10	9.79095	16	10.20905	9.89541	26	10.10459	10.10446	10	9.89554	50
11	.79111	17	.20889	.89567	26	.10433	.10456	10	.89544	49
12	.79128	16	.20872	.89593	26	.10407	.10466	10	.89534	48
13	.79144	16	.20856	.89619	26	.10381	.10476	10	.89524	47
14	.79160	16	.20840	.89645	26	.10355	.10486	10	.89514	46
15	9.79176	16	10.20824	9.89671	26	10.10329	10.10496	9	9.89504	45
16	.79192	16	.20808	.89697	26	.10303	.10505	10	.89495	44
17	.79208	16	.20792	.89723	26	.10277	.10515	10	.89485	43
18	.79224	16	.20776	.89749	26	.10251	.10525	10	.89475	42
19	.79240	16	.20760	.89775	26	.10225	.10535	10	.89465	41
20	9.79256	16	10.20744	9.89801	26	10.10199	10.10545	10	9.89455	40
21	.79272	16	.20728	.89827	26	.10173	.10555	10	.89445	39
22	.79288	16	.20712	.89853	26	.10147	.10565	10	.89435	38
23	.79304	15	.20696	.89879	26	.10121	.10575	10	.89425	37
24	.79319	16	.20681	.89905	26	.10095	.10585	10	.89415	36
25	9.79335	16	10.20665	9.89931	26	10.10069	10.10595	10	9.89405	35
26	.79351	16	.20649	.89957	26	.10043	.10605	10	.89395	34
27	.79367	16	.20633	.89983	26	.10017	.10615	10	.89385	33
28	.79383	16	.20617	.90009	26	.09991	.10625	11	.89375	32
29	.79399	16	.20601	.90035	26	.09965	.10636	10	.89364	31
30	9.79415	16	10.20585	9.90061	25	10.09939	10.10646	10	9.89354	30
31	.79431	16	.20569	.90086	26	.09914	.10656	10	.89344	29
32	.79447	16	.20553	.90112	26	.09888	.10666	10	.89334	28
33	.79463	15	.20537	.90138	26	.09862	.10676	10	.89324	27
34	.79478	16	.20522	.90164	26	.09836	.10686	10	.89314	26
35	9.79494	16	10.20506	9.90190	26	10.09810	10.10696	10	9.89304	25
36	.79510	16	.20490	.90216	26	.09784	.10706	10	.89294	24
37	.79526	16	.20474	.90242	26	.09758	.10716	10	.89284	23
38	.79542	16	.20458	.90268	26	.09732	.10726	10	.89274	22
39	.79558	15	.20442	.90294	26	.09706	.10736	10	.89264	21
40	9.79573	16	10.20427	9.90320	26	10.09680	10.10746	10	9.89254	20
41	.79589	16	.20411	.90346	25	.09654	.10756	11	.89244	19
42	.79605	16	.20395	.90371	26	.09629	.10767	10	.89233	18
43	.79621	15	.20379	.90397	26	.09603	.10777	10	.89223	17
44	.79636	16	.20364	.90423	26	.09577	.10787	10	.89213	16
45	9.79652	16	10.20348	9.90449	26	10.09551	10.10797	10	9.89203	15
46	.79668	16	.20332	.90475	26	.09525	.10807	10	.89193	14
47	.79684	15	.20316	.90501	26	.09499	.10817	10	.89183	13
48	.79699	16	.20301	.90527	26	.09473	.10827	11	.89173	12
49	.79715	16	.20285	.90553	25	.09447	.10838	10	.89162	11
50	9.79731	15	10.20269	9.90578	26	10.09422	10.10848	10	9.89152	10
51	.79746	16	.20254	.90604	26	.09396	.10858	10	.89142	9
52	.79762	16	.20238	.90630	26	.09370	.10868	10	.89132	8
53	.79778	15	.20222	.90656	26	.09344	.10878	10	.89122	7
54	.79793	16	.20207	.90682	26	.09318	.10888	11	.89112	6
55	9.79809	16	10.20191	9.90708	26	10.09292	10.10899	10	9.89101	5
56	.79825	15	.20175	.90734	25	.09266	.10909	10	.89091	4
57	.79840	16	.20160	.90759	26	.09241	.10919	10	.89081	3
58	.79856	16	.20144	.90785	26	.09215	.10929	11	.89071	2
59	.79872	15	.20128	.90811	26	.09189	.10940	10	.89060	1
60	9.79887		10.20113	9.90837		10.09163	10.10950		9.89050	0
128° →	cos	Diff. 1′	sec	cot	Diff. 1′	tan	csc	Diff. 1′	sin	← 51°

TABLE 5

Common Logarithms of Trigonometric Functions (offset +10)

39° → ′	sin	Diff. 1′	csc	tan	Diff. 1′	cot	sec	Diff. 1′	cos	← 140° ′
0	9.79887	16	10.20113	9.90837	26	10.09163	10.10950	10	9.89050	60
1	.79903	15	.20097	.90863	26	.09137	.10960	10	.89040	59
2	.79918	16	.20082	.90889	25	.09111	.10970	10	.89030	58
3	.79934	16	.20066	.90914	26	.09086	.10980	11	.89020	57
4	.79950	15	.20050	.90940	26	.09060	.10991	10	.89009	56
5	9.79965	16	10.20035	9.90966	26	10.09034	10.11001	10	9.88999	55
6	.79981	15	.20019	.90992	26	.09008	.11011	11	.88989	54
7	.79996	16	.20004	.91018	25	.08982	.11022	10	.88978	53
8	.80012	15	.19988	.91043	26	.08957	.11032	10	.88968	52
9	.80027	16	.19973	.91069	26	.08931	.11042	10	.88958	51
10	9.80043	15	10.19957	9.91095	26	10.08905	10.11052	11	9.88948	50
11	.80058	16	.19942	.91121	26	.08879	.11063	10	.88937	49
12	.80074	15	.19926	.91147	25	.08853	.11073	10	.88927	48
13	.80089	16	.19911	.91172	26	.08828	.11083	11	.88917	47
14	.80105	15	.19895	.91198	26	.08802	.11094	10	.88906	46
15	9.80120	16	10.19880	9.91224	26	10.08776	10.11104	10	9.88896	45
16	.80136	15	.19864	.91250	26	.08750	.11114	11	.88886	44
17	.80151	15	.19849	.91276	25	.08724	.11125	10	.88875	43
18	.80166	16	.19834	.91301	26	.08699	.11135	10	.88865	42
19	.80182	15	.19818	.91327	26	.08673	.11145	11	.88855	41
20	9.80197	16	10.19803	9.91353	26	10.08647	10.11156	10	9.88844	40
21	.80213	15	.19787	.91379	25	.08621	.11166	10	.88834	39
22	.80228	16	.19772	.91404	26	.08596	.11176	11	.88824	38
23	.80244	15	.19756	.91430	26	.08570	.11187	10	.88813	37
24	.80259	15	.19741	.91456	26	.08544	.11197	10	.88803	36
25	9.80274	16	10.19726	9.91482	25	10.08518	10.11207	11	9.88793	35
26	.80290	15	.19710	.91507	26	.08493	.11218	10	.88782	34
27	.80305	15	.19695	.91533	26	.08467	.11228	11	.88772	33
28	.80320	16	.19680	.91559	26	.08441	.11239	10	.88761	32
29	.80336	15	.19664	.91585	25	.08415	.11249	10	.88751	31
30	9.80351	15	10.19649	9.91610	26	10.08390	10.11259	11	9.88741	30
31	.80366	16	.19634	.91636	26	.08364	.11270	10	.88730	29
32	.80382	15	.19618	.91662	26	.08338	.11280	11	.88720	28
33	.80397	15	.19603	.91688	25	.08312	.11291	10	.88709	27
34	.80412	16	.19588	.91713	26	.08287	.11301	11	.88699	26
35	9.80428	15	10.19572	9.91739	26	10.08261	10.11312	10	9.88688	25
36	.80443	15	.19557	.91765	26	.08235	.11322	10	.88678	24
37	.80458	15	.19542	.91791	25	.08209	.11332	11	.88668	23
38	.80473	16	.19527	.91816	26	.08184	.11343	10	.88657	22
39	.80489	15	.19511	.91842	26	.08158	.11353	11	.88647	21
40	9.80504	15	10.19496	9.91868	25	10.08132	10.11364	10	9.88636	20
41	.80519	15	.19481	.91893	26	.08107	.11374	11	.88626	19
42	.80534	16	.19466	.91919	26	.08081	.11385	10	.88615	18
43	.80550	15	.19450	.91945	26	.08055	.11395	11	.88605	17
44	.80565	15	.19435	.91971	25	.08029	.11406	10	.88594	16
45	9.80580	15	10.19420	9.91996	26	10.08004	10.11416	11	9.88584	15
46	.80595	15	.19405	.92022	26	.07978	.11427	10	.88573	14
47	.80610	15	.19390	.92048	25	.07952	.11437	11	.88563	13
48	.80625	16	.19375	.92073	26	.07927	.11448	10	.88552	12
49	.80641	15	.19359	.92099	26	.07901	.11458	11	.88542	11
50	9.80656	15	10.19344	9.92125	25	10.07875	10.11469	10	9.88531	10
51	.80671	15	.19329	.92150	26	.07850	.11479	11	.88521	9
52	.80686	15	.19314	.92176	26	.07824	.11490	11	.88510	8
53	.80701	15	.19299	.92202	25	.07798	.11501	10	.88499	7
54	.80716	15	.19284	.92227	26	.07773	.11511	11	.88489	6
55	9.80731	15	10.19269	9.92253	26	10.07747	10.11522	10	9.88478	5
56	.80746	16	.19254	.92279	25	.07721	.11532	11	.88468	4
57	.80762	15	.19238	.92304	26	.07696	.11543	10	.88457	3
58	.80777	15	.19223	.92330	26	.07670	.11553	11	.88447	2
59	.80792	15	.19208	.92356	25	.07644	.11564	11	.88436	1
60	9.80807		10.19193	9.92381		10.07619	10.11575		9.88425	0
129° →	cos	Diff. 1′	sec	cot	Diff. 1′	tan	csc	Diff. 1′	sin	← 50°

TABLE 3

Common Logarithms of Trigonometric Functions (offset +10)

40°→ ′	sin	Diff. 1′	csc	tan	Diff. 1′	cot	sec	Diff. 1′	cos	←139° ′
0	9.80807	15	10.19193	9.92381	26	10.07619	10.11575	10	9.88425	60
1	.80822	15	.19178	.92407	26	.07593	.11585	11	.88415	59
2	.80837	15	.19163	.92433	25	.07567	.11596	10	.88404	58
3	.80852	15	.19148	.92458	26	.07542	.11606	11	.88394	57
4	.80867	15	.19133	.92484	26	.07516	.11617	11	.88383	56
5	9.80882	15	10.19118	9.92510	25	10.07490	10.11628	10	9.88372	55
6	.80897	15	.19103	.92535	26	.07465	.11638	11	.88362	54
7	.80912	15	.19088	.92561	26	.07439	.11649	11	.88351	53
8	.80927	15	.19073	.92587	25	.07413	.11660	10	.88340	52
9	.80942	15	.19058	.92612	26	.07388	.11670	11	.88330	51
10	9.80957	15	10.19043	9.92638	25	10.07362	10.11681	11	9.88319	50
11	.80972	15	.19028	.92663	26	.07337	.11692	10	.88308	49
12	.80987	15	.19013	.92689	26	.07311	.11702	11	.88298	48
13	.81002	15	.18998	.92715	25	.07285	.11713	11	.88287	47
14	.81017	15	.18983	.92740	26	.07260	.11724	10	.88276	46
15	9.81032	15	10.18968	9.92766	26	10.07234	10.11734	11	9.88266	45
16	.81047	14	.18953	.92792	25	.07208	.11745	11	.88255	44
17	.81061	15	.18939	.92817	26	.07183	.11756	10	.88244	43
18	.81076	15	.18924	.92843	25	.07157	.11766	11	.88234	42
19	.81091	15	.18909	.92868	26	.07132	.11777	11	.88223	41
20	9.81106	15	10.18894	9.92894	26	10.07106	10.11788	11	9.88212	40
21	.81121	15	.18879	.92920	25	.07080	.11799	10	.88201	39
22	.81136	15	.18864	.92945	26	.07055	.11809	11	.88191	38
23	.81151	15	.18849	.92971	25	.07029	.11820	11	.88180	37
24	.81166	14	.18834	.92996	26	.07004	.11831	11	.88169	36
25	9.81180	15	10.18820	9.93022	26	10.06978	10.11842	10	9.88158	35
26	.81195	15	.18805	.93048	25	.06952	.11852	11	.88148	34
27	.81210	15	.18790	.93073	26	.06927	.11863	11	.88137	33
28	.81225	15	.18775	.93099	25	.06901	.11874	11	.88126	32
29	.81240	14	.18760	.93124	26	.06876	.11885	10	.88115	31
30	9.81254	15	10.18746	9.93150	25	10.06850	10.11895	11	9.88105	30
31	.81269	15	.18731	.93175	26	.06825	.11906	11	.88094	29
32	.81284	15	.18716	.93201	26	.06799	.11917	11	.88083	28
33	.81299	15	.18701	.93227	25	.06773	.11928	11	.88072	27
34	.81314	14	.18686	.93252	26	.06748	.11939	10	.88061	26
35	9.81328	15	10.18672	9.93278	25	10.06722	10.11949	11	9.88051	25
36	.81343	15	.18657	.93303	26	.06697	.11960	11	.88040	24
37	.81358	14	.18642	.93329	25	.06671	.11971	11	.88029	23
38	.81372	15	.18628	.93354	26	.06646	.11982	11	.88018	22
39	.81387	15	.18613	.93380	26	.06620	.11993	11	.88007	21
40	9.81402	15	10.18598	9.93406	25	10.06594	10.12004	11	9.87996	20
41	.81417	14	.18583	.93431	26	.06569	.12015	10	.87985	19
42	.81431	15	.18569	.93457	25	.06543	.12025	11	.87975	18
43	.81446	15	.18554	.93482	26	.06518	.12036	11	.87964	17
44	.81461	14	.18539	.93508	25	.06492	.12047	11	.87953	16
45	9.81475	15	10.18525	9.93533	26	10.06467	10.12058	11	9.87942	15
46	.81490	15	.18510	.93559	25	.06441	.12069	11	.87931	14
47	.81505	14	.18495	.93584	26	.06416	.12080	11	.87920	13
48	.81519	15	.18481	.93610	26	.06390	.12091	11	.87909	12
49	.81534	15	.18466	.93636	25	.06364	.12102	11	.87898	11
50	9.81549	14	10.18451	9.93661	26	10.06339	10.12113	10	9.87887	10
51	.81563	15	.18437	.93687	25	.06313	.12123	11	.87877	9
52	.81578	14	.18422	.93712	26	.06288	.12134	11	.87866	8
53	.81592	15	.18408	.93738	25	.06262	.12145	11	.87855	7
54	.81607	15	.18393	.93763	26	.06237	.12156	11	.87844	6
55	9.81622	14	10.18378	9.93789	25	10.06211	10.12167	11	9.87833	5
56	.81636	15	.18364	.93814	26	.06186	.12178	11	.87822	4
57	.81651	14	.18349	.93840	25	.06160	.12189	11	.87811	3
58	.81665	15	.18335	.93865	26	.06135	.12200	11	.87800	2
59	.81680	14	.18320	.93891	25	.06109	.12211	11	.87789	1
60	9.81694		10.18306	9.93916		10.06084	10.12222		9.87778	0
130°→	cos	Diff. 1′	sec	cot	Diff. 1′	tan	csc	Diff. 1′	sin	←49°

TABLE 3

Common Logarithms of Trigonometric Functions (offset +10)

41°→ ′	sin	Diff. 1′	csc	tan	Diff. 1′	cot	sec	Diff. 1′	cos	←138° ′
0	9.81694	15	10.18306	9.93916	26	10.06084	10.12222	11	9.87778	60
1	.81709	14	.18291	.93942	25	.06058	.12233	11	.87767	59
2	.81723	15	.18277	.93967	26	.06033	.12244	11	.87756	58
3	.81738	14	.18262	.93993	25	.06007	.12255	11	.87745	57
4	.81752	15	.18248	.94018	26	.05982	.12266	11	.87734	56
5	9.81767	14	10.18233	9.94044	25	10.05956	10.12277	11	9.87723	55
6	.81781	15	.18219	.94069	26	.05931	.12288	11	.87712	54
7	.81796	14	.18204	.94095	25	.05905	.12299	11	.87701	53
8	.81810	15	.18190	.94120	26	.05880	.12310	11	.87690	52
9	.81825	14	.18175	.94146	25	.05854	.12321	11	.87679	51
10	9.81839	15	10.18161	9.94171	26	10.05829	10.12332	11	9.87668	50
11	.81854	14	.18146	.94197	25	.05803	.12343	11	.87657	49
12	.81868	14	.18132	.94222	26	.05778	.12354	11	.87646	48
13	.81882	15	.18118	.94248	25	.05752	.12365	11	.87635	47
14	.81897	14	.18103	.94273	26	.05727	.12376	11	.87624	46
15	9.81911	15	10.18089	9.94299	25	10.05701	10.12387	12	9.87613	45
16	.81926	14	.18074	.94324	26	.05676	.12399	11	.87601	44
17	.81940	15	.18060	.94350	25	.05650	.12410	11	.87590	43
18	.81955	14	.18045	.94375	26	.05625	.12421	11	.87579	42
19	.81969	14	.18031	.94401	25	.05599	.12432	11	.87568	41
20	9.81983	15	10.18017	9.94426	26	10.05574	10.12443	11	9.87557	40
21	.81998	14	.18002	.94452	25	.05548	.12454	11	.87546	39
22	.82012	14	.17988	.94477	26	.05523	.12465	11	.87535	38
23	.82026	15	.17974	.94503	25	.05497	.12476	11	.87524	37
24	.82041	14	.17959	.94528	26	.05472	.12487	12	.87513	36
25	9.82055	14	10.17945	9.94554	25	10.05446	10.12499	11	9.87501	35
26	.82069	15	.17931	.94579	25	.05421	.12510	11	.87490	34
27	.82084	14	.17916	.94604	26	.05396	.12521	11	.87479	33
28	.82098	14	.17902	.94630	25	.05370	.12532	11	.87468	32
29	.82112	14	.17888	.94655	26	.05345	.12543	11	.87457	31
30	9.82126	15	10.17874	9.94681	25	10.05319	10.12554	12	9.87446	30
31	.82141	14	.17859	.94706	26	.05294	.12566	11	.87434	29
32	.82155	14	.17845	.94732	25	.05268	.12577	11	.87423	28
33	.82169	15	.17831	.94757	26	.05243	.12588	11	.87412	27
34	.82184	14	.17816	.94783	25	.05217	.12599	11	.87401	26
35	9.82198	14	10.17802	9.94808	26	10.05192	10.12610	12	9.87390	25
36	.82212	14	.17788	.94834	25	.05166	.12622	11	.87378	24
37	.82226	14	.17774	.94859	25	.05141	.12633	11	.87367	23
38	.82240	15	.17760	.94884	26	.05116	.12644	11	.87356	22
39	.82255	14	.17745	.94910	25	.05090	.12655	11	.87345	21
40	9.82269	14	10.17731	9.94935	26	10.05065	10.12666	12	9.87334	20
41	.82283	14	.17717	.94961	25	.05039	.12678	11	.87322	19
42	.82297	14	.17703	.94986	26	.05014	.12689	11	.87311	18
43	.82311	15	.17689	.95012	25	.04988	.12700	12	.87300	17
44	.82326	14	.17674	.95037	25	.04963	.12712	11	.87288	16
45	9.82340	14	10.17660	9.95062	26	10.04938	10.12723	11	9.87277	15
46	.82354	14	.17646	.95088	25	.04912	.12734	11	.87266	14
47	.82368	14	.17632	.95113	26	.04887	.12745	12	.87255	13
48	.82382	14	.17618	.95139	25	.04861	.12757	11	.87243	12
49	.82396	14	.17604	.95164	26	.04836	.12768	11	.87232	11
50	9.82410	14	10.17590	9.95190	25	10.04810	10.12779	12	9.87221	10
51	.82424	15	.17576	.95215	25	.04785	.12791	11	.87209	9
52	.82439	14	.17561	.95240	26	.04760	.12802	11	.87198	8
53	.82453	14	.17547	.95266	25	.04734	.12813	12	.87187	7
54	.82467	14	.17533	.95291	26	.04709	.12825	11	.87175	6
55	9.82481	14	10.17519	9.95317	25	10.04683	10.12836	11	9.87164	5
56	.82495	14	.17505	.95342	26	.04658	.12847	12	.87153	4
57	.82509	14	.17491	.95368	25	.04632	.12859	11	.87141	3
58	.82523	14	.17477	.95393	25	.04607	.12870	11	.87130	2
59	.82537	14	.17463	.95418	26	.04582	.12881	12	.87119	1
60	9.82551		10.17449	9.95444		10.04556	10.12893		9.87107	0
131°→	cos	Diff. 1′	sec	cot	Diff. 1′	tan	csc	Diff. 1′	sin	←48°

Common Logarithms of Trigonometric Functions (offset +10)

42° → ↓	sin	Diff. 1′	csc	tan	Diff. 1′	cot	sec	Diff. 1′	cos	← 137° ↓
′										′
0	9.82551	14	10.17449	9.95444	25	10.04556	10.12893	11	9.87107	60
1	.82565	14	.17435	.95469	26	.04531	.12904	11	.87096	59
2	.82579	14	.17421	.95495	25	.04505	.12915	12	.87085	58
3	.82593	14	.17407	.95520	25	.04480	.12927	11	.87073	57
4	.82607	14	.17393	.95545	26	.04455	.12938	12	.87062	56
5	9.82621	14	10.17379	9.95571	25	10.04429	10.12950	11	9.87050	55
6	.82635	14	.17365	.95596	26	.04404	.12961	11	.87039	54
7	.82649	14	.17351	.95622	25	.04378	.12972	12	.87028	53
8	.82663	14	.17337	.95647	25	.04353	.12984	11	.87016	52
9	.82677	14	.17323	.95672	26	.04328	.12995	12	.87005	51
10	9.82691	14	10.17309	9.95698	25	10.04302	10.13007	11	9.86993	50
11	.82705	14	.17295	.95723	25	.04277	.13018	12	.86982	49
12	.82719	14	.17281	.95748	26	.04252	.13030	11	.86970	48
13	.82733	14	.17267	.95774	25	.04226	.13041	12	.86959	47
14	.82747	14	.17253	.95799	26	.04201	.13053	11	.86947	46
15	9.82761	14	10.17239	9.95825	25	10.04175	10.13064	12	9.86936	45
16	.82775	13	.17225	.95850	25	.04150	.13076	11	.86924	44
17	.82788	14	.17212	.95875	26	.04125	.13087	11	.86913	43
18	.82802	14	.17198	.95901	25	.04099	.13098	12	.86902	42
19	.82816	14	.17184	.95926	26	.04074	.13110	11	.86890	41
20	9.82830	14	10.17170	9.95952	25	10.04048	10.13121	12	9.86879	40
21	.82844	14	.17156	.95977	25	.04023	.13133	12	.86867	39
22	.82858	14	.17142	.96002	26	.03998	.13145	11	.86855	38
23	.82872	13	.17128	.96028	25	.03972	.13156	12	.86844	37
24	.82885	14	.17115	.96053	25	.03947	.13168	11	.86832	36
25	9.82899	14	10.17101	9.96078	26	10.03922	10.13179	12	9.86821	35
26	.82913	14	.17087	.96104	25	.03896	.13191	11	.86809	34
27	.82927	14	.17073	.96129	26	.03871	.13202	12	.86798	33
28	.82941	14	.17059	.96155	25	.03845	.13214	11	.86786	32
29	.82955	13	.17045	.96180	25	.03820	.13225	12	.86775	31
30	9.82968	14	10.17032	9.96205	26	10.03795	10.13237	11	9.86763	30
31	.82982	14	.17018	.96231	25	.03769	.13248	12	.86752	29
32	.82996	14	.17004	.96256	25	.03744	.13260	12	.86740	28
33	.83010	13	.16990	.96281	26	.03719	.13272	11	.86728	27
34	.83023	14	.16977	.96307	25	.03693	.13283	12	.86717	26
35	9.83037	14	10.16963	9.96332	25	10.03668	10.13295	11	9.86705	25
36	.83051	14	.16949	.96357	26	.03643	.13306	12	.86694	24
37	.83065	13	.16935	.96383	25	.03617	.13318	12	.86682	23
38	.83078	14	.16922	.96408	25	.03592	.13330	11	.86670	22
39	.83092	14	.16908	.96433	26	.03567	.13341	12	.86659	21
40	9.83106	14	10.16894	9.96459	25	10.03541	10.13353	12	9.86647	20
41	.83120	13	.16880	.96484	26	.03516	.13365	11	.86635	19
42	.83133	14	.16867	.96510	25	.03490	.13376	12	.86624	18
43	.83147	14	.16853	.96535	25	.03465	.13388	12	.86612	17
44	.83161	13	.16839	.96560	26	.03440	.13400	11	.86600	16
45	9.83174	14	10.16826	9.96586	25	10.03414	10.13411	12	9.86589	15
46	.83188	14	.16812	.96611	25	.03389	.13423	12	.86577	14
47	.83202	13	.16798	.96636	26	.03364	.13435	11	.86565	13
48	.83215	14	.16785	.96662	25	.03338	.13446	12	.86554	12
49	.83229	13	.16771	.96687	25	.03313	.13458	12	.86542	11
50	9.83242	14	10.16758	9.96712	26	10.03288	10.13470	12	9.86530	10
51	.83256	14	.16744	.96738	25	.03262	.13482	11	.86518	9
52	.83270	13	.16730	.96763	25	.03237	.13493	12	.86507	8
53	.83283	14	.16717	.96788	26	.03212	.13505	12	.86495	7
54	.83297	13	.16703	.96814	25	.03186	.13517	11	.86483	6
55	9.83310	14	10.16690	9.96839	25	10.03161	10.13528	12	9.86472	5
56	.83324	14	.16676	.96864	26	.03136	.13540	12	.86460	4
57	.83338	13	.16662	.96890	25	.03110	.13552	12	.86448	3
58	.83351	14	.16649	.96915	25	.03085	.13564	11	.86436	2
59	.83365	13	.16635	.96940	26	.03060	.13575	12	.86425	1
60	9.83378		10.16622	9.96966		10.03034	10.13587		9.86413	0
↑ 132° →	cos	Diff. 1′	sec	cot	Diff. 1′	tan	csc	Diff. 1′	sin	← 47° ↑

Common Logarithms of Trigonometric Functions (offset +10)

43° → ↓	sin	Diff. 1′	csc	tan	Diff. 1′	cot	sec	Diff. 1′	cos	← 136° ↓
′										′
0	9.83378	14	10.16622	9.96966	25	10.03034	10.13587	12	9.86413	60
1	.83392	13	.16608	.96991	25	.03009	.13599	12	.86401	59
2	.83405	14	.16595	.97016	26	.02984	.13611	12	.86389	58
3	.83419	13	.16581	.97042	25	.02958	.13623	11	.86377	57
4	.83432	14	.16568	.97067	25	.02933	.13634	12	.86366	56
5	9.83446	13	10.16554	9.97092	26	10.02908	10.13646	12	9.86354	55
6	.83459	14	.16541	.97118	25	.02882	.13658	12	.86342	54
7	.83473	13	.16527	.97143	25	.02857	.13670	12	.86330	53
8	.83486	14	.16514	.97168	25	.02832	.13682	12	.86318	52
9	.83500	13	.16500	.97193	26	.02807	.13694	11	.86306	51
10	9.83513	14	10.16487	9.97219	25	10.02781	10.13705	12	9.86295	50
11	.83527	13	.16473	.97244	25	.02756	.13717	12	.86283	49
12	.83540	14	.16460	.97269	26	.02731	.13729	12	.86271	48
13	.83554	13	.16446	.97295	25	.02705	.13741	12	.86259	47
14	.83567	14	.16433	.97320	25	.02680	.13753	12	.86247	46
15	9.83581	13	10.16419	9.97345	26	10.02655	10.13765	12	9.86235	45
16	.83594	14	.16406	.97371	25	.02629	.13777	12	.86223	44
17	.83608	13	.16392	.97396	25	.02604	.13789	11	.86211	43
18	.83621	13	.16379	.97421	26	.02579	.13800	12	.86200	42
19	.83634	14	.16366	.97447	25	.02553	.13812	12	.86188	41
20	9.83648	13	10.16352	9.97472	25	10.02528	10.13824	12	9.86176	40
21	.83661	13	.16339	.97497	26	.02503	.13836	12	.86164	39
22	.83674	14	.16326	.97523	25	.02477	.13848	12	.86152	38
23	.83688	13	.16312	.97548	25	.02452	.13860	12	.86140	37
24	.83701	14	.16299	.97573	25	.02427	.13872	12	.86128	36
25	9.83715	13	10.16285	9.97598	26	10.02402	10.13884	12	9.86116	35
26	.83728	13	.16272	.97624	25	.02376	.13896	12	.86104	34
27	.83741	14	.16259	.97649	25	.02351	.13908	12	.86092	33
28	.83755	13	.16245	.97674	26	.02326	.13920	12	.86080	32
29	.83768	13	.16232	.97700	25	.02300	.13932	12	.86068	31
30	9.83781	14	10.16219	9.97725	25	10.02275	10.13944	12	9.86056	30
31	.83795	13	.16205	.97750	26	.02250	.13956	12	.86044	29
32	.83808	13	.16192	.97776	25	.02224	.13968	12	.86032	28
33	.83821	13	.16179	.97801	25	.02199	.13980	12	.86020	27
34	.83834	14	.16166	.97826	25	.02174	.13992	12	.86008	26
35	9.83848	13	10.16152	9.97851	26	10.02149	10.14004	12	9.85996	25
36	.83861	13	.16139	.97877	25	.02123	.14016	12	.85984	24
37	.83874	13	.16126	.97902	25	.02098	.14028	12	.85972	23
38	.83887	14	.16113	.97927	26	.02073	.14040	12	.85960	22
39	.83901	13	.16099	.97953	25	.02047	.14052	12	.85948	21
40	9.83914	13	10.16086	9.97978	25	10.02022	10.14064	12	9.85936	20
41	.83927	13	.16073	.98003	26	.01997	.14076	12	.85924	19
42	.83940	14	.16060	.98029	25	.01971	.14088	12	.85912	18
43	.83954	13	.16046	.98054	25	.01946	.14100	12	.85900	17
44	.83967	13	.16033	.98079	25	.01921	.14112	12	.85888	16
45	9.83980	13	10.16020	9.98104	26	10.01896	10.14124	12	9.85876	15
46	.83993	13	.16007	.98130	25	.01870	.14136	13	.85864	14
47	.84006	14	.15994	.98155	25	.01845	.14149	12	.85851	13
48	.84020	13	.15980	.98180	26	.01820	.14161	12	.85839	12
49	.84033	13	.15967	.98206	25	.01794	.14173	12	.85827	11
50	9.84046	13	10.15954	9.98231	25	10.01769	10.14185	12	9.85815	10
51	.84059	13	.15941	.98256	25	.01744	.14197	12	.85803	9
52	.84072	13	.15928	.98281	26	.01719	.14209	12	.85791	8
53	.84085	13	.15915	.98307	25	.01693	.14221	13	.85779	7
54	.84098	14	.15902	.98332	25	.01668	.14234	12	.85766	6
55	9.84112	13	10.15888	9.98357	26	10.01643	10.14246	12	9.85754	5
56	.84125	13	.15875	.98383	25	.01617	.14258	12	.85742	4
57	.84138	13	.15862	.98408	25	.01592	.14270	12	.85730	3
58	.84151	13	.15849	.98433	25	.01567	.14282	12	.85718	2
59	.84164	13	.15836	.98458	26	.01542	.14294	13	.85706	1
60	9.84177		10.15823	9.98484		10.01516	10.14307		9.85693	0
↑ 133° →	cos	Diff. 1′	sec	cot	Diff. 1′	tan	csc	Diff. 1′	sin	← 46° ↑

TABLE 3

Common Logarithms of Trigonometric Functions (offset +10)

44°→ ↓ ′	sin	Diff. 1′	csc	tan	Diff. 1′	cot	sec	Diff. 1′	cos	←135° ↓ ′
0	9.84177	13	10.15823	9.98484	25	10.01516	10.14307	12	9.85693	60
1	.84190	13	.15810	.98509	25	.01491	.14319	12	.85681	59
2	.84203	13	.15797	.98534	26	.01466	.14331	12	.85669	58
3	.84216	13	.15784	.98560	25	.01440	.14343	12	.85657	57
4	.84229	13	.15771	.98585	25	.01415	.14355	13	.85645	56
5	9.84242	13	10.15758	9.98610	25	10.01390	10.14368	12	9.85632	55
6	.84255	14	.15745	.98635	26	.01365	.14380	12	.85620	54
7	.84269	13	.15731	.98661	25	.01339	.14392	12	.85608	53
8	.84282	13	.15718	.98686	25	.01314	.14404	13	.85596	52
9	.84295	13	.15705	.98711	26	.01289	.14417	12	.85583	51
10	9.84308	13	10.15692	9.98737	25	10.01263	10.14429	12	9.85571	50
11	.84321	13	.15679	.98762	25	.01238	.14441	12	.85559	49
12	.84334	13	.15666	.98787	25	.01213	.14453	13	.85547	48
13	.84347	13	.15653	.98812	26	.01188	.14466	12	.85534	47
14	.84360	13	.15640	.98838	25	.01162	.14478	12	.85522	46
15	9.84373	12	10.15627	9.98863	25	10.01137	10.14490	13	9.85510	45
16	.84385	13	.15615	.98888	25	.01112	.14503	12	.85497	44
17	.84398	13	.15602	.98913	26	.01087	.14515	12	.85485	43
18	.84411	13	.15589	.98939	25	.01061	.14527	13	.85473	42
19	.84424	13	.15576	.98964	25	.01036	.14540	12	.85460	41
20	9.84437	13	10.15563	9.98989	26	10.01011	10.14552	12	9.85448	40
21	.84450	13	.15550	.99015	25	.00985	.14564	13	.85436	39
22	.84463	13	.15537	.99040	25	.00960	.14577	12	.85423	38
23	.84476	13	.15524	.99065	25	.00935	.14589	12	.85411	37
24	.84489	13	.15511	.99090	26	.00910	.14601	13	.85399	36
25	9.84502	13	10.15498	9.99116	25	10.00884	10.14614	12	9.85386	35
26	.84515	13	.15485	.99141	25	.00859	.14626	13	.85374	34
27	.84528	12	.15472	.99166	25	.00834	.14639	12	.85361	33
28	.84540	13	.15460	.99191	26	.00809	.14651	12	.85349	32
29	.84553	13	.15447	.99217	25	.00783	.14663	13	.85337	31
30	9.84566	13	10.15434	9.99242	25	10.00758	10.14676	12	9.85324	30
31	.84579	13	.15421	.99267	26	.00733	.14688	13	.85312	29
32	.84592	13	.15408	.99293	25	.00707	.14701	12	.85299	28
33	.84605	13	.15395	.99318	25	.00682	.14713	13	.85287	27
34	.84618	12	.15382	.99343	25	.00657	.14726	12	.85274	26
35	9.84630	13	10.15370	9.99368	26	10.00632	10.14738	12	9.85262	25
36	.84643	13	.15357	.99394	25	.00606	.14750	13	.85250	24
37	.84656	13	.15344	.99419	25	.00581	.14763	12	.85237	23
38	.84669	13	.15331	.99444	25	.00556	.14775	13	.85225	22
39	.84682	12	.15318	.99469	26	.00531	.14788	12	.85212	21
40	9.84694	13	10.15306	9.99495	25	10.00505	10.14800	13	9.85200	20
41	.84707	13	.15293	.99520	25	.00480	.14813	12	.85187	19
42	.84720	13	.15280	.99545	25	.00455	.14825	13	.85175	18
43	.84733	12	.15267	.99570	26	.00430	.14838	12	.85162	17
44	.84745	13	.15255	.99596	25	.00404	.14850	13	.85150	16
45	9.84758	13	10.15242	9.99621	25	10.00379	10.14863	12	9.85137	15
46	.84771	13	.15229	.99646	26	.00354	.14875	13	.85125	14
47	.84784	12	.15216	.99672	25	.00328	.14888	12	.85112	13
48	.84796	13	.15204	.99697	25	.00303	.14900	13	.85100	12
49	.84809	13	.15191	.99722	25	.00278	.14913	13	.85087	11
50	9.84822	13	10.15178	9.99747	26	10.00253	10.14926	12	9.85074	10
51	.84835	12	.15165	.99773	25	.00227	.14938	13	.85062	9
52	.84847	13	.15153	.99798	25	.00202	.14951	12	.85049	8
53	.84860	13	.15140	.99823	25	.00177	.14963	13	.85037	7
54	.84873	12	.15127	.99848	26	.00152	.14976	12	.85024	6
55	9.84885	13	10.15115	9.99874	25	10.00126	10.14988	13	9.85012	5
56	.84898	13	.15102	.99899	25	.00101	.15001	13	.84999	4
57	.84911	12	.15089	.99924	25	.00076	.15014	12	.84986	3
58	.84923	13	.15077	.99949	26	.00051	.15026	13	.84974	2
59	.84936	13	.15064	.99975	25	.00025	.15039	12	.84961	1
60	9.84949		10.15051	10.00000		10.00000	10.15051		9.84949	0
↑ 134°→	cos	Diff. 1′	sec	cot	Diff. 1′	tan	csc	Diff. 1′	sin	← ↑ 45°

TABLE 4

1°

Traverse Table

359° | 001°
181° | 179°

Dist.	D. Lat.	Dep.	Dist.	D. Lat.	Dep.	Dist.	D. Lat.	Dep.	Dist.	D. Lat.	Dep.	Dist.	D. Lat.	Dep.
1	1.0	0.0	61	61.0	1.1	121	121.0	2.1	181	181.0	3.2	241	241.0	4.2
2	2.0	0.0	62	62.0	1.1	22	122.0	2.1	82	182.0	3.2	42	242.0	4.2
3	3.0	0.1	63	63.0	1.1	23	123.0	2.1	83	183.0	3.2	43	243.0	4.2
4	4.0	0.1	64	64.0	1.1	24	124.0	2.2	84	184.0	3.2	44	244.0	4.3
5	5.0	0.1	65	65.0	1.1	25	125.0	2.2	85	185.0	3.2	45	245.0	4.3
6	6.0	0.1	66	66.0	1.2	26	126.0	2.2	86	186.0	3.2	46	246.0	4.3
7	7.0	0.1	67	67.0	1.2	27	127.0	2.2	87	187.0	3.3	47	247.0	4.3
8	8.0	0.1	68	68.0	1.2	28	128.0	2.2	88	188.0	3.3	48	248.0	4.3
9	9.0	0.2	69	69.0	1.2	29	129.0	2.3	89	189.0	3.3	49	249.0	4.3
10	10.0	0.2	70	70.0	1.2	30	130.0	2.3	90	190.0	3.3	50	250.0	4.4
11	11.0	0.2	71	71.0	1.2	131	131.0	2.3	191	191.0	3.3	251	251.0	4.4
12	12.0	0.2	72	72.0	1.3	32	132.0	2.3	92	192.0	3.4	52	252.0	4.4
13	13.0	0.2	73	73.0	1.3	33	133.0	2.3	93	193.0	3.4	53	253.0	4.4
14	14.0	0.2	74	74.0	1.3	34	134.0	2.3	94	194.0	3.4	54	254.0	4.4
15	15.0	0.3	75	75.0	1.3	35	135.0	2.4	95	195.0	3.4	55	255.0	4.5
16	16.0	0.3	76	76.0	1.3	36	136.0	2.4	96	196.0	3.4	56	256.0	4.5
17	17.0	0.3	77	77.0	1.3	37	137.0	2.4	97	197.0	3.4	57	257.0	4.5
18	18.0	0.3	78	78.0	1.4	38	138.0	2.4	98	198.0	3.5	58	258.0	4.5
19	19.0	0.3	79	79.0	1.4	39	139.0	2.4	99	199.0	3.5	59	259.0	4.5
20	20.0	0.3	80	80.0	1.4	40	140.0	2.4	200	200.0	3.5	60	260.0	4.5
21	21.0	0.4	81	81.0	1.4	141	141.0	2.5	201	201.0	3.5	261	261.0	4.6
22	22.0	0.4	82	82.0	1.4	42	142.0	2.5	02	202.0	3.5	62	262.0	4.6
23	23.0	0.4	83	83.0	1.4	43	143.0	2.5	03	203.0	3.5	63	263.0	4.6
24	24.0	0.4	84	84.0	1.5	44	144.0	2.5	04	204.0	3.6	64	264.0	4.6
25	25.0	0.4	85	85.0	1.5	45	145.0	2.5	05	205.0	3.6	65	265.0	4.6
26	26.0	0.5	86	86.0	1.5	46	146.0	2.5	06	206.0	3.6	66	266.0	4.6
27	27.0	0.5	87	87.0	1.5	47	147.0	2.6	07	207.0	3.6	67	267.0	4.7
28	28.0	0.5	88	88.0	1.5	48	148.0	2.6	08	208.0	3.6	68	268.0	4.7
29	29.0	0.5	89	89.0	1.6	49	149.0	2.6	09	209.0	3.6	69	269.0	4.7
30	30.0	0.5	90	90.0	1.6	50	150.0	2.6	10	210.0	3.7	70	270.0	4.7
31	31.0	0.5	91	91.0	1.6	151	151.0	2.6	211	211.0	3.7	271	271.0	4.7
32	32.0	0.6	92	92.0	1.6	52	152.0	2.7	12	212.0	3.7	72	272.0	4.7
33	33.0	0.6	93	93.0	1.6	53	153.0	2.7	13	213.0	3.7	73	273.0	4.8
34	34.0	0.6	94	94.0	1.6	54	154.0	2.7	14	214.0	3.7	74	274.0	4.8
35	35.0	0.6	95	95.0	1.7	55	155.0	2.7	15	215.0	3.8	75	275.0	4.8
36	36.0	0.6	96	96.0	1.7	56	156.0	2.7	16	216.0	3.8	76	276.0	4.8
37	37.0	0.6	97	97.0	1.7	57	157.0	2.7	17	217.0	3.8	77	277.0	4.8
38	38.0	0.7	98	98.0	1.7	58	158.0	2.8	18	218.0	3.8	78	278.0	4.9
39	39.0	0.7	99	99.0	1.7	59	159.0	2.8	19	219.0	3.8	79	279.0	4.9
40	40.0	0.7	100	100.0	1.7	60	160.0	2.8	20	220.0	3.8	80	280.0	4.9
41	41.0	0.7	101	101.0	1.8	161	161.0	2.8	221	221.0	3.9	281	281.0	4.9
42	42.0	0.7	02	102.0	1.8	62	162.0	2.8	22	222.0	3.9	82	282.0	4.9
43	43.0	0.8	03	103.0	1.8	63	163.0	2.8	23	223.0	3.9	83	283.0	4.9
44	44.0	0.8	04	104.0	1.8	64	164.0	2.9	24	224.0	3.9	84	284.0	5.0
45	45.0	0.8	05	105.0	1.8	65	165.0	2.9	25	225.0	3.9	85	285.0	5.0
46	46.0	0.8	06	106.0	1.8	66	166.0	2.9	26	226.0	3.9	86	286.0	5.0
47	47.0	0.8	07	107.0	1.9	67	167.0	2.9	27	227.0	4.0	87	287.0	5.0
48	48.0	0.8	08	108.0	1.9	68	168.0	2.9	28	228.0	4.0	88	288.0	5.0
49	49.0	0.9	09	109.0	1.9	69	169.0	2.9	29	229.0	4.0	89	289.0	5.0
50	50.0	0.9	10	110.0	1.9	70	170.0	3.0	30	230.0	4.0	90	290.0	5.1
51	51.0	0.9	111	111.0	1.9	171	171.0	3.0	231	231.0	4.0	291	291.0	5.1
52	52.0	0.9	12	112.0	2.0	72	172.0	3.0	32	232.0	4.0	92	292.0	5.1
53	53.0	0.9	13	113.0	2.0	73	173.0	3.0	33	233.0	4.1	93	293.0	5.1
54	54.0	0.9	14	114.0	2.0	74	174.0	3.0	34	234.0	4.1	94	294.0	5.1
55	55.0	1.0	15	115.0	2.0	75	175.0	3.1	35	235.0	4.1	95	295.0	5.1
56	56.0	1.0	16	116.0	2.0	76	176.0	3.1	36	236.0	4.1	96	296.0	5.2
57	57.0	1.0	17	117.0	2.0	77	177.0	3.1	37	237.0	4.1	97	297.0	5.2
58	58.0	1.0	18	118.0	2.1	78	178.0	3.1	38	238.0	4.2	98	298.0	5.2
59	59.0	1.0	19	119.0	2.1	79	179.0	3.1	39	239.0	4.2	99	299.0	5.2
60	60.0	1.0	20	120.0	2.1	80	180.0	3.1	40	240.0	4.2	300	300.0	5.2
Dist.	Dep.	D. Lat.	Dist.	Dep.	D. Lat.	Dist.	Dep.	D. Lat.	Dist.	Dep.	D. Lat.	Dist.	Dep.	D. Lat.

271° | 089°
269° | 091°

89°

Dist.	D. Lat.	Dep.
N.	N x Cos.	N x Sin.
Hypotenuse	Side Adj.	Side Opp.

TABLE 4

1°

Traverse Table

359° | 001°
181° | 179°

Dist.	D. Lat.	Dep.	Dist.	D. Lat.	Dep.	Dist.	D. Lat.	Dep.	Dist.	D. Lat.	Dep.	Dist.	D. Lat.	Dep.
301	301.0	5.3	361	360.9	6.3	421	420.9	7.3	481	480.9	8.4	541	540.9	9.4
02	302.0	5.3	62	361.9	6.3	22	421.9	7.4	82	481.9	8.4	42	541.9	9.5
03	303.0	5.3	63	362.9	6.3	23	422.9	7.4	83	482.9	8.4	43	542.9	9.5
04	304.0	5.3	64	363.9	6.4	24	423.9	7.4	84	483.9	8.4	44	543.9	9.5
05	305.0	5.3	65	364.9	6.4	25	424.9	7.4	85	484.9	8.5	45	544.9	9.5
06	306.0	5.3	66	365.9	6.4	26	425.9	7.4	86	485.9	8.5	46	545.9	9.5
07	307.0	5.4	67	366.9	6.4	27	426.9	7.5	87	486.9	8.5	47	546.9	9.5
08	308.0	5.4	68	367.9	6.4	28	427.9	7.5	88	487.9	8.5	48	547.9	9.6
09	309.0	5.4	69	368.9	6.4	29	428.9	7.5	89	488.9	8.5	49	548.9	9.6
10	310.0	5.4	70	369.9	6.5	30	429.9	7.5	90	489.9	8.6	50	549.9	9.6
311	311.0	5.4	371	370.9	6.5	431	430.9	7.5	491	490.9	8.6	551	550.9	9.6
12	312.0	5.4	72	371.9	6.5	32	431.9	7.5	92	491.9	8.6	52	551.9	9.6
13	313.0	5.5	73	372.9	6.5	33	432.9	7.6	93	492.9	8.6	53	552.9	9.7
14	314.0	5.5	74	373.9	6.5	34	433.9	7.6	94	493.9	8.6	54	553.9	9.7
15	315.0	5.5	75	374.9	6.5	35	434.9	7.6	95	494.9	8.6	55	554.9	9.7
16	316.0	5.5	76	375.9	6.6	36	435.9	7.6	96	495.9	8.7	56	555.9	9.7
17	317.0	5.5	77	376.9	6.6	37	436.9	7.6	97	496.9	8.7	57	556.9	9.7
18	318.0	5.5	78	377.9	6.6	38	437.9	7.6	98	497.9	8.7	58	557.9	9.7
19	319.0	5.6	79	378.9	6.6	39	438.9	7.7	99	498.9	8.7	59	558.9	9.8
20	320.0	5.6	80	379.9	6.6	40	439.9	7.7	500	499.9	8.7	60	559.9	9.8
321	321.0	5.6	381	380.9	6.6	441	440.9	7.7	501	500.9	8.7	561	560.9	9.8
22	322.0	5.6	82	381.9	6.7	42	441.9	7.7	02	501.9	8.8	62	561.9	9.8
23	323.0	5.6	83	382.9	6.7	43	442.9	7.7	03	502.9	8.8	63	562.9	9.8
24	324.0	5.7	84	383.9	6.7	44	443.9	7.7	04	503.9	8.8	64	563.9	9.8
25	325.0	5.7	85	384.9	6.7	45	444.9	7.8	05	504.9	8.8	65	564.9	9.9
26	326.0	5.7	86	385.9	6.7	46	445.9	7.8	06	505.9	8.8	66	565.9	9.9
27	327.0	5.7	87	386.9	6.8	47	446.9	7.8	07	506.9	8.8	67	566.9	9.9
28	328.0	5.7	88	387.9	6.8	48	447.9	7.8	08	507.9	8.9	68	567.9	9.9
29	328.9	5.7	89	388.9	6.8	49	448.9	7.8	09	508.9	8.9	69	568.9	9.9
30	329.9	5.8	90	389.9	6.8	50	449.9	7.9	10	509.9	8.9	70	569.9	9.9
331	330.9	5.8	391	390.9	6.8	451	450.9	7.9	511	510.9	8.9	571	570.9	10.0
32	331.9	5.8	92	391.9	6.8	52	451.9	7.9	12	511.9	8.9	72	571.9	10.0
33	332.9	5.8	93	392.9	6.9	53	452.9	7.9	13	512.9	9.0	73	572.9	10.0
34	333.9	5.8	94	393.9	6.9	54	453.9	7.9	14	513.9	9.0	74	573.9	10.0
35	334.9	5.8	95	394.9	6.9	55	454.9	7.9	15	514.9	9.0	75	574.9	10.0
36	335.9	5.9	96	395.9	6.9	56	455.9	8.0	16	515.9	9.0	76	575.9	10.1
37	336.9	5.9	97	396.9	6.9	57	456.9	8.0	17	516.9	9.0	77	576.9	10.1
38	337.9	5.9	98	397.9	6.9	58	457.9	8.0	18	517.9	9.0	78	577.9	10.1
39	338.9	5.9	99	398.9	7.0	59	458.9	8.0	19	518.9	9.1	79	578.9	10.1
40	339.9	5.9	400	399.9	7.0	60	459.9	8.0	20	519.9	9.1	80	579.9	10.1
341	340.9	6.0	401	400.9	7.0	461	460.9	8.0	521	520.9	9.1	581	580.9	10.1
42	341.9	6.0	02	401.9	7.0	62	461.9	8.1	22	521.9	9.1	82	581.9	10.2
43	342.9	6.0	03	402.9	7.0	63	462.9	8.1	23	522.9	9.1	83	582.9	10.2
44	343.9	6.0	04	403.9	7.1	64	463.9	8.1	24	523.9	9.1	84	583.9	10.2
45	344.9	6.0	05	404.9	7.1	65	464.9	8.1	25	524.9	9.2	85	584.9	10.2
46	345.9	6.0	06	405.9	7.1	66	465.9	8.1	26	525.9	9.2	86	585.9	10.2
47	346.9	6.1	07	406.9	7.1	67	466.9	8.2	27	526.9	9.2	87	586.9	10.2
48	347.9	6.1	08	407.9	7.1	68	467.9	8.2	28	527.9	9.2	88	587.9	10.3
49	348.9	6.1	09	408.9	7.1	69	468.9	8.2	29	528.9	9.2	89	588.9	10.3
50	349.9	6.1	10	409.9	7.2	70	469.9	8.2	30	529.9	9.2	90	589.9	10.3
351	350.9	6.1	411	410.9	7.2	471	470.9	8.2	531	530.9	9.3	591	590.9	10.3
52	351.9	6.1	12	411.9	7.2	72	471.9	8.2	32	531.9	9.3	92	591.9	10.3
53	352.9	6.2	13	412.9	7.2	73	472.9	8.3	33	532.9	9.3	93	592.9	10.3
54	353.9	6.2	14	413.9	7.2	74	473.9	8.3	34	533.9	9.3	94	593.9	10.4
55	354.9	6.2	15	414.9	7.2	75	474.9	8.3	35	534.9	9.3	95	594.9	10.4
56	355.9	6.2	16	415.9	7.3	76	475.9	8.3	36	535.9	9.4	96	595.9	10.4
57	356.9	6.2	17	416.9	7.3	77	476.9	8.3	37	536.9	9.4	97	596.9	10.4
58	357.9	6.2	18	417.9	7.3	78	477.9	8.3	38	537.9	9.4	98	597.9	10.4
59	358.9	6.3	19	418.9	7.3	79	478.9	8.4	39	538.9	9.4	99	598.9	10.5
60	359.9	6.3	20	419.9	7.3	80	479.9	8.4	40	539.9	9.4	600	599.9	10.5
Dist.	Dep.	D. Lat.	Dist.	Dep.	D. Lat.	Dist.	Dep.	D. Lat.	Dist.	Dep.	D. Lat.	Dist.	Dep.	D. Lat.

Dist.	D. Lat.	Dep.
D Lo	Dep.	
	m	D Lo

89°

271° | 089°
269° | 091°

TABLE 4
Traverse 2° Table

358° | 002°
182° | 178°

Dist.	D. Lat.	Dep.	Dist.	D. Lat.	Dep.	Dist.	D. Lat.	Dep.	Dist.	D. Lat.	Dep.	Dist.	D. Lat.	Dep.
1	1.0	0.0	61	61.0	2.1	121	120.9	4.2	181	180.9	6.3	241	240.9	8.4
2	2.0	0.1	62	62.0	2.2	22	121.9	4.3	82	181.9	6.4	42	241.9	8.4
3	3.0	0.1	63	63.0	2.2	23	122.9	4.3	83	182.9	6.4	43	242.9	8.5
4	4.0	0.1	64	64.0	2.2	24	123.9	4.3	84	183.9	6.4	44	243.9	8.5
5	5.0	0.2	65	65.0	2.3	25	124.9	4.4	85	184.9	6.5	45	244.9	8.6
6	6.0	0.2	66	66.0	2.3	26	125.9	4.4	86	185.9	6.5	46	245.9	8.6
7	7.0	0.2	67	67.0	2.3	27	126.9	4.4	87	186.9	6.5	47	246.8	8.6
8	8.0	0.3	68	68.0	2.4	28	127.9	4.5	88	187.9	6.6	48	247.8	8.7
9	9.0	0.3	69	69.0	2.4	29	128.9	4.5	89	188.9	6.6	49	248.8	8.7
10	10.0	0.3	70	70.0	2.4	30	129.9	4.5	90	189.9	6.6	50	249.8	8.7
11	11.0	0.4	71	71.0	2.5	131	130.9	4.6	191	190.9	6.7	251	250.8	8.8
12	12.0	0.4	72	72.0	2.5	32	131.9	4.6	92	191.9	6.7	52	251.8	8.8
13	13.0	0.5	73	73.0	2.5	33	132.9	4.6	93	192.9	6.7	53	252.8	8.8
14	14.0	0.5	74	74.0	2.6	34	133.9	4.7	94	193.9	6.8	54	253.8	8.9
15	15.0	0.5	75	75.0	2.6	35	134.9	4.7	95	194.9	6.8	55	254.8	8.9
16	16.0	0.6	76	76.0	2.7	36	135.9	4.7	96	195.9	6.8	56	255.8	8.9
17	17.0	0.6	77	77.0	2.7	37	136.9	4.8	97	196.9	6.9	57	256.8	9.0
18	18.0	0.6	78	78.0	2.7	38	137.9	4.8	98	197.9	6.9	58	257.8	9.0
19	19.0	0.7	79	79.0	2.8	39	138.9	4.9	99	198.9	6.9	59	258.8	9.0
20	20.0	0.7	80	80.0	2.8	40	139.9	4.9	200	199.9	7.0	60	259.8	9.1
21	21.0	0.7	81	81.0	2.8	141	140.9	4.9	201	200.9	7.0	261	260.8	9.1
22	22.0	0.8	82	82.0	2.9	42	141.9	5.0	02	201.9	7.0	62	261.8	9.1
23	23.0	0.8	83	82.9	2.9	43	142.9	5.0	03	202.9	7.1	63	262.8	9.2
24	24.0	0.8	84	83.9	2.9	44	143.9	5.0	04	203.9	7.1	64	263.8	9.2
25	25.0	0.9	85	84.9	3.0	45	144.9	5.1	05	204.9	7.2	65	264.8	9.2
26	26.0	0.9	86	85.9	3.0	46	145.9	5.1	06	205.9	7.2	66	265.8	9.3
27	27.0	0.9	87	86.9	3.0	47	146.9	5.1	07	206.9	7.2	67	266.8	9.3
28	28.0	1.0	88	87.9	3.1	48	147.9	5.2	08	207.9	7.3	68	267.8	9.4
29	29.0	1.0	89	88.9	3.1	49	148.9	5.2	09	208.9	7.3	69	268.8	9.4
30	30.0	1.0	90	89.9	3.1	50	149.9	5.2	10	209.9	7.3	70	269.8	9.4
31	31.0	1.1	91	90.9	3.2	151	150.9	5.3	211	210.9	7.4	271	270.8	9.5
32	32.0	1.1	92	91.9	3.2	52	151.9	5.3	12	211.9	7.4	72	271.8	9.5
33	33.0	1.2	93	92.9	3.2	53	152.9	5.3	13	212.9	7.4	73	272.8	9.5
34	34.0	1.2	94	93.9	3.3	54	153.9	5.4	14	213.9	7.5	74	273.8	9.6
35	35.0	1.2	95	94.9	3.3	55	154.9	5.4	15	214.9	7.5	75	274.8	9.6
36	36.0	1.3	96	95.9	3.4	56	155.9	5.4	16	215.9	7.5	76	275.8	9.6
37	37.0	1.3	97	96.9	3.4	57	156.9	5.5	17	216.9	7.6	77	276.8	9.7
38	38.0	1.3	98	97.9	3.4	58	157.9	5.5	18	217.9	7.6	78	277.8	9.7
39	39.0	1.4	99	98.9	3.5	59	158.9	5.5	19	218.9	7.6	79	278.8	9.7
40	40.0	1.4	100	99.9	3.5	60	159.9	5.6	20	219.9	7.7	80	279.8	9.8
41	41.0	1.4	101	100.9	3.5	161	160.9	5.6	221	220.9	7.7	281	280.8	9.8
42	42.0	1.5	02	101.9	3.6	62	161.9	5.7	22	221.9	7.7	82	281.8	9.8
43	43.0	1.5	03	102.9	3.6	63	162.9	5.7	23	222.9	7.8	83	282.8	9.9
44	44.0	1.5	04	103.9	3.6	64	163.9	5.7	24	223.9	7.8	84	283.8	9.9
45	45.0	1.6	05	104.9	3.7	65	164.9	5.8	25	224.9	7.9	85	284.8	9.9
46	46.0	1.6	06	105.9	3.7	66	165.9	5.8	26	225.9	7.9	86	285.8	10.0
47	47.0	1.6	07	106.9	3.7	67	166.9	5.8	27	226.9	7.9	87	286.8	10.0
48	48.0	1.7	08	107.9	3.8	68	167.9	5.9	28	227.9	8.0	88	287.8	10.1
49	49.0	1.7	09	108.9	3.8	69	168.9	5.9	29	228.9	8.0	89	288.8	10.1
50	50.0	1.7	10	109.9	3.8	70	169.9	5.9	30	229.9	8.0	90	289.8	10.1
51	51.0	1.8	111	110.9	3.9	171	170.9	6.0	231	230.9	8.1	291	290.8	10.2
52	52.0	1.8	12	111.9	3.9	72	171.9	6.0	32	231.9	8.1	92	291.8	10.2
53	53.0	1.8	13	112.9	3.9	73	172.9	6.0	33	232.9	8.1	93	292.8	10.2
54	54.0	1.9	14	113.9	4.0	74	173.9	6.1	34	233.9	8.2	94	293.8	10.3
55	55.0	1.9	15	114.9	4.0	75	174.9	6.1	35	234.9	8.2	95	294.8	10.3
56	56.0	2.0	16	115.9	4.0	76	175.9	6.1	36	235.9	8.2	96	295.8	10.3
57	57.0	2.0	17	116.9	4.1	77	176.9	6.2	37	236.9	8.3	97	296.8	10.4
58	58.0	2.0	18	117.9	4.1	78	177.9	6.2	38	237.9	8.3	98	297.8	10.4
59	59.0	2.1	19	118.9	4.2	79	178.9	6.2	39	238.9	8.3	99	298.8	10.4
60	60.0	2.1	20	119.9	4.2	80	179.9	6.3	40	239.9	8.4	300	299.8	10.5
Dist.	Dep.	D. Lat.	Dist.	Dep.	D. Lat.	Dist.	Dep.	D. Lat.	Dist.	Dep.	D. Lat.	Dist.	Dep.	D. Lat.

272° | 088°
268° | 092°

88°

Dist.	D. Lat.	Dep.
N.	N x Cos.	N x Sin.
Hypotenuse	Side Adj.	Side Opp.

TABLE 4
Traverse 2° Table

358° | 002°
182° | 178°

Dist.	D. Lat.	Dep.	Dist.	D. Lat.	Dep.	Dist.	D. Lat.	Dep.	Dist.	D. Lat.	Dep.	Dist.	D. Lat.	Dep.
301	300.8	10.5	361	360.8	12.6	421	420.7	14.7	481	480.7	16.8	541	540.7	18.9
02	301.8	10.5	62	361.8	12.6	22	421.7	14.7	82	481.7	16.8	42	541.7	18.9
03	302.8	10.6	63	362.8	12.7	23	422.7	14.8	83	482.7	16.9	43	542.7	19.0
04	303.8	10.6	64	363.8	12.7	24	423.7	14.8	84	483.7	16.9	44	543.7	19.0
05	304.8	10.6	65	364.8	12.7	25	424.7	14.8	85	484.7	16.9	45	544.7	19.0
06	305.8	10.7	66	365.8	12.8	26	425.7	14.9	86	485.7	17.0	46	545.7	19.1
07	306.8	10.7	67	366.8	12.8	27	426.7	14.9	87	486.7	17.0	47	546.7	19.1
08	307.8	10.7	68	367.8	12.8	28	427.7	14.9	88	487.7	17.0	48	547.7	19.1
09	308.8	10.8	69	368.8	12.9	29	428.7	15.0	89	488.7	17.1	49	548.7	19.2
10	309.8	10.8	70	369.8	12.9	30	429.7	15.0	90	489.7	17.1	50	549.7	19.2
311	310.8	10.9	371	370.8	12.9	431	430.7	15.0	491	490.7	17.1	551	550.7	19.2
12	311.8	10.9	72	371.8	13.0	32	431.7	15.1	92	491.7	17.2	52	551.7	19.3
13	312.8	10.9	73	372.8	13.0	33	432.7	15.1	93	492.7	17.2	53	552.7	19.3
14	313.8	11.0	74	373.8	13.1	34	433.7	15.1	94	493.7	17.2	54	553.7	19.3
15	314.8	11.0	75	374.8	13.1	35	434.7	15.2	95	494.7	17.3	55	554.7	19.4
16	315.8	11.0	76	375.8	13.1	36	435.7	15.2	96	495.7	17.3	56	555.7	19.4
17	316.8	11.1	77	376.8	13.2	37	436.7	15.3	97	496.7	17.3	57	556.7	19.4
18	317.8	11.1	78	377.8	13.2	38	437.7	15.3	98	497.7	17.4	58	557.7	19.5
19	318.8	11.1	79	378.8	13.2	39	438.7	15.3	99	498.7	17.4	59	558.7	19.5
20	319.8	11.2	80	379.8	13.3	40	439.7	15.4	500	499.7	17.4	60	559.7	19.5
321	320.8	11.2	381	380.8	13.3	441	440.7	15.4	501	500.7	17.5	561	560.7	19.6
22	321.8	11.2	82	381.8	13.3	42	441.7	15.4	02	501.7	17.5	62	561.7	19.6
23	322.8	11.3	83	382.8	13.4	43	442.7	15.5	03	502.7	17.6	63	562.7	19.6
24	323.8	11.3	84	383.8	13.4	44	443.7	15.5	04	503.7	17.6	64	563.7	19.7
25	324.8	11.3	85	384.8	13.4	45	444.7	15.5	05	504.7	17.6	65	564.7	19.7
26	325.8	11.4	86	385.8	13.5	46	445.7	15.6	06	505.7	17.7	66	565.7	19.8
27	326.8	11.4	87	386.8	13.5	47	446.7	15.6	07	506.7	17.7	67	566.7	19.8
28	327.8	11.4	88	387.8	13.5	48	447.7	15.6	08	507.7	17.7	68	567.7	19.8
29	328.8	11.5	89	388.8	13.6	49	448.7	15.7	09	508.7	17.8	69	568.7	19.9
30	329.8	11.5	90	389.8	13.6	50	449.7	15.7	10	509.7	17.8	70	569.7	19.9
331	330.8	11.6	391	390.8	13.6	451	450.7	15.7	511	510.7	17.8	571	570.7	19.9
32	331.8	11.6	92	391.8	13.7	52	451.7	15.8	12	511.7	17.9	72	571.7	20.0
33	332.8	11.6	93	392.8	13.7	53	452.7	15.8	13	512.7	17.9	73	572.7	20.0
34	333.8	11.7	94	393.8	13.8	54	453.7	15.8	14	513.7	17.9	74	573.7	20.0
35	334.8	11.7	95	394.8	13.8	55	454.7	15.9	15	514.7	18.0	75	574.6	20.1
36	335.8	11.7	96	395.8	13.8	56	455.7	15.9	16	515.7	18.0	76	575.6	20.1
37	336.8	11.8	97	396.8	13.9	57	456.7	15.9	17	516.7	18.0	77	576.6	20.1
38	337.8	11.8	98	397.8	13.9	58	457.7	16.0	18	517.7	18.1	78	577.6	20.2
39	338.8	11.8	99	398.8	13.9	59	458.7	16.0	19	518.7	18.1	79	578.6	20.2
40	339.8	11.9	400	399.8	14.0	60	459.7	16.1	20	519.7	18.1	80	579.6	20.2
341	340.8	11.9	401	400.8	14.0	461	460.7	16.1	521	520.7	18.2	581	580.6	20.3
42	341.8	11.9	02	401.8	14.0	62	461.7	16.1	22	521.7	18.2	82	581.6	20.3
43	342.8	12.0	03	402.8	14.1	63	462.7	16.2	23	522.7	18.3	83	582.6	20.3
44	343.8	12.0	04	403.8	14.1	64	463.7	16.2	24	523.7	18.3	84	583.6	20.4
45	344.8	12.0	05	404.8	14.1	65	464.7	16.2	25	524.7	18.3	85	584.6	20.4
46	345.8	12.1	06	405.8	14.2	66	465.7	16.3	26	525.7	18.4	86	585.6	20.5
47	346.8	12.1	07	406.8	14.2	67	466.7	16.3	27	526.7	18.4	87	586.6	20.5
48	347.8	12.1	08	407.8	14.2	68	467.7	16.3	28	527.7	18.4	88	587.6	20.5
49	348.8	12.2	09	408.8	14.3	69	468.7	16.4	29	528.7	18.5	89	588.6	20.6
50	349.8	12.2	10	409.8	14.3	70	469.7	16.4	30	529.7	18.5	90	589.6	20.6
351	350.8	12.2	411	410.7	14.3	471	470.7	16.4	531	530.7	18.5	591	590.6	20.6
52	351.8	12.3	12	411.7	14.4	72	471.7	16.5	32	531.7	18.6	92	591.6	20.7
53	352.8	12.3	13	412.7	14.4	73	472.7	16.5	33	532.7	18.6	93	592.6	20.7
54	353.8	12.4	14	413.7	14.4	74	473.7	16.5	34	533.7	18.6	94	593.6	20.7
55	354.8	12.4	15	414.7	14.5	75	474.7	16.6	35	534.7	18.7	95	594.6	20.8
56	355.8	12.4	16	415.7	14.5	76	475.7	16.6	36	535.7	18.7	96	595.6	20.8
57	356.8	12.5	17	416.7	14.6	77	476.7	16.6	37	536.7	18.7	97	596.6	20.8
58	357.8	12.5	18	417.7	14.6	78	477.7	16.7	38	537.7	18.8	98	597.6	20.9
59	358.8	12.5	19	418.7	14.6	79	478.7	16.7	39	538.7	18.8	99	598.6	20.9
60	359.8	12.6	20	419.7	14.7	80	479.7	16.8	40	539.7	18.8	600	599.6	20.9
Dist.	Dep.	D. Lat.	Dist.	Dep.	D. Lat.	Dist.	Dep.	D. Lat.	Dist.	Dep.	D. Lat.	Dist.	Dep.	D. Lat.

Dist.	D. Lat.	Dep.
D Lo	Dep.	
	m	D Lo

88°

272° | 088°
268° | 092°

TABLE 4

Traverse **3°** Table

357° / 183°	003° / 177°

Dist.	D. Lat.	Dep.	Dist.	D. Lat.	Dep.	Dist.	D. Lat.	Dep.	Dist.	D. Lat.	Dep.	Dist.	D. Lat.	Dep.
1	1.0	0.1	61	60.9	3.2	121	120.8	6.3	181	180.8	9.5	241	240.7	12.6
2	2.0	0.1	62	61.9	3.2	22	121.8	6.4	82	181.8	9.5	42	241.7	12.7
3	3.0	0.2	63	62.9	3.3	23	122.8	6.4	83	182.7	9.6	43	242.7	12.7
4	4.0	0.2	64	63.9	3.3	24	123.8	6.5	84	183.7	9.6	44	243.7	12.8
5	5.0	0.3	65	64.9	3.4	25	124.8	6.5	85	184.7	9.7	45	244.7	12.8
6	6.0	0.3	66	65.9	3.5	26	125.8	6.6	86	185.7	9.7	46	245.7	12.9
7	7.0	0.4	67	66.9	3.5	27	126.8	6.6	87	186.7	9.8	47	246.7	12.9
8	8.0	0.4	68	67.9	3.6	28	127.8	6.7	88	187.7	9.8	48	247.7	13.0
9	9.0	0.5	69	68.9	3.6	29	128.8	6.8	89	188.7	9.9	49	248.7	13.0
10	10.0	0.5	70	69.9	3.7	30	129.8	6.8	90	189.7	9.9	50	249.7	13.1
11	11.0	0.6	71	70.9	3.7	131	130.8	6.9	191	190.7	10.0	251	250.7	13.1
12	12.0	0.6	72	71.9	3.8	32	131.8	6.9	92	191.7	10.0	52	251.7	13.2
13	13.0	0.7	73	72.9	3.8	33	132.8	7.0	93	192.7	10.1	53	252.7	13.2
14	14.0	0.7	74	73.9	3.9	34	133.8	7.0	94	193.7	10.2	54	253.7	13.3
15	15.0	0.8	75	74.9	3.9	35	134.8	7.1	95	194.7	10.2	55	254.7	13.3
16	16.0	0.8	76	75.9	4.0	36	135.8	7.1	96	195.7	10.3	56	255.6	13.4
17	17.0	0.9	77	76.9	4.0	37	136.8	7.2	97	196.7	10.3	57	256.6	13.5
18	18.0	0.9	78	77.9	4.1	38	137.8	7.2	98	197.7	10.4	58	257.6	13.5
19	19.0	1.0	79	78.9	4.1	39	138.8	7.3	99	198.7	10.4	59	258.6	13.6
20	20.0	1.0	80	79.9	4.2	40	139.8	7.3	200	199.7	10.5	60	259.6	13.6
21	21.0	1.1	81	80.9	4.2	141	140.8	7.4	201	200.7	10.5	261	260.6	13.7
22	22.0	1.2	82	81.9	4.3	42	141.8	7.4	02	201.7	10.6	62	261.6	13.7
23	23.0	1.2	83	82.9	4.3	43	142.8	7.5	03	202.7	10.6	63	262.6	13.8
24	24.0	1.3	84	83.9	4.4	44	143.8	7.5	04	203.7	10.7	64	263.6	13.8
25	25.0	1.3	85	84.9	4.4	45	144.8	7.6	05	204.7	10.7	65	264.6	13.9
26	26.0	1.4	86	85.9	4.5	46	145.8	7.6	06	205.7	10.8	66	265.6	13.9
27	27.0	1.4	87	86.9	4.6	47	146.8	7.7	07	206.7	10.8	67	266.6	14.0
28	28.0	1.5	88	87.9	4.6	48	147.8	7.7	08	207.7	10.9	68	267.6	14.0
29	29.0	1.5	89	88.9	4.7	49	148.8	7.8	09	208.7	10.9	69	268.6	14.1
30	30.0	1.6	90	89.9	4.7	50	149.8	7.9	10	209.7	11.0	70	269.6	14.1
31	31.0	1.6	91	90.9	4.8	151	150.8	7.9	211	210.7	11.0	271	270.6	14.2
32	32.0	1.7	92	91.9	4.8	52	151.8	8.0	12	211.7	11.1	72	271.6	14.2
33	33.0	1.7	93	92.9	4.9	53	152.8	8.0	13	212.7	11.1	73	272.6	14.3
34	34.0	1.8	94	93.9	4.9	54	153.8	8.1	14	213.7	11.2	74	273.6	14.3
35	35.0	1.8	95	94.9	5.0	55	154.8	8.1	15	214.7	11.3	75	274.6	14.4
36	36.0	1.9	96	95.9	5.0	56	155.8	8.2	16	215.7	11.3	76	275.6	14.4
37	36.9	1.9	97	96.9	5.1	57	156.8	8.2	17	216.7	11.4	77	276.6	14.5
38	37.9	2.0	98	97.9	5.1	58	157.8	8.3	18	217.7	11.4	78	277.6	14.5
39	38.9	2.0	99	98.9	5.2	59	158.8	8.3	19	218.7	11.5	79	278.6	14.6
40	39.9	2.1	100	99.9	5.2	60	159.8	8.4	20	219.7	11.5	80	279.6	14.7
41	40.9	2.1	101	100.9	5.3	161	160.8	8.4	221	220.7	11.6	281	280.6	14.7
42	41.9	2.2	02	101.9	5.3	62	161.8	8.5	22	221.7	11.6	82	281.6	14.8
43	42.9	2.3	03	102.9	5.4	63	162.8	8.5	23	222.7	11.7	83	282.6	14.8
44	43.9	2.3	04	103.9	5.4	64	163.8	8.6	24	223.7	11.7	84	283.6	14.9
45	44.9	2.4	05	104.9	5.5	65	164.8	8.6	25	224.7	11.8	85	284.6	14.9
46	45.9	2.4	06	105.9	5.5	66	165.8	8.7	26	225.7	11.8	86	285.6	15.0
47	46.9	2.5	07	106.9	5.6	67	166.8	8.7	27	226.7	11.9	87	286.6	15.0
48	47.9	2.5	08	107.9	5.7	68	167.8	8.8	28	227.7	11.9	88	287.6	15.1
49	48.9	2.6	09	108.9	5.7	69	168.8	8.8	29	228.7	12.0	89	288.6	15.1
50	49.9	2.6	10	109.8	5.8	70	169.8	8.9	30	229.7	12.0	90	289.6	15.2
51	50.9	2.7	111	110.8	5.8	171	170.8	8.9	231	230.7	12.1	291	290.6	15.2
52	51.9	2.7	12	111.8	5.9	72	171.8	9.0	32	231.7	12.1	92	291.6	15.3
53	52.9	2.8	13	112.8	5.9	73	172.8	9.1	33	232.7	12.2	93	292.6	15.3
54	53.9	2.8	14	113.8	6.0	74	173.8	9.1	34	233.7	12.2	94	293.6	15.4
55	54.9	2.9	15	114.8	6.0	75	174.8	9.2	35	234.7	12.3	95	294.6	15.4
56	55.9	2.9	16	115.8	6.1	76	175.8	9.2	36	235.7	12.4	96	295.6	15.5
57	56.9	3.0	17	116.8	6.1	77	176.8	9.3	37	236.7	12.4	97	296.6	15.5
58	57.9	3.0	18	117.8	6.2	78	177.8	9.3	38	237.7	12.5	98	297.6	15.6
59	58.9	3.1	19	118.8	6.2	79	178.8	9.4	39	238.7	12.5	99	298.6	15.6
60	59.9	3.1	20	119.8	6.3	80	179.8	9.4	40	239.7	12.6	300	299.6	15.7
Dist.	Dep.	D. Lat.	Dist.	Dep.	D. Lat.	Dist.	Dep.	D. Lat.	Dist.	Dep.	D. Lat.	Dist.	Dep.	D. Lat.

273° / 267°	087° / 093°

87°

Dist.	D. Lat.	Dep.
N.	N x Cos.	N x Sin.
Hypotenuse	Side Adj.	Side Opp.

TABLE 4

Traverse **3°** Table

357° / 183°	003° / 177°

Dist.	D. Lat.	Dep.	Dist.	D. Lat.	Dep.	Dist.	D. Lat.	Dep.	Dist.	D. Lat.	Dep.	Dist.	D. Lat.	Dep.
301	300.6	15.8	361	360.5	18.9	421	420.4	22.0	481	480.3	25.2	541	540.3	28.3
02	301.6	15.8	62	361.5	18.9	22	421.4	22.1	82	481.3	25.2	42	541.3	28.4
03	302.6	15.9	63	362.5	19.0	23	422.4	22.1	83	482.3	25.3	43	542.3	28.4
04	303.6	15.9	64	363.5	19.1	24	423.4	22.2	84	483.3	25.3	44	543.3	28.5
05	304.6	16.0	65	364.5	19.1	25	424.4	22.2	85	484.3	25.4	45	544.3	28.5
06	305.6	16.0	66	365.5	19.2	26	425.4	22.3	86	485.3	25.4	46	545.3	28.6
07	306.6	16.1	67	366.5	19.2	27	426.4	22.3	87	486.3	25.5	47	546.3	28.6
08	307.6	16.1	68	367.5	19.3	28	427.4	22.4	88	487.3	25.5	48	547.2	28.7
09	308.6	16.2	69	368.5	19.3	29	428.4	22.5	89	488.3	25.6	49	548.2	28.7
10	309.6	16.2	70	369.5	19.4	30	429.4	22.5	90	489.3	25.6	50	549.2	28.8
311	310.6	16.3	371	370.5	19.4	431	430.4	22.6	491	490.3	25.7	551	550.2	28.8
12	311.6	16.3	72	371.5	19.5	32	431.4	22.6	92	491.3	25.7	52	551.2	28.9
13	312.6	16.4	73	372.5	19.5	33	432.4	22.7	93	492.3	25.8	53	552.2	28.9
14	313.6	16.4	74	373.5	19.6	34	433.4	22.7	94	493.3	25.9	54	553.2	29.0
15	314.6	16.5	75	374.5	19.6	35	434.4	22.8	95	494.3	25.9	55	554.2	29.0
16	315.6	16.5	76	375.5	19.7	36	435.4	22.8	96	495.3	26.0	56	555.2	29.1
17	316.6	16.6	77	376.5	19.7	37	436.4	22.9	97	496.3	26.0	57	556.2	29.2
18	317.6	16.6	78	377.5	19.8	38	437.4	22.9	98	497.3	26.1	58	557.2	29.2
19	318.6	16.7	79	378.5	19.8	39	438.4	23.0	99	498.3	26.1	59	558.2	29.3
20	319.6	16.7	80	379.5	19.9	40	439.4	23.0	500	499.3	26.2	60	559.2	29.3
321	320.6	16.8	381	380.5	19.9	441	440.4	23.1	501	500.3	26.2	561	560.2	29.4
22	321.6	16.9	82	381.5	20.0	42	441.4	23.1	02	501.3	26.3	62	561.2	29.4
23	322.6	16.9	83	382.5	20.0	43	442.4	23.2	03	502.3	26.3	63	562.2	29.5
24	323.6	17.0	84	383.5	20.1	44	443.4	23.2	04	503.3	26.4	64	563.2	29.5
25	324.6	17.0	85	384.5	20.1	45	444.4	23.3	05	504.3	26.4	65	564.2	29.6
26	325.6	17.1	86	385.5	20.2	46	445.4	23.3	06	505.3	26.5	66	565.2	29.6
27	326.6	17.1	87	386.5	20.3	47	446.4	23.4	07	506.3	26.5	67	566.2	29.7
28	327.6	17.2	88	387.5	20.3	48	447.4	23.4	08	507.3	26.6	68	567.2	29.7
29	328.5	17.2	89	388.5	20.4	49	448.4	23.5	09	508.3	26.6	69	568.2	29.8
30	329.5	17.3	90	389.5	20.4	50	449.4	23.6	10	509.3	26.7	70	569.2	29.8
331	330.5	17.3	391	390.5	20.5	451	450.4	23.6	511	510.3	26.7	571	570.2	29.9
32	331.5	17.4	92	391.5	20.5	52	451.4	23.7	12	511.3	26.8	72	571.2	29.9
33	332.5	17.4	93	392.5	20.6	53	452.4	23.7	13	512.3	26.8	73	572.2	30.0
34	333.5	17.5	94	393.5	20.6	54	453.4	23.8	14	513.3	26.9	74	573.2	30.0
35	334.5	17.5	95	394.5	20.7	55	454.4	23.8	15	514.3	27.0	75	574.2	30.1
36	335.5	17.6	96	395.5	20.7	56	455.4	23.9	16	515.3	27.0	76	575.2	30.1
37	336.5	17.6	97	396.5	20.8	57	456.4	23.9	17	516.3	27.1	77	576.2	30.2
38	337.5	17.7	98	397.5	20.8	58	457.4	24.0	18	517.3	27.1	78	577.2	30.3
39	338.5	17.7	99	398.5	20.9	59	458.4	24.0	19	518.3	27.2	79	578.2	30.3
40	339.5	17.8	400	399.5	20.9	60	459.4	24.1	20	519.3	27.2	80	579.2	30.4
341	340.5	17.8	401	400.5	21.0	461	460.4	24.1	521	520.3	27.3	581	580.2	30.4
42	341.5	17.9	02	401.4	21.0	62	461.4	24.2	22	521.3	27.3	82	581.2	30.5
43	342.5	18.0	03	402.4	21.1	63	462.4	24.2	23	522.3	27.4	83	582.2	30.5
44	343.5	18.0	04	403.4	21.1	64	463.4	24.3	24	523.3	27.4	84	583.2	30.6
45	344.5	18.1	05	404.4	21.2	65	464.4	24.3	25	524.3	27.5	85	584.2	30.6
46	345.5	18.1	06	405.4	21.2	66	465.4	24.4	26	525.3	27.5	86	585.2	30.7
47	346.5	18.2	07	406.4	21.3	67	466.4	24.4	27	526.3	27.6	87	586.2	30.7
48	347.5	18.2	08	407.4	21.4	68	467.4	24.5	28	527.3	27.6	88	587.2	30.8
49	348.5	18.3	09	408.4	21.4	69	468.4	24.5	29	528.3	27.7	89	588.2	30.8
50	349.5	18.3	10	409.4	21.5	70	469.4	24.6	30	529.3	27.7	90	589.2	30.9
351	350.5	18.4	411	410.4	21.5	471	470.4	24.7	531	530.3	27.8	591	590.2	30.9
52	351.5	18.4	12	411.4	21.6	72	471.4	24.7	32	531.3	27.8	92	591.2	31.0
53	352.5	18.5	13	412.4	21.6	73	472.4	24.8	33	532.3	27.9	93	592.2	31.0
54	353.5	18.5	14	413.4	21.7	74	473.4	24.8	34	533.3	27.9	94	593.2	31.1
55	354.5	18.6	15	414.4	21.7	75	474.3	24.9	35	534.3	28.0	95	594.2	31.1
56	355.5	18.6	16	415.4	21.8	76	475.3	24.9	36	535.3	28.1	96	595.2	31.2
57	356.5	18.7	17	416.4	21.8	77	476.3	25.0	37	536.3	28.1	97	596.2	31.2
58	357.5	18.7	18	417.4	21.9	78	477.3	25.0	38	537.3	28.2	98	597.2	31.3
59	358.5	18.8	19	418.4	21.9	79	478.3	25.1	39	538.3	28.2	99	598.2	31.3
60	359.5	18.8	20	419.4	22.0	80	479.3	25.1	40	539.3	28.3	600	599.2	31.4
Dist.	Dep.	D. Lat.	Dist.	Dep.	D. Lat.	Dist.	Dep.	D. Lat.	Dist.	Dep.	D. Lat.	Dist.	Dep.	D. Lat.

Dist.	D. Lat.	Dep.
D Lo	Dep.	
	m	D Lo

87°

273° / 267°	087° / 093°

TABLE 4

4° Traverse Table

356° | 004° — 184° | 176°

Dist.	D. Lat.	Dep.	Dist.	D. Lat.	Dep.	Dist.	D. Lat.	Dep.	Dist.	D. Lat.	Dep.	Dist.	D. Lat.	Dep.
1	1.0	0.1	61	60.9	4.3	121	120.7	8.4	181	180.6	12.6	241	240.4	16.8
2	2.0	0.1	62	61.8	4.3	22	121.7	8.5	82	181.6	12.7	42	241.4	16.9
3	3.0	0.2	63	62.8	4.4	23	122.7	8.6	83	182.6	12.8	43	242.4	17.0
4	4.0	0.3	64	63.8	4.5	24	123.7	8.6	84	183.6	12.8	44	243.4	17.0
5	5.0	0.3	65	64.8	4.5	25	124.7	8.7	85	184.5	12.9	45	244.4	17.1
6	6.0	0.4	66	65.8	4.6	26	125.7	8.8	86	185.5	13.0	46	245.4	17.2
7	7.0	0.5	67	66.8	4.7	27	126.7	8.9	87	186.5	13.0	47	246.4	17.2
8	8.0	0.6	68	67.8	4.7	28	127.7	8.9	88	187.5	13.1	48	247.4	17.3
9	9.0	0.6	69	68.8	4.8	29	128.7	9.0	89	188.5	13.2	49	248.4	17.4
10	10.0	0.7	70	69.8	4.9	30	129.7	9.1	90	189.5	13.3	50	249.4	17.4
11	11.0	0.8	71	70.8	5.0	131	130.7	9.1	191	190.5	13.3	251	250.4	17.5
12	12.0	0.8	72	71.8	5.0	32	131.7	9.2	92	191.5	13.4	52	251.4	17.6
13	13.0	0.9	73	72.8	5.1	33	132.7	9.3	93	192.5	13.5	53	252.4	17.6
14	14.0	1.0	74	73.8	5.2	34	133.7	9.3	94	193.5	13.5	54	253.4	17.7
15	15.0	1.0	75	74.8	5.2	35	134.7	9.4	95	194.5	13.6	55	254.4	17.8
16	16.0	1.1	76	75.8	5.3	36	135.7	9.5	96	195.5	13.7	56	255.4	17.9
17	17.0	1.2	77	76.8	5.4	37	136.7	9.6	97	196.5	13.7	57	256.4	17.9
18	18.0	1.3	78	77.8	5.4	38	137.7	9.6	98	197.5	13.8	58	257.4	18.0
19	19.0	1.3	79	78.8	5.5	39	138.7	9.7	99	198.5	13.9	59	258.4	18.1
20	20.0	1.4	80	79.8	5.6	40	139.7	9.8	200	199.5	14.0	60	259.4	18.1
21	20.9	1.5	81	80.8	5.7	141	140.7	9.8	201	200.5	14.0	261	260.4	18.2
22	21.9	1.5	82	81.8	5.7	42	141.7	9.9	02	201.5	14.1	62	261.4	18.3
23	22.9	1.6	83	82.8	5.8	43	142.7	10.0	03	202.5	14.2	63	262.4	18.3
24	23.9	1.7	84	83.8	5.9	44	143.6	10.0	04	203.5	14.2	64	263.4	18.4
25	24.9	1.7	85	84.8	5.9	45	144.6	10.1	05	204.5	14.3	65	264.4	18.5
26	25.9	1.8	86	85.8	6.0	46	145.6	10.2	06	205.5	14.4	66	265.4	18.6
27	26.9	1.9	87	86.8	6.1	47	146.6	10.3	07	206.5	14.4	67	266.3	18.6
28	27.9	2.0	88	87.8	6.1	48	147.6	10.3	08	207.5	14.5	68	267.3	18.7
29	28.9	2.0	89	88.8	6.2	49	148.6	10.4	09	208.5	14.6	69	268.3	18.8
30	29.9	2.1	90	89.8	6.3	50	149.6	10.5	10	209.5	14.6	70	269.3	18.8
31	30.9	2.2	91	90.8	6.3	151	150.6	10.5	211	210.5	14.7	271	270.3	18.9
32	31.9	2.2	92	91.8	6.4	52	151.6	10.6	12	211.5	14.8	72	271.3	19.0
33	32.9	2.3	93	92.8	6.5	53	152.6	10.7	13	212.5	14.9	73	272.3	19.0
34	33.9	2.4	94	93.8	6.6	54	153.6	10.7	14	213.5	14.9	74	273.3	19.1
35	34.9	2.4	95	94.8	6.6	55	154.6	10.8	15	214.5	15.0	75	274.3	19.2
36	35.9	2.5	96	95.8	6.7	56	155.6	10.9	16	215.5	15.1	76	275.3	19.3
37	36.9	2.6	97	96.8	6.8	57	156.6	11.0	17	216.5	15.1	77	276.3	19.3
38	37.9	2.7	98	97.8	6.8	58	157.6	11.0	18	217.5	15.2	78	277.3	19.4
39	38.9	2.7	99	98.8	6.9	59	158.6	11.1	19	218.5	15.3	79	278.3	19.5
40	39.9	2.8	100	99.8	7.0	60	159.6	11.2	20	219.5	15.3	80	279.3	19.5
41	40.9	2.9	101	100.8	7.0	161	160.6	11.2	221	220.5	15.4	281	280.3	19.6
42	41.9	2.9	02	101.8	7.1	62	161.6	11.3	22	221.5	15.5	82	281.3	19.7
43	42.9	3.0	03	102.7	7.2	63	162.6	11.4	23	222.5	15.6	83	282.3	19.7
44	43.9	3.1	04	103.7	7.3	64	163.6	11.4	24	223.5	15.6	84	283.3	19.8
45	44.9	3.1	05	104.7	7.3	65	164.6	11.5	25	224.5	15.7	85	284.3	19.9
46	45.9	3.2	06	105.7	7.4	66	165.6	11.6	26	225.4	15.8	86	285.3	20.0
47	46.9	3.3	07	106.7	7.5	67	166.6	11.6	27	226.4	15.8	87	286.3	20.0
48	47.9	3.3	08	107.7	7.5	68	167.6	11.7	28	227.4	15.9	88	287.3	20.1
49	48.9	3.4	09	108.7	7.6	69	168.6	11.8	29	228.4	16.0	89	288.3	20.2
50	49.9	3.5	10	109.7	7.7	70	169.6	11.9	30	229.4	16.0	90	289.3	20.2
51	50.9	3.6	111	110.7	7.7	171	170.6	11.9	231	230.4	16.1	291	290.3	20.3
52	51.9	3.6	12	111.7	7.8	72	171.6	12.0	32	231.4	16.2	92	291.3	20.4
53	52.9	3.7	13	112.7	7.9	73	172.6	12.1	33	232.4	16.3	93	292.3	20.4
54	53.9	3.8	14	113.7	8.0	74	173.6	12.1	34	233.4	16.3	94	293.3	20.5
55	54.9	3.8	15	114.7	8.0	75	174.6	12.2	35	234.4	16.4	95	294.3	20.6
56	55.9	3.9	16	115.7	8.1	76	175.6	12.3	36	235.4	16.5	96	295.3	20.6
57	56.9	4.0	17	116.7	8.2	77	176.6	12.3	37	236.4	16.5	97	296.3	20.7
58	57.9	4.0	18	117.7	8.2	78	177.6	12.4	38	237.4	16.6	98	297.3	20.8
59	58.9	4.1	19	118.7	8.3	79	178.6	12.5	39	238.4	16.7	99	298.3	20.9
60	59.9	4.2	20	119.7	8.4	80	179.6	12.6	40	239.4	16.7	300	299.3	20.9
Dist.	Dep.	D. Lat.	Dist.	Dep.	D. Lat.	Dist.	Dep.	D. Lat.	Dist.	Dep.	D. Lat.	Dist.	Dep.	D. Lat.

274° | 086° — 266° | 094°

86°

Dist.	D. Lat.	Dep.
N.	N x Cos.	N x Sin.
Hypotenuse	Side Adj.	Side Opp.

TABLE 4

4° Traverse Table

356° | 004° — 184° | 176°

Dist.	D. Lat.	Dep.	Dist.	D. Lat.	Dep.	Dist.	D. Lat.	Dep.	Dist.	D. Lat.	Dep.	Dist.	D. Lat.	Dep.
301	300.3	21.0	361	360.1	25.2	421	420.0	29.4	481	479.8	33.6	541	539.7	37.7
02	301.3	21.1	62	361.1	25.3	22	421.0	29.4	82	480.8	33.6	42	540.7	37.8
03	302.3	21.1	63	362.1	25.3	23	422.0	29.5	83	481.8	33.7	43	541.7	37.9
04	303.3	21.2	64	363.1	25.4	24	423.0	29.6	84	482.8	33.8	44	542.7	37.9
05	304.3	21.3	65	364.1	25.5	25	424.0	29.6	85	483.8	33.8	45	543.7	38.0
06	305.3	21.3	66	365.1	25.5	26	425.0	29.7	86	484.8	33.9	46	544.7	38.1
07	306.3	21.4	67	366.1	25.6	27	426.0	29.8	87	485.8	34.0	47	545.7	38.2
08	307.2	21.5	68	367.1	25.7	28	427.0	29.9	88	486.8	34.0	48	546.7	38.2
09	308.2	21.6	69	368.1	25.7	29	428.0	29.9	89	487.8	34.1	49	547.7	38.3
10	309.2	21.6	70	369.1	25.8	30	429.0	30.0	90	488.8	34.2	50	548.7	38.4
311	310.2	21.7	371	370.1	25.9	431	430.0	30.1	491	489.8	34.3	551	549.7	38.4
12	311.2	21.8	72	371.1	25.9	32	430.9	30.1	92	490.8	34.3	52	550.7	38.5
13	312.2	21.8	73	372.1	26.0	33	431.9	30.2	93	491.8	34.4	53	551.7	38.6
14	313.2	21.9	74	373.1	26.1	34	432.9	30.3	94	492.8	34.5	54	552.7	38.6
15	314.2	22.0	75	374.1	26.2	35	433.9	30.3	95	493.8	34.5	55	553.6	38.7
16	315.2	22.0	76	375.1	26.2	36	434.9	30.4	96	494.8	34.6	56	554.6	38.8
17	316.2	22.1	77	376.1	26.3	37	435.9	30.5	97	495.8	34.7	57	555.6	38.9
18	317.2	22.2	78	377.1	26.4	38	436.9	30.6	98	496.8	34.7	58	556.6	38.9
19	318.2	22.3	79	378.1	26.4	39	437.9	30.6	99	497.8	34.8	59	557.6	39.0
20	319.2	22.3	80	379.1	26.5	40	438.9	30.7	500	498.8	34.9	60	558.6	39.1
321	320.2	22.4	381	380.1	26.6	441	439.9	30.8	501	499.8	34.9	561	559.6	39.1
22	321.2	22.5	82	381.1	26.6	42	440.9	30.8	02	500.8	35.0	62	560.6	39.2
23	322.2	22.5	83	382.1	26.7	43	441.9	30.9	03	501.8	35.1	63	561.6	39.3
24	323.2	22.6	84	383.1	26.8	44	442.9	31.0	04	502.8	35.2	64	562.6	39.3
25	324.2	22.7	85	384.1	26.9	45	443.9	31.0	05	503.8	35.2	65	563.6	39.4
26	325.2	22.7	86	385.1	26.9	46	444.9	31.1	06	504.8	35.3	66	564.6	39.5
27	326.2	22.8	87	386.1	27.0	47	445.9	31.2	07	505.8	35.4	67	565.6	39.6
28	327.2	22.9	88	387.1	27.1	48	446.9	31.3	08	506.8	35.4	68	566.6	39.6
29	328.2	22.9	89	388.1	27.1	49	447.9	31.3	09	507.8	35.5	69	567.6	39.7
30	329.2	23.0	90	389.0	27.2	50	448.9	31.4	10	508.8	35.6	70	568.6	39.8
331	330.2	23.1	391	390.0	27.3	451	449.9	31.5	511	509.8	35.6	571	569.6	39.8
32	331.2	23.2	92	391.0	27.3	52	450.9	31.5	12	510.8	35.7	72	570.6	39.9
33	332.2	23.2	93	392.0	27.4	53	451.9	31.6	13	511.8	35.8	73	571.6	40.0
34	333.2	23.3	94	393.0	27.5	54	452.9	31.7	14	512.7	35.9	74	572.6	40.0
35	334.2	23.4	95	394.0	27.6	55	453.9	31.7	15	513.7	35.9	75	573.6	40.1
36	335.2	23.4	96	395.0	27.6	56	454.9	31.8	16	514.7	36.0	76	574.6	40.2
37	336.2	23.5	97	396.0	27.7	57	455.9	31.9	17	515.7	36.1	77	575.6	40.2
38	337.2	23.6	98	397.0	27.8	58	456.9	31.9	18	516.7	36.1	78	576.6	40.3
39	338.2	23.6	99	398.0	27.8	59	457.9	32.0	19	517.7	36.2	79	577.6	40.4
40	339.2	23.7	400	399.0	27.9	60	458.9	32.1	20	518.7	36.3	80	578.6	40.5
341	340.2	23.8	401	400.0	28.0	461	459.9	32.2	521	519.7	36.3	581	579.6	40.5
42	341.2	23.9	02	401.0	28.0	62	460.9	32.2	22	520.7	36.4	82	580.6	40.6
43	342.2	23.9	03	402.0	28.1	63	461.9	32.3	23	521.7	36.5	83	581.6	40.7
44	343.2	24.0	04	403.0	28.2	64	462.9	32.4	24	522.7	36.6	84	582.6	40.7
45	344.2	24.1	05	404.0	28.3	65	463.9	32.4	25	523.7	36.6	85	583.6	40.8
46	345.2	24.1	06	405.0	28.3	66	464.9	32.5	26	524.7	36.7	86	584.6	40.9
47	346.2	24.2	07	406.0	28.4	67	465.9	32.6	27	525.7	36.8	87	585.6	40.9
48	347.2	24.3	08	407.0	28.5	68	466.9	32.6	28	526.7	36.8	88	586.6	41.0
49	348.1	24.3	09	408.0	28.5	69	467.9	32.7	29	527.7	36.9	89	587.6	41.1
50	349.1	24.4	10	409.0	28.6	70	468.9	32.8	30	528.7	37.0	90	588.6	41.2
351	350.1	24.5	411	410.0	28.7	471	469.9	32.9	531	529.7	37.0	591	589.6	41.2
52	351.1	24.6	12	411.0	28.7	72	470.9	32.9	32	530.7	37.1	92	590.6	41.3
53	352.1	24.6	13	412.0	28.8	73	471.8	33.0	33	531.7	37.2	93	591.6	41.4
54	353.1	24.7	14	413.0	28.9	74	472.8	33.1	34	532.7	37.2	94	592.6	41.4
55	354.1	24.8	15	414.0	28.9	75	473.8	33.1	35	533.7	37.3	95	593.6	41.5
56	355.1	24.8	16	415.0	29.0	76	474.8	33.2	36	534.7	37.4	96	594.5	41.6
57	356.1	24.9	17	416.0	29.1	77	475.8	33.3	37	535.7	37.5	97	595.5	41.6
58	357.1	25.0	18	417.0	29.2	78	476.8	33.3	38	536.7	37.5	98	596.5	41.7
59	358.1	25.0	19	418.0	29.2	79	477.8	33.4	39	537.7	37.6	99	597.5	41.8
60	359.1	25.1	20	419.0	29.3	80	478.8	33.5	40	538.7	37.7	600	598.5	41.9
Dist.	Dep.	D. Lat.	Dist.	Dep.	D. Lat.	Dist.	Dep.	D. Lat.	Dist.	Dep.	D. Lat.	Dist.	Dep.	D. Lat.

Dist.	D. Lat.	Dep.
D Lo	Dep.	
	m	D Lo

86°

274° | 086° — 266° | 094°

TABLE 4

Traverse 5° Table

355° | 005° — 185° | 175°

Dist.	D. Lat.	Dep.	Dist.	D. Lat.	Dep.	Dist.	D. Lat.	Dep.	Dist.	D. Lat.	Dep.	Dist.	D. Lat.	Dep.
1	1.0	0.1	61	60.8	5.3	121	120.5	10.5	181	180.3	15.8	241	240.1	21.0
2	2.0	0.2	62	61.8	5.4	22	121.5	10.6	82	181.3	15.9	42	241.1	21.1
3	3.0	0.3	63	62.8	5.5	23	122.5	10.7	83	182.3	15.9	43	242.1	21.2
4	4.0	0.3	64	63.8	5.6	24	123.5	10.8	84	183.3	16.0	44	243.1	21.3
5	5.0	0.4	65	64.8	5.7	25	124.5	10.9	85	184.3	16.1	45	244.1	21.4
6	6.0	0.5	66	65.7	5.8	26	125.5	11.0	86	185.3	16.2	46	245.1	21.4
7	7.0	0.6	67	66.7	5.8	27	126.5	11.1	87	186.3	16.3	47	246.1	21.5
8	8.0	0.7	68	67.7	5.9	28	127.5	11.2	88	187.3	16.4	48	247.1	21.6
9	9.0	0.8	69	68.7	6.0	29	128.5	11.2	89	188.3	16.5	49	248.1	21.7
10	10.0	0.9	70	69.7	6.1	30	129.5	11.3	90	189.3	16.6	50	249.0	21.8
11	11.0	1.0	71	70.7	6.2	131	130.5	11.4	191	190.3	16.6	251	250.0	21.9
12	12.0	1.0	72	71.7	6.3	32	131.5	11.5	92	191.3	16.7	52	251.0	22.0
13	13.0	1.1	73	72.7	6.4	33	132.5	11.6	93	192.3	16.8	53	252.0	22.1
14	13.9	1.2	74	73.7	6.4	34	133.5	11.7	94	193.3	16.9	54	253.0	22.1
15	14.9	1.3	75	74.7	6.5	35	134.5	11.8	95	194.3	17.0	55	254.0	22.2
16	15.9	1.4	76	75.7	6.6	36	135.5	11.9	96	195.3	17.1	56	255.0	22.3
17	16.9	1.5	77	76.7	6.7	37	136.5	11.9	97	196.3	17.2	57	256.0	22.4
18	17.9	1.6	78	77.7	6.8	38	137.5	12.0	98	197.2	17.3	58	257.0	22.5
19	18.9	1.7	79	78.7	6.9	39	138.5	12.1	99	198.2	17.3	59	258.0	22.6
20	19.9	1.7	80	79.7	7.0	40	139.5	12.2	200	199.2	17.4	60	259.0	22.7
21	20.9	1.8	81	80.7	7.1	141	140.5	12.3	201	200.2	17.5	261	260.0	22.7
22	21.9	1.9	82	81.7	7.1	42	141.5	12.4	02	201.2	17.6	62	261.0	22.8
23	22.9	2.0	83	82.7	7.2	43	142.5	12.5	03	202.2	17.7	63	262.0	22.9
24	23.9	2.1	84	83.7	7.3	44	143.5	12.6	04	203.2	17.8	64	263.0	23.0
25	24.9	2.2	85	84.7	7.4	45	144.4	12.6	05	204.2	17.9	65	264.0	23.1
26	25.9	2.3	86	85.7	7.5	46	145.4	12.7	06	205.2	18.0	66	265.0	23.2
27	26.9	2.4	87	86.7	7.6	47	146.4	12.8	07	206.2	18.0	67	266.0	23.3
28	27.9	2.4	88	87.7	7.7	48	147.4	12.9	08	207.2	18.1	68	267.0	23.4
29	28.9	2.5	89	88.7	7.8	49	148.4	13.0	09	208.2	18.2	69	268.0	23.4
30	29.9	2.6	90	89.7	7.8	50	149.4	13.1	10	209.2	18.3	70	269.0	23.5
31	30.9	2.7	91	90.7	7.9	151	150.4	13.2	211	210.2	18.4	271	270.0	23.6
32	31.9	2.8	92	91.6	8.0	52	151.4	13.2	12	211.2	18.5	72	271.0	23.7
33	32.9	2.9	93	92.6	8.1	53	152.4	13.3	13	212.2	18.6	73	272.0	23.8
34	33.9	3.0	94	93.6	8.2	54	153.4	13.4	14	213.2	18.7	74	273.0	23.9
35	34.9	3.1	95	94.6	8.3	55	154.4	13.5	15	214.2	18.7	75	274.0	24.0
36	35.9	3.1	96	95.6	8.4	56	155.4	13.6	16	215.2	18.8	76	274.9	24.1
37	36.9	3.2	97	96.6	8.5	57	156.4	13.7	17	216.2	18.9	77	275.9	24.1
38	37.9	3.3	98	97.6	8.5	58	157.4	13.8	18	217.2	19.0	78	276.9	24.2
39	38.9	3.4	99	98.6	8.6	59	158.4	13.9	19	218.2	19.1	79	277.9	24.3
40	39.8	3.5	100	99.6	8.7	60	159.4	13.9	20	219.2	19.2	80	278.9	24.4
41	40.8	3.6	101	100.6	8.8	161	160.4	14.0	221	220.2	19.3	281	279.9	24.5
42	41.8	3.7	02	101.6	8.9	62	161.4	14.1	22	221.2	19.3	82	280.9	24.6
43	42.8	3.7	03	102.6	9.0	63	162.4	14.2	23	222.2	19.4	83	281.9	24.7
44	43.8	3.8	04	103.6	9.1	64	163.4	14.3	24	223.1	19.5	84	282.9	24.8
45	44.8	3.9	05	104.6	9.2	65	164.4	14.4	25	224.1	19.6	85	283.9	24.8
46	45.8	4.0	06	105.6	9.2	66	165.4	14.5	26	225.1	19.7	86	284.9	24.9
47	46.8	4.1	07	106.6	9.3	67	166.4	14.6	27	226.1	19.8	87	285.9	25.0
48	47.8	4.2	08	107.6	9.4	68	167.4	14.6	28	227.1	19.9	88	286.9	25.1
49	48.8	4.3	09	108.6	9.5	69	168.4	14.7	29	228.1	20.0	89	287.9	25.2
50	49.8	4.4	10	109.6	9.6	70	169.4	14.8	30	229.1	20.0	90	288.9	25.3
51	50.8	4.4	111	110.6	9.7	171	170.3	14.9	231	230.1	20.1	291	289.9	25.4
52	51.8	4.5	12	111.6	9.8	72	171.3	15.0	32	231.1	20.2	92	290.9	25.4
53	52.8	4.6	13	112.6	9.8	73	172.3	15.1	33	232.1	20.3	93	291.9	25.5
54	53.8	4.7	14	113.6	9.9	74	173.3	15.2	34	233.1	20.4	94	292.9	25.6
55	54.8	4.8	15	114.6	10.0	75	174.3	15.3	35	234.1	20.5	95	293.9	25.7
56	55.8	4.9	16	115.6	10.1	76	175.3	15.3	36	235.1	20.6	96	294.9	25.8
57	56.8	5.0	17	116.6	10.2	77	176.3	15.4	37	236.1	20.7	97	295.9	25.9
58	57.8	5.1	18	117.6	10.3	78	177.3	15.5	38	237.1	20.7	98	296.9	26.0
59	58.8	5.1	19	118.5	10.4	79	178.3	15.6	39	238.1	20.8	99	297.9	26.1
60	59.8	5.2	20	119.5	10.5	80	179.3	15.7	40	239.1	20.9	300	298.9	26.1
Dist.	Dep.	D. Lat.	Dist.	Dep.	D. Lat.	Dist.	Dep.	D. Lat.	Dist.	Dep.	D. Lat.	Dist.	Dep.	D. Lat.

275° | 085° — 265° | 095°

85°

Dist.	D. Lat.	Dep.
N.	N x Cos.	N x Sin.
Hypotenuse	Side Adj.	Side Opp.

TABLE 4

Traverse 5° Table

355° | 005° — 185° | 175°

Dist.	D. Lat.	Dep.	Dist.	D. Lat.	Dep.	Dist.	D. Lat.	Dep.	Dist.	D. Lat.	Dep.	Dist.	D. Lat.	Dep.
301	299.9	26.2	361	359.6	31.5	421	419.4	36.7	481	479.2	41.9	541	538.9	47.2
02	300.9	26.3	62	360.6	31.6	22	420.4	36.8	82	480.2	42.0	42	539.9	47.2
03	301.8	26.4	63	361.6	31.6	23	421.4	36.9	83	481.2	42.1	43	540.9	47.3
04	302.8	26.5	64	362.6	31.7	24	422.4	37.0	84	482.2	42.2	44	541.9	47.4
05	303.8	26.6	65	363.6	31.8	25	423.4	37.0	85	483.2	42.3	45	542.9	47.5
06	304.8	26.7	66	364.6	31.9	26	424.4	37.1	86	484.2	42.4	46	543.9	47.6
07	305.8	26.8	67	365.6	32.0	27	425.4	37.2	87	485.1	42.4	47	544.9	47.7
08	306.8	26.8	68	366.6	32.1	28	426.4	37.3	88	486.1	42.5	48	545.9	47.8
09	307.8	26.9	69	367.6	32.2	29	427.4	37.4	89	487.1	42.6	49	546.9	47.8
10	308.8	27.0	70	368.6	32.2	30	428.4	37.5	90	488.1	42.7	50	547.9	47.9
311	309.8	27.1	371	369.6	32.3	431	429.4	37.6	491	489.1	42.8	551	548.9	48.0
12	310.8	27.2	72	370.6	32.4	32	430.4	37.7	92	490.1	42.9	52	549.9	48.1
13	311.8	27.3	73	371.6	32.5	33	431.4	37.7	93	491.1	43.0	53	550.9	48.2
14	312.8	27.4	74	372.6	32.6	34	432.3	37.8	94	492.1	43.1	54	551.9	48.3
15	313.8	27.5	75	373.6	32.7	35	433.3	37.9	95	493.1	43.1	55	552.9	48.4
16	314.8	27.5	76	374.6	32.8	36	434.3	38.0	96	494.1	43.2	56	553.9	48.5
17	315.8	27.6	77	375.6	32.9	37	435.3	38.1	97	495.1	43.3	57	554.9	48.5
18	316.8	27.7	78	376.6	32.9	38	436.3	38.2	98	496.1	43.4	58	555.9	48.6
19	317.8	27.8	79	377.6	33.0	39	437.3	38.3	99	497.1	43.5	59	556.9	48.7
20	318.8	27.9	80	378.6	33.1	40	438.3	38.3	500	498.1	43.6	60	557.9	48.8
321	319.8	28.0	381	379.6	33.2	441	439.3	38.4	501	499.1	43.7	561	558.9	48.9
22	320.8	28.1	82	380.5	33.3	42	440.3	38.5	02	500.1	43.8	62	559.9	49.0
23	321.8	28.2	83	381.5	33.4	43	441.3	38.6	03	501.1	43.8	63	560.9	49.1
24	322.8	28.2	84	382.5	33.5	44	442.3	38.7	04	502.1	43.9	64	561.9	49.2
25	323.8	28.3	85	383.5	33.6	45	443.3	38.8	05	503.1	44.0	65	562.9	49.2
26	324.8	28.4	86	384.5	33.6	46	444.3	38.9	06	504.1	44.1	66	563.8	49.3
27	325.8	28.5	87	385.5	33.7	47	445.3	39.0	07	505.1	44.2	67	564.8	49.4
28	326.8	28.6	88	386.5	33.8	48	446.3	39.0	08	506.1	44.3	68	565.8	49.5
29	327.7	28.7	89	387.5	33.9	49	447.3	39.1	09	507.1	44.4	69	566.8	49.6
30	328.7	28.8	90	388.5	34.0	50	448.3	39.2	10	508.1	44.4	70	567.8	49.7
331	329.7	28.8	391	389.5	34.1	451	449.3	39.3	511	509.1	44.5	571	568.8	49.8
32	330.7	28.9	92	390.5	34.2	52	450.3	39.4	12	510.1	44.6	72	569.8	49.9
33	331.7	29.0	93	391.5	34.3	53	451.3	39.5	13	511.0	44.7	73	570.8	49.9
34	332.7	29.1	94	392.5	34.3	54	452.3	39.6	14	512.0	44.8	74	571.8	50.0
35	333.7	29.2	95	393.5	34.4	55	453.3	39.7	15	513.0	44.9	75	572.8	50.1
36	334.7	29.3	96	394.5	34.5	56	454.3	39.7	16	514.0	45.0	76	573.8	50.2
37	335.7	29.4	97	395.5	34.6	57	455.3	39.8	17	515.0	45.1	77	574.8	50.3
38	336.7	29.5	98	396.5	34.7	58	456.3	39.9	18	516.0	45.1	78	575.8	50.4
39	337.7	29.5	99	397.5	34.8	59	457.3	40.0	19	517.0	45.2	79	576.8	50.5
40	338.7	29.6	400	398.5	34.9	60	458.2	40.1	20	518.0	45.3	80	577.8	50.6
341	339.7	29.7	401	399.5	34.9	461	459.2	40.2	521	519.0	45.4	581	578.8	50.6
42	340.7	29.8	02	400.5	35.0	62	460.2	40.3	22	520.0	45.5	82	579.8	50.7
43	341.7	29.9	03	401.5	35.1	63	461.2	40.4	23	521.0	45.6	83	580.8	50.8
44	342.7	30.0	04	402.5	35.2	64	462.2	40.4	24	522.0	45.7	84	581.8	50.9
45	343.7	30.1	05	403.5	35.3	65	463.2	40.5	25	523.0	45.8	85	582.8	51.0
46	344.7	30.2	06	404.5	35.4	66	464.2	40.6	26	524.0	45.8	86	583.8	51.1
47	345.7	30.2	07	405.5	35.5	67	465.2	40.7	27	525.0	45.9	87	584.8	51.2
48	346.7	30.3	08	406.4	35.6	68	466.2	40.8	28	526.0	46.0	88	585.8	51.2
49	347.7	30.4	09	407.4	35.6	69	467.2	40.9	29	527.0	46.1	89	586.8	51.3
50	348.7	30.5	10	408.4	35.7	70	468.2	41.0	30	528.0	46.2	90	587.8	51.4
351	349.7	30.6	411	409.4	35.8	471	469.2	41.1	531	529.0	46.3	591	588.8	51.5
52	350.7	30.7	12	410.4	35.9	72	470.2	41.1	32	530.0	46.4	92	589.7	51.6
53	351.7	30.8	13	411.4	36.0	73	471.2	41.2	33	531.0	46.5	93	590.7	51.7
54	352.7	30.9	14	412.4	36.1	74	472.2	41.3	34	532.0	46.5	94	591.7	51.8
55	353.6	30.9	15	413.4	36.2	75	473.2	41.4	35	533.0	46.6	95	592.7	51.9
56	354.6	31.0	16	414.4	36.3	76	474.2	41.5	36	534.0	46.7	96	593.7	51.9
57	355.6	31.1	17	415.4	36.3	77	475.2	41.6	37	535.0	46.8	97	594.7	52.0
58	356.6	31.2	18	416.4	36.4	78	476.2	41.7	38	536.0	46.9	98	595.7	52.1
59	357.6	31.3	19	417.4	36.5	79	477.2	41.7	39	536.9	47.0	99	596.7	52.2
60	358.6	31.4	20	418.4	36.6	80	478.2	41.8	40	537.9	47.1	600	597.7	52.3
Dist.	Dep.	D. Lat.	Dist.	Dep.	D. Lat.	Dist.	Dep.	D. Lat.	Dist.	Dep.	D. Lat.	Dist.	Dep.	D. Lat.

Dist.	D. Lat.	Dep.
D Lo	Dep.	
	m	D Lo

85°

275° | 085° — 265° | 095°

TABLE 4

Traverse 6° Table

354° | 006° / 186° | 174°

Dist.	D. Lat.	Dep.	Dist.	D. Lat.	Dep.	Dist.	D. Lat.	Dep.	Dist.	D. Lat.	Dep.	Dist.	D. Lat.	Dep.
1	1.0	0.1	61	60.7	6.4	121	120.3	12.6	181	180.0	18.9	241	239.7	25.2
2	2.0	0.2	62	61.7	6.5	22	121.3	12.8	82	181.0	19.0	42	240.7	25.3
3	3.0	0.3	63	62.7	6.6	23	122.3	12.9	83	182.0	19.1	43	241.7	25.4
4	4.0	0.4	64	63.6	6.7	24	123.3	13.0	84	183.0	19.2	44	242.7	25.5
5	5.0	0.5	65	64.6	6.8	25	124.3	13.1	85	184.0	19.3	45	243.7	25.6
6	6.0	0.6	66	65.6	6.9	26	125.3	13.2	86	185.0	19.4	46	244.7	25.7
7	7.0	0.7	67	66.6	7.0	27	126.3	13.3	87	186.0	19.5	47	245.6	25.8
8	8.0	0.8	68	67.6	7.1	28	127.3	13.4	88	187.0	19.7	48	246.6	25.9
9	9.0	0.9	69	68.6	7.2	29	128.3	13.5	89	188.0	19.8	49	247.6	26.0
10	9.9	1.0	70	69.6	7.3	30	129.3	13.6	90	189.0	19.9	50	248.6	26.1
11	10.9	1.1	71	70.6	7.4	131	130.3	13.7	191	190.0	20.0	251	249.6	26.2
12	11.9	1.3	72	71.6	7.5	32	131.3	13.8	92	190.9	20.1	52	250.6	26.3
13	12.9	1.4	73	72.6	7.6	33	132.3	13.9	93	191.9	20.2	53	251.6	26.4
14	13.9	1.5	74	73.6	7.7	34	133.3	14.0	94	192.9	20.3	54	252.6	26.6
15	14.9	1.6	75	74.6	7.8	35	134.3	14.1	95	193.9	20.4	55	253.6	26.7
16	15.9	1.7	76	75.6	7.9	36	135.3	14.2	96	194.9	20.5	56	254.6	26.8
17	16.9	1.8	77	76.6	8.0	37	136.2	14.3	97	195.9	20.6	57	255.6	26.9
18	17.9	1.9	78	77.6	8.2	38	137.2	14.4	98	196.9	20.7	58	256.6	27.0
19	18.9	2.0	79	78.6	8.3	39	138.2	14.5	99	197.9	20.8	59	257.6	27.1
20	19.9	2.1	80	79.6	8.4	40	139.2	14.6	200	198.9	20.9	60	258.6	27.2
21	20.9	2.2	81	80.6	8.5	141	140.2	14.7	201	199.9	21.0	261	259.6	27.3
22	21.9	2.3	82	81.6	8.6	42	141.2	14.8	02	200.9	21.1	62	260.6	27.4
23	22.9	2.4	83	82.5	8.7	43	142.2	14.9	03	201.9	21.2	63	261.6	27.5
24	23.9	2.5	84	83.5	8.8	44	143.2	15.1	04	202.9	21.3	64	262.6	27.6
25	24.9	2.6	85	84.5	8.9	45	144.2	15.2	05	203.9	21.4	65	263.5	27.7
26	25.9	2.7	86	85.5	9.0	46	145.2	15.3	06	204.9	21.5	66	264.5	27.8
27	26.9	2.8	87	86.5	9.1	47	146.2	15.4	07	205.9	21.6	67	265.5	27.9
28	27.8	2.9	88	87.5	9.2	48	147.2	15.5	08	206.9	21.7	68	266.5	28.0
29	28.8	3.0	89	88.5	9.3	49	148.2	15.6	09	207.9	21.8	69	267.5	28.1
30	29.8	3.1	90	89.5	9.4	50	149.2	15.7	10	208.8	22.0	70	268.5	28.2
31	30.8	3.2	91	90.5	9.5	151	150.2	15.8	211	209.8	22.1	271	269.5	28.3
32	31.8	3.3	92	91.5	9.6	52	151.2	15.9	12	210.8	22.2	72	270.5	28.4
33	32.8	3.4	93	92.5	9.7	53	152.2	16.0	13	211.8	22.3	73	271.5	28.5
34	33.8	3.6	94	93.5	9.8	54	153.2	16.1	14	212.8	22.4	74	272.5	28.6
35	34.8	3.7	95	94.5	9.9	55	154.2	16.2	15	213.8	22.5	75	273.5	28.7
36	35.8	3.8	96	95.5	10.0	56	155.1	16.3	16	214.8	22.6	76	274.5	28.8
37	36.8	3.9	97	96.5	10.1	57	156.1	16.4	17	215.8	22.7	77	275.5	29.0
38	37.8	4.0	98	97.5	10.2	58	157.1	16.5	18	216.8	22.8	78	276.5	29.1
39	38.8	4.1	99	98.5	10.3	59	158.1	16.6	19	217.8	22.9	79	277.5	29.2
40	39.8	4.2	100	99.5	10.5	60	159.1	16.7	20	218.8	23.0	80	278.5	29.3
41	40.8	4.3	101	100.4	10.6	161	160.1	16.8	221	219.8	23.1	281	279.5	29.4
42	41.8	4.4	02	101.4	10.7	62	161.1	16.9	22	220.8	23.2	82	280.5	29.5
43	42.8	4.5	03	102.4	10.8	63	162.1	17.0	23	221.8	23.3	83	281.4	29.6
44	43.8	4.6	04	103.4	10.9	64	163.1	17.1	24	222.8	23.4	84	282.4	29.7
45	44.8	4.7	05	104.4	11.0	65	164.1	17.2	25	223.8	23.5	85	283.4	29.8
46	45.7	4.8	06	105.4	11.1	66	165.1	17.4	26	224.8	23.6	86	284.4	29.9
47	46.7	4.9	07	106.4	11.2	67	166.1	17.5	27	225.8	23.7	87	285.4	30.0
48	47.7	5.0	08	107.4	11.3	68	167.1	17.6	28	226.8	23.8	88	286.4	30.1
49	48.7	5.1	09	108.4	11.4	69	168.1	17.7	29	227.7	23.9	89	287.4	30.2
50	49.7	5.2	10	109.4	11.5	70	169.1	17.8	30	228.7	24.0	90	288.4	30.3
51	50.7	5.3	111	110.4	11.6	171	170.1	17.9	231	229.7	24.1	291	289.4	30.4
52	51.7	5.4	12	111.4	11.7	72	171.1	18.0	32	230.7	24.3	92	290.4	30.5
53	52.7	5.5	13	112.4	11.8	73	172.1	18.1	33	231.7	24.4	93	291.4	30.6
54	53.7	5.6	14	113.4	11.9	74	173.0	18.2	34	232.7	24.5	94	292.4	30.7
55	54.7	5.7	15	114.4	12.0	75	174.0	18.3	35	233.7	24.6	95	293.4	30.8
56	55.7	5.9	16	115.4	12.1	76	175.0	18.4	36	234.7	24.7	96	294.4	30.9
57	56.7	6.0	17	116.4	12.2	77	176.0	18.5	37	235.7	24.8	97	295.4	31.0
58	57.7	6.1	18	117.4	12.3	78	177.0	18.6	38	236.7	24.9	98	296.4	31.1
59	58.7	6.2	19	118.3	12.4	79	178.0	18.7	39	237.7	25.0	99	297.4	31.3
60	59.7	6.3	20	119.3	12.5	80	179.0	18.8	40	238.7	25.1	300	298.4	31.4
Dist.	Dep.	D. Lat.	Dist.	Dep.	D. Lat.	Dist.	Dep.	D. Lat.	Dist.	Dep.	D. Lat.	Dist.	Dep.	D. Lat.

276° | 084° / 264° | 096°

84°

Dist.	D. Lat.	Dep.
N.	N x Cos.	N x Sin.
Hypotenuse	Side Adj.	Side Opp.

TABLE 4

Traverse 6° Table

354° | 006° / 186° | 174°

Dist.	D. Lat.	Dep.	Dist.	D. Lat.	Dep.	Dist.	D. Lat.	Dep.	Dist.	D. Lat.	Dep.	Dist.	D. Lat.	Dep.
301	299.4	31.5	361	359.0	37.7	421	418.7	44.0	481	478.4	50.3	541	538.0	56.5
02	300.3	31.6	62	360.0	37.8	22	419.7	44.1	82	479.4	50.4	42	539.0	56.7
03	301.3	31.7	63	361.0	37.9	23	420.7	44.2	83	480.4	50.5	43	540.0	56.8
04	302.3	31.8	64	362.0	38.0	24	421.7	44.3	84	481.3	50.6	44	541.0	56.9
05	303.3	31.9	65	363.0	38.2	25	422.7	44.4	85	482.3	50.7	45	542.0	57.0
06	304.3	32.0	66	364.0	38.3	26	423.7	44.5	86	483.3	50.8	46	543.0	57.1
07	305.3	32.1	67	365.0	38.4	27	424.7	44.6	87	484.3	50.9	47	544.0	57.2
08	306.3	32.2	68	366.0	38.5	28	425.7	44.7	88	485.3	51.0	48	545.0	57.3
09	307.3	32.3	69	367.0	38.6	29	426.6	44.8	89	486.3	51.1	49	546.0	57.4
10	308.3	32.4	70	368.0	38.7	30	427.6	44.9	90	487.3	51.2	50	547.0	57.5
311	309.3	32.5	371	369.0	38.8	431	428.6	45.1	491	488.3	51.3	551	548.0	57.6
12	310.3	32.6	72	370.0	38.9	32	429.6	45.2	92	489.3	51.4	52	549.0	57.7
13	311.3	32.7	73	371.0	39.0	33	430.6	45.3	93	490.3	51.5	53	550.0	57.8
14	312.3	32.8	74	372.0	39.1	34	431.6	45.4	94	491.3	51.6	54	551.0	57.9
15	313.3	32.9	75	372.9	39.2	35	432.6	45.5	95	492.3	51.7	55	552.0	58.0
16	314.3	33.0	76	373.9	39.3	36	433.6	45.6	96	493.3	51.8	56	553.0	58.1
17	315.3	33.1	77	374.9	39.4	37	434.6	45.7	97	494.3	52.0	57	553.9	58.2
18	316.3	33.2	78	375.9	39.5	38	435.6	45.8	98	495.3	52.1	58	554.9	58.3
19	317.3	33.3	79	376.9	39.6	39	436.6	45.9	99	496.3	52.2	59	555.9	58.4
20	318.2	33.4	80	377.9	39.7	40	437.6	46.0	500	497.3	52.3	60	556.9	58.5
321	319.2	33.6	381	378.9	39.8	441	438.6	46.1	501	498.3	52.4	561	557.9	58.6
22	320.2	33.7	82	379.9	39.9	42	439.6	46.2	02	499.2	52.5	62	558.9	58.7
23	321.2	33.8	83	380.9	40.0	43	440.6	46.3	03	500.2	52.6	63	559.9	58.8
24	322.2	33.9	84	381.9	40.1	44	441.6	46.4	04	501.2	52.7	64	560.9	59.0
25	323.2	34.0	85	382.9	40.2	45	442.6	46.5	05	502.2	52.8	65	561.9	59.1
26	324.2	34.1	86	383.9	40.3	46	443.6	46.6	06	503.2	52.9	66	562.9	59.2
27	325.2	34.2	87	384.9	40.5	47	444.6	46.7	07	504.2	53.0	67	563.9	59.3
28	326.2	34.3	88	385.9	40.6	48	445.5	46.8	08	505.2	53.1	68	564.9	59.4
29	327.2	34.4	89	386.9	40.7	49	446.5	46.9	09	506.2	53.2	69	565.9	59.5
30	328.2	34.5	90	387.9	40.8	50	447.5	47.0	10	507.2	53.3	70	566.9	59.6
331	329.2	34.6	391	388.9	40.9	451	448.5	47.1	511	508.2	53.4	571	567.9	59.7
32	330.2	34.7	92	389.9	41.0	52	449.5	47.2	12	509.2	53.5	72	568.9	59.8
33	331.2	34.8	93	390.8	41.1	53	450.5	47.4	13	510.2	53.6	73	569.9	59.9
34	332.2	34.9	94	391.8	41.2	54	451.5	47.5	14	511.2	53.7	74	570.9	60.0
35	333.2	35.0	95	392.8	41.3	55	452.5	47.6	15	512.2	53.8	75	571.9	60.1
36	334.2	35.1	96	393.8	41.4	56	453.5	47.7	16	513.2	53.9	76	572.8	60.2
37	335.2	35.2	97	394.8	41.5	57	454.5	47.8	17	514.2	54.0	77	573.8	60.3
38	336.1	35.3	98	395.8	41.6	58	455.5	47.9	18	515.2	54.1	78	574.8	60.4
39	337.1	35.4	99	396.8	41.7	59	456.5	48.0	19	516.2	54.3	79	575.8	60.5
40	338.1	35.5	400	397.8	41.8	60	457.5	48.1	20	517.2	54.4	80	576.8	60.6
341	339.1	35.6	401	398.8	41.9	461	458.5	48.2	521	518.1	54.5	581	577.8	60.7
42	340.1	35.7	02	399.8	42.0	62	459.5	48.3	22	519.1	54.6	82	578.8	60.8
43	341.1	35.9	03	400.8	42.1	63	460.5	48.4	23	520.1	54.7	83	579.8	60.9
44	342.1	36.0	04	401.8	42.2	64	461.5	48.5	24	521.1	54.8	84	580.8	61.0
45	343.1	36.1	05	402.8	42.3	65	462.5	48.6	25	522.1	54.9	85	581.8	61.1
46	344.1	36.2	06	403.8	42.4	66	463.4	48.7	26	523.1	55.0	86	582.8	61.3
47	345.1	36.3	07	404.8	42.5	67	464.4	48.8	27	524.1	55.1	87	583.8	61.4
48	346.1	36.4	08	405.8	42.6	68	465.4	48.9	28	525.1	55.2	88	584.8	61.5
49	347.1	36.5	09	406.8	42.8	69	466.4	49.0	29	526.1	55.3	89	585.8	61.6
50	348.1	36.6	10	407.8	42.9	70	467.4	49.1	30	527.1	55.4	90	586.8	61.7
351	349.1	36.7	411	408.7	43.0	471	468.4	49.2	531	528.1	55.5	591	587.8	61.8
52	350.1	36.8	12	409.7	43.1	72	469.4	49.3	32	529.1	55.6	92	588.8	61.9
53	351.1	36.9	13	410.7	43.2	73	470.4	49.4	33	530.1	55.7	93	589.8	62.0
54	352.1	37.0	14	411.7	43.3	74	471.4	49.5	34	531.1	55.8	94	590.7	62.1
55	353.1	37.1	15	412.7	43.4	75	472.4	49.7	35	532.1	55.9	95	591.7	62.2
56	354.0	37.2	16	413.7	43.5	76	473.4	49.8	36	533.1	56.0	96	592.7	62.3
57	355.0	37.3	17	414.7	43.6	77	474.4	49.9	37	534.1	56.1	97	593.7	62.4
58	356.0	37.4	18	415.7	43.7	78	475.4	50.0	38	535.1	56.2	98	594.7	62.5
59	357.0	37.5	19	416.7	43.8	79	476.4	50.1	39	536.0	56.3	99	595.7	62.6
60	358.0	37.6	20	417.7	43.9	80	477.4	50.2	40	537.0	56.4	600	596.7	62.7
Dist.	Dep.	D. Lat.	Dist.	Dep.	D. Lat.	Dist.	Dep.	D. Lat.	Dist.	Dep.	D. Lat.	Dist.	Dep.	D. Lat.

Dist.	D. Lat.	Dep.
D Lo	Dep.	
	m	D Lo

84°

276° | 084° / 264° | 096°

TABLE 4

7°

Traverse Table

353° | 007°
187° | 173°

Dist.	D. Lat.	Dep.	Dist.	D. Lat.	Dep.	Dist.	D. Lat.	Dep.	Dist.	D. Lat.	Dep.	Dist.	D. Lat.	Dep.
1	1.0	0.1	61	60.5	7.4	121	120.1	14.7	181	179.7	22.1	241	239.2	29.4
2	2.0	0.2	62	61.5	7.6	22	121.1	14.9	82	180.6	22.2	42	240.2	29.5
3	3.0	0.4	63	62.5	7.7	23	122.1	15.0	83	181.6	22.3	43	241.2	29.6
4	4.0	0.5	64	63.5	7.8	24	123.1	15.1	84	182.6	22.4	44	242.2	29.7
5	5.0	0.6	65	64.5	7.9	25	124.1	15.2	85	183.6	22.5	45	243.2	29.9
6	6.0	0.7	66	65.5	8.0	26	125.1	15.4	86	184.6	22.7	46	244.2	30.0
7	6.9	0.9	67	66.5	8.2	27	126.1	15.5	87	185.6	22.8	47	245.2	30.1
8	7.9	1.0	68	67.5	8.3	28	127.0	15.6	88	186.6	22.9	48	246.2	30.2
9	8.9	1.1	69	68.5	8.4	29	128.0	15.7	89	187.6	23.0	49	247.1	30.3
10	9.9	1.2	70	69.5	8.5	30	129.0	15.8	90	188.6	23.2	50	248.1	30.5
11	10.9	1.3	71	70.5	8.7	131	130.0	16.0	191	189.6	23.3	251	249.1	30.6
12	11.9	1.5	72	71.5	8.8	32	131.0	16.1	92	190.6	23.4	52	250.1	30.7
13	12.9	1.6	73	72.5	8.9	33	132.0	16.2	93	191.6	23.5	53	251.1	30.8
14	13.9	1.7	74	73.4	9.0	34	133.0	16.3	94	192.6	23.6	54	252.1	31.0
15	14.9	1.8	75	74.4	9.1	35	134.0	16.5	95	193.5	23.8	55	253.1	31.1
16	15.9	1.9	76	75.4	9.3	36	135.0	16.6	96	194.5	23.9	56	254.1	31.2
17	16.9	2.1	77	76.4	9.4	37	136.0	16.7	97	195.5	24.0	57	255.1	31.3
18	17.9	2.2	78	77.4	9.5	38	137.0	16.8	98	196.5	24.1	58	256.1	31.4
19	18.9	2.3	79	78.4	9.6	39	138.0	16.9	99	197.5	24.3	59	257.1	31.6
20	19.9	2.4	80	79.4	9.7	40	139.0	17.1	200	198.5	24.4	60	258.1	31.7
21	20.8	2.6	81	80.4	9.9	141	139.9	17.2	201	199.5	24.5	261	259.1	31.8
22	21.8	2.7	82	81.4	10.0	42	140.9	17.3	02	200.5	24.6	62	260.0	31.9
23	22.8	2.8	83	82.4	10.1	43	141.9	17.4	03	201.5	24.7	63	261.0	32.1
24	23.8	2.9	84	83.4	10.2	44	142.9	17.5	04	202.5	24.9	64	262.0	32.2
25	24.8	3.0	85	84.4	10.4	45	143.9	17.7	05	203.5	25.0	65	263.0	32.3
26	25.8	3.2	86	85.4	10.5	46	144.9	17.8	06	204.5	25.1	66	264.0	32.4
27	26.8	3.3	87	86.4	10.6	47	145.9	17.9	07	205.5	25.2	67	265.0	32.5
28	27.8	3.4	88	87.3	10.7	48	146.9	18.0	08	206.4	25.3	68	266.0	32.7
29	28.8	3.5	89	88.3	10.8	49	147.9	18.2	09	207.4	25.5	69	267.0	32.8
30	29.8	3.7	90	89.3	11.0	50	148.9	18.3	10	208.4	25.6	70	268.0	32.9
31	30.8	3.8	91	90.3	11.1	151	149.9	18.4	211	209.4	25.7	271	269.0	33.0
32	31.8	3.9	92	91.3	11.2	52	150.9	18.5	12	210.4	25.8	72	270.0	33.1
33	32.8	4.0	93	92.3	11.3	53	151.9	18.6	13	211.4	26.0	73	271.0	33.3
34	33.7	4.1	94	93.3	11.5	54	152.9	18.8	14	212.4	26.1	74	272.0	33.4
35	34.7	4.3	95	94.3	11.6	55	153.8	18.9	15	213.4	26.2	75	273.0	33.5
36	35.7	4.4	96	95.3	11.7	56	154.8	19.0	16	214.4	26.3	76	273.9	33.6
37	36.7	4.5	97	96.3	11.8	57	155.8	19.1	17	215.4	26.4	77	274.9	33.8
38	37.7	4.6	98	97.3	11.9	58	156.8	19.3	18	216.4	26.6	78	275.9	33.9
39	38.7	4.8	99	98.3	12.1	59	157.8	19.4	19	217.4	26.7	79	276.9	34.0
40	39.7	4.9	100	99.3	12.2	60	158.8	19.5	20	218.4	26.8	80	277.9	34.1
41	40.7	5.0	101	100.2	12.3	161	159.8	19.6	221	219.4	26.9	281	278.9	34.2
42	41.7	5.1	02	101.2	12.4	62	160.8	19.7	22	220.3	27.1	82	279.9	34.4
43	42.7	5.2	03	102.2	12.6	63	161.8	19.9	23	221.3	27.2	83	280.9	34.5
44	43.7	5.4	04	103.2	12.7	64	162.8	20.0	24	222.3	27.3	84	281.9	34.6
45	44.7	5.5	05	104.2	12.8	65	163.8	20.1	25	223.3	27.4	85	282.9	34.7
46	45.7	5.6	06	105.2	12.9	66	164.8	20.2	26	224.3	27.5	86	283.9	34.9
47	46.6	5.7	07	106.2	13.0	67	165.8	20.4	27	225.3	27.7	87	284.9	35.0
48	47.6	5.8	08	107.2	13.2	68	166.7	20.5	28	226.3	27.8	88	285.9	35.1
49	48.6	6.0	09	108.2	13.3	69	167.7	20.6	29	227.3	27.9	89	286.8	35.2
50	49.6	6.1	10	109.2	13.4	70	168.7	20.7	30	228.3	28.0	90	287.8	35.3
51	50.6	6.2	111	110.2	13.5	171	169.7	20.8	231	229.3	28.2	291	288.8	35.5
52	51.6	6.3	12	111.2	13.6	72	170.7	21.0	32	230.3	28.3	92	289.8	35.6
53	52.6	6.5	13	112.2	13.8	73	171.7	21.1	33	231.3	28.4	93	290.8	35.7
54	53.6	6.6	14	113.2	13.9	74	172.7	21.2	34	232.3	28.5	94	291.8	35.8
55	54.6	6.7	15	114.1	14.0	75	173.7	21.3	35	233.2	28.6	95	292.8	36.0
56	55.6	6.8	16	115.1	14.1	76	174.7	21.4	36	234.2	28.8	96	293.8	36.1
57	56.6	6.9	17	116.1	14.3	77	175.7	21.6	37	235.2	28.9	97	294.8	36.2
58	57.6	7.1	18	117.1	14.4	78	176.7	21.7	38	236.2	29.0	98	295.8	36.3
59	58.6	7.2	19	118.1	14.5	79	177.7	21.8	39	237.2	29.1	99	296.8	36.4
60	59.6	7.3	20	119.1	14.6	80	178.7	21.9	40	238.2	29.2	300	297.8	36.6
Dist.	Dep.	D. Lat.	Dist.	Dep.	D. Lat.	Dist.	Dep.	D. Lat.	Dist.	Dep.	D. Lat.	Dist.	Dep.	D. Lat.

277° | 083°
263° | 097°

83°

Dist.	D. Lat.	Dep.
N.	N x Cos.	N x Sin.
Hypotenuse	Side Adj.	Side Opp.

TABLE 4

7°

Traverse Table

353° | 007°
187° | 173°

Dist.	D. Lat.	Dep.	Dist.	D. Lat.	Dep.	Dist.	D. Lat.	Dep.	Dist.	D. Lat.	Dep.	Dist.	D. Lat.	Dep.
301	298.8	36.7	361	358.3	44.0	421	417.9	51.3	481	477.4	58.6	541	537.0	65.9
02	299.7	36.8	62	359.3	44.1	22	418.9	51.4	82	478.4	58.7	42	538.0	66.1
03	300.7	36.9	63	360.3	44.2	23	419.8	51.6	83	479.4	58.9	43	539.0	66.2
04	301.7	37.0	64	361.3	44.4	24	420.8	51.7	84	480.4	59.0	44	539.9	66.3
05	302.7	37.2	65	362.3	44.5	25	421.8	51.8	85	481.4	59.1	45	540.9	66.4
06	303.7	37.3	66	363.3	44.6	26	422.8	51.9	86	482.4	59.2	46	541.9	66.5
07	304.7	37.4	67	364.3	44.7	27	423.8	52.0	87	483.4	59.4	47	542.9	66.7
08	305.7	37.5	68	365.3	44.8	28	424.8	52.2	88	484.4	59.5	48	543.9	66.8
09	306.7	37.7	69	366.2	45.0	29	425.8	52.3	89	485.4	59.6	49	544.9	66.9
10	307.7	37.8	70	367.2	45.1	30	426.8	52.4	90	486.3	59.7	50	545.9	67.0
311	308.7	37.9	371	368.2	45.2	431	427.8	52.5	491	487.3	59.8	551	546.9	67.2
12	309.7	38.0	72	369.2	45.3	32	428.8	52.6	92	488.3	60.0	52	547.9	67.3
13	310.7	38.1	73	370.2	45.5	33	429.8	52.8	93	489.3	60.1	53	548.9	67.4
14	311.7	38.3	74	371.2	45.6	34	430.8	52.9	94	490.3	60.2	54	549.9	67.5
15	312.7	38.4	75	372.2	45.7	35	431.8	53.0	95	491.3	60.3	55	550.9	67.6
16	313.6	38.5	76	373.2	45.8	36	432.8	53.1	96	492.3	60.4	56	551.9	67.8
17	314.6	38.6	77	374.2	45.9	37	433.7	53.3	97	493.3	60.6	57	552.8	67.9
18	315.6	38.8	78	375.2	46.1	38	434.7	53.4	98	494.3	60.7	58	553.8	68.0
19	316.6	38.9	79	376.2	46.2	39	435.7	53.5	99	495.3	60.8	59	554.8	68.1
20	317.6	39.0	80	377.2	46.3	40	436.7	53.6	500	496.3	60.9	60	555.8	68.2
321	318.6	39.1	381	378.2	46.4	441	437.7	53.7	501	497.3	61.1	561	556.8	68.4
22	319.6	39.2	82	379.2	46.6	42	438.7	53.9	02	498.3	61.2	62	557.8	68.5
23	320.6	39.4	83	380.1	46.7	43	439.7	54.0	03	499.3	61.3	63	558.8	68.6
24	321.6	39.5	84	381.1	46.8	44	440.7	54.1	04	500.2	61.4	64	559.8	68.7
25	322.6	39.6	85	382.1	46.9	45	441.7	54.2	05	501.2	61.5	65	560.8	68.9
26	323.6	39.7	86	383.1	47.0	46	442.7	54.4	06	502.2	61.7	66	561.8	69.0
27	324.6	39.9	87	384.1	47.2	47	443.7	54.5	07	503.2	61.8	67	562.8	69.1
28	325.6	40.0	88	385.1	47.3	48	444.7	54.6	08	504.2	61.9	68	563.8	69.2
29	326.5	40.1	89	386.1	47.4	49	445.7	54.7	09	505.2	62.0	69	564.8	69.3
30	327.5	40.2	90	387.1	47.5	50	446.6	54.8	10	506.2	62.2	70	565.8	69.5
331	328.5	40.3	391	388.1	47.7	451	447.6	55.0	511	507.2	62.3	571	566.7	69.6
32	329.5	40.5	92	389.1	47.8	52	448.6	55.1	12	508.2	62.4	72	567.7	69.7
33	330.5	40.6	93	390.1	47.9	53	449.6	55.2	13	509.2	62.5	73	568.7	69.8
34	331.5	40.7	94	391.1	48.0	54	450.6	55.3	14	510.2	62.6	74	569.7	70.0
35	332.5	40.8	95	392.1	48.1	55	451.6	55.5	15	511.2	62.8	75	570.7	70.1
36	333.5	40.9	96	393.0	48.3	56	452.6	55.6	16	512.2	62.9	76	571.7	70.2
37	334.5	41.1	97	394.0	48.4	57	453.6	55.7	17	513.1	63.0	77	572.7	70.3
38	335.5	41.2	98	395.0	48.5	58	454.6	55.8	18	514.1	63.1	78	573.7	70.4
39	336.5	41.3	99	396.0	48.6	59	455.6	55.9	19	515.1	63.3	79	574.7	70.6
40	337.5	41.4	400	397.0	48.7	60	456.6	56.1	20	516.1	63.4	80	575.7	70.7
341	338.5	41.6	401	398.0	48.9	461	457.6	56.2	521	517.1	63.5	581	576.7	70.8
42	339.5	41.7	02	399.0	49.0	62	458.6	56.3	22	518.1	63.6	82	577.7	70.9
43	340.4	41.8	03	400.0	49.1	63	459.5	56.4	23	519.1	63.7	83	578.7	71.0
44	341.4	41.9	04	401.0	49.2	64	460.5	56.5	24	520.1	63.9	84	579.6	71.2
45	342.4	42.0	05	402.0	49.4	65	461.5	56.7	25	521.1	64.0	85	580.6	71.3
46	343.4	42.2	06	403.0	49.5	66	462.5	56.8	26	522.1	64.1	86	581.6	71.4
47	344.4	42.3	07	404.0	49.6	67	463.5	56.9	27	523.1	64.2	87	582.6	71.5
48	345.4	42.4	08	405.0	49.7	68	464.5	57.0	28	524.1	64.3	88	583.6	71.7
49	346.4	42.5	09	406.0	49.8	69	465.5	57.2	29	525.1	64.5	89	584.6	71.8
50	347.4	42.7	10	406.9	50.0	70	466.5	57.3	30	526.0	64.6	90	585.6	71.9
351	348.4	42.8	411	407.9	50.1	471	467.5	57.4	531	527.0	64.7	591	586.6	72.0
52	349.4	42.9	12	408.9	50.2	72	468.5	57.5	32	528.0	64.8	92	587.6	72.1
53	350.4	43.0	13	409.9	50.3	73	469.5	57.6	33	529.0	65.0	93	588.6	72.3
54	351.4	43.1	14	410.9	50.5	74	470.5	57.8	34	530.0	65.1	94	589.6	72.4
55	352.4	43.3	15	411.9	50.6	75	471.5	57.9	35	531.0	65.2	95	590.6	72.5
56	353.3	43.4	16	412.9	50.7	76	472.5	58.0	36	532.0	65.3	96	591.6	72.6
57	354.3	43.5	17	413.9	50.8	77	473.4	58.1	37	533.0	65.4	97	592.6	72.8
58	355.3	43.6	18	414.9	50.9	78	474.4	58.3	38	534.0	65.6	98	593.5	72.9
59	356.3	43.8	19	415.9	51.1	79	475.4	58.4	39	535.0	65.7	99	594.5	73.0
60	357.3	43.9	20	416.9	51.2	80	476.4	58.5	40	536.0	65.8	600	595.5	73.1
Dist.	Dep.	D. Lat.	Dist.	Dep.	D. Lat.	Dist.	Dep.	D. Lat.	Dist.	Dep.	D. Lat.	Dist.	Dep.	D. Lat.

Dist.	D. Lat.	Dep.
D Lo	Dep.	
	m	D Lo

83°

277° | 083°
263° | 097°

TABLE 4

Traverse 8° Table

352° 188°	008° 172°

Dist.	D. Lat.	Dep.	Dist.	D. Lat.	Dep.	Dist.	D. Lat.	Dep.	Dist.	D. Lat.	Dep.	Dist.	D. Lat.	Dep.
1	1.0	0.1	61	60.4	8.5	121	119.8	16.8	181	179.2	25.2	241	238.7	33.5
2	2.0	0.3	62	61.4	8.6	22	120.8	17.0	82	180.2	25.3	42	239.6	33.7
3	3.0	0.4	63	62.4	8.8	23	121.8	17.1	83	181.2	25.5	43	240.6	33.8
4	4.0	0.6	64	63.4	8.9	24	122.8	17.3	84	182.2	25.6	44	241.6	34.0
5	5.0	0.7	65	64.4	9.0	25	123.8	17.4	85	183.2	25.7	45	242.6	34.1
6	5.9	0.8	66	65.4	9.2	26	124.8	17.5	86	184.2	25.9	46	243.6	34.2
7	6.9	1.0	67	66.3	9.3	27	125.8	17.7	87	185.2	26.0	47	244.6	34.4
8	7.9	1.1	68	67.3	9.5	28	126.8	17.8	88	186.2	26.2	48	245.6	34.5
9	8.9	1.3	69	68.3	9.6	29	127.7	18.0	89	187.2	26.3	49	246.6	34.7
10	9.9	1.4	70	69.3	9.7	30	128.7	18.1	90	188.2	26.4	50	247.6	34.8
11	10.9	1.5	71	70.3	9.9	131	129.7	18.2	191	189.1	26.6	251	248.6	34.9
12	11.9	1.7	72	71.3	10.0	32	130.7	18.4	92	190.1	26.7	52	249.5	35.1
13	12.9	1.8	73	72.3	10.2	33	131.7	18.5	93	191.1	26.9	53	250.5	35.2
14	13.9	1.9	74	73.3	10.3	34	132.7	18.6	94	192.1	27.0	54	251.5	35.3
15	14.9	2.1	75	74.3	10.4	35	133.7	18.8	95	193.1	27.1	55	252.5	35.5
16	15.8	2.2	76	75.3	10.6	36	134.7	18.9	96	194.1	27.3	56	253.5	35.6
17	16.8	2.4	77	76.3	10.7	37	135.7	19.1	97	195.1	27.4	57	254.5	35.8
18	17.8	2.5	78	77.2	10.9	38	136.7	19.2	98	196.1	27.6	58	255.5	35.9
19	18.8	2.6	79	78.2	11.0	39	137.6	19.3	99	197.1	27.7	59	256.5	36.0
20	19.8	2.8	80	79.2	11.1	40	138.6	19.5	200	198.1	27.8	60	257.5	36.2
21	20.8	2.9	81	80.2	11.3	141	139.6	19.6	201	199.0	28.0	261	258.5	36.3
22	21.8	3.1	82	81.2	11.4	42	140.6	19.8	02	200.0	28.1	62	259.5	36.5
23	22.8	3.2	83	82.2	11.6	43	141.6	19.9	03	201.0	28.3	63	260.4	36.6
24	23.8	3.3	84	83.2	11.7	44	142.6	20.0	04	202.0	28.4	64	261.4	36.7
25	24.8	3.5	85	84.2	11.8	45	143.6	20.2	05	203.0	28.5	65	262.4	36.9
26	25.7	3.6	86	85.2	12.0	46	144.6	20.3	06	204.0	28.7	66	263.4	37.0
27	26.7	3.8	87	86.2	12.1	47	145.6	20.5	07	205.0	28.8	67	264.4	37.2
28	27.7	3.9	88	87.1	12.2	48	146.6	20.6	08	206.0	28.9	68	265.4	37.3
29	28.7	4.0	89	88.1	12.4	49	147.5	20.7	09	207.0	29.1	69	266.4	37.4
30	29.7	4.2	90	89.1	12.5	50	148.5	20.9	10	208.0	29.2	70	267.4	37.6
31	30.7	4.3	91	90.1	12.7	151	149.5	21.0	211	208.9	29.4	271	268.4	37.7
32	31.7	4.5	92	91.1	12.8	52	150.5	21.2	12	209.9	29.5	72	269.4	37.9
33	32.7	4.6	93	92.1	12.9	53	151.5	21.3	13	210.9	29.6	73	270.3	38.0
34	33.7	4.7	94	93.1	13.1	54	152.5	21.4	14	211.9	29.8	74	271.3	38.1
35	34.7	4.9	95	94.1	13.2	55	153.5	21.6	15	212.9	29.9	75	272.3	38.3
36	35.6	5.0	96	95.1	13.4	56	154.5	21.7	16	213.9	30.1	76	273.3	38.4
37	36.6	5.1	97	96.1	13.5	57	155.5	21.9	17	214.9	30.2	77	274.3	38.6
38	37.6	5.3	98	97.0	13.6	58	156.5	22.0	18	215.9	30.3	78	275.3	38.7
39	38.6	5.4	99	98.0	13.8	59	157.5	22.1	19	216.9	30.5	79	276.3	38.8
40	39.6	5.6	100	99.0	13.9	60	158.4	22.3	20	217.9	30.6	80	277.3	39.0
41	40.6	5.7	101	100.0	14.1	161	159.4	22.4	221	218.8	30.8	281	278.3	39.1
42	41.6	5.8	02	101.0	14.2	62	160.4	22.5	22	219.8	30.9	82	279.3	39.2
43	42.6	6.0	03	102.0	14.3	63	161.4	22.7	23	220.8	31.0	83	280.2	39.4
44	43.6	6.1	04	103.0	14.5	64	162.4	22.8	24	221.8	31.2	84	281.2	39.5
45	44.6	6.3	05	104.0	14.6	65	163.4	23.0	25	222.8	31.3	85	282.2	39.7
46	45.6	6.4	06	105.0	14.8	66	164.4	23.1	26	223.8	31.5	86	283.2	39.8
47	46.5	6.5	07	106.0	14.9	67	165.4	23.2	27	224.8	31.6	87	284.2	39.9
48	47.5	6.7	08	106.9	15.0	68	166.4	23.4	28	225.8	31.7	88	285.2	40.1
49	48.5	6.8	09	107.9	15.2	69	167.4	23.5	29	226.8	31.9	89	286.2	40.2
50	49.5	7.0	10	108.9	15.3	70	168.3	23.7	30	227.8	32.0	90	287.2	40.4
51	50.5	7.1	111	109.9	15.4	171	169.3	23.8	231	228.8	32.1	291	288.2	40.5
52	51.5	7.2	12	110.9	15.6	72	170.3	23.9	32	229.7	32.3	92	289.2	40.6
53	52.5	7.4	13	111.9	15.7	73	171.3	24.1	33	230.7	32.4	93	290.1	40.8
54	53.5	7.5	14	112.9	15.9	74	172.3	24.2	34	231.7	32.6	94	291.1	40.9
55	54.5	7.7	15	113.9	16.0	75	173.3	24.4	35	232.7	32.7	95	292.1	41.1
56	55.5	7.8	16	114.9	16.1	76	174.3	24.5	36	233.7	32.8	96	293.1	41.2
57	56.4	7.9	17	115.9	16.3	77	175.3	24.6	37	234.7	33.0	97	294.1	41.3
58	57.4	8.1	18	116.9	16.4	78	176.3	24.8	38	235.7	33.1	98	295.1	41.5
59	58.4	8.2	19	117.8	16.6	79	177.3	24.9	39	236.7	33.3	99	296.1	41.6
60	59.4	8.4	20	118.8	16.7	80	178.2	25.1	40	237.7	33.4	300	297.1	41.8
Dist.	Dep.	D. Lat.	Dist.	Dep.	D. Lat.	Dist.	Dep.	D. Lat.	Dist.	Dep.	D. Lat.	Dist.	Dep.	D. Lat.

278° 262°	082° 098°

82°

Dist.	D. Lat.	Dep.
N.	N x Cos.	N x Sin.
Hypotenuse	Side Adj.	Side Opp.

TABLE 4

Traverse 8° Table

352° 188°	008° 172°

Dist.	D. Lat.	Dep.	Dist.	D. Lat.	Dep.	Dist.	D. Lat.	Dep.	Dist.	D. Lat.	Dep.	Dist.	D. Lat.	Dep.
301	298.1	41.9	361	357.5	50.2	421	416.9	58.6	481	476.3	66.9	541	535.7	75.3
02	299.1	42.0	62	358.5	50.4	22	417.9	58.7	82	477.3	67.1	42	536.7	75.4
03	300.1	42.2	63	359.5	50.5	23	418.9	58.9	83	478.3	67.2	43	537.7	75.6
04	301.0	42.3	64	360.5	50.7	24	419.9	59.0	84	479.3	67.4	44	538.7	75.7
05	302.0	42.4	65	361.4	50.8	25	420.9	59.1	85	480.3	67.5	45	539.7	75.8
06	303.0	42.6	66	362.4	50.9	26	421.9	59.3	86	481.3	67.6	46	540.7	76.0
07	304.0	42.7	67	363.4	51.1	27	422.8	59.4	87	482.3	67.8	47	541.7	76.1
08	305.0	42.9	68	364.4	51.2	28	423.8	59.6	88	483.3	67.9	48	542.7	76.3
09	306.0	43.0	69	365.4	51.4	29	424.8	59.7	89	484.2	68.1	49	543.7	76.4
10	307.0	43.1	70	366.4	51.5	30	425.8	59.8	90	485.2	68.2	50	544.6	76.5
311	308.0	43.3	371	367.4	51.6	431	426.8	60.0	491	486.2	68.3	551	545.6	76.7
12	309.0	43.4	72	368.4	51.8	32	427.8	60.1	92	487.2	68.5	52	546.6	76.8
13	310.0	43.6	73	369.4	51.9	33	428.8	60.3	93	488.2	68.6	53	547.6	77.0
14	310.9	43.7	74	370.4	52.1	34	429.8	60.4	94	489.2	68.8	54	548.6	77.1
15	311.9	43.8	75	371.4	52.2	35	430.8	60.5	95	490.2	68.9	55	549.6	77.2
16	312.9	44.0	76	372.3	52.3	36	431.8	60.7	96	491.2	69.0	56	550.6	77.4
17	313.9	44.1	77	373.3	52.5	37	432.7	60.8	97	492.2	69.2	57	551.6	77.5
18	314.9	44.3	78	374.3	52.6	38	433.7	61.0	98	493.2	69.3	58	552.6	77.7
19	315.9	44.4	79	375.3	52.7	39	434.7	61.1	99	494.1	69.4	59	553.6	77.8
20	316.9	44.5	80	376.3	52.9	40	435.7	61.2	500	495.1	69.6	60	554.6	77.9
321	317.9	44.7	381	377.3	53.0	441	436.7	61.4	501	496.1	69.7	561	555.5	78.1
22	318.9	44.8	82	378.3	53.2	42	437.7	61.5	02	497.1	69.9	62	556.5	78.2
23	319.9	45.0	83	379.3	53.3	43	438.7	61.7	03	498.1	70.0	63	557.5	78.4
24	320.8	45.1	84	380.3	53.4	44	439.7	61.8	04	499.1	70.1	64	558.5	78.5
25	321.8	45.2	85	381.3	53.6	45	440.7	61.9	05	500.1	70.3	65	559.5	78.6
26	322.8	45.4	86	382.2	53.7	46	441.7	62.1	06	501.1	70.4	66	560.5	78.8
27	323.8	45.5	87	383.2	53.9	47	442.6	62.2	07	502.1	70.6	67	561.5	78.9
28	324.8	45.6	88	384.2	54.0	48	443.6	62.3	08	503.1	70.7	68	562.5	79.1
29	325.8	45.8	89	385.2	54.1	49	444.6	62.5	09	504.0	70.8	69	563.5	79.2
30	326.8	45.9	90	386.2	54.3	50	445.6	62.6	10	505.0	71.0	70	564.5	79.3
331	327.8	46.1	391	387.2	54.4	451	446.6	62.8	511	506.0	71.1	571	565.4	79.5
32	328.8	46.2	92	388.2	54.6	52	447.6	62.9	12	507.0	71.3	72	566.4	79.6
33	329.8	46.3	93	389.2	54.7	53	448.6	63.0	13	508.0	71.4	73	567.4	79.7
34	330.7	46.5	94	390.2	54.8	54	449.6	63.2	14	509.0	71.5	74	568.4	79.9
35	331.7	46.6	95	391.2	55.0	55	450.6	63.3	15	510.0	71.7	75	569.4	80.0
36	332.7	46.8	96	392.1	55.1	56	451.6	63.5	16	511.0	71.8	76	570.4	80.2
37	333.7	46.9	97	393.1	55.3	57	452.6	63.6	17	512.0	72.0	77	571.4	80.3
38	334.7	47.0	98	394.1	55.4	58	453.5	63.7	18	513.0	72.1	78	572.4	80.4
39	335.7	47.2	99	395.1	55.5	59	454.5	63.9	19	513.9	72.2	79	573.4	80.6
40	336.7	47.3	400	396.1	55.7	60	455.5	64.0	20	514.9	72.4	80	574.4	80.7
341	337.7	47.5	401	397.1	55.8	461	456.5	64.2	521	515.9	72.5	581	575.3	80.9
42	338.7	47.6	02	398.1	55.9	62	457.5	64.3	22	516.9	72.6	82	576.3	81.0
43	339.7	47.7	03	399.1	56.1	63	458.5	64.4	23	517.9	72.8	83	577.3	81.1
44	340.7	47.9	04	400.1	56.2	64	459.5	64.6	24	518.9	72.9	84	578.3	81.3
45	341.6	48.0	05	401.1	56.4	65	460.5	64.7	25	519.9	73.1	85	579.3	81.4
46	342.6	48.2	06	402.0	56.5	66	461.5	64.9	26	520.9	73.2	86	580.3	81.6
47	343.6	48.3	07	403.0	56.6	67	462.5	65.0	27	521.9	73.3	87	581.3	81.7
48	344.6	48.4	08	404.0	56.8	68	463.4	65.1	28	522.9	73.5	88	582.3	81.8
49	345.6	48.6	09	405.0	56.9	69	464.4	65.3	29	523.9	73.6	89	583.3	82.0
50	346.6	48.7	10	406.0	57.1	70	465.4	65.4	30	524.8	73.8	90	584.3	82.1
351	347.6	48.8	411	407.0	57.2	471	466.4	65.6	531	525.8	73.9	591	585.2	82.3
52	348.6	49.0	12	408.0	57.3	72	467.4	65.7	32	526.8	74.0	92	586.2	82.4
53	349.6	49.1	13	409.0	57.5	73	468.4	65.8	33	527.8	74.2	93	587.2	82.5
54	350.6	49.3	14	410.0	57.6	74	469.4	66.0	34	528.8	74.3	94	588.2	82.7
55	351.5	49.4	15	411.0	57.8	75	470.4	66.1	35	529.8	74.5	95	589.2	82.8
56	352.5	49.5	16	412.0	57.9	76	471.4	66.2	36	530.8	74.6	96	590.2	82.9
57	353.5	49.7	17	412.9	58.0	77	472.4	66.4	37	531.8	74.7	97	591.2	83.1
58	354.5	49.8	18	413.9	58.2	78	473.3	66.5	38	532.8	74.9	98	592.2	83.2
59	355.5	50.0	19	414.9	58.3	79	474.3	66.7	39	533.8	75.0	99	593.2	83.4
60	356.5	50.1	20	415.9	58.5	80	475.3	66.8	40	534.7	75.2	600	594.2	83.5
Dist.	Dep.	D. Lat.	Dist.	Dep.	D. Lat.	Dist.	Dep.	D. Lat.	Dist.	Dep.	D. Lat.	Dist.	Dep.	D. Lat.

Dist.	D. Lat.	Dep.
D Lo	Dep.	
	m	D Lo

82°

278° 262°	082° 098°

TABLE 4

9°

Traverse Table

351°	009°
189°	171°

Dist.	D. Lat.	Dep.	Dist.	D. Lat.	Dep.	Dist.	D. Lat.	Dep.	Dist.	D. Lat.	Dep.	Dist.	D. Lat.	Dep.
1	1.0	0.2	61	60.2	9.5	121	119.5	18.9	181	178.8	28.3	241	238.0	37.7
2	2.0	0.3	62	61.2	9.7	22	120.5	19.1	82	179.8	28.5	42	239.0	37.9
3	3.0	0.5	63	62.2	9.9	23	121.5	19.2	83	180.7	28.6	43	240.0	38.0
4	4.0	0.6	64	63.2	10.0	24	122.5	19.4	84	181.7	28.8	44	241.0	38.2
5	4.9	0.8	65	64.2	10.2	25	123.5	19.6	85	182.7	28.9	45	242.0	38.3
6	5.9	0.9	66	65.2	10.3	26	124.4	19.7	86	183.7	29.1	46	243.0	38.5
7	6.9	1.1	67	66.2	10.5	27	125.4	19.9	87	184.7	29.3	47	244.0	38.6
8	7.9	1.3	68	67.2	10.6	28	126.4	20.0	88	185.7	29.4	48	244.9	38.8
9	8.9	1.4	69	68.2	10.8	29	127.4	20.2	89	186.7	29.6	49	245.9	39.0
10	9.9	1.6	70	69.1	11.0	30	128.4	20.3	90	187.7	29.7	50	246.9	39.1
11	10.9	1.7	71	70.1	11.1	131	129.4	20.5	191	188.6	29.9	251	247.9	39.3
12	11.9	1.9	72	71.1	11.3	32	130.4	20.6	92	189.6	30.0	52	248.9	39.4
13	12.8	2.0	73	72.1	11.4	33	131.4	20.8	93	190.6	30.2	53	249.9	39.6
14	13.8	2.2	74	73.1	11.6	34	132.4	21.0	94	191.6	30.3	54	250.9	39.7
15	14.8	2.3	75	74.1	11.7	35	133.3	21.1	95	192.6	30.5	55	251.9	39.9
16	15.8	2.5	76	75.1	11.9	36	134.3	21.3	96	193.6	30.7	56	252.8	40.0
17	16.8	2.7	77	76.1	12.0	37	135.3	21.4	97	194.6	30.8	57	253.8	40.2
18	17.8	2.8	78	77.0	12.2	38	136.3	21.6	98	195.6	31.0	58	254.8	40.4
19	18.8	3.0	79	78.0	12.4	39	137.3	21.7	99	196.5	31.1	59	255.8	40.5
20	19.8	3.1	80	79.0	12.5	40	138.3	21.9	200	197.5	31.3	60	256.8	40.7
21	20.7	3.3	81	80.0	12.7	141	139.3	22.1	201	198.5	31.4	261	257.8	40.8
22	21.7	3.4	82	81.0	12.8	42	140.3	22.2	02	199.5	31.6	62	258.8	41.0
23	22.7	3.6	83	82.0	13.0	43	141.2	22.4	03	200.5	31.8	63	259.8	41.1
24	23.7	3.8	84	83.0	13.1	44	142.2	22.5	04	201.5	31.9	64	260.7	41.3
25	24.7	3.9	85	84.0	13.3	45	143.2	22.7	05	202.5	32.1	65	261.7	41.5
26	25.7	4.1	86	84.9	13.5	46	144.2	22.8	06	203.5	32.2	66	262.7	41.6
27	26.7	4.2	87	85.9	13.6	47	145.2	23.0	07	204.5	32.4	67	263.7	41.8
28	27.7	4.4	88	86.9	13.8	48	146.2	23.2	08	205.4	32.5	68	264.7	41.9
29	28.6	4.5	89	87.9	13.9	49	147.2	23.3	09	206.4	32.7	69	265.7	42.1
30	29.6	4.7	90	88.9	14.1	50	148.2	23.5	10	207.4	32.9	70	266.7	42.2
31	30.6	4.8	91	89.9	14.2	151	149.1	23.6	211	208.4	33.0	271	267.7	42.4
32	31.6	5.0	92	90.9	14.4	52	150.1	23.8	12	209.4	33.2	72	268.7	42.6
33	32.6	5.2	93	91.9	14.5	53	151.1	23.9	13	210.4	33.3	73	269.6	42.7
34	33.6	5.3	94	92.8	14.7	54	152.1	24.1	14	211.4	33.5	74	270.6	42.9
35	34.6	5.5	95	93.8	14.9	55	153.1	24.2	15	212.4	33.6	75	271.6	43.0
36	35.6	5.6	96	94.8	15.0	56	154.1	24.4	16	213.3	33.8	76	272.6	43.2
37	36.5	5.8	97	95.8	15.2	57	155.1	24.6	17	214.3	33.9	77	273.6	43.3
38	37.5	5.9	98	96.8	15.3	58	156.1	24.7	18	215.3	34.1	78	274.6	43.5
39	38.5	6.1	99	97.8	15.5	59	157.0	24.9	19	216.3	34.3	79	275.6	43.6
40	39.5	6.3	100	98.8	15.6	60	158.0	25.0	20	217.3	34.4	80	276.6	43.8
41	40.5	6.4	101	99.8	15.8	161	159.0	25.2	221	218.3	34.6	281	277.5	44.0
42	41.5	6.6	02	100.7	16.0	62	160.0	25.3	22	219.3	34.7	82	278.5	44.1
43	42.5	6.7	03	101.7	16.1	63	161.0	25.5	23	220.3	34.9	83	279.5	44.3
44	43.5	6.9	04	102.7	16.3	64	162.0	25.7	24	221.2	35.0	84	280.5	44.4
45	44.4	7.0	05	103.7	16.4	65	163.0	25.8	25	222.2	35.2	85	281.5	44.6
46	45.4	7.2	06	104.7	16.6	66	164.0	26.0	26	223.2	35.4	86	282.5	44.7
47	46.4	7.4	07	105.7	16.7	67	164.9	26.1	27	224.2	35.5	87	283.5	44.9
48	47.4	7.5	08	106.7	16.9	68	165.9	26.3	28	225.2	35.7	88	284.5	45.1
49	48.4	7.7	09	107.7	17.1	69	166.9	26.4	29	226.2	35.8	89	285.4	45.2
50	49.4	7.8	10	108.6	17.2	70	167.9	26.6	30	227.2	36.0	90	286.4	45.4
51	50.4	8.0	111	109.6	17.4	171	168.9	26.8	231	228.2	36.1	291	287.4	45.5
52	51.4	8.1	12	110.6	17.5	72	169.9	26.9	32	229.1	36.3	92	288.4	45.7
53	52.3	8.3	13	111.6	17.7	73	170.9	27.1	33	230.1	36.4	93	289.4	45.8
54	53.3	8.4	14	112.6	17.8	74	171.9	27.2	34	231.1	36.6	94	290.4	46.0
55	54.3	8.6	15	113.6	18.0	75	172.8	27.4	35	232.1	36.8	95	291.4	46.1
56	55.3	8.8	16	114.6	18.1	76	173.8	27.5	36	233.1	36.9	96	292.4	46.3
57	56.3	8.9	17	115.6	18.3	77	174.8	27.7	37	234.1	37.1	97	293.3	46.5
58	57.3	9.1	18	116.5	18.5	78	175.8	27.8	38	235.1	37.2	98	294.3	46.6
59	58.3	9.2	19	117.5	18.6	79	176.8	28.0	39	236.1	37.4	99	295.3	46.8
60	59.3	9.4	20	118.5	18.8	80	177.8	28.2	40	237.0	37.5	300	296.3	46.9
Dist.	Dep.	D. Lat.	Dist.	Dep.	D. Lat.	Dist.	Dep.	D. Lat.	Dist.	Dep.	D. Lat.	Dist.	Dep.	D. Lat.

279°	081°
261°	099°

81°

Dist.	D. Lat.	Dep.
N.	N x Cos.	N x Sin.
Hypotenuse	Side Adj.	Side Opp.

TABLE 4

9°

Traverse Table

351°	009°
189°	171°

Dist.	D. Lat.	Dep.	Dist.	D. Lat.	Dep.	Dist.	D. Lat.	Dep.	Dist.	D. Lat.	Dep.	Dist.	D. Lat.	Dep.
301	297.3	47.1	361	356.6	56.5	421	415.8	65.9	481	475.1	75.2	541	534.3	84.6
02	298.3	47.2	62	357.5	56.6	22	416.8	66.0	82	476.1	75.4	42	535.3	84.8
03	299.3	47.4	63	358.5	56.8	23	417.8	66.2	83	477.1	75.6	43	536.3	84.9
04	300.3	47.6	64	359.5	56.9	24	418.8	66.3	84	478.0	75.7	44	537.3	85.1
05	301.2	47.7	65	360.5	57.1	25	419.8	66.5	85	479.0	75.9	45	538.3	85.3
06	302.2	47.9	66	361.5	57.3	26	420.8	66.6	86	480.0	76.0	46	539.3	85.4
07	303.2	48.0	67	362.5	57.4	27	421.7	66.8	87	481.0	76.2	47	540.3	85.6
08	304.2	48.2	68	363.5	57.6	28	422.7	67.0	88	482.0	76.3	48	541.3	85.7
09	305.2	48.3	69	364.5	57.7	29	423.7	67.1	89	483.0	76.5	49	542.2	85.9
10	306.2	48.5	70	365.4	57.9	30	424.7	67.3	90	484.0	76.7	50	543.2	86.0
311	307.2	48.7	371	366.4	58.0	431	425.7	67.4	491	485.0	76.8	551	544.2	86.2
12	308.2	48.8	72	367.4	58.2	32	426.7	67.6	92	485.9	77.0	52	545.2	86.4
13	309.1	49.0	73	368.4	58.4	33	427.7	67.7	93	486.9	77.1	53	546.2	86.5
14	310.1	49.1	74	369.4	58.5	34	428.7	67.9	94	487.9	77.3	54	547.2	86.7
15	311.1	49.3	75	370.4	58.7	35	429.6	68.0	95	488.9	77.4	55	548.2	86.8
16	312.1	49.4	76	371.4	58.8	36	430.6	68.2	96	489.9	77.6	56	549.2	87.0
17	313.1	49.6	77	372.4	59.0	37	431.6	68.4	97	490.9	77.7	57	550.1	87.1
18	314.1	49.7	78	373.3	59.1	38	432.6	68.5	98	491.9	77.9	58	551.1	87.3
19	315.1	49.9	79	374.3	59.3	39	433.6	68.7	99	492.9	78.1	59	552.1	87.4
20	316.1	50.1	80	375.3	59.4	40	434.6	68.8	500	493.8	78.2	60	553.1	87.6
321	317.0	50.2	381	376.3	59.6	441	435.6	69.0	501	494.8	78.4	561	554.1	87.8
22	318.0	50.4	82	377.3	59.8	42	436.6	69.1	02	495.8	78.5	62	555.1	87.9
23	319.0	50.5	83	378.3	59.9	43	437.5	69.3	03	496.8	78.7	63	556.1	88.1
24	320.0	50.7	84	379.3	60.1	44	438.5	69.5	04	497.8	78.8	64	557.1	88.2
25	321.0	50.8	85	380.3	60.2	45	439.5	69.6	05	498.8	79.0	65	558.0	88.4
26	322.0	51.0	86	381.2	60.4	46	440.5	69.8	06	499.8	79.2	66	559.0	88.5
27	323.0	51.2	87	382.2	60.5	47	441.5	69.9	07	500.8	79.3	67	560.0	88.7
28	324.0	51.3	88	383.2	60.7	48	442.5	70.1	08	501.7	79.5	68	561.0	88.9
29	324.9	51.5	89	384.2	60.9	49	443.5	70.2	09	502.7	79.6	69	562.0	89.0
30	325.9	51.6	90	385.2	61.0	50	444.5	70.4	10	503.7	79.8	70	563.0	89.2
331	326.9	51.8	391	386.2	61.2	451	445.4	70.6	511	504.7	79.9	571	564.0	89.3
32	327.9	51.9	92	387.2	61.3	52	446.4	70.7	12	505.7	80.1	72	565.0	89.5
33	328.9	52.1	93	388.2	61.5	53	447.4	70.9	13	506.7	80.3	73	565.9	89.6
34	329.9	52.2	94	389.1	61.6	54	448.4	71.0	14	507.7	80.4	74	566.9	89.8
35	330.9	52.4	95	390.1	61.8	55	449.4	71.2	15	508.7	80.6	75	567.9	89.9
36	331.9	52.6	96	391.1	61.9	56	450.4	71.3	16	509.6	80.7	76	568.9	90.1
37	332.9	52.7	97	392.1	62.1	57	451.4	71.5	17	510.6	80.9	77	569.9	90.3
38	333.8	52.9	98	393.1	62.3	58	452.4	71.6	18	511.6	81.0	78	570.9	90.4
39	334.8	53.0	99	394.1	62.4	59	453.3	71.8	19	512.6	81.2	79	571.9	90.6
40	335.8	53.2	400	395.1	62.6	60	454.3	72.0	20	513.6	81.3	80	572.9	90.7
341	336.8	53.3	401	396.1	62.7	461	455.3	72.1	521	514.6	81.5	581	573.8	90.9
42	337.8	53.5	02	397.1	62.9	62	456.3	72.3	22	515.6	81.7	82	574.8	91.0
43	338.8	53.7	03	398.0	63.0	63	457.3	72.4	23	516.6	81.8	83	575.8	91.2
44	339.8	53.8	04	399.0	63.2	64	458.3	72.6	24	517.5	82.0	84	576.8	91.4
45	340.8	54.0	05	400.0	63.4	65	459.3	72.7	25	518.5	82.1	85	577.8	91.5
46	341.7	54.1	06	401.0	63.5	66	460.3	72.9	26	519.5	82.3	86	578.8	91.7
47	342.7	54.3	07	402.0	63.7	67	461.3	73.1	27	520.5	82.4	87	579.8	91.8
48	343.7	54.4	08	403.0	63.8	68	462.2	73.2	28	521.5	82.6	88	580.8	92.0
49	344.7	54.6	09	404.0	64.0	69	463.2	73.4	29	522.5	82.8	89	581.7	92.1
50	345.7	54.8	10	405.0	64.1	70	464.2	73.5	30	523.5	82.9	90	582.7	92.3
351	346.7	54.9	411	405.9	64.3	471	465.2	73.7	531	524.5	83.1	591	583.7	92.5
52	347.7	55.1	12	406.9	64.5	72	466.2	73.8	32	525.5	83.2	92	584.7	92.6
53	348.7	55.2	13	407.9	64.6	73	467.2	74.0	33	526.4	83.4	93	585.7	92.8
54	349.6	55.4	14	408.9	64.8	74	468.2	74.1	34	527.4	83.5	94	586.7	92.9
55	350.6	55.5	15	409.9	64.9	75	469.2	74.3	35	528.4	83.7	95	587.7	93.1
56	351.6	55.7	16	410.9	65.1	76	470.1	74.5	36	529.4	83.8	96	588.7	93.2
57	352.6	55.8	17	411.9	65.2	77	471.1	74.6	37	530.4	84.0	97	589.6	93.4
58	353.6	56.0	18	412.9	65.4	78	472.1	74.8	38	531.4	84.2	98	590.6	93.5
59	354.6	56.2	19	413.8	65.5	79	473.1	74.9	39	532.4	84.3	99	591.6	93.7
60	355.6	56.3	20	414.8	65.7	80	474.1	75.1	40	533.4	84.5	600	592.6	93.9
Dist.	Dep.	D. Lat.	Dist.	Dep.	D. Lat.	Dist.	Dep.	D. Lat.	Dist.	Dep.	D. Lat.	Dist.	Dep.	D. Lat.

Dist.	D. Lat.	Dep.
D Lo	Dep.	
	m	D Lo

81°

279°	081°
261°	099°

TABLE 4

10°

Traverse Table

350° | 010°
190° | 170°

Dist.	D. Lat.	Dep.	Dist.	D. Lat.	Dep.	Dist.	D. Lat.	Dep.	Dist.	D. Lat.	Dep.	Dist.	D. Lat.	Dep.
1	1.0	0.2	61	60.1	10.6	121	119.2	21.0	181	178.3	31.4	241	237.3	41.8
2	2.0	0.3	62	61.1	10.8	22	120.1	21.2	82	179.2	31.6	42	238.3	42.0
3	3.0	0.5	63	62.0	10.9	23	121.1	21.4	83	180.2	31.8	43	239.3	42.2
4	3.9	0.7	64	63.0	11.1	24	122.1	21.5	84	181.2	32.0	44	240.3	42.4
5	4.9	0.9	65	64.0	11.3	25	123.1	21.7	85	182.2	32.1	45	241.3	42.5
6	5.9	1.0	66	65.0	11.5	26	124.1	21.9	86	183.2	32.3	46	242.3	42.7
7	6.9	1.2	67	66.0	11.6	27	125.1	22.1	87	184.2	32.5	47	243.2	42.9
8	7.9	1.4	68	67.0	11.8	28	126.1	22.2	88	185.1	32.6	48	244.2	43.1
9	8.9	1.6	69	68.0	12.0	29	127.0	22.4	89	186.1	32.8	49	245.2	43.2
10	9.8	1.7	70	68.9	12.2	30	128.0	22.6	90	187.1	33.0	50	246.2	43.4
11	10.8	1.9	71	69.9	12.3	131	129.0	22.7	191	188.1	33.2	251	247.2	43.6
12	11.8	2.1	72	70.9	12.5	32	130.0	22.9	92	189.1	33.3	52	248.2	43.8
13	12.8	2.3	73	71.9	12.7	33	131.0	23.1	93	190.1	33.5	53	249.2	43.9
14	13.8	2.4	74	72.9	12.8	34	132.0	23.3	94	191.1	33.7	54	250.1	44.1
15	14.8	2.6	75	73.9	13.0	35	132.9	23.4	95	192.0	33.9	55	251.1	44.3
16	15.8	2.8	76	74.8	13.2	36	133.9	23.6	96	193.0	34.0	56	252.1	44.5
17	16.7	3.0	77	75.8	13.4	37	134.9	23.8	97	194.0	34.2	57	253.1	44.6
18	17.7	3.1	78	76.8	13.5	38	135.9	24.0	98	195.0	34.4	58	254.1	44.8
19	18.7	3.3	79	77.8	13.7	39	136.9	24.1	99	196.0	34.6	59	255.1	45.0
20	19.7	3.5	80	78.8	13.9	40	137.9	24.3	200	197.0	34.7	60	256.1	45.1
21	20.7	3.6	81	79.8	14.1	141	138.9	24.5	201	197.9	34.9	261	257.0	45.3
22	21.7	3.8	82	80.8	14.2	42	139.8	24.7	02	198.9	35.1	62	258.0	45.5
23	22.7	4.0	83	81.7	14.4	43	140.8	24.8	03	199.9	35.3	63	259.0	45.7
24	23.6	4.2	84	82.7	14.6	44	141.8	25.0	04	200.9	35.4	64	260.0	45.8
25	24.6	4.3	85	83.7	14.8	45	142.8	25.2	05	201.9	35.6	65	261.0	46.0
26	25.6	4.5	86	84.7	14.9	46	143.8	25.4	06	202.9	35.8	66	262.0	46.2
27	26.6	4.7	87	85.7	15.1	47	144.8	25.5	07	203.9	35.9	67	262.9	46.4
28	27.6	4.9	88	86.7	15.3	48	145.8	25.7	08	204.8	36.1	68	263.9	46.5
29	28.6	5.0	89	87.6	15.5	49	146.7	25.9	09	205.8	36.3	69	264.9	46.7
30	29.5	5.2	90	88.6	15.6	50	147.7	26.0	10	206.8	36.5	70	265.9	46.9
31	30.5	5.4	91	89.6	15.8	151	148.7	26.2	211	207.8	36.6	271	266.9	47.1
32	31.5	5.6	92	90.6	16.0	52	149.7	26.4	12	208.8	36.8	72	267.9	47.2
33	32.5	5.7	93	91.6	16.1	53	150.7	26.6	13	209.8	37.0	73	268.9	47.4
34	33.5	5.9	94	92.6	16.3	54	151.7	26.7	14	210.7	37.2	74	269.8	47.6
35	34.5	6.1	95	93.6	16.5	55	152.6	26.9	15	211.7	37.3	75	270.8	47.8
36	35.5	6.3	96	94.5	16.7	56	153.6	27.1	16	212.7	37.5	76	271.8	47.9
37	36.4	6.4	97	95.5	16.8	57	154.6	27.3	17	213.7	37.7	77	272.8	48.1
38	37.4	6.6	98	96.5	17.0	58	155.6	27.4	18	214.7	37.9	78	273.8	48.3
39	38.4	6.8	99	97.5	17.2	59	156.6	27.6	19	215.7	38.0	79	274.8	48.4
40	39.4	6.9	100	98.5	17.4	60	157.6	27.8	20	216.7	38.2	80	275.7	48.6
41	40.4	7.1	101	99.5	17.5	161	158.6	28.0	221	217.6	38.4	281	276.7	48.8
42	41.4	7.3	02	100.5	17.7	62	159.5	28.1	22	218.6	38.5	82	277.7	49.0
43	42.3	7.5	03	101.4	17.9	63	160.5	28.3	23	219.6	38.7	83	278.7	49.1
44	43.3	7.6	04	102.4	18.1	64	161.5	28.5	24	220.6	38.9	84	279.7	49.3
45	44.3	7.8	05	103.4	18.2	65	162.5	28.7	25	221.6	39.1	85	280.7	49.5
46	45.3	8.0	06	104.4	18.4	66	163.5	28.8	26	222.6	39.2	86	281.7	49.7
47	46.3	8.2	07	105.4	18.6	67	164.5	29.0	27	223.6	39.4	87	282.6	49.8
48	47.3	8.3	08	106.4	18.8	68	165.4	29.2	28	224.5	39.6	88	283.6	50.0
49	48.3	8.5	09	107.3	18.9	69	166.4	29.3	29	225.5	39.8	89	284.6	50.2
50	49.2	8.7	10	108.3	19.1	70	167.4	29.5	30	226.5	39.9	90	285.6	50.4
51	50.2	8.9	111	109.3	19.3	171	168.4	29.7	231	227.5	40.1	291	286.6	50.5
52	51.2	9.0	12	110.3	19.4	72	169.4	29.9	32	228.5	40.3	92	287.6	50.7
53	52.2	9.2	13	111.3	19.6	73	170.4	30.0	33	229.5	40.5	93	288.5	50.9
54	53.2	9.4	14	112.3	19.8	74	171.4	30.2	34	230.4	40.6	94	289.5	51.1
55	54.2	9.6	15	113.3	20.0	75	172.3	30.4	35	231.4	40.8	95	290.5	51.2
56	55.1	9.7	16	114.2	20.1	76	173.3	30.6	36	232.4	41.0	96	291.5	51.4
57	56.1	9.9	17	115.2	20.3	77	174.3	30.7	37	233.4	41.2	97	292.5	51.6
58	57.1	10.1	18	116.2	20.5	78	175.3	30.9	38	234.4	41.3	98	293.5	51.7
59	58.1	10.2	19	117.2	20.7	79	176.3	31.1	39	235.4	41.5	99	294.5	51.9
60	59.1	10.4	20	118.2	20.8	80	177.3	31.3	40	236.4	41.7	300	295.4	52.1
Dist.	Dep.	D. Lat.	Dist.	Dep.	D. Lat.	Dist.	Dep.	D. Lat.	Dist.	Dep.	D. Lat.	Dist.	Dep.	D. Lat.

280° | 080°
260° | 100°

80°

Dist.	D. Lat.	Dep.
N.	N x Cos.	N x Sin.
Hypotenuse	Side Adj.	Side Opp.

TABLE 4

10°

Traverse Table

350° | 010°
190° | 170°

Dist.	D. Lat.	Dep.	Dist.	D. Lat.	Dep.	Dist.	D. Lat.	Dep.	Dist.	D. Lat.	Dep.	Dist.	D. Lat.	Dep.
301	296.4	52.3	361	355.5	62.7	421	414.6	73.1	481	473.7	83.5	541	532.8	93.9
02	297.4	52.4	62	356.5	62.9	22	415.6	73.3	82	474.7	83.7	42	533.8	94.1
03	298.4	52.6	63	357.5	63.0	23	416.6	73.5	83	475.7	83.9	43	534.8	94.3
04	299.4	52.8	64	358.5	63.2	24	417.6	73.6	84	476.6	84.0	44	535.7	94.5
05	300.4	53.0	65	359.5	63.4	25	418.5	73.8	85	477.6	84.2	45	536.7	94.6
06	301.4	53.1	66	360.4	63.6	26	419.5	74.0	86	478.6	84.4	46	537.7	94.8
07	302.3	53.3	67	361.4	63.7	27	420.5	74.1	87	479.6	84.6	47	538.7	95.0
08	303.3	53.5	68	362.4	63.9	28	421.5	74.3	88	480.6	84.7	48	539.7	95.2
09	304.3	53.7	69	363.4	64.1	29	422.5	74.5	89	481.6	84.9	49	540.7	95.3
10	305.3	53.8	70	364.4	64.2	30	423.5	74.7	90	482.6	85.1	50	541.6	95.5
311	306.3	54.0	371	365.4	64.4	431	424.5	74.8	491	483.5	85.3	551	542.6	95.7
12	307.3	54.2	72	366.3	64.6	32	425.4	75.0	92	484.5	85.4	52	543.6	95.9
13	308.2	54.4	73	367.3	64.8	33	426.4	75.2	93	485.5	85.6	53	544.6	96.0
14	309.2	54.5	74	368.3	64.9	34	427.4	75.4	94	486.5	85.8	54	545.6	96.2
15	310.2	54.7	75	369.3	65.1	35	428.4	75.5	95	487.5	86.0	55	546.6	96.4
16	311.2	54.9	76	370.3	65.3	36	429.4	75.7	96	488.5	86.1	56	547.6	96.5
17	312.2	55.0	77	371.3	65.5	37	430.4	75.9	97	489.4	86.3	57	548.5	96.7
18	313.2	55.2	78	372.3	65.6	38	431.3	76.1	98	490.4	86.5	58	549.5	96.9
19	314.2	55.4	79	373.2	65.8	39	432.3	76.2	99	491.4	86.7	59	550.5	97.1
20	315.1	55.6	80	374.2	66.0	40	433.3	76.4	500	492.4	86.8	60	551.5	97.2
321	316.1	55.7	381	375.2	66.2	441	434.3	76.6	501	493.4	87.0	561	552.5	97.4
22	317.1	55.9	82	376.2	66.3	42	435.3	76.8	02	494.4	87.2	62	553.5	97.6
23	318.1	56.1	83	377.2	66.5	43	436.3	76.9	03	495.4	87.3	63	554.4	97.8
24	319.1	56.3	84	378.2	66.7	44	437.3	77.1	04	496.3	87.5	64	555.4	97.9
25	320.1	56.4	85	379.2	66.9	45	438.2	77.3	05	497.3	87.7	65	556.4	98.1
26	321.0	56.6	86	380.1	67.0	46	439.2	77.4	06	498.3	87.9	66	557.4	98.3
27	322.0	56.8	87	381.1	67.2	47	440.2	77.6	07	499.3	88.0	67	558.4	98.5
28	323.0	57.0	88	382.1	67.4	48	441.2	77.8	08	500.3	88.2	68	559.4	98.6
29	324.0	57.1	89	383.1	67.5	49	442.2	78.0	09	501.3	88.4	69	560.4	98.8
30	325.0	57.3	90	384.1	67.7	50	443.2	78.1	10	502.3	88.6	70	561.3	99.0
331	326.0	57.5	391	385.1	67.9	451	444.1	78.3	511	503.2	88.7	571	562.3	99.2
32	327.0	57.7	92	386.0	68.1	52	445.1	78.5	12	504.2	88.9	72	563.3	99.3
33	327.9	57.8	93	387.0	68.2	53	446.1	78.7	13	505.2	89.1	73	564.3	99.5
34	328.9	58.0	94	388.0	68.4	54	447.1	78.8	14	506.2	89.3	74	565.3	99.7
35	329.9	58.2	95	389.0	68.6	55	448.1	79.0	15	507.2	89.4	75	566.3	99.8
36	330.9	58.3	96	390.0	68.8	56	449.1	79.2	16	508.2	89.6	76	567.2	100.0
37	331.9	58.5	97	391.0	68.9	57	450.1	79.4	17	509.1	89.8	77	568.2	100.2
38	332.9	58.7	98	392.0	69.1	58	451.0	79.5	18	510.1	89.9	78	569.2	100.4
39	333.8	58.9	99	392.9	69.3	59	452.0	79.7	19	511.1	90.1	79	570.2	100.5
40	334.8	59.0	400	393.9	69.5	60	453.0	79.9	20	512.1	90.3	80	571.2	100.7
341	335.8	59.2	401	394.9	69.6	461	454.0	80.1	521	513.1	90.5	581	572.2	100.9
42	336.8	59.4	02	395.9	69.8	62	455.0	80.2	22	514.1	90.6	82	573.2	101.1
43	337.8	59.6	03	396.9	70.0	63	456.0	80.4	23	515.1	90.8	83	574.1	101.2
44	338.8	59.7	04	397.9	70.2	64	457.0	80.6	24	516.0	91.0	84	575.1	101.4
45	339.8	59.9	05	398.8	70.3	65	457.9	80.7	25	517.0	91.2	85	576.1	101.6
46	340.7	60.1	06	399.8	70.5	66	458.9	80.9	26	518.0	91.3	86	577.1	101.8
47	341.7	60.3	07	400.8	70.7	67	459.9	81.1	27	519.0	91.5	87	578.1	101.9
48	342.7	60.4	08	401.8	70.8	68	460.9	81.3	28	520.0	91.7	88	579.1	102.1
49	343.7	60.6	09	402.8	71.0	69	461.9	81.4	29	521.0	91.9	89	580.1	102.3
50	344.7	60.8	10	403.8	71.2	70	462.9	81.6	30	521.9	92.0	90	581.0	102.5
351	345.7	61.0	411	404.8	71.4	471	463.8	81.8	531	522.9	92.2	591	582.0	102.6
52	346.7	61.1	12	405.7	71.5	72	464.8	82.0	32	523.9	92.4	92	583.0	102.8
53	347.6	61.3	13	406.7	71.7	73	465.8	82.1	33	524.9	92.6	93	584.0	103.0
54	348.6	61.5	14	407.7	71.9	74	466.8	82.3	34	525.9	92.7	94	585.0	103.1
55	349.6	61.6	15	408.7	72.1	75	467.8	82.5	35	526.9	92.9	95	586.0	103.3
56	350.6	61.8	16	409.7	72.2	76	468.8	82.7	36	527.9	93.1	96	586.9	103.5
57	351.6	62.0	17	410.7	72.4	77	469.8	82.8	37	528.8	93.2	97	587.9	103.7
58	352.6	62.2	18	411.6	72.6	78	470.7	83.0	38	529.8	93.4	98	588.9	103.8
59	353.5	62.3	19	412.6	72.8	79	471.7	83.2	39	530.8	93.6	99	589.9	104.0
60	354.5	62.5	20	413.6	72.9	80	472.7	83.4	40	531.8	93.8	600	590.9	104.2
Dist.	Dep.	D. Lat.	Dist.	Dep.	D. Lat.	Dist.	Dep.	D. Lat.	Dist.	Dep.	D. Lat.	Dist.	Dep.	D. Lat.

Dist.	D. Lat.	Dep.
D Lo	Dep.	
	m	D Lo

80°

280° | 080°
260° | 100°

TABLE 4

Traverse 11° Table

349° | 011° — 191° | 169°

Dist.	D. Lat.	Dep.	Dist.	D. Lat.	Dep.	Dist.	D. Lat.	Dep.	Dist.	D. Lat.	Dep.	Dist.	D. Lat.	Dep.
1	1.0	0.2	61	59.9	11.6	121	118.8	23.1	181	177.7	34.5	241	236.6	46.0
2	2.0	0.4	62	60.9	11.8	22	119.8	23.3	82	178.7	34.7	42	237.6	46.2
3	2.9	0.6	63	61.8	12.0	23	120.7	23.5	83	179.6	34.9	43	238.5	46.4
4	3.9	0.8	64	62.8	12.2	24	121.7	23.7	84	180.6	35.1	44	239.5	46.6
5	4.9	1.0	65	63.8	12.4	25	122.7	23.9	85	181.6	35.3	45	240.5	46.7
6	5.9	1.1	66	64.8	12.6	26	123.7	24.0	86	182.6	35.5	46	241.5	46.9
7	6.9	1.3	67	65.8	12.8	27	124.7	24.2	87	183.6	35.7	47	242.5	47.1
8	7.9	1.5	68	66.8	13.0	28	125.6	24.4	88	184.5	35.9	48	243.4	47.3
9	8.8	1.7	69	67.7	13.2	29	126.6	24.6	89	185.5	36.1	49	244.4	47.5
10	9.8	1.9	70	68.7	13.4	30	127.6	24.8	90	186.5	36.3	50	245.4	47.7
11	10.8	2.1	71	69.7	13.5	131	128.6	25.0	191	187.5	36.4	251	246.4	47.9
12	11.8	2.3	72	70.7	13.7	32	129.6	25.2	92	188.5	36.6	52	247.4	48.1
13	12.8	2.5	73	71.7	13.9	33	130.6	25.4	93	189.5	36.8	53	248.4	48.3
14	13.7	2.7	74	72.6	14.1	34	131.5	25.6	94	190.4	37.0	54	249.3	48.5
15	14.7	2.9	75	73.6	14.3	35	132.5	25.8	95	191.4	37.2	55	250.3	48.7
16	15.7	3.1	76	74.6	14.5	36	133.5	26.0	96	192.4	37.4	56	251.3	48.8
17	16.7	3.2	77	75.6	14.7	37	134.5	26.1	97	193.4	37.6	57	252.3	49.0
18	17.7	3.4	78	76.6	14.9	38	135.5	26.3	98	194.4	37.8	58	253.3	49.2
19	18.7	3.6	79	77.5	15.1	39	136.4	26.5	99	195.3	38.0	59	254.2	49.4
20	19.6	3.8	80	78.5	15.3	40	137.4	26.7	200	196.3	38.2	60	255.2	49.6
21	20.6	4.0	81	79.5	15.5	141	138.4	26.9	201	197.3	38.4	261	256.2	49.8
22	21.6	4.2	82	80.5	15.6	42	139.4	27.1	02	198.3	38.5	62	257.2	50.0
23	22.6	4.4	83	81.5	15.8	43	140.4	27.3	03	199.3	38.7	63	258.2	50.2
24	23.6	4.6	84	82.5	16.0	44	141.4	27.5	04	200.3	38.9	64	259.1	50.4
25	24.5	4.8	85	83.4	16.2	45	142.3	27.7	05	201.2	39.1	65	260.1	50.6
26	25.5	5.0	86	84.4	16.4	46	143.3	27.9	06	202.2	39.3	66	261.1	50.8
27	26.5	5.2	87	85.4	16.6	47	144.3	28.0	07	203.2	39.5	67	262.1	50.9
28	27.5	5.3	88	86.4	16.8	48	145.3	28.2	08	204.2	39.7	68	263.1	51.1
29	28.5	5.5	89	87.4	17.0	49	146.3	28.4	09	205.2	39.9	69	264.1	51.3
30	29.4	5.7	90	88.3	17.2	50	147.2	28.6	10	206.1	40.1	70	265.0	51.5
31	30.4	5.9	91	89.3	17.4	151	148.2	28.8	211	207.1	40.3	271	266.0	51.7
32	31.4	6.1	92	90.3	17.6	52	149.2	29.0	12	208.1	40.5	72	267.0	51.9
33	32.4	6.3	93	91.3	17.7	53	150.2	29.2	13	209.1	40.6	73	268.0	52.1
34	33.4	6.5	94	92.3	17.9	54	151.2	29.4	14	210.1	40.8	74	269.0	52.3
35	34.4	6.7	95	93.3	18.1	55	152.2	29.6	15	211.0	41.0	75	269.9	52.5
36	35.3	6.9	96	94.2	18.3	56	153.1	29.8	16	212.0	41.2	76	270.9	52.7
37	36.3	7.1	97	95.2	18.5	57	154.1	30.0	17	213.0	41.4	77	271.9	52.9
38	37.3	7.3	98	96.2	18.7	58	155.1	30.1	18	214.0	41.6	78	272.9	53.0
39	38.3	7.4	99	97.2	18.9	59	156.1	30.3	19	215.0	41.8	79	273.9	53.2
40	39.3	7.6	100	98.2	19.1	60	157.1	30.5	20	216.0	42.0	80	274.9	53.4
41	40.2	7.8	101	99.1	19.3	161	158.0	30.7	221	216.9	42.2	281	275.8	53.6
42	41.2	8.0	02	100.1	19.5	62	159.0	30.9	22	217.9	42.4	82	276.8	53.8
43	42.2	8.2	03	101.1	19.7	63	160.0	31.1	23	218.9	42.6	83	277.8	54.0
44	43.2	8.4	04	102.1	19.8	64	161.0	31.3	24	219.9	42.7	84	278.8	54.2
45	44.2	8.6	05	103.1	20.0	65	162.0	31.5	25	220.9	42.9	85	279.8	54.4
46	45.2	8.8	06	104.1	20.2	66	163.0	31.7	26	221.8	43.1	86	280.7	54.6
47	46.1	9.0	07	105.0	20.4	67	163.9	31.9	27	222.8	43.3	87	281.7	54.8
48	47.1	9.2	08	106.0	20.6	68	164.9	32.1	28	223.8	43.5	88	282.7	55.0
49	48.1	9.3	09	107.0	20.8	69	165.9	32.2	29	224.8	43.7	89	283.7	55.1
50	49.1	9.5	10	108.0	21.0	70	166.9	32.4	30	225.8	43.9	90	284.7	55.3
51	50.1	9.7	111	109.0	21.2	171	167.9	32.6	231	226.8	44.1	291	285.7	55.5
52	51.0	9.9	12	109.9	21.4	72	168.8	32.8	32	227.7	44.3	92	286.6	55.7
53	52.0	10.1	13	110.9	21.6	73	169.8	33.0	33	228.7	44.5	93	287.6	55.9
54	53.0	10.3	14	111.9	21.8	74	170.8	33.2	34	229.7	44.6	94	288.6	56.1
55	54.0	10.5	15	112.9	21.9	75	171.8	33.4	35	230.7	44.8	95	289.6	56.3
56	55.0	10.7	16	113.9	22.1	76	172.8	33.6	36	231.7	45.0	96	290.6	56.5
57	56.0	10.9	17	114.9	22.3	77	173.7	33.8	37	232.6	45.2	97	291.5	56.7
58	56.9	11.1	18	115.8	22.5	78	174.7	34.0	38	233.6	45.4	98	292.5	56.9
59	57.9	11.3	19	116.8	22.7	79	175.7	34.2	39	234.6	45.6	99	293.5	57.1
60	58.9	11.4	20	117.8	22.9	80	176.7	34.3	40	235.6	45.8	300	294.5	57.2
Dist.	Dep.	D. Lat.	D.st.	Dep.	D. Lat.	Dist.	Dep.	D. Lat.	Dist.	Dep.	D. Lat.	Dist.	Dep.	D. Lat.

281° | 079° — 259° | 101°

79°

Dist.	D. Lat.	Dep.
N.	N x Cos.	N x Sin.
Hypotenuse	Side Adj.	Side Opp.

TABLE 4

Traverse 11° Table

349° | 011° — 191° | 169°

Dist.	D. Lat.	Dep.	Dist.	D. Lat.	Dep.	Dist.	D. Lat.	Dep.	Dist.	D. Lat.	Dep.	Dist.	D. Lat.	Dep.
301	295.5	57.4	361	354.4	68.9	421	413.3	80.3	481	472.2	91.8	541	531.1	103.2
02	296.5	57.6	62	355.3	69.1	22	414.2	80.5	82	473.1	92.0	42	532.0	103.4
03	297.4	57.8	63	356.3	69.3	23	415.2	80.7	83	474.1	92.2	43	533.0	103.6
04	298.4	58.0	64	357.3	69.5	24	416.2	80.9	84	475.1	92.4	44	534.0	103.8
05	299.4	58.2	65	358.3	69.6	25	417.2	81.1	85	476.1	92.5	45	535.0	104.0
06	300.4	58.4	66	359.3	69.8	26	418.2	81.3	86	477.1	92.7	46	536.0	104.2
07	301.4	58.6	67	360.3	70.0	27	419.2	81.5	87	478.1	92.9	47	537.0	104.4
08	302.3	58.8	68	361.2	70.2	28	420.1	81.7	88	479.0	93.1	48	537.9	104.6
09	303.3	59.0	69	362.2	70.4	29	421.1	81.9	89	480.0	93.3	49	538.9	104.8
10	304.3	59.2	70	363.2	70.6	30	422.1	82.0	90	481.0	93.5	50	539.9	104.9
311	305.3	59.3	371	364.2	70.8	431	423.1	82.2	491	482.0	93.7	551	540.9	105.1
12	306.3	59.5	72	365.2	71.0	32	424.1	82.4	92	483.0	93.9	52	541.9	105.3
13	307.2	59.7	73	366.1	71.2	33	425.0	82.6	93	483.9	94.1	53	542.8	105.5
14	308.2	59.9	74	367.1	71.4	34	426.0	82.8	94	484.9	94.3	54	543.8	105.7
15	309.2	60.1	75	368.1	71.6	35	427.0	83.0	95	485.9	94.5	55	544.8	105.9
16	310.2	60.3	76	369.1	71.7	36	428.0	83.2	96	486.9	94.6	56	545.8	106.1
17	311.2	60.5	77	370.1	71.9	37	429.0	83.4	97	487.9	94.8	57	546.8	106.3
18	312.2	60.7	78	371.1	72.1	38	430.0	83.6	98	488.9	95.0	58	547.7	106.5
19	313.1	60.9	79	372.0	72.3	39	430.9	83.8	99	489.8	95.2	59	548.7	106.7
20	314.1	61.1	80	373.0	72.5	40	431.9	84.0	500	490.8	95.4	60	549.7	106.9
321	315.1	61.2	381	374.0	72.7	441	432.9	84.1	501	491.8	95.6	561	550.7	107.0
22	316.1	61.4	82	375.0	72.9	42	433.9	84.3	02	492.8	95.8	62	551.7	107.2
23	317.1	61.6	83	376.0	73.1	43	434.9	84.5	03	493.8	96.0	63	552.7	107.4
24	318.0	61.8	84	376.9	73.3	44	435.8	84.7	04	494.7	96.2	64	553.6	107.6
25	319.0	62.0	85	377.9	73.5	45	436.8	84.9	05	495.7	96.4	65	554.6	107.8
26	320.0	62.2	86	378.9	73.7	46	437.8	85.1	06	496.7	96.5	66	555.6	108.0
27	321.0	62.4	87	379.9	73.8	47	438.8	85.3	07	497.7	96.7	67	556.6	108.2
28	322.0	62.6	88	380.9	74.0	48	439.8	85.5	08	498.7	96.9	68	557.6	108.4
29	323.0	62.8	89	381.9	74.2	49	440.8	85.7	09	499.6	97.1	69	558.5	108.6
30	323.9	63.0	90	382.8	74.4	50	441.7	85.9	10	500.6	97.3	70	559.5	108.8
331	324.9	63.2	391	383.8	74.6	451	442.7	86.1	511	501.6	97.5	571	560.5	109.0
32	325.9	63.3	92	384.8	74.8	52	443.7	86.2	12	502.6	97.7	72	561.5	109.1
33	326.9	63.5	93	385.8	75.0	53	444.7	86.4	13	503.6	97.9	73	562.5	109.3
34	327.9	63.7	94	386.8	75.2	54	445.7	86.6	14	504.6	98.1	74	563.5	109.5
35	328.8	63.9	95	387.7	75.4	55	446.6	86.8	15	505.5	98.3	75	564.4	109.7
36	329.8	64.1	96	388.7	75.6	56	447.6	87.0	16	506.5	98.5	76	565.4	109.9
37	330.8	64.3	97	389.7	75.8	57	448.6	87.2	17	507.5	98.6	77	566.4	110.1
38	331.8	64.5	98	390.7	75.9	58	449.6	87.4	18	508.5	98.8	78	567.4	110.3
39	332.8	64.7	99	391.7	76.1	59	450.6	87.6	19	509.5	99.0	79	568.4	110.5
40	333.8	64.9	400	392.7	76.3	60	451.5	87.8	20	510.4	99.2	80	569.3	110.7
341	334.7	65.1	401	393.6	76.5	461	452.5	88.0	521	511.4	99.4	581	570.3	110.9
42	335.7	65.3	02	394.6	76.7	62	453.5	88.2	22	512.4	99.6	82	571.3	111.1
43	336.7	65.4	03	395.6	76.9	63	454.5	88.3	23	513.4	99.8	83	572.3	111.2
44	337.7	65.6	04	396.6	77.1	64	455.5	88.5	24	514.4	100.0	84	573.3	111.4
45	338.7	65.8	05	397.6	77.3	65	456.5	88.7	25	515.4	100.2	85	574.3	111.6
46	339.6	66.0	06	398.5	77.5	66	457.4	88.9	26	516.3	100.4	86	575.2	111.8
47	340.6	66.2	07	399.5	77.7	67	458.4	89.1	27	517.3	100.6	87	576.2	112.0
48	341.6	66.4	08	400.5	77.9	68	459.4	89.3	28	518.3	100.7	88	577.2	112.2
49	342.6	66.6	09	401.5	78.0	69	460.4	89.5	29	519.3	100.9	89	578.2	112.4
50	343.6	66.8	10	402.5	78.2	70	461.4	89.7	30	520.3	101.1	90	579.2	112.6
351	344.6	67.0	411	403.4	78.4	471	462.3	89.9	531	521.2	101.3	591	580.1	112.8
52	345.5	67.2	12	404.4	78.6	72	463.3	90.1	32	522.2	101.5	92	581.1	113.0
53	346.5	67.4	13	405.4	78.8	73	464.3	90.3	33	523.2	101.7	93	582.1	113.1
54	347.5	67.5	14	406.4	79.0	74	465.3	90.4	34	524.2	101.9	94	583.1	113.3
55	348.5	67.7	15	407.4	79.2	75	466.3	90.6	35	525.2	102.1	95	584.1	113.5
56	349.5	67.9	16	408.4	79.4	76	467.3	90.8	36	526.2	102.3	96	585.0	113.7
57	350.4	68.1	17	409.3	79.6	77	468.2	91.0	37	527.1	102.5	97	586.0	113.9
58	351.4	68.3	18	410.3	79.8	78	469.2	91.2	38	528.1	102.7	98	587.0	114.1
59	352.4	68.5	19	411.3	79.9	79	470.2	91.4	39	529.1	102.8	99	588.0	114.3
60	353.4	68.7	20	412.3	80.1	80	471.2	91.6	40	530.1	103.0	600	589.0	114.5
Dist.	Dep.	D. Lat.	Dist.	Dep.	D. Lat.	Dist.	Dep.	D. Lat.	Dist.	Dep.	D. Lat.	Dist.	Dep.	D. Lat.

Dist.	D. Lat.	Dep.
D Lo	Dep.	
	m	D Lo

79°

281° | 079° — 259° | 101°

TABLE 4

Traverse **12°** Table

348° | 012° / 192° | 168°

Dist.	D. Lat.	Dep.	Dist.	D. Lat.	Dep.	Dist.	D. Lat.	Dep.	Dist.	D. Lat.	Dep.	Dist.	D. Lat.	Dep.
1	1.0	0.2	61	59.7	12.7	121	118.4	25.2	181	177.0	37.6	241	235.7	50.1
2	2.0	0.4	62	60.6	12.9	22	119.3	25.4	82	178.0	37.8	42	236.7	50.3
3	2.9	0.6	63	61.6	13.1	23	120.3	25.6	83	179.0	38.0	43	237.7	50.5
4	3.9	0.8	64	62.6	13.3	24	121.3	25.8	84	180.0	38.3	44	238.7	50.7
5	4.9	1.0	65	63.6	13.5	25	122.3	26.0	85	181.0	38.5	45	239.6	50.9
6	5.9	1.2	66	64.6	13.7	26	123.2	26.2	86	181.9	38.7	46	240.6	51.1
7	6.8	1.5	67	65.5	13.9	27	124.2	26.4	87	182.9	38.9	47	241.6	51.4
8	7.8	1.7	68	66.5	14.1	28	125.2	26.6	88	183.9	39.1	48	242.6	51.6
9	8.8	1.9	69	67.5	14.3	29	126.2	26.8	89	184.9	39.3	49	243.6	51.8
10	9.8	2.1	70	68.5	14.6	30	127.2	27.0	90	185.8	39.5	50	244.5	52.0
11	10.8	2.3	71	69.4	14.8	131	128.1	27.2	191	186.8	39.7	251	245.5	52.2
12	11.7	2.5	72	70.4	15.0	32	129.1	27.4	92	187.8	39.9	52	246.5	52.4
13	12.7	2.7	73	71.4	15.2	33	130.1	27.7	93	188.8	40.1	53	247.5	52.6
14	13.7	2.9	74	72.4	15.4	34	131.1	27.9	94	189.8	40.3	54	248.4	52.8
15	14.7	3.1	75	73.4	15.6	35	132.0	28.1	95	190.7	40.5	55	249.4	53.0
16	15.7	3.3	76	74.3	15.8	36	133.0	28.3	96	191.7	40.8	56	250.4	53.2
17	16.6	3.5	77	75.3	16.0	37	134.0	28.5	97	192.7	41.0	57	251.4	53.4
18	17.6	3.7	78	76.3	16.2	38	135.0	28.7	98	193.7	41.2	58	252.4	53.6
19	18.6	4.0	79	77.3	16.4	39	136.0	28.9	99	194.7	41.4	59	253.3	53.8
20	19.6	4.2	80	78.3	16.6	40	136.9	29.1	200	195.6	41.6	60	254.3	54.1
21	20.5	4.4	81	79.2	16.8	141	137.9	29.3	201	196.6	41.8	261	255.3	54.3
22	21.5	4.6	82	80.2	17.0	42	138.9	29.5	02	197.6	42.0	62	256.3	54.5
23	22.5	4.8	83	81.2	17.3	43	139.9	29.7	03	198.6	42.2	63	257.3	54.7
24	23.5	5.0	84	82.2	17.5	44	140.9	29.9	04	199.5	42.4	64	258.2	54.9
25	24.5	5.2	85	83.1	17.7	45	141.8	30.1	05	200.5	42.6	65	259.2	55.1
26	25.4	5.4	86	84.1	17.9	46	142.8	30.4	06	201.5	42.8	66	260.2	55.3
27	26.4	5.6	87	85.1	18.1	47	143.8	30.6	07	202.5	43.0	67	261.2	55.5
28	27.4	5.8	88	86.1	18.3	48	144.8	30.8	08	203.5	43.2	68	262.1	55.7
29	28.4	6.0	89	87.1	18.5	49	145.7	31.0	09	204.4	43.5	69	263.1	55.9
30	29.3	6.2	90	88.0	18.7	50	146.7	31.2	10	205.4	43.7	70	264.1	56.1
31	30.3	6.4	91	89.0	18.9	151	147.7	31.4	211	206.4	43.9	271	265.1	56.3
32	31.3	6.7	92	90.0	19.1	52	148.7	31.6	12	207.4	44.1	72	266.1	56.6
33	32.3	6.9	93	91.0	19.3	53	149.7	31.8	13	208.3	44.3	73	267.0	56.8
34	33.3	7.1	94	91.9	19.5	54	150.6	32.0	14	209.3	44.5	74	268.0	57.0
35	34.2	7.3	95	92.9	19.8	55	151.6	32.2	15	210.3	44.7	75	269.0	57.2
36	35.2	7.5	96	93.9	20.0	56	152.6	32.4	16	211.3	44.9	76	270.0	57.4
37	36.2	7.7	97	94.9	20.2	57	153.6	32.6	17	212.3	45.1	77	270.9	57.6
38	37.2	7.9	98	95.9	20.4	58	154.5	32.9	18	213.2	45.3	78	271.9	57.8
39	38.1	8.1	99	96.8	20.6	59	155.5	33.1	19	214.2	45.5	79	272.9	58.0
40	39.1	8.3	100	97.8	20.8	60	156.5	33.3	20	215.2	45.7	80	273.9	58.2
41	40.1	8.5	101	98.8	21.0	161	157.5	33.5	221	216.2	45.9	281	274.9	58.4
42	41.1	8.7	02	99.8	21.2	62	158.5	33.7	22	217.1	46.2	82	275.8	58.6
43	42.1	8.9	03	100.7	21.4	63	159.4	33.9	23	218.1	46.4	83	276.8	58.8
44	43.0	9.1	04	101.7	21.6	64	160.4	34.1	24	219.1	46.6	84	277.8	59.0
45	44.0	9.4	05	102.7	21.8	65	161.4	34.3	25	220.1	46.8	85	278.8	59.3
46	45.0	9.6	06	103.7	22.0	66	162.4	34.5	26	221.1	47.0	86	279.8	59.5
47	46.0	9.8	07	104.7	22.2	67	163.4	34.7	27	222.0	47.2	87	280.7	59.7
48	47.0	10.0	08	105.6	22.5	68	164.3	34.9	28	223.0	47.4	88	281.7	59.9
49	47.9	10.2	09	106.6	22.7	69	165.3	35.1	29	224.0	47.6	89	282.7	60.1
50	48.9	10.4	10	107.6	22.9	70	166.3	35.3	30	225.0	47.8	90	283.7	60.3
51	49.9	10.6	111	108.6	23.1	171	167.3	35.6	231	226.0	48.0	291	284.6	60.5
52	50.9	10.8	12	109.6	23.3	72	168.2	35.8	32	226.9	48.2	92	285.6	60.7
53	51.8	11.0	13	110.5	23.5	73	169.2	36.0	33	227.9	48.4	93	286.6	60.9
54	52.8	11.2	14	111.5	23.7	74	170.2	36.2	34	228.9	48.7	94	287.6	61.1
55	53.8	11.4	15	112.5	23.9	75	171.2	36.4	35	229.9	48.9	95	288.6	61.3
56	54.8	11.6	16	113.5	24.1	76	172.2	36.6	36	230.8	49.1	96	289.5	61.5
57	55.8	11.9	17	114.4	24.3	77	173.1	36.8	37	231.8	49.3	97	290.5	61.7
58	56.7	12.1	18	115.4	24.5	78	174.1	37.0	38	232.8	49.5	98	291.5	62.0
59	57.7	12.3	19	116.4	24.7	79	175.1	37.2	39	233.8	49.7	99	292.5	62.2
60	58.7	12.5	20	117.4	24.9	80	176.1	37.4	40	234.8	49.9	300	293.4	62.4
Dist.	Dep.	D. Lat.	Dist.	Dep.	D. Lat.	Dist.	Dep.	D. Lat.	Dist.	Dep.	D. Lat.	Dist.	Dep.	D. Lat.

282° | 078° / 258° | 102°

78°

Dist.	D. Lat.	Dep.
N.	N x Cos.	N x Sin.
Hypotenuse	Side Adj.	Side Opp.

TABLE 4

Traverse **12°** Table

348° | 012° / 192° | 168°

Dist.	D. Lat.	Dep.	Dist.	D. Lat.	Dep.	Dist.	D. Lat.	Dep.	Dist.	D. Lat.	Dep.	Dist.	D. Lat.	Dep.
301	294.4	62.6	361	353.1	75.1	421	411.8	87.5	481	470.5	100.0	541	529.2	112.5
02	295.4	62.8	62	354.1	75.3	22	412.8	87.7	82	471.5	100.2	42	530.2	112.7
03	296.4	63.0	63	355.1	75.5	23	413.8	87.9	83	472.4	100.4	43	531.1	112.9
04	297.4	63.2	64	356.0	75.7	24	414.7	88.2	84	473.4	100.6	44	532.1	113.1
05	298.3	63.4	65	357.0	75.9	25	415.7	88.4	85	474.4	100.8	45	533.1	113.3
06	299.3	63.6	66	358.0	76.1	26	416.7	88.6	86	475.4	101.0	46	534.1	113.5
07	300.3	63.8	67	359.0	76.3	27	417.7	88.8	87	476.4	101.3	47	535.0	113.7
08	301.3	64.0	68	360.0	76.5	28	418.6	89.0	88	477.3	101.5	48	536.0	113.9
09	302.2	64.2	69	360.9	76.7	29	419.6	89.2	89	478.3	101.7	49	537.0	114.1
10	303.2	64.5	70	361.9	76.9	30	420.6	89.4	90	479.3	101.9	50	538.0	114.4
311	304.2	64.7	371	362.9	77.1	431	421.6	89.6	491	480.3	102.1	551	539.0	114.6
12	305.2	64.9	72	363.9	77.3	32	422.6	89.8	92	481.2	102.3	52	539.9	114.8
13	306.2	65.1	73	364.8	77.6	33	423.5	90.0	93	482.2	102.5	53	540.9	115.0
14	307.1	65.3	74	365.8	77.8	34	424.5	90.2	94	483.2	102.7	54	541.9	115.2
15	308.1	65.5	75	366.8	78.0	35	425.5	90.4	95	484.2	102.9	55	542.9	115.4
16	309.1	65.7	76	367.8	78.2	36	426.5	90.6	96	485.2	103.1	56	543.9	115.6
17	310.1	65.9	77	368.8	78.4	37	427.5	90.9	97	486.1	103.3	57	544.8	115.8
18	311.1	66.1	78	369.7	78.6	38	428.4	91.1	98	487.1	103.5	58	545.8	116.0
19	312.0	66.3	79	370.7	78.8	39	429.4	91.3	99	488.1	103.7	59	546.8	116.2
20	313.0	66.5	80	371.7	79.0	40	430.4	91.5	500	489.1	104.0	60	547.8	116.4
321	314.0	66.7	381	372.7	79.2	441	431.4	91.7	501	490.1	104.2	561	548.7	116.6
22	315.0	66.9	82	373.7	79.4	42	432.3	91.9	02	491.0	104.4	62	549.7	116.8
23	315.9	67.2	83	374.6	79.6	43	433.3	92.1	03	492.0	104.6	63	550.7	117.1
24	316.9	67.4	84	375.6	79.8	44	434.3	92.3	04	493.0	104.8	64	551.7	117.3
25	317.9	67.6	85	376.6	80.0	45	435.3	92.5	05	494.0	105.0	65	552.7	117.5
26	318.9	67.8	86	377.6	80.3	46	436.3	92.7	06	494.9	105.2	66	553.6	117.7
27	319.9	68.0	87	378.5	80.5	47	437.2	92.9	07	495.9	105.4	67	554.6	117.9
28	320.8	68.2	88	379.5	80.7	48	438.2	93.1	08	496.9	105.6	68	555.6	118.1
29	321.8	68.4	89	380.5	80.9	49	439.2	93.4	09	497.9	105.8	69	556.6	118.3
30	322.8	68.6	90	381.5	81.1	50	440.2	93.6	10	498.9	106.0	70	557.5	118.5
331	323.8	68.8	391	382.5	81.3	451	441.1	93.8	511	499.8	106.2	571	558.5	118.7
32	324.7	69.0	92	383.4	81.5	52	442.1	94.0	12	500.8	106.5	72	559.5	118.9
33	325.7	69.2	93	384.4	81.7	53	443.1	94.2	13	501.8	106.7	73	560.5	119.1
34	326.7	69.4	94	385.4	81.9	54	444.1	94.4	14	502.8	106.9	74	561.5	119.3
35	327.7	69.7	95	386.4	82.1	55	445.1	94.6	15	503.7	107.1	75	562.4	119.5
36	328.7	69.9	96	387.3	82.3	56	446.0	94.8	16	504.7	107.3	76	563.4	119.8
37	329.6	70.1	97	388.3	82.5	57	447.0	95.0	17	505.7	107.5	77	564.4	120.0
38	330.6	70.3	98	389.3	82.7	58	448.0	95.2	18	506.7	107.7	78	565.4	120.2
39	331.6	70.5	99	390.3	83.0	59	449.0	95.4	19	507.7	107.9	79	566.3	120.4
40	332.6	70.7	400	391.3	83.2	60	449.9	95.6	20	508.6	108.1	80	567.3	120.6
341	333.5	70.9	401	392.2	83.4	461	450.9	95.8	521	509.6	108.3	581	568.3	120.8
42	334.5	71.1	02	393.2	83.6	62	451.9	96.1	22	510.6	108.5	82	569.3	121.0
43	335.5	71.3	03	394.2	83.8	63	452.9	96.3	23	511.6	108.7	83	570.3	121.2
44	336.5	71.5	04	395.2	84.0	64	453.9	96.5	24	512.5	108.9	84	571.2	121.4
45	337.5	71.7	05	396.1	84.2	65	454.8	96.7	25	513.5	109.2	85	572.2	121.6
46	338.4	71.9	06	397.1	84.4	66	455.8	96.9	26	514.5	109.4	86	573.2	121.8
47	339.4	72.1	07	398.1	84.6	67	456.8	97.1	27	515.5	109.6	87	574.2	122.0
48	340.4	72.4	08	399.1	84.8	68	457.8	97.3	28	516.5	109.8	88	575.2	122.3
49	341.4	72.6	09	400.1	85.0	69	458.8	97.5	29	517.4	110.0	89	576.1	122.5
50	342.4	72.8	10	401.0	85.2	70	459.7	97.7	30	518.4	110.2	90	577.1	122.7
351	343.3	73.0	411	402.0	85.5	471	460.7	97.9	531	519.4	110.4	591	578.1	122.9
52	344.3	73.2	12	403.0	85.7	72	461.7	98.1	32	520.4	110.6	92	579.1	123.1
53	345.3	73.4	13	404.0	85.9	73	462.7	98.3	33	521.4	110.8	93	580.0	123.3
54	346.3	73.6	14	405.0	86.1	74	463.6	98.6	34	522.3	111.0	94	581.0	123.5
55	347.2	73.8	15	405.9	86.3	75	464.6	98.8	35	523.3	111.2	95	582.0	123.7
56	348.2	74.0	16	406.9	86.5	76	465.6	99.0	36	524.3	111.4	96	583.0	123.9
57	349.2	74.2	17	407.9	86.7	77	466.6	99.2	37	525.3	111.6	97	584.0	124.1
58	350.2	74.4	18	408.9	86.9	78	467.6	99.4	38	526.2	111.9	98	584.9	124.3
59	351.2	74.6	19	409.8	87.1	79	468.5	99.6	39	527.2	112.1	99	585.9	124.5
60	352.1	74.8	20	410.8	87.3	80	469.5	99.8	40	528.2	112.3	600	586.9	124.7
Dist.	Dep.	D. Lat.	Dist.	Dep.	D. Lat.	Dist.	Dep.	D. Lat.	Dist.	Dep.	D. Lat.	Dist.	Dep.	D. Lat.

Dist.	D. Lat.	Dep.
D Lo	Dep.	
	m	D Lo

78°

282° | 078° / 258° | 102°

TABLE 4

Traverse **13°** Table

347° | 013°
193° | 167°

Dist.	D. Lat.	Dep.	Dist.	D. Lat.	Dep.	Dist.	D. Lat.	Dep.	Dist.	D. Lat.	Dep.	Dist.	D. Lat.	Dep.
1	1.0	0.2	61	59.4	13.7	121	117.9	27.2	181	176.4	40.7	241	234.8	54.2
2	1.9	0.4	62	60.4	13.9	22	118.9	27.4	82	177.3	40.9	42	235.8	54.4
3	2.9	0.7	63	61.4	14.2	23	119.8	27.7	83	178.3	41.2	43	236.8	54.7
4	3.9	0.9	64	62.4	14.4	24	120.8	27.9	84	179.3	41.4	44	237.7	54.9
5	4.9	1.1	65	63.3	14.6	25	121.8	28.1	85	180.3	41.6	45	238.7	55.1
6	5.8	1.3	66	64.3	14.8	26	122.8	28.3	86	181.2	41.8	46	239.7	55.3
7	6.8	1.6	67	65.3	15.1	27	123.7	28.6	87	182.2	42.1	47	240.7	55.6
8	7.8	1.8	68	66.3	15.3	28	124.7	28.8	88	183.2	42.3	48	241.6	55.8
9	8.8	2.0	69	67.2	15.5	29	125.7	29.0	89	184.2	42.5	49	242.6	56.0
10	9.7	2.2	70	68.2	15.7	30	126.7	29.2	90	185.1	42.7	50	243.6	56.2
11	10.7	2.5	71	69.2	16.0	131	127.6	29.5	191	186.1	43.0	251	244.6	56.5
12	11.7	2.7	72	70.2	16.2	32	128.6	29.7	92	187.1	43.2	52	245.5	56.7
13	12.7	2.9	73	71.1	16.4	33	129.6	29.9	93	188.1	43.4	53	246.5	56.9
14	13.6	3.1	74	72.1	16.6	34	130.6	30.1	94	189.0	43.6	54	247.5	57.1
15	14.6	3.4	75	73.1	16.9	35	131.5	30.4	95	190.0	43.9	55	248.5	57.4
16	15.6	3.6	76	74.1	17.1	36	132.5	30.6	96	191.0	44.1	56	249.4	57.6
17	16.6	3.8	77	75.0	17.3	37	133.5	30.8	97	192.0	44.3	57	250.4	57.8
18	17.5	4.0	78	76.0	17.5	38	134.5	31.0	98	192.9	44.5	58	251.4	58.0
19	18.5	4.3	79	77.0	17.8	39	135.4	31.3	99	193.9	44.8	59	252.4	58.3
20	19.5	4.5	80	77.9	18.0	40	136.4	31.5	200	194.9	45.0	60	253.3	58.5
21	20.5	4.7	81	78.9	18.2	141	137.4	31.7	201	195.8	45.2	261	254.3	58.7
22	21.4	4.9	82	79.9	18.4	42	138.4	31.9	02	196.8	45.4	62	255.3	58.9
23	22.4	5.2	83	80.9	18.7	43	139.3	32.2	03	197.8	45.7	63	256.3	59.2
24	23.4	5.4	84	81.8	18.9	44	140.3	32.4	04	198.8	45.9	64	257.2	59.4
25	24.4	5.6	85	82.8	19.1	45	141.3	32.6	05	199.7	46.1	65	258.2	59.6
26	25.3	5.8	86	83.8	19.3	46	142.3	32.8	06	200.7	46.3	66	259.2	59.8
27	26.3	6.1	87	84.8	19.6	47	143.2	33.1	07	201.7	46.6	67	260.2	60.1
28	27.3	6.3	88	85.7	19.8	48	144.2	33.3	08	202.7	46.8	68	261.1	60.3
29	28.3	6.5	89	86.7	20.0	49	145.2	33.5	09	203.6	47.0	69	262.1	60.5
30	29.2	6.7	90	87.7	20.2	50	146.2	33.7	10	204.6	47.2	70	263.1	60.7
31	30.2	7.0	91	88.7	20.5	151	147.1	34.0	211	205.6	47.5	271	264.1	61.0
32	31.2	7.2	92	89.6	20.7	52	148.1	34.2	12	206.6	47.7	72	265.0	61.2
33	32.2	7.4	93	90.6	20.9	53	149.1	34.4	13	207.5	47.9	73	266.0	61.4
34	33.1	7.6	94	91.6	21.1	54	150.1	34.6	14	208.5	48.1	74	267.0	61.6
35	34.1	7.9	95	92.6	21.4	55	151.0	34.9	15	209.5	48.4	75	268.0	61.9
36	35.1	8.1	96	93.5	21.6	56	152.0	35.1	16	210.5	48.6	76	268.9	62.1
37	36.1	8.3	97	94.5	21.8	57	153.0	35.3	17	211.4	48.8	77	269.9	62.3
38	37.0	8.5	98	95.5	22.0	58	154.0	35.5	18	212.4	49.0	78	270.9	62.5
39	38.0	8.8	99	96.5	22.3	59	154.9	35.8	19	213.4	49.3	79	271.8	62.8
40	39.0	9.0	100	97.4	22.5	60	155.9	36.0	20	214.4	49.5	80	272.8	63.0
41	39.9	9.2	101	98.4	22.7	161	156.9	36.2	221	215.3	49.7	281	273.8	63.2
42	40.9	9.4	02	99.4	22.9	62	157.8	36.4	22	216.3	49.9	82	274.8	63.4
43	41.9	9.7	03	100.4	23.2	63	158.8	36.7	23	217.3	50.2	83	275.7	63.7
44	42.9	9.9	04	101.3	23.4	64	159.8	36.9	24	218.3	50.4	84	276.7	63.9
45	43.8	10.1	05	102.3	23.6	65	160.8	37.1	25	219.2	50.6	85	277.7	64.1
46	44.8	10.3	06	103.3	23.8	66	161.7	37.3	26	220.2	50.8	86	278.7	64.3
47	45.8	10.6	07	104.3	24.1	67	162.7	37.6	27	221.2	51.1	87	279.6	64.6
48	46.8	10.8	08	105.2	24.3	68	163.7	37.8	28	222.2	51.3	88	280.6	64.8
49	47.7	11.0	09	106.2	24.5	69	164.7	38.0	29	223.1	51.5	89	281.6	65.0
50	48.7	11.2	10	107.2	24.7	70	165.6	38.2	30	224.1	51.7	90	282.6	65.2
51	49.7	11.5	111	108.2	25.0	171	166.6	38.5	231	225.1	52.0	291	283.5	65.5
52	50.7	11.7	12	109.1	25.2	72	167.6	38.7	32	226.1	52.2	92	284.5	65.7
53	51.6	11.9	13	110.1	25.4	73	168.6	38.9	33	227.0	52.4	93	285.5	65.9
54	52.6	12.1	14	111.1	25.6	74	169.5	39.1	34	228.0	52.6	94	286.5	66.1
55	53.6	12.4	15	112.1	25.9	75	170.5	39.4	35	229.0	52.9	95	287.4	66.4
56	54.6	12.6	16	113.0	26.1	76	171.5	39.6	36	230.0	53.1	96	288.4	66.6
57	55.5	12.8	17	114.0	26.3	77	172.5	39.8	37	230.9	53.3	97	289.4	66.8
58	56.5	13.0	18	115.0	26.5	78	173.4	40.0	38	231.9	53.5	98	290.4	67.0
59	57.5	13.3	19	116.0	26.8	79	174.4	40.3	39	232.9	53.8	99	291.3	67.3
60	58.5	13.5	20	116.9	27.0	80	175.4	40.5	40	233.8	54.0	300	292.3	67.5
Dist.	Dep.	D. Lat.	Dist.	Dep.	D. Lat.	Dist.	Dep.	D. Lat.	Dist.	Dep.	D. Lat.	Dist.	Dep.	D. Lat.

283° | 077°
257° | 103°

77°

Dist.	D. Lat.	Dep.
N.	N x Cos.	N x Sin.
Hypotenuse	Side Adj.	Side Opp.

TABLE 4

Traverse **13°** Table

347° | 013°
193° | 167°

Dist.	D. Lat.	Dep.	Dist.	D. Lat.	Dep.	Dist.	D. Lat.	Dep.	Dist.	D. Lat.	Dep.	Dist.	D. Lat.	Dep.
301	293.3	67.7	361	351.7	81.2	421	410.2	94.7	481	468.7	108.2	541	527.1	121.7
02	294.3	67.9	62	352.7	81.4	22	411.2	94.9	82	469.6	108.4	42	528.1	121.9
03	295.2	68.2	63	353.7	81.7	23	412.2	95.2	83	470.6	108.7	43	529.1	122.1
04	296.2	68.4	64	354.7	81.9	24	413.1	95.4	84	471.6	108.9	44	530.1	122.4
05	297.2	68.6	65	355.6	82.1	25	414.1	95.6	85	472.6	109.1	45	531.0	122.6
06	298.2	68.8	66	356.6	82.3	26	415.1	95.8	86	473.5	109.3	46	532.0	122.8
07	299.1	69.1	67	357.6	82.6	27	416.1	96.1	87	474.5	109.6	47	533.0	123.0
08	300.1	69.3	68	358.6	82.8	28	417.0	96.3	88	475.5	109.8	48	534.0	123.3
09	301.1	69.5	69	359.5	83.0	29	418.0	96.5	89	476.5	110.0	49	534.9	123.5
10	302.1	69.7	70	360.5	83.2	30	419.0	96.7	90	477.4	110.2	50	535.9	123.7
311	303.0	70.0	371	361.5	83.5	431	420.0	97.0	491	478.4	110.5	551	536.9	123.9
12	304.0	70.2	72	362.5	83.7	32	420.9	97.2	92	479.4	110.7	52	537.9	124.2
13	305.0	70.4	73	363.4	83.9	33	421.9	97.4	93	480.4	110.9	53	538.8	124.4
14	306.0	70.6	74	364.4	84.1	34	422.9	97.6	94	481.3	111.1	54	539.8	124.6
15	306.9	70.9	75	365.4	84.4	35	423.9	97.9	95	482.3	111.4	55	540.8	124.8
16	307.9	71.1	76	366.4	84.6	36	424.8	98.1	96	483.3	111.6	56	541.7	125.1
17	308.9	71.3	77	367.3	84.8	37	425.8	98.3	97	484.3	111.8	57	542.7	125.3
18	309.8	71.5	78	368.3	85.0	38	426.8	98.5	98	485.2	112.0	58	543.7	125.5
19	310.8	71.8	79	369.3	85.3	39	427.7	98.8	99	486.2	112.3	59	544.7	125.7
20	311.8	72.0	80	370.3	85.5	40	428.7	99.0	500	487.2	112.5	60	545.6	126.0
321	312.8	72.2	381	371.2	85.7	441	429.7	99.2	501	488.2	112.7	561	546.6	126.2
22	313.7	72.4	82	372.2	85.9	42	430.7	99.4	02	489.1	112.9	62	547.6	126.4
23	314.7	72.7	83	373.2	86.2	43	431.6	99.7	03	490.1	113.2	63	548.6	126.6
24	315.7	72.9	84	374.2	86.4	44	432.6	99.9	04	491.1	113.4	64	549.5	126.9
25	316.7	73.1	85	375.1	86.6	45	433.6	100.1	05	492.1	113.6	65	550.5	127.1
26	317.6	73.3	86	376.1	86.8	46	434.6	100.3	06	493.0	113.8	66	551.5	127.3
27	318.6	73.6	87	377.1	87.1	47	435.5	100.6	07	494.0	114.1	67	552.5	127.5
28	319.6	73.8	88	378.1	87.3	48	436.5	100.8	08	495.0	114.3	68	553.4	127.8
29	320.6	74.0	89	379.0	87.5	49	437.5	101.0	09	496.0	114.5	69	554.4	128.0
30	321.5	74.2	90	380.0	87.7	50	438.5	101.2	10	496.9	114.7	70	555.4	128.2
331	322.5	74.5	391	381.0	88.0	451	439.4	101.5	511	497.9	114.9	571	556.4	128.4
32	323.5	74.7	92	382.0	88.2	52	440.4	101.7	12	498.9	115.2	72	557.3	128.7
33	324.5	74.9	93	382.9	88.4	53	441.4	101.9	13	499.9	115.4	73	558.3	128.9
34	325.4	75.1	94	383.9	88.6	54	442.4	102.1	14	500.8	115.6	74	559.3	129.1
35	326.4	75.4	95	384.9	88.9	55	443.3	102.4	15	501.8	115.8	75	560.3	129.3
36	327.4	75.6	96	385.9	89.1	56	444.3	102.6	16	502.8	116.1	76	561.2	129.6
37	328.4	75.8	97	386.8	89.3	57	445.3	102.8	17	503.7	116.3	77	562.2	129.8
38	329.3	76.0	98	387.8	89.5	58	446.3	103.0	18	504.7	116.5	78	563.2	130.0
39	330.3	76.3	99	388.8	89.8	59	447.2	103.3	19	505.7	116.7	79	564.2	130.2
40	331.3	76.5	400	389.7	90.0	60	448.2	103.5	20	506.7	117.0	80	565.1	130.5
341	332.3	76.7	401	390.7	90.2	461	449.2	103.7	521	507.6	117.2	581	566.1	130.7
42	333.2	76.9	02	391.7	90.4	62	450.2	103.9	22	508.6	117.4	82	567.1	130.9
43	334.2	77.2	03	392.7	90.7	63	451.1	104.2	23	509.6	117.6	83	568.1	131.1
44	335.2	77.4	04	393.6	90.9	64	452.1	104.4	24	510.6	117.9	84	569.0	131.4
45	336.2	77.6	05	394.6	91.1	65	453.1	104.6	25	511.5	118.1	85	570.0	131.6
46	337.1	77.8	06	395.6	91.3	66	454.1	104.8	26	512.5	118.3	86	571.0	131.8
47	338.1	78.1	07	396.6	91.6	67	455.0	105.1	27	513.5	118.5	87	572.0	132.0
48	339.1	78.3	08	397.5	91.8	68	456.0	105.3	28	514.5	118.8	88	572.9	132.3
49	340.1	78.5	09	398.5	92.0	69	457.0	105.5	29	515.4	119.0	89	573.9	132.5
50	341.0	78.7	10	399.5	92.2	70	458.0	105.7	30	516.4	119.2	90	574.9	132.7
351	342.0	79.0	411	400.5	92.5	471	458.9	106.0	531	517.4	119.4	591	575.9	132.9
52	343.0	79.2	12	401.4	92.7	72	459.9	106.2	32	518.4	119.7	92	576.8	133.2
53	344.0	79.4	13	402.4	92.9	73	460.9	106.4	33	519.3	119.9	93	577.8	133.4
54	344.9	79.6	14	403.4	93.1	74	461.9	106.6	34	520.3	120.1	94	578.8	133.6
55	345.9	79.9	15	404.4	93.4	75	462.8	106.9	35	521.3	120.3	95	579.8	133.8
56	346.9	80.1	16	405.3	93.6	76	463.8	107.1	36	522.3	120.6	96	580.7	134.1
57	347.9	80.3	17	406.3	93.8	77	464.8	107.3	37	523.2	120.8	97	581.7	134.3
58	348.8	80.5	18	407.3	94.0	78	465.7	107.5	38	524.2	121.0	98	582.7	134.5
59	349.8	80.8	19	408.3	94.3	79	466.7	107.8	39	525.2	121.2	99	583.6	134.7
60	350.8	81.0	20	409.2	94.5	80	467.7	108.0	40	526.2	121.5	600	584.6	135.0
Dist.	Dep.	D. Lat.	Dist.	Dep.	D. Lat.	Dist.	Dep.	D. Lat.	Dist.	Dep.	D. Lat.	Dist.	Dep.	D. Lat.

Dist.	D. Lat.	Dep.
D Lo	Dep.	
	m	D Lo

77°

283° | 077°
257° | 103°

TABLE 4

14°

Traverse Table

346°	014°
194°	166°

Dist.	D. Lat.	Dep.	Dist.	D. Lat.	Dep.	Dist.	D. Lat.	Dep.	Dist.	D. Lat.	Dep.	Dist.	D. Lat.	Dep.
1	1.0	0.2	61	59.2	14.8	121	117.4	29.3	181	175.6	43.8	241	233.8	58.3
2	1.9	0.5	62	60.2	15.0	22	118.4	29.5	82	176.6	44.0	42	234.8	58.5
3	2.9	0.7	63	61.1	15.2	23	119.3	29.8	83	177.6	44.3	43	235.8	58.8
4	3.9	1.0	64	62.1	15.5	24	120.3	30.0	84	178.5	44.5	44	236.8	59.0
5	4.9	1.2	65	63.1	15.7	25	121.3	30.2	85	179.5	44.8	45	237.7	59.3
6	5.8	1.5	66	64.0	16.0	26	122.3	30.5	86	180.5	45.0	46	238.7	59.5
7	6.8	1.7	67	65.0	16.2	27	123.2	30.7	87	181.4	45.2	47	239.7	59.8
8	7.8	1.9	68	66.0	16.5	28	124.2	31.0	88	182.4	45.5	48	240.6	60.0
9	8.7	2.2	69	67.0	16.7	29	125.2	31.2	89	183.4	45.7	49	241.6	60.2
10	9.7	2.4	70	67.9	16.9	30	126.1	31.4	90	184.4	46.0	50	242.6	60.5
11	10.7	2.7	71	68.9	17.2	131	127.1	31.7	191	185.3	46.2	251	243.5	60.7
12	11.6	2.9	72	69.9	17.4	32	128.1	31.9	92	186.3	46.4	52	244.5	61.0
13	12.6	3.1	73	70.8	17.7	33	129.0	32.2	93	187.3	46.7	53	245.5	61.2
14	13.6	3.4	74	71.8	17.9	34	130.0	32.4	94	188.2	46.9	54	246.5	61.4
15	14.6	3.6	75	72.8	18.1	35	131.0	32.7	95	189.2	47.2	55	247.4	61.7
16	15.5	3.9	76	73.7	18.4	36	132.0	32.9	96	190.2	47.4	56	248.4	61.9
17	16.5	4.1	77	74.7	18.6	37	132.9	33.1	97	191.1	47.7	57	249.4	62.2
18	17.5	4.4	78	75.7	18.9	38	133.9	33.4	98	192.1	47.9	58	250.3	62.4
19	18.4	4.6	79	76.7	19.1	39	134.9	33.6	99	193.1	48.1	59	251.3	62.7
20	19.4	4.8	80	77.6	19.4	40	135.8	33.9	200	194.1	48.4	60	252.3	62.9
21	20.4	5.1	81	78.6	19.6	141	136.8	34.1	201	195.0	48.6	261	253.2	63.1
22	21.3	5.3	82	79.6	19.8	42	137.8	34.4	02	196.0	48.9	62	254.2	63.4
23	22.3	5.6	83	80.5	20.1	43	138.8	34.6	03	197.0	49.1	63	255.2	63.6
24	23.3	5.8	84	81.5	20.3	44	139.7	34.8	04	197.9	49.4	64	256.2	63.9
25	24.3	6.0	85	82.5	20.6	45	140.7	35.1	05	198.9	49.6	65	257.1	64.1
26	25.2	6.3	86	83.4	20.8	46	141.7	35.3	06	199.9	49.8	66	258.1	64.4
27	26.2	6.5	87	84.4	21.0	47	142.6	35.6	07	200.9	50.1	67	259.1	64.6
28	27.2	6.8	88	85.4	21.3	48	143.6	35.8	08	201.8	50.3	68	260.0	64.8
29	28.1	7.0	89	86.4	21.5	49	144.6	36.0	09	202.8	50.6	69	261.0	65.1
30	29.1	7.3	90	87.3	21.8	50	145.5	36.3	10	203.8	50.8	70	262.0	65.3
31	30.1	7.5	91	88.3	22.0	151	146.5	36.5	211	204.7	51.0	271	263.0	65.6
32	31.0	7.7	92	89.3	22.3	52	147.5	36.8	12	205.7	51.3	72	263.9	65.8
33	32.0	8.0	93	90.2	22.5	53	148.5	37.0	13	206.7	51.5	73	264.9	66.0
34	33.0	8.2	94	91.2	22.7	54	149.4	37.3	14	207.6	51.8	74	265.9	66.3
35	34.0	8.5	95	92.2	23.0	55	150.4	37.5	15	208.6	52.0	75	266.8	66.5
36	34.9	8.7	96	93.1	23.2	56	151.4	37.7	16	209.6	52.3	76	267.8	66.8
37	35.9	9.0	97	94.1	23.5	57	152.3	38.0	17	210.6	52.5	77	268.8	67.0
38	36.9	9.2	98	95.1	23.7	58	153.3	38.2	18	211.5	52.7	78	269.7	67.3
39	37.8	9.4	99	96.1	24.0	59	154.3	38.5	19	212.5	53.0	79	270.7	67.5
40	38.8	9.7	100	97.0	24.2	60	155.2	38.7	20	213.5	53.2	80	271.7	67.7
41	39.8	9.9	101	98.0	24.4	161	156.2	38.9	221	214.4	53.5	281	272.7	68.0
42	40.8	10.2	02	99.0	24.7	62	157.2	39.2	22	215.4	53.7	82	273.6	68.2
43	41.7	10.4	03	99.9	24.9	63	158.2	39.4	23	216.4	53.9	83	274.6	68.5
44	42.7	10.6	04	100.9	25.2	64	159.1	39.7	24	217.3	54.2	84	275.6	68.7
45	43.7	10.9	05	101.9	25.4	65	160.1	39.9	25	218.3	54.4	85	276.5	68.9
46	44.6	11.1	06	102.9	25.6	66	161.1	40.2	26	219.3	54.7	86	277.5	69.2
47	45.6	11.4	07	103.8	25.9	67	162.0	40.4	27	220.3	54.9	87	278.5	69.4
48	46.6	11.6	08	104.8	26.1	68	163.0	40.6	28	221.2	55.2	88	279.4	69.7
49	47.5	11.9	09	105.8	26.4	69	164.0	40.9	29	222.2	55.4	89	280.4	69.9
50	48.5	12.1	10	106.7	26.6	70	165.0	41.1	30	223.2	55.6	90	281.4	70.2
51	49.5	12.3	111	107.7	26.9	171	165.9	41.4	231	224.1	55.9	291	282.4	70.4
52	50.5	12.6	12	108.7	27.1	72	166.9	41.6	32	225.1	56.1	92	283.3	70.6
53	51.4	12.8	13	109.6	27.3	73	167.9	41.9	33	226.1	56.4	93	284.3	70.9
54	52.4	13.1	14	110.6	27.6	74	168.8	42.1	34	227.0	56.6	94	285.3	71.1
55	53.4	13.3	15	111.6	27.8	75	169.8	42.3	35	228.0	56.9	95	286.2	71.4
56	54.3	13.5	16	112.6	28.1	76	170.8	42.6	36	229.0	57.1	96	287.2	71.6
57	55.3	13.8	17	113.5	28.3	77	171.7	42.8	37	230.0	57.3	97	288.2	71.9
58	56.3	14.0	18	114.5	28.5	78	172.7	43.1	38	230.9	57.6	98	289.1	72.1
59	57.2	14.3	19	115.5	28.8	79	173.7	43.3	39	231.9	57.8	99	290.1	72.3
60	58.2	14.5	20	116.4	29.0	80	174.7	43.5	40	232.9	58.1	300	291.1	72.6
Dist.	Dep.	D. Lat.	Dist.	Dep.	D. Lat.	Dist.	Dep.	D. Lat.	Dist.	Dep.	D. Lat.	Dist.	Dep.	D. Lat.

284°	076°
256°	104°

76°

Dist.	D. Lat.	Dep.
N.	N x Cos.	N x Sin.
Hypotenuse	Side Adj.	Side Opp.

TABLE 4

14°

Traverse Table

346°	014°
194°	166°

Dist.	D. Lat.	Dep.	Dist.	D. Lat.	Dep.	Dist.	D. Lat.	Dep.	Dist.	D. Lat.	Dep.	Dist.	D. Lat.	Dep.
301	292.1	72.8	361	350.3	87.3	421	408.5	101.8	481	466.7	116.4	541	524.9	130.9
02	293.0	73.1	62	351.2	87.6	22	409.5	102.1	82	467.7	116.6	42	525.9	131.1
03	294.0	73.3	63	352.2	87.8	23	410.4	102.3	83	468.7	116.8	43	526.9	131.4
04	295.0	73.5	64	353.2	88.1	24	411.4	102.6	84	469.6	117.1	44	527.8	131.6
05	295.9	73.8	65	354.2	88.3	25	412.4	102.8	85	470.6	117.3	45	528.8	131.8
06	296.9	74.0	66	355.1	88.5	26	413.3	103.1	86	471.6	117.6	46	529.8	132.1
07	297.9	74.3	67	356.1	88.8	27	414.3	103.3	87	472.5	117.8	47	530.8	132.3
08	298.9	74.5	68	357.1	89.0	28	415.3	103.5	88	473.5	118.1	48	531.7	132.6
09	299.8	74.8	69	358.0	89.3	29	416.3	103.8	89	474.5	118.3	49	532.7	132.8
10	300.8	75.0	70	359.0	89.5	30	417.2	104.0	90	475.4	118.5	50	533.7	133.1
311	301.8	75.2	371	360.0	89.8	431	418.2	104.3	491	476.4	118.8	551	534.6	133.3
12	302.7	75.5	72	361.0	90.0	32	419.2	104.5	92	477.4	119.0	52	535.6	133.5
13	303.7	75.7	73	361.9	90.2	33	420.1	104.8	93	478.4	119.3	53	536.6	133.8
14	304.7	76.0	74	362.9	90.5	34	421.1	105.0	94	479.3	119.5	54	537.5	134.0
15	305.6	76.2	75	363.9	90.7	35	422.1	105.2	95	480.3	119.8	55	538.5	134.3
16	306.6	76.4	76	364.8	91.0	36	423.0	105.5	96	481.3	120.0	56	539.5	134.5
17	307.6	76.7	77	365.8	91.2	37	424.0	105.7	97	482.2	120.2	57	540.5	134.8
18	308.6	76.9	78	366.8	91.4	38	425.0	106.0	98	483.2	120.5	58	541.4	135.0
19	309.5	77.2	79	367.7	91.7	39	426.0	106.2	99	484.2	120.7	59	542.4	135.2
20	310.5	77.4	80	368.7	91.9	40	426.9	106.4	500	485.1	121.0	60	543.4	135.5
321	311.5	77.7	381	369.7	92.2	441	427.9	106.7	501	486.1	121.2	561	544.3	135.7
22	312.4	77.9	82	370.7	92.4	42	428.9	106.9	02	487.1	121.4	62	545.3	136.0
23	313.4	78.1	83	371.6	92.7	43	429.8	107.2	03	488.1	121.7	63	546.3	136.2
24	314.4	78.4	84	372.6	92.9	44	430.8	107.4	04	489.0	121.9	64	547.2	136.4
25	315.3	78.6	85	373.6	93.1	45	431.8	107.7	05	490.0	122.2	65	548.2	136.7
26	316.3	78.9	86	374.5	93.4	46	432.8	107.9	06	491.0	122.4	66	549.2	136.9
27	317.3	79.1	87	375.5	93.6	47	433.7	108.1	07	491.9	122.7	67	550.2	137.2
28	318.3	79.4	88	376.5	93.9	48	434.7	108.4	08	492.9	122.9	68	551.1	137.4
29	319.2	79.6	89	377.4	94.1	49	435.7	108.6	09	493.9	123.1	69	552.1	137.7
30	320.2	79.8	90	378.4	94.3	50	436.6	108.9	10	494.9	123.4	70	553.1	137.9
331	321.2	80.1	391	379.4	94.6	451	437.6	109.1	511	495.8	123.6	571	554.0	138.1
32	322.1	80.3	92	380.4	94.8	52	438.6	109.3	12	496.8	123.9	72	555.0	138.4
33	323.1	80.6	93	381.3	95.1	53	439.5	109.6	13	497.8	124.1	73	556.0	138.6
34	324.1	80.8	94	382.3	95.3	54	440.5	109.8	14	498.7	124.3	74	556.9	138.9
35	325.0	81.0	95	383.3	95.6	55	441.5	110.1	15	499.7	124.6	75	557.9	139.1
36	326.0	81.3	96	384.2	95.8	56	442.5	110.3	16	500.7	124.8	76	558.9	139.3
37	327.0	81.5	97	385.2	96.0	57	443.4	110.6	17	501.6	125.1	77	559.9	139.6
38	328.0	81.8	98	386.2	96.3	58	444.4	110.8	18	502.6	125.3	78	560.8	139.8
39	328.9	82.0	99	387.1	96.5	59	445.4	111.0	19	503.6	125.6	79	561.8	140.1
40	329.9	82.3	400	388.1	96.8	60	446.3	111.3	20	504.6	125.8	80	562.8	140.3
341	330.9	82.5	401	389.1	97.0	461	447.3	111.5	521	505.5	126.0	581	563.7	140.6
42	331.8	82.7	02	390.1	97.3	62	448.3	111.8	22	506.5	126.3	82	564.7	140.8
43	332.8	83.0	03	391.0	97.5	63	449.2	112.0	23	507.5	126.5	83	565.7	141.0
44	333.8	83.2	04	392.0	97.7	64	450.2	112.3	24	508.4	126.8	84	566.7	141.3
45	334.8	83.5	05	393.0	98.0	65	451.2	112.5	25	509.4	127.0	85	567.6	141.5
46	335.7	83.7	06	393.9	98.2	66	452.2	112.7	26	510.4	127.3	86	568.6	141.8
47	336.7	83.9	07	394.9	98.5	67	453.1	113.0	27	511.3	127.5	87	569.6	142.0
48	337.7	84.2	08	395.9	98.7	68	454.1	113.2	28	512.3	127.7	88	570.5	142.3
49	338.6	84.4	09	396.9	98.9	69	455.1	113.5	29	513.3	128.0	89	571.5	142.5
50	339.6	84.7	10	397.8	99.2	70	456.0	113.7	30	514.3	128.2	90	572.5	142.7
351	340.6	84.9	411	398.8	99.4	471	457.0	113.9	531	515.2	128.5	591	573.4	143.0
52	341.5	85.2	12	399.8	99.7	72	458.0	114.2	32	516.2	128.7	92	574.4	143.2
53	342.5	85.4	13	400.7	99.9	73	458.9	114.4	33	517.2	128.9	93	575.4	143.5
54	343.5	85.6	14	401.7	100.2	74	459.9	114.7	34	518.1	129.2	94	576.4	143.7
55	344.5	85.9	15	402.7	100.4	75	460.9	114.9	35	519.1	129.4	95	577.3	143.9
56	345.4	86.1	16	403.6	100.6	76	461.9	115.2	36	520.1	129.7	96	578.3	144.2
57	346.4	86.4	17	404.6	100.9	77	462.8	115.4	37	521.0	129.9	97	579.3	144.4
58	347.4	86.6	18	405.6	101.1	78	463.8	115.6	38	522.0	130.2	98	580.2	144.7
59	348.3	86.8	19	406.6	101.4	79	464.8	115.9	39	523.0	130.4	99	581.2	144.9
60	349.3	87.1	20	407.5	101.6	80	465.7	116.1	40	524.0	130.6	600	582.2	145.2
Dist.	Dep.	D. Lat.	Dist.	Dep.	D. Lat.	Dist.	Dep.	D. Lat.	Dist.	Dep.	D. Lat.	Dist.	Dep.	D. Lat.

Dist.	D. Lat.	Dep.
D Lo	Dep.	
	m	D Lo

76°

284°	076°
256°	104°

TABLE 4

Traverse 15° Table

345° | 015°
195° | 165°

Dist.	D. Lat.	Dep.	Dist.	D. Lat.	Dep.	Dist	D. Lat.	Dep.	Dist.	D. Lat.	Dep.	Dist.	D. Lat.	Dep.
1	1.0	0.3	61	58.9	15.8	121	116.9	31.3	181	174.8	46.8	241	232.8	62.4
2	1.9	0.5	62	59.9	16.0	22	117.8	31.6	82	175.8	47.1	42	233.8	62.6
3	2.9	0.8	63	60.9	16.3	23	118.8	31.8	83	176.8	47.4	43	234.7	62.9
4	3.9	1.0	64	61.8	16.6	24	119.8	32.1	84	177.7	47.6	44	235.7	63.2
5	4.8	1.3	65	62.8	16.8	25	120.7	32.4	85	178.7	47.9	45	236.7	63.4
6	5.8	1.6	66	63.8	17.1	26	121.7	32.6	86	179.7	48.1	46	237.6	63.7
7	6.8	1.8	67	64.7	17.3	27	122.7	32.9	87	180.6	48.4	47	238.6	63.9
8	7.7	2.1	68	65.7	17.6	28	123.6	33.1	88	181.6	48.7	48	239.5	64.2
9	8.7	2.3	69	66.6	17.9	29	124.6	33.4	89	182.6	48.9	49	240.5	64.4
10	9.7	2.6	70	67.6	18.1	30	125.6	33.6	90	183.5	49.2	50	241.5	64.7
11	10.6	2.8	71	68.6	18.4	131	126.5	33.9	191	184.5	49.4	251	242.4	65.0
12	11.6	3.1	72	69.5	18.6	32	127.5	34.2	92	185.5	49.7	52	243.4	65.2
13	12.6	3.4	73	70.5	18.9	33	128.5	34.4	93	186.4	50.0	53	244.4	65.5
14	13.5	3.6	74	71.5	19.2	34	129.4	34.7	94	187.4	50.2	54	245.3	65.7
15	14.5	3.9	75	72.4	19.4	35	130.4	34.9	95	188.4	50.5	55	246.3	66.0
16	15.5	4.1	76	73.4	19.7	36	131.4	35.2	96	189.3	50.7	56	247.3	66.3
17	16.4	4.4	77	74.4	19.9	37	132.3	35.5	97	190.3	51.0	57	248.2	66.5
18	17.4	4.7	78	75.3	20.2	38	133.3	35.7	98	191.3	51.2	58	249.2	66.8
19	18.4	4.9	79	76.3	20.4	39	134.3	36.0	99	192.2	51.5	59	250.2	67.0
20	19.3	5.2	80	77.3	20.7	40	135.2	36.2	200	193.2	51.8	60	251.1	67.3
21	20.3	5.4	81	78.2	21.0	141	136.2	36.5	201	194.2	52.0	261	252.1	67.6
22	21.3	5.7	82	79.2	21.2	42	137.2	36.8	02	195.1	52.3	62	253.1	67.8
23	22.2	6.0	83	80.2	21.5	43	138.1	37.0	03	196.1	52.5	63	254.0	68.1
24	23.2	6.2	84	81.1	21.7	44	139.1	37.3	04	197.0	52.8	64	255.0	68.3
25	24.1	6.5	85	82.1	22.0	45	140.1	37.5	05	198.0	53.1	65	256.0	68.6
26	25.1	6.7	86	83.1	22.3	46	141.0	37.8	06	199.0	53.3	66	256.9	68.8
27	26.1	7.0	87	84.0	22.5	47	142.0	38.0	07	199.9	53.6	67	257.9	69.1
28	27.0	7.2	88	85.0	22.8	48	143.0	38.3	08	200.9	53.8	68	258.9	69.4
29	28.0	7.5	89	86.0	23.0	49	143.9	38.6	09	201.9	54.1	69	259.8	69.6
30	29.0	7.8	90	86.9	23.3	50	144.9	38.8	10	202.8	54.4	70	260.8	69.9
31	29.9	8.0	91	87.9	23.6	151	145.9	39.1	211	203.8	54.6	271	261.8	70.1
32	30.9	8.3	92	88.9	23.8	52	146.8	39.3	12	204.8	54.9	72	262.7	70.4
33	31.9	8.5	93	89.8	24.1	53	147.8	39.6	13	205.7	55.1	73	263.7	70.7
34	32.8	8.8	94	90.8	24.3	54	148.8	39.9	14	206.7	55.4	74	264.7	70.9
35	33.8	9.1	95	91.8	24.6	55	149.7	40.1	15	207.7	55.6	75	265.6	71.2
36	34.8	9.3	96	92.7	24.8	56	150.7	40.4	16	208.6	55.9	76	266.6	71.4
37	35.7	9.6	97	93.7	25.1	57	151.7	40.6	17	209.6	56.2	77	267.6	71.7
38	36.7	9.8	98	94.7	25.4	58	152.6	40.9	18	210.6	56.4	78	268.5	72.0
39	37.7	10.1	99	95.6	25.6	59	153.6	41.2	19	211.5	56.7	79	269.5	72.2
40	38.6	10.4	100	96.6	25.9	60	154.5	41.4	20	212.5	56.9	80	270.5	72.5
41	39.6	10.6	101	97.6	26.1	161	155.5	41.7	221	213.5	57.2	281	271.4	72.7
42	40.6	10.9	02	98.5	26.4	62	156.5	41.9	22	214.4	57.5	82	272.4	73.0
43	41.5	11.1	03	99.5	26.7	63	157.4	42.2	23	215.4	57.7	83	273.4	73.2
44	42.5	11.4	04	100.5	26.9	64	158.4	42.4	24	216.4	58.0	84	274.3	73.5
45	43.5	11.6	05	101.4	27.2	65	159.4	42.7	25	217.3	58.2	85	275.3	73.8
46	44.4	11.9	06	102.4	27.4	66	160.3	43.0	26	218.3	58.5	86	276.3	74.0
47	45.4	12.2	07	103.4	27.7	67	161.3	43.2	27	219.3	58.8	87	277.2	74.3
48	46.4	12.4	08	104.3	28.0	68	162.3	43.5	28	220.2	59.0	88	278.2	74.5
49	47.3	12.7	09	105.3	28.2	69	163.2	43.7	29	221.2	59.3	89	279.2	74.8
50	48.3	12.9	10	106.3	28.5	70	164.2	44.0	30	222.2	59.5	90	280.1	75.1
51	49.3	13.2	111	107.2	28.7	171	165.2	44.3	231	223.1	59.8	291	281.1	75.3
52	50.2	13.5	12	108.2	29.0	72	166.1	44.5	32	224.1	60.0	92	282.1	75.6
53	51.2	13.7	13	109.1	29.2	73	167.1	44.8	33	225.1	60.3	93	283.0	75.8
54	52.2	14.0	14	110.1	29.5	74	168.1	45.0	34	226.0	60.6	94	284.0	76.1
55	53.1	14.2	15	111.1	29.8	75	169.0	45.3	35	227.0	60.8	95	284.9	76.4
56	54.1	14.5	16	112.0	30.0	76	170.0	45.6	36	228.0	61.1	96	285.9	76.6
57	55.1	14.8	17	113.0	30.3	77	171.0	45.8	37	228.9	61.3	97	286.9	76.9
58	56.0	15.0	18	114.0	30.5	78	171.9	46.1	38	229.9	61.6	98	287.8	77.1
59	57.0	15.3	19	114.9	30.8	79	172.9	46.3	39	230.9	61.9	99	288.8	77.4
60	58.0	15.5	20	115.9	31.1	80	173.9	46.6	40	231.8	62.1	300	289.8	77.6
Dist.	Dep.	D. Lat.	Dist.	Dep.	D. Lat.	Dist.	Dep.	D. Lat.	Dist.	Dep.	D. Lat.	Dist.	Dep.	D. Lat.

285° | 075°
255° | 105°

75°

Dist.	D. Lat.	Dep.
N.	N x Cos.	N x Sin.
Hypotenuse	Side Adj.	Side Opp.

TABLE 4

Traverse 15° Table

345° | 015°
195° | 165°

Dist.	D. Lat.	Dep.	Dist.	D. Lat.	Dep.	Dist.	D. Lat.	Dep.	Dist.	D. Lat.	Dep.	Dist.	D. Lat.	Dep.
301	290.7	77.9	361	348.7	93.4	421	406.7	109.0	481	464.6	124.5	541	522.6	140.0
02	291.7	78.2	62	349.7	93.7	22	407.6	109.2	82	465.6	124.8	42	523.5	140.3
03	292.7	78.4	63	350.6	94.0	23	408.6	109.5	83	466.5	125.0	43	524.5	140.5
04	293.6	78.7	64	351.6	94.2	24	409.6	109.7	84	467.5	125.3	44	525.5	140.8
05	294.6	78.9	65	352.6	94.5	25	410.5	110.0	85	468.5	125.5	45	526.4	141.1
06	295.6	79.2	66	353.5	94.7	26	411.5	110.3	86	469.4	125.8	46	527.4	141.3
07	296.5	79.5	67	354.5	95.0	27	412.5	110.5	87	470.4	126.0	47	528.4	141.6
08	297.5	79.7	68	355.5	95.2	28	413.4	110.8	88	471.4	126.3	48	529.3	141.8
09	298.5	80.0	69	356.4	95.5	29	414.4	111.0	89	472.3	126.6	49	530.3	142.1
10	299.4	80.2	70	357.4	95.8	30	415.3	111.3	90	473.3	126.8	50	531.3	142.4
311	300.4	80.5	371	358.4	96.0	431	416.3	111.6	491	474.3	127.1	551	532.2	142.6
12	301.4	80.8	72	359.3	96.3	32	417.3	111.8	92	475.2	127.3	52	533.2	142.9
13	302.3	81.0	73	360.3	96.5	33	418.2	112.1	93	476.2	127.6	53	534.2	143.1
14	303.3	81.3	74	361.3	96.8	34	419.2	112.3	94	477.2	127.9	54	535.1	143.4
15	304.3	81.5	75	362.2	97.1	35	420.2	112.6	95	478.1	128.1	55	536.1	143.6
16	305.2	81.8	76	363.2	97.3	36	421.1	112.8	96	479.1	128.4	56	537.1	143.9
17	306.2	82.0	77	364.2	97.6	37	422.1	113.1	97	480.1	128.6	57	538.0	144.2
18	307.2	82.3	78	365.1	97.8	38	423.1	113.4	98	481.0	128.9	58	539.0	144.4
19	308.1	82.6	79	366.1	98.1	39	424.0	113.6	99	482.0	129.2	59	540.0	144.7
20	309.1	82.8	80	367.1	98.4	40	425.0	113.9	500	483.0	129.4	60	540.9	144.9
321	310.1	83.1	381	368.0	98.6	441	426.0	114.1	501	483.9	129.7	561	541.9	145.2
22	311.0	83.3	82	369.0	98.9	42	426.9	114.4	02	484.9	129.9	62	542.9	145.5
23	312.0	83.6	83	369.9	99.1	43	427.9	114.7	03	485.9	130.2	63	543.8	145.7
24	313.0	83.9	84	370.9	99.4	44	428.9	114.9	04	486.8	130.4	64	544.8	146.0
25	313.9	84.1	85	371.9	99.6	45	429.8	115.2	05	487.8	130.7	65	545.7	146.2
26	314.9	84.4	86	372.8	99.9	46	430.8	115.4	06	488.8	131.0	66	546.7	146.5
27	315.9	84.6	87	373.8	100.2	47	431.8	115.7	07	489.7	131.2	67	547.7	146.8
28	316.8	84.9	88	374.8	100.4	48	432.7	116.0	08	490.7	131.5	68	548.6	147.0
29	317.8	85.2	89	375.7	100.7	49	433.7	116.2	09	491.7	131.7	69	549.6	147.3
30	318.8	85.4	90	376.7	100.9	50	434.7	116.5	10	492.6	132.0	70	550.6	147.5
331	319.7	85.7	391	377.7	101.2	451	435.6	116.7	511	493.6	132.3	571	551.5	147.8
32	320.7	85.9	92	378.6	101.5	52	436.6	117.0	12	494.6	132.5	72	552.5	148.0
33	321.7	86.2	93	379.6	101.7	53	437.6	117.2	13	495.5	132.8	73	553.5	148.3
34	322.6	86.4	94	380.6	102.0	54	438.5	117.5	14	496.5	133.0	74	554.4	148.6
35	323.6	86.7	95	381.5	102.2	55	439.5	117.8	15	497.5	133.3	75	555.4	148.8
36	324.6	87.0	96	382.5	102.5	56	440.5	118.0	16	498.4	133.6	76	556.4	149.1
37	325.5	87.2	97	383.5	102.8	57	441.4	118.3	17	499.4	133.8	77	557.3	149.3
38	326.5	87.5	98	384.4	103.0	58	442.4	118.5	18	500.3	134.1	78	558.3	149.6
39	327.4	87.7	99	385.4	103.3	59	443.4	118.8	19	501.3	134.3	79	559.3	149.9
40	328.4	88.0	400	386.4	103.5	60	444.3	119.1	20	502.3	134.6	80	560.2	150.1
341	329.4	88.3	401	387.3	103.8	461	445.3	119.3	521	503.2	134.8	581	561.2	150.4
42	330.3	88.5	02	388.3	104.0	62	446.3	119.6	22	504.2	135.1	82	562.2	150.6
43	331.3	88.8	03	389.3	104.3	63	447.2	119.8	23	505.2	135.4	83	563.1	150.9
44	332.3	89.0	04	390.2	104.6	64	448.2	120.1	24	506.1	135.6	84	564.1	151.2
45	333.2	89.3	05	391.2	104.8	65	449.2	120.4	25	507.1	135.9	85	565.1	151.4
46	334.2	89.6	06	392.2	105.1	66	450.1	120.6	26	508.1	136.1	86	566.0	151.7
47	335.2	89.8	07	393.1	105.3	67	451.1	120.9	27	509.0	136.4	87	567.0	151.9
48	336.1	90.1	08	394.1	105.6	68	452.1	121.1	28	510.0	136.7	88	568.0	152.2
49	337.1	90.3	09	395.1	105.9	69	453.0	121.4	29	511.0	136.9	89	568.9	152.4
50	338.1	90.6	10	396.0	106.1	70	454.0	121.6	30	511.9	137.2	90	569.9	152.7
351	339.0	90.8	411	397.0	106.4	471	455.0	121.9	531	512.9	137.4	591	570.9	153.0
52	340.0	91.1	12	398.0	106.6	72	455.9	122.2	32	513.9	137.7	92	571.8	153.2
53	341.0	91.4	13	398.9	106.9	73	456.9	122.4	33	514.8	138.0	93	572.8	153.5
54	341.9	91.6	14	399.9	107.2	74	457.8	122.7	34	515.8	138.2	94	573.8	153.7
55	342.9	91.9	15	400.9	107.4	75	458.8	122.9	35	516.8	138.5	95	574.7	154.0
56	343.9	92.1	16	401.8	107.7	76	459.8	123.2	36	517.7	138.7	96	575.7	154.3
57	344.8	92.4	17	402.8	107.9	77	460.7	123.5	37	518.7	139.0	97	576.7	154.5
58	345.8	92.7	18	403.8	108.2	78	461.7	123.7	38	519.7	139.2	98	577.6	154.8
59	346.8	92.9	19	404.7	108.4	79	462.7	124.0	39	520.6	139.5	99	578.6	155.0
60	347.7	93.2	20	405.7	108.7	80	463.6	124.2	40	521.6	139.8	600	579.6	155.3
Dist.	Dep.	D. Lat.	Dist.	Dep.	D. Lat.	Dist.	Dep.	D. Lat.	Dist.	Dep.	D. Lat.	Dist.	Dep.	D. Lat.

Dist.	D. Lat.	Dep.
D Lo	Dep.	
	m	D Lo

75°

285° | 075°
255° | 105°

TABLE 4

Traverse 16° Table

344° | 016° / 196° | 164°

Dist.	D. Lat.	Dep.	Dist.	D. Lat.	Dep.	Dist.	D. Lat.	Dep.	Dist.	D. Lat.	Dep.	Dist.	D. Lat.	Dep.
1	1.0	0.3	61	58.6	16.8	121	116.3	33.4	181	174.0	49.9	241	231.7	66.4
2	1.9	0.6	62	59.6	17.1	22	117.3	33.6	82	174.9	50.2	42	232.6	66.7
3	2.9	0.8	63	60.6	17.4	23	118.2	33.9	83	175.9	50.4	43	233.6	67.0
4	3.8	1.1	64	61.5	17.6	24	119.2	34.2	84	176.9	50.7	44	234.5	67.3
5	4.8	1.4	65	62.5	17.9	25	120.2	34.5	85	177.8	51.0	45	235.5	67.5
6	5.8	1.7	66	63.4	18.2	26	121.1	34.7	86	178.8	51.3	46	236.5	67.8
7	6.7	1.9	67	64.4	18.5	27	122.1	35.0	87	179.8	51.5	47	237.4	68.1
8	7.7	2.2	68	65.4	18.7	28	123.0	35.3	88	180.7	51.8	48	238.4	68.4
9	8.7	2.5	69	66.3	19.0	29	124.0	35.6	89	181.7	52.1	49	239.4	68.6
10	9.6	2.8	70	67.3	19.3	30	125.0	35.8	90	182.6	52.4	50	240.3	68.9
11	10.6	3.0	71	68.2	19.6	131	125.9	36.1	191	183.6	52.6	251	241.3	69.2
12	11.5	3.3	72	69.2	19.8	32	126.9	36.4	92	184.6	52.9	52	242.2	69.5
13	12.5	3.6	73	70.2	20.1	33	127.8	36.7	93	185.5	53.2	53	243.2	69.7
14	13.5	3.9	74	71.1	20.4	34	128.8	36.9	94	186.5	53.5	54	244.2	70.0
15	14.4	4.1	75	72.1	20.7	35	129.8	37.2	95	187.4	53.7	55	245.1	70.3
16	15.4	4.4	76	73.1	20.9	36	130.7	37.5	96	188.4	54.0	56	246.1	70.6
17	16.3	4.7	77	74.0	21.2	37	131.7	37.8	97	189.4	54.3	57	247.0	70.8
18	17.3	5.0	78	75.0	21.5	38	132.7	38.0	98	190.3	54.6	58	248.0	71.1
19	18.3	5.2	79	75.9	21.8	39	133.6	38.3	99	191.3	54.9	59	249.0	71.4
20	19.2	5.5	80	76.9	22.1	40	134.6	38.6	200	192.3	55.1	60	249.9	71.7
21	20.2	5.8	81	77.9	22.3	141	135.5	38.9	201	193.2	55.4	261	250.9	71.9
22	21.1	6.1	82	78.8	22.6	42	136.5	39.1	02	194.2	55.7	62	251.9	72.2
23	22.1	6.3	83	79.8	22.9	43	137.5	39.4	03	195.1	56.0	63	252.8	72.5
24	23.1	6.6	84	80.7	23.2	44	138.4	39.7	04	196.1	56.2	64	253.8	72.8
25	24.0	6.9	85	81.7	23.4	45	139.4	40.0	05	197.1	56.5	65	254.7	73.0
26	25.0	7.2	86	82.7	23.7	46	140.3	40.2	06	198.0	56.8	66	255.7	73.3
27	26.0	7.4	87	83.6	24.0	47	141.3	40.5	07	199.0	57.1	67	256.7	73.6
28	26.9	7.7	88	84.6	24.3	48	142.3	40.8	08	199.9	57.3	68	257.6	73.9
29	27.9	8.0	89	85.6	24.5	49	143.2	41.1	09	200.9	57.6	69	258.6	74.1
30	28.8	8.3	90	86.5	24.8	50	144.2	41.3	10	201.9	57.9	70	259.5	74.4
31	29.8	8.5	91	87.5	25.1	151	145.2	41.6	211	202.8	58.2	271	260.5	74.7
32	30.8	8.8	92	88.4	25.4	52	146.1	41.9	12	203.8	58.4	72	261.5	75.0
33	31.7	9.1	93	89.4	25.6	53	147.1	42.2	13	204.7	58.7	73	262.4	75.2
34	32.7	9.4	94	90.4	25.9	54	148.0	42.4	14	205.7	59.0	74	263.4	75.5
35	33.6	9.6	95	91.3	26.2	55	149.0	42.7	15	206.7	59.3	75	264.3	75.8
36	34.6	9.9	96	92.3	26.5	56	150.0	43.0	16	207.6	59.5	76	265.3	76.1
37	35.6	10.2	97	93.2	26.7	57	150.9	43.3	17	208.6	59.8	77	266.3	76.4
38	36.5	10.5	98	94.2	27.0	58	151.9	43.6	18	209.6	60.1	78	267.2	76.6
39	37.5	10.7	99	95.2	27.3	59	152.8	43.8	19	210.5	60.4	79	268.2	76.9
40	38.5	11.0	100	96.1	27.6	60	153.8	44.1	20	211.5	60.6	80	269.2	77.2
41	39.4	11.3	101	97.1	27.8	161	154.8	44.4	221	212.4	60.9	281	270.1	77.5
42	40.4	11.6	02	98.0	28.1	62	155.7	44.7	22	213.4	61.2	82	271.1	77.7
43	41.3	11.9	03	99.0	28.4	63	156.7	44.9	23	214.4	61.5	83	272.0	78.0
44	42.3	12.1	04	100.0	28.7	64	157.6	45.2	24	215.3	61.7	84	273.0	78.3
45	43.3	12.4	05	100.9	28.9	65	158.6	45.5	25	216.3	62.0	85	274.0	78.6
46	44.2	12.7	06	101.9	29.2	66	159.6	45.8	26	217.2	62.3	86	274.9	78.8
47	45.2	13.0	07	102.9	29.5	67	160.5	46.0	27	218.2	62.6	87	275.9	79.1
48	46.1	13.2	08	103.8	29.8	68	161.5	46.3	28	219.2	62.8	88	276.8	79.4
49	47.1	13.5	09	104.8	30.0	69	162.5	46.6	29	220.1	63.1	89	277.8	79.7
50	48.1	13.8	10	105.7	30.3	70	163.4	46.9	30	221.1	63.4	90	278.8	79.9
51	49.0	14.1	111	106.7	30.6	171	164.4	47.1	231	222.1	63.7	291	279.7	80.2
52	50.0	14.3	12	107.7	30.9	72	165.3	47.4	32	223.0	63.9	92	280.7	80.5
53	50.9	14.6	13	108.6	31.1	73	166.3	47.7	33	224.0	64.2	93	281.6	80.8
54	51.9	14.9	14	109.6	31.4	74	167.3	48.0	34	224.9	64.5	94	282.6	81.0
55	52.9	15.2	15	110.5	31.7	75	168.2	48.2	35	225.9	64.8	95	283.6	81.3
56	53.8	15.4	16	111.5	32.0	76	169.2	48.5	36	226.9	65.1	96	284.5	81.6
57	54.8	15.7	17	112.5	32.2	77	170.1	48.8	37	227.8	65.3	97	285.5	81.9
58	55.8	16.0	18	113.4	32.5	78	171.1	49.1	38	228.8	65.6	98	286.5	82.1
59	56.7	16.3	19	114.4	32.8	79	172.1	49.3	39	229.7	65.9	99	287.4	82.4
60	57.7	16.5	20	115.4	33.1	80	173.0	49.6	40	230.7	66.2	300	288.4	82.7
Dist.	Dep.	D. Lat.	Dist.	Dep.	D. Lat.	Dist.	Dep.	D. Lat.	Dist.	Dep.	D. Lat.	Dist.	Dep.	D. Lat.

286° | 074° / 254° | 106°

74°

Dist.	D. Lat.	Dep.
N.	N x Cos.	N x Sin.
Hypotenuse	Side Adj.	Side Opp.

TABLE 4

Traverse 16° Table

344° | 016° / 196° | 164°

Dist.	D. Lat.	Dep.	Dist.	D. Lat.	Dep.	Dist.	D. Lat.	Dep.	Dist.	D. Lat.	Dep.	Dist.	D. Lat.	Dep.
301	289.3	83.0	361	347.0	99.5	421	404.7	116.0	481	462.4	132.6	541	520.0	149.1
02	290.3	83.2	62	348.0	99.8	22	405.7	116.3	82	463.3	132.9	42	521.0	149.4
03	291.3	83.5	63	348.9	100.1	23	406.6	116.6	83	464.3	133.1	43	522.0	149.7
04	292.2	83.8	64	349.9	100.3	24	407.6	116.9	84	465.3	133.4	44	522.9	149.9
05	293.2	84.1	65	350.9	100.6	25	408.5	117.1	85	466.2	133.7	45	523.9	150.2
06	294.1	84.3	66	351.8	100.9	26	409.5	117.4	86	467.2	134.0	46	524.8	150.5
07	295.1	84.6	67	352.8	101.2	27	410.5	117.7	87	468.1	134.2	47	525.8	150.8
08	296.1	84.9	68	353.7	101.4	28	411.4	118.0	88	469.1	134.5	48	526.8	151.0
09	297.0	85.2	69	354.7	101.7	29	412.4	118.2	89	470.1	134.8	49	527.7	151.3
10	298.0	85.4	70	355.7	102.0	30	413.3	118.5	90	471.0	135.1	50	528.7	151.6
311	299.0	85.7	371	356.6	102.3	431	414.3	118.8	491	472.0	135.3	551	529.7	151.9
12	299.9	86.0	72	357.6	102.5	32	415.3	119.1	92	472.9	135.6	52	530.6	152.2
13	300.9	86.3	73	358.6	102.8	33	416.2	119.4	93	473.9	135.9	53	531.6	152.4
14	301.8	86.6	74	359.5	103.1	34	417.2	119.6	94	474.9	136.2	54	532.5	152.7
15	302.8	86.8	75	360.5	103.4	35	418.1	119.9	95	475.8	136.4	55	533.5	153.0
16	303.8	87.1	76	361.4	103.6	36	419.1	120.2	96	476.8	136.7	56	534.5	153.3
17	304.7	87.4	77	362.4	103.9	37	420.1	120.5	97	477.7	137.0	57	535.4	153.5
18	305.7	87.7	78	363.4	104.2	38	421.0	120.7	98	478.7	137.3	58	536.4	153.8
19	306.6	87.9	79	364.3	104.5	39	422.0	121.0	99	479.7	137.5	59	537.3	154.1
20	307.6	88.2	80	365.3	104.7	40	423.0	121.3	500	480.6	137.8	60	538.3	154.4
321	308.6	88.5	381	366.2	105.0	441	423.9	121.6	501	481.6	138.1	561	539.3	154.6
22	309.5	88.8	82	367.2	105.3	42	424.9	121.8	02	482.6	138.4	62	540.2	154.9
23	310.5	89.0	83	368.2	105.6	43	425.8	122.1	03	483.5	138.6	63	541.2	155.2
24	311.4	89.3	84	369.1	105.8	44	426.8	122.4	04	484.5	138.9	64	542.2	155.5
25	312.4	89.6	85	370.1	106.1	45	427.8	122.7	05	485.4	139.2	65	543.1	155.7
26	313.4	89.9	86	371.0	106.4	46	428.7	122.9	06	486.4	139.5	66	544.1	156.0
27	314.3	90.1	87	372.0	106.7	47	429.7	123.2	07	487.4	139.7	67	545.0	156.3
28	315.3	90.4	88	373.0	106.9	48	430.6	123.5	08	488.3	140.0	68	546.0	156.6
29	316.3	90.7	89	373.9	107.2	49	431.6	123.8	09	489.3	140.3	69	547.0	156.8
30	317.2	91.0	90	374.9	107.5	50	432.6	124.0	10	490.2	140.6	70	547.9	157.1
331	318.2	91.2	391	375.9	107.8	451	433.5	124.3	511	491.2	140.9	571	548.9	157.4
32	319.1	91.5	92	376.8	108.0	52	434.5	124.6	12	492.2	141.1	72	549.8	157.7
33	320.1	91.8	93	377.8	108.3	53	435.5	124.9	13	493.1	141.4	73	550.8	157.9
34	321.1	92.1	94	378.7	108.6	54	436.4	125.1	14	494.1	141.7	74	551.8	158.2
35	322.0	92.3	95	379.7	108.9	55	437.4	125.4	15	495.0	142.0	75	552.7	158.5
36	323.0	92.6	96	380.7	109.2	56	438.3	125.7	16	496.0	142.2	76	553.7	158.8
37	323.9	92.9	97	381.6	109.4	57	439.3	126.0	17	497.0	142.5	77	554.6	159.0
38	324.9	93.2	98	382.6	109.7	58	440.3	126.2	18	497.9	142.8	78	555.6	159.3
39	325.9	93.4	99	383.5	110.0	59	441.2	126.5	19	498.9	143.1	79	556.6	159.6
40	326.8	93.7	400	384.5	110.3	60	442.2	126.8	20	499.9	143.3	80	557.5	159.9
341	327.8	94.0	401	385.5	110.5	461	443.1	127.1	521	500.8	143.6	581	558.5	160.1
42	328.8	94.3	02	386.4	110.8	62	444.1	127.3	22	501.8	143.9	82	559.5	160.4
43	329.7	94.5	03	387.4	111.1	63	445.1	127.6	23	502.7	144.2	83	560.4	160.7
44	330.7	94.8	04	388.3	111.4	64	446.0	127.9	24	503.7	144.4	84	561.4	161.0
45	331.6	95.1	05	389.3	111.6	65	447.0	128.2	25	504.7	144.7	85	562.3	161.2
46	332.6	95.4	06	390.3	111.9	66	447.9	128.4	26	505.6	145.0	86	563.3	161.5
47	333.6	95.6	07	391.2	112.2	67	448.9	128.7	27	506.6	145.3	87	564.3	161.8
48	334.5	95.9	08	392.2	112.5	68	449.9	129.0	28	507.5	145.5	88	565.2	162.1
49	335.5	96.2	09	393.2	112.7	69	450.8	129.3	29	508.5	145.8	89	566.2	162.4
50	336.4	96.5	10	394.1	113.0	70	451.8	129.5	30	509.5	146.1	90	567.1	162.6
351	337.4	96.7	411	395.1	113.3	471	452.8	129.8	531	510.4	146.4	591	568.1	162.9
52	338.4	97.0	12	396.0	113.6	72	453.7	130.1	32	511.4	146.6	92	569.1	163.2
53	339.3	97.3	13	397.0	113.8	73	454.7	130.4	33	512.4	146.9	93	570.0	163.5
54	340.3	97.6	14	398.0	114.1	74	455.6	130.7	34	513.3	147.2	94	571.0	163.7
55	341.2	97.9	15	398.9	114.4	75	456.6	130.9	35	514.3	147.5	95	572.0	164.0
56	342.2	98.1	16	399.9	114.7	76	457.6	131.2	36	515.2	147.7	96	572.9	164.3
57	343.2	98.4	17	400.8	114.9	77	458.5	131.5	37	516.2	148.0	97	573.9	164.6
58	344.1	98.7	18	401.8	115.2	78	459.5	131.8	38	517.2	148.3	98	574.8	164.8
59	345.1	99.0	19	402.8	115.5	79	460.4	132.0	39	518.1	148.6	99	575.8	165.1
60	346.1	99.2	20	403.7	115.8	80	461.4	132.3	40	519.1	148.8	600	576.8	165.4
Dist.	Dep.	D. Lat.	Dist.	Dep.	D. Lat.	Dist.	Dep.	D. Lat.	Dist.	Dep.	D. Lat.	Dist.	Dep.	D. Lat.

Dist.	D. Lat.	Dep.
D Lo	Dep.	
	m	D Lo

74°

286° | 074° / 254° | 106°

TABLE 4

Traverse 17° Table

343° | 017°
197° | 163°

Dist.	D. Lat.	Dep.	Dist.	D. Lat.	Dep.	Dist.	D. Lat.	Dep.	Dist.	D. Lat.	Dep.	Dist.	D. Lat.	Dep.
1	1.0	0.3	61	58.3	17.8	121	115.7	35.4	181	173.1	52.9	241	230.5	70.5
2	1.9	0.6	62	59.3	18.1	22	116.7	35.7	82	174.0	53.2	42	231.4	70.8
3	2.9	0.9	63	60.2	18.4	23	117.6	36.0	83	175.0	53.5	43	232.4	71.0
4	3.8	1.2	64	61.2	18.7	24	118.6	36.3	84	176.0	53.8	44	233.3	71.3
5	4.8	1.5	65	62.2	19.0	25	119.5	36.5	85	176.9	54.1	45	234.3	71.6
6	5.7	1.8	66	63.1	19.3	26	120.5	36.8	86	177.9	54.4	46	235.3	71.9
7	6.7	2.0	67	64.1	19.6	27	121.5	37.1	87	178.8	54.7	47	236.2	72.2
8	7.7	2.3	68	65.0	19.9	28	122.4	37.4	88	179.8	55.0	48	237.2	72.5
9	8.6	2.6	69	66.0	20.2	29	123.4	37.7	89	180.7	55.3	49	238.1	72.8
10	9.6	2.9	70	66.9	20.5	30	124.3	38.0	90	181.7	55.6	50	239.1	73.1
11	10.5	3.2	71	67.9	20.8	131	125.3	38.3	191	182.7	55.8	251	240.0	73.4
12	11.5	3.5	72	68.9	21.1	32	126.2	38.6	92	183.6	56.1	52	241.0	73.7
13	12.4	3.8	73	69.8	21.3	33	127.2	38.9	93	184.6	56.4	53	241.9	74.0
14	13.4	4.1	74	70.8	21.6	34	128.1	39.2	94	185.5	56.7	54	242.9	74.3
15	14.3	4.4	75	71.7	21.9	35	129.1	39.5	95	186.5	57.0	55	243.9	74.6
16	15.3	4.7	76	72.7	22.2	36	130.1	39.8	96	187.4	57.3	56	244.8	74.8
17	16.3	5.0	77	73.6	22.5	37	131.0	40.1	97	188.4	57.6	57	245.8	75.1
18	17.2	5.3	78	74.6	22.8	38	132.0	40.3	98	189.3	57.9	58	246.7	75.4
19	18.2	5.6	79	75.5	23.1	39	132.9	40.6	99	190.3	58.2	59	247.7	75.7
20	19.1	5.8	80	76.5	23.4	40	133.9	40.9	200	191.3	58.5	60	248.6	76.0
21	20.1	6.1	81	77.5	23.7	141	134.8	41.2	201	192.2	58.8	261	249.6	76.3
22	21.0	6.4	82	78.4	24.0	42	135.8	41.5	02	193.2	59.1	62	250.6	76.6
23	22.0	6.7	83	79.4	24.3	43	136.8	41.8	03	194.1	59.4	63	251.5	76.9
24	23.0	7.0	84	80.3	24.6	44	137.7	42.1	04	195.1	59.6	64	252.5	77.2
25	23.9	7.3	85	81.3	24.9	45	138.7	42.4	05	196.0	59.9	65	253.4	77.5
26	24.9	7.6	86	82.2	25.1	46	139.6	42.7	06	197.0	60.2	66	254.4	77.8
27	25.8	7.9	87	83.2	25.4	47	140.6	43.0	07	198.0	60.5	67	255.3	78.1
28	26.8	8.2	88	84.2	25.7	48	141.5	43.3	08	198.9	60.8	68	256.3	78.4
29	27.7	8.5	89	85.1	26.0	49	142.5	43.6	09	199.9	61.1	69	257.2	78.6
30	28.7	8.8	90	86.1	26.3	50	143.4	43.9	10	200.8	61.4	70	258.2	78.9
31	29.6	9.1	91	87.0	26.6	151	144.4	44.1	211	201.8	61.7	271	259.2	79.2
32	30.6	9.4	92	88.0	26.9	52	145.4	44.4	12	202.7	62.0	72	260.1	79.5
33	31.6	9.6	93	88.9	27.2	53	146.3	44.7	13	203.7	62.3	73	261.1	79.8
34	32.5	9.9	94	89.9	27.5	54	147.3	45.0	14	204.6	62.6	74	262.0	80.1
35	33.5	10.2	95	90.8	27.8	55	148.2	45.3	15	205.6	62.9	75	263.0	80.4
36	34.4	10.5	96	91.8	28.1	56	149.2	45.6	16	206.6	63.2	76	263.9	80.7
37	35.4	10.8	97	92.8	28.4	57	150.1	45.9	17	207.5	63.4	77	264.9	81.0
38	36.3	11.1	98	93.7	28.7	58	151.1	46.2	18	208.5	63.7	78	265.9	81.3
39	37.3	11.4	99	94.7	28.9	59	152.1	46.5	19	209.4	64.0	79	266.8	81.6
40	38.3	11.7	100	95.6	29.2	60	153.0	46.8	20	210.4	64.3	80	267.8	81.9
41	39.2	12.0	101	96.6	29.5	161	154.0	47.1	221	211.3	64.6	281	268.7	82.2
42	40.2	12.3	02	97.5	29.8	62	154.9	47.4	22	212.3	64.9	82	269.7	82.4
43	41.1	12.6	03	98.5	30.1	63	155.9	47.7	23	213.3	65.2	83	270.6	82.7
44	42.1	12.9	04	99.5	30.4	64	156.8	47.9	24	214.2	65.5	84	271.6	83.0
45	43.0	13.2	05	100.4	30.7	65	157.8	48.2	25	215.2	65.8	85	272.5	83.3
46	44.0	13.4	06	101.4	31.0	66	158.7	48.5	26	216.1	66.1	86	273.5	83.6
47	44.9	13.7	07	102.3	31.3	67	159.7	48.8	27	217.1	66.4	87	274.5	83.9
48	45.9	14.0	08	103.3	31.6	68	160.7	49.1	28	218.0	66.7	88	275.4	84.2
49	46.9	14.3	09	104.2	31.9	69	161.6	49.4	29	219.0	67.0	89	276.4	84.5
50	47.8	14.6	10	105.2	32.2	70	162.6	49.7	30	220.0	67.2	90	277.3	84.8
51	48.8	14.9	111	106.1	32.5	171	163.5	50.0	231	220.9	67.5	291	278.3	85.1
52	49.7	15.2	12	107.1	32.7	72	164.5	50.3	32	221.9	67.8	92	279.2	85.4
53	50.7	15.5	13	108.1	33.0	73	165.4	50.6	33	222.8	68.1	93	280.2	85.7
54	51.6	15.8	14	109.0	33.3	74	166.4	50.9	34	223.8	68.4	94	281.2	86.0
55	52.6	16.1	15	110.0	33.6	75	167.4	51.2	35	224.7	68.7	95	282.1	86.2
56	53.6	16.4	16	110.9	33.9	76	168.3	51.5	36	225.7	69.0	96	283.1	86.5
57	54.5	16.7	17	111.9	34.2	77	169.3	51.7	37	226.6	69.3	97	284.0	86.8
58	55.5	17.0	18	112.8	34.5	78	170.2	52.0	38	227.6	69.6	98	285.0	87.1
59	56.4	17.2	19	113.8	34.8	79	171.2	52.3	39	228.6	69.9	99	285.9	87.4
60	57.4	17.5	20	114.8	35.1	80	172.1	52.6	40	229.5	70.2	300	286.9	87.7
Dist.	Dep.	D. Lat.	Dist.	Dep.	D. Lat.	Dist.	Dep.	D. Lat.	Dist.	Dep.	D. Lat.	Dist.	Dep.	D. Lat.

287° | 073°
253° | 107°

73°

Dist.	D. Lat.	Dep.
N.	N x Cos.	N x Sin.
Hypotenuse	Side Adj.	Side Opp.

TABLE 4

Traverse 17° Table

343° | 017°
197° | 163°

Dist.	D. Lat.	Dep.	Dist.	D. Lat.	Dep.	Dist.	D. Lat.	Dep.	Dist.	D. Lat.	Dep.	Dist.	D. Lat.	Dep.
301	287.8	88.0	361	345.2	105.5	421	402.6	123.1	481	460.0	140.6	541	517.4	158.2
02	288.8	88.3	62	346.2	105.8	22	403.6	123.4	82	460.9	140.9	42	518.3	158.5
03	289.8	88.6	63	347.1	106.1	23	404.5	123.7	83	461.9	141.2	43	519.3	158.8
04	290.7	88.9	64	348.1	106.4	24	405.5	124.0	84	462.9	141.5	44	520.2	159.1
05	291.7	89.2	65	349.1	106.7	25	406.4	124.3	85	463.8	141.8	45	521.2	159.3
06	292.6	89.5	66	350.0	107.0	26	407.4	124.6	86	464.8	142.1	46	522.1	159.6
07	293.6	89.8	67	351.0	107.3	27	408.3	124.8	87	465.7	142.4	47	523.1	159.9
08	294.5	90.1	68	351.9	107.6	28	409.3	125.1	88	466.7	142.7	48	524.1	160.2
09	295.5	90.3	69	352.9	107.9	29	410.3	125.4	89	467.6	143.0	49	525.0	160.5
10	296.5	90.6	70	353.8	108.2	30	411.2	125.7	90	468.6	143.3	50	526.0	160.8
311	297.4	90.9	371	354.8	108.5	431	412.2	126.0	491	469.5	143.6	551	526.9	161.1
12	298.4	91.2	72	355.7	108.8	32	413.1	126.3	92	470.5	143.8	52	527.9	161.4
13	299.3	91.5	73	356.7	109.1	33	414.1	126.6	93	471.5	144.1	53	528.8	161.7
14	300.3	91.8	74	357.7	109.3	34	415.0	126.9	94	472.4	144.4	54	529.8	162.0
15	301.2	92.1	75	358.6	109.6	35	416.0	127.2	95	473.4	144.7	55	530.7	162.3
16	302.2	92.4	76	359.6	109.9	36	416.9	127.5	96	474.3	145.0	56	531.7	162.6
17	303.1	92.7	77	360.5	110.2	37	417.9	127.8	97	475.3	145.3	57	532.7	162.9
18	304.1	93.0	78	361.5	110.5	38	418.9	128.1	98	476.2	145.6	58	533.6	163.1
19	305.1	93.3	79	362.4	110.8	39	419.8	128.4	99	477.2	145.9	59	534.6	163.4
20	306.0	93.6	80	363.4	111.1	40	420.8	128.6	500	478.2	146.2	60	535.5	163.7
321	307.0	93.9	381	364.4	111.4	441	421.7	128.9	501	479.1	146.5	561	536.5	164.0
22	307.9	94.1	82	365.3	111.7	42	422.7	129.2	02	480.1	146.8	62	537.4	164.3
23	308.9	94.4	83	366.3	112.0	43	423.6	129.5	03	481.0	147.1	63	538.4	164.6
24	309.8	94.7	84	367.2	112.3	44	424.6	129.8	04	482.0	147.4	64	539.4	164.9
25	310.8	95.0	85	368.2	112.6	45	425.6	130.1	05	482.9	147.6	65	540.3	165.2
26	311.8	95.3	86	369.1	112.9	46	426.5	130.4	06	483.9	147.9	66	541.3	165.5
27	312.7	95.6	87	370.1	113.1	47	427.5	130.7	07	484.8	148.2	67	542.2	165.8
28	313.7	95.9	88	371.0	113.4	48	428.4	131.0	08	485.8	148.5	68	543.2	166.1
29	314.6	96.2	89	372.0	113.7	49	429.4	131.3	09	486.8	148.8	69	544.1	166.4
30	315.6	96.5	90	373.0	114.0	50	430.3	131.6	10	487.7	149.1	70	545.1	166.7
331	316.5	96.8	391	373.9	114.3	451	431.3	131.9	511	488.7	149.4	571	546.1	166.9
32	317.5	97.1	92	374.9	114.6	52	432.2	132.2	12	489.6	149.7	72	547.0	167.2
33	318.4	97.4	93	375.8	114.9	53	433.2	132.4	13	490.6	150.0	73	548.0	167.5
34	319.4	97.7	94	376.8	115.2	54	434.2	132.7	14	491.5	150.3	74	548.9	167.8
35	320.4	97.9	95	377.7	115.5	55	435.1	133.0	15	492.5	150.6	75	549.9	168.1
36	321.3	98.2	96	378.7	115.8	56	436.1	133.3	16	493.5	150.9	76	550.8	168.4
37	322.3	98.5	97	379.7	116.1	57	437.0	133.6	17	494.4	151.2	77	551.8	168.7
38	323.2	98.8	98	380.6	116.4	58	438.0	133.9	18	495.4	151.4	78	552.7	169.0
39	324.2	99.1	99	381.6	116.7	59	438.9	134.2	19	496.3	151.7	79	553.7	169.3
40	325.1	99.4	400	382.5	116.9	60	439.9	134.5	20	497.3	152.0	80	554.7	169.6
341	326.1	99.7	401	383.5	117.2	461	440.9	134.8	521	498.2	152.3	581	555.6	169.9
42	327.1	100.0	02	384.4	117.5	62	441.8	135.1	22	499.2	152.6	82	556.6	170.2
43	328.0	100.3	03	385.4	117.8	63	442.8	135.4	23	500.1	152.9	83	557.5	170.5
44	329.0	100.6	04	386.3	118.1	64	443.7	135.7	24	501.1	153.2	84	558.5	170.7
45	329.9	100.9	05	387.3	118.4	65	444.7	136.0	25	502.1	153.5	85	559.4	171.0
46	330.9	101.2	06	388.3	118.7	66	445.6	136.2	26	503.0	153.8	86	560.4	171.3
47	331.8	101.5	07	389.2	119.0	67	446.6	136.5	27	504.0	154.1	87	561.4	171.6
48	332.8	101.7	08	390.2	119.3	68	447.6	136.8	28	504.9	154.4	88	562.3	171.9
49	333.8	102.0	09	391.1	119.6	69	448.5	137.1	29	505.9	154.7	89	563.3	172.2
50	334.7	102.3	10	392.1	119.9	70	449.5	137.4	30	506.8	155.0	90	564.2	172.5
351	335.7	102.6	411	393.0	120.2	471	450.4	137.7	531	507.8	155.2	591	565.2	172.8
52	336.6	102.9	12	394.0	120.5	72	451.4	138.0	32	508.8	155.5	92	566.1	173.1
53	337.6	103.2	13	395.0	120.7	73	452.3	138.3	33	509.7	155.8	93	567.1	173.4
54	338.5	103.5	14	395.9	121.0	74	453.3	138.6	34	510.7	156.1	94	568.0	173.7
55	339.5	103.8	15	396.9	121.3	75	454.2	138.9	35	511.6	156.4	95	569.0	174.0
56	340.4	104.1	16	397.8	121.6	76	455.2	139.2	36	512.6	156.7	96	570.0	174.3
57	341.4	104.4	17	398.8	121.9	77	456.2	139.5	37	513.5	157.0	97	570.9	174.5
58	342.4	104.7	18	399.7	122.2	78	457.1	139.8	38	514.5	157.3	98	571.9	174.8
59	343.3	105.0	19	400.7	122.5	79	458.1	140.0	39	515.4	157.6	99	572.8	175.1
60	344.3	105.3	20	401.6	122.8	80	459.0	140.3	40	516.4	157.9	600	573.8	175.4
Dist.	Dep.	D. Lat.	Dist.	Dep.	D. Lat.	Dist.	Dep.	D. Lat.	Dist.	Dep.	D. Lat.	Dist.	Dep.	D. Lat.

Dist.	D. Lat.	Dep.
D Lo	Dep.	
	m	D Lo

73°

287° | 073°
253° | 107°

TABLE 4

Traverse 18° Table

342° | 018°
198° | 162°

Dist.	D. Lat.	Dep.	Dist.	D. Lat.	Dep.	Dist.	D. Lat.	Dep.	Dist.	D. Lat.	Dep.	Dist.	D. Lat.	Dep.
1	1.0	0.3	61	58.0	18.9	121	115.1	37.4	181	172.1	55.9	241	229.2	74.5
2	1.9	0.6	62	59.0	19.2	22	116.0	37.7	82	173.1	56.2	42	230.2	74.8
3	2.9	0.9	63	59.9	19.5	23	117.0	38.0	83	174.0	56.6	43	231.1	75.1
4	3.8	1.2	64	60.9	19.8	24	117.9	38.3	84	175.0	56.9	44	232.1	75.4
5	4.8	1.5	65	61.8	20.1	25	118.9	38.6	85	175.9	57.2	45	233.0	75.7
6	5.7	1.9	66	62.8	20.4	26	119.8	38.9	86	176.9	57.5	46	234.0	76.0
7	6.7	2.2	67	63.7	20.7	27	120.8	39.2	87	177.8	57.8	47	234.9	76.3
8	7.6	2.5	68	64.7	21.0	28	121.7	39.6	88	178.8	58.1	48	235.9	76.6
9	8.6	2.8	69	65.6	21.3	29	122.7	39.9	89	179.7	58.4	49	236.8	76.9
10	9.5	3.1	70	66.6	21.6	30	123.6	40.2	90	180.7	58.7	50	237.8	77.3
11	10.5	3.4	71	67.5	21.9	131	124.6	40.5	191	181.7	59.0	251	238.7	77.6
12	11.4	3.7	72	68.5	22.2	32	125.5	40.8	92	182.6	59.3	52	239.7	77.9
13	12.4	4.0	73	69.4	22.6	33	126.5	41.1	93	183.6	59.6	53	240.6	78.2
14	13.3	4.3	74	70.4	22.9	34	127.4	41.4	94	184.5	59.9	54	241.6	78.5
15	14.3	4.6	75	71.3	23.2	35	128.4	41.7	95	185.5	60.3	55	242.5	78.8
16	15.2	4.9	76	72.3	23.5	36	129.3	42.0	96	186.4	60.6	56	243.5	79.1
17	16.2	5.3	77	73.2	23.8	37	130.3	42.3	97	187.4	60.9	57	244.4	79.4
18	17.1	5.6	78	74.2	24.1	38	131.2	42.6	98	188.3	61.2	58	245.4	79.7
19	18.1	5.9	79	75.1	24.4	39	132.2	43.0	99	189.3	61.5	59	246.3	80.0
20	19.0	6.2	80	76.1	24.7	40	133.1	43.3	200	190.2	61.8	60	247.3	80.3
21	20.0	6.5	81	77.0	25.0	141	134.1	43.6	201	191.2	62.1	261	248.2	80.7
22	20.9	6.8	82	78.0	25.3	42	135.1	43.9	02	192.1	62.4	62	249.2	81.0
23	21.9	7.1	83	78.9	25.6	43	136.0	44.2	03	193.1	62.7	63	250.1	81.3
24	22.8	7.4	84	79.9	26.0	44	137.0	44.5	04	194.0	63.0	64	251.1	81.6
25	23.8	7.7	85	80.8	26.3	45	137.9	44.8	05	195.0	63.3	65	252.0	81.9
26	24.7	8.0	86	81.8	26.6	46	138.9	45.1	06	195.9	63.7	66	253.0	82.2
27	25.7	8.3	87	82.7	26.9	47	139.8	45.4	07	196.9	64.0	67	253.9	82.5
28	26.6	8.7	88	83.7	27.2	48	140.8	45.7	08	197.8	64.3	68	254.9	82.8
29	27.6	9.0	89	84.6	27.5	49	141.7	46.0	09	198.8	64.6	69	255.8	83.1
30	28.5	9.3	90	85.6	27.8	50	142.7	46.4	10	199.7	64.9	70	256.8	83.4
31	29.5	9.6	91	86.5	28.1	151	143.6	46.7	211	200.7	65.2	271	257.7	83.7
32	30.4	9.9	92	87.5	28.4	52	144.6	47.0	12	201.6	65.5	72	258.7	84.1
33	31.4	10.2	93	88.4	28.7	53	145.5	47.3	13	202.6	65.8	73	259.6	84.4
34	32.3	10.5	94	89.4	29.0	54	146.5	47.6	14	203.5	66.1	74	260.6	84.7
35	33.3	10.8	95	90.4	29.4	55	147.4	47.9	15	204.5	66.4	75	261.5	85.0
36	34.2	11.1	96	91.3	29.7	56	148.4	48.2	16	205.4	66.7	76	262.5	85.3
37	35.2	11.4	97	92.3	30.0	57	149.3	48.5	17	206.4	67.1	77	263.4	85.6
38	36.1	11.7	98	93.2	30.3	58	150.3	48.8	18	207.3	67.4	78	264.4	85.9
39	37.1	12.1	99	94.2	30.6	59	151.2	49.1	19	208.3	67.7	79	265.3	86.2
40	38.0	12.4	100	95.1	30.9	60	152.2	49.4	20	209.2	68.0	80	266.3	86.5
41	39.0	12.7	101	96.1	31.2	161	153.1	49.8	221	210.2	68.3	281	267.2	86.8
42	39.9	13.0	02	97.0	31.5	62	154.1	50.1	22	211.1	68.6	82	268.2	87.1
43	40.9	13.3	03	98.0	31.8	63	155.0	50.4	23	212.1	68.9	83	269.1	87.5
44	41.8	13.6	04	98.9	32.1	64	156.0	50.7	24	213.0	69.2	84	270.1	87.8
45	42.8	13.9	05	99.9	32.4	65	156.9	51.0	25	214.0	69.5	85	271.1	88.1
46	43.7	14.2	06	100.8	32.8	66	157.9	51.3	26	214.9	69.8	86	272.0	88.4
47	44.7	14.5	07	101.8	33.1	67	158.8	51.6	27	215.9	70.1	87	273.0	88.7
48	45.7	14.8	08	102.7	33.4	68	159.8	51.9	28	216.8	70.5	88	273.9	89.0
49	46.6	15.1	09	103.7	33.7	69	160.7	52.2	29	217.8	70.8	89	274.9	89.3
50	47.6	15.5	10	104.6	34.0	70	161.7	52.5	30	218.7	71.1	90	275.8	89.6
51	48.5	15.8	111	105.6	34.3	171	162.6	52.8	231	219.7	71.4	291	276.8	89.9
52	49.5	16.1	12	106.5	34.6	72	163.6	53.2	32	220.6	71.7	92	277.7	90.2
53	50.4	16.4	13	107.5	34.9	73	164.5	53.5	33	221.6	72.0	93	278.7	90.5
54	51.4	16.7	14	108.4	35.2	74	165.5	53.8	34	222.5	72.3	94	279.6	90.9
55	52.3	17.0	15	109.4	35.5	75	166.4	54.1	35	223.5	72.6	95	280.6	91.2
56	53.3	17.3	16	110.3	35.8	76	167.4	54.4	36	224.4	72.9	96	281.5	91.5
57	54.2	17.6	17	111.3	36.2	77	168.3	54.7	37	225.4	73.2	97	282.5	91.8
58	55.2	17.9	18	112.2	36.5	78	169.3	55.0	38	226.4	73.5	98	283.4	92.1
59	56.1	18.2	19	113.2	36.8	79	170.2	55.3	39	227.3	73.9	99	284.4	92.4
60	57.1	18.5	20	114.1	37.1	80	171.2	55.6	40	228.3	74.2	300	285.3	92.7
Dist.	Dep.	D. Lat.	Dist.	Dep.	D. Lat.	Dist.	Dep.	D. Lat.	Dist.	Dep.	D. Lat.	Dist.	Dep.	D. Lat.

288° | 072°
252° | 108°

72°

Dist.	D. Lat.	Dep.
N.	N x Cos.	N x Sin.
Hypotenuse	Side Adj.	Side Opp.

TABLE 4

Traverse 18° Table

342° | 018°
198° | 162°

Dist.	D. Lat.	Dep.	Dist.	D. Lat.	Dep.	Dist.	D. Lat.	Dep.	Dist.	D. Lat.	Dep.	Dist.	D. Lat.	Dep.
301	286.3	93.0	361	343.3	111.6	421	400.4	130.1	481	457.5	148.6	541	514.5	167.2
02	287.2	93.3	62	344.3	111.9	22	401.3	130.4	82	458.4	148.9	42	515.5	167.5
03	288.2	93.6	63	345.2	112.2	23	402.3	130.7	83	459.4	149.3	43	516.4	167.8
04	289.1	93.9	64	346.2	112.5	24	403.2	131.0	84	460.3	149.6	44	517.4	168.1
05	290.1	94.3	65	347.1	112.8	25	404.2	131.3	85	461.3	149.9	45	518.3	168.4
06	291.0	94.6	66	348.1	113.1	26	405.2	131.6	86	462.2	150.2	46	519.3	168.7
07	292.0	94.9	67	349.0	113.4	27	406.1	132.0	87	463.2	150.5	47	520.2	169.0
08	292.9	95.2	68	350.0	113.7	28	407.1	132.3	88	464.1	150.8	48	521.2	169.3
09	293.9	95.5	69	350.9	114.0	29	408.0	132.6	89	465.1	151.1	49	522.1	169.7
10	294.8	95.8	70	351.9	114.3	30	409.0	132.9	90	466.0	151.4	50	523.1	170.0
311	295.8	96.1	371	352.8	114.6	431	409.9	133.2	491	467.0	151.7	551	524.0	170.3
12	296.7	96.4	72	353.8	115.0	32	410.9	133.5	92	467.9	152.0	52	525.0	170.6
13	297.7	96.7	73	354.7	115.3	33	411.8	133.8	93	468.9	152.3	53	525.9	170.9
14	298.6	97.0	74	355.7	115.6	34	412.8	134.1	94	469.8	152.7	54	526.9	171.2
15	299.6	97.3	75	356.6	115.9	35	413.7	134.4	95	470.8	153.0	55	527.8	171.5
16	300.5	97.6	76	357.6	116.2	36	414.7	134.7	96	471.7	153.3	56	528.8	171.8
17	301.5	98.0	77	358.5	116.5	37	415.6	135.0	97	472.7	153.6	57	529.7	172.1
18	302.4	98.3	78	359.5	116.8	38	416.6	135.3	98	473.6	153.9	58	530.7	172.4
19	303.4	98.6	79	360.5	117.1	39	417.5	135.7	99	474.6	154.2	59	531.6	172.7
20	304.3	98.9	80	361.4	117.4	40	418.5	136.0	500	475.5	154.5	60	532.6	173.0
321	305.3	99.2	381	362.4	117.7	441	419.4	136.3	501	476.5	154.8	561	533.5	173.4
22	306.2	99.5	82	363.3	118.0	42	420.4	136.6	02	477.4	155.1	62	534.5	173.7
23	307.2	99.8	83	364.3	118.4	43	421.3	136.9	03	478.4	155.4	63	535.4	174.0
24	308.1	100.1	84	365.2	118.7	44	422.3	137.2	04	479.3	155.7	64	536.4	174.3
25	309.1	100.4	85	366.2	119.0	45	423.2	137.5	05	480.3	156.1	65	537.3	174.6
26	310.0	100.7	86	367.1	119.3	46	424.2	137.8	06	481.2	156.4	66	538.3	174.9
27	311.0	101.0	87	368.1	119.6	47	425.1	138.1	07	482.2	156.7	67	539.2	175.2
28	311.9	101.4	88	369.0	119.9	48	426.1	138.4	08	483.1	157.0	68	540.2	175.5
29	312.9	101.7	89	370.0	120.2	49	427.0	138.7	09	484.1	157.3	69	541.2	175.8
30	313.8	102.0	90	370.9	120.5	50	428.0	139.1	10	485.0	157.6	70	542.1	176.1
331	314.8	102.3	391	371.9	120.8	451	428.9	139.4	511	486.0	157.9	571	543.1	176.4
32	315.8	102.6	92	372.8	121.1	52	429.9	139.7	12	486.9	158.2	72	544.0	176.8
33	316.7	102.9	93	373.8	121.4	53	430.8	140.0	13	487.9	158.5	73	545.0	177.1
34	317.7	103.2	94	374.7	121.8	54	431.8	140.3	14	488.8	158.8	74	545.9	177.4
35	318.6	103.5	95	375.7	122.1	55	432.7	140.6	15	489.8	159.1	75	546.9	177.7
36	319.6	103.8	96	376.6	122.4	56	433.7	140.9	16	490.7	159.5	76	547.8	178.0
37	320.5	104.1	97	377.6	122.7	57	434.6	141.2	17	491.7	159.8	77	548.8	178.3
38	321.5	104.4	98	378.5	123.0	58	435.6	141.5	18	492.6	160.1	78	549.7	178.6
39	322.4	104.8	99	379.5	123.3	59	436.5	141.8	19	493.6	160.4	79	550.7	178.9
40	323.4	105.1	400	380.4	123.6	60	437.5	142.1	20	494.5	160.7	80	551.6	179.2
341	324.3	105.4	401	381.4	123.9	461	438.4	142.5	521	495.5	161.0	581	552.6	179.5
42	325.3	105.7	02	382.3	124.2	62	439.4	142.8	22	496.5	161.3	82	553.5	179.8
43	326.2	106.0	03	383.3	124.5	63	440.3	143.1	23	497.4	161.6	83	554.5	180.2
44	327.2	106.3	04	384.2	124.8	64	441.3	143.4	24	498.4	161.9	84	555.4	180.5
45	328.1	106.6	05	385.2	125.2	65	442.2	143.7	25	499.3	162.2	85	556.4	180.8
46	329.1	106.9	06	386.1	125.5	66	443.2	144.0	26	500.3	162.5	86	557.3	181.1
47	330.0	107.2	07	387.1	125.8	67	444.1	144.3	27	501.2	162.9	87	558.3	181.4
48	331.0	107.5	08	388.0	126.1	68	445.1	144.6	28	502.2	163.2	88	559.2	181.7
49	331.9	107.8	09	389.0	126.4	69	446.0	144.9	29	503.1	163.5	89	560.2	182.0
50	332.9	108.2	10	389.9	126.7	70	447.0	145.2	30	504.1	163.8	90	561.1	182.3
351	333.8	108.5	411	390.9	127.0	471	447.9	145.5	531	505.0	164.1	591	562.1	182.6
52	334.8	108.8	12	391.8	127.3	72	448.9	145.9	32	506.0	164.4	92	563.0	182.9
53	335.7	109.1	13	392.8	127.6	73	449.8	146.2	33	506.9	164.7	93	564.0	183.2
54	336.7	109.4	14	393.7	127.9	74	450.8	146.5	34	507.9	165.0	94	564.9	183.6
55	337.6	109.7	15	394.7	128.2	75	451.8	146.8	35	508.8	165.3	95	565.9	183.9
56	338.6	110.0	16	395.6	128.6	76	452.7	147.1	36	509.8	165.6	96	566.8	184.2
57	339.5	110.3	17	396.6	128.9	77	453.7	147.4	37	510.7	165.9	97	567.8	184.5
58	340.5	110.6	18	397.5	129.2	78	454.6	147.7	38	511.7	166.3	98	568.7	184.8
59	341.4	110.9	19	398.5	129.5	79	455.6	148.0	39	512.6	166.6	99	569.7	185.1
60	342.4	111.2	20	399.4	129.8	80	456.5	148.3	40	513.6	166.9	600	570.6	185.4
Dist.	Dep.	D. Lat.	Dist.	Dep.	D. Lat.	Dist.	Dep.	D. Lat.	Dist.	Dep.	D. Lat.	Dist.	Dep.	D. Lat.

Dist.	D. Lat.	Dep.
D Lo	Dep.	
	m	D Lo

72°

288° | 072°
252° | 108°

TABLE 4

Traverse 19° Table

341°	019°
199°	161°

Dist.	D. Lat.	Dep.	Dist.	D. Lat.	Dep.	Dist.	D. Lat.	Dep.	Dist.	D. Lat.	Dep.	Dist.	D. Lat.	Dep.
1	0.9	0.3	61	57.7	19.9	121	114.4	39.4	181	171.1	58.9	241	227.9	78.5
2	1.9	0.7	62	58.6	20.2	22	115.4	39.7	82	172.1	59.3	42	228.8	78.8
3	2.8	1.0	63	59.6	20.5	23	116.3	40.0	83	173.0	59.6	43	229.8	79.1
4	3.8	1.3	64	60.5	20.8	24	117.2	40.4	84	174.0	59.9	44	230.7	79.4
5	4.7	1.6	65	61.5	21.2	25	118.2	40.7	85	174.9	60.2	45	231.7	79.8
6	5.7	2.0	66	62.4	21.5	26	119.1	41.0	86	175.9	60.6	46	232.6	80.1
7	6.6	2.3	67	63.3	21.8	27	120.1	41.3	87	176.8	60.9	47	233.5	80.4
8	7.6	2.6	68	64.3	22.1	28	121.0	41.7	88	177.8	61.2	48	234.5	80.7
9	8.5	2.9	69	65.2	22.5	29	122.0	42.0	89	178.7	61.5	49	235.4	81.1
10	9.5	3.3	70	66.2	22.8	30	122.9	42.3	90	179.6	61.9	50	236.4	81.4
11	10.4	3.6	71	67.1	23.1	131	123.9	42.6	191	180.6	62.2	251	237.3	81.7
12	11.3	3.9	72	68.1	23.4	32	124.8	43.0	92	181.5	62.5	52	238.3	82.0
13	12.3	4.2	73	69.0	23.8	33	125.8	43.3	93	182.5	62.8	53	239.2	82.4
14	13.2	4.6	74	70.0	24.1	34	126.7	43.6	94	183.4	63.2	54	240.2	82.7
15	14.2	4.9	75	70.9	24.4	35	127.6	44.0	95	184.4	63.5	55	241.1	83.0
16	15.1	5.2	76	71.9	24.7	36	128.6	44.3	96	185.3	63.8	56	242.1	83.3
17	16.1	5.5	77	72.8	25.1	37	129.5	44.6	97	186.3	64.1	57	243.0	83.7
18	17.0	5.9	78	73.8	25.4	38	130.5	44.9	98	187.2	64.5	58	243.9	84.0
19	18.0	6.2	79	74.7	25.7	39	131.4	45.3	99	188.2	64.8	59	244.9	84.3
20	18.9	6.5	80	75.6	26.0	40	132.4	45.6	200	189.1	65.1	60	245.8	84.6
21	19.9	6.8	81	76.6	26.4	141	133.3	45.9	201	190.0	65.4	261	246.8	85.0
22	20.8	7.2	82	77.5	26.7	42	134.3	46.2	02	191.0	65.8	62	247.7	85.3
23	21.7	7.5	83	78.5	27.0	43	135.2	46.6	03	191.9	66.1	63	248.7	85.6
24	22.7	7.8	84	79.4	27.3	44	136.2	46.9	04	192.9	66.4	64	249.6	85.9
25	23.6	8.1	85	80.4	27.7	45	137.1	47.2	05	193.8	66.7	65	250.6	86.3
26	24.6	8.5	86	81.3	28.0	46	138.0	47.5	06	194.8	67.1	66	251.5	86.6
27	25.5	8.8	87	82.3	28.3	47	139.0	47.9	07	195.7	67.4	67	252.5	86.9
28	26.5	9.1	88	83.2	28.6	48	139.9	48.2	08	196.7	67.7	68	253.4	87.3
29	27.4	9.4	89	84.2	29.0	49	140.9	48.5	09	197.6	68.0	69	254.3	87.6
30	28.4	9.8	90	85.1	29.3	50	141.8	48.8	10	198.6	68.4	70	255.3	87.9
31	29.3	10.1	91	86.0	29.6	151	142.8	49.2	211	199.5	68.7	271	256.2	88.2
32	30.3	10.4	92	87.0	30.0	52	143.7	49.5	12	200.4	69.0	72	257.2	88.6
33	31.2	10.7	93	87.9	30.3	53	144.7	49.8	13	201.4	69.3	73	258.1	88.9
34	32.1	11.1	94	88.9	30.6	54	145.6	50.1	14	202.3	69.7	74	259.1	89.2
35	33.1	11.4	95	89.8	30.9	55	146.6	50.5	15	203.3	70.0	75	260.0	89.5
36	34.0	11.7	96	90.8	31.3	56	147.5	50.8	16	204.2	70.3	76	261.0	89.9
37	35.0	12.0	97	91.7	31.6	57	148.4	51.1	17	205.2	70.6	77	261.9	90.2
38	35.9	12.4	98	92.7	31.9	58	149.4	51.4	18	206.1	71.0	78	262.9	90.5
39	36.9	12.7	99	93.6	32.2	59	150.3	51.8	19	207.1	71.3	79	263.8	90.8
40	37.8	13.0	100	94.6	32.6	60	151.3	52.1	20	208.0	71.6	80	264.7	91.2
41	38.8	13.3	101	95.5	32.9	161	152.2	52.4	221	209.0	72.0	281	265.7	91.5
42	39.7	13.7	02	96.4	33.2	62	153.2	52.7	22	209.9	72.3	82	266.6	91.8
43	40.7	14.0	03	97.4	33.5	63	154.1	53.1	23	210.9	72.6	83	267.6	92.1
44	41.6	14.3	04	98.3	33.9	64	155.1	53.4	24	211.8	72.9	84	268.5	92.5
45	42.5	14.7	05	99.3	34.2	65	156.0	53.7	25	212.7	73.3	85	269.5	92.8
46	43.5	15.0	06	100.2	34.5	66	157.0	54.0	26	213.7	73.6	86	270.4	93.1
47	44.4	15.3	07	101.2	34.8	67	157.9	54.4	27	214.6	73.9	87	271.4	93.4
48	45.4	15.6	08	102.1	35.2	68	158.8	54.7	28	215.6	74.2	88	272.3	93.8
49	46.3	16.0	09	103.1	35.5	69	159.8	55.0	29	216.5	74.6	89	273.3	94.1
50	47.3	16.3	10	104.0	35.8	70	160.7	55.3	30	217.5	74.9	90	274.2	94.4
51	48.2	16.6	111	105.0	36.1	171	161.7	55.7	231	218.4	75.2	291	275.1	94.7
52	49.2	16.9	12	105.9	36.5	72	162.6	56.0	32	219.4	75.5	92	276.1	95.1
53	50.1	17.3	13	106.8	36.8	73	163.6	56.3	33	220.3	75.9	93	277.0	95.4
54	51.1	17.6	14	107.8	37.1	74	164.5	56.6	34	221.3	76.2	94	278.0	95.7
55	52.0	17.9	15	108.7	37.4	75	165.5	57.0	35	222.2	76.5	95	278.9	96.0
56	52.9	18.2	16	109.7	37.8	76	166.4	57.3	36	223.1	76.8	96	279.9	96.4
57	53.9	18.6	17	110.6	38.1	77	167.4	57.6	37	224.1	77.2	97	280.8	96.7
58	54.8	18.9	18	111.6	38.4	78	168.3	58.0	38	225.0	77.5	98	281.8	97.0
59	55.8	19.2	19	112.5	38.7	79	169.2	58.3	39	226.0	77.8	99	282.7	97.3
60	56.7	19.5	20	113.5	39.1	80	170.2	58.6	40	226.9	78.1	300	283.7	97.7
Dist.	Dep.	D. Lat.	Dist.	Dep.	D. Lat.	Dist.	Dep.	D. Lat.	Dist.	Dep.	D. Lat.	Dist.	Dep.	D. Lat.

289°	071°
251°	109°

71°

Dist.	D. Lat.	Dep.
N.	N x Cos.	N x Sin.
Hypotenuse	Side Adj.	Side Opp.

TABLE 4

Traverse 19° Table

341°	019°
199°	161°

Dist.	D. Lat.	Dep.	Dist.	D. Lat.	Dep.	Dist.	D. Lat.	Dep.	Dist.	D. Lat.	Dep.	Dist.	D. Lat.	Dep.
301	284.6	98.0	361	341.3	117.5	421	398.1	137.1	481	454.8	156.6	541	511.5	176.1
02	285.5	98.3	62	342.3	117.9	22	399.0	137.4	82	455.7	156.9	42	512.5	176.5
03	286.5	98.6	63	343.2	118.2	23	400.0	137.7	83	456.7	157.2	43	513.4	176.8
04	287.4	99.0	64	344.2	118.5	24	400.9	138.0	84	457.6	157.6	44	514.4	177.1
05	288.4	99.3	65	345.1	118.8	25	401.8	138.4	85	458.6	157.9	45	515.3	177.4
06	289.3	99.6	66	346.1	119.2	26	402.8	138.7	86	459.5	158.2	46	516.3	177.8
07	290.3	99.9	67	347.0	119.5	27	403.7	139.0	87	460.5	158.6	47	517.2	178.1
08	291.2	100.3	68	348.0	119.8	28	404.7	139.3	88	461.4	158.9	48	518.1	178.4
09	292.2	100.6	69	348.9	120.1	29	405.6	139.7	89	462.4	159.2	49	519.1	178.7
10	293.1	100.9	70	349.8	120.5	30	406.6	140.0	90	463.3	159.5	50	520.0	179.1
311	294.1	101.3	371	350.8	120.8	431	407.5	140.3	491	464.2	159.9	551	521.0	179.4
12	295.0	101.6	72	351.7	121.1	32	408.5	140.6	92	465.2	160.2	52	521.9	179.7
13	295.9	101.9	73	352.7	121.4	33	409.4	141.0	93	466.1	160.5	53	522.9	180.0
14	296.9	102.2	74	353.6	121.8	34	410.4	141.3	94	467.1	160.8	54	523.8	180.4
15	297.8	102.6	75	354.6	122.1	35	411.3	141.6	95	468.0	161.2	55	524.8	180.7
16	298.8	102.9	76	355.5	122.4	36	412.2	141.9	96	469.0	161.5	56	525.7	181.0
17	299.7	103.2	77	356.5	122.7	37	413.2	142.3	97	469.9	161.8	57	526.7	181.3
18	300.7	103.5	78	357.4	123.1	38	414.1	142.6	98	470.9	162.1	58	527.6	181.7
19	301.6	103.9	79	358.4	123.4	39	415.1	142.9	99	471.8	162.5	59	528.5	182.0
20	302.6	104.2	80	359.3	123.7	40	416.0	143.2	500	472.8	162.8	60	529.5	182.3
321	303.5	104.5	381	360.2	124.0	441	417.0	143.6	501	473.7	163.1	561	530.4	182.6
22	304.5	104.8	82	361.2	124.4	42	417.9	143.9	02	474.7	163.4	62	531.4	183.0
23	305.4	105.2	83	362.1	124.7	43	418.9	144.2	03	475.6	163.8	63	532.3	183.3
24	306.3	105.5	84	363.1	125.0	44	419.8	144.6	04	476.5	164.1	64	533.3	183.6
25	307.3	105.8	85	364.0	125.3	45	420.8	144.9	05	477.5	164.4	65	534.2	183.9
26	308.2	106.1	86	365.0	125.7	46	421.7	145.2	06	478.4	164.7	66	535.2	184.3
27	309.2	106.5	87	365.9	126.0	47	422.6	145.5	07	479.4	165.1	67	536.1	184.6
28	310.1	106.8	88	366.9	126.3	48	423.6	145.9	08	480.3	165.4	68	537.1	184.9
29	311.1	107.1	89	367.8	126.6	49	424.5	146.2	09	481.3	165.7	69	538.0	185.2
30	312.0	107.4	90	368.8	127.0	50	425.5	146.5	10	482.2	166.0	70	538.9	185.6
331	313.0	107.8	391	369.7	127.3	451	426.4	146.8	511	483.2	166.4	571	539.9	185.9
32	313.9	108.1	92	370.6	127.6	52	427.4	147.2	12	484.1	166.7	72	540.8	186.2
33	314.9	108.4	93	371.6	127.9	53	428.3	147.5	13	485.1	167.0	73	541.8	186.6
34	315.8	108.7	94	372.5	128.3	54	429.3	147.8	14	486.0	167.3	74	542.7	186.9
35	316.7	109.1	95	373.5	128.6	55	430.2	148.1	15	486.9	167.7	75	543.7	187.2
36	317.7	109.4	96	374.4	128.9	56	431.2	148.5	16	487.9	168.0	76	544.6	187.5
37	318.6	109.7	97	375.4	129.3	57	432.1	148.8	17	488.8	168.3	77	545.6	187.9
38	319.6	110.0	98	376.3	129.6	58	433.0	149.1	18	489.8	168.6	78	546.5	188.2
39	320.5	110.4	99	377.3	129.9	59	434.0	149.4	19	490.7	169.0	79	547.5	188.5
40	321.5	110.7	400	378.2	130.2	60	434.9	149.8	20	491.7	169.3	80	548.4	188.8
341	322.4	111.0	401	379.2	130.6	461	435.9	150.1	521	492.6	169.6	581	549.3	189.2
42	323.4	111.3	02	380.1	130.9	62	436.8	150.4	22	493.6	169.9	82	550.3	189.5
43	324.3	111.7	03	381.0	131.2	63	437.8	150.7	23	494.5	170.3	83	551.2	189.8
44	325.3	112.0	04	382.0	131.5	64	438.7	151.1	24	495.5	170.6	84	552.2	190.1
45	326.2	112.3	05	382.9	131.9	65	439.7	151.4	25	496.4	170.9	85	553.1	190.5
46	327.1	112.6	06	383.9	132.2	66	440.6	151.7	26	497.3	171.2	86	554.1	190.8
47	328.1	113.0	07	384.8	132.5	67	441.6	152.0	27	498.3	171.6	87	555.0	191.1
48	329.0	113.3	08	385.8	132.8	68	442.5	152.4	28	499.2	171.9	88	556.0	191.4
49	330.0	113.6	09	386.7	133.2	69	443.4	152.7	29	500.2	172.2	89	556.9	191.8
50	330.9	113.9	10	387.7	133.5	70	444.4	153.0	30	501.1	172.6	90	557.9	192.1
351	331.9	114.3	411	388.6	133.8	471	445.3	153.3	531	502.1	172.9	591	558.8	192.4
52	332.8	114.6	12	389.6	134.1	72	446.3	153.7	32	503.0	173.2	92	559.7	192.7
53	333.8	114.9	13	390.5	134.5	73	447.2	154.0	33	504.0	173.5	93	560.7	193.1
54	334.7	115.3	14	391.4	134.8	74	448.2	154.3	34	504.9	173.9	94	561.6	193.4
55	335.7	115.6	15	392.4	135.1	75	449.1	154.6	35	505.9	174.2	95	562.6	193.7
56	336.6	115.9	16	393.3	135.4	76	450.1	155.0	36	506.8	174.5	96	563.5	194.0
57	337.6	116.2	17	394.3	135.8	77	451.0	155.3	37	507.7	174.8	97	564.5	194.4
58	338.5	116.6	18	395.2	136.1	78	452.0	155.6	38	508.7	175.2	98	565.4	194.7
59	339.4	116.9	19	396.2	136.4	79	452.9	155.9	39	509.6	175.5	99	566.4	195.0
60	340.4	117.2	20	397.1	136.7	80	453.8	156.3	40	510.6	175.8	600	567.3	195.3
Dist.	Dep.	D. Lat.	Dist.	Dep.	D. Lat.	Dist.	Dep.	D. Lat.	Dist.	Dep.	D. Lat.	Dist.	Dep.	D. Lat.

Dist.	D. Lat.	Dep.
D Lo	Dep.	
	m	D Lo

71°

289°	071°
251°	109°

TABLE 4

Traverse 20° Table

340° | 020°
200° | 160°

Dist.	D. Lat.	Dep.	Dist.	D. Lat.	Dep.	Dist.	D. Lat.	Dep.	Dist.	D. Lat.	Dep.	Dist.	D. Lat.	Dep.
1	0.9	0.3	61	57.3	20.9	121	113.7	41.4	181	170.1	61.9	241	226.5	82.4
2	1.9	0.7	62	58.3	21.2	22	114.6	41.7	82	171.0	62.2	42	227.4	82.8
3	2.8	1.0	63	59.2	21.5	23	115.6	42.1	83	172.0	62.6	43	228.3	83.1
4	3.8	1.4	64	60.1	21.9	24	116.5	42.4	84	172.9	62.9	44	229.3	83.5
5	4.7	1.7	65	61.1	22.2	25	117.5	42.8	85	173.8	63.3	45	230.2	83.8
6	5.6	2.1	66	62.0	22.6	26	118.4	43.1	86	174.8	63.6	46	231.2	84.1
7	6.6	2.4	67	63.0	22.9	27	119.3	43.4	87	175.7	64.0	47	232.1	84.5
8	7.5	2.7	68	63.9	23.3	28	120.3	43.8	88	176.7	64.3	48	233.0	84.8
9	8.5	3.1	69	64.8	23.6	29	121.2	44.1	89	177.6	64.6	49	234.0	85.2
10	9.4	3.4	70	65.8	23.9	30	122.2	44.5	90	178.5	65.0	50	234.9	85.5
11	10.3	3.8	71	66.7	24.3	131	123.1	44.8	191	179.5	65.3	251	235.9	85.8
12	11.3	4.1	72	67.7	24.6	32	124.0	45.1	92	180.4	65.7	52	236.8	86.2
13	12.2	4.4	73	68.6	25.0	33	125.0	45.5	93	181.4	66.0	53	237.7	86.5
14	13.2	4.8	74	69.5	25.3	34	125.9	45.8	94	182.3	66.4	54	238.7	86.9
15	14.1	5.1	75	70.5	25.7	35	126.9	46.2	95	183.2	66.7	55	239.6	87.2
16	15.0	5.5	76	71.4	26.0	36	127.8	46.5	96	184.2	67.0	56	240.6	87.6
17	16.0	5.8	77	72.4	26.3	37	128.7	46.9	97	185.1	67.4	57	241.5	87.9
18	16.9	6.2	78	73.3	26.7	38	129.7	47.2	98	186.1	67.7	58	242.4	88.2
19	17.9	6.5	79	74.2	27.0	39	130.6	47.5	99	187.0	68.1	59	243.4	88.6
20	18.8	6.8	80	75.2	27.4	40	131.6	47.9	200	187.9	68.4	60	244.3	88.9
21	19.7	7.2	81	76.1	27.7	141	132.5	48.2	201	188.9	68.7	261	245.3	89.3
22	20.7	7.5	82	77.1	28.0	42	133.4	48.6	02	189.8	69.1	62	246.2	89.6
23	21.6	7.9	83	78.0	28.4	43	134.4	48.9	03	190.8	69.4	63	247.1	90.0
24	22.6	8.2	84	78.9	28.7	44	135.3	49.3	04	191.7	69.8	64	248.1	90.3
25	23.5	8.6	85	79.9	29.1	45	136.3	49.6	05	192.6	70.1	65	249.0	90.6
26	24.4	8.9	86	80.8	29.4	46	137.2	49.9	06	193.6	70.5	66	250.0	91.0
27	25.4	9.2	87	81.8	29.8	47	138.1	50.3	07	194.5	70.8	67	250.9	91.3
28	26.3	9.6	88	82.7	30.1	48	139.1	50.6	08	195.5	71.1	68	251.8	91.7
29	27.3	9.9	89	83.6	30.4	49	140.0	51.0	09	196.4	71.5	69	252.8	92.0
30	28.2	10.3	90	84.6	30.8	50	141.0	51.3	10	197.3	71.8	70	253.7	92.3
31	29.1	10.6	91	85.5	31.1	151	141.9	51.6	211	198.3	72.2	271	254.7	92.7
32	30.1	10.9	92	86.5	31.5	52	142.8	52.0	12	199.2	72.5	72	255.6	93.0
33	31.0	11.3	93	87.4	31.8	53	143.8	52.3	13	200.2	72.9	73	256.5	93.4
34	31.9	11.6	94	88.3	32.1	54	144.7	52.7	14	201.1	73.2	74	257.5	93.7
35	32.9	12.0	95	89.3	32.5	55	145.7	53.0	15	202.0	73.5	75	258.4	94.1
36	33.8	12.3	96	90.2	32.8	56	146.6	53.4	16	203.0	73.9	76	259.4	94.4
37	34.8	12.7	97	91.2	33.2	57	147.5	53.7	17	203.9	74.2	77	260.3	94.7
38	35.7	13.0	98	92.1	33.5	58	148.5	54.0	18	204.9	74.6	78	261.2	95.1
39	36.6	13.3	99	93.0	33.9	59	149.4	54.4	19	205.8	74.9	79	262.2	95.4
40	37.6	13.7	100	94.0	34.2	60	150.4	54.7	20	206.7	75.2	80	263.1	95.8
41	38.5	14.0	101	94.9	34.5	161	151.3	55.1	221	207.7	75.6	281	264.1	96.1
42	39.5	14.4	02	95.8	34.9	62	152.2	55.4	22	208.6	75.9	82	265.0	96.4
43	40.4	14.7	03	96.8	35.2	63	153.2	55.7	23	209.6	76.3	83	265.9	96.8
44	41.3	15.0	04	97.7	35.6	64	154.1	56.1	24	210.5	76.6	84	266.9	97.1
45	42.3	15.4	05	98.7	35.9	65	155.0	56.4	25	211.4	77.0	85	267.8	97.5
46	43.2	15.7	06	99.6	36.3	66	156.0	56.8	26	212.4	77.3	86	268.8	97.8
47	44.2	16.1	07	100.5	36.6	67	156.9	57.1	27	213.3	77.6	87	269.7	98.2
48	45.1	16.4	08	101.5	36.9	68	157.9	57.5	28	214.2	78.0	88	270.6	98.5
49	46.0	16.8	09	102.4	37.3	69	158.8	57.8	29	215.2	78.3	89	271.6	98.8
50	47.0	17.1	10	103.4	37.6	70	159.7	58.1	30	216.1	78.7	90	272.5	99.2
51	47.9	17.4	111	104.3	38.0	171	160.7	58.5	231	217.1	79.0	291	273.5	99.5
52	48.9	17.8	12	105.2	38.3	72	161.6	58.8	32	218.0	79.3	92	274.4	99.9
53	49.8	18.1	13	106.2	38.6	73	162.6	59.2	33	218.9	79.7	93	275.3	100.2
54	50.7	18.5	14	107.1	39.0	74	163.5	59.5	34	219.9	80.0	94	276.3	100.6
55	51.7	18.8	15	108.1	39.3	75	164.4	59.9	35	220.8	80.4	95	277.2	100.9
56	52.6	19.2	16	109.0	39.7	76	165.4	60.2	36	221.8	80.7	96	278.1	101.2
57	53.6	19.5	17	109.9	40.0	77	166.3	60.5	37	222.7	81.1	97	279.1	101.6
58	54.5	19.8	18	110.9	40.4	78	167.3	60.9	38	223.6	81.4	98	280.0	101.9
59	55.4	20.2	19	111.8	40.7	79	168.2	61.2	39	224.6	81.7	99	281.0	102.3
60	56.4	20.5	20	112.8	41.0	80	169.1	61.6	40	225.5	82.1	300	281.9	102.6
Dist.	Dep.	D. Lat.	Dist.	Dep.	D. Lat.	Dist.	Dep.	D. Lat.	Dist.	Dep.	D. Lat.	Dist.	Dep.	D. Lat.

290° | 070°
250° | 110°

70°

Dist.	D. Lat.	Dep.
N.	N x Cos.	N x Sin.
Hypotenuse	Side Adj.	Side Opp.

TABLE 4

Traverse 20° Table

340° | 020°
200° | 160°

Dist.	D. Lat.	Dep.	Dist.	D. Lat.	Dep.	Dist.	D. Lat.	Dep.	Dist.	D. Lat.	Dep.	Dist.	D. Lat.	Dep.
301	282.8	102.9	361	339.2	123.5	421	395.6	144.0	481	452.0	164.5	541	508.4	185.0
02	283.8	103.3	62	340.2	123.8	22	396.6	144.3	82	452.9	164.9	42	509.3	185.4
03	284.7	103.6	63	341.1	124.2	23	397.5	144.7	83	453.9	165.2	43	510.3	185.7
04	285.7	104.0	64	342.0	124.5	24	398.4	145.0	84	454.8	165.5	44	511.2	186.1
05	286.6	104.3	65	343.0	124.8	25	399.4	145.4	85	455.8	165.9	45	512.1	186.4
06	287.5	104.7	66	343.9	125.2	26	400.3	145.7	86	456.7	166.2	46	513.1	186.7
07	288.5	105.0	67	344.9	125.5	27	401.2	146.0	87	457.6	166.6	47	514.0	187.1
08	289.4	105.3	68	345.8	125.9	28	402.2	146.4	88	458.6	166.9	48	515.0	187.4
09	290.4	105.7	69	346.7	126.2	29	403.1	146.7	89	459.5	167.2	49	515.9	187.8
10	291.3	106.0	70	347.7	126.5	30	404.1	147.1	90	460.4	167.6	50	516.8	188.1
311	292.2	106.4	371	348.6	126.9	431	405.0	147.4	491	461.4	167.9	551	517.8	188.5
12	293.2	106.7	72	349.6	127.2	32	405.9	147.8	92	462.3	168.3	52	518.7	188.8
13	294.1	107.1	73	350.5	127.6	33	406.9	148.1	93	463.3	168.6	53	519.7	189.1
14	295.1	107.4	74	351.4	127.9	34	407.8	148.4	94	464.2	169.0	54	520.6	189.5
15	296.0	107.7	75	352.4	128.3	35	408.8	148.8	95	465.1	169.3	55	521.5	189.8
16	296.9	108.1	76	353.3	128.6	36	409.7	149.1	96	466.1	169.6	56	522.5	190.2
17	297.9	108.4	77	354.3	128.9	37	410.6	149.5	97	467.0	170.0	57	523.4	190.5
18	298.8	108.8	78	355.2	129.3	38	411.6	149.8	98	468.0	170.3	58	524.3	190.8
19	299.8	109.1	79	356.1	129.6	39	412.5	150.1	99	468.9	170.7	59	525.3	191.2
20	300.7	109.4	80	357.1	130.0	40	413.5	150.5	500	469.8	171.0	60	526.2	191.5
321	301.6	109.8	381	358.0	130.3	441	414.4	150.8	501	470.8	171.4	561	527.2	191.9
22	302.6	110.1	82	359.0	130.7	42	415.3	151.2	02	471.7	171.7	62	528.1	192.2
23	303.5	110.5	83	359.9	131.0	43	416.3	151.5	03	472.7	172.0	63	529.0	192.6
24	304.5	110.8	84	360.8	131.3	44	417.2	151.9	04	473.6	172.4	64	530.0	192.9
25	305.4	111.2	85	361.8	131.7	45	418.2	152.2	05	474.5	172.7	65	530.9	193.2
26	306.3	111.5	86	362.7	132.0	46	419.1	152.5	06	475.5	173.1	66	531.9	193.6
27	307.3	111.8	87	363.7	132.4	47	420.0	152.9	07	476.4	173.4	67	532.8	193.9
28	308.2	112.2	88	364.6	132.7	48	421.0	153.2	08	477.4	173.7	68	533.7	194.3
29	309.2	112.5	89	365.5	133.0	49	421.9	153.6	09	478.3	174.1	69	534.7	194.6
30	310.1	112.9	90	366.5	133.4	50	422.9	153.9	10	479.2	174.4	70	535.6	195.0
331	311.0	113.2	391	367.4	133.7	451	423.8	154.3	511	480.2	174.8	571	536.6	195.3
32	312.0	113.6	92	368.4	134.1	52	424.7	154.6	12	481.1	175.1	72	537.5	195.6
33	312.9	113.9	93	369.3	134.4	53	425.7	154.9	13	482.1	175.5	73	538.4	196.0
34	313.9	114.2	94	370.2	134.8	54	426.6	155.3	14	483.0	175.8	74	539.4	196.3
35	314.8	114.6	95	371.2	135.1	55	427.6	155.6	15	483.9	176.1	75	540.3	196.7
36	315.7	114.9	96	372.1	135.4	56	428.5	156.0	16	484.9	176.5	76	541.3	197.0
37	316.7	115.3	97	373.1	135.8	57	429.4	156.3	17	485.8	176.8	77	542.2	197.3
38	317.6	115.6	98	374.0	136.1	58	430.4	156.6	18	486.8	177.2	78	543.1	197.7
39	318.6	115.9	99	374.9	136.5	59	431.3	157.0	19	487.7	177.5	79	544.1	198.0
40	319.5	116.3	400	375.9	136.8	60	432.3	157.3	20	488.6	177.9	80	545.0	198.4
341	320.4	116.6	401	376.8	137.2	461	433.2	157.7	521	489.6	178.2	581	546.0	198.7
42	321.4	117.0	02	377.8	137.5	62	434.1	158.0	22	490.5	178.5	82	546.9	199.1
43	322.3	117.3	03	378.7	137.8	63	435.1	158.4	23	491.5	178.9	83	547.8	199.4
44	323.3	117.7	04	379.6	138.2	64	436.0	158.7	24	492.4	179.2	84	548.8	199.7
45	324.2	118.0	05	380.6	138.5	65	437.0	159.0	25	493.3	179.6	85	549.7	200.1
46	325.1	118.3	06	381.5	138.9	66	437.9	159.4	26	494.3	179.9	86	550.7	200.4
47	326.1	118.7	07	382.5	139.2	67	438.8	159.7	27	495.2	180.2	87	551.6	200.8
48	327.0	119.0	08	383.4	139.5	68	439.8	160.1	28	496.2	180.6	88	552.5	201.1
49	328.0	119.4	09	384.3	139.9	69	440.7	160.4	29	497.1	180.9	89	553.5	201.4
50	328.9	119.7	10	385.3	140.2	70	441.7	160.7	30	498.0	181.3	90	554.4	201.8
351	329.8	120.0	411	386.2	140.6	471	442.6	161.1	531	499.0	181.6	591	555.4	202.1
52	330.8	120.4	12	387.2	140.9	72	443.5	161.4	32	499.9	182.0	92	556.3	202.5
53	331.7	120.7	13	388.1	141.3	73	444.5	161.8	33	500.9	182.3	93	557.2	202.8
54	332.7	121.1	14	389.0	141.6	74	445.4	162.1	34	501.8	182.6	94	558.2	203.2
55	333.6	121.4	15	390.0	141.9	75	446.4	162.5	35	502.7	183.0	95	559.1	203.5
56	334.5	121.8	16	390.9	142.3	76	447.3	162.8	36	503.7	183.3	96	560.1	203.8
57	335.5	122.1	17	391.9	142.6	77	448.2	163.1	37	504.6	183.7	97	561.0	204.2
58	336.4	122.4	18	392.8	143.0	78	449.2	163.5	38	505.6	184.0	98	561.9	204.5
59	337.3	122.8	19	393.7	143.3	79	450.1	163.8	39	506.5	184.3	99	562.9	204.9
60	338.3	123.1	20	394.7	143.6	80	451.1	164.2	40	507.4	184.7	600	563.8	205.2
Dist.	Dep.	D. Lat.	Dist.	Dep.	D. Lat.	Dist.	Dep.	D. Lat.	Dist.	Dep.	D. Lat.	Dist.	Dep.	D. Lat.

Dist.	D. Lat.	Dep.
D Lo	Dep.	
	m	D Lo

70°

290° | 070°
250° | 110°

TABLE 4

339° | 021°
201° | 159°

Traverse **21°** Table

Dist.	D. Lat.	Dep.	Dist.	D. Lat.	Dep.	Dist.	D. Lat.	Dep.	Dist.	D. Lat.	Dep.	Dist.	D. Lat.	Dep.
1	0.9	0.4	61	56.9	21.9	121	113.0	43.4	181	169.0	64.9	241	225.0	86.4
2	1.9	0.7	62	57.9	22.2	22	113.9	43.7	82	169.9	65.2	42	225.9	86.7
3	2.8	1.1	63	58.8	22.6	23	114.8	44.1	83	170.8	65.6	43	226.9	87.1
4	3.7	1.4	64	59.7	22.9	24	115.8	44.4	84	171.8	65.9	44	227.8	87.4
5	4.7	1.8	65	60.7	23.3	25	116.7	44.8	85	172.7	66.3	45	228.7	87.8
6	5.6	2.2	66	61.6	23.7	26	117.6	45.2	86	173.6	66.7	46	229.7	88.2
7	6.5	2.5	67	62.5	24.0	27	118.6	45.5	87	174.6	67.0	47	230.6	88.5
8	7.5	2.9	68	63.5	24.4	28	119.5	45.9	88	175.5	67.4	48	231.5	88.9
9	8.4	3.2	69	64.4	24.7	29	120.4	46.2	89	176.4	67.7	49	232.5	89.2
10	9.3	3.6	70	65.4	25.1	30	121.4	46.6	90	177.4	68.1	50	233.4	89.6
11	10.3	3.9	71	66.3	25.4	131	122.3	46.9	191	178.3	68.4	251	234.3	90.0
12	11.2	4.3	72	67.2	25.8	32	123.2	47.3	92	179.2	68.8	52	235.3	90.3
13	12.1	4.7	73	68.2	26.2	33	124.2	47.7	93	180.2	69.2	53	236.2	90.7
14	13.1	5.0	74	69.1	26.5	34	125.1	48.0	94	181.1	69.5	54	237.1	91.0
15	14.0	5.4	75	70.0	26.9	35	126.0	48.4	95	182.0	69.9	55	238.1	91.4
16	14.9	5.7	76	71.0	27.2	36	127.0	48.7	96	183.0	70.2	56	239.0	91.7
17	15.9	6.1	77	71.9	27.6	37	127.9	49.1	97	183.9	70.6	57	239.9	92.1
18	16.8	6.5	78	72.8	28.0	38	128.8	49.5	98	184.8	71.0	58	240.9	92.5
19	17.7	6.8	79	73.8	28.3	39	129.8	49.8	99	185.8	71.3	59	241.8	92.8
20	18.7	7.2	80	74.7	28.7	40	130.7	50.2	200	186.7	71.7	60	242.7	93.2
21	19.6	7.5	81	75.6	29.0	141	131.6	50.5	201	187.6	72.0	261	243.7	93.5
22	20.5	7.9	82	76.6	29.4	42	132.6	50.9	02	188.6	72.4	62	244.6	93.9
23	21.5	8.2	83	77.5	29.7	43	133.5	51.2	03	189.5	72.7	63	245.5	94.3
24	22.4	8.6	84	78.4	30.1	44	134.4	51.6	04	190.5	73.1	64	246.5	94.6
25	23.3	9.0	85	79.4	30.5	45	135.4	52.0	05	191.4	73.5	65	247.4	95.0
26	24.3	9.3	86	80.3	30.8	46	136.3	52.3	06	192.3	73.8	66	248.3	95.3
27	25.2	9.7	87	81.2	31.2	47	137.2	52.7	07	193.3	74.2	67	249.3	95.7
28	26.1	10.0	88	82.2	31.5	48	138.2	53.0	08	194.2	74.5	68	250.2	96.0
29	27.1	10.4	89	83.1	31.9	49	139.1	53.4	09	195.1	74.9	69	251.1	96.4
30	28.0	10.8	90	84.0	32.3	50	140.0	53.8	10	196.1	75.3	70	252.1	96.8
31	28.9	11.1	91	85.0	32.6	151	141.0	54.1	211	197.0	75.6	271	253.0	97.1
32	29.9	11.5	92	85.9	33.0	52	141.9	54.5	12	197.9	76.0	72	253.9	97.5
33	30.8	11.8	93	86.8	33.3	53	142.8	54.8	13	198.9	76.3	73	254.9	97.8
34	31.7	12.2	94	87.8	33.7	54	143.8	55.2	14	199.8	76.7	74	255.8	98.2
35	32.7	12.5	95	88.7	34.0	55	144.7	55.5	15	200.7	77.0	75	256.7	98.6
36	33.6	12.9	96	89.6	34.4	56	145.6	55.9	16	201.7	77.4	76	257.7	98.9
37	34.5	13.3	97	90.6	34.8	57	146.6	56.3	17	202.6	77.8	77	258.6	99.3
38	35.5	13.6	98	91.5	35.1	58	147.5	56.6	18	203.5	78.1	78	259.5	99.6
39	36.4	14.0	99	92.4	35.5	59	148.4	57.0	19	204.5	78.5	79	260.5	100.0
40	37.3	14.3	100	93.4	35.8	60	149.4	57.3	20	205.4	78.8	80	261.4	100.3
41	38.3	14.7	101	94.3	36.2	161	150.3	57.7	221	206.3	79.2	281	262.3	100.7
42	39.2	15.1	02	95.2	36.6	62	151.2	58.1	22	207.3	79.6	82	263.3	101.1
43	40.1	15.4	03	96.2	36.9	63	152.2	58.4	23	208.2	79.9	83	264.2	101.4
44	41.1	15.8	04	97.1	37.3	64	153.1	58.8	24	209.1	80.3	84	265.1	101.8
45	42.0	16.1	05	98.0	37.6	65	154.0	59.1	25	210.1	80.6	85	266.1	102.1
46	42.9	16.5	06	99.0	38.0	66	155.0	59.5	26	211.0	81.0	86	267.0	102.5
47	43.9	16.8	07	99.9	38.3	67	155.9	59.8	27	211.9	81.3	87	267.9	102.9
48	44.8	17.2	08	100.8	38.7	68	156.8	60.2	28	212.9	81.7	88	268.9	103.2
49	45.7	17.6	09	101.8	39.1	69	157.8	60.6	29	213.8	82.1	89	269.8	103.6
50	46.7	17.9	10	102.7	39.4	70	158.7	60.9	30	214.7	82.4	90	270.7	103.9
51	47.6	18.3	111	103.6	39.8	171	159.6	61.3	231	215.7	82.8	291	271.7	104.3
52	48.5	18.6	12	104.6	40.1	72	160.6	61.6	32	216.6	83.1	92	272.6	104.6
53	49.5	19.0	13	105.5	40.5	73	161.5	62.0	33	217.5	83.5	93	273.5	105.0
54	50.4	19.4	14	106.4	40.9	74	162.4	62.4	34	218.5	83.9	94	274.5	105.4
55	51.3	19.7	15	107.4	41.2	75	163.4	62.7	35	219.4	84.2	95	275.4	105.7
56	52.3	20.1	16	108.3	41.6	76	164.3	63.1	36	220.3	84.6	96	276.3	106.1
57	53.2	20.4	17	109.2	41.9	77	165.2	63.4	37	221.3	84.9	97	277.3	106.4
58	54.1	20.8	18	110.2	42.3	78	166.2	63.8	38	222.2	85.3	98	278.2	106.8
59	55.1	21.1	19	111.1	42.6	79	167.1	64.1	39	223.1	85.6	99	279.1	107.2
60	56.0	21.5	20	112.0	43.0	80	168.0	64.5	40	224.1	86.0	300	280.1	107.5
Dist.	Dep.	D. Lat.	Dist.	Dep.	D. Lat.	Dist.	Dep.	D. Lat.	Dist.	Dep.	D. Lat.	Dist.	Dep.	D. Lat.

291° | 069°
249° | 111°

69°

Dist.	D. Lat.	Dep.
N.	N x Cos.	N x Sin.
Hypotenuse	Side Adj.	Side Opp.

TABLE 4

339° | 021°
201° | 159°

Traverse **21°** Table

Dist.	D. Lat.	Dep.	Dist.	D. Lat.	Dep.	Dist.	D. Lat.	Dep.	Dist.	D. Lat.	Dep.	Dist.	D. Lat.	Dep.
301	281.0	107.9	361	337.0	129.4	421	393.0	150.9	481	449.1	172.4	541	505.1	193.9
02	281.9	108.2	62	338.0	129.7	22	394.0	151.2	82	450.0	172.7	42	506.0	194.2
03	282.9	108.6	63	338.9	130.1	23	394.9	151.6	83	450.9	173.1	43	506.9	194.6
04	283.8	108.9	64	339.8	130.4	24	395.8	151.9	84	451.9	173.5	44	507.9	195.0
05	284.7	109.3	65	340.8	130.8	25	396.8	152.3	85	452.8	173.8	45	508.8	195.3
06	285.7	109.7	66	341.7	131.2	26	397.7	152.7	86	453.7	174.2	46	509.7	195.7
07	286.6	110.0	67	342.6	131.5	27	398.6	153.0	87	454.7	174.5	47	510.7	196.0
08	287.5	110.4	68	343.6	131.9	28	399.6	153.4	88	455.6	174.9	48	511.6	196.4
09	288.5	110.7	69	344.5	132.2	29	400.5	153.7	89	456.5	175.2	49	512.5	196.7
10	289.4	111.1	70	345.4	132.6	30	401.4	154.1	90	457.5	175.6	50	513.5	197.1
311	290.3	111.5	371	346.4	133.0	431	402.4	154.5	491	458.4	176.0	551	514.4	197.5
12	291.3	111.8	72	347.3	133.3	32	403.3	154.8	92	459.3	176.3	52	515.3	197.8
13	292.2	112.2	73	348.2	133.7	33	404.2	155.2	93	460.3	176.7	53	516.3	198.2
14	293.1	112.5	74	349.2	134.0	34	405.2	155.5	94	461.2	177.0	54	517.2	198.5
15	294.1	112.9	75	350.1	134.4	35	406.1	155.9	95	462.1	177.4	55	518.1	198.9
16	295.0	113.2	76	351.0	134.7	36	407.0	156.2	96	463.1	177.8	56	519.1	199.3
17	295.9	113.6	77	352.0	135.1	37	408.0	156.6	97	464.0	178.1	57	520.0	199.6
18	296.9	114.0	78	352.9	135.5	38	408.9	157.0	98	464.9	178.5	58	520.9	200.0
19	297.8	114.3	79	353.8	135.8	39	409.8	157.3	99	465.9	178.8	59	521.9	200.3
20	298.7	114.7	80	354.8	136.2	40	410.8	157.7	500	466.8	179.2	60	522.8	200.7
321	299.7	115.0	381	355.7	136.5	441	411.7	158.0	501	467.7	179.5	561	523.7	201.0
22	300.6	115.4	82	356.6	136.9	42	412.6	158.4	02	468.7	179.9	62	524.7	201.4
23	301.5	115.8	83	357.6	137.3	43	413.6	158.8	03	469.6	180.3	63	525.6	201.8
24	302.5	116.1	84	358.5	137.6	44	414.5	159.1	04	470.5	180.6	64	526.5	202.1
25	303.4	116.5	85	359.4	138.0	45	415.4	159.5	05	471.5	181.0	65	527.5	202.5
26	304.3	116.8	86	360.4	138.3	46	416.4	159.8	06	472.4	181.3	66	528.4	202.8
27	305.3	117.2	87	361.3	138.7	47	417.3	160.2	07	473.3	181.7	67	529.3	203.2
28	306.2	117.5	88	362.2	139.0	48	418.2	160.5	08	474.3	182.1	68	530.3	203.6
29	307.1	117.9	89	363.2	139.4	49	419.2	160.9	09	475.2	182.4	69	531.2	203.9
30	308.1	118.3	90	364.1	139.8	50	420.1	161.3	10	476.1	182.8	70	532.1	204.3
331	309.0	118.6	391	365.0	140.1	451	421.0	161.6	511	477.1	183.1	571	533.1	204.6
32	309.9	119.0	92	366.0	140.5	52	422.0	162.0	12	478.0	183.5	72	534.0	205.0
33	310.9	119.3	93	366.9	140.8	53	422.9	162.3	13	478.9	183.8	73	534.9	205.3
34	311.8	119.7	94	367.8	141.2	54	423.8	162.7	14	479.9	184.2	74	535.9	205.7
35	312.7	120.1	95	368.8	141.6	55	424.8	163.1	15	480.8	184.6	75	536.8	206.1
36	313.7	120.4	96	369.7	141.9	56	425.7	163.4	16	481.7	184.9	76	537.7	206.4
37	314.6	120.8	97	370.6	142.3	57	426.6	163.8	17	482.7	185.3	77	538.7	206.8
38	315.6	121.1	98	371.6	142.6	58	427.6	164.1	18	483.6	185.6	78	539.6	207.1
39	316.5	121.5	99	372.5	143.0	59	428.5	164.5	19	484.5	186.0	79	540.5	207.5
40	317.4	121.8	400	373.4	143.3	60	429.4	164.8	20	485.5	186.4	80	541.5	207.9
341	318.4	122.2	401	374.4	143.7	461	430.4	165.2	521	486.4	186.7	581	542.4	208.2
42	319.3	122.6	02	375.3	144.1	62	431.3	165.6	22	487.3	187.1	82	543.3	208.6
43	320.2	122.9	03	376.2	144.4	63	432.2	165.9	23	488.3	187.4	83	544.3	208.9
44	321.2	123.3	04	377.2	144.8	64	433.2	166.3	24	489.2	187.8	84	545.2	209.3
45	322.1	123.6	05	378.1	145.1	65	434.1	166.6	25	490.1	188.1	85	546.1	209.6
46	323.0	124.0	06	379.0	145.5	66	435.0	167.0	26	491.1	188.5	86	547.1	210.0
47	324.0	124.4	07	380.0	145.9	67	436.0	167.4	27	492.0	188.9	87	548.0	210.4
48	324.9	124.7	08	380.9	146.2	68	436.9	167.7	28	492.9	189.2	88	548.9	210.7
49	325.8	125.1	09	381.8	146.6	69	437.8	168.1	29	493.9	189.6	89	549.9	211.1
50	326.8	125.4	10	382.8	146.9	70	438.8	168.4	30	494.8	189.9	90	550.8	211.4
351	327.7	125.8	411	383.7	147.3	471	439.7	168.8	531	495.7	190.3	591	551.7	211.8
52	328.6	126.1	12	384.6	147.6	72	440.6	169.1	32	496.7	190.7	92	552.7	212.2
53	329.6	126.5	13	385.6	148.0	73	441.6	169.5	33	497.6	191.0	93	553.6	212.5
54	330.5	126.9	14	386.5	148.4	74	442.5	169.9	34	498.5	191.4	94	554.5	212.9
55	331.4	127.2	15	387.4	148.7	75	443.5	170.2	35	499.5	191.7	95	555.5	213.2
56	332.4	127.6	16	388.4	149.1	76	444.4	170.6	36	500.4	192.1	96	556.4	213.6
57	333.3	127.9	17	389.3	149.4	77	445.3	170.9	37	501.3	192.4	97	557.3	213.9
58	334.2	128.3	18	390.2	149.8	78	446.3	171.3	38	502.3	192.8	98	558.3	214.3
59	335.2	128.7	19	391.2	150.2	79	447.2	171.7	39	503.2	193.2	99	559.2	214.7
60	336.1	129.0	20	392.1	150.5	80	448.1	172.0	40	504.1	193.5	600	560.1	215.0
Dist.	Dep.	D. Lat.	Dist.	Dep.	D. Lat.	Dist.	Dep.	D. Lat.	Dist.	Dep.	D. Lat.	Dist.	Dep.	D. Lat.

Dist.	D. Lat.	Dep.
D Lo	Dep.	
	m	D Lo

69°

291° | 069°
249° | 111°

TABLE 4
Traverse 22° Table

338° | 022°
202° | 158°

Dist.	D. Lat.	Dep.	Dist.	D. Lat.	Dep.	Dist.	D. Lat.	Dep.	Dist.	D. Lat.	Dep.	Dist.	D. Lat.	Dep.
1	0.9	0.4	61	56.6	22.9	121	112.2	45.3	181	167.8	67.8	241	223.5	90.3
2	1.9	0.7	62	57.5	23.2	22	113.1	45.7	82	168.7	68.2	42	224.4	90.7
3	2.8	1.1	63	58.4	23.6	23	114.0	46.1	83	169.7	68.6	43	225.3	91.0
4	3.7	1.5	64	59.3	24.0	24	115.0	46.5	84	170.6	68.9	44	226.2	91.4
5	4.6	1.9	65	60.3	24.3	25	115.9	46.8	85	171.5	69.3	45	227.2	91.8
6	5.6	2.2	66	61.2	24.7	26	116.8	47.2	86	172.5	69.7	46	228.1	92.2
7	6.5	2.6	67	62.1	25.1	27	117.8	47.6	87	173.4	70.1	47	229.0	92.5
8	7.4	3.0	68	63.0	25.5	28	118.7	47.9	88	174.3	70.4	48	229.9	92.9
9	8.3	3.4	69	64.0	25.8	29	119.6	48.3	89	175.2	70.8	49	230.9	93.3
10	9.3	3.7	70	64.9	26.2	30	120.5	48.7	90	176.2	71.2	50	231.8	93.7
11	10.2	4.1	71	65.8	26.6	131	121.5	49.1	191	177.1	71.5	251	232.7	94.0
12	11.1	4.5	72	66.8	27.0	32	122.4	49.4	92	178.0	71.9	52	233.7	94.4
13	12.1	4.9	73	67.7	27.3	33	123.3	49.8	93	178.9	72.3	53	234.6	94.8
14	13.0	5.2	74	68.6	27.7	34	124.2	50.2	94	179.9	72.7	54	235.5	95.2
15	13.9	5.6	75	69.5	28.1	35	125.2	50.6	95	180.8	73.0	55	236.4	95.5
16	14.8	6.0	76	70.5	28.5	36	126.1	50.9	96	181.7	73.4	56	237.4	95.9
17	15.8	6.4	77	71.4	28.8	37	127.0	51.3	97	182.7	73.8	57	238.3	96.3
18	16.7	6.7	78	72.3	29.2	38	128.0	51.7	98	183.6	74.2	58	239.2	96.6
19	17.6	7.1	79	73.2	29.6	39	128.9	52.1	99	184.5	74.5	59	240.1	97.0
20	18.5	7.5	80	74.2	30.0	40	129.8	52.4	200	185.4	74.9	60	241.1	97.4
21	19.5	7.9	81	75.1	30.3	141	130.7	52.8	201	186.4	75.3	261	242.0	97.8
22	20.4	8.2	82	76.0	30.7	42	131.7	53.2	02	187.3	75.7	62	242.9	98.1
23	21.3	8.6	83	77.0	31.1	43	132.6	53.6	03	188.2	76.0	63	243.8	98.5
24	22.3	9.0	84	77.9	31.5	44	133.5	53.9	04	189.1	76.4	64	244.8	98.9
25	23.2	9.4	85	78.8	31.8	45	134.4	54.3	05	190.1	76.8	65	245.7	99.3
26	24.1	9.7	86	79.7	32.2	46	135.4	54.7	06	191.0	77.2	66	246.6	99.6
27	25.0	10.1	87	80.7	32.6	47	136.3	55.1	07	191.9	77.5	67	247.6	100.0
28	26.0	10.5	88	81.6	33.0	48	137.2	55.4	08	192.9	77.9	68	248.5	100.4
29	26.9	10.9	89	82.5	33.3	49	138.2	55.8	09	193.8	78.3	69	249.4	100.8
30	27.8	11.2	90	83.4	33.7	50	139.1	56.2	10	194.7	78.7	70	250.3	101.1
31	28.7	11.6	91	84.4	34.1	151	140.0	56.6	211	195.6	79.0	271	251.3	101.5
32	29.7	12.0	92	85.3	34.5	52	140.9	56.9	12	196.6	79.4	72	252.2	101.9
33	30.6	12.4	93	86.2	34.8	53	141.9	57.3	13	197.5	79.8	73	253.1	102.3
34	31.5	12.7	94	87.2	35.2	54	142.8	57.7	14	198.4	80.2	74	254.0	102.6
35	32.5	13.1	95	88.1	35.6	55	143.7	58.1	15	199.3	80.5	75	255.0	103.0
36	33.4	13.5	96	89.0	36.0	56	144.6	58.4	16	200.3	80.9	76	255.9	103.4
37	34.3	13.9	97	89.9	36.3	57	145.6	58.8	17	201.2	81.3	77	256.8	103.8
38	35.2	14.2	98	90.9	36.7	58	146.5	59.2	18	202.1	81.7	78	257.8	104.1
39	36.2	14.6	99	91.8	37.1	59	147.4	59.6	19	203.1	82.0	79	258.7	104.5
40	37.1	15.0	100	92.7	37.5	60	148.3	59.9	20	204.0	82.4	80	259.6	104.9
41	38.0	15.4	101	93.6	37.8	161	149.3	60.3	221	204.9	82.8	281	260.5	105.3
42	38.9	15.7	02	94.6	38.2	62	150.2	60.7	22	205.8	83.2	82	261.5	105.6
43	39.9	16.1	03	95.5	38.6	63	151.1	61.1	23	206.8	83.5	83	262.4	106.0
44	40.8	16.5	04	96.4	39.0	64	152.1	61.4	24	207.7	83.9	84	263.3	106.4
45	41.7	16.9	05	97.4	39.3	65	153.0	61.8	25	208.6	84.3	85	264.2	106.8
46	42.7	17.2	06	98.3	39.7	66	153.9	62.2	26	209.5	84.7	86	265.2	107.1
47	43.6	17.6	07	99.2	40.1	67	154.8	62.6	27	210.5	85.0	87	266.1	107.5
48	44.5	18.0	08	100.1	40.5	68	155.8	62.9	28	211.4	85.4	88	267.0	107.9
49	45.4	18.4	09	101.1	40.8	69	156.7	63.3	29	212.3	85.8	89	268.0	108.3
50	46.4	18.7	10	102.0	41.2	70	157.6	63.7	30	213.3	86.2	90	268.9	108.6
51	47.3	19.1	111	102.9	41.6	171	158.5	64.1	231	214.2	86.5	291	269.8	109.0
52	48.2	19.5	12	103.8	42.0	72	159.5	64.4	32	215.1	86.9	92	270.7	109.4
53	49.1	19.9	13	104.8	42.3	73	160.4	64.8	33	216.0	87.3	93	271.7	109.8
54	50.1	20.2	14	105.7	42.7	74	161.3	65.2	34	217.0	87.7	94	272.6	110.1
55	51.0	20.6	15	106.6	43.1	75	162.3	65.6	35	217.9	88.0	95	273.5	110.5
56	51.9	21.0	16	107.6	43.5	76	163.2	65.9	36	218.8	88.4	96	274.4	110.9
57	52.8	21.4	17	108.5	43.8	77	164.1	66.3	37	219.7	88.8	97	275.4	111.3
58	53.8	21.7	18	109.4	44.2	78	165.0	66.7	38	220.7	89.2	98	276.3	111.6
59	54.7	22.1	19	110.3	44.6	79	166.0	67.1	39	221.6	89.5	99	277.2	112.0
60	55.6	22.5	20	111.3	45.0	80	166.9	67.4	40	222.5	89.9	300	278.2	112.4
Dist.	Dep.	D. Lat.	Dist.	Dep.	D. Lat.	Dist.	Dep.	D. Lat.	Dist.	Dep.	D. Lat.	Dist.	Dep.	D. Lat.

292° | 068°
248° | 112°

68°

Dist.	D. Lat.	Dep.
N.	N x Cos.	N x Sin.
Hypotenuse	Side Adj.	Side Opp.

TABLE 4
Traverse 22° Table

338° | 022°
202° | 158°

Dist.	D. Lat.	Dep.	Dist.	D. Lat.	Dep.	Dist.	D. Lat.	Dep.	Dist.	D. Lat.	Dep.	Dist.	D. Lat.	Dep.
301	279.1	112.8	361	334.7	135.2	421	390.3	157.7	481	446.0	180.2	541	501.6	202.7
02	280.0	113.1	62	335.6	135.6	22	391.3	158.1	82	446.9	180.6	42	502.5	203.0
03	280.9	113.5	63	336.6	136.0	23	392.2	158.5	83	447.8	180.9	43	503.5	203.4
04	281.9	113.9	64	337.5	136.4	24	393.1	158.8	84	448.8	181.3	44	504.4	203.8
05	282.8	114.3	65	338.4	136.7	25	394.1	159.2	85	449.7	181.7	45	505.3	204.2
06	283.7	114.6	66	339.3	137.1	26	395.0	159.6	86	450.6	182.1	46	506.2	204.5
07	284.6	115.0	67	340.3	137.5	27	395.9	160.0	87	451.5	182.4	47	507.2	204.9
08	285.6	115.4	68	341.2	137.9	28	396.8	160.3	88	452.5	182.8	48	508.1	205.3
09	286.5	115.8	69	342.1	138.2	29	397.8	160.7	89	453.4	183.2	49	509.0	205.7
10	287.4	116.1	70	343.1	138.6	30	398.7	161.1	90	454.3	183.6	50	510.0	206.0
311	288.4	116.5	371	344.0	139.0	431	399.6	161.5	491	455.2	183.9	551	510.9	206.4
12	289.3	116.9	72	344.9	139.4	32	400.5	161.8	92	456.2	184.3	52	511.8	206.8
13	290.2	117.3	73	345.8	139.7	33	401.5	162.2	93	457.1	184.7	53	512.7	207.2
14	291.1	117.6	74	346.8	140.1	34	402.4	162.6	94	458.0	185.1	54	513.7	207.5
15	292.1	118.0	75	347.7	140.5	35	403.3	163.0	95	459.0	185.4	55	514.6	207.9
16	293.0	118.4	76	348.6	140.9	36	404.3	163.3	96	459.9	185.8	56	515.5	208.3
17	293.9	118.8	77	349.5	141.2	37	405.2	163.7	97	460.8	186.2	57	516.4	208.7
18	294.8	119.1	78	350.5	141.6	38	406.1	164.1	98	461.7	186.6	58	517.4	209.0
19	295.8	119.5	79	351.4	142.0	39	407.0	164.5	99	462.7	186.9	59	518.3	209.4
20	296.7	119.9	80	352.3	142.4	40	408.0	164.8	500	463.6	187.3	60	519.2	209.8
321	297.6	120.2	381	353.3	142.7	441	408.9	165.2	501	464.5	187.7	561	520.2	210.2
22	298.6	120.6	82	354.2	143.1	42	409.8	165.6	02	465.4	188.1	62	521.1	210.5
23	299.5	121.0	83	355.1	143.5	43	410.7	166.0	03	466.4	188.4	63	522.0	210.9
24	300.4	121.4	84	356.0	143.8	44	411.7	166.3	04	467.3	188.8	64	522.9	211.3
25	301.3	121.7	85	357.0	144.2	45	412.6	166.7	05	468.2	189.2	65	523.9	211.7
26	302.3	122.1	86	357.9	144.6	46	413.5	167.1	06	469.2	189.6	66	524.8	212.0
27	303.2	122.5	87	358.8	145.0	47	414.5	167.4	07	470.1	189.9	67	525.7	212.4
28	304.1	122.9	88	359.7	145.3	48	415.4	167.8	08	471.0	190.3	68	526.6	212.8
29	305.0	123.2	89	360.7	145.7	49	416.3	168.2	09	471.9	190.7	69	527.6	213.2
30	306.0	123.6	90	361.6	146.1	50	417.2	168.6	10	472.9	191.0	70	528.5	213.5
331	306.9	124.0	391	362.5	146.5	451	418.2	168.9	511	473.8	191.4	571	529.4	213.9
32	307.8	124.4	92	363.5	146.8	52	419.1	169.3	12	474.7	191.8	72	530.3	214.3
33	308.8	124.7	93	364.4	147.2	53	420.0	169.7	13	475.6	192.2	73	531.3	214.6
34	309.7	125.1	94	365.3	147.6	54	420.9	170.1	14	476.6	192.5	74	532.2	215.0
35	310.6	125.5	95	366.2	148.0	55	421.9	170.4	15	477.5	192.9	75	533.1	215.4
36	311.5	125.9	96	367.2	148.3	56	422.8	170.8	16	478.4	193.3	76	534.1	215.8
37	312.5	126.2	97	368.1	148.7	57	423.7	171.2	17	479.4	193.7	77	535.0	216.1
38	313.4	126.6	98	369.0	149.1	58	424.7	171.6	18	480.3	194.0	78	535.9	216.5
39	314.3	127.0	99	369.9	149.5	59	425.6	171.9	19	481.2	194.4	79	536.8	216.9
40	315.2	127.4	400	370.9	149.8	60	426.5	172.3	20	482.1	194.8	80	537.8	217.3
341	316.2	127.7	401	371.8	150.2	461	427.4	172.7	521	483.1	195.2	581	538.7	217.6
42	317.1	128.1	02	372.7	150.6	62	428.4	173.1	22	484.0	195.5	82	539.6	218.0
43	318.0	128.5	03	373.7	151.0	63	429.3	173.4	23	484.9	195.9	83	540.5	218.4
44	319.0	128.9	04	374.6	151.3	64	430.2	173.8	24	485.8	196.3	84	541.5	218.8
45	319.9	129.2	05	375.5	151.7	65	431.1	174.2	25	486.8	196.7	85	542.4	219.1
46	320.8	129.6	06	376.4	152.1	66	432.1	174.6	26	487.7	197.0	86	543.3	219.5
47	321.7	130.0	07	377.4	152.5	67	433.0	174.9	27	488.6	197.4	87	544.3	219.9
48	322.7	130.4	08	378.3	152.8	68	433.9	175.3	28	489.6	197.8	88	545.2	220.3
49	323.6	130.7	09	379.2	153.2	69	434.8	175.7	29	490.5	198.2	89	546.1	220.6
50	324.5	131.1	10	380.1	153.6	70	435.8	176.1	30	491.4	198.5	90	547.0	221.0
351	325.4	131.5	411	381.1	154.0	471	436.7	176.4	531	492.3	198.9	591	548.0	221.4
52	326.4	131.9	12	382.0	154.3	72	437.6	176.8	32	493.3	199.3	92	548.9	221.8
53	327.3	132.2	13	382.9	154.7	73	438.6	177.2	33	494.2	199.7	93	549.8	222.1
54	328.2	132.6	14	383.9	155.1	74	439.5	177.6	34	495.1	200.0	94	550.7	222.5
55	329.2	133.0	15	384.8	155.5	75	440.4	177.9	35	496.0	200.4	95	551.7	222.9
56	330.1	133.4	16	385.7	155.8	76	441.3	178.3	36	497.0	200.8	96	552.6	223.3
57	331.0	133.7	17	386.6	156.2	77	442.3	178.7	37	497.9	201.2	97	553.5	223.6
58	331.9	134.1	18	387.6	156.6	78	443.2	179.1	38	498.8	201.5	98	554.5	224.0
59	332.9	134.5	19	388.5	157.0	79	444.1	179.4	39	499.8	201.9	99	555.4	224.4
60	333.8	134.9	20	389.4	157.3	80	445.0	179.8	40	500.7	202.3	600	556.3	224.8
Dist.	Dep.	D. Lat.	Dist.	Dep.	D. Lat.	Dist.	Dep.	D. Lat.	Dist.	Dep.	D. Lat.	Dist.	Dep.	D. Lat.

Dist.	D. Lat.	Dep.
D Lo	Dep.	
	m	D Lo

68°

292° | 068°
248° | 112°

TABLE 4

Traverse 23° Table

337° | 023° / 203° | 157°

Dist.	D. Lat.	Dep.	D.st.	D. Lat.	Dep.	Dist.	D. Lat.	Dep.	Dist.	D. Lat.	Dep.	Dist.	D. Lat.	Dep.
1	0.9	0.4	61	56.2	23.8	121	111.4	47.3	181	166.6	70.7	241	221.8	94.2
2	1.8	0.8	62	57.1	24.2	22	112.3	47.7	82	167.5	71.1	42	222.8	94.6
3	2.8	1.2	63	58.0	24.6	23	113.2	48.1	83	168.5	71.5	43	223.7	94.9
4	3.7	1.6	64	58.9	25.0	24	114.1	48.5	84	169.4	71.9	44	224.6	95.3
5	4.6	2.0	65	59.8	25.4	25	115.1	48.8	85	170.3	72.3	45	225.5	95.7
6	5.5	2.3	66	60.8	25.8	26	116.0	49.2	86	171.2	72.7	46	226.4	96.1
7	6.4	2.7	67	61.7	26.2	27	116.9	49.6	87	172.1	73.1	47	227.4	96.5
8	7.4	3.1	68	62.6	26.6	28	117.8	50.0	88	173.1	73.5	48	228.3	96.9
9	8.3	3.5	69	63.5	27.0	29	118.7	50.4	89	174.0	73.8	49	229.2	97.3
10	9.2	3.9	70	64.4	27.4	30	119.7	50.8	90	174.9	74.2	50	230.1	97.7
11	10.1	4.3	71	65.4	27.7	131	120.6	51.2	191	175.8	74.6	251	231.0	98.1
12	11.0	4.7	72	66.3	28.1	32	121.5	51.6	92	176.7	75.0	52	232.0	98.5
13	12.0	5.1	73	67.2	28.5	33	122.4	52.0	93	177.7	75.4	53	232.9	98.9
14	12.9	5.5	74	68.1	28.9	34	123.3	52.4	94	178.6	75.8	54	233.8	99.2
15	13.8	5.9	75	69.0	29.3	35	124.3	52.7	95	179.5	76.2	55	234.7	99.6
16	14.7	6.3	76	70.0	29.7	36	125.2	53.1	96	180.4	76.6	56	235.6	100.0
17	15.6	6.6	77	70.9	30.1	37	126.1	53.5	97	181.3	77.0	57	236.6	100.4
18	16.6	7.0	78	71.8	30.5	38	127.0	53.9	98	182.3	77.4	58	237.5	100.8
19	17.5	7.4	79	72.7	30.9	39	128.0	54.3	99	183.2	77.8	59	238.4	101.2
20	18.4	7.8	80	73.6	31.3	40	128.9	54.7	200	184.1	78.1	60	239.3	101.6
21	19.3	8.2	81	74.6	31.6	141	129.8	55.1	201	185.0	78.5	261	240.3	102.0
22	20.3	8.6	82	75.5	32.0	42	130.7	55.5	02	185.9	78.9	62	241.2	102.4
23	21.2	9.0	83	76.4	32.4	43	131.6	55.9	03	186.9	79.3	63	242.1	102.8
24	22.1	9.4	84	77.3	32.8	44	132.6	56.3	04	187.8	79.7	64	243.0	103.2
25	23.0	9.8	85	78.2	33.2	45	133.5	56.7	05	188.7	80.1	65	243.9	103.5
26	23.9	10.2	86	79.2	33.6	46	134.4	57.0	06	189.6	80.5	66	244.9	103.9
27	24.9	10.5	87	80.1	34.0	47	135.3	57.4	07	190.5	80.9	67	245.8	104.3
28	25.8	10.9	88	81.0	34.4	48	136.2	57.8	08	191.5	81.3	68	246.7	104.7
29	26.7	11.3	89	81.9	34.8	49	137.2	58.2	09	192.4	81.7	69	247.6	105.1
30	27.6	11.7	90	82.8	35.2	50	138.1	58.6	10	193.3	82.1	70	248.5	105.5
31	28.5	12.1	91	83.8	35.6	151	139.0	59.0	211	194.2	82.4	271	249.5	105.9
32	29.5	12.5	92	84.7	35.9	52	139.9	59.4	12	195.1	82.8	72	250.4	106.3
33	30.4	12.9	93	85.6	36.3	53	140.8	59.8	13	196.1	83.2	73	251.3	106.7
34	31.3	13.3	94	86.5	36.7	54	141.8	60.2	14	197.0	83.6	74	252.2	107.1
35	32.2	13.7	95	87.4	37.1	55	142.7	60.6	15	197.9	84.0	75	253.1	107.5
36	33.1	14.1	96	88.4	37.5	56	143.6	61.0	16	198.8	84.4	76	254.1	107.8
37	34.1	14.5	97	89.3	37.9	57	144.5	61.3	17	199.7	84.8	77	255.0	108.2
38	35.0	14.8	98	90.2	38.3	58	145.4	61.7	18	200.7	85.2	78	255.9	108.6
39	35.9	15.2	99	91.1	38.7	59	146.4	62.1	19	201.6	85.6	79	256.8	109.0
40	36.8	15.6	100	92.1	39.1	60	147.3	62.5	20	202.5	86.0	80	257.7	109.4
41	37.7	16.0	101	93.0	39.5	161	148.2	62.9	221	203.4	86.4	281	258.7	109.8
42	38.7	16.4	02	93.9	39.9	62	149.1	63.3	22	204.4	86.7	82	259.6	110.2
43	39.6	16.8	03	94.8	40.2	63	150.0	63.7	23	205.3	87.1	83	260.5	110.6
44	40.5	17.2	04	95.7	40.6	64	151.0	64.1	24	206.2	87.5	84	261.4	111.0
45	41.4	17.6	05	96.7	41.0	65	151.9	64.5	25	207.1	87.9	85	262.3	111.4
46	42.3	18.0	06	97.6	41.4	66	152.8	64.9	26	208.0	88.3	86	263.3	111.7
47	43.3	18.4	07	98.5	41.8	67	153.7	65.3	27	209.0	88.7	87	264.2	112.1
48	44.2	18.8	08	99.4	42.2	68	154.6	65.6	28	209.9	89.1	88	265.1	112.5
49	45.1	19.1	09	100.3	42.6	69	155.6	66.0	29	210.8	89.5	89	266.0	112.9
50	46.0	19.5	10	101.3	43.0	70	156.5	66.4	30	211.7	89.9	90	266.9	113.3
51	46.9	19.9	111	102.2	43.4	171	157.4	66.8	231	212.6	90.3	291	267.9	113.7
52	47.9	20.3	12	103.1	43.8	72	158.3	67.2	32	213.6	90.6	92	268.8	114.1
53	48.8	20.7	13	104.0	44.2	73	159.2	67.6	33	214.5	91.0	93	269.7	114.5
54	49.7	21.1	14	104.9	44.5	74	160.2	68.0	34	215.4	91.4	94	270.6	114.9
55	50.6	21.5	15	105.9	44.9	75	161.1	68.4	35	216.3	91.8	95	271.5	115.3
56	51.5	21.9	16	106.8	45.3	76	162.0	68.8	36	217.2	92.2	96	272.5	115.7
57	52.5	22.3	17	107.7	45.7	77	162.9	69.2	37	218.2	92.6	97	273.4	116.0
58	53.4	22.7	18	108.6	46.1	78	163.8	69.6	38	219.1	93.0	98	274.3	116.4
59	54.3	23.1	19	109.5	46.5	79	164.8	69.9	39	220.0	93.4	99	275.2	116.8
60	55.2	23.4	20	110.5	46.9	80	165.7	70.3	40	220.9	93.8	300	276.2	117.2
Dist.	Dep.	D. Lat.	Dist.	Dep.	D. Lat.	Dist.	Dep.	D. Lat.	Dist.	Dep.	D. Lat.	Dist.	Dep.	D. Lat.

293° | 067° / 247° | 113°

67°

Dist.	D. Lat.	Dep.
N.	N x Cos.	N x Sin.
Hypotenuse	Side Adj.	Side Opp.

TABLE 4

Traverse 23° Table

337° | 023° / 203° | 157°

Dist.	D. Lat.	Dep.	Dist.	D. Lat.	Dep.	Dist.	D. Lat.	Dep.	Dist.	D. Lat.	Dep.	Dist.	D. Lat.	Dep.
301	277.1	117.6	361	332.3	141.1	421	387.5	164.5	481	442.8	187.9	541	498.0	211.4
02	278.0	118.0	62	333.2	141.4	22	388.5	164.9	82	443.7	188.3	42	498.9	211.8
03	278.9	118.4	63	334.1	141.8	23	389.4	165.3	83	444.6	188.7	43	499.8	212.2
04	279.8	118.8	64	335.1	142.2	24	390.3	165.7	84	445.5	189.1	44	500.8	212.6
05	280.8	119.2	65	336.0	142.6	25	391.2	166.1	85	446.4	189.5	45	501.7	212.9
06	281.7	119.6	66	336.9	143.0	26	392.1	166.5	86	447.4	189.9	46	502.6	213.3
07	282.6	120.0	67	337.8	143.4	27	393.1	166.8	87	448.3	190.3	47	503.5	213.7
08	283.5	120.3	68	338.7	143.8	28	394.0	167.2	88	449.2	190.7	48	504.4	214.1
09	284.4	120.7	69	339.7	144.2	29	394.9	167.6	89	450.1	191.1	49	505.4	214.5
10	285.4	121.1	70	340.6	144.6	30	395.8	168.0	90	451.0	191.5	50	506.3	214.9
311	286.3	121.5	371	341.5	145.0	431	396.7	168.4	491	452.0	191.8	551	507.2	215.3
12	287.2	121.9	72	342.4	145.4	32	397.7	168.8	92	452.9	192.2	52	508.1	215.7
13	288.1	122.3	73	343.3	145.7	33	398.6	169.2	93	453.8	192.6	53	509.0	216.1
14	289.0	122.7	74	344.3	146.1	34	399.5	169.6	94	454.7	193.0	54	510.0	216.5
15	290.0	123.1	75	345.2	146.5	35	400.4	170.0	95	455.6	193.4	55	510.9	216.9
16	290.9	123.5	76	346.1	146.9	36	401.3	170.4	96	456.6	193.8	56	511.8	217.2
17	291.8	123.9	77	347.0	147.3	37	402.3	170.7	97	457.5	194.2	57	512.7	217.6
18	292.7	124.3	78	348.0	147.7	38	403.2	171.1	98	458.4	194.6	58	513.6	218.0
19	293.6	124.6	79	348.9	148.1	39	404.1	171.5	99	459.3	195.0	59	514.6	218.4
20	294.6	125.0	80	349.8	148.5	40	405.0	171.9	500	460.3	195.4	60	515.5	218.8
321	295.5	125.4	381	350.7	148.9	441	405.9	172.3	501	461.2	195.8	561	516.4	219.2
22	296.4	125.8	82	351.6	149.3	42	406.9	172.7	02	462.1	196.1	62	517.3	219.6
23	297.3	126.2	83	352.6	149.7	43	407.8	173.1	03	463.0	196.5	63	518.2	220.0
24	298.2	126.6	84	353.5	150.0	44	408.7	173.5	04	463.9	196.9	64	519.2	220.4
25	299.2	127.0	85	354.4	150.4	45	409.6	173.9	05	464.9	197.3	65	520.1	220.8
26	300.1	127.4	86	355.3	150.8	46	410.5	174.3	06	465.8	197.7	66	521.0	221.2
27	301.0	127.8	87	356.2	151.2	47	411.5	174.7	07	466.7	198.1	67	521.9	221.5
28	301.9	128.2	88	357.2	151.6	48	412.4	175.0	08	467.6	198.5	68	522.8	221.9
29	302.8	128.6	89	358.1	152.0	49	413.3	175.4	09	468.5	198.9	69	523.8	222.3
30	303.8	128.9	90	359.0	152.4	50	414.2	175.8	10	469.5	199.3	70	524.7	222.7
331	304.7	129.3	391	359.9	152.8	451	415.1	176.2	511	470.4	199.7	571	525.6	223.1
32	305.6	129.7	92	360.8	153.2	52	416.1	176.6	12	471.3	200.1	72	526.5	223.5
33	306.5	130.1	93	361.8	153.6	53	417.0	177.0	13	472.2	200.4	73	527.4	223.9
34	307.4	130.5	94	362.7	153.9	54	417.9	177.4	14	473.1	200.8	74	528.4	224.3
35	308.4	130.9	95	363.6	154.3	55	418.8	177.8	15	474.1	201.2	75	529.3	224.7
36	309.3	131.3	96	364.5	154.7	56	419.8	178.2	16	475.0	201.6	76	530.2	225.1
37	310.2	131.7	97	365.4	155.1	57	420.7	178.6	17	475.9	202.0	77	531.1	225.5
38	311.1	132.1	98	366.4	155.5	58	421.6	179.0	18	476.8	202.4	78	532.1	225.8
39	312.1	132.5	99	367.3	155.9	59	422.5	179.3	19	477.7	202.8	79	533.0	226.2
40	313.0	132.8	400	368.2	156.3	60	423.4	179.7	20	478.7	203.2	80	533.9	226.6
341	313.9	133.2	401	369.1	156.7	461	424.4	180.1	521	479.6	203.6	581	534.8	227.0
42	314.8	133.6	02	370.0	157.1	62	425.3	180.5	22	480.5	204.0	82	535.7	227.4
43	315.7	134.0	03	371.0	157.5	63	426.2	180.9	23	481.4	204.4	83	536.7	227.8
44	316.7	134.4	04	371.9	157.9	64	427.1	181.3	24	482.3	204.7	84	537.6	228.2
45	317.6	134.8	05	372.8	158.2	65	428.0	181.7	25	483.3	205.1	85	538.5	228.6
46	318.5	135.2	06	373.7	158.6	66	429.0	182.1	26	484.2	205.5	86	539.4	229.0
47	319.4	135.6	07	374.6	159.0	67	429.9	182.5	27	485.1	205.9	87	540.3	229.4
48	320.3	136.0	08	375.6	159.4	68	430.8	182.9	28	486.0	206.3	88	541.3	229.7
49	321.3	136.4	09	376.5	159.8	69	431.7	183.3	29	486.9	206.7	89	542.2	230.1
50	322.2	136.8	10	377.4	160.2	70	432.6	183.6	30	487.9	207.1	90	543.1	230.5
351	323.1	137.1	411	378.3	160.6	471	433.6	184.0	531	488.8	207.5	591	544.0	230.9
52	324.0	137.5	12	379.2	161.0	72	434.5	184.4	32	489.7	207.9	92	544.9	231.3
53	324.9	137.9	13	380.2	161.4	73	435.4	184.8	33	490.6	208.3	93	545.9	231.7
54	325.9	138.3	14	381.1	161.8	74	436.3	185.2	34	491.5	208.7	94	546.8	232.1
55	326.8	138.7	15	382.0	162.2	75	437.2	185.6	35	492.5	209.0	95	547.7	232.5
56	327.7	139.1	16	382.9	162.5	76	438.2	186.0	36	493.4	209.4	96	548.6	232.9
57	328.6	139.5	17	383.9	162.9	77	439.1	186.4	37	494.3	209.8	97	549.5	233.3
58	329.5	139.9	18	384.8	163.3	78	440.0	186.8	38	495.2	210.2	98	550.5	233.7
59	330.5	140.3	19	385.7	163.7	79	440.9	187.2	39	496.2	210.6	99	551.4	234.0
60	331.4	140.7	20	386.6	164.1	80	441.8	187.6	40	497.1	211.0	600	552.3	234.4
Dist.	Dep.	D. Lat.	Dist.	Dep.	D. Lat.	Dist.	Dep.	D. Lat.	Dist.	Dep.	D. Lat.	Dist.	Dep.	D. Lat.

Dist.	D. Lat.	Dep.
D Lo	Dep.	
	m	D Lo

67°

293° | 067° / 247° | 113°

TABLE 4

Traverse 24° Table

336° | 024° / 204° | 156°

Dist.	D. Lat.	Dep.	Dist.	D. Lat.	Dep.	Dist.	D. Lat.	Dep.	Dist.	D. Lat.	Dep.	Dist.	D. Lat.	Dep.
1	0.9	0.4	61	55.7	24.8	121	110.5	49.2	181	165.4	73.6	241	220.2	98.0
2	1.8	0.8	62	56.6	25.2	22	111.5	49.6	82	166.3	74.0	42	221.1	98.4
3	2.7	1.2	63	57.6	25.6	23	112.4	50.0	83	167.2	74.4	43	222.0	98.8
4	3.7	1.6	64	58.5	26.0	24	113.3	50.4	84	168.1	74.8	44	222.9	99.2
5	4.6	2.0	65	59.4	26.4	25	114.2	50.8	85	169.0	75.2	45	223.8	99.7
6	5.5	2.4	66	60.3	26.8	26	115.1	51.2	86	169.9	75.7	46	224.7	100.1
7	6.4	2.8	67	61.2	27.3	27	116.0	51.7	87	170.8	76.1	47	225.6	100.5
8	7.3	3.3	68	62.1	27.7	28	116.9	52.1	88	171.7	76.5	48	226.6	100.9
9	8.2	3.7	69	63.0	28.1	29	117.8	52.5	89	172.7	76.9	49	227.5	101.3
10	9.1	4.1	70	63.9	28.5	30	118.8	52.9	90	173.6	77.3	50	228.4	101.7
11	10.0	4.5	71	64.9	28.9	131	119.7	53.3	191	174.5	77.7	251	229.3	102.1
12	11.0	4.9	72	65.8	29.3	32	120.6	53.7	92	175.4	78.1	52	230.2	102.5
13	11.9	5.3	73	66.7	29.7	33	121.5	54.1	93	176.3	78.5	53	231.1	102.9
14	12.8	5.7	74	67.6	30.1	34	122.4	54.5	94	177.2	78.9	54	232.0	103.3
15	13.7	6.1	75	68.5	30.5	35	123.3	54.9	95	178.1	79.3	55	233.0	103.7
16	14.6	6.5	76	69.4	30.9	36	124.2	55.3	96	179.1	79.7	56	233.9	104.1
17	15.5	6.9	77	70.3	31.3	37	125.2	55.7	97	180.0	80.1	57	234.8	104.5
18	16.4	7.3	78	71.3	31.7	38	126.1	56.1	98	180.9	80.5	58	235.7	104.9
19	17.4	7.7	79	72.2	32.1	39	127.0	56.5	99	181.8	80.9	59	236.6	105.3
20	18.3	8.1	80	73.1	32.5	40	127.9	56.9	200	182.7	81.3	60	237.5	105.8
21	19.2	8.5	81	74.0	32.9	141	128.8	57.3	201	183.6	81.8	261	238.4	106.2
22	20.1	8.9	82	74.9	33.4	42	129.7	57.8	02	184.5	82.2	62	239.3	106.6
23	21.0	9.4	83	75.8	33.8	43	130.6	58.2	03	185.4	82.6	63	240.3	107.0
24	21.9	9.8	84	76.7	34.2	44	131.6	58.6	04	186.4	83.0	64	241.2	107.4
25	22.8	10.2	85	77.7	34.6	45	132.5	59.0	05	187.3	83.4	65	242.1	107.8
26	23.8	10.6	86	78.6	35.0	46	133.4	59.4	06	188.2	83.8	66	243.0	108.2
27	24.7	11.0	87	79.5	35.4	47	134.3	59.8	07	189.1	84.2	67	243.9	108.6
28	25.6	11.4	88	80.4	35.8	48	135.2	60.2	08	190.0	84.6	68	244.8	109.0
29	26.5	11.8	89	81.3	36.2	49	136.1	60.6	09	190.9	85.0	69	245.7	109.4
30	27.4	12.2	90	82.2	36.6	50	137.0	61.0	10	191.8	85.4	70	246.7	109.8
31	28.3	12.6	91	83.1	37.0	151	137.9	61.4	211	192.8	85.8	271	247.6	110.2
32	29.2	13.0	92	84.0	37.4	52	138.9	61.8	12	193.7	86.2	72	248.5	110.6
33	30.1	13.4	93	85.0	37.8	53	139.8	62.2	13	194.6	86.6	73	249.4	111.0
34	31.1	13.8	94	85.9	38.2	54	140.7	62.6	14	195.5	87.0	74	250.3	111.4
35	32.0	14.2	95	86.8	38.6	55	141.6	63.0	15	196.4	87.4	75	251.2	111.9
36	32.9	14.6	96	87.7	39.0	56	142.5	63.5	16	197.3	87.9	76	252.1	112.3
37	33.8	15.0	97	88.6	39.5	57	143.4	63.9	17	198.2	88.3	77	253.1	112.7
38	34.7	15.5	98	89.5	39.9	58	144.3	64.3	18	199.2	88.7	78	254.0	113.1
39	35.6	15.9	99	90.4	40.3	59	145.3	64.7	19	200.1	89.1	79	254.9	113.5
40	36.5	16.3	100	91.4	40.7	60	146.2	65.1	20	201.0	89.5	80	255.8	113.9
41	37.5	16.7	101	92.3	41.1	161	147.1	65.5	221	201.9	89.9	281	256.7	114.3
42	38.4	17.1	02	93.2	41.5	62	148.0	65.9	22	202.8	90.3	82	257.6	114.7
43	39.3	17.5	03	94.1	41.9	63	148.9	66.3	23	203.7	90.7	83	258.5	115.1
44	40.2	17.9	04	95.0	42.3	64	149.8	66.7	24	204.6	91.1	84	259.4	115.5
45	41.1	18.3	05	95.9	42.7	65	150.7	67.1	25	205.5	91.5	85	260.4	115.9
46	42.0	18.7	06	96.8	43.1	66	151.6	67.5	26	206.5	91.9	86	261.3	116.3
47	42.9	19.1	07	97.7	43.5	67	152.6	67.9	27	207.4	92.3	87	262.2	116.7
48	43.9	19.5	08	98.7	43.9	68	153.5	68.3	28	208.3	92.7	88	263.1	117.1
49	44.8	19.9	09	99.6	44.3	69	154.4	68.7	29	209.2	93.1	89	264.0	117.5
50	45.7	20.3	10	100.5	44.7	70	155.3	69.1	30	210.1	93.5	90	264.9	118.0
51	46.6	20.7	111	101.4	45.1	171	156.2	69.6	231	211.0	94.0	291	265.8	118.4
52	47.5	21.2	12	102.3	45.6	72	157.1	70.0	32	211.9	94.4	92	266.8	118.8
53	48.4	21.6	13	103.2	46.0	73	158.0	70.4	33	212.9	94.8	93	267.7	119.2
54	49.3	22.0	14	104.1	46.4	74	159.0	70.8	34	213.8	95.2	94	268.6	119.6
55	50.2	22.4	15	105.1	46.8	75	159.9	71.2	35	214.7	95.6	95	269.5	120.0
56	51.2	22.8	16	106.0	47.2	76	160.8	71.6	36	215.6	96.0	96	270.4	120.4
57	52.1	23.2	17	106.9	47.6	77	161.7	72.0	37	216.5	96.4	97	271.3	120.8
58	53.0	23.6	18	107.8	48.0	78	162.6	72.4	38	217.4	96.8	98	272.2	121.2
59	53.9	24.0	19	108.7	48.4	79	163.5	72.8	39	218.3	97.2	99	273.2	121.6
60	54.8	24.4	20	109.6	48.8	80	164.4	73.2	40	219.3	97.6	300	274.1	122.0
Dist.	Dep.	D. Lat.	Dist.	Dep.	D. Lat.	Dist.	Dep.	D. Lat.	Dist.	Dep.	D. Lat.	Dist.	Dep.	D. Lat.

294° | 066° / 246° | 114°

66°

Dist.	D. Lat.	Dep.
N.	N x Cos.	N x Sin.
Hypotenuse	Side Adj.	Side Opp.

TABLE 4

Traverse 24° Table

336° | 024° / 204° | 156°

Dist.	D. Lat.	Dep.	Dist.	D. Lat.	Dep.	Dist.	D. Lat.	Dep.	Dist.	D. Lat.	Dep.	Dist.	D. Lat.	Dep.
301	275.0	122.4	361	329.8	146.8	421	384.6	171.2	481	439.4	195.6	541	494.2	220.0
02	275.9	122.8	62	330.7	147.2	22	385.5	171.6	82	440.3	196.0	42	495.1	220.5
03	276.8	123.2	63	331.6	147.6	23	386.4	172.0	83	441.2	196.5	43	496.1	220.9
04	277.7	123.6	64	332.5	148.1	24	387.3	172.5	84	442.2	196.9	44	497.0	221.3
05	278.6	124.1	65	333.4	148.5	25	388.3	172.9	85	443.1	197.3	45	497.9	221.7
06	279.5	124.5	66	334.4	148.9	26	389.2	173.3	86	444.0	197.7	46	498.8	222.1
07	280.5	124.9	67	335.3	149.3	27	390.1	173.7	87	444.9	198.1	47	499.7	222.5
08	281.4	125.3	68	336.2	149.7	28	391.0	174.1	88	445.8	198.5	48	500.6	222.9
09	282.3	125.7	69	337.1	150.1	29	391.9	174.5	89	446.7	198.9	49	501.5	223.3
10	283.2	126.1	70	338.0	150.5	30	392.8	174.9	90	447.6	199.3	50	502.5	223.7
311	284.1	126.5	371	338.9	150.9	431	393.7	175.3	491	448.6	199.7	551	503.4	224.1
12	285.0	126.9	72	339.8	151.3	32	394.7	175.7	92	449.5	200.1	52	504.3	224.5
13	285.9	127.3	73	340.8	151.7	33	395.6	176.1	93	450.4	200.5	53	505.2	224.9
14	286.9	127.7	74	341.7	152.1	34	396.5	176.5	94	451.3	200.9	54	506.1	225.3
15	287.8	128.1	75	342.6	152.5	35	397.4	176.9	95	452.2	201.3	55	507.0	225.7
16	288.7	128.5	76	343.5	152.9	36	398.3	177.3	96	453.1	201.7	56	507.9	226.1
17	289.6	128.9	77	344.4	153.3	37	399.2	177.7	97	454.0	202.1	57	508.8	226.6
18	290.5	129.3	78	345.3	153.7	38	400.1	178.2	98	454.9	202.6	58	509.8	227.0
19	291.4	129.7	79	346.2	154.2	39	401.0	178.6	99	455.9	203.0	59	510.7	227.4
20	292.3	130.2	80	347.1	154.6	40	402.0	179.0	500	456.8	203.4	60	511.6	227.8
321	293.2	130.6	381	348.1	155.0	441	402.9	179.4	501	457.7	203.8	561	512.5	228.2
22	294.2	131.0	82	349.0	155.4	42	403.8	179.8	02	458.6	204.2	62	513.4	228.6
23	295.1	131.4	83	349.9	155.8	43	404.7	180.2	03	459.5	204.6	63	514.3	229.0
24	296.0	131.8	84	350.8	156.2	44	405.6	180.6	04	460.4	205.0	64	515.2	229.4
25	296.9	132.2	85	351.7	156.6	45	406.5	181.0	05	461.3	205.4	65	516.2	229.8
26	297.8	132.6	86	352.6	157.0	46	407.4	181.4	06	462.3	205.8	66	517.1	230.2
27	298.7	133.0	87	353.5	157.4	47	408.4	181.8	07	463.2	206.2	67	518.0	230.6
28	299.6	133.4	88	354.5	157.8	48	409.3	182.2	08	464.1	206.6	68	518.9	231.0
29	300.6	133.8	89	355.4	158.2	49	410.2	182.6	09	465.0	207.0	69	519.8	231.4
30	301.5	134.2	90	356.3	158.6	50	411.1	183.0	10	465.9	207.4	70	520.7	231.8
331	302.4	134.6	391	357.2	159.0	451	412.0	183.4	511	466.8	207.8	571	521.6	232.2
32	303.3	135.0	92	358.1	159.4	52	412.9	183.8	12	467.7	208.2	72	522.5	232.7
33	304.2	135.4	93	359.0	159.8	53	413.8	184.3	13	468.6	208.7	73	523.5	233.1
34	305.1	135.9	94	359.9	160.3	54	414.7	184.7	14	469.6	209.1	74	524.4	233.5
35	306.0	136.3	95	360.9	160.7	55	415.7	185.1	15	470.5	209.5	75	525.3	233.9
36	307.0	136.7	96	361.8	161.1	56	416.6	185.5	16	471.4	209.9	76	526.2	234.3
37	307.9	137.1	97	362.7	161.5	57	417.5	185.9	17	472.3	210.3	77	527.1	234.7
38	308.8	137.5	98	363.6	161.9	58	418.4	186.3	18	473.2	210.7	78	528.0	235.1
39	309.7	137.9	99	364.5	162.3	59	419.3	186.7	19	474.1	211.1	79	528.9	235.5
40	310.6	138.3	400	365.4	162.7	60	420.2	187.1	20	475.0	211.5	80	529.9	235.9
341	311.5	138.7	401	366.3	163.1	461	421.1	187.5	521	476.0	211.9	581	530.8	236.3
42	312.4	139.1	02	367.2	163.5	62	422.1	187.9	22	476.9	212.3	82	531.7	236.7
43	313.3	139.5	03	368.2	163.9	63	423.0	188.3	23	477.8	212.7	83	532.6	237.1
44	314.3	139.9	04	369.1	164.3	64	423.9	188.7	24	478.7	213.1	84	533.5	237.5
45	315.2	140.3	05	370.0	164.7	65	424.8	189.1	25	479.6	213.5	85	534.4	237.9
46	316.1	140.7	06	370.9	165.1	66	425.7	189.5	26	480.5	213.9	86	535.3	238.3
47	317.0	141.1	07	371.8	165.5	67	426.6	189.9	27	481.4	214.4	87	536.3	238.8
48	317.9	141.5	08	372.7	165.9	68	427.5	190.4	28	482.4	214.8	88	537.2	239.2
49	318.8	142.0	09	373.6	166.4	69	428.5	190.8	29	483.3	215.2	89	538.1	239.6
50	319.7	142.4	10	374.6	166.8	70	429.4	191.2	30	484.2	215.6	90	539.0	240.0
351	320.7	142.8	411	375.5	167.2	471	430.3	191.6	531	485.1	216.0	591	539.9	240.4
52	321.6	143.2	12	376.4	167.6	72	431.2	192.0	32	486.0	216.4	92	540.8	240.8
53	322.5	143.6	13	377.3	168.0	73	432.1	192.4	33	486.9	216.8	93	541.7	241.2
54	323.4	144.0	14	378.2	168.4	74	433.0	192.8	34	487.8	217.2	94	542.6	241.6
55	324.3	144.4	15	379.1	168.8	75	433.9	193.2	35	488.7	217.6	95	543.6	242.0
56	325.2	144.8	16	380.0	169.2	76	434.8	193.6	36	489.7	218.0	96	544.5	242.4
57	326.1	145.2	17	380.9	169.6	77	435.8	194.0	37	490.6	218.4	97	545.4	242.8
58	327.0	145.6	18	381.9	170.0	78	436.7	194.4	38	491.5	218.8	98	546.3	243.2
59	328.0	146.0	19	382.8	170.4	79	437.6	194.8	39	492.4	219.2	99	547.2	243.6
60	328.9	146.4	20	383.7	170.8	80	438.5	195.2	40	493.3	219.6	600	548.1	244.0
Dist.	Dep.	D. Lat.	Dist.	Dep.	D. Lat.	Dist.	Dep.	D. Lat.	Dist.	Dep.	D. Lat.	Dist.	Dep.	D. Lat.

Dist.	D. Lat.	Dep.
D Lo	Dep.	
	m	D Lo

66°

294° | 066° / 246° | 114°

TABLE 4

Traverse 25° Table

335° | 025°
205° | 155°

Dist.	D. Lat.	Dep.	Dist.	D. Lat.	Dep.	Dist.	D. Lat.	Dep.	Dist.	D. Lat.	Dep.	Dist.	D. Lat.	Dep.
1	0.9	0.4	61	55.3	25.8	121	109.7	51.1	181	164.0	76.5	241	218.4	101.9
2	1.8	0.8	62	56.2	26.2	22	110.6	51.6	82	164.9	76.9	42	219.3	102.3
3	2.7	1.3	63	57.1	26.6	23	111.5	52.0	83	165.9	77.3	43	220.2	102.7
4	3.6	1.7	64	58.0	27.0	24	112.4	52.4	84	166.8	77.8	44	221.1	103.1
5	4.5	2.1	65	58.9	27.5	25	113.3	52.8	85	167.7	78.2	45	222.0	103.5
6	5.4	2.5	66	59.8	27.9	26	114.2	53.2	86	168.6	78.6	46	223.0	104.0
7	6.3	3.0	67	60.7	28.3	27	115.1	53.7	87	169.5	79.0	47	223.9	104.4
8	7.3	3.4	68	61.6	28.7	28	116.0	54.1	88	170.4	79.5	48	224.8	104.8
9	8.2	3.8	69	62.5	29.2	29	116.9	54.5	89	171.3	79.9	49	225.7	105.2
10	9.1	4.2	70	63.4	29.6	30	117.8	54.9	90	172.2	80.3	50	226.6	105.7
11	10.0	4.6	71	64.3	30.0	131	118.7	55.4	191	173.1	80.7	251	227.5	106.1
12	10.9	5.1	72	65.3	30.4	32	119.6	55.8	92	174.0	81.1	52	228.4	106.5
13	11.8	5.5	73	66.2	30.9	33	120.5	56.2	93	174.9	81.6	53	229.3	106.9
14	12.7	5.9	74	67.1	31.3	34	121.4	56.6	94	175.8	82.0	54	230.2	107.3
15	13.6	6.3	75	68.0	31.7	35	122.4	57.1	95	176.7	82.4	55	231.1	107.8
16	14.5	6.8	76	68.9	32.1	36	123.3	57.5	96	177.6	82.8	56	232.0	108.2
17	15.4	7.2	77	69.8	32.5	37	124.2	57.9	97	178.5	83.3	57	232.9	108.6
18	16.3	7.6	78	70.7	33.0	38	125.1	58.3	98	179.4	83.7	58	233.8	109.0
19	17.2	8.0	79	71.6	33.4	39	126.0	58.7	99	180.4	84.1	59	234.7	109.5
20	18.1	8.5	80	72.5	33.8	40	126.9	59.2	200	181.3	84.5	60	235.6	109.9
21	19.0	8.9	81	73.4	34.2	141	127.8	59.6	201	182.2	84.9	261	236.5	110.3
22	19.9	9.3	82	74.3	34.7	42	128.7	60.0	02	183.1	85.4	62	237.5	110.7
23	20.8	9.7	83	75.2	35.1	43	129.6	60.4	03	184.0	85.8	63	238.4	111.1
24	21.8	10.1	84	76.1	35.5	44	130.5	60.9	04	184.9	86.2	64	239.3	111.6
25	22.7	10.6	85	77.0	35.9	45	131.4	61.3	05	185.8	86.6	65	240.2	112.0
26	23.6	11.0	86	77.9	36.3	46	132.3	61.7	06	186.7	87.1	66	241.1	112.4
27	24.5	11.4	87	78.8	36.8	47	133.2	62.1	07	187.6	87.5	67	242.0	112.8
28	25.4	11.8	88	79.8	37.2	48	134.1	62.5	08	188.5	87.9	68	242.9	113.3
29	26.3	12.3	89	80.7	37.6	49	135.0	63.0	09	189.4	88.3	69	243.8	113.7
30	27.2	12.7	90	81.6	38.0	50	135.9	63.4	10	190.3	88.7	70	244.7	114.1
31	28.1	13.1	91	82.5	38.5	151	136.9	63.8	211	191.2	89.2	271	245.6	114.5
32	29.0	13.5	92	83.4	38.9	52	137.8	64.2	12	192.1	89.6	72	246.5	115.0
33	29.9	13.9	93	84.3	39.3	53	138.7	64.7	13	193.0	90.0	73	247.4	115.4
34	30.8	14.4	94	85.2	39.7	54	139.6	65.1	14	193.9	90.4	74	248.3	115.8
35	31.7	14.8	95	86.1	40.1	55	140.5	65.5	15	194.9	90.9	75	249.2	116.2
36	32.6	15.2	96	87.0	40.6	56	141.4	65.9	16	195.8	91.3	76	250.1	116.6
37	33.5	15.6	97	87.9	41.0	57	142.3	66.4	17	196.7	91.7	77	251.0	117.1
38	34.4	16.1	98	88.8	41.4	58	143.2	66.8	18	197.6	92.1	78	252.0	117.5
39	35.3	16.5	99	89.7	41.8	59	144.1	67.2	19	198.5	92.6	79	252.9	117.9
40	36.3	16.9	100	90.6	42.3	60	145.0	67.6	20	199.4	93.0	80	253.8	118.3
41	37.2	17.3	101	91.5	42.7	161	145.9	68.0	221	200.3	93.4	281	254.7	118.8
42	38.1	17.7	02	92.4	43.1	62	146.8	68.5	22	201.2	93.8	82	255.6	119.2
43	39.0	18.2	03	93.3	43.5	63	147.7	68.9	23	202.1	94.2	83	256.5	119.6
44	39.9	18.6	04	94.3	44.0	64	148.6	69.3	24	203.0	94.7	84	257.4	120.0
45	40.8	19.0	05	95.2	44.4	65	149.5	69.7	25	203.9	95.1	85	258.3	120.4
46	41.7	19.4	06	96.1	44.8	66	150.4	70.2	26	204.8	95.5	86	259.2	120.9
47	42.6	19.9	07	97.0	45.2	67	151.4	70.6	27	205.7	95.9	87	260.1	121.3
48	43.5	20.3	08	97.9	45.6	68	152.3	71.0	28	206.6	96.4	88	261.0	121.7
49	44.4	20.7	09	98.8	46.1	69	153.2	71.4	29	207.5	96.8	89	261.9	122.1
50	45.3	21.1	10	99.7	46.5	70	154.1	71.8	30	208.5	97.2	90	262.8	122.6
51	46.2	21.6	111	100.6	46.9	171	155.0	72.3	231	209.4	97.6	291	263.7	123.0
52	47.1	22.0	12	101.5	47.3	72	155.9	72.7	32	210.3	98.0	92	264.6	123.4
53	48.0	22.4	13	102.4	47.8	73	156.8	73.1	33	211.2	98.5	93	265.5	123.8
54	48.9	22.8	14	103.3	48.2	74	157.7	73.5	34	212.1	98.9	94	266.5	124.2
55	49.8	23.2	15	104.2	48.6	75	158.6	74.0	35	213.0	99.3	95	267.4	124.7
56	50.8	23.7	16	105.1	49.0	76	159.5	74.4	36	213.9	99.7	96	268.3	125.1
57	51.7	24.1	17	106.0	49.4	77	160.4	74.8	37	214.8	100.2	97	269.2	125.5
58	52.6	24.5	18	106.9	49.9	78	161.3	75.2	38	215.7	100.6	98	270.1	125.9
59	53.5	24.9	19	107.9	50.3	79	162.2	75.6	39	216.6	101.0	99	271.0	126.4
60	54.4	25.4	20	108.8	50.7	80	163.1	76.1	40	217.5	101.4	300	271.9	126.8
Dist.	Dep.	D. Lat.	Dist.	Dep.	D. Lat.	Dist.	Dep.	D. Lat.	Dist.	Dep.	D. Lat.	Dist.	Dep.	D. Lat.

295° | 065°
245° | 115°

65°

Dist.	D. Lat.	Dep.
N.	N x Cos.	N x Sin.
Hypotenuse	Side Adj.	Side Opp.

TABLE 4

Traverse 25° Table

335° | 025°
205° | 155°

Dist.	D. Lat.	Dep.	Dist.	D. Lat.	Dep.	Dist.	D. Lat.	Dep.	Dist.	D. Lat.	Dep.	Dist.	D. Lat.	Dep.
301	272.8	127.2	361	327.2	152.6	421	381.6	177.9	481	435.9	203.3	541	490.3	228.6
02	273.7	127.6	62	328.1	153.0	22	382.5	178.3	82	436.8	203.7	42	491.2	229.1
03	274.6	128.1	63	329.0	153.4	23	383.4	178.8	83	437.7	204.1	43	492.1	229.5
04	275.5	128.5	64	329.9	153.8	24	384.3	179.2	84	438.7	204.5	44	493.0	229.9
05	276.4	128.9	65	330.8	154.3	25	385.2	179.6	85	439.6	205.0	45	493.9	230.3
06	277.3	129.3	66	331.7	154.7	26	386.1	180.0	86	440.5	205.4	46	494.8	230.7
07	278.2	129.7	67	332.6	155.1	27	387.0	180.5	87	441.4	205.8	47	495.8	231.2
08	279.1	130.2	68	333.5	155.5	28	387.9	180.9	88	442.3	206.2	48	496.7	231.6
09	280.0	130.6	69	334.4	155.9	29	388.8	181.3	89	443.2	206.7	49	497.6	232.0
10	281.0	131.0	70	335.3	156.4	30	389.7	181.7	90	444.1	207.1	50	498.5	232.4
311	281.9	131.4	371	336.2	156.8	431	390.6	182.1	491	445.0	207.5	551	499.4	232.9
12	282.8	131.9	72	337.1	157.2	32	391.5	182.6	92	445.9	207.9	52	500.3	233.3
13	283.7	132.3	73	338.1	157.6	33	392.4	183.0	93	446.8	208.4	53	501.2	233.7
14	284.6	132.7	74	339.0	158.1	34	393.3	183.4	94	447.7	208.8	54	502.1	234.1
15	285.5	133.1	75	339.9	158.5	35	394.2	183.8	95	448.6	209.2	55	503.0	234.6
16	286.4	133.5	76	340.8	158.9	36	395.2	184.3	96	449.5	209.6	56	503.9	235.0
17	287.3	134.0	77	341.7	159.3	37	396.1	184.7	97	450.4	210.0	57	504.8	235.4
18	288.2	134.4	78	342.6	159.7	38	397.0	185.1	98	451.3	210.5	58	505.7	235.8
19	289.1	134.8	79	343.5	160.2	39	397.9	185.5	99	452.2	210.9	59	506.6	236.2
20	290.0	135.2	80	344.4	160.6	40	398.8	186.0	500	453.2	211.3	60	507.5	236.7
321	290.9	135.7	381	345.3	161.0	441	399.7	186.4	501	454.1	211.7	561	508.4	237.1
22	291.8	136.1	82	346.2	161.4	42	400.6	186.8	02	455.0	212.2	62	509.3	237.5
23	292.7	136.5	83	347.1	161.9	43	401.5	187.2	03	455.9	212.6	63	510.3	237.9
24	293.6	136.9	84	348.0	162.3	44	402.4	187.6	04	456.8	213.0	64	511.2	238.4
25	294.6	137.4	85	348.9	162.7	45	403.3	188.1	05	457.7	213.4	65	512.1	238.8
26	295.5	137.8	86	349.8	163.1	46	404.2	188.5	06	458.6	213.8	66	513.0	239.2
27	296.4	138.2	87	350.7	163.6	47	405.1	188.9	07	459.5	214.3	67	513.9	239.6
28	297.3	138.6	88	351.6	164.0	48	406.0	189.3	08	460.4	214.7	68	514.8	240.0
29	298.2	139.0	89	352.6	164.4	49	406.9	189.8	09	461.3	215.1	69	515.7	240.5
30	299.1	139.5	90	353.5	164.8	50	407.8	190.2	10	462.2	215.5	70	516.6	240.9
331	300.0	139.9	391	354.4	165.2	451	408.7	190.6	511	463.1	216.0	571	517.5	241.3
32	300.9	140.3	92	355.3	165.7	52	409.7	191.0	12	464.0	216.4	72	518.4	241.7
33	301.8	140.7	93	356.2	166.1	53	410.6	191.4	13	464.9	216.8	73	519.3	242.2
34	302.7	141.2	94	357.1	166.5	54	411.5	191.9	14	465.8	217.2	74	520.2	242.6
35	303.6	141.6	95	358.0	166.9	55	412.4	192.3	15	466.7	217.6	75	521.1	243.0
36	304.5	142.0	96	358.9	167.4	56	413.3	192.7	16	467.7	218.1	76	522.0	243.4
37	305.4	142.4	97	359.8	167.8	57	414.2	193.1	17	468.6	218.5	77	522.9	243.9
38	306.3	142.8	98	360.7	168.2	58	415.1	193.6	18	469.5	218.9	78	523.8	244.3
39	307.2	143.3	99	361.6	168.6	59	416.0	194.0	19	470.4	219.3	79	524.8	244.7
40	308.1	143.7	400	362.5	169.0	60	416.9	194.4	20	471.3	219.8	80	525.7	245.1
341	309.1	144.1	401	363.4	169.5	461	417.8	194.8	521	472.2	220.2	581	526.6	245.5
42	310.0	144.5	02	364.3	169.9	62	418.7	195.2	22	473.1	220.6	82	527.5	246.0
43	310.9	145.0	03	365.2	170.3	63	419.6	195.7	23	474.0	221.0	83	528.4	246.4
44	311.8	145.4	04	366.1	170.7	64	420.5	196.1	24	474.9	221.5	84	529.3	246.8
45	312.7	145.8	05	367.1	171.2	65	421.4	196.5	25	475.8	221.9	85	530.2	247.2
46	313.6	146.2	06	368.0	171.6	66	422.3	196.9	26	476.7	222.3	86	531.1	247.7
47	314.5	146.6	07	368.9	172.0	67	423.2	197.4	27	477.6	222.7	87	532.0	248.1
48	315.4	147.1	08	369.8	172.4	68	424.2	197.8	28	478.5	223.1	88	532.9	248.5
49	316.3	147.5	09	370.7	172.9	69	425.1	198.2	29	479.4	223.6	89	533.8	248.9
50	317.2	147.9	10	371.6	173.3	70	426.0	198.6	30	480.3	224.0	90	534.7	249.3
351	318.1	148.3	411	372.5	173.7	471	426.9	199.1	531	481.2	224.4	591	535.6	249.8
52	319.0	148.8	12	373.4	174.1	72	427.8	199.5	32	482.2	224.8	92	536.5	250.2
53	319.9	149.2	13	374.3	174.5	73	428.7	199.9	33	483.1	225.3	93	537.4	250.6
54	320.8	149.6	14	375.2	175.0	74	429.6	200.3	34	484.0	225.7	94	538.3	251.0
55	321.7	150.0	15	376.1	175.4	75	430.5	200.7	35	484.9	226.1	95	539.3	251.5
56	322.6	150.5	16	377.0	175.8	76	431.4	201.2	36	485.8	226.5	96	540.2	251.9
57	323.6	150.9	17	377.9	176.2	77	432.3	201.6	37	486.7	226.9	97	541.1	252.3
58	324.5	151.3	18	378.8	176.7	78	433.2	202.0	38	487.6	227.4	98	542.0	252.7
59	325.4	151.7	19	379.7	177.1	79	434.1	202.4	39	488.5	227.8	99	542.9	253.1
60	326.3	152.1	20	380.6	177.5	80	435.0	202.9	40	489.4	228.2	600	543.8	253.6
Dist.	Dep.	D. Lat.	Dist.	Dep.	D. Lat.	Dist.	Dep.	D. Lat.	Dist.	Dep.	D. Lat.	Dist.	Dep.	D. Lat.

Dist.	D. Lat.	Dep.
D Lo	Dep.	
	m	D Lo

65°

295° | 065°
245° | 115°

TABLE 4

Traverse 26° Table

334° 206° | 026° 154°

Dist.	D. Lat.	Dep.	Dist.	D. Lat.	Dep.	Dist.	D. Lat.	Dep.	Dist.	D. Lat.	Dep.	Dist.	D. Lat.	Dep.
1	0.9	0.4	61	54.8	26.7	121	108.8	53.0	181	162.7	79.3	241	216.6	105.6
2	1.8	0.9	62	55.7	27.2	22	109.7	53.5	82	163.6	79.8	42	217.5	106.1
3	2.7	1.3	63	56.6	27.6	23	110.6	53.9	83	164.5	80.2	43	218.4	106.5
4	3.6	1.8	64	57.5	28.1	24	111.5	54.4	84	165.4	80.7	44	219.3	107.0
5	4.5	2.2	65	58.4	28.5	25	112.3	54.8	85	166.3	81.1	45	220.2	107.4
6	5.4	2.6	66	59.3	28.9	26	113.2	55.2	86	167.2	81.5	46	221.1	107.8
7	6.3	3.1	67	60.2	29.4	27	114.1	55.7	87	168.1	82.0	47	222.0	108.3
8	7.2	3.5	68	61.1	29.8	28	115.0	56.1	88	169.0	82.4	48	222.9	108.7
9	8.1	3.9	69	62.0	30.2	29	115.9	56.5	89	169.9	82.9	49	223.8	109.2
10	9.0	4.4	70	62.9	30.7	30	116.8	57.0	90	170.8	83.3	50	224.7	109.6
11	9.9	4.8	71	63.8	31.1	131	117.7	57.4	191	171.7	83.7	251	225.6	110.0
12	10.8	5.3	72	64.7	31.6	32	118.6	57.9	92	172.6	84.2	52	226.5	110.5
13	11.7	5.7	73	65.6	32.0	33	119.5	58.3	93	173.5	84.6	53	227.4	110.9
14	12.6	6.1	74	66.5	32.4	34	120.4	58.7	94	174.4	85.0	54	228.3	111.3
15	13.5	6.6	75	67.4	32.9	35	121.3	59.2	95	175.3	85.5	55	229.2	111.8
16	14.4	7.0	76	68.3	33.3	36	122.2	59.6	96	176.2	85.9	56	230.1	112.2
17	15.3	7.5	77	69.2	33.8	37	123.1	60.1	97	177.1	86.4	57	231.0	112.7
18	16.2	7.9	78	70.1	34.2	38	124.0	60.5	98	178.0	86.8	58	231.9	113.1
19	17.1	8.3	79	71.0	34.6	39	124.9	60.9	99	178.9	87.2	59	232.8	113.5
20	18.0	8.8	80	71.9	35.1	40	125.8	61.4	200	179.8	87.7	60	233.7	114.0
21	18.9	9.2	81	72.8	35.5	141	126.7	61.8	201	180.7	88.1	261	234.6	114.4
22	19.8	9.6	82	73.7	35.9	42	127.6	62.2	02	181.6	88.6	62	235.5	114.9
23	20.7	10.1	83	74.6	36.4	43	128.5	62.7	03	182.5	89.0	63	236.4	115.3
24	21.6	10.5	84	75.5	36.8	44	129.4	63.1	04	183.4	89.4	64	237.3	115.7
25	22.5	11.0	85	76.4	37.3	45	130.3	63.6	05	184.3	89.9	65	238.2	116.2
26	23.4	11.4	86	77.3	37.7	46	131.2	64.0	06	185.2	90.3	66	239.1	116.6
27	24.3	11.8	87	78.2	38.1	47	132.1	64.4	07	186.1	90.7	67	240.0	117.0
28	25.2	12.3	88	79.1	38.6	48	133.0	64.9	08	186.9	91.2	68	240.9	117.5
29	26.1	12.7	89	80.0	39.0	49	133.9	65.3	09	187.8	91.6	69	241.8	117.9
30	27.0	13.2	90	80.9	39.5	50	134.8	65.8	10	188.7	92.1	70	242.7	118.4
31	27.9	13.6	91	81.8	39.9	151	135.7	66.2	211	189.6	92.5	271	243.6	118.8
32	28.8	14.0	92	82.7	40.3	52	136.6	66.6	12	190.5	92.9	72	244.5	119.2
33	29.7	14.5	93	83.6	40.8	53	137.5	67.1	13	191.4	93.4	73	245.4	119.7
34	30.6	14.9	94	84.5	41.2	54	138.4	67.5	14	192.3	93.8	74	246.3	120.1
35	31.5	15.3	95	85.4	41.6	55	139.3	67.9	15	193.2	94.2	75	247.2	120.6
36	32.4	15.8	96	86.3	42.1	56	140.2	68.4	16	194.1	94.7	76	248.1	121.0
37	33.3	16.2	97	87.2	42.5	57	141.1	68.8	17	195.0	95.1	77	249.0	121.4
38	34.2	16.7	98	88.1	43.0	58	142.0	69.3	18	195.9	95.6	78	249.9	121.9
39	35.1	17.1	99	89.0	43.4	59	142.9	69.7	19	196.8	96.0	79	250.8	122.3
40	36.0	17.5	100	89.9	43.8	60	143.8	70.1	20	197.7	96.4	80	251.7	122.7
41	36.9	18.0	101	90.8	44.3	161	144.7	70.6	221	198.6	96.9	281	252.6	123.2
42	37.7	18.4	02	91.7	44.7	62	145.6	71.0	22	199.5	97.3	82	253.5	123.6
43	38.6	18.8	03	92.6	45.2	63	146.5	71.5	23	200.4	97.8	83	254.4	124.1
44	39.5	19.3	04	93.5	45.6	64	147.4	71.9	24	201.3	98.2	84	255.3	124.5
45	40.4	19.7	05	94.4	46.0	65	148.3	72.3	25	202.2	98.6	85	256.2	124.9
46	41.3	20.2	06	95.3	46.5	66	149.2	72.8	26	203.1	99.1	86	257.1	125.4
47	42.2	20.6	07	96.2	46.9	67	150.1	73.2	27	204.0	99.5	87	258.0	125.8
48	43.1	21.0	08	97.1	47.3	68	151.0	73.6	28	204.9	99.9	88	258.9	126.3
49	44.0	21.5	09	98.0	47.8	69	151.9	74.1	29	205.8	100.4	89	259.8	126.7
50	44.9	21.9	10	98.9	48.2	70	152.8	74.5	30	206.7	100.8	90	260.7	127.1
51	45.8	22.4	111	99.8	48.7	171	153.7	75.0	231	207.6	101.3	291	261.5	127.6
52	46.7	22.8	12	100.7	49.1	72	154.6	75.4	32	208.5	101.7	92	262.4	128.0
53	47.6	23.2	13	101.6	49.5	73	155.5	75.8	33	209.4	102.1	93	263.3	128.4
54	48.5	23.7	14	102.5	50.0	74	156.4	76.3	34	210.3	102.6	94	264.2	128.9
55	49.4	24.1	15	103.4	50.4	75	157.3	76.7	35	211.2	103.0	95	265.1	129.3
56	50.3	24.5	16	104.3	50.9	76	158.2	77.2	36	212.1	103.5	96	266.0	129.8
57	51.2	25.0	17	105.2	51.3	77	159.1	77.6	37	213.0	103.9	97	266.9	130.2
58	52.1	25.4	18	106.1	51.7	78	160.0	78.0	38	213.9	104.3	98	267.8	130.6
59	53.0	25.9	19	107.0	52.2	79	160.9	78.5	39	214.8	104.8	99	268.7	131.1
60	53.9	26.3	20	107.9	52.6	80	161.8	78.9	40	215.7	105.2	300	269.6	131.5
Dist.	Dep.	D. Lat.	Dist.	Dep.	D. Lat.	Dist.	Dep.	D. Lat.	Dist.	Dep.	D. Lat.	Dist.	Dep.	D. Lat.

296° 244° | 064° 116°

64°

Dist.	D. Lat.	Dep.
N.	N x Cos.	N x Sin.
Hypotenuse	Side Adj.	Side Opp.

TABLE 4

Traverse 26° Table

334° 206° | 026° 154°

Dist.	D. Lat.	Dep.	Dist.	D. Lat.	Dep.	Dist.	D. Lat.	Dep.	Dist.	D. Lat.	Dep.	Dist.	D. Lat.	Dep.
301	270.5	131.9	361	324.5	158.3	421	378.4	184.6	481	432.3	210.9	541	486.2	237.2
02	271.4	132.4	62	325.4	158.7	22	379.3	185.0	82	433.2	211.3	42	487.1	237.6
03	272.3	132.8	63	326.3	159.1	23	380.2	185.4	83	434.1	211.7	43	488.0	238.0
04	273.2	133.3	64	327.2	159.6	24	381.1	185.9	84	435.0	212.2	44	488.9	238.5
05	274.1	133.7	65	328.1	160.0	25	382.0	186.3	85	435.9	212.6	45	489.8	238.9
06	275.0	134.1	66	329.0	160.4	26	382.9	186.7	86	436.8	213.0	46	490.7	239.4
07	275.9	134.6	67	329.9	160.9	27	383.8	187.2	87	437.7	213.5	47	491.6	239.8
08	276.8	135.0	68	330.8	161.3	28	384.7	187.6	88	438.6	213.9	48	492.5	240.2
09	277.7	135.5	69	331.7	161.8	29	385.6	188.1	89	439.5	214.4	49	493.4	240.7
10	278.6	135.9	70	332.6	162.2	30	386.5	188.5	90	440.4	214.8	50	494.3	241.1
311	279.5	136.3	371	333.5	162.6	431	387.4	188.9	491	441.3	215.2	551	495.2	241.5
12	280.4	136.8	72	334.4	163.1	32	388.3	189.4	92	442.2	215.7	52	496.1	242.0
13	281.3	137.2	73	335.3	163.5	33	389.2	189.8	93	443.1	216.1	53	497.0	242.4
14	282.2	137.6	74	336.1	164.0	34	390.1	190.3	94	444.0	216.6	54	497.9	242.9
15	283.1	138.1	75	337.0	164.4	35	391.0	190.7	95	444.9	217.0	55	498.8	243.3
16	284.0	138.5	76	337.9	164.8	36	391.9	191.1	96	445.8	217.4	56	499.7	243.7
17	284.9	139.0	77	338.8	165.3	37	392.8	191.6	97	446.7	217.9	57	500.6	244.2
18	285.8	139.4	78	339.7	165.7	38	393.7	192.0	98	447.6	218.3	58	501.5	244.6
19	286.7	139.8	79	340.6	166.1	39	394.6	192.4	99	448.5	218.7	59	502.4	245.0
20	287.6	140.3	80	341.5	166.6	40	395.5	192.9	500	449.4	219.2	60	503.3	245.5
321	288.5	140.7	381	342.4	167.0	441	396.4	193.3	501	450.3	219.6	561	504.2	245.9
22	289.4	141.2	82	343.3	167.5	42	397.3	193.8	02	451.2	220.1	62	505.1	246.4
23	290.3	141.6	83	344.2	167.9	43	398.2	194.2	03	452.1	220.5	63	506.0	246.8
24	291.2	142.0	84	345.1	168.3	44	399.1	194.6	04	453.0	220.9	64	506.9	247.2
25	292.1	142.5	85	346.0	168.8	45	400.0	195.1	05	453.9	221.4	65	507.8	247.7
26	293.0	142.9	86	346.9	169.2	46	400.9	195.5	06	454.8	221.8	66	508.7	248.1
27	293.9	143.3	87	347.8	169.6	47	401.8	196.0	07	455.7	222.3	67	509.6	248.6
28	294.8	143.8	88	348.7	170.1	48	402.7	196.4	08	456.6	222.7	68	510.5	249.0
29	295.7	144.2	89	349.6	170.5	49	403.6	196.8	09	457.5	223.1	69	511.4	249.4
30	296.6	144.7	90	350.5	171.0	50	404.5	197.3	10	458.4	223.6	70	512.3	249.9
331	297.5	145.1	391	351.4	171.4	451	405.4	197.7	511	459.3	224.0	571	513.2	250.3
32	298.4	145.5	92	352.3	171.8	52	406.3	198.1	12	460.2	224.4	72	514.1	250.7
33	299.3	146.0	93	353.2	172.3	53	407.2	198.6	13	461.1	224.9	73	515.0	251.2
34	300.2	146.4	94	354.1	172.7	54	408.1	199.0	14	462.0	225.3	74	515.9	251.6
35	301.1	146.9	95	355.0	173.2	55	409.0	199.5	15	462.9	225.8	75	516.8	252.1
36	302.0	147.3	96	355.9	173.6	56	409.9	199.9	16	463.8	226.2	76	517.7	252.5
37	302.9	147.7	97	356.8	174.0	57	410.7	200.3	17	464.7	226.6	77	518.6	252.9
38	303.8	148.2	98	357.7	174.5	58	411.6	200.8	18	465.6	227.1	78	519.5	253.4
39	304.7	148.6	99	358.6	174.9	59	412.5	201.2	19	466.5	227.5	79	520.4	253.8
40	305.6	149.0	400	359.5	175.3	60	413.4	201.7	20	467.4	228.0	80	521.3	254.3
341	306.5	149.5	401	360.4	175.8	461	414.3	202.1	521	468.3	228.4	581	522.2	254.7
42	307.4	149.9	02	361.3	176.2	62	415.2	202.5	22	469.2	228.8	82	523.1	255.1
43	308.3	150.4	03	362.2	176.7	63	416.1	203.0	23	470.1	229.3	83	524.0	255.6
44	309.2	150.8	04	363.1	177.1	64	417.0	203.4	24	471.0	229.7	84	524.9	256.0
45	310.1	151.2	05	364.0	177.5	65	417.9	203.8	25	471.9	230.1	85	525.8	256.4
46	311.0	151.7	06	364.9	178.0	66	418.8	204.3	26	472.8	230.6	86	526.7	256.9
47	311.9	152.1	07	365.8	178.4	67	419.7	204.7	27	473.7	231.0	87	527.6	257.3
48	312.8	152.6	08	366.7	178.9	68	420.6	205.2	28	474.6	231.5	88	528.5	257.8
49	313.7	153.0	09	367.6	179.3	69	421.5	205.6	29	475.5	231.9	89	529.4	258.2
50	314.6	153.4	10	368.5	179.7	70	422.4	206.0	30	476.4	232.3	90	530.3	258.6
351	315.5	153.9	411	369.4	180.2	471	423.3	206.5	531	477.3	232.8	591	531.2	259.1
52	316.4	154.3	12	370.3	180.6	72	424.2	206.9	32	478.2	233.2	92	532.1	259.5
53	317.3	154.7	13	371.2	181.0	73	425.1	207.3	33	479.1	233.7	93	533.0	260.0
54	318.2	155.2	14	372.1	181.5	74	426.0	207.8	34	480.0	234.1	94	533.9	260.4
55	319.1	155.6	15	373.0	181.9	75	426.9	208.2	35	480.9	234.5	95	534.8	260.8
56	320.0	156.1	16	373.9	182.4	76	427.8	208.7	36	481.8	235.0	96	535.7	261.3
57	320.9	156.5	17	374.8	182.8	77	428.7	209.1	37	482.7	235.4	97	536.6	261.7
58	321.8	156.9	18	375.7	183.2	78	429.6	209.5	38	483.6	235.8	98	537.5	262.1
59	322.7	157.4	19	376.6	183.7	79	430.5	210.0	39	484.4	236.3	99	538.4	262.6
60	323.6	157.8	20	377.5	184.1	80	431.4	210.4	40	485.3	236.7	600	539.3	263.0
Dist.	Dep.	D. Lat.	Dist.	Dep.	D. Lat.	Dist.	Dep.	D. Lat.	Dist.	Dep.	D. Lat.	Dist.	Dep.	D. Lat.

Dist.	D. Lat.	Dep.
D Lo	Dep.	
	m	D Lo

64°

296° 244° | 064° 116°

TABLE 4

Traverse 27° Table

333° 207°	027° 153°													
Dist.	D. Lat.	Dep.	Dist.	D. Lat.	Dep.	Dist.	D. Lat.	Dep.	Dist.	D. Lat.	Dep.	Dist.	D. Lat.	Dep.
1	0.9	0.5	61	54.4	27.7	121	107.8	54.9	181	161.3	82.2	241	214.7	109.4
2	1.8	0.9	62	55.2	28.1	22	108.7	55.4	82	162.2	82.6	42	215.6	109.9
3	2.7	1.4	63	56.1	28.6	23	109.6	55.8	83	163.1	83.1	43	216.5	110.3
4	3.6	1.8	64	57.0	29.1	24	110.5	56.3	84	163.9	83.5	44	217.4	110.8
5	4.5	2.3	65	57.9	29.5	25	111.4	56.7	85	164.8	84.0	45	218.3	111.2
6	5.3	2.7	66	58.8	30.0	26	112.3	57.2	86	165.7	84.4	46	219.2	111.7
7	6.2	3.2	67	59.7	30.4	27	113.2	57.7	87	166.6	84.9	47	220.1	112.1
8	7.1	3.6	68	60.6	30.9	28	114.0	58.1	88	167.5	85.4	48	221.0	112.6
9	8.0	4.1	69	61.5	31.3	29	114.9	58.6	89	168.4	85.8	49	221.9	113.0
10	8.9	4.5	70	62.4	31.8	30	115.8	59.0	90	169.3	86.3	50	222.8	113.5
11	9.8	5.0	71	63.3	32.2	131	116.7	59.5	191	170.2	86.7	251	223.6	114.0
12	10.7	5.4	72	64.2	32.7	32	117.6	59.9	92	171.1	87.2	52	224.5	114.4
13	11.6	5.9	73	65.0	33.1	33	118.5	60.4	93	172.0	87.6	53	225.4	114.9
14	12.5	6.4	74	65.9	33.6	34	119.4	60.8	94	172.9	88.1	54	226.3	115.3
15	13.4	6.8	75	66.8	34.0	35	120.3	61.3	95	173.7	88.5	55	227.2	115.8
16	14.3	7.3	76	67.7	34.5	36	121.2	61.7	96	174.6	89.0	56	228.1	116.2
17	15.1	7.7	77	68.6	35.0	37	122.1	62.2	97	175.5	89.4	57	229.0	116.7
18	16.0	8.2	78	69.5	35.4	38	123.0	62.7	98	176.4	89.9	58	229.9	117.1
19	16.9	8.6	79	70.4	35.9	39	123.8	63.1	99	177.3	90.3	59	230.8	117.6
20	17.8	9.1	80	71.3	36.3	40	124.7	63.6	200	178.2	90.8	60	231.7	118.0
21	18.7	9.5	81	72.2	36.8	141	125.6	64.0	201	179.1	91.3	261	232.6	118.5
22	19.6	10.0	82	73.1	37.2	42	126.5	64.5	02	180.0	91.7	62	233.4	118.9
23	20.5	10.4	83	74.0	37.7	43	127.4	64.9	03	180.9	92.2	63	234.3	119.4
24	21.4	10.9	84	74.8	38.1	44	128.3	65.4	04	181.8	92.6	64	235.2	119.9
25	22.3	11.3	85	75.7	38.6	45	129.2	65.8	05	182.7	93.1	65	236.1	120.3
26	23.2	11.8	86	76.6	39.0	46	130.1	66.3	06	183.5	93.5	66	237.0	120.8
27	24.1	12.3	87	77.5	39.5	47	131.0	66.7	07	184.4	94.0	67	237.9	121.2
28	24.9	12.7	88	78.4	40.0	48	131.9	67.2	08	185.3	94.4	68	238.8	121.7
29	25.8	13.2	89	79.3	40.4	49	132.8	67.6	09	186.2	94.9	69	239.7	122.1
30	26.7	13.6	90	80.2	40.9	50	133.7	68.1	10	187.1	95.3	70	240.6	122.6
31	27.6	14.1	91	81.1	41.3	151	134.5	68.6	211	188.0	95.8	271	241.5	123.0
32	28.5	14.5	92	82.0	41.8	52	135.4	69.0	12	188.9	96.2	72	242.4	123.5
33	29.4	15.0	93	82.9	42.2	53	136.3	69.5	13	189.8	96.7	73	243.2	123.9
34	30.3	15.4	94	83.8	42.7	54	137.2	69.9	14	190.7	97.2	74	244.1	124.4
35	31.2	15.9	95	84.6	43.1	55	138.1	70.4	15	191.6	97.6	75	245.0	124.8
36	32.1	16.3	96	85.5	43.6	56	139.0	70.8	16	192.5	98.1	76	245.9	125.3
37	33.0	16.8	97	86.4	44.0	57	139.9	71.3	17	193.3	98.5	77	246.8	125.8
38	33.9	17.3	98	87.3	44.5	58	140.8	71.7	18	194.2	99.0	78	247.7	126.2
39	34.7	17.7	99	88.2	44.9	59	141.7	72.2	19	195.1	99.4	79	248.6	126.7
40	35.6	18.2	100	89.1	45.4	60	142.6	72.6	20	196.0	99.9	80	249.5	127.1
41	36.5	18.6	101	90.0	45.9	161	143.5	73.1	221	196.9	100.3	281	250.4	127.6
42	37.4	19.1	02	90.9	46.3	62	144.3	73.5	22	197.8	100.8	82	251.3	128.0
43	38.3	19.5	03	91.8	46.8	63	145.2	74.0	23	198.7	101.2	83	252.2	128.5
44	39.2	20.0	04	92.7	47.2	64	146.1	74.5	24	199.6	101.7	84	253.0	128.9
45	40.1	20.4	05	93.6	47.7	65	147.0	74.9	25	200.5	102.1	85	253.9	129.4
46	41.0	20.9	06	94.4	48.1	66	147.9	75.4	26	201.4	102.6	86	254.8	129.8
47	41.9	21.3	07	95.3	48.6	67	148.8	75.8	27	202.3	103.1	87	255.7	130.3
48	42.8	21.8	08	96.2	49.0	68	149.7	76.3	28	203.1	103.5	88	256.6	130.7
49	43.7	22.2	09	97.1	49.5	69	150.6	76.7	29	204.0	104.0	89	257.5	131.2
50	44.6	22.7	10	98.0	49.9	70	151.5	77.2	30	204.9	104.4	90	258.4	131.7
51	45.4	23.2	111	98.9	50.4	171	152.4	77.6	231	205.8	104.9	291	259.3	132.1
52	46.3	23.6	12	99.8	50.8	72	153.3	78.1	32	206.7	105.3	92	260.2	132.6
53	47.2	24.1	13	100.7	51.3	73	154.1	78.5	33	207.6	105.8	93	261.1	133.0
54	48.1	24.5	14	101.6	51.8	74	155.0	79.0	34	208.5	106.2	94	262.0	133.5
55	49.0	25.0	15	102.5	52.2	75	155.9	79.4	35	209.4	106.7	95	262.8	133.9
56	49.9	25.4	16	103.4	52.7	76	156.8	79.9	36	210.3	107.1	96	263.7	134.4
57	50.8	25.9	17	104.2	53.1	77	157.7	80.4	37	211.2	107.6	97	264.6	134.8
58	51.7	26.3	18	105.1	53.6	78	158.6	80.8	38	212.1	108.0	98	265.5	135.3
59	52.6	26.8	19	106.0	54.0	79	159.5	81.3	39	213.0	108.5	99	266.4	135.7
60	53.5	27.2	20	106.9	54.5	80	160.4	81.7	40	213.8	109.0	300	267.3	136.2
Dist.	Dep.	D. Lat.	Dist.	Dep.	D. Lat.	Dist.	Dep.	D. Lat.	Dist.	Dep.	D. Lat.	Dist.	Dep.	D. Lat.
297° 243°	063° 117°													

63°

Dist.	D. Lat.	Dep.
N.	N x Cos.	N x Sin.
Hypotenuse	Side Adj.	Side Opp.

TABLE 4

Traverse 27° Table

333° 207°	027° 153°													
Dist.	D. Lat.	Dep.	Dist.	D. Lat.	Dep.	Dist.	D. Lat.	Dep.	Dist.	D. Lat.	Dep.	Dist.	D. Lat.	Dep.
301	268.2	136.7	361	321.7	163.9	421	375.1	191.1	481	428.6	218.4	541	482.0	245.6
02	269.1	137.1	62	322.5	164.3	22	376.0	191.6	82	429.5	218.8	42	482.9	246.1
03	270.0	137.6	63	323.4	164.8	23	376.9	192.0	83	430.4	219.3	43	483.8	246.5
04	270.9	138.0	64	324.3	165.3	24	377.8	192.5	84	431.2	219.7	44	484.7	247.0
05	271.8	138.5	65	325.2	165.7	25	378.7	192.9	85	432.1	220.2	45	485.6	247.4
06	272.6	138.9	66	326.1	166.2	26	379.6	193.4	86	433.0	220.6	46	486.5	247.9
07	273.5	139.4	67	327.0	166.6	27	380.5	193.9	87	433.9	221.1	47	487.4	248.3
08	274.4	139.8	68	327.9	167.1	28	381.4	194.3	88	434.8	221.5	48	488.3	248.8
09	275.3	140.3	69	328.8	167.5	29	382.2	194.8	89	435.7	222.0	49	489.2	249.2
10	276.2	140.7	70	329.7	168.0	30	383.1	195.2	90	436.6	222.5	50	490.1	249.7
311	277.1	141.2	371	330.6	168.4	431	384.0	195.7	491	437.5	222.9	551	490.9	250.1
12	278.0	141.6	72	331.5	168.9	32	384.9	196.1	92	438.4	223.4	52	491.8	250.6
13	278.9	142.1	73	332.3	169.3	33	385.8	196.6	93	439.3	223.8	53	492.7	251.1
14	279.8	142.6	74	333.2	169.8	34	386.7	197.0	94	440.2	224.3	54	493.6	251.5
15	280.7	143.0	75	334.1	170.2	35	387.6	197.5	95	441.0	224.7	55	494.5	252.0
16	281.6	143.5	76	335.0	170.7	36	388.5	197.9	96	441.9	225.2	56	495.4	252.4
17	282.4	143.9	77	335.9	171.2	37	389.4	198.4	97	442.8	225.6	57	496.3	252.9
18	283.3	144.4	78	336.8	171.6	38	390.3	198.8	98	443.7	226.1	58	497.2	253.3
19	284.2	144.8	79	337.7	172.1	39	391.2	199.3	99	444.6	226.5	59	498.1	253.8
20	285.1	145.3	80	338.6	172.5	40	392.0	199.8	500	445.5	227.0	60	499.0	254.2
321	286.0	145.7	381	339.5	173.0	441	392.9	200.2	501	446.4	227.4	561	499.9	254.7
22	286.9	146.2	82	340.4	173.4	42	393.8	200.7	02	447.3	227.9	62	500.7	255.1
23	287.8	146.6	83	341.3	173.9	43	394.7	201.1	03	448.2	228.4	63	501.6	255.6
24	288.7	147.1	84	342.1	174.3	44	395.6	201.6	04	449.1	228.8	64	502.5	256.1
25	289.6	147.5	85	343.0	174.8	45	396.5	202.0	05	450.0	229.3	65	503.4	256.5
26	290.5	148.0	86	343.9	175.2	46	397.4	202.5	06	450.8	229.7	66	504.3	257.0
27	291.4	148.5	87	344.8	175.7	47	398.3	202.9	07	451.7	230.2	67	505.2	257.4
28	292.3	148.9	88	345.7	176.1	48	399.2	203.4	08	452.6	230.6	68	506.1	257.9
29	293.1	149.4	89	346.6	176.6	49	400.1	203.8	09	453.5	231.1	69	507.0	258.3
30	294.0	149.8	90	347.5	177.1	50	401.0	204.3	10	454.4	231.5	70	507.9	258.8
331	294.9	150.3	391	348.4	177.5	451	401.8	204.7	511	455.3	232.0	571	508.8	259.2
32	295.8	150.7	92	349.3	178.0	52	402.7	205.2	12	456.2	232.4	72	509.7	259.7
33	296.7	151.2	93	350.2	178.4	53	403.6	205.7	13	457.1	232.9	73	510.5	260.1
34	297.6	151.6	94	351.1	178.9	54	404.5	206.1	14	458.0	233.4	74	511.4	260.6
35	298.5	152.1	95	351.9	179.3	55	405.4	206.6	15	458.9	233.8	75	512.3	261.0
36	299.4	152.5	96	352.8	179.8	56	406.3	207.0	16	459.8	234.3	76	513.2	261.5
37	300.3	153.0	97	353.7	180.2	57	407.2	207.5	17	460.7	234.7	77	514.1	262.0
38	301.2	153.4	98	354.6	180.7	58	408.1	207.9	18	461.5	235.2	78	515.0	262.4
39	302.1	153.9	99	355.5	181.1	59	409.0	208.4	19	462.4	235.6	79	515.9	262.9
40	302.9	154.4	400	356.4	181.6	60	409.9	208.8	20	463.3	236.1	80	516.8	263.3
341	303.8	154.8	401	357.3	182.1	461	410.8	209.3	521	464.2	236.5	581	517.7	263.8
42	304.7	155.3	02	358.2	182.5	62	411.6	209.7	22	465.1	237.0	82	518.6	264.2
43	305.6	155.7	03	359.1	183.0	63	412.5	210.2	23	466.0	237.4	83	519.5	264.7
44	306.5	156.2	04	360.0	183.4	64	413.4	210.7	24	466.9	237.9	84	520.3	265.1
45	307.4	156.6	05	360.9	183.9	65	414.3	211.1	25	467.8	238.3	85	521.2	265.6
46	308.3	157.1	06	361.7	184.3	66	415.2	211.6	26	468.7	238.8	86	522.1	266.0
47	309.2	157.5	07	362.6	184.8	67	416.1	212.0	27	469.6	239.3	87	523.0	266.5
48	310.1	158.0	08	363.5	185.2	68	417.0	212.5	28	470.5	239.7	88	523.9	266.9
49	311.0	158.4	09	364.4	185.7	69	417.9	212.9	29	471.3	240.2	89	524.8	267.4
50	311.9	158.9	10	365.3	186.1	70	418.8	213.4	30	472.2	240.6	90	525.7	267.9
351	312.7	159.4	411	366.2	186.6	471	419.7	213.8	531	473.1	241.1	591	526.6	268.3
52	313.6	159.8	12	367.1	187.0	72	420.6	214.3	32	474.0	241.5	92	527.5	268.8
53	314.5	160.3	13	368.0	187.5	73	421.4	214.7	33	474.9	242.0	93	528.4	269.2
54	315.4	160.7	14	368.9	188.0	74	422.3	215.2	34	475.8	242.4	94	529.3	269.7
55	316.3	161.2	15	369.8	188.4	75	423.2	215.6	35	476.7	242.9	95	530.1	270.1
56	317.2	161.6	16	370.7	188.9	76	424.1	216.1	36	477.6	243.3	96	531.0	270.6
57	318.1	162.1	17	371.5	189.3	77	425.0	216.6	37	478.5	243.8	97	531.9	271.0
58	319.0	162.5	18	372.4	189.8	78	425.9	217.0	38	479.4	244.2	98	532.8	271.5
59	319.9	163.0	19	373.3	190.2	79	426.8	217.5	39	480.3	244.7	99	533.7	271.9
60	320.8	163.4	20	374.2	190.7	80	427.7	217.9	40	481.1	245.2	600	534.6	272.4
Dist.	Dep.	D. Lat.	Dist.	Dep.	D. Lat.	Dist.	Dep.	D. Lat.	Dist.	Dep.	D. Lat.	Dist.	Dep.	D. Lat.
												297° 243°	063° 117°	

Dist.	D. Lat.	Dep.
D Lo	Dep.	
	m	D Lo

63°

TABLE 4
Traverse 28° Table

332° | 028° / 208° | 152°

Dist.	D. Lat.	Dep.	Dist.	D. Lat.	Dep.	Dist.	D. Lat.	Dep.	Dist.	D. Lat.	Dep.	Dist.	D. Lat.	Dep.
1	0.9	0.5	61	53.9	28.6	121	106.8	56.8	181	159.8	85.0	241	212.8	113.1
2	1.8	0.9	62	54.7	29.1	22	107.7	57.3	82	160.7	85.4	42	213.7	113.6
3	2.6	1.4	63	55.6	29.6	23	108.6	57.7	83	161.6	85.9	43	214.6	114.1
4	3.5	1.9	64	56.5	30.0	24	109.5	58.2	84	162.5	86.4	44	215.4	114.6
5	4.4	2.3	65	57.4	30.5	25	110.4	58.7	85	163.3	86.9	45	216.3	115.0
6	5.3	2.8	66	58.3	31.0	26	111.3	59.2	86	164.2	87.3	46	217.2	115.5
7	6.2	3.3	67	59.2	31.5	27	112.1	59.6	87	165.1	87.8	47	218.1	116.0
8	7.1	3.8	68	60.0	31.9	28	113.0	60.1	88	166.0	88.3	48	219.0	116.4
9	7.9	4.2	69	60.9	32.4	29	113.9	60.6	89	166.9	88.7	49	219.9	116.9
10	8.8	4.7	70	61.8	32.9	30	114.8	61.0	90	167.8	89.2	50	220.7	117.4
11	9.7	5.2	71	62.7	33.3	131	115.7	61.5	191	168.6	89.7	251	221.6	117.8
12	10.6	5.6	72	63.6	33.8	32	116.5	62.0	92	169.5	90.1	52	222.5	118.3
13	11.5	6.1	73	64.5	34.3	33	117.4	62.4	93	170.4	90.6	53	223.4	118.8
14	12.4	6.6	74	65.3	34.7	34	118.3	62.9	94	171.3	91.1	54	224.3	119.2
15	13.2	7.0	75	66.2	35.2	35	119.2	63.4	95	172.2	91.5	55	225.2	119.7
16	14.1	7.5	76	67.1	35.7	36	120.1	63.8	96	173.1	92.0	56	226.0	120.2
17	15.0	8.0	77	68.0	36.1	37	121.0	64.3	97	173.9	92.5	57	226.9	120.7
18	15.9	8.5	78	68.9	36.6	38	121.8	64.8	98	174.8	93.0	58	227.8	121.1
19	16.8	8.9	79	69.8	37.1	39	122.7	65.3	99	175.7	93.4	59	228.7	121.6
20	17.7	9.4	80	70.6	37.6	40	123.6	65.7	200	176.6	93.9	60	229.6	122.1
21	18.5	9.9	81	71.5	38.0	141	124.5	66.2	201	177.5	94.4	261	230.4	122.5
22	19.4	10.3	82	72.4	38.5	42	125.4	66.7	02	178.4	94.8	62	231.3	123.0
23	20.3	10.8	83	73.3	39.0	43	126.3	67.1	03	179.2	95.3	63	232.2	123.5
24	21.2	11.3	84	74.2	39.4	44	127.1	67.6	04	180.1	95.8	64	233.1	123.9
25	22.1	11.7	85	75.1	39.9	45	128.0	68.1	05	181.0	96.2	65	234.0	124.4
26	23.0	12.2	86	75.9	40.4	46	128.9	68.5	06	181.9	96.7	66	234.9	124.9
27	23.8	12.7	87	76.8	40.8	47	129.8	69.0	07	182.8	97.2	67	235.7	125.3
28	24.7	13.1	88	77.7	41.3	48	130.7	69.5	08	183.7	97.7	68	236.6	125.8
29	25.6	13.6	89	78.6	41.8	49	131.6	70.0	09	184.5	98.1	69	237.5	126.3
30	26.5	14.1	90	79.5	42.3	50	132.4	70.4	10	185.4	98.6	70	238.4	126.8
31	27.4	14.6	91	80.3	42.7	151	133.3	70.9	211	186.3	99.1	271	239.3	127.2
32	28.3	15.0	92	81.2	43.2	52	134.2	71.4	12	187.2	99.5	72	240.2	127.7
33	29.1	15.5	93	82.1	43.7	53	135.1	71.8	13	188.1	100.0	73	241.0	128.2
34	30.0	16.0	94	83.0	44.1	54	136.0	72.3	14	189.0	100.5	74	241.9	128.6
35	30.9	16.4	95	83.9	44.6	55	136.9	72.8	15	189.8	100.9	75	242.8	129.1
36	31.8	16.9	96	84.8	45.1	56	137.7	73.2	16	190.7	101.4	76	243.7	129.6
37	32.7	17.4	97	85.6	45.5	57	138.6	73.7	17	191.6	101.9	77	244.6	130.0
38	33.6	17.8	98	86.5	46.0	58	139.5	74.2	18	192.5	102.3	78	245.5	130.5
39	34.4	18.3	99	87.4	46.5	59	140.4	74.6	19	193.4	102.8	79	246.3	131.0
40	35.3	18.8	100	88.3	46.9	60	141.3	75.1	20	194.2	103.3	80	247.2	131.5
41	36.2	19.2	101	89.2	47.4	161	142.2	75.6	221	195.1	103.8	281	248.1	131.9
42	37.1	19.7	02	90.1	47.9	62	143.0	76.1	22	196.0	104.2	82	249.0	132.4
43	38.0	20.2	03	90.9	48.4	63	143.9	76.5	23	196.9	104.7	83	249.9	132.9
44	38.8	20.7	04	91.8	48.8	64	144.8	77.0	24	197.8	105.2	84	250.8	133.3
45	39.7	21.1	05	92.7	49.3	65	145.7	77.5	25	198.7	105.6	85	251.6	133.8
46	40.6	21.6	06	93.6	49.8	66	146.6	77.9	26	199.5	106.1	86	252.5	134.3
47	41.5	22.1	07	94.5	50.2	67	147.5	78.4	27	200.4	106.6	87	253.4	134.7
48	42.4	22.5	08	95.4	50.7	68	148.3	78.9	28	201.3	107.0	88	254.3	135.2
49	43.3	23.0	09	96.2	51.2	69	149.2	79.3	29	202.2	107.5	89	255.2	135.7
50	44.1	23.5	10	97.1	51.6	70	150.1	79.8	30	203.1	108.0	90	256.1	136.1
51	45.0	23.9	111	98.0	52.1	171	151.0	80.3	231	204.0	108.4	291	256.9	136.6
52	45.9	24.4	12	98.9	52.6	72	151.9	80.7	32	204.8	108.9	92	257.8	137.1
53	46.8	24.9	13	99.8	53.1	73	152.7	81.2	33	205.7	109.4	93	258.7	137.6
54	47.7	25.4	14	100.7	53.5	74	153.6	81.7	34	206.6	109.9	94	259.6	138.0
55	48.6	25.8	15	101.5	54.0	75	154.5	82.2	35	207.5	110.3	95	260.5	138.5
56	49.4	26.3	16	102.4	54.5	76	155.4	82.6	36	208.4	110.8	96	261.4	139.0
57	50.3	26.8	17	103.3	54.9	77	156.3	83.1	37	209.3	111.3	97	262.2	139.4
58	51.2	27.2	18	104.2	55.4	78	157.2	83.6	38	210.1	111.7	98	263.1	139.9
59	52.1	27.7	19	105.1	55.9	79	158.0	84.0	39	211.0	112.2	99	264.0	140.4
60	53.0	28.2	20	106.0	56.3	80	158.9	84.5	40	211.9	112.7	300	264.9	140.8
Dist.	Dep.	D. Lat.	Dist.	Dep.	D. Lat.	Dist.	Dep.	D. Lat.	Dist.	Dep.	D. Lat.	Dist.	Dep.	D. Lat.

298° | 062° / 242° | 118°

62°

Dist.	D. Lat.	Dep.
N.	N x Cos.	N x Sin.
Hypotenuse	Side Adj.	Side Opp.

TABLE 4
Traverse 28° Table

332° | 028° / 208° | 152°

Dist.	D. Lat.	Dep.	Dist.	D. Lat.	Dep.	Dist.	D. Lat.	Dep.	Dist.	D. Lat.	Dep.	Dist.	D. Lat.	Dep.
301	265.8	141.3	361	318.7	169.5	421	371.7	197.6	481	424.7	225.8	541	477.7	254.0
02	266.7	141.8	62	319.6	169.9	22	372.6	198.1	82	425.6	226.3	42	478.6	254.5
03	267.5	142.2	63	320.5	170.4	23	373.5	198.6	83	426.5	226.8	43	479.4	254.9
04	268.4	142.7	64	321.4	170.9	24	374.4	199.1	84	427.3	227.2	44	480.3	255.4
05	269.3	143.2	65	322.3	171.4	25	375.3	199.5	85	428.2	227.7	45	481.2	255.9
06	270.2	143.7	66	323.2	171.8	26	376.1	200.0	86	429.1	228.2	46	482.1	256.3
07	271.1	144.1	67	324.0	172.3	27	377.0	200.5	87	430.0	228.6	47	483.0	256.8
08	271.9	144.6	68	324.9	172.8	28	377.9	200.9	88	430.9	229.1	48	483.9	257.3
09	272.8	145.1	69	325.8	173.2	29	378.8	201.4	89	431.8	229.6	49	484.7	257.7
10	273.7	145.5	70	326.7	173.7	30	379.7	201.9	90	432.6	230.0	50	485.6	258.2
311	274.6	146.0	371	327.6	174.2	431	380.6	202.3	491	433.5	230.5	551	486.5	258.7
12	275.5	146.5	72	328.5	174.6	32	381.4	202.8	92	434.4	231.0	52	487.4	259.1
13	276.4	146.9	73	329.3	175.1	33	382.3	203.3	93	435.3	231.4	53	488.3	259.6
14	277.2	147.4	74	330.2	175.6	34	383.2	203.8	94	436.2	231.9	54	489.2	260.1
15	278.1	147.9	75	331.1	176.1	35	384.1	204.2	95	437.1	232.4	55	490.0	260.6
16	279.0	148.4	76	332.0	176.5	36	385.0	204.7	96	437.9	232.9	56	490.9	261.0
17	279.9	148.8	77	332.9	177.0	37	385.8	205.2	97	438.8	233.3	57	491.8	261.5
18	280.8	149.3	78	333.8	177.5	38	386.7	205.6	98	439.7	233.8	58	492.7	262.0
19	281.7	149.8	79	334.6	177.9	39	387.6	206.1	99	440.6	234.3	59	493.6	262.4
20	282.5	150.2	80	335.5	178.4	40	388.5	206.6	500	441.5	234.7	60	494.5	262.9
321	283.4	150.7	381	336.4	178.9	441	389.4	207.0	501	442.4	235.2	561	495.3	263.4
22	284.3	151.2	82	337.3	179.3	42	390.3	207.5	02	443.2	235.7	62	496.2	263.8
23	285.2	151.6	83	338.2	179.8	43	391.1	208.0	03	444.1	236.1	63	497.1	264.3
24	286.1	152.1	84	339.1	180.3	44	392.0	208.4	04	445.0	236.6	64	498.0	264.8
25	287.0	152.6	85	339.9	180.7	45	392.9	208.9	05	445.9	237.1	65	498.9	265.3
26	287.8	153.0	86	340.8	181.2	46	393.8	209.4	06	446.8	237.6	66	499.7	265.7
27	288.7	153.5	87	341.7	181.7	47	394.7	209.9	07	447.7	238.0	67	500.6	266.2
28	289.6	154.0	88	342.6	182.2	48	395.6	210.3	08	448.5	238.5	68	501.5	266.7
29	290.5	154.5	89	343.5	182.6	49	396.4	210.8	09	449.4	239.0	69	502.4	267.1
30	291.4	154.9	90	344.3	183.1	50	397.3	211.3	10	450.3	239.4	70	503.3	267.6
331	292.3	155.4	391	345.2	183.6	451	398.2	211.7	511	451.2	239.9	571	504.2	268.1
32	293.1	155.9	92	346.1	184.0	52	399.1	212.2	12	452.1	240.4	72	505.0	268.5
33	294.0	156.3	93	347.0	184.5	53	400.0	212.7	13	453.0	240.8	73	505.9	269.0
34	294.9	156.8	94	347.9	185.0	54	400.9	213.1	14	453.8	241.3	74	506.8	269.5
35	295.8	157.3	95	348.8	185.4	55	401.7	213.6	15	454.7	241.8	75	507.7	269.9
36	296.7	157.7	96	349.6	185.9	56	402.6	214.1	16	455.6	242.2	76	508.6	270.4
37	297.6	158.2	97	350.5	186.4	57	403.5	214.5	17	456.5	242.7	77	509.5	270.9
38	298.4	158.7	98	351.4	186.8	58	404.4	215.0	18	457.4	243.2	78	510.3	271.4
39	299.3	159.2	99	352.3	187.3	59	405.3	215.5	19	458.2	243.7	79	511.2	271.8
40	300.2	159.6	400	353.2	187.8	60	406.2	216.0	20	459.1	244.1	80	512.1	272.3
341	301.1	160.1	401	354.1	188.3	461	407.0	216.4	521	460.0	244.6	581	513.0	272.8
42	302.0	160.6	02	354.9	188.7	62	407.9	216.9	22	460.9	245.1	82	513.9	273.2
43	302.9	161.0	03	355.8	189.2	63	408.8	217.4	23	461.8	245.5	83	514.8	273.7
44	303.7	161.5	04	356.7	189.7	64	409.7	217.8	24	462.7	246.0	84	515.6	274.2
45	304.6	162.0	05	357.6	190.1	65	410.6	218.3	25	463.5	246.5	85	516.5	274.6
46	305.5	162.4	06	358.5	190.6	66	411.5	218.8	26	464.4	246.9	86	517.4	275.1
47	306.4	162.9	07	359.4	191.1	67	412.3	219.2	27	465.3	247.4	87	518.3	275.6
48	307.3	163.4	08	360.2	191.5	68	413.2	219.7	28	466.2	247.9	88	519.2	276.0
49	308.1	163.8	09	361.1	192.0	69	414.1	220.2	29	467.1	248.4	89	520.1	276.5
50	309.0	164.3	10	362.0	192.5	70	415.0	220.7	30	468.0	248.8	90	520.9	277.0
351	309.9	164.8	411	362.9	193.0	471	415.9	221.1	531	468.8	249.3	591	521.8	277.5
52	310.8	165.3	12	363.8	193.4	72	416.8	221.6	32	469.7	249.8	92	522.7	277.9
53	311.7	165.7	13	364.7	193.9	73	417.6	222.1	33	470.6	250.2	93	523.6	278.4
54	312.6	166.2	14	365.5	194.4	74	418.5	222.5	34	471.5	250.7	94	524.5	278.9
55	313.4	166.7	15	366.4	194.8	75	419.4	223.0	35	472.4	251.2	95	525.4	279.3
56	314.3	167.1	16	367.3	195.3	76	420.3	223.5	36	473.3	251.6	96	526.2	279.8
57	315.2	167.6	17	368.2	195.8	77	421.2	223.9	37	474.1	252.1	97	527.1	280.3
58	316.1	168.1	18	369.1	196.2	78	422.0	224.4	38	475.0	252.6	98	528.0	280.7
59	317.0	168.5	19	370.0	196.7	79	422.9	224.9	39	475.9	253.0	99	528.9	281.2
60	317.9	169.0	20	370.8	197.2	80	423.8	225.3	40	476.8	253.5	600	529.8	281.7
Dist.	Dep.	D. Lat.	Dist.	Dep.	D. Lat.	Dist.	Dep.	D. Lat.	Dist.	Dep.	D. Lat.	Dist.	Dep.	D. Lat.

Dist.	D. Lat.	Dep.
D Lo	Dep.	
	m	D Lo

62°

298° | 062° / 242° | 118°

TABLE 4

Traverse 29° Table

331°	029°
209°	151°

Dist.	D. Lat.	Dep.	Dist.	D. Lat.	Dep.	Dist.	D. Lat.	Dep.	Dist.	D. Lat.	Dep.	Dist.	D. Lat.	Dep.
1	0.9	0.5	61	53.4	29.6	121	105.8	58.7	181	158.3	87.8	241	210.8	116.8
2	1.7	1.0	62	54.2	30.1	22	106.7	59.1	82	159.2	88.2	42	211.7	117.3
3	2.6	1.5	63	55.1	30.5	23	107.6	59.6	83	160.1	88.7	43	212.5	117.8
4	3.5	1.9	64	56.0	31.0	24	108.5	60.1	84	160.9	89.2	44	213.4	118.3
5	4.4	2.4	65	56.9	31.5	25	109.3	60.6	85	161.8	89.7	45	214.3	118.8
6	5.2	2.9	66	57.7	32.0	26	110.2	61.1	86	162.7	90.2	46	215.2	119.3
7	6.1	3.4	67	58.6	32.5	27	111.1	61.6	87	163.6	90.7	47	216.0	119.7
8	7.0	3.9	68	59.5	33.0	28	112.0	62.1	88	164.4	91.1	48	216.9	120.2
9	7.9	4.4	69	60.3	33.5	29	112.8	62.5	89	165.3	91.6	49	217.8	120.7
10	8.7	4.8	70	61.2	33.9	30	113.7	63.0	90	166.2	92.1	50	218.7	121.2
11	9.6	5.3	71	62.1	34.4	131	114.6	63.5	191	167.1	92.6	251	219.5	121.7
12	10.5	5.8	72	63.0	34.9	32	115.4	64.0	92	167.9	93.1	52	220.4	122.2
13	11.4	6.3	73	63.8	35.4	33	116.3	64.5	93	168.8	93.6	53	221.3	122.7
14	12.2	6.8	74	64.7	35.9	34	117.2	65.0	94	169.7	94.1	54	222.2	123.1
15	13.1	7.3	75	65.6	36.4	35	118.1	65.4	95	170.6	94.5	55	223.0	123.6
16	14.0	7.8	76	66.5	36.8	36	118.9	65.9	96	171.4	95.0	56	223.9	124.1
17	14.9	8.2	77	67.3	37.3	37	119.8	66.4	97	172.3	95.5	57	224.8	124.6
18	15.7	8.7	78	68.2	37.8	38	120.7	66.9	98	173.2	96.0	58	225.7	125.1
19	16.6	9.2	79	69.1	38.3	39	121.6	67.4	99	174.0	96.5	59	226.5	125.6
20	17.5	9.7	80	70.0	38.8	40	122.4	67.9	200	174.9	97.0	60	227.4	126.1
21	18.4	10.2	81	70.8	39.3	141	123.3	68.4	201	175.8	97.4	261	228.3	126.5
22	19.2	10.7	82	71.7	39.8	42	124.2	68.8	02	176.7	97.9	62	229.2	127.0
23	20.1	11.2	83	72.6	40.2	43	125.1	69.3	03	177.5	98.4	63	230.0	127.5
24	21.0	11.6	84	73.5	40.7	44	125.9	69.8	04	178.4	98.9	64	230.9	128.0
25	21.9	12.1	85	74.3	41.2	45	126.8	70.3	05	179.3	99.4	65	231.8	128.5
26	22.7	12.6	86	75.2	41.7	46	127.7	70.8	06	180.2	99.9	66	232.6	129.0
27	23.6	13.1	87	76.1	42.2	47	128.6	71.3	07	181.0	100.4	67	233.5	129.4
28	24.5	13.6	88	77.0	42.7	48	129.4	71.8	08	181.9	100.8	68	234.4	129.9
29	25.4	14.1	89	77.8	43.1	49	130.3	72.2	09	182.8	101.3	69	235.3	130.4
30	26.2	14.5	90	78.7	43.6	50	131.2	72.7	10	183.7	101.8	70	236.1	130.9
31	27.1	15.0	91	79.6	44.1	151	132.1	73.2	211	184.5	102.3	271	237.0	131.4
32	28.0	15.5	92	80.5	44.6	52	132.9	73.7	12	185.4	102.8	72	237.9	131.9
33	28.9	16.0	93	81.3	45.1	53	133.8	74.2	13	186.3	103.3	73	238.8	132.4
34	29.7	16.5	94	82.2	45.6	54	134.7	74.7	14	187.2	103.7	74	239.6	132.8
35	30.6	17.0	95	83.1	46.1	55	135.6	75.1	15	188.0	104.2	75	240.5	133.3
36	31.5	17.5	96	84.0	46.5	56	136.4	75.6	16	188.9	104.7	76	241.4	133.8
37	32.4	17.9	97	84.8	47.0	57	137.3	76.1	17	189.8	105.2	77	242.3	134.3
38	33.2	18.4	98	85.7	47.5	58	138.2	76.6	18	190.7	105.7	78	243.1	134.8
39	34.1	18.9	99	86.6	48.0	59	139.1	77.1	19	191.5	106.2	79	244.0	135.3
40	35.0	19.4	100	87.5	48.5	60	139.9	77.6	20	192.4	106.7	80	244.9	135.7
41	35.9	19.9	101	88.3	49.0	161	140.8	78.1	221	193.3	107.1	281	245.8	136.2
42	36.7	20.4	02	89.2	49.5	62	141.7	78.5	22	194.2	107.6	82	246.6	136.7
43	37.6	20.8	03	90.1	49.9	63	142.6	79.0	23	195.0	108.1	83	247.5	137.2
44	38.5	21.3	04	91.0	50.4	64	143.4	79.5	24	195.9	108.6	84	248.4	137.7
45	39.4	21.8	05	91.8	50.9	65	144.3	80.0	25	196.8	109.1	85	249.3	138.2
46	40.2	22.3	06	92.7	51.4	66	145.2	80.5	26	197.7	109.6	86	250.1	138.7
47	41.1	22.8	07	93.6	51.9	67	146.1	81.0	27	198.5	110.1	87	251.0	139.1
48	42.0	23.3	08	94.5	52.4	68	146.9	81.4	28	199.4	110.5	88	251.9	139.6
49	42.9	23.8	09	95.3	52.8	69	147.8	81.9	29	200.3	111.0	89	252.8	140.1
50	43.7	24.2	10	96.2	53.3	70	148.7	82.4	30	201.2	111.5	90	253.6	140.6
51	44.6	24.7	111	97.1	53.8	171	149.6	82.9	231	202.0	112.0	291	254.5	141.1
52	45.5	25.2	12	98.0	54.3	72	150.4	83.4	32	202.9	112.5	92	255.4	141.6
53	46.4	25.7	13	98.8	54.8	73	151.3	83.9	33	203.8	113.0	93	256.3	142.0
54	47.2	26.2	14	99.7	55.3	74	152.2	84.4	34	204.7	113.4	94	257.1	142.5
55	48.1	26.7	15	100.6	55.8	75	153.1	84.8	35	205.5	113.9	95	258.0	143.0
56	49.0	27.1	16	101.5	56.2	76	153.9	85.3	36	206.4	114.4	96	258.9	143.5
57	49.9	27.6	17	102.3	56.7	77	154.8	85.8	37	207.3	114.9	97	259.8	144.0
58	50.7	28.1	18	103.2	57.2	78	155.7	86.3	38	208.2	115.4	98	260.6	144.5
59	51.6	28.6	19	104.1	57.7	79	156.6	86.8	39	209.0	115.9	99	261.5	145.0
60	52.5	29.1	20	105.0	58.2	80	157.4	87.3	40	209.9	116.4	300	262.4	145.4
Dist.	Dep.	D. Lat.	Dist.	Dep.	D. Lat.	Dist.	Dep.	D. Lat.	Dist.	Dep.	D. Lat.	Dist.	Dep.	D. Lat.

299°	061°
241°	119°

61°

Dist.	D. Lat.	Dep.
N.	N x Cos.	N x Sin.
Hypotenuse	Side Adj.	Side Opp.

TABLE 4

Traverse 29° Table

331°	029°
209°	151°

Dist.	D. Lat.	Dep.	Dist.	D. Lat.	Dep.	Dist.	D. Lat.	Dep.	Dist.	D. Lat.	Dep.	Dist.	D. Lat.	Dep.
301	263.3	145.9	361	315.7	175.0	421	368.2	204.1	481	420.7	233.2	541	473.2	262.3
02	264.1	146.4	62	316.6	175.5	22	369.1	204.6	82	421.6	233.7	42	474.0	262.8
03	265.0	146.9	63	317.5	176.0	23	370.0	205.1	83	422.4	234.2	43	474.9	263.3
04	265.9	147.4	64	318.4	176.5	24	370.8	205.6	84	423.3	234.6	44	475.8	263.7
05	266.8	147.9	65	319.2	177.0	25	371.7	206.0	85	424.2	235.1	45	476.7	264.2
06	267.6	148.4	66	320.1	177.4	26	372.6	206.5	86	425.1	235.6	46	477.5	264.7
07	268.5	148.8	67	321.0	177.9	27	373.5	207.0	87	425.9	236.1	47	478.4	265.2
08	269.4	149.3	68	321.9	178.4	28	374.3	207.5	88	426.8	236.6	48	479.3	265.7
09	270.3	149.8	69	322.7	178.9	29	375.2	208.0	89	427.7	237.1	49	480.2	266.2
10	271.1	150.3	70	323.6	179.4	30	376.1	208.5	90	428.6	237.6	50	481.0	266.6
311	272.0	150.8	371	324.5	179.9	431	377.0	209.0	491	429.4	238.0	551	481.9	267.1
12	272.9	151.3	72	325.4	180.3	32	377.8	209.4	92	430.3	238.5	52	482.8	267.6
13	273.8	151.7	73	326.2	180.8	33	378.7	209.9	93	431.2	239.0	53	483.7	268.1
14	274.6	152.2	74	327.1	181.3	34	379.6	210.4	94	432.1	239.5	54	484.5	268.6
15	275.5	152.7	75	328.0	181.8	35	380.5	210.9	95	432.9	240.0	55	485.4	269.1
16	276.4	153.2	76	328.9	182.3	36	381.3	211.4	96	433.8	240.5	56	486.3	269.6
17	277.3	153.7	77	329.7	182.8	37	382.2	211.9	97	434.7	241.0	57	487.2	270.0
18	278.1	154.2	78	330.6	183.3	38	383.1	212.3	98	435.6	241.4	58	488.0	270.5
19	279.0	154.7	79	331.5	183.7	39	384.0	212.8	99	436.4	241.9	59	488.9	271.0
20	279.9	155.1	80	332.4	184.2	40	384.8	213.3	500	437.3	242.4	60	489.8	271.5
321	280.8	155.6	381	333.2	184.7	441	385.7	213.8	501	438.2	242.9	561	490.7	272.0
22	281.6	156.1	82	334.1	185.2	42	386.6	214.3	02	439.1	243.4	62	491.5	272.5
23	282.5	156.6	83	335.0	185.7	43	387.5	214.8	03	439.9	243.9	63	492.4	272.9
24	283.4	157.1	84	335.9	186.2	44	388.3	215.3	04	440.8	244.3	64	493.3	273.4
25	284.3	157.6	85	336.7	186.7	45	389.2	215.7	05	441.7	244.8	65	494.2	273.9
26	285.1	158.0	86	337.6	187.1	46	390.1	216.2	06	442.6	245.3	66	495.0	274.4
27	286.0	158.5	87	338.5	187.6	47	391.0	216.7	07	443.4	245.8	67	495.9	274.9
28	286.9	159.0	88	339.4	188.1	48	391.8	217.2	08	444.3	246.3	68	496.8	275.4
29	287.7	159.5	89	340.2	188.6	49	392.7	217.7	09	445.2	246.8	69	497.7	275.9
30	288.6	160.0	90	341.1	189.1	50	393.6	218.2	10	446.1	247.3	70	498.5	276.3
331	289.5	160.5	391	342.0	189.6	451	394.5	218.6	511	446.9	247.7	571	499.4	276.8
32	290.4	161.0	92	342.9	190.0	52	395.3	219.1	12	447.8	248.2	72	500.3	277.3
33	291.2	161.4	93	343.7	190.5	53	396.2	219.6	13	448.7	248.7	73	501.2	277.8
34	292.1	161.9	94	344.6	191.0	54	397.1	220.1	14	449.6	249.2	74	502.0	278.3
35	293.0	162.4	95	345.5	191.5	55	398.0	220.6	15	450.4	249.7	75	502.9	278.8
36	293.9	162.9	96	346.3	192.0	56	398.8	221.1	16	451.3	250.2	76	503.8	279.3
37	294.7	163.4	97	347.2	192.5	57	399.7	221.6	17	452.2	250.6	77	504.7	279.7
38	295.6	163.9	98	348.1	193.0	58	400.6	222.0	18	453.1	251.1	78	505.5	280.2
39	296.5	164.4	99	349.0	193.4	59	401.5	222.5	19	453.9	251.6	79	506.4	280.7
40	297.4	164.8	400	349.8	193.9	60	402.3	223.0	20	454.8	252.1	80	507.3	281.2
341	298.2	165.3	401	350.7	194.4	461	403.2	223.5	521	455.7	252.6	581	508.2	281.7
42	299.1	165.8	02	351.6	194.9	62	404.1	224.0	22	456.6	253.1	82	509.0	282.2
43	300.0	166.3	03	352.5	195.4	63	404.9	224.5	23	457.4	253.6	83	509.9	282.6
44	300.9	166.8	04	353.3	195.9	64	405.8	225.0	24	458.3	254.0	84	510.8	283.1
45	301.7	167.3	05	354.2	196.3	65	406.7	225.4	25	459.2	254.5	85	511.7	283.6
46	302.6	167.7	06	355.1	196.8	66	407.6	225.9	26	460.0	255.0	86	512.5	284.1
47	303.5	168.2	07	356.0	197.3	67	408.4	226.4	27	460.9	255.5	87	513.4	284.6
48	304.4	168.7	08	356.8	197.8	68	409.3	226.9	28	461.8	256.0	88	514.3	285.1
49	305.2	169.2	09	357.7	198.3	69	410.2	227.4	29	462.7	256.5	89	515.2	285.6
50	306.1	169.7	10	358.6	198.8	70	411.1	227.9	30	463.5	256.9	90	516.0	286.0
351	307.0	170.2	411	359.5	199.3	471	411.9	228.3	531	464.4	257.4	591	516.9	286.5
52	307.9	170.7	12	360.3	199.7	72	412.8	228.8	32	465.3	257.9	92	517.8	287.0
53	308.7	171.1	13	361.2	200.2	73	413.7	229.3	33	466.2	258.4	93	518.6	287.5
54	309.6	171.6	14	362.1	200.7	74	414.6	229.8	34	467.0	258.9	94	519.5	288.0
55	310.5	172.1	15	363.0	201.2	75	415.4	230.3	35	467.9	259.4	95	520.4	288.5
56	311.4	172.6	16	363.8	201.7	76	416.3	230.8	36	468.8	259.9	96	521.3	288.9
57	312.2	173.1	17	364.7	202.2	77	417.2	231.3	37	469.7	260.3	97	522.1	289.4
58	313.1	173.6	18	365.6	202.7	78	418.1	231.7	38	470.5	260.8	98	523.0	289.9
59	314.0	174.0	19	366.5	203.1	79	418.9	232.2	39	471.4	261.3	99	523.9	290.4
60	314.9	174.5	20	367.3	203.6	80	419.8	232.7	40	472.3	261.8	600	524.8	290.9
Dist.	Dep.	D. Lat.	Dist.	Dep.	D. Lat.	Dist.	Dep.	D. Lat.	Dist.	Dep.	D. Lat.	Dist.	Dep.	D. Lat.

Dist.	D. Lat.	Dep.
D Lo	Dep.	
	m	D Lo

61°

299°	061°
241°	119°

TABLE 4

30° Traverse Table

330° | 030° — 210° | 150°

Dist.	D. Lat.	Dep.	Dist.	D. Lat.	Dep.	Dist.	D. Lat.	Dep.	Dist.	D. Lat.	Dep.	Dist.	D. Lat.	Dep.
1	0.9	0.5	61	52.8	30.5	121	104.8	60.5	181	156.8	90.5	241	208.7	120.5
2	1.7	1.0	62	53.7	31.0	22	105.7	61.0	82	157.6	91.0	42	209.6	121.0
3	2.6	1.5	63	54.6	31.5	23	106.5	61.5	83	158.5	91.5	43	210.4	121.5
4	3.5	2.0	64	55.4	32.0	24	107.4	62.0	84	159.3	92.0	44	211.3	122.0
5	4.3	2.5	65	56.3	32.5	25	108.3	62.5	85	160.2	92.5	45	212.2	122.5
6	5.2	3.0	66	57.2	33.0	26	109.1	63.0	86	161.1	93.0	46	213.0	123.0
7	6.1	3.5	67	58.0	33.5	27	110.0	63.5	87	161.9	93.5	47	213.9	123.5
8	6.9	4.0	68	58.9	34.0	28	110.9	64.0	88	162.8	94.0	48	214.8	124.0
9	7.8	4.5	69	59.8	34.5	29	111.7	64.5	89	163.7	94.5	49	215.6	124.5
10	8.7	5.0	70	60.6	35.0	30	112.6	65.0	90	164.5	95.0	50	216.5	125.0
11	9.5	5.5	71	61.5	35.5	131	113.4	65.5	191	165.4	95.5	251	217.4	125.5
12	10.4	6.0	72	62.4	36.0	32	114.3	66.0	92	166.3	96.0	52	218.2	126.0
13	11.3	6.5	73	63.2	36.5	33	115.2	66.5	93	167.1	96.5	53	219.1	126.5
14	12.1	7.0	74	64.1	37.0	34	116.0	67.0	94	168.0	97.0	54	220.0	127.0
15	13.0	7.5	75	65.0	37.5	35	116.9	67.5	95	168.9	97.5	55	220.8	127.5
16	13.9	8.0	76	65.8	38.0	36	117.8	68.0	96	169.7	98.0	56	221.7	128.0
17	14.7	8.5	77	66.7	38.5	37	118.6	68.5	97	170.6	98.5	57	222.6	128.5
18	15.6	9.0	78	67.5	39.0	38	119.5	69.0	98	171.5	99.0	58	223.4	129.0
19	16.5	9.5	79	68.4	39.5	39	120.4	69.5	99	172.3	99.5	59	224.3	129.5
20	17.3	10.0	80	69.3	40.0	40	121.2	70.0	200	173.2	100.0	60	225.2	130.0
21	18.2	10.5	81	70.1	40.5	141	122.1	70.5	201	174.1	100.5	261	226.0	130.5
22	19.1	11.0	82	71.0	41.0	42	123.0	71.0	02	174.9	101.0	62	226.9	131.0
23	19.9	11.5	83	71.9	41.5	43	123.8	71.5	03	175.8	101.5	63	227.8	131.5
24	20.8	12.0	84	72.7	42.0	44	124.7	72.0	04	176.7	102.0	64	228.6	132.0
25	21.7	12.5	85	73.6	42.5	45	125.6	72.5	05	177.5	102.5	65	229.5	132.5
26	22.5	13.0	86	74.5	43.0	46	126.4	73.0	06	178.4	103.0	66	230.4	133.0
27	23.4	13.5	87	75.3	43.5	47	127.3	73.5	07	179.3	103.5	67	231.2	133.5
28	24.2	14.0	88	76.2	44.0	48	128.2	74.0	08	180.1	104.0	68	232.1	134.0
29	25.1	14.5	89	77.1	44.5	49	129.0	74.5	09	181.0	104.5	69	233.0	134.5
30	26.0	15.0	90	77.9	45.0	50	129.9	75.0	10	181.9	105.0	70	233.8	135.0
31	26.8	15.5	91	78.8	45.5	151	130.8	75.5	211	182.7	105.5	271	234.7	135.5
32	27.7	16.0	92	79.7	46.0	52	131.6	76.0	12	183.6	106.0	72	235.6	136.0
33	28.6	16.5	93	80.5	46.5	53	132.5	76.5	13	184.5	106.5	73	236.4	136.5
34	29.4	17.0	94	81.4	47.0	54	133.4	77.0	14	185.3	107.0	74	237.3	137.0
35	30.3	17.5	95	82.3	47.5	55	134.2	77.5	15	186.2	107.5	75	238.2	137.5
36	31.2	18.0	96	83.1	48.0	56	135.1	78.0	16	187.1	108.0	76	239.0	138.0
37	32.0	18.5	97	84.0	48.5	57	136.0	78.5	17	187.9	108.5	77	239.9	138.5
38	32.9	19.0	98	84.9	49.0	58	136.8	79.0	18	188.8	109.0	78	240.8	139.0
39	33.8	19.5	99	85.7	49.5	59	137.7	79.5	19	189.7	109.5	79	241.6	139.5
40	34.6	20.0	100	86.6	50.0	60	138.6	80.0	20	190.5	110.0	80	242.5	140.0
41	35.5	20.5	101	87.5	50.5	161	139.4	80.5	221	191.4	110.5	281	243.4	140.5
42	36.4	21.0	02	88.3	51.0	62	140.3	81.0	22	192.3	111.0	82	244.2	141.0
43	37.2	21.5	03	89.2	51.5	63	141.2	81.5	23	193.1	111.5	83	245.1	141.5
44	38.1	22.0	04	90.1	52.0	64	142.0	82.0	24	194.0	112.0	84	246.0	142.0
45	39.0	22.5	05	90.9	52.5	65	142.9	82.5	25	194.9	112.5	85	246.8	142.5
46	39.8	23.0	06	91.8	53.0	66	143.8	83.0	26	195.7	113.0	86	247.7	143.0
47	40.7	23.5	07	92.7	53.5	67	144.6	83.5	27	196.6	113.5	87	248.5	143.5
48	41.6	24.0	08	93.5	54.0	68	145.5	84.0	28	197.5	114.0	88	249.4	144.0
49	42.4	24.5	09	94.4	54.5	69	146.4	84.5	29	198.3	114.5	89	250.3	144.5
50	43.3	25.0	10	95.3	55.0	70	147.2	85.0	30	199.2	115.0	90	251.1	145.0
51	44.2	25.5	111	96.1	55.5	171	148.1	85.5	231	200.1	115.5	291	252.0	145.5
52	45.0	26.0	12	97.0	56.0	72	149.0	86.0	32	200.9	116.0	92	252.9	146.0
53	45.9	26.5	13	97.9	56.5	73	149.8	86.5	33	201.8	116.5	93	253.7	146.5
54	46.8	27.0	14	98.7	57.0	74	150.7	87.0	34	202.6	117.0	94	254.6	147.0
55	47.6	27.5	15	99.6	57.5	75	151.6	87.5	35	203.5	117.5	95	255.5	147.5
56	48.5	28.0	16	100.5	58.0	76	152.4	88.0	36	204.4	118.0	96	256.3	148.0
57	49.4	28.5	17	101.3	58.5	77	153.3	88.5	37	205.2	118.5	97	257.2	148.5
58	50.2	29.0	18	102.2	59.0	78	154.2	89.0	38	206.1	119.0	98	258.1	149.0
59	51.1	29.5	19	103.1	59.5	79	155.0	89.5	39	207.0	119.5	99	258.9	149.5
60	52.0	30.0	20	103.9	60.0	80	155.9	90.0	40	207.8	120.0	300	259.8	150.0
Dist.	Dep.	D. Lat.	Dist.	Dep.	D. Lat.	Dist.	Dep.	D. Lat.	Dist.	Dep.	D. Lat.	Dist.	Dep.	D. Lat.

300° | 060° — 240° | 120°

60°

Dist.	D. Lat.	Dep.
N.	N x Cos.	N x Sin.
Hypotenuse	Side Adj.	Side Opp.

TABLE 4

30° Traverse Table

330° | 030° — 210° | 150°

Dist.	D. Lat.	Dep.	Dist.	D. Lat.	Dep.	Dist.	D. Lat.	Dep.	Dist.	D. Lat.	Dep.	Dist.	D. Lat.	Dep.
301	260.7	150.5	361	312.6	180.5	421	364.6	210.5	481	416.6	240.5	541	468.5	270.5
02	261.5	151.0	62	313.5	181.0	22	365.5	211.0	82	417.4	241.0	42	469.4	271.0
03	262.4	151.5	63	314.4	181.5	23	366.3	211.5	83	418.3	241.5	43	470.3	271.5
04	263.3	152.0	64	315.2	182.0	24	367.2	212.0	84	419.2	242.0	44	471.1	272.0
05	264.1	152.5	65	316.1	182.5	25	368.1	212.5	85	420.0	242.5	45	472.0	272.5
06	265.0	153.0	66	317.0	183.0	26	368.9	213.0	86	420.9	243.0	46	472.8	273.0
07	265.9	153.5	67	317.8	183.5	27	369.8	213.5	87	421.8	243.5	47	473.7	273.5
08	266.7	154.0	68	318.7	184.0	28	370.7	214.0	88	422.6	244.0	48	474.6	274.0
09	267.6	154.5	69	319.6	184.5	29	371.5	214.5	89	423.5	244.5	49	475.4	274.5
10	268.5	155.0	70	320.4	185.0	30	372.4	215.0	90	424.4	245.0	50	476.3	275.0
311	269.3	155.5	371	321.3	185.5	431	373.3	215.5	491	425.2	245.5	551	477.2	275.5
12	270.2	156.0	72	322.2	186.0	32	374.1	216.0	92	426.1	246.0	52	478.0	276.0
13	271.1	156.5	73	323.0	186.5	33	375.0	216.5	93	427.0	246.5	53	478.9	276.5
14	271.9	157.0	74	323.9	187.0	34	375.9	217.0	94	427.8	247.0	54	479.8	277.0
15	272.8	157.5	75	324.8	187.5	35	376.7	217.5	95	428.7	247.5	55	480.6	277.5
16	273.7	158.0	76	325.6	188.0	36	377.6	218.0	96	429.5	248.0	56	481.5	278.0
17	274.5	158.5	77	326.5	188.5	37	378.5	218.5	97	430.4	248.5	57	482.4	278.5
18	275.4	159.0	78	327.4	189.0	38	379.3	219.0	98	431.3	249.0	58	483.2	279.0
19	276.3	159.5	79	328.2	189.5	39	380.2	219.5	99	432.1	249.5	59	484.1	279.5
20	277.1	160.0	80	329.1	190.0	40	381.1	220.0	500	433.0	250.0	60	485.0	280.0
321	278.0	160.5	381	330.0	190.5	441	381.9	220.5	501	433.9	250.5	561	485.8	280.5
22	278.9	161.0	82	330.8	191.0	42	382.8	221.0	02	434.7	251.0	62	486.7	281.0
23	279.7	161.5	83	331.7	191.5	43	383.6	221.5	03	435.6	251.5	63	487.6	281.5
24	280.6	162.0	84	332.6	192.0	44	384.5	222.0	04	436.5	252.0	64	488.4	282.0
25	281.5	162.5	85	333.4	192.5	45	385.4	222.5	05	437.3	252.5	65	489.3	282.5
26	282.3	163.0	86	334.3	193.0	46	386.2	223.0	06	438.2	253.0	66	490.2	283.0
27	283.2	163.5	87	335.2	193.5	47	387.1	223.5	07	439.1	253.5	67	491.0	283.5
28	284.1	164.0	88	336.0	194.0	48	388.0	224.0	08	439.9	254.0	68	491.9	284.0
29	284.9	164.5	89	336.9	194.5	49	388.8	224.5	09	440.8	254.5	69	492.8	284.5
30	285.8	165.0	90	337.7	195.0	50	389.7	225.0	10	441.7	255.0	70	493.6	285.0
331	286.7	165.5	391	338.6	195.5	451	390.6	225.5	511	442.5	255.5	571	494.5	285.5
32	287.5	166.0	92	339.5	196.0	52	391.4	226.0	12	443.4	256.0	72	495.4	286.0
33	288.4	166.5	93	340.3	196.5	53	392.3	226.5	13	444.3	256.5	73	496.2	286.5
34	289.3	167.0	94	341.2	197.0	54	393.2	227.0	14	445.1	257.0	74	497.1	287.0
35	290.1	167.5	95	342.1	197.5	55	394.0	227.5	15	446.0	257.5	75	498.0	287.5
36	291.0	168.0	96	342.9	198.0	56	394.9	228.0	16	446.9	258.0	76	498.8	288.0
37	291.9	168.5	97	343.8	198.5	57	395.8	228.5	17	447.7	258.5	77	499.7	288.5
38	292.7	169.0	98	344.7	199.0	58	396.6	229.0	18	448.6	259.0	78	500.6	289.0
39	293.6	169.5	99	345.5	199.5	59	397.5	229.5	19	449.5	259.5	79	501.4	289.5
40	294.4	170.0	400	346.4	200.0	60	398.4	230.0	20	450.3	260.0	80	502.3	290.0
341	295.3	170.5	401	347.3	200.5	461	399.2	230.5	521	451.2	260.5	581	503.2	290.5
42	296.2	171.0	02	348.1	201.0	62	400.1	231.0	22	452.1	261.0	82	504.0	291.0
43	297.0	171.5	03	349.0	201.5	63	401.0	231.5	23	452.9	261.5	83	504.9	291.5
44	297.9	172.0	04	349.9	202.0	64	401.8	232.0	24	453.8	262.0	84	505.8	292.0
45	298.8	172.5	05	350.7	202.5	65	402.7	232.5	25	454.7	262.5	85	506.6	292.5
46	299.6	173.0	06	351.6	203.0	66	403.6	233.0	26	455.5	263.0	86	507.5	293.0
47	300.5	173.5	07	352.5	203.5	67	404.4	233.5	27	456.4	263.5	87	508.4	293.5
48	301.4	174.0	08	353.3	204.0	68	405.3	234.0	28	457.3	264.0	88	509.2	294.0
49	302.2	174.5	09	354.2	204.5	69	406.2	234.5	29	458.1	264.5	89	510.1	294.5
50	303.1	175.0	10	355.1	205.0	70	407.0	235.0	30	459.0	265.0	90	511.0	295.0
351	304.0	175.5	411	355.9	205.5	471	407.9	235.5	531	459.9	265.5	591	511.8	295.5
52	304.8	176.0	12	356.8	206.0	72	408.8	236.0	32	460.7	266.0	92	512.7	296.0
53	305.7	176.5	13	357.7	206.5	73	409.6	236.5	33	461.6	266.5	93	513.6	296.5
54	306.6	177.0	14	358.5	207.0	74	410.5	237.0	34	462.5	267.0	94	514.4	297.0
55	307.4	177.5	15	359.4	207.5	75	411.4	237.5	35	463.3	267.5	95	515.3	297.5
56	308.3	178.0	16	360.3	208.0	76	412.2	238.0	36	464.2	268.0	96	516.2	298.0
57	309.2	178.5	17	361.1	208.5	77	413.1	238.5	37	465.1	268.5	97	517.0	298.5
58	310.0	179.0	18	362.0	209.0	78	414.0	239.0	38	465.9	269.0	98	517.9	299.0
59	310.9	179.5	19	362.9	209.5	79	414.8	239.5	39	466.8	269.5	99	518.7	299.5
60	311.8	180.0	20	363.7	210.0	80	415.7	240.0	40	467.7	270.0	600	519.6	300.0
Dist.	Dep.	D. Lat.	Dist.	Dep.	D. Lat.	Dist.	Dep.	D. Lat.	Dist.	Dep.	D. Lat.	Dist.	Dep.	D. Lat.

Dist.	D. Lat.	Dep.
D Lo	Dep.	
	m	D Lo

60°

300° | 060° — 240° | 120°

TABLE 4

Traverse 31° Table

329°	031°
211°	149°

Dist.	D. Lat.	Dep.	Dist.	D. Lat.	Dep.	Dist.	D. Lat.	Dep.	Dist.	D. Lat.	Dep.	Dist.	D. Lat.	Dep.
1	0.9	0.5	61	52.3	31.4	121	103.7	62.3	181	155.1	93.2	241	206.6	124.1
2	1.7	1.0	62	53.1	31.9	22	104.6	62.8	82	156.0	93.7	42	207.4	124.6
3	2.6	1.5	63	54.0	32.4	23	105.4	63.3	83	156.9	94.3	43	208.3	125.2
4	3.4	2.1	64	54.9	33.0	24	106.3	63.9	84	157.7	94.8	44	209.1	125.7
5	4.3	2.6	65	55.7	33.5	25	107.1	64.4	85	158.6	95.3	45	210.0	126.2
6	5.1	3.1	66	56.6	34.0	26	108.0	64.9	86	159.4	95.8	46	210.9	126.7
7	6.0	3.6	67	57.4	34.5	27	108.9	65.4	87	160.3	96.3	47	211.7	127.2
8	6.9	4.1	68	58.3	35.0	28	109.7	65.9	88	161.1	96.8	48	212.6	127.7
9	7.7	4.6	69	59.1	35.5	29	110.6	66.4	89	162.0	97.3	49	213.4	128.2
10	8.6	5.2	70	60.0	36.1	30	111.4	67.0	90	162.9	97.9	50	214.3	128.8
11	9.4	5.7	71	60.9	36.6	131	112.3	67.5	191	163.7	98.4	251	215.1	129.3
12	10.3	6.2	72	61.7	37.1	32	113.1	68.0	92	164.6	98.9	52	216.0	129.8
13	11.1	6.7	73	62.6	37.6	33	114.0	68.5	93	165.4	99.4	53	216.9	130.3
14	12.0	7.2	74	63.4	38.1	34	114.9	69.0	94	166.3	99.9	54	217.7	130.8
15	12.9	7.7	75	64.3	38.6	35	115.7	69.5	95	167.1	100.4	55	218.6	131.3
16	13.7	8.2	76	65.1	39.1	36	116.6	70.0	96	168.0	100.9	56	219.4	131.8
17	14.6	8.8	77	66.0	39.7	37	117.4	70.6	97	168.9	101.5	57	220.3	132.4
18	15.4	9.3	78	66.9	40.2	38	118.3	71.1	98	169.7	102.0	58	221.1	132.9
19	16.3	9.8	79	67.7	40.7	39	119.1	71.6	99	170.6	102.5	59	222.0	133.4
20	17.1	10.3	80	68.6	41.2	40	120.0	72.1	200	171.4	103.0	60	222.9	133.9
21	18.0	10.8	81	69.4	41.7	141	120.9	72.6	201	172.3	103.5	261	223.7	134.4
22	18.9	11.3	82	70.3	42.2	42	121.7	73.1	02	173.1	104.0	62	224.6	134.9
23	19.7	11.8	83	71.1	42.7	43	122.6	73.7	03	174.0	104.6	63	225.4	135.5
24	20.6	12.4	84	72.0	43.3	44	123.4	74.2	04	174.9	105.1	64	226.3	136.0
25	21.4	12.9	85	72.9	43.8	45	124.3	74.7	05	175.7	105.6	65	227.1	136.5
26	22.3	13.4	86	73.7	44.3	46	125.1	75.2	06	176.6	106.1	66	228.0	137.0
27	23.1	13.9	87	74.6	44.8	47	126.0	75.7	07	177.4	106.6	67	228.9	137.5
28	24.0	14.4	88	75.4	45.3	48	126.9	76.2	08	178.3	107.1	68	229.7	138.0
29	24.9	14.9	89	76.3	45.8	49	127.7	76.7	09	179.1	107.6	69	230.6	138.5
30	25.7	15.5	90	77.1	46.4	50	128.6	77.3	10	180.0	108.2	70	231.4	139.1
31	26.6	16.0	91	78.0	46.9	151	129.4	77.8	211	180.9	108.7	271	232.3	139.6
32	27.4	16.5	92	78.9	47.4	52	130.3	78.3	12	181.7	109.2	72	233.1	140.1
33	28.3	17.0	93	79.7	47.9	53	131.1	78.8	13	182.6	109.7	73	234.0	140.6
34	29.1	17.5	94	80.6	48.4	54	132.0	79.3	14	183.4	110.2	74	234.9	141.1
35	30.0	18.0	95	81.4	48.9	55	132.9	79.8	15	184.3	110.7	75	235.7	141.6
36	30.9	18.5	96	82.3	49.4	56	133.7	80.3	16	185.1	111.2	76	236.6	142.2
37	31.7	19.1	97	83.1	50.0	57	134.6	80.9	17	186.0	111.8	77	237.4	142.7
38	32.6	19.6	98	84.0	50.5	58	135.4	81.4	18	186.9	112.3	78	238.3	143.2
39	33.4	20.1	99	84.9	51.0	59	136.3	81.9	19	187.7	112.8	79	239.1	143.7
40	34.3	20.6	100	85.7	51.5	60	137.1	82.4	20	188.6	113.3	80	240.0	144.2
41	35.1	21.1	101	86.6	52.0	161	138.0	82.9	221	189.4	113.8	281	240.9	144.7
42	36.0	21.6	02	87.4	52.5	62	138.9	83.4	22	190.3	114.3	82	241.7	145.2
43	36.9	22.1	03	88.3	53.0	63	139.7	84.0	23	191.1	114.9	83	242.6	145.8
44	37.7	22.7	04	89.1	53.6	64	140.6	84.5	24	192.0	115.4	84	243.4	146.3
45	38.6	23.2	05	90.0	54.1	65	141.4	85.0	25	192.9	115.9	85	244.3	146.8
46	39.4	23.7	06	90.9	54.6	66	142.3	85.5	26	193.7	116.4	86	245.1	147.3
47	40.3	24.2	07	91.7	55.1	67	143.1	86.0	27	194.6	116.9	87	246.0	147.8
48	41.1	24.7	08	92.6	55.6	68	144.0	86.5	28	195.4	117.4	88	246.9	148.3
49	42.0	25.2	09	93.4	56.1	69	144.9	87.0	29	196.3	117.9	89	247.7	148.8
50	42.9	25.8	10	94.3	56.7	70	145.7	87.6	30	197.1	118.5	90	248.6	149.4
51	43.7	26.3	111	95.1	57.2	171	146.6	88.1	231	198.0	119.0	291	249.4	149.9
52	44.6	26.8	12	96.0	57.7	72	147.4	88.6	32	198.9	119.5	92	250.3	150.4
53	45.4	27.3	13	96.9	58.2	73	148.3	89.1	33	199.7	120.0	93	251.2	150.9
54	46.3	27.8	14	97.7	58.7	74	149.1	89.6	34	200.6	120.5	94	252.0	151.4
55	47.1	28.3	15	98.6	59.2	75	150.0	90.1	35	201.4	121.0	95	252.9	151.9
56	48.0	28.8	16	99.4	59.7	76	150.9	90.6	36	202.3	121.5	96	253.7	152.5
57	48.9	29.4	17	100.3	60.3	77	151.7	91.2	37	203.1	122.1	97	254.6	153.0
58	49.7	29.9	18	101.1	60.8	78	152.6	91.7	38	204.0	122.6	98	255.4	153.5
59	50.6	30.4	19	102.0	61.3	79	153.4	92.2	39	204.9	123.1	99	256.3	154.0
60	51.4	30.9	20	102.9	61.8	80	154.3	92.7	40	205.7	123.6	300	257.2	154.5
Dist.	Dep.	D. Lat.	Dist.	Dep.	D. Lat.	Dist.	Dep.	D. Lat.	Dist.	Dep.	D. Lat.	Dist.	Dep.	D. Lat.

301°	059°
239°	121°

59°

Dist.	D. Lat.	Dep.
N.	N x Cos.	N x Sin.
Hypotenuse	Side Adj.	Side Opp.

TABLE 4

Traverse 31° Table

329°	031°
211°	149°

Dist.	D. Lat.	Dep.	Dist.	D. Lat.	Dep.	Dist.	D. Lat.	Dep.	Dist.	D. Lat.	Dep.	Dist.	D. Lat.	Dep.
301	258.0	155.0	361	309.4	185.9	421	360.9	216.8	481	412.3	247.7	541	463.7	278.6
02	258.9	155.5	62	310.3	186.4	22	361.7	217.3	82	413.2	248.2	42	464.6	279.2
03	259.7	156.1	63	311.2	187.0	23	362.6	217.9	83	414.0	248.8	43	465.4	279.7
04	260.6	156.6	64	312.0	187.5	24	363.4	218.4	84	414.9	249.3	44	466.3	280.2
05	261.4	157.1	65	312.9	188.0	25	364.3	218.9	85	415.7	249.8	45	467.2	280.7
06	262.3	157.6	66	313.7	188.5	26	365.2	219.4	86	416.6	250.3	46	468.0	281.2
07	263.2	158.1	67	314.6	189.0	27	366.0	219.9	87	417.4	250.8	47	468.9	281.7
08	264.0	158.6	68	315.4	189.5	28	366.9	220.4	88	418.3	251.3	48	469.7	282.2
09	264.9	159.1	69	316.3	190.0	29	367.7	221.0	89	419.2	251.9	49	470.6	282.8
10	265.7	159.7	70	317.2	190.6	30	368.6	221.5	90	420.0	252.4	50	471.4	283.3
311	266.6	160.2	371	318.0	191.1	431	369.4	222.0	491	420.9	252.9	551	472.3	283.8
12	267.4	160.7	72	318.9	191.6	32	370.3	222.5	92	421.7	253.4	52	473.2	284.3
13	268.3	161.2	73	319.7	192.1	33	371.2	223.0	93	422.6	253.9	53	474.0	284.8
14	269.2	161.7	74	320.6	192.6	34	372.0	223.5	94	423.4	254.4	54	474.9	285.3
15	270.0	162.2	75	321.4	193.1	35	372.9	224.0	95	424.3	254.9	55	475.7	285.8
16	270.9	162.8	76	322.3	193.7	36	373.7	224.6	96	425.2	255.5	56	476.6	286.4
17	271.7	163.3	77	323.2	194.2	37	374.6	225.1	97	426.0	256.0	57	477.4	286.9
18	272.6	163.8	78	324.0	194.7	38	375.4	225.6	98	426.9	256.5	58	478.3	287.4
19	273.4	164.3	79	324.9	195.2	39	376.3	226.1	99	427.7	257.0	59	479.2	287.9
20	274.3	164.8	80	325.7	195.7	40	377.2	226.6	500	428.6	257.5	60	480.0	288.4
321	275.2	165.3	381	326.6	196.2	441	378.0	227.1	501	429.4	258.0	561	480.9	288.9
22	276.0	165.8	82	327.4	196.7	42	378.9	227.6	02	430.3	258.5	62	481.7	289.5
23	276.9	166.4	83	328.3	197.3	43	379.7	228.2	03	431.2	259.1	63	482.6	290.0
24	277.7	166.9	84	329.2	197.8	44	380.6	228.7	04	432.0	259.6	64	483.4	290.5
25	278.6	167.4	85	330.0	198.3	45	381.4	229.2	05	432.9	260.1	65	484.3	291.0
26	279.4	167.9	86	330.9	198.8	46	382.3	229.7	06	433.7	260.6	66	485.2	291.5
27	280.3	168.4	87	331.7	199.3	47	383.2	230.2	07	434.6	261.1	67	486.0	292.0
28	281.2	168.9	88	332.6	199.8	48	384.0	230.7	08	435.4	261.6	68	486.9	292.5
29	282.0	169.4	89	333.4	200.3	49	384.9	231.3	09	436.3	262.2	69	487.7	293.1
30	282.9	170.0	90	334.3	200.9	50	385.7	231.8	10	437.2	262.7	70	488.6	293.6
331	283.7	170.5	391	335.2	201.4	451	386.6	232.3	511	438.0	263.2	571	489.4	294.1
32	284.6	171.0	92	336.0	201.9	52	387.4	232.8	12	438.9	263.7	72	490.3	294.6
33	285.4	171.5	93	336.9	202.4	53	388.3	233.3	13	439.7	264.2	73	491.2	295.1
34	286.3	172.0	94	337.7	202.9	54	389.2	233.8	14	440.6	264.7	74	492.0	295.6
35	287.2	172.5	95	338.6	203.4	55	390.0	234.3	15	441.4	265.2	75	492.9	296.1
36	288.0	173.1	96	339.4	204.0	56	390.9	234.9	16	442.3	265.8	76	493.7	296.7
37	288.9	173.6	97	340.3	204.5	57	391.7	235.4	17	443.2	266.3	77	494.6	297.2
38	289.7	174.1	98	341.2	205.0	58	392.6	235.9	18	444.0	266.8	78	495.4	297.7
39	290.6	174.6	99	342.0	205.5	59	393.4	236.4	19	444.9	267.3	79	496.3	298.2
40	291.4	175.1	400	342.9	206.0	60	394.3	236.9	20	445.7	267.8	80	497.2	298.7
341	292.3	175.6	401	343.7	206.5	461	395.2	237.4	521	446.6	268.3	581	498.0	299.2
42	293.2	176.1	02	344.6	207.0	62	396.0	237.9	22	447.4	268.8	82	498.9	299.8
43	294.0	176.7	03	345.4	207.6	63	396.9	238.5	23	448.3	269.4	83	499.7	300.3
44	294.9	177.2	04	346.3	208.1	64	397.7	239.0	24	449.2	269.9	84	500.6	300.8
45	295.7	177.7	05	347.2	208.6	65	398.6	239.5	25	450.0	270.4	85	501.4	301.3
46	296.6	178.2	06	348.0	209.1	66	399.4	240.0	26	450.9	270.9	86	502.3	301.8
47	297.4	178.7	07	348.9	209.6	67	400.3	240.5	27	451.7	271.4	87	503.2	302.3
48	298.3	179.2	08	349.7	210.1	68	401.2	241.0	28	452.6	271.9	88	504.0	302.8
49	299.2	179.7	09	350.6	210.7	69	402.0	241.6	29	453.4	272.5	89	504.9	303.4
50	300.0	180.3	10	351.4	211.2	70	402.9	242.1	30	454.3	273.0	90	505.7	303.9
351	300.9	180.8	411	352.3	211.7	471	403.7	242.6	531	455.2	273.5	591	506.6	304.4
52	301.7	181.3	12	353.2	212.2	72	404.6	243.1	32	456.0	274.0	92	507.4	304.9
53	302.6	181.8	13	354.0	212.7	73	405.4	243.6	33	456.9	274.5	93	508.3	305.4
54	303.4	182.3	14	354.9	213.2	74	406.3	244.1	34	457.7	275.0	94	509.2	305.9
55	304.3	182.8	15	355.7	213.7	75	407.2	244.6	35	458.6	275.5	95	510.0	306.4
56	305.2	183.4	16	356.6	214.3	76	408.0	245.2	36	459.4	276.1	96	510.9	307.0
57	306.0	183.9	17	357.4	214.8	77	408.9	245.7	37	460.3	276.6	97	511.7	307.5
58	306.9	184.4	18	358.3	215.3	78	409.7	246.2	38	461.2	277.1	98	512.6	308.0
59	307.7	184.9	19	359.2	215.8	79	410.6	246.7	39	462.0	277.6	99	513.4	308.5
60	308.6	185.4	20	360.0	216.3	80	411.4	247.2	40	462.9	278.1	600	514.3	309.0
Dist.	Dep.	D. Lat.	Dist.	Dep.	D. Lat.	Dist.	Dep.	D. Lat.	Dist.	Dep.	D. Lat.	Dist.	Dep.	D. Lat.

Dist.	D. Lat.	Dep.
D Lo	Dep.	
	m	D Lo

59°

301°	059°
239°	121°

TABLE 4

Traverse **32°** Table

328° | 032°
212° | 148°

Dist.	D. Lat.	Dep.	Dist.	D. Lat.	Dep.	Dist.	D. Lat.	Dep.	Dist.	D. Lat.	Dep.	Dist.	D. Lat.	Dep.
1	0.8	0.5	61	51.7	32.3	121	102.6	64.1	181	153.5	95.9	241	204.4	127.7
2	1.7	1.1	62	52.6	32.9	22	103.5	64.7	82	154.3	96.4	42	205.2	128.2
3	2.5	1.6	63	53.4	33.4	23	104.3	65.2	83	155.2	97.0	43	206.1	128.8
4	3.4	2.1	64	54.3	33.9	24	105.2	65.7	84	156.0	97.5	44	206.9	129.3
5	4.2	2.6	65	55.1	34.4	25	106.0	66.2	85	156.9	98.0	45	207.8	129.8
6	5.1	3.2	66	56.0	35.0	26	106.9	66.8	86	157.7	98.6	46	208.6	130.4
7	5.9	3.7	67	56.8	35.5	27	107.7	67.3	87	158.6	99.1	47	209.5	130.9
8	6.8	4.2	68	57.7	36.0	28	108.6	67.8	88	159.4	99.6	48	210.3	131.4
9	7.6	4.8	69	58.5	36.6	29	109.4	68.4	89	160.3	100.2	49	211.2	131.9
10	8.5	5.3	70	59.4	37.1	30	110.2	68.9	90	161.1	100.7	50	212.0	132.5
11	9.3	5.8	71	60.2	37.6	131	111.1	69.4	191	162.0	101.2	251	212.9	133.0
12	10.2	6.4	72	61.1	38.2	32	111.9	69.9	92	162.8	101.7	52	213.7	133.5
13	11.0	6.9	73	61.9	38.7	33	112.8	70.5	93	163.7	102.3	53	214.6	134.1
14	11.9	7.4	74	62.8	39.2	34	113.6	71.0	94	164.5	102.8	54	215.4	134.6
15	12.7	7.9	75	63.6	39.7	35	114.5	71.5	95	165.4	103.3	55	216.3	135.1
16	13.6	8.5	76	64.5	40.3	36	115.3	72.1	96	166.2	103.9	56	217.1	135.7
17	14.4	9.0	77	65.3	40.8	37	116.2	72.6	97	167.1	104.4	57	217.9	136.2
18	15.3	9.5	78	66.1	41.3	38	117.0	73.1	98	167.9	104.9	58	218.8	136.7
19	16.1	10.1	79	67.0	41.9	39	117.9	73.7	99	168.8	105.5	59	219.6	137.2
20	17.0	10.6	80	67.8	42.4	40	118.7	74.2	200	169.6	106.0	60	220.5	137.8
21	17.8	11.1	81	68.7	42.9	141	119.6	74.7	201	170.5	106.5	261	221.3	138.3
22	18.7	11.7	82	69.5	43.5	42	120.4	75.2	02	171.3	107.0	62	222.2	138.8
23	19.5	12.2	83	70.4	44.0	43	121.3	75.8	03	172.2	107.6	63	223.0	139.4
24	20.4	12.7	84	71.2	44.5	44	122.1	76.3	04	173.0	108.1	64	223.9	139.9
25	21.2	13.2	85	72.1	45.0	45	123.0	76.8	05	173.8	108.6	65	224.7	140.4
26	22.0	13.8	86	72.9	45.6	46	123.8	77.4	06	174.7	109.2	66	225.6	141.0
27	22.9	14.3	87	73.8	46.1	47	124.7	77.9	07	175.5	109.7	67	226.4	141.5
28	23.7	14.8	88	74.6	46.6	48	125.5	78.4	08	176.4	110.2	68	227.3	142.0
29	24.6	15.4	89	75.5	47.2	49	126.4	79.0	09	177.2	110.8	69	228.1	142.5
30	25.4	15.9	90	76.3	47.7	50	127.2	79.5	10	178.1	111.3	70	229.0	143.1
31	26.3	16.4	91	77.2	48.2	151	128.1	80.0	211	178.9	111.8	271	229.8	143.6
32	27.1	17.0	92	78.0	48.8	52	128.9	80.5	12	179.8	112.3	72	230.7	144.1
33	28.0	17.5	93	78.9	49.3	53	129.8	81.1	13	180.6	112.9	73	231.5	144.7
34	28.8	18.0	94	79.7	49.8	54	130.6	81.6	14	181.5	113.4	74	232.4	145.2
35	29.7	18.5	95	80.6	50.3	55	131.4	82.1	15	182.3	113.9	75	233.2	145.7
36	30.5	19.1	96	81.4	50.9	56	132.3	82.7	16	183.2	114.5	76	234.1	146.3
37	31.4	19.6	97	82.3	51.4	57	133.1	83.2	17	184.0	115.0	77	234.9	146.8
38	32.2	20.1	98	83.1	51.9	58	134.0	83.7	18	184.9	115.5	78	235.8	147.3
39	33.1	20.7	99	84.0	52.5	59	134.8	84.3	19	185.7	116.1	79	236.6	147.8
40	33.9	21.2	100	84.8	53.0	60	135.7	84.8	20	186.6	116.6	80	237.5	148.4
41	34.8	21.7	101	85.7	53.5	161	136.5	85.3	221	187.4	117.1	281	238.3	148.9
42	35.6	22.3	02	86.5	54.1	62	137.4	85.8	22	188.3	117.6	82	239.1	149.4
43	36.5	22.8	03	87.3	54.6	63	138.2	86.4	23	189.1	118.2	83	240.0	150.0
44	37.3	23.3	04	88.2	55.1	64	139.1	86.9	24	190.0	118.7	84	240.8	150.5
45	38.2	23.8	05	89.0	55.6	65	139.9	87.4	25	190.8	119.2	85	241.7	151.0
46	39.0	24.4	06	89.9	56.2	66	140.8	88.0	26	191.7	119.8	86	242.5	151.6
47	39.9	24.9	07	90.7	56.7	67	141.6	88.5	27	192.5	120.3	87	243.4	152.1
48	40.7	25.4	08	91.6	57.2	68	142.5	89.0	28	193.4	120.8	88	244.2	152.6
49	41.6	26.0	09	92.4	57.8	69	143.3	89.6	29	194.2	121.4	89	245.1	153.1
50	42.4	26.5	10	93.3	58.3	70	144.2	90.1	30	195.1	121.9	90	245.9	153.7
51	43.3	27.0	111	94.1	58.8	171	145.0	90.6	231	195.9	122.4	291	246.8	154.2
52	44.1	27.6	12	95.0	59.4	72	145.9	91.1	32	196.7	122.9	92	247.6	154.7
53	44.9	28.1	13	95.8	59.9	73	146.7	91.7	33	197.6	123.5	93	248.5	155.3
54	45.8	28.6	14	96.7	60.4	74	147.6	92.2	34	198.4	124.0	94	249.3	155.8
55	46.6	29.1	15	97.5	60.9	75	148.4	92.7	35	199.3	124.5	95	250.2	156.3
56	47.5	29.7	16	98.4	61.5	76	149.3	93.3	36	200.1	125.1	96	251.0	156.9
57	48.3	30.2	17	99.2	62.0	77	150.1	93.8	37	201.0	125.6	97	251.9	157.4
58	49.2	30.7	18	100.1	62.5	78	151.0	94.3	38	201.8	126.1	98	252.7	157.9
59	50.0	31.3	19	100.9	63.1	79	151.8	94.9	39	202.7	126.7	99	253.6	158.4
60	50.9	31.8	20	101.8	63.6	80	152.6	95.4	40	203.5	127.2	300	254.4	159.0
Dist.	Dep.	D. Lat.	Dist.	Dep.	D. Lat.	Dist.	Dep.	D. Lat.	Dist.	Dep.	D. Lat.	Dist.	Dep.	D. Lat.

302° | 058°
238° | 122°

58°

Dist.	D. Lat.	Dep.
N.	N x Cos.	N x Sin.
Hypotenuse	Side Adj.	Side Opp.

TABLE 4

Traverse **32°** Table

328° | 032°
212° | 148°

Dist.	D. Lat.	Dep.	Dist.	D. Lat.	Dep.	Dist.	D. Lat.	Dep.	Dist.	D. Lat.	Dep.	Dist.	D. Lat.	Dep.
301	255.3	159.5	361	306.1	191.3	421	357.0	223.1	481	407.9	254.9	541	458.8	286.7
02	256.1	160.0	62	307.0	191.8	22	357.9	223.6	82	408.8	255.4	42	459.6	287.2
03	257.0	160.6	63	307.8	192.4	23	358.7	224.2	83	409.6	256.0	43	460.5	287.7
04	257.8	161.1	64	308.7	192.9	24	359.6	224.7	84	410.5	256.5	44	461.3	288.3
05	258.7	161.6	65	309.5	193.4	25	360.4	225.2	85	411.3	257.0	45	462.2	288.8
06	259.5	162.2	66	310.4	194.0	26	361.3	225.7	86	412.2	257.5	46	463.0	289.3
07	260.4	162.7	67	311.2	194.5	27	362.1	226.3	87	413.0	258.1	47	463.9	289.9
08	261.2	163.2	68	312.1	195.0	28	363.0	226.8	88	413.8	258.6	48	464.7	290.4
09	262.0	163.7	69	312.9	195.5	29	363.8	227.3	89	414.7	259.1	49	465.6	290.9
10	262.9	164.3	70	313.8	196.1	30	364.7	227.9	90	415.5	259.7	50	466.4	291.5
311	263.7	164.8	371	314.6	196.6	431	365.5	228.4	491	416.4	260.2	551	467.3	292.0
12	264.6	165.3	72	315.5	197.1	32	366.4	228.9	92	417.2	260.7	52	468.1	292.5
13	265.4	165.9	73	316.3	197.7	33	367.2	229.5	93	418.1	261.3	53	469.0	293.0
14	266.3	166.4	74	317.2	198.2	34	368.1	230.0	94	418.9	261.8	54	469.8	293.6
15	267.1	166.9	75	318.0	198.7	35	368.9	230.5	95	419.8	262.3	55	470.7	294.1
16	268.0	167.5	76	318.9	199.2	36	369.7	231.0	96	420.6	262.8	56	471.5	294.6
17	268.8	168.0	77	319.7	199.8	37	370.6	231.6	97	421.5	263.4	57	472.4	295.2
18	269.7	168.5	78	320.6	200.3	38	371.4	232.1	98	422.3	263.9	58	473.2	295.7
19	270.5	169.0	79	321.4	200.8	39	372.3	232.6	99	423.2	264.4	59	474.1	296.2
20	271.4	169.6	80	322.3	201.4	40	373.1	233.2	500	424.0	265.0	60	474.9	296.8
321	272.2	170.1	381	323.1	201.9	441	374.0	233.7	501	424.9	265.5	561	475.8	297.3
22	273.1	170.6	82	324.0	202.4	42	374.8	234.2	02	425.7	266.0	62	476.6	297.8
23	273.9	171.2	83	324.8	203.0	43	375.7	234.8	03	426.6	266.5	63	477.5	298.3
24	274.8	171.7	84	325.7	203.5	44	376.5	235.3	04	427.4	267.1	64	478.3	298.9
25	275.6	172.2	85	326.5	204.0	45	377.4	235.8	05	428.3	267.6	65	479.1	299.4
26	276.5	172.8	86	327.3	204.5	46	378.2	236.3	06	429.1	268.1	66	480.0	299.9
27	277.3	173.3	87	328.2	205.1	47	379.1	236.9	07	430.0	268.7	67	480.8	300.5
28	278.2	173.8	88	329.0	205.6	48	379.9	237.4	08	430.8	269.2	68	481.7	301.0
29	279.0	174.3	89	329.9	206.1	49	380.8	237.9	09	431.7	269.7	69	482.5	301.5
30	279.9	174.9	90	330.7	206.7	50	381.6	238.5	10	432.5	270.3	70	483.4	302.1
331	280.7	175.4	391	331.6	207.2	451	382.5	239.0	511	433.4	270.8	571	484.2	302.6
32	281.6	175.9	92	332.4	207.7	52	383.3	239.5	12	434.2	271.3	72	485.1	303.1
33	282.4	176.5	93	333.3	208.3	53	384.2	240.1	13	435.0	271.8	73	485.9	303.6
34	283.2	177.0	94	334.1	208.8	54	385.0	240.6	14	435.9	272.4	74	486.8	304.2
35	284.1	177.5	95	335.0	209.3	55	385.9	241.1	15	436.7	272.9	75	487.6	304.7
36	284.9	178.1	96	335.8	209.8	56	386.7	241.6	16	437.6	273.4	76	488.5	305.2
37	285.8	178.6	97	336.7	210.4	57	387.6	242.2	17	438.4	274.0	77	489.3	305.8
38	286.6	179.1	98	337.5	210.9	58	388.4	242.7	18	439.3	274.5	78	490.2	306.3
39	287.5	179.6	99	338.4	211.4	59	389.3	243.2	19	440.1	275.0	79	491.0	306.8
40	288.3	180.2	400	339.2	212.0	60	390.1	243.8	20	441.0	275.6	80	491.9	307.4
341	289.2	180.7	401	340.1	212.5	461	391.0	244.3	521	441.8	276.1	581	492.7	307.9
42	290.0	181.2	02	340.9	213.0	62	391.8	244.8	22	442.7	276.6	82	493.6	308.4
43	290.9	181.8	03	341.8	213.6	63	392.6	245.4	23	443.5	277.1	83	494.4	308.9
44	291.7	182.3	04	342.6	214.1	64	393.5	245.9	24	444.4	277.7	84	495.3	309.5
45	292.6	182.8	05	343.5	214.6	65	394.3	246.4	25	445.2	278.2	85	496.1	310.0
46	293.4	183.4	06	344.3	215.1	66	395.2	246.9	26	446.1	278.7	86	497.0	310.5
47	294.3	183.9	07	345.2	215.7	67	396.0	247.5	27	446.9	279.3	87	497.8	311.1
48	295.1	184.4	08	346.0	216.2	68	396.9	248.0	28	447.8	279.8	88	498.7	311.6
49	296.0	184.9	09	346.9	216.7	69	397.7	248.5	29	448.6	280.3	89	499.5	312.1
50	296.8	185.5	10	347.7	217.3	70	398.6	249.1	30	449.5	280.9	90	500.3	312.7
351	297.7	186.0	411	348.5	217.8	471	399.4	249.6	531	450.3	281.4	591	501.2	313.2
52	298.5	186.5	12	349.4	218.3	72	400.3	250.1	32	451.2	281.9	92	502.0	313.7
53	299.4	187.1	13	350.2	218.9	73	401.1	250.7	33	452.0	282.4	93	502.9	314.2
54	300.2	187.6	14	351.1	219.4	74	402.0	251.2	34	452.9	283.0	94	503.7	314.8
55	301.1	188.1	15	351.9	219.9	75	402.8	251.7	35	453.7	283.5	95	504.6	315.3
56	301.9	188.7	16	352.8	220.4	76	403.7	252.2	36	454.6	284.0	96	505.4	315.8
57	302.8	189.2	17	353.6	221.0	77	404.5	252.8	37	455.4	284.6	97	506.3	316.4
58	303.6	189.7	18	354.5	221.5	78	405.4	253.3	38	456.2	285.1	98	507.1	316.9
59	304.4	190.2	19	355.3	222.0	79	406.2	253.8	39	457.1	285.6	99	508.0	317.4
60	305.3	190.8	20	356.2	222.6	80	407.1	254.4	40	457.9	286.2	600	508.8	318.0
Dist.	Dep.	D. Lat.	Dist.	Dep.	D. Lat.	Dist.	Dep.	D. Lat.	Dist.	Dep.	D. Lat.	Dist.	Dep.	D. Lat.

Dist.	D. Lat.	Dep.
D Lo	Dep.	
	m	D Lo

58°

302° | 058°
238° | 122°

TABLE 4

Traverse 33° Table

327° | 033°
213° | 147°

Dist.	D. Lat.	Dep.	Dist.	D. Lat.	Dep.	Dist.	D. Lat.	Dep.	Dist.	D. Lat.	Dep.	Dist.	D. Lat.	Dep.
1	0.8	0.5	61	51.2	33.2	121	101.5	65.9	181	151.8	98.6	241	202.1	131.3
2	1.7	1.1	62	52.0	33.8	22	102.3	66.4	82	152.6	99.1	42	203.0	131.8
3	2.5	1.6	63	52.8	34.3	23	103.2	67.0	83	153.5	99.7	43	203.8	132.3
4	3.4	2.2	64	53.7	34.9	24	104.0	67.5	84	154.3	100.2	44	204.6	132.9
5	4.2	2.7	65	54.5	35.4	25	104.8	68.1	85	155.2	100.8	45	205.5	133.4
6	5.0	3.3	66	55.4	35.9	26	105.7	68.6	86	156.0	101.3	46	206.3	134.0
7	5.9	3.8	67	56.2	36.5	27	106.5	69.2	87	156.8	101.8	47	207.2	134.5
8	6.7	4.4	68	57.0	37.0	28	107.3	69.7	88	157.7	102.4	48	208.0	135.1
9	7.5	4.9	69	57.9	37.6	29	108.2	70.3	89	158.5	102.9	49	208.8	135.6
10	8.4	5.4	70	58.7	38.1	30	109.0	70.8	90	159.3	103.5	50	209.7	136.2
11	9.2	6.0	71	59.5	38.7	131	109.9	71.3	191	160.2	104.0	251	210.5	136.7
12	10.1	6.5	72	60.4	39.2	32	110.7	71.9	92	161.0	104.6	52	211.3	137.2
13	10.9	7.1	73	61.2	39.8	33	111.5	72.4	93	161.9	105.1	53	212.2	137.8
14	11.7	7.6	74	62.1	40.3	34	112.4	73.0	94	162.7	105.7	54	213.0	138.3
15	12.6	8.2	75	62.9	40.8	35	113.2	73.5	95	163.5	106.2	55	213.9	138.9
16	13.4	8.7	76	63.7	41.4	36	114.1	74.1	96	164.4	106.7	56	214.7	139.4
17	14.3	9.3	77	64.6	41.9	37	114.9	74.6	97	165.2	107.3	57	215.5	140.0
18	15.1	9.8	78	65.4	42.5	38	115.7	75.2	98	166.1	107.8	58	216.4	140.5
19	15.9	10.3	79	66.3	43.0	39	116.6	75.7	99	166.9	108.4	59	217.2	141.1
20	16.8	10.9	80	67.1	43.6	40	117.4	76.2	200	167.7	108.9	60	218.1	141.6
21	17.6	11.4	81	67.9	44.1	141	118.3	76.8	201	168.6	109.5	261	218.9	142.2
22	18.5	12.0	82	68.8	44.7	42	119.1	77.3	02	169.4	110.0	62	219.7	142.7
23	19.3	12.5	83	69.6	45.2	43	119.9	77.9	03	170.3	110.6	63	220.6	143.2
24	20.1	13.1	84	70.4	45.7	44	120.8	78.4	04	171.1	111.1	64	221.4	143.8
25	21.0	13.6	85	71.3	46.3	45	121.6	79.0	05	171.9	111.7	65	222.2	144.3
26	21.8	14.2	86	72.1	46.8	46	122.4	79.5	06	172.8	112.2	66	223.1	144.9
27	22.6	14.7	87	73.0	47.4	47	123.3	80.1	07	173.6	112.7	67	223.9	145.4
28	23.5	15.2	88	73.8	47.9	48	124.1	80.6	08	174.4	113.3	68	224.8	146.0
29	24.3	15.8	89	74.6	48.5	49	125.0	81.2	09	175.3	113.8	69	225.6	146.5
30	25.2	16.3	90	75.5	49.0	50	125.8	81.7	10	176.1	114.4	70	226.4	147.1
31	26.0	16.9	91	76.3	49.6	151	126.6	82.2	211	177.0	114.9	271	227.3	147.6
32	26.8	17.4	92	77.2	50.1	52	127.5	82.8	12	177.8	115.5	72	228.1	148.1
33	27.7	18.0	93	78.0	50.7	53	128.3	83.3	13	178.6	116.0	73	229.0	148.7
34	28.5	18.5	94	78.8	51.2	54	129.2	83.9	14	179.5	116.6	74	229.8	149.2
35	29.4	19.1	95	79.7	51.7	55	130.0	84.4	15	180.3	117.1	75	230.6	149.8
36	30.2	19.6	96	80.5	52.3	56	130.8	85.0	16	181.2	117.6	76	231.5	150.3
37	31.0	20.2	97	81.4	52.8	57	131.7	85.5	17	182.0	118.2	77	232.3	150.9
38	31.9	20.7	98	82.2	53.4	58	132.5	86.1	18	182.8	118.7	78	233.2	151.4
39	32.7	21.2	99	83.0	53.9	59	133.3	86.6	19	183.7	119.3	79	234.0	152.0
40	33.5	21.8	100	83.9	54.5	60	134.2	87.1	20	184.5	119.8	80	234.8	152.5
41	34.4	22.3	101	84.7	55.0	161	135.0	87.7	221	185.3	120.4	281	235.7	153.0
42	35.2	22.9	02	85.5	55.6	62	135.9	88.2	22	186.2	120.9	82	236.5	153.6
43	36.1	23.4	03	86.4	56.1	63	136.7	88.8	23	187.0	121.5	83	237.3	154.1
44	36.9	24.0	04	87.2	56.6	64	137.5	89.3	24	187.9	122.0	84	238.2	154.7
45	37.7	24.5	05	88.1	57.2	65	138.4	89.9	25	188.7	122.5	85	239.0	155.2
46	38.6	25.1	06	88.9	57.7	66	139.2	90.4	26	189.5	123.1	86	239.9	155.8
47	39.4	25.6	07	89.7	58.3	67	140.1	91.0	27	190.4	123.6	87	240.7	156.3
48	40.3	26.1	08	90.6	58.8	68	140.9	91.5	28	191.2	124.2	88	241.5	156.9
49	41.1	26.7	09	91.4	59.4	69	141.7	92.0	29	192.1	124.7	89	242.4	157.4
50	41.9	27.2	10	92.3	59.9	70	142.6	92.6	30	192.9	125.3	90	243.2	157.9
51	42.8	27.8	111	93.1	60.5	171	143.4	93.1	231	193.7	125.8	291	244.1	158.5
52	43.6	28.3	12	93.9	61.0	72	144.3	93.7	32	194.6	126.4	92	244.9	159.0
53	44.4	28.9	13	94.8	61.5	73	145.1	94.2	33	195.4	126.9	93	245.7	159.6
54	45.3	29.4	14	95.6	62.1	74	145.9	94.8	34	196.2	127.4	94	246.6	160.1
55	46.1	30.0	15	96.4	62.6	75	146.8	95.3	35	197.1	128.0	95	247.4	160.7
56	47.0	30.5	16	97.3	63.2	76	147.6	95.9	36	197.9	128.5	96	248.2	161.2
57	47.8	31.0	17	98.1	63.7	77	148.4	96.4	37	198.8	129.1	97	249.1	161.8
58	48.6	31.6	18	99.0	64.3	78	149.3	96.9	38	199.6	129.6	98	249.9	162.3
59	49.5	32.1	19	99.8	64.8	79	150.1	97.5	39	200.4	130.2	99	250.8	162.8
60	50.3	32.7	20	100.6	65.4	80	151.0	98.0	40	201.3	130.7	300	251.6	163.4
Dist.	Dep.	D. Lat.	Dist.	Dep.	D. Lat.	Dist.	Dep.	D. Lat.	Dist.	Dep.	D. Lat.	Dist.	Dep.	D. Lat.

303° | 057°
237° | 123°

57°

Dist.	D. Lat.	Dep.
N.	N x Cos.	N x Sin.
Hypotenuse	Side Adj.	Side Opp.

TABLE 4

Traverse 33° Table

327° | 033°
213° | 147°

Dist.	D. Lat.	Dep.	Dist.	D. Lat.	Dep.	Dist.	D. Lat.	Dep.	Dist.	D. Lat.	Dep.	Dist.	D. Lat.	Dep.
301	252.4	163.9	361	302.8	196.6	421	353.1	229.3	481	403.4	262.0	541	453.7	294.6
02	253.3	164.5	62	303.6	197.2	22	353.9	229.8	82	404.2	262.5	42	454.6	295.2
03	254.1	165.0	63	304.4	197.7	23	354.8	230.4	83	405.1	263.1	43	455.4	295.7
04	255.0	165.6	64	305.3	198.2	24	355.6	230.9	84	405.9	263.6	44	456.2	296.3
05	255.8	166.1	65	306.1	198.8	25	356.4	231.5	85	406.8	264.1	45	457.1	296.8
06	256.6	166.7	66	307.0	199.3	26	357.3	232.0	86	407.6	264.7	46	457.9	297.4
07	257.5	167.2	67	307.8	199.9	27	358.1	232.6	87	408.4	265.2	47	458.8	297.9
08	258.3	167.7	68	308.6	200.4	28	359.0	233.1	88	409.3	265.8	48	459.6	298.5
09	259.1	168.3	69	309.5	201.0	29	359.8	233.7	89	410.1	266.3	49	460.4	299.0
10	260.0	168.8	70	310.3	201.5	30	360.6	234.2	90	410.9	266.9	50	461.3	299.6
311	260.8	169.4	371	311.1	202.1	431	361.5	234.7	491	411.8	267.4	551	462.1	300.1
12	261.7	169.9	72	312.0	202.6	32	362.3	235.3	92	412.6	268.0	52	462.9	300.6
13	262.5	170.5	73	312.8	203.2	33	363.1	235.8	93	413.5	268.5	53	463.8	301.2
14	263.3	171.0	74	313.7	203.7	34	364.0	236.4	94	414.3	269.1	54	464.6	301.7
15	264.2	171.6	75	314.5	204.2	35	364.8	236.9	95	415.1	269.6	55	465.5	302.3
16	265.0	172.1	76	315.3	204.8	36	365.7	237.5	96	416.0	270.1	56	466.3	302.8
17	265.9	172.7	77	316.2	205.3	37	366.5	238.0	97	416.8	270.7	57	467.1	303.4
18	266.7	173.2	78	317.0	205.9	38	367.3	238.6	98	417.7	271.2	58	468.0	303.9
19	267.5	173.7	79	317.9	206.4	39	368.2	239.1	99	418.5	271.8	59	468.8	304.5
20	268.4	174.3	80	318.7	207.0	40	369.0	239.6	500	419.3	272.3	60	469.7	305.0
321	269.2	174.8	381	319.5	207.5	441	369.9	240.2	501	420.2	272.9	561	470.5	305.5
22	270.1	175.4	82	320.4	208.1	42	370.7	240.7	02	421.0	273.4	62	471.3	306.1
23	270.9	175.9	83	321.2	208.6	43	371.5	241.3	03	421.9	274.0	63	472.2	306.6
24	271.7	176.5	84	322.0	209.1	44	372.4	241.8	04	422.7	274.5	64	473.0	307.2
25	272.6	177.0	85	322.9	209.7	45	373.2	242.4	05	423.5	275.0	65	473.8	307.7
26	273.4	177.6	86	323.7	210.2	46	374.0	242.9	06	424.4	275.6	66	474.7	308.3
27	274.2	178.1	87	324.6	210.8	47	374.9	243.5	07	425.2	276.1	67	475.5	308.8
28	275.1	178.6	88	325.4	211.3	48	375.7	244.0	08	426.0	276.7	68	476.4	309.4
29	275.9	179.2	89	326.2	211.9	49	376.6	244.5	09	426.9	277.2	69	477.2	309.9
30	276.8	179.7	90	327.1	212.4	50	377.4	245.1	10	427.7	277.8	70	478.0	310.4
331	277.6	180.3	391	327.9	213.0	451	378.2	245.6	511	428.6	278.3	571	478.9	311.0
32	278.4	180.8	92	328.8	213.5	52	379.1	246.2	12	429.4	278.9	72	479.7	311.5
33	279.3	181.4	93	329.6	214.0	53	379.9	246.7	13	430.2	279.4	73	480.6	312.1
34	280.1	181.9	94	330.4	214.6	54	380.8	247.3	14	431.1	279.9	74	481.4	312.6
35	281.0	182.5	95	331.3	215.1	55	381.6	247.8	15	431.9	280.5	75	482.2	313.2
36	281.8	183.0	96	332.1	215.7	56	382.4	248.4	16	432.8	281.0	76	483.1	313.7
37	282.6	183.5	97	333.0	216.2	57	383.3	248.9	17	433.6	281.6	77	483.9	314.3
38	283.5	184.1	98	333.8	216.8	58	384.1	249.4	18	434.4	282.1	78	484.8	314.8
39	284.3	184.6	99	334.6	217.3	59	384.9	250.0	19	435.3	282.7	79	485.6	315.3
40	285.1	185.2	400	335.5	217.9	60	385.8	250.5	20	436.1	283.2	80	486.4	315.9
341	286.0	185.7	401	336.3	218.4	461	386.6	251.1	521	436.9	283.8	581	487.3	316.4
42	286.8	186.3	02	337.1	218.9	62	387.5	251.6	22	437.8	284.3	82	488.1	317.0
43	287.7	186.8	03	338.0	219.5	63	388.3	252.2	23	438.6	284.8	83	488.9	317.5
44	288.5	187.4	04	338.8	220.0	64	389.1	252.7	24	439.5	285.4	84	489.8	318.1
45	289.3	187.9	05	339.7	220.6	65	390.0	253.3	25	440.3	285.9	85	490.6	318.6
46	290.2	188.4	06	340.5	221.1	66	390.8	253.8	26	441.1	286.5	86	491.5	319.2
47	291.0	189.0	07	341.3	221.7	67	391.7	254.3	27	442.0	287.0	87	492.3	319.7
48	291.9	189.5	08	342.2	222.2	68	392.5	254.9	28	442.8	287.6	88	493.1	320.2
49	292.7	190.1	09	343.0	222.8	69	393.3	255.4	29	443.7	288.1	89	494.0	320.8
50	293.5	190.6	10	343.9	223.3	70	394.2	256.0	30	444.5	288.7	90	494.8	321.3
351	294.4	191.2	411	344.7	223.8	471	395.0	256.5	531	445.3	289.2	591	495.7	321.9
52	295.2	191.7	12	345.5	224.4	72	395.9	257.1	32	446.2	289.7	92	496.5	322.4
53	296.1	192.3	13	346.4	224.9	73	396.7	257.6	33	447.0	290.3	93	497.3	323.0
54	296.9	192.8	14	347.2	225.5	74	397.5	258.2	34	447.9	290.8	94	498.2	323.5
55	297.7	193.3	15	348.0	226.0	75	398.4	258.7	35	448.7	291.4	95	499.0	324.1
56	298.6	193.9	16	348.9	226.6	76	399.2	259.2	36	449.5	291.9	96	499.8	324.6
57	299.4	194.4	17	349.7	227.1	77	400.0	259.8	37	450.4	292.5	97	500.7	325.1
58	300.2	195.0	18	350.6	227.7	78	400.9	260.3	38	451.2	293.0	98	501.5	325.7
59	301.1	195.5	19	351.4	228.2	79	401.7	260.9	39	452.0	293.6	99	502.4	326.2
60	301.9	196.1	20	352.2	228.7	80	402.6	261.4	40	452.9	294.1	600	503.2	326.8
Dist.	Dep.	D. Lat.	Dist.	Dep.	D. Lat.	Dist.	Dep.	D. Lat.	Dist.	Dep.	D. Lat.	Dist.	Dep.	D. Lat.

Dist.	D. Lat.	Dep.
D Lo	Dep.	
	m	D Lo

57°

303° | 057°
237° | 123°

TABLE 4

34°

Traverse Table

326° | 034°
214° | 146°

Dist.	D. Lat.	Dep.	Dist.	D. Lat.	Dep.	Dist.	D. Lat.	Dep.	Dist.	D. Lat.	Dep.	Dist.	D. Lat.	Dep.
1	0.8	0.6	61	50.6	34.1	121	100.3	67.7	181	150.1	101.2	241	199.8	134.8
2	1.7	1.1	62	51.4	34.7	22	101.1	68.2	82	150.9	101.8	42	200.6	135.3
3	2.5	1.7	63	52.2	35.2	23	102.0	68.8	83	151.7	102.3	43	201.5	135.9
4	3.3	2.2	64	53.1	35.8	24	102.8	69.3	84	152.5	102.9	44	202.3	136.4
5	4.1	2.8	65	53.9	36.3	25	103.6	69.9	85	153.4	103.5	45	203.1	137.0
6	5.0	3.4	66	54.7	36.9	26	104.5	70.5	86	154.2	104.0	46	203.9	137.6
7	5.8	3.9	67	55.5	37.5	27	105.3	71.0	87	155.0	104.6	47	204.8	138.1
8	6.6	4.5	68	56.4	38.0	28	106.1	71.6	88	155.9	105.1	48	205.6	138.7
9	7.5	5.0	69	57.2	38.6	29	106.9	72.1	89	156.7	105.7	49	206.4	139.2
10	8.3	5.6	70	58.0	39.1	30	107.8	72.7	90	157.5	106.2	50	207.3	139.8
11	9.1	6.2	71	58.9	39.7	131	108.6	73.3	191	158.3	106.8	251	208.1	140.4
12	9.9	6.7	72	59.7	40.3	32	109.4	73.8	92	159.2	107.4	52	208.9	140.9
13	10.8	7.3	73	60.5	40.8	33	110.3	74.4	93	160.0	107.9	53	209.7	141.5
14	11.6	7.8	74	61.3	41.4	34	111.1	74.9	94	160.8	108.5	54	210.6	142.0
15	12.4	8.4	75	62.2	41.9	35	111.9	75.5	95	161.7	109.0	55	211.4	142.6
16	13.3	8.9	76	63.0	42.5	36	112.7	76.1	96	162.5	109.6	56	212.2	143.2
17	14.1	9.5	77	63.8	43.1	37	113.6	76.6	97	163.3	110.2	57	213.1	143.7
18	14.9	10.1	78	64.7	43.6	38	114.4	77.2	98	164.1	110.7	58	213.9	144.3
19	15.8	10.6	79	65.5	44.2	39	115.2	77.7	99	165.0	111.3	59	214.7	144.8
20	16.6	11.2	80	66.3	44.7	40	116.1	78.3	200	165.8	111.8	60	215.5	145.4
21	17.4	11.7	81	67.2	45.3	141	116.9	78.8	201	166.6	112.4	261	216.4	145.9
22	18.2	12.3	82	68.0	45.9	42	117.7	79.4	02	167.5	113.0	62	217.2	146.5
23	19.1	12.9	83	68.8	46.4	43	118.6	80.0	03	168.3	113.5	63	218.0	147.1
24	19.9	13.4	84	69.6	47.0	44	119.4	80.5	04	169.1	114.1	64	218.9	147.6
25	20.7	14.0	85	70.5	47.5	45	120.2	81.1	05	170.0	114.6	65	219.7	148.2
26	21.6	14.5	86	71.3	48.1	46	121.0	81.6	06	170.8	115.2	66	220.5	148.7
27	22.4	15.1	87	72.1	48.6	47	121.9	82.2	07	171.6	115.8	67	221.4	149.3
28	23.2	15.7	88	73.0	49.2	48	122.7	82.8	08	172.4	116.3	68	222.2	149.9
29	24.0	16.2	89	73.8	49.8	49	123.5	83.3	09	173.3	116.9	69	223.0	150.4
30	24.9	16.8	90	74.6	50.3	50	124.4	83.9	10	174.1	117.4	70	223.8	151.0
31	25.7	17.3	91	75.4	50.9	151	125.2	84.4	211	174.9	118.0	271	224.7	151.5
32	26.5	17.9	92	76.3	51.4	52	126.0	85.0	12	175.8	118.5	72	225.5	152.1
33	27.4	18.5	93	77.1	52.0	53	126.8	85.6	13	176.6	119.1	73	226.3	152.7
34	28.2	19.0	94	77.9	52.6	54	127.7	86.1	14	177.4	119.7	74	227.2	153.2
35	29.0	19.6	95	78.8	53.1	55	128.5	86.7	15	178.2	120.2	75	228.0	153.8
36	29.8	20.1	96	79.6	53.7	56	129.3	87.2	16	179.1	120.8	76	228.8	154.3
37	30.7	20.7	97	80.4	54.2	57	130.2	87.8	17	179.9	121.3	77	229.6	154.9
38	31.5	21.2	98	81.2	54.8	58	131.0	88.4	18	180.7	121.9	78	230.5	155.5
39	32.3	21.8	99	82.1	55.4	59	131.8	88.9	19	181.6	122.5	79	231.3	156.0
40	33.2	22.4	100	82.9	55.9	60	132.6	89.5	20	182.4	123.0	80	232.1	156.6
41	34.0	22.9	101	83.7	56.5	161	133.5	90.0	221	183.2	123.6	281	233.0	157.1
42	34.8	23.5	02	84.6	57.0	62	134.3	90.6	22	184.0	124.1	82	233.8	157.7
43	35.6	24.0	03	85.4	57.6	63	135.1	91.1	23	184.9	124.7	83	234.6	158.3
44	36.5	24.6	04	86.2	58.2	64	136.0	91.7	24	185.7	125.3	84	235.4	158.8
45	37.3	25.2	05	87.0	58.7	65	136.8	92.3	25	186.5	125.8	85	236.3	159.4
46	38.1	25.7	06	87.9	59.3	66	137.6	92.8	26	187.4	126.4	86	237.1	159.9
47	39.0	26.3	07	88.7	59.8	67	138.4	93.4	27	188.2	126.9	87	237.9	160.5
48	39.8	26.8	08	89.5	60.4	68	139.3	93.9	28	189.0	127.5	88	238.8	161.0
49	40.6	27.4	09	90.4	61.0	69	140.1	94.5	29	189.8	128.1	89	239.6	161.6
50	41.5	28.0	10	91.2	61.5	70	140.9	95.1	30	190.7	128.6	90	240.4	162.2
51	42.3	28.5	111	92.0	62.1	171	141.8	95.6	231	191.5	129.2	291	241.2	162.7
52	43.1	29.1	12	92.9	62.6	72	142.6	96.2	32	192.3	129.7	92	242.1	163.3
53	43.9	29.6	13	93.7	63.2	73	143.4	96.7	33	193.2	130.3	93	242.9	163.8
54	44.8	30.2	14	94.5	63.7	74	144.3	97.3	34	194.0	130.9	94	243.7	164.4
55	45.6	30.8	15	95.3	64.3	75	145.1	97.9	35	194.8	131.4	95	244.6	165.0
56	46.4	31.3	16	96.2	64.9	76	145.9	98.4	36	195.7	132.0	96	245.4	165.5
57	47.3	31.9	17	97.0	65.4	77	146.7	99.0	37	196.5	132.5	97	246.2	166.1
58	48.1	32.4	18	97.8	66.0	78	147.6	99.5	38	197.3	133.1	98	247.1	166.6
59	48.9	33.0	19	98.7	66.5	79	148.4	100.1	39	198.1	133.6	99	247.9	167.2
60	49.7	33.6	20	99.5	67.1	80	149.2	100.7	40	199.0	134.2	300	248.7	167.8
Dist.	Dep.	D. Lat.	Dist.	Dep.	D. Lat.	Dist.	Dep.	D. Lat.	Dist.	Dep.	D. Lat.	Dist.	Dep.	D. Lat.

304° | 056°
236° | 124°

56°

Dist.	D. Lat.	Dep.
N.	N x Cos.	N x Sin.
Hypotenuse	Side Adj.	Side Opp.

TABLE 4

34°

Traverse Table

326° | 034°
214° | 146°

Dist.	D. Lat.	Dep.	Dist.	D. Lat.	Dep.	Dist.	D. Lat.	Dep.	Dist.	D. Lat.	Dep.	Dist.	D. Lat.	Dep.
301	249.5	168.3	361	299.3	201.9	421	349.0	235.4	481	398.8	269.0	541	448.5	302.5
02	250.4	168.9	62	300.1	202.4	22	349.9	236.0	82	399.6	269.5	42	449.3	303.1
03	251.2	169.4	63	300.9	203.0	23	350.7	236.5	83	400.4	270.1	43	450.2	303.6
04	252.0	170.0	64	301.8	203.5	24	351.5	237.1	84	401.3	270.6	44	451.0	304.2
05	252.9	170.6	65	302.6	204.1	25	352.3	237.7	85	402.1	271.2	45	451.8	304.8
06	253.7	171.1	66	303.4	204.7	26	353.2	238.2	86	402.9	271.8	46	452.7	305.3
07	254.5	171.7	67	304.3	205.2	27	354.0	238.8	87	403.7	272.3	47	453.5	305.9
08	255.3	172.2	68	305.1	205.8	28	354.8	239.3	88	404.6	272.9	48	454.3	306.4
09	256.2	172.8	69	305.9	206.3	29	355.7	239.9	89	405.4	273.4	49	455.1	307.0
10	257.0	173.3	70	306.7	206.9	30	356.5	240.5	90	406.2	274.0	50	456.0	307.6
311	257.8	173.9	371	307.6	207.5	431	357.3	241.0	491	407.1	274.6	551	456.8	308.1
12	258.7	174.5	72	308.4	208.0	32	358.1	241.6	92	407.9	275.1	52	457.6	308.7
13	259.5	175.0	73	309.2	208.6	33	359.0	242.1	93	408.7	275.7	53	458.5	309.2
14	260.3	175.6	74	310.1	209.1	34	359.8	242.7	94	409.5	276.2	54	459.3	309.8
15	261.1	176.1	75	310.9	209.7	35	360.6	243.2	95	410.4	276.8	55	460.1	310.4
16	262.0	176.7	76	311.7	210.3	36	361.5	243.8	96	411.2	277.4	56	460.9	310.9
17	262.8	177.3	77	312.5	210.8	37	362.3	244.4	97	412.0	277.9	57	461.8	311.5
18	263.6	177.8	78	313.4	211.4	38	363.1	244.9	98	412.9	278.5	58	462.6	312.0
19	264.5	178.4	79	314.2	211.9	39	363.9	245.5	99	413.7	279.0	59	463.4	312.6
20	265.3	178.9	80	315.0	212.5	40	364.8	246.0	500	414.5	279.6	60	464.3	313.1
321	266.1	179.5	381	315.9	213.1	441	365.6	246.6	501	415.3	280.2	561	465.1	313.7
22	267.0	180.1	82	316.7	213.6	42	366.4	247.2	02	416.2	280.7	62	465.9	314.3
23	267.8	180.6	83	317.5	214.2	43	367.3	247.7	03	417.0	281.3	63	466.7	314.8
24	268.6	181.2	84	318.4	214.7	44	368.1	248.3	04	417.8	281.8	64	467.6	315.4
25	269.4	181.7	85	319.2	215.3	45	368.9	248.8	05	418.7	282.4	65	468.4	315.9
26	270.3	182.3	86	320.0	215.8	46	369.8	249.4	06	419.5	283.0	66	469.2	316.5
27	271.1	182.9	87	320.8	216.4	47	370.6	250.0	07	420.3	283.5	67	470.1	317.1
28	271.9	183.4	88	321.7	217.0	48	371.4	250.5	08	421.2	284.1	68	470.9	317.6
29	272.8	184.0	89	322.5	217.5	49	372.2	251.1	09	422.0	284.6	69	471.7	318.2
30	273.6	184.5	90	323.3	218.1	50	373.1	251.6	10	422.8	285.2	70	472.6	318.7
331	274.4	185.1	391	324.2	218.6	451	373.9	252.2	511	423.6	285.7	571	473.4	319.3
32	275.2	185.7	92	325.0	219.2	52	374.7	252.8	12	424.5	286.3	72	474.2	319.9
33	276.1	186.2	93	325.8	219.8	53	375.6	253.3	13	425.3	286.9	73	475.0	320.4
34	276.9	186.8	94	326.6	220.3	54	376.4	253.9	14	426.1	287.4	74	475.9	321.0
35	277.7	187.3	95	327.5	220.9	55	377.2	254.4	15	427.0	288.0	75	476.7	321.5
36	278.6	187.9	96	328.3	221.4	56	378.0	255.0	16	427.8	288.5	76	477.5	322.1
37	279.4	188.4	97	329.1	222.0	57	378.9	255.6	17	428.6	289.1	77	478.4	322.7
38	280.2	189.0	98	330.0	222.6	58	379.7	256.1	18	429.4	289.7	78	479.2	323.2
39	281.0	189.6	99	330.8	223.1	59	380.5	256.7	19	430.3	290.2	79	480.0	323.8
40	281.9	190.1	400	331.6	223.7	60	381.4	257.2	20	431.1	290.8	80	480.8	324.3
341	282.7	190.7	401	332.4	224.2	461	382.2	257.8	521	431.9	291.3	581	481.7	324.9
42	283.5	191.2	02	333.3	224.8	62	383.0	258.3	22	432.8	291.9	82	482.5	325.5
43	284.4	191.8	03	334.1	225.4	63	383.8	258.9	23	433.6	292.5	83	483.3	326.0
44	285.2	192.4	04	334.9	225.9	64	384.7	259.5	24	434.4	293.0	84	484.2	326.6
45	286.0	192.9	05	335.8	226.5	65	385.5	260.0	25	435.2	293.6	85	485.0	327.1
46	286.8	193.5	06	336.6	227.0	66	386.3	260.6	26	436.1	294.1	86	485.8	327.7
47	287.7	194.0	07	337.4	227.6	67	387.2	261.1	27	436.9	294.7	87	486.6	328.2
48	288.5	194.6	08	338.2	228.2	68	388.0	261.7	28	437.7	295.3	88	487.5	328.8
49	289.3	195.2	09	339.1	228.7	69	388.8	262.3	29	438.6	295.8	89	488.3	329.4
50	290.2	195.7	10	339.9	229.3	70	389.6	262.8	30	439.4	296.4	90	489.1	329.9
351	291.0	196.3	411	340.7	229.8	471	390.5	263.4	531	440.2	296.9	591	490.0	330.5
52	291.8	196.8	12	341.6	230.4	72	391.3	263.9	32	441.0	297.5	92	490.8	331.0
53	292.7	197.4	13	342.4	230.9	73	392.1	264.5	33	441.9	298.0	93	491.6	331.6
54	293.5	198.0	14	343.2	231.5	74	393.0	265.1	34	442.7	298.6	94	492.4	332.2
55	294.3	198.5	15	344.1	232.1	75	393.8	265.6	35	443.5	299.2	95	493.3	332.7
56	295.1	199.1	16	344.9	232.6	76	394.6	266.2	36	444.4	299.7	96	494.1	333.3
57	296.0	199.6	17	345.7	233.2	77	395.5	266.7	37	445.2	300.3	97	494.9	333.8
58	296.8	200.2	18	346.5	233.7	78	396.3	267.3	38	446.0	300.8	98	495.8	334.4
59	297.6	200.8	19	347.4	234.3	79	397.1	267.9	39	446.9	301.4	99	496.6	335.0
60	298.5	201.3	20	348.2	234.9	80	397.9	268.4	40	447.7	302.0	600	497.4	335.5
Dist.	Dep.	D. Lat.	Dist.	Dep.	D. Lat.	Dist.	Dep.	D. Lat.	Dist.	Dep.	D. Lat.	Dist.	Dep.	D. Lat.

Dist.	D. Lat.	Dep.
D Lo	Dep.	
	m	D Lo

56°

304° | 056°
236° | 124°

TABLE 4
Traverse 35° Table

325° | 035°
215° | 145°

Dist.	D. Lat.	Dep.	Dist.	D. Lat.	Dep.	Dist.	D. Lat.	Dep.	Dist.	D. Lat.	Dep.	Dist.	D. Lat.	Dep.
1	0.8	0.6	61	50.0	35.0	121	99.1	69.4	181	148.3	103.8	241	197.4	138.2
2	1.6	1.1	62	50.8	35.6	22	99.9	70.0	82	149.1	104.4	42	198.2	138.8
3	2.5	1.7	63	51.6	36.1	23	100.8	70.5	83	149.9	105.0	43	199.1	139.4
4	3.3	2.3	64	52.4	36.7	24	101.6	71.1	84	150.7	105.5	44	199.9	140.0
5	4.1	2.9	65	53.2	37.3	25	102.4	71.7	85	151.5	106.1	45	200.7	140.5
6	4.9	3.4	66	54.1	37.9	26	103.2	72.3	86	152.4	106.7	46	201.5	141.1
7	5.7	4.0	67	54.9	38.4	27	104.0	72.8	87	153.2	107.3	47	202.3	141.7
8	6.6	4.6	68	55.7	39.0	28	104.9	73.4	88	154.0	107.8	48	203.1	142.2
9	7.4	5.2	69	56.5	39.6	29	105.7	74.0	89	154.8	108.4	49	204.0	142.8
10	8.2	5.7	70	57.3	40.2	30	106.5	74.6	90	155.6	109.0	50	204.8	143.4
11	9.0	6.3	71	58.2	40.7	131	107.3	75.1	191	156.5	109.6	251	205.6	144.0
12	9.8	6.9	72	59.0	41.3	32	108.1	75.7	92	157.3	110.1	52	206.4	144.5
13	10.6	7.5	73	59.8	41.9	33	108.9	76.3	93	158.1	110.7	53	207.2	145.1
14	11.5	8.0	74	60.6	42.4	34	109.8	76.9	94	158.9	111.3	54	208.1	145.7
15	12.3	8.6	75	61.4	43.0	35	110.6	77.4	95	159.7	111.8	55	208.9	146.3
16	13.1	9.2	76	62.3	43.6	36	111.4	78.0	96	160.6	112.4	56	209.7	146.8
17	13.9	9.8	77	63.1	44.2	37	112.2	78.6	97	161.4	113.0	57	210.5	147.4
18	14.7	10.3	78	63.9	44.7	38	113.0	79.2	98	162.2	113.6	58	211.3	148.0
19	15.6	10.9	79	64.7	45.3	39	113.9	79.7	99	163.0	114.1	59	212.2	148.6
20	16.4	11.5	80	65.5	45.9	40	114.7	80.3	200	163.8	114.7	60	213.0	149.1
21	17.2	12.0	81	66.4	46.5	141	115.5	80.9	201	164.6	115.3	261	213.8	149.7
22	18.0	12.6	82	67.2	47.0	42	116.3	81.4	02	165.5	115.9	62	214.6	150.3
23	18.8	13.2	83	68.0	47.6	43	117.1	82.0	03	166.3	116.4	63	215.4	150.9
24	19.7	13.8	84	68.8	48.2	44	118.0	82.6	04	167.1	117.0	64	216.3	151.4
25	20.5	14.3	85	69.6	48.8	45	118.8	83.2	05	167.9	117.6	65	217.1	152.0
26	21.3	14.9	86	70.4	49.3	46	119.6	83.7	06	168.7	118.2	66	217.9	152.6
27	22.1	15.5	87	71.3	49.9	47	120.4	84.3	07	169.6	118.7	67	218.7	153.1
28	22.9	16.1	88	72.1	50.5	48	121.2	84.9	08	170.4	119.3	68	219.5	153.7
29	23.8	16.6	89	72.9	51.0	49	122.1	85.5	09	171.2	119.9	69	220.4	154.3
30	24.6	17.2	90	73.7	51.6	50	122.9	86.0	10	172.0	120.5	70	221.2	154.9
31	25.4	17.8	91	74.5	52.2	151	123.7	86.6	211	172.8	121.0	271	222.0	155.4
32	26.2	18.4	92	75.4	52.8	52	124.5	87.2	12	173.7	121.6	72	222.8	156.0
33	27.0	18.9	93	76.2	53.3	53	125.3	87.8	13	174.5	122.2	73	223.6	156.6
34	27.9	19.5	94	77.0	53.9	54	126.1	88.3	14	175.3	122.7	74	224.4	157.2
35	28.7	20.1	95	77.8	54.5	55	127.0	88.9	15	176.1	123.3	75	225.3	157.7
36	29.5	20.6	96	78.6	55.1	56	127.8	89.5	16	176.9	123.9	76	226.1	158.3
37	30.3	21.2	97	79.5	55.6	57	128.6	90.1	17	177.8	124.5	77	226.9	158.9
38	31.1	21.8	98	80.3	56.2	58	129.4	90.6	18	178.6	125.0	78	227.7	159.5
39	31.9	22.4	99	81.1	56.8	59	130.2	91.2	19	179.4	125.6	79	228.5	160.0
40	32.8	22.9	100	81.9	57.4	60	131.1	91.8	20	180.2	126.2	80	229.4	160.6
41	33.6	23.5	101	82.7	57.9	161	131.9	92.3	221	181.0	126.8	281	230.2	161.2
42	34.4	24.1	02	83.6	58.5	62	132.7	92.9	22	181.9	127.3	82	231.0	161.7
43	35.2	24.7	03	84.4	59.1	63	133.5	93.5	23	182.7	127.9	83	231.8	162.3
44	36.0	25.2	04	85.2	59.7	64	134.3	94.1	24	183.5	128.5	84	232.6	162.9
45	36.9	25.8	05	86.0	60.2	65	135.2	94.6	25	184.3	129.1	85	233.5	163.5
46	37.7	26.4	06	86.8	60.8	66	136.0	95.2	26	185.1	129.6	86	234.3	164.0
47	38.5	27.0	07	87.6	61.4	67	136.8	95.8	27	185.9	130.2	87	235.1	164.6
48	39.3	27.5	08	88.5	61.9	68	137.6	96.4	28	186.8	130.8	88	235.9	165.2
49	40.1	28.1	09	89.3	62.5	69	138.4	96.9	29	187.6	131.3	89	236.7	165.8
50	41.0	28.7	10	90.1	63.1	70	139.3	97.5	30	188.4	131.9	90	237.6	166.3
51	41.8	29.3	111	90.9	63.7	171	140.1	98.1	231	189.2	132.5	291	238.4	166.9
52	42.6	29.8	12	91.7	64.2	72	140.9	98.7	32	190.0	133.1	92	239.2	167.5
53	43.4	30.4	13	92.6	64.8	73	141.7	99.2	33	190.9	133.6	93	240.0	168.1
54	44.2	31.0	14	93.4	65.4	74	142.5	99.8	34	191.7	134.2	94	240.8	168.6
55	45.1	31.5	15	94.2	66.0	75	143.4	100.4	35	192.5	134.8	95	241.6	169.2
56	45.9	32.1	16	95.0	66.5	76	144.2	100.9	36	193.3	135.4	96	242.5	169.8
57	46.7	32.7	17	95.8	67.1	77	145.0	101.5	37	194.1	135.9	97	243.3	170.4
58	47.5	33.3	18	96.7	67.7	78	145.8	102.1	38	195.0	136.5	98	244.1	170.9
59	48.3	33.8	19	97.5	68.3	79	146.6	102.7	39	195.8	137.1	99	244.9	171.5
60	49.1	34.4	20	98.3	68.8	80	147.4	103.2	40	196.6	137.7	300	245.7	172.1
Dist.	Dep.	D. Lat.	Dist.	Dep.	D. Lat.	Dist.	Dep.	D. Lat.	Dist.	Dep.	D. Lat.	Dist.	Dep.	D. Lat.

305° | 055°
235° | 125°

55°

Dist.	D. Lat.	Dep.
N.	N x Cos.	N x Sin.
Hypotenuse	Side Adj.	Side Opp.

TABLE 4
Traverse 35° Table

325° | 035°
215° | 145°

Dist.	D. Lat.	Dep.	Dist.	D. Lat.	Dep.	Dist.	D. Lat.	Dep.	Dist.	D. Lat.	Dep.	Dist.	D. Lat.	Dep.
301	246.6	172.6	361	295.7	207.1	421	344.9	241.5	481	394.0	275.9	541	443.2	310.3
02	247.4	173.2	62	296.5	207.6	22	345.7	242.0	82	394.8	276.5	42	444.0	310.9
03	248.2	173.8	63	297.4	208.2	23	346.5	242.6	83	395.7	277.0	43	444.8	311.5
04	249.0	174.4	64	298.2	208.8	24	347.3	243.2	84	396.5	277.6	44	445.6	312.0
05	249.8	174.9	65	299.0	209.4	25	348.1	243.8	85	397.3	278.2	45	446.4	312.6
06	250.7	175.5	66	299.8	209.9	26	349.0	244.3	86	398.1	278.8	46	447.3	313.2
07	251.5	176.1	67	300.6	210.5	27	349.8	244.9	87	398.9	279.3	47	448.1	313.7
08	252.3	176.7	68	301.4	211.1	28	350.6	245.5	88	399.7	279.9	48	448.9	314.3
09	253.1	177.2	69	302.3	211.6	29	351.4	246.1	89	400.6	280.5	49	449.7	314.9
10	253.9	177.8	70	303.1	212.2	30	352.2	246.6	90	401.4	281.1	50	450.5	315.5
311	254.8	178.4	371	303.9	212.8	431	353.1	247.2	491	402.2	281.6	551	451.4	316.0
12	255.6	179.0	72	304.7	213.4	32	353.9	247.8	92	403.0	282.2	52	452.2	316.6
13	256.4	179.5	73	305.5	213.9	33	354.7	248.4	93	403.8	282.8	53	453.0	317.2
14	257.2	180.1	74	306.4	214.5	34	355.5	248.9	94	404.7	283.3	54	453.8	317.8
15	258.0	180.7	75	307.2	215.1	35	356.3	249.5	95	405.5	283.9	55	454.6	318.3
16	258.9	181.3	76	308.0	215.7	36	357.2	250.1	96	406.3	284.5	56	455.4	318.9
17	259.7	181.8	77	308.8	216.2	37	358.0	250.7	97	407.1	285.1	57	456.3	319.5
18	260.5	182.4	78	309.6	216.8	38	358.8	251.2	98	407.9	285.6	58	457.1	320.1
19	261.3	183.0	79	310.5	217.4	39	359.6	251.8	99	408.8	286.2	59	457.9	320.6
20	262.1	183.5	80	311.3	218.0	40	360.4	252.4	500	409.6	286.8	60	458.7	321.2
321	262.9	184.1	381	312.1	218.5	441	361.2	252.9	501	410.4	287.4	561	459.5	321.8
22	263.8	184.7	82	312.9	219.1	42	362.1	253.5	02	411.2	287.9	62	460.4	322.3
23	264.6	185.3	83	313.7	219.7	43	362.9	254.1	03	412.0	288.5	63	461.2	322.9
24	265.4	185.8	84	314.6	220.3	44	363.7	254.7	04	412.9	289.1	64	462.0	323.5
25	266.2	186.4	85	315.4	220.8	45	364.5	255.2	05	413.7	289.7	65	462.8	324.1
26	267.0	187.0	86	316.2	221.4	46	365.3	255.8	06	414.5	290.2	66	463.6	324.6
27	267.9	187.6	87	317.0	222.0	47	366.2	256.4	07	415.3	290.8	67	464.5	325.2
28	268.7	188.1	88	317.8	222.5	48	367.0	257.0	08	416.1	291.4	68	465.3	325.8
29	269.5	188.7	89	318.7	223.1	49	367.8	257.5	09	416.9	292.0	69	466.1	326.4
30	270.3	189.3	90	319.5	223.7	50	368.6	258.1	10	417.8	292.5	70	466.9	326.9
331	271.1	189.9	391	320.3	224.3	451	369.4	258.7	511	418.6	293.1	571	467.7	327.5
32	272.0	190.4	92	321.1	224.8	52	370.3	259.3	12	419.4	293.7	72	468.6	328.1
33	272.8	191.0	93	321.9	225.4	53	371.1	259.8	13	420.2	294.2	73	469.4	328.7
34	273.6	191.6	94	322.7	226.0	54	371.9	260.4	14	421.0	294.8	74	470.2	329.2
35	274.4	192.1	95	323.6	226.6	55	372.7	261.0	15	421.9	295.4	75	471.0	329.8
36	275.2	192.7	96	324.4	227.1	56	373.5	261.6	16	422.7	296.0	76	471.8	330.4
37	276.1	193.3	97	325.2	227.7	57	374.4	262.1	17	423.5	296.5	77	472.7	331.0
38	276.9	193.9	98	326.0	228.3	58	375.2	262.7	18	424.3	297.1	78	473.5	331.5
39	277.7	194.4	99	326.8	228.9	59	376.0	263.3	19	425.1	297.7	79	474.3	332.1
40	278.5	195.0	400	327.7	229.4	60	376.8	263.8	20	426.0	298.3	80	475.1	332.7
341	279.3	195.6	401	328.5	230.0	461	377.6	264.4	521	426.8	298.8	581	475.9	333.2
42	280.1	196.2	02	329.3	230.6	62	378.4	265.0	22	427.6	299.4	82	476.7	333.8
43	281.0	196.7	03	330.1	231.2	63	379.3	265.6	23	428.4	300.0	83	477.6	334.4
44	281.8	197.3	04	330.9	231.7	64	380.1	266.1	24	429.2	300.6	84	478.4	335.0
45	282.6	197.9	05	331.8	232.3	65	380.9	266.7	25	430.1	301.1	85	479.2	335.5
46	283.4	198.5	06	332.6	232.9	66	381.7	267.3	26	430.9	301.7	86	480.0	336.1
47	284.2	199.0	07	333.4	233.4	67	382.5	267.9	27	431.7	302.3	87	480.8	336.7
48	285.1	199.6	08	334.2	234.0	68	383.4	268.4	28	432.5	302.8	88	481.7	337.3
49	285.9	200.2	09	335.0	234.6	69	384.2	269.0	29	433.3	303.4	89	482.5	337.8
50	286.7	200.8	10	335.9	235.2	70	385.0	269.6	30	434.2	304.0	90	483.3	338.4
351	287.5	201.3	411	336.7	235.7	471	385.8	270.2	531	435.0	304.6	591	484.1	339.0
52	288.3	201.9	12	337.5	236.3	72	386.6	270.7	32	435.8	305.1	92	484.9	339.6
53	289.2	202.5	13	338.3	236.9	73	387.5	271.3	33	436.6	305.7	93	485.8	340.1
54	290.0	203.0	14	339.1	237.5	74	388.3	271.9	34	437.4	306.3	94	486.6	340.7
55	290.8	203.6	15	339.9	238.0	75	389.1	272.4	35	438.2	306.9	95	487.4	341.3
56	291.6	204.2	16	340.8	238.6	76	389.9	273.0	36	439.1	307.4	96	488.2	341.9
57	292.4	204.8	17	341.6	239.2	77	390.7	273.6	37	439.9	308.0	97	489.0	342.4
58	293.3	205.3	18	342.4	239.8	78	391.6	274.2	38	440.7	308.6	98	489.9	343.0
59	294.1	205.9	19	343.2	240.3	79	392.4	274.7	39	441.5	309.2	99	490.7	343.6
60	294.9	206.5	20	344.0	240.9	80	393.2	275.3	40	442.3	309.7	600	491.5	344.1
Dist.	Dep.	D. Lat.	Dist.	Dep.	D. Lat.	Dist.	Dep.	D. Lat.	Dist.	Dep.	D. Lat.	Dist.	Dep.	D. Lat.

Dist.	D. Lat.	Dep.
D Lo	Dep.	
	m	D Lo

55°

305° | 055°
235° | 125°

TABLE 4
Traverse 36° Table

324° 216° | 036° 144°

Dist.	D. Lat.	Dep.	Dist.	D. Lat.	Dep.	Dist.	D. Lat.	Dep.	Dist.	D. Lat.	Dep.	Dist.	D. Lat.	Dep.
1	0.8	0.6	61	49.4	35.9	121	97.9	71.1	181	146.4	106.4	241	195.0	141.7
2	1.6	1.2	62	50.2	36.4	22	98.7	71.7	82	147.2	107.0	42	195.8	142.2
3	2.4	1.8	63	51.0	37.0	23	99.5	72.3	83	148.1	107.6	43	196.6	142.8
4	3.2	2.4	64	51.8	37.6	24	100.3	72.9	84	148.9	108.2	44	197.4	143.4
5	4.0	2.9	65	52.6	38.2	25	101.1	73.5	85	149.7	108.7	45	198.2	144.0
6	4.9	3.5	66	53.4	38.8	26	101.9	74.1	86	150.5	109.3	46	199.0	144.6
7	5.7	4.1	67	54.2	39.4	27	102.7	74.6	87	151.3	109.9	47	199.8	145.2
8	6.5	4.7	68	55.0	40.0	28	103.6	75.2	88	152.1	110.5	48	200.6	145.8
9	7.3	5.3	69	55.8	40.6	29	104.4	75.8	89	152.9	111.1	49	201.4	146.4
10	8.1	5.9	70	56.6	41.1	30	105.2	76.4	90	153.7	111.7	50	202.3	146.9
11	8.9	6.5	71	57.4	41.7	131	106.0	77.0	191	154.5	112.3	251	203.1	147.5
12	9.7	7.1	72	58.2	42.3	32	106.8	77.6	92	155.3	112.9	52	203.9	148.1
13	10.5	7.6	73	59.1	42.9	33	107.6	78.2	93	156.1	113.4	53	204.7	148.7
14	11.3	8.2	74	59.9	43.5	34	108.4	78.8	94	156.9	114.0	54	205.5	149.3
15	12.1	8.8	75	60.7	44.1	35	109.2	79.4	95	157.8	114.6	55	206.3	149.9
16	12.9	9.4	76	61.5	44.7	36	110.0	79.9	96	158.6	115.2	56	207.1	150.5
17	13.8	10.0	77	62.3	45.3	37	110.8	80.5	97	159.4	115.8	57	207.9	151.1
18	14.6	10.6	78	63.1	45.8	38	111.6	81.1	98	160.2	116.4	58	208.7	151.6
19	15.4	11.2	79	63.9	46.4	39	112.5	81.7	99	161.0	117.0	59	209.5	152.2
20	16.2	11.8	80	64.7	47.0	40	113.3	82.3	200	161.8	117.6	60	210.3	152.8
21	17.0	12.3	81	65.5	47.6	141	114.1	82.9	201	162.6	118.1	261	211.2	153.4
22	17.8	12.9	82	66.3	48.2	42	114.9	83.5	02	163.4	118.7	62	212.0	154.0
23	18.6	13.5	83	67.1	48.8	43	115.7	84.1	03	164.2	119.3	63	212.8	154.6
24	19.4	14.1	84	68.0	49.4	44	116.5	84.6	04	165.0	119.9	64	213.6	155.2
25	20.2	14.7	85	68.8	50.0	45	117.3	85.2	05	165.8	120.5	65	214.4	155.8
26	21.0	15.3	86	69.6	50.5	46	118.1	85.8	06	166.7	121.1	66	215.2	156.4
27	21.8	15.9	87	70.4	51.1	47	118.9	86.4	07	167.5	121.7	67	216.0	156.9
28	22.7	16.5	88	71.2	51.7	48	119.7	87.0	08	168.3	122.3	68	216.8	157.5
29	23.5	17.0	89	72.0	52.3	49	120.5	87.6	09	169.1	122.8	69	217.6	158.1
30	24.3	17.6	90	72.8	52.9	50	121.4	88.2	10	169.9	123.4	70	218.4	158.7
31	25.1	18.2	91	73.6	53.5	151	122.2	88.8	211	170.7	124.0	271	219.2	159.3
32	25.9	18.8	92	74.4	54.1	52	123.0	89.3	12	171.5	124.6	72	220.1	159.9
33	26.7	19.4	93	75.2	54.7	53	123.8	89.9	13	172.3	125.2	73	220.9	160.5
34	27.5	20.0	94	76.0	55.3	54	124.6	90.5	14	173.1	125.8	74	221.7	161.1
35	28.3	20.6	95	76.9	55.8	55	125.4	91.1	15	173.9	126.4	75	222.5	161.6
36	29.1	21.2	96	77.7	56.4	56	126.2	91.7	16	174.7	127.0	76	223.3	162.2
37	29.9	21.7	97	78.5	57.0	57	127.0	92.3	17	175.6	127.5	77	224.1	162.8
38	30.7	22.3	98	79.3	57.6	58	127.8	92.9	18	176.4	128.1	78	224.9	163.4
39	31.6	22.9	99	80.1	58.2	59	128.6	93.5	19	177.2	128.7	79	225.7	164.0
40	32.4	23.5	100	80.9	58.8	60	129.4	94.0	20	178.0	129.3	80	226.5	164.6
41	33.2	24.1	101	81.7	59.4	161	130.3	94.6	221	178.8	129.9	281	227.3	165.2
42	34.0	24.7	02	82.5	60.0	62	131.1	95.2	22	179.6	130.5	82	228.1	165.8
43	34.8	25.3	03	83.3	60.5	63	131.9	95.8	23	180.4	131.1	83	229.0	166.3
44	35.6	25.9	04	84.1	61.1	64	132.7	96.4	24	181.2	131.7	84	229.8	166.9
45	36.4	26.5	05	84.9	61.7	65	133.5	97.0	25	182.0	132.3	85	230.6	167.5
46	37.2	27.0	06	85.8	62.3	66	134.3	97.6	26	182.8	132.8	86	231.4	168.1
47	38.0	27.6	07	86.6	62.9	67	135.1	98.2	27	183.6	133.4	87	232.2	168.7
48	38.8	28.2	08	87.4	63.5	68	135.9	98.7	28	184.5	134.0	88	233.0	169.3
49	39.6	28.8	09	88.2	64.1	69	136.7	99.3	29	185.3	134.6	89	233.8	169.9
50	40.5	29.4	10	89.0	64.7	70	137.5	99.9	30	186.1	135.2	90	234.6	170.5
51	41.3	30.0	111	89.8	65.2	171	138.3	100.5	231	186.9	135.8	291	235.4	171.0
52	42.1	30.6	12	90.6	65.8	72	139.2	101.1	32	187.7	136.4	92	236.2	171.6
53	42.9	31.2	13	91.4	66.4	73	140.0	101.7	33	188.5	137.0	93	237.0	172.2
54	43.7	31.7	14	92.2	67.0	74	140.8	102.3	34	189.3	137.5	94	237.9	172.8
55	44.5	32.3	15	93.0	67.6	75	141.6	102.9	35	190.1	138.1	95	238.7	173.4
56	45.3	32.9	16	93.8	68.2	76	142.4	103.5	36	190.9	138.7	96	239.5	174.0
57	46.1	33.5	17	94.7	68.8	77	143.2	104.0	37	191.7	139.3	97	240.3	174.6
58	46.9	34.1	18	95.5	69.4	78	144.0	104.6	38	192.5	139.9	98	241.1	175.2
59	47.7	34.7	19	96.3	69.9	79	144.8	105.2	39	193.4	140.5	99	241.9	175.7
60	48.5	35.3	20	97.1	70.5	80	145.6	105.8	40	194.2	141.1	300	242.7	176.3
Dist.	Dep.	D. Lat.	Dist.	Dep.	D. Lat.	Dist.	Dep.	D. Lat.	Dist.	Dep.	D. Lat.	Dist.	Dep.	D. Lat.

306° 234° | 054° 126°

54°

Dist.	D. Lat.	Dep.
N.	N x Cos.	N x Sin.
Hypotenuse	Side Adj.	Side Opp.

TABLE 4
Traverse 36° Table

324° 216° | 036° 144°

Dist.	D. Lat.	Dep.	Dist.	D. Lat.	Dep.	Dist.	D. Lat.	Dep.	Dist.	D. Lat.	Dep.	Dist.	D. Lat.	Dep.
301	243.5	176.9	361	292.1	212.2	421	340.6	247.5	481	389.1	282.7	541	437.7	318.0
02	244.3	177.5	62	292.9	212.8	22	341.4	248.0	82	389.9	283.3	42	438.5	318.6
03	245.1	178.1	63	293.7	213.4	23	342.2	248.6	83	390.8	283.9	43	439.3	319.2
04	245.9	178.7	64	294.5	214.0	24	343.0	249.2	84	391.6	284.5	44	440.1	319.8
05	246.8	179.3	65	295.3	214.5	25	343.8	249.8	85	392.4	285.1	45	440.9	320.3
06	247.6	179.9	66	296.1	215.1	26	344.6	250.4	86	393.2	285.7	46	441.7	320.9
07	248.4	180.5	67	296.9	215.7	27	345.5	251.0	87	394.0	286.3	47	442.5	321.5
08	249.2	181.0	68	297.7	216.3	28	346.3	251.6	88	394.8	286.8	48	443.3	322.1
09	250.0	181.6	69	298.5	216.9	29	347.1	252.2	89	395.6	287.4	49	444.2	322.7
10	250.8	182.2	70	299.3	217.5	30	347.9	252.7	90	396.4	288.0	50	445.0	323.3
311	251.6	182.8	371	300.1	218.1	431	348.7	253.3	491	397.2	288.6	551	445.8	323.9
12	252.4	183.4	72	301.0	218.7	32	349.5	253.9	92	398.0	289.2	52	446.6	324.5
13	253.2	184.0	73	301.8	219.2	33	350.3	254.5	93	398.8	289.8	53	447.4	325.0
14	254.0	184.6	74	302.6	219.8	34	351.1	255.1	94	399.7	290.4	54	448.2	325.6
15	254.8	185.2	75	303.4	220.4	35	351.9	255.7	95	400.5	291.0	55	449.0	326.2
16	255.6	185.7	76	304.2	221.0	36	352.7	256.3	96	401.3	291.5	56	449.8	326.8
17	256.5	186.3	77	305.0	221.6	37	353.5	256.9	97	402.1	292.1	57	450.6	327.4
18	257.3	186.9	78	305.8	222.2	38	354.3	257.4	98	402.9	292.7	58	451.4	328.0
19	258.1	187.5	79	306.6	222.8	39	355.2	258.0	99	403.7	293.3	59	452.2	328.6
20	258.9	188.1	80	307.4	223.4	40	356.0	258.6	500	404.5	293.9	60	453.0	329.2
321	259.7	188.7	381	308.2	223.9	441	356.8	259.2	501	405.3	294.5	561	453.9	329.7
22	260.5	189.3	82	309.0	224.5	42	357.6	259.8	02	406.1	295.1	62	454.7	330.3
23	261.3	189.9	83	309.9	225.1	43	358.4	260.4	03	406.9	295.7	63	455.5	330.9
24	262.1	190.4	84	310.7	225.7	44	359.2	261.0	04	407.7	296.2	64	456.3	331.5
25	262.9	191.0	85	311.5	226.3	45	360.0	261.6	05	408.6	296.8	65	457.1	332.1
26	263.7	191.6	86	312.3	226.9	46	360.8	262.2	06	409.4	297.4	66	457.9	332.7
27	264.5	192.2	87	313.1	227.5	47	361.6	262.7	07	410.2	298.0	67	458.7	333.3
28	265.4	192.8	88	313.9	228.1	48	362.4	263.3	08	411.0	298.6	68	459.5	333.9
29	266.2	193.4	89	314.7	228.6	49	363.2	263.9	09	411.8	299.2	69	460.3	334.4
30	267.0	194.0	90	315.5	229.2	50	364.1	264.5	10	412.6	299.8	70	461.1	335.0
331	267.8	194.6	391	316.3	229.8	451	364.9	265.1	511	413.4	300.4	571	461.9	335.6
32	268.6	195.1	92	317.1	230.4	52	365.7	265.7	12	414.2	300.9	72	462.8	336.2
33	269.4	195.7	93	317.9	231.0	53	366.5	266.3	13	415.0	301.5	73	463.6	336.8
34	270.2	196.3	94	318.8	231.6	54	367.3	266.9	14	415.8	302.1	74	464.4	337.4
35	271.0	196.9	95	319.6	232.2	55	368.1	267.4	15	416.6	302.7	75	465.2	338.0
36	271.8	197.5	96	320.4	232.8	56	368.9	268.0	16	417.5	303.3	76	466.0	338.6
37	272.6	198.1	97	321.2	233.4	57	369.7	268.6	17	418.3	303.9	77	466.8	339.2
38	273.4	198.7	98	322.0	233.9	58	370.5	269.2	18	419.1	304.5	78	467.6	339.7
39	274.3	199.3	99	322.8	234.5	59	371.3	269.8	19	419.9	305.1	79	468.4	340.3
40	275.1	199.8	400	323.6	235.1	60	372.1	270.4	20	420.7	305.6	80	469.2	340.9
341	275.9	200.4	401	324.4	235.7	461	373.0	271.0	521	421.5	306.2	581	470.0	341.5
42	276.7	201.0	02	325.2	236.3	62	373.8	271.6	22	422.3	306.8	82	470.8	342.1
43	277.5	201.6	03	326.0	236.9	63	374.6	272.1	23	423.1	307.4	83	471.7	342.7
44	278.3	202.2	04	326.8	237.5	64	375.4	272.7	24	423.9	308.0	84	472.5	343.3
45	279.1	202.8	05	327.7	238.1	65	376.2	273.3	25	424.7	308.6	85	473.3	343.9
46	279.9	203.4	06	328.5	238.6	66	377.0	273.9	26	425.5	309.2	86	474.1	344.4
47	280.7	204.0	07	329.3	239.2	67	377.8	274.5	27	426.4	309.8	87	474.9	345.0
48	281.5	204.5	08	330.1	239.8	68	378.6	275.1	28	427.2	310.4	88	475.7	345.6
49	282.3	205.1	09	330.9	240.4	69	379.4	275.7	29	428.0	310.9	89	476.5	346.2
50	283.2	205.7	10	331.7	241.0	70	380.2	276.3	30	428.8	311.5	90	477.3	346.8
351	284.0	206.3	411	332.5	241.6	471	381.0	276.8	531	429.6	312.1	591	478.1	347.4
52	284.8	206.9	12	333.3	242.2	72	381.9	277.4	32	430.4	312.7	92	478.9	348.0
53	285.6	207.5	13	334.1	242.8	73	382.7	278.0	33	431.2	313.3	93	479.7	348.6
54	286.4	208.1	14	334.9	243.3	74	383.5	278.6	34	432.0	313.9	94	480.6	349.1
55	287.2	208.7	15	335.7	243.9	75	384.3	279.2	35	432.8	314.5	95	481.4	349.7
56	288.0	209.3	16	336.6	244.5	76	385.1	279.8	36	433.6	315.1	96	482.2	350.3
57	288.8	209.8	17	337.4	245.1	77	385.9	280.4	37	434.4	315.6	97	483.0	350.9
58	289.6	210.4	18	338.2	245.7	78	386.7	281.0	38	435.3	316.2	98	483.8	351.5
59	290.4	211.0	19	339.0	246.3	79	387.5	281.5	39	436.1	316.8	99	484.6	352.1
60	291.2	211.6	20	339.8	246.9	80	388.3	282.1	40	436.9	317.4	600	485.4	352.7
Dist.	Dep.	D. Lat.	Dist.	Dep.	D. Lat.	Dist.	Dep.	D. Lat.	Dist.	Dep.	D. Lat.	Dist.	Dep.	D. Lat.

Dist.	D. Lat.	Dep.
D Lo	Dep.	
	m	D Lo

54°

306° 234° | 054° 126°

323° | 037° / 217° | 143°

TABLE 4
Traverse 37° Table

Dist.	D. Lat.	Dep.	Dist.	D. Lat.	Dep.	Dist.	D. Lat.	Dep.	Dist.	D. Lat.	Dep.	Dist.	D. Lat.	Dep.
1	0.8	0.6	61	48.7	36.7	121	96.6	72.8	181	144.6	108.9	241	192.5	145.0
2	1.6	1.2	62	49.5	37.3	22	97.4	73.4	82	145.4	109.5	42	193.3	145.6
3	2.4	1.8	63	50.3	37.9	23	98.2	74.0	83	146.2	110.1	43	194.1	146.2
4	3.2	2.4	64	51.1	38.5	24	99.0	74.6	84	146.9	110.7	44	194.9	146.8
5	4.0	3.0	65	51.9	39.1	25	99.8	75.2	85	147.7	111.3	45	195.7	147.4
6	4.8	3.6	66	52.7	39.7	26	100.6	75.8	86	148.5	111.9	46	196.5	148.0
7	5.6	4.2	67	53.5	40.3	27	101.4	76.4	87	149.3	112.5	47	197.3	148.6
8	6.4	4.8	68	54.3	40.9	28	102.2	77.0	88	150.1	113.1	48	198.1	149.3
9	7.2	5.4	69	55.1	41.5	29	103.0	77.6	89	150.9	113.7	49	198.9	149.9
10	8.0	6.0	70	55.9	42.1	30	103.8	78.2	90	151.7	114.3	50	199.7	150.5
11	8.8	6.6	71	56.7	42.7	131	104.6	78.8	191	152.5	114.9	251	200.5	151.1
12	9.6	7.2	72	57.5	43.3	32	105.4	79.4	92	153.3	115.5	52	201.3	151.7
13	10.4	7.8	73	58.3	43.9	33	106.2	80.0	93	154.1	116.2	53	202.1	152.3
14	11.2	8.4	74	59.1	44.5	34	107.0	80.6	94	154.9	116.8	54	202.9	152.9
15	12.0	9.0	75	59.9	45.1	35	107.8	81.2	95	155.7	117.4	55	203.7	153.5
16	12.8	9.6	76	60.7	45.7	36	108.6	81.8	96	156.5	118.0	56	204.5	154.1
17	13.6	10.2	77	61.5	46.3	37	109.4	82.4	97	157.3	118.6	57	205.2	154.7
18	14.4	10.8	78	62.3	46.9	38	110.2	83.1	98	158.1	119.2	58	206.0	155.3
19	15.2	11.4	79	63.1	47.5	39	111.0	83.7	99	158.9	119.8	59	206.8	155.9
20	16.0	12.0	80	63.9	48.1	40	111.8	84.3	200	159.7	120.4	60	207.6	156.5
21	16.8	12.6	81	64.7	48.7	141	112.6	84.9	201	160.5	121.0	261	208.4	157.1
22	17.6	13.2	82	65.5	49.3	42	113.4	85.5	02	161.3	121.6	62	209.2	157.7
23	18.4	13.8	83	66.3	50.0	43	114.2	86.1	03	162.1	122.2	63	210.0	158.3
24	19.2	14.4	84	67.1	50.6	44	115.0	86.7	04	162.9	122.8	64	210.8	158.9
25	20.0	15.0	85	67.9	51.2	45	115.8	87.3	05	163.7	123.4	65	211.6	159.5
26	20.8	15.6	86	68.7	51.8	46	116.6	87.9	06	164.5	124.0	66	212.4	160.1
27	21.6	16.2	87	69.5	52.4	47	117.4	88.5	07	165.3	124.6	67	213.2	160.7
28	22.4	16.9	88	70.3	53.0	48	118.2	89.1	08	166.1	125.2	68	214.0	161.3
29	23.2	17.5	89	71.1	53.6	49	119.0	89.7	09	166.9	125.8	69	214.8	161.9
30	24.0	18.1	90	71.9	54.2	50	119.8	90.3	10	167.7	126.4	70	215.6	162.5
31	24.8	18.7	91	72.7	54.8	151	120.6	90.9	211	168.5	127.0	271	216.4	163.1
32	25.6	19.3	92	73.5	55.4	52	121.4	91.5	12	169.3	127.6	72	217.2	163.7
33	26.4	19.9	93	74.3	56.0	53	122.2	92.1	13	170.1	128.2	73	218.0	164.3
34	27.2	20.5	94	75.1	56.6	54	123.0	92.7	14	170.9	128.8	74	218.8	164.9
35	28.0	21.1	95	75.9	57.2	55	123.8	93.3	15	171.7	129.4	75	219.6	165.5
36	28.8	21.7	96	76.7	57.8	56	124.6	93.9	16	172.5	130.0	76	220.4	166.1
37	29.5	22.3	97	77.5	58.4	57	125.4	94.5	17	173.3	130.6	77	221.2	166.7
38	30.3	22.9	98	78.3	59.0	58	126.2	95.1	18	174.1	131.2	78	222.0	167.3
39	31.1	23.5	99	79.1	59.6	59	127.0	95.7	19	174.9	131.8	79	222.8	167.9
40	31.9	24.1	100	79.9	60.2	60	127.8	96.3	20	175.7	132.4	80	223.6	168.5
41	32.7	24.7	101	80.7	60.8	161	128.6	96.9	221	176.5	133.0	281	224.4	169.1
42	33.5	25.3	02	81.5	61.4	62	129.4	97.5	22	177.3	133.6	82	225.2	169.7
43	34.3	25.9	03	82.3	62.0	63	130.2	98.1	23	178.1	134.2	83	226.0	170.3
44	35.1	26.5	04	83.1	62.6	64	131.0	98.7	24	178.9	134.8	84	226.8	170.9
45	35.9	27.1	05	83.9	63.2	65	131.8	99.3	25	179.7	135.4	85	227.6	171.5
46	36.7	27.7	06	84.7	63.8	66	132.6	99.9	26	180.5	136.0	86	228.4	172.1
47	37.5	28.3	07	85.5	64.4	67	133.4	100.5	27	181.3	136.6	87	229.2	172.7
48	38.3	28.9	08	86.3	65.0	68	134.2	101.1	28	182.1	137.2	88	230.0	173.3
49	39.1	29.5	09	87.1	65.6	69	135.0	101.7	29	182.9	137.8	89	230.8	173.9
50	39.9	30.1	10	87.8	66.2	70	135.8	102.3	30	183.7	138.4	90	231.6	174.5
51	40.7	30.7	111	88.6	66.8	171	136.6	102.9	231	184.5	139.0	291	232.4	175.1
52	41.5	31.3	12	89.4	67.4	72	137.4	103.5	32	185.3	139.6	92	233.2	175.7
53	42.3	31.9	13	90.2	68.0	73	138.2	104.1	33	186.1	140.2	93	234.0	176.3
54	43.1	32.5	14	91.0	68.6	74	139.0	104.7	34	186.9	140.8	94	234.8	176.9
55	43.9	33.1	15	91.8	69.2	75	139.8	105.3	35	187.7	141.4	95	235.6	177.5
56	44.7	33.7	16	92.6	69.8	76	140.6	105.9	36	188.5	142.0	96	236.4	178.1
57	45.5	34.3	17	93.4	70.4	77	141.4	106.5	37	189.3	142.6	97	237.2	178.7
58	46.3	34.9	18	94.2	71.0	78	142.2	107.1	38	190.1	143.2	98	238.0	179.3
59	47.1	35.5	19	95.0	71.6	79	143.0	107.7	39	190.9	143.8	99	238.8	179.9
60	47.9	36.1	20	95.8	72.2	80	143.8	108.3	40	191.7	144.4	300	239.6	180.5
Dist.	Dep.	D. Lat.	Dist.	Dep.	D. Lat.	Dist.	Dep.	D. Lat.	Dist.	Dep.	D. Lat.	Dist.	Dep.	D. Lat.

307° | 053° / 233° | 127°

53°

Dist.	D. Lat.	Dep.
N.	N x Cos.	N x Sin.
Hypotenuse	Side Adj.	Side Opp.

323° | 037° / 217° | 143°

TABLE 4

Traverse 37° Table

Dist.	D. Lat.	Dep.	Dist.	D. Lat.	Dep.	Dist.	D. Lat.	Dep.	Dist.	D. Lat.	Dep.	Dist.	D. Lat.	Dep.
301	240.4	181.1	361	288.3	217.3	421	336.2	253.4	481	384.1	289.5	541	432.1	325.6
02	241.2	181.7	62	289.1	217.9	22	337.0	254.0	82	384.9	290.1	42	432.9	326.2
03	242.0	182.3	63	289.9	218.5	23	337.8	254.6	83	385.7	290.7	43	433.7	326.8
04	242.8	183.0	64	290.7	219.1	24	338.6	255.2	84	386.5	291.3	44	434.5	327.4
05	243.6	183.6	65	291.5	219.7	25	339.4	255.8	85	387.3	291.9	45	435.3	328.0
06	244.4	184.2	66	292.3	220.3	26	340.2	256.4	86	388.1	292.5	46	436.1	328.6
07	245.2	184.8	67	293.1	220.9	27	341.0	257.0	87	388.9	293.1	47	436.9	329.2
08	246.0	185.4	68	293.9	221.5	28	341.8	257.6	88	389.7	293.7	48	437.7	329.8
09	246.8	186.0	69	294.7	222.1	29	342.6	258.2	89	390.5	294.3	49	438.5	330.4
10	247.6	186.6	70	295.5	222.7	30	343.4	258.8	90	391.3	294.9	50	439.2	331.0
311	248.4	187.2	371	296.3	223.3	431	344.2	259.4	491	392.1	295.5	551	440.0	331.6
12	249.2	187.8	72	297.1	223.9	32	345.0	260.0	92	392.9	296.1	52	440.8	332.2
13	250.0	188.4	73	297.9	224.5	33	345.8	260.6	93	393.7	296.7	53	441.6	332.8
14	250.8	189.0	74	298.7	225.1	34	346.6	261.2	94	394.5	297.3	54	442.4	333.4
15	251.6	189.6	75	299.5	225.7	35	347.4	261.8	95	395.3	297.9	55	443.2	334.0
16	252.4	190.2	76	300.3	226.3	36	348.2	262.4	96	396.1	298.5	56	444.0	334.6
17	253.2	190.8	77	301.1	226.9	37	349.0	263.0	97	396.9	299.1	57	444.8	335.2
18	254.0	191.4	78	301.9	227.5	38	349.8	263.6	98	397.7	299.7	58	445.6	335.8
19	254.8	192.0	79	302.7	228.1	39	350.6	264.2	99	398.5	300.3	59	446.4	336.4
20	255.6	192.6	80	303.5	228.7	40	351.4	264.8	500	399.3	300.9	60	447.2	337.0
321	256.4	193.2	381	304.3	229.3	441	352.2	265.4	501	400.1	301.5	561	448.0	337.6
22	257.2	193.8	82	305.1	229.9	42	353.0	266.0	02	400.9	302.1	62	448.8	338.2
23	258.0	194.4	83	305.9	230.5	43	353.8	266.6	03	401.7	302.7	63	449.6	338.8
24	258.8	195.0	84	306.7	231.1	44	354.6	267.2	04	402.5	303.3	64	450.4	339.4
25	259.6	195.6	85	307.5	231.7	45	355.4	267.8	05	403.3	303.9	65	451.2	340.0
26	260.4	196.2	86	308.3	232.3	46	356.2	268.4	06	404.1	304.5	66	452.0	340.6
27	261.2	196.8	87	309.1	232.9	47	357.0	269.0	07	404.9	305.1	67	452.8	341.2
28	262.0	197.4	88	309.9	233.5	48	357.8	269.6	08	405.7	305.7	68	453.6	341.8
29	262.8	198.0	89	310.7	234.1	49	358.6	270.2	09	406.5	306.3	69	454.4	342.4
30	263.5	198.6	90	311.5	234.7	50	359.4	270.8	10	407.3	306.9	70	455.2	343.0
331	264.3	199.2	391	312.3	235.3	451	360.2	271.4	511	408.1	307.5	571	456.0	343.6
32	265.1	199.8	92	313.1	235.9	52	361.0	272.0	12	408.9	308.1	72	456.8	344.2
33	265.9	200.4	93	313.9	236.5	53	361.8	272.6	13	409.7	308.7	73	457.6	344.8
34	266.7	201.0	94	314.7	237.1	54	362.6	273.2	14	410.5	309.3	74	458.4	345.4
35	267.5	201.6	95	315.5	237.7	55	363.4	273.8	15	411.3	309.9	75	459.2	346.0
36	268.3	202.2	96	316.3	238.3	56	364.2	274.4	16	412.1	310.5	76	460.0	346.6
37	269.1	202.8	97	317.1	238.9	57	365.0	275.0	17	412.9	311.1	77	460.8	347.2
38	269.9	203.4	98	317.9	239.5	58	365.8	275.6	18	413.7	311.7	78	461.6	347.8
39	270.7	204.0	99	318.7	240.1	59	366.6	276.2	19	414.5	312.3	79	462.4	348.5
40	271.5	204.6	400	319.5	240.7	60	367.4	276.8	20	415.3	312.9	80	463.2	349.1
341	272.3	205.2	401	320.3	241.3	461	368.2	277.4	521	416.1	313.5	581	464.0	349.7
42	273.1	205.8	02	321.1	241.9	62	369.0	278.0	22	416.9	314.1	82	464.8	350.3
43	273.9	206.4	03	321.9	242.5	63	369.8	278.6	23	417.7	314.7	83	465.6	350.9
44	274.7	207.0	04	322.6	243.1	64	370.6	279.2	24	418.5	315.4	84	466.4	351.5
45	275.5	207.6	05	323.4	243.7	65	371.4	279.8	25	419.3	316.0	85	467.2	352.1
46	276.3	208.2	06	324.2	244.3	66	372.2	280.4	26	420.1	316.6	86	468.0	352.7
47	277.1	208.8	07	325.0	244.9	67	373.0	281.0	27	420.9	317.2	87	468.8	353.3
48	277.9	209.4	08	325.8	245.5	68	373.8	281.6	28	421.7	317.8	88	469.6	353.9
49	278.7	210.0	09	326.6	246.1	69	374.6	282.3	29	422.5	318.4	89	470.4	354.5
50	279.5	210.6	10	327.4	246.7	70	375.4	282.9	30	423.3	319.0	90	471.2	355.1
351	280.3	211.2	411	328.2	247.3	471	376.2	283.5	531	424.1	319.6	591	472.0	355.7
52	281.1	211.8	12	329.0	247.9	72	377.0	284.1	32	424.9	320.2	92	472.8	356.3
53	281.9	212.4	13	329.8	248.5	73	377.8	284.7	33	425.7	320.8	93	473.6	356.9
54	282.7	213.0	14	330.6	249.2	74	378.6	285.3	34	426.5	321.4	94	474.4	357.5
55	283.5	213.6	15	331.4	249.8	75	379.4	285.9	35	427.3	322.0	95	475.2	358.1
56	284.3	214.2	16	332.2	250.4	76	380.2	286.5	36	428.1	322.6	96	476.0	358.7
57	285.1	214.8	17	333.0	251.0	77	380.9	287.1	37	428.9	323.2	97	476.8	359.3
58	285.9	215.4	18	333.8	251.6	78	381.7	287.7	38	429.7	323.8	98	477.6	359.9
59	286.7	216.1	19	334.6	252.2	79	382.5	288.3	39	430.5	324.4	99	478.4	360.5
60	287.5	216.7	20	335.4	252.8	80	383.3	288.9	40	431.3	325.0	600	479.2	361.1
Dist.	Dep.	D. Lat.	Dist.	Dep.	D. Lat.	Dist.	Dep.	D. Lat.	Dist.	Dep.	D. Lat.	Dist.	Dep.	D. Lat.

Dist.	D. Lat.	Dep.
D Lo	Dep.	
	m	D Lo

53°

307° | 053° / 233° | 127°

TABLE 4

Traverse 38° Table

322° | 038° — 218° | 142°

Dist.	D. Lat.	Dep.	Dist.	D. Lat.	Dep.	Dist.	D. Lat.	Dep.	Dist.	D. Lat.	Dep.	Dist.	D. Lat.	Dep.
1	0.8	0.6	61	48.1	37.6	121	95.3	74.5	181	142.6	111.4	241	189.9	148.4
2	1.6	1.2	62	48.9	38.2	22	96.1	75.1	82	143.4	112.1	42	190.7	149.0
3	2.4	1.8	63	49.6	38.8	23	96.9	75.7	83	144.2	112.7	43	191.5	149.6
4	3.2	2.5	64	50.4	39.4	24	97.7	76.3	84	145.0	113.3	44	192.3	150.2
5	3.9	3.1	65	51.2	40.0	25	98.5	77.0	85	145.8	113.9	45	193.1	150.8
6	4.7	3.7	66	52.0	40.6	26	99.3	77.6	86	146.6	114.5	46	193.9	151.5
7	5.5	4.3	67	52.8	41.2	27	100.1	78.2	87	147.4	115.1	47	194.6	152.1
8	6.3	4.9	68	53.6	41.9	28	100.9	78.8	88	148.1	115.7	48	195.4	152.7
9	7.1	5.5	69	54.4	42.5	29	101.7	79.4	89	148.9	116.4	49	196.2	153.3
10	7.9	6.2	70	55.2	43.1	30	102.4	80.0	90	149.7	117.0	50	197.0	153.9
11	8.7	6.8	71	55.9	43.7	131	103.2	80.7	191	150.5	117.6	251	197.8	154.5
12	9.5	7.4	72	56.7	44.3	32	104.0	81.3	92	151.3	118.2	52	198.6	155.1
13	10.2	8.0	73	57.5	44.9	33	104.8	81.9	93	152.1	118.8	53	199.4	155.8
14	11.0	8.6	74	58.3	45.6	34	105.6	82.5	94	152.9	119.4	54	200.2	156.4
15	11.8	9.2	75	59.1	46.2	35	106.4	83.1	95	153.7	120.1	55	200.9	157.0
16	12.6	9.9	76	59.9	46.8	36	107.2	83.7	96	154.5	120.7	56	201.7	157.6
17	13.4	10.5	77	60.7	47.4	37	108.0	84.3	97	155.2	121.3	57	202.5	158.2
18	14.2	11.1	78	61.5	48.0	38	108.7	85.0	98	156.0	121.9	58	203.3	158.8
19	15.0	11.7	79	62.3	48.6	39	109.5	85.6	99	156.8	122.5	59	204.1	159.5
20	15.8	12.3	80	63.0	49.3	40	110.3	86.2	200	157.6	123.1	60	204.9	160.1
21	16.5	12.9	81	63.8	49.9	141	111.1	86.8	201	158.4	123.7	261	205.7	160.7
22	17.3	13.5	82	64.6	50.5	42	111.9	87.4	02	159.2	124.4	62	206.5	161.3
23	18.1	14.2	83	65.4	51.1	43	112.7	88.0	03	160.0	125.0	63	207.2	161.9
24	18.9	14.8	84	66.2	51.7	44	113.5	88.7	04	160.8	125.6	64	208.0	162.5
25	19.7	15.4	85	67.0	52.3	45	114.3	89.3	05	161.5	126.2	65	208.8	163.2
26	20.5	16.0	86	67.8	52.9	46	115.0	89.9	06	162.3	126.8	66	209.6	163.8
27	21.3	16.6	87	68.6	53.6	47	115.8	90.5	07	163.1	127.4	67	210.4	164.4
28	22.1	17.2	88	69.3	54.2	48	116.6	91.1	08	163.9	128.1	68	211.2	165.0
29	22.9	17.9	89	70.1	54.8	49	117.4	91.7	09	164.7	128.7	69	212.0	165.6
30	23.6	18.5	90	70.9	55.4	50	118.2	92.3	10	165.5	129.3	70	212.8	166.2
31	24.4	19.1	91	71.7	56.0	151	119.0	93.0	211	166.3	129.9	271	213.6	166.8
32	25.2	19.7	92	72.5	56.6	52	119.8	93.6	12	167.1	130.5	72	214.3	167.5
33	26.0	20.3	93	73.3	57.3	53	120.6	94.2	13	167.8	131.1	73	215.1	168.1
34	26.8	20.9	94	74.1	57.9	54	121.4	94.8	14	168.6	131.8	74	215.9	168.7
35	27.6	21.5	95	74.9	58.5	55	122.1	95.4	15	169.4	132.4	75	216.7	169.3
36	28.4	22.2	96	75.6	59.1	56	122.9	96.0	16	170.2	133.0	76	217.5	169.9
37	29.2	22.8	97	76.4	59.7	57	123.7	96.7	17	171.0	133.6	77	218.3	170.5
38	29.9	23.4	98	77.2	60.3	58	124.5	97.3	18	171.8	134.2	78	219.1	171.2
39	30.7	24.0	99	78.0	61.0	59	125.3	97.9	19	172.6	134.8	79	219.9	171.8
40	31.5	24.6	100	78.8	61.6	60	126.1	98.5	20	173.4	135.4	80	220.6	172.4
41	32.3	25.2	101	79.6	62.2	161	126.9	99.1	221	174.2	136.1	281	221.4	173.0
42	33.1	25.9	02	80.4	62.8	62	127.7	99.7	22	174.9	136.7	82	222.2	173.6
43	33.9	26.5	03	81.2	63.4	63	128.4	100.4	23	175.7	137.3	83	223.0	174.2
44	34.7	27.1	04	82.0	64.0	64	129.2	101.0	24	176.5	137.9	84	223.8	174.8
45	35.5	27.7	05	82.7	64.6	65	130.0	101.6	25	177.3	138.5	85	224.6	175.5
46	36.2	28.3	06	83.5	65.3	66	130.8	102.2	26	178.1	139.1	86	225.4	176.1
47	37.0	28.9	07	84.3	65.9	67	131.6	102.8	27	178.9	139.8	87	226.2	176.7
48	37.8	29.6	08	85.1	66.5	68	132.4	103.4	28	179.7	140.4	88	226.9	177.3
49	38.6	30.2	09	85.9	67.1	69	133.2	104.0	29	180.5	141.0	89	227.7	177.9
50	39.4	30.8	10	86.7	67.7	70	134.0	104.7	30	181.2	141.6	90	228.5	178.5
51	40.2	31.4	111	87.5	68.3	171	134.7	105.3	231	182.0	142.2	291	229.3	179.2
52	41.0	32.0	12	88.3	69.0	72	135.5	105.9	32	182.8	142.8	92	230.1	179.8
53	41.8	32.6	13	89.0	69.6	73	136.3	106.5	33	183.6	143.4	93	230.9	180.4
54	42.6	33.2	14	89.8	70.2	74	137.1	107.1	34	184.4	144.1	94	231.7	181.0
55	43.3	33.9	15	90.6	70.8	75	137.9	107.7	35	185.2	144.7	95	232.5	181.6
56	44.1	34.5	16	91.4	71.4	76	138.7	108.4	36	186.0	145.3	96	233.3	182.2
57	44.9	35.1	17	92.2	72.0	77	139.5	109.0	37	186.8	145.9	97	234.0	182.9
58	45.7	35.7	18	93.0	72.6	78	140.3	109.6	38	187.5	146.5	98	234.8	183.5
59	46.5	36.3	19	93.8	73.3	79	141.1	110.2	39	188.3	147.1	99	235.6	184.1
60	47.3	36.9	20	94.6	73.9	80	141.8	110.8	40	189.1	147.8	300	236.4	184.7
Dist.	Dep.	D. Lat.	Dist.	Dep.	D. Lat.	Dist.	Dep.	D. Lat.	Dist.	Dep.	D. Lat.	Dist.	Dep.	D. Lat.

308° | 052° — 232° | 128°

52°

Dist.	D. Lat.	Dep.
N.	N x Cos.	N x Sin.
Hypotenuse	Side Adj.	Side Opp.

TABLE 4

Traverse 38° Table

322° | 038° — 218° | 142°

Dist.	D. Lat.	Dep.	Dist.	D. Lat.	Dep.	Dist.	D. Lat.	Dep.	Dist.	D. Lat.	Dep.	Dist.	D. Lat.	Dep.
301	237.2	185.3	361	284.5	222.3	421	331.8	259.2	481	379.0	296.1	541	426.3	333.1
02	238.0	185.9	62	285.3	222.9	22	332.5	259.8	82	379.8	296.7	42	427.1	333.7
03	238.8	186.5	63	286.0	223.5	23	333.3	260.4	83	380.6	297.4	43	427.9	334.3
04	239.6	187.2	64	286.8	224.1	24	334.1	261.0	84	381.4	298.0	44	428.7	334.9
05	240.3	187.8	65	287.6	224.7	25	334.9	261.7	85	382.2	298.6	45	429.5	335.5
06	241.1	188.4	66	288.4	225.3	26	335.7	262.3	86	383.0	299.2	46	430.3	336.2
07	241.9	189.0	67	289.2	225.9	27	336.5	262.9	87	383.8	299.8	47	431.0	336.8
08	242.7	189.6	68	290.0	226.6	28	337.3	263.5	88	384.5	300.4	48	431.8	337.4
09	243.5	190.2	69	290.8	227.2	29	338.1	264.1	89	385.3	301.1	49	432.6	338.0
10	244.3	190.9	70	291.6	227.8	30	338.8	264.7	90	386.1	301.7	50	433.4	338.6
311	245.1	191.5	371	292.4	228.4	431	339.6	265.4	491	386.9	302.3	551	434.2	339.2
12	245.9	192.1	72	293.1	229.0	32	340.4	266.0	92	387.7	302.9	52	435.0	339.8
13	246.6	192.7	73	293.9	229.6	33	341.2	266.6	93	388.5	303.5	53	435.8	340.5
14	247.4	193.3	74	294.7	230.3	34	342.0	267.2	94	389.3	304.1	54	436.6	341.1
15	248.2	193.9	75	295.5	230.9	35	342.8	267.8	95	390.1	304.8	55	437.3	341.7
16	249.0	194.5	76	296.3	231.5	36	343.6	268.4	96	390.9	305.4	56	438.1	342.3
17	249.8	195.2	77	297.1	232.1	37	344.4	269.0	97	391.6	306.0	57	438.9	342.9
18	250.6	195.8	78	297.9	232.7	38	345.1	269.7	98	392.4	306.6	58	439.7	343.5
19	251.4	196.4	79	298.7	233.3	39	345.9	270.3	99	393.2	307.2	59	440.5	344.2
20	252.2	197.0	80	299.4	234.0	40	346.7	270.9	500	394.0	307.8	60	441.3	344.8
321	253.0	197.6	381	300.2	234.6	441	347.5	271.5	501	394.8	308.4	561	442.1	345.4
22	253.7	198.2	82	301.0	235.2	42	348.3	272.1	02	395.6	309.1	62	442.9	346.0
23	254.5	198.9	83	301.8	235.8	43	349.1	272.7	03	396.4	309.7	63	443.7	346.6
24	255.3	199.5	84	302.6	236.4	44	349.9	273.4	04	397.2	310.3	64	444.4	347.2
25	256.1	200.1	85	303.4	237.0	45	350.7	274.0	05	397.9	310.9	65	445.2	347.8
26	256.9	200.7	86	304.2	237.6	46	351.5	274.6	06	398.7	311.5	66	446.0	348.5
27	257.7	201.3	87	305.0	238.3	47	352.2	275.2	07	399.5	312.1	67	446.8	349.1
28	258.5	201.9	88	305.7	238.9	48	353.0	275.8	08	400.3	312.8	68	447.6	349.7
29	259.3	202.6	89	306.5	239.5	49	353.8	276.4	09	401.1	313.4	69	448.4	350.3
30	260.0	203.2	90	307.3	240.1	50	354.6	277.0	10	401.9	314.0	70	449.2	350.9
331	260.8	203.8	391	308.1	240.7	451	355.4	277.7	511	402.7	314.6	571	450.0	351.5
32	261.6	204.4	92	308.9	241.3	52	356.2	278.3	12	403.5	315.2	72	450.7	352.2
33	262.4	205.0	93	309.7	242.0	53	357.0	278.9	13	404.2	315.8	73	451.5	352.8
34	263.2	205.6	94	310.5	242.6	54	357.8	279.5	14	405.0	316.4	74	452.3	353.4
35	264.0	206.2	95	311.3	243.2	55	358.5	280.1	15	405.8	317.1	75	453.1	354.0
36	264.8	206.9	96	312.1	243.8	56	359.3	280.7	16	406.6	317.7	76	453.9	354.6
37	265.6	207.5	97	312.8	244.4	57	360.1	281.4	17	407.4	318.3	77	454.7	355.2
38	266.3	208.1	98	313.6	245.0	58	360.9	282.0	18	408.2	318.9	78	455.5	355.9
39	267.1	208.7	99	314.4	245.6	59	361.7	282.6	19	409.0	319.5	79	456.3	356.5
40	267.9	209.3	400	315.2	246.3	60	362.5	283.2	20	409.8	320.1	80	457.0	357.1
341	268.7	209.9	401	316.0	246.9	461	363.3	283.8	521	410.6	320.8	581	457.8	357.7
42	269.5	210.6	02	316.8	247.5	62	364.1	284.4	22	411.3	321.4	82	458.6	358.3
43	270.3	211.2	03	317.6	248.1	63	364.8	285.1	23	412.1	322.0	83	459.4	358.9
44	271.1	211.8	04	318.4	248.7	64	365.6	285.7	24	412.9	322.6	84	460.2	359.5
45	271.9	212.4	05	319.1	249.3	65	366.4	286.3	25	413.7	323.2	85	461.0	360.2
46	272.7	213.0	06	319.9	250.0	66	367.2	286.9	26	414.5	323.8	86	461.8	360.8
47	273.4	213.6	07	320.7	250.6	67	368.0	287.5	27	415.3	324.5	87	462.6	361.4
48	274.2	214.3	08	321.5	251.2	68	368.8	288.1	28	416.1	325.1	88	463.4	362.0
49	275.0	214.9	09	322.3	251.8	69	369.6	288.7	29	416.9	325.7	89	464.1	362.6
50	275.8	215.5	10	323.1	252.4	70	370.4	289.4	30	417.6	326.3	90	464.9	363.2
351	276.6	216.1	411	323.9	253.0	471	371.2	290.0	531	418.4	326.9	591	465.7	363.9
52	277.4	216.7	12	324.7	253.7	72	371.9	290.6	32	419.2	327.5	92	466.5	364.5
53	278.2	217.3	13	325.4	254.3	73	372.7	291.2	33	420.0	328.1	93	467.3	365.1
54	279.0	217.9	14	326.2	254.9	74	373.5	291.8	34	420.8	328.8	94	468.1	365.7
55	279.7	218.6	15	327.0	255.5	75	374.3	292.4	35	421.6	329.4	95	468.9	366.3
56	280.5	219.2	16	327.8	256.1	76	375.1	293.1	36	422.4	330.0	96	469.7	366.9
57	281.3	219.8	17	328.6	256.7	77	375.9	293.7	37	423.2	330.6	97	470.4	367.5
58	282.1	220.4	18	329.4	257.3	78	376.7	294.3	38	423.9	331.2	98	471.2	368.2
59	282.9	221.0	19	330.2	258.0	79	377.5	294.9	39	424.7	331.8	99	472.0	368.8
60	283.7	221.6	20	331.0	258.6	80	378.2	295.5	40	425.5	332.5	600	472.8	369.4
Dist.	Dep.	D. Lat.	Dist.	Dep.	D. Lat.	Dist.	Dep.	D. Lat.	Dist.	Dep.	D. Lat.	Dist.	Dep.	D. Lat.

Dist.	D. Lat.	Dep.
D Lo	Dep.	
	m	D Lo

52°

308° | 052° — 232° | 128°

TABLE 4

Traverse 39° Table

321° | 039°
219° | 141°

Dist.	D. Lat.	Dep.	Dist.	D. Lat.	Dep.	Dist.	D. Lat.	Dep.	Dist.	D. Lat.	Dep.	Dist.	D. Lat.	Dep.
1	0.8	0.6	61	47.4	38.4	121	94.0	76.1	181	140.7	113.9	241	187.3	151.7
2	1.6	1.3	62	48.2	39.0	22	94.8	76.8	82	141.4	114.5	42	188.1	152.3
3	2.3	1.9	63	49.0	39.6	23	95.6	77.4	83	142.2	115.2	43	188.8	152.9
4	3.1	2.5	64	49.7	40.3	24	96.4	78.0	84	143.0	115.8	44	189.6	153.6
5	3.9	3.1	65	50.5	40.9	25	97.1	78.7	85	143.8	116.4	45	190.4	154.2
6	4.7	3.8	66	51.3	41.5	26	97.9	79.3	86	144.5	117.1	46	191.2	154.8
7	5.4	4.4	67	52.1	42.2	27	98.7	79.9	87	145.3	117.7	47	192.0	155.4
8	6.2	5.0	68	52.8	42.8	28	99.5	80.6	88	146.1	118.3	48	192.7	156.1
9	7.0	5.7	69	53.6	43.4	29	100.3	81.2	89	146.9	118.9	49	193.5	156.7
10	7.8	6.3	70	54.4	44.1	30	101.0	81.8	90	147.7	119.6	50	194.3	157.3
11	8.5	6.9	71	55.2	44.7	131	101.8	82.4	191	148.4	120.2	251	195.1	158.0
12	9.3	7.6	72	56.0	45.3	32	102.6	83.1	92	149.2	120.8	52	195.8	158.6
13	10.1	8.2	73	56.7	45.9	33	103.4	83.7	93	150.0	121.5	53	196.6	159.2
14	10.9	8.8	74	57.5	46.6	34	104.1	84.3	94	150.8	122.1	54	197.4	159.8
15	11.7	9.4	75	58.3	47.2	35	104.9	85.0	95	151.5	122.7	55	198.2	160.5
16	12.4	10.1	76	59.1	47.8	36	105.7	85.6	96	152.3	123.3	56	198.9	161.1
17	13.2	10.7	77	59.8	48.5	37	106.5	86.2	97	153.1	124.0	57	199.7	161.7
18	14.0	11.3	78	60.6	49.1	38	107.2	86.8	98	153.9	124.6	58	200.5	162.4
19	14.8	12.0	79	61.4	49.7	39	108.0	87.5	99	154.7	125.2	59	201.3	163.0
20	15.5	12.6	80	62.2	50.3	40	108.8	88.1	200	155.4	125.9	60	202.1	163.6
21	16.3	13.2	81	62.9	51.0	141	109.6	88.7	201	156.2	126.5	261	202.8	164.3
22	17.1	13.8	82	63.7	51.6	42	110.4	89.4	02	157.0	127.1	62	203.6	164.9
23	17.9	14.5	83	64.5	52.2	43	111.1	90.0	03	157.8	127.8	63	204.4	165.5
24	18.7	15.1	84	65.3	52.9	44	111.9	90.6	04	158.5	128.4	64	205.2	166.1
25	19.4	15.7	85	66.1	53.5	45	112.7	91.3	05	159.3	129.0	65	205.9	166.8
26	20.2	16.4	86	66.8	54.1	46	113.5	91.9	06	160.1	129.6	66	206.7	167.4
27	21.0	17.0	87	67.6	54.8	47	114.2	92.5	07	160.9	130.3	67	207.5	168.0
28	21.8	17.6	88	68.4	55.4	48	115.0	93.1	08	161.6	130.9	68	208.3	168.7
29	22.5	18.3	89	69.2	56.0	49	115.8	93.8	09	162.4	131.5	69	209.1	169.3
30	23.3	18.9	90	69.9	56.6	50	116.6	94.4	10	163.2	132.2	70	209.8	169.9
31	24.1	19.5	91	70.7	57.3	151	117.3	95.0	211	164.0	132.8	271	210.6	170.5
32	24.9	20.1	92	71.5	57.9	52	118.1	95.7	12	164.8	133.4	72	211.4	171.2
33	25.6	20.8	93	72.3	58.5	53	118.9	96.3	13	165.5	134.0	73	212.2	171.8
34	26.4	21.4	94	73.1	59.2	54	119.7	96.9	14	166.3	134.7	74	212.9	172.4
35	27.2	22.0	95	73.8	59.8	55	120.5	97.5	15	167.1	135.3	75	213.7	173.1
36	28.0	22.7	96	74.6	60.4	56	121.2	98.2	16	167.9	135.9	76	214.5	173.7
37	28.8	23.3	97	75.4	61.0	57	122.0	98.8	17	168.6	136.6	77	215.3	174.3
38	29.5	23.9	98	76.2	61.7	58	122.8	99.4	18	169.4	137.2	78	216.0	175.0
39	30.3	24.5	99	76.9	62.3	59	123.6	100.1	19	170.2	137.8	79	216.8	175.6
40	31.1	25.2	100	77.7	62.9	60	124.3	100.7	20	171.0	138.5	80	217.6	176.2
41	31.9	25.8	101	78.5	63.6	161	125.1	101.3	221	171.7	139.1	281	218.4	176.8
42	32.6	26.4	02	79.3	64.2	62	125.9	101.9	22	172.5	139.7	82	219.2	177.5
43	33.4	27.1	03	80.0	64.8	63	126.7	102.6	23	173.3	140.3	83	219.9	178.1
44	34.2	27.7	04	80.8	65.4	64	127.5	103.2	24	174.1	141.0	84	220.7	178.7
45	35.0	28.3	05	81.6	66.1	65	128.2	103.8	25	174.9	141.6	85	221.5	179.4
46	35.7	28.9	06	82.4	66.7	66	129.0	104.5	26	175.6	142.2	86	222.3	180.0
47	36.5	29.6	07	83.2	67.3	67	129.8	105.1	27	176.4	142.9	87	223.0	180.6
48	37.3	30.2	08	83.9	68.0	68	130.6	105.7	28	177.2	143.5	88	223.8	181.2
49	38.1	30.8	09	84.7	68.6	69	131.3	106.4	29	178.0	144.1	89	224.6	181.9
50	38.9	31.5	10	85.5	69.2	70	132.1	107.0	30	178.7	144.7	90	225.4	182.5
51	39.6	32.1	111	86.3	69.9	171	132.9	107.6	231	179.5	145.4	291	226.1	183.1
52	40.4	32.7	12	87.0	70.5	72	133.7	108.2	32	180.3	146.0	92	226.9	183.8
53	41.2	33.4	13	87.8	71.1	73	134.4	108.9	33	181.1	146.6	93	227.7	184.4
54	42.0	34.0	14	88.6	71.7	74	135.2	109.5	34	181.9	147.3	94	228.5	185.0
55	42.7	34.6	15	89.4	72.4	75	136.0	110.1	35	182.6	147.9	95	229.3	185.6
56	43.5	35.2	16	90.1	73.0	76	136.8	110.8	36	183.4	148.5	96	230.0	186.3
57	44.3	35.9	17	90.9	73.6	77	137.6	111.4	37	184.2	149.1	97	230.8	186.9
58	45.1	36.5	18	91.7	74.3	78	138.3	112.0	38	185.0	149.8	98	231.6	187.5
59	45.9	37.1	19	92.5	74.9	79	139.1	112.6	39	185.7	150.4	99	232.4	188.2
60	46.6	37.8	20	93.3	75.5	80	139.9	113.3	40	186.5	151.0	300	233.1	188.8
Dist.	Dep.	D. Lat.	Dist.	Dep.	D. Lat.	Dist.	Dep.	D. Lat.	Dist.	Dep.	D. Lat.	Dist.	Dep.	D. Lat.

309° | 051°
231° | 129°

51°

Dist.	D. Lat.	Dep.
N.	N x Cos.	N x Sin.
Hypotenuse	Side Adj.	Side Opp.

TABLE 4

Traverse 39° Table

321° | 039°
219° | 141°

Dist.	D. Lat.	Dep.	Dist.	D. Lat.	Dep.	Dist.	D. Lat.	Dep.	Dist.	D. Lat.	Dep.	Dist.	D. Lat.	Dep.
301	233.9	189.4	361	280.5	227.2	421	327.2	264.9	481	373.8	302.7	541	420.4	340.5
02	234.7	190.1	62	281.3	227.8	22	328.0	265.6	82	374.6	303.3	42	421.2	341.1
03	235.5	190.7	63	282.1	228.4	23	328.7	266.2	83	375.4	304.0	43	422.0	341.7
04	236.3	191.3	64	282.9	229.1	24	329.5	266.8	84	376.1	304.6	44	422.8	342.4
05	237.0	191.9	65	283.7	229.7	25	330.3	267.5	85	376.9	305.2	45	423.5	343.0
06	237.8	192.6	66	284.4	230.3	26	331.1	268.1	86	377.7	305.8	46	424.3	343.6
07	238.6	193.2	67	285.2	231.0	27	331.8	268.7	87	378.5	306.5	47	425.1	344.2
08	239.4	193.8	68	286.0	231.6	28	332.6	269.3	88	379.2	307.1	48	425.9	344.9
09	240.1	194.5	69	286.8	232.2	29	333.4	270.0	89	380.0	307.7	49	426.7	345.5
10	240.9	195.1	70	287.5	232.8	30	334.2	270.6	90	380.8	308.4	50	427.4	346.1
311	241.7	195.7	371	288.3	233.5	431	334.9	271.2	491	381.6	309.0	551	428.2	346.8
12	242.5	196.3	72	289.1	234.1	32	335.7	271.9	92	382.4	309.6	52	429.0	347.4
13	243.2	197.0	73	289.9	234.7	33	336.5	272.5	93	383.1	310.3	53	429.8	348.0
14	244.0	197.6	74	290.7	235.4	34	337.3	273.1	94	383.9	310.9	54	430.5	348.6
15	244.8	198.2	75	291.4	236.0	35	338.1	273.8	95	384.7	311.5	55	431.3	349.3
16	245.6	198.9	76	292.2	236.6	36	338.8	274.4	96	385.5	312.1	56	432.1	349.9
17	246.4	199.5	77	293.0	237.3	37	339.6	275.0	97	386.2	312.8	57	432.9	350.5
18	247.1	200.1	78	293.8	237.9	38	340.4	275.6	98	387.0	313.4	58	433.6	351.2
19	247.9	200.8	79	294.5	238.5	39	341.2	276.3	99	387.8	314.0	59	434.4	351.8
20	248.7	201.4	80	295.3	239.1	40	341.9	276.9	500	388.6	314.7	60	435.2	352.4
321	249.5	202.0	381	296.1	239.8	441	342.7	277.5	501	389.4	315.3	561	436.0	353.0
22	250.2	202.6	82	296.9	240.4	42	343.5	278.2	02	390.1	315.9	62	436.8	353.7
23	251.0	203.3	83	297.6	241.0	43	344.3	278.8	03	390.9	316.5	63	437.5	354.3
24	251.8	203.9	84	298.4	241.7	44	345.1	279.4	04	391.7	317.2	64	438.3	354.9
25	252.6	204.5	85	299.2	242.3	45	345.8	280.0	05	392.5	317.8	65	439.1	355.6
26	253.3	205.2	86	300.0	242.9	46	346.6	280.7	06	393.2	318.4	66	439.9	356.2
27	254.1	205.8	87	300.8	243.5	47	347.4	281.3	07	394.0	319.1	67	440.6	356.8
28	254.9	206.4	88	301.5	244.2	48	348.2	281.9	08	394.8	319.7	68	441.4	357.5
29	255.7	207.0	89	302.3	244.8	49	348.9	282.6	09	395.6	320.3	69	442.2	358.1
30	256.5	207.7	90	303.1	245.4	50	349.7	283.2	10	396.3	321.0	70	443.0	358.7
331	257.2	208.3	391	303.9	246.1	451	350.5	283.8	511	397.1	321.6	571	443.8	359.3
32	258.0	208.9	92	304.6	246.7	52	351.3	284.5	12	397.9	322.2	72	444.5	360.0
33	258.8	209.6	93	305.4	247.3	53	352.0	285.1	13	398.7	322.8	73	445.3	360.6
34	259.6	210.2	94	306.2	248.0	54	352.8	285.7	14	399.5	323.5	74	446.1	361.2
35	260.3	210.8	95	307.0	248.6	55	353.6	286.3	15	400.2	324.1	75	446.9	361.9
36	261.1	211.5	96	307.7	249.2	56	354.4	287.0	16	401.0	324.7	76	447.6	362.5
37	261.9	212.1	97	308.5	249.8	57	355.2	287.6	17	401.8	325.4	77	448.4	363.1
38	262.7	212.7	98	309.3	250.5	58	355.9	288.2	18	402.6	326.0	78	449.2	363.7
39	263.5	213.3	99	310.1	251.1	59	356.7	288.9	19	403.3	326.6	79	450.0	364.4
40	264.2	214.0	400	310.9	251.7	60	357.5	289.5	20	404.1	327.2	80	450.7	365.0
341	265.0	214.6	401	311.6	252.4	461	358.3	290.1	521	404.9	327.9	581	451.5	365.6
42	265.8	215.2	02	312.4	253.0	62	359.0	290.7	22	405.7	328.5	82	452.3	366.3
43	266.6	215.9	03	313.2	253.6	63	359.8	291.4	23	406.4	329.1	83	453.1	366.9
44	267.3	216.5	04	314.0	254.2	64	360.6	292.0	24	407.2	329.8	84	453.9	367.5
45	268.1	217.1	05	314.7	254.9	65	361.4	292.6	25	408.0	330.4	85	454.6	368.2
46	268.9	217.7	06	315.5	255.5	66	362.2	293.3	26	408.8	331.0	86	455.4	368.8
47	269.7	218.4	07	316.3	256.1	67	362.9	293.9	27	409.6	331.7	87	456.2	369.4
48	270.4	219.0	08	317.1	256.8	68	363.7	294.5	28	410.3	332.3	88	457.0	370.0
49	271.2	219.6	09	317.9	257.4	69	364.5	295.2	29	411.1	332.9	89	457.7	370.7
50	272.0	220.3	10	318.6	258.0	70	365.3	295.8	30	411.9	333.5	90	458.5	371.3
351	272.8	220.9	411	319.4	258.7	471	366.0	296.4	531	412.7	334.2	591	459.3	371.9
52	273.6	221.5	12	320.2	259.3	72	366.8	297.0	32	413.4	334.8	92	460.1	372.6
53	274.3	222.2	13	321.0	259.9	73	367.6	297.7	33	414.2	335.4	93	460.8	373.2
54	275.1	222.8	14	321.7	260.5	74	368.4	298.3	34	415.0	336.1	94	461.6	373.8
55	275.9	223.4	15	322.5	261.2	75	369.1	298.9	35	415.8	336.7	95	462.4	374.4
56	276.7	224.0	16	323.3	261.8	76	369.9	299.6	36	416.6	337.3	96	463.2	375.1
57	277.4	224.7	17	324.1	262.4	77	370.7	300.2	37	417.3	337.9	97	464.0	375.7
58	278.2	225.3	18	324.8	263.1	78	371.5	300.8	38	418.1	338.6	98	464.7	376.3
59	279.0	225.9	19	325.6	263.7	79	372.3	301.4	39	418.9	339.2	99	465.5	377.0
60	279.8	226.6	20	326.4	264.3	80	373.0	302.1	40	419.7	339.8	600	466.3	377.6
Dist.	Dep.	D. Lat.	Dist.	Dep.	D. Lat.	Dist.	Dep.	D. Lat.	Dist.	Dep.	D. Lat.	Dist.	Dep.	D. Lat.

Dist.	D. Lat.	Dep.
D Lo	Dep.	
	m	D Lo

51°

309° | 051°
231° | 129°

TABLE 4

Traverse 40° Table

320° | 040°
220° | 140°

Dist.	D. Lat.	Dep.	Dist.	D. Lat.	Dep.	Dist.	D. Lat.	Dep.	Dist.	D. Lat.	Dep.	Dist.	D. Lat.	Dep.
1	0.8	0.6	61	46.7	39.2	121	92.7	77.8	181	138.7	116.3	241	184.6	154.9
2	1.5	1.3	62	47.5	39.9	22	93.5	78.4	82	139.4	117.0	42	185.4	155.6
3	2.3	1.9	63	48.3	40.5	23	94.2	79.1	83	140.2	117.6	43	186.1	156.2
4	3.1	2.6	64	49.0	41.1	24	95.0	79.7	84	141.0	118.3	44	186.9	156.8
5	3.8	3.2	65	49.8	41.8	25	95.8	80.3	85	141.7	118.9	45	187.7	157.5
6	4.6	3.9	66	50.6	42.4	26	96.5	81.0	86	142.5	119.6	46	188.4	158.1
7	5.4	4.5	67	51.3	43.1	27	97.3	81.6	87	143.3	120.2	47	189.2	158.8
8	6.1	5.1	68	52.1	43.7	28	98.1	82.3	88	144.0	120.8	48	190.0	159.4
9	6.9	5.8	69	52.9	44.4	29	98.8	82.9	89	144.8	121.5	49	190.7	160.1
10	7.7	6.4	70	53.6	45.0	30	99.6	83.6	90	145.5	122.1	50	191.5	160.7
11	8.4	7.1	71	54.4	45.6	131	100.4	84.2	191	146.3	122.8	251	192.3	161.3
12	9.2	7.7	72	55.2	46.3	32	101.1	84.8	92	147.1	123.4	52	193.0	162.0
13	10.0	8.4	73	55.9	46.9	33	101.9	85.5	93	147.8	124.1	53	193.8	162.6
14	10.7	9.0	74	56.7	47.6	34	102.6	86.1	94	148.6	124.7	54	194.6	163.3
15	11.5	9.6	75	57.5	48.2	35	103.4	86.8	95	149.4	125.3	55	195.3	163.9
16	12.3	10.3	76	58.2	48.9	36	104.2	87.4	96	150.1	126.0	56	196.1	164.6
17	13.0	10.9	77	59.0	49.5	37	104.9	88.1	97	150.9	126.6	57	196.9	165.2
18	13.8	11.6	78	59.8	50.1	38	105.7	88.7	98	151.7	127.3	58	197.6	165.8
19	14.6	12.2	79	60.5	50.8	39	106.5	89.3	99	152.4	127.9	59	198.4	166.5
20	15.3	12.9	80	61.3	51.4	40	107.2	90.0	200	153.2	128.6	60	199.2	167.1
21	16.1	13.5	81	62.0	52.1	141	108.0	90.6	201	154.0	129.2	261	199.9	167.8
22	16.9	14.1	82	62.8	52.7	42	108.8	91.3	02	154.7	129.8	62	200.7	168.4
23	17.6	14.8	83	63.6	53.4	43	109.5	91.9	03	155.5	130.5	63	201.5	169.1
24	18.4	15.4	84	64.3	54.0	44	110.3	92.6	04	156.3	131.1	64	202.2	169.7
25	19.2	16.1	85	65.1	54.6	45	111.1	93.2	05	157.0	131.8	65	203.0	170.3
26	19.9	16.7	86	65.9	55.3	46	111.8	93.8	06	157.8	132.4	66	203.8	171.0
27	20.7	17.4	87	66.6	55.9	47	112.6	94.5	07	158.6	133.1	67	204.5	171.6
28	21.4	18.0	88	67.4	56.6	48	113.4	95.1	08	159.3	133.7	68	205.3	172.3
29	22.2	18.6	89	68.2	57.2	49	114.1	95.8	09	160.1	134.3	69	206.1	172.9
30	23.0	19.3	90	68.9	57.9	50	114.9	96.4	10	160.9	135.0	70	206.8	173.6
31	23.7	19.9	91	69.7	58.5	151	115.7	97.1	211	161.6	135.6	271	207.6	174.2
32	24.5	20.6	92	70.5	59.1	52	116.4	97.7	12	162.4	136.3	72	208.4	174.8
33	25.3	21.2	93	71.2	59.8	53	117.2	98.3	13	163.2	136.9	73	209.1	175.5
34	26.0	21.9	94	72.0	60.4	54	118.0	99.0	14	163.9	137.6	74	209.9	176.1
35	26.8	22.5	95	72.8	61.1	55	118.7	99.6	15	164.7	138.2	75	210.7	176.8
36	27.6	23.1	96	73.5	61.7	56	119.5	100.3	16	165.5	138.8	76	211.4	177.4
37	28.3	23.8	97	74.3	62.4	57	120.3	100.9	17	166.2	139.5	77	212.2	178.1
38	29.1	24.4	98	75.1	63.0	58	121.0	101.6	18	167.0	140.1	78	213.0	178.7
39	29.9	25.1	99	75.8	63.6	59	121.8	102.2	19	167.8	140.8	79	213.7	179.3
40	30.6	25.7	100	76.6	64.3	60	122.6	102.8	20	168.5	141.4	80	214.5	180.0
41	31.4	26.4	101	77.4	64.9	161	123.3	103.5	221	169.3	142.1	281	215.3	180.6
42	32.2	27.0	02	78.1	65.6	62	124.1	104.1	22	170.1	142.7	82	216.0	181.3
43	32.9	27.6	03	78.9	66.2	63	124.9	104.8	23	170.8	143.3	83	216.8	181.9
44	33.7	28.3	04	79.7	66.8	64	125.6	105.4	24	171.6	144.0	84	217.6	182.6
45	34.5	28.9	05	80.4	67.5	65	126.4	106.1	25	172.4	144.6	85	218.3	183.2
46	35.2	29.6	06	81.2	68.1	66	127.2	106.7	26	173.1	145.3	86	219.1	183.8
47	36.0	30.2	07	82.0	68.8	67	127.9	107.3	27	173.9	145.9	87	219.9	184.5
48	36.8	30.9	08	82.7	69.4	68	128.7	108.0	28	174.7	146.6	88	220.6	185.1
49	37.5	31.5	09	83.5	70.1	69	129.5	108.6	29	175.4	147.2	89	221.4	185.8
50	38.3	32.1	10	84.3	70.7	70	130.2	109.3	30	176.2	147.8	90	222.2	186.4
51	39.1	32.8	111	85.0	71.3	171	131.0	109.9	231	177.0	148.5	291	222.9	187.1
52	39.8	33.4	12	85.8	72.0	72	131.8	110.6	32	177.7	149.1	92	223.7	187.7
53	40.6	34.1	13	86.6	72.6	73	132.5	111.2	33	178.5	149.8	93	224.5	188.3
54	41.4	34.7	14	87.3	73.3	74	133.3	111.8	34	179.3	150.4	94	225.2	189.0
55	42.1	35.4	15	88.1	73.9	75	134.1	112.5	35	180.0	151.1	95	226.0	189.6
56	42.9	36.0	16	88.9	74.6	76	134.8	113.1	36	180.8	151.7	96	226.7	190.3
57	43.7	36.6	17	89.6	75.2	77	135.6	113.8	37	181.6	152.3	97	227.5	190.9
58	44.4	37.3	18	90.4	75.8	78	136.4	114.4	38	182.3	153.0	98	228.3	191.6
59	45.2	37.9	19	91.2	76.5	79	137.1	115.1	39	183.1	153.6	99	229.0	192.2
60	46.0	38.6	20	91.9	77.1	80	137.9	115.7	40	183.9	154.3	300	229.8	192.8
Dist.	Dep.	D. Lat.	Dist.	Dep.	D. Lat.	Dist.	Dep.	D. Lat.	Dist.	Dep.	D. Lat.	Dist.	Dep.	D. Lat.

310° | 050°
230° | 130°

50°

Dist.	D. Lat.	Dep.
N.	N x Cos.	N x Sin.
Hypotenuse	Side Adj.	Side Opp.

TABLE 4

Traverse 40° Table

320° | 040°
220° | 140°

Dist.	D. Lat.	Dep.	Dist.	D. Lat.	Dep.	Dist.	D. Lat.	Dep.	Dist.	D. Lat.	Dep.	Dist.	D. Lat.	Dep.
301	230.6	193.5	361	276.5	232.0	421	322.5	270.6	481	368.5	309.2	541	414.4	347.7
02	231.3	194.1	62	277.3	232.7	22	323.3	271.3	82	369.2	309.8	42	415.2	348.4
03	232.1	194.8	63	278.1	233.3	23	324.0	271.9	83	370.0	310.5	43	416.0	349.0
04	232.9	195.4	64	278.8	234.0	24	324.8	272.5	84	370.8	311.1	44	416.7	349.7
05	233.6	196.1	65	279.6	234.6	25	325.6	273.2	85	371.5	311.8	45	417.5	350.3
06	234.4	196.7	66	280.4	235.3	26	326.3	273.8	86	372.3	312.4	46	418.3	351.0
07	235.2	197.3	67	281.1	235.9	27	327.1	274.5	87	373.1	313.0	47	419.0	351.6
08	235.9	198.0	68	281.9	236.5	28	327.9	275.1	88	373.8	313.7	48	419.8	352.2
09	236.7	198.6	69	282.7	237.2	29	328.6	275.8	89	374.6	314.3	49	420.6	352.9
10	237.5	199.3	70	283.4	237.8	30	329.4	276.4	90	375.4	315.0	50	421.3	353.5
311	238.2	199.9	371	284.2	238.5	431	330.2	277.0	491	376.1	315.6	551	422.1	354.2
12	239.0	200.5	72	285.0	239.1	32	330.9	277.7	92	376.9	316.3	52	422.9	354.8
13	239.8	201.2	73	285.7	239.8	33	331.7	278.3	93	377.7	316.9	53	423.6	355.5
14	240.5	201.8	74	286.5	240.4	34	332.5	279.0	94	378.4	317.5	54	424.4	356.1
15	241.3	202.5	75	287.3	241.0	35	333.2	279.6	95	379.2	318.2	55	425.2	356.7
16	242.1	203.1	76	288.0	241.7	36	334.0	280.3	96	380.0	318.8	56	425.9	357.4
17	242.8	203.8	77	288.8	242.3	37	334.8	280.9	97	380.7	319.5	57	426.7	358.0
18	243.6	204.4	78	289.6	243.0	38	335.5	281.5	98	381.5	320.1	58	427.5	358.7
19	244.4	205.0	79	290.3	243.6	39	336.3	282.2	99	382.3	320.8	59	428.2	359.3
20	245.1	205.7	80	291.1	244.3	40	337.1	282.8	500	383.0	321.4	60	429.0	360.0
321	245.9	206.3	381	291.9	244.9	441	337.8	283.5	501	383.8	322.0	561	429.8	360.6
22	246.7	207.0	82	292.6	245.5	42	338.6	284.1	02	384.6	322.7	62	430.5	361.2
23	247.4	207.6	83	293.4	246.2	43	339.4	284.8	03	385.3	323.3	63	431.3	361.9
24	248.2	208.3	84	294.2	246.8	44	340.1	285.4	04	386.1	324.0	64	432.0	362.5
25	249.0	208.9	85	294.9	247.5	45	340.9	286.0	05	386.9	324.6	65	432.8	363.2
26	249.7	209.5	86	295.7	248.1	46	341.7	286.7	06	387.6	325.3	66	433.6	363.8
27	250.5	210.2	87	296.5	248.8	47	342.4	287.3	07	388.4	325.9	67	434.3	364.5
28	251.3	210.8	88	297.2	249.4	48	343.2	288.0	08	389.2	326.5	68	435.1	365.1
29	252.0	211.5	89	298.0	250.0	49	344.0	288.6	09	389.9	327.2	69	435.9	365.7
30	252.8	212.1	90	298.8	250.7	50	344.7	289.3	10	390.7	327.8	70	436.6	366.4
331	253.6	212.8	391	299.5	251.3	451	345.5	289.9	511	391.4	328.5	571	437.4	367.0
32	254.3	213.4	92	300.3	252.0	52	346.3	290.5	12	392.2	329.1	72	438.2	367.7
33	255.1	214.0	93	301.1	252.6	53	347.0	291.2	13	393.0	329.8	73	438.9	368.3
34	255.9	214.7	94	301.8	253.3	54	347.8	291.8	14	393.7	330.4	74	439.7	369.0
35	256.6	215.3	95	302.6	253.9	55	348.6	292.5	15	394.5	331.0	75	440.5	369.6
36	257.4	216.0	96	303.4	254.5	56	349.3	293.1	16	395.3	331.7	76	441.2	370.2
37	258.2	216.6	97	304.1	255.2	57	350.1	293.8	17	396.0	332.3	77	442.0	370.9
38	258.9	217.3	98	304.9	255.8	58	350.8	294.4	18	396.8	333.0	78	442.8	371.5
39	259.7	217.9	99	305.7	256.5	59	351.6	295.0	19	397.6	333.6	79	443.5	372.2
40	260.5	218.5	400	306.4	257.1	60	352.4	295.7	20	398.3	334.2	80	444.3	372.8
341	261.2	219.2	401	307.2	257.8	461	353.1	296.3	521	399.1	334.9	581	445.1	373.5
42	262.0	219.8	02	307.9	258.4	62	353.9	297.0	22	399.9	335.5	82	445.8	374.1
43	262.8	220.5	03	308.7	259.0	63	354.7	297.6	23	400.6	336.2	83	446.6	374.7
44	263.5	221.1	04	309.5	259.7	64	355.4	298.3	24	401.4	336.8	84	447.4	375.4
45	264.3	221.8	05	310.2	260.3	65	356.2	298.9	25	402.2	337.5	85	448.1	376.0
46	265.1	222.4	06	311.0	261.0	66	357.0	299.5	26	402.9	338.1	86	448.9	376.7
47	265.8	223.0	07	311.8	261.6	67	357.7	300.2	27	403.7	338.7	87	449.7	377.3
48	266.6	223.7	08	312.5	262.3	68	358.5	300.8	28	404.5	339.4	88	450.4	378.0
49	267.3	224.3	09	313.3	262.9	69	359.3	301.5	29	405.2	340.0	89	451.2	378.6
50	268.1	225.0	10	314.1	263.5	70	360.0	302.1	30	406.0	340.7	90	452.0	379.2
351	268.9	225.6	411	314.8	264.2	471	360.8	302.8	531	406.8	341.3	591	452.7	379.9
52	269.6	226.3	12	315.6	264.8	72	361.6	303.4	32	407.5	342.0	92	453.5	380.5
53	270.4	226.9	13	316.4	265.5	73	362.3	304.0	33	408.3	342.6	93	454.3	381.2
54	271.2	227.5	14	317.1	266.1	74	363.1	304.7	34	409.1	343.2	94	455.0	381.8
55	271.9	228.2	15	317.9	266.8	75	363.9	305.3	35	409.8	343.9	95	455.8	382.5
56	272.7	228.8	16	318.7	267.4	76	364.6	306.0	36	410.6	344.5	96	456.6	383.1
57	273.5	229.5	17	319.4	268.0	77	365.4	306.6	37	411.4	345.2	97	457.3	383.7
58	274.2	230.1	18	320.2	268.7	78	366.2	307.3	38	412.1	345.8	98	458.1	384.4
59	275.0	230.8	19	321.0	269.3	79	366.9	307.9	39	412.9	346.5	99	458.9	385.0
60	275.8	231.4	20	321.7	270.0	80	367.7	308.5	40	413.7	347.1	600	459.6	385.7
Dist.	Dep.	D. Lat.	Dist.	Dep.	D. Lat.	Dist.	Dep.	D. Lat.	Dist.	Dep.	D. Lat.	Dist.	Dep.	D. Lat.

Dist.	D. Lat.	Dep.
D Lo	Dep.	
	m	D Lo

50°

310° | 050°
230° | 130°

TABLE 4

Traverse 41° Table

319° | 041°
221° | 139°

Dist.	D. Lat.	Dep.	Dist.	D. Lat.	Dep.	Dist.	D. Lat.	Dep.	Dist.	D. Lat.	Dep.	Dist.	D. Lat.	Dep.
1	0.8	0.7	61	46.0	40.0	121	91.3	79.4	181	136.6	118.7	241	181.9	158.1
2	1.5	1.3	62	46.8	40.7	22	92.1	80.0	82	137.4	119.4	42	182.6	158.8
3	2.3	2.0	63	47.5	41.3	23	92.8	80.7	83	138.1	120.1	43	183.4	159.4
4	3.0	2.6	64	48.3	42.0	24	93.6	81.4	84	138.9	120.7	44	184.1	160.1
5	3.8	3.3	65	49.1	42.6	25	94.3	82.0	85	139.6	121.4	45	184.9	160.7
6	4.5	3.9	66	49.8	43.3	26	95.1	82.7	86	140.4	122.0	46	185.7	161.4
7	5.3	4.6	67	50.6	44.0	27	95.8	83.3	87	141.1	122.7	47	186.4	162.0
8	6.0	5.2	68	51.3	44.6	28	96.6	84.0	88	141.9	123.3	48	187.2	162.7
9	6.8	5.9	69	52.1	45.3	29	97.4	84.6	89	142.6	124.0	49	187.9	163.4
10	7.5	6.6	70	52.8	45.9	30	98.1	85.3	90	143.4	124.7	50	188.7	164.0
11	8.3	7.2	71	53.6	46.6	131	98.9	85.9	191	144.1	125.3	251	189.4	164.7
12	9.1	7.9	72	54.3	47.2	32	99.6	86.6	92	144.9	126.0	52	190.2	165.3
13	9.8	8.5	73	55.1	47.9	33	100.4	87.3	93	145.7	126.6	53	190.9	166.0
14	10.6	9.2	74	55.8	48.5	34	101.1	87.9	94	146.4	127.3	54	191.7	166.6
15	11.3	9.8	75	56.6	49.2	35	101.9	88.6	95	147.2	127.9	55	192.5	167.3
16	12.1	10.5	76	57.4	49.9	36	102.6	89.2	96	147.9	128.6	56	193.2	168.0
17	12.8	11.2	77	58.1	50.5	37	103.4	89.9	97	148.7	129.2	57	194.0	168.6
18	13.6	11.8	78	58.9	51.2	38	104.1	90.5	98	149.4	129.9	58	194.7	169.3
19	14.3	12.5	79	59.6	51.8	39	104.9	91.2	99	150.2	130.6	59	195.5	169.9
20	15.1	13.1	80	60.4	52.5	40	105.7	91.8	200	150.9	131.2	60	196.2	170.6
21	15.8	13.8	81	61.1	53.1	141	106.4	92.5	201	151.7	131.9	261	197.0	171.2
22	16.6	14.4	82	61.9	53.8	42	107.2	93.2	02	152.5	132.5	62	197.7	171.9
23	17.4	15.1	83	62.6	54.5	43	107.9	93.8	03	153.2	133.2	63	198.5	172.5
24	18.1	15.7	84	63.4	55.1	44	108.7	94.5	04	154.0	133.8	64	199.2	173.2
25	18.9	16.4	85	64.2	55.8	45	109.4	95.1	05	154.7	134.5	65	200.0	173.9
26	19.6	17.1	86	64.9	56.4	46	110.2	95.8	06	155.5	135.1	66	200.8	174.5
27	20.4	17.7	87	65.7	57.1	47	110.9	96.4	07	156.2	135.8	67	201.5	175.2
28	21.1	18.4	88	66.4	57.7	48	111.7	97.1	08	157.0	136.5	68	202.3	175.8
29	21.9	19.0	89	67.2	58.4	49	112.5	97.8	09	157.7	137.1	69	203.0	176.5
30	22.6	19.7	90	67.9	59.0	50	113.2	98.4	10	158.5	137.8	70	203.8	177.1
31	23.4	20.3	91	68.7	59.7	151	114.0	99.1	211	159.2	138.4	271	204.5	177.8
32	24.2	21.0	92	69.4	60.4	52	114.7	99.7	12	160.0	139.1	72	205.3	178.4
33	24.9	21.6	93	70.2	61.0	53	115.5	100.4	13	160.8	139.7	73	206.0	179.1
34	25.7	22.3	94	70.9	61.7	54	116.2	101.0	14	161.5	140.4	74	206.8	179.8
35	26.4	23.0	95	71.7	62.3	55	117.0	101.7	15	162.3	141.1	75	207.5	180.4
36	27.2	23.6	96	72.5	63.0	56	117.7	102.3	16	163.0	141.7	76	208.3	181.1
37	27.9	24.3	97	73.2	63.6	57	118.5	103.0	17	163.8	142.4	77	209.1	181.7
38	28.7	24.9	98	74.0	64.3	58	119.2	103.7	18	164.5	143.0	78	209.8	182.4
39	29.4	25.6	99	74.7	64.9	59	120.0	104.3	19	165.3	143.7	79	210.6	183.0
40	30.2	26.2	100	75.5	65.6	60	120.8	105.0	20	166.0	144.3	80	211.3	183.7
41	30.9	26.9	101	76.2	66.3	161	121.5	105.6	221	166.8	145.0	281	212.1	184.4
42	31.7	27.6	02	77.0	66.9	62	122.3	106.3	22	167.5	145.6	82	212.8	185.0
43	32.5	28.2	03	77.7	67.6	63	123.0	106.9	23	168.3	146.3	83	213.6	185.7
44	33.2	28.9	04	78.5	68.2	64	123.8	107.6	24	169.1	147.0	84	214.3	186.3
45	34.0	29.5	05	79.2	68.9	65	124.5	108.2	25	169.8	147.6	85	215.1	187.0
46	34.7	30.2	06	80.0	69.5	66	125.3	108.9	26	170.6	148.3	86	215.8	187.6
47	35.5	30.8	07	80.8	70.2	67	126.0	109.6	27	171.3	148.9	87	216.6	188.3
48	36.2	31.5	08	81.5	70.9	68	126.8	110.2	28	172.1	149.6	88	217.4	188.9
49	37.0	32.1	09	82.3	71.5	69	127.5	110.9	29	172.8	150.2	89	218.1	189.6
50	37.7	32.8	10	83.0	72.2	70	128.3	111.5	30	173.6	150.9	90	218.9	190.3
51	38.5	33.5	111	83.8	72.8	171	129.1	112.2	231	174.3	151.5	291	219.6	190.9
52	39.2	34.1	12	84.5	73.5	72	129.8	112.8	32	175.1	152.2	92	220.4	191.6
53	40.0	34.8	13	85.3	74.1	73	130.6	113.5	33	175.8	152.9	93	221.1	192.2
54	40.8	35.4	14	86.0	74.8	74	131.3	114.2	34	176.6	153.5	94	221.9	192.9
55	41.5	36.1	15	86.8	75.4	75	132.1	114.8	35	177.4	154.2	95	222.6	193.5
56	42.3	36.7	16	87.5	76.1	76	132.8	115.5	36	178.1	154.8	96	223.4	194.2
57	43.0	37.4	17	88.3	76.8	77	133.6	116.1	37	178.9	155.5	97	224.1	194.8
58	43.8	38.1	18	89.1	77.4	78	134.3	116.8	38	179.6	156.1	98	224.9	195.5
59	44.5	38.7	19	89.8	78.1	79	135.1	117.4	39	180.4	156.8	99	225.7	196.2
60	45.3	39.4	20	90.6	78.7	80	135.8	118.1	40	181.1	157.5	300	226.4	196.8
Dist.	Dep.	D. Lat.	Dist.	Dep.	D. Lat.	Dist.	Dep.	D. Lat.	Dist.	Dep.	D. Lat.	Dist.	Dep.	D. Lat.

311° | 049°
229° | 131°

49°

Dist.	D. Lat.	Dep.
N.	N x Cos.	N x Sin.
Hypotenuse	Side Adj.	Side Opp.

TABLE 4

Traverse 41° Table

319° | 041°
221° | 139°

Dist.	D. Lat.	Dep.	Dist.	D. Lat.	Dep.	Dist.	D. Lat.	Dep.	Dist.	D. Lat.	Dep.	Dist.	D. Lat.	Dep.
301	227.2	197.5	361	272.5	236.8	421	317.7	276.2	481	363.0	315.6	541	408.3	354.9
02	227.9	198.1	62	273.2	237.5	22	318.5	276.9	82	363.8	316.2	42	409.1	355.6
03	228.7	198.8	63	274.0	238.1	23	319.2	277.5	83	364.5	316.9	43	409.8	356.2
04	229.4	199.4	64	274.7	238.8	24	320.0	278.2	84	365.3	317.5	44	410.6	356.9
05	230.2	200.1	65	275.5	239.5	25	320.8	278.8	85	366.0	318.2	45	411.3	357.6
06	230.9	200.8	66	276.2	240.1	26	321.5	279.5	86	366.8	318.8	46	412.1	358.2
07	231.7	201.4	67	277.0	240.8	27	322.3	280.1	87	367.5	319.5	47	412.8	358.9
08	232.5	202.1	68	277.7	241.4	28	323.0	280.8	88	368.3	320.2	48	413.6	359.5
09	233.2	202.7	69	278.5	242.1	29	323.8	281.4	89	369.1	320.8	49	414.3	360.2
10	234.0	203.4	70	279.2	242.7	30	324.5	282.1	90	369.8	321.5	50	415.1	360.8
311	234.7	204.0	371	280.0	243.4	431	325.3	282.8	491	370.6	322.1	551	415.8	361.5
12	235.5	204.7	72	280.8	244.1	32	326.0	283.4	92	371.3	322.8	52	416.6	362.1
13	236.2	205.3	73	281.5	244.7	33	326.8	284.1	93	372.1	323.4	53	417.4	362.8
14	237.0	206.0	74	282.3	245.4	34	327.5	284.7	94	372.8	324.1	54	418.1	363.5
15	237.7	206.7	75	283.0	246.0	35	328.3	285.4	95	373.6	324.7	55	418.9	364.1
16	238.5	207.3	76	283.8	246.7	36	329.1	286.0	96	374.3	325.4	56	419.6	364.8
17	239.2	208.0	77	284.5	247.3	37	329.8	286.7	97	375.1	326.1	57	420.4	365.4
18	240.0	208.6	78	285.3	248.0	38	330.6	287.4	98	375.8	326.7	58	421.1	366.1
19	240.8	209.3	79	286.0	248.6	39	331.3	288.0	99	376.6	327.4	59	421.9	366.7
20	241.5	209.9	80	286.8	249.3	40	332.1	288.7	500	377.4	328.0	60	422.6	367.4
321	242.3	210.6	381	287.5	250.0	441	332.8	289.3	501	378.1	328.7	561	423.4	368.0
22	243.0	211.3	82	288.3	250.6	42	333.6	290.0	02	378.9	329.3	62	424.1	368.7
23	243.8	211.9	83	289.1	251.3	43	334.3	290.6	03	379.6	330.0	63	424.9	369.4
24	244.5	212.6	84	289.8	251.9	44	335.1	291.3	04	380.4	330.7	64	425.7	370.0
25	245.3	213.2	85	290.6	252.6	45	335.8	291.9	05	381.1	331.3	65	426.4	370.7
26	246.0	213.9	86	291.3	253.2	46	336.6	292.6	06	381.9	332.0	66	427.2	371.3
27	246.8	214.5	87	292.1	253.9	47	337.4	293.3	07	382.6	332.6	67	427.9	372.0
28	247.5	215.2	88	292.8	254.6	48	338.1	293.9	08	383.4	333.3	68	428.7	372.6
29	248.3	215.8	89	293.6	255.2	49	338.9	294.6	09	384.1	333.9	69	429.4	373.3
30	249.1	216.5	90	294.3	255.9	50	339.6	295.2	10	384.9	334.6	70	430.2	374.0
331	249.8	217.2	391	295.1	256.5	451	340.4	295.9	511	385.7	335.2	571	430.9	374.6
32	250.6	217.8	92	295.8	257.2	52	341.1	296.5	12	386.4	335.9	72	431.7	375.3
33	251.3	218.5	93	296.6	257.8	53	341.9	297.2	13	387.2	336.6	73	432.4	375.9
34	252.1	219.1	94	297.4	258.5	54	342.6	297.9	14	387.9	337.2	74	433.2	376.6
35	252.8	219.8	95	298.1	259.1	55	343.4	298.5	15	388.7	337.9	75	434.0	377.2
36	253.6	220.4	96	298.9	259.8	56	344.1	299.2	16	389.4	338.5	76	434.7	377.9
37	254.3	221.1	97	299.6	260.5	57	344.9	299.8	17	390.2	339.2	77	435.5	378.5
38	255.1	221.7	98	300.4	261.1	58	345.7	300.5	18	390.9	339.8	78	436.2	379.2
39	255.8	222.4	99	301.1	261.8	59	346.4	301.1	19	391.7	340.5	79	437.0	379.9
40	256.6	223.1	400	301.9	262.4	60	347.2	301.8	20	392.4	341.2	80	437.7	380.5
341	257.4	223.7	401	302.6	263.1	461	347.9	302.4	521	393.2	341.8	581	438.5	381.2
42	258.1	224.4	02	303.4	263.7	62	348.7	303.1	22	394.0	342.5	82	439.2	381.8
43	258.9	225.0	03	304.1	264.4	63	349.4	303.8	23	394.7	343.1	83	440.0	382.5
44	259.6	225.7	04	304.9	265.0	64	350.2	304.4	24	395.5	343.8	84	440.8	383.1
45	260.4	226.3	05	305.7	265.7	65	350.9	305.1	25	396.2	344.4	85	441.5	383.8
46	261.1	227.0	06	306.4	266.4	66	351.7	305.7	26	397.0	345.1	86	442.3	384.5
47	261.9	227.7	07	307.2	267.0	67	352.4	306.4	27	397.7	345.7	87	443.0	385.1
48	262.6	228.3	08	307.9	267.7	68	353.2	307.0	28	398.5	346.4	88	443.8	385.8
49	263.4	229.0	09	308.7	268.3	69	354.0	307.7	29	399.2	347.1	89	444.5	386.4
50	264.1	229.6	10	309.4	269.0	70	354.7	308.3	30	400.0	347.7	90	445.3	387.1
351	264.9	230.3	411	310.2	269.6	471	355.5	309.0	531	400.8	348.4	591	446.0	387.7
52	265.7	230.9	12	310.9	270.3	72	356.2	309.7	32	401.5	349.0	92	446.8	388.4
53	266.4	231.6	13	311.7	271.0	73	357.0	310.3	33	402.3	349.7	93	447.5	389.0
54	267.2	232.2	14	312.4	271.6	74	357.7	311.0	34	403.0	350.3	94	448.3	389.7
55	267.9	232.9	15	313.2	272.3	75	358.5	311.6	35	403.8	351.0	95	449.1	390.4
56	268.7	233.6	16	314.0	272.9	76	359.2	312.3	36	404.5	351.6	96	449.8	391.0
57	269.4	234.2	17	314.7	273.6	77	360.0	312.9	37	405.3	352.3	97	450.6	391.7
58	270.2	234.9	18	315.5	274.2	78	360.8	313.6	38	406.0	353.0	98	451.3	392.3
59	270.9	235.5	19	316.2	274.9	79	361.5	314.3	39	406.8	353.6	99	452.1	393.0
60	271.7	236.2	20	317.0	275.5	80	362.3	314.9	40	407.5	354.3	600	452.8	393.6
Dist.	Dep.	D. Lat.	Dist.	Dep.	D. Lat.	Dist.	Dep.	D. Lat.	Dist.	Dep.	D. Lat.	Dist.	Dep.	D. Lat.

Dist.	D. Lat.	Dep.
D Lo	Dep.	
	m	D Lo

49°

311° | 049°
229° | 131°

TABLE 4

Traverse 42° Table

318° | 042° — 222° | 138°

Dist.	D. Lat.	Dep.	Dist.	D. Lat.	Dep.	Dist.	D. Lat.	Dep.	Dist.	D. Lat.	Dep.	Dist.	D. Lat.	Dep.
1	0.7	0.7	61	45.3	40.8	121	89.9	81.0	181	134.5	121.1	241	179.1	161.3
2	1.5	1.3	62	46.1	41.5	22	90.7	81.6	82	135.3	121.8	42	179.8	161.9
3	2.2	2.0	63	46.8	42.2	23	91.4	82.3	83	136.0	122.5	43	180.6	162.6
4	3.0	2.7	64	47.6	42.8	24	92.1	83.0	84	136.7	123.1	44	181.3	163.3
5	3.7	3.3	65	48.3	43.5	25	92.9	83.6	85	137.5	123.8	45	182.1	163.9
6	4.5	4.0	66	49.0	44.2	26	93.6	84.3	86	138.2	124.5	46	182.8	164.6
7	5.2	4.7	67	49.8	44.8	27	94.4	85.0	87	139.0	125.1	47	183.6	165.3
8	5.9	5.4	68	50.5	45.5	28	95.1	85.6	88	139.7	125.8	48	184.3	165.9
9	6.7	6.0	69	51.3	46.2	29	95.9	86.3	89	140.5	126.5	49	185.0	166.6
10	7.4	6.7	70	52.0	46.8	30	96.6	87.0	90	141.2	127.1	50	185.8	167.3
11	8.2	7.4	71	52.8	47.5	131	97.4	87.7	191	141.9	127.8	251	186.5	168.0
12	8.9	8.0	72	53.5	48.2	32	98.1	88.3	92	142.7	128.5	52	187.3	168.6
13	9.7	8.7	73	54.2	48.8	33	98.8	89.0	93	143.4	129.1	53	188.0	169.3
14	10.4	9.4	74	55.0	49.5	34	99.6	89.7	94	144.2	129.8	54	188.8	170.0
15	11.1	10.0	75	55.7	50.2	35	100.3	90.3	95	144.9	130.5	55	189.5	170.6
16	11.9	10.7	76	56.5	50.9	36	101.1	91.0	96	145.7	131.1	56	190.2	171.3
17	12.6	11.4	77	57.2	51.5	37	101.8	91.7	97	146.4	131.8	57	191.0	172.0
18	13.4	12.0	78	58.0	52.2	38	102.6	92.3	98	147.1	132.5	58	191.7	172.6
19	14.1	12.7	79	58.7	52.9	39	103.3	93.0	99	147.9	133.2	59	192.5	173.3
20	14.9	13.4	80	59.5	53.5	40	104.0	93.7	200	148.6	133.8	60	193.2	174.0
21	15.6	14.1	81	60.2	54.2	141	104.8	94.3	201	149.4	134.5	261	194.0	174.6
22	16.3	14.7	82	60.9	54.9	42	105.5	95.0	02	150.1	135.2	62	194.7	175.3
23	17.1	15.4	83	61.7	55.5	43	106.3	95.7	03	150.9	135.8	63	195.4	176.0
24	17.8	16.1	84	62.4	56.2	44	107.0	96.4	04	151.6	136.5	64	196.2	176.7
25	18.6	16.7	85	63.2	56.9	45	107.8	97.0	05	152.3	137.2	65	196.9	177.3
26	19.3	17.4	86	63.9	57.5	46	108.5	97.7	06	153.1	137.8	66	197.7	178.0
27	20.1	18.1	87	64.7	58.2	47	109.2	98.4	07	153.8	138.5	67	198.4	178.7
28	20.8	18.7	88	65.4	58.9	48	110.0	99.0	08	154.6	139.2	68	199.2	179.3
29	21.6	19.4	89	66.1	59.6	49	110.7	99.7	09	155.3	139.8	69	199.9	180.0
30	22.3	20.1	90	66.9	60.2	50	111.5	100.4	10	156.1	140.5	70	200.6	180.7
31	23.0	20.7	91	67.6	60.9	151	112.2	101.0	211	156.8	141.2	271	201.4	181.3
32	23.8	21.4	92	68.4	61.6	52	113.0	101.7	12	157.5	141.9	72	202.1	182.0
33	24.5	22.1	93	69.1	62.2	53	113.7	102.4	13	158.3	142.5	73	202.9	182.7
34	25.3	22.8	94	69.9	62.9	54	114.4	103.0	14	159.0	143.2	74	203.6	183.3
35	26.0	23.4	95	70.6	63.6	55	115.2	103.7	15	159.8	143.9	75	204.4	184.0
36	26.8	24.1	96	71.3	64.2	56	115.9	104.4	16	160.5	144.5	76	205.1	184.7
37	27.5	24.8	97	72.1	64.9	57	116.7	105.1	17	161.3	145.2	77	205.9	185.3
38	28.2	25.4	98	72.8	65.6	58	117.4	105.7	18	162.0	145.9	78	206.6	186.0
39	29.0	26.1	99	73.6	66.2	59	118.2	106.4	19	162.7	146.5	79	207.3	186.7
40	29.7	26.8	100	74.3	66.9	60	118.9	107.1	20	163.5	147.2	80	208.1	187.4
41	30.5	27.4	101	75.1	67.6	161	119.6	107.7	221	164.2	147.9	281	208.8	188.0
42	31.2	28.1	02	75.8	68.3	62	120.4	108.4	22	165.0	148.5	82	209.6	188.7
43	32.0	28.8	03	76.5	68.9	63	121.1	109.1	23	165.7	149.2	83	210.3	189.4
44	32.7	29.4	04	77.3	69.6	64	121.9	109.7	24	166.5	149.9	84	211.1	190.0
45	33.4	30.1	05	78.0	70.3	65	122.6	110.4	25	167.2	150.6	85	211.8	190.7
46	34.2	30.8	06	78.8	70.9	66	123.4	111.1	26	168.0	151.2	86	212.5	191.4
47	34.9	31.4	07	79.5	71.6	67	124.1	111.7	27	168.7	151.9	87	213.3	192.0
48	35.7	32.1	08	80.3	72.3	68	124.8	112.4	28	169.4	152.6	88	214.0	192.7
49	36.4	32.8	09	81.0	72.9	69	125.6	113.1	29	170.2	153.2	89	214.8	193.4
50	37.2	33.5	10	81.7	73.6	70	126.3	113.8	30	170.9	153.9	90	215.5	194.0
51	37.9	34.1	111	82.5	74.3	171	127.1	114.4	231	171.7	154.6	291	216.3	194.7
52	38.6	34.8	12	83.2	74.9	72	127.8	115.1	32	172.4	155.2	92	217.0	195.4
53	39.4	35.5	13	84.0	75.6	73	128.6	115.8	33	173.2	155.9	93	217.7	196.1
54	40.1	36.1	14	84.7	76.3	74	129.3	116.4	34	173.9	156.6	94	218.5	196.7
55	40.9	36.8	15	85.5	77.0	75	130.1	117.1	35	174.6	157.2	95	219.2	197.4
56	41.6	37.5	16	86.2	77.6	76	130.8	117.8	36	175.4	157.9	96	220.0	198.1
57	42.4	38.1	17	86.9	78.3	77	131.5	118.4	37	176.1	158.6	97	220.7	198.7
58	43.1	38.8	18	87.7	79.0	78	132.3	119.1	38	176.9	159.3	98	221.5	199.4
59	43.8	39.5	19	88.4	79.6	79	133.0	119.8	39	177.6	159.9	99	222.2	200.1
60	44.6	40.1	20	89.2	80.3	80	133.8	120.4	40	178.4	160.6	300	222.9	200.7
Dist.	Dep.	D. Lat.	Dist.	Dep.	D. Lat.	Dist.	Dep.	D. Lat.	Dist.	Dep.	D. Lat.	Dist.	Dep.	D. Lat.

312° | 048° — 228° | 132°

48°

Dist.	D. Lat.	Dep.
N.	N x Cos.	N x Sin.
Hypotenuse	Side Adj.	Side Opp.

TABLE 4

Traverse 42° Table

318° | 042° — 222° | 138°

Dist.	D. Lat.	Dep.	Dist.	D. Lat.	Dep.	Dist.	D. Lat.	Dep.	Dist.	D. Lat.	Dep.	Dist.	D. Lat.	Dep.
301	223.7	201.4	361	268.3	241.6	421	312.9	281.7	481	357.5	321.9	541	402.0	362.0
02	224.4	202.1	62	269.0	242.2	22	313.6	282.4	82	358.2	322.5	42	402.8	362.7
03	225.2	202.7	63	269.8	242.9	23	314.4	283.0	83	358.9	323.2	43	403.5	363.3
04	225.9	203.4	64	270.5	243.6	24	315.1	283.7	84	359.7	323.9	44	404.3	364.0
05	226.7	204.1	65	271.2	244.2	25	315.8	284.4	85	360.4	324.5	45	405.0	364.7
06	227.4	204.8	66	272.0	244.9	26	316.6	285.0	86	361.2	325.2	46	405.8	365.3
07	228.1	205.4	67	272.7	245.6	27	317.3	285.7	87	361.9	325.9	47	406.5	366.0
08	228.9	206.1	68	273.5	246.2	28	318.1	286.4	88	362.7	326.5	48	407.2	366.7
09	229.6	206.8	69	274.2	246.9	29	318.8	287.1	89	363.4	327.2	49	408.0	367.4
10	230.4	207.4	70	275.0	247.6	30	319.6	287.7	90	364.1	327.9	50	408.7	368.0
311	231.1	208.1	371	275.7	248.2	431	320.3	288.4	491	364.9	328.5	551	409.5	368.7
12	231.9	208.8	72	276.4	248.9	32	321.0	289.1	92	365.6	329.2	52	410.2	369.4
13	232.6	209.4	73	277.2	249.6	33	321.8	289.7	93	366.4	329.9	53	411.0	370.0
14	233.3	210.1	74	277.9	250.3	34	322.5	290.4	94	367.1	330.6	54	411.7	370.7
15	234.1	210.8	75	278.7	250.9	35	323.3	291.1	95	367.9	331.2	55	412.4	371.4
16	234.8	211.4	76	279.4	251.6	36	324.0	291.7	96	368.6	331.9	56	413.2	372.0
17	235.6	212.1	77	280.2	252.3	37	324.8	292.4	97	369.3	332.6	57	413.9	372.7
18	236.3	212.8	78	280.9	252.9	38	325.5	293.1	98	370.1	333.2	58	414.7	373.4
19	237.1	213.5	79	281.7	253.6	39	326.2	293.7	99	370.8	333.9	59	415.4	374.0
20	237.8	214.1	80	282.4	254.3	40	327.0	294.4	500	371.6	334.6	60	416.2	374.7
321	238.5	214.8	381	283.1	254.9	441	327.7	295.1	501	372.3	335.2	561	416.9	375.4
22	239.3	215.5	82	283.9	255.6	42	328.5	295.8	02	373.1	335.9	62	417.6	376.1
23	240.0	216.1	83	284.6	256.3	43	329.2	296.4	03	373.8	336.6	63	418.4	376.7
24	240.8	216.8	84	285.4	256.9	44	330.0	297.1	04	374.5	337.2	64	419.1	377.4
25	241.5	217.5	85	286.1	257.6	45	330.7	297.8	05	375.3	337.9	65	419.9	378.1
26	242.3	218.1	86	286.9	258.3	46	331.4	298.4	06	376.0	338.6	66	420.6	378.7
27	243.0	218.8	87	287.6	259.0	47	332.2	299.1	07	376.8	339.2	67	421.4	379.4
28	243.8	219.5	88	288.3	259.6	48	332.9	299.8	08	377.5	339.9	68	422.1	380.1
29	244.5	220.1	89	289.1	260.3	49	333.7	300.4	09	378.3	340.6	69	422.8	380.7
30	245.2	220.8	90	289.8	261.0	50	334.4	301.1	10	379.0	341.3	70	423.6	381.4
331	246.0	221.5	391	290.6	261.6	451	335.2	301.8	511	379.7	341.9	571	424.3	382.1
32	246.7	222.2	92	291.3	262.3	52	335.9	302.4	12	380.5	342.6	72	425.1	382.7
33	247.5	222.8	93	292.1	263.0	53	336.6	303.1	13	381.2	343.3	73	425.8	383.4
34	248.2	223.5	94	292.8	263.6	54	337.4	303.8	14	382.0	343.9	74	426.6	384.1
35	249.0	224.2	95	293.5	264.3	55	338.1	304.5	15	382.7	344.6	75	427.3	384.8
36	249.7	224.8	96	294.3	265.0	56	338.9	305.1	16	383.5	345.3	76	428.1	385.4
37	250.4	225.5	97	295.0	265.6	57	339.6	305.8	17	384.2	345.9	77	428.8	386.1
38	251.2	226.2	98	295.8	266.3	58	340.4	306.5	18	384.9	346.6	78	429.5	386.8
39	251.9	226.8	99	296.5	267.0	59	341.1	307.1	19	385.7	347.3	79	430.3	387.4
40	252.7	227.5	400	297.3	267.7	60	341.8	307.8	20	386.4	347.9	80	431.0	388.1
341	253.4	228.2	401	298.0	268.3	461	342.6	308.5	521	387.2	348.6	581	431.8	388.8
42	254.2	228.8	02	298.7	269.0	62	343.3	309.1	22	387.9	349.3	82	432.5	389.4
43	254.9	229.5	03	299.5	269.7	63	344.1	309.8	23	388.7	350.0	83	433.3	390.1
44	255.6	230.2	04	300.2	270.3	64	344.8	310.5	24	389.4	350.6	84	434.0	390.8
45	256.4	230.9	05	301.0	271.0	65	345.6	311.1	25	390.2	351.3	85	434.7	391.4
46	257.1	231.5	06	301.7	271.7	66	346.3	311.8	26	390.9	352.0	86	435.5	392.1
47	257.9	232.2	07	302.5	272.3	67	347.0	312.5	27	391.6	352.6	87	436.2	392.8
48	258.6	232.9	08	303.2	273.0	68	347.8	313.2	28	392.4	353.3	88	437.0	393.4
49	259.4	233.5	09	303.9	273.7	69	348.5	313.8	29	393.1	354.0	89	437.7	394.1
50	260.1	234.2	10	304.7	274.3	70	349.3	314.5	30	393.9	354.6	90	438.5	394.8
351	260.8	234.9	411	305.4	275.0	471	350.0	315.2	531	394.6	355.3	591	439.2	395.5
52	261.6	235.5	12	306.2	275.7	72	350.8	315.8	32	395.4	356.0	92	439.9	396.1
53	262.3	236.2	13	306.9	276.4	73	351.5	316.5	33	396.1	356.6	93	440.7	396.8
54	263.1	236.9	14	307.7	277.0	74	352.3	317.2	34	396.8	357.3	94	441.4	397.5
55	263.8	237.5	15	308.4	277.7	75	353.0	317.8	35	397.6	358.0	95	442.2	398.1
56	264.6	238.2	16	309.1	278.4	76	353.7	318.5	36	398.3	358.7	96	442.9	398.8
57	265.3	238.9	17	309.9	279.0	77	354.5	319.2	37	399.1	359.3	97	443.7	399.5
58	266.0	239.5	18	310.6	279.7	78	355.2	319.8	38	399.8	360.0	98	444.4	400.1
59	266.8	240.2	19	311.4	280.4	79	356.0	320.5	39	400.6	360.7	99	445.1	400.8
60	267.5	240.9	20	312.1	281.0	80	356.7	321.2	40	401.3	361.3	600	445.9	401.5
Dist.	Dep.	D. Lat.	Dist.	Dep.	D. Lat.	Dist.	Dep.	D. Lat.	Dist.	Dep.	D. Lat.	Dist.	Dep.	D. Lat.

Dist.	D. Lat.	Dep.
D Lo	Dep.	
	m	D Lo

48°

312° | 048° — 228° | 132°

TABLE 4
Traverse 43° Table

317° | 043°
223° | 137°

Dist.	D. Lat.	Dep.	Dist.	D. Lat.	Dep.	Dist.	D. Lat.	Dep.	Dist.	D. Lat.	Dep.	Dist.	D. Lat.	Dep.
1	0.7	0.7	61	44.6	41.6	121	88.5	82.5	181	132.4	123.4	241	176.3	164.4
2	1.5	1.4	62	45.3	42.3	22	89.2	83.2	82	133.1	124.1	42	177.0	165.0
3	2.2	2.0	63	46.1	43.0	23	90.0	83.9	83	133.8	124.8	43	177.7	165.7
4	2.9	2.7	64	46.8	43.6	24	90.7	84.6	84	134.6	125.5	44	178.5	166.4
5	3.7	3.4	65	47.5	44.3	25	91.4	85.2	85	135.3	126.2	45	179.2	167.1
6	4.4	4.1	66	48.3	45.0	26	92.2	85.9	86	136.0	126.9	46	179.9	167.8
7	5.1	4.8	67	49.0	45.7	27	92.9	86.6	87	136.8	127.5	47	180.6	168.5
8	5.9	5.5	68	49.7	46.4	28	93.6	87.3	88	137.5	128.2	48	181.4	169.1
9	6.6	6.1	69	50.5	47.1	29	94.3	88.0	89	138.2	128.9	49	182.1	169.8
10	7.3	6.8	70	51.2	47.7	30	95.1	88.7	90	139.0	129.6	50	182.8	170.5
11	8.0	7.5	71	51.9	48.4	131	95.8	89.3	191	139.7	130.3	251	183.6	171.2
12	8.8	8.2	72	52.7	49.1	32	96.5	90.0	92	140.4	130.9	52	184.3	171.9
13	9.5	8.9	73	53.4	49.8	33	97.3	90.7	93	141.2	131.6	53	185.0	172.5
14	10.2	9.5	74	54.1	50.5	34	98.0	91.4	94	141.9	132.3	54	185.8	173.2
15	11.0	10.2	75	54.9	51.1	35	98.7	92.1	95	142.6	133.0	55	186.5	173.9
16	11.7	10.9	76	55.6	51.8	36	99.5	92.8	96	143.3	133.7	56	187.2	174.6
17	12.4	11.6	77	56.3	52.5	37	100.2	93.4	97	144.1	134.4	57	188.0	175.3
18	13.2	12.3	78	57.0	53.2	38	100.9	94.1	98	144.8	135.0	58	188.7	176.0
19	13.9	13.0	79	57.8	53.9	39	101.7	94.8	99	145.5	135.7	59	189.4	176.6
20	14.6	13.6	80	58.5	54.6	40	102.4	95.5	200	146.3	136.4	60	190.2	177.3
21	15.4	14.3	81	59.2	55.2	141	103.1	96.2	201	147.0	137.1	261	190.9	178.0
22	16.1	15.0	82	60.0	55.9	42	103.9	96.8	02	147.7	137.8	62	191.6	178.7
23	16.8	15.7	83	60.7	56.6	43	104.6	97.5	03	148.5	138.4	63	192.3	179.4
24	17.6	16.4	84	61.4	57.3	44	105.3	98.2	04	149.2	139.1	64	193.1	180.0
25	18.3	17.0	85	62.2	58.0	45	106.0	98.9	05	149.9	139.8	65	193.8	180.7
26	19.0	17.7	86	62.9	58.7	46	106.8	99.6	06	150.7	140.5	66	194.5	181.4
27	19.7	18.4	87	63.6	59.3	47	107.5	100.3	07	151.4	141.2	67	195.3	182.1
28	20.5	19.1	88	64.4	60.0	48	108.2	100.9	08	152.1	141.9	68	196.0	182.8
29	21.2	19.8	89	65.1	60.7	49	109.0	101.6	09	152.9	142.5	69	196.7	183.5
30	21.9	20.5	90	65.8	61.4	50	109.7	102.3	10	153.6	143.2	70	197.5	184.1
31	22.7	21.1	91	66.6	62.1	151	110.4	103.0	211	154.3	143.9	271	198.2	184.8
32	23.4	21.8	92	67.3	62.7	52	111.2	103.7	12	155.0	144.6	72	198.9	185.5
33	24.1	22.5	93	68.0	63.4	53	111.9	104.3	13	155.8	145.3	73	199.7	186.2
34	24.9	23.2	94	68.7	64.1	54	112.6	105.0	14	156.5	145.9	74	200.4	186.9
35	25.6	23.9	95	69.5	64.8	55	113.4	105.7	15	157.2	146.6	75	201.1	187.5
36	26.3	24.6	96	70.2	65.5	56	114.1	106.4	16	158.0	147.3	76	201.9	188.2
37	27.1	25.2	97	70.9	66.2	57	114.8	107.1	17	158.7	148.0	77	202.6	188.9
38	27.8	25.9	98	71.7	66.8	58	115.6	107.8	18	159.4	148.7	78	203.3	189.6
39	28.5	26.6	99	72.4	67.5	59	116.3	108.4	19	160.2	149.4	79	204.0	190.3
40	29.3	27.3	100	73.1	68.2	60	117.0	109.1	20	160.9	150.0	80	204.8	191.0
41	30.0	28.0	101	73.9	68.9	161	117.7	109.8	221	161.6	150.7	281	205.5	191.6
42	30.7	28.6	02	74.6	69.6	62	118.5	110.5	22	162.4	151.4	82	206.2	192.3
43	31.4	29.3	03	75.3	70.2	63	119.2	111.2	23	163.1	152.1	83	207.0	193.0
44	32.2	30.0	04	76.1	70.9	64	119.9	111.8	24	163.8	152.8	84	207.7	193.7
45	32.9	30.7	05	76.8	71.6	65	120.7	112.5	25	164.6	153.4	85	208.4	194.4
46	33.6	31.4	06	77.5	72.3	66	121.4	113.2	26	165.3	154.1	86	209.2	195.1
47	34.4	32.1	07	78.3	73.0	67	122.1	113.9	27	166.0	154.8	87	209.9	195.7
48	35.1	32.7	08	79.0	73.7	68	122.9	114.6	28	166.7	155.5	88	210.6	196.4
49	35.8	33.4	09	79.7	74.3	69	123.6	115.3	29	167.5	156.2	89	211.4	197.1
50	36.6	34.1	10	80.4	75.0	70	124.3	115.9	30	168.2	156.9	90	212.1	197.8
51	37.3	34.8	111	81.2	75.7	171	125.1	116.6	231	168.9	157.5	291	212.8	198.5
52	38.0	35.5	12	81.9	76.4	72	125.8	117.3	32	169.7	158.2	92	213.6	199.1
53	38.8	36.1	13	82.6	77.1	73	126.5	118.0	33	170.4	158.9	93	214.3	199.8
54	39.5	36.8	14	83.4	77.7	74	127.3	118.7	34	171.1	159.6	94	215.0	200.5
55	40.2	37.5	15	84.1	78.4	75	128.0	119.3	35	171.9	160.3	95	215.7	201.2
56	41.0	38.2	16	84.8	79.1	76	128.7	120.0	36	172.6	161.0	96	216.5	201.9
57	41.7	38.9	17	85.6	79.8	77	129.4	120.7	37	173.3	161.6	97	217.2	202.6
58	42.4	39.6	18	86.3	80.5	78	130.2	121.4	38	174.1	162.3	98	217.9	203.2
59	43.1	40.2	19	87.0	81.2	79	130.9	122.1	39	174.8	163.0	99	218.7	203.9
60	43.9	40.9	20	87.8	81.8	80	131.6	122.8	40	175.5	163.7	300	219.4	204.6
Dist.	Dep.	D. Lat.	Dist.	Dep.	D. Lat.	Dist.	Dep.	D. Lat.	Dist.	Dep.	D. Lat.	Dist.	Dep.	D. Lat.

313° | 047°
227° | 133°

47°

Dist.	D. Lat.	Dep.
N.	N x Cos.	N x Sin.
Hypotenuse	Side Adj.	Side Opp.

TABLE 4
Traverse 43° Table

317° | 043°
223° | 137°

Dist.	D. Lat.	Dep.	Dist.	D. Lat.	Dep.	Dist.	D. Lat.	Dep.	Dist.	D. Lat.	Dep.	Dist.	D. Lat.	Dep.
301	220.1	205.3	361	264.0	246.2	421	307.9	287.1	481	351.8	328.0	541	395.7	369.0
02	220.9	206.0	62	264.8	246.9	22	308.6	287.8	82	352.5	328.7	42	396.4	369.6
03	221.6	206.6	63	265.5	247.6	23	309.4	288.5	83	353.2	329.4	43	397.1	370.3
04	222.3	207.3	64	266.2	248.2	24	310.1	289.2	84	354.0	330.1	44	397.9	371.0
05	223.1	208.0	65	266.9	248.9	25	310.8	289.8	85	354.7	330.8	45	398.6	371.7
06	223.8	208.7	66	267.7	249.6	26	311.6	290.5	86	355.4	331.5	46	399.3	372.4
07	224.5	209.4	67	268.4	250.3	27	312.3	291.2	87	356.2	332.1	47	400.1	373.1
08	225.3	210.1	68	269.1	251.0	28	313.0	291.9	88	356.9	332.8	48	400.8	373.7
09	226.0	210.7	69	269.9	251.7	29	313.8	292.6	89	357.6	333.5	49	401.5	374.4
10	226.7	211.4	70	270.6	252.3	30	314.5	293.3	90	358.4	334.2	50	402.2	375.1
311	227.5	212.1	371	271.3	253.0	431	315.2	293.9	491	359.1	334.9	551	403.0	375.8
12	228.2	212.8	72	272.1	253.7	32	315.9	294.6	92	359.8	335.5	52	403.7	376.5
13	228.9	213.5	73	272.8	254.4	33	316.7	295.3	93	360.6	336.2	53	404.4	377.1
14	229.6	214.1	74	273.5	255.1	34	317.4	296.0	94	361.3	336.9	54	405.2	377.8
15	230.4	214.8	75	274.3	255.7	35	318.1	296.7	95	362.0	337.6	55	405.9	378.5
16	231.1	215.5	76	275.0	256.4	36	318.9	297.4	96	362.8	338.3	56	406.6	379.2
17	231.8	216.2	77	275.7	257.1	37	319.6	298.0	97	363.5	339.0	57	407.4	379.9
18	232.6	216.9	78	276.5	257.8	38	320.3	298.7	98	364.2	339.6	58	408.1	380.6
19	233.3	217.6	79	277.2	258.5	39	321.1	299.4	99	364.9	340.3	59	408.8	381.2
20	234.0	218.2	80	277.9	259.2	40	321.8	300.1	500	365.7	341.0	60	409.6	381.9
321	234.8	218.9	381	278.6	259.8	441	322.5	300.8	501	366.4	341.7	561	410.3	382.6
22	235.5	219.6	82	279.4	260.5	42	323.3	301.4	02	367.1	342.4	62	411.0	383.3
23	236.2	220.3	83	280.1	261.2	43	324.0	302.1	03	367.9	343.0	63	411.8	384.0
24	237.0	221.0	84	280.8	261.9	44	324.7	302.8	04	368.6	343.7	64	412.5	384.6
25	237.7	221.6	85	281.6	262.6	45	325.5	303.5	05	369.3	344.4	65	413.2	385.3
26	238.4	222.3	86	282.3	263.3	46	326.2	304.2	06	370.1	345.1	66	413.9	386.0
27	239.2	223.0	87	283.0	263.9	47	326.9	304.9	07	370.8	345.8	67	414.7	386.7
28	239.9	223.7	88	283.8	264.6	48	327.6	305.5	08	371.5	346.5	68	415.4	387.4
29	240.6	224.4	89	284.5	265.3	49	328.4	306.2	09	372.3	347.1	69	416.1	388.1
30	241.3	225.1	90	285.2	266.0	50	329.1	306.9	10	373.0	347.8	70	416.9	388.7
331	242.1	225.7	391	286.0	266.7	451	329.8	307.6	511	373.7	348.5	571	417.6	389.4
32	242.8	226.4	92	286.7	267.3	52	330.6	308.3	12	374.5	349.2	72	418.3	390.1
33	243.5	227.1	93	287.4	268.0	53	331.3	308.9	13	375.2	349.9	73	419.1	390.8
34	244.3	227.8	94	288.2	268.7	54	332.0	309.6	14	375.9	350.5	74	419.8	391.5
35	245.0	228.5	95	288.9	269.4	55	332.8	310.3	15	376.6	351.2	75	420.5	392.1
36	245.7	229.2	96	289.6	270.1	56	333.5	311.0	16	377.4	351.9	76	421.3	392.8
37	246.5	229.8	97	290.3	270.8	57	334.2	311.7	17	378.1	352.6	77	422.0	393.5
38	247.2	230.5	98	291.1	271.4	58	335.0	312.4	18	378.8	353.3	78	422.7	394.2
39	247.9	231.2	99	291.8	272.1	59	335.7	313.0	19	379.6	354.0	79	423.5	394.9
40	248.7	231.9	400	292.5	272.8	60	336.4	313.7	20	380.3	354.6	80	424.2	395.6
341	249.4	232.6	401	293.3	273.5	461	337.2	314.4	521	381.0	355.3	581	424.9	396.2
42	250.1	233.2	02	294.0	274.2	62	337.9	315.1	22	381.8	356.0	82	425.6	396.9
43	250.9	233.9	03	294.7	274.8	63	338.6	315.8	23	382.5	356.7	83	426.4	397.6
44	251.6	234.6	04	295.5	275.5	64	339.3	316.4	24	383.2	357.4	84	427.1	398.3
45	252.3	235.3	05	296.2	276.2	65	340.1	317.1	25	384.0	358.0	85	427.8	399.0
46	253.0	236.0	06	296.9	276.9	66	340.8	317.8	26	384.7	358.7	86	428.6	399.7
47	253.8	236.7	07	297.7	277.6	67	341.5	318.5	27	385.4	359.4	87	429.3	400.3
48	254.5	237.3	08	298.4	278.3	68	342.3	319.2	28	386.2	360.1	88	430.0	401.0
49	255.2	238.0	09	299.1	278.9	69	343.0	319.9	29	386.9	360.8	89	430.8	401.7
50	256.0	238.7	10	299.9	279.6	70	343.7	320.5	30	387.6	361.5	90	431.5	402.4
351	256.7	239.4	411	300.6	280.3	471	344.5	321.2	531	388.3	362.1	591	432.2	403.1
52	257.4	240.1	12	301.3	281.0	72	345.2	321.9	32	389.1	362.8	92	433.0	403.7
53	258.2	240.7	13	302.0	281.7	73	345.9	322.6	33	389.8	363.5	93	433.7	404.4
54	258.9	241.4	14	302.8	282.3	74	346.7	323.3	34	390.5	364.2	94	434.4	405.1
55	259.6	242.1	15	303.5	283.0	75	347.4	323.9	35	391.3	364.9	95	435.2	405.8
56	260.4	242.8	16	304.2	283.7	76	348.1	324.6	36	392.0	365.6	96	435.9	406.5
57	261.1	243.5	17	305.0	284.4	77	348.9	325.3	37	392.7	366.2	97	436.6	407.2
58	261.8	244.2	18	305.7	285.1	78	349.6	326.0	38	393.5	366.9	98	437.3	407.8
59	262.6	244.8	19	306.4	285.8	79	350.3	326.7	39	394.2	367.6	99	438.1	408.5
60	263.3	245.5	20	307.2	286.4	80	351.0	327.4	40	394.9	368.3	600	438.8	409.2
Dist.	Dep.	D. Lat.	Dist.	Dep.	D. Lat.	Dist.	Dep.	D. Lat.	Dist.	Dep.	D. Lat.	Dist.	Dep.	D. Lat.

Dist.	D. Lat.	Dep.
D Lo	Dep.	
	m	D Lo

47°

313° | 047°
227° | 133°

TABLE 4

Traverse **44°** Table

316° | 044°
224° | 136°

Dist.	D. Lat.	Dep.	Dist.	D. Lat.	Dep.	Dist.	D. Lat.	Dep.	Dist.	D. Lat.	Dep.	Dist.	D. Lat.	Dep.
1	0.7	0.7	61	43.9	42.4	121	87.0	84.1	181	130.2	125.7	241	173.4	167.4
2	1.4	1.4	62	44.6	43.1	22	87.8	84.7	82	130.9	126.4	42	174.1	168.1
3	2.2	2.1	63	45.3	43.8	23	88.5	85.4	83	131.6	127.1	43	174.8	168.8
4	2.9	2.8	64	46.0	44.5	24	89.2	86.1	84	132.4	127.8	44	175.5	169.5
5	3.6	3.5	65	46.8	45.2	25	89.9	86.8	85	133.1	128.5	45	176.2	170.2
6	4.3	4.2	66	47.5	45.8	26	90.6	87.5	86	133.8	129.2	46	177.0	170.9
7	5.0	4.9	67	48.2	46.5	27	91.4	88.2	87	134.5	129.9	47	177.7	171.6
8	5.8	5.6	68	48.9	47.2	28	92.1	88.9	88	135.2	130.6	48	178.4	172.3
9	6.5	6.3	69	49.6	47.9	29	92.8	89.6	89	136.0	131.3	49	179.1	173.0
10	7.2	6.9	70	50.4	48.6	30	93.5	90.3	90	136.7	132.0	50	179.8	173.7
11	7.9	7.6	71	51.1	49.3	131	94.2	91.0	191	137.4	132.7	251	180.6	174.4
12	8.6	8.3	72	51.8	50.0	32	95.0	91.7	92	138.1	133.4	52	181.3	175.1
13	9.4	9.0	73	52.5	50.7	33	95.7	92.4	93	138.8	134.1	53	182.0	175.7
14	10.1	9.7	74	53.2	51.4	34	96.4	93.1	94	139.6	134.8	54	182.7	176.4
15	10.8	10.4	75	54.0	52.1	35	97.1	93.8	95	140.3	135.5	55	183.4	177.1
16	11.5	11.1	76	54.7	52.8	36	97.8	94.5	96	141.0	136.2	56	184.2	177.8
17	12.2	11.8	77	55.4	53.5	37	98.5	95.2	97	141.7	136.8	57	184.9	178.5
18	12.9	12.5	78	56.1	54.2	38	99.3	95.9	98	142.4	137.5	58	185.6	179.2
19	13.7	13.2	79	56.8	54.9	39	100.0	96.6	99	143.1	138.2	59	186.3	179.9
20	14.4	13.9	80	57.5	55.6	40	100.7	97.3	200	143.9	138.9	60	187.0	180.6
21	15.1	14.6	81	58.3	56.3	141	101.4	97.9	201	144.6	139.6	261	187.7	181.3
22	15.8	15.3	82	59.0	57.0	42	102.1	98.6	02	145.3	140.3	62	188.5	182.0
23	16.5	16.0	83	59.7	57.7	43	102.9	99.3	03	146.0	141.0	63	189.2	182.7
24	17.3	16.7	84	60.4	58.4	44	103.6	100.0	04	146.7	141.7	64	189.9	183.4
25	18.0	17.4	85	61.1	59.0	45	104.3	100.7	05	147.5	142.4	65	190.6	184.1
26	18.7	18.1	86	61.9	59.7	46	105.0	101.4	06	148.2	143.1	66	191.3	184.8
27	19.4	18.8	87	62.6	60.4	47	105.7	102.1	07	148.9	143.8	67	192.1	185.5
28	20.1	19.5	88	63.3	61.1	48	106.5	102.8	08	149.6	144.5	68	192.8	186.2
29	20.9	20.1	89	64.0	61.8	49	107.2	103.5	09	150.3	145.2	69	193.5	186.9
30	21.6	20.8	90	64.7	62.5	50	107.9	104.2	10	151.1	145.9	70	194.2	187.6
31	22.3	21.5	91	65.5	63.2	151	108.6	104.9	211	151.8	146.6	271	194.9	188.3
32	23.0	22.2	92	66.2	63.9	52	109.3	105.6	12	152.5	147.3	72	195.7	188.9
33	23.7	22.9	93	66.9	64.6	53	110.1	106.3	13	153.2	148.0	73	196.4	189.6
34	24.5	23.6	94	67.6	65.3	54	110.8	107.0	14	153.9	148.7	74	197.1	190.3
35	25.2	24.3	95	68.3	66.0	55	111.5	107.7	15	154.7	149.4	75	197.8	191.0
36	25.9	25.0	96	69.1	66.7	56	112.2	108.4	16	155.4	150.0	76	198.5	191.7
37	26.6	25.7	97	69.8	67.4	57	112.9	109.1	17	156.1	150.7	77	199.3	192.4
38	27.3	26.4	98	70.5	68.1	58	113.7	109.8	18	156.8	151.4	78	200.0	193.1
39	28.1	27.1	99	71.2	68.8	59	114.4	110.5	19	157.5	152.1	79	200.7	193.8
40	28.8	27.8	100	71.9	69.5	60	115.1	111.1	20	158.3	152.8	80	201.4	194.5
41	29.5	28.5	101	72.7	70.2	161	115.8	111.8	221	159.0	153.5	281	202.1	195.2
42	30.2	29.2	02	73.4	70.9	62	116.5	112.5	22	159.7	154.2	82	202.9	195.9
43	30.9	29.9	03	74.1	71.5	63	117.3	113.2	23	160.4	154.9	83	203.6	196.6
44	31.7	30.6	04	74.8	72.2	64	118.0	113.9	24	161.1	155.6	84	204.3	197.3
45	32.4	31.3	05	75.5	72.9	65	118.7	114.6	25	161.9	156.3	85	205.0	198.0
46	33.1	32.0	06	76.3	73.6	66	119.4	115.3	26	162.6	157.0	86	205.7	198.7
47	33.8	32.6	07	77.0	74.3	67	120.1	116.0	27	163.3	157.7	87	206.5	199.4
48	34.5	33.3	08	77.7	75.0	68	120.8	116.7	28	164.0	158.4	88	207.2	200.1
49	35.2	34.0	09	78.4	75.7	69	121.6	117.4	29	164.7	159.1	89	207.9	200.8
50	36.0	34.7	10	79.1	76.4	70	122.3	118.1	30	165.4	159.8	90	208.6	201.5
51	36.7	35.4	111	79.8	77.1	171	123.0	118.8	231	166.2	160.5	291	209.3	202.1
52	37.4	36.1	12	80.6	77.8	72	123.7	119.5	32	166.9	161.2	92	210.0	202.8
53	38.1	36.8	13	81.3	78.5	73	124.4	120.2	33	167.6	161.9	93	210.8	203.5
54	38.8	37.5	14	82.0	79.2	74	125.2	120.9	34	168.3	162.6	94	211.5	204.2
55	39.6	38.2	15	82.7	79.9	75	125.9	121.6	35	169.0	163.2	95	212.2	204.9
56	40.3	38.9	16	83.4	80.6	76	126.6	122.3	36	169.8	163.9	96	212.9	205.6
57	41.0	39.6	17	84.2	81.3	77	127.3	123.0	37	170.5	164.6	97	213.6	206.3
58	41.7	40.3	18	84.9	82.0	78	128.0	123.6	38	171.2	165.3	98	214.4	207.0
59	42.4	41.0	19	85.6	82.7	79	128.8	124.3	39	171.9	166.0	99	215.1	207.7
60	43.2	41.7	20	86.3	83.4	80	129.5	125.0	40	172.6	166.7	300	215.8	208.4
Dist.	Dep.	D. Lat.	Dist.	Dep.	D. Lat.	Dist.	Dep.	D. Lat.	Dist.	Dep.	D. Lat.	Dist.	Dep.	D. Lat.

314° | 046°
226° | 134°

46°

Dist.	D. Lat.	Dep.
N.	N x Cos.	N x Sin.
Hypotenuse	Side Adj.	Side Opp.

TABLE 4

Traverse **44°** Table

316° | 044°
224° | 136°

Dist.	D. Lat.	Dep.	Dist.	D. Lat.	Dep.	Dist.	D. Lat.	Dep.	Dist.	D. Lat.	Dep.	Dist.	D. Lat.	Dep.
301	216.5	209.1	361	259.7	250.8	421	302.8	292.5	481	346.0	334.1	541	389.2	375.8
02	217.2	209.8	62	260.4	251.5	22	303.6	293.1	82	346.7	334.8	42	389.9	376.5
03	218.0	210.5	63	261.1	252.2	23	304.3	293.8	83	347.4	335.5	43	390.6	377.2
04	218.7	211.2	64	261.8	252.9	24	305.0	294.5	84	348.2	336.2	44	391.3	377.9
05	219.4	211.9	65	262.6	253.6	25	305.7	295.2	85	348.9	336.9	45	392.0	378.6
06	220.1	212.6	66	263.3	254.2	26	306.4	295.9	86	349.6	337.6	46	392.8	379.3
07	220.8	213.3	67	264.0	254.9	27	307.2	296.6	87	350.3	338.3	47	393.5	380.0
08	221.6	214.0	68	264.7	255.6	28	307.9	297.3	88	351.0	339.0	48	394.2	380.7
09	222.3	214.6	69	265.4	256.3	29	308.6	298.0	89	351.8	339.7	49	394.9	381.4
10	223.0	215.3	70	266.2	257.0	30	309.3	298.7	90	352.5	340.4	50	395.6	382.1
311	223.7	216.0	371	266.9	257.7	431	310.0	299.4	491	353.2	341.1	551	396.4	382.8
12	224.4	216.7	72	267.6	258.4	32	310.8	300.1	92	353.9	341.8	52	397.1	383.5
13	225.2	217.4	73	268.3	259.1	33	311.5	300.8	93	354.6	342.5	53	397.8	384.1
14	225.9	218.1	74	269.0	259.8	34	312.2	301.5	94	355.4	343.2	54	398.5	384.8
15	226.6	218.8	75	269.8	260.5	35	312.9	302.2	95	356.1	343.9	55	399.2	385.5
16	227.3	219.5	76	270.5	261.2	36	313.6	302.9	96	356.8	344.6	56	400.0	386.2
17	228.0	220.2	77	271.2	261.9	37	314.4	303.6	97	357.5	345.2	57	400.7	386.9
18	228.8	220.9	78	271.9	262.6	38	315.1	304.3	98	358.2	345.9	58	401.4	387.6
19	229.5	221.6	79	272.6	263.3	39	315.8	305.0	99	359.0	346.6	59	402.1	388.3
20	230.2	222.3	80	273.3	264.0	40	316.5	305.6	500	359.7	347.3	60	402.8	389.0
321	230.9	223.0	381	274.1	264.7	441	317.2	306.3	501	360.4	348.0	561	403.5	389.7
22	231.6	223.7	82	274.8	265.4	42	317.9	307.0	02	361.1	348.7	62	404.3	390.4
23	232.3	224.4	83	275.5	266.1	43	318.7	307.7	03	361.8	349.4	63	405.0	391.1
24	233.1	225.1	84	276.2	266.7	44	319.4	308.4	04	362.5	350.1	64	405.7	391.8
25	233.8	225.8	85	276.9	267.4	45	320.1	309.1	05	363.3	350.8	65	406.4	392.5
26	234.5	226.5	86	277.7	268.1	46	320.8	309.8	06	364.0	351.5	66	407.1	393.2
27	235.2	227.2	87	278.4	268.8	47	321.5	310.5	07	364.7	352.2	67	407.9	393.9
28	235.9	227.8	88	279.1	269.5	48	322.3	311.2	08	365.4	352.9	68	408.6	394.6
29	236.7	228.5	89	279.8	270.2	49	323.0	311.9	09	366.1	353.6	69	409.3	395.3
30	237.4	229.2	90	280.5	270.9	50	323.7	312.6	10	366.9	354.3	70	410.0	396.0
331	238.1	229.9	391	281.3	271.6	451	324.4	313.3	511	367.6	355.0	571	410.7	396.6
32	238.8	230.6	92	282.0	272.3	52	325.1	314.0	12	368.3	355.7	72	411.5	397.3
33	239.5	231.3	93	282.7	273.0	53	325.9	314.7	13	369.0	356.4	73	412.2	398.0
34	240.3	232.0	94	283.4	273.7	54	326.6	315.4	14	369.7	357.1	74	412.9	398.7
35	241.0	232.7	95	284.1	274.4	55	327.3	316.1	15	370.5	357.7	75	413.6	399.4
36	241.7	233.4	96	284.9	275.1	56	328.0	316.8	16	371.2	358.4	76	414.3	400.1
37	242.4	234.1	97	285.6	275.8	57	328.7	317.5	17	371.9	359.1	77	415.1	400.8
38	243.1	234.8	98	286.3	276.5	58	329.5	318.2	18	372.6	359.8	78	415.8	401.5
39	243.9	235.5	99	287.0	277.2	59	330.2	318.8	19	373.3	360.5	79	416.5	402.2
40	244.6	236.2	400	287.7	277.9	60	330.9	319.5	20	374.1	361.2	80	417.2	402.9
341	245.3	236.9	401	288.5	278.6	461	331.6	320.2	521	374.8	361.9	581	417.9	403.6
42	246.0	237.6	02	289.2	279.3	62	332.3	320.9	22	375.5	362.6	82	418.7	404.3
43	246.7	238.3	03	289.9	279.9	63	333.1	321.6	23	376.2	363.3	83	419.4	405.0
44	247.5	239.0	04	290.6	280.6	64	333.8	322.3	24	376.9	364.0	84	420.1	405.7
45	248.2	239.7	05	291.3	281.3	65	334.5	323.0	25	377.7	364.7	85	420.8	406.4
46	248.9	240.4	06	292.1	282.0	66	335.2	323.7	26	378.4	365.4	86	421.5	407.1
47	249.6	241.0	07	292.8	282.7	67	335.9	324.4	27	379.1	366.1	87	422.3	407.8
48	250.3	241.7	08	293.5	283.4	68	336.7	325.1	28	379.8	366.8	88	423.0	408.5
49	251.0	242.4	09	294.2	284.1	69	337.4	325.8	29	380.5	367.5	89	423.7	409.2
50	251.8	243.1	10	294.9	284.8	70	338.1	326.5	30	381.3	368.2	90	424.4	409.8
351	252.5	243.8	411	295.6	285.5	471	338.8	327.2	531	382.0	368.9	591	425.1	410.5
52	253.2	244.5	12	296.4	286.2	72	339.5	327.9	32	382.7	369.6	92	425.8	411.2
53	253.9	245.2	13	297.1	286.9	73	340.2	328.6	33	383.4	370.3	93	426.6	411.9
54	254.6	245.9	14	297.8	287.6	74	341.0	329.3	34	384.1	370.9	94	427.3	412.6
55	255.4	246.6	15	298.5	288.3	75	341.7	330.0	35	384.8	371.6	95	428.0	413.3
56	256.1	247.3	16	299.2	289.0	76	342.4	330.7	36	385.6	372.3	96	428.7	414.0
57	256.8	248.0	17	300.0	289.7	77	343.1	331.4	37	386.3	373.0	97	429.4	414.7
58	257.5	248.7	18	300.7	290.4	78	343.8	332.0	38	387.0	373.7	98	430.2	415.4
59	258.2	249.4	19	301.4	291.1	79	344.6	332.7	39	387.7	374.4	99	430.9	416.1
60	259.0	250.1	20	302.1	291.8	80	345.3	333.4	40	388.4	375.1	600	431.6	416.8
Dist.	Dep.	D. Lat.	Dist.	Dep.	D. Lat.	Dist.	Dep.	D. Lat.	Dist.	Dep.	D. Lat.	Dist.	Dep.	D. Lat.

Dist.	D. Lat.	Dep.
D Lo	Dep.	
	m	D Lo

46°

314° | 046°
226° | 134°

TABLE 4

315° | 045° / 225° | 135°

Traverse **45°** Table

Dist.	D. Lat.	Dep.	Dist.	D. Lat.	Dep.	Dist.	D. Lat.	Dep.	Dist.	D. Lat.	Dep.	Dist.	D. Lat.	Dep.
1	0.7	0.7	61	43.1	43.1	121	85.6	85.6	181	128.0	128.0	241	170.4	170.4
2	1.4	1.4	62	43.8	43.8	22	86.3	86.3	82	128.7	128.7	42	171.1	171.1
3	2.1	2.1	63	44.5	44.5	23	87.0	87.0	83	129.4	129.4	43	171.8	171.8
4	2.8	2.8	64	45.3	45.3	24	87.7	87.7	84	130.1	130.1	44	172.5	172.5
5	3.5	3.5	65	46.0	46.0	25	88.4	88.4	85	130.8	130.8	45	173.2	173.2
6	4.2	4.2	66	46.7	46.7	26	89.1	89.1	86	131.5	131.5	46	173.9	173.9
7	4.9	4.9	67	47.4	47.4	27	89.8	89.8	87	132.2	132.2	47	174.7	174.7
8	5.7	5.7	68	48.1	48.1	28	90.5	90.5	88	132.9	132.9	48	175.4	175.4
9	6.4	6.4	69	48.8	48.8	29	91.2	91.2	89	133.6	133.6	49	176.1	176.1
10	7.1	7.1	70	49.5	49.5	30	91.9	91.9	90	134.4	134.4	50	176.8	176.8
11	7.8	7.8	71	50.2	50.2	131	92.6	92.6	191	135.1	135.1	251	177.5	177.5
12	8.5	8.5	72	50.9	50.9	32	93.3	93.3	92	135.8	135.8	52	178.2	178.2
13	9.2	9.2	73	51.6	51.6	33	94.0	94.0	93	136.5	136.5	53	178.9	178.9
14	9.9	9.9	74	52.3	52.3	34	94.8	94.8	94	137.2	137.2	54	179.6	179.6
15	10.6	10.6	75	53.0	53.0	35	95.5	95.5	95	137.9	137.9	55	180.3	180.3
16	11.3	11.3	76	53.7	53.7	36	96.2	96.2	96	138.6	138.6	56	181.0	181.0
17	12.0	12.0	77	54.4	54.4	37	96.9	96.9	97	139.3	139.3	57	181.7	181.7
18	12.7	12.7	78	55.2	55.2	38	97.6	97.6	98	140.0	140.0	58	182.4	182.4
19	13.4	13.4	79	55.9	55.9	39	98.3	98.3	99	140.7	140.7	59	183.1	183.1
20	14.1	14.1	80	56.6	56.6	40	99.0	99.0	200	141.4	141.4	60	183.8	183.8
21	14.8	14.8	81	57.3	57.3	141	99.7	99.7	201	142.1	142.1	261	184.6	184.6
22	15.6	15.6	82	58.0	58.0	42	100.4	100.4	02	142.8	142.8	62	185.3	185.3
23	16.3	16.3	83	58.7	58.7	43	101.1	101.1	03	143.5	143.5	63	186.0	186.0
24	17.0	17.0	84	59.4	59.4	44	101.8	101.8	04	144.2	144.2	64	186.7	186.7
25	17.7	17.7	85	60.1	60.1	45	102.5	102.5	05	145.0	145.0	65	187.4	187.4
26	18.4	18.4	86	60.8	60.8	46	103.2	103.2	06	145.7	145.7	66	188.1	188.1
27	19.1	19.1	87	61.5	61.5	47	103.9	103.9	07	146.4	146.4	67	188.8	188.8
28	19.8	19.8	88	62.2	62.2	48	104.7	104.7	08	147.1	147.1	68	189.5	189.5
29	20.5	20.5	89	62.9	62.9	49	105.4	105.4	09	147.8	147.8	69	190.2	190.2
30	21.2	21.2	90	63.6	63.6	50	106.1	106.1	10	148.5	148.5	70	190.9	190.9
31	21.9	21.9	91	64.3	64.3	151	106.8	106.8	211	149.2	149.2	271	191.6	191.6
32	22.6	22.6	92	65.1	65.1	52	107.5	107.5	12	149.9	149.9	72	192.3	192.3
33	23.3	23.3	93	65.8	65.8	53	108.2	108.2	13	150.6	150.6	73	193.0	193.0
34	24.0	24.0	94	66.5	66.5	54	108.9	108.9	14	151.3	151.3	74	193.7	193.7
35	24.7	24.7	95	67.2	67.2	55	109.6	109.6	15	152.0	152.0	75	194.5	194.5
36	25.5	25.5	96	67.9	67.9	56	110.3	110.3	16	152.7	152.7	76	195.2	195.2
37	26.2	26.2	97	68.6	68.6	57	111.0	111.0	17	153.4	153.4	77	195.9	195.9
38	26.9	26.9	98	69.3	69.3	58	111.7	111.7	18	154.1	154.1	78	196.6	196.6
39	27.6	27.6	99	70.0	70.0	59	112.4	112.4	19	154.9	154.9	79	197.3	197.3
40	28.3	28.3	100	70.7	70.7	60	113.1	113.1	20	155.6	155.6	80	198.0	198.0
41	29.0	29.0	101	71.4	71.4	161	113.8	113.8	221	156.3	156.3	281	198.7	198.7
42	29.7	29.7	02	72.1	72.1	62	114.6	114.6	22	157.0	157.0	82	199.4	199.4
43	30.4	30.4	03	72.8	72.8	63	115.3	115.3	23	157.7	157.7	83	200.1	200.1
44	31.1	31.1	04	73.5	73.5	64	116.0	116.0	24	158.4	158.4	84	200.8	200.8
45	31.8	31.8	05	74.2	74.2	65	116.7	116.7	25	159.1	159.1	85	201.5	201.5
46	32.5	32.5	06	75.0	75.0	66	117.4	117.4	26	159.8	159.8	86	202.2	202.2
47	33.2	33.2	07	75.7	75.7	67	118.1	118.1	27	160.5	160.5	87	202.9	202.9
48	33.9	33.9	08	76.4	76.4	68	118.8	118.8	28	161.2	161.2	88	203.6	203.6
49	34.6	34.6	09	77.1	77.1	69	119.5	119.5	29	161.9	161.9	89	204.4	204.4
50	35.4	35.4	10	77.8	77.8	70	120.2	120.2	30	162.6	162.6	90	205.1	205.1
51	36.1	36.1	111	78.5	78.5	171	120.9	120.9	231	163.3	163.3	291	205.8	205.8
52	36.8	36.8	12	79.2	79.2	72	121.6	121.6	32	164.0	164.0	92	206.5	206.5
53	37.5	37.5	13	79.9	79.9	73	122.3	122.3	33	164.8	164.8	93	207.2	207.2
54	38.2	38.2	14	80.6	80.6	74	123.0	123.0	34	165.5	165.5	94	207.9	207.9
55	38.9	38.9	15	81.3	81.3	75	123.7	123.7	35	166.2	166.2	95	208.6	208.6
56	39.6	39.6	16	82.0	82.0	76	124.5	124.5	36	166.9	166.9	96	209.3	209.3
57	40.3	40.3	17	82.7	82.7	77	125.2	125.2	37	167.6	167.6	97	210.0	210.0
58	41.0	41.0	18	83.4	83.4	78	125.9	125.9	38	168.3	168.3	98	210.7	210.7
59	41.7	41.7	19	84.1	84.1	79	126.6	126.6	39	169.0	169.0	99	211.4	211.4
60	42.4	42.4	20	84.9	84.9	80	127.3	127.3	40	169.7	169.7	300	212.1	212.1
Dist.	Dep.	D. Lat.	Dist.	Dep.	D. Lat.	Dist.	Dep.	D. Lat.	Dist.	Dep.	D. Lat.	Dist.	Dep.	D. Lat.

315° | 045° / 225° | 135°

45°

Dist.	D. Lat.	Dep.
N.	N x Cos.	N x Sin.
Hypotenuse	Side Adj.	Side Opp.

TABLE 4

315° | 045° / 225° | 135°

Traverse **45°** Table

Dist.	D. Lat.	Dep.	Dist.	D. Lat.	Dep.	Dist.	D. Lat.	Dep.	Dist.	D. Lat.	Dep.	Dist.	D. Lat.	Dep.
301	212.8	212.8	361	255.3	255.3	421	297.7	297.7	481	340.1	340.1	541	382.5	382.5
02	213.5	213.5	62	256.0	256.0	22	298.4	298.4	82	340.8	340.8	42	383.3	383.3
03	214.3	214.3	63	256.7	256.7	23	299.1	299.1	83	341.5	341.5	43	384.0	384.0
04	215.0	215.0	64	257.4	257.4	24	299.8	299.8	84	342.2	342.2	44	384.7	384.7
05	215.7	215.7	65	258.1	258.1	25	300.5	300.5	85	342.9	342.9	45	385.4	385.4
06	216.4	216.4	66	258.8	258.8	26	301.2	301.2	86	343.7	343.7	46	386.1	386.1
07	217.1	217.1	67	259.5	259.5	27	301.9	301.9	87	344.4	344.4	47	386.8	386.8
08	217.8	217.8	68	260.2	260.2	28	302.6	302.6	88	345.1	345.1	48	387.5	387.5
09	218.5	218.5	69	260.9	260.9	29	303.3	303.3	89	345.8	345.8	49	388.2	388.2
10	219.2	219.2	70	261.6	261.6	30	304.1	304.1	90	346.5	346.5	50	388.9	388.9
311	219.9	219.9	371	262.3	262.3	431	304.8	304.8	491	347.2	347.2	551	389.6	389.6
12	220.6	220.6	72	263.0	263.0	32	305.5	305.5	92	347.9	347.9	52	390.3	390.3
13	221.3	221.3	73	263.8	263.8	33	306.2	306.2	93	348.6	348.6	53	391.0	391.0
14	222.0	222.0	74	264.5	264.5	34	306.9	306.9	94	349.3	349.3	54	391.7	391.7
15	222.7	222.7	75	265.2	265.2	35	307.6	307.6	95	350.0	350.0	55	392.4	392.4
16	223.4	223.4	76	265.9	265.9	36	308.3	308.3	96	350.7	350.7	56	393.2	393.2
17	224.2	224.2	77	266.6	266.6	37	309.0	309.0	97	351.4	351.4	57	393.9	393.9
18	224.9	224.9	78	267.3	267.3	38	309.7	309.7	98	352.1	352.1	58	394.6	394.6
19	225.6	225.6	79	268.0	268.0	39	310.4	310.4	99	352.8	352.8	59	395.3	395.3
20	226.3	226.3	80	268.7	268.7	40	311.1	311.1	500	353.6	353.6	60	396.0	396.0
321	227.0	227.0	381	269.4	269.4	441	311.8	311.8	501	354.3	354.3	561	396.7	396.7
22	227.7	227.7	82	270.1	270.1	42	312.5	312.5	02	355.0	355.0	62	397.4	397.4
23	228.4	228.4	83	270.8	270.8	43	313.2	313.2	03	355.7	355.7	63	398.1	398.1
24	229.1	229.1	84	271.5	271.5	44	314.0	314.0	04	356.4	356.4	64	398.8	398.8
25	229.8	229.8	85	272.2	272.2	45	314.7	314.7	05	357.1	357.1	65	399.5	399.5
26	230.5	230.5	86	272.9	272.9	46	315.4	315.4	06	357.8	357.8	66	400.2	400.2
27	231.2	231.2	87	273.7	273.7	47	316.1	316.1	07	358.5	358.5	67	400.9	400.9
28	231.9	231.9	88	274.4	274.4	48	316.8	316.8	08	359.2	359.2	68	401.6	401.6
29	232.6	232.6	89	275.1	275.1	49	317.5	317.5	09	359.9	359.9	69	402.3	402.3
30	233.3	233.3	90	275.8	275.8	50	318.2	318.2	10	360.6	360.6	70	403.1	403.1
331	234.1	234.1	391	276.5	276.5	451	318.9	318.9	511	361.3	361.3	571	403.8	403.8
32	234.8	234.8	92	277.2	277.2	52	319.6	319.6	12	362.0	362.0	72	404.5	404.5
33	235.5	235.5	93	277.9	277.9	53	320.3	320.3	13	362.7	362.7	73	405.2	405.2
34	236.2	236.2	94	278.6	278.6	54	321.0	321.0	14	363.5	363.5	74	405.9	405.9
35	236.9	236.9	95	279.3	279.3	55	321.7	321.7	15	364.2	364.2	75	406.6	406.6
36	237.6	237.6	96	280.0	280.0	56	322.4	322.4	16	364.9	364.9	76	407.3	407.3
37	238.3	238.3	97	280.7	280.7	57	323.1	323.1	17	365.6	365.6	77	408.0	408.0
38	239.0	239.0	98	281.4	281.4	58	323.9	323.9	18	366.3	366.3	78	408.7	408.7
39	239.7	239.7	99	282.1	282.1	59	324.6	324.6	19	367.0	367.0	79	409.4	409.4
40	240.4	240.4	400	282.8	282.8	60	325.3	325.3	20	367.7	367.7	80	410.1	410.1
341	241.1	241.1	401	283.5	283.5	461	326.0	326.0	521	368.4	368.4	581	410.8	410.8
42	241.8	241.8	02	284.3	284.3	62	326.7	326.7	22	369.1	369.1	82	411.5	411.5
43	242.5	242.5	03	285.0	285.0	63	327.4	327.4	23	369.8	369.8	83	412.2	412.2
44	243.2	243.2	04	285.7	285.7	64	328.1	328.1	24	370.5	370.5	84	413.0	413.0
45	244.0	244.0	05	286.4	286.4	65	328.8	328.8	25	371.2	371.2	85	413.7	413.7
46	244.7	244.7	06	287.1	287.1	66	329.5	329.5	26	371.9	371.9	86	414.4	414.4
47	245.4	245.4	07	287.8	287.8	67	330.2	330.2	27	372.6	372.6	87	415.1	415.1
48	246.1	246.1	08	288.5	288.5	68	330.9	330.9	28	373.4	373.4	88	415.8	415.8
49	246.8	246.8	09	289.2	289.2	69	331.6	331.6	29	374.1	374.1	89	416.5	416.5
50	247.5	247.5	10	289.9	289.9	70	332.3	332.3	30	374.8	374.8	90	417.2	417.2
351	248.2	248.2	411	290.6	290.6	471	333.0	333.0	531	375.5	375.5	591	417.9	417.9
52	248.9	248.9	12	291.3	291.3	72	333.8	333.8	32	376.2	376.2	92	418.6	418.6
53	249.6	249.6	13	292.0	292.0	73	334.5	334.5	33	376.9	376.9	93	419.3	419.3
54	250.3	250.3	14	292.7	292.7	74	335.2	335.2	34	377.6	377.6	94	420.0	420.0
55	251.0	251.0	15	293.4	293.4	75	335.9	335.9	35	378.3	378.3	95	420.7	420.7
56	251.7	251.7	16	294.2	294.2	76	336.6	336.6	36	379.0	379.0	96	421.4	421.4
57	252.4	252.4	17	294.9	294.9	77	337.3	337.3	37	379.7	379.7	97	422.1	422.1
58	253.1	253.1	18	295.6	295.6	78	338.0	338.0	38	380.4	380.4	98	422.8	422.8
59	253.9	253.9	19	296.3	296.3	79	338.7	338.7	39	381.1	381.1	99	423.6	423.6
60	254.6	254.6	20	297.0	297.0	80	339.4	339.4	40	381.8	381.8	600	424.3	424.3
Dist.	Dep.	D. Lat.	Dist.	Dep.	D. Lat.	Dist.	Dep.	D. Lat.	Dist.	Dep.	D. Lat.	Dist.	Dep.	D. Lat.

Dist.	D. Lat.	Dep.
D Lo	Dep.	
	m	D Lo

45°

315° | 045° / 225° | 135°

TABLE 5
Natural and Numerical Chart Scales

Natural Scale	Miles Per Inch		Inches Per Mile		Feet Per Inch
	Nautical	Statute	Nautical	Statute	
1:500	0.007	0.008	145.83	126.72	41.67
1:600	0.008	0.009	121.52	105.60	50.00
1:1,000	0.014	0.016	72.91	63.36	83.33
1:1,200	0.016	0.019	60.76	52.80	100.00
1:1,500	0.021	0.024	48.61	42.24	125.00
1:2,000	0.027	0.032	36.46	31.68	166.67
1:2,400	0.033	0.038	30.38	26.40	200.00
1:2,500	0.034	0.039	29.17	25.34	208.33
1:3,000	0.041	0.047	24.30	21.12	250.00
1:3,600	0.049	0.057	20.25	17.60	300.00
1:4,000	0.055	0.063	18.23	15.84	333.33
1:4,800	0.066	0.076	15.19	13.20	400.00
1:5,000	0.069	0.079	14.58	12.67	416.67
1:6,000	0.082	0.095	12.15	10.56	500.00
1:7,000	0.096	0.110	10.42	9.05	583.33
1:7,200	0.099	0.114	10.13	8.80	600.00
1:7,920	0.109	0.125	9.21	8.00	660.00
1:8,000	0.110	0.126	9.11	7.92	666.67
1:8,400	0.115	0.133	8.68	7.54	700.00
1:9,000	0.123	0.142	8.10	7.04	750.00
1:9,600	0.132	0.152	7.60	6.60	800.00
1:10,000	0.137	0.158	7.29	6.34	833.33
1:10,800	0.148	0.170	6.75	5.87	900.00
1:12,000	0.165	0.189	6.08	5.28	1,000.00
1:13,200	0.181	0.208	5.52	4.80	1,100.00
1:14,400	0.197	0.227	5.06	4.40	1,200.00
1:15,000	0.206	0.237	4.86	4.22	1,250.00
1:15,600	0.214	0.246	4.67	4.06	1,300.00
1:15,840	0.217	0.250	4.60	4.00	1,320.00
1:16,000	0.219	0.253	4.56	3.96	1,333.33
1:16,800	0.230	0.265	4.34	3.77	1,400.00
1:18,000	0.247	0.284	4.05	3.52	1,500.00
1:19,200	0.263	0.303	3.80	3.30	1,600.00
1:20,000	0.274	0.316	3.65	3.17	1,666.67
1:20,400	0.280	0.322	3.57	3.11	1,700.00
1:21,120	0.290	0.333	3.45	3.00	1,760.00
1:21,600	0.296	0.341	3.38	2.93	1,800.00
1:22,800	0.313	0.360	3.20	2.78	1,900.00
1:24,000	0.329	0.379	3.04	2.64	2,000.00
1:25,000	0.343	0.395	2.92	2.53	2,083.33
1:40,000	0.549	0.631	1.82	1.58	3,333.33
1:48,000	0.658	0.758	1.52	1.32	4,000.00
1:50,000	0.686	0.789	1.46	1.27	4,166.67
1:62,500	0.857	0.986	1.17	1.01	5,208.33
1:63,360	0.869	1.000	1.15	1.00	5,280.00
1:75,000	1.029	1.184	0.97	0.85	6,250.00
1:80,000	1.097	1.263	0.91	0.79	6,666.67
1:100,000	1.371	1.578	0.73	0.63	8,333.33
1:125,000	1.714	1.973	0.58	0.51	10,416.67
1:200,000	2.743	3.157	0.36	0.32	16,666.67
1:250,000	3.429	3.946	0.29	0.25	20,833.33
1:400,000	5.486	6.313	0.18	0.16	33,333.33
1:500,000	6.857	7.891	0.15	0.13	41,666.67
1:750,000	10.286	11.837	0.10	0.08	62,500.00
1:1,000,000	13.715	15.783	0.07	0.06	83,333.33
FORMULAS	$\frac{\text{SCALE}}{72{,}913.39}$	$\frac{\text{SCALE}}{63{,}360}$	$\frac{72{,}913.39}{\text{SCALE}}$	$\frac{63{,}360}{\text{SCALE}}$	$\frac{\text{SCALE}}{12}$

TABLE 6

Meridional Parts

Lat.	0°	1°	2°	3°	4°	5°	6°	7°	8°	9°	Lat.
′											′
0	0.0	59.6	119.2	178.9	238.6	298.4	358.3	418.2	478.4	538.6	0
1	01.0	60.6	20.2	79.9	39.6	299.4	59.3	19.2	79.4	39.6	1
2	02.0	61.6	21.2	80.9	40.6	300.4	60.3	20.2	80.4	40.6	2
3	03.0	62.6	22.2	81.9	41.6	01.4	61.3	21.2	81.4	41.7	3
4	04.0	63.6	23.2	82.9	42.6	02.4	62.2	22.2	82.4	42.7	4
5	5.0	64.6	124.2	183.9	243.6	303.4	363.2	423.2	483.4	543.7	5
6	06.0	65.6	25.2	84.8	44.6	04.4	64.2	24.2	84.4	44.7	6
7	07.0	66.6	26.2	85.8	45.6	05.4	65.2	25.3	85.4	45.7	7
8	07.9	67.5	27.2	86.8	46.6	06.4	66.2	26.3	86.4	46.7	8
9	08.9	68.5	28.2	87.8	47.6	07.4	67.2	27.3	87.4	47.7	9
10	9.9	69.5	129.2	188.8	248.5	308.3	368.2	428.3	488.4	548.7	10
11	10.9	70.5	30.2	89.8	49.5	09.3	69.2	29.3	89.4	49.7	11
12	11.9	71.5	31.1	90.8	50.5	10.3	70.2	30.3	90.4	50.7	12
13	12.9	72.5	32.1	91.8	51.5	11.3	71.2	31.3	91.4	51.7	13
14	13.9	73.5	33.1	92.8	52.5	12.3	72.2	32.3	92.4	52.7	14
15	14.9	74.5	134.1	193.8	253.5	313.3	373.2	433.3	493.4	553.7	15
16	15.9	75.5	35.1	94.8	54.5	14.3	74.2	34.3	94.4	54.7	16
17	16.9	76.5	36.1	95.8	55.5	15.3	75.2	35.3	95.4	55.7	17
18	17.9	77.5	37.1	96.8	56.5	16.3	76.2	36.3	96.4	56.7	18
19	18.9	78.5	38.1	97.8	57.5	17.3	77.2	37.3	97.4	57.8	19
20	19.9	79.5	139.1	198.8	258.5	318.3	378.2	438.3	498.4	558.8	20
21	20.9	80.5	40.1	199.8	59.5	19.3	79.2	39.3	499.4	59.8	21
22	21.9	81.5	41.1	200.8	60.5	20.3	80.2	40.3	500.4	60.8	22
23	22.8	82.5	42.1	01.8	61.5	21.3	81.2	41.3	01.4	61.8	23
24	23.8	83.4	43.1	02.8	62.5	22.3	82.2	42.3	02.5	62.8	24
25	24.8	84.4	144.1	203.8	263.5	323.3	383.2	443.3	503.5	563.8	25
26	25.8	85.4	45.1	04.7	64.5	24.3	84.2	44.3	04.5	64.8	26
27	26.8	86.4	46.1	05.7	65.5	25.3	85.2	45.3	05.5	65.8	27
28	27.8	87.4	47.1	06.7	66.5	26.3	86.2	46.3	06.5	66.8	28
29	28.8	88.4	48.0	07.7	67.5	27.3	87.2	47.3	07.5	67.8	29
30	29.8	89.4	149.0	208.7	268.5	328.3	388.2	448.3	508.5	568.8	30
31	30.8	90.4	50.0	09.7	69.5	29.3	89.2	49.3	09.5	69.8	31
32	31.8	91.4	51.0	10.7	70.5	30.3	90.2	50.3	10.5	70.8	32
33	32.8	92.4	52.0	11.7	71.5	31.3	91.2	51.3	11.5	71.9	33
34	33.8	93.4	53.0	12.7	72.5	32.3	92.2	52.3	12.5	72.9	34
35	34.8	94.4	154.0	213.7	273.5	333.3	393.2	453.3	513.5	573.9	35
36	35.8	95.4	55.0	14.7	74.5	34.3	94.2	54.3	14.5	74.9	36
37	36.8	96.4	56.0	15.7	75.4	35.3	95.2	55.3	15.5	75.9	37
38	37.7	97.4	57.0	16.7	76.4	36.3	96.2	56.3	16.5	76.9	38
39	38.7	98.4	58.0	17.7	77.4	37.3	97.2	57.3	17.5	77.9	39
40	39.7	99.3	159.0	218.7	278.4	338.3	398.2	458.3	518.5	578.9	40
41	40.7	100.3	60.0	19.7	79.4	39.3	399.2	59.3	19.5	79.9	41
42	41.7	01.3	61.0	20.7	80.4	40.3	400.2	60.3	20.5	80.9	42
43	42.7	02.3	62.0	21.7	81.4	41.3	01.2	61.3	21.5	81.9	43
44	43.7	03.3	63.0	22.7	82.4	42.3	02.2	62.3	22.5	82.9	44
45	44.7	104.3	164.0	223.7	283.4	343.3	403.2	463.3	523.6	583.9	45
46	45.7	05.3	65.0	24.7	84.4	44.3	04.2	64.3	24.6	85.0	46
47	46.7	06.3	65.9	25.6	85.4	45.3	05.2	65.3	25.6	86.0	47
48	47.7	07.3	66.9	26.6	86.4	46.3	06.2	66.3	26.6	87.0	48
49	48.7	08.3	67.9	27.6	87.4	47.3	07.2	67.3	27.6	88.0	49
50	49.7	109.3	168.9	228.6	288.4	348.3	408.2	468.3	528.6	589.0	50
51	50.7	10.3	69.9	29.6	89.4	49.3	09.2	69.3	29.6	90.0	51
52	51.7	11.3	70.9	30.6	90.4	50.3	10.2	70.3	30.6	91.0	52
53	52.6	12.3	71.9	31.6	91.4	51.3	11.2	71.3	31.6	92.0	53
54	53.6	13.3	72.9	32.6	92.4	52.3	12.2	72.3	32.6	93.0	54
55	54.6	114.3	173.9	233.6	293.4	353.3	413.2	473.3	533.6	594.0	55
56	55.6	15.2	74.9	34.6	94.4	54.3	14.2	74.4	34.6	95.0	56
57	56.6	16.2	75.9	35.6	95.4	55.3	15.2	75.4	35.6	96.0	57
58	57.6	17.2	76.9	36.6	96.4	56.3	16.2	76.4	36.6	97.1	58
59	58.6	18.2	77.9	37.6	97.4	57.3	17.2	77.4	37.6	98.1	59
60	59.6	119.2	178.9	238.6	298.4	358.3	418.2	478.4	538.6	599.1	60
Lat.	0°	1°	2°	3°	4°	5°	6°	7°	8°	9°	Lat.

TABLE 6

Meridional Parts

Lat.	10°	11°	12°	13°	14°	15°	16°	17°	18°	19°	Lat.
′											′
0	599.1	659.7	720.5	781.6	842.9	904.5	966.4	1028.6	1091.1	1154.0	0
1	600.1	60.7	21.6	82.6	43.9	05.5	67.4	29.6	92.1	55.0	1
2	01.1	61.7	22.6	83.6	45.0	06.6	68.4	30.7	93.2	56.1	2
3	02.1	62.7	23.6	84.7	46.0	07.6	69.5	31.7	94.2	57.1	3
4	03.1	63.7	24.6	85.7	47.0	08.6	70.5	32.7	95.3	58.2	4
5	604.1	664.8	725.6	786.7	848.0	909.6	971.6	1033.8	1096.3	1159.3	5
6	05.1	65.8	26.6	87.7	49.1	10.7	72.6	34.8	97.4	60.3	6
7	06.1	66.8	27.6	88.7	50.1	11.7	73.6	35.9	98.4	61.4	7
8	07.1	67.8	28.7	89.8	51.1	12.7	74.7	36.9	1099.5	62.4	8
9	08.2	68.8	29.7	90.8	52.1	13.8	75.7	37.9	1100.5	63.5	9
10	609.2	669.8	730.7	791.8	853.2	914.8	976.7	1039.0	1101.6	1164.5	10
11	10.2	70.8	31.7	92.8	54.2	15.8	77.8	40.0	02.6	65.6	11
12	11.2	71.9	32.7	93.8	55.2	16.9	78.8	41.1	03.7	66.6	12
13	12.2	72.9	33.7	94.9	56.2	17.9	79.8	42.1	04.7	67.7	13
14	13.2	73.9	34.8	95.9	57.3	18.9	80.9	43.1	05.7	68.7	14
15	614.2	674.9	735.8	796.9	858.3	919.9	981.9	1044.2	1106.8	1169.8	15
16	15.2	75.9	36.8	97.9	59.3	21.0	82.9	45.2	07.8	70.8	16
17	16.2	76.9	37.8	798.9	60.3	22.0	84.0	46.3	08.9	71.9	17
18	17.2	77.9	38.8	800.0	61.4	23.0	85.0	47.3	09.9	72.9	18
19	18.3	78.9	39.8	01.0	62.4	24.1	86.0	48.3	11.0	74.0	19
20	619.3	680.0	740.9	802.0	863.4	925.1	987.1	1049.4	1112.0	1175.0	20
21	20.3	81.0	41.9	03.0	64.4	26.1	88.1	50.4	13.1	76.1	21
22	21.3	82.0	42.9	04.1	65.5	27.2	89.1	51.5	14.1	77.1	22
23	22.3	83.0	43.9	05.1	66.5	28.2	90.2	52.5	15.2	78.2	23
24	23.3	84.0	44.9	06.1	67.5	29.2	91.2	53.5	16.2	79.3	24
25	624.3	685.0	746.0	807.1	868.5	930.2	992.3	1054.6	1117.3	1180.3	25
26	25.3	86.0	47.0	08.1	69.6	31.3	93.3	55.6	18.3	81.4	26
27	26.3	87.1	48.0	09.2	70.6	32.3	94.3	56.7	19.4	82.4	27
28	27.3	88.1	49.0	10.2	71.6	33.3	95.4	57.7	20.4	83.5	28
29	28.4	89.1	50.0	11.2	72.6	34.4	96.4	58.8	21.5	84.5	29
30	629.4	690.1	751.0	812.2	873.7	935.4	997.4	1059.8	1122.5	1185.6	30
31	30.4	91.1	52.1	13.2	74.7	36.4	98.5	60.8	23.5	86.6	31
32	31.4	92.1	53.1	14.3	75.7	37.5	999.5	61.9	24.6	87.7	32
33	32.4	93.1	54.1	15.3	76.8	38.5	1000.5	62.9	25.6	88.7	33
34	33.4	94.1	55.1	16.3	77.8	39.5	01.6	64.0	26.7	89.8	34
35	634.4	695.2	756.1	817.3	878.8	940.6	1002.6	1065.0	1127.7	1190.9	35
36	35.4	96.2	57.1	18.4	79.8	41.6	03.7	66.0	28.8	91.9	36
37	36.4	97.2	58.2	19.4	80.9	42.6	04.7	67.1	29.8	93.0	37
38	37.4	98.2	59.2	20.4	81.9	43.7	05.7	68.1	30.9	94.0	38
39	38.5	99.2	60.2	21.4	82.9	44.7	06.8	69.2	31.9	95.1	39
40	639.5	700.2	761.2	822.5	883.9	945.7	1007.8	1070.2	1133.0	1196.1	40
41	40.5	01.2	62.2	23.5	85.0	46.8	08.8	71.3	34.0	97.2	41
42	41.5	02.3	63.3	24.5	86.0	47.8	09.9	72.3	35.1	98.2	42
43	42.5	03.3	64.3	25.5	87.0	48.8	10.9	73.4	36.1	1199.3	43
44	43.5	04.3	65.3	26.5	88.1	49.9	12.0	74.4	37.2	1200.4	44
45	644.5	705.3	766.3	827.6	889.1	950.9	1013.0	1075.4	1138.2	1201.4	45
46	45.5	06.3	67.3	28.6	90.1	51.9	14.0	76.5	39.3	02.5	46
47	46.5	07.3	68.4	29.6	91.1	52.9	15.1	77.5	40.3	03.5	47
48	47.6	08.4	69.4	30.6	92.2	54.0	16.1	78.6	41.4	04.6	48
49	48.6	09.4	70.4	31.7	93.2	55.0	17.1	79.6	42.4	05.6	49
50	649.6	710.4	771.4	832.7	894.2	956.0	1018.2	1080.7	1143.5	1206.7	50
51	50.6	11.4	72.4	33.7	95.2	57.1	19.2	81.7	44.5	07.7	51
52	51.6	12.4	73.4	34.7	96.3	58.1	20.3	82.7	45.6	08.8	52
53	52.6	13.4	74.5	35.8	97.3	59.1	21.3	83.8	46.6	09.9	53
54	53.6	14.4	75.5	36.8	98.3	60.2	22.3	84.8	47.7	10.9	54
55	654.6	715.5	776.5	837.8	899.4	961.2	1023.4	1085.9	1148.7	1212.0	55
56	55.7	16.5	77.5	38.8	900.4	62.2	24.4	86.9	49.8	13.0	56
57	56.7	17.5	78.5	39.8	01.4	63.3	25.5	88.0	50.8	14.1	57
58	57.7	18.5	79.6	40.9	02.4	64.3	26.5	89.0	51.9	15.2	58
59	58.7	19.5	80.6	41.9	03.5	65.3	27.5	90.1	52.9	16.2	59
60	659.7	720.5	781.6	842.9	904.5	966.4	1028.6	1091.1	1154.0	1217.3	60
Lat.	10°	11°	12°	13°	14°	15°	16°	17°	18°	19°	Lat.

TABLE 6
Meridional Parts

Lat.	20°	21°	22°	23°	24°	25°	26°	27°	28°	29°	Lat.
′											′
0	1217.3	1280.9	1345.1	1409.6	1474.7	1540.3	1606.4	1673.1	1740.4	1808.3	0
1	18.3	82.0	46.1	10.7	75.8	41.4	07.5	74.2	41.5	09.4	1
2	19.4	83.1	47.2	11.8	76.9	42.5	08.6	75.3	42.6	10.5	2
3	20.4	84.1	48.3	12.9	78.0	43.6	09.7	76.4	43.7	11.7	3
4	21.5	85.2	49.4	14.0	79.0	44.7	10.8	77.5	44.9	12.8	4
5	1222.6	1286.3	1350.4	1415.0	1480.1	1545.8	1611.9	1678.6	1746.0	1814.0	5
6	23.6	87.3	51.5	16.1	81.2	46.9	13.0	79.8	47.1	15.1	6
7	24.7	88.4	52.6	17.2	82.3	48.0	14.1	80.9	48.2	16.2	7
8	25.7	89.5	53.6	18.3	83.4	49.0	15.2	82.0	49.4	17.4	8
9	26.8	90.5	54.7	19.4	84.5	50.1	16.3	83.1	50.5	18.5	9
10	1227.9	1291.6	1355.8	1420.4	1485.6	1551.2	1617.5	1684.2	1751.6	1819.7	10
11	28.9	92.7	56.9	21.5	86.7	52.3	18.6	85.4	52.8	20.8	11
12	30.0	93.7	57.9	22.6	87.8	53.4	19.7	86.5	53.9	21.9	12
13	31.0	94.8	59.0	23.7	88.9	54.5	20.8	87.6	55.0	23.1	13
14	32.1	95.9	60.1	24.8	89.9	55.6	21.9	88.7	56.1	24.2	14
15	1233.1	1296.9	1361.2	1425.9	1491.0	1556.7	1623.0	1689.8	1757.3	1825.4	15
16	34.2	98.0	62.2	26.9	92.1	57.8	24.1	90.9	58.4	26.5	16
17	35.3	1299.1	63.3	28.0	93.2	58.9	25.2	92.1	59.5	27.6	17
18	36.3	1300.1	64.4	29.1	94.3	60.0	26.3	93.2	60.7	28.8	18
19	37.4	01.2	65.5	30.2	95.4	61.1	27.4	94.3	61.8	29.9	19
20	1238.4	1302.3	1366.5	1431.3	1496.5	1562.2	1628.5	1695.4	1762.9	1831.1	20
21	39.5	03.3	67.6	32.3	97.6	63.3	29.7	96.5	64.1	32.2	21
22	40.6	04.4	68.7	33.4	98.7	64.4	30.8	97.7	65.2	33.3	22
23	41.6	05.5	69.8	34.5	1499.8	65.5	31.9	98.8	66.3	34.5	23
24	42.7	06.5	70.8	35.6	1500.9	66.6	33.0	1699.9	67.4	35.6	24
25	1243.7	1307.6	1371.9	1436.7	1502.0	1567.7	1634.1	1701.0	1768.6	1836.8	25
26	44.8	08.7	73.0	37.8	03.0	68.8	35.2	02.1	69.7	37.9	26
27	45.9	09.7	74.1	38.8	04.1	69.9	36.3	03.3	70.8	39.1	27
28	46.9	10.8	75.1	39.9	05.2	71.0	37.4	04.4	72.0	40.2	28
29	48.0	11.9	76.2	41.0	06.3	72.2	38.5	05.5	73.1	41.3	29
30	1249.1	1312.9	1377.3	1442.1	1507.4	1573.3	1639.6	1706.6	1774.2	1842.5	30
31	50.1	14.0	78.4	43.2	08.5	74.4	40.8	07.8	75.4	43.6	31
32	51.2	15.1	79.4	44.3	09.6	75.5	41.9	08.9	76.5	44.8	32
33	52.2	16.2	80.5	45.4	10.7	76.6	43.0	010.0	77.6	45.9	33
34	53.3	17.2	81.6	46.4	11.8	77.7	44.1	11.1	78.8	47.1	34
35	1254.4	1318.3	1382.7	1447.5	1512.9	1578.8	1645.2	1712.2	1779.9	1848.2	35
36	55.4	19.4	83.7	48.6	14.0	79.9	46.3	13.4	81.0	49.3	36
37	56.5	20.4	84.8	49.7	15.1	81.0	47.4	14.5	82.2	50.5	37
38	57.5	21.5	85.9	50.8	16.2	82.1	48.5	15.6	83.3	51.6	38
39	58.6	22.6	87.0	51.9	17.3	83.2	49.7	16.7	84.4	52.8	39
40	1259.7	1323.6	1388.1	1453.0	1518.3	1584.3	1650.8	1717.9	1785.6	1853.9	40
41	60.7	24.7	89.1	54.0	19.4	85.4	51.9	19.0	86.7	55.1	41
42	61.8	25.8	90.2	55.1	20.5	86.5	53.0	20.1	87.8	56.2	42
43	62.9	26.9	91.3	56.2	21.6	87.6	54.1	21.2	89.0	57.4	43
44	63.9	27.9	92.4	57.3	22.7	88.7	55.2	22.3	90.1	58.5	44
45	1265.0	1329.0	1393.4	1458.4	1523.8	1589.8	1656.3	1723.5	1791.2	1859.7	45
46	66.1	30.1	94.5	59.5	24.9	90.9	57.5	24.6	92.4	60.8	46
47	67.1	31.1	95.6	60.6	26.0	92.0	58.6	25.7	93.5	61.9	47
48	68.2	32.2	96.7	61.6	27.1	93.1	59.7	26.8	94.6	63.1	48
49	69.2	33.3	97.8	62.7	28.2	94.2	60.8	28.0	95.8	64.2	49
50	1270.3	1334.3	1398.8	1463.8	1529.3	1595.3	1661.9	1729.1	1796.9	1865.4	50
51	71.4	35.4	1399.9	64.9	30.4	96.4	63.0	30.2	98.0	66.5	51
52	72.4	36.5	1401.0	66.0	31.5	97.5	64.1	31.3	1799.2	67.7	52
53	73.5	37.6	02.1	67.1	32.6	98.6	65.3	32.5	1800.3	68.8	53
54	74.6	38.6	03.2	68.2	33.7	1599.7	66.4	33.6	01.5	70.0	54
55	1275.6	1339.7	1404.2	1469.3	1534.8	1600.8	1667.5	1734.7	1802.6	1871.1	55
56	76.7	40.8	05.3	70.3	35.9	02.0	68.6	35.8	03.7	72.3	56
57	77.8	41.8	06.4	71.4	37.0	03.1	69.7	37.0	04.9	73.4	57
58	78.8	42.9	07.5	72.5	38.1	04.2	70.8	38.1	06.0	74.6	58
59	79.9	44.0	08.6	73.6	39.2	05.3	71.9	39.2	07.1	75.7	59
60	1280.9	1345.1	1409.6	1474.7	1540.3	1606.4	1673.1	1740.4	1808.3	1876.9	60
Lat.	20°	21°	22°	23°	24°	25°	26°	27°	28°	29°	Lat.

TABLE 6
Meridional Parts

Lat.	30°	31°	32°	33°	34°	35°	36°	37°	38°	39°	Lat.
′											′
0	1876.9	1946.2	2016.2	2087.0	2158.6	2231.1	2304.5	2378.8	2454.1	2530.4	0
1	78.0	47.3	17.4	88.2	59.8	32.3	05.7	80.0	55.3	31.7	1
2	79.2	48.5	18.5	89.4	61.0	33.5	06.9	81.3	56.6	33.0	2
3	80.3	49.6	19.7	90.5	62.2	34.7	08.1	82.5	57.9	34.3	3
4	81.5	50.8	20.9	91.7	63.4	35.9	09.4	83.8	59.1	35.6	4
5	1882.6	1952.0	2022.1	2092.9	2164.6	2237.2	2310.6	2385.0	2460.4	2536.8	5
6	83.8	53.1	23.2	94.1	65.8	38.4	11.8	86.3	61.7	38.1	6
7	84.9	54.3	24.4	95.3	67.0	39.6	13.1	87.5	62.9	39.4	7
8	86.1	55.4	25.6	96.5	68.2	40.8	14.3	88.8	64.2	40.7	8
9	87.2	56.6	26.8	97.7	69.4	42.0	15.5	90.0	65.5	42.0	9
10	1888.4	1957.8	2027.9	2098.9	2170.6	2243.2	2316.8	2391.3	2466.7	2543.3	10
11	89.5	58.9	29.1	2100.1	71.8	44.5	18.0	92.5	68.0	44.5	11
12	90.7	60.1	30.3	01.2	73.0	45.7	19.2	93.8	69.3	45.8	12
13	91.8	61.3	31.5	02.4	74.2	46.9	20.5	95.0	70.5	47.1	13
14	93.0	62.4	32.6	03.6	75.4	48.1	21.7	96.3	71.8	48.4	14
15	1894.1	1963.6	2033.8	2104.8	2176.6	2249.3	2322.9	2397.5	2473.1	2549.7	15
16	95.3	64.8	35.0	06.0	77.8	50.6	24.2	2398.8	74.3	51.0	16
17	96.4	65.9	36.2	07.2	79.0	51.8	25.4	2400.0	75.6	52.3	17
18	97.6	67.1	37.3	08.4	80.3	53.0	26.6	01.3	76.9	53.6	18
19	98.7	68.2	38.5	09.6	81.5	54.2	27.9	02.5	78.1	54.8	19
20	1899.9	1969.4	2039.7	2110.8	2182.7	2255.4	2329.1	2403.8	2479.4	2556.1	20
21	1901.0	70.6	40.9	12.0	83.9	56.7	30.4	05.0	80.7	57.4	21
22	02.2	71.7	42.1	13.1	85.1	57.9	31.6	06.3	82.0	58.7	22
23	03.3	72.9	43.2	14.3	86.3	59.1	32.8	07.5	83.2	60.0	23
24	04.5	74.1	44.4	15.5	87.5	60.3	34.1	08.8	84.5	61.3	24
25	1905.6	1975.2	2045.6	2116.7	2188.7	2261.5	2335.3	2410.0	2485.8	2562.6	25
26	06.8	76.4	46.8	17.9	89.9	62.8	36.5	11.3	87.0	63.9	26
27	08.0	77.6	47.9	19.1	91.1	64.0	37.8	12.5	88.3	65.1	27
28	09.1	78.7	49.1	20.3	92.3	65.2	39.0	13.8	89.6	66.4	28
29	10.3	79.9	50.3	21.5	93.5	66.4	40.3	15.0	90.9	67.7	29
30	1911.4	1981.1	2051.5	2122.7	2194.7	2267.6	2341.5	2416.3	2492.1	2569.0	30
31	12.6	82.2	52.7	23.9	95.9	68.9	42.7	17.6	93.4	70.3	31
32	13.7	83.4	53.8	25.1	97.1	70.1	44.0	18.8	94.7	71.6	32
33	14.9	84.6	55.0	26.3	98.4	71.3	45.2	20.1	95.9	72.9	33
34	16.0	85.7	56.2	27.5	2199.6	72.5	46.4	21.3	97.2	74.2	34
35	1917.2	1986.9	2057.4	2128.7	2200.8	2273.8	2347.7	2422.6	2498.5	2575.5	35
36	18.4	88.1	58.6	29.9	02.0	75.0	48.9	23.8	2499.8	76.8	36
37	19.5	89.2	59.7	31.1	03.2	76.2	50.2	25.1	2501.0	78.1	37
38	20.7	90.4	60.9	32.2	04.4	77.4	51.4	26.3	02.3	79.4	38
39	21.8	91.6	62.1	33.4	05.6	78.7	52.6	27.6	03.6	80.6	39
40	1923.0	1992.8	2063.3	2134.6	2206.8	2279.9	2353.9	2428.9	2504.9	2581.9	40
41	24.1	93.9	64.5	35.8	08.0	81.1	55.1	30.1	06.1	83.2	41
42	25.3	95.1	65.7	37.0	09.2	82.3	56.4	31.4	07.4	84.5	42
43	26.4	96.3	66.8	38.2	10.5	83.6	57.6	32.6	08.7	85.8	43
44	27.6	97.4	68.0	39.4	11.7	84.8	58.9	33.9	010.0	87.1	44
45	1928.8	1998.6	2069.2	2140.6	2212.9	2286.0	2360.1	2435.2	2511.2	2588.4	45
46	29.9	1999.8	70.4	41.8	14.1	87.2	61.3	36.4	12.5	89.7	46
47	31.1	2000.9	71.6	43.0	15.3	88.5	62.6	37.7	13.8	91.0	47
48	32.2	02.1	72.8	44.2	16.5	89.7	63.8	38.9	15.1	92.3	48
49	33.4	03.3	73.9	45.4	17.7	90.9	65.1	40.2	16.4	93.6	49
50	1934.6	2004.5	2075.1	2146.6	2218.9	2292.2	2366.3	2441.5	2517.6	2594.9	50
51	35.7	05.6	76.3	47.8	20.1	93.4	67.6	42.7	18.9	96.2	51
52	36.9	06.8	77.5	49.0	21.4	94.6	68.8	44.0	20.2	97.5	52
53	38.0	08.0	78.7	50.2	22.6	95.8	70.0	45.2	21.5	2598.8	53
54	39.2	09.1	79.9	51.4	23.8	97.1	71.2	46.5	22.8	2600.1	54
55	1940.4	2010.3	2081.1	2152.6	2225.0	2298.3	2372.5	2447.8	2524.0	2601.4	55
56	41.5	11.5	82.2	53.8	26.2	2299.5	73.8	49.0	25.3	02.7	56
57	42.7	12.7	83.4	55.0	27.4	2300.8	75.0	50.3	26.6	04.0	57
58	43.8	13.8	84.6	56.2	28.6	02.0	76.3	51.6	27.9	05.3	58
59	45.0	15.0	85.8	57.4	29.9	03.2	77.5	52.8	29.2	06.6	59
60	1946.2	2016.2	2087.0	2158.6	2231.1	2304.5	2378.8	2454.1	2530.4	2607.9	60
Lat.	30°	31°	32°	33°	34°	35°	36°	37°	38°	39°	Lat.

TABLE 6

Meridional Parts

Lat. ′	40°	41°	42°	43°	44°	45°	46°	47°	48°	49°	Lat. ′
0	2607.9	2686.5	2766.3	2847.4	2929.8	3013.6	3099.0	3185.9	3274.4	3364.7	0
1	09.2	87.8	67.6	48.7	31.2	15.1	3100.4	87.3	75.9	66.2	1
2	10.5	89.1	69.0	50.1	32.6	16.5	01.8	88.8	77.4	67.7	2
3	11.8	90.4	70.3	51.5	34.0	17.9	03.3	90.2	78.9	69.3	3
4	13.1	91.8	71.7	52.8	35.4	19.3	04.7	91.7	80.4	70.8	4
5	2614.4	2693.1	2773.0	2854.2	2936.7	3020.7	3106.2	3193.2	3281.9	3372.3	5
6	15.7	94.4	74.3	55.6	38.1	22.1	07.6	94.6	83.4	73.8	6
7	17.0	95.7	75.7	56.9	39.5	23.5	09.0	96.1	84.8	75.3	7
8	18.3	97.1	77.0	58.3	40.9	24.9	10.5	97.6	86.3	76.9	8
9	19.6	98.4	78.4	59.7	42.3	26.4	11.9	3199.0	87.8	78.4	9
10	2620.9	2699.7	2779.7	2861.0	2943.7	3027.8	3113.3	3200.5	3289.3	3379.9	10
11	22.2	2701.0	81.1	62.4	45.1	29.2	14.8	02.0	90.8	81.4	11
12	23.5	02.3	82.4	63.8	46.5	30.6	16.2	03.4	92.3	83.0	12
13	24.8	03.7	83.8	65.1	47.9	32.0	17.7	04.9	93.8	84.5	13
14	26.1	05.0	85.1	66.5	49.2	33.4	19.1	06.4	95.3	86.0	14
15	2627.4	2706.3	2786.4	2867.9	2950.6	3034.8	3120.5	3207.8	3296.8	3387.5	15
16	28.7	07.6	87.8	69.2	52.0	36.3	22.0	09.3	98.3	89.1	16
17	30.0	09.0	89.1	70.6	53.4	37.7	23.4	10.8	3299.8	90.6	17
18	31.3	10.3	90.5	72.0	54.8	39.1	24.9	12.2	3301.3	92.1	18
19	32.6	11.6	91.8	73.3	56.2	40.5	26.3	13.7	02.8	93.7	19
20	2634.0	2713.0	2793.2	2874.7	2957.6	3041.9	3127.8	3215.2	3304.3	3395.2	20
21	35.3	14.3	94.5	76.1	59.0	43.3	29.2	16.7	05.8	96.7	21
22	36.6	15.6	95.9	77.4	60.4	44.8	30.6	18.1	07.3	98.3	22
23	37.9	16.9	97.2	78.8	61.8	46.2	32.1	19.6	08.8	3399.8	23
24	39.2	18.3	98.6	80.2	63.2	47.6	33.5	21.1	10.3	3401.3	24
25	2640.5	2719.6	2799.9	2881.6	2964.6	3049.0	3135.0	3222.5	3311.8	3402.8	25
26	41.8	20.9	2801.3	82.9	66.0	50.4	36.4	24.0	13.3	04.4	26
27	43.1	22.2	02.6	84.3	67.4	51.9	37.9	25.5	14.8	05.9	27
28	44.4	23.6	04.0	85.7	68.8	53.3	39.3	27.0	16.3	07.4	28
29	45.7	24.9	05.3	87.1	70.2	54.7	40.8	28.4	17.8	09.0	29
30	2647.0	2726.2	2806.7	2888.4	2971.5	3056.1	3142.2	3229.9	3319.3	3410.5	30
31	48.3	27.6	08.0	89.8	72.9	57.5	43.7	31.4	20.8	12.1	31
32	49.7	28.9	09.4	91.2	74.3	59.0	45.1	32.9	22.3	13.6	32
33	51.0	30.2	10.7	92.5	75.7	60.4	46.6	34.4	23.8	15.1	33
34	52.3	31.6	12.1	93.9	77.1	61.8	48.0	35.8	25.3	16.7	34
35	2653.6	2732.9	2813.4	2895.3	2978.5	3063.2	3149.5	3237.3	3326.9	3418.2	35
36	54.9	34.2	14.8	96.7	79.9	64.7	50.9	38.8	28.4	19.7	36
37	56.2	35.6	16.1	98.1	81.3	66.1	52.4	40.3	29.9	21.3	37
38	57.5	36.9	17.5	2899.4	82.7	67.5	53.8	41.7	31.4	22.8	38
39	58.8	38.2	18.9	2900.8	84.1	68.9	55.3	43.2	32.9	24.4	39
40	2660.2	2739.6	2820.2	2902.2	2985.5	3070.4	3156.7	3244.7	3334.4	3425.9	40
41	61.5	40.9	21.6	03.6	86.9	71.8	58.2	46.2	35.9	27.4	41
42	62.8	42.2	22.9	04.9	88.3	73.2	59.6	47.7	37.4	29.0	42
43	64.1	43.6	24.3	06.3	89.7	74.6	61.1	49.1	38.9	30.5	43
44	65.4	44.9	25.6	07.7	91.1	76.1	62.5	50.6	40.4	32.1	44
45	2666.7	2746.2	2827.0	2909.1	2992.6	3077.5	3164.0	3252.1	3342.0	3433.6	45
46	68.0	47.6	28.3	10.5	94.0	78.9	65.4	53.6	43.5	35.2	46
47	69.4	48.9	29.7	11.8	95.4	80.4	66.9	55.1	45.0	36.7	47
48	70.7	50.2	31.1	13.2	96.8	81.8	68.4	56.6	46.5	38.2	48
49	72.0	51.6	32.4	14.6	98.2	83.2	69.8	58.0	48.0	39.8	49
50	2673.3	2752.9	2833.8	2916.0	2999.6	3084.6	3171.3	3259.5	3349.5	3441.3	50
51	74.6	54.2	35.1	17.4	3001.0	86.1	72.7	61.0	51.0	42.9	51
52	75.9	55.6	36.5	18.7	02.4	87.5	74.2	62.5	52.5	44.4	52
53	77.3	56.9	37.9	20.1	03.8	88.9	75.6	64.0	54.1	46.0	53
54	78.6	58.3	39.2	21.5	05.2	90.4	77.1	65.5	55.6	47.5	54
55	2679.9	2759.6	2840.6	2922.9	3006.6	3091.8	3178.6	3267.0	3357.1	3449.1	55
56	81.2	60.9	41.9	24.3	08.0	93.2	80.0	68.4	58.6	50.6	56
57	82.5	62.3	43.3	25.7	09.4	94.7	81.5	69.9	60.1	52.2	57
58	83.8	63.6	44.7	27.0	10.8	96.1	82.9	71.4	61.7	53.7	58
59	85.2	65.0	46.0	28.4	12.2	97.5	84.4	72.9	63.2	55.3	59
60	2686.5	2766.3	2847.4	2929.8	3013.6	3099.0	3185.9	3274.4	3364.7	3456.8	60
Lat.	40°	41°	42°	43°	44°	45°	46°	47°	48°	49°	Lat.

TABLE 6

Meridional Parts

Lat. ′	50°	51°	52°	53°	54°	55°	56°	57°	58°	59°	Lat. ′
0	3456.8	3550.9	3647.0	3745.4	3846.0	3949.1	4054.8	4163.3	4274.8	4389.4	0
1	58.4	52.5	48.7	47.0	47.7	50.8	56.6	65.1	76.6	91.3	1
2	59.9	54.1	50.3	48.7	49.4	52.6	58.4	67.0	78.5	93.3	2
3	61.5	55.7	51.9	50.3	51.1	54.3	60.2	68.8	80.4	95.2	3
4	63.0	57.2	53.5	52.0	52.8	56.1	61.9	70.6	82.3	97.1	4
5	3464.6	3558.8	3655.1	3753.7	3854.5	3957.8	4063.7	4172.5	4284.2	4399.1	5
6	66.1	60.4	56.8	55.3	56.2	59.5	65.5	74.3	86.1	4401.0	6
7	67.7	62.0	58.4	57.0	57.9	61.3	67.3	76.1	88.0	03.0	7
8	69.2	63.6	60.0	58.6	59.6	63.0	69.1	78.0	89.8	04.9	8
9	70.8	65.2	61.6	60.3	61.3	64.8	70.9	79.8	91.7	06.9	9
10	3472.4	3566.8	3663.3	3762.0	3863.0	3966.5	4072.7	4181.7	4293.6	4408.8	10
11	73.9	68.4	64.9	63.6	64.7	68.3	74.5	83.5	95.5	10.8	11
12	75.5	70.0	66.5	65.3	66.4	70.0	76.3	85.3	97.4	12.7	12
13	77.0	71.5	68.1	67.0	68.1	71.8	78.1	87.2	4299.3	14.7	13
14	78.6	73.1	69.8	68.6	69.8	73.5	79.9	89.0	4301.2	16.6	14
15	3480.2	3574.7	3671.4	3770.3	3871.5	3975.3	4081.7	4190.9	4303.1	4418.6	15
16	81.7	76.3	73.0	72.0	73.2	77.0	83.4	92.7	05.0	20.5	16
17	83.3	77.9	74.7	73.6	74.9	78.8	85.2	94.6	06.9	22.5	17
18	84.8	79.5	76.3	75.3	76.7	80.5	87.0	96.4	08.8	24.4	18
19	86.4	81.1	77.9	77.0	78.4	82.3	88.8	98.3	10.7	26.4	19
20	3488.0	3582.7	3679.6	3778.6	3880.1	3984.0	4090.6	4200.1	4312.6	4428.3	20
21	89.5	84.3	81.2	80.3	81.8	85.8	92.4	02.0	14.5	30.3	21
22	91.1	85.9	82.8	82.0	83.5	87.5	94.2	03.8	16.4	32.3	22
23	92.6	87.5	84.5	83.7	85.2	89.3	96.0	05.7	18.3	34.2	23
24	94.2	89.1	86.1	85.3	86.9	91.0	97.8	07.5	20.2	36.2	24
25	3495.8	3590.7	3687.7	3787.0	3888.6	3992.8	4099.7	4209.4	4322.1	4438.1	25
26	97.3	92.3	89.4	88.7	90.4	94.6	4101.5	11.2	24.0	40.1	26
27	3498.9	93.9	91.0	90.3	92.1	96.3	03.3	13.1	25.9	42.1	27
28	3500.5	95.5	92.6	92.0	93.8	98.1	05.1	14.9	27.8	44.0	28
29	02.0	97.1	94.3	93.7	95.5	3999.8	06.9	16.8	29.7	46.0	29
30	3503.6	3598.7	3695.9	3795.4	3897.2	4001.6	4108.7	4218.6	4331.7	4448.0	30
31	05.2	3600.3	97.6	97.1	3898.9	03.4	10.5	20.5	33.6	49.9	31
32	06.7	01.9	3699.2	3798.7	3900.7	05.1	12.3	22.4	35.5	51.9	32
33	08.3	03.5	3700.8	3800.4	02.4	06.9	14.1	24.2	37.4	53.9	33
34	09.9	05.1	02.5	02.1	04.1	08.7	15.9	26.1	39.3	55.8	34
35	3511.5	3606.7	3704.1	3803.8	3905.8	4010.4	4117.7	4227.9	4341.2	4457.8	35
36	13.0	08.3	05.8	05.4	07.5	12.2	19.5	29.8	43.1	59.8	36
37	14.6	09.9	07.4	07.1	09.3	14.0	21.4	31.7	45.1	61.7	37
38	16.2	11.5	09.0	08.8	11.0	15.7	23.2	33.5	47.0	63.7	38
39	17.7	13.1	10.7	10.5	12.7	17.5	25.0	35.4	48.9	65.7	39
40	3519.3	3614.8	3712.3	3812.2	3914.4	4019.3	4126.8	4237.3	4350.8	4467.7	40
41	20.9	16.4	14.0	13.9	16.2	21.0	28.6	39.1	52.7	69.6	41
42	22.5	18.0	15.6	15.5	17.9	22.8	30.4	41.0	54.6	71.6	42
43	24.0	19.6	17.3	17.2	19.6	24.6	32.3	42.9	56.6	73.6	43
44	25.6	21.2	18.9	18.9	21.3	26.3	34.1	44.7	58.5	75.6	44
45	3527.2	3622.8	3720.6	3820.6	3923.1	4028.1	4135.9	4246.6	4360.4	4477.6	45
46	28.8	24.4	22.2	22.3	24.8	29.9	37.7	48.5	62.3	79.5	46
47	30.3	26.0	23.9	24.0	26.5	31.7	39.5	50.3	64.3	81.5	47
48	31.9	27.6	25.5	25.7	28.3	33.4	41.4	52.2	66.2	83.5	48
49	33.5	29.2	27.2	27.4	30.0	35.2	43.2	54.1	68.1	85.5	49
50	3535.1	3630.9	3728.8	3829.1	3931.7	4037.0	4145.0	4256.0	4370.0	4487.5	50
51	36.7	32.5	30.5	30.7	33.5	38.8	46.8	57.8	72.0	89.5	51
52	38.2	34.1	32.1	32.4	35.2	40.5	48.7	59.7	73.9	91.5	52
53	39.8	35.7	33.8	34.1	36.9	42.3	50.5	61.6	75.8	93.5	53
54	41.4	37.3	35.4	35.8	38.7	44.1	52.3	63.5	77.8	95.4	54
55	3543.0	3638.9	3737.1	3837.5	3940.4	4045.9	4154.1	4265.3	4379.7	4497.4	55
56	44.6	40.6	38.7	39.2	42.1	47.7	56.0	67.2	81.6	4499.4	56
57	46.1	42.2	40.4	40.9	43.9	49.4	57.8	69.1	83.6	4501.4	57
58	47.7	43.8	42.0	42.6	45.6	51.2	59.6	71.0	85.5	03.4	58
59	49.3	45.4	43.7	44.3	47.3	53.0	61.5	72.9	87.4	05.4	59
60	3550.9	3647.0	3745.4	3846.0	3949.1	4054.8	4163.3	4274.8	4389.4	4507.4	60
Lat.	50°	51°	52°	53°	54°	55°	56°	57°	58°	59°	Lat.

TABLE 6
Meridional Parts

Lat.	60°	61°	62°	63°	64°	65°	66°	67°	68°	69°	Lat.
′											′
0	4507.4	4629.1	4754.6	4884.4	5018.7	5157.9	5302.4	5452.8	5609.4	5773.0	0
1	09.4	31.1	56.8	86.6	21.0	60.3	04.9	55.3	12.1	75.8	1
2	11.4	33.2	58.9	88.8	23.3	62.6	07.4	57.9	14.8	78.6	2
3	13.4	35.2	61.0	91.0	25.5	65.0	09.8	60.5	17.4	81.4	3
4	15.4	37.3	63.1	93.2	27.8	67.4	12.3	63.0	20.1	84.2	4
5	4517.4	4639.4	4765.3	4895.4	5030.1	5169.7	5314.7	5465.6	5622.8	5787.0	5
6	19.4	41.4	67.4	97.6	32.4	72.1	17.2	68.2	25.5	89.8	6
7	21.4	43.5	69.5	4899.8	34.7	74.5	19.7	70.7	28.2	92.6	7
8	23.4	45.6	71.7	4902.0	37.0	76.9	22.1	73.3	30.8	95.4	8
9	25.4	47.6	73.8	04.2	39.3	79.2	24.6	75.9	33.5	5798.2	9
10	4527.4	4649.7	4776.0	4906.5	5041.6	5181.6	5327.1	5478.4	5636.2	5801.0	10
11	29.4	51.8	78.1	08.7	43.8	84.0	29.6	81.0	38.9	03.8	11
12	31.4	53.9	80.2	10.9	46.1	86.4	32.0	83.6	41.6	06.6	12
13	33.4	55.9	82.4	13.1	48.4	88.8	34.5	86.2	44.3	09.5	13
14	35.5	58.0	84.5	15.3	50.7	99.1	37.0	88.7	47.0	12.3	14
15	4537.5	4660.1	4786.7	4917.5	5053.0	5193.5	5339.5	5491.3	5649.7	5815.1	15
16	39.5	62.2	88.8	19.8	55.3	95.9	42.0	93.9	52.4	17.9	16
17	41.5	64.2	91.0	22.0	57.6	5198.3	44.4	96.5	55.1	20.7	17
18	43.5	66.3	93.1	24.2	59.9	5200.7	46.9	5499.1	57.8	23.6	18
19	45.5	68.4	95.3	26.4	62.2	03.1	49.4	5501.7	60.5	26.4	19
20	4547.5	4670.5	4797.4	4928.6	5064.5	5205.5	5351.9	5504.3	5663.2	5829.2	20
21	49.6	72.6	4799.6	30.9	66.8	07.9	54.4	06.9	65.9	32.1	21
22	51.6	74.6	4801.7	33.1	69.2	10.3	56.9	09.5	68.6	34.9	22
23	53.6	76.7	03.9	35.3	71.5	12.7	59.4	12.1	71.3	37.7	23
24	55.6	78.8	06.0	37.6	73.8	15.1	61.9	14.7	74.0	40.6	24
25	4557.6	4680.9	4808.2	4939.8	5076.1	5217.5	5364.4	5517.3	5676.7	5843.4	25
26	59.7	83.0	10.3	42.0	78.4	19.9	66.9	19.9	79.4	46.2	26
27	61.7	85.1	12.5	44.2	80.7	22.3	69.4	22.5	82.2	49.1	27
28	63.7	87.2	14.6	46.5	83.0	24.7	71.9	25.1	84.9	51.9	28
29	65.7	89.2	16.8	48.7	85.3	27.1	74.4	27.7	87.6	54.8	29
30	4567.8	4691.3	4819.0	4951.0	5087.7	5229.5	5376.9	5530.3	5690.3	5857.6	30
31	69.8	93.4	21.1	53.2	90.0	31.9	79.4	32.9	93.1	60.5	31
32	71.8	95.5	23.3	55.4	92.3	34.3	81.9	35.5	95.8	63.3	32
33	73.8	97.6	25.5	57.7	94.6	36.7	84.4	38.1	5698.5	66.2	33
34	75.9	4699.7	27.6	59.9	97.0	39.1	86.9	40.7	5701.2	69.1	34
35	4577.9	4701.8	4829.8	4962.2	5099.3	5241.6	5389.4	5543.6	5704.0	5871.9	35
36	79.9	03.9	32.0	64.4	5101.6	44.0	91.9	46.0	06.7	74.8	36
37	82.0	06.0	34.1	66.7	03.9	46.4	94.4	48.6	09.5	77.7	37
38	84.0	08.1	36.3	68.9	06.3	48.8	97.0	51.2	12.2	80.5	38
39	86.1	10.2	38.5	71.2	08.6	51.2	5399.5	53.9	14.9	83.4	39
40	4588.1	4712.3	4840.7	4973.4	5110.9	5253.7	5402.0	5556.5	5717.7	5886.3	40
41	90.1	14.4	42.8	75.7	13.3	56.1	04.5	59.1	20.4	89.2	41
42	92.2	16.5	45.0	77.9	15.6	58.5	07.0	61.7	23.2	92.0	42
43	94.2	18.6	47.2	80.2	17.9	60.9	09.6	64.4	25.9	94.9	43
44	96.3	20.7	49.4	82.4	20.3	63.4	12.1	67.0	28.7	5897.8	44
45	4598.3	4722.9	4851.5	4984.7	5122.6	5265.8	5414.6	5569.7	5731.4	5900.7	45
46	4600.3	25.0	53.7	86.9	25.0	68.2	17.2	72.3	34.2	03.6	46
47	02.4	27.1	55.9	89.2	27.3	70.7	19.7	74.9	37.0	06.5	47
48	04.4	29.2	58.1	91.5	29.7	73.1	22.2	77.6	39.7	09.4	48
49	06.5	31.3	60.3	93.7	32.0	75.5	24.8	80.2	42.5	12.3	49
50	4608.5	4733.4	4862.5	4996.0	5134.4	5278.0	5427.3	5582.9	5745.3	5915.2	50
51	10.6	35.5	64.6	4998.3	36.7	80.4	29.8	85.5	48.0	18.1	51
52	12.6	37.6	66.8	5000.5	39.1	82.9	32.4	88.2	50.8	21.0	52
53	14.7	39.8	69.0	02.8	41.4	85.3	34.9	90.8	53.6	23.9	53
54	16.7	41.9	71.2	05.1	43.8	87.7	37.5	93.5	56.3	26.8	54
55	4618.8	4744.0	4873.4	5007.3	5146.1	5290.2	5440.0	5596.1	5759.1	5929.7	55
56	20.8	46.1	75.6	09.6	48.5	92.6	42.6	5598.8	61.9	32.6	56
57	22.9	48.3	77.8	11.9	50.8	95.1	45.1	5601.4	64.7	35.5	57
58	24.9	50.4	80.0	14.1	53.2	5297.5	47.7	04.1	67.5	38.4	58
59	27.0	52.5	82.2	16.4	55.5	5300.0	50.2	06.8	70.2	41.3	59
60	4629.1	4754.6	4884.4	5018.7	5157.9	5302.4	5452.8	5609.4	5773.0	5944.2	60
Lat.	60°	61°	62°	63°	64°	65°	66°	67°	68°	69°	Lat.

TABLE 6
Meridional Parts

Lat.	70°	71°	72°	73°	74°	75°	76°	77°	78°	79°	Lat.
′											′
0	5944.2	6123.9	6312.9	6512.4	6723.6	6948.1	7187.7	7444.7	7722.0	8023.1	0
1	47.2	27.0	16.1	15.8	27.2	51.9	91.8	49.2	26.8	28.3	1
2	50.1	30.0	19.4	19.2	30.8	55.8	7196.0	53.6	31.6	33.6	2
3	53.0	33.1	22.6	22.6	34.5	59.7	7200.1	58.1	36.5	38.8	3
4	56.0	36.2	25.9	26.1	38.1	63.5	04.3	62.6	41.3	44.1	4
5	5958.9	6139.3	6329.1	6529.5	6741.8	6967.4	7208.4	7467.0	7746.1	8049.4	5
6	61.8	42.4	32.4	32.9	45.4	71.3	12.6	71.5	51.0	54.6	6
7	64.8	45.4	35.6	36.4	49.0	75.2	16.7	76.0	55.8	59.9	7
8	67.7	48.5	38.9	39.8	52.7	79.1	20.9	80.5	60.7	65.2	8
9	70.6	51.6	42.1	43.3	56.4	83.0	25.1	85.0	65.6	70.5	9
10	5973.6	6154.7	6345.4	6546.7	6760.0	6986.9	7229.3	7489.5	7770.4	8075.9	10
11	76.5	57.8	48.7	50.2	63.7	90.8	33.4	94.0	75.3	81.2	11
12	79.5	60.9	51.9	53.6	67.4	94.7	37.6	7498.5	80.2	86.5	12
13	82.4	64.0	55.2	57.1	71.0	6998.6	41.8	7503.0	85.1	91.9	13
14	85.4	67.1	58.5	60.5	74.7	7002.5	46.0	07.5	90.0	8097.2	14
15	5988.3	6170.2	6361.7	6564.0	6778.4	7006.5	7250.2	7512.0	7794.9	8102.6	15
16	91.3	73.3	65.0	67.5	82.1	10.4	54.4	16.6	7799.8	07.9	16
17	94.3	76.5	68.3	71.0	85.8	14.3	58.6	21.1	7804.7	13.3	17
18	5997.2	79.6	71.6	74.4	89.4	18.3	62.9	25.6	09.6	18.7	18
19	6000.2	82.7	74.9	77.9	93.1	22.2	67.1	30.2	14.6	24.1	19
20	6003.2	6185.8	6378.2	6581.4	6796.8	7026.2	7271.3	7534.8	7819.5	8129.5	20
21	06.1	88.9	81.5	84.9	6800.5	30.1	75.5	39.3	24.5	34.9	21
22	09.1	92.1	84.8	88.4	04.3	34.1	79.8	43.9	29.4	40.3	22
23	12.1	95.2	88.1	91.9	08.0	38.0	84.0	48.5	34.4	45.7	23
24	15.0	6198.3	91.4	95.4	11.7	42.0	88.3	53.0	39.3	51.1	24
25	6018.0	6201.5	6394.7	6598.9	6815.4	7045.9	7292.5	7557.6	7844.3	8156.6	25
26	21.0	04.6	6398.0	6602.4	19.1	49.9	7296.8	62.2	49.3	62.0	26
27	24.0	07.7	6401.3	05.9	22.8	53.9	7301.1	66.8	54.3	67.5	27
28	27.0	10.9	04.6	09.4	26.6	57.9	05.3	71.4	59.3	72.9	28
29	30.0	14.0	07.9	12.9	30.3	61.9	09.6	76.0	64.3	78.4	29
30	6033.0	6217.2	6411.3	6616.4	6834.1	7065.9	7313.9	7580.6	7869.3	8183.9	30
31	36.0	20.3	14.6	19.9	37.8	69.8	18.2	85.3	74.3	89.4	31
32	39.0	23.5	17.9	23.5	41.5	73.8	22.4	89.9	79.3	8194.9	32
33	42.0	26.6	21.2	27.0	45.3	77.9	26.7	94.5	84.4	8200.4	33
34	45.0	29.8	24.6	30.5	49.0	81.9	31.0	7599.2	89.4	05.9	34
35	6048.0	6233.0	6427.9	6634.1	6852.8	7085.9	7335.4	7603.8	7894.5	8211.4	35
36	51.0	36.1	31.3	37.6	56.6	89.9	39.7	08.5	7899.5	17.0	36
37	54.0	39.3	34.6	41.1	60.3	93.9	44.0	13.1	7904.6	22.5	37
38	57.0	42.5	37.9	44.7	64.1	7097.9	48.3	17.8	09.7	28.1	38
39	60.0	45.6	41.3	48.2	67.9	7102.0	52.6	22.5	14.7	33.6	39
40	6063.0	6248.8	6444.6	6651.8	6871.7	7106.0	7357.0	7627.1	7919.8	8239.2	40
41	66.0	52.0	48.0	55.3	75.4	10.0	61.3	31.8	24.9	44.8	41
42	69.1	55.2	51.4	58.9	79.2	14.1	65.6	36.5	30.0	50.4	42
43	72.1	58.3	54.7	62.5	83.0	18.1	70.0	41.2	35.1	56.0	43
44	75.1	61.5	58.1	66.0	86.8	22.2	74.3	45.9	40.2	61.6	44
45	6078.1	6264.7	6461.5	6669.6	6890.6	7126.2	7378.7	7650.6	7945.3	8267.2	45
46	81.2	67.9	64.8	73.2	94.4	30.3	83.1	55.3	50.5	72.8	46
47	84.2	71.1	68.2	76.7	6898.2	34.4	87.4	60.0	55.6	78.4	47
48	87.3	74.3	71.6	80.3	6902.0	38.5	91.8	64.8	60.7	84.1	48
49	90.3	77.5	75.0	83.9	05.8	42.5	7396.2	69.5	65.9	89.7	49
50	6093.3	6280.7	6478.3	6687.5	6909.7	7146.6	7400.6	7674.2	7971.1	8295.4	50
51	96.4	83.9	81.7	91.1	13.5	50.7	05.0	79.0	76.2	8301.0	51
52	6099.4	87.1	85.1	94.7	17.3	54.8	09.4	83.7	81.4	06.7	52
53	6102.5	90.3	88.5	6698.3	21.1	58.9	13.8	88.5	86.6	12.4	53
54	05.5	93.6	91.9	6701.9	25.0	63.0	18.2	93.3	91.8	18.1	54
55	6108.6	6296.8	6495.3	6705.5	6928.8	7167.1	7422.6	7698.0	7997.0	8323.8	55
56	11.6	6300.0	6498.7	09.1	32.6	71.2	27.0	7702.8	8002.2	29.5	56
57	14.7	03.2	6502.1	12.7	36.5	75.3	31.4	07.6	07.4	35.3	57
58	17.8	06.4	05.5	16.3	40.3	79.4	35.9	12.4	12.6	41.0	58
59	20.8	09.7	09.0	20.0	44.2	83.6	40.3	17.2	17.8	46.7	59
60	6123.9	6312.9	6512.4	6723.6	6948.1	7187.7	7444.7	7722.0	8023.1	8352.5	60
Lat.	70°	71°	72°	73°	74°	75°	76°	77°	78°	79°	Lat.

TABLE 6
Meridional Parts

Lat.	80°	81°	82°	83°	84°	85°	86°	87°	88°	89°	Lat.
′											′
0	8352.5	8716.3	9122.6	9582.9	10113.9	10741.6	11509.5	12499.1	13893.4	16276.5	0
1	58.2	22.7	29.8	91.1	123.5	753.1	523.9	518.2	922.2	334.3	1
2	64.0	29.1	37.0	9599.4	133.1	764.7	538.3	537.5	951.2	393.0	2
3	69.8	35.5	44.2	9607.6	142.8	776.2	552.8	556.9	13980.4	452.8	3
4	75.6	41.9	51.5	15.9	152.4	787.8	567.3	576.4	14009.9	513.7	4
5	8381.4	8748.4	9158.7	9624.2	10162.1	10799.5	11581.9	12596.0	14039.7	16575.6	5
6	87.2	54.8	66.0	32.5	171.8	811.2	596.6	615.7	069.7	638.7	6
7	93.0	61.3	73.3	40.8	181.6	822.9	611.3	635.5	100.0	703.0	7
8	8398.9	67.8	80.6	49.2	191.3	834.7	626.1	655.4	130.6	768.5	8
9	8404.7	74.3	87.9	57.6	201.1	846.5	641.0	675.5	161.4	835.2	9
10	8410.5	8780.8	9195.2	9666.0	10211.0	10858.3	11655.9	12695.7	14192.6	16903.3	10
11	16.4	87.3	9202.6	74.4	220.8	870.2	670.9	715.9	224.0	16972.8	11
12	22.3	8793.8	09.9	82.8	230.7	882.1	686.0	736.4	255.6	17043.6	12
13	28.1	8800.4	17.3	91.3	240.6	894.1	701.1	756.9	287.6	116.0	13
14	34.0	06.9	24.7	9699.7	250.5	906.1	716.3	777.5	319.9	189.9	14
15	8439.9	8813.5	9232.1	9708.2	10260.5	10918.2	11731.5	12798.3	14352.5	17265.5	15
16	45.8	20.1	39.5	16.8	270.5	930.3	746.9	819.2	385.4	342.8	16
17	51.8	26.7	47.0	25.3	280.5	942.4	762.3	840.3	418.6	421.8	17
18	57.7	33.3	54.4	33.9	290.6	954.6	777.7	861.4	452.2	502.7	18
19	63.6	39.9	61.9	42.4	300.7	966.8	793.2	882.7	486.0	585.5	19
20	8469.6	8846.5	9269.4	9751.0	10310.8	10979.1	11808.8	12904.1	14520.3	17670.4	20
21	75.5	53.2	76.9	59.7	320.9	10991.4	824.5	925.7	554.8	757.5	21
22	81.5	59.8	84.4	68.3	331.1	11003.8	840.3	947.4	589.7	846.8	22
23	87.5	66.5	91.9	77.0	341.3	016.2	856.1	969.2	625.0	17938.4	23
24	93.5	73.2	9299.5	85.7	351.5	028.6	872.0	12991.2	660.6	18032.6	24
25	8499.5	8879.9	9307.0	9794.4	10361.8	11041.1	11887.9	13013.3	14696.6	18129.5	25
26	8505.5	86.6	14.6	9803.1	372.1	053.6	904.0	035.6	733.0	229.1	26
27	11.5	8893.3	22.2	11.9	382.4	066.2	920.1	058.0	769.8	331.8	27
28	17.5	8900.0	29.8	20.6	392.7	078.8	936.3	080.5	806.9	437.6	28
29	23.6	06.8	37.5	29.4	403.1	091.5	952.5	103.2	844.5	546.7	29
30	8529.6	8913.5	9345.1	9838.3	10413.6	11104.2	11968.9	13126.1	14882.5	18659.4	30
31	35.7	20.3	52.8	47.1	424.0	117.0	11985.3	149.1	920.9	776.0	31
32	41.8	27.1	60.5	56.0	434.5	129.8	12001.8	172.2	959.8	18896.6	32
33	47.9	33.9	68.2	64.9	445.0	142.7	018.4	195.6	14999.1	19021.6	33
34	54.0	40.7	75.9	73.8	455.5	155.6	035.0	219.0	15038.8	151.4	34
35	8560.1	8947.5	9383.7	9882.7	10466.1	11168.6	12051.8	13242.7	15079.0	19286.2	35
36	66.2	54.3	91.4	9891.7	476.7	181.6	068.6	266.5	119.7	426.5	36
37	72.3	61.2	9399.2	9900.6	487.4	194.6	085.5	290.4	160.9	572.9	37
38	78.4	68.1	9407.0	09.7	498.0	207.7	102.5	314.6	202.6	725.7	38
39	84.6	74.9	14.8	18.7	508.7	220.9	119.5	338.9	244.7	19885.6	39
40	8590.7	8981.8	9422.6	9927.7	10519.5	11234.1	12136.7	13363.3	15287.5	20053.3	40
41	8596.9	88.7	30.4	36.8	530.3	247.4	153.9	388.0	330.7	229.7	41
42	8603.1	8995.7	38.3	45.9	541.1	260.7	171.3	412.8	374.5	415.5	42
43	09.3	9002.6	46.2	55.0	551.9	274.0	188.7	437.8	418.9	612.0	43
44	15.5	09.5	54.1	64.2	562.8	287.5	206.2	463.0	463.8	20820.4	44
45	8621.7	9016.5	9462.0	9973.4	10573.7	11300.9	12223.8	13488.4	15509.3	21042.3	45
46	27.9	23.5	69.9	82.6	584.6	314.4	241.5	513.9	555.5	279.5	46
47	34.2	30.5	77.9	9991.8	595.6	328.0	259.2	539.7	602.3	534.2	47
48	40.4	37.5	85.8	10001.0	606.6	341.6	277.1	565.7	649.7	21809.4	48
49	46.7	44.5	9493.8	010.3	617.7	355.3	295.1	591.8	697.8	22108.5	49
50	8652.9	9051.5	9501.8	10019.6	10628.8	11369.1	12313.1	13618.2	15746.5	22436.2	50
51	59.2	58.6	09.9	028.9	639.9	382.8	331.3	644.7	796.0	22798.4	51
52	65.5	65.6	17.9	038.3	651.1	396.7	349.5	671.5	846.2	23203.3	52
53	71.8	72.7	26.0	047.6	662.3	410.6	367.9	698.4	897.1	23662.4	53
54	78.1	79.8	34.0	057.0	673.5	424.6	386.3	725.6	15948.8	24192.3	54
55	8684.5	9086.9	9542.1	10066.4	10684.8	11438.6	12404.8	13753.0	16001.3	24819.1	55
56	90.8	9094.0	50.3	075.9	696.1	452.6	423.5	780.6	054.6	25586.2	56
57	8697.2	9101.1	58.4	085.4	707.4	466.8	442.2	808.5	108.8	26575.1	57
58	8703.5	08.3	66.6	094.9	718.8	481.0	461.1	836.5	163.8	27969.8	58
59	09.9	15.4	74.7	10104.4	730.2	495.2	480.0	864.8	219.7	30351.6	59
60	8716.3	9122.6	9582.9	10113.9	10741.6	11509.5	12499.1	13893.4	16276.5	- - - - -	60
Lat.	80°	81°	82°	83°	84°	85°	86°	87°	88°	89°	Lat.

TABLE 7

Length of a Degree of Latitude and Longitude

Lat.	Degree of latitude				Degree of longitude				Lat.
	Nautical Miles	Statute Miles	Feet	Meters	Nautical Miles	Statute Miles	Feet	Meters	
°									°
0	59. 705	68. 708	362 776	110 574	60. 108	69. 171	365 221	111 319	0
1	. 706	. 708	778	575	60. 099	69. 160	365 166	111 303	1
2	. 706	. 709	781	576	60. 071	69. 129	365 000	111 252	2
3	. 707	. 710	786	577	60. 026	69. 077	364 724	111 168	3
4	. 708	. 711	794	580	59. 962	69. 003	364 338	111 050	4
5	59. 710	68. 713	362 804	110 583	59. 880	68. 909	363 841	110 899	5
6	. 712	. 715	816	586	59. 781	68. 794	363 234	110 714	6
7	. 714	. 718	831	591	59. 663	68. 659	362 517	110 495	7
8	. 717	. 721	847	596	59. 527	68. 502	361 690	110 243	8
9	. 720	. 725	866	601	59. 373	68. 325	360 754	109 958	9
10	59. 723	68. 728	362 886	110 608	59. 201	68. 127	359 709	109 639	10
11	. 727	. 733	909	615	59. 011	67. 908	358 555	109 288	11
12	. 731	. 738	934	622	58. 803	67. 669	357 292	108 903	12
13	. 736	. 743	961	630	58. 577	67. 409	355 921	108 485	13
14	. 740	. 748	990	639	58. 334	67. 129	354 442	108 034	14
15	59. 746	68. 754	363 021	110 649	58. 073	66. 829	352 856	107 550	15
16	. 751	. 760	053	659	57. 794	66. 508	351 163	107 034	16
17	. 757	. 767	088	669	57. 498	66. 167	349 363	106 486	17
18	. 763	. 774	125	680	57. 184	65. 806	347 457	105 905	18
19	. 769	. 781	163	692	56. 853	65. 425	345 446	105 292	19
20	59. 776	68. 788	363 203	110 704	56. 505	65. 025	343 330	104 647	20
21	. 782	. 796	245	717	56. 140	64. 604	341 110	103 970	21
22	. 790	. 805	288	730	55. 757	64. 164	338 786	103 262	22
23	. 797	. 813	333	744	55. 358	63. 705	336 360	102 523	23
24	. 805	. 822	380	758	54. 942	63. 226	333 831	101 752	24
25	59. 813	68. 831	363 428	110 773	54. 509	62. 727	331 201	100 950	25
26	. 821	. 840	478	788	54. 059	62. 210	328 470	100 118	26
27	. 829	. 850	529	804	53. 593	61. 674	325 639	99 255	27
28	. 838	. 860	581	819	53. 111	61. 119	322 709	98 362	28
29	. 847	. 870	634	836	52. 613	60. 546	319 681	97 439	29
30	59. 856	68. 881	363 689	110 852	52. 098	59. 954	316 556	96 486	30
31	. 865	. 891	745	869	51. 568	59. 344	313 334	95 504	31
32	. 874	. 902	802	887	51. 022	58. 715	310 017	94 493	32
33	. 884	. 913	860	904	50. 461	58. 069	306 605	93 453	33
34	. 893	. 924	919	922	49. 884	57. 405	303 100	92 385	34
35	59. 903	68. 935	363 978	110 941	49. 292	56. 724	299 502	91 288	35
36	. 913	. 947	364 039	959	48. 684	56. 025	295 813	90 164	36
37	. 923	. 958	100	978	48. 062	55. 309	292 033	89 012	37
38	. 933	. 970	162	996	47. 426	54. 577	288 164	87 832	38
39	. 944	. 982	224	111 015	46. 774	53. 827	284 207	86 626	39
40	59. 954	68. 994	364 287	111 035	46. 109	53. 061	280 163	85 394	40
41	. 964	69. 006	350	054	45. 429	52. 279	276 034	84 135	41
42	. 975	. 018	414	073	44. 736	51. 481	271 820	82 851	42
43	. 985	. 030	477	093	44. 029	50. 667	267 523	81 541	43
44	. 996	. 042	541	112	43. 308	49. 838	263 144	80 206	44
45	60. 006	69. 054	364 605	111 132	42. 574	48. 993	258 684	78 847	45

TABLE 7

Length of a Degree of Latitude and Longitude

Lat.	Degree of latitude				Degree of longitude				Lat.
	Nautical Miles	Statute Miles	Feet	Meters	Nautical Miles	Statute Miles	Feet	Meters	
°									°
45	60. 006	69. 054	364 605	111 132	42. 574	48. 993	258 684	78 847	45
46	. 017	. 066	670	151	41. 827	48. 133	254 145	77 463	46
47	. 027	. 078	734	171	41. 067	47. 259	249 527	76 056	47
48	. 038	. 090	798	190	40. 294	46. 370	244 834	74 625	48
49	. 048	. 103	861	210	39. 510	45. 467	240 065	73 172	49
50	60. 059	69. 115	364 925	111 229	38. 713	44. 550	235 222	71 696	50
51	. 069	. 126	988	248	37. 904	43. 619	230 307	70 198	51
52	. 080	. 138	365 050	267	37. 083	42. 675	225 321	68 678	52
53	. 090	. 150	112	286	36. 251	41. 717	220 266	67 137	53
54	. 100	. 162	174	305	35. 408	40. 747	215 144	65 576	54
55	60. 110	69. 173	365 235	111 323	34. 554	39. 764	209 954	63 994	55
56	. 120	. 185	295	342	33. 689	38. 769	204 701	62 393	56
57	. 129	. 196	354	360	32. 814	37. 762	199 384	60 772	57
58	. 139	. 207	412	378	31. 929	36. 743	194 005	59 133	58
59	. 149	. 218	469	395	31. 034	35. 713	188 567	57 475	59
60	60. 158	69. 228	365 526	111 412	30. 130	34. 672	183 071	55 800	60
61	. 167	. 239	581	429	29. 216	33. 621	177 518	54 107	61
62	. 176	. 249	635	446	28. 293	32. 559	171 910	52 398	62
63	. 184	. 259	688	462	27. 361	31. 487	166 249	50 673	63
64	. 193	. 269	739	477	26. 421	30. 405	160 537	48 932	64
65	60. 201	69. 278	365 789	111 493	25. 473	29. 314	154 775	47 176	65
66	. 209	. 287	838	507	24. 517	28. 213	148 966	45 405	66
67	. 217	. 296	885	522	23. 553	27. 104	143 110	43 620	67
68	. 224	. 305	931	536	22. 582	25. 987	137 210	41 822	68
69	. 232	. 313	975	549	21. 604	24. 861	131 267	40 010	69
70	60. 239	69. 321	366 017	111 562	20. 619	23. 728	125 284	38 187	70
71	. 245	. 329	058	574	19. 628	22. 587	119 262	36 351	71
72	. 252	. 336	096	586	18. 631	21. 440	113 203	34 504	72
73	. 258	. 343	133	597	17. 628	20. 286	107 109	32 647	73
74	. 264	. 350	169	608	16. 619	19. 125	100 981	30 779	74
75	60. 269	69. 356	366 202	111 618	15. 606	17. 959	94 823	28 902	75
76	. 274	. 362	233	628	14. 587	16. 787	88 635	27 016	76
77	. 279	. 368	262	637	13. 564	15. 610	82 419	25 121	77
78	. 284	. 373	290	645	12. 537	14. 428	76 178	23 219	78
79	. 288	. 378	315	653	11. 506	13. 241	69 913	21 310	79
80	60. 292	69. 382	366 338	111 660	10. 472	12. 051	63 627	19 393	80
81	. 295	. 386	359	666	9. 434	10. 856	57 321	17 471	81
82	. 298	. 390	378	672	8. 393	9. 658	50 997	15 544	82
83	. 301	. 393	395	677	7. 350	8. 458	44 657	13 611	83
84	. 303	. 396	409	682	6. 304	7. 254	38 303	11 675	84
85	60. 305	69. 398	366 422	111 685	5. 256	6. 049	31 937	9 735	85
86	. 307	. 400	432	688	4. 207	4. 841	25 562	7 791	86
87	. 308	. 401	440	691	3. 156	3. 632	19 178	5 846	87
88	. 309	. 403	445	693	2. 105	2. 422	12 789	3 898	88
89	. 310	. 403	449	694	1. 053	1. 211	6 395	1 949	89
90	60. 310	69. 403	366 450	111 694	0. 000	0. 000	0	0	90

TABLE 8

Conversion Table for Meters, Feet, and Fathoms

Meters	Feet	Fathoms	Meters	Feet	Fathoms	Feet	Meters	Feet	Meters	Fathoms	Meters	Fathoms	Meters
1	3.28	0.55	61	200.13	33.36	1	0.30	61	18.59	1	1.83	61	111.56
2	6.56	1.09	62	203.41	33.90	2	0.61	62	18.90	2	3.66	62	113.39
3	9.84	1.64	63	206.69	34.45	3	0.91	63	19.20	3	5.49	63	115.21
4	13.12	2.19	64	209.97	35.00	4	1.22	64	19.51	4	7.32	64	117.04
5	16.40	2.73	65	213.25	35.54	5	1.52	65	19.81	5	9.14	65	118.87
6	19.69	3.28	66	216.54	36.09	6	1.83	66	20.12	6	10.97	66	120.70
7	22.97	3.83	67	219.82	36.64	7	2.13	67	20.42	7	12.80	67	122.53
8	26.25	4.37	68	223.10	37.18	8	2.44	68	20.73	8	14.63	68	124.36
9	29.53	4.92	69	226.38	37.73	9	2.74	69	21.03	9	16.46	69	126.19
10	32.81	5.47	70	229.66	38.28	10	3.05	70	21.34	10	18.29	70	128.02
11	36.09	6.01	71	232.94	38.82	11	3.35	71	21.64	11	20.12	71	129.84
12	39.37	6.56	72	236.22	39.37	12	3.66	72	21.95	12	21.95	72	131.67
13	42.65	7.11	73	239.50	39.92	13	3.96	73	22.25	13	23.77	73	133.50
14	45.93	7.66	74	242.78	40.46	14	4.27	74	22.56	14	25.60	74	135.33
15	49.21	8.20	75	246.06	41.01	15	4.57	75	22.86	15	27.43	75	137.16
16	52.49	8.75	76	249.34	41.56	16	4.88	76	23.16	16	29.26	76	138.99
17	55.77	9.30	77	252.62	42.10	17	5.18	77	23.47	17	31.09	77	140.82
18	59.06	9.84	78	255.91	42.65	18	5.49	78	23.77	18	32.92	78	142.65
19	62.34	10.39	79	259.19	43.20	19	5.79	79	24.08	19	34.75	79	144.48
20	65.62	10.94	80	262.47	43.74	20	6.10	80	24.38	20	36.58	80	146.30
21	68.90	11.48	81	265.75	44.29	21	6.40	81	24.69	21	38.40	81	148.13
22	72.18	12.03	82	269.03	44.84	22	6.71	82	24.99	22	40.23	82	149.96
23	75.46	12.58	83	272.31	45.38	23	7.01	83	25.30	23	42.06	83	151.79
24	78.74	13.12	84	275.59	45.93	24	7.32	84	25.60	24	43.89	84	153.62
25	82.02	13.67	85	278.87	46.48	25	7.62	85	25.91	25	45.72	85	155.45
26	85.30	14.22	86	282.15	47.03	26	7.92	86	26.21	26	47.55	86	157.28
27	88.58	14.76	87	285.43	47.57	27	8.23	87	26.52	27	49.38	87	159.11
28	91.86	15.31	88	288.71	48.12	28	8.53	88	26.82	28	51.21	88	160.93
29	95.14	15.86	89	291.99	48.67	29	8.84	89	27.13	29	53.04	89	162.76
30	98.43	16.40	90	295.28	49.21	30	9.14	90	27.43	30	54.86	90	164.59
31	101.71	16.95	91	298.56	49.76	31	9.45	91	27.74	31	56.69	91	166.42
32	104.99	17.50	92	301.84	50.31	32	9.75	92	28.04	32	58.52	92	168.25
33	108.27	18.04	93	305.12	50.85	33	10.06	93	28.35	33	60.35	93	170.08
34	111.55	18.59	94	308.40	51.40	34	10.36	94	28.65	34	62.18	94	171.91
35	114.83	19.14	95	311.68	51.95	35	10.67	95	28.96	35	64.01	95	173.74
36	118.11	19.69	96	314.96	52.49	36	10.97	96	29.26	36	65.84	96	175.56
37	121.39	20.23	97	318.24	53.04	37	11.28	97	29.57	37	67.67	97	177.39
38	124.67	20.78	98	321.52	53.59	38	11.58	98	29.87	38	69.49	98	179.22
39	127.95	21.33	99	324.80	54.13	39	11.89	99	30.18	39	71.32	99	181.05
40	131.23	21.87	100	328.08	54.68	40	12.19	100	30.48	40	73.15	100	182.88
41	134.51	22.42	101	331.36	55.23	41	12.50	101	30.78	41	74.98	101	184.71
42	137.80	22.97	102	334.65	55.77	42	12.80	102	31.09	42	76.81	102	186.54
43	141.08	23.51	103	337.93	56.32	43	13.11	103	31.39	43	78.64	103	188.37
44	144.36	24.06	104	341.21	56.87	44	13.41	104	31.70	44	80.47	104	190.20
45	147.64	24.61	105	344.49	57.41	45	13.72	105	32.00	45	82.30	105	192.02
46	150.92	25.15	106	347.77	57.96	46	14.02	106	32.31	46	84.12	106	193.85
47	154.20	25.70	107	351.05	58.51	47	14.33	107	32.61	47	85.95	107	195.68
48	157.48	26.25	108	354.33	59.06	48	14.63	108	32.92	48	87.78	108	197.51
49	160.76	26.79	109	357.61	59.60	49	14.94	109	33.22	49	89.61	109	199.34
50	164.04	27.34	110	360.89	60.15	50	15.24	110	33.53	50	91.44	110	201.17
51	167.32	27.89	111	364.17	60.70	51	15.54	111	33.83	51	93.27	111	203.00
52	170.60	28.43	112	367.45	61.24	52	15.85	112	34.14	52	95.10	112	204.83
53	173.88	28.98	113	370.73	61.79	53	16.15	113	34.44	53	96.93	113	206.65
54	177.17	29.53	114	374.02	62.34	54	16.46	114	34.75	54	98.76	114	208.48
55	180.45	30.07	115	377.30	62.88	55	16.76	115	35.05	55	100.58	115	210.31
56	183.73	30.62	116	380.58	63.43	56	17.07	116	35.36	56	102.41	116	212.14
57	187.01	31.17	117	383.86	63.98	57	17.37	117	35.66	57	104.24	117	213.97
58	190.29	31.71	118	387.14	64.52	58	17.68	118	35.97	58	106.07	118	215.80
59	193.57	32.26	119	390.42	65.07	59	17.98	119	36.27	59	107.90	119	217.63
60	196.85	32.81	120	393.70	65.62	60	18.29	120	36.58	60	109.73	120	219.46

TABLE 9

Conversion Table for Nautical and Statute Miles

1 nautical mile = 6,076.11548 . . . feet 1 statute mile = 5,280 feet

Nautical miles to statute miles				Statute miles to nautical miles			
Nautical miles	Statute miles	Nautical miles	Statute miles	Statute miles	Nautical miles	Statute miles	Nautical miles
1	1.151	51	58.690	1	0.869	51	44.318
2	2.302	52	59.841	2	1.738	52	45.187
3	3.452	53	60.991	3	2.607	53	46.056
4	4.603	54	62.142	4	3.476	54	46.925
5	5.754	55	63.293	5	4.345	55	47.794
6	6.905	56	64.444	6	5.214	56	48.663
7	8.055	57	65.594	7	6.083	57	49.532
8	9.206	58	66.745	8	6.952	58	50.401
9	10.357	59	67.896	9	7.821	59	51.270
10	11.508	60	69.047	10	8.690	60	52.139
11	12.659	61	70.198	11	9.559	61	53.008
12	13.809	62	71.348	12	10.428	62	53.877
13	14.960	63	72.499	13	11.297	63	54.746
14	16.111	64	73.650	14	12.166	64	55.614
15	17.262	65	74.801	15	13.035	65	56.483
16	18.412	66	75.951	16	13.904	66	57.352
17	19.563	67	77.102	17	14.773	67	58.221
18	20.714	68	78.253	18	15.642	68	59.090
19	21.865	69	79.404	19	16.511	69	59.959
20	23.016	70	80.555	20	17.380	70	60.828
21	24.166	71	81.705	21	18.249	71	61.697
22	25.317	72	82.856	22	19.117	72	62.566
23	26.468	73	84.007	23	19.986	73	63.435
24	27.619	74	85.158	24	20.855	74	64.304
25	28.769	75	86.308	25	21.724	75	65.173
26	29.920	76	87.459	26	22.593	76	66.042
27	31.071	77	88.610	27	23.462	77	66.911
28	32.222	78	89.761	28	24.331	78	67.780
29	33.373	79	90.912	29	25.200	79	68.649
30	34.523	80	92.062	30	26.069	80	69.518
31	35.674	81	93.213	31	26.938	81	70.387
32	36.825	82	94.364	32	27.807	82	71.256
33	37.976	83	95.515	33	28.676	83	72.125
34	39.127	84	96.665	34	29.545	84	72.994
35	40.277	85	97.816	35	30.414	85	73.863
36	41.428	86	98.967	36	31.283	86	74.732
37	42.579	87	100.118	37	32.152	87	75.601
38	43.730	88	101.269	38	33.021	88	76.470
39	44.880	89	102.419	39	33.890	89	77.339
40	46.031	90	103.570	40	34.759	90	78.208
41	47.182	91	104.721	41	35.628	91	79.077
42	48.333	92	105.872	42	36.497	92	79.946
43	49.484	93	107.022	43	37.366	93	80.815
44	50.634	94	108.173	44	38.235	94	81.684
45	51.785	95	109.324	45	39.104	95	82.553
46	52.936	96	110.475	46	39.973	96	83.422
47	54.087	97	111.626	47	40.842	97	84.291
48	55.237	98	112.776	48	41.711	98	85.160
49	56.388	99	113.927	49	42.580	99	86.029
50	57.539	100	115.078	50	43.449	100	86.898

TABLE 10

Speed Table for Measured Mile

Sec.	Minutes 1	2	3	4	5	6	7	8	9	10	11	12	Sec.
	Knots	*Knots*	*Knots*	*Knots*	*Knots*	*Knots*	*Knots*	*Knots*	*Knots*	*Knots*	*Knots*	*Knots*	
0	60.000	30.000	20.000	15.000	12.000	10.000	8.571	7.500	6.667	6.000	5.455	5.000	0
1	59.016	29.752	19.890	14.938	11.960	9.972	8.551	7.484	6.654	5.990	5.446	4.993	1
2	58.065	29.508	19.780	14.876	11.921	9.945	8.531	7.469	6.642	5.980	5.438	4.986	2
3	57.143	29.268	19.672	14.815	11.881	9.917	8.511	7.453	6.630	5.970	5.430	4.979	3
4	56.250	29.032	19.565	14.754	11.842	9.890	8.491	7.438	6.618	5.960	5.422	4.972	4
5	55.385	28.800	19.459	14.694	11.803	9.863	8.471	7.423	6.606	5.950	5.414	4.966	5
6	54.545	28.571	19.355	14.634	11.765	9.836	8.451	7.407	6.593	5.941	5.405	4.959	6
7	53.731	28.346	19.251	14.575	11.726	9.809	8.431	7.392	6.581	5.931	5.397	4.952	7
8	52.941	28.125	19.149	14.516	11.688	9.783	8.411	7.377	6.569	5.921	5.389	4.945	8
9	52.174	27.907	19.048	14.458	11.650	9.756	8.392	7.362	6.557	5.911	5.381	4.938	9
10	51.429	27.692	18.947	14.400	11.613	9.730	8.372	7.347	6.545	5.902	5.373	4.932	10
11	50.704	27.481	18.848	14.343	11.576	9.704	8.353	7.332	6.534	5.892	5.365	4.925	11
12	50.000	27.273	18.750	14.286	11.538	9.677	8.333	7.317	6.522	5.882	5.357	4.918	12
13	49.315	27.068	18.653	14.229	11.502	9.651	8.314	7.302	6.510	5.873	5.349	4.911	13
14	48.649	26.866	18.557	14.173	11.465	9.626	8.295	7.287	6.498	5.863	5.341	4.905	14
15	48.000	26.667	18.462	14.118	11.429	9.600	8.276	7.273	6.486	5.854	5.333	4.898	15
16	47.368	26.471	18.367	14.062	11.392	9.574	8.257	7.258	6.475	5.844	5.325	4.891	16
17	46.753	26.277	18.274	14.008	11.356	9.549	8.238	7.243	6.463	5.835	5.318	4.885	17
18	46.154	26.087	18.182	13.953	11.321	9.524	8.219	7.229	6.452	5.825	5.310	4.878	18
19	45.570	25.899	18.090	13.900	11.285	9.499	8.200	7.214	6.440	5.816	5.302	4.871	19
20	45.000	25.714	18.000	13.846	11.250	9.474	8.182	7.200	6.429	5.806	5.294	4.865	20
21	44.444	25.532	17.910	13.793	11.215	9.449	8.163	7.186	6.417	5.797	5.286	4.858	21
22	43.902	25.352	17.822	13.740	11.180	9.424	8.145	7.171	6.406	5.788	5.279	4.852	22
23	43.373	25.175	17.734	13.688	11.146	9.399	8.126	7.157	6.394	5.778	5.271	4.845	23
24	42.857	25.000	17.647	13.636	11.111	9.375	8.108	7.143	6.383	5.769	5.263	4.839	24
25	42.353	24.828	17.561	13.585	11.077	9.351	8.090	7.129	6.372	5.760	5.255	4.832	25
26	41.860	24.658	17.476	13.534	11.043	9.326	8.072	7.115	6.360	5.751	5.248	4.826	26
27	41.379	24.490	17.391	13.483	11.009	9.302	8.054	7.101	6.349	5.742	5.240	4.819	27
28	40.909	24.324	17.308	13.433	10.976	9.278	8.036	7.087	6.338	5.732	5.233	4.813	28
29	40.449	24.161	17.225	13.383	10.942	9.254	8.018	7.073	6.327	5.723	5.225	4.806	29
30	40.000	24.000	17.143	13.333	10.909	9.231	8.000	7.059	6.316	5.714	5.217	4.800	30
31	39.560	23.841	17.062	13.284	10.876	9.207	7.982	7.045	6.305	5.705	5.210	4.794	31
32	39.130	23.684	16.981	13.235	10.843	9.184	7.965	7.031	6.294	5.696	5.202	4.787	32
33	38.710	23.529	16.901	13.187	10.811	9.160	7.947	7.018	6.283	5.687	5.195	4.781	33
34	38.298	23.377	16.822	13.139	10.778	9.137	7.930	7.004	6.272	5.678	5.187	4.775	34
35	37.895	23.226	16.744	13.091	10.746	9.114	7.912	6.990	6.261	5.669	5.180	4.768	35
36	37.500	23.077	16.667	13.043	10.714	9.091	7.895	6.977	6.250	5.660	5.172	4.762	36
37	37.113	22.930	16.590	12.996	10.682	9.068	7.877	6.963	6.239	5.651	5.165	4.756	37
38	36.735	22.785	16.514	12.950	10.651	9.045	7.860	6.950	6.228	5.643	5.158	4.749	38
39	36.364	22.642	16.438	12.903	10.619	9.023	7.843	6.936	6.218	5.634	5.150	4.743	39
40	36.000	22.500	16.364	12.857	10.588	9.000	7.826	6.923	6.207	5.625	5.143	4.737	40
41	35.644	22.360	16.290	12.811	10.557	8.978	7.809	6.910	6.196	5.616	5.136	4.731	41
42	35.294	22.222	16.216	12.766	10.526	8.955	7.792	6.897	6.186	5.607	5.128	4.724	42
43	34.951	22.086	16.143	12.721	10.496	8.933	7.775	6.883	6.175	5.599	5.121	4.718	43
44	34.615	21.951	16.071	12.676	10.465	8.911	7.759	6.870	6.164	5.590	5.114	4.712	44
45	34.286	21.818	16.000	12.632	10.435	8.889	7.742	6.857	6.154	5.581	5.106	4.706	45
46	33.962	21.687	15.929	12.587	10.405	8.867	7.725	6.844	6.143	5.573	5.099	4.700	46
47	33.645	21.557	15.859	12.544	10.375	8.845	7.709	6.831	6.133	5.564	5.092	4.694	47
48	33.333	21.429	15.789	12.500	10.345	8.824	7.692	6.818	6.122	5.556	5.085	4.688	48
49	33.028	21.302	15.721	12.457	10.315	8.802	7.676	6.805	6.112	5.547	5.078	4.681	49
50	32.727	21.176	15.652	12.414	10.286	8.780	7.660	6.792	6.102	5.538	5.070	4.675	50
51	32.432	21.053	15.584	12.371	10.256	8.759	7.643	6.780	6.091	5.530	5.063	4.669	51
52	32.143	20.930	15.517	12.329	10.227	8.738	7.627	6.767	6.081	5.521	5.056	4.663	52
53	31.858	20.809	15.451	12.287	10.198	8.717	7.611	6.754	6.071	5.513	5.049	4.657	53
54	31.579	20.690	15.385	12.245	10.169	8.696	7.595	6.742	6.061	5.505	5.042	4.651	54
55	31.304	20.571	15.319	12.203	10.141	8.675	7.579	6.729	6.050	5.496	5.035	4.645	55
56	31.034	20.455	15.254	12.162	10.112	8.654	7.563	6.716	6.040	5.488	5.028	4.639	56
57	30.769	20.339	15.190	12.121	10.084	8.633	7.547	6.704	6.030	5.479	5.021	4.633	57
58	30.508	20.225	15.126	12.081	10.056	8.612	7.531	6.691	6.020	5.471	5.014	4.627	58
59	30.252	20.112	15.063	12.040	10.028	8.592	7.516	6.679	6.010	5.463	5.007	4.621	59
60	30.000	20.000	15.000	12.000	10.000	8.571	7.500	6.667	6.000	5.455	5.000	4.615	60
Sec.	1	2	3	4	5	6	7	8	9	10	11	12	Sec.

TABLE 11
Speed, Time, and Distance

Minutes	0.5	1.0	1.5	2.0	2.5	3.0	3.5	4.0	4.5	5.0	5.5	6.0	6.5	7.0	7.5	8.0	Minutes
	Miles	*Miles*	*Miles*	*Miles*	*Miles*	*Miles*	*Miles*	*Miles*	*Miles*	*Miles*	*Miles*	*Miles*	*Miles*	*Miles*	*Miles*	*Miles*	
1	0.0	0.0	0.0	0.0	0.0	0.0	0.1	0.1	0.1	0.1	0.1	0.1	0.1	0.1	0.1	0.1	1
2	0.0	0.0	0.0	0.1	0.1	0.1	0.1	0.1	0.2	0.2	0.2	0.2	0.2	0.2	0.2	0.3	2
3	0.0	0.0	0.1	0.1	0.1	0.2	0.2	0.2	0.2	0.2	0.3	0.3	0.3	0.4	0.4	0.4	3
4	0.0	0.1	0.1	0.1	0.2	0.2	0.2	0.3	0.3	0.3	0.4	0.4	0.4	0.5	0.5	0.5	4
5	0.0	0.1	0.1	0.2	0.2	0.2	0.3	0.3	0.4	0.4	0.5	0.5	0.5	0.6	0.6	0.7	5
6	0.0	0.1	0.2	0.2	0.2	0.3	0.4	0.4	0.4	0.5	0.6	0.6	0.6	0.7	0.8	0.8	6
7	0.1	0.1	0.2	0.2	0.3	0.4	0.4	0.5	0.5	0.6	0.6	0.7	0.8	0.8	0.9	0.9	7
8	0.1	0.1	0.2	0.3	0.3	0.4	0.5	0.5	0.6	0.7	0.7	0.8	0.9	0.9	1.0	1.1	8
9	0.1	0.2	0.2	0.3	0.4	0.4	0.5	0.6	0.7	0.8	0.8	0.9	1.0	1.0	1.1	1.2	9
10	0.1	0.2	0.2	0.3	0.4	0.5	0.6	0.7	0.8	0.8	0.9	1.0	1.1	1.2	1.2	1.3	10
11	0.1	0.2	0.3	0.4	0.5	0.6	0.6	0.7	0.8	0.9	1.0	1.1	1.2	1.3	1.4	1.5	11
12	0.1	0.2	0.3	0.4	0.5	0.6	0.7	0.8	0.9	1.0	1.1	1.2	1.3	1.4	1.5	1.6	12
13	0.1	0.2	0.3	0.4	0.5	0.6	0.8	0.9	1.0	1.1	1.2	1.3	1.4	1.5	1.6	1.7	13
14	0.1	0.2	0.4	0.5	0.6	0.7	0.8	0.9	1.0	1.2	1.3	1.4	1.5	1.6	1.8	1.9	14
15	0.1	0.2	0.4	0.5	0.6	0.8	0.9	1.0	1.1	1.2	1.4	1.5	1.6	1.8	1.9	2.0	15
16	0.1	0.3	0.4	0.5	0.7	0.8	0.9	1.1	1.2	1.3	1.5	1.6	1.7	1.9	2.0	2.1	16
17	0.1	0.3	0.4	0.6	0.7	0.8	1.0	1.1	1.3	1.4	1.6	1.7	1.8	2.0	2.1	2.3	17
18	0.2	0.3	0.4	0.6	0.8	0.9	1.0	1.2	1.4	1.5	1.6	1.8	2.0	2.1	2.2	2.4	18
19	0.2	0.3	0.5	0.6	0.8	1.0	1.1	1.3	1.4	1.6	1.7	1.9	2.1	2.2	2.4	2.5	19
20	0.2	0.3	0.5	0.7	0.8	1.0	1.2	1.3	1.5	1.7	1.8	2.0	2.2	2.3	2.5	2.7	20
21	0.2	0.4	0.5	0.7	0.9	1.0	1.2	1.4	1.6	1.8	1.9	2.1	2.3	2.4	2.6	2.8	21
22	0.2	0.4	0.6	0.7	0.9	1.1	1.3	1.5	1.6	1.8	2.0	2.2	2.4	2.6	2.8	2.9	22
23	0.2	0.4	0.6	0.8	1.0	1.2	1.3	1.5	1.7	1.9	2.1	2.3	2.5	2.7	2.9	3.1	23
24	0.2	0.4	0.6	0.8	1.0	1.2	1.4	1.6	1.8	2.0	2.2	2.4	2.6	2.8	3.0	3.2	24
25	0.2	0.4	0.6	0.8	1.0	1.2	1.5	1.7	1.9	2.1	2.3	2.5	2.7	2.9	3.1	3.3	25
26	0.2	0.4	0.6	0.9	1.1	1.3	1.5	1.7	2.0	2.2	2.4	2.6	2.8	3.0	3.2	3.5	26
27	0.2	0.4	0.7	0.9	1.1	1.4	1.6	1.8	2.0	2.2	2.5	2.7	2.9	3.2	3.4	3.6	27
28	0.2	0.5	0.7	0.9	1.2	1.4	1.6	1.9	2.1	2.3	2.6	2.8	3.0	3.3	3.5	3.7	28
29	0.2	0.5	0.7	1.0	1.2	1.4	1.7	1.9	2.2	2.4	2.7	2.9	3.1	3.4	3.6	3.9	29
30	0.2	0.5	0.8	1.0	1.2	1.5	1.8	2.0	2.2	2.5	2.8	3.0	3.2	3.5	3.8	4.0	30
31	0.3	0.5	0.8	1.0	1.3	1.6	1.8	2.1	2.3	2.6	2.8	3.1	3.4	3.6	3.9	4.1	31
32	0.3	0.5	0.8	1.1	1.3	1.6	1.9	2.1	2.4	2.7	2.9	3.2	3.5	3.7	4.0	4.3	32
33	0.3	0.6	0.8	1.1	1.4	1.6	1.9	2.2	2.5	2.8	3.0	3.3	3.6	3.8	4.1	4.4	33
34	0.3	0.6	0.8	1.1	1.4	1.7	2.0	2.3	2.6	2.8	3.1	3.4	3.7	4.0	4.2	4.5	34
35	0.3	0.6	0.9	1.2	1.5	1.8	2.0	2.3	2.6	2.9	3.2	3.5	3.8	4.1	4.4	4.7	35
36	0.3	0.6	0.9	1.2	1.5	1.8	2.1	2.4	2.7	3.0	3.3	3.6	3.9	4.2	4.5	4.8	36
37	0.3	0.6	0.9	1.2	1.5	1.8	2.2	2.5	2.8	3.1	3.4	3.7	4.0	4.3	4.6	4.9	37
38	0.3	0.6	1.0	1.3	1.6	1.9	2.2	2.5	2.8	3.2	3.5	3.8	4.1	4.4	4.8	5.1	38
39	0.3	0.6	1.0	1.3	1.6	2.0	2.3	2.6	2.9	3.2	3.6	3.9	4.2	4.6	4.9	5.2	39
40	0.3	0.7	1.0	1.3	1.7	2.0	2.3	2.7	3.0	3.3	3.7	4.0	4.3	4.7	5.0	5.3	40
41	0.3	0.7	1.0	1.4	1.7	2.0	2.4	2.7	3.1	3.4	3.8	4.1	4.4	4.8	5.1	5.5	41
42	0.4	0.7	1.0	1.4	1.8	2.1	2.4	2.8	3.2	3.5	3.8	4.2	4.6	4.9	5.2	5.6	42
43	0.4	0.7	1.1	1.4	1.8	2.2	2.5	2.9	3.2	3.6	3.9	4.3	4.7	5.0	5.4	5.7	43
44	0.4	0.7	1.1	1.5	1.8	2.2	2.6	2.9	3.3	3.7	4.0	4.4	4.8	5.1	5.5	5.9	44
45	0.4	0.8	1.1	1.5	1.9	2.2	2.6	3.0	3.4	3.8	4.1	4.5	4.9	5.2	5.6	6.0	45
46	0.4	0.8	1.2	1.5	1.9	2.3	2.7	3.1	3.4	3.8	4.2	4.6	5.0	5.4	5.8	6.1	46
47	0.4	0.8	1.2	1.6	2.0	2.4	2.7	3.1	3.5	3.9	4.3	4.7	5.1	5.5	5.9	6.3	47
48	0.4	0.8	1.2	1.6	2.0	2.4	2.8	3.2	3.6	4.0	4.4	4.8	5.2	5.6	6.0	6.4	48
49	0.4	0.8	1.2	1.6	2.0	2.4	2.9	3.3	3.7	4.1	4.5	4.9	5.3	5.7	6.1	6.5	49
50	0.4	0.8	1.2	1.7	2.1	2.5	2.9	3.3	3.8	4.2	4.6	5.0	5.4	5.8	6.2	6.7	50
51	0.4	0.8	1.3	1.7	2.1	2.6	3.0	3.4	3.8	4.2	4.7	5.1	5.5	6.0	6.4	6.8	51
52	0.4	0.9	1.3	1.7	2.2	2.6	3.0	3.5	3.9	4.3	4.8	5.2	5.6	6.1	6.5	6.9	52
53	0.4	0.9	1.3	1.8	2.2	2.6	3.1	3.5	4.0	4.4	4.9	5.3	5.7	6.2	6.6	7.1	53
54	0.4	0.9	1.4	1.8	2.2	2.7	3.2	3.6	4.1	4.5	5.0	5.4	5.8	6.3	6.8	7.2	54
55	0.5	0.9	1.4	1.8	2.3	2.8	3.2	3.7	4.1	4.6	5.0	5.5	6.0	6.4	6.9	7.3	55
56	0.5	0.9	1.4	1.9	2.3	2.8	3.3	3.7	4.2	4.7	5.1	5.6	6.1	6.5	7.0	7.5	56
57	0.5	1.0	1.4	1.9	2.4	2.8	3.3	3.8	4.3	4.8	5.2	5.7	6.2	6.6	7.1	7.6	57
58	0.5	1.0	1.4	1.9	2.4	2.9	3.4	3.9	4.4	4.8	5.3	5.8	6.3	6.8	7.2	7.7	58
59	0.5	1.0	1.5	2.0	2.5	3.0	3.4	3.9	4.4	4.9	5.4	5.9	6.4	6.9	7.4	7.9	59
60	0.5	1.0	1.5	2.0	2.5	3.0	3.5	4.0	4.5	5.0	5.5	6.0	6.5	7.0	7.5	8.0	60

Speed in knots

TABLE 11
Speed, Time, and Distance

Minutes	8.5	9.0	9.5	10.0	10.5	11.0	11.5	12.0	12.5	13.0	13.5	14.0	14.5	15.0	15.5	16.0	Minutes
	Miles	*Miles*	*Miles*	*Miles*	*Miles*	*Miles*	*Miles*	*Miles*	*Miles*	*Miles*	*Miles*	*Miles*	*Miles*	*Miles*	*Miles*	*Miles*	
1	0.1	0.2	0.2	0.2	0.2	0.2	0.2	0.2	0.2	0.2	0.2	0.2	0.2	0.2	0.3	0.3	1
2	0.3	0.3	0.3	0.3	0.4	0.4	0.4	0.4	0.4	0.4	0.4	0.5	0.5	0.5	0.5	0.5	2
3	0.4	0.4	0.5	0.5	0.5	0.6	0.6	0.6	0.6	0.6	0.7	0.7	0.7	0.8	0.8	0.8	3
4	0.6	0.6	0.6	0.7	0.7	0.7	0.8	0.8	0.8	0.9	0.9	0.9	1.0	1.0	1.0	1.1	4
5	0.7	0.8	0.8	0.8	0.9	0.9	1.0	1.0	1.0	1.1	1.1	1.2	1.2	1.2	1.3	1.3	5
6	0.9	0.9	1.0	1.0	1.0	1.1	1.2	1.2	1.2	1.3	1.4	1.4	1.4	1.5	1.6	1.6	6
7	1.0	1.0	1.1	1.2	1.2	1.3	1.3	1.4	1.5	1.5	1.6	1.6	1.7	1.8	1.8	1.9	7
8	1.1	1.2	1.3	1.3	1.4	1.5	1.5	1.6	1.7	1.7	1.8	1.9	1.9	2.0	2.1	2.1	8
9	1.3	1.4	1.4	1.5	1.6	1.6	1.7	1.8	1.9	2.0	2.0	2.1	2.2	2.2	2.3	2.4	9
10	1.4	1.5	1.6	1.7	1.8	1.8	1.9	2.0	2.1	2.2	2.2	2.3	2.4	2.5	2.6	2.7	10
11	1.6	1.6	1.7	1.8	1.9	2.0	2.1	2.2	2.3	2.4	2.5	2.6	2.7	2.8	2.8	2.9	11
12	1.7	1.8	1.9	2.0	2.1	2.2	2.3	2.4	2.5	2.6	2.7	2.8	2.9	3.0	3.1	3.2	12
13	1.8	2.0	2.1	2.2	2.3	2.4	2.5	2.6	2.7	2.8	2.9	3.0	3.1	3.2	3.4	3.5	13
14	2.0	2.1	2.2	2.3	2.4	2.6	2.7	2.8	2.9	3.0	3.2	3.3	3.4	3.5	3.6	3.7	14
15	2.1	2.2	2.4	2.5	2.6	2.8	2.9	3.0	3.1	3.2	3.4	3.5	3.6	3.8	3.9	4.0	15
16	2.3	2.4	2.5	2.7	2.8	2.9	3.1	3.2	3.3	3.5	3.6	3.7	3.9	4.0	4.1	4.3	16
17	2.4	2.5	2.7	2.8	3.0	3.1	3.3	3.4	3.5	3.7	3.8	4.0	4.1	4.2	4.4	4.5	17
18	2.6	2.7	2.8	3.0	3.2	3.3	3.4	3.6	3.8	3.9	4.0	4.2	4.4	4.5	4.6	4.8	18
19	2.7	2.8	3.0	3.2	3.3	3.5	3.6	3.8	4.0	4.1	4.3	4.4	4.6	4.8	4.9	5.1	19
20	2.8	3.0	3.2	3.3	3.5	3.7	3.8	4.0	4.2	4.3	4.5	4.7	4.8	5.0	5.2	5.3	20
21	3.0	3.2	3.3	3.5	3.7	3.8	4.0	4.2	4.4	4.6	4.7	4.9	5.1	5.2	5.4	5.6	21
22	3.1	3.3	3.5	3.7	3.8	4.0	4.2	4.4	4.6	4.8	5.0	5.1	5.3	5.5	5.7	5.9	22
23	3.3	3.4	3.6	3.8	4.0	4.2	4.4	4.6	4.8	5.0	5.2	5.4	5.6	5.8	5.9	6.1	23
24	3.4	3.6	3.8	4.0	4.2	4.4	4.6	4.8	5.0	5.2	5.4	5.6	5.8	6.0	6.2	6.4	24
25	3.5	3.8	4.0	4.2	4.4	4.6	4.8	5.0	5.2	5.4	5.6	5.8	6.0	6.2	6.5	6.7	25
26	3.7	3.9	4.1	4.3	4.6	4.8	5.0	5.2	5.4	5.6	5.8	6.1	6.3	6.5	6.7	6.9	26
27	3.8	4.0	4.3	4.5	4.7	5.0	5.2	5.4	5.6	5.8	6.1	6.3	6.5	6.8	7.0	7.2	27
28	4.0	4.2	4.4	4.7	4.9	5.1	5.4	5.6	5.8	6.1	6.3	6.5	6.8	7.0	7.2	7.5	28
29	4.1	4.4	4.6	4.8	5.1	5.3	5.6	5.8	6.0	6.3	6.5	6.8	7.0	7.2	7.5	7.7	29
30	4.2	4.5	4.8	5.0	5.2	5.5	5.8	6.0	6.2	6.5	6.8	7.0	7.2	7.5	7.8	8.0	30
31	4.4	4.6	4.9	5.2	5.4	5.7	5.9	6.2	6.5	6.7	7.0	7.2	7.5	7.8	8.0	8.3	31
32	4.5	4.8	5.1	5.3	5.6	5.9	6.1	6.4	6.7	6.9	7.2	7.5	7.7	8.0	8.3	8.5	32
33	4.7	5.0	5.2	5.5	5.8	6.0	6.3	6.6	6.9	7.2	7.4	7.7	8.0	8.2	8.5	8.8	33
34	4.8	5.1	5.4	5.7	6.0	6.2	6.5	6.8	7.1	7.4	7.6	7.9	8.2	8.5	8.8	9.1	34
35	5.0	5.2	5.5	5.8	6.1	6.4	6.7	7.0	7.3	7.6	7.9	8.2	8.5	8.8	9.0	9.3	35
36	5.1	5.4	5.7	6.0	6.3	6.6	6.9	7.2	7.5	7.8	8.1	8.4	8.7	9.0	9.3	9.6	36
37	5.2	5.6	5.9	6.2	6.5	6.8	7.1	7.4	7.7	8.0	8.3	8.6	8.9	9.2	9.6	9.9	37
38	5.4	5.7	6.0	6.3	6.8	7.0	7.3	7.6	7.9	8.2	8.6	8.9	9.2	9.5	9.8	10.1	38
39	5.5	5.8	6.2	6.5	6.8	7.2	7.5	7.8	8.1	8.4	8.8	9.1	9.4	9.8	10.1	10.4	39
40	5.7	6.0	6.3	6.7	7.0	7.3	7.7	8.0	8.3	8.7	9.0	9.3	9.7	10.0	10.3	10.7	40
41	5.8	6.2	6.5	6.8	7.2	7.5	7.9	8.2	8.5	8.9	9.2	9.6	9.9	10.2	10.6	10.9	41
42	6.0	6.3	6.6	7.0	7.4	7.7	8.0	8.4	8.8	9.1	9.4	9.8	10.2	10.5	10.9	11.2	42
43	6.1	6.4	6.8	7.2	7.5	7.9	8.2	8.6	9.0	9.3	9.7	10.0	10.4	10.8	11.1	11.5	43
44	6.2	6.6	7.0	7.3	7.7	8.1	8.4	8.8	9.2	9.5	9.9	10.3	10.6	11.0	11.4	11.7	44
45	6.4	6.8	7.1	7.5	7.9	8.2	8.6	9.0	9.4	9.8	10.1	10.5	10.9	11.2	11.6	12.0	45
46	6.5	6.9	7.3	7.7	8.0	8.4	8.8	9.2	9.6	10.0	10.4	10.7	11.1	11.5	11.9	12.3	46
47	6.7	7.0	7.4	7.8	8.2	8.6	9.0	9.4	9.8	10.2	10.6	11.0	11.4	11.8	12.1	12.5	47
48	6.8	7.2	7.6	8.0	8.4	8.8	9.2	9.6	10.0	10.4	10.8	11.2	11.6	12.0	12.4	12.8	48
49	6.9	7.3	7.8	8.2	8.6	9.0	9.4	9.8	10.2	10.6	11.0	11.4	11.8	12.2	12.7	13.1	49
50	7.1	7.5	7.9	8.3	8.8	9.2	9.6	10.0	10.4	10.8	11.2	11.7	12.1	12.5	12.9	13.3	50
51	7.2	7.6	8.1	8.5	8.9	9.4	9.8	10.2	10.6	11.1	11.5	11.9	12.3	12.8	13.2	13.6	51
52	7.4	7.8	8.2	8.7	9.1	9.5	10.0	10.4	10.8	11.3	11.7	12.1	12.6	13.0	13.4	13.9	52
53	7.5	8.0	8.4	8.8	9.3	9.7	10.2	10.6	11.0	11.5	11.9	12.4	12.8	13.2	13.7	14.1	53
54	7.6	8.1	8.6	9.0	9.4	9.9	10.4	10.8	11.2	11.7	12.1	12.6	13.0	13.5	14.0	14.4	54
55	7.8	8.2	8.7	9.2	9.6	10.1	10.5	11.0	11.5	11.9	12.4	12.8	13.3	13.8	14.2	14.7	55
56	7.9	8.4	8.9	9.3	9.8	10.3	10.7	11.2	11.7	12.1	12.6	13.1	13.5	14.0	14.5	14.9	56
57	8.1	8.6	9.0	9.5	10.0	10.4	10.9	11.4	11.9	12.4	12.8	13.3	13.8	14.2	14.7	15.2	57
58	8.2	8.7	9.2	9.7	10.2	10.6	11.1	11.6	12.1	12.6	13.0	13.5	14.0	14.5	15.0	15.5	58
59	8.4	8.9	9.3	9.8	10.3	10.8	11.3	11.8	12.3	12.8	13.3	13.8	14.3	14.8	15.2	15.7	59
60	8.5	9.0	9.5	10.0	10.5	11.0	11.5	12.0	12.5	13.0	13.5	14.0	14.5	15.0	15.5	16.0	60

Speed in knots

TABLE 11
Speed, Time, and Distance

Minutes	16.5	17.0	17.5	18.0	18.5	19.0	19.5	20.0	20.5	21.0	21.5	22.0	22.5	23.0	23.5	24.0	Minutes
	Miles	Miles	Miles	Miles	Miles	Miles	Miles	Miles	Miles	Miles	Miles	Miles	Miles	Miles	Miles	Miles	
1	0.3	0.3	0.3	0.3	0.3	0.3	0.3	0.3	0.3	0.4	0.4	0.4	0.4	0.4	0.4	0.4	1
2	0.6	0.6	0.6	0.6	0.6	0.6	0.6	0.7	0.7	0.7	0.7	0.7	0.8	0.8	0.8	0.8	2
3	0.8	0.8	0.9	0.9	0.9	1.0	1.0	1.0	1.0	1.0	1.1	1.1	1.1	1.1	1.2	1.2	3
4	1.1	1.1	1.2	1.2	1.2	1.3	1.3	1.3	1.4	1.4	1.4	1.5	1.5	1.5	1.6	1.6	4
5	1.4	1.4	1.5	1.5	1.5	1.6	1.6	1.7	1.7	1.8	1.8	1.8	1.9	1.9	2.0	2.0	5
6	1.6	1.7	1.8	1.8	1.8	1.9	2.0	2.0	2.0	2.1	2.2	2.2	2.2	2.3	2.4	2.4	6
7	1.9	2.0	2.0	2.1	2.2	2.2	2.3	2.3	2.4	2.4	2.5	2.6	2.6	2.7	2.7	2.8	7
8	2.2	2.3	2.3	2.4	2.5	2.5	2.6	2.7	2.7	2.8	2.9	2.9	3.0	3.1	3.1	3.2	8
9	2.5	2.6	2.6	2.7	2.8	2.8	2.9	3.0	3.1	3.2	3.2	3.3	3.4	3.4	3.5	3.6	9
10	2.8	2.8	2.9	3.0	3.1	3.2	3.2	3.3	3.4	3.5	3.6	3.7	3.8	3.8	3.9	4.0	10
11	3.0	3.1	3.2	3.3	3.4	3.5	3.6	3.7	3.8	3.8	3.9	4.0	4.1	4.2	4.3	4.4	11
12	3.3	3.4	3.5	3.6	3.7	3.8	3.9	4.0	4.1	4.2	4.3	4.4	4.5	4.6	4.7	4.8	12
13	3.6	3.7	3.8	3.9	4.0	4.1	4.2	4.3	4.4	4.6	4.7	4.8	4.9	5.0	5.1	5.2	13
14	3.8	4.0	4.1	4.2	4.3	4.4	4.6	4.7	4.8	4.9	5.0	5.1	5.2	5.4	5.5	5.6	14
15	4.1	4.2	4.4	4.5	4.6	4.8	4.9	5.0	5.1	5.2	5.4	5.5	5.6	5.8	5.9	6.0	15
16	4.4	4.5	4.7	4.8	4.9	5.1	5.2	5.3	5.5	5.6	5.7	5.9	6.0	6.1	6.3	6.4	16
17	4.7	4.8	5.0	5.1	5.2	5.4	5.5	5.7	5.8	6.0	6.1	6.2	6.4	6.5	6.7	6.8	17
18	5.0	5.1	5.2	5.4	5.6	5.7	5.8	6.0	6.2	6.3	6.4	6.6	6.8	6.9	7.0	7.2	18
19	5.2	5.4	5.5	5.7	5.9	6.0	6.2	6.3	6.5	6.6	6.8	7.0	7.1	7.3	7.4	7.6	19
20	5.5	5.7	5.8	6.0	6.2	6.3	6.5	6.7	6.8	7.0	7.2	7.3	7.5	7.7	7.8	8.0	20
21	5.8	6.0	6.1	6.3	6.5	6.6	6.8	7.0	7.2	7.4	7.5	7.7	7.9	8.0	8.2	8.4	21
22	6.0	6.2	6.4	6.6	6.8	7.0	7.2	7.3	7.5	7.7	7.9	8.1	8.2	8.4	8.6	8.8	22
23	6.3	6.5	6.7	6.9	7.1	7.3	7.5	7.7	7.9	8.0	8.2	8.4	8.6	8.8	9.0	9.2	23
24	6.6	6.8	7.0	7.2	7.4	7.6	7.8	8.0	8.2	8.4	8.6	8.8	9.0	9.2	9.4	9.6	24
25	6.9	7.1	7.3	7.5	7.7	7.9	8.1	8.3	8.5	8.8	9.0	9.2	9.4	9.6	9.8	10.0	25
26	7.2	7.4	7.6	7.8	8.0	8.2	8.4	8.7	8.9	9.1	9.3	9.5	9.8	10.0	10.2	10.4	26
27	7.4	7.6	7.9	8.1	8.3	8.6	8.8	9.0	9.2	9.4	9.7	9.9	10.1	10.4	10.6	10.8	27
28	7.7	7.9	8.2	8.4	8.6	8.9	9.1	9.3	9.6	9.8	10.0	10.3	10.5	10.7	11.0	11.2	28
29	8.0	8.2	8.5	8.7	8.9	9.2	9.4	9.7	9.9	10.2	10.4	10.6	10.9	11.1	11.4	11.6	29
30	8.2	8.5	8.8	9.0	9.2	9.5	9.8	10.0	10.2	10.5	10.8	11.0	11.2	11.5	11.8	12.0	30
31	8.5	8.8	9.0	9.3	9.6	9.8	10.1	10.3	10.6	10.8	11.1	11.4	11.6	11.9	12.1	12.4	31
32	8.8	9.1	9.3	9.6	9.9	10.1	10.4	10.7	10.9	11.2	11.5	11.7	12.0	12.3	12.5	12.8	32
33	9.1	9.4	9.6	9.9	10.2	10.4	10.7	11.0	11.3	11.6	11.8	12.1	12.4	12.6	12.9	13.2	33
34	9.4	9.6	9.9	10.2	10.5	10.8	11.0	11.3	11.6	11.9	12.2	12.5	12.8	13.0	13.3	13.6	34
35	9.6	9.9	10.2	10.5	10.8	11.1	11.4	11.7	12.0	12.2	12.5	12.8	13.1	13.4	13.7	14.0	35
36	9.9	10.2	10.5	10.8	11.1	11.4	11.7	12.0	12.3	12.6	12.9	13.2	13.5	13.8	14.1	14.4	36
37	10.2	10.5	10.8	11.1	11.4	11.7	12.0	12.3	12.6	13.0	13.3	13.6	13.9	14.2	14.5	14.8	37
38	10.4	10.8	11.1	11.4	11.7	12.0	12.4	12.7	13.0	13.3	13.6	13.9	14.2	14.6	14.9	15.2	38
39	10.7	11.0	11.4	11.7	12.0	12.4	12.7	13.0	13.3	13.6	14.0	14.3	14.6	15.0	15.3	15.6	39
40	11.0	11.3	11.7	12.0	12.3	12.7	13.0	13.3	13.7	14.0	14.3	14.7	15.0	15.3	15.7	16.0	40
41	11.3	11.6	12.0	12.3	12.6	13.0	13.3	13.7	14.0	14.4	14.7	15.0	15.4	15.7	16.1	16.4	41
42	11.6	11.9	12.2	12.6	13.0	13.3	13.6	14.0	14.4	14.7	15.0	15.4	15.8	16.1	16.4	16.8	42
43	11.8	12.2	12.5	12.9	13.3	13.6	14.0	14.3	14.7	15.0	15.4	15.8	16.1	16.5	16.8	17.2	43
44	12.1	12.5	12.8	13.2	13.6	13.9	14.3	14.7	15.0	15.4	15.8	16.1	16.5	16.9	17.2	17.6	44
45	12.4	12.8	13.1	13.5	13.9	14.2	14.6	15.0	15.4	15.8	16.1	16.5	16.9	17.2	17.6	18.0	45
46	12.6	13.0	13.4	13.8	14.2	14.6	15.0	15.3	15.7	16.1	16.5	16.9	17.2	17.6	18.0	18.4	46
47	12.9	13.3	13.7	14.1	14.5	14.9	15.3	15.7	16.1	16.4	16.8	17.2	17.6	18.0	18.4	18.8	47
48	13.2	13.6	14.0	14.4	14.8	15.2	15.6	16.0	16.4	16.8	17.2	17.6	18.0	18.4	18.8	19.2	48
49	13.5	13.9	14.3	14.7	15.1	15.5	15.9	16.3	16.7	17.2	17.6	18.0	18.4	18.8	19.2	19.6	49
50	13.8	14.2	14.6	15.0	15.4	15.8	16.2	16.7	17.1	17.5	17.9	18.3	18.8	19.2	19.6	20.0	50
51	14.0	14.4	14.9	15.3	15.7	16.2	16.6	17.0	17.4	17.8	18.3	18.7	19.1	19.6	20.0	20.4	51
52	14.3	14.7	15.2	15.6	16.0	16.5	16.9	17.3	17.8	18.2	18.6	19.1	19.5	19.9	20.4	20.8	52
53	14.6	15.0	15.5	15.9	16.3	16.8	17.2	17.7	18.1	18.6	19.0	19.4	19.9	20.3	20.8	21.2	53
54	14.8	15.3	15.8	16.2	16.6	17.1	17.6	18.0	18.4	18.9	19.4	19.8	20.2	20.7	21.2	21.6	54
55	15.1	15.6	16.0	16.5	17.0	17.4	17.9	18.3	18.8	19.2	19.7	20.2	20.6	21.1	21.5	22.0	55
56	15.4	15.9	16.3	16.8	17.3	17.7	18.2	18.7	19.1	19.6	20.1	20.5	21.0	21.5	21.9	22.4	56
57	15.7	16.2	16.6	17.1	17.6	18.0	18.5	19.0	19.5	20.0	20.4	20.9	21.4	21.8	22.3	22.8	57
58	16.0	16.4	16.9	17.4	17.9	18.4	18.8	19.3	19.8	20.3	20.8	21.3	21.8	22.2	22.7	23.2	58
59	16.2	16.7	17.2	17.7	18.2	18.7	19.2	19.7	20.2	20.6	21.1	21.6	22.1	22.6	23.1	23.6	59
60	16.5	17.0	17.5	18.0	18.5	19.0	19.5	20.0	20.5	21.0	21.5	22.0	22.5	23.0	23.5	24.0	60

Speed in knots

TABLE 11
Speed, Time, and Distance

Minutes	24.5	25.0	25.5	26.0	26.5	27.0	27.5	28.0	28.5	29.0	29.5	30.0	30.5	31.0	31.5	32.0	Minutes
	Miles	Miles	Miles	Miles	Miles	Miles	Miles	Miles	Miles	Miles	Miles	Miles	Miles	Miles	Miles	Miles	
1	0.4	0.4	0.4	0.4	0.4	0.4	0.5	0.5	0.5	0.5	0.5	0.5	0.5	0.5	0.5	0.5	1
2	0.8	0.8	0.8	0.9	0.9	0.9	0.9	0.9	1.0	1.0	1.0	1.0	1.0	1.0	1.0	1.1	2
3	1.2	1.2	1.3	1.3	1.3	1.4	1.4	1.4	1.4	1.4	1.5	1.5	1.5	1.6	1.6	1.6	3
4	1.6	1.7	1.7	1.7	1.8	1.8	1.8	1.9	1.9	1.9	2.0	2.0	2.0	2.1	2.1	2.1	4
5	2.0	2.1	2.1	2.2	2.2	2.2	2.3	2.3	2.4	2.4	2.5	2.5	2.5	2.6	2.6	2.7	5
6	2.4	2.5	2.6	2.6	2.6	2.7	2.8	2.8	2.8	2.9	3.0	3.0	3.0	3.1	3.2	3.2	6
7	2.9	2.9	3.0	3.0	3.1	3.2	3.2	3.3	3.3	3.4	3.4	3.5	3.6	3.6	3.7	3.7	7
8	3.3	3.3	3.4	3.5	3.5	3.6	3.7	3.7	3.8	3.9	3.9	4.0	4.1	4.1	4.2	4.3	8
9	3.7	3.8	3.8	3.9	4.0	4.0	4.1	4.2	4.3	4.4	4.4	4.5	4.6	4.6	4.7	4.8	9
10	4.1	4.2	4.2	4.3	4.4	4.5	4.6	4.7	4.8	4.8	4.9	5.0	5.1	5.2	5.2	5.3	10
11	4.5	4.6	4.7	4.8	4.9	5.0	5.0	5.1	5.2	5.3	5.4	5.5	5.6	5.7	5.8	5.9	11
12	4.9	5.0	5.1	5.2	5.3	5.4	5.5	5.6	5.7	5.8	5.9	6.0	6.1	6.2	6.3	6.4	12
13	5.3	5.4	5.5	5.6	5.7	5.8	6.0	6.1	6.2	6.3	6.4	6.5	6.6	6.7	6.8	6.9	13
14	5.7	5.8	6.0	6.1	6.2	6.3	6.4	6.5	6.6	6.8	6.9	7.0	7.1	7.2	7.4	7.5	14
15	6.1	6.2	6.4	6.5	6.6	6.8	6.9	7.0	7.1	7.2	7.4	7.5	7.6	7.8	7.9	8.0	15
16	6.5	6.7	6.8	6.9	7.1	7.2	7.3	7.5	7.6	7.7	7.9	8.0	8.1	8.3	8.4	8.5	16
17	6.9	7.1	7.2	7.4	7.5	7.6	7.8	7.9	8.1	8.2	8.4	8.5	8.6	8.8	8.9	9.1	17
18	7.4	7.5	7.6	7.8	8.0	8.1	8.2	8.4	8.6	8.7	8.8	9.0	9.2	9.3	9.4	9.6	18
19	7.8	7.9	8.1	8.2	8.4	8.6	8.7	8.9	9.0	9.2	9.3	9.5	9.7	9.8	10.0	10.1	19
20	8.2	8.3	8.5	8.7	8.8	9.0	9.2	9.3	9.5	9.7	9.8	10.0	10.2	10.3	10.5	10.7	20
21	8.6	8.8	8.9	9.1	9.3	9.4	9.6	9.8	10.0	10.2	10.3	10.5	10.7	10.8	11.0	11.2	21
22	9.0	9.2	9.4	9.5	9.7	9.9	10.1	10.3	10.4	10.6	10.8	11.0	11.2	11.4	11.6	11.7	22
23	9.4	9.6	9.8	10.0	10.2	10.4	10.5	10.7	10.9	11.1	11.3	11.5	11.7	11.9	12.1	12.3	23
24	9.8	10.0	10.2	10.4	10.6	10.8	11.0	11.2	11.4	11.6	11.8	12.0	12.2	12.4	12.6	12.8	24
25	10.2	10.4	10.6	10.8	11.0	11.2	11.5	11.7	11.9	12.1	12.3	12.5	12.7	12.9	13.1	13.3	25
26	10.6	10.8	11.0	11.3	11.5	11.7	11.9	12.1	12.4	12.6	12.8	13.0	13.2	13.4	13.6	13.9	26
27	11.0	11.2	11.5	11.7	11.9	12.2	12.4	12.6	12.8	13.0	13.3	13.5	13.7	14.0	14.2	14.4	27
28	11.4	11.7	11.9	12.1	12.4	12.6	12.8	13.1	13.3	13.5	13.8	14.0	14.2	14.5	14.7	14.9	28
29	11.8	12.1	12.3	12.6	12.8	13.0	13.3	13.5	13.8	14.0	14.3	14.5	14.7	15.0	15.2	15.5	29
30	12.2	12.5	12.8	13.0	13.2	13.5	13.8	14.0	14.2	14.5	14.8	15.0	15.2	15.5	15.8	16.0	30
31	12.7	12.9	13.2	13.4	13.7	14.0	14.2	14.5	14.7	15.0	15.2	15.5	15.8	16.0	16.3	16.5	31
32	13.1	13.3	13.6	13.9	14.1	14.4	14.7	14.9	15.2	15.5	15.7	16.0	16.3	16.5	16.8	17.1	32
33	13.5	13.8	14.0	14.3	14.6	14.8	15.1	15.4	15.7	16.0	16.2	16.5	16.8	17.0	17.3	17.6	33
34	13.9	14.2	14.4	14.7	15.0	15.3	15.6	15.9	16.2	16.4	16.7	17.0	17.3	17.6	17.8	18.1	34
35	14.3	14.6	14.9	15.2	15.5	15.8	16.0	16.3	16.6	16.9	17.2	17.5	17.8	18.1	18.4	18.7	35
36	14.7	15.0	15.3	15.6	15.9	16.2	16.5	16.8	17.1	17.4	17.7	18.0	18.3	18.6	18.9	19.2	36
37	15.1	15.4	15.7	16.0	16.3	16.6	17.0	17.3	17.6	17.9	18.2	18.5	18.8	19.1	19.4	19.7	37
38	15.5	15.8	16.2	16.5	16.8	17.1	17.4	17.7	18.0	18.4	18.7	19.0	19.3	19.6	20.0	20.3	38
39	15.9	16.2	16.6	16.9	17.2	17.6	17.9	18.2	18.5	18.8	19.2	19.5	19.8	20.2	20.5	20.8	39
40	16.3	16.7	17.0	17.3	17.7	18.0	18.3	18.7	19.0	19.3	19.7	20.0	20.3	20.7	21.0	21.3	40
41	16.7	17.1	17.4	17.8	18.1	18.4	18.8	19.1	19.5	19.8	20.2	20.5	20.8	21.2	21.5	21.9	41
42	17.2	17.5	17.8	18.2	18.6	18.9	19.2	19.6	20.0	20.3	20.6	21.0	21.4	21.7	22.0	22.4	42
43	17.6	17.9	18.3	18.6	19.0	19.4	19.7	20.1	20.4	20.8	21.1	21.5	21.9	22.2	22.6	22.9	43
44	18.0	18.3	18.7	19.1	19.4	19.8	20.2	20.5	20.9	21.3	21.6	22.0	22.4	22.7	23.1	23.5	44
45	18.4	18.8	19.1	19.5	19.9	20.2	20.6	21.0	21.4	21.8	22.1	22.5	22.9	23.2	23.6	24.0	45
46	18.8	19.2	19.6	19.9	20.3	20.7	21.1	21.5	21.8	22.2	22.6	23.0	23.4	23.8	24.2	24.5	46
47	19.2	19.6	20.0	20.4	20.8	21.2	21.5	21.9	22.3	22.7	23.1	23.5	23.9	24.3	24.7	25.1	47
48	19.6	20.0	20.4	20.8	21.2	21.6	22.0	22.4	22.8	23.2	23.6	24.0	24.4	24.8	25.2	25.6	48
49	20.0	20.4	20.8	21.2	21.6	22.0	22.5	22.9	23.3	23.7	24.1	24.5	24.9	25.3	25.7	26.1	49
50	20.4	20.8	21.2	21.7	22.1	22.5	22.9	23.3	23.8	24.2	24.6	25.0	25.4	25.8	26.2	26.7	50
51	20.8	21.2	21.7	22.1	22.5	23.0	23.4	23.8	24.2	24.6	25.1	25.5	25.9	26.4	26.8	27.2	51
52	21.2	21.7	22.1	22.5	23.0	23.4	23.8	24.3	24.7	25.1	25.6	26.0	26.4	26.9	27.3	27.7	52
53	21.6	22.1	22.5	23.0	23.4	23.8	24.3	24.7	25.2	25.6	26.1	26.5	26.9	27.4	27.8	28.3	53
54	22.0	22.5	23.0	23.4	23.8	24.3	24.8	25.2	25.6	26.1	26.6	27.0	27.4	27.9	28.4	28.8	54
55	22.5	22.9	23.4	23.8	24.3	24.8	25.2	25.7	26.1	26.6	27.0	27.5	28.0	28.4	28.9	29.3	55
56	22.9	23.3	23.8	24.3	24.7	25.2	25.7	26.1	26.6	27.1	27.5	28.0	28.5	28.9	29.4	29.9	56
57	23.3	23.8	24.2	24.7	25.2	25.6	26.1	26.6	27.1	27.6	28.0	28.5	29.0	29.4	29.9	30.4	57
58	23.7	24.2	24.6	25.1	25.6	26.1	26.6	27.1	27.6	28.0	28.5	29.0	29.5	30.0	30.4	30.9	58
59	24.1	24.6	25.1	25.6	26.1	26.6	27.0	27.5	28.0	28.5	29.0	29.5	30.0	30.5	31.0	31.5	59
60	24.5	25.0	25.5	26.0	26.5	27.0	27.5	28.0	28.5	29.0	29.5	30.0	30.5	31.0	31.5	32.0	60

Speed in knots

TABLE 11
Speed, Time, and Distance

Minutes	Speed in knots																Minutes
	32.5	33.0	33.5	34.0	34.5	35.0	35.5	36.0	36.5	37.0	37.5	38.0	38.5	39.0	39.5	40.0	
	Miles	*Miles*	*Miles*	*Miles*	*Miles*	*Miles*	*Miles*	*Miles*	*Miles*	*Miles*	*Miles*	*Miles*	*Miles*	*Miles*	*Miles*	*Miles*	
1	0.5	0.6	0.6	0.6	0.6	0.6	0.6	0.6	0.6	0.6	0.6	0.6	0.6	0.6	0.7	0.7	1
2	1.1	1.1	1.1	1.1	1.1	1.2	1.2	1.2	1.2	1.2	1.2	1.3	1.3	1.3	1.3	1.3	2
3	1.6	1.6	1.7	1.7	1.7	1.8	1.8	1.8	1.8	1.8	1.9	1.9	1.9	2.0	2.0	2.0	3
4	2.2	2.2	2.2	2.3	2.3	2.3	2.4	2.4	2.4	2.5	2.5	2.5	2.6	2.6	2.6	2.7	4
5	2.7	2.8	2.8	2.8	2.9	2.9	3.0	3.0	3.0	3.1	3.1	3.2	3.2	3.2	3.3	3.3	5
6	3.2	3.3	3.4	3.4	3.4	3.5	3.6	3.6	3.6	3.7	3.8	3.8	3.8	3.9	4.0	4.0	6
7	3.8	3.8	3.9	4.0	4.0	4.1	4.1	4.2	4.3	4.3	4.4	4.4	4.5	4.6	4.6	4.7	7
8	4.3	4.4	4.5	4.5	4.6	4.7	4.7	4.8	4.9	4.9	5.0	5.1	5.1	5.2	5.3	5.3	8
9	4.9	5.0	5.0	5.1	5.2	5.2	5.3	5.4	5.5	5.6	5.6	5.7	5.8	5.8	5.9	6.0	9
10	5.4	5.5	5.6	5.7	5.8	5.8	5.9	6.0	6.1	6.2	6.2	6.3	6.4	6.5	6.6	6.7	10
11	6.0	6.0	6.1	6.2	6.3	6.4	6.5	6.6	6.7	6.8	6.9	7.0	7.1	7.2	7.2	7.3	11
12	6.5	6.6	6.7	6.8	6.9	7.0	7.1	7.2	7.3	7.4	7.5	7.6	7.7	7.8	7.9	8.0	12
13	7.0	7.2	7.3	7.4	7.5	7.6	7.7	7.8	7.9	8.0	8.1	8.2	8.3	8.4	8.6	8.7	13
14	7.6	7.7	7.8	7.9	8.0	8.2	8.3	8.4	8.5	8.6	8.8	8.9	9.0	9.1	9.2	9.3	14
15	8.1	8.2	8.4	8.5	8.6	8.8	8.9	9.0	9.1	9.2	9.4	9.5	9.6	9.8	9.9	10.0	15
16	8.7	8.8	8.9	9.1	9.2	9.3	9.5	9.6	9.7	9.9	10.0	10.1	10.3	10.4	10.5	10.7	16
17	9.2	9.4	9.5	9.6	9.8	9.9	10.1	10.2	10.3	10.5	10.6	10.8	10.9	11.0	11.2	11.3	17
18	9.8	9.9	10.1	10.2	10.4	10.5	10.6	10.8	11..0	11.1	11.2	11.4	11.6	11.7	11.8	12.0	18
19	10.3	10.4	10.6	10.8	10.9	11.1	11.2	11.4	11.6	11.7	11.9	12.0	12.2	12.4	12.5	12.7	19
20	10.8	11.0	11.2	11.3	11.5	11.7	11.8	12.0	12.2	12.3	12.5	12.7	12.8	13.0	13.2	13.3	20
21	11.4	11.6	11.7	11.9	12.1	12.2	12.4	12.6	12.8	13.0	13.1	13.3	13.5	13.6	13.8	14.0	21
22	11.9	12.1	12.3	12.5	12.6	12.8	13.0	13.2	13.4	13.6	13.8	13.9	14.1	14.3	14.5	14.7	22
23	12.5	12.6	12.8	13.0	13.2	13.4	13.6	13.8	14.0	14.2	14.4	14.6	14.8	15.0	15.1	15.3	23
24	13.0	13.2	13.4	13.6	13.8	14.0	14.2	14.4	14.6	14.8	15.0	15.2	15.4	15.6	15.8	16.0	24
25	13.5	13.8	14.0	14.2	14.4	14.6	14.8	15.0	15.2	15.4	15.6	15.8	16.0	16.2	16.5	16.7	25
26	14.1	14.3	14.5	14.7	15.0	15.2	15.4	15.6	15.8	16.0	16.2	16.5	16.7	16.9	17.1	17.3	26
27	14.6	14.8	15.1	15.3	15.5	15.8	16.0	16.2	16.4	16.6	16.9	17.1	17.3	17.6	17.8	18.0	27
28	15.2	15.4	15.6	15.9	16.1	16.3	16.6	16.8	17.0	17.3	17.5	17.7	18.0	18.2	18.4	18.7	28
29	15.7	16.0	16.2	16.4	16.7	16.9	17.2	17.4	17.6	17.9	18.1	18.4	18.6	18.9	19.1	19.3	29
30	16.2	16.5	16.8	17.0	17.2	17.5	17.8	18.0	18.2	18.5	18.8	19.0	19.2	19.5	19.8	20.0	30
31	16.8	17.0	17.3	17.6	17.8	18.1	18.3	18.6	18.9	19.1	19.4	19.6	19.9	20.2	20.4	20.7	31
32	17.3	17.6	17.9	18.1	18.4	18.7	18.9	19.2	19.5	19.7	20.0	20.3	20.5	20.8	21.1	21.3	32
33	17.9	18.2	18.4	18.7	19.0	19.2	19.5	19.8	20.1	20.4	20.6	20.9	21.2	21.4	21.7	22.0	33
34	18.4	18.7	19.0	19.3	19.6	19.8	20.1	20.4	20.7	21.0	21.2	21.5	21.8	22.1	22.4	22.7	34
35	19.0	19.2	19.5	19.8	20.1	20.4	20.7	21.0	21.3	21.6	21.9	22.2	22.5	22.8	23.0	23.3	35
36	19.5	19.8	20.1	20.4	20.7	21.0	21.3	21.6	21.9	22.2	22.5	22.8	23.1	23.4	23.7	24.0	36
37	20.0	20.4	20.7	21.0	21.3	21.6	21.9	22.2	22.5	22.8	23.1	23.4	23.7	24.0	24.4	24.7	37
38	20.6	20.9	21.2	21.5	21.8	22.2	22.5	22.8	23.1	23.4	23.8	24.1	24.4	24.7	25.0	25.3	38
39	21.1	21.4	21.8	22.1	22.4	22.8	23.1	23.4	23.7	24.0	24.4	24.7	25.0	25.4	25.7	26.0	39
40	21.7	22.0	22.3	22.7	23.0	23.3	23.7	24.0	24.3	24.7	25.0	25.3	25.7	26.0	26.3	26.7	40
41	22.2	22.6	22.9	23.2	23.6	23.9	24.3	24.6	24.9	25.3	25.6	26.0	26.3	26.6	27.0	27.3	41
42	22.8	23.1	23.4	23.8	24.2	24.5	24.8	25.2	25.6	25.9	26.2	26.6	27.0	27.3	27.6	28.0	42
43	23.3	23.6	24.0	24.4	24.7	25.1	25.4	25.8	26.2	26.5	26.9	27.2	27.6	28.0	28.3	28.7	43
44	23.8	24.2	24.6	24.9	25.3	25.7	26.0	26.4	26.8	27.1	27.5	27.9	28.2	28.6	29.0	29.3	44
45	24.4	24.8	25.1	25.5	25.9	26.2	26.6	27.0	27.4	27.8	28.1	28.5	28.9	29.2	29.6	30.0	45
46	24.9	25.3	25.7	26.1	26.4	26.8	27.2	27.6	28.0	28.4	28.8	29.1	29.5	29.9	30.3	30.7	46
47	25.5	25.8	26.2	26.6	27.0	27.4	27.8	28.2	28.6	29.0	29.4	29.8	30.2	30.6	30.9	31.3	47
48	26.0	26.4	26.8	27.2	27.6	28.0	28.4	28.8	29.2	29.6	30.0	30.4	30.8	31.2	31.6	32.0	48
49	26.5	27.0	27.4	27.8	28.2	28.6	29.0	29.4	29.8	30.2	30.6	31.0	31.4	31.8	32.3	32.7	49
50	27.1	27.5	27.9	28.3	28.8	29.2	29.6	30.0	30.4	30.8	31.2	31.7	32.1	32.5	32.9	33.3	50
51	27.6	28.0	28.5	28.9	29.3	29.8	30.2	30.6	31.0	31.4	31.9	32.3	32.7	33.2	33.6	34.0	51
52	28.2	28.6	29.0	29.5	29.9	30.3	30.8	31.2	31.6	32.1	32.5	32.9	33.4	33.8	34.2	34.7	52
53	28.7	29.2	29.6	30.0	30.5	30.9	31.4	31.8	32.2	32.7	33.1	33.6	34.0	34.4	34.9	35.3	53
54	29.2	29.7	30.2	30.6	31.0	31.5	32.0	32.4	32.8	33.3	33.8	34.2	34.6	35.1	35.6	36.0	54
55	29.8	30.2	30.7	31.2	31.6	32.1	32.5	33.0	33.5	33.9	34.4	34.8	35.3	35.8	36.2	36.7	55
56	30.3	30.8	31.3	31.7	32.2	32.7	33.1	33.6	34.1	34.5	35.0	35.5	35.9	36.4	36.9	37.3	56
57	30.9	31.4	31.8	32.3	32.8	33.2	33.7	34.2	34.7	35.2	35.6	36.1	36.6	37.0	37.5	38.0	57
58	31.4	31.9	32.4	32.9	33.4	33.8	34.3	34.8	35.3	35.8	36.2	36.7	37.2	37.7	38.2	38.7	58
59	32.0	32.4	32.9	33.4	33.9	34.4	34.9	35.4	35.9	36.4	36.9	37.4	37.9	38.4	38.8	39.3	59
60	32.5	33.0	33.5	34.0	34.5	35.0	35.5	36.0	36.5	37.0	37.5	38.0	38.5	39.0	39.5	40.0	60

TABLE 12
Distance of the Horizon

Height Feet	Nautical Miles	Statute Miles	Height *meters*	Height Feet	Nautical Miles	Statute Miles	Height *meters*
1	1.2	1.3	.30	120	12.8	14.7	36.58
2	1.7	1.9	.61	125	13.1	15.1	38.10
3	2.0	2.3	.91	130	13.3	15.4	39.62
4	2.3	2.7	1.22	135	13.6	15.6	41.15
5	2.6	3.0	1.52	140	13.8	15.9	42.67
6	2.9	3.3	1.83	145	14.1	16.2	44.20
7	3.1	3.6	2.13	150	14.3	16.5	45.72
8	3.3	3.8	2.44	160	14.8	17.0	48.77
9	3.5	4.0	2.74	170	15.3	17.6	51.82
10	3.7	4.3	3.05	180	15.7	18.1	54.86
11	3.9	4.5	3.35	190	16.1	18.6	57.91
12	4.1	4.7	3.66	200	16.5	19.0	60.96
13	4.2	4.9	3.96	210	17.0	19.5	64.01
14	4.4	5.0	4.27	220	17.4	20.0	67.06
15	4.5	5.2	4.57	230	17.7	20.4	70.10
16	4.7	5.4	4.88	240	18.1	20.9	73.15
17	4.8	5.6	5.18	250	18.5	21.3	76.20
18	5.0	5.7	5.49	260	18.9	21.7	79.25
19	5.1	5.9	5.79	270	19.2	22.1	82.30
20	5.2	6.0	6.10	280	19.6	22.5	85.34
21	5.4	6.2	6.40	290	19.9	22.9	88.39
22	5.5	6.3	6.71	300	20.3	23.3	91.44
23	5.6	6.5	7.01	310	20.6	23.7	94.49
24	5.7	6.6	7.32	320	20.9	24.1	97.54
25	5.9	6.7	7.62	330	21.3	24.5	100.58
26	6.0	6.9	7.92	340	21.6	24.8	103.63
27	6.1	7.0	8.23	350	21.9	25.2	106.68
28	6.2	7.1	8.53	360	22.2	25.5	109.73
29	6.3	7.3	8.84	370	22.5	25.9	112.78
30	6.4	7.4	9.14	380	22.8	26.2	115.82
31	6.5	7.5	9.45	390	23.1	26.6	118.87
32	6.6	7.6	9.75	400	23.4	26.9	121.92
33	6.7	7.7	10.06	410	23.7	27.3	124.97
34	6.8	7.9	10.36	420	24.0	27.6	128.02
35	6.9	8.0	10.67	430	24.3	27.9	131.06
36	7.0	8.1	10.97	440	24.5	28.2	134.11
37	7.1	8.2	11.28	450	24.8	28.6	137.16
38	7.2	8.3	11.58	460	25.1	28.9	140.21
39	7.3	8.4	11.89	470	25.4	29.2	143.26
40	7.4	8.5	12.19	480	25.6	29.5	146.30
41	7.5	8.6	12.50	490	25.9	29.8	149.35
42	7.6	8.7	12.80	500	26.2	30.1	152.40
43	7.7	8.8	13.11	510	26.4	30.4	155.45
44	7.8	8.9	13.41	520	26.7	30.7	158.50
45	7.8	9.0	13.72	530	26.9	31.0	161.54
46	7.9	9.1	14.02	540	27.2	31.3	164.59
47	8.0	9.2	14.33	550	27.4	31.6	167.64
48	8.1	9.3	14.63	560	27.7	31.9	170.69
49	8.2	9.4	14.94	570	27.9	32.1	173.74
50	8.3	9.5	15.24	580	28.2	32.4	176.78
55	8.7	10.0	16.76	590	28.4	32.7	179.83
60	9.1	10.4	18.29	600	28.7	33.0	182.88
65	9.4	10.9	19.81	620	29.1	33.5	188.98
70	9.8	11.3	21.34	640	29.5	34.1	195.07
75	10.1	11.7	22.86	660	30.1	34.6	201.17
80	10.5	12.0	24.38	680	30.5	35.1	207.26
85	10.8	12.4	25.91	700	31.0	35.6	213.36
90	11.1	12.8	27.43	720	31.4	36.1	219.46
95	11.4	13.1	28.96	740	31.8	36.6	225.55
100	11.7	13.5	30.48	760	32.3	37.1	231.65
105	12.0	13.8	32.00	780	32.7	37.6	237.74
110	12.3	14.1	33.53	800	33.1	38.1	243.84
115	12.5	14.4	35.05	820	33.5	38.6	249.94

TABLE 13
Geographic Range

Object Height		Height of eye of observer in feet and meters										Object Height	
Feet		7	10	13	16	20	23	26	30	33	36		Feet
	Meters	2	3	4	5	6	7	8	9	10	11	Meters	
		Miles	Miles	Miles	Miles	Miles	Miles	Miles	Miles	Miles	Miles		
0	0	3.1	3.7	4.2	4.7	5.2	5.6	6.0	6.4	6.7	7.0	0	**0**
3	1	5.1	5.7	6.2	6.7	7.3	7.6	8.0	8.4	8.7	9.0	1	**3**
7	2	6.2	6.8	7.3	7.8	8.3	8.7	9.1	9.5	9.8	10.1	2	**7**
10	3	6.8	7.4	7.9	8.4	8.9	9.3	9.7	10.1	10.4	10.7	3	**10**
13	4	7.3	7.9	8.4	8.9	9.5	9.8	10.2	10.6	10.9	11.2	4	**13**
16	5	7.8	8.4	8.9	9.4	9.9	10.3	10.6	11.1	11.4	11.7	5	**16**
20	6	8.3	8.9	9.5	9.9	10.5	10.8	11.2	11.6	12.0	12.3	6	**20**
23	7	8.7	9.3	9.8	10.3	10.8	11.2	11.6	12.0	12.3	12.6	7	**23**
26	8	9.1	9.7	10.2	10.6	11.2	11.6	11.9	12.4	12.7	13.0	8	**26**
30	9	9.5	10.1	10.6	11.1	11.6	12.0	12.4	12.8	13.1	13.4	9	**30**
33	10	9.8	10.4	10.9	11.4	12.0	12.3	12.7	13.1	13.4	13.7	10	**33**
36	11	10.1	10.7	11.2	11.7	12.3	12.6	13.0	13.4	13.7	14.0	11	**36**
39	12	10.4	11.0	11.5	12.0	12.5	12.9	13.3	13.7	14.0	14.3	12	**39**
43	13	10.8	11.4	11.9	12.4	12.9	13.3	13.6	14.1	14.4	14.7	13	**43**
46	14	11.0	11.6	12.2	12.6	13.2	13.5	13.9	14.3	14.7	15.0	14	**46**
49	15	11.3	11.9	12.4	12.9	13.4	13.8	14.2	14.6	14.9	15.2	15	**49**
52	16	11.5	12.1	12.7	13.1	13.7	14.0	14.4	14.8	15.2	15.5	16	**52**
56	17	11.9	12.5	13.0	13.4	14.0	14.4	14.7	15.2	15.5	15.8	17	**56**
59	18	12.1	12.7	13.2	13.7	14.2	14.6	15.0	15.4	15.7	16.0	18	**59**
62	19	12.3	12.9	13.4	13.9	14.4	14.8	15.2	15.6	15.9	16.2	19	**62**
66	20	12.6	13.2	13.7	14.2	14.7	15.1	15.5	15.9	16.2	16.5	20	**66**
72	22	13.0	13.6	14.1	14.6	15.2	15.5	15.9	16.3	16.6	16.9	22	**72**
79	24	13.5	14.1	14.6	15.1	15.6	16.0	16.4	16.8	17.1	17.4	24	**79**
85	26	13.9	14.5	15.0	15.5	16.0	16.4	16.8	17.2	17.5	17.8	26	**85**
92	28	14.3	14.9	15.4	15.9	16.5	16.8	17.2	17.6	17.9	18.2	28	**92**
98	30	14.7	15.3	15.8	16.3	16.8	17.2	17.5	18.0	18.3	18.6	30	**98**
115	35	15.6	16.2	16.8	17.2	17.8	18.2	18.5	19.0	19.3	19.6	35	**115**
131	40	16.5	17.1	17.6	18.1	18.6	19.0	19.4	19.8	20.1	20.4	40	**131**
148	45	17.3	17.9	18.5	18.9	19.5	19.8	20.2	20.6	21.0	21.3	45	**148**
164	50	18.1	18.7	19.2	19.7	20.2	20.6	20.9	21.4	21.7	22.0	50	**164**
180	55	18.8	19.4	19.9	20.4	20.9	21.3	21.7	22.1	22.4	22.7	55	**180**
197	60	19.5	20.1	20.6	21.1	21.7	22.0	22.4	22.8	23.1	23.4	60	**197**
213	65	20.2	20.8	21.3	21.8	22.3	22.7	23.0	23.5	23.8	24.1	65	**213**
230	70	20.8	21.4	22.0	22.4	23.0	23.4	23.7	24.2	24.5	24.8	70	**230**
246	75	21.4	22.1	22.6	23.0	23.6	24.0	24.3	24.8	25.1	25.4	75	**246**
262	80	22.0	22.6	23.2	23.6	24.2	24.5	24.9	25.3	25.7	26.0	80	**262**
279	85	22.6	23.2	23.8	24.2	24.8	25.2	25.5	26.0	26.3	26.6	85	**279**
295	90	23.2	23.8	24.3	24.8	25.3	25.7	26.1	26.5	26.8	27.1	90	**295**
312	95	23.8	24.4	24.9	25.3	25.9	26.3	26.6	27.1	27.4	27.7	95	**312**
328	100	24.3	24.9	25.4	25.9	26.4	26.8	27.2	27.6	27.9	28.2	100	**328**
361	110	25.3	25.9	26.4	26.9	27.5	27.8	28.2	28.6	29.0	29.3	110	**361**
394	120	26.3	26.9	27.4	27.9	28.5	28.8	29.2	29.6	29.9	30.2	120	**394**
427	130	27.3	27.9	28.4	28.9	29.4	29.8	30.1	30.6	30.9	31.2	130	**427**
459	140	28.2	28.8	29.3	29.7	30.3	30.7	31.0	31.5	31.8	32.1	140	**459**
492	150	29.0	29.7	30.2	30.6	31.2	31.6	31.9	32.4	32.7	33.0	150	**492**
525	160	29.9	30.5	31.0	31.5	32.0	32.4	32.8	33.2	33.5	33.8	160	**525**
558	170	30.7	31.3	31.9	32.3	32.9	33.2	33.6	34.0	34.4	34.7	170	**558**
591	180	31.5	32.1	32.7	33.1	33.7	34.1	34.4	34.9	35.2	35.5	180	**591**
623	190	32.3	32.9	33.4	33.9	34.4	34.8	35.2	35.6	35.9	36.2	190	**623**
656	200	33.1	33.7	34.2	34.6	35.2	35.6	35.9	36.4	36.7	37.0	200	**656**
722	220	34.5	35.1	35.7	36.1	36.7	37.0	37.4	37.8	38.2	38.5	220	**722**
787	240	35.9	36.5	37.0	37.5	38.1	38.4	38.8	39.2	39.5	39.8	240	**787**
853	260	37.3	37.9	38.4	38.9	39.4	39.8	40.1	40.6	40.9	41.2	260	**853**
919	280	38.6	39.2	39.7	40.1	40.7	41.1	41.4	41.9	42.2	42.5	280	**919**
984	300	39.8	40.4	40.9	41.4	41.9	42.3	42.7	43.1	43.4	43.7	300	**984**

TABLE 13
Geographic Range

Object Height		Height of eye of observer in feet and meters										Object Height	
Feet		39	43	46	49	52	56	59	62	66	69		Feet
	Meters	12	13	14	15	16	17	18	19	20	21	Meters	
		Miles	Miles	Miles	Miles	Miles	Miles	Miles	Miles	Miles	Miles		
0	0	7.3	7.7	7.9	8.2	8.4	8.8	9.0	9.2	9.5	9.7	0	**0**
3	1	9.3	9.7	10.0	10.2	10.5	10.8	11.0	11.2	11.5	11.7	1	**3**
7	2	10.4	10.8	11.0	11.3	11.5	11.9	12.1	12.3	12.6	12.8	2	**7**
10	3	11.0	11.4	11.6	11.9	12.1	12.5	12.7	12.9	13.2	13.4	3	**10**
13	4	11.5	11.9	12.2	12.4	12.7	13.0	13.2	13.4	13.7	13.9	4	**13**
16	5	12.0	12.4	12.6	12.9	13.1	13.4	13.7	13.9	14.2	14.4	5	**16**
20	6	12.5	12.9	13.2	13.4	13.7	14.0	14.2	14.4	14.7	15.0	6	**20**
23	7	12.9	13.3	13.5	13.8	14.0	14.4	14.6	14.8	15.1	15.3	7	**23**
26	8	13.3	13.6	13.9	14.2	14.4	14.7	15.0	15.2	15.5	15.7	8	**26**
30	9	13.7	14.1	14.3	14.6	14.8	15.2	15.4	15.6	15.9	16.1	9	**30**
33	10	14.0	14.4	14.7	14.9	15.2	15.5	15.7	15.9	16.2	16.4	10	**33**
36	11	14.3	14.7	15.0	15.2	15.5	15.8	16.0	16.2	16.5	16.7	11	**36**
39	12	14.6	15.0	15.2	15.5	15.7	16.1	16.3	16.5	16.8	17.0	12	**39**
43	13	15.0	15.3	15.6	15.9	16.1	16.4	16.7	16.9	17.2	17.4	13	**43**
46	14	15.2	15.6	15.9	16.1	16.4	16.7	16.9	17.1	17.4	17.7	14	**46**
49	15	15.5	15.9	16.1	16.4	16.6	16.9	17.2	17.4	17.7	17.9	15	**49**
52	16	15.7	16.1	16.4	16.6	16.9	17.2	17.4	17.6	17.9	18.2	16	**52**
56	17	16.1	16.4	16.7	16.9	17.2	17.5	17.7	18.0	18.3	18.5	17	**56**
59	18	16.3	16.7	16.9	17.2	17.4	17.7	18.0	18.2	18.5	18.7	18	**59**
62	19	16.5	16.9	17.1	17.4	17.6	18.0	18.2	18.4	18.7	18.9	19	**62**
66	20	16.8	17.2	17.4	17.7	17.9	18.3	18.5	18.7	19.0	19.2	20	**66**
72	22	17.2	17.6	17.9	18.1	18.4	18.7	18.9	19.1	19.4	19.6	22	**72**
79	24	17.7	18.1	18.3	18.6	18.8	19.2	19.4	19.6	19.9	20.1	24	**79**
85	26	18.1	18.5	18.7	19.0	19.2	19.5	19.8	20.0	20.3	20.5	26	**85**
92	28	18.5	18.9	19.2	19.4	19.7	20.0	20.2	20.4	20.7	20.9	28	**92**
98	30	18.9	19.3	19.5	19.8	20.0	20.3	20.6	20.8	21.1	21.3	30	**98**
115	35	19.9	20.2	20.5	20.7	21.0	21.3	21.5	21.8	22.1	22.3	35	**115**
131	40	20.7	21.1	21.3	21.6	21.8	22.1	22.4	22.6	22.9	23.1	40	**131**
148	45	21.5	21.9	22.2	22.4	22.7	23.0	23.2	23.4	23.7	24.0	45	**148**
164	50	22.3	22.7	22.9	23.2	23.4	23.7	24.0	24.2	24.5	24.7	50	**164**
180	55	23.0	23.4	23.6	23.9	24.1	24.5	24.7	24.9	25.2	25.4	55	**180**
197	60	23.7	24.1	24.4	24.6	24.9	25.2	25.4	25.6	25.9	26.1	60	**197**
213	65	24.4	24.7	25.0	25.3	25.5	25.8	26.1	26.3	26.6	26.8	65	**213**
230	70	25.1	25.4	25.7	25.9	26.2	26.5	26.7	27.0	27.2	27.5	70	**230**
246	75	25.7	26.0	26.3	26.5	26.8	27.1	27.3	27.6	27.9	28.1	75	**246**
262	80	26.2	26.6	26.9	27.1	27.4	27.7	27.9	28.2	28.4	28.7	80	**262**
279	85	26.8	27.2	27.5	27.7	28.0	28.3	28.5	28.8	29.0	29.3	85	**279**
295	90	27.4	27.8	28.0	28.3	28.5	28.9	29.1	29.3	29.6	29.8	90	**295**
312	95	28.0	28.3	28.6	28.9	29.1	29.4	29.7	29.9	30.2	30.4	95	**312**
328	100	28.5	28.9	29.1	29.4	29.6	29.9	30.2	30.4	30.7	30.9	100	**328**
361	110	29.5	29.9	30.2	30.4	30.7	31.0	31.2	31.4	31.7	31.9	110	**361**
394	120	30.5	30.9	31.2	31.4	31.7	32.0	32.2	32.4	32.7	32.9	120	**394**
427	130	31.5	31.8	32.1	32.4	32.6	32.9	33.2	33.4	33.7	33.9	130	**427**
459	140	32.4	32.7	33.0	33.3	33.5	33.8	34.1	34.3	34.6	34.8	140	**459**
492	150	33.3	33.6	33.9	34.1	34.4	34.7	34.9	35.2	35.5	35.7	150	**492**
525	160	34.1	34.5	34.7	35.0	35.2	35.6	35.8	36.0	36.3	36.5	160	**525**
558	170	34.9	35.3	35.6	35.8	36.1	36.4	36.6	36.9	37.1	37.4	170	**558**
591	180	35.7	36.1	36.4	36.6	36.9	37.2	37.4	37.7	37.9	38.2	180	**591**
623	190	36.5	36.9	37.1	37.4	37.6	38.0	38.2	38.4	38.7	38.9	190	**623**
656	200	37.3	37.6	37.9	38.2	38.4	38.7	39.0	39.2	39.5	39.7	200	**656**
722	220	38.7	39.1	39.4	39.6	39.9	40.2	40.4	40.7	40.9	41.2	220	**722**
787	240	40.1	40.5	40.8	41.0	41.3	41.6	41.8	42.0	42.3	42.5	240	**787**
853	260	41.5	41.8	42.1	42.4	42.6	42.9	43.2	43.4	43.7	43.9	260	**853**
919	280	42.8	43.1	43.4	43.7	43.9	44.2	44.5	44.7	45.0	45.2	280	**919**
984	300	44.0	44.4	44.6	44.9	45.1	45.5	45.7	45.9	46.2	46.4	300	**984**

TABLE 13
Geographic Range

Object Height		Height of eye of observer in feet and meters										Object Height	
Feet		72	75	79	82	85	89	92	95	98	115		Feet
	Meters	22	23	24	25	26	27	28	29	30	35	Meters	
		Miles	Miles	Miles	Miles	Miles	Miles	Miles	Miles	Miles	Miles		
0	0	9.9	10.2	10.4	10.6	10.8	11.0	11.2	11.4	11.6	12.5	0	**0**
3	1	12.0	12.2	12.4	12.6	12.8	13.1	13.2	13.4	13.6	14.6	1	**3**
7	2	13.0	13.3	13.5	13.7	13.9	14.1	14.3	14.5	14.7	15.6	2	**7**
10	3	13.6	13.9	14.1	14.3	14.5	14.7	14.9	15.1	15.3	16.2	3	**10**
13	4	14.1	14.4	14.6	14.8	15.0	15.3	15.4	15.6	15.8	16.8	4	**13**
16	5	14.6	14.9	15.1	15.3	15.5	15.7	15.9	16.1	16.3	17.2	5	**16**
20	6	15.2	15.4	15.6	15.8	16.0	16.3	16.5	16.6	16.8	17.8	6	**20**
23	7	15.5	15.8	16.0	16.2	16.4	16.6	16.8	17.0	17.2	18.2	7	**23**
26	8	15.9	16.2	16.4	16.6	16.8	17.0	17.2	17.4	17.5	18.5	8	**26**
30	9	16.3	16.6	16.8	17.0	17.2	17.4	17.6	17.8	18.0	19.0	9	**30**
33	10	16.6	16.9	17.1	17.3	17.5	17.8	17.9	18.1	18.3	19.3	10	**33**
36	11	16.9	17.2	17.4	17.6	17.8	18.1	18.2	18.4	18.6	19.6	11	**36**
39	12	17.2	17.5	17.7	17.9	18.1	18.3	18.5	18.7	18.9	19.8	12	**39**
43	13	17.6	17.9	18.1	18.3	18.5	18.7	18.9	19.1	19.3	20.2	13	**43**
46	14	17.9	18.1	18.3	18.5	18.7	19.0	19.2	19.3	19.5	20.5	14	**46**
49	15	18.1	18.4	18.6	18.8	19.0	19.2	19.4	19.6	19.8	20.7	15	**49**
52	16	18.4	18.6	18.8	19.0	19.2	19.5	19.7	19.8	20.0	21.0	16	**52**
56	17	18.7	19.0	19.2	19.4	19.5	19.8	20.0	20.2	20.3	21.3	17	**56**
59	18	18.9	19.2	19.4	19.6	19.8	20.0	20.2	20.4	20.6	21.5	18	**59**
62	19	19.1	19.4	19.6	19.8	20.0	20.3	20.4	20.6	20.8	21.8	19	**62**
66	20	19.4	19.7	19.9	20.1	20.3	20.5	20.7	20.9	21.1	22.0	20	**66**
72	22	19.9	20.1	20.3	20.5	20.7	21.0	21.2	21.3	21.5	22.5	22	**72**
79	24	20.3	20.6	20.8	21.0	21.2	21.4	21.6	21.8	22.0	22.9	24	**79**
85	26	20.7	21.0	21.2	21.4	21.6	21.8	22.0	22.2	22.4	23.3	26	**85**
92	28	21.2	21.4	21.6	21.8	22.0	22.3	22.4	22.6	22.8	23.8	28	**92**
98	30	21.5	21.8	22.0	22.2	22.4	22.6	22.8	23.0	23.2	24.1	30	**98**
115	35	22.5	22.7	22.9	23.1	23.3	23.6	23.8	24.0	24.1	25.1	35	**115**
131	40	23.3	23.6	23.8	24.0	24.2	24.4	24.6	24.8	25.0	25.9	40	**131**
148	45	24.2	24.4	24.6	24.8	25.0	25.3	25.5	25.6	25.8	26.8	45	**148**
164	50	24.9	25.2	25.4	25.6	25.8	26.0	26.2	26.4	26.6	27.5	50	**164**
180	55	25.6	25.9	26.1	26.3	26.5	26.7	26.9	27.1	27.3	28.2	55	**180**
197	60	26.3	26.6	26.8	27.0	27.2	27.5	27.6	27.8	28.0	29.0	60	**197**
213	65	27.0	27.3	27.5	27.7	27.9	28.1	28.3	28.5	28.7	29.6	65	**213**
230	70	27.7	27.9	28.1	28.3	28.5	28.8	29.0	29.1	29.3	30.3	70	**230**
246	75	28.3	28.6	28.7	28.9	29.1	29.4	29.6	29.7	29.9	30.9	75	**246**
262	80	28.9	29.1	29.3	29.5	29.7	30.0	30.1	30.3	30.5	31.5	80	**262**
279	85	29.5	29.7	29.9	30.1	30.3	30.6	30.8	30.9	31.1	32.1	85	**279**
295	90	30.0	30.3	30.5	30.7	30.9	31.1	31.3	31.5	31.7	32.6	90	**295**
312	95	30.6	30.9	31.1	31.3	31.4	31.7	31.9	32.1	32.2	33.2	95	**312**
328	100	31.1	31.4	31.6	31.8	32.0	32.2	32.4	32.6	32.8	33.7	100	**328**
361	110	32.2	32.4	32.6	32.8	33.0	33.3	33.5	33.6	33.8	34.8	110	**361**
394	120	33.2	33.4	33.6	33.8	34.0	34.3	34.4	34.6	34.8	35.8	120	**394**
427	130	34.1	34.4	34.6	34.8	35.0	35.2	35.4	35.6	35.8	36.7	130	**427**
459	140	35.0	35.3	35.5	35.7	35.9	36.1	36.3	36.5	36.6	37.6	140	**459**
492	150	35.9	36.2	36.4	36.5	36.7	37.0	37.2	37.3	37.5	38.5	150	**492**
525	160	36.7	37.0	37.2	37.4	37.6	37.8	38.0	38.2	38.4	39.4	160	**525**
558	170	37.6	37.8	38.0	38.2	38.4	38.7	38.9	39.0	39.2	40.2	170	**558**
591	180	38.4	38.6	38.8	39.0	39.2	39.5	39.7	39.8	40.0	41.0	180	**591**
623	190	39.1	39.4	39.6	39.8	40.0	40.2	40.4	40.6	40.8	41.8	190	**623**
656	200	39.9	40.2	40.4	40.6	40.8	41.0	41.2	41.4	41.5	42.5	200	**656**
722	220	41.4	41.6	41.8	42.0	42.2	42.5	42.7	42.8	43.0	44.0	220	**722**
787	240	42.8	43.0	43.2	43.4	43.6	43.9	44.0	44.2	44.4	45.4	240	**787**
853	260	44.1	44.4	44.6	44.8	45.0	45.2	45.4	45.6	45.8	46.7	260	**853**
919	280	45.4	45.7	45.9	46.1	46.3	46.5	46.7	46.9	47.1	48.0	280	**919**
984	300	46.6	46.9	47.1	47.3	47.5	47.7	47.9	48.1	48.3	49.2	300	**984**

TABLE 14

Dip of the Sea Short of the Horizon

Dis-tance	5 (1.5)	10 (3.0)	15 (4.6)	20 (6.1)	25 (7.6)	30 (9.1)	35 (10.7)	40 (12.2)	45 (13.7)	50 (15.2)	Dis-tance
	Height of eye above the sea, in feet and (meters)										
Miles	′	′	′	′	′	′	′	′	′	′	*Miles*
0.2	14.2	28.4	42.5	56.7	70.8	84.9	99.1	113.2	127.3	141.5	0.2
0.3	9.6	19.0	28.4	37.8	47.3	56.7	66.1	75.6	85.0	94.4	0.3
0.4	7.2	14.3	21.4	28.5	35.5	42.6	49.7	56.7	63.8	70.9	0.4
0.5	5.9	11.5	17.2	22.8	28.5	34.2	39.8	45.5	51.1	56.8	0.5
0.6	5.0	9.7	14.4	19.1	23.8	28.5	33.3	38.0	42.7	47.4	0.6
0.7	4.3	8.4	12.4	16.5	20.5	24.5	28.6	32.6	36.7	40.7	0.7
0.8	3.9	7.4	10.9	14.5	18.0	21.5	25.1	28.6	32.2	35.7	0.8
0.9	3.5	6.7	9.8	12.9	16.1	19.2	22.4	25.5	28.7	31.8	0.9
1.0	3.2	6.1	8.9	11.7	14.6	17.4	20.2	23.0	25.9	28.7	1.0
1.1	3.0	5.6	8.2	10.7	13.3	15.9	18.5	21.0	23.6	26.2	1.1
1.2	2.9	5.2	7.6	9.9	12.3	14.6	17.0	19.4	21.7	24.1	1.2
1.3	2.7	4.9	7.1	9.2	11.4	13.6	15.8	17.9	20.1	22.3	1.3
1.4	2.6	4.6	6.6	8.7	10.7	12.7	14.7	16.7	18.8	20.8	1.4
1.5	2.5	4.4	6.3	8.2	10.1	11.9	13.8	15.7	17.6	19.5	1.5
1.6	2.4	4.2	6.0	7.7	9.5	11.3	13.0	14.8	16.6	18.3	1.6
1.7	2.4	4.0	5.7	7.4	9.0	10.7	12.4	14.0	15.7	17.3	1.7
1.8	2.3	3.9	5.5	7.0	8.6	10.2	11.7	13.3	14.9	16.5	1.8
1.9	2.3	3.8	5.3	6.7	8.2	9.7	11.2	12.7	14.2	15.7	1.9
2.0	2.2	3.7	5.1	6.5	7.9	9.3	10.7	12.1	13.6	15.0	2.0
2.1	2.2	3.6	4.9	6.3	7.6	9.0	10.3	11.7	13.0	14.3	2.1
2.2	2.2	3.5	4.8	6.1	7.3	8.6	9.9	11.2	12.5	13.8	2.2
2.3	2.2	3.4	4.6	5.9	7.1	8.3	9.6	10.8	12.0	13.3	2.3
2.4	2.2	3.4	4.5	5.7	6.9	8.1	9.2	10.4	11.6	12.8	2.4
2.5	2.2	3.3	4.4	5.6	6.7	7.8	9.0	10.1	11.2	12.4	2.5
2.6	2.2	3.3	4.3	5.4	6.5	7.6	8.7	9.8	10.9	12.0	2.6
2.7	2.2	3.2	4.3	5.3	6.4	7.4	8.5	9.5	10.6	11.6	2.7
2.8	2.2	3.2	4.2	5.2	6.2	7.2	8.2	9.2	10.3	11.3	2.8
2.9	2.2	3.2	4.1	5.1	6.1	7.1	8.0	9.0	10.0	11.0	2.9
3.0	2.2	3.1	4.1	5.0	6.0	6.9	7.8	8.8	9.7	10.7	3.0
3.1	2.2	3.1	4.0	4.9	5.9	6.8	7.7	8.6	9.5	10.4	3.1
3.2	2.2	3.1	4.0	4.9	5.8	6.6	7.5	8.4	9.3	10.2	3.2
3.3	2.2	3.1	3.9	4.8	5.7	6.5	7.4	8.2	9.1	9.9	3.3
3.4	2.2	3.1	3.9	4.7	5.6	6.4	7.2	8.1	8.9	9.7	3.4
3.5	2.2	3.1	3.9	4.7	5.5	6.3	7.1	7.9	8.7	9.5	3.5
3.6	2.2	3.1	3.9	4.6	5.4	6.2	7.0	7.8	8.6	9.4	3.6
3.7	2.2	3.1	3.8	4.6	5.4	6.1	6.9	7.7	8.4	9.2	3.7
3.8	2.2	3.1	3.8	4.6	5.3	6.0	6.8	7.5	8.3	9.0	3.8
3.9	2.2	3.1	3.8	4.5	5.2	6.0	6.7	7.4	8.2	8.9	3.9
4.0	2.2	3.1	3.8	4.5	5.2	5.9	6.6	7.3	8.0	8.7	4.0
4.1	2.2	3.1	3.8	4.5	5.2	5.8	6.5	7.2	7.9	8.6	4.1
4.2	2.2	3.1	3.8	4.4	5.1	5.8	6.5	7.1	7.8	8.5	4.2
4.3	2.2	3.1	3.8	4.4	5.1	5.7	6.4	7.1	7.7	8.4	4.3
4.4	2.2	3.1	3.8	4.4	5.0	5.7	6.3	7.0	7.6	8.3	4.4
4.5	2.2	3.1	3.8	4.4	5.0	5.6	6.3	6.9	7.5	8.2	4.5
4.6	2.2	3.1	3.8	4.4	5.0	5.6	6.2	6.8	7.4	8.1	4.6
4.7	2.2	3.1	3.8	4.4	5.0	5.6	6.2	6.8	7.4	8.0	4.7
4.8	2.2	3.1	3.8	4.4	4.9	5.5	6.1	6.7	7.3	7.9	4.8
4.9	2.2	3.1	3.8	4.3	4.9	5.5	6.1	6.7	7.2	7.8	4.9
5.0	2.2	3.1	3.8	4.3	4.9	5.5	6.0	6.6	7.2	7.7	5.0
5.5	2.2	3.1	3.8	4.3	4.9	5.4	5.9	6.4	6.9	7.4	5.5
6.0	2.2	3.1	3.8	4.3	4.9	5.3	5.8	6.3	6.7	7.2	6.0
6.5	2.2	3.1	3.8	4.3	4.9	5.3	5.7	6.2	6.6	7.1	6.5
7.0	2.2	3.1	3.8	4.3	4.9	5.3	5.7	6.1	6.5	7.0	7.0
7.5	2.2	3.1	3.8	4.3	4.9	5.3	5.7	6.1	6.5	6.9	7.5
8.0	2.2	3.1	3.8	4.3	4.9	5.3	5.7	6.1	6.5	6.9	8.0
8.5	2.2	3.1	3.8	4.3	4.9	5.3	5.7	6.1	6.5	6.9	8.5
9.0	2.2	3.1	3.8	4.3	4.9	5.3	5.7	6.1	6.5	6.9	9.0
9.5	2.2	3.1	3.8	4.3	4.9	5.3	5.7	6.1	6.5	6.9	9.5
10.0	2.2	3.1	3.8	4.3	4.9	5.3	5.7	6.1	6.5	6.9	10.0

TABLE 14

Dip of the Sea Short of the Horizon

Dis-tance	55 (16.8)	60 (18.3)	65 (19.8)	70 (21.3)	75 (22.9)	80 (24.4)	85 (25.9)	90 (27.4)	95 (29.0)	100 (30.5)	Dis-tance
	Height of eye above the sea, in feet and (meters)										
Miles	′	′	′	′	′	′	′	′	′	′	*Miles*
0.2	155.6	169.7	183.3	197.9	212.0	226.1	240.2	254.2	268.3	282.3	0.2
0.3	103.8	113.3	122.7	132.1	141.6	151.0	160.4	169.9	179.3	188.7	0.3
0.4	77.9	85.0	92.1	99.2	106.2	113.3	120.3	127.4	134.5	141.5	0.4
0.5	62.4	68.1	73.8	79.4	85.1	90.7	96.4	102.0	107.7	113.3	0.5
0.6	52.1	56.8	61.5	66.3	71.0	75.7	80.4	85.1	89.8	94.5	0.6
0.7	44.7	48.8	52.8	56.9	60.9	64.9	69.0	73.0	77.1	81.1	0.7
0.8	39.2	42.8	46.3	49.8	53.4	56.9	60.4	64.0	67.5	71.1	0.8
0.9	34.9	38.1	41.2	44.4	47.5	50.7	53.8	56.9	60.1	63.2	0.9
1.0	31.5	34.4	37.2	40.0	42.8	45.7	48.5	51.3	54.2	57.0	1.0
1.1	28.7	31.3	33.9	36.5	39.0	41.6	44.2	46.7	49.3	51.9	1.1
1.2	26.4	28.8	31.1	33.5	35.9	38.2	40.6	42.9	45.3	47.6	1.2
1.3	24.5	26.7	28.8	31.0	33.2	35.4	37.5	39.7	41.9	44.1	1.3
1.4	22.8	24.8	26.8	28.9	30.9	32.9	34.9	37.0	39.0	41.0	1.4
1.5	21.4	23.3	25.1	27.0	28.9	30.8	32.7	34.6	36.5	38.3	1.5
1.6	20.1	21.9	23.6	25.4	27.2	29.0	30.7	32.5	34.3	36.0	1.6
1.7	19.0	20.7	22.3	24.0	25.7	27.3	29.0	30.7	32.3	34.0	1.7
1.8	18.0	19.6	21.2	22.8	24.3	25.9	27.5	29.0	30.6	32.2	1.8
1.9	17.2	18.7	20.1	21.6	23.1	24.6	26.1	27.6	29.1	30.6	1.9
2.0	16.4	17.8	19.2	20.6	22.0	23.5	24.9	26.3	27.7	29.1	2.0
2.1	15.7	17.0	18.4	19.7	21.1	22.4	23.8	25.1	26.5	27.8	2.1
2.2	15.1	16.3	17.6	18.9	20.2	21.5	22.8	24.1	25.3	26.6	2.2
2.3	14.5	15.7	16.9	18.2	19.4	20.6	21.9	23.1	24.3	25.6	2.3
2.4	14.0	15.1	16.3	17.5	18.7	19.9	21.0	22.2	23.4	24.6	2.4
2.5	13.5	14.6	15.8	16.9	18.0	19.1	20.3	21.4	22.5	23.7	2.5
2.6	13.0	14.1	15.2	16.3	17.4	18.5	19.6	20.7	21.8	22.8	2.6
2.7	12.6	13.7	14.7	15.8	16.8	17.9	18.9	20.0	21.0	22.1	2.7
2.8	12.3	13.3	14.3	15.3	16.3	17.3	18.3	19.3	20.4	21.4	2.8
2.9	11.9	12.9	13.9	14.9	15.8	16.8	17.8	18.8	19.7	20.7	2.9
3.0	11.6	12.6	13.5	14.4	15.4	16.3	17.3	18.2	19.2	20.1	3.0
3.1	11.3	12.2	13.2	14.1	15.0	15.9	16.8	17.7	18.6	19.5	3.1
3.2	11.1	11.9	12.8	13.7	14.6	15.5	16.4	17.2	18.1	19.0	3.2
3.3	10.8	11.7	12.5	13.4	14.2	15.1	15.9	16.8	17.7	18.5	3.3
3.4	10.6	11.4	12.2	13.1	13.9	14.7	15.6	16.4	17.2	18.1	3.4
3.5	10.3	11.2	12.0	12.8	13.6	14.4	15.2	16.0	16.8	17.6	3.5
3.6	10.1	10.9	11.7	12.5	13.3	14.1	14.9	15.6	16.4	17.2	3.6
3.7	9.9	10.7	11.5	12.2	13.0	13.8	14.5	15.3	16.1	16.8	3.7
3.8	9.8	10.5	11.3	12.0	12.7	13.5	14.2	15.0	15.7	16.5	3.8
3.9	9.6	10.3	11.1	11.8	12.5	13.2	14.0	14.7	15.4	16.1	3.9
4.0	9.4	10.1	10.9	11.6	12.3	13.0	13.7	14.4	15.1	15.8	4.0
4.1	9.3	10.0	10.7	11.4	12.1	12.7	13.4	14.1	14.8	15.5	4.1
4.2	9.2	9.8	10.5	11.2	11.8	12.5	13.2	13.9	14.5	15.2	4.2
4.3	9.0	9.7	10.3	11.0	11.7	12.3	13.0	13.6	14.3	14.9	4.3
4.4	8.9	9.5	10.2	10.8	11.5	12.1	12.8	13.4	14.0	14.7	4.4
4.5	8.8	9.4	10.0	10.7	11.3	11.9	12.6	13.2	13.8	14.4	4.5
4.6	8.7	9.3	9.9	10.5	11.1	11.8	12.4	13.0	13.6	14.2	4.6
4.7	8.6	9.2	9.8	10.4	11.0	11.6	12.2	12.8	13.4	14.0	4.7
4.8	8.5	9.1	9.7	10.2	10.8	11.4	12.0	12.6	13.2	13.8	4.8
4.9	8.4	9.0	9.5	10.1	10.7	11.3	11.9	12.4	13.0	13.6	4.9
5.0	8.3	8.9	9.4	10.0	10.6	11.1	11.7	12.3	12.8	13.4	5.0
5.5	7.9	8.5	9.0	9.5	10.0	10.5	11.0	11.5	12.1	12.6	5.5
6.0	7.7	8.2	8.6	9.1	9.6	10.0	10.5	11.0	11.5	11.9	6.0
6.5	7.5	7.9	8.4	8.8	9.2	9.7	10.1	10.5	11.0	11.4	6.5
7.0	7.4	7.8	8.2	8.6	9.0	9.4	9.8	10.2	10.6	11.0	7.0
7.5	7.3	7.6	8.0	8.4	8.8	9.2	9.5	9.9	10.3	10.7	7.5
8.0	7.2	7.6	7.9	8.3	8.6	9.0	9.3	9.7	10.0	10.4	8.0
8.5	7.2	7.5	7.9	8.2	8.5	8.9	9.2	9.5	9.9	10.2	8.5
9.0	7.2	7.5	7.8	8.1	8.5	8.8	9.1	9.4	9.7	10.0	9.0
9.5	7.2	7.5	7.8	8.1	8.4	8.7	9.0	9.3	9.6	9.9	9.5
10.0	7.2	7.5	7.8	8.1	8.4	8.7	9.0	9.3	9.5	9.8	10.0

TABLE 15

Distance by Vertical Angle

Measured Between Sea Horizon and Top of Object Beyond Sea Horizon

Angle	Difference in feet between height of object and height of eye of observer										Angle
	25	30	35	40	45	50	60	70	80	90	
° ′	*Miles*	*Miles*	*Miles*	*Miles*	*Miles*	*Miles*	*Miles*	*Miles*	*Miles*	*Miles*	° ′
−0 04	12.4	12.8	13.2	13.6	14.0	14.4	15.0	15.7	16.3	16.9	−0 04
−0 03	10.5	10.9	11.4	11.8	12.2	12.6	13.3	14.0	14.6	15.2	−0 03
−0 02	8.7	9.2	9.7	10.2	10.6	11.0	11.8	12.5	13.1	13.7	−0 02
−0 01	7.2	7.7	8.2	8.7	9.1	9.5	10.3	11.0	11.7	12.3	−0 01
0 00	5.8	6.4	6.9	7.4	7.8	8.2	9.0	9.7	10.4	11.1	0 00
0 01	4.8	5.3	5.8	6.3	6.7	7.1	7.9	8.6	9.3	9.9	0 01
0 02	3.9	4.4	4.9	5.4	5.8	6.2	6.9	7.6	8.3	8.9	0 02
0 03	3.3	3.7	4.2	4.6	5.0	5.4	6.1	6.8	7.4	8.0	0 03
0 04	2.8	3.2	3.6	4.0	4.4	4.7	5.4	6.1	6.7	7.3	0 04
0 05	2.4	2.8	3.1	3.5	3.9	4.2	4.8	5.5	6.0	6.6	0 05
0 06	2.1	2.4	2.8	3.1	3.4	3.7	4.3	4.9	5.5	6.0	0 06
0 07	1.8	2.2	2.5	2.8	3.1	3.4	3.9	4.5	5.0	5.5	0 07
0 08	1.6	1.9	2.2	2.5	2.8	3.1	3.6	4.1	4.6	5.0	0 08
0 09	1.5	1.7	2.0	2.3	2.5	2.8	3.3	3.8	4.2	4.7	0 09
0 10	1.3	1.6	1.8	2.1	2.3	2.6	3.0	3.5	3.9	4.3	0 10
0 15	0.9	1.1	1.3	1.5	1.6	1.8	2.1	2.5	2.8	3.1	0 15
0 20	0.7	0.8	1.0	1.1	1.2	1.4	1.6	1.9	2.2	2.4	0 20
0 25	0.6	0.7	0.8	0.9	1.0	1.1	1.3	1.5	1.8	2.0	0 25
0 30	0.5	0.6	0.7	0.7	0.8	0.9	1.1	1.3	1.5	1.7	0 30
0 35		0.5	0.6	0.6	0.7	0.8	1.0	1.1	1.3	1.4	0 35
0 40			0.5	0.6	0.6	0.7	0.8	1.0	1.1	1.3	0 40
0 45				0.5	0.6	0.6	0.7	0.9	1.0	1.1	0 45
0 50				0.5	0.5	0.6	0.7	0.8	0.9	1.0	0 50
0 55					0.5	0.5	0.6	0.7	0.8	0.9	0 55
1 00						0.5	0.6	0.7	0.8	0.8	1 00
1 10							0.5	0.6	0.6	0.7	1 10
1 20								0.5	0.6	0.6	1 20
1 30									0.5	0.6	1 30
1 40									0.5	0.5	1 40
1 50										0.5	1 50

TABLE 15

Distance by Vertical Angle

Measured Between Sea Horizon and Top of Object Beyond Sea Horizon

Angle	Difference in feet between height of object and height of eye of observer											Angle
	100	120	140	160	180	200	250	300	350	400	450	
° ′	*Miles*	*Miles*	*Miles*	*Miles*	*Miles*	*Miles*	*Miles*	*Miles*	*Miles*	*Miles*	*Miles*	° ′
0 00	11.7	12.8	13.8	14.8	15.7	16.5	18.4	20.2	21.8	23.3	24.7	0 00
0 01	10.5	11.6	12.7	13.6	14.5	15.3	17.3	19.0	20.7	22.2	23.6	0 01
0 02	9.5	10.6	11.6	12.5	13.4	14.3	16.2	17.9	19.6	21.0	22.5	0 02
0 03	8.6	9.7	10.7	11.6	12.5	13.3	15.2	16.9	18.5	20.0	21.4	0 03
0 04	7.8	8.8	9.8	10.7	11.6	12.4	14.3	16.0	17.5	19.0	20.4	0 04
0 05	7.1	8.1	9.0	9.9	10.8	11.5	13.4	15.1	16.6	18.1	19.5	0 05
0 06	6.5	7.5	8.4	9.2	10.0	10.8	12.6	14.2	15.8	17.2	18.6	0 06
0 07	6.0	6.9	7.7	8.6	9.4	10.1	11.9	13.5	15.0	16.4	17.7	0 07
0 08	5.5	6.4	7.2	8.0	8.8	9.5	11.2	12.8	14.2	15.6	16.9	0 08
0 09	5.1	5.9	6.7	7.5	8.2	8.9	10.6	12.1	13.5	14.9	16.2	0 09
0 10	4.7	5.5	6.3	7.0	7.7	8.4	10.0	11.5	12.9	14.2	15.5	0 10
0 11	4.4	5.2	5.9	6.6	7.3	7.9	9.5	10.9	12.3	13.6	14.8	0 11
0 12	4.1	4.8	5.5	6.2	6.9	7.5	9.0	10.4	11.7	13.0	14.2	0 12
0 13	3.9	4.6	5.2	5.9	6.5	7.1	8.5	9.9	11.2	12.5	13.6	0 13
0 14	3.6	4.3	4.9	5.6	6.2	6.7	8.1	9.5	10.7	11.9	13.1	0 14
0 15	3.4	4.1	4.7	5.3	5.8	6.4	7.8	9.0	10.3	11.5	12.6	0 15
0 20	2.7	3.2	3.7	4.2	4.6	5.1	6.3	7.4	8.4	9.5	10.5	0 20
0 25	2.2	2.6	3.0	3.4	3.8	4.2	5.2	6.2	7.1	8.0	8.9	0 25
0 30	1.8	2.2	2.6	2.9	3.2	3.6	4.4	5.3	6.1	6.9	7.7	0 30
0 35	1.6	1.9	2.2	2.5	2.8	3.1	3.9	4.6	5.3	6.0	6.7	0 35
0 40	1.4	1.7	1.9	2.2	2.5	2.8	3.4	4.1	4.7	5.4	6.0	0 40
0 45	1.2	1.5	1.7	2.0	2.2	2.5	3.1	3.6	4.2	4.8	5.4	0 45
0 50	1.1	1.3	1.6	1.8	2.0	2.2	2.8	3.3	3.8	4.4	4.9	0 50
0 55	1.0	1.2	1.4	1.6	1.8	2.0	2.5	3.0	3.5	4.0	4.5	0 55
1 00	0.9	1.1	1.3	1.5	1.7	1.9	2.3	2.8	3.2	3.7	4.1	1 00
1 10	0.8	1.0	1.1	1.3	1.4	1.6	2.0	2.4	2.8	3.2	3.6	1 10
1 20	0.7	0.8	1.0	1.1	1.3	1.4	1.8	2.1	2.4	2.8	3.1	1 20
1 30	0.6	0.8	0.9	1.0	1.1	1.2	1.6	1.9	2.2	2.5	2.8	1 30
1 40	0.6	0.7	0.8	0.9	1.0	1.1	1.4	1.7	2.0	2.2	2.5	1 40
1 50	0.5	0.6	0.7	0.8	0.9	1.0	1.3	1.5	1.8	2.0	2.3	1 50
2 00	0.5	0.6	0.7	0.8	0.8	0.9	1.2	1.4	1.6	1.9	2.1	2 00
2 30		0.5	0.5	0.6	0.7	0.8	0.9	1.1	1.3	1.5	1.7	2 30
3 00				0.5	0.6	0.6	0.8	0.9	1.1	1.3	1.4	3 00
3 30					0.5	0.5	0.7	0.8	0.9	1.1	1.2	3 30
4 00						0.5	0.6	0.7	0.8	0.9	1.1	4 00
4 30							0.5	0.6	0.7	0.8	0.9	4 30
5 00							0.5	0.6	0.7	0.8	0.8	5 00
6 00								0.5	0.5	0.6	0.7	6 00
7 00									0.5	0.5	0.6	7 00
8 00										0.5	0.5	8 00
10 00												10 00

TABLE 15

Distance by Vertical Angle

Measured Between Sea Horizon and Top of Object Beyond Sea Horizon

Angle	Difference in feet between height of object and height of eye of observer											Angle
	500	600	700	800	900	1000	1200	1400	1600	1800	2000	
° ′	*Miles*	*Miles*	*Miles*	*Miles*	*Miles*	*Miles*	*Miles*	*Miles*	*Miles*	*Miles*	*Miles*	° ′
0 05	20. 8	23. 2	25. 4	27. 5	29. 5	31. 4	34. 8	38. 0	41. 0	43. 8	46. 5	0 05
0 06	19. 8	22. 3	24. 5	26. 6	28. 5	30. 4	33. 8	37. 0	40. 0	42. 8	45. 4	0 06
0 07	19. 0	21. 4	23. 6	25. 6	27. 6	29. 4	32. 9	36. 0	39. 0	41. 8	44. 4	0 07
0 08	18. 2	20. 5	22. 7	24. 7	26. 7	28. 5	31. 9	35. 1	38. 0	40. 8	43. 4	0 08
0 09	17. 4	19. 7	21. 9	23. 9	25. 8	27. 6	31. 0	34. 1	37. 0	39. 8	42. 5	0 09
0 10	16. 7	19. 0	21. 1	23. 1	25. 0	26. 8	30. 1	33. 2	36. 2	38. 9	41. 5	0 10
0 11	16. 0	18. 3	20. 4	22. 3	24. 2	26. 0	29. 3	32. 4	35. 3	38. 0	40. 6	0 11
0 12	15. 4	17. 6	19. 6	21. 6	23. 4	25. 2	28. 5	31. 5	34. 4	37. 1	39. 7	0 12
0 13	14. 8	16. 9	19. 0	20. 9	22. 7	24. 4	27. 7	30. 7	33. 6	36. 3	38. 8	0 13
0 14	14. 2	16. 3	18. 3	20. 2	22. 0	23. 7	26. 9	30. 0	32. 8	35. 4	38. 0	0 14
0 15	13. 7	15. 8	17. 7	19. 6	21. 3	23. 0	26. 2	29. 2	32. 0	34. 6	37. 2	0 15
0 17	12. 7	14. 7	16. 6	18. 4	20. 1	21. 7	24. 8	27. 8	30. 5	33. 1	35. 6	0 17
0 20	11. 4	13. 3	15. 1	16. 8	18. 4	20. 0	23. 0	25. 8	28. 4	31. 0	33. 4	0 20
0 25	9. 7	11. 4	13. 0	14. 6	16. 1	17. 5	20. 3	22. 9	25. 4	27. 8	30. 1	0 25
0 30	8. 4	9. 9	11. 4	12. 8	14. 2	15. 5	18. 1	20. 5	22. 9	25. 2	27. 4	0 30
0 35	7. 4	8. 8	10. 1	11. 4	12. 6	13. 9	16. 3	18. 5	20. 7	22. 9	24. 9	0 35
0 40	6. 6	7. 8	9. 0	10. 2	11. 4	12. 5	14. 7	16. 9	18. 9	20. 9	22. 9	0 40
0 45	6. 0	7. 1	8. 2	9. 3	10. 3	11. 4	13. 4	15. 4	17. 3	19. 2	21. 1	0 45
0 50	5. 4	6. 4	7. 5	8. 5	9. 4	10. 4	12. 3	14. 2	16. 0	17. 7	19. 5	0 50
0 55	5. 0	5. 9	6. 8	7. 8	8. 7	9. 6	11. 4	13. 1	14. 8	16. 5	18. 1	0 55
1 00	4. 6	5. 5	6. 3	7. 2	8. 0	8. 9	10. 5	12. 2	13. 8	15. 3	16. 9	1 00
1 10	3. 9	4. 7	5. 5	6. 2	7. 0	7. 7	9. 2	10. 6	12. 1	13. 5	14. 9	1 10
1 20	3. 5	4. 2	4. 8	5. 5	6. 2	6. 8	8. 1	9. 4	10. 7	12. 0	13. 2	1 20
1 30	3. 1	3. 7	4. 3	4. 9	5. 5	6. 1	7. 3	8. 5	9. 6	10. 8	11. 9	1 30
1 40	2. 8	3. 3	3. 9	4. 4	5. 0	5. 5	6. 6	7. 7	8. 7	9. 8	10. 8	1 40
1 50	2. 5	3. 0	3. 6	4. 1	4. 5	5. 0	6. 0	7. 0	8. 0	9. 0	9. 9	1 50
2 00	2. 3	2. 8	3. 3	3. 7	4. 2	4. 6	5. 5	6. 5	7. 4	8. 2	9. 1	2 00
2 30	1. 9	2. 2	2. 6	3. 0	3. 4	3. 7	4. 5	5. 2	5. 9	6. 7	7. 4	2 30
3 00	1. 6	1. 9	2. 2	2. 5	2. 8	3. 1	3. 7	4. 4	5. 0	5. 6	6. 2	3 00
3 30	1. 3	1. 6	1. 9	2. 1	2. 4	2. 7	3. 2	3. 7	4. 3	4. 8	5. 3	3 30
4 00	1. 2	1. 4	1. 6	1. 9	2. 1	2. 3	2. 8	3. 3	3. 7	4. 2	4. 7	4 00
5 00	0. 9	1. 1	1. 3	1. 5	1. 7	1. 9	2. 3	2. 6	3. 0	3. 4	3. 7	5 00
6 00	0. 8	0. 9	1. 1	1. 3	1. 4	1. 6	1. 9	2. 2	2. 5	2. 8	3. 1	6 00
7 00	0. 7	0. 8	0. 9	1. 1	1. 2	1. 3	1. 6	1. 9	2. 1	2. 4	2. 7	7 00
8 00	0. 6	0. 7	0. 8	0. 9	1. 1	1. 2	1. 4	1. 6	1. 9	2. 1	2. 3	8 00
10 00	0. 5	0. 6	0. 7	0. 7	0. 8	0. 9	1. 1	1. 3	1. 5	1. 7	1. 9	10 00
12 00		0. 5	0. 5	0. 6	0. 7	0. 8	0. 9	1. 1	1. 2	1. 4	1. 5	12 00
15 00				0. 5	0. 6	0. 6	0. 7	0. 9	1. 0	1. 1	1. 2	15 00
20 00						0. 5	0. 5	0. 6	0. 7	0. 8	0. 9	20 00
25 00								0. 5	0. 6	0. 6	0. 7	25 00
30 00									0. 5	0. 5	0. 6	30 00

TABLE 16
Distance by Vertical Angle
Measured Between Waterline at Object and Top of Object

Angle	Height of object above the sea, in **feet** and (meters)										Angle
	10 (3.0)	15 (4.6)	20 (6.1)	25 (7.6)	30 (9.1)	35 (10.7)	40 (12.2)	45 (13.7)	50 (15.2)	55 (16.8)	
° ′	*Miles*	*Miles*	*Miles*	*Miles*	*Miles*	*Miles*	*Miles*	*Miles*	*Miles*	*Miles*	° ′
0 10	0.57	0.85	1.13	1.41	1.70	1.98	2.26	2.55	2.83	3.11	0 10
0 11	0.51	0.77	1.03	1.29	1.54	1.80	2.06	2.31	2.57	2.83	0 11
0 12	0.47	0.71	0.94	1.18	1.41	1.65	1.89	2.12	2.36	2.59	0 12
0 13	0.44	0.65	0.87	1.09	1.31	1.52	1.74	1.96	2.18	2.39	0 13
0 14	0.40	0.61	0.81	1.01	1.21	1.41	1.62	1.82	2.02	2.22	0 14
0 15	0.38	0.57	0.75	0.94	1.18	1.32	1.51	1.70	1.89	2.07	0 15
0 20	0.28	0.42	0.57	0.71	0.85	0.99	1.13	1.27	1.41	1.56	0 20
0 25	0.23	0.34	0.45	0.57	0.68	0.79	0.91	1.02	1.13	1.24	0 25
0 30	0.19	0.28	0.38	0.47	0.57	0.66	0.75	0.85	0.94	1.04	0 30
0 35	0.16	0.24	0.32	0.40	0.46	0.57	0.65	0.73	0.81	0.89	0 35
0 40	0.14	0.21	0.28	0.35	0.42	0.50	0.57	0.64	0.71	0.78	0 40
0 45	0.13	0.19	0.25	0.31	0.38	0.44	0.50	0.57	0.63	0.69	0 45
0 50	0.11	0.17	0.23	0.28	0.34	0.40	0.45	0.51	0.57	0.62	0 50
0 55	0.10	0.15	0.21	0.26	0.31	0.36	0.41	0.46	0.51	0.57	0 55
1 00		0.14	0.19	0.24	0.28	0.33	0.38	0.42	0.47	0.52	1 00
1 10		0.12	0.16	0.20	0.24	0.28	0.32	0.36	0.40	0.44	1 10
1 20		0.11	0.14	0.18	0.21	0.25	0.28	0.32	0.35	0.39	1 20
1 30		0.09	0.13	0.16	0.19	0.22	0.25	0.28	0.31	0.35	1 30
1 40			0.11	0.14	0.17	0.20	0.23	0.25	0.28	0.31	1 40
1 50			0.10	0.13	0.15	0.18	0.21	0.23	0.26	0.28	1 50
2 00				0.12	0.14	0.16	0.19	0.21	0.24	0.26	2 00
2 15				0.10	0.13	0.15	0.17	0.19	0.21	0.23	2 15
2 30					0.11	0.13	0.15	0.17	0.19	0.21	2 30
2 45					0.10	0.12	0.14	0.15	0.17	0.19	2 45
3 00						0.11	0.13	0.14	0.16	0.17	3 00
3 20						0.10	0.11	0.13	0.14	0.16	3 20
3 40							0.10	0.12	0.13	0.14	3 40
4 00								0.11	0.12	0.13	4 00
4 20								0.10	0.11	0.12	4 20
4 40									0.10	0.11	4 40
5 00										0.10	5 00

TABLE 16
Distance by Vertical Angle
Measured Between Waterline at Object and Top of Object

Angle	Height of object above the sea, in **feet** and (meters)										Angle
	60 (18.3)	65 (19.8)	70 (21.3)	75 (22.9)	80 (24.4)	85 (25.9)	90 (27.4)	95 (29.0)	100 (30.5)	105 (32.0)	
° ′	*Miles*	*Miles*	*Miles*	*Miles*	*Miles*	*Miles*	*Miles*	*Miles*	*Miles*	*Miles*	° ′
0 10	3.39	3.68	3.96	4.24	4.53	4.81					0 10
0 11	3.09	3.34	3.60	3.86	4.11	4.37	4.63	4.89			0 11
0 12	2.83	3.06	3.30	3.54	3.77	4.01	4.24	4.48	4.71	4.95	0 12
0 13	2.61	2.83	3.05	3.26	3.48	3.70	3.92	4.13	4.35	4.57	0 13
0 14	2.42	2.63	2.83	3.03	3.23	3.44	3.64	3.84	4.04	4.24	0 14
0 15	2.26	2.45	2.64	2.83	3.02	3.21	3.39	3.58	3.77	3.96	0 15
0 20	1.70	1.84	1.98	2.12	2.26	2.40	2.55	2.69	2.83	2.97	0 20
0 25	1.36	1.47	1.58	1.70	1.81	1.92	2.04	2.15	2.26	2.38	0 25
0 30	1.13	1.23	1.32	1.41	1.51	1.60	1.70	1.79	1.89	1.98	0 30
0 35	0.97	1.05	1.13	1.21	1.29	1.37	1.45	1.54	1.62	1.70	0 35
0 40	0.85	0.92	0.99	1.06	1.13	1.20	1.27	1.34	1.41	1.49	0 40
0 45	0.75	0.82	0.88	0.94	1.01	1.07	1.13	1.19	1.26	1.32	0 45
0 50	0.68	0.74	0.79	0.85	0.91	0.96	1.02	1.07	1.13	1.19	0 50
0 55	0.62	0.67	0.72	0.77	0.82	0.87	0.93	0.98	1.03	1.08	0 55
1 00	0.57	0.61	0.66	0.71	0.75	0.80	0.85	0.90	0.94	0.99	1 00
1 10	0.48	0.53	0.57	0.61	0.65	0.69	0.73	0.77	0.81	0.85	1 10
1 20	0.42	0.46	0.49	0.53	0.57	0.60	0.64	0.67	0.71	0.74	1 20
1 30	0.38	0.41	0.44	0.47	0.50	0.53	0.57	0.60	0.63	0.66	1 30
1 40	0.34	0.37	0.40	0.42	0.45	0.48	0.51	0.54	0.57	0.59	1 40
1 50	0.31	0.33	0.36	0.39	0.41	0.44	0.46	0.49	0.51	0.54	1 50
2 00	0.28	0.31	0.33	0.35	0.38	0.40	0.42	0.45	0.47	0.49	2 00
2 15	0.25	0.27	0.29	0.31	0.34	0.36	0.38	0.40	0.42	0.44	2 15
2 30	0.23	0.25	0.26	0.28	0.30	0.32	0.34	0.36	0.38	0.40	2 30
2 45	0.21	0.22	0.24	0.26	0.27	0.29	0.31	0.33	0.34	0.36	2 45
3 00	0.19	0.20	0.22	0.24	0.25	0.27	0.28	0.30	0.31	0.33	3 00
3 20	0.17	0.18	0.20	0.21	0.23	0.24	0.25	0.27	0.28	0.30	3 20
3 40	0.15	0.17	0.18	0.19	0.21	0.22	0.23	0.24	0.26	0.27	3 40
4 00	0.14	0.15	0.16	0.18	0.19	0.20	0.21	0.22	0.24	0.25	4 00
4 20	0.13	0.14	0.15	0.16	0.17	0.18	0.20	0.21	0.22	0.23	4 20
4 40	0.12	0.13	0.14	0.15	0.16	0.17	0.18	0.19	0.20	0.21	4 40
5 00	0.11	0.12	0.13	0.14	0.15	0.16	0.17	0.18	0.19	0.20	5 00
5 20	0.11	0.11	0.12	0.13	0.14	0.15	0.16	0.17	0.18	0.19	5 20
5 40	0.10	0.11	0.12	0.12	0.13	0.14	0.15	0.16	0.17	0.17	5 40
6 00		0.10	0.11	0.12	0.13	0.13	0.14	0.15	0.16	0.16	6 00
6 20			0.10	0.11	0.12	0.13	0.13	0.14	0.15	0.16	6 20
6 40				0.11	0.11	0.12	0.13	0.13	0.14	0.15	6 40
7 00				0.10	0.11	0.11	0.12	0.13	0.13	0.14	7 00
7 20					0.10	0.11	0.12	0.12	0.13	0.13	7 20
7 40						0.10	0.11	0.12	0.12	0.13	7 40
8 00							0.11	0.11	0.12	0.12	8 00
8 20							0.10	0.11	0.11	0.12	8 20
8 40								0.10	0.11	0.11	8 40
9 00									0.10	0.11	9 00
9 30										0.10	9 30
10 00											10 00

TABLE 16

Distance by Vertical Angle
Measured Between Waterline at Object and Top of Object

Angle	Height of object above the sea, in **feet** and (meters)										Angle
	110 (33.5)	115 (35.1)	120 (36.6)	125 (38.1)	130 (39.6)	135 (41.1)	140 (42.7)	145 (44.2)	150 (45.7)	155 (47.2)	
° ′	*Miles*	*Miles*	*Miles*	*Miles*	*Miles*	*Miles*	*Miles*	*Miles*	*Miles*	*Miles*	° ′
0 10											0 10
0 11											0 11
0 12											0 12
0 13	4.79	5.00									0 13
0 14	4.45	4.65	4.85								0 14
0 15	4.15	4.34	4.53	4.71	4.90						0 15
0 20	3.11	3.25	3.39	3.54	3.68	3.82	3.96	4.10	4.24	4.38	0 20
0 25	2.49	2.60	2.72	2.83	2.94	3.06	3.17	3.28	3.30	3.51	0 25
0 30	2.07	2.17	2.26	2.36	2.45	2.55	2.64	2.73	2.83	2.92	0 30
0 35	1.78	1.86	1.94	2.02	2.10	2.18	2.26	2.34	2.42	2.51	0 35
0 40	1.56	1.63	1.70	1.77	1.84	1.91	1.98	2.05	2.12	2.19	0 40
0 45	1.30	1.45	1.51	1.57	1.63	1.70	1.76	1.82	1.89	1.95	0 45
0 50	1.24	1.30	1.36	1.41	1.47	1.53	1.58	1.64	1.70	1.75	0 50
0 55	1.13	1.18	1.23	1.29	1.34	1.39	1.44	1.49	1.54	1.59	0 55
1 00	1.04	1.08	1.13	1.18	1.23	1.27	1.32	1.37	1.41	1.46	1 00
1 10	0.89	0.93	0.97	1.01	1.05	1.09	1.13	1.17	1.21	1.25	1 10
1 20	0.78	0.81	0.85	0.88	0.92	0.95	0.99	1.03	1.06	1.10	1 20
1 30	0.69	0.72	0.75	0.79	0.82	0.85	0.88	0.91	0.94	0.97	1 30
1 40	0.62	0.65	0.66	0.71	0.74	0.76	0.79	0.82	0.85	0.88	1 40
1 50	0.57	0.59	0.62	0.64	0.67	0.69	0.72	0.75	0.77	0.80	1 50
2 00	0.52	0.54	0.57	0.59	0.61	0.64	0.66	0.68	0.71	0.73	2 00
2 15	0.46	0.48	0.50	0.52	0.54	0.57	0.59	0.61	0.63	0.65	2 15
2 30	0.41	0.43	0.45	0.47	0.49	0.51	0.53	0.55	0.57	0.58	2 30
2 45	0.38	0.39	0.41	0.43	0.45	0.46	0.48	0.50	0.51	0.53	2 45
3 00	0.35	0.36	0.38	0.39	0.41	0.42	0.44	0.46	0.47	0.49	3 00
3 20	0.31	0.32	0.34	0.35	0.37	0.38	0.40	0.41	0.42	0.44	3 20
3 40	0.28	0.30	0.31	0.32	0.33	0.35	0.36	0.37	0.39	0.40	3 40
4 00	0.26	0.27	0.28	0.29	0.31	0.32	0.33	0.34	0.35	0.36	4 00
4 20	0.24	0.25	0.26	0.27	0.28	0.29	0.30	0.31	0.33	0.34	4 20
4 40	0.22	0.23	0.24	0.25	0.26	0.27	0.28	0.29	0.30	0.31	4 40
5 00	0.21	0.22	0.23	0.24	0.24	0.25	0.26	0.27	0.28	0.29	5 00
5 20	0.19	0.20	0.21	0.22	0.23	0.24	0.25	0.26	0.26	0.27	5 20
5 40	0.18	0.19	0.20	0.21	0.22	0.22	0.23	0.24	0.25	0.26	5 40
6 00	0.17	0.18	0.19	0.20	0.20	0.21	0.22	0.23	0.23	0.24	6 00
6 20	0.16	0.17	0.18	0.19	0.19	0.20	0.21	0.22	0.22	0.23	6 20
6 40	0.15	0.17	0.17	0.18	0.18	0.19	0.20	0.20	0.21	0.22	6 40
7 00	0.15	0.15	0.16	0.17	0.17	0.18	0.19	0.19	0.20	0.21	7 00
7 20	0.14	0.15	0.15	0.16	0.17	0.17	0.18	0.19	0.19	0.20	7 20
7 40	0.13	0.14	0.15	0.15	0.16	0.17	0.17	0.18	0.18	0.19	7 40
8 00	0.13	0.13	0.14	0.15	0.15	0.16	0.16	0.17	0.18	0.18	8 00
8 20	0.12	0.13	0.13	0.14	0.15	0.15	0.16	0.16	0.17	0.17	8 20
8 40	0.12	0.12	0.13	0.13	0.14	0.15	0.15	0.16	0.16	0.17	8 40
9 00	0.11	0.12	0.12	0.13	0.14	0.14	0.15	0.15	0.16	0.16	9 00
9 30	0.11	0.11	0.12	0.13	0.13	0.13	0.14	0.14	0.15	0.15	9 30
10 00	0.10	0.11	0.11	0.12	0.12	0.13	0.13	0.14	0.14	0.14	10 00
10 30		0.10	0.11	0.11	0.12	0.12	0.12	0.13	0.13	0.14	10 30
11 00			0.10	0.11	0.11	0.11	0.12	0.12	0.13	0.13	11 00
11 30				0.10	0.11	0.11	0.11	0.12	0.12	0.13	11 30
12 00					0.10	0.10	0.11	0.11	0.12	0.12	12 00
12 30							0.10	0.11	0.11	0.12	12 30
13 00								0.10	0.11	0.11	13 00
13 30									0.10	0.11	13 30
14 00										0.10	14 00

TABLE 16

Distance by Vertical Angle
Measured Between Waterline at Object and Top of Object

Angle	Height of object above the sea, in **feet** and (meters)										Angle
	160 (48.8)	165 (50.3)	175 (53.3)	185 (56.4)	195 (59.4)	200 (61.0)	225 (68.6)	250 (76.2)	275 (83.8)	300 (91.4)	
° ′	*Miles*	*Miles*	*Miles*	*Miles*	*Miles*	*Miles*	*Miles*	*Miles*	*Miles*	*Miles*	° ′
0 15											0 15
0 20	4.53	4.67	4.95								0 20
0 25	3.62	3.73	3.96	4.19	4.41	4.53					0 25
0 30	3.02	3.11	3.30	3.49	3.68	3.77	4.24	4.71			0 30
0 35	2.59	2.67	2.83	2.99	3.15	3.23	3.64	4.04	4.45	4.85	0 35
0 40	2.26	2.33	2.48	2.62	2.76	2.83	3.18	3.54	3.89	4.24	0 40
0 45	2.01	2.07	2.20	2.33	2.45	2.51	2.83	3.14	3.46	3.77	0 45
0 50	1.81	1.87	1.98	2.09	2.21	2.26	2.55	2.83	3.11	3.39	0 50
0 55	1.65	1.70	1.80	1.90	2.01	2.06	2.31	2.57	2.83	3.09	0 55
1 00	1.51	1.56	1.65	1.74	1.84	1.89	2.12	2.36	2.59	2.83	1 00
1 10	1.20	1.33	1.41	1.50	1.58	1.62	1.82	2.02	2.22	2.42	1 10
1 20	1.13	1.17	1.24	1.31	1.38	1.41	1.59	1.77	1.94	2.12	1 20
1 30	1.01	1.04	1.10	1.16	1.23	1.26	1.41	1.57	1.73	1.89	1 30
1 40	0.91	0.93	0.99	1.05	1.10	1.13	1.27	1.41	1.56	1.70	1 40
1 50	0.82	0.85	0.90	0.95	1.00	1.03	1.16	1.29	1.41	1.54	1 50
2 00	0.75	0.78	0.82	0.87	0.92	0.94	1.06	1.18	1.30	1.41	2 00
2 15	0.67	0.69	0.73	0.77	0.82	0.84	0.94	1.05	1.15	1.26	2 15
2 30	0.60	0.62	0.66	0.70	0.74	0.75	0.85	0.94	1.04	1.13	2 30
2 45	0.55	0.57	0.60	0.63	0.67	0.69	0.77	0.86	0.94	1.03	2 45
3 00	0.50	0.52	0.55	0.58	0.61	0.63	0.71	0.79	0.86	0.94	3 00
3 20	0.45	0.47	0.49	0.52	0.55	0.57	0.64	0.71	0.78	0.85	3 20
3 40	0.41	0.42	0.45	0.48	0.50	0.51	0.58	0.64	0.71	0.77	3 40
4 00	0.38	0.39	0.41	0.44	0.46	0.47	0.53	0.59	0.65	0.71	4 00
4 20	0.35	0.36	0.38	0.40	0.42	0.43	0.49	0.54	0.60	0.65	4 20
4 40	0.32	0.33	0.35	0.37	0.39	0.40	0.45	0.50	0.55	0.60	4 40
5 00	0.30	0.31	0.33	0.35	0.37	0.38	0.42	0.47	0.52	0.56	5 00
5 20	0.28	0.29	0.31	0.33	0.34	0.35	0.40	0.44	0.48	0.53	5 20
5 40	0.27	0.27	0.29	0.31	0.32	0.33	0.37	0.41	0.46	0.50	5 40
6 00	0.25	0.26	0.27	0.29	0.31	0.31	0.35	0.39	0.43	0.47	6 00
6 20	0.24	0.24	0.26	0.27	0.29	0.30	0.33	0.37	0.41	0.44	6 20
6 40	0.23	0.23	0.25	0.26	0.27	0.28	0.32	0.35	0.39	0.42	6 40
7 00	0.21	0.22	0.23	0.25	0.26	0.27	0.30	0.34	0.37	0.40	7 00
7 20	0.20	0.21	0.22	0.24	0.25	0.26	0.29	0.32	0.35	0.38	7 20
7 40	0.20	0.20	0.21	0.23	0.24	0.24	0.28	0.31	0.34	0.37	7 40
8 00	0.19	0.19	0.20	0.22	0.23	0.23	0.26	0.29	0.32	0.35	8 00
8 20	0.18	0.19	0.20	0.21	0.22	0.22	0.25	0.28	0.31	0.34	8 20
8 40	0.17	0.19	0.19	0.20	0.21	0.22	0.24	0.27	0.30	0.32	8 40
9 00	0.17	0.17	0.18	0.19	0.20	0.21	0.23	0.26	0.29	0.31	9 00
9 30	0.16	0.16	0.17	0.18	0.19	0.20	0.22	0.25	0.27	0.30	9 30
10 00	0.15	0.15	0.16	0.17	0.18	0.19	0.21	0.23	0.26	0.28	10 00
10 30	0.14	0.15	0.16	0.16	0.17	0.18	0.20	0.22	0.24	0.27	10 30
11 00	0.14	0.14	0.15	0.16	0.17	0.17	0.19	0.21	0.23	0.25	11 00
11 30	0.13	0.13	0.14	0.15	0.16	0.16	0.18	0.20	0.22	0.24	11 30
12 00	0.12	0.13	0.14	0.14	0.15	0.15	0.17	0.19	0.21	0.23	12 00
12 30	0.12	0.12	0.13	0.14	0.14	0.15	0.17	0.19	0.20	0.22	12 30
13 00	0.11	0.11	0.12	0.13	0.14	0.14	0.16	0.18	0.20	0.21	13 00
13 30	0.11	0.11	0.12	0.13	0.13	0.14	0.15	0.17	0.19	0.21	13 30
14 00	0.11	0.11	0.12	0.12	0.13	0.13	0.15	0.17	0.18	0.20	14 00
14 30	0.10	0.10	0.11	0.12	0.12	0.13	0.14	0.16	0.18	0.19	14 30
15 00			0.11	0.11	0.12	0.12	0.14	0.15	0.17	0.18	15 00
16 00			0.10	0.11	0.11	0.11	0.13	0.14	0.16	0.17	16 00
17 00				0.10	0.10	0.11	0.12	0.13	0.15	0.16	17 00
18 00						0.10	0.11	0.13	0.14	0.15	18 00
19 00							0.11	0.12	0.13	0.14	19 00
20 00							0.10	0.11	0.12	0.14	20 00

TABLE 17
Distance by Vertical Angle
Measured Between Waterline at Object and Sea Horizon Beyond Object

Distance	Height of eye above the sea, in feet										Distance
	5	10	15	20	25	30	35	40	45	50	
Yards	° ′	° ′	° ′	° ′	° ′	° ′	° ′	° ′	° ′	° ′	*Yards*
100	0 55	1 52	2 48	3 45	4 41	5 37	6 34	7 30	8 26	9 21	100
200	27	0 54	1 22	1 50	2 18	2 46	3 15	3 43	4 11	4 39	200
300	17	35	0 54	1 12	1 31	1 49	2 08	2 27	2 45	3 04	300
400	12	26	39	0 53	1 07	1 21	1 35	1 49	2 02	2 16	400
500	9	20	31	42	0 53	1 04	1 15	1 26	1 37	1 48	500
600		16	25	34	43	0 52	1 01	1 10	1 20	1 29	600
700		13	21	29	36	44	0 52	0 59	1 07	1 15	700
800		11	18	24	31	38	45	51	0 58	1 05	800
900		10	16	21	27	33	39	45	51	0 57	900
1, 000			14	19	24	29	35	40	45	51	1, 000
1, 100			12	17	21	26	31	36	41	45	1, 100
1, 200			11	15	19	24	28	32	37	41	1, 200
1, 300			10	14	17	21	25	29	33	37	1, 300
1, 400				12	16	20	23	27	31	34	1, 400
1, 500				11	15	18	21	25	28	32	1, 500
1, 600				10	13	17	20	23	26	29	1, 600
1, 700					12	15	18	21	24	27	1, 700
1, 800					11	14	17	20	23	25	1, 800
1, 900					11	13	16	18	21	24	1, 900
2, 000					10	12	15	17	20	22	2, 000
2, 100						11	14	16	18	21	2, 100
2, 200						11	13	15	17	20	2, 200
2, 300						10	12	14	16	19	2, 300
2, 400							11	13	15	18	2, 400
2, 500							11	13	15	17	2, 500
2, 600							10	12	14	16	2, 600
2, 700								11	13	15	2, 700
2, 800								11	12	14	2, 800
2, 900								10	12	14	2, 900
3, 000									11	13	3, 000
3, 100									11	12	3, 100
3, 200									10	12	3, 200
3, 300										11	3, 300
3, 400										11	3, 400
3, 500										10	3, 500

TABLE 17
Distance by Vertical Angle
Measured Between Waterline at Object and Sea Horizon Beyond Object

Distance	Height of eye above the sea, in feet										Distance
	55	60	65	70	75	80	85	90	95	100	
Yards	° ′	° ′	° ′	° ′	° ′	° ′	° ′	° ′	° ′	° ′	*Yards*
100	10 16	11 11	12 06	13 00	13 54	14 48	15 41	16 34	17 26	18 17	100
200	5 07	5 35	6 03	6 31	6 59	7 27	7 55	8 23	8 51	9 18	200
300	3 23	3 41	4 00	4 19	4 38	4 56	5 15	5 34	5 52	6 11	300
400	2 30	2 44	2 58	3 12	3 26	3 40	3 54	4 08	4 22	4 36	400
500	1 59	2 10	2 21	2 32	2 43	2 55	3 06	3 17	3 28	3 39	500
600	1 38	1 47	1 56	2 06	2 15	2 24	2 33	2 43	2 52	3 01	600
700	1 23	1 31	1 39	1 47	1 54	2 02	2 10	2 18	2 26	2 34	700
800	1 12	1 19	1 25	1 32	1 39	1 46	1 53	2 00	2 07	2 14	800
900	1 03	1 09	1 15	1 21	1 27	1 33	1 39	1 46	1 52	1 58	900
1, 000	0 56	1 01	1 07	1 12	1 18	1 23	1 29	1 34	1 40	1 45	1, 000
1, 100	50	0 55	1 00	1 05	1 10	1 15	1 20	1 25	1 30	1 35	1, 100
1, 200	46	50	0 55	0 59	1 03	1 08	1 12	1 17	1 22	1 26	1, 200
1, 300	42	46	50	54	0 58	1 02	1 06	1 10	1 15	1 19	1, 300
1, 400	38	42	46	49	53	0 57	1 01	1 05	1 09	1 12	1, 400
1, 500	35	39	42	46	49	53	0 56	1 00	1 03	1 07	1, 500
1, 600	33	36	39	42	46	49	52	0 56	0 59	1 02	1, 600
1, 700	30	33	36	39	43	46	49	52	55	0 58	1, 700
1, 800	28	31	34	37	40	43	46	48	51	54	1, 800
1, 900	26	29	32	35	37	40	43	45	48	51	1, 900
2, 000	25	27	30	32	35	38	40	43	45	48	2, 000
2, 100	23	26	28	31	33	35	38	40	43	45	2, 100
2, 200	22	24	27	29	31	33	36	38	40	43	2, 200
2, 300	21	23	25	27	29	32	34	36	38	41	2, 300
2, 400	20	22	24	26	28	30	32	34	36	39	2, 400
2, 500	19	21	23	25	27	29	31	33	35	37	2, 500
2, 600	18	19	21	23	25	27	29	31	33	35	2, 600
2, 700	17	19	20	22	24	26	28	30	31	33	2, 700
2, 800	16	18	19	21	23	25	26	28	30	32	2, 800
2, 900	15	17	18	20	22	24	25	27	29	30	2, 900
3, 000	14	16	18	19	21	23	24	26	27	29	3, 000
3, 100	14	15	17	18	20	22	23	25	26	28	3, 100
3, 200	13	15	16	18	19	21	22	24	25	27	3, 200
3, 300	13	14	15	17	18	20	21	23	24	26	3, 300
3, 400	12	13	15	16	18	19	20	22	23	25	3, 400
3, 500	12	13	14	16	17	18	20	21	22	24	3, 500
3, 600	11	12	14	15	16	18	19	20	22	23	3, 600
3, 700	11	12	13	14	16	17	18	19	21	22	3, 700
3, 800	10	11	13	14	15	16	17	19	20	21	3, 800
3, 900		11	12	13	14	16	17	18	19	21	3, 900
4, 000		11	12	13	14	15	16	17	19	20	4, 000
4, 100		10	11	12	13	15	16	17	18	19	4, 100
4, 200			11	12	13	14	15	16	17	18	4, 200
4, 300			10	11	12	14	15	16	17	18	4, 300
4, 400			10	11	12	13	14	15	16	17	4, 400
4, 500			10	11	12	13	14	15	16	17	4, 500
4, 600				10	11	12	13	14	15	16	4, 600
4, 700					11	12	13	14	15	16	4, 700
4, 800					11	11	12	13	14	15	4, 800
4, 900					10	11	12	13	14	15	4, 900
5, 000					10	11	12	12	13	14	5, 000

TABLE 18
Distance of an Object by Two Bearings

Difference between the course and second bearing °	20°		22°		24°		26°		28°		30°		32°	
	Difference between the course and first bearing													
30	1.97	0.98												
32	1.64	0.87	2.16	1.14										
34	1.41	0.79	1.80	1.01	2.34	1.31								
36	1.24	0.73	1.55	0.91	1.96	1.15	2.52	1.48						
38	1.11	0.68	1.36	0.84	1.68	1.04	2.11	1.30	2.70	1.66				
40	1.00	0.64	1.21	0.78	1.48	0.95	1.81	1.16	2.26	1.45	2.88	1.85		
42	0.91	0.61	1.10	0.73	1.32	0.88	1.59	1.06	1.94	1.30	2.40	1.61	3.05	2.04
44	0.84	0.58	1.00	0.69	1.19	0.83	1.42	0.98	1.70	1.18	2.07	1.44	2.55	1.77
46	0.78	0.56	0.92	0.66	1.09	0.78	1.28	0.92	1.52	1.09	1.81	1.30	2.19	1.58
48	0.73	0.54	0.85	0.64	1.00	0.74	1.17	0.87	1.37	1.02	1.62	1.20	1.92	1.43
50	0.68	0.52	0.80	0.61	0.93	0.71	1.08	0.83	1.25	0.96	1.46	1.12	1.71	1.31
52	0.65	0.51	0.75	0.59	0.87	0.68	1.00	0.79	1.15	0.91	1.33	1.05	1.55	1.22
54	0.61	0.49	0.71	0.57	0.81	0.66	0.93	0.76	1.07	0.87	1.23	0.99	1.41	1.14
56	0.58	0.48	0.67	0.56	0.77	0.64	0.88	0.73	1.00	0.83	1.14	0.95	1.30	1.08
58	0.56	0.47	0.64	0.54	0.73	0.62	0.83	0.70	0.94	0.80	1.07	0.90	1.21	1.03
60	0.53	0.46	0.61	0.53	0.69	0.60	0.78	0.68	0.89	0.77	1.00	0.87	1.13	0.98
62	0.51	0.45	0.58	0.51	0.66	0.58	0.75	0.66	0.84	0.74	0.94	0.83	1.06	0.94
64	0.49	0.44	0.56	0.50	0.63	0.57	0.71	0.64	0.80	0.72	0.89	0.80	1.00	0.90
66	0.48	0.43	0.54	0.49	0.61	0.56	0.68	0.62	0.76	0.70	0.85	0.78	0.95	0.87
68	0.46	0.43	0.52	0.48	0.59	0.54	0.66	0.61	0.73	0.68	0.81	0.75	0.90	0.84
70	0.45	0.42	0.50	0.47	0.57	0.53	0.63	0.59	0.70	0.66	0.78	0.73	0.86	0.81
72	0.43	0.41	0.49	0.47	0.55	0.52	0.61	0.58	0.68	0.64	0.75	0.71	0.82	0.78
74	0.42	0.41	0.48	0.46	0.53	0.51	0.59	0.57	0.65	0.63	0.72	0.69	0.79	0.76
76	0.41	0.40	0.46	0.45	0.52	0.50	0.57	0.56	0.63	0.61	0.70	0.67	0.76	0.74
78	0.40	0.39	0.45	0.44	0.50	0.49	0.56	0.54	0.61	0.60	0.67	0.66	0.74	0.72
80	0.39	0.39	0.44	0.44	0.49	0.48	0.54	0.53	0.60	0.59	0.65	0.64	0.71	0.70
82	0.39	0.38	0.43	0.43	0.48	0.47	0.53	0.52	0.58	0.57	0.63	0.63	0.69	0.69
84	0.38	0.38	0.42	0.42	0.47	0.47	0.52	0.51	0.57	0.56	0.62	0.61	0.67	0.67
86	0.37	0.37	0.42	0.42	0.46	0.46	0.51	0.50	0.55	0.55	0.60	0.60	0.66	0.65
88	0.37	0.37	0.41	0.41	0.45	0.45	0.50	0.50	0.54	0.54	0.59	0.59	0.64	0.64
90	0.36	0.36	0.40	0.40	0.45	0.45	0.49	0.49	0.53	0.53	0.58	0.58	0.62	0.62
92	0.36	0.36	0.40	0.40	0.44	0.44	0.48	0.48	0.52	0.52	0.57	0.57	0.61	0.61
94	0.36	0.35	0.39	0.39	0.43	0.43	0.47	0.47	0.51	0.51	0.56	0.55	0.60	0.60
96	0.35	0.35	0.39	0.39	0.43	0.43	0.47	0.46	0.51	0.50	0.55	0.54	0.59	0.59
98	0.35	0.35	0.39	0.38	0.42	0.42	0.46	0.46	0.50	0.50	0.54	0.53	0.58	0.57
100	0.35	0.34	0.38	0.38	0.42	0.41	0.46	0.45	0.49	0.49	0.53	0.52	0.57	0.56
102	0.35	0.34	0.38	0.37	0.42	0.41	0.45	0.44	0.49	0.48	0.53	0.51	0.56	0.55
104	0.34	0.33	0.38	0.37	0.41	0.40	0.45	0.43	0.48	0.47	0.52	0.50	0.56	0.54
106	0.34	0.33	0.38	0.36	0.41	0.39	0.45	0.43	0.48	0.46	0.52	0.50	0.55	0.53
108	0.34	0.32	0.38	0.36	0.41	0.39	0.44	0.42	0.48	0.45	0.51	0.49	0.55	0.52
110	0.34	0.32	0.37	0.35	0.41	0.38	0.44	0.41	0.47	0.44	0.51	0.48	0.54	0.51
112	0.34	0.32	0.37	0.35	0.41	0.38	0.44	0.41	0.47	0.44	0.50	0.47	0.54	0.50
114	0.34	0.31	0.37	0.34	0.41	0.37	0.44	0.40	0.47	0.43	0.50	0.46	0.54	0.49
116	0.34	0.31	0.38	0.34	0.41	0.37	0.44	0.39	0.47	0.42	0.50	0.45	0.53	0.48
118	0.35	0.31	0.38	0.33	0.41	0.36	0.44	0.39	0.47	0.41	0.50	0.44	0.53	0.47
120	0.35	0.30	0.38	0.33	0.41	0.36	0.44	0.38	0.47	0.41	0.50	0.43	0.53	0.46
122	0.35	0.30	0.38	0.32	0.41	0.35	0.44	0.37	0.47	0.40	0.50	0.42	0.53	0.45
124	0.35	0.29	0.38	0.32	0.41	0.34	0.44	0.37	0.47	0.39	0.50	0.42	0.53	0.44
126	0.36	0.29	0.39	0.31	0.42	0.34	0.45	0.36	0.47	0.38	0.50	0.41	0.53	0.43
128	0.36	0.28	0.39	0.31	0.42	0.33	0.45	0.35	0.48	0.38	0.50	0.40	0.53	0.42
130	0.36	0.28	0.39	0.30	0.42	0.32	0.45	0.35	0.48	0.37	0.51	0.39	0.54	0.41
132	0.37	0.27	0.40	0.30	0.43	0.32	0.46	0.34	0.48	0.36	0.51	0.38	0.54	0.40
134	0.37	0.27	0.40	0.29	0.43	0.31	0.46	0.33	0.49	0.35	0.52	0.37	0.54	0.39
136	0.38	0.26	0.41	0.28	0.44	0.30	0.47	0.32	0.49	0.34	0.52	0.36	0.55	0.38
138	0.39	0.26	0.42	0.28	0.45	0.30	0.47	0.32	0.50	0.33	0.53	0.35	0.55	0.37
140	0.39	0.25	0.42	0.27	0.45	0.29	0.48	0.31	0.51	0.33	0.53	0.34	0.56	0.36
142	0.40	0.25	0.43	0.27	0.46	0.28	0.49	0.30	0.51	0.32	0.54	0.33	0.56	0.35
144	0.41	0.24	0.44	0.26	0.47	0.28	0.50	0.29	0.52	0.31	0.55	0.32	0.57	0.34
146	0.42	0.24	0.45	0.25	0.48	0.27	0.51	0.28	0.53	0.30	0.56	0.31	0.58	0.32
148	0.43	0.23	0.46	0.25	0.49	0.26	0.52	0.27	0.54	0.29	0.57	0.30	0.59	0.31
150	0.45	0.22	0.48	0.24	0.50	0.25	0.53	0.26	0.55	0.28	0.58	0.29	0.60	0.30
152	0.46	0.22	0.49	0.23	0.52	0.24	0.54	0.25	0.57	0.27	0.59	0.28	0.61	0.29
154	0.48	0.21	0.50	0.22	0.53	0.23	0.56	0.24	0.58	0.25	0.60	0.26	0.62	0.27
156	0.49	0.20	0.52	0.21	0.55	0.22	0.57	0.23	0.60	0.24	0.62	0.25	0.64	0.26
158	0.51	0.19	0.54	0.20	0.57	0.21	0.59	0.22	0.61	0.23	0.63	0.24	0.66	0.25
160	0.53	0.18	0.56	0.19	0.59	0.20	0.61	0.21	0.63	0.22	0.65	0.22	0.67	0.23

TABLE 18
Distance of an Object by Two Bearings

Difference between the course and second bearing °	34°		36°		38°		40°		42°		44°		46°	
	Difference between the course and first bearing													
44	3.22	2.24												
46	2.69	1.93	3.39	2.43										
48	2.31	1.72	2.83	2.10	3.55	2.63								
50	2.03	1.55	2.43	1.86	2.96	2.27	3.70	2.84						
52	1.81	1.43	2.13	1.68	2.54	2.01	3.09	2.44	3.85	3.04				
54	1.63	1.32	1.90	1.54	2.23	1.81	2.66	2.15	3.22	2.60	4.00	3.24		
56	1.49	1.24	1.72	1.42	1.99	1.65	2.33	1.93	2.77	2.29	3.34	2.77	4.14	3.43
58	1.37	1.17	1.57	1.33	1.80	1.53	2.08	1.76	2.43	2.06	2.87	2.44	3.46	2.93
60	1.28	1.10	1.45	1.25	1.64	1.42	1.88	1.63	2.17	1.88	2.52	2.18	2.97	2.57
62	1.19	1.05	1.34	1.18	1.51	1.34	1.72	1.52	1.96	1.73	2.25	1.98	2.61	2.30
64	1.12	1.01	1.25	1.13	1.40	1.26	1.58	1.42	1.79	1.61	2.03	1.83	2.33	2.09
66	1.06	0.96	1.18	1.07	1.31	1.20	1.47	1.34	1.65	1.51	1.85	1.69	2.10	1.92
68	1.00	0.93	1.11	1.03	1.23	1.14	1.37	1.27	1.53	1.42	1.71	1.58	1.92	1.78
70	0.95	0.89	1.05	0.99	1.16	1.09	1.29	1.21	1.43	1.34	1.58	1.49	1.77	1.66
72	0.91	0.86	1.00	0.95	1.10	1.05	1.21	1.15	1.34	1.27	1.48	1.41	1.64	1.56
74	0.87	0.84	0.95	0.92	1.05	1.01	1.15	1.10	1.26	1.21	1.39	1.34	1.53	1.47
76	0.84	0.81	0.91	0.89	1.00	0.97	1.09	1.06	1.20	1.16	1.31	1.27	1.44	1.40
78	0.80	0.79	0.88	0.86	0.96	0.94	1.04	1.02	1.14	1.11	1.24	1.22	1.36	1.33
80	0.78	0.77	0.85	0.83	0.92	0.91	1.00	0.98	1.09	1.07	1.18	1.16	1.28	1.27
82	0.75	0.75	0.82	0.81	0.89	0.88	0.96	0.95	1.04	1.03	1.13	1.12	1.22	1.21
84	0.73	0.73	0.79	0.79	0.86	0.85	0.93	0.92	1.00	0.99	1.08	1.07	1.17	1.16
86	0.71	0.71	0.77	0.77	0.83	0.83	0.89	0.89	0.96	0.96	1.04	1.04	1.12	1.12
88	0.69	0.69	0.75	0.75	0.80	0.80	0.86	0.86	0.93	0.93	1.00	1.00	1.08	1.07
90	0.67	0.67	0.73	0.73	0.78	0.78	0.84	0.84	0.90	0.90	0.97	0.97	1.04	1.04
92	0.66	0.66	0.71	0.71	0.76	0.76	0.82	0.82	0.87	0.87	0.93	0.93	1.00	1.00
94	0.65	0.64	0.69	0.69	0.74	0.74	0.79	0.79	0.85	0.85	0.91	0.90	0.97	0.97
96	0.63	0.63	0.68	0.67	0.73	0.72	0.78	0.77	0.83	0.82	0.88	0.88	0.94	0.93
98	0.62	0.62	0.67	0.66	0.71	0.70	0.76	0.75	0.81	0.80	0.86	0.85	0.91	0.90
100	0.61	0.60	0.65	0.64	0.70	0.69	0.74	0.73	0.79	0.78	0.84	0.83	0.89	0.88
102	0.60	0.59	0.64	0.63	0.68	0.67	0.73	0.71	0.77	0.76	0.82	0.80	0.87	0.85
104	0.60	0.58	0.63	0.61	0.67	0.65	0.72	0.69	0.76	0.74	0.80	0.78	0.85	0.82
106	0.59	0.57	0.63	0.60	0.66	0.64	0.70	0.68	0.74	0.72	0.79	0.76	0.83	0.80
108	0.58	0.55	0.62	0.59	0.66	0.62	0.69	0.66	0.73	0.70	0.77	0.74	0.81	0.77
110	0.58	0.54	0.61	0.57	0.65	0.61	0.68	0.64	0.72	0.68	0.76	0.71	0.80	0.75
112	0.57	0.53	0.61	0.56	0.64	0.59	0.68	0.63	0.71	0.66	0.75	0.69	0.79	0.73
114	0.57	0.52	0.60	0.55	0.63	0.58	0.67	0.61	0.70	0.64	0.74	0.68	0.78	0.71
116	0.56	0.51	0.60	0.54	0.63	0.57	0.66	0.60	0.70	0.63	0.73	0.66	0.77	0.69
118	0.56	0.50	0.59	0.52	0.63	0.55	0.66	0.58	0.69	0.61	0.72	0.64	0.76	0.67
120	0.56	0.49	0.59	0.51	0.62	0.54	0.65	0.57	0.68	0.59	0.72	0.62	0.75	0.65
122	0.56	0.47	0.59	0.50	0.62	0.53	0.65	0.55	0.68	0.58	0.71	0.60	0.74	0.63
124	0.56	0.46	0.59	0.49	0.62	0.51	0.65	0.54	0.68	0.56	0.71	0.58	0.74	0.61
126	0.56	0.45	0.59	0.48	0.62	0.50	0.64	0.52	0.67	0.54	0.70	0.57	0.73	0.59
128	0.56	0.44	0.59	0.46	0.62	0.49	0.64	0.51	0.67	0.53	0.70	0.55	0.73	0.57
130	0.56	0.43	0.59	0.45	0.62	0.47	0.64	0.49	0.67	0.51	0.70	0.53	0.72	0.55
132	0.56	0.42	0.59	0.44	0.62	0.46	0.64	0.48	0.67	0.50	0.70	0.52	0.72	0.54
134	0.57	0.41	0.59	0.43	0.62	0.45	0.64	0.46	0.67	0.48	0.69	0.50	0.72	0.52
136	0.57	0.40	0.60	0.41	0.62	0.43	0.65	0.45	0.67	0.47	0.70	0.48	0.72	0.50
138	0.58	0.39	0.60	0.40	0.63	0.42	0.65	0.43	0.67	0.45	0.70	0.47	0.72	0.48
140	0.58	0.37	0.61	0.39	0.63	0.40	0.65	0.42	0.68	0.43	0.70	0.45	0.72	0.46
142	0.59	0.36	0.61	0.38	0.63	0.39	0.66	0.41	0.68	0.42	0.70	0.43	0.72	0.45
144	0.60	0.35	0.62	0.36	0.64	0.38	0.66	0.39	0.68	0.40	0.71	0.41	0.73	0.43
146	0.60	0.34	0.63	0.35	0.65	0.36	0.67	0.37	0.69	0.39	0.71	0.40	0.73	0.41
148	0.61	0.32	0.63	0.34	0.66	0.35	0.68	0.36	0.70	0.37	0.72	0.38	0.74	0.39
150	0.62	0.31	0.64	0.32	0.66	0.33	0.68	0.34	0.70	0.35	0.72	0.36	0.74	0.37
152	0.63	0.30	0.65	0.31	0.67	0.32	0.69	0.33	0.71	0.33	0.73	0.34	0.75	0.35
154	0.65	0.28	0.67	0.29	0.68	0.30	0.70	0.31	0.72	0.32	0.74	0.32	0.76	0.33
156	0.66	0.27	0.68	0.28	0.70	0.28	0.72	0.29	0.73	0.30	0.75	0.30	0.77	0.31
158	0.67	0.25	0.69	0.26	0.71	0.27	0.73	0.27	0.74	0.28	0.76	0.28	0.78	0.29
160	0.69	0.24	0.71	0.24	0.73	0.25	0.74	0.25	0.76	0.26	0.77	0.26	0.79	0.27

TABLE 18

Distance of an Object by Two Bearings

Difference between the course and second bearing °	Difference between the course and first bearing 48°		50°		52°		54°		56°		58°		60°	
58	4.28	3.63												
60	3.57	3.10	4.41	3.82										
62	3.07	2.71	3.68	3.25	4.54	4.01								
64	2.70	2.42	3.17	2.85	3.79	3.41	4.66	4.19						
66	2.40	2.20	2.78	2.54	3.26	2.98	3.89	3.55	4.77	4.36				
68	2.17	2.01	2.48	2.30	2.86	2.65	3.34	3.10	3.99	3.71	4.88	4.53		
70	1.98	1.86	2.24	2.10	2.55	2.39	2.94	2.76	3.43	3.22	4.08	3.83	4.99	4.69
72	1.83	1.74	2.04	1.94	2.30	2.19	2.62	2.49	3.01	2.86	3.51	3.33	4.17	3.96
74	1.70	1.63	1.88	1.81	2.10	2.02	2.37	2.27	2.68	2.58	3.08	2.96	3.58	3.44
76	1.58	1.54	1.75	1.70	1.94	1.88	2.16	2.10	2.42	2.35	2.74	2.66	3.14	3.05
78	1.49	1.45	1.63	1.60	1.80	1.76	1.99	1.95	2.21	2.16	2.48	2.43	2.80	2.74
80	1.40	1.38	1.53	1.51	1.68	1.65	1.85	1.82	2.04	2.01	2.26	2.23	2.53	2.49
82	1.33	1.32	1.45	1.43	1.58	1.56	1.72	1.71	1.89	1.87	2.08	2.06	2.31	2.29
84	1.26	1.26	1.37	1.36	1.49	1.48	1.62	1.61	1.77	1.76	1.93	1.92	2.13	2.12
86	1.21	1.20	1.30	1.30	1.41	1.41	1.53	1.52	1.66	1.65	1.81	1.80	1.98	1.97
88	1.16	1.16	1.24	1.24	1.34	1.34	1.45	1.45	1.56	1.56	1.70	1.70	1.84	1.84
90	1.11	1.11	1.19	1.19	1.28	1.28	1.38	1.38	1.48	1.48	1.60	1.60	1.73	1.73
92	1.07	1.07	1.14	1.14	1.23	1.23	1.31	1.31	1.41	1.41	1.52	1.52	1.63	1.63
94	1.03	1.03	1.10	1.10	1.18	1.17	1.26	1.26	1.35	1.34	1.44	1.44	1.55	1.54
96	1.00	0.99	1.06	1.06	1.13	1.13	1.21	1.20	1.29	1.28	1.38	1.37	1.47	1.47
98	0.97	0.96	1.03	1.02	1.10	1.08	1.16	1.15	1.24	1.23	1.32	1.31	1.41	1.39
100	0.94	0.93	1.00	0.98	1.06	1.04	1.12	1.11	1.19	1.18	1.27	1.25	1.35	1.33
102	0.92	0.90	0.97	0.95	1.03	1.01	1.09	1.06	1.15	1.13	1.22	1.19	1.29	1.27
104	0.90	0.87	0.95	0.92	1.00	0.97	1.06	1.02	1.12	1.08	1.18	1.14	1.25	1.21
106	0.88	0.84	0.92	0.89	0.97	0.94	1.03	0.99	1.09	1.04	1.14	1.10	1.20	1.16
108	0.86	0.82	0.90	0.86	0.95	0.90	1.00	0.95	1.05	1.00	1.11	1.05	1.17	1.11
110	0.84	0.79	0.88	0.83	0.93	0.87	0.98	0.92	1.02	0.96	1.08	1.01	1.13	1.06
112	0.83	0.77	0.87	0.80	0.91	0.84	0.95	0.88	1.00	0.93	1.05	0.97	1.10	1.02
114	0.81	0.74	0.85	0.78	0.89	0.82	0.93	0.85	0.98	0.89	1.02	0.93	1.07	0.98
116	0.80	0.72	0.84	0.75	0.88	0.79	0.92	0.82	0.96	0.85	1.00	0.90	1.04	0.94
118	0.79	0.70	0.83	0.73	0.86	0.76	0.90	0.79	0.94	0.83	0.98	0.86	1.02	0.90
120	0.78	0.68	0.82	0.71	0.85	0.74	0.89	0.77	0.91	0.80	0.96	0.83	1.00	0.87
122	0.77	0.66	0.81	0.68	0.84	0.71	0.87	0.74	0.90	0.77	0.95	0.80	0.98	0.83
124	0.77	0.63	0.80	0.66	0.83	0.69	0.86	0.71	0.90	0.74	0.93	0.77	0.96	0.80
126	0.76	0.61	0.79	0.64	0.82	0.66	0.85	0.69	0.88	0.71	0.91	0.74	0.95	0.77
128	0.75	0.59	0.78	0.62	0.81	0.64	0.84	0.66	0.87	0.69	0.90	0.71	0.93	0.74
130	0.75	0.57	0.78	0.60	0.81	0.62	0.83	0.64	0.86	0.66	0.89	0.68	0.92	0.71
132	0.75	0.56	0.77	0.57	0.80	0.59	0.83	0.61	0.85	0.64	0.88	0.66	0.91	0.68
134	0.74	0.54	0.77	0.55	0.80	0.57	0.82	0.59	0.85	0.61	0.87	0.63	0.90	0.65
136	0.74	0.52	0.77	0.53	0.80	0.55	0.82	0.57	0.84	0.58	0.87	0.60	0.89	0.62
138	0.74	0.50	0.77	0.51	0.79	0.53	0.81	0.54	0.84	0.56	0.86	0.58	0.89	0.59
140	0.74	0.48	0.77	0.49	0.79	0.51	0.81	0.52	0.83	0.54	0.86	0.55	0.88	0.57
142	0.74	0.46	0.77	0.47	0.79	0.49	0.81	0.50	0.83	0.51	0.85	0.52	0.87	0.54
144	0.75	0.44	0.77	0.45	0.79	0.46	0.81	0.48	0.83	0.49	0.85	0.50	0.87	0.51
146	0.75	0.42	0.77	0.43	0.79	0.44	0.81	0.45	0.83	0.46	0.85	0.47	0.87	0.49
148	0.76	0.40	0.77	0.41	0.79	0.42	0.81	0.43	0.83	0.44	0.85	0.45	0.87	0.46
150	0.76	0.38	0.78	0.39	0.80	0.40	0.81	0.41	0.83	0.42	0.85	0.42	0.87	0.43
152	0.77	0.36	0.78	0.37	0.80	0.38	0.82	0.38	0.83	0.39	0.85	0.40	0.87	0.41
154	0.77	0.34	0.79	0.35	0.81	0.35	0.82	0.36	0.84	0.37	0.85	0.37	0.87	0.38
156	0.78	0.32	0.80	0.32	0.81	0.33	0.83	0.34	0.84	0.34	0.86	0.35	0.87	0.35
158	0.79	0.30	0.81	0.30	0.82	0.31	0.83	0.31	0.85	0.32	0.86	0.32	0.87	0.33
160	0.80	0.27	0.82	0.28	0.83	0.28	0.84	0.29	0.85	0.29	0.86	0.30	0.88	0.30

TABLE 18

Distance of an Object by Two Bearings

Difference between the course and second bearing °	Difference between the course and first bearing 62°		64°		66°		68°		70°		72°		74°		76°	
72	5.08	4.84														
74	4.25	4.08	5.18	4.98												
76	3.65	3.54	4.32	4.19	5.26	5.10										
78	3.20	3.13	3.72	3.63	4.39	4.30	5.34	5.22								
80	2.86	2.81	3.26	3.21	3.78	3.72	4.46	4.39	5.41	5.33						
82	2.58	2.56	2.91	2.88	3.31	3.28	3.83	3.80	4.52	4.48	5.48	5.42				
84	2.36	2.34	2.63	2.61	2.96	2.94	3.36	3.35	3.88	3.86	4.57	4.55	5.54	5.51		
86	2.17	2.17	2.40	2.39	2.67	2.66	3.00	2.99	3.41	3.40	3.93	3.92	4.62	4.61	5.59	5.57
88	2.01	2.01	2.21	2.21	2.44	2.44	2.71	2.71	3.04	3.04	3.45	3.45	3.97	3.97	4.67	4.66
90	1.88	1.88	2.05	2.05	2.25	2.25	2.48	2.48	2.75	2.75	3.08	3.08	3.49	3.49	4.01	4.01
92	1.77	1.76	1.91	1.91	2.08	2.08	2.28	2.28	2.51	2.51	2.78	2.78	3.11	3.11	3.52	3.52
94	1.67	1.66	1.80	1.79	1.95	1.94	2.12	2.11	2.31	2.30	2.54	2.53	2.81	2.80	3.14	3.13
96	1.58	1.57	1.70	1.69	1.83	1.82	1.97	1.96	2.14	2.13	2.34	2.33	2.57	2.55	2.84	2.82
98	1.50	1.49	1.61	1.59	1.72	1.71	1.85	1.84	2.00	1.98	2.17	2.15	2.36	2.34	2.59	2.56
100	1.43	1.41	1.53	1.51	1.63	1.61	1.75	1.72	1.88	1.85	2.03	2.00	2.19	2.16	2.39	2.35
102	1.37	1.34	1.46	1.43	1.55	1.52	1.66	1.62	1.77	1.73	1.90	1.86	2.05	2.00	2.21	2.16
104	1.32	1.28	1.40	1.36	1.48	1.44	1.58	1.53	1.68	1.63	1.79	1.74	1.92	1.87	2.07	2.01
106	1.27	1.22	1.34	1.29	1.42	1.37	1.51	1.45	1.60	1.54	1.70	1.63	1.81	1.74	1.94	1.87
108	1.23	1.17	1.29	1.23	1.37	1.30	1.44	1.37	1.53	1.45	1.62	1.54	1.72	1.63	1.83	1.74
110	1.19	1.12	1.25	1.17	1.32	1.24	1.39	1.30	1.46	1.37	1.54	1.45	1.64	1.54	1.74	1.63
112	1.15	1.07	1.21	1.12	1.27	1.18	1.33	1.24	1.40	1.30	1.48	1.37	1.56	1.45	1.65	1.53
114	1.12	1.02	1.17	1.07	1.23	1.12	1.29	1.18	1.35	1.24	1.42	1.30	1.50	1.37	1.58	1.44
116	1.09	0.98	1.14	1.03	1.19	1.07	1.25	1.12	1.31	1.17	1.37	1.23	1.44	1.29	1.51	1.36
118	1.07	0.94	1.11	0.98	1.16	1.02	1.21	1.07	1.26	1.12	1.32	1.17	1.38	1.22	1.45	1.28
120	1.04	0.90	1.08	0.94	1.13	0.98	1.18	1.02	1.23	1.06	1.28	1.11	1.34	1.16	1.40	1.21
122	1.02	0.86	1.06	0.90	1.10	0.93	1.15	0.97	1.19	1.01	1.24	1.05	1.29	1.10	1.35	1.14
124	1.00	0.83	1.04	0.86	1.08	0.89	1.12	0.93	1.16	0.96	1.21	1.00	1.25	1.04	1.31	1.08
126	0.98	0.79	1.02	0.82	1.05	0.85	1.09	0.88	1.13	0.92	1.18	0.95	1.22	0.99	1.27	1.02
128	0.97	0.76	1.00	0.79	1.03	0.82	1.07	0.84	1.11	0.87	1.15	0.90	1.19	0.94	1.23	0.97
130	0.95	0.73	0.98	0.75	1.02	0.78	1.05	0.80	1.09	0.83	1.12	0.86	1.16	0.89	1.20	0.92
132	0.94	0.70	0.97	0.72	1.00	0.74	1.03	0.77	1.06	0.79	1.10	0.82	1.13	0.84	1.17	0.87
134	0.93	0.67	0.96	0.69	0.99	0.71	1.01	0.73	1.04	0.75	1.08	0.77	1.11	0.80	1.14	0.82
136	0.92	0.64	0.95	0.66	0.97	0.68	1.00	0.69	1.03	0.71	1.06	0.74	1.09	0.76	1.12	0.78
138	0.91	0.61	0.94	0.63	0.96	0.64	0.99	0.66	1.01	0.68	1.04	0.70	1.07	0.72	1.10	0.74
140	0.90	0.58	0.93	0.60	0.95	0.61	0.97	0.63	1.00	0.64	1.03	0.66	1.05	0.68	1.08	0.69
142	0.90	0.55	0.92	0.57	0.94	0.58	0.96	0.59	0.99	0.61	1.01	0.62	1.04	0.64	1.06	0.65
144	0.89	0.52	0.91	0.54	0.93	0.55	0.96	0.56	0.98	0.57	1.00	0.59	1.02	0.60	1.05	0.62
146	0.89	0.50	0.91	0.51	0.93	0.52	0.95	0.53	0.97	0.54	0.99	0.55	1.01	0.57	1.03	0.58
148	0.89	0.47	0.90	0.48	0.92	0.49	0.94	0.50	0.96	0.51	0.98	0.52	1.00	0.53	1.02	0.54
150	0.88	0.44	0.90	0.45	0.92	0.46	0.94	0.47	0.95	0.48	0.97	0.49	0.99	0.50	1.01	0.50
152	0.88	0.41	0.90	0.42	0.92	0.43	0.93	0.44	0.95	0.45	0.97	0.45	0.98	0.46	1.00	0.47
154	0.88	0.39	0.90	0.39	0.91	0.40	0.93	0.41	0.94	0.41	0.96	0.42	0.98	0.43	0.99	0.43
156	0.89	0.36	0.90	0.37	0.91	0.37	0.93	0.38	0.94	0.38	0.96	0.39	0.97	0.39	0.99	0.40
158	0.89	0.33	0.90	0.34	0.91	0.34	0.93	0.35	0.94	0.35	0.95	0.36	0.97	0.36	0.98	0.37
160	0.89	0.30	0.90	0.31	0.91	0.31	0.93	0.32	0.94	0.32	0.95	0.33	0.96	0.33	0.98	0.33

TABLE 18

Distance of an Object by Two Bearings

Difference between the course and second bearing	Difference between the course and first bearing															
	78°		80°		82°		84°		86°		88°		90°		92°	
°																
88	5.63	5.63														
90	4.70	4.70	5.67	5.67												
92	4.04	4.04	4.74	4.73	5.70	5.70										
94	3.55	3.54	4.07	4.06	4.76	4.75	5.73	5.71								
96	3.17	3.15	3.57	3.55	4.09	4.07	4.78	4.76	5.74	5.71						
98	2.86	2.83	3.19	3.16	3.59	3.56	4.11	4.07	4.80	4.75	5.76	5.70				
100	2.61	2.57	2.88	2.84	3.20	3.16	3.61	3.55	4.12	4.06	4.81	4.73	5.76	5.67		
102	2.40	2.35	2.63	2.57	2.90	2.83	3.22	3.15	3.62	3.54	4.13	4.04	4.81	4.70	5.76	5.63
104	2.23	2.16	2.42	2.35	2.64	2.56	2.91	2.82	3.23	3.13	3.63	3.52	4.13	4.01	4.81	4.66
106	2.08	2.00	2.25	2.16	2.43	2.34	2.65	2.55	2.92	2.80	3.23	3.11	3.63	3.49	4.13	3.97
108	1.96	1.86	2.10	2.00	2.26	2.15	2.45	2.33	2.66	2.53	2.92	2.78	3.24	3.08	3.63	3.45
110	1.85	1.73	1.97	1.85	2.11	1.98	2.27	2.13	2.45	2.31	2.67	2.51	2.92	2.75	3.23	3.04
112	1.75	1.62	1.86	1.72	1.98	1.83	2.12	1.96	2.28	2.11	2.46	2.28	2.67	2.48	2.92	2.71
114	1.66	1.52	1.76	1.61	1.87	1.71	1.99	1.82	2.12	1.94	2.28	2.08	2.46	2.25	2.67	2.44
116	1.59	1.43	1.68	1.51	1.77	1.59	1.88	1.69	2.00	1.79	2.13	1.91	2.28	2.05	2.46	2.21
118	1.52	1.34	1.60	1.41	1.68	1.49	1.78	1.57	1.88	1.66	2.00	1.76	2.13	1.88	2.28	2.01
120	1.46	1.27	1.53	1.33	1.61	1.39	1.69	1.47	1.78	1.54	1.89	1.63	2.00	1.73	2.13	1.84
122	1.41	1.19	1.47	1.25	1.54	1.31	1.62	1.37	1.70	1.44	1.79	1.52	1.89	1.60	2.00	1.70
124	1.36	1.13	1.42	1.18	1.48	1.23	1.55	1.28	1.62	1.34	1.70	1.41	1.79	1.48	1.89	1.56
126	1.32	1.06	1.37	1.11	1.43	1.15	1.48	1.20	1.55	1.26	1.62	1.31	1.70	1.38	1.79	1.45
128	1.28	1.01	1.33	1.04	1.38	1.08	1.43	1.13	1.49	1.17	1.55	1.23	1.62	1.28	1.70	1.34
130	1.24	0.95	1.29	0.98	1.33	1.02	1.38	1.06	1.44	1.10	1.49	1.14	1.56	1.19	1.62	1.24
132	1.21	0.90	1.25	0.93	1.29	0.96	1.34	0.99	1.39	1.03	1.44	1.07	1.49	1.11	1.55	1.16
134	1.18	0.85	1.22	0.88	1.26	0.90	1.30	0.93	1.34	0.97	1.39	1.00	1.44	1.04	1.49	1.07
136	1.15	0.80	1.19	0.83	1.22	0.85	1.26	0.88	1.30	0.90	1.34	0.93	1.39	0.97	1.44	1.00
138	1.13	0.76	1.16	0.78	1.19	0.80	1.23	0.82	1.27	0.85	1.30	0.87	1.35	0.90	1.39	0.93
140	1.11	0.71	1.14	0.73	1.17	0.75	1.20	0.77	1.23	0.79	1.27	0.82	1.31	0.84	1.34	0.86
142	1.09	0.67	1.12	0.69	1.14	0.70	1.17	0.72	1.20	0.74	1.24	0.76	1.27	0.78	1.30	0.80
144	1.07	0.63	1.10	0.64	1.12	0.66	1.15	0.67	1.18	0.69	1.21	0.71	1.24	0.73	1.27	0.75
146	1.05	0.59	1.08	0.60	1.10	0.62	1.13	0.63	1.15	0.64	1.18	0.66	1.21	0.67	1.24	0.69
148	1.04	0.55	1.06	0.56	1.08	0.57	1.11	0.59	1.13	0.60	1.15	0.61	1.18	0.62	1.21	0.64
150	1.03	0.51	1.05	0.52	1.07	0.53	1.09	0.54	1.11	0.55	1.13	0.57	1.15	0.58	1.18	0.59
152	1.02	0.48	1.04	0.49	1.05	0.49	1.07	0.50	1.09	0.51	1.11	0.52	1.13	0.53	1.15	0.54
154	1.01	0.44	1.02	0.45	1.04	0.46	1.06	0.46	1.08	0.47	1.09	0.48	1.11	0.49	1.13	0.50
156	1.00	0.41	1.01	0.41	1.03	0.42	1.05	0.43	1.06	0.43	1.08	0.44	1.09	0.45	1.11	0.45
158	0.99	0.37	1.01	0.38	1.02	0.38	1.03	0.39	1.05	0.39	1.06	0.40	1.08	0.40	1.09	0.41
160	0.99	0.34	1.00	0.34	1.01	0.35	1.02	0.35	1.04	0.35	1.05	0.36	1.06	0.36	1.08	0.37
	94°		96°		98°		100°		102°		104°		106°		108°	
°																
104	5.74	5.57														
106	4.80	4.61	5.73	5.51												
108	4.12	3.92	4.78	4.55	5.70	5.42										
110	3.62	3.40	4.11	3.86	4.76	4.48	5.67	5.33								
112	3.23	2.99	3.61	3.35	4.09	3.80	4.74	4.40	5.63	5.22						
114	2.92	2.66	3.22	2.94	3.59	3.28	4.07	3.72	4.70	4.30	5.59	5.10				
116	2.66	2.39	2.91	2.61	3.20	2.88	3.57	3.21	4.04	3.63	4.67	4.19	5.54	4.98		
118	2.45	2.17	2.65	2.34	2.90	2.56	3.19	2.81	3.55	3.13	4.01	3.54	4.62	4.08	5.48	4.84
120	2.28	1.97	2.45	2.12	2.64	2.29	2.88	2.49	3.17	2.74	3.52	3.05	3.97	3.44	4.57	3.96
122	2.12	1.80	2.27	1.92	2.43	2.06	2.63	2.23	2.86	2.43	3.14	2.66	3.49	2.96	3.93	3.33
124	2.00	1.65	2.12	1.76	2.26	1.87	2.42	2.01	2.61	2.16	2.84	2.35	3.11	2.58	3.45	2.86
126	1.88	1.52	1.99	1.61	2.11	1.71	2.25	1.82	2.40	1.95	2.59	2.10	2.81	2.27	3.08	2.49
128	1.78	1.41	1.88	1.48	1.98	1.56	2.10	1.65	2.23	1.76	2.39	1.88	2.57	2.02	2.78	2.19
130	1.70	1.30	1.78	1.36	1.87	1.43	1.97	1.51	2.08	1.60	2.21	1.70	2.36	1.81	2.54	1.94
132	1.62	1.20	1.69	1.26	1.77	1.32	1.86	1.38	1.96	1.45	2.07	1.54	2.19	1.63	2.34	1.74
134	1.55	1.12	1.62	1.16	1.68	1.21	1.76	1.27	1.85	1.33	1.94	1.40	2.05	1.47	2.17	1.56
136	1.49	1.04	1.55	1.07	1.61	1.12	1.68	1.16	1.75	1.22	1.83	1.27	1.92	1.34	2.03	1.41
138	1.44	0.96	1.49	0.99	1.54	1.03	1.60	1.07	1.66	1.11	1.74	1.16	1.81	1.21	1.90	1.27
140	1.39	0.89	1.43	0.92	1.48	0.95	1.53	0.98	1.59	1.02	1.65	1.06	1.72	1.10	1.79	1.15
142	1.34	0.83	1.38	0.85	1.43	0.88	1.47	0.91	1.52	0.94	1.58	0.97	1.64	1.01	1.70	1.05
144	1.30	0.77	1.34	0.79	1.38	0.81	1.42	0.83	1.46	0.86	1.51	0.89	1.56	0.92	1.62	0.95
146	1.27	0.71	1.30	0.73	1.33	0.75	1.37	0.77	1.41	0.79	1.45	0.81	1.50	0.84	1.54	0.86
148	1.23	0.65	1.26	0.67	1.29	0.69	1.33	0.70	1.36	0.72	1.40	0.74	1.44	0.76	1.48	0.78
150	1.20	0.60	1.23	0.61	1.26	0.63	1.29	0.64	1.32	0.66	1.35	0.67	1.38	0.69	1.42	0.71
152	1.18	0.55	1.20	0.56	1.22	0.57	1.25	0.59	1.28	0.60	1.31	0.61	1.34	0.63	1.37	0.64
154	1.15	0.50	1.17	0.51	1.19	0.52	1.22	0.53	1.24	0.54	1.27	0.56	1.29	0.57	1.32	0.58
156	1.13	0.46	1.15	0.47	1.17	0.47	1.19	0.48	1.21	0.49	1.23	0.50	1.25	0.51	1.28	0.52
158	1.11	0.42	1.13	0.42	1.14	0.43	1.16	0.44	1.18	0.44	1.20	0.45	1.22	0.46	1.24	0.47
160	1.09	0.37	1.11	0.38	1.12	0.38	1.14	0.39	1.15	0.39	1.17	0.40	1.19	0.41	1.21	0.41

TABLE 18

Distance of an Object by Two Bearings

Difference between the course and second bearing	Difference between the course and first bearing													
	110°		112°		114°		116°		118°		120°		122°	
°														
120	5.41	4.69												
122	4.52	3.83	5.34	4.53										
124	3.88	3.22	4.46	3.70	5.26	4.36								
126	3.41	2.76	3.83	3.10	4.39	3.55	5.18	4.19						
128	3.04	2.40	3.36	2.65	3.78	2.98	4.32	3.41	5.08	4.01				
130	2.75	2.10	3.00	2.30	3.31	2.54	3.72	2.85	4.25	3.25	4.99	3.82		
132	2.51	1.86	2.71	2.01	2.96	2.20	3.26	2.42	3.65	2.71	4.17	3.10	4.88	3.63
134	2.31	1.66	2.48	1.78	2.67	1.92	2.91	2.09	3.20	2.30	3.58	2.57	4.08	2.93
136	2.14	1.49	2.28	1.58	2.44	1.69	2.63	1.83	2.86	1.98	3.14	2.18	3.51	2.44
138	2.00	1.34	2.12	1.42	2.25	1.50	2.40	1.61	2.58	1.73	2.80	1.88	3.08	2.06
140	1.88	1.21	1.97	1.27	2.08	1.34	2.21	1.42	2.36	1.52	2.53	1.63	2.74	1.76
142	1.77	1.09	1.85	1.14	1.95	1.20	2.05	1.26	2.17	1.34	2.31	1.42	2.48	1.53
144	1.68	0.99	1.75	1.03	1.83	1.07	1.91	1.13	2.01	1.18	2.13	1.25	2.26	1.33
146	1.60	0.89	1.66	0.93	1.72	0.96	1.80	1.01	1.88	1.05	1.98	1.10	2.08	1.17
148	1.53	0.81	1.58	0.84	1.63	0.87	1.70	0.90	1.77	0.94	1.84	0.98	1.93	1.03
150	1.46	0.73	1.51	0.75	1.55	0.78	1.61	0.80	1.67	0.83	1.73	0.87	1.81	0.90
152	1.40	0.66	1.44	0.68	1.48	0.70	1.53	0.72	1.58	0.74	1.63	0.77	1.70	0.80
154	1.35	0.59	1.39	0.61	1.42	0.62	1.46	0.64	1.50	0.66	1.55	0.68	1.60	0.70
156	1.31	0.53	1.33	0.54	1.37	0.56	1.40	0.57	1.43	0.58	1.47	0.60	1.52	0.62
158	1.26	0.47	1.29	0.48	1.32	0.49	1.34	0.50	1.37	0.51	1.41	0.53	1.44	0.54
160	1.23	0.42	1.25	0.43	1.27	0.43	1.29	0.44	1.32	0.45	1.35	0.46	1.38	0.47
	124°		126°		128°		130°		132°		134°		136°	
134	4.77	3.43												
136	3.99	2.77	4.66	3.24										
138	3.43	2.29	3.89	2.60	4.54	3.04								
140	3.01	1.93	3.34	2.15	3.79	2.44	4.41	2.84						
142	2.68	1.65	2.94	1.81	3.26	2.01	3.68	2.27	4.28	2.63				
144	2.42	1.42	2.62	1.54	2.86	1.68	3.17	1.86	3.57	2.10	4.14	2.43		
146	2.21	1.24	2.37	1.32	2.55	1.43	2.78	1.55	3.07	1.72	3.46	1.93	4.00	2.24
148	2.04	1.08	2.16	1.14	2.30	1.22	2.48	1.31	2.70	1.43	2.97	1.58	3.34	1.77
150	1.89	0.95	1.99	0.99	2.10	1.05	2.24	1.12	2.40	1.20	2.61	1.30	2.87	1.44
152	1.77	0.83	1.85	0.87	1.94	0.91	2.04	0.96	2.17	1.02	2.33	1.09	2.52	1.18
154	1.66	0.73	1.72	0.76	1.80	0.79	1.88	0.83	1.98	0.87	2.10	0.92	2.25	0.99
156	1.56	0.64	1.62	0.66	1.68	0.68	1.75	0.71	1.83	0.74	1.92	0.78	2.03	0.83
158	1.48	0.56	1.53	0.57	1.58	0.59	1.63	0.61	1.70	0.64	1.77	0.66	1.85	0.69
160	1.41	0.48	1.45	0.49	1.49	0.51	1.53	0.52	1.58	0.54	1.64	0.56	1.71	0.58
	138°		140°		142°		144°		146°		148°		150°	
148	3.85	2.04												
150	3.22	1.61	3.70	1.85										
152	2.77	1.30	3.09	1.45	3.55	1.66								
154	2.43	1.06	2.66	1.16	2.96	1.30	3.38	1.48						
156	2.17	0.88	2.33	0.95	2.54	1.04	2.83	1.15	3.22	1.31				
158	1.96	0.73	2.08	0.78	2.23	0.84	2.43	0.91	2.69	1.01	3.05	1.14		
160	1.79	0.61	1.88	0.64	1.99	0.68	2.13	0.73	2.31	0.79	2.55	0.87	2.88	0.98

TABLE 19
Table of Offsets

	DISTANCE ALONG POSITION LINE FROM INTERCEPT										
	00′	05′	10′	15′	20′	25′	30′	35′	40′	45′	
ALT.	**OFFSETS**										***ALT.***
0°	0′.0	0′.0	0′.0	0′.0	0′.0	0′.0	0′.0	0′.0	0′.0	0′.0	0°
30	0. 0	0. 0	0. 0	0. 0	0. 0	0. 1	0. 1	0. 1	0. 1	0. 2	30
40	0. 0	0. 0	0. 0	0. 0	0. 1	0. 1	0. 1	0. 2	0. 2	0. 3	40
50	0. 0	0. 0	0. 0	0. 0	0. 1	0. 1	0. 2	0. 2	0. 3	0. 3	50
55	0. 0	0. 0	0. 0	0. 0	0. 1	0. 1	0. 2	0. 3	0. 3	0. 4	55
60	0. 0	0. 0	0. 0	0. 1	0. 1	0. 2	0. 2	0. 3	0. 4	0. 5	60
62	0. 0	0. 0	0. 0	0. 1	0. 1	0. 2	0. 2	0. 3	0. 4	0. 5	62
64	0. 0	0. 0	0. 0	0. 1	0. 1	0. 2	0. 3	0. 4	0. 5	0. 6	64
66	0. 0	0. 0	0. 0	0. 1	0. 1	0. 2	0. 3	0. 4	0. 5	0. 7	66
68	0. 0	0. 0	0. 0	0. 1	0. 1	0. 2	0. 3	0. 4	0. 6	0. 7	68
70	0. 0	0. 0	0. 0	0. 1	0. 2	0. 2	0. 4	0. 5	0. 6	0. 8	70
71	0. 0	0. 0	0. 0	0. 1	0. 2	0. 3	0. 4	0. 5	0. 7	0. 9	71
72	0. 0	0. 0	0. 0	0. 1	0. 2	0. 3	0. 4	0. 5	0. 7	0. 9	72
73	0. 0	0. 0	0. 0	0. 1	0. 2	0. 3	0. 4	0. 6	0. 8	1. 0	73
74	0. 0	0. 0	0. 1	0. 1	0. 2	0. 3	0. 5	0. 6	0. 8	1. 0	74
75	0. 0	0. 0	0. 1	0. 1	0. 2	0. 3	0. 5	0. 7	0. 9	1. 1	75
76	0. 0	0. 0	0. 1	0. 1	0. 2	0. 4	0. 5	0. 7	0. 9	1. 2	76
77	0. 0	0. 0	0. 1	0. 1	0. 3	0. 4	0. 6	0. 8	1. 0	1. 3	77
78	0. 0	0. 0	0. 1	0. 2	0. 3	0. 4	0. 6	0. 8	1. 1	1. 4	78
79	0. 0	0. 0	0. 1	0. 2	0. 3	0. 5	0. 7	0. 9	1. 2	1. 5	79
80 . 0	0. 0	0. 0	0. 1	0. 2	0. 3	0. 5	0. 7	1. 0	1. 3	1. 7	80 . 0
80 . 5	0. 0	0. 0	0. 1	0. 2	0. 3	0. 5	0. 8	1. 1	1. 4	1. 8	80 . 5
81 . 0	0. 0	0. 0	0. 1	0. 2	0. 4	0. 6	0. 8	1. 1	1. 5	1. 9	81 . 0
81 . 5	0. 0	0. 0	0. 1	0. 2	0. 4	0. 6	0. 9	1. 2	1. 6	2. 0	81 . 5
82 . 0	0. 0	0. 0	0. 1	0. 2	0. 4	0. 6	0. 9	1. 3	1. 7	2. 1	82 . 0
82 . 5	0. 0	0. 0	0. 1	0. 2	0. 4	0. 7	1. 0	1. 4	1. 8	2. 2	82 . 5
83 . 0	0. 0	0. 0	0. 1	0. 3	0. 5	0. 7	1. 1	1. 5	1. 9	2. 4	83 . 0
83 . 5	0. 0	0. 0	0. 1	0. 3	0. 5	0. 8	1. 2	1. 6	2. 0	2. 6	83 . 5
84 . 0	0. 0	0. 0	0. 1	0. 3	0. 5	0. 9	1. 2	1. 7	2. 2	2. 8	84 . 0
84 . 5	0. 0	0. 0	0. 2	0. 3	0. 6	1. 0	1. 4	1. 9	2. 4	3. 1	84 . 5
85 . 0	0. 0	0. 0	0. 2	0. 4	0. 7	1. 0	1. 5	2. 1	2. 7	3. 4	85 . 0
85 . 5	0. 0	0. 0	0. 2	0. 4	0. 7	1. 2	1. 7	2. 3	3. 0	3. 8	85 . 5
86 . 0	0. 0	0. 1	0. 2	0. 5	0. 8	1. 3	1. 9	2. 6	3. 4	4. 3	86 . 0
86 . 5	0. 0	0. 1	0. 2	0. 5	1. 0	1. 5	2. 2	2. 9	3. 8	4. 9	86 . 5
87 . 0	0. 0	0. 1	0. 3	0. 6	1. 1	1. 7	2. 5	3. 4	4. 5	5. 7	87 . 0
87 . 5	0. 0	0. 1	0. 3	0. 8	1. 3	2. 1	3. 0	4. 1	5. 4	6. 9	87 . 5
88 . 0	0. 0	0. 1	0. 4	0. 9	1. 7	2. 7	3. 8	5. 2	6. 9	8. 8	88 . 0
88 . 5	0. 0	0. 2	0. 6	1. 3	2. 3	3. 5	5. 1	7. 1	9. 4	12. 1	88 . 5
89 . 0	0. 0	0. 3	0. 8	1. 9	3. 4	5. 5	8. 0	11. 3	15. 3	20. 3	89 . 0

TABLE 20

Meridian Angle and Altitude of a Body on the Prime Vertical Circle

Latitude	Declination (same name as Latitude)												Latitude
	0°		1°		2°		3°		4°		5°		
	t	Alt.	t	Alt.	t	Alt.	t	Alt.	t	Alt.	t	Alt.	
°	°	°	°	°	°	°	°	°	°	°	°	°	°
0	—	—	90.0	0.0	90.0	0.0	90.0	0.0	90.0	0.0	90.0	0.0	0
1	90.0	0.0	0.0	90.0	60.0	30.0	70.5	19.5	75.5	14.5	78.5	11.6	1
2	90.0	0.0	60.0	30.0	0.0	90.0	48.2	41.8	60.0	30.0	66.5	23.6	2
3	90.0	0.0	70.5	19.5	48.2	41.8	0.0	90.0	41.5	48.6	53.2	36.9	3
4	90.0	0.0	75.5	14.5	60.0	30.0	41.5	48.6	0.0	90.0	36.9	53.2	4
5	90.0	0.0	78.5	11.6	66.5	23.6	53.2	36.9	36.9	53.2	0.0	90.0	5
6	90.0	0.0	80.4	9.6	70.6	19.5	60.1	30.0	48.3	41.9	33.7	56.5	6
7	90.0	0.0	81.8	8.2	73.5	16.6	64.7	25.4	55.3	34.9	44.6	45.7	7
8	90.0	0.0	82.9	7.2	75.6	14.5	68.1	22.1	60.2	30.1	51.5	38.8	8
9	90.0	0.0	83.7	6.4	77.3	12.9	70.7	19.5	63.8	26.5	56.5	33.9	9
10	90.0	0.0	84.3	5.8	78.6	11.6	72.7	17.5	66.6	23.7	60.3	30.1	10
11	90.0	0.0	84.8	5.2	79.7	10.5	74.4	15.9	68.9	21.4	63.3	27.2	11
12	90.0	0.0	85.3	4.8	80.5	9.7	75.7	14.6	70.8	19.6	65.7	24.8	12
13	90.0	0.0	85.7	4.4	81.3	8.9	76.9	13.5	72.4	18.1	67.7	22.8	13
14	90.0	0.0	86.0	4.1	81.9	8.3	77.9	12.5	73.7	16.8	69.5	21.1	14
15	90.0	0.0	86.3	3.9	82.5	7.7	78.7	11.7	74.9	15.6	70.9	19.7	15
16	90.0	0.0	86.5	3.6	83.0	7.3	79.5	10.9	75.9	14.7	72.2	18.4	16
17	90.0	0.0	86.7	3.4	83.4	6.9	80.1	10.3	76.8	13.8	73.4	17.3	17
18	90.0	0.0	86.9	3.2	83.8	6.5	80.7	9.8	77.6	13.0	74.4	16.4	18
19	90.0	0.0	87.1	3.1	84.2	6.2	81.2	9.3	78.3	12.4	75.3	15.5	19
20	90.0	0.0	87.3	2.9	84.5	5.9	81.7	8.8	78.9	11.8	76.1	14.8	20
21	90.0	0.0	87.4	2.8	84.8	5.6	82.2	8.4	79.5	11.2	76.8	14.1	21
22	90.0	0.0	87.5	2.7	85.0	5.3	82.5	8.0	80.0	10.7	77.5	13.5	22
23	90.0	0.0	87.6	2.6	85.3	5.1	82.9	7.7	80.5	10.3	78.1	12.9	23
24	90.0	0.0	87.8	2.5	85.5	4.9	83.2	7.4	81.0	9.9	78.7	12.4	24
25	90.0	0.0	87.9	2.4	85.7	4.7	83.5	7.1	81.4	9.5	79.2	11.9	25
26	90.0	0.0	87.9	2.3	85.9	4.6	83.8	6.9	81.8	9.2	79.7	11.5	26
27	90.0	0.0	88.0	2.2	86.1	4.4	84.1	6.6	82.1	8.8	80.1	11.1	27
28	90.0	0.0	88.1	2.1	86.2	4.3	84.3	6.4	82.4	8.5	80.5	10.7	28
29	90.0	0.0	88.2	2.1	86.4	4.1	84.6	6.2	82.8	8.3	80.9	10.4	29
30	90.0	0.0	88.3	2.0	86.5	4.0	84.8	6.0	83.0	8.0	81.3	10.0	30
31	90.0	0.0	88.3	1.9	86.7	3.9	85.0	5.8	83.3	7.8	81.6	9.7	31
32	90.0	0.0	88.4	1.9	86.8	3.8	85.2	5.7	83.6	7.6	82.0	9.5	32
33	90.0	0.0	88.5	1.8	86.9	3.7	85.4	5.5	83.8	7.4	82.3	9.2	33
34	90.0	0.0	88.5	1.8	87.0	3.6	85.5	5.4	84.0	7.2	82.5	9.0	34
35	90.0	0.0	88.6	1.7	87.1	3.5	85.7	5.2	84.3	7.0	82.8	8.7	35
36	90.0	0.0	88.6	1.7	87.2	3.4	85.9	5.1	84.5	6.8	83.1	8.5	36
37	90.0	0.0	88.7	1.7	87.3	3.3	86.0	5.0	84.7	6.7	83.3	8.3	37
38	90.0	0.0	88.7	1.6	87.4	3.2	86.2	4.9	84.9	6.5	83.6	8.1	38
39	90.0	0.0	88.8	1.6	87.5	3.2	86.3	4.8	85.0	6.4	83.8	8.0	39
40	90.0	0.0	88.8	1.6	87.6	3.1	86.4	4.7	85.2	6.2	84.0	7.8	40
41	90.0	0.0	88.8	1.5	87.7	3.0	86.5	4.6	85.4	6.1	84.2	7.6	41
42	90.0	0.0	88.9	1.5	87.8	3.0	86.7	4.5	85.5	6.0	84.4	7.5	42
43	90.0	0.0	88.9	1.5	87.9	2.9	86.8	4.4	85.7	5.9	84.6	7.3	43
44	90.0	0.0	89.0	1.4	87.9	2.9	86.9	4.3	85.8	5.8	84.8	7.2	44
45	90.0	0.0	89.0	1.4	88.0	2.8	87.0	4.2	86.0	5.7	85.0	7.1	45
46	90.0	0.0	89.0	1.4	88.1	2.8	87.1	4.2	86.1	5.6	85.2	7.0	46
47	90.0	0.0	89.1	1.4	88.1	2.7	87.2	4.1	86.3	5.5	85.3	6.8	47
48	90.0	0.0	89.1	1.3	88.2	2.7	87.3	4.0	86.4	5.4	85.5	6.7	48
49	90.0	0.0	89.1	1.3	88.3	2.7	87.4	4.0	86.5	5.3	85.6	6.6	49
50	90.0	0.0	89.2	1.3	88.3	2.6	87.5	3.9	86.6	5.2	85.8	6.5	50
52	90.0	0.0	89.2	1.3	88.4	2.5	87.7	3.8	86.9	5.1	86.1	6.4	52
54	90.0	0.0	89.3	1.2	88.5	2.5	87.8	3.7	87.1	4.9	86.4	6.2	54
56	90.0	0.0	89.3	1.2	88.7	2.4	88.0	3.6	87.3	4.8	86.6	6.0	56
58	90.0	0.0	89.4	1.2	88.7	2.4	88.1	3.5	87.5	4.7	86.9	5.9	58
60	90.0	0.0	89.4	1.2	88.8	2.3	88.3	3.5	87.7	4.6	87.1	5.8	60
65	90.0	0.0	89.5	1.1	89.1	2.2	88.6	3.3	88.1	4.4	87.7	5.5	65
70	90.0	0.0	89.6	1.1	89.3	2.1	88.9	3.2	88.5	4.3	88.2	5.3	70
75	90.0	0.0	89.7	1.0	89.5	2.1	89.2	3.1	88.9	4.1	88.7	5.2	75
80	90.0	0.0	89.8	1.0	89.6	2.0	89.5	3.0	89.3	4.1	89.1	5.1	80
85	90.0	0.0	89.9	1.0	89.8	2.0	89.7	3.0	89.6	4.0	89.6	5.0	85

Numbers in *italics* indicate nearest approach to prime vertical

TABLE 20

Meridian Angle and Altitude of a Body on the Prime Vertical Circle

Latitude	Declination (same name as Latitude)												Latitude
	6°		7°		8°		9°		10°		11°		
	t	Alt.	t	Alt.	t	Alt.	t	Alt.	t	Alt.	t	Alt.	
°	°	°	°	°	°	°	°	°	°	°	°	°	°
0	90.0	0.0	90.0	0.0	90.0	0.0	90.0	0.0	90.0	0.0	90.0	0.0	0
1	80.4	9.6	81.8	8.2	82.9	7.2	83.7	6.4	84.3	5.8	84.8	5.2	1
2	70.6	19.5	73.5	16.6	75.6	14.5	77.3	12.9	78.6	11.6	79.7	10.5	2
3	60.1	30.0	64.7	25.4	68.1	22.1	70.7	19.5	72.7	17.5	74.4	15.9	3
4	48.3	41.9	55.3	34.9	60.2	30.1	63.8	26.5	66.6	23.7	68.9	21.4	4
5	33.7	56.5	44.6	45.7	51.5	38.8	56.5	33.9	60.3	30.1	63.3	27.2	5
6	0.0	90.0	31.1	59.1	41.6	48.7	48.4	41.9	53.4	37.0	57.3	33.2	6
7	31.1	59.1	0.0	90.0	29.1	61.1	39.2	51.2	45.9	44.6	50.8	39.7	7
8	41.6	48.7	29.1	61.1	0.0	90.0	27.5	62.8	37.2	53.3	43.7	46.8	8
9	48.4	41.9	39.2	51.2	27.5	62.8	0.0	90.0	26.1	64.3	35.4	55.1	9
10	53.4	37.0	45.9	44.6	37.2	53.3	26.1	64.3	0.0	90.0	24.9	65.5	10
11	57.3	33.2	50.8	39.7	43.7	46.8	35.4	55.1	24.9	65.5	0.0	90.0	11
12	60.4	30.2	54.7	35.9	48.6	42.0	41.8	48.8	33.9	56.6	23.9	66.6	12
13	62.9	27.7	57.9	32.8	52.5	38.2	46.7	44.1	40.2	50.5	32.7	58.0	13
14	65.1	25.6	60.5	30.2	55.7	35.1	50.6	40.3	45.0	45.9	38.8	52.1	14
15	66.9	23.8	62.7	28.1	58.4	32.5	53.8	37.2	48.8	42.1	43.5	47.5	15
16	68.5	22.3	64.6	26.2	60.7	30.3	56.5	34.6	52.1	39.0	47.3	43.8	16
17	69.9	20.9	66.3	24.6	62.6	28.4	58.8	32.3	54.8	36.4	50.5	40.7	17
18	71.1	19.8	67.8	23.2	64.4	26.8	60.8	30.4	57.1	34.2	53.3	38.1	18
19	72.2	18.7	69.1	22.0	65.9	25.3	62.6	28.7	59.2	32.2	55.6	35.9	19
20	73.2	17.8	70.3	20.9	67.3	24.0	64.2	27.2	61.0	30.5	57.7	33.9	20
21	74.1	17.0	71.3	19.9	68.5	22.9	65.6	25.9	62.7	29.0	59.6	32.2	21
22	74.9	16.2	72.3	19.0	69.6	21.8	66.9	24.7	64.1	27.6	61.2	30.6	22
23	75.7	15.5	73.2	18.2	70.7	20.9	68.1	23.6	65.5	26.4	62.7	29.2	23
24	76.3	14.9	74.0	17.4	71.6	20.0	69.2	22.6	66.7	25.3	64.1	28.0	24
25	77.0	14.3	74.7	16.8	72.5	19.2	70.1	21.7	67.8	24.3	65.4	26.8	25
26	77.6	13.8	75.4	16.1	73.3	18.5	71.1	20.9	68.8	23.3	66.5	25.8	26
27	78.1	13.3	76.1	15.6	74.0	17.9	71.9	20.2	69.8	22.5	67.6	24.9	27
28	78.6	12.9	76.6	15.0	74.7	17.2	72.7	19.5	70.6	21.7	68.6	24.0	28
29	79.1	12.5	77.2	14.6	75.3	16.7	73.4	18.8	71.5	21.0	69.5	23.2	29
30	79.5	12.1	77.7	14.1	75.9	16.2	74.1	18.2	72.2	20.3	70.3	22.4	30
31	79.9	11.7	78.2	13.7	76.5	15.7	74.7	17.7	72.9	19.7	71.1	21.7	31
32	80.3	11.4	78.7	13.3	77.0	15.2	75.3	17.2	73.6	19.1	71.9	21.1	32
33	80.7	11.1	79.1	12.9	77.5	14.8	75.9	16.7	74.2	18.6	72.6	20.5	33
34	81.0	10.8	79.5	12.6	78.0	14.4	76.4	16.2	74.8	18.1	73.3	20.0	34
35	81.4	10.5	79.9	12.3	78.4	14.0	76.9	15.8	75.4	17.6	73.9	19.4	35
36	81.7	10.2	80.3	12.0	78.8	13.7	77.4	15.4	76.0	17.2	74.5	18.9	36
37	82.0	10.0	80.6	11.7	79.3	13.4	77.9	15.1	76.5	16.8	75.1	18.5	37
38	82.3	9.8	81.0	11.4	79.6	13.1	78.3	14.7	77.0	16.4	75.6	18.1	38
39	82.5	9.6	81.3	11.2	80.0	12.8	78.7	14.4	77.4	16.0	76.1	17.6	39
40	82.8	9.4	81.6	10.9	80.4	12.5	79.1	14.1	77.9	15.7	76.6	17.3	40
41	83.1	9.2	81.9	10.7	80.7	12.2	79.5	13.8	78.3	15.3	77.1	16.9	41
42	83.3	9.0	82.2	10.5	81.0	12.0	79.9	13.5	78.7	15.0	77.5	16.6	42
43	83.5	8.8	82.4	10.3	81.3	11.8	80.2	13.3	79.1	14.8	78.0	16.2	43
44	83.8	8.7	82.7	10.1	81.6	11.6	80.6	13.0	79.5	14.5	78.4	15.9	44
45	84.0	8.5	82.9	9.9	81.9	11.4	80.9	12.8	79.8	14.2	78.8	15.7	45
46	84.2	8.4	83.2	9.8	82.2	11.2	81.2	12.6	80.2	14.0	79.2	15.4	46
47	84.4	8.2	83.4	9.6	82.5	11.0	81.5	12.4	80.5	13.7	79.6	15.1	47
48	84.6	8.1	83.7	9.4	82.7	10.8	81.8	12.2	80.9	13.5	79.9	14.9	48
49	84.8	8.0	83.9	9.3	83.0	10.6	82.1	12.0	81.2	13.3	80.3	14.6	49
50	84.9	7.8	84.1	9.2	83.2	10.5	82.4	11.8	81.5	13.1	80.6	14.4	50
52	85.3	7.6	84.5	8.9	83.7	10.2	82.9	11.5	82.1	12.7	81.3	14.0	52
54	85.6	7.4	84.9	8.7	84.1	9.9	83.4	11.1	82.6	12.4	81.9	13.6	54
56	85.9	7.2	85.2	8.5	84.6	9.7	83.9	10.9	83.2	12.1	82.5	13.3	56
58	86.2	7.1	85.6	8.3	85.0	9.4	84.3	10.6	83.7	11.8	83.0	13.0	58
60	86.5	6.9	85.9	8.1	85.3	9.2	84.8	10.4	84.2	11.6	83.6	12.7	60
65	87.2	6.6	86.7	7.7	86.2	8.8	85.8	9.9	85.3	11.0	84.8	12.2	65
70	87.8	6.4	87.4	7.5	87.1	8.5	86.7	9.6	86.3	10.6	85.9	11.7	70
75	88.4	6.2	88.1	7.2	87.8	8.3	87.6	9.3	87.3	10.4	87.0	11.4	75
80	88.9	6.1	88.8	7.1	88.6	8.1	88.4	9.1	88.2	10.2	88.0	11.2	80
85	89.5	6.0	89.4	7.0	89.3	8.0	89.2	9.0	89.1	10.0	89.0	11.0	85

Numbers in *italics* indicate nearest approach to prime vertical

TABLE 20

Meridian Angle and Altitude of a Body on the Prime Vertical Circle

Latitude	12° t	12° Alt.	13° t	13° Alt.	14° t	14° Alt.	15° t	15° Alt.	16° t	16° Alt.	17° t	17° Alt.	Latitude
	Declination (same name as Latitude)												
°	°	°	°	°	°	°	°	°	°	°	°	°	°
0	90.0	0.0	90.0	0.0	90.0	0.0	90.0	0.0	90.0	0.0	90.0	0.0	0
1	85.3	4.8	85.7	4.4	86.0	4.1	86.3	3.9	86.5	3.6	86.7	3.4	1
2	80.5	9.7	81.3	8.9	81.9	8.3	82.5	7.7	83.0	7.3	83.4	6.9	2
3	75.7	14.6	76.9	13.5	77.9	12.5	78.7	11.7	79.5	10.9	80.1	10.3	3
4	70.8	19.6	72.4	18.1	73.7	16.8	74.9	15.6	75.9	14.7	76.8	13.8	4
5	65.7	24.8	67.7	22.8	69.5	21.1	70.9	19.7	72.2	18.4	73.4	17.3	5
6	60.4	30.2	62.9	27.7	65.1	25.6	66.9	23.8	68.5	22.3	69.9	20.9	6
7	54.7	35.9	57.9	32.8	60.5	30.2	62.7	28.1	64.6	26.2	66.3	24.6	7
8	48.6	42.0	52.5	38.2	55.7	35.1	58.4	32.5	60.7	30.3	62.6	28.4	8
9	41.8	48.8	46.7	44.1	50.6	40.3	53.8	37.2	56.5	34.6	58.8	32.3	9
10	33.9	56.6	40.2	50.5	45.0	45.9	48.8	42.1	52.1	39.0	54.8	36.4	10
11	23.9	66.6	32.7	58.0	38.8	52.1	43.5	47.5	47.3	43.8	50.5	40.7	11
12	0.0	90.0	23.0	67.6	31.5	59.3	37.5	53.4	42.2	49.0	46.0	45.3	12
13	23.0	67.6	0.0	90.0	22.2	68.4	30.5	60.4	36.4	54.7	41.0	50.3	13
14	31.5	59.3	22.2	68.4	0.0	90.0	21.5	69.2	29.6	61.4	35.4	55.8	14
15	37.5	53.4	30.5	60.4	21.5	69.2	0.0	90.0	20.9	69.9	28.8	62.3	15
16	42.2	49.0	36.4	54.7	29.6	61.4	20.9	69.9	0.0	90.0	20.3	70.5	16
17	46.0	45.3	41.0	50.3	35.4	55.8	28.8	62.3	20.3	70.5	0.0	90.0	17
18	49.1	42.3	44.7	46.7	39.9	51.5	34.4	56.9	28.1	63.1	19.8	71.1	18
19	51.9	39.7	47.9	43.7	43.6	48.0	38.9	52.7	33.6	57.8	27.4	63.9	19
20	54.3	37.4	50.6	41.1	46.8	45.0	42.6	49.2	38.0	53.7	32.9	58.7	20
21	56.4	35.5	53.0	38.9	49.5	42.5	45.7	46.2	41.7	50.3	37.2	54.7	21
22	58.3	33.7	55.2	36.9	51.9	40.2	48.5	43.7	44.8	47.4	40.8	51.3	22
23	60.0	32.1	57.1	35.1	54.0	38.3	50.9	41.5	47.5	44.9	43.9	48.4	23
24	61.5	30.7	58.8	33.6	55.9	36.5	53.0	39.5	49.9	42.7	46.6	46.0	24
25	62.9	29.5	60.3	32.2	57.7	34.9	54.9	37.8	52.1	40.7	49.0	43.8	25
26	64.2	28.3	61.7	30.9	59.3	33.5	56.7	36.2	54.0	39.0	51.2	41.8	26
27	65.3	27.3	63.1	29.7	60.7	32.2	58.3	34.8	55.8	37.4	53.1	40.1	27
28	66.4	26.3	64.3	28.6	62.0	31.0	59.7	33.5	57.4	36.0	54.9	38.5	28
29	67.5	25.4	65.4	27.6	63.3	29.9	61.1	32.3	58.8	34.6	56.5	37.1	29
30	68.4	24.6	66.4	26.7	64.4	28.9	62.3	31.2	60.2	33.5	58.0	35.8	30
31	69.3	23.8	67.4	25.9	65.5	28.0	63.5	30.2	61.5	32.4	59.4	34.6	31
32	70.1	23.1	68.3	25.1	66.5	27.2	64.6	29.2	62.7	31.3	60.7	33.5	32
33	70.9	22.4	69.2	24.4	67.4	26.4	65.6	28.4	63.8	30.4	61.9	32.5	33
34	71.6	21.8	70.0	23.7	68.3	25.6	66.6	27.6	64.8	29.5	63.0	31.5	34
35	72.3	21.3	70.7	23.1	69.1	24.9	67.5	26.8	65.8	28.7	64.1	30.6	35
36	73.0	20.7	71.5	22.5	69.9	24.3	68.4	26.1	66.8	28.0	65.1	29.8	36
37	73.6	20.2	72.2	21.9	70.7	23.7	69.2	25.5	67.6	27.3	66.1	29.1	37
38	74.2	19.7	72.8	21.4	71.4	23.1	69.9	24.9	68.5	26.6	67.0	28.4	38
39	74.8	19.3	73.4	20.9	72.1	22.6	70.7	24.3	69.3	26.0	67.8	27.7	39
40	75.3	18.9	74.0	20.5	72.7	22.1	71.4	23.7	70.0	25.4	68.6	27.1	40
41	75.8	18.5	74.6	20.1	73.3	21.6	72.0	23.2	70.7	24.8	69.4	26.5	41
42	76.3	18.1	75.1	19.6	73.9	21.2	72.7	22.8	71.4	24.3	70.2	25.9	42
43	76.8	17.7	75.7	19.3	74.5	20.8	73.3	22.3	72.1	23.8	70.9	25.4	43
44	77.3	17.4	76.2	18.9	75.0	20.4	73.9	21.9	72.7	23.4	71.5	24.9	44
45	77.7	17.1	76.7	18.5	75.6	20.0	74.5	21.5	73.3	22.9	72.2	24.4	45
46	78.2	16.8	77.1	18.2	76.1	19.7	75.0	21.1	73.9	22.5	72.8	24.0	46
47	78.6	16.5	77.6	17.9	76.6	19.3	75.5	20.7	74.5	22.1	73.4	23.6	47
48	79.0	16.2	78.0	17.6	77.0	19.0	76.0	20.4	75.0	21.8	74.0	23.2	48
49	79.4	16.0	78.4	17.3	77.5	18.7	76.5	20.1	75.6	21.4	74.6	22.8	49
50	79.7	15.7	78.8	17.1	77.9	18.4	77.0	19.7	76.1	21.1	75.1	22.4	50
52	80.4	15.3	79.6	16.6	78.8	17.9	77.9	19.2	77.1	20.5	76.2	21.8	52
54	81.1	14.9	80.3	16.1	79.6	17.4	78.8	18.7	78.0	19.9	77.2	21.2	54
56	81.8	14.5	81.0	15.7	80.3	17.0	79.6	18.2	78.8	19.4	78.1	20.7	56
58	82.4	14.2	81.7	15.4	81.0	16.6	80.4	17.8	79.7	19.0	79.0	20.2	58
60	83.0	13.9	82.3	15.1	81.7	16.2	81.1	17.4	80.5	18.6	79.8	19.7	60
65	84.3	13.3	83.8	14.4	83.3	15.5	82.8	16.6	82.3	17.7	81.8	18.8	65
70	85.6	12.8	85.2	13.8	84.8	14.9	84.4	16.0	84.0	17.1	83.6	18.1	70
75	86.7	12.4	86.5	13.5	86.2	14.5	85.9	15.5	85.6	16.6	85.3	17.6	75
80	87.9	12.2	87.7	13.2	87.5	14.2	87.3	15.2	87.1	16.3	86.9	17.3	80
85	88.9	12.0	88.8	13.1	88.8	14.1	88.7	15.1	88.6	16.1	88.5	17.1	85

Numbers in *italics* indicate nearest approach to prime vertical

TABLE 20

Meridian Angle and Altitude of a Body on the Prime Vertical Circle

Latitude	18° t	18° Alt.	19° t	19° Alt.	20° t	20° Alt.	21° t	21° Alt.	22° t	22° Alt.	23° t	23° Alt.	Latitude
	Declination (same name as Latitude)												
°	°	°	°	°	°	°	°	°	°	°	°	°	°
0	90.0	0.0	90.0	0.0	90.0	0.0	90.0	0.0	90.0	0.0	90.0	0.0	0
1	86.9	3.2	87.1	3.1	87.3	2.9	87.4	2.8	87.5	2.7	87.6	2.6	1
2	83.8	6.5	84.2	6.2	84.5	5.9	84.8	5.6	85.0	5.3	85.3	5.1	2
3	80.7	9.8	81.2	9.3	81.7	8.8	82.2	8.4	82.5	8.0	82.9	7.7	3
4	77.6	13.0	78.3	12.4	78.9	11.8	79.5	11.2	80.0	10.7	80.5	10.3	4
5	74.4	16.4	75.3	15.5	76.1	14.8	76.8	14.1	77.5	13.5	78.1	12.9	5
6	71.1	19.8	72.2	18.7	73.2	17.8	74.1	17.0	74.9	16.2	75.7	15.5	6
7	67.8	23.2	69.1	22.0	70.3	20.9	71.3	19.9	72.3	19.0	73.2	18.2	7
8	64.4	26.8	65.9	25.3	67.3	24.0	68.5	22.9	69.6	21.8	70.7	20.9	8
9	60.8	30.4	62.6	28.7	64.2	27.2	65.6	25.9	66.9	24.7	68.1	23.6	9
10	57.1	34.2	59.2	32.2	61.0	30.5	62.7	29.0	64.1	27.6	65.5	26.4	10
11	53.3	38.1	55.6	35.9	57.7	33.9	59.6	32.2	61.2	30.6	62.7	29.2	11
12	49.1	42.3	51.9	39.7	54.3	37.4	56.4	35.5	58.3	33.7	60.0	32.1	12
13	44.7	46.7	47.9	43.7	50.6	41.1	53.0	38.9	55.2	36.9	57.1	35.1	13
14	39.9	51.5	43.6	48.0	46.8	45.0	49.5	42.5	51.9	40.2	54.0	38.3	14
15	34.4	56.9	38.9	52.7	42.6	49.2	45.7	46.2	48.5	43.7	50.9	41.5	15
16	28.1	63.1	33.6	57.8	38.0	53.7	41.7	50.3	44.8	47.4	47.5	44.9	16
17	19.8	71.1	27.4	63.9	32.9	58.7	37.2	54.7	40.8	51.3	43.9	48.4	17
18	0.0	90.0	19.3	71.7	26.8	64.6	32.2	59.6	36.5	55.6	40.1	52.3	18
19	19.3	71.7	0.0	90.0	18.9	72.2	26.2	65.3	31.5	60.4	35.8	56.4	19
20	26.8	64.6	18.9	72.2	0.0	90.0	18.5	72.6	25.7	65.9	31.0	61.1	20
21	32.2	59.6	26.2	65.3	18.5	72.6	0.0	90.0	18.2	73.1	25.3	66.5	21
22	36.5	55.6	31.5	60.4	25.7	65.9	18.2	73.1	0.0	90.0	17.9	73.5	22
23	40.1	52.3	35.8	56.4	31.0	61.1	25.3	66.5	17.9	73.5	0.0	90.0	23
24	43.1	49.4	39.3	53.2	35.2	57.2	30.4	61.8	24.8	67.1	17.6	73.9	24
25	45.8	47.0	42.4	50.4	38.7	54.0	34.6	58.0	30.0	62.4	24.5	67.6	25
26	48.2	44.8	45.1	48.0	41.7	51.3	38.1	54.8	34.1	58.7	29.5	63.0	26
27	50.4	42.9	47.5	45.8	44.4	48.9	41.1	52.1	37.5	55.6	33.6	59.4	27
28	52.3	41.2	49.6	43.9	46.8	46.8	43.8	49.8	40.5	52.9	37.0	56.3	28
29	54.1	39.6	51.6	42.2	49.0	44.9	46.2	47.7	43.2	50.6	40.0	53.7	29
30	55.8	38.2	53.4	40.6	50.9	43.2	48.3	45.8	45.6	48.5	42.7	51.4	30
31	57.3	36.9	55.0	39.2	52.7	41.6	50.3	44.1	47.7	46.7	45.1	49.3	31
32	58.7	35.7	56.6	37.9	54.4	40.2	52.1	42.6	49.7	45.0	47.2	47.5	32
33	60.0	34.6	58.0	36.7	55.9	38.9	53.8	41.1	51.5	43.5	49.2	45.8	33
34	61.2	33.5	59.3	35.6	57.3	37.7	55.3	39.9	53.2	42.1	51.0	44.3	34
35	62.4	32.6	60.5	34.6	58.7	36.6	56.8	38.7	54.8	40.8	52.7	42.9	35
36	63.4	31.7	61.7	33.6	59.9	35.6	58.1	37.6	56.2	39.6	54.3	41.7	36
37	64.5	30.9	62.8	32.8	61.1	34.6	59.4	36.5	57.6	38.5	55.7	40.5	37
38	65.4	30.1	63.9	31.9	62.2	33.7	60.6	35.6	58.9	37.5	57.1	39.4	38
39	66.3	29.4	64.8	31.2	63.3	32.9	61.7	34.7	60.1	36.5	58.4	38.4	39
40	67.2	28.7	65.8	30.4	64.3	32.1	62.8	33.9	61.2	35.6	59.6	37.4	40
41	68.1	28.1	66.7	29.8	65.2	31.4	63.8	33.1	62.3	34.8	60.8	36.6	41
42	68.8	27.5	67.5	29.1	66.2	30.7	64.8	32.4	63.3	34.0	61.9	35.7	42
43	69.6	26.9	68.3	28.5	67.0	30.1	65.7	31.7	64.3	33.3	62.9	35.0	43
44	70.3	26.4	69.1	27.9	67.9	29.5	66.6	31.1	65.3	32.6	63.9	34.2	44
45	71.0	25.9	69.9	27.4	68.7	28.9	67.4	30.5	66.2	32.0	64.9	33.5	45
46	71.7	25.4	70.6	26.9	69.4	28.4	68.2	29.9	67.0	31.4	65.8	32.9	46
47	72.4	25.0	71.3	26.4	70.2	27.9	69.0	29.3	67.9	30.8	66.7	32.3	47
48	73.0	24.6	71.9	26.0	70.9	27.4	69.8	28.8	68.7	30.3	67.5	31.7	48
49	73.6	24.2	72.6	25.6	71.6	26.9	70.5	28.3	69.4	29.8	68.3	31.2	49
50	74.2	23.8	73.2	25.2	72.2	26.5	71.2	27.9	70.2	29.3	69.1	30.7	50
52	75.3	23.1	74.4	24.4	73.5	25.7	72.5	27.1	71.6	28.4	70.6	29.7	52
54	76.3	22.5	75.5	23.7	74.7	25.0	73.8	26.3	72.9	27.6	72.0	28.9	54
56	77.3	21.9	76.6	23.1	75.8	24.4	75.0	25.6	74.2	26.9	73.4	28.1	56
58	78.3	21.4	77.6	22.6	76.9	23.8	76.1	25.0	75.4	26.2	74.6	27.4	58
60	79.2	20.9	78.5	22.1	77.9	23.3	77.2	24.4	76.5	25.6	75.8	26.8	60
65	81.3	19.9	80.8	21.1	80.2	22.2	79.7	23.3	79.1	24.4	78.6	25.5	65
70	83.2	19.2	82.8	20.3	82.4	21.3	82.0	22.4	81.5	23.5	81.1	24.6	70
75	85.0	18.7	84.7	19.7	84.4	20.7	84.1	21.8	83.8	22.8	83.5	23.9	75
80	86.7	18.3	86.5	19.3	86.3	20.3	86.1	21.3	85.9	22.4	85.7	23.4	80
85	88.4	18.1	88.3	19.1	88.2	20.1	88.1	21.1	88.0	22.1	87.9	23.1	85

Numbers in *italics* indicate nearest approach to prime vertical

TABLE 21
Latitude and Longitude Factors

f, the change of latitude for a unit change in longitude
F, the change of longitude for a unit change in latitude

Azimuth angle	0° f	0° F	2° f	2° F	4° f	4° F	6° f	6° F	8° f	8° F	Azimuth angle
°											°
0	0.00	—	0.00	—	0.00	—	0.00	—	0.00	—	180
1	0.02	57.29	0.02	57.32	0.02	57.43	0.02	57.61	0.02	57.85	179
2	0.03	28.64	0.03	28.65	0.03	28.71	0.03	28.79	0.03	28.92	178
3	0.05	19.08	0.05	19.09	0.05	19.13	0.05	19.19	0.05	19.27	177
4	0.07	14.30	0.07	14.31	0.07	14.34	0.07	14.38	0.07	14.44	176
5	0.09	11.43	0.09	11.44	0.09	11.46	0.09	11.49	0.09	11.54	175
6	0.11	9.51	0.11	9.52	0.10	9.54	0.10	9.57	0.10	9.61	174
7	0.12	8.14	0.12	8.15	0.12	8.16	0.12	8.19	0.12	8.22	173
8	0.14	7.12	0.14	7.12	0.14	7.13	0.14	7.15	0.14	7.18	172
9	0.16	6.31	0.16	6.32	0.16	6.33	0.16	6.35	0.16	6.38	171
10	0.18	5.67	0.18	5.68	0.18	5.69	0.18	5.70	0.17	5.73	170
12	0.21	4.70	0.21	4.71	0.21	4.72	0.21	4.73	0.21	4.75	168
14	0.25	4.01	0.25	4.01	0.25	4.02	0.25	4.03	0.25	4.05	166
16	0.29	3.49	0.29	3.49	0.29	3.50	0.28	3.51	0.28	3.52	164
18	0.32	3.08	0.32	3.08	0.32	3.08	0.32	3.10	0.32	3.11	162
20	0.36	2.75	0.36	2.75	0.36	2.75	0.36	2.76	0.36	2.77	160
22	0.40	2.48	0.40	2.48	0.40	2.48	0.40	2.49	0.40	2.50	158
24	0.45	2.25	0.44	2.25	0.44	2.25	0.44	2.26	0.44	2.27	156
26	0.49	2.05	0.49	2.06	0.49	2.06	0.49	2.06	0.48	2.07	154
28	0.53	1.88	0.53	1.88	0.53	1.88	0.53	1.89	0.53	1.90	152
30	0.58	1.73	0.58	1.73	0.57	1.74	0.57	1.74	0.57	1.75	150
32	0.62	1.60	0.62	1.60	0.62	1.60	0.62	1.61	0.62	1.62	148
34	0.67	1.48	0.67	1.48	0.67	1.49	0.67	1.49	0.67	1.50	146
36	0.73	1.38	0.73	1.38	0.72	1.38	0.72	1.38	0.72	1.39	144
38	0.78	1.28	0.78	1.28	0.78	1.28	0.78	1.29	0.78	1.29	142
40	0.84	1.19	0.84	1.19	0.84	1.19	0.83	1.20	0.83	1.20	140
42	0.90	1.11	0.90	1.11	0.90	1.11	0.90	1.12	0.89	1.12	138
44	0.97	1.04	0.97	1.04	0.96	1.04	0.96	1.04	0.96	1.05	136
46	1.04	0.97	1.04	0.97	1.03	0.97	1.03	0.97	1.03	0.98	134
48	1.11	0.90	1.11	0.90	1.11	0.90	1.11	0.90	1.10	0.91	132
50	1.19	0.84	1.19	0.84	1.19	0.84	1.19	0.84	1.18	0.85	130
52	1.28	0.78	1.28	0.78	1.28	0.78	1.27	0.79	1.27	0.79	128
54	1.38	0.73	1.38	0.73	1.37	0.73	1.37	0.73	1.36	0.73	126
56	1.48	0.67	1.48	0.67	1.48	0.68	1.47	0.68	1.47	0.68	124
58	1.60	0.62	1.60	0.63	1.60	0.63	1.59	0.63	1.58	0.63	122
60	1.73	0.58	1.73	0.58	1.73	0.58	1.72	0.58	1.72	0.58	120
62	1.88	0.53	1.88	0.53	1.88	0.53	1.87	0.53	1.86	0.54	118
64	2.05	0.49	2.05	0.49	2.05	0.49	2.04	0.49	2.03	0.49	116
66	2.25	0.45	2.24	0.45	2.24	0.45	2.23	0.45	2.22	0.45	114
68	2.48	0.40	2.47	0.40	2.47	0.40	2.46	0.40	2.45	0.41	112
70	2.75	0.36	2.75	0.36	2.74	0.36	2.73	0.37	2.72	0.37	110
72	3.08	0.32	3.08	0.33	3.07	0.33	3.06	0.33	3.05	0.33	108
74	3.49	0.29	3.49	0.29	3.48	0.29	3.47	0.29	3.45	0.29	106
76	4.01	0.25	4.01	0.25	4.00	0.25	3.99	0.25	3.97	0.25	104
78	4.70	0.21	4.70	0.21	4.69	0.21	4.68	0.21	4.66	0.21	102
80	5.67	0.18	5.67	0.18	5.66	0.18	5.64	0.18	5.62	0.18	100
81	6.31	0.16	6.31	0.16	6.30	0.16	6.28	0.16	6.25	0.16	99
82	7.12	0.14	7.11	0.14	7.10	0.14	7.07	0.14	7.05	0.14	98
83	8.14	0.12	8.14	0.12	8.12	0.12	8.10	0.12	8.07	0.12	97
84	9.51	0.11	9.51	0.11	9.49	0.11	9.46	0.11	9.42	0.11	96
85	11.43	0.09	11.42	0.09	11.40	0.09	11.37	0.09	11.32	0.09	95
86	14.30	0.07	14.29	0.07	14.27	0.07	14.22	0.07	14.16	0.07	94
87	19.08	0.05	19.07	0.05	19.03	0.05	18.98	0.05	18.91	0.05	93
88	28.64	0.03	28.62	0.03	28.57	0.03	28.48	0.03	28.36	0.03	92
89	57.29	0.02	57.26	0.02	57.15	0.02	56.98	0.02	56.73	0.02	91
90	—	0.00	—	0.00	—	0.00	—	0.00	—	0.00	90
	0°		2°		4°		6°		8°		

Correction to latitude = f × error in longitude — Correction to longitude = F × error in latitude

TABLE 21
Latitude and Longitude Factors

f, the change of latitude for a unit change in longitude
F, the change of longitude for a unit change in latitude

Azimuth angle	10° f	10° F	12° f	12° F	14° f	14° F	16° f	16° F	18° f	18° F	Azimuth angle
°											°
0	0.00	—	0.00	—	0.00	—	0.00	—	0.00	—	180
1	0.02	58.17	0.02	58.57	0.02	59.04	0.02	59.60	0.02	60.24	179
2	0.03	29.08	0.03	29.28	0.03	29.51	0.03	29.79	0.03	30.11	178
3	0.05	19.38	0.05	19.51	0.05	19.67	0.05	19.85	0.05	20.06	177
4	0.07	14.52	0.07	14.62	0.07	14.74	0.07	14.88	0.07	15.04	176
5	0.09	11.61	0.09	11.69	0.08	11.78	0.08	11.89	0.08	12.02	175
6	0.10	9.66	0.10	9.73	0.10	9.81	0.10	9.90	0.10	10.00	174
7	0.12	8.27	0.12	8.33	0.12	8.39	0.12	8.47	0.12	8.56	173
8	0.14	7.22	0.14	7.27	0.14	7.33	0.14	7.40	0.13	7.48	172
9	0.16	6.41	0.15	6.45	0.15	6.51	0.15	6.57	0.15	6.64	171
10	0.17	5.76	0.17	5.80	0.17	5.85	0.17	5.90	0.17	5.96	170
12	0.21	4.78	0.21	4.81	0.21	4.85	0.20	4.89	0.20	4.95	168
14	0.25	4.07	0.24	4.10	0.24	4.13	0.24	4.17	0.24	4.22	166
16	0.28	3.54	0.28	3.56	0.28	3.59	0.28	3.63	0.27	3.67	164
18	0.32	3.13	0.32	3.15	0.32	3.17	0.31	3.20	0.31	3.24	162
20	0.36	2.79	0.36	2.81	0.35	2.83	0.35	2.86	0.35	2.89	160
22	0.40	2.51	0.40	2.53	0.39	2.55	0.39	2.57	0.38	2.60	158
24	0.44	2.28	0.44	2.30	0.43	2.32	0.43	2.34	0.42	2.36	156
26	0.48	2.08	0.48	2.10	0.47	2.11	0.47	2.13	0.46	2.16	154
28	0.52	1.91	0.52	1.92	0.52	1.94	0.51	1.96	0.51	1.98	152
30	0.57	1.76	0.56	1.77	0.56	1.78	0.56	1.80	0.55	1.82	150
32	0.62	1.63	0.61	1.64	0.61	1.65	0.60	1.66	0.59	1.68	148
34	0.66	1.50	0.66	1.52	0.65	1.53	0.65	1.54	0.64	1.56	146
36	0.72	1.40	0.71	1.41	0.70	1.42	0.70	1.43	0.69	1.45	144
38	0.77	1.30	0.76	1.31	0.76	1.32	0.75	1.33	0.74	1.35	142
40	0.83	1.21	0.82	1.22	0.81	1.23	0.81	1.24	0.80	1.25	140
42	0.88	1.13	0.88	1.14	0.88	1.14	0.87	1.15	0.85	1.17	138
44	0.95	1.05	0.94	1.06	0.94	1.07	0.93	1.08	0.92	1.09	136
46	1.02	0.98	1.01	0.99	1.01	1.00	1.00	1.01	0.99	1.02	134
48	1.10	0.91	1.09	0.92	1.08	0.93	1.07	0.94	1.06	0.95	132
50	1.17	0.85	1.17	0.86	1.16	0.87	1.15	0.87	1.13	0.88	130
52	1.26	0.79	1.25	0.80	1.24	0.80	1.23	0.81	1.22	0.82	128
54	1.36	0.74	1.35	0.74	1.34	0.75	1.32	0.76	1.31	0.76	126
56	1.46	0.68	1.45	0.69	1.44	0.69	1.43	0.70	1.41	0.71	124
58	1.58	0.63	1.57	0.64	1.55	0.64	1.54	0.65	1.52	0.66	122
60	1.71	0.59	1.69	0.59	1.68	0.60	1.67	0.60	1.65	0.61	120
62	1.85	0.54	1.84	0.54	1.83	0.55	1.81	0.55	1.79	0.56	118
64	2.02	0.50	2.01	0.50	1.99	0.50	1.97	0.51	1.95	0.51	116
66	2.21	0.45	2.20	0.46	2.18	0.46	2.16	0.46	2.14	0.47	114
68	2.44	0.41	2.42	0.41	2.40	0.42	2.38	0.42	2.35	0.42	112
70	2.71	0.37	2.69	0.37	2.67	0.37	2.64	0.38	2.61	0.38	110
72	3.03	0.33	3.01	0.33	2.99	0.33	2.96	0.34	2.93	0.34	108
74	3.43	0.29	3.41	0.29	3.38	0.30	3.35	0.30	3.32	0.30	106
76	3.95	0.25	3.92	0.25	3.89	0.26	3.86	0.26	3.81	0.26	104
78	4.63	0.22	4.60	0.22	4.56	0.22	4.52	0.22	4.47	0.22	102
80	5.59	0.18	5.55	0.18	5.50	0.18	5.45	0.18	5.39	0.18	100
81	6.22	0.16	6.18	0.16	6.13	0.16	6.07	0.16	6.01	0.17	99
82	7.01	0.14	6.96	0.14	6.90	0.14	6.84	0.15	6.77	0.15	98
83	8.02	0.12	7.97	0.13	7.90	0.13	7.83	0.13	7.75	0.13	97
84	9.37	0.11	9.31	0.11	9.23	0.11	9.15	0.11	9.05	0.11	96
85	11.25	0.09	11.18	0.09	11.09	0.09	10.99	0.09	10.87	0.09	95
86	14.08	0.07	13.99	0.07	13.88	0.07	13.75	0.07	13.60	0.07	94
87	18.79	0.05	18.66	0.05	18.51	0.05	18.34	0.05	18.15	0.05	93
88	28.20	0.03	28.01	0.04	27.79	0.04	27.53	0.04	27.23	0.04	92
89	56.42	0.02	56.04	0.02	55.59	0.02	55.07	0.02	54.49	0.02	91
90	—	0.00	—	0.00	—	0.00	—	0.00	—	0.00	90
	10°		12°		14°		16°		18°		

Correction to latitude = f × error in longitude — Correction to longitude = F × error in latitude

TABLE 21

Latitude and Longitude Factors

f, the change of latitude for a unit change in longitude
F, the change of longitude for a unit change in latitude

Azimuth angle	Latitude 20° f	20° F	22° f	22° F	24° f	24° F	26° f	26° F	28° f	28° F	Azimuth angle
°											°
0	0.00	—	0.00	—	0.00	—	0.00	—	0.00	—	180
1	0.02	60.97	0.02	61.79	0.02	62.71	0.02	63.74	0.02	64.88	179
2	0.03	30.47	0.03	30.89	0.03	31.35	0.03	31.86	0.03	32.43	178
3	0.05	20.31	0.05	20.58	0.05	20.89	0.05	21.23	0.05	21.61	177
4	0.07	15.22	0.06	15.42	0.06	15.65	0.06	15.91	0.06	16.20	176
5	0.08	12.16	0.08	12.33	0.08	12.51	0.08	12.72	0.08	12.95	175
6	0.10	10.12	0.10	10.26	0.10	10.41	0.09	10.59	0.09	10.78	174
7	0.12	8.67	0.11	8.78	0.11	8.91	0.11	9.06	0.11	9.22	173
8	0.13	7.57	0.13	7.67	0.13	7.79	0.13	7.92	0.12	8.06	172
9	0.15	6.72	0.15	6.81	0.14	6.91	0.14	7.02	0.14	7.15	171
10	0.17	6.03	0.16	6.12	0.16	6.21	0.16	6.31	0.16	6.42	170
12	0.20	5.01	0.20	5.07	0.19	5.15	0.19	5.23	0.19	5.33	168
14	0.23	4.27	0.23	4.33	0.23	4.39	0.22	4.46	0.22	4.54	166
16	0.27	3.71	0.27	3.76	0.26	3.82	0.26	3.88	0.25	3.95	164
18	0.30	3.28	0.30	3.32	0.30	3.37	0.29	3.42	0.29	3.49	162
20	0.34	2.92	0.34	2.96	0.33	3.01	0.33	3.06	0.32	3.11	160
22	0.38	2.63	0.38	2.67	0.37	2.71	0.36	2.75	0.36	2.80	158
24	0.42	2.39	0.41	2.42	0.41	2.46	0.40	2.50	0.39	2.54	156
26	0.46	2.18	0.45	2.21	0.45	2.24	0.44	2.28	0.43	2.32	154
28	0.50	2.00	0.49	2.03	0.49	2.06	0.48	2.09	0.47	2.13	152
30	0.54	1.84	0.53	1.87	0.53	1.90	0.52	1.93	0.51	1.96	150
32	0.59	1.70	0.58	1.73	0.57	1.75	0.56	1.78	0.55	1.81	148
34	0.63	1.58	0.63	1.60	0.62	1.62	0.61	1.65	0.60	1.68	146
36	0.68	1.47	0.67	1.48	0.66	1.51	0.65	1.53	0.64	1.56	144
38	0.74	1.36	0.72	1.38	0.71	1.40	0.70	1.42	0.69	1.45	142
40	0.79	1.27	0.78	1.28	0.77	1.30	0.75	1.33	0.74	1.35	140
42	0.85	1.18	0.83	1.20	0.82	1.22	0.81	1.24	0.80	1.26	138
44	0.91	1.10	0.90	1.12	0.88	1.13	0.87	1.15	0.85	1.17	136
46	0.97	1.03	0.96	1.04	0.95	1.06	0.93	1.07	0.91	1.09	134
48	1.04	0.96	1.03	0.97	1.02	0.99	1.00	1.00	0.98	1.02	132
50	1.12	0.89	1.10	0.91	1.09	0.92	1.07	0.93	1.05	0.95	130
52	1.20	0.83	1.19	0.84	1.17	0.85	1.15	0.87	1.13	0.88	128
54	1.29	0.77	1.28	0.78	1.26	0.79	1.24	0.81	1.22	0.82	126
56	1.39	0.72	1.38	0.73	1.35	0.74	1.33	0.75	1.31	0.76	124
58	1.50	0.66	1.48	0.67	1.46	0.68	1.44	0.70	1.41	0.71	122
60	1.63	0.61	1.61	0.62	1.58	0.63	1.56	0.64	1.53	0.65	120
62	1.77	0.57	1.74	0.57	1.72	0.58	1.69	0.59	1.66	0.60	118
64	1.93	0.52	1.90	0.53	1.87	0.53	1.84	0.54	1.81	0.55	116
66	2.11	0.47	2.08	0.48	2.05	0.49	2.02	0.50	1.98	0.50	114
68	2.33	0.43	2.30	0.44	2.26	0.44	2.23	0.45	2.18	0.46	112
70	2.58	0.39	2.55	0.39	2.51	0.40	2.47	0.40	2.43	0.41	110
72	2.89	0.35	2.85	0.35	2.81	0.36	2.77	0.36	2.72	0.37	108
74	3.28	0.31	3.23	0.31	3.19	0.31	3.14	0.32	3.08	0.33	106
76	3.77	0.27	3.72	0.27	3.66	0.27	3.61	0.28	3.54	0.28	104
78	4.42	0.23	4.36	0.23	4.30	0.23	4.23	0.24	4.15	0.24	102
80	5.33	0.19	5.26	0.19	5.18	0.19	5.10	0.20	5.01	0.20	100
81	5.93	0.17	5.86	0.17	5.77	0.17	5.68	0.18	5.58	0.18	99
82	6.69	0.15	6.60	0.15	6.50	0.15	6.40	0.16	6.28	0.16	98
83	7.65	0.13	7.55	0.13	7.44	0.13	7.32	0.14	7.19	0.14	97
84	8.94	0.11	8.82	0.11	8.69	0.12	8.55	0.12	8.40	0.12	96
85	10.74	0.09	10.60	0.09	10.44	0.10	10.26	0.10	10.09	0.10	95
86	13.44	0.07	13.26	0.08	13.07	0.08	12.86	0.08	12.63	0.08	94
87	17.93	0.06	17.69	0.06	17.43	0.06	17.15	0.06	16.85	0.06	93
88	26.91	0.04	26.55	0.04	26.16	0.04	25.74	0.04	25.28	0.04	92
89	53.84	0.02	53.12	0.02	52.33	0.02	51.50	0.02	50.58	0.02	91
90	—	0.00	—	0.00	—	0.00	—	0.00	—	0.00	90
	20°		22°		24°		26°		28°		

Correction to latitude = f × error in longitude | Correction to longitude = F × error in latitude

TABLE 21

Latitude and Longitude Factors

f, the change of latitude for a unit change in longitude
F, the change of longitude for a unit change in latitude

Azimuth angle	Latitude 30° f	30° F	32° f	32° F	34° f	34° F	36° f	36° F	38° f	38° F	Azimuth angle
°											°
0	0.00	—	0.00	—	0.00	—	0.00	—	0.00	—	180
1	0.02	66.15	0.01	67.56	0.01	69.10	0.01	70.81	0.01	72.70	179
2	0.03	33.07	0.03	33.77	0.03	34.54	0.03	35.40	0.03	36.34	178
3	0.05	22.03	0.05	22.50	0.04	23.02	0.04	23.59	0.04	24.21	177
4	0.06	16.51	0.06	16.86	0.06	17.25	0.06	17.68	0.06	18.15	176
5	0.08	13.20	0.07	13.48	0.07	13.79	0.07	14.13	0.07	14.50	175
6	0.09	10.99	0.09	11.22	0.09	11.48	0.09	11.76	0.08	12.07	174
7	0.11	9.40	0.10	9.60	0.10	9.82	0.10	10.07	0.10	10.34	173
8	0.12	8.22	0.12	8.39	0.12	8.58	0.11	8.79	0.11	9.03	172
9	0.14	7.29	0.13	7.45	0.13	7.62	0.13	7.80	0.12	8.01	171
10	0.15	6.55	0.15	6.69	0.15	6.84	0.14	7.01	0.14	7.20	170
12	0.18	5.43	0.18	5.55	0.18	5.67	0.17	5.82	0.17	5.97	168
14	0.22	4.63	0.21	4.73	0.21	4.84	0.20	4.96	0.20	5.09	166
16	0.25	4.03	0.24	4.11	0.24	4.21	0.23	4.31	0.23	4.43	164
18	0.28	3.55	0.28	3.63	0.27	3.71	0.26	3.80	0.26	3.91	162
20	0.32	3.17	0.31	3.24	0.30	3.31	0.29	3.40	0.29	3.49	160
22	0.35	2.86	0.34	2.92	0.34	2.99	0.33	3.06	0.32	3.14	158
24	0.39	2.59	0.38	2.65	0.37	2.71	0.36	2.78	0.35	2.85	156
26	0.42	2.37	0.41	2.42	0.40	2.47	0.40	2.53	0.38	2.60	154
28	0.46	2.17	0.45	2.22	0.44	2.27	0.43	2.32	0.42	2.39	152
30	0.50	2.00	0.49	2.04	0.48	2.09	0.47	2.14	0.45	2.20	150
32	0.54	1.85	0.53	1.89	0.52	1.93	0.51	1.98	0.49	2.03	148
34	0.58	1.71	0.57	1.75	0.56	1.79	0.55	1.83	0.53	1.88	146
36	0.63	1.59	0.62	1.62	0.60	1.66	0.59	1.70	0.57	1.75	144
38	0.68	1.48	0.66	1.51	0.65	1.54	0.63	1.58	0.62	1.62	142
40	0.72	1.38	0.71	1.41	0.69	1.44	0.68	1.47	0.66	1.51	140
42	0.78	1.28	0.76	1.31	0.75	1.34	0.73	1.37	0.71	1.41	138
44	0.84	1.20	0.82	1.22	0.80	1.25	0.78	1.28	0.76	1.31	136
46	0.90	1.11	0.88	1.14	0.86	1.16	0.84	1.19	0.82	1.23	134
48	0.96	1.04	0.94	1.06	0.92	1.09	0.90	1.11	0.88	1.14	132
50	1.03	0.97	1.01	0.99	0.99	1.01	0.96	1.04	0.94	1.06	130
52	1.11	0.90	1.09	0.92	1.06	0.94	1.04	0.97	1.01	0.99	128
54	1.19	0.84	1.16	0.86	1.14	0.88	1.11	0.90	1.08	0.92	126
56	1.28	0.78	1.26	0.79	1.23	0.81	1.20	0.83	1.17	0.86	124
58	1.39	0.72	1.36	0.74	1.33	0.75	1.30	0.77	1.26	0.79	122
60	1.49	0.67	1.47	0.68	1.44	0.70	1.40	0.71	1.37	0.73	120
62	1.63	0.61	1.59	0.63	1.56	0.64	1.52	0.66	1.48	0.67	118
64	1.78	0.56	1.74	0.57	1.70	0.59	1.66	0.60	1.62	0.62	116
66	1.95	0.51	1.91	0.52	1.85	0.54	1.82	0.55	1.77	0.56	114
68	2.14	0.47	2.10	0.48	2.05	0.49	2.00	0.50	1.95	0.51	112
70	2.38	0.42	2.33	0.43	2.28	0.44	2.22	0.45	2.17	0.46	110
72	2.67	0.38	2.61	0.38	2.55	0.39	2.50	0.40	2.43	0.41	108
74	3.02	0.33	2.96	0.34	2.89	0.35	2.82	0.35	2.75	0.36	106
76	3.47	0.29	3.40	0.29	3.33	0.30	3.25	0.31	3.16	0.32	104
78	4.07	0.24	3.99	0.25	3.90	0.26	3.81	0.26	3.71	0.27	102
80	4.91	0.20	4.81	0.21	4.70	0.21	4.59	0.22	4.47	0.22	100
81	5.47	0.18	5.35	0.19	5.24	0.19	5.11	0.20	4.98	0.20	99
82	6.16	0.16	6.03	0.17	5.90	0.17	5.76	0.17	5.61	0.18	98
83	7.05	0.14	6.91	0.14	6.75	0.15	6.59	0.15	6.42	0.16	97
84	8.24	0.12	8.07	0.12	7.89	0.13	7.70	0.13	7.50	0.13	96
85	9.90	0.10	9.69	0.10	9.48	0.11	9.25	0.11	9.01	0.11	95
86	12.39	0.08	12.13	0.08	11.86	0.08	11.57	0.09	11.27	0.09	94
87	16.52	0.06	16.18	0.06	15.82	0.06	15.44	0.06	15.04	0.07	93
88	24.80	0.04	24.28	0.04	23.74	0.04	23.17	0.04	22.57	0.04	92
89	49.61	0.02	48.58	0.02	47.50	0.02	46.36	0.02	45.14	0.02	91
90	—	0.00	—	0.00	—	0.00	—	0.00	—	0.00	90
	30°		32°		34°		36°		38°		

Correction to latitude = f × error in longitude | Correction to longitude = F × error in latitude

TABLE 21

Latitude and Longitude Factors

f, the change of latitude for a unit change in longitude
F, the change of longitude for a unit change in latitude

Azimuth angle	Latitude 40° f	40° F	42° f	42° F	44° f	44° F	46° f	46° F	48° f	48° F	Azimuth angle
°											°
0	0.00	—	0.00	—	0.00	—	0.00	—	0.00	—	180
1	0.01	74.79	0.01	77.09	0.01	79.64	0.01	82.47	0.01	85.62	179
2	0.03	37.38	0.03	38.53	0.03	39.81	0.02	41.22	0.02	42.80	178
3	0.04	24.91	0.04	25.68	0.04	26.53	0.04	27.47	0.03	28.52	177
4	0.05	18.67	0.05	19.24	0.05	19.88	0.05	20.59	0.05	21.37	176
5	0.07	14.92	0.07	15.38	0.06	15.89	0.06	16.45	0.06	17.08	175
6	0.08	12.42	0.08	12.80	0.08	13.23	0.07	13.70	0.07	14.22	174
7	0.09	10.63	0.09	10.96	0.09	11.32	0.08	11.72	0.08	12.17	173
8	0.11	9.29	0.10	9.57	0.10	9.89	0.10	10.24	0.09	10.63	172
9	0.12	8.24	0.12	8.50	0.11	8.78	0.11	9.09	0.11	9.44	171
10	0.14	7.40	0.13	7.63	0.13	7.88	0.12	8.16	0.12	8.48	170
12	0.16	6.14	0.16	6.33	0.15	6.54	0.15	6.77	0.14	7.03	168
14	0.19	5.24	0.19	5.40	0.18	5.58	0.17	5.77	0.17	5.99	166
16	0.22	4.55	0.21	4.69	0.21	4.85	0.20	5.02	0.19	5.21	164
18	0.25	4.02	0.24	4.14	0.23	4.28	0.23	4.43	0.22	4.60	162
20	0.28	3.59	0.27	3.70	0.26	3.82	0.25	3.95	0.24	4.11	160
22	0.31	3.23	0.30	3.33	0.29	3.44	0.28	3.56	0.27	3.70	158
24	0.34	2.93	0.33	3.02	0.32	3.12	0.31	3.23	0.30	3.36	156
26	0.37	2.68	0.36	2.76	0.35	2.85	0.34	2.95	0.33	3.06	154
28	0.41	2.45	0.40	2.53	0.38	2.61	0.37	2.71	0.36	2.81	152
30	0.44	2.26	0.43	2.33	0.41	2.41	0.40	2.49	0.39	2.59	150
32	0.48	2.09	0.46	2.15	0.45	2.22	0.43	2.30	0.42	2.39	148
34	0.52	1.93	0.50	1.99	0.49	2.06	0.47	2.13	0.45	2.22	146
36	0.56	1.80	0.54	1.85	0.52	1.91	0.50	1.98	0.49	2.06	144
38	0.60	1.67	0.58	1.72	0.56	1.78	0.54	1.84	0.52	1.91	142
40	0.64	1.56	0.63	1.60	0.60	1.66	0.58	1.71	0.56	1.78	140
42	0.69	1.45	0.67	1.49	0.65	1.54	0.63	1.60	0.60	1.66	138
44	0.74	1.35	0.72	1.39	0.69	1.44	0.67	1.49	0.65	1.55	136
46	0.79	1.26	0.77	1.30	0.74	1.34	0.72	1.39	0.69	1.44	134
48	0.85	1.17	0.83	1.21	0.80	1.25	0.77	1.30	0.74	1.35	132
50	0.91	1.09	0.88	1.13	0.86	1.17	0.83	1.21	0.80	1.25	130
52	0.98	1.02	0.95	1.05	0.92	1.09	0.89	1.12	0.86	1.17	128
54	1.05	0.95	1.02	0.98	0.99	1.01	0.96	1.05	0.92	1.09	126
56	1.14	0.88	1.10	0.91	1.07	0.94	1.03	0.97	0.99	1.01	124
58	1.23	0.82	1.19	0.84	1.15	0.87	1.11	0.90	1.07	0.93	122
60	1.33	0.75	1.29	0.78	1.25	0.80	1.20	0.83	1.16	0.86	120
62	1.44	0.69	1.40	0.72	1.35	0.74	1.31	0.77	1.26	0.79	118
64	1.57	0.64	1.52	0.66	1.48	0.68	1.42	0.70	1.37	0.73	116
66	1.72	0.58	1.67	0.60	1.62	0.62	1.56	0.64	1.50	0.66	114
68	1.90	0.53	1.84	0.54	1.78	0.56	1.72	0.58	1.66	0.60	112
70	2.10	0.47	2.04	0.49	1.98	0.51	1.91	0.52	1.84	0.54	110
72	2.36	0.42	2.29	0.44	2.21	0.45	2.14	0.47	2.06	0.49	108
74	2.67	0.37	2.59	0.39	2.51	0.40	2.42	0.41	2.33	0.43	106
76	3.07	0.32	2.98	0.34	2.89	0.35	2.79	0.36	2.68	0.37	104
78	3.60	0.28	3.50	0.29	3.38	0.29	3.27	0.31	3.15	0.32	102
80	4.34	0.23	4.22	0.24	4.08	0.24	3.94	0.25	3.80	0.26	100
81	4.84	0.21	4.69	0.21	4.54	0.22	4.39	0.23	4.23	0.24	99
82	5.45	0.18	5.29	0.19	5.12	0.20	4.94	0.20	4.76	0.21	98
83	6.24	0.16	6.05	0.16	5.86	0.17	5.66	0.18	5.45	0.18	97
84	7.29	0.14	7.07	0.14	6.84	0.15	6.61	0.15	6.37	0.16	96
85	8.75	0.11	8.49	0.12	8.22	0.12	7.94	0.13	7.65	0.13	95
86	10.95	0.09	10.63	0.09	10.29	0.10	9.94	0.10	9.57	0.10	94
87	14.62	0.07	14.18	0.07	13.73	0.07	13.26	0.08	12.77	0.08	93
88	21.94	0.05	21.28	0.05	20.60	0.05	19.89	0.05	19.16	0.05	92
89	43.98	0.02	42.58	0.02	41.21	0.02	39.80	0.02	38.34	0.03	91
90	—	0.00	—	0.00	—	0.00	—	0.00	—	0.00	90
	40°		42°		44°		46°		48°		

Correction to latitude = f × error in longitude | Correction to longitude = F × error in latitude

TABLE 21

Latitude and Longitude Factors

f, the change of latitude for a unit change in longitude
F, the change of longitude for a unit change in latitude

Azimuth angle	Latitude 50° f	50° F	52° f	52° F	54° f	54° F	56° f	56° F	58° f	58° F	Azimuth angle
°											°
0	0.00	—	0.00	—	0.00	—	0.00	—	0.00	—	180
1	0.01	89.13	0.01	93.05	0.01	97.47	0.01	102.45	0.01	108.11	179
2	0.02	44.55	0.02	46.51	0.02	48.72	0.02	51.21	0.02	54.04	178
3	0.03	29.68	0.03	30.99	0.03	32.46	0.03	34.12	0.03	36.01	177
4	0.04	22.25	0.04	23.23	0.04	24.33	0.04	25.57	0.04	26.99	176
5	0.06	17.78	0.05	18.57	0.05	19.45	0.05	20.44	0.05	21.57	175
6	0.07	14.80	0.06	15.45	0.06	16.19	0.06	17.01	0.06	17.95	174
7	0.08	12.67	0.08	13.23	0.07	13.86	0.07	14.56	0.06	15.37	173
8	0.09	11.07	0.08	11.56	0.08	12.11	0.08	12.72	0.07	13.43	172
9	0.10	9.82	0.10	10.26	0.09	10.74	0.09	11.29	0.08	11.91	171
10	0.11	8.82	0.11	9.21	0.10	9.65	0.10	10.14	0.09	10.70	170
12	0.14	7.32	0.13	7.64	0.13	8.00	0.12	8.41	0.11	8.88	168
14	0.16	6.24	0.15	6.51	0.15	6.82	0.14	7.17	0.13	7.57	166
16	0.18	5.42	0.18	5.66	0.17	5.93	0.16	6.24	0.15	6.58	164
18	0.21	4.79	0.20	5.00	0.19	5.24	0.18	5.50	0.17	5.81	162
20	0.23	4.27	0.22	4.46	0.21	4.67	0.20	4.91	0.19	5.19	160
22	0.26	3.85	0.25	4.02	0.24	4.21	0.23	4.43	0.21	4.67	158
24	0.29	3.49	0.27	3.65	0.26	3.82	0.25	4.02	0.24	4.24	156
26	0.31	3.19	0.30	3.33	0.29	3.49	0.27	3.66	0.26	3.87	154
28	0.34	2.93	0.33	3.05	0.31	3.20	0.30	3.36	0.28	3.55	152
30	0.37	2.69	0.36	2.81	0.34	2.95	0.32	3.10	0.31	3.27	150
32	0.40	2.49	0.38	2.60	0.37	2.72	0.35	2.86	0.33	3.02	148
34	0.43	2.31	0.42	2.41	0.40	2.52	0.38	2.65	0.36	2.80	146
36	0.47	2.14	0.45	2.24	0.43	2.34	0.41	2.46	0.39	2.60	144
38	0.50	1.99	0.48	2.08	0.46	2.18	0.44	2.29	0.41	2.41	142
40	0.54	1.85	0.52	1.94	0.49	2.03	0.47	2.13	0.44	2.25	140
42	0.58	1.73	0.56	1.80	0.53	1.89	0.50	1.99	0.48	2.09	138
44	0.62	1.61	0.59	1.68	0.57	1.76	0.54	1.85	0.51	1.95	136
46	0.67	1.50	0.64	1.57	0.61	1.64	0.58	1.73	0.55	1.82	134
48	0.71	1.40	0.68	1.46	0.65	1.53	0.62	1.61	0.59	1.70	132
50	0.77	1.31	0.73	1.36	0.70	1.43	0.67	1.50	0.63	1.58	130
52	0.82	1.22	0.79	1.27	0.75	1.33	0.72	1.40	0.68	1.47	128
54	0.88	1.13	0.85	1.18	0.81	1.23	0.77	1.30	0.73	1.37	126
56	0.95	1.05	0.91	1.10	0.87	1.15	0.83	1.21	0.79	1.27	124
58	1.03	0.97	0.99	1.01	0.94	1.06	0.89	1.12	0.85	1.18	122
60	1.11	0.90	1.07	0.94	1.02	0.98	0.97	1.03	0.92	1.09	120
62	1.21	0.83	1.16	0.86	1.11	0.90	1.05	0.95	1.00	1.00	118
64	1.32	0.76	1.26	0.79	1.20	0.83	1.15	0.87	1.09	0.92	116
66	1.44	0.69	1.38	0.72	1.32	0.76	1.26	0.79	1.19	0.84	114
68	1.59	0.63	1.52	0.65	1.45	0.69	1.38	0.72	1.31	0.76	112
70	1.77	0.57	1.69	0.59	1.61	0.62	1.54	0.65	1.45	0.68	110
72	1.98	0.51	1.89	0.53	1.81	0.55	1.72	0.58	1.63	0.61	108
74	2.24	0.45	2.15	0.46	2.05	0.49	1.95	0.51	1.85	0.54	106
76	2.58	0.39	2.47	0.40	2.36	0.42	2.24	0.45	2.13	0.47	104
78	3.02	0.33	2.90	0.34	2.77	0.36	2.63	0.38	2.49	0.40	102
80	3.65	0.27	3.49	0.29	3.33	0.30	3.17	0.31	3.01	0.33	100
81	4.06	0.25	3.89	0.26	3.71	0.27	3.53	0.28	3.35	0.30	99
82	4.57	0.22	4.38	0.23	4.18	0.24	3.98	0.25	3.77	0.26	98
83	5.24	0.19	5.01	0.20	4.79	0.21	4.55	0.22	4.32	0.23	97
84	6.12	0.16	5.86	0.17	5.59	0.18	5.32	0.19	5.04	0.20	96
85	7.35	0.14	7.04	0.14	6.72	0.15	6.39	0.16	6.06	0.16	95
86	9.19	0.11	8.81	0.11	8.41	0.12	8.00	0.12	7.58	0.13	94
87	12.27	0.08	11.75	0.08	11.22	0.09	10.67	0.09	10.11	0.10	93
88	18.41	0.05	17.63	0.06	16.83	0.06	16.01	0.06	15.17	0.07	92
89	36.83	0.03	35.27	0.03	33.68	0.03	32.04	0.03	30.36	0.03	91
90	—	0.00	—	0.00	—	0.00	—	0.00	—	0.00	90
	50°		52°		54°		56°		58°		

Correction to latitude = f × error in longitude | Correction to longitude = F × error in latitude

TABLE 21

Latitude and Longitude Factors

f, the change of latitude for a unit change in longitude
F, the change of longitude for a unit change in latitude

Azimuth angle	Latitude 60° f	60° F	62° f	62° F	64° f	64° F	66° f	66° F	68° f	68° F	Azimuth angle
°											°
0	0.00	—	0.00	—	0.00	—	0.00	—	0.00	—	180
1	0.01	114.58	0.01	122.03	0.01	130.69	0.01	140.85	0.01	152.93	179
2	0.02	57.27	0.02	61.00	0.02	65.32	0.01	70.40	0.01	76.44	178
3	0.03	38.16	0.02	40.64	0.02	43.53	0.02	46.91	0.02	50.94	177
4	0.03	28.60	0.03	30.46	0.03	32.62	0.03	35.16	0.03	38.18	176
5	0.04	22.86	0.04	24.35	0.04	26.07	0.04	28.10	0.03	30.51	175
6	0.05	19.03	0.05	20.27	0.05	21.70	0.04	23.39	0.04	25.40	174
7	0.06	16.29	0.06	17.35	0.05	18.58	0.05	20.02	0.05	21.74	173
8	0.07	14.23	0.07	15.16	0.06	16.23	0.06	17.49	0.05	18.99	172
9	0.08	12.63	0.07	13.45	0.07	14.40	0.06	15.52	0.06	16.85	171
10	0.09	11.34	0.08	12.08	0.08	12.94	0.07	13.94	0.07	15.14	170
12	0.11	9.41	0.10	10.02	0.09	10.73	0.09	11.57	0.08	12.56	168
14	0.12	8.02	0.12	8.54	0.11	9.15	0.10	9.86	0.09	10.71	166
16	0.14	6.97	0.13	7.43	0.13	7.96	0.12	8.57	0.11	9.31	164
18	0.16	6.15	0.15	6.56	0.14	7.02	0.13	7.57	0.12	8.22	162
20	0.18	5.49	0.17	5.85	0.16	6.27	0.15	6.75	0.14	7.33	160
22	0.20	4.95	0.19	5.27	0.18	5.65	0.16	6.09	0.15	6.61	158
24	0.22	4.49	0.21	4.78	0.20	5.12	0.18	5.52	0.17	6.00	156
26	0.24	4.10	0.23	4.37	0.21	4.68	0.20	5.04	0.18	5.47	154
28	0.27	3.76	0.25	4.01	0.23	4.29	0.22	4.62	0.20	5.02	152
30	0.29	3.46	0.27	3.69	0.25	3.95	0.23	4.26	0.22	4.62	150
32	0.31	3.20	0.29	3.41	0.27	3.65	0.25	3.93	0.23	4.27	148
34	0.34	2.96	0.32	3.16	0.30	3.38	0.27	3.65	0.25	3.96	146
36	0.36	2.75	0.34	2.93	0.32	3.14	0.30	3.38	0.27	3.67	144
38	0.39	2.56	0.37	2.73	0.34	2.92	0.32	3.15	0.29	3.42	142
40	0.42	2.38	0.39	2.54	0.37	2.72	0.34	2.93	0.31	3.18	140
42	0.45	2.22	0.42	2.37	0.39	2.53	0.37	2.73	0.34	2.96	138
44	0.48	2.07	0.45	2.21	0.42	2.36	0.39	2.55	0.36	2.76	136
46	0.52	1.93	0.49	2.06	0.45	2.20	0.42	2.37	0.39	2.58	134
48	0.56	1.80	0.52	1.92	0.49	2.05	0.45	2.21	0.42	2.40	132
50	0.60	1.68	0.56	1.79	0.52	1.91	0.48	2.06	0.45	2.24	130
52	0.64	1.56	0.60	1.66	0.56	1.78	0.52	1.92	0.48	2.09	128
54	0.69	1.45	0.65	1.55	0.60	1.66	0.56	1.79	0.52	1.94	126
56	0.74	1.35	0.70	1.44	0.65	1.54	0.60	1.66	0.56	1.80	124
58	0.80	1.25	0.75	1.33	0.70	1.43	0.65	1.54	0.60	1.67	122
60	0.87	1.15	0.81	1.23	0.76	1.32	0.70	1.42	0.65	1.54	120
62	0.94	1.06	0.88	1.13	0.82	1.21	0.76	1.31	0.70	1.42	118
64	1.03	0.97	0.96	1.04	0.90	1.11	0.83	1.20	0.77	1.30	116
66	1.12	0.89	1.05	0.95	0.98	1.02	0.91	1.09	0.84	1.19	114
68	1.24	0.81	1.16	0.86	1.09	0.92	1.01	0.99	0.93	1.08	112
70	1.37	0.73	1.29	0.78	1.20	0.83	1.12	0.89	1.03	0.97	110
72	1.54	0.65	1.44	0.69	1.35	0.74	1.25	0.80	1.15	0.87	108
74	1.74	0.57	1.64	0.61	1.53	0.65	1.42	0.70	1.31	0.77	106
76	2.01	0.50	1.88	0.53	1.76	0.57	1.63	0.61	1.50	0.67	104
78	2.35	0.42	2.21	0.45	2.06	0.48	1.91	0.52	1.76	0.57	102
80	2.84	0.35	2.66	0.38	2.49	0.40	2.31	0.43	2.12	0.47	100
81	3.16	0.32	2.96	0.34	2.77	0.36	2.57	0.39	2.37	0.42	99
82	3.56	0.28	3.34	0.30	3.12	0.32	2.89	0.35	2.67	0.38	98
83	4.07	0.25	3.82	0.26	3.57	0.28	3.31	0.30	3.05	0.33	97
84	4.76	0.21	4.47	0.22	4.17	0.24	3.87	0.26	3.56	0.28	96
85	5.72	0.17	5.37	0.19	5.01	0.20	4.65	0.22	4.28	0.23	95
86	7.15	0.14	6.71	0.15	6.27	0.16	5.82	0.17	5.36	0.19	94
87	9.54	0.10	8.96	0.11	8.36	0.12	7.76	0.13	7.15	0.14	93
88	14.32	0.07	13.44	0.07	12.55	0.08	11.65	0.09	10.73	0.09	92
89	28.65	0.03	26.90	0.04	25.11	0.04	23.30	0.04	21.46	0.05	91
90	—	0.00	—	0.00	—	0.00	—	0.00	—	0.00	90
	60°		62°		64°		66°		68°		

Correction to latitude = f × error in longitude

Correction to longitude = F × error in latitude

TABLE 22
Amplitudes

Latitude	Declination 0°0	0°5	1°0	1°5	2°0	2°5	3°0	3°5	4°0	4°5	5°0	5°5	6°0	Latitude
°	°	°	°	°	°	°	°	°	°	°	°	°	°	°
0	0.0	0.5	1.0	1.5	2.0	2.5	3.0	3.5	4.0	4.5	5.0	5.5	6.0	0
10	0.0	0.5	1.0	1.5	2.0	2.5	3.0	3.6	4.1	4.6	5.1	5.6	6.1	10
15	0.0	0.5	1.0	1.6	2.1	2.6	3.1	3.6	4.1	4.7	5.2	5.7	6.2	15
20	0.0	0.5	1.1	1.6	2.1	2.7	3.2	3.7	4.3	4.8	5.3	5.9	6.4	20
25	0.0	0.6	1.1	1.7	2.2	2.8	3.3	3.9	4.4	5.0	5.5	6.1	6.6	25
30	0.0	0.6	1.2	1.7	2.3	2.9	3.5	4.0	4.6	5.2	5.8	6.4	6.9	30
32	0.0	0.6	1.2	1.8	2.4	2.9	3.5	4.1	4.7	5.3	5.9	6.5	7.1	32
34	0.0	0.6	1.2	1.8	2.4	3.0	3.6	4.2	4.8	5.4	6.0	6.6	7.2	34
36	0.0	0.6	1.2	1.9	2.5	3.1	3.7	4.3	4.9	5.6	6.2	6.8	7.4	36
38	0.0	0.6	1.3	1.9	2.5	3.2	3.8	4.4	5.1	5.7	6.4	7.0	7.6	38
40	0.0	0.7	1.3	2.0	2.6	3.3	3.9	4.6	5.2	5.9	6.5	7.2	7.8	40
42	0.0	0.7	1.3	2.0	2.7	3.4	4.0	4.7	5.4	6.1	6.7	7.4	8.1	42
44	0.0	0.7	1.4	2.1	2.8	3.5	4.2	4.9	5.6	6.3	7.0	7.7	8.4	44
46	0.0	0.7	1.4	2.2	2.9	3.6	4.3	5.0	5.8	6.5	7.2	7.9	8.7	46
48	0.0	0.7	1.5	2.2	3.0	3.7	4.5	5.2	6.0	6.7	7.5	8.2	9.0	48
50	0.0	0.8	1.6	2.3	3.1	3.9	4.7	5.4	6.2	7.0	7.8	8.6	9.4	50
51	0.0	0.8	1.6	2.4	3.2	4.0	4.8	5.6	6.4	7.2	8.0	8.8	9.6	51
52	0.0	0.8	1.6	2.4	3.2	4.1	4.9	5.7	6.5	7.3	8.1	9.0	9.8	52
53	0.0	0.8	1.7	2.5	3.3	4.2	5.0	5.8	6.7	7.5	8.3	9.2	10.0	53
54	0.0	0.9	1.7	2.6	3.4	4.3	5.1	6.0	6.8	7.7	8.5	9.4	10.2	54
55	0.0	0.9	1.7	2.6	3.5	4.4	5.2	6.1	7.0	7.9	8.7	9.6	10.5	55
56	0.0	0.9	1.8	2.7	3.6	4.5	5.4	6.3	7.2	8.1	9.0	9.9	10.8	56
57	0.0	0.9	1.8	2.8	3.7	4.6	5.5	6.4	7.4	8.3	9.2	10.1	11.1	57
58	0.0	0.9	1.9	2.8	3.8	4.7	5.7	6.6	7.6	8.5	9.5	10.4	11.4	58
59	0.0	1.0	1.9	2.9	3.9	4.9	5.8	6.8	7.8	8.8	9.7	10.7	11.7	59
60	0.0	1.0	2.0	3.0	4.0	5.0	6.0	7.0	8.0	9.0	10.0	11.1	12.1	60
61	0.0	1.0	2.1	3.1	4.1	5.2	6.2	7.2	8.3	9.3	10.3	11.4	12.5	61
62	0.0	1.1	2.1	3.2	4.3	5.3	6.4	7.5	8.5	9.6	10.7	11.8	12.9	62
63	0.0	1.1	2.2	3.3	4.4	5.5	6.6	7.7	8.8	10.0	11.1	12.2	13.3	63
64	0.0	1.1	2.3	3.4	4.6	5.7	6.9	8.0	9.2	10.3	11.5	12.6	13.8	64
65.0	0.0	1.2	2.4	3.6	4.7	5.9	7.1	8.3	9.5	10.7	11.9	13.1	14.3	65.0
65.5	0.0	1.2	2.4	3.6	4.8	6.0	7.3	8.5	9.7	10.9	12.1	13.4	14.6	65.5
66.0	0.0	1.2	2.5	3.7	4.9	6.2	7.4	8.6	9.9	11.1	12.4	13.6	14.9	66.0
66.5	0.0	1.3	2.5	3.8	5.0	6.3	7.5	8.8	10.1	11.3	12.6	13.9	15.2	66.5
67.0	0.0	1.3	2.6	3.8	5.1	6.4	7.7	9.0	10.3	11.6	12.9	14.2	15.5	67.0
67.5	0.0	1.3	2.6	3.9	5.2	6.5	7.9	9.2	10.5	11.8	13.2	14.5	15.9	67.5
68.0	0.0	1.3	2.7	4.0	5.3	6.7	8.0	9.4	10.7	12.1	13.5	14.8	16.2	68.0
68.5	0.0	1.4	2.7	4.1	5.5	6.8	8.2	9.6	11.0	12.4	13.8	15.2	16.6	68.5
69.0	0.0	1.4	2.8	4.2	5.6	7.0	8.4	9.8	11.2	12.6	14.1	15.5	17.0	69.0
69.5	0.0	1.4	2.9	4.3	5.7	7.2	8.6	10.0	11.5	12.9	14.4	15.9	17.4	69.5
70.0	0.0	1.5	2.9	4.4	5.9	7.3	8.8	10.3	11.8	13.3	14.8	16.3	17.8	70.0
70.5	0.0	1.5	3.0	4.5	6.0	7.5	9.0	10.5	12.1	13.6	15.1	16.7	18.2	70.5
71.0	0.0	1.5	3.1	4.6	6.2	7.7	9.3	10.8	12.4	13.9	15.5	17.1	18.7	71.0
71.5	0.0	1.6	3.2	4.7	6.3	7.9	9.5	11.1	12.7	14.3	15.9	17.6	19.2	71.5
72.0	0.0	1.6	3.2	4.9	6.5	8.1	9.8	11.4	13.0	14.7	16.4	18.1	19.8	72.0
72.5	0.0	1.7	3.3	5.0	6.7	8.3	10.0	11.7	13.4	15.1	16.8	18.6	20.3	72.5
73.0	0.0	1.7	3.4	5.1	6.9	8.6	10.3	12.1	13.8	15.6	17.3	19.1	20.9	73.0
73.5	0.0	1.8	3.5	5.3	7.1	8.8	10.6	12.4	14.2	16.0	17.9	19.7	21.6	73.5
74.0	0.0	1.8	3.6	5.4	7.3	9.1	10.9	12.8	14.7	16.5	18.4	20.3	22.3	74.0
74.5	0.0	1.9	3.7	5.6	7.5	9.4	11.3	13.2	15.1	17.1	19.0	21.0	23.0	74.5
75.0	0.0	1.9	3.9	5.8	7.7	9.7	11.7	13.6	15.6	17.6	19.7	21.7	23.8	75.0
75.5	0.0	2.0	4.0	6.0	8.0	10.0	12.1	14.1	16.2	18.3	20.4	22.5	24.7	75.5
76.0	0.0	2.1	4.1	6.2	8.3	10.4	12.5	14.6	16.8	18.9	21.1	23.3	25.6	76.0
76.5	0.0	2.1	4.3	6.4	8.6	10.8	13.0	15.2	17.4	19.6	21.9	24.2	26.6	76.5
77.0	0.0	2.2	4.4	6.7	8.9	11.2	13.5	15.7	18.1	20.4	22.8	25.2	27.7	77.0

TABLE 22
Amplitudes

Latitude	Declination 6°0	6°5	7°0	7°5	8°0	8°5	9°0	9°5	10°0	10°5	11°0	11°5	12°0	Latitude
°	°	°	°	°	°	°	°	°	°	°	°	°	°	°
0	6.0	6.5	7.0	7.5	8.0	8.5	9.0	9.5	10.0	10.5	11.0	11.5	12.0	0
10	6.1	6.6	7.1	7.6	8.1	8.6	9.1	9.6	10.2	10.7	11.2	11.7	12.2	10
15	6.2	6.7	7.2	7.8	8.3	8.8	9.3	9.8	10.4	10.9	11.4	11.9	12.4	15
20	6.4	6.9	7.5	8.0	8.5	9.0	9.6	10.1	10.6	11.2	11.7	12.2	12.8	20
25	6.6	7.2	7.7	8.3	8.8	9.4	9.9	10.5	11.0	11.6	12.2	12.7	13.3	25
30	6.9	7.5	8.1	8.7	9.2	9.8	10.4	11.0	11.6	12.1	12.7	13.3	13.9	30
32	7.1	7.7	8.3	8.9	9.4	10.0	10.6	11.2	11.8	12.4	13.0	13.6	14.2	32
34	7.2	7.8	8.5	9.1	9.7	10.3	10.9	11.5	12.1	12.7	13.3	13.9	14.5	34
36	7.4	8.0	8.7	9.3	9.9	10.5	11.1	11.8	12.4	13.0	13.6	14.3	14.9	36
38	7.6	8.3	8.9	9.5	10.2	10.8	11.5	12.1	12.7	13.4	14.0	14.7	15.3	38
40	7.8	8.5	9.2	9.8	10.5	11.1	11.8	12.4	13.1	13.8	14.4	15.1	15.7	40
42	8.1	8.8	9.4	10.1	10.8	11.5	12.1	12.8	13.5	14.2	14.9	15.6	16.2	42
44	8.4	9.1	9.8	10.5	11.2	11.9	12.6	13.3	14.0	14.7	15.4	16.1	16.8	44
46	8.7	9.4	10.1	10.8	11.6	12.3	13.0	13.7	14.5	15.2	15.9	16.7	17.4	46
48	9.0	9.7	10.5	11.2	12.0	12.8	13.5	14.3	15.0	15.8	16.6	17.3	18.1	48
50	9.4	10.1	10.9	11.7	12.5	13.3	14.1	14.9	15.7	16.5	17.3	18.1	18.9	50
51	9.6	10.4	11.2	12.0	12.8	13.6	14.4	15.2	16.0	16.8	17.7	18.5	19.3	51
52	9.8	10.6	11.4	12.2	13.1	13.9	14.7	15.6	16.4	17.2	18.1	18.9	19.7	52
53	10.0	10.8	11.7	12.5	13.4	14.2	15.1	15.9	16.8	17.6	18.5	19.3	20.2	53
54	10.2	11.1	12.0	12.8	13.7	14.6	15.4	16.3	17.2	18.1	18.9	19.8	20.7	54
55	10.5	11.4	12.3	13.2	14.0	14.9	15.8	16.7	17.6	18.5	19.4	20.3	21.3	55
56	10.8	11.7	12.6	13.5	14.4	15.3	16.2	17.2	18.1	19.0	20.0	20.9	21.8	56
57	11.1	12.0	12.9	13.9	14.8	15.7	16.7	17.6	18.6	19.6	20.5	21.5	22.4	57
58	11.4	12.3	13.3	14.3	15.2	16.2	17.2	18.1	19.1	20.1	21.1	22.1	23.1	58
59	11.7	12.7	13.7	14.7	15.7	16.7	17.7	18.7	19.7	20.7	21.7	22.8	23.8	59
60	12.1	13.1	14.1	15.1	16.2	17.2	18.2	19.3	20.3	21.4	22.4	23.5	24.6	60
61	12.5	13.5	14.6	15.6	16.7	17.8	18.8	19.9	21.0	22.1	23.2	24.3	25.4	61
62	12.9	14.0	15.0	16.1	17.2	18.4	19.5	20.6	21.7	22.8	24.0	25.1	26.3	62
63	13.3	14.4	15.6	16.7	17.9	19.0	20.2	21.3	22.5	23.7	24.9	26.0	27.3	63
64	13.8	15.0	16.2	17.3	18.5	19.7	20.9	22.1	23.3	24.6	25.8	27.1	28.3	64
65.0	14.3	15.5	16.8	18.0	19.2	20.5	21.7	23.0	24.3	25.5	26.8	28.1	29.5	65.0
65.5	14.6	15.8	17.1	18.3	19.6	20.9	22.2	23.5	24.8	26.1	27.4	28.7	30.1	65.5
66.0	14.9	16.2	17.4	18.7	20.0	21.3	22.6	23.9	25.3	26.6	28.0	29.4	30.7	66.0
66.5	15.2	16.5	17.8	19.1	20.4	21.8	23.1	24.5	25.8	27.2	28.6	30.0	31.4	66.5
67.0	15.5	16.8	18.2	19.5	20.9	22.2	23.6	25.0	26.4	27.8	29.2	30.7	32.1	67.0
67.5	15.9	17.2	18.6	19.9	21.3	22.7	24.1	25.5	27.0	28.4	29.9	31.4	32.9	67.5
68.0	16.2	17.6	19.0	20.4	21.8	23.2	24.7	26.1	27.6	29.1	30.6	32.2	33.7	68.0
68.5	16.6	18.0	19.4	20.9	22.3	23.8	25.3	26.8	28.3	29.8	31.4	33.0	34.6	68.5
69.0	17.0	18.4	19.9	21.4	22.9	24.4	25.9	27.4	29.0	30.6	32.2	33.8	35.5	69.0
69.5	17.4	18.9	20.4	21.9	23.4	25.0	26.5	28.1	29.7	31.4	33.0	34.7	36.4	69.5
70.0	17.8	19.3	20.9	22.4	24.0	25.6	27.2	28.9	30.5	32.2	33.9	35.7	37.4	70.0
70.5	18.2	19.8	21.4	23.0	24.6	26.3	27.9	29.6	31.3	33.1	34.9	36.7	38.5	70.5
71.0	18.7	20.3	22.0	23.6	25.3	27.0	28.7	30.5	32.2	34.0	35.9	37.8	39.7	71.0
71.5	19.2	20.9	22.6	24.3	26.0	27.8	29.5	31.3	33.2	35.1	37.0	38.9	40.9	71.5
72.0	19.8	21.5	23.2	25.0	26.8	28.6	30.4	32.3	34.2	36.1	38.1	40.2	42.3	72.0
72.5	20.3	22.1	23.9	25.7	27.6	29.4	31.3	33.3	35.3	37.3	39.4	41.5	43.7	72.5
73.0	20.9	22.8	24.6	26.5	28.4	30.4	32.3	34.4	36.4	38.6	40.7	43.0	45.3	73.0
73.5	21.6	23.5	25.4	27.4	29.3	31.4	33.4	35.5	37.7	39.9	42.2	44.6	47.1	73.5
74.0	22.3	24.2	26.2	28.3	30.3	32.4	34.6	36.8	39.0	41.4	43.8	46.3	49.0	74.0
74.5	23.0	25.1	27.1	29.3	31.4	33.6	35.8	38.1	40.5	43.0	45.6	48.2	51.1	74.5
75.0	23.8	25.9	28.1	30.3	32.5	34.8	37.2	39.6	42.1	44.8	47.5	50.4	53.4	75.0
75.5	24.7	26.9	29.1	31.4	33.8	36.2	38.7	41.2	43.9	46.7	49.6	52.8	56.1	75.5
76.0	25.6	27.9	30.2	32.7	35.1	37.7	40.3	43.0	45.9	48.9	52.1	55.5	59.3	76.0
76.5	26.6	29.0	31.5	34.0	36.6	39.3	42.1	45.0	48.1	51.3	54.8	58.7	63.0	76.5
77.0	27.7	30.2	32.8	35.5	38.2	41.1	44.1	47.2	50.5	54.1	58.0	62.4	67.6	77.0

TABLE 22
Amplitudes

Latitude	Declination 12°0	12°5	13°0	13°5	14°0	14°5	15°0	15°5	16°0	16°5	17°0	17°5	18°0	Latitude
°	°	°	°	°	°	°	°	°	°	°	°	°	°	°
0	12.0	12.5	13.0	13.5	14.0	14.5	15.0	15.5	16.0	16.5	17.0	17.5	18.0	0
10	12.2	12.7	13.2	13.7	14.2	14.7	15.2	15.7	16.3	16.8	17.3	17.8	18.3	10
15	12.4	12.9	13.5	14.0	14.5	15.0	15.5	16.1	16.6	17.1	17.6	18.1	18.7	15
20	12.8	13.3	13.9	14.4	14.9	15.5	16.0	16.5	17.1	17.6	18.1	18.7	19.2	20
25	13.3	13.8	14.4	14.9	15.5	16.0	16.6	17.1	17.7	18.3	18.8	19.4	19.9	25
30	13.9	14.5	15.1	15.6	16.2	16.8	17.4	18.0	18.6	19.1	19.7	20.3	20.9	30
32	14.2	14.8	15.4	16.0	16.6	17.2	17.8	18.4	19.0	19.6	20.2	20.8	21.4	32
34	14.5	15.1	15.7	16.4	17.0	17.6	18.2	18.8	19.4	20.0	20.7	21.3	21.9	34
36	14.9	15.5	16.1	16.8	17.4	18.0	18.7	19.3	19.9	20.6	21.2	21.8	22.5	36
38	15.3	15.9	16.6	17.2	17.9	18.5	19.2	19.8	20.5	21.1	21.8	22.4	23.1	38
40	15.7	16.4	17.1	17.7	18.4	19.1	19.7	20.4	21.1	21.8	22.4	23.1	23.8	40
41	16.0	16.7	17.3	18.0	18.7	19.4	20.1	20.8	21.4	22.1	22.8	23.5	24.2	41
42	16.2	16.9	17.6	18.3	19.0	19.7	20.4	21.1	21.8	22.5	23.2	23.9	24.6	42
43	16.5	17.2	17.9	18.6	19.3	20.0	20.7	21.4	22.1	22.9	23.6	24.3	25.0	43
44	16.8	17.5	18.2	18.9	19.7	20.4	21.1	21.8	22.5	23.3	24.0	24.7	25.4	44
45	17.1	17.8	18.5	19.3	20.0	20.7	21.5	22.2	22.9	23.7	24.4	25.2	25.9	45
46	17.4	18.2	18.9	19.6	20.4	21.1	21.9	22.6	23.4	24.1	24.9	25.7	26.4	46
47	17.7	18.5	19.3	20.0	20.8	21.5	22.3	23.1	23.8	24.6	25.4	26.2	26.9	47
48	18.1	18.9	19.6	20.4	21.2	22.0	22.8	23.5	24.3	25.1	25.9	26.7	27.5	48
49	18.5	19.3	20.1	20.8	21.6	22.4	23.2	24.0	24.8	25.7	26.5	27.3	28.1	49
50	18.9	19.7	20.5	21.3	22.1	22.9	23.7	24.6	25.4	26.2	27.1	27.9	28.7	50
51	19.3	20.1	20.9	21.8	22.6	23.4	24.3	25.1	26.0	26.8	27.7	28.5	29.4	51
52	19.7	20.6	21.4	22.3	23.1	24.0	24.9	25.7	26.6	27.5	28.3	29.2	30.1	52
53	20.2	21.1	21.9	22.8	23.7	24.6	25.5	26.4	27.3	28.2	29.1	30.0	30.9	53
54	20.7	21.6	22.5	23.4	24.3	25.2	26.1	27.0	28.0	28.9	29.8	30.8	31.7	54
55	21.3	22.2	23.1	24.0	24.9	25.9	26.8	27.8	28.7	29.7	30.6	31.6	32.6	55
56	21.8	22.8	23.7	24.7	25.6	26.6	27.6	28.5	29.5	30.5	31.5	32.5	33.5	56
57	22.4	23.4	24.4	25.4	26.4	27.4	28.4	29.4	30.4	31.4	32.5	33.5	34.6	57
58	23.1	24.1	25.1	26.1	27.2	28.2	29.2	30.3	31.3	32.4	33.5	34.6	35.7	58
59	23.8	24.8	25.9	27.0	28.0	29.1	30.2	31.3	32.4	33.5	34.6	35.7	36.9	59
60	24.6	25.7	26.7	27.8	28.9	30.1	31.2	32.3	33.5	34.6	35.8	37.0	38.2	60
61	25.4	26.5	27.6	28.8	29.9	31.1	32.3	33.5	34.6	35.9	37.1	38.3	39.6	61
62	26.3	27.5	28.6	29.8	31.0	32.2	33.5	34.7	36.0	37.2	38.5	39.8	41.2	62
63	27.3	28.5	29.7	30.9	32.2	33.5	34.8	36.1	37.4	38.7	40.1	41.5	42.9	63
64	28.3	29.6	30.9	32.2	33.5	34.8	36.2	37.6	39.0	40.4	41.8	43.3	44.8	64
65.0	29.5	30.8	32.2	33.5	34.9	36.3	37.8	39.2	40.7	42.2	43.8	45.4	47.0	65.0
65.5	30.1	31.5	32.9	34.3	35.7	37.1	38.6	40.1	41.7	43.2	44.8	46.5	48.2	65.5
66.0	30.7	32.1	33.6	35.0	36.5	38.0	39.5	41.1	42.7	44.3	46.0	47.7	49.4	66.0
66.5	31.4	32.9	34.3	35.8	37.3	38.9	40.5	42.1	43.7	45.4	47.2	48.9	50.8	66.5
67.0	32.1	33.6	35.1	36.7	38.3	39.9	41.5	43.2	44.9	46.6	48.4	50.3	52.3	67.0
67.5	32.9	34.4	36.0	37.6	39.2	40.9	42.6	44.3	46.1	47.9	49.8	51.8	53.9	67.5
68.0	33.7	35.3	36.9	38.6	40.2	41.9	43.7	45.5	47.4	49.3	51.3	53.4	55.6	68.0
68.5	34.6	36.2	37.9	39.6	41.3	43.1	44.9	46.8	48.8	50.8	52.9	55.1	57.5	68.5
69.0	35.5	37.2	38.9	40.6	42.5	44.3	46.2	48.2	50.3	52.4	54.7	57.0	59.6	69.0
69.5	36.4	38.2	40.0	41.8	43.7	45.6	47.7	49.7	51.9	54.2	56.6	59.2	61.9	69.5
70.0	37.4	39.3	41.1	43.0	45.0	47.1	49.2	51.4	53.7	56.1	58.7	61.5	64.6	70.0
70.5	38.5	40.4	42.4	44.4	46.4	48.6	50.8	53.2	55.7	58.3	61.1	64.3	67.8	70.5
71.0	39.7	41.7	43.7	45.8	48.0	50.3	52.7	55.2	57.8	60.7	63.9	67.5	71.7	71.0
71.5	40.9	43.0	45.1	47.4	49.7	52.1	54.7	57.4	60.3	63.5	67.1	71.4	76.9	71.5
72.0	42.3	44.5	46.7	49.1	51.5	54.1	56.9	59.9	63.1	66.8	71.1	76.7	90.0	72.0
72.5	43.7	46.0	48.4	50.9	53.6	56.4	59.4	62.7	66.4	70.8	76.5	90.0		72.5
73.0	45.3	47.8	50.3	53.0	55.8	58.9	62.3	66.1	70.5	76.3	90.0			73.0
73.5	47.1	49.6	52.4	55.3	58.4	61.8	65.7	70.2	76.0	90.0				73.5
74.0	49.0	51.7	54.7	57.9	61.4	65.3	69.9	75.8	90.0					74.0
74.5	51.1	54.1	57.3	60.9	64.9	69.5	75.6	90.0						74.5

TABLE 22
Amplitudes

Latitude	Declination 18°0	18°5	19°0	19°5	20°0	20°5	21°0	21°5	22°0	22°5	23°0	23°5	24°0	Latitude
°	°	°	°	°	°	°	°	°	°	°	°	°	°	°
0	18.0	18.5	19.0	19.5	20.0	20.5	21.0	21.5	22.0	22.5	23.0	23.5	24.0	0
10	18.3	18.8	19.3	19.8	20.3	20.8	21.3	21.8	22.4	22.9	23.4	23.9	24.4	10
15	18.7	19.2	19.7	20.2	20.7	21.3	21.8	22.3	22.8	23.3	23.9	24.4	24.9	15
20	19.2	19.7	20.3	20.8	21.3	21.9	22.4	23.0	23.5	24.0	24.6	25.1	25.6	20
25	19.9	20.5	21.1	21.6	22.2	22.7	23.3	23.9	24.4	25.0	25.5	26.1	26.7	25
30	20.9	21.5	22.1	22.7	23.3	23.9	24.4	25.0	25.6	26.2	26.8	27.4	28.0	30
32	21.4	22.0	22.6	23.2	23.8	24.4	25.0	25.6	26.2	26.8	27.4	28.0	28.7	32
34	21.9	22.5	23.1	23.7	24.4	25.0	25.6	26.2	26.9	27.5	28.1	28.7	29.4	34
36	22.5	23.1	23.7	24.4	25.0	25.7	26.3	26.9	27.6	28.2	28.9	29.5	30.2	36
38	23.1	23.7	24.4	25.1	25.7	26.4	27.1	27.7	28.4	29.1	29.7	30.4	31.1	38
40	23.8	24.5	25.2	25.8	26.5	27.2	27.9	28.6	29.3	30.0	30.7	31.4	32.1	40
41	24.2	24.9	25.6	26.3	26.9	27.6	28.3	29.1	29.8	30.5	31.2	31.9	32.6	41
42	24.6	25.3	26.0	26.7	27.4	28.1	28.8	29.5	30.3	31.0	31.7	32.5	33.2	42
43	25.0	25.7	26.4	27.2	27.9	28.6	29.3	30.1	30.8	31.6	32.3	33.0	33.8	43
44	25.4	26.2	26.9	27.6	28.4	29.1	29.9	30.6	31.4	32.1	32.9	33.7	34.4	44
45	25.9	26.7	27.4	28.2	28.9	29.7	30.5	31.2	32.0	32.8	33.5	34.3	35.1	45
46	26.4	27.2	27.9	28.7	29.5	30.3	31.1	31.8	32.6	33.4	34.2	35.0	35.8	46
47	26.9	27.7	28.5	29.3	30.1	30.9	31.7	32.5	33.3	34.1	35.0	35.8	36.6	47
48	27.5	28.3	29.1	29.9	30.7	31.6	32.4	33.2	34.0	34.9	35.7	36.6	37.4	48
49	28.1	28.9	29.8	30.6	31.4	32.3	33.1	34.0	34.8	35.7	36.6	37.4	38.3	49
50	28.7	29.6	30.4	31.3	32.1	33.0	33.9	34.8	35.6	36.5	37.4	38.3	39.3	50
51	29.4	30.3	31.2	32.0	32.9	33.8	34.7	35.6	36.5	37.5	38.4	39.3	40.3	51
52	30.1	31.0	31.9	32.8	33.7	34.7	35.6	36.5	37.5	38.4	39.4	40.4	41.3	52
53	30.9	31.8	32.8	33.7	34.6	35.6	36.5	37.5	38.5	39.5	40.5	41.5	42.5	53
54	31.7	32.7	33.6	34.6	35.6	36.6	37.6	38.6	39.6	40.6	41.7	42.7	43.8	54
55	32.6	33.6	34.6	35.6	36.6	37.6	38.7	39.7	40.8	41.9	42.9	44.0	45.2	55
56	33.5	34.6	35.6	36.7	37.7	38.8	39.9	41.0	42.1	43.2	44.3	45.5	46.7	56
57	34.6	35.6	36.7	37.8	38.9	40.0	41.1	42.3	43.5	44.6	45.8	47.1	48.3	57
58	35.7	36.8	37.9	39.1	40.2	41.4	42.6	43.8	45.0	46.2	47.5	48.8	50.1	58
59	36.9	38.0	39.2	40.4	41.6	42.8	44.1	45.4	46.7	48.0	49.3	50.7	52.2	59
60.0	38.2	39.4	40.6	41.9	43.2	44.5	45.8	47.1	48.5	49.9	51.4	52.9	54.4	60.0
60.5	38.9	40.1	41.4	42.7	44.0	45.3	46.7	48.1	49.5	51.0	52.5	54.1	55.7	60.5
61.0	39.6	40.9	42.2	43.5	44.9	46.3	47.7	49.1	50.6	52.1	53.7	55.3	57.0	61.0
61.5	40.4	41.7	43.0	44.4	45.8	47.2	48.7	50.2	51.7	53.3	55.0	56.7	58.5	61.5
62.0	41.2	42.5	43.9	45.3	46.8	48.2	49.8	51.3	52.9	54.6	56.3	58.1	60.0	62.0
62.5	42.0	43.4	44.8	46.3	47.8	49.3	50.9	52.5	54.2	56.0	57.8	59.7	61.7	62.5
63.0	42.9	44.3	45.8	47.3	48.9	50.5	52.1	53.8	55.6	57.5	59.4	61.4	63.6	63.0
63.5	43.8	45.3	46.9	48.4	50.0	51.7	53.4	55.2	57.1	59.1	61.1	63.4	65.7	63.5
64.0	44.8	46.4	48.0	49.6	51.3	53.0	54.8	56.7	58.7	60.8	63.0	65.5	68.1	64.0
64.5	45.9	47.5	49.1	50.8	52.6	54.4	56.3	58.4	60.5	62.7	65.2	67.9	70.9	64.5
65.0	47.0	48.7	50.4	52.2	54.0	56.0	58.0	60.1	62.4	64.9	67.6	70.7	74.2	65.0
65.5	48.2	49.9	51.7	53.6	55.6	57.6	59.8	62.1	64.6	67.3	70.4	74.1	78.8	65.5
66.0	49.4	51.3	53.2	55.2	57.2	59.4	61.8	64.3	67.1	70.2	73.9	78.6	90.0	66.0
66.5	50.8	52.7	54.7	56.8	59.1	61.4	64.0	66.8	70.0	73.7	78.5	90.0		66.5
67.0	52.3	54.3	56.4	58.7	61.1	63.7	66.5	69.7	73.5	78.4	90.0			67.0
67.5	53.9	56.0	58.3	60.7	63.3	66.2	69.5	73.3	78.2	90.0				67.5
68.0	55.6	57.9	60.4	63.0	65.9	69.2	73.1	78.1	90.0					68.0
68.5	57.5	60.0	62.7	65.6	68.9	72.9	77.9	90.0						68.5
69.0	59.6	62.3	65.3	68.7	72.6	77.7	90.0							69.0
69.5	61.9	65.0	68.4	72.4	77.6	90.0								69.5
70.0	64.6	68.1	72.2	77.4	90.0									70.0
70.5	67.8	71.9	77.2	90.0										70.5
71.0	71.7	77.1	90.0											71.0
71.5	76.9	90.0												71.5
72.0	90.0													72.0

TABLE 23

Correction of Amplitude as Observed on the Visible Horizon

Latitude	Declination													Latitude
	0°	2°	4°	6°	8°	10°	12°	14°	16°	18°	20°	22°	24°	
°	°	°	°	°	°	°	°	°	°	°	°	°	°	°
0	0.0	0.0	0.0	0.0	0.0	0.0	0.0	0.0	0.0	0.0	0.0	0.0	0.0	0
10	0.1	0.1	0.1	0.1	0.1	0.1	0.1	0.1	0.1	0.1	0.1	0.1	0.1	10
15	0.2	0.2	0.2	0.2	0.2	0.2	0.2	0.2	0.2	0.2	0.2	0.2	0.2	15
20	0.3	0.3	0.3	0.3	0.3	0.3	0.3	0.3	0.3	0.3	0.3	0.3	0.3	20
25	0.3	0.3	0.3	0.3	0.3	0.4	0.3	0.3	0.3	0.3	0.3	0.3	0.3	25
30	0.4	0.4	0.4	0.4	0.5	0.4	0.4	0.4	0.4	0.4	0.4	0.5	0.5	30
32	0.4	0.4	0.4	0.4	0.5	0.4	0.4	0.4	0.4	0.4	0.5	0.5	0.5	32
34	0.5	0.5	0.5	0.5	0.5	0.5	0.5	0.5	0.5	0.5	0.5	0.5	0.5	34
36	0.5	0.5	0.5	0.5	0.5	0.5	0.5	0.5	0.6	0.5	0.6	0.6	0.6	36
38	0.6	0.6	0.6	0.6	0.6	0.6	0.6	0.6	0.6	0.6	0.6	0.6	0.6	38
40	0.6	0.6	0.6	0.6	0.6	0.6	0.6	0.6	0.6	0.6	0.7	0.7	0.7	40
42	0.6	0.6	0.6	0.6	0.7	0.7	0.7	0.7	0.7	0.7	0.7	0.7	0.7	42
44	0.7	0.7	0.7	0.6	0.6	0.7	0.7	0.7	0.8	0.8	0.8	0.8	0.9	44
46	0.7	0.7	0.7	0.7	0.7	0.8	0.8	0.8	0.8	0.8	0.8	0.9	0.9	46
48	0.8	0.8	0.8	0.8	0.8	0.8	0.8	0.8	0.9	0.9	1.0	1.0	1.0	48
50	0.8	0.8	0.8	0.8	0.9	0.9	0.9	0.9	0.9	1.0	1.0	1.1	1.0	50
51	0.8	0.8	0.8	0.8	0.9	0.9	0.9	0.9	0.9	1.0	1.1	1.1	1.1	51
52	0.9	0.9	0.9	0.9	0.9	0.9	1.0	1.0	1.0	1.1	1.1	1.1	1.3	52
53	0.9	0.9	0.9	0.9	0.9	0.9	1.0	1.0	1.0	1.1	1.2	1.2	1.3	53
54	1.0	1.0	1.0	1.0	1.0	1.0	1.1	1.1	1.1	1.2	1.2	1.3	1.3	54
55	1.0	1.0	1.0	1.0	1.1	1.1	1.0	1.2	1.2	1.2	1.3	1.3	1.4	55
56	1.0	1.0	1.0	1.0	1.1	1.1	1.2	1.2	1.2	1.3	1.3	1.4	1.5	56
57	1.1	1.1	1.1	1.1	1.1	1.1	1.2	1.2	1.3	1.3	1.4	1.5	1.7	57
58	1.1	1.1	1.1	1.1	1.2	1.2	1.2	1.2	1.4	1.4	1.5	1.6	1.8	58
59	1.2	1.2	1.2	1.2	1.2	1.2	1.3	1.3	1.3	1.4	1.6	1.7	1.9	59
60	1.2	1.2	1.2	1.2	1.2	1.3	1.3	1.4	1.4	1.5	1.7	1.9	2.2	60
61	1.3	1.3	1.3	1.3	1.3	1.3	1.4	1.5	1.6	1.7	1.8	2.0	2.4	61
62	1.3	1.3	1.3	1.3	1.4	1.4	1.5	1.6	1.6	1.7	1.9	2.3	2.6	62
63	1.3	1.4	1.4	1.4	1.4	1.5	1.5	1.6	1.7	1.9	2.1	2.5	3.3	63
64	1.4	1.4	1.4	1.5	1.5	1.6	1.7	1.7	1.8	2.1	2.3	2.9	4.3	64
65.0	1.5	1.5	1.5	1.6	1.6	1.6	1.7	1.9	2.0	2.2	2.7	3.5	7.2	65.0
65.5	1.5	1.5	1.5	1.6	1.6	1.7	1.8	1.9	2.1	2.3	2.8	3.9		65.5
66.0	1.6	1.6	1.6	1.6	1.7	1.7	1.9	2.0	2.1	2.5	3.1	4.4		66.0
66.5	1.6	1.6	1.6	1.7	1.7	1.8	1.9	2.1	2.3	2.6	3.3	5.4		66.5
67.0	1.7	1.7	1.7	1.7	1.7	1.8	2.0	2.1	2.3	2.8	3.6	7.5		67.0
67.5	1.7	1.7	1.7	1.7	1.8	1.9	2.0	2.2	2.5	2.9	4.1			67.5
68.0	1.7	1.8	1.8	1.8	1.9	2.0	2.1	2.3	2.6	3.2	4.7			68.0
68.5	1.8	1.8	1.8	1.8	2.0	2.0	2.2	2.4	2.8	3.5	5.7			68.5
69.0	1.8	1.9	1.9	1.9	1.9	2.1	2.2	2.5	2.9	3.8	7.9			69.0
69.5	1.9	1.9	1.9	1.9	2.1	2.2	2.4	2.6	3.2	4.3				69.5
70.0	1.9	1.9	1.9	2.0	2.1	2.3	2.5	2.8	3.4	5.0				70.0
70.5	2.0	2.0	2.0	2.2	2.2	2.4	2.6	3.0	3.6	6.0				70.5
71.0	2.0	2.0	2.1	2.2	2.3	2.5	2.7	3.1	4.1	8.3				71.0
71.5	2.1	2.1	2.2	2.3	2.4	2.5	2.9	3.3	4.6					71.5
72.0	2.2	2.2	2.3	2.3	2.4	2.6	3.0	3.6	5.3					72.0
72.5	2.2	2.2	2.3	2.4	2.5	2.7	3.2	3.9	6.4					72.5
73.0	2.3	2.3	2.4	2.5	2.7	2.9	3.4	4.4	8.9					73.0
73.5	2.4	2.4	2.5	2.6	2.8	3.0	3.6	4.9						73.5
74.0	2.4	2.4	2.5	2.7	2.9	3.3	3.8	5.6						74.0
74.5	2.5	2.6	2.7	2.8	3.0	3.4	4.2	6.8						74.5
75.0	2.6	2.7	2.8	2.9	3.2	3.7	4.7	9.3						75.0
75.5	2.7	2.8	2.8	3.0	3.3	3.9	5.3							75.5
76.0	2.8	2.8	2.9	3.2	3.5	4.2	5.6							76.0
76.5	2.9	3.0	3.1	3.3	3.7	4.5	7.3							76.5
77.0	3.0	3.1	3.2	3.5	4.0	5.1	10.2							77.0

For the **sun**, a **planet**, or a **star**, apply the correction to the observed amplitude in the direction **away** from the elevated pole. For the **moon** apply **half** the correction **toward** the elevated pole.

TABLE 24
Altitude Factor

a, the change of altitude in one minute from meridian transit.

Lati-tude	Declination **contrary** name to latitude, **upper** transit: **add** correction to observed altitude												Lati-tude
	0°	1°	2°	3°	4°	5°	6°	7°	8°	9°	10°	11°	
°	″	″	″	″	″	″	″	″	″	″	″	″	°
0					28.1	22.4	18.7	16.0	14.0	12.4	11.1	10.1	0
1				28.1	22.4	18.7	16.0	14.0	12.4	11.2	10.1	9.3	1
2			28.1	22.4	18.7	16.0	14.0	12.5	11.2	10.2	9.3	8.6	2
3		28.1	22.4	18.7	16.0	14.0	12.5	11.2	10.2	9.3	8.6	8.0	3
4	28.1	22.4	18.7	16.0	14.0	12.5	11.2	10.2	9.3	8.6	8.0	7.4	4
5	22.4	18.7	16.0	14.0	12.5	11.2	10.2	9.3	8.6	8.0	7.4	7.0	5
6	18.7	16.0	14.0	12.5	11.2	10.2	9.3	8.6	8.0	7.5	7.0	6.6	6
7	16.0	14.0	12.4	11.2	10.2	9.3	8.6	8.0	7.5	7.0	6.6	6.2	7
8	14.0	12.4	11.2	10.2	9.3	8.6	8.0	7.5	7.0	6.6	6.2	5.9	8
9	12.4	11.2	10.2	9.3	8.6	8.0	7.5	7.0	6.6	6.2	5.9	5.6	9
10	11.1	10.1	9.3	8.6	8.0	7.4	7.0	6.6	6.2	5.9	5.6	5.3	10
11	10.1	9.3	8.6	8.0	7.4	7.0	6.6	6.2	5.9	5.6	5.3	5.1	11
12	9.2	8.5	7.9	7.4	7.0	6.5	6.2	5.9	5.6	5.3	5.0	4.8	12
13	8.5	7.9	7.4	6.9	6.5	6.2	5.8	5.6	5.3	5.0	4.8	4.6	13
14	7.9	7.4	6.9	6.5	6.2	5.8	5.5	5.3	5.0	4.8	4.6	4.4	14
15	7.3	6.9	6.5	6.1	5.8	5.5	5.3	5.0	4.8	4.6	4.4	4.2	15
16	6.8	6.5	6.1	5.8	5.5	5.2	5.0	4.8	4.6	4.4	4.2	4.1	16
17	6.4	6.1	5.8	5.5	5.2	5.0	4.8	4.6	4.4	4.2	4.1	3.9	17
18	6.0	5.7	5.5	5.2	5.0	4.8	4.6	4.4	4.2	4.1	3.9	3.8	18
19	5.7	5.4	5.2	4.9	4.7	4.5	4.4	4.2	4.0	3.9	3.8	3.6	19
20	5.4	5.1	4.9	4.7	4.5	4.3	4.2	4.0	3.9	3.8	3.6	3.5	20
21	5.1	4.9	4.7	4.5	4.3	4.2	4.0	3.9	3.7	3.6	3.5	3.4	21
22	4.9	4.7	4.5	4.3	4.1	4.0	3.9	3.7	3.6	3.5	3.4	3.3	22
23	4.6	4.4	4.3	4.1	4.0	3.8	3.7	3.6	3.5	3.4	3.3	3.2	23
24	4.4	4.2	4.1	3.9	3.8	3.7	3.6	3.5	3.4	3.3	3.2	3.1	24
25	4.2	4.1	3.9	3.8	3.7	3.5	3.4	3.3	3.2	3.1	3.1	3.0	25
26	4.0	3.9	3.8	3.6	3.5	3.4	3.3	3.2	3.1	3.0	3.0	2.9	26
27	3.9	3.7	3.6	3.5	3.4	3.3	3.2	3.1	3.0	2.9	2.9	2.8	27
28	3.7	3.6	3.5	3.4	3.3	3.2	3.1	3.0	2.9	2.8	2.8	2.7	28
29	3.5	3.4	3.3	3.2	3.1	3.1	3.0	2.9	2.8	2.8	2.7	2.6	29
30	3.4	3.3	3.2	3.1	3.0	3.0	2.9	2.8	2.7	2.7	2.6	2.5	30
31	3.3	3.2	3.1	3.0	2.9	2.9	2.8	2.7	2.6	2.6	2.5	2.5	31
32	3.2	3.1	3.0	2.9	2.8	2.8	2.7	2.6	2.6	2.5	2.5	2.4	32
33	3.0	2.9	2.9	2.8	2.7	2.7	2.6	2.5	2.5	2.4	2.4	2.3	33
34	2.9	2.8	2.8	2.7	2.6	2.6	2.5	2.5	2.4	2.4	2.3	2.3	34
35	2.8	2.7	2.7	2.6	2.5	2.5	2.4	2.4	2.3	2.3	2.2	2.2	35
36	2.7	2.6	2.6	2.5	2.5	2.4	2.4	2.3	2.3	2.2	2.2	2.1	36
37	2.6	2.5	2.5	2.4	2.4	2.3	2.3	2.2	2.2	2.2	2.1	2.1	37
38	2.5	2.5	2.4	2.4	2.3	2.3	2.2	2.2	2.1	2.1	2.1	2.0	38
39	2.4	2.4	2.3	2.3	2.2	2.2	2.1	2.1	2.1	2.0	2.0	2.0	39
40	2.3	2.3	2.2	2.2	2.2	2.1	2.1	2.0	2.0	2.0	1.9	1.9	40
41	2.3	2.2	2.2	2.1	2.1	2.1	2.0	2.0	1.9	1.9	1.9	1.8	41
42	2.2	2.1	2.1	2.1	2.0	2.0	2.0	1.9	1.9	1.9	1.8	1.8	42
43	2.1	2.1	2.0	2.0	2.0	1.9	1.9	1.9	1.8	1.8	1.8	1.7	43
44	2.0	2.0	2.0	1.9	1.9	1.9	1.8	1.8	1.8	1.7	1.7	1.7	44
45	2.0	1.9	1.9	1.9	1.8	1.8	1.8	1.7	1.7	1.7	1.7	1.6	45
46	1.9	1.9	1.8	1.8	1.8	1.7	1.7	1.7	1.7	1.6	1.6	1.6	46
47	1.8	1.8	1.8	1.7	1.7	1.7	1.7	1.6	1.6	1.6	1.6	1.6	47
48	1.8	1.7	1.7	1.7	1.7	1.6	1.6	1.6	1.6	1.6	1.5	1.5	48
49	1.7	1.7	1.7	1.6	1.6	1.6	1.6	1.5	1.5	1.5	1.5	1.5	49
50	1.6	1.6	1.6	1.6	1.6	1.5	1.5	1.5	1.5	1.5	1.4	1.4	50
51	1.6	1.6	1.6	1.5	1.5	1.5	1.5	1.5	1.4	1.4	1.4	1.4	51
52	1.5	1.5	1.5	1.5	1.5	1.4	1.4	1.4	1.4	1.4	1.4	1.3	52
53	1.5	1.5	1.4	1.4	1.4	1.4	1.4	1.4	1.3	1.3	1.3	1.3	53
54	1.4	1.4	1.4	1.4	1.4	1.3	1.3	1.3	1.3	1.3	1.3	1.3	54
55	1.4	1.4	1.3	1.3	1.3	1.3	1.3	1.3	1.3	1.2	1.2	1.2	55
56	1.3	1.3	1.3	1.3	1.3	1.3	1.2	1.2	1.2	1.2	1.2	1.2	56
57	1.3	1.3	1.3	1.2	1.2	1.2	1.2	1.2	1.2	1.2	1.1	1.1	57
58	1.2	1.2	1.2	1.2	1.2	1.2	1.2	1.1	1.1	1.1	1.1	1.1	58
59	1.2	1.2	1.2	1.2	1.1	1.1	1.1	1.1	1.1	1.1	1.1	1.1	59
60	1.1	1.1	1.1	1.1	1.1	1.1	1.1	1.1	1.0	1.0	1.0	1.0	60
Lati-tude	0°	1°	2°	3°	4°	5°	6°	7°	8°	9°	10°	11°	Lati-tude
	Declination **contrary** name to latitude, **upper** transit: **add** correction to observed altitude												

TABLE 24
Altitude Factor

a, the change of altitude in one minute from meridian transit.

Lati-tude	Declination **same** name to latitude, **upper** transit: **add** correction to observed altitude												Lati-tude
	0°	1°	2°	3°	4°	5°	6°	7°	8°	9°	10°	11°	
°	″	″	″	″	″	″	″	″	″	″	″	″	°
0					28.1	22.4	18.7	16.0	14.0	12.4	11.1	10.1	0
1						28.0	22.4	18.6	16.0	13.9	12.4	11.1	1
2							28.0	22.3	18.6	15.9	13.9	12.3	2
3								27.9	22.3	18.5	15.8	13.8	3
4	28.1								27.8	22.2	18.5	15.8	4
5	22.4	28.0								27.7	22.1	18.4	5
6	18.7	22.4	28.0								27.6	22.0	6
7	16.0	18.6	22.3	27.9								27.4	7
8	14.0	16.0	18.6	22.3	27.8								8
9	12.4	13.9	15.9	18.5	22.2	27.7							9
10	11.1	12.4	13.9	15.8	18.5	22.1	27.6						10
11	10.1	11.1	12.3	13.8	15.8	18.4	22.0	27.4					11
12	9.2	10.1	11.1	12.3	13.8	15.7	18.3	21.9	27.3				12
13	8.5	9.2	10.0	11.0	12.2	13.7	15.6	18.2	21.7	27.1			13
14	7.9	8.5	9.2	10.0	10.9	12.1	13.6	15.5	18.0	21.6	26.9		14
15	7.3	7.8	8.4	9.1	9.9	10.9	12.1	13.5	15.4	17.9	21.4	26.7	15
16	6.8	7.3	7.8	8.4	9.1	9.8	10.8	12.0	13.4	15.3	17.8	21.3	16
17	6.4	6.8	7.2	7.8	8.3	9.0	9.8	10.7	11.9	13.3	15.2	17.6	17
18	6.0	6.4	6.8	7.2	7.7	8.3	8.9	9.7	10.6	11.8	13.2	15.0	18
19	5.7	6.0	6.3	6.7	7.2	7.6	8.2	8.9	9.6	10.6	11.7	13.1	19
20	5.4	5.7	6.0	6.3	6.7	7.1	7.6	8.1	8.8	9.5	10.5	11.6	20
21	5.1	5.4	5.6	5.9	6.3	6.6	7.0	7.5	8.1	8.7	9.5	10.4	21
22	4.9	5.1	5.3	5.6	5.9	6.2	6.6	7.0	7.5	8.0	8.6	9.4	22
23	4.6	4.8	5.0	5.3	5.5	5.8	6.1	6.5	6.9	7.4	7.9	8.5	23
24	4.4	4.6	4.8	5.0	5.2	5.5	5.8	6.1	6.4	6.8	7.3	7.8	24
25	4.2	4.4	4.6	4.7	5.0	5.2	5.4	5.7	6.0	6.4	6.8	7.2	25
26	4.0	4.2	4.3	4.5	4.7	4.9	5.1	5.4	5.7	6.0	6.3	6.7	26
27	3.9	4.0	4.1	4.3	4.5	4.7	4.9	5.1	5.3	5.6	5.9	6.2	27
28	3.7	3.8	4.0	4.1	4.3	4.4	4.6	4.8	5.0	5.3	5.5	5.8	28
29	3.5	3.7	3.8	3.9	4.1	4.2	4.4	4.6	4.7	5.0	5.2	5.5	29
30	3.4	3.5	3.6	3.7	3.9	4.0	4.2	4.3	4.5	4.7	4.9	5.1	30
31	3.3	3.4	3.5	3.6	3.7	3.8	4.0	4.1	4.3	4.4	4.6	4.8	31
32	3.1	3.2	3.3	3.4	3.5	3.7	3.8	3.9	4.1	4.2	4.4	4.6	32
33	3.0	3.1	3.2	3.3	3.4	3.5	3.6	3.7	3.9	4.0	4.2	4.3	33
34	2.9	3.0	3.1	3.2	3.2	3.3	3.4	3.6	3.7	3.8	3.9	4.1	34
35	2.8	2.9	3.0	3.0	3.1	3.2	3.3	3.4	3.5	3.6	3.7	3.9	35
36	2.7	2.8	2.8	2.9	3.0	3.1	3.2	3.3	3.4	3.5	3.6	3.7	36
37	2.6	2.7	2.7	2.8	2.9	2.9	3.0	3.1	3.2	3.3	3.4	3.5	37
38	2.5	2.6	2.6	2.7	2.8	2.8	2.9	3.0	3.0	3.2	3.2	3.3	38
39	2.4	2.5	2.5	2.6	2.7	2.7	2.8	2.9	2.9	3.0	3.1	3.2	39
40	2.3	2.4	2.4	2.5	2.6	2.6	2.7	2.7	2.8	2.9	3.0	3.0	40
41	2.3	2.3	2.4	2.4	2.5	2.5	2.6	2.6	2.7	2.8	2.8	2.9	41
42	2.2	2.2	2.3	2.3	2.4	2.4	2.5	2.5	2.6	2.6	2.7	2.8	42
43	2.1	2.1	2.2	2.2	2.3	2.3	2.4	2.4	2.5	2.5	2.6	2.7	43
44	2.0	2.1	2.1	2.1	2.2	2.2	2.3	2.3	2.4	2.4	2.5	2.5	44
45	2.0	2.0	2.0	2.1	2.1	2.2	2.2	2.2	2.3	2.3	2.4	2.4	45
46	1.9	1.9	2.0	2.0	2.0	2.1	2.1	2.2	2.2	2.2	2.3	2.3	46
47	1.8	1.9	1.9	1.9	2.0	2.0	2.0	2.1	2.1	2.1	2.2	2.2	47
48	1.8	1.8	1.8	1.9	1.9	1.9	2.0	2.0	2.0	2.1	2.1	2.1	48
49	1.7	1.7	1.8	1.8	1.8	1.8	1.9	1.9	1.9	2.0	2.0	2.1	49
50	1.6	1.7	1.7	1.7	1.8	1.8	1.8	1.8	1.9	1.9	1.9	2.0	50
51	1.6	1.6	1.6	1.7	1.7	1.7	1.7	1.8	1.8	1.8	1.9	1.9	51
52	1.5	1.6	1.6	1.6	1.6	1.6	1.7	1.7	1.7	1.8	1.8	1.8	52
53	1.5	1.5	1.5	1.5	1.6	1.6	1.6	1.6	1.7	1.7	1.7	1.7	53
54	1.4	1.4	1.5	1.5	1.5	1.5	1.5	1.6	1.6	1.6	1.6	1.7	54
55	1.4	1.4	1.4	1.4	1.5	1.5	1.5	1.5	1.5	1.6	1.6	1.6	55
56	1.3	1.3	1.4	1.4	1.4	1.4	1.4	1.4	1.5	1.5	1.5	1.5	56
57	1.3	1.3	1.3	1.3	1.3	1.4	1.4	1.4	1.4	1.4	1.4	1.5	57
58	1.2	1.2	1.3	1.3	1.3	1.3	1.3	1.3	1.3	1.4	1.4	1.4	58
59	1.2	1.2	1.2	1.2	1.2	1.3	1.3	1.3	1.3	1.3	1.3	1.3	59
60	1.1	1.1	1.2	1.2	1.2	1.2	1.2	1.2	1.2	1.2	1.3	1.3	60
Lati-tude	0°	1°	2°	3°	4°	5°	6°	7°	8°	9°	10°	11°	Lati-tude
	Declination **same** name to latitude, **upper** transit: **add** correction to observed altitude												

TABLE 24
Altitude Factor

a, the change of altitude in one minute from meridian transit.

Latitude	12°	13°	14°	15°	16°	17°	18°	19°	20°	21°	22°	23°	24°	Latitude
	Declination **same** name as latitude, **upper** transit: **add** correction to observed altitude													
°	″	″	″	″	″	″	″	″	″	″	″	″	″	°
0	9.2	8.5	7.9	7.3	6.8	6.4	6.0	5.7	5.4	5.1	4.9	4.6	4.4	0
1	10.1	9.2	8.5	7.8	7.3	6.8	6.4	6.0	5.7	5.4	5.1	4.8	4.6	1
2	11.1	10.0	9.2	8.4	7.8	7.2	6.8	6.3	6.0	5.6	5.3	5.0	4.8	2
3	12.3	11.0	10.0	9.1	8.4	7.8	7.2	6.7	6.3	5.9	5.6	5.3	5.0	3
4	13.8	12.2	10.9	9.9	9.1	8.3	7.7	7.2	6.7	6.3	5.9	5.5	5.2	4
5	15.7	13.7	12.1	10.9	9.8	9.0	8.3	7.6	7.1	6.6	6.2	5.8	5.5	5
6	18.3	15.6	13.6	12.1	10.8	9.8	8.9	8.2	7.6	7.0	6.6	6.1	5.8	6
7	21.9	18.2	15.5	13.5	12.0	10.7	9.7	8.9	8.1	7.5	7.0	6.5	6.1	7
8	27.3	21.7	18.0	15.4	13.4	11.9	10.6	9.6	8.8	8.1	7.5	6.9	6.4	8
9		27.1	21.6	17.9	15.3	13.3	11.8	10.6	9.5	8.7	8.0	7.4	6.8	9
10			26.9	21.4	17.8	15.2	13.2	11.7	10.5	9.5	8.6	7.9	7.3	10
11				26.7	21.3	17.6	15.0	13.1	11.6	10.4	9.4	8.5	7.8	11
12					26.5	21.1	17.5	14.9	13.0	11.5	10.3	9.3	8.4	12
13						26.2	20.9	17.3	14.8	12.8	11.3	10.1	9.2	13
14							26.0	20.7	17.1	14.6	12.7	11.2	10.0	14
15								25.7	20.4	16.9	14.4	12.5	11.1	15
16	26.5								25.4	20.2	16.7	14.3	12.4	16
17	21.1	26.2								25.1	20.0	16.5	14.1	17
18	17.5	20.9	26.0								24.8	19.7	16.3	18
19	14.9	17.3	20.7	25.7								24.5	19.5	19
20	13.0	14.8	17.1	20.4	25.4								24.2	20
21	11.5	12.8	14.6	16.9	20.2	25.1								21
22	10.3	11.3	12.7	14.4	16.7	20.0	24.8							22
23	9.3	10.1	11.2	12.5	14.3	16.5	19.7	24.5						23
24	8.4	9.2	10.0	11.1	12.4	14.1	16.3	19.5	24.2					24
25	7.7	8.3	9.0	9.9	10.9	12.2	13.9	16.1	19.2	23.8				25
26	7.1	7.6	8.2	8.9	9.8	10.8	12.1	13.7	15.9	18.9	23.5			26
27	6.6	7.0	7.5	8.1	8.8	9.6	10.6	11.9	13.5	15.6	18.6	23.1		27
28	6.2	6.5	7.0	7.4	8.0	8.7	9.5	10.5	11.7	13.3	15.4	18.3	22.7	28
29	5.7	6.1	6.4	6.9	7.3	7.9	8.6	9.4	10.3	11.5	13.1	15.1	18.0	29
30	5.4	5.7	6.0	6.4	6.8	7.2	7.8	8.4	9.2	10.1	11.3	12.8	14.9	30
31	5.1	5.3	5.6	5.9	6.3	6.7	7.1	7.7	8.3	9.0	10.0	11.1	12.6	31
32	4.8	5.0	5.2	5.5	5.8	6.2	6.5	7.0	7.5	8.1	8.9	9.8	10.9	32
33	4.5	4.7	4.9	5.1	5.4	5.7	6.1	6.4	6.9	7.4	8.0	8.7	9.6	33
34	4.3	4.4	4.6	4.8	5.1	5.3	5.6	5.9	6.3	6.8	7.3	7.8	8.6	34
35	4.0	4.2	4.4	4.5	4.7	5.0	5.2	5.5	5.8	6.2	6.6	7.1	7.7	35
36	3.8	4.0	4.1	4.3	4.5	4.7	4.9	5.1	5.4	5.7	6.1	6.5	7.0	36
37	3.6	3.8	3.9	4.0	4.2	4.4	4.6	4.8	5.0	5.3	5.6	6.0	6.4	37
38	3.4	3.6	3.7	3.8	4.0	4.1	4.3	4.5	4.7	4.9	5.2	5.5	5.8	38
39	3.3	3.4	3.5	3.6	3.8	3.9	4.0	4.2	4.4	4.6	4.8	5.1	5.4	39
40	3.1	3.2	3.3	3.4	3.6	3.7	3.8	4.0	4.1	4.3	4.5	4.7	5.0	40
41	3.0	3.1	3.2	3.3	3.4	3.5	3.6	3.7	3.9	4.0	4.2	4.4	4.6	41
42	2.9	2.9	3.0	3.1	3.2	3.3	3.4	3.5	3.7	3.8	4.0	4.1	4.3	42
43	2.7	2.8	2.9	3.0	3.0	3.1	3.2	3.3	3.5	3.6	3.7	3.9	4.0	43
44	2.6	2.7	2.7	2.8	2.9	3.0	3.1	3.2	3.3	3.4	3.5	3.6	3.8	44
45	2.5	2.6	2.6	2.7	2.8	2.8	2.9	3.0	3.1	3.2	3.3	3.4	3.5	45
46	2.4	2.4	2.5	2.6	2.6	2.7	2.8	2.8	2.9	3.0	3.1	3.2	3.3	46
47	2.3	2.3	2.4	2.4	2.5	2.6	2.6	2.7	2.8	2.9	2.9	3.0	3.1	47
48	2.2	2.2	2.3	2.3	2.4	2.4	2.5	2.6	2.6	2.7	2.8	2.9	3.0	48
49	2.1	2.1	2.2	2.2	2.3	2.3	2.4	2.4	2.5	2.6	2.6	2.7	2.8	49
50	2.0	2.0	2.1	2.1	2.2	2.2	2.3	2.3	2.4	2.4	2.5	2.6	2.6	50
51	1.9	2.0	2.0	2.0	2.1	2.1	2.2	2.2	2.3	2.3	2.4	2.4	2.5	51
52	1.8	1.9	1.9	1.9	2.0	2.0	2.1	2.1	2.1	2.2	2.2	2.3	2.4	52
53	1.8	1.8	1.8	1.9	1.9	1.9	2.0	2.0	2.0	2.1	2.1	2.2	2.2	53
54	1.7	1.7	1.7	1.8	1.8	1.8	1.9	1.9	1.9	2.0	2.0	2.1	2.1	54
55	1.6	1.6	1.7	1.7	1.7	1.8	1.8	1.8	1.9	1.9	1.9	2.0	2.0	55
56	1.5	1.6	1.6	1.6	1.6	1.7	1.7	1.7	1.8	1.8	1.8	1.9	1.9	56
57	1.5	1.5	1.5	1.5	1.6	1.6	1.6	1.6	1.7	1.7	1.7	1.8	1.8	57
58	1.4	1.4	1.5	1.5	1.5	1.5	1.5	1.6	1.6	1.6	1.6	1.7	1.7	58
59	1.4	1.4	1.4	1.4	1.4	1.5	1.5	1.5	1.5	1.5	1.6	1.6	1.6	59
60	1.3	1.3	1.3	1.3	1.4	1.4	1.4	1.4	1.4	1.5	1.5	1.5	1.5	60
Latitude	12°	13°	14°	15°	16°	17°	18°	19°	20°	21°	22°	23°	24°	Latitude
	Declination **same** name as latitude, **upper** transit: **add** correction to observed altitude													

TABLE 24
Altitude Factor

a, the change of altitude in one minute from meridian transit.

Latitude	12°	13°	14°	15°	16°	17°	18°	19°	20°	21°	22°	23°	24°	Latitude
	Declination **contrary** name to latitude, **upper** transit: **add** correction to observed altitude													
°	″	″	″	″	″	″	″	″	″	″	″	″	″	°
0	9.2	8.5	7.9	7.3	6.8	6.4	6.0	5.7	5.4	5.1	4.9	4.6	4.4	0
1	8.5	7.9	7.4	6.9	6.5	6.1	5.7	5.4	5.1	4.9	4.7	4.4	4.2	1
2	7.9	7.4	6.9	6.5	6.1	5.8	5.5	5.2	4.9	4.7	4.5	4.3	4.1	2
3	7.4	6.9	6.5	6.1	5.8	5.5	5.2	4.9	4.7	4.5	4.3	4.1	3.9	3
4	7.0	6.5	6.2	5.8	5.5	5.2	5.0	4.7	4.5	4.3	4.1	4.0	3.8	4
5	6.5	6.2	5.8	5.5	5.2	5.0	4.8	4.5	4.3	4.2	4.0	3.8	3.7	5
6	6.2	5.8	5.5	5.3	5.0	4.8	4.6	4.4	4.2	4.0	3.9	3.7	3.6	6
7	5.9	5.6	5.3	5.0	4.8	4.6	4.4	4.2	4.0	3.9	3.7	3.6	3.5	7
8	5.6	5.3	5.0	4.8	4.6	4.4	4.2	4.0	3.9	3.7	3.6	3.5	3.4	8
9	5.3	5.0	4.8	4.6	4.4	4.2	4.1	3.9	3.8	3.6	3.5	3.4	3.3	9
10	5.0	4.8	4.6	4.4	4.2	4.1	3.9	3.8	3.6	3.5	3.4	3.3	3.2	10
11	4.8	4.6	4.4	4.2	4.1	3.9	3.8	3.6	3.5	3.4	3.3	3.2	3.1	11
12	4.6	4.4	4.3	4.1	3.9	3.8	3.7	3.5	3.4	3.3	3.2	3.1	3.0	12
13	4.4	4.3	4.1	3.9	3.8	3.7	3.5	3.4	3.3	3.2	3.1	3.0	2.9	13
14	4.2	4.1	3.9	3.8	3.7	3.5	3.4	3.3	3.2	3.1	3.0	2.9	2.8	14
15	4.1	3.9	3.8	3.7	3.5	3.4	3.3	3.2	3.1	3.0	2.9	2.8	2.8	15
16	3.9	3.8	3.7	3.5	3.4	3.3	3.2	3.1	3.0	2.9	2.8	2.8	2.7	16
17	3.8	3.7	3.5	3.4	3.3	3.2	3.1	3.0	2.9	2.8	2.8	2.7	2.6	17
18	3.7	3.5	3.4	3.3	3.2	3.1	3.0	2.9	2.9	2.8	2.7	2.6	2.5	18
19	3.5	3.4	3.3	3.2	3.1	3.0	2.9	2.9	2.8	2.7	2.6	2.6	2.5	19
20	3.4	3.3	3.2	3.1	3.0	2.9	2.9	2.8	2.7	2.6	2.6	2.5	2.4	20
21	3.3	3.2	3.1	3.0	2.9	2.8	2.8	2.7	2.6	2.6	2.5	2.4	2.4	21
22	3.2	3.1	3.0	2.9	2.8	2.8	2.7	2.6	2.6	2.5	2.4	2.4	2.3	22
23	3.1	3.0	2.9	2.8	2.8	2.7	2.6	2.6	2.5	2.4	2.4	2.3	2.3	23
24	3.0	2.9	2.8	2.8	2.7	2.6	2.5	2.5	2.4	2.4	2.3	2.3	2.2	24
25	2.9	2.8	2.7	2.7	2.6	2.5	2.5	2.4	2.4	2.3	2.3	2.2	2.2	25
26	2.8	2.7	2.7	2.6	2.5	2.5	2.4	2.4	2.3	2.3	2.2	2.1	2.1	26
27	2.7	2.7	2.6	2.5	2.5	2.4	2.4	2.3	2.2	2.2	2.1	2.1	2.1	27
28	2.6	2.6	2.5	2.5	2.4	2.3	2.3	2.2	2.2	2.1	2.1	2.1	2.0	28
29	2.6	2.5	2.4	2.4	2.3	2.3	2.2	2.2	2.1	2.1	2.0	2.0	2.0	29
30	2.5	2.4	2.4	2.3	2.3	2.2	2.2	2.1	2.1	2.0	2.0	2.0	1.9	30
31	2.4	2.4	2.3	2.3	2.2	2.2	2.1	2.1	2.0	2.0	2.0	1.9	1.9	31
32	2.3	2.3	2.2	2.2	2.2	2.1	2.1	2.0	2.0	1.9	1.9	1.9	1.8	32
33	2.3	2.2	2.2	2.1	2.1	2.1	2.0	2.0	1.9	1.9	1.9	1.8	1.8	33
34	2.2	2.2	2.1	2.1	2.0	2.0	2.0	1.9	1.9	1.9	1.8	1.8	1.8	34
35	2.2	2.1	2.1	2.0	2.0	2.0	1.9	1.9	1.8	1.8	1.8	1.7	1.7	35
36	2.1	2.1	2.0	2.0	1.9	1.9	1.9	1.8	1.8	1.8	1.7	1.7	1.7	36
37	2.0	2.0	2.0	1.9	1.9	1.9	1.8	1.8	1.8	1.7	1.7	1.7	1.6	37
38	2.0	1.9	1.9	1.9	1.8	1.8	1.8	1.7	1.7	1.7	1.7	1.6	1.6	38
39	1.9	1.9	1.9	1.8	1.8	1.8	1.7	1.7	1.7	1.6	1.6	1.6	1.6	39
40	1.9	1.8	1.8	1.8	1.7	1.7	1.7	1.7	1.6	1.6	1.6	1.6	1.5	40
41	1.8	1.8	1.8	1.7	1.7	1.7	1.6	1.6	1.6	1.6	1.5	1.5	1.5	41
42	1.8	1.7	1.7	1.7	1.7	1.6	1.6	1.6	1.6	1.5	1.5	1.5	1.5	42
43	1.7	1.7	1.7	1.6	1.6	1.6	1.6	1.5	1.5	1.5	1.5	1.4	1.4	43
44	1.7	1.6	1.6	1.6	1.6	1.5	1.5	1.5	1.5	1.5	1.4	1.4	1.4	44
45	1.6	1.6	1.6	1.5	1.5	1.5	1.5	1.5	1.4	1.4	1.4	1.4	1.4	45
46	1.6	1.6	1.5	1.5	1.5	1.5	1.4	1.4	1.4	1.4	1.4	1.3	1.3	46
47	1.5	1.5	1.5	1.5	1.4	1.4	1.4	1.4	1.4	1.3	1.3	1.3	1.3	47
48	1.5	1.5	1.4	1.4	1.4	1.4	1.4	1.3	1.3	1.3	1.3	1.3	1.3	48
49	1.4	1.4	1.4	1.4	1.4	1.3	1.3	1.3	1.3	1.3	1.3	1.2	1.2	49
50	1.4	1.4	1.4	1.3	1.3	1.3	1.3	1.3	1.3	1.3	1.2	1.2	1.2	50
51	1.4	1.3	1.3	1.3	1.3	1.3	1.3	1.2	1.2	1.2	1.2	1.2	1.2	51
52	1.3	1.3	1.3	1.3	1.3	1.3	1.2	1.2	1.2	1.2	1.2	1.1	1.1	52
53	1.3	1.3	1.3	1.2	1.2	1.2	1.2	1.2	1.2	1.2	1.1	1.1	1.1	53
54	1.2	1.2	1.2	1.2	1.2	1.2	1.2	1.1	1.1	1.1	1.1	1.1	1.1	54
55	1.2	1.2	1.2	1.2	1.1	1.1	1.1	1.1	1.1	1.1	1.1	1.1	1.1	55
56	1.2	1.1	1.1	1.1	1.1	1.1	1.1	1.1	1.1	1.1	1.0	1.0	1.0	56
57	1.1	1.1	1.1	1.1	1.1	1.1	1.1	1.0	1.0	1.0	1.0	1.0	1.0	57
58	1.1	1.1	1.1	1.1	1.0	1.0	1.0	1.0	1.0	1.0	1.0	1.0	1.0	58
59	1.0	1.0	1.0	1.0	1.0	1.0	1.0	1.0	1.0	1.0	1.0	0.9	0.9	59
60	1.0	1.0	1.0	1.0	1.0	1.0	1.0	0.9	0.9	0.9	0.9	0.9	0.9	60
Latitude	12°	13°	14°	15°	16°	17°	18°	19°	20°	21°	22°	23°	24°	Latitude
	Declination **contrary** name to latitude, **upper** transit: **add** correction to observed altitude													

TABLE 24
Altitude Factor

a. the change of altitude in one minute from meridian transit.

Lati- tude	Declination **same** name as latitude, **upper** transit: **add** correction to observed altitude													Lati- tude
	25°	26°	27°	28°	29°	30°	31°	32°	33°	34°	35°	36°	37°	
°	″	″	″	″	″	″	″	″	″	″	″	″	″	°
0	4.2	4.0	3.9	3.7	3.5	3.4	3.3	3.1	3.0	2.9	2.8	2.7	2.6	0
1	4.4	4.2	4.0	3.8	3.7	3.5	3.4	3.2	3.1	3.0	2.9	2.8	2.7	1
2	4.6	4.3	4.1	4.0	3.8	3.6	3.5	3.3	3.2	3.1	3.0	2.8	2.7	2
3	4.7	4.5	4.3	4.1	3.9	3.7	3.6	3.4	3.3	3.2	3.0	2.9	2.8	3
4	5.0	4.7	4.5	4.3	4.1	3.9	3.7	3.5	3.4	3.2	3.1	3.0	2.9	4
5	5.2	4.9	4.7	4.4	4.2	4.0	3.8	3.7	3.5	3.3	3.2	3.1	3.0	5
6	5.4	5.1	4.9	4.6	4.4	4.2	4.0	3.8	3.6	3.5	3.3	3.2	3.0	6
7	5.7	5.4	5.1	4.8	4.6	4.3	4.1	3.9	3.7	3.6	3.4	3.3	3.1	7
8	6.0	5.7	5.3	5.0	4.8	4.5	4.3	4.1	3.9	3.7	3.5	3.4	3.2	8
9	6.4	6.0	5.6	5.3	5.0	4.7	4.4	4.2	4.0	3.8	3.6	3.5	3.3	9
10	6.8	6.3	5.9	5.5	5.2	4.9	4.6	4.4	4.2	3.9	3.7	3.6	3.4	10
11	7.2	6.7	6.2	5.8	5.5	5.1	4.8	4.6	4.3	4.1	3.9	3.7	3.5	11
12	7.7	7.1	6.6	6.2	5.8	5.4	5.1	4.8	4.5	4.3	4.0	3.8	3.6	12
13	8.3	7.6	7.1	6.5	6.1	5.7	5.3	5.0	4.7	4.4	4.2	4.0	3.8	13
14	9.1	8.2	7.6	7.0	6.4	6.0	5.6	5.2	4.9	4.6	4.4	4.1	3.9	14
15	9.9	8.9	8.1	7.4	6.9	6.4	5.9	5.5	5.2	4.8	4.5	4.3	4.0	15
16	10.9	9.8	8.8	8.0	7.3	6.8	6.3	5.8	5.4	5.1	4.8	4.5	4.2	16
17	12.2	10.8	9.6	8.7	7.9	7.2	6.7	6.2	5.7	5.3	5.0	4.7	4.4	17
18	13.9	12.1	10.6	9.5	8.6	7.8	7.1	6.6	6.1	5.6	5.2	4.9	4.6	18
19	16.1	13.7	11.9	10.5	9.4	8.4	7.7	7.0	6.4	6.0	5.5	5.1	4.8	19
20	19.2	15.9	13.5	11.7	10.3	9.2	8.3	7.5	6.9	6.3	5.8	5.4	5.0	20
21	23.8	18.9	15.6	13.3	11.5	10.1	9.1	8.2	7.4	6.8	6.2	5.7	5.3	21
22		23.5	18.6	15.4	13.1	11.3	10.0	8.9	8.0	7.3	6.6	6.1	5.6	22
23			23.1	18.3	15.1	12.8	11.1	9.8	8.7	7.9	7.1	6.5	6.0	23
24				22.7	18.0	14.9	12.6	10.9	9.6	8.6	7.7	7.0	6.4	24
25					22.3	17.7	14.6	12.4	10.7	9.4	8.4	7.5	6.8	25
26						21.9	17.4	14.3	12.1	10.5	9.2	8.2	7.4	26
27							21.5	17.0	14.0	11.9	10.3	9.1	8.1	27
28								21.1	16.7	13.8	11.7	10.1	8.9	28
29	22.3								20.6	16.3	13.5	11.4	9.9	29
30	17.7	21.9								20.2	16.0	13.2	11.1	30
31	14.6	17.4	21.5								19.8	15.6	12.9	31
32	12.4	14.3	17.0	21.1								19.3	15.3	32
33	10.7	12.1	14.0	16.7	20.6								18.9	33
34	9.4	10.5	11.9	13.8	16.3	20.2								34
35	8.4	9.2	10.3	11.7	13.5	16.0	19.8							35
36	7.5	8.2	9.1	10.1	11.4	13.2	15.6	19.3						36
37	6.8	7.4	8.1	8.9	9.9	11.1	12.9	15.3	18.9					37
38	6.2	6.7	7.2	7.9	8.7	9.6	10.9	12.6	14.9	18.4				38
39	5.7	6.1	6.5	7.1	7.7	8.5	9.4	10.6	12.2	14.5	17.9			39
40	5.3	5.6	6.0	6.4	6.9	7.5	8.2	9.2	10.4	11.9	14.1	17.4		40
41	4.9	5.2	5.5	5.8	6.2	6.7	7.3	8.0	8.9	10.1	11.6	13.8	17.0	41
42	4.5	4.8	5.0	5.3	5.7	6.1	6.6	7.1	7.8	8.7	9.8	11.3	13.4	42
43	4.2	4.4	4.6	4.9	5.2	5.5	5.9	6.4	6.9	7.6	8.5	9.5	11.0	43
44	3.9	4.1	4.3	4.5	4.8	5.1	5.4	5.8	6.2	6.7	7.4	8.2	9.3	44
45	3.7	3.8	4.0	4.2	4.4	4.7	4.9	5.2	5.6	6.0	6.6	7.2	8.0	45
46	3.5	3.6	3.7	3.9	4.1	4.3	4.5	4.8	5.1	5.4	5.9	6.4	7.0	46
47	3.3	3.4	3.5	3.6	3.8	4.0	4.2	4.4	4.6	4.9	5.3	5.7	6.2	47
48	3.1	3.2	3.3	3.4	3.5	3.7	3.9	4.0	4.3	4.5	4.8	5.1	5.5	48
49	2.9	3.0	3.1	3.2	3.3	3.4	3.6	3.7	3.9	4.1	4.4	4.6	5.0	49
50	2.7	2.8	2.9	3.0	3.1	3.2	3.3	3.5	3.6	3.8	4.0	4.2	4.5	50
51	2.6	2.6	2.7	2.8	2.9	3.0	3.1	3.2	3.4	3.5	3.7	3.9	4.1	51
52	2.4	2.5	2.6	2.6	2.7	2.8	2.9	3.0	3.1	3.2	3.4	3.6	3.7	52
53	2.3	2.3	2.4	2.5	2.5	2.6	2.7	2.8	2.9	3.0	3.1	3.3	3.4	53
54	2.2	2.2	2.3	2.3	2.4	2.5	2.5	2.6	2.7	2.8	2.9	3.0	3.2	54
55	2.0	2.1	2.1	2.2	2.2	2.3	2.4	2.4	2.5	2.6	2.7	2.8	2.9	55
56	1.9	2.0	2.0	2.1	2.1	2.2	2.2	2.3	2.4	2.4	2.5	2.6	2.7	56
57	1.8	1.9	1.9	2.0	2.0	2.0	2.1	2.2	2.2	2.3	2.3	2.4	2.5	57
58	1.7	1.8	1.8	1.8	1.9	1.9	2.0	2.0	2.1	2.1	2.2	2.3	2.3	58
59	1.6	1.7	1.7	1.7	1.8	1.8	1.9	1.9	1.9	2.0	2.0	2.1	2.2	59
60	1.6	1.6	1.6	1.6	1.7	1.7	1.7	1.8	1.8	1.9	1.9	2.0	2.0	60
Lati- tude	25°	26°	27°	28°	29°	30°	31°	32°	33°	34°	35°	36°	37°	Lati- tude
	Declination **same** name as latitude, **upper** transit: **add** correction to observed altitude													

TABLE 24
Altitude Factor

a, the change of altitude in one minute from meridian transit.

Lati- tude	Declination **contrary** name to latitude **upper** transit: **add** correction to observed altitude													Lati- tude
	25°	26°	27°	28°	29°	30°	31°	32°	33°	34°	35°	36°	37°	
°	″	″	″	″	″	″	″	″	″	″	″	″	″	°
0	4.2	4.0	3.9	3.7	3.5	3.4	3.3	3.1	3.0	2.9	2.8	2.7	2.6	0
1	4.1	3.9	3.7	3.6	3.4	3.3	3.2	3.1	2.9	2.8	2.7	2.6	2.6	1
2	3.9	3.8	3.6	3.5	3.3	3.2	3.1	3.0	2.9	2.8	2.7	2.6	2.5	2
3	3.8	3.6	3.5	3.4	3.2	3.1	3.0	2.9	2.8	2.7	2.6	2.5	2.4	3
4	3.7	3.5	3.4	3.3	3.2	3.0	2.9	2.8	2.7	2.6	2.5	2.5	2.4	4
5	3.6	3.4	3.3	3.2	3.1	3.0	2.9	2.8	2.7	2.6	2.5	2.4	2.3	5
6	3.4	3.3	3.2	3.1	3.0	2.9	2.8	2.7	2.6	2.5	2.4	2.4	2.3	6
7	3.3	3.2	3.1	3.0	2.9	2.8	2.7	2.6	2.5	2.5	2.4	2.3	2.2	7
8	3.2	3.1	3.0	2.9	2.8	2.7	2.7	2.6	2.5	2.4	2.3	2.3	2.2	8
9	3.1	3.0	2.9	2.8	2.8	2.7	2.6	2.5	2.4	2.4	2.3	2.2	2.2	9
10	3.1	3.0	2.9	2.8	2.7	2.6	2.5	2.5	2.4	2.3	2.2	2.2	2.1	10
11	3.0	2.9	2.8	2.7	2.6	2.5	2.5	2.4	2.3	2.3	2.2	2.1	2.1	11
12	2.9	2.8	2.7	2.6	2.6	2.5	2.4	2.3	2.3	2.2	2.2	2.1	2.0	12
13	2.8	2.7	2.7	2.6	2.5	2.4	2.4	2.3	2.2	2.2	2.1	2.1	2.0	13
14	2.7	2.7	2.6	2.5	2.4	2.4	2.3	2.3	2.2	2.2	2.1	2.0	2.0	14
15	2.7	2.6	2.5	2.5	2.4	2.3	2.3	2.2	2.1	2.1	2.0	2.0	1.9	15
16	2.6	2.5	2.5	2.4	2.3	2.3	2.2	2.2	2.1	2.0	2.0	1.9	1.9	16
17	2.5	2.5	2.4	2.3	2.3	2.2	2.2	2.1	2.1	2.0	2.0	1.9	1.9	17
18	2.5	2.4	2.4	2.3	2.2	2.2	2.1	2.1	2.0	2.0	1.9	1.9	1.8	18
19	2.4	2.4	2.3	2.2	2.2	2.1	2.1	2.0	2.0	1.9	1.9	1.8	1.8	19
20	2.4	2.3	2.3	2.2	2.1	2.1	2.0	2.0	1.9	1.9	1.9	1.8	1.8	20
21	2.3	2.3	2.2	2.1	2.1	2.0	2.0	2.0	1.9	1.9	1.8	1.8	1.7	21
22	2.3	2.2	2.2	2.1	2.1	2.0	2.0	1.9	1.9	1.8	1.8	1.7	1.7	22
23	2.2	2.2	2.1	2.1	2.0	2.0	1.9	1.9	1.8	1.8	1.8	1.7	1.7	23
24	2.2	2.1	2.1	2.0	2.0	1.9	1.9	1.8	1.8	1.8	1.7	1.7	1.6	24
25	2.1	2.1	2.0	2.0	1.9	1.9	1.8	1.8	1.8	1.7	1.7	1.6	1.6	25
26	2.1	2.0	2.0	1.9	1.9	1.8	1.8	1.8	1.7	1.7	1.7	1.6	1.6	26
27	2.0	2.0	1.9	1.9	1.9	1.8	1.8	1.7	1.7	1.7	1.6	1.6	1.6	27
28	2.0	1.9	1.9	1.8	1.8	1.8	1.7	1.7	1.7	1.6	1.6	1.6	1.5	28
29	1.9	1.9	1.9	1.8	1.8	1.7	1.7	1.7	1.6	1.6	1.6	1.5	1.5	29
30	1.9	1.8	1.8	1.8	1.7	1.7	1.7	1.6	1.6	1.6	1.5	1.5	1.5	30
31	1.8	1.8	1.8	1.7	1.7	1.7	1.6	1.6	1.6	1.5	1.5	1.5	1.5	31
32	1.8	1.8	1.7	1.7	1.7	1.6	1.6	1.6	1.5	1.5	1.5	1.5	1.4	32
33	1.8	1.7	1.7	1.7	1.6	1.6	1.6	1.5	1.5	1.5	1.5	1.4	1.4	33
34	1.7	1.7	1.7	1.6	1.6	1.6	1.5	1.5	1.5	1.5	1.4	1.4	1.4	34
35	1.7	1.7	1.6	1.6	1.6	1.5	1.5	1.5	1.5	1.4	1.4	1.4	1.4	35
36	1.6	1.6	1.6	1.6	1.5	1.5	1.5	1.5	1.4	1.4	1.4	1.4	1.3	36
37	1.6	1.6	1.6	1.5	1.5	1.5	1.5	1.4	1.4	1.4	1.4	1.3	1.3	37
38	1.6	1.5	1.5	1.5	1.5	1.5	1.4	1.4	1.4	1.4	1.3	1.3	1.3	38
39	1.5	1.5	1.5	1.5	1.4	1.4	1.4	1.4	1.4	1.3	1.3	1.3	1.3	39
40	1.5	1.5	1.5	1.4	1.4	1.4	1.4	1.3	1.3	1.3	1.3	1.3	1.2	40
41	1.5	1.4	1.4	1.4	1.4	1.4	1.3	1.3	1.3	1.3	1.3	1.2	1.2	41
42	1.4	1.4	1.4	1.4	1.4	1.3	1.3	1.3	1.3	1.2	1.2	1.2	1.2	42
43	1.4	1.4	1.4	1.3	1.3	1.3	1.3	1.3	1.2	1.2	1.2	1.2	1.2	43
44	1.4	1.4	1.3	1.3	1.3	1.3	1.3	1.2	1.2	1.2	1.2	1.2	1.2	44
45	1.3	1.3	1.3	1.3	1.3	1.2	1.2	1.2	1.2	1.2	1.2	1.1	1.1	45
46	1.3	1.3	1.3	1.3	1.2	1.2	1.2	1.2	1.2	1.2	1.1	1.1	1.1	46
47	1.3	1.3	1.2	1.2	1.2	1.2	1.2	1.2	1.1	1.1	1.1	1.1	1.1	47
48	1.2	1.2	1.2	1.2	1.2	1.2	1.1	1.1	1.1	1.1	1.1	1.1		48
49	1.2	1.2	1.2	1.2	1.2	1.1	1.1	1.1	1.1	1.1	1.1			49
50	1.2	1.2	1.2	1.1	1.1	1.1	1.1	1.1	1.1	1.1				50
51	1.2	1.1	1.1	1.1	1.1	1.1	1.1	1.1	1.0					51
52	1.1	1.1	1.1	1.1	1.1	1.1	1.0	1.0						52
53	1.1	1.1	1.1	1.1	1.0	1.0	1.0							53
54	1.1	1.1	1.0	1.0	1.0	1.0								54
55	1.0	1.0	1.0	1.0	1.0									55
56	1.0	1.0	1.0	1.0										56
57	1.0	1.0	1.0											57
58	1.0	0.9												58
59	0.9												0.8	59
60												0.8	0.8	60
Lati- tude	25°	26°	27°	28°	29°	30°	31°	32°	33°	34°	35°	36°	37°	Lati- tude
	Declination **contrary** name to latitude, **upper** transit: **add** correction to observed altitude													

TABLE 24
Altitude Factor

a, the change of altitude in one minute from meridian transit.

Latitude	38°	39°	40°	41°	42°	43°	44°	45°	46°	47°	48°	49°	50°	Latitude
	Declination **same** name as latitude, **upper** transit: **add** correction to observed altitude													
°	″	″	″	″	″	″	″	″	″	″	″	″	″	°
0	2.5	2.4	2.3	2.3	2.2	2.1	2.0	2.0	1.9	1.8	1.8	1.7	1.7	0
1	2.6	2.5	2.4	2.3	2.2	2.2	2.1	2.0	1.9	1.9	1.8	1.7	1.7	1
2	2.6	2.5	2.4	2.4	2.3	2.2	2.1	2.0	2.0	1.9	1.8	1.8	1.7	2
3	2.7	2.6	2.5	2.4	2.3	2.2	2.2	2.1	2.0	1.9	1.9	1.8	1.7	3
4	2.8	2.7	2.6	2.5	2.4	2.3	2.2	2.1	2.0	2.0	1.9	1.8	1.8	4
5	2.8	2.7	2.6	2.5	2.4	2.3	2.2	2.2	2.1	2.0	1.9	1.9	1.8	5
6	2.9	2.8	2.7	2.6	2.5	2.4	2.3	2.2	2.1	2.0	2.0	1.9	1.8	6
7	3.0	2.9	2.7	2.6	2.5	2.4	2.3	2.2	2.2	2.1	2.0	1.9	1.8	7
8	3.1	2.9	2.8	2.7	2.6	2.5	2.4	2.3	2.2	2.1	2.0	1.9	1.9	8
9	3.2	3.0	2.9	2.8	2.6	2.5	2.4	2.3	2.2	2.2	2.1	2.0	1.9	9
10	3.2	3.1	3.0	2.8	2.7	2.6	2.5	2.4	2.3	2.2	2.1	2.0	1.9	10
11	3.4	3.2	3.1	2.9	2.8	2.7	2.6	2.4	2.3	2.2	2.1	2.1	2.0	11
12	3.5	3.3	3.1	3.0	2.9	2.7	2.6	2.5	2.4	2.3	2.2	2.1	2.0	12
13	3.6	3.4	3.2	3.1	2.9	2.8	2.7	2.6	2.4	2.3	2.2	2.1	2.0	13
14	3.7	3.5	3.3	3.2	3.0	2.9	2.7	2.6	2.5	2.4	2.3	2.2	2.1	14
15	3.8	3.6	3.4	3.3	3.1	3.0	2.8	2.7	2.6	2.4	2.3	2.2	2.1	15
16	4.0	3.8	3.6	3.4	3.2	3.0	2.9	2.8	2.6	2.5	2.4	2.3	2.2	16
17	4.1	3.9	3.7	3.5	3.3	3.1	3.0	2.8	2.7	2.6	2.4	2.3	2.2	17
18	4.3	4.1	3.8	3.6	3.4	3.2	3.1	2.9	2.8	2.6	2.5	2.4	2.3	18
19	4.5	4.2	4.0	3.7	3.5	3.3	3.2	3.0	2.8	2.7	2.6	2.4	2.3	19
20	4.7	4.4	4.1	3.9	3.7	3.5	3.3	3.1	2.9	2.8	2.6	2.5	2.4	20
21	4.9	4.6	4.3	4.0	3.8	3.6	3.4	3.2	3.0	2.9	2.7	2.6	2.4	21
22	5.2	4.8	4.5	4.2	4.0	3.7	3.5	3.3	3.1	2.9	2.8	2.6	2.5	22
23	5.5	5.1	4.7	4.4	4.1	3.9	3.6	3.4	3.2	3.0	2.9	2.7	2.6	23
24	5.8	5.4	5.0	4.6	4.3	4.0	3.8	3.5	3.3	3.1	3.0	2.8	2.6	24
25	6.2	5.7	5.3	4.9	4.5	4.2	3.9	3.7	3.5	3.3	3.1	2.9	2.7	25
26	6.7	6.1	5.6	5.2	4.8	4.4	4.1	3.8	3.6	3.4	3.2	3.0	2.8	26
27	7.2	6.5	6.0	5.5	5.0	4.6	4.3	4.0	3.7	3.5	3.3	3.1	2.9	27
28	7.9	7.1	6.4	5.8	5.3	4.9	4.5	4.2	3.9	3.6	3.4	3.2	3.0	28
29	8.7	7.7	6.9	6.2	5.7	5.2	4.8	4.4	4.1	3.8	3.5	3.3	3.1	29
30	9.6	8.5	7.5	6.7	6.1	5.5	5.1	4.7	4.3	4.0	3.7	3.4	3.2	30
31	10.9	9.4	8.2	7.3	6.6	5.9	5.4	4.9	4.5	4.2	3.9	3.6	3.3	31
32	12.6	10.6	9.2	8.0	7.1	6.4	5.8	5.2	4.8	4.4	4.0	3.7	3.5	32
33	14.9	12.2	10.4	8.9	7.8	6.9	6.2	5.6	5.1	4.6	4.3	3.9	3.6	33
34	18.4	14.5	11.9	10.1	8.7	7.6	6.7	6.0	5.4	4.9	4.5	4.1	3.8	34
35		17.9	14.1	11.6	9.8	8.5	7.4	6.6	5.9	5.3	4.8	4.4	4.0	35
36			17.4	13.8	11.3	9.5	8.2	7.2	6.4	5.7	5.1	4.6	4.2	36
37				17.0	13.4	11.0	9.3	8.0	7.0	6.2	5.5	5.0	4.5	37
38					16.5	13.0	10.7	9.0	7.7	6.8	6.0	5.3	4.8	38
39						16.0	12.6	10.3	8.7	7.5	6.5	5.8	5.1	39
40							15.5	12.2	10.0	8.4	7.2	6.3	5.6	40
41								15.0	11.8	9.7	8.1	7.0	6.1	41
42	16.5								14.5	11.4	9.3	7.9	6.7	42
43	13.0	16.0								14.0	11.0	9.0	7.6	43
44	10.6	12.6	15.5								13.6	10.6	8.7	44
45	9.0	10.3	12.2	15.0								13.1	10.2	45
46	7.7	8.7	10.0	11.8	14.5								12.6	46
47	6.8	7.5	8.4	9.7	11.4	14.0								47
48	6.0	6.5	7.2	8.1	9.3	11.0	13.6							48
49	5.3	5.8	6.3	7.0	7.9	9.0	10.6	13.1						49
50	4.8	5.1	5.6	6.1	6.7	7.6	8.7	10.2	12.6					50
51	4.3	4.6	5.0	5.4	5.9	6.5	7.3	8.4	9.9	12.1				51
52	3.9	4.2	4.5	4.8	5.2	5.7	6.3	7.0	8.0	9.5	11.6			52
53	3.6	3.8	4.0	4.3	4.6	5.0	5.4	6.0	6.7	7.7	9.1	11.1		53
54	3.3	3.5	3.7	3.9	4.1	4.4	4.8	5.2	5.8	6.5	7.4	8.7	10.6	54
55	3.0	3.2	3.3	3.5	3.7	4.0	4.3	4.6	5.0	5.5	6.2	7.1	8.3	55
56	2.8	2.9	3.1	3.2	3.4	3.6	3.8	4.1	4.4	4.8	5.3	5.9	6.8	56
57	2.6	2.7	2.8	2.9	3.1	3.2	3.4	3.6	3.9	4.2	4.6	5.0	5.6	57
58	2.4	2.5	2.6	2.7	2.8	2.9	3.1	3.3	3.5	3.7	4.0	4.4	4.8	58
59	2.2	2.3	2.4	2.5	2.6	2.7	2.8	3.0	3.1	3.3	3.6	3.8	4.2	59
60	2.1	2.1	2.2	2.3	2.4	2.5	2.6	2.7	2.8	3.0	3.2	3.4	3.6	60
Latitude	38°	39°	40°	41°	42°	43°	44°	45°	46°	47°	48°	49°	50°	Latitude
	Declination **same** name as latitude, **upper** transit: **add** correction to observed altitude													

TABLE 24
Altitude Factor

a, the change of altitude in one minute from meridian transit.

Latitude	38°	39°	40°	41°	42°	43°	44°	45°	46°	47°	48°	49°	50°	Latitude
	Declination **contrary** name to latitude, **upper** transit: **add** correction to observed altitude													
°	″	″	″	″	″	″	″	″	″	″	″	″	″	°
0	2.5	2.4	2.3	2.3	2.2	2.1	2.0	2.0	1.9	1.8	1.8	1.7	1.7	0
1	2.5	2.4	2.3	2.2	2.1	2.1	2.0	1.9	1.9	1.8	1.7	1.7	1.6	1
2	2.4	2.3	2.3	2.2	2.1	2.0	2.0	1.9	1.8	1.8	1.7	1.7	1.6	2
3	2.4	2.3	2.2	2.1	2.1	2.0	1.9	1.9	1.8	1.8	1.7	1.6	1.6	3
4	2.3	2.2	2.2	2.1	2.0	2.0	1.9	1.8	1.8	1.7	1.7	1.6	1.6	4
5	2.3	2.2	2.1	2.1	2.0	1.9	1.9	1.8	1.8	1.7	1.6	1.6	1.5	5
6	2.2	2.2	2.1	2.0	2.0	1.9	1.8	1.8	1.7	1.7	1.6	1.6	1.5	6
7	2.2	2.1	2.0	2.0	1.9	1.9	1.8	1.8	1.7	1.6	1.6	1.5	1.5	7
8	2.1	2.1	2.0	1.9	1.9	1.8	1.8	1.7	1.7	1.6	1.6	1.5	1.5	8
9	2.1	2.0	2.0	1.9	1.9	1.8	1.8	1.7	1.6	1.6	1.6	1.5	1.5	9
10	2.1	2.0	1.9	1.9	1.8	1.8	1.7	1.7	1.6	1.6	1.5	1.5	1.4	10
11	2.0	2.0	1.9	1.8	1.8	1.7	1.7	1.6	1.6	1.6	1.5	1.5	1.4	11
12	2.0	1.9	1.9	1.8	1.8	1.7	1.7	1.6	1.6	1.5	1.5	1.4	1.4	12
13	1.9	1.9	1.8	1.8	1.7	1.7	1.6	1.6	1.6	1.5	1.5	1.4	1.4	13
14	1.9	1.9	1.8	1.8	1.7	1.7	1.6	1.6	1.5	1.5	1.4	1.4	1.4	14
15	1.9	1.8	1.8	1.7	1.7	1.6	1.6	1.6	1.5	1.5	1.4	1.4	1.4	15
16	1.8	1.8	1.7	1.7	1.7	1.6	1.6	1.5	1.5	1.4	1.4	1.4	1.3	16
17	1.8	1.8	1.7	1.7	1.6	1.6	1.5	1.5	1.5	1.4	1.4	1.4	1.3	17
18	1.8	1.7	1.7	1.6	1.6	1.6	1.5	1.5	1.4	1.4	1.4	1.3	1.3	18
19	1.7	1.7	1.7	1.6	1.6	1.5	1.5	1.5	1.4	1.4	1.4	1.3	1.3	19
20	1.7	1.7	1.6	1.6	1.6	1.5	1.5	1.4	1.4	1.4	1.3	1.3	1.3	20
21	1.7	1.6	1.6	1.6	1.5	1.5	1.5	1.4	1.4	1.4	1.3	1.3	1.3	21
22	1.7	1.6	1.6	1.5	1.5	1.5	1.4	1.4	1.4	1.3	1.3	1.3	1.2	22
23	1.6	1.6	1.6	1.5	1.5	1.4	1.4	1.4	1.3	1.3	1.3	1.3	1.2	23
24	1.6	1.6	1.5	1.5	1.5	1.4	1.4	1.4	1.3	1.3	1.3	1.2	1.2	24
25	1.6	1.5	1.5	1.5	1.4	1.4	1.4	1.3	1.3	1.3	1.2	1.2	1.2	25
26	1.6	1.5	1.5	1.5	1.4	1.4	1.4	1.3	1.3	1.3	1.2	1.2	1.2	26
27	1.5	1.5	1.5	1.4	1.4	1.4	1.3	1.3	1.3	1.2	1.2	1.2	1.2	27
28	1.5	1.5	1.4	1.4	1.4	1.3	1.3	1.3	1.3	1.2	1.2	1.2	1.1	28
29	1.5	1.4	1.4	1.4	1.4	1.3	1.3	1.3	1.2	1.2	1.2	1.2	1.1	29
30	1.5	1.4	1.4	1.4	1.3	1.3	1.3	1.2	1.2	1.2	1.2	1.1	1.1	30
31	1.4	1.4	1.4	1.3	1.3	1.3	1.3	1.2	1.2	1.2	1.2	1.1	1.1	31
32	1.4	1.4	1.3	1.3	1.3	1.3	1.2	1.2	1.2	1.2	1.1	1.1	1.1	32
33	1.4	1.4	1.3	1.3	1.3	1.2	1.2	1.2	1.2	1.1	1.1	1.1	1.1	33
34	1.4	1.3	1.3	1.3	1.3	1.2	1.2	1.2	1.2	1.1	1.1	1.1	1.1	34
35	1.3	1.3	1.3	1.3	1.2	1.2	1.2	1.2	1.1	1.1	1.1	1.1		35
36	1.3	1.3	1.3	1.2	1.2	1.2	1.2	1.1	1.1	1.1	1.1			36
37	1.3	1.3	1.2	1.2	1.2	1.2	1.2	1.1	1.1	1.1				37
38	1.3	1.2	1.2	1.2	1.2	1.2	1.1	1.1	1.1					38
39	1.2	1.2	1.2	1.2	1.2	1.1	1.1	1.1						39
40	1.2	1.2	1.2	1.2	1.1	1.1	1.1							40
41	1.2	1.2	1.2	1.1	1.1	1.1								41
42	1.2	1.2	1.1	1.1	1.1									42
43	1.2	1.1	1.1	1.1										43
44	1.1	1.1	1.1											44
45	1.1	1.1												45
46	1.1												0.9	46
47												0.9	0.9	47
48											0.9	0.9	0.9	48
49										0.9	0.9	0.9	0.8	49
50									0.9	0.9	0.9	0.8	0.8	50
51								0.9	0.9	0.9	0.8	0.8	0.8	51
52							0.9	0.9	0.9	0.8	0.8	0.8	0.8	52
53						0.9	0.9	0.8	0.8	0.8	0.8	0.8	0.8	53
54					0.9	0.9	0.8	0.8	0.8	0.8	0.8	0.8	0.8	54
55				0.9	0.8	0.8	0.8	0.8	0.8	0.8	0.8	0.8	0.7	55
56			0.8	0.8	0.8	0.8	0.8	0.8	0.8	0.8	0.8	0.7	0.7	56
57		0.8	0.8	0.8	0.8	0.8	0.8	0.8	0.8	0.8	0.7	0.7	0.7	57
58	0.8	0.8	0.8	0.8	0.8	0.8	0.8	0.8	0.8	0.7	0.7	0.7	0.7	58
59	0.8	0.8	0.8	0.8	0.8	0.8	0.8	0.7	0.7	0.7	0.7	0.7	0.7	59
60	0.8	0.8	0.8	0.8	0.8	0.7	0.7	0.7	0.7	0.7	0.7	0.7	0.7	60
Latitude	38°	39°	40°	41°	42°	43°	44°	45°	46°	47°	48°	49°	50°	Latitude
	Declination **contrary** name to latitude, **upper** transit: **add** correction to observed altitude													

TABLE 24
Altitude Factor

a, the change of altitude in one minute from meridian transit.

Lati-tude	51°	52°	53°	54°	55°	56°	57°	58°	59°	60°	61°	62°	63°	Lati-tude
	Declination **same** name as latitude, **upper** transit: **add** correction to observed altitude													
°	″	″	″	″	″	″	″	″	″	″	″	″	″	°
0	1.6	1.5	1.5	1.4	1.4	1.3	1.3	1.2	1.2	1.1	1.1	1.0	1.0	0
1	1.6	1.6	1.5	1.4	1.4	1.3	1.3	1.2	1.2	1.2	1.1	1.1	1.0	1
2	1.6	1.6	1.5	1.5	1.4	1.4	1.3	1.3	1.2	1.2	1.1	1.1	1.0	2
3	1.7	1.6	1.5	1.5	1.4	1.4	1.3	1.3	1.2	1.2	1.1	1.1	1.0	3
4	1.7	1.6	1.6	1.5	1.5	1.4	1.3	1.3	1.2	1.2	1.1	1.1	1.0	4
5	1.7	1.7	1.6	1.5	1.5	1.4	1.4	1.3	1.3	1.2	1.1	1.1	1.1	5
6	1.7	1.7	1.6	1.5	1.5	1.4	1.4	1.3	1.3	1.2	1.2	1.1	1.1	6
7	1.8	1.7	1.6	1.6	1.5	1.4	1.4	1.3	1.3	1.2	1.2	1.1	1.1	7
8	1.8	1.7	1.7	1.6	1.5	1.5	1.4	1.4	1.3	1.2	1.2	1.1	1.1	8
9	1.8	1.8	1.7	1.6	1.6	1.5	1.4	1.4	1.3	1.3	1.2	1.1	1.1	9
10	1.9	1.8	1.7	1.6	1.6	1.5	1.4	1.4	1.3	1.3	1.2	1.2	1.1	10
11	1.9	1.8	1.7	1.7	1.6	1.5	1.5	1.4	1.3	1.3	1.2	1.2	1.1	11
12	1.9	1.8	1.8	1.7	1.6	1.6	1.5	1.4	1.4	1.3	1.2	1.2	1.1	12
13	2.0	1.9	1.8	1.7	1.6	1.6	1.5	1.4	1.4	1.3	1.3	1.2	1.1	13
14	2.0	1.9	1.8	1.7	1.7	1.6	1.5	1.5	1.4	1.3	1.3	1.2	1.2	14
15	2.0	1.9	1.9	1.8	1.7	1.6	1.5	1.5	1.4	1.3	1.3	1.2	1.2	15
16	2.1	2.0	1.9	1.8	1.7	1.6	1.6	1.5	1.4	1.4	1.3	1.2	1.2	16
17	2.1	2.0	1.9	1.8	1.8	1.7	1.6	1.5	1.5	1.4	1.3	1.3	1.2	17
18	2.2	2.1	2.0	1.9	1.8	1.7	1.6	1.5	1.5	1.4	1.3	1.3	1.2	18
19	2.2	2.1	2.0	1.9	1.8	1.7	1.6	1.6	1.5	1.4	1.3	1.3	1.2	19
20	2.3	2.1	2.0	1.9	1.9	1.8	1.7	1.6	1.5	1.4	1.4	1.3	1.2	20
21	2.3	2.2	2.1	2.0	1.9	1.8	1.7	1.6	1.5	1.5	1.4	1.3	1.2	21
22	2.4	2.2	2.1	2.0	1.9	1.8	1.7	1.6	1.6	1.5	1.4	1.3	1.3	22
23	2.4	2.3	2.2	2.1	2.0	1.9	1.8	1.7	1.6	1.5	1.4	1.4	1.3	23
24	2.5	2.4	2.2	2.1	2.0	1.9	1.8	1.7	1.6	1.5	1.5	1.4	1.3	24
25	2.6	2.4	2.3	2.2	2.0	1.9	1.8	1.7	1.6	1.6	1.5	1.4	1.3	25
26	2.6	2.5	2.3	2.2	2.1	2.0	1.9	1.8	1.7	1.6	1.5	1.4	1.3	26
27	2.7	2.6	2.4	2.3	2.1	2.0	1.9	1.8	1.7	1.6	1.5	1.4	1.4	27
28	2.8	2.6	2.5	2.3	2.2	2.1	2.0	1.8	1.7	1.6	1.5	1.5	1.4	28
29	2.9	2.7	2.5	2.4	2.3	2.1	2.0	1.9	1.8	1.7	1.6	1.5	1.4	29
30	3.0	2.8	2.6	2.5	2.3	2.2	2.0	1.9	1.8	1.7	1.6	1.5	1.4	30
31	3.1	2.9	2.7	2.5	2.4	2.2	2.1	2.0	1.9	1.7	1.6	1.5	1.4	31
32	3.2	3.0	2.8	2.6	2.4	2.3	2.2	2.0	1.9	1.8	1.7	1.6	1.5	32
33	3.4	3.1	2.9	2.7	2.5	2.4	2.2	2.1	1.9	1.8	1.7	1.6	1.5	33
34	3.5	3.2	3.0	2.8	2.6	2.4	2.3	2.1	2.0	1.9	1.7	1.6	1.5	34
35	3.7	3.4	3.1	2.9	2.7	2.5	2.3	2.2	2.0	1.9	1.8	1.7	1.6	35
36	3.9	3.6	3.3	3.0	2.8	2.6	2.4	2.3	2.1	2.0	1.8	1.7	1.6	36
37	4.1	3.7	3.4	3.2	2.9	2.7	2.5	2.3	2.2	2.0	1.9	1.7	1.6	37
38	4.3	3.9	3.6	3.3	3.0	2.8	2.6	2.4	2.2	2.1	1.9	1.8	1.7	38
39	4.6	4.2	3.8	3.5	3.2	2.9	2.7	2.5	2.3	2.1	2.0	1.8	1.7	39
40	5.0	4.5	4.0	3.7	3.3	3.1	2.8	2.6	2.4	2.2	2.0	1.9	1.8	40
41	5.4	4.8	4.3	3.9	3.5	3.2	2.9	2.7	2.5	2.3	2.1	1.9	1.8	41
42	5.9	5.2	4.6	4.1	3.7	3.4	3.1	2.8	2.6	2.4	2.2	2.0	1.9	42
43	6.5	5.7	5.0	4.4	4.0	3.6	3.2	2.9	2.7	2.5	2.3	2.1	1.9	43
44	7.3	6.3	5.4	4.8	4.3	3.8	3.4	3.1	2.8	2.6	2.3	2.2	2.0	44
45	8.4	7.0	6.0	5.2	4.6	4.1	3.6	3.3	3.0	2.7	2.4	2.2	2.0	45
46	9.9	8.0	6.7	5.8	5.0	4.4	3.9	3.5	3.1	2.8	2.6	2.3	2.1	46
47	12.1	9.5	7.7	6.5	5.5	4.8	4.2	3.7	3.3	3.0	2.7	2.4	2.2	47
48		11.6	9.1	7.4	6.2	5.3	4.6	4.0	3.6	3.2	2.8	2.6	2.3	48
49			11.1	8.7	7.1	5.9	5.0	4.4	3.8	3.4	3.0	2.7	2.4	49
50				10.6	8.3	6.8	5.6	4.8	4.2	3.6	3.2	2.9	2.6	50
51					10.2	7.9	6.4	5.4	4.6	4.0	3.5	3.0	2.7	51
52						9.7	7.6	6.1	5.1	4.3	3.8	3.3	2.9	52
53							9.2	7.2	5.9	4.9	4.1	3.6	3.1	53
54								8.8	6.8	5.5	4.6	3.9	3.4	54
55	10.2								8.3	6.5	5.3	4.3	3.7	55
56	7.9	9.7								7.9	6.1	5.0	4.1	56
57	6.4	7.6	9.2								7.4	5.8	4.7	57
58	5.4	6.1	7.2	8.8								7.0	5.4	58
59	4.6	5.1	5.9	6.8	8.3								6.6	59
60	4.0	4.3	4.9	5.5	6.5	7.9								60
Lati-tude	51°	52°	53°	54°	55°	56°	57°	58°	59°	60°	61°	62°	63°	Lati-tude
	Declination **same** name as latitude, **upper** transit: **add** correction to observed altitude													

TABLE 24
Altitude Factor

a, the change of altitude in one minute from meridian transit.

Lati-tude	51°	52°	53°	54°	55°	56°	57°	58°	59°	60°	61°	62°	63°	Lati-tude
	Declination **contrary** name to latitude, **upper** transit: **add** correction to observed altitude													
°	″	″	″	″	″	″	″	″	″	″	″	″	″	°
0	1.6	1.5	1.5	1.4	1.4	1.3	1.3	1.2	1.2	1.1	1.1	1.0	1.0	0
1	1.6	1.5	1.5	1.4	1.4	1.3	1.3	1.2	1.2	1.1	1.1	1.0	1.0	1
2	1.5	1.5	1.4	1.4	1.3	1.3	1.3	1.2	1.2	1.1	1.1	1.0	1.0	2
3	1.5	1.5	1.4	1.4	1.3	1.3	1.2	1.2	1.1	1.1	1.1	1.0	1.0	3
4	1.5	1.5	1.4	1.4	1.3	1.3	1.2	1.2	1.1	1.1	1.1	1.0	1.0	4
5	1.5	1.4	1.4	1.3	1.3	1.3	1.2	1.2	1.1	1.1	1.0	1.0	1.0	5
6	1.5	1.4	1.4	1.3	1.3	1.2	1.2	1.2	1.1	1.1	1.0	1.0	1.0	6
7	1.4	1.4	1.4	1.3	1.3	1.2	1.2	1.1	1.1	1.1	1.0	1.0	0.9	7
8	1.4	1.4	1.3	1.3	1.3	1.2	1.2	1.1	1.1	1.1	1.0	1.0	0.9	8
9	1.4	1.4	1.3	1.3	1.2	1.2	1.2	1.1	1.1	1.0	1.0	1.0	0.9	9
10	1.4	1.4	1.3	1.3	1.2	1.2	1.1	1.1	1.1	1.0	1.0	1.0	0.9	10
11	1.4	1.3	1.3	1.3	1.2	1.2	1.1	1.1	1.1	1.0	1.0	1.0	0.9	11
12	1.4	1.3	1.3	1.2	1.2	1.2	1.1	1.1	1.1	1.0	1.0	0.9	0.9	12
13	1.3	1.3	1.3	1.2	1.2	1.2	1.1	1.1	1.0	1.0	1.0	0.9	0.9	13
14	1.3	1.3	1.3	1.2	1.2	1.1	1.1	1.1	1.0	1.0	1.0	0.9	0.9	14
15	1.3	1.3	1.2	1.2	1.2	1.1	1.1	1.1	1.0	1.0	1.0	0.9	0.9	15
16	1.3	1.3	1.2	1.2	1.1	1.1	1.1	1.0	1.0	1.0	0.9	0.9	0.9	16
17	1.3	1.2	1.2	1.2	1.1	1.1	1.1	1.0	1.0	1.0	0.9	0.9	0.9	17
18	1.3	1.2	1.2	1.2	1.1	1.1	1.1	1.0	1.0	1.0	0.9	0.9	0.9	18
19	1.2	1.2	1.2	1.1	1.1	1.1	1.0	1.0	1.0	1.0	0.9	0.9	0.9	19
20	1.2	1.2	1.2	1.1	1.1	1.1	1.0	1.0	1.0	0.9	0.9	0.9	0.8	20
21	1.2	1.2	1.2	1.1	1.1	1.1	1.0	1.0	1.0	0.9	0.9	0.9	0.8	21
22	1.2	1.2	1.1	1.1	1.1	1.0	1.0	1.0	1.0	0.9	0.9	0.9		22
23	1.2	1.2	1.1	1.1	1.1	1.0	1.0	1.0	0.9	0.9	0.9			23
24	1.2	1.1	1.1	1.1	1.1	1.0	1.0	1.0	0.9	0.9				24
25	1.2	1.1	1.1	1.1	1.0	1.0	1.0	1.0	0.9					25
26	1.1	1.1	1.1	1.1	1.0	1.0	1.0	0.9						26
27	1.1	1.1	1.1	1.0	1.0	1.0	1.0							27
28	1.1	1.1	1.1	1.0	1.0	1.0								28
29	1.1	1.1	1.0	1.0	1.0									29
30	1.1	1.1	1.0	1.0										30
31	1.1	1.0	1.0											31
32	1.1	1.0												32
33	1.1												0.8	33
34												0.8	0.7	34
35											0.8	0.8	0.7	35
36										0.8	0.8	0.8	0.7	36
37									0.8	0.8	0.8	0.7	0.7	37
38								0.8	0.8	0.8	0.8	0.7	0.7	38
39							0.8	0.8	0.8	0.8	0.8	0.7	0.7	39
40						0.8	0.8	0.8	0.8	0.8	0.8	0.7	0.7	40
41					0.9	0.8	0.8	0.8	0.8	0.8	0.7	0.7	0.7	41
42				0.9	0.8	0.8	0.8	0.8	0.8	0.8	0.7	0.7	0.7	42
43			0.9	0.9	0.8	0.8	0.8	0.8	0.8	0.7	0.7	0.7	0.7	43
44		0.9	0.9	0.8	0.8	0.8	0.8	0.8	0.8	0.7	0.7	0.7	0.7	44
45	0.9	0.9	0.8	0.8	0.8	0.8	0.8	0.8	0.7	0.7	0.7	0.7	0.7	45
46	0.9	0.9	0.8	0.8	0.8	0.8	0.8	0.8	0.7	0.7	0.7	0.7	0.7	46
47	0.9	0.8	0.8	0.8	0.8	0.8	0.8	0.7	0.7	0.7	0.7	0.7	0.6	47
48	0.8	0.8	0.8	0.8	0.8	0.8	0.7	0.7	0.7	0.7	0.7	0.7	0.6	48
49	0.8	0.8	0.8	0.8	0.8	0.7	0.7	0.7	0.7	0.7	0.7	0.6	0.6	49
50	0.8	0.8	0.8	0.8	0.7	0.7	0.7	0.7	0.7	0.7	0.7	0.6	0.6	50
51	0.8	0.8	0.8	0.8	0.7	0.7	0.7	0.7	0.7	0.7	0.7	0.6	0.6	51
52	0.8	0.8	0.8	0.7	0.7	0.7	0.7	0.7	0.7	0.7	0.6	0.6	0.6	52
53	0.8	0.8	0.7	0.7	0.7	0.7	0.7	0.7	0.7	0.6	0.6	0.6	0.6	53
54	0.8	0.7	0.7	0.7	0.7	0.7	0.7	0.7	0.6	0.6	0.6	0.6	0.6	54
55	0.7	0.7	0.7	0.7	0.7	0.7	0.7	0.7	0.6	0.6	0.6	0.6	0.6	55
56	0.7	0.7	0.7	0.7	0.7	0.7	0.7	0.6	0.6	0.6	0.6	0.6	0.6	56
57	0.7	0.7	0.7	0.7	0.7	0.7	0.6	0.6	0.6	0.6	0.6	0.6	0.6	57
58	0.7	0.7	0.7	0.7	0.7	0.6	0.6	0.6	0.6	0.6	0.6	0.6	0.6	58
59	0.7	0.7	0.7	0.6	0.6	0.6	0.6	0.6	0.6	0.6	0.6	0.6	0.5	59
60	0.7	0.7	0.6	0.6	0.6	0.6	0.6	0.6	0.6	0.6	0.6	0.6	0.5	60
Lati-tude	51°	52°	53°	54°	55°	56°	57°	58°	59°	60°	61°	62°	63°	Lati-tude
	Declination **contrary** name to latitude, **upper** transit: **add** correction to observed altitude													

TABLE 25

Change of Altitude in Given Time from Meridian Transit

a (table 24)	t, meridian angle														a (table 24)
	5′	10′	15′	20′	25′	30′	35′	40′	45′	50′	55′	1° 00′	1° 05′	1° 10′	
	$0^m 20^s$	$0^m 40^s$	$1^m 00^s$	$1^m 20^s$	$1^m 40^s$	$2^m 00^s$	$2^m 20^s$	$2^m 40^s$	$3^m 00^s$	$3^m 20^s$	$3^m 40^s$	$4^m 00^s$	$4^m 20^s$	$4^m 40^s$	
″	′	′	′	′	′	′	′	′	′	′	′	′	′	′	″
0.1	0.0	0.0	0.0	0.0	0.0	0.0	0.0	0.0	0.0	0.0	0.0	0.0	0.0	0.0	0.1
0.2	0.0	0.0	0.0	0.0	0.0	0.0	0.0	0.0	0.0	0.0	0.0	0.1	0.1	0.1	0.2
0.3	0.0	0.0	0.0	0.0	0.0	0.0	0.0	0.0	0.0	0.1	0.1	0.1	0.1	0.1	0.3
0.4	0.0	0.0	0.0	0.0	0.0	0.0	0.0	0.0	0.1	0.1	0.1	0.1	0.1	0.1	0.4
0.5	0.0	0.0	0.0	0.0	0.0	0.0	0.0	0.1	0.1	0.1	0.1	0.1	0.2	0.2	0.5
0.6	0.0	0.0	0.0	0.0	0.0	0.0	0.1	0.1	0.1	0.1	0.1	0.2	0.2	0.2	0.6
0.7	0.0	0.0	0.0	0.0	0.0	0.0	0.1	0.1	0.1	0.1	0.2	0.2	0.2	0.3	0.7
0.8	0.0	0.0	0.0	0.0	0.0	0.1	0.1	0.1	0.1	0.1	0.2	0.2	0.3	0.3	0.8
0.9	0.0	0.0	0.0	0.0	0.0	0.1	0.1	0.1	0.1	0.2	0.2	0.2	0.3	0.3	0.9
1.0	0.0	0.0	0.0	0.0	0.0	0.1	0.1	0.1	0.2	0.2	0.2	0.3	0.3	0.4	1.0
2.0	0.0	0.0	0.0	0.1	0.1	0.1	0.2	0.2	0.3	0.4	0.4	0.5	0.6	0.7	2.0
3.0	0.0	0.0	0.1	0.1	0.1	0.2	0.3	0.4	0.4	0.6	0.7	0.8	0.9	1.1	3.0
4.0	0.0	0.0	0.1	0.1	0.2	0.3	0.4	0.5	0.6	0.7	0.9	1.1	1.3	1.5	4.0
5.0	0.0	0.0	0.1	0.1	0.2	0.3	0.5	0.6	0.8	0.9	1.1	1.3	1.6	1.8	5.0
6.0	0.0	0.0	0.1	0.2	0.3	0.4	0.5	0.7	0.9	1.1	1.3	1.6	1.9	2.2	6.0
7.0	0.0	0.1	0.1	0.2	0.3	0.5	0.6	0.8	1.0	1.3	1.6	1.9	2.2	2.5	7.0
8.0	0.0	0.1	0.1	0.2	0.4	0.5	0.7	0.9	1.2	1.5	1.8	2.1	2.5	2.9	8.0
9.0	0.0	0.1	0.2	0.3	0.4	0.6	0.8	1.1	1.4	1.7	2.0	2.4	2.8	3.3	9.0
10.0	0.0	0.1	0.2	0.3	0.5	0.7	0.9	1.2	1.5	1.9	2.2	2.7	3.1	3.6	10.0
11.0	0.0	0.1	0.2	0.3	0.5	0.7	1.0	1.3	1.6	2.0	2.5	2.9	3.4	4.0	11.0
12.0	0.0	0.1	0.2	0.4	0.6	0.8	1.1	1.4	1.8	2.2	2.7	3.2	3.8	4.4	12.0
13.0	0.0	0.1	0.2	0.4	0.6	0.9	1.2	1.5	2.0	2.4	2.9	3.5	4.1	4.7	13.0
14.0	0.0	0.1	0.2	0.4	0.6	0.9	1.3	1.7	2.1	2.6	3.1	3.7	4.4	5.1	14.0
15.0	0.0	0.1	0.3	0.4	0.7	1.0	1.4	1.8	2.2	2.8	3.4	4.0	4.7	5.4	15.0
16.0	0.0	0.1	0.3	0.5	0.7	1.1	1.5	1.9	2.4	3.0	3.6	4.3	5.0	5.8	16.0
17.0	0.0	0.1	0.3	0.5	0.8	1.1	1.5	2.0	2.6	3.1	3.8	4.5	5.3	6.2	17.0
18.0	0.0	0.1	0.3	0.5	0.8	1.2	1.6	2.1	2.7	3.3	4.0	4.8	5.6	6.5	18.0
19.0	0.0	0.1	0.3	0.6	0.9	1.3	1.7	2.3	2.8	3.5	4.3	5.1	5.9	6.9	19.0
20.0	0.0	0.1	0.3	0.6	0.9	1.3	1.8	2.4	3.0	3.7	4.5	5.3	6.3	7.3	20.0
21.0	0.0	0.2	0.4	0.6	1.0	1.4	1.9	2.5	3.2	3.9	4.7	5.6	6.6	7.6	21.0
22.0	0.0	0.2	0.4	0.7	1.0	1.5	2.0	2.6	3.3	4.1	4.9	5.9	6.9	8.0	22.0
23.0	0.0	0.2	0.4	0.7	1.1	1.5	2.1	2.7	3.4	4.3	5.2	6.1	7.2	8.3	23.0
24.0	0.0	0.2	0.4	0.7	1.1	1.6	2.2	2.8	3.6	4.4	5.4	6.4	7.5	8.7	24.0
25.0	0.0	0.2	0.4	0.7	1.2	1.7	2.3	3.0	3.8	4.6	5.6	6.7	7.8	9.1	25.0
26.0	0.0	0.2	0.4	0.8	1.2	1.7	2.4	3.1	3.9	4.8	5.8	6.9	8.1	9.4	26.0
27.0	0.0	0.2	0.4	0.8	1.2	1.8	2.4	3.2	4.0	5.0	6.0	7.2	8.5	9.8	27.0
28.0	0.1	0.2	0.5	0.8	1.3	1.9	2.5	3.3	4.2	5.2	6.3	7.5	8.8	10.2	28.0

Caution. —If this table is entered with the meridian angle of the Moon in arc units, such units should correspond to the meridian angle in time units as given in the Increments and Corrections section of the *Nautical Almanac.*

TABLE 25

Change of Altitude in Given Time from Meridian Transit

a (table 24)	t, meridian angle														a (table 24)
	1° 15′	1° 20′	1° 25′	1° 30′	1° 35′	1° 40′	1° 45′	1° 50′	1° 55′	2° 00′	2° 05′	2° 10′	2° 15′	2° 20′	
	$5^m 00^s$	$5^m 20^s$	$5^m 40^s$	$6^m 00^s$	$6^m 20^s$	$6^m 40^s$	$7^m 00^s$	$7^m 20^s$	$7^m 40^s$	$8^m 00^s$	$8^m 20^s$	$8^m 40^s$	$9^m 00^s$	$9^m 20^s$	
″	′	′	′	′	′	′	′	′	′	′	′	′	′	′	″
0.1	0.0	0.0	0.1	0.1	0.1	0.1	0.1	0.1	0.1	0.1	0.1	0.1	0.1	0.1	0.1
0.2	0.1	0.1	0.1	0.1	0.1	0.1	0.2	0.2	0.2	0.2	0.2	0.3	0.3	0.3	0.2
0.3	0.1	0.1	0.2	0.2	0.2	0.2	0.2	0.3	0.3	0.3	0.3	0.4	0.4	0.4	0.3
0.4	0.2	0.2	0.2	0.2	0.3	0.3	0.3	0.4	0.4	0.4	0.5	0.5	0.5	0.6	0.4
0.5	0.2	0.2	0.3	0.3	0.3	0.4	0.4	0.5	0.5	0.5	0.6	0.6	0.7	0.7	0.5
0.6	0.2	0.3	0.3	0.4	0.4	0.4	0.5	0.5	0.6	0.6	0.7	0.8	0.8	0.9	0.6
0.7	0.3	0.3	0.4	0.4	0.5	0.5	0.6	0.6	0.7	0.7	0.8	0.9	0.9	1.0	0.7
0.8	0.3	0.4	0.4	0.5	0.5	0.6	0.7	0.7	0.8	0.9	0.9	1.0	1.1	1.2	0.8
0.9	0.4	0.4	0.5	0.5	0.6	0.7	0.7	0.8	0.9	1.0	1.0	1.1	1.2	1.3	0.9
1.0	0.4	0.5	0.5	0.6	0.7	0.7	0.8	0.9	1.0	1.1	1.2	1.3	1.4	1.5	1.0
2.0	0.8	0.9	1.1	1.2	1.3	1.5	1.6	1.8	2.0	2.1	2.3	2.5	2.7	2.9	2.0
3.0	1.2	1.4	1.6	1.8	2.0	2.2	2.4	2.7	2.9	3.2	3.5	3.8	4.0	4.4	3.0
4.0	1.7	1.9	2.1	2.4	2.7	3.0	3.3	3.6	3.9	4.3	4.6	5.0	5.4	5.8	4.0
5.0	2.1	2.4	2.7	3.0	3.3	3.7	4.1	4.5	4.9	5.3	5.8	6.3	6.8	7.3	5.0
6.0	2.5	2.8	3.2	3.6	4.0	4.4	4.9	5.4	5.9	6.4	6.9	7.5	8.1	8.7	6.0
7.0	2.9	3.3	3.7	4.2	4.7	5.2	5.7	6.3	6.9	7.5	8.1	8.8	9.4	10.2	7.0
8.0	3.3	3.8	4.3	4.8	5.3	5.9	6.5	7.2	7.8	8.5	9.3	10.0	10.8	11.6	8.0
9.0	3.8	4.3	4.8	5.4	6.0	6.7	7.4	8.1	8.8	9.6	10.4	11.3	12.2	13.1	9.0
10.0	4.2	4.7	5.4	6.0	6.7	7.4	8.2	9.0	9.8	10.7	11.6	12.5	13.5	14.5	10.0
11.0	4.6	5.2	5.9	6.6	7.4	8.1	9.0	9.9	10.8	11.7	12.7	13.8	14.8	16.0	11.0
12.0	5.0	5.7	6.4	7.2	8.0	8.9	9.8	10.8	11.8	12.8	13.9	15.0	16.2	17.4	12.0
13.0	5.4	6.2	7.0	7.8	8.7	9.6	10.6	11.7	12.7	13.9	15.0	16.3	17.6	18.9	13.0
14.0	5.8	6.6	7.5	8.4	9.4	10.4	11.4	12.5	13.7	14.9	16.2	17.5	18.9	20.3	14.0
15.0	6.2	7.1	8.0	9.0	10.0	11.1	12.2	13.4	14.7	16.0	17.4	18.8	20.2	21.8	15.0
16.0	6.7	7.6	8.6	9.6	10.7	11.9	13.1	14.3	15.7	17.1	18.5	20.0	21.6	23.2	16.0
17.0	7.1	8.1	9.1	10.2	11.4	12.6	13.9	15.2	16.7	18.1	19.7	21.3	23.0	24.7	17.0
18.0	7.5	8.5	9.6	10.8	12.0	13.3	14.7	16.1	17.6	19.2	20.8	22.5	24.3	26.1	18.0
19.0	7.9	9.0	10.2	11.4	12.7	14.1	15.5	17.0	18.6	20.3	22.0	23.8			19.0
20.0	8.3	9.5	10.7	12.0	13.4	14.8	16.3	17.9	19.6	21.3	23.1				20.0
21.0	8.8	10.0	11.2	12.6	14.0	15.6	17.2	18.8	20.6						21.0
22.0	9.2	10.4	11.8	13.2	14.7	16.3	18.0	19.7	21.6						22.0
23.0	9.6	10.9	12.3	13.8	15.4	17.0	18.8	20.6							23.0
24.0	10.0	11.4	12.8	14.4	16.0	17.8	19.6	21.5							24.0
25.0	10.4	11.9	13.4	15.0	16.7	18.5	20.4								25.0
26.0	10.8	12.3	13.9	15.6	17.4	19.3									26.0
27.0	11.2	12.8	14.4	16.2	18.0	20.0									27.0

Caution. —If this table is entered with the meridian angle of the Moon in arc units, such units should correspond to the meridian angle in time units as given in the Increments and Corrections section of the *Nautical Almanac.*

TABLE 25

Change of Altitude in Given Time from Meridian Transit

a (table 24)	t, meridian angle														a (table 24)
	2° 25′	2° 30′	2° 35′	2° 40′	2° 45′	2° 50′	2° 55′	3° 00′	3° 05′	3° 10′	3° 15′	3° 20′	3° 25′	3° 30′	
	$9^m 40^s$	$10^m 00^s$	$10^m 20^s$	$10^m 40^s$	$11^m 00^s$	$11^m 20^s$	$11^m 40^s$	$12^m 00^s$	$12^m 20^s$	$12^m 40^s$	$13^m 00^s$	$13^m 20^s$	$13^m 40^s$	$14^m 00^s$	
″	′	′	′	′	′	′	′	′	′	′	′	′	′	′	″
0.1	0.2	0.2	0.2	0.2	0.2	0.2	0.2	0.2	0.3	0.3	0.3	0.3	0.3	0.3	0.1
0.2	0.3	0.3	0.4	0.4	0.4	0.4	0.5	0.5	0.5	0.5	0.6	0.6	0.6	0.7	0.2
0.3	0.5	0.5	0.5	0.6	0.6	0.6	0.7	0.7	0.8	0.8	0.8	0.9	0.9	1.0	0.3
0.4	0.6	0.7	0.7	0.8	0.8	0.9	0.9	1.0	1.0	1.1	1.1	1.2	1.2	1.3	0.4
0.5	0.8	0.8	0.9	0.9	1.0	1.1	1.1	1.2	1.3	1.3	1.4	1.5	1.6	1.6	0.5
0.6	0.9	1.0	1.1	1.1	1.2	1.3	1.4	1.4	1.5	1.6	1.7	1.8	1.9	2.0	0.6
0.7	1.1	1.2	1.2	1.3	1.4	1.5	1.6	1.7	1.8	1.9	2.0	2.1	2.2	2.3	0.7
0.8	1.2	1.3	1.4	1.5	1.6	1.7	1.8	1.9	2.0	2.1	2.3	2.4	2.5	2.6	0.8
0.9	1.4	1.5	1.6	1.7	1.8	1.9	2.0	2.2	2.3	2.4	2.5	2.7	2.8	2.9	0.9
1.0	1.6	1.7	1.8	1.9	2.0	2.1	2.3	2.4	2.5	2.7	2.8	3.0	3.1	3.3	1.0
2.0	3.1	3.3	3.6	3.8	4.0	4.3	4.5	4.8	5.1	5.3	5.6	5.9	6.2	6.5	2.0
3.0	4.7	5.0	5.3	5.7	6.0	6.4	6.8	7.2	7.6	8.0	8.4	8.9	9.3	9.8	3.0
4.0	6.2	6.7	7.1	7.6	8.1	8.6	9.1	9.6	10.1	10.7	11.3	11.9	12.5	13.1	4.0
5.0	7.8	8.3	8.9	9.5	10.1	10.7	11.3	12.0	12.7	13.4	14.1	14.8	15.6	16.3	5.0
6.0	9.3	10.0	10.7	11.4	12.1	12.8	13.6	14.4	15.2	16.0	16.9	17.8	18.7	19.6	6.0
7.0	10.9	11.7	12.5	13.3	14.1	15.0	15.9	16.8	17.7	18.7	19.7	20.7	21.8	22.9	7.0
8.0	12.5	13.3	14.2	15.2	16.1	17.1	18.1	19.2	20.3	21.4	22.5	23.7	24.9	26.1	8.0
9.0	14.0	15.0	16.0	17.1	18.2	19.3	20.4	21.6	22.8	24.1	25.4	26.7	28.0	29.4	9.0
10.0	15.6	16.7	17.8	19.0	20.2	21.4	22.7	24.0	25.4	26.7	28.2	29.6			10.0
11.0	17.1	18.3	19.6	20.9	22.2	23.5	25.0	26.4	27.9	29.4					11.0
12.0	18.7	20.0	21.4	22.8	24.2	25.7	27.2	28.8							12.0
13.0	20.2	21.7	23.1	24.7	26.2	27.8	29.5								13.0
14.0	21.8	23.3	24.9	26.6	28.2	30.0									14.0
15.0	23.4	25.0	26.7	28.5	30.2										15.0
16.0	24.9	26.7	28.5	30.3											16.0
17.0	26.5	28.3	30.3												17.0

a (table 24)	t, meridian angle														a (table 24)
	3° 35′	3° 40′	3° 45′	3° 50′	3° 55′	4° 00′	4° 05′	4° 10′	4° 15′	4° 20′	4° 25′	4° 30′	4° 35′	4° 40′	
	$14^m 20^s$	$14^m 40^s$	$15^m 00^s$	$15^m 20^s$	$15^m 40^s$	$16^m 00^s$	$16^m 20^s$	$16^m 40^s$	$17^m 00^s$	$17^m 20^s$	$17^m 40^s$	$18^m 00^s$	$18^m 20^s$	$18^m 40^s$	
″	′	′	′	′	′	′	′	′	′	′	′	′	′	′	″
0.1	0.3	0.4	0.4	0.4	0.4	0.4	0.4	0.5	0.5	0.5	0.5	0.5	0.6	0.6	0.1
0.2	0.7	0.7	0.8	0.8	0.8	0.9	0.9	0.9	1.0	1.0	1.0	1.1	1.1	1.2	0.2
0.3	1.0	1.1	1.1	1.2	1.2	1.3	1.3	1.4	1.4	1.5	1.6	1.6	1.7	1.7	0.3
0.4	1.4	1.4	1.5	1.6	1.6	1.7	1.8	1.9	1.9	2.0	2.1	2.2	2.2	2.3	0.4
0.5	1.7	1.8	1.9	2.0	2.0	2.1	2.2	2.3	2.4	2.5	2.6	2.7	2.8	2.9	0.5
0.6	2.1	2.2	2.2	2.4	2.5	2.6	2.7	2.8	2.9	3.0	3.1	3.2	3.4	3.5	0.6
0.7	2.4	2.5	2.6	2.7	2.9	3.0	3.1	3.2	3.4	3.5	3.6	3.8	3.9	4.1	0.7
0.8	2.7	2.9	3.0	3.1	3.3	3.4	3.6	3.7	3.9	4.0	4.2	4.3	4.5	4.6	0.8
0.9	3.1	3.2	3.4	3.5	3.7	3.8	4.0	4.2	4.3	4.5	4.7	4.9	5.0	5.2	0.9
1.0	3.4	3.6	3.8	3.9	4.1	4.3	4.4	4.6	4.8	5.0	5.2	5.4	5.6	5.8	1.0
2.0	6.8	7.2	7.5	7.8	8.2	8.5	8.9	9.3	9.6	10.0	10.4	10.8	11.2	11.6	2.0
3.0	10.3	10.8	11.3	11.8	12.3	12.8	13.3	13.9	14.4	15.0	15.6	16.2	16.8	17.4	3.0
4.0	13.7	14.3	15.0	15.7	16.4	17.1	17.8	18.5	19.3	20.0	20.8	21.6	22.4	23.2	4.0
5.0	17.1	17.9	18.8	19.6	20.5	21.3	22.2	23.1	24.1	25.0	26.0	27.0	28.0	29.0	5.0
6.0	20.5	21.5	22.5	23.5	24.5	25.6	26.7	27.8							6.0
7.0	24.0	25.1	26.3	27.4											7.0
8.0	27.4	28.7	30.0												8.0

Caution. —If this table is entered with the meridian angle of the Moon in arc units, such units should correspond to the meridian angle in time units as given in the Increments and Corrections section of the *Nautical Almanac.*

TABLE 25

Change of Altitude in Given Time from Meridian Transit

a (table 24)	t, meridian angle														a (table 24)
	4° 45′	4° 50′	4° 55′	5° 00′	5° 05′	5° 10′	5° 15′	5° 20′	5° 25′	5° 30′	5° 35′	5° 40′	5° 45′	5° 50′	
	$19^m 00^s$	$19^m 20^s$	$19^m 40^s$	$20^m 00^s$	$20^m 20^s$	$20^m 40^s$	$21^m 00^s$	$21^m 20^s$	$21^m 40^s$	$22^m 00^s$	$22^m 20^s$	$22^m 40^s$	$23^m 00^s$	$23^m 20^s$	
″	′	′	′	′	′	′	′	′	′	′	′	′	′	′	″
0.1	0.6	0.6	0.6	0.7	0.7	0.7	0.7	0.8	0.8	0.8	0.8	0.9	0.9	0.9	0.1
0.2	1.2	1.2	1.3	1.3	1.4	1.4	1.5	1.5	1.6	1.6	1.7	1.7	1.8	1.8	0.2
0.3	1.8	1.9	1.9	2.0	2.1	2.1	2.2	2.3	2.3	2.4	2.5	2.6	2.6	2.7	0.3
0.4	2.4	2.5	2.6	2.7	2.8	2.8	2.9	3.0	3.1	3.2	3.3	3.4	3.5	3.6	0.4
0.5	3.0	3.1	3.2	3.3	3.4	3.6	3.7	3.8	3.9	4.0	4.2	4.3	4.4	4.5	0.5
0.6	3.6	3.7	3.9	4.0	4.1	4.3	4.4	4.6	4.7	4.8	5.0	5.1	5.3	5.4	0.6
0.7	4.2	4.4	4.5	4.7	4.8	5.0	5.1	5.3	5.5	5.6	5.8	6.0	6.2	6.4	0.7
0.8	4.8	5.0	5.2	5.3	5.5	5.7	5.9	6.1	6.3	6.5	6.7	6.9	7.1	7.3	0.8
0.9	5.4	5.6	5.8	6.0	6.2	6.4	6.6	6.8	7.0	7.3	7.5	7.7	7.9	8.2	0.9
1.0	6.0	6.2	6.4	6.7	6.9	7.1	7.4	7.6	7.8	8.1	8.3	8.6	8.8	9.1	1.0
2.0	12.0	12.5	12.9	13.3	13.8	14.2	14.7	15.2	15.6	16.1	16.6	17.1	17.6	18.1	2.0
3.0	18.0	18.7	19.3	20.0	20.7	21.4	22.0	22.8	23.5	24.2	24.9	25.7	26.4	27.2	3.0
4.0	24.1	24.9	25.8	26.7	27.6	28.5	29.4	30.3	31.3						4.0

a (table 24)	t, meridian angle														a (table 24)
	5° 55′	6° 00′	6° 05′	6° 10′	6° 15′	6° 20′	6° 25′	6° 30′	6° 35′	6° 40′	6° 45′	6° 50′	6° 55′	7° 00′	
	$23^m 40^s$	$24^m 00^s$	$24^m 20^s$	$24^m 40^s$	$25^m 00^s$	$25^m 20^s$	$25^m 40^s$	$26^m 00^s$	$26^m 20^s$	$26^m 40^s$	$27^m 00^s$	$27^m 20^s$	$27^m 40^s$	$28^m 20^s$	
″	′	′	′	′	′	′	′	′	′	′	′	′	′	′	″
0.1	0.9	1.0	1.0	1.0	1.0	1.1	1.1	1.1	1.2	1.2	1.2	1.2	1.3	1.3	0.1
0.2	1.9	1.9	2.0	2.0	2.1	2.1	2.2	2.3	2.3	2.4	2.4	2.5	2.6	2.6	0.2
0.3	2.8	2.9	3.0	3.0	3.1	3.2	3.3	3.4	3.5	3.6	3.6	3.7	3.8	3.9	0.3
0.4	3.7	3.8	3.9	4.1	4.2	4.3	4.4	4.5	4.6	4.7	4.9	5.0	5.1	5.2	0.4
0.5	4.7	4.8	4.9	5.1	5.2	5.3	5.5	5.6	5.8	5.9	6.1	6.2	6.4	6.5	0.5
0.6	5.6	5.8	5.9	6.1	6.2	6.4	6.6	6.8	6.9	7.1	7.3	7.5	7.7	7.8	0.6
0.7	6.5	6.7	6.9	7.1	7.3	7.5	7.7	7.9	8.1	8.3	8.5	8.7	8.9	9.1	0.7
0.8	7.5	7.7	7.9	8.1	8.3	8.6	8.8	9.0	9.2	9.5	9.7	10.0	10.2	10.5	0.8
0.9	8.4	8.6	8.9	9.1	9.4	9.6	9.9	10.1	10.4	10.7	10.9	11.2	11.5	11.8	0.9
1.0	9.3	9.6	9.9	10.1	10.4	10.7	11.0	11.3	11.6	11.9	12.2	12.5	12.8	13.1	1.0
2.0	18.7	19.2	19.7	20.3	20.8	21.4	22.0	22.5	23.1	23.7	24.3	24.9	25.5	26.1	2.0
3.0	28.0	28.8	29.6	30.4											3.0

Caution. —If this table is entered with the meridian angle of the Moon in arc units, such units should correspond to the meridian angle in time units as given in the Increments and Corrections section of the *Nautical Almanac.*

TABLE 26

Time Zones, Zone Descriptions, and Suffixes

ZONE	ZD	SUFFIX	ZONE	ZD	SUFFIX
7½° W to 7½° E.	0	Z	7½° W. to 22½° W.	+ 1	N
7½° E. to 22½° E.	– 1	A	22½° W. to 37½° W.	+ 2	O
22½° E. to 37½° E.	– 2	B	37½° W. to 52½° W.	+ 3	P
37½° E. to 52½° E.	– 3	C	52½° W. to 67½° W.	+ 4	Q
52½° E. to 67½° E.	– 4	D	67½° W. to 82½° W.	+ 5	R
67½° E. to 82½° E.	– 5	E	82½° W. to 97½° W.	+ 6	S
82½° E. to 97½° E.	– 6	F	97½° W. to 112½° W.	+ 7	T
97½° E. to 112½° E.	– 7	G	112½° W. to 127½° W.	+ 8	U
112½° E. to 127½° E.	– 8	H	127½° W. to 142½° W.	+ 9	V
127½° E. to 142½° E.	– 9	I	142½° W. to 157½° W.	+ 10	W
142½° E. to 157½° E.	– 10	K	157½° W. to 172½° W.	+ 11	X
157½° E. to 172½° E.	– 11	L	172½° W. to 180°	+ 12	Y
172½° E. to 180 °E.	– 12	M			

NOTE. – G M T is indicated by suffix Z. Standard times as kept in various places or countries are listed in *The Nautical Almanac* and *The Air Almanac*.

TABLE 27

Altitude Correction for Air Temperature

Altitude	Temperature—degrees Fahrenheit								Altitude
	− 40	− 30	− 20	− 10	0	+ 10	+ 20	+ 30	
° ′	′	′	′	′	′	′	′	′	° ′
− 0 10	− 7.9	− 6.8	− 5.8	− 4.9	− 4.0	− 3.1	− 2.3	− 1.5	− 0 10
0 00	7.4	6.4	5.5	4.6	3.8	2.9	2.2	1.4	0 00
+ 0 10	6.9	6.0	5.2	4.3	3.5	2.8	2.0	1.3	+ 0 10
0 20	6.6	5.7	4.9	4.1	3.3	2.6	1.9	1.2	0 20
0 30	6.1	5.3	4.6	3.8	3.1	2.4	1.8	1.2	0 30
+ 0 45	− 5.7	− 4.9	− 4.2	− 3.5	− 2.9	− 2.2	− 1.6	− 1.1	+ 0 45
1 00	5.2	4.5	3.9	3.2	2.6	2.1	1.5	1.0	1 00
1 20	4.7	4.1	3.5	2.9	2.4	1.9	1.4	0.9	1 20
1 40	4.3	3.7	3.2	2.7	2.2	1.7	1.2	0.8	1 40
2 00	3.9	3.4	2.9	2.4	2.0	1.6	1.1	0.7	2 00
+ 2 30	− 3.4	− 3.0	− 2.6	− 2.1	− 1.8	− 1.4	− 1.0	− 0.7	+ 2 30
3 00	3.1	2.7	2.3	1.9	1.6	1.2	0.9	0.6	3 00
4	2.5	2.2	1.9	1.6	1.3	1.0	0.7	0.5	4
5	2.1	1.8	1.6	1.3	1.1	0.8	0.6	0.4	5
6	1.8	1.6	1.4	1.1	0.9	0.7	0.5	0.3	6
+ 7	− 1.6	− 1.4	− 1.2	− 1.0	− 0.8	− 0.6	− 0.5	− 0.3	+ 7
8	1.4	1.2	1.0	0.9	0.7	0.6	0.4	0.3	8
9	1.3	1.1	0.9	0.8	0.6	0.5	0.4	0.2	9
10	1.1	1.0	0.8	0.7	0.6	0.5	0.3	0.2	10
15	0.8	0.7	0.6	0.5	0.4	0.3	0.2	0.1	15
+ 20	− 0.6	− 0.5	− 0.4	− 0.3	− 0.3	− 0.2	− 0.2	− 0.1	+ 20
30	0.4	0.3	0.3	0.2	0.2	0.1	0.1	0.1	30
50	0.2	0.1	0.1	0.1	0.1	0.1	0.0	0.0	50
70	0.1	0.1	0.1	0.1	0.0	0.0	0.0	0.0	70
+ 90	0.0	0.0	0.0	0.0	0.0	0.0	0.0	0.0	+ 90

Altitude	Temperature—degrees Fahrenheit								Altitude
	+ 40	+ 50	+ 60	+ 70	+ 80	+ 90	+ 100	+ 110	
° ′	′	′	′	′	′	′	′	′	° ′
− 0 10	− 0.7	0.0	+ 0.7	+ 1.4	+ 2.0	+ 2.7	+ 3.3	+ 3.9	− 0 10
0 00	0.7	0.0	0.7	1.3	1.9	2.5	3.1	3.6	0 00
+ 0 10	0.6	0.0	0.6	1.2	1.8	2.4	2.9	3.4	+ 0 10
0 20	0.6	0.0	0.6	1.2	1.7	2.2	2.7	3.2	0 20
0 30	0.6	0.0	0.6	1.1	1.6	2.1	2.6	3.0	0 30
+ 0 45	− 0.5	0.0	+ 0.5	+ 1.0	+ 1.5	+ 1.9	+ 2.4	+ 2.8	+ 0 45
1 00	0.5	0.0	0.5	0.9	1.4	1.8	2.2	2.6	1 00
1 20	0.4	0.0	0.4	0.8	1.2	1.6	2.0	2.3	1 20
1 40	0.4	0.0	0.4	0.8	1.1	1.5	1.8	2.1	1 40
2 00	0.4	0.0	0.4	0.7	1.0	1.3	1.6	1.9	2 00
+ 2 30	− 0.3	0.0	+ 0.3	+ 0.6	+ 0.9	+ 1.2	+ 1.4	+ 1.7	+ 2 30
3 00	0.3	0.0	0.3	0.5	0.8	1.0	1.3	1.5	3 00
4	0.2	0.0	0.2	0.4	0.7	0.9	1.1	1.2	4
5	0.2	0.0	0.2	0.4	0.6	0.7	0.9	1.0	5
6	0.2	0.0	0.2	0.3	0.5	0.6	0.8	0.9	6
+ 7	− 0.1	0.0	+ 0.1	+ 0.3	+ 0.4	+ 0.5	+ 0.7	+ 0.8	+ 7
8	0.1	0.0	0.1	0.2	0.4	0.5	0.6	0.7	8
9	0.1	0.0	0.1	0.2	0.3	0.4	0.5	0.6	9
10	0.1	0.0	0.1	0.2	0.3	0.4	0.5	0.6	10
15	0.1	0.0	0.1	0.1	0.2	0.3	0.3	0.4	15
+ 20	− 0.1	0.0	+ 0.1	+ 0.1	+ 0.1	+ 0.2	+ 0.2	+ 0.3	+ 20
30	0.0	0.0	0.0	0.1	0.1	0.1	0.2	0.2	30
50	0.0	0.0	0.0	0.0	0.0	0.1	0.1	0.1	50
70	0.0	0.0	0.0	0.0	0.0	0.0	0.0	0.0	70
+ 90	0.0	0.0	0.0	0.0	0.0	0.0	0.0	0.0	+ 90

TABLE 28

Altitude Correction for Atmospheric Pressure

Altitude	Pressure in inches or millibars— **Subtract** correction from sextant or rectified altitude								Altitude
	31.2	31.0	30.8	30.6	30.4	30.2	30.0	29.8	
	1056.56	1049.78	1043.01	1036.24	1029.46	1022.69	1015.92	1009.15	
° ′	′	′	′	′	′	′	′	′	° ′
− 0 10	− 1.7	− 1.4	− 1.2	− 1.0	− 0.7	− 0.5	− 0.2	0.0	− 0 10
0 00	1.6	1.4	1.1	0.9	0.7	0.4	0.2	0.0	0 00
+ 0 10	1.5	1.3	1.1	0.8	0.6	0.4	0.2	0.0	+ 0 10
0 20	1.4	1.2	1.0	0.8	0.6	0.4	0.2	0.0	0 20
0 30	1.3	1.1	0.9	0.7	0.6	0.4	0.2	0.0	0 30
+ 0 45	− 1.2	− 1.0	− 0.9	− 0.7	− 0.5	− 0.3	− 0.2	− 0.0	+ 0 45
1 00	1.1	1.0	0.8	0.6	0.5	0.3	0.1	0.0	1 00
1 20	1.0	0.9	0.7	0.6	0.4	0.3	0.1	0.0	1 20
1 40	0.9	0.8	0.7	0.5	0.4	0.3	0.1	0.0	1 40
2 00	0.8	0.7	0.6	0.5	0.4	0.2	0.1	0.0	2 00
+ 2 30	− 0.7	− 0.6	− 0.5	− 0.4	− 0.3	− 0.2	− 0.1	0.0	+ 2 30
3 00	0.7	0.6	0.5	0.4	0.3	0.2	0.1	0.0	3 00
4	0.5	0.5	0.4	0.3	0.2	0.1	0.1	0.0	4
5	0.5	0.4	0.3	0.3	0.2	0.1	0.0	0.0	5
6	0.4	0.3	0.3	0.2	0.2	0.1	0.0	0.0	6
+ 7	− 0.3	− 0.3	− 0.2	− 0.2	− 0.1	− 0.1	− 0.0	− 0.0	+ 7
8	0.3	0.3	0.2	0.2	0.1	0.1	0.0	0.0	8
9	0.3	0.2	0.2	0.2	0.1	0.1	0.0	0.0	9
10	0.2	0.2	0.2	0.1	0.1	0.1	0.0	0.0	10
15	0.2	0.1	0.1	0.1	0.1	0.0	0.0	0.0	15
+ 20	− 0.1	− 0.1	− 0.1	− 0.1	− 0.1	− 0.0	− 0.0	− 0.0	+ 20
30	0.1	0.1	0.1	0.0	0.0	0.0	0.0	0.0	30
50	0.0	0.0	0.0	0.0	0.0	0.0	0.0	0.0	50
70	0.0	0.0	0.0	0.0	0.0	0.0	0.0	0.0	70
+ 90	0.0	0.0	0.0	0.0	0.0	0.0	0.0	0.0	+ 90

Altitude	Pressure in inches or millibars— **Add** correction to sextant or rectified altitude								Altitude
	29.6	29.4	29.2	29.0	28.8	28.6	28.4	28.2	
	1002.37	995.60	988.83	982.05	975.28	968.51	961.74	954.96	
° ′	′	′	′	′	′	′	′	′	° ′
− 0 10	+ 0.3	+ 0.5	+ 0.8	+ 1.0	+ 1.3	+ 1.5	+ 1.8	+ 2.0	− 0 10
0 00	0.3	0.5	0.7	1.0	1.2	1.4	1.6	1.9	0 00
+ 0 10	0.2	0.5	0.7	0.9	1.1	1.3	1.5	1.8	+ 0 10
0 20	0.2	0.4	0.6	0.8	1.1	1.3	1.5	1.7	0 20
0 30	0.2	0.4	0.6	0.8	1.0	1.2	1.4	1.6	0 30
+ 0 45	+ 0.2	+ 0.4	+ 0.6	+ 0.7	+ 0.9	+ 1.1	+ 1.3	+ 1.4	+ 0 45
1 00	0.2	0.3	0.5	0.7	0.8	1.0	1.2	1.3	1 00
1 20	0.2	0.3	0.5	0.6	0.8	0.9	1.1	1.2	1 20
1 40	0.2	0.3	0.4	0.6	0.7	0.8	1.0	1.1	1 40
2 00	0.1	0.3	0.4	0.5	0.6	0.8	0.9	1.0	2 00
+ 2 30	+ 0.1	+ 0.2	+ 0.3	+ 0.4	+ 0.6	+ 0.7	+ 0.8	+ 0.9	+ 2 30
3 00	0.1	0.2	0.3	0.4	0.5	0.6	0.7	0.8	3 00
4	0.1	0.2	0.2	0.3	0.4	0.5	0.6	0.6	4
5	0.1	0.1	0.2	0.3	0.3	0.4	0.5	0.5	5
6	0.1	0.1	0.2	0.2	0.3	0.3	0.4	0.5	6
+ 7	+ 0.1	+ 0.1	+ 0.2	+ 0.2	+ 0.3	+ 0.3	+ 0.4	+ 0.4	+ 7
8	0.0	0.1	0.1	0.2	0.2	0.3	0.3	0.4	8
9	0.0	0.1	0.1	0.2	0.2	0.2	0.3	0.3	9
10	0.0	0.1	0.1	0.1	0.2	0.2	0.3	0.3	10
15	0.0	0.1	0.1	0.1	0.1	0.1	0.2	0.2	15
+ 20	+ 0.0	+ 0.0	+ 0.1	+ 0.1	+ 0.1	+ 0.1	+ 0.1	+ 0.1	+ 20
30	0.0	0.0	0.0	0.0	0.1	0.1	0.1	0.1	30
50	0.0	0.0	0.0	0.0	0.0	0.0	0.0	0.0	50
70	0.0	0.0	0.0	0.0	0.0	0.0	0.0	0.0	70
+ 90	0.0	0.0	0.0	0.0	0.0	0.0	0.0	0.0	+ 90

TABLE 29

Conversion Tables for Thermometer Scales

F = Fahrenheit, C = Celsius (centigrade), K = Kelvin

F	C	K	F	C	K	C	F	K	K	F	C
°	°	°	°	°	°	°	°	°	°	°	°
−20	−28.9	244.3	+40	+4.4	277.6	−25	−13.0	248.2	250	−9.7	−23.2
19	28.3	244.8	41	5.0	278.2	24	11.2	249.2	251	7.9	22.2
18	27.8	245.4	42	5.6	278.7	23	9.4	250.2	252	6.1	21.2
17	27.2	245.9	43	6.1	279.3	22	7.6	251.2	253	4.3	20.2
16	26.7	246.5	44	6.7	279.8	21	5.8	252.2	254	2.5	19.2
−15	−26.1	247.0	+45	+7.2	280.4	−20	−4.0	253.2	255	−0.7	−18.2
14	25.6	247.6	46	7.8	280.9	19	2.2	254.2	256	+1.1	17.2
13	25.0	248.2	47	8.3	281.5	18	−0.4	255.2	257	2.9	16.2
12	24.4	248.7	48	8.9	282.0	17	+1.4	256.2	258	4.7	15.2
11	23.9	249.3	49	9.4	282.6	16	3.2	257.2	259	6.5	14.2
−10	−23.3	249.8	+50	+10.0	283.2	−15	+5.0	258.2	260	+8.3	−13.2
9	22.8	250.4	51	10.6	283.7	14	6.8	259.2	261	10.1	12.2
8	22.2	250.9	52	11.1	284.3	13	8.6	260.2	262	11.9	11.2
7	21.7	251.5	53	11.7	284.8	12	10.4	261.2	263	13.7	10.2
6	21.1	252.0	54	12.2	285.4	11	12.2	262.2	264	15.5	9.2
−5	−20.6	252.6	+55	+12.8	285.9	−10	+14.0	263.2	265	+17.3	−8.2
4	20.0	253.2	56	13.3	286.5	9	15.8	264.2	266	19.1	7.2
3	19.4	253.7	57	13.9	287.0	8	17.6	265.2	267	20.9	6.2
2	18.9	254.3	58	14.4	287.6	7	19.4	266.2	268	22.7	5.2
−1	18.3	254.8	59	15.0	288.2	6	21.2	267.2	269	24.5	4.2
0	−17.8	255.4	+60	+15.6	288.7	−5	+23.0	268.2	270	+26.3	−3.2
+1	17.2	255.9	61	16.1	289.3	4	24.8	269.2	271	28.1	2.2
2	16.7	256.5	62	16.7	289.8	3	26.6	270.2	272	29.9	1.2
3	16.1	257.0	63	17.2	290.4	2	28.4	271.2	273	31.7	−0.2
4	15.6	257.6	64	17.8	290.9	−1	30.2	272.2	274	33.5	+0.8
+5	−15.0	258.2	+65	+18.3	291.5	0	+32.0	273.2	275	+35.3	+1.8
6	14.4	258.7	66	18.9	292.0	+1	33.8	274.2	276	37.1	2.8
7	13.9	259.3	67	19.4	292.6	2	35.6	275.2	277	38.9	3.8
8	13.3	259.8	68	20.0	293.2	3	37.4	276.2	278	40.7	4.8
9	12.8	260.4	69	20.6	293.7	4	39.2	277.2	279	42.5	5.8
+10	−12.2	260.9	+70	+21.1	294.3	+5	+41.0	278.2	280	+44.3	+6.8
11	11.7	261.5	71	21.7	294.8	6	42.8	279.2	281	46.1	7.8
12	11.1	262.0	72	22.2	295.4	7	44.6	280.2	282	47.9	8.8
13	10.6	262.6	73	22.8	295.9	8	46.4	281.2	283	49.7	9.8
14	10.0	263.2	74	23.3	296.5	9	48.2	282.2	284	51.5	10.8
+15	−9.4	263.7	+75	+23.9	297.0	+10	+50.0	283.2	285	+53.3	+11.8
16	8.9	264.3	76	24.4	297.6	11	51.8	284.2	286	55.1	12.8
17	8.3	264.8	77	25.0	298.2	12	53.6	285.2	287	56.9	13.8
18	7.8	265.4	78	25.6	298.7	13	55.4	286.2	288	58.7	14.8
19	7.2	265.9	79	26.1	299.3	14	57.2	287.2	289	60.5	15.8
+20	−6.7	266.5	+80	+26.7	299.8	+15	+59.0	288.2	290	+62.3	+16.8
21	6.1	267.0	81	27.2	300.4	16	60.8	289.2	291	64.1	17.8
22	5.6	267.6	82	27.8	300.9	17	62.6	290.2	292	65.9	18.8
23	5.0	268.2	83	28.3	301.5	18	64.4	291.2	293	67.7	19.8
24	4.4	268.7	84	28.9	302.0	19	66.2	292.2	294	69.5	20.8
+25	−3.9	269.3	+85	+29.4	302.6	+20	+68.0	293.2	295	+71.3	+21.8
26	3.3	269.8	86	30.0	303.2	21	69.8	294.2	296	73.1	22.8
27	2.8	270.4	87	30.6	303.7	22	71.6	295.2	297	74.9	23.8
28	2.2	270.9	88	31.1	304.3	23	73.4	296.2	298	76.7	24.8
29	1.7	271.5	89	31.7	304.8	24	75.2	297.2	299	78.5	25.8
+30	−1.1	272.0	+90	+32.2	305.4	+25	+77.0	298.2	300	+80.3	+26.8
31	0.6	272.6	91	32.8	305.9	26	78.8	299.2	301	82.1	27.8
32	0.0	273.2	92	33.3	306.5	27	80.6	300.2	302	83.9	28.8
33	+0.6	273.7	93	33.9	307.0	28	82.4	301.2	303	85.7	29.8
34	1.1	274.3	94	34.4	307.6	29	84.2	302.2	304	87.5	30.8
+35	+1.7	274.8	+95	+35.0	308.2	+30	+86.0	303.2	305	+89.3	+31.8
36	2.2	275.4	96	35.6	308.7	31	87.8	304.2	306	91.1	32.8
37	2.8	275.9	97	36.1	309.3	32	89.6	305.2	307	92.9	33.8
38	3.3	276.5	98	36.7	309.8	33	91.4	306.2	308	94.7	34.8
39	3.9	277.0	99	37.2	310.4	34	93.2	307.2	309	96.5	35.8
+40	+4.4	277.6	+100	+37.8	310.9	+35	+95.0	308.2	310	+98.3	+36.8

TABLE 30

Direction and Speed of True Wind in Units of Ship's Speed

Apparent wind speed	0°		10°		20°		30°		40°		Apparent wind speed
	Difference between the heading and apparent wind direction										
	°		°		°		°		°		
0.0	180	1.00	180	1.00	180	1.00	180	1.00	180	1.00	0.0
0.1	180	0.90	179	0.90	178	0.91	177	0.91	176	0.93	0.1
0.2	180	0.80	178	0.80	175	0.81	173	0.83	171	0.86	0.2
0.3	180	0.70	176	0.71	172	0.73	169	0.76	166	0.79	0.3
0.4	180	0.60	173	0.61	168	0.64	163	0.68	160	0.74	0.4
0.5	180	0.50	170	0.51	162	0.56	156	0.62	152	0.70	0.5
0.6	180	0.40	166	0.42	155	0.48	148	0.57	144	0.66	0.6
0.7	180	0.30	159	0.33	145	0.42	138	0.53	136	0.65	0.7
0.8	180	0.20	147	0.25	132	0.37	128	0.50	127	0.64	0.8
0.9	180	0.10	126	0.19	117	0.34	116	0.50	118	0.66	0.9
1.0	calm	0.00	95	0.17	100	0.35	105	0.52	110	0.68	1.0
1.1	0	0.10	66	0.21	85	0.38	95	0.55	103	0.72	1.1
1.2	0	0.20	49	0.28	73	0.43	86	0.60	96	0.78	1.2
1.3	0	0.30	39	0.36	64	0.50	79	0.66	90	0.84	1.3
1.4	0	0.40	33	0.45	57	0.57	73	0.73	85	0.90	1.4
1.5	0	0.50	29	0.54	51	0.66	68	0.81	81	0.98	1.5
1.6	0	0.60	26	0.64	47	0.74	64	0.89	78	1.05	1.6
1.7	0	0.70	24	0.74	44	0.83	61	0.97	75	1.13	1.7
1.8	0	0.80	22	0.83	42	0.93	58	1.06	72	1.22	1.8
1.9	0	0.90	21	0.93	40	1.02	56	1.15	70	1.30	1.9
2.0	0	1.00	20	1.03	38	1.11	54	1.24	68	1.39	2.0
2.5	0	1.50	17	1.52	32	1.60	47	1.71	60	1.85	2.5
3.0	0	2.00	15	2.02	29	2.09	43	2.19	56	2.32	3.0
3.5	0	2.50	14	2.52	28	2.58	41	2.68	53	2.81	3.5
4.0	0	3.00	13	3.02	26	3.08	39	3.17	51	3.30	4.0
4.5	0	3.50	13	3.52	25	3.58	38	3.67	50	3.79	4.5
5.0	0	4.00	12	4.02	25	4.08	37	4.16	49	4.28	5.0
6.0	0	5.00	12	5.02	24	5.07	36	5.16	47	5.27	6.0
7.0	0	6.00	12	6.02	23	6.07	35	6.15	46	6.27	7.0
8.0	0	7.00	11	7.02	23	7.07	34	7.15	45	7.26	8.0
9.0	0	8.00	11	8.02	22	8.07	34	8.15	44	8.26	9.0
10.0	0	9.00	11	9.02	22	9.06	33	9.15	44	9.26	10.0

Apparent wind speed	50°		60°		70°		80°		90°		Apparent wind speed
	°		°		°		°		°		
0.0	180	1.00	180	1.00	180	1.00	180	1.00	180	1.00	0.0
0.1	175	0.94	175	0.95	174	0.97	174	0.99	174	1.00	0.1
0.2	170	0.88	169	0.92	169	0.95	168	0.99	169	1.02	0.2
0.3	164	0.84	163	0.89	163	0.94	163	0.99	163	1.04	0.3
0.4	158	0.80	157	0.87	156	0.94	157	1.01	158	1.08	0.4
0.5	151	0.78	150	0.87	150	0.95	152	1.04	153	1.12	0.5
0.6	143	0.77	143	0.87	145	0.97	147	1.07	149	1.17	0.6
0.7	136	0.77	137	0.89	139	1.01	142	1.12	145	1.22	0.7
0.8	128	0.78	131	0.92	134	1.05	138	1.17	141	1.28	0.8
0.9	121	0.81	125	0.95	129	1.09	134	1.22	138	1.35	0.9
1.0	115	0.85	120	1.00	125	1.15	130	1.29	135	1.41	1.0
1.1	109	0.89	115	1.05	121	1.21	127	1.35	132	1.49	1.1
1.2	104	0.95	111	1.11	118	1.27	124	1.42	130	1.56	1.2
1.3	99	1.01	107	1.18	114	1.34	121	1.50	128	1.64	1.3
1.4	95	1.08	104	1.25	112	1.42	119	1.57	126	1.72	1.4
1.5	92	1.15	101	1.32	109	1.49	117	1.65	124	1.80	1.5
1.6	89	1.23	98	1.40	107	1.57	115	1.73	122	1.89	1.6
1.7	86	1.31	96	1.48	105	1.65	113	1.82	120	1.97	1.7
1.8	84	1.39	94	1.56	103	1.73	111	1.90	119	2.06	1.8
1.9	81	1.47	92	1.65	101	1.82	110	1.99	118	2.15	1.9
2.0	79	1.56	90	1.73	100	1.91	108	2.07	117	2.24	2.0
2.5	72	2.01	83	2.18	94	2.35	103	2.53	112	2.69	2.5
3.0	68	2.48	79	2.65	89	2.82	99	2.99	108	3.16	3.0
3.5	65	2.96	76	3.12	87	3.29	96	3.47	106	3.64	3.5
4.0	63	3.44	74	3.61	84	3.78	94	3.95	104	4.12	4.0
4.5	61	3.93	72	4.09	83	4.26	93	4.44	103	4.61	4.5
5.0	60	4.42	71	4.58	81	4.75	92	4.93	101	5.10	5.0
6.0	58	5.41	69	5.57	79	5.74	90	5.91	99	6.08	6.0
7.0	57	6.40	68	6.56	78	6.72	88	6.90	98	7.07	7.0
8.0	56	7.40	67	7.55	77	7.72	87	7.89	97	8.06	8.0
9.0	55	8.39	66	8.54	76	8.71	86	8.88	96	9.06	9.0
10.0	55	9.39	65	9.54	76	9.70	86	9.88	96	10.01	10.0

TABLE 30

Direction and Speed of True Wind in Units of Ship's Speed

Apparent wind speed	90°		100°		110°		120°		130°		Apparent wind speed
	Difference between the heading and apparent wind direction										
	°		°		°		°		°		
0.0	180	1.00	180	1.00	180	1.00	180	1.00	180	1.00	0.0
0.1	174	1.00	174	1.02	175	1.04	175	1.05	176	1.07	0.1
0.2	169	1.02	169	1.05	170	1.08	171	1.11	172	1.14	0.2
0.3	163	1.04	164	1.09	166	1.14	167	1.18	169	1.21	0.3
0.4	158	1.08	160	1.14	162	1.20	164	1.25	166	1.29	0.4
0.5	153	1.12	156	1.19	158	1.26	161	1.32	164	1.38	0.5
0.6	149	1.17	152	1.25	155	1.33	158	1.40	162	1.46	0.6
0.7	145	1.22	148	1.32	152	1.40	156	1.48	160	1.55	0.7
0.8	141	1.28	145	1.38	149	1.48	154	1.56	158	1.63	0.8
0.9	138	1.35	143	1.46	147	1.56	152	1.65	156	1.72	0.9
1.0	135	1.41	140	1.53	145	1.64	150	1.73	155	1.81	1.0
1.1	132	1.49	138	1.61	143	1.72	148	1.82	154	1.90	1.1
1.2	130	1.56	136	1.69	141	1.81	147	1.91	153	2.00	1.2
1.3	128	1.64	134	1.77	140	1.89	146	2.00	152	2.09	1.3
1.4	126	1.72	132	1.86	138	1.98	145	2.09	151	2.18	1.4
1.5	124	1.80	130	1.94	137	2.07	143	2.18	150	2.28	1.5
1.6	122	1.89	129	2.03	136	2.16	142	2.27	149	2.37	1.6
1.7	120	1.97	128	2.12	135	2.25	141	2.36	148	2.46	1.7
1.8	119	2.06	127	2.21	134	2.34	141	2.46	147	2.56	1.8
1.9	118	2.15	125	2.30	133	2.43	140	2.55	147	2.66	1.9
2.0	117	2.24	124	2.39	132	2.52	139	2.65	146	2.75	2.0
2.5	112	2.69	120	2.85	128	2.99	136	3.12	144	3.23	2.5
3.0	108	3.16	117	3.32	126	3.47	134	3.61	142	3.72	3.0
3.5	106	3.64	115	3.80	124	3.96	132	4.09	140	4.21	3.5
4.0	104	4.12	113	4.29	122	4.44	131	4.58	139	4.71	4.0
4.5	103	4.61	112	4.78	121	4.93	130	5.07	138	5.20	4.5
5.0	101	5.10	111	5.27	120	5.42	129	5.57	138	5.69	5.0
6.0	99	6.08	109	6.25	118	6.41	128	6.56	137	6.69	6.0
7.0	98	7.07	108	7.24	117	7.40	127	7.55	136	7.68	7.0
8.0	97	8.06	107	8.23	116	8.39	126	8.54	135	8.68	8.0
9.0	96	9.06	106	9.23	116	9.39	125	9.54	135	9.67	9.0
10.0	96	10.01	106	10.22	115	10.39	125	10.54	134	10.67	10.0

Apparent wind speed	140°		150°		160°		170°		180°		Apparent wind speed
	°		°		°		°		°		
0.0	180	1.00	180	1.00	180	1.00	180	1.00	180	1.00	0.0
0.1	177	1.08	177	1.09	178	1.09	179	1.10	180	1.10	0.1
0.2	174	1.16	175	1.18	177	1.19	178	1.20	180	1.20	0.2
0.3	171	1.24	173	1.27	175	1.29	178	1.30	180	1.30	0.3
0.4	169	1.33	172	1.36	174	1.38	177	1.40	180	1.40	0.4
0.5	167	1.42	170	1.45	173	1.48	177	1.50	180	1.50	0.5
0.6	165	1.51	169	1.55	173	1.58	176	1.60	180	1.60	0.6
0.7	164	1.60	168	1.64	172	1.68	176	1.69	180	1.70	0.7
0.8	162	1.69	167	1.74	171	1.77	176	1.79	180	1.80	0.8
0.9	161	1.79	166	1.84	171	1.87	175	1.89	180	1.90	0.9
1.0	160	1.88	165	1.93	170	1.97	175	1.99	180	2.00	1.0
1.1	159	1.97	164	2.03	170	2.07	175	2.09	180	2.10	1.1
1.2	158	2.07	164	2.13	169	2.17	175	2.19	180	2.20	1.2
1.3	157	2.16	163	2.22	169	2.27	174	2.29	180	2.30	1.3
1.4	157	2.26	162	2.32	168	2.36	174	2.39	180	2.40	1.4
1.5	156	2.36	162	2.42	168	2.46	174	2.49	180	2.50	1.5
1.6	155	2.45	161	2.52	168	2.56	174	2.59	180	2.60	1.6
1.7	155	2.55	161	2.61	167	2.66	174	2.69	180	2.70	1.7
1.8	154	2.65	161	2.71	167	2.76	174	2.79	180	2.80	1.8
1.9	154	2.74	160	2.81	167	2.86	173	2.89	180	2.90	1.9
2.0	153	2.84	160	2.91	167	2.96	173	2.99	180	3.00	2.0
2.5	151	3.33	158	3.40	166	3.46	173	3.49	180	3.50	2.5
3.0	150	3.82	157	3.90	165	3.95	172	3.99	180	4.00	3.0
3.5	149	4.31	157	4.39	164	4.45	172	4.49	180	4.50	3.5
4.0	148	4.81	156	4.89	164	4.95	172	4.99	180	5.00	4.0
4.5	147	5.31	155	5.39	164	5.45	172	5.49	180	5.50	4.5
5.0	146	5.80	155	5.89	163	5.95	172	5.99	180	6.00	5.0
6.0	145	6.80	154	6.88	163	6.95	171	6.99	180	7.00	6.0
7.0	145	7.79	154	7.88	162	7.95	171	7.99	180	8.00	7.0
8.0	144	8.79	153	8.88	162	8.95	171	8.99	180	9.00	8.0
9.0	144	9.79	153	9.88	162	9.95	171	9.99	180	10.00	9.0
10.0	143	10.78	153	10.88	162	10.95	171	10.98	180	11.00	10.0

TABLE 31
Correction of Barometer Reading for Height Above Sea Level

All barometers. All values positive.

Height in Feet	Outside temperature in degrees Fahrenheit													Height in Feet
	−20°	−10°	0°	10°	20°	30°	40°	50°	60°	70°	80°	90°	100°	
	Inches	*Inches*	*Inches*	*Inches*	*Inches*	*Inches*	*Inches*	*Inches*	*Inches*	*Inches*	*Inches*	*Inches*	*Inches*	
5	0.01	0.01	0.01	0.01	0.01	0.01	0.01	0.01	0.01	0.01	0.01	0.01	0.01	5
10	0.01	0.01	0.01	0.01	0.01	0.01	0.01	0.01	0.01	0.01	0.01	0.01	0.01	10
15	0.02	0.02	0.02	0.02	0.02	0.02	0.02	0.02	0.02	0.02	0.02	0.02	0.02	15
20	0.03	0.02	0.02	0.02	0.02	0.02	0.02	0.02	0.02	0.02	0.02	0.02	0.02	20
25	0.03	0.03	0.03	0.03	0.03	0.03	0.03	0.03	0.03	0.03	0.03	0.03	0.03	25
30	0.04	0.04	0.04	0.04	0.04	0.03	0.03	0.03	0.03	0.03	0.03	0.03	0.03	30
35	0.04	0.04	0.04	0.04	0.04	0.04	0.04	0.04	0.04	0.04	0.04	0.04	0.04	35
40	0.05	0.05	0.05	0.05	0.05	0.05	0.04	0.04	0.04	0.04	0.04	0.04	0.04	40
45	0.06	0.06	0.05	0.05	0.05	0.05	0.05	0.05	0.05	0.05	0.05	0.05	0.05	45
50	0.06	0.06	0.06	0.06	0.06	0.06	0.06	0.06	0.05	0.05	0.05	0.05	0.05	50
55	0.07	0.07	0.07	0.07	0.06	0.06	0.06	0.06	0.06	0.06	0.06	0.06	0.06	55
60	0.08	0.07	0.07	0.07	0.07	0.07	0.07	0.07	0.06	0.06	0.06	0.06	0.06	60
65	0.08	0.08	0.08	0.08	0.08	0.07	0.07	0.07	0.07	0.07	0.07	0.07	0.07	65
70	0.09	0.09	0.09	0.08	0.08	0.08	0.08	0.08	0.08	0.07	0.07	0.07	0.07	70
75	0.10	0.09	0.09	0.09	0.09	0.09	0.08	0.08	0.08	0.08	0.08	0.08	0.08	75
80	0.10	0.10	0.10	0.10	0.09	0.09	0.09	0.09	0.09	0.08	0.08	0.08	0.08	80
85	0.11	0.11	0.10	0.10	0.10	0.10	0.10	0.09	0.09	0.09	0.09	0.09	0.09	85
90	0.11	0.11	0.11	0.11	0.11	0.10	0.10	0.10	0.10	0.10	0.09	0.09	0.09	90
95	0.12	0.12	0.12	0.11	0.11	0.11	0.11	0.10	0.10	0.10	0.10	0.10	0.10	95
100	0.13	0.12	0.12	0.12	0.12	0.11	0.11	0.11	0.11	0.11	0.10	0.10	0.10	100
105	0.13	0.13	0.13	0.13	0.12	0.12	0.12	0.12	0.11	0.11	0.11	0.11	0.11	105
110	0.14	0.14	0.13	0.13	0.13	0.13	0.12	0.12	0.12	0.12	0.11	0.11	0.11	110
115	0.15	0.14	0.14	0.14	0.13	0.13	0.13	0.13	0.12	0.12	0.12	0.12	0.12	115
120	0.15	0.15	0.15	0.14	0.14	0.14	0.13	0.13	0.13	0.13	0.12	0.12	0.12	120
125	0.16	0.16	0.15	0.15	0.15	0.14	0.14	0.14	0.13	0.13	0.13	0.13	0.12	125

TABLE 32
Correction of Barometer Reading for Gravity

Mercurial barometers only.

Latitude	Correction	Latitude	Correction	Latitude	Correction	Latitude	Correction
°	*Inches*	°	*Inches*	°	*Inches*	°	*Inches*
0	− 0.08	25	− 0.05	50	+ 0.01	75	+ 0.07
5	− 0.08	30	− 0.04	55	+ 0.03	80	+ 0.07
10	− 0.08	35	− 0.03	60	+ 0.04	85	+ 0.08
15	− 0.07	40	− 0.02	65	+ 0.05	90	+ 0.08
20	− 0.06	45	0.00	70	+ 0.06		

TABLE 33
Correction of Barometer Reading for Temperature

Mercurial barometers only.

Temp. F	Height of Barometers in inches								Temp. F
	27.5	28.0	28.5	29.0	29.5	30.0	30.5	31.0	
°	*Inches*	*Inches*	*Inches*	*Inches*	*Inches*	*Inches*	*Inches*	*Inches*	°
− 20	+ 0.12	+ 0.12	+ 0.13	+ 0.13	+ 0.13	+ 0.13	+ 0.14	+ 0.14	− 20
18	0.12	0.12	0.12	0.12	0.13	0.13	0.13	0.13	18
16	0.11	0.11	0.12	0.12	0.12	0.12	0.12	0.13	16
14	0.11	0.11	0.11	0.11	0.11	0.12	0.12	0.12	14
12	0.10	0.10	0.11	0.11	0.11	0.11	0.11	0.11	12
− 10	+ 0.10	+ 0.10	+ 0.10	+ 0.10	+ 0.10	+ 0.11	+ 0.11	+ 0.11	− 10
8	0.09	0.09	0.10	0.10	0.10	0.10	0.10	0.10	8
6	0.09	0.09	0.09	0.09	0.09	0.09	0.10	0.10	6
4	0.08	0.08	0.08	0.09	0.09	0.09	0.09	0.09	4
− 2	0.08	0.08	0.08	0.08	0.08	0.08	0.09	0.09	− 2
0	+ 0.07	+ 0.07	+ 0.07	+ 0.08	+ 0.08	+ 0.08	+ 0.08	+ 0.08	0
+ 2	0.07	0.07	0.07	0.07	0.07	0.07	0.07	0.08	+ 2
4	0.06	0.06	0.06	0.07	0.07	0.07	0.07	0.07	4
6	0.06	0.06	0.06	0.06	0.06	0.06	0.06	0.06	6
8	0.05	0.05	0.05	0.05	0.06	0.06	0.06	0.06	8
+ 10	+ 0.05	+ 0.05	+ 0.05	+ 0.05	+ 0.05	+ 0.05	+ 0.05	+ 0.05	+ 10
12	0.04	0.04	0.04	0.04	0.04	0.05	0.05	0.05	12
14	0.04	0.04	0.04	0.04	0.04	0.04	0.04	0.04	14
16	0.03	0.03	0.03	0.03	0.03	0.03	0.03	0.04	16
18	0.03	0.03	0.03	0.03	0.03	0.03	0.03	0.03	18
+ 20	+ 0.02	+ 0.02	+ 0.02	+ 0.02	+ 0.02	+ 0.02	+ 0.02	+ 0.02	+ 20
22	0.02	0.02	0.02	0.02	0.02	0.02	0.02	0.02	22
24	0.01	0.01	0.01	0.01	0.01	0.01	0.01	0.01	24
26	+ 0.01	+ 0.01	+ 0.01	+ 0.01	+ 0.01	+ 0.01	+ 0.01	+ 0.01	26
28	0.00	0.00	0.00	0.00	0.00	0.00	0.00	0.00	28
+ 30	0.00	0.00	0.00	0.00	0.00	0.00	0.00	0.00	+ 30
32	− 0.01	− 0.01	− 0.01	− 0.01	− 0.01	− 0.01	− 0.01	− 0.01	32
34	0.01	0.01	0.01	0.01	0.01	0.01	0.01	0.02	34
36	0.02	0.02	0.02	0.02	0.02	0.02	0.02	0.02	36
38	0.02	0.02	0.02	0.02	0.03	0.03	0.03	0.03	38
+ 40	− 0.03	− 0.03	− 0.03	− 0.03	− 0.03	− 0.03	− 0.03	− 0.03	+ 40
42	0.03	0.03	0.03	0.04	0.04	0.04	0.04	0.04	42
44	0.04	0.04	0.04	0.04	0.04	0.04	0.04	0.04	44
46	0.04	0.04	0.04	0.05	0.05	0.05	0.05	0.05	46
48	0.05	0.05	0.05	0.05	0.05	0.05	0.05	0.05	48
+ 50	− 0.05	− 0.05	− 0.06	− 0.06	− 0.06	− 0.06	− 0.06	− 0.06	+ 50
52	0.06	0.06	0.06	0.06	0.06	0.06	0.06	0.07	52
54	0.06	0.06	0.07	0.07	0.07	0.07	0.07	0.07	54
56	0.07	0.07	0.07	0.07	0.07	0.07	0.08	0.08	56
58	0.07	0.07	0.08	0.08	0.08	0.08	0.08	0.08	58
+ 60	− 0.08	− 0.08	− 0.08	− 0.08	− 0.08	− 0.09	− 0.09	− 0.09	+ 60
62	0.08	0.08	0.09	0.09	0.09	0.09	0.09	0.09	62
64	0.09	0.09	0.09	0.09	0.09	0.10	0.10	0.10	64
66	0.09	0.09	0.10	0.10	0.10	0.10	0.10	0.10	66
68	0.10	0.10	0.10	0.10	0.11	0.11	0.11	0.11	68
+ 70	− 0.10	− 0.10	− 0.11	− 0.11	− 0.11	− 0.11	− 0.11	− 0.12	+ 70
72	0.11	0.11	0.11	0.11	0.12	0.12	0.12	0.12	72
74	0.11	0.11	0.12	0.12	0.12	0.12	0.13	0.13	74
76	0.12	0.12	0.12	0.12	0.13	0.13	0.13	0.13	76
78	0.12	0.12	0.13	0.13	0.13	0.13	0.14	0.14	78
+ 80	− 0.13	− 0.13	− 0.13	− 0.13	− 0.14	− 0.14	− 0.14	− 0.14	+ 80
82	0.13	0.14	0.14	0.14	0.14	0.14	0.15	0.15	82
84	0.14	0.14	0.14	0.15	0.15	0.15	0.15	0.16	84
86	0.14	0.15	0.15	0.15	0.15	0.16	0.16	0.16	86
88	0.15	0.15	0.15	0.16	0.16	0.16	0.16	0.17	88
+ 90	− 0.15	− 0.16	− 0.16	− 0.16	− 0.16	− 0.17	− 0.17	− 0.17	+ 90
92	0.16	0.16	0.16	0.17	0.17	0.17	0.17	0.18	92
94	0.16	0.17	0.17	0.17	0.17	0.18	0.18	0.18	94
96	0.17	0.17	0.17	0.18	0.18	0.18	0.19	0.19	96
98	0.17	0.18	0.18	0.18	0.18	0.19	0.19	0.19	98
100	− 0.18	− 0.18	− 0.18	− 0.19	− 0.19	− 0.19	− 0.20	− 0.20	100

TABLE 34

Conversion Table for hecto-Pascals (millibars), Inches of Mercury, and Millimeters of Mercury

hPa	Inches	Millimeters	hPa	Inches	Millimeters	hPa	Inches	Millimeters
900	26.58	675.1	960	28.35	720.1	1020	30.12	765.1
901	26.61	675.8	961	28.38	720.8	1021	30.15	765.8
902	26.64	676.6	962	28.41	721.6	1022	30.18	766.6
903	26.67	677.3	963	28.44	722.3	1023	30.21	767.3
904	26.70	678.1	964	28.47	723.1	1024	30.24	768.1
905	26.72	678.8	965	28.50	723.8	1025	30.27	768.8
906	26.75	679.6	966	28.53	724.6	1026	30.30	769.6
907	26.78	680.3	967	28.56	725.3	1027	30.33	770.3
908	26.81	681.1	968	28.58	726.1	1028	30.36	771.1
909	26.84	681.8	969	28.61	726.8	1029	30.39	771.8
910	26.87	682.6	970	28.64	727.6	1030	30.42	772.6
911	26.90	683.3	971	28.67	728.3	1031	30.45	773.3
912	26.93	684.1	972	28.70	729.1	1032	30.47	774.1
913	26.96	684.8	973	28.73	729.8	1033	30.50	774.8
914	26.99	685.6	974	28.76	730.6	1034	30.53	775.6
915	27.02	686.3	975	28.79	731.3	1035	30.56	776.3
916	27.05	687.1	976	28.82	732.1	1036	30.59	777.1
917	27.08	687.8	977	28.85	732.8	1037	30.62	777.8
918	27.11	688.6	978	28.88	733.6	1038	30.65	778.6
919	27.14	689.3	979	28.91	734.3	1039	30.68	779.3
920	27.17	690.1	980	28.94	735.1	1040	30.71	780.1
921	27.20	690.8	981	28.97	735.8	1041	30.74	780.8
922	27.23	691.6	982	29.00	736.6	1042	30.77	781.6
923	27.26	692.3	983	29.03	737.3	1043	30.80	782.3
924	27.29	693.1	984	29.06	738.1	1044	30.83	783.1
925	27.32	693.8	985	29.09	738.8	1045	30.86	783.8
926	27.34	694.6	986	29.12	739.6	1046	30.89	784.6
927	27.37	695.3	987	29.15	740.3	1047	30.92	785.3
928	27.40	696.1	988	29.18	741.1	1048	30.95	786.1
929	27.43	696.8	989	29.21	741.8	1049	30.98	786.8
930	27.46	697.6	990	29.23	742.6	1050	31.01	787.6
931	27.49	698.3	991	29.26	743.3	1051	31.04	788.3
932	27.52	699.1	992	29.29	744.1	1052	31.07	789.1
933	27.55	699.8	993	29.32	744.8	1053	31.10	789.8
934	27.58	700.6	994	29.35	745.6	1054	31.12	790.6
935	27.61	701.3	995	29.38	746.3	1055	31.15	791.3
936	27.64	702.1	996	29.41	747.1	1056	31.18	792.1
937	27.67	702.8	997	29.44	747.8	1057	31.21	792.8
938	27.70	703.6	998	29.47	748.6	1058	31.24	793.6
939	27.73	704.3	999	29.50	749.3	1059	31.27	794.3
940	27.76	705.1	1000	29.53	750.1	1060	31.30	795.1
941	27.79	705.8	1001	29.56	750.8	1061	31.33	795.8
942	27.82	706.6	1002	29.59	751.6	1062	31.36	796.6
943	27.85	707.3	1003	29.62	752.3	1063	31.39	797.3
944	27.88	708.1	1004	29.65	753.1	1064	31.42	798.1
945	27.91	708.8	1005	29.68	753.8	1065	31.45	798.8
946	27.94	709.6	1006	29.71	754.6	1066	31.48	799.6
947	27.96	710.3	1007	29.74	755.3	1067	31.51	800.3
948	27.99	711.1	1008	29.77	756.1	1068	31.54	801.1
949	28.02	711.8	1009	29.80	756.8	1069	31.57	801.8
950	28.05	712.6	1010	29.83	757.6	1070	31.60	802.6
951	28.08	713.3	1011	29.85	758.3	1071	31.63	803.3
952	28.11	714.1	1012	29.88	759.1	1072	31.66	804.1
953	28.14	714.8	1013	29.91	759.8	1073	31.69	804.8
954	28.17	715.6	1014	29.94	760.6	1074	31.72	805.6
955	28.20	716.3	1015	29.97	761.3	1075	31.74	806.3
956	28.23	717.1	1016	30.00	762.1	1076	31.77	807.1
957	28.26	717.8	1017	30.03	762.8	1077	31.80	807.8
958	28.29	718.6	1018	30.06	763.6	1078	31.83	808.6
959	28.32	719.3	1019	30.09	764.3	1079	31.86	809.3
960	28.35	720.1	1020	30.12	765.1	1080	31.89	810.1

TABLE 35
Relative Humidity

Dry-bulb temp. F	1°	2°	3°	4°	5°	6°	7°	8°	9°	10°	11°	12°	13°	14°	Dry-bulb temp. F
°	%	%	%	%	%	%	%	%	%	%	%	%	%	%	°
− 20	7														− 20
18	14														18
16	21														16
14	27														14
12	32														12
− 10	37														− 10
8	41	2													8
6	45	9													6
4	49	16													4
− 2	52	22													− 2
0	56	28													0
+ 2	59	33	7												+ 2
4	62	37	14												4
6	64	42	20												6
8	67	46	25	5											8
+ 10	69	50	30	11											+ 10
12	71	53	35	17											12
14	73	56	40	23	7										14
16	76	60	44	28	13										16
18	77	62	48	33	19	4									18
+ 20	79	65	51	37	24	10									+ 20
22	81	68	55	42	29	16	4								22
24	83	70	58	45	33	21	10								24
26	85	73	61	49	38	26	15	4							26
28	86	75	64	53	42	31	20	10							28
+ 30	88	77	66	56	45	35	25	15	6						+ 30
32	89	79	69	59	49	39	30	20	11	2					32
34	90	81	71	62	52	43	34	25	16	8					34
36	91	82	73	64	55	47	38	29	21	13	5				36
38	91	83	74	66	58	50	42	33	25	18	10	2			38
+ 40	92	84	76	68	60	52	45	37	30	22	15	7			+ 40
42	92	84	77	69	62	54	47	40	33	26	19	12	5		42
44	92	85	78	70	63	56	49	43	36	29	23	17	10	4	44
46	93	86	79	72	65	58	52	45	39	32	26	20	14	8	46
48	93	86	79	73	66	60	54	47	41	35	29	24	18	12	48
+ 50	93	87	80	74	68	61	55	49	44	38	32	27	21	16	+ 50
52	94	87	81	75	69	63	57	51	46	40	35	29	24	19	52
54	94	88	82	76	70	64	59	53	48	42	37	32	27	22	54
56	94	88	82	77	71	65	60	55	50	44	39	35	30	25	56
58	94	88	83	77	72	67	61	56	51	46	42	37	32	28	58
+ 60	94	89	83	78	73	68	63	58	53	48	43	39	34	30	+ 60
62	95	89	84	79	74	69	64	59	54	50	45	41	37	32	62
64	95	89	84	79	74	70	65	60	56	51	47	43	38	34	64
66	95	90	85	80	75	71	66	61	57	53	49	44	40	36	66
68	95	90	85	81	76	71	67	63	58	54	50	46	42	38	68
+ 70	95	90	86	81	77	72	68	64	59	55	51	48	44	40	+ 70
72	95	91	86	82	77	73	69	65	61	57	53	49	45	42	72
74	95	91	86	82	78	74	69	65	62	58	54	50	47	43	74
76	95	91	87	82	78	74	70	66	63	59	55	51	48	45	76
78	96	91	87	83	79	75	71	67	63	60	56	53	49	46	78
+ 80	96	91	87	83	79	75	72	68	64	61	57	54	50	47	+ 80
82	96	92	88	84	80	76	72	69	65	62	58	55	52	48	82
84	96	92	88	84	80	76	73	69	66	62	59	56	53	49	84
86	96	92	88	84	81	77	73	70	67	63	60	57	54	51	86
88	96	92	88	85	81	77	74	71	67	64	61	58	55	52	88
+ 90	96	92	89	85	81	78	74	71	68	65	61	58	55	52	+ 90
92	96	92	89	85	82	78	75	72	68	65	62	59	56	53	92
94	96	93	89	85	82	79	75	72	69	66	63	60	57	54	94
96	96	93	89	86	82	79	76	73	70	67	64	61	58	55	96
98	96	93	89	86	83	79	76	73	70	67	64	61	59	56	98
+ 100	96	93	90	86	83	80	77	74	71	68	65	62	59	57	+ 100

(Column headings: Difference between dry-bulb and wet-bulb temperatures)

TABLE 35
Relative Humidity

Dry-bulb temp. F	15°	16°	17°	18°	19°	20°	21°	22°	23°	24°	25°	26°	27°	28°	Dry-bulb temp. F
°	%	%	%	%	%	%	%	%	%	%	%	%	%	%	°
+ 46	2														+ 46
48	7	1													48
+ 50	10	5													+ 50
52	14	9	4												52
54	17	12	7	3											54
56	20	16	11	7	2										56
58	23	19	14	10	6	2									58
+ 60	26	21	17	13	9	5	1								+ 60
62	28	24	20	16	12	8	4	1							62
64	30	26	22	19	15	11	8	4							64
66	32	29	25	21	17	14	10	7	4						66
68	34	31	27	23	20	16	13	10	7	3					68
+ 70	36	33	29	26	22	19	16	12	9	6	3				+ 70
72	38	34	31	28	24	21	18	15	12	9	6	3			72
74	40	36	33	30	26	23	20	17	14	11	8	6	3		74
76	41	38	35	31	28	25	22	19	16	14	11	8	5	3	76
78	43	39	36	33	30	27	24	21	18	16	13	10	8	5	78
+ 80	44	41	38	35	32	29	26	23	20	18	15	13	10	8	+ 80
82	45	42	39	36	33	30	28	25	22	20	17	15	12	10	82
84	46	43	40	38	35	32	29	27	24	21	19	17	14	12	84
86	48	45	42	39	36	33	31	28	26	23	21	18	16	14	86
88	49	46	43	40	37	35	32	30	27	25	22	20	18	16	88
+ 90	50	47	44	41	39	36	34	31	29	26	24	22	19	17	+ 90
92	51	48	45	42	40	37	35	32	30	28	25	23	21	19	92
94	51	49	46	44	41	39	36	34	31	29	27	25	23	20	94
96	52	50	47	45	42	40	37	35	33	30	28	26	24	22	96
98	53	51	48	45	43	41	38	36	34	32	29	27	25	23	98
+ 100	54	51	49	46	44	42	39	37	35	33	31	29	27	25	+ 100

(Column headings: Difference between dry-bulb and wet-bulb temperatures)

Dry-bulb temp. F	29°	30°	31°	32°	33°	34°	35°	36°	37°	38°	39°	40°	41°	42°	Dry-bulb temp. F
°	%	%	%	%	%	%	%	%	%	%	%	%	%	%	°
+ 78	3														+ 78
+ 80	5	3													+ 80
82	7	5	3	1											82
84	10	7	5	3	1										84
86	11	9	7	5	3	1									86
88	13	11	9	7	5	3	1								88
+ 90	15	13	11	9	7	5	3	1							+ 90
92	17	15	13	11	9	7	5	3	1						92
94	18	16	14	12	11	9	7	5	3	2					94
96	20	18	16	14	12	10	9	7	5	4	2				96
98	21	19	17	16	14	12	10	9	7	5	4	2	1		98
+ 100	23	21	19	17	15	14	12	10	9	7	5	4	2	1	+ 100

(Column headings: Difference between dry-bulb and wet-bulb temperatures)

TABLE 36
Dew Point

Dry-bulb temp. F	Difference between dry-bulb and wet-bulb temperatures														Dry-bulb temp. F
	1°	2°	3°	4°	5°	6°	7°	8°	9°	10°	11°	12°	13°	14°	
°	°	°	°	°	°	°	°	°	°	°	°	°	°	°	°
−20															−20
18	−52														18
16	45														16
14	39														14
12	34														12
−10	−29														−10
8	25	−75													8
6	22	50													6
4	18	39													4
−2	15	32													−2
0	−12	−26													0
+2	9	21	−49												+2
4	6	16	35												4
6	3	12	27												6
8	−1	9	20	−50											8
+10	+2	−5	−15	−34											+10
12	5	−2	10	24											12
14	7	+1	6	17											14
16	10	4	−2	11											16
18	12	7	+1	6											18
+20	+15	+10	+5	−2											+20
22	17	13	8	+2											22
24	20	16	11	6											24
26	22	18	14	10	+4	−4									26
28	24	21	17	13	8	+1	−8	−22							28
+30	+27	+24	+20	+16	+11	+6	−1	−12	−31						+30
32	29	26	23	19	15	10	+4	−4	16	−47					32
34	32	29	26	22	18	14	9	+2	−7	22					34
36	34	31	28	25	22	18	13	7	0	11	−30				36
38	36	33	31	28	25	21	17	12	+6	−2	14	−42			38
+40	+38	+35	+33	+30	+27	+24	+20	+16	+11	+4	−4	−18	−79		+40
42	40	38	35	33	30	27	23	19	15	10	+3	−7	23		42
44	42	40	37	35	32	29	26	23	19	14	9	+2	−9	−29	44
46	44	42	40	37	35	32	29	26	22	18	13	7	0	11	46
48	46	44	42	40	37	35	32	29	26	22	18	13	+6	−2	48
+50	+48	+46	+44	+42	+40	+37	+35	+32	+29	+25	+21	+17	+12	+5	+50
52	50	48	46	44	42	40	37	35	32	29	25	21	17	11	52
54	52	50	49	47	44	42	40	37	35	32	28	25	21	16	54
56	54	53	51	49	47	45	42	40	37	35	32	28	25	21	56
58	56	55	53	51	49	47	45	43	40	38	35	32	28	25	58
+60	+58	+57	+55	+53	+51	+49	+47	+45	+43	+40	+38	+35	+32	+28	+60
62	60	59	57	55	54	52	50	48	45	43	41	38	35	32	62
64	62	61	59	57	56	54	52	50	48	46	43	41	38	35	64
66	64	63	61	60	58	56	54	52	50	48	46	44	41	39	66
68	67	65	63	62	60	58	57	55	53	51	49	46	44	42	68
+70	+69	+67	+66	+64	+62	+61	+59	+57	+55	+53	+51	+49	+47	+45	+70
72	71	69	68	66	64	63	61	59	58	56	54	52	50	47	72
74	73	71	70	68	67	65	63	62	60	58	56	54	52	50	74
76	75	73	72	70	69	67	66	64	62	61	59	57	55	53	76
78	77	75	74	72	71	69	68	66	65	63	61	59	57	55	78
+80	+79	+77	+76	+74	+73	+72	+70	+68	+67	+65	+64	+62	+60	+58	+80
82	81	79	78	77	75	74	72	71	69	67	66	64	62	61	82
84	83	81	80	79	77	76	74	73	71	70	68	67	65	63	84
86	85	83	82	81	79	78	76	75	74	72	70	69	67	66	86
88	87	85	84	83	81	80	79	77	76	74	73	71	70	68	88
+90	+89	+87	+86	+85	+84	+82	+81	+79	+78	+76	+75	+73	+72	+70	+90
92	91	89	88	87	86	84	83	82	80	79	77	76	74	73	92
94	93	92	90	89	88	86	85	84	82	81	79	78	76	75	94
96	95	94	92	91	90	88	87	86	84	83	82	80	79	77	96
98	97	96	94	93	92	91	89	88	87	85	84	82	81	80	98
+100	+99	+98	+96	+95	+94	+93	+91	+90	+89	+87	+86	+85	+83	+82	+100

TABLE 36
Dew Point

Dry-bulb temp. F	Difference between dry-bulb and wet-bulb temperatures														Dry-bulb temp. F
	15°	16°	17°	18°	19°	20°	21°	22°	23°	24°	25°	26°	27°	28°	
°	°	°	°	°	°	°	°	°	°	°	°	°	°	°	°
+46	−36														+46
48	14	−45													48
+50	−3	−17	−78												+50
52	+4	−5	21												52
54	10	+3	−7	−25											54
56	16	10	+2	−8	−29										56
58	20	16	10	+2	−10	−34									58
+60	+25	+20	+15	+9	+1	−11	−39								+60
62	29	25	20	15	9	+1	−12	−45							62
64	32	29	25	20	15	9	0	−13	−52						64
66	36	33	29	25	21	15	+9	0	−14	−59					66
68	39	36	33	29	25	21	16	+9	0	−14	−68				68
+70	+42	+39	+36	+33	+30	+26	+21	+16	+9	0	−14	−76			+70
72	45	43	40	37	34	30	26	22	16	+10	+1	−14	−77		72
74	48	46	43	40	37	34	31	27	22	17	10	+1	−13	−70	74
76	51	48	46	44	41	38	35	31	27	23	17	11	+2	−12	76
78	53	51	49	47	44	41	38	35	32	28	23	18	11	+3	78
+80	+56	+54	+52	+50	+47	+45	+42	+39	+36	+32	+28	+24	+19	+12	+80
82	59	57	55	53	50	48	45	43	40	37	33	29	25	20	82
84	61	59	57	55	53	51	49	46	43	41	37	34	30	26	84
86	64	62	60	58	56	54	52	49	47	44	41	38	35	31	86
88	66	64	63	61	59	57	55	52	50	48	45	42	39	36	88
+90	+69	+67	+65	+63	+62	+60	+58	+55	+53	+51	+48	+46	+43	+40	+90
92	71	69	68	66	64	62	60	58	56	54	52	49	47	44	92
94	73	72	70	68	67	65	63	61	59	57	55	52	50	47	94
96	76	74	73	71	69	67	66	64	62	60	58	56	53	51	96
98	78	77	75	73	72	70	68	67	65	63	61	59	57	54	98
+100	+80	+79	+77	+76	+74	+73	+71	+69	+67	+66	+64	+62	+60	+57	+100

Dry-bulb temp. F	Difference between dry-bulb and wet-bulb temperatures														Dry-bulb temp. F
	29°	30°	31°	32°	33°	34°	35°	36°	37°	38°	39°	40°	41°	42°	
°	°	°	°	°	°	°	°	°	°	°	°	°	°	°	°
+76	−61														+76
78	−11	−53													78
+80	+4	−10	−45												+80
82	13	+5	−8	−39											82
84	20	14	+6	−6	−33										84
86	27	21	15	+7	−4	−28									86
88	32	27	22	16	+9	−2	−23								88
+90	+36	+33	+28	+24	+18	+10	0	−18							+90
92	41	37	34	30	25	19	+12	+2	−14						92
94	45	42	38	35	31	26	20	13	+4	−10					94
96	48	46	43	39	36	32	27	22	15	+6	−7	−43			96
98	52	49	47	44	40	37	33	28	23	17	+9	−4	−30		98
+100	+55	+53	+50	+47	+45	+41	+38	+34	+30	+25	+19	+11	0	−21	+100

GLOSSARY OF MARINE NAVIGATION

A

abaft., *adv.* In a direction farther aft in a ship than a specified reference position, such as abaft the mast. See also ABAFT THE BEAM, AFT, ASTERN.

abaft the beam.. Any direction between broad on the beam and astern. See also FORWARD OF THE BEAM.

abampere., *n.* The unit of current in the centimeter gram-second electromagnetic system. The abampere is 10 amperes.

abeam., *adv.* In a line approximately at right angle to the ship's keel opposite the waist or middle part of a ship. See also BROAD ON THE BEAM.

aberration., *n.* 1. The apparent displacement of a celestial body in the direction of motion of the earth in its orbit caused by the motion of the earth combined with the finite velocity of light. When, in addition to the combined effect of the velocity of light and the motion of the earth, account is taken of the motion of the celestial body in space during the interval that the light is traveling to the earth from the luminous body, as in the case of planets, the phenomenon is termed planetary aberration. The aberration due to the rotation of the earth on its axis is termed diurnal aberration or daily aberration. The aberration due to the revolution of the earth about the sun is termed annual aberration. The aberration due to the motion of the center of mass of the solar system in space is termed secular aberration but is not taken into account in practical astronomy. See also CONSTANT OF ABERRATION. 2. The convergence to different foci, by a lens or mirror, of parallel rays of light. In a single lens having spherical surfaces, aberration may be caused by differences in the focal lengths of the various parts of the lens: rays passing through the outer part of the lens come to a focus nearer the lens than do rays passing through its central part. This is termed spherical aberration and, being due to the faulty figure of the lens, is eliminated by correcting that figure. A lens so corrected is called an aplanatic lens. Aberration may also result from differences in the wavelengths of light of different colors: light of the shorter wavelengths (violet end of the spectrum) comes to a focus nearer the lens than light of the longer wavelengths (red end of the spectrum). This is termed chromatic aberration, and is practically eliminated over a moderate range of wavelengths by using a composite lens, called an achromatic lens, composed of parts having different dispersive powers.

aberration constant.. See CONSTANT OF ABERRATION.

ablation., *n.* Wasting of snow or ice by melting or evaporation.

abnormal., *adj.* Deviating from normal.

abrasion., *n.* Rubbing or wearing away, or the result of such action.

abroholos., *n.* A squall frequent from May through August between Cabo de Sao Tome and Cabo Frio on the coast of Brazil.

abrupt., *adv.* Steep, precipitous. See also BOLD.

abscissa., *n.* The horizontal coordinate of a set of rectangular coordinates. Also used in a similar sense in connection with oblique coordinates.

absolute.. Pertaining to measurement relative to a universal constant or natural datum.

absolute accuracy.. The ability of a navigation or positioning system to define an exact location in relation to a coordinate system.

absolute gain.. See ISOTROPIC GAIN (of an antenna).

absolute humidity.. The mass of water vapor per unit volume of air.

absolute motion.. Motion relative to a fixed point. If the earth were stationary in space, any change in the position of another body, relative to the earth, would be due only to the motion of that body. This would be absolute motion, or motion relative to a fixed point. Actual motion is motion of an object relative to the earth.

absolute temperature.. Temperature measured from absolute zero which is zero on the Kelvin scale, 273.16°C on the Celsius scale, and 459.69°F on the Fahrenheit scale. The sizes of the Kelvin and Celsius degree are equal. The size of a degree on the Fahrenheit scale equals that on the Rankine scale.

absolute value.. The value of a real number without regard to sign. Thus, the absolute value of +8 or -8 is |8|. Vertical lines on each side of a number indicate that its absolute value is intended.

absorption.. The process by which radiant energy is absorbed and converted to other forms of energy. See ATTENUATION.

absolute zero.. The theoretical temperature at which molecular motion ceases, 459.69°F or -273.16°C.

abyss., *n.* A very deep area of the ocean. The term is used to refer to a particular deep part of the ocean, or to any part below 300 fathoms.

abyssal plain.. See under PLAIN.

accelerate., *v., t.* To move or cause to move with increasing velocity.

acceleration., *n.* 1. The rate of change of velocity. 2. The act or process of accelerating, or the state of being accelerated. Negative acceleration is called DECELERATION.

acceleration error.. The error resulting from change in velocity (either speed or direction); specifically, deflection of the apparent vertical, as indicated by an artificial horizon, due to acceleration. Also called BUBBLE ACCELERATION ERROR when applied to an instrument using a bubble as an artificial horizon.

accelerometer., *n.* A device used to measure the accelerations of a craft, resulting from the craft's acceleration with respect to the earth, acceleration of gravity, and Coriolis acceleration.

accidental error.. See RANDOM ERROR. An error of accidental nature. (Not to be confused with MISTAKE.)

accretion., *n.* Accumulation of material on the surface of an object.

accuracy., *n.* 1. In navigation, a measure of the difference between the position indicated by measurement and the true position. Some expressions of accuracy are defined in terms of probability. 2. A measure of how close the outcome of a series of observations or measurements approaches the true value of a desired quantity. The degree of exactness with which the true value of the quantity is determined from observations is limited by the presence of both systematic and random errors. Accuracy should not be confused with PRECISION, which is a measure of the repeatability of the observations. Observations may be of high precision due to the quality of the observing instrument, the skill of the observer and the resulting small random errors, but inaccurate due to the presence of large systematic errors. Accuracy implies precision, but precision does not imply accuracy. See also ERROR, RADIAL ERROR, ABSOLUTE ACCURACY, PREDICTABLE ACCURACY, RELATIVE ACCURACY, REPEATABLE ACCURACY.

achromatic lens.. See under ABERRATION, definition 2.

aclinal., *adj.* Without dip; horizontal.

aclinic., *adj.* Without magnetic dip.

aclinic line.. The magnetic equator; the line on the surface of the earth connecting all points of zero magnetic dip.

acoustic depth finder.. See ECHO SOUNDER.

acoustic navigation.. See SONIC NAVIGATION.

acoustics., *n.* 1. That branch of physics dealing with sound. 2. The sound characteristics of a room, auditorium, etc., which determine its quality with respect to distinct hearing.

acoustic sounding.. See ECHO SOUNDING.

acquisition., *n.* The selection of those targets or satellites requiring a tracking procedure and the initiation of their tracking.

acre., *n.* A unit of area equal to 43,560 square feet.

across-the-scope echo. . See CLASSIFICATION OF RADAR ECHOES.

active satellite. . 1. An artificial satellite which transmits an electromagnetic signal. A satellite with the capability to transmit, repeat, or retransmit electromagnetic information, as contrasted with PASSIVE SATELLITE. 2. As defined by International Telecommunications Union (ITU), an earth satellite carrying a station intended to transmit or re transmit radio communication signals.

active tracking system. . A satellite tracking system which operates by transmission of signals to and receipt of responses from the satellite.

actual motion. . Motion of an object relative to the earth. See also MOTION.

acute angle. . An angle less than 90°.

additional secondary phase factor correction. . A correction in addition to the secondary phase factor correction for the additional time (or phase delay) for transmission of a low frequency signal over a composite land-water path when the signal transit time is based on the free-space velocity.

ADF reversal. . The swinging of the needle on the direction indicator of an automatic direction finder through 180°, indicating that the station to which the direction finder is tuned has been passed.

adiabatic. , *adj.* Referring to a thermodynamic change of state of a system in which there is no transfer of heat or mass across the boundaries of the system. In an adiabatic process, compression causes warming, expansion causes cooling.

adjacent angles. . Two angles having a common vertex and lying at opposite ends of a common side.

adjustment. , *n.* The determination and application of corrections to observations, for the purpose of reducing errors or removing internal inconsistencies in derived results.

admiralty. . Pertaining to the body of law that governs maritime affairs.

adrift. , *adj. & adv.* Afloat and unattached to the shore or the sea bottom, and without propulsive power. See also UNDERWAY.

advance. , *n.* 1. The distance a vessel moves in its initial direction from the point where the rudder is started over until the heading has changed 90°. 2. The distance a vessel moves in the initial direction for heading changes of less than 90°. See also TRANSFER.

advance. , *v., t. & i.* To move forward, as to move a line of position forward, parallel to itself, along a course line to obtain a line of position at a later time. The opposite is RETIRE.

advanced line of position. . A line of position which has been moved forward along the course line to allow for the run since the line was established. The opposite is RETIRED LINE OF POSITION.

advection. , *n.* Transport of atmospheric properties solely by mass motion of the atmosphere. WIND refers to air motion, while ADVECTION refers more specifically to the transfer of any property of the atmosphere (temperature, humidity, etc.) from one area to another.

advection fog. . A type of fog caused by the advection of moist air over a cold surface, and the consequent cooling of that air to below its dew point. SEA FOG is a very common advection fog that is caused by moist air in transport over a cold body of water.

aero light. . Short for AERONAUTICAL LIGHT.

aeromarine light. . A marine light having part of its beam deflected to an angle of 10° to 15° above the horizon for use by aircraft.

aeromarine radiobeacon. . A radiobeacon established for use by both mariners and airmen.

aeronautical. , *adj.* Of or pertaining to the operation or navigation of aircraft.

aeronautical beacon. . A visual aid to navigation, displaying flashes of white or colored light or both, used to indicate the location of airports, landmarks, and certain points of the Federal airways in mountainous terrain and to mark hazards.

aeronautical chart. . See under CHART.

aeronautical light. . A luminous or lighted aid to navigation intended primarily for air navigation. Often shortened to AERO LIGHT.

aeronautical radiobeacon. . A radiobeacon whose service is intended primarily for aircraft.

aestival. , *adj.* Pertaining to summer. The corresponding adjectives for fall, winter, and spring are autumnal, hibernal and vernal.

affluent. , *n.* A stream flowing into a larger stream or lake; a tributary.

afloat. , *adj. & adv.* Floating on the water; water-borne. See also SURFACED, UNCOVERED, AGROUND, ASHORE.

aft. , *adv.* Near, toward, or at the stern of a craft. See also ABAFT, ASTERN.

afterglow. , *n.* 1. The slowly decaying luminescence of the screen of the cathode-ray tube after excitation by an electron beam has ceased. See also PERSISTENCE. 2. A broad, high arch of radiance or glow seen occasionally in the western sky above the highest clouds in deepening twilight, caused by the scattering effect of very fine particles of dust suspended in the upper atmosphere.

aged ridge. . A ridge of ice forced up by pressure which has undergone considerable weathering.

age of diurnal inequality. . The time interval between the maximum semimonthly north or south declination of the moon and the maximum effect of the declination upon the range of tide or the speed of the tidal current; this effect is manifested chiefly by an increase in the height or speed difference between the two high (low) waters or flood (ebb) currents during the day. The tides occurring at this time are called TROPIC TIDES. Also called DIURNAL AGE.

age of parallax inequality. . The time interval between perigee of the moon and the maximum effect of parallax upon the range of tide or the speed of the tidal current. See also PARALLAX INEQUALITY.

age of phase inequality. . The time interval between new or full moon and the maximum effect of these phases upon the range of tide or the speed of the tidal current. Also called AGE OF TIDE.

age of the moon. . The elapsed time, usually expressed in days, since the last new moon. See also PHASES OF THE MOON.

age of tide. . See AGE OF PHASE INEQUALITY.

Ageton. . *n.* 1. A divided triangle method of sight reduction in which a perpendicular is dropped from the GP of the body to the meridian of the observer. 2. Rear Admiral Arthur A. Ageton, USN, inventor of the Ageton method.

agger. , *n.* See DOUBLE TIDE.

agonic line. . A line joining points of no magnetic variation, a special case of an isogonic line.

agravic. , *adj.* Of or pertaining to a condition of no gravitation.

aground. , *adj. & adv.* Resting or lodged on the bottom.

Agulhas Current. . A generally southwestward flowing ocean current of the Indian Ocean, one of the swiftest ocean currents. To the south of latitude 30°S the Agulhas Current is a well-defined and narrow current that extends less than 100 km from the coast of South Africa. To the south of South Africa the greatest volume of its water bends sharply to the south and then toward the east, thus returning to the Indian Ocean.

ahead. , *adv.* Bearing approximately 000° relative. The term is often used loosely for DEAD AHEAD or bearing exactly 000° relative. The opposite is ASTERN.

ahead reach. . The distance traveled by a vessel proceeding ahead at full power from the time the engines are reversed until she is at full stop.

ahull. . The condition of a vessel making no way in a storm, allowing wind and sea to determine the position of the ship. Sailing vessels lying ahull lash the helm alee, and may carry storm sails.

aid. , *n.* Short for AID TO NAVIGATION.

aid to navigation. . A device or structure external to a craft, designed to assist in determination of position, to define a safe course, or to warn of dangers or obstructions. If the information is transmitted by light waves, the device is called a visual aid to navigation; if by sound waves, an audible aid to navigation; if by radio waves; a radio aid to navigation. Any aid to navigation using electronic equipment, whether or not radio waves are involved, may be called an electronic aid to navigation. Compare with NAVIGATIONAL AID, meaning an instrument, device, chart, method, etc., intended to assist in the navigation of a craft.

air. , *n.* 1. The mixture of gases comprising the earth's atmosphere. It is composed of about 78% nitrogen, 21% oxygen, 1% other gases, and a variable amount of impurities such as water vapor, suspended dust particles, smoke, etc. See also ATMOSPHERE. 2. Wind of force 1 (1-3 knots or 1-3 miles per hour) on the Beaufort wind scale, called LIGHT AIR.

air almanac. . 1. A periodical publication of astronomical data designed primarily for air navigation, but often used in marine navigation. See also ALMANAC FOR COMPUTERS. 2. ***Air Almanac***, a joint publication of the U.S. Naval Observatory and H. M. Nautical Almanac Office, Royal Greenwich Observatory, designed primarily for air navigation. In general the information is similar to that of the *Nautical Almanac*, but is given to a precision of 1' of arc and 1s of time, at intervals of 10m (values for the sun and Aries are given to a precision of 0.1').

air defense identification zone (ADIZ). . Airspace of defined dimensions within which the ready identification location, and control of aircraft are required.

air mass. . An extensive body of air with fairly uniform (horizontal) physical properties, especially temperature and humidity. In its incipient stage the properties of the air mass are determined by the characteristics of the region in which it forms. It is a cold or warm air mass if it is colder or warmer than the surrounding air.

air-mass classification. . Air masses are classified according to their source regions. Four such regions are generally recognized- (1) equatorial (E), the doldrum area between the north and south trades; (2) tropical (T), the trade wind and lower temperate regions, (3) polar (P), the higher temperate latitudes; and (4) Arctic or Antarctic (A), the north or south polar regions of ice and snow. This classification is a general indication of relative temperature, as well as latitude of origin. Air masses are further classified as maritime (m) or continental (c), depending upon whether they form over water or land. This classification is an indication of the relative moisture content of the air mass. A third classification sometimes applied to tropical and polar air masses indicates whether the air mass is warm (w) or cold (k) relative to the underlying surface. The w and k classifications are primarily indications of stability, cold air being more stable.

air temperature correction. . A correction due to nonstandard air temperature, particularly the sextant altitude correction due to changes in refraction caused by difference between the actual temperature and the standard temperature used in the computation of the refraction table. The *Nautical Almanac* refraction table is based upon an air temperature of 50°F (10°C) at the surface of the earth. Refraction is greater at lower temperatures, and less at higher temperatures. The correction for air temperature varies with the temperature of the air and the altitude of the celestial body, and applies to all celestial bodies, regardless of the method of observation. It is not applied in normal navigation.

Alaska Current. . A North Pacific Ocean current flowing counterclockwise in the Gulf of Alaska. It is the northward flowing division of the Aleutian Current.

Alaska-Hawaii standard time. . See STANDARD TIME.

albedo. , *n.* The ratio of radiant energy reflected to that received by a surface, usually expressed as a percentage; reflectivity. The term generally refers to energy within a specific frequency range, as the visible spectrum. Its most frequent application in navigation is to the light reflected by a celestial body.

alert. , *n.* See ALERT TIME CALCULATIONS.

alert time calculations. . Computations of times and-altitudes of available satellite passes in a given period of time at a given location, based on orbital data transmitted from satellite memory. Sometimes called ALERT.

Aleutian Current. . An eastward flowing North Pacific Ocean current which lies north of the North Pacific Current. As it approaches the coast of North America it divides to form the northward-flowing ALASKA CURRENT, and the southward-flowing CALIFORNIA CURRENT. Also called SUBARCTIC CURRENT.

alga. *(pl. algae)*, *n.* A plant of simple structure which grows chiefly in water, such as the various forms of seaweed. It ranges in size from a microscopic plant, large numbers of which sometimes cause discoloration of water, to the giant kelp which may extend for more than 600 feet in length. The Red Sea owes its name to red algae, as does the "red tide."

algorithm. . A defined procedure or routine used for solving a specific mathematical problem.

alidade. , *n.* The part of an optical measuring instrument comprising the optical system, indicator, vernier, etc. In modern practice the term is used principally in connection with a bearing circle fitted with a telescope to facilitate observation of bearings. Also called TELESCOPIC ALIDADE.

align. , *v., t.* To place objects in line.

alignment. , *n.* 1. The placing of objects in a line. 2. The process of orienting the measuring axes of the inertial components of inertial navigation equipment with respect to the coordinate system in which the equipment is to be used.

Allard's law. . A formula relating the illuminance produced on a normal surface at a given distance from a point source of light, the intensity of the light, and the degree of transparency of the atmosphere, assumed to be uniform. See OMNIDIRECTIONAL LIGHT.

all-weather. , *adj.* Designed or equipped to perform by day or night under any weather conditions.

almanac. , *n.* A periodical publication of ephemeral astronomical data. If information is given in a form and to a precision suitable for marine navigation, it is called a nautical almanac. See also nautical almanac; if designed primarily for air navigation, it is called an air almanac. See also EPHEMERIS, ASTRONOMICAL ALMANAC.

almucantar. , *n.* A small circle on the celestial sphere paralleled to the horizon. Also called CIRCLE OF EQUAL ALTITUDE, PARALLEL OF ALTITUDE.

almucantar staff. . An ancient instrument formerly used for amplitude observations.

alnico. , *n.* An alloy composed principally of aluminum, nickel, cobalt, and iron; used for permanent magnets.

aloft. . Up in the rigging of a ship.

alongshore current. . See LONGSHORE CURRENT.

alphanumeric. . Referring to a set of computer characters consisting of alphabetic and numeric symbols.

alphanumeric grid. . See ATLAS GRID.

alternate blanking. . See under DUAL-RATE BLANKING.

alternating current. . An electric current that continually changes in magnitude and periodically reverses polarity.

alternating. . Referring to periodic changes in color of a lighted aid to navigation.

alternating fixed and flashing light. . A fixed light varied at regular intervals by a single flash of greater luminous intensity, with color variations in either the fixed light or flash, or both. See ALTERNATING LIGHT.

alternating fixed and group flashing light. . A fixed light varied at regular intervals by a group of two or more flashes of greater luminous intensity, with color variations in either the fixed light or flashes or both.

alternating flashing light. . A light showing a single flash with color variations at regular intervals, the duration of light being shorter than that of darkness. See also FLASHING LIGHT.

alternating group flashing light. . A group flashing light which shows periodic color change.

alternating group occulting light. . A group occulting light which shows periodic color change.

alternating occulting light. . A light totally eclipsed at regular intervals, the duration of light always being longer than the duration of darkness, which shows periodic color change. See also ALTERNATING LIGHT.

alternating light. . A light showing different colors alternately.

altitude. , *n.* Angular distance above the horizon; the arc of a vertical circle between the horizon and a point on the celestial sphere, measured upward from the horizon. Angular distance below the horizon is called negative altitude or depression. Altitude indicated by a sextant is called sextant altitude. Sextant altitude corrected only for inaccuracies in the reading (instrument, index, and personal errors, as applicable) and inaccuracies in the reference level (principally dip) is called apparent or rectified altitude. After all corrections are applied, it is called corrected sextant altitude or observed altitude. An altitude taken directly from a table, before interpolation, is called tabulated altitude. After interpolation, or if determined by calculation, mechanical device, or graphics, it is called computed altitude. If the altitude of a celestial body is computed before observation, and sextant altitude corrections are applied with reversed sign, the result is called precomputed altitude. The difference between computed and observed altitudes (corrected sextant altitudes), or between precomputed and sextant altitudes, is called altitude intercept or altitude difference. An altitude determined by inexact means, as by estimation or star finder, is called an approximate altitude. The altitude of a celestial body on the celestial meridian is called meridian altitude. The expression ex-meridian altitude is applied to the altitude of a celestial body near the celestial meridian, to which a correction is to be applied to determine the meridian altitude. A parallel of altitude is a circle of the celestial sphere parallel to the horizon, connecting all points of equal altitude. See also EQUAL ALTITUDES.

altitude azimuth.. An azimuth determined by solution of the navigational triangle with altitude, declination, and latitude given. A time azimuth is computed with meridian angle, declination, and latitude given. A time and altitude azimuth is computed with meridian angle, declination, and altitude given.

altitude circle.. See PARALLEL OF ALTITUDE.

altitude difference.. 1. See ALTITUDE INTERCEPT. 2. The change in the altitude of a celestial body occurring with change in declination, latitude, or hour angle, for example the *first difference* between successive tabulations of altitude in a latitude column of *Pub. No. 229*, Sight Reduction Tables for Marine Navigation.

altitude intercept.. The difference in minutes of arc between the computed and the observed altitude (corrected sextant altitude), or between precomputed and sextant altitudes. It is labeled T (toward) or A (away) as the observed (or sextant) altitude is greater or smaller than the computed (or precomputed) altitude. Also called ALTITUDE DIFFERENCE, INTERCEPT.

altitude intercept method.. See ST. HILAIRE METHOD.

altitude of the apogee.. As defined by the International Telecommunication Union (ITU), the altitude of the apogee above a specified reference surface serving to represent the surface of the earth.

altitude of the perigee.. As defined by the International Telecommunication Union (ITU), the altitude of the perigee above a specified reference surface serving to represent the surface of the earth.

altitude tints.. See HYPSOMETRIC TINTING.

alto-.. A prefix used in cloud classification to indicate the middle level. See also CIRRO-.

altocumulus., *n.* Clouds within the middle level (mean height 6,500-20,000 ft.) composed of flattened globular masses, the smallest elements of the regularly arranged layers being fairly thin, with or without shading. These elements are arranged in groups, in lines, or waves, following one or two directions, and are sometimes so close together that their edges join. See also CLOUD CLASSIFICATION.

altostratus., *n.* A sheet of gray or bluish cloud within the middle level (mean height 6,500-20,000 ft.). Sometimes the sheet is composed of a compact mass of dark, thick, gray clouds of fibrous structure; at other times the sheet is thin and through it the sun or moon can be seen dimly. See also CLOUD CLASSIFICATION.

A.M.. Abbreviation for Ante Meridian; before noon in zone time.

ambient temperature.. The temperature of the air or other medium surrounding an object. See also FREE-AIR TEMPERATURE.

ambiguity., *n.* In navigation, the condition obtained when a given set of observations defines more than one point, direction, line of position, or surface of position.

ambiguous., *adj.* Having two or more possible meanings or values.

American Ephemeris and Nautical Almanac.. See ASTRONOMICAL ALMANAC.

American Practical Navigator, The.. A navigational text and reference book published by the National Imagery and Mapping Agency (NIMA); originally by Nathaniel Bowditch (1773-1838). Popularly called BOWDITCH.

amidships., *adv.* At, near, or toward the middle of a ship.

ampere., *n.* The base unit of electric current in the International System of Units; it is that constant current which, if maintained in two straight parallel conductors of infinite length, of negligible circular cross section, and placed 1 meter apart in vacuum, would produce between these conductors a force equal to 2×10^{-7} newton per meter of length.

ampere per meter.. The derived unit of magnetic field strength in the International System of Units.

amphidromic point.. Point on a tidal chart where the cotidal lines meet.

amphidromic region.. An area surrounding a no-tide point from which the radiating cotidal lines progress through all hours of the tidal cycle.

amplification., *n.* 1. An increase in signal magnitude from one point to another, or the process causing this increase. 2. Of a transducer, the scalar ratio of the signal output to the signal input.

amplifier., *n.* A device which enables an input signal to control power from a source independent of the signal and thus be capable of delivering an output which is greater than the input signal.

amplitude., *n.* 1. Angular distance of a celestial body north or south of the prime vertical circle; the arc of the horizon or the angle at the zenith between the prime vertical circle and a vertical circle through the celestial body measured north or south from the prime vertical to the vertical circle. The term is customarily used only with reference to bodies whose centers are on the celestial horizon, and is prefixed E or W, as the body is rising or setting, respectively; and suffixed N or S to agree with the declination. The prefix indicates the origin and the suffix the direction of measurement. Amplitude is designated as true, magnetic, compass, or grid as the reference direction is true, magnetic, compass, or grid east or west, respectively. 2. The maximum value of the displacement of a wave, or other periodic phenomenon, from the zero position. 3. One-half the range of a constituent tide. By analogy, it may be applied also to the maximum speed of a constituent current.

amplitude compass.. A compass intended primarily for measuring amplitude. It is graduated from 0° at east and west to 90° at north and south. Seldom used on modern vessels.

amplitude distortion.. Distortion occurring in an amplifier or other device when the output amplitude is not a linear function of the input amplitude.

amplitude modulation.. The process of changing the amplitude of a carrier wave in accordance with the variations of a modulating wave. See also MODULATION.

Amver System.. Operated by the U.S. Coast Guard, the Amver System is a maritime mutual-assistance program that aids coordination of search and rescue efforts by maintaining a worldwide computerized DR plot of participating vessels.

anabatic wind.. Any wind blowing up an incline. A KATABATIC WIND blows down an incline.

analemma., *n.* A graduated scale of the declination of the sun and the equation of time for each day of the year located in the Torrid Zone on the terrestrial globe.

analog., *adj.* Referring to the processing and/or transfer of information via physical means such as waves, fluids, or mechanical devices.

analog computer.. A computer in which quantities are represented by physical variables. Problem parameters are translated into equivalent mechanical or electrical circuits as an analog for the physical phenomenon being investigated without the use of a machine language. An analog computer measures continuously; a digital computer counts discretely. See DIGITAL.

anchorage., *n.* An area where vessels may anchor, either because of suitability or designation.

anchorage buoy.. A buoy which marks the limits of an anchorage, not to be confused with a MOORING BUOY.

anchorage chart.. A nautical chart showing prescribed or recommended anchorages.

anchorage mark.. A navigation mark which indicates an anchorage area or defines its limits.

anchor., *n.* A device used to secure a ship to the sea floor.

anchor., *v,t.* To use the anchor to secure a ship to the sea floor. If more than one anchor is used the ship is moored.

anchor buoy.. A buoy marking the position of an anchor on the bottom, usually painted green for the starboard anchor and red for the port anchor, and secured to the crown of the anchor by a buoy rope.

anchor ice.. Submerged ice attached or anchored to the bottom, irrespective of the nature of its formation.

anchor light.. A light shown from a vessel or aircraft to indicate its position when riding at anchor. Also called RIDING LIGHT.

anemometer., *n.* An instrument for measuring the speed of the wind. Some instruments also indicate the direction from which it is blowing. See also VANE, definition l; WIND INDICATOR.

aneroid barometer.. An instrument which determines atmospheric pressure by the effect of such pressure on a thin-metal cylinder from which the air has been partly exhausted. See also MERCURIAL BAROMETER.

angel.. A radar echo caused by a physical phenomenon which cannot be seen.

angle., *n.* The inclination to each other of two intersecting lines, measured by the arc of a circle intercepted between the two lines forming the angle, the center of the circle being the point of intersection. An acute angle is less than 90°; a right angle, 90° an obtuse angle, more than 90° but less than 180°- a straight angle 180°; a reflex angle, more than 180° but less than 360°; a perigon, 360°. Any angle not a multiple of 90 is an oblique angle. If the sum of two angles is 90°, they are complementary angles; if 180°, supplementary angles; if 360°, explementary angles. Two adjacent angles have a common vertex and lie on opposite sides of a common side. A dihedral angle is the angle between two intersecting planes. A spherical angle is the angle between two intersecting great circles.

angle of cut. . The smaller angular difference of two bearings or lines of position.

angle of depression. . The angle in a vertical plane between the horizontal and a descending line. Also called DEPRESSION ANGLE. See ANGLE OF ELEVATION.

angle of deviation. . The angle through which a ray is bent by refraction.

angle of elevation. . The angle in a vertical plane between the horizontal and an ascending line, as from an observer to an object. A negative angle of elevation is usually called an ANGLE OF DEPRESSION. Also called ELEVATION ANGLE.

angle of incidence. . The angle between the line of motion of a ray of radiant energy and the perpendicular to a surface, at the point of impingement. This angle is numerically equal to the ANGLE OF REFLECTION.

angle of reflection. . The angle between the line of motion of a ray of reflected radiant energy and the perpendicular to a surface, at the point of reflection. This angle is numerically equal to the ANGLE OF INCIDENCE.

angle of refraction. . The angle between a refracted ray and the perpendicular to the refracting surface.

angle of roll. . The angle between the transverse axis of a craft and the horizontal. Also called ROLL ANGLE.

angle of uncertainty. . The horizontal angle of the region of indefinite characteristic near the boundaries of a sector of a sector light. Also called ARC OF UNCERTAINTY.

angstrom. , *n.* A unit of length, used especially in expressing the length of light waves, equal to one ten-thousandth of a micron or one hundred millionth of a centimeter.

angular. , *adj.* Of or pertaining to an angle or angles.

angular distance. . 1. The angular difference between two directions, numerically equal to the angle between two lines extending in the given directions. 2. The arc of the great circle joining two points, expressed in angular units. 3. Distance between two points, expressed in angular units of a specified frequency. It is equal to the number of waves between the points multiplied by 2π if expressed in radians, or multiplied by 360° if measured in degrees.

angular distortion. . Distortion in a map projection because of non-conformity.

angular momentum. . The quantity obtained by multiplying the moment of inertia of a body by its angular speed.

angular rate. . See ANGULAR SPEED.

angular rate of the earth's rotation. . Time rate of change of angular displacement of the earth relative to the fixed stars equal to 0.729211×10^{-4} radian per second.

angular resolution. . See BEARING RESOLUTION.

angular speed. . Change of direction per unit time. Also called ANGULAR RATE. See also LINEAR SPEED.

anneal. , *v., t.* To heat to a high temperature and then allow to cool slowly, for the purpose of softening, making less brittle, or removing permanent magnetism. When Flinders bars or quadrantal correctors acquire permanent magnetism which decreases their effectiveness as compass correctors, they are annealed.

annotation. , *n.* Any marking on illustrative material for the purpose of clarification such as numbers, letters, symbols, and signs.

annual. , *adj.* Of or pertaining to a year; yearly.

annual aberration. . See under ABERRATION, definition 1.

annual inequality. . Seasonal variation in water level or tidal current speed, more or less periodic due chiefly to meteorological causes.

annual parallax. . See HELIOCENTRIC PARALLAX.

annular. , *adj.* Ring-shaped.

annular eclipse. . An eclipse in which a thin ring of the source of light appears around the obscuring body. Annular solar eclipses occur, but never annular lunar eclipses.

annulus. , *n.* A ring-shaped band.

anode. , *n.* 1. A positive electrode; the plate of a vacuum tube; the electrode of an electron tube through which a principal stream of electrons leaves the inter-electrode space. 2. The positive electrode of an electrochemical device, such as a primary or secondary cell, toward which the negative ions are drawn. See also CATHODE.

anomalistic. , *adj.* Pertaining to the periodic return of the moon to its perigee, or of the earth to its perihelion.

anomalistic month. . The average period of revolution of the moon from perigee to perigee, a period of 27 days, 13 hours, 18 minutes, and 33.2 seconds in 1900. The secular variation does not exceed a few hundredths of a second per century. anomalistic period. The interval between two successive passes of a satellite through perigee. Also called PERIGEE-TO-PERIGEE PERIOD RADIAL PERIOD. See also ORBITAL PERIOD.

anomalistic year. . The period of one revolution of the earth around the sun, from perihelion to perihelion, averaging 365 days, 6 hours, 13 minutes, 53.0 seconds in 1900, and increasing at the rate of 0.26 second per century.

anomaly. , *n.* 1. Departure from the strict characteristics of the type, pattern, scheme, etc. 2. An angle used in the mathematical description of the orbit of one body about another. It is the angle between the radius vector of the body and the line of apsides and is measured from pericenter in the direction of motion. When the radius vector is from the center of the primary to the orbiting body, the angle is called true anomaly. When the radius vector is from the center of the primary to a fictitious body moving with a uniform angular velocity in such a way that its period is equal to that of the actual body, the angle is called mean anomaly. When the radius vector is from the center of the elliptical orbit to the point of intersection of the circle defined by the semimajor axis with the line perpendicular to the semimajor axis and passing through the orbiting body, the angle is called eccentric anomaly or eccentric angle. 3. Departure of the local mean value of a meteorological element from the mean value for the latitude. See also MAGNETIC ANOMALY.

antarctic. , *adj.* referring to the Antarctic region.

Antarctic. , *n.* The region within the Antarctic Circle, or, loosely, the extreme southern regions of the earth.

antarctic air. . A type of air whose characteristics are developed in an Antarctic region. Antarctic air appears to be colder at the surface in all seasons, and at all levels in fall and winter, than ARCTIC AIR.

Antarctic Circle. . The parallel of latitude at about 66° 33'S, marking the northern limit of the south Frigid Zone. This latitude is the complement of the sun's greatest southerly declination, and marks the approximate northern limit at which the sun becomes circumpolar. The actual limit is extended somewhat by the combined effect of refraction, semidiameter of the sun, parallax, and the height of the observer's eye above the surface of the earth. A similar circle marking the southern limit of the north Frigid Zone is called ARCTIC or NORTH POLAR CIRCLE. Also called SOUTH POLAR CIRCLE.

Antarctic Circumpolar Current. . See WEST WIND DRIFT.

antarctic front. . The semi-permanent, semi-continuous front between the Antarctic air of the Antarctic Continent and the polar air of the southern oceans; generally comparable to the **arctic front** of the Northern Hemisphere.

antarctic whiteout. . The obliteration of contrast between surface features in the Antarctic when a covering of snow obscuring all landmarks is accompanied by an overcast sky, resulting in an absence of shadows and an unrelieved expanse of white, the earth and sky blending so that the horizon is not distinguishable. A similar occurrence in the Arctic is called ARCTIC WHITEOUT.

ante meridian (AM). . Before noon, or the period of time between midnight (0000) and noon (1200). The period between noon and midnight is called POST MERIDIAN.

antenna. , *n.* A structure or device used to collect or radiate electromagnetic waves.

antenna array. . A combination of antennas with suitable spacing and with all elements excited to make the radiated fields from the individual elements add in the desired direction, i.e., to obtain directional characteristics.

antenna assembly. . The complete equipment associated with an antenna, including, in addition to the antenna, the base, switches, lead-in wires, revolving mechanism, etc.

antenna bearing. . The generated bearing of the antenna of a radar set, as delivered to the indicator.

antenna coupler. . 1. A radio-frequency transformer used to connect an antenna to a transmission line or to connect a transmission line to a radio receiver. 2. A radio-frequency transformer, link circuit, or tuned line used to transfer radio-frequency energy from the final plate-tank circuit of a transmitter to the transmitter to the transmission line feeding the antenna.

antenna directivity diagram. . See DIRECTIVITY DIAGRAM.

antenna effect. . A spurious effect, in a loop antenna, resulting from the capacitance of the loop to ground.

antenna feed. . The component of an antenna of mirror or lens type that irradiates, or receives energy from, the mirror or lens. See also HORN ANTENNA.

antenna radiation pattern. . See RADIATION PATTERN.

anthelion. , *n.* A rare kind of halo, which appears as a bright spot at the same altitude as the sun and 180° from it in azimuth. See also PARHELION.

anti-clutter gain control. . See SENSITIVITY TIME CONTROL.

anti-clutter rain. . See FAST TIME CONSTANT CIRCUIT.

anti-clutter sea. . See SENSITIVITY TIME CONTROL.

anticorona. , *n.* A diffraction phenomenon very similar to but complementary to the corona, appearing at a point directly opposite to the sun or moon from the observer. Also called BROKEN BOW, GLORY.

anticrepuscular arch. . See ANTITWILIGHT.

anti-crepuscular rays. . Extensions of crepuscular rays, converging toward a point 180° from the sun.

anticyclone. , *n.* An approximately circular portion of the atmosphere, having relatively high atmospheric pressure and winds which blow clockwise around the center in the Northern Hemisphere and counterclockwise in the Southern Hemisphere. An anticyclone is characterized by good weather. Also called HIGH. See also CYCLONE.

anticyclonic winds. . The winds associated with a high pressure area and constituting part of an anticyclone.

Antilles Current. . This current originates in the vicinity of the Leeward Islands as part of the Atlantic North Equatorial Current. It flows along the northern side of the Greater Antilles. The Antilles Current eventually joins the Florida Current (north of Grand Bahama Island) to form the Gulf Stream.

antilogarithm. , *n.* The number corresponding to a given logarithm. Also called INVERSE LOGARITHM.

antinode. , *n.* Either of the two points on an orbit where a line in the orbit plane, perpendicular to the line of nodes, and passing through the focus, intersects the orbit.

antipodal effects. . See as LONG PATH INTERFERENCE under MULTIPATH ERROR.

antipode. , *n.* Anything exactly opposite to something else. Particularly, that point on the earth 180° from a given place.

antisolar point. . The point on the celestial sphere 180° from the sun.

antitrades. , *n., pl.* The prevailing western winds which blow over and in the opposite direction to the trade winds. Also called COUNTERTRADES.

anti-TR tube. . See TR TUBE.

antitwilight. , *n.* The pink or purplish zone of illumination bordering the shadow of the earth in the dark part of the sky opposite the sun after sunset or before sunrise. Also called ANTI CREPUSCULAR ARCH.

anvil cloud. . Heavy cumulus or cumulonimbus having an anvil-like upper part.

apastron. , *n.* The point of the orbit of one member of a double star system at which the stars are farthest apart. That point at which they are nearest together is called PERIASTRON.

aperiodic. , *adj.* Without a period; of irregular occurrence.

aperiodic compass. . Literally "a compass without a period," or a compass that, after being deflected, returns by one direct movement to its proper reading without oscillation. Also called DEADBEAT COMPASS.

aperture. , *n.* 1. An opening; particularly, the opening in the front of a camera through which light rays pass when a picture is taken. 2. The diameter of the objective of a telescope or other optical instrument, usually expressed in inches, but sometimes as the angle between lines from the principal focus to opposite ends of a diameter of the objective. 3. Of a directional antenna, that portion of nearby plane surface that is perpendicular to the direction of maximum radiation and through which the major part of the radiation passes.

aperture antenna. . An antenna in which the beam width is determined by the dimensions of a horn, lens, or reflector.

aperture ratio. . The ratio of the diameter of the objective to the focal length of an optical instrument.

apex. , *n.* The highest point of something, as of a cone or triangle, or the maximum latitude (vertex) of a great circle.

aphelion. , *n.* That point in the elliptical orbit of a body about the sun farthest from the sun. That point nearest the sun is called PERIHELION.

aphylactic map projection. . A map projection which is neither conformal nor equal area. Also called ARBITRARY MAP PROJECTION.

aplanatic lens. . See under ABERRATION, definition 2.

apoapsis. , *n.* See APOCENTER.

apocenter. , *n.* In an elliptical orbit, the point in the orbit which is the farthest distance from the focus, where the attracting mass is located. The apocenter is at one end of the major axis of the orbital ellipse. The opposite is PERICENTER, PERIFOCUS, PERIAPSIS. Also called APOAPSIS, APOFOCUS.

apofocus. , *n.* See APOCENTER.

apogean range. . The average semidiurnal range of the tide occurring at the time of apogean tides. It is smaller than the mean range, where the type of tide is either semidiurnal or mixed, and is of no practical significance where the type of tide is diurnal.

apogean tidal currents. . Tidal currents of decreased speed occurring monthly as the result of the moon being at apogee (farthest from the earth).

apogean tides. . Tides of decreased range occurring monthly as the result of the moon being at apogee (farthest from the earth).

apogee. , *n.* That orbital point of a non-circular orbit farthest from the center of attraction. Opposite is PERIGEE. See APOCENTER, PERICENTER.

apparent altitude. . Sextant altitude corrected for inaccuracies in the reading (instrument, index, and personal errors) and inaccuracies in the reference level (principally dip or Coriolis/acceleration), but not for other errors. Apparent altitude is used in obtaining a more accurate refraction correction than would be obtained with an uncorrected sextant altitude. Also called RECTIFIED ALTITUDE. See also OBSERVED ALTITUDE, SEXTANT ALTITUDE.

apparent horizon. . See VISIBLE HORIZON.

apparent motion. . Motion relative to a specified or implied reference point which may itself be in motion. The expression usually refers to movement of celestial bodies as observed from the earth. Usually called RELATIVE MOVEMENT when applied to the motion of one vessel relative to that of another. Also called RELATIVE MOTION.

apparent noon. . Twelve o'clock apparent time, or the instant the apparent sun is over the upper branch of the meridian. Apparent noon may be either local or Greenwich depending upon the reference meridian. High noon is local apparent noon.

apparent place. . The position on the celestial sphere at which a celestial body would be seen if the effects of refraction, diurnal aberration, and geocentric parallax were removed; the position at which the object would actually be seen from the center of the earth. Also called APPARENT POSITION.

apparent position. . See APPARENT PLACE.

apparent precession. . Apparent change in the direction of the axis of rotation of a spinning body, such as a gyroscope, due to rotation of the earth. As a result of gyroscopic inertia or rigidity in space, to an observer on the rotating earth a gyroscope appears to turn or precess.

apparent secular trend. . The non-periodic tendency of sea level to rise, fall and/or remain stationary with time. Technically, it is frequently defined as the slope of a least-squares line of regression through a relatively long series of yearly mean sea level values. The word apparent is used since it is often not possible to know whether a trend is truly non periodic or merely a segment of a very long oscillation.

apparent shoreline. . A line drawn on the chart in lieu of the mean high water line or the mean water level line in areas where either may be obscured by marsh, mangrove, cypress, or other marine vegetation.

This line represents the intersection of the appropriate datum with the outer limits of vegetation and appears to the navigator as the shoreline.

apparent sidereal time.. See under SIDEREAL TIME.

apparent solar day.. The duration of one rotation of the earth on its axis, with respect to the apparent sun. It is measured by successive transits of the apparent sun over the lower branch of a meridian. The length of the apparent solar day is 24 hours of apparent time and averages the length of the mean solar day, but varies somewhat from day to day.

apparent sun.. The actual sun as it appears in the sky. Also called TRUE SUN. See also MEAN SUN, DYNAMICAL MEAN SUN.

apparent time.. Time based upon the rotation of the earth relative to the apparent or true sun. This is the time shown by a sun dial. Apparent time may be designated as either **local** or **Greenwich**, as the local or Greenwich meridian is used as the reference. Also called TRUE SOLAR TIME. See also EQUATION OF TIME.

apparent wind.. The speed and true direction from which the wind appears to blow with reference to a moving point. Sometimes called RELATIVE WIND. See also TRUE WIND.

application program.. A computer program designed to do a specific task or group of tasks.

approach chart.. A chart used to approach a harbor. See CHART CLASSIFICATION BY SCALE.

approximate altitude.. An altitude determined by inexact means, as by estimation or by a star finder or star chart.

approximate coefficients.. The six coefficients used in the analysis of the magnetic properties of a vessel in the course of magnetic compass adjustment. The values of these coefficients are determined from deviations of an unadjusted compass. See also COEFFICIENT A, COEFFICIENT B, COEFFICIENT C, COEFFICIENT D, COEFFICIENT E, COEFFICIENT J.

appulse., *n.* 1. The near approach of one celestial body to another on the celestial sphere, as in occultation, conjunction, etc. 2. The penumbral eclipse of the moon.

apron., *n.* 1. On the sea floor a gentle slope, with a generally smooth surface, particularly as found around groups of islands or sea mounts. Sometimes called ARCHIPELAGIC APRON. 2. The area of wharf or quay for handling cargo. 3. A sloping underwater extension of an iceberg. 4. An outwash plain along the front of a glacier.

apse line.. See LINE OF APSIDES.

apsis. *(pl. apsides), n.* Either of the two orbital points nearest or farthest from the center of attraction, the perihelion and aphelion in the case of an orbit about the sun, and the perigee and apogee in the case of an orbit about the earth. The line connecting these two points is called LINE OF APSIDES.

aqueduct., *n.* A conduit or artificial channel for the conveyance of water, often elevated, especially one for the conveyance of a large quantity of water that flows by gravitation.

arbitrary map projection.. See APHYLACTIC MAP PROJECTION.

arc., *n.* 1. A part of a curved line, as of a circle. See also ANGULAR DISTANCE. 2. The semi-circular graduated scale of an instrument for measuring angles. See also EXCESS OF ARC.

arched squall.. A squall which is relatively high in the center, tapering off on both sides.

archipelagic apron.. See APRON, definition 1.

archipelago., *n.* 1. A sea or broad expanse of water containing many islands or groups of islands. 2. A group of such islands.

arc of uncertainty.. See ANGLE OF UNCERTAINTY.

arc of visibility.. The arc of a light sector, designated by its limiting bearings as observed from seaward.

Arcs of Lowitz.. Oblique, rare, downward extensions of the parhelia of 22°, concave toward the sun, and with red inner borders. They are formed by refraction by ice crystals oscillating about the vertical, such as with snowflakes.

arctic., *adj.* Of or pertaining to the arctic, or intense cold.

Arctic., *n.* The region within the Arctic Circle, or, loosely, northern regions in general, characterized by very low temperatures.

arctic air.. A type of air which develops mostly in winter over the arctic. Arctic air is cold aloft and extends to great heights, but the surface temperatures are often higher than those of POLAR AIR. For 2 or 3 months in summer arctic air masses are shallow and rapidly lose the characteristics as they move southward. See also ANTARCTIC AIR.

Arctic Circle.. The parallel of latitude at about 66° 33'N, marking the southern limit of the north Frigid Zone. This latitude is the complement of the sun's greatest northerly declination and marks the approximate southern limit at which the sun becomes circumpolar. The actual limit is extended somewhat by the combined effect of refraction, semidiameter of the sun, parallax, and the height of the observer's eye above the surface of the earth. A similar circle marking the northern limit of the south Frigid Zone is called ANTARCTIC or SOUTH POLAR CIRCLE. Also called NORTH POLAR CIRCLE.

arctic front.. The semi-permanent, semi-continuous front between the deep, cold arctic air and the shallower, generally less cold polar air of northern latitudes; generally comparable to the ANTARCTIC FRONT of the Southern Hemisphere.

arctic sea smoke.. Steam fog, but often specifically applied to steam fog rising from small areas of open water within sea ice. See also FROST SMOKE.

arctic smoke.. See STEAM FOG.

arctic whiteout.. The obliteration of contrast between surface features in the Arctic when a covering of snow obscuring all landmarks is accompanied by an overcast sky, resulting in an absence of shadows and an unrelieved expanse of white, the earth and sky blending so that the horizon is not distinguishable. A similar occurrence in the Antarctic is called ANTARCTIC WHITEOUT.

arc to chord correction.. See CONVERSION ANGLE.

areal feature.. A topographic feature, such as sand, swamp, vegetation, etc., which extends over an area. It is represented on the published map or chart by a solid or screened color, by a prepared pattern of symbols, or by a delimiting line.

area to be avoided.. A ship routing measure comprising an area with defined limits which should be avoided by all ships, or certain classes of ships; instituted to protect natural features or to define a particularly hazardous area for navigation. See also PRECAUTIONARY AREA, ROUTING SYSTEM.

argument., *n.* One of the values used for entering a table or diagram.

argument of latitude.. The angular distance measured in the orbital plane from the ascending node to the orbiting body; the sum of the argument of pericenter and the true anomaly.

argument of pericenter.. The angle at the center of attraction from the ascending node to the pericenter point, measured in the direction of motion of the orbiting body. Also called ARGUMENT OF PERIFOCUS.

argument of perifocus.. See ARGUMENT OF PERICENTER.

argument of perigee.. The angle at the center of attraction from the ascending node to the perigee point, measured in the direction of motion of the orbiting body.

Aries., *n.* 1. Vernal equinox. Also called FIRST POINT OF ARIES. 2. The first sign of the zodiac.

arithmetic mean.. See MEAN.

arm., *v., t.* To place tallow or other substance in the recess at the lower end of a sounding lead for obtaining a sample of the bottom.

Armco., *n.* The registered trade name for a high purity, low carbon iron, used for Flinders bars, quadrantal correctors, etc., to correct magnetic compass errors resulting from induced magnetism.

arming., *n.* Tallow or other substance placed in the recess at the lower end of a sounding lead, for obtaining a sample of the bottom.

array., *n.* See as ANTENNA ARRAY.

articulated light.. An offshore aid to navigation consisting of a pipe attached to a mooring by a pivoting or universal joint; more accurate in position than a buoy but less than a fixed light.

artificial antenna.. See DUMMY ANTENNA.

artificial asteroid.. A man-made object placed in orbit about the sun.

artificial earth satellite.. A man-made earth satellite, as distinguished from the moon. Often shortened to ARTIFICIAL SATELLITE.

artificial harbor.. A harbor where the desired protection from wind and sea is obtained from breakwaters, moles, jetties, or other man-made works. See also NATURAL HARBOR.

artificial horizon.. A device for indicating the horizontal, such as a bubble, gyroscope, pendulum, or the surface of a liquid.

artificial magnet.. A magnet produced by artificial means, either by placing magnetic material in the field of another magnet or by means of an electric current, as contrasted with a NATURAL MAGNET occurring in nature.

artificial range.. A range formed by two objects such as buildings, towers, etc., not designed as aids to navigation. See also NATURAL RANGE.

artificial satellite.. See ARTIFICIAL EARTH SATELLITE.

ascending node.. That point at which a planet, planetoid, or comet crosses the ecliptic from south to north, or a satellite crosses the plane of the equator of its primary from south to north. Also called NORTHBOUND NODE. The opposite is called DESCENDING NODE.

ASCII.. Acronym for American Standard Code for Information Interchange, a standard method of representing alphanumeric characters with numbers in a computer.

ash breeze.. Expression referring to rowing a sailing vessel in a calm, usually from ship's boats which tow the ship. (Oars are commonly made of ash wood.)

ashore., *adj.* & *adv*. On the shore; on land; aground. See also AFLOAT.

aspect., *n*. The relative bearing of own ship from the target ship, measured 0° to 180° port (red) or starboard (green). See also TARGET ANGLE.

aspects., *n., pl.* The apparent positions of celestial bodies relative to one another; particularly the apparent positions of the moon or a planet relative to the sun.

assigned frequency.. The center of the frequency band assigned to a radio station. Sometimes called CHANNEL FREQUENCY, CENTER FREQUENCY.

assigned frequency band.. The frequency band whose center coincides with the frequency assigned to the station and whose width equals the necessary bandwidth plus twice the absolute value of the frequency tolerance.

assumed latitude.. The latitude at which an observer is assumed to be located for an observation or computation, as the latitude of an assumed position or the latitude used for determining the longitude of time sight. Also called CHOSEN LATITUDE.

assumed longitude.. The longitude at which an observer is assumed to be located for an observation or computation, as the longitude of an assumed position or the longitude used for determining the latitude by meridian altitude. Also called CHOSEN LONGITUDE.

assumed position.. A point at which a craft is assumed to be located, particularly one used as a preliminary to establishing certain navigational data, as that point on the surface of the earth for which the computed altitude is determined in the solution of a celestial observation, also called CHOSEN POSITION.

astern., *adv*. Bearing approximately 180° relative. The term is often used loosely for DEAD ASTERN, or bearing exactly 180° relative. The opposite is AHEAD.

asteroid., *n*. A minor planet, one of the many small celestial bodies revolving around the sun, most of the orbits being between those of Mars and Jupiter. Also called PLANETOID, MINOR PLANET. See under PLANET.

astigmatism., *n*. A defect of a lens which causes the image of a point to appear as a line, rather than a point.

astigmatizer., *n*. A lens which introduces astigmatism into an optical system. Such a lens is so arranged that it can be placed in or removed from the optical path at will. In a sextant, an astigmatizer may be used to elongate the image of a celestial body into a horizontal line.

astre fictif.. Any of several fictitious stars which are assumed to move along the celestial equator at uniform rates corresponding to the speeds of the several harmonic constituents of the tide producing force. Each astre fictif crosses the meridian at a time corresponding to the maximum of the constituent that it represents.

astro.. A prefix meaning *star* or *stars* and, by extension, sometimes used as the equivalent of *celestial*.

astrodynamics., *n*. The practical application of celestial mechanics, astroballistics, propulsion theory, and allied fields to the problem of planning and directing the trajectories of space vehicles.

astrograph., *n*. A device for projecting a set of precomputed altitude curves onto a chart or plotting sheet, the curves moving with time such that if they are properly adjusted, they will remain in the correct position on the chart or plotting sheet.

astrolabe., *n*. An instrument which measures altitudes of celestial bodies, used for determining an accurate astronomical position, usually while ashore in survey work. Originally, the astrolabe consisted of a disk with an arm pivoted at the center, the whole instrument being hung by a ring at the top to establish the vertical.

astrometry., *n*. The branch of astronomy dealing with the geometrical relations of the celestial bodies and their real and apparent motions.

astronomical., *adj*. Of or pertaining to astronomy.

Astronomical Almanac, The.. An annual publication prepared jointly by the Nautical Almanac Office, U.S. Naval Observatory, and H.M. Nautical Almanac Office, Royal Greenwich Observatory. With the exception of certain introductory pages, the publication as printed in the United Kingdom is identical to that printed in the United States. This ephemeris gives high precision, detailed information on a large number of celestial bodies. It is arranged to suit the convenience of the astronomer for whom it is primarily intended and is not intended for ordinary purposes of navigation. But it does contain some information of general interest to the navigator, such as various astronomical constants, details of eclipses, information on planetary configurations, and miscellaneous phenomena. Prior to 1981 this publication was entitled *American Ephemeris and Nautical Almanac*. See also NAUTICAL ALMANAC.

astronomical day.. Prior to January 1, 1925, a mean solar day which began at mean noon, 12 hours later than the beginning of the calendar day of the same date. Since 1925 the astronomical day agrees with the civil day.

astronomical equator.. A line connecting points having 0° astronomical latitude. Because the deflection of the vertical varies from point to point, the astronomical equator is not a plane curve. But since the verticals through all points on it are parallel, the zenith at any point on the astronomical equator lies in the plane of the celestial equator. When the astronomical equator is corrected for station error, it becomes the GEODETIC EQUATOR. Sometimes called TERRESTRIAL EQUATOR.

astronomical latitude.. Angular distance between the plumb line at a station and the plane of the celestial equator It is the latitude which results directly from observations of celestial bodies, uncorrected for deflection of the vertical which, in the United States, may amount to as much as 25". Astronomical latitude applies only to positions on the earth, and is reckoned from the astronomical equator (0°), north and south through 90°. Also called ASTRONOMIC LATITUDE and sometimes GEOGRAPHIC LATITUDE. See also GEODETIC LATITUDE.

astronomical longitude.. Angular distance between the plane of the celestial meridian at a station and the plane of the celestial meridian at Greenwich. It is the longitude which results directly from observations of celestial bodies, uncorrected for deflection of the vertical, the prime vertical component of which, in the United States, may amount to more than 18". Astronomical longitude applies only to positions on the earth, and is reckoned from the Greenwich meridian (0°) east and west through 180°. Also called ASTRONOMIC LONGITUDE and sometimes GEOGRAPHIC LONGITUDE. See also GEODETIC LONGITUDE.

astronomical mean sun.. See MEAN SUN.

astronomical meridian.. A line connecting points having the same astronomical longitude. Because the deflection of the vertical (station error) varies from point to point, the astronomical meridian is not a plane curve. When the astronomical meridian is corrected for station error, it becomes the GEODETIC MERIDIAN. Also called TERRESTRIAL MERIDIAN and sometimes called GEOGRAPHIC MERIDIAN.

astronomical parallel.. A line connecting points having the same astronomical latitude. Because the deflection of the vertical varies from point to point, the astronomical parallel is an irregular line not lying in a single plane. When the astronomical parallel is corrected for station error, it becomes the GEODETIC PARALLEL. Sometimes called GEOGRAPHIC PARALLEL.

astronomical position.. 1. A point on the earth whose coordinates have been determined as a result of observation of celestial bodies. The expression is usually used in connection with positions on land determined with great accuracy for survey purposes. 2. A point on

astronomical refraction. . Atmospheric refraction of a ray of radiant energy passing through the atmosphere from outer space, as contrasted with TERRESTRIAL REFRACTION of a ray emanating from a point on or near the surface of the earth. See also REFRACTION.

astronomical tide. . The tide without constituents having their origin in the daily or seasonal variations in weather conditions which may occur with some degree of periodicity. See also METEOROLOGICAL TIDES.

astronomical time. . Time used with the astronomical day which prior to 1926 began at noon of the civil day of same date. The hours of the day were numbered consecutively from 0 (noon) to 23 (11 AM of the following morning).

astronomical triangle. . The navigational triangle either terrestrial or celestial, used in the solution of celestial observations.

astronomical twilight. . The period of incomplete darkness when the center of the sun is more than 12° but not more than 18° below the celestial horizon. See also CIVIL TWILIGHT, NAUTICAL TWILIGHT.

astronomical unit. . 1. The mean distance between the earth and the sun, approximately 92,960,000 miles. 2. The astronomical unit is often used as a unit of measurement for distances within the solar system. In the system of astronomical constants of the International Astronomical Union the adopted value for it is 1 AU = $149{,}600 \times 10^6$ meters.

astronomical year. . See TROPICAL YEAR.

astronomic latitude. . See ASTRONOMICAL LATITUDE.

astronomic longitude. . See ASTRONOMICAL LONGITUDE.

astronomy. , *n.* The science which deals with the size, constitution, motions, relative position, etc. of celestial bodies, including the earth. That part of astronomy of direct use to a navigator, comprising principally celestial coordinates, time, and the apparent motions of celestial bodies is called navigational or nautical astronomy.

astro-tracker. . A navigation equipment which automatically acquires and continuously tracks a celestial body in azimuth and altitude.

asymmetrical. , *adj.* Not symmetrical.

asymptote. , *n.* A straight line or curve which a curve of infinite length approaches but never quite reaches.

Atlantic Equatorial Counter Current. . An ocean current that flows eastward between the westward flowing Atlantic North and South Equatorial Currents. The counter current is most prominent during August and September, when it extends from about 52° W to 10° W and joins the GUINEA CURRENT. In October it narrows and separates into two parts at about latitude 7° N, longitude 35° W. The western part, which appears to be a region where the counter current probably sinks and flows eastward beneath the equatorial currents, gradually diminishes in size to the west-northwest, while the eastern part diminishes to the east-southeast. The greatest separation occurs during March; during April the western part of the counter current disappears, but in May it reappears in the vicinity of latitude 0°, longitude 40° W. The two segments progress west-northwestward without much change in size. They merge at about latitude 6°N, longitude 43°W during August and continue their flow eastward uninterrupted through September.

Atlantic North Equatorial Current. . A broad, slow, westward flowing ocean current generated mainly by the northeast trade winds. The current originates near longitude 26° W between about latitude 15° N and 30° N and flows across the ocean past longitude 60° W. It forms the ANTILLES CURRENT in the vicinity of the Leeward Islands. The part of the current between 12° N and 15° N joins the Guiana Current and forms the CARIBBEAN CURRENT.

Atlantic South Equatorial Current. . The major part of this westward flowing ocean current is located south of the equator, the central portion extending to about latitude 20° S. The northern part expands northward during January, February, and March when the Atlantic Equatorial Counter current dissipates and is least evident. On approaching the coast of South America one part turns northwestward as the GUIANA CURRENT; the other part turns below Natal and flows southwestward along the coast of Brazil as the BRAZIL CURRENT. Of the two equatorial currents in the Atlantic, the Atlantic South Equatorial Current is the stronger and more extensive.

Atlantic standard time. . See STANDARD TIME.

atlas. , *n.* A collection of charts or maps kept loose or bound in a volume.

atlas grid. . A reference system that permits the designation of the location of a point or an area on a map, photograph, or other graphic in terms of numbers and letters. Also called ALPHANUMERIC GRID.

atmosphere. , *n.* 1. The envelope of air surrounding the earth and bound to it more or less permanently by gravity. The earth's atmosphere extends from the surface of the earth to an indefinite height, its density asymptotically approaching that of interplanetary space. At heights of the order of 80 kilometers (50 miles) the atmosphere is barely dense enough to scatter sunlight to a visible degree. The atmosphere may be subdivided vertically into a number of atmospheric layers, but the most common basic subdivision is that which recognizes a troposphere from the surface to about 10 kilometers, a stratosphere from about 10 kilometers to about 80 kilometers, and an ionosphere above 80 kilometers. See also STANDARD ATMOSPHERE. 2. The gaseous envelope surrounding any celestial body, including the Earth.

atmospheric absorption. . The loss of power in transmission of radiant energy by dissipation in the atmosphere.

atmospheric drag. . A major cause of perturbations of close artificial satellite orbits caused by the resistance of the atmosphere. The secular effects are decreasing magnitudes of eccentricity, major axis, and period. Sometimes shortened to DRAG.

atmospheric noise. . See ATMOSPHERIC RADIO NOISE.

atmospheric pressure. . The pressure exerted by the weight of the earth's atmosphere, about 14.7 pounds per square inch. See also STANDARD ATMOSPHERE, definition 1; BAROMETRIC PRESSURE.

atmospheric radio noise. . In radio reception noise or static due to natural causes such as thunderstorm activity. Sometimes shortened to ATMOSPHERIC NOISE. See also MAN-MADE NOISE, RADIO INTERFERENCE.

atmospheric refraction. . Refraction resulting when a ray of radiant energy passes obliquely through the atmosphere. It may be called astronomical refraction if the ray enters the atmosphere from outer space, or terrestrial refraction if it emanates from a point on or near the surface of the earth.

atoll. , *n.* A ring-shaped coral reef which has closely spaced islands or islets on it enclosing a central area or lagoon. The diameter may vary from less than a mile to 80 or more.

atollon. , *n.* A large reef ring in the Maldive Islands consisting of many smaller reef rings. The word ATOLL was derived from this name.

atomic clock. . A precision clock that depends for its operation upon an electrical oscillator regulated by an atomic system. The basic principle of the clock is that electromagnetic waves of a particular frequency are emitted when an atomic transition occurs.

atomic second. . See SECOND, definition 1.

Atomic Time. . A fundamental kind of time based on transitions in the atom. International Atomic Time (TAI) is the time reference coordinate established by the Bureau International de l'Heure (BIH) on the basis of the readings of atomic clocks functioning in various establishments in accordance with the definition of the atomic second, the unit of time in the International System of Units (SI). The Atomic Time scales maintained in the United States by the National Institute of Standards and Technology and the U.S. Naval Observatory constitute approximately 37 1/2 percent of the stable reference information used in maintaining a stable TAI scale by the BIH.

A-trace. . The first trace of an oscilloscope having more than one displayed.

ATR tube. . See ANTI-TR TUBE.

attenuation. , *n.* 1. A lessening in amount, particularly the reduction of the amplitude of a wave with distance from the origin. 2. The decrease in the strength of a radar wave resulting from absorption, scattering, and reflection by the medium through which it passes (wave guide, atmosphere) and by obstructions in its path. Also attenuation of the wave may be the result of artificial means, such as the inclusion of an attenuator in the circuitry or by placing an absorbing device in the path of the wave.

attitude. , *n.* The position of a body as determined by the inclination of the axes to some other frame of reference. If not otherwise specified, this frame of reference is fixed to the earth.

atto-. . A prefix meaning one-quintillionth (10^{-18}).

audible. , *adj.* Capable of being translated into sound by the human ear.

audible aid to navigation. . An aid to navigation which uses sound waves.

audio frequency. . A frequency within the audible range, about 20 to 20,000 hertz. Also called SONIC FREQUENCY.

augmentation. , *n.* The apparent increase in the semidiameter of a celestial body as its altitude increases, due to the reduced distance from the observer. The term is used principally in reference to the moon.

augmentation correction. . A correction due to augmentation, particularly that sextant altitude correction due to the apparent increase in the semidiameter of a celestial body as its altitude increases.

augmenting factor. . A factor used in connection with the harmonic analysis of tides or tidal currents to allow for the difference between the times of hourly tabulation and the corresponding constituent hours.

aural. , *adj.* Of or pertaining to the ear or sense of hearing.

aural null. . A null detected by listening for the minimum or the absence of an audible signal.

aureole. , *n.* A poorly developed corona, characterized by a bluish-white disk immediately around the luminary and a reddish-brown outer edge. An aureole, rather than a corona, is produced when the cloud responsible for this diffraction effect is composed of droplets distributed over a wide size-range. The diffracted rays approach the observer from a wide variety of angles, in contrast to the relative uniform diffraction produced by a cloud of more limited drop-size range. In as much as most clouds exhibit rather broad drop-size distributions, aureoles are observed much more frequently than coronas.

aurora. , *n.* A luminous phenomenon due to electrical discharges in the atmosphere, probably confined to the thin air high above the surface of the earth It is most commonly seen in high latitudes where it is most frequent during periods of greatest sunspot activity. If it occurs in the Northern Hemisphere, it is called aurora borealis or northern lights; and if in the Southern, aurora Australis.

aurora Australis. . The aurora in the Southern Hemisphere.

aurora borealis. . The aurora in the Northern Hemisphere. Also called NORTHERN LIGHTS.

auroral zone. . The area of maximum auroral activity. Two such areas exist, each being a 10° wide annulus centered at an average distance of 23° from a geomagnetic pole.

aurora polaris. . A high latitude aurora borealis.

austral. , *adj.* Of or pertaining to south.

authalic map projection. . See EQUAL-AREA MAP PROJECTION.

Automated Mutual-assistance Vessel Rescue System. . Operated by the United States Coast Guard, the AMVER System is a maritime mutual assistance program that aids coordination of search and rescue efforts in the oceans of the world, by maintaining a computerized worldwide merchant vessel plot.

automatic direction finder. . A radio direction finder in which the bearing to the transmitter is indicated automatically and continuously, in contrast with a MANUAL RADIO DIRECTION FINDER which requires manual operation. Also called AUTOMATIC RADIO DIRECTION FINDER (ADF).

automatic frequency control. . The technique of automatically maintaining, or a circuit or device which automatically maintains, the frequency of a receiver within specified limits.

automatic gain control. . A feature involving special circuitry designed to maintain the output of a radio, radar, or television receiver essentially constant, or to prevent its exceeding certain limits, regardless of variations in the strength of the incoming signal.

automatic radar plotting aid. . A computer-assisted radar data processing system which generates predicted ship vectors based on the recent plotted positions. For such a system to meet the specifications of the Inter Governmental Maritime Consultative Organization (IMCO), it must satisfy requirements with respect to detection, acquisition, tracking, display, warnings, data display, and trial maneuvers.

automatic radio direction finder. . See AUTOMATIC DIRECTION FINDER.

automatic tide gage. . An instrument that automatically registers the rise and fall of the tide. In some instruments, the registration is accomplished by recording the heights at regular intervals in digital format, in others by a continuous graph in which the height versus corresponding time is recorded.

auto pilot. , *n.* A device which steers a vessel unattended along a given bearing. See GYRO PILOT.

autumn. , *n.* The season between summer and winter. In the Northern Hemisphere autumn begins astronomically at the autumnal equinox and ends at the winter solstice. In the Southern Hemisphere the limits are the vernal equinox and the summer solstice. The meteorological limits vary with the locality and the year. Also called FALL.

autumnal. , *adj.* Pertaining to fall (autumn). The corresponding adjectives for winter, spring, and summer are *hibernal, vernal,* and *aestival.*

autumnal equinox. . 1. That point of intersection of the ecliptic and the celestial equator occupied by the sun as it changes from north to south declination, on or about September 23. Also called SEPTEMBER EQUINOX, FIRST POINT OF LIBRA. 2. The instant the sun reaches the point of zero declination when crossing the celestial equator from north to south.

auxiliary lights. . See under VERTICAL LIGHTS.

average. , *adj.* Equaling or approximating a mean.

average. , *n.* See MEAN.

average. , *v., t.* To determine a mean.

avoirdupois pound. . See POUND.

avulsion. , *n.* The rapid erosion of shore land by waves during a storm.

awash. , *adj.* & *adv.* Situated so that the top is intermittently washed by waves or tidal action. The term applies both to fixed objects such as rocks, and to floating objects with their tops flush with or slightly above the surface of the water. See also ROCK AWASH, SUBMERGED, UNCOVERED.

axial. , *adj.* Of or pertaining to an axis.

axis. , *n. (pl. axes).* 1. A straight line about which a body rotates, or around which a plane figure may rotate to produce a solid; a line of symmetry. A polar axis is the straight line connecting the poles of a body. The major axis of an ellipse or ellipsoid is its longest diameter; the minor axis, its shortest diameter. 2. One of a set of reference lines for certain systems of coordinates. 3. The principal line about which anything may extend, as the axis of a channel or compass card axis. 4. A straight line connecting two related points.

axis of freedom. . An axis about which the gimbal of a gyro provides a degree-of-freedom of movement.

azimuth. , *n.* The horizontal direction or bearing of a celestial point from a terrestrial point, expressed as the angular distance from a reference direction. It is usually measured from 000° at the reference direction clockwise through 360°. An azimuth is often designated as true, magnetic, compass grid, or relative as the reference direction is true, magnetic, compass, or grid north, or heading, respectively. Unless otherwise specified, the term is generally understood to apply to true azimuth, which may be further defined as the arc of the horizon, or the angle at the zenith, between the north part of the celestial meridian or principal vertical circle and a vertical circle, measured from 000° at the north part of the principal vertical circle clockwise through 360°. Azimuth taken directly from a table, before interpolation, is called tabulated azimuth. After interpolation, or, if determined by calculation, mechanical device, or graphics, it is called computed azimuth. When the angle is measured in either direction from north or south, and labeled accordingly, it is properly called azimuth angle; when measured either direction from east or west, and labeled accordingly, it is called amplitude. An azimuth determined by solution of the navigational triangle with altitude, declination, and latitude then is called an altitude azimuth; if meridian angle, declination, and latitude are given, it is called a time azimuth; if meridian angle, declination and altitude are given, it is called a time and altitude azimuth. See also BACK AZIMUTH, BEARING.

azimuthal. , *adj.* Of or pertaining to azimuth.

azimuthal chart. . A chart on an azimuthal map projection. Also called ZENITHAL CHART.

azimuthal equidistant chart. . A chart on the azimuthal equidistant map projection.

azimuthal equidistant map projection. . An azimuthal map projection on which straight lines radiating from the center or pole of projection represent great circles in their true azimuths from that center,

azimuthal map projection. . A map projection on which the azimuths or directions of all lines radiating from a central point or pole are the same as the azimuths or directions of the corresponding lines on the ellipsoid. This classification includes the gnomonic, stereographic, orthographic, and the azimuthal equidistant map projections. Also called ZENITHAL MAP PROJECTION.

azimuthal orthomorphic projection. . See STEREOGRAPHIC MAP PROJECTION.

azimuth angle. . Azimuth measured from 0° at the north or south reference direction clockwise or counterclockwise through 90° or 180". It is labeled with the reference direction as a prefix and the direction of measurement from the reference direction as a suffix. When azimuth angle is measured through 180°, it is labeled N or S to agree with the latitude and E or W to agree with the meridian angle.

azimuth bar. . An instrument for measuring azimuths, particularly a device consisting of a slender bar with a vane at each end, and designed to fit over a central pivot in the glass cover of a magnetic compass. See also BEARING BAR.

azimuth circle. . A ring designed to fit snugly over a compass or compass repeater, and provided with means for observing compass bearings and azimuths. A similar ring without the means for observing azimuths of the sun is called a BEARING CIRCLE.

azimuth instrument. . An instrument for measuring azimuths, particularly a device which fits over a central pivot in the glass cover of a magnetic compass.

azimuth stabilized display. . See as STABILIZED IN AZIMUTH under STABILIZATION OF RADARSCOPE DISPLAY.

azimuth tables. . Publications providing tabulated azimuths or azimuth angles of celestial bodies for various combinations of declination, latitude and hour angle. Great circle course angles can also be obtained by substitution of values.

Azores Current. . A slow but fairly constant southeast branch of the North Atlantic Current and part of the Gulf Stream System. Its mean speed is only 0.4 knot, and the mean maximum speed computed from all observations above 1 knot in the prevailing direction is 1.3 knots. There is no discernible seasonal fluctuation. The speed and direction of the current is easily influenced for short periods by changing winds. The Azores Current is an inner part of the general clockwise oceanic circulation of the North Atlantic Ocean. Also called SOUTHEAST DRIFT CURRENT.

B

back. , *adj.* Reciprocal.

back. , *v., i.* 1. A change in wind direction in reverse of the normal pattern, or counterclockwise in the Northern Hemisphere and clockwise in the Southern Hemisphere. Change in the opposite direction is called *veer*. See also HAUL. 2. To go stern first, or to operate the engines in reverse. 3. To brace the yard of a square sail so as to bring the wind on the forward side.

back azimuth. . An azimuth 180° from a given azimuth.

back echo. . The effect on a radar display produced by a back lobe of a radar antenna. See also SIDE ECHO.

backlash. , *n.* 1. The amount which a gear or other part of a machine, instrument, etc., can be moved without moving an adjoining part, resulting from loose fit. See also LOST MOTION. 2. The tangle resulting when a reel of line or cable revolves faster than line is being stripped off.

back lobe. . The lobe of the radiation pattern of a directional antenna which makes an angle of approximately 180° with the direction of the axis of the main lobe.

back range. . A range observed astern, particularly one used as guidance for a craft moving away from the objects forming the range.

backrush. , *n.* The seaward return of water following the uprush onto the foreshore. See also RIP CURRENT, UNDERTOW.

backshore. , *n.* That part of a beach which is usually dry, being reached only by the highest tides, and by extension, a narrow strip of relatively flat coast bordering the sea. See also FORESHORE.

back sight. . A marine sextant observation of a celestial body made by facing away from the body, measuring an angle of more than 90°.

backstaff. , *n.* A forerunner of the sextant, consisting essentially of a graduated arc and a single mirror. To use the instrument it was necessary to face away from the body being observed. Also called QUADRANT WITH TWO ARCS, SEA QUADRANT.

backstays of the sun. . Crepuscular rays extending downward toward the horizon.

backwash. , *n.* Water or waves thrown back by an obstruction such as a seaward, breakwater, cliff, etc.

backwater. , *n.* Water held back from the main flow, as that which overflows the land and collects in low places or that forming an inlet approximately parallel to the main body and connected thereto by a narrow outlet.

bad-bearing sector. . Relative to a radio direction finder station or radiobeacon, a sector within which bearings are known to be liable to significant errors of unknown magnitudes.

baguio. , *n.* Local term in the Philippines for a tropical cyclone.

balancer. , *n.* A device used with a radio direction finder to balance out antenna effect and thus produce a sharper reading.

balancing. , *n.* The process of neutralizing antenna effect in order to improve the definition of the observed bearing. See also BALANCER.

Bali wind. . A strong east wind at the eastern end of Java.

ball. , *n.* 1. A spherical identifying mark placed at the top of a perch. 2. A time ball.

ballast ground. . A designated area for discharging solid ballast before entering harbor.

ballistic damping error. . A temporary oscillatory error of a gyrocompass introduced during changes of course or speed as a result of the means used to damp the oscillations of the spin axis.

ballistic deflection error. . A temporary oscillatory error of a gyrocompass introduced when the north-south component of the speed changes, as by speed or course change. An accelerating force acts upon the compass, causing a surge of mercury from one part of the system to another in the case of the non pendulous compass, or a deflection (along the meridian) of a mass in the case of a pendulous compass. In either case, a precessing force introduces a temporary ballistic deflection error in the reading of the compass unless it is corrected.

band. , *n.* A specific section or range of anything. See also FREQUENCY BAND.

band of error. . An area either side of a line of position, within which, for a stated level of probability, the true position is considered to lie.

bandwidth. , *n.* 1. The range of frequencies of a device within which its performance, in respect to some characteristic, conforms to a specified standard. 2. The range within the limits of a frequency band.

bank. , *n.* 1. An elevation of the sea floor typically located on a shelf, over which the depth of water is relatively shallow. Reefs or shoals, dangerous to surface navigation, may rise above the general depths of a bank. 2. A shallow area of shifting sand, gravel, mud, etc., such as a *sand bank, mud bank,* etc. 3. A ridge of any material such as earth, rock, snow, etc., or anything resembling such a ridge, as a *fog bank* or *cloud bank*. 4. The edge of a cut or fill. 5. The margin of a watercourse. 6. A number of similar devices connected so as to be used as a single device in common.

bank cushion. . In a restricted channel, especially one with steep banks, bank cushion tends to force the bow away from the bank due to the increase in the bow wave on the near side.

bank suction. . The bodily movement of a ship toward the near bank due to a decrease in pressure as a result of increased velocity of flow of water past the hull in a restricted channel.

banner cloud. . A banner like cloud streaming off from a mountain peak in a strong wind. See also CAP CLOUD.

bar. , *n.* 1. A ridge or mound of sand, gravel, or other unconsolidated material below the high water level, especially at the mouth of a river or estuary, or lying a short distance from and usually parallel to the beach, and which may obstruct navigation. 2. A unit accepted temporarily for use with the International System of Units; 1 bar is equal to 100,000 pascals.

barat., *n.* A heavy northwest squall in Manado Bay on the north coast of the island of Celebes, prevalent from December to February.

barber., *n.*1. A strong wind carrying damp snow or sleet and spray that freezes upon contact with objects, especially the beard and hair. 2. See FROST SMOKE, definition 2.

bar buoy.. A buoy marking the location of a bar at the mouth of a river on approach to a harbor.

bare ice.. Ice without snow cover.

bare rock.. A rock that extends above the mean high water datum in tidal areas or above the low water datum in the Great Lakes. See also ROCK AWASH, SUBMERGED ROCK.

barogram., *n.* The record made by a barograph.

barograph., *n.* A recording barometer. A highly sensitive barograph may be called a microbarograph.

barometer., *n.* An instrument for measuring atmospheric pressure. A **mercurial barometer** employs a column of mercury supported by the atmosphere. An aneroid barometer has a partly exhausted, thin metal cylinder somewhat compressed by atmospheric pressure.

barometric pressure.. Atmospheric pressure as indicated by a barometer.

barometric pressure correction.. A correction due to nonstandard barometric pressure, particularly the sextant altitude correction due to changes in refraction caused by difference between the actual barometric pressure and the standard barometric pressure used in the computation of the refraction table.

barometric tendency.. See PRESSURE TENDENCY.

barothermogram., *n.* The record made by a barothermograph.

barothermograph., *n.* An instrument which automatically records pressure and temperature.

barothermohygrogram., *n.* The record made by a barothermohygrograph.

barothermohygrograph., *n.* An instrument which automatically records pressure, temperature and humidity of the atmosphere.

barrel., *n.* A unit of volume or weight, the U.S. petroleum value being 42 U.S. gallons.

barrel buoy.. A buoy having the shape of a barrel or cylinder floating horizontally, usually for special purposes, including mooring.

barrier beach.. A bar essentially parallel to the shore, the crest of which is above high water.

barrier reef.. A coral reef which roughly parallels land but is some distance offshore, with deeper water adjacent to the land, as contrasted with a FRINGING REEF closely attached to the shore.

bar scale.. A line or series of lines on a chart, subdivided and labeled with the distances represented on the chart. Also called GRAPHIC SCALE. See also SCALE.

barycenter., *n.* The center of mass of a system of masses; the common point about which two or more celestial bodies revolve.

base chart.. See BASE MAP.

base course up.. One of the three basic orientations of display of relative or true motion on a radarscope. In the BASE COURSE UP orientation, the target pips are painted at their measured distances and in their directions relative to a preset base course of own ship maintained UP in relation to the display. This orientation is most often used with automated radar plotting systems. Also called COURSE UP. See also HEAD UP, NORTH UP.

base line.. 1. The reference used to position limits of the territorial sea and the contiguous zone. 2. One side of a series of connected survey triangles, the length of which is measured with prescribed accuracy and precision, and from which the lengths of the other triangle sides are obtained by computation. Important factors in the accuracy and precision of base measurements are the use of standardized invar tapes, controlled conditions of support and tension, and corrections for temperatures, inclination, and alignment. Base lines in triangulation are classified according to the character of the work they are intended to control, and the instruments and methods used in their measurement are such that prescribed probable errors for each class are not exceeded. These probable errors, expressed in terms of the lengths, are as follows: first order, 1 part in 1,000,000; second order, 1 part in 500,000; and third order, 1 part in 250,000. 3. The line along the surface of the earth between two radio navigation stations operating in conjunction for the determination of a line of position.

baseline delay.. The time interval needed for the signal from a master station of a hyperbolic radionavigation system to travel the length of the baseline, introduced as a delay between transmission of the master and slave (or secondary) signals to make it possible to distinguish between the signals and to permit measurement of time differences.

baseline extension.. The extension of the baseline in both directions beyond the transmitters of a pair of radio stations operating in conjunction for determination of a line of position.

base map.. 1. A map or chart showing certain fundamental information, used as a base upon which additional data of specialized nature are compiled or overprinted. 2. A map containing all the information from which maps showing specialized information can be prepared. Also called BASE CHART in nautical charting.

base map symbol.. A symbol used on a base map or chart as opposed to one used on an overprint to the base map or chart. Also called BASE SYMBOL.

base symbol.. See BASE MAP SYMBOL.

base units.. See under INTERNATIONAL SYSTEM OF UNITS.

basin., *n.* 1. A depression of the sea floor approximately equidimensional in plan view and of variable extent. 2. An area of water surrounded by quay walls, usually created or enlarged by excavation, large enough to receive one or more ships for a specific purpose. See also GRAVING DOCK, HALF. TIDE BASIN, NON-TIDAL BASIN, SCOURING BASIN, TIDAL BASIN, TURNING BASIN. 3. An area of land which drains into a lake or sea through a river and its tributaries. 4. A nearly land-locked area of water leading off an inlet, firth, or sound.

bathyal., *adj.* Pertaining to ocean depths between 100 and 2,000 fathoms; also to the ocean bottom between those depths, sometimes identical with the continental slope environment.

bathymeter., *n.* An instrument for measuring depths of water.

bathymetric., *adj.* Of or pertaining to bathymetry.

bathymetric chart.. A topographic chart of the seabed of a body of water, or a part of it. Generally, bathymetric charts show depths by contour lines and gradient tints.

bathymetry., *n.* The science of measuring water depths (usually in the ocean) in order to determine bottom topography.

bathysphere., *n.* A spherical chamber in which persons are lowered for observation and study of ocean depths.

bathythermogram., *n.* The record made by a bathythermograph.

bathythermograph., *n.* An instrument which automatically draws a graph showing temperature as a function of depth when lowered in the sea.

batture., *n.* An elevation of the bed of a river under the surface of the water; sometimes used to signify the same elevation when it has risen above the surface.

baud.. A measure of the speed of computer data transmission in bits per second.

bay., *n.* A recess in the shore, on an inlet of a sea or lake between two capes or headlands, that may vary greatly in size but is usually smaller than a gulf but larger than a cove.

bayamo., *n.* A violent blast of wind, accompanied by vivid lightning, blowing from the land on the south coast of Cuba, especially near the Bight of Bayamo.

Bayer's letter.. The Greek (or Roman) letter used in a Bayer's name.

Bayer's name.. The Greek (or Roman) letter and the possessive form of the Latin name of a constellation, used as a star name.

baymouth bar.. A bar extending partially or entirely across the mouth of a bay.

bayou., *n.* A minor, sluggish waterway or estuaries creek, generally tidal or with a slow or imperceptible current, and with its course generally through lowlands or swamps, tributary to or connecting with other bodies of water. Various specific meanings have been implied in different parts of the southern United States. Sometimes called SLOUGH.

beach., *n.* The zone of unconsolidated material that extends landward from the low water line to the place where there is a marked change in material or physiographic form, or to the line of permanent vegetation (usually the effective limit of storm waves). A beach includes foreshore and backshore. The beach along the margin of the sea may be called SEABEACH. Also called STRAND, especially when the beach is composed of sand. See also TIDELAND.

beach., *v., t. & i.* To intentionally run a craft ashore.

beach berm. . See BERM.

beach erosion. . The carrying away of beach materials by wave action, tidal or littoral currents, or wind.

beacon. , *n*. A fixed artificial navigation mark. See also MARK, definition 1; DAYBEACON; DAYMARK; LIGHTED BEACON; RADIOBEACON.

beaconage. , *n*. A system of fixed aids to navigation comprised of beacons and minor lights. See also BUOYAGE.

beacon buoy. . See PILLAR BUOY.

beacon tower. . A beacon which is a major structure, having a support as distinctive as the topmark. See also LATTICE BEACON, REFUGE BEACON.

beam. , *n*. 1. A directed flow of electromagnetic radiation from an antenna. See also MAIN BEAM under LOBE, BEAM WIDTH. 2. A group of nearly parallel rays, as a *light beam*.

beam compass. . Compass for drawing circles of large diameter. In its usual form it consists of a bar with sliding holders for points, pencils, or pens which can be set at any desired position.

beam sea. . Waves moving in a direction approximately 90° from the vessel's heading. Those moving in a direction approximately opposite to the heading are called HEAD SEA, those moving in the general direction of the heading are called FOLLOWING SEA, and those moving in a direction approximately 45° from the heading (striking the quarter) are called QUARTERING SEA. See also CROSS SEA.

beam tide. . A tidal current setting in a direction approximately 90° from the heading of a vessel One setting in a direction approximately 90° from the course is called a CROSS TIDE. In common usage these two expressions are usually used synonymously. One setting in a direction approximately opposite to the heading is called a HEAD TIDE. One setting in such a direction as to increase the speed of a vessel is called a FAIR TIDE.

beam width. . The angular measure of the transverse section of a beam (usually in the main lobe) Lying within directions corresponding to specified values of field strength relative to the maximum (e.g., half field strength beam width and half power beam width). The beam width is usually measured in one or more specified planes containing the axis of the beam. See also HORIZONTAL BEAM WIDTH, VERTICAL BEAM WIDTH.

beam-width error. . An azimuth or bearing distortion on a radar display caused by the width of the radar beam. See also BEAM WIDTH, PULSE LENGTH ERROR.

beam wind. . Wind blowing in a direction approximately 90° from the heading. One blowing in a direction approximately 90° from the course is called a CROSS WIND. In common usage these two expressions are usually used synonymously, BEAM WIND being favored by mariners, and CROSS WIND by aviators. One blowing from ahead is called a HEAD WIND. One blowing from astern is called a FOLLOWING WIND by mariners and a TAIL WIND by aviators. See also FAIR WIND, FAVORABLE WIND, UNFAVORABLE WIND.

bear. , *v., i*. To be situated as to direction, as, the light bears 165°.

bear down. . To approach from windward.

bearing. , *n*. The horizontal direction of one terrestrial point from another, expressed as the angular distance from a reference direction. It is usually measured from 000° at the reference direction clockwise through 360°. The terms BEARING and AZIMUTH are sometimes used interchangeably, but in navigation the former customarily applies to terrestrial objects and the latter to the direction of a point on the celestial sphere from a point on the earth. A bearing is often designated as true, magnetic, compass, grid, or relative as the reference direction is true, magnetic, compass, or grid north, or heading, respectively. The angular distance between a reference direction and the initial direction of a great circle through two terrestrial points is called great-circle bearing. The angular distance between a reference direction and the rhumb line through two terrestrial points is called rhumb or Mercator bearing. A bearing differing by 180°, or one measured in the opposite direction, from a given bearing is called a reciprocal bearing. The maximum or minimum bearing of a point for safe passage of an off-lying danger is called a danger bearing. A relative bearing of 045° or 315° is sometimes called a four-point bearing. Successive relative bearings (right or left) of 45° and 90° taken on a fixed object to obtain a running fix are often called bow and beam bearings. Two or more bearings used as intersecting lines of position for fixing the position of a craft are called cross bearings. The bearing of a radio transmitter from a receiver, as determined by a radio direction finder, is called a radio bearing. A bearing obtained by radar is called a radar bearing. A bearing obtained by visual observation is called a visual bearing. A constant bearing maintained while the distance between two craft is decreasing is called a collision bearing. See also CURVE OF EQUAL BEARING.

bearing angle. . Bearing measured from 0° at the reference direction clockwise or counterclockwise through 90° or 180°. It is labeled with the reference direction as a prefix and the direction of measurement from the reference direction as a suffix. Thus, bearing angle N37°W is 37° west of north, or true bearing 323°.

bearing bar. . An instrument for measuring bearings, particularly a device consisting of a slender bar with a vane at each end, and designed to fit over a central pivot in the glass cover of a magnetic compass. See also AZIMUTH BAR.

bearing book. . A log for the recording of visual bearings.

bearing calibration. . The determination of bearing corrections of a radiodirection finder by observations of a radiobeacon, particularly a calibration radiobeacon, of known visual bearing, observations being taken over 360° of swing of the observing vessel.

bearing circle. . A ring designed to fit snugly over a compass or compass repeater, and provided with vanes for observing compass bearings. A similar ring provided with means for observing azimuths of the sun is called an AZIMUTH CIRCLE.

bearing compass. . A compass intended primarily for use in observing bearings.

bearing cursor. . The radial line on a radar set inscribed on a transparent disk which can be rotated manually about an axis coincident with the center of the PPI. It is used for bearing determination. Also called MECHANICAL BEARING CURSOR.

bearing light. . A navigation light using two superimposed optical systems which provides an approximate bearing without the use of a compass.

bearing line. . A line extending in the direction of a bearing.

bearing repeater. . A compass repeater used primarily for observing bearings.

bearing resolution. . See as RESOLUTION IN BEARING under RESOLUTION, definition 2. Also called ANGULAR RESOLUTION.

beat frequency. . Either of the two additional frequencies obtained when signals of two frequencies are combined, equal to the sum or difference, respectively, of the original frequencies.

Beaufort wind scale. . A numerical scale for indicating wind speed, devised by Admiral Sir Francis Beaufort in 1805. Beaufort numbers (or forces) range from force 0 (calm) to force 12 (hurricane).

bed. , *n*. The ground upon which a body of water rests. The term is usually used with a modifier to indicate the type of water body, as river bed or sea bed. See also BOTTOM.

before the wind. . In the direction of the wind. The expression applies particularly to a sailing vessel having the wind well aft. See also DOWNWIND.

bell. , *n*. A device for producing a distinctive sound by the vibration of a hollow, cup-shaped metallic vessel which gives forth a ringing sound when struck.

bell book. . The log of ordered engine speeds and directions.

bell buoy. . A buoy with a skeleton tower in which a bell is fixed.

belt. , *n*. A band of pack ice from 1 km to more than 100 km in width.

bench. , *n*. On the sea floor, a small terrace.

bench mark. . A fixed physical object used as reference for a vertical datum. A tidal bench mark is one near a tide station to which the tide staff and tidal datums are referred. A primary tidal bench mark is the principal (or only) mark of a group of tidal bench marks to which the tide staff and tidal datum's are referred. A geodetic bench mark identifies a surveyed point in the National Geodetic Vertical Network. Geodetic bench mark disks contain the inscription VERTICAL CONTROL MARK, NATIONAL GEODETIC SURVEY with other individual identifying information. Bench mark disks of either type may, on occasion, serve simultaneously to reference both tidal and geodetic datum's. Numerous bench marks, both tidal and geodetic, still bear the inscription U.S. COAST & GEODETIC SURVEY.

beneaped. , *adj.* See NEAPED.

Benguela Current. . A slow-moving ocean current flowing generally northwestward along the west coast of Africa. It is caused mainly by the prevailing southeast trade winds. Near the equator the current flows westward and becomes the ATLANTIC SOUTH EQUATORIAL CURRENT.

bentu de soli. . An east wind on the coast of Sardinia.

berg. , *n.* Short for ICEBERG.

bergy bit. . A large piece of floating glacier ice, generally showing less than 5 meters above sea level but more than 1 meter and normally about 100 to 300 square meters in area. It is smaller than an ICEBERG but larger than a GROWLER. A typical bergy bit is about the size of a small house.

Bering Current. . A northward flowing current through the eastern half of the Bering Sea, through Bering Strait, and in the eastern Chukchi Sea. The current speed in the Bering Sea is estimated to be usually 0.5 knot or less but at times as high as 1.0 knot. In the Bering Strait, current speeds frequently reach 2 knots. However, in the eastern half of the strait, currents are even stronger and usually range between 1.0 and 2.5 knots. Strong southerly winds may increase current speeds in the strait to 3 knots, and up to 4 knots in the eastern part. Persistent, strong northerly winds during autumn may cause the current to reverse direction for short periods. During winter a southward flow may occur in the western part of the strait. After flowing through Bering Strait, the current widens, and part continues toward Point Barrow, where it turns northwestward. Along the Alaska coast, current t speeds have been observed to range between 0.1 and 1.5 knots and increase to 2.0 or 2.5 knots with southerly winds. In the western part of the Chukchi Sea, currents are considerably weaker and do not usually exceed 0.5 knot.

berm. , *n.* A nearly horizontal portion of a beach or backshore having an abrupt fall and formed by wave deposition of material and marking the limit of ordinary high tides. Also called BEACH BERM.

berm crest. . The seaward limit of a berm. Also called BERM EDGE.

berm edge. . See BERM CREST.

berth. , *n., v., t.* 1. A place for securing a vessel. 2. To secure a vessel at a berth. See also FOUL BERTH, MUD BERTH.

beset. , *adj.* State of a vessel surrounded by ice and unable to move. If the ice forcibly squeezes the hull, the vessel is said to be NIPPED.

Bessel ellipsoid of 1841. . The reference ellipsoid of which the semimajor axis is 6,377,397.155 meters, the semiminor axis is 6,356,078.963 meters and the flattening or ellipticity equals 1/299.1528. Also called BESSEL SPHEROID OF 1841.

Besselian year. . See FICTITIOUS YEAR.

Bessel spheroid of 1841. . See BESSEL ELLIPSOID OF 1841.

bias error. . See CONSTANT ERROR.

bifurcation. , *n.* A division into two branches.

bifurcation buoy. . A buoy which indicates the place at which a channel divides into two. See also JUNCTION BUOY.

bifurcation mark. . A navigation mark which indicates the place at which the channel divides into two. See also JUNCTION MARK.

big floe. . See under FLOE.

bight. , *n.* 1. A long and gradual bend or recess in the coastline which forms a large open receding bay. 2. A bend in a river or mountain range. 3. An extensive crescent-shaped indentation in the ice edge.

bill. , *n.* A narrow promontory.

bi-margin format. . The format of a map or chart on which the cartographic detail is extended to two edges of the sheet, thus leaving only two margins. See also BLEED.

binary notation. . Referring to a system of numbers with a base of 2; used extensively in computers, which use electronic on-off storage devices to represent the numbers 0 and 1.

binary star. . A system of two stars that revolve about their common center of mass. See also DOUBLE STAR.

binnacle. , *n.* The stand in which a compass is mounted. For a magnetic compass it is usually provided with means of mounting various correctors for adjustment and compensation of the compass.

binocular. , *n., adj.* 1. An optical instrument for use with both eyes simultaneously. 2. Referring to vision with two eyes.

bioluminescence. , *n.* The production of light by living organisms in the sea. Generally, these displays are stimulated by surface wave action, ship movement, subsurface waves, up welling, eddies, physical changes in sea water, surfs, and rip tides.

bisect. , *v., t.* To divide into two equal parts.

bit. *(from binary digit).* The smallest unit of information in a computer. Bits are grouped together into bytes, which represent characters or other information.

bit-map. . A type of computerized display which consists of a single layer of data; individual elements cannot be manipulated. See VECTOR, RASTER.

bivariate error distribution. . A two-dimensional error distribution.

blackbody. , *n.* An ideal emitter which radiates energy at the maximum possible rate per unit area at each wavelength for any given temperature. A blackbody also absorbs all the radiant energy in the near visible spectrum incident upon it. No actual substance behaves as a true blackbody.

black light. . Ultraviolet or infrared radiant energy. It is neither black nor light.

blanket. , *v, t.* To blank out or obscure weak radio signals by a stronger signal.

blanketing. , *n.* The blanking out or obscuring of weak radio signals by a stronger signal.

blanking. , *n.* See as DUAL-RATE BLANKING.

blank tube. . A marine sextant accessory consisting of a tubular sighting vane, the function of which is to keep the line of vision parallel to the frame of the instrument when observing horizontal sextant angles.

blather. , *n.* Very wet mud of such nature that a weight will rapidly sink into it. See also QUICKSAND.

bleed. , *n.* The edge of a map or chart on which cartographic detail is extended to the edge of the sheet. Also called BLEEDING EDGE.

bleeding edge. . See BLEED.

blind lead. . A lead with only one outlet.

blind pilotage. . *British terminology.* The task of conducting the passage of a ship in pilot waters using means available to the navigator in low visibility.

blind rollers. . Long, high swells which have increased in height, almost to the breaking point, as they pass over shoals or run in shoaling water. Also called BLIND SEAS.

blind seas. . See BLIND ROLLERS.

blind sector. . A sector on the radarscope in which radar echoes cannot be received because of an obstruction near the antenna. See also SHADOW SECTOR.

blink. , *n.* A glare on the underside of extensive cloud areas, created by light reflected from snow or ice-covered surfaces.

snow blink. . Blink caused by a snow-covered surface, which is whitish and brighter than the yellowish-white glare of ice blink. See also LAND SKY, WATER SKY, SKY MAP.

blinking. , *n.* A means of providing information in radionavigation systems of the pulse type by modifying the signal at its source so that the signal presentation alternately appears and disappears or shifts along the time base. In Loran, blinking is used to indicate that a station is malfunctioning.

blip. , *n.* On a radarscope, a deflection or spot of contrasting luminescence caused by an echo, i.e., the radar signal reflected back to the antenna by an object. Also called PIP, ECHO, RETURN.

blip scan ratio. . The ratio of the number of paints from a target to the maximum possible number of paints for a given number of revolutions of the radar antenna. The maximum number of paints is usually equivalent to the number of revolutions of the antenna.

blister. , *n.* See BORDER BREAK.

blizzard. , *n.* A severe weather condition characterized by low temperatures and by strong winds bearing a great amount of snow (mostly fine, dry snow picked up from the ground). The National Weather Service specifies the following conditions for a blizzard: a wind of 32 miles per hour or higher, low temperatures, and sufficient snow in the air to reduce visibility to less than 500 feet; for a severe blizzard, it specifies wind speeds exceeding 45 miles per hour, temperature near or below 10°F, and visibility reduced by snow to near zero. In popular usage in the United States, the term is often used for any heavy snowstorm accompanied by strong winds.

block. , *n.* See CHARTLET, definition 2.

block correction. . See CHARTLET, definition 2.

blocky iceberg. . An iceberg with steep sides and a flat top. The length-to-height ratio is less than 5:1. See also TABULAR ICEBERG.

Blondel-Rey effect. . The effect that the flashing of a light has on reducing its apparent intensity as compared to the intensity of the same light when operated continuously or fixed.

blooming. , *n.* Expansion of the spot produced by a beam of electrons striking the face of a cathode-ray indicator, caused by maladjustment.

blowing snow. . Snow raised from the ground and carried by the wind to such a height that both vertical and horizontal visibility are considerably reduced. The expression DRIFTING SNOW is used when only the horizontal visibility is reduced.

blue ice. . The oldest and hardest form of glacier ice, distinguished by a slightly bluish or greenish color.

blue magnetism. . The magnetism displayed by the south-seeking end of a freely suspended magnet. This is the magnetism of the earth's north magnetic pole.

bluff. , *n.* A headland or stretch of cliff having a broad nearly perpendicular face. See also CLIFF.

blunder. , *n.* See MISTAKE.

Board of Geographic Names. . An agency of the U.S Government, first established by Executive Order in 1890 and currently functioning under Public Law 242-80, 25 July 1947. Twelve departments and agencies have Board membership. The board provides for "uniformity in geographic nomenclature and orthography throughout the Federal Government." It develops policies and romanization systems under which names are derived and it standardizes geographic names for use on maps and in textual materials.

boat. , *n.* A small vessel. The term is often modified to indicate the means of propulsion, such as motorboat, rowboat, steamboat, sailboat, and sometimes to indicate the intended use, such as lifeboat, fishing boat, etc. See also SHIP.

boat compass. . A small compass mounted in a box for small craft. use.

boat harbor. . A sheltered area in a harbor set aside for the use of boats, usually with docks, moorings, etc.

boat sheet. . The work sheet used in the field for plotting details of a hydrographic survey as it progresses.

bobbing a light. . Quickly lowering the height of eye and raising it again when a navigational light is first sighted to determine if the observer is at the geographic range of the light.

bold. , *adj.* Rising steeply from the sea; as a bold coast. See also ABRUPT.

bolide. , *n.* A meteor having a magnitude brighter than 4 magnitude. Bolides are observed with much less frequency than shooting stars. Light bursts, spark showers, or splitting of the luminous trail are sometimes seen along their trails. The luminous trails persist for minutes and may persist up to an hour in exceptional cases. Also called FIREBALL. See also METEOR.

bollard. , *n.* A post (usually steel or reinforced concrete) firmly secured on a wharf, quay, etc., for mooring vessels with lines.

bombing range. . An area of land or water, and the air space above, designated for use as a bombing practice area.

boom. , *n.* A floating barrier used for security, shelter, or environmental cleanup.

boot. . To start a computer, which initiates a series of internal checks and programs which ready the computer for use.

bora. , *n.* A cold, northerly wind blowing from the Hungarian basin into the Adriatic Sea. See also FALL WIND.

borasco. , *n.* A thunderstorm or violent squall, especially in the Mediterranean.

border break. . A cartographic technique used when it is required to extend cartographic detail of a map or chart beyond the neatline into the margin, which eliminates the necessity of producing an additional sheet. Also called BLISTER.

borderland. , *n.* A region bordering a continent, normally occupied by or bordering a shelf that is highly irregular with depths well in excess of those typical of a shelf.

bore. , *n.* See TIDAL BORE.

boring. , *n.* Forcing a vessel under power through ice, by breaking a lead.

borrow. , *v., t.* To approach closer to the shore or wind.

bottom. , *n.* The ground under a body of water. The terms FLOOR, and BOTTOM have nearly the same meaning, but BED refers more specifically to the whole hollowed area supporting a body of water, FLOOR refers to the essential horizontal surface constituting the principal level of the ground under a body of water, and BOTTOM refers to any ground covered with water.

bottom characteristics. . Designations used on surveys and nautical charts to indicate the consistency, color, and classification of the sea bottom. Also called NATURE OF THE BOTTOM, CHARACTER OF THE BOTTOM.

bottom contour chart. . A chart designed for surface and sub-surface bathymetric navigation seaward of the 10 fathom contour. Bottom configuration is portrayed by depth contours and selected soundings.

bottom sample. . A portion of the material forming the bottom, brought up for inspection.

bottom sampler. . A device for obtaining a portion of the bottom for inspection.

Bouguer's halo. . An infrequently observed, faint, white. circular arc or complete ring of light which has a radius of about 39°, and is centered on the antisolar point. When observed, it usually is in the form of a separate outer ring around an anticorona. Also called ULLOA'S RING. See also FOGBOW.

boulder. , *n.* A detached water-rounded stone more than 256 millimeters in diameter, i.e., larger than a man's head. See also COBBLE.

boundary disclaimer. . A statement on a map or chart that the status and/or alignment of international or administrative boundaries is not necessarily recognized by the government of the publishing nation.

boundary lines of inland waters. . Lines dividing the high seas from rivers, harbors, and inland waters. The waters inshore of the lines are "inland waters" and upon them the Inland Rules of the Road or Pilot Rules apply. The waters outside of the lines are the high seas and upon them the International Rules apply.

boundary monument. . A material object placed on or near a boundary line to preserve and identify the location of the boundary line on the ground.

bow. , *n.* The forward part of a ship, craft, aircraft, or float.

bow and beam bearings. . Successive relative bearings (right or left) of 45° and 90° taken on a fixed object to obtain a running fix. The length of the run between such bearings is equal to the distance of the craft from the object at the time the object is broad on the beam., neglecting current.

Bowditch. , *n.* Popular title for Pub. No. 9, The American Practical Navigator.

bow wave. . 1. The wave set up by the bow of a vessel moving through the water. Also called WAVE OF DISPLACEMENT. 2. A shock wave in front of a body such as an airfoil.

boxing the compass. . Stating in order the names of the points (and sometimes the half and quarter points) of the compass.

brackish. , *adj.* Containing salt to a moderate degree, such as sea water which has been diluted by fresh water, such as near the mouth of a river. The salinity values of brackish water range from approximately 0.50 to 17.00 parts per thousand.

branch. , *n.* 1. A creek or brook, as used locally in the southern U.S. 2. One of the bifurcations of a stream.

brash ice. . Accumulations of floating ice made up of fragments not more than 2 meters across, the wreckage of other forms of ice.

brave west winds. . The strong, often stormy, winds from the west-northwest and northwest which blow at all seasons of the year between latitudes 40° S and 60°S. See also ROARING FORTIES.

Brazil Current. . The ocean current flowing southwestward along the Brazilian coast. Its origin is in the westward flowing Atlantic South Equatorial Current, part of which turns south-and flows along the South American coast as the Brazil Current. The mean speed of the current along its entire length is about 0.6 knot. Off Uruguay at about 35° S, it meets the Falkland Current, the two turning eastward to join the South Atlantic Current.

break-circuit chronometer. . A chronometer equipped with an electrical contact assembly and program wheel which automatically makes or breaks an electric circuit at precise intervals, the sequence and duration of circuit-open circuit closed conditions being recorded on a chronograph. The program sequence is controlled by the design of the program wheel installed. Various programs of make or break sequence, up to 60 seconds, are possible. In some chronometers the breaks occur every other second, on the even seconds, and a break occurs also on the 59th second to identify the beginning of the minute; in other chronometers, breaks occur every second except at the beginning of the minute. By recording the occurrence of events (such as star transits) on a chronograph sheet along with the chronometer breaks, the chronometer times of those occurrences are obtained.

breaker. , *n.* A wave which breaks, either because it becomes unstable, usually when it reaches shallow water, or because it dashes against an obstacle. Instability is caused by an increase in wave height and a decrease in the speed of the trough of the wave in shallow water. The momentum of the crest, often aided by the wind, causes the upper part of the wave to move forward faster than the lower part. The crest of a wave which becomes unstable in deep water and topples over or "breaks" is called a WHITECAP.

breakwater. , *n.* A line of rocks, concrete, pilings, or other material which breaks the force of the sea at a particular place, forming a protected area. Often an artificial embankment built to protect the entrance to a harbor or to form an artificial harbor. See also JETTY.

breasting float. . See CAMEL.

breeze. , *n.* 1. Wind of force 2 to 6 (4-31 miles per hour or 4-27 knots) on the Beaufort wind scale. Wind of force 2 (4-7 miles per hour or 4-6 knots) is classified as a light breeze; wind of force 3 (8-12 miles per hour or 7-10 knots), a gentle breeze; wind of force 4 (13-18 miles per hour or 11-16 knots), a moderate breeze; wind, of force 5 (19-24 miles per hour or 17-21 knots), a fresh breeze; and wind of force 6 (25-31 miles per hour or 22-27 knots), a strong breeze. See also LIGHT AIR. 2. Any light wind.

bridge. , *n.* 1. An elevated structure extending across or over the weather deck of a vessel, or part of such a structure. The term is sometimes modified to indicate the intended use, such as *navigating bridge* or *signal bridge*. 2. A structure erected over a depression or an obstacle such as a body of water, railroad, etc. to provide a roadway for vehicles or pedestrians. See also CAUSEWAY, VIADUCT.

bridge resource management. The study of the resources available to the navigator and the exploitation of them in order to achieve the goal of safe and efficient voyages.

Briggsian logarithm. . See COMMON LOGARITHM.

bright display. . A radar display capable of being used under relatively high ambient light levels.

brisa, briza. , *n.* 1. A northeast wind which blows on the coast of South America or an east wind which blows on Puerto Rico during the trade wind season. 2. The northeast monsoon in the Philippines.

brisote. , *n.* The northeast trade wind when it is blowing stronger than usual on Cuba.

Broadcast Notice to Mariners. . Notices to mariners disseminated by radio broadcast, generally of immediate interest to navigators.

broad on the beam. . Bearing 090° relative *(broad on the starboard beam)* or 270° relative *(broad on the port beam)*. If the bearings are approximate, the expression ON THE BEAM or ABEAM should be used.

broad on the bow. . Bearing 045° relative *(broad on the starboard bow)* or 315° relative *(broad on the port bow)*. If the bearings are approximate, the expression ON THE BOW should be used.

broad on the quarter. . Bearing 135° relative *(broad on the starboard quarter)* or 225° relative *(broad on the port quarter)*. If the bearings are approximate, the expression ON THE QUARTER should be used.

broadside on. . Beam on, such as to the wind or sea.

broad tuning. . Low selectivity, usually resulting in simultaneous reception of signals of different frequencies (spill-over). The opposite is SHARP TUNING.

Broken bow. . See ANTICORONA.

broken water. . An area of small waves and eddies occurring in what otherwise is a calm sea.

brook. , *n.* A very small natural stream; a rivulet. Also called RUN, RUNNEL. See also CREEK, definition 2.

brubu. , *n.* A name for a squall in the East Indies.

B-trace. . The second trace of an oscilloscope having more than one displayed.

bubble acceleration error. . The error of a bubble sextant observation caused by displacement of the bubble by acceleration or deceleration resulting from motion of a craft. Also called ACCELERATION ERROR.

bubble horizon. . An artificial horizon parallel to the celestial horizon, established by means of a bubble level.

bubble sextant. . A sextant with a bubble or spirit level to indicate the horizontal.

bucket temperature. . Temperature of surface sea water trapped and measured in a bucket or similar receptacle.

buffer. . In computers, a temporary storage area used when incoming data cannot be processed as fast as it is transmitted.

building. , *n.* A label on a nautical chart which is used when the entire structure is the landmark, rather than an individual feature of it. Also labeled HOUSE.

bull's eye squall. . A squall forming in fair weather, characteristic of the ocean off the coast of South Africa. It is named for the peculiar appearance of the small isolated cloud marking the top of the invisible vortex of the storm.

bull the buoy. . To bump into a buoy.

bummock. , *n.* A downward projection from the underside of an ice field; the counterpart of a HUMMOCK.

bund. , *n.* An embankment or embanked thoroughfare along a body of water. The term is used particularly for such structures in the Far East.

buoy. , *n.* An unmanned floating device moored or anchored to the bottom as an aid to navigation. Buoys may be classified according to shape, as spar, cylindrical or can, conical, nun, spherical, barrel, or pillar buoy. They may also be classified according to the color scheme as a red, green, striped, banded, or checkered buoy. A buoy fitted with a characteristic shape at the top to aid in its identification is called a topmark buoy. A sound buoy is one equipped with a characteristic sound signal, and may be further classified according to the manner in which the sound is produced, as a bell, gong, horn, trumpet, or whistle buoy. A lighted buoy is one with a light having definite characteristics for detection and identification during darkness. A buoy equipped with a marker radiobeacon is called a radiobeacon buoy. A buoy with equipment for automatically transmitting a radio signal when triggered by an underwater sound signal is called a sonobuoy. A combination buoy has more than one means of conveying information; it may be called a lighted sound buoy if it is a lighted buoy provided with a sound signal. Buoys may be classified according to location, as channel mid channel, middle ground, turning, fairway junction, junction, or sea buoy. A bar buoy marks the location of a bar. A buoy marking a hazard to navigation may be classified according to the nature of the hazard, such as obstruction, wreck, telegraph, cable, fish net, dredging, or spoil ground buoys. Buoys used for particular purposes may be classified according to their use, as anchor, anchorage, quarantine, mooring, marker, station, watch, or position buoy. A light-weight buoy especially designed to withstand strong currents is called a river buoy. An ice buoy is a sturdy one used to replace a more easily damaged buoy during a period when heavy ice is anticipated.

buoyage. , *n.* A system of buoys. One in which the buoys are assigned shape, color, and number distinction in accordance with location relative to the nearest obstruction is called a cardinal system. One in which buoys are assigned shape, color, and number distinction as a means of indicating navigable waters is called a lateral system. See also IALA MARITIME BUOYAGE SYSTEM.

buoy station. . The established (charted) location of a buoy.

buoy tender. . A vessel designed for, and engaged in, servicing aids to navigation, particularly buoys.

butte. , *n.* An isolated flat-topped hill, similar to but smaller than a MESA.

Buys Ballot's law. . A rule useful in locating the center of cyclones and anticyclones. It states that, facing away from the wind in the northern hemisphere, the low pressure lies to the left. Facing away from the wind in the southern hemisphere, it is to the right; named after Dutch meteorologist C. H. D. Buys Ballot, who published it in 1857.

byte. . Basic unit of measurement of computer memory. A byte usually consists of 8 BITS; each ASCII character is represented by 1 byte.

by the head. . See DOWN BY THE HEAD.

by the stern. . See DOWN BY THE STERN.

C

C/A code. , *n*. The coarse acquisition, or "civilian code," modulated on the GPS L1 signal.

cable. , *n*. 1. A unit of distance equal to one-tenth of a sea mile. Sometimes called CABLE LENGTH. 2. A chain or very strong fiber or wire rope used to anchor or moor vessels or buoys. 3. A stranded conductor or an assembly of two or more electric conductors insulated from each other, but laid up together with a strong, waterproof covering. A coaxial cable consists of two concentric conductors insulated from each other.

cable buoy. . 1. A buoy used to mark one end of a cable being worked by a cable ship. 2. A floating support of a submarine cable.

cable length. . See CABLE, definition 1.

cage. , *n*. The upper part of the buoy built on top of the body of the buoy and used as a daymark or part thereof, usually to support a light, topmark and/or radar reflector. Also called SUPERSTRUCTURE.

cage. , *v., t*. To erect a gyro or lock it in place by means of a caging mechanism.

caging mechanism. . A device for erecting a gyroscope or locking it in position.

cairn. , *n*. A mound of rough stones or concrete, particularly one intended to serve as a landmark or message location. The stones are customarily piled in a pyramidal or beehive shape.

caisson. , *n*. A watertight gate for a lock, basin, etc.

calcareous. , *adj*. Containing or composed of calcium or one of its compounds.

calculated altitude. . See under COMPUTED ALTITUDE.

calculator. . A device for mathematical computations; originally mechanical, modern ones are exclusively electronic, and able to run simple programs. A navigational calculator contains ephemeral data and algorithms for the solution of navigation problems. Compare with computers, which can be used for many other applications and run complex programs.

caldera. , *n*. A volcanic crater.

calendar. , *n*. A graphic or printed record of time, usually of days, weeks, months, etc., used to refer to future events. The Gregorian calendar is in common use today. See also JULIAN DAY.

calendar day. . The period from midnight to midnight. The calendar day is 24 hours of mean solar time in length and coincides with the civil day unless a time change occurs during a day.

calendar line. . *British terminology*. See DATE LINE.

calendar month. . The month of the calendar, varying from 28 to 31 days in length.

calendar year. . The year of the calendar. Common years have 365 days and leap years 366 days. Each year exactly divisible by 4 is a leap year, except century years (1800, 1900, etc.), which must be exactly divisible by 400 (2000, 2400, etc.) to be leap years. The calendar year is based on the tropical year. Also called CIVIL YEAR.

calibrate. , *n*. To determine or rectify the scale graduations of an instrument.

calibration card. . See under CALIBRATION TABLE.

calibration correction. . The value to be added to or subtracted from the reading of an instrument to obtain the correct reading.

calibration error. . The error in an instrument due to imperfection of calibration or maladjustment of its parts. Also called SCALE ERROR.

calibration radiobeacon. . A special radiobeacon operated primarily for calibrating shipboard radio direction finders. These radiobeacons transmit either continuously during scheduled hours or upon request.

calibration table. . A list of calibration corrections or calibrated values. A card having such a table on it is called a CALIBRATION CARD.

California Current. . A North Pacific Ocean current flowing southeastward along the west coast of North America from a point west of Vancouver Island to the west of Baja (Lower) California where it gradually widens and curves southward and southwestward, to continue as the westerly flowing PACIFIC NORTH EQUATORIAL CURRENT. The California Current is the southern branch of the Aleutian Current, augmented by the North Pacific Current, and forms the eastern part of the general clockwise oceanic circulation of the North Pacific Ocean. Although usually described as a permanent ocean current, the California Current is actually a poorly defined and variable flow easily influenced by the winds. See also MEXICO CURRENT.

California Norther. . See NORTHER.

Callipic cycle. . A period of four Meteoric cycles equal to 76 Julian years of 27759 days. Devised by Callipus, a Greek astronomer, about 350 B.C., as a suggested improvement on the Meteoric cycle for a period in which new and full moon would recur on the same day of the year. Taking the length of the synodical month as 29.530588 days, there are 940 lunations in the Callipic cycle with about 0.25 day remaining.

calm. , *adj*. In a state of calm; without motion.

calm. , *n*. 1. Absence of appreciable wind; specifically, force 0 (less than 1 knot or 1 mile per hour) on the Beaufort wind scale. 2. The state of the sea when there are no waves.

calm belt. . 1. The doldrum sides of the trade winds, called *calms of Cancer* and *calms of Capricorn*, respectively.

calving. , *n*. The breaking away of a mass of ice from an ice wall, ice front, or iceberg.

camanchaca. , *n*. See GARUA.

camel. , *n*. A float used as a fender. Also called BREASTING FLOAT.

canal. , *n*. 1. An artificial waterway for navigation. 2. A long, fairly straight natural channel with steep sloping sides. 3. Any watercourse or channel. 4. A sluggish coastal stream, as used locally on the Atlantic coast of the U.S.

Canary Current. . The southern branch of the North Atlantic Current (which divides on the eastern side of the ocean); it moves south past Spain and southwestward along the Northwest coast of Africa and past the Canary islands. In the vicinity of the Cape Verde Islands, it divides into two branches, the western branch augmenting the Atlantic North Equatorial Current and the Eastern branch curving southward and continuing as the GUINEA CURRENT. The Canary Current forms the southeastern part of the general clockwise oceanic circulation of the North Atlantic Ocean. Also called the Canaries Current.

can buoy. . An unlighted buoy of which the upper part of the body (above the waterline), or the larger part of the superstructure has the shape of a cylinder or nearly so. Also called CYLINDRICAL BUOY.

candela. , *n*. The base unit of luminous intensity in the International System of Units (SI). It is the luminous intensity, in the perpendicular direction, of a surface of 1/600,000 square meter of a blackbody at the temperature of freezing platinum, under a pressure of 101,325 newtons per square meter. The definition was adopted by the Thirteenth General Conference on Weights and Measures (1967).

candela per square meter. . The derived unit of luminance in the International System of Units.

candlepower. , *n*. Luminous intensity expressed in candelas.

canyon. , *n*. On the sea floor, a relatively narrow, deep depression with steep sides, the bottom of which generally has a continuous slope.

cap cloud. . 1. A cloud resting on the top of an isolated mountain peak. The cloud appears stationary, but actually is being continually formed to windward and dissipated to leeward. A similar cloud over a mountain ridge is called a CREST CLOUD. See also BANNER CLOUD. 2. False cirrus over a towering cumulus, in the form of a cap or hood. See also SCARF CLOUD.

cape. , *n*. A relatively extensive land area jutting seaward from a continent, or large island, which prominently marks a change in or interrupts notably the coastal trend.

Cape Breton Current. . Originating in the Gulf of St. Lawrence, the Cape Breton Current flows southeastward in the southwestern half of Cabot Strait, and merges with the Labrador Current Extension. It may be augmented by a branch of the constant but tide influenced Gaspe' Current to the northwest.

cape doctor. . The strong southeast wind which blows on the South African coast. Also called DOCTOR.

Cape Horn Current. . An ocean current that flows continuously eastward close to the tip of South America. It enters Drake Passage, at about longitude 70° W, in a 150-mile-wide band, with observed surface speeds to 2.4 knots. The current veers north-northeastward; when it crosses longitude 65° W, the current has narrowed to a width of about 85 miles, and its speed has decreased considerably. The current continues as the FALKLAND CURRENT.

card. . An element of a computer consisting of the hard surface on which components are mounted. A completed card performs one or more specific functions, such as graphics.

cardinal heading. . A heading in the direction of any of the cardinal points of the compass. See also INTERCARDINAL HEADING.

cardinal mark. . An IALA aid to navigation intended to show the location of a danger to navigation based on its position relative to the danger. Its distinguishing features are black double-cone topmarks and black and yellow horizontal bands.

cardinal point. . Any of the four principal directions; north, east, south, or west. Directions midway between cardinal points are called INTERCARDINAL POINTS.

cardinal system. . A system of aids to navigation in which the shape, color, and number distinction are assigned in accordance with location relative to the nearest obstruction. The cardinal points delineate the sectors for aid location. The cardinal system is particularly applicable to a region having numerous small islands and isolated dangers. In the LATERAL SYSTEM, used in United States waters, the aids are assigned shape, color, and number distinction as a means of indicating navigable waters.

cardioid. , *n.* The figure traced by a point on a circle which rolls around an equal fixed circle.

cargo transfer area. . See under CARGO TRANSSHIPMENT AREAS.

cargo transshipment area. . An area generally outside port limits that is specifically designated as suitable for the transshipment of oil or other materials from large ships to smaller ones. As the purpose of transshipment is usually to reduce the draft of the larger vessel to allow her to proceed to port, the operation is often known as lightening and the area may be called lightening area or cargo transfer area.

Caribbean Current. . An ocean current flowing westward through the Caribbean Sea to the Yucatan Channel. It is formed by the comingling of part of the waters of the Atlantic North Equatorial Current with those of the Guiana Current.

carrier. , *n.* 1. A radio wave having at least one characteristic which may be varied from a known reference value by modulation. 2. The part of a modulated wave that corresponds in a specified manner to the unmodulated wave. 3. In a frequency stabilized system, the sinusoidal component of a modulated wave; or the output of a transmitter when the modulating wave is made zero; or a wave generated at a point in the transmitting system and subsequently modulated by the signal; or a wave generated locally at the receiving terminal which, when combined with the sidebands in a suitable detector, produces the modulating wave. Also called CARRIER WAVE.

carrier frequency. . 1. The frequency of the unmodulated fundamental output of a radio transmitter. 2. In a periodic carrier, the reciprocal of its period. The frequency of a periodic pulse carrier often is called PULSE REPETITION FREQUENCY.

carrier power. . See under POWER (OF A RADIO TRANSMITTER).

carrier wave. . See CARRIER.

cartesian coordinates. . Magnitudes defining a point relative to two intersecting lines, called AXES. The magnitudes indicate the distance from each axis, measured along a parallel to the other axis. If the axes are perpendicular, the coordinates are rectangular; if not perpendicular, they are oblique coordinates.

cartographer. , *n.* One who designs and constructs charts or maps.

cartographic feature. . A natural or cultural object shown on a map or chart by a symbol or line. See also TOPOGRAPHY.

cartography. , *n.* The art and science of making charts or maps.

cartometer. , *n.* A device consisting of a small wheel and a calibrated dial used to measure distances on a map by following the desired route.

cartouche. , *n.* A panel of a map, often with decoration, enclosing the title, scale, publishing information, and other notes.

cask buoy. . A buoy in the shape of a cask.

Cassegrainian telescope. . A reflecting telescope in which the incoming light is reflected from the primary mirror onto a secondary mirror and back through a small central aperture in the primary mirror. See also NEWTONIAN TELESCOPE.

cast. , *n., t.* 1. To turn a ship in her own length. 2. To turn a ship to a desired direction without gaining headway or sternway. 3. To take a sounding with the lead.

catamaran. , *n.* 1. A double-hulled vessel. 2. A raft consisting of a rectangular frame attached to two parallel cylindrical floats and which may be used for working alongside a ship. See also CAMEL.

catenary. , *n.* The curve formed by a uniform cable supported only at its ends. Navigators are concerned with the catenary of overhead cables which determines clearance underneath, and the catenary of the anchor rode, which in part determines holding power and swing circle.

cathode. , *n.* 1. The electrode through which a primary stream of electrons enters the interelectrode space. 2. The general term for a negative electrode. See also ANODE.

cathode ray. . A stream of electrons emitted from the cathode of any vacuum tube, but normally used in reference to special purpose tubes designed to provide a visual display.

cathode-ray tube (CRT). . A vacuum tube in which the instantaneous position of a sharply focused electron beam, deflected by means of electrostatic or electromagnetic fields, is indicated by a spot of light produced by impact of the electrons on a fluorescent screen at the end of the tube opposite the cathode. Used in radar displays.

catoptric light. . A light concentrated into a parallel beam by means of one or more reflectors. One so concentrated by means of refracting lens or prisms is a DIOPTRIC LIGHT.

cat's paw. . A puff of wind; a light breeze affecting a small area, as one that causes patches of ripples on the surface the water.

causeway. , *n.* A raised earthen road across wet ground or water. See also BRIDGE definition 2; VIADUCT.

cautionary characteristic. . Of a light, a unique characteristic which can be recognized as imparting a special cautionary significance e.g., a quick flashing characteristic phase indicating a sharp turn in a channel.

cautionary note. . Information calling special attention to some fact, usually a danger area, shown on a map or chart.

caver, kaver. , *n.* A gentle breeze in the Hebrides.

cavitation. . The formation of bubbles in a liquid which occurs when the static pressure becomes less than the fluid vapor pressure; it usually occurs from rotating propellers and is acoustically very noisy.

cay, kay. , *n.* A low, flat, tropical or sub-tropical island of sand and coral built up on a reef lying slightly above high water. Also called KEY.

C-band. . A radiofrequency band of 3,900 to 6,200 megahertz. This band overlaps the S- and X-bands. See also FREQUENCY.

ceiling. , *n.* The height above the earth's surface of the lowest layer of generally solid clouds, not classified as thin or partial.

celestial. , *adj.* Of or pertaining to the heavens.

celestial body. . Any aggregation of matter in space constituting a unit for astronomical study, as the sun, moon, a planet, comet, star, nebula, etc. Also called HEAVENLY BODY.

celestial concave. . See CELESTIAL SPHERE.

celestial coordinates. . Any set of coordinates used to define a point on the celestial sphere. The horizon, celestial equator, and the ecliptic systems of celestial coordinates are based on the celestial horizon, celestial equator, and the ecliptic, respectively, as the primary great circle.

celestial equator. . The primary great circle of the celestial sphere, everywhere 90° from the celestial poles; the intersection of the extended plane of the equator and the celestial sphere. Also called EQUINOCTIAL.

celestial equator system of coordinates. . A set of celestial coordinates based on the celestial equator as the primary great circle. Also called EQUINOCTIAL SYSTEM OF COORDINATES.

celestial fix. . A fix established by means of two or more celestial bodies.

celestial globe. . See STAR GLOBE.

celestial horizon. . That circle of the celestial sphere formed by the intersection of the celestial sphere and a plane through the center of the earth and perpendicular to the zenith-nadir line. Also called RATIONAL HORIZON. See also HORIZON.

celestial latitude. . Angular distance north or south of the ecliptic; the arc of a circle of latitude between the ecliptic and a point on the celestial sphere, measured northward or southward from the ecliptic through 90°, and labeled N or S indicate the direction of measurement.

celestial line of position. . A line of position determined by means of a celestial body.

celestial longitude. . Angular distance east of the vernal equinox, along the ecliptic; the arc of the ecliptic or the angle at the ecliptic pole between the circle of latitude of the vernal equinox at the circle of latitude of a point on the celestial sphere, measured eastward from the circle of latitude of the vernal equinox, through 360°.

celestial mechanics. . The study of the motions of celestial bodies under the influence of gravitational fields.

celestial meridian. . A great circle of the celestial sphere, through the celestial poles and the zenith. The expression usually refers to the upper branch, that half from pole to pole which passes through the zenith; the other half being called the lower branch. The celestial meridian coincides with the hour circle through the zenith and the vertical circle through the elevated pole.

celestial navigation. . Navigation by celestial bodies.

celestial observation. . Observation of celestial phenomena. The expression is applied in navigation principally to the measurement of the altitude of a celestial body, and sometimes to measurement of azimuth, or to both altitude azimuth. The expression may also be applied to the data obtained by such measurement. Also called SIGHT in navigation usage.

celestial parallel. . See PARALLEL OF DECLINATION.

celestial pole. . Either of the two points of intersection section of the celestial sphere and the extended axis of the earth, labeled N or S to indicate whether the north celestial pole or the south celestial pole.

celestial sphere. . An imaginary sphere of infinite radius concentric with the earth, on which all celestial bodies except the earth are imagined to be projected.

celestial triangle. . A spherical triangle on the celestial sphere, especially the navigational triangle.

Celsius temperature. . The designation given to the temperature measured on the International Practical Temperature Scale with the zero taken as 0.01° below the triple point of water. Normally called CENTIGRADE TEMPERATURE, but the Ninth General Conference of Weights and Measures, held in October 1948, adopted the name *Celsius* in preference to *centigrade,* to be consistent with naming other temperature scales after their inventors, and to avoid the use of different names in different countries. On the original Celsius scale, invented in 1742 by a Swedish astronomer named Andres Celsius, the numbering was the reverse of the modern scale, 0°C representing the boiling point of water, and 100° C its freezing point.

center frequency. . See ASSIGNED FREQUENCY.

centering control. . On a radar indicator, a control used to place the sweep origin at the center of the plan position indicator.

centering error. . Error in an instrument due to inaccurate pivoting of a moving part, as the index arm of a marine sextant. Also called ECCENTRIC ERROR.

center line. . 1. The locus of points equidistant from two reference points or lines. 2. *(Usually centerline)* The line separating the port and starboard sides of a vessel, center of buoyancy. The geometric center of the immersed portion of the hull and appendages of a floating vessel All buoyant forces may be resolved into one resultant force acting upwards at this point.

center of gravity. . The point in any body at which the force of gravity may be considered to be concentrated. Same as CENTER OF MASS in a uniform gravitational field.

center of mass. . The point at which all the given mass of a body or bodies may be regarded as being concentrated as far as motion is concerned. Commonly called CENTER OF GRAVITY.

centi-. . A prefix meaning *one-hundredth.*

centibar. , *n.* One-hundredth of a bar; 10 millibars.

centigrade temperature. . See under CELSIUS TEMPERATURE.

centimeter. , *n.* One-hundredth of a meter.

centimeter-gram-second system. . A system of units based on the centimeter as the unit of length, the gram as the unit of mass, and the mean solar second as the unit of time. Its units with special names include the erg, the dyne, the gauss, and the oersted. See also INTERNATIONAL SYSTEM OF UNITS.

centimetric wave. . A super high frequency radio wave, approximately 0.01 to 0.1 meter in length (3 to 30 gigaHertz). See also ULTRA SHORT WAVE.

central force. . A force which for purposes of computation can be considered to be concentrated at one central point with its intensity at any other point being a function of the distance from the central point. Gravitation is considered as a central force in celestial mechanics.

central force field. . The spatial distribution of the influence of a central force.

central force orbit. . The theoretical orbit achieved by a particle of negligible mass moving in the vicinity of a point mass with no other forces acting; an unperturbed orbit.

central processing unit (CPU). . The computer chip which is the brain of a computer, which runs PROGRAMS and processes DATA; also the container in which the CPU is located, along with many other associated devices such as the power supply, disk drives, etc., distinct from the MONITOR and other peripherals.

central standard time. . See STANDARD TIME.

centrifugal force. . The force acting on a body or part of a body moving under constraint along a curved path, tending to force it outward from the center of revolution or rotation. The opposite is CENTRIPETAL FORCE.

centripetal force. . The force directed toward the center of curvature, which constrains a body to move in a curved path. The opposite is CENTRIFUGAL FORCE.

chain. , *n.* A group of associated stations of a radionavigation system. A Loran C chain consists of a master station and two to four secondary stations.

chains. . The platform or station from which soundings are taken with a hand lead.

chain signature. . See under GROUP REPETITION INTERVAL.

chalk. , *n.* Soft earthy sandstone of marine origin, composed chiefly of minute shells. It is white, gray, or buff in color. Part of the ocean bed and shores and composed of chalk, notably the "white cliffs of Dover," England.

challenge. , *n.* A signal transmitted by a interrogator.

challenge. , *v. t.* To cause an interrogator to transmit a signal which puts a transponder into operation.

challenger. , *n.* See INTERROGATOR.

chance error. . See RANDOM ERROR.

change of the moon. . The time of new moon. See also PHASES OF THE MOON.

change of tide. . A reversal of the direction of motion (rising or falling) of a tide. The expression is also sometimes applied somewhat loosely to a reversal in the set of a tidal current. Also called TURN OF THE TIDE.

channel. , *n.* 1. The part of a body of water deep enough for navigation through an area otherwise not suitable. It is usually marked by a single or double line of buoys and sometimes by ranges. 2. The deepest part of a stream, bay, or strait, through which the main current flows. 3. A name given to certain large straits, such as the English Channel. 4. A hollow bed through which water may run. 5. A band of radio frequencies within which a radio station must maintain its modulated carrier frequency to prevent interference with stations on adjacent channels. Also called FREQUENCY CHANNEL.

channel buoy. . A buoy marking a channel.

channel light. . A light either on a fixed support or on a buoy, marking the limit of a navigable channel. In French, the term FEU DE RIVE is commonly used for a channel light on a fixed support.

characteristic. , *n.* 1. The color and shape of a daymark or buoy or the color and period of a light used for identifying the aid. See also CHARACTERISTIC COLOR, CHARACTERISTIC PHASE. 2. The identifying signal transmitted by a radiobeacon. 3. That part of a logarithm (base 10) to the left of the decimal point. That part of a logarithm (base 10) to the right of the decimal point is called the MANTISSA. 4. A quality, attribute, or distinguishing property of anything.

characteristic color. . The unique identifying color of a light.

characteristic frequency. . A frequency which can be easily identified and measured in a given emission.

characteristic phase.. Of a light, the sequence and length of light and dark periods by which a navigational light is identified, i.e., whether fixed, flashing, interrupted quick flashing, etc. See also CAUTIONARY CHARACTERISTIC.

characteristics of a light.. The sequence and length of light and dark periods and the color or colors by which a navigational light is identified.

character of the bottom.. See BOTTOM CHARACTERISTICS.

chart., *n.* A map intended primarily for navigational use by aircraft or vessels.

chart amendment patch.. See CHARTLET, definition 2.

chart catalog.. A list or enumeration of navigational charts, sometimes with index charts indicating the extent of coverage of the various navigational charts.

chart classification by scale.. 1. Charts are constructed on many different scales, ranging from about 1:2,500 to 1:14,000,000 (and even smaller for some world charts). Small-scale charts are used for voyage planning and offshore navigation. Charts of larger scale are used as the vessel approaches land. Several methods of classifying charts according to scale are in use in various nations. The following classifications of nautical charts are those used by the National Ocean Survey: Sailing charts are the smallest scale charts used for planning, fixing position at sea, and for plotting while proceeding on a long voyage. The scale is generally smaller than 1:600,000. The shoreline and topography are generalized and only offshore soundings, the principal navigational lights, outer buoys, and landmarks visible at considerable distances are shown. General charts are intended for coastwise navigation outside of outlying reefs and shoals. The scales range from about 1:150,000 to 1:600,000. Coast (coastal) charts are intended for inshore coastwise navigation where the course may lie inside outlying reefs and shoals, for entering or leaving bays and harbors of considerable width, and for navigating large inland waterways. The scales range from about 1:50,000 to 1:150,000. Harbor charts are intended for navigation and anchorage in harbors and small waterways. The scale is generally larger than 1:50,000. 2. The classification system used by the National Imagery and Mapping Agency differs from the system in definition 1 above in that the sailing charts are incorporated in the general charts classification (smaller than about 1:150,000); those coast charts especially useful for approaching more confined waters (bays, harbors) are classified as approach charts.

chart comparison unit.. An optical device used to superimpose the plan position indicator radar picture on a navigational chart.

chart convergence.. Convergence of the meridians as shown on a chart.

chart datum.. See CHART SOUNDING DATUM.

chart desk.. A flat surface on which charts are spread out, usually with stowage space for charts and other navigating equipment below the plotting surface. One without stowage space is called a CHART TABLE.

charted depth.. The vertical distance from the chart sounding datum to the bottom.

charthouse.. A room, usually adjacent to or on the bridge, where charts and other navigational equipment are stored, and where navigational computations, plots, etc., may be made. Also called CHARTROOM.

chartlet., *n.* 1. A corrected reproduction of a small area of a nautical chart which is pasted to the chart for which it is issued. These chartlets are disseminated in *Notice to Mariners* when the corrections are too numerous or of such detail as not to be feasible in printed form. Also called BLOCK, BLOCK CORRECTION, CHART AMENDMENT PATCH.

chart portfolio.. A systematic grouping of nautical charts covering a specific geographical area.

chart projection.. See MAP PROJECTION.

chart reading.. Interpretation of the symbols, lines, abbreviations, and terms appearing on charts. May be called MAP READING when applied to maps generally.

chartroom., *n.* See CHARTHOUSE.

chart scale.. The ratio between a distance on a chart and the corresponding distance represented as a ratio such as 1:80,000 (natural scale), or 30 miles to an inch (numerical scale). May be called MAP SCALE when applied to any map. See also REPRESENTATIVE FRACTION.

chart sounding datum.. The tidal datum to which soundings and drying heights on a chart are referred. It is usually taken to correspond to a low water stage of the tide. Often shortened to CHART DATUM, especially when it is clear that reference is not being made to a geodetic datum.

chart symbol.. A character, letter, or similar graphic representation used on a chart to indicate some object, characteristic, etc. May be called MAP SYMBOL when applied to any map.

chart table.. A flat surface on which charts are spread out, particularly one without stowage space below the plotting surface. One provided with stowage space is usually called a CHART DESK.

Charybdis., *n.* See GALOFARO.

chasm., *n.* A deep breach in the earth's surface; an abyss; a gorge; a deep canyon.

check bearing.. An additional bearing, using a charted object other than those used to fix the position, observed and plotted in order to insure that the fix is not the result of a blunder.

cheese antenna.. An antenna consisting of a mirror in the shape of part of a parabolic cylinder bounded by two parallel plates normal to the cylinder axis, and of an antenna feed placed on or near the focal point.

Chile Current.. See under PERU CURRENT.

chimney., *n.* A label on a nautical chart which indicates a relatively small smokestack.

chip., *n.* 1. An integrated circuit. 2. The length of time to transmit a "0" or "1" in a binary pulse code.

chip log.. A historical speed measuring device consisting of a weighted wooden quadrant (quarter of a circle) attached to a bridle in such a manner that it will float in a vertical position, and a line with equally spaced knots, usually each 47 feet 3 inches apart. Speed is measured by casting the quadrant overboard and counting the number of knots paid out in a unit of time, usually 28 seconds.

chip rate., *n.* The number of chips per second. See CHIP.

chopped response.. See CHOPPING.

chopping., *n.* The rapid and regular on and off switching of a transponder, for recognition purposes.

choppy., *adj.* description of short, breaking waves.

chord., *n.* A straight line connecting two points on a curve.

chromatic aberration.. See under ABERRATION, definition 2.

chromosphere., *n.* A thin layer of relatively transparent gases above the photosphere of the sun.

chromospheric eruption.. See SOLAR FLARE.

chronograph., *n.* An instrument for producing a graphical record of time as shown by a clock or other device. The chronograph produces a double record: the first is made by the associated clock and forms a continuous time scale with significant marks indicating periodic beats of the time keepers; the second is made by some external agency, human or mechanical, and records the occurrence of an event or a series of events. The time interval of such occurrences are read on the time scale made by the clock. See also BREAK-CIRCUIT CHRONOMETER.

chronogram., *n.* The record of a chronograph.

chronometer., *n.* A timepiece with a nearly constant rate. It is customarily used for comparison of watches and clocks to determine their errors. A chronometer is usually set approximately to Greenwich mean time and not reset as the craft changes time zones. A hack chronometer is one which has failed to meet the exacting requirements of a standard chronometer, and is used for timing observations of celestial bodies. Hack chronometers are seldom used in modern practice, any chronometer failing to meet the requirements being rejected. See also CHRONOMETER WATCH.

chronometer correction.. The amount that must be added algebraically to the chronometer time to obtain the correct time. Chronometer correction is numerically equal to the chronometer error, but of opposite sign.

chronometer error.. The amount by which chronometer time differs from the correct time to which it was set, usually Greenwich mean time. It is usually expressed to an accuracy of 1s and labeled fast (F) or slow (S) as the chronometer time is later or earlier, respectively, than the correct time. CHRONOMETER ERROR and CHRONOMETER CORRECTION are numerically the same, but of opposite sign. See also WATCH ERROR.

chronometer rate. . The amount gained or lost by a chronometer in a unit of time. It is usually expressed in seconds per 24 hours, to an accuracy of 0.1s, and labeled gaining or losing, as appropriate, when it is sometimes called DAILY RATE.

chronometer time. . The hour of the day as indicated by a chronometer. Shipboard chronometers are generally set to Greenwich mean time. Unless the chronometer has a 24-hour dial, chronometer time is usually expressed on a 12-hour cycle and labeled AM or PM.

chronometer watch. . A small chronometer, especially one with an enlarged watch-type movement.

chubasco. , *n.* A very violent wind and rain squall attended by thunder and vivid lightning often encountered during the rainy season along the west coast of Central America.

churada. , *n.* A severe rain squall in the Mariana Islands during the northeast monsoon. They occur from November to April or May, especially from January through March.

cierzo. , *n.* See MISTRAL.

cinders. , *n., pl.* See SCORIAE.

circle. , *n.* 1. A plane closed curve all points of which are equidistant from a point within, called the center. A great circle is the intersection of a sphere and a plane through its center; it is the largest circle that can be drawn on a sphere. A small circle is the intersection of a sphere and a plane which does not pass through its center. See also PARALLEL OF ALTITUDE, PARALLEL OF DECLINATION, PARALLEL OF LATITUDE; AZIMUTH CIRCLE, BEARING CIRCLE, DIURNAL CIRCLE, EQUATOR, HOUR CIRCLE, PARASELENIC CIRCLES, POSITION CIRCLE, SPEED CIRCLE, VERTICAL CIRCLE. 2. A section of a plane, bounded by a curve all points of which are equidistant from a point within, called the center.

circle of declination. . See HOUR CIRCLE.

circle of equal altitude. . A circle on the surface of the earth, on every point of which the altitude of a given celestial body is the same at a given instant. The center of this circle is the geographical position of the body, and the great circle distance from this pole to the circle is the zenith distance of the body. See PARALLEL OF ALTITUDE.

circle of equal declination. . See PARALLEL OF DECLINATION.

circle of equivalent probability. . A circle with the same center as an error ellipse of specified probability and of such radius that the probability of being located within the circle is the same as the probability of being located within the ellipse. See also CIRCULAR ERROR PROBABLE.

circle of latitude. . A great circle of the celestial sphere through the ecliptic poles and along which celestial latitude is measured.

circle of longitude. . See PARALLEL OF LATITUDE, definition 2.

circle of perpetual apparition. . The circle of the celestial sphere, centered on the polar axis and having a polar distance from the elevated pole approximately equal to the latitude of the observer, within which celestial bodies do not set. The circle within which bodies do not rise is called the CIRCLE OF PERPETUAL OCCULTATION.

circle of perpetual occultation. . The circle of the celestial sphere, centered on the polar axis and having a polar distance from the depressed pole approximately equal to the latitude of the observer, within which celestial bodies do not rise. The circle within which bodies do not set Is called the CIRCLE OF PERPETUAL APPARITION.

circle of position. . A circular line of position. The expression is most frequently used with reference to the circle of equal altitude surrounding the geographical position of a celestial body. Also called POSITION CIRCLE.

circle of right ascension. . See HOUR CIRCLE.

circle of uncertainty. . A circle having as its center a given position and as its radius the maximum likely error of the position—a circle within which a vessel is considered to be located. See also CIRCLE OF EQUAL PROBABILITY, CIRCLE OF POSITION, POSITION CIRCLE.

circle of visibility. . The circle surrounding an aid to navigation in which the aid is visible. See also VISUAL RANGE (OF A LIGHT).

circle sheet. . A chart with curves enabling a graphical solution of the three-point problem rather than using a three-arm protractor. Also called SEXTANT CHART, STANDARD CIRCLE SHEET.

circuit. , *n.* 1. An electrical path between two or more points. 2. Conductors connected together for the purpose of carrying an electric current. 3. A connected assemblage of electrical components, such as resistors, capacitors, and inductors.

circular error probable. . 1. In a circular normal distribution (the magnitudes of the two one-dimensional input errors are equal and the angle of cut is 90°), the radius of the circle containing 50 percent of the individual measurements being made, or the radius of the circle inside of which there is a 50 percent probability of being located. 2. The radius of a circle inside of which there is a 50 percent probability of being located even though the actual error figure is an ellipse. That is, it is the radius of a circle of equivalent probability when the probability is specified as 50 percent. See also ERROR ELLIPSE, CIRCLE OF EQUIVALENT PROBABILITY. Also called CIRCULAR PROBABLE ERROR.

circular fix. . The designation of any one of the erroneous fix positions obtained with a revolver or swinger.

circularly polarized wave. . An electromagnetic wave which can be resolved into two plane polarized waves which are perpendicular to each other and which propagate in the same direction. The amplitudes of the two waves are equal and in time-phase quadrature. The tip of the component of the electric field vector in the plane normal to the direction of propagation describes a circle. See also ELLIPTICALLY POLARIZED WAVE.

circular normal distribution. . A two-dimensional error distribution defined by two equal single axis normal distributions, the axes being perpendicular. The error figure is a circle.

circular probable error. . See CIRCULAR ERROR PROBABLE.

circular radiobeacon. . See under RADIOBEACON.

circular velocity. . The magnitude of the velocity required of a body at a given point in a gravitational field which will result in the body following a circular orbital path about the center of the field. With respect to circular velocities characteristic of the major bodies of the solar system, this is defined for a circular orbit at the surface of the body in question. Circular velocity equals escape velocity divided by the square root of 2.

circumference. , *n.* 1. The boundary line of a circle or other closed plane curve or the outer limits of a sphere or other round body. 2. The length of the boundary line of a circle or closed plane curve or of the outer limits of a sphere or other rounded body. The circumference of a sphere is the circumference of any great circle on the sphere.

circumlunar. , *adj.* Around the moon, generally applied to trajectories.

circummeridian altitude. . See EX-MERIDIAN ALTITUDE.

circumpolar. , *adj.* Revolving about the elevated pole without setting. A celestial body is circumpolar when its polar distance is approximately equal to or less than the latitude of the observer. The actual limit is extended somewhat by the combined effect of refraction, semidiameter parallax, and the height of the observer's eye above the horizon.

circumscribed halo. . A halo formed by the junction of the upper and lower tangent arcs of the halo of 22°.

circumzenithal arc. . A brilliant rainbow-colored arc of about a quarter of a circle with its center at the zenith and about 46° above the sun. It is produced by refraction and dispersion of the sun's light striking the top of prismatic ice crystals in the atmosphere. It usually lasts for only a few minutes. See also HALO.

cirriform. , *adj.* Like cirrus; more generally, descriptive of clouds composed of small particles, mostly ice crystals, which are fairly widely dispersed, usually resulting in relative transparency and whiteness, and often producing halo phenomena not observed with other cloud forms. Irisation may also be observed. Cirriform clouds are high clouds. As a result, when near the horizon, their reflected light traverses a sufficient thickness of air to cause them often to take on a yellow or orange tint even during the midday period. On the other hand, cirriform clouds near the zenith always appear whiter than any other clouds in that part of the sky. With the sun on the horizon, this type of cloud is whitish, while other clouds may be tinted with yellow or orange; when the sun sets a little below the horizon, cirriform clouds become yellow, then pink or red- and when the sun is well below the horizon, they are gray. All species and varieties of cirrus, cirrocumulus, and cirrostratus clouds are cirriform in nature. See also CUMULIFORM, STRATIFORM.

cirro-. . A prefix used in cloud classification to indicate the highest of three levels generally recognized. See also ALTO-.

cirrocumulus. , *n.* A principal cloud type (cloud genus), appearing as a thin, white patch of cloud without shadows, composed of very small elements in the form of grains, ripples, etc. The elements may be merged or separate, and more or less regularly arranged; they subtend an angle of less than 1° when observed at an angle of more than 30° above the horizon. Holes or rifts often occur in a sheet of cirrocumulus. Cirrocumulus may be composed of highly super cooled water droplets, as well as small ice crystals, or a mixture of both; usually, the droplets are rapidly replaced by ice crystals. Sometimes corona or irisation may be observed. Mamma may appear. Small virga may fall, particularly from cirrocumulus castellanus and floccus. Cirrocumulus, as well as altocumulus, often forms in a layer of cirrus and/or cirrostratus. In middle and high latitudes, cirrocumulus is usually associated in space and time with cirrus and/or cirrostratus; this association occurs less often in low latitudes. Cirrocumulus differs from these other cirriform clouds in that it is not on the whole fibrous, or both silky and smooth; rather, it is rippled and subdivided into little cloudlets. Cirrocumulus is most often confused with altocumulus. It differs primarily in that its constituent elements are very small and are without shadows. The term *cirrocumulus* is not used for incompletely developed small elements such as those on the margin of a sheet of altocumulus, or in separate patches at that level. See also CIRRIFORM, CLOUD CLASSIFICATION.

cirrostratus. , *n.* A principal cloud type (cloud genus), appearing as a whitish veil, usually fibrous but sometimes smooth, which may totally cover the sky, and which often produces halo phenomena, either partial or complete. Sometimes a banded aspect may appear, but the intervals between the bands are filled with thinner cloud veil. The edge of a veil of cirrostratus may be straight and clear-cut, but more often it is irregular and fringed with cirrus. Some of the ice crystals which comprise the cloud are large enough to fall, and thereby produce a fibrous aspect. Cirrostratus occasionally may be so thin and transparent as to render it nearly indiscernible, especially through haze or at night. At such times, the existence of a halo may be the only revealing feature. The angle of incidence of illumination upon a cirrostratus layer is an important consideration in evaluating the identifying characteristics. When the sun is high (generally above 50° altitude), cirrostratus never prevents the casting of shadows by terrestrial objects, and a halo might be completely circular. At progressively lower altitudes of the sun, halos become fragmentary and light intensity noticeably decreases. Cirrostratus may be produced by the merging of elements of cirrus; from cirrocumulus; from the thinning of altostratus; or from the anvil of cumulonimbus. Since cirrostratus and altostratus form from each other, it frequently is difficult to delineate between the two. In general, altostratus does not cause halo phenomena, is thicker than cirrostratus, appears to move more rapidly, and has a more even optical thickness. When near the horizon, cirrostratus may be impossible to distinguish from cirrus. See also CIRRIFORM, CLOUD CLASSIFICATION.

cirrus. , *n.* A principal cloud type (cloud genus) composed of detached cirriform elements in the form of delicate filaments or white (or mostly white) patches, or of narrow bands. These clouds have a fibrous aspect and/or a silky sheen. Many of the ice crystal particles of cirrus are sufficiently large to acquire an appreciable speed of fall; therefore, the cloud elements have a considerable vertical extent. Wind shear and variations in particle size usually cause these fibrous trails to be slanted or irregularly curved. For this reason, cirrus does not usually tend, as do other clouds, to appear horizontal when near the horizon. Because cirrus elements are too narrow, they do not produce a complete circular halo. Cirrus often evolves from virga of cirrocumulus or altocumulus, or from the upper part of cumulonimbus. Cirrus may also result from the transformation of cirrostratus of uneven optical thickness, the thinner parts of which dissipate. It may be difficult at times to distinguish cirrus from cirrostratus (often impossible when near the horizon); cirrostratus has a much more continuous structure, and if subdivided, its bands are wider. Thick cirrus (usually cirrus spissatus) is differentiated from patches of altostratus by its lesser extension and white color. The term *cirrus* is frequently used for all types of cirriform clouds. See also CIRRIFORM, CLOUD CLASSIFICATION.

cirrus spissatus. . See FALSE CIRRUS.

cislunar. , *adj.* Of or pertaining to phenomena, projects, or activity in the space between the earth and moon, or between the earth and the moon's orbit.

civil day. . A mean solar day beginning at midnight. See also CALENDAR DAY.

civil noon. . United States terminology from 1925 through 1952. See MEAN NOON.

civil time. . United States terminology from 1925 through 1952. See MEAN TIME.

civil twilight. . The period of incomplete darkness when the upper limb of the sun is below the visible horizon, and the center of the sun is not more than 6° below the celestial horizon.

civil year. . A year of the Gregorian calendar of 365 days in common years, or 366 days in leap years.

clamp screw. . A screw for holding a moving part in place, as during an observation or reading, particularly such a device used in connection with the tangent screw of a marine sextant.

clamp screw sextant. . A marine sextant having a clamp screw for controlling the position of the tangent screw.

clapper. , *n.* A heavy pendulum suspended inside a bell which sounds the bell by striking it.

Clarke ellipsoid of 1866. . The reference ellipsoid adopted by the U.S. Coast and Geodetic Survey in 1880 for charting North America. This ellipsoid is not to be confused with the Clarke ellipsoid of 1880, which was the estimate of the size and shape of the earth at that time by the English geodesist A. R. Clarke. For the Clarke ellipsoid of 1866, the semimajor axis is 6,378,206.4 meters, the semiminor axis is 6,356,583.8 meters, and the flattening or ellipticity is 1/294.98. Also called CLARKE SPHEROID OF 1866.

Clarke ellipsoid of 1880. . The reference ellipsoid of which the semimajor axis is 6,378,249.145 meters, the semiminor axis is 6,356,514.870 meters and the flattening or ellipticity is 1/293.65. This ellipsoid should not be confused with the CLARKE ELLIPSOID OF 1866. Also called CLARKE SPHEROID OF 1880.

Clarke spheroid of 1866. . See CLARKE ELLIPSOID OF 1866.

Clarke spheroid of 1880. . See CLARKE ELLIPSOID OF 1880.

classification of radar echoes. . When observing a radarscope having a stabilized relative motion display, the echoes (targets) may be classified as follows as an aid in rapid predictions of effects of evasive action on the compass direction of relative movement: an up-the-scope echo is an echo whose direction of relative movement differs by less than 90° from own ship's heading; a down-the-scope echo is an echo whose direction of relative movement differs by more than 90° from own ship's heading; an across-the scope (limbo) echo is an echo whose direction of relative movement differs by 90° from own ship's heading, i.e., the echo's tail is perpendicular to own ship's heading flasher.

clay. , *n.* See under MUD.

clean. , *adj.* Free from obstructions, unevenness, imperfections, as a clean anchorage.

clear. , *v., t.* To leave port or pass safely by an obstruction.

clearance. , *n.* The clear space between two objects, such as the nearest approach of a vessel to a navigational light, hazard to navigation, or other vessel.

clear berth. . A berth in which a vessel may swing at anchor without striking or fouling another vessel or an obstruction. See also FOUL BERTH.

cliff. , *n.* Land arising abruptly for a considerable distance above water or surrounding land. See also BLUFF.

climate. , *n.* The prevalent or characteristic meteorological conditions of a place or region, in contrast with weather, the state of the atmosphere at any time. A marine climate is characteristic of coastal areas, islands, and the oceans, the distinctive features being small annual and daily temperature range and high relative humidity, in contrast with continental climate, which is characteristic of the interior of a large land mass, and the distinctive features of which are large annual and daily temperature range and dry air with few clouds.

climatology. , *n.* 1. The study of climate. 2. An account of the climate of a particular place or region.

clinometer. , *n.* An instrument for indicating the degree of the angle of heel, roll, or pitch of a vessel; may be of the pivot arm or bubble type, usually indicating in whole degrees.

clock. , *n.* A timepiece not meant to be carried on the person. See also CHRONOMETER.

clock speed. . The speed with which a computer performs operations, commonly measured in mega- or gigaHertz.

clockwise. , *adv.* In the direction of rotation of the hands of a clock.

close. , *v., i.* To move or appear to move together. An order is sometimes given by a flagship for a vessel to close to yards, or miles. When a craft moves onto a range, the objects forming the range appear to move closer together or close. The opposite is OPEN.

close aboard. . Very near.

closed. , *adj.* Said of a manned aid to navigation that has been temporarily discontinued for the winter season. See also COMMISSIONED, WITHDRAWN.

closed sea. . 1. A part of the ocean enclosed by headlands, within narrow straits, etc. 2. A part of the ocean within the territorial jurisdiction of a country. The opposite is OPEN SEA. See also HIGH SEAS, INLAND SEA.

close pack ice. . Pack ice in which the concentration is 7/10 to 8/10, composed of floes mostly in contact.

closest approach. . 1. The event that occurs when two planets or other bodies are nearest to each other as they orbit about the primary body. 2. The place or time of the event in definition 1. 3. The time or place where an orbiting earth satellite is closest to the observer. Also called CLOSEST POINT OF APPROACH.

cloud. , *n.* 1. A hydrometeor consisting of a visible aggregate of minute water and/or ice particles in the atmosphere above the earth's surface. Cloud differs from fog only in that the latter is, by definition, in contact with the earth's surface. Clouds form in the free atmosphere as a result of condensation of water vapor in rising currents of air, or by the evaporation of the lowest stratum of fog. For condensation to occur at the point of saturation or a low degree of supersaturation, there must be an abundance of condensation nuclei for water clouds, or ice nuclei for ice-crystal clouds. The size of cloud drops varies from one cloud to another, and within any given cloud there always exists a finite range of sizes. In general, cloud drops range between 1 and 100 microns in diameter and hence are very much smaller than rain drops. See also CLOUD CLASSIFICATION. 2. Any collection of particulate matter in the atmosphere dense enough to be perceptible to the eye, such as a dust cloud or smoke cloud.

cloud bank. . A fairly well defined mass of clouds observed at a distance; it covers an appreciable portion of the horizon sky, but does not extend overhead.

cloud base. . For a given cloud or cloud layer, that lowest level in the atmosphere at which the air contains a perceptible quantity of cloud particles.

cloudburst. , *n.* In popular terminology, any sudden and heavy fall of rain. An unofficial criterion sometimes used specifies a rate of fall equal to or greater than 100 millimeters (3.94 inches) per hour. Also called RAIN GUSH, RAIN GUST.

cloud classification. . 1. A scheme of distinguishing and grouping clouds according to their appearance and, where possible, to their process of formation. The one in general use, based on a classification system introduced by Luke Howard in 1803, is that adopted by the World Meteorological Organization and published in the International Cloud Atlas (1956). This classification is based on the determination of (a) genera, the main characteristic forms of clouds; (b) species, the peculiarities in shape and differences in internal structure of clouds; (c) varieties, special characteristics of arrangement and transparency of clouds; (d) supplementary features and accessory clouds, appended and associated minor clouds forms; and (e) mother-clouds, the origin of clouds if formed from other clouds. The ten cloud genera are cirrus, cirrocumulus, cirrostratus, altocumulus, altostratus, nimbostratus, stratocumulus, stratus, cumulus, and cumulonimbus. The fourteen cloud species are fibratus, uncinus, spissatus, castellanus, floccus, stratiformis, nebulous, lenticularis, fractus, humilis, mediocris, congestus, calvus, and capillatus. The nine cloud varieties are intortus, vertebratus, undulatus, radiatus, lacunosis, duplicatus, translucidus, perlucidus, and opacus. The nine supplementary features and accessory clouds are inclus, mamma, virga, praecipitatio, arcus, tuba, pileus, velum, and pannus. Note that although these are Latin words, it is proper convention to use only the singular endings, e.g., more than one cirrus cloud are, collectively, cirrus, not cirri. 2. A scheme of classifying clouds according to their usual altitudes. Three classes are distinguished: high, middle, and low. High clouds include cirrus, cirrocumulus, cirrostratus, occasionally altostratus and the tops of cumulonimbus. The middle clouds are altocumulus, altostratus, nimbostratus, and portions of cumulus and cumulonimbus. The low clouds are stratocumulus, stratus, most cumulus and cumulonimbus bases, and sometimes nimbostratus. 3. A scheme of classifying clouds according to their particulate composition; namely water clouds, ice-crystal clouds, and mixed clouds. The first are composed entirely of water droplets (ordinary and/or super cooled), the second entirely of ice crystals, and the third a combination of the first two. Of the cloud genera, only cirrostratus and cirrus are always ice-crystal clouds; cirrocumulus can also be mixed; and only cumulonimbus is always mixed. Altostratus nearly always is mixed, but occasionally can be ice crystal. All the rest of the genera are usually water clouds, occasionally mixed: altocumulus, cumulus, nimbostratus and stratocumulus.

cloud cover. . That portion of the sky cover which is attributed to clouds, usually measured in tenths of sky covered.

cloud deck. . The upper surface of a cloud.

cloud height. . In weather observations, the height of the cloud base above local terrain.

cloud layer. . An array of clouds, not necessarily all of the same type, whose bases are at approximately the same level. It may be either continuous or composed of detached elements.

club. , *v., i.* To drift in a current with an anchor dragging to provide control. Usually used with the word down, ie. club down.

clutter. , *n.* Unwanted radar echoes reflected from heavy rain, snow, waves, etc., which may obscure relatively large areas on the radarscope. See also RAIN CLUTTER, SEA RETURN.

co-. . A prefix meaning 90° minus the value with which it is used. Thus, if the latitude is 30° the colatitude is 90° - 30° = 60°. The cofunction of an angle is the function of its complement.

coalsack. , *n.* Any of several dark areas in the Milky Way, especially, when capitalized, a prominent one near the Southern Cross.

coaltitude. , *n.* Ninety degrees minus the altitude. The term has significance only when used in connection with altitude measured from the celestial horizon, when it is synonymous with ZENITH DISTANCE.

coast. , *n.* The general region of indefinite width that extends from the sea inland to the first major change in terrain features. Sometimes called SEACOAST. See also SEABOARD.

coastal aid. . See COASTAL MARK.

coastal area. . The land and sea area bordering the shoreline.

coastal boundary. . A general term for the boundary defined as the line (or measured from the line or points thereon) used to depict the intersection of the ocean surface and the land at an elevation of a particular datum, excluding one established by treaty or by the U.S. Congress.

coastal chart. . See under CHART CLASSIFICATION BY SCALE.

coastal current. . An ocean current flowing roughly parallel to a coast, outside the surf zone. See also LONGSHORE CURRENT.

coastal mark. . A navigation mark placed on the coast to assist coastal navigation. Particularly used with reference to marks placed on a long straight coastline devoid of many natural landmarks. Also called COASTAL AID.

coastal marsh. . An area of salt-tolerant vegetation in brackish and/or salt-water habitats subject to tidal inundation.

coastal plain. . Any plain which has its margin on the shore of a large body of water, particularly the sea, and generally represents a strip of recently emerged sea bottom.

coastal refraction. . The bending of the wave front of a radio wave traveling parallel to a coastline or crossing it at an acute angle due to the differences in the conducting and reflective properties of the land and water over which the wave travels. This refraction affects the accuracy of medium frequency radio direction finding systems. Also called COAST REFRACTION.

Coast and Geodetic Survey. . Mapping, charting, and surveying arm of the National Ocean Service (NOS), a component of the National Oceanic and Atmospheric Administration (NOAA). The organization was known as: The Survey of the Coast from its founding in 1807 to 1836, Coast Survey from 1836 to 1878, and Coast and Geodetic Survey from 1878 to 1970, when it became the Office of Charting and Geodetic Services under the newly formed NOAA. In 1991 the name Coast and Geodetic Survey was reinstated.

Coast Earth Station (CES).. A station which receives communications from an earth orbiting satellite for retransmission via landlines, and vice versa.

coast chart.. See under CHART CLASSIFICATION BY SCALE.

coasting., *n.* Proceeding approximately parallel to a coastline (headland to headland) in sight of land, or sufficiently often in sight of land to fix the ship's position by observations of land features.

coasting lead.. A light deep sea lead (30 to 50 pounds), used for sounding in water 20 to 60 fathoms.

coastline., *n.* The configuration made by the meeting of land and sea.

Coast Pilot.. See UNITED STATES COAST PILOT.

coast refraction.. See COASTAL REFRACTION.

coastwise., *adv.* & *adj.* By way of the coast; moving along the coast. coastwise navigation. Navigation in the vicinity of a coast, in contrast with OFFSHORE NAVIGATION at a distance from a coast. See also COASTING.

coaxial cable.. A transmission cable consisting of two concentric conductors insulated from each other.

cobble., *n.* A stone particle between 64 and 256 millimeters (about 2.5 to 10 inches) in diameter. See also STONE.

cocked hat.. Error triangle formed by lines of position which do not cross at a common point.

cockeyed bob.. A colloquial term in western Australia for a squall, associated with thunder, on the northwest coast in Southern Hemisphere summer.

code beacon.. A beacon that flashes a characteristic signal by which it may be recognized.

codeclination., *n.* Ninety degrees minus the declination. When the declination and latitude are of the same name, codeclination is the same as POLAR DISTANCE measured from the elevated pole.

coding delay.. An arbitrary time delay in the transmission of pulse signals. In hyperbolic radionavigation systems of the pulse type, the coding delay is inserted between the transmission of the master and slave (or secondary) signals to prevent zero or small readings, and thus aid in distinguishing between master and slave (or secondary) station signals.

coefficient., *n.* 1. A number indicating the amount of some change under certain specified conditions, often expressed as a ratio. For example, the coefficient of linear expansion of a substance is the ratio of its change in length to the original length for a unit change of temperature, from a standard. 2. A constant in an algebraic equation. 3. One of several parts which combine to make a whole, as the maximum deviation produced by each of several causes. See also APPROXIMATE COEFFICIENTS.

coefficient A.. A component of magnetic compass deviation of constant value with compass heading resulting from mistakes in calculations, compass and pelorus misalignment, and unsymmetrical arrangements of horizontal soft iron. See also APPROXIMATE COEFFICIENTS.

coefficient B.. A component of magnetic compass deviation, varying with the sine function of the compass heading, resulting from the fore-and-aft component of the craft's permanent magnetic field and induced magnetism in unsymmetrical vertical iron forward or abaft the compass. See also APPROXIMATE COEFFICIENTS.

coefficient C.. A component of magnetic compass deviation, varying with the cosine function of the compass heading, resulting from the athwartship component of the craft's permanent magnetic field and induced magnetism in unsymmetrical vertical iron port or starboard of the compass. See also APPROXIMATE COEFFICIENTS.

coefficient D.. A component of magnetic compass deviation, varying with the sine function of twice the compass heading, resulting from induced magnetism in all symmetrical arrangements of the craft's horizontal soft iron. See also APPROXIMATE COEFFICIENTS.

coefficient E.. A component of magnetic compass deviation varying with the cosine function of twice the compass heading, resulting from induced magnetism in all unsymmetrical arrangements of the craft's horizontal soft iron. See also APPROXIMATE COEFFICIENTS.

coefficient J.. A change in magnetic compass deviation, varying with the cosine function of the compass heading for a given value of J, where J is the change of deviation for a heel of 1° on compass heading 000°. See also APPROXIMATE COEFFICIENTS.

coercive force.. The opposing magnetic intensity that must be applied to a magnetic substance to remove the residual magnetism.

COGARD., *n.* Acronym for U.S. Coast Guard usually used in radio messages.

coherence., *n.* The state of there being correlation between the phases of two or more waves, as is necessary in making phase comparisons in radionavigation.

coincidence., *n.* The condition of occupying the same position as regards location, time, etc.

col., *n.* 1. A neck of relative low pressure between two anticyclones. 2. A depression in the summit line of a mountain range. Also called PASS.

colatitude., *n.* Ninety degrees minus the latitude, the angle between the polar axis and the radius vector locating a point.

cold air mass.. An air mass that is colder than surrounding air. The expression implies that the air mass is colder than the surface over which it is moving.

cold front.. Any non-occluded front, or portion thereof, that moves so that the colder air replaces the warmer air, i.e., the leading edge of a relatively cold air mass. While some occluded fronts exhibit this characteristic, they are more properly called COLD OCCLUSIONS.

cold occlusion.. See under OCCLUDED FRONT.

cold wave.. Unseasonably low temperatures extending over a period of a day or longer, particularly during the cold season of the year.

collada., *n.* A strong wind (35 to 50 miles per hour or stronger) blowing from the north or northwest in the northern part of the Gulf of California and from the northeast in the southern part of the Gulf of California.

collimate., *v., t.* 1. To render parallel, as rays of light. 2. To adjust the line of sight of an optical instrument, such as a theodolite, in proper relation to other parts of the instrument.

collimation error.. The angle by which the line of sight of an optical instrument differs from its collimation axis. Also called ERROR OF COLLIMATION.

collimator., *n.* An optical device which renders rays of light parallel. One of the principal navigational uses of a collimator is to determine the index error of a bubble sextant.

collision bearing.. A constant bearing maintained while the distance between two craft is decreasing.

collision course.. A course which, if followed, will bring two craft together.

cologarithm., *n.* The logarithm of the reciprocal of a number, or the negative logarithm. The sum of the logarithm and cologarithm of the same number is zero. The addition of a cologarithm accomplishes the same result as the subtraction of a logarithm.

colored light.. An aid to navigation exhibiting a light of a color other than white.

color gradients.. See HYPSOMETRIC TINTING.

COLREGS., *n.* Acronym for International Regulations for Prevention of Collisions at Sea.

COLREGS Demarcation Lines.. Lines delineating the waters upon which mariners must comply with the International Regulations for Preventing Collisions at Sea 1972 (72 COLREGS) and those waters upon which mariners must comply with the Navigation Rules for Harbors, Rivers, and Inland Waters (Inland Rules). The waters outside the lines are COLREGS waters. For specifics concerning COLREGS Demarcation Lines see U.S. Code of Federal Regulations, Title 33, Navigation and Navigable Waters; Part 82, COLREGS Demarcation Lines.

column., *n.* A vertical line of anything, such as a column of air, a column of figures in a table, etc.

colure., *n.* A great circle of the celestial sphere through the celestial poles and either the equinoxes or solstices, called, respectively, the equinoctial colure or the solstitial colure.

coma., *n.* The foggy envelope surrounding the nucleus of a comet.

combat chart.. A special-purpose chart of a land-sea area using the characteristics of a map to represent the land area and a chart to represent the sea area, with special features to make the chart useful in naval operations, particularly amphibious operations. Also called MAP CHART.

comber. , *n.* A deep water wave whose crest is pushed forward by a strong wind and is much larger than a whitecap. A long spilling breaker. See ROLLER.

comet. , *n.* A luminous member of the solar system composed of a head or coma, at the center of which a nucleus of many small solid particles is sometimes situated, and often with a spectacular gaseous tail extending a great distance from the head. The orbits of comets are highly elliptical and present no regularity as to their angle to the plane of the ecliptic.

command and control. . The facilities, equipment, communications, procedures, and personnel essential to a commander for planning, locating, directing, and controlling operations of assigned forces pursuant to the missions assigned. In many cases, a locating or position fixing capability exists in, or as a by-product to, command and control systems.

commissioned. , *adj.* Officially placed in operation. In navigation, most commonly used to describe seasonal aids to navigation, which are *decommissioned* in the fall or winter, *commissioned* in spring.

common establishment. . See under ESTABLISHMENT OF THE PORT.

common logarithm. . A logarithm to the base 10. Also called BRIGGSIAN LOGARITHM.

common-user. , *adj.* Having the characteristics of being planned, operated or used to provide services for both military and civil applications. The availability of a system having such characteristics is not dependent on tactical military operations or use.

common year. . A calendar year of 365 days. One of 366 days is called a LEAP YEAR.

communication. , *n.* The transfer of intelligence between points. If by wire, radio, or other electromagnetic means, it may be called telecommunication; if by radio, radiocommunication.

commutation. , *n.* A method by means of which the transmissions from a number of stations of a radionavigation system are time shared on the same frequency.

compact disk. . A type of computer storage media which records data using bubbles melted into the surface of a disk. It cannot be erased and is therefore called Read Only Memory (ROM).

compacted ice edge. . A close, clear-cut ice edge compacted by wind or current. It is usually on the windward side of an area of pack ice.

compacting. , *adj.* Pieces of sea ice are said to be compacting when they are subjected to a converging motion, which increases ice concentration and/or produces stresses which may result in ice deformations.

compact pack ice. . Pack ice in which the concentration is 10/10 and no water is visible.

comparing watch. . A watch used for timing observations of celestial bodies. Generally its error is determined by comparison with a chronometer, hence its name. A comparing watch normally has a large sweep second hand to facilitate reading time to the nearest second. Sometimes called HACK WATCH. See also SPLIT-SECOND TIMER.

comparison frequency. . In the Decca Navigator System, the common frequency to which the incoming signals are converted in order that their phase relationships may be compared.

comparison of simultaneous observations. . A reduction process in which a short series of tide or tidal current observations at any place is compared with simultaneous observations at a control station where tidal or tidal current constants have previously been determined from a long series of observations. For tides, it is usually used to adjust constants from a subordinate station to the equivalent of that which would be obtained from a 19-year series.

compass. , *adj.* Of or pertaining to a compass or related to compass north.

compass. , *n.* An instrument for indicating a horizontal reference direction relative to the earth. Compasses used for navigation are equipped with a graduated compass card for direct indication of any horizontal direction. A magnetic compass depends for its directive force upon the attraction of the magnetism of the earth for a magnet free to turn in any horizontal direction. A compass having one or more gyroscopes as the directive element, and tending to indicate true north is called a gyrocompass. A compass intended primarily for use in observing bearings is called a bearing compass; one intended primarily for measuring amplitudes, an amplitude compass. A directional gyro is a gyroscopic device used to indicate a selected horizontal direction for a limited time. A remote-indicating compass is equipped with one or more indicators, called compass repeaters, to repeat at a distance the readings of a master compass. A compass designated as the standard for a vessel is called a standard compass; one by which a craft is steered is called a steering compass. A liquid, wet, or spirit compass is a magnetic compass having a bowl completely filled with liquid; a magnetic compass without liquid is called a dry compass. An aperiodic or deadbeat compass, after being deflected, returns by one direct movement to its proper reading, without oscillation. A small compass mounted in a box for convenient use in small water craft is called a boat compass. A pelorus is sometimes called a dumb compass. A radio direction finder was formerly called a radio compass.

compass adjustment. . The process of neutralizing undesired magnetic effects on a magnetic compass. Permanent magnets and soft iron correctors are arranged about the binnacle so that their effects are about equal and opposite to the magnetic material in the craft, thus reducing the deviations and eliminating the sectors of sluggishness and unsteadiness. See also COMPASS COMPENSATION.

compass adjustment buoy. . See SWINGING BUOY.

compass amplitude. . Amplitude relative to compass east or west.

compass azimuth. . Azimuth relative to compass north.

compass bearing. . Bearing relative to compass north.

compass bowl. . The housing in which the compass card is mounted, usually filled with liquid.

compass card. . The part of a compass on which the direction graduations are placed. It is usually in the form of a thin disk or annulus graduated in degrees, clockwise from 0° at the reference direction to 360°, and sometimes also in compass points. A similar card on a pelorus is called a PELORUS CARD.

compass card axis. . The line joining 0° and 180° on a compass card. Extended, this line is sometimes called COMPASS MERIDIAN.

compass compensation. . The process of neutralizing the effects of degaussing currents on a marine magnetic compass. The process of neutralizing the magnetic effects the vessel itself exerts on a magnetic compass is properly called COMPASS ADJUSTMENT, but the expression COMPASS COMPENSATION is often used for this process, too.

compass course. . Course relative to compass north.

compass direction. . Horizontal direction expressed as angular distance from compass north.

compass error. . The angle by which a compass direction differs from the true direction; the algebraic sum of variation and deviation; the angle between the true meridian and the axis of the compass card, expressed in degrees east or west to indicate the direction of compass north with respect to true north. See also ACCELERATION ERROR, GAUSSIN ERROR, GYRO ERROR, HEELING ERROR, LUBBER'S LINE ERROR, QUADRANTAL ERROR, RETENTIVE ERROR, SWIRL ERROR.

compasses. , *n.* An instrument for drawing circles. In its most common form it consists of two legs joined by a pivot, one leg carrying a pen or pencil and the other leg being pointed. An instrument for drawing circles of large diameter, usually consisting of a bar with sliding holders for points, pencils, or pens is called beam compasses. If both legs are pointed, the instrument is called DIVIDERS and is used principally for measuring distances or coordinates.

compass heading. . Heading relative to compass north.

compass meridian. . A line through the north-south points of a magnetic compass. The COMPASS CARD AXIS lies in the compass meridian.

compass north. . The direction north as indicated by a magnetic compass; the reference direction for measurement of compass directions.

compass points. . The 32 divisions of a compass, at intervals of 11 1/4°. Each division is further divided into quarter points. Stating in order the names of the points (and sometimes the half and quarter points) is called BOXING THE COMPASS.

compass prime vertical. . The vertical circle through the compass east and west points of the horizon.

compass repeater. . That part of a remote-indicating compass system which repeats at a distance the indications of the master compass. One used primarily for observing bearings may be called a bearing repeater. Also called REPEATER COMPASS. See also GYRO REPEATER.

compass rose.. A circle graduated in degrees, clockwise from 0° at the reference direction to 360°, and sometimes also in compass points. Compass roses are placed at convenient locations on the Mercator chart or plotting sheet to facilitate measurement of direction. See also PROTRACTOR.

compass track.. The direction of the track relative to compass north.

compass transmitter.. The part of a remote-indicating compass system which sends the direction indications to the repeaters.

compensate., *v., t.* To counteract an error; to counterbalance.

compensated loop radio direction finder.. A loop antenna radio direction finder for bearing determination, which incorporates a second antenna system designed to reduce the effect of polarization and radiation error.

compensating coils.. The coils placed near a magnetic compass to neutralize the effect of the vessel's degaussing system on the compass. See also COMPASS COMPENSATION.

compensating error.. An error that tends to offset a companion error and thus obscure or reduce the effect of each.

compensator., *n.* 1. A corrector used in the compensation of a magnetic compass. 2. The part of a radio direction finder which applies all or part of the necessary correction to the direction indication.

compile.. To assemble various elements of a system into a whole.

compiler.. 1. One who compiles. 2. Computer software which translates programs into machine language which a computer can use.

complement., *n.* An angle equal to 90° minus a given angle. See also EXPLEMENT, SUPPLEMENT.

complementary angles.. Two angles whose sum is 90°.

component., *n.* 1. See CONSTITUENT. 2. The part of a tidal force of tidal current velocity which, by resolution into orthogonal vectors, is found to act in a specified direction. 3. One of the parts into which a vector quantity can be divided. For example, the earth's magnetic force at any point can be divided into *horizontal* and *vertical components*.

composite., *adj.* Composed of two or more separate parts.

composite group flashing light.. A light similar to a group flashing light except that successive groups in a single period have different numbers of flashes.

composite group occulting light.. A group occulting light in which the occultations are combined in successive groups of different numbers of occultations.

composite sailing.. A modification of great-circle sailing used when it is desired to limit the highest latitude. The composite track consists of a great circle from the point of departure and tangent to the limiting parallel, a course line along the parallel, and a great circle tangent to the limiting parallel to the destination. Composite sailing applies only when the vertex lies between the point of departure and destination.

composite track.. A modified great-circle track consisting of an initial great circle track from the point of departure with its vertex on a limiting parallel of latitude, a parallel-sailing track from this vertex along the limiting parallel to the vertex of a final great-circle track to the destination.

composition of vectors.. See VECTOR ADDITION.

compound harmonic motion.. The projection of two or more uniform circular motions on a diameter of the circle of such motion. The projection of a simple uniform circular motion is called SIMPLE HARMONIC MOTION.

compound tide.. A tidal constituent with a speed equal to the sum or difference of the speeds of two or more elementary constituents. Compound tides are usually the result of shallow water.

compressed-air horn.. See DIAPHRAGM HORN.

compression., *n.* See FLATTENING.

computed altitude.. 1. Tabulated altitude interpolated for increments of latitude, declination, or hour angle. If no interpolation is required, the tabulated altitude and computed altitude are identical. 2. Altitude determined by computation, table, mechanical computer, or graphics, particularly such an altitude of the center of a celestial body measured as an arc on a vertical circle of the celestial sphere from the celestial horizon. Also called CALCULATED ALTITUDE.

computed azimuth.. Azimuth determined by computation, table, mechanical device, or graphics for a given place and time. See also TABULATED AZIMUTH.

computed azimuth angle.. Azimuth angle determined by computation, table, mechanical device, or graphics for a given place and time. See also TABULATED AZIMUTH ANGLE.

computed point.. In the construction of the line of position by the Marcq St. Hilaire method, the foot of the perpendicular from the assumed position to the line of position. Also called SUMNER POINT.

concave., *adj.* Curving and hollow, such as the inside of a circle or sphere. The opposite is CONVEX.

concave., *n.* A concave line or surface.

concentration., *n.* The ratio, expressed in tenths, of the sea surface actually covered by ice to the total area of sea surface, both ice-covered and ice-free, at a specific location or over a defined area.

concentration boundary.. The transition between two areas of pack ice with distinctly different concentrations.

concentric., *adj.* Having the same center. The opposite is ECCENTRIC.

concurrent line.. A line on a map or chart passing through places having the same current hour.

condensation., *n.* The physical process by which a vapor becomes a liquid or solid. The opposite is EVAPORATION.

conduction., *n.* Transmission of electricity, heat, or other form of energy from one point to another along a conductor, or transference of heat from particle to particle through a substance, such as air, without any obvious motion. Heat is also transferred by CONVECTION and RADIATION.

conductivity., *n.* The ability to transmit, as electricity, heat, sound, etc. Conductivity is the opposite of RESISTIVITY.

conductor., *n.* A substance which transmits electricity, heat, sound, etc.

cone., *n.* 1. A solid having a plane base bounded by a closed curve and a surface formed by lines from every point on the circumference of the base to a common point or APEX. 2. A surface generated by a straight line of indefinite length, one point of which is fixed and another point of which follows a fixed curve. Also called a CONICAL SURFACE.

configuration., *n.* 1. The position or disposition of various parts, or the figure or pattern so formed. 2. A geometric figure, usually consisting principally of points and connecting lines.

conformal., *adj.* Having correct angular representation.

conformal chart.. A chart using a conformal projection; also called orthomorphic chart.

conformal map projection.. A map projection in which all angles around any point are correctly represented, In such a projection the scale is the same in all directions about any point. Very small shapes are correctly represented, resulting in an orthomorphic projection. The terms *conformal* and *orthomorphic* are used synonymously since neither characteristic can exist without the other.

confusion region.. The region surrounding a radar target within which the radar echo from the target cannot be distinguished from other echoes.

conic., *adj.* Pertaining to a cone.

conical buoy.. See NUN BUOY.

conical surface.. See CONE, definition 2.

conic chart.. A chart on a conic projection.

conic chart with two standard parallels.. A chart on the conic projection with two standard parallels. Also called SECANT CONIC CHART. See also LAMBERT CONFORMAL CHART.

conic map projection.. A map projection in which the surface of a sphere or spheroid, such as the earth, is conceived as projected onto a tangent or secant cone which is then developed into a plane. In a simple conic map projection the cone is tangent to the sphere or spheroid, in a conic map projection with two standard parallels the cone intersects the sphere or spheroid along two chosen parallels, and in a polyconic map projection a series of cones are tangent to the sphere or spheroid. See also LAMBERT CONFORMAL CONIC MAP PROJECTION, MODIFIED LAMBERT CONFORMAL MAP PROJECTION.

conic map projection with two standard parallels.. A conic map projection in which the surface of a sphere or spheroid is conceived as developed on a cone which intersects the sphere or spheroid along two standard parallels, the cone being spread out to form a plane. The Lambert conformal map projection is an example. Also called SECANT CONIC MAP PROJECTION.

conic section. . Any plane curve which is the locus of a point which moves so that the ratio of its distance from a fixed point to its distance from a fixed line is constant. The ratio is called the eccentricity; the fixed point is the focus; the fixed line is the directrix. When the eccentricity is equal to unity, the conic section is a parabola; when less than unity an ellipse; and when greater than unity, a hyperbola. They are so called because they are formed by the intersection of a plane and a right circular cone.

conjunction. , *n.* The situation of two celestial bodies having either the same celestial longitude or the same sidereal hour angle. A planet is at superior conjunction if the sun is between it and the earth; at inferior conjunction if it is between the sun and the earth. The situation of two celestial bodies having either celestial longitudes or sidereal hour angles differing by 180° is called OPPOSITION.

conn. , *v., t.* 1. To direct the course and speed of a vessel. The person giving orders to the helmsman (not just relaying orders) is said to have the conn or to be conning the ship. 2. *n.* Control of the maneuvering of a ship.

Consol. , *n.* A long range, obsolete azimuthal radionavigation system of low accuracy operated primarily for air navigation.

console. , *n.* The housing of the main operating unit of electronic equipment, in which indicators and general controls are located. The term is popularly limited to large housings resting directly on the deck, as contrasted with smaller cabinets such as rack or bracket-mounted units.

consolidated pack ice. . Pack ice in which the concentration is 10/10 and the floes are frozen together.

consolidated ridge. . A ridge (a line or wall of ice forced up by pressure) in which the base has frozen together.

Consol station. . A short baseline directional antenna system used to generate Consol signals.

constant. , *n.* A fixed quantity; one that does not change.

constant bearing, decreasing range. . See STEADY BEARING.

constant deviation. . Deviation which is the same on any heading, as that which may result from certain arrangements of asymmetrical horizontal soft iron.

constant error. . A systematic error of unchanging magnitude and sign throughout a given series of observations. Also called BIAS ERROR.

constant of aberration. . The measure of the maximum angle between the true direction and the apparent direction of a celestial body as observed from earth due to aberration. It has a value of 20.496 seconds of arc. The aberration angle depends upon the ratio of the velocity of the earth in its orbit and the velocity of light in addition to the angle between the direction of the light and the direction of motion of the observing telescope. The maximum value is obtained when the celestial body is at the pole of the ecliptic. Also called ABERRATION CONSTANT.

constant of the cone. . The chart convergence factor for a conic projection. See also CONVERGENCE FACTOR.

constant-pressure chart. . The synoptic chart for any constant-pressure surface, usually containing plotted data and analyses of the distribution of, e.g., height of the surface, wind, temperature, and humidity. Constant-pressure charts are most commonly known by their pressure value; for example the *1000-millibar chart*. Also called ISOBARIC CHART.

constant-pressure surface. . In meteorology, an imaginary surface along which the atmospheric pressure is everywhere equal at a given instant. Also called ISOBARIC SURFACE.

constellation. , *n.* A group of stars which appear close together, regardless of actual distances, particularly if the group forms a striking configuration. Among astronomers a constellation is now considered a region of the sky having precise boundaries so arranged that all of the sky is covered, without overlap. The ancient Greeks recognized 48 constellations covering only certain groups of stars. Modern astronomers recognize 88 constellations.

constituent. , *n.* One of the harmonic elements in a mathematical expression for the tide-producing force and in corresponding formulas for the tide or tidal current. Each constituent represents a periodic change or variation in the relative positions of the earth, moon, and sun. Also called HARMONIC CONSTITUENT, TIDAL CONSTITUENT, COMPONENT.

constituent day. . The duration of one rotation of the earth on its axis, with respect to an astre fictif, a fictitious star representing one of the periodic elements in tidal forces. It approximates the length of a lunar or solar day. The expression is not applicable to a long period.

constituent, constituent hour. . One twenty-fourth part of a constituent day.

contact. , *n.* Any echo detected on the radarscope and not evaluated as clutter or as a false echo. Although the term *contact is* often used interchangeably with *target,* the latter term specifically indicates that the echo is from an object about which information is being sought.

conterminous. . U.S. Forty-eight states and the District of Columbia, i.e., the United States before January 3, 1959 (excluding Alaska and Hawaii).

contiguous zone. . The band of water outside or beyond the territorial sea in which a coastal nation may exercise customs control and enforce public health and other regulations.

continent. , *n.* An expanse of continuous land constituting one of the major divisions of the land surface of the earth.

continental borderland. . A region adjacent to a continent, normally occupied by or bordering a shelf, that is highly irregular with depths well in excess of those typical of a shelf. See also INSULAR BORDERLAND.

continental climate. . The type of climate characteristic of the interior of a large land mass, the distinctive features of which are large annual and daily temperature range and dry air with few clouds, in contrast with MARINE CLIMATE.

continental polar air. . See under AIR-MASS CLASSIFICATION.

continental rise. . A gentle slope rising from oceanic depths toward the foot of a continental slope.

continental shelf. . A zone adjacent to a continent that extends from the low water line to a depth at which there is usually a marked increase of slope towards oceanic depths. See also INSULAR SHELF.

continental tropical air. . See under AIR-MASS CLASSIFICATION.

Continental United States. . United States territory, including the adjacent territorial waters, located within the North American continent between Canada and Mexico. See also CONTERMINOUS U.S.

continuous carrier radiobeacon. . A radiobeacon whose carrier wave is unbroken but which is modulated with the identification signal. The continuous carrier wave signal is not audible to the operator of an aural null direction finder not having a beat frequency oscillator. The use of the continuous carrier wave improves the performance of automatic direction finders. The marine radiobeacons on the Atlantic and Pacific coasts of the U.S. are of this type. See also DUAL CARRIER RADIOBEACON.

continuous quick light. . A quick flashing light (flashing 50-80 times per minute) which operates continuously with no eclipses.

continuous system. . A classification of a navigation system with respect to availability. A continuous system gives the capability to determine position at any time.

continuous ultra quick light. . An ultra quick light (flashing not less than 160 flashes per minute) with no eclipses.

continuous very quick light. . A very quick light (flashing 80-160 times per minute) with no eclipses.

continuous wave. . 1. Electromagnetic radiation of a constant amplitude and frequency. 2. Radio waves, the successive sinusoidal oscillations of which are identical under steady-state conditions.

contour. , *n.* The imaginary line on the ground, all points of which are at the same elevation above or below a specified datum.

contour interval. . The difference in elevation between two adjacent contours.

contour line. . A line connecting points of equal elevation or equal depth. One connecting points of equal depth is usually called a depth contour, but if depth is expressed in fathoms, it may be called a fathom curve or fathom line. See also FORM LINES.

contour map. . A topographic map showing relief by means of contour lines.

contrary name. . A name opposite or contrary to that possessed by something else, as declination has a name *contrary* to that of latitude if one is north and the other south. If both are north or both are south, they are said to be of SAME NAME.

contrastes., *n., pl.* Winds a short distance apart blowing from opposite quadrants, frequent in the spring and fall in the western Mediterranean.

contrast threshold.. The minimum contrast at the eye of a given observer at which an object can be detected. The contrast threshold is a property of the eye of the individual observer. See METEOROLOGICAL VISIBILITY, VISUAL RANGE.

control., *n.* 1. The coordinated and correlated dimensional data used in geodesy and cartography to determine the positions and elevations of points on the earth's surface or on a cartographic representation of that surface. 2. A collective term for a system of marks or objects on the earth or on a map or a photograph, whose positions and/or elevations have been or will be determined.

control current station.. A current station at which continuous velocity observations have been made over a minimum of 29 days. Its purpose is to provide data for computing accepted values of the harmonic and nonharmonic constants essential to tidal current predictions and circulatory studies. The data series from this station serves as the control for the reduction of relatively short series from subordinate current stations through the method of comparison of simultaneous observations. See also CURRENT STATION, SUBORDINATE CURRENT STATION.

controlled air space.. An airspace of defined dimensions within which air traffic control service is provided.

controlling depth.. 1. The least depth in the approach or channel to an area, such as a port or anchorage, governing the maximum draft of vessels that can enter. 2. The least depth within the limits of a channel; it restricts the safe use of the channel to drafts of less than that depth. The centerline controlling depth of a channel applies only to the channel centerline; lesser depths may exist in the remainder of the channel. The mid-channel controlling depth of a channel is the controlling depth of only the middle half of the channel. See also FEDERAL PROJECT DEPTH.

control station.. See PRIMARY CONTROL TIDE STATION, SECONDARY CONTROL TIDE STATION, CONTROL CURRENT STATION.

convection., *n.* Circulation in a fluid of nonuniform temperature, due to the differences in density and the action of gravity. In the atmosphere, convection takes place on a large scale. It is essential to the formation of many clouds, especially those of the cumulus type. Heat is transferred by CONVECTION and also by ADVECTION, CONDUCTION, and RADIATION.

convention., *n.* A body of regulations adopted by the IMO which regulate one aspect of maritime affairs. See also GEOGRAPHIC SIGN CONVENTIONS.

conventional direction of buoyage.. 1. The general direction taken by the mariner when approaching a harbor, river, estuary or other waterway from seaward, or 2. The direction determined by the proper authority. In general it follows a clockwise direction around land masses.

converge., *v., i.* To tend to come together.

converged beam.. See under FAN BEAM.

convergence constant.. The angle at a given latitude between meridians 1° apart. Sometimes loosely called CONVERGENCY. On a map or chart having a convergence constant of 1.0, the true direction of a straight line on the map or chart changes 1° for each 1° of longitude that the line crosses; the true direction of a straight line on a map or chart having a convergence constant of 0.785 changes 0.785° for each 1° of longitude the line crosses. Also called CONVERGENCE FACTOR. See also CONVERGENCE OF MERIDIANS.

convergence factor.. See CONVERGENCE CONSTANT.

convergence of meridians.. The angular drawing together of the geographic meridians in passing from the Equator to the poles, At the Equator all meridians are mutually parallel; passing from the Equator, they converge until they meet at the poles, intersecting at angles that are equal to their differences of longitude. See also CONVERGENCE CONSTANT.

convergency., *n.* See under CONVERGENCE CONSTANT.

conversion., *n.* Determination of the rhumb line direction of one point from another when the initial great circle direction is known, or vice versa. The difference between the two directions is the conversion angle, and is used in great circle sailing.

conversion angle.. The angle between the rhumb line and the great circle between two points. Also called ARC TO CHORD CORRECTION. See also HALF-CONVERGENCY.

conversion scale.. A scale for the conversion of units of one measurement to equivalent units of another measurement. See NOMOGRAM.

conversion table.. A table for the conversion of units of one measurement to equivalent units of another measurement. See NOMOGRAM.

convex., *adj.* Curving away from, such as the outside of a circle or sphere. The opposite is CONCAVE.

convex., *n.* A convex line or surface.

coordinate., *n.* One of a set of magnitudes defining a point in space. If the point is known to be on a given line, only one coordinate is needed; if on a surface, two are required; if in space, three. Cartesian coordinates define a point relative to two intersecting lines, called AXES. If the axes are perpendicular, the coordinates are rectangular; if not perpendicular, they are oblique coordinates. A three-dimensional system of Cartesian coordinates is called space coordinates. Polar coordinates define a point by its distance and direction from a fixed point called the POLE. Direction is given as the angle between a reference radius vector and a radius vector to the point. If three dimensions are involved, two angles are used to locate the radius vector. Space-polar coordinates define a point on the surface of a sphere by (1) its distance from a fixed point at the center, called the POLE (2) the COLATITUDE or angle between the POLAR AXIS (a reference line through the pole) and the RADIUS VECTOR (a straight line connecting the pole and the point)- and (3) the LONGITUDE or angle between a reference plane through the polar axis and a plane through the radius vector and the polar axis. Spherical coordinates define a point on a sphere or spheroid by its angular distances from a primary great circle and from a reference secondary great circle. Geographical or terrestrial coordinates define a point on the surface of the earth. Celestial coordinates define a point on the celestial sphere. The horizon, celestial equator and the ecliptic systems of celestial coordinates are based on the celestial horizon, celestial equator, and the ecliptic, respectively, as the primary great circle.

coordinate conversion.. Changing the coordinate values from one system to those of another.

Coordinated Universal Time (UTC).. The time scale that is available from most broadcast time signals. It differs from International Atomic Time (TAI) by an integral number of seconds. UTC is maintained within 1 second of UT1 by the introduction of 1-second steps (leap seconds) when necessary, normally at the end of December. DUT1, an approximation to the difference UT1 minus UTC, is transmitted in code on broadcast time signals.

coordinate paper.. Paper ruled with lines to aid in the plotting of coordinates. In its most common form, it has two sets of parallel lines, usually at right angles to each other, when it is also called CROSS-SECTION PAPER. A type ruled with two sets of mutually-perpendicular, parallel lines spaced according to the logarithms of consecutive numbers is called logarithmic coordinate papa or semilogarithmic coordinate paper as both or only one set of lines is spaced logarithmically. A type ruled with concentric circles and radial lines from the common center is called polar coordinate paper. Also called GRAPH PAPER.

coplanar., *adj.* Lying in the same plane.

coprocessor.. A microprocessor chip which performs numerical functions for the CPU, freeing it for other tasks.

coral., *n.* The hard skeleton of certain tiny sea animals; or the stony, solidified mass of a number of such skeletons.

coral head.. A large mushroom or pillar shaped coral growth.

coral reef.. A reef made up of coral, fragments of coral and other organisms, and the limestone resulting from their consolidation. Coral may constitute less than half of the reef material.

corange line.. A line passing through places of equal tidal range.

cordillera., *n.* On the sea floor, an entire mountain system including all the subordinate ranges, interior plateaus, and basins.

cordonazo., *n.* The "Lash of St. Francis." Name applied locally to southerly hurricane winds along the west coast of Mexico. The cordonazo is associated with tropical cyclones in the southeastern North Pacific Ocean. These storms may occur from May to November, but ordinarily affect the coastal areas most severely near or after the Feast of St. Francis, October 4.

Coriolis acceleration. . An acceleration of a body in motion in a relative (moving) coordinate system. The total acceleration of the body, as measured in an inertial coordinate system, may be expressed as the sum of the acceleration within the relative system, the acceleration of the relative system itself, and the Coriolis acceleration. In the case of the earth, moving with angular velocity Ω, a body moving relative to the earth with velocity V has the Coriolis acceleration $252 \times \Omega$. If Newton's laws are to be applied in the relative system, the Coriolis acceleration and the acceleration of the relative system must be treated as forces. See also CORIOLIS FORCE.

Coriolis correction. . 1. A correction applied to an assumed position, celestial line of position, celestial fix, or to a computed or observed altitude to allow for Coriolis acceleration. 2. In inertial navigation equipment, an acceleration correction which must be applied to measurements of acceleration with respect to a coordinate system in translation to compensate for the effect of any angular motion of the coordinate system with respect to inertial space.

Coriolis force. . An inertial force acting on a body in motion, due to rotation of the earth, causing deflection to the right in the Northern Hemisphere and to the left in the Southern Hemisphere. It affects air (wind), water (current), etc. and introduces an error in bubble sextant observations made from a moving craft due to the liquid in the bubble being deflected, the effect increasing with higher latitude and greater speed of the craft.

corner reflector. . A radar reflector consisting of three mutually perpendicular flat reflecting surfaces designed to return incident electromagnetic radiation toward its source. The reflector is used to render objects such as buoys and sailboats more conspicuous to radar observations. Since maximum effectiveness is obtained when the incident beam coincides with the axis of symmetry of the reflector, clusters of reflectors are sometimes used to insure that the object will be a good reflector in all directions. See also RADAR REFLECTOR. Also called TRIHEDRAL REFLECTOR.

coromell. , *n*. A night land breeze prevailing from November to May at La Paz, near the southern extremity of the Gulf of California.

corona. , *n*. 1. The luminous envelope surrounding the sun but visible only during a total eclipse. 2. A luminous discharge due to ionization of the air surrounding an electric conductor. 3. A set of one or more rainbow-colored rings of small radii surrounding the sun, moon, or other source of light covered by a thin cloud veil. It is caused by diffraction of the light by tiny droplets in the atmosphere, and hence the colors are in the reverse order to those of a HALO caused by refraction. 4. A circle of light occasionally formed by the apparent convergency of the beams of the aurora.

corona discharge. . Luminous and often audible discharge of electricity intermediate between a spark and a point discharge. See ST. ELMO'S FIRE.

corposant. , *n*. See CORONA DISCHARGE, ST. ELMO'S FIRE.

corrasion. , *n*. The wearing away of the earth's surface by the abrasive action of material transported by glacier, water, or air; a process of erosion.

corrected compass course. . Compass course with deviation applied; magnetic course.

corrected compass heading. . Compass heading with deviation applied; magnetic heading.

corrected current. . A relatively short series of current observations from a subordinate station to which a factor is applied to adjust the current to a more representative value, based on a relatively long series from a nearby control station. See also CURRENT, definition 1; TOTAL CURRENT.

corrected establishment. . See under ESTABLISHMENT OF THE PORT.

corrected sextant altitude. . Sextant altitude corrected for index error, height of eye, parallax, refraction, etc. Also called OBSERVED ALTITUDE, TRUE ALTITUDE.

correcting. , *n*. The process of applying corrections, particularly the process of converting compass to magnetic direction, or compass, magnetic, or gyro to true direction. The opposite is UNCORRECTING.

correction. , *n*. That which is added to or subtracted from a reading, as of an instrument, to eliminate the effect of an error, or to reduce an observation to an arbitrary standard.

correction of soundings. . The adjustment of soundings for any departure from true depth because of the method of sounding or any fault in the measuring apparatus. See also REDUCTION OF SOUNDINGS.

corrector. , *n*. A magnet, piece of soft iron, or device used in the adjustment of a magnetic compass. See also FLINDERS BAR, HEELING MAGNET, QUADRANTAL CORRECTORS.

corrosion. , *n*. The wearing or wasting away by chemical action, usually by oxidation. A distinction is usually made between CORROSION and EROSION, the latter referring to the wearing away of the earth's surface primarily by non-chemical action. See also CORRASION.

cosecant. , *n*. The ratio of the hypotenuse of a plane right triangle to the side opposite one of the acute angles of the triangle, equal to l/sin. The expression NATURAL COSECANT is sometimes used to distinguish the cosecant from its logarithm (called LOGARITHMIC COSECANT).

cosine. , *n*. The ratio of the side adjacent to an acute angle of a plane right triangle to the hypotenuse. The expression NATURAL COSINE is sometimes used to distinguish the cosine from its logarithm (called LOGARITHMIC COSINE).

COSPAS/SARSAT. . A cooperative search and rescue satellite system operated by the U.S. and Russia which provides worldwide coverage by sensing the signals of Emergency Position Indicating Radiobeacons (EPIRB's).

cotangent. , *n*. The ratio of the shorter side adjacent to an acute angle of a plane right triangle to the side opposite the same angle, equal to l/tan. The expression NATURAL COTANGENT is sometimes used to distinguish the cotangent from its logarithm (called LOGARITHMIC COTANGENT).

cotidal. , *adj*. Having tides occurring at the same time.

cotidal chart. . A chart showing cotidal lines.

cotidal hour. . The average interval between the moon's transit over the meridian of Greenwich and the time of the following high water at any place, expressed in either mean solar or lunar time units. When expressed in solar time, it is the same as the Greenwich high water interval. When expressed in lunar time, it is equal to the Greenwich high water interval multiplied by the factor 0.966.

cotidal line. . A line on a map or chart passing through places having the same cotidal hour.

coulomb. , *n*. A derived unit of quantity of electricity in the International System of Units; it is the quantity of electricity carried in 1 second by a current of 1 ampere.

counterclockwise. , *adv*. In a direction of rotation opposite to that of the hands of a clock.

countercurrent. , *n*. A current usually setting in a direction opposite to that of a main current.

counterglow. , *n*. See GEGENSCHEIN.

countertrades. , *n., pl*. See ANTITRADES.

coupler. , *n*. See as ANTENNA COUPLER.

course. , *n*. The direction in which a vessel is steered or intended to be steered, expressed as angular distance from north, usually from 000° at north, clockwise through 360°. Strictly, the term applies to direction through the water, not the direction intended to be made good over the ground. The course is often designated as true, magnetic, compass, or grid as the reference direction is true, magnetic compass, or grid north, respectively. TRACK MADE GOOD is the single resultant direction from the point of departure to point of arrival at any given time. The use of this term to indicate a single resultant direction is preferred to the use of the misnomer course made good. A course line is a line, as drawn on a chart, extending in the direction of a course. See also COURSE ANGLE, COURSE OF ADVANCE, COURSE OVER GROUND. HEADING. TRACK.

course angle. . Course measured from 0° at the reference direction clockwise or counterclockwise through 90° or 180°. It is labeled with the reference direction as a prefix and the direction of measurement from the reference direction as a suffix.

course beacon. . A directional radiobeacon which gives an "on course" signal in the receiver of a vessel which is on, or in close proximity to, the prescribed course line and "off course" signals in sectors adjacent to this line.

course board. . A board located on the navigation bridge used to display the course to steer, track, drift angle, leeway angle, compass error, etc.

course line. . 1. The graphic representation of a ship's course, usually with respect to true north. 2. A line of position approximately parallel to the course line (definition 1), thus providing a check as to deviating left or right of the track. See also SPEED LINE.

course made good. . A misnomer indicating the resultant direction from a point of departure to a point of arrival at any given time. See also COURSE, COURSE OVER GROUND, TRACK MADE GOOD.

course of advance. . An expression sometimes used to indicate the direction intended to be made good over the ground. The preferred term is TRACK, definition 1. This is a misnomer in that courses are directions steered or intended to be steered through the water with respect to a reference meridian. See also COURSE, COURSE OVER GROUND.

course over ground. . The direction of the path over the ground actually followed by a vessel. The preferred term is TRACK, definition 1. It is normally a somewhat irregular line. This is a misnomer in that courses are directions steered or intended to be steered through the water with respect to a reference meridian. See also COURSE, COURSE MADE GOOD.

course recorder. . A device which makes an automatic graphic record of the headings of a vessel vs. time. See also DEAD RECKONING TRACER.

course up. . See BASE COURSE UP.

cove. , *n.* A small sheltered recess or indentation in a shore or coast, generally inside a larger embayment.

coverage diagram. . A chart which depicts the area serviced by a radionavigation system.

crab. , *v., t.* To drift sideways while in forward motion.

crack line. , *n.* Any fracture (in ice) which has not parted.

creek. , *n.* 1. A stream of less volume than a river but larger than a brook. 2. A small tidal channel through a coastal marsh. 3. A wide arm of a river or bay, as used locally in Maryland and Virginia.

crepuscular rays. . Literally, "twilight rays," alternating lighter and darker bands (rays and shadows) which appear to diverge in fanlike array from the sun's position at about twilight. This term is applied to two quite different phenomena: a. It refers to shadows cast across the purple light, a true twilight phenomenon, by cloud tops that are high enough and far enough away from the observer to intercept some of the sunlight that would ordinarily produce the purple light. b. A more common occurrence is that of shadows and rays made visible by haze in the lower atmosphere. Towering clouds produce this effect also, but they may be fairly close to the observer and the sun need not be below the horizon. The apparent divergence of crepuscular rays is merely a perspective effect. When they continue across the sky to the antisolar point, these extensions are called ANTICREPUSCULAR RAYS. Also called SHADOW BANDS.

crescent. , *adj.* Bounded by a convex and a concave curve. Originally, the term applied only to the "increasing" moon, from which the word was derived. By extension, it is now generally applied to the moon between last quarter and new as well as between new and first quarter, and to any other celestial body presenting a similar appearance, or any similarly shaped object. See also PHASES OF THE MOON.

crest. , *n.* The highest part of a wave or swell; or terrestrially, a hill or ridge.

crest cloud. . A type of cloud over a mountain ridge, similar to a cap cloud over an isolated peak. The cloud is apparently stationary, but actually is continually being formed to windward and dissipated to leeward.

crevasse. , *n.* A deep fissure or rift in a glacier.

critical angle. . 1. The maximum angle at which a radio wave may be emitted from an antenna, in respect to the plane of the earth, and still be returned to the earth by refraction or reflection by an ionospheric layer. 2. The angle at which radiation, about to pass from a medium of greater density into one of lesser density, is refracted along the surface of the denser medium.

critical table. . A single entering argument table in which values of the quantity to be found are tabulated for limiting values of the entering argument. In such a table interpolation is avoided through dividing the argument into intervals so chosen that successive intervals correspond to successive values of the required quantity, called the respondent. For any value of the argument within these intervals, the respondent can be extracted from the table without interpolation. The lower and upper limits (critical values) of the argument correspond to half-way values of the respondent and, by convention, are chosen so that when the argument is equal to one of the critical values, the respondent corresponding to the preceding (upper) interval is to be used.

critical temperature. . The temperature above which a substance cannot exist in the liquid state, regardless of pressure.

cross-band Racon. . A Racon which transmits at a frequency not within the marine radar frequency band. To be able to use this type of Racon, the ship's radar receiver must be capable of being tuned to the frequency of the crossband Racon or special accessory equipment is required. In either case, normal radar echoes will not be painted on the radarscope. This is an experimental type of Racon. See also INBAND RACON.

cross-band transponder. . A transponder which responds on a frequency different from that of the interrogating signal.

cross bearings. . Two or more bearings used as intersecting lines of position for fixing the position of a craft.

cross hair. . A hair, thread, or wire constituting part of a reticle.

cross sea. . A series of waves imposed across the prevailing waves. It is called CROSS SWELL when the imposed waves are the longer swell waves.

cross-section paper. . Paper ruled with two sets of parallel lines, useful as an aid in plotting Cartesian coordinates. Usually, the two sets are mutually perpendicular. See also COORDINATE PAPER.

cross-staff. , *n.* A forerunner of the modern sextant used for measuring altitudes of celestial bodies, consisting of a wooden rod with one or more perpendicular cross pieces free to slide along the main rod. Also called FORESTAFF, JACOB'S STAFF.

cross swell. . See under CROSS SEA.

cross tide. . A tidal current setting in a direction approximately 90° from the course of a vessel One setting in a direction approximately 90° from the heading is called a BEAM TIDE. In common usage these two expressions are usually used synonymously. One setting from ahead is called a HEAD TIDE. One setting from aft is called a FAIR TIDE.

cross wind. . See under BEAM WIND.

cruising radius. . The distance a craft can travel at cruising speed without refueling. Also called CRUISING RANGE.

cruising range. . See CRUISING RADIUS.

cryogenics. , *n.* 1. The study of the methods of producing very low temperatures. 2. The study of the behavior of materials and processes at cryogenic temperatures.

cryogenic temperature. . In general, a temperature range below the boiling point of nitrogen (-195°C); more particularly, temperatures within a few degrees of absolute zero.

crystal. , *n.* A crystalline substance which allows electric current to pass in only one direction.

crystal clock. . See QUARTZ CRYSTAL CLOCK.

cube. , *n.* 1. A solid bounded by six equal square sides. 2. The third power of a quantity.

cubic meter. . The derived unit of volume in the International System of Units.

cul-de-sac. , *n.* An inlet with a single small opening.

culmination. , *n.* See MERIDIAN TRANSIT.

culture. , *n.* 1. The man-made features of a map or chart, including roads, rails, cables, etc.; boundary lines, latitude and longitude lines, isogonic lines, etc. are also properly classified as culture.

cumuliform. , *adj.* Like cumulus; generally descriptive of all clouds, the principal characteristic of which is vertical development in the form of rising mounds, domes, or towers. This is the contrasting form to the horizontally extended STRATIFORM types. See also CIRRIFORM.

cumulonimbus. , *n.* An exceptionally dense cloud of great vertical development, occurring either as an isolated cloud or one of a line or wall of clouds with separated upper portions. These clouds appear as mountains or huge towers, at least a part of the upper portions of which are usually smooth, fibrous, striated, and almost flattened. This part often spreads out in the form of an anvil or plume. Under the base of cumulonimbus, which often is very dark, there frequently exists virga, precipitation, and low, ragged clouds, either merged with it or not. Its precipitation is often heavy and always of a showery nature. The usual occurrence of lightning and thunder within or from this cloud leads to its being popularly called THUN-

DERCLOUD and THUNDERHEAD. The latter term usually refers to only the upper portion of the cloud. See also CLOUD CLASSIFICATION.

cumulus. , *n.* A cloud type in the form of individual, detached elements which are generally dense and possess sharp non-fibrous outlines. These elements develop vertically, appearing as rising mounds, domes, or towers, the upper parts of which often resemble a cauliflower. The sunlit parts of these clouds are mostly brilliant white; their bases are relatively dark and nearly horizontal. Near the horizon the vertical development of cumulus often causes the individual clouds to appear merged. If precipitation occurs, it is usually of a showery nature. Various effects of wind, illumination, etc. may modify many of the above characteristics. Strong winds may shred the clouds, often tearing away the cumulus tops to form the species *fractus*. See also CLOUD CLASSIFICATION.

cupola. , *n.* A label on a nautical chart which indicates a small dome-shaped tower or turret rising from a building.

current. , *n.* 1. A horizontal movement of water. Currents may be classified as tidal and nontidal. Tidal currents are caused by gravitational interactions between the sun, moon, and earth and are a part of the same general movement of the sea that is manifested in the vertical rise and fall, called TIDE. Tidal currents are periodic with a net velocity of zero over the tidal cycle. Nontidal currents include the permanent currents in the general circulatory systems of the sea as well as temporary currents arising from more pronounced meteorological variability. The SET of a current is the direction toward which it flows; the DRIFT is its speed. In British usage, tidal current is called TIDAL STREAM, and nontidal current is called CURRENT.

current chart. . A chart on which current data are graphically depicted. See also TIDAL CURRENT CHARTS.

current constants. . Tidal current relations that remain practically constant for any particular locality. Current constants are classified as **harmonic** and **nonharmonic**. The harmonic constants consist of the amplitudes and epochs of the harmonic constituents, and the nonharmonic constants include the velocities and intervals derived directly from the current observations.

current curve. . A graphic representation of the flow of the current. In the reversing type of tidal current, the curve is referred to rectangular coordinates with time represented by the abscissas and the speed of the current by the ordinates, the flood speeds being considered as positive' and the ebb speeds as negative. In general, the current curve for a reversing tidal current approximates a cosine curve.

current cycle. . A complete set of tidal current conditions, as those occurring during a tidal day, lunar month, or Metonic cycle.

current diagram. . A graphic table showing the speeds of the flood and ebb currents and the times of slack and strength over a considerable stretch of the channel of a tidal waterway, the times being referred to tide or tidal current phases at some reference station.

current difference. . The difference between the time of slack water (or minimum current) or strength of current in any locality and the time of the corresponding phase of the tidal current at a reference station, for which predictions are given in the *Tidal Current Tables*.

current direction. . The direction toward which a current is flowing, called the SET of the current.

current ellipse. . A graphic representation of a rotary current in which the velocity of the current at different hours of the tidal cycle is represented by radius vectors and vectorial angles. A line joining the extremities of the radius vectors will form a curve roughly approximating an ellipse. The cycle is completed in one half tidal day or in a whole tidal day according to whether the tidal current is of the semidiurnal or the diurnal type. A current of the mixed type will give a curve of two unequal loops each tidal day.

current hour. . The mean interval between the transit of the moon over the meridian of Greenwich and the time of strength of flood, modified by the times of slack water (or minimum current) and strength of ebb. In computing the mean current hour an average is obtained of the intervals for the following phases: flood strength, slack (or minimum) before flood increased by 3.10 hours (one-fourth of tidal cycle), slack (or minimum) after flood decreased by 3.10 hours, and ebb strength increased or decreased by 6.21 hours (one-half of tidal cycle). Before taking the average, the four phases are made comparable by the addition or rejection of such multiples of 12.42 hours as may be necessary. The current hour is usually expressed in solar time, but if the use of lunar time is desired the solar hour should be multiplied by the factor 0.966.

current line. . A graduated line attached to a CURRENT POLE, used in measuring the velocity of the current. The line is marked so that the speed of the current, expressed in knots and tenths, is indicated directly by the length of line carried out by the current pole in a specified interval of time. When marked for a 60 second run, the principal divisions for the whole knots are spaced 101.33 feet and the subdivisions for tenths of knots are spaced at 10.13 feet. Also called LOG LINE.

current meter. . An instrument for measuring the speed and direction or just speed of a current. The measurements are usually Eulerian since the meter is most often fixed or moored at a specific location.

current pole. . A pole used in observing the velocity of the current. In use, the pole, which is weighted at one end so as to float upright, is attached to the current line but separated from the graduated portion by an ungraduated section of approximately 100 feet, known as the *stray line*. As the pole is carried out from an observing vessel by the current, the amount of line passing from the vessel during a specific time interval indicates the speed of the current. The set is obtained from a bearing from the vessel to the pole.

current rips. . See RIPS.

current sailing. . The process of allowing for current when predicting the track to be made good or of determining the effect of a current on the direction of motion of a vessel. The expression is better avoided, as the process is not strictly a sailing.

current station. . The geographic location at which current observations are conducted. Also, the facilities used to make current observations. These may include a buoy, ground tackle, current meters, recording mechanism, and radio transmitter. See also CONTROL CURRENT STATION, SUBORDINATE CURRENT STATION.

current tables. . See TIDAL CURRENT TABLES.

cursor. , *n.* A device used with an instrument to provide a moveable reference. A symbol indicating the location in a file of the data entry point of a computer.

curve of constant bearing. . See CURVE OF EQUAL BEARING.

curve of equal bearing. . A curve connecting all points at which the great-circle bearing of a given point is the same. Also called CURVE OF CONSTANT BEARING.

curvilinear. , *adj.* Consisting of or bounded by a curve.

curvilinear triangle. . A closed figure having three curves as sides.

cusp. , *n.* One of the horns or pointed ends of the crescent moon or other luminary.

cut. , *n.* 1. A notch or depression produced by excavation or erosion. 2. The intersection of lines of position, constituting a fix, with particular reference to the angle of intersection.

cut in. . To observe and plot lines of position locating an object or craft, particularly by bearings.

cut-off. , *n.* 1. A new and relatively short channel formed when a stream cuts through the neck of an oxbow or horseshoe bend. 2. An artificial straightening or short-cut in a channel.

Cyclan. , *n.* The designation of Loran C in its earliest stage of development but later superseded by the term CYTAC.

cycle. , *n.* One complete train of events or phenomena that recur sequentially. When used in connection with sound or radio the term refers to one complete wave, or to a frequency of one wave per second. See also KILOCYCLE, MEGACYCLE, CALLIPPIC CYCLE, CURRENT CYCLE, DUTY CYCLE, LUNAR CYCLE, METONIC CYCLE, TIDAL CYCLE.

cycle match. . In Loran C, the comparison, in time difference, between corresponding carrier cycles contained in the rise times of a master and secondary station pulse. The comparison is refined to a determination of the phase difference between these two cycles. See also ENVELOPE MATCH.

cyclic. , *adj.* Of or pertaining to a cycle or cycles.

cyclogenesis., *n.* A development or strengthening of cyclonic circulation in the atmosphere. The opposite is CYCLOLYSIS. The term is applied to the development of cyclonic circulation where previously it did not exist, as well as to the intensification of existing cyclonic flow. While cyclogenesis usually occurs with a deepening (a decrease in atmospheric pressure), the two terms should not be used synonymously.

cyclolysis., *n.* Any weakening of cyclonic circulation in the atmosphere. The opposite is CYCLOGENESIS. While cyclolysis usually occurs with a filling (an increase in atmospheric pressure), the two terms should not be used synonymously.

cyclone., *n.* 1. A meteorological phenomena characterized by relatively low atmospheric pressure and winds which blow counterclockwise around the center in the Northern Hemisphere and clockwise in the Southern Hemisphere. 2. The name by which a tropical storm having winds of 34 knots or greater is known in the South Indian Ocean. See TROPICAL CYCLONE.

cyclonic storm.. See under TROPICAL CYCLONE.

cyclonic winds.. The winds associated with a low pressure area and constituting part of a cyclone.

cylinder., *n.* 1. A solid figure having two parallel plane bases bounded by closed congruent curves, and a surface formed by parallel lines connecting similar points on the two curves. 2. A surface formed by a straight line moving parallel to itself and constantly intersecting a curve. Also called CYLINDRICAL SURFACE.

cylindrical., *adj.* Of or pertaining to a cylinder.

cylindrical buoy.. See CAN BUOY.

cylindrical chart.. A chart on a cylindrical map projection.

cylindrical map projection.. A map projection in which the surface of a sphere or spheroid, such as the earth, is conceived as developed on a tangent cylinder, which is then spread out to form a plane. See also MERCATOR MAP PROJECTION, RECTANGULAR MAP PROJECTION, EQUATORIAL MAP PROJECTION, OBLIQUE MAP PROJECTION, OBLIQUE MERCATOR MAP PROJECTION, TRANSVERSE MAP PROJECTION.

cylindrical surface.. A surface formed by a straight line moving parallel to itself and constantly intersecting a curve. Also called a CYLINDER.

Cytac., *n.* The designation of Loran C in an earlier stage of development. See also CYCLAN.

D

daily aberration.. See under ABERRATION, definition 1.

Daily Memorandum.. An electronic file of the National Imagery and Mapping Agency's Maritime Safety Information System web site, containing HYDROLANTS, HYDROPACS, and NAVAREA Warnings from NAVAREAS IV and XII. The HYDROLANTS, HYDROPACS, and NAVAREA Warnings are broadcast messages restricted to the more important marine incidents or navigational changes for which a delay in disseminating the information to mariners would adversely affect navigational safety.

daily rate.. See CHRONOMETER RATE, WATCH RATE.

dale., *n.* A vale or small valley.

dam., *n.* A barrier to check or confine anything in motion; particularly a bank of earth, masonry, etc., across a watercourse to keep back moving water.

damped wave.. 1. A wave such that, at every point, the amplitude of each sinusoidal component is a decreasing function of time. 2. A wave in which the amplitudes of successive peaks (crests) progressively diminish.

damp haze.. See under HAZE.

damping., *n.* 1. The reduction of energy in a mechanical or electrical system by absorption or radiation. 2. The act of reducing the amplitude of the oscillations of an oscillatory system; hindering or preventing oscillation or vibration; diminishing the sharpness of resonance of the natural frequency of a system.

damping error.. See as BALLISTIC DAMPING ERROR.

dan buoy.. A buoy consisting of a ballasted float carrying a staff which supports a flag or light. Dan buoys are used principally in minesweeping, and by fisherman to mark the position of deepsea fishing lines or nets.

danger angle.. The maximum (or minimum) angle between two points, as observed from a craft indicating the limit of safe approach to an offlying danger. A horizontal danger angle is measured between points shown on the chart. A vertical danger angle is measured between the top and bottom of an object of known height.

danger area.. A specified area above, below, or within which there may exist potential danger. See also PROHIBITED AREA, RESTRICTED AREA.

danger bearing.. The maximum or minimum bearing of a point for safe passage of an off-lying danger. As a vessel proceeds along a coast, the bearing of a fixed point on shore, such as a lighthouse, is measured frequently. As long as the bearing does not exceed the limit of the predetermined danger bearing, the vessel is on a safe course.

danger buoy.. A buoy marking an isolated danger to navigation, such as a rock, shoal or sunken wreck.

danger line.. 1. A line drawn on a chart to indicate the limits of safe navigation for a vessel of specific draft. 2. A line of small dots used to draw the navigator's attention to a danger which would not stand out clearly enough if it were represented on the chart solely by the specific symbols. This line of small dots is also used to delimit areas containing numerous dangers, through which it is unsafe to navigate.

dangerous semicircle.. The half of a cyclonic Storm in which the rotary and forward motions of the storm reinforce each other and the winds tend to blow a vessel into the storm track. In the Northern Hemisphere this is to the right of the storm center (when facing the direction the storm is moving) and in the Southern Hemisphere it is to the left. The opposite is the LESS DANGEROUS or NAVIGABLE SEMICIRCLE.

danger sounding.. A minimum sounding chosen for a vessel of specific draft in a given area to indicate the limit of safe navigation.

dark nilas.. Nilas which is under 5 centimeters in thickness and is very dark in color.

dark-trace tube.. A cathode-ray tube having a specially coated screen which changes color but does not necessarily luminesce when struck by the electron beam. It shows a dark trace on a bright background.

data.. Factual information.

data-acquisition station.. A ground station used for performing the various functions necessary to control satellite operations and to obtain data from the satellite.

data base.. A uniform, organized set of data.

data processing.. Changing data from one form or format to another by application of specified routines or algorithms.

data reduction.. The process of transforming raw data into more ordered data.

data smoothing.. The process of fitting dispersed data points to a smooth or uniform curve or line.

date., *n.* A designated mark or point on a time scale.

date line.. The line coinciding approximately with the 180th meridian, at which each calendar day first begins; the boundary between the -12 and +12 time zones. The date on each side of this line differs by 1 day, but the time is the same in these two zones. When crossing this line on a westerly course, the date must be advanced 1 day; when crossing on an easterly course, the date must be put back 1 day. Sometimes called INTERNATIONAL DATE LINE.

datum., *n.* Any numerical or geometrical quantity or set of such quantities which may serve as reference or base for other quantities. In navigation two types of datums are used: horizontal and vertical. See also HORIZONTAL GEODETIC DATUM, VERTICAL GEODETIC DATUM. CHART SOUNDING DATUM, VERTICAL DATUM.

datum-centered ellipsoid.. The reference ellipsoid that gives the best fit to the astrogeodetic network of a particular datum, and hence does not necessarily have its center at the center of the earth.

datum plane.. A misnomer for collection of datums used in mapping, charting, and geodesy which are not strictly planar. This term should not be used.

datum transformation.. The systematic elimination of discrepancies between adjoining or overlapping triangulation networks from different datums by moving the origins, rotating, and stretching the networks to fit each other.

Davidson Current.. A seasonal North Pacific Ocean countercurrent flowing northwestward along the west coast of North America from north of 32° N to at least latitude 48° N, inshore of the southeasterly-flowing California Current. This current occurs generally between November and April, but is best established in January. Strong opposing winds may cause the current to reverse. Also called WINTER COASTAL COUNTERCURRENT.

Davidson Inshore Current.. See DAVIDSON CURRENT.

dawn., *n*. The first appearance of light in the eastern sky before sunrise; daybreak. See also DUSK, TWILIGHT.

day., *n*. 1. The duration of one rotation of a celestial body on its axis. It is measured by successive transits of a reference point on the celestial sphere over the meridian, and each type takes its name from the reference used. Thus, for a solar day on earth the reference is the sun; a mean solar day uses the mean sun; and an apparent solar day uses the apparent sun. For a lunar day the reference is the moon; for a sidereal day the vernal equinox; for a constituent day an astre fictif or fictitious star representing one of the periodic elements in the tidal forces. The expression lunar day refers also to the duration of one rotation of the moon with respect to the sun. A Julian day begins at Greenwich mean noon and the days are consecutively numbered from January 1, 4713 B.C. 2. A period of 24 hours beginning at a specified time, as the civil day beginning at midnight, or the astronomical day beginning at noon, which was used up to 1925 by astronomers. 3. A specified time or period, usually of approximately 24-hours duration. A calendar day extends from midnight to midnight, and is of 24-hours duration unless a time change occurs during the day. A tidal day is either the same as a lunar day (on the earth), or the period of the daily cycle of the tides, differing slightly from the lunar day because of priming and lagging. 4. The period of daylight, as distinguished from night.

daybeacon., *n*. An unlighted beacon. A daybeacon is identified by its color and the color, shape and number of its daymark. The simplest form of daybeacon consists of a single pile with a daymark affixed at or near its top. See also DAYMARK.

daybreak., *n*. See DAWN.

daylight control.. A photoelectric device that automatically lights and extinguishes a navigation light, usually lighting it at or about sunset and extinguishing it at or about sunrise. Also called SUN RELAY, SUN SWITCH, SUN VALVE.

daylight saving meridian.. The meridian used for reckoning daylight saving time. This is generally 15° east of the ZONE or STANDARD MERIDIAN.

daylight saving noon.. Twelve o'clock daylight saving time, or the instant the mean sun is over the upper branch of the daylight saving meridian. Also called SUMMER NOON, especially in Europe. See also MEAN NOON.

daylight saving time.. A variation of standard time in order to make better use of daylight. In the United States the "Uniform Time Act of 1966" (Public Law 99-359 Sect. 2) establishes the annual advancement and retardation of standard time by 1 hour at 2 AM on the first Sunday of April and last Sunday of October, respectively, except in those states which have by law exempted themselves from the observance of daylight saving time. Also called SUMMER TIME, especially in Europe.

daylight signal light.. A signal light exhibited by day and also, usually with reduced intensity by night. The reduction of intensity is made in order to avoid glare. Daylight signals may be used to indicate whether or not the entrance to a lock is free.

daymark., *n*. 1. The daytime identifying characteristics of an aid to navigation. See also DAYBEACON. 2. An unlighted navigation mark. 3. The shaped signals used to identify vessels engaged in special operations during daytime, more properly known as day shapes.

day's run.. The distance traveled by a vessel in 1 day, usually reckoned from noon to noon.

dead ahead.. Bearing 000° relative. If the bearing is approximate, the term AHEAD should be used.

dead astern.. Bearing 180° relative. If the bearing is approximate, the term ASTERN should be used. Also called RIGHT ASTERN.

deadbeat., *adj*. Aperiodic, or without a period.

deadbeat compass.. See APERIODIC COMPASS.

deadhead., *n*. 1. A block of wood used as an anchor buoy. 2. A bollard, particularly one of wood set in the ground.

deadman.. Timber or other long sturdy object buried in ice or ground to which ship's mooring lines are attached.

dead reckoning.. Determining the position of a vessel by adding to the last fix the ship's course and speed for a given time. The position so obtained is called a DEAD RECKONING POSITION. Comparison of the dead reckoning position with the fix for the same time indicates the sum of currents, winds, and other forces acting on the vessel during the intervening period.

Dead Reckoning Altitude and Azimuth Table.. See *H.O. PUB. NO. 211*.

dead reckoning equipment.. A device that continuously indicates the dead reckoning position of a vessel. It may also provide, on a dead reckoning tracer, a graphical record of the dead reckoning. See also COURSE RECORDER.

dead reckoning plot.. The graphic plot of the dead reckoning, suitably labeled with time, direction, and speed. See also NAVIGATIONAL PLOT.

dead reckoning position.. See under DEAD RECKONING.

dead reckoning tracer.. A device that automatically provides a graphic record of the dead reckoning. It may be part of dead reckoning equipment. See also COURSE RECORDER.

dead water.. The water carried along with a ship as it moves through the water. It is maximum at the waterline and decreases with depth. It increases in a direction towards the stern.

deca-.. A prefix meaning *ten*.

decameter., *n*. Ten meters.

Decca., *n*. See as DECCA NAVIGATOR SYSTEM.

Decca chain.. A group of associated stations of the Decca Navigator System. A Decca chain normally consists of one master and three slave stations. Each slave station is called by the color of associated pattern of hyperbolic lines as printed on the chart, i.e., red slave, green slave, purple slave. See also CHAIN.

Decca Navigator System.. A short to medium range low frequency (70-130 kHz) radionavigation system which yields a hyperbolic line of position of high accuracy. The system is an arrangement of fixed, phase locked, continuous wave transmitters operating on harmonically related frequencies and special receiving and display equipment carried on a vessel or other craft. The operation of the system depends on phase comparison of the signals from the transmitters brought to a common comparison frequency within the receiver.

decelerate., *v., t*. To cause to more slower. *v. i*. To decrease speed.

deceleration., *n*. Negative acceleration.

December solstice.. Winter solstice in the Northern Hemisphere.

deci-.. A prefix meaning one-tenth. decibar, *n*. One-tenth of a bar; 100 millibars.

decibel., *n*. A dimensionless unit used for expressing the ratio between widely different powers. It is 10 times the logarithm to the base 10 of the power ratio.

decimeter., *n*. One-tenth of a meter.

deck log.. See LOG, definition 2.

declination., *n*. 1. Angular distance north or south of the celestial equator; the arc of an hour circle between the celestial equator and a point on the celestial sphere, measured northward or southward from the celestial equator through 90°, and labeled N or S (+ or -) to indicate the direction of measurement. 2. Short for MAGNETIC DECLINATION.

declinational inequality.. See DIURNAL INEQUALITY.

declinational reduction.. A processing of observed high and low waters or flood and ebb tidal currents to obtain quantities depending upon changes in the declination of the moon; such as tropic ranges or speeds, height or speed inequalities, and tropic intervals.

declination difference.. The difference between two declinations, particularly between the declination of a celestial body and the value used as an argument for entering a table.

declinometer. , *n.* An instrument for measuring magnetic declination. See also MAGNETOMETER.

Decometer. , *n.* A phase meter used in the Decca Navigator System.

decrement. , *n.* 1. A decrease in the value of a variable. 2. *v.* To decrease a variable in steps. See also INCREMENT.

deep. , *n.* 1. An unmarked fathom point on a lead line. 2. A relatively small area of exceptional depth found in a depression of the ocean floor. The term is generally restricted to depths greater than 3,000 fathoms. If it is very limited in area, it is referred to as a HOLE. 3. A relatively deep channel in a strait or estuary.

deepening. , *n.* Decrease in atmospheric pressure, particularly within a low. Increase in pressure is called FILLING. See also CYCLOGENESIS.

deep sea lead. . A heavy sounding lead (about 30 to 100 pounds), usually having a line 100 fathoms or more in length. A light deep sea lead is sometimes called a COASTING LEAD. Sometimes called DIPSEY LEAD.

deep water route. . A route for deep draft vessels within defined limits which has been accurately surveyed for clearance of sea bottom and submerged obstacles as indicated on the chart. See also ROUTING SYSTEM.

definition. , *n.* The clarity and fidelity of the detail of radar images on the radarscope. A combination of good resolution and focus is required for good definition.

definitive orbit. . An orbit that is defined in a highly precise manner with due regard taken for accurate constants and observational data, and precision computational techniques including perturbations.

deflection of the plumb line. . See under DEFLECTION OF THE VERTICAL.

deflection of the vertical. . The angular difference at any place, between the direction of a plumb line (the vertical) and the perpendicular to the reference ellipsoid. This difference seldom exceeds 30". Often expressed in two components, meridian and prime vertical. Also called STATION ERROR.

deflection of the vertical correction. . The correction due to deflection of the vertical resulting from irregularities in the density and form of the earth. Deflection of the vertical affects the accuracy of sextant altitudes.

deflector. , *n.* An instrument for measuring the directive force acting on a magnetic compass. It is used for adjusting a compass when ordinary methods of determining deviation are not available, and operates on the theory that when the directive force is the same on all cardinal headings, the compass is approximately adjusted.

deformed ice. . A general term for ice which has been squeezed together and in places forced forwards (and downwards). Subdivisions are RAFTED ICE, RIDGED ICE, and HUMMOCKED ICE.

degaussing. , *n.* Neutralization of the strength of the magnetic field of a vessel, using electric coils permanently installed in the vessel. See also DEPERMING.

degaussing cable. . A cable carrying an electric current for degaussing a vessel.

degaussing range. . An area for determining magnetic signatures of ships and other marine craft. Such signatures are used to determine required degaussing coil current settings and other required corrective actions. Sensing instruments and cables are installed on the sea bed in the range, and there are cables leading from the range to a control position ashore.

degree. , *n.* 1. A unit of circular measure equal to 1/360th of a circle. 2. A unit of measurement of temperature.

degree-of-freedom. . The number of orthogonal axes of a gyroscope about which the spin axis is free to rotate, the spin axis freedom not being counted. This is not a universal convention. For example, the free gyro is frequently referred to as a three-degree-of-freedom gyro, the spin axis being counted.

deka-. . A prefix meaning ten (10).

delayed plan position indicator. . A plan position indicator on which the start of the sweep is delayed so that the center represents a selected range. This allows distant targets to be displayed on a larger-scale presentation.

delayed sweep. . Short for DELAYED TIME BASE SWEEP.

delayed time base. . Short for DELAYED TIME BASE SWEEP.

delayed time base sweep. . A sweep, the start of which is delayed, usually to provide an expanded scale for a particular part. Usually shortened to DELAYED SWEEP, and sometimes to DELAYED TIME BASE.

delta. , *n.* 1. The low alluvial land, deposited in a more or less triangular form, as the Greek letter delta, at the mouth of a river, which is often cut by several distributaries of the main stream. 2. A change in a variable quantity, such as a change in the value of the declination of a celestial body.

demagnetize. , *v., t.* To remove magnetism. The opposite is MAGNETIZE.

demodulation. , *n.* The process of obtaining a modulating wave from a modulated carrier. The opposite is MODULATION.

departure. , *n.* 1. The distance between two meridians at any given parallel of latitude, expressed in linear units, usually nautical miles; the distance to the east or west made good by a craft in proceeding from one point to another. 2. The point at which reckoning of a voyage begins. It is usually established by bearings of prominent landmarks as the vessel clears a harbor and proceeds to sea. When a navigator establishes this point, he is said to take departure. Also called POINT OF DEPARTURE. 3. Act of departing or leaving. 4. The amount by which the value of a meteorological element differs from the normal value.

dependent surveillance. . Position determination requiring the cooperation of the tracked craft.

deperming. , *n.* The process of changing the magnetic condition of a vessel by wrapping a large conductor around it a number of times in a vertical plane, athwartships, and energizing the coil thus formed. If a single coil is placed horizontally around the vessel and energized, the process is called FLASHING if the coil remains stationary, and WIPING if it is moved up and down. See also DEGAUSSING.

depressed pole. . The celestial pole below the horizon, of opposite name to the latitude. The celestial pole above the horizon is called ELEVATED POLE.

depression. , *n.* 1. See NEGATIVE ALTITUDE. 2. A developing cyclonic area, or low pressure area.

depression angle. . See ANGLE OF DEPRESSION.

depth. , *n.* The vertical distance from a given water level to the sea bottom. The charted depth is the vertical distance from the tidal datum to the bottom. The least depth in the approach or channel to an area, such as a port or anchorage, governing the maximum draft of vessels that can enter is called the controlling depth. See also CHART SOUNDING DATUM.

depth contour. . A line connecting points of equal depth below the sounding datum. It may be called FATHOM CURVE or FATHOM LINE if depth is expressed in fathoms. Also called DEPTH CURVE, ISOBATH.

depth curve. . See DEPTH CONTOUR.

depth finder. . See ECHO SOUNDER.

depth of water. . The vertical distance from the surface of the water to the bottom. See also SOUNDING.

depth perception. . The ability to estimate depth or distance between points in the field of vision.

derelict. , *n.* Any property abandoned at sea, often large enough to constitute a menace to navigation; especially an abandoned vessel. See also JETTISON, WRECK.

derived units. . See under INTERNATIONAL SYSTEM OF UNITS.

descending node. . The point at which a planet, planetoid, or comet crosses the ecliptic from north to south, or a satellite crosses the plane of the equator of its primary from north to south. Also called SOUTHBOUND NODE. The opposite is ASCENDING NODE.

destination. , *n.* The port of intended arrival. Also called POINT OF DESTINATION. See also POINT OF ARRIVAL.

detection. , *n.* 1. The process of extracting information from an electromagnetic wave. 2. In the use of radar, the recognition of the presence of a target.

detritus. , *n.* An accumulation of the fragments resulting from the disintegration of rocks.

developable. , *adj.* Capable of being flattened without distortion. The opposite is UNDEVELOPABLE.

developable surface. . A curved surface that can be spread out in a plane without distortion, e.g., the cone and the cylinder.

deviascope. , *n.* A device for demonstration of various forms of deviation and compass adjustment, or compass compensation.

deviation. , *n.* 1. The angle between the magnetic meridian and the axis of a compass card, expressed in degrees east or west to indicate the direction in which the northern end of the compass card is offset from magnetic north. Deviation is caused by disturbing magnetic influences in the immediate vicinity of the compass. Semicircular deviation changes sign (E or W) approximately each 180° change of heading; quadrantal deviation changes sign approximately each 90° change of heading; constant deviation is the same on any heading. Deviation of a magnetic compass after adjustment or compensation is RESIDUAL DEVIATION. Called MAGNETIC DEVIATION when a distinction is needed to prevent possible ambiguity. 2. Given a series of observations or measurements of a given quantity, the deviation of a single observation is the algebraic difference between the single observation and the mean or average value of the series of observations. See also RANDOM ERROR.

deviation table. . A table of the deviation of a magnetic compass on various headings, magnetic or compass. Also called MAGNETIC COMPASS TABLE. See also NAPIER DIAGRAM.

dew point. . The temperature to which air must be cooled at constant pressure and constant water vapor content to reach saturation. Any further cooling usually results in the formation of dew or frost.

DGPS. . Differential Global Positioning System; a method of increasing the accuracy of GPS positions by transmitting corrections generated by precisely surveyed reference stations.

diagram on the plane of the celestial equator. . See TIME DIAGRAM.

diagram on the plane of the celestial meridian. . A theoretical orthographic view of the celestial sphere from a point outside the sphere and over the celestial equator. The great circle appearing as the outer limit is the local celestial meridian; other celestial meridians appear as ellipses. The celestial equator appears as a diameter 90° from the poles. Parallels of declination appear as straight lines parallel to the equator. The celestial horizon appears as a diameter 90° from the zenith.

diagram on the plane of the equinoctial. . See TIME DIAGRAM.

diameter. , *n.* Any chord passing through the center of a figure, as a circle, ellipse, sphere, etc., or the length of such chord. See also RADIUS.

diaphone. , *n.* A sound signal emitter operating on the principle of periodic release of compressed air controlled by the reciprocating motion of a piston operated by compressed air. The diaphone usually emits a powerful sound of low pitch which often concludes with a brief sound of lowered pitch called the GRUNT. The emitted signal of a TWO-TONE DIAPHONE consists of two tones of different pitch, in which case the second tone is of lower pitch.

diaphragm horn. . A sound signal emitter comprising a resonant horn excited at its throat by impulsive emissions of compressed air regulated by an elastic diaphragm. Duplex or triplex horn units of different pitch produce a chime signal. Also called COMPRESSED-AIR HORN.

diatom. , *n.* A microscopic alga with an external skeleton of silica, found in both fresh and salt water. Part of the ocean bed is composed of a sedimentary ooze consisting principally of large collections of the skeletal remains of diatoms.

dichroic mirror. . A glass surface coated with a special metallic film that permits some colors of light to pass through the glass while reflecting certain other colors of light. Also called SEMIREFLECTING MIRROR.

dichroism. , *n.* The optical property of exhibiting two colors, as one color in transmitted light and another in reflected light. See also DICHROIC MIRROR.

dielectric reflector. . A device composed of dielectric material which returns the greater part of the incident electromagnetic waves parallel to the direction of incidence. See also RADAR REFLECTOR.

difference of latitude. . The shorter arc of any meridian between the parallels of two places, expressed in angular measure.

difference of longitude. . The smaller angle at the pole or the shorter arc of a parallel between the meridians of two places, expressed in angular measure.

difference of meridional parts. . See MERIDIONAL DIFFERENCE.

differential. . Relating to the technology of increasing the accuracy of an electronic navigation system by monitoring the system error from a known, fixed location and transmitting corrections to vessels using the system. Differential GPS is in operation. Differential Loran has been in an experimental phase.

differentiator. , *n.* See FAST TIME CONSTANT CIRCUIT.

diffraction. , *n.* 1. The bending of the rays of radiant energy around the edges of an obstacle or when passing near the edges of an opening, or through a small hole or slit, resulting in the formation of a spectrum. See also REFLECTION REFRACTION. 2. The bending of a wave as it passes an obstruction.

diffuse ice edge. . A poorly defined ice edge limiting an area of dispersed ice. It is usually on the leeward side of an area of pack ice.

diffuse reflection. . A reflection process in which the reflected radiation is sent out in many directions usually bearing no simple relationship to the angle of incidence. It results from reflection from a rough surface with small irregularities. See also SPECULAR REFLECTION.

diffusion. , *n.* See DIFFUSE REFLECTION.

digit. , *n.* A single character representing an integer.

digital. . Referring to the use of discreet expressions to represent variables. See ANALOG.

digital calculator. . In navigation, a small electronic device which does arithmetical calculations by applying mathematical formulas (ALGORITHMS) to user-entered values. A navigational calculator has preloaded programs to solve navigational problems.

digital computer. . An electronic device larger and more sophisticated than a calculator which can operate a variety of software programs. In navigation, computers are used to run celestial sight reduction programs, tide computing programs, electronic chart programs, ECDIS, and for a number of other tasks in ship management.

digital nautical chart (DNC). . The electronic chart data base used in the U.S. Navy's NAVSSI.

digital selective calling (DSC). . A communications technique using coded digitized signals which allows transmitters and receivers to manage message traffic, accepting or rejecting messages according to certain variables.

digital tide gage. . See AUTOMATIC TIDE GAGE.

digitize. . To convert analog data to digital data.

dihedral angle. . The angle between two intersecting planes.

dihedral reflector. . A radar reflector consisting of two flat surfaces intersecting mutually at right angles. Incident radar waves entering the aperture so formed with a direction of incidence perpendicular to the edge, are returned parallel to their direction of incidence. Also called RIGHT ANGLE REFLECTOR.

dike. , *n.* A bank of earth or stone used to form a barrier, which restrains water outside of an area that is normally flooded. See LEVEE.

dioptric light. . A light concentrated into a parallel beam by means of refracting lenses or prisms. One so concentrated by means of a reflector is a CATOPTRIC LIGHT.

dip. , *n.* 1. The vertical angle, at the eye of an observer, between the horizontal and the line of sight to the visible horizon. Altitudes of celestial bodies measured from the visible sea horizon as a reference are too great by the amount of dip. Since dip arises from and varies with the elevation of the eye of the observer above the surface of the earth, the correction for dip is sometimes called HEIGHT OF EYE CORRECTION. Dip is smaller than GEOMETRICAL DIP by the amount of terrestrial refraction. Also called DIP OF THE HORIZON. 2. The angle between the horizontal and the lines of force of the earth's magnetic field at any point. Also called MAGNETIC DIP, MAGNETIC LATITUDE, MAGNETIC INCLINATION. 3. The first detectable decrease in the altitude of a celestial body after reaching its maximum altitude on or near meridian transit.

dip. , *v., i.* To begin to descend in altitude after reaching a maximum on or near meridian transit.

dip circle. . An instrument for measuring magnetic dip. It consists of a DIP NEEDLE, or magnetic needle, suspended in such manner as to be free to rotate about a horizontal axis.

dip correction. . The correction to sextant altitude due to dip of the horizon. Also called HEIGHT OF EYE CORRECTION.

dip needle. . A magnetic needle suspended so as to be free to rotate about a horizontal axis. An instrument using such a needle to measure magnetic dip is called a DIP CIRCLE. A dip needle with a sliding weight that can be moved along one of its arms to balance the magnetic force is called a HEELING ADJUSTER.

dip of the horizon. . See DIP, *n.*, definition 1.

dipole antenna. , *n.* A straight center-fed one-half wavelength antenna. Horizontally polarized it produces a figure eight radiation pattern, with maximum radiation at right angles to the plane of the antenna. Also called DOUBLET ANTENNA.

dip pole. . See as MAGNETIC DIP POLE.

dipsey lead. (led). See DEEP SEA LEAD.

direct indicating compass. . A compass in which the dial, scale, or index is carried on the sensing element.

direction. , *n.* The position of one point in space relative to another without reference to the distance between them. Direction may be either three-dimensional or two-dimensional, the horizontal being the usual plane of the latter. Direction is not an angle but is often indicated in terms of its angular distance from a REFERENCE DIRECTION. Thus, a horizontal direction may be specified as compass, magnetic, true, grid or relative. A Mercator or rhumb direction is the horizontal direction of a rhumb line, expressed as angular distance from a reference direction, while great circle direction is the horizontal direction of a great circle, similarly expressed. See also CURRENT DIRECTION, SWELL DIRECTION, WAVE DIRECTION, WIND DIRECTION.

directional antenna. . An antenna designed so that the radiation pattern is largely concentrated in a single lobe.

directional gyro. . A gyroscopic device used to indicate a selected horizontal direction for a limited time.

directional gyro mode. . The mode of operation of a gyrocompass in which the compass operates as a free gyro with the spin axis oriented to grid north.

directional radiobeacon. . See under RADIOBEACON. Also see as COURSE BEACON.

direction finder. . See RADIO DIRECTION FINDER.

direction finder deviation. . The angular difference between a bearing observed by a radio direction finder and the correct bearing, caused by disturbances due to the characteristics of the receiving craft or station.

direction finder station. . See RADIO DIRECTION FINDER STATION.

direction light. . A light illuminating a sector of very narrow angle and intended to mark a direction to be followed. A direction light bounded by other sectors of different characteristics which define its margins with small angles of uncertainty is called a SINGLE STATION RANGE LIGHT.

direction of current. . The direction toward which a current is flowing, called the SET of the current.

direction of force of gravity. . The direction indicated by a plumb line. It is perpendicular (normal) to the surface of the geoid. Also called DIRECTION OF GRAVITY.

direction of gravity. . See DIRECTION OF FORCE OF GRAVITY.

direction of relative movement. . The direction of motion relative to a reference point, itself usually in motion.

direction of waves or swell. . The direction from which waves or swell are moving.

direction of wind. . The direction from which a wind is blowing.

directive force. . The force tending to cause the directive element of a compass to line up with the reference direction. Also, the value of this force. Of a magnetic compass, it is the intensity of the horizontal component of the earth's magnetic field.

directive gain. . Four times the ratio of the radiation intensity of an antenna for a given direction to the total power radiated by the antenna. Also called GAIN FUNCTION.

directivity. , *n.* 1. The characteristic of an antenna which makes it radiate or receive more efficiently in some directions than in others. 2. An expression of the value of the directive gain of an antenna in the direction of its maximum gain. Also called POWER GAIN (OF AN ANTENNA).

directivity diagram. . See RADIATION PATTERN.

direct motion. . The apparent motion of a planet eastward among the stars. Apparent motion westward is called RETROGRADE MOTION. The usual motion of planets is direct.

directory. . A list of files in a computer.

direct wave. , 1. A radio wave that travels directly from the transmitting to the receiving antenna without reflections from any object or layer of the ionosphere. The path may be curved as a result of refraction. 2. A radio wave that is propagated directly through space; it is not influenced by the ground. Also called SPACE WAVE.

discontinued. , *adj.* Said of a previously authorized aid to navigation that has been removed from operation (permanent or temporary).

discontinuity. , *n.* 1. A zone of the atmosphere within which there is a comparatively rapid transition of any meteorological element. 2. A break in sequence of continuity of anything.

discrepancy. , *n.* 1. Failure of an aid to navigation to maintain its position or function exactly as prescribed in the *Light List*. 2. The difference between two or more observations or measurements of a given quantity.

discrepancy buoy. . An easily transportable buoy used to temporarily replace a buoy missing, damaged or otherwise not watching properly.

disk. . A type of computer data storage which consists of a plastic or metallic disk which rotates to provide access to the stored data. Data is stored in discreet areas of the disk known as tracks and sectors.

Disk Operating System (DOS). . A collection of computer programs which enables an operator to use a computer.

dismal. , *n.* A swamp bordering on, or near the sea. Also called POCOSIN.

dispersion. , *n.* The separation of light into its component colors by its passage through a diffraction grating or by refraction such as that provided by a prism.

display. , *n.* 1. The visual presentation of radar echoes or electronic charts. 2. The equipment for the visual display.

disposal area. . Area designated by the Corps of Engineers for depositing dredged material where existing depths indicate that the intent is not to cause sufficient shoaling to create a danger to surface navigation. Disposal areas are shown on nautical charts. See also DUMPING GROUND, DUMP SITE, SPOIL AREA.

disposition of lights. . The arrangement, order, etc., of navigational lights in an area.

distance circles. . Circles concentric to the center of a formation of ships, designated by their radii in thousands of yards.

distance finding station. . An attended light station or lightship emitting simultaneous radio and sound signals as a means of determining distance from the source of sound, by measuring the difference in the time of reception of the signals. The sound may be transmitted through either air or water or both and either from the same location as the radio signal or a location remote from it. Very few remain in use.

distance of relative movement. . The distance traveled relative to a reference point, itself usually in motion.

distance resolution. . See RANGE RESOLUTION.

Distances Between Ports. . See PUB. 151.

Distances Between United States Ports. . A publication of the National Ocean Survey providing calculated distances in nautical miles over water areas between United States ports. A similar publication published by NIMA for foreign waters is entitled *Distances Between Ports*.

dithering. , *n.* The introduction of digital noise intended to slightly degrade the accuracy of the civilian code in order to apply Selective Availability (SA).

diurnal. , *adj.* Having a period or cycle of approximately 1 day. The tide is said to be diurnal when only one high water and one low water occur during a tidal day, and the tidal current is said to be diurnal when there is a single flood and single ebb period in the tidal day. A rotary current is diurnal if it changes its direction through 360° once each tidal day. A diurnal constituent is one which has a single period in the constituent day. See also STATIONARY WAVE THEORY, TYPE OF TIDE.

diurnal aberration. . See under ABERRATION definition 1.

diurnal age. . See AGE OF DIURNAL INEQUALITY.

diurnal circle. . The apparent daily path of a celestial body, approximating a PARALLEL OF DECLINATION.

diurnal current. . Tidal current in which the tidal day current cycle consists of one flood current and one ebb current, separated by slack water; or a change in direction of 360° of a rotary current. A SEMIDIURNAL CURRENT is one in which two floods and two ebbs, or two changes of 360°, occur each tidal day.

diurnal inequality. . The difference in height of the two high waters or of the two low waters of each tidal day; the difference in speed between the two flood tidal currents or the two ebb tidal currents of each tidal day. The difference changes with the declination of the moon and to a lesser extent with declination of the sun. In general, the inequality tends to increase with an increasing declination, either north or south. Mean diurnal high water inequality is one-half the average difference between the two high waters of each day observed over a specific 19-year Metonic cycle (the National Tidal Datum Epoch). It is obtained by subtracting the mean of all high waters from the mean of the higher high waters. Mean diurnal low water inequality is one-half the average difference between the two low waters of each day observed over a specific 19-year Metonic cycle (the National Tidal Datum Epoch). It is obtained by subtracting the mean of the lower low waters from the mean of all low waters. Tropic high water inequality is the average difference between the two high waters of the day at the times of the tropic tides. Tropic low water inequality is the average difference between the two low waters of the day at the times of the tropic tides. Mean and tropic inequalities as defined above are applicable only when the type of tide is either semidiurnal or mixed. Sometimes called DECLINATIONAL INEQUALITY.

diurnal motion. . The apparent daily motion of a celestial body.

diurnal parallax. . See GEOCENTRIC PARALLAX.

diurnal range. . See GREAT DIURNAL RANGE.

diurnal tide. . See under TYPE OF TIDE; DIURNAL, *adj.*

dive. , *n.* Submergence with one end foremost.

dive. , *v., i.* To submerge with one end foremost.

diverged beam. . See under FAN BEAM.

dividers. , *n.* An instrument consisting two pointed legs joined by a pivot, used principally for measuring distances or coordinates on charts. If the legs are pointed at both ends and provided with an adjustable pivot in the middle of the legs, the instrument is called proportional dividers. An instrument having one pointed leg and one leg carrying a pen or pencil is called COMPASSES.

D-layer. , *n.* The lowest of the ionized layers in the upper atmosphere, or ionosphere. It is present only during daylight hours, and its density is proportional to the altitude of the sun. The D-layer's only significant effect upon radio waves is its tendency to absorb their energy, particularly at frequencies below 3 megahertz. High angle radiation and signals of a frequency greater than 3 megahertz may penetrate the D-layer and be refracted or reflected by the somewhat higher E-layer.

dock. , *n.* 1. The slip or waterway between two piers, or cut into the land for the berthing of ships. A PIER is sometimes erroneously called a DOCK. Also called SLIP. See also JETTY; LANDING, definition 1; QUAY; WHARF. 2. A basin or enclosure for reception of vessels, provided with means for controlling the water level. A wet dock is one in which water can be maintained at various levels by closing a gate when the water is at the desired level. A dry dock is a dock providing support for a ship, and means of removing the water so that the bottom of the ship can be exposed. A dry dock consisting of an artificial basin is called a graving dock; one consisting of a floating structure is called a floating dock. 3. Used in the plural, a term used to describe area of the docks, wharves, basins, quays, etc.

dock. , *v., t.* To place in a dock.

docking signals. . See TRAFFIC CONTROL SIGNALS.

dock sill. . The foundation at the bottom of the entrance to a dry dock or lock against which the caisson or gates close. The depth of water controlling the use of the dock or lock is measured from the sill to the surface.

dockyard. , *n. British terminology.* Shipyard.

doctor. , *n.* 1. A cooling sea breeze in the Tropics. 2. See HARMATTAN. 3. The strong southeast wind which blows on the south African coast. Usually called CAPE DOCTOR.

dog days. . The period of greatest heat in the summer.

doldrums. , *n., pl.* The equatorial belt of calms or light variable winds, lying between the two trade wind belts. Also called EQUATORIAL CALM S.

dolphin. , *n.* A post or group of posts, used for mooring or warping a vessel. The dolphin may be in the water, on a wharf, or on the beach. See PILE DOLPHIN.

dome. , *n.* A label on a nautical chart which indicates a large, rounded, hemispherical structure rising from a building or a roof.

dome-shaped iceberg. . A solid type iceberg with a large, round, smooth top.

doppler effect. . First described by Christian Johann Doppler in 1842, an effect observed as a frequency shift which results from relative motion between a transmitter and receiver or reflector of acoustic or electromagnetic energy. The effect on electromagnetic energy is used in doppler satellite navigation to determine an observer's position relative to a satellite. The effect on ultrasonic energy is used in doppler sonar speed logs to measure the relative motion between the vessel and the reflective sea bottom (for bottom return mode) or suspended particulate matter in the seawater itself (for volume reverberation mode). The velocity so obtained and integrated with respect to time is used in doppler sonar navigators to determine position with respect to a start point. The doppler effect is also used in docking aids which provide precise speed measurements. Also called DOPPLER SHIFT.

doppler navigation. . The use of the doppler effect in navigation. See also DOPPLER SONAR NAVIGATION, DOPPLER SATELLITE NAVIGATION.

doppler radar. . Any form of radar which detects radial motion of a distant object relative to a radar apparatus by means of the change of the radio frequency of the echo signal due to motion.

doppler satellite navigation. . The use of a navigation system which determines positions based on the doppler effect of signals received from an artificial satellite.

doppler shift. . See DOPPLER EFFECT.

doppler sonar navigation. . The use of the doppler effect observed as a frequency shift resulting from relative motion between a transmitter and receiver of ultrasonic energy to measure the relative motion between the vessel and the reflective sea bottom (for bottom return mode) or suspended particulate matter in the seawater itself (for volume reverberation mode) to determine the vessel's velocity. The velocity so obtained by a doppler sonar speed log may be integrated with respect to time to determine distance traveled. This integration of velocity with time is correlated with direction of travel in a doppler sonar navigator to determine position with respect to a start point. The doppler effect is also used in docking aids to provide precise speed measurements.

double. , *v., t.* To travel around with a near reversal of course. See also ROUND.

double altitudes. . See EQUAL ALTITUDES.

double ebb. . An ebb tidal current having two maxima of speed separated by a lesser ebb speed.

double flood. . A flood tidal current having two maxima of speed separated by a lesser flood speed.

double interpolation. . Interpolation when there are two arguments or variables.

double sextant. . A sextant designed to enable the observer to simultaneously measure the left and right horizontal sextant angles of the three-point problem.

double stabilization. . See under STABILIZATION Of RADARSCOPE DISPLAY.

double star. . Two stars appearing close together. If they appear close because they are in nearly the same line of sight but differ greatly in distance from the observer, they are called an optical double star; if in nearly the same line of sight and at approximately the same distance from the observer, they are called a physical double star. If they revolve about their common center of mass, they are called a binary star.

double summer time. . See under SUMMER TIME.

doublet antenna. . See DIPOLE ANTENNA.

double tide. . A high water consisting of two maxima of nearly the same height separated by a relatively small depression, or a low water consisting of two minima separated by a relatively small elevation. Sometimes called AGGER. See also GULDER.

doubling the angle on the bow. . A method of obtaining a running fix by measuring the distance a vessel travels on a steady course while the relative bearing (right or left) of a fixed object doubles. The distance from the object at the time of the second bearing is equal to the run between bearings, neglecting drift.

doubly stabilized. . See under STABILIZATION OF RADARSCOPE DISPLAY.

doubtful. , *adj.* Of questionable accuracy. APPROXIMATE or SECOND CLASS may be used with the same meaning.

doubtful sounding. . Of uncertain depth. The expression, as abbreviated, is used principally on charts to indicate a position where the depth may be less than indicated, the position not being in doubt.

down. , *n.* 1. See DUNE. 2. An area of high, treeless ground, usually undulating and covered with grass.

down by the head. . Having greater draft at the bow than at the stern. The opposite is DOWN BY THE STERN or BY THE STERN. Also called BY THE HEAD.

down by the stern. . Having greater draft at the stern than at the bow. The opposite is DOWN BY THE HEAD or BY THE HEAD. Also called BY THE STERN. See DRAG *n.*, definition 3.

downstream. , *adj.* & *adv.* In the direction of flow of a current or stream. The opposite is UPSTREAM.

down-the-scope echo. . See CLASSIFICATION OF RADAR ECHOES.

downwind. , *adj.* & *adv.* In the direction toward which the wind is blowing. The term applies particularly to the situation of moving in this direction, whether desired or not. BEFORE THE WIND implies assistance from the wind in making progress in a desired direction. LEEWARD applies to the direction toward which the wind blows, without implying motion. The opposite is UPWIND.

draft. , *n.* The depth to which a vessel is submerged. Draft is customarily indicated by numerals called DRAFT MARKS at the bow and stern. It may also be determined by means of a DRAFT GAUGE.

draft gauge. . A hydrostatic instrument installed in the side of a vessel, below the light load line, to indicate the depth to which a vessel is submerged.

drafting machine. . See PARALLEL MOTION PROTRACTOR.

draft marks. . Numerals placed on the sides of a vessel, customarily at the bow and stern, to indicate the depth to which a vessel is submerged.

drag. , *n.* 1. See SEA ANCHOR. 2. Short for WIRE DRAG. 3. The designed difference between the draft forward and aft when a vessel is down by the stern. See also TRIM, definition 1. 4. The retardation of a ship when in shallow water. 5. Short for ATMOSPHERIC DRAG.

drag. , *v., t.* 1. To tow a line or object below the surface, to determine the least depth in an area or to insure that a given area is free from navigational dangers to a certain depth. DRAG and SWEEP have nearly the same meanings. DRAG refers particularly to the location of obstructions, or the determination that obstructions do not exist. SWEEP may include, additionally, the removal of any obstructions located. 2. To pull along the bottom, as in dragging anchor.

dragging. , *n.* 1. The process of towing a wire or horizontally set bar below the surface, to determine the least depth in an area or to insure that a given area is free from navigational dangers to a certain depth. 2. The process of pulling along the bottom, as in dragging anchor.

draw. , *v., i.* 1. To be immersed to a specified draft. 2. To change relative bearing forward or aft, or to port or starboard.

dredge. , *n.* A vessel used to dredge an area.

dredge. , *v., t.* To remove solid matter from the bottom of a water area.

dredging area. . An area where dredging vessels may be encountered dredging material for construction. Channels dredged to provide an adequate depth of water for navigation are not considered as dredging areas.

dredging buoy. . A buoy marking the limit of an area where dredging is being performed. See also SPOIL GROUND BUOY.

dried ice. . Sea ice from the surface of which meltwater has disappeared after the formation of cracks and thaw holes. During the period of drying, the surface whitens.

drift. , *n.* 1. The speed of a current as defined in CURRENT, definition 1. 2. The speed of the current as defined in CURRENT, definition 2. 3. The distance a craft is moved by current and wind. 4. Downwind or downcurrent motion of airborne or waterborne objects due to wind or current. 5. Material moved from one place and deposited in another, as sand by a river, rocks by a glacier, material washed ashore and left stranded, snow or sand piled up by wind. Rock material deposited by a glacier is also called ERRATIC. 6. The horizontal component of real precession or apparent precession, or the algebraic sum of the two. When it is desired to differentiate between the sum and its components, the sum is called total drift.

drift. , *v., i.* To move by action of wind or current without control. drift angle. 1. The angle between the tangent-to the turning circle and the centerline of the vessel during a turn. 2. The angular difference between a vessel's ground track and the water track. See also LEEWAY ANGLE.

drift axis. . On a gyroscope, the axis about which drift occurs. In a directional gyro with the spin axis mounted horizontally the drift axis is the vertical axis. See also SPIN AXIS, TOPPLE AXIS.

drift bottle. . An identifiable float allowed to drift with ocean currents to determine their sets and drifts.

drift current. . A wide, slow-moving ocean current principally caused by prevailing winds.

drifting snow. . Snow raised from the ground and carried by the wind to such a height that the horizontal visibility is considerably reduced but the vertical visibility is not materially diminished. The expression BLOWING SNOW is used when both the horizontal and vertical visibility are considerably reduced.

drift lead. . A lead placed on the bottom to indicate movement of a vessel. At anchor the lead line is usually secured to the rail with a little slack and if the ship drags anchor, the line tends forward. A drift lead is also used to indicate when a vessel coming to anchor is dead in the water or when it is moving astern. A drift lead can be used to indicate current if a ship is dead in the water.

drilling rig. . A term used solely to indicate a mobile drilling structure. A drilling rig is not charted except in the rare cases where it is converted to a permanent production platform.

drizzle. , *n.* Very small, numerous, and uniformly dispersed water drops that may appear to float while following air currents. Unlike fog droplets, drizzle falls to the ground. It usually falls from low stratus clouds and is frequently accompanied by low visibility and fog. See also MIST.

drogue. , *n.* 1. See SEA ANCHOR. 2. A current measuring assembly consisting of a weighted parachute and an attached surface buoy.

drought. , *n.* A protracted period of dry weather.

droxtal. , *n.* A very small ice particle (about 10 to 20 microns in diameter) formed by the direct freezing of supercooled water droplets at temperatures below –30°C. Droxtals cause most of the restriction to visibility in ice fog.

dry-bulb temperature. . The temperature of the air, as indicated by the dry-bulb thermometer of a psychrometer.

dry-bulb thermometer. . A thermometer with an uncovered bulb, used with a wet-bulb thermometer to determine atmosphere humidity. The two thermometers constitute the essential parts of a PSYCHROMETER.

dry compass. . A compass without a liquid-filled bowl, particularly a magnetic compass having a very light compass card. Such a magnetic compass is seldom, if ever, used in marine applications. See also LIQUID COMPASS.

dry dock. . A dock providing support for a vessel, and means for removing the water so that the bottom of the vessel can be exposed. A dry dock consisting of an artificial basin is called a graving dock; one consisting of a floating structure is called a floating dock. See also MARINE RAILWAY.

dry-dock. , *v., t.* To place in a dry dock.

drydock iceberg. . An iceberg eroded in such manner that a large U-shaped slot is formed with twin columns. The slot extends into or near the waterline.

dry fog. . A fog that does not moisten exposed surfaces.

dry harbor. . A small harbor which either dries at low water or has insufficient depths to keep vessels afloat during all states of the tide. Vessels using it must be prepared to take the ground on the falling tide.

dry haze. . See under HAZE.

drying heights. . Heights above chart sounding datum of those features which are periodically covered and exposed by the rise and fall of the tide.

dual-carrier radiobeacon. . A continuous carrier radiobeacon in which identification is accomplished by means of a keyed second carrier. The frequency difference between the two carriers is made equal to the desired audio frequency. The object of the system is to reduce the bandwidth of the transmission.

dual-rate blanking. . To provide continuous service from one Loran C chain to the next, some stations are operated as members of two chains and radiate signals at both rates. Such a station is faced periodically with an impossible requirement to radiate two overlapping pulse groups at the same time. During the time of overlap, the subordinate signal is blanked or suppressed. Blanking is accomplished in one of two ways: priority blanking in which case one rate is always superior or alternate blanking in which case the two rates alternate in the superior and subordinate roll.

duct. , *n*. See as TROPOSPHERIC RADIO DUCT.

dumb compass. . See PELORUS.

dummy antenna. . A substantially non-radiating device used to simulate an antenna with respect to input impedance over some specified range of frequencies. Also called ARTIFICIAL ANTENNA.

dumping ground. . An area used for the disposal of dredge spoil. Although shown on nautical charts as dumping grounds in United States waters, the Federal regulations for these areas have been revoked and their use for dumping discontinued. These areas will continue to be shown on nautical charts until they are no longer considered to be a danger to navigation. See also DUMP SITE, SPOIL AREA, DISPOSAL AREA.

dump site. . Area established by Federal regulation in which dumping of dredged and fill material and other nonbuoyant objects is allowed with the issuance of a permit. Dump sites are shown on nautical charts. See also DISPOSAL AREA, DUMPING GROUND, SPOIL AREA.

dune. , *n*. A mound ridge, or hill of sand piled up by the wind on the shore or in a desert. Also called SAND DUNE.

duplex. . Concurrent transmission and reception of radio signals, electronic data, or other information.

duplexer. , *n*. A device which permits a single antenna system to be used for both transmitting and receiving.

duration of flood, duration of ebb. . Duration of flood is the interval of time in which a tidal current is flooding, and the duration of ebb is the interval in which it is ebbing; these intervals being reckoned from the middle of the intervening slack waters or minimum currents. Together they cover, on an average, a period of 12.42 hours for a semidiurnal tidal current or a period of 24.84 hours for a diurnal current. In a normal semidiurnal tidal current, the duration of flood and duration of ebb will each be approximately equal to 6.21 hours, but the times may be modified greatly by the presence of a nontidal flow. In a river the duration of ebb is usually longer than the duration of flood because of the fresh water discharge, especially during the spring months when snow and ice melt are the predominant influences. See also DURATION OF RISE, DURATION OF FALL.

duration of rise, duration of fall. . Duration of rise is the interval from low water to high water, and duration of fall is the interval from high water to low water. Together they cover, on an average, a period of 12.4 2 hours for a semidiurnal tide or a period of 24.84 hours for a diurnal tide. In a normal semidiurnal tide, the duration of' rise and duration of fall will each be approximately equal to 6.21 hours, but in shallow waters and in rivers there is a tendency for a decrease in the duration of rise and a corresponding increase in the duration of fall. See also DURATION OF FLOOD, DURATION OF EBB.

dusk. , *n*. The darker part of twilight; that part of twilight between complete darkness and the darker limit of civil twilight, both morning and evening.

dust devil. . A well-developed dust whirl, a small but vigorous whirlwind, usually of short duration, rendered visible by dust, sand, and debris picked up from the ground. Diameters of dust devils range from about 10 feet to greater than 100 feet; their average height is about 600 feet, but a few have been observed as high as several thousand feet. They have been observed to rotate anticyclonically as well as cyclonically. Dust devils are best developed on a hot, calm afternoon with clear skies, in a dry region when intense surface heating causes a very steep lapse rate of temperature in the lower few hundred feet of the atmosphere.

dust storm. , *n*. An unusual, frequently severe weather condition characterized by strong winds and dust-filled air over an extensive area. Prerequisite to a dust storm is a period of drought over an area of normally arable land, thus providing very fine particles of dust which distinguish it from the much more common SANDSTORM.

dust whirl. . A rapidly rotating column of air (whirlwind) over a dry and dusty or sandy area, carrying dust, leaves, and other light material picked up from the ground. When well developed it is called DUST DEVIL.

Dutchman's log. . A buoyant object thrown overboard to determine the speed of a vessel. The time required for a known length of the vessel to pass the object is measured.

duty cycle. . An expression of the fraction of the total time of pulse radar that radio-frequency energy is radiated. It is the ratio of pulse length to pulse repetition time.

dynamical mean sun. . A fictitious sun conceived to move eastward along the ecliptic at the average rate of the apparent sun. The dynamical mean sun and the apparent sun occupy the same position when the earth is at perihelion in January. See also MEAN SUN.

dyne. , *n*. A force which imparts an acceleration of 1 centimeter per second to a mass of 1 gram. The dyne is the unit of force in the centimeter-gram-second system. It corresponds to 10^{-5} newton in the International System of Units.

E

earth-centered ellipsoid. . A reference ellipsoid whose geometric center coincides with the earth's center of gravity and whose semiminor axis coincides with the earth's rotational axis.

earth-fixed coordinate system. . Any coordinate system in which the axes are stationary with respect to the earth. See also INERTIAL COORDINATE SYSTEM.

earthlight. , *n*. The faint illumination of the dark part of the moon by sunlight reflected from the earth. Also called EARTHSHINE.

earth rate. . The angular velocity or rate of the earth's rotation. See also EARTH-RATE CORRECTION, HORIZONTAL EARTH RATE, VERTICAL EARTH RATE.

earth-rate correction. . A rate applied to a gyroscope to compensate for the apparent precession of the spin axis caused by the rotation of the earth. See also EARTH RATE, HORIZONTAL EARTH RATE, VERTICAL EARTH RATE.

earth satellite. . A body that orbits about the earth. See also ARTIFICIAL EARTH SATELLITE.

earthshine. , *n*. See EARTHLIGHT.

earth tide. . Periodic movement of the earth's crust caused by the gravitational interactions between the sun, moon, and earth.

east. , *n*. The direction 90° to the right of north. See also CARDINAL POINT.

East Africa Coastal Current. . An Indian Ocean current which originates mainly from the part of the Indian South Equatorial Current which turns northward off the northeast coast of Africa in the vicinity of latitude 10°S. The current appears to vary considerably in speed and direction from month to month. The greatest changes coincide with the period of the opposing northeast monsoon during November through March. This coastal current is most persistent in a north or northeast direction and strongest during the southwest monsoon from May through September, particularly during August. Speed and frequency begin to decrease during the transition month of October. In November at about latitude 4°N a part of the current begins to reverse; this part expands northward and southward until February. The region of reverse flow begins to diminish in March and disappear in April, when the northward set again predominates. Also called SOMALI CURRENT. See also MONSOON.

East Australia Current. . A South Pacific Ocean current flowing southward along the east coast of Australia, from the Coral Sea to a point northeast of Tasmania, where it turns to join the northeastward flow through the Tasman Sea. It is formed by that part of the Pacific South Equatorial Current that turns south east of Australia. In the southern hemisphere summer, a small part of this current flows westward along the south coast of Australia into the Indian Ocean. The East Australia Current forms the western part of the general counterclockwise oceanic circulation of the South Pacific Ocean.

eastern standard time. . See STANDARD TIME.

East Greenland Current.. An ocean current flowing southward along the east coast of Greenland carrying water of low salinity and low temperature. The East Greenland Current is joined by most of the water of the Irminger Current. The greater part of the current continues through Denmark Strait between Iceland and Greenland, but one branch turns to the east and forms a portion of the counterclockwise circulation in the southern part of the Norwegian Sea. Some of the East Greenland Current curves to the right around the tip of Greenland, flowing northward into Davis Strait as the WEST GREENLAND CURRENT. The main discharge of the Arctic Ocean is via the East Greenland Current.

easting., *n.* The distance a craft makes good to the east. The opposite is WESTING.

East Siberian Coastal Current.. An ocean current in the Chukchi Sea which joins the northward flowing Bering Current north of East Cape.

ebb., *n.* Tidal current moving away from land or down a tidal stream. The opposite is FLOOD. Sometimes the terms EBB and FLOOD are also used with reference to vertical tidal movement, but for this vertical movement the expressions FALLING TIDE and RISING TIDE are preferable. Also called EBB CURRENT.

ebb axis.. The average direction of current at strength of ebb.

ebb current.. The movement of a tidal current away from shore or down a tidal river or estuary. In the mixed type of reversing tidal current, the terms *greater ebb* and *lesser ebb* are applied respectively to the ebb tidal currents of greater and lesser speed of each day. The terms maximum *ebb* and minimum *ebb* are applied to the maximum and minimum speeds of a current running continuously. The expression *maximum ebb* is also applicable to any ebb current at the time of greatest speed. The opposite is FLOOD CURRENT.

ebb interval.. Short for STRENGTH OF EBB INTERVAL. The interval between the transit of the moon over the meridian of a place and the time of the following strength of ebb. See also LUNICURRENT INTERVAL.

ebb strength.. Phase of the ebb tidal current at the time of maximum velocity. Also, the velocity at this time. Also called STRENGTH OF EBB.

eccentric., *adj.* Not having the same center. The opposite is CONCENTRIC.

eccentric angle.. See under ANOMALY, definition 2.

eccentric anomaly.. See under ANOMALY, definition 2.

eccentric error.. See CENTERING ERROR.

eccentricity., *n.* 1. Degree of deviating from a center. 2. The ratio of the distance between foci of an ellipse to the length of the major axis, or the ratio of the distance between the center and a focus to the length of the semimajor axis. 3. The ratio of the distances from any point of a conic section to a focus and the corresponding directrix.

eccentricity component.. That part of the equation of time due to the ellipticity of the orbit and known as the eccentricity component is the difference, in mean solar time units, between the hour angles of the apparent (true) sun and the dynamical mean sun. It is also the difference in the right ascensions of these two suns.

echo., *n.* 1. A wave which has been reflected or otherwise returned with sufficient magnitude and delay to be perceived. 2. A signal reflected by a target to a radar antenna. Also called RETURN. 3. The deflection or indication on a radarscope representing a target. Also called PIP, BLIP, RETURN.

echo box.. A resonant cavity, energized by part of the transmitted pulse of a radar set, which produces an artificial target signal for tuning or testing the overall performance of a radar set. Also called PHANTOM TARGET.

echo box performance monitor.. See under PERFORMANCE MONITOR.

echogram., *n.* A graphic record of depth measurements obtained by an echo sounder. See also FATHOGRAM.

echo ranging.. The determination of distance by measuring the time interval between transmission of a radiant energy signal and the return of its echo. Since echo ranging equipment is usually provided with means for determining direction as well as distance, both functions are generally implied. The expression is customarily applied only to ranging by utilization of the travel of sonic or ultrasonic signals through water. See also RADIO ACOUSTIC RANGING, SONAR.

echo sounder.. An instrument used to determine water depth by measuring the time interval for sound waves to go from a source of sound near the surface to the bottom and back again. Also called DEPTH FINDER, ACOUSTIC DEPTH FINDER.

echo sounding.. Determination of the depth of water by measuring the time interval between emission of a sonic or ultrasonic signal and the return of its echo from the bottom. The instrument used for this purpose is called an ECHO SOUNDER. Also called ACOUSTIC SOUNDING.

eclipse., *n.* 1. Obscuring of a source of light by the intervention of an object. When the moon passes between the earth and the sun, casting a shadow on the earth, a **solar eclipse** takes place within the shadow. When the moon enters the earth's shadow, a **lunar eclipse** occurs. When the moon enters only the penumbra of the earth's shadow, a **penumbral lunar eclipse** occurs. A solar eclipse is partial if the sun is partly obscured and total if the entire surface is obscured; or **annular** if a thin ring of the sun's surface appears around the obscuring body. A lunar eclipse can be either total or partial. 2. An interval of darkness between flashes of a navigation light.

eclipse year.. The interval between two successive conjunctions of the sun with the same node of the moon's orbit, averaging 346 days, 14 hours, 52 minutes 50.7 seconds in 1900, and increasing at the rate of 2.8 seconds per century.

ecliptic., *n.* The apparent annual path of the sun among the stars; the intersection of the plane of the earth's orbit with the celestial sphere. This is a great circle of the celestial sphere inclined at an angle of about 23°27' to the celestial equator. See also ZODIAC.

ecliptic diagram.. A diagram of the zodiac, indicating the positions of certain celestial bodies in this region.

ecliptic pole.. On the celestial sphere, either of the two points 90° from the ecliptic.

ecliptic system of coordinates.. A set of celestial coordinates based on the ecliptic as the primary great circle; celestial latitude and celestial longitude.

eddy., *n.* A quasi-circular movement of water whose area is relatively small in comparison to the current with which it is associated. Eddies may be formed between two adjacent currents flowing counter to each other and where currents pass obstructions, especially on the downstream side. See also WHIRLPOOL.

effective radiated power.. The power supplied to the antenna multiplied by the relative gain of the antenna in a given direction.

effective radius of the earth.. The radius of a hypothetical earth for which the distance to the radio horizon, assuming rectilinear propagation, is the same as that for the actual earth with an assumed uniform vertical gradient of a refractive index. For the standard atmosphere, the effective radius is 4/3 that of the actual earth.

Ekman spiral.. A logarithmic spiral (when projected on a horizontal plane) formed by current velocity vectors at increasing depth intervals. The current vectors become progressively smaller with depth. They spiral to the right (looking in the direction of flow) in the Northern Hemisphere and to the left in the Southern with increasing depth. Theoretically, the surface current vector sets 45° from the direction toward which the wind is blowing. Flow opposite to the surface current occurs at the depth of frictional resistance. The phenomenon occurs in wind drift currents in which only the Coriolis and frictional forces are significant. Named for Vagn Walfrid Ekman who, assuming a constant eddy viscosity, steady wind stress, and unlimited depth and extent, published the effect in 1905.

E-layer., *n.* From the standpoint of its effect upon radio wave propagation, the lowest useful layer of the Kennelly-Heaviside radiation region. Its average height is about 70 miles, and its density is greatest about local apparent noon. For practical purposes, the layer disappears during the hours of darkness.

elbow., *n.* A sharp change in direction of a coast line, a channel, river, etc.

electrical distance.. A distance expressed in terms of the duration of travel of an electromagnetic wave in a given medium between two points.

electrically suspended gyro.. A gyroscope in which the main rotating element is suspended by a magnetic field or any other similar electrical phenomenon. See also GYRO, ELECTROSTATIC GYRO.

electrical storm.. See THUNDERSTORM.

electric field.. That region in space which surrounds an electrically charged object and in which the forces due to this charge are detectable. See also ELECTRIC VECTOR.

electric tape gage.. A tide gage consisting of a monel metal tape on a metal reel (with supporting frame), voltmeter, and battery. The tape is graduated with numbers increasing toward the unattached end. Tidal heights can be measured directly by unreeling the tape into its stilling well. When contact is made with the water's surface, the circuit is completed and the voltmeter needle moves. At that moment, the length of tape is read against an index mark, the mark having a known elevation relative to the tidal bench marks. Used at many long term control stations in place of the tide staff.

electric vector.. The component of the electromagnetic field associated with electromagnetic radiation which is of the nature of an electric field. The electric vector is considered to coexist with, but to act at right angles to, the magnetic vector.

electrode., *n.* A terminal at which electricity passes from one medium into another. The positive electrode is called the anode; the negative electrode is called the cathode.

electromagnetic., *adj.* Of, pertaining to, or produced by electromagnetism.

electromagnetic energy.. All forms of radiant energy, such as radio waves, light waves, X-rays, heat waves, gamma rays, and cosmic rays.

electromagnetic field.. 1. The field of influence which an electric current produces around the conductor through which it flows. 2. A rapidly moving electric field and its associated magnetic field located at right angles to both electric lines of force and to their direction of motion. 3. The magnetic field resulting from the flow of electricity.

electromagnetic log.. A log containing an electromagnetic sensing element extended below the hull of the vessel, which produces a voltage directly proportional to speed through the water.

electromagnetic waves.. Waves of associated electric and magnetic fields characterized by variations of the fields. The electric and magnetic fields are at right angles to each other and to the direction of propagation. The waves are propagated at the speed of light and are known as radio (Hertzian) waves, infrared rays, light, ultraviolet rays, X-rays, etc., depending on their frequencies.

electromagnetism., *n.* 1. Magnetism produced by an electric current. 2. The science dealing with the physical relations between electricity and magnetism.

electron., *n.* A negatively-charged particle of matter constituting a part of an atom. Its electric charge is the most elementary unit of negative electricity.

electron gun.. A group of electrodes which produces an electron beam of controllable intensity. By extension, the expression is often used to include, also, the elements which focus and deflect the beam.

electronic aid to navigation.. An aid to navigation using electronic equipment. If the navigational information is transmitted by radio waves, the device may be called a RADIO AID TO NAVIGATION.

electronic bearing cursor.. The bright rotatable radial line on the display of a marine radar set, used for bearing determination.

electronic chart (EC).. A chart displayed on a video terminal, usually integrated with other navigational aids.

electronic chart data base (ECDB).. The master electronic chart data base for the electronic navigation chart held in digital form by the hydrographic authority.

electronic chart display and information system (ECDIS).. An electronic chart system which complies with IMO guidelines and is the legal equivalent of a paper chart.

electronic navigation chart (ENC).. The standardized electronic data base, a subset of the ECDB, issued by a hydrographic authority for use with an ECDIS.

electronic cursor.. Short for ELECTRONIC BEARING CURSOR.

electronic distance measuring devices.. Instruments that measure the phase differences between transmitted and reflected or retransmitted electromagnetic waves of known frequency, or that measure the round-trip transit time of a pulsed signal, from which distance is computed.

electronic navigation.. Navigation by means of electronic equipment. The expression ELECTRONIC NAVIGATION is more inclusive than RADIONAVIGATION, since it includes navigation involving any electronic device or instrument.

electronics., *n.* The science and technology relating to the emission, flow, and effects of electrons in a vacuum or through a semiconductor such as a gas, and to systems using devices in which this action takes place.

electronic telemeter.. An electronic device that measures the phase difference or transit time between a transmitted electromagnetic impulse of known frequency and speed and its return.

electrostatic gyro.. A gyroscope in which a small ball rotor is electrically suspended within an array of electrodes in a vacuum inside a ceramic envelope. See also GYRO, ELECTRICALLY SUSPENDED GYRO.

elements of a fix.. The specific values of the coordinates used to define a position.

elephanta., *n.* A strong southerly or southeasterly wind which blows on the Malabar coast of India during the months of September and October and marks the end of the southwest monsoon.

elevated duct.. A tropospheric radio duct of which the lower boundary is above the surface of the earth.

elevated pole.. The celestial pole above the horizon, agreeing in name with the latitude. The celestial pole below the horizon is called DEPRESSED POLE.

elevation., *n.* 1. Vertical distance of a point above a datum, usually mean sea level. Elevation usually applies to a point on the surface of the earth. The term HEIGHT is used for points on or above the surface. See also SPOT ELEVATION. 2. An area higher than its surroundings, as a hill.

elevation angle.. See ANGLE OF ELEVATION.

elevation tints.. See HYPSOMETRIC TINTING.

elimination., *n.* One of the final processes in the harmonic analysis of tides in which preliminary values of the harmonic constants of a number of constituents are cleared of residual effects of each other.

E-link.. A bracket attached to one of the arms of a binnacle to permit the mounting of a quadrantal corrector in an intermediate position between the fore-and-aft and athwartship lines through a magnetic compass.

ellipse., *n.* A plane curve constituting the locus of all points the sum of whose distances from two fixed points called FOCI is constant; an elongated circle. The orbits of planets, satellites, planetoids, and comets are ellipses with the center of attraction at one focus. See also CONIC SECTION, CURRENT ELLIPSE.

ellipsoid., *n.* A surface whose plane sections (cross-sections) are all ellipses or circles, or the solid enclosed by such a surface. Also called ELLIPSOID OF REVOLUTION, SPHEROID.

ellipsoidal height.. The height above the reference ellipsoid, measured along the ellipsoidal outer normal through the point in question. Also called GEODETIC HEIGHT.

ellipsoid of reference.. See REFERENCE ELLIPSOID.

ellipsoid of revolution.. A term used for an ellipsoid which can be formed by revolving an ellipse about one of its axes. Also called ELLIPSOID OF ROTATION.

ellipsoid of rotation.. See ELLIPSOID OF REVOLUTION.

elliptically polarized wave.. An electromagnetic wave which can be resolved into two plane polarized waves which are perpendicular to each other and which propagate in the same direction. The amplitudes of the waves may be equal or unequal and of arbitrary time-phase. The tip of the component of the electric field vector in the plane normal to the direction of propagation describes an ellipse. See also CIRCULARLY POLARIZED WAVE.

ellipticity., *n.* The amount by which a spheroid differs from a sphere or an ellipse differs from a circle, found by dividing the difference in the lengths of the semiaxes of the ellipse by the length of the semimajor axis. See also FLATTENING.

elongation., *n.* The angular distance of a body of the solar system from the sun; the angle at the earth between lines to the sun and another celestial body of the solar system. The greatest elongation is the maximum angular distance of an inferior planet from the sun before it starts back toward conjunction. The direction of the body east or west of the sun is usually specified, as greatest elongation east (or west).

embayed., *adj.* 1. Formed into or having bays. 2. Unable to put to sea safely because of wind, current, or sea conditions.

embayment., *n.* Any indentation of a coast regardless of width at the entrance or depth of penetration into the land. See also ESTUARY.

emergency light.. A light put into service in an emergency when the permanent or standby light has failed. It often provides reduced service in comparison with the permanent light.

Emergency Position Indicating Radiobeacon.. A small portable radiobeacon carried by vessels and aircraft which transmits radio signals which can be used by search and rescue authorities to locate a marine emergency.

emergency position indicating radiobeacon station.. As defined by the International Telecommunication Union (ITU), a station in the mobile service whose emissions are intended to facilitate search and rescue operations.

emission delay.. 1. A delay in the transmission of a pulse signal from a slave (or secondary) station of a hyperbolic radionavigation system, introduced as an aid in distinguishing between master and slave (or secondary) station signals. 2. In Loran C the time interval between the master station's transmission and the secondary station's transmission in the same group repetition interval (GRI). The GRI is selected of sufficient duration to provide time for each station to transmit its pulse group and additional time between each pulse group so that signals from two or more stations cannot overlap in time anywhere within the coverage area. In general, emission delays are kept as small as possible to allow the use of the smallest GRI.

empirical., *adj*. Derived by observation or experience rather than by rules or laws.

endless tangent screw.. A tangent screw which can be moved over its entire range without resetting.

endless tangent screw sextant.. A marine sextant having an endless tangent screw for controlling the position of the index arm and the vernier or micrometer drum. The index arm may be moved over the entire arc without resetting, by means of the endless tangent screw.

enhanced group call (EGC).. A global automated satellite communications service capable of addressing messages to specific areas or specific groups of vessels

entrance., *n*. The seaward end of channel, harbor, etc.

entrance lock.. A lock between the tideway and an enclosed basin when their water levels vary. By means of the lock, which has two sets of gates vessels can pass either way at all states of the tide. Also called TIDAL LOCK. See also NONTIDAL BASIN.

envelope match.. In Loran C, the comparison, in time difference, between the leading edges of the demodulated and filtered pulses from a master and secondary station. The pulses are superimposed and matched manually or automatically. See also CYCLE MATCH.

envelope to cycle difference.. The time relationship between the phase of the Loran C carrier and the time origin of the envelope waveform. Zero envelope to cycle difference (ECD) is defined as the signal condition occurring when the 30 microsecond point of the Loran C pulse envelope is in time coincidence with the third positive-going zero crossing of the 100 kHz carrier.

envelope to cycle discrepancy.. An error in a Loran C time difference measurement which results from upsetting the precise relationship between the shape of the pulse envelope and the phase of the carrier wave necessary for an accurate measurement due to some of the large number of frequencies (90-110 kHz) governing the envelope shape being transmitted more readily than others because of the medium over which the groundwave propagates.

ephemeris. *(pl. ephemerides), n*. 1. A periodical publication tabulating the predicted positions of celestial bodies at regular intervals, such as daily, and containing other data of interest to astronomers and navigators. The *Astronomical Almanac* is an ephemeris. See also ALMANAC. 2. A statement, not necessarily in a publication, presenting a correlation of time and position of celestial bodies or artificial satellites.

ephemeris day.. See under EPHEMERIS SECOND.

ephemeris second.. The ephemeris second is defined as 1/31,556,925.9747 of the tropical year for 1900 January 0^d 12^h ET. The ephemeris day is 86,400 ephemeris seconds. See also EPHEMERIS TIME.

Ephemeris Time.. The time scale used by astronomers as the tabular argument of the precise fundamental ephemerides of the sun, moon and planets. It is the independent variable in the gravitational theories of the solar system. It is determined in arrears from astronomical observations and extrapolated into the future, based on International Atomic Time.

epicenter., *n*. The point on the earth's surface directly above the focus of an earthquake.

epoch., *n*. 1. A particular instant of time or a date for which values of data, which vary with time, are given. 2. A given period of time during which a series of related acts or events takes place. 3. Angular retardation of the maximum of a constituent of the observed tide behind the corresponding maximum of the same constituent of the hypothetical equilibrium. Also called PHASE LAG, TIDAL EPOCH. 4. As used in tidal datum determinations, a 19-year Metonic cycle over which tidal height observations are meaned in order to establish the various datums.

equal altitudes.. Two altitudes numerically the same. The expression applies particularly to the practice of determining the instant of local apparent noon by observing the altitude of the sun a short time before it reaches the meridian and again at the same altitude after transit, the time of local apparent noon being midway between the times of the two observations, if the second is corrected as necessary for the run of the ship. Also called DOUBLE ALTITUDES.

equal-area map projection.. A map projection having a constant area scale. Such a projection is not conformal and is not used for navigation. Also called AUTHALIC MAP PROJECTION, EQUIVALENT MAP PROJECTION.

equal interval light.. A navigation light having equal periods of light and darkness. Also called ISOPHASE LIGHT.

equation of time.. The difference at any instant between apparent time and local mean time. It is a measure of the difference of the hour angles of the apparent (true) sun and the mean (fictitious) sun. The curve drawn for the equation of time during a year has two maxima: February 12 (+14.3^m) and July 27 (+6.3^m) and two minima: May 15 (-3.7^m) and November 4 (-16.4^m). The curve crosses the zero line on April 15, June 14, September 1, and December 24. The equation of time is tabulated in the *Nautical Almanac*, without sign, for 00^h and 12^h GMT on each day. To obtain apparent time, apply the equation of time to mean time with a positive sign when GHA sun at 00^h GMT exceeds 180°, or at 12^h exceeds 0°, corresponding to a meridian passage of the sun before 12^h GMT; otherwise apply with a negative sign.

equator., *n*. The primary great circle of a sphere or spheroid, such as the earth, perpendicular to the polar axis, or a line resembling or approximating such a circle. The terrestrial equator is 90° from the earth's geographical poles, the celestial equator or equinoctial is 90° from the celestial poles. The astronomical equator is a line connecting points having 0° astronomical latitude, the geodetic equator connects points having 0° geodetic latitude. The expression terrestrial equator is sometimes applied to the astronomical equator. The equator shown on charts is the geodetic equator. A fictitious equator is a reference line serving as the origin for measurement of fictitious latitude. A transverse or inverse equator is a meridian the plane of which is perpendicular to the axis of a transverse projection. An oblique equator is a great circle the plane of which is perpendicular to the axis of an oblique projection. A grid equator is a line perpendicular to a prime grid meridian at the origin. The magnetic equator or aclinic line is the line on the surface of the earth connecting all points at which the magnetic dip is zero. The geomagnetic equator is the great circle 90° from the geomagnetic poles of the earth.

equatorial., *adj*. Of or pertaining to the equator.

equatorial air.. See under AIR-MASS CLASSIFICATION.

equatorial bulge.. The excess of the earth's equatorial diameter over the polar diameter.

equatorial calms.. See DOLDRUMS.

equatorial chart.. 1. A chart of equatorial areas. 2. A chart on an equatorial map projection.

equatorial countercurrent.. An oceanic current flowing between and counter to the EQUATORIAL CURRENTS. See ATLANTIC EQUATORIAL COUNTERCURRENT, PACIFIC EQUATORIAL COUNTERCURRENT, INDIAN EQUATORIAL COUNTERCURRENT.

equatorial current.. See NORTH EQUATORIAL CURRENT, SOUTH EQUATORIAL CURRENT.

equatorial cylindrical orthomorphic chart.. See MERCATOR CHART.

equatorial cylindrical orthomorphic map projection.. See MERCATOR MAP PROJECTION.

equatorial gravity value.. The mean acceleration of gravity at the equator, approximately equal to 978.03 centimeters per second per second.

equatorial map projection.. A map projection centered on the equator.

equatorial node.. Either of the two points where the orbit of the satellite intersects the equatorial plane of its primary.

equatorial satellite.. A satellite whose orbital plane coincides, or almost coincides, with the earth's equatorial plane.

equatorial tidal currents.. Tidal currents occurring semimonthly as a result of the moon being over the equator. At these times the tendency of the moon to produce a diurnal inequality in the tidal current is at a minimum.

equatorial tides.. Tides occurring semimonthly as the result of the moon being over the equator. At these times the tendency of the moon to produce a diurnal inequality in the tide is at a minimum.

equiangular., *adj.* Having equal angles.

equilateral., *adj.* Having equal sides.

equilateral triangle.. A triangle having all of its sides equal. An equilateral triangle is necessarily equiangular.

equilibrium., *n.* A state of balance between forces. A body is said to be in equilibrium when the vector sum or all forces acting upon it is zero.

equilibrium argument.. The theoretical phase of a constituent of the equilibrium tide.

equilibrium theory.. A model under which it is assumed that the waters covering the face of the earth instantly respond to the tide-producing forces of the moon and sun, and form a surface of equilibrium under the action of these forces. The model disregards friction and inertia and the irregular distribution of the land masses of the earth. The theoretical tide formed under these conditions is called EQUILIBRIUM TIDE.

equilibrium tide.. Hypothetical tide due to the tide producing forces under the equilibrium theory. Also called GRAVITATIONAL TIDE.

equinoctial., *adj.* Of or pertaining to an equinox or the equinoxes.

equinoctial., *n.* See CELESTIAL EQUATOR.

equinoctial colure.. The great circle of the celestial sphere through the celestial poles and the equinoxes; the hour circle of the vernal equinox. See also SOLSTITIAL COLURE.

equinoctial point.. One of the two points of intersection of the ecliptic and the celestial equator. Also called EQUINOX.

equinoctial system of coordinates.. See CELESTIAL EQUATOR SYSTEM OF COORDINATES.

equinoctial tides.. Tides occurring near the times of the equinoxes, when the spring range is greater than average.

equinoctial year.. See TROPICAL YEAR.

equinox., *n.* 1. One of the two points of intersection of the ecliptic and celestial equator, occupied by the sun when its declination is 0°. The point occupied on or about March 21, when the sun's declination changes from south to north, is called vernal equinox, March equinox, or first point of Aries; the point occupied on or about September 23, when the declination changes from north to south, is called autumnal equinox, September equinox, or first point of Libra. Also called EQUINOCTIAL POINT. 2. The instant the sun occupies one of the equinoctial points.

equiphase zone.. The region in space within which there is no difference in phase between two radio signals.

equipotential surface.. A surface having the same potential of gravity at every point. See also GEOID.

equisignal., *adj.* Pertaining to two signals of equal intensity.

equisignal., *n.* See under CONSOL STATION.

equisignal zone.. The region in space within which the difference in amplitude of two radio signals (usually emitted by a signal station) is indistinguishable.

equivalent echoing area.. See RADAR CROSS SECTION.

equivalent map projection.. See EQUAL-AREA MAP PROJECTION.

erect image.. See under IMAGE, definition 1.

erecting telescope.. A telescope with which the observer sees objects right side up as opposed to the upside down view provided by the INVERTING TELESCOPE. The eyepiece in the optical system of an erecting telescope usually has four lenses, and the eyepiece in the optical system of an inverting telescope has two lenses.

erg., *n.* The work performed by a force of 1 dyne acting through a distance of 1 centimeter. The erg is the unit of energy or work in the centimeter-gram-second system. It corresponds to 10^{-7} joule in the International System of Units.

ergonomics.. The science of making mechanical and electronic devices easily usable by humans; human factors engineering.

error., *n.* The difference between the value of a quantity determined by observation, measurement or calculation and the true, correct, accepted, adopted or standard value of that quantity. Usually, the true value of the quantity cannot be determined with exactness due to insufficient knowledge of the errors encountered in the observations. Exceptions occur (1) when the value is mathematically determinable, or (2) when the value is an adopted or standard value established by authority. In order to analyze the exactness with which the true value of a quantity has been determined from observations, errors are classified into two categories, random and systematic errors. For the purpose of error analysis, blunders or mistakes are not classified as errors. The significant difference between the two categories is that random errors must be treated by means of statistical and probability methods due to their accidental or chance nature whereas systematic errors are usually expressible in terms of a unique mathematical formula representing some physical law or phenomenon. See also ACCURACY.

error budget.. A correlated set of individual major error sources with statements of the percentage of the total system error contributed by each source.

error ellipse.. The contour of equal probability density centered on the intersection of two straight lines of position which results from the one-dimensional normal error distribution associated with each line. For the 50 percent error ellipse, there is a 50 percent probability that a fix will lie within such ellipse. If the angle of cut is 90° and the standard deviations are equal, the error figure is a circle.

error of collimation.. See COLLIMATION ERROR.

error of perpendicularity.. That error in the reading of a marine sextant due to non-perpendicularity of the index mirror to the frame.

escape velocity., *n.* The minimum velocity required of a body at a given point in a gravitational field which will permit the body to escape from the field. The orbit followed is a parabola and the body arrives at an infinite distance from the center of the field with zero velocity. With respect to escape velocities characteristic of the major bodies of the solar system, this is defined as escape from the body's gravitational field from the surface of the body in question. Escape velocity equals circular velocity times the square root of 2. Also called PARABOLIC VELOCITY.

escarpment., *n.* An elongated and comparatively steep slope separating flat or gently sloping areas. Also called SCARP.

established direction of traffic flow.. A traffic flow pattern indicating the directional movement of traffic as established within a traffic separation scheme. See also RECOMMENDED DIRECTION OF TRAFFIC FLOW.

establishment of the port.. Average high water interval on days of the new and full moon. This interval is also sometimes called the COMMON or VULGAR ESTABLISHMENT to distinguish it from the CORRECTED ESTABLISHMENT, the latter being the mean of all high water intervals. The latter is usually 10 to 15 minutes less than the common establishment. Also called HIGH WATER FULL AND CHANGE.

estimate., *v., t.* To determine roughly or with incomplete information.

estimated position.. The most probable position of a craft determined from incomplete data or data of questionable accuracy. Such a position might be determined by applying a correction to the dead reckoning position, as for estimated current; by plotting a line of soundings; or by plotting lines of position of questionable accuracy. If no better information is available, a dead reckoning position is an estimated position, but the expression *estimated position* is not customarily used in this case. The distinction between an estimated position and a fix or running fix is a matter of judgment. See also MOST PROBABLE POSITION.

estimated time of arrival.. The predicted time of reaching a destination or waypoint.

estimated time of departure. . The predicted time of leaving a place.

estimation. , *n*. A mathematical method or technique of making a decision concerning the approximate value of a desired quantity when the decision is weighted or influenced by all available information.

estuarine sanctuary. . A research area which may include any part or all of an estuary, adjoining transitional areas, and adjacent uplands, constituting to the extent feasible a natural unit, set aside to provide scientists and students the opportunity to examine over a period of time the ecological relationships within the area. See also MARINE SANCTUARY.

estuary. , *n*. 1. An embayment of the coast in which fresh river water entering at its head mixes with the relatively saline ocean water. When tidal action is the dominant mixing agent, it is usually called TIDAL ESTUARY. 2. the lower reaches and mouth of a river emptying directly into the sea where tidal mixing takes place. Sometimes called RIVER ESTUARY. 3. A drowned river mouth due to sinking of the land near the coast.

etesian. , *n*. A refreshing northerly summer wind of the Mediterranean, especially over the Aegean Sea.

Eulerian current measurement. . The direct observation of the current speed or direction, or both, during a period of time as it flows past a recording instrument such as the Ekman or Roberts current meter. See also LAGRANGIAN CURRENT MEASUREMENT.

Eulerian motion. . A slight wobbling of the earth about its axis of rotation, often called polar motion, and sometimes wandering of the poles. This motion which does not exceed 40 feet from the mean position, produces slight variation of latitude and longitude of places on the earth.

European Datum. . The origin of this datum is at Potsdam, Germany. Numerous national systems have been joined in a large datum based upon the International Ellipsoid 1924 which was oriented by a modified astrogeodetic method. European, African, and Asian triangulation chains were connected. African arc measurements from Cairo to Cape Town were completed. Thus, all Europe, Africa, and Asia are molded into one great system. Through common survey stations, it was possible to convert data from the Russian Pulkova 1932 system to the European Datum, and as a result the European Datum includes triangulation as far east as the 84th meridian. Additional ties across the Middle East have permitted connection of the Indian and European Datums.

evaporation. , *n*. The physical process by which a liquid or solid is transformed to the gaseous state. The opposite is CONDENSATION. In meteorology, the term evaporation is usually restricted in use to the change of water vapor from liquid to gas, while SUBLIMATION is used for the change from solid to gas as well as from gas to solid. Energy is lost by an evaporating liquid, and when no heat is added externally, the liquid always cools. The heat thus removed is called LATENT HEAT OF VAPORIZATION.

evection. , *n*. A perturbation of the moon depending upon the alternate increase or decrease of the eccentricity of its orbit, which is always a maximum when the sun is passing the moon's line of apsides and at minimum when the sun is at right angles to it.

evening star. . The brightest planet appearing in the western sky during evening twilight.

evening twilight. . The period of time between sunset and darkness.

everglade. , *n*. 1. A tract of swampy land covered mostly with tall grass. 2. A swamp or inundated tract of low land, as used locally in the southern U.S.

excess of arc. . That part of a sextant arc beginning at zero and extending in the direction opposite to that part usually considered positive. See also ARC, definition 2.

existence doubtful. . Of uncertain existence. The expression is used principally on charts to indicate the possible existence of a rock, shoal, etc., the actual existence of which has not been established. See also VIGIA.

ex-meridian altitude. . An altitude of a celestial body near the celestial meridian of the observer to which a correction must be applied to determine the meridian altitude. Also called CIRCUM-MERIDIAN ALTITUDE.

ex-meridian observation. . Measurement of the altitude of a celestial body near the celestial meridian of the observer, for conversion to a meridian altitude; or the altitude so measured.

expanded center PPI display. . A plan position indicator display on which zero range corresponds to a ring around the center of the display. expanded sweep. Short for EXPANDED TIME BASE SWEEP.

expanded time base. . A time base having a selected part of increased speed. Particularly an EXPANDED TIME BASE SWEEP.

expanded time base sweep. . A sweep in which the sweep speed is increased during a selected part of the cycle. Usually shortened to EXPANDED SWEEP, and sometimes to EXPANDED TIME BASE.

explement. , *n*. An angle equal to 360° minus a given angle. See also COMPLEMENT, SUPPLEMENT.

explementary angles. . Two angles whose sum is 360°.

explosive fog signal. . A fog signal consisting of short reports produced by detonating explosive charges.

exponent. , *n*. A number which indicates the power to which another number is to be raised.

external noise. . In radio reception, atmospheric radio noise and man-made noise, singly or in combination. Internal noise is produced in the receiver circuits.

extragalactic nebula. . An aggregation of matter beyond our galaxy, large enough to occupy a perceptible area but which has not been resolved into individual stars.

extrapolation. , *n*. The process of estimating the value of a quantity beyond the limits of known values by assuming that the rate or system of change between the last few known values continues.

extratropical cyclone. . Any cyclonic-scale storm that is not a tropical cyclone, usually referring only to the migratory frontal cyclones of middle and high latitudes. Also called EXTRATROPICAL LOW.

extratropical low. . See EXTRATROPICAL CYCLONE.

extreme high water. . The highest elevation reached by the sea as recorded by a tide gage during a given period. The National Ocean Survey routinely documents monthly and yearly extreme high waters for its control stations. See also EXTREME LOW WATER.

extreme low water. . The lowest elevation reached by the sea as recorded by a tide gage during a given period. The National Ocean Survey routinely documents monthly and yearly extreme low water for its control stations. See also EXTREME HIGH WATER.

extremely high frequency. . Radio frequency of 30,000 to 300,000 megahertz.

eye guard. . A guard or shield on an eyepiece of an optical system, to protect the eye from stray light, wind, etc., and to maintain proper eye distance. Also called EYE SHIELD, EYE SHADE, SHADE.

eye of the storm. . The center of a tropical cyclone marked by relatively light winds, confused seas, rising temperature, lowered relative humidity, and often by clear skies. The general area of lowest atmospheric pressure of a cyclone is called STORM CENTER.

eye of the wind. . Directly into the wind; the point or direction from which the wind is blowing. See also IN THE WIND.

eyepiece. , *n*. In an optical device, the lens group which is nearest the eye and with which the image formed by the preceding elements is viewed.

eye shade. . See EYE GUARD.

eye shield. . See EYE GUARD.

F

facsimile. , *n*. The process of transmission of images electronically. The hard-copy result of a facsimile transmission.

fading. , *n*. The fluctuation in intensity or relative phase of any or all of the frequency components of a received radio signal due to changes in the characteristics of the propagation path. See also SELECTIVE FADING.

Fahrenheit temperature. . Temperature based on a scale in which, under standard atmospheric pressure, water freezes at 32° and boils at 212° above zero.

fair. , *adj*. Not stormy; good; fine; clear.

fair tide. . A tidal current setting in such a direction as to increase the speed of a vessel. One setting in a direction approximately opposite to the heading is called a HEAD TIDE. One abeam is called a BEAM TIDE. One approximately 90° from the course is called a CROSS TIDE.

fairway. , *n.* 1. The main thoroughfare of shipping in a harbor or channel. 2. The middle of a channel.

fairway buoy. . A buoy marking a fairway, with safe water on either side. Its color is red and white vertical stripes. Also called MIDCHANNEL BUOY.

fair wind. . A wind which aids a craft in making progress in a desired direction. Used chiefly in connection with sailing vessels, when it refers to a wind which permits the vessel to proceed in the desired direction without tacking. See also FOLLOWING WIND.

Falkland Current. . Originating mainly from the Cape Horn Current in the north part of Drake Passage, the Falkland Current flows northward between the continent and the Falkland Islands after passing through the strait. The current follows the coast of South America until it joins the BRAZIL CURRENT at about latitude 36° S near the entrance to Rio de la Plata. Also called MALVIN CURRENT.

fall. , *n.* 1. See AUTUMN. 2. Decrease in a value, such as a fall of temperature. 3. Sinking, subsidence, etc., as the rise and fall of the sea due to tidal action or when waves or swell are present. See also WATERFALL.

fall equinox. . See AUTUMNAL EQUINOX.

falling star. . See METEOR.

falling tide. . The portion of the tide cycle between high water and the following low water in which the depth of water is decreasing. Sometimes the term EBB is used as an equivalent, but since ebb refers primarily to horizontal rather than vertical movement, falling tide is considered more appropriate. The opposite is RISING TIDE.

fall streaks. . See VIRGA.

fall wind. . A cold wind blowing down a mountain slope. It is warmed by its descent, but is still cool relative to surrounding air. A warm wind blowing down a mountain slope is called a FOEHN. The bora, mistral, papagayo, and vardar are examples of fall winds. See also KATABATIC WIND.

false cirrus. . A cloud species unique to the genus cirrus, of such optical thickness as to appear grayish on the side away from the sun, and to veil the sun, conceal its outline, or even hide it. These often originate from the upper part of a cumulonimbus, and are often so dense that they suggest clouds of the middle level. Also called THUNDERSTORM CIRRUS, CIRRUS SPISSATUS.

false echo. . See INDIRECT ECHO, PHANTOM TARGET.

false horizon. . A line resembling the VISIBLE HORIZON but above or below it.

false light. . A light which is unavoidably exhibited by an aid to navigation and which is not intended to be a part of the proper characteristic of the light. Reflections from storm panes come under this category.

false relative motion. . False indications of the movement of a target relative to own ship on a radar display that is unstabilized in azimuth due to continuous reorientation of the display as own ship's heading changes. See also STABILIZATION OF RADARSCOPE DISPLAY.

fan. , *n.* On the sea floor, a relatively smooth feature normally sloping away from the lower termination of a canyon or canyon system.

fan beam. . A beam in which the radiant energy is concentrated in and about a single plane. The angular spread in the plane of concentration may be any amount to 360°. This type beam is most widely used for navigational lights. A converged beam is a fan beam in which the angular spread is decreased laterally to increase the intensity of the remaining beam over all or part of its arc; a diverged beam is a fan beam formed by increasing the divergence of a pencil beam in one plane only.

farad. , *n.* A derived unit of capacitance in the International System of Units; it is the capacitance of a capacitor between the plates of which there appears a potential difference of 1 volt when it is charged by a quantity of electricity of 1 coulomb.

far vane. . That instrument sighting vane on the opposite side of the instrument from the observer's eye. The opposite is NEAR VANE.

fast ice. . Sea ice which forms and remains attached to the shore, to an ice wall, to an ice front, between shoals or grounded icebergs. Vertical fluctuations may be observed during changes of sea level. Fast ice may be formed in situ from the sea water or by freezing of pack ice of any age to the shore, and it may extend a few meters or several hundred kilometers from the coast. Fast ice may be more than 1 year old and may then be prefixed with the appropriate age category (old, second-year or multi-year). If it is thicker than about 2 meters above sea level, it is called an ICE SHELF.

fast-ice boundary. . The ice boundary at any given time between fast ice and pack ice.

fast-ice edge. . The demarcation at any given time between fast ice and open water.

fast-sweep racon. . See under SWEPT-FREQUENCY RACON.

fast time constant circuit. . A type of coupling circuit, with high pass frequency characteristics used in radar receivers to permit discrimination against received pulses of duration longer than the transmitted pulse. With the fast time constant (FTC) circuit in operation, only the leading edge of an echo having a long time duration is displayed on the radarscope. The use of this circuit tends to reduce saturation of the scope which could be caused by clutter. Also called ANTI-CLUTTER, RAIN, DIFFERENTIATOR.

fata morgana. . A complex mirage, characterized by marked distortion, generally in the vertical. It may cause objects to appear towering, magnified, and at times even multiplied.

fathogram. , *n.* A graphic record of depth measurements obtained by a fathometer. See also ECHOGRAM.

fathom. , *n.* A unit of length equal to 6 feet. This unit of measure is used principally as a measure of depth of water and the length of lead lines, anchor chains, and cordage. See also CABLE, definition 1.

fathom curve, fathom line. . A depth contour, with depths expressed in fathoms.

Fathometer. , *n.* The registered trade name for a widely-used echo sounder.

favorable current. . A current flowing in such a direction as to increase the speed of a vessel over the ground. The opposite is UNFAVORABLE CURRENT.

favorable wind. . A wind which aids a craft in making progress in a desired direction. Usually used in connection with sailing vessels. A wind which delays the progress of a craft is called an UNFAVORABLE WIND. Also called FAIR WIND. See also FOLLOWING WIND.

feasibility orbit. . An orbit that can be rapidly and inexpensively computed on the basis of simplifying assumptions (e.g., two-body motion, circular orbit, rectilinear orbit, three-body motion approximated by two two-body orbits, etc.) and yields an indication of the general feasibility of a system based upon the orbit without having to carry out a full-blown definitive orbit computation.

federal project depth. . The design dredging depth of a channel constructed by the Corps of Engineers, U.S. Army; the project depth may or may not be the goal of maintenance dredging after completion of the channel. For this reason federal project depth must not be confused with CONTROLLING DEPTH.

feel the bottom. . The effect on a ship underway in shallow water which tends to reduce her speed, make her slow in answering the helm, and often make her sheer off course. The speed reduction is largely due to increased wave making resistance resulting from higher pressure differences due to restriction of flow around the hull. The increased velocity of the water flowing past the hull results in an increase in squat. Also called SMELL THE BOTTOM.

femto-. . A prefix meaning one-quadrillionth (10^{-15})

fen. , *n.* A low-lying tract of land, wholly or partly covered with water at times.

fetch. , *n.* 1. An area of the sea surface over which seas are generated by a wind having a constant direction and speed. Also called GENERATING AREA. 2. The length of the fetch area, measured in the direction of the wind, in which the seas are generated.

fictitious equator. . A reference line serving as the origin for measurement of fictitious latitude. A transverse or inverse equator is a meridian the plane of which is perpendicular to the axis of a transverse map projection. An oblique equator is a great circle the plane of which is perpendicular to the axis of an oblique map projection. A grid equator is a line perpendicular to a prime grid meridian, at the origin.

fictitious graticule. . The network of lines representing fictitious parallels and fictitious meridians on a map, chart, or plotting sheet. It may be either a transverse graticule or an oblique graticule depending upon the kind of projection; a fictitious graticule may also be a GRID. See also OBLIQUE GRATICULE, TRANSVERSE GRATICULE.

fictitious latitude. . Angular distance from a fictitious equator. It may be called transverse, oblique, or grid latitude depending upon the type of fictitious equator.

fictitious longitude. . The arc of the fictitious equator between the prime fictitious meridian and any given fictitious meridian. It may be called transverse, oblique, or grid longitude depending upon the type of fictitious meridian.

fictitious loxodrome. . See FICTITIOUS RHUMB LINE.

fictitious loxodromic curve. . See FICTITIOUS RHUMB LINE.

fictitious meridian. . One of a series of great circles or lines used in place of a meridian for certain purposes. A transverse meridian is a great circle perpendicular to a transverse equator; an oblique meridian is a great circle perpendicular to an oblique equator; a grid meridian is one of the grid lines extending in a grid north-south direction. The reference meridian (real or fictitious) used as the origin for measurement of fictitious longitude is called prime fictitious meridian.

fictitious parallel. . A circle or line parallel to a fictitious equator, connecting all points of equal fictitious latitude. It may be called transverse, oblique, or grid parallel depending upon the type of fictitious equator.

fictitious pole. . One of the two points 90° from a fictitious equator. It may be called the transverse or oblique pole depending upon the type of fictitious equator.

fictitious rhumb. . See FICTITIOUS RHUMB LINE.

fictitious rhumb line. . A line making the same oblique angle with all fictitious meridians. It may be called transverse, oblique, or grid rhumb line depending upon the type of fictitious meridian. The expression OBLIQUE RHUMB LINE applies also to any rhumb line, real or fictitious, which makes an oblique angle with its meridians; as distinguished from parallels and meridians real or fictitious, which may be consider special cases of the rhumb line. Also called FICTITIOUS RHUMB, FICTITIOUS LOXODROME, FICTITIOUS LOXODROMIC CURVE.

fictitious ship. . An imaginary craft used in the solution of certain maneuvering problems, as when a ship to be intercepted is expected to change course or speed during the interception run.

fictitious sun. . An imaginary sun conceived to move eastward along the celestial equator at a rate equal to the average rate of the apparent sun or to move eastward along the ecliptic at the average rate of the apparent sun. See also DYNAMICAL MEAN SUN, MEAN SUN.

fictitious year. . The period between successive returns of the sun to a sidereal hour angle of 80° (about January 1). The length of the fictitious year is the same as that of the tropical year, since both are based upon the position of the sun with respect to the vernal equinox. Also called BESSELIAN YEAR.

fidelity. , *n.* The accuracy to which an electrical system, such as a radio, reproduces at its output the essential characteristics of its input signal.

field glass. . A telescopic binocular.

field lens. . A lens at or near the plane of a real image, to collect and redirect the rays into another part of the optical system; particularly, the eyepiece lens nearest the object, to direct the rays into the eye lens.

field of view. . The maximum angle of vision, particularly of an optical instrument.

figure of the earth. . See GEOID.

filling. , *n.* Increase in atmospheric pressure, particularly within a low. Decrease in pressure is called DEEPENING.

final diameter. . The diameter of the circle traversed by a vessel after turning through 360° and maintaining the same speed and rudder angle. This diameter is always less than the tactical diameter. It is measured perpendicular to the original course and between the tangents at the points where 180° and 360° of the turn have been completed.

final great circle course. . The direction, at the destination, of the great circle through that point and the point of departure, expressed as the angular distance from a reference direction, usually north, to that part of the great circle extending beyond the destination. See also INITIAL GREAT CIRCLE COURSE.

finger rafted ice. . The type of rafted ice in which floes thrust "fingers" alternately over and under the other.

finger rafting. . A type of rafting whereby interlocking thrusts are formed, each floe thrusting "fingers" alternately over and under the other. Finger rafting is common in NILAS and GRAY ICE.

finite. , *adj.* Having limits. The opposite is INFINITE.

fireball. , *n.* See BOLIDE.

firn. , *n.* Old snow which has recrystallized into a dense material. Unlike snow, the particles are to some extent joined together; but, unlike ice, the air spaces in it still connect with each other.

first estimate-second estimate method. . The process of determining the value of a variable quantity by trial and error. The expression applies particularly to the method of determining time of meridian transit (especially local apparent noon) at a moving craft. The time of transit is computed for an estimated longitude of the craft, the longitude estimate is then revised to agree with the time determined by the first estimate, and a second computation is made. The process is repeated as many times as necessary to obtain an answer of the desired precision.

first light. . The beginning of morning nautical twilight, i.e., when the center of the morning sun is 12° below the horizon.

first point of Aries. . See VERNAL EQUINOX.

first point of Cancer. . See SUMMER SOLSTICE.

first point of Capricornus. . See WINTER SOLSTICE.

first point of Libra. . See AUTUMNAL EQUINOX.

first quarter. . The phase of the moon when it is near east quadrature, when the western half of it is visible to an observer on the earth. See also PHASES OF THE MOON.

first-year ice. . Sea ice of not more than one winter's growth, developing from young ice, with a thickness of 30 centimeters to 2 meters. First-year ice may be subdivided into THIN FIRST YEAR ICE, WHITE ICE, MEDIUM FIRST YEAR ICE, and THICK FIRST YEAR ICE.

firth. , *n.* A long, narrow arm of the sea.

Fischer ellipsoid of 1960. . The reference ellipsoid of which the semimajor axis is 6,378,166.000 meters, the semiminor axis is 6,356,784.298 meters, and the flattening or ellipticity is 1/298.3. Also called FISCHER SPHEROID OF 1960.

Fischer ellipsoid of 1968. . The reference ellipsoid of which the semimajor axis is 6,378,150 meters, the semiminor axis is 6,356,768.337 meters, and the flattening or ellipticity is 1/298.3. Also called FISCHER SPHEROID OF 1968.

Fischer spheroid of 1960. . See FISCHER ELLIPSOID OF 1960.

Fischer spheroid of 1968. . See FISCHER ELLIPSOID OF 1968.

fish. , *n.* Any towed sensing device.

fishery conservation zone. . See under FISHING ZONE.

fish havens. . Areas established by private interests, usually sport fishermen, to simulate natural reefs and wrecks that attract fish. The reefs are constructed by dumping assorted junk in areas which may be of very small extent or may stretch a considerable distance along a depth contour. Fish havens are outlined and labeled on charts. Also called FISHERY REEFS.

fishing zone. . The offshore zone in which exclusive fishing rights and management are held by the coastal nation. The U.S. fishing zone, known as the fishery conservation zone, is defined under P.L. 94-265. The law states, "The inner boundary of the fishery conservation zone is a line conterminous with the seaward boundary of catch of the coastal states, and the outer boundary of such zone is a line drawn in such manner that each point on it is 200 nautical miles from the baseline from which the territorial sea is measured."

fish lead. . A type of sounding lead used without removal from the water between soundings.

fish stakes. . Poles or stakes placed in shallow water to outline fishing grounds or to catch fish.

fish trap areas. . Areas established by the Corps of Engineers in which traps may be built and maintained according to established regulations. The fish stakes which may exist in these areas are obstructions to navigation and may be dangerous. The limits of fish trap areas and a cautionary note are usually charted.

fix. , *n.* A position determined without reference to any former position; the common intersection of two or more lines of position obtained from simultaneous observations. Fixes obtained from electronic systems are often given as lat./long. coordinates determined by algorithms in the system software. See also RUNNING FIX.

fixed. . A light which is continuously on.

fixed and flashing light. . A light in which a fixed light is combined with a flashing light of higher luminous intensity. The aeronautical light equivalent is called UNDULATING LIGHT.

fixed and group flashing light. . A fixed light varied at regular intervals by a group of two or more flashes of greater intensity.

fixed and variable parameters of satellite orbit. . The fixed parameters are those parameters which describe a satellite's approximate orbit and which are used over a period of hours. The variable parameters describe the fine structure of the orbit as a function of time and are correct only for the time at which they are transmitted by the satellite.

fixed antenna radio direction finder. . A radio direction finder whose use does not require the rotation of the antenna system.

fixed light. . A light which appears continuous and steady. The term is sometimes loosely used for a light supported on a fixed structure, as distinct from a light on a floating support.

fixed mark. . A navigation mark fixed in position.

fixed satellite. . See GEOSTATIONARY SATELLITE.

fixed star. . A star whose apparent position relative to surrounding stars appears to be unvarying or fixed for long periods of time.

fjord. , *n*. A long, deep, narrow arm of the sea between high land. A fjord often has a relatively shallow sill across its entrance.

flag alarm. . A semaphore-type flag in the indicator of an instrument, to serve as a signal, usually to warn that the indications are unreliable.

flagpole. , *n*. A label on a nautical chart which indicates a single pole from which flags are displayed. The term is used when the pole is not attached to a building. The label flagstaff is used for a flagpole rising from a building.

flagstaff. , *n*. See under FLAGPOLE.

Flamsteed's number. . A number sometimes used with the possessive form of the Latin name of the constellation to identify a star.

flash. , *n*. A relatively brief appearance of a light, in comparison with the longest interval of darkness in the period of the light. See also OCCULTATION.

flasher. , *n*. An electrical device which controls the characteristic of a lighted aid to navigation by regulating power to the lamp according to a certain pattern.

flashing. , *n*. The process of reducing the amount of permanent magnetism in a vessel by placing a single coil horizontally around the vessel and energizing it. If the energized coil is moved up and down along the sides of the vessel, the process is called WIPING. See also DEPERMING.

flashing light. . A navigation light in which the total duration of light in a cycle is shorter than the total duration of darkness. The term is commonly used for a SINGLE-FLASHING LIGHT, a flashing light in which a flash is regularly repeated at a rate of less then 50 flashes per minute. See also GROUP-FLASHING LIGHT, COMPOSITE GROUP-FLASHING LIGHT LONG-FLASHING LIGHT, QUICK LIGHT.

flat. , *n*. 1. A large flat area attached to the shore consisting usually of mud, but sometimes of sand and rock. Also called TIDAL FLATS. See also SALT MARSH, SLOUGH, TIDAL MARSH. 2. On the sea floor, a small level or nearly level area.

flattening. , *n*. The ratio of the difference between the equatorial and polar radii of the earth to its equatorial radius. The flattening of the earth is the ellipticity of the spheroid. The magnitude of the flattening is sometimes expressed as the numerical value of the reciprocal of the flattening. Also called COMPRESSION.

flaw. , *n*. A narrow separation zone between pack ice and fast ice, where the pieces of ice are in a chaotic state. The flaw forms when pack ice shears under the effect of a strong wind or current along the fast-ice boundary. See also SHEARING.

flaw lead. . A passage-way between pack ice and fast ice which is navigable by surface vessels.

flaw polynya. . A polynya between pack ice and fast ice.

F-layer. , *n*. The second principal layer of ionization in the Kennelly-Heaviside region (the E-layer is the first principal layer; the D-layer is of minor significance except for a tendency to absorb energy from radio waves in the medium frequency range). Situated about 175 miles above the earth's surface, the F-layer exists as a single layer only during the hours of darkness. It divides into two separate layers during daylight hours.

F1-layer. , *n*. The lower of the two layers into which the F-layer divides during daylight hours. Situated about 140 miles above the earth's surface, it reaches its maximum density at noon. Since its density varies with the extent of the sun's radiation, it is subject to daily and seasonal variations. It may disappear completely at some point during the winter months.

F2-layer. , *n*. The higher of the two layers into which the F-layer divides during daylight hours. It reaches its maximum density at noon and, over the continental U.S., varies in height from about 185 miles in winter to 250 miles in the summer. The F2-layer normally has a greater influence on radio wave propagation than the F1-layer.

FleetNET. . INMARSAT broadcast service for commercial traffic.

***Fleet Guide*.** . One of a series of port information booklets for United States naval bases prepared for U.S. Navy use only.

Flinders bar. . A bar of soft unmagnetized iron placed vertically near a magnetic compass to counteract deviation caused by magnetic induction in vertical soft iron of the craft.

float chamber. . A sealed, hollow part attached to the compass card of a magnetic compass as part of the compass card assembly, to provide buoyancy to reduce the friction on the pivot bearing.

floating aid. . A buoy serving as an aid to navigation secured in its charted position by a mooring.

floating breakwater. . A moored assembly of floating objects used for protection of vessels riding at anchor.

floating dock. . A form of dry dock consisting of a floating structure of one or more sections, which can be partly submerged by controlled flooding to receive a vessel, then raised by pumping out the water so that the vessel's bottom can be exposed. See also GRAVING DOCK.

floating ice. . Any form of ice found floating in water. The principal kinds of floating ice are lake ice, river ice and sea ice which form by the freezing of water at the surface, and glacier ice (ice of land origin) formed on land or in an ice shelf. The concept includes ice that is stranded or grounded.

floating mark. . A navigation mark carried on a floating body such as a lightship or buoy.

float pipe. . A pipe used as a float well.

float well. . A vertical pipe or box with a relatively small opening (orifice) in the bottom. It is used as a tide gage installation to dampen the wind waves while freely admitting the tide to actuate a float which, in turn, operates the gage. Also called STILLING WELL.

floe. , *n*. Any relatively flat piece of sea ice 20 meters or more across. Floes are subdivided according to horizontal extent. A giant flow is over 5.4 nautical miles across; a vast floe is 1.1 to 5.4 nautical miles across; a big floe is 500 to 2000 meters across; a medium floe is 100 to 500 meters across; and a small floe is 20 to 100 meters across.

floeberg. , *n*. A massive piece of sea ice composed of a hummock, or a group of hummocks frozen together, and separated from any ice surroundings. It may float showing up to 5 meters above sea level.

flood. , *n*. Tidal current moving toward land or up a tidal stream. The opposite is EBB. Also called FLOOD CURRENT.

flood axis. . Average direction of tidal current at strength of flood.

flood current. . The movement of a tidal current toward the shore or up a tidal river or estuary. In the mixed type of reversing current, the terms *greater flood* and *lesser flood* are applied respectively to the flood currents of greater and lesser speed of each day. The terms *maximum flood* and *minimum flood* are applied to the maximum and minimum speeds of a flood current, the speed of which alternately increases and decreases without coming to a slack or reversing. The expression maximum flood is also applicable to any flood current at the time of greatest velocity. The opposite is EBB CURRENT.

flooded ice. . Sea ice which has been flooded by melt-water or river water and is heavily loaded by water and wet snow.

floodgate. , *n*. A gate for shutting out, admitting, or releasing a body of water, a sluice.

flood interval. . Short for STRENGTH OF FLOOD INTERVAL. The interval between the transit of the moon over the meridian of a place and the time of the following strength of flood. See also LUNICURRENT INTERVAL.

flood plain.. The belt of low flat ground bordering a stream or river channel that is flooded when runoff exceeds the capacity of the stream channel.

flood strength.. Phase of the flood current at time of maximum speed. Also, the speed at this time. Also called STRENGTH OF FLOOD.

floor., *n*. The ground under a body of water. See also BOTTOM.

floppy disk.. A type of magnetic computer data storage media consisting of a thin circular plastic disk enclosed in a rigid or semi-rigid housing.

Florida Current.. A swift ocean current that flows through the Straits of Florida from the Gulf of Mexico to the Atlantic Ocean. It shows a gradual increase in speed and persistency as it flows northeastward and then northward along the Florida coast. In summer, the part of the surface current south of latitude 25° N moves farther south of its mean position, with a mean speed of 2.0 knots and a maximum speed of about 6.0 knots; the part of the current north of latitude 25° N moves farther west of its mean position, with a mean speed of 2.9 knots and a maximum speed of 6.5 knots. In winter the shift of position is in the opposite direction, and speeds are somewhat less by about 0.2 to 0.5 knot. The flow prevails throughout the year, with no significant changes in direction; the speed, however, varies slightly from one season to another. North of Grand Bahama Island, it merges with the Antilles Current to form the GULF STREAM. The Florida Current is part of the GULF STREAM SYSTEM.

flotsam.. *n*. Floating articles, particularly those that are thrown overboard to lighten a vessel in distress. See also JETSAM, JETTISON, LAGAN.

flow., *n. British terminology*. Total current or the combination of tidal current and nontidal current. In British usage, tidal current is called TIDAL STREAM and nontidal current is called CURRENT.

fluorescence., *n*. Emission of light or other radiant energy as a result of and only during absorption of radiation from some other source.

fluorescent chart.. A chart reproduced with fluorescent ink or on fluorescent paper, which enables the user to read the chart under ultraviolet light.

flurry., *n*. See SNOW FLURRY.

flux-gate.. The magnetic direction-sensitive element of a flux-gate compass. Also called FLUX VALVE.

fluxmeter., *n*. An instrument for measuring the intensity of a magnetic field.

flux valve.. See FLUX GATE.

focal length.. The distance between the optical center of a lens, or the surface of a mirror, and its focus.

focal plane.. A plane parallel to the plane of a lens or mirror and passing through the focus.

focal point.. See FOCUS.

focus. *(pl. foci), n*. 1. The point at which parallel rays of light meet after being refracted by a lens or reflected by a mirror. Also called FOCAL PO-I NT. 2. A point having specific significance relative to a geometrical figure. See under ELLIPSE, HYPERBOLA, PARABOLA. 3. The true center of an earthquake, within which the strain energy is first converted to elastic wave energy.

focus., *v., t*. The process of adjusting an optical instrument, projector, cathode-ray tube, etc., to produce a clear and well-defined image.

foehn., *n*. A warm, dry, wind blowing down the leeward slope of a mountain and across a valley floor or plain.

fog., *n*. A visible accumulation of tiny droplets of water, formed by condensation of water vapor in the air, with the base at the surface of the earth. It reduces visibility below 1 kilometer (0.54 nautical mile). If this is primarily the result of movement of air over a surface of lower temperature, it is called advection fog; if primarily the result of cooling of the surface of the earth and the adjacent layer of atmosphere by radiational cooling, it is called radiation fog. An advection fog occurring as monsoon circulation transports warm moist air over a colder surface is called a monsoon fog. A fog that hides less than six-tenths of the sky, and does not extend to the base of any clouds is called a ground fog. Fog formed at sea, usually when air from a warm-water surface moves to a cold-water surface, is called sea fog. Fog produced by apparent steaming of a relatively warm sea in the presence of very cold air is called steam fog, steam mist, frost smoke, sea smoke, arctic sea smoke, arctic smoke, or water smoke. Fog composed of suspended particles of ice, partly ice crystals 20 to 100 microns in diameter but chiefly, especially when dense, droxtals 12 to 20 microns in diameter is called ice fog. A rare simulation of true fog by anomalous atmospheric refraction is called mock fog. A dry fog is a fog that does not moisten exposed surfaces.

fog bank.. A well defined mass of fog observed at a distance, most commonly at sea.

fogbound., *adj*. Surrounded by fog. The term is used particularly with reference to vessels which are unable to proceed because of the fog.

fogbow., *n*. A faintly colored circular arc similar to a RAINBOW but formed on fog layers containing drops whose diameters are of the order of 100 microns or less. See also BOUGUER'S HALO.

fog detector.. A device used to automatically determine conditions of visibility which warrant sounding a fog signal.

fog signal.. See under SOUND SIGNAL.

following sea.. A sea in which the waves move in the general direction of the heading. The opposite is HEAD SEA. Those moving in a direction approximately 90° from the heading are called BEAM SEA, and those moving in a direction approximately 45° from the heading (striking the quarter) are called QUARTERING SEA.

following wind.. Wind blowing in the general direction of a vessel's course. The equivalent aeronautical expression is TAIL Wind. Wind blowing in the opposite direction is called a HEAD WIND. Wind blowing in a direction approximately 90° from the heading is called a BEAM WIND. One blowing in a direction approximately 90° from the course is called a CROSS WIND. See also FAIR WIND, FAVORABLE WIND, UNFAVORABLE WIND.

foot., *n*. Twelve inches or 30.48 centimeters. The latter value was adopted in 1959 by Australia, Canada, New Zealand, South Africa, the United Kingdom, and the United States. See also U.S. SURVEY FOOT. 2. The bottom of a slope, grade, or declivity.

foraminifera., *n., pl*. Small, single-cell, jellylike marine animals with hard shells of many chambers. In some areas the shells of dead foraminifera are so numerous they cover the ocean bottom.

Forbes log.. A log consisting of a small rotator in a tube projecting below the bottom of a vessel, and suitable registering devices.

forced wave.. A wave generated and maintained by a continuous force, in contrast with a FREE WAVE that continues to exist after the generating force has ceased to act.

foreland., *n*. See PROMONTORY, HEADLAND.

foreshore., *n*. That part of the shore or beach which lies between the low water mark and the upper limit of normal wave action. See also BACKSHORE.

forestaff., *n*. See CROSS-STAFF.

fork., *n*. On the sea floor, a branch of a canyon or valley.

format., *v., t*. To prepare a computer disk for data storage; formatting defines tracks and sectors, sets up a directory, and performs other functions before a new disk can be used.

form lines.. Broken lines resembling contour lines but representing no actual elevations, which have been sketched from visual observation or from inadequate or unreliable map sources, to show collectively the shape of the terrain rather than the elevation.

formation axis.. An arbitrarily selected direction within a formation of ships from which all bearings used designation of station are measured; bearings are always expressed in true direction from the center.

formation center.. An arbitrary point around which a formation of ships is centered, designated "station zero."

formation guide.. A ship designated by the OTC as the reference vessel upon which all ships in a formation maintain position.

forward., *adj*. In a direction towards the bow of a vessel. See also AHEAD, ABAFT.

forward of the beam.. Any direction between broad on the beam and ahead. See also ABAFT THE BEAM.

foul berth.. A berth in which a vessel cannot swing to her anchor or moorings without fouling another vessel or striking an obstruction. See also FOUL GROUND, CLEAR BERTH.

foul bottom.. A term used to describe the bottom of a vessel when encrusted with marine growth.

foul ground.. An area unsuitable for anchoring or fishing due to rocks, boulders, coral or other obstructions. See also FOUL BERTH.

four-point bearing.. A relative bearing of 045° or 315°. See also BOW AND BEAM BEARINGS.

fractional scale.. See REPRESENTATIVE FRACTION.

fracto-. . A prefix used with the name of a basic cloud form to indicate a torn, ragged, and scattered appearance caused by strong winds. See also SCUD.

fracture. , *n*. A break or rupture through very close pack ice, compact pack ice, consolidated pack ice, fast ice, or a single floe resulting from deformation processes. Fractures may contain brash ice and/or be covered with nilas and/or young ice. The length of a fracture may vary from a few meters to many miles. A large fracture is more than 500 meters wide- a medium fracture is 200 to 500 meters wide- a small fracture is 50 to 200 meters wide, and a very small fracture is 0 to 50 meters wide.

fracture zone. . 1. An extensive linear zone of irregular topography of the sea floor characterized by steep-sided or asymmetrical ridges, troughs, or escarpments. 2. An ice area which has a great number of fractures. See also FRACTURE.

fracturing. , *n*. The pressure process whereby ice is permanently deformed, and rupture occurs. The term is most commonly used to describe breaking across very close pack ice, compact pack ice, and consolidated pack ice.

Franklin continuous radar plot technique. . A method of providing continuous correlation of a small fixed radar-conspicuous object with own ship's position and movement relative to a planned track. Named for QMCM Byron Franklin, USN.

Franklin piloting technique. . A method of finding the most probable position of a ship from three lines of position which do not intersect in a point.

frazil ice. . Fine spicules or plates of ice, suspended in water.

free-air temperature. . Temperature of the atmosphere, obtained by a thermometer located so as to avoid as completely as practicable the effects of extraneous heating. See also AMBIENT TEMPERATURE, WET-BULB TEMPERATURE.

freeboard. , *n*. The vertical distance from the uppermost complete, watertight deck of a vessel to the surface of the water, usually measured amidships. Minimum permissible freeboards may be indicated by LOAD LINE MARKS.

free gyro. . A two-degree-of-freedom gyro or a gyro the spin axis of which may be oriented in any specified altitude. The rotor of this gyro has freedom to spin on its axis, freedom to tilt about its horizontal axis, and freedom to turn about its vertical axis. Also called FREE GYROSCOPE. See also DEGREE-OF-FREEDOM.

free gyroscope. . See FREE GYRO.

free wave. . A wave that continues to exist after the generating force has ceased to act, in contrast with a FORCED WAVE that is generated and maintained by a continuous force.

freezing drizzle. . Drizzle that falls in liquid form but freezes upon impact to form a coating of glaze upon the ground and exposed objects.

freezing fog. . A fog whose droplets freeze upon contact with exposed objects and form a coating of rime and/or glaze. See also FREEZING PRECIPITATION.

freezing precipitation. . Precipitation which falls to the earth in a liquid state and then freezes to exposed surfaces. Such precipitation is called freezing rain if it consists of relatively large drops of water, and freezing drizzle if of smaller drops. See also GLAZE.

freezing rain. . Rain that falls in liquid form but freezes upon impact to form a coating of ice on the ground and exposed objects.

frequency. , *n*. The rate at which a cycle is repeated. See also AUDIO FREQUENCY, RADIO FREQUENCY.

frequency band. . 1. A specified segment of the frequency spectrum. 2. One of two or more segments of the total frequency coverage of a radio receiver or transmitter, each segment being selectable by means of a band change switch. 3. Any range of frequencies extending from a specified lower to a specified upper limit.

frequency channel. . The assigned frequency band commonly referred to by number, letter, symbol, or some salient frequency within the band.

frequency-modulated radar. . A type of radar in which the radiated wave is frequency modulated and the frequency of an echo is compared with the frequency of the transmitted wave at the instant of reception, thus enabling range to be measured.

frequency modulation. . Angle modulation of a sinewave carrier in which the instantaneous frequency of the modulated wave differs from the carrier frequency by an amount proportional to the instantaneous value of the modulating.

frequency tolerance. . The maximum permissible departure by the center frequency of the frequency band occupied by an emission from the assigned frequency, or by the characteristic frequency of an emission from the reference frequency. The frequency tolerance is expressed in parts in 106 or in hertz.

fresh breeze. . Wind of force 5 (17 to 21 knots or 19 to 24 miles per hour) on the Beaufort wind scale.

freshen. , *v., i*. To become stronger applied particularly to wind.

freshet. ,n. A sudden increased flow of fresh water, as from a flood, emptying from a river into a larger body of salt or brackish water.

fresh gale. A term once used by seamen to what is now called GALE on the Beaufort wind scale.

fresh-water marsh. . A tract of low wet ground, usually miry and covered with rank vegetation.

friction. , *n*. Resistance to motion due to interaction between the surface of a body and anything in contact with it.

friction error. . The error of an instrument reading due to friction in the moving parts of the instrument.

friction layer. . See SURFACE BOUNDARY LAYER.

friendly ice. . From the point of view of the submariner, an ice canopy containing many large skylights or other features which permit a submarine to surface. There must be more than 10 such features per 30 nautical miles along the submarine's track.

frigid zones. . Either of the two zones between the polar circles and the poles, called the north frigid zone and the south frigid zone.

fringing reef. . A reef attached directly to the shore of an island or continental landmass. Its outer margin is submerged and often consists of algal limestone, coral rock, and living coral. See also BARRIER REEF.

front. , *n*. Generally, the interface or transition zone between two air masses of different density. Since the temperature distribution is the most important regulator of atmospheric density, a front almost invariably separates air masses of different temperature. Along with the basic density criterion and the common temperature criterion, many other features may distinguish a front, such as a pressure trough, a change in wind direction, a moisture discontinuity, and certain characteristic cloud and precipitation forms. The term front is used ambiguously for: frontal zone, the three-dimensional zone or layer of large horizontal density gradient, bounded by frontal surfaces across which the horizontal density gradient is discontinuous (frontal surface usually refers specifically to the warmer side of the frontal zone); and surface front, the line of intersection of a frontal surface or frontal zone with the earth's surface or less frequently, with a specified constant-pressure surface. See also POLAR FRONT, ARCTIC FRONT, COLD FRONT, WARM FRONT, OCCLUDED FRONT.

frontal. , *adj*. Of or pertaining to a front.

frontal cyclone. . In general, any cyclone associated with a front; often used synonymously with WAVE CYCLONE or with EXTRATROPICAL CYCLONE (as opposed to tropical cyclones, which are non-frontal).

frontal occlusion. . See OCCLUDED Front; OCCLUSION, definition 2.

frontal surface. . See under FRONT.

frontal zone. . See under FRONT.

front light. . The closer of two range lights. It is the lowest of the lights of an established range. Also called LOW LIGHT.

frontogenesis. , *n*. 1. The initial formation of a front or frontal zone. 2. In general, an increase in the horizontal gradient of an air mass property, principally density, and the development of the accompanying features of the wind field that characterize a front.

frontolysis. , *n*. 1 The dissipation of a front or frontal zone. 2. In general, a decrease in the horizontal gradient of an air mass property, principally density, and the dissipation of the accompanying features of the wind field.

frost. , *n*. 1. A deposit of interlocking ice crystals formed by direct sublimation on objects, usually those of small diameter freely exposed to the air. The deposition is similar to the process in which dew is formed, except that the temperature of the object must be below freezing. It forms when air with a dew point below freezing is brought to saturation by cooling. It is more fluffy and feathery than rime which in turn is lighter than glaze. Also called HOAR, HOARFROST. 2. The condition which exists when the temperature of the earth's surface and earthbound objects falls below 0°C or 32°F. Temperatures below the freezing point of water are sometimes expressed as "degrees of frost."

frost smoke. . 1. Fog like clouds due to contact of cold air with relatively warm water, which can appear over openings in the ice, or leeward of the ice edge, and which may persist while ice is forming. 2. A rare type of fog formed in the same manner as a steam fog but at lower temperatures. It is composed of ice particles or droxtals instead of liquid water as is steam fog. Thus, it is a type of ice fog. Sometimes called BARBER. 3. See STEAM FOG.

frozen precipitation. . Any form of precipitation that reaches the ground in frozen form; i.e., snow, snow pellets, snow grains, ice crystals, ice pellets, and hail.

frustum, frustrum. , *n*. That part of a solid figure between the base and a parallel intersecting plane; or between any two intersecting planes, generally parallel.

full depiction of detail. . Since even on charts of the largest scale full depiction of detail is impossible because all features are symbolized to an extent which is partly determined by scale and partly by the conventions of charting practice, the term *full depiction of detail* is used to indicate that over the greater part of a chart nothing essential to navigation is omitted. See also GENERALIZATION OF DETAIL, MINIMAL DEPICTION OF DETAIL.

full moon. . The moon at opposition, when it appears as a round disk to an observer on the earth because the illuminated side is toward him. See also PHASES OF THE MOON.

function. , *n*. A magnitude so related to another magnitude that for any value of one there is a corresponding value of the other. See also TRIGONOMETRIC FUNCTIONS.

fundamental circle. . See PRIMARY GREAT CIRCLE.

fundamental frequency. . In the Decca Navigator System, the frequency from which other frequencies in a chain are derived by harmonic multiplication.

fundamental star places. . The apparent right ascensions and declinations of 1,535 standard comparison stars obtained by leading observatories and published annually under the auspices of the International Astronomical Union.

funnel cloud. . A cloud column or inverted cloud cone, pendant from a cloud base. This supplementary feature occurs mostly with cumulus and cumulonimbus; when it reaches the earth's surface, it constitutes a tornado or waterspout. Also called TUBA, TORNADO CLOUD.

furrow. , *n*. On the sea floor, a closed, linear, narrow, shallow depression.

fusion. , *n*. The phase transition of a substance passing from the solid to the liquid state; melting. In meteorology, fusion is almost always understood to refer to the melting of ice, which, if the ice is pure and subjected to 1 standard atmosphere of pressure, takes place at the ice point of 0°C or 32°F. Additional heat at the melting point is required to fuse any substance. This quantity of heat is called LATENT HEAT OF FUSION; in the case of ice, it is approximately 80 calories per gram.

G

G. , *n*. An acceleration equal to the acceleration of gravity, approximately 32.2 feet per second per second at sea level.

gain. , *n*. The ratio of output voltage, current, or power to input voltage, current, or power in electronic instruments.

gain control. . See RECEIVER GAIN CONTROL.

gain function. . See DIRECTIVE GAIN.

gain of an antenna. . An expression of radiation effectiveness, it is the ratio of the power required at the input of a reference antenna to the power supplied to the input of the given antenna to produce, in a given direction, the same field at the same distance. When not specified otherwise, the figure expressing the gain of an antenna refers to the gain in the direction of the radiation main lobe. In services using scattering modes of propagation, the full gain of an antenna may not be realizable in practice and the apparent gain may vary with time.

gain referred to a short vertical antenna. . The gain of an antenna in a given direction when the reference antenna is a perfect vertical antenna, much shorter than one quarter of the wavelength, placed on the surface of a perfectly conducting plane earth.

gal. , *n*. A special unit employed in geodesy and geophysics to express the acceleration due to gravity. The gal is a unit accepted temporarily for use with the International System of Units; 1 gal is equal to 1 centimeter per second, per second.

galactic nebula. . An aggregation of matter within our galaxy but beyond the solar system, large enough to occupy a perceptible area but which has not been resolved into individual stars.

galaxy. , *n*. A vast assemblage of stars, planets, nebulae, and other bodies composing a distinct group in the universe. The sun and its family of planets is part of a galaxy commonly called the MILKY WAY.

gale. , *n*. Wind of force 8 on the Beaufort wind scale (34 to 40 knots or 39 to 46 miles per hour) is classified as a gale. Wind of force 9 (41 to 47 knots or 47 to 54 miles per hour) is classified as a strong gale. Wind of force 7 (28 to 33 knots or 32 to 38 miles per hour) is classified as a near gale. See also MODERATE GALE, FRESH GALE WHOLE GALE.

gallon. , *n*. A unit of volume equal to 4 quarts or 231 cubic inches.

Galofaro. , *n*. A whirlpool in the Strait of Messina; formerly called CHARYBDIS.

galvanometer. , *n*. An instrument for measuring the magnitude of a small electric current or for detecting the presence or direction of such a current by means of motion of an indicator in a magnetic field.

gap. , *n*. On the sea floor, a narrow break in a ridge or rise.

garua. , *n*. A thick, damp fog on the coasts of Ecuador, Peru, and Chile. Also called CAMANCHACA.

gas. , *n*. A fluid without shape or volume, which tends to expand indefinitely, or to completely fill a closed container of any size.

gas buoy. . A buoy having a gas light. See also LIGHTED BUOY.

gat. , *n*. A natural or artificial passage or channel extending inland through shoals or steep banks. See also OPENING.

gather way. . To begin to move.

gauge, gage. , *n*. An instrument for measuring the size or state of anything.

gauge, gage. , *v., t*. To determine the size or state of anything.

gauss. , *n*. The centimeter-gram-second electromagnetic unit of magnetic induction. It corresponds to 10^{-4} tesla in the International System.

Gaussian distribution. . See normal DISTRIBUTION.

Gaussin error. . Deviation of a magnetic compass due to transient magnetism caused by eddy currents set up by a changing number of lines of force through soft iron as the ship changes heading. Due to these eddy currents, the induced magnetism on a given heading does not arrive at its normal value until about 2 minutes after change to the heading. This error should not be confused with RETENTIVE ERROR.

gazeteer. , *n*. An alphabetical list of place names giving geographic coordinates.

Gegenschein. , *n*. A faint light area of the sky always opposite the position of the sun on the celestial sphere. It is believed to be the reflection of sunlight from particles moving beyond the earth's orbit. Also called COUNTERGLOW.

general chart. . See CHART CLASSIFICATION BY SCALE.

generalization. . The process of selectively removing less important features of charts as scale becomes smaller, to avoid over-crowding charts. See also FULL DEPICTION OF DETAIL, MINIMAL DEPICTION OF DETAIL.

general precession. . The resultant motion of the components causing precession of the equinoxes westward along the ecliptic at the rate of about 50.3" per year, completing the cycle in about 25,800 years. The effect of the sun and moon, called lunisolar precession, is to produce a westward motion of the equinoxes along the ecliptic. The effect of other planets, called planetary precession, tends to produce a much smaller motion eastward along the ecliptic. The component t of general precession along the celestial equator, called precession in right ascension, is about 46.1" per year; and the component along a celestial meridian, called precession in declination, is about 20.0" per year.

General Prudential Rule.. Rule 2(b) of the International Rules and Inland Rules. Rule 2(b) states "In construing and complying with these Rules due regard shall be had to all dangers of navigation and collision and to any special circumstances, including the limitations of the vessels involved, which may make a departure from these Rules necessary to avoid immediate danger."

generating area.. The area in which ocean waves are generated by the wind. Also called FETCH.

gentle breeze.. Wind of force 3 (7 to 10 knots or 8 to 12 miles per hour) on the Beaufort wind scale.

geo., *n.* A narrow coastal inlet bordered by steep cliffs. Also called GIO.

geo-.. A prefix meaning earth.

geocentric., *adj.* Relative to the earth as a center; measured from the center of the earth.

geocentric latitude.. The angle at the center of the reference ellipsoid between the celestial equator and a radius vector to a point on the ellipsoid. This differs from the geographic latitude by a maximum of 11.6' of arc at Lat. 45°.

geocentric parallax.. The difference in apparent direction of a celestial body from a point on the surface of the earth and from the center of the earth. This difference varies with the body's altitude and distance from the earth. Also called DIURNAL PARALLAX. See also HELIOCENTRIC PARALLAX.

geodesic., *adj.* Of or pertaining to geodesy; geodetic.

geodesic., *n.* See GEODESIC LINE.

geodesic line.. A line of shortest distance between any two points on any mathematically defined surface. A geodesic line is a line of double curvature and usually lies between the two normal section lines which the two points determine. If the two terminal points are in nearly the same latitude, the geodesic line may cross one of the normal section lines It should be noted that, except along the equator and along the meridians, the geodesic line is not a plane curve and cannot be sighted over directly. Also called GEODESIC, GEODETIC LINE.

geodesy., *n.* The science of the determination of the size and shape of the earth.

geodetic., *adj.* Of or pertaining to geodesy; geodesic.

geodetic bench mark.. See under BENCH MARK.

geodetic datum.. See DATUM, HORIZONTAL GEODETIC DATUM, VERTICAL GEODETIC DATUM.

geodetic equator.. The line of zero geodetic latitude; the great circle described by the semimajor axis of the reference ellipsoid as it is rotated about the minor axis. See also ASTRONOMICAL EQUATOR.

geodetic height.. See ELLIPSOIDAL HEIGHT.

geodetic latitude.. The angle which the normal to the ellipsoid at a station makes with the plane of the geodetic equator. It differs from the corresponding astronomical latitude by the amount of the meridional component of the local deflection of the vertical. Also called TOPOGRAPHICAL LATITUDE and sometimes GEOGRAPHIC LATITUDE.

geodetic line.. See GEODESIC LINE.

geodetic longitude.. The angle between the plane of the geodetic meridian at a station and the plane of the geodetic meridian at Greenwich. A geodetic longitude differs from the corresponding astronomical longitude by the amount of the prime vertical component of the local deflection of the vertical divided by the cosine of the latitude. Sometimes called GEOGRAPHIC LONGITUDE.

geodetic meridian.. A line on a reference ellipsoid which has the same geodetic longitude at every point. Sometimes called GEOGRAPHIC MERIDIAN.

geodetic parallel.. A line on a reference ellipsoid which has the same geodetic latitude of every point. A geodetic parallel, other than the equator, is not a geodesic line. In form, it is a small circle whose plane is parallel with the plane of the geodetic equator. See also ASTRONOMICAL PARALLEL.

geodetic position.. A position of a point on the surface of the earth expressed in terms of geodetic latitude and geodetic longitude. A geodetic position implies an adopted geodetic datum.

geodetic satellite.. Any satellite whose orbit and payload render it useful for geodetic purposes.

geodetic survey.. A survey that takes into account the shape and size of the earth. It is applicable for large areas and long lines and is used for the precise location of basic points suitable for controlling other surveys.

geographic, geographical., *adj.* Of or pertaining to geography.

geographical coordinates.. Spherical coordinates defining a point on the surface of the earth, usually latitude and longitude. Also called TERRESTRIAL COORDINATES.

geographical mile.. The length of 1 minute of arc of the equator, or 6,087.08 feet. This approximates the length of the nautical mile.

geographical plot.. A plot of the movements of one or more craft relative to the surface of the earth. Also called TRUE PLOT. See also NAVIGATIONAL PLOT.

geographical pole.. Either of the two points of intersection of the surface of the earth with its axis, where all meridians meet, labeled N or S to indicate whether the north geographical pole or the south geographical pole.

geographical position.. 1. That point on the earth at which a given celestial body is in the zenith at a specified time. The geographical position of the sun is also called the sub solar point, of the moon the sublunar point, and of a star the substellar or subastral point. 2. Any position on the earth defined by means of its geographical coordinates either astronomical or geodetic.

geographic graticule.. The system of coordinates of latitude and longitude used to define the position of a point on the surface of the earth with respect to the reference ellipsoid.

geographic latitude.. A general term applying to astronomic and geodetic latitudes.

geographic longitude.. A general term applying to astronomic and geodetic longitudes.

geographic meridian.. A general term applying to astronomical and geodetic meridians.

geographic number.. The number assigned to an aid to navigation for identification purposes in accordance with the lateral system of numbering.

geographic parallel.. A general term applying to astronomical and geodetic parallels.

geographic range.. The maximum distance at which the curvature of the earth and terrestrial refraction permit an aid to navigation to be seen from a particular height of eye without regard to the luminous intensity of the light. The geographic range sometimes printed on charts or tabulated in light lists is the maximum distance at which the curvature of the earth and terrestrial refraction permit a light to be seen from a height of eye of 15 feet above the water when the elevation of the light is taken above the height datum of the largest scale chart of the locality. Therefore, this range is a nominal geographic range. See also VISUAL RANGE (OF A LIGHT).

geographic sign conventions.. In mapping, charting, and geodesy, the inconsistent application of algebraic sign to geographical references and the angular reference of azimuthal systems is a potential trouble area in scientific data collection. The following conventions have wide use in the standardization of scientific notation: Longitude references are positive eastward of the Greenwich meridian to 180°, and negative westward of Greenwich. Latitude references are positive to the north of the equator and negative to the south. Azimuths are measured clockwise, using South as the origin and continuing to 360°. Bearings are measured clockwise, using North as the origin and continuing to 360°. Tabulated coordinates, or individual coordinates, are annotated N, S, E, W, as appropriate.

geoid., *n.* The equipotential surface in the gravity field of the earth; the surface to which the oceans would conform over the entire earth if free to adjust to the combined effect of the earth's mass attraction and the centrifugal force of the earth's rotation. As a result of the uneven distribution of the earth's mass, the geoidal surface is irregular. The geoid is a surface along which the gravity potential is everywhere equal (equipotential surface) and to which the direction of gravity is always perpendicular. Also called FIGURE OF THE EARTH.

geoidal height.. The distance of the geoid above (positive) or below (negative) the mathematical reference ellipsoid. Also called GEOIDAL SEPARATION, GEOIDAL UNDULATION, UNDULATION OF THE GEOID.

geoidal horizon.. The circle of the celestial sphere formed by the intersection of the celestial sphere and a plane through a point on the sea level surface of the earth, and perpendicular to the zenith-nadir line. See also HORIZON.

geoidal separation.. See GEOIDAL HEIGHT.

geoidal undulation.. See GEOIDAL HEIGHT.

geological oceanography.. The study of the floors and margins of the oceans, including description of submarine relief features, chemical and physical composition of bottom materials, interaction of sediments and rocks with air and seawater, and action of various forms of wave energy in the submarine crust of the earth.

geomagnetic., *adj.* Of or pertaining to geomagnetism.

geomagnetic equator.. The terrestrial great circle everywhere 90° from the geomagnetic poles. GEOMAGNETIC EQUATOR is not the same as the MAGNETIC EQUATOR, the line connecting all points of zero magnetic dip.

geomagnetic latitude.. Angular distance from the geomagnetic equator, measured northward or southward on the geomagnetic meridian through 90° and labeled N or S to indicate the direction of measurement. GEOMAGNETIC LATITUDE should not be confused with MAGNETIC LATITUDE.

geomagnetic pole.. Either of two antipodal points marking the intersection of the earth's surface with the extended axis of a bar magnet assumed to be located at the center of the earth and approximating the source of the actual magnetic field of the earth. The pole in the Northern Hemisphere (at about lat. 78.5° N, long. 69° W) is designated north geomagnetic pole, and the pole in the Southern Hemisphere (at about lat. 78°S, long. 111° E) is designated south.

geomagnetic pole.. The great circle midway between these poles is called GEOMAGNETIC EQUATOR. The expression GEOMAGNETIC POLE should not be confused with MAGNETIC POLE, which relates to the actual magnetic field of the earth. See also GEOMAGNETIC LATITUDE.

geomagnetism., *n.* Magnetic phenomena, collectively considered, exhibited by the earth and its atmosphere. Also called TERRESTRIAL MAGNETISM.

geometrical dip.. The vertical angle between the horizontal and a straight line tangent to the surface of the earth. It is larger than DIP by the amount of terrestrial refraction.

geometrical horizon.. Originally, the celestial horizon; now more commonly the intersection of the celestial sphere and an infinite number of straight lines tangent to the earth's surface, and radiating from the eye of the observer. If there were no terrestrial refraction, GEOMETRICAL and VISIBLE HORIZONS would coincide. See also RADIO HORIZON.

geometric dilution.. See GEOMETRIC DILUTION OF PRECISION.

geometric dilution of precision.. All geometric factors that degrade the accuracy of position fixes derived from externally referenced navigation systems. Often shortened to GEOMETRIC DILUTION.

geometric map projection.. See PERSPECTIVE MAP PROJECTION.

geometric projection.. See PERSPECTIVE PROJECTION.

geomorphology., *n.* A branch of both geography and geology that deals with the form of the earth, the general configuration of its surface, and the changes that take place in the evolution of land forms.

geo-navigation., *n.* Navigation by means of reference points on the earth. The term is obsolete.

geophysics., *n.* The study of the composition and physical phenomena of the earth and its liquid and gaseous envelopes; it embraces the study of terrestrial magnetism, atmospheric electricity, and gravity; and it includes seismology, volcanology, oceanography, meteorology, and related sciences.

geopotential., *n.* The gravity potential of the actual earth. It is the sum of the gravitational (attraction) potential and the potential of the centrifugal force.

Georef., *n.* See WORLD GEOGRAPHIC REFERENCE SYSTEM.

geosphere., *n.* The portion of the earth, including land (lithosphere) and water (hydrosphere), but excluding the atmosphere.

geostationary satellite.. An earth satellite moving eastward in an equatorial, circular orbit at an altitude (approximately 35,900 kilometers) such that its period of revolution is exactly equal to and synchronous with the rotational period of the earth. Such a satellite will remain fixed over a point on the earth's equator. Although geostationary satellites are frequently called GEOSYNCHRONOUS or SYNCHRONOUS SATELLITES, the orbit of an eastward moving synchronous satellite must be equatorial if the satellite is to remain fixed over a point on the equator. Otherwise, the satellite moves daily in a figure eight pattern relative to the earth. Also called FIXED SATELLITE. See also STATIONARY ORBIT.

geostrophic wind.. The horizontal wind velocity for which the Coriolis force exactly balances the horizontal pressure force. See also GRADIENT WIND.

geosynchronous satellite.. An earth satellite whose period of rotation is equal to the period of rotation of the earth about its axis. The orbit of a geosynchronous satellite must be equatorial if the satellite is to remain fixed over a point on the earth's equator. Also called TWENTY-FOUR HOUR SATELLITE. See also SYNCHRONOUS SATELLITE, GEOSTATIONARY SATELLITE.

ghost., *n.* 1. An unwanted image appearing on a radarscope caused by echoes which experience multiple reflections before reaching the receiver. See also SECOND-TRACE ECHO, MULTIPLE ECHOES, INDIRECT ECHO. 2. An image appearing on a radarscope the origin of which cannot readily be determined.

giant floe.. See under FLOE.

gibbous., *adj.* Bounded by convex curves. The term is used particularly in reference to the moon when it is between first quarter and full or between full and last quarter, or to other celestial bodies when they present a similar appearance. See also PHASES OF THE MOON.

giga-.. A prefix meaning one billion (10^9).

gigahertz., *n.* One thousand megahertz, or one billion cycles per second.

gimbal freedom.. The maximum angular displacement of a gyro about the output axis of a gimbal.

gimballess inertial navigation equipment.. See STRAPPED-DOWN INERTIAL NAVIGATION EQUIPMENT.

gimballing error.. That error introduced in a gyro-compass by the tilting of the gimbal mounting system of the compass due to horizontal acceleration caused by motion of the vessel, such as rolling.

gimbal lock.. A condition of a two-degree-of-freedom gyro wherein the alignment of the spin axis with an axis of freedom deprives the gyro of a degree-of-freedom and therefore its useful properties.

gimbals., *n., pl.* A device for supporting anything, such as an instrument, in such a manner that it will remain horizontal when the support tilts. It consists of a ring inside which the instrument is supported at two points 180° apart, the ring being similarly supported at two points 90° from the instrument supports.

gio., *n.* See GEO.

glacial., *adj.* Of or pertaining to a glacier.

glacier., *n.* A mass of snow and ice continuously moving from higher to lower ground or, if afloat, continuously spreading. The principal forms of glacier are INLAND ICE SHEETS, ICE SHELVES, ICE STREAMS, ICE CAPS, ICE PIEDMONTS, CIRQUE GLACIERS, and various types of mountain (valley) glaciers.

glacier berg.. An irregularly shaped iceberg. Also called WEATHERED BERG.

glacier ice.. Ice in, or originating from, a glacier, whether on land or floating on the sea as icebergs, bergy bits, or growlers.

glacier tongue.. The seaward projecting extension of a glacier, usually afloat. In the Antarctic, glacier tongues may extend many tens of kilo-meters.

glare., *n.* Dazzling brightness of the atmosphere caused by excessive reflection and scattering of light by particles in the line of sight.

glaze., *n.* A coating of ice, generally clear and smooth but usually containing some air pockets, formed on exposed objects by the freezing of a film of super cooled water deposited by rain, drizzle, fog, or possibly condensed from super cooled water vapor. Glaze is denser, harder and more transparent than either rime or hoarfrost Also called GLAZE ICE, GLAZED FROST VERGLAS.

glazed frost.. See GLAZE.

glaze ice.. See GLAZE.

glint., *n.* The pulse-to-pulse variation in amplitude of reflected radar signals due to rapid change of the reflecting surface, as in the case of the propeller of an aircraft in flight.

Global Positioning System.. See as NAVSTAR GLOBAL POSITIONING SYSTEM.

globigerina . *(pl. globlgerinae), n.* A very small marine animal of the foraminifera order, with a chambered shell; or the shell of such an animal. In large areas of the ocean the calcareous shells of these animals are very numerous, being the principal constituent of a soft mud or globigerina *ooze* forming the ocean bed.

GLONASS. . A satellite navigation system operated by Russia, analogous to the U.S. Global Positioning System (GPS).

gloom. , *n.* The condition existing when daylight is very much reduced by dense cloud or smoke accumulation above the surface, the surface visibility not being materially reduced.

glory. , *n.* See ANTICORONA.

gnomon. , *n.* Any object the shadow of which serves as an indicator, as the SHADOW PIN on a sun.

gnomonic. , *adj.* Of or pertaining to a gnomon.

gnomonic chart. . A chart constructed on the gnomonic projection and often used as an adjunct for transferring a great circle to a Mercator chart. Commonly called GREAT CIRCLE CHART.

gnomonic map projection. . A perspective azimuthal map projection in which points on the surface of a sphere or spheroid, such as the earth, are conceived as projected by radials from the center to a tangent plane. Great circles project as straight lines. For this reason the projection is used principally for charts for great circle sailing. The projection is neither conformal nor equal area.

gong. , *n.* A sound signal producing a sound by the vibration of a resonant disc struck by a clapper.

gong buoy. . A buoy fitted with a group of saucer shaped bells of different tones as an audible signal.

goniometer. , 1. An instrument for measuring angles. 2. A pick-up coil which eliminates the necessity of having to rotate a radio direction finder antenna to determine direction.

gore. , *n.* A lune-shaped map which may be fitted to the surface of a globe with a negligible amount of distortion.

gorge. , *n.* 1. A narrow opening between mountains, especially one with steep, rocky walls. 2. A collection of solid matter obstructing a channel, river, etc., as *ice gorge*.

GPS. , *n.* Global Positioning System; the US Department of Defense-operated world-wide satellite positioning system.

gradient. , *n.* 1. A rate of rise or fall of a quantity against horizontal distance expressed as a ratio, decimal, fraction, percentage, or the tangent of the angle of inclination. 2. The rate of increase or decrease of one quantity with respect to another. 3. A term used in radionavigation to refer to the spacing between consecutive hyperbolas of a family of hyperbolas per unit time difference. If the gradient is high, a relatively small time-difference error in determining a hyperbolic line of position will result in a relatively high position error. See also GEOMETRIC DILUTION OF PRECISION.

gradient current. . An ocean current associated with horizontal pressure gradients in the ocean and determined by the condition that the pressure force due to the distribution of mass balances the Coriolis force due to the earth's rotation. See also OCEAN CURRENT.

gradient tints. . See HYPSOMETRIC TINTING.

gradient wind. . Any horizontal wind velocity tangent to the contour line of a constant pressure surface (or to the isobar of a geopotential surface) at the point in question. At such points where the wind is gradient, the Coriolis force and the centrifugal force together exactly balance the horizontal pressure force. See also GEOSTROPHIC WIND.

graduation error. . Inaccuracy in the graduations of the scale of an instrument.

graduations. , *n., pl.* The marks on a scale.

grain noise. . See SNOW, definition 2.

gram. , *n.* One one-thousandth of a kilogram.

granular snow. . See SNOW GRAINS.

graph. , *n.* A diagram indicating the relationship between two or more variables.

graph. , *v., t.* To represent by a graph.

graphic scale. . See BAR SCALE.

graticule. , *n.* 1. The network of lines representing parallels and meridians on a map, chart, or plotting sheet. A fictitious graticule represents fictitious parallels and fictitious meridians. See also GRID, *n.* 2. A scale at the focal plane of an optical instrument to aid in the measurement of objects. See also RETICULE.

graupel. , *n.* See SNOW PELLETS.

gravel. , *n.* See under STONES.

graving dock. . A form of dry dock consisting of an artificial basin fitted with a gate or caisson, into which vessels can be floated and the water pumped out to expose the vessels' bottoms. The term is derived from the term used to describe the process of burning barnacles and other accretions from a ship's bottom. See also FLOATING DOCK.

gravisphere. , *n.* The spherical extent in which the force of a given celestial body's gravity is predominant in relation to that of other celestial bodies.

gravitation. , *n.* 1. The force of attraction between two bodies. According to Newton, gravitation is directly proportional to the product of the masses of two bodies and inversely proportional to the square of the distance between them. 2. The acceleration produced by the mutual attraction of two masses, directed along the line joining their centers of mass, and of magnitude inversely proportional to the square of the distance between the two centers of mass.

gravitational disturbance. . See GRAVITY DISTURBANCE.

gravitational gradient. . The change in the gravitational acceleration per unit distance.

gravitational perturbations. . Perturbations caused by body forces due to nonspherical terrestrial effects, lunisolar effect, tides, and the effect of relativity.

gravitational tide. . See EQUILIBRIUM TIDE.

gravity. , *n.* The force of attraction of the earth, or another body, on nearby objects.

gravity anomaly. . The difference between the observed gravity value properly reduced to sea level and the theoretical gravity obtained from gravity formula. Also called OBSERVED GRAVITY ANOMALY.

gravity anomaly map. . A map showing the positions and magnitudes of gravity anomalies. Also, a map on which contour lines are used to represent points at which the gravity anomalies are equal.

gravity data. . Information concerning that acceleration which attracts bodies and is expressed as observations or in the form of gravity anomaly charts or spherical harmonics for spatial representation of the earth and other celestial bodies.

gravity disturbance. . The difference between the observed gravity and the normal gravity at the same point (the vertical gradient of the disturbing potential) as opposed to GRAVITY ANOMALY which uses corresponding points on two different surfaces. Because the centrifugal force is the same when both are taken at the same point, it can also be called GRAVITATIONAL DISTURBANCE.

gravity field of the earth. . The field of force arising from a combination of the mass attraction and rotation of the earth. The field is normally expressed in terms of point values, mean area values, and/or series expansion for the potential of the field.

gravity network. . A network of gravity stations.

gravity reduction. . A combination of gravity corrections to obtain reduced gravity on the geoid.

gravity reference stations. . Stations which serve as reference values for a gravity survey, i.e., with respect to which the differences at the other stations are determined in a relative survey. The absolute value of gravity may or may not be known at the reference stations.

gravity station. . A station at which observations are made to determine the value of gravity.

gravity wind. . A wind blowing down an incline. Also called KATABATIC WIND.

grease ice. . Ice at that stage of freezing when the crystals have coagulated to form a soupy layer on the surface. Grease ice is at a later stage of freezing than *frazil ice* and reflects little light, giving the sea a matte appearance.

great circle. . The intersection of a sphere and a plane through its center. The intersection of a sphere and a plane which does not pass through its center is called a small circle. Also called ORTHODROME, ORTHODROMIC CURVE.

great circle bearing. . The initial direction of a great circle through two terrestrial points, expressed as angular distance from a reference direction. It is usually measured from 000° at the reference direction clockwise through 360°. Bearings obtained by any form of radiant energy are great circle bearings.

great circle chart. . A chart on which a great circle appears as a straight line or approximately so, particularly a chart on the gnomonic map projection.

great circle course. . The direction of the great circle through the point of departure and the destination, expressed as the angular distance from a reference direction, usually north, to the direction of the great circle. The angle varies from point to point along the great circle. At the point of departure it is called initial great circle course; at the destination it is called final great circle course.

great circle direction. . Horizontal direction of a great circle, expressed as angular distance from a reference direction.

great circle distance. . The length of the shorter arc of the great circle joining two points. It is usually expressed in nautical miles.

great circle sailing. . Any method of solving the various problems involving courses, distance, etc., as they are related to a great circle track.

great circle track. . The track of a vessel following a great circle, or a great circle which it is intended that a vessel follow approximately.

great diurnal range. . The difference in height between mean higher high water and mean lower low water. Often shortened to DIURNAL RANGE. The difference in height between mean lower high water and mean higher low water is called SMALL DIURNAL RANGE.

greater ebb. . See under EBB CURRENT.

greater flood. . See under FLOOD CURRENT.

greatest elongation. . The maximum angular distance of an inferior planet from the sun before it starts back toward conjunction, as observed from the earth. The direction of the body east or west of the sun is usually specified, as *greatest elongation east* (or *west*). See also ELONGATION.

great tropic range. . The difference in height between tropic higher high water and tropic lower low water. Often shortened to TROPIC RANGE. See also MEAN TROPIC RANGE, SMALL TROPIC RANGE.

great year. . The period of one complete cycle of the equinoxes around the ecliptic, about 25,800 years. Also called PLATONIC YEAR. See also PRECESSION OF THE EQUINOXES.

green flash. . A brilliant green coloring of the upper edge of the sun as it appears at sunrise or disappears at sunset when there is a clear, distinct horizon. It is due to refraction by the atmosphere, which disperses the first (or last) spot of light into a spectrum and causes the colors to appear (or disappear) in the order of refrangibility. The green is bent more than red or yellow and hence is visible sooner at sunrise and later at sunset.

green house effect. . The heating phenomenon due to shorter wavelengths of insolation passing through the atmosphere to the earth, which radiates longer wavelength infrared radiation that is trapped by the atmosphere. Some of this trapped radiation is reradiated to the earth. This causes a higher earth temperature than would occur from direct insolation alone.

Greenwich apparent noon. . Local apparent noon at the Greenwich meridian; 12 o'clock Greenwich apparent time, or the instant the apparent sun is over the upper branch of the Greenwich meridian.

Greenwich apparent time. . Local apparent time at the Greenwich meridian; the arc of the celestial equator, or the angle at the celestial pole between the lower branch of the Greenwich celestial meridian and the hour circle of the apparent or true sun, measured westward from the lower branch of the Greenwich celestial meridian through 24 hours, Greenwich hour angle of the apparent or true sun, expressed in time units, plus 12 hours.

Greenwich civil time. . United States terminology from 1925 through 1952. See GREENWICH MEAN TIME.

Greenwich hour angle. . Angular distance west of the Greenwich celestial meridian; the arc of the celestial equator, or the angle at the celestial pole, between the upper branch of the Greenwich celestial meridian and the hour circle of a point on the celestial sphere, measured westward from the Greenwich celestial meridian through 360°; local hour angle at the Greenwich meridian.

Greenwich interval. . An interval based on the moon's transit of the Greenwich celestial meridian, as distinguished from a local interval based on the moon's transit of the local celestial meridian.

Greenwich lunar time. . Local lunar time at the Greenwich meridian; the arc of the celestial equator, or the angle at the celestial pole, between the lower branch of the Greenwich celestial meridian and the hour circle of the moon, measured westward from the lower branch of the Greenwich celestial meridian through 24 hours; Greenwich hour angle of the moon expressed in time units, plus 12 hours.

Greenwich mean noon. . Local mean noon at the Greenwich meridian, 12 o'clock Greenwich mean time, or the instant the mean sun is over the upper branch of the Greenwich meridian.

Greenwich mean time. . Local mean time at the Greenwich meridian; the arc of the celestial equator, or the angle at the celestial pole, between the lower branch of the Greenwich celestial meridian and the hour circle of the mean sun, measured westward from the lower branch of the Greenwich celestial meridian through 24 hours; Greenwich hour angle of the mean sun expressed in time units, plus 12 hours. Also called UNIVERSAL TIME, ZULU TIME.

Greenwich meridian. . The meridian through Greenwich, England, serving as the reference for Greenwich time, in contrast with LOCAL MERIDIAN. It is accepted almost universally as the PRIME MERIDIAN, or the origin of measurement of longitude.

Greenwich noon. . Noon at the Greenwich meridian.

Greenwich sidereal noon. . Local sidereal noon at the Greenwich meridian; zero hours Greenwich sidereal time, or the instant the vernal equinox is over the upper branch of the Greenwich meridian.

Greenwich sidereal time. . Local sidereal time at the Greenwich meridian; the arc of the celestial equator, or the angle at the celestial pole, between the upper branch of the Greenwich celestial meridian and the hour circle of the vernal equinox, measured westward from the upper branch of the Greenwich celestial meridian through 24 hours; Greenwich hour angle of the vernal equinox expressed in time units.

Greenwich time. . Time based upon the Greenwich meridian as reference.

gregale. , *n.* A strong northeast wind of the central Mediterranean.

Gregorian calendar. . The calendar now in almost universal use for civil purposes in which each year has 365 days, except leap years which have 366 days. Leap years are those years which are divisible by 4, and in the case of centurial years, those years divisible by 400. This calendar, a modification of the Julian calendar, was not adopted in Great Britain and the English colonies in North America until 1752. The calendar was instituted in 1582 by Pope Gregory XIII to keep calendar days in adjustment with the tropical year for the purpose of regulating the date of Easter and the civil and ecclesiastical calendars.

gray ice. . A subdivision of YOUNG ICE 10 to 15 centimeters thick. Gray ice is less elastic than nilas and breaks in swells. It usually rafts under pressure.

gray-white ice. . A subdivision of YOUNG ICE 15 to 30 centimeters thick. Gray-white ice under pressure is more likely to ridge than to raft.

grid. , *adj.* Pertaining to a grid or related to grid north.

grid. , *n.* 1. A series of lines, usually straight and parallel, superimposed on a chart or plotting sheet to serve as a directional reference for navigation. See also FICTITIOUS GRATICULE, GRATICULE, definition 1. 2. Two sets of mutually perpendicular lines dividing a map or chart into squares or rectangles to permit location of any point by a system of rectangular coordinates. Also called REFERENCE GRID. See also MILITARY GRID, UNIVERSAL POLAR STENOGRAPHIC GRID, UNIVERSAL TRANSVERSE MERCATOR GRID, WORLD GEOGRAPHIC REFERENCING SYSTEM.

grid amplitude. . Amplitude relative to grid east or west.

grid azimuth. . Azimuth relative to grid north.

grid bearing. . Bearing relative to grid north.

grid convergence. . The angular difference in direction between grid north and true north. It is measured east or west from true north.

grid course. . Course relative to grid north.

grid declination. . The angular difference between grid north and true north.

grid direction. . Horizontal direction expressed as angular distance from grid north. Grid direction is measured from grid north, clockwise through 360°.

grid equator. . A line perpendicular to a prime grid meridian, at the origin. For the usual orientation in polar regions the grid equator is the 90°W - 90°E meridian forming the basic grid parallel, from which grid latitude is measured. See also FICTITIOUS EQUATOR.

grid heading. . Heading relative to grid north.

grid latitude. . Angular distance from a grid equator. See also FICTITIOUS LATITUDE.

grid line. . One of the lines of a grid.

grid longitude. . Angular distance between a prime grid meridian and any given grid meridian. See also FICTITIOUS LONGITUDE.

grid magnetic angle. . Angular difference in direction between grid north and magnetic north. It is measured east or west from grid north. Grid magnetic angle is sometimes called GRID VARIATION or GRIVATION.

grid meridian. . One of the grid lines extending in a grid north-south direction. The reference grid meridian is called prime grid meridian. In polar regions the prime grid meridian is usually the 180° - 0° geographic meridian. See also FICTITIOUS MERIDIAN.

grid navigation. . Navigation by the use of grid directions.

grid north. . 1. An arbitrary reference direction used with grid navigation. The direction of the 180th geographical meridian from the north pole is used almost universally as grid north. 2. The northerly or zero direction indicated by the grid datum of directional reference.

grid parallel. . A line parallel to a grid equator, connecting all points of equal grid latitude. See also FICTITIOUS PARALLEL.

grid prime vertical. . The vertical circle through the grid east and west points of the horizon.

grid rhumb line. . A line making the same oblique angle with all grid meridians. Grid parallels and meridians may be considered special cases of the grid rhumb line. See also FICTITIOUS RHUMB LINE.

grid track. . The direction of the track relative to grid north.

grid variation. . See GRID MAGNETIC ANGLE.

grivation. , *n.* See GRID MAGNETIC ANGLE.

groin. , *n.* A structure (usually one of a group) extending approximately perpendicular from a shore to protect the shore from erosion by tides currents, or waves or to trap sand for making a beach. See also JETTY, definition 1.

ground. , *n.* A conducting connection between an electric circuit and the earth or some other conducting body of zero potential with respect to the earth.

ground. , *v., t. & i.* To touch bottom or run aground. *v., t.* To connect an electric circuit with the earth or some other conducting body, such that the earth or body serves as part of the circuit.

ground absorption. . The dissipation of energy in radio waves because of absorption by the ground over which the waves are transmitted.

ground-based duct. . See SURFACE DUCT.

ground chain. . Heavy chain used with permanent moorings and connecting the various legs or bridles.

grounded hummock. . Hummocked grounded ice formation. There are single grounded hummocks and lines (or chains) of grounded hummocks.

grounded ice. . Floating ice which is aground in shoal water. See also STRANDED ICE, FLOATING ICE.

ground fog. . A fog that obscures less than six tenths of the sky, and does not extend to the base of any clouds.

grounding. , *n.* The touching of the bottom by a vessel. A serious grounding is called a stranding.

ground log. . A device for determining the course and speed over the ground in shallow water consisting of a lead or weight attached to a line. The lead is thrown overboard and allowed to rest on the bottom. The course over ground is indicated by the direction the line tends and the speed by the amount of line paid out in a unit of time.

ground swell. . A long, deep swell or undulation of the ocean often caused by a long-continued gale and sometimes a seismic disturbance and felt even at a remote distance. In shallow water the swell rises to a prominent height. See SWELL definition 1.

ground tackle. . The anchors, anchor chains, fittings etc., used for anchoring a vessel.

ground track. . 1. See under TRACK, definition 2. 2. See under TRUE TRACK OF TARGET.

groundwave. . A radio wave that is propagated over the earth and is ordinarily influenced by the presence of the ground and the troposphere. Except for ionospheric and tropospheric waves, the groundwave includes all components of a radio wave.

group flashing light. . A flashing light in which the flashes are combined in groups, each group having the same number of flashes, and in which the groups are repeated at regular intervals. The eclipses separating the flashes within each group are of equal duration and this duration is clearly shorter than the duration of the eclipse between two successive groups.

group occulting light. . An occulting light in which the occultations are combined in groups, each group including the same number of occultations, and in which the groups are repeated at regular intervals. The intervals of light separating the occultations within each group are of equal duration and this duration is clearly shorter than the duration of the interval of light between two successive groups.

group quick light. . A quick flashing light in which a specified group of flashes is regularly repeated. See also CONTINUOUS QUICK LIGHT, INTERRUPTED QUICK LIGHT.

group repetition interval. . The specified time interval of a Loran C chain for all stations of the chain to transmit their pulse groups. For each chain a minimum group repetition interval (GRI) is selected of sufficient duration to provide time for each station to transmit its pulse group and additional time between each pulse group so that signals from two or more stations cannot overlap in time anywhere within the coverage area. The GRI is normally stated in terms of tens of microseconds; i.e., the GRI having a duration of 79,900 microseconds is stated as 7900.

group repetition interval code. . The group repetition interval in microseconds divided by 10.

group very quick light. . A very quick flashing light in which a specified group of flashes is regularly repeated. See also CONTINUOUS VERY QUICK LIGHT, INTERRUPTED VERY QUICK LIGHT.

growler. , *n.* A piece of ice smaller than a BERGY BIT or FLOEBERG, often transparent but appearing green or almost black in color. It extends less than 1 meter above the sea surface and its length is less than 20 feet (6 meters). A growler is large enough to be a hazard to shipping but small enough that it may escape visual or radar detection.

grunt. , *n.* See under DIAPHONE.

Guiana Current. . An ocean current flowing northwestward along the northeast coast of South America. The Guiana Current is an extension of the Atlantic South Equatorial Current, which crosses the equator and approaches the coast of South America. Eventually, it is joined by part of the Atlantic North Equatorial Current and becomes, successively, the CARIBBEAN ISLANDS, and the FLORIDA CURRENT. Also called NORTH BRAZIL CURRENT.

Guinea Current. . A North Atlantic Ocean current flowing eastward along the south coast of northwest Africa into the Gulf of Guinea. The Guinea Current is the continuation of the Atlantic Equatorial Countercurrent augmented by the eastern branch of the Canary Current.

gulder. , *n.* Local name given to double low water occurring on the south coast of England. See DOUBLE TIDE.

gulf. , *n.* A major indentation of the sea into the land, usually larger than a bay.

Gulf Coast Low Water Datum. . Gulf Coast Low Water Datum (GCLWD) is defined as mean lower low water when the type of tide is mixed, and mean low water when the type of tide is diurnal. GCLWD was used as chart tidal datum from November 14, 1977, to November 28, 1980, for the coastal waters of the gulf coast of the United States.

Gulf Stream. . A warm, well defined, swift, relatively narrow ocean current which originates where the Florida Current and the Antilles Current meet north of Grand Bahama Island. It gains its impetus from the large volume of water that flows through the Straits of Florida. Near the edge of the Grand Banks of Newfoundland extensions of the Gulf Stream and the Labrador Current continue as the NORTH ATLANTIC CURRENT, which fans outward and widens in a northeastward to eastward flow across the ocean. The Florida Current, the Gulf Stream, and the North Atlantic Current together form the GULF STREAM SYSTEM. Sometimes the entire system is referred to as the Gulf Stream The Gulf Stream forms the western and northwestern part of the general clockwise oceanic circulation of the North Atlantic Ocean.

Gulf Stream System. . A system of ocean currents comprised of the Florida Current, the Gulf Stream, and the North Atlantic Current.

gulfweed. , *n.* See SARGASSUM.

gully. , *n.* 1. A small ravine, especially one cut by running water, but through which water flows only after a rain. 2. On the sea floor, a small valley-like feature.

gust. , *n.* 1. A sudden brief increase in the speed of the wind of more transient character than a squall, and followed by a lull or slackening of the wind. 2. The violet wind or squall that accompanies a thunderstorm.

gut., *n.* A narrow passage or contracted strait connecting two bodies of water.

guyot., *n.* See TABLEMOUNT.

gyre., *n.* A closed circulatory system, but larger than a whirlpool or eddy.

gyro., *n.* Short for GYROSCOPE.

gyrocompass., *n.* A compass having one or more gyroscopes as the directive element, and which is north-seeking. Its operation depends upon four natural phenomena, namely gyroscopic inertia, gyroscopic precession, the earth's rotation, and gravity. When such a compass controls remote indicators, called GYRO REPEATERS, it is called a master gyrocompass. See also DIRECTIONAL GYRO MODE.

gyro error.. The error in the reading of the gyrocompass, expressed in degrees east or west to indicate the direction in which the axis of the compass is offset from true north. See also BALLISTIC DAMPING ERROR, BALLISTIC DEFLECTION ERROR, COMPASS ERROR, GIMBALLING ERROR, INTERCARDINAL ROLLING ERROR, LUBBER'S LINE ERROR SPEED ERROR.

gyro log.. A written record of the performance of a gyrocompass.

gyropilot., *n.* An automatic device for steering a vessel by means of control signals received from a gyrocompass. Also called AUTOPILOT.

gyro repeater.. A device which displays at a different location the indications of the master gyrocompass. See also COMPASS REPEATER.

gyroscope., *n.* A rapidly rotating mass free to move about one or both axes perpendicular to the axis of rotation and to each other. It is characterized by GYROSCOPIC INERTIA and PRECESSION. Usually shortened to GYRO. The term also refers colloquially to the GYROCOMPASS. See also DIRECTIONAL GYRO, FREE GYRO.

gyroscopic drift.. The horizontal rotation of the spin axis of a gyroscope about the vertical axis.

gyroscopic inertia.. The property of a gyroscope of resisting any force which tends to change its axis of rotation. A gyroscope tends to maintain the direction of its axis of rotation in space. Also called RIGIDITY IN SPACE.

gyro sextant.. A sextant provided with a gyroscope to indicate the horizontal.

H

haar., *n.* A wet sea fog or very fine drizzle which drifts in from the sea in coastal districts of eastern Scotland and northeast England, especially in summer.

habitat sanctuary.. A marine sanctuary established for the preservation, protection and management of essential or specialized habitats representative of important marine systems. See also MARINE SANCTUARY.

hachules., *n. pl.* 1. Short lines on topographic maps or nautical charts to indicate the slope of the ground or the submarine bottom. They usually follow the direction of the slope. 2. Inward-pointing short lines or "ticks" around the circumference of a closed contour indicating a depression or a minimum.

hack., *n.* A chronometer which has failed to meet the exacting requirements of a standard chronometer, and is used for timing observations of celestial bodies, regulating ship's clocks, etc. A comparing watch, which may be of high quality, is normally used for timing celestial observations, the watch being compared with the chronometer, preferably both before and after observations. Sometimes called HACK CHRONOMETER.

hack chronometer.. See HACK.

hack watch.. See COMPARING WATCH.

hail., *n.* Frozen precipitation consisting of ice balls or irregular lumps of ice of varying size, ranging from that of a raindrop to an inch or considerably more. They are composed of clear ice or of alternate layers of ice and snow, and may fall detached or frozen together into irregular lumps. Hail is usually associated with thunderstorms. A hailstone is a single unit of hail. Small hail consists of snow pellets surrounded by a very thin ice covering. See also SNOW PELLETS.

hailstone., *n.* See under HAIL.

hail storm.. See under STORM, definition 2.

half-power points.. Power ratios used to define the angular width of a radar beam. One convention defines beam width as the angular width between points at which the field strength is 71 percent of its maximum value. Expressed in terms of power ratio, this convention defines beam width as the angular width between half-power points. A second convention defines beam width as the angular width between points at which the field strength is 50 percent of its maximum value. Expressed in terms of power ratio, the latter convention defines beam width as the angular width between quarter-power points.

half tide.. The condition or time of the tide when midway between high and low.

half-tide basin.. A lock of very large size and usually of irregular shape, the gates of which are kept open for several hours after high tide so that vessels may enter as long as there is sufficient depth over the sill. Vessels remain in the half-tide basin until the ensuing flood tide before they may pass through the gate to the inner harbor. If entry to the inner harbor is required before this time, water must be admitted to the half-tide basin from some external source. See also TIDAL BASIN, NON-TIDAL BASIN.

half-tide level.. A tidal datum midway between mean high water and mean low water. Mean sea level may coincide with half-tide level, but seldom does; the variation is generally about 3 centimeters and rarely exceeds 6 centimeters. Also called MEAN TIDE LEVEL. See also MID-EXTREME TIDE.

halo., *n.* Any of a group of optical phenomena caused by refraction or reflection of light by ice crystals in the atmosphere. The most common form is a ring of light of radius 22° or 46° around the sun or moon. See also CORONA, PARHELION, CIRCUMSCRIBED HALO, PARHELIC CIRCLE, SUN CROSS, SUN PILLAR, CIRCUMZENITHAL ARC, ANTHELION, PARANTHELION, HAVELIAN HALO, TANGENT ARC.

halving., *n.* The process of adjusting magnetic compass correctors so as to remove half of the deviation on the opposite cardinal or adjacent intercardinal headings to those on which adjustment was originally made when all deviation was removed. This is done to equalize the error on opposite headings.

***Handbook of Magnetic Compass Adjustment*.**. See PUB. NO. 226. (No longer in print)

hand lead.. A light sounding lead (7 to 14 pounds), usually having a line of not more than 25 fathoms.

hanging compass.. See INVERTED COMPASS.

harbor., *n.* 1. A body of water providing protection for vessels and, generally, anchorage and docking facilities. 2. A haven or space of deep water so sheltered by the adjacent land as to afford a safe anchorage for ships. See also NATURAL HARBOR, ARTIFICIAL HARBOR.

harbor chart.. See under CHART CLASSIFICATION BY SCALE.

harbor line.. The line beyond which wharves and other structures cannot be extended.

harbor reach.. See under REACH.

hard beach.. A portion of a beach especially prepared with a hard surface extending into the water, employed for the purpose of loading or unloading directly into landing ships or landing craft.

hard disk.. Rigid computer data storage in disk form.

hard iron.. Iron or steel which is not readily magnetized by induction, but which retains a high percentage of the magnetism acquired. The opposite is SOFT IRON.

hardware.. The physical parts of a computer system; compare with SOFTWARE, the programs which accomplish work.

harmattan., *n.* The dry, dusty trade wind blowing off the Sahara Desert across the Gulf of Guinea and the Cape Verde Islands. Sometimes called the DOCTOR, because of its supposed healthful properties.

harmful interference.. Any emission, radiation or induction which endangers the functioning of a radionavigation service or of other safety services or seriously degrades, obstructs or repeatedly interrupts a radio-communication service operating in accordance with the International Telecommunications Union Regulations.

harmonic., *n.* 1. A sinusoidal quantity having a frequency that is an integral multiple of the frequency of a periodic quantity to which it is related. 2. A signal having a frequency which is an integral multiple of the fundamental frequency.

harmonic analysis.. The process by which the observed tide or tidal current at any place is separated into basic harmonic constituents. Also called HARMONIC REDUCTION.

harmonic analyzer.. A machine designed for the resolution of a periodic curve into its harmonic constituents. Now performed by computer.

harmonic component.. Any of the simple sinusoidal components into which a periodic quantity may be resolved.

harmonic constants.. The amplitudes and epochs of the harmonic constituents of the tide or tidal current at any place.

harmonic constituent.. See CONSTITUENT.

harmonic expressions.. Trigonometric terms of an infinite series used to approximate irregular curves in two or three dimensions.

harmonic function.. Any real function that satisfies a certain equation. In its simplest form, as used in tide and tidal current predictions, it is a quantity that varies as the cosine of an angle that increases uniformly with time.

harmonic motion.. The projection of circular motion on a diameter of the circle of such motion. Simple harmonic motion is produced if the circular motion is of constant speed. The combination of two or more simple harmonic motions results in compound harmonic motion.

harmonic prediction. *(tidal).* Method of predicting tides and tidal currents by combining the harmonic constituents into a single tide curve, usually performed by computer.

harmonic reduction.. See HARMONIC ANALYSIS.

harmonic tide plane.. See INDIAN SPRING LOW WATER.

harpoon log.. A log which consists of a rotator and distance registering device combined in a single unit, which is towed through the water. The TAFFRAIL LOG is similar except that the registering device is located at the taffrail, with only the rotator in the water.

harvest moon.. The full moon occurring nearest the autumnal equinox. See also PHASES OF THE MOON.

haul., *v., i.* 1. A counterclockwise change in direction of the wind. 2. A shift in the direction of the wind forward. The opposite is to VEER. 2. *v., t.* To change the course of a sailing vessel to bring the wind farther forward, usually used with up, such as *haul up*.

haven., *n.* A place of safety for vessels.

haze., *n.* Fine dust or salt particles in the air, too small to be individually apparent but in sufficient number to reduce horizontal visibility and give the atmosphere a characteristic hazy appearance which casts a bluish or yellowish veil over the landscape, subduing its colors. This is sometimes called a dry haze to distinguish it from damp haze, small water droplets or very hygroscopic particles in the air, smaller and more scattered than light fog.

head., *n.* See HEADLAND.

heading., *n.* The horizontal direction in which a ship actually points or heads at any instant, expressed in angular units from a reference direction, usually from 000° at the reference direction clockwise through 360°. Heading is often designated as true, magnetic, compass, or grid. Heading should not be confused with COURSE, which is the intended direction of movement through the water. At a specific instant the heading may or may not coincide with the course. The heading of a ship is also called SHIP'S HEAD.

heading angle.. Heading measured from 0° at the reference direction clockwise or counterclockwise through 90° or 180°. It is labeled with the reference direction as a prefix and the direction of measurement from the reference direction as a suffix.

heading flasher.. An illuminated radial line on the radar for indicating own ship's heading on the bearing dial. Also called HEADING MARKER.

heading line.. The line extending in the direction of a heading.

heading marker.. See HEADING FLASHER.

headland., *n.* A comparatively high promontory having a steep face. Usually called HEAD when coupled with a specific name. Also called FORELAND.

head sea.. A sea in which the waves move in a direction approximately opposite to the heading. The opposite is FOLLOWING SEA.

head tide.. A tidal current setting in a direction approximately opposite to the heading of a vessel. One setting in such a direction as to increase the speed of a vessel is called a FAIR TIDE. One abeam is called a BEAM TIDE. One approximately 90° from the course is called a CROSS TIDE.

head up, heading upward.. One of the three basic orientations of display of relative or true motion on a radarscope. In the HEAD UP orientation, the target pips are painted at their measured distances and in their directions relative to own ship's heading maintained UP in relation to the display and so indicated by the HEADING FLASHER. See also NORTH UP, BASE COURSE UP.

headwaters., *n., pl.* The source of a stream or river.

headway., *n.* Motion in a forward direction. Motion in the opposite direction is called STERNWAY.

head wind.. Wind from ahead of the vessel.

heat lightning.. A flash of light from an electric discharge, without thunder, believed to be the reflection by haze or clouds of a distant flash of lightning, too far away for the thunder to be audible.

heat wave.. Unseasonably high temperatures extending over a period of a day or longer, particularly during the warm season of the year.

heave., *n.* The oscillatory vertical rise and fall, due to the entire hull being lifted by the force of the sea. Also called HEAVING. See also SHIP MOTIONS.

heavenly body.. See CELESTIAL BODY.

heave the lead.. To take a sounding with a lead.

heaving., *n.* See HEAVE.

Heaviside layer.. See under KENNELLY-HEAVISIDE REGION.

hecto-.. A prefix meaning one hundred (10^2).

hectometer., *n.* One hundred meters.

heel., *n.* Lateral inclination of a vessel. See also LIST, *n.*

heel., *v., t., i.* To incline or be inclined to one side. See also LIST, *n.*

heeling adjuster.. A dip needle with a sliding weight that can be moved along one of its arms to balance magnetic force, used to determine the correct position of a heeling magnet. Also called HEELING ERROR INSTRUMENT, VERTICAL FORCE INSTRUMENT. See also HEELING ERROR.

heeling error.. The change in the deviation of a magnetic compass when a craft heels, due to the change in the position of the magnetic influences of the craft relative to the earth's magnetic field and to the compass.

heeling error instrument.. Heeling adjuster. Also called VERTICAL FORCE INSTRUMENT.

heeling magnet.. A permanent magnet placed vertically in a tube under the center of a marine magnetic compass, to correct for heeling error.

height., *n.* Vertical distance above a datum.

height of eye correction.. The correction to sextant altitude due to dip of the horizon. Also called DIP CORRECTION.

height of tide.. Vertical distance from the chart sounding datum to the water surface at any stage of the tide. It is positive if the water level is higher than the chart sounding datum. The vertical distance from the chart sounding datum to a high water datum is called RISE OF TIDE.

heliocentric., *adj.* Relative to the sun as a center.

heliocentric parallax.. The difference in the apparent direction or positions of a celestial body outside the solar system, as observed from the earth and sun. Also called STELLAR PARALLAX, ANNUAL PARALLAX. See also GEOCENTRIC PARALLAX.

helm., *n.* The apparatus by which a vessel is steered; the tiller or wheel.

hemisphere., *n.* Half of a sphere.

henry., *n.* A derived unit of electric inductance in the International System of Units; it is the inductance of a closed circuit in which an electromotive force of 1 volt is produced when the electric current in the circuit varies uniformly at a rate of 1 ampere per second.

hertz., *n.* The special name for the derived unit of frequency in the International System of Units, it is one cycle per second.

Hertzian waves.. See RADIO WAVES.

heterodyne reception.. Radio reception in which an audio frequency is derived by beating the signal frequency with that produced by a local oscillator, followed by detection. Also called BEAT RECEPTION.

Hevellan halo. . A faint white halo consisting of a ring occasionally seen 90° from the sun, and probably caused by the refraction and internal reflection of the sun's light by bi-pyramidal ice crystals.

hexagon. , *n*. A closed plane figure having six sides.

hibernal. , *adj*. Pertaining to winter. The corresponding adjectives for spring, summer, and fall are vernal, aestival, and autumnal.

high. , *n*. An area of high pressure. Since a high is, on a synoptic chart, always associated with anticyclonic circulation, the term is used interchangeably with ANTICYCLONE. See also LOW.

high altitude method. . The establishing of a circular line of position from the observation of the altitude of a celestial body by means of the geographical position and zenith distance of the body. The line of position is a circle having the geographical position as its center and a radius equal to the zenith distance. The method is normally used only for bodies at high altitudes having small zenith distances. See also ST. HILAIRE METHOD, SUMNER METHOD LONGITUDE METHOD.

high clouds. . Types of clouds the mean lower level of which is above 20,000 feet. The principal clouds in this group are cirrus, cirrocumulus, and cirrostratus.

higher high water. . The higher of the two high waters of any tidal day.

higher high water interval. . See under LUNITIDAL INTERVAL.

higher lower water. . The higher of the two low waters of any tidal day.

higher low water interval. . See under LUNITIDAL INTERVAL.

high fidelity. . The ability to reproduce modulating waves at various audio frequencies without serious distortion.

high focal plane buoy. . A type of lighted buoy in which the light is mounted exceptionally high above the surface of the sea.

high frequency. . Radio frequency of 3 to 30 megahertz.

high light. . The rear light of a lighted range. See REAR LIGHT.

high noon. . See LOCAL APPARENT NOON.

high sea, high seas. . All water beyond the outer limit of the territorial sea. Although the high seas are in part coextensive with the waters of the contiguous zone, the fishing zone, and those over the continental shelf, freedom of the seas is not invalidated by the zonal overlap.

high tide. . See under HIGH WATER.

high water. . The maximum height reached by a rising tide. The height may be due solely to the periodic tidal forces or it may have superimposed upon it the effects of prevailing meteorological conditions. Use of the synonymous term HIGH TIDE is discouraged.

high water full and change. . See ESTABLISHMENT OF THE PORT.

high water inequality. . The difference between the heights of the two high waters during a tidal day. See under DIURNAL INEQUALITY.

high water interval. . See under LUNITIDAL INTERVAL.

high water line. . 1. The intersection of the land with the water surface at an elevation of high water. 2. The line along the shore to which the waters normally reach at high water.

high water mark. . A line or mark left upon tide flats, beach, or alongshore objects indicating the elevation of the intrusion of high water. It should not be confused with the MEAN HIGH WATER LINE or MEAN HIGHER HIGH WATER LINE.

high water neaps. . See under NEAP TIDES.

high water springs. . Short for MEAN HIGH WATER SPRINGS.

high water stand. . The condition at high water when there is no sensible change in the height of the water. A similar condition at low water is called LOW WATER STAND. See also STAND.

hill. , *n*. 1. A relatively low, rounded elevation of the earth's surface. 2. On the sea floor, an elevation rising generally less than 500 meters.

hillock. , *n*. A small hill.

hoar. , *n*. See FROST, definition 1.

hoarfrost. , *n*. See FROST, definition 1.

holding ground. . The bottom ground of an anchorage. The expression is usually used with a modifying adjective to indicate the quality of the holding power of the material constituting the bottom.

hole. , *n*. 1. A small depression of the sea floor. 2. An opening through a piece of sea ice, or an open space between ice cakes. 3. A small bay, particularly in New England.

homing. , *n*. Navigation toward a point by following a signal from that point. Radiobeacons are commonly used for homing.

homogenous. , *adj*. Uniform throughout, or composed of parts which are similar in every detail.

hood. , *n*. A shield placed over a radarscope, to eliminate extraneous light and thus make the radar picture appear clearly.

hook. , *n*. A feature resembling a hook in shape, particularly, a. a spit or narrow cape of sand or gravel which turns landward at the outer end; or b. a sharp bend or curve, as in a stream.

hooked spit. . See RECURVED SPIT.

hop. , *n*. Travel of a radio wave to the ionosphere and back to earth. The number of hops a radio signal has experienced is usually designated by the expression one-hop, two-hop, multihop, etc.

H.O. Pub. No. 208. ., *Navigation Tables for Mariners and Aviators*; a sight reduction table first published in 1928 by the U.S. Navy Hydrographic Office but discontinued on 31 December 1970 by the successor, the U.S. Naval Oceanographic Office. The method was devised by Lieutenant Commander J. Y. Dreisonstok USN. It is based upon a navigational triangle divided by dropping a perpendicular from the zenith The table has been published commercially. Popularly called DREISONSTOK.

H.O. Pub. No. 211. . *Dead Reckoning Altitude and Azimuth Table*; a sight reduction table first published by the U.S. Navy Hydrographic Office in 1931 but discontinued as a separate publication on 31 December 1972 by the successor, the Defense Mapping Agency Hydrographic/Topographic Center. The method was devised by Lieutenant Arthur A. Ageton, USN. It is based upon a navigational triangle divided by dropping a perpendicular from the GP of the body. The table was republished in 1975 by the Defense Mapping Agency Hydrographic/Topographic Center as table 35 of *Volume II: American Practical Navigator*, but is no longer included. Popularly called the AGETON method.

H.O. Pub. No. 214. . *Tables of Computed Altitude and Azimuth;* a nine-volume set of sight reduction tables of the inspection type published between 1936 and 1946 by the U.S. Navy Hydrographic Office, and reprinted from time to time until discontinued on 31 December 1973. These tables were superseded by *Pub. No. 229, Sight Reduction Tables for Marine Navigation*.

horizon. , *n*. The great circle of the celestial sphere midway between the zenith and nadir, or a line resembling or approximating such a circle. The line where earth and sky appear to meet, and the projection of this line upon the celestial sphere, is called the visible or apparent horizon. A line resembling the visible horizon but above or below it is called a false horizon. The circle of the celestial sphere-formed by the intersection of the celestial sphere and a plane perpendicular to the zenith-nadir line is called sensible horizon if the plane is through any point, such as the eye of an observer; geoidal horizon if through any sea-level point; and celestial or rational horizon if through the center of the earth. The geometrical horizon was originally considered identical with the celestial horizon, but the expression is now more commonly used to refer to the intersection of the celestial sphere and an infinite number of straight lines tangent to the earth's surface, and radiating from the eye of the observer. If there were no terrestrial refraction, GEOMETRICAL AND VISIBLE HORIZONS would coincide. An artificial horizon is a device for indicating the horizontal. A radio horizon is the line at which direct rays from a transmitting antenna become tangent to the earth's surface. A radar horizon is the radio horizon of a radar antenna.

horizon glass. . The glass of a marine sextant, attached to the frame, through which the horizon is observed. The half of this glass nearer the frame is silvered to form the HORIZON MIRROR for reflecting the image of a celestial body; the other half is clear.

horizon mirror. . The mirror part of the horizon glass. The expression is sometimes used somewhat loosely to refer to the horizon glass.

horizon prism. . A prism which can be inserted in the optical path of an instrument, such as a bubble sextant, to permit observation of the visible horizon.

horizon system of coordinates. . A set of celestial coordinates based on the celestial horizon as the primary great circle; usually altitude and azimuth or azimuth angle.

horizontal. , *adj*. Parallel to the plane of the horizon; perpendicular to the direction of gravity.

horizontal. , *n*. A horizontal line, plane, etc. horizontal beam width. The beam width measured in a horizontal plane.

horizontal control datum. . See HORIZONTAL GEODETIC DATUM.

horizontal danger angle. . The maximum or minimum angle between two points on a chart, as observed from a vessel, indicating the limit of safe approach to an off-lying danger. See also DANGER ANGLE.

horizontal datum. . See HORIZONTAL GEODETIC DATUM.

horizontal earth rate. . The rate at which the spin axis of a gyroscope must be tilted about the horizontal axis to remain parallel to the earth's surface. Horizontal earth rate is maximum at the equator, zero at the poles, and varies as the cosine of the latitude. See also EARTH RATE, VERTICAL EARTH RATE.

horizontal force instrument. . An instrument used to make a comparison between the intensity of the horizontal component of the earth's magnetic field and the magnetic field at the compass location on board. Basically, it consists of a magnetized needle pivoted in a horizontal plane, as a dry card compass. It will settle in some position which will indicate the direction of the resultant magnetic field. If the needle is started swinging, it will be damped down with a certain period of oscillation dependent upon the strength of the magnetic field. Also called HORIZONTAL VIBRATING NEEDLE. See also DEFLECTOR.

horizontal geodetic datum. . The basis for computations of horizontal control surveys in which the curvature of the earth is considered It consists of the astronomical and geodetic latitude and the astronomical and geodetic longitude of an initial point (origin); an azimuth of a line from this point; the parameters (radius and flattening) of the reference ellipsoid; and the geoidal separation at the origin. A change in any of these quantities affects every point on the datum. For this reason, while positions within a system are directly and accurately relatable, those points from different datums must be transformed to a common datum for consistency. The horizontal geodetic datum may extend over a continent or be limited to a small area. See also DATUM. Also called HORIZONTAL DATUM, HORIZONTAL CONTROL DATUM.

horizontal intensity of the earth's magnetic field. . The strength of the horizontal component of the earth's magnetic field.

horizontally polarized wave. . A plane polarized electromagnetic wave in which the electric field vector is in a horizontal plane.

horizontal parallax. . The geocentric parallax when a body is on the horizon. The expression is usually used only in connection with the moon, for which the tabulated horizontal parallax is given for an observer on the equator. The parallax at any altitude is called PARALLAX IN ALTITUDE.

horizontal vibrating needle. . See HORIZONTAL FORCE INSTRUMENT.

horn. , *n*. 1. A flared tube designed to match the acoustic impedance to the impedance of the atmosphere; it can behave as a resonator and can influence the directivity; the narrow end is called the throat and the large end the mouth. Also called TRUMPET. 2. See HORN ANTENNA.

horn antenna. . An antenna consisting of a waveguide the cross-sectional area of which increases toward the open end. Often shortened to HORN.

horse latitudes. . The regions of calms and variable winds coinciding with the subtropical high pressure belts on the poleward sides of the trade winds. The expression is generally applied only to the northern of these two regions in the North Atlantic Ocean, or to the portion of it near Bermuda.

hostile ice. . An ice canopy containing no large sky lights or other features which permit a submarine to surface.

hour. , *n*. 1. A 24th part of a day. 2. A specified interval. See also COTIDAL HOUR, CURRENT HOUR.

hour angle. . Angular distance west of a celestial meridian or hour circle; the arc of the celestial equator, or the angle at the celestial pole, between the upper branch of a celestial meridian or hour circle and the hour circle of a celestial body or the vernal equinox, measured westward through 360°. It is usually further designated as local, Greenwich, or sidereal as the origin of measurement is the local or Greenwich celestial meridian or the hour circle of the vernal equinox. See also MERIDIAN ANGLE.

hour angle difference. . See MERIDIAN ANGLE DIFFERENCE.

hour circle. . On the celestial sphere, a great circle through the celestial poles. An hour circle through the zenith is called a celestial meridian Also called CIRCLE OF DECLINATION, CIRCLE OF RIGHT ASCENSION.

hour-glass effect. . A radarscope phenomenon which appears as a constriction or expansion of the display near the center of the plan position indicator, which can be caused by a nonlinear time base or the sweep plot starting on the radar indicator at the same instant as the transmission of the pulse. The phenomenon is most apparent when in narrow rivers or close to shore.

hug. , *v., t.* To remain close to, as to *hug the land*.

Humboldt Current. . See PERU CURRENT.

humidity. , *n*. The amount of water vapor in the air. The mass of water vapor per unit volume of air is called absolute humidity. The mass of water vapor per unit mass of moist air is called specific humidity. The ratio of the actual vapor pressure to the vapor pressure corresponding to saturation at the prevailing temperature is called relative humidity.

hummock. , *n*. 1. A hillock of broken ice which has been forced upwards by pressure. It may be fresh or weathered. The submerged volume of broken ice under the hummocks, forced downwards by pressure, is called a BUMMOCK; 2. A natural elevation of the earth's surface resembling a hillock, but smaller and lower.

hummocked ice. . Sea ice piled haphazardly one piece over another to form an uneven surface. When weathered, hummocked ice has the appearance of smooth hillocks.

hummocking. , *n*. The pressure process by which sea ice is forced into hummocks. When the floes rotate in the process, it is called SCREWING.

hunter's moon. . The full moon next following the harvest moon. See also PHASES OF THE MOON.

hunting. , *n*. Fluctuation about a mid-point due to instability, as oscillations of the needle of an instrument about the zero point.

hurricane. , *n*. 1. See under TROPICAL CYCLONE. 2. Wind of force 12 (64 knots and higher or 73 miles per hour and higher) on the Beaufort wind scale.

hydraulic current. . A current in a channel caused by a difference in the surface level at the two ends. Such a current may be expected in a strait connecting two bodies of water in which the tides differ in time or range. The current in the East River, N.Y., connecting Long Island Sound and New York Harbor, is an example.

hydrographer. , *n*. One who studies and practices the science of hydrography.

hydrographic. , *adj*. Of or pertaining to hydrography.

hydrographic datum. . A datum used for referencing depths of water or the heights of predicted tides. See also DATUM.

hydrographic sextant. . A surveying sextant similar to those used for celestial navigation but smaller and lighter, constructed so that the maximum angle that can be read on it is slightly greater than that on the navigating sextant. Usually the angles can be read only to the nearest minute by means of a vernier. It is fitted with a telescope with a large object glass and field of view. Although the ordinary navigating sextant may be used in place of the hydrographic sextant, it is not entirely satisfactory for use in observing objects ashore which are difficult to see. Hydrographic sextants are either not provided with shade glasses or they are removed before use. Also called SOUNDING SEXTANT, SURVEYING SEXTANT.

hydrographic survey. . The survey of a water area, with particular reference to submarine relief, and any adjacent land. See also OCEANOGRAPHIC SURVEY.

hydrography. , *n*. The science that deals with the measurement and description of the physical features of the oceans, seas, lakes, rivers, and their adjoining coastal areas, with particular reference to their use for navigation.

HYDROLANT. , *n*. A radio message disseminated by the National Imagery and Mapping Agency and restricted to important marine incidents or navigational changes which affect navigational safety. The HYDROLANT broadcast covers those water areas outside and eastward of NAVAREA IV in the Atlantic Ocean. HYDROLANTS constitute part of the U.S. long range radio navigational warning system. The text of HYDROLANTS issued during a week which are in effect are available through NAVINFONET and are printed in the weekly *Notice to Mariners*.

hydrology. , *n*. The scientific study of the waters of the earth, especially with relation to the effects of precipitation and evaporation upon the occurrence and character of ground water.

hydrometeor. , *n*. Any product of the condensation or sublimation of atmospheric water vapor whether formed in the free atmosphere or at the earth's surface, also any water particles blown by the wind from the earth s surface. See also LITHOMETEOR.

HYDROPAC.. A radio message disseminated by the National Imagery and mapping Agency and restricted to important marine incidents or navigational changes which affect navigational safety. The HYDROPAC broadcast covers those water areas outside of NAVAREA XII in the Pacific Ocean. HYDROPACS constitute part of the U.S. long range radio navigational warning system. The text of HYDROPACS issued during a week which is in effect are available through NAVINFONET and are printed in the weekly *Notice to Mariners*.

hydrophone., *n.* A listening device for receiving underwater sounds.

hydrosphere., *n.* The water portion of the earth as distinguished from the solid part, called the LITHOSPHERE, and from the gaseous outer envelope, called the ATMOSPHERE.

hyetal., *adj.* Of or pertaining to rain.

hygrometer., *n.* An instrument for measuring the humidity of the air. The most common type is a psychrometer consisting of drybulb and wet-bulb thermometers.

hygroscope., *n.* An instrument which indicates variation in atmospheric moisture.

hygroscopic., *adj.* Able to absorb moisture.

hyperbola., *n.* An open curve with two parts, all points of which have a constant difference in distance from two fixed points called FOCI.

hyperbolic., *adj.* Of or pertaining to a hyperbola.

hyperbolic lattice.. A pattern formed by two or more families of intersecting hyperbolas.

hyperbolic line of position.. A line of position in the shape of a hyperbola, determined by measuring the difference in distance to two fixed points. Loran C lines of position are an example.

hyperbolic navigation.. Radionavigation based on the measurement of the time differences in the reception of signals from several pairs of synchronized transmitters. For each pair of transmitters the isochrones are substantially hyperbolic. The combination of isochrones for two or more pairs of transmitters forms a hyperbolic lattice within which position can be determined according to the measured time differences.

hypersonic., *adj.* Of or pertaining to high supersonic speed, of the order of five times the speed of sound, or greater.

hypotenuse., *n.* The side of a plane right triangle opposite the right angle; the longest side of a plane right triangle.

hypsographic detail.. The features pertaining to relief or elevation of terrain.

hypsographic map.. A map showing land or submarine bottom relief in terms of height above, or below, a datum by any method, such as contours, hachures, shading, or hypsometric tinting. Also called HYPSOMETRIC MAP, RELIEF MAP.

hypsography., *n.* 1. The science or art of describing elevations of land surfaces with reference to a datum, usually sea level. 2. That part of topography dealing with relief or elevation of terrain.

hypsometer., *n.* An instrument for measuring height by determining the boiling temperature of a liquid. Its operation depends on the principle that boiling temperature is dependent on pressure, which normally varies with height.

hypsometric map.. See HYPSOGRAPHIC MAP.

hypsometric tinting.. A method of showing relief on maps and charts by coloring, in different shades, those parts which lie between different levels. Also called ALTITUDE TINTS, COLOR GRADIENTS, ELEVATION TINTS, GRADIENT TINTS, LAYER TINTS. See also HYPSOMETRIC TINT SCALE.

hypsometric tint scale.. A graphic scale in the margin of maps and charts which indicates heights or depths by graduated shades of color. See also HYPSOMETRIC TINTING.

hysteresis., *n.* The lagging of the effect caused by change of a force acting on anything.

hysteresis error.. That error in the reading of an instrument due to hysteresis.

I

IALA Maritime Buoyage System.. A uniform system of maritime buoyage which is now implemented by most maritime nations. Within the single system there are two buoyage *regions*, designated as Region A and Region B, where lateral marks differ only in the colors of port and starboard hand marks. In Region A, red is to port on entering; in Region B, red is to starboard on entering. The system is a combined cardinal and lateral system, and applies to all fixed and floating marks, other than lighthouses, sector lights, leading lights and marks, lightships and large navigational buoys.

ice., *n.* Frozen water, the solid form of H_2O.

ice anchor.. An anchor designed for securing a vessel to ice.

ice atlas.. A publication containing a series of ice charts showing geographic distribution of ice, usually by seasons or months.

iceberg., *n.* A massive piece of ice greatly varying in shape, showing more than 5 meters above the sea surface, which has broken away from a glacier, and which may be afloat or aground. Icebergs may be described as tabular, dome shaped, pinnacled, drydock, glacier or weathered, blocky, tilted blocky, or drydock icebergs. For reports to the International Ice Patrol they are described with respect to size as small, medium, or large icebergs.

iceberg tongue.. A major accumulation of icebergs projecting from the coast, held in place by grounding and joined together by fast ice.

ice-blink.. A whitish glare on low clouds above an accumulation of distant ice.

ice-bound., *adj.* Pertaining to a harbor, inlet, etc. when entry or exit is prevented by ice, except possibly with the assistance of an icebreaker.

ice boundary.. The demarcation at any given time between fast ice and pack ice or between areas of pack ice of different concentrations. See also ICE EDGE.

ice breccia.. Ice pieces of different age frozen together.

ice bridge., *n.* 1. Surface river ice of sufficient thickness to impede or prevent navigation. 2. An area of fast ice between the mainland and nearby inhabited islands used in winter as a means of travel.

ice buoy.. A sturdy buoy, usually a metal spar, used to replace a more easily damaged buoy during a period when heavy ice is anticipated.

ice cake.. Any relatively flat piece of sea ice less than 20 meters across. See also SMALL ICE CAKE.

ice canopy.. From the point of view of the submariner, PACK ICE.

ice-cap.. A perennial cover of ice and snow over an extensive portion of the earth's surface. The largest ice caps are those in Antarctica and Greenland. Arctic Ocean ice is seasonal and in motion, and is not considered an ice cap.

ice cover.. The ratio, expressed in tenths, of the amount of ice to the total area of sea surface in a defined area; this locale may be global, hemispheric, or a specific geographic entity.

ice crystal.. Any one of a number of macroscopic crystalline forms in which ice appears.

ice-crystal haze.. A type of very light ice fog composed only of ice crystals (no droxtals). It is usually associated with precipitation of ice crystals.

ice crystals.. A type of precipitation composed of slowly falling, very small, unbranched crystals of ice which often seem to float in the air. It may fall from a cloud or from a cloudless sky. It is visible only in direct sunlight or in an artificial light beam, and does not appreciably reduce visibility. The latter quality helps to distinguish it from ice fog, which is composed largely of droxtals.

ice edge.. The demarcation at any given time between the open sea and sea ice of any kind, whether fast or drifting. See also COMPACTED ICE EDGE, DIFFUSE ICE EDGE, ICE BOUNDARY.

ice field.. An area of pack ice consisting of floes of any size, which is greater than 10 kilometers (5.4 nautical miles) across. Ice fields are subdivided according to areal extent. A large ice field is over 11 nautical miles across; a medium ice field is 8 to 11 nautical miles across; a small ice field is 5.4 to 8 nautical miles across.

ice fog. . Fog composed of suspended particles of ice, partly ice crystals 20 to 100 microns in diameter but chiefly, especially when dense, droxtals 12 to 20 microns in diameter. It occurs at very low temperatures, and usually in clear, calm weather in high latitudes. The sun is usually visible and may cause halo phenomena. Ice fog is rare at temperatures warmer than -30° C or -20°F. Also called RIME FOG. See also FREEZING FOG.

icefoot. , *n.* A narrow fringe of ice attached to the coast, unmoved by tides and remaining after the fast ice has moved away.

ice-free. , *adj.* Referring to a locale with no sea ice; there may be some ice of land origin present.

ice front. . The vertical cliff forming the seaward face of an ice shelf or other floating glacier varying in height from 2 to 50 meters above sea level. See also ICE WALL.

ice island. . A large piece of floating ice showing about 5 meters above the sea surface, which has broken away from an ice shelf, having a thickness of 30 to 50 meters and an area of from a few thousand square meters to 150 square nautical miles or more; usually characterized by a regularly undulating surface which gives it a ribbed appearance from the air.

ice jam. . An accumulation of broken river ice or sea ice caught in a narrow channel.

ice keel. . A downward projecting ridge on the underside of the ICE CANOPY, the counterpart of a RIDGE. An ice keel may extend as much as 50 meters below sea level.

ice limit. . The climatological term referring to the extreme minimum or extreme maximum extent of the ice edge in any given month or period based on observations over a number of years. The term should be preceded by minimum or maximum, as appropriate. See also MEAN ICE EDGE.

ice massif. . A concentration of sea ice covering an area of hundreds of kilometers, which is found in the same region every summer.

ice needle. . A long, thin ice crystal whose cross-section is typically hexagonal. The expression ICE NEEDLE should not be confused with NEEDLE ICE.

ice of land origin. . Ice formed on land or in an ice shelf, found floating in water, including ice that is stranded or grounded.

ice patch. . An area of pack ice less than 5.4 nautical miles (10 kilometers) across.

ice pellets. . A type of precipitation consisting of transparent or translucent pellets of ice, 5 millimeters or less in diameter. The pellets may be spherical, irregular, or (rarely) conical in shape. They usually bounce when hitting hard ground, and make a sound upon impact. Ice pellets includes two basically different types of precipitation, those which are known in the United States as SLEET and SMALL HAIL. Sleet is generally transparent, globular, solid grains of ice which have formed from the freezing of raindrops or the refreezing of largely melted snowflakes when falling through a below-freezing layer of air near the earth's surface. Small hail is generally translucent particles, consisting of snow pellets encased in a thin layer of ice. The ice layer may form either by the accretion of droplets upon the snow pellet, or by the melting and refreezing of the surface of the snow pellet.

ice port. . An embayment in an ice front, often of a temporary nature, where ships can moor alongside and unload directly onto the ice shelf.

ice rind. . A brittle shiny crust of ice formed on a quiet surface by direct freezing or from grease ice, usually in water of low salinity. Of thickness to about 5 centimeters, ice rind is easily broken by wind or swell, commonly breaking into rectangular pieces.

ice sheet. . Continuous ice overlaying a large land area.

ice shelf. . A floating ice sheet attached to the coast and of considerable thickness, showing 20 to 50 meters or more above sea level. Usually of great horizontal extent and with a level or gently undulating surface, the ice shelf is augmented by annual snow accumulation and often also by the seaward extension of land glaciers. Limited areas of the ice shelf may be aground. The seaward edge is called ICE FRONT.

ice storm. A storm characterized by a fall of freezing precipitation with significant buildup of ice on exposed surfaces.

ice stream. . The part of an inland ice sheet in which the ice flows more rapidly and not necessarily in the same direction as the surrounding ice. The margins are sometimes clearly marked by a change in direction of the surface slope, but may be indistinct.

ice under pressure. . Ice in which deformation processes are actively occurring; hence the ice is a potential impediment or danger to shipping.

ice wall. . An ice cliff forming the seaward margin of a glacier which is not afloat. An ice wall is aground with the underlying land at or below sea level. See also ICE FRONT.

ice-worn. , *adj.* Abraded by ice.

icicle. , *n.* A hanging mass of ice, usually conical, formed by the freezing of dripping water.

illuminance. , *n.* The luminous flux per unit of area. The derived unit of illuminance in the International System of Units is the LUX.

image. , *n.* 1. The optical counterpart of an object. A real image is actually produced and is capable of being shown on a surface, as in a camera; while a virtual image cannot be shown on a surface, but is visible, as in a mirror. 2. A visual representation, as on a radarscope.

improved channels. . Dredged channels under the jurisdiction of the U.S Army Corps of Engineers, and maintained to provide an assigned CONTROLLING DEPTH. Symbolized on National Ocean Survey charts by black, broken lines to represent side limits, with the controlling depth and date of the survey given together with a tabulation of more detailed information.

impulse train. . See PULSE TRAIN.

in-band racon. . A racon which transmits in the marine radar frequency band. There are two types of in-band racons, swept-frequency racons and experimental fixed-frequency racons. The transmitter of the swept-frequency racon sweeps through a range of frequencies within the band to insure that a radar receiver tuned to a particular frequency within the band will be able to detect the signal. The fixed-frequency racon transmits on a fixed frequency at the band edge. It is therefore necessary that the radar set be tuned to the racon's transmitting frequency or that auxiliary receiving equipment be used. When the radar is tuned to the fixed-frequency racon, normal radar echoes are not painted on the radarscope. See also CROSS-BAND RACON.

incandescence. , *n.* Emission of light due to high temperature. Any other emission of light is called LUMINESCENCE.

inch. , *n.* A unit of length equal to one-twelfth of foot, or 2.54 centimeters.

incidence. , *n.* 1. Partial coincidence, as a circle and a tangent line. 2. The impingement of a ray on a surface.

incident ray. . A ray impinging on a surface.

incineration area. . An officially designated offshore area for the burning of chemical waste by specially equipped vessels. The depiction of incineration areas on charts (in conjunction with radio warnings) is necessary to insure that passing vessels do not mistake the burning of waste for a vessel on fire.

inclination. , *n.* 1. The angle which a line or surface makes with the vertical, horizontal, or with another line or surface. 2. One of the orbital elements (parameters) that specifies the orientation of an orbit. It is the angle between the orbital plane and a reference plane, the plane of the celestial equator for geocentric orbits and the ecliptic for heliocentric orbits. See also ORBITAL ELEMENTS, ORBITAL PARAMETERS OF ARTIFICIAL EARTH SATELLITE.

inclination of an orbit. . 1. See INCLINATION, definition 2. 2. As defined by the International Telecommunication Union (ITU), the angle determined by the plane containing an orbit and the plane of the earth's equator.

increment. , *n.* A change in the value of a variable. A negative increment is also called DECREMENT.

independent surveillance. , Position determination by means requiring no cooperation from the craft or vehicle.

index. *(pl. indices or indexes), n.* 1. A mark on the scale of an instrument, diagram, etc., to indicate the origin of measurement 2. A pointer or part of an instrument which points to a value, like the needle of a gage. 3. A list or diagram serving as a guide to a book, set of charts, etc. 4. A ratio or value used as a basis for comparison of other values.

index arm. . A slender bar carrying an index; particularly the bar which pivots at the center of curvature of the arc of a marine sextant and carries the index and the vernier or micrometer.

index chart. . An outline chart showing the limits and identifying designations of navigational charts, volumes of sailing directions, etc.

index correction. . The correction due to index error.

index error. . The error in the reading of an instrument equal to the difference between the zero of the scale and the zero of the index. In a marine sextant it is due primarily to lack of parallelism of the index mirror and the horizon glass at zero reading.

index glass. . See INDEX MIRROR.

index mirror. . The mirror attached to the index arm of a marine sextant. The bubble or pendulum sextant counterpart is called INDEX PRISM. Also called INDEX GLASS.

index prism. . A sextant prism which can be rotated to any angle corresponding to altitudes between established limits. It is the bubble or pendulum sextant counterpart of the INDEX MIRROR of a marine sextant.

Indian Equatorial Countercurrent. . A complex Indian Ocean current which is influenced by the monsoons and the circulations of the Arabian Sea and the Bay of Bengal. At times it is easily distinguishable; at other times it is not evident. During December through March, the countercurrent has a marked tendency to migrate southward and to become narrower. In December the northern and southern boundaries are at 2° N and 4° S, respectively, moving southward to 3° S and 6° S by February. The northern boundary of Indian Equatorial Countercurrent is easily discernible at this time due to the generally westward current flow in the region immediately north. During May through July the cell, within which the Indian Equatorial Countercurrent and the Monsoon Drift flow clockwise, moves toward the west side of the region. In June and July the southeastward flowing currents prevail in the region between the Bay of Bengal and the Indian South Equatorial Current; only traces of the countercurrent remain. During August through November eastward flowing currents prevail north of the Indian Equatorial Countercurrent. As a result, the northern boundary of the countercurrent is difficult to distinguish from the eastward drift currents. See also MONSOON.

Indian South Equatorial Current. . An Indian Ocean current that flows westward throughout the year, controlled by the southeast trade winds. Its northern and southern boundaries are at approximately 10° S and 25° S, respectively. The northern boundary of the current fluctuates seasonally between 9° S and 11° S, being at its northernmost limit during the southwest monsoon and at its southernmost limit during the northeast monsoon. The current flows westward toward the east coast of Madagascar to the vicinity of Tamatave and Ile Sainte-Marie, where it divides; one part turns northward, flows past the northern tip of the island with speeds up to 3.3 knots, and then flows westward and northwestward toward the African coast. The northern branch of the current divides upon reaching the coast of Africa near Cabo Delgado; one part turns and flows northward, the other turns and flows southward in the western part of the Mozambique Channel and forms the AGULHAS CURRENT. See also MONSOON.

Indian spring low water. . A tidal datum originated by G.H. Darwin when investigating the tides of India. It is an elevation depressed below mean sea level by an amount equal to the sum of the amplitudes of certain constituents as given in the *Tide and Current Glossary* published by the National Ocean Survey. Also called INDIAN TIDE PLANE, HARMONIC TIDE PLANE.

Indian summer. . An indefinite and irregular period of mild, calm, hazy weather often occurring in autumn or early winter, especially in the United States and Canada.

Indian tide plane. . See INDIAN SPRING LOW WATER.

indicator. , *n.* See RADAR INDICATOR.

indirect echo. . A radar echo which is caused by the electromagnetic energy being transmitted to the target by an indirect path and returned as an echo along the same path. An indirect echo may appear on the radar display when the main lobe of the radar beam is reflected off part of the structure of the ship (the stack for example) from which it is reflected to the target. Returning to own ship by the same indirect path, the echo appears on the PPI at the bearing of the reflecting surface. Assuming that the additional distance by the indirect path is negligible, the indirect echo appears on the PPI at the same range as the direct echo received. Also called FALSE ECHO.

indirect wave. . A radio wave which reaches a given reception point by a path from the transmitting point other than the direct line path between the two. An example is the SKYWAVE received after reflection from one of the layers of the ionosphere.

induced magnetism. . The magnetism acquired by soft iron while it is in a magnetic field. Soft iron will lose its induced magnetism when it is removed from a magnetic field. The strength and polarity of the induced magnetism will alter immediately as its magnetic latitude, or its orientation in a magnetic field, is changed. The induced magnetism has an immediate effect upon the magnetic compass as the magnetic latitude or heading of a craft changes. See also PERMANENT MAGNETISM, SUBPERMANENT MAGNETISM.

induced precession. . See REAL PRECESSION.

inequality . *(tidal), n.* A systematic departure from the mean value of a tidal quantity.

inertia. , *n.* The tendency of a body at rest to remain at rest and of a body in motion to remain in motion, unless acted upon by another force. See also GYROSCOPIC INERTIA.

inertial alignment. . The process of orienting the measuring axes of the inertial components of inertial navigation equipment with respect to the coordinate system in which the equipment is to be used.

inertial coordinate system. . A coordinate system in which the axes do not rotate with respect to the "fixed stars" and in which dynamic behavior can be described using Newton's laws of motion. See also EARTH-FIXED COORDINATE.

inertial force. . A force in a given coordinate system arising from the inertia of a mass moving with respect to another coordinate system.

inertial navigation. . The process of measuring a craft's velocity, attitude, and displacement from a known start point through sensing the accelerations acting on it in known directions using devices that mechanize Newton's laws of motion. Inertial navigation is described as self-contained because it is independent of external aids to navigation, and passive because no energy is emitted to obtain information. The basic principle of inertial navigation is the measurement of the accelerations acting on a craft, other than those not associated with its orientation or motion with respect to the earth, and the double integration of these accelerations along known directions to obtain the displacement from the start point. Due to increasing position errors with time, an inertial system must be reset from time to time using another navigation system.

in extremis. . Condition in which either course or speed changes or both are required on the part of both ships if the ships are to avoid collision.

inferior conjunction. . The conjunction of an inferior planet and the sun when the planet is between the earth and the sun.

inferior planets. . The planets with orbits smaller than that of the earth; Mercury and Venus. See also PLANET.

inferior transit. . See LOWER TRANSIT.

infinite. , *adj.* Without limits. The opposite is FINITE.

infinitesimal. , *adj.* 1. Immeasurably small. 2. Approaching zero as a limit.

infinity. , *n.* Beyond finite limits. In navigation, a source of light is regarded as at infinity if it is at such a great distance that rays from it can be considered parallel. The sun, planets, and stars can be considered at infinity without serious error. See also PARALLAX.

inflection, inflexion. , *n.* Reversal of direction of curvature. A point at which reversal takes place is called POINT OF INFLECTION.

infrared. , *adj.* Having a frequency immediately beyond the red end of the visible spectrum; rays of longer wavelength than visible light, but shorter than radio waves.

infrasonic. , *adj.* Having a frequency below the audible range. Frequencies above the audible range are called ULTRASONIC.

initial great circle course. . The direction, at the point of departure, of the great circle through that point and the destination, expressed as the angular distance from a reference direction, usually north, to that part of the great circle extending toward the designation. Also called INITIAL GREAT CIRCLE DIRECTION. See also FINAL GREAT CIRCLE COURSE.

initial great circle direction. . See INITIAL GREAT CIRCLE COURSE.

injection messages. . Messages periodically transmitted to artificial satellites for storage in satellite memory.

Inland Rules of the Road.. Officially the Inland Navigation Rules; Rules to be followed by all vessels while navigating upon certain defined inland waters of the United States. See also COLREGS DEMARCATION LINES, RULES OF THE ROAD.

inland sea.. A body of water nearly or completely surrounded by land, especially if very large or composed of salt water. If completely surrounded by land, it is usually called a LAKE. This should not be confused with CLOSED SEA, that part of the ocean enclosed by headlands, within narrow straits, etc., or within the territorial jurisdiction of a country.

inlet., *n.* A narrow body of water extending into the land from a larger body of water. A long, narrow inlet with gradually decreasing depth inward is called a ria. Also called ARM, TONGUE.

inner harbor.. The part of a harbor most remote from the sea, as contrasted with the OUTER HARBOR. These expressions are usually used only in a harbor that is clearly divided into two parts by a narrow passageway or man-made structures.

inner planets.. The four planets nearest the sun; Mercury, Venus, Earth, and Mars.

inoperative., *adj.* Said of a sound signal or radionavigation aid out of service due to a malfunction.

in phase.. The condition of two or more cyclic motions which are at the same part of their cycles at the same instant. Two or more cyclic motions which are not at the same part of their cycles at the same instant are said to be OUT OF PHASE.

input axis.. The axis of applied torque of a gyroscope. See also OUTPUT AXIS, PRECESSION.

inshore., *adj., adv.* Near or toward the shore.

inshore., *n.* The zone of variable width between the shore face and the seaward limit of the breaker zone.

inshore traffic zone.. A routing measure comprising a designated area between the landward boundary of a traffic separation scheme and the adjacent coast, intended for local traffic.

in situ.. A Latin term meaning "in place"; in the natural or original position.

insolation., *n.* Solar radiation received, or the rate of delivery of such radiation.

instability., *n.* The state or property of submitting to change or of tending to increase the departure from original conditions after being disturbed. The opposite is STABILITY.

instability line.. Any non-frontal line or band of convective activity in the atmosphere. This is the general term and includes the developing, mature, and dissipating stages. However, when the mature stage consists of a line of active thunderstorms, it is properly called SQUALL LINE; therefore, in practice, *instability line* often refers only to the less active phases. Instability lines are usually hundreds of miles long (not necessarily continuous), 10 to 50 miles wide, and are most often formed in the warm sectors of wave cyclones. Unlike true fronts, they are transitory in character, ordinarily developing to maximum intensity in less than 12 hours and then dissipating in about the same time. Maximum intensity is usually attained in late afternoon.

instrument correction.. That correction due to instrument error.

instrument error.. The inaccuracy of an instrument due to imperfections within the instrument. See CALIBRATION ERROR, CENTERING ERROR, FRICTION ERROR, GRADUATION ERROR, HYSTERESIS ERROR, LAG ERROR, PRISMATIC ERROR, SECULAR ERROR, TEMPERATURE ERROR, VERNIER ERROR.

instrument shelter.. A cage or screen in which a thermometer and sometimes other instrument are placed to shield them from the direct rays of the sun and from other conditions that would interfere with registration of true conditions. It is usually a small wooden structure with louvered sides.

insular., *adj.* Of or pertaining to an island or islands.

insular borderland.. A region around an island normally occupied by or bordering a shelf, that is highly irregular with depths well in excess of those typical of a shelf. See also CONTINENTAL BORDERLAND.

insular shelf.. A zone around an island that extends from the low water line to a depth at which there is usually a marked increase of slope towards oceanic depths. See also CONTINENTAL SHELF.

insulate., *v., t.* To separate or isolate a conducting body from its surroundings, by means of a nonconductor, as to prevent transfer of electricity, heat, or sound.

insulator., *n.* A non conducting substance or one offering high resistance to passage of energy.

integer., *n.* A whole number; a number that is not a fraction.

integral., *adj.* Of or pertaining to an integer.

integral Doppler navigation.. Navigation by means of integrating the Doppler frequency shift that occurs over a specific interval of time as the distance between a navigational satellite and navigator is changing to determine the time rate of change of range of the satellite from the navigator for the same interval. See also DOPPLER SATELLITE NAVIGATION BASIC PRINCIPLES, NAVY NAVIGATION SATELLITE SYSTEM.

integrated navigation system.. A navigation system which comprises two or more positioning systems combined in such manner as to achieve performance better than each constituent system.

integrating accelerometer.. An instrument which senses the component of specific acceleration along an axis known as the sensitive axis of the accelerometer, and produces an output equal to the time integral of that quantity. Also called VELOCITY METER.

intended track.. See TRACK, definition 2.

intercalary day.. A day inserted or introduced among others in a calendar, such as February 29 during leap years.

intercardinal heading.. A heading in the direction of any of the intercardinal points. See also CARDINAL HEADING.

intercardinal point.. Any of the four directions midway between the cardinal points; northeast, southeast, southwest, or northwest. Also called QUADRANTAL POINT.

intercardinal rolling error.. See under QUADRANTAL ERROR.

intercept., *n.* See ALTITUDE INTERCEPT, ALTITUDE INTERCEPT METHOD.

interference., *n.* 1. Unwanted and confusing signals or patterns produced by nearby electrical equipment or machinery, or by atmospheric phenomena. 2. The variation of wave amplitude with distance or time, caused by superposition of two or more waves. Sometimes called WAVE INTERFERENCE.

interferometer., *n.* An apparatus used to produce and measure interference from two or more coherent wave trains from the same source. Used to measure wavelengths, to measure angular width of sources, to determine the angular position of sources (as in satellite tracking), and for other purposes. See also RADIO INTERFEROMETER.

interlaced.. Referring to a computer monitor which displays data by scanning alternate lines instead of each line sequentially.

intermediate frequency.. In super heterodyne reception, the frequency which is derived by mixing the signal-carrying frequency with the local oscillator frequency. If there are more than one such mixing process, the successive intermediate frequencies are known as the first, second, etc. intermediate frequency.

intermediate light.. The middle light of the three-light range.

intermediate orbit.. A central force orbit that is tangent to the real (or disturbed) orbit at some point. A fictitious satellite traveling in the intermediate orbit would have the same position, but not the same velocity, as the real satellite at the point of tangency.

internal noise.. In radio reception, the noise which is produced in the receiver circuits. Internal noise is in addition to external noise.

internal tide.. A tidal wave propagating along a sharp density discontinuity, such as at a thermocline, or in an area of gradual changing density (vertically).

International Atomic Time.. See under ATOMIC TIME.

International Bureau of Weights and Measures.. The International Bureau of Weights and Measures (BIPM) insures worldwide unification of physical measurements. It is responsible for establishing the fundamental standards and scales for measurement of the principal physical quantities and maintaining the international prototypes, carrying out comparisons of national and international standards insuring coordination of corresponding measuring techniques; and carrying out and coordinating the determinations relating to the fundamental physical constants.

international call sign.. An alpha-numeric symbol assigned in accordance with the provisions of the International Telecommunications Union to identify a radio station. The nationality or the radio station is identified by the first three characters; also referred to as call letters or signal letters.

international chart. . One of a coordinated series of small-scale charts for planning and long range navigation. The charts are prepared and published by different Member States of the International Hydrographic Organization using the same specifications.

Intentional Code of Signals. , See PUB. 102.

international date line. . See DATE LINE.

International ellipsoid of reference. . The reference ellipsoid of which the semimajor axis is 6,378 388.0 meters, the semiminor axis is 6,356 911.9 meters, and the flattening or ellipticity is 1/297. Also called INTERNATIONAL SPHEROID OF REFERENCE.

International Great Lakes Datum (1955). . Mean water level at Pointe-au-Pere, Quebec, on the Gulf of St. Lawrence over the period 1941-1956, from which dynamic elevations throughout the Great Lakes region are measured. The term is often used to mean the entire system of dynamic elevations rather than just the referenced water level.

International Hydrographic Bulletin. . A publication, published monthly by the International Hydrographic Bureau for the International Hydrographic Organization, which contains information of current hydrographic interest.

International Hydrographic Bureau. . The Directors and administrative staff of the International Hydrographic Organization, based in Monaco.

International Hydrographic Organization. . An institution formed in 1921, consisting of representatives of maritime nations organized for the purpose of coordinating the hydrographic work of the participating governments.

international low water. . A hydrographic datum originally suggested for international use at the International Hydrographic Conference in London in 1919 and later discussed at the Monaco Conference in 1926. The proposed datum, which has not yet been generally adopted, was to be "a plane so low that the tide will but seldom fall below it." This datum was the subject of the International Hydrographic Bureau's Special Publications No. 5 (March 1925) and No. 10 (January 1926), reproduced in the *Hydrographic Reviews* for May 1925 and July 1926.

International Maritime Organization (IMO). . A Specialized Agency of the United Nations responsible for maritime safety and efficiency of navigation. The IMO provides for cooperation among governments in the field of governmental regulations and practices relating to technical matters of all kinds affecting shipping engaged in international trade: to encourage the general adoption of the highest practicable standards in matters concerning maritime safety, efficiency of navigation, and the prevention and control of marine pollution from ships, and to deal with legal matters related to the purposes set out in Article 1 of the Convention.

International Nautical Mile. . A unit of length equal to 1,852 meters, exactly. See also NAUTICAL MILE.

international number. . The number of a navigational light, assigned in accordance with the Resolution adopted at the Fifth International Hydrographic Conference in 1949 by Member Nations of the International Hydrographic Bureau (now the International Hydrographic Organization). This number is in italic type and under the light list number in the light list.

International spheroid of reference. . See INTERNATIONAL ELLIPSOID OF REFERENCE.

International System of Units. . A modern form of the metric system adopted in 1960 by the General Conference of Weights and Measures (CGPM). The units of the International System of Units (SI) are divided into three classes. The first class of SI units are the base units or the seven well defined units which by convention are regarded as dimensionally independent: the meter the kilogram, the second, the ampere, the kelvin, the mole, and the candela. The second class of SI units are the derived units, i.e., the units that can be formed by combining base units according to the algebraic relations linking the corresponding quantities. Several of these algebraic expressions in terms of base units can be replaced by special names and symbols which can themselves be used to form other derived units. The third class of SI units are the supplementary units, those units not yet classified by the CGPM as either base units or derived units. In 1969 the International Committee of Weights and Measures (CIPM) recognized that users of SI units will wish to employ with it certain units not part of SI, but which are important and ale widely used. These are the minute, the hour, the day, the degree of arc, the minute of arc, the second of arc, the liter, and the tonne. Outside the International System are some other units useful in specialized fields. Their value expressed in SI units must be obtained by experiment, and are therefore not known exactly These are the electron-volt, the unified atomic mass unit, the astronomical unit, and the parsec. Other temporary units are the nautical mile, the knot, the angstrom, the arc, the hectare, the barn, the bar, the standard atmosphere, the gal, the curie, the röntgen, and the rod.

interpolation. , *n.* The process of determining intermediate values between given values in accordance with some known or assumed rate or system of change. Linear interpolation assumes that changes of tabulated values are proportional to changes in entering arguments. Interpolation is designated as single, double, or triple if there are one, two, or three arguments or variables respectively. The extension of the process of interpolation beyond the limits of known value is called EXTRAPOLATION.

interpolation table. . An auxiliary table used for interpolating. See also PROPORTIONAL PART.

interrogating signal. . The signal emitted by interrogator to trigger a transponder.

interrogation. , *n.* The transmission of a radio frequency pulse, or combination of pulses, intended to trigger a transponder or group of transponder.

interrogator. , *n.* A radar transmitter which sends out a pulse that triggers a transponder. An interrogator may be combined in a single unit with a responsor, which receives the reply from a transponder and produces an output suitable for feeding a display system; the combined unit is called INTERROGATOR-RESPONDER. Also called CHALLENGER.

interrogator-responder. , *n.* A radar transmitter and receiver combined to interrogate a transponder and display the resulting replies. Often shortened to INTERROGATOR and sometimes called CHALLENGER.

interrupted quick flashing light. . A quick flashing light (50-80 flashes per minute) is interrupted at regular intervals by eclipses of long duration. See also QUICK FLASHING LIGHT, VERY QUICK FLASHING LIGHT.

interrupted quick light. . A quick light in which the sequence of flashes is interrupted by regularly repeated eclipses of constant and long duration. See also CONTINUOUS QUICK LIGHT, GROUP QUICK LIGHT.

interrupted very quick light. . A very quick light (80-160 flashes per minute) in which the sequence of flashes is interrupted by regularly repeated eclipses of long duration. See also CONTINUOUS VERY QUICK LIGHT, GROUP VERY QUICK LIGHT.

interscan. , *n.* See INTER-TRACE DISPLAY.

intersect. , *v., t. & i.* To cut or cross. For example, two non parallel lines in a plane intersect in a point, and a plane intersects a sphere in a circle.

inter-trace display. . A technique for presenting additional information, in the form of alphanumerics, markers, cursors, etc., on a radar display, by using the intervals between the normal presentation scans. Also called INTER-SCAN.

Intracoastal Waterway. . An inside protected route for small craft and small commercial vessels extending through New Jersey; from Norfolk, Virginia to Key West, Florida; across Florida from St. Lucie Inlet to Fort Myers, Charlotte Harbor, Tampa Bay, and Tarpon Springs; and from Carabelle, Florida, to Brownsville, Texas. Some portions are in exposed waters; some portions are very limited in depth.

Invar. , *n.* The registered trade name for an alloy of nickel and iron, containing about 36% nickel. Its coefficient of expansion is extremely small over a wide range of temperature.

inverse chart. . See TRANSVERSE CHART.

inverse cylindrical orthomorphic chart.. See TRANSVERSE MERCATOR CHART.

inverse cylindrical orthomorphic map projection.. See TRANSVERSE MERCATOR MAP Projection.

inverse equator.. See TRANSVERSE EQUATOR.

inverse latitude.. See TRANSVERSE LATITUDE.

inverse logarithm.. See ANTILOGARITHM.

inverse longitude.. See TRANSVERSE LONGITUDE.

inverse Mercator chart.. See TRANSVERSE MERCATOR CHART.

inverse Mercator map projection.. See TRANSVERSE MERCATOR MAP PROJECTION.

inverse meridian.. See TRANSVERSE MERIDIAN.

inverse parallel.. See TRANSVERSE PARALLEL.

inverse rhumb line.. See TRANSVERSE RHUMB LINE.

inversion., *n.* In meteorology, a departure from the usual decrease or increase with altitude of the value of an atmospheric property. This term is almost always used to refer to a temperature inversion, an atmospheric condition in which the temperature increases with increasing altitude.

inverted compass.. A marine magnetic compass designed and installed for observation from below the compass card. Frequently used as a telltale compass. Also called HANGING COMPASS, OVERHEAD COMPASS.

inverted image.. An image that appears upside down in relation to the object.

inverter., *n.* A device for changing direct current to alternating current. A device for changing alternating current to direct current is called a CONVERTER if a rotary device and a RECTIFIER if a static device.

inverting telescope.. An instrument with the optics so arranged that the light rays entering the objective of the lens meet at the crosshairs and appear inverted when viewed through the eyepiece without altering the orientation of the image. See also ERECTING TELESCOPE.

inward bound.. Heading toward the land or up a harbor away from the open sea. The opposite is OUTWARD BOUND.

ion., *n.* An atom or group of atoms which has become electrically charged, either positively or negatively, by the loss or gain of one or more electrons.

ionization., *n.* The process by which neutral atoms or groups of atoms become electrically charged either positively or negatively, by the loss or gain of electrons; or the state of a substance whose atoms or groups of atoms have become thus charged.

ionized layers.. Layers of charged particles existing in the upper reaches of the atmosphere as a result of solar radiation.

ionosphere., *n.* 1. The region of the atmosphere extending from about 40 to 250 miles above the earth's surface, in which there is appreciable ionization. The presence of charged particles in this region profoundly affects the propagation of certain electromagnetic radiation. 2. A region composed of highly ionized layers at varying heights above the surface of the earth which may cause the return to the earth of radio waves originating below these layers. See also D-LAYER, E-LAYER, F-LAYER, F1-LAYER, F2-LAYER.

ionospheric correction.. A correction for ionospheric refraction, a major potential source of error in all satellite radionavigation systems. Navigation errors can result from the effect of refraction on the measurement of the doppler shift and from the errors in the satellite's orbit if refraction is not accurately accounted for in the satellite tracking. The refraction contribution can be eliminated by the proper mixing of the received Doppler shift from two harmonically related frequencies to yield an accurate estimate of the vacuum doppler shift. Also called REFRACTION CORRECTION.

ionospheric delay.. The delay experienced by a wave or signal as it passes through the ionosphere.

ionospheric disturbance.. A sudden outburst of ultraviolet light on the sun, known as a SOLAR FLARE or CHROMOSPHERIC ERUPTION, which produces abnormally high ionization in the region of the D-layer. The result is a sudden increase in radio wave absorption, with particular severity in the upper medium frequencies and lower high frequencies. It has negligible effects on the heights of the reflecting/refracting layers and, consequently, upon critical frequencies, but enormous transmission losses may occur. See also SUDDEN IONOSPHERIC DISTURBANCE.

ionospheric error.. The total systematic and random error resulting from the reception of a navigation signal after ionospheric reflections. It may be due to variations in transmission paths, non-uniform height of the ionosphere, or non-uniform propagation within the ionosphere. Also called IONOSPHERIC-PATH ERROR, SKYWAVE ERROR.

ionospheric-path error.. See IONOSPHERIC ERROR.

ionospheric storm.. An ionospheric disturbance characterized by wide variations from normal in the state of the ionosphere, such as turbulence in the F-region, absorption increase, height increase, and ionization density decreases. The effects are most marked in high magnetic latitudes and are associated with abnormal solar activity.

ionospheric refraction.. Change in the propagation speed of a signal as it passes through the ionosphere.

ionospheric wave.. See SKYWAVE.

iridescence., *n.* Changing-color appearance, such as of a soap bubble, caused by interference of colors in a thin film or by diffraction.

iridescent clouds.. Ice-crystal clouds which exhibit brilliant spots or borders of colors, usually red and green, observed up to about 30° from the sun.

irisation., *n.* The coloration exhibited by iridescent clouds.

Irminger Current.. A North Atlantic Ocean current, one of the terminal branches of the Gulf Stream System (part of the northern branch of the North Atlantic Current); it flows toward the west off the southwest coast of Iceland. A small portion of the water of the Irminger Current bends around the west coast of Iceland but the greater amount turns south and becomes more or less mixed with the water of the East Greenland Current.

ironbound., *adj.* Rugged, rocky, as an *ironbound coast.*

irradiation., *n.* The apparent enlargement of a bright surface against a darker background.

irradiation correction.. A correction due to irradiation, particularly that sextant altitude correction caused by the apparent enlargement of the bright surface of a celestial body against the darker background of the sky.

irregular error.. See RANDOM ERROR.

irregular iceberg.. See PINNACLE ICEBERG.

isallobar., *n.* A line of equal change in atmospheric pressure during a specified time interval.

isallotherm., *n.* A line connecting points having the same anomalies of temperature, pressure, etc.

isanomal., *n.* A line connecting points of equal variations from a normal value.

island., *n.* An area of land not a continent, surrounded by water.

islet., *n.* A very small and minor island.

iso-. .A prefix meaning equal.

isobar., *n.* A line connecting points having the same atmospheric pressure reduced to a common datum, usually sea level.

isobaric., *adj.* Having the same pressure.

isobaric chart.. See CONSTANT-PRESSURE CHART.

isobaric surface.. See CONSTANT PRESSURE SURFACE.

isobath., *n.* See DEPTH CONTOUR.

isobathic., *adj.* Having equal depth.

isobathytherm., *n.* A line on the earth's surface connecting points at which the same temperature occurs at some specified depth.

isobront., *n.* A line connecting points at which some specified phase of a thunderstorm occurs at the same time.

isoceraunic, isokeraunic., *adj.* Indicating or having equal frequency or intensity of thunderstorms.

isochasm., *n.* A line connecting points having the same average frequency of auroras.

isochronal., *adj.* Of equal time; recurring at equal intervals of time. Also called ISOCHRONOUS.

isochrone., *n.* A line connecting points having the same time or time difference relationship, as a line representing all points having the same time difference in the reception of signals from two radio stations such as the master and slave stations of a Loran rate.

isochronize., *v., t.* To render isochronal.

isochronon., *n.* A clock designed to keep very accurate time.

isochronous., *adj.* See ISOCHRONAL.

isoclinal., *adj.* Of or pertaining to equal magnetic dip.

isoclinal., *n.* See ISOCLINIC LINE.

isoclinal chart.. See ISOCLINIC CHART.

isoclinic chart.. A chart of which the chief feature is a system of isoclinic lines. Also called ISOCLINAL CHART.

isoclinic line.. A line drawn through all points on the earth's surface having the same magnetic dip. The particular isoclinic line drawn through points of zero dip is called ACLINIC LINE. Also called ISOCLINAL.

isodynamic chart.. A chart showing isodynamic lines. See also MAGNETIC CHART.

isodynamic line.. A line connecting points of equal magnetic intensity, either the total or any component.

isogonal., *adj*. Having equal angles; isogonic.

isogonic., *adj*. Having equal angles; isogonal.

isogonic., *n*. A line connecting points of equal magnetic variation. Also called ISOGONIC LINE, ISOGONAL.

isogonic chart.. A chart showing magnetic variation with isogonic lines and the annual rate of change in variation with isoporic lines. See also MAGNETIC CHART.

isogonic line.. See ISOGONIC, *n*.

isogram., *n*. That line, on a chart or diagram, connecting points of equal value of some phenomenon.

isogriv., *n*. A line drawn on a map or chart joining points of equal grivation.

isogriv chart.. A chart showing isogrivs. See also MAGNETIC CHART.

isohaline, isohalsine., *n*. A line connecting points of equal salinity in the ocean.

isolated danger mark (or buoy).. An IALA navigation aid marking a danger with clear water all around; it has a double ball topmark and is black with at least one red band. If lighted its characteristic is Fl(2).

isosceles., *adj*. Having two equal sides.

isosceles triangle.. A triangle having two of its sides equal.

isomagnetic., *adj;*. Of or pertaining to lines connecting points of equality in some magnetic element *t*.

isomagnetic., *n*. A line connecting points of equality in some magnetic element. Also called ISOMAGNETIC LINE.

isomagnetic chart.. A chart showing isomagnetics. See also MAGNETIC CHART.

isomagnetic line.. See ISOMAGNETIC, *n*.

isometric., *n*. Of or pertaining to equal measure.

isophase., *adj*. Referring to a light having a characteristic of equal intervals of light and darkness.

isopleth., *n*. 1. An isogram indicating the variation of an element with respect to two variables, one of which is usually the time of year. The other may be time of day, altitude, or some other variable. 2. A line on a map depicting points of constant value of a variable. Examples are contours, isobars, and isogons.

isopor., *n*. See ISOPORIC LINE.

isoporic chart.. A chart with lines connecting points of equal annual rate of change of any magnetic element. See also ISOPORIC LINE.

isoporic line.. A line connecting points of equal annual rate of change of any magnetic element. Also called ISOPOR. See also ISOGONIC.

isostasy., *n*. A supposed equality existing in vertical sections of the earth, whereby the weight of any column from the surface of the earth to a constant depth is approximately the same as that of any other column of equal area, the equilibrium being maintained by plastic flow of material from one part of the earth to another.

isotropic antenna.. A hypothetical antenna which radiates or receives equally well in all directions. Although such an antenna does not physically exist, it provides a convenient reference for expressing the directional properties of actual antennas. Also called UNIPOLE.

isotropic gain of an antenna.. The gain of an antenna in a given direction when the reference antenna is an isotropic antenna isolated in space. Also called ABSOLUTE GAIN OF AN ANTENNA.

isthmus., *n*. A narrow strip of land connecting two larger portions of land. A submarine elevation joining two land areas and separating two basins or depressions by a depth less than that of the basins is called a submarine isthmus.

J

Jacob's staff.. See CROSS-STAFF.

jamming., *n*. Intentional transmission or re-radiation of radio signals in such a way as to interfere with reception of desired signals by the intended receiver.

Janus configuration.. A term describing orientations of the beams of acoustic or electromagnetic energy employed with doppler navigation systems. The Janus configuration normally used with doppler sonar speed logs, navigators, and docking aids employs four beams of ultrasonic energy, displaced laterally 90° from each other, and each directed obliquely (30° from the vertical) from the ship's bottom, to obtain true ground speed in the fore and aft and athwartship directions. These speeds are measured as doppler frequency shifts in the reflected beams. Certain errors in data extracted from one beam tend to cancel the errors associated with the oppositely directed beam.

Japan Current.. See KUROSHIO.

jetsam., *n*. Articles that sink when thrown overboard, particularly those jettisoned for the purpose of lightening a vessel in distress. See also FLOTSAM, JETTISON, LAGAN.

jet stream.. Relatively strong winds (50 knots or greater) concentrated in a narrow stream in the atmosphere. It usually refers only to a quasi-horizontal stream of maximum winds imbedded in the middle latitude westerlies, and concentrated in the high troposphere.

jettison., *n*. To throw objects overboard, especially to lighten a craft in distress. Jettisoned objects that float are termed FLOTSAM; those that sink JETSAM; and heavy articles that are buoyed for future recovery, LAGAN. See also DERELICT.

jetty., *n*. A structure built out into the water to restrain or direct currents, usually to protect a river mouth or harbor entrance from silting, etc. See also GROIN; MOLE, definition 1.

jitter., *n*. A term used to describe the short-time instability of a signal. The instability may be in amplitude, phase, or both. The term is applied especially to signals reproduced on the screen of a cathode-ray tube.

joule., *n*. A derived unit of energy of work in the International System of Units; it is the work done when the point of application of 1 newton (that force which gives to a mass of 1 kilogram an acceleration of 1 meter per second, per second) moves a distance of 1 meter in the direction of the force.

Julian calendar.. A revision of the ancient calendar of the city of Rome, instituted in the Roman Empire by Julius Caesar in 46 BC, which reached its final form in about 8 A.D. It consisted of years of 365 days, with an intercalary day every fourth year. The current Gregorian calendar is the same as the Julian calendar except that October 5, 1582, of the Julian calendar became October 15, 1582 of the Gregorian calendar and of the centurial years, only those divisible by 400 are leap years.

Julian day.. The number of each day, as reckoned consecutively since the beginning of the present Julian period on January 1, 4713 BC. It is used primarily by astronomers to avoid confusion due to the use of different calendars at different times and places. The Julian day begins at noon, 12 hours later than the corresponding civil day. The day beginning at noon January 1, 1968, was Julian day 2,439,857.

junction buoy.. A buoy which, when viewed from a vessel approaching from the open sea or in the same direction as the main stream of flood current, or in the direction established by appropriate authority, indicates the place at which two channels meet. See also BIFURCATION BUOY.

junction mark.. A navigation mark which, when viewed from a vessel approaching from the open sea or in the same direction as the main stream of flood current, or in the direction established by appropriate authority, indicates the place at which two channels meet. See also BIFURCATION MARK.

June solstice.. Summer solstice in the Northern Hemisphere.

Jupiter., *n*. The navigational planet whose orbit lies between those of Mars and Saturn. Largest of the known planets.

Jutland Current.. A narrow and localized nontidal current off the coast of Denmark between longitudes 8°30'E and 10°30'E. It originates partly from the resultant counterclockwise flow in the tidal North Sea. The main cause, however, appears to be the winds which prevail from south through west to northwest over 50 percent of the time throughout the year and the transverse flows from the English coast toward the Skaggerak. The current retains the characteristics of a major nontidal current and flows northeastward along the northwest coast of Denmark at speeds ranging between 1.5 to 2.0 knots 75 to 100 percent of the time.

K

Kaléma. , *n.* A very heavy surf breaking on the Guinea coast during the winter, even when there is no wind.

Kalman filtering. . A statistical method for estimating the parameters of a dynamic system, using recursive techniques of estimation, measurement, weighting, and correction. Weighting is based on variances of the measurements and of the estimates. The filter acts to reduce the variance of the estimate with each measurement cycle. In navigation, the technique is used to refine the positions given by one or more electronic systems.

katabatic wind. . Any wind blowing down an incline. If the wind is warm, it is called a foehn; if cold, a fall wind. An ANABATIC WIND blows up an incline. Also called GRAVITY WIND.

kaver. , *n.* See CAVER.

kay. , *n.* See CAY.

K-band. . A radio-frequency band of 10,900 to 36,000 megahertz. See also FREQUENCY, FREQUENCY BAND.

kedge. , *v., t.* To move a vessel by carrying out an anchor, letting it go, and winching the ship to the anchor. See also WARP.

keeper. , *n.* A piece of magnetic material placed across the poles of a permanent magnet to assist in the maintenance of magnetic strength.

kelp. , *n.* 1. A family of seaweed found in cool to cold waters along rocky coasts, characterized by its extreme length. 2. Any large seaweed. 3. The ashes of seaweed.

kelvin. , *n.* The base unit of thermodynamic temperature in the International System of Units; it is the fraction 1/273.16 of the thermodynamic temperature of the triple point of water, which is -273.16K°.

Kelvin temperature. . Temperature based upon a thermodynamic scale with its zero point at absolute zero (-273.16°C) and using Celsius degrees. Rankine temperature is based upon the Rankine scale starting at absolute zero (-459.69° F) and using Fahrenheit degrees.

Kennelly-Heaviside layer. . See under KENNELLY-HEAVISIDE REGION.

Kennelly-Heaviside region. . The region of the ionosphere, extending from approximately 40 to 250 miles above the earth's surface within which ionized layers form which may affect radio wave propagation. The E-layer, which is the lowest useful layer from the standpoint of wave propagation, is sometimes called KENNELLY-HEAVISIDE LAYER or, in some instances, simply the HEAVISIDE LAYER.

Kepler's laws. . The three empirical laws describing the motions of the planets in their orbits. These are: (1) The orbits of the planets are ellipses, with the sun at a common focus; (2) As a planet moves in its orbit, the line joining the planet and sun sweeps over equal areas in equal intervals of time; (3) The squares of the periods of revolution of any two planets are proportional to the cubes of their mean distances from the sun. Also called KEPLER'S PLANETARY LAWS.

Kepler's planetary laws. . See KEPLER'S LAWS.

key. , *n.* See CAY.

kick. , *n.* 1. The distance a ship moves sidewise from the original course away from the direction of turn after the rudder is first put over. 2. The swirl of water toward the inside of the turn when the rudder is put over to begin the turn.

kilo-. . A prefix meaning one thousand(10^3).

kilobyte. . One thousand bytes of information in a computer.

kilocycle. , *n.* One thousand cycles, the term is often used as the equivalent of one thousand cycles per second.

kilogram. , *n.* 1. The base unit of mass in the International System of Units; it is equal to the mass of the international prototype of the kilogram, which is made of platinum-iridium and kept at the International Bureau of Weights and Measures. 2. One thousand grams exactly, or 2.204623 pounds, approximately.

kilometer. , *n.* One thousand meters; about 0.54 nautical mile, 0.62 U.S. Survey mile, or 3,281 feet.

kinetic energy. . Energy possessed by a body by virtue of its motion, in contrast with POTENTIAL ENERGY, that possessed by virtue of its position.

klaxon. , *n.* A diaphragm horn similar to a nautophone, but smaller, and sometimes operated by hand.

knik wind. . A strong southeast wind in the vicinity of Palmer, Alaska, most frequent in the winter.

knoll. , *n.* 1. On the sea floor, an elevation rising generally more than 500 meters and less than 1,000 meters and of limited extent across the summit. 2. A small rounded hill.

knot. , *n.* A unit of speed equal to 1 nautical mile per hour.

kona storm. . A storm over the Hawaiian Islands, characterized by strong southerly or southwesterly winds and heavy rains.

Krassowski ellipsoid of 1938. . A reference ellipsoid of which the semimajor axis is 6,378,245 meters and the flattening of ellipticity equals 1/298.3.

Kuroshio. , *n.* A North Pacific Ocean current flowing northeastward from Taiwan to the Ryukyu Islands and close to the coast of Japan. The Kuroshio is the northward flowing part of the Pacific North Equatorial Current (which divides east of the Philippines). The Kuroshio divides near Yaku Shima, the weaker branch flowing northward through the Korea Strait and the stronger branch flowing through Tokara Kaikyo and then along the south coast of Shikoku. There are light seasonal variations in speed; the Kuroshio is usually strongest in summer, weakens in autumn, strengthens in winter, and weakens in spring. Strong winds can accelerate or retard the current but seldom change its direction. Beyond latitude 35°N on the east coast of Japan, the current turns east-northeastward to form the transitional KUROSHIO EXTENSION. The Kuroshio is part of the KUROSHIO SYSTEM. Also called JAPAN CURRENT.

Kuroshio Extension. . The transitional, eastward flowing ocean current that connects the Kuroshio and the North Pacific Current.

Kuroshio System. . A system of ocean currents which includes part of the Pacific North Equatorial Current, the Tsushima Current, the Kuroshio, and the Kuroshio Extension.

kymatology. , *n.* The science of waves and wave motion.

L

L-1 Signal. , The primary L-band signal transmitted by each GPS satellite at 1572.42 MHz. It is modulated with the C/A and P codes and the navigation message.

L-2 Signal. , The second L-band signal of the GPS satellite, transmitted at 1227.60 MHz, modulated with the P-code and navigation message.

labor. , *v., i.* To pitch and roll heavily under conditions which subject the ship to unusually heavy stresses caused by confused or turbulent seas or unstable stowage of cargo.

Labrador Current. . Originating from cold arctic water flowing southeastward through Davis Strait at speeds of 0.2 to 0.5 knot and from a westward branching of the warmer West Greenland Current, the Labrador Current flows south eastward along the shelf of the Canadian coast. Part of the current flows into Hudson Strait along its north shore. The outflow of fresh water along the south shore of the strait augments the part of the current flowing along the Labrador coast. The current also appears to be influenced by surface outflow from inlets and fjords along the Labrador coast. The mean speed is about 0.5 knot, but current speed at times may reach 1.5 to 2.0 knots.

Labrador Current Extension. . A name sometimes given to the nontidal current flowing southwestward along the northeast coast of the United States. This coastal current originates from part of the Labrador Current flowing clockwise around the southeastern tip of Newfoundland. Its speeds are fairly constant throughout the year and average about 0.6 knot. The greatest seasonal fluctuation appears to be in the width of the current. The current is widest during winter between Newfoundland and Cape Cod. Southwest of Cape Cod to Cape Hatteras the current shows very little seasonal change. The current narrows considerably during summer and flows closest to shore in the vicinity of Cape Sable, Nova Scotia and between Cape Cod and Long Island in July and August. The current in some places encroaches on tidal regions.

lagan., *n.* A heavy object thrown overboard and buoyed to mark its location for future recovery. See also JETTISON.

lag error.. Error in the reading of an instrument due to lag.

lagging of tide.. The periodic retardation in the time of occurrence of high and low water due to changes in the relative positions of the moon and the sun. See also PRIMING OF TIDE.

lagoon., *n.* 1. A shallow sound, pond, or lake generally separated from the open sea. 2. A body of water enclosed by the reefs and islands of an atoll.

Lagrangian current measurement.. The direct observation of the current speed or direction, or both, by a recording device such as a parachute drogue which follows the movement of a water mass through the ocean. See also EULERIAN CURRENT MEASUREMENT.

lake., *n.* 1. A standing body of inland water, generally of considerable size. There are exceptions such as the lakes in Louisiana which are open to or connect with the Gulf of Mexico. Occasionally a lake is called a SEA, especially if very large and composed of salt water. 2. An expanded part of a river.

lake ice.. Ice formed on a lake.

Lambert conformal chart.. A chart on the Lambert conformal projection. See also CONIC CHART WITH TWO STANDARD PARALLELS, MODIFIED LAMBERT CONFORMAL CHART.

Lambert conformal map projection.. A conformal map projection of the conic type, on which all geographic meridians are represented by straight lines which meet in a common point outside the limits of the map, and the geographic parallels are represented by a series of arcs of circles having this common point for a center. Meridians and parallels intersect at right angles, and angles on the earth are correctly represented on the projection. This projection may have one standard parallel along which the scale is held exact; or there may be two such standard parallels, both maintaining exact scale. At any point on the map, the scale is the same in every direction. The scale changes along the meridians and is constant along each parallel. Where there are two standard parallels, the scale between those parallels is too small; beyond them, too large. Also called LAMBERT CONFORMAL MAP PROJECTION. See also MODIFIED LAMBERT CONFORMAL MAP PROJECTION.

laminar flow.. See under STREAMLINE FLOW.

land., *v., t. & i.* To bring a vessel to a landing.

land breeze.. A breeze blowing from the land to the sea. It usually blows by night, when the sea is warmer than the land, and alternates with a SEA BREEZE, which blows in the opposite direction by day. See also OFFSHORE WIND.

landfall., *n.* The first sighting of land when approached from seaward. By extension, the term is sometimes used to refer to the first contact with land by any means, as by radar.

landfall buoy.. See SEA BUOY.

landfall light.. See PRIMARY SEACOAST LIGHT.

landing., *n.* 1. A place where boats receive or discharge passengers, freight, etc. See also LANDING STAGE, WHARF. 2. Bringing of a vessel to a landing.

landing compass.. A compass taken ashore so as to be unaffected by deviation. If reciprocal bearings of the landing compass and the magnetic compass on board are observed, the deviation of the latter can be determined.

landing stage.. A platform attached to the shore for landing or embarking passengers or cargo. In some cases the outer end of the landing stage is floating. Ships can moor alongside larger landing stages.

landmark., *n.* A conspicuous artificial feature on land, other than an established aid to navigation, which can be used as an aid to navigation. See also SEA MARK.

land mile.. See U.S. SURVEY MILE.

land sky.. Dark streaks or patches or a grayness on the underside of extensive cloud areas, due to the absence of reflected light from bare ground. Land sky is not as dark as WATER SKY. The clouds above ice or snow covered surfaces have a white or yellowish white glare called ICE BLINK. See also SKY MAP.

lane., *n.* In any continuous wave phase comparison system, the distance between two successive equiphase lines, taken as 0°–360°, in a system of hyperbolic or circular coordinates.

lane count.. An automatic method of counting and totaling the number of hyperbolic or circular lanes traversed by a moving vessel.

language.. A set of characters and rules which allow human interface with the computer, allowing PROGRAMS to be written.

lapse rate.. The rate of decrease of temperature in the atmosphere with height, or, sometimes, the rate of change of any meteorological element with height.

large fracture.. See under FRACTURE.

large iceberg.. For reports to the International Ice Patrol, an iceberg that extends more than 150 feet (45 meters) above the sea surface and which has a length of more than 400 feet (122 meters). See also SMALL ICEBERG, MEDIUM ICEBERG.

large ice field.. See under ICE FIELD.

large navigational buoy (LNB).. A large buoy designed to take the place of a lightship where construction of an offshore light station is not feasible. These buoys may show secondary lights from heights of about 30–40 feet above the water. In addition to the light, they may mount a radiobeacon and provide sound signals. A station buoy may be moored nearby.

large scale.. A scale involving a relatively small reduction in size. A large-scale chart is one covering a small area. The opposite is SMALL SCALE. See also REPRESENTATIVE FRACTION.

large-scale chart.. See under CHART. See also LARGE SCALE.

last quarter.. The phase of the moon when it is near west quadrature, when the eastern half of it is visible to an observer on the earth. See also PHASES OF THE MOON.

latent heat of fusion.. See under FUSION.

latent heat of vaporization.. See under EVAPORATION.

lateral., *adj.* Of or pertaining to the side, such as lateral motion.

lateral drifting.. See SWAY.

lateral mark.. A navigation aid intended to mark the sides of a channel or waterway. See CARDINAL MARKS.

lateral sensitivity.. The property of a range which determines the rapidity with which the two lights of the range open up as a vessel moves laterally from the range line, indicating to the mariner that he is off the center line.

lateral system.. A system of aids to navigation in which the shape, color, and number are assigned in accordance with their location relative to navigable waters. When used to mark a channel, they are assigned colors to indicate the side they mark and numbers to indicate their sequence along the channel. In the CARDINAL SYSTEM the aids are assigned shape, color, and number distinction in accordance with location relative to obstructions.

latitude., *n.* Angular distance from a primary great circle or plane. Terrestrial latitude is angular distance from the equator, measured northward or southward through 90° and labeled N or S to indicate the direction of measurement; astronomical latitude at a station is angular distance between the plumb line and the plane of the celestial equator; geodetic or topographical latitude at a station is angular distance between the plane of the geodetic equator and a normal to the ellipsoid; geocentric latitude is the angle at the center of the reference ellipsoid between the celestial equator and a radius vector to a point on the ellipsoid. Geodetic and sometimes astronomical latitude are also called geographic latitude. Geodetic latitude is used for charts. Assumed (or chosen) latitude is the latitude at which an observer is assumed to be located for an observation or computation. Observed latitude is determined by one or more lines of position extending in a generally east-west direction. Fictitious latitude is angular distance from a fictitious equator. Grid latitude is angular distance from a grid equator. Transverse or inverse latitude is angular distance from a transverse equator. Oblique latitude is angular distance from an oblique equator. Middle or mid latitude is the latitude at which the arc length of the parallel separating the meridians passing through two specific points is exactly equal to the departure in proceeding from one point to the other by middle-latitude sailing. Mean latitude is half the arithmetical sum of the latitude of two places on the same side of the equator. The mean latitude is usually used in middle-latitude sailing for want of a practical means of determining middle latitude. Difference of latitude is the shorter arc of any meridian between the parallels of two places, expressed in angular measure. Magnetic latitude, magnetic inclination, or magnetic dip is angular distance between the horizontal and the direction of a line of force of the earth's magnetic field at any point. Geomagnetic latitude is angular distance from the geomagnetic equator. A parallel of latitude is a circle (or approximation of a circle) of the earth, parallel to the equator, and connecting points of equal latitude- or a circle of the celestial sphere, parallel to the ecliptic. Celestial latitude is angular distance north or south of the ecliptic. See also VARIATION OF LATITUDE.

latitude factor. . The change in latitude along a celestial line of position per 1' change in longitude. The change in longitude for a 1' change in latitude is called LONGITUDE FACTOR.

latitude line. . A line of position extending in a generally east-west direction. Sometimes called OBSERVED LATITUDE. See also LONGITUDE LINE; COURSE LINE, definition 2; SPEED LINE.

lattice. , *n.* A pattern formed by two or more families of intersecting lines, such as that pattern formed by two or more families of hyperbolas representing, for example, curves of equal time difference associated with a hyperbolic radionavigation system. Sometimes the term pattern is used to indicate curves of equal time difference, with the term lattice being used to indicate its representation on the chart. See also PATTERN, definition 2.

lattice beacon. . A beacon or daymark in the form of a lattice. See also BEACON TOWER, REFUGE BEACON.

laurence. , *n.* A shimmering seen over a hot surface on a calm, cloudless day, caused by the unequal refraction of light by innumerable convective air columns of different temperatures and densities.

lava. , *n.* Rock in the fluid state, or such material after it has solidified. Lava is formed at very high temperature and issues from the earth through volcanoes. Part of the ocean bed is composed of lava.

law of equal areas. . Kepler's second law.

layer tints. . See HYPSOMETRIC TINTING.

L-band. . A radio-frequency band of 390 to 1,550 megahertz. See also FREQUENCY, FREQUENCY BAND.

lead. , *n.* A fracture or passage-way through ice which is navigable by surface vessels.

lead. , *n.* A weight attached to a line. A sounding lead is used for determining depth of water. A hand lead is a light sounding lead (7 to 14 pounds), usually having a line of not more than 25 fathoms. A deep sea lead is a heavy sounding lead (about 30 to 100 pounds), usually having a line 100 fathoms or more in length. A light deep sea lead (30 to 50 pounds), used for sounding depths of 20 to 60 fathoms is called a coasting lead. A type of sounding lead used without removal from the water between soundings is called a fish lead. A drift lead is one placed on the bottom to indicate movement of a vessel.

leader cable. . A cable carrying an electric current, signals from or the magnetic influence of which indicates the path to be followed by a craft equipped with suitable instruments.

leading lights. . See RANGE LIGHTS.

leading line. . On a nautical chart, a straight line, drawn through leading marks. A ship moving along such line will clear certain dangers or remain in the best channel. See also CLEARING LINE, RANGE, definition 1.

leading marks. . See RANGE, *n.* definition 1.

lead line. . A line, graduated with attached marks and fastened to a sounding lead, used for determining the depth of water when making soundings by hand. The lead line is usually used in depths of less than 25 fathoms. Also called SOUNDING LINE.

leadsman. , *n.* A person using a sounding lead to determine depth of water.

leap second. . A step adjustment to Coordinated Universal Time (UTC) to maintain it within 0.95^{s} of UT1. The 1 second adjustments, when necessary, are normally made at the end of June or December. Because of the variations in the rate of rotation of the earth, the occurrences of the leap second adjustments are not predictable in detail.

leap year. . A calendar year having 366 days as opposed to the COMMON YEAR having 365 days. Each year exactly divisible by 4 is a leap year, except century years (1800, 1900, etc.) which must be exactly divisible by 400 (2000, 2400, etc.) to be leap years.

least squares adjustment. . A statistical method of adjusting observations in which the sum of the squares of all the deviations or residuals derived in fitting the observations to a mathematical model is made a minimum.

ledge. , *n.* On the sea floor, a rocky, projection or datum outcrop, commonly linear and near shore.

lee. , *adj.* Referring to the downwind, or sheltered side of an object.

lee. , *n.* The sheltered area on the downwind side of an object.

lee shore. . As observed from a ship, the shore towards which the wind is blowing. See also WEATHER SHORE.

lee side. . That side of a craft which is away from the wind and therefore sheltered.

lee tide. . See LEEWARD TIDAL CURRENT.

leeward. , *adj.* & *adv.* Toward the lee, or in the general direction toward which the wind is blowing. The opposite is WINDWARD.

leeward. , *n.* The lee side. The opposite is WINDWARD.

leeward tidal current. . A tidal current setting in the same direction as that in which the wind is blowing. Also called LEE TIDE, LEEWARD TIDE.

leeward tide. . See LEEWARD TIDAL CURRENT.

leeway. , *n.* The leeward motion of a vessel due to wind. See also LEEWAY ANGLE.

leeway angle. . The angular difference between a vessel's course and the track due to the effect of wind in moving a vessel bodily to leeward. See also DRIFT ANGLE, definition 2.

left bank. . The bank of a stream or river on the left of an observer facing downstream.

leg. , *n.* A part of a ship's track line that can be represented by a single course line.

legend. , *n.* A title or explanation on a chart, diagram, illustration, etc.

lens. , *n.* A piece of glass or transparent material with plane, convex, or concave surfaces adapted for changing the direction of light rays to enlarge or reduce the apparent size of objects. See also EYEPIECE; FIELD LENS MENISCUS, definition 2, OBJECTIVE.

lenticular, lenticularis. , *adj.* In the shape of a lens, used to refer to an apparently stationary cloud resembling a lens, being broad in its middle and tapering at the ends and having a smooth appearance. Actually, the cloud continually forms to windward and dissipates to leeward.

lesser ebb. . See under EBB CURRENT.

lesser flood. . See under FLOOD CURRENT.

leste. , *n.* A hot, dry, easterly wind of the Madeira and Canary Islands.

levanter. , *n.* A strong easterly wind of the Mediterranean, especially in the Strait of Gibraltar, attended by cloudy, foggy, and sometimes rainy weather especially in winter.

levantera. , *n.* A persistent east wind of the Adriatic, usually accompanied by cloudy weather.

levanto. , *n.* A hot southeasterly wind which blows over the Canary Islands.

leveche. , *n.* A warm wind in Spain, either a foehn or a hot southerly wind in advance of a low pressure area moving from the Sahara Desert. Called a SIROCCO in other parts of the Mediterranean area.

levee. , *n.* 1. An artificial bank confining a stream channel or limiting adjacent areas subject to flooding. 2. on the sea floor, an embankment bordering a canyon, valley, or sea channel.

level ice. . Sea ice which is unaffected by deformation.

leveling. , *n.* A survey operation in which heights of objects are determined relative to a specified datum.

libration. , *n.* A real or apparent oscillatory motion, particularly the apparent oscillation of the moon, which results in more than half of the moon's surface being revealed to an observer on the earth, even though the same side of the moon is always toward the earth because of the moon's periods of rotation and revolution are the same.

light. , *adj.* 1. Of or pertaining to low speed, such as light air, force 1 (1-3 miles per hour or 1-3 knots) on the Beaufort scale or light breeze, force 2 (4-7 miles per hour or 4-6 knots) on the Beaufort scale. 2. Of or pertaining to low intensity, as light rain, light fog, etc.

light. , *n.* 1. Luminous energy. 2. An apparatus emitting light of distinctive character for use as an aid to navigation.

light air. . Wind of force 1 (1 to 3 knots or 1 to 3 miles per hour) on the Beaufort wind scale.

light attendant station. . A shore unit established for the purpose of servicing minor aids to navigation within an assigned area.

light-beacon. , *n.* See LIGHTED BEACON.

light breeze. . Wind of force 2 (4 to 6 knots or 4 to 7 miles per hour) on the Beaufort wind scale.

lighted beacon. . A beacon exhibiting a light. Also called LIGHT-BEACON.

lighted buoy. . A buoy exhibiting a light.

lighted sound buoy. . See under SOUND BUOY.

lightering area. . An area designated for handling ship's cargo by barge or lighter.

light-float. , *n.* A buoy having a boat-shaped body. Light-floats are usually unmanned and are used instead of smaller lighted buoys in waters where strong currents are experienced.

lighthouse. , *n.* A distinctive structure exhibiting a major navigation light.

light list. . 1. A publication giving detailed information regarding lighted navigational aids and fog signals. In the United States, light lists are published by the U.S. Coast Guard as USCG Light Lists and by the National Imagery and Mapping Agency as List of Lights.

light list number. . The sequential number used to identify a navigational light in the light list. This may or may not be the same as the INTERNATIONAL NUMBER, which is an identifying number assigned by the International Hydrographic Organization. The international number is in italic type and is located under the light list number in the list.

light nilas. . Nilas which is more than 5 centimeters in thickness and somewhat lighter in color than dark nilas.

light sector. . As defined by bearings from seaward, the sector in which a navigational light is visible or in which it has a distinctive color different from that of adjoining sectors, or in which it is obscured. See also SECTOR LIGHT.

lightship. , *n.* A distinctively marked vessel providing aids to navigation services similar to a light station, i.e., a light of high intensity and reliability, sound signal, and radiobeacon, and moored at a station where erection of a fixed structure is not feasible. Most lightships are anchored to a very long scope of chain and, as a result, the radius of their swinging circle is considerable. The chart symbol represents the approximate location of the anchor. Also called LIGHT VESSEL. See also LIGHT-FLOAT.

lights in line. . Two or more lights so situated that when observed in transit they define the alignment of a submarine cable, the limit of an area, an alignment for use in anchoring, etc. Not to be confused with RANGE LIGHTS which mark a direction to be followed. See also RANGE, definition 1.

light station. . A manned station providing a light usually of high intensity and reliability. It may also provide sound signal and radiobeacon services.

light valve. . See SUN VALVE.

light vessel. . See LIGHTSHIP.

light-year. , *n.* A unit of length equal to the distance light travels in 1 year, equal to about $5.88X\ 10^{12}$ miles. This unit is used as a measure of stellar distances.

liman. , *n.* A shallow coastal lagoon or embayment with a muddy bottom; also a region of mud or slime deposited near a stream mouth.

Liman Current. . Formed by part of the Tsushima Current and river discharge in Tatar Strait, the coastal Liman Current flows southward in the western part of the Sea of Japan. During winter, it may reach as far south as 35°N. See also under TSUSHIMA CURRENT.

limb. , *n.* 1. The graduated curved part of an instrument for measuring angles, such as the part of a marine sextant carrying the altitude scale, or ARC. 2. The circular outer edge of a celestial body, usually referred to with the designation upper or lower.

limbo echo. . See CLASSIFICATION OF RADAR ECHOES.

line. , *n.* 1. A series of related points, the path of a moving point. A line has only one dimension; length. 2. A row of letters, numbers, etc. 3. A mark of division or demarcation, as a *boundary line*.

linear. , *adj.* 1. Of or pertaining to a line. 2. Having a relation such that a change in one quantity is accompanied by an exactly proportional change in a related quantity.

linear interpolation. . Interpolation in which changes of tabulated values are assumed to be proportional to changes in entering arguments.

linear light. . A luminous signal having perceptible length, as contrasted with a POINT LIGHT, which does not have perceptible length.

linearly polarized wave. . A transverse electromagnetic wave the electric field vector of which lies along a fixed line at all times.

linear scale. . A scale graduated at uniform intervals.

linear speed. . Rate of motion in a straight line. See also ANGULAR RATE.

linear sweep. . Short for LINEAR TIME BASE SWEEP.

linear time base. . A time base having a constant speed, particularly a linear time base sweep.

linear time base sweep. . A sweep having a constant sweep speed before retrace. Usually shortened to LINEAR SWEEP, and sometimes to LINEAR TIME BASE.

line blow. . A strong wind on the equator side of an anticyclone, probably so called because there is little shifting of wind direction during the blow, as contrasted with the marked shifting which occurs with a cyclonic windstorm.

line of apsides. . The line connecting the two points of an orbit that are nearest and farthest from the center of attraction, such as the perigee and apogee of the moon or the perihelion and aphelion of a planet. Also called APSE LINE.

line of force. . A line indicating the direction in which a force acts, as in a magnetic field.

line of nodes. . The straight line connecting the two points of intersection of the orbit of a planet, planetoid, or comet and the ecliptic; or the line of intersection of the planes of the orbits of a satellite and the equator of its primary.

line of position. . A plotted line on which a vessel is located, determined by observation or measurement. Also called POSITION LINE.

line of sight. . The straight line between two points, which does not follow the curvature of the earth.

line of soundings. . A series of soundings obtained by a vessel underway, usually at regular intervals. In piloting, this information may be used to determine an estimated position, by recording the soundings at appropriate intervals (to the scale of the chart) along a line drawn on transparent paper or plastic, to represent the track, and then fitting the plot to the chart, by trial and error. A vessel obtaining soundings along a course line, for use in making or improving a chart, is said to run *a line of soundings*.

line of total force. . The direction of a freely suspended magnetic needle when acted upon by the earth's magnetic field alone.

line squall. . A squall that occurs along a squall line.

lipper. , *n.* 1. Slight ruffling or roughness on a water surface. 2. Light spray from small waves.

liquid compass. . A magnetic compass of which the bowl mounting the compass card is completely filled with liquid. Nearly all modern magnetic compasses are of this type. An older liquid compass using a solution of alcohol and water is sometimes called a SPIRIT COMPASS. Also called WET COMPASS. See also DRY COMPASS.

list. , *n.* Inclination to one side. LIST generally implies equilibrium in an inclined condition caused by uneven distribution of mass aboard the vessel itself, while HEEL implies either a continuing or momentary inclination caused by an outside force, such as the wind. The term ROLL refers to the oscillatory motion of a vessel rather than its inclined condition.

list. , *v., t. & i.* To incline or be inclined to one side.

lithometeor. , *n.* The general term for dry atmospheric suspensoids, including dust, haze, smoke, and sand. See also HYDROMETEOR.

little brother. . A secondary tropical cyclone sometimes following a more severe disturbance.

littoral. , *adj. & n.* 1. A littoral region. 2. The marine environment influenced by a land mass. 3. Of or pertaining to a shore, especially a seashore. See also SEABOARD.

load line marks. . Markings stamped and painted amidships on the side of a vessel, to indicate the minimum permissible freeboard. Also called PLIMSOLL MARKS. See also DRAFT MARKS.

lobe. , *n.* 1. The portion of the overall radiation pattern of a directional antenna which is contained within a region bounded by adjacent minima. The main beam is the beam in the lobe containing the direction of maximum radiation (main lobe) lying within specified values of field strength relative to the maximum field strength. See also BACK LOBE, SIDE LOBE, BEAM WIDTH 2. The radiation within the region of definition 1.

local apparent noon. Twelve o'clock local apparent time, or the instant the apparent sun is over the upper branch of the local meridian. Local apparent noon at the Greenwich meridian is called Greenwich apparent noon. Sometimes called HIGH NOON.

local apparent time. The arc of the celestial equator, or the angle at the celestial pole, between the lower branch of the local celestial meridian and the hour circle of the apparent or true sun, measured westward from the lower branch of the local celestial meridian through 24 hours; local hour angle of the apparent or true sun, expressed in time units, plus 12 hours. Local apparent time at the Greenwich meridian is called Greenwich apparent time.

local attraction. See LOCAL MAGNETIC DISTURBANCE.

local civil noon. *United States terminology from 1925 through 1952*. See LOCAL MEAN NOON.

local civil time. United States terminology from 1925 through 1952. See LOCAL MEAN TIME.

local hour angle (LHA). Angular distance west of the local celestial meridian; the arc of the celestial equator, or the angle at the celestial pole, between the upper branch of the local celestial meridian and the hour circle of a point on the celestial sphere, measured westward from the local celestial meridian through 360°. The local hour angle at longitude 0° is called Greenwich hour angle.

local knowledge. The term applied to specialized, detailed knowledge of a port, harbor, or other navigable water considered necessary for safe navigation. Local knowledge extends beyond that available in charts and publications, being more detailed, intimate, and current.

local lunar time. The arc of the celestial equator, or the angle at the celestial pole, between the lower branch of the local celestial meridian and the hour circle of the moon, measured westward from the lower branch of the local celestial meridian through 24 hours; local hour angle of the moon, expressed in time units, plus 12 hours. Local lunar time at the Greenwich meridian is called Greenwich lunar time.

local magnetic disturbance. An anomaly of the magnetic field of the earth, extending over a relatively small area, due to local magnetic influences. Also called LOCAL ATTRACTION, MAGNETIC ANOMALY.

local mean noon. Twelve o'clock local mean time, or the instant the mean sun is over the upper branch of the local meridian. Local mean noon at the Greenwich meridian is called Greenwich mean noon.

local mean time. The arc of the celestial equator, or the angle at the celestial pole, between the lower branch of the local celestial meridian and the hour circle of the mean sun, measured westward from the lower branch of the local celestial meridian through 24 hours; local hour angle of the mean sun, expressed in time units, plus 12 hours. Local mean time at the Greenwich meridian is called Greenwich mean time, or Universal Time.

local meridian. The meridian through any particular place of observer, serving as the reference for local time, in contrast with GREENWICH MERIDIAN.

local noon. Noon at the local meridian.

Local Notice to Mariners. A notice issued by each U.S. Coast Guard District to disseminate important information affecting navigational safety within the District. The *Local Notice* reports changes to and deficiencies in aids to navigation maintained by and under the authority of the U.S. Coast Guard. Other information includes channel depths, new charts, naval operations, regattas, etc. Since temporary information, known or expected to be of short duration, is not included in the weekly *Notice to Mariners* published by the Defense Mapping Agency Hydrographic/ Topographic Center, the appropriate *Local Notice to Mariners* may be the only source of such information. Much of the information contained in the *Local Notice to Mariners* is included in the weekly *Notice to Mariners*. The *Local Notice to Mariners* is published as often as required; usually weekly. It may be obtained, free of charge, the appropriate Coast Guard District Commander.

local oscillator. An oscillator used to drive an intermediate frequency by beating with the signal carrying frequency in superheterodyne reception.

local sidereal noon. Zero hours local sidereal time, or the instant the vernal equinox is over the upper branch of the local meridian. Local sidereal noon at the Greenwich meridian is called Greenwich sidereal noon.

local sidereal time. Local hour angle of the vernal equinox, expressed in time units; the arc of the celestial equator, or the angle at the celestial pole, between the upper branch of the local celestial meridian and the hour circle of the vernal equinox, measured westward from the upper branch of the local celestial meridian through 24 hours. Local sidereal time at the Greenwich meridian is called Greenwich sidereal time.

local time. 1. Time based upon the local meridian as reference, as contrasted with that based upon a standard meridian. Local time was in general use in the United States until 1883, when standard time was adopted. 2. Any time kept locally.

local vertical. The direction of the acceleration of gravity as opposed to the normal to the reference ellipsoid. It is in the direction of the resultant of the gravitational and centrifugal accelerations of the earth at the location of the observer. Also called PLUMB-BOB VERTICAL. See also MASS ATTRACTION VERTICAL.

loch, *n*. 1. A lake. 2. An arm of the sea, especially when nearly landlocked.

lock, *n*. 1. A basin in a waterway with caissons or gates at each end by means of which vessels are passed from one water level to another.

lock, *v. t.* To pass through a lock, referred to as *locking through*.

lock on. To identify and begin to continuously track a target in one or more coordinates (e.g., range, bearing, elevation).

locus, *n*. All possible positions of a point or curve satisfying stated conditions.

log, *n*. 1. An instrument for measuring the speed or distance or both traveled by a vessel. A chip log (ancient) consists essentially of a weighted wooden quadrant (quarter of a circle) attached to a bridle in such a manner that it will float in a vertical position, and a line with equally spaced knots. A mechanical means of determining speed or distance is called a patent log. A harpoon log consists essentially of a combined rotator and distance registering device towed through the water. This has been largely replaced by the taffrail log, a somewhat similar device but with the registering unit secured at the taffrail. A Pitometer log consists essentially of a Pitot tube projecting into the water, and suitable registering devices. An electromagnetic log consists of suitable registering devices and an electromagnetic sensing element, extended below the hull of a vessel, which produces a voltage directly proportional to speed through the water. A Forbes log consists of a small rotator in a tube projecting below the bottom of the vessel, and suitable registering devices. A Dutchman's log is a buoyant object thrown overboard, the speed of a vessel being determined by noting the time required for a known length of the vessel to pass the object. 2. A written record of the movements of a craft, with regard to courses, speeds, positions, and other information of interest to navigators, and of important happenings aboard the craft. The book in which the log is kept is called a LOG BOOK. Also called DECK LOG. See also NIGHT ORDER BOOK 3. A written record of specific related information, as that concerning performance of an instrument. See GYRO LOG.

logarithm, *n*. The power to which a fixed number, called the base, usually 10 or *e* (2.7182818), must be raised to produce the value to which the logarithm corresponds. A logarithm (base 10) consists of two parts: the characteristic is that part to the left of the decimal point and the mantissa is that part to the right of the decimal point. An ANTILOGARITHM or INVERSE LOGARITHM is the value corresponding to a given logarithm. Logarithms are used to multiply or divide numbers, the sum or difference of the logarithms of two numbers being the logarithm of the product or quotient, respectively, of the two numbers. A COLOGARITHM is the logarithm of the reciprocal of a number. Logarithms to the base 10 are called common or Briggsian and those to the base *e* are called natural or Napierian logarithms.

logarithmic, *adj*. Having to do with a logarithm, used with the name of a trigonometric function to indicate that the value given is the logarithm of that function, rather than the function itself which is called the natural trigonometric function.

logarithmic coordinate paper. Paper ruled with two sets of mutually-perpendicular, parallel lines spaced according to the logarithms of consecutive numbers, rather than the numbers themselves. On SEMILOGARITHMIC COORDINATE PAPER one set of lines is

logarithmic scale. A scale graduated in the logarithms of uniformly-spaced consecutive numbers.

logarithmic tangent. See under TANGENT, definition 1.

logarithmic trigonometric function. See under TRIGONOMETRIC FUNCTIONS.

log book. See LOG, definition 2.

log chip. The wooden quadrant forming part of a chip log. Also called LOG SHIP.

log glass. A small hour glass used to time a chip log. The period most frequently used is 28 seconds.

log line. 1. A graduated line used to measure the speed of a vessel through the water or to measure the speed of a current, the line may be called a CURRENT LINE. 2. The line secured to a log.

long flashing light. A navigation light with a duration of flash of not less than 2 seconds.

longitude, *n*. Angular distance, along a primary great circle, from the adopted reference point. Terrestrial longitude is the arc of a parallel, or the angle at the pole, between the prime meridian and the meridian of a point on the earth measured eastward or westward from the prime meridian through 180°, and labeled E or W to indicate the direction of measurement. Astronomical longitude is the angle between the plane of the prime meridian and the plane of the celestial meridian; geodetic longitude is the angle between the plane of the geodetic meridian and a station and the plane of the geodetic meridian at Greenwich. Geodetic and sometimes astronomical longitude are also called geographic longitude. Geodetic longitude is used in charting. Assumed longitude is the longitude at which an observer is assumed to be located for an observation or computation. Observed longitude is determined by one or more lines of position extending in a generally north-south direction. Difference of longitude is the smaller angle at the pole or the shorter arc of a parallel between the meridians of two places, expressed in angular measure. Fictitious longitude is the arc of the fictitious equator between the prime fictitious meridian and any given fictitious meridian. Grid longitude is angular distance between a prime grid meridian and any given grid meridian. Oblique longitude is angular distance between a prime oblique meridian and any given oblique meridian. Transverse or inverse longitude is angular distance between a prime transverse meridian and any given meridian. Celestial longitude is angular distance east of the vernal equinox, along the ecliptic.

longitude factor. The change in longitude along a celestial line of position per 1' change in latitude. The change in latitude for a 1' change in longitude is called LATITUDE FACTOR.

longitude line. A line of position extending in a generally north-south direction. Sometimes called OBSERVED LONGITUDE. See also LATITUDE LINE; COURSE LINE, definition 2; SPEED LINE.

longitude method. The establishing of a line of position from the observation of the latitude of a celestial body by assuming a latitude (or longitude), and calculating the longitude (or latitude) through which the line of position passes, and the azimuth. The line of position is drawn through the point thus found, perpendicular to the azimuth. See also ST. HILAIRE METHOD, SUMNER METHOD, HIGH ALTITUDE METHOD.

longitude of Greenwich at time of perigee. See RIGHT ASCENSION OF GREENWICH AT TIME OF PERIGEE.

longitude of pericenter. An orbital element that specifies the orientation of an orbit; it is a broken angle consisting of the angular distance in the ecliptic from the vernal equinox to the ascending node of the orbit plus the angular distance in the orbital plane from the ascending node to the pericenter, i.e. the sum of the longitude of the ascending node and the argument of pericenter.

longitude of the ascending node. 1. The angular distance in the ecliptic from the vernal equinox to the ascending node of the orbit. See also LONGITUDE OF PERICENTER, RIGHT ASCENSION OF THE ASCENDING NODE. 2. The angular distance, always measured eastward, in the plane of the celestial equator from Greenwich through 360°.

longitude of the moon's nodes. The angular distance along the ecliptic of the moon's nodes from the vernal equinox; the nodes have a retrograde motion, and complete a cycle of 360° in approximately 19 years.

longitudinal axis. The fore-and-aft line through the center of gravity of a craft, around which it rolls.

longitudinal wave. A wave in which the vibration is in the direction of propagation, as in sound waves. This is in contrast with a TRANSVERSE WAVE, in which the vibration is perpendicular to the direction of propagation.

long path interference. See under MULTIPATH ERROR.

long period constituent. A tidal or tidal current constituent with a period that is independent of the rotation of the earth but which depends upon the orbital movement of the moon or of the earth. The principal lunar long period constituents have periods approximating the month and half-month, and the principal solar long period constituents have periods approximating the year and half-year.

long period perturbations. Periodic eccentricities in the orbit of a planet or satellite which require more than one orbital period to execute one complete periodic variation.

long range systems. Radionavigation systems providing positioning capability on the high seas. Loran C is an example. See also SHORT RANGE SYSTEMS.

longshore current. A current paralleling the shore largely within the surf zone. It is caused by the excess water brought to the zone by the small net mass transport of wind waves. Longshore currents feed into rip currents.

look angles. The elevation and azimuth at which a particular satellite is predicted to be found at a specified time.

lookout station. A label on a nautical chart which indicates a tower surmounted by a small house from which a watch is kept regularly.

loom, *n*. The diffused glow observed from a light below the horizon, due to atmospheric scattering.

looming, *n*. 1. An apparent elevation of distant terrestrial objects by abnormal atmospheric refraction. Because of looming, objects below the horizon are sometimes visible. The opposite is SINKING. 2. The appearance indistinctly of an object during a period of low visibility.

loop antenna. A closed circuit antenna in the form of a loop, lying in the same plane, or of several loops lying in parallel planes.

loop of stationary wave. See under STATIONARY WAVE.

Loran, *n*. The general designation of a type of radionavigation system by which a hyperbolic line of position is determined through measuring the difference in the times of reception of synchronized signals from two fixed transmitters. The name Loran is derived from the words <u>lo</u>ng <u>r</u>ange <u>n</u>avigation.

Loran A, *n*. A long range medium frequency (1850 to 1950 kHz) radionavigation system by which a hyperbolic line of position of medium accuracy was obtained. System operation in U.S. waters was terminated on 31 December 1980. See also LORAN, HYPERBOLIC NAVIGATION.

Loran C, *n*. A long range, low frequency (90-110 kHz) radionavigation system by which a hyperbolic line of position of high accuracy is obtained by measuring the difference in the times of arrival of signals radiated by a pair of synchronized transmitters (master station and secondary station) which are separated by several hundred miles. See also LORAN, HYPERBOLIC NAVIGATION.

Loran C plotting chart. See under Plotting CHART.

Loran C reliability diagram. One of a series of charts which depict the following data for the area covered: (1) for each station of the chain, predicted maximum usable groundwave signal limits for signal-to-noise ratios of 1:3 and 1:10, and (2) contours which indicate the regions within which positions can be fixed with repeatable accuracies of 500, 750, or 1500 feet or better on a 95 percent probability basis. See also COVERAGE DIAGRAM.

Loran C Table. See PUB. 221. LORAN C TABLE.

Loran rate. See RATE, definition 2.

Lorhumb line. A line along which the rates of change of the values of two families of hyperbolae are constants.

lost motion. Mechanical motion which is not transmitted to connected or related parts, due to loose fit. See also BACKLASH.

low, *n*. Short for area of low pressure. Since a low is, on a synoptic chart, always associated with cyclonic circulation, the term is used interchangeably with CYCLONE. See also HIGH.

low clouds. Types of clouds the mean level of which is between the surface and 6,500 feet. The principal clouds in this group are stratocumulus, stratus, and nimbostratus.

lower branch. The half of a meridian or celestial meridian from pole to pole which passes through the antipode or nadir of a place. See also UPPER BRANCH.

lower culmination. See LOWER TRANSIT.

lower high water. The lower of the two high waters of any tidal day.

lower high water interval. See under LUNITIDAL INTERVAL.

lower limb. The lower edge (closest to the horizon) of a celestial body having measurable diameter; opposite is the UPPER LIMB, or the upper edge.

lower low water. The lower of the two low waters of any tidal day.

lower low water datum. An approximation of mean lower low water that has been adopted as a standard reference for a limited area, and is retained for an indefinite period regardless of the fact that it may differ slightly from a better determination of mean lower low water from a subsequent series of observations. Used primarily for river and harbor engineering purposes. Columbia River lower low water datum is an example.

lower low water interval. See under LUNITIDAL INTERVAL.

lower transit. Transit of the lower branch of the celestial meridian. Transit of the upper branch is called UPPER TRANSIT. Also called INFERIOR TRANSIT, LOWER CULMINATION.

low frequency. Radio frequency of 30 to 300 kilohertz.

low light. See FRONT LIGHT.

low tide. See under LOW WATER.

low water. The minimum height reached by a falling tide. The height may be due solely to the periodic tidal forces or it may have superimposed upon it the effects of meteorological conditions.

low water datum. 1. The dynamic elevation for each of the Great Lakes, Lake St. Clair, and the corresponding sloping surfaces of the St. Marys, St. Clair, Detroit, Niagara, and St. Lawrence Rivers to which are referred the depths shown on the navigation charts and the authorized depths for navigation improvement projects. Elevations of these planes are referred to International Great Lakes Datum (1955) and are: Lake Superior - 600.0 feet, Lakes Michigan and Huron - 576.8 feet, Lake St. Clair - 571.7 feet, Lake Erie - 568.6 feet, and Lake Ontario- 242.8 feet. 2. An approximation of mean low water that has been adopted as a standard reference for a limited area and is retained for an indefinite period regardless of the fact that it may differ slightly from a better determination of mean low water from a subsequent series of observations. Used primarily for river and harbor engineering purposes.

low water equinoctial springs. Low water spring tides near the times of the equinoxes. Expressed in terms of the harmonic constituents, it is an elevation depressed below mean sea level by an amount equal to the sum of the amplitudes of certain constituents as given in the *Tide and Current Glossary* published by the National Ocean Survey.

low water inequality. See under DIURNAL INEQUALITY.

low water interval. See under LUNITIDAL INTERVAL.

low water line. The intersection of the land with the water surface at an elevation of low water.

low water neaps. See under NEAP TIDES.

low water springs. Short for MEAN LOW WATER SPRINGS.

low water stand. The condition at low water when there is no sensible change in the height of the tide. A similar condition at high water is called HIGH WATER STAND. See also STAND.

loxodrome, *n*. See RHUMB LINE. See also ORTHODROME.

loxodromic curve. See RHUMB LINE.

lubber's line. A reference line on a compass marking the reading which coincides with the heading.

lubber's line error. The angular difference between the heading as indicated by a lubber's line, and the actual heading; the horizontal angle, at the center of an instrument, between a line through the lubber's line and one parallel to the keel.

lull, *n*. A momentary decrease in the speed of the wind.

lumen, *n*. The derived unit of luminous flux in the International System of Units; it is the luminous flux emitted within unit solid angle (1 steradian) by a point source having a uniform luminous intensity of 1 candela.

luminance, *n*. In a given direction, at a point on the surface of a source or receptor, or at a point on the path of a beam, the quotient of the luminous flux leaving, arriving at, or passing through an element of surface at this point and propagated in directions defined by an elementary cone containing the given directions, by the product of the solid angle of the cone and the area of the orthogonal projection of the element of surface on a plane perpendicular to the given direction. The derived unit of luminance in the International System of Units is the CANDELA PER SQUARE METER.

luminescence, *n*. Emission of light other than incandescence, as in bioluminescence; emission as a result of and only during absorption of radiation from some other source is called FLUORESCENCE; continued emission after absorption of radiation has ceased is called PHOSPHORESCENCE.

luminous, *adj*. Emitting or reflecting light.

luminous flux. The quantity characteristic of radiant flux which expresses its capacity to produce a luminous sensation, evaluated according to the values of spectral luminous efficiency. Unless otherwise indicated, the luminous flux relates to photopic vision, and is connected with the radiant flux in accordance with the formula adopted in 1948 by the International Commission on Illumination. The derived unit of luminous flux in the International System of Units is the LUMEN.

luminous range. See under VISUAL RANGE (OF A LIGHT).

luminous Range Diagram. A diagram used to convert the nominal range of a light to its luminous range under existing conditions.

lunar, *adj*. Of or pertaining to the moon.

lunar cycle. An ambiguous expression which has been applied to various cycles associated with the moon's motion, including CALLIPPIC CYCLE, METONIC CYCLE, NODE CYCLE, SYNODICAL MONTH or LUNATION.

lunar day. 1. The duration of one rotation of the earth on its axis, with respect to the moon. Its average length is about $24^h 50^m$ of mean solar time. Also called TIDAL DAY. 2. The duration of one rotation of the moon on its axis, with respect to the sun.

lunar distance. The angle, at an observer on the earth, between the moon and another celestial body. This was the basis of a method formerly used to determine longitude at sea.

lunar eclipse. An eclipse of the moon. When the moon enters the shadow of the earth, it appears eclipsed to an observer on the earth. A lunar eclipse is penumbral when it enters only the penumbra of the earth's shadow, partial when part of its surface enters the umbra of the earth's shadow, and total if its entire surface is obscured by the umbra.

lunar inequality. 1. Variation in the moon's motion in its orbit, due to attraction by other bodies of the solar system. See also EVECTION, PERTURBATIONS. 2. A minute fluctuation of a magnetic needle from its mean position, caused by the moon.

lunar interval. The difference in time between the transit of the moon over the Greenwich meridian and a local meridian. The lunar interval equals the difference between the Greenwich and local intervals of a tide or current phase.

lunar month. The period of revolution of the moon about the earth, especially a synodical month.

lunar node. A node of the moon's orbit. See also LINE OF NODES.

lunar noon. The instant at which the sun is over the upper branch of any meridian of the moon.

lunar parallax. Parallax of the moon.

lunar rainbow. See MOON BOW.

lunar tide. That part of the tide due solely to the tide-producing force of the moon. That part due to the tide-producing force of the sun is called SOLAR TIDE.

lunar time. Time based upon the rotation of the earth relative to the moon. Lunar time may be designated as local or Greenwich according to whether the local or Greenwich meridian is used as the reference.

lunation, *n*. See SYNODICAL MONTH.

lune, *n*. The part of the surface of a sphere bounded by halves of two great circles.

lunicurrent internal. The interval between the moon's transit (upper or lower) over the local or Greenwich meridian and a specified phase of the tidal current following the transit. Examples are strength of flood interval and strength of ebb interval, which may be abbreviated to flood interval and ebb interval, respectively. The interval is described as local or Greenwich according to whether the reference is to the moon's transit over the local or Greenwich meridian. When not otherwise specified, the reference is assumed to be local. See also LUNITIDAL INTERVAL.

lunisolar effect. Gravitational effects caused by the attractions of the moon and of the sun.

lunisolar perturbation. Perturbations of the orbits of artificial earth satellites due to the attractions of the sun and the moon. The most important effects are secular variations in the mean anomaly, in the right ascension of the ascending node, and in the argument of perigee.

lunisolar precession. That component of general precession caused by the combined effect of the sun and moon on the equatorial protuberance of the earth, producing a westward motion of the equinoxes along the ecliptic. See also PRECESSION OF THE EQUINOXES.

lunitidal interval. The interval between the moon's transit (upper or lower) over the local or Greenwich meridian and the following high or low water. The average of all high water intervals for all phases of the moon is known as mean high water lunitidal interval and is abbreviated to high water interval. Similarly the mean low water lunitidal interval is abbreviated to low water interval. The interval is described as local or Greenwich according to whether the reference is to the transit over the local or Greenwich meridian. When not otherwise specified, the reference is assumed to be local. When there is considerable diurnal inequality in the tide separate intervals may be obtained for the higher high waters, the lower high waters, the higher low waters and the lower low waters. These are designated respectively as higher high water interval, lower high water interval higher low water interval, and lower low water interval. In such cases, and also when the tide is diurnal, it is necessary to distinguish between the upper and lower transit of the moon with reference to its declination.

lux, *n*. The derived unit of illuminance in the International System of Units; it is equal to 1 lumen per square meter.

M

mackerel sky. An area of sky with a formation of rounded and isolated cirrocumulus or altocumulus resembling the pattern of scales on the back of a mackerel.

macroscopic, *adj*. Large enough to be seen by the unaided eye.

madrepore, *n*. A branching or stag-horn coral, or any perforated stone coral.

maelstrom, *n*. A whirlpool similar to the Maelstrom off the west coast of Norway.

maestro, *n*. A northwesterly wind with fine weather which blows, especially in summer, in the Adriatic. It is most frequent on the western shore. This wind is also found on the coasts of Corsica and Sardinia.

magnet, *n*. A body which produces a magnetic field around itself. It has the property of attracting certain materials capable of being magnetized. A magnet occurring in nature is called a natural magnet in contrast with a man-made artificial magnet. See also HEELING MAGNET, KEEPER.

magnetic, *adj*. Of or pertaining to a magnet or related to magnetic north.

magnetic amplitude. Amplitude relative to magnetic east or west.

magnetic annual change. The amount of secular change in the earth's magnetic field which occurs in 1 year. magnetic annual variation; the small systematic temporal variation in the earth's magnetic field which occurs after the trend for secular change has been removed from the average monthly values.

magnetic anomaly. See LOCAL MAGNETIC DISTURBANCE.

magnetic azimuth. Azimuth relative to magnetic north.

magnetic bay. A small magnetic disturbance whose magnetograph resembles an indentation of a coastline. On earth, magnetic bays occur mainly in the polar regions and have duration of a few hours.

magnetic bearing. Bearing relative to magnetic north; compass bearing corrected for deviation.

magnetic chart. A chart showing magnetic information. If it shows lines of equality in one or more magnetic elements, it may be called an isomagnetic chart. It is an isoclinal or isoclinic chart if it shows lines of equal magnetic dip, an isodynamic chart if it shows lines of equal magnetic intensity, an isogonic chart if it shows lines of equal magnetic variation, an isogriv chart if it shows lines of equal grid variation, an isoporic chart if it shows lines of equal rate or change of a magnetic element.

magnetic circle. A sphere of specified radius about the magnetic compass location to be kept free of any magnetic or electrical equipment which would interfere with the compass.

magnetic compass. A compass depending for its directive force upon the attraction of the horizontal component of the earth's magnetic field for a magnetized needle or sensing element free to turn in a horizontal direction.

magnetic course. Course relative to magnetic north; compass course corrected for deviation. magnetic daily variation. See MAGNETIC DIURNAL VARIATION.

magnetic declination. See VARIATION, definition 1.

magnetic deviation. See DEVIATION, definition 1.

magnetic dip. Angular distance between the horizontal and the direction of a line of force of the earth's magnetic field at any point. Also called DIP, MAGNETIC INCLINATION.

magnetic dip pole. See MAGNETIC POLE, definition 1.

magnetic direction. Horizontal direction expressed as angular distance from magnetic north. magnetic diurnal variation. Oscillations of the earth's magnetic field which have a periodicity of about a day and which depend to a close approximation only on local time and geographic latitude. Also called MAGNETIC DAILY VARIATION.

magnetic element. 1. Variation, dip, or magnetic intensity. 2. The part of an instrument producing or influenced by magnetism.

magnetic equator. The line on the surface of the earth connecting all points at which the magnetic dip is zero. Also called ACLINIC LINE. See also GEOMAGNETIC EQUATOR.

magnetic field. Any space or region in which magnetic forces are present, as in the earth's magnetic field, or in or about a magnet, or in or about an electric current. See also MAGNETIC VECTOR.

magnetic force. The strength of a magnetic field. Also called MAGNETIC INTENSITY.

magnetic heading. Heading relative to magnetic north; compass heading corrected for deviation.

magnetic inclination. See MAGNETIC DIP.

magnetic induction. The act or process by which material becomes magnetized when placed in a magnetic field.

magnetic intensity. The strength of a magnetic field. Also called MAGNETIC FORCE.

magnetic latitude. Angular distance north or south of the magnetic equator. The angle is equal to an angle, the tangent of which is equal to half the tangent of the magnetic dip at the point.

magnetic lines of force. Closed lines indicating by their direction the direction of magnetic influence.

magnetic meridian. A line of horizontal magnetic force of the earth. A compass needle without deviation lies in the magnetic meridian.

magnetic moment. The quantity obtained by multiplying the distance between two magnetic poles by the average strength of the poles.

magnetic needle. A small, slender, magnetized bar which tends to align itself with magnetic lines of force.

magnetic north. The direction indicated by the north seeking pole of a freely suspended magnetic needle, influenced only by the earth's magnetic field.

magnetic observation. Measurement of any of the magnetic elements.

magnetic parallel. An isoclinal; a line connecting points of equal magnetic dip.

magnetic pole. 1. Either of the two places on the surface of the earth where the magnetic dip is 90°, that in the Northern Hemisphere being designated north magnetic pole, and that in the Southern Hemisphere being designated south magnetic pole. Also called MAGNETIC DIP POLE. See also MAGNETIC LATITUDE, GEOMAGNETIC POLE, MAGNETIC LATITUDE. 2. Either of those two points of a magnet where the magnetic force is greatest.

magnetic prime vertical. The vertical circle through the magnetic east and west points of the horizon.

magnetic range. A range oriented in a given magnetic direction and used to assist in the determination of the deviation of a magnetic compass.

magnetic retentivity. The ability to retain magnetism after removal of the magnetizing force.

magnetic secular change. The gradual variation in the value of a magnetic element which occurs over a period of years.

magnetic storm. A disturbance in the earth's magnetic field, associated with abnormal solar activity, and capable of seriously affecting both radio and wire transmission.

magnetic temporal variation. Any change in the earth's magnetic field which is a function of time.

magnetic track. The direction of the track relative to magnetic north.

magnetic variation. See VARIATION, definition 1.

magnetic vector. The component of the electromagnetic field associated with electromagnetic radiation which is of the nature of a magnetic field. The magnetic vector is considered to coexist with, but to act at right angles to, the electric vector.

magnetism, *n*. The phenomena associated with magnetic fields and their effects upon magnetic materials, notably iron and steel. The magnetism of the north-seeking end of a freely suspended magnet is called red magnetism; the magnetism of the south-seeking end is called blue magnetism. Magnetism acquired by a piece of magnetic material while it is in a magnetic field is called induced magnetism. Permanent magnetism is retained for long periods without appreciable reduction, unless the magnet is subjected to a demagnetizing force. The magnetism in the intermediate iron of a ship which tends to change as the result of vibration, aging, or cruising in the same direction for a long period but does not alter immediately so as to be properly termed induced magnetism is called sub permanent magnetism. Magnetism which remains after removal of the magnetizing force may be called residual magnetism. The magnetism of the earth is called terrestrial magnetism or geomagnetism.

magnetize, *v., t*. To produce magnetic properties. The opposite is DEMAGNETIZE.

magnetometer, *n*. An instrument for measuring the intensity and direction of the earth's magnetic field. See also DECLINOMETER.

magnetron, *n*. An electron tube characterized by the interaction of electrons with the electric field of circuit element in crossed steady electric and magnetic fields to produce an alternating current power output. It is used to generate high power output in the ultra-high and super-high frequency bands.

magnification, *n*. The apparent enlargement of anything.

magnifying power. The ratio of the apparent length of a linear dimension as seen through an optical instrument to that seen by the unaided eye. See POWER.

magnitude, *n*. 1. Relative brightness of a celestial body. The smaller (algebraically) the number indicating magnitude, the brighter the body. The expression first magnitude is often used somewhat loosely to refer to all bodies of magnitude 1.5 or brighter, including negative magnitudes. 2. Amount; size; greatness.

magnitude ratio. The ratio of relative brightness of two celestial bodies differing in magnitude by 1.0. This ratio is 2.512, the 5th root of 100. A body of magnitude 1.0 is 2.512 times as bright as a body of magnitude 2.0, etc.

main beam. See under LOBE.

mainland, *n*. The principal portion of a large land area. The term is used loosely to contrast a principal land mass from outlying islands and sometimes peninsulas.

main light. The principal light of two or more lights situated on the same support or neighboring supports.

main lobe. The lobe of the radiation pattern of a directional antenna which contains the direction of maximum radiation.

major axis. The longest diameter of an ellipse or ellipsoid. Opposite is MINOR AXIS.

major datum. See PREFERRED DATUM.

major light. A light of high intensity and reliability exhibited from a fixed structure or on marine site (except range lights). Major lights include primary seacoast lights and secondary lights. See also MINOR LIGHT.

major planets. See under PLANET.

make the land. To sight and approach or reach land from seaward.

make way. To progress through the water.

making way. Progressing through the water. See also UNDERWAY.

Malvin Current. See FALKLAND CURRENT.

mamma, *n*. Hanging protuberances, like pouches on the under surface of a cloud. This supplementary cloud feature occurs mostly with cirrus, cirrocumulus, altocumulus, altostratus, stratocumulus, and cumulonimbus; in the case of cumulonimbus, mamma generally appear on the under side of the anvil.

mammatus, *n*. See MAMMA.

maneuvering board. A polar coordinate plotting sheet devised to facilitate solution of problems involving relative movement.

Maneuvering Board Manual. See PUB. NO. 217.

man-made noise. In radio reception, noise due entirely to unwanted transmissions from electrical or electronic apparatus, which has been insufficiently suppressed.

manned light. A light which is operated and maintained by full-time resident personnel.

mantissa, *n*. The part of a logarithm (base 10) to the right of the decimal point. The part of a logarithm (base 10) to the left of the decimal point is called the CHARACTERISTIC.

manual, *adj*. By hand, in contrast with AUTOMATIC.

manual radio direction finder. A radio direction finder which requires manual operation of the antenna and determination of the aural null by speaker or headphones.

map, *n*. A representation, usually on a plane surface, of all or part of the surface of the earth, celestial sphere, or other area; showing relative size and position, according to a given projection, of the features represented. Such a representation intended primarily for navigational use is called a chart. A planimetric map indicates only the horizontal positions of features; a topographic map both horizontal and vertical positions. The pattern on the underside of extensive cloud areas, created by the varying amounts of light reflected from the earth's surface, is called a sky map. A chart which shows the distribution of meteorological conditions over an area at a given moment may be called a weather map.

map accuracy standards. See UNITED STATES NATIONAL MAP ACCURACY STANDARDS.

map chart. See COMBAT CHART.

mapping, charting and geodesy. The collection, transformation, generation, dissemination, and storing of geodetic, geomagnetic, gravimetric, aeronautical, topographic, hydrographic, cultural, and toponymic data. These data may be used for military planning, training, and operations including aeronautical, nautical, and land navigation, as well as for weapon orientation and target positioning. Mapping, charting and geodesy (MC&G) also includes the evaluation of topographic, hydrographic, or aeronautical features for their effect on military operations or intelligence. The data may be presented in the form of topographic, planimetric, relief, or thematic maps and graphics; nautical and aeronautical charts and publications, and in simulated, photographic, digital, or computerized formats.

map projection. A systematic drawing of lines on a plane surface to represent the parallels of latitude and the meridians of longitude of the earth or a section of the earth. A map projection may be established by analytical computation or may be constructed geometrically.

map symbol. A character, letter, or similar graphic representation used on a map to indicate some object, characteristic, etc. May be called a CHART SYMBOL when applied to a chart.

March equinox. See VERNAL EQUINOX.

mare's tails. Long, slender, well-defined streaks of cirrus cloud which resemble horse's tails.

marigram, *n*. A graphic record of the rise and fall of the tide. The record is in the form of a curve, in which time is generally represented on the abscissa and the height of the tide on the ordinate.

marina, *n*. A harbor facility for small boats, yachts, etc., where supplies, repairs, and various services are available.

marine, *adj*. Of or pertaining to the sea. See also NAUTICAL.

marine chart. See NAUTICAL CHART.

marine climate. The type of climate characteristic of coastal areas, islands, and the oceans, the distinctive features of which are small annual and daily temperature range and high relative humidity in contrast with CONTINENTAL CLIMATE, which is characteristic of the interior of a large landmass, and the distinctive features of which are large annual and daily temperature range and dry air with few clouds.

marine light. A luminous or lighted aid to navigation intended primarily for marine navigation. One intended primarily for air navigation is called an AERONAUTICAL LIGHT.

marine parade. See MARINE REGATTA.

marine radiobeacon. A radiobeacon whose service is intended primarily for the benefit of ships.

marine railway. A track, a wheeled cradle, and winching mechanism for hauling vessels out of the water so that the bottom can be exposed.

marine regatta. An organized race or other public water event, conducted according to a prearranged schedule, noted in the Local Notice to Mariners. Also called MARINE PARADE.

marine sanctuary. An area established under provisions of the Marine Protection, Research, and Sanctuaries Act of 1972, Public Law 92-532 (86 Stat. 1052), for the preservation and restoration of its conservation, recreational, ecological, or esthetic values. Such an area may lie in ocean waters as far seaward as the outer edge of the continental shelf, in coastal waters where the tide ebbs and flows, or in the Great Lakes and connecting waters, and may be classified as a habitat, species, research, recreational and esthetic, or unique area.

marine sextant. A sextant designed primarily for marine navigation. On a clamp screw sextant the position of the tangent screw is controlled by a clamp screw; on an endless tangent screw sextant the position of the index arm and the vernier or micrometer drum is controlled by an endless tangent screw. A vernier sextant provides a precise reading by means of a vernier used directly with the arc, and may have either a clamp screw or an endless tangent screw for controlling the position of the tangent screw or the index arm. A micrometer drum sextant provides a precise reading by means of a micrometer drum attached to the index arm, and has an endless tangent screw for controlling the position of the index arm. See also SEXTANT.

maritime, *adj*. Bordering on, concerned with, or related to the sea. See also NAUTICAL.

maritime polar air. See under AIR-MASS CLASSIFICATION.

maritime position. The location of a seaport or other point along a coast.

Maritime Safety Information (MSI). Designation of the IHO/IMO referring to navigational information of immediate importance to mariners, affecting the safety of life and/or property at sea.

maritime tropical air. See under AIR-MASS CLASSIFICATION.

mark, *n*. 1. An artificial or natural object of easily recognizable shape or color, or both, situated in such a position that it may be identified on a chart. A fixed artificial navigation mark is often called a BEACON. This may be lighted or unlighted. Also called NAVIGATION MARK; SEAMARK. See also CLEARING MARKS. 2. A major design or redesign of an instrument, denoted by a number. Minor changes are designated MODIFICATIONS. 3. One of the bits of leather, cloth, etc., indicating a specified length of a lead line. 4. An indication intended as a datum or reference, such as a bench mark.

mark, *v., i*. "Now" or "at this moment." A call used when simultaneous observations are being made, to indicate to the second person the moment a reading is to be made, as when the time of a celestial observation is to be noted; or the moment a reading is a prescribed value, as when the heading of a vessel is exactly a desired value.

marker beacon. 1. See MARKER RADIOBEACON. 2. As defined by the International Telecommunication Union (ITU), a transmitter in the aeronautical radionavigation service which radiates vertically a distinctive pattern for providing position information to aircraft.

marker buoy. A small, brightly painted moored float used to temporarily mark a location on the water while placing a buoy on station.

marker radiobeacon. A low powered radiobeacon used primarily to mark a specific location such as the end of a jetty. Usually used primarily for homing bearings. Also called MARKER BEACON.

marl, *n*. A crumbling, earthy deposit, particularly one of clay mixed with sand, lime, decomposed shells, etc. Sometimes a layer of marl becomes quite compact.

Mars, *n*. The navigational planet whose orbit lies between the orbits of the Earth and Jupiter.

marsh, *n*. An area of soft wet land. Flat land periodically flooded by salt water is called a salt marsh. Sometimes called SLOUGH.

mascaret, *n*. See TIDAL BORE.

mass, *n*. The measure of a body's inertia, or the amount of material it contains. This term should not be confused with WEIGHT.

mass attraction vertical. The normal to any surface of constant geopotential. On the earth this vertical is a function only of the distribution of mass and is unaffected by forces resulting from the motions of the earth.

master, *n*. Short for MASTER STATION.

master compass. The main part of a remote-indicating compass system which determines direction for transmission to various repeaters.

master gyrocompass. See under GYROCOMPASS.

master station. In a radionavigation system, the station of a chain which provides a reference by which the emissions of other (slave or secondary) stations are controlled.

masthead light. A fixed running light placed on the centerline of a vessel showing an unbroken white light over an arc of the horizon from dead ahead to 22.5° abaft the beam on either side of the vessel.

Matanuska wind. A strong, gusty, northeast wind which occasionally occurs during the winter in the vicinity of Palmer, Alaska.

maximum ebb. See under EBB CURRENT.

maximum flood. See under FLOOD CURRENT.

maximum thermometer. A thermometer which automatically registers the highest temperature occurring since its last setting. One which registers the lowest temperature is called a MINIMUM THERMOMETER.

mean, *adj*. Occupying a middle position.

mean, *n*. The average of a number of quantities, obtained by adding the values and dividing the sum by the number of quantities involved. Also called AVERAGE, ARITHMETIC MEAN. See also MEDIAN.

mean anomaly. See under ANOMALY, definition 2.

mean diurnal high water inequality. See under DIURNAL INEQUALITY.

mean diurnal low water inequality. See under DIURNAL INEQUALITY.

mean elements. Elements of an adopted reference orbit that approximates the actual, perturbed orbit. Mean elements serve as the basis for calculating perturbations. See also ORBITAL ELEMENTS.

mean higher high water. A tidal datum that is the average of the highest high water height of each tidal day observed over the National Tidal Datum Epoch. For stations with shorter series, simultaneous observational comparisons are made with a control tide station in order to derive the equivalent of a 19-year datum. See also HIGH WATER.

mean higher high water line. The intersection of the land with the water surface at the elevation of mean higher high water.

mean high tide. See under MEAN HIGH WATER.

mean high water. A tidal datum, the average of all the high water heights observed over the National Tidal Datum Epoch. For stations with shorter series, simultaneous observational comparisons are made with a control tide station in order to derive the equivalent of a 19-year datum. See also HIGH WATER.

mean high water line. The intersection of the land with the water surface at the elevation of mean high water. See also SHORELINE.

mean high water lunitidal interval. See under LUNITIDAL INTERVAL. mean high water neaps. See as NEAP HIGH WATER or HIGH WATER NEAPS under NEAP TIDES.

mean high water springs. See under SPRING TIDES.

mean ice edge. The average position of the ice edge in any given month or period based on observations over a number of years. Other terms which may be used are mean maximum ice edge and mean minimum ice edge. See also ICE LIMIT.

mean latitude. Half the arithmetical sum of the latitudes of two places on the same side of the equator. Mean latitude is labeled N or S to indicate whether it is north or south of the equator. The expression is occasionally used with reference to two places on opposite sides of the equator, but this usage is misleading as it lacks the significance usually associated with the expression. When the places are on opposite sides of the equator, two mean latitudes are generally used, the mean of each latitude north and south of the equator. The mean latitude is usually used in middle-latitude sailing for want of a practicable means of determining the middle latitude. See also MIDDLE LATITUDE, MIDDLE-LATITUDE SAILING.

mean lower low water. A tidal datum that is the average of the lowest low water height of each tidal day observed over the National Tidal Datum Epoch. For station with shorter series, simultaneous observational comparisons are made with a control tide station in order to derive the equivalent of a 19-year datum. See also LOW WATER.

mean lower low water line. The intersection of the land with the water surface at the elevation of mean lower low water.

mean low water. A tidal datum that is the average of all the low water heights observed over the National Tidal Datum Epoch. For stations with shorter series, simultaneous observational comparisons are made with a control tide station in order to derive the equivalent of a 19-year datum. See also LOW WATER.

mean low water line. The intersection of the land with the water surface at the elevation of mean low water.

mean low water lunitidal interval. See under LUNITIDAL INTERVAL.

mean low water neaps. See as NEAP LOW WATER or LOW WATER NEAPS under NEAP TIDES.

mean low water springs. 1. A tidal datum that is the arithmetic mean of the low waters occurring at the time of the spring tides observed over a specific 19-year Metonic cycle (the National Tidal Datum Epoch). It is usually derived by taking an elevation depressed below the halftide level by an amount equal to one-half the spring range of tide, necessary corrections being applied to reduce the result to a mean value. This datum is used, to a considerable extent, for hydrographic work outside of the United States and is the level of reference for the Pacific approaches to the Panama Canal. Often shortened to SPRING LOW WATER. See also DATUM. 2. See under SPRING TIDES.

mean motion. In undisturbed elliptic motion, the constant angular speed required for a body of a specified mass to complete one revolution in an orbit of a specified semimajor axis.

mean noon. Twelve o'clock mean time, or the instant the mean sun is over the upper branch of the meridian. Mean noon may be either local or Greenwich depending upon the reference meridian. Zone, standard, daylight saving or summer noon are also forms of mean noon, the mean sun being over the upper branch of the zone, standard, daylight saving or summer reference meridian, respectively.

mean power. See under POWER (OF A RADIO TRANSMITTER).

mean range. The average difference in the extreme values of a variable quantity, as the mean range of tide.

mean range of tide. The difference in height between mean high water and mean low water.

mean rise interval. The average interval between the meridian transit of the moon and the middle of the period of the rise of the tide. It may be computed by adding the half of the duration of rise to the mean low water interval, rejecting the semidiurnal tidal period of 12.42 hours when greater than this amount. The mean rise interval may be either local or Greenwich according to whether it is referred to the local or Greenwich meridian.

mean rise of tide. The height of mean high water above the reference or chart sounding datum.

mean river level. A tidal datum that is the average height of the surface of a tidal river at any point for all stages of the tide observed over a 19-year Metonic cycle (the National Tidal Datum Epoch) usually determined from hourly height readings. In rivers subject to occasional freshets, the river level may undergo wide variations, and for practical purposes certain months of the year may be excluded in the determination of tidal datums. For charting purposes, tidal datums for rivers are usually based on observations during selected periods when the river is at or near low water state. See also DATUM.

mean sea level. A tidal datum that is the arithmetic mean of hourly water elevations observed over a specific 19-year Metonic cycle (the National Tidal Datum Epoch). Shorter series are specified in the name, e.g., monthly mean sea level and yearly mean sea level. See also DATUM; EPOCH, definition 2.

mean sidereal time. See under SIDEREAL TIME.

mean solar day. The duration of one rotation of the earth on its axis, with respect to the mean sun. The length of the mean solar day is 24 hours of mean solar time or $24^h 03^m 56.555^s$ of mean sidereal time. See also CALENDAR DAY.

mean solar time. See MEAN TIME, the term usually used.

mean sun. A fictitious sun conceived to move eastward along the celestial equator at a rate that provides a uniform measure of time equal to the average apparent time. It is used as a reference for reckoning mean time, zone time, etc. Also called ASTRONOMICAL MEAN SUN. See also DYNAMICAL MEAN SUN.

mean tide level. See HALF-TIDE LEVEL.

mean time. Time based upon the rotation of the earth relative to the mean sun. Mean time may be designated as local or Greenwich as the local or Greenwich meridian is the reference. Greenwich mean time is also called UNIVERSAL TIME. Zone, standard, daylight saving or summer time are also variations of mean time, specified meridians being used as the reference. See also EQUATION OF TIME, MEAN SIDEREAL TIME.

mean tropic range. The mean between the great tropic tidal range and the small tropic range. The small tropic range and the mean tropic range are applicable only when the type of tide is semidiurnal or mixed. See also GREAT TROPIC RANGE.

mean water level. The mean surface elevation as determined by averaging the heights of the water at equal intervals of time, usually hourly.

mean water level line. The line formed by the intersection of the land with the water surface at an elevation of mean water level.

measured mile. A length of 1 nautical mile, the limits of which have been accurately measured and are indicated by ranges ashore. It is used by vessels to calibrate logs, engine revolution counters, etc., and to determine speed.

measured-mile buoy. A buoy marking the end of a measured mile.

mechanical scanning. Scanning effected by moving all or part of the antenna.

median, *n*. A value in a group of quantities below and above which fall an equal number of quantities. Of the group 60, 75, 80, 95, and 100, the median is 80. If there is no middle quantity in the group, the median is the value interpolated between the two middle quantities. The median of the group 6, 10, 20, and 31 is 15. See also MEAN.

median valley. The axial depression of the midoceanic ridge system.

medium. A method of electronic data storage and physical transfer, commonly relying on the properties of electromagnetic coatings on tape, disks, or other surfaces, or on the effects of laser light on light-sensitive surfaces.

medium first-year ice. First-year ice 70 to 120 centimeters thick.

medium floe. See under FLOE.

medium fracture. See under FRACTURE.

medium frequency. Radio frequency of 300 to 3,000 kilohertz.

medium iceberg. For reports to the International Ice Patrol, an iceberg that extends 51 to 150 feet (16 to 45 meters) above the sea surface and which has a length of 201 to 400 feet (61 to 122 meters). See also SMALL ICEBERG, LARGE ICEBERG.

medium ice field. See under ICE FIELD.

medium range systems. Those radionavigation systems providing positioning capability beyond the range of short range systems, but their use is generally limited to ranges permitting reliable positioning for about 1 day prior to making landfall; Decca is an example.

mega-. A prefix meaning one million (10^6).

megabyte. One million bytes of information in a computer.

megacycle, *n*. One million cycles; one thousand kilocycles. The term is often used as the equivalent of one million cycles per second.

megahertz, *n*. One million hertz or one million cycles per second.

megaripple, *n*. See SAND WAVE.

meniscus, *n*. 1. The curved upper surface of a liquid in a tube. 2. A type of lens.

mensuration, *n*. 1. The act, process, or art of measuring. 2. That branch of mathematics dealing with determination of length, area, or volume.

Mentor Current. Originating mainly from the easternmost extension of the South Pacific Current at about latitude 40°S, longitude 90°W, the Mentor Current flows first northward and then northwestward. It has the characteristic features of a WIND DRIFT in that it is a broad, slow-moving flow that extends about 900 miles westward from the Peru Current to about longitude 90°W at its widest section and tends to be easily influenced by winds. It joins the westward flowing Pacific South Equatorial Current and forms the eastern part of the general counterclockwise oceanic circulation of the South Pacific Ocean. The speed in the central part of the current at about latitude 26°S, longitude 80°W, may at times reach about 0.9 knot. Also called PERU OCEANIC CURRENT.

Mercator bearing. See RHUMB BEARING.

Mercator chart. A chart on the Mercator projection. This is the chart commonly used for marine navigation. Also called EQUATORIAL CYLINDRICAL ORTHOMORPHIC CHART.

Mercator course. See RHUMB-LINE COURSE.

Mercator direction. Horizontal direction of a rhumb line, expressed as angular distance from a reference direction. Also called RHUMB DIRECTION.

Mercator map projection. A conformal cylindrical map projection in which the surface of a sphere or spheroid, such as the earth, is developed on a cylinder tangent along the equator. Meridians appear as equally spaced vertical lines and parallels as horizontal lines drawn farther apart as the latitude increases, such that the correct relationship between latitude and longitude scales at any point is maintained. The expansion at any point is equal to the secant of the latitude of that point, with a small correction for the ellipticity of the earth. The Mercator is not a perspective projection. Since rhumb lines appear as straight lines and directions can be measured directly, this projection is widely used in navigation. If the cylinder is tangent along a meridian. a transverse Mercator map projection results; if the cylinder is tangent along an oblique great circle, an oblique Mercator map projection results. Also called EQUATORIAL CYLINDRICAL ORTHOMORPHIC MAP PROJECTION.

Mercator sailing. A method of solving the various problems involving course, distance, difference of latitude, difference of longitude, and departure by considering them in the relation in which they are plotted on a Mercator chart. It is similar to plane sailing, but uses meridional difference and difference of longitude in place of difference of latitude and departure, respectively.

mercurial barometer. An instrument which determines atmospheric pressure by measuring the height of a column of mercury which the atmosphere will support. See also ANEROID BAROMETER.

mercury ballistic. A system of reservoirs and connecting tubes containing mercury used with a type of non-pendulous gyrocompass. The action of gravity on this system provides the torques and resultant precessions required to convert the gyroscope into a compass.

meridian, *n*. A north-south reference line, particularly a great circle through the geographical poles of the earth. The term usually refers to the upper branch, the half, from pole to pole, which passes through a given place; the other half being called the lower branch. An astronomical (terrestrial) meridian is a line connecting points having the same astronomical longitude. A geodetic meridian is a line connecting points of equal geodetic longitude. Geodetic and sometime astronomical meridians are also called geographic meridians. Geodetic meridians are shown on charts. The prime meridian passes through longitude 0°. Sometimes designated TRUE MERIDIAN to distinguish it from magnetic meridian, compass meridian, or grid meridian, the north-south lines relative to magnetic, compass, or grid direction, respectively. A fictitious meridian is one of a series of great circles or lines used in place of a meridian for certain purposes. A transverse or inverse meridian is a great circle perpendicular to a transverse equator. An oblique meridian is a great circle perpendicular to an oblique equator. Any meridian used as a reference for reckoning time is called a time meridian. The meridian used for reckoning standard zone, daylight saving, or war time is called standard, zone, daylight saving, or war meridian respectively. The meridian through any particular place or observer, serving as the reference for local time, is called local meridian, in contrast with the Greenwich meridian, the reference for Greenwich time. A celestial meridian is a great circle of the celestial sphere, through the celestial poles and the zenith. Also called CIRCLE OF LATITUDE. See also ANTE MERIDIAN, POST MERIDIAN.

meridian altitude. The altitude of a celestial body when it is on the celestial meridian of the observer, bearing 000° or 180° true.

meridian angle. Angular distance east or west of the local celestial meridian; the arc of the celestial equator, or the angle at the celestial pole, between the upper branch of the local celestial meridian and the hour circle of a celestial body measured eastward or westward from the local celestial meridian through 180°, and labeled E or W to indicate the direction of measurement. See also HOUR ANGLE.

meridian angle difference. The difference between two meridian angles, particularly between the meridian angle of a celestial body and the value used as an argument for entering a table. Also called HOUR ANGLE DIFFERENCE.

meridian observation. Measurement of the altitude of a celestial body on the celestial meridian of the observer, or the altitude so measured.

meridian passage. See MERIDIAN TRANSIT.

meridian sailing. Following a true course of 000° or 180°, sailing along a meridian. Under these conditions the dead reckoning latitude is assumed to change 1 minute for each mile run and the dead reckoning longitude remains unchanged.

meridian transit. The passage of a celestial body across a celestial meridian. Upper transit, the crossing of the upper branch of the celestial meridian, is understood unless lower transit, the crossing of the lower branch, is specified. Also called TRANSIT, MERIDIAN PASSAGE, CULMINATION.

meridional difference. The difference between the meridional parts of any two given parallels. This difference is found by subtraction if the two parallels are on the same side of the equator and by addition if on opposite sides. Also called DIFFERENCE OF MERIDIONAL PARTS.

meridional parts. The length of the arc of a meridian between the equator and a given parallel on a Mercator chart, expressed in units of 1 minute of longitude at the equator.

metacenter, *n*. For small angles of inclination of a ship, the instantaneous center of a very small increment of the curved path of the center of buoyancy locus. Or, for small angles of inclination, the point of intersection of the lines of action of the buoyant force and the original vertical through the center of buoyancy.

meteor, *n*. The phenomenon occurring when a solid particle from space enters the earth's atmosphere and is heated to incandescence by friction of the air. A meteor whose brightness does not exceed that of Venus (magnitude -4) is popularly called SHOOTING STAR or FALLING STAR. A shooting star results from the entrance into the atmosphere of a particle having a diameter between a few centimeters and just visible to the naked eye. Shooting stars are observed first as a light source, similar to a star, which suddenly appears in the sky and moves along a long or short path to a point where it just as suddenly disappears. The brighter shooting stars may leave a trail which remains luminous for a short time. Meteors brighter than magnitude -4 are called BOLIDES or FIREBALLS. Light bursts, spark showers, or splitting of the trail are sometimes seen along their luminous trails which persist for minutes and for an hour in exceptional cases. The intensity of any meteor is dependent upon the size of the particle which enters the atmosphere. A particle 10 centimeters in diameter can produce a bolide as bright as the full moon. See also METEORITE.

meteorite, *n*. 1. The solid particle which causes the phenomenon known as a METEOR. 2. The remnant of the solid particle, causing the meteor, which reaches the earth.

meteorological optical range. The length of path in the atmosphere required to reduce the luminous flux in a collimated beam from an incandescent lamp at a color temperature of 2,700°K to 0.05 of its original value, the luminous flux being evaluated by means of the curve of spectral luminous efficiencies for photopic vision given by the International Commission on Illumination. The quantity so defined corresponds approximately to the distance in the atmosphere required to reduce the contrast of an object against its background to 5 percent of the value it would have at zero distance, for daytime observation. See also METEOROLOGICAL VISIBILITY.

Meteorological Optical Range Table. A table from the International Visibility Code which gives the code number of meteorological visibility and the meteorological visibility for several weather conditions.

meteorological tide. A change in water level caused by local meteorological conditions, in contrast to an ASTRONOMICAL TIDE, caused by the attractions of the sun and moon. See also SEICHE, STORM SURGE.

meteorological tides. Tidal constituents having origin in the daily or seasonal variations in weather conditions which may occur with some degree of periodicity. See also STORM SURGE.

meteorological visibility. The greatest distance at which a black object of suitable dimensions can be seen and recognized by day against the horizon sky, or, in the case of night observations, could be seen and recognized if the general illumination were raised to the normal daylight level. It has been established that the object may be seen and recognized if the contrast threshold is 0.05 or higher. The term may express the visibility in a single direction or the prevailing visibility in all directions. See also VISIBILITY, METEOROLOGICAL OPTICAL RANGE, CONTRAST THRESHOLD.

meteor swarm. The scattered remains of comets that have broken up.

meter, *n*. 1. The base unit of length in the International System of Units, equal to 1,650,763.73 wavelengths in vacuum of the radiation corresponding to the transition between the levels $2p_{10}$ and $5p_5$ of the krypton-86 atom. It is equal to 39.37008 inches, approximately, or approximately one ten-millionth of the distance from the equator to the North or South Pole. The old international prototype of the meter is still kept at the International Bureau of Weights and Measures under the conditions specified in 1889. 2. A device for measuring, and usually indicating, some quantity.

method of bisectors. As applied to celestial lines of position, the movement of each of three or four intersecting lines of position an equal amount, in the same direction toward or away from the celestial bodies, so as to bring them as nearly as possible to a common intersection. When there are more than four lines of position, the lines of position in the same general direction are combined to reduce the data to not more than four lines of position. See also OUTSIDE FIX.

Metonic cycle. A period of 19 years or 235 lunations, devised by Meton, an Athenian astronomer who lived in the fifth century B.C., for the purpose of obtaining a period in which new and full moon would recur on the same day of the year. Taking the Julian year of 365.25 days and the synodic month as 29.53058 days, we have the 19-year period of 6939.75 days as compared with the 235 lunations of 6939.69 days, a difference of only 0.06 days. See also CALLIPPIC CYCLE.

meter per second. The derived unit of speed in the International System of Units.

meter per second squared. The derived unit of acceleration in the International System of Units.

metric system. A decimal system of weights and measures based on the meter as the unit of length and the kilogram as a unit mass. See also INTERNATIONAL SYSTEM OF UNITS.

Mexico Current. From late October through April an extension of the California Current, known as the Mexico Current, flows southeastward along the coast to the vicinity of longitude 95°W where it usually turns west, but at times extends southward as far as Honduras with speeds from 0.5 to 1 knot. During the remainder of the year, this current flows northwestward along the Mexican coast as far as Cabo Corrientes, where it turns westward and becomes a part of the Pacific North Equatorial Current.

micro-. A prefix meaning one-millionth (10^{-6}).

micrometer, *n*. An auxiliary device to provide measurement of very small angles or dimensions by an instrument such as a telescope.

micrometer drum. A cylinder carrying an auxiliary scale and sometimes a vernier, for precise measurement, as in certain type sextants.

micrometer drum sextant. A marine sextant providing a precise reading by means of a micrometer drum attached to the index arm, and having an endless tangent screw for controlling the position of the index arm. The micrometer drum may include a vernier to enable a more precise reading. On a vernier sextant the vernier is directly on the arc.

micron, *n*. A unit of length equal to one-millionth of a meter.

microprocessor. An integrated circuit in a computer which executes machine-language instructions.

microsecond, *n*. One-millionth of a second.

microwave, *n*. A very short electromagnetic wave, usually considered to be about 30 centimeters to 1 millimeter in length. While the limits are not clearly defined, it is generally considered as the wavelength of radar operation.

microwave frequency. Radio frequency of 1,000 to 300,000 megahertz, having wavelengths of 30 centimeters to 1 millimeter.

mid-channel buoy. See FAIRWAY BUOY.

mid-channel mark. A navigation mark serving to indicate the middle of a channel, which can be passed on either side safely.

middle clouds. Types of clouds the mean level of which is between 6,500 and 20,000 feet. The principal clouds in this group are altocumulus and altostratus.

middle ground. A shoal in a fairway having a channel on either side.

middle ground buoy. One of the buoys placed at each end of a middle ground. See BIFURCATION BUOY, JUNCTION BUOY.

middle latitude. The latitude at which the arc length of the parallel separating the meridians passing through two specific points is exactly equal to the departure in proceeding from one point to the other by middle-latitude sailing. Also called MID-LATITUDE. See also MEAN LATITUDE, MIDDLE-LATITUDE SAILING.

middle-latitude sailing. A method that combines plane sailing and parallel sailing. Plane sailing is used to find difference of latitude and departure when course and distance are known, or vice versa. Parallel sailing is used to inter-convert departure and difference of longitude. The mean latitude is normally used for want of a practicable means of determining the middle latitude, the latitude at which the arc length of the parallel separating the meridians passing through two specific points is exactly equal to the departure in proceeding from one point to the other. See also MEAN LATITUDE.

mid-extreme tide. An elevation midway between the extreme high water and the extreme low water occurring in any locality. See also HALFTIDE LEVEL.

mid-latitude. See MIDDLE LATITUDE.

midnight, *n*. Twelve hours from noon, or the instant the time reference crosses the lower branch of the reference celestial meridian.

midnight sun. The sun when it is visible at midnight. This occurs during the summer in high latitudes, poleward of the circle at which the latitude is approximately equal to the polar distance of the sun.

mill, *n*. 1. A unit of angular measurement equal to an angle having a tangent of 0.001. 2. A unit of angular measurement equal to an angle subtended by an arc equal to 1/6,400th part of the circumference of a circle.

mile, *n*. A unit of distance. The nautical mile, or sea mile, is used primarily in navigation. Nearly all maritime nations have adopted the International Nautical Mile of 1,852 meters proposed in 1929 by the International Hydrographic Bureau. The U.S. Departments of Defense and Commerce adopted this value on July 1, 1954. Using the yard-meter conversion factor effective July 1, 1959, (1 yard = 0.9144 meter, exactly) the International Nautical Mile is equivalent to 6076.11549 feet, approximately. The geographical mile is the length of 1 minute of arc of the equator considered to be 6,087.08 feet. The U.S. Survey mile or land mile (5,280 feet in the United States) is commonly used for navigation on rivers and lakes, notably the Great Lakes of North America. See also CABLE, MEASURED MILE.

mileage number. A number assigned to aids to navigation which gives the distance in sailing miles along the river from a reference point to the aid. The number is used principally in the Mississippi and other river systems.

miles of relative movement. The distance, in miles, traveled relative to a reference point which is usually in motion.

military grid. Two sets of parallel lines intersecting at right angles and forming squares; the grid is superimposed on maps, charts, and other similar representations of the earth's surface in an accurate and consistent manner to permit identification of ground locations with respect to other locations and the computation of direction and distance to other points. See also MILITARY GRID REFERENCE SYSTEM, UNIVERSAL POLAR STEREOGRAPHIC GRID, UNIVERSAL TRANSVERSE MERCATOR GRID, WORLD GEOGRAPHIC REFERENCE SYSTEM.

military grid reference system. A system which uses a standard-scaled grid square, based on a point of origin on a map projection of the earth's surface in an accurate and consistent manner to permit either position referencing or the computation of direction and distance between grid positions. See also MILITARY GRID.

Milky Way. The galaxy of which the sun and its family of planets are a part. It appears as an irregular band of misty light across the sky. Through a telescope, it is seen to be composed of numerous individual stars. See also COALSACK.

milli-. A prefix meaning one-thousandth.

millibar, *n*. A unit of pressure equal to 1,000 dynes per square centimeter, or 1/1,000th of a bar. The millibar is used as a unit of measure of atmospheric pressure, a standard atmosphere being equal to 1,013.25 millibars or 29.92 inches of mercury.

milligal, *n*. A unit of acceleration equal to 1/1,000th of a gal, or 1/1,000 centimeter per second per second. This unit is used in gravity measurements, being approximately one-millionth of the average gravity at the earth's surface.

millimeter, *n*. One thousandth of a meter- one tenth of a centimeter;.03937008 inch.

millisecond, *n*. One-thousandth of a second.

minaret, *n*. A tall, slender tower attached to a mosque and surrounded by one or more projecting balconies; frequently charted as landmarks.

minimal depiction of detail. A term used to indicate the extreme case of generalization of detail on a chart. In the extreme case most features are omitted even through there is space to show at least some of them. The practice is most frequently used for semi-enclosed areas such as estuaries and harbors on smaller-scale charts, where use of a larger scale chart is essential.

minimum distance (of a navigational system). The minimum distance at which a navigational system will function within its prescribed tolerances.

minimum ebb. See under EBB CURRENT.

minimum flood. See under FLOOD CURRENT.

minimum signal. The smallest signal capable of satisfactorily operating an equipment, e.g., the smallest signal capable of triggering a racon.

minimum thermometer. A thermometer which automatically registers the lowest temperature occurring since its last setting. One which registers the highest temperature is called a MAXIMUM THERMOMETER.

minor axis. The shortest diameter of an ellipse or ellipsoid.

minor light. An automatic unmanned light on a fixed structure usually showing low to moderate intensity. Minor lights are established in harbors, along channels, along rivers, and in isolated locations. See also MAJOR LIGHT.

minor planets. See under PLANET.

minute, *n*. 1. The sixtieth part of a degree of arc. 2. The sixtieth part of an hour.

mirage, *n*. An optical phenomenon in which objects appear distorted, displaced (raised or lowered), magnified, multiplied, or inverted due to varying atmospheric refraction when a layer of air near the earth's surface differs greatly in density from surrounding air. See also TOWERING, STOOPING, LOOMING, SINKING, FATA MORGANA.

mirror reelection. See SPECULAR REFLECTION.

missing, *adj*. Said of a floating aid to navigation which is not on station with its whereabouts unknown.

mist, *n*. An aggregate of very small water droplets suspended in the atmosphere. It produces a thin, grayish veil over the landscape. It reduces visibility to a lesser extent than fog. The relative humidity with mist is often less than 95 percent. Mist is intermediate in all respects between haze (particularly damp haze) and fog. See also DRIZZLE.

mistake, *n*. The result of carelessness or of a mistake. For the purpose of error analysis, a mistake is not classified as an error. Also called BLUNDER.

mistral, *n*. A cold, dry wind blowing from the north over the northwest coast of the Mediterranean Sea, particularly over the Gulf of Lions. Also called CIERZO. See also FALL WIND.

mixed current. Type of tidal current characterized by a conspicuous speed difference between the two floods and/or ebbs usually occurring each tidal day. See also TYPE OF TIDE.

mixed tide. Type of tide with a large inequality in either the high and/or low water heights, with two high waters and two low waters usually occurring each tidal day. All tides are mixed, but the name is usually applied to the tides intermediate to those predominantly semidiurnal and those predominantly diurnal. See also TYPE OF TIDE.

moat, *n*. An annular depression that may not be continuous, located at the base of many sea mounts, islands, and other isolated elevations of the sea floor, analogous to the moat around a castle.

mobile service. As defined by the International Telecommunication Union (ITU), a service of radiocommunication between mobile and land stations, or between mobile stations.

mobile offshore drilling unit (MODU). A movable drilling platform used in offshore oil exploration and production. It is kept stationary by vertically movable legs or by mooring with several anchors. After drilling for oil it may be replaced by a production platform or a submerged structure.

mock fog. A rare simulation of true fog by anomalous atmospheric refraction.

mock moon. See PARASALENE.

mock sun. See PARHELION.

mock-sun ring. See PARHELIC CIRCLE.

modal interference. Omega signals propagate in the earth-ionosphere wave guide. This waveguide can support many different electromagnetic field configurations, each of which can be regarded as an identifiable signal component or mode having the same signal frequency, but with slightly different phase velocity. Modal interference is a special form of signal interference wherein two or more waveguide modes interfere with each other and irregularities appear in the phase pattern. This type of interference occurs predominantly under nighttime conditions when most of the propagation path is not illuminated and the boundary conditions of the waveguide are unstable. It is most severe for signals originating at stations located close to the geomagnetic equator. During all daylight path conditions, the only region of modal interference is a more-less circular area of radius 500-1000 kilometers immediately surrounding a transmitting station.

model atmosphere. Any theoretical representation of the atmosphere, particularly of vertical temperature distribution. See also STANDARD ATMOSPHERE.

modem. An electronic device which converts digital information to analog signals and vice-versa, used in computer file transfer over telephone lines; derived from MOdulator-DEModulator.

moderate breeze. Wind of force 4 (11 to 16 knots or 13 to 18 miles per hour) on the Beaufort wind scale.

moderate gale. A term once used by seamen for what is now called NEAR GALE on the Beaufort wind scale.

modification, *n*. An instrument design resulting from a minor change, and indicated by number. A design resulting from a major change is called a MARK.

modified Julian day. An abbreviated form of the Julian day which requires fewer digits and translates the beginning of each day from Greenwich noon to Greenwich midnight; obtained by subtracting 2400000.5 from Julian days.

modified Lambert conformal chart. A chart on the modified Lambert conformal map projection. Also called NEY'S CHART.

modified Lambert conformal map projection. A modification of the Lambert conformal projection for use in polar regions, one of the standard parallels being at latitude 89°59'58" and the other at latitude 71° or 74°, and the parallels being expanded slightly to form complete concentric circles. Also called NEY'S MAP PROJECTION.

modified refractive index. For a given height above sea level, the sum of the refractive index of the air at this height and the ratio of the height to the radius of the earth.

modulated wave. A wave which varies in some characteristic in accordance with the variations of a modulating wave. See also CONTINUOUS WAVE.

modulating wave. A wave which modulates a carrier wave.

modulation, *n*. A variation of some characteristic of a radio wave, called the CARRIER WAVE in accordance with instantaneous values of another wave called the MODULATING WAVE. These variations can be amplitude, frequency, phase, or pulse.

modulator, *n*. The component in pulse radar which generates a succession of short pulses of energy which in turn cause a transmitter tube to oscillate during each pulse.

mole, *n*. 1. A structure, usually massive, on the seaward side of a harbor for its protection against current and wave action, drift ice, wind, etc. Sometimes it may be suitable for the berthing of ships. See also JETTY, definition 1; QUAY. 2. The base unit of amount of substance in the International System of Units; it is the amount of substance of a system which contains as many elementary entities as there are atoms in 0.012 kilogram of carbon atom 12. When the mole is used, the elementary entities must be specified and may be atoms, molecules, ions, electrons, other particles, or specified groups of such particles.

moment, *n*. The tendency or degree of tendency to produce motion about an axis. Numerically it is the quantity obtained by multiplying the force, speed, or mass by the distance from the point of application or center of gravity to the axis. See also MAGNETIC MOMENT.

moment of inertia. The quantity obtained by multiplying the mass of each small part of a body by the square of its distance from an axis, and adding all the results.

momentum, *n*. The quantity of motion. Linear momentum is the quantity obtained by multiplying the mass of a body by its linear speed. Angular momentum is the quantity obtained by multiplying the moment of inertia of a body by its angular speed.

monitor, *v. t.* In radionavigation, to receive the signals of a system in order to check its operation and performance.

monitor, *n*. The video display portion of a computer system.

monitoring, *n*. In radionavigation, the checking of the operation and performance of a system through reception of its signals.

monsoon, *n*. A name for seasonal winds first applied to the winds over the Arabian Sea, which blow for 6 months from the northeast (northeast monsoon) and for 6 months from the southwest (southwest monsoon). The primary cause is the much greater annual variation of temperature over large land areas compared with the neighboring ocean surfaces, causing an excess of pressure over the continents in winter and a deficit in summer, but other factors such as the relief features of the land have a considerable effect. In India the term is popularly applied chiefly to the southwest monsoon and by extension, to the rain which it brings.

monsoon current. A seasonal wind-driven current occurring in the northern part of the Indian Ocean and the northwest Pacific Ocean. See also MONSOON DRIFT.

Monsoon Drift. A drift current of the northeast Indian Ocean located north of the Indian Equatorial Countercurrent and south of the Bay of Bengal. During February and March when the northeast monsoon decreases in intensity, the monsoon drift is formed from the outflow of the Strait of Malacca and a small amount of northwestward flow along the upper southwest coast of Sumatra. Off the southwest coast of Sumatra, a current generally sets southeast during all months. It is strongest during October through April. The monsoon drift broadens as it flows westward and divides off the east coast of Sri Lanka, part joining the circulation of the Bay of Bengal and part joining the flow from the Arabian Sea. During April, the transition period between monsoons, the monsoon drift is ill-defined. A counterclockwise circulation exists between Sumatra and Sri Lanka. During May through October, the monsoon drift flows east to southeast. During November and December part of the monsoon drift is deflected into the Bay of Bengal and the remainder turns clockwise and flows southeastward. See also MONSOON.

monsoon fog. An advection fog occurring as a monsoon circulation transports warm moist air over a colder surface.

month, *n*. 1. The period of the revolution of the moon around the earth. The month is designated as sidereal, tropical, anomalistic, nodical or synodical, according to whether the revolution is relative to the stars, the vernal equinox, the perigee, the ascending node, or the sun. 2. The calendar month, which is a rough approximation to the synodical month.

month of the phases. See SYNODICAL MONTH.

moon, *n*. The astronomical satellite of the earth.

moonbow, *n*. A rainbow formed by light from the moon. Colors in a moonbow are usually very difficult to detect. Also called LUNAR RAINBOW.

moon dog. See PARASELENE.

moonrise, *n*. The crossing of the visible horizon by the upper limb of the ascending moon.

moonset, *n*. The crossing of the visible horizon by the upper limb of the descending moon.

moor, *v., t.* To secure a vessel to land by tying to a pier, wharf or other land-based structure, or to anchor with two or more anchors.

mooring, *n*. 1. The act of securing a craft to the ground, a wharf, pier, quay, etc., other than anchoring with a single anchor. 2. The place where a craft may be moored. 3. Chains, bridles, anchors, etc. used in securing a craft to the ground.

mooring buoy. A buoy secured to the bottom by permanent moorings and provided with means for mooring a vessel by use of its anchor chain or mooring lines.

morning star. The brightest planet appearing in the eastern sky during morning twilight.

morning twilight. The period of time between darkness and sunrise.

Morse code light. A navigation light which flashes one or more characters in Morse code.

motion, *n*. The act, process, or instance of change of position. Absolute motion is motion relative to a fixed point. Actual motion is motion of an object relative to the earth. Apparent or relative motion is change of position as observed from a reference point which may itself be in motion. Diurnal motion is the apparent daily motion of a celestial body. Direct motion is the apparent motion of a planet eastward among the stars; retrograde motion, the apparent motion westward among the stars. Motion of a celestial body through space is called space motion, which is composed of two components: proper motion, that component perpendicular to the line of sight; and radial motion, that component in the direction of the line of sight. Also called MOVEMENT, especially when used in connection with problems involving the motion of one vessel relative to another.

mound, *n*. On the sea floor, a low, isolated, rounded hill.

mountain breeze. A breeze that blows down a mountain slope due to the gravitational flow of cooled air. See also KATABATIC WIND, VALLEY BREEZE.

mountains, *n., pl.* On the sea floor, a well delineated subdivision of a large and complex positive feature, generally part of a cordillera.

movement, *n*. See MOTION.

moving havens. Moving restricted areas established to prevent mutual interference of Naval vessels in transit.

moving target indication. A radar presentation in which stationary targets are wholly or partially suppressed.

Mozambique Current. The part of the Indian South Equatorial Current that turns and flows along the African coast in the Mozambique Channel. It is considered part of the AGULHAS CURRENT.

mud, *n*. A general term applied to mixtures of sediments in water. Where the grains are less than 0.002 millimeter in diameter, the mixture is called clay. Where the grains are between 0.002 and 0.0625 millimeter in diameter, the mixture is called silt. See also SAND; STONES; ROCK, definition 2.

mud berth. A berth where a vessel rests on the bottom at low water.

mud flat. A tidal flat composed of mud.

mud pilot. A person who pilots a vessel by visually observing changes in the color of the water as the depth of the water increases or decreases.

multihop transmission. See MULTIPLE-HOP TRANSMISSION.

multipath error. Interference between radio waves which have traveled between the transmitter and the receiver by two paths of different lengths, which may cause fading or phase changes at the receiving point due to the vector addition of the signals, making it difficult to obtain accurate information.

multipath propagation. Radio propagation from the transmitter to the receiver by two or more paths simultaneously. Also called MULTIPATH TRANSMISSION.

multipath transmission. See MULTIPATH PROPAGATION.

multiple echoes. Radar echoes which may occur when a strong echo is received from another ship at close range. A second or third or more echoes may be observed on the radarscope at double triple, or other multiples of the actual range of the radar target, resulting from the echo's being reflected by own ship back to the target and received once again as an echo at a multiple of the preceding range to the target. This term should not be confused with MULTIPLE-TRACE ECHO. See also SECOND-TRACE ECHO.

multiple-hop transmission. Radio wave transmission in which the waves traveling between transmitter and receiver undergo multiple reflections and refractions between the earth and ionosphere. Also called MULTIHOP TRANSMISSION.

multiple ranges. A group of two ranges, having one of the range marks (either front or rear) in common.

multiple star. A group of three or more stars so close together that they appear as a single star, whether through physical closeness or as a result of lying in approximately the same direction. See also STAR CLUSTER.

multiple tide staff. A succession of tide staffs on a sloping shore so placed that the vertical graduations on the several staffs will form a continuous scale referred to the same datum.

multiple-trace echo. See SECOND-TRACE ECHO.

multi-year ice. Old ice up to 3 meters or more thick which has survived at least two summer's melt. Hummocks are even smoother than in second-year ice. The ice is almost salt-free. The color, where bare, is usually blue. The melt pattern consists of large interconnecting irregular puddles and a well-developed drainage system.

Mumetal, *n*. The registered trade name for an alloy of about 75% nickel and 25% iron, having high magnetic permeability and low hysteresis.

N

nadir, *n*. The point on the celestial sphere vertically below the observer, or 180° from the zenith.

name, *n*. The label of a numerical value, used particularly to refer to the N (north) or S (south) label of latitude and declination. When latitude and declination are both N or both S, they are said to be of same name, but if one is N and the other S, they are said to be of contrary name.

nano-. A prefix meaning one-billionth (10^{-9}).

nanosecond, *n*. One-billionth of a second.

Napier diagram. A diagram on which compass deviation is plotted for various headings, and the points connected by a smooth curve, permitting deviation problems to be solved quickly without interpolation. It consists of a vertical line, usually in two parts, each part being graduated for 180° of heading, and two additional sets of lines at an angle of 60° to each other and to the vertical lines. See also DEVIATION TABLE.

Napierian logarithm. A logarithm to the base e (2.7182818). Also called NATURAL LOGARITHM. See also COMMON LOGARITHM.

narrows, *n*. A navigable narrow part of a bay, strait, river, etc.

nashi, n'aschi, *n*. A northeast wind which occurs in winter on the Iranian coast of the Persian Gulf, especially near the entrance to the gulf, and also on the Makran coast. It is probably associated with an outflow from the central Asiatic anticyclone which extends over the high land of Iran. It is similar in character but less severe than the BORA.

National Geodetic Vertical Datum. A fixed reference once adopted as a standard geodetic datum for heights in the United States. The geodetic datum now in use in the United States is the North American Vertical Datum of 1988. The geodetic datum is fixed and does not take into account the changing stands of sea level. Because there are many variables affecting sea level, and because the geodetic datum represents a best fit over a broad area, the relationship between the geodetic datum and local mean sea level is not consistent from one location to another in either time or space. For this reason the National Geodetic Vertical Datum should not be confused with MEAN SEA LEVEL.

National Tidal Datum Epoch. The specific 19-year cycle adopted by the National Ocean Survey as the official time segment over which tide observations are taken and reduced to obtain mean values (e.g., mean lower low water, etc.) for tidal datums. It is necessary for standardization because of apparent periodic and apparent secular trends in sea level. The present National Tidal Datum Epoch is 1960 through 1978.

National Water Level Observation Network. (National Tidal Datum Control Network). A network composed of the primary control tide stations of the National Ocean Service. This network of coastal observation stations provides the basic tidal datums for coastal boundaries and chart datums of the United States. Tidal datums obtained at secondary control tide stations and tertiary tide stations are referenced to the Network.

natural, *adj*. 1. Occurring in nature; not artificial. 2. Not logarithmic-used with the name of a trigonometric function to distinguish it from its logarithm (called LOGARITHMIC TRIGONOMETRIC FUNCTION).

natural frequency. The lowest resonant frequency of a body or system.

natural harbor. A harbor where the configuration of the coast provides the necessary protection See also ARTIFICIAL HARBOR.

natural logarithm. See NAPIERIAN LOGARITHM.

natural magnet. A magnet occurring m nature, as contrasted with an ARTIFICIAL MAGNET, produced by artificial means.

natural period. The period of the natural frequency of a body or system.

natural range. A range formed by natural objects such as rocks, peaks, etc. See also ARTIFICIAL RANGE.

natural scale. See REPRESENTATIVE FRACTION.

natural tangent. See under TANGENT, definition 1.

natural trigonometric function. See under TRIGONOMETRIC FUNCTIONS.

natural year. See TROPICAL YEAR.

nature of the bottom. See BOTTOM CHARACTERISTICS.

nautical, *adj*. Of or pertaining to ships, marine navigation, or seamen.

nautical almanac. 1. A periodical publication of astronomical data designed primarily for marine navigation. Such a publication designed primarily for air navigation is called an AIR ALMANAC. 2. *Nautical Almanac*; a joint annual publication of the U.S. Naval Observatory and the Nautical Almanac Office, Royal Greenwich Observatory listing the Greenwich hour angle and declination of various celestial bodies to a precision of 0.1' at hourly intervals; time of sunrise, sunset, moon rise, moonset; and other astronomical information useful to navigators.

nautical astronomy. See NAVIGATIONAL ASTRONOMY.

nautical chart. A representation of a portion of the navigable waters of the earth and adjacent coastal areas on a specified map projection, designed specifically to meet requirements of marine navigation.

nautical day. Until January 1, 1925, a day that began at noon, 12 hours earlier than the calendar day, or 24 hours earlier than the astronomical day of the same date.

nautical mile. A unit of distance used principally in navigation. For practical consideration it is usually considered the length of 1 minute of any great circle of the earth, the meridian being the great circle most commonly used. Because of various lengths of the nautical mile in use throughout the world, due to differences in definition and the assumed size and shape of the earth, the International Hydrographic Bureau in 1929 proposed a standard length of 1,852 meters, which is known as the International Nautical Mile. This has been adopted by nearly all maritime nations. The U.S. Departments of Defense and Commerce adopted this value on July 1, 1954. With the yard-meter relationship then in use, the International Nautical Mile was equivalent to 6076.10333 feet, approximately. Using the yard-meter conversion factor effective July 1, 1959, (1 yard = 0.9144 meter, exactly) the International Nautical Mile is equivalent to 6076.11549 feet, approximately. See also SEA MILE.

nautical twilight. The time of incomplete darkness which begins (morning) or ends (evening) when the center of the sun is 12° below the celestial horizon. The times of nautical twilight are tabulated in the *Nautical Almanac*; at the times given the horizon is generally not visible and it is too dark for marine sextant observations. See also FIRST LIGHT.

nautophone, *n*. A sound signal emitter comprising an electrically oscillated diaphragm. It emits a signal similar in power and tone to that of a REED HORN.

Naval Vessel Lights Act. Authorized departure from the rules of the road for character and position of navigation lights for certain naval ships. Such modifications are published in *Notice to Mariners*.

NAVAREA. A geographical subdivision of the Long Range Radio Broadcast Service.

NAVAREA Warnings. Broadcast messages containing information which may affect the safety of navigation on the high seas. In accordance with international obligations, the Defense Mapping

Agency Hydrographic/Topographic Center is responsible for disseminating navigation information for ocean areas designated as NAVAREAS IV and XII of the World Wide Navigational Warning Service. NAVAREA IV broadcasts cover the waters contiguous to North America from the Atlantic coast eastward to 35°W and between latitudes 7°N and 67°N. NAVAREA XII broadcasts cover the waters contiguous to North America extending westward to the International Date Line and from 67°N to the equator east of 120°W, south to 3°25′S, thence east to the coast. Other countries are responsible for disseminating navigational information for the remaining NAVAREAS. NAVAREA Warnings may be superseded by a numbered paragraph in *Notice to Mariners*. The text of effective warnings for NAVAREAS IV and XII is available through NAVINFONET and is printed in the weekly *Notice to Mariners*.

navigable, *adj*. Affording passage to a craft; capable of being navigated.

navigable semicircle (less dangerous semicircle). The half of a cyclonic storm area in which the rotary and forward motions of the storm tend to counteract each other and the winds are in such a direction as to tend to blow a vessel away from the storm track. In the Northern Hemisphere this is to the left of the storm center and in the Southern Hemisphere it is to the right. The opposite is DANGEROUS SEMICIRCLE.

navigable waters. Waters usable, with or without improvements, as routes for commerce in the customary means of travel on water.

navigating sextant. A sextant designed and used for observing the altitudes of celestial bodies, as opposed to a hydrographic sextant.

navigation, *n*. The process of planning, recording, and controlling the movement of a craft or vehicle from one place to another. The word navigate is from the Latin navigatus, the past participle of the verb navigere, which is derived from the words navis, meaning "ship," and agere meaning "to move" or "to direct." Navigation of water craft is called marine navigation to distinguish it from navigation of aircraft, called air navigation. Navigation of a vessel on the surface is sometimes called surface navigation to distinguish it from navigation of a submarine. Navigation of vehicles across land or ice is called land navigation. The expression polar navigation refers to navigation in the regions near the geographical poles of the earth, where special techniques are employed.

navigational aid. An instrument, tool, system, device, chart, method, etc., intended to assist in navigation. This expression is not the same as AID TO NAVIGATION, which refers to devices external to a craft such as lights and buoys.

navigational astronomy. Astronomy of direct use to a navigator, comprising principally celestial coordinates, time, and the apparent motions of celestial bodies. Also called NAUTICAL ASTRONOMY.

navigational planets. The four planets commonly used for celestial observations: Venus, Mars Jupiter, and Saturn.

navigational plot. A graphic plot of the movements of a craft. A dead reckoning plot is the graphic plot of the dead reckoning, suitably labeled with respect to time, direction, and speed; a geographical plot is one relative to the surface of the earth.

navigational triangle. The spherical triangle solved in computing altitude and azimuth and great circle sailing problems. The celestial triangle is formed on the celestial sphere by the great circles connecting the elevated pole, zenith of the assumed position of the observer, and a celestial body. The terrestrial triangle is formed on the earth by the great circles connecting the pole and two places on the earth; the assumed position of the observer and geographical position of the body for celestial observations, and the point of departure and destination for great circle sailing problems. The expression astronomical triangle applies to either the celestial or terrestrial triangle used for solving celestial observations.

navigation, head of. A transshipment point at the end of a waterway where loads are transferred between water carriers and land carriers; also the point at which a river is no longer navigable due to rapids or falls.

navigation lights. Statutory, required lights shown by vessels during the hours between sunset and sunrise, in accordance with international agreements.

navigation mark. See MARK.

navigation/positioning system. A system capable of being used primarily for navigation or position fixing. It includes the equipment, its operators, the rules and procedures governing their actions and, to some extent, the environment which affects the craft or vehicle being navigated.

navigation satellite. An artificial satellite used in a system which determines positions based upon signals received from the satellite.

Navigation Sensor System Interface (NAVSSI). The U.S. Naval version of the electronic chart display and information system (ECDIS). It is integrated with command and control, weapons, and other systems.

Navigation Tables for Mariners and Aviators. See H.O. PUB. NO. 208.

navigator, *n*. 1. A person who navigates or is directly responsible for the navigation of a craft. 2. A book of instructions on navigation, such as the *The American Practical Navigator (Bowditch)*.

NAVSTAR Global Positioning System. A satellite navigation system developed by the Department of Defense. The system is provides highly accurate position and velocity information in three dimensions and precise time and time interval on a global basis continuously, to an unlimited number of users. It is unaffected by weather and provides a worldwide common grid reference system. The objective of the program is to provide very precise position information for a wide spectrum of military missions. In addition, current policy calls for civil availability with a slight degradation in system accuracy required to protect U.S. national security interests.

NAVTEX. A medium frequency radiocommunications system intended for the broadcast of navigational information up to 200 miles at sea, which uses narrow band direct printing technology to print out MSI and safety messages aboard vessels, without operator monitoring.

Navy Navigation Satellite System. A satellite navigation system of the United States conceived and developed by the Applied Physics Laboratory of the Johns Hopkins University. It is an all-weather, worldwide, and passive system which provides two-dimensional positioning from low-altitude satellites in near-polar orbits. The Transit launch program ended in 1988, and the system is scheduled for termination in 1996, replaced by GPS.

neaped, *adj*. Left aground following a spring high tide. Also called BENEAPED.

neap high water. See under NEAP TIDES.

neap low water. See under NEAP TIDES.

neap range. See under NEAP TIDES.

neap rise. The height of neap high water above the elevation of reference or datum of chart.

neap tidal currents. Tidal currents of decreased speed occurring semimonthly as the result of the moon being in quadrature. See also NEAP TIDES.

neap tides. Tides of decreased range occurring semimonthly as the result of the moon being in quadrature. The neap range of the tide is the average semidiurnal range occurring at the time of neap tides and is most conveniently computed from the harmonic constants. It is smaller than the mean range where the type of tide is either semidiurnal or mixed and is of no practical significance where the type of tide is diurnal. The average height of the high waters of the neap tides is called neap high water or high water neaps and the average height of the corresponding low waters is called neap low water or low water neaps.

nearest approach. The least distance between two objects having relative motion with respect to each other.

near gale. Wind of force 8 (28 to 33 knots or 32 to 38 miles per hour) on the Beaufort wind scale. See also GALE.

nearshore current system. The current system caused by wave action in or near the surf zone. The nearshore current system consists of four parts: the shoreward mass transport of water; longshore currents; rip currents; the longshore movement of expanding heads of rip currents.

near vane. That instrument sighting vane on the same side of the instrument as the observer's eye. The opposite is FAR VANE.

neatline, *n*. That border line which indicates the limit of the body of a map or chart. Also called SHEET LINE.

nebula *(pl. nebulae)*, *n*. 1. An aggregation of matter outside the solar system, large enough to occupy a perceptible area but which has not been resolved into individual stars. One within our galaxy is called a galactic nebula and one beyond is called an extragalactic nebula. If a nebula is resolved into numerous individual stars, it is called a STAR CLUSTER. 2. A galaxy.

necessary bandwidth. As defined by the International Telecommunication Union (ITU) for a given class of emission, the minimum value of the occupied bandwidth sufficient to ensure the transmission of information at the rate and with the quality required for the system employed, under specified conditions. Emissions useful for the good functioning of the receiving equipment as, for example, the emission corresponding to the carrier of reduced carrier systems, shall be included in the necessary bandwidth.

neck, *n*. 1. A narrow isthmus, cape or promontory. 2. The land areas between streams flowing into a sound or bay. 3. A narrow strip of land which connects a peninsula with the mainland. 4. A narrow body of water between two larger bodies; a strait.

negative altitude. Angular distance below the horizon. Also called DEPRESSION.

Network Coordinating Station. An INMARSAT COAST EARTH STATION (CES) equipped to process messages in the EGC SafetyNET system.

neutral occlusion. See under OCCLUDED FRONT.

new ice. A general term for recently formed ice which includes frazil ice, grease ice, slush, and shuga. These types of ice are composed of ice crystals which are only weakly frozen together (if at all) and have definite form only while they are afloat.

new moon. The moon at conjunction, when little or none of it is visible to an observer on the earth because the illuminated side is away from him. Also called CHANGE OF THE MOON. See also PHASES OF THE MOON.

new ridge. A newly formed ice ridge with sharp peaks, the slope of the sides usually being about 40°. Fragments are visible from the air at low altitude.

newton, *n*. The special name for the derived unit of force in the International System of Units; it is that force which gives to a mass of 1 kilogram an acceleration of 1 meter per second, per second.

Newtonian telescope. A reflecting telescope in which a small plane mirror reflects the convergent beam from the speculum to an eyepiece at one side of the telescope. After the second reflection the rays travel approximately perpendicular to the longitudinal axis of the telescope. See also CASSEGRAINIAN TELESCOPE.

newton per square meter. The derived unit of pressure in the International System of Units. See also PASCAL.

Newton's laws of motion. Universal laws governing all motion, formulated by Isaac Newton. These are: (1) Every body continues in a state of rest or of uniform motion in a straight line unless acted upon by a force; (2) When a body is acted upon by a force, its acceleration is directly proportional to the force and inversely proportional to the mass of the body, and the acceleration takes place in the direction in which the force acts; (3) To every action there is always an equal and opposite reaction; or, the mutual actions of two bodies are always equal and oppositely directed.

Ney's chart. See MODIFIED LAMBERT CONFORMAL CHART.

Ney's map projection. See MODIFIED LAMBERT CONFORMAL MAP PROJECTION.

night, *n*. The part of the solar day when the sun is below the visible horizon, especially the period between dusk and dawn.

night effect. See under POLARIZATION ERROR.

night error. See under POLARIZATION ERROR.

night order book. A notebook in which the commanding officer of a ship writes orders with respect to courses and speeds, any special precautions concerning the speed and navigation of the ship, and all other orders for the night for the officer of the deck.

nilas, *n*. A thin elastic crust of ice, easily bending on waves and swell and under pressure, thrusting in a pattern of interlocking "fingers." Nilas has a matte surface and is up to 10 centimeters in thickness. It may be subdivided into DARK NILAS and LIGHT NILAS. See also FINGER RAFTING.

nimbostratus, *n*. A dark, low shapeless cloud layer (mean upper level below 6,500 ft.) usually nearly uniform; the typical rain cloud. When precipitation falls from nimbostratus, it is in the form of continuous or intermittent rain or snow, as contrasted with the showery precipitation of cumulonimbus.

nimbus, *n*. A characteristic rain cloud. The term is not used in the international cloud classification except as a combining term, as cumulonimbus.

nipped, *adj*. Beset in the ice with the surrounding ice forcibly pressing against the hull.

nipping, *n*. The forcible closing of ice around a vessel such that it is held fast by ice under pressure. See also BESET, ICE-BOUND.

no-bottom sounding. A sounding in which the bottom is not reached.

nocturnal, *n*. An old navigation instrument which consisted of two arms pivoted at the enter of a disk graduated for date, time and arc. The nocturnal was used for determining time during the night and for obtaining a correction to be applied to an altitude observation of Polaris for finding latitude.

nodal, *adj*. Related to or located at or near a node or nodes.

nodal line. A line in an oscillating body of water along which there is a minimum or no rise and fall of the tide.

nodal point. 1. See NODE, definition 1. 2. The no-tide point in an amphidromic region.

node, *n*. 1. One of the two points of intersection of the orbit of a planet, planetoid, or comet with the ecliptic, or of the orbit of a satellite with the plane of the orbit of its primary. That point at which the body crosses to the north side of the reference plane is called the ascending node; the other, the descending node. The line connecting the nodes is called LINE OF NODES. Also called NODAL POINT. See also REGRESSION OF THE NODES. 2. A zero point in any stationary wave system.

node cycle. The period of approximately 18.61 Julian years required for the regression of the moon's nodes to complete a circuit of 360° of longitude. It is accompanied by a corresponding cycle of changing inclination of the moon's orbit relative to the plane of the earth's equator, with resulting inequalities in the rise and fall of the tide and speed of the tidal current.

node factor. A factor depending upon the longitude of the moon's node which, when applied to the mean coefficient of a tidal constituent, will adapt the same to a particular year for which predictions are to be made.

nodical, *adj*. Of or pertaining to astronomical nodes; measured from node to node.

nodical month. The average period of revolution of the moon about the earth with respect to the moon's ascending node, a period of 27 days, 5 hours, 5 minutes, 35.8 seconds.

nodical period. The interval between two successive passes of a satellite through the ascending node. See also ORBITAL PERIOD.

nominal orbit. The true or ideal orbit in which an artificial satellite is expected to travel. See also NORMAL ORBIT.

nominal range. See under VISUAL RANGE (OF A LIGHT).

nomogram, *n*. A diagram showing, to scale, the relationship between several variables in such manner that the value of one which corresponds to known values of the others can be determined graphically. Also called NOMOGRAPH.

nomograph, *n*. See NOMOGRAM.

non-dangerous wreck. A term used to describe a wreck having more than 20 meters of water over it. This term excludes a FOUL GROUND, which is frequently covered by the remains of a wreck and is a hazard only for anchoring, taking the ground, or bottom fishing.

nongravitational perturbations. Perturbations caused by surface forces due to mechanical drag of the atmosphere (in case of low flying satellites), electromagnetism, and solar radiation pressure.

nonharmonic constants. Tidal constants such as lunitidal intervals, ranges, and inequalities which may be derived directly from high and low water observations without regard to the harmonic constituents of the tide. Also applicable to tidal currents.

non-standard buoys. The general classification of all lighted and unlighted buoys built to specifications other than modern standard designs.

non-tidal basin. An enclosed basin separated from tidal waters by a caisson or flood gates. Ships are moved into the dock near high tide. The dock is closed when the tide begins to fall. If necessary, ships are kept afloat by pumping water into the dock to maintain the desired level. Also called WET DOCK. See also BASIN, definition 2.

nontidal current. See under CURRENT.

noon, *n*. The instant at which a time reference is over the upper branch of the reference meridian. Noon may be solar or sidereal as the sun or vernal equinox is over the upper branch of the reference meridian. Solar noon may be further classified as mean or apparent as the mean or apparent sun is the reference. Noon may also be classified according to the reference meridian, either the local or Greenwich meridian or additionally in the case of mean noon, a designated zone meridian. Standard, daylight saving or summer noon are variations of zone noon. The instant the sun is over the upper branch of any meridian of the moon is called lunar noon. Local apparent noon may also be called high noon.

noon constant. A predetermined value added to a meridian or ex-meridian sextant altitude to determine the latitude.

noon interval. The predicted time interval between a given instant, usually the time of a morning observation, and local apparent noon. This is used to predict the time for observing the sun on the celestial meridian.

noon sight. Measurement of the altitude of the sun at local apparent noon, or the altitude so measured.

normal, *adj*. Perpendicular. A line is normal to another line or a plane when it is perpendicular to it. A line is normal to a curve or curved surface when it is perpendicular to the tangent line or plane at the point of tangency.

normal, *n*. 1. A straight line perpendicular to a surface or to another line. 2. In geodesy, the straight line perpendicular to the surface of the reference ellipsoid. 3. The average, regular, or expected value of a quantity.

normal curve. Short for NORMAL DISTRIBUTION CURVE.

normal distribution. A mathematical law which predicts the probability that the random error of any given observation of a series of observations of a certain quantity will lie within certain bounds. The law can be derived from the following properties of random errors: (1) positive and negative errors of the same magnitude are about equal in number, (2) small errors occur more frequently than large errors, and (3) extremely large errors rarely occur. One immediate consequence of these properties is that the average or mean value of a large number of observations of a given quantity is zero. Also called GAUSSIAN DISTRIBUTION. See also SINGLE-AXIS NORMAL DISTRIBUTION, CIRCULAR NORMAL DISTRIBUTION, STANDARD DEVIATION.

normal distribution curve. The graph of the normal distribution. Often shortened to NORMAL CURVE.

normal orbit. The orbit of a spherical satellite about a spherical primary during which there are no disturbing elements present due to other celestial bodies, or to some physical phenomena. Also called UNPERTURBED ORBIT, UNDISTURBED ORBIT.

normal section line. A line on the surface of a reference ellipsoid, connecting two points on that surface, and traced by a plane containing the normal at one point and passing through the other point.

normal tide. A non technical term synonymous with tide, i.e., the rise and fall of the ocean due to the gravitational interactions of the sun, moon, and earth alone.

norte, *n*. A strong cold northeasterly wind which blows in Mexico and on the shores of the Gulf of Mexico. It results from an outbreak of cold air from the north. It is the Mexican extension of a norther.

north, *n*. The primary reference direction relative to the earth; the direction indicated by 000° in any system other than relative. True north is the direction of the north geographical pole; magnetic north the direction north as determined by the earth's magnetic compass; grid north an arbitrary reference direction used with grid navigation. See also CARDINAL POINT.

North Africa Coast Current. A nontidal current in the Mediterranean Sea that flows eastward along the African coast from the Strait of Gibraltar to the Strait of Sicily. It is the most permanent current in the Mediterranean Sea. The stability of the current is indicated by the proportion of no current observations, which averages less than 1 percent. The current is most constant just after it passes through the Strait of Gibraltar; in this region, west of longitude 3°W, 65 percent of all observations show an eastward set, with a mean speed of 1.1 knots and a mean maximum speed of 3.5 knots. Although the current is weaker between longitudes 3°W and 11°E, it remains constant, the speed averaging 0.7 knot through its length and its maximum speed being about 2.5 knots.

North American Datum of 1927. The geodetic datum the origin of which is located at Meade's Ranch, Kansas. Based on the Clarke spheroid of 1866, the geodetic position of triangulation station Meades Ranch and azimuth from that station to station Waldo are as follows: Latitude of Meades Ranch: 39° 13' 25.686"N; Longitude of Meades Ranch: 98° 32' 30.506"W Azimuth to Waldo: 75° 28' 09.64" The geoidal height at Meades Ranch is assumed to be zero.

North American Datum of 1983. The modern geodetic datum for North America; it is the functional equivalent of the World Geodetic System (WGS). It is based on the GRS 80 ellipsoid, which fits the size and shape of the earth more closely, and has its origin at the earth's center of mass.

North Atlantic Current. An ocean current which results from extensions of the Gulf Stream and the Labrador Current near the edge of the Grand Banks of Newfoundland. As the current fans outward and widens in a northeastward through eastward flow, it decreases sharply in speed and persistence. Some influence of the Gulf Stream is noticeable near the extreme southwestern boundary of the current. The North Atlantic Current is a sluggish, slow-moving flow that can easily be influenced by opposing or augmenting winds. There is some evidence that the weaker North Atlantic Current may consist of separate eddies or branches which are frequently masked by a shallow, wind-driven surface now called the NORTH ATLANTIC DRIFT. A branch of the North Atlantic Current flows along the west coasts of the British Isles at speeds up to 0.6 knot and enters the Norwegian Sea as the NORWAY CURRENT mainly through the east side of the Faeroe–Shetland Channel. A small portion of this current to the west of the Faeroe Islands mixes with part of the southeastward flow from the north coast of Iceland; these two water masses join and form a clockwise circulation around the Faeroe Islands. The very weak nontidal current in the Irish Sea, which averages only about 0.1 knot, depends on the wind. The part of the North Atlantic Current that flows eastward into the western approaches to the English Channel tends to increase or decrease the speed of the reversing tidal currents. The southern branch of the North Atlantic Current turns southward near the Azores to become the CANARY CURRENT.

North Atlantic Drift. See under NORTH ATLANTIC CURRENT.

northbound node. See ASCENDING NODE.

North Brazil Current. See GUIANA CURRENT.

North Cape Current. An Arctic Ocean current flowing northeastward and eastward around northern Norway, and curving northeastward into the Barents Sea. The North Cape Current is the continuation of the northeastern branch of the NORWAY CURRENT.

northeaster, nor'easter, *n*. A northeast wind, particularly a strong wind or gale associated with cold rainy weather. In the U.S., nor'easters generally occur on the north side of late-season low pressure systems which pass off the Atlantic seaboard, bringing onshore gales to the region north of the low. Combined with high tides, they can be very destructive.

northeast monsoon. See under MONSOON.

north equatorial current. See ATLANTIC NORTH EQUATORIAL CURRENT, PACIFIC NORTH EQUATORIAL CURRENT.

norther, *n*. A northerly wind. In the southern United States, especially in Texas (Texas norther) in the Gulf of Mexico, in the Gulf of Panama away from the coast, and in central America (the norte), the norther is a strong cold wind from the northeast to northwest. It occurs between November and April, freshening during the afternoon and decreasing at night. It is a cold air outbreak associated with the southward movement of a cold anticyclone. It is usually preceded by a warm and cloudy or rainy spell with southerly winds. The norther comes as a rushing blast and brings a sudden drop of temperature of as much as 25°F in 1 hour or 50°F in 3 hours in winter. The California norther is a strong, very dry, dusty, northerly wind which blows in late spring, summer and early fall in the valley of California or on the west coast when pressure is high over the mountains to the north. It lasts from 1 to 4 days. The dryness is due to adiabatic warming during descent. In summer it is very hot. The Portuguese norther is the beginning of the trade wind west of Portugal. The term is used for a strong north wind on the coast of Chile which blows occasionally in summer. In southeast Australia, a hot dry wind from the desert is called a norther.

northern lights. See AURORA BOREALIS.

north frigid zone. That part of the earth north o the Arctic Circle.

north geographical pole. The geographical pole in the Northern Hemisphere, at lat. 90°N.

north geomagnetic pole. The geomagnetic pole in the Northern Hemisphere. This term should not be confused with NORTH MAGNETIC POLE. See also GEOMAGNETIC POLE.

northing, *n*. The distance a craft makes good to the north. The opposite is SOUTHING.

north magnetic pole. The magnetic pole in the Northern Hemisphere. This term should not be confused with NORTH GEOMAGNETIC POLE. See also GEOMAGNETIC POLE.

North Pacific Current. Flowing eastward from the eastern limit of the Kuroshio Extension (about longitude 170° E), the North Pacific Current forms the northern part of the general clockwise oceanic circulation of the North Pacific Ocean.

north polar circle. See ARCTIC CIRCLE.

North Pole. 1. The north geographical pole. See also MAGNETIC POLE GEOMAGNETIC POLE. 2. The north-seeking end of a magnet. See also RED MAGNETISM.

north temperate zone. That part of the earth between the Tropic of Cancer and the Arctic Circle.

north up, north upward. One of the three basic orientations of display of relative or true motion on a radarscope or electronic chart. In the NORTH UP orientation, the presentation is in true (gyrocompass) directions from own ship, north being maintained UP or at the top of the radarscope. See also HEAD UP, BASE COURSE UP.

northwester, nor'wester, *n*. A northwesterly wind.

Norway Coastal Current. Originating mainly from Oslofjord outflow, counterclockwise return flow of the Jutland Current within the Skaggerak, and outflow from the Kattegat, the Norway Coastal Current begins at about latitude 59°N, longitude 10°E and follows the coast of Norway, and is about 20 miles in width. Speeds are strongest off the southeast coast of Norway, where they frequently range between 1 and 2 knots. Along the remainder of the coast the current gradually weakens. It may widen to almost 30 miles at about latitude 63°N, where it joins the NORWAY CURRENT. South of latitude 62°N the current speed usually ranges between 0.4 and 0.9 knots. Speeds are generally stronger in spring and summer, when the flow is augmented by increased discharge from fjords.

Norway Current. An Atlantic Ocean current flowing northeastward along the northwest coast of Norway, and gradually branching and continuing as the SPITZBERGEN ATLANTIC CURRENT and the NORTH CAPE CURRENT. The Norway Current is the continuation of part of the northern branch of the North Atlantic Current. Also called NORWEGIAN CURRENT.

Norwegian Current. See NORWAY CURRENT.

notch filter. An arrangement of electronic components designed to attenuate or reject a specific frequency band with a sharp cut-off at either end.

notice board. A signboard used to indicate speed restrictions, cable landings, etc.

notice to mariners. A periodic publication used by the navigator to correct charts and publications.

Notice to Mariners. A weekly publication of the Defense Mapping Agency Hydrographic/Topographic Center prepared jointly with the National Ocean Survey and the U.S. Coast Guard giving information on changes in aids to navigation, dangers to navigation, selected items from the *Local Notice to Mariners*, important new soundings, changes in channels, harbor construction, radionavigation information, new and revised charts and publications, special warnings and notices, pertinent HYDROLANT, HYDROPAC, NAVAREA IV and XII messages and corrections to charts, manuals, catalogs, sailing directions (pilots), etc. The *Notice to Mariners* should be used routinely for updating the latest editions of nautical charts and related publications.

nova *(pl. novae), n*. A star which suddenly becomes many times brighter than previously, and then gradually fades. Novae are believed to be exploding stars.

nucleus, *n*. The central, massive part of anything, such as an atom or comet.

numerical scale. A statement of that distance on the earth shown in one unit (usually an inch) on the chart, or vice versa. See also REPRESENTATIVE FRACTION.

nun buoy. An unlighted buoy of which the upper part of the body (above the waterline), or the larger part of the superstructure, has a cone shape with vertex upwards.

nutation, *n*. Irregularities in the precessional motion of the equinoxes due chiefly to regression of the nodes.

O

object glass. See OBJECTIVE.

objective, *n*. The lens or combination of lenses which receives light rays from an object, and refracts them to form an image in the focal plane of the eyepiece of an optical instrument, such as a telescope. Also called OBJECT GLASS.

oblate spheroid. An ellipsoid of revolution, the shorter axis of which is the axis of revolution. An ellipsoid of revolution, the longer axis of which is the axis of revolution, is called a PROLATE SPHEROID. The earth is approximately an oblate spheroid.

oblique, *adj*. Neither perpendicular nor parallel; slanting.

oblique angle. Any angle not a multiple of 90°.

oblique ascension. The arc of the celestial equator, or the angle at the celestial pole, between the hour circle of the vernal equinox and the hour circle through the intersection of the celestial equator and the eastern horizon at the instant a point on the oblique sphere rises, measured eastward from the hour circle of the vernal equinox through 24h. The expression is not used in modern navigation.

oblique chart. A chart on an oblique map projection.

oblique coordinates. Magnitudes defining a point relative to two intersecting non-perpendicular lines, called AXES. The magnitudes indicate the distance from each axis, measured along a parallel to the other axis. The horizontal distance is called the abscissa and the other distance the ordinate. This is a form of CARTESIAN COORDINATES.

oblique cylindrical orthomorphic chart. See OBLIQUE MERCATOR CHART.

oblique cylindrical orthomorphic map projection. See OBLIQUE MERCATOR MAP PROJECTION oblique equator. A great circle the plane of which is perpendicular to the axis of an oblique projection. An oblique equator serves as the origin for measurement of oblique latitude. On an oblique Mercator map projection, the oblique equator is the tangent great circle. See also FICTITIOUS EQUATOR.

oblique graticule. A fictitious graticule based upon an oblique map projection.

oblique latitude. Angular distance from an oblique equator. See also FICTITIOUS LATITUDE.

oblique longitude. Angular distance between a prime oblique meridian and any given oblique meridian. See also FICTITIOUS LONGITUDE.

oblique map projection. A map projection with an axis inclined at an oblique angle to the plane of the equator.

oblique Mercator chart. A chart on the oblique Mercator map projection. Also called OBLIQUE CYLINDRICAL ORTHOMORPHIC CHART. See also MERCATOR CHART.

oblique Mercator map projection. A conformal cylindrical map projection in which points on the surface of a sphere or spheroid, such as the earth, are developed by Mercator principles on a cylinder tangent along an oblique great circle. Also called OBLIQUE CYLINDRICAL ORTHOMORPHIC MAP PROJECTION. See also MERCATOR MAP PROJECTION.

oblique meridian. A great circle perpendicular to an oblique equator. The reference oblique meridian is called prime oblique meridian. See also FICTITIOUS MERIDIAN.

oblique parallel. A circle or line parallel to an oblique equator, connecting all points of equal oblique latitude. See also FICTITIOUS PARALLEL.

oblique pole. One of the two points 90° from an oblique equator.

oblique rhumb line. 1. A line making the same oblique angle with all fictitious meridians of an oblique Mercator map projection. Oblique parallels and meridians may be considered special cases of the oblique rhumb line. 2. Any rhumb line, real or fictitious, making an oblique angle with its meridians. In this sense the expression is used to distinguish such rhumb lines from parallels and meridians, real or fictitious, which may be included in the expression rhumb line. See also FICTITIOUS RHUMB LINE.

oblique sphere. The celestial sphere as it appears to an observer between the equator and the pole, where celestial bodies appear to rise obliquely to the horizon.

obliquity factor. A factor in an expression for a constituent tide or tidal current involving the angle of the inclination of the moon's orbit to the plane of the earth's equator.

obliquity of the ecliptic. The acute angle between the plane of the ecliptic and the plane of the celestial equator, about 23° 27'.

obscuration, *n*. The designation for the sky cover when the sky is completely hidden by obscuring phenomena in contact with, or extending to the surface.

obscuring phenomenon. Any atmospheric phenomenon, not including clouds, which restricts the vertical or slant visibility.

observed altitude. Corrected sextant altitude; angular distance of the center of a celestial body above the celestial horizon of an observer measured along a vertical circle, through 90°. Occasionally called TRUE ALTITUDE. See also ALTITUDE INTERCEPT, APPARENT ALTITUDE, SEXTANT ALTITUDE.

observed gravity anomaly. See GRAVITY ANOMALY.

observed latitude. See LATITUDE LINE.

observed longitude. See LONGITUDE LINE.

obstruction, *n*. Anything that hinders or prevents movement, particularly anything that endangers or prevents passage of a vessel or aircraft. The term is usually used to refer to an isolated danger to navigation, such as a submerged rock or reef in the case of marine navigation, and a tower, tall building, mountain peak, etc., in the case of air navigation.

obstruction buoy. A buoy used to indicate a dangerous obstruction. See ISOLATED DANGER BUOY.

obstruction light. A light indicating a radio tower or other obstruction to aircraft.

obstruction mark. A navigation mark used to indicate a dangerous obstruction. See ISOLATED DANGER MARK.

obtuse angle. An angle greater than 90° and less than 180°.

occasional light. A light put into service only on demand.

occluded front. A composite of two fronts, formed when a cold front overtakes a warm front or stationary front. This is common in the late stages of wave-cyclone development, but is not limited to occurrence within a wave-cyclone. There are three basic types of occluded front, determined by the relative coldness of the air behind the original cold front to the air ahead of the warm (or stationary) front. A cold occlusion results when the coldest air is behind the cold front. The cold front undercuts the warm front and, at the earth's surface, cold air replaces less-cold air. When the coldest air lies ahead of the warm front, a warm occlusion is formed in which case the original cold front is forced aloft at the warm-front surface. At the earth's surface, cold air is replaced by less-cold air. A third and frequent type, a neutral occlusion, results when there is no appreciable temperature difference between the cold air masses of the cold and warm fronts. In this case frontal characteristics at the earth's surface consist mainly of a pressure trough, a wind-shift line, and a band of cloudiness and precipitation. Commonly called OCCLUSION. Also called FRONTAL OCCLUSION.

occlusion, *n*. 1. See OCCLUDED FRONT. 2. The process of formation of an occluded front. Also called FRONTAL OCCLUSION.

occultation, *n*. 1. The concealment of a celestial body by another which crosses the line of view. Thus, the moon occults a star when it passes between the observer and the star. 2. The interval of darkness in the period of the light. See also FLASH.

occulting light. A light totally eclipsed at regular intervals, with the duration of light always longer than the intervals of darkness called OCCULTATIONS. The term is commonly used for a SINGLE OCCULTING LIGHT, an occulting light exhibiting only single occultations which are repeated at regular intervals.

occupied bandwidth. As defined by the International Telecommunication Union (ITU) the frequency bandwidth such that, below its lower and above its upper frequency limits, the mean powers radiated are each equal to 0.5 percent of the total mean power radiated by a given emission. In some cases, for example multichannel frequency-division systems, the percentage of 0.5 percent may lead to certain difficulties in the practical application of the definitions of occupied and necessary bandwidth; in such cases a different percentage may prove useful.

ocean, *n*. 1. The major area of salt water covering the greater part of the earth. 2. One of the major divisions of the expanse of salt water covering the earth.

ocean current. A movement of ocean water characterized by regularity, either of a cyclic nature, or as a continuous stream flowing along a definable path. Three general classes may be distinguished, by cause: (a) currents associated with horizontal pressure gradients, comprising the various types of gradient current; (b) wind-driven currents, which are those directly produced by the stress exerted by the wind upon the ocean surface; (c) currents produced by long-wave motions. The latter are principally tidal currents, but may also include currents associated with internal waves, tsunamis and seiches. The major ocean currents are of continuous, stream-flow character, and are of first-order importance in the maintenance of the earth's thermodynamic balance.

oceanic, *adj*. Of or pertaining to the ocean.

oceanographic, *adj*. Of or pertaining to oceanography, or knowledge of the oceans.

oceanographic survey. The study or examination of conditions in the ocean or any part of it. with reference to zoology, chemistry, geology, or other scientific discipline. See also HYDROGRAPHIC SURVEY.

oceanography, *n*. The study of the sea, embracing and integrating all knowledge pertaining to the sea's physical boundaries, the chemistry and physics of sea water, and marine biology. Strictly, oceanography is the description of the marine environment, whereas OCEANOLOGY is the study of the oceans.

oceanology, *n*. The study of the ocean. See also OCEANOGRAPHY.

Ocean Passages for the World. A British publication relating to the planning and conduct of ocean passages. Published by the Hydrographer of the Navy, *Ocean Passages for the World* addresses those areas which lie mainly out side the areas covered in detail by Admiralty Sailing Directions. It is kept up-to-date by periodical supplements. The publication should not be used without reference to the latest supplement and those *Notices to Mariners* published to correct Sailing Directions.

ocean waters. For application to the provisions of the Marine Protection, Research, and Sanctuaries Act of 1972, those waters of the open sea lying seaward of the base line from which the territorial sea is measured.

octagon, *n*. A closed plane figure having 8 sides.

octahedral cluster. An arrangement of eight corner reflectors with common faces designed to give substantially uniform response in all directions. The octahedral cluster is formed by mounting three rectangular plates mutually at right angles with the geometric centers of the plates coincident. See also PENTAGONAL CLUSTER.

octant, *n*. A double-reflecting instrument for measuring angles, used primarily for measuring altitude of celestial bodies. It has a range of 90°, with the graduated arc subtending 45°, or 1/8 of a circle, hence the term octant; a precursor of the sextant, whose arc subtends 60° or 1/6 of a circle.

octant altitude. See SEXTANT ALTITUDE.

Odessey protractor. A device used in conjunction with a plotting sheet having equally spaced concentric circles (range circles) drawn about two or more stations of a radio determination system being operated in the ranging mode.

oe, *n*. A whirlwind off the Faeroe Islands.

oersted, *n*. The centimeter-gram-second electromagnetic system unit of magnetic field strength. It corresponds to $1000/4\pi$ ampere per meter.

off-center PPI display. A plan position indicator display in which the center about which the sweep rotates is offset from the center of the radarscope.

offing, *n*. The part of the visible sea a considerable distance from the shore, or that part just beyond the limits of the area in which a pilot is needed.

offshore, *adj*. & *adv*. Away from the shore.

offshore, *n*. The comparatively flat zone of variable width which extends from the outer margin of the rather steeply sloping shore face to the edge of the shelf.

offshore light stations. Manned light stations built on exposed marine sites to replace lightships.

offshore navigation. Navigation at a distance from a coast, in contrast with COASTWISE NAVIGATION in the vicinity of a coast.

offshore water. Water adjacent to land in which the physical properties are slightly influenced by continental conditions.

offshore wind. Wind blowing from the land toward the sea. An ONSHORE WIND blows in the opposite direction. See also LAND BREEZE.

off soundings. Navigating beyond the 100-fathom curve. In earlier times, said of a vessel in water deeper than could be sounded with the sounding lead.

off station. Not in charted position.

ogival buoy. A buoy with a pointed-arch shaped vertical cross-section. Used in the cardinal system.

ohm, *n*. A derived unit of electrical resistance in the International System of Units; it is the electrical resistance between two points of a conductor when a constant potential difference of 1 volt, applied to these points, produces in the conductor a current of 1 ampere, the conductor not being the seat of an electromotive force.

old ice. Sea ice which has survived at least one summer's melt. Most topographic features are smoother than on first-year ice. Old ice may be subdivided into SECOND-YEAR ICE and MULTI YEAR ICE.

Omega Navigation System. A worldwide. continuous, radionavigation system of medium accuracy which provides hyperbolic lines of position through phase comparisons of VLF (10-14kHz) continuous wave signals transmitted on a common frequency on a time-shared basis. The full system is comprised of eight transmitting stations.

Omega plotting chart. See under PLOTTING CHART.

Omega Table. *See PUB. 224.*

omni-. A prefix meaning all.

omniazimuthal antenna. See OMNIDIRECTIONAL ANTENNA.

omnidirectional antenna. An antenna whose radiating or receiving properties at any instant are the same on all bearings. Also called OMNIAZIMUTHAL ANTENNA. See also DIRECTIONAL ANTENNA.

omnidirectional light. A light which presents the same characteristic over the whole horizon of interest to marine navigation. Also called ALL-ROUND LIGHT.

omnidirectional radiobeacon. A radiobeacon transmitting a signal in all directions. A circular radiobeacon is an omnidirectional beacon which transmits in all horizontal directions simultaneously. A rotating radiobeacon is an omnidirectional beacon with one or more beams that rotate. A DIRECTIONAL RADIOBEACON is a beacon which beams its signals in one or several prescribed directions.

onshore wind. Wind blowing from the sea towards the land. An OFFSHORE WIND blows in the opposite direction. See also SEA BREEZE.

on soundings. Navigating within the 100-fathom curve. In earlier times, said of a vessel in water sufficiently shallow for sounding by sounding lead.

on the beam. Bearing approximately 090° relative (on the starboard beam) or 270° relative (on the port beam). The expression is often used loosely for BROAD ON THE BEAM, or bearing exactly 090° or 270° relative. Also called ABEAM.

on the bow. Bearing approximately 045° relative (on the starboard bow) or 315° relative (on the port bow). The expression is often used loosely for BROAD ON THE BOW, or bearing exactly 045° or 315° relative.

on the quarter. Bearing approximately 135° relative (on the starboard quarter) or 225° relative (on the port quarter). The expression is often used loosely for BROAD ON THE QUARTER, or bearing exactly 135° or 225° relative.

ooze, *n*. A soft, slimy, organic sediment covering part of the ocean bottom, composed principally of shells or other hard parts of minute organisms.

open, *v., i*. To move or appear to move apart, such as when range lights appear to separate as the vessel moves off the channel centerline. The opposite is CLOSE.

open basin. See TIDAL BASIN.

open berth. An anchorage berth in an open roadstead.

open coast. A coast that is not sheltered from the sea.

open harbor. An unsheltered harbor exposed to the sea.

opening, *n*. A break in a coastline or a passage between shoals, etc. See also GAT.

open pack ice. Pack ice in which the concentration is 4/10 to 6/10, with many leads and polynyas, and the floes generally not in contact with one another.

open roadstead. A roadstead with relatively little protection from the sea.

open sea. 1. The part of the ocean not enclosed by headlands, within narrow straits, etc. 2. The part of the ocean outside the territorial jurisdiction of any country. The opposite is CLOSED SEA. See also HIGH SEAS.

open water. A large area of freely navigable water in which sea ice is present in concentration less than 1/10. When there is no sea ice present, the area should be described as ICE FREE, even though icebergs may be present.

operating area chart. A base chart with overprints of various operating areas necessary to control fleet exercise activities. Submarine Transit Lanes, Surface and Sub-surface Operating Areas, Air Space Warning Areas, Controlled Air Spaces, and other restricted areas are portrayed.

operating system. The portion of a computer's software devoted to running programs and providing for operator interface.

opposition, *n*. The situation of two celestial bodies having either celestial longitudes or sidereal hour angles differing by 180°. The term is usually used only in relation to the position of a superior planet or the moon with reference to the sun. The situation of two celestial bodies having either the same celestial longitude or the same sidereal hour angle is called conjunction.

optic, *adj*. Of or pertaining to vision.

optical, *adj*. Of or pertaining to optics or to vision.

optical double star. Two stars in nearly the same line of sight but differing greatly in distance from the observer, as distinguished from a PHYSICAL DOUBLE STAR (two stars in nearly the same line of sight and at approximately the same distance from the observer).

optical glass. Glass of which the composition and molding are carefully controlled in order to insure uniform refractive index and high transmission factor.

optical path. The path followed by a ray of light through an optical system.

optical system. A series of lenses, apertures, prisms, mirrors, etc., so arranged as to perform a definite optical function.

optics, *n*. The science dealing with light, lenses, etc.

Optimum Track Ship Routing. See under SHIP WEATHER ROUTING.

orbit, *n*. 1. The path of a body or particle under the influence of a gravitational or other force. See also CENTRAL FORCE ORBIT, INERTIAL ORBIT, INTERMEDIATE ORBIT, NOMINAL ORBIT, NORMAL ORBIT, OSCULATING ORBIT, PERTURBED ORBIT, POLAR ORBIT, STATIONARY ORBIT.

orbital altitude. The mean altitude of the orbit of a satellite above the surface of the parent body.

orbital elements. Parameters that specify the position and motion of a body in orbit. The elliptical orbit of a satellite attracted by an exactly central gravitational force is specified by a set of six parameters as follows: Two parameters, the semimajor axis and eccentricity of the ellipse, establish the size and shape of the elliptical orbit. A third parameter, time of perifocal passage, enables determination of the location of the satellite in its orbit at any instant. The three remaining parameters establish the orientation of the orbit in space. These are the inclination of the orbital plane to a reference plane, the right ascension of the ascending node of the satellite, and the argument of pericenter. See also ORBITAL PARAMETERS OF ARTIFICIAL SATELLITE, MEAN ELEMENTS, OSCULATING ELEMENTS.

orbital inclination. See as INCLINATION, definition 2.

orbital mode. A method for determining the position of an unknown station position when the unknown position cannot be viewed simultaneously with known positions. The arc of the satellite orbit is extrapolated from the ephemeris of the satellite determined by the known stations which permits the determination of the position of the unknown station dependent completely on the satellite's orbital parameters.

orbital motion. Continuous motion in a closed path about and as a direct result of a source of gravitational attraction.

orbital parameters of artificial earth satellite. The precessing elliptical orbit of an artificial earth satellite is unambiguously specified by the following set of parameters: semimajor axis, eccentricity, time of perigee, inclination of the orbital plane to the plane of the reference plane (celestial equator), the right ascension of the ascending node of the satellite at time of perigee, the argument of perigee at time of perigee, right ascension of Greenwich at time of perigee, mean motion (rate of change of mean anomaly), rate of change of argument of perigee, and rate of change of right ascension of the ascending node at time of perigee. With the inclination expressed as the sine and cosine of the orbital inclination, the parameters number 11. See also ORBITAL ELEMENTS.

orbital path. One of the tracks on a primary body's surface traced by the subpoint of a satellite that orbits about it several times in a direction other than normal to the primary body's axis of rotation. Each track is displaced in a direction opposite and by an amount equal to the degrees of rotation between each satellite orbit and of the nodical precession of the plane of the orbit. Also called SUBTRACK. See also WESTWARD MOTION.

orbital period. If the orbit is unchanging and ideal, the in travel between successive passages of a satellite through the same point in its orbit. If the orbit is not ideal, the point must be specified. When the perigee is specified it is called radial or anomalistic period. When the ascending node is specified, it is called nodical period. When the same geocentric right ascension is specified, it is called sidereal period. Also called PERIOD OF SATELLITE.

orbital plane. The plane of the ellipse defined by a central force orbit.

orbital velocity. The velocity of an earth satellite or other orbiting body at any given point in its orbit.

ordinary, *adj*. With respect to tides, the use of this non technical term has, for the most part, been determined to be synonymous with mean. The use of the term ordinary in tidal terms is discouraged.

ordinate, *n*. The vertical coordinate of a set of rectangular coordinates. Also used in a similar sense in connection with oblique coordinates.

orient, *v., t*. 1. To line up or adjust with respect to a reference. 2. To obtain a mental grasp of the existing situation.

orientability of a sound signal. The property of a sound signal by virtue of which a listener can estimate the direction of the location of the signal.

orographic rain. Rain resulting when moist air is forced upward by a mountain range.

orthodrome, *n*. See GREAT CIRCLE.

orthodromic curve. See GREAT CIRCLE.

orthogonal, *adj*. Right angled, rectangular.

orthogonal map projection. See ORTHOGRAPHIC MAP PROJECTION.

orthographic, *adj*. Of or pertaining to right angles or perpendicular lines.

orthographic chart. A chart on the orthographic map projection.

orthographic map projection. A perspective azimuthal projection in which the projecting lines, emanating from a point at infinity, are perpendicular to a tangent plane. The projection is used chiefly in navigational astronomy for inter converting coordinates of the celestial equator and horizon systems. Also called ORTHOGONAL PROJECTION.

orthomorphic, *adj*. Preserving the correct shape. See also CONFORMAL MAP PROJECTION.

orthomorphic chart. A chart on which very small shapes are correctly represented. See also CONFORMAL MAP PROJECTION.

orthomorphic map projection. A projection in which very small shapes are correctly represented. See also CONFORMAL MAP PROJECTION.

oscar satellite. A general term for one of the operational satellites of the Navy Navigation Satellite System, except for satellite 30110 called TRANSAT, placed in orbit prior to 1981. The improved satellites placed in orbit beginning in 1981 are called NOVA.

oscillation, *n*. 1. Fluctuation or vibration to each side of a mean value or position. 2. Half an oscillatory cycle, consisting of fluctuation or vibration in one direction; half a vibration.

oscillator, *n*. A sound signal emitter comprising a resonant diaphragm maintained in vibrating motion by electromagnetic action.

oscillatory wave. A wave in which only the form advances, the individual particles of the medium moving in closed orbits, as ocean waves in deep water; in contrast with a WAVE OF TRANSLATION, in which the individual particles are shifted in the direction of wave travel, as ocean waves in shoal water.

oscilloscope, *n*. An instrument for producing a visual representation of oscillations or changes in an electric current. The face of the cathode-ray tube used for this representation is called a SCOPE or SCREEN.

osculating elements. A set of parameters that specifies the instantaneous position and velocity of a celestial body, or artificial satellite in a perturbed orbit. Osculating elements describe the unperturbed (two-body) orbit (osculating orbit) that the body would follow if perturbations were to cease instantaneously.

osculating orbit. The ellipse that a satellite would follow after a specific time "t" (the epoch of osculation) if all forces other than central force ceased to act from "t" on. An osculating orbit is tangent to the real, perturbed, orbit and has the same velocity at the point of tangency. See also OSCULATING ELEMENTS.

outage, *n*. The failure of an aid to navigation to function exactly as described in the light list.

outer harbor. See under INNER HARBOR.

outfall, *n*. The discharge end of a narrow street sewer, drain, etc.

outfall buoy. A buoy marking the position where a sewer or other drain discharges.

outline chart. A chart with only a generally presentation of the landmass with little or no culture or relief. See also PLOT CHART.

output axis. The axis of precession of a gyroscope. See also INPUT AXIS, PRECESSION.

outside fix. A term describing the fix position determined by the method of bisectors when the lines of position result from observations of objects or celestial bodies lying within a 180° arc of the horizon. See also METHOD OF BISECTORS.

outward bound. Heading for the open sea. The opposite is INWARD BOUND. See also HOMEWARD BOUND.

overcast, *adj*. Pertaining to a sky cover of 95% or more.

overcast, *n*. A cloud cover.

overfalls, *n. pl*. Breaking waves caused by the meeting of currents or by waves moving against the current. See also RIPS.

overhead cable effect. A radar phenomenon which may occur in the vicinity of an overhead power cable. The echo from the cable appears on the plan position indicator as a single echo, the echo being returned from that part of cable where the radar beam is at right angles to the cable. If this phenomenon is not recognized, the echo can be wrongly identified as the echo from a ship on a steady bearing. Evasive action results in the echo remaining on a constant bearing and moving to the same side of the channel as the ship altering course. This phenomenon is particularly apparent for the power cable spanning the Straits of Messina.

overhead compass. See INVERTED COMPASS.

overhead constraints. The elevation angle limitations between which usable navigation data may be obtained from a satellite in the doppler mode.

overlay, *n*. A printing or drawing on a transparent or translucent medium at the same scale as a map, chart, etc., to show details not appearing on the original.

overprint, *n*. New material printed on a map or chart to show data of importance or special value in addition to that originally printed.

overtide, *n*. A harmonic tidal or tidal current constituent with a speed that is an exact multiple of the speed of one of the fundamental constituents derived from the development of the tide-producing force. The presence of overtides is usually attributed to shallow water conditions.

Oyashio, *n*. A cold ocean current flowing from the Bering Sea southwestward along the coast of Kamchatka, past the Kuril Islands to meet the Kuroshio off the coast of Honshu. The Oyashio turns and continues eastward, eventually joining the Aleutian Current.

P

Pacific Equatorial Countercurrent. A Pacific Ocean current that flows eastward, counter to and between the westward flowing Pacific North and South Equatorial Currents, between latitudes 3°N and 10°N. East of the Philippines it is joined by the southern part of the Pacific North Equatorial Current.

Pacific North Equatorial Current. A North Pacific Ocean current that flows westward between latitudes 10°N and 20°N. East of the Philippines, it divides, part turning south to join the Pacific Equatorial Counter current and part turning north to flow along the coast of Japan as the KUROSHIO.

Pacific South Equatorial Current. A Pacific Ocean current that flows westward between latitudes 3°N and 10°S. In mid ocean, much of it turns south to form a large whirl. The portion that continues across the ocean divides as it approaches Australia, part flowing north toward New Guinea and part turning south along the east coast of Australia as the EAST AUSTRALIA CURRENT.

Pacific standard time. See STANDARD TIME.

pack ice. The term used in a wide sense to include any area of sea ice, other than fast ice, no matter what form it takes or how it is disposed.

pagoda, *n*. As a landmark, a tower having a number of stories and a characteristic architecture, used as a place of worship or as a memorial, primarily in Japan, China, and India.

paint, *n*. The bright area on the phosphorescent plan position indicator screen resulting from the brightening of the sweep by the echoes.

paint, *v., t & i*. To brighten the phosphorescent plan position indicator screen through the effects of the echoes on the sweep.

painted mark. A navigation mark formed simply by painting a cliff, wall, rock, etc.

pancake ice. Predominantly circular pieces of ice from 30 centimeters to 3 meters in diameter, and up to about 10 centimeters in thickness with raised rims due to pieces striking against one another. It may be formed on a slight swell from grease ice, shuga, or slush or as a result of the breaking of ice rind, nilas, or under severe conditions of swell or waves, of gray ice. It also sometimes forms at some depth, at an interface between water bodies of different physical characteristics, from where it floats to the surface; its appearance may rapidly cover wide areas of water.

pantograph, *n*. An instrument for copying maps, drawings, or other graphics at a predetermined scale.

papagayo, *n*. A violet northeasterly fall wind on the Pacific coast of Nicaragua and Guatemala. It consists of the cold air mass of a *norte* which has overridden the mountains of Central America. See also TEHUANTEPECER.

parabola, *n*. An open curve all points of which are equidistant from a fixed point, called the FOCUS, and a straight line. The limiting case occurs when the point is on the line, in which case the parabola becomes a straight line.

parabolic reflector. A reflecting surface having the cross section along the axis in the shape of a parabola. Parallel rays striking the reflector are brought to a focus at a point, or if the source of the rays is placed at the focus, the reflected rays are parallel. See also CORNER REFLECTION RADAR REFLECTOR, SCANNER.

parabolic velocity. See ESCAPE VELOCITY.

parallactic angle. That angle at the navigational triangle at the celestial body; the angle between a body's hour circle and its vertical circle. Also called POSITION ANGLE.

parallax, *n*. The difference in apparent direction or position of an object when viewed from different points. For bodies of the solar system, parallax is the difference in the direction of the body due to the displacement of the observer from the center of the earth, and is called geocentric parallax, varying with the body's altitude and distance from the earth. The geocentric parallel when a body is in the horizon is called horizontal parallax, as contrasted with the parallax at any altitude, called parallax in altitude. Parallax of the moon is called lunar parallax. In marine navigation it is customary to apply a parallax correction to sextant altitudes of the sun, moon, Venus, and Mars. For stars, parallax is the angle at the star subtended by the semimajor axis of the earth's orbit and is called heliocentric or stellar parallax, which is too small to be significant as a sextant error.

parallax correction. A correction due to parallax, particularly that sextant altitude correction due to the difference between the apparent direction from a point on the surface of the earth to celestial body and the apparent direction from the center of the earth to the same body.

parallax in altitude. Geocentric parallax of a body at any altitude. The expression is used to distinguish the parallax at the given altitude from the horizontal parallax when the body is in the horizon. See also PARALLAX.

parallax inequality. The variation in the range of tide or in the speed of a tidal current due to changes in the distance of the moon from the earth. The range of tide and speed of the current tend alternately to increase and decrease as the moon approaches its perigee and apogee, respectively, the complete cycle being the anomalistic month. There is a similar but relatively unimportant inequality due to the sun; this cycle is the anomalistic year. The parallax has little direct effect upon the lunitidal intervals but tends to modify the phase effect. When the moon is in perigee, the priming and lagging of the tide due to the phase is diminished and when in apogee the priming and lagging is increased.

parallax reduction. Processing of observed high and low waters to obtain quantities depending upon changes in the distance of the moon, such as perigean and apogean ranges.

parallel, *adj*. Everywhere equidistant, as of lines or surfaces.

parallel, *n*. See PARALLEL OF LATITUDE, definition 1.

parallel indexing. The use of rotating parallel lines overlayed on a radar display to aid in piloting.

parallel motion protractor. An instrument consisting of a protractor and one or more arms attached to a parallel motion device, so that the movement of the arms is everywhere parallel. The protractor can be rotated and set at any position so that it can be oriented to a chart. Also called DRAFTING MACHINE.

parallel of altitude. A circle of the celestial sphere parallel to the horizon, connecting all points of equal altitude. Also called ALTITUDE CIRCLE, ALMUCANTAR. See also CIRCLE OF EQUAL ALTITUDE.

parallel of declination. A circle of the celestial sphere parallel to the celestial equator. Also called CELESTIAL PARALLEL, CIRCLE OF EQUAL DECLINATION. See also DIURNAL CIRCLE.

parallel of latitude. 1. A circle (or approximation of a circle) on the surface of the earth, parallel to the equator, and connecting points of equal latitude. Also called a PARALLEL. 2. A circle of the celestial sphere, parallel to the ecliptic, and connecting points of equal celestial latitude. Also called CIRCLE OF LONGITUDE.

parallelogram, *n*. A four-sided figure with both pairs of opposite sides parallel. A right-angled parallelogram is a rectangle; a rectangle with sides of equal length is a square. A parallelogram with oblique angles is a rhomboid; a rhomboid with sides of equal length is a rhombus.

parallel rulers. An instrument for transferring a line parallel to itself. In its most common form it consists of two parallel bars or rulers connected in such manner that when one is held in place, the other may be moved, remaining parallel to its original position.

parallel sailing. A method of converting departure into difference of longitude, or vice versa, when the true course is 090° or 270°.

parallel sphere. The celestial sphere as it appears to an observer at the pole, where celestial bodies appear to move parallel to the horizon.

parameter, *n*. 1. A quantity which remains constant within the limits of a given case or situation. 2. One of the components into which a craft's magnetic field is assumed to be resolved for the purpose of compass adjustment. The field caused by permanent magnetism is resolved into orthogonal components or parameters: Parameter P,

Parameter Q, and Parameter R. The field caused by induced magnetism is resolved into that magnetism induced in 9 imaginary soft iron bars or rods. With respect to the axis of a craft, these parameters lie in a fore-and-aft direction, an athwart ships direction, and in a vertical direction. See also ROD, definition 2.

paranthelion, *n*. A phenomenon similar to a PARHELION but occurring generally at a distance of 120° (occasionally 90° or 140°) from the sun.

paraselene *(pl. paraselenae), n*. A form of halo consisting of an image of the moon at the same altitude as the moon and some distance from it, usually about 22°, but occasionally about 46°. Similar phenomena may occur about 90°, 120°, 140°, or 180° from the moon. A similar phenomenon in relation to the sun is called a PARHELION, SUN DOG, or MOCK SUN. Also called MOCK MOON.

paraselenic circle. A halo consisting of a faint white circle through the moon and parallel to the horizon. It is produced by reflection of moonlight from vertical faces of ice crystals. A similar circle through the sun is called a PARHELIC CIRCLE.

parhelic circle. A halo consisting of a faint white circle through the sun and parallel to the horizon. It is produced by reflection of sunlight from vertical faces of ice crystals. A similar circle through the moon is called a PARASELENIC CIRCLE. Also called MOCK SUN RING.

parhelion *(pl. parhelia), n*. A form of halo, consisting of an image of the sun at the same altitude as the sun and some distance from usually about 22°, but occasionally about 40°. A similar phenomenon occurring at a distance of 90°, 120°, or 140° from the sun is called a PARANTHELION, and if occurring at a distance of 180° from the sun, an ANTHELION. A similar phenomenon in relation to the moon is called PARASELENE, MOON DOG, or MOCK MOON. The term PARHELION should not be confused with PERIHELION, the orbital point near the sun when the sun is the center of attraction. Also called SUN DOG, MOCK SUN.

parsec, *n*. The distance at which 1 astronomical unit subtends an angle of 1 second of arc. One parsec equals about 206,265 astronomical units or 30,857 × 10^{12} meters or 3.26 light years. The name parsec is derived from parallax second.

partial eclipse. An eclipse in which only part of the source of light is obscured. See ECLIPSE.

pascal, *n*. The special name for the derived unit of pressure and stress in the International System of Units; it is 1 newton per square meter.

pass, *n*. 1. A navigable channel leading to a harbor or river. Sometimes called PASSAGE. 2. A break in a mountain range, permitting easier passage from one side of the range to the other; also called COL. 3. A narrow opening through a barrier reef atoll, or sand bar. 4. A single circuit of the earth by a satellite. See also ORBIT. 5. The period of time a satellite is within telemetry range of a data acquisition station.

passage, *n*. 1. A navigable channel, especially one through reefs or islands. Also called PASS. 2. A transit from one place to another; one leg of a voyage.

passing light. A low intensity light which may be mounted on the structure of another light to enable the mariner to keep the latter light in sight when he passes out of its beam. See also SUBSIDIARY LIGHT.

passive satellite. 1. A satellite which contains power source to augment the output signal (i.e., reflected only) as contrasted with ACTIVE SATELLITE; a satellite which is a passive reflector. 2. As defined by the International Telecommunications Union (ITU), an earth satellite intended to transmit radiocommunication signals by reflection.

passive system. A term used to describe a navigation system whose operation does not require the user to transmit a signal.

patent log. A mechanical log, particularly a TAFFRAIL LOG.

patent slip. See MARINE RAILWAY.

path, *n*. See as ORBITAL PATH.

pattern, *n*. 1. See under LATTICE. 2. In a hyperbolic radionavigation system, the family of hyperbolas associated with a single pair of stations, usually the master station and a slave (secondary) station.

P-band. A radio-frequency band of 225 to 390 megahertz. See also FREQUENCY, FREQUENCY BAND.

P-code. The precise code of the GPS signal, used by military receivers.

polar cap anomaly. See under POLAR CAP DISTURBANCE.

peak, *n*. 1. On the sea floor, a prominent elevation, part of a larger feature, either pointed or of very limited extent across the summit. 2. A pointed mountain summit. 3. An individual or conspicuous mountain with a single conspicuous summit, as Pikes Peak. 4. The summit of a mountain. 5. A term sometimes used for a headland or promontory.

peak envelope power. See under POWER (OF A RADIO TRANSMITTER).

pebble, *n*. See under STONES.

pelorus, *n*. A dumb compass, or a compass card (called a PELORUS CARD) without a directive element, suitably mounted and provided with vanes to permit observation of relative bearings unless used in conjunction with a compass to give true or magnetic bearings.

pelorus card. The part of a pelorus on which the direction graduations are placed. It is usually in the form of a thin disk or annulus graduated in degrees, clockwise, from 0° at the reference direction to 360°.

pendulous gyroscope. A gyroscope with its axis of rotation constrained by a suitable weight to remain horizontal. The pendulous gyroscope is the basis of one type of gyrocompass.

peninsula, *n*. A section of land nearly surrounded by water. Frequently, but not necessarily, a peninsula is connected to a larger body of land by a neck or isthmus.

pentagon, *n*. A closed plane figure having five sides.

pentagonal cluster. An arrangement of five corner reflectors, mounted so as to give their maximum response in a horizontal direction, and equally spaced on the circumference of a circle. The response is substantially uniform in all horizontal directions. See also OCTAHEDRAL CLUSTER.

penumbra, *n*. 1. That part of a shadow in which light is partly cut off by an intervening object. The penumbra surrounds the darker UMBRA in which light is completely cut off. 2. The lighter part of a sun spot, surrounding the darker UMBRA.

penumbral lunar eclipse. The eclipse of the moon when the moon passes only through the penumbra of the earth's shadow.

performance monitor. A device used to check the performance of the transmitter and receiver of a radar set. Such device does not provide any indication of performance as it might be affected by the propagation of the radar waves through the atmosphere. An echo box is used in one type of performance monitor called an echo box performance monitor.

per gyrocompass (PGC). Relating to or from the gyrocompass.

periapsis, *n*. See PERICENTER.

periastron, *n*. That point of the orbit of one member of a double star system at which the stars are nearest together. That point at which they are farthest apart is called APASTRON.

pericenter, *n*. In an elliptical orbit, the point in the orbit which is the nearest distance from the focus where the attracting mass is located. the pericenter is at one end of the major axis of the orbital ellipse. The opposite is APOAPSIS, APOCENTER. Also called PERIAPSIS, PERIFOCUS.

perifocus, *n*. See PERICENTER.

perigean range. See under PERIGEAN TIDES.

perigean tidal currents. Tidal currents of increased speed occurring monthly as the result of the moon being in perigee or nearest the earth.

perigean tides. Tides of increased range occurring monthly as the result of the moon being in perigee or nearest the earth. The perigean range of tide is the average semidiurnal range occurring at the time of perigean tides and is most conveniently computed from the harmonic constants. It is larger than the mean range where the type of tide is either semidiurnal or mixed and is of no practical significance where the type of tide is diurnal.

perigee, *n*. The orbital point nearest the earth when the earth is the center of attraction. The orbital point farthest from the earth is called APOGEE. See also APOCENTER, PERICENTER.

perigee-to-perigee period. See ANOMALISTIC PERIOD.

perigon, *n*. An angle of 360°.

perihelion, *n*. That orbital point nearest the sun when the sun is the center of attraction. That point farthest from the sun is called APHELION.

perimeter, *n*. 1. The length of a closed plane curve or the sum of the sides of a polygon. 2. The boundary of a plane figure. Also called PERIPHERY.

period, *n*. 1. The interval needed to complete a cycle. See also NATURAL PERIOD, SIDEREAL PERIOD, SYNODIC PERIOD, WAVE PERIOD). 2. The interval of time between the commencement of two identical successive cycles of the characteristic of the light.

periodic, *adj*. Of or pertaining to a period.

periodic error. An error whose amplitude and direction vary systematically with time.

periodic perturbations. Perturbations to the orbit of a satellite which change direction in regular or periodic manner in time, such that the average effect over a long period of time is zero.

periodic terms. In the mathematical expression of the orbit of a satellite, terms which vary with time in both magnitude and direction in a periodic manner. See also SECULAR TERMS.

period of satellite. 1. See ORBITAL PERIOD. 2. As defined by the International Telecommunication Union (ITU), the time elapsing between two consecutive passages of a satellite or planet through a characteristic point on its orbit.

periphery, *n*. See PERIMETER.

periplus, *n*. The early Greek name for SAILING DIRECTIONS. The literal meaning of the term is "a sailing round."

periscope, *n*. An optical instrument which displaces the line of sight parallel to itself, to permit a view which may otherwise be obstructed.

periscope sextant. A sextant designed to be used in conjunction with the periscope of a submarine.

permafrost, *n*. Permanently frozen subsoil. Any soil or other deposit, including rock, the temperature of which has been below freezing continuously for 2 years or more is considered permafrost.

Permalloy, *n*. The trade name for an alloy of about 80% nickel and 20% iron, which is very easily magnetized and demagnetized.

permanent current. A current that runs fairly continuously and is independent of tides and other temporary causes.

permanent echo. An echo from an object whose position relative to the radar set is fixed.

permanent light. A light used in regular service.

permanent magnetism. The magnetism which is acquired by hard iron, which is not readily magnetized by induction, but which retains a high percentage of magnetism acquired unless subjected to a demagnetizing force. The strength and polarity of this magnetism in a craft depends upon the heading, magnetic latitude, and building stresses imposed during construction. See also INDUCED MAGNETISM, SUBPERMANENT MAGNETISM.

permeability, *n*. 1. The ability to transmit magnetism; magnetic conductivity. 2. The ability to permit penetration or passage. In this sense the term is applied particularly to substances which permit penetration or passage of fluids.

perpendicular, *adj*. At right angles; normal.

perpendicular, *n*. A perpendicular line, plane, etc. A distinction is sometimes made between PERPENDICULAR and NORMAL, the former applying to a line at right angles to a straight line or plane, and the latter referring to a line at right angles to a curve or curved surface.

persistence, *n*. A measure of the time of decay of the luminescence of the face of the cathode ray tube after excitation by the stream of electrons has ceased. Relatively slow decay is indicative of high persistence. Persistence is the length of time during which phosphorescence takes place. See also AFTERGLOW, definition 1.

personal correction. A correction due to personal error. Also called PERSONAL EQUATION.

personal equation. A term used for both PERSONAL ERROR and PERSONAL CORRECTION.

personal error. A systematic error in the observation of a quantity due to the personal idiosyncrasies of the observer. Also called PERSONAL EQUATION.

perspective chart. A chart on a perspective map projection.

perspective map projection. A map projection produced by the direct projection of the points of the ellipsoid (used to represent the earth) by straight lines drawn through them from some given point. The projection is usually made upon a plane tangent to the ellipsoid at the end of the diameter joining the point of projection and the center of the ellipsoid. The plane of projection is usually tangent to the ellipsoid at the center of the area being mapped. he analytical expressions that determine the elements of the projection. If the point of projection is at the center of the ellipsoid, a gnomonic map projection results; if it is at the point opposite the plane's point of tangency a stereographic map projection; and if at infinity (the projecting lines being parallel to each other), an orthographic map projection. Most map projections are not perspective. Also called GEOMETRIC MAP PROJECTION.

perspective map projection upon a tangent cylinder. A cylindrical map projection upon a cylinder tangent to the ellipsoid produced by perspective projection from the ellipsoid's center. The geographic meridians are represented by a family of equally spaced parallel straight lines, perpendicular to a second family of parallel straight lines which represent the geographic parallels of latitude. The spacing, with respect to the equator of the lines which represent the parallels of latitude, increases as the tangent function of the latitude; the line representing 90° latitude is at an infinite distance from the line which represents the equator. Not to be confused with MERCATOR MAP PROJECTION to which it bears a general resemblance.

perspective projection. The representation of a figure on a surface, either plane or curved, by means of projecting lines emanating from a single point, which may be infinity. Also called GEOMETRIC PROJECTION. See also PERSPECTIVE MAP PROJECTION.

per standard compass. Relating to the standard magnetic compass.

per steering compass. Relating to the magnetic steering compass.

perturbations, *n. (pl.)*. In celestial mechanics differences of the actual orbit from a central force orbit, arising from some external force such as a third body attracting the other two; a resisting medium (atmosphere); failure of the parent body to act as a point mass, and so forth. Also the forces that cause differences between the actual and reference (central force) orbits. See also GRAVITATIONAL PERTURBATIONS, LONG PERIOD PERTURBATIONS, LUNISOLAR PERTURBATIONS, NONGRAVITATIONAL PERTURBATIONS, PERIODIC PERTURBATIONS, SECULAR PERTURBATIONS, SHORT PERIOD PERTURBATIONS, TERRESTRIAL PERTURBATIONS.

perturbed orbit. The orbit of a satellite differing from its normal orbit due to various disturbing effects, such as nonsymmetrical gravitational effects, atmospheric drag, radiation pressure, and so forth. See also PERTURBATIONS.

perturbing factor. In celestial mechanics, any factor that acts on an orbiting body to change its orbit from a central force orbit. Also called PERTURBING FORCE.

perturbing force. See PERTURBING FACTOR.

Peru Coastal Current. See PERU CURRENT.

Peru Current. A narrow, fairly stable ocean current that flows northward close to the South American coast. It originates off the coast of Chile at about latitude 40°S and flows past Peru and Ecuador to the southwest extremity of Colombia. The southern portion of the Peru Current is sometimes called the CHILE CURRENT. It has sometimes been called the HUMBOLDT CURRENT because an early record of its temperature was taken by the German scientist Alexander von Humboldt in 1802. The name Corriente del Peru was adopted by a resolution of the Ibero-American Oceanographic Conference at its Madrid-Malaga meeting in April 1935. Also called PERU COASTAL CURRENT.

Peru Oceanic Current. See MENTOR CURRENT.

phantom, *n*. That part of a gyrocompass carrying the compass card.

phantom bottom. A false bottom indicated by an echo sounder, some distance above the actual bottom. Such an indication, quite common in the deeper parts of the ocean, is due to large quantities of small organisms.

phantom echo. See PHANTOM TARGET.

phantom target. 1. An indication of an object on a radar display that does not correspond to the presence of an actual object at the point indicated. Also called PHANTOM ECHO. 2. See ECHO BOX.

phase, *n*. The amount by which a cycle has progressed from a specified origin. For most purposes it is stated in circular measure, a complete cycle being considered 360°. See also PHASES OF THE MOON.

phase angle. The angle at a celestial body between the sun and earth.

phase coding. In Loran C, the shifting in a fixed sequence of the relative phase of the carrier cycles between certain pulses of a group. This shifting facilitates automatic synchronization in identical sequence within the group of eight pulses that are transmitted during each group repetition interval. It also minimizes the effect of unusually long skywave transmissions causing one pulse to interfere with the succeeding pulse in the group received by groundwave.

phase inequality. Variations in the tides or tidal currents due to changes in the phase of the moon. At the times of new and full moon the tide-producing forces of the moon and sun act in conjunction, causing the range of tide and speed of the tidal current to be greater than the average, the tides at these times being known as spring tides. At the time of quadrature of the moon these forces are opposed to each other, causing the neap tides with diminished range and current speed.

phase lag. See EPOCH, definition 3.

phase lock. The technique whereby the phase of an oscillator signal is made to follow exactly the phase of a reference signal by first comparing the phases of the two signals and then using the resulting phase difference signal to adjust the reference oscillator frequency to eliminate phase difference when the two signals are next compared.

phase meter. An instrument for measuring the difference in phase of two waves of the same frequency.

phase modulation. The process of changing the phase of a carrier wave in accordance with the variations of a modulating wave. See also MODULATION.

phase reduction. Processing of observed high and low waters to obtain quantities depending upon the phase of the moon, such as the spring and neap ranges of tide. Formerly this process was known as SECOND REDUCTION. Also applicable to tidal currents.

phases of the moon. The various appearances of the moon during different parts of the synodical month. The cycle begins with new moon or change of the moon at conjunction. The visible part of the waxing moon increases in size during the first half of the cycle until full moon appears at opposition, after which the visible part of the waning moon decreases for the remainder of the cycle. First quarter occurs when the waxing moon is at east quadrature; last quarter when the waning moon is at west quadrature. From last quarter to new and from new to first quarter the moon is crescent; from first quarter to full and from full to last quarter it is gibbous. The elapsed time, usually expressed in days, since the last new moon is called age of the moon. The full moon occurring nearest the autumnal equinox is called harvest moon; the next full moon, hunter's moon.

phase synchronized. A term used to indicate that radio wave transmissions have the same phase at their sources at any instant of time.

phenomenon *(pl. phenomena), n.* 1. An occurrence or event capable of being explained scientifically, particularly one relating to the unusual. 2. A rare or unusual event.

phonetic alphabet. A list of standard words used to identify letters in a message transmitted by radio or telephone.

phosphor, *n.* A phosphorescent substance which emits light when excited by radiation, as on the scope of a cathode-ray tube.

phosphorescence, *n.* Emission of light without sensible heat, particularly as a result of but continuing after absorption of radiation from some other source. PERSISTENCE is the length of time during which phosphorescence takes place. The emission of light or other radiant energy as a result of and only during absorption of radiation from some other source is called FLUORESCENCE.

photogrammetry, *n.* 1. The science of obtaining reliable measurements from photographic images. 2. The science of preparing charts and maps from aerial photographs using stereoscopic equipment and methods.

photosphere, *n.* The bright portion of the sun visible to the unaided eye.

physical double star. Two stars in nearly the same line of sight and at approximately the same distance from the observer, as distinguished from an OPTICAL DOUBLE STAR (two stars in nearly the same line of sight but differing greatly in distance from the observer). If they revolve about their common center of mass, they are called a **binary star**.

pico-. A prefix meaning one-trillionth (10^{-12}).

piedmont, *n.* An area of hills situated at the base of a range of mountains.

pier, *n.* 1. A structure extending into the water from a shore or a bank which provides berthing for ships, or use as a promenade or fishing pier. See also WHARF. 2. A support for the spans of a bridge.

pierhead, *n.* The outer end of a pier or jetty.

pile, *n.* A long, heavy timber or section of steel, concrete, etc., forced into the earth to serve as a support, as for a pier, or to resist lateral pressure.

pile beacon. A beacon formed of one or more piles.

pile dolphin. A minor light structure consisting of a number of piles driven into the bottom in a circular pattern and drawn together with a light mounted at the top. Referred to in the *Light List* as a DOLPHIN.

pillar buoy. A buoy composed of a tall central structure mounted on a broad flat base.

pilot, *n.* 1. A person who directs the movement of a vessel through pilot waters, usually a person who has demonstrated extensive knowledge of channels, aids to navigation, dangers to navigation, etc., in a particular area and is licensed in that area. See also LOCAL KNOWLEDGE. 2. A book of sailing directions. For waters the United States and its possessions, They are prepared by the National Ocean Survey, and are called COAST PILOTS.

pilotage, *n.* 1. The services of especially qualified navigators having local knowledge who assist in the navigation of vessels in particular areas. Also called PILOTAGE SERVICE. 2. A term loosely used for piloting.

pilotage service. See PILOTAGE, definition 1.

pilotage waters. See PILOT WATERS.

pilot boat. A small vessel used by the pilot to go or from a vessel employing his services. Also called PILOT VESSEL.

pilot chart. A chart of a major ocean area which presents in graphic form averages obtained from weather, wave, ice, and other marine data gathered over many years in meteorology and oceanography to aid the navigator in selecting the quickest and safest routes; published by the Defense Mapping Agency Hydrographic/Topographic Center from data provided by the U.S. Naval Oceanographic Office and the Environmental Data and Information Service of the National Oceanic and Atmospheric Administration.

piloting, *n.* Navigation involving frequent or continuous determination of position relative to observed geographical points, to a high order of accuracy; directing the movements of a vessel near a coast by means of terrestrial reference points is called coast piloting. Sometimes called PILOTAGE. See also PILOTAGE, definition 1.

pilot rules. Regulations supplementing the Inland Rules of the Road, superseded by the adoption of the Inland Navigation Rules in 1980 (1983 on the Great Lakes).

pilot station. The office or headquarters of pilots; the place where the services of a pilot may be obtained.

pilot vessel. See PILOT BOAT.

pilot waters. 1. Areas in which the services of a marine pilot are essential. 2. Waters in which navigation is by piloting. Also called PILOTAGE WATERS.

pinnacle, *n.* A high tower or spire-shaped pillar of rock or coral on the sea floor, alone or cresting a summit. It may or may not be a hazard to surface navigation. Due to the steep rise from the sea floor no warning is given by sounding.

pinnacled iceberg. An iceberg weathered in such manner as to produce spires or pinnacles. Also called PYRAMIDAL ICEBERG, IRREGULAR ICEBERG.

pip, *n.* See BLIP.

pitch, *n.* 1. Oscillation of a vessel about the transverse axis due to the vessel's bow and stern being raised or lowered on passing through successive crests and troughs of waves. Also called PITCHING. See also SHIP MOTIONS. 2. The distance a propeller would advance longitudinally in one revolution if there were no slip.

pitch, *v., i.* To oscillate about the transverse axis. See also SHIP MOTIONS.

pitching, *n.* See PITCH, definition 1.

pivot point. The point on the centerline between the bow and the center of gravity at which the resultant of the velocities of rotation and translation is directed along the centerline, after a ship has assumed its drift angle in a turn. To an observer on board, the ship appears to rotate about this point.

pixel. The smallest area of phosphors on a video terminal that can be excited to form a picture element.

place name. See TOPONYM.

plain, *n*. On the sea floor, a flat, gently sloping or nearly level region. Sometimes called ABYSSAL PLAIN in very deep water.

plan, *n*. 1. An orthographic drawing or view on a horizontal plane, as of an instrument, a horizontal section, or a layout. 2. A large-scale map or chart of a small area, generally showing at increased scale a portion of the chart on which it is placed.

planar, *adj*. Lying in a plane.

plane, *n*. A surface without curvature, such that a straight line joining any two of its points lies wholly on the surface.

plane of polarization. With respect to a plane polarized wave, the plane containing the electric field vector and the direction of propagation.

plane polarized wave. An electromagnetic wave the electric field vector of which lies at all times in a fixed plane which contains the direction of propagation.

plane sailing. A method of solving the various problems involving a single course and distance, difference of latitude, and departure, in which the earth, or that part traversed. is considered as a plane surface.

planet, *n*. A celestial body of a solar system, in orbit around the sun or a star and shining by reflected light. The larger of such bodies are sometimes called major planets to distinguish them from minor planets (asteroids) which are very much smaller. Larger planets may have satellites. In the solar system an inferior planet has an orbit smaller than that of the earth; a superior planet has an orbit larger than that of the earth. The four planets commonly used for celestial observations are called navigational planets. The word planet is of Greek origin, meaning, literally, wanderer, applied because the planets appear to move relative to the stars.

planetary, *adj*. Of a planet or the planets; terrestrial; worldwide.

planetary aberration. See under ABERRATION definition 1.

planetary configurations. Apparent positions of the planets relative to each other and to other bodies of the solar system, as seen from the earth.

planetary precession. The component of general precession caused by the effect of other planets on the equatorial protuberance of the earth producing an eastward motion of the equinoxes along the ecliptic. See also PRECESSION OF THE EQUINOXES.

planetoid, *n*. See ASTEROID.

plane triangle. A closed plane figure having three straight lines as sides.

planimetric map. A map indicating only the horizontal positions of features, without regard to elevation, in contrast with a TOPOGRAPHIC MAP, which indicates both horizontal and vertical positions.

planisphere, *n*. A representation on a plane of the celestial sphere, especially one on a polar projection, with means provided for making certain measurements such as altitude and azimuth. See also STAR FINDER.

plankton, *n*. Floating, drifting, or feebly swimming plant and animal organisms of the sea. These are usually microscopic or very small, although jellyfish are included.

planning chart. A chart designed for use in planning voyages or flight operations or investigating areas of marine or aviation activities.

plan position indicator. An intensity-modulated radar display in which the radial sweep rotates on the cathode-ray tube in synchronism with the rotating antenna. The display presents a maplike representation of the positions of echo-producing objects. It is generally one of two main types: RELATIVE MOTION DISPLAY or TRUE MOTION DISPLAY.

plastic relief map. A topographic map printed on plastic and molded into a three-dimensional form.

plateau, *n*. On the sea floor, a comparatively flat-topped feature of considerable extent, dropping off abruptly on one or more sides.

plate glass. A fine quality sheet glass obtained by rolling, grinding, and polishing.

platform erection. In the alignment of inertial navigation equipment, the alignment of the stable platform vertical axis with the local vertical.

platform tide. See STAND.

Platonic year. See GREAT YEAR.

Plimsoll mark. A mark on a ship's side indicating how deeply she may be loaded.

plot, *n*. A drawing consisting of lines and points representing certain conditions graphically, as the progress of a craft. See also NAVIGATIONAL PLOT.

plot, *v., t*. To draw lines and points to represent certain conditions graphically, as the various lines and points on a chart or plotting sheet representing the progress of a vessel, a curve of magnetic azimuths vs. time or of altitude vs. time, or a graphical solution of a problem, such as a relative motion solution.

plotter, *n*. An instrument used for plotting straight lines and measuring angles on a chart or plotting sheet. See also PROTRACTOR.

plotting chart. An outline chart on a specific scale and projection, usually showing a graticule and compass rose, designed to be used ancillary to a standard nautical chart, and produced either as an independent chart or part of a coordinated series. See also POSITION PLOTTING SHEET.

plotting head. See REFLECTION PLOTTER.

plumb bob. A conical device, usually of brass and suspended by a chord, by means of which a point can be projected vertically into space over relatively short distances.

plumb-bob vertical. See LOCAL VERTICAL.

plumb line. 1. A line in the direction of gravity. 2. A cord with a weight at one end for determining the direction of gravity.

pluvial, *adj*. Of or pertaining to rain. The expression pluvial period is often used to designate an extended period or age of heavy rainfall.

P.M. Abbreviation for Post Meridian; after noon in zone time.

pocosin, *n*. See DISMAL.

point, *n*. 1. A place having position, but no extent. 2. A tapering piece of land projecting into a body of water. It is generally less prominent than a CAPE. 3. One thirty-second of a circle, or 11 1/4°. Also called COMPASS POINT when used in reference to compass directions. See also FOUR-POINT BEARING.

point designation grid. A system of lines, having no relation to the actual scale or orientation, drawn on a map, chart, or air photograph, dividing it into squares so that points can be more readily located.

point light. A luminous signal without perceptible length, as contrasted with a LINEAR LIGHT which has perceptible length.

point of arrival. The position at which a craft is assumed to have reached or will reach after following specified courses for specified distance from a point of departure. See also DESTINATION.

point of departure. The point from which the initial course to reach the destination begins. It is usually established by bearings of prominent landmarks as the vessel clears a harbor and proceeds to sea. When a person establishes this point, he is said to take departure. Also called the DEPARTURE.

point of destination. See DESTINATION.

point of inflection. The point at which a reverse in direction of curvature takes place.

polar, *adj*. Of or pertaining to a pole or the poles.

polar air. A type of air whose characteristics are developed over high latitudes, especially Within the subpolar highs. Continental polar air has low surface temperature, low moisture content, and especially in its source regions, has great stability in the lower layers. It is shallow in comparison with arctic air. Maritime polar air initially possesses similar properties to those of continental polar air, but in passing over warmer water it becomes unstable with a higher moisture content.

polar axis. 1. The straight line connecting the poles of a body 2. A reference line for one of the spherical coordinates.

polar cap absorption. See under POLAR DISTURBANCE.

polar cap disturbance. An ionospheric disturbance (which does not refer to the ice cap in the polar regions). It is a result of the focusing effect that the earth's magnetic field has on particles released from the sun during a solar proton event. The effect concentrates high-energy particles in the region of the magnetic pole with the result that normal very low frequency Omega propagation is disrupted. The effect on radio waves is known as POLAR CAP ABSORPTION (PCA). Historically, polar cap disturbances (PCDs) produced large or total absorption of high frequency radio waves crossing the polar region, hence the term POLAR CAP ABSORPTION. A transmission path which is entirely outside the polar region is unaffected by a PCD. The PCDs, often called PCA EVENTS (PCAs), may persist for a week or more, but duration of only a few days is more common. The PCD can cause line of position errors about 6 to 8 nautical miles. The *Omega Propagation Correction Tables* make no allowance for this phenomenon since it is not predictable. However, the frequency of the phenomenon increases during those years of peak solar activity. See also SUDDEN IONOSPHERIC DISTURBANCE, MODAL INTERFERENCE.

polar chart. 1. A chart of polar areas. 2. A chart on a polar projection. The projections most used for polar charts are the gnomonic, stereographic, azimuthal equidistant, transverse Mercator, and modified Lambert conformal.

polar circles. The minimum latitudes, north and south, at which the sun becomes circumpolar.

polar continental air. Air of an air mass that originates over land or frozen ocean areas in polar regions. Polar continental air is characterized by low temperature, stability, low specific humidity, and shallow vertical extent.

polar coordinates. A system of coordinates defining a point by its distance and direction from a fixed point, called the POLE. Direction is given as the angle between a reference radius vector and a radius vector to the point. If three dimensions are involved, two angles are used to locate the radius vector. See also SPACE-POLAR COORDINATES.

polar distance. Angular distance from a celestial pole; the arc of an hour circle between a celestial pole, usually the elevated pole, and a point on the celestial sphere, measured from the celestial pole through 180°. See also CODECLINATION.

polar front. The semi-permanent, semi-continuous front separating air masses of tropical and polar origin. This is the major front in terms of air mass contrast and susceptibility to cyclonic disturbance.

Polaris correction. A correction to be applied to the corrected sextant altitude of Polaris to obtain latitude. This correction for the offset of Polaris from the north celestial pole varies with the local hour angle of Aries, latitude, and date. See Q-CORRECTION.

polarization, *n*. The attribute of an electromagnetic wave which describes the direction of the electric field vector.

polarization error. An error in a radio direction finder bearing or the course indicated by a radiobeacon because of a change in the polarization of the radio waves between the transmitter and receiver on being reflected and refracted from the ionosphere. Because the medium frequency radio direction finder normally operates with vertically polarized waves, a change to horizontal polarization in the process of reflection and refraction of the waves from the ionosphere can have a serious effect on bearing measurements. If the horizontally polarized skywaves are of higher signal strength than the vertically polarized groundwaves, the null position for the loop antenna cannot be obtained. If the skywaves are of lower signal strength than the groundwaves, the null position is made less distinct. Before the cause of the error was understood, it was called NIGHT EFFECT or NIGHT ERROR because it occurs principally during the night, and especially during twilight when rapid changes are occurring in the ionosphere.

polar map projection. A map projection centered on a pole.

polar maritime air. An air mass that originates in the polar regions and is then modified by passing over a relatively warm ocean surface. It is characterized by moderately low temperature, moderately high surface specific humidity, and a considerable degree of vertical instability. When the air is colder than the sea surface, it is further characterized by gusts and squalls, showery precipitation, variable sky, and good visibility between showers.

polar motion. See EULERIAN MOTION.

polar navigation. Navigation in polar regions, where unique considerations and techniques are applied. No definite limit for these regions is recognized but polar navigation techniques are usually used from about latitude 70°N.

polar orbit. An earth satellite orbit that has an inclination of about 90° and, hence, passes over or near the earth's poles.

polar orthographic map projection. An orthographic map projection having the plane of the projection perpendicular to the axis of rotation of the earth, in this projection, the geographic parallels are full circles, true to scale, and the geographic meridians are straight lines.

polar regions. The regions near the geographic poles. No definite limit for these regions is recognized.

polar satellite. A satellite that passes over or near the earth's poles, i.e., a satellite whose orbital plane has an inclination of about 90° to the plane of the earth's equator.

polar stereographic map projection. A stereographic map projection having the center of the projection located at a pole of the sphere.

pole, *n*. 1. Either of the two points of intersection of the surface of a sphere or spheroid and its axis, labeled N or S to indicate whether the north pole or south pole. The two points of intersection of the surface of the earth with its axis are called geographical poles. The two points of intersection of the celestial sphere and the extended axis of the earth are called celestial poles. The celestial pole above the horizon is called the elevated pole; that below the horizon the depressed pole. The ecliptic poles are 90° from the ecliptic. Also, one of a pair of similar points on the surface of a sphere or spheroid, as a magnetic pole, definition l; a geomagnetic pole; or a fictitious pole. 2. A magnetic pole, definition 2. 3. The origin of measurement of distance in polar or spherical coordinates. 4. Any point around which something centers.

pole beacon. A vertical spar fixed in the ground or in the sea bed or a river bed to show as a navigation mark. Sometimes called SPINDLE BEACON or SINGLE-PILE BEACON in the United States.

polyconic, *adj*. Consisting of or related to many cones.

polyconic chart. A chart on the polyconic map projection.

polyconic map projection. A conic map projection in which the surface of a sphere or spheroid, such as the earth, is conceived as developed on a series of tangent cones, which are then spread out to form a plane. A separate cone is used for each small zone. This projection is widely used for maps but seldom used for charts, except for survey purposes. It is not conformal.

polygon, *n*. A closed plane figure bounded by straight lines. See also HEXAGON, OCTAGON, PARALLELOGRAM, PENTAGON, QUADRILATERAL, RECTANGLE, SQUARE, TRAPEZOID, TRIANGLE.

polynya, *n*. A non-linear shaped area of water enclosed by ice. Polynyas may contain brash ice and/or be covered with new ice, nilas, or young ice; submariners refer to these as SKYLIGHTS. Sometimes the POLYNYA is limited on one side by the coast and is called a SHORE POLYNYA or by fast ice and is called a FLAW POLYNYA. If it recurs in the same position every year, it is called a RECURRING POLYNYA.

polyzoa, *n., pl*. Very small marine animals which reproduce by budding, many generations often being permanently connected by branchlike structures. These animals are often very numerous and in some areas they cover the bottom. Also called BRYOZOA.

pond, *n*. A relatively small body of water, usually surrounded on all sides by land. A larger body of water is called a LAKE.

pontoon, *n*. A float or low, flat-bottomed vessel to float machinery such as cranes, capstans, etc. or to support weights such as floating bridges boat landings, etc.

pool, *n*. 1. A small body of water, usually smaller than a pond, especially one that is quite deep. One left by an ebb tide is called a **tide pool**. 2. A small and comparatively still, deep part of a larger body of water such as a river or harbor.

poop, *n*. A short enclosed structure at the stern of a vessel, extending from side to side. It is covered by the poop deck, which is surrounded by the poop rail.

pooped. To have shipped a sea or wave over the stern.

pororoca, *n*. See TIDAL BORE.

port, *n*. 1. A place provided with moorings and transfer facilities for loading and discharging cargo or passengers, usually located in a harbor. 2. The left side of a craft, facing forward. The opposite is STARBOARD.

portfolio, *n*. A portable case for carrying papers. See also CHART PORTFOLIO.

port hand buoy. A buoy which is to be left to the port side when approaching from the open sea or proceeding in the direction of the main stream of flood current, or in the direction established by appropriate authority.

port of call. A port visited by a ship.

Portugal Current. A slow-moving current that is the prevailing southward flow off the Atlantic coasts of Spain and Portugal. Its speed averages only about 0.5 knot during both winter and summer. The maximum speed seldom exceeds 2.0 knots north of latitude 40°N and 2.5 knots south of 40°N. It is easily influenced by winds.

Portuguese norther. See under NORTHER.

position, *n*. A point defined by stated or implied coordinates, particularly one on the surface of the earth. A fix is a relatively accurate position determined without reference to any former position. A running fix is a position determined by crossing lines of position obtained at different times and advanced or retired to a common time. An estimated position is determined from incomplete data or data of questionable accuracy. A dead reckoning position is determined by advancing a previous position for courses and distances. A most probable position is a position judged to be most accurate when an element of doubt exists as to the true position. It may be a fix, running fix, estimated position, or dead reckoning position depending upon the information upon which it is based. An assumed position is a point at which a craft is assumed to be located. A geographical position is that point on the earth at which a given celestial body is in the zenith at a specified time, or any position defined by means of its geographical coordinates. A geodetic position is a point on the earth the coordinates of which have been determined by triangulation from an accurately known initial station, or one defined in terms of geodetic latitude and longitude. An astronomical position is a point on the earth whose coordinates have been determined as a result of observation of celestial bodies, or one defined in terms of astronomical latitude and longitude. A maritime position is the location of a seaport or other point along a coast. A relative position is one defined with reference to another position, either fixed or moving. See also PINPOINT, LINE OF POSITION, BAND OF POSITION, SURFACE OF POSITION.

position angle. See PARALLACTIC ANGLE.

position approximate. Of inexact position. The expression is used principally on charts to indicate that the position of a wreck, shoal, etc., has not been accurately determined or does not remain fixed.

position buoy. An object towed astern to assist a following vessel in maintaining the desired or prescribed distance, particularly in conditions of low visibility.

position circle. 1. The chart symbol denoting the position of a buoy. 2. See CIRCLE OF POSITION.

position doubtful. Of uncertain position. The expression is used principally on charts to indicate that a wreck, shoal, etc., has been reported in various positions and not definitely determined in any. See also VIGIA.

positioning, *n*. The process of determining, at a particular point in time, the precise physical location of a craft, vehicle, person or site.

position line. See LINE OF POSITION.

position plotting sheet. A blank chart, usually on the Mercator projection, showing only the graticule and a compass rose. The meridians are usually unlabeled by the publisher so that they can be appropriately labeled when the chart is used in any longitude. It is designed and intended for use in conjunction with the standard nautical chart. See also SMALL AREA PLOTTING SHEET, UNIVERSAL PLOTTING SHEET, PLOTTING CHART.

post meridian (PM). After noon, or the period of time between noon (1200) and midnight (2400). The period between midnight and noon is called ANTE MERIDIAN.

potential, *n*. The difference in voltage at two points in a circuit.

potential energy. Energy possessed by a body by virtue of its position, in contrast with KINETIC ENERGY, that possessed by virtue of its motion.

pound, *n*. A unit of mass equal to 0.45359237 kilograms. Also called AVOIRDUPOIS POUND.

pound, *v., i*. To strike oncoming waves repeatedly or heavily.

pounding, *n*. A series of shocks received by a pitching vessel as it repeatedly or heavily strikes the water in a heavy sea. The shocks can be felt over the entire vessel and each one is followed by a short period of vibration.

power, *n*. 1. Rate of doing work. 2. Luminous intensity. 3. The number of times an object is magnified by an optical system, such as a telescope. Usually called MAGNIFYING POWER. 4. The result of multiplying a number by itself a given number of times. See also EXPONENT.

power gain (of an antenna). See DIRECTIVITY, definition 2.

power gain (of a transmitter). The ratio of the output power delivered to a specified load by an amplifier to the power absorbed by its input circuit.

power (of a radio transmitter), *n*. The power of a radio transmitter is expressed in one of the following forms: The peak envelope power is the average power supplied to the antenna transmission line by a transmitter during one radio frequency cycle at the highest crest of the modulation envelope, taken under conditions of normal operation. The mean power is the power supplied to the antenna transmission line by a transmitter during normal operation, averaged over a time sufficiently long compared with the period of the lowest frequency encountered in the modulation. The carrier power is the average power supplied to the antenna transmission line by a transmitter during one radio frequency cycle under conditions of no modulation. This definition does not apply to pulse modulated emissions.

PPI display. See as PLAN POSITION INDICATOR.

PPI repeater. See RADAR REPEATER.

precautionary area. A routing measure comprising an area within defined limits where ships must navigate with particular caution and within which the direction of traffic flow may be recommended. See also ROUTING SYSTEM.

precession, *n*. The change in the direction of the axis of rotation of a spinning body, as a gyroscope, when acted upon by a torque. The direction of motion of the axis is such that it causes the direction of spin of the gyroscope to tend to coincide with that of the impressed torque. The horizontal component of precession is called drift, and the vertical component is called topple. Also called INDUCED PRECESSION, REAL PRECESSION. See also APPARENT PRECESSION, PRECESSION OF THE EQUINOXES.

precession in declination. The component of general precession along a celestial meridian, amounting to about 20.0" per year.

precession in right ascension. The component of general precession along the celestial equator, amounting to about 46.1" per year.

precession of the equinoxes. The conical motion of the earth's axis about the vertical to the plane of the ecliptic, caused by the attractive force of the sun, moon, and other planets on the equatorial protuberance of the earth. The effect of the sun and moon, called lunisolar precession, is to produce a westward motion of the equinoxes along the ecliptic. The effect of other planets, called planetary precession, tends to produce a much smaller motion eastward along the ecliptic. The resultant motion, called general precession, is westward along the ecliptic at the rate of about 50.3" per year. The component of general precession along the celestial equator, called precession in right ascension, is about 46.1" per year and the component along a celestial meridian, called precession in declination, is about 20.0" per year.

precipice, *n*. A high and very steep cliff.

precipitation, *n*. 1. Any or all forms of water particles, whether liquid or solid, that fall from the atmosphere and reach the ground. It is distinguished from cloud, fog, dew, rime, frost, etc., in that it must fall; and it is distinguished from cloud and virga in that it must reach the ground. Precipitation includes drizzle, rain, snow, snow pellets, snow grains, ice crystals, ice pellets, and hail. 2. The amount usually expressed in inches of liquid water depth, of the water substance that has fallen at a given point over a specified period of time.

precipitation static. A type of interference experienced in a radio receiver, during snow storms, rain storms, and dust storms, caused by the impact of dust particles against the antenna. It may also be caused by the existence of induction fields created by nearby corona discharges.

precipitation trails. See VIRGA.

precision, *n*. A measure of how close the outcome of a series of observations or measurement cluster about some estimated value of a desired quantity. Precision implies repeatability of the observations within some specified limit and depends upon the random errors encountered due to the quality of the observing instrument, the skill of the observer and randomly fluctuating conditions such as temperature, pressure, refraction, etc. Precision should not be confused with ACCURACY. Observations may be of high precision but inaccurate due to the presence of systematic errors. For a quantity to be accurately measured, both systematic and random errors should be small. For a quantity to be known with high precision, only the random errors due to irregular effects need to be small. See ERROR.

precision graphic recorder. A device used with the standard hydrographic echo sounder in ocean depths where soundings cannot be recorded on the expanded scale of the standard recorder. It provides a sounding record with a scale expansion and high accuracy. Commonly called a PGR.

precision index. A measure of the magnitude of the random errors of a series of observations of some given quantity. If the precision index is large, most of the random errors of the observations are small. The precision index appears as a parameter in the normal (Gaussian) distribution law. While making a series of observations, the standard deviation can be calculated. The precision index is then calculated using a formula and a measure of the precision of the observing instrument is obtained. See also RANDOM ERROR, NORMAL DISTRIBUTION, PRECISION, STANDARD DEVIATION.

Precise Positioning Service. The most accurate military positioning service of the Global Positioning System.

precomputation, *n*. The process of making navigational solutions in advance; applied particularly to the determination of computed altitude and azimuth before making a celestial observation for a line of position. When this is done, the observation must be made at the time used for the computation, or a correction applied.

precomputed altitude. The altitude of a celestial body computed before observation, and with the sextant altitude corrections applied with reversed sign. When a precomputed altitude has been calculated, the altitude difference can be determined by comparison with the sextant altitude.

precomputed curve, A graphical representation of the azimuth or altitude of a celestial body plotted against time for a given assumed position, computed for use with celestial observations.

predictability, *n*. In a navigation system, the measure of the accuracy with which the system can define the position in terms of geographical coordinates. See also REPEATABILITY, definition 2.

predicable accuracy. The accuracy of predicting position with respect to precise space and surface coordinates. See also REPEATABLE ACCURACY.

predicted tides. The times and heights of the tide as given in the Tide Tables in advance of their occurrence.

predicting machine. See TIDE PREDICTING MACHINE.

preferred datum. A geodetic datum selected as a base for consolidation of local independent datums within a geographical area. Also called MAJOR DATUM.

pressure, *n*. Force per unit area. The pressure exerted by the weight of the earth's atmosphere is called atmospheric or, if indicated by a barometer, barometric pressure. Pressure exerted by the vapor of a liquid is called vapor pressure. The pressure exerted by a fluid as a result of its own weight or position is called static pressure. Pressure exerted by radiant energy is called radiation pressure.

pressure gage. A tide gage that is operated by the change in pressure at the bottom of a body of water due to rise and fall of the tide.

pressure tendency. The character and amount of atmospheric pressure change for a 3-hour or other specified period ending at the time of observation. Also called BAROMETRIC TENDENCY.

prevailing westerlies. The prevailing westerly winds on the poleward sides of the sub-tropical high-pressure belts.

prevailing wind. The average or characteristic wind at any place.

primary, *n*. See PRIMARY BODY.

primary body. The celestial body or central force field about which a satellite orbits, or from which it is escaping, or towards which it is falling. The primary body of the earth is the sun, the primary body of the moon is the earth. Usually shortened to PRIMARY.

primary circle. See PRIMARY GREAT CIRCLE.

primary control tide station. A tide station at which continuous observations have been made over a minimum of a 19-year Metonic cycle. Its purpose is to provide data for computing accepted values of the harmonic and non harmonic constants essential to tide predictions and to the determination of tidal datums for charting and coastal boundaries. The data series from this station serves as a primary control for the reduction of relatively short series from subordinate tide stations through the method of comparisons of simultaneous observations, and for monitoring long-period sea-level trends and variations. See also TIDE STATION; SUBORDINATE TIDE STATION, definition 1; SECONDARY CONTROL TIDE STATION; TEMPORARY TIDE STATION.

primary great circle. A great circle used as the origin of measurement of a coordinate; particularly such a circle 90° from the poles of a SYSTEM of spherical coordinates, as the equator. Also called PRIMARY CIRCLE, FUNDAMENTAL CIRCLE.

primary radar. 1. Radar which transmits a SIGNAL and receives the incident energy reflected from an object to detect the object. 2. As defined by the International Telecommunications Union (ITU), a radio-determination system based on the comparison of reference signals with radio signals reflected from a position to be determined.

primary seacoast light. A light established for purpose of making landfall or coastwise past from headland to headland. Also called LAND FALL LIGHT.

primary tidal bench mark. See under BENCH MARK.

primary tide station. See PRIMARY CONTROL TIDE STATION.

prime fictitious meridian. The reference meridian (real or fictitious) used as the origin for measurement of fictitious longitude. Prime grid meridian is the reference meridian of a grid; prime transverse or prime inverse meridian is the reference meridian of a transverse graticule; prime oblique meridian is the reference fictitious meridian of an oblique graticule.

prime grid meridian. The reference meridian of a grid. In polar regions it is usually the 180°-0° geographic meridian, used as the origin for measuring grid longitude.

prime inverse meridian. See PRIME TRANSVERSE MERIDIAN.

prime meridian. The 0° meridian of longitude, used as the origin for measurement of longitude The meridian of Greenwich, England, is almost universally used for this purpose. See also PRIME FICTITIOUS MERIDIAN.

prime oblique meridian. The reference fictitious meridian of an oblique graticule.

prime transverse meridian. The reference meridian of a transverse graticule. Also called PRIME INVERSE MERIDIAN.

prime vertical. See PRIME VERTICAL CIRCLE.

prime vertical circle. The vertical circle perpendicular to the principal vertical circle. The intersections of the prime vertical circle with the horizon define the east and west points of the horizon. Often shortened to PRIME VERTICAL; Sometimes called TRUE PRIME VERTICAL to distinguish from magnetic, compass, or grid prime vertical, defined as the vertical circle passing through the magnetic, compass, or grid east and west points of the horizon, respectively.

priming of tide. The periodic acceleration in the time of occurrence of high and low waters due changes in the relative positions of the moon and the sun. Priming occurs when the moon between new and first quarter and between full and third quarter. High tide occurs before transit of the moon. Lagging occurs when the moon is between first quarter and full and between third quarter and new. High tide occurs after transit of the moon. See also LAGGING OF TIDE.

principal vertical circle. The vertical circle passing through the north and south celestial poles. The intersection of the principal vertical circle with the horizon defines the north and south points of the horizon.

priority blanking. See DUAL-RATE BLANKING.

prism, *n*. A solid having parallel, similar, equal, plane geometric figures as bases, and parallelograms as sides. By extension, the term is also applied to a similar solid having nonparallel bases, and trapezoids or a combination of trapezoids and parallelograms as sides. Prisms are used for changing the direction of motion of a ray of light and for forming spectra.

prismatic error. That error due to lack of parallelism of the two faces of an optical element, such as a mirror or a shade glass. See also SHADE ERROR.

private aids to navigation. In United States waters, those aids to navigation not established and maintained by the U.S. Coast Guard. Private aids include those established by other federal agencies with prior U.S. Coast Guard approval, aids to navigation on marine structures or other works which the owners are legally obligated to establish, maintain, and operate as prescribed by the U.S. Coast Guard, and those aids which are merely desired, for one reason or another, by the individual corporation, state or local government or other body that has established the aid with U.S. Coast Guard approval.

probable error. A measure of the dispersion or spread of a series of observations about some value, usually the mean or average value of all the observations. See also CIRCULAR ERROR PROBABLE.

processor. The brain of a computer, which executes programs to do work. Also known more correctly as the CENTRAL PROCESSING UNIT (CPU).

production platform. A term used to indicate a permanent offshore structure equipped to control the flow of oil or gas. For charting purposes, the use of the term is extended to include all permanent platforms associated with oil or gas production, e.g. field terminal, drilling and accommodation platforms, and "booster" platforms sited at intervals along some pipelines. It does not include entirely submarine structures.

prognostic chart. A chart showing, principally, the expected pressure pattern of a given synoptic chart at a specified future time. Usually, positions of fronts are also included, and the forecast values of other meteorological elements may be superimposed.

program. A set of instructions which a computer executes to perform work. Programs are written in one of many LANGUAGES, which translate the instructions into MACHINE LANGUAGE used by the PROCESSOR.

progressive wave. In the ocean, a wave that advances in distance along the sea surfaces or at some intermediate depth. Although the wave form itself travels significant distances, the water particles that make up the wave merely describe circular (in relatively deep water) or elliptical (in relatively shallow water) orbits. With high, steep, wind waves, a small overlap in the orbit motion becomes significant. This overlapping gives rise to a small net transport.

prohibited area. 1. An area shown on nautical charts within which navigation and/or anchoring is prohibited except as authorized by appropriate authority. 2. A specified area within the land areas of a state or territorial waters adjacent thereto over which the flight of aircraft is prohibited. See also DANGER AREA, RESTRICTED AREA.

projection, *n*. The extension of lines or planes to intersect a given surface; the transfer of a point from one surface to a corresponding position on another surface by graphical or analytical means. See also MAP PROJECTION.

projector compass. A magnetic compass in which the lubber's line and compass card, or a portion thereof, are viewed as an image projected through a system of lenses upon a screen adjacent to the helmsman's position. See also REFLECTOR COMPASS.

prolate cycloid. See TROCHOID.

prolate spheroid. An ellipsoid of revolution, the longer axis of which is the axis of revolution. An ellipsoid of revolution, the shorter axis of which is the axis of REVOLUTION, is called an OBLATE SPHEROID.

promontory, *n*. High land extending into a large body of water beyond the line of the coast. Called HEADLAND when the promontory is comparatively high and has a steep face. Also called FORELAND.

propagation, *n*. The travel of waves of energy through or along a medium other than a specially constructed path such as an electrical circuit.

proper motion. The component of the space motion of a celestial body perpendicular to line of sight, resulting in the change of a stars apparent position relative to other stars. Proper motion is expressed in angular units.

proportional dividers. An instrument consisting in its simple form of two legs pointed at both ends and provided with an adjustable pivot, so that for any given pivot setting, the distance between one set of pointed ends always bears the same ratio to the distance between the other set. A change in the pivot changes the ratio. The dividers are used in transferring measurements between charts or other graphics which are not the same scale.

proportional parts. Numbers in the same proportion as a set of given numbers. Such numbers are used in an auxiliary interpolation table based on the assumption that the tabulated quantity and entering arguments differ in the same proportion. For each intermediate argument a "proportional part" or number is given to be applied the preceding tabulated value in the main table.

protractor, *n*. An instrument for measuring angles on a surface; an angular scale. In its most usual form it consists of a circle or part of one (usually a semicircle) graduated in degrees. See also COMPASS ROSE, THREE-ARM PROTRACTOR.

province, *n*. On the sea floor, a region identifiable by a group of similar physiographic features whose characteristics are markedly in contrast with surrounding areas.

pseudo-independent surveillance. Position determination that relies on craft or vehicle cooperation but is not subject to craft or vehicle navigational errors (e.g., secondary radar).

pseudo-random noise. An apparently random but reproducible sequence of binary code used in the GPS signal.

pseudo-range. Measure of distance from GPS satellite to receiver, uncorrected for synchronization errors between satellite and receiver clocks.

psychrometer, *n*. A type of hygrometer (an instrument for determining atmospheric humidity) consisting of dry-bulb and wet-bulb thermometers. The dry-bulb thermometer indicates the temperature of the air, and the wet bulb thermometer the lowest temperature to which air can be cooled by evaporating water into it at constant pressure. With the information obtained from a psychrometer, the humidity, dew point, and vapor pressure for any atmospheric pressure can be obtained by means of appropriate tables.

psychrometric chart. A nomogram for graphically determining relative humidity, absolute humidity, and dew point from wet- and dry-bulb thermometer readings.

pteropod *(pl. pteropoda)*, *n*. A small marine animal with or without a shell and having two thin, winglike feet. These animals are often so numerous they may cover the surface of the sea for miles. In some areas, their shells cover the bottom.

Pub. No. 9. The American Practical Navigator. A publication of the Defense Mapping Agency Hydrographic/Topographic Center, originally by Nathaniel Bowditch (1773-1838) and first published in 1802, comprising a complete manual of navigation with tables for solution of navigational problems. Popularly called BOWDITCH.

Pub. No. 102. International Code of Signals. A publication of the Defense Mapping Agency Hydrographic/Topographic Center intended primarily for communication at sea in situations involving safety of life at sea and navigational safety, especially when language difficulties arise between ships or stations of different nationalities. The Code is suitable for transmission by all means of communication, including radiotelephony, radiotelegraphy, sound, flashing light, and flags.

Pub. 117. Radio Navigational Aids. A publication of the Defense Mapping Agency Hydrographic/Topographic Center which contains data on radio aids to navigation services provided to mariners. Information on radio direction finder and radar stations, radio time signals, radio navigational warnings, distress signals, stations transmitting medical advice, long range radionavigation systems, emergency procedures and communications instructions, listed in text and tabular format.

Pub. 150. World Port Index. A publication of the Defense Mapping Agency Hydrographic/Topographic Center listing the location, characteristics, known facilities, and available services of ports, shipping facilities and oil terminals throughout the world. The applicable chart and Sailing Direction volume is given for each place listed. A code indicates certain types of information.

Pub. 151. Distances Between Ports. A publication of the Defense Mapping Agency Hydrographic/Topographic Center providing calculated distances in nautical miles over water areas between most of the seaports of the world. A similar publication published by the National Ocean Service of United States waters is entitled *Distances between United States Ports*.

Pub. 217. Maneuvering Board Manual. A publication of the Defense Mapping Agency Hydrographic/Topographic Center providing explanations and examples of various problems involved in maneuvering and in relative movement.

Pub. 221. Loran C Table. A series of tables published by the Defense Mapping Agency Hydrographic/Topographic Center, published primarily for manufacturers who use computers to correct Loran C time differences to geographic coordinates. The tables also correct time differences for ASF.

Pub. 224. Omega Tables. A series of tables published by the Defense Mapping Agency Hydrographic/Topographic Center providing the tabular counterpart of the Omega chart. With the appropriate charting coordinate or lattice table, Omega lines of position can be plotted on suitable a plotting sheet or chart having a scale large as 1:800,000. 2. *Omega Propagation Correction Tables*; a series of tables published by the Defense Mapping Agency Hydrographic/Topographic Center providing necessary data for correcting Omega Navigation System receiver readouts affected by the prevailing propagation conditions, to the standard conditions on which all Omega hyperbolic charts and lattice tables are based.

Pub. No. 226. Handbook of Magnetic Compass Adjustment. A publication of the Defense Mapping Agency Hydrographic/Topographic Center, providing information for adjustment of marine magnetic compasses.

Pub. No. 229. Sight Reduction Tables for Marine Navigation. A publication of the Defense Mapping Agency Hydrographic/Topographic Center, in six volumes each of which includes two 8° zones of latitude. An overlap of 1° of latitude occurs between volumes. The six volumes cover latitude bands 0°-15°, 15°-30°, 30°-45°, 45°-60°, 60°-75°, and 75°-90°. For entering arguments of integral degrees of latitude, declination, and local hour angle, altitudes and their differences are tabulated to the nearest tenth of a minute, azimuth angles to the nearest tenth of a degree. The tables are designed for precise interpolation of altitude for declination only by means of interpolation tables which facilitate linear interpolation and provide additionally for the effect of second differences. The data are applicable to the solutions of sights of all celestial bodies; there are no limiting values of altitude, latitude, hour angle, or declination.

Pub. No. 249. Sight Reduction Tables for Air Navigation. A publication of the Defense Mapping Agency Hydrographic/Topographic Center, in three volumes, with volume 1 containing tabulated altitudes and azimuths of selected stars, the entering arguments being latitude, local hour angle of the vernal equinox, and the name of the star; and volumes 2 and 3 containing tabulated altitudes and azimuth angles of any body within the limits of the entering arguments, which are latitude, local hour angle, and declination (0°-29°) of the body.

Pub. 1310. Radar Navigation Manual. A publication of the Defense Mapping Agency Hydrographic/Topographic Center which explains the fundamentals of shipboard radar, radar operation collision avoidance, radar navigation, and radar-assisted vessel traffic systems in the U.S.

puddles *n*. An accumulation of melt-water on ice, mainly due to melting snow, but in the more advanced stages also due to the melting of ice.

pulse, *n*. A short burst of electromagnetic energy, such as emitted by a radar.

pulse decay time. The interval of time required for the trailing edge of a pulse to decay from 90 percent to 10 percent of the pulse amplitude.

pulse duration. The time interval during which the amplitude of a pulse is at or greater than a specified value, usually stated in terms of a fraction or percentage of the maximum value.

pulse duration error. A range distortion of a radar return caused by the duration of the pulse. See also SPOT-SIZE ERROR.

pulse group. See PULSE TRAIN.

pulse interval. See PULSE SPACING.

pulse length. See PULSE DURATION.

pulse-modulated radar. The type of radar generally used for shipboard navigational applications. The radio-frequency energy transmitted by a pulse-modulated radar consists of a series of equally spaced short pulses having a pulse duration of about 1 microsecond or less. The distance to the target is determined by measuring the transmit time of a pulse and its return to the source as a reflected echo. Also called PULSE RADAR.

pulse modulation. 1. The modulation of a carrier wave by a pulse train. In this sense, the term describes the process of generating carrier-frequency pulses. 2. The modulation of one or more characteristics of a pulse carrier. In this sense, the term describes methods of transmitting information on a pulse carrier.

pulse radar. See PULSE-MODULATED RADAR.

pulse repetition frequency. The pulse repetition rate of a periodic pulse train.

pulse repetition rate. The average number pulses per unit of time. See also PULSE REPETITION FREQUENCY.

pulse rise time. The interval of time required for the leading edge of a pulse to rise from 10 to 90 percent of the pulse amplitude.

pulse spacing. The interval between corresponding points on consecutive pulses. Also called PULSE INTERVAL.

pulse train. A series of pulses of similar characteristics. Also called PULSE GROUP, IMPULSE TRAIN.

pulse width. See PULSE DURATION.

pumice, *n*. Cooled volcanic glass with a great number of minute cavities caused by the expulsion of water vapor at high temperature, resulting in a very light rocky material.

pumping, *n*. Unsteadiness of the mercury in a barometer, caused by fluctuations of the air pressure produced by a gusty wind or due to the motion of a vessel.

pure sound. See PURE TONE.

pure tone. A sound produced by a sinusoidal acoustic oscillation. Also called PURE SOUND.

purple light. The faint purple glow observed on clear days over a large region of the western sky after sunset and over the eastern sky before sunrise.

put to sea. To leave a sheltered area and head out to sea.

pyramidal iceberg. See PINNACLED ICEBERG.

Q

Q-band. A radio-frequency band 36 to 46 gigahertz. See also FREQUENCY, FREQUENCY BAND.

Q-correction. The Polaris correction as tabulated in the *Air Almanac*.

Q signals. Conventional code signals used in radiotelegraphy, each signal of three letters beginning with Q and representing a complete sentence.

quadrant, *n*. 1. A quarter of a circle; either an arc of 90° or the area bounded by such an arc and two radii. 2. A double-reflecting instrument for measuring angles used primarily for measuring altitudes of celestial bodies.

quadrantal correctors. Masses of soft iron placed near a magnetic compass to correct for quadrantal deviation. Spherical quadrantal correctors are called quadrantal spheres.

quadrantal deviation. Deviation which changes its sign (E or W) approximately each 90° change of heading. It is caused by induced magnetism in horizontal soft iron.

quadrantal error. An error which changes sign (plus or minus) each 90°. Also called INTERCARDINAL ROLLING ERROR when related to a gyrocompass.

quadrantal point. See INTERCARDINAL POINT.

quadrantal spheres. Two hollow spheres of soft iron placed near a magnetic compass to correct for quadrantal deviation. See also QUADRANTAL CORRECTORS.

quadrant with two arcs. See BACKSTAFF.

quadrature, *n*. An elongation of 90° usually specified as east or west in accordance with the direction of the body from the sun. The moon is at quadrature at first and last quarters.

quadrilateral, *adj*. Having four sides.

quadrilateral, *n*. A closed plane figure having four sides. See also PARALLELOGRAM, TRAPEZOID.

quarantine anchorage. An area where a vessel anchors while satisfying quarantine regulations.

quarantine buoy. A buoy marking the location of a quarantine anchorage. In U.S. waters a quarantine buoy is yellow.

quarantine mark. A navigation mark indicating a quarantine anchorage area for shipping, or defining its limits.

quartering sea. Waves striking the vessel on the quarter, or relative bearings approximately 045°, 135°, 225°, and 315°.

quarter-power points. See under HALF-POWER POINTS.

quartz, *n*. Crystalline form of silica. In its most common form it is colorless and transparent, but it takes a large variety of forms of varying degrees of opaqueness and color. It is the most common solid mineral.

quartz clock. See QUARTZ CRYSTAL CLOCK.

quartz crystal clock. A precision timepiece, consisting of a current generator of constant frequency controlled by a resonator made of quartz crystal with suitable methods for producing continuous rotation to operate time-indicating and related mechanisms. See also QUARTZ CRYSTAL MARINE CHRONOMETER.

quartz crystal marine chronometer. A quartz crystal clock intended for marine use. The degree of accuracy is such that it requires no chronometer rate, but can be reset electrically if necessary.

quasi-stationary front. See STATIONARY FRONT.

quay, *n*. A structure of solid construction along a shore or bank which provides berthing for ships and which usually provides cargo handling facilities. A similar facility of open construction is called WHARF. See also MOLE, definition 1.

quick flashing light. A light flashing 50-80 flashes per minute. See also CONTINUOUS QUICK LIGHT, GROUP QUICK LIGHT, INTERRUPTED QUICK LIGHT.

quick light. See QUICK FLASHING LIGHT.

quicksand, *n*. A loose mixture of sand and water that yields to the pressure of heavy objects. Such objects are difficult to extract once they begin sinking.

quiet sun. The sun when it is free from unusual radio wave or thermal radiation such as that associated with sun spots.

quintant, *n*. A double-reflecting instrument for measuring angles, used primarily for measuring altitudes of celestial bodies, having an arc of 72°.

R

race, *n*. A rapid current or a constricted channel in which such a current flows. The term is usually used only in connection with a tidal current, when it may be called a TIDE RACE.

racon, *n*. As defined by the International Telecommunication Union (ITU), in the maritime radionavigation service, a receiver-transmitter device which, when triggered by a surface search radar, automatically returns a distinctive signal which can appear on the display of the triggering radar, providing range, bearing and identification information. See also IN-BAND RACON, CROSS BAND RACON, SWEPT-FREQUENCY RACON, RAMARK. Also called RADAR TRANSPONDER BEACON.

radar, *n*. 1. (from radio detection and ranging) A radio system which measures distance and usually direction by a comparison of reference signals with the radio signals reflected or retransmitted from the target whose position is to be determined. Pulse-modulated radar is used for shipboard navigational applications. In this type of radar the distance to the target is determined by measuring the time required for an extremely short burst or pulse of radio-frequency energy to travel to the target and return to its source as a reflected echo. Directional antennas allow determination of the direction of the target echo from the source. 2. As defined by the International Telecommunication Union (ITU) a radiodetermination system based on the comparison of reference signals with radio signals reflected, or re-transmitted, from the position to be determined.

radar beacon. A radar transmitter whose emissions enable a ship to determine its direction and frequently position relative to the transmitter using the ship's radar equipment. There are two general types of radar beacons: one type, the RACON, must be triggered by the ship's radar emissions; the other type, the RAMARK transmits continuously and provides bearings only. See also TRANSPONDER.

radar bearing. A bearing obtained by radar.

radar buoy. A buoy having corner reflectors designed into the superstructure, the characteristic shape of the buoy being maintained. This is to differentiate from a buoy on which a corner reflector is mounted.

radar conspicuous object. An object which return a strong radar echo which can be identified with a high degree of certainty.

radar cross section. The area of a plane element situated at the position of an object and normal to the direction of the radar transmitter, which would be traversed by a power such that, if the power were re-radiated equally in all directions with suitable polarization, it would give an echo of the same power as that given by the object itself. Also called EQUIVALENT ECHOING AREA.

radar echo. See ECHO, definition 3.

radar fix. A fix established by means of radar.

radar horizon. The sensible horizon of a radar antenna.

radar indicator. A unit of a radar set which provides a visual indication of radar echoes received using a cathode-ray tube or video monitor. Besides the cathode-ray tube, the radar indicator is comprised of sweep and calibration circuit; and associated power supplies. Often shortened to INDICATOR.

radar link. A means by which the information from a radar set is reproduced at a distance by use of a radio link or cable. Also called RADAR RELAY SYSTEM.

radar nautical mile. The time interval required for the electromagnetic energy of a radar pulse to travel 1 nautical mile and the echo to return; approximately 12.4 microseconds.

radar picture. See DISPLAY, definition 1.

radar range. 1. The distance of a target as measured by radar. 2. The maximum distance at which a radar is effective in detecting targets. Radar range depends upon variables such as the weather, transmitted power, antenna height, pulse duration, receiver sensitivity, target size, target shape, etc.

radar receiver. A unit of a radar set which demodulates received radar echoes, amplifies the echoes and delivers them to the radar indicator. A radar receiver differs from the usual superheterodyne communications receiver in that its sensitivity is much greater; it has a better signal noise ratio, and it is designed to pass a pulse-type signal.

radar reference line. A mid-channel line on a chart which corresponds to a line incorporated in harbor radar display for the purpose of providing a reference for informing a vessel of its position. In some cases the line may be coincident with the recommended track. The line may be broken into sections of specified length having assigned names or numbers.

radar reflector. A device arranged so that incident electromagnetic energy reflects back to its source. See also CORNER REFLECTOR, PENTAGONAL CLUSTER, OCTAHEDRAL CLUSTER, DIHEDRAL REFLECTOR, DIELECTRIC REFLECTOR, REFLECTOR.

radar relay system. See RADAR LINK.

radar repeater. A unit which duplicates the radar display at a location remote from the main radar indicator installation. Also called PPI REPEATER, REMOTE PPI.

radar return. See ECHO, definition 2.

radar scan. The motion of a radar beam through space in searching for an echo.

radar scanning. The process or action of directing a radar beam through a search pattern.

radarscope, *n*. The cathode-ray tube or video monitor in the indicator of a radar set which displays the received echo to indicate range and bearing. Often shortened to SCOPE. See also PLAN POSITION INDICATOR.

radar set. An electronic apparatus consisting of a transmitter, antenna, receiver, and indicator for sending out radio-frequency energy and receiving and displaying reflected energy so as to indicate the range and bearing of the reflecting object. See also RADAR.

radar shadow. The area shielded from radar signals because of an intervening obstruction or absorbing medium. The shadow region appears as an area void of targets.

radar target. See as TARGET.

radar transponder beacon. See RACON.

radial, *adj*. Of or pertaining to a ray or radius; extending in a straight line outward from a center.

radial, *n*. A straight line extending outward from a center.

radial error. In a two-dimensional or elliptical error distribution, the measure of error as the radius of a circle of equivalent probability derived from the error ellipse. The error, expressed as 1 d_{rms}, is the square root of the sum of the error components along the major and minor axes of the probability ellipse. The use of radial error or d_{rms} error as a measure of error is somewhat confusing because the term does not correspond to a fixed value of probability for a given value of the error measure.

radial motion. Motion along a radius, or a component in such a direction, particularly the component of space motion of a celestial body in the direction of the line of sight.

radial period. See ANOMALISTIC PERIOD.

radian, *n*. The supplementary unit of plane angle in the International System of Units; it is the plane angle subtended at the center of a circle by an arc equal in length to the radius of the circle. It is equal to 360 Ö 2π, or approximately 57°17'48.8".

radian per second. The derived unit of angular velocity in the International System of Units.

radian per second squared. The derived unit of angular acceleration in the International System of Units.

radiant, *adj*. Of, pertaining to, or transmitted by radiation.

radiant energy. Energy consisting of electromagnetic waves.

radiate, *v., t. & i*. To send out in rays or straight lines from a center.

radiation, *n*. 1. The process of emitting energy in the form of electromagnetic waves. 2. The energy radiated in definition 1 above.

radiational cooling. The cooling of the earth's surface and adjacent air, occurring mainly at night whenever the earth's surface suffers a net loss of heat due to terrestrial radiation.

radiational tides. Periodic variations in sea level primarily related to meteorological changes such as the semi-daily (solar) cycle in barometric pressure, daily (solar) land and sea breezes, and seasonal (annual) changes in temperature. Only changes in sea level due to meteorological changes that are random in phase are not considered radiational tides.

radiation fog. A major type of fog, produced over land when radiational cooling reduces the temperature to or below its dew point. Radiation fog is a nighttime occurrence although it may begin to form by evening twilight and often does not dissipate until aft sunrise.

radiation pattern. A curve representing, in polar or Cartesian coordinates, the relative amounts of energy radiated in various directions. Also called DIRECTIVITY DIAGRAM.

radiatus, *adj*. Radial. A term used to refer to clouds in parallel bands which, owing to perspective, appear to converge toward a point on the horizon, or two opposite points if the bands cross the sky.

radio, *n*. A general term applied to the use of radio waves.

radio acoustic ranging. Determining distance by a combination of radio and sound, the radio being used to determine the instant of transmission or reception of the sound, and distance being determined by the time of transit of sound usually in water. See also ECHO RANGING.

radio aid to navigation. An aid to navigation transmitting information by radio waves. See also ELECTRONIC AID TO NAVIGATION.

radio altimeter. As defined by the International Telecommunications Union (ITU), a radionavigation device for aircraft, which uses reflected radio waves from the ground to determine the height of the aircraft above the ground.

radiobeacon, *n*. A radio transmitting station which emits a distinctive or characteristic signal so a navigator can determine the direction of the source using a radio direction finder, providing a line of position. The most common type of marine radiobeacon transmits radio waves of approximately uniform strength in all directions. These omnidirectional beacons are called circular radiobeacons. A radiobeacon some or all of the emissions of which are directional so that the signal characteristic changes according to the vessel's bearing from the beacon is called a directional radiobeacon. A radiobeacon all or part of the emissions of which is concentrated in a beam which rotates is called a rotating radiobeacon. See also CONTINUOUS CARRIER RADIOBEACON, DUAL-CARRIER RADIOBEACON, SEQUENCED RADIOBEACON, ROTATING PATTERN RADIOBEACON, COURSE BEACON.

radiobeacon characteristic. The description of the complete cycle of transmission of a radiobeacon in a given period of time, inclusive of any silent period.

radiobeacon station. As defined by the International Telecommunications Union (ITU), a station in the radionavigation service the emissions of which are intended to enable a mobile station to determine its bearing or direction from the radiobeacon station.

radio bearing. The bearing of a radio transmitter from a receiver, as determined by a radio direction finder.

radio compass. The name by which the radio direction finder was formerly known.

radiodetermination, *n*. As defined by the International Telecommunication Union (ITU), the determination of position using propagation properties of radio waves.

radiodetermination-satellite service. As defined by the International Telecommunication Union (ITU), a radiocommunication service involving the use of radiodetermination and the use of one or more space stations.

radio direction finder. A radio receiver system used for radio direction finding. Also called DIRECTION FINDER. Formerly called RADIO COMPASS. See also AUTOMATIC DIRECTION FINDER.

radio direction finder station. A radio station equipped with special apparatus for determining the direction of radio signals transmitted by ships and other stations. The bearing taken by a radio direction finder station, and reported to a ship, is corrected for all determinable errors except conversion angle. Also called DIRECTION FINDER STATION.

radio direction finding. As defined by the International Telecommunication Union (ITU), radiodetermination using the reception of radio waves to determine the direction of a station or object.

radio direction-finding station. As defined by the International Telecommunication Union (ITU), a radiodetermination station using radio direction finding.

radio fix. A navigational position determined by radio direction finder.

radio frequency. Any electromagnetic wave occurring within that segment of the spectrum normally associated with some form of radio propagation.

radio guard. A ship, aircraft, or radio station designated to listen for and record transmissions, and to handle traffic on a designated frequency for a certain unit or units.

radio horizon. The locus of points at which direct rays from a transmitting antenna become tangent to the earth's surface, taking into account the curvature due to refraction. Its distance from the transmitting antenna is greater than that of the visible horizon, and increases with decreasing frequency.

radio interference. Interference due to unwanted signals from other radio transmitting stations operating on the same or adjacent frequencies.

radio interferometer. An interferometer operating at radio frequencies; used in radio astronomy and in satellite tracking.

radiolarian *(pl. radiolaria), n*. A minute sea animal with a siliceous outer shell. The skeletons of such animals are very numerous, covering the ocean bottom in certain areas, principally in the tropics.

radiolocation, *n*. As defined by the International Telecommunication Union (ITU), radiodetermination used for purposes other than navigation.

radio mast. A label on a nautical chart which indicates a pole or structure for elevating radio antennas, usually found in groups.

radionavigation, *n*. 1. The determination of position, or the obtaining of information relating to position, for the purposes of navigation by means of the propagation properties of radio waves. 2. As defined by the International Telecommunication Union (ITU), radiodetermination used for the purposes of navigation, including obstruction warning. See also RADIODETERMINATION, RADIOLOCATION.

Radio Navigational Aids. See *PUB. 117*.

radio navigational warning. A radio-transmitted message affecting the safe navigation of vessels or aircraft. See also HYDROLANT, HYDROPAC, NAVAREA WARNINGS, WORLD WIDE NAVIGATIONAL WARNING SERVICE.

radionavigation-satellite service. As defined by the International Telecommunication Union (ITU) a radiodetermination-satellite service used for the same purposes as the radionavigation service; in certain cases this service includes transmission or retransmission of supplementary information necessary for the operation of radionavigation systems.

radio receiver. An electronic device connected to an antenna or other receptor of radio signals which receives and processes the signals for use.

radio silence. A period during which all or certain radio equipment capable of radiation is kept inoperative.

radio spectrum. The range of electromagnetic radiation useful for communication by radio (approximately 10 kilohertz to 300,000 megahertz).

radio station. A place equipped with one or more transmitters or receivers and accessory equipment for carrying on a radiocommunication service.

radio tower. A label on a nautical chart which indicates a tall pole or structure for elevating radio antennas.

radio transmitter. Equipment for generation and modulation of radio-frequency energy for the purpose of radiocommunication.

radio wave propagation. The transfer of energy by electromagnetic radiation at radio frequencies.

radio waves. Electromagnetic waves of frequencies lower than 3,000 gHz propagated in space without artificial guide. The practicable limits of radio frequency are approximately 10 kHz to 100 GHz. Also called HERTZIAN WAVES.

radius, *n*. A straight line from the center of a circle, arc, or sphere to its circumference, or the length of such a line. Also called SEMIDIAMETER for a circle or sphere. See also DIAMETER.

radius of action. The maximum distance a ship, aircraft, or vehicle can travel away from its base along a given course with normal combat load and return without refueling, allowing for all safety and operating factors.

radius vector. A straight line connecting a fixed reference point or center with a second point, which may be moving. In astronomy the expression is usually used to refer to the straight line connecting a celestial body with another which revolves around it. See also POLAR COORDINATES, SPHERICAL COORDINATES.

radome, *n*. A dome-shaped structure used to enclose radar apparatus.

rafted ice. A type of deformed ice formed by one piece of ice overriding another. See also FINGER RAFTING.

rain, *n*. Liquid precipitation consisting of drops of water larger than those which comprise DRIZZLE. Orographic rain results when moist air is forced upward by a mountain range. See also FREEZING RAIN.

rainbow, *n*. A circular arc of concentric spectrally colored bands formed by the refraction of light in drops of water. One seen in ocean spray is called a marine or sea rainbow. See also FOGBOW, MOONBOW.

rain clutter. Clutter on the radarscope which is the result of the radar signal being reflected by rain or other forms of precipitation.

rain gush. See CLOUDBURST.

rain gust. See CLOUDBURST.

rain shadow. The condition of diminished rainfall on the lee side of a mountain or mountain range, where the rainfall is noticeably less than on the windward side.

rain storm. See under STORM, definition 2.

raise. To cause to appear over the horizon or higher above the horizon by approaching closer.

ram, *n*. An underwater ice projection from an ice wall, ice front, iceberg, or floe. Its formation is usually due to a more intensive melting and erosion of the unsubmerged part.

ramark, *(from radar marker) n*. A radar beacon which continuously transmits a signal appearing as a radial line on the radar display, indicating the direction of the beacon from the ship. For identification purposes, the radial line may be formed by a series of dots or dashes. The radial line appears even if the beacon is outside the range for which the radar is set, as long as the radar receiver is within the power range of the beacon. Unlike the RACON, the ramark does not provide the range to the beacon.

ramming, *n*. In ice navigation, the act of an icebreaker at full power striking ice to break a track through it.

ramp, *n*. On the sea floor, a gentle slope connecting areas of different elevations.

random access memory (RAM). Type of computer memory used for temporary storage and processing of data, as opposed to permanent storage of data. RAM is volatile, meaning it is unable to store data without a constant source of power. See READ ONLY MEMORY(ROM).

random error. One of the two categories of errors of observation and measurement, the other category being systematic error. Random errors are the errors which occur when irregular, randomly occurring conditions affect the observing instrument, the observer and the environment, and the quantity being observed so that observations of the same quantity made with the same equipment and observer under the same observing conditions result in different values of the observed quantity. Random errors depend upon (1) the quality of the observing instrument. (2) the skill of the observer, particularly, the ability to estimate the fraction of the smallest division or graduation on the observing instrument, and (3) randomly fluctuating conditions such as temperature, pressure, refraction, etc. For many types of observations, random errors are characterized by the following properties: (1) positive and negative errors of the same magnitude are about equal in number, (2) small errors occur more frequently than large errors. and (3) extremely large errors rarely occur. These properties of random errors permit the use of a mathematical law called the Gaussian or normal distribution of errors to calculate the probability that the random error of any given observation of a series of observations will lie within certain limits. Random error might more properly be called deviation since mathematically, the random error of an individual observation is calculated as the difference or deviation between the actual observation and an improved or adjusted value of the observation obtained by some mathematical technique such as averaging all the observations. Also called ACCIDENTAL ERROR, CHANCE ERROR, IRREGULAR ERROR, STATISTICAL ERROR. See also ERROR, PRECISION, PRECISION INDEX, STANDARD DEVIATION.

range, *n*. 1. Two or more objects in line. Such objects are said to be in range. An observer having them in range is said to be on the range. Two beacons are frequently located for the specific purpose of forming a range to indicate a safe route or the centerline of a channel. See also BACK RANGE, LEADING LINE, MAGNETIC RANGE, MULTIPLE RANGES. 2. Distance in a single direction or along a great circle. 3. The extreme distance at which an object or light can be seen is called VISUAL RANGE. When the extreme distance is limited by the curvature of the earth and the heights of the object and the observer, this is called geographic range; when the range of a light is limited only by its intensity, clearness of the atmosphere, and sensitiveness of the observer's eyes, it is called luminous range. 4. The extreme distance at which a signal can be detected or used. The maximum distance at which reliable service is provided is called operating range. The spread of ranges in which there is an element of uncertainty of interpretation is called critical range. 5. The distance a vessel can travel at cruising speed without refueling is called CRUISING RADIUS. 6. The difference in extreme values of a variable quantity. See also RANGE OF TIDE. 7. A series of mountains or mountain ridges is called MOUNTAIN RANGE. 8. A predetermined line along which a craft moves while certain data are recorded by instruments usually placed below the line, or the entire station at which such information is determined. See also DEGAUSSING RANGE. 9. An area where practice firing of ordnance equipment is authorized is a firing range. See also BOMBING RANGE. 10. On the sea floor, a series of ridges or seamounts.

range, *v., t*. 1. To place in line. 2 To determine the distance to an object. 3 To move along or approximately parallel to something, as to range along coast.

range daymark. 1. One of a pair of unlighted structures used to mark a definite line of bearing. See also RANGE, definition 1. 2. A daymark on a range light.

range finder. An optical instrument for measuring the distance to an object. See also STADIMETER.

range lights. Two or more lights at different elevations so situated to form a range (leading line) when brought into transit. The one nearest the observer is the front light and the one farthest from the observer is the rear light. The front light is at a lower elevation than the rear light.

range marker. A visual presentation on a radar display for measuring the range or for calibrating the time base. See also VARIABLE RANGE MARKER, RANGE RING.

range (of a light). See VISUAL RANGE (OF A LIGHT).

range of tide. The difference in height between consecutive high and low waters. The mean range is the difference in height between mean high water and mean low water. The great diurnal range or diurnal range is the difference in height between mean higher high water and mean lower low water. Where the type of tide is diurnal the mean range is the same as the diurnal range. For other ranges see APOGEAN TIDES, NEAP TIDES, PERIGEAN TIDES, SPRING TIDES, TROPIC TIDES.

range-range mode. See RANGING MODE.

range rate. Rate of change in range between satellite and receiver, measured by determining the Doppler shift of the satellite carrier signal.

range resolution. See as RESOLUTION IN RANGE under RESOLUTION, definition 2. Also called DISTANCE RESOLUTION.

range ring. One of a set of equally spaced concentric rings, centered on own ship's position, providing a visual presentation of range on a radar display. See also VARIABLE RANGE MARKER.

ranging mode. A mode of operation of a radionavigation system in which the times for the radio signals to travel from each transmitting station to the receiver are measured rather than their *differences* as in the HYPERBOLIC MODE. Also called RHO-RHO MODE, RANGE-RANGE MODE.

Rankine temperature. Temperature based upon a scale starting at absolute zero (–459.69°F) and using Fahrenheit degrees.

rapids, *n*. A portion of a stream in swift, disturbed motion, but without cascade or waterfall.

raster. A type of computerized display which consists of a single undifferentiated data file, analogous to a picture. See BIT-MAP, VECTOR.

ratan, *n*. An experimental short-range aid to navigation, not operational, in which radar harbor surveillance information is transmitted to the user by television.

rate, *n*. 1. Quantity or amount per unit of something else, usually time. See also ANGULAR RATE, CHRONOMETER RATE, PULSE REPETITION RATE, REPETITION RATE, WATCH RATE. 2. With respect to Loran C, the term rate, implying the number of pulses per unit time, is used for the character designation, and also the station pair, their signals, and the resulting hyperbolic lines of position and the tables and curves by which they are represented.

rate gyro. A single-degree-of-freedom gyro having primarily elastic restraint of its spin axis about the output axis. In this gyro, an output signal is produced by gimbal angular displacement, relative to the base, which is proportional to the angular rate of the base about the input axis. See also RATE INTEGRATING GYRO.

rate integrating gyro. A single-degree-of-freedom gyro having restraint of its spin axis about the output axis. In this gyro an output signal is produced by gimbal angular displacement, relative to the base, which is proportional to the integral of the angular rate of the base about the input axis. See also RATE GYRO.

ratio, *n*. The relation of one magnitude to another of the same kind, the quotient obtained by dividing one magnitude by another of the same kind. See also MAGNITUDE RATIO.

rational horizon. See CELESTIAL HORIZON.

ratio of ranges. The ratio of the ranges of tide at two places. It is used in the tide tables where the times and heights of all high and low tides are given for a relatively few places, called REFERENCE STATIONS. The tides at other places called SUBORDINATE TIDE STATIONS, are found by applying corrections to the values given for the reference stations. One of these corrections is the ratio of ranges, or the ratio between the height of the tide at the subordinate station and its reference station.

ratio of rise. The ratio of the height of tide at two places.

ravine, *n*. 1. A gulch; a small canyon or gorge, the sides of which have comparatively uniform slopes. 2. On the sea floor, a small canyon.

read only memory (ROM). Computer memory used for permanent storage of data. It retains the data without a source of power. See RANDOM ACCESS MEMORY (RAM).

reach, *n*. A comparatively straight segment of a river or channel between two bends.

reach ahead. The distance traveled from the time a new speed is ordered to the time the new speed is being made.

real image. An image actually produced and capable of being shown on a surface, as in a camera.

real precession. Precession of a gyroscope resulting from an applied torque such as that resulting from friction and dynamic unbalance as opposed to APPARENT PRECESSION. Also called INDUCED PRECESSION, PRECESSION.

rear-light. The range light which is farthest from the observer. It is the highest of the lights of an established range. Also called HIGH LIGHT.

receiver, *n*. A person who or a device which receives anything, particularly a radio receiver.

receiver gain control. An operating control on a radar indicator used to increase or decrease the sensitivity of the receiver. The control regulates the intensity of the echoes displayed on the radarscope.

receiver monitor. See under PERFORMANCE MONITOR.

reciprocal, *adj*. In a direction 180° from a given direction. Also called BACK.

reciprocal, *n*. 1. A direction 180° from a given direction 2. The quotient of 1 divided by a given number.

reciprocal bearing. A bearing differing by 180° or one measured in the opposite direction, from a given bearing.

recommended direction of traffic flow. A traffic flow pattern indicating a recommended directional movement of traffic in a routing system within which it is impractical or unnecessary to adopt an established direction of traffic flow.

recommended track. A route which has been examined to ensure that it is free of dangers and along which vessels are advised to navigate. See also ROUTING SYSTEM.

rectangle, *n*. A four-sided figure with its opposite sides parallel and its angles 90°, a -right-angle parallelogram.

rectangular chart. A chart on the rectangular projection.

rectangular coordinates. Magnitudes defining a point relative to two perpendicular lines, called AXES. The magnitudes indicate the perpendicular distance from each axis. The vertical distance is called the ordinate and the horizontal distance the abscissa. This is a form of CARTESIAN COORDINATES.

rectangular error. An error which results from rounding off values prior to their inclusion in table or which results from the fact that an instrument cannot be read closer than a certain value The error is so called because of the shape of its plot. For example: if the altitudes tabulated in a sight reduction table are stated to the nearest 01', the error in the altitude as extracted from the table might have any value from (+) 0.05' to (-) 0.05', and any value within these limits is as likely to occur as another value having similar decimals. See also SIMILAR DECIMALS.

rectangular projection. A cylindrical map projection with uniform spacing of the parallels. This projection is used for the star chart in the *Air Almanac*.

rectified altitude. See APPARENT ALTITUDE.

rectilinear, *adj*. Moving in or characterized by straight line.

rectilinear current. See REVERSING CURRENT.

recurring decimal. See REPEATING DECIMAL.

recurring polynya. See under POLYNYA.

recurved spit. A hook developed when the end or spit is turned toward the shore by current deflection or by opposing action of two or more currents. Also called HOOK, HOOKED SPIT.

red magnetism. The magnetism of the northseeking end of a freely suspended magnet. This is the magnetism of the earth's south magnetic pole.

red sector. A sector of the circle of visibility of a navigational light in which a red light is exhibited. Such sectors are designated by their limiting bearings, as observed from a vessel. Red sectors are often located to warn of dangers.

red shift. In astronomy, the displacement of observed spectral lines toward the longer wavelengths of the red end of the spectrum. The red shift in the spectrum of distant galaxies has been interpreted as evidence that the universe is expanding.

red snow. Snow colored red by the presence in it either of minute algae or of red dust particles.

reduction, *n*. The process of substituting for an observed value one derived from it; often referring specifically to the adjustment of soundings to the selected chart datum. Usually the term reduction of soundings does not pertain to corrections other than those for height of tide. See also CORRECTION OF SOUNDINGS.

reduction of tidal current. The processing of observed tidal current data to obtain mean values of tidal current constants. See also REDUCTION OF TIDES.

reduction of tides. The processing of observed tidal data to obtain mean values of tidal constants. See also REDUCTION OF TIDAL CURRENTS.

reduction tables. See SIGHT REDUCTION TABLES.

reduction to the meridian. The process of applying a correction to an altitude observed when a body is near the celestial meridian of the observer, to find the altitude at meridian transit. The altitude at the time of such an observation is called an EX-MERIDIAN ALTITUDE.

reed, *n*. A steel tongue which is designed to vibrate when air is passed across its unsupported end.

reed horn. A sound signal emitter comprising a resonant horn excited by a jet of air which is modulated by a vibrating reed. The signal is a high-pitched note. See also REED, HORN.

reef, *n*. 1. An offshore consolidated rock hazard to navigation with a depth of 16 fathoms (or 30 meters) or less over it. See also SHOAL. 2. Sometimes used as a term for a low rocky or coral area some of which is above water. See BARRIER REEF, CORAL REEF, FRINGING REEF.

reef flat. A flat expanse of dead reef rock which is partly or entirely dry at low tide. Shallow pools, potholes, gullies, and patches of coral debris and sand are features of the reef flat.

reference datum. A general term applied to any datum, plane, or surface used as a reference or base from which other quantities can be measured.

reference ellipsoid. A theoretical figure whose dimensions closely approach the dimensions of the geoid; the exact dimensions of the ellipsoid are determined by various considerations of the section of the earth's surface of concern. Also called REFERENCE SPHEROID, SPHEROID OF REFERENCE, ELLIPSOID OF REFERENCE.

reference frequency. A frequency having a fixed and specified position with respect to the assigned frequency. The displacement of this frequency, with respect to the assigned frequency, has the same absolute value and sign that the displacement of the characteristic frequency has with respect to the center of the frequency band occupied by the emission.

reference grid. See GRID, definition 2.

reference orbit. An orbit, usually but not exclusively, the best two-body orbit available, on the basis of which the perturbations are computed.

reference ship. The ship to which the movement of other ships is referred.

reference spheroid. See REFERENCE ELLIPSOID.

reference station. A tide or current station for which independent daily predictions are given in the Tide Tables and Tidal Current Tables, and from which corresponding predictions obtained for subordinate stations by means differences and ratios. Also called STANDARD STATION. See also SUBORDINATE CURRENT STATION, SUBORDINATE TIDE STATION.

reflecting prism. A prism that deviates a light beam by internal reflection.

reflecting telescope. A telescope which collects light by means of a concave mirror. All telescopes more than 40 inches in diameter arc of this type. See also CASSEGRAINIAN TELESCOPE, NEWTONIAN TELESCOPE.

reflection, *n*. The return or the change in direction of travel of radiation by a surface without change of frequency of the monochromal components of which the radiation is composed. The radiation does not enter the substance providing the reflecting surface. If reflecting surface is smooth, specular reflection occurs; if the reflecting surface is rough with small irregularities, diffuse reflection occurs.

reflection plotter. An attachment fitted to a radar display which provides a plotting surface permitting plotting without parallax errors. Marks made on the plotting surface are reflected on the radarscope directly below. Also called PLOTTING HEAD.

reflectivity, *n*. The ratio of the radiant energy reflected by a surface to that incident upon it.

reflector, *n*. A reflecting surface situated behind the primary radiator, an array of primary radiators or a feed for the purpose of increasing forward and reducing backward radiation from antenna. See also RADAR REFLECTOR.

reflector compass. A magnetic compass in which the image of the compass card is viewed by direct reflection in a mirror adjacent to helmsman's position. See also PROJECTOR COMPASS.

reflex angle. An angle greater than 180° and less than 360°.

reflex reflection. See RETRO-REFLECTION.

reflex-reflector, *n*. See RETRO-REFLECTOR.

refracted ray. A ray extending onward from point of refraction.

refracting prism. A prism that deviates a beam light by refraction. The angular deviation is function of the wavelength of light; therefore if the beam is composed of white light, the prism will spread the beam into a spectrum.

refracting telescope. A telescope which collects light by means of a lens or system of lenses.

refraction, *n*. The change in direction of motion of a ray of radiant energy as it passes obliquely from one medium into another in which the speed of propagation is different. Atmospheric refraction is caused by the atmosphere and may be further designated astronomical refraction if the ray enters from outside the atmosphere or terrestrial refraction if it emanates from a point on or near the surface of the earth. Super-refraction is greater than normal and sub-refraction is less than normal. See also DIFFRACTION, REFLECTION.

refraction correction. 1. A correction due to refraction, particularly such a correction to a sextant altitude, due to atmospheric refraction. 2. See IONOSPHERIC CORRECTION.

refractive index. The ratio of the velocity of light in vacuum to the velocity of light in a medium. This index is equal to the ratio of the sines of the angles of incidence and refraction when a ray crosses the surface separating vacuum and medium.

refractive modulus. One million times the amount by which the modified refractive index exceeds unity.

refrangible, *adj*. Capable of being refracted.

regelation, *n*. The melting of ice under pressure and the subsequent refreezing when the pressure is reduced or removed.

region. One of the major subdivisions of the earth based on the DMAHTC chart numbering system.

regression of the nodes. Precessional motion of a set of nodes. The expression is used principally with respect to the moon, the nodes of which make a complete westerly revolution in approximately 18.6 years.

regular error. See SYSTEMATIC ERROR.

regular reflection. See SPECULAR REFLECTION.

relative, *adj*. Having relationship. In navigation the term has several specific applications: a. related to a moving point; apparent, as relative wind, relative movement; b. related to or measured from the heading, as relative bearing; c. related or proportional to a variable, as relative humidity. See also TRUE.

relative accuracy. The accuracy with which a user can measure current position relative to that of another user of the same navigation system at the same time. Hence, a system with high relative accuracy provides good rendezvous capability for the users of the system. The correlation between the geographical coordinates and the system coordinates is not relevant. See also PREDICTABLE ACCURACY, REPEATABLE ACCURACY.

relative azimuth. Azimuth relative to heading.

relative bearing. Bearing relative to heading of a vessel, expressed as the angular difference between the heading and the direction. It is usually measured from 000° at the heading clockwise through 360°, but is sometimes measured from 0° at the heading either clockwise or counterclockwise through 180°, when it is designated right or left.

relative course. Misnomer for DIRECTION OF RELATIVE MOVEMENT.

relative direction. Horizontal direction expressed as angular distance from heading.

relative distance. Distance relative to a specified reference point, usually one in motion.

relative gain of an antenna. The gain of an antenna in a given direction when the reference antenna is a half-wave loss-free dipole isolated in space, the equatorial plane of which contains the given direction.

relative humidity. See under HUMIDITY.

relative motion. See RELATIVE MOVEMENT.

relative motion display. A type of radarscope display in which the position of own ship is fixed, usually at the center of the display, and all detected targets move relative own ship. See also TRUE MOTION DISPLAY.

relative movement. Motion of one object relative to another. The expression is usually used in connection with problems involving motion of one vessel to another, the direction such motion being called DIRECTION RELATIVE MOVEMENT and the speed of the motion being called SPEED OF RELATIVE MOVEMENT or RELATIVE SPEED. Distance relative to a specified reference point, usually one in motion, is called RELATIVE DISTANCE. Usually called APPARENT MOTION applied to the change of position of a celestial body as observed from the earth. Also called RELATIVE MOTION.

relative plot. A plot of the successive positions of a craft relative to a reference point, which is usually in motion. A line connecting successive relative positions of a maneuvering ship relative to a reference ship is called a RELATIVE MOVEMENT LINE. A relative plot includes relative movement lines and the position of the reference ship.

relative position. A point defined with reference to another position, either fixed or moving coordinates of such a point are usually between true or relative, and distance from an identified reference point.

relative speed. See SPEED OF RELATIVE MOVEMENT.

relative wind. The wind with reference to a moving point. Sometimes called APPARENT WIND. See also APPARENT WIND, TRUE WIND.

release, *n*. A device for holding or releasing a mechanism, particularly the device by which the tangent screw of a sextant is engaged or disengaged from the limb.

reliability diagram. See LORAN C RELIABILITY DIAGRAM.

relief, *n*. 1. The elevations of a land surface; represented graphics by contours, hypsometric tints, spot elevations, hachures, etc. Similar representation of the ocean floor is called SUBMARINE RELIEF. 2. The removal of a buoy (formerly also referred to lightships) from station and provision of another buoy having the operating characteristics authorized for that station.

relief map. See HYPSOGRAPHIC MAP.

relief model. Any three-dimensional representation of an object or geographic area, modeled in any size or medium. See also PLASTIC RELIEF MAP.

relieved, *adj*. Said of a buoy that has been removed from a station and replaced by another having the proper operating characteristics.

relighted, *adj*. Said of an extinguished aid to navigation returned to its advertised light characteristic.

relocated, *adj*. Said of aid to navigation that has been permanently moved from one position to another.

reluctance, *n*. Magnetic resistance.

remanence, *n*. Ability to retain magnetism after removal of the magnetizing force. Also See RETENTIVITY.

remote-indicating compass. A compass equipped with one or more indicators to repeat at a distance the readings of the master compass. The directive element and controls are called a master compass to distinguish this part of the system from the repeaters, or remote indicators. Most marine gyrocompass installations are of this type. Also called REMOTE-READING COMPASS.

remotely controlled light. A light which is operated by personnel at a considerable distance from the light, through electrical or radio links.

remote PPI. See RADAR REPEATER.

remote-reading compass. See REMOTE-INDICATING COMPASS.

repaired, *adj*. Said of a sound signal or radionavigation aid previously INOPERATIVE, placed back in operation, or of a structure previously DAMAGED, that has been restored as an effective aid to navigation.

repeatability, *n*. 1. A measure of the variation in the accuracy of an instrument when identical tests are made under fixed conditions. 2. In a navigation system, the measure of the accuracy with which the system permits the user to return to a specified point as defined only in terms of the coordinates peculiar to that system. See also PREDICTABILITY.

repeatable accuracy. In a navigation system, the measure of the accuracy with which the system permits the user to return to a position as defined only in terms of the coordinates peculiar to that system. For example, the distance specified for the repeatable accuracy of a system such as Loran C is the distance between two Loran C positions established using the same stations and time-difference readings at different times. The correlation between the geographical coordinates and the system coordinates may or may not be known. See also PREDICTABLE ACCURACY, RELATIVE ACCURACY.

repeater, *n*. A device for repeating at a distance the indications of an instrument or device. See also COMPASS REPEATER, GYRO REPEATER, RADAR REPEATER, STEERING REPEATER.

repeating decimal. A decimal in which all the digits after a certain digit consist of a set of one or more digits repeated and infinitum. Also called RECURRING DECIMAL.

replaced, *adj*. Said of an aid to navigation previously OFF STATION, ADRIFT or MISSING that has been restored by another aid of the same type and characteristic.

representative fraction. The scale of a map or chart expressed as a fraction or ratio that relates unit distance on the map to distance measured in the same unit on the ground. Also called NATURAL SCALE, FRACTIONAL SCALE. See also NUMERICAL SCALE.

reradiation, *n*. 1. The scattering of incident radiation. Reradiation from metallic objects in proximity to either the transmitting or receiving antennas can introduce unwanted effects. This is particularly true on a vessel having a number of metallic structures or wires in the vicinity of an antenna. Where such structures are permanent, the effects can sometimes be allowed for by calibration. Also called SECONDARY RADIATION. 2. Radiation from a radio receiver due to poor isolation between the antenna circuit and the local oscillator within the receiver, causing unwanted interference in other receivers.

research sanctuary. A marine sanctuary established for scientific research in support of management programs, and to establish ecological baselines. See also MARINE SANCTUARY.

reset, *adj*. Said of a floating aid to navigation previously OFF STATION, ADRIFT, or MISSING that has been returned to its station.

residual deviation. Deviation of a magnetic compass after adjustment or compensation. The values on various headings are called RESIDUALS.

residual magnetism. Magnetism which remains after removal of the magnetizing force.

residuals, *n., pl*. The remaining deviation of a magnetic compass on various headings after adjustment or compensation. See also DEVIATION TABLE.

resistance, *n*. Opposition, particularly to the flow of electric current.

resistivity, *n*. The amount of resistance in a system. Resistivity is the reciprocal of CONDUCTIVITY.

resolution, *n*. 1. The ability of an optical system to distinguish between individual objects; the degree of ability to make such a separation, called RESOLVING POWER, is expressed as the minimum distance between two objects that can be separated. 2. The degree of ability of a radar set to indicate separately the echoes of two targets in range, bearing, and elevation. Resolution in range is the minimum range difference between separate targets at the same bearing which will allow both to appear separately; Resolution in bearing is the minimum horizontal angular separation between two targets at the same range which will allow both to appear separately. Resolution in elevation is the minimum separation in the vertical plane between two contacts at the same range and bearing which will allow both to appear as distinct echoes.

resolution of vectors. The resolving of a vector into two or more components. The opposite is called VECTOR ADDITION.

resolving power. The degree of ability of an optical system to distinguish between objects close together. See also RESOLUTION.

resolving time. 1. The minimum time interval between two events which permits one event to be distinguishable from the other. 2. In computers, the shortest permissible period between trigger pulses for reliable operation of a binary cell.

resonance, *n*. Re-enforcement or prolongation any wave motion, such as sound, radio waves etc., resulting when the natural frequency of a body or system in vibration is equal to that of an impressed vibration.

resonant frequency. Any frequency at which a body or system vibrates most readily. The lowest resonant frequency is the natural frequency of the body or system.

responsor, *n*. A unit which receives the response emitted by a transponder.

restricted area. 1. An area (land, sea, or air) in which there are special restrictive measures employed to prevent or minimize interference between friendly forces. 2. An area under military jurisdiction in which special security measures are employed to prevent unauthorized entry. See also DANGER AREA, PROHIBITED AREA.

restricted waters. Areas which for navigational reasons such as the presence of shoals or other dangers confine the movements of shipping within narrow limits.

resultant, *n*. The sum of two or more vectors.

retard, *v., t & i*. To delay. This term is sometimes used as the equivalent of RETIRE (meaning "to move back"), but this usage is not appropriate.

retarded line of position. See RETIRED LINE OF POSITION.

retentive error. Deviation of a magnetic compass due to the tendency of a vessel's structure to retain some of the induced magnetic effects for short periods of time. For example, a vessel on a northerly course for several days, especially if pounding in heavy seas, will tend to retain some fore-and-aft magnetism gained through induction. Although this effect is not large and generally decays within a few hours, it may cause incorrect observations or adjustments, if neglected. This error should not be confused with GAUSSIN ERROR.

retentivity, *n*. See REMANENCE.

reticle, *n*. A system of lines, wires, etc., placed in the focal plane of an optical instrument to serve as a reference. A cross hair is a hair, thread, or wire constituting part of a reticle. See also GRATICULE, definition 2.

retire, *v., t. & i*. To move back, as to move a line of position back, parallel to itself, along a course line to obtain a line of position at an earlier time. The term RETARD (meaning "to delay") is sometimes used as an equivalent, but the term RETIRE (meaning "to move back") is more appropriate. The opposite is ADVANCE.

retired line of position. A line of position which has been moved backward along the course line to correspond with a time previous to that at which the line was established. The opposite is ADVANCED LINE OF POSITION.

retrace, *n*. The path of the visible dot from the end of one sweep to the start of the next sweep across the face of a cathode-ray tube.

retract, *v., t. & i*. The opposite of BEACH, *v., t & i*.

retrograde motion. The apparent motion of a planet westward among the stars. Apparent motion eastward, called DIRECT MOTION, is more common. Also called RETROGRESSION.

retrogression, *n*. See RETROGRADE MOTION.

retro-reflecting material. A material which produces retro-reflection over a wide range of angles of incidence of a light beam, by use of a large number of very small reflecting and refracting elements, usually very small beads.

retro-reflection, *n*. Reflection in which light is returned in directions close to the direction from which it came over wide variations of the direction of the incident light. Also called REFLEX REFLECTION.

retro-reflector, *n*. A device intended to produce retro-reflection. It may comprise one or more retro-reflecting optical units, for example, comer reflectors or special lens units of glass or plastic. Such devices may be installed generally on unlighted buoys or other aids to navigation to increase the range at which they may be seen at night. Also called REFLEX REFLECTOR.

return, *n*. See BLIP; ECHO, definition 2.

reverberation, *n*. Continuation of radiant energy, particularly sound, by multiple reflection.

reversing current. A tidal current which flows alternately in approximately opposite directions with a slack water at each reversal of direction. Currents of this type usually occur in rivers and straits where the direction of flow is somewhat restricted to certain channels. When the movement is towards the shore or up a stream the current is said to be flooding, and when in the opposite direction it is said to be ebbing. The combined flood and ebb movement including the slack water covers, on an average, 12.4. hours for the semidiurnal current. If unaffected by a nontidal flow, the flood and ebb movements will each last about 6 hours, but when combined with such a flow, the durations of flood and ebb may be quite unequal. During the low in each direction the speed of the current will vary from zero at the time of slack water to a maximum about midway between the slacks. Also called RECTILINEAR CURRENT.

reversing falls. Falls which flow alternately in opposite directions in a narrow channel in the St. John River, New Brunswick, Canada, due to the large range of tide and a constriction in the river. The direction of flow is upstream or downstream according to whether it is high or low water on the outside, the falls disappearing at the half-tide level.

revolution, *n*. Circular motion about an axis usually external to the body. The terms REVOLUTION and ROTATION are often used interchangeably but, with reference to the motions of a celestial body, REVOLUTION refers to the motion in an orbit or about an axis external to the body while ROTATION refers to motion about axis within the body. Thus, the earth revolves about the sun annually and rotates about its axis daily.

revolution counter, revolution indicator. An instrument for registering the number of revolutions of a shaft, particularly a propeller shaft of a vessel (when it may be called ENGINE REVOLUTION COUNTER). This information is useful in estimating a vessel's speed through the water.

revolution table. A table listing the number of shaft revolutions corresponding to various speeds of a vessel.

revolver, *n*. The pair of horizontal angles between three points, as observed at any place on the circle defined by the three points. This is the only situation in which such angles do not establish a fix. Also called SWINGER.

revolving light. See ROTATING LIGHT.

revolving storm. A cyclonic storm, or one in which the wind revolves about a central low pressure area.

rheostat, *n*. A variable resistor for changing the amount of current in an electrical circuit.

rhomboid, *n*. A parallelogram with oblique angles. A rhomboid with sides of equal length is rhombus.

rhombus, *n*. A rhomboid with sides of equal length.

Rho-Rho mode. See RANGING MODE.

rho-theta navigation. Navigation by means measuring ranges and bearings of a known position.

rhumb, *n*. Short for RHUMB LINE.

rhumb bearing. The direction of a rhumb line through two terrestrial points, expressed angular distance from a reference direction. It is usually measured from 000° at the reference direction clockwise through 360°. Also called MERCATOR BEARING.

rhumb direction. See MERCATOR DIRECTION.

rhumb line. A line on the surface of the earth making the same oblique angle with all meridians; a loxodrome or loxodromic curve spirals toward the poles in a constant true direction. Parallels and meridians, which also maintain constant true directions, may be considered special cases of the rhumb line. A rhumb line is a straight line on a Mercator projection. Sometimes shortened to RHUMB. See also FICTITIOUS RHUMB LINE.

rhumb-line course. The direction of the rhumb line from the point of departure to the destination, expressed as the angular distance from a reference direction, usually north. Also called MERCATOR COURSE.

rhumb-line distance. Distance point to point along a rhumb line, usually expressed in nautical miles.

rhumb-line sailing. Any method of solving the various problems involving course, distance, difference of latitude, difference of longitude, and departure as they are related to a rhumb line.

rhythmic light. A light showing intermittently with a regular periodicity.

ria, *n*. A long, narrow inlet with gradually decreasing depth inward.

ridge, *n*. 1. On the sea floor, a long, narrow elevation with steep sides. 2. A line or wall of broken ice forced up by pressure. The ridge may be fresh or weathered. See also AGED RIDGE. 3. In meteorology, an elongated area of relatively high atmospheric pressure, almost always associated with and most clearly identified as an area of maximum anticyclonic curvature of wind flow. The opposite of a ridge is called TROUGH. Sometimes called WEDGE.

ridged ice. Ice piled haphazardly one piece over another in the form of ridges or walls; usually found in first-year ice.

ridged-ice zone. An area in which much ridged ice with similar characteristics has formed.

ridging, *n*. The pressure process by which sea ice is forced into ridges.

riding light. See ANCHOR LIGHT.

rift, *n*. An opening made by splitting; a crevasse; usually in the earth.

right angle. An angle of 90°.

right angle reflector. See DIHEDRAL REFLECTOR.

right ascension. Angular distance east of the vernal equinox; the arc of the celestial equator, or the angle at the celestial pole, between the hour circle of the vernal equinox and the hour circle of a point on the celestial sphere, measured eastward from the hour circle of the vernal equinox through 24 hours. Angular distance west of the vernal equinox, through 360°, is SIDEREAL HOUR ANGLE.

right astern. See DEAD ASTERN.

right bank. The bank of a stream or river on the right of the observer when he is facing in the direction of flow, or downstream. See also LEFT BANK.

right circular cone. A cone having a circular base perpendicular to the axis of the cone. Often shortened to RIGHT CONE.

right cone. Short for RIGHT CIRCULAR CONE.

right sphere. The celestial sphere as it appears to an observer at the equator, where celestial bodies appear to rise vertically above the horizon.

right triangle. A triangle one angle of which is 90°.

rigidity in space. See GYROSCOPIC INERTIA.

rime, *n*. A white or milky and opaque granular deposit of ice formed by the rapid freezing of supercooled water drops as they impinge on an exposed object. It is denser and harder than frost, but lighter, softer, and less transparent than glaze.

rime fog. See ICE FOG.

ring time. The time, reckoned from the end of pulse transmitted by a radar set, during which the output of an echo box produces a visible signal on the display.

rip current. A narrow intense current setting seaward through the surf zone. It removes excess water brought to the zone by the small net mass transport of waves, and is fed by longshore currents. Rip currents usually occur at points groins, jetties, etc., of irregular beaches, and at regular intervals along straight, uninterrupted beaches. See also RIPS.

riprap, *n*. Stones or broken rock thrown together without order to provide a revetment.

riprap mounds. Mounds of riprap maintained at certain light structures to protect the structures against ice damage and scouring action. Submerged portions present a hazard to vessels attempting to pass very close aboard.

rips, *n. pl*. Agitation of water caused by the meeting of currents or by a rapid current setting over an irregular bottom. Called TIDE RIPS when the tidal current is involved. See also OVERFALLS, RIP CURRENT.

rise, *n*. A broad elevation that rises gently and generally smoothly from the sea floor. See also CONTINENTAL RISE.

rise, *v., i*. To ascend past the visible horizon. The opposite is SET.

rise of tide. Vertical distance from the chart sounding datum to a higher water datum. Mean rise of tide is the height of mean high water above the chart sounding datum. Spring rise and neap rise are the heights of spring high water and neap high water, respectively, above the chart sounding datum; while mean spring rise and mean neap rise are the heights of mean high water springs and mean high water neaps, respectively above the chart sounding datum. Also called TIDAL RISE. See also HEIGHT OF TIDE.

rising tide. A tide in which the depth of water is increasing. Sometimes the term FLOOD is used as an equivalent, but since flood refers primarily to horizontal rather than vertical movement RISING TIDE is more appropriate. The opposite is FALLING TIDE.

river, *n*. A natural stream of water, of greater volume than a creek or rivulet, flowing in a more or less permanent bed or channel, between defined banks or walls, with a current which may either be continuous in one direction or affected by the ebb and flow of the tidal current.

river buoy. A lightweight nun or can buoy especially designed to withstand strong currents.

river estuary. See ESTUARY, definition 2.

river ice. Ice formed on a river, regardless of observed location.

river radar. A marine radar set especially designated for river pilotage, generally characterized by high degree of resolution and a wide selection of range scales.

rivulet, *n*. A small stream; a brook.

road, *n*. An open anchorage affording less protection than a harbor. Some protection may be afforded by reefs, shoals, etc. Often used in the plural. Also called ROADSTEAD.

roadstead, *n*. See ROAD.

roaring forties. The area of the oceans between 40° and 50° south latitude, where strong westerly winds prevail. See also BRAVE WEST WIND.

roche moutonnée. A rock worn into a rounded shape by a glacier.

rock, *n*. 1. An isolated rocky formation or single large stone, usually one constituting a danger navigation. It may be always submerged, always uncovered, or alternately covered and uncovered by the tide. A pinnacle is a sharp-pointed rock rising from the bottom. 2. The naturally occurring material that forms the firm, hard, and solid masses of the ocean floor. Also, rock is a collective term for hard material generally not smaller than 256 millimeters.

rock awash. A rock that becomes exposed, or nearly so, between chart sounding datum and mean high water. In the Great Lakes, the rock awash symbol is used on charts for rocks that are awash, or nearly so, at low water datum. See also BARE ROCK, SUBMERGED ROCK.

rocking the sextant. See SWINGING THE ARC.

rod, *n*. 1. A unit of length equal to 5.5 yards or 16.5 feet. Also called POLE, PERCH. 2. One of the imaginary slender soft iron bars which are assumed to be components or parameters of a craft's magnetic field caused by magnetism induced in soft iron.

roll, *n*. Oscillation of a craft about its longitudinal axis. Also called ROLLING. See also LIST, *n*.; SHIP MOTIONS.

roll, *v., t. & i*. To oscillate or be oscillated about the longitudinal axis.

roll angle. See ANGLE OF ROLL.

rollers, *n*. Amongst the islands of the West Indies, the South Atlantic and the South Indian Ocean, swell waves which after moving into shallow water have grown to such height as to be destructive. See also COMBER.

rolling, *n*. See ROLL, *n*.

root mean square. The square root of the arithmetical mean of the squares of a group of numbers.

root mean square error. For the one-dimensional error distribution, this term has the same meaning as STANDARD DEVIATION or STANDARD ERROR. For the two-dimensional error distribution, this term has the same meaning as RADIAL (d_{rms}) ERROR. However, such use of the term is deprecated. Root mean square error is commonly called RMS ERROR.

rotary current. A tidal current that flows continually, with the direction of flow changing through 360° during the tidal period. Rotary currents are usually found offshore where the direction of flow is not restricted by any barriers. The tendency for rotation is due to the Coriolis force and, unless modified by local conditions, is clockwise in the Northern Hemisphere and counterclockwise in the Southern Hemisphere. The speed of the current usually varies throughout the tidal cycle, passing through the two maxima in approximately opposite directions and the two minima with the direction of the current at approximately 90° from the direction at time of maximum speed.

rotating light. A light with one or more beams that rotate. Sometimes called REVOLVING LIGHT.

rotation, *n*. Turning of a body about an axis within the body, such as the daily rotation of the earth. See also REVOLUTION.

rotten ice. Sea ice which has become honeycombed and is in an advanced state of disintegration.

round, *v., t*. To pass and alter direction of travel, as a vessel ROUNDS A CAPE. If the course is nearly reversed, the term DOUBLE may be used.

roundabout, *n*. A routing measure comprising a separation point or circular separation zone and a circular traffic lane within defined limits. Traffic within the roundabout moves in a counterclockwise direction around the separation point or zone. See also ROUTING SYSTEM, TRAFFIC SEPARATION SCHEME.

round of bearings. A group of bearings observed together for plotting as a fix.
round of sights. A group of celestial observations made together for plotting a fix.
round wind. A wind that gradually changes direction through approximately 180° during the daylight hours. See also LAND BREEZE.
route chart. A chart showing routes between various places, usually with distances indicated.
routing system. Any system of one or more defined tracks and/or traffic control measures for reducing the risk of casualties; it includes traffic separation schemes, two-way routes, recommended tracks, areas to be avoided, inshore traffic zones, roundabouts, precautionary areas, and deep water routes.
rubble, *n*. 1. Fragments of hard sea ice, roughly spherical and up to 5 feet in diameter, resulting from the disintegration of larger ice formations. When afloat, commonly called BRASH ICE. 2. Loose angular rock fragments.
Rude Star Finder. A star finder previously published by the U.S. Navy Hydrographic Office, and named for Captain Gilbert T. Rude, U.S. Coast and Geodetic Survey. This star finder preceded No. 2102-D Star Finder and Identifier.
rugged, *adj*. Rock-bound; craggy.
rules of navigation. Rules of the road.
rules of the road. The *International Regulations for Prevention of Collisions at Sea*, commonly called *International Rules of the Road*, and the *Inland Navigation Rules*, to be followed by all vessels while navigating upon certain inland waters of the United States. Also called RULES OF NAVIGATION.
run, *n*. 1. A brook, or small creek. 2. A small, swift watercourse. 3. The distance traveled by a craft during any given time interval, or since leaving a designated place. See also DAY'S RUN.
run a line of soundings. To obtain soundings along a course line, for use in making or improving a chart.
run before the wind. To steer a course downwind, especially under sail.
run down a coast. To sail approximately parallel with the coast.
runnel, *n*. The smallest of natural streams; a brook or run.
running fix. A position determined by crossing lines of position obtained at different times and advanced or retired to a common time. However in celestial navigation or when using long-range electronic aids, a position determined by crossing lines of position obtained within a few minutes is considered a FIX; the expression RUNNING FIX is applied to a position determined by advancing or retiring a line over a considerable period of time. There is no sharp dividing line between a fix and a running fix in this case.
running light. See NAVIGATION LIGHTS.
run-off, *n*. That portion of precipitation which is discharged from the area of fall as surface water in streams.
run of the coast. The directional trend of a coast.
run-up. The rush of water up a structure on the breaking of a wave. The amount of run-up is the vertical height above the still water level that the rush of water reaches. Also called UPRUSH.

S

saddle, *n*. A low part of the sea floor resembling in shape a saddle, in a ridge or between contiguous seamounts.
safety lanes. Specified sea lanes designated for use by submarines and surface ships in transit to prevent attack by friendly forces. They may be called SUBMARINE SAFETY LANES when designated for use by submarines in transit.
safe water mark. See under IALA MARITIME BUOYAGE SYSTEM.
SafetyNET. The INMARSAT broadcast service for MARITIME SAFETY INFORMATION (MSI).
sailing, *n*. A method of solving the various problems involving course, distance, difference of latitude, difference of longitude, and departure. The various methods are collectively spoken of as the sailings. Plane sailing considers the earth as a plane. Traverse sailing applies the principles of plane sailing to determine the equivalent course and distance made good by a craft following a track consisting of a series of rhumb lines. Any of the sailings which considers the spherical or spheroidal shape of the earth is called spherical sailing. Middle-latitude sailing is a method of converting departure into difference of longitude, or vice versa, by assuming that such a course is steered at the middle or mean latitude; if the course is 090° or 270° true, it is called parallel sailing. Mercator sailing applies when the various elements are considered in their relation on a Mercator chart. Meridian sailing is used when the course is 000° or 180° true. Rhumb-line sailing is used when a rhumb line is involved; great-circle sailing when a great circle track is involved. Composite sailing is a modification of great circle sailing used when it is desired to limit the highest latitude. The expression current sailing is occasionally used to refer to the process of allowing for current in determining the predicted course made good, or of determining the effect of a current on the direction of motion of a vessel.
sailing chart. See under CHART CLASSIFICATION BY SCALE.
sailing directions. 1. A descriptive book for the use of mariners, containing detailed information of coastal waters, harbor facilities, etc. of an area. For waters of the United States and its possessions, they are published by the National Ocean Survey and are called UNITED STATES COAST PILOTS. Sailing directions, as well as light lists, provide the information that cannot be shown graphically on the nautical chart and that is not readily available elsewhere. See also UNITED STATES COAST PILOT.
St. Elmo's fire. A luminous discharge of electricity from pointed objects such as the masts and arms of ships, lightning rods, steeples, etc. occurring when there is a considerable atmospheric difference in potential. Also called CORPOSANT, CORONA DISCHARGE.
St. Hilaire method. Establishing a line position from observation of the altitude of a celestial body by using an assumed position, the difference between the observed and computed altitudes, and the azimuth. The method was devised by Marcq St. Hilaire, a French naval officer, in 1874. See also SUMNER METHOD, LONGITUDE METHOD, HIGH ALTITUDE METHOD. Also see ALTITUDE INTERCEPT METHOD.
sallying ship. Producing rolling motion of a ship by having the crew run in unison from to side. This is usually done to help float a ship which is aground or to assist it to make way when it is beset by ice.
salt marsh. A flat coastal area flooded by most high tides, characterized by various species of marsh grasses and animal life.
salt-water wedge. The intrusion of a tidal estuary by sea water in the form of a wedge underneath the less dense fresh water.
same name. A name the same as that possessed by something else, as declination has the same name as latitude if both are north or both south. They are of CONTRARY NAME if one is north and the other south.
sand, *n*. Sediment consisting of small but distinguishable separate grains between 0.0625 and 2.0 millimeters in diameter. It is called very fine sand if the grains are between 0.0625 and 0.125 millimeter in diameter, fine sand between 0.125 and 0.25 millimeter, medium sand if between 0.25 and 0.50 millimeters, coarse sand if between 0.50 and 1.0 millimeters, and very coarse sand if between 1.0 and 2.0 millimeters. See also MUD, STONES, ROCK definition 2.
sand dune. See DUNE.
sandstorm, *n*. A strong wind carrying sand through the air, the diameter of most of the particles ranging from 0.08 to 1.0 millimeter. In contrast to a DUST STORM, the sand particles are mostly confined to the lowest 10 feet, and rarely rise more than 50 feet above the ground.
sandwave, *n*. A large wavelike sea-floor sediment feature in very shallow water and composed of sand. The wavelength may reach 100 meters, the amplitude is about 0.5 meter. Also called MEGARIPPLE.
Santa Ana. A strong, dust-laden foehn occurring in Southern California near the mouth of the Santa Ana pass and river.
Sargasso Sea. The west central region of the subtropical gyre of the North Atlantic Ocean. It is bounded by the North Atlantic, Canary, Atlantic North Equatorial, and Antilles Currents, and the Gulf Stream. It is characterized by the absence of well-marked currents and by large quantities of drifting Sargassum, or gulfweed.
sargasso weed. See SARGASSUM.

sargassum, *n*. A genus of brown algae characterized by a bushy form, a substantial holdfast when attached, and a yellowish brown, greenish yellow, or orange color. Species of the group have a large variety of forms and are widely distributed in warm seas as attached and free floating plants. Two species (S. *fluitans* and *S. matans)* make up 99 percent of the macroscopic vegetation in the Sargasso Sea. Also called SARGASSO WEED, GULFWEED.

Saros, *n*. A period of 223 synodic months corresponding approximately to 19 eclipse years or 18.03 Julian years, and is a cycle in which solar and lunar eclipses repeat themselves under approximately the same conditions.

sastrugi, *(sing. sastruga), n., pl.* Sharp, irregular ridges formed on a snow surface by wind erosion and deposition. On mobile floating ice, the ridges are parallel to the direction of the prevailing wind at the time they were formed.

satellite, *n*. 1. A body, natural or man-made, that orbits about another body, the primary body. The moon is a satellite of the earth, the primary body. 2. As defined by the International Telecommunication Union (ITU), a body which revolves around another body of preponderant mass and which has a motion primarily and permanently determined by the force of attraction of that other body. See also ACTIVE SATELLITE, EARTH SATELLITE, EQUATORIAL SATELLITE, GEODETIC SATELLITE, NAVIGATION SATELLITE, PASSIVE SATELLITE, POLAR SATELLITE, SNYCHRONOUS SATELLITE, TWENTY-FOUR HOUR SATELLITE.

satellite geodesy. The discipline which employs observations of an earth satellite to extract geodetic information.

satellite triangulation. The determination of the angular relationships between two or more stations by the simultaneous observation of an earth satellite from these stations.

satellite triangulation stations. Triangulation stations whose angular positions relative to one another are determined by the simultaneous observations of an earth satellite from two or more of them.

saturable system. A term used to describe a navigation system whose use is limited to a single user or a limited number of users on a time-shared basis.

saturation, *n*. Complete impregnation under given conditions, such as the condition that exists in the atmosphere when no additional water vapor can added at the prevailing temperature without condensation or supersaturation occurring.

Saturn, *n*. The navigational planet whose orbit lies outside that of Jupiter.

santanna, *n*. A plain with low vegetation, especially in the sub-tropical latitudes.

S-band. A radio-frequency band of 1,550 to 5,200 megahertz. See also FREQUENCY, FREQUENCY BAND.

scalar, *adj*. Having magnitude only.

scalar, *n*. Any physical quantity whose field can be described by a single numerical value at each point in space. A scalar quantity is distinguished from a VECTOR quantity by the fact that scalar quantity possesses only magnitude, where as, a vector quantity possesses both magnitude and direction.

scale, *n*. 1. A series of marks or graduations at definite intervals. A linear scale is a scale graduated at uniform intervals; a logarithmic scale is graduated in the logarithms of uniformly-spaced consecutive numbers. 2. The ratio between the linear dimensions of chart, map drawing, etc. and the actual dimensions. See also CONVERSION SCALE, BAR SCALE, REPRESENTATIVE FRACTION, SMALL SCALE, LARGE SCALE.

scale error. See CALIBRATION ERROR.

scan, *v., t.* In the use of radar, to search or investigate an area or space by varying the direction of the radar antenna and thus the beam. Normally scanning is done by continuous rotation of the antenna.

scanner, *n*. 1. A unit of a radar set consisting of the antenna and drive assembly for rotating the antenna. 2. A computerized electronic device which digitizes printed images.

scarf cloud. A thin cirrus-like cloud sometimes observed above a developing cumulus. See also CAP CLOUD.

scarp, *n*. See ESCARPMENT.

scatter reflections. Reflections from portions of the ionosphere having different virtual height which mutually interfere and cause rapid fading.

Schuler frequency. The natural frequency of simple pendulum with a length equal to the earth's radius. The corresponding period is 84 minutes.

Schuler loop. The portion of the inertial navigator in which the instrumental local vertical is established.

Schuler tuned. The condition wherein gyroscopic devices should be insensitive to applied accelerations. M. Schuler determined that if gyroscopic devices were not to be affected by the motions of the craft in which installed, the devices should have a natural period of oscillation of about 84.4 minutes. This period is equal to the product of 2π and the square root of the quotient: radius of the earth divided by the acceleration of gravity.

scintillation, *n*. Twinkling; emission of sparks or quick flashes; shimmer.

scope, *n*. Short for RADARSCOPE.

scoria *(pl. scoriae), n.* Volcanic rock fragments usually of basic composition, characterized by marked vesicularity, dark color, high density and a partly crystalline structure. Scoria is a constituent of certain marine sediments.

scouring basin. A basin containing impounded water which is released at about low water in order to maintain the desired depth in the entrance channel by scouring the bottom. Also called SLUICING POND.

screen, *n*. The chemically coated inside surface of the large end of a cathode-ray tube which becomes luminous when struck by an electron beam.

scud, *n*. Shreds or small detached masses of cloud moving rapidly before the wind, often below a layer of lighter clouds. See also FRACTO.

scud, *v., i.* To run before a storm.

sea, *n*. 1. A body of salt water more or less confined by continuous land or chains of islands and forming a distinct region. 2. A body of water nearly or completely surrounded by land, especially if very large or composed of salt water. Sometimes called INLAND SEA. See also LAKE. 3. Ocean areas in general, including major indentations in the coast line, such as gulfs. See also CLOSED SEA, OPEN SEA, HIGH SEA. 4. Waves generated or sustained by winds within their fetch as opposed to SWELL. 5. The character of a water surface, particularly the height, length (period), and direction of travel of waves generated locally. A smooth sea has waves no higher than ripples or small wavelets. A short sea has short, irregular, and broken waves. A confused sea has a highly disturbed surface without a single, well-defined direction of travel, as when waves from different directions meet following a sudden shift in the direction of the wind. A cross sea is a series of waves imposed across the prevailing waves. A sea may be designated as head, beam, quartering, or following. See also SWELL definition 1.

Sea Area. A defined area under the Global Maritime Distress and Safety System (GMDSS) which regulates certain safety and communication equipment necessary according to the area of the ship's operations. Sea Area A-1 is within coverage of VHF coast radio stations (25-30 miles) providing digital selective calling. Sea Area A-2 is within range of the medium frequency coast radio stations (to approximately 300 miles). Sea Area A-3 is within the footprint of the geostationary INMARSAT communications satellites, covering the rest of the open seas except the poles. Sea Area A-4 covers the rest of the earth, chiefly the polar areas. The areas do not overlap.

sea-air temperature difference correction. A correction due to a difference in the temperature of the sea and air, particularly the sextant altitude correction caused by abnormal terrestrial refraction occurring when there is a nonstandard density lapse rate in the atmosphere due to a difference in the temperature of the water and air at the surface.

sea anchor. An object towed by a vessel, usually a small one, to keep the vessel end-on to a heavy sea or surf or to reduce the drift. Also called DRAG, DROGUE.

seabeach, *n*. See under BEACH.

seaboard, *n*. The region of land bordering the sea. The terms SEABOARD, COAST, and LITTORAL have nearly the same meanings. SEABOARD is a general term used somewhat loosely to indicate a rather extensive region bordering the sea. COAST is the region of indefinite width that extends from the sea inland to the first major change in terrain features. LITTORAL applies more specifically to the various parts of a region bordering the sea, including the coast, foreshore, backshore, beach, etc.

sea breeze. A breeze blowing from the sea to adjacent land. It usually blows by day, when the land is warmer than the sea, and alternates with a LAND BREEZE, which blows in the opposite direction by night. See also ONSHORE WIND.

sea buoy. The outermost buoy marking the entrance to a channel or harbor.

seachannel, *n*. On the sea floor, a continuously sloping, elongated depression commonly found in fans or plains and usually bordered by levees on one or two sides.

sea clutter. See SEA RETURN.

seacoast, *n*. See COAST.

sea fog. A type of advection fog formed when air that has been lying over a warm water surface is transported over colder water, resulting in cooling of the lower layer of air below its dew point. See also HAAR.

sea gate. 1. A gate which serves to protect a harbor tidal basin from the sea, such as one of a pair of supplementary gates at the entrance to a tidal basin exposed to the sea. 2. A movable gate which protects the main deck of a ferry from waves and sea spray.

seagirt, *adj*. Surrounded by sea. Also called SEA BOUND.

sea ice. Any form of ice found at sea which has originated from the freezing of sea water.

sea-ice nomenclature. See WMO SEA-ICE NOMENCLATURE.

sea kindliness. A measure of the ease of motion of a vessel in heavy seas, particularly in regard to rolling, pitching, and shipping water. It is not to be confused with seaworthiness which implies that the vessel is able to sustain heavy rolling, pitching, etc., without structural damage or impaired stability.

sea level. Height of the surface of the sea at any time.

sea manners. Understood by seamen to mean consideration for the other vessel and the exercise of good judgment under certain condition when vessels meet.

seamark, *n*. See MARK, *n*., definition 1.

sea mile. An approximate mean value of the nautical mile equal to 6,080 feet; the length of a minute of arc along the meridian at latitude 48°.

sea mist. See STEAM FOG.

seamount, *n*. On the sea floor, an elevation rising generally more than 1,000 meters and of limited extent across the summit.

sea quadrant. See BACKSTAFF.

search and rescue chart. A chart designed primarily for directing and conducting search and rescue operations.

search and rescue radar transponder (SART). An electronic device which transmits a homing signal on the radar frequency used by rescue ships and aircraft.

sea reach. The reach of a channel entering a harbor from seaward.

sea return. Clutter on the radarscope which is the result of the radar signal being reflected from the sea, especially near the ship. Also called SEA CLUTTER. See also CLUTTER.

sea room. Space in which to maneuver without danger of grounding or colliding.

seashore, *n*. A loose term referring to the general area in close proximity to the sea.

season, *n*. 1. One of the four principal divisions of the year: spring, summer, autumn, and winter. 2. An indefinite part of the year, such as the rainy season.

seasonal current. An ocean current which changes in speed or direction due to seasonal winds.

sea-temperature difference correction. A correction due to a difference in the temperature of the sea and air, particularly the sextant altitude correction caused by abnormal terrestrial refraction occurring when there is a nonstandard density lapse rate in the atmosphere due to a difference in the temperature of the water and air at the surface.

seaward, *adj*. In a direction away from the land; toward the sea.

seaward, *adv*. Away from the land; toward the sea.

seaward boundary. Limits of any area or zone offshore from the mean low, or mean lower low water line and established by an act of the U.S. Congress.

seaway, *n*. 1. A moderately rough sea. Used chiefly in the expression in a seaway. 2. The sea as a route of travel from one place to another; a shipping lane.

secant, *n*. 1. The ratio of the hypotenuse of a plane right triangle to the side adjacent to one of the acute angles of the triangle, equal to 1/cos. The expression NATURAL SECANT is sometimes used to distinguish the secant from its logarithm (called LOGARITHMIC SECANT). 2. A line that intersects another, especially a straight line intersecting a curve at two or more points.

secant conic chart. See CONIC CHART WITH TWO STANDARD PARALLELS.

secant conic map projection. See CONIC MAP PROJECTION WITH TWO STANDARD PARALLELS.

second, *n*. 1. The base unit of time in the International System of Units (SI). In 1967 the second was defined by the Thirteenth General Conference on Weights and Measures as the duration of 9,192,631,770 periods of the radiation corresponding to the transition between two hyperfine levels of the ground state of the cesium-133 atom. This value was established to agree as closely as possible with the ephemeris second. Also called ATOMIC SECOND. See also ATOMIC TIME. 2. A sixtieth part of a minute in either time or arc.

secondary, *n*. A small low pressure area accompanying a large or primary one. The secondary often grows at the expense of the primary, eventually replacing it.

secondary circle. See SECONDARY GREAT CIRCLE.

secondary control tide station. A tide station at which continuous observations have been made over a minimum period of 1 year but less than a 19-year Metonic cycle. The series is reduced by comparison with simultaneous observations from a primary control tide station. This station provides for a 365-day harmonic analysis including the seasonal fluctuation of sea level. See also PRIMARY CONTROL TIDE STATION; SUBORDINATE TIDE STATION, definition 1; TERTIARY TIDE STATION; TIDE STATION.

secondary great circle. A great circle perpendicular to a primary great circle, as a meridian. Also called SECONDARY CIRCLE.

secondary light. A major light, other than a primary seacoast light, established at harbor entrances and other locations where high intensity and reliability are required. See also MINOR LIGHT.

secondary phase factor correction. A correction for additional time (or phase delay) for transmission of a low frequency signal over an all seawater path when the signal transit time is based on the free-space velocity. The Loran C lattices as tabulated in tables or overprinted on the nautical chart normally include compensation for secondary phase factor. See also ADDITIONAL SECONDARY PHASE FACTOR CORRECTION.

secondary radar. 1. Radar in which the target is fitted with a transponder and in which the target retransmits automatically on the interrogating frequency, or a different frequency. The response may be coded. See also PRIMARY RADAR, RACON, RAMARK. 2. As defined by the International Telecommunication Union (ITU), a radiodetermination system based on the comparison of reference signals with radio signals re-transmitted from the position to be determined.

secondary radiation. See RERADIATION, definition 2.

secondary station. In a radionavigation system, the station of a chain whose emissions are made with reference to the emissions of a master station without being triggered by the emissions of such station, as in Loran C. See also SLAVE STATION.

secondary tide station. See as SECONDARY CONTROL TIDE STATION.

second reduction. See PHASE REDUCTION.

second-trace echo. A radar echo received from a target after the following pulse has been transmitted. Second-trace echoes are unusual except under abnormal atmospheric conditions, or conditions under which super-refraction is present, and are received from targets at actual ranges greater than the radar range scale setting. They may be recognized through changes in their position on the radarscope on changing the pulse repetition rate; their hazy, streaky or distorted shape; and their erratic movements on plotting. Also called MULTIPLE-TRACE ECHO.

second-year ice. Old ice which has survived only one summer's melt. Because it is thicker and less dense than first-year ice, it stands higher out of the water. In contrast to multi-year ice, summer melting produces a regular pattern of numerous small puddles. Bare patches and puddles are usually greenish-blue.

sector, *n.* 1. Part of a circle bounded by two radii and an arc. See also RED SECTOR. 2. Something resembling the sector of a circle, as a warm sector between the warm and cold fronts of a cyclone.
sector display. A radar display in which a high persistence screen is excited only when the radar beam is within a narrow sector which can be selected at will.
sector light. A light having sectors of different colors or the same color in specific sectors separated by dark sectors.
sector scanning. In the use of radar, the process of scanning within a sector as opposed to scanning around the horizon.
secular, *adj.* Of or pertaining to a long period of time.
secular aberration. See under ABERRATION, definition 1.
secular error. That error in the reading of an instrument due to secular change within the materials of the instrument.
secular perturbations. Perturbations of the orbit of a planet or satellite that continue to act in one direction without limit, in contrast to periodic perturbations which change direction in a regular manner.
secular terms. In the mathematical expression of the orbit of a satellite, terms which are proportional to time, resulting in secular perturbations. See also PERIODIC TERMS.
secular trend. See APPARENT SECULAR TREND.
seiche, *n.* A stationary wave usually caused by strong winds and/or changes in barometric pressure. It is usually found in lakes and semi-enclosed bodies of water. It may also be found in areas of the open ocean. See also STANDING WAVE.
Seismic sea wave. See as TSUNAMI.
selective availability. A Department of Defense program which degrades the accuracy of the pseudorange measurement of the GPS signal by dithering the clock time and ephemerides data, providing a less accurate fix for civilian users. It can be turned on or off at will by DoD.
selective fading. 1. Fading of the skywave in which the carrier and various sideband frequencies fade at different rates, causing audio-frequency distortion. 2. Fading that affects the different frequencies within a specified band unequally. 3. Fading in which the variation in the received signal strength is not the same for all frequencies in the frequency band of the received signal. See also FADING.
selectivity, *n.* 1. The characteristic of a radio receiver which enables it to differentiate between the desired signal and those of other frequencies. 2. The ability of a receiver to reject transmissions other than the one to which tuned. 3. The degree to which a radio receiver can accept the signals of one station while rejecting those of stations on adjacent channels. See also SENSITIVITY.
selenographic, *adj.* Of or pertaining to the physical geography of the moon.
semaphore, *n.* A device using visual signals, usually bodies of defined shapes or positions or both, by which information can be transmitted.
semi-. A prefix meaning half.
semicircle, *n.* Half of a circle. See also DANGEROUS SEMICIRCLE, LESS DANGEROUS SEMICIRCLE, NAVIGABLE SEMICIRCLE.
semicircular deviation. Deviation which changes sign (E or W) approximately each 180° change of heading.
semidiameter, *n.* 1. Half the angle at the observer subtended by the visible disk of a celestial body. Sextant altitudes of the sun and moon should be corrected for semidiameter unless the center is observed. 2. The radius of a circle or sphere.
semidiameter correction. A correction due to semidiameter, particularly that sextant altitude correction, when applied to the observation of the upper or lower limb of a celestial body, determines the altitude of the center of that body.
semidiurnal, *adj.* Having a period or cycle of approximately one-half of a day. The predominating type of tide throughout the world is semidiurnal, with two high waters and two low waters each tidal day. The tidal current is said to be semidiurnal when there are two flood and two ebb periods each tidal day. A semidiurnal constituent has two maxima and minima each constituent day. See also TYPE OF TIDE.
semidiurnal current. Tidal current in which tidal day current cycle consists of two flood currents and two ebb currents, separated by slack water; or two changes in direction, 360° of a rotary current. This is the most common type of tidal current throughout the world.
semidiurnal tide. See under TYPE OF TIDE, SEMIDIURNAL, *adj.*
semilogarithmic coordinate paper. Paper ruled with two sets of mutually-perpendicular parallel lines, one set being spaced according to the logarithms of consecutive numbers, and the other set uniformly spaced.
semimajor axis. One-half of the longest diameter of an ellipse.
semiminor axis. One-half of the shortest diameter of an ellipse.
semi-reflecting mirror. See DICHROIC MIRROR.
sense, *n.* The solution of the 180° ambiguity present in some radio direction finding systems.
sense antenna. An antenna used to resolve a 180° ambiguity in a directional antenna.
sense finding. The process of eliminating 180° ambiguity from the bearing indication some types of radio direction finder.
sensibility, *n.* The ability of a magnetic compass card to align itself with the magnetic meridian after deflection.
sensible horizon. The circle of the celestial sphere formed by the intersection of the celestial sphere and a plane through any point, such as the eye of an observer, and perpendicular to the zenith-nadir line. See also HORIZON.
sensitive axis. 1. The axis Of an accelerometer along which specific acceleration is measured. 2. See also INPUT AXIS.
sensitivity, *n.* The minimum input signal required to produce a specified output signal from a radio or similar device, having a specific signal-to-noise ratio. See also SELECTIVITY.
sensitivity time control. An electronic circuit designed to reduce automatically the sensitivity of the radar receiver to nearby targets. Also called SWEPT GAIN, ANTI-CLUTTER GAIN CONTROL, ANTI-CLUTTER SEA.
separation line. A line separating the traffic lanes in which ships are proceeding in opposite or nearly opposite directions, or separating a traffic lane from the adjacent inshore traffic zone. See also ROUTING SYSTEM, SEPARATION ZONE.
separation zone. A defined zone which separates traffic lanes in which ships are proceeding in opposite directions, or which separates traffic lanes from the adjacent inshore traffic zone. See also ROUTING SYSTEM, SEPARATION LINE.
September equinox. See AUTUMNAL EQUINOX.
sequenced radiobeacon. One of a group of marine radiobeacons in the same geographical area, except those operating continuously, that transmit on a single frequency. Each radiobeacon transmits for 1 minute of each period in sequence with other beacons of the group. If less than six radiobeacons are assigned to a group, one or more of the beacons may transmit during two 1-minute periods.
sequence of current. The order of occurrence of the four tidal current strengths of a day, with special reference as to whether the greater flood immediately precedes or follows the greater ebb.
sequence of tide. The order in which the four tides of a day occur, with special reference as to whether the higher high water immediately precedes or follows the lower low water.
service area. The area within which a navigational aid is of use. This may be divided into primary and secondary service areas having different degrees of accuracy.
service area diagram. See RELIABILITY DIAGRAM.
service period. The number of days that an automatic light or buoy is expected to operate without requiring recharging.
set, *n.* The direction towards which a current flows.
set, *v., i.* Of a celestial body, to cross the visible horizon while descending. The opposite is RISE.
set, *v., t.* To establish, as to set a course.
set screw. A screw for locking a movable part of an instrument or device.
setting a buoy. The act of placing a buoy on station in the water.
settled, *adj.* Pertaining to weather, devoid of storms for a considerable period. See also UNSETTLED.
seven-eighths rule. A rule of thumb which states that the approximate distance to an object broad on the beam equals 7/8 of the distance traveled by a craft while the relative bearing (right or left) changes from 30° or 60° or from 120° to 150°, neglecting current and wind.
seven seas. Figuratively, all the waters or oceans of the world. Applied generally to the seven oceans - Arctic, Antarctic, North Atlantic, South Atlantic, North Pacific, South Pacific, and Indian.
seven-tenths rule. A rule of thumb which states that the approximate distance to an object broad on the beam equals 7/10 of the distance traveled by a craft while the relative bearing (right or left) changes from 22.5° to 45° or from 135° to 157.5°, neglecting current and wind.

seven-thirds rule. A rule of thumb which states that the approximate distance to an object broad on the beam equals 7/3 of the distance traveled by a craft while the relative bearing (right or left) changes from 22.5° to 26.5°, 67.5° to 90°, 90° to 112.5°, or 153.5° to 157.5°, neglecting current and wind.

sexagesimal system. A system of notation by increments of 60°, such as the division of the circle into 360°, each degree into 60 minutes, and each minute into 60 seconds.

sextant, *n*. A double-reflecting instrument for measuring angles, primarily altitudes of celestial bodies. As originally used, the term applied only to instruments having an arc of 60°, a sixth of a circle, from which the instrument derived its name. Such an instrument had a range of 120°. In modern practice the term applies to a similar instrument, regardless of its range, very few modern instruments being sextants in the original sense. Thus, an octant, having a range of 90°; a quintant, having a range of 144°; and a quadrant, having a range of 180°, may be called sextants. A marine sextant is designed primarily for marine navigation. See also MARINE SEXTANT.

sextant adjustment. The process of checking the accuracy of a sextant and removing or reducing its error.

sextant altitude. Altitude as indicated by a sextant or similar instrument, before corrections are applied. See also OBSERVED ALTITUDE, APPARENT ALTITUDE.

sextant altitude correction. Any of several corrections applied to a sextant altitude in the process of converting it to observed altitude. See also ACCELERATION CORRECTION, AIR TEMPERATURE CORRECTION, AUGMENTATION CORRECTION, BAROMETRIC PRESSURE CORRECTION, CORIOLIS CORRECTION, DEFLECTION OF THE VERTICAL CORRECTION, DIP CORRECTION, HEIGHT OF EYE CORRECTION, INDEX CORRECTION, INSTRUMENT CORRECTION, IRRADIATION CORRECTION, PARALLAX CORRECTION, PERSONAL CORRECTION, REFRACTION CORRECTION, SEA-AIR TEMPERATURE DIFFERENCE CORRECTION, SEMI-DIAMETER CORRECTION, TIDE CORRECTION, TILT CORRECTION, WAVE HEIGHT CORRECTION.

sextant chart. See CIRCLE SHEET.

sextant error. The error in reading a sextant, due either to lack of proper adjustment or imperfection of manufacture. See CALIBRATION ERROR, CENTERING ERROR, COLLIMATION ERROR, ERROR OF PERPENDICULARITY, GRADUATION ERROR, INDEX ERROR, INSTRUMENT ERROR, PRISMATIC ERROR, SHADE ERROR, SIDE ERROR, VERNIER ERROR.

shade, *n*. See SHADE GLASS.

shaded relief. A cartographic technique that provides an apparent three-dimensional configuration of the terrain on maps and charts by the use of graded shadows that would be cast if light were shining from the northwest. Shaded relief is usually used in combination with contours.

shade error. The error of an optical instrument due to refraction in the shade glasses. If this effect is due to lack of parallelism of the faces it is usually called PRISMATIC ERROR.

shade glass. A darkened transparent glass that can be moved into the line of sight of an optical instrument, such as a sextant, to reduce the intensity of light reaching the eye. Also called SHADE.

shadow, *n*. 1. Darkness in a region, caused by an obstruction between the source of light and the region. By extension, the term is applied to similar condition when any form of radiant energy is cut off by an obstruction, as in a radar shadow. The darkest part of a shadow in which light is completely cut off is called the UMBRA; the lighter part surrounding the umbra in which the light is only partly cut off is called the PENUMBRA. 2. A region of diminished rainfall on the lee side of a mountain or mountain range, where the rainfall is noticeably less than on the windward side. Usually called RAIN SHADOW.

shadow bands. See CREPUSCULAR RAYS.

shadow bar. A rod or bar used to cast a shadow, such as on the sighting assembly of an astro compass.

shadow pin. A small rod or pin used to cast a shadow on an instrument, such as a magnetic compass or sun compass, to determine the direction of the luminary; a GNOMON.

shadow region. A region shielded from radar signals because of an intervening obstruction or absorbing medium. This region appears as an area void of targets on a radar display such as a plan position indicator. The phenomenon is called RADAR SHADOW. See also SHADOW SECTOR, BLIND SECTOR.

shadow sector. A sector on the radarscope in which the appearance of radar echoes is improbable because of an obstruction near the antenna. While both blind and shadow sectors have the same basic cause, blind sectors generally occur within the larger angles subtended by the obstruction. See also SHADOW REGION.

shallow, *adj*. Having little depth; shoal.

shallow, *n*. An area where the depth of water is relatively slight.

shallow water constituent. A short-period harmonic term introduced into the formula of tidal (or tidal current) constituents to take account of the change in the form of a tide wave resulting from shallow water conditions. Shallow water constituents include the overtides and compound tides.

shallow water wave. A wave is classified as a shallow water wave whenever the ratio of the depth (the vertical distance of the still water level from the bottom) to the wave length (the horizontal distance between crests) is less than 0.04. Tidal waves are shallow water waves.

shamal, *n*. A northwesterly wind blowing over Iraq and the Persian Gulf, in summer, often strong during the day, but decreasing during the night.

sharki, *n*. A southeasterly wind which sometimes blows in the Persian Gulf.

shearing, *n*. An area of pack ice is subject to shear when the ice motion varies significantly in the direction normal to the motion, subjecting the ice to rotational forces. These forces may result in phenomena similar to a FLAW.

sheet line. See NEATLINE.

shelf, *n*. A zone adjacent to a continent, or around an island, that extends from the low water line to a depth at which there is usually a marked increase of slope towards oceanic depths.

shelf valley. A valley on the shelf, generally the shoreward extension of a canyon.

shield, *n*. A metal housing around an electrical or magnetic element to eliminate or reduce the effect of its electric or magnetic field, or to reduce the effect of an exterior field on the element.

shielding factor. The ratio of the strength of the magnetic field at a compass to the strength if there were no disturbing material nearby; usually expressed as a decimal. Because of the metal of a vessel, the strength of the earth's magnetic field is reduced somewhat at a compass location aboard ship. The shielding factor is one minus the percentage of reduction.

shimmer, *v., i*. To appear tremulous or wavering due to varying atmospheric refraction in the line of sight.

shingle, *n*. See under STONES.

ship, *n*. Originally a sailing vessel with three or more masts, square-rigged on all. The term is now generally applied to any large, ocean-going vessel, except submarines which are called boats regardless of size.

ship earth station (SES). An INMARSAT satellite system installed aboard a vessel.

ship error. The error in radio direction finder bearings due to reradiation of radio waves by the metal of the ship.

ship motions. Surge is the bodily motion of a ship forward and backward along the longitudinal axis, caused by the force of the sea acting alternately on the bow and stern; heave is the oscillatory rise and fall due to the entire hull being lifted by the force of the sea; sway is the side-to-side bodily motion, independent of rolling caused by uniform pressure being exerted all along one side of the hull; yaw is the oscillation about a vertical axis approximately through the center of gravity of the vessel; roll is the oscillation about the longitudinal axis; and pitch is oscillation about the transverse axis, due to the bow and stern being raised or lowered on passing through successive crests and troughs of waves.

shipping lane. An established route traversed by ocean shipping.

ship's emergency transmitter. As defined by the International Telecommunication Union (ITU) a ship's transmitter to be used exclusively on a distress frequency for distress, urgency or safety purposes.

ship's head. Heading of a vessel.

ship simulator. A computerized system which uses video projection techniques to simulate navigational and shiphandling situations. A full capability system includes a completely equipped ship's bridge and can duplicate almost any aspect of ship operation; partial systems focus on a particular function, such as radar collision avoidance or nighttime navigation.

Ships' Routing. A publication of the International Maritime Organization (IMO) which describes the general provisions of ships' routing, traffic separation schemes, deep water routes and areas to be avoided, which have been adopted by IMO. All details of routing systems are promulgated through Notices to Mariners and Sailing Directions and are depicted on charts.

ship weather routing. A procedure whereby an optimum route is developed based on the forecasts of weather and seas and the ship's characteristics for a particular transit. Within specified limits of weather and sea conditions, ship weather routing seeks maximum safety and crew comfort, minimum fuel consumption, minimum time underway, or any desired combination of these factors.

shoal, *adj*. Shallow.

shoal, *n*. An offshore hazard to navigation on which there is a depth of 16 fathoms or 30 meters or less, composed of unconsolidated material. See also REEF.

shoal, *v., i*. To become less deep.

shoal, *v., t*. To cause to become less deep.

shoal patches. Individual and scattered elevations of the bottom, with depths of 16 fathoms (or 30 meters) or less, but composed of any material except rock or coral.

shoal water. Shallow water; water over a shoal.

shoot, *v., t*. To observe the altitude of (a celestial body).

shooting star. See METEOR.

shore, *n*. That part of the land in immediate contact with a body of water including the area between high and low water lines. The term SHORE is usually used with reference to the body of water and COAST with reference to the land, as the east coast of the United States is part of the western shore of Atlantic Ocean. The term SHORE usually refers to a narrow strip of land in immediate contact with any body of water, while COAST refers to a general region in proximity to the sea. A shore bordering the sea may be called a SEASHORE. See also FORESHORE, BACKSHORE.

shoreface, *n*. The narrow zone seaward from the low tide shoreline, permanently covered by water, over which the beach sands and gravels actively oscillate with changing wave conditions.

shore lead. A lead between pack ice and the shore or between pack ice and an ice front.

shoreline, *n*. The intersection of the land with the water surface. The shoreline shown on charts represents the line of contact between the land and a selected water elevation.

shore polynya. See under POLYNYA.

short period perturbations. Periodic perturbations in the orbit of a planet or satellite which execute one complete periodic variation in the time of one orbital period or less.

short range systems. Radionavigation systems limited in their positioning capability to coastal regions, or those systems limited to making landfall. See also MEDIUM RANGE SYSTEMS, LONG RANGE SYSTEMS.

short sea. A sea in which the waves are short, irregular, and broken.

short wave. A radio wave shorter than those of the standard broadcast band. See also WAVE, definition 2.

shower, *n*. Precipitation from a convective cloud. Showers are characterized by the suddenness with which they start and stop, by the rapid changes of intensity, and usually by rapid changes in the appearance of the sky. In weather observing practice, showers are always reported in terms of the basic type of precipitation that is falling, i.e., rain showers, snow showers, sleet showers.

shuga, *n*. An accumulation of spongy white ice lumps, a few centimeters across, the lumps are formed from grease ice or slush and sometimes from anchor ice rising to the surface.

side echo. The effect on a radar display by a side lobe of a radar antenna. See also ECHO.

side error. The error in the reading of a sextant due to nonperpendicularity of horizon glass to the frame.

side lights. Running lights placed on the sides of a vessel, green to starboard and red to port, showing an unbroken light over an arc of the horizon from dead ahead to 22.5° abaft the beam.

side lobe. Any lobe of the radiation pattern of a directional antenna other than the main or lobe.

sidereal, *adj*. Of or pertaining to the stars, though SIDEREAL generally refers to the stars and TROPICAL to the vernal equinox, sidereal time and the sidereal day are based upon position of the vernal equinox relative the meridian. The SIDEREAL YEAR is based on the stars.

sidereal day. See under SIDEREAL TIME.

sidereal hour angle. Angular distance west of the vernal equinox; the arc of the celestial equator or the angle at the celestial pole between the hour circle of the vernal equinox and the hour circle of a point on the celestial sphere, measured westward from the hour circle of the equinox through 360°. Angular distance east of the vernal equinox, through 24 hours, is RIGHT ASCENSION.

sidereal month. The average period of revolution of the moon with respect to the stars, a period of 27 days, 7 hours, 43 minutes, 11.5 seconds.

sidereal noon. See under SIDEREAL TIME.

sidereal period. 1. The length of time required for one revolution of a celestial body about a primary, with respect to the stars. 2. The interval between two successive returns of an artificial earth satellite in orbit to the same geocentric right ascension.

sidereal time. Time defined by the daily rotation of the earth with respect to the vernal equinox of the first point of Aries. Sidereal time is numerically measured by the hour angle of the equinox, which represents the position of the equinox in the daily rotation. The period of one rotation of the equinox in hour angle, between two successive upper meridian transits, is a sidereal day. It is divided into 24 sidereal hours, reckoned at upper transit which is known as sidereal noon. The true equinox is at the intersection of the true celestial equator of date with the ecliptic of date; the time measured by its daily rotation is apparent sidereal time. The position of the equinox is affected by the nutation of the axis of rotation of the earth, and the nutation consequently introduces irregular periodic inequities into the apparent sidereal time and the length of the sidereal day. The time measured by the motion of the mean equinox of date, affected only by the secular inequalities due to the precession of the axis, is mean sidereal time. The maximum difference between apparent mean sidereal times is only a little over a second and its greatest daily change is a little more than a hundredth of a second. Because of its variable rate, apparent sidereal time is used by astronomers only as a measure of epoch; it is not used for time interval. Mean sidereal time is deduced from apparent sidereal time by applying the equation of equinoxes.

sidereal year. The period of one apparent rotation of the earth around the sun, with relation to a fixed point, or a distant star devoid of proper motion, being 365 days, 6 hours, 9 days and 9.5 seconds in 1900, and increasing at a rate of rate of 0.0001 second annually. Because of the precession of the equinoxes this is about 20 minutes longer than a tropical year.

sight, *n*. Observation of the altitude, and sometimes also the azimuth, of a celestial body for a line of position; or the data obtained by such observation. An observation of a celestial body made by facing 180° from the azimuth of the body is called a back sight. See also NOON SIGHT, TIME SIGHT.

sighting vane. See VANE, definition 2.

sight reduction. The process of deriving from a sight the information needed for establishing a line of position.

sight reduction tables. Tables for performing sight reduction, particularly those for comparison with the observed altitude of a celestial body to determine the altitude difference for establishing a line of position.

Sight Reduction Tables for Air Navigation. See *PUB. NO. 249*.

Sight Reduction Tables for Marine Navigation. See *PUB. NO. 229*.

signal, *n*. 1. As applied to electronics, any transmitted electrical impulse 2. That which conveys intelligence in any form of communication, such as a time signal or a distress signal.

signal-to-noise ratio. The ratio of the magnitude of the signal to that of the noise, often expressed in decibels.

signature, *n*. The graphic record of the magnetic or acoustic properties of a vessel.

sign conventions. See as GEOGRAPHIC SIGN CONVENTIONS.

significant digits. Those digits of a number which have a significance, zeros at the left and sometimes those at the right being excluded.

sikussak, *n*. Very old ice trapped in fjords. Sikussak resembles glacier ice, since it is formed partly from snow.

sill, *n*. On the sea floor, the low part of a gap or saddle separating basins. See also DOCK SILL.

sill depth. The depth over a sill.

silt, *n*. See under MUD.

similar decimals. Decimals having the same number of decimal places, as 3.141 and 0.789. Decimals can be made similar by adding the appropriate number of zeros. For example, 0.789 can be made similar to 3.1416 by stating it as 0.7890. See also REPEATING DECIMAL, SIGNIFICANT DIGITS.

simple conic chart. A chart on a simple conic projection.

simple conic map projection. A conic map projection in which the surface of a sphere or spheroid, such as the earth, is conceived as developed on a tangent cone, which is then spread out to form a plane.

simple harmonic motion. The projection of uniform circular motion on a diameter of the circle of such motion. The combination of two or more simple harmonic motions results in COMPOUND HARMONIC MOTION.

simultaneous altitudes. Altitudes of two or more celestial bodies observed at the same time.

simultaneous observations (of a satellite). Observations of a satellite that are made from two or more distinct points or tracking stations at exactly the same time.

sine, *n*. The ratio of the side opposite an angle of a plane right triangle to the hypotenuse. The expression NATURAL SINE is used to distinguish the sine from its logarithm (called LOGARITHMIC SINE).

sine curve. Characteristic simple wave pattern; a curve which represents the plotted values of sines of angles, with the sine as the ordinate and the angle as the abscissa. The curve starts at 0 amplitude at the origin, increases to a maximum at 90°, decreases to 0 at 180°, increases negatively to a maximum negative amplitude at 270°, and returns to 0 at 360°, to repeat the cycle. Also called SINUSOID.

sine wave. A simple wave in the form of curve.

single astronomic station datum orientation. Orientation of a geodetic datum by accepting the astronomically determined coordinates of the origin and the azimuth to one other station without any correction.

single-axis normal distribution. A one-time normal distribution along an axis perpendicular to a line of position. Two single-axis normal distributions may be used to establish the error ellipse and the corresponding circle of equivalent probability when the error distribution is two-dimensional or bivariate.

single-degree-of-freedom gyro. A gyroscope, the spin axis of which is free to rotate about one of the orthogonal axes, the spin axis not being counted. See also DEGREE-FREEDOM, RATE GYRO.

single-flashing light. See under FLASHING LIGHT.

single interpolation. Interpolation with only one argument or variable.

single-occulting light. See under OCCULTING LIGHT.

single-sideband transmission. A method of transmission in which the frequencies produced by the process of modulation on one side of the carrier are transmitted and those on the other side are suppressed. The carrier frequency may either be transmitted or suppressed. With this method, less power is required for the effective signal at the receiver, a narrower frequency band can be used, and the signal is less subject to man-made interference or selective fading.

single station range light. A directional light bound by other sectors of different characteristic which define its margins with small angular uncertainty. Most commonly the bounding sectors are of different colors (red and green).

sinking, *n*. An apparent lowering of distant terrestrial objects by abnormal atmospheric refraction. Because of sinking, objects normally visible near the horizon sometimes disappear below the horizon. The opposite is LOOMING.

sinusoid, *n*. See SINE CURVE.

sinusoidal, *adj*. Of or pertaining to a sine wave or sinusoid.

siren, *n*. A sound signal emitter using the periodic escape of compressed air through a rotary shutter.

sirocco, *n*. A warm wind of the Mediterranean area, either a foehn or a hot southerly wind in advance of a low pressure area moving from the Sahara or Arabian deserts. Called LEVECHE in Spain.

skeleton tower. A tower, usually of steel and often used for navigation aids, constructed of open legs with various horizontal and diagonal bracing members.

skip distance. The least distance from a transmitting antenna at which a skywave can normally be received at a given frequency.

skip zone. The area between the outer limit of reception of groundwaves and the inner limit of reception of skywaves, where no signal is received.

sky diagram. A diagram of the heavens, indicating the apparent position of various celestial bodies with reference to the horizon system of coordinates.

skylight, *n*. Thin places in the ice canopy, usually less than 1 meter thick and appearing from below as relatively light, translucent patches in dark surroundings. The under-surface of a skylight is normally flat, but may have ice keels below. Skylights are called large if big enough for a submarine to attempt to surface through them, or small if not.

sky map. The pattern on the underside of extensive cloud areas, created by the varying amounts of light reflected from the earth's surface. Snow surfaces produce a white glare (SNOW BLINK) and ice surfaces produce a yellowish-white glare (ICE BLINK). Bare land reflects relatively little light (LAND SKY) and open water even less (WATER SKY).

skywave, *n*. A radio wave that is propagated by way of the ionosphere. Also called IONOSPHERIC WAVE.

skywave correction. The correction to be applied to the time difference reading of signals received via the ionosphere to convert it to the equivalent groundwave reading. The correction for a particular place is established on the basis of an average height of the ionosphere.

skywave error. See IONOSPHERIC ERROR.

skywave transmission delay. The amount by which the time of transit from transmitter to receiver of a pulse carried by skywaves reflected once from the E-layer exceeds the time of transit of the same pulse carried by groundwaves.

slack water. The state of a tidal current when its speed is near zero, especially the moment when a reversing current changes direction and its speed is zero. The term is also applied to the entire period of low speed near the time of turning of the current when it is too weak to be of any practical importance in navigation. The relation of the time of slack water to the tidal phases varies in different localities. For standing tidal waves, slack water occurs near the times of high and low water, while for progressive tidal waves, slack water occurs midway between high and low water.

slant range. The line-of-sight distance between two points not at the same elevation.

slave, *n*. Short for SLAVE STATION.

slaved gyro magnetic compass. A directional gyro compass with an input from a flux valve to keep the gyro oriented to magnetic north.

slave station. In a radionavigation system, the station of a chain whose emissions are made with reference to the emissions of a master station, its emissions being triggered by the emissions of the master station. See also SECONDARY STATION.

sleet, *n*. See under ICE PELLETS; colloquially some parts of the United States, precipitation the form of a mixture of rain and snow.

slewing, *n*. In ice navigation, the act of forcing a ship through ice by pushing apart adjoining ice floes.

slick, *n*. A smooth area of water, such as one caused by the sweep of a vessel's stern during a turn, or by a film of oil on the water.

slime, *n*. Soft, fine, oozy mud or other substance of similar consistency.

slip, *n*. 1. A berthing space between two piers. Also called DOCK. 2. The difference between the distance a propeller would travel longitudinally in one revolution if operating in a solid and the distance it travels through a fluid.

slope, *n*. On the sea floor, the slope seaward from the shelf edge to the beginning of a continental or insular rise or the point where there is a general reduction in slope.

slot radiator. A slot in the wall of a slotted wave guide antenna which acts as a radiating element.

slotted guide antenna. See SLOTTED WAVE GUIDE ANTENNA.

slotted wave guide antenna. An antenna consisting of a metallic waveguide in the walls of which are cut one or more slot radiators.

slough (sloo), *n.* A minor marshland or tidal waterway which usually connects other tidal areas; often more or less equivalent to a bayou occasionally applied to the sea level portion of a creek on the U.S. West Coast.

slow-sweep racon. See under SWEPT-FREQUENCY RACON.

slue, *n.* A slough or swamp.

sluice, *n.* A floodgate. sluicing pond. See SCOURING BASIN.

slush, *n.* Snow which is saturated and mixed with water on land or ice surfaces, or which is viscous floating mass in water after a heavy snow fall.

small area plotting sheet. For a relatively small area, a good approximation of a Mercator position plotting sheet, constructed by the navigator by either of two methods based upon graphical solution of the secant of the latitude which approximates the expansion. A partially completed small area plotting sheet printed in advance for later rapid completion according to requirements is called UNIVERSAL PLOTTING SHEET.

small circle. The intersection of a sphere and plane which does not pass through its center.

small diurnal range. The difference in height between mean lower high water and mean higher low water. Applicable only when the type of tide is either semidiurnal or mixed. See also TROPIC RANGES.

small floe. See under FLOE.

small fracture. See under FRACTURE.

small hail. See under ICE PELLETS.

small iceberg. For reports to the International Ice Patrol, an iceberg that extends 4 to 50 feet (1 to 15 meters) above the sea surface and which has a length of 20 to 200 feet (6 to 60 meters). See also MEDIUM ICEBERG, LARGE ICEBERG.

small ice cake. A flat piece of ice less than 2 meters across.

small ice field. See under ICE FIELD.

small scale. A scale involving a relatively large reduction in size. A small-scale chart usually covers a large area. The opposite is LARGE SCALE, which covers a small area. See also REPRESENTATIVE FRACTION.

small-scale chart. See under CHART. See also SMALL SCALE.

small tropic range. The difference in height between tropic lower high water and tropic higher low water. Applicable only when the type of tide is either semidiurnal or mixed. See also MEAN TROPIC RANGE, GREAT TROPIC RANGE.

smell the bottom. See FEEL THE BOTTOM.

smog, *n.* Originally a natural fog contaminated by industrial pollutants, or a mixture of smoke and fog. Today, smog is a common term applied to visible air pollution with or without fog.

smoke, *n.* Small particles of carbon and other solid matter, resulting from incomplete combustion, suspended in the air. When it settles, it is called SOOT.

smokes, *n., pl.* Dense white haze and dust clouds common in the dry season on the Guinea coast of Africa, particularly at the approach of the harmattan.

smooth sea. Sea with waves no higher than ripples or small wavelets.

snow, *n.* 1. Frozen precipitation consisting of translucent or white ice crystals which fall either separately or in loose clusters called snowflakes. Very fine, simple crystals, or minute branched, star-like snowflakes are called snow grains. Snow pellets are white, opaque, roundish grains which are crisp and easily compressible, and may rebound or burst when striking a hard surface. Snow is called brown, red, or yellow when it is colored by the presence of brown dust, red dust or algae, or pine or cypress pollen, respectively. See also BLOWING SNOW, DRIFTING SNOW. 2. The speckled background on the plan position indicator or video display due to electrical noise.

snow barchan. See under SNOWDRIFT.

snow blink. A white glare on the underside of extensive cloud areas, created by light reflected from snow-covered surfaces. Snow blink is brighter than the yellowish-white glare of ICE BLINK. Clouds above bare land or open water have no glare. See also LAND SKY, WATER SKY, SKY MAP.

snowdrift, *n.* An accumulation of wind-blown snow deposited in the lee of obstructions or heaped by wind eddies. A crescent-shaped snowdrift, with ends pointing downwind, is called a SNOW BARCHAN.

snowflake, *n.* A loose cluster if ice crystals, or rarely, a single crystal.

snow flurry. A popular term for SNOW SHOWER, particularly of a very light and brief nature.

snow grains. Frozen precipitation consisting of very fine, single crystals, or of minute, branched star-like snowflakes. Snow grains are the solid equivalent of drizzle. Also called GRANULAR SNOW.

snow pellets. Frozen precipitation consisting of small, white, opaque, roundish grains of snowlike structure which are crisp and easily compressible, and may rebound or burst when striking a hard surface. Also called SOFT HAIL, GRAUPEL. See also SMALL HAIL.

snow storm. See under STORM, definition 2.

soft hail. See SNOW PELLETS.

soft iron. Iron or steel which is easily magnetized by induction, but loses its magnetism when the magnetic field is removed. The opposite is HARD IRON.

solar, *adj.* Of or pertaining to the sun.

solar day. 1. The duration of one rotation of the earth on its axis, with respect to the sun. This may be either a mean solar day, or an apparent solar day, as the reference is the mean or apparent sun, respectively. 2. The duration of one apparent rotation of the sun.

solar eclipse. An eclipse of the sun. When the moon passes between the sun and the earth, the sun appears eclipsed to an observer in the moon's shadow. A solar eclipse is partial if the sun is partly obscured; total if the entire surface is obscured, or annular if a thin ring of the sun's surface appears around the obscuring body.

solar flare. A bright eruption from the sun's chromosphere. Solar flares may appear within minutes and fade within an hour.

solar noon. Twelve o'clock solar time, or the instant the sun is over the upper branch of the reference meridian. Solar noon may be classified as mean if the mean sun is the reference, or as apparent if the apparent sun is the reference. It may be further classified according to the reference meridian, either the local or Greenwich meridian or additionally in the case of mean noon, a designated zone meridian. Standard, daylight saving or summer noon are variations of zone noon. Local apparent noon may also be called high noon.

solar-radiation pressure. A cause of perturbations of high flying artificial satellites of large diameter. The greater part is directly from the sun, a minor part is from the earth, which is usually divided into direct (reflected) and indirect terrestrial (radiated) radiation pressures.

solar system. The sun and other celestial bodies within its gravitational influence, including planets, planetoids, satellites, comets, and meteors.

solar tide. 1. The part of the tide that is due to the tide-producing force of the sun. See also LUNAR TIDE. 2. The observed tide in areas where the solar tide is dominant. This condition provides for phase repetition at about the same time each solar day.

solar time. Time based upon the rotation of the earth relative to the sun. Solar time may be classified as mean if the mean sun is the reference; or as apparent if the apparent sun is the reference. The difference between mean and apparent time is called EQUATION OF TIME. Solar time may be further classified according to the reference meridian, either the local or Greenwich meridian or additionally in the case of mean time, a designated zone meridian. Standard and daylight saving or summer time are variations of zone time. Time may also be designated according to the timepiece, as chronometer time or watch time, the time indicated by these instruments.

solar year. See TROPICAL YEAR.

solid color buoy. A buoy which is painted only one color above the water line.

solitary wave. A wave of translation consisting of a single crest rising above the undisturbed water level, without any accompanying trough, in contrast with a WAVE TRAIN. The rate of advance of a solitary wave depends upon the depth of water.

solstice, *n.* 1. One of the two points of the ecliptic farthest from the celestial equator; one of the two points on the celestial sphere occupied by the sun at maximum declination. That in the Northern Hemisphere is called the summer solstice and that in the Southern Hemisphere the winter solstice. Also called SOLSTITIAL POINT. 2. That instant at which the sun reaches one of the solstices about June 21 (summer solstice) or December 22 (winter solstice).

solstitial colure. The great circle of the celestial sphere through the celestial poles and the solstices.

solstitial point. One of the two points on the ecliptic at the greatest distance from the celestial equator. Also called SOLSTICE.

solstitial tides. Tides occurring near the times of the solstices. The tropic range may be expected to be especially large at these times.

Somali Current. See EAST AFRICA COASTAL CURRENT.

sonar, *n*. A system which determines distance and/or direction of an underwater object by measuring the interval of time between transmission of an underwater sonic or ultrasonic signal and the return of its echo. The name sonar is derived from the words sound navigation and ranging. See also ECHO RANGING.

sonic, *adj*. Of, or pertaining to, the speed of sound.

sonic depth finder. A direct-reading instrument which determines the depth of water by measuring the time interval between the emission of a sound and the return of its echo from the bottom. A similar instrument utilizing signals above audible range is called an ULTRASONIC DEPTH FINDER. Both instruments are also called ECHO SOUNDERS.

sonic frequency. See AUDIO FREQUENCY.

sonic navigation. Navigation by means of sound waves whether or not they are within the audible range. Also called ACOUSTIC NAVIGATION.

sonne, *n*. A German forerunner of the CONSOL navigation system.

sonobuoy, *n*. A buoy with equipment for automatically transmitting a radio signal when triggered by an underwater sound signal.

sound, *n*. 1. A relatively long arm of the sea or ocean forming a channel between an island and a mainland or connecting two larger bodies of water, as a sea and the ocean, or two parts of the same body but usually wider and more extensive than a strait. The term has been applied to many features which do not fit the accepted definition. Many are very large bodies of water such as Mississippi Sound and Prince William Sound, others are mere salt water ponds or small passages between islands. 2. A vibratory disturbance in air or some other elastic medium, capable of being heard by the human ear, and generally of a frequency between about 20 and 20,000 cycles per second.

sound, *v., i*. To measure the depth of the water.

sound, *v., t*. For a whale or other large sea mammal to dive for an extended period of time.

sound buoy. A buoy equipped with a gong, bell, whistle, or horn.

sounding, *n*. Measured or charted depth of water, or the measurement of such depth. A minimum sounding chosen for a vessel of specific draft in a given area to indicate the limit of safe navigation is called a danger sounding. See also ECHO SOUNDING, LINE OF SOUNDINGS.

sounding datum. Short for CHART SOUNDING DATUM.

sounding lead. See under LEAD.

sounding machine. An instrument for measuring depth of water, consisting essentially of a reel of wire to one end of which is attached a weight which carries a device for recording the depth. A crank or motor is provided for reeling in the wire.

sounding sextant. See HYDROGRAPHIC SEXTANT.

sound signal. A sound transmitted in order to convey information.

sound signal station. An attended station whose function is to operate a sound signal.

sound wave. An audio-frequency wave in any material medium, in which vibration is in the direction of travel, resulting in alternate compression and rarefaction of the medium, or, by extension, a similar wave outside the audible range.

south, *n*. The direction 180° from north. See also CARDINAL POINT.

South Atlantic Current. An eastward flowing current of the South Atlantic Ocean that is continuous with the northern edge of the WEST WIND DRIFT. It appears to originate mainly from the Brazil Current and partly from the northernmost flow of the West Wind Drift west of longitude 40°W. The current is under the influence of the prevailing westerly trade winds; the constancy and speed increase from the northern boundary to about latitude 40°S, where the current converges with the West Wind Drift. The mean speed varies from about 0.5 to 0.7 knot.

southbound node. See DESCENDING NODE.

Southeast Drift Current. See AZORES CURRENT.

southeaster, sou'easter, *n*. A southeasterly wind, particularly a strong wind or gale.

south equatorial current. See ATLANTIC SOUTH EQUATORIAL CURRENT, PACIFIC SOUTH EQUATORIAL CURRENT, INDIAN SOUTH EQUATORIAL CURRENT.

south frigid zone. That part of the earth south of the Antarctic Circle.

south geographical pole. The geographical pole in the Southern Hemisphere, at lat. 90°S.

south geomagnetic pole. The geomagnetic pole in the Southern Hemisphere. This term should not be confused with SOUTH MAGNETIC POLE. See also GEOMAGNETIC POLE.

South Indian Current. An eastward flowing current of the Indian Ocean that is continuous with the northern edge of the WEST WIND DRIFT.

southing, *n*. The distance a craft makes good to the south. The opposite is NORTHING.

south magnetic pole. The magnetic pole in the Southern Hemisphere. This term should not be confused with SOUTH GEOMAGNETIC POLE. See also GEOMAGNETIC POLE.

South Pacific Current. An eastward flowing current of the South Pacific Ocean that is continuous with the northern edge of the WEST WIND DRIFT.

south polar circle. See ANTARCTIC CIRCLE.

South Pole. 1. The south geographical pole. See also MAGNETIC POLE, GEOMAGNETIC POLE. 2. The south-seeking end of a magnet. See also BLUE MAGNETISM.

south temperate zone. The part of the earth between the Tropic of Capricorn and the Antarctic Circle.

southwester, sou'wester, *n*. A southwest wind, particularly a strong wind or gale.

southwest monsoon. See under MONSOON.

space coordinates. A three-dimensional system of Cartesian coordinates by which a point is located by three magnitudes indicating distance from three planes which intersect at a point.

spacecraft, *n*. Devices, manned and unmanned which are designed to be placed into an orbit about the earth or into a trajectory to another celestial body.

space motion. Motion of a celestial body through space. The component perpendicular to the line of sight is called proper motion and that component in the direction of the line of sight is called radial motion.

space-polar coordinates. A system of coordinates by which a point on the surface of a sphere is located in space by (1) its distance from a fixed point at the center, called the POLE; (2) the COLATITUDE or angle between the POLAR AXIS (a reference line through the pole) and the RADIUS VECTOR (a straight line connecting the pole and the point); and (3) the LONGITUDE or angle between a reference plane through the polar axis and a plane through the radius vector and polar axis. See also POLAR COORDINATES, SPHERICAL COORDINATES.

space wave. See DIRECT WAVE, definition 2.

spar buoy. A buoy in the shape of a spar, or tapered pole, floating nearly vertically. See also SPINDLE BUOY.

special mark. See under IALA MARITIME BUOYAGE SYSTEM.

Special Notice To Mariners. These notices contain important information of interest to all mariners such as cautions on the use of foreign charts; warning on use of floating aids; use of the Automated Mutual-Assistance Vessel Rescue (AMVER) system; rules, regulations, and proclamations issued by foreign governments; oil pollution regulations, etc. *Special Notice to Mariners* is published annually in *Notice to Mariners No. 1* by the Defense Mapping Agency Hydrographic/Topographic Center.

special purpose buoy. A buoy used to indicate a special meaning to the mariner and having no lateral significance, such as one used to mark a quarantine or anchorage area.

Special Warnings. Messages originated by the U.S. government which promulgate official warning of dangers to navigation, generally involving political situations. They remain active until canceled, and are published in *Notice to Mariners No. 1* issued by DMAHTC.

species of constituent. A classification depending upon the period of a constituent. The principal species are semidiurnal, diurnal, and long period.

species sanctuary. A sanctuary established for the conservation of marine life. See also MARINE SANCTUARY.

specific humidity. See HUMIDITY.

spectral, *adj*. Of or pertaining to a spectrum.

spectroscope, *n*. An optical instrument for forming spectra, very useful in studying the characteristics of celestial bodies.

spectrum *(pl. spectra)*, *n*. 1. A series of images formed when a beam of radiant energy is separated into its various wavelength components. 2. The entire range of electromagnetic radiation, or any part of it used for a specific purpose, such as the radio spectrum (10 kilohertz to 300 gigahertz).

specular reflection. Reflection without diffusion in accordance with the laws of optical reflection, such as in a mirror. Also called REGULAR REFLECTION, MIRROR REFLECTION.

speculum, *n*. An optical instrument reflector of polished metal or of glass with a film of metal.

speed, *n*. Rate of motion. The terms SPEED and VELOCITY are often used interchangeably but SPEED is a scalar, having magnitude only while VELOCITY is a vector quantity, having both magnitude and direction. Rate of motion in a straight line is called linear speed, while change of direction per unit time is called angular velocity. Subsonic, sonic, and supersonic refer to speeds respectively less than, equal to, greater than the speed of sound in standard air at sea level. Transonic speeds are those in the range in which flow patterns change from subsonic to supersonic, or vice versa.

speed circle. A circle having a radius equal to a given speed and drawn about a specified center. The expression is used chiefly in connection with relative movement problems.

speed-course-latitude error. See SPEED ERROR.

speed error. An error in both pendulous and nonpendulous type gyrocompasses resulting from movement of the gyrocompass in other than an east-west direction. The error is westerly if any component of the ship's course is north, and easterly if south. Its magnitude is proportional to the course, speed, and latitude of the ship. Sometimes called SPEED-COURSE-LATITUDE ERROR.

speed line. A line of position approximately perpendicular to the course line, thus providing a check on the speed of advance. See also COURSE LINE.

speed made good. The speed estimated by dividing the distance between the last fix and an EP by the time between the fix and the EP.

speed of advance. 1. The speed intended to be made good along the track. 2. The average speed in knots which must be maintained during a passage to arrive at a destination at an appointed time.

speed of relative movement. Speed relative to a reference point, usually itself in motion.

speed over ground. The vessel's actual speed, determined by dividing the distance between successive fixes by the time between the fixes.

speed triangle. See under VECTOR DIAGRAM.

spending beach. In a wave basin, the beach on which the entering waves spend themselves, except for the small remainder entering the inner harbor.

sphere, *n*. 1. A curved surface all points of which are equidistant from a fixed point within, called the center. The celestial sphere is an imaginary sphere of infinite radius concentric with the earth, on which all celestial bodies except the earth are imagined to be projected. The celestial sphere as it appears to an observer at the equator, where celestial bodies appear to rise vertically above the horizon, is called a right sphere; at the pole, where bodies appear to move parallel to the horizon, it is called a parallel sphere; between the equator and pole, where bodies appear to rise obliquely to the horizon, it is called an oblique sphere. Half a sphere is called a HEMISPHERE. 2. A body or the space bounded by a spherical surface. For most practical problems of navigation, the earth is considered a sphere, called the terrestrial sphere.

spherical, *adj*. Of or pertaining to a sphere.

spherical aberration. See under ABERRATION, definition 2.

spherical angle. The angle between two intersecting great circles.

spherical buoy. A buoy of which the upper part of the body (above the waterline), or the larger part of the superstructure, is spherical.

spherical coordinates. A system of coordinates defining a point on a sphere or spheroid by its angular distances from a primary great circle and from a reference secondary great circle, as latitude and longitude. See also CELESTIAL COORDINATES, POLAR COORDINATES.

spherical excess. The amount by which the sum of the three angles of a spherical triangle exceeds 180°.

spherical harmonics. Trigonometric terms of an infinite series used to approximate a two- or three-dimensional function of locations on or above the earth.

spherical sailing. Any of the sailings which solve the problems of course, distance, difference of latitude, difference of longitude, and departure by considering the spherical or spheroidal shape of the earth.

spherical triangle. A closed figure having arcs of three great circles as sides.

spherical wave. A wave with a spherical wave front.

spheroid, *n*. An ellipsoid; a figure resembling a sphere. Also called ELLIPSOID or ELLIPSOID OF REVOLUTION, from the fact that it can be formed by revolving an ellipse about one of its axes. If the shorter axis is used as the axis of revolution, an oblate spheroid results, and if the longer axis is used, a prolate spheroid results. The earth is approximately an oblate spheroid.

spheroidal excess. The amount by which the sum of the three angles on the surface of a spheroid exceeds 180°.

spheroid of reference. See REFERENCE ELLIPSOID.

spin axis. The axis of rotation of a gyroscope.

spindle buoy. A buoy having a spindle-like shape floating nearly vertically. See also SPAR BUOY.

spire, *n*. A pointed structure extending above a building, often charted with the symbol of a position circle. The spire is seldom less than two-thirds of the entire height of the structure, and its tines are rarely broken by stages or other features.

spirit compass. A magnetic compass of which the bowl mounting the compass card is filled with a solution of alcohol and water.

spit, *n*. A small tongue of land or a long narrow shoal (usually sand) extending from the shore into a body of water. Generally the tongue of land continues in a long narrow shoal for some distance from the shore.

Spitzbergen Atlantic Current. An ocean current flowing northward and westward from a point south of Spitzbergen, and gradually merging with the EAST GREENLAND CURRENT in the Greenland Sea. The Spitzbergen Atlantic Current is the continuation of the northwestern branch of the NORWAY CURRENT. Also called SPITZBERGEN CURRENT.

Spitzbergen Current. See SPITZBERGEN ATLANTIC CURRENT.

split fix. A fix by horizontal sextant angles obtained by measuring two angles between four charted features, with no common center object observed.

split-second timer. A watch with two sweep second hands which can be started and stopped together with one push button.

spoil area. Area for the purpose of disposing dredged material, usually near dredged channels. Spoil areas are usually a hazard to navigation and navigators should avoid crossing these areas. Spoil areas are shown on nautical charts. See also DISPOSAL AREA, DUMPING GROUND DUMP SITE. Also called SPOIL GROUND.

spoil ground. See SPOIL AREA.

spoil ground buoy. A buoy which marks a spoil ground.

spoil ground mark. A navigation mark indicating an area used for deposition of dredge spoil.

sporadic E-ionization. Ionization that appears at E-layer heights, is more noticeable toward the polar regions, and is caused by particle radiation from the sun. It may occur at any time of day. A sporadic E-layer sometimes breaks away from the normal E-layer and exhibits especially erratic characteristics.

spot elevation. A point on a map or chart where height above a specified datum is noted, usually by a dot and the height value.

spot-size error. The distortion of the radar return on the radarscope caused by the diameter of the electron beam which displays the returns on the scope and the lateral radiation across the scope of part of the glow produced when the electron beam strikes the phosphorescent coating of the cathode-ray tube. See also PULSE-DURATION ERROR.

spring, *n*. The season in the Northern Hemisphere which begins astronomically at the vernal equinox and ends at the summer solstice. In the Southern Hemisphere the limits are the autumnal equinox and the winter solstice.

spring high water. See under SPRING TIDES.

spring low water. See under SPRING TIDES.

spring range. See under SPRING TIDES.

spring tidal currents. Tidal currents of increased speed occurring semimonthly as the result of the moon being new or full. See also SPRING TIDES.

spring tides. Tides of increased range occurring semimonthly as the result of the moon being new or full. The spring range of tide is the average semidiurnal range occurring at the time of spring tides and is most conveniently computed from the harmonic constants. It is larger than the mean range where the type of tide is either semidiurnal or mixed, and is of no practical significance where the type of tide is diurnal. The average height of the high waters of the spring tides is called spring high water or mean high water springs and the average height of the corresponding low waters is called spring low water or mean low water springs. See also SPRING TIDAL CURRENTS.

spur, *n*. A terrestrial or bathymetric feature consisting of a subordinate elevation, ridge, or rise projecting outward from a larger feature.

spurious disk. The round image of perceptible diameter of a star as seen through a telescope, due to diffraction of light in the telescope.

spurious emission. Emission on a frequency or frequencies which are outside the necessary band, the level of which may be reduced without affecting the corresponding transmission of information. Spurious emissions include harmonic emissions, parasitic emissions and intermodulation products, but exclude emissions in the immediate vicinity of the necessary band, which are a result of the modulation process for the transmission of information.

squall, *n*. A wind of considerable intensity caused by atmospheric instability. It forms and dissipates relatively quickly, and is often accompanied by thunder, lightning, and precipitation, when it may be called a thundersquall. An arched squall is one relatively high in the center, tapering off on both sides. A bull's eye squall is one formed in fair weather, characteristic of the ocean off the coast of South Africa. See also GUST, LINE SQUALL, SQUALL LINE, WHITE SQUALL.

squall cloud. A small eddy cloud sometimes formed below the leading edge of a thunderstorm cloud, between the upward and downward currents.

squall line. A non-frontal line or narrow band of active thunderstorms (with or without squalls); a mature instability line.

squally, *adj*. Having or threatening numerous squalls.

squamish, *n*. A strong and often violent wind occurring in many of the fjords of British Columbia. Squamishes occur in those fjords oriented in a northeast-southwest or east-west direction where cold polar air can be funneled westward. They are notable in Jervis, Toba, and Bute inlets and in Dean Channel and Portland Canal. Squamishes lose their strength when free of the confining fjords and are not noticeable 15 to 20 miles offshore.

square, *n*. 1. A four-sided geometrical figure with all sides equal and all angles 90°; a rectangle or right-angled parallelogram with sides of equal length. 2. The second power of a quantity.

square meter. The derived unit of area in the International System of Units.

squat, *n*. For a vessel underway, the bodily sinkage and change of trim which are caused by the pressure distribution on the hull due to the relative motion of water and hull. The effect begins to increase significantly at depth-to-draft ratios less than 2.5. It increases rapidly with speed and is augmented in narrow channels.

stability, *n*. The state or property of resisting change or of tending to return to original conditions after being disturbed. The opposite is INSTABILITY.

stabilization of radarscope display. Orientation of the radar display to some reference direction. A radarscope display is said to be STABILIZED IN AZIMUTH when the orientation of the display is fixed to an unchanging reference (usually north). The NORTH UP orientation is an example. A radarscope display is said to be UNSTABILIZED IN AZIMUTH when the orientation of the display changes with changes in own ship's heading. The HEAD UP orientation is an example. A radarscope display is said to be DOUBLY STABILIZED or to have DOUBLE STABILIZATION when the basic orientation of the display is fixed to an unchanging reference (usually north) but the radarscope is rotated to keep own ship's heading or heading flasher up on the radarscope.

stabilized in azimuth. See under STABILIZATION OF RADARSCOPE DISPLAY.

stabilized platform. A gimbal-mounted platform, usually containing gyros and accelerometers, the purpose of which is to maintain a desired orientation in inertial space independent of craft motion. Also called STABLE PLATFORM.

stable platform. See STABILIZED PLATFORM.

stack, *n*. A label on a nautical chart which indicates a tall smokestack or chimney. The term is used when the stack is more prominent as a landmark than the accompanying buildings.

stadimeter, *n*. An instrument for determining the distance to an object of known height by measuring the vertical angle subtended by the object. The instrument is graduated directly in distance. See also RANGE FINDER.

stand, *n*. The state of the tide at high or low water when there is no sensible change in the height of the tide. The water level is stationary at high and low water for only an instant, but the change in level near these times is so slow that it is not usually perceptible. In general, the duration of the apparent stand will depend upon the range of tide, being longer for a small range than for a large range, but where there is a tendency for a double tide the stand may last for several hours, even with a large range of tide. It may be called high water stand if it occurs at the time of high water, and low water stand if it occurs at low water. Sometimes called PLATFORM TIDE.

standard, *n*. 1. Something established by custom, agreement, or authority as a basis for comparison. 2. A physical embodiment of a unit. In general it is not independent of physical conditions, and it is a true embodiment of the unit only under specified conditions.

standard acceleration of gravity. The value adopted in the International Service of Weights and Measures for the standard acceleration due to gravity is 980.665 centimeters per second, per second. See also WEIGHT.

standard atmosphere. 1. A unit accepted temporarily for use with the International System of Units; 1 standard atmosphere is equal to 101,325 pascals. 2. A hypothetical vertical distribution of atmospheric temperature, pressure, and density which is taken to be representative of the atmosphere for various purposes.

standard chronometer. See CHRONOMETER.

standard circle sheet. See CIRCLE SHEET.

standard compass. A magnetic compass designated as the standard for a vessel. It is normally located in a favorable position with respect to magnetic influences.

standard deviation. A measure of the dispersion of random errors about the mean value. If a large number of measurements or observations of the same quantity are made, the standard deviation is the square root of the sum of the squares of deviations from the mean value divided by the number of observations less one. The square of the standard deviation is called the VARIANCE. Also called RMS ERROR. See also ROOT MEAN SQUARE ERROR.

standard error. See under STANDARD DEVIATION.

standard meridian. 1. The meridian used for reckoning standard time. Throughout most of the world the standard meridians are those whose longitudes are exactly divisible by 15°. The DAYLIGHT SAVING MERIDIAN is usually 15° east of the standard meridian. 2. A meridian of a map projection, along which the scale is as stated.

standard noon. Twelve o'clock standard time, or the instant the mean sun is over the upper branch of the standard meridian. DAYLIGHT SAVING or SUMMER NOON usually occurs 1 hour later than standard noon.

standard parallel. 1. A parallel of latitude which is used as a control line in the computation of a map projection. 2. A parallel of latitude on a map or chart along which the scale is as stated for that map or chart.

standard propagation. The propagation of radio waves over a smooth spherical earth of uniform electrical characteristics, under conditions of standard refraction in the atmosphere.

standard positioning service (SPS). GPS service provided to non-military users using the single-frequency C/A code. Accuracy is 100 meters 95% (2 drms) of the time with SA turned on.

standard radio atmosphere. An atmosphere having the standard refractive modulus gradient.

standard radio horizon. The radio horizon corresponding to propagation through the standard radio atmosphere.

standard refraction. The refraction which would occur in a standard atmosphere.

standard refractive modulus gradient. The uniform variation of refractive modulus with height above the earth's surface which is regarded as a standard for comparison. The gradient considered as normal has a value of 0.12M unit per meter. The M unit is the unit in terms of which the refractive modulus is expressed.

standard station. Use of this term is discouraged. See REFERENCE STATION.

standard tactical diameter. A prescribed tactical diameter used by different types of vessels, or by vessels of the same formation in maneuvers.

standard time. The legally established time for a given zone. The United States and its possessions are, by law, divided into eight time zones. The limits of each time zone are defined by the Secretary of Transportation in Part 71, Title 49 of the *Code of Federal Regulations*. The standard time within each zone is the local mean time at the standard meridian that passes approximately through the center of the zone. Since the standard meridians are the same as those used with ZONE TIME, standard time conforms generally with the zone time for a given area. The standard time zone boundary may vary considerably from the zone time limits (71/2° in longitude on each side of the standard meridian) to conform to political or geographic boundaries or both. The standard times used in various countries and places are tabulated in the *Air Almanac* and the *Nautical Almanac* and are displayed on Chart 76, *Standard Time Zone Chart of the World*.

standard type buoy. The general classification of lighted and unlighted buoys in U.S. waters built to modern (1962) specifications.

standby lamp. A lamp brought into service in the event of failure of the lamp in regular service.

standby light. A permanently installed navigation light used in the event of failure of the main light; it is usually of lesser intensity.

standing floe. A separate floe standing vertically or inclined and enclosed by rather smooth ice.

standing wave. See STATIONARY WAVE.

stand on. To proceed on the same course.

standpipe, *n*. A label on a nautical chart which indicates a tall cylindrical structure in a waterworks system.

star, *n*. A large self-luminous celestial body. Stars are generally at such great distances from the earth that they appear to the eye to be fixed in space relative to each other. Comets, meteors, and nebulae may also be self-luminous, but are much smaller. Two stars appearing close together are called a double star, an optical double star if they appear close because they are in nearly the same line of sight but differ greatly in distance from the observer, a physical double star if in nearly the same line of sight and at approximately the same distance from the observer. A system of two stars that revolve about their common center of mass is called a binary star. A group of three or more stars so close together that they appear as a single star is called a multiple star. A group of stars physically close together is called a star cluster. A variable star changes in magnitude. A star which suddenly becomes many times brighter than previously, and then gradually fades, is called a nova. The brightest planet appearing in the western sky during evening twilight is called evening star, and the brightest one appearing in the eastern sky during morning twilight is called morning star. A shooting star or meteor is a solid particle too small to be seen until it enters the earth's atmosphere, when it is heated to incandescence by friction of the air. See also GALAXY, MILKY WAY.

starboard, *n*. The right side of a craft, facing forward. The opposite is PORT.

starboard hand buoy. A buoy which is to be left to the starboard side when approaching from seaward or in the general direction of buoyage, or in the direction established by the appropriate authority.

star chain. A radionavigation transmitting system comprised of a master station about which three (or more) slave (secondary) stations are more or less symmetrically located.

star chart. A representation, on a flat surface, of the celestial sphere or a part of it, showing the positions of the stars and sometimes other features of the celestial sphere.

star cloud. A large number of stars close together, forming a congested part of a galaxy.

star cluster. A group of stars physically close together. See also MULTIPLE STAR.

star finder. A device to facilitate the identification of stars. Sometimes called a STAR IDENTIFIER. See also PLANISPHERE.

Star Finder and Identifier (No. 2102-D). A circular star finder and identifier formerly published by the U.S. Navy Hydrographic Office and later by the U.S. Naval Oceanographic Office. It consists of a white opaque base with an azimuthal equidistant projection of most of the celestial sphere on each side, one side having the north celestial pole at the center and the other side having the south celestial pole at the center, and a series of transparent templates, at 10° intervals of latitude, each template having a family of altitude and azimuth curves.

star globe. A small globe representing the celestial sphere, on which the apparent positions of the stars are indicated. It is usually provided with graduated arcs and a suitable mount for determining the approximate altitude and azimuth of the stars, to serve as a star finder. Star globes are more commonly used by the British than by Americans. Also called CELESTIAL GLOBE.

star identifier. See STAR FINDER.

Star Sight Reduction and Identification Table. See under STAR SIGHT REDUCTION TABLES FOR 42 STARS.

Star Sight Reduction Tables for 42 Stars. A sight reduction table which provides for the reduction of 42 selected stars by the assumed altitude method. Of the 42 stars included in the table, 21 are above the observer's horizon at any time and are so tabulated in each column for integral values of latitude and altitude. This large number of star tabulations is particularly useful when clouds make identification difficult or obscure stars. Since the tabulations are for a given epoch, provision is made for precession and nutation corrections.

star telescope. An accessory of the marine navigational sextant designed primarily for star observations. It has a large object glass to give a greater field of view and increased illumination. It is an erect telescope, i.e., the object viewed is seen erect as opposed to the inverting telescope in which the object viewed is inverted. The latter type telescope requires one less lens than the erect telescope, consequently for the same size object glass, it has greater illumination. The telescope may be used for all observations.

static, *adj*. Having a fixed, nonvarying condition.

static, *n*. 1. Radio wave interference caused by natural electrical disturbances in the atmosphere, or the electromagnetic phenomena capable of causing such interference 2. Noise heard in a radio receiver caused by electrical disturbances in the atmosphere, such as lightning, northern lights, etc.

station, *n*. 1. The authorized location of an aid to navigation. 2. One or more transmitters or receivers, or a combination of transmitters and receivers, including the accessory equipment necessary at one location, for carrying on a radiocommunication service.

stationary front. A front which is stationary or nearly so. A front which is moving at a speed less than about 5 knots is generally considered to be stationary. In synoptic chart analysis, a stationary front is one that has not moved appreciably from its position on the last previous synoptic chart (3 or 6 hours before). Also called QUASI-STATIONARY FRONT.

stationary orbit. An equatorial orbit in which the satellite revolves about the primary at the angular rate at which the primary rotates on its axis. From the primary, the satellite appears to be stationary over a point on the primary's equator. See also GEOSTATIONARY SATELLITE.

stationary wave. A wave that oscillates without progressing. One-half of such a wave may be illustrated by the oscillation of the water in a pan that has been tilted. Near the axis, which is called the node or nodal line, there is no vertical rise and fall of the water. The ends of the wave are called loops and at these places the vertical rise and fall is at a maximum. The current is maximum near the node and minimum at the loops. The period of a stationary wave depends upon the length and depth of the body of water. A stationary wave may be resolved into two progressive waves of equal amplitude and equal speeds moving in opposite directions. Also called STANDING WAVE.

stationary wave theory. An assumption that the basic tidal movement in the open ocean consists of a system of stationary wave oscillations, any progressive wave movement being of secondary importance except as the tide advances into tributary waters. The continental masses divide the sea into irregular basins, which, although not completely enclosed, are capable of sustaining oscillations which are more or less independent. The tide-producing force consists principally of two parts, a semidiurnal force with a period approximating the half-day and a diurnal force with a period of a whole day. Insofar as the free period of oscillation of any part of the ocean, as determined by its dimensions and depth, is in accord with the semidiurnal or diurnal tide producing forces, there will he built up corresponding oscillations of considerable amplitude which will be manifested in the rise and fall of the tide. The diurnal oscillations, superimposed upon the semidiurnal oscillations, cause the inequalities in the heights of the two high and the two low waters of each day. Although the tidal movement as a whole is somewhat complicated by the overlapping of oscillating areas, the theory is consistent with observational data.

station buoy. An unlighted buoy established in the vicinity of a lightship or an important lighted buoy as a reference point in case the lightship or buoy should be dragged off station. Also called WATCH BUOY.

station error. See DEFLECTION OF THE VERTICAL.

statistical error. See RANDOM ERROR.

U.S. Survey mile. A unit of distance equal to 5,280 feet. This mile is generally used on land, and is sometimes called LAND MILE. It is commonly used to express navigational distances by navigators of river and lake vessels, particularly those navigating the Great Lakes.

steady bearing. A bearing line to another vessel or object, which does not change over time. An approaching or closing craft is said to be on a steady bearing if the compass bearing does not change and risk of collision therefore exists. Also called CONSTANT BEARING, DECREASING RANGE (CBDR).

steam fog. Fog formed when water vapor is added to air which is much colder than the source of the vapor. It may be formed when very cold air drifts across relatively warm water. At temperatures below about-20°F, ice particles or droxtals may be formed in the air producing a type of ice fog known as frost smoke. See also ARCTIC SEA SMOKE, FROST SMOKE. Also called ARCTIC SMOKE, SEA MIST, STEAM MIST, WATER SMOKE, ARCTIC SEA SMOKE, FROST SMOKE.

steam mist. See STEAM FOG.

steep-to, *adj.* Precipitous. The term is applied particularly to a shore, bank, or shoal that descends steeply to the sea.

steerage way, *n.* The condition wherein a ship has sufficient way on to respond to rudder movements to maintain a desired course.

steering compass. A compass by which a craft is steered, generally meaning the magnetic compass at the helm. See STEERING REPEATER.

steering repeater. A compass repeater by which a craft is steered. Sometimes loosely called a STEERING COMPASS.

stellar, *adj.* Of or pertaining to stars.`

stellar observation. See CELESTIAL OBSERVATION.

stellar parallax. See HELIOCENTRIC PARALLAX.

stem, *v., t.* To make headway against a current.

steradian, *n.* The supplementary unit of solid angle in the International System of Units, which, having its vertex in the center of a sphere, cuts off an area on the surface of the sphere equal to that of a square with sides of length equal to the radius of the sphere.

stereographic, *adj.* Of or pertaining to stereography, the art of representing the forms of solid bodies on a plane.

stereographic chart. A chart on the stereographic map projection.

stereographic map projection. A perspective, conformal, azimuthal map projection in which points on the surface of a sphere or spheroid, such as the earth, are conceived as projected by radial lines from any point on the surface to a plane tangent to the antipode of the point of projection. Circles project as circles except for great circles through the point of tangency, which project as straight lines. The principal navigational use of the projection is for charts of the polar regions. Also called AZIMUTHAL ORTHOMORPHIC MAP PROJECTION.

sternboard, *n.* Making way through the water in a direction opposite to the heading. Also called STERNWAY, though the term STERNBOARD is sometimes used to refer to the beginning of motion astern and STERNWAY is used as the vessel picks up speed. Motion in the forward direction is called HEADWAY.

stern light. A running light placed on the centerline of a vessel showing a continuous white light from dead astern to 67.5° to either side.

sternway, *n.* Making way through the water in a direction opposite to the heading. Motion in the forward direction is called HEADWAY. See also STERNBOARD.

stilling well. See FLOAT WELL.

still water level. The level that the sea surface would assume in the absence of wind waves not to be confused with MEAN SEA LEVEL or HALF TIDE LEVEL.

stippling, *n.* Graduation of shading by numerous separate dots or marks. Shallow areas on charts, for instance, are sometimes indicated by numerous dots decreasing in density as the depth increases.

stones, *n., pl.* A general term for rock fragments ranging in size from 2 to 256 millimeters. An individual water-rounded stone is called a cobble if between 64 to 256 millimeters (size of clenched fist to size of man's head), a pebble if between 4 and 64 millimeters (size of small pea to size of clenched fist), and gravel if between 2 and 4 millimeters (thickness of standard pencil lead to size of small pea). An aggregate of stones ranging from 16 to 256 millimeters is called shingle. See also MUD; SAND; ROCK, definition 2.

stooping, *n.* Apparent decrease in the vertical dimension of an object near the horizon, due to large inequality of atmospheric refraction in the line of sight to the top and bottom of the object. The opposite is TOWERING.

stop watch. A watch that can be started, stopped, and reset at will, to indicate elapsed time.

storm, *n.* 1. Wind of force 10 (48 to 55 knots or 55 to 63 miles per hour) on the Beaufort wind scale. See also VIOLENT STORM. 2. Any disturbed state of the atmosphere implying severe weather. In synoptic meteorology, a storm is a complete individual disturbance identified on synoptic charts as a complex of pressure, wind, clouds, precipitation, etc., or identified by such means as radar. Thus, storms range in scale from tornadoes and thunderstorms, through tropical cyclones, to widespread extra tropical cyclones. From a local and special interest viewpoint, a storm is a transient occurrence identified by its most destructive or spectacular aspect. Examples are rain storms, wind storms, hail storms, snow storms, etc. Notable special cases are blizzards, ice storms, sandstorms, and dust storms. 3. A term once used by seamen for what is now called VIOLENT STORM on the Beaufort wind scale.

storm center. The area of lowest atmospheric pressure of a cyclone. This is a more general expression than EYE OF THE STORM, which refers only to the center of a well-developed tropical cyclone, in which there is a tendency of the skies to clear.

storm surge. Increase or decrease in sea level by strong winds such as those accompanying a hurricane or other intense storm. Reduced atmospheric pressure often contributes to the decrease in height during hurricanes. It is potentially catastrophic, especially in deltaic regions with onshore winds at the time of high water and extreme wind wave heights. Also called STORM TIDE, STORM WAVE, TIDAL WAVE.

storm tide. See STORM SURGE.

storm track. The horizontal component of the path followed or expected to be followed by a storm CENTER.

storm wave. See STORM SURGE.

straight angle. An angle of 180°.

strait, *n.* A relatively narrow waterway connecting two larger bodies of water.

strand, *n.* See BEACH.

strand, *v., t. & i.* To run hard aground. The term STRAND usually refers to a serious grounding, while the term GROUND refers to any grounding, however slight.

stranded ice. Ice which has been floating and has been deposited on the shore by retreating high water.

stranding, *n.* The grounding of a vessel so that it is not easily refloated; a serious grounding.

strapped-down inertial navigation equipment. Inertial navigation equipment in which a stable platform and gimbal system are not utilized. The inertial devices are attached or strapped directly to the carrier. A computer utilizing gyro information resolves accelerations sensed along the carrier axes and refers these accelerations to an inertial frame of reference. Also called GIMBALLESS INERTIAL NAVIGATION EQUIPMENT. See also INERTIAL NAVIGATION.

stratiform, *adj.* Descriptive of clouds of extensive horizontal development, as contrasted to the vertically developed CUMULIFORM types. See also CIRRIFORM.

stratocumulus, *n.* A principal cloud type (cloud genus), predominantly stratiform, in the form of a gray and/or whitish layer or patch, which nearly always has dark parts and is non-fibrous (except for virga). Its elements are tessellated, rounded, roll-shaped, etc.; they may or may not be merged, and usually are arranged in orderly groups, lines or undulations, giving the appearance of a simple (or occasionally a cross-pattern) wave system. These elements are generally flat-topped, smooth and large; observed at an angle of more than 30° above the horizon, the individual stratocumulus element subtends an angle of greater than 5°. Stratocumulus is composed of small water droplets, sometimes accompanied by larger droplets, soft hail, and (rarely) by snowflakes. When the cloud is not very thick, the diffraction phenomena corona and irisation appear. Precipitation rarely occurs with stratocumulus. Stratocumulus frequently forms in clear air. It may also form from the rising of stratus, and by the convective or undulatory transformation of stratus, or nimbostratus, with or without change of height. Since stratocumulus may be transformed directly from or into altocumulus, stratus, and nimbostratus, all transitional stages may be observed. When the base of stratocumulus is rendered diffuse by precipitation, the cloud becomes nimbostratus. See also STRATIFORM, CLOUD CLASSIFICATION.

stratosphere, *n.* The atmospheric shell extending upward from the tropopause to the height where the temperature begins to increase in the 20- to 25-kilometer region.

stratus, *n.* A low cloud (mean upper level below 6,500 ft.) in a uniform layer, resembling fog but not resting on the surface.

stray line. Ungraduated portion of line connected with a current pole used in taking current observations The stray line is usually about 100 feet long and permits the pole to acquire the velocity of the current at some distance from the disturbed waters in the immediate vicinity of the observing vessel before the current velocity is read from the graduated portion of the current line.

stream, *v., t.* To place overboard and tow, as to stream a log or stream a sea anchor.

stream current. A relatively narrow, deep, fast-moving ocean current. The opposite is DRIFT CURRENT.

streamline, *n.* The path followed by a particle of fluid flowing past an obstruction. The term generally excludes the path of a particle in an eddy current.

streamline flow. Fluid motion in which the fluid moves uniformly without eddies or turbulence. If it moves in thin layers, it is called laminar flow. The opposite is TURBULENT FLOW.

stream the log. To throw the log overboard and secure it in place for taking readings.

strength of current. Phase of tidal current in which the speed is a maximum; also the speed at this time.

strength of ebb. See EBB STRENGTH.

strength of ebb interval. See EBB INTERVAL. See also LUNICURRENT INTERVAL.

strength of flood. See FLOOD STRENGTH.

strength of flood interval. See FLOOD INTERVAL. See also LUNICURRENT INTERVAL.

strip, *n.* A long narrow area of pack ice, about 1 kilometer or less in width, usually composed of small fragments detached from the main mass of ice, and run together under the influence of wind, swell, or current.

stripes, *n.* In navigation terminology, stripes are vertically arranged areas of color, such as the red and white stripes on a safe-water buoy. Horizontal areas are called bands.

strong breeze. Wind of force 6 (22 to 27 knots or 25 to 31 miles per hour) on the Beaufort wind scale.

strong fix. A fix determined from horizontal sextant angles between objects so situated as to give very accurate results.

strong gale. Wind of force 9 (41 to 47 knots or 47 to 54 miles per hour) on the Beaufort wind scale See also GALE.

sub-. A prefix meaning under, less, or marginal. The opposite is SUPER-.

Subarctic Current. See ALEUTIAN CURRENT.

subastral point. See SUBSTELLAR POINT.

sublimation, *n.* The transition of a substance directly from the solid state to the vapor state, or vice versa, without passing through the intermediate liquid state. See also CONDENSATION, EVAPORATION, FUSION.

sublunar point. The geographical position of the moon; the point on the earth at which the moon is in the zenith.

submarine bell. See under BELL.

submarine cable. A submarine conductor or fiber-optic conduit for electric current or communications.

submarine havens. Specified sea areas for submarine operations established by the submarine commander in which no friendly ASW attack may be launched. Compare with MOVING HAVENS, which are designed to prevent collisions.

submarine relief. Variations in elevation of the sea bed, or their representation by depth contours, hypsometric tints, or soundings.

submarine safety lanes. See SAFETY LANES.

submarine site. The site of a structure when located below the surface of the water.

submerge, *v., i.* To descend below the surface The opposite is SURFACE. See also DIVE.

submerged, *adj. & adv.* 1. Under water. The opposite is UNCOVERED. See also AWASH. 2. Having descended below the surface. The opposite is SURFACED.

submerged breakwater. A breakwater with its top below the still water level. When this structure is struck by a wave, part of the wave energy is reflected seaward. The remaining energy is largely dissipated in a breaker, transmitted shoreward as a multiple crest system, or as a simple wave system.

submerged lands. Lands covered by water at any stage of the tide, as distinguished from tidelands which are attached to the mainland or an island and cover and uncover with the tide. Tidelands presuppose a highwater line as the upper boundary; submerged lands do not.

submerged production well. An oil or gas well that is a seabed installation only, i.e., the installation does not include a permanent production platform. See also WELLHEAD.

submerged rock. A rock covered at the chart sounding datum and considered to be potentially dangerous to navigation. See also BARE ROCK, ROCK AWASH.

submerged screw log. A type of electric log which is actuated by the flow of water past a propeller.

subordinate current station. 1. A current station from which a relatively short series of observations is reduced by comparison with simultaneous observations from a control current station. 2. A station listed in the *Tidal Current Tables* for which predictions are to be obtained by means of differences and ratios applied to the full predictions at a reference station. See also CURRENT STATION, CONTROL CURRENT STATION. REFERENCE STATION.

subordinate tide station. 1. A tide station from which a relatively short series of observations is reduced by comparison with simultaneous observations from a tide station with a relatively long series of observations. 2. A station listed in the *Tide Tables* for which predictions are to be obtained by means of differences and ratios applied to the full predictions at a reference station. See also PRIMARY CONTROL TIDE STATION, REFERENCE STATION, SECONDARY CONTROL TIDE STATION, TERTIARY TIDE STATION.

subpermanent magnetism. The magnetism in the intermediate iron of a ship which tends to change as a result of vibration, aging, or cruising in the same direction for a long period, but does not alter immediately so as to be properly termed induced magnetism. This magnetism is the principal cause of deviation changes of a magnetic compass. At any instant this magnetism is recognized as part of the ship's permanent magnetism, and consequently must be corrected as such by means of permanent magnet correctors. See also MAGNETISM.

sub-refraction, *n*. Less-than-normal refraction, particularly as related to the atmosphere. Greater than normal refraction is called SUPER-REFRACTION.

subregion. One of the subdivisions of the earth based on the DMAHTC chart numbering system.

subsatellite point. The point at which a line from the satellite perpendicular to the ellipsoid intersects the surface of the earth.

subsidence, *n*. Decrease in the elevation of land without removal of surface material due to tectonic, seismic, or artificial forces.

subsidiary light. A light placed on or near the support of a main light and having a special use in navigation. See also PASSING LIGHT.

subsolar point. The geographical position of the sun; the point on the earth at which the sun is in the zenith at a specified time.

substellar point. The geographical position of a star; that point on the earth at which the star is in the zenith at a specified time. Also called SUBASTRAL POINT.

substratosphere, *n*. A region of indefinite lower limit just below the stratosphere.

subsurface current. An underwater current which is not present at the surface. See also SURFACE CURRENT, UNDERCURRENT, UNDERTOW.

subtend, *v., t*. To be opposite, as an arc of a circle subtends an angle at the center of the circle, the angle being formed by the radii joining the ends of the arc with the center.

subtrack, *n*. See ORBITAL PATH.

subtropical anticyclones. High pressure belts which prevail on the poleward sides of the trade winds characterized by calms, light breezes, and dryness.

sudden ionospheric disturbances (SID's). Sudden increases in the ionization density in the lower part of the ionosphere caused by very sudden and large increases in X-ray flux emitted from the sun, usually during a solar flare. SID's also occur during flares called X-ray flares that produce large X-ray flux, but which have no components in the visible light spectrum. The effect, which is restricted to sunlit propagation paths, causes a phase advance in certain radionavigation systems and is known as a SUDDEN PHASE ANOMALY (SPA). The SID effects are related to solar zenith angle, and consequently, occur mostly in lower latitude regions. Usually there is a phase advance over a period of 5 to 10 minutes followed by a recovery over a period of 30 to 60 minutes. See also POLAR CAP DISTURBANCE, MODAL INTERFERENCE.

sudden phase anomaly. See under SUDDEN IONOSPHERIC DISTURBANCES.

Suestado, *n*. A storm with southeast gales, caused by intense cyclonic activity off the coasts of Argentina and Uruguay, which affects the southern part of the coast of Brazil in the winter.

sugarloaf sea. A sea characterized by waves that rise into sugarloaf (conical) shapes, with little wind, resulting from intersecting waves.

sugg, *v., i*. To roll with the action of the sea when aground.

sumatra, *n*. A squall with violent thunder, lightning, and rain, which blows at night in the Malacca Straits, especially during the southwest monsoon. It is intensified by strong mountain breezes.

Summary of Corrections. A cumulative summary of corrections to charts, *Sailing Directions*, and *United States Coast Pilots* previously published in *Notice to Mariners*, published by the Defense Mapping Agency Hydrographic/Topographic Center.

summer, *n*. In the Northern Hemisphere summer begins astronomically at the summer solstice and ends at the autumnal equinox. In the Southern Hemisphere the limits are the winter solstice and the vernal equinox. The meteorological limits vary with the locality and the year. See also INDIAN SUMMER.

summer noon. Daylight saving noon. The expression applies where summer time is used, particularly in Europe.

summer solstice. 1. The point on the ecliptic occupied by the sun at maximum northerly declination. Sometimes called JUNE SOLSTICE, FIRST POINT OF CANCER. 2. That instant at which the sun reaches the point of maximum northerly declination, about June 21.

summer time. A variation of standard time in which the clocks are advanced 1 hour. The variation when the clocks are advanced 2 hours is called double summer time. The expression is used principally in Europe. See also DAYLIGHT SAVING TIME.

Sumner line. A line of position established by the Sumner method or, loosely, any celestial line of position.

Sumner method. The establishing of a line of position from the observation of the altitude of a celestial body by assuming two latitudes (or longitudes) and calculating the longitudes (or latitudes) through which the line of position passes. The line of position is the straight line connecting these two points (extended if necessary). This method, discovered by Thomas H. Sumner, an American sea captain, is seldom used by modern navigators, an adaptation of it, called ST. HILAIRE METHOD, being favored. See also LONGITUDE METHOD, HIGH ALTITUDE METHOD.

Sumner point. See COMPUTED POINT.

sun, *n*. The luminous celestial body at the center of the solar system, around which the planets asteroids, and comets revolve. It is an average star in terms of size and age. The sun visible in the sky is called apparent or true sun. A fictitious sun conceived to move eastward along the celestial equator at a rate that provides a uniform measure of time equal to the average apparent time is called mean sun or astronomical mean sun; a fictitious sun conceived to move eastward along the ecliptic at the average rate of the apparent sun is called dynamical mean sun. When the sun is observable at midnight, in high latitudes, it is called midnight sun.

sun cross. A rare halo phenomenon in which horizontal and vertical shafts of light intersect at the sun. It is probably due to the simultaneous occurrence of a sun pillar and a parhelic circle.

sun dog. See PARHELION.

sun line, *n*. A line of position determined from a sextant observation of the sun.

sun pillar. A glittering shaft of light, white or reddish, extending above and below the sun, most frequently observed at sunrise or sunset. If a parhelic circle is observed at the same time, a SUN CROSS results. See also HALO.

sun relay. See DAYLIGHT CONTROL.

sunrise, *n*. The crossing of the visible horizon by the upper limb of the rising sun.

sunset, *n*. The crossing of the visible horizon by the upper limb of the setting sun.

sunspot, *n*. Dark spots on the sun's surface. These spots are apparently magnetic in character and exert a disturbing influence on radio propagation on the earth.

sun's way. The path of the solar system through space.

sun switch. See DAYLIGHT CONTROL.

super-. A prefix meaning over, more, greater. The opposite is SUB-.

super-buoy. A very large buoy, generally more than 5 meters in diameter, used for navigation, offshore mooring, or data acquisition.

superheterodyne receiver. A receiver in which the incoming radio frequency signals are normally amplified before being fed into a mixer (first detector) for conversion into a fixed, lower carrier (the intermediate frequency). The intermediate frequency signals undergo very high amplification in the intermediate frequency amplifier stages and are then fed into a detector (second detector) for demodulation. The resulting audio or video signals are then usually further amplified before use.

super high frequency. Radio frequency of 3,000 to 30,000 megahertz.

superior conjunction. The conjunction of an inferior planet and the sun when the sun is between the earth and the other planet.

superior planets. The planets with orbits outside that of the Earth: Mars, Jupiter, Saturn Uranus, Neptune, and Pluto. See also PLANET.

superior transit. See UPPER TRANSIT.

super-refraction, *n*. Greater than normal refraction, particularly as related to the atmosphere. Less than normal refraction is called SUB-REFRACTION.

supersaturation, *n*. Beyond the usual point of saturation. As an example, if saturated air is cooled, condensation takes place only if nuclei are present. If they are not present, the air continues to hold more water than required for saturation until the temperature is increased or until a nucleus is introduced.

supersonic, *adj*. Faster than sound. Formerly this term was also applied to a frequency above the audible range, but in this usage it has been replaced by the term ULTRASONIC.

superstructure, *n*. See CAGE.

supplement, *n*. An angle equal to 180° minus a given angle. Two angles which equal 180° supplementary. See also COMPLEMENT, EXPLEMENT.

supplementary angles. Two angles whose sum is 180°.

supplementary units. See under INTERNATIONAL SYSTEM OF UNITS.

surf, *n*. The region of breaking waves near a beach or over a detached reef.

surface, *v., i*. To rise to the surface. The opposite is SUBMERGE.

surface boundary layer. That thin layer of air adjacent to the earth's surface extending up to a level of about 10 to 100 meters. Within this layer the wind distribution is determined largely by the vertical temperature gradient and the nature and contours of the underlying surface; shearing stresses are approximately constant. Also called FRICTION LAYER.

surface chart. Short for SYNOPTIC SURFACE CHART.

surface current. A current which does not extend more than about 3 meters below the surface. See also SUBSURFACE CURRENT, UNDERCURRENT, UNDERTOW.

surfaced, *adj*. & *adv*. Having come to the surface from below the water. The opposite is SUBMERGED. See also AFLOAT, UNCOVERED.

surface duct. A tropospheric radio duct in which the lower boundary is the surface of the earth. Also called GROUND-BASED DUCT.

surface front. See under FRONT.

surface of position. A surface on some point of which a craft is located. See also LINE OPPOSITION, FIX.

surface wave. A radio wave which is propagated along the boundary between two media in a manner determined by the properties of the two media in the vicinity of the boundary.

surf zone. The area between the outermost limit of breakers and the limit of wave uprush.

surge, *n*. 1. The bodily motion of a vessel in a seaway forward and backward along the longitudinal axis, caused by the force of the sea acting alternately on the bow and stern. Also called SURGING. See also SHIP MOTIONS. 2. See as STORM SURGE.

surging, *n*. See SURGE, *n*., definition.

surveillance, *n*. The observation of an area or space for the purpose of determining the position and movements of craft or vehicles in that area or space. Surveillance can be either dependent, independent, or pseudo-independent.

surveillance radar. A primary radar installation at a land station used to display at that station the position of vessels within its range, usually for advisory purposes.

survey, *n*. 1. The act or operation of making measurements for determining the relative positions of points on, above, or beneath the earth's surface. 2. The results of operations as in definition 1. 3. An organization for making surveys. See also GEODETIC SURVEY, HYDROGRAPHIC SURVEY, OCEANOGRAPHIC SURVEY, TOPOGRAPHIC SURVEY.

surveying, *n*. The branch of applied mathematics which teaches the art of determining accurately the area of any part of the earth's surface, the lengths and directions of bounding lines, the contour of the surface, etc., and accurately delineating the whole on a map or chart for a specified datum.

surveying sextant. See HYDROGRAPHIC SEXTANT.

swamp, *n*. An area of spongy land saturated with water. It may have a shallow covering of water, usually with a considerable amount of vegetation appearing above the surface. Sometimes called SLOUGH.

swash, *n*. 1. A narrow channel or sound within a sand bank, or between a sand bank and the shore. 2. A bar over which the sea washes. 3. The rush of water up onto the beach following the breaking of a wave.

sway, *n*. The side-to-side bodily motion of a vessel in a seaway, independent of rolling, caused by uniform pressure being exerted all along one side of the hull. Also called LATERAL DRIFTING, SWAYING. See also SHIP MOTIONS.

swaying, *n*. See SWAY.

sweep, *v., t*. To tow a line or object below the surface, to determine the least depth in an area or to insure that a given area is free from navigational dangers to a certain depth; or the removal of such dangers. See also DRAG, *v., t*.

sweep (of radarscope), *n*. As determined by the time base or range calibration, the radial movement of the stream of electrons impinging on the face of the cathode-ray tube.

sweeping, *n*. 1. The process of towing a line or object below the surface, to determine whether an area is free from isolated submerged dangers to vessels and to determine the position of any dangers that exist, or to determine the least depth of an area. 2. The process of clearing an area or channel of mines or other dangers to navigation.

sweep rate. The number of times a radar radiation pattern rotates during 1 minute of time. Sometimes expressed as the duration of one complete rotation in seconds of time.

swell, *n*. A relatively long wind wave, or series of waves, that has traveled out of the generating area. In contrast the term SEA is applied to the waves while still in the generating area. As these waves travel away from the area in which they are formed, the shorter ones die out. The surviving waves exhibit a more regular and longer period with flatter crests. When these waves reach shoal water, they become more prominent in height and of decreased wave length and are then known as ground swell.

swell direction. The direction from which swell is moving.

swept-frequency racon. An in-band racon which sweeps through the marine radar band (2920-3100 MHz in the 10-centimeter band and 9220-9500 MHz in the 3-centimeter band) in order that it may be triggered at the frequency of the interrogating radar transmitting at a given frequency within the band. Almost all such racons operate in the 3-centimeter band only. There are two types of swept-frequency racons: the slow-sweep racon sweeps through the 180 MHz frequency band in 10s of seconds (1.5 to 3.0 MHz per second); the fast-sweep racon sweeps through the band in microseconds.

swept gain. See SENSITIVITY TIME CONTROL.

swinger, *n*. See REVOLVER.

swinging buoy. A buoy placed at a favorable location to assist a vessel to adjust its compass or swing ship. The bow of the vessel is made fast to one buoy and the vessel is swung by means of lines to a tug or to additional buoys. Also called COMPASS ADJUSTMENT BUOY.

swinging ship. The process of placing a vessel on various headings and comparing magnetic compass readings with the corresponding magnetic directions, to determine deviation. This usually follows compass adjustment or compass compensation, and is done to obtain information for making a deviation table.

swinging the arc. The process of rotating a sextant about the line of sight to the horizon to determine the foot of the vertical circle through a body being observed. Also called ROCKING THE SEXTANT.

swirl error. The additional error in the reading of a magnetic compass during a turn, due to friction in the compass liquid.

symmetrical, *adj*. Being equal or identical on each side of a center line or middle value. The opposite is ASYMMETRICAL.

synchronism, *n*. The relationship between two or more periodic quantities of the same frequency when the phase difference between them is zero or constant at a predetermined value.

synchronization error. In radionavigation, the error due to imperfect timing of two operations.

synchronize, *v., t*. To bring into synchronization.

synchronous, *adj*. Coincident in time, phase, rate, etc.

synchronous lights. Two or more lights the characteristics of which are in synchronism.

synchronous satellite. A satellite whose period of rotation is equal to the period of rotation of the primary about its axis. The orbit of a synchronous satellite must be equatorial if the satellite is to remain fixed over a point on the primary's equator. See also GEOSYNCHRONOUS SATELLITE, GEOSTATIONARY SATELLITE.

synodical month. The average period of revolution of the moon about the earth with respect to the sun, a period of 29 days, 12 hours, 44 minutes, 2.8 seconds. This is sometimes called the MONTH OF THE PHASES, since it extends from new moon to the next new moon. Also called LUNATION.

synodical period. See SYNODIC PERIOD.

synodic period. The interval of time between any planetary configuration of a celestial body, with respect to the sun, and the next successive same configuration of that body, as from inferior conjunction to inferior conjunction. Also called SYNODICAL PERIOD.

synoptic chart. In meteorology, any chart or map on which data and analyses are presented that describe the state of the atmosphere over a large area at a given moment of time. A synoptic surface chart is an analyzed synoptic chart of surface weather observations.

synoptic surface chart. See under SYNOPTIC CHART.

system accuracy. The expected accuracy of a navigation system expressed in d_{rms} units, not including errors which may be introduced by the user, or geodetic or cartographic errors.

systematic error. One of the two categories of errors of observation, measurement and calculation, the other category being random error. Systematic errors are characterized by an orderly trend, and are usually predictable once the cause is known. They are divided into three classes: (1) errors resulting from changing or nonstandard natural physical conditions, sometimes called theoretical errors, (2) personal (nonaccidental) errors, and (3) instrument errors. Also called REGULAR ERROR. See also ERROR.

system electronic navigation chart. The electronic chart data base actually accessed aboard ship for the display of electronic charts. It is developed from the ENC provided by hydrographic authorities, but is specific to the shipboard system. When corrected, it is the equivalent of a paper chart.

syzygy, *n*. 1. A point of the orbit of a planet or satellite at which it is in conjunction or opposition. The term is used chiefly in connection with the moon at its new and full phase. 2. A west wind on the seas between New Guinea and Australia preceding the summer northwest monsoon.

T

table, *n*. An orderly, condensed arrangement of numerical or other information, usually in parallel rows or columns. A table in which values of the quantity to be found are tabulated for limiting values of the entering argument is called critical table. See also CALIBRATION TABLE, CONVERSION TABLE, CURRENT TABLES, TIDE TABLES, TRAVERSE TABLE.

tablemount, *n*. A seamount having a comparatively smooth, flat top. Also called GUYOT.

Tables of Computed Altitude and Azimuth. See H.O. PUB. NO. 214.

tabular altitude. See TABULATED ALTITUDE.

tabular azimuth. See TABULATED AZIMUTH.

tabular azimuth angle. See TABULATED AZIMUTH ANGLE.

tabular iceberg. A flat-topped iceberg with length-to-height ratio greater than 5:1. Most tabular bergs form by calving from an ice shelf and show horizontal banding. See also ICE ISLAND, BLOCKY ICEBERG.

tabulated altitude. In navigational sight reduction tables, the altitude taken directly from a table for the entering arguments. After interpolation for argument increments, i.e., the difference between each entering argument and the actual value, it is called COMPUTED ALTITUDE. Also called TABULAR ALTITUDE.

tabulated azimuth. Azimuth taken directly from a table, before interpolation. After interpolation, it becomes COMPUTED AZIMUTH.

tabulated azimuth angle. Azimuth angle taken directly from a table, before interpolation. After interpolation, it becomes COMPUTED AZIMUTH ANGLE.

Tacan, *n*. An ultra high frequency aeronautical radionavigation system which provides a continuous indication of bearing and distance to a Tacan station. The term is derived from Tactical Air Navigation.

tactical diameter. The distance gained to the right or left of the original course when a turn of 180° with a constant rudder angle has been completed. See also STANDARD TACTICAL DIAMETER.

taffrail, *n*. The after rail at the stern of a vessel.

taffrail log. A log consisting of a rotator towed through the water by a braided log line attached to a distance-registering device usually secured at the taffrail. Also called PATENT LOG.

tail wind. A wind from behind the vessel. See FOLLOWING WIND.

take departure. See under DEPARTURE, definition 2.

take the ground. To become stranded by the tide.

Taku wind. A strong, gusty, east-northeast wind, occurring in the vicinity of Juneau, Alaska, between October and March. At the mouth of the Taku River, after which it is named, it sometimes attains hurricane force.

tangent, *adj*. Touching at a single point.

tangent, *n*. 1. The ratio of the side opposite an acute angle of a plane right triangle to the shorter side adjacent to the same angle. The expression NATURAL TANGENT is sometimes used to distinguish the tangent from its logarithm (called LOGARITHMIC TANGENT). 2. A straight line, curve, or surface touching a curve or surface at one point.

tangent arc. 1. An arc touching a curve or surface at one point. 2. A halo tangent to a circular halo.

tangent latitude error. On a nonpendulous gyrocompass where damping is accomplished by offsetting the point of application of the force of a mercury ballistic, the angle between the local meridian and the settling position or spin axis. Where the offset of the point of application of a mercury ballistic is to the east of the vertical axis of the gyrocompass, the settling position is to the east of the meridian in north latitudes and to the west of the meridian in south latitudes. The error is so named because it is approximately proportional to the tangent of the latitude in which the gyrocompass is operating. The tangent latitude error varies from zero at the equator to a maximum at high northern and southern latitudes.

tank, *n*. An elevated water tank, indicated on a chart by a position circle.

tape gage. See ELECTRIC TAPE GAGE.

tapper, *n*. A heavy pendulum suspended outside a bell which rings it.

target, *n*. In navigation, an object observed on a radar screen. See also CONTACT.

target angle. The relative bearing of own ship from a target vessel, measured clockwise through 360°. See also ASPECT.

target tail. The display of diminishing luminance seen to follow a target on a radar display which results from afterglow and the progress of the target between successive scans of the radar. Also called TARGET TRAIL.

target trail. See TARGET TAIL.

tehuantepecer, *n*. A violent squally wind from north or north-northeast in the Gulf of Tehuantepec (south of southern Mexico) in winter. It originates in the Gulf of Mexico as a norther which crosses the isthmus and blows through the gap between the Mexican and Guatamalan mountains. It may be felt up to 100 miles out to sea. See also PAPAGAYO.

telecommunication, *n*. Any transmission, emission, sound, or intelligence of any nature by wire, radio, or other electromagnetic system. If the transfer is by radio, it may be called radiocommunication.

telegraph buoy. A buoy used to mark the position of a submarine telegraph cable.

telemeter, *n*. The complete equipment for measuring any quantity, transmitting the results electrically to a distant point, and there recording the values measured.

telemetry, *n*. The science of measuring a quantity or quantities, transmitting the measured value to a distant station, and there interpreting, indicating, or recording the quantities measured.

telemotor, *n*. A device for controlling the application of power at a distance, especially one by which the steering gear of a vessel is controlled from the wheel house.

telescope, *n*. An optical instrument used as an aid in viewing or photographing distant objects, particularly celestial objects. A reflecting telescope collects light by means of a concave mirror; a refracting telescope by means of a lens or system of lenses. A Cassegrainian telescope is a reflecting telescope in which the immergent light is reflected from the main mirror onto a secondary mirror, where it is reflected through a hole in the main mirror to an eyepiece; a Newtonian telescope is a reflecting telescope in which the immergent beam is reflected from the main mirror onto a small plane mirror, and from there to an eyepiece at the side of the telescope.

telescopic alidade. See ALIDADE.

telescopic meteor. See under METEOR.

telltale compass. A marine magnetic compass, usually of the inverted type, frequently installed in the master's cabin for his convenience.

temperate zone. Either of the two zones between the frigid and torrid zones, called the north temperate zone and the south temperate zone.

temperature, *n*. Intensity or degree of heat. Fahrenheit temperature is based upon a scale in which water freezes at 32°F and boils at about 212°F; Celsius temperature upon a scale in which water freezes at 0°C and boils at 100°C. Absolute temperature is measured from absolute zero which is zero on the Kelvin scale, –273.16° on the Celsius scale, and 459.69°F on the Fahrenheit scale. Absolute temperature based upon degrees Fahrenheit is called Rankine temperature and that based upon degrees Celsius is called Kelvin temperature.

temperature error. That instrument error due to nonstandard temperature of the instrument.

temperature inversion. An atmospheric condition in which the usual lapse rate is inverted, i.e., the temperature increases with increasing altitude.

temporal, *adj*. Pertaining to or limited by time.

temporary light. A light put into service for a limited period.

temporary units. See under INTERNATIONAL SYSTEM OF UNITS.

tend, *v., i*. To extend in a stated direction, as an anchor cable.

tera-. A prefix meaning one trillion (10^{12}).

terdiurnal, *adj*. Occurring three times per day. A terdiurnal tidal constituent has three periods in a constituent day.

terminator, *n*. The line separating illuminated and dark portions of a non-self-luminous body, as the moon.

terrace, *n*. On the sea floor, a relatively flat horizontal or gently inclined surface, sometimes long and narrow, which is bounded by a steeper ascending slope on one side and by a steeper descending slope on the opposite side.

terrestrial, *adj*. Of or pertaining to the earth.

terrestrial coordinates. See GEOGRAPHICAL COORDINATES.

terrestrial equator. 1. The earth's equator, 90° from its geographical poles. 2. See ASTRONOMICAL EQUATOR.

terrestrial latitude. Latitude on the earth; angular distance from the equator, measured northward or southward through 90° and labeled N or S to indicate the direction of measurement. See also LATITUDE.

terrestrial longitude. Longitude on the earth, the arc of a parallel, or the angle at the pole, between the prime meridian and the meridian of a point on the earth, measured eastward or westward from the prime meridian through 180°, and labeled E or W to indicate the direction of measurement. See also LONGITUDE.

terrestrial magnetism. See GEOMAGNETISM.

terrestrial meridian. See ASTRONOMICAL MERIDIAN.

terrestrial perturbations. The largest gravitational perturbations of artificial satellites which are caused by the fact that the gravity field of the earth is not spherically symmetrical.

terrestrial pole. One of the poles of the earth. See also GEOGRAPHICAL POLE, GEOMAGNETIC POLE, MAGNETIC POLE.

terrestrial radiation. The total infrared radiation emitted from the earth's surface.

terrestrial refraction. Atmospheric refraction of a ray of radiant energy emanating from a point on or near the surface of the earth, as contrasted with ASTRONOMICAL REFRACTION of a ray passing through the earth's atmosphere from outer space.

terrestrial sphere. The earth.

terrestrial triangle. A triangle on the surface of the earth, especially the navigational triangle.

territorial sea. The zone off the coast of a nation immediately seaward from a base line. Sovereignty is maintained over this coastal zone by the coastal nation, subject to the right of innocent passage to the ships of all nations. The United States recognizes this zone as extending 4.8 kilometers from the base line. See also FISHING ZONE, FISHERY CONSERVATION ZONE.

tertiary tide station. A tide station at which continuous observations have been made over a minimum period of 30 days but less than 1 year. The series is reduced by comparison with simultaneous observations from a secondary control tide station. This station provides for a 29-day harmonic analysis. See also PRIMARY CONTROL TIDE STATION; SECONDARY CONTROL TIDE STATION; SUBORDINATE TIDE STATION, definition 2; TIDE STATION.

tesla, *n*. The derived unit of magnetic flux density in the International System of Units; it is equal to 1 weber per square meter.

Texas norther. See under NORTHER.

thaw holes. Vertical holes in sea ice formed when surface puddles melt through to the underlying water.

thematic map. See TOPICAL MAP.

theoretical error. See under SYSTEMATIC ERROR.

thermometer, *n*. An instrument for measuring temperature. A maximum thermometer automatically registers the highest temperature and a minimum thermometer the lowest temperature since the last thermometer setting.

thermostat, *n*. A device for automatically regulating temperature or detecting temperature changes.

thick first-year ice. First-year ice over 120 centimeters thick.

thick weather. Condition of greatly reduced visibility, as by fog, snow, rain, etc.

thin first-year ice. First-year ice 30 to 70 centimeters thick. Also called WHITE ICE.

thin overcast. An overcast sky cover which is predominantly transparent.

thorofare, *n*. This shortened form of thoroughfare has become standard for a natural waterway in marshy areas. It is the same type of feature as a slough or bayou.

thoroughfare, *n*. A public waterway such as a river or strait. See also THOROFARE.

three-arm protractor. An instrument consisting of a circle graduated in degrees, to which is attached one fixed arm and two arms pivoted at the center and provided with clamps so that they can be set at any angle to the fixed arm, within the limits of the instrument. It is used for finding a ship's position when the horizontal angles between three fixed and known points are measured.

three-point problem. From the observation of two horizontal angles between three objects or points of known (charted) positions, to determine the position of the point of observation. The problem is solved graphically by means of the three-arm protractor and analytically by trigonometrical calculation.

threshold signal. The smallest signal capable of being detected above the background noise level.

threshold speed. The minimum speed of current at which a particular current meter will measure at its rated reliability.

thundercloud, *n*. See CUMULONIMBUS.

thunderhead, *n*. See CUMULONIMBUS.

thundersquall, *n*. Strictly, the combined occurrence of a thunderstorm and a squall, the squall usually being associated with the downrush phenomenon typical of a well-developed thunderstorm.

thunderstorm, *n*. A local storm invariably produced by a cumulonimbus cloud and always accompanied by lightning and thunder, usually with strong gusts of wind, heavy rain, and sometimes with hail. It is usually of short duration. Sometimes called ELECTRICAL STORM.

thunderstorm cirrus. See FALSE CIRRUS.

thundery sky. A sky with an overcast and chaotic aspect, a general absence of wind except during showers, a mammatus appearance of the lower clouds, and dense cirrostratus and altocumulus above.

tick, *n*. A short, audible sound or beat, as that of a clock. A time signal in the form of one or more ticks is called a TIME TICK.

tickle, *n*. A narrow channel, as used locally in the Arctic and Newfoundland.

tidal, *adj*. Of or pertaining to tides.

tidal amplitude. One-half the range of a constituent tide.

tidal basin. A basin without a caisson or gate in which the level of water rises and falls with the tides. Also called OPEN BASIN. See also TIDAL HARBOR, NON-TIDAL BASIN.

tidal bench mark. See under BENCH MARK.

tidal bench mark description. A published, concise description of the location, stamped number of designation, date established, and elevation (referred to a tidal datum) of a specific bench mark.

tidal bench mark state index map. A state map which indicates the locations for which tidal datums and tidal bench mark descriptions are available.

tidal bore. A tidal wave that propagates up a relatively shallow and sloping estuary or river in a solitary wave. The leading edge presents an abrupt rise in level, frequently with continuous breaking and often immediately followed by several large undulations. An uncommon phenomenon, the tidal bore is usually associated with very large ranges in tide as well as wedge-shaped and rapidly shoaling entrances. Also called EAGRE, EAGER, MASCARET, POROROCA, BORE.

tidal constants. Tidal relations that remain practically constant for any particular locality. Tidal constants are classified as harmonic and nonharmonic. The harmonic constants consist of the amplitudes and epochs of the harmonic constituents, and the nonharmonic constants include the ranges and intervals derived directly from the high and low water observations.

tidal constituent. See CONSTITUENT.

tidal current. A horizontal movement of the water caused by gravitational interactions between the sun, moon, and earth. The horizontal component of the particulate motion of a tidal wave. Part of the same general movement of the sea that is manifested in the vertical rise and fall, called tide. Also called TIDAL STREAM. See also CURRENT, TIDAL WAVE, TIDE.

tidal current charts. 1. Charts on which tidal current data are depicted graphically. 2. *Tidal Current Chart,* as published by the National Ocean Survey, part of a set of charts which depict, by means of arrows and figures, the direction and velocity of the tidal current for each hour of the tidal cycle. The charts, which may be used for any year, present a comprehensive view of the tidal current movement in the respective waterways as a whole and also supply a means for readily determining for any time the direction and velocity of the current at various localities throughout the water area covered.

tidal current constants. See CURRENT CONSTANTS.

tidal current diagrams. Monthly diagrams which are used with tidal current charts to provide a convenient method to determine the current flow on a particular day.

tidal current station. See CURRENT STATION.

tidal current tables. 1. Tables which give the predicted times of slack water and the predicted times and velocities of maximum current flood and ebb for each day of the year at a number of reference stations, together with time differences and velocity ratios for obtaining predictions at subordinate stations. 2. *Tidal Current Tables,* published annually by the National Ocean Survey.

tidal cycle. A complete set of tidal conditions as those occurring during a tidal day, lunar month, or Metonic cycle.

tidal datum. See VERTICAL DATUM.

tidal day. See LUNAR DAY, definition 1.

tidal difference. Difference in time or height of a high or low water at a subordinate station and at a reference station for which predictions are given in the *Tide Tables*. The difference, when applied according to sign to the prediction at the reference station, gives the corresponding time or height for the subordinate station.

tidal epoch. See EPOCH, definition 3.

tidal estuary. See under ESTUARY, definition 1.

tidal flats. See FLAT.

tidal harbor. A harbor affected by the tides, distinct from a harbor in which the water level is maintained by caissons or gates. See also NON-TIDAL BASIN.

tidal lights. Lights shown at the entrance of a harbor, to indicate tide and tidal current conditions within the harbor.

tidal lock. See ENTRANCE LOCK.

tidal marsh. Any marsh the surface of which is covered and uncovered by tidal flow. See also FLAT.

tidal platform ice foot. An ice foot between high and low water levels, produced by the rise and fall of the tide.

tidal quay. A quay in an open harbor or basin with sufficient depth alongside to enable ships lying alongside to remain afloat at any state of the tide.

tidal range. See RANGE OF TIDE.

tidal rise. See RISE OF TIDE.

tidal stream. See TIDAL CURRENT.

tidal water. Any water subject to tidal action. See also TIDEWATER.

tidal wave. 1. A wave caused by the gravitational interactions between the sun, moon and earth. Essentially, high water is the crest of a tidal wave and low water is the trough. Tide is the vertical component of the particulate motion and tidal current is the horizontal component. The observed tide and tidal current can be considered the result of the combination of several tidal waves, each of which may vary from nearly pure progressive to nearly pure standing and with differing periods, heights, phase relationships, and directions. 2. Any unusually high and destructive water level along a shore. It usually refers to either a storm surge or tsunami.

tide, *n*. The periodic rise and fall of the water resulting from gravitational interactions between the sun, moon, and earth. The vertical component of the particulate motion of a tidal wave. Although the accompanying horizontal movement of the water is part of the same phenomenon, it is preferable to designate this motion as TIDAL CURRENT. See also TIDAL WAVE definition 1.

tide-bound, *adj*. Unable to proceed because of insufficient depth of water due to tidal action.

tide crack. A crack at the line of junction between an immovable icefoot or ice wall and fast ice the latter subject to rise and fall of the tide.

tide curve. A graphic representation of the rise and fall of the tide in which time is usually represented by the abscissa and height by the ordinate of the graph. For a normal tide the graphic representation approximates a cosine curve. See also MARIGRAM.

tide datum. See VERTICAL DATUM.

tide gage. An instrument for measuring the rise and fall of the tide. See also AUTOMATIC TIDE GAGE, ELECTRIC TAPE GAGE, PRESSURE GAGE, TIDE STAFF.

tide gate. 1. A restricted passage through which water runs with great speed due to tidal action. 2. An opening through which water may flow freely when the tide sets in one direction, but which closes automatically and prevents the water from flowing in the other direction when the direction of flow is reversed.

tidehead, *n*. Inland limit of water affected by a tide.

tide hole. A hole made in ice to observe the height of the tide.

tide indicator. The part of a tide gage which indicates the height of tide at any time. The indicator may be in the immediate vicinity of the tidal water or at some distance from it.

tideland, *n*. Land which is under water at high tide and uncovered at low tide.

tidemark, *n*. 1. A high water mark left by tidal water. 2. The highest point reached by a high tide. 3. A mark placed to indicate the highest point reached by a high tide, or, occasionally, any specified state of tide.

tide notes. Notes included on nautical charts which give information on the mean range or the diurnal range of the tide, mean tide level, and extreme low water at key places on the chart.

tide pole. A graduated spar used for measuring the rise and fall of the tide. Also called TIDE STAFF.

tide pool. A pool left by an ebb tide.

tide predicting machine. A mechanical analog machine especially designed to handle the great quantity of constituent summations required in the harmonic method. William Ferrel's Maxima and Minima Tide Predictor was the first such machine used in the United States. Summing only 19 constituents, but giving direct readings of the predicted times and heights of the high and low waters, the Ferrel machine was used for the predictions of 1885 through 1914. A second machine was used for the predictions of 1912 through 1965. Predictions are now prepared using a computer.

tide-producing force. The part of the gravitational attraction of the moon and sun which is effective in producing the tides on the earth. The force varies approximately as the mass of the attracting body and inversely as the cube of its distance. The tide-producing force exerted by the sun is a little less than one-half as great as that of the moon.

tide producing potential. Tendency for particles on the earth to change their positions as a result of the gravitational interactions between the sun, moon, and earth. Although the gravitational attraction varies inversely as the square of the distance of the tide-producing body, the resulting potential varies inversely as the cube of the distance.

tide race. A very rapid tidal current through a comparatively narrow channel. Also called RACE.

tide rips. Small waves formed on the surface of water by the meeting of opposing tidal currents or by a tidal current crossing an irregular bottom. Vertical oscillation, rather than progressive waves, is characteristic of tide rips. See also RIPS.

tide rode. The condition of a ship at anchor heading into the tidal current. See also WIND RODE.

tide signals. Signals showing to navigators the state or change of the tide according to a prearranged code, or by direct display on a scale.

tide staff. A tide gage consisting of a vertical graduated staff from which the height of the tide can be read directly. See also ELECTRIC TAPE GAGE.

tide station. The geographic location at which tidal observations are conducted. Also, the facilities used to make tidal observations. These may include a tide house, tide gage, tide staff, and tidal bench marks. See also PRIMARY CONTROL TIDE STATION, SECONDARY CONTROL TIDE STATION, SUBORDINATE TIDE STATION, TERTIARY TIDE STATION.

tide tables. 1. Tables which give the predicted times and heights of high and low water for every day in the year for a number of reference stations, and tidal differences and ratios by which additional predictions can be obtained for subordinate stations. From these values it is possible to interpolate by a simple procedure the height of the tide at any hour of the day. See also TIDAL CURRENT TABLES.

tidewater, *n*. Water affected by tides or sometimes that part of it which covers the tideland. The term is sometimes used broadly to designate the seaboard. See also TIDAL WATER.

tide wave. See TIDAL WAVE, definition 1.

tideway, *n*. A channel through which a tidal current runs.

tilt, *n*. The angle which anything makes with the horizontal.

tilted blocky iceberg. A blocky iceberg which has tilted to present a triangular shape from the side.

tilt correction. The correction due to tilt error.

tilt error. The error introduced in the reading of an instrument when it is tilted, as a marine sextant held so that its frame is not perpendicular to the horizon.

time, *n*. 1. The interval between two events. 2. The date or other designated mark on a time scale. See also TIME SCALE, APPARENT TIME MEAN TIME, SIDEREAL TIME.

time and altitude azimuth. An azimuth determined by solution of the navigational triangle with meridian angle, declination, and altitude given. A TIME AZIMUTH is computed with meridian angle, declination, and latitude given. An ALTITUDE AZIMUTH is computed with altitude, declination, and latitude given.

time azimuth. An azimuth determined by solution of the navigational triangle, with meridian angle, declination, and latitude given. An ALTITUDE AZIMUTH is computed with altitude, declination, and latitude given. A TIME AND ALTITUDE AZIMUTH is computed with meridian angle, declination, and altitude given.

time ball. A visual time signal in the form of a ball. Before the widespread use of radio time signals, time balls were dropped, usually at local noon, from conspicuously-located masts in various ports. The accuracy of the signal was usually controlled by a telegraphic time signal from an observatory.

time base. A motion, of known but not necessarily of constant speed, used for measuring time intervals, particularly the sweep of a cathode-ray tube. In a linear time base the speed is constant in an expanded time base a selected part is of increased speed, and in a delayed time base the start is delayed. See also SWEEP.

time diagram. A diagram in which the celestial equator appears as a circle, and celestial meridians and hour circles as radial lines; used to facilitate solution of time problems and others involving arcs of the celestial equator or angles at the pole, by indicating relations between various quantities involved. Conventionally the relationships are given as viewed from a point over the south pole westward direction being counterclockwise. Also called DIAGRAM ON THE PLANE OF THE CELESTIAL EQUATOR, DIAGRAM ON THE PLANE OF THE EQUINOCTIAL.

time line. A line joining the heads of two vectors which represent successive courses and speeds of a ship in passing from one point to another in a known time via a specified intermediate point.

time meridian. Any meridian used as a reference for reckoning time, particularly a zone or standard meridian.

timepiece, *n*. An instrument for measuring time. See also CHRONOMETER, CLOCK, WATCH.

time scale. A system of assigning dates to events. There are three fundamental scales: Ephemeris Time, time based upon the rotation of the earth, and atomic time or time obtained by counting the cycles of a signal in resonance with certain kinds of atoms. Ephemeris Time (ET), the independent variable in the gravitational theories of the solar system, is the scale used by astronomers as the tabular argument of the precise, fundamental ephemerides of the sun, moon, and planets. Universal Time (UT1), time based on the rotation of the earth, is the scale used by astronomers as the tabular argument for most other ephemerides, e.g., the *Nautical Almanac*. Although ET and UT1 differ in concept, both are determined in arrears from astronomical observations and are extrapolated into the future based on International Atomic Time (TAI). Coordinated Universal Time (UTC) is the scale disseminated by most broadcast time services; it differs from TAI by an integral number of seconds.

time sight. Originally, an observation of the altitude of a celestial body, made for the purpose of determining longitude. Now, the expression is applied primarily to the common method of reducing such an observation.

time signal. An accurate signal marking a specified time or time interval. It is used primarily for determining errors of timepieces; usually sent from an observatory by radio. As defined by the International Telecommunications Union (ITU), a radiocommunication service for the transmission of time signals of stated high precision, intended for general reception.

time switch. A device for lighting or extinguishing a light at predetermined times, controlled by a timing device.

time tick. A time signal consisting of one or more short audible sounds or beats.

time zone. An area in all parts of which the same time is kept. In general, each zone is 15° of longitude in width with the Greenwich meridian (0° longitude) designated as the central meridian of zone 0 and the remaining zones centered on a meridian whose longitude is exactly divisible by 15. The zone boundary may vary considerably to conform to political and geographic boundaries. See also STANDARD TIME.

Tokyo datum. A geodetic datum that has its origin in Tokyo. It is defined in terms of the Bessel ellipsoid and is oriented by means of a single astronomic station. Using triangulation ties through Korea, the Tokyo datum is connected with the Manchurian datum. Unfortunately, since Tokyo is situated on a steep geoidal slope, the single station orientation has resulted in large systematic geoidal separations as the system is extended from its initial point.

tombolo, *n*. An islet and a shoal connecting it to a larger land area.

tonnage. A measure of the weight, size or capacity of a vessel. Deadweight tonnage refers to the number of tons of 2240 lbs. that a vessel will carry in salt water loaded to summer marks. It may also be considered the difference between loaded and light displacement tonnage. Displacement tonnage refers to the amount of water displaced by a vessel afloat, and is thus a measure of actual weight. Gross tonnage or gross register tonnage refers to the total measured cubic volume (100 cubic feet per ton of 2240 lbs.), based on varying formulas. Net tonnage or net registered tonnage refers to the gross tonnage minus spaces generally not used for cargo, according to varying formulas. Register tonnage is the tonnage listed on the ship's registration certificate, usually gross and/or net. Cargo tonnage refers to the weight of the cargo, independent of the vessel. Merchant ships are normally referred to by their gross or deadweight tonnage, warships by their displacement tonnage.

tongue, *n*. 1. A projection of the ice edge up to several kilometers in length, caused by wind or current. 2. An elongated extension of flat sea floor into an adjacent higher feature.

topical map. A map portraying a special subject. Also called SPECIAL SUBJECT MAP, THEMATIC MAP.

topmark, *n*. One or more objects of characteristic shape and color placed on top of a beacon or buoy to aid in its identification.

topographical latitude. See GEODETIC LATITUDE.

topographic feature. See under TOPOGRAPHY definition 1.

topographic map. A map which presents the vertical position of features in measurable form as well as their horizontal positions.

topography, *n*. 1. The configuration of the surface of the earth, including its relief and the position of features on it; the earth's natural and physical features collectively. 2. The science of delineation of natural and man-made features of a place or region especially in a way to show their positions and elevations.

toponym, *n*. A name applied to a physical or cultural topographic feature. For U.S. Government usage, policies and decisions governing place names on earth are established by the Board on Geographic Names. Also called PLACE NAME.

toponymy, *n*. 1. The study and treatment of toponyms. 2. A body of toponyms.

topple, *n*. 1. The vertical rotation of the spin axis of a gyroscope about the topple axis. 2. The vertical component of real precession or apparent precession, or the algebraic sum of the two. See also DRIFT, *n*. definition 6; TOTAL DRIFT.

topple axis. Of a gyroscope, the horizontal axis perpendicular to the horizontal spin axis, around which topple occurs. See also DRIFT AXIS, SPIN AXIS.

tornado, *n*. A violently rotating column of air, pendant from a cumulonimbus cloud, and nearly always observable as a funnel cloud. On a local scale, it is the most destructive of all atmospheric phenomena. Its vortex, commonly several hundreds of yards in diameter, whirls usually cyclonically with wind speeds estimated at 100 to more than 200 miles per hour. Its general direction of travel is governed by the motion of its parent cloud. Tornadoes occur on all continents, but are most common in Australia and the United States where the average number is 140 to 150 per year. They occur throughout the year and at any time of day, but are most frequent in spring and in middle and late afternoon. In the United States, tornadoes often develop several hundred miles southeast of a deep low centered in the central or north-central states. However, they may appear in any sector of the low, and/or be associated with fronts, instability lines, troughs, and even form within high-pressure ridges. A distinction sometimes is made between cyclonic tornadoes and convective tornadoes, the former occurring within the circulation of a well-developed parent cyclone, and the latter referring to all others. A tornado over water is called WATERSPOUT.

tornado cloud. See FUNNEL CLOUD.

torque, *n*. That which effects or tends to effect rotation or torsion and which is measured by the product of the applied force and the perpendicular distance from the line of action of the force to the axis of rotation.

torrid zone. The region of the earth between the Tropic of Cancer and the Tropic of Capricorn. Also called the TROPICS.

total current. The combination of the tidal and nontidal current. See also CURRENT.

total drift. The algebraic sum of drift due to real precession and that due to apparent precession.

total eclipse. An eclipse in which the entire source of light is obscured.

tower, *n*. A tall, slender structure, which may be charted with a position circle.

towering, *n*. Apparent increase in the vertical dimension of an object near the horizon, due to large inequality of atmospheric refraction in the line of sight to the top and bottom of the object. The opposite is STOOPING.

towing light. A yellow light having the same characteristics as a STERN LIGHT.

trace, *n*. The luminous line resulting from the radial movement of the points of impingement of the electron stream on the face of the cathode-ray tube of a radar indicator. See also SWEEP.

track, *n*. 1. The intended or desired horizontal direction of travel with respect to the earth. The track as expressed in degrees of the compass may be different from the course due to such factors as making allowance for current or sea or steering to resume the TRACK, definition 2. 2. The path of intended travel with respect to the earth as drawn on the chart. Also called INTENDED TRACK, TRACK-LINE. 3. The actual path of a vessel over the ground, such as may be determined by tracking.

track, *v., t*. To follow the movements of an object such as by radar or an optical system.

track angle. See TRACK, definition 1.

track chart. A chart showing recommended, required, or established tracks, and usually indicating turning points, courses, and distances. A distinction is sometimes made between a TRACK CHART and a ROUTE CHART, the latter generally showing less specific information, and sometimes only the area for some distance each side of the great circle or rhumb line connecting two terminals.

tracking, *n*. In the operation of automated radar plotting aids, the process of observing the sequential changes in the position of a target to establish its motion.

track-line, *n*. See TRACK, definition 2.

track made good. The single resultant direction from a point of departure to a point of arrival at any given time. The use of this term to indicate a single resultant direction is preferred to the use of the misnomer course made good. See also COURSE, TRACK.

trade winds. Relatively permanent winds on each side of the equatorial doldrums, blowing from the northeast in the Northern Hemisphere and from the southeast in the Southern Hemisphere. See also ANTI-TRADES.

traffic control signals. Visual signals placed in a harbor or waterway to indicate to shipping the movements authorized or prohibited at the time at which they are shown. Also called DOCKING SIGNALS.

traffic lane. An area of defined limits in which one-way traffic is established. See also TWO-WAY ROUTE, ROUTING SYSTEM.

traffic separation scheme. A routing measure designed for separating opposing streams of traffic in congested areas by the establishment of traffic lanes, precautionary areas, and other measures. See also ROUTING SYSTEM.

train, *v., t*. To control motion in bearing.

training wall. A wall, bank, or jetty, often submerged, built to direct or confine the flow of a river or tidal current.

tramontana, *n*. A northeasterly or northerly wind occurring in winter off the west coast of Italy. It is a fresh wind of the fine weather mistral type.

transceiver, *n*. A combination transmitter and receiver in a single housing, with some components being used by both parts. See also TRANSPONDER.

transducer, *n*. A device that converts one type of energy to another, such as the part of a depth sounder that changes electrical energy into acoustical energy.

transfer, *n*. 1. The distance a vessel moves perpendicular to its initial direction in making a turn of 90° with a constant rudder angle. 2. The distance a vessel moves perpendicular to its initial direction for turns of less than 90°. See also ADVANCE.

transit, *n*. 1. The passage of a celestial body across a celestial meridian, usually called MERIDIAN TRANSIT. 2. The apparent passage of a celestial body across the face of another celestial body or across any point, area, or line. 3. An instrument used by an astronomer to determine the exact instant of meridian transit of a celestial body. 4. A reversing instrument used by a surveyor for accurately measuring horizontal and vertical angles; a theodolite which can be reversed in its supports without being lifted from them.

transit, *v., t*. To cross. In navigation the term is generally used with reference to the passage of a celestial body over a meridian, across the face of another celestial body, or across the reticle of an optical instrument.

TRANSIT, *n*. See NAVY NAVIGATION SATELLITE SYSTEM.

transition buoy. A buoy indicating the transition between the lateral and cardinal systems of buoyage.

transition mark. A navigation mark indicating the transition between the lateral and cardinal systems of marking.

translocation, *n*. The determination of the relative positions of two points by simultaneous Doppler satellite observations from each point.

translunar, *adj*. Of or pertaining to space outside the moon's orbit about the earth.

transmit-receive tube. See as TR TUBE.

transponder, *n.* A component of a secondary radar system capable of accepting the interrogating signal, received from a radar set or interrogator, and in response automatically transmitting a signal which enables the transponder to be identified by the interrogating station. Also called TRANSPONDER BEACON. See also RADAR BEACON, RACON.

transponder beacon. See TRANSPONDER.

transpose, *v., t.* To change the relative place or position of, as to move a term from one side of an equation to the other with a change of sign.

transverse bar. A bar which extends approximately normal to the shoreline.

transverse chart. A chart on a transverse map projection. Also called INVERSE CHART.

transverse cylindrical orthomorphic chart. See TRANSVERSE MERCATOR CHART.

transverse cylindrical orthomorphic projection. See TRANSVERSE MERCATOR MAP PROJECTION.

transverse equator. The plane which is perpendicular to the axis of a transverse map projection. Also called INVERSE EQUATOR. See also FICTITIOUS EQUATOR.

transverse graticule. A fictitious graticule based upon a transverse map projection.

transverse latitude. Angular distance from a transverse equator. Also called INVERSE LATITUDE. See also FICTITIOUS LATITUDE.

transverse longitude. Angular distance between a prime transverse meridian and any given transverse meridian. Also called INVERSE LONGITUDE. See also FICTITIOUS LONGITUDE.

transverse map projection. A map projection with its axis in the plane of the equator.

transverse Mercator chart. A chart on the transverse Mercator projection. Also called TRANSVERSE CYLINDRICAL ORTHOMORPHIC CHART, INVERSE MERCATOR CHART, INVERSE CYLINDRICAL ORTHOMORPHIC CHART. See also MERCATOR CHART.

transverse Mercator map projection. A conformal cylindrical map projection, being in principle equivalent to the regular Mercator map projection turned (transversed) 90° in azimuth. In this projection, the central meridian is represented by a straight line, corresponding to the line which represents the equator on the regular Mercator projection. Neither the geographic meridians (except the central meridian) nor the geodetic parallels (except the equator) are represented by straight lines. Also called INVERSE MERCATOR MAP PROJECTION, TRANSVERSE CYLINDRICAL ORTHOMORPHIC MAP PROJECTION, INVERSE CYLINDRICAL ORTHOMORPHIC MAP PROJECTION. See also MERCATOR MAP PROJECTION.

transverse meridian. A great circle perpendicular to a transverse equator. The reference transverse meridian is called prime transverse meridian. Also called INVERSE MERIDIAN. See also FICTITIOUS MERIDIAN.

transverse parallel. A circle or line parallel to a transverse equator connecting all points of equal transverse latitude. Also called INVERSE PARALLEL. See also FICTITIOUS PARALLEL.

transverse pole. One of the two points 90° from a transverse equator.

transverse rhumb line. A line making the same oblique angle with all fictitious meridians of a transverse Mercator map projection. Transverse parallels and meridians may be considered special cases of the transverse rhumb line. Also called INVERSE RHUMB LINE. See also FICTITIOUS RHUMB LINE.

transverse wave. A wave in which the vibration is perpendicular to the direction of propagation, as in light waves. This is in contrast with a LONGITUDINAL WAVE, in which the vibration is in the direction of propagation.

trapezoid, *n.* A quadrilateral having two parallel sides and two nonparallel sides.

traverse, *n.* A series of directions and distances, such as when a sailing vessel beats into the wind, a steam vessel zigzags, or a surveyor makes measurements for determination of position.

traverse sailing. A method of determining the equivalent course and distance made good by a craft following a track consisting of a series of rhumb lines. The solution is usually made by means of traverse tables.

traverse table. A table giving relative values of various parts of plane right triangles, for use in solving such triangles, particularly in connection with various sailings.

TR box. See TR SWITCH.

trench, *n.* A long, narrow, characteristically very deep and asymmetrical depression of the sea floor, with relatively steep sides. See also TROUGH.

triad, *n.* Three radionavigation stations operated as a group for the determination of positions. Also called TRIPLET. See also STAR CHAIN.

triangle, *n.* A closed figure having three sides. The triangle is plane, spherical, or curvilinear as the sides are straight lines, arcs of great circles, or curves, respectively. See also EQUILATERAL TRIANGLE, ISOSCELES TRIANGLE, NAVIGATIONAL TRIANGLE, RIGHT TRIANGLE.

triangulation, *n.* A method of surveying in which the stations are points on the ground, located on the vertices of a chain or network of triangles. The angles of the triangles are measured instrumentally, and the sides are derived by computation from selected sides which are called BASE LINES, the lengths of which are obtained from direction measurements on the ground. See also TRILATERATION.

triaxial ellipsoid. A reference ellipsoid having three unequal axes; the shortest is the polar axis, and the two longer ones lie in the plane of the equator.

tributary, *n.* A stream that flows into another stream or a lake.

tributary. Any body of water that flows into a larger body, i.e., a creek in relation to a river, or a river in relation to a bay.

trigger, *n.* In a radar set, a sharp voltage pulse which is applied to the modulator tubes to fire the transmitter, applied simultaneously to the sweep generator to start the electron beam moving radially from the sweep origin to the edge of the face of the cathode-ray tube.

triggering, *n.* The process of causing a transponder to respond.

trigonometric functions. The ratios of the sides of a plane right triangle, as related to one of its angles. If a is the side opposite an acute angle, b the adjacent side, and c the hypotenuse the trigonometric functions are: sine = a/c, cosine = b/c, tangent = a/b, cotangent = b/a, secant = c/b, cosecant = c/a. The expression NATURAL TRIGONOMETRIC FUNCTION is sometimes used to distinguish a trigonometric function from its logarithm (called LOGARITHMIC TRIGONOMETRIC FUNCTION).

trihedral reflector. See CORNER REFLECTOR.

trilateration, *n.* A method of surveying wherein the lengths of the triangle sides are measured, usually by electronic methods, and the angles are computed from the measured lengths. See also TRIANGULATION.

trim, *n.* The relation of the draft of a vessel at the bow and stern. See also DOWN BY THE HEAD; DOWN BY THE STERN; DRAG, *n.*, definition 3; SQUAT, *n.*

triple interpolation. Interpolation when there are three arguments or variables.

triples, *n.* See TRIAD.

trochoid, *n.* In relation to wave motion, a curve described by a point on a radius of a circle that rolls along a straight line. Also called PROLATE CYCLOID.

tropic, *adj.* Of or pertaining to a tropic or the tropics.

tropic, *n.* Either of the two parallels of declination (north or south), approximately 23°27' from the celestial equator, reached by the sun at its maximum declination, or the corresponding parallels on the earth. The northern of these is called the TROPIC OF CANCER and the southern, the TROPIC OF CAPRICORN. The region of the earth between these two parallels is called the TORRID ZONE, or often the TROPICS.

tropical, *adj.* 1. Of or pertaining to the vernal equinox. See also SIDEREAL. 2. Of or pertaining to the Tropics.

tropical air. Warm air of an air mass originating in subtropical anticyclones, further classified as tropical continental air and tropical maritime air, as it originates over land or sea, respectively.

tropical continental air. Air of an air mass originating over a land area in low latitudes, such as the Sahara desert. Tropical continental air is characterized by high surface temperature and low specific humidity.

tropical cyclone. The general term for cyclones originating in the tropics or subtropics. These cyclones are classified by form and intensity as follows: A tropical disturbance is a discrete system of apparently organized convection generally 100 to 300 miles in diameter, having a nonfrontal migratory character, having maintained its identity for 24 hours or more. It may or may not be associated with a detectable perturbation of the wind field. It has no strong winds and no closed isobars, i.e., isobars that completely enclose the low. In successive stages of intensification, the tropical cyclone are classified as tropical disturbance, tropical depression, tropical storm, and hurricane or typhoon. The tropical depression has one or more closed isobars and some rotary circulation at the surface. The highest sustained (l-minute mean) surface wind speed is 33 knots. The tropical storm has closed isobars and a distinct rotary circulation. The highest sustained (1-minute mean) surface wind speed is 34 to 63 knots. The hurricane or typhoon has closed isobars, a strong and very pronounced rotary circulation, and a sustained (1-minute mean) surface wind speed of 64 knots or higher. Tropical cyclones occur almost entirely in six rather distinct areas, four in the Northern Hemisphere and two in the Southern Hemisphere. The name by which the tropical cyclone is commonly known varies somewhat with locality as follows: North Atlantic: A tropical cyclone with winds of 64 knots or greater is called a HURRICANE. Eastern North Pacific: The name HURRICANE is used as in the North Atlantic. Western North Pacific: A fully developed storm with winds of 64 knots or greater is called a TYPHOON or, locally in the Philippines, a BAGUIO. North Indian Ocean: A tropical cyclone with winds of 34 knots or greater is called a CYCLONIC STORM. South Indian Ocean: A tropical storm with winds of 34 knots or greater is called a CYCLONE. Southwest Pacific and Australian Area: The name CYCLONE is used as in the South Indian Ocean. A severe tropical cyclone originating in the Timor Sea and moving southwestward and then southeastward across the interior of northwestern Australia is called a WILLY-WILLY. Tropical cyclones have not been observed in the South Atlantic Ocean or in the South Pacific Ocean east of longitude 140°W.

tropical depression. See under TROPICAL CYCLONE.

tropical disturbance. See under TROPICAL CYCLONE.

tropical maritime air. Air of an air mass originating over an ocean area in low latitudes. Tropical maritime air is characterized by high surface temperature and high specific humidity.

tropical month. The average period of the revolution of the moon about the earth with respect to the vernal equinox, a period of 27 days, 7 hours, 43 minutes, 4.7 seconds. This is almost the same length as the sidereal month.

tropical storm. See under TROPICAL CYCLONE.

tropical year. The period of one revolution of the earth around the sun, with respect to the vernal equinox. Because of precession of the equinoxes, this is not 360° with respect to the stars, but 50.3" less. A tropical year is about 20 minutes shorter than a sidereal year, averaging 365 days, 5 hours, 48 minutes, and 46 seconds in 1900, decreasing at the rate of 0.00530 second annually. Also called ASTRONOMICAL, EQUINOCTIAL, NATURAL, or SOLAR YEAR.

tropic currents. Tidal currents occurring semimonthly when the effect of the moon's maximum declination is greatest. At these times the tendency of the moon to produce a diurnal inequality in the current is at a maximum.

tropic higher high water. The higher high water of tropic tides. See also TROPIC TIDES.

tropic higher high water interval. The lunitidal interval pertaining to the higher high waters at the time of the tropic tides. See also TROPIC LOWER LOW WATER INTERVAL.

tropic higher low water. The higher low water of tropic tides. See also TROPIC TIDES.

tropic high water inequality. The average difference between the two high waters of the day at the times of the tropic tides. Applicable only when the tide is semidiurnal or mixed. See also TROPIC TIDES, TROPIC LOW WATER INEQUALITY.

tropic inequalities. See TROPIC HIGH WATER INEQUALITY, TROPIC LOW WATER INEQUALITY.

tropic intervals. See TROPIC HIGH WATER INTERVAL, TROPIC LOWER LOW WATER INTERVAL.

tropic lower high water. The lower high water of tropic tides. See also TROPIC TIDES.

tropic lower low water. The lower low water of tropic tides. See also TROPIC TIDES.

tropic lower low water interval. The lunitidal interval pertaining to the lower low waters at the time of tropic tides. See also TROPIC HIGHER HIGH WATER INTERVAL.

tropic low water inequality. The average difference between the two low waters of the day at the times of the tropic tides. Applicable only when the type of tide is semidiurnal or mixed. See also TROPIC TIDES, TROPIC HIGH WATER INEQUALITY.

Tropic of Cancer. The northern parallel of declination, approximately 23°27' from the celestial equator, reached by the sun at its maximum northerly declination, or the corresponding parallel on the earth. It is named for the sign of the zodiac in which the sun reached its maximum northerly declination at the time the parallel was so named.

Tropic of Capricorn. The southern parallel of declination, approximately 23°27' from the celestial equator, reached by the sun at its maximum southerly declination, or the corresponding parallel on the earth. It is named for the sign of the zodiac in which the sun reached its maximum southerly declination at the time the parallel was so named.

tropic ranges. See GREAT TROPIC RANGE, MEAN TROPIC RANGE, SMALL TROPIC RANGE.

tropics, *n*. See TORRID ZONE.

tropic speed. The greater flood or greater ebb speed at the time of tropic currents.

tropic tides. Tides occurring semimonthly when the effect of the moon's maximum declination is greatest. At these times there is a tendency for an increase in the diurnal range. The tidal datums pertaining to the tropic tides are designated as tropic higher high water, tropic lower high water, tropic higher low water, and tropic lower low water.

tropopause, *n*. The boundary between the troposphere and the stratosphere.

troposphere, *n*. The portion of the atmosphere from the earth's surface to the tropopause, i.e., the lowest 10 to 20 kilometers of the atmosphere. It is characterized by decreasing temperature with height, appreciable vertical wind motion, appreciable water vapor content, and variable weather.

tropospheric radio duct. A quasi-horizontal layer in the troposphere between the boundaries of which radio energy of sufficiently high frequency is substantially confined and propagated with abnormally low attenuation. The duct may be formed in the lower portion of the atmosphere when there is a marked temperature inversion or a sharp decrease in water vapor with increased height. See also SURFACE DUCT, ELEVATED DUCT.

tropospheric wave. A radio wave traveling between points on or near the surface of the earth by one or more paths lying wholly within the troposphere. The propagation of this wave is determined primarily by the distribution of the refractive index in the troposphere.

trough, *n*. 1. A long depression of the sea floor, characteristically flat bottomed and steep sided, and normally shallower than a trench. 2. In meteorology, an elongated area of relatively low pressure. The opposite of a trough is called RIDGE. The term trough is commonly used to distinguish the above elongated area from the closed circulation of a low (or cyclone). But a large-scale trough may include one or more lows. 3. The lowest part of a wave between two crests.

TR switch *(from transmit/receive)*. A switch used to automatically decouple the receiver from the antenna during transmission when there is a common transmitting and receiving antenna. Also called TR BOX.

TR tube. An electronic switch capable of rapid switching between transmit and receive functions, used to protect the receiver from damage from energy generated by the transmitter. Another device called the anti-TR tube is used to block the passage of echoes to the receiver during the relatively long periods when the transmitter is inactive. See also TR SWITCH, ATR TUBE.

true, *adj*. 1. Related to true north. 2. Actual, as contrasted with fictitious, such as the true sun. 3. Related to a fixed point, either on the earth or in space, such as true wind, in contrast with RELATIVE, which is related to a moving point. 4. Corrected, as in the term true altitude.

true altitude. See OBSERVED ALTITUDE.

true amplitude. Amplitude relative to true east or west.

true anomaly. See under ANOMALY, definition 2.

true azimuth. Azimuth relative to true north.

true bearing. Bearing relative to true north; compass bearing corrected for compass error.

true course. Course relative to true north.

true direction. Horizontal direction expressed as angular distance from true north.

true heading. Heading relative to true north.

true meridian. A meridian through the geographical pole; compare with MAGNETIC MERIDIAN, COMPASS MERIDIAN, or GRID MERIDIAN, the north-south lines according to magnetic, compass, or grid direction, respectively.

true motion display. A type of radarscope display in which own ship and other moving targets move on the plan position indicator in accordance with their true courses and speeds. All fixed targets appear as stationary echoes. However, uncompensated set and drift of own ship may result in some movement of the echoes of stationary targets. This display is similar to a navigational (geographical) plot. See also RELATIVE MOTION DISPLAY.

true motion radar. A radar set which provides a true motion display as opposed to the relative motion display most commonly used. The true motion radar requires own ship's speed input, either log or manual, in addition to own ship's course input.

true north. The direction of the north geographical pole; the reference direction for measurement of true directions.

true plot. See GEOGRAPHICAL PLOT.

true prime vertical. See under PRIME VERTICAL CIRCLE.

true solar time. See APPARENT TIME.

true sun. The actual sun as it appears in the sky. Usually called APPARENT SUN. See also MEAN SUN, DYNAMICAL MEAN SUN.

true track of target. The motion of a radar target on a true motion display. When the true motion display is ground stabilized, i.e., allowance is made for the set and drift of current, the motion displayed is called GROUND TRACK. Without such stabilization the motion displayed is called WATER TRACK.

true wind. Wind relative to a fixed point on the earth. Wind relative to a moving point is called APPARENT or RELATIVE WIND.

trumpet, *n*. See HORN.

tsunami, *n*. A long-period sea wave, potentially catastrophic, produced by a submarine earthquake or volcanic eruption. It may travel unnoticed across the ocean for thousands of miles from its point of origin, building up to great heights over shoal water. Also called SEISMIC SEA WAVE, TIDAL WAVE.

Tsushima Current. That part of the Kuroshio flowing northeastward through Korea Strait and along the Japanese coast in the Japan Sea; it flows strongly eastward through Tsugaru Strait at speeds to 7 knots. The Tsushima Current is strong most of the time, averaging about 1 knot; however, it may weaken somewhat during autumn. In Western Channel, between Tsushima and southeastern Korea, tidal currents retard the general northeastward flowing Tsushima Current during the southwest-setting flood and reinforce it during the northeast-setting ebb. Resultant current speeds range from 1/4 knot during flood to 3 knots during ebb. In the strait between Tsushima and Kyushu, the current flows northeastward throughout the year. Current speeds in Korea Strait also are affected by the seasonal variations of the monsoons. The strongest currents usually occur from July through November. The Tsushima Current divides after flowing through Korea Strait, a small branch flowing northward along the east coast of Korea as far as Vladivostok in summer. During this season the current is strongest and overcomes the weak southward flowing, coastal Liman Current. When the current combines with the ebb current, the resultant speed may reach 2 knots. During winter this branch of the Tsushima Current is weakest and is influenced by the stronger southward flowing Liman Current which normally extends as far south as 39°N, with speeds from 1/4 to 3/4 knot. The main body of the Tsushima Current flows northeastward off the northeast coast of Honshu. In summer, after entering the Japan Sea, its speed is about 1/2 to 1 knot. In winter the current is relatively weak, although near the islands and headlands speeds may exceed 1 knot, especially after northwesterly gales.

tuba, *n*. See FUNNEL CLOUD.

tufa, *n*. A porous rocky deposit formed in streams and in the ocean near the mouths of rivers.

tumble, *v., i.* The tendency of a gyroscope to precess suddenly and to an extreme extent as a result of exceeding its operating limits of bank or pitch.

tune, *v., t.* To adjust the frequency of a circuit or system to obtain optimum performance, commonly to adjust to resonance.

turbidity, *n*. A measure of the amount of suspended material in water.

turbulent, *n*. Agitated or disturbed fluid motion, not flowing smoothly or uniformly.

turbulent flow. Fluid motion in which random motions of parts of the fluid are superimposed upon a simple pattern of flow. All or nearly all fluid flow displays some degree of turbulence. The opposite is STREAMLINE FLOW.

turning basin. A water area, usually dredged to well-defined limits, used for turning vessels.

turning buoy. A buoy marking a turn in a channel.

turning circle. The path described by the pivot point of the vessel as it makes a turn of 360° with constant rudder and speed.

turn of the tide. See CHANGE OF TIDE.

twenty-four hour satellite. See GEOSYNCHRONOUS SATELLITE.

twilight, *n*. The period of incomplete darkness following sunset (evening twilight) or preceding sunrise (morning twilight). Twilight is designated as civil, nautical, or astronomical, as the darker limit occurs when the center of the sun is 6°, 12°, or 18° below the celestial horizon, respectively. See also DAWN, DUSK.

twinkle, *v., i.* To flicker randomly, or vary in intensity.

two-body orbit. The motion of a point mass in the presence of the gravitational attraction of another point mass, and in the absence of other forces. This orbit is usually an ellipse, but may be a parabola or hyperbola.

two-degree-of-freedom gyro. A gyroscope the spin axis of which is free to rotate about two orthogonal axes, not counting the spin axis. See also DEGREE-OF-FREEDOM.

two-tone diaphone. See under DIAPHONE.

two-way route. A route within defined limits in which two-way traffic is established, aimed at providing safe passage of ships through waters where navigation is difficult or dangerous. See also ROUTING SYSTEM.

tyfon, *n*. See TYPHON.

type of tide. A classification based on characteristic forms of a tide curve. Qualitatively, when the two high waters and two low waters of each tidal day are approximately equal in height, the tide is said to be semidiurnal; when there is a relatively large diurnal inequality in the high or low waters or both, it said to be mixed; and when there is only one high water and one low water in each tidal day, it is said to be diurnal.

typhon, *n*. A diaphragm horn which operates under the influence of compressed air or steam. Also called TYFON.

typhoon, *n*. See under TROPICAL CYCLONE.

U

Ulloa's ring. See BOUGUER'S HALO.

ultra high frequency. Radio frequency of 300 to 3,000 megahertz.

ultra quick light. A navigation light flashing at a rate of not less than 160 flashes per minute. See also CONTINUOUS ULTRA QUICK LIGHT, INTERRUPTED ULTRA QUICK LIGHT.

ultrashort wave. A radio wave shorter than 10 meters. A wave shorter than 1 meter is called a MICROWAVE. See also WAVE.

ultrasonic, *adj*. Having a frequency above the audible range. Frequencies below the audible range are called INFRASONIC. See also SUPERSONIC.

ultrasonic depth finder. A direct-reading instrument which determines the depth of water by measuring the time interval between the emission of an ultrasonic signal and the return of its echo from the bottom. A similar instrument utilizing signals within the audible range is called a SONIC DEPTH FINDER. Both instruments are also called ECHO SOUNDERS.

umbra, *n*. 1. The darkest part of a shadow in which light is completely cut off by an intervening object. A lighter part surrounding the umbra, in which the light is only partly cut off, is called the PENUMBRA. 2. The darker central portion of a sun spot, surrounded by the lighter PENUMBRA.

uncorrecting, *n*. The process of converting true to magnetic, compass, or gyro direction, or magnetic to compass direction. The opposite is CORRECTING.

uncovered, *adj*. & *adv*. Above water. The opposite is SUBMERGED. See also AFLOAT; AWASH.

undercurrent, *n*. A current below the surface, particularly one flowing in a direction or at a speed differing from the surface current. See UNDERTOW, SUBSURFACE CURRENT, SURFACE CURRENT.

under the lee. To leeward.

undertow, *n*. Receding water below the surface of breakers on a beach. See also UNDERCURRENT, SUBSURFACE CURRENT, SURFACE CURRENT, BACKRUSH, RIP CURRENT.

underway, under way, *adv*. Not moored or anchored. See also ADRIFT. See also MAKING WAY.

undevelopable, *adj*. A surface not capable of being flattened without distortion. The opposite is DEVELOPABLE.

undisturbed orbit. See NORMAL ORBIT.

undulating, *adj*. Having the form of more or less regular waves.

undulating light. See under FIXED AND FLASHING LIGHT.

undulation of the geoid. See GEOIDAL HEIGHT.

undulatus, *adj*. Having undulations, referring to a cloud composed of elongated and parallel elements resembling ocean waves.

unfavorable current. A current flowing in such a direction as to decrease the speed of a vessel over the ground. The opposite is FAVORABLE CURRENT.

unfavorable wind. A wind which delays the progress of a craft in a desired direction. Usually used in plural and chiefly in connection with sailing vessels. A wind which aids the progress of a craft is called a FAIR or FAVORABLE WIND. See also FOLLOWING WIND, HEAD WIND.

Uniform State Waterway Marking System. An aids to navigation system developed jointly by the U.S. Coast Guard and state boating administrators to assist the small craft operator in inland state waters marked by states. It consists of two categories of aids to navigation. One is a system of aids to navigation, generally compatible with the Federal lateral system of buoyage, to supplement the federal system in state waters The other is a system of regulatory markers to warn the small craft operator of dangers or to provide general information and directions.

unipole antenna, *n*. See ISOTROPIC ANTENNA.

unique sanctuary. A marine sanctuary established to protect a unique geologic, oceanographic, or living feature. See also MARINE SANCTUARY.

unit, *n*. A value, quantity, or magnitude in terms of which other values, quantities, or magnitudes are expressed. In general, a unit is fixed by definition and is independent of such physical conditions as temperature. See also STANDARD, definition 2; INTERNATIONAL SYSTEM OF UNITS.

United States Coast Pilot. One of a series of SAILING DIRECTIONS published by the National Ocean Service, that cover a wide variety of information important to navigators of U.S. coastal and intracoastal waters, and waters of the Great Lakes. Most of this information cannot be shown graphically on the standard nautical charts and is not readily available elsewhere. This information includes navigation regulations, outstanding landmarks, channel and anchorage peculiarities, dangers, weather, ice, currents, and port facilities. Each *Coast Pilot* is corrected through the dates of *Notices to Mariners* shown on the title page and should not be used without reference to the *Notices to Mariners* issued subsequent to those dates.

United States National Map Accuracy Standards. A set of standards which define the accuracy with which features of U.S. maps are to be portrayed. 1. Horizontal accuracy: For maps at publication scales larger than 1:20,000, 90 percent of all well-defined features, with the exception of those unavoidably displaced by exaggerated symbolization, will be located within 0.85 mm of their geographic positions as referred to the map projection; for maps at publication scales of 1:20,000 or smaller, 0.50 mm. 2. Vertical accuracy: 90 percent of all contours will be accurate within one-half of the basic contour interval. Discrepancies in the accuracy of contours and elevations beyond this tolerance may be decreased by assuming a horizontal displacement within 0.50 mm. Also called MAP ACCURACY STANDARDS.

universal plotting sheet. See under SMALL AREA PLOTTING SHEET.

Universal Polar Stereographic grid. A military grid system based on the polar stereographic map projection, applied to maps of the earth's polar regions north of 84° N and south of 80° S.

Universal Time. Conceptually, time as determined from the apparent diurnal motion of a fictitious mean sun which moves uniformly along the celestial equator at the average rate of the apparent sun. Actually, Universal Time (UT) is related to the rotation of the earth through its definition in terms of sidereal time. Universal Time at any instant is derived from observations of the diurnal motions of the stars. The time scale determined directly from such observations is slightly dependent on the place of observation; this scale is designated UT0. By removing from UT0 the effect of the variation of the observer's meridian due to the observed motion of the geographic pole, the scale UT1 is established. A scale designated UT2 results from applying to UT1 an adopted formula for the seasonal variation in the rate of the earth's rotation. UT1 and UT2 are independent of the location of the observer. UT1 is the same as Greenwich mean time used in navigation. See also TIME SCALE.

Universal Transverse Mercator (UTM) grid. A military grid system based on the transverse Mercator map projection, applied to maps of the earth's surface extending to 84°N and 80°S.

unlighted buoy. A buoy not fitted with a light, whose shape and color are the defining features; may have a sound signal.

unlighted sound buoy. See under SOUND BUOY.

unmanned light. A light which is operated automatically and may be maintained in service automatically for extended periods of time, but with routine visits for maintenance purposes. Also called UNWATCHED LIGHT.

unperturbed orbit. See NORMAL ORBIT.

unsettled, *adj*. Pertaining to fair weather which may at any time become rainy, cloudy, or stormy. See also SETTLED.

unstabilized display. A radarscope display in which the orientation of the relative motion presentation is set to the ship's heading and changes with it.

unstabilized in azimuth. See under STABILIZATION OF RADARSCOPE DISPLAY.

unwatched light. See UNMANNED LIGHT.

upper branch. That half of a meridian or celestial meridian from pole to pole which passes through a place or its zenith.

upper culmination. See UPPER TRANSIT.

upper limb. The upper edge of a celestial body, in contrast with the LOWER LIMB, the lower edge.

upper transit. Transit of the upper branch of the celestial meridian. Transit of the lower branch is called LOWER TRANSIT. Also called SUPERIOR TRANSIT, UPPER CULMINATION.

uprush, *n*. 1. The rush of the water onto the foreshore following the breaking of a wave. 2. See RUN-UP.

upstream, *adj*. & *adv*. Toward the source of a stream. The opposite is DOWNSTREAM.

up-the-scope echo. See CLASSIFICATION OF RADAR ECHOES.

upwelling, *n*. The process by which water rises from a lower to a higher depth, usually as a result of divergence and offshore currents. Upwelling is most prominent where persistent wind blows parallel to a coastline so that the resultant wind-driven current sets away from the coast. Over the open ocean, upwelling occurs whenever the wind circulation is cyclonic, but is appreciable only in areas where that circulation is relatively permanent. It is also observable when the southern trade winds cross the equator.

upwind, *adj.* & *adv.* In the direction from which the wind is blowing. The opposite is DOWNWIND.

U.S. Survey foot. The foot used by the National Ocean Service in which 1 inch is equal to 2.540005 centimeters. The foot equal to 0.3048 meter, exactly, adopted by Australia, Canada, New Zealand, South Africa, the United Kingdom, and the United States in 1959 was not adopted by the National Ocean Service because of the extensive revisions which would be necessary to their charts and measurement records.

UTC, *n.* See under COORDINATED UNIVERSAL TIME.

UT0, *n.* See under UNIVERSAL TIME.

UT1, *n.* See under UNIVERSAL TIME.

UT2, *n.* See under UNIVERSAL. TIME.

V

vacuum, *n.* A space containing no matter.

valley, *n.* On the sea floor, a relatively shallow, wide depression, the bottom of which usually has a continuous gradient. This term is generally not used for features that have canyon-like characteristics for a significant portion of their extent.

valley breeze. A gentle wind blowing up a valley or mountain slope in the absence of cyclonic or anticyclonic winds, caused by the warming of the mountainside and valley floor before the sun. See also KATABATIC WIND, MOUNTAIN BREEZE.

Van Allen Radiation Belts. Popular term for regions of high energy charged particles trapped in the earth's magnetic field. Definition of size and shape of these belts depends on selection of an arbitrary standard of radiation intensity and the predominant particle component. Belts known to exist are: a proton region centered at about 2,000 miles altitude at the geomagnetic equator; an electron region centered at about 12,000 miles altitude at the geomagnetic equator; overlapping electron and proton regions centered at about 20,000 miles altitude at the geomagnetic equator. Trapped radiation regions from artificial sources also exist. These belts were first reported by Dr. James A. Van Allen of Iowa State University.

vane, *n.* 1. A device to sense or indicate the direction from which the wind blows. Also called WEATHER VANE, WIND VANE. See also ANEMOMETER. 2. A sight on an instrument used for observing bearings, as on a pelorus, azimuth circle, etc. That vane nearest the observer's eye is called near vane and that on the opposite side is called far vane. Also called SIGHTING VANE. 3. In current measurements, a device to indicate the direction toward which the current flows.

vanishing tide. In a mixed tide with very large diurnal inequality, the lower high water (or higher low water) frequently becomes indistinct (or vanishes) at time of extreme declinations. During these periods the diurnal tide has such overriding dominance that the semidiurnal tide, although still present, cannot be readily seen on the tide curve.

vapor pressure. 1. The pressure exerted by the vapor of a volatile liquid. Each component of a mixed-gas vapor has its own pressure, called partial pressure.

vardar, *n.* A cold fall wind blowing from the northwest down the Vardar valley in Greece to the Gulf of Salonica. It occurs when atmospheric pressure over eastern Europe is higher than over the Aegean Sea, as is often the case in winter. Also called VARDARAC.

vardarac, *n.* See VARDAR.

variable, *n.* A quantity to which a number of values can be assigned.

variable parameters of satellite orbit. See under FIXED AND VARIABLE PARAMETERS OF SATELLITE ORBIT.

variable range marker. An adjustable range ring on the radar display.

variable star. A star which is not of constant magnitude.

variance, *n.* The square of the standard deviation.

variation, *n.* 1. The angle between the magnetic and geographic meridians at any place, expressed in degrees and minutes east or west to indicate the direction of magnetic north from true north. The angle between magnetic and grid meridians is called GRID MAGNETIC ANGLE, GRID VARIATION, or GRIVATION. Called MAGNETIC VARIATION when a distinction is needed to prevent possible ambiguity. Also called MAGNETIC DECLINATION. 2. Change or difference from a given value.

variation of latitude. A small change in the astronomical latitude of points on the earth due to polar motion.

variation of the poles. See POLAR MOTION.

variometer, *n.* An instrument for comparing magnetic forces, especially of the earth's magnetic field.

vast floe. See under FLOE.

V-band. A radio-frequency band of 46.0 to 56.0 kilomegahertz. See also FREQUENCY, FREQUENCY BAND.

vector, *n.* Any quantity, such as a force, velocity, or acceleration, which has both magnitude and direction, as opposed to a SCALAR which has magnitude only. Such a quantity may be represented geometrically by an arrow of length proportional to its magnitude, pointing in the given direction.

vector, *adj.* A type of computerized display which consists of layers of differentiated data, each with discreet features. Individual data files can be independently manipulated. See RASTER, BIT-MAP.

vector addition. The combining of two or more vectors in such manner as to determine the equivalent single vector. The opposite is RESOLUTION OF VECTORS. Also called COMPOSITION OF VECTORS.

vector diagram. A diagram of more than one vector drawn to the same scale and reference direction and in correct position relative to each other. A vector diagram composed of vectors representing the actual courses and speeds of two craft and the relative motion vector of either one in relation to the other may be called a SPEED TRIANGLE.

vector quantity. A quantity having both magnitude and direction and hence capable of being represented by a vector. A quantity having magnitude only is called a SCALAR.

veer, *v., i.* 1. For the wind to change direction in a clockwise direction in the Northern Hemisphere and a counterclockwise direction in the Southern Hemisphere. Change in the opposite direction is called BACK. 2. Of the wind, to shift aft. The opposite motion is to HAUL forward.

veer, *v., t.* To pay or let out, as to veer anchor chain.

vehicle location monitoring. A service provided to maintain the orderly and safe movement of platforms or vehicles. It encompasses the systematic observation of airspace, surface, or subsurface areas by electronic, visual, and other means to locate, identify, and control the movement of vehicles.

velocity, *n.* A vector quantity equal to speed in a given direction.

velocity meter. See INTEGRATING ACCELEROMETER.

velocity of current. Speed and set of the current.

velocity ratio. The ratio of two speeds, particularly the ratio of the speed of tidal current at a subordinate station to the speed of the corresponding current at the reference station.

Venus, *n.* The planet whose orbit is next nearer the sun than that of the earth.

verglas, *n.* See GLAZE.

vernal, *adj.* Pertaining to spring. The corresponding adjectives for summer, fall, and winter are aestival, autumnal, and hibernal.

vernal equinox. 1. The point of intersection of the ecliptic and the celestial equator, occupied by the sun as it changes from south to north declination, on or about March 21. Also called MARCH EQUINOX, FIRST POINT OF ARIES. 2. That instant the sun reaches the point of zero declination when crossing the celestial equator from south to north.

vernier, *n*. A short, auxiliary scale situated alongside the graduated scale of an instrument, by which fractional parts of the smallest division of the primary scale can be measured with greater accuracy by a factor of ten. If 10 graduations on a vernier equal 9 graduations on the micrometer drum of a sextant, when the zero on the vernier lies one-tenth of a graduation beyond zero on the micrometer drum, the first graduation beyond zero on the vernier coincides with a graduation on the micrometer drum. Likewise, when the zero on the vernier lies five-tenths of a graduation beyond zero on the micrometer drum, the fifth graduation beyond zero on the vernier coincides with a graduation on the micrometer drum.

vernier error. Inaccuracy in the graduations of the scale of a vernier.

vernier sextant. A marine sextant providing a precise reading by means of a vernier used directly with the arc, and having either a clamp screw or an endless tangent screw for controlling the position of the index arm. The micrometer drum on a micrometer drum sextant may include a vernier to enable a more precise reading.

vertex *(pl. vertices), n*. The highest point. See also APEX.

vertical, *adj*. In the direction of gravity, or perpendicular to the plane of the horizon.

vertical, *n*. A vertical line, plane, etc.

vertical axis. The line through the center of gravity of a craft, perpendicular to both the longitudinal and lateral axes, around which it yaws.

vertical beam width. The beam width measured in a vertical plane.

vertical circle. A great circle of the celestial sphere through the zenith and nadir. Vertical circles are perpendicular to the horizon. The prime vertical circle or prime vertical passes through the east and west points of the horizon. The principal vertical circle passes through the north and south points of the horizon and coincides with the celestial meridian.

vertical control datum. See VERTICAL GEODETIC DATUM.

vertical danger angle. The maximum or minimum angle between the top and bottom of an object of known height, as observed from a craft, indicating the limit of safe approach to an offlying danger. See also DANGER ANGLE.

vertical datum. 1. A base elevation used as a reference from which to reckon heights or depths. It is called TIDAL DATUM when defined by a certain phase of the tide. Tidal datums are local datums and should not be extended into areas which have differing topographic features without substantiating measurements. In order that they may be recovered when needed, such datums are referenced to fixed points known as bench marks. See also CHART SOUNDING DATUM. 2. See VERTICAL GEODETIC DATUM.

vertical earth rate. To compensate for the effect of earth rate, the rate at which a gyroscope must be turned about its vertical axis for the spin axis to remain in the meridian. Vertical earth rate is maximum at the poles, zero at the equator and varies as the sine of the latitude. See also EARTH RATE, HORIZONTAL EARTH RATE.

vertical force instrument. See HEELING ADJUSTER.

vertical geodetic datum. A surface derived by geodetic means and taken as a surface of reference from which to reckon geodetic elevations. See also DATUM. Also called VERTICAL DATUM, VERTICAL CONTROL DATUM.

vertical intensity of the earth's magnetic field. The strength of the vertical component of the earth's magnetic field.

vertical lights. Two or more lights disposed vertically, or geometrically to form a triangle, square or other figure. If the individual lights serve different purposes, those of lesser importance are called AUXILIARY LIGHTS.

vertically polarized wave. A plane polarized electromagnetic wave in which the electric field vector is in a vertical plane.

very close pack ice. Pack ice in which the concentration is 9/10 to less than 10/10.

very high frequency. Radio frequency of 30 to 300 megahertz.

very low frequency. Radio frequency below 30 kilohertz.

very open pack ice. Pack ice in which the concentration is 1/10 to 3/10.

very quick flashing light. A navigation light flashing 80-160 flashes per minute. See also CONTINUOUS VERY QUICK LIGHT, GROUP VERY QUICK LIGHT, INTERRUPTED VERY QUICK LIGHT.

very small fracture. See under FRACTURE.

very weathered ridge. A ridge with tops very rounded, the slopes of the sides usually being about 20° to 30°.

vessel, *n*. Any type of craft which can be used for transportation on water.

Vessel Traffic Services. A system of regulations, communications, and monitoring facilities established to provide active position monitoring, collision avoidance services, and navigational advice for vessels in confined and busy waterways. There are two main types of VTS, surveilled and non-surveilled. Surveilled systems consist of one or more land-based radar sites which output their signals to a central location where operators monitor and to a certain extent control traffic flows. Non-surveilled systems consist of one or more calling-in points at which ships are required to report their identity, course, speed, and other data to the monitoring authority.

viaduct, *n*. A type of bridge which carries a roadway or railway across a ravine; distinct from an aquaduct, which carries water over a ravine. See also BRIDGE, definition 2; CAUSEWAY.

vibrating needle. A magnetic needle used in compass adjustment to find the relative intensity of the horizontal components of the earth's magnetic field and the magnetic field at the compass location. Also called HORIZONTAL FORCE INSTRUMENT.

vibration, *n*. 1. Periodic motion of an elastic body or medium in alternately opposite directions from equilibrium; oscillation. 2. The motion of a vibrating body during one complete cycle; two oscillations.

video, *n*. In the operation of a radar set, the demodulated receiver output that is applied to the indicator. Video contains the relevant radar information after removal of the carrier frequency.

violent storm. Wind of force 11 (56 to 63 knots or 64 to 72 miles per hour) on the Beaufort wind scale. See also STORM, definition 1.

virga, *n*. Wisps or streaks of water or ice particles falling out of a cloud but evaporating before reaching the earth's surface as precipitation. Virga is frequently seen trailing from altocumulus and altostratus clouds, but also is discernible below the bases of high-level cumuliform clouds from which precipitation is falling into a dry subcloud layer. It typically exhibits a hooked form in which the streaks descend nearly vertically just under the precipitation source but appear to be almost horizontal at their lower extremities. Such curvature of virga can be produced simply by effects of strong vertical windshear, but ordinarily it results from the fact that droplet or crystal evaporation decreases the particle terminal fall velocity near the ends of the streaks. Also called FALL STREAKS, PRECIPITATION TRAILS.

virtual image. An image that cannot be shown on a surface but is visible, as in a mirror.

virtual meridian. The meridian in which the spin axis of a gyrocompass will settle as a result of speed-course-latitude error.

visibility, *n*. A measure of the ability of an observer to see objects at a distance through the atmosphere. A measure of this property is expressed in units of distance. This term should not be confused with VISUAL RANGE. See also METEOROLOGICAL VISIBILITY.

visible horizon. The line where earth and sky appear to meet, and the projection of this line upon the celestial sphere. If there were no terrestrial refraction, VISIBLE and GEOMETRICAL HORIZONS would coincide. Also called APPARENT HORIZON.

visual aid to navigation. An aid to navigation which transmits information through its visible characteristics. It may be lighted or unlighted.

visual bearing. A bearing obtained by visual observation.

visual range. The maximum distance at which a given object can be seen, limited by the atmospheric transmission. The distance is such that the contrast of the object with its background is reduced by the atmosphere to the contrast threshold value for the observer. This term should not be confused with VISIBILITY. See also CONTRAST THRESHOLD, VISUAL RANGE OF A LIGHT.

visual range of light. The predicted range at which a light can be observed. The predicted range is the lesser of either the luminous range or the geographic range. If the luminous range is less than the geographic range, the luminous range must be taken as the limiting range. The luminous range is the maximum distance at which a light can be seen under existing visibility conditions. This luminous range takes no account of the elevation of the light, the observer's height of eye, the curvature of the earth, or interference from background lighting. The luminous range is determined from the nominal range and the existing visibility conditions, using the Luminous Range Diagram. The nominal range is the maximum distance at which a light can be seen in clear weather as defined by the International Visibility Code (meteorological visibility of 10 nautical miles). The geographic range is the maximum distance at which the curvature of the earth and terrestrial refraction permit a light to be seen from a particular height of eye without regard to the luminous intensity of the light. The geographic range sometimes printed on charts or tabulated in light lists is the maximum distance at which the curvature of the earth and refraction permit a light to be seen from a height of eye of 15 feet above the water when the elevation of the light is taken above the height datum of the largest scale chart of the locality.) See also VISUAL RANGE, CONTRAST THRESHOLD.

volcano, *n*. An opening in the earth from which hot gases, smoke, and molten material issue, or a hill or mountain composed of volcanic material. A volcano is characteristically conical in shape with a crater in the top.

volt, *n*. A derived unit of electric potential in the International System of Units, it is the difference of electric potential between two points of a conducting wire carrying a constant current of 1 ampere, when the power dissipated between these points is equal to 1 watt.

volt per meter. The derived unit of electric field strength in the International System of Units.

volume, *n*. 1. A measure of the amount of space contained within a solid. 2. Loudness of a sound, usually measured in decibels.

voyage, *n*. 1. A trip by sea.

vulgar establishment. See under ESTABLISHMENT OF THE PORT.

W

wandering of the poles. See EULERIAN MOTION.

waning moon. The moon between full and new when its visible part is decreasing. See also PHASES OF THE MOON.

warble tone. A tone whose frequency varies periodically about a mean value.

warm air mass. An air mass that is warmer than surrounding air. The expression implies that the air mass is warmer than the surface over which it is moving.

warm braw. A foehn in the Schouten Islands north of New Guinea.

warm front. Any non-occluded front, or portion thereof, which moves in such a way that warmer air replaces colder air. While some occluded fronts exhibit this characteristic, they are more properly called WARM OCCLUSIONS.

warm occlusion. See under OCCLUDED FRONT.

warm sector. An area at the earth's surface bounded by the warm and cold fronts of a cyclone.

warning beacon. See WARNING RADIOBEACON.

warning radiobeacon. An auxiliary radiobeacon located at a lightship to warn vessels of their proximity to the lightship. It is of short range and sounds a warbling note for 1 minute immediately following the main radiobeacon on the same frequency. Also called WARNING BEACON.

warp, *v., t*. To move, as a vessel, from one place to another by means of lines fastened to an object, such as a buoy, wharf, etc., secured to the ground. See also KEDGE.

warp, *n*. A heavy line used in warping or mooring.

warping buoy. A buoy located so that lines to it can be used for the movement of ships.

wash, *n*. The dry channel of an intermittent stream.

watch, *n*. A small timepiece of a size convenient to be carried on the person. A hack or comparing watch is used for timing observations of celestial bodies. A stop watch can be started, stopped, and reset at will, to indicate elapsed time. A chronometer watch is a small chronometer, especially one with an enlarged watch-type movement.

watch buoy. See STATION BUOY.

watch error. The amount by which watch time differs from the correct time. It is usually expressed to an accuracy of 1 second and labeled fast (F) or slow (S) as the watch time is later or earlier, respectively, than the correct time. See also CHRONOMETER ERROR.

watching properly. The state of an aid to navigation on charted position and exhibiting its proper characteristics.

watch rate. The amount gained or lost by a watch or clock in a unit of time. It is usually expressed in seconds per 24 hours, to an accuracy of 0.1^s, and labeled gaining or losing, as appropriate, when it is sometimes called DAILY RATE.

watch time. The hour of the day as indicated by a watch or clock. Watches and clocks are generally set approximately to zone time. Unless a watch or clock has a 24-hour dial, watch time is usually expressed on a 12-hour cycle and labeled AM or PM.

watch tower. See LOOKOUT STATION.

water-borne, *adj*. Floating on water; afloat. See also SEA-BORNE.

watercourse, *n*. 1. A stream of water. 2. A natural channel through which water runs. See also GULLY, WASH.

waterfall, *n*. A perpendicular or nearly perpendicular descent of river or stream water.

waterline, *n*. The line marking the junction of water and land. See also HIGH WATER LINE, LOW WATER LINE, SHORELINE.

water sky. Dark streaks on the underside of low clouds, indicating the presence of water features in the vicinity of sea ice.

water smoke. See STEAM FOG.

waterspout, *n*. 1. A tornado occurring over water; most common over tropical and subtropical waters. 2. A whirlwind over water comparable in intensity to a dust devil over land.

water tower. A structure erected to store water at an elevation above the surrounding terrain; often charted with a position circle and label.

water track. 1. See under TRACK, definition 2. 2. See under TRUE TRACK OF TARGET.

waterway, *n*. A water area providing a means of transportation from one place to another, principally one providing a regular route for water traffic, such as a bay, channel, passage, or the regularly traveled parts of the open sea. The terms WATERWAY, FAIRWAY, and THOROUGHFARE have nearly the same meanings. WATERWAY refers particularly to the navigable part of a water area. FAIRWAY refers to the main traveled part of a waterway. A THOROUGHFARE is a public waterway. See also CANAL.

watt, *n*. A derived unit of power in the International System of Units; it is that power which in 1 second gives rise to energy of 1 joule.

wave, *n*. 1. An undulation or ridge on the surface of a fluid. See also STORM SURGE, TIDAL WAVE, TSUNAMI. 2. A disturbance propagated in such a manner that it may progress from point to point. See also ELECTROMAGNETIC WAVES, RADIO WAVES, SKYWAVE, GROUNDWAVE, DIRECT WAVE, INDIRECT WAVE, MODULATED WAVE, MICROWAVE, SPHERICAL WAVE, TRANSVERSE WAVE, LONGITUDINAL WAVE.

wave basin. A basin close to the inner entrance of a harbor in which the waves from the outer entrance are absorbed, thus reducing the size of the waves entering the inner harbor. See also WAVE TRAP.

wave crest. The highest part of a wave.

wave cyclone. A cyclone which forms and moves along a front. The circulation about the cyclone center tends to produce a wavelike deformation of the front. The wave cyclone is the most frequent form of extratropical cyclone (or low). Also called WAVE DEPRESSION. See also FRONTAL CYCLONE.

wave depression. See WAVE CYCLONE.

wave direction. The direction from which waves are coming.

waveguide, *n*. A transmission line for electromagnetic waves consisting of a hollow conducting tube within which electromagnetic waves may be propagated; or a solid dielectric or dielectric-filled conductor designed for the same purpose.

wave height. The distance from the trough to the crest of a wave, equal to double the amplitude, and measured perpendicular to the direction of advance.

wave height correction. A correction due to the elevation of parts of the sea surface by wave action, particularly such a correction to a sextant altitude because of altered dip.

wave interference. See INTERFERENCE, definition 2.

wavelength, *n*. The distance between corresponding points in consecutive cycles in a wave train, measured in the direction of propagation at any instant.

wave of translation. A wave in which the individual particles of the medium are shifted in the direction of wave travel, as ocean waves in shoal waters; in contrast with an OSCILLATORY WAVE, in which only the form advances, the individual particles moving in closed orbits, as ocean waves in deep water.

wave period. The time interval between passage of successive wave crests at a fixed point.

wave train. A series of waves moving in the same direction. See also SOLITARY WAVE.

wave trap. Breakwaters situated close within the entrance used to reduce the size of waves from sea or swell which enter a harbor before they penetrate into the harbor. See also WAVE BASIN.

wave trough. The lowest part of a wave form between successive wave crests.

waxing moon. The moon between new and full when its visible part is increasing. See also PHASES OF THE MOON.

waypoint, *n*. A reference point on the track.

weak fix. A fix determined from horizontal sextant angles between objects poorly located.

weather, *adj*. Pertaining to the windward side, or the side in the direction from which the wind is blowing. LEE pertains to the leeward or sheltered side.

weather, *n*. 1. The state of the atmosphere as defined by various meteorological elements, such as temperature, pressure, wind speed and direction, humidity, cloudiness, precipitation, etc. This is in contrast with CLIMATE, the prevalent or characteristic meteorological conditions of a place or region. 2. Bad weather. See also THICK WEATHER.

weathered, *adj*. Eroded by action of the weather.

weathered berg. An irregularly shaped iceberg. Also called GLACIER BERG.

weathered ridge. An ice ridge with peaks slightly rounded, the slopes of the sides usually being about 30° to 40°. Individual fragments are not discernible.

weathering, *n*. Processes of ablation and accumulation which gradually eliminate irregularities in an ice surface.

weather map. See under SYNOPTIC CHART.

weather shore. As observed from a vessel, the shore lying in the direction from which the wind is blowing. See also LEE SHORE.

weather side. The side of a ship exposed to the wind or weather.

weather vane. A device to indicate the direction from which the wind blows. Also called WIND DIRECTION INDICATOR, WIND VANE. See also ANEMOMETER.

weber, *n*. A derived unit of magnetic flux in the International System of Units; it is that magnetic flux which, linking a circuit of one turn, would produce in it an electromotive force of 1 volt if it were reduced to zero at a uniform rate in 1 second.

wedge. See RIDGE, definition 3.

weight, *n*. A quantity of the same nature as a force; the weight of a body is the product of its mass and the acceleration due to gravity; in particular, the standard weight of a body is the product of its mass and the standard acceleration due to gravity. The value adopted in the International Service of Weights and Measures for the standard acceleration due to gravity is 980.665 centimeters per second, per second.

weighted mean. A value obtained by multiplying each of a series of values by its assigned weight and dividing the sum of those products by the sum of the weights. See also WEIGHT OF OBSERVATION.

weight of observation. The relative value of an observation, source, or quantity when compared with other observations, sources, or quantities of the same or related quantities. The value determined by the most reliable method is assigned the greatest weight. See also WEIGHTED MEAN.

wellhead, *n*. A submarine structure projecting some distance above the seabed and capping a temporarily abandoned or suspended oil or gas well. See also SUBMERGED PRODUCTION WELL.

west, *n*. The direction 90° to the left or 270° to the right of north. See also CARDINAL POINT.

West Australia Current. An Indian Ocean current which generally first flows northward and then northwestward off the west coast of Australia. This current varies seasonally with the strength of the wind and is most stable during November, December, and January, and least stable during May, June, and July, when it may set in any direction. North of 20°S the main part of this current flows northwestward into the Indian South Equatorial Current.

westerlies, *n., pl*. Winds blowing from the west on the poleward sides of the subtropical high-pressure belts.

West Greenland Current. The ocean current flowing northward along the west coast of Greenland into Davis Strait. It is a continuation of the East Greenland Current. Part of the West Greenland Current turns around when approaching the Davis Strait and joins the Labrador Current; the rest rapidly loses its character as a warm current as it continues into Baffin Bay.

westing, *n*. The distance a craft makes good to the west. The opposite is EASTING.

westward motion. The motion in a westerly direction of the subtrack of a satellite, including the motion due to the earth's rotation and the nodical precession of the orbital plane.

West Wind Drift. An ocean current that flows eastward through all the oceans around the Antarctic Continent, under the influence of the prevailing west winds. On its northern edge it is continuous with the South Atlantic Current, the South Pacific Current, and the South Indian Current. Also called ANTARCTIC CIRCUMPOLAR CURRENT.

wet-bulb temperature. The lowest temperature to which air can be cooled at any given time by evaporating water into it at constant pressure, when the heat required for evaporation is supplied by the cooling of the air. This temperature is indicated by a well-ventilated wet-bulb thermometer. See also FREE-AIR TEMPERATURE.

wet-bulb thermometer. A thermometer having the bulb covered with a cloth, usually muslin or cambric, saturated with water. See also PSYCHROMETER.

wet compass. See LIQUID COMPASS.

wet dock. See NON-TIDAL BASIN.

wharf, *n*. A structure of open pilings covered with a deck along a shore or a bank which provides berthing for ships and which generally provides cargo-handling facilities. A similar facility of solid construction is called QUAY. See also PIER, definition 1; DOCK; LANDING; MOLE, definition 1.

whirlpool, *n*. Water in rapid rotary motion. See also EDDY.

whirlwind, *n*. A general term for a small-scale, rotating column of air. More specific terms include DUST WHIRL, DUST DEVIL, WATERSPOUT, and TORNADO.

whirly, *n*. A small violent storm, a few yards to 100 yards or more in diameter, frequent in Antarctica near the time of the equinoxes.

whistle, *n*. A sound signal emitter comprising a resonator having an orifice of suitable shape such that when a jet of air is passed through the orifice the turbulence produces a sound.

whistle buoy. A sound buoy equipped with a whistle operated by wave action. The whistle makes a loud moaning sound as the buoy rises and falls in the sea.

whitecap, *n*. A crest of a wave which becomes unstable in deep water, toppling over or "breaking." The instability is caused by the too rapid addition of energy from a strong wind. A wave which becomes unstable due shallow water is called a BREAKER.

white ice. See THIN FIRST-YEAR ICE.

white squall. A sudden, strong gust of wind coming up without warning, noted by whitecaps or white, broken water; usually seen in whirlwind form in clear weather in the tropics.

white water. 1. Frothy water as in whitecaps or breakers. 2. Light-colored water over a shoal.

whole gale. A term once used by seamen for what is now called STORM on the Beaufort wind scale.

wide berth. A generous amount of room given to a navigational danger.

williwaw, *n*. A sudden blast of wind descending from a mountainous coast to the sea, especially in the vicinity of either the Strait of Magellan or the Aleutian Islands.

willy-willy, *n*. See under TROPICAL CYCLONE.

wind. Air in horizontal motion over the earth.

wind cone. See WIND SOCK.

wind direction. The direction from which wind blows.

wind direction indicator. See WEATHER VANE.

wind drift current. See DRIFT CURRENT.

wind driven current. A current created by the action of the wind.

wind indicator. A device to indicate the direction or speed of the wind. See also ANEMOMETER.

wind rode. A ship riding at anchor is said to be wind rode when it is heading into the wind. See also TIDE RODE.

wind rose. A diagram showing the relative frequency and sometimes the average speed of the winds blowing from different directions in a specified region.

winds aloft. Wind speeds and directions at various levels beyond the domain of surface weather observations.

wind shear. A change in wind direction or speed in a short distance, resulting in a shearing effect. It can act in a horizontal or vertical direction and, occasionally, in both. The degree of turbulence increases as the amount of wind shear increases.

wind-shift line. In meteorology, a line or narrow zone along which there is an abrupt change of wind direction.

wind sock. A tapered fabric sleeve mounted so as to catch and swing with the wind, thus indicating the wind direction. Also called WIND CONE.

wind speed. The rate of motion of air. See also ANEMOMETER.

wind storm. See under STORM, definition 2.

wind vane. See WEATHER VANE.

wind velocity. The speed and direction of wind.

windward, *adj.* & *adv.* In the general direction from which the wind blows; in the wind; on the weather side. The opposite is LEEWARD.

windward, *n.* The weather side. The opposite is LEEWARD.

windward tide. A tidal current setting to windward. One setting in the opposite direction is called a LEEWARD TIDE or LEE TIDE.

wind wave. A wave generated by friction between wind and a fluid surface. Ocean waves are produced principally in this way.

winged headland. A seacliff with two bays or spits, one on either side.

winter, *n.* The coldest season of the year. In the Northern Hemisphere, winter begins astronomically at the winter solstice and ends at the vernal equinox. In the Southern Hemisphere the limits are the summer solstice and the autumnal equinox. The meteorological limits vary with the locality and the year.

winter buoy. An unlighted buoy which is maintained in certain areas during winter months when other aids to navigation are temporarily removed or extinguished.

Winter Coastal Countercurrent. See DAVIDSON CURRENT.

winter light. A light which is in service during the winter months when the regular light is out of service. It has lower intensity than the regular light but usually has the same characteristic.

winter marker. An unlighted buoy or small lighted buoy which is established as a replacement during the winter months when other aids are out of service or withdrawn.

winter solstice. The point on the ecliptic occupied by the sun at maximum southerly declination. Sometimes called DECEMBER SOLSTICE, FIRST POINT OF CAPRICORNUS.

wiping, *n.* The process of reducing the amount of permanent magnetism in a vessel by placing a single coil horizontally around the vessel and moving it, while energized, up and down along the sides of the vessel. If the coil remains stationary, the process is called FLASHING. See also DEPERMING.

wire drag. An apparatus for surveying rock areas where the normal sounding methods are insufficient to insure the discovery of all existing obstructions above a given depth, or for determining the least depth of an area. It consists of a buoyed wire towed at the desired depth by two vessels. Often shortened to DRAG. See also DRAG, *v., t.*

withdrawn, *adj.* Removed from service during severe ice conditions or for the winter season. Compare with the term disestablished, which means permanently removed. See also CLOSED, COMMISSIONED.

WMO Sea-Ice Nomenclature (WMO/OMM/BMO No. 259. TP. 145). A publication of the World Meteorological Organization which is comprised of sea-ice terminology, ice reporting codes, and an illustrated glossary. This publication results from international cooperation in the standardization of ice terminology.

working, *n.* In sea ice navigation, making headway through an ice pack by boring, breaking, and slewing.

World Geographic Reference System. A worldwide position reference system that may be applied to any map or chart graduated in latitude and longitude (with Greenwich as prime meridian) regardless of projection. It is a method of expressing latitude and longitude in a form suitable for rapid reporting and plotting. Commonly referred to by use of the acronym GEOREF.

World Geodetic System. A consistent set of parameters describing the size and shape of the earth, the positions of a network of points with respect to the center of mass of the earth, transformations from major geodetic datums, and the potential of the earth (usually in terms of harmonic coefficients). It forms the common geodetic reference system for modern charts on which positions from electronic navigation systems can be plotted directly without correction.

Worldwide Marine Weather Broadcasts. A joint publication of the National Weather Service and the Naval Weather Service Command providing information on marine weather broadcasts in all areas of the world. In general, English language broadcasts (or foreign language broadcasts repeated in English) are included in the publication. For areas where English language broadcasts are not available foreign language transmissions are also included.

World Meteorological Organization. A specialized agency of the United Nations which seeks to facilitate world-wide cooperation in the establishment of stations for meteorological and related geophysical observations of centers providing meteorological services, of systems of rapid exchange of weather information; and to promote the standardization and publication of meteorological and hydrometeorological observations and statistics; to further the application of meteorology to aviation, shipping, agriculture, and other related activities; to encourage research and training in meteorology and their international coordination.

World Port Index. See *PUB. 150.*

World Wide Navigational Warning Service. Established through the joint efforts of the International Hydrographic Organization (IHO) and the Intergovernmental Maritime Consultative Organization (IMCO) now called the International Maritime Organization (IMO), the World Wide Navigational Warning Service (WWNWS) is a coordinated global service for the promulgation by radio of information on hazards to navigation which might endanger international shipping. The basic objective of the WWNWS is the timely promulgation by radio of information of concern to the ocean-going navigator. Such information includes failure and or changes to major navigational aids, newly discovered wrecks or natural hazards in or near main shipping lanes; areas where search and rescue, antipollution operations, cable-laying or other underway activities are taking place. For WWNWS purposes, the world is divided into 16 NAVAREAS. Within each NAVAREA one national authority, designated the Area Coordinator, has assumed responsibility for the coordination and promulgation of warnings. Designated "National Coordinators" of other coastal states in a NAVAREA are responsible for collecting and forwarding information to the Area Coordinator. In the Baltic, a Sub-Area Coordinator has been established to filter information prior to passing to the Area Coordinator. Coordinators are responsible for the exchange of information as appropriate with other coordinators, including that which should be further promulgated by charting authorities in *Notice to Mariners*. The language used is English, although warnings may also be transmitted in one or more of the official languages of the United Nations. Broadcast schedules appear in an Annex to the International Telecommunication Union *List of Radiodetermination and Special Service Stations Volume II*, and in the lists of radio signals published by various hydrographic authorities (for the U.S., *Pub 117, Radio Navigational Aids*.) Transmissions usually occur frequently enough during day to fall within at least one normal radio watch period, and the information is repeated with varying frequency as time passes until either the danger has passed or the information on it has appeared as a notice to mariners.

worldwide system. A term used to describe a navigation system providing positioning capability wherever the observer may be located. Also

wreck, *n*. The ruined remains of a vessel which has been rendered useless, usually by violent action by the sea and weather, on a stranded or sunken vessel. In hydrography the term is limited to a wrecked vessel, either submerged or visible, which is attached to or foul of the bottom or cast up on the shore. In nautical cartography wrecks are designated visible, dangerous, or non-dangerous according to whether they are above tidal datum, less than, or more than 20 meters (66 feet; 11 fathoms) below tidal datum, respectively.

wreck buoy. A buoy marking the position of a wreck. It is usually placed on the seaward or channel side of the wreck and as near to the wreck as conditions will permit. To avoid confusion in some situations, two buoys may be used to mark the wreck.

wreck mark. A navigation mark which marks the position of a wreck.

X-Y-Z

X-band. A radio-frequency band of 5,200 to 10,900 megahertz. See also FREQUENCY, FREQUENCY BAND.

yard, *n*. A unit of length equal to 3 feet, 36 inches, or 0.9144 meter.

yaw, *n*. The oscillation of a vessel in a seaway about a vertical axis approximately through the center of gravity.

Y-code, *n*. The encrypted version of the P-code.

yawing, *n*. See YAW.

year, *n*. A period of one revolution of a planet around the sun. The period of one revolution of the earth with respect to the vernal equinox, averaging 365 days, 5 hours, 48 minutes, 46 seconds in 1900, is called a tropical, astronomical, equinoctial, or solar year. The period with respect to the stars, averaging 365 days, 6 hours, 9 minutes, 9.5 seconds in 1900, is called a sidereal year. The period of revolution from perihelion to perihelion, averaging 365 days, 6 hours, 13 minutes, 53.0 seconds in 1900, is an anomalistic year. The period between successive returns of the sun to a sidereal hour angle of 80° is called a fictitious or Besselian year. A civil year is the calendar year of 365 days in common years, or 366 days in leap years. A light-year is a unit of length equal to the distance light travels in 1 year, about 5.88×10^{12} miles. The term year is occasionally applied to other intervals such as an eclipse year, the interval between two successive conjunctions of the sun with the same node of the moon's orbit, a period averaging 346 days, 14 hours, 52 minutes, 50.7 seconds in 1900, or a great or Platonic year, the period of one complete cycle of the equinoxes around the ecliptic, about 25,800 years.

young coastal ice. The initial stage of fast ice formation consisting of nilas or young ice, its width varying from a few meters up to 100 to 200 meters from the shoreline.

young ice. Ice in the transition stage between nilas and first-year ice, 10 to 30 centimeters in thickness. Young ice may be subdivided into GRAY ICE and GRAY-WHITE ICE.

zenith, *n*. The point on the celestial sphere vertically overhead. The point 180° from the zenith is called the NADIR.

zenithal, *adj*. Of or pertaining to the zenith.

zenithal chart. See AZIMUTHAL CHART.

zenithal map projection. See AZIMUTHAL MAP PROJECTION.

zenith distance. Angular distance from the zenith; the arc of a vertical circle between the zenith and a point on the celestial sphere, measured from the zenith through 90°, for bodies above the horizon. This is the same as COALTITUDE with reference to the celestial horizon.

zephyr, *n*. A warm, gentle breeze, especially one from the west.

zodiac, *n*. The band of the sky extending 9° either side of the ecliptic. The sun, moon, and navigational planets are always within this band, with the occasional exception of Venus. The zodiac is divided into 12 equal parts, called signs, each part being named for the principal constellation originally within it.

zodiacal light. A faint cone of light which extends upward from the horizon along the ecliptic after sunset or before sunrise, seen best in the tropics and believed to be the reflection of sunlight by extraterrestrial particles in the zodiac.

zone, *n*. 1. A defined area or region. The surface of the earth is divided into climatic zones by the polar circles and the tropics; the parts between the poles and polar circles are called the north and south frigid zones; the parts between the polar circles and the tropics are the north and south temperate zones; the part between the two tropics is the torrid zone. 2. A time zone, within which the same time is kept.

zone description. The number, with its sign, that must be added to or subtracted from the zone time to obtain the Greenwich mean time. The zone description is usually a whole number of hours.

zone meridian. The meridian used for reckoning zone time. This is generally the nearest meridian whose longitude is exactly divisible by 15°. The DAYLIGHT SAVING MERIDIAN is usually 15° east of the zone meridian.

zone noon. Twelve o'clock zone time, or the instant the mean sun is over the upper branch of the zone meridian. Standard noon is 12 o'clock standard time.

zone time. The local mean time of a reference or zone meridian whose time is kept throughout a designated zone. The zone meridian is usually the nearest meridian whose longitude is exactly divisible by 15°. Standard time is a variation of zone time with irregular but well-defined zone limits. Daylight saving or summer time is usually 1 hour later than zone or standard time. See ZONE DESCRIPTION.

zulu. See GREENWICH MEAN TIME.

GLOSSARY OF ABBREVIATIONS AND ACRONYMS

A

A	amplitude; augmentation; away (altitude intercept); Arctic/Antarctic (air mass).
a	semimajor axis.
a	altitude intercept (Ho~Hc); altitude factor (change of altitude in 1 minute of time from meridian transit); assumed.
ABAND	abandoned.
AC	alternating current; altocumulus.
ACC	Antarctic Circumpolar Current.
add'l	additional.
ADF	automatic direction finder.
ADIZ	air defense identification zone.
AEB	acquisition exclusion boundary.
AERO	aeronautical.
AF	audio frequency.
AFC	automatic frequency control.
AGC	automatic gain control.
AISM	Association Internationale de Signalisation Maritime (International Association of Lighthouse Authorities).
aL	assumed latitude.
Al., Alt,	alternating (light).
A.L.R.S.	*Admiralty List of Radio Signals.*
am	amber.
AM	amplitude modulation.
AM	ante meridian (before noon).
Anch	anchorage.
antilog	antilogarithm.
AP	assumed position.
approx.	approximate, approximately.
ARPA	automatic radar plotting aid.
ASAM	anti-ship action message.
ASF	Additional Secondary Phase Factor.
AT	atomic time.
AU	astronomical unit.
AUSREP	Australian Ships Reporting System.
al	assumed longitude.

B

B	atmospheric pressure correction (altitude); bearing, bearing angle.
Bdy Mon	boundary monument.
BFO	beat frequency oscillator.
BIH	Bureau Internationale de l'Heure.
BIPM	International Bureau of Weights and Measures.
bk	broken.
bkw	breakwater.
bl	blue.
BM	bench mark.
Bn	beacon.
Bpgc	bearing per gyrocompass.
br	breakers.
Brg.	bearing (as distinguished from bearing angle).
bu	blue.

C

C	Celsius (centigrade); chronometer time; compass (direction); correction; course, course angle; can; cylindrical; cove.
CALM	catenary anchor leg mooring.
CB	compass bearing.
CBDR	constant bearing, decreasing range.
CC	compass course; chronometer correction.
CCIR	International Radio Consultative Committee.
CCU	Consultative Committee for Units of the International Committee of Weights and Measures (CIPM).
CCZ	Coastal Confluence Zone.
cd	candela, candelas.
CD	chart datum.
CD-ROM	compact disk-read only memory.
CG	Coast Guard.
CE	chronometer error- compass error.
CFR	Code of Federal Regulations.
cec	centicycle.
cel	centilane.
CEP	circular probable error.
CES	coast earth station.
CFR	Code of Federal Regulations.
CGPM	General Conference of Weights and Measures.
CH	compass heading.
CIPM	International Committee of Weights and Measures.
Cl	clearance.

cm	centimeter(s).
CMG	course made good.
Cn	course (as distinguished from course angle).
co	coral.
co-	the complement of (90° minus).
COA	course of advance.
COE	Committee on ECDIS (IHO).
COG	course over ground.
coL	colatitude.
colog	cologarithm.
corr.	correction.
cos	cosine.
cot	cotangent.
cov	coversine.
CPA	closest point of approach.
CPE	circular probable error.
Cpgc	course per gyrocompass.
cps	cycles per second.
Cpsc	course per standard compass.
Cp stg c	course per steering compass.
CPU	central processing unit.
crs	course.
CRT	cathode-ray tube.
csc	cosecant.
cup	cupola.
Cus Ho	customs house.
CW	continuous wave.
CZn	compass azimuth.

D

D	deviation; dip (of horizon); distance; destroyed.
d	declination (astronomical); altitude difference.
d	declination change in 1 hour.
dA	difference of longitude (time units).
DC	direct current.
deg.	degree(s).
Dec.	declination.
Dec. Inc.	declination increment.
Dep.	departure.
destr	destroyed.
Dev.	deviation.
DG	degaussing.
DGPS	differential global positioning system.
DHQ	mean diurnal high water inequality.
Dia	diaphone.
diff.	difference.
Dist.	distance.
D. Lat.	difference of latitude.
DLo	difference of longitude (arc units).
DLQ	mean diurnal low water inequality.
dm	decimeters.
DMAHTC	Defense Mapping Agency Hydrographic/Topographic Center.
DNC	digital navigation chart.
dol	dolphin.
DR	dead reckoning; dead reckoning position.
DRE	dead reckoning equipment.
DRM	direction of relative movement.
DRT	dead reckoning tracer.
Ds	dip short of horizon.
DSC	digital selective calling.
DSD	double second difference.
DSVL	doppler sonar velocity log.
dur.	duration.
DW	Deep Water Route.
DZ	danger zone.

E

E	east.
e	base of Naperian logarithms; origin of own ship's true vector.
e	eccentricity.
EBL	electronic bearing line.
ECD	envelope to cycle difference; envelope to cycle discrepancy.
EC	electronic chart.
ECDB	electronic chart data base.
ECDIS	electronic chart display and information system.
ED	existence doubtful.
EDD	estimated date of departure.
EEZ	exclusive economic zone.
EGC	enhanced group calling.
EHF	extremely high frequency.
E. Int.	equal interval; isophase.
EM	electromagnetic (underwater log).
em	other ship's true vector.
ENC	electronic navigation chart.
ENCDB	electronic navigation chart data base.
EP	estimated position.
EPIRB	Emergency Position Indicating Radiobeacon.
EPROM	erasable programmable read only memory.
Eq.T	equation of time.
er	own ship's true vector.
ET	Ephemeris Time.
ETA	estimated time of arrival.
ETD	estimated time of departure.
Exting	extinguished.

F

F	Fahrenheit; fast; longitude factor; phase correction (altitude); fixed (light)
f	latitude factor.

f	flattening or ellipticity.
F.Fl.	fixed and flashing.
Fl.	flashing (light).
Fl. (2)	group flashing (light).
Fl. (2+1)	composite group flashing (light).
fm(s)	fathom(s).
FM	frequency modulation.
Fog Det.	fog detector.
Fog Sig.	fog signal.
ft.	foot, feet.
FTC	fast time constant.

G

G	Greenwich; Greenwich meridian (upper branch); grid (direction); gravel; green.
g	acceleration due to gravity; Greenwich meridian (lower branch).
GAT	Greenwich apparent time.
GB	grid bearing.
GC	grid course.
GCLWD	Gulf Coast Low Water Datum.
GDOP	geometric dilution of precision.
GE	gyro error.
GH	grid heading.
GHA	Greenwich hour angle.
GMDSS	Global Maritime Distress and Safety System.
GMT	Greenwich mean time.
Gp. Fl.	group flashing.
GP	geographical position.
GPS	Global Positioning System.
Gr.	Greenwich.
GRI	group repetition interval.
GST	Greenwich sidereal time.
GV	grid variation.
GZn	grid azimuth.

H

h	altitude (astronomical); height above sea level; hours.
ha	apparent altitude.
Hc	computed altitude.
Hdg.	heading.
HE	heeling error; height of eye.
HF	high frequency.
hf	height above sea level in feet.
HHW	higher high water.
HHWI	higher high water interval.
Hk	hulk.
HLW	higher low water.
HLWI	higher low water interval.
hm	height above sea level in meters.
Ho	observed altitude.
Hor	horizontal.
Hor Cl	horizontal clearance.
HP	horizontal parallax.
Hp	precomputed altitude.
Hpgc	heading per gyrocompass.
Hpsc	heading per standard compass.
Hp stg c	heading per steering compass.
hr	rectified (apparent) altitude.
hr.	hour, hrs., hours.
hs	sextant altitude.
HSD	high speed data.
ht	tabulated altitude.
HW	high water.
H.W.F.&C.	high water full and change.
HWI	high water interval, mean high water lunitidal interval.
HWQ	tropic high water inequality.
Hz	Hertz.

I

I	instrument correction.
i	inclination (of satellite orbit).
IALA	International Association of Lighthouse Authorities.
IAU	International Astronomical Union.
IC	index correction.
ICW	Intracoastal Waterway.
IGLD	International Great Lakes Datum.
IHB	International Hydrographic Bureau.
IHO	International Hydrographic Organization.
IMO	International Maritime Organization.
in.	inch, inches.
INM	International Nautical Mile.
INMARSAT	International Maritime Satellite Organization.
INS	inertial navigation system.
int.	interval.
Int. Qk.	Interrupted quick flashing.
ION	Institute of Navigation.
I.Q.	interrupted quick flashing.
IR	interference rejection.
IRP	image-retaining panel.
ISLW	Indian spring low water.
ISO	International Order of Standardization; isophase (light).
ITU	International Telecommunications Union.
IUGG	International Union of Geodesy and Geophysics.
I.U.Q.	interrupted ultra quick flashing.
I.V.Q.	interrupted very quick flashing.
IWW	Intracoastal Waterway.

J – K – L

J	irradiation correction (altitude).
K	Kelvin (temperature).
kHz	kilohertz.
km	kilometer, kilometers.
kn	knot, knots.
L	latitude; lower limb correction for moon.
l	difference of latitude; logarithm, logarithmic.
LAN	local apparent noon.
LANBY	large automatic navigational buoy.
LASH	lighter aboard ship.
LAT	local apparent time.
lat.	latitude.
LF	low frequency.
L.Fl.	long flashing.
LHA	local hour angle.
LHW	lower high water.
LHWI	lower high water interval.
LL	lower limb.
LLW	lower low water.
LLWD	lower low water datum.
LLWI	lower low water interval.
Lm	middle latitude; mean latitude.
LMT	local mean time.
LNB	large navigational buoy.
LNG	liquified natural gas.
LPG	liquified petroleum gas.
Log	logarithm, logarithmic.
Loge	natural logarithm (to the base e).
Log10	common logarithm (to the base 10).
Long.	longitude.
LOP	line of position.
LST	local sidereal time.
Lt.	light.
Lt Ho	light house.
Lt V	light vessel.
LW	low water.
LWD	low water datum.
LWI	low water interval; mean low water lunitidal interval.
LWQ	tropic low water inequality.

M

M	celestial body; meridian (upper branch); magnetic (direction); meridional parts; nautical mile, miles; other ship.
m	meridian (lower branch); meridional difference; meter,(s); U.S. survey mile, miles; end of other ship's true vector; minutes.
mag.	magnetic; magnitude.
MARAD	United States Maritime Administration.
MB	magnetic bearing.
mb	millibar(s).
MC	magnetic course.
mc	megacycle, megacycles; megacycles per second.
MC&G	mapping, charting and geodesy.
MCPA	minutes to closest point of approach.
Mer. Pass.	meridian passage.
MF	medium frequency.
MGRS	military grid reference system.
MH	magnetic heading.
MHHW	mean higher high water.
MHHWL	mean higher high water line.
MHW	mean high water.
MHWI	mean high water lunitidal interval.
MHWL	mean high water line.
MHWN	neap high water or high water neaps.
MHWS	mean high water springs.
MHz	megahertz.
mi.	mile, miles.
mid	middle.
min.	minute(s).
MLLW	mean lower low water.
MLLWL	mean lower low water line.
MLW	mean low water.
MLWI	mean low water lunitidal interval.
MLWL	mean low water line.
MLWN	neap low water or low water neaps.
MLWS	mean low water springs.
mm	millimeters.
Mn	mean range of tide.
mo(s)	month(s).
MODU	mobile offshore drilling unit.
Mon	monument.
Mo.(U)	Morse Uniform (light).
mph	miles per hour.
MPP	most probable position.
MRI	mean rise interval.
MRM	miles of relative movement.
ms	millisecond(s).
MSC	Military Sealift Command.
MSI	maritime safety information.
MSL	mean sea level.
MTI	moving target indication.
MTL	mean tide level.
MWL	mean water level.
MWLL	mean water level line.
MZn	magnetic azimuth.

N

N	north; nun.
n	natural (trigonometric function).
Na	nadir.
NAD	North American Datum.
NASA	National Aeronautics and Space Administration.

NAUTO nautophone.
NAVSAT Navy Navigation Satellite System.
NAVSSI navigation sensor system interface.
NBDP narrow band direct printing.
NBS National Bureau of Standards.
NCS network coordination station.
NESS National Earth Satellite Service.
NGVD National Geodetic Vertical Datum.
NLT not less than (used with danger bearing).
n. mi. nautical mile(s).
NM nautical mile, miles; notice to mariners.
NMEA National Marine Electronics Association.
NMT not more than (used with danger bearing).
NNSS Navy Navigation Satellite System.
NOAA National Oceanic and Atmospheric Administration.
NOS National Ocean Service.
NRML new relative movement line.
NtM notice to mariners.
NWS National Weather Service.

O

Obsc obscured.
Obs Spot Observation spot.
Obstr obstruction.
Oc. occulting.
Oc.(2) group occulting.
Oc.(2+1) composite group occulting.
Occas occasional.
ODAS oceanographic data acquisition systems.
Or orange.
OTC officer in tactical command.
OTSR Optimum Track Ship Routing.

P

P atmospheric pressure; parallax; planet; pole; pillar.
p departure; polar distance.
PA position approximate.
PC personal correction; personal computer.
PD position doubtful.
PCA polar cap absorption.
PCD polar cap disturbance.
PCP potential point of collision.
pgc per gyrocompass.
P in A parallax in altitude.
PM pulse modulation.
PM post meridian (after noon).
Pn north pole; north celestial pole.
PPC predicted propagation correction.
PPDB point positioning data base.
PPI plan position indicator.
PRF pulse repetition frequency.
Priv private; privately.
PROHIB prohibited.
PRR pulse repetition rate.
Ps south pole; south celestial pole.
psc per standard compass.
p stg c per steering compass.
Pub. publication.
PV prime vertical.
Pyl pylon.

Q

Q quick flashing.
Q(3) group quick flashing (3 flashes).
Q(6)+L.Fl. group quick flashing (6 flashes) plus a long flash.
Q Polaris correction.
QQ' celestial equator.

R

r end of own ship's true vector.
R Rankine (temperature); refraction; own ship; red; rocky; coast radio station.
RA right ascension.
RACON radar transponder beacon.
rad radian(s).
RB relative bearing.
R Bn radiobeacon.
RCC Rescue Coordination Center.
RDF radio direction finder, RDF station.
Rep. reported.
rev. reversed.
RF (rf) radio frequency.
R Fix running fix.
rk rock, rocky.
RLG ring laser gyro.
rm relative DRM-SRM vector.
R Mast radio mast.
RORO roll-on/roll-off.
RML relative movement line.
RMS root mean square.
RSS root sum square.
RTCM Radio Technical Commission for Maritime Services.
RZn relative azimuth.

S

s second(s).
S sea-air temperature difference correction; slow; south; set; speed; sand.

SALM	single anchor leg mooring.
SAM	system area monitor.
SAR	search and rescue.
SART	search and rescue radar transponder.
SBM	single buoy mooring.
SD	semidiameter; sounding doubtful.
sec	secant.
sec.	second, seconds.
semidur.	semiduration.
SENC	system electronic navigation chart.
SES	ship earth station.
SF	Secondary Phase Factor.
SH	ship's head (heading).
SHA	sidereal hour angle.
SHF	super high frequency.
SI	International System of Units.
SID	sudden ionospheric disturbance.
sin	sine.
SINS	Ships Inertial Navigation System.
SLD	sea level datum.
SMG	speed made good.
SNR	signal-to-noise ratio.
SOA	speed of advance.
SOG	speed over ground.
SOLAS	Safety of Life at Sea Convention.
SP	spire; spherical.
SPA	sudden phase anomaly.
SPM	single point mooring.
SRM	speed of relative movement.
SS	signal station.
U.S. Sur M	U.S. Survey mile(s).
sub, subm	submerged.

T

T	air temperature correction (altitude); table; temperature; time; toward (altitude intercept); true (direction).
t	dry-bulb temperature; elapsed time; meridian angle.
t'	wet-bulb temperature.
tab.	table.
TAI	International Atomic Time.
tan	tangent.
TB	true bearing; turning bearing; air temperature atmospheric pressure correction (altitude).
TC	true course.
TCA	time of satellite closest approach.
TCPA	time to closest point of approach.
TcHHW	tropic higher high water.
TcHHWI	tropic higher high water interval.
TcHLW	tropic higher low water.
TcLHW	tropic lower high water.
TcLLW	tropic lower low water.
TcLLWI	tropic lower low water interval.
TD	time difference (Loran C).
Tel	telephone; telegraph.
TG	time difference of groundwaves from master and secondary (slave) stations (Loran).
TGS	time difference of groundwave from master and skywave from secondary (slave) station (Loran).
TH	true heading.
TMG	track made good.
TOD	time of day (clock).
Tk	tank.
TR	track.
Tr	transit; tower.
Ts	time difference of skywaves from master and secondary (slave) stations (Loran).
TSG	time difference of skywave from master and groundwave from secondary (slave) station (Loran).
TSS	traffic separation scheme.
TZn	true azimuth.

U

U	upper limb correction for moon.
UHF	ultra high frequency.
UL	upper limb.
Uncov	uncovers.
UPS	Universal Polar Stereographic.
U.Q.	ultra quick flashing.
USGS	United States Geodetic Survey.
USWMS	Uniform State Waterway Marking System.
UT	Universal Time.
UT0	Universal Time 0.
UTl	Universal Time 1.
UT2	Universal Time 2.
UTC	Coordinated Universal Time.
UTM	Universal Transverse Mercator.

V

V	variation; vertex.
v	excess of GHA change from adopted value for 1 hour.
var.	variation.
vel	velocity.
Ver	vertical.
VHF	very high frequency.
VHSD	very high speed data.
Vi	violet.
VLCC	very large crude carrier.
VLF	very low frequency.
VPF	vector product format.

vol	volcano; volcanic.
VPF	vector product format.
V.Q.	very quick flashing.
V.Q.(3)	group very quick flashing.
VRM	variable range marker.
VTS	vessel traffic service.

W

W	west; white.
WARC	World Administrative Radio Council.
WE	watch error.
WGS	World Geodetic System.
Wk	wreck.
WMO	World Meteorological Organization.
WWNWS	World Wide Navigational Warning Service.
WT	watch time.

X – Y – Z

X	parallactic angle.
XMTR	transmitter.
y.	yellow.
yd(s).	yard(s).
yr(s).	year(s).
z	zenith distance.
Z	azimuth angle; zenith.
ZD	zone description.
Z Diff.	azimuth angle difference.
Zn	azimuth.
Znpgc	azimuth per gyrocompass.

INDEX

A

B

C

D

E

F

G

H

I

J

K

L

M

N

O

P

Q

R

S

T

U

V

W

X

Y

Z